MW01553564

# LINDEN'S
# HANDBOOK OF
# BATTERIES

## ABOUT THE EDITOR

**Kirby W. Beard** graduated from Lehigh University with a B.S. in Chemical Engineering (magna cum laude, Tau Beta Pi). His career includes extensive experience in lead-acid and lithium batteries. In addition, he has worked in the composites, plastics, cement, and engineering materials industries.

At Gates Energy Products (now EnerSys), Mr. Beard designed and qualified the first military lead-acid aircraft batteries in the early 1980s for the U.S. Navy's F/A-18A and Harrier AV-8B aircraft. Over the next 15 years at Honeywell/Alliant Techsystems, Lithium Technology Corp., and Yuasa/EnerSys, he developed various lithium secondary battery technologies, including the first patents for an internal safety disconnect mechanism and a commercially viable lithium-ion cell design (with a priority date the year before the market introduction of the Sony cell).

Recent achievements and patents include an advanced membrane/battery separator, novel composite materials, aqueous lithium-ion cell technology, and a high-voltage electrolyte based on supercritical fluids. Mr. Beard is a founder of a number of technology companies based on unique materials and process engineering. Serving in executive management, technology, and operational roles, he helped these companies achieve technical and manufacturing successes. Battery separators were commercialized that demonstrated exceptional performance capabilities, leading to the purchase of Mr. Beard's original patent portfolio by the leading consumer electronics company in Asia.

Additionally, he completed graduate studies in Engineering Management at the masters (Wichita State University, 1985) and doctorate (Warren National University, 2007) levels. He has also taught environmental engineering and engineering design courses at Drexel and Penn State universities.

Mr. Beard continues his consulting/entrepreneurial career, and in his free time renovates old homes, fabricates unique architectural building products (U.S. Patent, 2012), and studies family genealogy.

# LINDEN'S HANDBOOK OF BATTERIES

**Kirby W. Beard**  Editor

**Thomas B. Reddy**  Editor Emeritus

**Fifth Edition**

New York   Chicago   San Francisco   Athens   London
Madrid   Mexico City   Milan   New Delhi
Singapore   Sydney   Toronto

Library of Congress Control Number: 2019934406

**Linden's Handbook of Batteries, Fifth Edition**

2 3 4 5 6 7 8 9   LCR   24 23 22 21 20

ISBN 978-1-260-11592-5
MHID 1-260-11592-5

This book is printed on acid-free paper.

**Sponsoring Editor**
Lara Zoble

**Editing Supervisor**
Stephen M. Smith

**Production Supervisor**
Lynn M. Messina

**Acquisitions Coordinator**
Elizabeth M. Houde

**Project Manager**
Tania Andrabi,
Cenveo® Publisher Services

**Copy Editors**
Farah Naseem and Smriti Verma,
Cenveo Publisher Services

**Proofreader**
Alekha Chandra Jena

**Indexer**
Rahul Sachdeva

**Art Director, Cover**
Jeff Weeks

**Composition**
Cenveo Publisher Services

# CONTENTS

## Part 3   Battery Product Overview

## Part 4   Electrochemical Cell Designs (Platform Technologies)

## Part 5    Battery Applications

## Part 6    Battery Industry Infrastructure

# Appendices

# CONTRIBUTORS

**Daniel Abraham**   Vice President, Science and Business Strategy, MPEG LA LLC
Affiliation(s): Columbia Univ., General Electric
Career Synopsis: Strategy and licensing specialist.
Education: Ph.D., Univ. of California

**Vaidevutis Alminauskas**   Contributor to the Fourth Edition

**Terrill B. Atwater**   Senior Research Chemical Engineer/Leader, Materials Science Team, U.S. Army
Affiliation(s): U.S. Army, Communications Electronics Research, Development and Engineering Center
Career Synopsis: 35-yr electrochemistry career, Army Greatest Invention Top Ten (2003, BA-8180/U Zn-Air; 2010, Soldier Wearable Power System), 2006 R&D Achievement Award (Soldier Portable Power), 19 patents.
Education: Ph.D., Materials Science & Engr., Rutgers Univ.

**Rouse Roby Bailey**   Stanley Black & Decker (retired)
Affiliation(s): Black & Decker, Stanley Black & Decker, DeWALT Power Tools
Career Synopsis: Engineering Director/VP for power tool manufacturers with focus on cordless system design, efficiency optimization, cell technology and testing, brushless motors and controls, mechanicals.
Education: M.S., Tech. Mgmt., Johns Hopkins

**Gary A. Bayles**   Sr. Systems Engineer, Scientific Application International Corp.
Affiliation(s): Northrop Grumman Corporation, Westinghouse
Education: Ph.D., Chem. Engr., Univ. of Pittsburgh

**Kirby W. Beard**   Editor, *Linden's Handbook of Batteries*, Fifth Edition

**Trevor D. Beard**   Owner, TEEBS R&D, LLC
Affiliation(s): Porous Power Technologies, E3BV, Contract Consultant/Entrepreneur (TEEBS R&D)
Career Synopsis: Independent inventor/technology licensing of aqueous and Li ion electrodes/separator, filter materials and absorbent media; invited presenter, NAATBatt Emerging Technologies 2017.
Education: B.S., Business/Management, Thomas Edison State College

**Seth Ayliffe Binfield**   Member, Technical Staff, DfR Solutions
Affiliation(s): DfR Solutions
Career Synopsis: Over 10 years in failure analysis of electronics hardware, completing hundreds of projects using a wide variety of analytical tools.
Education: B.S., Chem., Univ. of Maryland

**Daniel Borneo**   Engineering Program/Project Lead, Sandia National Laboratories
Affiliation(s): U.S. DoE, OE, ESS projects team lead
Career Synopsis: U.S. DoE Grid Modernization Initiative; Devices and Integrated Systems Testing and Energy Storage Demonstrations lead with states, utilities, and storage providers analyzing benefits and challenges.
Education: B.S./M.S., EE, Univ. of New Mexico

**Jack N. Brill**   Contributor to the Fourth Edition

**Ralph J. Brodd**   Contributor to the Fourth Edition

**Andrew F. Burke**   Univ. of California, Davis

**George E. Bye**   Founder/CEO, Bye Aerospace

Affiliation(s): U.S. Air Force (Desert Storm veteran); Founder/CEO, ATG Aircraft Co.

Career Synopsis: T-38 instructor pilot, C-141B Instructor Aircraft Commander, ATP rated pilot with over 4000 flying hours, and designer for the Sun Flyer, StratoAirNet, Solesa, Javelin and Silent Falcon.

Education: B. Engr., Univ. of Washington

**Raymond H. Byrne**   Sandia National Laboratories

**Joseph A. Carcone**   Vice President/GM, Power Glory Tech.

Affiliation(s): General Electric Co., SANYO Electric Co., Powergenic Co., Contour Energy Co.

Career Synopsis: Strategic global management (VP, Sales and Marketing/COO/GM) for leading rechargeable battery companies; founding member of RBRC (Recycling).

Education: M.S., Rutgers Univ.

**Denis Carpenter**   Contributor to the Fourth Edition

**Babu R. Chalamala**   Department Head, Energy Storage Technology and Systems, Sandia National Laboratories

Affiliation(s): MEMC Electronic Materials, Indocel Technologies, Motorola, Semiconductor Products

Career Synopsis: IEEE Fellow; MEMC Corporate Fellow; James Eads Award, Academy of Sciences; IEEE Morton Antler Award Lecture; President, Indocel; Manager External Research, Motorola.

Education: Ph.D., Univ. of North Texas

**Vidyu Challa**   Technical Director, DfR Solutions

Affiliation(s): FlexEl, Danaher

Career Synopsis: Battery expert/consultant at DfR Solutions, helping *Fortune* 500 customers with a range of battery and electronic system reliability challenges.

Education: Ph.D., Univ. of Maryland

**Allen Charkey**   Board Advisor, ZAF Systems

Affiliation(s): Evercel Corp., Energy Research Corp. (now Fuel Cell Energy), Yardney Electric Corp.

Career Synopsis: Executive VP and CTO, VP, Battery Division in advanced alkaline cells (Ni-Zn, Zn-air), 25 U.S. patents.

Education: M.S., Chem., NYU; M.B.A., St. Johns Univ.

**Dennis A. Corrigan**   President, DC Energy Consulting LLC

Affiliation(s): General Motors, ECD-Ovonics, Cobasys, Wayne State Univ., TARDEC and XALT Energy

Career Synopsis: Over three decades in R&D and engineering of batteries for electric and hybrid vehicles, author of hundreds of papers and presentations, 19 U.S. patents, GM EV1 team.

Education: Ph.D., Chem., Univ. of Wisconsin

**Fausto Croce**   Professor, Università "G.d'Annunzio"

Affiliation(s): Dipartimento di Farmacia—Università "G.d'Annunzio" (Italy)

Career Synopsis: Professor of physical chemistry and electrochemistry, battery consulting expertise.

Education: Ph.D., Chem., Univ. 'La Sapienza' Roma (Italy)

**Jeff Dahn**   Professor of Physics, Professor of Chemistry, Dalhousie Univ., Canada

Affiliation(s): NSERC/3M (1996–2016) and NSERC/Tesla (current) Canada Industrial Research Chairs

Career Synopsis: Project Leader (1985–1987) and Director of Research (1987–1990), Moli Energy Ltd.; awarded the Gerhard Herzberg Gold Medal (Canada's top science prize) in 2017.

Education: Ph.D., Univ. of British Columbia

**Josef Daniel-Ivad**   President and CTO, Blizzard Technology

Affiliation(s): Electrochemical Society

Career Synopsis: Contributor to past edition of *Linden's Handbook*.

Education: Dipl. Ing./Dr. Techn., Technical Univ. Graz, Austria

**Steven M. Davis**   Senior Scientist, Abbott Laboratories

Affiliation(s): Greatbatch Ltd., Greatbatch Inc.

Career Synopsis: 23+ years in implantable power source technology (R&D and predictive modeling) as Sr. and Principal Scientist; 6 U.S. patents and 15 publications.

Education: B.A./M.A., Chem., SUNY Buffalo; M.S., Computational Science, SUNY Brockport

**Andreas de Vries**   Strategy and Transformation Consultant, Saudi Aramco

Affiliation(s): Qatar Petroleum and ExxonMobil (Advisor and Strategic Planning Manager)

Career Synopsis: 20+ years in finance and business strategy for oil companies; leading analyst on energy market trends, keynote speaker on energy strategy and transformation management.

Education: M.Economics, State Univ. of Groningen

**Arthur Dobley**   Principal Scientist, EaglePicher Technologies

Affiliation(s): Yardney Technical Products

Career Synopsis: Studied new battery materials (Prof. Stan Whittingham, advisor) and now developing lithium-ion, metal/air, and other new energy technology systems with a focus on safety.

Education: Ph.D., Materials Chem., Binghamton Univ.

**Daniel H. Doughty**   President, Battery Safety Consulting Inc.

Affiliation(s): Sandia National Laboratories, Sion Power Corporation

Career Synopsis: Battery expert witness for U.S. Dept. of Justice vs. British Petroleum on Deepwater Horizon oil spill litigation.

Education: Ph.D., Inorganic Chem., Univ. of Minnesota

**Nancy J. Dudney**   Group Leader, Oak Ridge National Laboratory (ORNL)

Affiliation(s): ORNL, Materials Science and Technology Division

Career Synopsis: ORNL Corporate Fellow.

Education: Ph.D., MIT

**Grant M. Ehrlich**   Partner, Cantor Colburn LLP

Affiliation(s): Pratt & Whitney, UTC Fuel Cells, Yardney Technical Products

Career Synopsis: Practicing patent law (multiple IPR petitions directed to lithium-ion batteries) and executive in Connecticut IP Law Association.

Education: Ph.D., Cornell Univ.; J.D., Univ. of Connecticut

**Christopher R. Feger**   Director, Electrochemical Engineering, Genesee Northern Research, LLC

Affiliation(s): St. Jude Medical

Career Synopsis: Passive and active component design, test development, and failure analysis for medical battery/electronic products; 14 issued U.S. patents.

Education: Ph.D., Chem., Ohio State Univ.

**Antonio L. Ferreira**   Battery Scientist (retired)

Career Synopsis: Professional career entirely in the field of lead-acid batteries, including both battery manufacturing and supply chain.

Education: B.S., Honours, Univ. of Waterloo, Canada

**Summer R. Ferreira**   Principal Member of Technical Staff, Sandia National Laboratories

Affiliation(s): Sandia National Laboratories

Career Synopsis: Grid energy storage safety research thrust lead at Sandia National Laboratories, investigating the interplay between battery cell performance, age, and reliability.

Education: Ph.D., Matl. Sci./Engr., Univ. of Illinois

**Michael Fetcenko**   BASF

**Daniel D. Friel**   Principal, Coulomb Consulting; New Business Development Manager, VARTA Microbattery

Affiliation(s): Duracell, Microchip Tech., Texas Instruments, Leyden Energy, ElectricFilm, PowerPrecise Solutions

Career Synopsis: Coauthor, *Smart Battery System/SMBus Specifications* (Intel/Duracell 1995).

Education: B.S.E.E., Purdue Univ.; M.B.A., Boston College

**Salman Ghouri**   Independent Energy Consultant

Affiliation(s): Qatar Petroleum and Oil & Gas Development Corp. (Econ./Sr. Advisor to QP/OGDCL Chair)

Career Synopsis: 35+ years advising IOCs and NOCs executives and investment bankers on energy markets, long-term market assessment, and energy policies; invited to speak in top energy conferences around world.

Education: Ph.D., Mineral Economics, Colorado School of Mines

**H. Frank Gibbard**  President and CEO, WattJoule Corp.

Affiliation(s): Gould, Inc., Power Conversion Inc., Duracell, H Power Corp., Altairnano

Career Synopsis: Cofounder WattJoule with 40 years of experience in batteries and fuel cells; Research Fellow, VP R&D; CEO at five companies; 100+ publications; Chair, 4th International Meeting on Lithium Batteries.

Education: Ph.D., MIT

**Imre Gyuk**  U.S. Department of Energy

**Michael A. Howard**  VP, Operations, DfR Solutions

Affiliation(s): U.S. Air Force, ARINC Engr. Services, Amer. Inst. of Aeronautics/Astronautics, INCOSE

Career Synopsis: Leads the reliability engineering services business at DfR Solutions, helping solve complex reliability and durability challenges with electronic systems, assemblies, and components.

Education: ME, Mech., Univ. of Maryland

**Nathan D. (Ned) Isaacs**  Contributor to the Fourth Edition

**Alexander P. Karpinski**  Director, Product Development/Engineering

Affiliation(s): Yardney Technical Products, Inc., EaglePicher Technologies, LLC

Career Synopsis: Contributor to past editions of *Linden's Handbook* and for multiple technology areas.

Education: B. Engr., Univ. of Connecticut; M.B.A., Rensselaer Polytech. Inst.

**Lisa Michelle King**  Engineering Director, Stanley Black & Decker

Affiliation(s): Mine Safety Appliances, Teledyne Energy Systems, Stanley Black & Decker

Career Synopsis: 25+ years of experience in battery research and development; cell design, applications testing, and supplier development across a wide range of primary and secondary battery chemistries and formats.

Education: B.S., Chem. Engr., Univ. of Maryland

**John Koch**  BASF

**David Linden**  Editor, *Handbook of Batteries*, First to Third Editions

**Eivind Listerud**  Vice President of Engineering, ZAF Energy Systems

Affiliation(s): EaglePicher Technologies, LLC

Career Synopsis: Principal Engineer in R&D and New Product Development, Registered Professional Engineer.

Education: Ph.D., Aerospace Engineering, Missouri Univ. of Science and Technology

**R. David Lucero**  General Manager—U.S. Space, Enersys

Affiliation(s): EaglePicher Technologies, LLC; Electro Energy Inc.

Career Synopsis: 25+ years in the aerospace, defense, medical, and grid energy storage battery markets; responsibilities as VP/GM in management, operations, program management, and engineering.

Education: B.S., Colorado State Univ.; M.B.A., Regis Univ.

**Timothy R. Marshall**  Senior Scientist, Abbott Laboratories

Career Synopsis: 21+ years of research in electrolytic capacitors and batteries for use in medical devices; 12+ patents.

Education: Ph.D., Clemson Univ.

**Alvaro Masias**  Ford Motor Company

**Geoffrey J. May**  Consulting Engineer, Focus Consulting

Affiliation(s): FIAMM SpA, Hawker Batteries Ltd (now EnerSys)

Career Synopsis: Former CTO/Group Technical Director; now managing a consulting business, FOCUS Consulting, which provides expert services to battery manufacturers, users, and investors.

Education: Ph.D., Univ. of Cambridge

**Rodney McKenzie**  Contributor to the Fourth Edition

**Benjamin M. Meyer**  R&D Manager, MaxPower, Inc.

Education: Ph.D., Matl. Science and Engr., Iowa State Univ.

**Ronald T. Moelker**   North American Motive Power Field Product Support Manager, Enersys

Affiliation(s): Enersys

Career Synopsis: 40+ years of connecting motive power products to actual field applications and fostering technical feedback to engineering, design, and manufacturing entities.

Education: Grand Rapids Community College

**John C. Nardi**   Contributor to the Fourth Edition

**John Olson**   President, Jolson Technologies LLC

Affiliation(s): Optima Batteries

Career Synopsis: Ph.D. electrochemist with industry experience in both manufacturing and R&D; provides consulting and lab services for a variety of battery technologies, including lead-acid and lithium-ion batteries.

Education: Ph.D., Univ. of Colorado

**Joseph Passaniti**   Contributor to the Fourth Edition

**Zhigang Qi**   CTO, Innoreagen Power Technology Co., Ltd.

Affiliation(s): H Power Corp., Plug Power Inc., MTI Micro Fuel Cells Inc.

Career Synopsis: Expertise/pioneer in commercializing fuel cells, author of *Proton Exchange Membrane Fuel Cells*, numerous invites to presentations, fuel cell standard documents with 19 issued patents.

Education: Ph.D., McGill Univ.

**Matthew Rappaport**   Founder and President, IP Checkups, Inc., ABC PatentEdge, PatentCAM

Affiliation(s): IP Checkups, PatentCAM

Career Synopsis: Patent Strategist, Intellectual Asset Management, Patent Categorization, Archiving, Monitoring Software.

**Thomas B. Reddy**   Editor, *Linden's Handbook of Batteries*, Third and Fourth Editions

**Charles M. Richard**   VP, New Products, Rebling, Inc.

Affiliation(s): AMP Inc. (VP Engineering), EBY Inc. (Director Engineering)

Career Synopsis: Led the teams that designed the connectors in Apple's and Dell's first computers, currently defining the products to enable winners of the battery revolution to rise to the top.

Education: B.S., E.E., Lehigh Univ.; M.B.A., Johns Hopkins Univ.

**Michael J. Root**   Science Fellow, Boston Scientific Corp.

Affiliation(s): Rayovac Corp. (now Spectrum Brands, Inc.)

Career Synopsis: Battery electrochemist with 30 years battery R&D, including primary and secondary cells for medical, consumer, military use; 20+ patents; numerous technical papers and book chapters.

Education: Ph.D., Univ. of Cincinnati

**Rose E. Ruther**   R&D Staff Scientist, Oak Ridge National Laboratory (ORNL)

Affiliation(s): ORNL, Oregon State Univ. Center for Sustainable Materials Chemistry

Career Synopsis: Fulbright Fellow at the Max Planck Institute for Microstructure Physics; co-chair of the 2016 Gordon Research Seminar on Batteries in Ventura, CA.

Education: Ph.D., Chem., Univ. of Wisconsin

**Mark Salomon**   Senior Scientist

Affiliation(s): Consultant, MaxPower, Inc.

Education: Ph.D., Univ. of Ottawa

**Miguel Sandoval**   Vice President Business Development, Maccor

Affiliation(s): Maccor, Inc.

Career Synopsis: 25+ years of experience in the battery industry with a primary focus on test and measurement equipment, also designs and develops test equipment hardware and software.

Education: B.S.E.E., Univ. of Tulsa

**Shriram Santhanagopalan**  Team Leader, Vehicle Electrification Group, National Renewable Energy Lab (NREL)

Affiliation(s): Celgard, NREL

Career Synopsis: Team Leader for Diagnostics and Characterization, Vehicle Electrification Group, NREL, and codeveloper of the single-particle model used widely for state estimation.

Education: Ph.D., Univ. of South Carolina

**Paul F. Schisselbauer**  Director of Engineering, Munitions Batteries, EnerSys Advanced Systems

Affiliation(s): Alliant Techsystems (Engineering Fellow), Honeywell Power Sources Center

Career Synopsis: 30+ years in the design, development, and manufacture of lithium batteries for military, space, and medical applications; several U.S. patent awards; multiple industry publications/presentations.

Education: M.E., Engr. Sci., Penn State Univ.

**Brooke Schumm, Jr.**  Contributor to the Fourth Edition

**Nicholas Shuster**  Director, Science and Technology, EnerSys Advanced Systems, Tampa

Affiliation(s): Westinghouse, Naval Systems Div.; Northrop Grumman; ENSER Corporation

Career Synopsis: Developed/commercialized Li-Si/Co-S, G. Westinghouse Innovation Award, Signature Award of Excellence, Westinghouse President's Quality Achievement Award, 16 U.S. patents.

Education: M.S., Chem. Engr., Case Western Reserve Univ.

**David Simm**  Spacecraft Power System Engineer (retired)

Education: Univ. of Maryland

**Patrick J. Spellman**  Contributor to the Fourth Edition

**Monica V. Stoka**  Sales Manager, EnerSys Advanced Systems, Munitions

Affiliation(s): Alliant Techsystems, Power Sources Center

Career Synopsis: Design engineering and technical sales.

Education: B. Mech. Engr., Villanova Univ.; M.B.A., Temple Univ.

**Thomas F. Strange**  Director R&D, Abbott Power Systems

Affiliation(s): Abbott Research Fellow

Career Synopsis: Led team to establish company to manufacture HV capacitors for ICDs, hybrids for pulse generators, and a range of battery types (now Abbott Power Components).

Education: B.S., Physics, Univ. of South Carolina

**Anthony Sudano**  Principal, Sudano Consulting

Affiliation(s): Avestor

Career Synopsis: 30+ years in process development/machine design/fabrication in battery and super-cap manufacturing/commercialization, 5 U.S. patents on electrode and cell design/manufacturing.

Education: B. Engr., Mech., Concordia Univ.

**Rob Sweney**  Director, Advanced Vehicle Development, Alta Motors

Affiliation(s): Volkswagen of America, Honeywell Aerospace, Atieva

Career Synopsis: Career focus has been on creating battery systems and developing new electric vehicles at both established companies and startups, including the Alta RedShift motorcycle battery pack.

Education: B.S., Engineering, Harvey Mudd College

**Travis Thompson**  Research Scientist, Univ. of Michigan

Affiliation(s): Univ. of Michigan

Career Synopsis: Advanced battery cell developer, solid-state prototype process development and materials scale-up, commercialization lead, *Forbes* "30 under 30" for Energy 2017 recognition award.

Education: Ph.D., Matl. Science/Engr., Michigan State Univ.

**Ron Turi**  Owner-Principal, Element 3 Battery Venture

Affiliation(s): Rexam Corp.; Lithium Technology Corp.; Advisor, SiNode and Momentum Dynamics

Career Synopsis: Expertise in key phases of Li-ion cells (materials, design, and manufacturing); consulting work includes business strategy/marketing, supply chain, and competitor/IP/investment analyses.

Education: M.S., Chem. Engr., Carnegie-Mellon/Univ. of Massachusetts

**Lawrence Edward Weinstein**    Contract Engineer

Affiliation(s): FlexEl, Y-Carbon

Career Synopsis: Battery design, materials, and fabrication; supercapacitors.

Education: M.S., Materials Sci. and Engr., Rutgers

**Adam Weisenstein**    Chief Technology Officer, ZAF Energy Systems

Affiliation(s): Montana State Univ.

Career Synopsis: Expertise in R&D and engineering of zinc battery systems.

Education: Ph.D., Engineering, Montana State Univ.

**Ralph E. White**    Professor, Univ. of South Carolina

Affiliation(s): Univ. of South Carolina

Career Synopsis: Distinguished Scientist and Professor (Dept. of Chem. Engr., Univ. of South Carolina), mathematical modeling of batteries and fuel cells.

Education: Ph.D., Univ. of California, Berkley

**Chase B. Whitman**    Enersys

**Steven Wicelinski**    Ph.D.

Affiliation(s): DURACELL, Global Regulatory Leader

Career Synopsis: Product Safety and Regulatory Affairs Executive.

Education: Ph.D., Louisiana State Univ.

**John A. Wozniak**    President, Energy Storage and Power Consulting

Affiliation(s): Energizer Power Systems, Hewlett-Packard (Distinguished Technologist), Microsoft

Career Synopsis: Prior to a lengthy career with HP and others, Dr. Wozniak was a 15-year academician; consulting since 2014, he drives quality manufacturing throughout Asia.

Education: Ph.D., Engineering, Univ. of Florida

**Kang Xu**    Team Leader, ARL Fellow

Affiliation(s): U.S. Army Research Lab (ARL)

Career Synopsis: 2017 IBA Technology, 2018 ECS Battery Research, 2017 DoD Scientist awards.

Education: Ph.D., Arizona State Univ.

**Kwo Young**    Chief Scientist, BASF

Affiliation(s): Ovonic Battery Co.

Career Synopsis: Battery research on solid-state proton conduction, high-energy silicone-based metal hydride, and high-power bipolar cells.

Education: Ph.D., Princeton Univ.

**Michael Zelinsky**    BASF

**George Zguris**    Chief Scientist at Hollingsworth and Vose (retired)

# PREFACE

## MAKING THE CASE FOR "STEWARDSHIP"

Early in my battery career, I was tasked with developing an improved lithium primary cell for an implantable medical device. I proposed a design that used lithium metal anodes with a "pre-charged" $LiCoO_2$ cathode (see U.S. Patent 5667660A). The concept was similar to the techniques used to activate lead-acid batteries, where electrodes are "formation charged" in an off-line process prior to transfer into the final battery assembly.

At the time this work was progressing, my company was actively seeking technology transfer partnerships. Hence, I was asked to present the concept to one of these potential partners (a pioneering medical device company that had expertise in both batteries and therapeutic, implantable electronic devices). I began my talk by launching into the technical details. However, about a minute into my review, the chief scientist for this visiting entourage interrupted to ask me, "What is your philosophy on batteries?"

Medical batteries, of course, need to have high levels of safety, performance, reliability, and quality, but such attributes are obvious. My presentation eventually got back on track, but I felt that I should have had a more impressive, clear, and precise answer. Indeed, defining one's philosophy might be good to do for any project, company mission statement, national policy, or other major, important undertaking.

Subsequently, many years later I identified a simple concept that functions well as a working philosophy: *stewardship*. Stewardship has a variety of narrow usages. Environmental stewardship is a well-used phrase that typically refers to conserving, recycling, pollution reduction, etc. However, a broader definition for stewardship that applies across the full spectrum of technical endeavors may be expressed as "Stewardship of the Energy Domain."

Energy concerns apply across many industries, but are a prime, universal factor for suppliers and users of electrical energy (including the battery industry). To achieve optimal outcomes for all stakeholders, energy issues must be meticulously and persistently analyzed. However, with few exceptions, energy considerations are typically limited to rather narrow issues (i.e., the cost of a barrel of oil, the tons of carbon dioxide emissions, plant and capital equipment costs, etc.). These approaches fail to include many critical factors.

In response to these shortcomings, I reasoned that the basic laws of thermodynamics dictate, at the most fundamental level, the consequences of all purposeful actions.[a] Just as all natural phenomena are governed by these constraints, the decisions we, as individuals, organizations, and societies, make on a constant and ongoing basis are impacted by the laws of thermodynamics. Additionally, and perhaps somewhat surprisingly, a number of dedicated philosophers have even made it their mission to contemplate the philosophy of energy transformation, conservation, and usage.[b]

To me, this sounds like a pretty good starting point for developing a "philosophy on batteries." In essence, battery development should be based on careful thermodynamic analyses of the conversion of chemicals into electrical energy (and vice versa), and most importantly, the magnitude and efficiency at which that is done. An optimal battery design and application will always be best benchmarked by the underlying thermodynamics of the process. And while thermodynamic calculations can be quite complex, the first and second laws of thermodynamics (i.e., conservation and entropy laws, respectively) will still serve to establish guiding principles for developing any new technology.

By using this philosophical approach, decision-making processes for implementing new battery technologies may be improved. Rather than only looking at the typical criteria, such as costs, carbon footprint,

---

[a]"Decision Models Based on Philosophical Considerations of Thermodynamic Principles," Doctoral Dissertation, Engineering Management, School of Engineering and Computer Science, Warren National University, Cheyenne, WY. Kirby W. Beard. ©2007.

[b]Robert D. Handscombe and Eann A. Patterson, *The Entropy Vector: Connecting Science and Business,* World Scientific, Singapore. ©2004.

recyclability, energy density, etc., the calculation of thermodynamic inefficiencies associated with any particular electrochemistry may offer a better means of comparison. In particular, though, the analyses should include all aspects of the proposed battery technology—the total supply chain, the performance characteristics, and disposal. While costs will remain a key factor, would it not provide valuable perspective to better understand all the energy conversions needed to produce and implement a battery technology, extending from before the cradle to after the grave?

For example, in the case of lithium-ion cells, the thermodynamic calculations would thus need to include the energy expended on infrastructure to mine and transport cobalt from the Democratic Republic of the Congo or lithium salts from the Atacama Desert of Bolivia. And how much energy is expended in building complex massive battery manufacturing plants? But also, without efficient collection and recycling systems in place, as is now largely the case, what resources are lost? For example, how do lithium batteries compare to lead-acid or natural gas on the basis of total system energy expenditures and "waste heat"?

Is a primary or a secondary cell potentially more thermodynamically superior for any specific application? The infrastructure necessary to charge cells is quite burdensome. Power plants, solar arrays, battery control, and monitoring systems are all significant expenditures of resources that may prove less efficient than a primary cell made from cheap, common materials and for which an efficient recycling program exists. Or would a mechanically rechargeable battery (i.e., replaceable cathodes and anodes) be even more logical as a means to optimize thermodynamic efficiency and reduce waste heat energy losses?

One problem, however, in using thermodynamic criteria to aid in battery technology development is that such analyses are often probabilistic (i.e., based on random, uncertain outcomes) and somewhat subjective (i.e., trade-offs of disparate metrics). Thermodynamic calculations provide a reference point, not a guaranteed outcome nor a precise and comprehensive measurement. Most electrochemists and battery engineers avoid such esoteric descriptors, focusing on hard data instead. Entrepreneurs and start-up companies, on the other hand, often avoid the hard facts, preferring to only espouse grandiose visions as a means of generating investor interest or market hype. Hence, it is not easy to find references that link the use of thermodynamics to rational strategic planning. Tying a solid field of science (thermodynamics) to a series of speculative outcomes based on imprecise decision analyses is problematic.

A case study on the ways that thermodynamic calculations intersect with battery selection criteria is presented in Chap. 1 (Sec. 1.12). However, two examples of the use of a thermodynamic perspective in analyzing battery technologies are presented below.

## Example 1: Total Energy Usage in the United States

Thermodynamic calculations are, of course, used to provide data on electrochemical reactions, but hardly ever to decide a business strategy. Fortunately, there appears to be an awakening. Lawrence Livermore National Laboratory (LLNL) has been publishing macroscale energy flow diagrams for at least a decade.[c] Different versions of these charts exist, but one of the more useful ones is reproduced below as Exhibit A. This diagram indicates that a great and immediate beneficial impact on energy use in the United States could be achieved by both improved energy conversion efficiency and reductions in consumption in certain sectors (e.g., transportation).

Specifically, this chart shows the massive scope of energy use in the United States, but more importantly, reveals that there is a penalty to pay for all energy conversions. First, the chart shows the United States used 97.3 quads of energy in 2016. A staggering two-thirds of this energy is listed as "rejected energy," also often referenced as "waste heat." This heat is dissipated energy that serves no purpose and is lost forever. Electricity generation consumes a third of all energy in the United States, but also loses two-thirds as waste heat. While the chart does not clarify which sources of electrical energy are most inefficient, clearly the burning of fossil fuels is a major culprit.

To make a major impact on energy consumption, the transportation section is a good target for improvement. Transportation consumes the largest single share of energy (~29%) but also does that in a most unequivocally inefficient manner: nearly 80% of energy consumed is rejected as waste heat. This loss is a huge problem. Development of renewable energy resources for electrical power generation and the expanded use of

---

[c]*Credit:* Lawrence Livermore National Laboratory and Department of Energy, Energy Information Administration, 2017 (https://www.eia.gov).

# Estimated U.S. Energy Consumption in 2016: 97.3 Quads

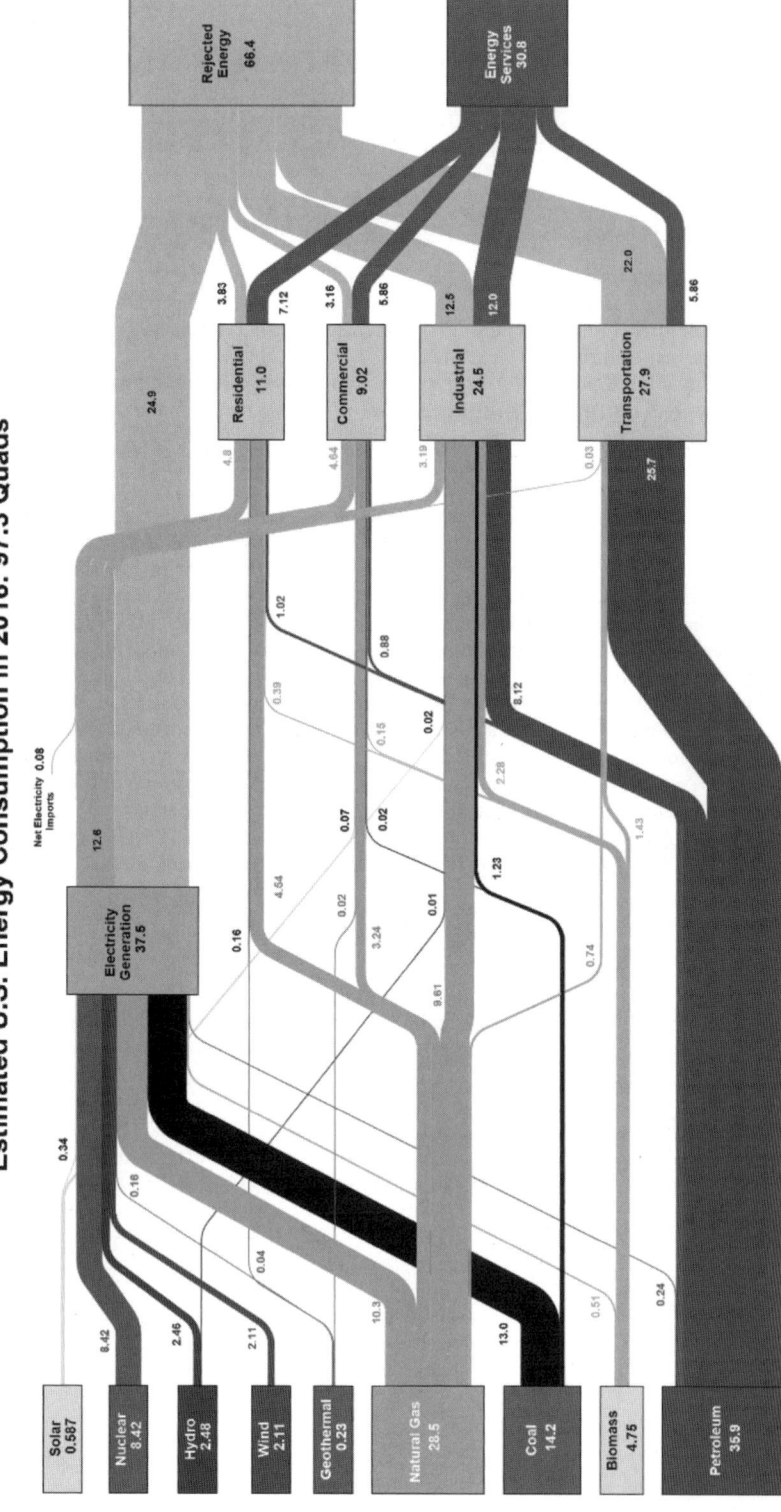

**EXHIBIT A** Macroscale energy flow diagram. (Source: LLNL, March 2017. Data is based on DOE/EIA MER (2016) by Lawrence Livermore National Laboratory and the Department of Energy, under whose auspices the work was performed. This chart was revised in 2017 to reflect changes made in mid-2016 to the Energy Information Administration's analysis methodology and reporting. The efficiency of electricity production is calculated as the total retail electricity delivered divided by the primary energy input into electricity generation. End-use energy efficiency estimates are 65% for residential and commercial, 21% for transportation, and 49% for industrial sectors.)

electric vehicles are examples of potential ways to combat inefficiencies. While these technologies may be popular, they are also desirable options to consider due to their superior thermodynamics (i.e., the high energy efficiency of solar arrays and electric motors). A technology should be selected, not because it is the latest craze, but because of favorable thermodynamic metrics:

1. Higher conversion efficiencies for electric vehicles relative to combustion engines
2. Diversion of solar energy typically lost as waste heat to the environment into stored energy

To help determine the best transportation option(s), all potential technologies should be analyzed based on both total, overall energy usage and conversion efficiencies. The following example provides details on one such effort.

### Example 2: Analysis of a Battery Supply Chain Based on Thermodynamic Calculations

An example of the beneficial use of thermodynamic analysis for comparing battery technology options comes from Argonne National Laboratory (ANL). ANL researchers presented a paper at a recent battery conference comparing the energy used to recycle lithium batteries to that used to build brand new ones.[d] While limited in scope, this analysis is exactly on-target for showing how to improve decision-making capabilities for selecting optimal technologies. While the paper showed recycling to be a lower-cost option (up to 27% lower $/kg cell produced), it is the emphasis on energy efficiency that is most intriguing (40% less MJ energy usage per kg cell produced). Clearly, thermodynamic efficiencies, not just cost, are critical to this analysis.

The most energy-efficient choice will not always have more favorable costs, but the use of thermodynamic calculations provides a "first principles" approach to making more logical long-term decisions. Specifically, while the lowest-cost option may seem better at first, the sustainability of that cost advantage may deteriorate over time. The advantage for the most energy-efficient technology, on the other hand, should not change. The use of large amounts of energy in an inefficient fashion is not a preferable choice even if it is the lowest-cost option. The selection of technologies based mainly on costs is potentially a missed opportunity for a more sustainable future society. An option with the smallest energy depletion and the highest efficiencies should reasonably be expected to be the best choice overall if all sourcing and life cycle factors are considered.

## HANDBOOK ORGANIZATION

This edition of the Handbook is built on four past editions that had grown magnificently in scope and collegiality. However, a number of changes were deemed necessary to improve conciseness and utility. The overall arrangement of the Handbook has undergone, perhaps, the biggest change, as detailed in Chap. 1 and as listed below.

PART 1: INTRODUCTION

Chapter 1: Overview/Prologue

Chapter 2: Raw Materials

Chapter 3: Battery Components

PART 2: ELECTROCHEMICAL/ANALYTICAL TECHNIQUES

Chapter 4: Electrochemical Principles and Reactions

Chapter 5: Battery Performance

Chapter 6: Mathematical Modeling

---

[d]"Comparison of Lithium-Ion Battery Recycling Process Using the ReCell Model." J. Spangenberger, L. Gaines, and Q. Dai, 35th Annual International Battery Seminar and Exhibit, Fort Lauderdale, FL, March 26–29, 2018.

## *SUMMARY AND CONCLUSION*

While the use of thermodynamic analyses as a primary criterion for determining the future of the world's energy supply is not without complications, decisions on our energy options are becoming increasingly fractious, zero-sum undertakings: more random, more convoluted, and less sustainable. A vision for the future based on thermodynamic analysis to validate the most fundamentally sound and most straightforward path is only logical.

Clearly, the battery industry has reached a critical mass where more thoughtful perspectives are needed. Social media, artificial intelligence, and other high-tech endeavors have attained extraordinary levels of sophistication. Is it not now an appropriate time for the battery industry to get ahead of the curve also and anticipate the challenges to come?

In summary, this Handbook provides the reader with a variety of claims, perspectives, and conjectures. The use of thermodynamic analyses potentially offers a powerful alternative means for properly assessing the technology.

---

[e]*Note:* Each electrochemistry chapter in Part 4 is now organized in the same fashion, so that various battery technologies are more readily compared, as follows:

    Section 1: Introduction/Overview/Background

    Section 2: Chemistry/Electrochemical Reactions

    Section 3: Cell Components/Designs/Constructions (and relevant technology)

    Section 4: Battery Applications and Performance

Additional sections are included as necessary in each chapter to cover special topics, such as maintenance, disposal, failure analysis, etc. This organization is different from the prior editions, but is intended to provide better clarity and easier data access.

[f]*Note:* This part is greatly expanded and fully updated with numerous examples of growth opportunities in the battery markets.

[g]*Note:* These chapters include brand new content on the miscellaneous factors that enable the battery industry to function: manufacturing, charging equipment, and miscellaneous support services, such as intellectual property services, battery safety analysis, electronic control/diagnostic equipment, and business/strategic analysis.

## *ACKNOWLEDGMENTS*

First, this Handbook would not have evolved into the comprehensive professional manual that exists today without the dedication of David Linden and Thomas Reddy over the last four editions and three decades. Additionally, the valuable contributions of the myriad of contributors required to make this text a versatile and effective tool cannot be understated. And of course, moral support and critical analyses from family, friends, and trusted associates are always necessary and beneficial for optimal outcomes. New technologies will always be used if they serve a truly useful purpose. Hopefully, this Handbook will help battery designers, marketing professionals, financial and business leaders, and others make wise decisions on energy technology development and use (i.e., Stewardship of the Energy Domain).

*Kirby W. Beard*
*Skippack, Pennsylvania*

# LINDEN'S
# HANDBOOK OF
# BATTERIES

# P · A · R · T · 1

# OVERVIEW AND INTRODUCTION

# CHAPTER 1
# ELECTRICITY, ELECTROCHEMISTRY, AND BATTERIES: PROLOGUE AND EXPOSITION

**David Linden, Thomas B. Reddy, Kirby W. Beard**

## 1.0  OVERVIEW

To best describe any field of human endeavor is to reveal the collective wisdom behind the creation and continued viability of the enterprise. The field of electrochemical cells and battery systems has at its core the ability to store chemical energy and then provide electrical energy derived from chemical processes. The battery industry is based on unique electrochemical energy transformations where the electrons, which are normally obscured by chemical reactions, are now separated from the reactants, collected by a current-carrying circuit and delivered, as a unit of commerce, to electronic equipment or an energy conversion device. Specifically, special types of oxidation-reduction (redox) reactions, or in some cases, charge separation processes, both involving electron transfer, are present, whereby the electrons are diverted through an external electrical circuit to do useful work. The force and volume of electron transfer from the high-potential to the low-potential electrode are proportional to the type and magnitude of the redox reactions. Ion flux and/or accumulation provide the other essential mechanism to achieve a practical working electrochemical energy source.

## 1.1  HANDBOOK CONTENT AND ORGANIZATION

The first several editions of *Linden's Handbook of Batteries* were an assembly of chapters targeted to summarize the large and growing body of work on various electrochemical systems developed up through the mid-1980s. Many new systems had been identified in preceding decades, and some were making inroads into commercial markets. David Linden's initial editions brought organization and clarity to a diversity of technologies. With the addition of Thomas Reddy to the editorial staff, sections on more modern systems (e.g., lithium primary cells) were expanded and the first series of battery application chapters were introduced.

This new 5th edition preserves the basics of the prior compendium while streamlining the content to allow coverage of the many innovations that have proliferated over the past 10 years. The Handbook is based on a process flow concept. The first two technical chapters are new additions that highlight battery raw materials (Chap. 2) and manufactured battery components (Chap. 3). Chapters 4 to 8 revisit the electroanalytical techniques, described in the prior edition, necessary for understanding the building blocks of batteries. Chapters 9 to 22 are updated chapter versions, also from the prior edition, that provide specific discussion of both new and traditional electrochemical systems. Battery applications, covering the evolving marketplace, are updated and supplemented with new content in Chaps. 23 to 29. The Handbook concludes with three new chapters intended to provide an added perspective on the battery industry as a whole:

Chapter 30: Manufacturing (a summary of current battery industry production concepts)

Chapter 31: Charging (a general review of charging strategies and related electronics)

Chapter 32: Ancillary and supporting services (a collection of discussions from an array of sources that contribute to the viability of the industry as a whole)

Figure 1.1 provides a schematic flow diagram of the 5th edition's organization. The organization of these topics/chapters will, ideally, provide two main benefits for this new edition. First, by reading through the entire text in order, one can gain a good overview of what makes the battery industry a unique and compelling field of study. But second, a reader who is interested in just a single or a few topics will be able to easily locate and quickly focus on the chapter(s) of interest.

The Handbook does not, however, provide a detailed, comprehensive treatment of the various specific electrochemical cells. Entire textbooks have been and can be written on each of these separate chapters. This Handbook is simply intended to provide a survey of the battery field and background knowledge for pursuing further investigation of any specific technology, market application, or strategic action plan. This introductory chapter provides an overview of the "battery basics." Further details for these various items are reviewed in Parts 1 to 4, as noted in Fig. 1.1.

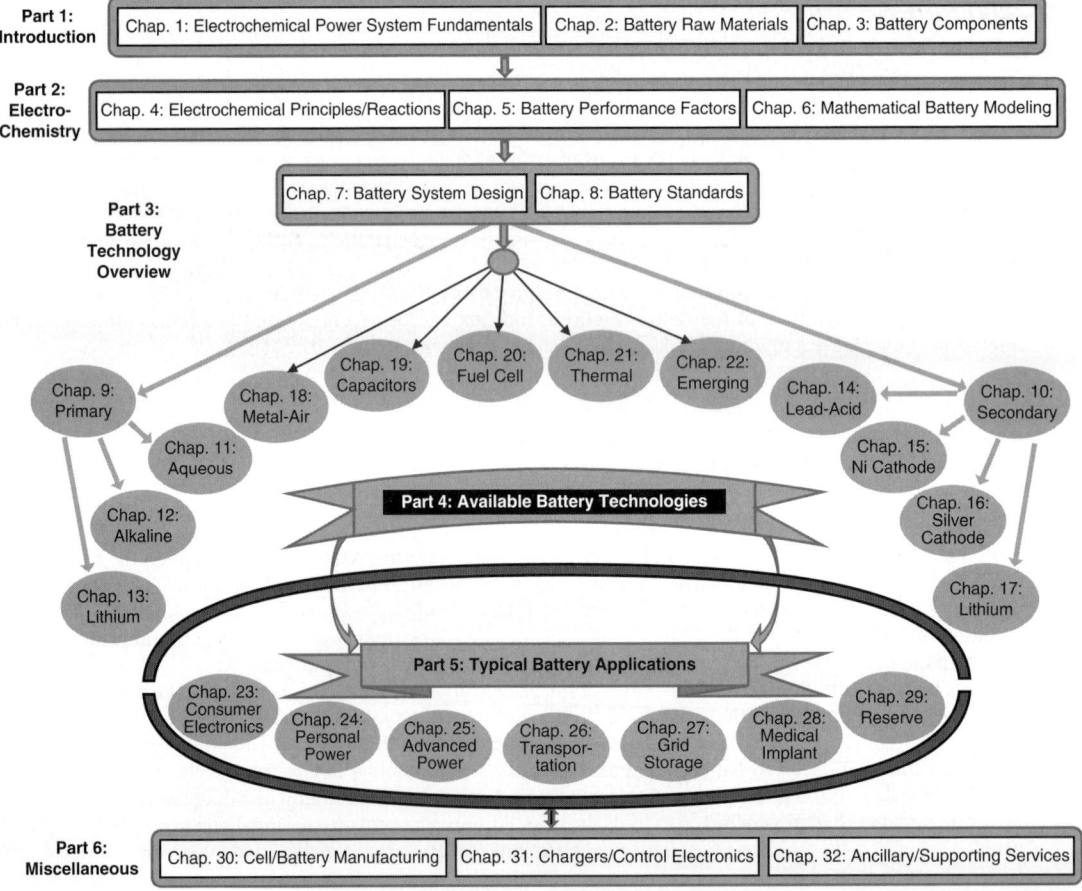

**FIGURE 1.1** Handbook content and organization.

## 1.2  *THE AGE OF ELECTROCHEMISTRY*

The invention of the electrochemical cell did not come easy. First, electricity needed to be better understood. In the 18th century, various electrical phenomena were known: lightning, Leyden jar static electricity (a type of capacitor), and Galvani's "animal electricity" discovery. Benjamin Franklin (Fig. 1.2) was the most prominent proponent for setting the record straight. He deduced that lightning was electricity and coined the term "battery," based on the resemblance of aligned, connected glass plates (or Leyden jars) to an artillery battery.

**FIGURE 1.2**  Franklin's burial site, Christ Church, Philadelphia, PA. (*Image courtesy of Kirby W. Beard.*)

However, it was Alessandro Volta (see Fig. 1.3) who actually built and documented the first functioning battery. Volta surmised that Galvani misunderstood the cause for movement in a frog's leg created by contact with a metal probe. Volta realized that the chemical reaction of the metal probe in the saline environment of the animal tissue was actually a corrosion reaction (now known as galvanic corrosion) that created the electrical stimulus. By interleaving metal plates between layers of cloth soaked in brine solution, the voltaic pile was invented.

Equally important to the invention of the battery was the ability to document the findings. Without a communication network (letters to The Royal Society, London), the scientific musings and experimentations of these early electrochemistry pioneers might have been lost for another century. But, it still took another 50 to 100 years before many practical batteries with useful applications succeeded. Communication through battery textbooks/e-books, web pages, conferences, and sponsored research reports are critical to continued advancements in the field.

**FIGURE 1.3**  Volta Memorial Plaque, Como, Italy. (*Image courtesy of Kirby W. Beard.*)

## 1.3    *BATTERY FUNCTIONING*

Batteries convert the chemical energy contained in materials directly into electric energy by means of electro-chemical oxidation-reduction (redox) reactions between two or more "ingredients." Primary cells have full capacity, as built, but in rechargeable systems, the battery must be recharged by a reversal of the process. The transfer of electrons from one material to another through an electric circuit differs from nonelectrochemical redox reaction, such as rusting or burning, where the transfer of electrons occurs directly between the reactants, producing thermal energy only and no electrical power. As the battery electrochemically converts chemical energy into electricity, it is not subjected, as are combustion or heat engines, to the limitations of the Carnot cycle dictated by the second law of thermodynamics.

Batteries, therefore, represent a revolutionary invention, capable of having higher energy conversion effi-ciencies than prior engines of the industrial revolution (i.e., steam, internal combustion, gas turbines, etc.). It is especially noteworthy to state, as noted in the Preface, that elimination of combustion and heat engines is key to solving society needs for energy supply. While environmental concerns are an important consideration, it is the thermodynamic inefficiency of the combustion or oxidation process that is the real culprit. A fuel cell can con-vert hydrocarbons to electrical energy with nearly an order of magnitude less wasted energy than a gasoline-powered engine. Energy efficiency (or waste heat) should be the prime metric to use in any energy study.

## 1.4    *BATTERY TERMINOLOGY*

While the term "battery" is often used interchangeably with "cell," the basic electrochemical unit is the "cell," and a battery consists of one or more connected cells. Cells may be connected in series or parallel, or both, depending on the desired output voltage and capacity. A *cell* consists of an assembly of electrodes, separators, electrolyte, container, and terminals. A *battery* consists of cells and any control circuitry and other ancillary components (e.g., fuses, diodes), case, terminals, etc. Popular usage considers the "battery" to be the product that is sold or provided to the "user." While the term "cell" may often be used for the single electrochemical unit and "battery" for the finished product, the chapters of this book may use the terms differently depending on the preferences of the chapter writers. Performance characteristics will also vary between single cells and battery packs that must be noted in comparing data.

## 1.5    *BATTERY COMPONENTS*

The cell consists of three major components:

1. *The anode or negative electrode:* the reducing or fuel electrode, which gives up electrons to the external circuit and is oxidized during the electrochemical reaction.

2. *The cathode or positive electrode:* the oxidizing electrode, which accepts electrons from the external circuit and is reduced during the electrochemical reaction.

3. *The electrolyte:* the ionic conductor, which provides the medium for transfer of charged species (ions) inside the cell between electrodes.

The electrolyte is typically a liquid, such as water or other solvents, with dissolved salts, acids, or alkalis to impart ionic conductivity. Some batteries use ionically conductive solids or gel-type polymers. The most advan-tageous combinations of anode and cathode materials are those that will be lightest and give a high cell voltage and capacity. Such combinations may not always be practical, however, due to reactivity with other cell compo-nents, kinetic limitations, processing issues, high cost, and other deficiencies.

Anodes are selected for efficiency as a reducing agent, coulombic output (Ah/g), conductivity (mS/cm), stability, ease of fabrication, and cost. Hydrogen is an exceptional anode material but requires containment vessels or absorption materials (see Chap. 15). Metals are a major category of anode materials. Zinc and lithium have been dominant in primary cells (Chaps. 11–13) because of favorable properties. Lithium, the lightest metal, with a high value of electrochemical equivalence, is a very attractive anode but requires electrolytes and cell designs to accommodate its high reducing potential. Lithium ion anodes have been developed that use lithiated carbons to temper the reactivity of lithium metal (see Chap. 17). The cathode must be an efficient oxidizing agent, stable in contact with the electrolyte, and should have sufficient working voltage. Oxygen is a good cathode material that can be extracted from air for use in zinc/air or similar batteries (Chap. 18). However, most of the common cathode materials are metallic oxides (see Chap. 3A). Other cathode materials, such as the halogens and the oxyhalides, and sulfur and its oxides, are also used for special battery systems.

The electrolyte must have good ionic conductivity but not be electronically conductive, as this would cause internal short-circuiting. Other important characteristics are electrode compatibility, tolerance to temperature variations, safety, and low cost. Traditional electrolytes are aqueous solutions, but thermal and lithium anode batteries use nonaqueous electrolytes due to reactivity of water with the anodes. Anode and cathode electrodes are electronically isolated in the cell to prevent internal short-circuiting. The electrodes are immersed in electrolyte with a permeable, ionically conductive separator material typically used to mechanically separate the electrodes. Various alternative designs do exist and are discussed in individual chapters. Examples include certain liquids that serve as both cathode and electrolyte and, hence, allow direct contact of the active materials.

The cell designs include many variations:

1. Cell shape—cylindrical, button, flat, and prismatic
2. Electrode configurations—series or parallel arrays, bipolar plates, etc.
3. Component modifications tailored to fit cell design or applications
4. Electrically conductive grid structures or additives that reduce internal resistance

The cells are sealed in a variety of ways to prevent leakage and dry-out. Some cells are provided with venting devices or other means to allow accumulated gases to escape. Suitable cases or containers, means for terminal connection, and labeling are added to complete the fabrication of the cell and battery.

## 1.6   CLASSIFICATION OF CELLS AND BATTERIES

Electrochemical cells and batteries are identified as primary (nonrechargeable) or secondary (rechargeable), depending on their capability of being electrically recharged. However, other classifications or subcategories may apply based on particular structures or designs and user preferences.

Also, while batteries and fuel cells both employ redox reactions, capacitors represent another type of electrochemical device, but not one that typically involves chemical reactions. These include electrostatic capacitors (such as the Leyden jar), which are not detailed in this Handbook, as well as electrolytic systems, where ions flow within the electrolyte and accumulate at the electrodes but do not participate in redox reactions. A new type of capacitor, referenced as a hybrid, does have a change in valence state at one of the electrodes but does not undergo permanent oxidation or reduction (see Chap. 20, including discussions on ultra- or super-capacitors).

### 1.6.1   Primary Cells or Batteries

While all electrochemical cells may conceptually be capable of charging after being depleted, there are practical limitations to recharge: activation energy, deleterious side reactions, safety issues, energy efficiency, costs, etc. Hence, batteries that are not suitable for recharge are discharged once and discarded.

The primary battery is a convenient, usually inexpensive, lightweight source of packaged energy for portable electronic and electromechanical devices and a myriad of other applications. The general advantages of primary batteries are good shelf life, high energy density at low to moderate discharge rates, little, if any, maintenance, and ease of use. Although large high-capacity primary batteries are used in military applications, signaling, and standby power, the vast majority of primary batteries are small, single-cell cylindrical cans (AAA-D size), flat button batteries or multicell stacks of these components (9-V transistor radio battery).

## 1.6.2    Secondary or Rechargeable Cells or Batteries

Secondary batteries can be recharged electrically, after discharge, to nearly original condition by passing current through the electrodes in the opposite direction of the discharge. They are storage devices for electric energy and are also known as "storage batteries" or "accumulators." In some cases, though, such as lead acid and lithium-ion cells, the batteries are manufactured in a depleted state and must actually be charged (called formation) before being able to function.

The applications of secondary batteries fall into two main categories:

1. *Energy storage for backup power or occasional use.* Batteries are electrically connected to and kept charged by a prime energy source, delivering energy to the load to supplement the main power system. Examples are automotive (Chap. 14), aircraft (Chap. 25B), emergency and standby power (uninterrupted power supply [UPS]), hybrid electric vehicles (HEVs; Chap. 26A), and electrical grid reserve energy storage systems (Chap. 27).

2. *Primary energy supply.* The secondary battery is used or discharged essentially as a primary battery, but recharged after use rather than being discarded. Examples include portable consumer electronics (Chap. 23), power tools/hand-held equipment (Chap. 24), etc., where life cycle costs are favorable compared to primary cells or where power drains exceed primary battery capabilities. Electric vehicles (EVs), plug-in hybrid electric vehicles (PHEVs; Chap. 26A), and light EVs (Chap. 25A) also fall into this category.

Secondary batteries are often characterized by high-power density/discharge rate, stable voltage levels, and good low-temperature performance. Compared to primary cells, most secondary cells have lower energy density and poorer charge retention. A few special-purpose batteries can be "mechanically" recharged by replacement of the discharged or depleted electrode, usually the metal anode or a liquid/gaseous fuel (see Chaps. 18 and 19).

## 1.6.3    Reserve Batteries

These special-purpose primary cells isolate the active materials or electrolyte from contact until power is needed. Self-discharge is essentially eliminated, and the battery is capable of long-term storage. Thermal batteries store the electrolyte in a frozen state (i.e., a solidified salt) until needed, whereupon, the salt melts and becomes conductive.

The reserve battery designs replace both primary and secondary cells where extreme environments or prolonged storage are encountered, such as in missiles, torpedoes, and other weapon systems (see Chap. 29).

## 1.6.4    Fuel Cells

Fuel cells, like batteries, are electrochemical galvanic cells that convert chemical energy directly into electrical energy and are not subject to the Carnot cycle limitations of heat engines. Fuel cells are similar to batteries except that the active materials are fed into the device from an external source when power is desired. The fuel cell produces electrical energy as long as the active materials are fed to the electrodes.

The electrode materials of the fuel cell do not react during cell operation but function as catalysts in the redox reactions of the active materials. Most fuel cells have gaseous or liquid anode materials (compared to most batteries with metal or solid anodes). The anode materials are more like the conventional fuels used in heat engines; hence, the term "fuel cell" is used. Oxygen or air is the predominant oxidant and is fed into the cathode side of the fuel cell.

Fuel cells have been known for almost two centuries and continue to be of interest as a more efficient and less polluting means for converting hydrogen and carbonaceous or fossil fuels to electricity compared to internal combustion engines. Hydrogen/oxygen fuel cells, using cryogenic fuels, have been used in spacecraft for about 70 years (see Chap. 15F). Fuel cell designs can vary but include two basic types: directly fueled and hydrogen fueled via chemical conversion of hydrocarbons. Fuel cell configurations vary from small portable devices to large power plants. The designs and electrolytes vary (i.e., direct methanol fuel cells, molten carbonate, proton exchange membrane, etc.) and can even include metal air systems where the anodes are replenished regularly. Applications include utility power, load leveling, remote electric generators, EVs, and potential replacements for batteries in consumer electronics (see Chap. 19).

## *1.7   OPERATION OF A CELL*

### 1.7.1   Discharge

A generic discharge reaction is shown schematically in Fig. 1.4. Under electrical load, electrons flow from the anode, which is oxidized, through the external circuit to the cathode, which is reduced. The electric current flow is balanced by the flow of anions (negative ions) and cations (positive ions) to the anode and cathode, respectively, in the electrolyte.

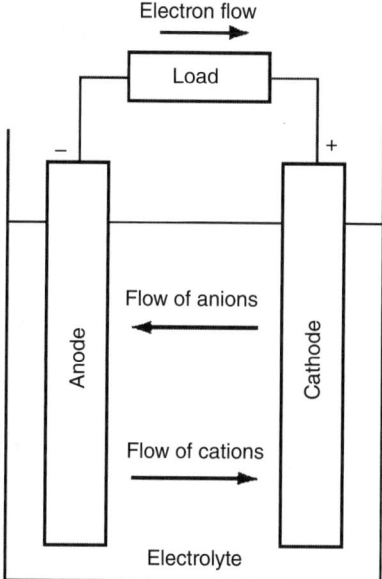

**FIGURE 1.4**   Electrochemical operation of a cell (discharge).

The discharge reaction can be generically written as follows:

*Negative electrode:* anodic reaction (oxidation, loss of electrons)

$$M \rightarrow M^+ + e^-$$

*Positive electrode:* cathodic reaction (reduction, gain of electrons)

$$X + e^- \rightarrow X^-$$

*Overall reaction (discharge):*

$$M + X \rightarrow M^+ + X^- (\rightarrow MX \downarrow)$$

where $M$ is typically a metal and $X$ is an oxidant, such as an oxide or a halogen, which react to form ions and may be a solid compound as shown.

## 1.7.2   Charge

During the recharge, the current flow is reversed and oxidation takes place at the positive electrode and reduction at the negative electrode. By definition, the oxidizing electrode is the anode and the reducing one, the cathode. Hence, the positive electrode is now the anode and the negative the cathode. (Note: To avoid confusion, battery practitioners typically do not switch the electrode names during recharge.)

In an example where zinc and chlorine have reacted to form $ZnCl_2$ discharge product, the reaction on charge of the solvated ions can be written as follows:

*Negative electrode:* cathodic reaction (reduction, gain of electrons)

$$Zn^{2+} + 2e \rightarrow Zn$$

*Positive electrode:* anodic reaction (oxidation, loss of electrons)

$$2Cl^- \rightarrow Cl_2 + 2e$$

*Overall reaction (charge):*

$$Zn^{2+} + 2Cl^- \rightarrow Zn + Cl_2$$

## 1.7.3   Specific Example (Nickel-Cadmium)

The chemical electrode reactions that create electron flow can be illustrated using the nickel-cadmium (NiCd) cell. At the anode (negative electrode), the discharge reaction is the oxidation of cadmium metal to cadmium hydroxide with the release of two electrons:

$$Cd + 2OH^- \rightarrow Cd(OH)_2 + 2e$$

At the cathode, nickel oxide (or more accurately, nickel oxyhydroxide) is reduced to nickel hydroxide with the acceptance of an electron:

$$NiOOH + H_2O + e \rightarrow OH^- + Ni(OH)_2$$

When these two "half-cell" reactions occur (via electron flow between electrodes through the external electrical circuit), the overall cell reaction converts cadmium to cadmium hydroxide at the anode and nickel oxyhydroxide to nickel hydroxide at the cathode:

$$Cd + 2NiOOH + 2H_2O \rightarrow Cd(OH)_2 + 2Ni(OH)_2$$

In primary nonrechargeable cells, the device would be discarded at the end of discharge, but in secondary (rechargeable) system, such as the NiCd cells, the reactions are reversed by electrical recharge. At the negative electrode, the reaction is:

$$Cd(OH)_2 + 2e \rightarrow Cd + 2OH^-$$

At the positive electrode, the reaction is:

$$Ni(OH)_2 + OH^- \rightarrow NiOOH + H_2O + e$$

Each electrochemical couple has its own unique discharge/charge reactions and characteristics. The chapters in Part 4 detail a variety of these variations.

### 1.7.4 Fuel Cell

One of the most basic electrochemical systems is the hydrogen/oxygen fuel cell. Hydrogen is oxidized at the anode and oxygen is reduced at the cathode, both electrocatalyzed with platinum or platinum alloys. The simplified anodic reaction is:

$$2H_2 \rightarrow 4H^+ + 4e$$

while the cathodic reaction is:

$$O_2 + 4H^+ + 4e \rightarrow 2H_2O$$

The overall reaction is the oxidation of hydrogen by oxygen, with water as the reaction product:

$$2H_2 + O_2 \rightarrow 2H_2O$$

This system can then be electrically (charged) or mechanically (replacement of the gases) reactivated (i.e., recharged).

## 1.8  THEORETICAL ELECTROCHEMISTRY FUNDAMENTALS

The theoretical voltage (i.e., the half-cell potentials) and capacity (i.e., the coulombs) of a cell are a function of the anode and cathode materials. Key concepts are introduced next (see Chap. 4 for details).

### 1.8.1 Free Energy

Gibbs free energy is known throughout science as a measure of the spontaneity of a reaction. Whenever a reaction occurs, there is a decrease in the free energy of the system, which is expressed as:

$$\Delta G^0 = -nFE^0$$

where  $F$ = constant known as the Faraday ($\approx$96,500 C or 26.8 Ah)
  $n$ = number of electrons involved in stoichiometric reaction
  $E^0$ = standard potential, V

Electrochemical systems may be analyzed by the use of this equation.

### 1.8.2 Theoretical Voltage

The standard potential (or theoretical voltage) of the cell is determined by the type of active materials and is calculated from free-energy data or obtained experimentally. A listing of electrode potentials (reduction potentials) under standard conditions is given in Table 1.1. A more complete list is presented in App. B.

**TABLE 1.1**    Characteristics of Typical Electrode Materials[*]

| Material | Atomic or molecular weight, g | Standard reduction potential at 25°C, V | Valence change | Melting point, °C | Density, g/cm$^3$ | Electrochemical equivalents | | |
|---|---|---|---|---|---|---|---|---|
| | | | | | | Ah/g | g/Ah | Ah/cm$^3$ |
| Anode materials | | | | | | | | |
| H$_2$ | 2.01 | 0 | 2 | — | — | 26.59 | 0.037 | — |
| | | −0.83[†] | | | | | | |
| Li | 6.94 | −3.01 | 1 | 180 | 0.54 | 3.86 | 0.259 | 2.06 |
| Na | 23.0 | −2.71 | 1 | 98 | 0.97 | 1.16 | 0.858 | 1.14 |
| Mg | 24.3 | −2.38 | 2 | 650 | 1.74 | 2.20 | 0.454 | 3.8 |
| | | −2.69[†] | | | | | | |
| Al | 26.9 | −1.66 | 3 | 659 | 2.69 | 2.98 | 0.335 | 8.1 |
| Ca | 40.1 | −2.84 | 2 | 851 | 1.54 | 1.34 | 0.748 | 2.06 |
| | | −2.35[†] | | | | | | |
| Fe | 55.8 | −0.44 | 2 | 1528 | 7.85 | 0.96 | 1.04 | 7.5 |
| | | −0.88[†] | | | | | | |
| Zn | 65.4 | −0.76 | 2 | 419 | 7.14 | 0.82 | 1.22 | 5.8 |
| | | −1.25[†] | | | | | | |
| Cd | 112.4 | −0.40 | 2 | 321 | 8.65 | 0.48 | 2.10 | 4.1 |
| | | −0.81[†] | | | | | | |
| Pb | 207.2 | −0.13 | 2 | 327 | 11.34 | 0.26 | 3.87 | 2.9 |
| (Li)C$_6$[§] | 72.06 | ~−2.8 | 1 | — | 2.25 | 0.372 | 2.69 | 0.837 |
| MH[¶] | | −0.83[†] | 2 | — | — | 0.305 | 3.28 | — |
| CH$_3$OH | 32.04 | — | 6 | — | — | 5.02 | 0.20 | — |
| Cathode materials | | | | | | | | |
| CuF$_2$ | 101.5 | 3.55 | 2 | | | 0.528 | 1.89 | |
| O$_2$ | 32.0 | 1.23 | 4 | — | — | 3.35 | 0.30 | |
| | | 0.40[†] | | | | | | |
| Cl$_2$ | 71.0 | 1.36 | 2 | — | — | 0.756 | 1.32 | |
| SO$_2$ | 64.0 | — | 1 | — | — | 0.419 | 2.38 | |
| MnO$_2$ | 86.9 | 1.28[‡] | 1 | — | 5.0 | 0.308 | 3.24 | 1.54 |
| NiOOH | 91.7 | 0.49 | 1 | — | 7.4 | 0.292 | 3.42 | 2.16 |
| CuCl | 99.0 | 0.14 | 1 | — | 3.5 | 0.270 | 3.69 | 0.95 |
| FeS$_2$ | 119.9 | — | 4 | — | — | 0.89 | 1.12 | 4.35 |
| AgO | 123.8 | 0.57[†] | 2 | — | 7.4 | 0.432 | 2.31 | 3.20 |
| Br$_2$ | 159.8 | 1.07 | 2 | — | — | 0.335 | 2.98 | |
| HgO | 216.6 | 0.10[†] | 2 | — | 11.1 | 0.247 | 4.05 | 2.74 |
| Ag$_2$O | 231.7 | 0.35[†] | 2 | — | 7.1 | 0.231 | 4.33 | 1.64 |
| PbO$_2$ | 239.2 | 1.69 | 2 | — | 9.4 | 0.224 | 4.45 | 2.11 |
| LiFePO$_4$ | 163.8 | ~0.42 | 1 | — | 3.44 | 0.160 | 6.25 | 0.554 |
| LiMn$_2$O$_4$ (spinel) | 148.8 | ~1.2 | 1 | — | 4.1 | 0.120 | 8.33 | 0.492 |
| Li$_x$CoO$_2$ | 98 | ~1.25 | 0.5 | — | 5.05 | 0.155 | 6.45 | 0.782 |
| I$_2$ | 253.8 | 0.54 | 2 | — | 4.94 | 0.211 | 4.73 | 1.04 |

[*]See also Apps. B and C.
[†]Basic electrolyte; all others, aqueous acid or nonaqueous electrolytes.
[‡]Based on density values shown.
[§]Calculations based only on weight of carbon.
[¶]Based on type AB$_5$ alloy.

The standard potential of a cell can be calculated from the standard electrode potentials as follows (the oxidation potential is the negative value of the reduction potential):

Anode (oxidation potential) + cathode (reduction potential) = standard cell potential

For example, in the reaction $Zn + Cl_2 \rightarrow ZnCl_2$, the standard cell potential is:

$$
\begin{array}{ll}
Zn \rightarrow Zn^2 + 2e & -(-0.76\ V) \\
Cl_2 \rightarrow 2Cl^- - 2e & \underline{\quad 1.36\ V} \\
E^O = & \quad 2.12\ V
\end{array}
$$

The cell voltage is also dependent on other factors, including concentration and temperature, as expressed by the Nernst equation (also see Chap. 4).

### 1.8.3 Theoretical Capacity (Coulombs of Stored/Generated Electricity)

The theoretical capacity of a cell is determined by the type and amount of active materials in the cell and is expressed in coulombs or ampere-hours (Ah) as the total quantity of electricity derived from the electrochemical reactions. Battery capacity is based on the ability of one (1) gram-equivalent weight of material to deliver 96,487 C or 26.8 Ah of electricity for each valence state change. (Gram-equivalent weight is the atomic or molecular weight of the active material in grams divided by the number of electrons, valence changes, involved in the reaction.) Electrochemical equivalence values are listed in Table 1.1 and App. C.

The theoretical capacity of an electrochemical cell, based only on the active materials participating in the electrochemical reaction, is calculated from the equivalent weight of the reactants. Hence, the theoretical capacity of the $Zn/Cl_2$ cell is 0.394 Ah/g, as follows:

$$
\begin{array}{llll}
Zn & + & Cl_2 & \longrightarrow & ZnCl_2 \\
(0.82\ Ah/g) & & (0.76\ Ah/g) & & \\
\end{array}
$$
1.22 g/Ah  +  1.32 g/Ah = 2.54 g/Ah or 0.394 Ah/g

Similarly, the ampere-hour capacity on a volume basis can be calculated using the appropriate data for ampere-hours per cubic centimeter as listed in Table 1.1.

The theoretical voltages and capacities of a number of the major electrochemical systems are given in Table 1.2. These theoretical values are based on active anode and cathode materials only. Water, electrolyte, or any other materials that may be involved in the cell reaction are not included in the calculation.

### 1.8.4 Theoretical Energy[a]

Cell output can also be considered on an energy (i.e., watt-hour) basis by factoring in the voltage level at which a quantity of electricity is provided as follows:

Watt-hour (Wh) = voltage (V) × ampere-hour (Ah)

---

[a]The energy output of a cell or battery is often expressed as a ratio of its weight or size. The preferred terminology for this ratio on a weight basis, e.g., watt-hours/kilogram (Wh/kg), is "specific energy." On a volume basis, e.g., watt-hours/liter (Wh/L), it is "energy density." Commonly, the term "energy density" may be used to refer to either ratio but should still specify a weight or volume basis.

**TABLE 1.2** Voltage, Capacity, and Specific Energy of Major Battery Systems—Theoretical and Practical Values

| Battery type | Anode | Cathode | Reaction mechanism | Theoretical values[a] | | | | | Practical battery[b] | |
|---|---|---|---|---|---|---|---|---|---|---|
| | | | | V | g/Ah | Ah/kg | Specific energy Wh/kg | Nominal voltage V | Specific energy Wh/kg | Energy density Wh/L |
| *Primary batteries* | | | | | | | | | | |
| Leclanché | Zn | $MnO_2$ | $Zn + 2MnO_2 \rightarrow ZnO \cdot Mn_2O_3$ | 1.6 | 4.46 | 224 | 358 | 1.5 | 85[f] | 165[f] |
| Magnesium | Mg | $MnO_2$ | $Mg + 2MnO_2 + H_2O \rightarrow Mn_2O_3 + Mg(OH)_2$ | 2.8 | 3.69 | 271 | 759 | 1.7 | 100[f] | 195[f] |
| Alkaline $MnO_2$ | Zn | $MnO_2$ | $Zn + 2MnO_2 \rightarrow ZnO + Mn_2O_3$ | 1.5 | 4.46 | 224 | 358 | 1.5 | 154[f] | 461[f] |
| Mercury | Zn | HgO | $Zn + HgO \rightarrow ZnO + Hg$ | 1.34 | 5.27 | 190 | 255 | 1.35 | 100[h] | 470[h] |
| Mercad | Cd | HgO | $Cd + HgO + H_2O \rightarrow Cd(OH)_2 + Hg$ | 0.91 | 6.15 | 163 | 148 | 0.9 | 55[h] | 230[h] |
| Silver oxide | Zn | $Ag_2O$ | $Zn + Ag_2O + H_2O \rightarrow Zn(OH)_2 + 2Ag$ | 1.6 | 5.55 | 180 | 288 | 1.6 | 135[h] | 525[h] |
| Zinc/$O_2$ | Zn | $O_2$ | $Zn + \frac{1}{2}O_2 \rightarrow ZnO$ | 1.65 | 1.52 | 658 | 1085 | — | — | — |
| Zinc/air | Zn | Ambient air | $Zn + (\frac{1}{2} O_2) \rightarrow ZnO$ | 1.65 | 1.22 | 820 | 1353 | 1.5 | 415[h] | 1350[h] |
| $Li/SOCl_2$ | Li | $SOCl_2$ | $4Li + 2SOCl_2 \rightarrow 4LiCl + S + SO_2$ | 3.65 | 3.25 | 403 | 1471 | 3.6 | 590[f] | 1100[f] |
| $Li/SO_2$ | Li | $SO_2$ | $2Li + 2SO_2 \rightarrow Li_2S_2O_4$ | 3.1 | 2.64 | 379 | 1175 | 3.0 | 260[g] | 415[g] |
| $LiMnO_2$ | Li | $MnO_2$ | $Li + Mn^{IV}O_2 \rightarrow Mn^{IV}O_2(Li^+)$ | 3.5 | 3.50 | 286 | 1001 | 3.0 | 260[g] | 546[g] |
| $Li/FeS_2$ | Li | $FeS_2$ | $4Li + FeS_2 \rightarrow 2Li_2S + Fe$ | 1.8 | 1.38 | 726 | 1307 | 1.5 | 310[g] | 560[g] |
| $Li/CF_x$ | Li | $CF_x$ | $xLi + CF_x \rightarrow xLiF + xC$ | 3.1 | 1.42 | 706 | 2189 | 3.0 | 360[g] | 540[g] |
| $Li/I_2$[e] | Li | $I_2(P2VP)$ | $Li + \frac{1}{2}I_2 \rightarrow LiI$ | 2.8 | 4.99 | 200 | 560 | 2.8 | 245 | 900 |
| *Reserve batteries* | | | | | | | | | | |
| Cuprous chloride | Mg | CuCl | $Mg + Cu_2Cl_2 \rightarrow MgCl_2 + 2Cu$ | 1.6 | 4.14 | 241 | 386 | 1.3 | 60[i] | 80[i] |
| Zinc/silver oxide | Zn | AgO | $Zn + AgO + H_2O \rightarrow Zn(OH)_2 + Hg$ | 1.81 | 3.53 | 283 | 512 | 1.5 | 30[j] | 75[j] |
| Thermal[d] | Li | $FeS_2$ | See Section 36.3.1 | 2.1–1.6 | 1.38 | 726 | 1307 | 2.1–1.6 | 40[k] | 100[k] |
| *Secondary batteries* | | | | | | | | | | |
| Lead-acid | Pb | $PbO_2$ | $Pb + PbO_2 + 2H_2SO_4 \rightarrow 2PbSO_4 + 2H_2O$ | 2.1 | 8.32 | 120 | 252 | 2.0 | 35 | 70[l] |
| Edison | Fe | Ni oxide | $Fe + 2NiOOH + 2H_2O \rightarrow 2Ni(OH)_2 + Fe(OH)_2$ | 1.4 | 4.46 | 224 | 314 | 1.2 | 30 | 55[l] |
| Nickel-cadmium | Cd | Ni oxide | $Cd + 2NiOOH + 2H_2O \rightarrow 2Ni(OH)_2 + Cd(OH)_2$ | 1.35 | 5.52 | 181 | 244 | 1.2 | 40 | 135[g] |
| Nickel-zinc | Zn | Ni oxide | $Zn + 2NiOOH + 2H_2O \rightarrow 2Ni(OH)_2 + Zn(OH)_2$ | 1.73 | 4.64 | 215 | 372 | 1.6 | 90 | 185 |
| Nickel-hydrogen | $H_2$ | Ni oxide | $H_2 + 2NiOOH \rightarrow 2Ni(OH)_2$ | 1.5 | 3.46 | 289 | 434 | 1.2 | 55 | 60 |
| Nickel-metal hydride | MH[c] | Ni oxide | $MH + NiOOH \rightarrow M + Ni(OH)_2$ | 1.35 | 5.63 | 178 | 240 | 1.2 | 100 | 235[g] |

| Battery | Anode | Cathode | Reaction | | | | | | | |
|---|---|---|---|---|---|---|---|---|---|---|
| Silver-zinc | Zn | AgO | $Zn + AgO + H_2O \rightarrow Zn(OH)_2 + Ag$ | 1.85 | 3.53 | 283 | 524 | 1.5 | 105 | 180[l] |
| Silver-cadmium | Cd | AgO | $Cd + AgO + H_2O \rightarrow Cd(OH)_2 + Ag$ | 1.4 | 4.41 | 227 | 318 | 1.1 | 70 | 120[l] |
| Zinc/chlorine | Zn | $Cl_2$ | $Zn + Cl_2 \rightarrow ZnCl_2$ | 2.12 | 2.54 | 394 | 835 | — | — | — |
| Zinc/bromine | Zn | $Br_2$ | $Zn + Br_2 \rightarrow ZnBr_2$ | 1.85 | 4.17 | 309 | 572 | 1.6 | 70 | 60 |
| Lithium-ion | $Li_xC_6$ | $Li_{(1-x)}CoO_2$ | $Li_xC_6 + Li_{(1-x)}CoO_2 \rightarrow LiCoO_2 + C_6$ | 4.1 | 9.14 | 109 | 448 | 3.8 | 200 | 570[g] |
| Lithium/manganese dioxide | Li | $MnO_2$ | $Li + Mn^{IV}O_2 \rightarrow Mn^{IV}O_2(Li^+)$ | 3.5 | 3.50 | 286 | 1001 | 3.0 | 120 | 265 |
| Lithium/iron disulfide[d] | Li(Al) | $FeS_2$ | $2Li(Al) + FeS_2 \rightarrow Li_2FeS_2 + 2Al$ | 1.73 | 3.50 | 285 | 493 | 1.7 | 180[m] | 350[m] |
| Sodium/sulfur[d] | Na | S | $2Na + 3S \rightarrow Na_2S_3$ | 2.1 | 2.65 | 377 | 792 | 2.0 | 170[m] | 345[m] |
| Sodium/nickel chloride[d] | Na | $NiCl_2$ | $2Na + NiCl_2 \rightarrow 2NaCl + Ni$ | 2.58 | 3.28 | 305 | 787 | 2.6 | 115[m] | 190[m] |
| Fuel cells | | | | | | | | | | |
| $H_2/O_2$ | $H_2$ | $O_2$ | $H_2 + \frac{1}{2}O_2 \rightarrow H_2O$ | 1.23 | 0.336 | 2975 | 3660 | | | |
| $H_2$/air | $H_2$ | Ambient air | $H_2 + (\frac{1}{2}O_2) \rightarrow H_2O$ | 1.23 | 0.037 | 26,587 | 32,702 | | | |
| Methanol/$O_2$ | $CH_3OH$ | $O_2$ | $CH_3OH + {}^3/_2 O_2 \rightarrow CO_2 + 2H_2O$ | 1.24 | 0.50 | 2000 | 2480 | — | — | — |
| Methanol/air | $CH_3OH$ | Ambient air | $CH_3OH + ({}^3/_2 O_2) \rightarrow CO_2 + 2H_2O$ | 1.24 | 0.20 | 5020 | 6225 | — | — | — |

[a]Based on active anode and cathode materials only, including $O_2$ but not air (electrolyte not included).

[b]These values are for single-cell batteries based on identified design and at discharge rates optimized for energy density, using midpoint voltage. More specific values are given in chapters on each battery system.

[c]MH = metal hydride, data based on type $AB_5$ alloy.

[d]High temperature batteries.

[e]Solid electrolyte battery ($Li/I_2$ [P2VP]).

[f]Cylindrical bobbin-type batteries.

[g]Cylindrical spiral-wound batteries.

[h]Button type batteries.

[i]Water-activated.

[j]Automatically activated 2- to 10-min rate.

[k]With lithium anodes.

[l]Prismatic batteries.

[m]Value based on cell performance, see appropriate chapter for details.

This energy value is the maximum that can be delivered by a specific electrochemical system since it is based on theoretical voltages, which are always greater than actual discharge voltages. In the Zn/Cl$_2$ cell example, if the standard potential is taken as 2.12 V, the theoretical watt-hour capacity per gram of active material (theoretical gravimetric specific energy or theoretical gravimetric energy density) is

$$\text{Specific energy (watt-hours/gram)} = 2.12 \text{ V} \times 0.394 \text{ Ah/g} = 0.835 \text{ Wh/g or 835 Wh/kg}$$

Table 1.2 also lists the theoretical specific energy of various electrochemical systems.

## 1.9    SPECIFIC ENERGY AND ENERGY DENSITY OF PRACTICAL BATTERIES

In summary, the maximum energy that can be delivered by an electrochemical system is based on the composition of active materials that are used, which determines voltage, and on the amount of the active materials that are used, which further determines ampere-hour capacity. In practice, only a fraction of the theoretical energy of the battery is realized, and this output is further diluted on an energy density basis by electrolyte and nonreactive components (containers, separators, electrodes), as illustrated in Fig. 1.5. This figure, which was modeled after an assumed (i.e., nonexistent) lithium-ion cell design, shows that active materials consume a relatively small fraction of the total battery and do not even contribute to the majority of cell costs. Clearly, then, cell designs are dominated by the peripheral parts, which decrease specific energy and increase costs. Also, the average cell discharge voltage is less than theoretical and a cell is never, practically, discharged to zero. Active materials in a practical battery are usually not stoichiometrically balanced, creating an excess amount of one of the active materials, which further impacts the delivered specific energy/energy density.

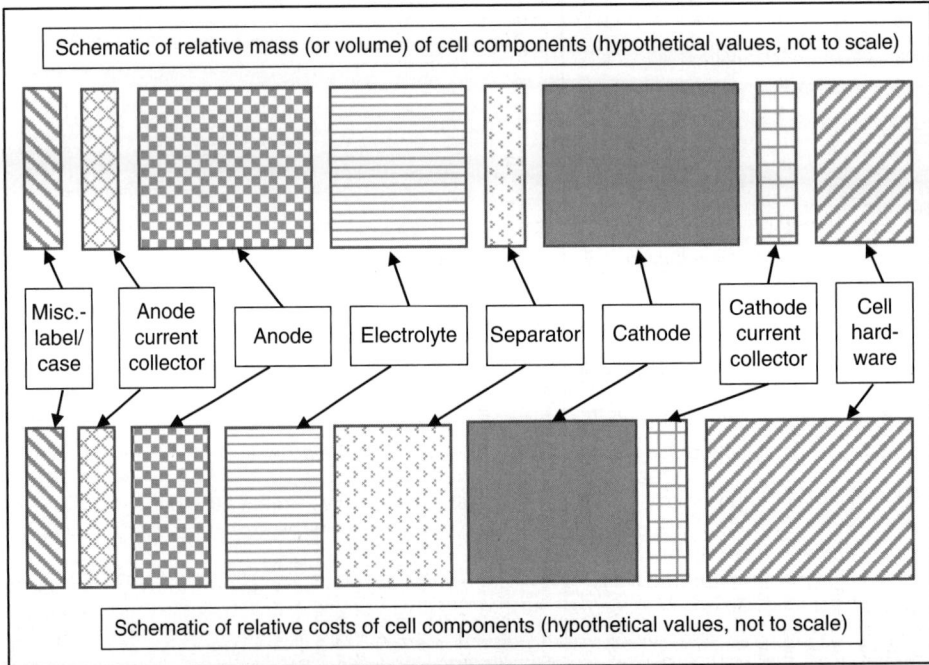

**FIGURE 1.5**    Illustrative example of cell component building blocks.

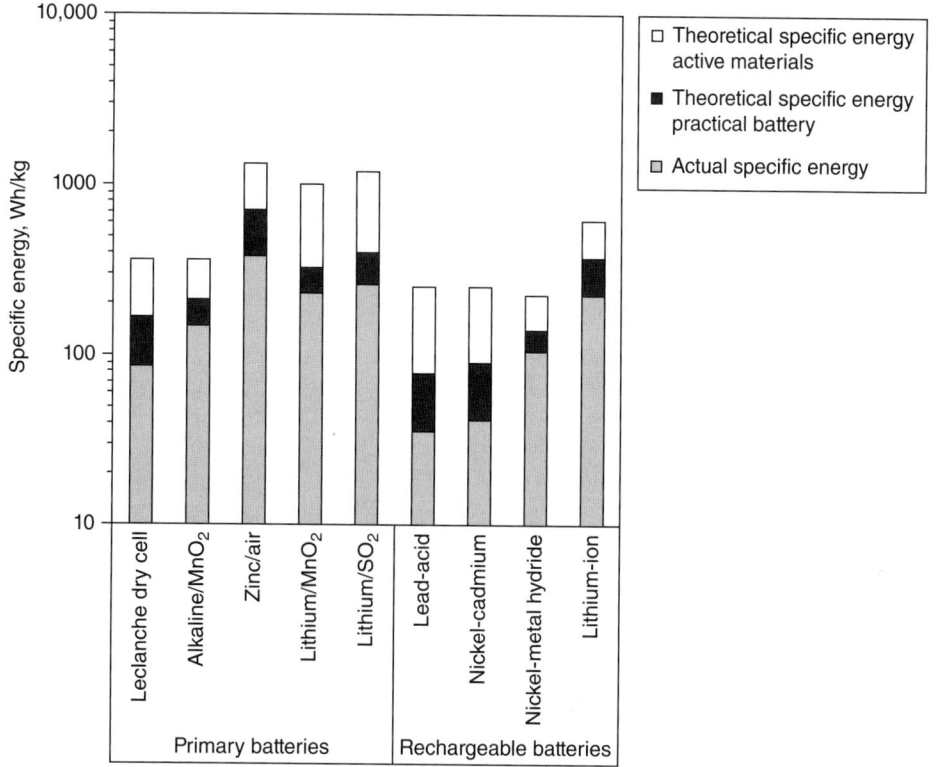

**FIGURE 1.6**    Theoretical and actual specific energy of battery systems.

In Fig. 1.6, rating factors, as follows, are plotted for various batteries:

1. The theoretical specific energy (based on the active anode and cathode materials only)
2. The theoretical specific energy of a practical battery (accounting for the electrolyte and nonreactive components)
3. The actual specific energy of these batteries when discharged at 20°C under optimal discharge conditions

These data show the following energy density deratings:

- A 50% reduction due to materials of construction
- A 50 to 75% reduction due to actual cell test inefficiencies (even under ideal conditions)

Thus, the actual energy that is available from a battery under even benign discharge conditions is only about 25 to 35% of the theoretical energy of the active materials. Chapter 5 covers the performance of batteries when used under more stringent conditions.

Table 1.2 summarizes these data along with the characteristics and performance of practical batteries. The specific energy (Wh/kg) and energy density (Wh/L) delivered by the major battery systems are also plotted in Fig. 1.7a for primary batteries and 1.7b for rechargeable batteries. In these figures, the energy storage capability is shown as a field, rather than as a single optimum value, to illustrate the spread in performance of that battery system under different conditions of use.

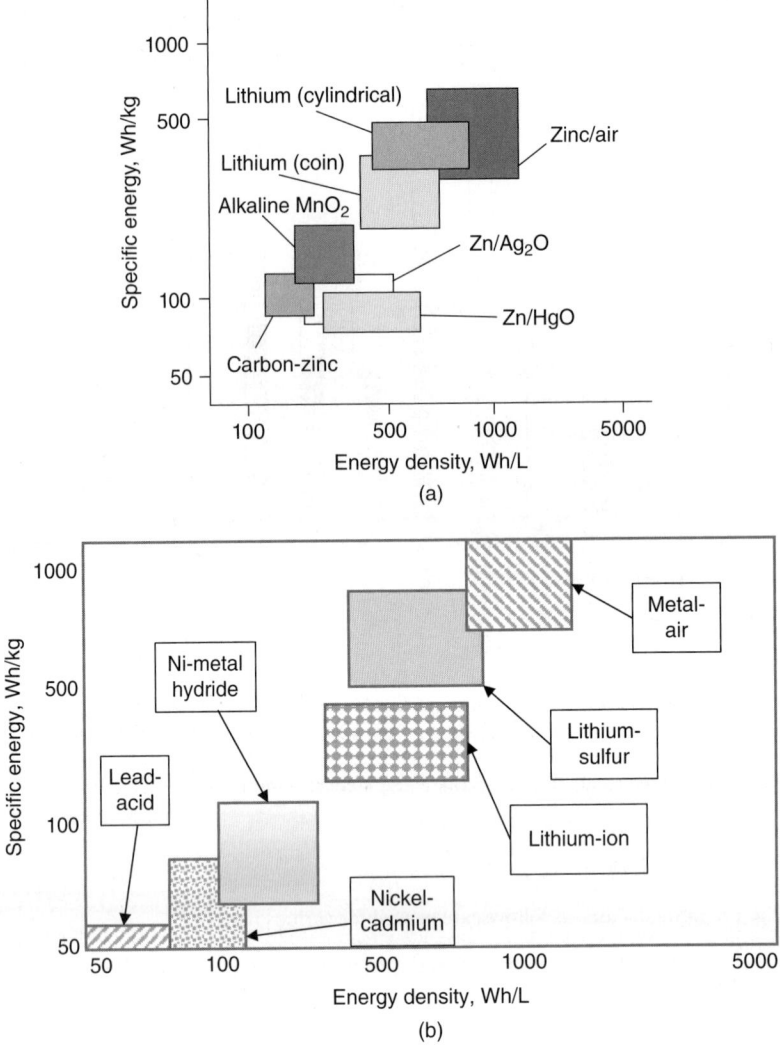

**FIGURE 1.7**   Comparison of the energy storage capability of various battery systems: (*a*) primary batteries, (*b*) secondary batteries.

## 1.10   *LIMITS OF SPECIFIC ENERGY AND ENERGY DENSITY*

Many advances have been made in battery technology over the years, as illustrated in Fig. 1.8, both through continued improvement of a specific electrochemical system and through the development and introduction of new battery chemistries. However, batteries, unlike electronic devices, consume materials when delivering electrical energy and, as discussed in Secs. 1.8 and 1.9, there are theoretical limits to the amount of electrical energy that can be delivered electrochemically by the available materials.

As shown in Table 1.2, except for some of the ambient air-breathing systems and the hydrogen/oxygen fuel cell, where the weight of the cathode active material is not included in the calculation, the values for the

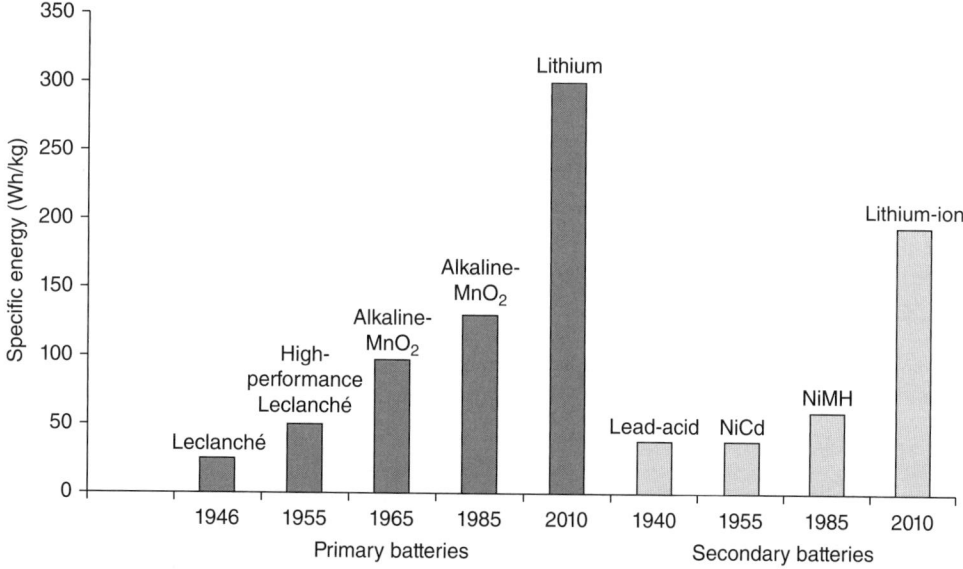

**FIGURE 1.8**    Advances in battery performance for portable applications.

theoretical energy density do not exceed 1500 Wh/kg. Even the values for hydrogen/air and liquid fuel cells are impacted since the weight and volume of suitable containers for these fuels must be included.

The data in Table 1.2 also show that the specific energy delivered by these batteries, based on the actual performance when discharged under optimum conditions, does not exceed 600 Wh/kg or 1300 Wh/L energy density. Rechargeable systems are more negatively impacted than primary batteries, due in part to a more limited selection of materials that can be recharged practically and the need for designs to facilitate recharging and cycle life.

## 1.11  MARKET TRENDS

The size and diversity of the battery industry are massive and growing. The supply of energy by conventional sources is simply inadequate to meet consumer expectations. The evolving statistics for specific electrochemical systems will be addressed in other chapters of this Handbook, but Table 1.3 presents an overview of the markets as a whole, based on recent analyses by The Freedonia Group.

**TABLE 1.3**    Battery Supply and Demand (million dollars)[*]

| Item | 2005 | 2010 | 2015 | 2020 | 2025 |
|---|---|---|---|---|---|
| Gross domestic product (bil $) | 13,094 | 14,964 | 18,037 | 22,200 | 27,500 |
| $ batteries/000$ GDP | 0.66 | 0.77 | 0.79 | 0.80 | 0.79 |
| Battery demand | 8590 | 11,450 | 14,200 | 17,650 | 21,750 |
| Single-use | 2895 | 3300 | 3250 | 3500 | 3750 |
| Rechargeable | 5695 | 8150 | 10,950 | 14,150 | 18,000 |
| −Net imports | −1185 | −950 | −1350 | −1000 | −250 |
| Battery shipments | 7405 | 10,500 | 12,850 | 16,650 | 21,500 |
| Price deflator (2009 = 100) | 76.8 | 102.6 | 107.0 | 116.3 | 126.2 |
| Battery shipments (mil 2009$) | 9642 | 10,234 | 12,009 | 14,316 | 17,036 |

[*]Table III-1, Battery Markets in the U.S. by Application and End User, published December 2016 by The Freedonia Group.

Table 1.3 reveals that batteries will have grown from less than 0.066% of the U.S. economy 15 years ago to nearly 0.08% of the gross domestic product (GDP) within 10 years—a remarkable increase, considering that rapid growth in other parts of the economy dilutes the growth of batteries. GDP nearly doubles over this 20-year period while batteries grow 2½ times with secondary batteries increasing threefold. Two other key facts are that primary cells are basically flat and that imported batteries are projected to nearly disappear (down >80% from a 2015 peak).

## 1.12   CASE STUDY: BATTERY-POWERED RIVER BARGE FOR HAULING COAL

As an example of the perspectives necessary to search out and implement new battery technologies in this new era of heightened energy awareness, a case study warrants consideration. Specifically, the development and commissioning of a battery-powered river barge was recently announced in China.[b] The use of electricity, rather than diesel fuel, to deliver cargo is an accomplishment that is certainly commendable and seemingly worthy of implementation. However, one critical circumstance is that the barge will be used to ship coal from the mine to the power plant down river. Examining this situation based on thermodynamic calculations brings up a major potential issue: What is the overall efficiency of this battery power barge system, as opposed to the alternative options? Locating the power plant at the mine is a typical practice, but one that is not always possible. Certainly, solar, wind, or even river currents would be more efficient (i.e., preferable) energy sources, but they are not very reliable. Other critics might say it is best to eliminate coal-fired power plants altogether and switch to solar or hydroelectric power. These options are certainly desirable, but not always practical.

Hence, assuming coal-fired power plants are still necessary to generate electricity in the near term, what is the best method for shipping coal? A complete thermodynamic efficiency analysis should compare the use of diesel engines (or diesel-electric power) to the use of batteries. There are two basic areas to consider in this analysis of the use of batteries:

1. The use of resources to produce the barge batteries (ideally prorated over battery life)

2. The use of electricity to recharge the batteries each trip

In the first category, electricity and other forms of energy, including human power, are used to mine the battery materials, construct the battery factory, and build each battery module and all related equipment. Additionally, energy is also expended to mine and ship the coal that is used to generate the electricity to produce the batteries. How this energy is derived and where it is used are not trivial issues. The energy consumed and the efficiencies (i.e., waste heat) for this entire portfolio of equipment, operating over the full battery product life cycle, deserve proper thermodynamic analyses (and comparison to fossil fuels) if the correct decision on barge power is to be made.

For instance, if not for the pollution, would it perhaps be preferable (more energy efficient) to simply use a coal-fired steam boiler on the barge? The coal mine then provides all the power needed to ship the coal. No resources to build the battery are consumed, no battery electrical storage is required, and no electrical energy generation is needed for either of these two tasks.

However, the heat (i.e., combustion) engine is, as a whole, the most inefficient power source available. Therefore, a complete analysis of energy efficiency of all potential power sources for the river barge is the only truly viable approach to making the proper selection. These analyses will ideally include current technologies as well as new or evolving power systems. Battery power may or may not be the right solution for this application. Once the thermodynamic calculations are completed and after other peripheral issues are considered (capital costs, operating costs, pollution, performance capabilities, etc.), then (and only then) can an informed decision be made.

Ultimately, though, the best choice for this situation (i.e., coal shipment from the mine to the power plant) seems fairly logical: Use fuel cells (or redox flow cells) based on new technology that can accept pulverized coal slurries as the fuel and air as the oxidant. Such electrochemical systems could have extremely high thermodynamic efficiencies and zero emissions (no particulates, no sulfurous or nitrous compounds, and no carbon dioxide). A fuel cell system can be engineered for long life with minimal use of strategic materials. The by-products, ash and $CO_2$, would be contained and easily redirected to other useful applications.

---

[b]https://cleantechnica.com/2017/12/02/china-launches-worlds-first-electric-cargo-ship-will-use-haul-coal/ (extracted from web 12/2/2017).

## 1.13   A SYSTEM OF SYSTEMS

New battery systems with significantly higher energy output and lower costs will be increasingly difficult to engineer and commercialize at this nexus of market forces and technical and financial barriers. The major goals for the near term will likely center on materials availability, cost, safety, and environmental acceptability.

Perfecting the use of lithium metal- or silicon-based anodes, enhancing cathode capabilities to match the anodes, improving the ratio of active to inactive components, increasing conversion efficiency and recharging ability, maximizing performance under more stringent operating conditions, and enhancing safety will be subjects of much R&D in the future. Alternative electrochemistry and components, such as fuel cells, redox flow cells, solid-state and hybrid electrolytes, wireless charging, etc., also offer opportunities for improving the state-of-the-art technology.

However, beyond such technical issues, attention needs to be paid to business models for battery development and equipment deployment. Intellectual property issues, supply chain complexities (including the need for recycling), and manufacturing strategies will play a major role in the future of the world's energy supply. Clearly, batteries are a system of systems, in which overall success will stem from unprecedented levels of cooperation and communication, just as was the case with Franklin and Volta at the dawn of the era of electrochemistry.

## ACKNOWLEDGMENT

Table 1.3 is provided by The Freedonia Group.

## REFERENCES

1. D. Linden and T. B. Reddy, *Battery Power and Products Technology*, vol. 5, no. 2, pp. 10–12, March/April 2008.
2. M. Winter and R. Brodd, *Chemical Reviews*, vol. 104, 4245–4270, 2004.

# CHAPTER 2
# RAW MATERIALS

## 2.0  OVERVIEW

The first step in establishing a viable battery industry, beyond the identification of a suitable electrochemistry, is sourcing of the raw materials. While the raw materials must be incorporated into batteries as engineered components, as detailed in Chap. 3, technical and commercial success will not be achieved unless a reliable, cost-efficient supply chain of raw materials is first determined. The primary categories of materials used in typical electrochemical devices are listed below and are further detailed in the referenced sections of this chapter.

Chapter 2A: Active Materials

Chapter 2B: Metals and Minerals

Chapter 2C: Polymers and Organic Materials

Chapter 2D: Ceramics and Inorganics

Chapter 2E: Carbon and Graphite

Many battery designs exist that use a wide variety of materials, but this list details the mainstream battery material chain of supply. By analyzing the supply and costs of raw materials used in various electrochemical devices, a better understanding of the potential long-term success of specific battery designs can be better projected. A recent paper detailed some of the supply and cost issues associated with various raw materials.[1]

Note that the active materials (i.e., as listed in Chap. 2A) depend on the supply of a variety of precursors, such as elemental metals, metal oxides, and carbon/graphite, which are obtained from independent sources as detailed in Chaps. 2B, 2D, and 2E. These precursor materials depend overwhelmingly on, and derive from, mining or other earth or brine extraction and refining processes. Once the metals, oxides, caustic/acid solutions, salts, etc., are produced, the active materials listed in Chap. 2A are then synthesized by various industrial chemical processing operations.

While polymer production may use some metals, inorganics, or ceramics (i.e., catalysts, etc.), the main source of raw materials is, of course, the petrochemical industry. Natural gas is a prime material source for manufacture of most commercial plastics. Recent, large shale gas discoveries have buoyed the prospects for sustainable supply of economical polymer precursor material. Petrochemicals are also, of course, the main source for organic solvents, a key raw material for many nonaqueous battery electrolytes and a key material used in the processing of electrodes and other solvent-based manufacturing technologies (i.e., lithium cell recycling, etc.).

In summary, the cost and supply of raw materials is a good starting point for the analysis of the battery industry. Safety, toxicity, and recycling ability are the other main concerns.

### Reference

1. A. Chagnes, Challenges for the development of sustainable lithium-ion batteries, *35th Annual International Battery Seminar and Exhibit*, Ft. Lauderdale, FL, March 26–29, 2018.

# SECTION A

# ACTIVE MATERIALS

**Ron Turi**

## 2A.1    INTRODUCTION

Electrochemical energy devices have negative (electron donor) and positive (electron receptor) electrodes when discharged in a typical reduction-oxidation reaction. The negative battery electrode (anode) contains a negative active material (NAM) for the reduction reaction, and the positive electrode (cathode) contains a positive active material (PAM) for the oxidation reaction. Anodes are often metals while cathode materials are often oxides of reducible metals or oxidizing nonmetals. Recent focus has been on active material development for rechargeable batteries.

### 2A.1.1    Properties of Battery Active Materials Are Contextual

Specific capacity, a measure of charge that a material can hold per unit mass, often expressed in units of mAh/g, is a key. This theoretical metric is an intrinsic property of a material based on chemical formula and the Faraday constant. However, capacity may be limited by the extractable range. Hence, specific capacity may be less than theoretical since the remaining charge is not accessed. Additionally, high internal resistance depresses output capacity on discharge and prevents full recharge, thereby causing a lower, observed specific capacity value.

Also, in many cases, active material capacity exceeds a cell's nominal design capacity to ensure that the capacity of one electrode is in excess—perhaps for safety concerns or to counteract depletion of charge as the cell ages. Additionally, if a cell is over-discharged, the loss of ions may cause a physical instability of an active material, leading to loss of capacity. Evaluating the specific capacity of active materials is system and end-use dependent.

### 2A.1.2    Cell Energy Properties Are Also Contextual

The contribution of active materials to cell energy density (Wh/L) and specific energy (Wh/kg) depends on total system context—average cell voltage, which depends on cell operating voltage range and rate of charge/discharge. Materials have relative potentials to each other, based on half-cell potentials, which determine cell operating voltage. Energy density also depends on specific gravity and bulk density of active materials in practical electrodes. For example, an active material in powder form has an effective (bulk) density proportional to the packing density. Some compressed powders have low void volume and low binder volume fractions, while others need more porosity or more binder optimal electrode properties.

The electronic and ionic conductivity of active materials contributes to power and capacity but infrequently possesses superior levels of both properties. However, conductivities can be controlled by extrinsic factors such as particle size, calendering, and electrode thickness. Also, high-power batteries often incorporate additive materials such as carbon black/nanotubes and graphene to boost electronic conductivity.

The reversibility of active materials in rechargeable cells is another contextual property. Active materials undergo structural changes when charged/discharged. The degree of reversibility depends on extensive factors—often including particle size, mechanical support from interstitial binders, the disruption and ability to repair any passivation layers, and the disruption and ability to repair any networks of electronic conductors. Also, exceeding optimal charge and discharge rates may cause stresses to the active materials and mechanical breakdowns that shorten cell cycle life. Environmental factors—temperature extremes, vibration/shock—also tend to accelerate the capacity loss of active materials either by direct degradation or by side reactions.

## 2A.2    *LITHIUM ION BATTERY ACTIVE MATERIALS (SEE CHAP. 17A)*

### 2A.2.1    Lithium Ion Cathode PAM

Lithium ion PAM development has advanced at a rapid rate in recent years with the aim of doubling materials utilization and improving batteries in parallel with advancements in electric vehicles (EVs). Some leading manufacturers of lithium ion PAM are listed in Table 2A.1.[1]

**TABLE 2A.1**    Leading Lithium Ion PAM Manufacturers (E3BV)

| |
|---|
| Umicore |
| Mitsubishi Chemical |
| Sumitomo Corporation |
| L&F Company |
| Nichia Corporation |
| Tianjin B&M Technology Company Limited |
| Toda Kogyo Corporation |
| Ningbo Shanshan Co., Ltd. |
| BASF SE |
| Ube-Dow |
| Beijing Easpring Material Technology Company, Ltd. |
| Johnson Matthey PLC |
| Ganfeng Lithium |
| Reshine |
| Tosoh (LMO) |
| Phostech Lithium (LFP) |

Most listed companies produce layered transition metal oxides. Battery companies that use lithium iron phosphate (LFP) PAM—e.g., BYD, Wanxiang-A123, Valence Technology Corporation—typically have in-house supply capabilities. Some EV companies have established partnerships with active material and battery producers, such as Sumitomo-Panasonic-Tesla or SKI-BESK-BAIC.

### 2A.2.2    Lithiated, Layered Transition Metal Oxide PAM

Commercial PAMs for lithium ion are typically first-row transition metal oxides that form layered crystals with lithium ion incorporated using lithium carbonate or lithium hydroxide. Initial compositions had one transition metal, but evolved to formulations of two or three metals, as shown in Table 2A.2.

**TABLE 2A.2**    Lithiated, Layered Transition Metal Oxide PAM (E3BV)

| Type | Description | Generic formula | Nominal charge voltage (first charge specific capacity, mAh/g) |
|---|---|---|---|
| LCO | Lithium cobalt oxide | $LiCoO_2$ | 4.2 V (~165 mAh/g) |
| LNO | Lithium nickel oxide | $LiNiO_2$ | 4.2 V (~200 mAh/g) |
| NCA | Nickel cobalt aluminum oxide | $LiNi_{0.8}Co_{0.15}Al_{0.05}O_2$ | 4.2 V (~200 mAh/g) |
| NMC-XYZ | Nickel manganese cobalt oxide | $LiNi_XMn_YCo_ZO_2$ | 4.2 V, XYZ = 111 (~175 mAh/g), XYZ = 811 (~200 mAh/g) |
| Layered-layered | Manganese oxide, nickel oxide | $xLi_aMn_bO_2$-$yLiNi_cO_2$ | 4.6 V (~270 mAh/g) |

Many commercial formulations also include Nb and other second-row transition metals as dopants that disrupt the regular structure of the metal oxides to enhance charge storage and charge mobility, improving energy density, power, and the stability of lithium ion cells. Some layered transition metal oxide PAMs are produced without prelithiation (e.g., $V_2O_5$), which can be made into cathode electrodes that cycle against lithiated anodes or lithium metal foils.

***Goals.*** Lithium ion cathodes, electrolyte systems (see Chap. 3B), and anodes (see Sec. 3A.2.2) are all targeting to increase capacity and reduce cell raw materials cost below 100 USD/kWh for EV applications. Cost, performance, and safety considerations weigh heavily in material selection and development. However, concurrent improvements in battery management—thermal management in particular—as well as electrolyte development contributed to selection criteria for commercial lithium ion cells.

***Initial Lithium Ion PAMs.*** The first major PAM was lithium cobalt oxide (LCO), following its use by Sony in the first commercial lithium ion battery applications. Although LCO has a specific capacity of 268 mAh/g, only 137 mAh/g was originally used. Currently, 165 mAh/g (<62% of the lithium content) is discharged to ensure stability of the cobalt oxide layers, which would cause collapse and prevent lithium recharge. Also, initial commercial lithium ion cells typically lost another 15 to 20 mAh/g in the first few cycles due to SEI (solid electrolyte interface) anode losses and to impurities, which limit reversible capacity of LCO to 145 to 150 mAh/g for conventional graphitic anodes with a standard 4.2 V charge. Higher voltage operation results in higher specific capacity but much shorter cycle life, even with improved anodes, due to irreversible damage to the LCO in the cathode.

Lithium nickel oxide (LNO), another first-generation commercial PAM, provides an initial 200 mAh/g out of a 268 mAh/g theoretical specific capacity and a reversible 180 mAh/g. The specific capacity of LNO exceeds that of LCO and the rate capability is also higher. However, the Ni-based material has a lower autothermal point around 160 to 170°C, which presents a significant risk for thermal runaway compared to LCO with an autothermal point of 190°C. Concerns over safety prevailed, and LNO was not commercially successful despite its superior performance.

***Ni-Rich PAM.*** By 2000, PAM manufacturers introduced mixed transition metal layered oxides. Umicore and Fuji Chemical—acquired by Toda Kogyo—developed nickel cobalt aluminum (NCA) cathode materials where the 85% Ni in the metal content of the crystal layers is stabilized by 15% Co and 5% Al. Despite the fact that Al is not electroactive, the NCA delivered 180 mAh/g of reversible specific capacity against graphitic anodes. The NCA was safer than LNO, but its higher price prevented widespread adoption due to a required process that achieves dispersion of Co and Al around Ni that enhances stability and safety. A new PAM based on layered, mixed transition metal oxide materials with 1:1:1 molar ratio of nickel, manganese, and cobalt entered the market. The dispersion of Mn and Co around Ni and the absence of Al precursors probably simplified the manufacturing of this PAM compared to NCA and provided 160 to 170 mAh/g reversible capacity at lower price. In addition, the Mn content raised the autothermal point making it comparable in safety to LCO. Electronic device batteries with blends of NMC-111 and LCO powders had improved energy density without increased safety hazards. The NMC-111 content in these cells evolved from <10% of the PAM in 2000 to ~90% of the PAM in most devices by 2015. As adoption of NMC-111 increased, the price for the material decreased due to economies of scale. But cobalt price pressures also decreased in turn.

***PAM for Electric Vehicles.*** In the late 2000s, as interest in EVs increased, a renewed focus on battery materials evolved. A decade earlier, General Motors (GM) was reported to have experimented with many types of battery chemistries from lead acid to NiMH to lithium chemistries, including NCA among other PAMs. But the cost of this battery chemistry was not deemed viable for the large format batteries needed to supply long-range all electric drive for EVs. Tesla, on the other hand, subsequently adopted NCA for the Roadster—its first EV model targeting a high-end, price-tolerant customer base. Tesla has stayed with NCA, a Ni-rich (80%) material for its batteries through Model S/X and Model X vehicles.

Meanwhile, GM contemplated LFP in the years preceding release of the Chevy Volt—a material with 290°C autothermal point and prospects for very low cost. However, by 2010, GM and Nissan selected lithium manganese spinel (LMO)—a lower cost material with a long history of safety testing (230°C autothermal point), but with material degradation issues exacerbated by high temperature. To counteract the issues with LMO, GM equipped the Chevy Volt with a liquid cooling (and heating) system to modulate battery temperature in a narrow and optimal operating range. Nissan opted for an air-cooled system for its LMO battery. Subsequently, GM switched the 16 kWh Chevy Volt plug-in hybrid electric vehicle (PHEV) battery to NMC-111 that was proven successful in portable consumer electronic devices. Nissan switched as well after some 24 kWh Nissan Leaf EV batteries experienced issues.

Tesla additionally demonstrated that the high Ni content NCA lithium ion batteries could operate in EVs without safety failures, although a liquid cooling system was included. In addition, Tesla used a monitoring/control system that collected data and provided feedback for improving battery performance. Originally, the NCA PAM was supplied by two leading producers, Umicore and Toda Kogy, and used by various companies such as Sony and Molicel. Sumitomo, partnered with Tesla and Panasonic, subsequently displaced Umicore and Toda as the leading NCA supplier and created even more limited market access to NCA supply. Tesla not only became the leading global customer for NCA by 2016, but it also helped verify that Ni-rich PAM was safe for large format EV batteries with proper battery management system (BMS) safeguards. With the strong, competitive vertical supply chain, the Ni-rich PAM proved to be viable in the high-end market for the Tesla Model S/X. A new Tesla EV, the Model 3 EV, was introduced with a lower price point, dependent on the build of a gigafactory (i.e., GWh/year battery capacity) and on leveraging larger economies of scale for the NCA PAM.

Various EV manufacturers (xEV OEMs) have emerged as listed in Table 2A.3 and are influencing the growth of the lithium ion battery industry.

**TABLE 2A.3**   Leading xEV Battery Manufacturers (E3BV)

Tesla-Panasonic
BYD Co., Ltd.
BESK—JV of Beijing Electric (BAIC) and SK Corporation
GM-LG Chemical
Nissan/Renault (formerly AESC) Mitsubishi/GS-Yuasa
BMW-Kreisel-Samsung
"CATL"—Contemporary Amperex Technology, Ltd.
"Guoxuan"—Guoxuan High-Tech Co., Ltd.
"Lishen"—Tianjin Lishen Battery Joint-Stock Co., Ltd.
Northvolt AB
Boston Power, Inc.

***Mn-Rich PAM.***   In 2006, Argonne National Laboratory (ANL)—a U.S. DOE National Laboratory—offered layered-layered materials with a high Ni-content in one layer and a second Mn-rich layer that provided very efficient lithium ion storage along with the intrinsic safety that comes with the higher autothermal point of Mn compounds. The resulting two-layered composite PAM provides a 268-mAh/g specific capacity, which can be even higher since the Mn-rich rock salt layer can hold more than one lithium ion per manganese atom in the oxide.

However, achieving these high specific capacity levels requires that cells operate to a charge voltage of 4.6 V using a graphitic anode. Cathodes at this level of delithiation cause the electrochemical breakdown of liquid-based electrolytes that use conventional organic carbonates. To reach the full benefit of layered-layered PAM will require the development of stable solid-state electrolytes (SSEs) or the use of another electrochemically stable ion conducting medium for contact with the cathode PAM. To meet this need, fluorinated organic solvents have become the subject of present battery industry research and development.

BASF and Toda Kogyo became prime licensees of the layered-layered PAM technology from ANL in 2010 and developed commercial scale production of their own layered-layered formulations until 2018, when the two companies formed the joint venture company, BTA. Dalhousie University developed very similar materials to the ANL layered-layered PAM and licensed the technology to 3M that later sold the license to Umicore. After settling legal matters concerning intellectual property rights, Umicore, BASF, and Toda Kogyo offer this class of PAM as of 2018. However, the large increase in performance relies on higher voltage operation and SSE or advanced liquid solution electrolyte.

***Ni-Rich PAM Improvement.***   The PAM in the lithium ion batteries for the vast majority of EVs sold in 2017 used NCA (e.g., Tesla), NMC-111 (e.g., GM, Nissan-Renault, Beijing Electric), or LFP (e.g., BYD). In China, subsidies for New Energy Vehicles (NEV)—i.e., electric and other green vehicles—had been based on meeting battery safety criteria up until 2017, when the NEV's subsidy policy was reformulated to qualify vehicles based on energy efficiency criteria. This change in China favors chemistries with higher specific capacity active materials (i.e., typically, with higher PAM utilization). While NMC-111 produces higher reversible specific capacity than LFP (135 mAh/g), Ni-rich NMC- and NCA-based lithium ion cells remained the highest in materials utilization and lowest in Co content in recent years. Despite the success of NCA use in Tesla vehicles, most other EV OEMs leveraged the early success of NMC-111 used in portable consumer electronics. PAM formulations were adjusted to decrease Co content and boost Ni content, such as NMC-532. While the increase in energy density that results from increased utilization is itself attractive for large format battery applications, the additional reduction of Co content decreases the dependency on this scarcer and more expensive element. Compared to NMC-111, the NMC-532 PAM reduces Co content to near the level of NCA, increases the reversible specific capacity to 175 mAh/g in lithium ion cells with graphitic anodes, and maintains the autothermal point.

Ni-rich NMC-532 cathode chemistry has already replaced NMC-111 in many EV battery applications, and commercial evaluations of PAM with higher Ni content are underway, including NMC-622 and NMC-811 (SK Innovation). There is also a grade of PAM with nearly all nickel, branded "eLNO" by Johnson Matthey, where the stability of the Ni may derive from stoichiometry and crystal structure as affected via processing technology and perhaps dopants. The aim of Ni-rich PAM is to increase material utilization and thereby decrease cost on a USD/kWh basis, while not affecting operational safety or battery life. These PAM compositions are sometimes viewed as an alternative or improvement compared to NCA PAM. Some notional comparisons are suggested in Fig. 2A.1.

Nevertheless, announcements by BTA in 2018 included the option to produce NCA as well as Ni-rich NMC grades. Similarly, LG Chemical announced a pair of joint ventures—one with Huajin New Energy Material and the other with Leyou New Energy Material—to produce either NCA or Ni-rich NMC grades in China.

Hedging between NCA and Ni-rich NMC not only reflects confidence in NCA PAM due to the successful track record of its use by Tesla, but also uncertainty in the performance of high-Ni PAM. Umicore and other cathode manufacturers have presented data that advises caution in adopting Ni-rich NMC based on projections of poor cycle life—especially at the higher end of the operating temperature range for automotive applications, 40°C and above.[2]

The layered-layered materials are Mn-rich with thinner Ni-rich "engine" layers. While the high temperature stability of Mn compounds is affected by valence state disproportionation and subsequent dissolution or diffusion, the Ni layers offer some protection. Also, Ni oxides have lower autothermal points than Mn oxides, which may stabilize the two-phase composite. The lower cost of Mn precursors is additional incentive to boost the rock layer content in these materials. This material is both promising and complex, which may explain why commercial development of layered-layered has taken so many years. Layered-layered development started prior to 2000 with licensing offered in 2006 and started in 2010 for implementation by 2025.

Cycle life is very important as the ultimate selection criterion is annualized total cost of ownership. Cycle life improved only after prolonged PAM development, involving materials synthesis, cell assembly processes, and battery-level design and control. At some point, battery cost is leveraged most by raw material costs. If the materials' cost for each of the PAM chemistries does level out over time, utilization of the materials becomes the dominant feature that influences cost as shown in Fig. 2A.2, where a 3.7-V potential versus a graphitic anode is assumed for all materials except layered-layered PAM, which is 4.1 V.

The hidden benefits of PAM utilization include greater utilization of all other materials in a cell. In addition, the use of SSE that enables layered-layered concurrently reduces or eliminates first charge SEI formation losses that consume 8% to 10% of cell capacity in conventional lithium ion liquid electrolyte systems. SSE development also improves the prospect of using higher capacity NAM than conventional graphite. Also, SSE will allow the use of silicon NAM or lithium metal foil without the need for high cost, ceramic-coated microporous polyolefin separator. The net amount of lithium (i.e., the lithium carbonate equivalent or LCE) needed—in terms of kg LCE/kWh—would decrease, since SSE improves Li utilization and reduces Li losses in the cells. If all these prospective gains are achievable, the overall lithium ion cell materials cost could easily go below 100 USD/kWh. A higher utilization PAM also increases energy density in terms of Wh/L, which is important in EV applications, since most xEVs are volume constrained.

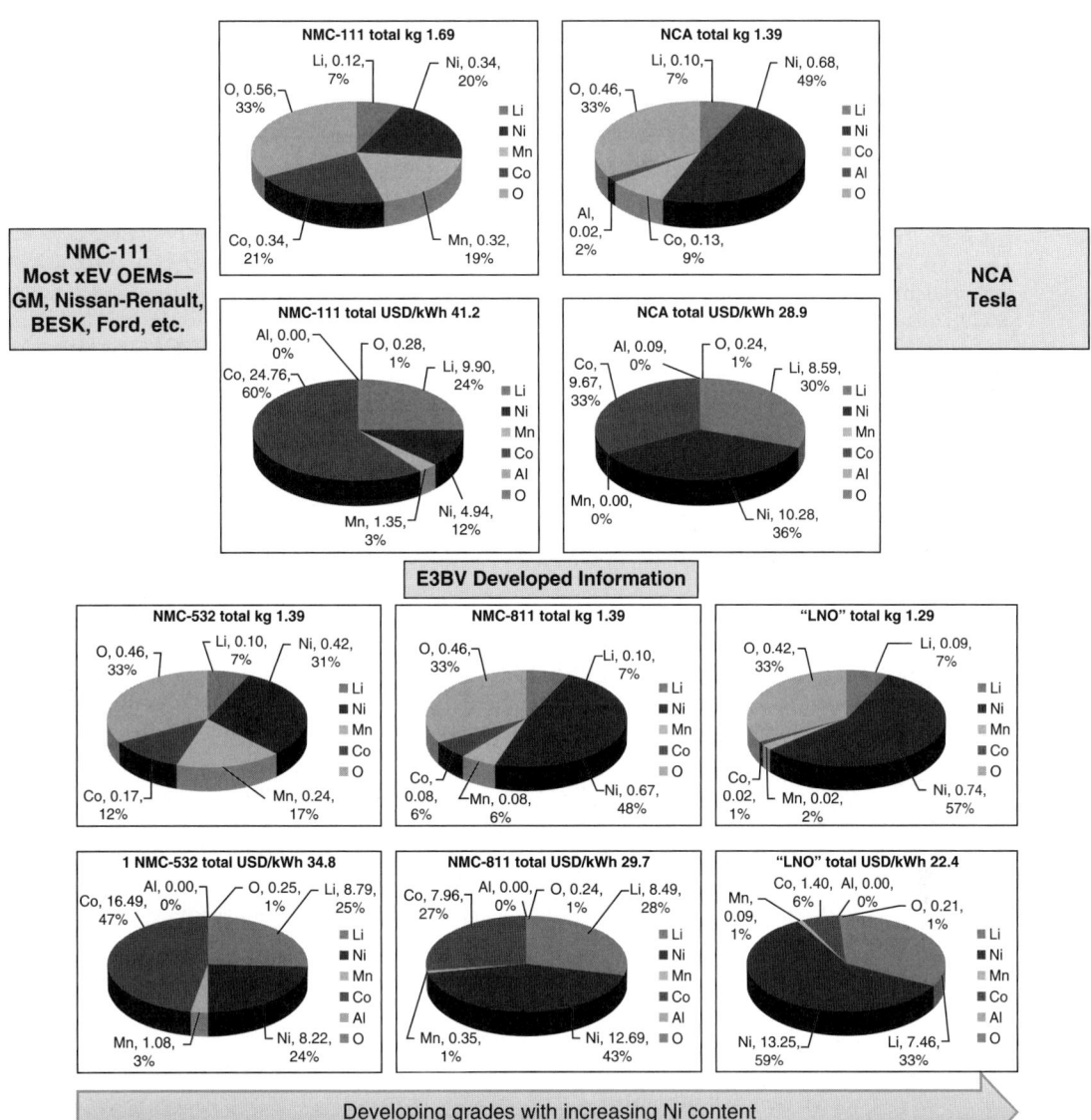

**FIGURE 2A.1**  Metal content and cost of lithium ion layered transition metal oxides (E3BV).

***Advanced Layered Transition Metal Oxide Production.***  Equipment cost depends on core process technology and target product offering. Equipment may include either low-cost, liquid-based calcination reactors geared to make cathode for portable electronic uses or high-cost, complex gas-phase furnace reactors geared to make Ni-rich cathode for EV uses. For prior generation, single-transition metal oxides, such as LCO, manufacturing technology is simpler. First-generation PAM produced for portable consumer electronic devices met less demanding product requirements so that in-house production of cathode was easier to include as part of the battery company technology. As multiple transition metal oxides evolved—via Umicore, OMG, Fuji-Toda, Seimi, FMC—the processes grew complex, the equipment cost increased, and smaller batch production became unviable.

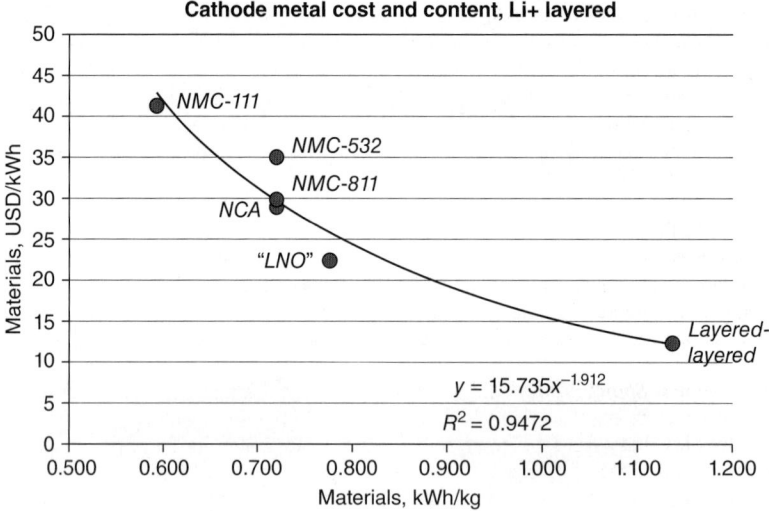

**FIGURE 2A.2**   Lithium ion PAM cost and content (E3BV).

***Lithiated, Layered Transition Metal Oxide PAM Outlook.***   This class of materials is projected to bring down materials' cost for lithium ion cells and to enable lithium metal chemistries, allowing lithium battery chemistries to grow not only in automotive applications but in all storage battery applications. While it may take years to reach full commercialization, if progress continues, these advances in PAM and the enabling SSE materials could change the battery industry landscape by the year 2030.

### 2A.2.3   Lithium Transition Metal Phosphate and Other Lithium PAM

The greater margin of safety of LFP and LMO helped launch many early lithium ion battery applications, and these chemistries also competed for use against conventional lead acid applications. In automotive, for example, these chemistries could replace 12 V starting-lighting-ignition batteries for ICE (internal combustion engines) vehicles and start-stop vehicles—as well as traction and/or auxiliary batteries for 48 V mild hybrid EV platforms.

***Lithium Iron Phosphate PAM.***   The promise of low-cost iron and phosphate precursors and a high 290°C autothermal temperature launched the interest in LFP, $LiFePO_4$. In addition, LFP cells with graphitic anodes reach close to full charge by 3.65 V, which is well within the safe operating limit of conventional lithium ion liquid electrolyte systems and a fact that renders LFP cells tolerant of overvoltage during battery charging. Another favorable feature of conventional LFP cells with graphitic anodes is that the nominal voltage of 3.3 V is easily made into batteries of 12 V and its multiples. These points, taken together, position LFP as a battery chemistry that could contend with lead acid battery chemistry. LFP cells are built to very large capacities and combine in series-parallel arrays with relative ease and minimal BMS requirements compared to lithium ion chemistries with layered oxide PAM.

LFP battery commercialization also received a boost in the early years from EV development in China. Battery safety failures prompted the government in China to base subsidies for EVs (NEV) on demonstrated battery safety, which favored the use of LFP PAM for lithium ion batteries. Similar concerns for other EVs promoted the use of LFP chemistry, which began to flourish around the year 2010. These factors helped to propel BYD and its use of LFP batteries in its line of EVs as well as other makers of LFP batteries—e.g., CALB, Winston-Thundersky, Sinopoly, BAK, and A123-Wanxiang.

However, LFP has a lower specific capacity than layered transition metal oxide PAM. Lower capacity and lower cell voltage, compared to the layered oxides, limit the energy density in conventional LFP cells with graphitic anodes. In addition, the cost of producing LFP is tied to the cost of producing phosphate precursors with reducible iron and to the cost-effectiveness of introducing electronic conductivity to the LFP PAM—either by adding conductive carbon particles to the manufacturing process or by using a carbothermal reduction process (licensed by Valence Technology Corporation). These manufacturing issues have impeded progress in realizing very low-cost LFP PAM.

The shift in 2016 of the NEV subsidy criteria in China favored energy efficiency for EVs. As a result, PAM with higher specific capacity threatens to displace LFP in the EV market in China.

There is interest in high-voltage phosphates based on Ni, Co, and Mn in place of Fe in the transition metal content of lithiated transition metal phosphates—including mixtures of these metals in the composition. At full charge, these PAMs with graphitic anodes operate very near to 5 V, which requires stable electrolytes. As with other high-voltage PAM compositions, SSE materials could enable the feasibility of high-voltage transition metal phosphate PAM.

*Lithium Manganese Spinel PAM.*    The LMO has a high autothermal point of 235°C, and the three-dimensional spinel crystal structure enables rapid movement of lithium ion in the PAM, providing LMO cells with high-rate capability. But LMO has a low specific capacity of ~90 mAh/g as it holds only one lithium atom for every two Mn atoms in the formula, $LiMn_2O_4$. Also, LMO is susceptible to dissolving in the electrolyte solutions, especially since the $Mn^{3+}$ ions disproportionate into $Mn^{4+}$ and $Mn^{2+}$ ions, and the latter are very soluble in electrolyte solvents. LMO was replaced with NMC PAM after brief use of LMO in early model years of Chevy Volt and Nissan Leaf batteries. Lessons from the long history of LMO technology contributed to the design of Mn-rich rock salt layers in layered-layered PAM, and LMO stability should improve in cells with SSE systems. The low cost and inherent safety of LMO remain attractive features despite a low specific capacity compared to other lithium ion PAM.

## 2A.2.4    Lithium Ion Anode NAM

*Graphitic Carbon NAM.*    Conventional lithium ion chemistry uses carbon in graphitic form to store lithium ion between the graphene layers of crystalline regions. Graphite with intercalated lithium ion has a formula of $LiC_6$ that has a theoretic capacity of 372 mAh/g. The specific capacity of graphitic NAM varies according to the crystalline content of each NAM and other variations such as particle surface area and impurity levels. Typical grades and specific capacities are listed in Table 2A.4.

**TABLE 2A.4**    Graphitic Carbon NAM for Conventional Lithium Ion Cells (E3BV)

| Type of graphitic PAM | Specific capacity, mAh/g (before SEI losses) |
| --- | --- |
| Synthetic, porous, two-layered graphite | 320 |
| Synthetic, crystalline | 340 |
| Mesocarbon | 290 |
| Purified natural graphite | 340 |

In liquid electrolyte systems, the particle surface area determines the extent to which capacity is lost due to the formation of SEI passivation layers as conventional lithium ion cells are charged for the first few times. The surface area of the graphite influences reversible capacity of a cell and the reversible specific capacity of the graphitic material itself, which is often 320 mAh/g or less.

The SEI layer on graphitic carbon NAM in lithium ion cells is meta-stable and susceptible to dissolution into liquid electrolyte systems at higher than normal cell operating temperatures. The SEI reaction is exothermic and can lead to anode thermal runaway if anodes reach around 150°C, at which point the SEI reaction proceeds as fast as the SEI dissolves into the electrolyte. Anode thermal runaway can lead to cathode thermal runaway and cell thermal runaway results. Attempts to avoid or reduce the extent of the SEI reaction include raising the thermal and electrochemical voltage breakdown thresholds of electrolyte solutions by using fluorinated carbonate solvents. Another method is to include carbonate solvents with unsaturated bonds, such as vinyl carbonate (VC), in place of part of the ethylene carbonate (EC) solvent present in most electrolytes in order to promote formation of SEI layers on the graphite that is more resilient to solvation, even at elevated temperatures. Similarly, vinyl ethyl carbonate (VEC) can be used in place of some diethyl carbonate (DEC) in the electrolyte solvent blend. Other attempts involve the use of ionic liquid systems. However, further improvement of SEI stability for graphite materials remains an area of commercial interest.

The synthetic, porous two-layer graphite produced by Hitachi (e.g., MAG-10) is a composite of crystalline and mesocarbon that enables both high-density storage of charge and absorption of high-rate charge influx for short durations as occurs during a braking event in EV applications and other high-rate application. A blend of synthetic crystalline graphite powder with mesocarbon microbead powder accomplishes a similar result. Synthetic materials are often produced using pure precursor materials such as acetylene gas or other processed petroleum products, which reduces the risk of impurities that is a major concern for batteries used in long service life applications such as EVs and grid energy storage systems. Some leading manufacturers of graphitic NAM are listed in Table 2A.5.

**TABLE 2A.5**   Manufacturers of Graphitic NAM (E3BV)

| |
|---|
| Hitachi Chemical |
| BTR |
| JFE Chemical |
| Nippon Chemical |
| Sumitomo Bakelite |
| Mitsubishi Chemical |
| Shanshan |
| Showa Denko |
| POSCO-Chemtech |
| Osaka Gas Chemical |
| Kureha |
| Tokai Carbon |
| Conoco-Phillips |
| Lonza/Timcal |
| Superior Graphite |

Commercial use of natural graphite tends toward applications where long device life is not a critical factor. However, there are purer grades that do work for xEV and for grid ESS applications. Higher purity is achieved by chemical washing, although some graphite mineral deposits of higher purity have been reported. Synthetic graphite from petroleum precursors (e.g., Kureha, Hitachi materials) have fewer concerns about impurity levels but tend to cost much more than natural graphite grades.

*Silicon-Based NAM.*   Lithiated silicon holds 4.4 lithium ions, which results in specific capacities of ~4200 mAh/g Si and ~2000 mAh/g $Li_{4.4}Si$. The more than threefold expansion of silicon for full lithiation, however, fractures elemental silicon crystals and thereby disrupts both the SEI passivation layers on silicon particles and the connectivity of electronically conductive additives surrounding the silicon. Contraction during discharge also means that the decrease in volume scatters the delithiated PAM across its original volume. This effect has limited practical use of silicon in commercial practice.

However, 3M produces a silicon NAM that incorporates first-row transition metals that stabilize the silicon but decrease the specific capacity of the material to ~600 mAh/g. Blending a few percent by mass with this material increases overall anode capacity in lithium ion cells, although the specific capacity of the silicon portion decreases more rapidly than the graphite portion. Liquid electrolyte solutions that contain silane-modified additives in the solvent provide some improvement in cycling. Cycling of silicon in general remains a key subject of materials development.

One route for the commercial development of the next generation of silicon NAM involves structuring silicon particles on a nanoscale to enable flexible, nondestructive expansion of lithiated crystals. Porous and peninsular substructures within a particle (or fibril materials) limit the absolute extent of linear expansion to provide stress relief to the local crystalline material. Although this nanoscale morphology increases the effective volume of the particle, this volume change could be designed to match the volume increase caused by lithiation. Companies such as Amprius and Sila Technologies are advancing solutions of this type. The materials are often produced in additive, vapor deposition processes with some work on subtractive processes to obtain the necessary particle morphology.

Another route for commercial development of silicon NAM involves preserving the macrostructure of fractured silicon within an electrode. Wrapping graphene layers around silicon particles to form graphene bags serves to hold fractured silicon pieces together, while the graphene also provides electronic conductivity for the NAM. SiNode Systems is a company advancing solutions of this type.

An alternative silicon NAM development effort involves the use of silicon monoxide or other nonstoichiometric ratios of silicon to oxygen that reduce specific capacity but stabilize the NAM during expansion and contraction as the cell cycles.

## 2A.3    *LEAD ACID ACTIVE MATERIALS (SEE CHAP. 14)*

Although the fundamental active materials that cycle in lead acid batteries—lead metal NAM and $PbO_2$ PAM—have not changed since 1859, mixtures of precursor materials in formulating electrode pastes have evolved. Extensive documentation exists in the public domain. One strong and comprehensive source is the website for the Institute of Electrochemistry and Energy Systems at the Bulgarian Academy of Sciences.[3]

In addition to conventional lead acid batteries with lead metal NAM, carbon NAM holds promise in providing longer cycle life in deep discharge and high-rate applications—such as 48 V mild-HEV platforms and other emerging applications with similar duty cycles and cost constraints that may compete with candidate lithium ion batteries.

## 2A.4    *NICKEL-METAL HYDRIDE ACTIVE MATERIALS (SEE CHAP. 15D)*

The volume of global production of nickel-metal hydride (NiMH) batteries is flat or declining, as the high nickel content of both NAM and PAM and limited specific capacity of these materials reduces cost competitiveness compared to other rechargeable battery chemistries. Nevertheless, BASF and FJK (formerly the NiMH business of Sanyo Electric) continue to develop NAM using nickel oxyhydroxide and intermetallic NAM with lanthanum added to the nickel metal mix. Recent development includes variations from conventional NiMH active materials such as using other transition metal oxyhydroxide PAM—e.g., cobalt oxyhydroxide—and other heavy, low LUMO (lowest unoccupied molecular orbital) transition metals added to the intermetallic NAM.

## 2A.5    *PRIMARY BATTERY ACTIVE MATERIALS (SEE CHAPS. 11 TO 13)*

A number of primary—nonrechargeable—cells provide high-energy density compared to secondary cells, although often at lower discharge rates. A few example chemistries are listed in Table 2A.6.

**TABLE 2A.6**    Primary Battery Chemistry Materials (E3BV)

| Chemistry | NAM-PAM |
|---|---|
| Alkaline cell | $Zn\text{-}MnO_2$ |
| Lithium iron sulfide | $Li^0\text{-}Fe_2S_3$ |
| Lithium manganese | $Li^0\text{-}MnO_2$ |
| Lithium thionyl chloride | $Li^0\text{-}SOCl_2$ |
| CFX | $Li^0\text{-}CF_x$ |
| Silver zinc | $Ag\text{-}Zn$ |
| Zinc air | $Zn\text{-}air$ |

Alkaline cells with low-cost Zn metal slurries and electrolytic manganese dioxide (EMD) dominate consumer retail use, although the use of primary lithium cells with iron sulfide PAM is growing. Many issues associated with lithium foil passivation are not as relevant in lithium primary cells as these are in secondary cells, since there is no charging issue and rates are often low, much less than a 1 hour discharge rate. Some of the highest energy density commercial batteries are based on $CF_x$ PAM, where graphitic carbon hosts the oxidation reaction of intercalated fluorine with lithium ion.

## 2A.6   "BEYOND LITHIUM ION" SECONDARY BATTERY MATERIALS

### 2A.6.1   Elemental Sulfur PAM (See Chap. 17B)

The lithium metal-sulfur ($Li^0$-$S_8$) battery is considered one of the "beyond lithium ion" battery chemistries that have received interest for commercial development due to the high specific capacity of sulfur, 1670 mAh/g S. Issues related to managing sulfur in the cathode coupled with issues managing the lithium metal anode have presented substantial barriers to commercialization. As sulfur discharges, it converts into lithium salt in the form of polysulfide chains that dissolve and improve ion transport in the cathode. At the same time, the polysulfides can migrate across separator materials and react with lithium in the anode to lose cell capacity. Advances in SSE may mitigate this undesirable effect, as may additives such as other PAMs that serve to immobilize or limit polysulfide movement. Another issue is the difficulty in providing electronic conductivity to sulfur PAM or to the cathode in general, without blocking the sulfur or displacing a fair fraction of it as the active ingredient.

Using lithium metal foils as the NAM introduces the theoretical 3.86 mAh/g specific capacity to the cell. However, the practical issues that challenge every lithium metal NAM remain.

### 2A.6.2   Sodium Sulfur Battery

The sodium-sulfur ($Na^0$-$S_8$) battery operates at temperatures above ~300°C to ensure that both Na metal NAM and sulfur PAM are molten liquids. The high specific capacity of sulfur along with the mobility of its molten state allows operation with a minimal amount of additives for electronic and ionic conductivity. The abundance of low-cost Na in the form of soda ash (<350 USD/ton) and S (<150 USD/ton) makes this chemistry attractive for commercial use, although thermal management and prediction of ceramic separator life remain practical issues for many applications. The leading manufacturer of this type of battery, NGK, makes beta alumina-based ceramic separators that permit $Na^+$ permeation but inhibit sulfide transport. The related sodium-nickel chloride ($Na^0$-$NiCl_2$) battery uses a similar type of separator and operates in the same temperature range.

### 2A.6.3   Magnesium-Based Battery

Magnesium metal cycles well as a reversible NAM, operates near lithium metal voltage in comparable cells, and has 2.2 mAh/g theoretical capacity, making Mg foil an attractive battery material. On the other hand, an equally suitable PAM that intercalates $Mg^+$ remains a commercial development objective. Pellion Technologies demonstrated a $Mg^0$-$MoS_2$ battery, although the development of lighter-weight and lower-cost PAM remains a critical research goal.

### 2A.6.4   Vanadium Redox Flow Battery (See Chap. 22B)

The multiple valence states of vanadium enable the use of two electrolyte solutions at different valences with ~1.4 to 2 V between them, depending on the acid used in the electrolytes. The more electronegative anolyte cycles between $V^{2+}$ and $V^{3+}$ while the more electropositive catholyte cycles between $V^{5+}$ and $V^{4+}$, although the precursor material used to produce both NAM and PAM is $V_2O_5$. Vanadium is a minor metal and developers of VRFB technology face supply challenges for precursor materials. However, the trend by mining companies and steelmakers to refine elements such as vanadium, cobalt, nickel, and other minor metals from tailings, coupled with the development of cost-effective, modular refining plants with electrowinning unit operations stands to improve the availability of vanadium for VRFB applications.

## REFERENCES

1. Tables provided courtesy of Element 3 Battery Venture (E3BV).

2. *Umicore downplay of high Ni NMC,* http://cii-resource.com/cet/AABE-03-17/Presentations/BTMT/Levasseur_Stephane.pdf.

3. www.labatscience.com.

# SECTION B

# METALS AND MINERALS

**Kirby W. Beard**

## 2B.1   INTRODUCTION

The mining industry is, undoubtedly, the most significant source of raw materials used in batteries, providing metals and metal compounds. Mining is a truly global enterprise, encompassing locations below sea level and above 4000 m elevations in both desert and artic regions. Mining locations are often in remote, politically unstable, or economically disadvantaged areas and involve both huge surface excavations (open pits) and deep shafts (down to 3 to 4000 m depths with dozens of lateral tunnels extending many kilometers). Some operations are based on liquids pumped from lakes, oceans, and deep wells, and others are even now targeting the moon, Mars, and asteroids. Deep sea manganese crust and nodules are also cited as a potential key source for nickel, cobalt, etc.[1,2]

However, despite the criticality of metals and minerals to batteries plus a wide range of other applications, there are on-going, concerted public efforts to shut down many mining operations. Most readily accessible surface deposits have been fairly well depleted and future operations will require more intrusive and extensive underground operations. Clearly, despite efforts to recycle materials, the battery industry will still be vulnerable to limited supplies (and high costs) of metals and minerals. If a large battery industry is to be sustained, companies, industries, and nations will need to assess the issues and enact policies and plans that ensure order and reliability in the supply of metals and minerals and proper closure/disposition of depleted deposits (see map coordinates: 45.773141, −71.952335).

## 2B.2   CLASSIFICATIONS

Materials extracted from the earth include a variety of elements and compounds that come in various forms: metal flakes, ores, mineral bodies, sands, brines, etc. Concentrates are processed to extract and purify the desired component(s). The materials may be used directly in battery constructions (i.e., copper in current collectors, lead as an active material, aluminum for battery housings, etc.), but are often converted to other chemical compounds also used in batteries. For example, manganese dioxide ($MnO_2$) is used in a wide variety of cells, including lithium and alkaline primary cells. However, $MnO_2$ can be used directly as mined or can be converted to more active forms by both chemical and electrolytic processes. In addition, other grades of manganese oxides, such as $LiMn_2O_4$ for lithium ion cells, can be synthesized from scratch. Various categories of materials used in batteries are listed next along with some general details of exemplary metals and minerals under each section. These lists are not all-inclusive, but are intended to convey a sampling of the variety of battery raw materials dependent on the mining industry.

The methodology used in these sections to present details on the various commodities is based on four issues:

1. Production levels
2. Costs
3. Reserves[a]
4. Trends over time

---

[a]Reserves refer to estimates of economically recoverable amounts that remain and are defined as that part of the reserve base which could be recovered even if extraction facilities are not in place or operative. (https://minerals.usgs.gov/minerals/pubs/mcs/2010/mcsapp2010.pdf [extracted May 11, 2018].)

Other factors are, of course, significant, but this set of data, along with details on the global distribution of the materials, is felt to serve as a good starting point for anyone interested in gauging the supply of raw materials critical to the battery industry. The selection and use of metals that are globally distributed and available in large quantities at low costs will be essential to sustained commercial success.[3] For instance, supply and costs of lithium and cobalt have created great consternation recently.[4] Meanwhile, materials such as lead and zinc present a more desirable supply choice for battery development and production. The data in the sections below are available online on the U.S. Geological Survey's website.[5]

### 2B.2.1    Pure Metals (and Alloys)

These materials include active materials, such as lithium foil, as well as inert battery components, such as current collectors. Additionally, metals may be used as (or converted to) oxides or other compounds as described in Chaps. 2A and 2C prior to use in batteries.

*Copper.*[6]    Copper is highly valued in electronics and other industries. It is widely available but requires massive operations to efficiently mine and purify. Costs can fluctuate greatly due to global supply and demand variations. Copper is used directly in lithium ion cell current collectors but finds other uses throughout the battery industry (particularly for electrical circuitry where conductivity and corrosion resistance is vital). Scrap copper is easily recycled. Table 2B.1 presents summary data on copper from the U.S. Geological Survey (USGS).

**TABLE 2B.1**    Copper Supply

| Year | Production: U.S./global (metric ton, 000s) | Estimated price (USD/lb) | Reserves: U.S./global (metric ton, 000s) |
|---|---|---|---|
| 2009 | 1190/15,800 | 2.30 | 35,000/540,000 |
| 2017 | 1270/19,700 | 2.80 | 45,000/790,000 |

An interesting depiction of the global deposits of copper, provided by the U.S. Geological Survey, is shown in Fig. 2B.1.[7]

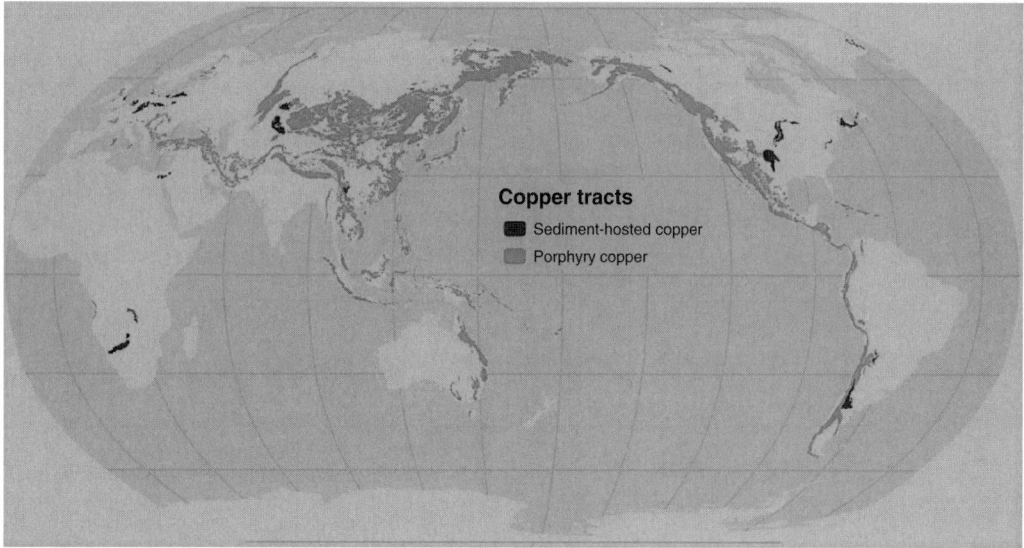

**FIGURE 2B.1**    Global distribution of copper deposits.

Being a key material in virtually all electronic industries, including batteries, it is quite fortuitous that copper is widely available. Other commodities are limited to maybe only a few economically viable deposits. For example, most cobalt comes from the Democratic Republic of Congo with no other country supplying >10% of world production.

*Nickel.*[8]   Similarly to copper, nickel is a reasonably available metal with many uses. A principal use is in metal alloys. In the battery industry, nickel is key to many alkaline-based cells (i.e., nickel cadmium, NiMH, nickel-zinc, etc.) but is also used in various battery components, including current collectors, steel plating, terminal pins, cases, etc. Nickel cadmium and even NiMH batteries are not seeing high levels of growth similar to lithium ion cells and therefore the demand for nickel in the battery industry has lessened. Table 2B.2 details key nickel supply figures.

**TABLE 2B.2**   Nickel Supply

| Year | Production: U.S./global (metric ton, 000s) | Estimated price (USD/lb) | Reserves: U.S./global (metric ton, 000s) |
|---|---|---|---|
| 2009 | ~0/1430 | 6.78 | NA/71,000 |
| 2017 | 23/2100 | 4.60 | 130/74,000 |

*Aluminum.*[9]   Aluminum is the second most abundant material in the earth's crust and is the most widely used metal next to iron. However, it is typically bounded to oxygen, silica, or other atoms, making it difficult to refine. Bauxite is a main source of aluminum metal.[b] A major use of aluminum is in lithium ion cathode current collectors, but various active materials and battery hardware are also made from aluminum due to its excellent electrical conductivity, high strength, light weight, and corrosion resistance (Table 2B.3).

**TABLE 2B.3**   Aluminum Supply

| Year | Production: U.S./global (metric ton, 000s) | Estimated price (USD/lb) | Reserves: U.S./global (metric ton, 000s) based on bauxite |
|---|---|---|---|
| 2009 | 1710/36,900 | 0.78 | 5000/6,750,000 |
| 2017 | 740/60,000 | 0.99 | 5000/7,500,000 |

*Lead.*[10]   Lead is available worldwide and finds principal use in lead-acid batteries. Lead is used for both negative and positive plate grids and as negative and positive active material (as lead powder and lead oxide powders, respectively). Recycling of lead-acid batteries serves as a model for the entire battery industry with an estimated 90% or better recovery rate. In the United States, recent surveys indicate a near 100% recycling rate for lead-acid batteries.[11] Lead use in paints, plumbing, pottery, and other such consumer applications has virtually disappeared due to toxic exposure issues. Given the high rates and lower cost to recycle batteries, reduced levels of lead mining are needed. Table 2B.4 details lead supply figures.

**TABLE 2B.4**   Lead Supply

| Year | Production: U.S./global (metric ton, 000s) | Estimated price (USD/lb) | Reserves: U.S./global (metric ton, 000s) |
|---|---|---|---|
| 2009 | 400/3900 | 0.74 | 7700/79,000 |
| 2017 | 313/4700 | 1.08 | 5000/88,000 |

Also, lead is often recovered as a by-product of zinc mining, and supply and prices may fluctuate accordingly.

---

[b]As a general rule, 4 tons of dried bauxite is required to produce 2 tons of alumina, which, in turn, produces 1 ton of aluminum.

***Zinc.***[12]  Zinc is fairly abundant and readily accessed. Its uses range across a wide spectrum of industries (especially metal alloys and plating), but has been prevalent in many primary alkaline cells (Chap. 11) over the years. Table 2B.5 provides supply details.

**TABLE 2B.5**  Zinc Supply

| Year | Production: U.S./global (metric ton, 000s) | Estimated price (USD/lb) | Reserves: U.S./global (metric ton, 000s) |
|---|---|---|---|
| 2009 | 690/11,100 | 0.76 | 14,000/200,000 |
| 2017 | 730/13,200 | 1.30 | 9700/230,000 |

***Lithium.***[13]  The interest in lithium has been intense due to its use in lithium cell anodes, cathodes, and electrolytes. The future of the entire electric vehicle (EV) industry has often been tied to the supply and pricing of lithium metal (see Table 2B.6). Depending on the latest market research projections, estimated future demand and costs can vary greatly. Although lithium is available from a variety of ores (e.g., spodumene), recovery of high-purity lithium for batteries has recently been focused on subsurface brine solutions in high-altitude deserts of Chile and Argentina. Batteries now use about half of the annual global lithium production, but other uses of lithium include ceramics, lubricant, and alloys.

**TABLE 2B.6**  Lithium Supply

| Year | Production: U.S./global (metric ton, 000s) | Estimated price—battery grade lithium carbonate (USD/lb) | Reserves: U.S./global (metric ton, 000s) |
|---|---|---|---|
| 2009 | NA/18 | 2.35 (2010) | 38/9900 |
| 2017 | 5.5/43 | 6.30 | 35/16,000 |

Recycling of lithium is minimal, but has been proposed as a top priority for the battery industry due to the expected intense expansion of EVs. According to a recent article, by 2040 EVs will exceed 50% of all car sales, and lower battery costs (undoubtedly based on recycling) will make EVs less expensive than internal combustion engine vehicles before 2030.[14]

Further details are available from a variety of sources, including special reports issued by the U.S. Department of Interior, U.S. Geological Survey.[15]

***Cobalt.***[16]  Cobalt is critical for use in metal alloys, pigments, catalysts, ceramics, and magnetics. However, cobalt has become a focal point of discussion (along with lithium metal) for the lithium ion battery industry and is deemed key to the future commercial success of EVs and grid storage applications. Table 2B.7 details recent production figures.

**TABLE 2B.7**  Cobalt Supply

| Year | Production: U.S./global (metric ton, 000s) | Estimated price (USD/lb) | Reserves: U.S./global (metric ton, 000s) |
|---|---|---|---|
| 2009 | ~0/62 | 18.00 | 33/6600 |
| 2017 | 0.65/110 | 24.70 | 23/7100 |

China has been the world's leading user of cobalt with nearly 80% of its consumption for rechargeable batteries. Cobalt is not easily substituted in many applications, but concerted efforts are underway to partially substitute cobalt with nickel, aluminum, and manganese in lithium ion cell cathodes (such as 8:1:1 Ni-Mn-Co). Cobalt prices have fluctuated widely over the last decade as supply, demand, and speculation intervene.

## 2B.2.2  Minerals/Compounds

*Iron Ore.*[17]   While iron is not a key component of many batteries, it does find some uses, such as in steel battery containers. Also, as the largest mined commodity, iron provides an interesting comparison to other battery metals as per Table 2B.8.

**TABLE 2B.8**   Iron Ore Supply

| Year | Production: U.S./global (million metric ton ore) | Estimated price—iron ore (USD/lb) | Reserves: U.S./global (million metric ton ore) |
|---|---|---|---|
| 2009 | 26/2300 | 0.032 | 6900/160,000 |
| 2017 | 46/2400 | 0.034 | 2900/170,000 |

*Manganese Ores.*[18]   Most of the manganese used in batteries is in the form of $MnO_2$. However, given its wide range of oxidation states, special lithiated manganese oxide compounds have been developed for lithium ion cells. The ore has other use in metal alloys and is widely available and very cheap (Table 2B.9).

**TABLE 2B.9**   Manganese Ore Supply

| Year | Production: U.S./global (metric ton, 000s) | Estimated price—~47% metric Mn ore (USD/mtu) | Reserves: U.S./global (metric ton, 000s) |
|---|---|---|---|
| 2009 | 0/9600 | 5.77–6.61 | 0/540,000 |
| 2017 | 0/16,000 | 4.40–5.88 | 0/680,000 |

*Sands (Titanium Oxide).*[19]   While metals, such as titanium, zirconium, and silicon, are not major components in batteries, it is important to mention the value of "sand" deposits as a source of these metals. Specifically, in certain areas (Florida, Western Australia, etc.) high purity sand deposits of various oxides can be quarried and purified to produce valuable metals such as titanium. Table 2B.10 shows details for one form of titanium oxide (ilmenite).

**TABLE 2B.10**   Ilmenite (Ti) Supply

| Year | Production: U.S./global (metric ton, 000s)—includes U.S. rutile data | Estimated price—54% ilmenite ore (USD/mtu) | Reserves: U.S./global (metric ton, 000s)—includes U.S. rutile data |
|---|---|---|---|
| 2009 | 200/5190 | 70 | 6000/680,000 |
| 2017 | 100/6200 | 170 | 2000/870,000 |

*Salts (Sodium Chloride).*[20]   A variety of salts are used in batteries. Many are synthesized from pure elements, but some salts may be directly mined or simply purified from brines (i.e., NaCl, LiCl, etc.). The electrolytic conversion of brines has enabled an entire industry (chlor-alkali production) that benefits the production of both plastics and inorganic chemicals that are both critical to the battery industry. However, a more important derivative obtained from brines includes chemicals such as caustic soda (NaOH). Sodium, potassium, and lithium hydroxide are key components in many battery electrolytes. Data for the production of table salt are shown in Table 2B.11.

**TABLE 2B.11**   Salt (NaCl) Supply

| Year | Production: U.S./global (metric ton, 000s) | Estimated price (USD/mtu) | Reserves: U.S./global (metric ton, 000s) |
|---|---|---|---|
| 2009 | 46,000/260,000 | 165 | Unlimited |
| 2017 | 73,000/280,000 | 190 | Unlimited |

### 2B.2.3 Strategic Minerals and Precious Metals

Other metals and minerals worth noting include:

1. Rare earth elements (e.g., yttrium) that are critical to technologies such as nickel-metal hydride cells and integrated circuits, and
2. Corrosion-resistant (precious) metals such as silver, gold, platinum, etc.

Both types of materials have important niche application in the battery industry. Additionally, a number of other metals are also being used in limited quantities (or are being evaluated for use) in new battery systems and may someday find widespread application. These include boron, tungsten, vanadium, tin, calcium, magnesium, and molybdenum among others. On the other hand, efforts are continuing to reduce or eliminate the use of toxic metals (mercury, cadmium, beryllium, arsenic, etc.). And while lead, antimony, and other heavy metals pose risks, the battery industry has attempted to limit human exposure and environmental hazards from the use of such materials.

## 2B.3 SUMMARY

Metals, minerals, and mining are key factors in the battery material supply chain. Companies and nations are engaging in numerous technical, political, and cultural efforts to secure supplies and satisfy the needs of all stakeholders. Any battery technology that is able to use common, low-cost, safe, environmentally sound raw materials will undoubtedly have a great advantage.[3] Finally, while the supply of metal and minerals is highly dependent on the mining industry, other types of materials are also largely obtained from mining operations, such as ceramics (Chap. 2D) and carbon/graphite (Chap. 2E).

## REFERENCES

1. https://worldoceanreview.com/en/wor-3/mineral-resources/manganese-nodules/ (extracted May 11, 2018).
2. J. R. Hein, "Manganese Nodules," in J. Harff, M. Meschede, S. Petersen, J. Thiede (eds.), *Encyclopedia of Marine Geosciences*, Encyclopedia of Earth Sciences Series, Springer, Dordrecht, 2016.
3. D. Rolison, J. Parker, J. Long, Dendrite-free rechargeable zinc-based batteries: Solving a chronic impediment through architectural design, 35th Annual International Battery Seminar and Exhibit, Ft. Lauderdale, FL, March 26–29, 2018, pp. 3–5, 13–15.
4. A. Ramkumar, A hunger for lithium juices deals. *Wall Street Journal*, May 18, 2018.
5. https://minerals.usgs.gov/minerals/pubs/commodity/myb/ (extracted May 11, 2018).
6. https://minerals.usgs.gov/minerals/pubs/commodity/copper/index.html#myb (extracted May 11, 2018).
7. https://www.usgs.gov/news/technical-announcement-usgs-puts-global-copper-assessments-map (extracted May 11, 2018).
8. https://minerals.usgs.gov/minerals/pubs/commodity/nickel/index.html#myb (extracted May 11, 2018).
9. https://minerals.usgs.gov/minerals/pubs/commodity/aluminum/index.html#myb (extracted May 11, 2018).
10. https://minerals.usgs.gov/minerals/pubs/commodity/lead/index.html#myb (extracted May 11, 2018).
11. http://www.recyclingtoday.com/article/battery-council-international-lead-battery-recycling/ (extracted May 11, 2018).
12. https://minerals.usgs.gov/minerals/pubs/commodity/zinc/index.html#myb (extracted May 11, 2018).
13. https://minerals.usgs.gov/minerals/pubs/commodity/lithium/index.html#myb (extracted May 11, 2018).
14. https://about.bnef.com/electric-vehicle-outlook/ (extracted May 11, 2018).
15. T. G. Goonan, Lithium Use in Batteries, U.S. Geological Survey Circular 1371, 2012, 14 p, http://pubs.usgs.gov/circ/1371/ (extracted May 11, 2018).
16. https://minerals.usgs.gov/minerals/pubs/commodity/cobalt/index.html#myb (extracted May 11, 2018).

17.  https://minerals.usgs.gov/minerals/pubs/commodity/iron_ore/index.html#myb (extracted May 11, 2018).

18.  https://minerals.usgs.gov/minerals/pubs/commodity/manganese/index.html#myb (extracted May 11, 2018).

19.  https://minerals.usgs.gov/minerals/pubs/commodity/titanium/index.html#myb (extracted May 11, 2018).

20.  https://minerals.usgs.gov/minerals/pubs/commodity/salt/index.html#myb (extracted May 11, 2018).

# SECTION C

# POLYMERS AND ORGANIC MATERIALS

**Kirby W. Beard**

## 2C.1 INTRODUCTION

Without organic materials, few practical batteries would likely exist. Organic molecules, starting with methane, cellulose, coal tar/pitch, and similar basic materials, are converted by the petrochemical and biochemical industries into numerous materials used directly within electrochemical cells, as packaging/insulating/support structures or as process aids for creating battery components. The categories include the following:

1. Solvents (organic liquids such as methanol, acetone, glycols, etc.)
   a. Polymer solutions for producing battery electrodes and other components
   b. Electrolyte solutions (i.e., nonaqueous lithium cells, ionic liquids, etc.)
   c. Process solvents for extracting, leaching, solvating, purifying, cleaning, etc.
2. Monomers (polymer synthesis)
3. Polymers (thermoset, thermoplastic, elastomers)
4. Miscellaneous (coatings, sealants, colorants, gases, etc.)

Hence, while organics are not used as active materials (except for a few exceptions, such as methane or methanol fuel cells, lithium-carbon monofluoride cells, etc.), organic materials are indispensable for achieving low-cost production, ruggedness, and long life. Organic chemistry is an extensive field and further references should be consulted for specifics.[1] Some general descriptions and uses of organic materials are listed next.

## 2C.2 MAIN CLASSIFICATIONS

While thousands of organic materials may be used in battery production and battery products, there are just a few key categories related to the battery industry. Battery scientists and engineers must be aware of the choices and scrupulous in selecting any given material for any particular use. The sections that follow discuss these main categories:

1. Polymers for binders/coatings/separators/insulators
2. Solvents for liquid battery electrolytes
3. Solvents used as processing media
4. Natural polymers such as cellulose used for casings and separators and for cellulosics

### 2C.2.1 Polymers

Polymers consist of many natural and synthetic molecules and are detailed in various reference books.[2] The fundamental characteristic of most polymers is the presence of long chains of carbon molecules. The two main polymer groups and several subgroups are summarized next.

***Thermoset.***   These polymers are created by reaction of short chain liquids (monomers) to form a cross-linked polymer that is typically rigid, temperature resistant, and durable in various solvents (depending on the exact polymer type). The more well-known materials include epoxy, polyester, phenolic, some urethanes, and silicones. Thermosets find limited but important uses in batteries such as electric terminal and case sealants and potting compounds.

***Thermoplastic.***   These materials consist of long chain, intertwined molecules, creating a fairly rigid solid but also allowing for flexibility (especially when heated and molded). Thermoplastics are used extensively in batteries, especially for the following applications:

1. Binders for electrodes
2. Separators
3. Electrical insulators (sheets, wire coverings, grommets, etc.) for both internal and external current-carrying parts
4. Barrier layers and spacers inside and between cells
5. Packaging for cells and batteries (including cell housings and battery compartments)

Thermoplastics represent a multibillion dollar expenditure for the battery industry that ranges from commodity polymers, such as polyethylene and polypropylene used to make lead-acid battery cases, to engineering plastics, marketed under tradenames such as Lexan®, Kapton®, and Teflon®, and used for high temperatures, ridged parts, electrical insulation, etc. One important distinction among thermoplastics is the degree of amorphous and crystalline structures (and the related glass transition temperature).

***Miscellaneous Polymers.***   Other special types of polymer used in the battery industry include elastomer sealants, cellulose paper and celluloid separators, polymer electrolytes (Chaps. 22A and 22C), etc. Various applications of these materials are discussed in more detail in the battery technology chapters (especially, Parts 4A and 4B).

The decision to use polymers for a specific component, rather than metal, glass, ceramic, etc., or to use one particular type of polymer over another, is typically based on performance (strength, stiffness, part weight/volume, etc.), processing ability, durability (corrosion, heat, wear, etc.), costs, and other intangibles (recycling, visual appearance, etc.). For some battery applications, such as military, medical, and aerospace, polymer selection may take years and involve multimillion dollar test programs. Despite these precautions, polymer failures of cell components (cell casings, separators, electrical insulators, etc.) from fatigue, creep/yield, oxidation, etc., still may occur.

***Polymer Characteristics.***   Each different polymer has its own range of properties. However, within each basic polymer group that has a common repeating structural unit, the properties will be somewhat similar. The basic structures and a few facts on each are presented as follows:

1. *Polyalkanes* ($-CH_2CH_2-$): These polymers are made from basic linear carbon chain molecules with single bonds (with or without branching side chains) and are derived from methane, ethane, etc. This group includes the polyolefins and are typically low cost and resistant to many chemicals. But as a result of good solvent resistance, they are not well suited for solvent processing. Strength and temperature limits tend to be fairly low. The bulk of all separators (for both aqueous and lithium cells), cell casings, and battery housings are either polyethylene or polypropylene, formed by melt extrusion or injection molding. And in the case of separators, the use of special stretching processes imparts high linear strength.

2. *Cyclic compounds (e.g., containing benzene ring structures)*: These polymers often have very large, complex repeating units with circular (aromatic) carbon structures and may also contain halogens and nitrogen (or even sulfur) atoms. Polystyrene and PET (polyethylene terephthalate) are prime examples. Properties may vary, but they are usually more rigid and robust than linear (aliphatic) chain polymers.

3. *Carbon-oxygen polymers* ($\diagdown \!C\!=\!O$, $-O\!-\!C\!=\!O$): These polymers are synthesized from carbonyl and carboxyl groups, polyacrylic acid, etc., and include polymers with ketone, ether, ester, carbonate, and similar structures. Alkanes form polymers such as the polyketones, and cyclic carbon rings are used to make polycarbonates such as Lexan. Polymers, such as acrylics, can have good transparency and hardness.

4. *Halogenated polymers* ($-CH_2CHCl-$, $-CH_2CF_2-$, $-CF_2CF_2-$, etc.): These materials include the various polymer types listed above but with substituted chlorine, fluorine, or both. They include polyvinyl chloride (PVC), Saran®, and specialty chlorofluoropolymers. Such polymers have found success as lead-acid separators (PVC) and lithium ion electrode binders (polyvinylidene fluoride [PVDF]). Fluoropolymers, such as Teflon and Tefzel®, have been used in batteries as electrode binders, and heat- and chemical-resistant insulators. Some of these materials exhibit very little softening upon heating and will survive to extreme temperatures.

5. *Nitrogen polymers:* These materials may be based on many of the structures listed previously, but incorporate nitrogen atoms. They include Nylon®, polyacrylonitrile, and many specialty polymers. Nylon has been used in many alkaline cells due to superior chemical stability. This list also includes the polyurethanes, polyamides/imides (Kevlar® and Torlon®), and polyamines (epoxy hardener).

6. *Miscellaneous:* Other more exotic polymers, such as polyalkenes, and silicone- and sulfur-based polymers (e.g., polysulfone and polyether sulfone, or Udel®), have found some use in battery applications, but typically only for limited, specialized applications. The cost and processing ability are sometimes impediments to more widespread use.

Also, a most important characteristic for polymers is their solution properties. Specifically, if a polymer component cannot be produced properly by melt processing (i.e., heating and molding, extruding, pressing, etc.), then the polymer must be dissolved in a solvent (typically, organic liquids or sometimes water) and cast as a solution into a sheet, membrane, coating, etc. Section 2C.2.3 discusses various process solvents, but the interactions of solvents and polymers require extensive analysis to achieve optimal production processes.

## 2C.2.2   Electrolytes

With the rapid growth of lithium ion batteries, the field of nonaqueous electrolytes has expanded greatly. While aqueous and inorganic solvents are still used in many cells, organic solvents used in lithium cells represent the overwhelming majority of new applications. As noted in Chap. 17, the costs of these organic solvent electrolytes are significant. Most of these solutions use inorganic salts, but some success has been achieved with organic salts and ionic liquids (see Chap. 3B for details on organic electrolytes).

## 2C.2.3   Process Solvents

Given the extensive use of polymers in the fabrication of battery components, the use of solvent processes is expected. While polymer extrusion (i.e., a polymer melt) can be used to produce many of these components, the need to incorporate solid particulates (powders, fibers, etc.), especially at high-volume loadings, often precludes the use of extrusion or thermoforming and requires the use of solvent mixing/casting. Slurries of dissolved polymers and fillers are mixed, deposited as a film, and dried to form both electrodes and ceramic-based separators for lithium cells. The solvents, which may be alcohols, ketones, aliphatic hydrocarbons, or other types, are fairly costly and are typically used at high levels. Hence, solvents removed from the cast films by heating/drying are often reclaimed by carbon bed absorption, distillation, and condensation processes, rather than incinerated. Other uses of organic solvents may include synthesis of active materials and cleaning of metal parts prior to cell build, welding, etc. Details on solvent properties can be found in various sources.[2,3]

## 2C.2.4   Cellulose

Early lead-acid battery designs used wood shingles for the separator. Today lead-acid cells still use wood-based products as electrode additives (Chap. 14). In addition, cellulose paper, which was used 100 years ago as the casing for dry cell batteries, is still used today for separators in various primary cells (Chaps. 11 and 12). A number of cellulose derivatives, such as carboxymethyl cellulose and cellophane, are also used in modern cells.

## 2C.3 PROPERTIES

An extensive array of material properties can be found in numerous reference books on polymers, solvents, and other organic materials. When selecting a material for use in a battery, these references may provide good information to make an initial selection. However, given the uniqueness of electrochemical cells, additional testing will likely be required. A polymer binder or an electrolyte solvent may be totally stable when exposed to the active materials or metals used in a cell. But when placed in the highly oxidizing or reducing environments found at the electrodes, unexpected results may occur. As noted in Chaps. 3B and 17, very few lithium cell electrolytes are stable from 0 to 4.5 V. Hence, materials must be verified for use in each applicable cell environment and under all conditions of use and abuse. Typical properties to be checked for polymers used as binders, separators, or insulators include:

1. *Physical:* density, tribology, surface energy
2. *Mechanical:* strength, modulus, creep, fatigue, yield
3. *Thermal:* glass transition, softening point, melt point
4. *Chemical:* acid/alkali/solvent/water resistance, solubility parameters
5. *Electrochemical:* cyclic voltammetry, ionic conductivity
6. *Electrical:* voltage breakdown, dielectric strength

Materials used for battery housings, wiring, or other external components will potentially require additional testing based on the environmental conditions (radiation, atomic oxygen, ozone, UV, etc.).

Once these properties are determined, the final step will always likely include complete cell and battery performance and safety testing, along with long-term storage and cycling tests over the full temperature range.

## REFERENCES

1. https://en.wikibooks.org/wiki/Organic_Chemistry (extracted June 16, 2018); J. G. Smith, *Organic Chemistry*, 4th ed., McGraw-Hill, New York, NY, 2014.

2. J. Brandrup, E. H. Immergut, *Polymer Handbook*, 3rd ed., Wiley-Interscience Publication, John Wiley & Sons, New York, 1989.

3. J. A. Riddick, W. B. Bunger, T. K. Sakano, *Organic Solvents—Physical Properties and Methods of Purification*, 4th ed., Wiley-Interscience Publication, John Wiley & Sons, New York, NY, 1986.

# SECTION D

# CERAMICS AND INORGANICS

## Kirby W. Beard

## 2D.1  INTRODUCTION

Ceramics and similar inert, inorganic compounds are another important category of materials used in the battery industry. Many of these materials include metallic elements, and most metals in fact are manufactured from metal oxides, sulfides, carbonates, etc. Ceramics typically do not participate in the electrochemical reactions (i.e., except for solid-state lithium cells and similar systems), but instead serve as insulators, fillers, and such. Inorganic compounds include a wide variety of material types and are typically used to synthesize active materials or formulate electrolytes. Some of the raw material types discussed next (i.e., bauxite, $TiO_2$, etc.) were detailed in Chaps. 2A and 2B in connection with the production of metals or with their use as active materials. The purpose of this section is to provide supplemental details on alternate forms and uses of such materials not fully covered in the prior sections. Additional references should be consulted on specific materials or applications. Much of the data in the following sections are available online on the USGS's website.[1]

## 2D.2  GENERIC DESCRIPTIONS

### 2D.2.1  Ceramics and Glass

Ceramics and glass are generally characterized by high temperature and electrical stability and have a fairly wide range of chemical resistance. While some special types of ceramic materials may function as active materials or solid-state electrolytes (see Chap. 22), the typical ceramic or glass is extremely inert, although strong alkali solutions and lithium metals may sometimes react or corrode certain compositions. However, except for such special cases, ceramics and glass are chosen for use in battery cells and packs for their ability to electrically insulate or isolate current-carrying parts or to provide high temperature and electrochemical stability. The use of these materials is preferred when polymers are unable to meet temperature or electrical requirements or are subject to chemical attack from acid, alkali, or organic solvent solutions.

### 2D.2.2  Inorganic Compounds

This category refers to the various generic inorganic chemicals that are consumed by the battery industry in the production of battery components. For instance, most active materials are synthesized by chemical reactions using ingredients such as sulfuric or nitric acid, calcium or ammonium hydroxide, chlorine or fluorine gas, bicarbonates or carbonates, etc. In other cases, the materials may be used directly in electrolyte solutions. For example, lithium carbonate or hydroxide is sometimes added to both alkaline primary and secondary cell electrolytes as well as lithium cells. The use of these materials in batteries is not always obvious, but details on costs, quantities, and sourcing need to be considered when analyzing large-scale production requirements. In recent years, materials such as potash (potassium carbonate) and helium (used as a processing media in certain battery technologies) have both had episodes of supply shortages.

### 2D.2.3   Physical Form

One major distinction among ceramics is based on physical forms. Ceramics and glasses can be used as fine powders (chemical reactants, precursors, fillers, etc.) or processed (i.e., fired, fluxed, bonded, etc.) into fibers, fibrous mats/cloth, solid bodies, porous sheets, etc. The intended use of the material will dictate the processing methods and final form. However, one of the bigger, more recent challenges facing battery component producers is the development of technology to process ultrafine particles (i.e., submicron and nanoparticles). Finer powders are now required in separators and electrodes to allow for enhanced performance, but poor packing density, repulsive forces, and other factors often cause issues in the use of fine powders.

The following sections highlight a random sampling of ceramics, glasses, and other inorganic raw materials relevant to the battery industry. Part 4 provides details on the use of ceramics and inorganics in various electrochemical cells.

### 2D.2.4   Oxides and Silicates

These materials may be used directly as active materials or fillers. Or they may be processed into bulk forms or converted to other compounds. Hundreds of applications are possible.

***Refractories (Alumina, Aluminosilicates, etc.).***[2]   Bauxite was referenced in Chap. 2B for use in aluminum production, but it is also important in manufacturing of alumina, an important component in ceramic production (see Table 2B.3 for supply details). Aluminosilicates are another important source of raw material for refractory products. Kyanite, mullite, and other minerals are major sources of the raw materials used to make high-temperature ceramics as detailed in Table 2D.1.

**TABLE 2D.1**   Aluminosilicate Supply

| Year | Production: U.S./global (metric ton, 000s)—various types | Estimated price—kyanite (USD/mtu) | Reserves: U.S./global (metric ton, 000s)—various types |
|---|---|---|---|
| 2009 | 80/440 | 250 (350 calcined) | NA (large) |
| 2017 | 90/NA | 270 (420 calcined) | NA (large) |

***Magnesium Oxide.***   Magnesium metal is used in a few types of cells and can be obtained from rock minerals, brines, seawater, and other sources. Also, sulfates, hydrates, and carbonate forms find some uses in the battery industry. However, one of the larger uses of this metal is in refractory products in the form of MgO. Details on sources, supplies, costs, etc., are available on the USGS's website.[3]

### 2D.2.5   Carbonates and Hydroxides

***Lime.***[4]   Lime is composed of calcium oxides and hydroxides, but calcium can also be obtained from carbonate rock (limestone, coral, etc.). While calcium ions are present in some electrochemistries, the real value of lime is for use in chemical synthesis. Supply estimates are shown in Table 2D.2.

**TABLE 2D.2**   Lime Supply

| Year | Production: U.S./global (metric ton, 000s)—various types | Estimated price—quick lime (USD/mtu) | Reserves: U.S./global (metric ton, 000s)—various types |
|---|---|---|---|
| 2009 | 15,000/280,000 | 101 (136-hydrate) | Large |
| 2017 | 18,000/350,000 | 123 (149-hydrate) | Large |

*Potash.*[5]    Potash refers to potassium carbonate. Rock deposits are the main source, but wood ashes were historically used to make potash. The value of potash is, of course, as a source of potassium, which is used in many electrolytes (i.e., KOH). Costs are under USD 1000/metric ton with nearly 4 million metric tons of reserves. However, the main use of potash is in fertilizer production. It is virtually irreplaceable for agricultural applications and that market will dictate any trends for supply and cost.

### 2D.2.6  Miscellaneous

*Phosphates.*[6]    Phosphates are another key ingredient in fertilizers and are an indispensable product for agricultural use. Hence, these uses dictate supply and costs. Phosphates are only available from phosphate rock (typically as $P_2O_5$) and mostly mined in the United States. Elemental phosphorous, phosphate salts, and phosphoric acid have been used in various electrodes, plastics, and other applications relevant to batteries, but the current most important use is in lithium ion electrolyte ($LiPF_6$). Data for mined phosphate rock are shown in Table 2D.3.

**TABLE 2D.3**  Phosphate Rock Supply

| Year | Production: U.S./global (metric ton, 000s)—various types | Estimated price—phosphate rock (USD/mtu) | Reserves: U.S./global (metric ton, 000s)—various types |
|---|---|---|---|
| 2009 | 27,200/158,000 | 50 | 1,100,000/16,000,000 |
| 2017 | 27,700/263,000 | 75 | 1,000,000/70,000,000 |

*Sulfur.*[7]    Sulfur has a number of uses in batteries. Undoubtedly, the key use is in lead-acid batteries as a constituent of sulfuric acid. But sulfur is also used as electrodes ($FeS_2$), electrolytes ($SO_2$), salts (lithium triflate), and many other peripheral products. The largest new potential use for sulfur may be in elemental form as an active cathode for new types of lithium secondary cells. The biggest source of sulfur is as a by-product of oil and gas production. Supply and costs can vary greatly based on oil production and competition from other uses such as phosphate rock processing for fertilizers. Table 2D.4 details the supply figures.

**TABLE 2D.4**  Sulfur Supply

| Year | Production: U.S./global (metric ton, 000s)—various sources | Estimated price—(USD/mtu) | Reserves: U.S./global (metric ton, 000s)—various sources |
|---|---|---|---|
| 2009 | 9800/70,300 | 10 | Large (oil reserves) |
| 2017 | 9660/83,000 | 60 | Large (oil reserves) |

*Silica and Silicon.*[8]    Silica is the common name for silicon dioxide ($SiO_2$) and is the base material in quartz, comprising 25% of the earth's crust as silicon. Silica is readily available everywhere and costs very little, depending on local availability of silica sand. $SiO_2$ is widely used for glass-making and for glass fibers, fillers, and additives. Battery applications of silica include glass mat separators, fumed silica for gelled electrolytes, fillers for plastics, and many other uses.

Another emerging application for silicon, as the elemental metal, is for potential use as anode material in lithium ion cells. The refined metal is fairly low cost (USD 100/metric ton silicon), but when used in lithium batteries, ultra-high-purity grades may be required. Until silicon anodes are perfected, the most important battery application for silicon metal will be in the integrated circuits used for charging electronics.

## 2D.3  SUMMARY

Mining, extraction, and processing of various inorganic materials for use in ceramic products, chemical operations, metallurgical plants, and other industries is a huge and diversified business. Battery manufacturing depends on this existing infrastructure. As suggested in this chapter, the supply chain of raw materials for battery manufacturing can be complex, diverse, and at times erratic.

## *REFERENCES*

1. https://minerals.usgs.gov/minerals/pubs/commodity/myb/ (extracted May 11, 2018).
2. https://minerals.usgs.gov/minerals/pubs/commodity/kyanite/index.html#myb (extracted May 11, 2018).
3. https://minerals.usgs.gov/minerals/pubs/commodity/magnesium/index.html#myb (extracted May 11, 2018).
4. https://minerals.usgs.gov/minerals/pubs/commodity/lime/index.html#myb (extracted May 11, 2018).
5. https://minerals.usgs.gov/minerals/pubs/commodity/potash/index.html#myb (extracted May 11, 2018).
6. https://minerals.usgs.gov/minerals/pubs/commodity/phosphate_rock/index.html#myb (extracted May 11, 2018).
7. https://minerals.usgs.gov/minerals/pubs/commodity/sulfur/index.html#myb (extracted May 11, 2018).
8. https://minerals.usgs.gov/minerals/pubs/commodity/silicon/index.html#myb (extracted May 11, 2018).

# SECTION E
# CARBON AND GRAPHITE

## Kirby W. Beard

## 2E.1   INTRODUCTION

There are two basic types of elemental carbon: amorphous and crystalline. Pioneering research identified the existence of purely crystalline forms of carbon, designated as fullerenes and graphene.[1,2] While graphene and fullerene offer promise for the energy storage industry, the supply is based on processing of carbon-based materials. The costs and volumes will depend on many factors that are not yet optimized. Hence, details on sourcing of sheets, tubes, spheres, etc., of these carbons are not discussed at length in this section.

Graphite has both metal and nonmetal properties and is considered both an organic and a nonorganic. The metallic properties benefit electrical and thermal conductivity, and the nonmetallic properties include inertness and thermal stability.[3]

Similarly, while carbon black and natural graphite have been used for decades in battery components, recent development efforts have led to purified and hybridized structures that have found great success in many batteries. However, again, these materials are derived from modifying natural products or from synthesizing new forms from scratch. The supply and costs vary accordingly.

Overall, then, the two basic carbon materials are produced by either petrochemical synthesis or mining of ores. By subjecting hydrocarbons to partial oxidation (i.e., incomplete combustion), carbon in the form of soot or coke is produced. This form of carbon (also called carbon black) has use as fillers for plastics, paint, ink, etc.; in steel production; as lubricants; and for other purposes. However, batteries have used various carbon species for over a century as an electrode component that enhances electrical conductivity and/or provides a surface for electrochemical reactions. These carbons are basically inert and do not participate in the redox reactions. Some carbon types, such as graphite, have also been used as conductive diluents, but have been modified to allow intercalation with lithium ions, making such materials a major component of the lithium ion battery industry. Modified graphite is made by various chemical and thermal processing operations, but the use of natural (i.e., mined) graphite is a primary source.

## 2E.2   GRAPHITE SUPPLY[3]

The supply of carbon is virtually unlimited since any fossil fuel or carbohydrate can be used to create carbon black. In fact, some of the preferred materials for capacitor applications come from materials such as carbonized coconut shells. The supply of graphitic forms of carbon is more limited, but it still may approach 800 million tons. Supply estimates are shown in Table 2E.1.

**TABLE 2E.1**   Graphite Supply

| Year | Production: U.S./global (metric ton, 000s) | Estimated price (USD/lb) | Reserves: U.S./global (metric ton, 000s) |
|---|---|---|---|
| 2009 | ~0/1130 | 866–2580 various, 256 amorphous | ~0/71,000 |
| 2017 | ~0/1200 | 1400–1840 various, 392 amorphous | 0/270,000 |

While supplies appear adequate for now, the USGS reports that one battery plant in the United States may soon require 93,000 tons of flake graphite (annually) that will be converted to 35,200 tons of spherical graphite. Largest reserves are in Turkey, Brazil, and China, but Canada has opened new mines recently. Fuel cells are also potentially a major use for graphite that could exceed all other uses combined.[3]

## 2E.3    SUMMARY

The role of carbon and graphite in batteries is evolving as new techniques provided new structures with new functionalities. Currently, lithium anode materials are being produced with various post-processing techniques that provide improved performance. Another new concept includes the use of carbon in a lead-acid cell, where activated carbon replaces the active lead anode material and serves as a site for hydrogen storage.[4,5] A key requirement for any graphite deposit includes the purity level and the ability to purify the ore body.

## REFERENCES

1. Nobel Prize in Chemistry "for their discovery of fullerene," R. F. Curl, Jr., H. W. Kroto, R. E. Smalley, https://www.nobelprize.org/nobel_prizes/chemistry/laureates/1996/ (extracted May 15, 2018).

2. Nobel Prize in Physics "for groundbreaking experiments regarding the two dimensional material graphene." A. Geim, K. Novoselov, https://www.nobelprize.org/nobel_prizes/physics/laureates/2010/ (extracted May 15, 2018).

3. https://minerals.usgs.gov/minerals/pubs/commodity/graphite/index.html#myb (extracted May 15, 2018).

4. Axion Power, www.axionpower.com.

5. https://businessjournaldaily.com/axion-power-enters-new-era-as-rd-company/ (extracted May 15, 2018).

# CHAPTER 3
# BATTERY COMPONENTS

## 3.0 OVERVIEW

The various raw materials incorporated directly into batteries or used indirectly in the production stream were reviewed at a high level in Chap. 2. The next step in the supply chain is to convert raw materials into battery components or peripheral items. As described in Chap. 1, the main components of typical electrochemical devices are listed below and are further detailed in the referenced sections of this chapter.

Chapter 3A: Electrodes

Chapter 3B: Battery Electrolytes

Chapter 3C: Separators

Chapter 3D: Electrical Connections from Cell to End User

Chapter 3E: Cell and Battery Packaging

This list is not definitive, as there are many unique battery designs that may use other components or even use alternative versions of these components. The main purpose of listing and analyzing the components used in an electrochemical device is to help provide a better understanding of cost and supply issues. For example, at a recent battery conference a paper was presented that broke costs and production volumes into various categories of components.[1] The analyses included estimates for NCA (Ni-Co-Al type) lithium-ion cells based on cost categories (labor vs. materials), the size of the production operation (up to 10 GWh annual capacity), cell configuration (cylindrical vs. prismatic), historical trends, etc. Current total costs, based on unit energy, are $280/kWh. While material costs clearly dominate (~$120/kWh), the cost of equipment, presumably depreciated over the life of the factory, was the second highest category (~$65/kWh). Hence, the complexity of the component manufacturing operations has a great influence on the final battery costs.

Labor and overhead also contribute a fair share to overall costs, but modifications to the components will be needed to meet the U.S. Department of Energy 2022 goals ($125/kWh). An interesting statistic in this analysis is that the cell hardware actually dominates material costs (~50% higher than cathode material). Separator, electrolyte, metal foil, and anode materials followed, respectively. Therefore, to make a greater impact on material (and total cell) costs, the implication is that battery architecture and packaging concepts should be a key focus.

## Reference

1. R. Ciez and J.F. Whitacre, "The costs and environmental impacts of lithium-ion battery production and recycling," *35th Annual International Battery Seminar and Exhibit*, Fort Lauderdale, FL, March 26–29, 2018.

# SECTION A

# ELECTRODES

**Trevor D. Beard, Kirby W. Beard**

## 3A.1   *INTRODUCTION*

Aside from differences in materials, electrodes will vary in design, physical properties, and functioning, depending on the specific battery electrochemistry and application. For various types of electrodes, the processing techniques and production equipment may also vary. Typically, electrode manufacturing is the most complex, expensive, and enabling technology in any battery. While electrode winding or stacking and cell assembly are also major complex operations, those processes are mainly concerned with mechanical issues. Cost and safety are negatively affected by poor assembly operations, but electrode processing has mechanical, electrochemical, and physical constraints that greatly impact performance as well.

In all cases, however, electrodes must serve the dual functions of transporting electrons and providing sites for oxidation-reduction (redox) electrochemical reactions—except in the case of some devices, such as capacitors, where the reactions involve charge separation only. In all cases, electrons flow into the electrode through electrically conductive pathways, while ions migrate through the electrolyte or active material (often starting from a source at the opposing electrode). The factors that benefit electron conduction often detract from ion transport. Hence, electrode designs are intrinsically a compromise of conflicting goals. And worse, what benefits one electrode may impede the performance of the opposite one. Electrodes, therefore, must be optimized both individually and as electrochemical pairs. This effort must also include striking a balance of properties, especially the specification of the relative rate capability and capacity of the individual electrodes. Various cell types may require precisely balanced properties, while others may purposefully have mismatched electrodes that achieve special results.

Various electrode types and associated production methods are summarized in the sections below. Greater details may also be found in the chapters on specific electrochemical cells in Part 4 of this Handbook.

## 3A.2   *GENERIC ELECTRODE TYPES*

Electrodes may be designed and categorized in four generic ways:

1. By void content (porosity) and pore structure
2. By electrical current collection design
3. By the bulk physical parameters (thickness, aspect ratio, etc.)
4. By the unit processes/operations used in electrode production

Additionally, various specialized or unique electrode designs, such as bipolar, interpenetrating three-dimensional printed,[1] bulk liquid tank storage (i.e., redox flow cells), etc., are feasible. (These various types of electrodes are covered in Part 4.)

1. *Porosity:* In this first category, the main issue is whether the electrode is a nonporous material (i.e., solid, gel, or liquid) or a porous structure. Solid electrodes rely on surface reactions or on ionic diffusion through the solid. Porous electrodes will typically involve transport of ions through a network of pores (whether macro-, micro-, or nanoporous). The pores are typically filled with liquid electrolyte solutions, but other configurations

(solid electrolytes) may be employed. As detailed in Chaps. 4 and 6, optimization of ion activation and flux can be quite challenging. Improvements rely on understanding difficult calculations or on conducting massive multivariate experimental designs. Solid electrodes also have issues, such as the evenness of active material utilization and ability to replate or redeposit materials on the surface during recharge of secondary cells.

2. *Electrical conductors:* Current collection is also a challenging design parameter. While some active materials readily conduct electrons, most electrodes still require a fine, highly dispersed, but connected network of conductive members, particles, or filaments. Carbon black and graphite are used in lithium-ion cells, but lead-acid and nickel-cathode cells rely mostly on metal current conductors, which have much better intrinsic electrical conductivity. Additionally, while the use of conductive diluents (i.e., carbon) is acceptable for localized current collection, the effective transport of electrons from within a large plate to a single tab is much more difficult. Lead-acid plates often have lead grids with a radial pattern that is embedded throughout the thickness and plane of the electrode that helps minimize the resistive losses as the current flow increases in closer proximity to the tab. Lithium-ion electrodes typically use only a thin metal foil current collector (a much less optimized design for tab connection).

3. *Physical parameters:* This fundamental design consideration (i.e., thickness, density, length, width, flat/bobbin/folded/wound styles, etc.) is a main factor in cell performance as well as manufacturing economics. Thinner electrodes, for example, typically have better rate capability and potentially even higher cycle life but are expensive to produce. Not only is machine throughput reduced with thinner electrodes, but also higher relative amounts of current collector and separator are consumed, further contributing to high cell costs and also reducing energy density. The history of many battery technologies has been a story based on electrode thickness and density (i.e., porosity) optimization.

4. *Unit processes:* In the chemical process industry, production operations are often broken into discrete steps, called unit operations.[2] Electrode manufacturing follows this basic model by employing a number of known chemical and material processing techniques in one or more sequences to yield a finished electrode. Some of the basic unit operations employed in electrode manufacturing include the following:

   a. Grinding/classification of powders

   b. Particle coating/drying/blending

   c. Mixing (heated/cooled)—pastes/slurries

   d. Deposition—coating/spraying/molding/filtering

   e. Densification (heated/cooled)—pressing/calendaring/rolling/die forming/centrifuging

   f. Cutting/stamping/slitting

   g. Advanced processes (i.e., three-dimensional printing)

Other than metal foil (Li), liquid ($SO_2$), or gaseous ($H_2$) electrodes, which simply use a base material in a mostly unadulterated form, a typical electrode will require at least a half dozen steps to produce a finished product. Many of these operations rely on empirically determined techniques and know-how to achieve optimal results. Some of the equipment is off-the-shelf, but much of the time customized machinery and process modifications are used. The quality of the electrodes may be screened by the use of physical measurements, but often only full-scale cell testing will validate the acceptability of the product. Building and testing of cells as a quality assurance measure is fraught with difficulties and still does not provide absolute guarantees of long-term performance. A summary of the various typical types of electrodes is listed next.

## 3A.3    SPECIFIC ELECTRODE TYPES

### 3A.3.1    Electrodes Based on Metal Sheet/Foil/Mesh Active Materials

The most common active materials for use as metallic electrodes are lead (Chap. 14), zinc (Chaps. 11, 15, etc.), and lithium (Chaps. 13, 17, etc.). Other metal electrodes have been researched and commercialized over the years but to a far lesser extent. These include cadmium (which is far less common now than 20 years ago), nickel, magnesium, aluminum, iron, copper and even sodium, tin, and a handful of other exotic metals or alloys. Many of these metals have reasonable costs and availability and are easily processed into foils or bonded powders. Such electrodes can then be readily incorporated into cells and will often have excellent specific capacity and reactivity (rate capability).

A current collector of a different composition may be used in these cells (i.e., nickel grids with lithium foil anodes). Additionally, while metals typically have good electrical conductivity, the main challenges in using metal electrodes are purity levels and surface contamination during processing or from electrolyte impurities. In summary, though, metal electrodes have been extensively studied and so much of the new research has shifted to other types of electrodes (see below).

### 3A.3.2   Electrodes Based on Powdered or Solid Inorganic Materials

This group includes the typical metal oxides common to a majority of electrochemistries. Some of the more typical materials include manganese dioxide, lead dioxide, nickel oxyhydroxide, and lithium cobalt dioxide, or a variety of similar transition metal lithium-ion insertion compounds. And of course, there are numerous other electrodes that have been used experimentally or are in limited production, including a variety of oxides, sulfides, or nitrides based on iron, titanium, tin, aluminum, magnesium, copper, mercury, and the like.

Another key material, applicable to this category, is carbon (see Chap. 2E). Carbon, which may be classified as organic, serves as the base material for most lithium-ion anodes, but it is also finding new uses in lead-acid cells (see Sec. 3A.5). Additionally, carbon is being engineered to function in concert with other materials to improve energy and stability. For instance, silicon, which is another new type of inorganic lithium-ion battery anode under development, is being combined with carbon to yield an improved active lithium anode material.[3]

Another inorganic material that has proven viable for use in lithium batteries is sulfur (in elemental form and in various sulfide compounds). First seriously considered in the 1990s for the construction of electrodes, lithium-sulfur batteries are now considered a major contender for the next generation of lithium batteries (Chap. 17B). One innovative example specifies the use of microbeads with $TiO_2$ shells and sulfur cores.[4]

Additionally, fluorocarbon polymers ($CF_x$) have been used commercially for decades in high-capacity lithium primary cells (Chap. 13). Lithium reacts directly to form carbon and LiF and is perhaps the highest energy density cell now available. Work has also been done and continues on developing ceramics and glasses (powders and sheets/films) as active materials, especially for use at high temperatures.

The key requirement in using most of these inorganic materials to produce electrodes is that the active materials, which are typically small particulates, must be fused, bonded, or entrapped/confined to function as a battery electrode. Some compounds, such as lead oxide, can be bonded together and onto a grid or current collector simply by reacting with moisture, which allows crystals to grow and interpenetrate until a stable electrode mass is created. Other powders can be fused by heat or pressure (i.e., sintered nickel-based cathode cells as per Chap. 15) to produce rugged electrodes.

However, a common technique to produce electrodes from particulates is via the use of a polymer binder. In general, polymer binders are used to produce electrodes in different ways as follows:

1. Mixing of dry polymer powders or liquid dispersions with the active materials to create a dough-like mixture that is kneaded and pressed or rolled into electrodes

2. Dissolving the polymers in a solvent to form a solution to which active materials and conductive diluents are added, followed by film casting and drying, usually using a foil current collector substrate

3. Drying polymer/active material mixtures produced by method 1 or 2 listed above into agglomerates that are then spread as dry powders into films, sheets, or coupons and compressed or bonded by other means

Further details on various inorganic active material processing techniques are discussed in the electrochemistry chapters (Part 4).

### 3A.3.3    Electrodes Based on Polymer Active Materials

A new innovative concept, identified in 1977, for the use of polymers as active materials in electrochemical devices is credited to work by MacDiarmid, Shirakawa, and Heeger. Their invention, for which they were awarded a Nobel Prize in Chemistry in 2000,[5] was based on ion doping of various organic polymers (e.g., poly-acetylene). Previously, it was not known that polymers could be "charged and discharged" and therefore crafted into a functional battery or capacitor. The electrodes were simply cast from solvent mixtures of the polymers and various ionic species. While these particular couples did not offer a commercially viable product at the time, interest in polymer electrodes has continued. Recent studies have identified the use of porphyrins, rubeanic acid, and trioxotriangulene as potential active cathode materials.[6,7] Such materials may have lower material costs and provide processing benefits when compared to metal-based materials.

### 3A.3.4    Polymer Composite Electrodes

A second wave of polymer research in the 1980s (starting with Armand in 1978)[8] centered on the concept of using solid (or gelled) conductive polymers as a binder for inorganic active material particulates (Sec. 3A.3.2). The potential ability to use lower-cost polymers and existing polymer processing equipment and technology to produce electrodes was considered a primary benefit. Secondly, the use of thermoplastic polymer binders had potential to benefit recycling of the electrodes. But also, by using conductive polymers as both an electrode binder and an ionically conductive separator, interfacial resistances could potentially be greatly reduced.

Initial trials used polyethylene oxide, compounded with an organic lithium salt (lithium triflate). This material was extruded into thin cathodes with added active material (i.e., $V_6O_{13}$) and attached to a current collector. A second extrusion of pure polymer electrolyte was then produced for use as a separator by laminating to the cathode. Later versions of polymer composite electrodes (and separators) were produced in a similar fashion but relied on the absorption of liquid salt solutions to provide improved conductivity. This technology was developed by Bellcore (later called Telcordia) but was sold to Valence Technology in 2000.[9] Current efforts by others include the use of solid polymers and ionic liquids as low volatility polymer gelling agents.[10,11]

### 3A.3.5    Liquids and Gases

Liquid electrodes, sometimes known as catholytes (or anolytes), have existed for decades. The unique feature of such electrodes is that the liquid active material also serves as the solvent for anions and cations. Sulfur dioxide and thionyl chloride are two of the more well-known examples (Chap. 13). More recently, redox flow cells (a type of battery-fuel cell hybrid detailed in Chap. 22B) have employed liquid (or slurry) electrodes.[12] Gases, such as hydrogen and methane, have also been used for active materials in certain types of electrochemical devices (i.e., fuel cells).

The key to the use of liquids and gases is that these active materials must be separated in one of two typical ways:

1. Physically separated by a selective polymer membrane or a conductive ceramic composite (typically used when both electrodes in the cell are liquids)
2. Separated by creation of a passivation layer on the opposing electrode (typically used in lithium metal anode cells)

Details for a variety of these electrochemical systems are covered in subsequent chapters (see Part 4).

The production of these electrodes is, seemingly, simply a matter of mixing the ingredients and injecting into a cell container. However, the nuances of having too much or too little active material can be catastrophic. Complex, detailed analyses of numerous events, such as thermal expansion of the liquid, pressurization at elevated temperature during normal use or abuse, evolution of gaseous decomposition products, crystallization over time or at reduced temperature, etc., can contribute to sudden death of the cell or spontaneous catastrophic failure. Hence, while liquid and gaseous electrodes at first seem like a simple, uncomplicated, and very desirable type of electrode, these problems must be fully analyzed and extensively tested.

## *3A.4   NEW ELECTRODE INNOVATIONS*

### 3A.4.1   Governing Principles

The incentives for research and development efforts to improve electrode technology are due to industry needs for more environmentally responsible, cost-effective, and technologically enabling processes. For instance, hazardous solvents used to create slurries for lithium-ion electrode production have been replaced by some battery manufacturers with water-soluble polymers, such as SBR (styrene-butadiene rubber)-type thermoplastic elastomers. The replacement in recent years of expensive hazardous solvents, such as N-methyl-3-pyrrolidone, is due in large part to increased government regulations for more eco-friendly and safer manufacturing prac-tices, and also due to potential cost benefits.

However, material cost savings are not the only cost advantage possible. The development of innovative electrode manufacturing practices may also help reduce actual production costs. To date, manufacturers of lithium-ion cells have primarily focused on lowering material costs, given its major contribution to direct vari-able costs. So, while reductions in material costs are helpful, any savings that can be achieved from improved process efficiencies provide an additional competitive edge. For instance, higher throughput electrode produc-tion (i.e., greater electrode ampere-hours per unit of machine time) equates to lower labor-hours and lower indirect costs per kWh of battery produced.

Additional savings can also come from reduction or elimination of solvent recovery or disposal costs. Dry polymer processing techniques, such as the use of kneaded, fibrillated polytetrafluoroethylene (PTFE) dough compounds, have gone through extensive development efforts due to their obvious potential benefits (Sec. 3A.3.2). However, the results have not been fully satisfactory due to either scale-up issues or inade-quate performance compared to solvent process electrodes. Hence, further innovations, such as dry powder electrostatic spray coatings, are being studied.[13]

Additionally, the development of new active and conductive materials has altered the landscape for electrode manufacturing. Sulfur is one such example of an inexpensive and potentially high-energy cathode that is under development and could benefit from an enhanced manufacturing process. Silicon-based anodes for lithium-ion cells provide another example of where an improved electrode manufacturing process could pay dividends as detailed below.

### 3A.4.2   Limitations of Current Electrode Production Processes

The incorporation of larger volumes of smaller particle sizes into battery components is a major trend in lithium-battery chemistries. As an example, sulfur cathodes need to be much thicker and should incorporate a high level of conductive carbon to improve electron transport. High loadings of carbon nanotubes or graphene nanoflakes are being evaluated in cathodes as a means to counteract the nonconductive nature of elemental sulfur. Specifically, porous polymer-carbon composite films, based on solvent coating and web processing oper-ations, are produced for subsequent impregnation with molten sulfur. However, these solvent cast cathode sub-strates have proven inadequate. Current solvent coating methods do not utilize the polymer binders efficiently. Volume loadings of nanosized carbon particles are minimal, and the ability to create thicker electrodes is also impacted. Most solvent coating processes are not able to produce electrodes above 250 micron with acceptable trade-offs of porosity/density/mechanical integrity.

As another example, the use of silicon particles in lithium-ion anodes has been advocated as a means of greatly increasing specific capacity. However, the intercalation of four lithium ions per silicon atom during charge causes excessive expansion of the silicon and fracturing of the particles. Repeated fracturing during each cycle disrupts the solid electrolyte interface (SEI) layer and potentially causes loss of structural integrity of the entire anode. To overcome the strain within the silicon, the use of nanosized silicon particles has been proposed.[14] Once again, however, current solvent coating techniques are inadequate for achieving anodes with high ratios of silicon to polymer binder. Clearly, advanced technologies beyond those traditionally used in lithium-ion battery manufacture are required.

The inherent problem with bonding smaller and smaller particles of active (silicon, nickel-manganese-cobalt [NMC], etc.), conductive (carbon), or heat resistant (ceramics) materials is that particle volume is inversely proportional to the square of the surface area. This effect causes the need for a disproportionate amount of polymer in order to bond each and every tiny particle. The properties of both electrode and separator components often then become less optimized by the use of smaller particles. Smaller particle sizes equate to less exposure of particle surface area within the matrix.

Traditional coating and drying techniques can capture only a minimal amount of these extremely small, high–surface area particles. Loose particles that are disconnected from the body of the electrode matrix do not fully participate in the electrochemical reactions. And if polymer binder levels are increased, the matrix becomes filled with too much of a dielectric material and is thereby less able to provide adequate rate capability. A need exists for a battery electrode manufacturing method that can more efficiently utilize the polymer to bond high levels of micron or submicron particles. Additionally, any process that allows for lower costs, greater film thickness, and higher particle volume loadings with optimized pore structure and better structural integrity would be especially beneficial. The case study below details work underway that offers significant improvements to electrode manufacture, compared with standard solvent coating and drying operations. This new invention is also highly applicable to the manufacture of polymer separators with small ceramic filler particles (see Chap. 3C).

## 3A.5    CASE STUDY: NEW TYPE OF PHASE INVERSION

Teebs R&D[15] has developed a novel method for producing electrode components that addresses many of the shortcomings highlighted in Sec. 3A.4. The proprietary process allows for the creation of nanostructured polymer networks that can bond very high levels of micron or submicron particles. For instance, a robust membrane, using polyvinylidene difluoride (PVDF) polymer, was created even with 95% particle loading (by weight or volume). In comparison, traditional solvent coating techniques, using this same formulation, would lose film structural integrity at around 50% loadings.

The Teebs R&D phase inversion process can create nanosized discrete polymer structures (i.e., filaments or whiskers) that are intimately bonded to and among multiple active or conductive particles. The polymer creates a three-dimensional structural network with high–surface area active and conductive materials held together by even finer polymer filaments. Figure 3A.1 shows polymer bonded microparticles.

**FIGURE 3A.1**    Polymer bonded microparticles composite film. (*Courtesy of Teebs R&D, LLC.*)

The proprietary process provides for reduced costs as well, especially since greater than 98% of the solvent is removed by filtering/decanting and not by drying. Specifically, the electrode production technique is based on the use of standard web handling equipment (i.e., the same type equipment used to produce nonwoven webs or cellulose paper). The filtering operation (referenced as "dewatering" in the paper industry) occurs quickly and requires virtually no energy (i.e., no hot air drying), as typically required by solvent coating techniques used to make standard electrode or separator films. Recovery of the solvents after filtering is much more efficient (higher throughput) and less costly than solvent drying.

However, lower costs can be further realized due to several other factors:

1. Higher throughput capabilities are possible compared with standard solvent coating and drying operations, which are limited by how fast the solvents can be dried to form the membrane.

2. Thicker electrodes are possible since the kinetics of the drying operation ("sheeting out" of the solvent slurry) are no longer a concern (i.e., liquids are removed after deposition of the film and thickness is dictated by the levels of material deposition and compression in a simple secondary process step).

3. Solvents are easily drained, collected, and recycled from the filter cake without the need for massive condensing units or large carbon absorption beds.

Ongoing tests of various battery electrochemistries have confirmed the performance capabilities achieved by using electrodes produced with these enhanced polymer-filler structures resulting from the Teebs R&D process. Scanning electron micrographs have shown that the active and conductive materials have well over 95% particle surface area exposure within these new electrodes. High specific surface area exposure allows for excellent interparticle contact (i.e., high electrical conductivity) and optimal active material utilization (i.e., high ionic reactivity with the electrolyte).

## REFERENCES

1. L. Pan, "Printing 3D gel polymer electrolyte in lithium ion microbattery using stereolithography," *35th Annual International Battery Seminar and Exhibit*, Fort Lauderdale, FL, March 26–29, 2018.

2. A.S. Foust, L.A. Wenzel, C.W. Clump, L. Maus, and L.B. Andersen, *Principles of Unit Operations*, John Wiley & Sons, Inc., New York, 1960.

3. J.E. Doninger, "Electrochemical performance of silicon enhance Lac Knife graphite for next-generation Li-ion batteries," *35th Annual International Battery Seminar and Exhibit*, Fort Lauderdale, FL, March 26–29, 2018.

4. Y. Nishi, "Past, present and future of LIB. Can new technologies open new horizons?" *35th Annual International Battery Seminar and Exhibit*, Fort Lauderdale, FL, March 26–29, 2018, pp. 45–48.

5. "The Nobel Prize in Chemistry 2000—Advanced Information." *Nobelprize.org.* Nobel Media AB 2014. May 10, 2018. http://www.nobelprize.org/nobel_prizes/chemistry/laureates/2000/advanced.html.

6. M. Fichtner, "Stabilized porphyrins as a new class of ultrafast storage materials with high capacity," *35th Annual International Battery Seminar and Exhibit*, Fort Lauderdale, FL, March 26–29, 2018.

7. Y. Nishi, "Past, present and future of LIB. Can new technologies open new horizons?" *35th Annual International Battery Seminar and Exhibit*, Fort Lauderdale, FL, March 26–29, 2018, pp. 33–44.

8. U.S. Patent 4303748A.

9. PRNewswire, "Valence Technology Completes Bellcore Battery Technology Acquisition," http://evworld.com/news.cfm?newsid=294 (web: May 17, 2018).

10. Ionic Materials, Inc. website, 2016, http://ionicmaterials.com/ (web: May 17, 2018).

11. M. Panzer, "Design of polymer-supported, low volatility, gel electrolytes," *35th Annual International Battery Seminar and Exhibit*, Fort Lauderdale, FL, March 26–29, 2018.

12. L. Zhang, "Annulated dialkoxybezenes as catholyte materials for non-aqueous redox flow batteries," *35th Annual International Battery Seminar and Exhibit*, Fort Lauderdale, FL, March 26–29, 2018.

13. Y.T. Cheng, M. Al-Shroofy, T. Chen, M. Wang, "Working towards making better and cheaper lithium-ion batteries," *35th Annual International Battery Seminar and Exhibit*, Fort Lauderdale, FL, March 26–29, 2018.

14. X. Xiao, M.W. Verbrugge, Q. Zhang, B. Sheldon, H. Gao, Y. Qi, Y-T. Cheng, and Z. Cheng, "Advanced silicon based electrodes: from fundamental understanding to practical applications," *35th Annual International Battery Seminar and Exhibit*, Fort Lauderdale, FL, March 26–29, 2018.

15. Teebs R&D, LLC is an early stage development company owned and operated by Trevor D. Beard.

# SECTION B

# BATTERY ELECTROLYTES

**Travis Thompson**
**(Emeritus Contributor: George E. Blomgren)**

## 3B.1  INTRODUCTION

Electrolytes, similar to electrodes, have an enabling role in any battery technology. The chapters in Part 4 on electrochemical systems include further details for specific types of cells. This chapter provides a more generic summary of various categories of electrolytes.

Most batteries of the last two centuries utilized aqueous electrolytes. This chapter, therefore, begins with a summary of aqueous electrolytes, including alkaline, neutral, and acid solutions. Electrolytes with good stability toward lithium metal were only commercialized in the last 50 years, starting with lithium primary cells, which became available in the 1970s. Efforts to commercialize rechargeable lithium metal anode batteries were unsuccessful due to safety issues. Hence, focus shifted to the lithium-ion battery, using lithiated carbon as the negative electrode, which made its first appearance in the early 1990s. An overview of the electrolytes used in both types of lithium batteries is included next. In addition, newly developed ionic liquid electrolytes, which offer the possibility of low flammability to enhance safety, are also discussed.

Solid electrolytes have been under development since the late 1950s first with silver iodide (Ag ions) and more recently (up to the 1980s) with sodium "β" alumina.[1-5] While rechargeable lithium metal batteries did not originally achieve success using liquid electrolytes, ion conductive solids have enabled the development of safer, long-life lithium metal secondary cells (see Chap. 17B). A brief summary of solid inorganic electrolytes is also included below. Additionally, several new approaches to solid electrolytes are presented (see Chaps. 22A and 22C also).

## 3B.2  BASIC FUNCTIONS OF AN ELECTROLYTE

The role of the electrolyte is to ionically connect and electronically isolate the chemical reaction that occurs heterogeneously in the cell at the surface of each electrode. As such, an electrolyte is to:

1. *Provide a pathway for the ions participating in the reaction with sufficient ionic conductivity.* This first consideration relates to the practicality of making cells with usable current output in high energy density (and power density) configurations for ambient operation. The conductivity of the electrolyte should be at least 1 mS/cm to make practical electrolytes, although higher values are desirable and allow thicker electrodes to be used. Many liquid and solid electrolytes exceed this conductivity at room temperature.

2. *Block electronic leakage.* The second consideration relates to shelf life and safety. If the electronic conductivity of the electrolyte is not sufficiently low, high rates of self-discharge will occur, since an internal electronic pathway exists. If the time for self-discharge is within an acceptable window for the application, there may be minimal impact on the operation of the device. These internal electronic shorts through the electrolyte could also lead to excessive joule heating with fire and current collector corrosion.

3. *Establish sufficient electrode interface contact to support charge transfer.* This consideration is perhaps too often overlooked. Poor electrolyte wetting of either the electrode or the separator can lead to high device impedance and poor power performance. Both electrode-electrolyte interfaces contribute to battery impedance in addition to the contribution due to the electrolyte itself beyond the region of the electrical double layer. Interfacial impedance becomes especially important under high-current conditions when mass

transport of ions through the electrolyte frequently becomes the limiting process as modeled by the Nernst equation relating the concentration of ions at the electrode interface to the electrode polarization. This consideration is especially important for solid electrolytes and is discussed further in Sec. 3B.5.

4. *Provide chemical and physical stability under the expected operating conditions of the device (i.e., freezing/ boiling or unwanted dissolution of components).* This fourth consideration is of practical importance and has fostered the use of solvent mixtures or additives to satisfy both high and low temperature requirements. Even so, extreme high or low temperature operation has necessitated special purpose electrolytes, which do not perform well at the opposite extremes. If a component has a high vapor pressure, it is usually found in modest concentration in the solvent mixture, providing enough conductivity to the electrolyte for improved low-temperature performance. Stability must also be viewed in context to the cell environment. One of the most common salts used in lithium-ion cells, $LiPF_6$, is thermally, as well as photochemically, unstable in many solutions, but in the cell environment maintains stability over many years of operation.

5. *Have a sufficiently wide electrochemical stability window in the presence of the electrodes.* The fifth consideration dictates the voltage window where the electrolyte is stable and directly impacts the energy density of the system. Figure 3B.1 shows an energy level schematic of an electrolyte in the presence of electrodes.

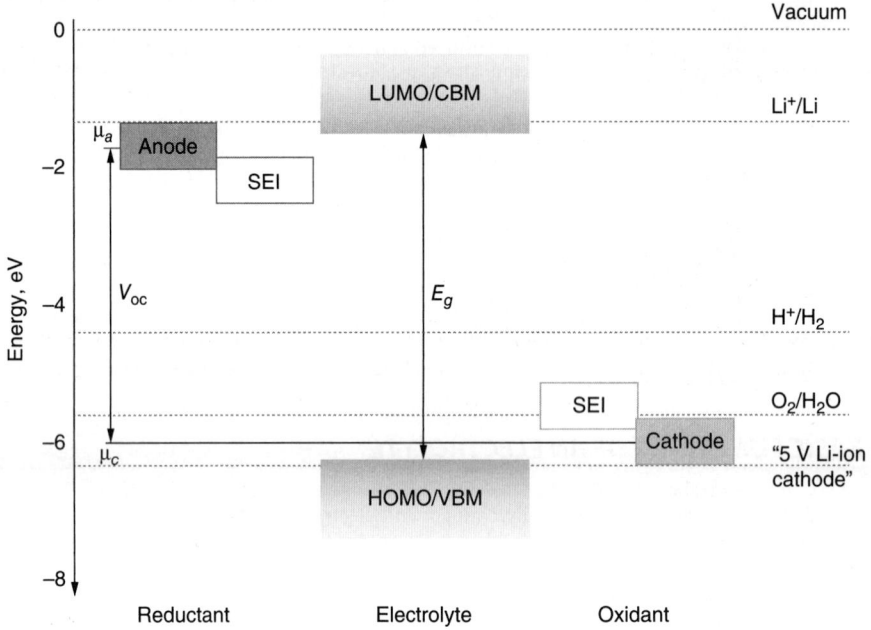

**FIGURE 3B.1**  Energy level diagram of a cell. Dashed lines are included to show the approximate location of relevant reactions for aqueous and Li-ion chemistries. The difference in the chemical potential of the anode ($\mu_a$) and the cathode ($\mu_c$) defines the open-circuit voltage ($V_{oc}$). The electrochemical window width ($E_g$) and center location of the electrolyte relative to the location of $\mu_a$ and $\mu_c$ are important considerations to prevent unwanted spontaneous reactions of the electrolyte.

For example, aqueous electrolytes undergo electrolysis of water to produce $O_2$ gas at the cathode and $H_2$ gas at the anode (i.e., the electrochemical stability window) if the cell exceeds ~1.23 V (zero pH vs. hydrogen). If the Faradaic reactions at the electrodes fall below the lowest unoccupied molecular orbital (LUMO)/conduction band minimum (CBM; anode) or above the highest occupied molecular orbital (HOMO)/valence band maximum (VBM; cathode), then decomposition of the electrolyte will spontaneously occur instead because of thermodynamic considerations (refer to Fig. 3B.1). However, in real systems, the ideal conditions are never met and (fortunately) kinetic limitations can be used as an advantage. An important example of this effect occurs in lithium-ion cells, due to the interplay of the electrochemical window and solid electrolyte interphase (SEI) formation. The SEI can effectively widen the electrochemical stability window so long as the film that is formed is self-limiting. It is important to note that evaluation of the electrochemical stability window of an electrolyte should take into account catalytic effects from the electrode surface in the actual cell environment.

If any one of the above conditions is violated, then the electrolyte fails to meet necessary requirements, and cell performance will suffer in some way. It is the role of the battery engineer to consider trade-offs.

## 3B.3 AQUEOUS ELECTROLYTES

Aqueous electrolytes can be classified by pH into alkaline, neutral (or mildly acidic), and strong-acid electrolytes. The alkaline electrolytes are usually very strong with pH values close to 13. Neutral electrolytes are generally composed of salts of strong acids and bases. Additives of weaker bases cause lower pH values. For example, in the Leclanché electrolyte, zinc chloride is an important electrolyte component, and its complex equilibria in aqueous media shift the pH to mildly acidic conditions. Likewise, the ubiquitous presence of carbon dioxide dissolved in aqueous solutions shifts the pH to mild-acid conditions. It must always be borne in mind that the electrochemical stability window for aqueous solutions is approximately 1.2 V (as noted above). Thus, much of the electrolyte work in aqueous media deals with the twin problems of anode corrosion and cathode gassing and the means to control these deleterious reactions. Rechargeable batteries have even more onerous requirements due to the increased potential that occurs at both negative and positive electrodes during charge.

### 3B.3.1 Alkaline Electrolytes

Alkaline electrolytes are utilized in a large variety of primary and rechargeable batteries. The most commonly used primary batteries are the dry cells (Chap. 11A) and the alkaline manganese dioxide battery (see Chap. 12A). Other common primary batteries with neutral or alkaline electrolytes include zinc-silver oxide (Chap. 12C) and zinc-air (Chap. 18A) batteries. Rechargeable cells using alkaline electrolytes include the nickel cathode cells with various anodes (see Chap. 15), such as Ni-cadmium (NiCd) and Ni-metal hydride.

Alkaline electrolytes generally have considerably higher conductivity than neutral electrolytes because of the enhanced proton conductance of high pH electrolytes. For example, 20% to 40% solutions of NaOH or KOH are frequently used in batteries, giving pH values near 14. KOH is generally preferred to NaOH because of its higher conductivity at a given concentration and lower freezing points in the eutectic region.[6] Figure 3B.2 shows the conductivity relationships of KOH and NaOH as a function of the weight percent of the hydroxide at 15 and 25°C.[7] Figure 3B.3 shows the effect of the dissolution of ZnO on the conductivity of both KOH and NaOH.[8]

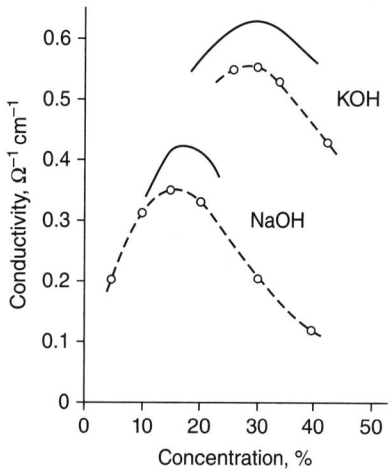

**FIGURE 3B.2** Specific conductance of NaOH and KOH aqueous solutions. Solid line is 25°C, dotted line is 15°C, Ref. 7.

The effect of ZnO dissolution is to diminish the conductivity due to the removal of hydroxide ion from the solution in the reaction,

$$ZnO + 2MOH + H_2O \rightarrow M_2Zn(OH)_4 \qquad (3B.1)$$

where M is either sodium or potassium.

This reaction (shown in Eq. 3B.1) is very important for batteries with zinc anodes because zinc ions, formed as a result of the anodic reaction, are converted to zincate ions, $Zn(OH)_4^{-2}$, until saturation of the solution occurs.[9] Thereafter, the product is ZnO or $Zn(OH)_2$ in the solid phase, although supersaturation of the solution with ZnO is a frequent occurrence along with many complications in the species and solution structure.[10]

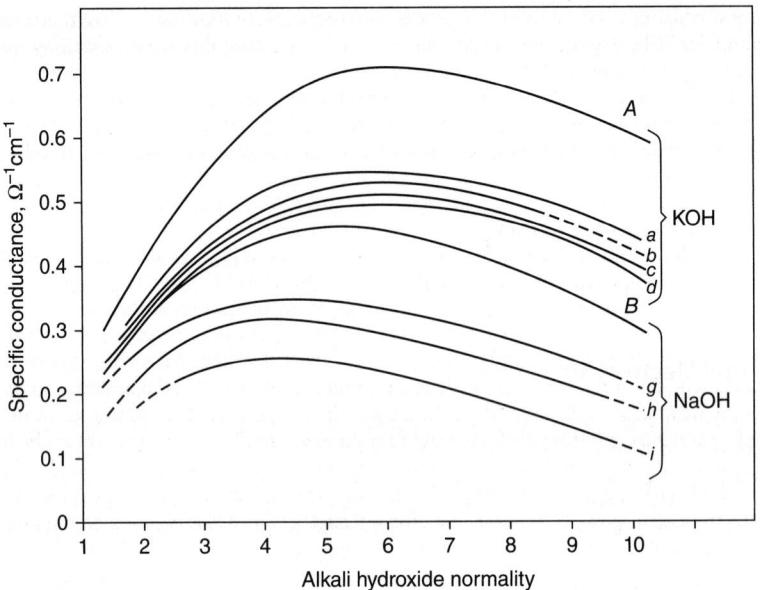

**FIGURE 3B.3**   Specific conductance of NaOH and KOH containing different ratios of dissolved ZnO to alkali hydroxide at 30°C, Ref. 6.

**A. KOH only**
  a. 1 mole ZnO:4.33 mole KOH
  b. 1 mole ZnO:3.71 mole KOH
  c. 1 mole ZnO:3.37 mole KOH
  d. 1 mole ZnO:3.00 mole KOH
**B. NaOH only**
  g. 1 mole ZnO:4.05 mole NaOH
  h. 1 mole ZnO:3.03 mole NaOH
  i. 1 mole ZnO:1.76 mole NaOH

Thus, the electrolyte changes throughout the early part of the discharge (unless it is pre-saturated with ZnO) until the zinc compound is precipitated, mainly in the anode compartment. The relatively high concentration of zincate ions in the electrolyte has important ramifications for the rechargeable zinc electrode, such as used in the nickel-zinc, silver oxide-zinc, or the $MnO_2$-Zn battery. Zinc deposition occurs mainly from the soluble species during charge and, as is common with deposition from soluble species of high concentration, this deposition can take a number of morphologies depending on the current density, zinc concentration, and other factors.

The electrolyte for primary alkaline batteries usually has a gel-forming polymer added to the solution, although the earlier wet cells such as Zn-air and Zn-CuO cells used the liquid without gelation. The gel-forming polymer must be chemically and electrochemically stable (see Points 4 and 5, Sec. 3B.2). For many years, the favorite polymer was sodium carboxymethylcellulose, first used in the zinc-mercuric oxide primary cell to immobilize the electrolyte and convert it from a wet cell to a dry cell. This material is still used in some alkaline batteries but is subject to oxidation by high-potential cathodes. Polymers such as other cellulosic or starch derivatives, polyacrylates, or ethylene maleic anhydride copolymers are used in some alkaline batteries. The manufacturers generally keep their anode gel formulation as trade secrets. An effect of the use of gelling agents is to lower the ionic conductivity of the electrolyte phase (Sec. 3B.2, Point 1), which must be taken into account in the cell design. Also, the removal of mercury from alkaline-$MnO_2$ cells places greater importance on electrolyte purity, particularly with regard to iron and chloride ions since these materials are corrosion accelerants (Sec. 3B.2, Point 4). To address the electrochemical corrosion reaction, certain organic additives, which have corrosion inhibition properties, are included in the electrolyte formulation. Mercuric oxide cells, which have declined substantially, are regarded as one of the few stable aqueous electrolyte batteries (even at high temperature) due to an operating voltage of only about 0.9 V.

Alkaline electrolytes are also used widely in secondary cells. The overpotential during charging can cause oxygen gassing with metal oxide positive electrodes and polymer gelling agent reactions. For example, with nickel oxyhydroxide positive electrodes, organic electrolyte compounds that lessen oxygen gassing are not used if high cycle life is needed.

Alkaline electrolyte rechargeable cells may also use additives, such as lithium hydroxide (for NiCd cells) and cobalt salts (for various NiOOH cathode cells).[11] The history of nickel-zinc secondary cells is essentially that of various electrolyte additives, which are used to suppress the solubility of zinc in the alkaline electrolyte. Otherwise, zinc dendrites form on charge and severely limit the cycle life (see Chap. 15E).

### 3B.3.2 Neutral Electrolytes

The main battery type that uses neutral or slightly acidic electrolytes is the Leclanché or carbon-zinc battery. There are two main versions of this cell, namely the Leclanché electrolyte (about 26% $NH_4Cl$, 9% $ZnCl_2$, and 65% water) and the zinc chloride cell (about 30–40% $ZnCl_2$ and 60–70% water) with small amount of corrosion inhibitors (Sec. 3B.2, Point 4). The conductivity of zinc chloride solutions reaches a maximum of 0.107 S/cm at 3.7 M $ZnCl_2$ with a relatively slow decrease with increasing or decreasing concentrations.[12] The addition of ammonium chloride increases the conductance substantially. The actual conductance in the cell is modified, however, by the inclusion of a gelling agent, such as starch, which immobilizes the electrolyte and minimizes leakage and orientation effects (see Chap. 11A).

Other neutral systems such as zinc-bromide secondary cells and magnesium and aluminum primary cells have been developed in the past, but they are not currently major commercial products. Some efforts are continuing, however, to look at near neutral pH electrolytes as a safer, cheaper option to alkaline or acid systems.

### 3B.3.3 Acid Electrolytes

The main electrolyte in this category is sulfuric acid, which, despite having a long history in batteries, is now mainly employed in the lead-acid or more recently in the carbon-lead-acid cells (see Chap. 14). It can be argued that this electrolyte is the most important one because of the widespread use of this battery type and its worldwide economic importance. The lead-acid cell was invented in 1859 by Gaston Planté and utilized dilute sulfuric acid. Today, typical cells use 37% by weight sulfuric acid solution, based on fully charged conditions. As is the case for alkaline batteries, the electrolyte concentration varies during charge and discharge according to the electrode reactions as given by

$$\textit{Positive:}\ \ PbSO_4 + 5H_2O \rightleftharpoons PbO_2 + 3H_3O^+ + HSO_4^- + 2e^- \qquad E° = 1.685\ \text{V} \qquad (3B.2)$$

$$\textit{Negative:}\ \ PbSO_4 + H_3O^+ + 2e^- \rightleftharpoons Pb + HSO_4^- + H_2O \qquad E° = -0.356\ \text{V} \qquad (3B.3)$$

$$\textit{Cell reaction:}\ \ 2PbSO_4 + 4H_2O \rightleftharpoons Pb + PbO_2 + 2H_3O^+ + 2HSO_4^- \qquad E° = -2.041\ \text{V} \qquad (3B.4)$$

where the cell reaction shows that 2 moles of sulfuric acid are used during discharge to produce 2 moles of water. This utilization of sulfuric acid contributes substantially to the weight and volume of the battery and also results in electrolyte properties changing during the course of the discharge as the electrolyte concentration changes. The conductivity of sulfuric acid at 35% concentration is about 800 mS/cm, one of the most conductive room-temperature battery electrolytes. Also, the individual electrode potentials given in Eqs. (3B.2) and (3B.3) are not stable with respect to oxygen evolution [Eq. (3B.2)] and hydrogen evolution [Eq. (3B.3)] in the fully charged condition (Sec. 3B.2, Point 5). These reactions cause a steady decline in capacity of the cell as gassing occurs, which is one of the principal problem areas of the lead-acid battery. In a sealed cell, hydrogen can cause increased pressure, while the oxygen evolution is recombined to some degree with the negative electrode material.

In some lead-acid cells, a catalyst is included in the head space to reform water. Otherwise, the cell simply vents the gases, thus causing a decrease in the amount of water in the cell (increasing the sulfuric acid concentration and depleting the volume of electrolyte). Hydrogen gas at a 4% level in air can cause explosions. Some newer types of lead-acid batteries include activated carbon in the lead negative electrode leading to a combined double-layer, Faradaic redox process that functions more like an aqueous capacitor (see Chap. 20).

Sulfuric acid has also been used in the vanadium and other redox types of batteries. These cells may have two different solutions, separated by a microporous membrane, although sometimes an ion-specific separator is used (proton conductive). Even in these systems, one must be concerned about gas evolution since the overpotential of the charging regime can carry the cell into the gassing range (see Chap. 22B).

## 3B.4   NONAQUEOUS ELECTROLYTES

This section is separated into organic solvent-based electrolytes, inorganic solvent-based electrolytes, and ionic liquids. The greatest development has been with organic solvent-based electrolytes as they have been used in numerous primary lithium batteries (Chap. 13) and many variations of secondary cells (Chap. 17). Polymeric materials may be used to cause gelation of the electrolyte (mainly with organic solvents) but typically have little effect on electrolyte properties such as conductivity and diffusivity. Inorganic solvent-based electrolytes have been used mainly in liquid cathode primary batteries (see Chap. 13). Ionic liquids are still in the development stage, but are discussed briefly here and elsewhere.

### 3B.4.1   Organic-Solvent Electrolytes

The greatest electrochemical use for organic-solvent electrolytes has been in the fields of primary lithium batteries and secondary lithium-ion batteries, which are the focus of this section. Successful primary lithium battery electrolytes predate the rechargeable lithium batteries by several decades, even though work was instituted on both primary and secondary batteries at about the same time in the early 1960s. Techniques for handling and purifying the organic liquids had to be developed first, with special attention on water contamination and other impurities. More than a decade of work was required to understand the importance of reducing impurities to the parts per million (ppm) levels in both salts and solvents. Cathode materials and other cell components often had much more adsorbed water than was commonly realized, requiring methods to remove moisture from all components. A good compilation of the techniques required to study nonaqueous electrochemistry is given in Ref. 13. In addition, it was necessary to develop an understanding of the stability of purified solvents and salts with active materials of both electrodes, especially the negative electrode. Studies of the electrochemical window (Sec. 3B.2, Point 5) on materials, such as conductive diamond, platinum, or glassy carbon, during cyclic voltammetry sweeps in the negative direction (cathodic scans) required much interpretation because the film-forming properties of these materials were very different from that of lithium metal (or more recently, carbon or graphite). Likewise, anodic scans on inert substrates were found to depend on both the salt and solvent, and most interpretations of oxidation reaction mechanisms are still unresolved.[14]

Chapter 17 further refines the guidelines (Sec. 3B.2) for electrolytes that are to be used specifically for lithium anode batteries (i.e., for Li-ion cells, Chap. 17A and lithium metal cells, Chap. 17B). However, because of the reactivity when using a lithium metal anode, the importance of the electrochemical stability window cannot be underscored enough. Polar solvents generally useful for electrolytes are at best only metastable to the lithium metal anode. Thermodynamic calculations show that even propylene and ethylene carbonates are capable of highly exothermic reactions with lithium metal to produce lithium carbonate and the corresponding alkene.[15] A recognition of the importance of the passivating film (or the SEI layer) is demonstrated in the book *Lithium Batteries: Solid Electrolyte Interphase*[16] and the cited references. Additives to the electrolyte have frequently been employed to improve the SEI.[17]

Reactions of solvents with the cathode are also of great concern to the electrochemist studying lithium and lithium-ion batteries. Again, the solvents are only metastable thermodynamically as demonstrated by calculations of simple reactions,[15] and the mechanisms are poorly understood. However, as discussed in Sec. 3B.2, the resulting SEI can form kinetically limiting film, and thus can stabilize the electrode/electrolyte interface.

The importance of additives to form stable SEIs on cathode surfaces for lithium-ion chemistries has become more important, as the voltage of the cathode has been increased toward 5 V.

Properties of the most common solvents for lithium primary batteries are given in Chap. 13, and the most common solvents for lithium rechargeable batteries are given in Chap. 17. A more complete listing of solvents can be found in Ref. 17. Mixtures of two or more solvents are usually employed to obtain the best combination of properties for the intended application. A high dielectric constant, high-viscosity solvent, such as propylene carbonate (permittivity = 64.4, viscosity = 2.5 cP), can be mixed with a low dielectric constant, low-viscosity solvent, such as dimethoxyethane (permittivity = 7.2, viscosity = 0.455 cP), to give a mixed solvent of intermediate dielectric constant and viscosity with good solvation properties toward lithium salts. Ideal mixing rules adapted from traditional physical chemistry generally predict the mixed solvent properties to within a few percent for aprotic solvents.[18] Certain empirical parameters, such as donor and acceptor numbers, have been usefully employed to help choose cosolvents to improve properties, while concepts such as ion association from physical chemical studies may also be useful to understand the conductivity and viscosity of electrolyte solutions.[19]

With the use of organic solvent electrolytes in lithium metal and lithium-ion chemistries, the choice of electrolyte salt has important considerations that must be balanced. It is not widely recognized, but many times the anodic window of an electrolyte is determined by the reactivity of the anion, which can play a catalytic role in the oxidation of the solution. Thus, a one-electron oxidation of the anion at the positive electrode (to form a neutral free radical) leads to attack of a solvent molecule in a chain reaction. Chain termination usually results from radical combination, but often great damage is done to the electrolyte, which becomes evident on further cycling. These reactions are especially important for rechargeable systems.[14] The salt may also be unstable at the negative electrode. For example, the tetrachloroaluminate anion may deposit aluminum in an exchange reaction with lithium. Furthermore, the salt may affect the SEI of the graphite anode in a beneficial way as exemplified by the LiBOB (lithium bis-oxalatoborate) salt in some lithium-ion battery electrolytes.

Another consideration is the impact of the salt on the collector substrate. For example with lithium-ion batteries, aluminum is the most commonly used collector. The short-term pitting corrosion for a group of salts may be similar, but over longer durations only a few salts, such as $LiPF_6$ and $LiBF_4$, are stable with aluminum.[20]

Thermal stability is affected by the salt as well as the solvent choice. One of the critical thermal aspects is related to the onset temperature of the dissolution of the SEI layer. These matters are discussed in Ref. 16. In summary, rechargeable cells are the most sensitive to the choice of salt, with $LiPF_6$ often preferred (see Chap. 17). A wider choice is available for primary batteries, in part because the cathode is never charged, and thus the collector is not exposed to high overpotentials. $LiCF_3SO_3$, $LiPF_6$, $LiBF_4$, LiBr, LiI, $LiN(CF_3SO_2)_2$, and $LiClO_4$ have all been used in primary lithium batteries (see Chap. 13).

However, in all cases, the conductivity of electrolytes with organic solvents (as well as the inorganic solvent systems discussed below) are at least an order of magnitude poorer than aqueous electrolyte solutions, especially those with enhanced proton conductance such as acids and bases. This has a profound effect on cell design for the nonaqueous systems. For high-rate cells with organic electrolytes, electrodes (and ideally separators) must be thinner with the corresponding electrode (and separator) surface area, much greater. Such cell designs are more expensive to manufacture than aqueous cells of similar rate capability. However, because of the high voltage or high capacity of the chosen nonaqueous systems, the cost of a given cell size per watthour of energy is mitigated.

Many electrolyte additives have been developed for lithium-ion batteries for the purpose of improving safety, extending the calendar life, and extending cycle life of cells. Low levels of additive are generally used (1% or less of the electrolyte) unless the additive is for flame retardancy, where ≥5% levels are used. Furthermore, extensive chemical testing before and after cycling are required, which complicates the analyses. Additive use in batteries is discussed further in Ref. 16.

### 3B.4.2    Inorganic-Solvent Electrolytes

A class of liquid cathode primary batteries has been developed that uses purely inorganic solvents. These cells have high energy density, in part because the electrolyte carries out the dual role of electrolyte solution and cathode active material. The main representatives are thionyl chloride, $SOCl_2$, and sulfuryl chloride, $SO_2Cl_2$, based on lithium chemistries, although a number of other solvents have appeared in the patent literature.

It is somewhat surprising that the conductivities of the electrolytes are relatively high (1 M $LiAlCl_4$ solution in thionyl chloride = 14.6 mS/cm; in sulfuryl chloride = 7.4 mS/cm), since the dielectric constants of these solvents are low (permittivity of thionyl chloride = 9.25; of sulfuryl chloride = 9.15).[21] The preferred salt in all of these solutions is $LiAlCl_4$, although some work has been carried out with $LiGaCl_4$ to show improved conductivity and reduced passivation. Both types have been modified by the inclusion of additives in the electrolyte, BrCl in the thionyl chloride electrolyte and $Cl_2$ in the sulfuryl chloride electrolyte. The effect on the energy output of the cells due to the additives is small, but there are some advantages in resistance of the cells to abuse conditions. Generally, the shelf life of the cells is adversely affected by the additives. These systems are surprisingly stable even though the oxidizing liquids are in direct contact with lithium metal, and one would expect at least a strong reaction. Extensive study of the lithium surface has shown, however, that a tight, compact layer of lithium chloride SEI is formed that impedes further reaction. The shelf life is accordingly very long (better than many organic solvent systems), although a delayed response in current is often shown after a long period of storage, as the layer must be at least somewhat disrupted in order for current to pass (see Chap. 13 and Refs. 22 and 23 for more details).

The lithium-sulfur dioxide liquid cathode system is very important in military and industrial applications. The electrolyte phase is a mixture of an organic solvent with condensed-phase sulfur dioxide. Acetonitrile is the typical organic solvent because of its high solubility for and stability with sulfur dioxide. Acetonitrile has a moderately high dielectric constant (35.95) and very low viscosity (0.341 cP). This combination (30/70 by volume) with a 1 M LiBr salt gives a conductivity of about 52 mS/cm at ambient temperature, approaching the conductivity of aqueous solutions. As with oxyhalide liquid cathode cells, a compact protective SEI (lithium dithionite, $Li_2S_2O_4$) on the lithium metal is the enabling feature of the electrolyte. The cell has lower energy content than oxyhalide cells in part because of dilution of the electrolyte with acetonitrile. It also has to contain pressurized sulfur dioxide, which has a boiling point of $-10°C$. This electrolyte allows $LiSO_2$ cell to provide excellent low temperature performance to $-40°C$.

### 3B.4.3   Ionic Liquids

Ionic liquids are defined as liquids that are primarily dissociated into ions even though they have complex polyatomic structures of each of the ions. Many of these are liquid at room temperature and below and also support the dissolution of lithium salts at reasonable concentrations, such as 1 M. Because they have very low vapor pressures in general, they offer flame-retardant properties that few other electrolytes accomplish. Also, because of their high concentration of ions, the conductivities are comparable to many organic solvent systems, even though the viscosities tend to be much higher. The high viscosity can cause problems in filling cells in short time periods as well as creating wetting problems of electrodes and separators. An electrochemical difficulty with most ionic liquids occurs due to reduction of onium cations at potentials more positive than lithium deposition or intercalation in graphite. Also, the SEI formed with these materials is frequently unstable due to dissolution in these excellent solvents. Therefore, the efficiency of lithium deposition or intercalation is reduced and the cycle life of the cell is correspondingly reduced. Additives that improve the SEI, such as vinylene chloride (VC), also help the cycle life, but not enough as yet to allow commercial production.[24] The reader is referred to Ref. 25 for further information.

## 3B.5   SOLID ELECTROLYTES

Solid electrolyte development for lithium batteries is relatively new, but other systems, including solid oxide fuel cells and especially sodium chemistries, have been studied since the 1970s (see Ref. 26 for solid electrolyte development for sodium batteries). While the focus is now on lithium systems, the findings from other technologies are also useful.

Just as with liquids, all the Sec. 3B.2 requirements must be met for solid electrolytes. Sufficient ionic conductivity (Sec. 3B.2, Point 1) has been demonstrated with several material systems with the possibility of a transference number of unity (compared to 0.5–0.6 for nonaqueous liquid electrolytes). Intimate atomic contact with the electrode surfaces (Sec. 3B.2, Point 3), readily achieved with liquid electrolytes, must be carefully orchestrated with solid systems. Solids are a condensed form of matter, however, so electrolyte freezing (Sec. 3B.2, Point 4) is not a concern. Similarly, vaporization, venting, and fire may not be an issue with inorganic materials. However, lithium can chemically reduce many elements imparting electronic conduction (Sec. 3B.2, Points 2 and 3). The band-gap of the solid has a similar interpretation to the electrochemical stability window.[27] Several solid materials demonstrate sufficiently wide electrochemical stability windows or kinetically limiting SEI films (Sec. 3B.2, Point 5). The primary advantage of a solid electrolyte in a lithium chemistry battery is the rechargeability of a lithium metal anode. Large numbers of cycles have been demonstrated, but the well-known dendrite penetration problem still exists. Precise mechanisms are still unknown and under investigation.[28-30]

### 3B.5.1  Solid Polymer Electrolytes

Batteries using solid polymer electrolytes (SPE) are discussed elsewhere (Chaps. 17, 22A, 22C, etc.). The initial solid electrolytes were simple polymers of polyethylene oxide (PEO) with salts such as $LiClO_4$ and more recently lithium bis(trifluoromethane)sulfonimide (LiTFSI). Instead of relying on a solvent, the ethereal oxygen atoms in the polymer allow for the solvation of the salt at reasonable concentrations (i.e., 1 M). Although extensively studied as a model system, the conductivity of the SPE based on PEO is too low at room temperature to be of use in a practical battery (of the order of $10^{-7}$ S/cm). Many efforts were made to increase the conductivity, including dissolution of a "plasticizer" solvent, such as carbonates or glymes; incorporation of inorganic nanoscopic fillers; and backbone modification, including isomerization and the formation of block copolymers.[31-33] Unfortunately, no practical solution has been found to increase the ionic conductivity at room temperature for the PEO system, but several efforts continue in the field. The incentive is due to the possibility of using lithium metal as the anode material, since PEO is electrochemically stable. However, it is unclear if any SPE will have sufficient mechanical properties to suppress lithium dendrite penetration upon cycling.[28] Work with SPE to discover a new materials system is ongoing.

### 3B.5.2  Inorganic Solid Electrolytes

Electrolytes of this type have, until recently, been employed primarily in thin-film batteries for lithium chemistries (about 10 μm stack thickness fabricated by vapor deposition processes; see Chaps. 17, 22A, and 22C). Glassy lithium phosphorous oxynitride (LiPON) is by far the most prevalent material used in thin-film batteries. This electrolyte has a wide electrochemical stability window and can cycle 5-V cathode chemistries over >10,000 cycles.[34] However, LiPON has poor conductivity (in the range of $10^{-5}$ S/cm at room temperature)[35] but is still adequate when thin layers are employed. Although thin-film batteries are now commercialized, most methods for making the electrolytes and the electrodes have been relatively slow and expensive.[36] Several companies are trying to find alternative bulk-scale processing approaches. To this end, several inorganic solid electrolyte material systems that do not require vapor-based deposition processing have been identified, including sulfide glasses, lithium NaSICONs, perovskites, anti-perovskites, thio-LiSICONs, and garnets.[37-40] Studies for each material system are underway to evaluate all the basic considerations an electrolyte must meet and weigh potential trade-offs. Additionally, alternative bulk processing routes are being explored.[41]

## *ACKNOWLEDGMENT*

Major content for this chapter was provided in whole or in part by George E. Blomgren, Chap. 7, Battery Electrolytes, in *Linden's Handbook of Batteries*, 4th ed., T. B. Reddy and D. Linden, eds., McGraw-Hill, 2011.

## *REFERENCES*

1. J.N. Mrgudich, *J. Electrochem. Soc.* **107**, 475–479 (1960).

2. K. Lehovec and J. Broder, *J. Electrochem. Soc.* **101**, 208–209 (1954).

3. S.M. Whittingham and R.A. Huggins, *J. Chem. Phys.* **54**, 414–416 (1971).

4. R.C. Galloway, *J. Electrochem. Soc.* **134**, 256–257 (1987).

5. B. Dunn and G.C. Farrington, *Mater. Res. Bull.* **15**, 1773–1777 (1980).

6. E.A. Schumacher, *The Primary Battery*, vol. 1, G.W. Heise and N.C. Cahoon (Eds.), John Wiley, New York, 1971, p. 179.

7. S.A. Megahed, J. Passaniti, and J.C. Springstead, *Handbook of Batteries*, 3rd ed., D. Linden and T.B. Reddy (Eds.), McGraw-Hill, New York, 2002, p. 12.9.

8. E.A. Schumacher, *The Primary Battery*, vol. 1, G.W. Heise and N.C. Cahoon (Eds.), John Wiley, New York, 1971, p. 180.

9. K.J. Cain, C.A. Melendres, and V.A. Maroni, *J. Electrochem. Soc.* **134**, 519–524 (1987) and references therein.

10. C. Debiemme-Chouvy, J. Vedel, M. Bellissent-Funel, and R. Cortes, *J. Electrochem. Soc.* **142**, 1359–1364 (1995) and references therein.

11. F. Beck and P. Ruetschi, *Electrochim. Acta* **145**, 2467–2482 (2000).

12. B.K. Thomas and D.J. Fray, *J. Applied Electrochem.* **12**, 1–5 (1982).

13. D. Aurbach and A. Zaban, Chap. 3 in *Nonaqueous Electrochemistry*, D. Aurbach (Ed.), Marcel Dekker, Inc., New York, 1999, pp. 81–136.

14. D. Aurbach and Y. Gofer, Chap. 4 in *Nonaqueous Electrochemistry*, D. Aurbach (Ed.), Marcel Dekker, Inc., New York, 1999, pp. 137–212.

15. G.E. Blomgren, Chap. 2 in *Lithium Batteries*, J.P. Gabano (Ed.), Academic Press, New York, 1983, pp. 13–42.

16. P.B. Balbuena and Y. Wang (Eds.), *Lithium Batteries: Solid Electrolyte Interphase*, Imperial College Press, London, 2004.

17. M. Winter, K.-C. Moeller, and J.O. Besenhard, Chap. 5 in *Lithium Batteries: Science and Technology*, G.A. Nazri and G. Pistoia (Eds.), Springer Science—Business Media, New York, 2009, pp. 144–194.

18. G.E. Blomgren, *J. Power Sources* **14**, 39–44 (1985).

19. G.E. Blomgren, Chap. 2 in *Nonaqueous Electrochemistry*, D. Aurbach (Ed.), Marcel Dekker, Inc., New York, 1999, pp. 53–80.

20. S.S. Zhang and T.R. Jow, *J. Power Sources* **109**, 458–464 (2002).

21. M.L. Kronenberg and G.E. Blomgren, Chap. 8 in *Comprehensive Treatise of Electrochemistry*, vol. 3, J. O'M. Bockris, B.E. Conway, E. Yeager, and R.E. White (Eds.), Plenum Press, New York, 1981, pp. 247–278.

22. E. Peled, Chap. 3 in *Lithium Batteries*, J-P. Gabano (Ed.), Academic Press, New York, 1983, pp. 43–72.

23. C.R. Schlaikjer, Chap. 13, in *Lithium Batteries*, J-P. Gabano (Ed.), Academic Press, New York, 1983, pp. 304–370.

24. A. Guerfi, M. Dontigny, P. Charest, M. Petitclerc, M. Lagacé, A. Vijh, and K. Zaghib, *J. Power Sources* **195**, 845–852 (2010).

25. A. Webber and G.E. Blomgren, Chap. 6 in *Advances in Lithium-Ion Batteries*, W.A. van Schalkwijk and B. Scrosati (Eds.), Kluwer Academic/Plenum Publishers, New York, 2002, pp. 185–232.

26. X. Lu, J.P. Lemmon, V. Sprenkle, and Z. Yang, *JOM* **62**, 31–36 (2010).

27. T. Thompson, S. Yu, L. Williams, R.D. Schmidt, R. Garcia-Mendez, J. Wolfenstine, J.L. Allen, E. Kioupakis, D.J. Siegel, and J. Sakamoto, *ACS Energy Lett.* **2**, 462–468 (2017).

28. C. Monroe and J. Newman, *J. Electrochem. Soc.* **150**, A1377–A1384 (2003).

29. R. Raj and J. Wolfenstine, *J. Power Sources* **343**, 119–126 (2017).

30. L. Porz, T. Swamy, B.W. Sheldon, D. Rettenwander, T. Frömling, H.L. Thaman, S. Berendts, R. Uecker, W.C. Carter, and Yet-Ming Chiang, *Adv. Energy Mater.* **7**, 20 (2017).

31. P.E. Trapa, Y-Y. Won, S.C. Mui, E.A. Olivetti, B. Huang, D.R. Sadoway, A.M. Mayes, and S. Dallek, *J. Electrochem. Soc.* **152**, A1–A5 (2005).

32. M. Singh, O. Odusanya, G.M. Wilmes, H.B. Etouni, E.D. Gomez, A.J. Patel, V.L. Chen, M.J. Park, P. Fragouli, H. Iatrou, N. Hadjichristidis, D. Cookson, and N.P. Balsara, *Macromolecules* **40**, 4578–4585 (2007).

33. M.A. Meador, V.A. Cubon, D.A. Schelman, and W.R. Bennett, *Chem. Materials* **15**, 3018–3025 (2003).

34. J. Li, C. Ma, M. Chi, C. Liang, and N.J. Dudney, *Adv. Energy Mater.* **5**, 1401408–1401414 (2015).

35. X. Yu, J.B. Bates, G.E. Jellison, Jr., and F.X. Hart, *J. Electrochem. Soc.* **144**, 524–532 (1997).

36. R. Salot, S. Martin, S. Oukassi, M. Bedjaoui, and J. Ubrig, *Appl. Surf. Sci.* **256**, S54–S57 (2009).

37. T. Thompson, J. Wolfenstine, J.L. Allen, M. Johannes, A. Huq, I.N. David, and J. Sakamoto, *J. Mat. Chem.* **33**, 13431–13436 (2014).

38. J.W. Fergus, *J. Power Sources* **195**, 4554–4569 (2010).

39. P. Knauth, *Solid State Ionics* **180**, 911–916 (2009).

40. Y. Kato, S. Hori, T. Saito, K. Suzuki, M. Hirayama, A. Mitsui, M. Yonemura, H. Iba, and R. Kanno, *Nat. Energy* **1**, 16030 (2016).

41. J. Schnell, T. Günther, T. Knoche, C. Vieider, L. Köhler, A. Just, M. Keller, S. Passerini, and G. Reinhart, *J. Power Sources* **382**, 160–175 (2018).

# SECTION C

# SEPARATORS

## Kirby W. Beard

## 3C.1  INTRODUCTION

Battery separators primarily serve to prevent direct contact and electrical shorting of active materials within an electrochemical cell. Typical lithium-ion cells are especially susceptible to this problem, where internal shorts will cause severe heating or even fire. However, in some other cases the active materials are actually required to be in direct contact for the cell to function. These exceptions occur when the electrolyte also serves as an active material.

Lead-acid cells are one such example where the sulfuric acid electrolyte, an active material, must contact the electrodes. Hence, rather than isolating the acid electrolyte from the electrodes, the separator's main role is to prevent contact of the current-carrying parts. While electrolyte must thoroughly wet and penetrate both electrodes, contact between electrically conductive active materials or current collectors is not acceptable.

An additional function of the separator in a cell with a reactive electrolyte is to serve as a reservoir for one of the cell's reactants. The sulfuric acid in a lead-acid cell participates in reactions at both positive and negative plates. While the electrodes also contain acid in their pore structures, the additional acid amounts held by the separator help to increase cell capacity.

Certain types of lithium primary cells also have electrolytes that serve as active materials. In these cells (i.e., Li/sulfur dioxide, Li/thionyl chloride, Li/iodine, etc.) the electrolyte is an active cathode material that must directly come in contact with the lithium anode for the cell to function. Common sense would normally suggest that such cells might fail catastrophically; however, in these systems the formation of a passivation layer on the lithium anode is key to stabilizing the cells from shorting. The cathode/electrolyte materials do not react with the lithium anode to any significant extent until after the external electrical circuit is closed.

Overall, then, the purpose of battery separators may vary and the design of the products may be altered accordingly. In the case of lithium-ion secondary cells, separators are thin with higher porosity for higher rate capability or lower porosity for greater robustness and safety. For lead-acid or lithium primary cell separators, where electrolytes are a key reactant, the separators will typically have high void levels but can vary in thickness and pore size depending on product application, electrode design, product type (i.e., sealed vs. flooded lead-acid cells), and performance requirements.

The usual, but not always optimal, strategy for selecting a battery separator is to first design the cell with a generic type of separator and then search for alternatives if the performance (i.e., rate capability, storage losses, cycle life, etc.) or safety of the cells are inadequate. Cooperative development efforts between separator vendors and battery companies are clearly a better option for optimizing the overall cell design from the start.

The next section lists various categories of separators, and the subsequent section provides a case study on evaluating separators. Various review articles on general separator technology are also available and should be consulted for further details.[1] Also, since each type of electrochemistry has its own unique features, the separator manufacturers should be consulted for design assistance. For instance, fuel cells, capacitors, thermal batteries, and similar unique electrochemical devices will each have their own special separator requirements. The appropriate chapters in Part 4 of this Handbook detail many of these various separator options.

## 3C.2 OVERVIEW OF SEPARATOR TECHNOLOGY

A wide variety of separator types are available that use various materials and processing techniques. Several popular options, used in a wide range of electrochemistries and applications, include microporous polyolefin membranes and nonwoven webs (made with either glass or polymer fibers). Often a single type of separator can be used in a variety of applications (i.e., for primary and secondary cells or for aqueous and organic electrolytes). In other cases, the separators have a single unique function (i.e., a solid-state conductive glass electrolyte 'separator' for lithium-ion transfer or Nafion® for water transmission).

An overview of various separator characteristics is listed in Table 3C.1. Separators are chosen (or developed) by matching the requirements for the proposed electrochemistry to separators with the optimal combination of these various characteristics.

**TABLE 3C.1**  Separator Design Parameters

| Separator material | Material form | Electrolyte type | Pore size | Porosity | Mechanical requirements |
|---|---|---|---|---|---|
| Nonconductive polymer | Microporous membrane | Acid | Nonporous (ion or water conductive) | Solid films (~0% voids) | Absorbent media with no structural requirements |
| Nonconductive ceramic/glass | Nonwoven web (fibrous types) | Alkaline | | Limited/controlled levels | |
| Solid-state ion conductors— polymer/glass | Solid ion conductive film/sheet | Organic solvent/ ionic liquid | Nanoporous | Full/open absorbent types | Resistant to puncture, shorts, compression |
| Composites/ multilayered | Passivation film layer | Molten salt | Microporous Macroporous | | Prevention of dendritic shorts |
| | | | | | Resistant to tearing, folding, abrasion |

Other options also exist, including recent proposals for eliminating separator as a distinct, discrete battery component (i.e., relying on air gaps or special passivation layers, perhaps).

The qualification of battery separators is a complex, expensive, and highly uncertain undertaking in many cases. Recent failures of lithium batteries in commercial airliners, electric vehicles, cell phones, hover boards, etc., provide an indication of the scope of the difficulties. The following section provides an overview of a specification that was recently developed to standardize test methods. Other standards are available that not only provide additional test methods but also establish minimum requirements for separator performance in various applications. Any separator product needs to be thoroughly tested and fully qualified before a battery system is fielded.

## 3C.3 LITHIUM BATTERY SEPARATOR TEST METHODS

Battery separator specifications are usually set by an individual battery manufacturer. The battery customers will typically specify battery performance but may also require separators meeting specific requirements or passing certain tests. As detailed in Chaps. 7 and 8 and elsewhere, various standards organizations (i.e., testing or regulating agencies) have also published guidelines or criteria for battery separators. In this section, SAE International's Recommended Practice for Determining Material Properties of Li-Battery Separator (J2983) will be used as an example of the process behind qualifying a separator for a specific battery design/application. Note that SAE J2983 only describes the testing procedures; the user must still set acceptable limits or ranges for each test method. In other cases, performance thresholds will be predetermined by the standards organization. The general categories included in SAE J2983 are listed in Table 3C.2

**TABLE 3C.2**    General Lithium Battery Separator Test Protocols

| Category | Parameters | Goals/purpose |
|---|---|---|
| Physical | Bulk dimensions, weight, internal properties, density | Verify proper thickness and porosity, and ensure quality and consistency for production/use |
| Mechanical | Tensile, puncture, tear, compression, fatigue, yield | Determine robustness/durability required for manufacture, performance, and long safe life |
| Thermal | Melt, heat shrinkage/integrity, embrittlement | Establish temperature effects/limits on performance, safety, and service life |
| Electrochemical | Wetting, solubility, chem./elec. stability, transport/diffusivity, breakdown | Confirm parameters related to processing, performance, and ultimate life |

## 3C.3.1    Physical Properties

Thickness, basis weight, and porosity (or void volume) are typically the most basic parameters stipulated. Other measures of the physical structure may include pore size, pore length, and tortuosity. Figure 3C.1 is a photomicrograph of a microporous separator showing pore structure as well as a unique adaptation: ceramic filler inclusions embedded in the membrane. Table 3C.3 exhibits SAE test results for a variety of separators.

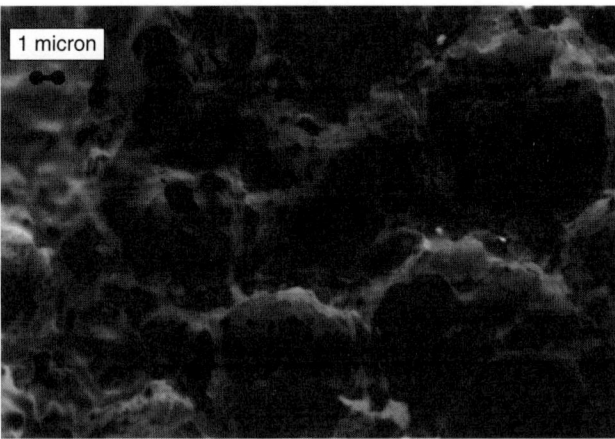

**FIGURE 3C.1**    Photomicrograph of microporous, polymer lithium battery separator with fillers.

## 3C.3.2    Mechanical Properties

The mechanical strength of lithium battery separators is fundamental to successful manufacturing. High-speed winding and electrode wrapping and stacking operations are extremely abusive toward separators. A combination of superior properties is required to build cells without failures and to ensure safety and long service life. Table 3C.4 covers these tests.

**TABLE 3C.3**    SAE J2983 Recommended Practice: Physical Property Tests

| J2983 Section | Parameter | Unit | Test method | Wet process polyolefin (12 μm nom.) | Wet process polyolefin (20 μm nom.) | Dry process polyolefin monolayer (25 μm nom.) | Dry process polyolefin shutdown trilayer (20 μm nom.) | Phase inversion membrane (reinforced with nonwoven web) | Phase inversion membrane (reinforced with nonwoven web and ceramics) |
|---|---|---|---|---|---|---|---|---|---|
| 4.1 | Thickness | μm | ASTM D5947 | 15 | 21 | 23 | 20 | 25 | 27 |
| 4.2 | Basis weight | g/m$^2$ | NA | 6.3 | 11.7 | 9.5 | 10.8 | 17.1 | 16.9 |
| 4.3 | Air permeability | s/100 mL air | JIS P 8117 | 190–275 | 450–640 | 190–250 | 480–600 | 45–60 | 12–18 |
| 4.4.1 | Porosity | % | NA | 56 | 46 | 58 | 48 | 60–65 | 65–70 |
| 5.1.2 | Mean pore size | μm | ASTM F316 | 0.045 | 0.025 | 0.036 | 0.016 | 0.17–0.19 | 0.3–0.6 |

### 3C.3.3    Thermal Parameter Testing

Exposure to high temperatures can be especially damaging to the polymer separators used in lithium cells. Most suppliers have actively pursued the use of filler additives or layers to help prevent thermal run-away and fire/explosion when abused. Measuring the ability of separators to withstand temperature extremes has been a difficult process. Table 3C.5 details these tests.

### 3C.3.4    Effects of Chemicals on Separators

Typical lithium-ion battery electrolytes are based on organic solvents that can wet, swell, and then dissolve many polymers. The ability to readily wet out a separator is a positive factor, allowing cells to be filled rapidly on high-speed production lines. However, if excessive swelling or softening occurs, the mechanical durability of the separators will be compromised. See Table 3C.6 for separator wetting and solubility testing results.

**TABLE 3C.4**    SAE J2983 Recommended Practice: Mechanical Property Tests

| J2983 Section | Parameter | Unit | Test method | Wet process polyolefin (12 μm) | Wet process polyolefin (20 μm) | Dry process polyolefin monolayer (25 μm) | Dry process polyolefin shutdown trilayer (20 μm) | Phase inversion membrane (reinforced with nonwoven web) | Phase inversion membrane (reinforced with nonwoven web and ceramics) |
|---|---|---|---|---|---|---|---|---|---|
| 5.2 | Ultimate tensile strength (MD) | N/mm$^2$ | ASTM D882 | ~60 | ~85 | ~120 | ~200 | ~20 | ~20 |
| 5.2 | Linear tensile strength (MD) | N/cm | ASTM D882 | ~9 | ~18 | ~28 | ~40 | ~5 | ~5 |
| 5.5 | Puncture | g | UL 2591 | ~200 | ~470 | ~340 | ~380 | ~100 | ~100 |
| 5.6 | Skew | | | Testing applies to slit roll good as a measure of straightness. | | | | | |
| 6.2.2 | Penetration by needle | g | 20 μm radius, 1 mm/min | 55 | 49 | 46 | 48 | 36 | 97 |
| 6.3 | Cycle fatigue | | Special method | Must repeat tests such as pore flow, yield, and tensile after cyclic fatigue testing. (Best results will show little/no change after fatigue cycling for various durations.) | | | | | |

**TABLE 3C.5**  SAE J2983 Recommended Practice for Determining Thermal Properties

| J2983 Section | Parameter | Unit | Test method | Wet process polyolefin (12 μm) | Wet process polyolefin (20 μm) | Dry process polyolefin monolayer (25 μm) | Dry process polyolefin shutdown trilayer (20 μm) | Phase inversion membrane (reinforced with nonwoven web) | Phase inversion membrane (reinforced with nonwoven web and ceramics) |
|---|---|---|---|---|---|---|---|---|---|
| 5.4 | Shutdown | | | | | | NA | | |
| 5.3.1 | Unrestrained shrinkage (130°C) MD | % | ASTM D1204 | 25.8 | 25.4 | 13.8 | 35.2 | 2.9 | 0.6 |
| 5.3.1 | Unrestrained shrinkage (130°C) TD | % | ASTM D1204 | 25.2 | 29.3 | 0.0 | 0.0 | 0.8 | 0.0 |
| 5.3.2 | Restrained shrinkage (130°C) | Visual | UL 2591 | No damage | No damage | No damage | No damage | No damage | No damage |
| 5.3.2 | Restrained shrinkage (130°C) | % inc. in air perm. | UL 2591 | Within range | 295 | Within range | 7 | Within range | Within range |
| 5.3.3 | Local shrinkage/ thermal penetration: 160°C | NA | | | | | Visual results. | | |
| 5.9 | Melt integrity | | Special | NA | | | | | |
| 5.10 | Melt temperature | °C | DSC | 120–140 | 120–142 | 154–175 | 120–167 | 140–257 | 140–257 (ceramics unaffected) |

## 3C.3.5  Separator Electrical Testing

Similar to separator exposure with solvents, electrical testing has two opposing purposes, also. First, on the positive side separators must be able to support high levels of ionic current flow. Conversely, separators subjected to high voltages have potential to degrade (arc) and fail. A variety of tests, as shown in Table 3C.7, can be used to analyze the conductivity and the breakdown of a separator. Figures 3C.2 and 3C.3 show one typical series of tests used to measure the voltage breakdown and temperature rise in test cells saturated with battery electrolyte for various separators.

**TABLE 3C.6**  SAE J2983 Chemical Exposure Testing

| J2983 Section | Parameter | Unit | Test method | Wet process polyolefin (12 μm) | Wet process polyolefin (20 μm) | Dry process polyolefin monolayer (25 μm) | Dry process polyolefin shutdown trilayer (20 μm) | Phase inversion membrane (reinforced with nonwoven web) | Phase inversion membrane (reinforced with nonwoven web and ceramics) |
|---|---|---|---|---|---|---|---|---|---|
| 5.6.1 | Wettability (with propylene carbonate) | Time for 1 drop to fully absorb, min. | NA | >30 | >30 | Did not fully absorb | Did not fully absorb | 3 | 4.5 |
| 5.6.3 | Rate of wetting (with propylene carbonate) | Height after 10 min wicking, mm | NA | 3 | 2 | 1 | <1 | 10 | 30 |
| 5.7 | Chemical Stability | Visual | NA | Stable | Stable | Stable | Stable | Stable | Stable |

**TABLE 3C.7** SAE J2983 Electrical Parameters

| J2983 Section | Parameter | Unit | Test method | Wet process polyolefin (12 μm) | Wet process polyolefin (20 μm) | Dry process polyolefin monolayer (25 μm) | Dry process polyolefin shutdown trilayer (20 μm) | Phase inversion membrane (reinforced with nonwoven web) | Phase inversion membrane (reinforced with nonwoven web and ceramics) |
|---|---|---|---|---|---|---|---|---|---|
| 5.11 | Electrical impedance | Ω | EIS (0.2–100 kHz) | 0.0723 | 0.1043 | 0.0640 | 0.1406 | 0.0867 | 0.0542 |
| | | | | | A full analysis requires AC electrical impedance testing. | | | | |
| 5.11.1 | McMullin # | No unit | EIS (0.2–100 kHz) | 5.4 | 5.6 | 3.1 | 7.9 | 3.9 | 2.3 |
| 5.12 | HiPot | V (Av. breakdown) | ASTM D3755 | 509 | >1100 | 955 | >1100 | 814 | 685 |
| 6.1 | Voltage stability | Time to failure (s) | ASTM D149/ D3755 | NA | 67 | NA | 51 | 102 | 78 |
| | | | | | Refer to voltage stability plots for more complete details. | | | | |

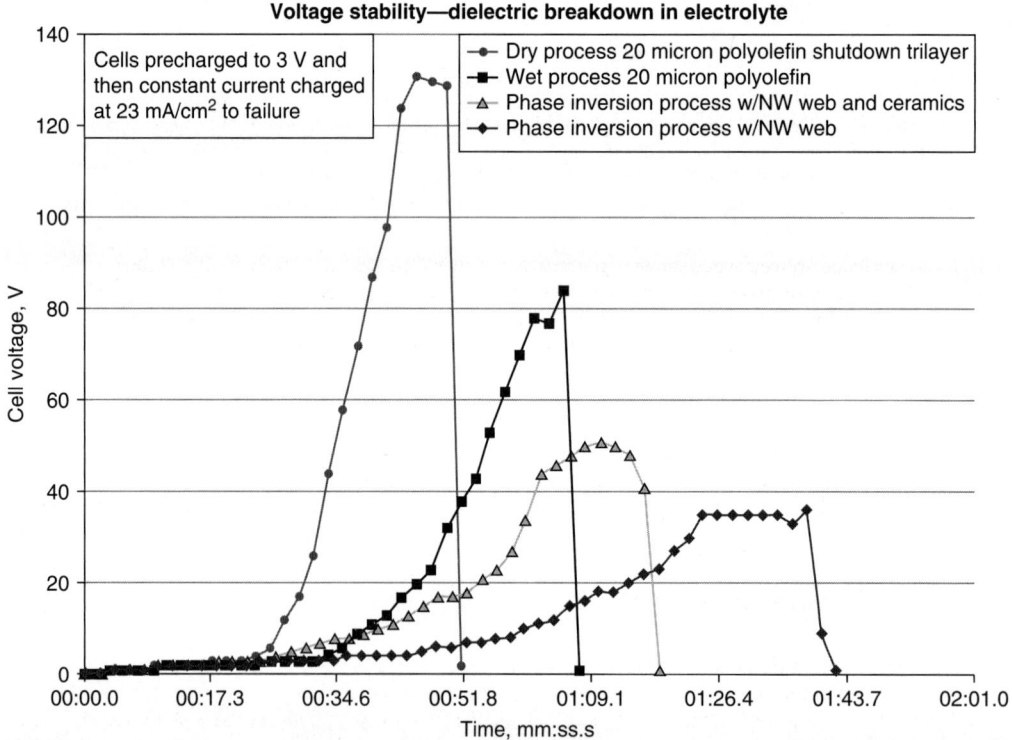

**FIGURE 3C.2**   SAE J2983 Electrical testing—Dielectric breakdown voltage.

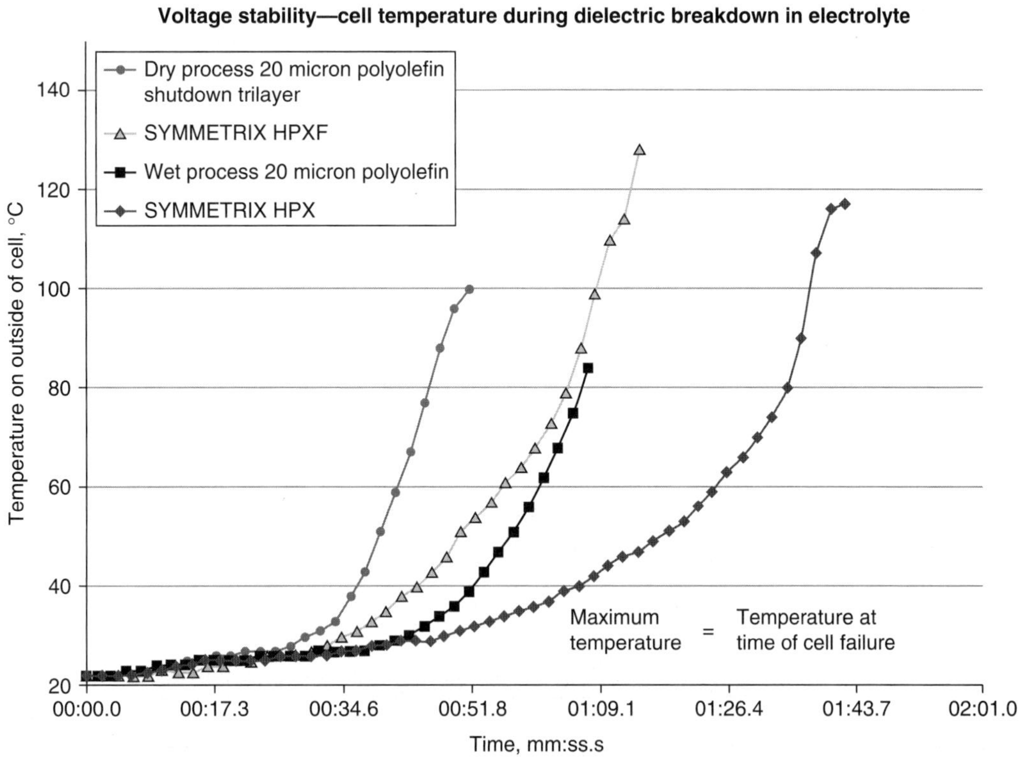

**FIGURE 3C.3**   SAE J2983 Dielectric breakdown testing—Thermal response.

### 3C.3.6   Test Parameter Summary

The tests listed above cover a range of parameters. However, each new battery design requires a thorough analysis for any and all possible separator failure modes. Numerous examples have been detailed in recent years where separators either failed to prevent or directly caused catastrophic battery failures. Hence, clearly, further development of separator test methods and qualification procedures are still required.

Also, while not discussed above, a wide range of other different test protocols are, of course, used for aqueous systems. The nature of acid, alkaline, and other aqueous systems entails completely different types of separators. Lead-acid separators, for example, are designed principally to hold a maximum amount of electrolyte. Issues such as separator shorting, melting, tearing, etc., are lesser concerns due to less energetic failure modes in lead-acid cells. On the other hand, gas transfer (recombination) is a major concern in sealed cells. In addition, all battery separator development programs should include complete life testing in full-up battery packs.

## 3C.4   *NEW SEPARATOR TECHNOLOGIES*

One of the most significant trends in lithium-ion batteries has been the effort to make separators both thinner *and* more resistant to shorting failures. The primary approach to these opposing goals has been to add ceramic fillers onto or into typical microporous polymer membranes.[2] Another approach is to use high-temperature polymer fibers to control the integrity of the films under abuse conditions.[3] The other recent major effort in the lithium battery field has been to use solid-state materials.[4] Work has focused on both polymer and ceramic/glass electrolytes (see Chap. 22C). In addition, hybrid systems with combinations of solid state, liquid, and/or gel components are now being developed (see Chap. 22A). But an even more far-reaching effort has been proposed that includes the use of electrical signals to detect battery shorts.[5]

Improved separators for other battery systems are also being developed. For instance, one of the biggest obstacles to the success of the alkaline rechargeable batteries with zinc electrodes has been the inability to adequately prevent the growth of dendrites through the separator. New efforts are discovering that the use of new anode structures (i.e., three-dimensional configurations), rather than modified separators, may prevent these internal shorts.[6]

Also, beyond various material improvements, the introduction of new process technology that reduces separator manufacturing costs and improves battery performance is being explored. One such example of an improved process technology that was detailed in Chap. 3A has also been found to be applicable to separator manufacture.[7] The processing technique could be used to produce both lead-acid and lithium battery separators with and without ceramic or other fillers. The technique can produce very porous and highly fibrous polymer films or mats with nanosized fibers and filler particles. Membranes have been made with filler levels exceeding 90% by volume in a low-cost, simple process. An example of polymer separator web made by this new process is shown in Fig. 3C.4.

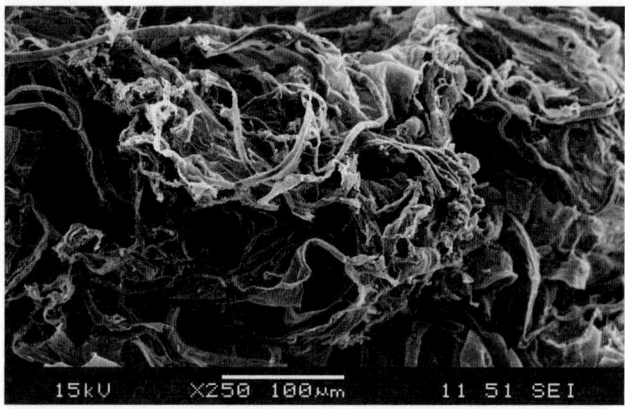

**FIGURE 3C.4**    Polymer separator web structure. (*Courtesy of Teebs R&D, LLC.*)

## REFERENCES

1. A. Pankaj and Z. Zhengming, "Battery Separators," *Chem. Rev.* **104** (10), 4419–4462 (2004).

2. L. Hock, "Separate from the Rest," R&D magazine, https://www.rdmag.com/award-winners/2013/08/separate-rest (web: May 18, 2018).

3. R. Clark, "From Energy Generation to Protection from Harm-Evolving Fiber Materials and their Applications Inside and Outsider Lithium ion Batteries," *35th Annual International Battery Seminar and Exhibit*, Fort Lauderdale, FL, March 26–29, 2018, pp. 13–15.

4. J. Voss, A. Luntz, S. Stegmaier, and H. Heenen, "Promise and Challenges of Practical High-Power Density Solid-State Batteries," *35th Annual International Battery Seminar and Exhibit*, Fort Lauderdale, FL, March 26–29, 2018, pp. 13–15.

5. Y. Barsukov, "Cell Internal Shorts as Next Frontier of Battery Safety: Types, Prevention and Detection Inside Battery Pack," *35th Annual International Battery Seminar and Exhibit*, Fort Lauderdale, FL, March 26–29, 2018, pp. 13–15.

6. D. Rolison, J. Parker, and J. Long, "Dendrite-Free Rechargeable Zinc-Based Batteries: Solving a Chronic Impediment through Architectural Design, pp. 6–11," *35th Annual International Battery Seminar and Exhibit*, Fort Lauderdale, FL, March 26–29, 2018, pp. 13–15.

7. Teebs R&D, LLC is an early stage development company owned and operated by Trevor D. Beard.

# SECTION D

# ELECTRICAL CONNECTIONS FROM CELL TO END USER

## Charles M. Richard

## 3D.1 OVERVIEW

The ultimate goal of most engineers and designers is to provide products that increase company sales and enhance market position. To best accomplish these goals in the battery industry, a designer must progress from the end user backward to the cell chemistry. Ideally, a design will start by conforming to the physical size and shape requirements of the end user's device. However, in doing so, the battery designer is also responsible for the electrical packaging of the battery module, which involves a little more than making a series of electrical connections.

Each step from the end user to the final cell relies upon electrical connections (i.e., the electrical packaging). While most end users do not basically care about cell chemistry, the battery-powered device must still function properly and be safe and reliable. The end user's demands dictate the features of the device, including attributes such as physical size, weight, portability, cost, charging speed, operating temperature range, etc. In turn, the features of a device dictate the design of the electrical packaging and subsequently the cell chemistry.

Successful designers follow this path when designing and packaging their batteries resulting in products such as the cylindrical-shaped D-cell battery that enables the user to comfortably wrap their fingers around a flashlight, a 9-V battery that could fit into a transistor radio, and a hearing aid battery that could be hidden inside the ear of a style-conscious user. Hence, fat smart phones, stuffed with three AA batteries, do not exist, and the 85-kWh battery pack of a Tesla Model S is not 2 m$^3$ in size and does not weigh 38,000 kg. Effective battery module designers must also be conscious of how their electrical packaging choices affect the manufacturer of the end user's device. If the designer specifies that a $10 connector be attached to the outside of the battery module, forcing the device manufacturer to purchase the $100 mating connector that also requires the purchase of a $15,000 crimping tool, the battery module will not enjoy widespread market acceptance.

In the section that follows, the design methodology best used to devise the network of electrical connections in a battery module is described. These guidelines are intended to help the battery designer determine the best options for the electrical connections in the battery pack. However, these steps should only be viewed as suggestions that need to be verified and tested to ensure safety and performance.

## 3D.2 CONNECTOR DESIGN CONSIDERATIONS

Before exploring the topic of electrical packaging for battery modules, some definitions of terms are needed:

- Electrical connection: A junction between two or more components of an electrical device that allows electrons to flow between those components.
- Bond: An electrical connection, not intended to be disconnected, created by joining conductive materials together using processes such as welding, soldering, crimping, etc.
- Cell pack: Two or more individual electrochemical cells connected together in series and/or parallel.
- Connector: An electrical connection that is designed to be easily connected and disconnected.
- Battery module: The smallest energy storage container in an end user' device that might need to be replaced if the device malfunctions (i.e., loss of energy output).

- Ampacity: The amount of electrical current that an electrical connection can conduct without exceeding the maximum temperature level desired by the battery module's designer or by the end user (i.e., the temperature rise caused by joule heating).
- Current profile: A graphical representation showing how the amplitude of the current flowing through an electrical connection changes with respect to time.

The multiple-pronged plug that enables a table lamp to be connected to and disconnected from an electrical outlet on the wall is an example of a connector, whereas the welded joint that connects one of the prongs of the lamp's plug to one of the wires in the lamp's power cord is an example of a bond. As is true in this table lamp example, a connector is always larger, more expensive, and heavier, and has a higher electrical resistance than a bond.

The designer is responsible for deciding which of the battery module's electrical connections should be connectors rather than bonds, based on the requirements of the end user. Due to a connector's disadvantages of cost, size, weight, and resistance, a battery module designer should favor the use of bonds for electrical connections unless the benefit of using a connector outweighs its drawbacks. The five situations when the use of a connector is recommended over the use of a bond are as follows:

1. Frequent disconnects—if the battery module needs to be connected and disconnected repeatedly.
2. Reduced manufacturing cost—if use of a connector results in a reduction in manufacturing labor cost, which, despite the added cost of the connector, reduces the total manufacturing cost of the battery module.
3. Unreliable components—if there is a high probability that, after the manufacturing process has been completed, a faulty component will be discovered and need to be replaced.
4. Simplify final assembly—if the final step in attaching the battery module to the end user's device is performed by someone less technically competent than the battery module's manufacturing team.
5. Simplify maintenance—if, after the device is in service, maintenance is performed by someone less technically competent than the battery module's manufacturing team.

Over the next 10 years, pressured by demand for greater density in energy and power, the size, nature, and visibility of the electrical connections in battery modules will change dramatically, mimicking the evolutionary path of the semiconductor industry. When transistors were in their infancy, each transistor was large (the size of a garden pea), packaged separately, and of dubious quality, which forced designers to include quick-disconnect sockets. These sockets, being the size of four garden peas, were included in their designs so that faulty transistors could be quickly and easily replaced when they inevitably failed. Eventually, as reliability and manufacturing processes improved, over 100 million transistors could fit into the volume of one garden pea. Due to the improvement in reliability (i.e., there was no longer a need to be able to remove a faulty individual transistor), the number of electrical connections required to connect 100 million transistors fell by a factor of 4 and the size of each connection decreased by a factor of 100 million.

## 3D.3   BATTERY CONNECTION DESIGN SPECIFICS

In this section, specific requirements for electrical connections in battery modules are described based on present-day size and quantity. How these connections evolve during the next 10 years will be interesting to witness. Also, notice that the terms "battery designer" and "battery module designer" seem to be used interchangeably. Around 100 years ago, when 20-kg automobile batteries were only one of a few energy storage devices capable of being recharged, and when the charger was located 1 m from the battery, a battery designer only had to worry about the relative proportions of lead and acid and whether the battery would leak if knocked over. Today, as batteries rise in power, shrink in size, and increase in complexity, the battery designer's responsibilities have grown to encompass battery management systems, charging systems, and all electrical connections between the cell's anode and cathode and the external environment.

While considering the electrical packaging requirements, envision the concept of a 15-V battery module, consisting of four lithium-ion cells, enclosed together with a battery management system and wireless charging receiver all packaged inside a $10 \times 50 \times 80$ mm sealed plastic shell. The end user, the device manufacturer, and the battery module manufacturer now clearly consider the battery module as a single component, not a collection of components. Thus, the demarcation lines that used to separate the responsibilities of a cell designer, battery designer, charger designer, electrical packaging designer, and a battery module designer have begun to disappear.

When designing the electrical connections in a battery module, the two most important considerations are resistance and ampacity. The lower the resistance, the lower the voltage drop per electrical connection, which maximizes the percentage of the cell's voltage delivered to the end user's device. However, a lower resistance connection is always more expensive than a higher resistance one. For example, connecting 10 lithium-ion cells in series using copper tabs each having a 50 mm$^2$ cross-sectional area will yield a lower resistance than that obtained by using 4 mm$^2$ aluminum tabs, but the copper connection is 30 times more expensive.

The designer's responsibility regarding a connection's ampacity follows the same pattern and is affected by the same trade-offs as resistance. Ampacity was defined earlier as "the amount of current that an electrical connection can conduct without exceeding the maximum temperature level desired by the battery module's designer or by the end user." This is a more useful definition than the one frequently cited online, which says that ampacity is "the amount of current a component can conduct before sustaining damage." In keeping with the concept of working backward from the end user's device, today's battery module designer must understand and quantify the maximum desirable temperature for the battery module as well as the end user's device over the current profile that is anticipated. An internal temperature that is too high can damage the battery's electrochemical cells, and an external temperature that is too high can cause discomfort or injury to the end user.

If the design of a battery module's connections are based solely upon the battery's expected 10-hour discharge rate (0.1 C), the connections will be under-designed for the end user's requirement based on the need for the same connections to recharge the battery module in a 15-min (4 C) charging cycle. Such under-designed connections would cause excessive heat, which could burn an end user who touches the cell phone while it is charging.

Since ampacity is a quantification of an electrical connection's ability to remain cool, and since the ability of a component to shed heat is affected by its ability to release thermal energy through conduction, convection, and radiation, there is no simple mathematical formula that can compare the ampacities of electrical conductors or electrical connections comprised of different materials in different configurations. In the absence of a mathematical model, the designer must rely upon the performance specifications provided by the components' manufacturers. Choose component manufacturers who express their component's ampacity performance based upon a current profile and a specific temperature rise, not simply a rated current.

## 3D.4   BATTERY CONNECTION ARCHITECTURE AND REQUIREMENTS

Whether designing a 3.8-V, 8-Wh battery module for a mobile phone, which has a total of four positive and four negative DC electrical connections between the single lithium cells and the power grid (see Fig. 3D.1), or a 650-V, 120,000-Wh microgrid energy storage system with 16,000 connections between the 2100 lithium cells and the power grid, the battery module designer is responsible for all the following levels of connections:

- Single cell
- Cell to cell
- Cell pack to battery module
- Battery module to device
- Device to power grid

Figure 3D.1 shows the wide range and deployment of connectors used within a typical cell phone.

Specifics for the wide array of connections used in a battery-powered electronic device are detailed below.

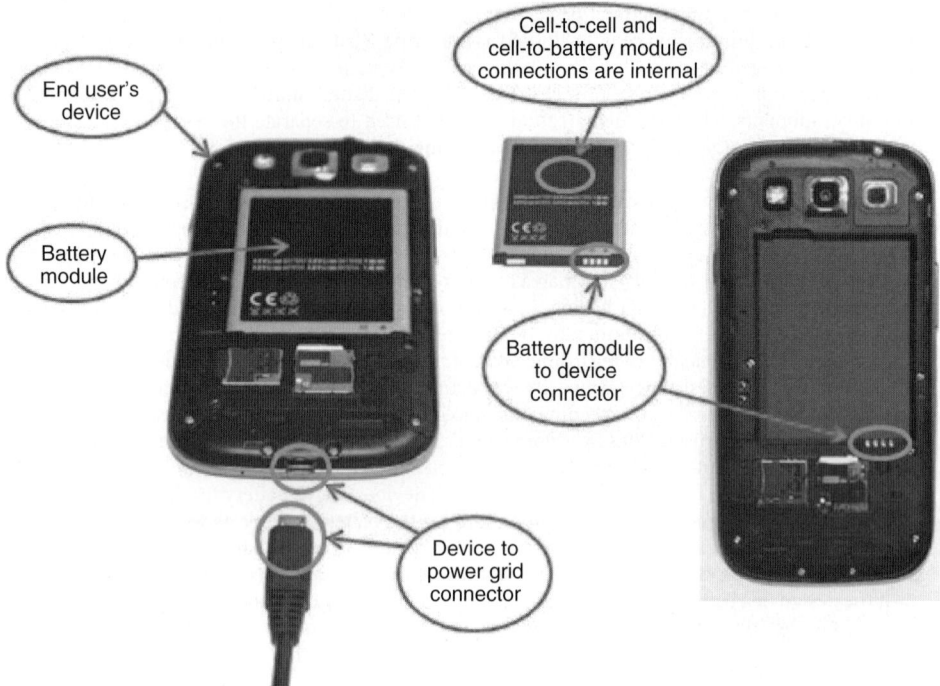

**FIGURE 3D.1**  Typical configuration connections for a cell phone.

### 3D.4.1   Single Cell and Internal Battery Module Connections

The most fundamental electrical connections of a battery cell are its anode and cathode. Connections to the battery plates are the initial mechanisms for moving electrons into and out of the electrochemical cell. Historically, the design of anodes and cathodes has been the domain of chemists and material scientists who select materials that are physically stable and facilitate the next higher electrical connection. However, a second type of basic cell connection may also be required.

Specifically, if the voltage and energy storage capacity of a single cell is sufficient for the end user's device, then no cell-to-cell connections are necessary inside the battery module. However, with today's available cell chemistries, any battery module with a voltage of over 4 V will also have internal connections joining multiple cells. The 12-V lead-acid automotive battery and the 9-V carbon-zinc transistor radio battery, both introduced in the 1950s, consist of six individual cells connected in series. Figure 3D.2 shows these two types of batteries that have both electrode and intercell connections within the internal environment of a cell.

More details on the options for making connections to electrode plates and to the terminals or tabs internal to the cells are discussed in the chapters under Part 4 of this Handbook. In general, though, any connection within a cell that is exposed to electrochemically active materials must resist chemical and galvanic corrosion and tolerate various conditions of use and abuse (temperature extremes, shock and vibration, fatigue, etc.). Techniques for making these connections typically fall under the following generic categories:

1. *Compression:* The use of physical contact, only, between the current-carrying parts. This technique is not commonly preferred due to potential interfacial resistance problems, but sometimes it is used in certain designs such as button/coin cells. Also, in some cells a spring-loaded tab or a socket type connector might be employed.

**FIGURE 3D.2**    Battery types that require series-connected electrode and intercell connections.

2. *Bonding material:* The use of a supplemental metal or other conductive material to join parts together. This method is distinguished in this list from other bonding techniques in that the electrical junction is created by materials with dissimilar compositions. The use of two or more materials is not an ideal technique, as interfacial reactions may contribute to defective circuits. Techniques such as soldering fall into this category. Additionally, other adhesive bonding systems have been developed based on conductive carbon-filled epoxies and such.

3. *Fusion:* The use of molten metals, similar in composition to the current-carrying parts, to create an essentially uniform, continuous electrical pathway upon solidification. This technique includes two main categories:

   - *Casting:* In lead-acid cells, for example, molten lead is used to bond the tabs of each plate to an internal lead bus-bar and then to cast lead terminals. As another option, some electrochemistries may use cast metals of varied compositions, but unlike the bonding method described above, the metals are capable of forming alloys with each other—creating a stable junction.
   - *Welding:* Joints are created by melting the electrical component parts, followed by fusing them together. Welding techniques include laser welding, percussive arc welding, ultrasonic welding, tungsten inert gas (TIG) and metal inert gas (MIG) welding, and a variety of other advanced methods.

Fusion joining methods are considered the most advanced, reliable, and cost-effective methods of connecting the internal cell current conductors, but require a high level of expertise, control, and precision to implement. Many battery failures have been attributed to the failure of cast or welded joints.

### 3D.4.2    Cell-to-Cell Connections in the External Environment

Cell-to-cell connections (i.e., intercell connections), with electrical pathways external to the cell or battery module, have different requirements. These connections are not typically directly exposed to the cell chemicals and need not typically be designed for such exposure. However, many cells still suffer from leakage and venting problems (lead acid, dry cells, etc.), and the external connectors must be designed to deal with these lapses.

In general, external intercell connections are made in the same manner as described for the internal cell connections. Compression, bonding, and fusion are all potentially acceptable methods. The decision is based on end user requirements and cost/performance analyses. While the fabrication of the external connections between cells is less stringent than internal connections, there is still a great need for care. Overheating of a terminal or tab during soldering or welding may damage the inner parts of the cell, causing shorts, decreased life, or even catastrophic failures. Excessive mechanical forces during a welding or crimping operation could also jeopardize cell integrity.

One final caveat in designing and making intercell connections is that the resultant connection will be sufficiently durable for actual use. An intercell connection may fail from issues such as physical damage, inadvertent electrical shorting, mechanical strain from thermal events, or corrosion from external media (moisture, heat, etc.).

### 3D.4.3    Cell Pack to Battery Module Connections

The method of connecting a cell pack to the battery module's external electrical connection differs according to the cell chemistry and the battery module's total weight. In 99% of lead-acid battery modules, the cell pack is connected to the battery module's external electrical connection with a bond. In battery modules based upon newer, higher energy density chemistries, 95% of those applications where the battery module weighs over 2 kg, the junction between the cell pack and the battery module's external electrical connection will be a connector. In 75% of battery modules that weigh less than 2 kg, the junction between the cells and the battery module is a bond. These electrical junctions exist inside the battery module's case and are connected to the "behind the bulkhead" half of the battery module to device connector, which is highlighted next.

### 3D.4.4    Battery Module to Device Connections

Almost all (99%) disposable batteries are joined to the end user's device using an electrical connector, allowing the device to resume operation once the discharged battery has been replaced with a fresh one. Additionally, 85% of the rechargeable battery modules are connected to their devices using electrical connectors. This percentage has dropped during the past 5 years due to the increased reliability of rechargeable battery modules. The proportional cost between the device and the battery module is a major determinant of whether a bond or a connector is used. Electric toothbrushes and electric shavers utilize bonds to connect their rechargeable battery modules to their devices, whereas 100% of mobile phones, laptops, electric vehicles, microgrid energy storage systems, and battery modules that weigh over 1 kg utilize electrical connectors.

When designing a battery module that has a peak charge or discharge current under 50 A, a wide range of connectors are offered by manufacturers such as TE, Delphi, Thomas & Betts, Molex, and Kostal. In fact, at currents under 50 A, you can select connectors that combine signal and power contacts in the same housing, enabling you to incorporate battery management and monitoring signals with power transmission in a single pluggable housing.

However, with battery module designs having a peak charge or discharge current above 100 A (including 90% of cutting edge applications such as self-driving cars, microgrid storage systems, electric buses, etc.), connector options are significantly more limited. Historically, 95% of connector manufacturing companies have focused upon high-volume applications, and high-volume applications have always been signal connectors that carry less than 1 A.

Connector companies that specialize in connectors above 100 A include Rebling, Anderson, Meltric, and Staubli. A typical high-power connector is shown in Fig. 3D.3

Choosing the right electrical connector for mating of the battery module to the electronic device has a significant impact upon a battery module's success. Choosing a connector that is low-cost, small, and easy to attach (see Fig. 3D.3) can help to ensure market acceptance and global success of a new battery module.

If an application requires toolless maintenance or a rapid battery swap, then a double-pole, quick disconnect, similar to the one shown in Fig. 3D.4, may be a good solution. If a battery module connector is specified that forces the end user or the device manufacturer to purchase a costly or complex mating connector that also requires expensive crimping equipment, the choice of the connector can significantly restrict widespread acceptance.

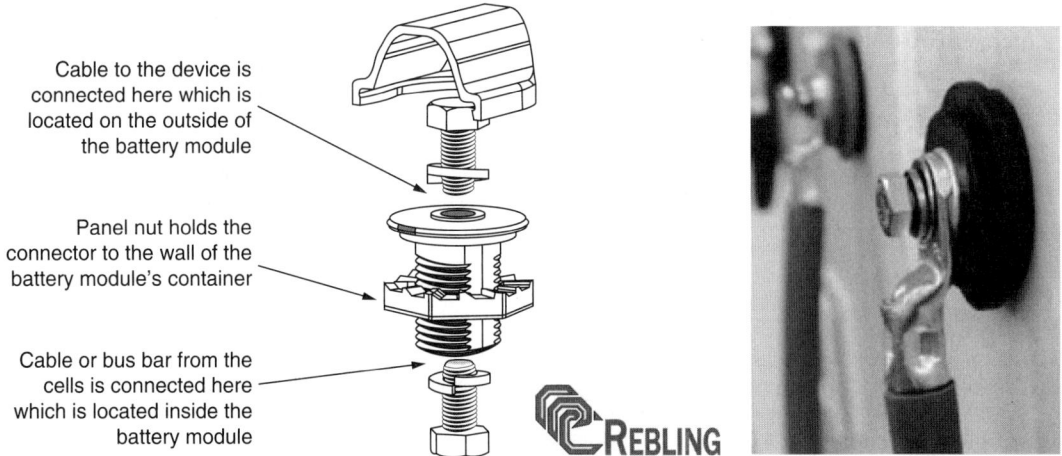

Cable to the device is connected here which is located on the outside of the battery module

Panel nut holds the connector to the wall of the battery module's container

Cable or bus bar from the cells is connected here which is located inside the battery module

**FIGURE 3D.3**   Lithium battery terminal. (*Courtesy of Rebling Corporation.*)

### 3D.4.5   Device to Power Grid Connections

Since the majority of new energy storage devices are rechargeable, they must be connected to some power source to be charged. Of all the connections between a cell and a power grid, the battery designer has historically had the least amount of involvement in specifying the connection between the device and the power grid (such as a universal serial bus [USB] connector that couples a cell phone to its charger or such as the pistol-grip connector used to recharge an electric vehicle). As devices shrink in size and as wireless charging grows in popularity, the battery designer's role in defining the device to power grid connection will increase.

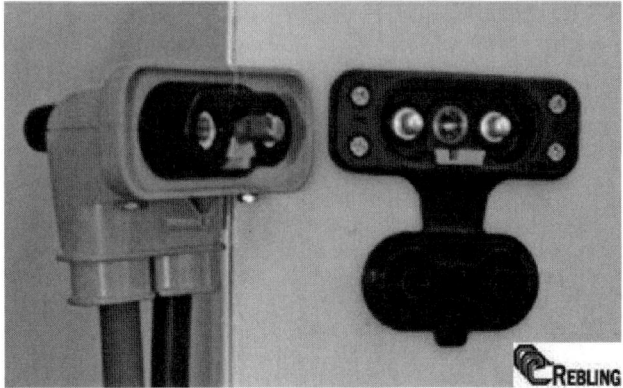

**FIGURE 3D.4**   Double-pole quick disconnect. (*Courtesy of Rebling Corporation.*)

## 3D.5   SUMMARY

The next 10 years hold great promise for the battery module industry and great challenges for battery module designers. The market demand for greater power density will drive miniaturization and, mirroring the path of the semiconductor industry, designers will need to surmount novel problems involving size, temperature, and electrical packaging. Here is an example of the challenges being faced today:

Imagine an electric vehicle that requires a battery module with an external shell composed of 3-mm thick aluminum, measuring 200 mm × 300 mm × 900 mm, containing a cell pack consisting of hundreds of lithium-ion cells, producing a total of 400 V. The module's steady state discharge current is 220 A, and superimposed upon that current is a repeating pattern of 750-A pulses, each 20 s in duration, that occur every 15 min. A serpentine cooling system reliant upon thermal conduction between the surfaces of the lithium cells and the cooling tubes could not provide sufficient cooling; so the battery module designer found it necessary to flood the entire module with a non-conductive coolant. The thermal expansion of the coolant due to the heat produced by the lithium cells creates a significant pressure inside the battery module.

The designer's challenge is to find a battery module connector that can be attached to the 3-mm aluminum shell with an ampacity to handle the current profile, a voltage rating suitable for the applied charge, and a seal to prevent pressurized coolant from escaping. The designer is at a disadvantage because 98% of connector manufacturers have designed their connectors to meet the sealing requirements of the IP (ingress protection) Code (IEC [International Electrotechnical Commission] Standard 60529), which rates a connector's ability to prevent the ingress of cold or warm water, not the egress of hot chemical solutions.

Table 3D.1 can assist battery module designers in selecting the optimal connector for attaching the battery module to the end user's device. It is important for the designer to remember that each additional feature adds cost and complexity to a connector. Increased cost and increased complexity are two characteristics that oppose the market success of any battery module. The lower the cost and the simpler the connection, the greater is the acceptance of a new battery module by the manufacturers of end user devices.

**TABLE 3D.1**   Connector Selection Guidelines

| Battery connector considerations | Guidance (Assumes battery module-to-device connectors for 100$^+$ A applications) |
|---|---|
| How important is the speed and ease of disconnection? | A connector that can be disconnected in 3 s without a tool is always more expensive than one requiring a tool and 20 s |
| How many power contacts must be disconnected in one motion? | Single-pole connectors, compared to double- or triple-pole, are less expensive per pole, dissipate heat more effectively, and give the device manufacturer more flexibility in routing cables in the device |
| What are the values for:<br>• The continuous discharge current?<br>• The peak discharge current and peak duration?<br>• The frequency of peak recurrences?<br>• Charging and discharging current magnitudes?<br>• Cable or bus bar sizes?<br>• Maximum temperature rise of cable and bus?<br>• Maximum temperature rise of the connector? | It is imperative that maximum temperature rise and cable size are specified along with the expected current profile. If not, an overly expensive connector with an excessively high ampacity might be selected in lieu of a more economical one that contributes a maximum acceptable 30°C temperature rise. Even worse, a connector that does not specify its current profile performance may contribute a temperature rise of >200°C to the battery module |
| What exposures are expected:<br>• Fluid ingress protection?<br>• Fluid egress prevention and pressure?<br>• Chemical contact with the connector? | Certain simple connectors can meet ingress and egress requirements at economical costs, whereas more complex connectors can only meet such demands at great expense. It is important to know whether the connector needs to prevent water ingress into the battery module from a rainstorm or from a 3 m continuous submersion |
| What environmental condition must the connector resist? | The more conditions a connector must resist, the more expensive the conductors and insulators and the more complex the design |
| Does the connector require EMI shielding? | A connector's ability to attenuate incoming or outgoing electromagnetic interference energy has a huge impact upon its cost, the expense of crimp tooling, and the labor costs of the assembly |
| What is the required operating temperature range? | The higher the maximum operating temperature, the more expensive and less flexible is the plastic that comprises the connector |
| What flammability resistance is required? | The higher the UL94 flammability rating, the more expensive the connector. If the battery module's shell is composed of a plastic with an HB rating, it makes little sense to specify a connector with the higher V-0 rating or the ultimate 5 VA rating |
| What is the nominal operating voltage? | Today's trend toward higher voltage battery modules renders obsolete those connectors which were designed during an era when 12, 24, and 48 V were dominant |

# SECTION E

# CELL AND BATTERY PACKAGING

**Kirby W. Beard**

## 3E.1 INTRODUCTION

When Alessandro Volta constructed the first electrochemical power source in 1799, the choices for packaging of the electrodes and electrolyte were limited. Initial cells comprised cups with metal coupons suspended in electrolyte solutions. The voltaic pile was made from stacked copper and zinc plates with interleafed organic fiber separators soaked in salt water that wicked electrolyte from a reservoir. The entire bipolar cell was simply stacked among a series of tie rods between two wooden end plates, as shown in Volta's letter to the London Royal Society (Fig. 3E.1) and in Volta's original model at the Tempio Voltiano, Como, Italy.[1]

While this battery did not need a container (since all electrolyte was absorbed by the separator and since the plates were held in alignment by tie rods), it was not long before more mechanically robust designs were needed and before flooded cells were common or water loss became a problem in long-term use or storage. The only viable containers available at the time for packaging a battery, aside from metals (which were often chemically reactive), were glass or ceramics. Eventually, glass jars were replaced with compatible, corrosion-resistant metal containers and modern-day polymer housings. The choice, then and now, was dictated by cost, electrochemical compatibility, material weight and volume, mechanical strength, the ability to preserve battery functioning over extended time periods, and extreme environmental conditions.

In particular, both electrolyte loss (evaporation) and contamination of the cell with moisture, oxygen, or other gases may be quite detrimental to many types of electrochemistries. For modern cells, commercial success is determined to a large degree by the product packaging method. The ability to properly insert the cell stacks, fill the electrolyte, close the container, and seal around the electric terminal(s) may all be impacted by the selection of packaging materials and assembly/closure processes. An entire industry has been created that has provided enabling technologies for accomplishing these functions across the full range of electrochemistries.

However, in addition to specifying individual cell containers, the battery designer must also take great care in assembling cells into packs. If the battery module or overall battery pack is not properly specified and constructed, damage may occur to the individual cells. Issues such as mechanical abuse, shock/vibration, thermal expansion/contraction, heat dissipation, venting of hydrogen gas, etc., are a few of the issues. While the battery pack housings are not subjected directly to the electrolyte or the electrochemical reactions, there is still the likelihood in many cell types for corrosion from vapors, leakage, or spillage. A flooded lead-acid battery compartment or storage rack must be resistant to acid, for example, to reduce potential for a collapse of the structures.

To provide a low-cost, high-performance battery system, cell packaging and battery compartments or housings must be fully analyzed and tested prior to fielding of the product. A number of historic battery failures have been traced to packaging issues. Various test standards and compliance requirements must normally be consulted to ensure product safety and life. Each application, whether electric vehicles, implantable medical devices, consumer electronics, commercial aircraft, or space satellites, will have its own set of packaging guidelines or parameters that apply. The sections below provide a high-level review of these issues.

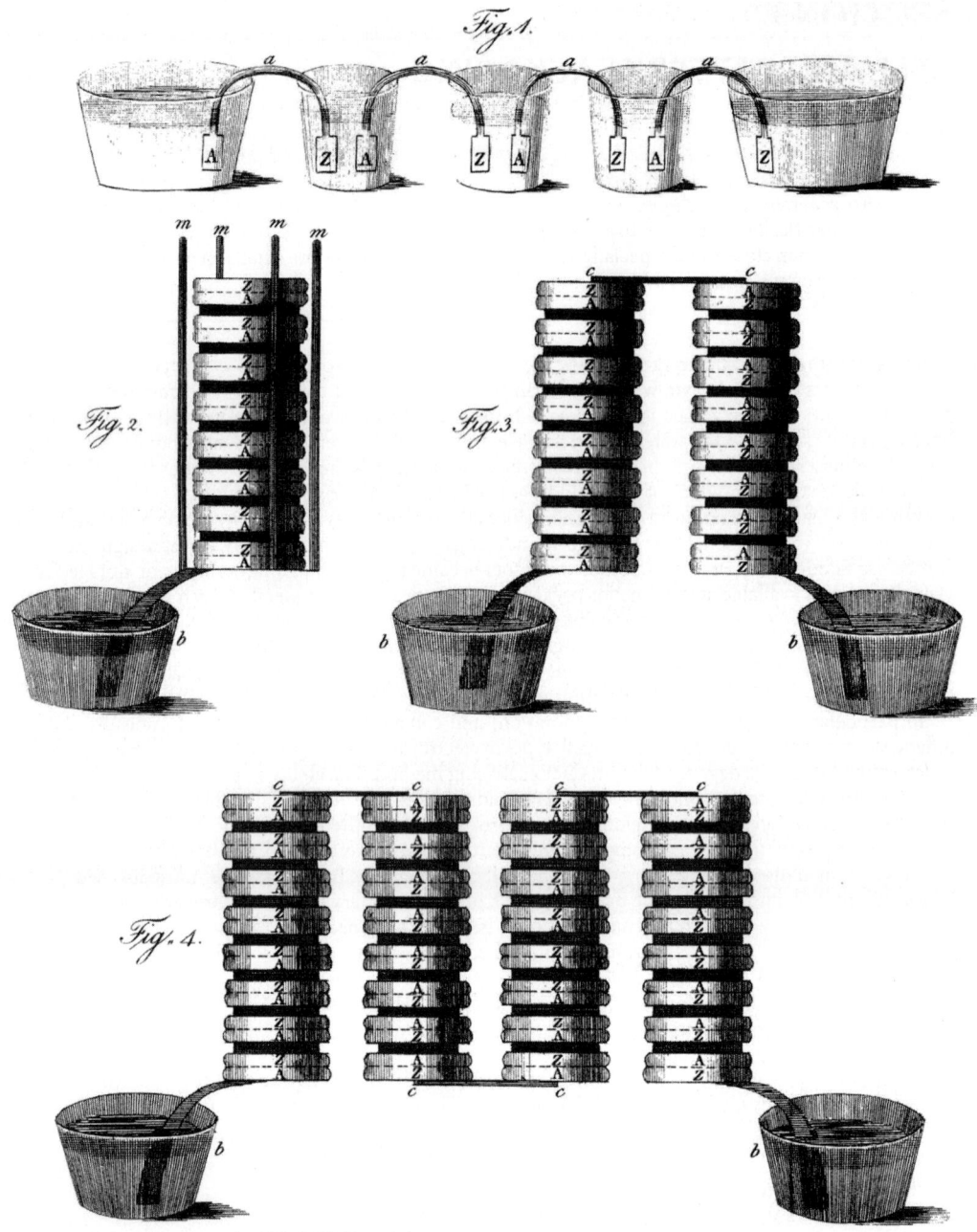

**FIGURE 3E.1**    The voltaic pile. (*U.S. Public Domain.*)

## 3E.2   ORIGINAL BATTERY CONTAINERS

The first battery cells used either glass or metal cans, or jars or boxes to contain the cell stack and retain the electrolyte. Dry cells (Chap. 11A) originally used paper cylinders. Some containers had lids or covers, with some sealed by asphalt or rubber gaskets, but overall such cells were not able to fully prevent liquid vapor losses or gas and moisture infiltration. This type of packaging was amenable to electrochemical systems that could tolerate some compositional changes and various side reactions. Certain other cells even require venting of excess moisture or gases generated during storage or charge. Hence, these particular cells (i.e., lead-acid, nickel-cadmium, etc.) may be specifically designed with a venting system. Venting devices would never likely be fully effective in sealing the cell when closed, and especially after venting and resealing. More details on packaging for these various types of legacy systems can be found in Part 4.

## 3E.3   HERMETIC AND SEALED PACKAGING

For the most demanding applications or the most sensitive electrochemistries, the use of hermetic cell containers is required. Hermeticity refers to the ability of the packaging to prevent the transmission of liquids or gases into or out of the cell. A hermetic container must have a casing material, a method of case closure, and an electric feedthrough system, which together make the cell totally "impermeable". The typical method of determining proper sealing is a helium gas leakage rate test. While vacuum or pressure checks can also be useful to detect gross issues, a truly hermetic cell must retard transmission on an atomic scale. The two materials that have found widespread use in hermetic cells are metal and glass. While ceramics may also work well, these materials are not common.

The common feature of hermetic cells is that both the container closures and the terminal feedthrough must be constructed of the materials with the same degree of hermeticity as the container itself. Hence, the lids are typically sealed by welds or solder, and the terminals sealed via molten glass placed between the container lid and the terminal. While one battery terminal may be attached to the metal container, which then serves as either the positive or negative lead attachment, the other terminal must be electrically isolated from the container to prevent a direct short. It is possible to use a sealed glass container with penetrating metal terminal pins, but these designs are not very practical and typically are confined to cells used in research.

A variety of other cell designs, while not truly hermetic, provide for high levels of impermeability. Lids and terminals may be sealed by crimping/compression with flexible or compressible gaskets with good results. Also, epoxy or other inert thermoset polymers have potential to function as cover and terminal seals in a few limited applications. However, regardless of the sealing method, the cell lid closure and terminal seals must be able to survive the conditions of use as discussed next.

### 3E.3.1   Glass Containers and Glass-to-Metal (GTM) Seals

Glass working was a common trade in the 1800s and allowed for the manufacture of various glass battery container designs long before advanced metals were available. Glass technology was also then used to create sealed electric terminal penetrations, leading to the modern glass-to-metal seal used extensively in various electronics such as the electric light bulbs and vacuum tubes, but more recently in corrosive electrolytes, such as thionyl chloride lithium batteries.[2]

In another application, glass ampoules, but without any GTMs, are still used extensively in reserve cells to provide decades of hermetic electrolyte storage. However, even this technology is quite complex. In lithium reserve cells with nonaqueous electrolytes, even trace amounts of moisture can be damaging. The organic solvents and lithium salts create an autocatalytic reaction that leads to gas evolution, ampoule pressurization, and eventual rupture. Extensive decades-long testing revealed some issues and provided some solutions. For instance, certain types of batteries will have a lithium chip added to the ampoule to counteract the moisture contamination problem.

Another example of a packaging issue relates to lithium-ion cells. While sulfur dioxide and thionyl chloride lithium primary cells require the use of a GTM, lithium-ion secondary cells have not typically been able to use this type of terminal seal. Specifically, the manufacture of a GTM relies on specific thermal response properties between the glass and the metal pin. The ideal design creates compressive forces on the terminal pin as the glass billet in the GTM cools after heat fluxing. The use of aluminum in a GTM does not provide compressive forces. Researchers have made progress on the use of aluminum in lithium-ion cell packaging, and some progress has been reported.[3]

### 3E.3.2 Metal Containers

A variety of metals and alloys are used for cell containers. The cans and lids are made by traditional metal working techniques (stamping, deep drawing, welded fabrications, etc.). In some cases plated metals may be used either to provide electrochemical reaction sites or to protect against any corrosion reactions. Polymer liners may also be used for corrosion protection. Cell closure occurs by crimping, sealing with elastomers/polymers, or by welding/brazing/soldering if the cell is intended for a hermetic application.

Metal cans may also be made from thin, cheap steel, as is common in alkaline cells, or from costly stainless steel alloys, titanium, nickel, etc., such as used for advanced battery applications. The technology for closure welds is extremely complex and includes advanced welding techniques and extensive analytical capabilities to insure proper welds and prevent heat damage to the cell components. Examples of metal cell containers and closure systems are shown in Figs. 3E.2 and 3E.3. When such cells are used in applications such as space craft or implanted medical cells, extensive long-term testing is needed to verify safety and performance. Examples of advanced hermetic cell packaging can be seen in Chap. 28 and other battery application chapters.

### 3E.3.3 Polymer Jars/Sleeves/Trays

The most common types of battery containers are based on thermoplastic moldings. Materials such as the polyolefins (polyethylene [PE] and polypropylene [PP]) have excellent chemical resistance and a fair range of mechanical and temperature capabilities. Other materials such as nylon, acrylics, and fluoropolymers have also found use in battery containers

**FIGURE 3E.2** Cylindrical metal battery cans and cover (AA to 2D sizes).

**FIGURE 3E.3** Prismatic metal battery can and header (with GTMs, rupture discs, fill hole).

(see Chap. 2C). While cost for injection molding dies is fairly high, the technology is well established and cost-effective in high-volume applications. Often, cell containers and pack assemblies will have multiple parts such as lids, insulating inserts, terminal holders, bottom bases, etc. The use of multicavity molds can be cost-effective in casting sets of parts. And the use of thermoplastics allows for component assembly by known plastic processing techniques such as solvent gluing, ultrasonic bonding, and thermal fusion. Figure 3E.4 provides an example of an innovative plastic case design that was proposed in the past for producing small lead-acid cells.

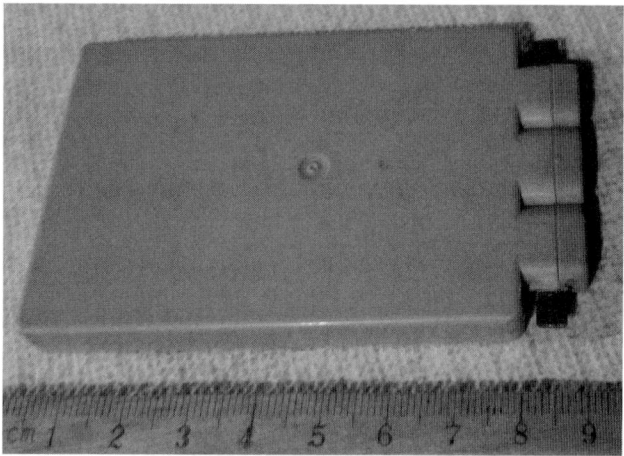

**FIGURE 3E.4**  Flat plastic case for single electrode pair (sealed lead-acid cell).

Container materials and molding techniques have been perfected for the different electrochemical systems over the years. Additionally, various peripheral equipment, such as battery trays and module compartments, can be made from a variety of other common thermoplastic or thermoset materials and production methods (i.e., thermoforming, rotational molding, shrink wrapping, etc.). Advancements in polymer science are also constantly being implemented into battery packaging systems, based on appropriate qualification testing and tradeoffs of cost and performance.

A network of resin suppliers, plastic compounders, and custom molders can be helpful in selecting the materials and setting up production. Additionally, new processing techniques, such as three-dimensional printing, can prove very useful for prototyping and battery qualification testing. Overall, though, packaging materials and methods will need to evolve as new cell chemistries and electrode stack designs progress. An example of this trend is the development of flexible cells that can bend to take the shape of the body in a wearable battery pack.

### 3E.3.4  Laminate Foil Pouches

A more recent battery packaging material that has found a great level of acceptance in the battery industry is the laminated foil pouch cell. Pouch cells were produced in the past by Polaroid (alkaline battery film packs), but they are now being used extensively for lithium-ion cells. The newest pouch materials are made from a multi-layer construction consisting of at least three layers:

1. An internal heat bonding polymer adhesive that allows for sealing of the cell by simple thermal fusion equipment (i.e., vacuum packaging heat sealers)

2. A solid metal foil layer that ensures the hermeticity of the cell (prevents moisture and oxygen ingress and electrolyte solvent losses)

3. An outer protective layer that prevents damage to the thin metal foil (generally consisting of durable, low-cost plastic films such as polyester)

Other layers may be present to add more robustness or functionality to the pouch.

To create a pouch cell, two oversized layers of film are positioned with the heat seal layer face-to-face on either side of a prebuilt electrode stack construction. Sometimes thick and more rigid multilayer foil pouch sheets are used by preforming a tray to accept the full thickness of the cell stack or flat winding. The cell assembly can be sealed and electrolyte filled in various ways. Some systems preseal three sides of the pouch and others may seal the entire assembly at once. Separate or combined processes are used to remove gases from the pouch and inject electrolyte, typically while applying a vacuum to the pouch. However, the pouch cell is almost always cut opened subsequently and then vacuum heat-sealed a second time due to gas evolution from contaminates and/or the formation process. A variety of specialized equipment, including vacuum heat seal chambers and electrolyte filling pumps, are used for these steps.

However, the most critical feature in a pouch cell is the sealing of the tabs that extend from the electrodes to the external electrical contacts. The bonding force between the inner adhesive layer and the battery tabs must be sufficient to prevent debonding or leakage but should not be overly aggressive such that the underlying metal foil layer is damaged, which can lead to electrolyte dry-out and even cell shorting if both tabs penetrate through to the metal foil layer. A number of common and proprietary techniques, involving tab coatings or protective tape layers, have been used to improve bond without damage to the foil pouch material itself. As noted elsewhere, any new or revised design will require extensive testing in various environments (high altitude, extreme temperature cycling, etc.). And while the foil material or the edge seals may fail, it is typical that the tab seal area will be most problematic. While foil pouches are considered less robust, the cost, thickness, and weight advantages are significant over metal cans or injection molded thermoplastics. Additionally, this type of packaging has benefited from years of development and experience from packaging food, medicines, electronics, and other critical items.

## 3E.4   SUMMARY

Continued development of advanced materials and multilayer composite films will undoubtedly help benefit cell packaging. Reactive fillers may also prove essential in blocking oxygen or moisture penetration or solvent loss. Modifications in electrochemistry could be significant in making cells last longer with even less robust packaging, too. The use of solid electrolytes is one example of such possibilities. Also, the trend to combine battery packaging with the electronics housing is very prevalent today. Overall, efforts to reduce cost, extend cell life, and enhance safety will continue.

## REFERENCES

1. https://commons.wikimedia.org/wiki/File:VoltaBattery.JPG (extracted April 1, 2018).

2. U.S. Patent 4556613A.

3. M. Moorthi, "Glass to metal seal with aluminum cans and lids." https://www.linkedin.com/pulse/glass-metal-seal-aluminum-cans-lids-mumu-moorthi-ph-d-mba (extracted May 22, 2018).

# PRINCIPLES OF ELECTROCHEMICAL CELL OPERATIONS

# CHAPTER 4
# ELECTROCHEMICAL PRINCIPLES AND REACTIONS

**Fausto Croce, Mark Salomon**

## 4.1 INTRODUCTION

Batteries and fuel cells are electrochemical devices that convert chemical energy into electrical energy by electrochemical oxidation and reduction reactions, which occur at the electrodes. A cell consists of an anode where oxidation takes place during discharge, a cathode where reduction takes place, and an electrolyte that conducts ions within the cell.[a]

The maximum electrical energy that can be delivered by the chemicals that are stored within or supplied to the electrodes in the cell depends on the change in Gibbs energy $\Delta G$ of the electrochemical couple, as shown in Eq. (4.5) and discussed in Sec. 4.2.

It would be desirable if during the discharge all of this energy could be converted to useful electric current. However, losses due to "polarization" occur when a load current $i$, accompanying the electrochemical reactions, flows through the electrodes. These losses include (1) activation polarization, which drives the electrochemical reaction at the electrode surface, and (2) concentration polarization, which arises from the concentration differences of the reactants and products at the electrode surface and in the bulk as a result of mass transfer.

These polarization effects consume part of the energy, which is given off as waste heat, and thus not all of the theoretically available energy stored in electrodes is fully converted into useful electrical energy.

In principle, activation polarization and concentration polarization can be calculated from several theoretical equations, as described in later sections of this chapter, as long as certain electrochemical parameters and mass-transfer conditions are known. However, in practice it is difficult to determine the values for both because of the complicated physical structure of the electrodes. As covered in Sec. 4.5, most battery and fuel cell electrodes are composite bodies made of active material, binder, performance-enhancing additives, and electrically conductive filler. Also, electrodes usually have a porous structure and are of limited thickness. Complex mathematical modeling with computer calculations is required to estimate the polarization components.

The internal impedance of the cell (a measure of resistance to electrical current flow) is another important factor that strongly affects the performance or rate capability. Impedance losses cause a voltage drop during operation, which also consumes part of the useful energy as waste heat. The voltage drop due to internal impedance is usually referred to as "ohmic polarization" or $IR$ drop and is proportional to the current drawn from the system. The total internal impedance of a cell is the sum of the ionic resistance of the electrolyte (within the separator and the porous electrodes); the electronic resistances of the active mass, the current

---

[a]Within the battery industry, oxidation and reduction are referenced relative to the discharge reactions. Also, anodes are sometimes called negative plates and cathodes, positive (e.g., lead-acid cells). In certain other electrochemical industries (e.g., electroplating), the opposite designations are used for anode/cathode and negative/positive.

collectors, and electrical tabs of both electrodes; and the contact resistance between the active mass and the current collector. These are ohmic resistances, following Ohm's law, with a linear relationship between current and voltage drop.

When connected to an external resistive load $R$, the cell voltage $E$ can be expressed as

$$E = E_o - [(\eta_{ct})_a + (\eta_c)_a] - [(\eta_{ct})_c + (\eta_c)_c] - iR_i = iR \tag{4.1}$$

where $E_o$ is the electromotive force or open-circuit voltage of cell; $(\eta_{ct})_a$, $(\eta_{ct})_c$ are the activation polarization or charge-transfer overvoltage at anode and cathode, respectively; $(\eta_c)_a$, $(\eta_c)_c$ are the concentration polarization at anode and cathode, respectively; $i$ is the operating current of cell on load; and $R_i$ is the internal resistance of cell.

As shown in Eq. (4.1), the useful voltage delivered by the cell is reduced by polarization and the internal $IR$ drop. It is only at very low operating currents, where polarization and the $IR$ drop are small, that the cell may operate close to the open-circuit voltage and deliver most of the theoretically available energy. Figure 4.1 shows the relation between cell polarization and discharge current.

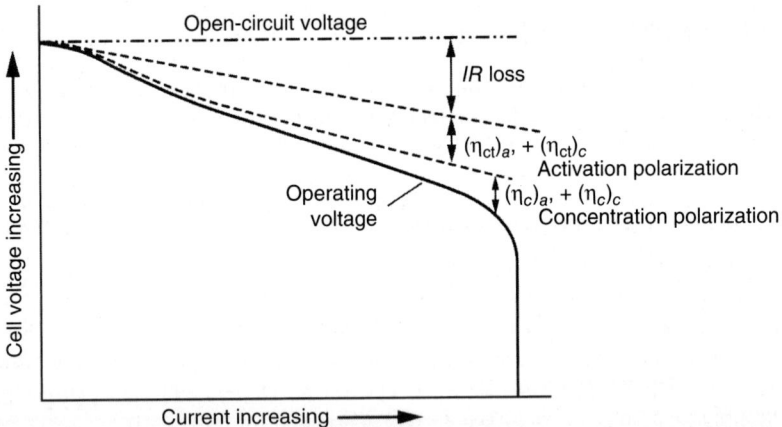

**FIGURE 4.1**    Cell polarization as a function of operating current.

Although the available energy of a battery or a fuel cell depends on the basic electrochemical reactions at both electrodes, there are many factors that affect the magnitude of the charge-transfer reaction, diffusion rates, and, subsequently, the energy loss. These factors include electrode formulation and design, electrolyte conductivity, and nature of the separators, among others. To achieve high operating efficiency with minimal loss of energy, battery and fuel cell designs should follow some essential rules, based on the electrochemical principles, as follows:

1. The ionic conductivity of the electrolyte should be high enough that the $IR$ polarization is not excessively large for practical operation. Table 4.1 shows the typical ranges of specific conductivities for various electrolyte systems used in batteries. Batteries are usually designed for specific drain rate applications ranging from microamperes to several hundred amperes. For a given electrolyte, a cell may be designed with a higher electrode interfacial area and thinner separator in order to reduce the $IR$ drop due to electrolyte resistance and to improve rate capability. Cells with a spirally wound electrode design are typical examples.

2. Electrolyte salt and solvents should have chemical stability to avoid direct chemical reaction with the anode or cathode materials.

**TABLE 4.1**   Conductivity Ranges of Various Electrolytes at Ambient Temperature

| Electrolyte system | Conductivity/S cm$^{-1}$ |
|---|---|
| Aqueous electrolytes | 0.1–0.55 |
| Molten salts | $\sim 10^{-1}$ |
| Inorganic electrolytes | $10^{-2}$–$10^{-1}$ |
| Organic electrolytes | $10^{-1}$–$10^{-2}$ |
| Ionic liquids | $10^{-4}$–$10^{-2}$ |
| Polymer electrolytes | $10^{-7}$–$10^{-3}$ |
| Inorganic solid electrolytes | $10^{-8}$–$10^{-5}$ |

3. The rate of electrode reaction at both the anode and the cathode should be sufficiently fast so that the activation or charge-transfer polarization is not too high, making the cell inoperable. A common method of minimizing the charge-transfer polarization is to use a porous electrode design. The porous electrode structure provides a high electrode surface area within given geometric dimensions for the electrode and reduces the local current density for a given total operating current.

4. In most battery and fuel cell systems, part or all of the reactants are supplied by the electrodes with part or all of the reaction products transported by diffusion away from the electrode surface. The cell should have electrolyte with adequate ionic transport to facilitate the mass transfer and avoid building up excessive concentration polarization. Adequate porosity and pore size of the electrode, proper separator thickness and structure, and sufficient concentrations of the reactants in the electrolyte are very important to ensure functionality of the cell. Mass-transfer limitations should be avoided for normal operation of the cell.

5. Rechargeable cells, on the other hand, will preferably yield reaction products that remain at the electrode surface to facilitate the reversible charge and discharge reactions. The reaction products should be stable mechanically as well as chemically with the electrolyte.

6. Current collector or substrate materials should be compatible with the electrode material and the electrolyte and should not be prone to corrosion. The design of the current collector should provide a uniform current distribution and low contact resistance to minimize electrode polarization during operation.

In general, the principles and various electrochemical techniques described in this chapter can be used to study all the important electrochemical aspects of a battery or fuel cell. These include the rate of electrode reaction; the existence of intermediate reaction steps; the stability of the electrolyte, the current collector, and the electrode materials; the mass-transfer conditions; the value of the limiting current; the formation of resistive films on the electrode surface; the impedance characteristics of the electrode or cell; and the existence of the rate-limiting processes.

## 4.2   THERMODYNAMIC BACKGROUND

In a cell, reactions essentially take place at two areas or sites in the device. These reaction sites are the two electrode-electrolyte interfaces. In generalized terms, the reaction at one electrode (reduction in forward direction) can be represented by

$$a\text{A} + ne \rightleftharpoons c\text{C} \tag{4.2}$$

where $a$ molecules of A take up $n$ electrons (e) to form $c$ molecules of C. At the other electrode, the reaction (oxidation in forward direction) can be represented by

$$b\text{B} \rightleftharpoons d\text{D} + ne \tag{4.3}$$

where the terms are equivalent to Eq. (4.2).

The overall reaction in the cell is given by addition of these two half-cell reactions

$$a\mathrm{A} + b\mathrm{B} \rightleftharpoons c\mathrm{C} + d\mathrm{D} \tag{4.4}$$

The change in the standard Gibbs energy $\Delta G°$ of this reaction is expressed as

$$\Delta G° = -nFE° \tag{4.5}$$

where $F$ is Faraday's constant (96,487 coulombs equiv$^{-1}$) and $E°$ is the standard electromotive force. (Selected values of standard electrode potentials are given in App. B.)

When conditions are other than in the standard state, the voltage $E$ of a cell is given by the Nernst equation,

$$E = E° - \frac{RT}{nF} \ln \frac{a_C^c a_D^d}{a_A^a a_B^b} \tag{4.6}$$

where $a_i$ is the activity of relevant species, $R$ is the gas constant (8.314 J K$^{-1}$ mol$^{-1}$), and $T$ is the absolute temperature in degrees Kelvin.

The change in Gibbs energy $\Delta G°$ of a cell reaction defines the driving force that enables a battery to deliver electrical energy to an external circuit. The measurement of the electromotive force, incidentally, also provides data for calculating changes in free energy, entropies, and enthalpies, together with activity coefficients, equilibrium constants, and solubility products.

Direct measurement of single (absolute) electrode potentials is considered practically impossible.[1] To establish a scale of half-cell or standard potentials, a reference potential "zero" must be established against which single-electrode potentials can be measured. By convention, the standard potential of the $H_2/H^+$ (aq.) reaction is taken as zero and all standard potentials are referenced to this value. (See App. B for various anode and cathode materials.)

## 4.3   ELECTRODE PROCESSES

Reactions at an electrode are characterized by both chemical and electrical changes that are of a heterogeneous type. Electrode reactions may be as simple as the reduction of a metal ion and incorporation of the resultant atom onto or into the electrode structure. Despite the apparent simplicity of the reaction, the mechanism of the overall process may be relatively complex and often involves several steps. Electroactive species must be transported to the electrode surface by migration or diffusion prior to the electron transfer step. Adsorption of electroactive material may be involved both prior to and after the electron transfer step. Chemical reactions may also be involved in the overall electrode reaction. As in any reaction, the overall rate of the electrochemical process is determined by the rate of the slowest step in the whole sequence of reactions.

The thermodynamic treatment of electrochemical processes presented in Sec. 4.2 describes the equilibrium condition of a system but does not present information on nonequilibrium conditions such as current flow resulting from electrode polarization (overvoltage) imposed to effect electrochemical reactions. Experimental determination of the current-voltage characteristics of many electrochemical systems has shown that there is an exponential relation between current and applied voltage. The generalized expression describing this relationship is called the Tafel equation,

$$\eta = a + b \log i \text{ for an anode reaction} \tag{4.7}$$

$$\eta = a - b \log i \text{ for a cathode reaction} \tag{4.8}$$

where $\eta$ is the overvoltage, $i$ is the current, and $a$ and $b$ are constants.

Typically, the constant $b$ is referred to as the Tafel slope. The Tafel relationship holds for a large number of electrochemical systems over a wide range of overpotential. At low values of overvoltage, however, the relationship breaks down and results in curvature in plots of $\eta$ versus $\log i$. Figure 4.2 is a schematic presentation of a Tafel plot, showing curvature at low values of overvoltage.

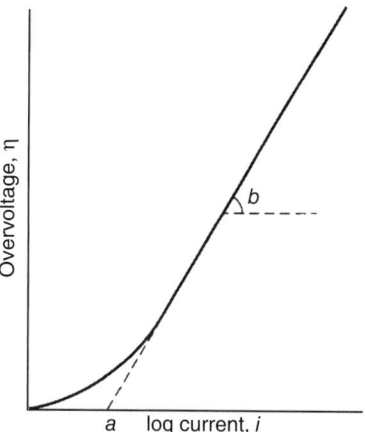

**FIGURE 4.2**  Schematic representation of a Tafel plot showing curvature at low overvoltage.

Success of the Tafel equation's fit to many experimental systems encouraged the quest for a kinetic theory of electrode processes. Since the range of validity of the Tafel relationship applies to high overvoltages, it is reasonable to assume that the expression does not apply to equilibrium situations but represents the current-voltage relationship of a unidirectional process. In an oxidation process, this means that there is a negligible contribution from reduction processes and vice versa. Combining and rearranging Eq. (4.7) into exponential form, we have

$$i = \exp\left(\pm\frac{a}{b}\right)\exp\frac{\eta}{b} \tag{4.9}$$

To consider a general theory, one must consider both forward and backward reactions of the electroreduction process, shown in simplified form in Fig. 4.3.

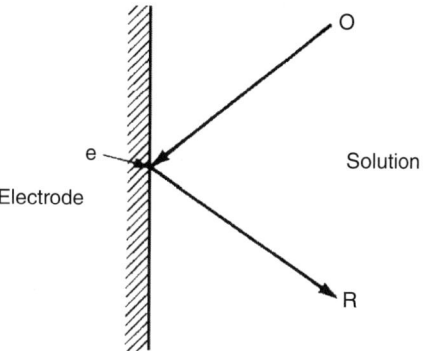

**FIGURE 4.3**  Simplified representation of electroreduction at an electrode.

The reaction is represented by the equation

$$O + ne \rightleftharpoons R \tag{4.10}$$

where O is the oxidized species, R is the reduced species, and $n$ is the number of electrons involved in electrode process.

The forward and backward reactions can be described by heterogeneous rate constants $k_f$ and $k_b$, respectively. The rates of the forward and backward reactions are then given by the products of these rate constants and the relevant concentrations, which typically are those at the electrode surface. As will be shown later, the concentrations of electroactive species at the electrode surface often are dissimilar from the bulk concentration in solution. The rate of the forward reaction is $k_f C_O$ and that for the backward reaction is $k_b C_R$. For convenience, these rates are usually expressed in terms of currents $i_f$ and $i_b$ for the forward and backward reactions, respectively,

$$i_f = nFAk_f C_O \tag{4.11}$$

$$i_b = nFAk_b C_R \tag{4.12}$$

where $A$ is the area of the electrode and $F$ is Faraday's constant.

Establishing these expressions is merely the result of applying the law of mass action to the forward and backward electrochemical processes. The role of electrons in the process is established by assuming that the magnitudes of the rate constants depend on the electrode potential. The dependence is usually described by assuming that a fraction of the electrode potential $\alpha E$ is involved in driving the reduction process, while the fraction $(1 - \alpha)E$ is effective in making the reoxidation process more difficult. Mathematically, these potential-dependent rate constants are expressed as

$$k_f = k_f^o \exp\left(\frac{-\alpha nFE}{RT}\right) \tag{4.13}$$

$$k_b = k_b^o \exp\left(\frac{(1 - \alpha)nFE}{RT}\right) \tag{4.14}$$

where $\alpha$ is the transfer coefficient and $E$ is the electrode potential relative to a suitable reference potential.

Since the meaning of the transfer coefficient ($\alpha$ or the symmetry factor $\beta$ as defined in some texts) is not implicit in the kinetic derivation, a discussion in mechanistic terms is appropriate.[2] Specifically, the transfer coefficient determines the fraction of electric energy affected by the rate of electrochemical transformation due to an associated reduction in cell potential below the equilibrium value. To understand the function of the transfer coefficient $\alpha$ it is necessary to describe a potential energy diagram for the reduction-oxidation process. Figure 4.4 shows an approximate potential energy curve (Morse curve) for an oxidized species (e.g., solvated $H^+$) approaching an electrode surface together with the potential energy curve for the resultant reduced species (e.g., H adsorbed on a metal surface).

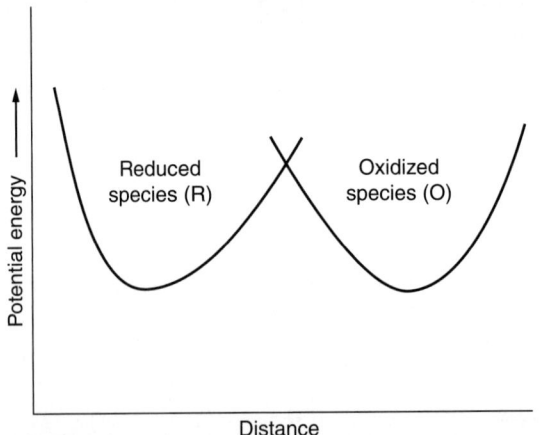

**FIGURE 4.4**  Potential energy diagram for a reduction-oxidation process taking place at an electrode.

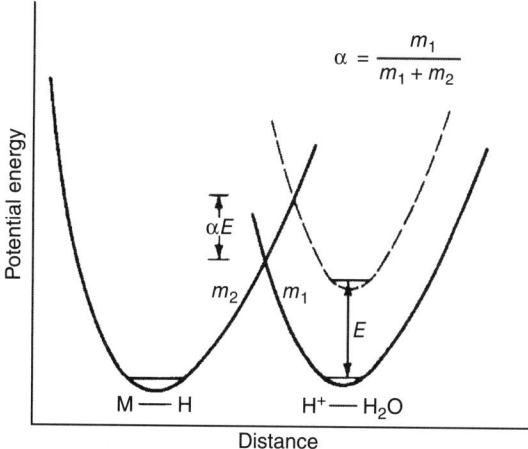

**FIGURE 4.5** Potential energy diagram for reduction of $H^+$ at an electrode such as platinum (Pt). The dashed line represents the potential energy in the absence of an electrode field, which is lowered when the ion is subjected to a potential $E$. However, the activation energy is lowered by an amount $\alpha E$ (usually half), as discussed in the text.

For convenience, consider the hydrogen ion reduction at a solid electrode as the model for a typical electro-reduction. According to Horiuti and Polanyi,[3] the potential energy diagram for reduction of the hydrogen ion (nominally $H_3O^+$) can be represented by Fig. 4.5, where the oxidized species O is the hydrated hydrogen ion and the reduced species R is a hydrogen atom bonded to the metal (electrode) surface. The effect of changing the electrode potential by a value of $E$ is to raise the potential energy of the Morse curve of the hydrogen ion. The intersection of the two Morse curves forms an energy barrier, the height of which is $\alpha E$. If the slope of the two Morse curves is approximately constant at the point of intersection, then $\alpha$ is defined by the ratio of the slope of the Morse curves at the point of intersection as follows:

$$\alpha = \frac{m_1}{m_1 + m_2} \tag{4.15}$$

where $m_1$ and $m_2$ are the slopes of the potential energy curves of the hydrated hydrogen ion and the hydrogen atom, respectively.

There are inadequacies in the theory of transfer coefficients. The theory assumes that $\alpha$ is constant and independent of $E$. At present there are no data to prove or disprove this assumption. The other main weakness is that the concept is used to describe processes involving a variety of different species such as (1) redox changes at an inert electrode ($Fe^{2+}/Fe^{3+}$ at Hg), (2) reactant and product soluble in different phases [$Cd^{2+}/Cd(Hg)$], and (3) electrodeposition ($Cu^{2+}/Cu$). Despite these inadequacies, the concept and application of the theory are appropriate in many cases and represent the best understanding of electrode processes at the present time. Examples for various electrode reactions on selected metals are given in Table 4.2.[4]

**TABLE 4.2** Values of Transfer Coefficients $\alpha$ at 25°C for Selected Systems[4]

| Electrode reaction | Metal | Electrolyte solution | $\alpha$ |
|---|---|---|---|
| $H^+ + e \rightleftharpoons \frac{1}{2}H_2$ | Pt (smooth) | 1.0 mol dm$^{-3}$ HCl | 2.0 |
| $H^+ + e \rightleftharpoons \frac{1}{2}H_2$ | Ni | 0.12 mol dm$^{-3}$ NaOH | 0.58 |
| $H^+ + e \rightleftharpoons \frac{1}{2}H_2$ | Hg | 10.0 mol dm$^{-3}$ HCl | 0.61 |
| $O_2 + 4H^+ + 4e \rightleftharpoons 2H_2O$ | Pt | 0.1 mol dm$^{-3}$ H$_2$SO$_4$ | 0.49 |
| $O_2 + 2H_2O + 4e \rightleftharpoons 4OH^-$ | Pt | 0.1 mol dm$^{-3}$ NaOH | 1.0 |
| $Cd^{2+} + 2e \rightleftharpoons Cd$ | Cd/Hg | $10^{-3}$ mol dm$^{-3}$ Cd(NO$_3$)$_2$ in 1 mol dm$^{-3}$ KNO$_3$ | 5.0 |
| $Cu^{2+} + 2e \rightleftharpoons Cu$ | Cu | 1 mol dm$^{-3}$ CuSO$_4$ | 0.5 |

From Eqs. (4.13) and (4.14) we can derive parameters useful for evaluating and describing an electrochemical system. Equations (4.13) and (4.14) are compatible both with the Nernst equation [Eq. (4.6)] for equilibrium conditions and with the Tafel relationships [Eqs. (4.7) and (4.8)] for unidirectional processes. Under equilibrium conditions both forward and backward currents exist, but because the system is at equilibrium the rates are equal, and thus there is no net current flow; hence,

$$i_f = i_b = i_o \tag{4.16}$$

where $i_o$ is the exchange current.

From Eqs. (4.11) to (4.14), together with Eq. (4.16), the following relationship is established:

$$C_o k_f^o \exp\left(\frac{\alpha n F E_o}{RT}\right) = C_R k_b^o \exp\left(\frac{(1-\alpha)n F E_e}{RT}\right) \tag{4.17}$$

where $E_e$ is the equilibrium potential. Rearranging,

$$E_e = \frac{RT}{nF}\ln\left(\frac{k_f^o}{k_b^o}\right) + \frac{RT}{nF}\ln\left(\frac{C_o}{C_R}\right) \tag{4.18}$$

From this equation we can establish the definition of formal standard potential $E_C^o$ where concentrations are used rather than activities,

$$E_C^o = \frac{RT}{nF}\ln\left(\frac{k_f^o}{k_b^o}\right) \tag{4.19}$$

For convenience, the formal standard potential is often taken as the reference point of the potential scale in reversible systems. Combining Eqs. (4.18) and (4.19), we can show consistency with the Nernst equation,

$$E_e = E_C^o + \frac{RT}{nF}\ln\left(\frac{C_o}{C_R}\right) \tag{4.20}$$

except that this expression is written in terms of concentrations rather than activities. From Eqs. (4.11) and (4.13), at equilibrium conditions,

$$i_o = i_f = nFAC_o k_f^o \exp\left(\frac{-\alpha n F E_e}{RT}\right) \tag{4.21}$$

The exchange current as defined in Eq. (4.16) is a parameter of interest to researchers in the battery field. This parameter may be conveniently expressed in terms of the rate constant $k$ by combining Eqs. (4.11), (4.13), (4.18), and (4.21),

$$i_o = nFAkC_o^{(t-u)}C_R^u \tag{4.22}$$

The exchange current $i_o$ is a measure of the rate of exchange of charge between oxidized and reduced species at any equilibrium potential without net overall change. The rate constant $k$, however, has been defined for a particular potential, the formal standard potential of the system. It is not in itself sufficient to characterize the system unless the transfer coefficient is also known. However, Eq. (4.22) can be used in the elucidation of the electrode reaction mechanism. The value of the transfer coefficient can be determined by measuring the exchange current density as a function of the concentration of the reduction species at constant concentration of the oxidation species or as a function of oxidation species at a constant reduction species concentration. A schematic representation of the forward and backward currents as a function of overvoltage, $\eta = E - E_e$, is shown in Fig. 4.6, where the net current (the middle curve) is the sum of the two components.

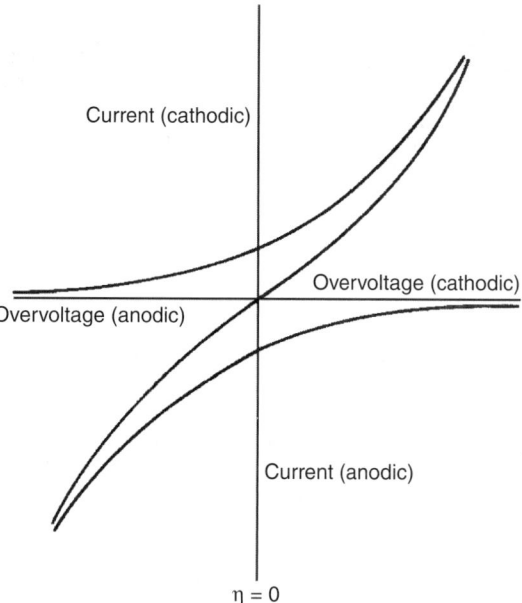

**FIGURE 4.6**  Schematic representation of relationship between overvoltage and current.

For situations where the net current is not zero, that is, where the potential is sufficiently different from the equilibrium potential, the net current approaches the net forward current (or, for anodic overvoltages, the backward current). One can then write

$$i = nFAC_o k \exp\left(\frac{-\alpha nF\eta}{RT}\right)$$

(4.23)

Now when $\eta = 0$, $i = i_o$, then

$$i = i_o \exp\left(\frac{-\alpha nF\eta}{RT}\right)$$

(4.24)

and

$$\eta = \frac{RT}{\alpha nF}\ln i_o - \frac{RT}{\alpha nF}\ln i$$

(4.25)

which is the Tafel equation introduced earlier in a generalized form as Eqs. (4.7) and (4.8). It can now be seen that the kinetic treatment here is self-consistent with both the Nernst equation (for equilibrium conditions) and the Tafel relationship (for unidirectional processes). To present the kinetic treatment in its most useful form, a transformation into a net current flow form is appropriate. Using the equation

$$i = i_f - i_b$$

(4.26)

and substituting Eqs. (4.11) to (4.14), and (4.19),

$$i = nFAk\left\{C_O \exp\left(\frac{-\alpha nFE_e^o}{RT}\right) - C_R \exp\left(\frac{(1-\alpha)nFE_e^o}{RT}\right)\right\}$$

(4.27)

When this equation is applied in practice, it is very important to remember that $C_O$ and $C_R$ are concentrations at the surface of the electrode (i.e., the effective concentrations) and not necessarily the same as the bulk concentrations. Concentrations at the interface are often (almost always) modified by differences in electric potential between the surface and the bulk solution. The effects of potential differences that are manifest at the electrode-electrolyte interface are given in the following section.

## 4.4   ELECTRICAL DOUBLE-LAYER CAPACITY AND IONIC ADSORPTION

When an electrode (metal surface) is immersed in an electrolyte, the electronic charge on the metal attracts ions of opposite charge and orients the solvent dipoles. There exist a layer of charge in the metal and a layer of charge in the electrolyte. This charge separation establishes what is commonly known as the "electrical double layer." The formation of the double layer is nonfaradaic, meaning that no electron transfer reactions (oxidation or reduction) occur. For simplicity and clarity, the different features of the electrical double layer are described below.

Consider a negatively charged electrode in a solution of electrolyte. The orientation of solvent molecules forms an initial layer. Using water for the sake of this discussion, the orientation is shown in Fig. 4.7. The water dipoles are oriented, as shown in the figure, so that the majority of the dipoles are oriented with their positive ends (arrow heads) toward the surface of the electrode. This layer is called the inner Helmholtz layer. Since the representation is statistical, not all dipoles are oriented the same way. Some dipoles are more influenced by dipole-dipole interactions than by dipole-electrode interactions.

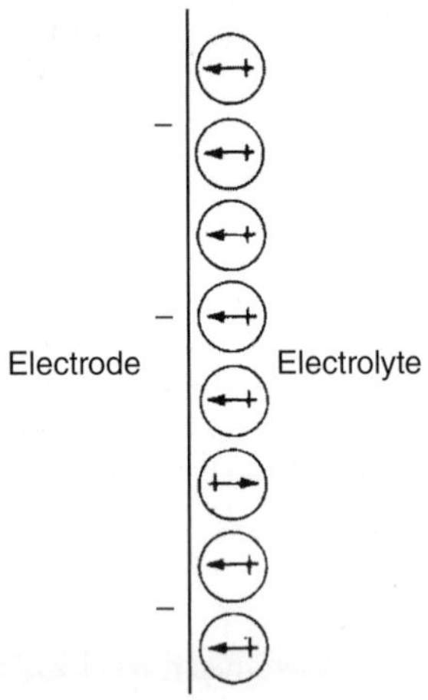

Next, consider the approach of a cation to the vicinity of the negative dipole end of the inner Helmholtz layer. The majority of cations are strongly solvated by water dipoles and maintain a sheath of water dipoles around them despite the orienting effect of the inner layer. With a few exceptions, cations do not penetrate beyond the "closest approach" to the electrode surface (see below) but remain outside the primary layer of solvent molecules and usually retain their solvation sheaths. Figure 4.8 shows a typical example of a cation in the electrical double layer. This layer is called the outer Helmholtz layer. The establishment of this particular typical cation configuration comes partly from experimental AC impedance measurements (see below) of mixed electrolytes but mainly from calculations of the free energy of approach of an ion to the electrode surface. In considering aqueous solutions with ion-electrode and ion-water interactions, the Gibbs energy of approach of a cation to an electrode surface is strongly influenced by the hydration of the cation. In rare cases, cations can specifically absorb (contact) onto the electrode surface, provided they have very large radius (e.g., $Cs^+$) and poor solvation.

**FIGURE 4.7**   Orientation of water molecules in the electrical double layer at a negatively charged electrode.

In analyzing the free-energy balance of the anion system at a positively charged electrode, it is found that anion-electrode contact solvated anions predominate. Figure 4.9 shows the generalized orientation of anion adsorption on a positively charged electrode

Extending out into solution from the electrical double layer is a continuous repetition of the layering effect, but with diminishing magnitude. This extension toward the bulk solution is known as the Gouy-Chapman diffuse double layer. Its effect on electrode kinetics and the concentration of electroactive species at the electrode surface is manifest when supporting electrolyte concentrations are low or zero. The establishment of an electrical double layer and various types of ion contact adsorption not only directly influences the real (actual) concentration of electroactive species at an electrode surface, but also indirectly modifies the potential gradient at the site of electron transfer. In this respect, it is important to understand the influence of the electrical double layer and adjust the models where and when appropriate.

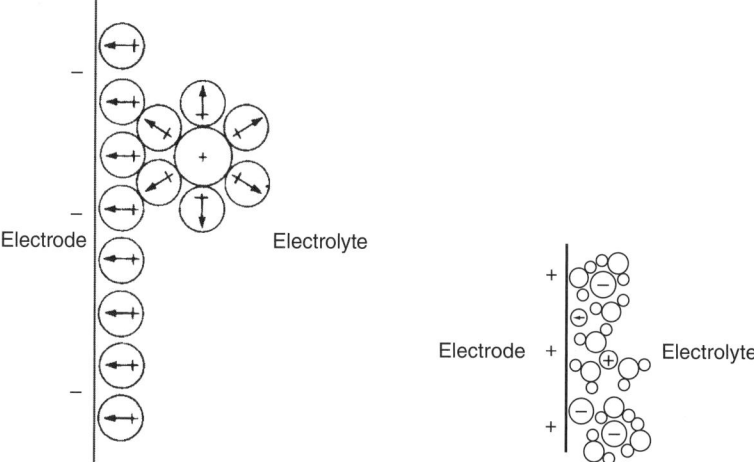

**FIGURE 4.8**  Typical cation situated in the electrical double layer at a negatively charged electrode.

**FIGURE 4.9**  Typical anions situated in the electrical double layer at a positively charged electrode.

The potential distribution near a negative electrode is shown schematically in Fig. 4.10. The inner Helmholtz plane (IHP) corresponds to the plane of the innermost layer of water molecules with, if present, specifically adsorbed cations. The electrode potential is defined as $\phi^i$ with the potential of the bulk solution assumed to be zero. The locus center of the solvated ions in the outer Helmholtz layer is defined as the outer Helmholtz plane (OHP) or as the plane of closest approach for those ions that do not contact-adsorb but approach the electrode with a sheath of solvated water molecules surrounding them. The potential at the OHP is defined as $\phi^o$ and again refers to the potential of the bulk solution. In some texts $\phi^i$ is defined as $\phi^1$ and $\phi^o$ as $\phi^2$.

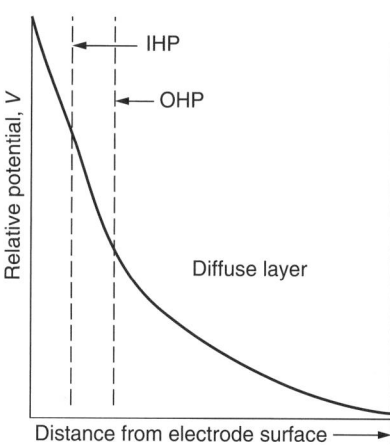

**FIGURE 4.10**  Potential distribution at a negatively charged electrode.

As mentioned previously, the bulk concentration of an electroactive species is often not the value to be used in kinetic equations. Species that are in the electrical double layer are in a different energy state from those in bulk solution. At equilibrium, the concentration $C^e$ of an ion or species that is about to take part in the charge-transfer process at the electrode is related to the bulk concentration by

$$C^e = C^B \exp\left(\frac{-zF\phi^e}{RT}\right) \tag{4.28}$$

where $z$ is the charge on the ion and $\phi^e$ is the potential of closest approach of the species to the electrode. For many species the plane of closet approach is the OHP, thus allowing the value of $\phi^e$ to often be equated to $\phi^o$. However, as noted in a few special cases, the plane of closest approach can be the IHP; the value of $\phi^e$ in these cases would be the same as $\phi^i$. A judgment has to be made as to what value of $\phi^e$ is best for any given situation.

The potential that is most effective in driving the electrode reaction occurs when a species is between closest approach and at the potential of the electrode. If $E$ is the potential of the electrode, then the driving force is $(E - \phi^e)$. Using this relationship together with Eqs. (4.26) and (4.27), we have

$$\begin{aligned}
\frac{i}{nFAk} &= C_O \exp\left(\frac{-z_O E\phi^e}{RT}\right) \exp\left(\frac{-\alpha nF(E - \phi^e)}{RT}\right) \\
&- C_R \exp\left(\frac{-z_R F\phi^e}{RT}\right) \exp\left(\frac{(1-\alpha)nF(E - \phi^e)}{RT}\right)
\end{aligned} \tag{4.29}$$

where $z_O$ and $z_R$ are the charges (with sign) of the oxidized and reduced species, respectively. Rearranging Eq. (4.29) and using

$$z_O - n = z_R \tag{4.30}$$

yields

$$\frac{i}{nFAk} = \exp\left(\frac{(\alpha n - z_O)F\phi^e}{RT}\right)\left\{C_O \exp\left(\frac{-\alpha nFE}{RT}\right) - C_R \exp\left(\frac{(1-\alpha)nFE}{RT}\right)\right\} \tag{4.31}$$

In experimental determination, the use of Eq. (4.27) will provide an apparent rate constant $k$, which does not take into account the effects of the electrical double layer. Taking into account the effects to the approach of a species to the plane of nearest (closest) approach,

$$k_{\mathrm{app}} = k \exp\left(\frac{(\alpha n - z_O)F\phi^e}{RT}\right) \tag{4.32}$$

For the exchange current the same applies,

$$(i_o)_{\mathrm{app}} = i_o \exp\left(\frac{(\alpha n - z_O)F\phi^e}{RT}\right) \tag{4.33}$$

In addition to the influence of the double layer on electrode kinetics as discussed above, the capacity of the double layer $C$ (farads), along with the cell voltage, defines the energy $E$ (watt-hours), of the electrochemical double-layer capacitor (EDLC) according to

$$E = \frac{1}{2}C(\Delta V)^2 \tag{4.34}$$

For solid electrodes such as platinum (Pt) and glassy carbon (GC) and for porous electrodes such as activated carbons, the double-layer capacity can be determined by various techniques such as linear sweep voltammetry, impedance spectroscopy, and potentiostatic intermittent titration (PITT) as discussed in Secs. 4.6.1, 4.6.3, and 4.6.4, respectively.

## 4.5 MASS TRANSPORT TO THE ELECTRODE SURFACE

The sections above have considered the thermodynamics of electrochemical processes, studied the kinetics of electrode processes, and investigated the properties of the electrochemical double layer and its effects on kinetic parameters. An understanding of these relationships is an important aspect of battery technology research. Another very important area of study that has a major impact on battery research is the evaluation of mass transport processes to and from electrode surfaces.

Mass transport to or from an electrode can occur by three processes: (1) convection and stirring, (2) electrical migration in an electric potential gradient, and (3) diffusion in a concentration gradient. The first of these processes can be handled relatively easily both mathematically and experimentally. If stirring is required, flow systems can be established, while if complete stagnation is an experimental necessity, this can also be imposed by careful design. In most cases, if stirring and convection are present or imposed, they can be handled mathematically.

The migration component of mass transport can also be handled experimentally (reduced to almost zero or occasionally increased in special cases) and described mathematically, provided certain parameters such as transport numbers or migration currents are known. Migration of electroactive species in an electric potential gradient can be reduced to near zero by addition of an excess of inert "supporting electrolyte," which effectively reduces the potential gradient to zero, and thus eliminates the electric field that produces migration. Enhancement of migration is more difficult, requiring the electric field to be increased so that the movement of charged species is increased. Electrode geometry design can increase migration slightly by altering electrode curvature. The fields at convex surfaces are greater than those at flat or concave surfaces, and thus migration is enhanced at convex curved surfaces.

The third process, diffusion in a concentration gradient, is the most important of the three processes and is the one that is typically dominant in mass transport in batteries. The analysis of diffusion uses the basic equation, given by Fick,[5] which defines the flux of material crossing a plane at distance $x$ and time $t$. The flux is proportional to the concentration gradient and is represented by the expression:

$$q = D\frac{\delta C}{\delta x} \tag{4.35}$$

where $q$ is the flux (mass flow rate per unit area), $D$ is the diffusion coefficient, and $C$ is the concentration.

The rate of change of concentration with time is defined by

$$\frac{\delta C}{\delta x} = D\frac{\delta^2 C}{\delta x^2} \tag{4.36}$$

This expression is referred to as Fick's second law of diffusion.[5] Solution of Eqs. (4.35) and (4.36) requires that boundary conditions be imposed. These are chosen according to the electrode's expected "discharge" regime dictated by battery performance or boundary conditions imposed by relevant electrochemical techniques.[6] Several electrochemical techniques are discussed in Sec. 4.6.

### 4.5.1 Concentration Polarization

Diffusion processes are the typical mass-transfer processes operative in the majority of battery systems where the transport of species to and from reaction sites is required for maintenance of electrical current flow. Enhancement and improvement of diffusion processes are an appropriate research approach for improving battery performance parameters. Equation (4.35) may be written in an approximate, yet more practical form, remembering that $i = nFq$, where $q$ is the flux through a plane of unit area, as follows, where symbols are defined as before and

$$i = nF\frac{DA(C_B - C_E)}{\delta} \tag{4.37}$$

where $C_B$ is the bulk concentration of electroactive species, $C_E$ is the concentration at electrode, $A$ is the electrode area, and $\delta$ is the boundary-layer thickness (i.e., the layer at the electrode surface where the majority of the concentration gradient exists; see Fig. 4.11). When $C_E = 0$, this expression defines the maximum diffusion current $i_L$ that can be sustained in solution under a given set of conditions,

$$i_L = nF\frac{DAC_B}{\delta_L} \tag{4.38}$$

where $\delta_L$ is the boundary-layer thickness at the limiting condition. Hence, to increase $i_L$, the bulk concentration, the electrode area, or the diffusion coefficient needs to be increased. In the design of a battery, an understanding of the implication of this expression is important. Specific cases can be analyzed quickly by applying Eq. (4.38), allowing parameters such as discharge rate and likely power densities of new systems to be estimated.

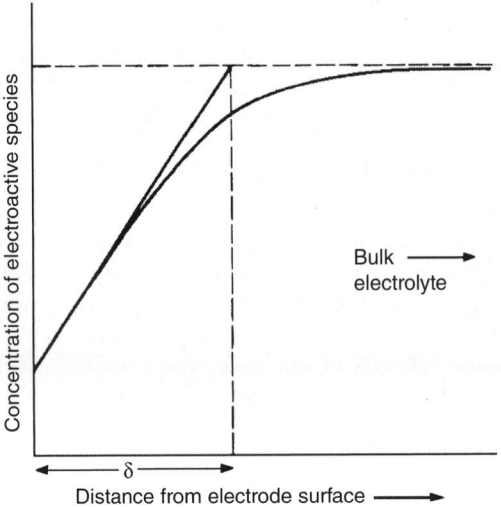

**FIGURE 4.11**   Boundary-layer thickness at an electrode surface.

Assuming that the thickness of the diffusion boundary layer does not change greatly with concentration, then $\delta_L = \delta$ and Eq. (4.37) may be rewritten as

$$i = \left(1 - \frac{C_E}{C_B}\right)i_L \tag{4.39}$$

The difference in concentration existing between the electrode surface and the bulk of the electrolyte results in a concentration polarization. According to the Nernst equation, the concentration polarization or overpotential produced from the change of concentration across the diffusion layer may be written as

$$\eta_c = \frac{RT}{nF} \ln\frac{C_E}{C_B} \tag{4.40}$$

From Eq. (4.39) we have

$$\eta_c = \frac{RT}{nF} \ln \frac{i_L}{i_L - i}$$

(4.41)

This gives the relationship of concentration polarization and current for mass transfer by diffusion. Equation (4.41) indicates that as $i$ approaches the limiting current $i_L$, the overpotential should theoretically increase to infinity. However, in a real process the potential will increase only to a point where another electrochemical reaction will occur, as illustrated in Fig. 4.12.

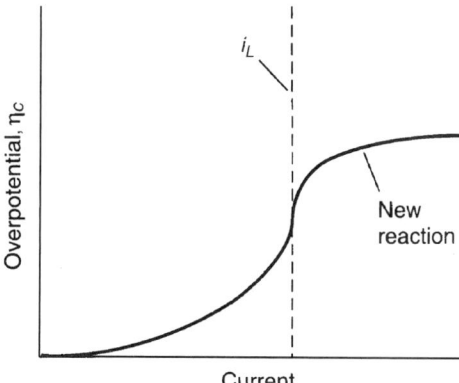

**FIGURE 4.12**   Plot of overpotential $\eta_c$ versus current $i$.

### 4.5.2   Porous Electrodes

Electrochemical reactions are heterogeneous reactions that occur at the electrode-electrolyte interface. In fuel cell systems, the reactants are supplied from the electrolyte phase to a catalytic electrode surface. In battery systems, the electrodes are usually composites made of reactants (active materials), binder, and conductive filler. In order to minimize the energy loss due to both activation and concentration polarizations at the electrode surface and to increase the electrode efficiency or utilization, it is preferred to have a large electrode surface area. This can be accomplished with the use of a porous electrode design. A porous electrode can provide an interfacial area per unit volume several orders of magnitude greater than that of a solid electrode (such as $10^4$ cm$^{-1}$).

A porous electrode consists of porous matrices of solids and void spaces. The electrolyte penetrates the void spaces of the porous matrix. In such an active porous mass, the mass-transfer condition in conjunction with the electrochemical reaction occurring at the interface is very complicated. Over a given time period of cell operation, the rate of reaction within the pores may vary significantly depending on the location. The distribution of current density within the porous electrode depends on the physical structure (such as tortuosity and pore sizes); the conductivity of the solid matrix, and the electrolyte; and the kinetic parameters of the electrochemical processes. A detailed treatment of such complex porous electrode systems can be found in Newman.[7]

## 4.6   *ELECTROCHEMICAL TECHNIQUES*

Many steady-state and transient electrochemical techniques are available to the experimentalist for determining electrochemical parameters and for assisting in both the improvement of existing battery systems and the evaluation of electrochemical couples as potential candidates for new batteries. In studying the mechanisms at a given electrode, the techniques described in this section all require a three-electrode cell: anode and cathode with a stable reference electrode.[1,6] It is preferable to employ a thermodynamically stable reference (i.e., a reference electrode reversible to an ion in solution, such as Li/Li$^+$ in a solution containing a lithium salt).

### 4.6.1   Cyclic and Linear Sweep Voltammetry

Of the electrochemical techniques, cyclic voltammetry (or linear sweep voltammetry) is probably one of the more versatile techniques available to the electrochemist. Specifically, the technique applies a linearly changing voltage (i.e., ramp voltage) to the working electrode.[1,6] The scan of voltage might be ±3 to 5 V from an appropriate resting potential such that most electrode reactions of interest would be observed at the working electrode.

To describe the principles behind cyclic voltammetry, Eq. (4.10) can be restated to describe the reversible reduction of an oxidized species, O,

$$O + ne \rightleftharpoons R \tag{4.10}$$

where O is the oxidized species, R is the reduced species, and $n$ is the number of electrons involved in electrode process.

In cyclic voltammetry, the initial potential sweep is represented by

$$E = E_i - \nu t \tag{4.42}$$

where $E_i$ is the initial potential, $t$ is the time, and $\nu$ is the rate of potential change or sweep rate (V/s).

The reverse sweep of the cycle is defined by

$$E = E_i + \nu' t \tag{4.43}$$

where $\nu'$ is often the same value as $\nu$. By combining Eq. (4.43) with the appropriate form of the Nernst equation [Eq. (4.6)] and with Fick's laws of diffusion [Eqs. (4.35) and (4.36)], an expression can be derived that describes the flux of species to the electrode surface. This expression is a complex differential equation and can be solved by the summation of an integral in small successive increments.

As the applied voltage approaches that of the reversible potential for the electrode process, a small current flows, the magnitude of which increases rapidly but later becomes limited at a potential slightly beyond the standard potential due to the subsequent depletion of reactants. This depletion of reactants establishes concentration profiles that spread out into the solution, as shown in Fig. 4.13. As the concentration profiles extend into solution, the rate of diffusive transport at the electrode surface decreases and with it the observed current.

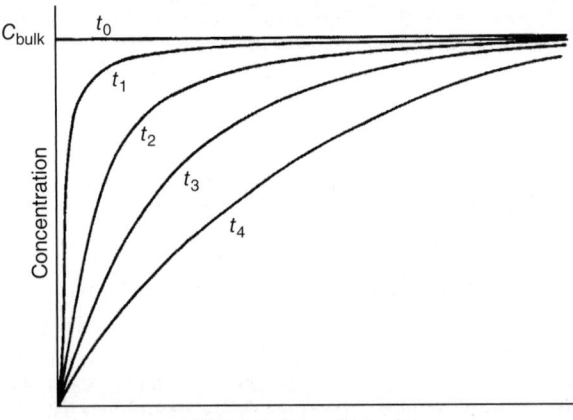

**FIGURE 4.13**   Concentration profiles for reduction of a species in cyclic voltammetry, $t_4 > t_0$.

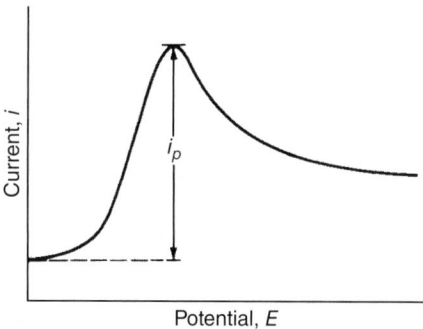

**FIGURE 4.14** Cyclic voltammetry peak current for reversible reduction of an electroactive species.

The current is thus seen to pass through a well-defined maximum, as illustrated in Fig. 4.14. The peak current of the reversible reduction [Eq. (4.10)] is defined by

$$i_p = \frac{0.447 F^{3/2} A n^{3/2} D^{1/2} C_0 v^{1/2}}{R^{1/2} T^{1/2}} \qquad (4.44)$$

The symbols have the same identity as before, while $i_p$ is the peak current and $A$ is the electrode area. It may be noted that the value of the constant varies slightly from one text or publication to another because the derivation of peak current height is performed numerically, as previously detailed.

A word of caution is due regarding the interpretation of the value of the peak current. Specifically, prior discussion of the effects of the electrical double layer on electrode kinetics revealed that there is a capacitance effect at an electrode-electrolyte interface. Consequently, the "true" electrode potential is modified by both the capacitance effect and the ohmic resistance of the solution. Equation (4.42) should ideally be written in a form that described these two components. Equation (4.45) shows such a modification,

$$E = E_i - vt + r(i_f + i_c) \qquad (4.45)$$

where $r$ is the cell resistance, $i_f$ is the faradaic current, and $i_c$ is the capacity current.

At small values of voltage sweep rate, typically below 1 mV/s, the capacity effects are small and in most cases can be ignored. At greater values of sweep rate, a correction needs to be applied to interpretations of $i_p$, as described by Nicholson and Shain.[8] To correct for a solution's ohmic drop, careful cell design and positive feedback compensation circuitry in the electronic instrumentation is typically adequate.

Cyclic voltammetry provides both qualitative and quantitative information on electrode processes. A reversible, diffusion-controlled reaction such as presented by Eq. (4.10) exhibits an approximately symmetrical pair of current peaks, as shown in Fig. 4.15. The voltage separation $\Delta E$ of these peaks is independent of voltage sweep rates and is expressed as

$$\Delta E = \frac{2.3 RT}{nF} \qquad (4.46)$$

In the case of the electrodeposition of an insoluble film, which can subsequently be reversibly reoxidized and which is not governed by diffusion to and from the electrode surface, the value of $\Delta E$ is considerably less than that given by Eq. (4.46), as shown in Fig. 4.16. In the ideal case, the value of $\Delta E$ for this system is close to zero. For quasi-reversible processes the current peaks are more separated, and the shape of the peak is less sharp at its summit and generally more rounded, as shown in Fig. 4.17. The voltage of the current peak is dependent on the voltage sweep rate with the voltage separation much greater than that given by Eq. (4.46). A completely irreversible electrode process produces a single peak, as shown in Fig. 4.18.

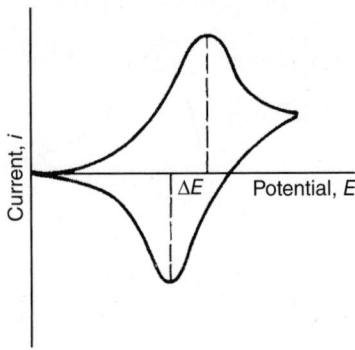

**FIGURE 4.15**    Cyclic voltammogram of a reversible diffusion-controlled process.

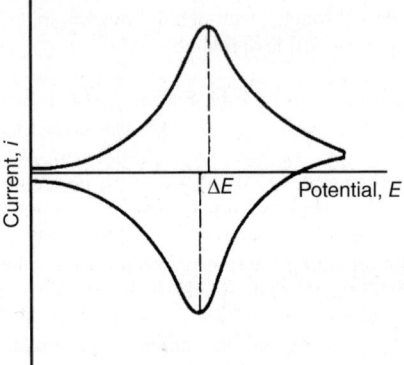

**FIGURE 4.16**    Cyclic voltammogram of electro-reduction and reoxidation of a deposited, insoluble film.

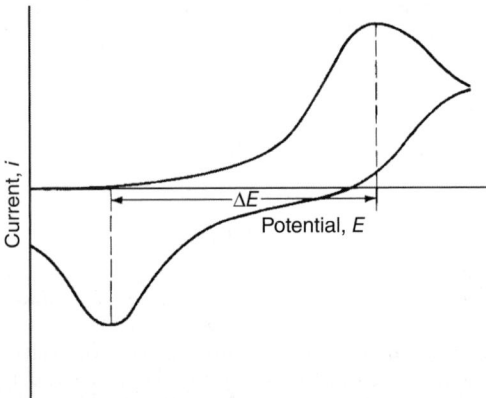

**FIGURE 4.17**    Cyclic voltammogram of a quasi-reversible process.

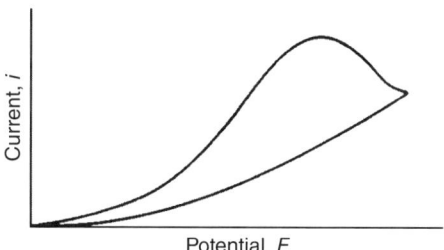

**FIGURE 4.18** Cyclic voltammogram of an irreversible process.

Again the voltage of the peak current is sweep-rate dependent, and in the case of an irreversible charge-transfer process for which the back reaction is negligible, the rate constant and transfer coefficient can be determined. With negligible back reaction, the expression for peak current as a function of peak potential $E_m$

$$i_p = 0.22 nFC_0 k_{app} \exp\left[-\alpha \frac{nF}{RT}(E_m - E^o)\right] \tag{4.47}$$

where the symbols are as noted before and $E_m$ is the potential of the current peak. A plot of $E_m$ versus $\ln i_p$, for different values of concentration, gives a slope that yields the transfer coefficient $\alpha$ and an intercept that yields the apparent rate constant $k_{app}$. Although both $\alpha$ and $k_{app}$ can be obtained by analyzing $E_m$ as a function of voltage sweep rate $\nu$ by a reiterative calculation, analysis by Eq. (4.47) (which is independent of $\nu$) is much more convenient.

Cyclic and linear voltammograms of electrochemical systems can often be much more complicated than the profiles presented here. It often takes some ingenuity and persistence to determine which peaks belong to which species or processes. Despite this drawback, cyclic and linear sweep voltammetric techniques are a versatile and relatively sensitive electrochemical method appropriate to the analysis of systems of interest in battery and EDLC development. This technique will identify reversible couples (desirable for secondary batteries and double-layer regions for EDLCs). The anodic sweep at a Pt electrode over the potential range of ~0.03 to ~1.3 V versus a reversible hydrogen electrode (RHE) is shown in Fig. 4.19. The figure clearly shows the faradaic regions for hydrogen oxidation and oxide formation, and the nonfaradaic double-layer region.[9]

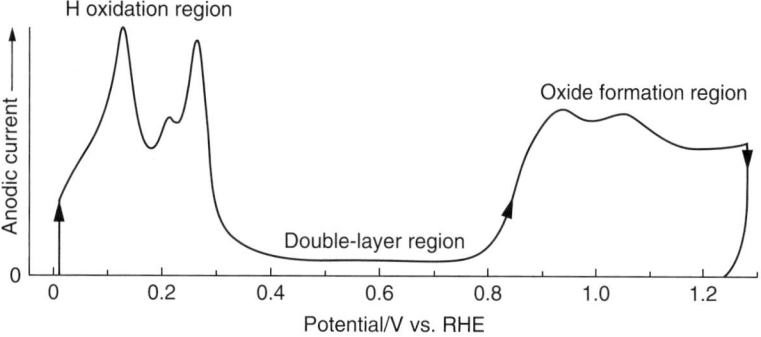

**FIGURE 4.19** Anodic linear sweep (0.1 V sec$^{-1}$) at Pt in 0.5 mol dm$^{-3}$ H$_2$SO$_4$.

The total capacity $Q$ (in farads) can be determined for any selected potential range $\Delta V$ from

$$Q = \int i_F \, dt + \int i_{nF} \, dt = Q_F + Q_{nF} \tag{4.48}$$

where $i_F$ is the faradaic current and $i_{nF}$ is the nonfaradaic current.

The potential range for nonfaradaic reactions in aqueous acidic solution are shown in Fig. 4.19. Useful energy is only derived where $\Delta V$ is ~0.4 to ~0.8 V versus the RHE—limiting the energy of an aqueous EDLC to a small $\Delta V$ range, as per Eq. (4.34).

$$E = \frac{1}{2}C(\Delta V)^2 \tag{4.34}$$

However, according to this equation, $E$ can be significantly increased by employing a nonaqueous electrolyte solution in which the $\Delta V$ can be significantly increased in the diffuse double-layer region. Using a three-electrode cell with carbon as the working electrode, Pt as the counter, and Li as the reference, Xu et al.,[10] used cyclic voltammetry in an electrolyte solution of $Et_3MeNPF_6$ in EC/DMC to determine a stable $\Delta V$ range of 6.9 V versus Li/Li$^+$.

### 4.6.2 Chronopotentiometry

Chronopotentiometry involves the study of voltage transients at an electrode upon which is imposed a constant current. It is sometimes alternately known as galvanostatic voltammetry. In this technique, a constant current is applied to an electrode, and its voltage response indicates the changes in electrode processes occurring at the interface. Consider, for example, the reduction of a species O, as expressed by Eq. (4.10). As the constant current is passed through the system, the concentration of O in the vicinity of the electrode surface begins to decrease. As a result of this depletion, O diffuses from the bulk solution into the depleted layer, and a concentration gradient grows out from the electrode surface into the solution. As the electrode process continues, the concentration profile extends further into the bulk solution, as shown in Fig. 4.20. When the surface concentration of O falls to zero (at time $t_6$ in Fig. 4.20), the electrode process can no longer be supported by electroreduction of O. An additional cathodic reaction must be brought into play and an abrupt change in potential occurs. The period of time between the commencement of electoreduction and the sudden change in potential is called the transition time $\tau$. The transition time for electroreduction of a species in the presence of excess supporting electrolyte was first quantified by Sand in 1901 who showed that the transition time $\tau$ was related to the diffusion coefficient of the electroactive species,

$$\tau^{1/2} = \frac{\pi^{1/2}nFC_oD^{1/2}}{2i} \tag{4.49}$$

where $D$ is the diffusion coefficient of species O, and the other symbols have their usual meanings.

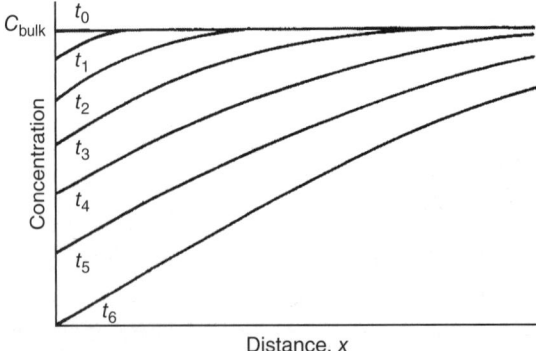

**FIGURE 4.20** Concentration profiles extending into bulk solution during constant current depletion of species at an electrode surface for $t_6 > t_0$.

Unlike cyclic voltammetry, solutions for Fick's diffusion equations [Eqs. (4.35) and (4.36)] can be obtained by chronopotentiometry as an exact expression by applying appropriate boundary conditions. For a reversible reduction of an electroactive species [Eq. (4.10)], the potential-time relationship has been derived by Delahay[11] for the case where O and R are free to diffuse to and from the electrode surface, including the case where R diffuses into a mercury electrode,

$$E = E_{\tau/4} + \frac{RT}{nf} \ln \frac{\tau^{1/2} - t^{1/2}}{t^{1/2}} \tag{4.50}$$

In this equation $E_{\tau/4}$ is the potential at the one-quarter transition time and $t$ is any time from zero to the transition time. The trace represented by this expression is shown in Fig. 4.21.

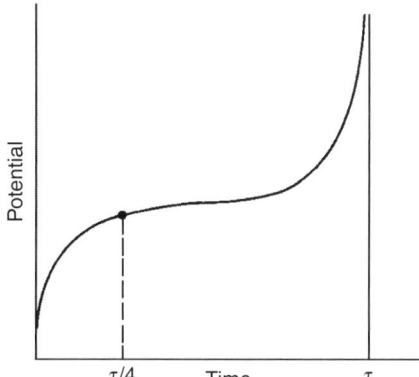

**FIGURE 4.21** Potential curve at constant current for reversible reduction of an electroactive species.

The corresponding expression for an irreversible process[12] with one rate-determining step is

$$E = \frac{RT}{\alpha n_a F} \ln\left[\frac{nFC_0 k_{app}}{i}\right] + \frac{RT}{\alpha n_a F} \ln\left[1 - \left(\frac{t}{\tau}\right)^{1/2}\right] \tag{4.51}$$

where $k_{app}$ is the apparent rate constant and $n_a$ is the number of electrons involved in the rate-determining step (often the same as $n$, the overall number of electrons involved in the total reaction) with the other symbols having their usual meanings. A plot of the logarithmic term versus potential yields both the transfer coefficient and the apparent rate constant.

In a practical system, the chronopotentiogram often has a potential trace with less than ideal shape. To accommodate variations in chronopotentiometric traces, measurement of the transition time can be assisted by use of a construction technique, as shown in Fig. 4.22. The transition time is measured at the potential of $E_{\tau/4}$.

To analyze two or more independent reactions separated by a potential sufficient to define individual transition times, the situation is slightly more complicated than with cyclic voltammetry. Analysis of the transition time of the reduction of the $n$th species has been derived elsewhere[13,14] and is

$$(\tau_1 + \tau_2 + \cdots + \tau_n)^{1/2} - (\tau_1 + \tau_2 + \cdots + \tau_{n-1})^{1/2} = \frac{\pi^{1/2} n F D_n^{1/2} C_n}{2i} \tag{4.52}$$

Clearly, this expression is somewhat cumbersome.

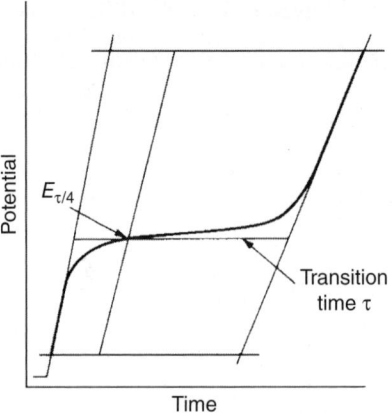

**FIGURE 4.22**    Construction of transition time τ for a chronopotentiogram.

An advantage of this technique is that it can be readily used to evaluate systems with high resistance. The trace conveniently displays segments due to the *IR* component, the charging of the double layer, and the onset of the faradaic process. Figure 4.23 shows these different features of the chronopotentiogram for solutions with significant resistance. If the solution is also one which does not contain an excess of supporting electrolyte to suppress the migration current, it is possible to describe the transition time of an electroreduction process in terms of the transport number of the electroactive species

$$\tau^{1/2} = \frac{\pi^{1/2} n F C_0 D_s^{1/2}}{2i(1-t_0)} \tag{4.53}$$

where $D_s$ is the diffusion coefficient of the salt (not the ion) and $t_0$ is the transport number of the electroactive species.

This expression can be useful in battery research since many battery systems do not have supporting electrolyte.

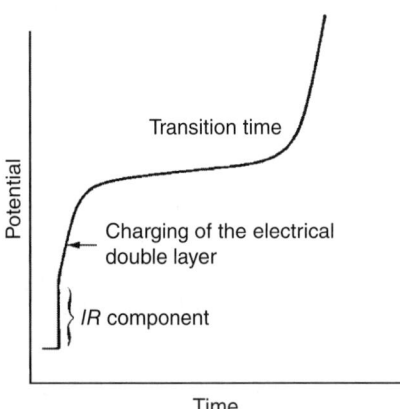

**FIGURE 4.23**    Chronopotentiogram of a system with significant resistance.

### 4.6.3   Intermittent Titration Techniques

The steady-state electrochemical measurement techniques reviewed above yield basic information derived from variables such as concentration, diffusion, and kinetics. However, transient measurements such as the galvanostatic intermittent titration technique (GITT)[15] and potentiostatic intermittent titration technique (PITT)[16] offer more direct determination of kinetic and thermodynamic properties of single and multiphase electrodes (e.g., alloys and Li-insertion materials). Both methods enable one to determine the phase diagram of an electrode material as a function of capacity and voltage. For example, in a two-phase material at equilibrium at constant temperature and pressure, the potential of the electrode material will be independent of composition. However, when a phase transition occurs such as from a two-phase to single-phase domain, the electrode potential will now vary as a function of composition. In addition, both GITT and PITT methods are convenient for determining the diffusion coefficient of the migrating ion in the various phases of the solid-state material. Solid-state diffusion coefficients can be determined from the Warburg impedance in electrochemical impedance spectroscopy (EIS) measurements as described below, and agreement is dependent on the equivalent circuit model used in the deconvolution of the EIS data. Thus, determination of the phases and diffusion coefficients by GITT and PITT are helpful in modeling equivalent circuits. Essentials of the GITT and PITT techniques are given below.

*Galvanostatic Intermittent Titration Technique (GITT).*   The GITT method[15] is a form of chronotpotentiometry discussed above, but it is simpler for determining diffusion coefficients for ions intercalating and deintercalating into and out of composite electrode materials, i.e., materials based on at least two components. In this transient method, a constant current pulse is sequentially applied to the electrode for a given time $t$ to remove or insert about 2% to 5% of the electrode's total capacity, e.g., $x$ in $Li_xCoO_2$. The capacity inserted or deinserted is simply $i$ times $t$ (i.e., Ah), and the total number of constant current steps required to fully cover the stable range of $x$ will depend on the amount of capacity added to or removed from the electrode material from each constant current step. After each defined constant current pulse, the electrode potential is allowed to rest until it reaches a new equilibrium before the next current pulse is applied. Figure 4.24 is a schematic of a single-step pulse for GITT,

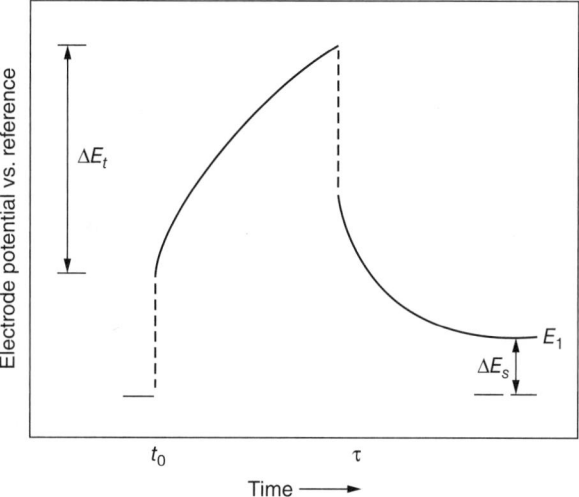

**FIGURE 4.24**   Typical single-step constant current pulse for galvanostatic intermittent titration technique (GITT).

where $\tau$ is the time over which the constant current pulse is applied starting from $t_0$, $\Delta E_t$ is the transit voltage change during the pulse (without the $IR$ drop), $\Delta E_s$ is change in the steady-state voltage due to the current pulse, and $E_1$ is the new OCV resulting from the insertion or deinsertion reaction ($E_1$ will of course not change for a two-phase material).

An example of determining phase transitions for $Li_xCoO_2$ from $x \approx 0.2$ to $1.0$ is shown in Fig. 4.25 based on data published by Plichta et al.[17] There are clearly three major phases.

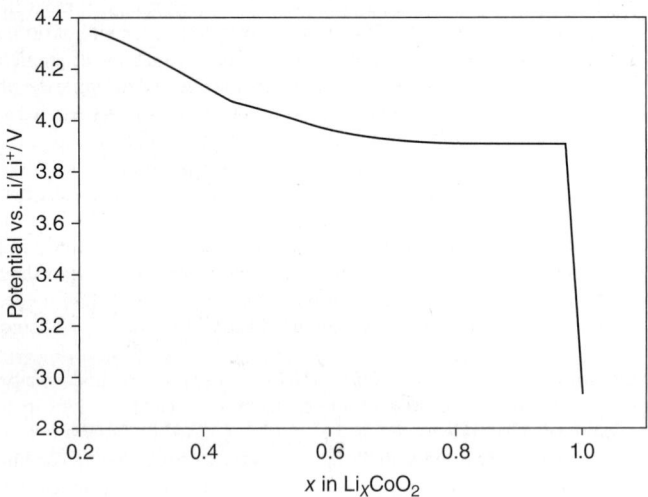

**FIGURE 4.25** Potential versus $x$ in $Li_xCoO_2$ (based on data from Ref. 17).

The figure shows that between ~0.2 and ~0.6 V versus Li/Li$^+$, there is a single-phase domain where the OCV varies with $x$ in $Li_xCoO_2$, followed by a transition to a two-phase domain at around $x$ ~0.6 to ~0.9 where the OCV is independent of $x$. Finally, at $x$ ~0.9 to 1.0, the OCV rapidly decreases indicating a transition to a single-phase domain. The voltage versus capacity relationship varies somewhat between various authors, but the important phase regions identified using GITT are quite clear. However, the precise voltages at which these phase transitions occur is best determined by the PITT method as described below. For designing a battery based on the rate and capacity ability of various intercalating materials, the available capacity from each phase over the major stable region for $x$ (Li$^+$ in the above example) is of interest to battery developers, but so are the diffusion rates of the intercalating ion, which is important for batteries specifically designed for high charge and discharge rates. A simple equation that can be used to calculate the chemical diffusion coefficient $D$ is[15]

$$D = \frac{4}{\pi\tau}\left(\frac{m_b V_M}{M_b S}\right)^2 \left(\frac{\Delta E_s}{\Delta E_t}\right)^2 \tag{4.54}$$

where $m_b$ is the mass of the active material, $V_M$ is the molar volume of the active material, $M_b$ is the molecular mass of the active material, and $S$ is the surface area of the electrode.

The remaining terms in Eq. (4.54) are defined above.

Note that the first term in parentheses in Eq. (4.54) is the thickness $L$ of the electrode, i.e.,

$$L = \frac{m_b V_M}{M_b S} \tag{4.55}$$

***Potentiostatic Intermittent Titration Technique (PITT).*** The PITT method[16] involves the application of a small amplitude voltage step to an electrode, typically around 10 mV, after which the current is recorded as a function of time, as shown in Fig. 4.26. Starting at an initial equilibrium voltage of $E_o$, the current decays to zero or very near zero (negligible), reaching a new equilibrium potential $E_1$. Subsequent potential pulses are imposed on the electrode to determine the incremental charge in coulombs $Q$ associated with each pulse covering the

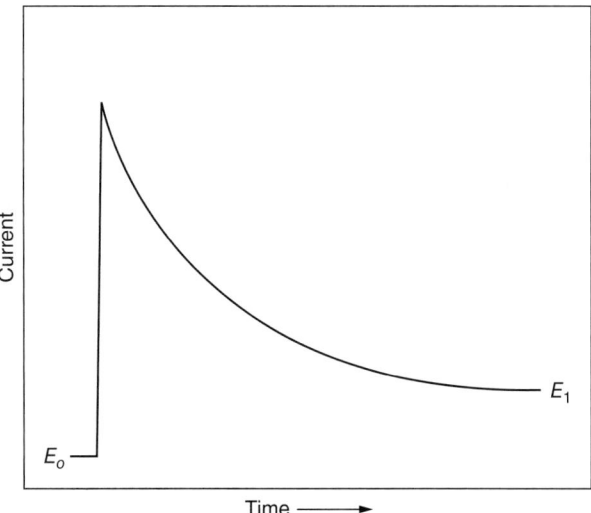

**FIGURE 4.26**  Typical single-step constant potential pulse for potentiostatic intermittent titration technique (PITT).

whole range of capacity defining the phase diagram for the material under investigation. The total current $i$ is recorded and the incremental charge (or differential charge) for each potential pulse is given by

$$Q = \int_0^t idt \tag{4.56}$$

Plotting the rest potential (the OCV) versus the composition of the electrode material (the amount of $x$ in $Li_xCoO_2$) results in a figure representing the phase diagram (i.e., similar to Fig. 4.25). Plots of the differential charge $Q/E$ versus potential $E$ result in sharp peaks precisely identifying the peak potentials for phase transitions. An example of this plot is shown in Fig. 4.27 for $Li_xCoO_2$.[17] Figure 4.27 is essentially identical to that obtained by cyclic voltammetry (CV) (Sec. 4.6.1), but there are important differences. The peak potentials for each phase are very sharp compared to what is observed in fast CV sweeps where the peak potentials are not precisely defined. By slowing down the sweep rate to ~0.001 mV/s, a cyclic voltammogram similar to Fig. 4.27 can be obtained, but because of the very low currents for such a low sweep rate, determination of capacity is difficult and not nearly as accurate as those obtained by the PITT method.

As demonstrated for the GITT method, the PITT method is also useful for determining the chemical diffusion coefficient for the insertion or removal of an ion from its host material. Two approaches are based on the thickness of the electrode $L$ [see Eq. (4.54)] and the time $t$ required to reach the equilibrium potential after each potential pulse.[16] The relations from Ref. 16 are given in Eqs. (4.57) and (4.58).

$$I(t) = \frac{QD^{1/2}}{L\pi^{1/2}}\left(\frac{1}{t^{1/2}}\right) \qquad \text{if } t << L^2/D \tag{4.57}$$

$$I(t) = \frac{2QD}{L^2}\exp\left(\frac{-\pi^2 Dt}{4L^2}\right) \qquad \text{if } t >> L^2/D \tag{4.58}$$

For very short times after a voltage step, the chemical diffusion coefficient can be determined from the slope of a plot of current versus ($1/t^{1/2}$) [Eq. (4.56)], and for large times, the chemical diffusion coefficient can be determined from the slope of a plot of the logarithm of the current versus time $t$ [Eq. (4.57)].

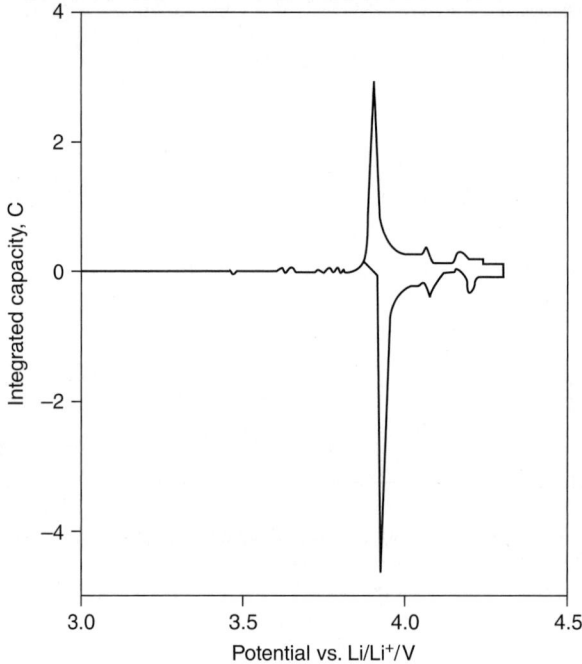

**FIGURE 4.27**    Plot of differential capacity $Q$ versus potential for $Li_xCoO_2$ (based on data from Ref. 17).

### 4.6.4    Electrochemical Impedance Spectroscopy (EIS) Methods

The two preceding electrochemical techniques (one in which the measured value was the current during imposition of a potential scan and the other a potential response under an imposed constant current) owe their electrical response to the change in impedance at the electrode-electrolyte interface. A more direct technique for studying electrode processes is to measure the change in the electrical impedance of an electrode by EIS. In this method, a small alternating current (AC) signal of ~5 to 10 mV is superimposed on an electrochemical cell with an applied finite direct current (DC) bias potential or OCV, and the impedance $Z$ (the equivalence of resistance $R$ in DC measurements) is determined over a wide frequency range, normally between 0.01 Hz and 1 MHz. The resulting wave forms for current $I$ and potential $E$ are sinusoidal, as shown in Fig. 4.28. The two wave forms in Fig. 4.28 differ in magnitude as well as phase. If the system is purely resistive, i.e., without capacitive and other elements, the two wave forms will be in-phase.

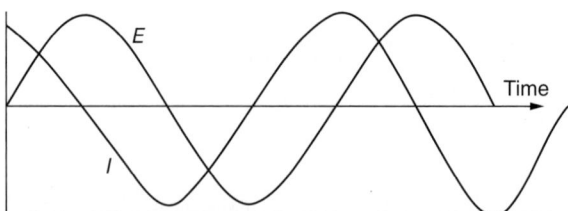

**FIGURE 4.28**    Sinusoidal current and potential wave forms at an electrode held at a specific DC (bias) potential or at OCV.

The potential sine wave and the current sine wave can be described, respectively, by

$$E_t = E_o \sin(\omega t) \tag{4.59}$$

and

$$I_t = I_o \sin(\omega t + \phi) \tag{4.60}$$

where $E_t$ and $E_o$ are the potentials at time $t$ and 0, respectively; $I_t$ and $I_o$ are the currents at time $t$ and 0, respectively; and $\omega$ is the frequency in rad/s (equal to $2\pi f$, where $f$ units are Hz).

Using the Euler's formula for trigonometric and complex functions, the impedance of the system can be represented by the following complex relation[25,26]

$$Z(\omega) = \frac{E}{I} = Z_o \exp(j\phi) = Z_o(\cos\phi + \sin\phi) \tag{4.61}$$

where $j = \sqrt{-1}$.

The components of $Z(\omega)$ thus consist of an imaginary part referred to as $Z_i$ and a real part referred to as $Z_r$. When plotting $Z_i$ versus $Z_r$, one obtains a semicircle called a "Nyquist" plot as described below. Note that in the absence of capacitance and inductance, the Nyquist plot for a simple resistor would be a simple point on the $Z_r$ axis representing the value of the resistor in ohms. For complex systems representing electrodes in cells and batteries, the Nyquist plots can be interpreted in terms of various electrode-electrolyte parameters such as solution resistance, kinetics (charge transfer), and capacitance. Inductive effects are generally not observed in these electrochemical systems. To relate the complex impedance of the electrode-electrolyte interface to electrochemical parameters, it is necessary to *model an equivalent circuit* to represent the dynamic characteristics of the interface. The model consists of a number of *impedance elements* in networks based on series, parallel, or series/parallel combinations. For example, the total impedance for $n$ elements in series is given by

$$Z_{total} = Z_1 + Z_2 + Z_3 + \cdots + Z_n \tag{4.62}$$

For $n$ elements in parallel, the impedance will be given by

$$\frac{1}{Z_{total}} = \frac{1}{Z_1} + \frac{1}{Z_2} + \frac{1}{Z_3} + \cdots + \frac{1}{Z_n} \tag{4.63}$$

The important elements to be considered in modeling EIS data for the electrode of interest are summarized in Table 4.3. A realistic model will then allow one to determine the electrochemical parameters for an electrode-electrolyte interface, which is discussed below.

**TABLE 4.3**  Equivalent Circuit Elements

| Circuit element | Impedance |
|---|---|
| Resistance $R$ | $R$ |
| Capacitance $C$ | $1/Cj\omega$ |
| Constant phase element $Q$ (CPE) | $1/Q(j\omega)^{\alpha}$ |
| Warburg impedance $W$ (infinite) | $1/Y(j\omega)^{1/2}$ |
| Warburg impedance $W$ (finite) | $\tan[\delta D^{-1/2}(j\omega)^{1/2}]/Y(j\omega)^{1/2}$ |
| Inductance $L$ | $j\omega L$ |

In all of the equivalent circuits and Nyquist figures shown below, the double-layer capacity is represented by $C$, which is the symbol used for a pure (ideal) capacitor. However, due to surface inhomogeneity and faradaic current (i.e., a "leaky" capacitor), the double layer rarely behaves as an ideal capacitor for most electrochemical systems. In this case, the capacity $C$ is replaced by a constant phase element (CPE) in which the impedance is given by (see Table 4.3)

$$Z = \frac{1}{Q(j\omega)^\alpha} \tag{4.64}$$

where $\alpha$ is an adjustable parameter.

When $\alpha = 1$, the CPE acts as an ideal capacitor, i.e., $Q = C$, and when $\alpha = 0$, the CPE is equivalent to a pure resistor. As indicated in Table 4.3, there are two ways to represent the Warburg impedance based on an infinite or finite diffusion layer thickness. For the latter, the relation for the impedance contains the thickness of the diffusion layer ($\delta$) and the diffusion coefficient ($D$) for the diffusing species.

In modeling the electrode-electrolyte interface for a single electrode (i.e., using a three-electrode cell comprised of a working electrode, a reference electrode, and a counter electrode), the adjustable (fitting) parameters to be evaluated include $R$, $C$, $Q$, $Y$, $L$, and $\alpha$. It is therefore important to select a realistic model for the analyses of EIS data. The equivalent circuit for the basic model of the electrode-electrolyte originally proposed by Randles[18] is shown in Fig. 4.29, and the basic Nyquist plot for this equivalent circuit is shown in Fig. 4.30,

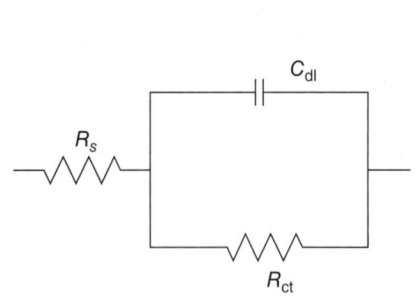

FIGURE 4.29   Randles' basic equivalent circuit for an electrode-electrolyte interface.

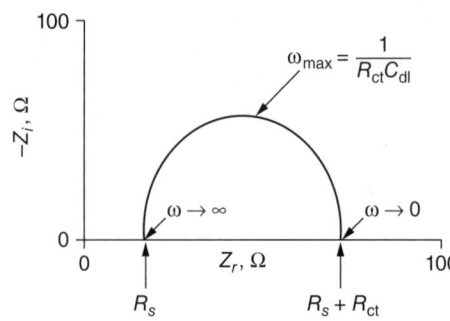

FIGURE 4.30   Schematic Nyquist plot for the Randles' circuit in Fig. 4.29.

where $R_s$ is the electrolyte solution resistance, $C_{dl}$ is the double-layer capacitance, and $R_{ct}$ is the charge-transfer resistance from which the exchange current density can be calculated.[19,20]

If the system exhibits diffusion control, this can be accounted for by the circuit shown in Fig. 4.31 in which the Warburg impedance is added in series with $R_{ct}$. The corresponding Nyquist plot is shown in Fig. 4.32 where the Warburg impedance appears at low frequencies as a straight line with a slope of 45°.

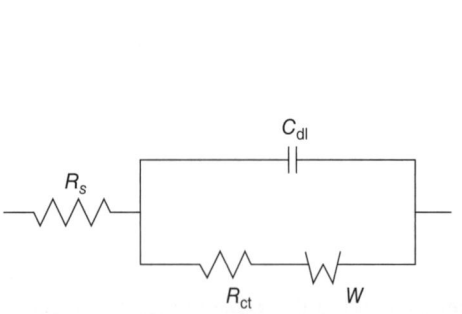

FIGURE 4.31   Randles' equivalent circuit for an electrode-electrolyte interface including the Warburg impedance.

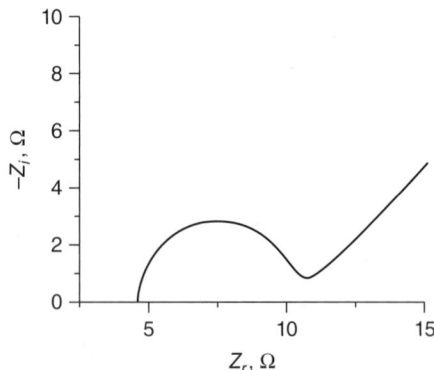

FIGURE 4.32   Schematic Nyquist plot for the equivalent circuit in Fig. 4.31.

Finally, for illustrative purposes, consider a system such as one based on a metallic Li anode or a $LiC_6$ anode in a Li-ion cell that reacts with the electrolyte solution to form a solid electrolyte interface (SEI) layer. The equivalent circuit for this system is given in Fig. 4.33, where $C_f$ and $R_f$ represent, respectively, the capacity of the SEI film and the impedance of the SEI film. In this case, experimentally the Nyquist plot shows either two distinct time constants (two symmetric semicircles) in the spectra, or an unsymmetrical semicircle in which the two semicircles overlap, as shown in Fig. 4.34.

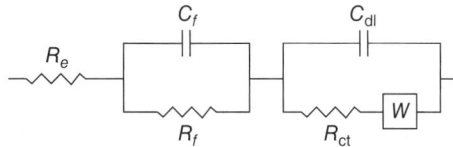

**FIGURE 4.33** Equivalent circuit accounting for solid electrolyte interface (SEI) formation.

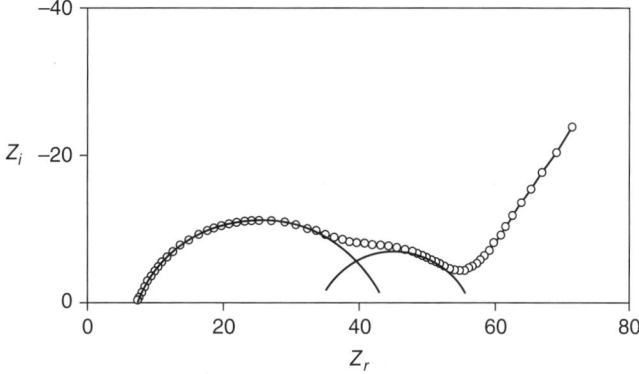

**FIGURE 4.34** Schematic Nyquist plot for the equivalent circuit in Fig. 4.33.

Selecting a reasonable model for the electrode-electrolyte interface is the first step in determining the parameters for a given system. The model can be any combination of impedances in series [Eq. (4.62)] or parallel [Eq. (4.63)] or any series and parallel combination of the elements given in Table 4.3. The next step is to select an equivalent circuit (e.g., Figs. 4.29, 4.31, or 4.33) and deconvolute the experimental data in each figure ($Z_r$, $Z_i$, and frequency) for each bias potential to obtain the numerical value of each element in the model selected. Mathematical methods used for the deconvolution process are conveniently accomplished by methods developed by Boukamp[21,22] and texts by Macdonald[19] and Orazem.[20] Many manufacturers of EIS instruments also offer software packages for EIS basics and/or deconvolution programs on their websites.

### 4.6.5 Transference Numbers

The transference (or transport) number of an ion is defined as the fraction of the current carried by an ion during charge or discharge of an electrochemical cell. According to this definition, the sum of transference numbers for the cation $t_+$ and anion $t_-$ is unity, as given in Eq. (4.65), i.e.,

$$1 = t_+ + t_- \qquad (4.65)$$

When cells are discharged at high rates, concentration polarization can be very severe and high value of the transference number of the ion of interest serves to mitigate this polarization, e.g., for $Li^+$ in Li-based batteries as discussed below.

Historically, transference numbers have been obtained by various time-consuming methods based on conductivity (emf measurements in liquid electrolyte solutions). A more rapid, simplified, and precise method to obtain transference numbers for Li$^+$ ions in nonaqueous solid polymer electrolytes was used by Bruce et al.[23,24] using a potentiostatic polarization technique with nonblocking Li cells of the type

$$\text{Li/electrolyte/Li} \tag{4.66}$$

Applying a small potential of ~10 mV to this cell, anions will accumulate at the anode and deplete at the cathode over a time period during which a steady-state condition will be reached, i.e., where the net anion flux is reduced to zero and only the Li$^+$ ions carry the current. At this time, the Li$^+$ transport number is given by

$$t_+ = \frac{i_{ss}(\Delta V - i_0 R_0)}{i_0(\Delta V - i_{ss} R_{ss})} \tag{4.67}$$

where the subscripts 0 and ss refer to initial and steady-state values, respectively, $R_0$ is the sum of the charge-transfer resistance $R_{ct}$ and the passivating film resistance $R_{film}$, $\Delta V$ is the applied voltage step, and $i$ is the current.

Figure 4.35 shows, schematically, a typical chronoamperogram used to obtain both $i_0$ and $i_{ss}$. Both $R_0$ and $R_{ss}$ are obtained from Nyquist plots (see Fig. 4.32) before and after electrode polarization. The equivalent circuit used to fit the impedance data to obtain the resistances $R_0$ and $R_{ss}$, shown schematically in Fig. 4.36, is a resistance series model of the electrolyte resistance $R_e$ and two subunits of a resistance and a CPE. This procedure originally developed by Bruce et al. for solid-polymer electrolytes has also been shown to be applicable to ionic liquid-based gelled electrolytes at temperatures of 50 to −20°C[25] and to nonaqueous liquid electrolyte solutions.[26]

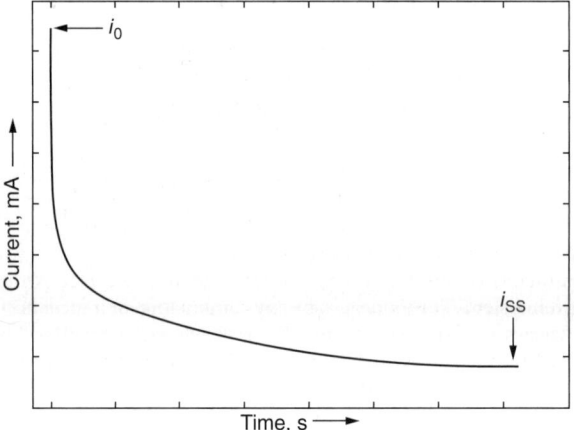

**FIGURE 4.35**    Schematic of a chronoamperogram, where $i_0$ is the initial current and $i_{ss}$ is the steady-state current.

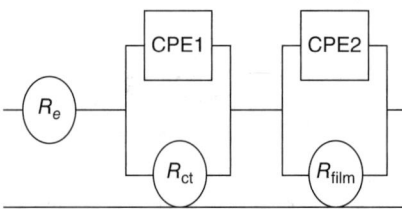

**FIGURE 4.36**    Equivalent circuit for deconvolution of the electrochemical impedance spectroscopy (EIS) spectra shown in Fig. 4.34.

## 4.6.6    Reference Electrodes

In all of the above techniques for determining mechanisms and properties of individual electrodes, a three-electrode cell is required. The three-electrode cell contains a working electrode (the electrode of interest); a counter-electrode at which oxidation or reduction products have no effect on the mechanisms at the working electrode; and a stable, nonpolarizable reference electrode to monitor the working electrode as a function of voltage, time, and current density. Figure 4.37 is a schematic of a three-electrode cell.

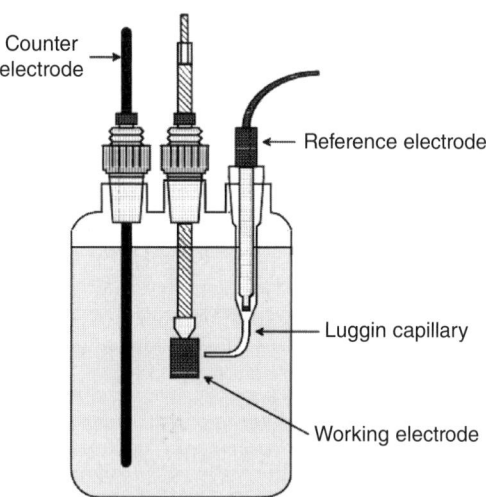

**FIGURE 4.37**    Three-electrode electrochemical cell.

A Luggin capillary, shown in Fig. 4.37, can be used to reduce uncompensated $IR$ drop when current is flowing between the counter and working electrodes. If a Luggin capillary is used, the reference electrode is placed inside the Luggin, as shown in Fig. 4.37, and the electrolyte solution inside the Luggin may be the same as the electrolyte solution in the cell or it may be a different electrolyte solution containing ions necessary to establish equilibrium at the reference electrode. For aqueous solutions, typical reference electrode systems such as $Ag/AgCl$, $Hg/Hg_2Cl_2$, and $Hg/HgO$ are often used. For comprehensive treatises on the subject of reference electrodes in aqueous solutions, the reader is referred to two excellent texts by Ives and Janz[27] and Bard et al.[28]

For studies in nonaqueous solvents where the preferred reference material reacts with the electrolyte solution, e.g., a metallic Li reference, a "pseudoreference" is sometimes used. Pseudoreference electrodes typically used in solutions reactive with Li are metal wires (e.g., Al, Pt, or Ag) immersed in the electrolyte solution. Although such electrodes often provide a near constant potential, they are not reversible and depend on factors that add to instability such as electrolyte concentration, impurities, and liquid junction potentials, and thus have no thermodynamic significance. Instead, a pseudoreference, such as a reversible and thermodynamically stable reference related to metallic lithium, is used. A metal oxide, such as $Li_4Ti_5O_{12}$ that has a wide range of stability at a potential of 1.55 V[29] positive to $Li/Li^+$, is an example. In any electrolyte solution containing only a lithium electrolyte, this reference has the advantage that the potential of the $Li/Li_4Ti_5O_{12}$ couple is constant and is independent of the $Li^+$ concentration in solution.

## *REFERENCES*

1. J. O'M. Bockris and A.K.N. Reddy, *Modern Electrochemistry,* Plenum, New York, 1970.

2. H.H. Bauer, *J. Electroanal. Chem.* **16**, 419 (1968).

3. J. Horiuti and M. Polanyi, *Acta Physicochim. U.S.S.R.* **2**, 505 (1935).

4. J. O'M. Bockris and A.K.N. Reddy, op. cit., p. 918; see also J. O'M. Bockris, "Electrode Kinetics" in *Modern Aspects of Electrochemistry*. J. O'M. Bockris and B.E. Conway, Editors, Butterworths, Lonmdon, 1954, Chap. 2.

5. A. Fick, *Ann. Phvs.* **94**, 59 (1855).

6. A.J. Bard and L.R. Faulkner, *Electrohemical Methods: Fundamentals and Applications*, John Wiley, NY, 1987.

7. J.S. Newman, *Electrochemical Systems,* 2d edition, Prentice-Hall, Englewood Cliffs, NJ, 1991.

8. R.S. Nicholson and I. Shain, *Anal. Chem.* **36**, 706 (1964).

9. H. Angerstein-Kozlowska, B.E. Conway, and W.B.A. Sharp. *J. Electroanal. Chem.* **43**, 9 (1973).

10. K. Xu, M.S. Ding, and T.R. Jow, *Electrochim. Acta* **46**, 1823 (2001).

11. P. Delahay, *New Instrumental Methods in Electrochemistry*, Interscience, NY, 1954.

12. P. Delahay, *J. Am. Chem. Soc.* **75**, 1190 (1953).

13. C.N. Reilley, G. W. Everett, and R. H. Johns, *Anal. Chem.* **27**, 483 (1955).

14. T. Kambara and I. Tachi, *J. Phys. Chem.* **61**, 405 (1957).

15. W. Weppner and R.A. Huggins, *J. Electrochem. Soc.* **124**, 1569 (1977).

16. C. John Wen, B.A. Boukamp, and R.A. Huggins, *J. Electrochem. Soc.* **126**, 2558 (1979).

17. E. Plichta, S. Slane, M. Uchiyami, M. Salomon, D. Chua, W.B. Ebner, and H.-p. Lin, *J. Electrochem. Soc.* **137**, 1865 (1989).

18. J.E.B. Randles, *Disc. Faraday Soc.* **1**, 11 (1947).

19. J.R. MacDonald, *Impedance Spectroscopy, Emphasizing Solid Materials and Systems,* Wiley, New York, 1987.

20. M.E. Orazem and B. Tribollet, *Electrochemical Impedance Spectroscopy,* The ECS Series of Texts and Monographs, 2nd edition, Wiley-Blackwell, 2017.

21. B.A. Boukamp, *Solid State Ionics* **18**, 136 (1986).

22. B.A. Boukamp, *Solid State Ionics* **20**, 31 (1986).

23. J. Evans, C.A. Vincent, and P.G. Bruce, *Polymer* **28**, 2324 (1987).

24. P.G. Bruce and C.A. Vincent, *J. Electroanal. Chem. Interfacial Electrochem.* **225**, 1 (1987).

25. D. Bansal, F. Cassel, F. Croce, M. Hendrickson, E. Plichta, and M. Salomon, *J. Phys. Chem. B.* **109**, 4492 (2005).

26. V. Mauro, A. D'Aprano, F. Croce, and M. Salomon, *J. Power Sources* **141**, 167 (2005).

27. D.J.G. Ives and G.J. Janz, *Reference Electrodes, Theory and Practice*, Academic, New York, 1961.

28. A.J. Bard, R. Parsons, and J. Jordan, *Standard Potentials in Aqueous Solutions*, Marcelle Dekker, NY, 1985.

29. T. Ohzuku, A. Ueda, and N. Yamamoto, *J. Electrochem. Soc.* **142**, 1431 (1995).

## BIBLIOGRAPHY

### General

A.J. Bard and L.R. Faulkner, *Electrochemical Methods: Fundamentals and Applications*, John Wiley, NY, 1980.

J. O'M. Bockris and A.K.N. Reddy, *Modern Electrochemistry*, vols. 1 and 2, Plenum, New York, 1970.

V.S. Bogotsky, *Fundamentals of Electrochemistry*, Wiley-Interscience, NY, 2006.

B.E. Conway, *Theory and Principles of Electrode Processes*, Ronald Press, New York, 1965.

B.E. Conway, *Electrochemical Supercapacitors: Scientific Fundamentals and Technological Applications,* Kluwer Academic/ Plenum Publishing, New York, 1999.

E. Gileadi, E. Kirowa-Eisner, and J. Penciner, *Interfacial Electrochemistry,* Addison-Wesley, Reading, Mass., 1975.

C.A. Vincent and B. Scrosati, *Modern Batteries*, 2nd edition, Butterworth-Heinemann, Oxford, 1997.

### Transfer Coefficient (Symmetry Factor)

J. O'M. Bockris and A.K.N. Reddy, *Modern Electrochemistry*, Plenum, New York, 1970.

B.E. Conway, *Theory and Principles of Electrode Processes,* Ronald Press, New York, 1965.

## Electrical Double Layer

J. O'M. Bockris and A.K.N. Reddy, *Modern Electrochemistry*, Plenum, New York, 1970.

V.S. Bogotsky, *Fundamentals of Electrochemistry*, Wiley-Interscience, NY, 2006.

P. Delahay, *Double Layer and Electrode Kinetics,* Interscience, New York, 1965.

D.C. Grahame, *Chemical Reviews* **41**, 441 (1947).

R. Parsons, "Equilibrium Properties of Electrified Interphases" in *Modern Aspects of Electrochemistry*, J. O'M. Bockris and B.E. Conway, Editors, vol. 1, Butterworths, London, 1954, pp. 103–179.


## Electrochemical Techniques

P. Delahay, *New Instrumental Methods in Electrochemistry,* Interscience, New York, 1954.

D.T. Sawyer, A. Sobkowiak, and J.L. Roberts, *Experimental Electrochemistry for Chemists,* 2nd edition, Wiley, New York, 1995.

E.B. Yeager and J. Kuta, *Techniques for the Study of Electrode Processes,* in *Physical Chemistry,* vol. IXA, *Electrochemistry,* Academic, New York, 1970.


## Reference Electrodes

A.J. Bard, R. Parsons, and J. Jordan, *Standard Potentials in Aqueous Solutions*, Marcelle Dekker, NY, 1985.

J.N. Butler, "Reference Electrodes in Aprotic Organic Solvents" in *Advances in Electrochemistry and Electrochemical Engineering,* P. Delahay, Editor, vol. 7, 1970, pp. 77–175.

D.J.D. Ives and G.J. Janz, *Reference Electrodes, Theory and Practice,* Academic, New York, 1961.


## Organic Electrode Reactions

O. Hammerich and B. Speiser, *Organic Electrochemistry,* 5th edition, CRC Press, Boca Raton, Florida, 2016.

L. Meites and P. Zuman, *Electrochemical Data,* Wiley, New York, 1974.


## AC Impedance Techniques

E. Barsoukov and J.R. Macdonald, *Impedance Spectroscopy: Theory, Experiment and Applications,* Wiley-Interscience, NY, 2005.

J.R. MacDonald, *Impedance Spectroscopy, Emphasizing Solid Materials and Systems,* Wiley, New York, 1987.


## Transference Numbers

H.S. Harnad and B.B. Owen, *The Physical Chemistry of Electrolyte Solutions*, 3rd edition, Reinhold, NY, 1965.

R.A. Robinson and R.H. Stokes, *Electrolyte Solutions*, Butterworths, London, 1959.

S. Zugmann, M. Fleischmann, M. Ameareller, R.M. Gschwind, H.D. Wiemhofear, and H.J. Gores, *Electrochim. Acta* **56**, 3926, (2011).

# CHAPTER 5
# FACTORS AFFECTING BATTERY PERFORMANCE

**David Linden**

## 5.1 BASELINE CHARACTERISTICS

The fundamental metric for assessing any battery system is specific energy or, alternatively, energy density (i.e., volumetric and gravimetric unit energy, respectively). Table 1.2 provides various theoretical specific energy values derived from half-cell reaction calculations. While these values establish the maximum potential energy output of each battery system, the actual performance of the battery may be significantly different. Specifically, if a battery is discharged under less thermodynamically favorable conditions (e.g., cold temperatures or faster rates), the performance of the battery will likely suffer. Additionally, all battery systems include electrochemically inactive components (packaging, terminals, current collectors, etc.) that further decrease the calculated unit energy output. To properly analyze performance for any given application, a battery system must be tested with actual cell hardware and under the intended specific conditions of use. As indicated in Chap. 1 and further detailed in the sections below, a variety of factors depress output voltage, reduce gravimetric and volumetric energy densities, and shorten the life of actual working batteries.

## 5.2 OVERVIEW

Among the many factors that influence the performance of batteries, the following design parameters and operational characteristics warrant special consideration:

1. Cell design (Sec. 5.3)
2. Battery pack design (Sec. 5.4)
3. Voltage response (Sec. 5.5)
4. Discharge current (Sec. 5.6)
5. Discharge profiles/regimes (Sec. 5.7)
6. Temperature effects (Sec. 5.8)
7. Service life (Sec. 5.9)
8. Battery charging (Sec. 5.10)
9. Battery storage (Sec. 5.11)

The effects of these factors on battery performance are discussed in subsequent sections. However, the generalizations presented below may be modified by various interactions and stray influences. For example, high-rate discharge following high-temperature storage can be especially debilitating. Capacity loss will usually be greater under heavy discharge currents for aged cells compared to fresh ones. Similarly, high-rate discharge at cold temperatures will typically cause higher observed loss of capacity than predicted by considering the combination of these conditions from separate data sets. Battery specifications and standards ideally list specific test and operational conditions to eliminate the undue influence of uncontrolled or random conditions on battery performance.

Furthermore, battery performance may differ subtly or even dramatically based on seemingly minor variations in cell materials, battery designs, manufacturing sources, or production lots. As with any manufactured product, the performance variability depends on the materials and process controls as well as even slight deviations in the application conditions. Manufacturers' data should be consulted to obtain specific performance characteristics. References 1 and 2 detail various concepts for improved battery performance.

## 5.3    EFFECT OF CELL DESIGN

Cell electrochemistry and construction will strongly influence battery performance characteristics. The type of cell (primary alkaline, rechargeable lithium, lead-acid, fuel cells, etc.) will clearly impact the capabilities. However, within any given category of cell, specific features such as electrode design or cell shape and capacity will also be critical influences.

### 5.3.1    Electrode Design

Cell designs are often based on goals for either capacity or rate capability. To achieve high capacity, the electrodes will be designed to contain maximum quantities of active material but usually at the expense of rate capability. High-rate cells, on the other hand, are designed with thin electrodes, having large surface area and high reactivity but reduced capacity. In addition, current collectors, tabs, and terminals can be manipulated to minimize internal resistance and enhance current density (amperes per area of electrode surface) or to maximize capacity.

For example, cylindrical cells may be designed with either a bobbin construction, typical for zinc-carbon cells as well as some alkaline-manganese dioxide cells, or with a spirally wound construction, common for many small rechargeable cells and some high-rate primary cells. In bobbin cells, a solid cylindrical core is used for one electrode and a concentric hollow cylinder for the second electrode with an interposing separator in the annular space between electrodes (Fig. 5.1a). This design maximizes the amount of active material that can be placed into the cylindrical can, but at the expense of interfacial electrode surface area necessary for fast electrochemical reaction kinetics.

The spirally wound (jelly roll) electrode construction is shown in Fig. 5.1b. Positive and negative electrodes are prepared as thin strips, rolled, with two layers of intervening separator, into a cylindrical shape and then inserted into a round can. The high electrode surface area of this design enhances high-rate performance, but at the expense of active material and capacity.

Another popular cell construction is based on flat plate electrodes and is typically used in many lead-acid cell designs, including SLI (starting-lighting-ignition) and large storage batteries (Fig. 5.1c). Flat plate cells can be designed with varied electrode thickness and surface area that will optimize the trade-off between capacity and rate capability. Another version of the flat plate cell is based on the use of flat, round electrode discs. Known as button, coin, or watch cells, this cell type is generally quite small (1- to 3-cm diameter and less than 1-cm thick).

Another key design feature of electrode pairs and stacks is the relative dimensional (i.e., length/width) and/or thickness (i.e., capacity) ratios of the anode to the cathode. The two electrodes can be of identical size and rated capacity, or they can vary by 5% to 10% when compared to each other. Significant consequences, either beneficial or harmful, result from these design choices. In general, anodes and cathodes are rarely designed with exactly the same dimensions or theoretical capacity. Many factors, specific to the cell electrochemistry, and extensive testing are often needed in selecting the optimal electrode designs.

### 5.3.2    Miscellaneous Designs

Some additional cell electrode configurations include the following:

1.  Bipolar cells: made by stacking anode and cathode electrodes back to back that share a common impervious but electrically conductive divider plate or foil (Fig. 5.1d and Chap. 14).

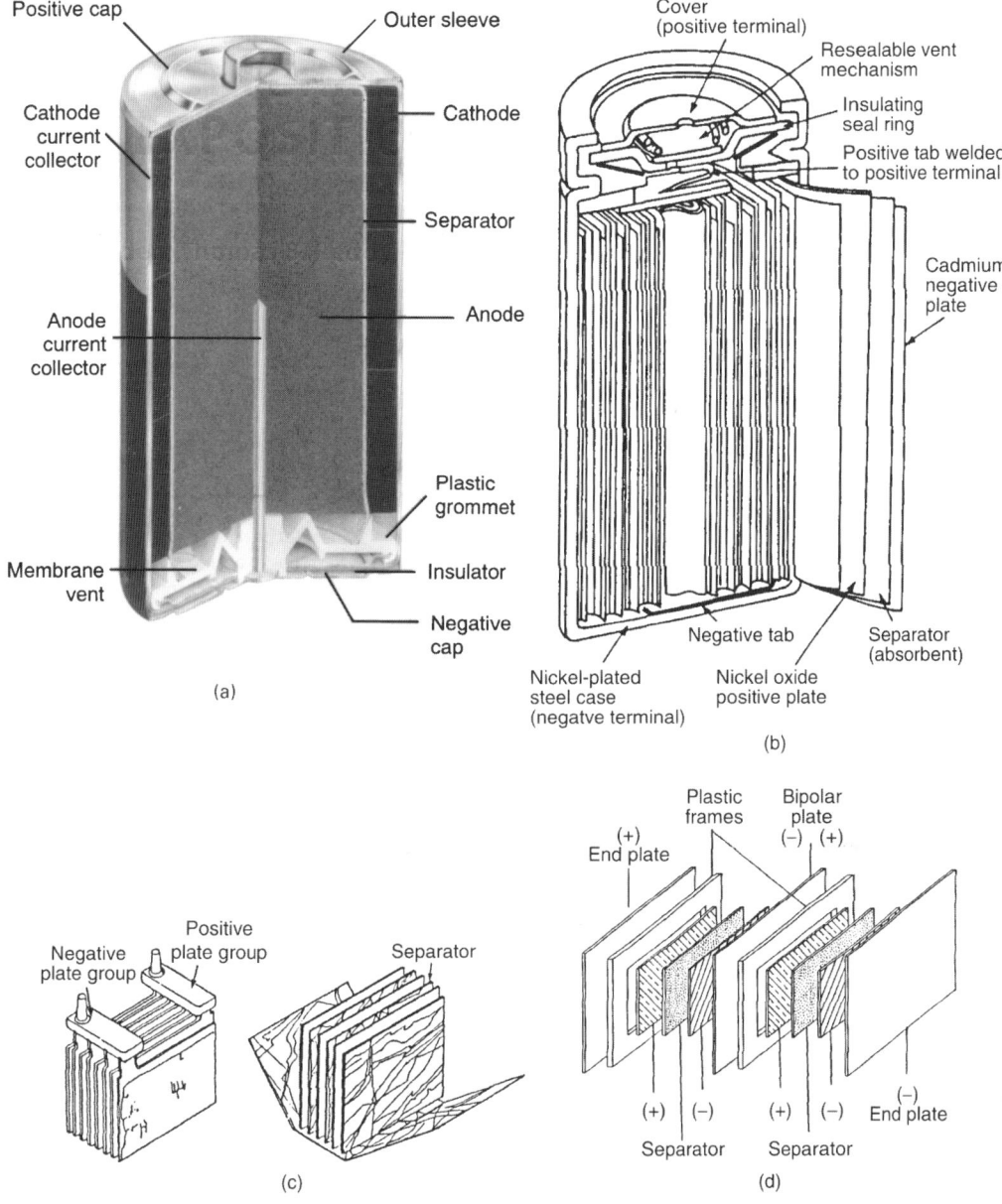

**FIGURE 5.1**  Cell design; typical internal configurations. (*a*) Bobbin construction. (*b*) Spiral wound construction. (*c*) Flat-plate construction. (*d*) Bipolar-plate construction.

2. Oval ("flat wound") cells: made similar to cylindrical spirally wound cells but with a flat center winding mandrel/tab that yields a nearly prismatic final cell shape (Chap. 17).

3. Air/water cathode cells: made using an anode with a membrane layer that is exposed to the exterior cell environment to allow reaction with ambient air (i.e., oxygen) or water as the cathode (Chap. 18).

4. Fuel cells: made with containers that store anode and cathode active materials in either liquid or gaseous form that are transferred simultaneously into the two halves of a reaction chamber each housing current collectors, segregated by a separator membrane with the reaction products then exhausted externally or in other cases converted back to reactants (Chap. 19).

5. Integrated multilayer constructions (including solid state cells): made by advanced techniques such as vacuum vapor deposition and three-dimensional printing (Chap. 22A) and including features such as interpenetrating/overlapping electrodes.

6. Flow batteries: made with reservoirs that store anode and cathode active materials in liquid or slurry form that are pumped through the two halves of a reaction chamber each housing current collectors, segregated by a separator membrane (Chap. 22B) and that are then collected in retention containers for later electrical recharge or physical replacement.

7. Thin bonded/laminated flexible electrode sandwiches (Chap. 22C).

The choice of the cell electrode configuration depends on electrochemistry issues, manufacturing preferences, and desired cost and performance goals for the applications and markets of interest.

### 5.3.3   Dual Systems

An ideal cell will combine high energy with high-rate capability. However, if a single power source cannot properly meet the electrical requirements, dual power sources may be used. For example, a high-energy battery may be coupled with electrochemical capacitors that better meet peak power requirements. Other hybrid power systems may include high-energy/high-power battery combinations, batteries with fuel cells, internal combustion engines with electrochemical power sources (common in hybrid electric vehicles), etc.

### 5.3.4   Cell Shape

Cell geometry will also influence the battery capabilities due to effects on internal resistance and heat dissipation. For example, bobbin cells with large length to diameter ratio will generally have lower internal resistance and higher discharge rate capability. In addition, heat dissipation also will be better for long, narrow bobbin cells with higher surface-to-volume ratio.

### 5.3.5   Internal Component Packaging Efficiency

Improving component packaging within the internal volume of the cell will benefit the specific energy output of the cell. Generally, volumetric energy density (watt-hours per liter) decreases with decreasing cell volume as the percentage of active materials is reduced for the smaller cells. This relationship is illustrated for several button-type cells in Fig. 5.2. Note that diameter also influences the volumetric energy efficiency as the relative volume for the seal and other inert cell materials increases for smaller cell diameter.

### 5.3.6   Effect of Current Density

For a given current drain, a higher capacity cell will have improved voltage characteristics over a lower capacity cell of similar design. Specifically, smaller cells will have higher current flow per unit area of electrode interface area (defined as current density, $mA/cm^2$) that depresses cell discharge voltage. For a fixed discharge current (mA) small cells may have a discharge profile similar to curve 2 in Fig. 5.3, but a higher capacity cell would experience lower current density, resulting in a discharge curve similar to curve 1 (or under very low rate, similar to the "ideal" curve). The cell discharge current relative to the size of the battery (i.e., current density) is a key factor to consider.

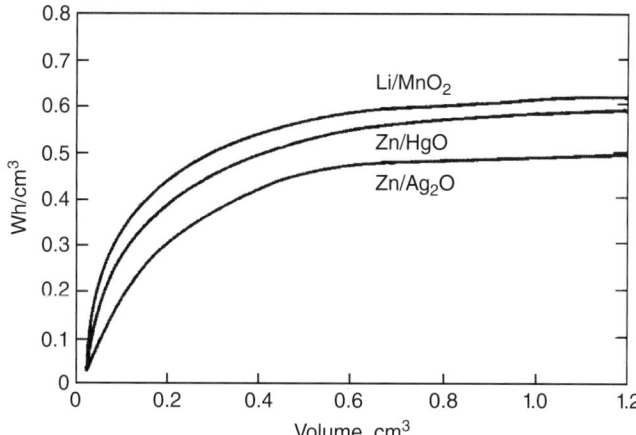

**FIGURE 5.2**   Energy density, watt-hour per cubic centimeter, of button batteries as a function of cell volume. (See Ref. 3.)

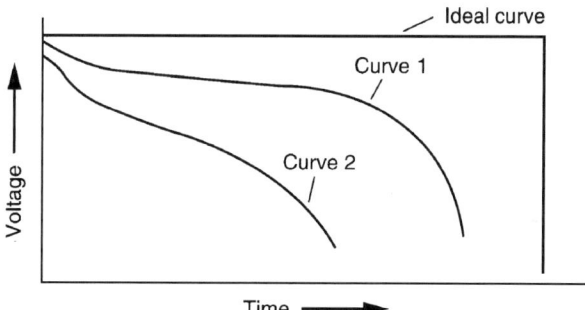

**FIGURE 5.3**   Characteristic discharge curves.

Additionally, rather than using oversized cells or parallel-connected small cells, battery designers have used various other techniques, such as series-connected cells with a voltage converter, to improve performance. Costs, reliability, voltage regulation, and other pertinent factors must all be considered.

## 5.4   EFFECT OF BATTERY DESIGN

A multicell battery will often perform differently than individual cells. First, cells will not be identical, but even if identical cells were used, each cell would experience different environments within the battery pack.

Second, battery pack designs and hardware (such as cell array configurations, spacing between cells, container material, electrical insulation, potting compound, fuses and other electronic controls, etc.) affect the environment, especially the temperature, of the individual cells. All cells generate heat during discharge and recharge. Internal heating of the cells can actually improve battery performance at low temperatures, but excessive heat generation and retention can reduce cell life and create safety issues. Ideally, battery packs should be designed to maintain a uniform internal temperature and avoid "hot spots."

Additionally, battery packaging materials add to size and weight, further reducing specific energy or energy density relative to component cells. Accordingly, when comparing specific energy of various cells, values must be specified based on single cells, single-cell batteries, or multicell battery packs (with or without charger/ electronics) as is appropriate for the situation. Finally, in a rechargeable battery pack individual cells will inevitably become "unbalanced," leading to differences in voltage levels, capacity, rate capability, or other characteristics. Performance of the battery pack will suffer or safety problems may arise. Advanced electronics and "smart" control algorithms are now used extensively to reduce these cell variations.

## 5.5   VOLTAGE RESPONSE

The initial voltage of an activated cell derives from thermodynamic mechanisms. Complex and rigorous experimental electrochemical techniques (Chap. 4) are used to establish half-cell reduction potentials (App. B) from which "standard" cell voltages are derived. However, the actual/measured cell voltage is not a fixed or precise value and depends on the cell environment and operating conditions as follows:

1. The *theoretical voltage* is a function of the anode and cathode materials, the composition of the electrolyte, and the temperature (usually stated at 25°C).

2. The *open-circuit voltage* is the voltage under a no-load condition and is usually a close approximation of the theoretical voltage.

3. The *closed-circuit voltage* is the voltage under a load condition.

4. The *nominal voltage* is one that is generally accepted as typical of the operating voltage of the battery as, for example, 1.5 V for a zinc-manganese dioxide battery.

5. The *working voltage* is more representative of the actual operating voltage of the battery under load and will be lower than the open-circuit voltage.

6. The *average voltage* is the time-weighted measured voltage during discharge.

7. The *midpoint voltage* is the voltage when half discharged.

8. The *end* or *cutoff voltage* is designated as the end of the discharge. Usually it is the voltage above which most of the capacity of the cell or battery has been delivered. The end voltage may also be dependent on the application requirements.

9. The charge (or float) voltage is the applied voltage necessary to overcome thermodynamic forces needed to restore the discharge products back to initial states.

Using the lead-acid battery as an example, the theoretical and open-circuit voltages are ~2.1 V, the nominal voltage is 2.0 V, the working voltage is between 1.8 and 2.0 V, and the end voltage is typically 1.75 V on moderate and low-drain discharges and 1.5 V for engine-cranking loads. On charge, the voltage may range from 2.3 to 2.8 V.

When a cell or battery is discharged, its voltage is lower than the theoretical voltage due to ohmic losses (i.e., the voltage drop calculated by Ohm's law) as well as polarization effects at both electrodes (Chap. 4). In the idealized case, the discharge of the battery proceeds at the theoretical voltage until the active materials are consumed and the capacity is fully utilized. The voltage then drops to zero. Under actual conditions voltage under a discharge load drops during discharge as the cell resistance increases due to the accumulation of discharge products, activation and concentration polarization, and related factors (Chap. 4). As the cell resistance or the discharge current is increased, the discharge voltage decreases and the discharge shows a more sloping profile.

The greatest impact from decreased cell working voltage is the reduction in delivered specific energy due to the inability to extract full capacity at the theoretical voltage level.

The actual delivered specific energy (volts × amps × hours on a unit basis) is reduced proportionate to the difference between the theoretical cell voltage and the actual operating discharge voltage (i.e., the integral area under the voltage-time curve at the specified current).

Discharge profiles will vary depending on the electro-chemical system, cell design, and discharge parameters as shown in Fig. 5.4 as follows:

1. *Flat discharge (curve 1):* thermodynamically efficient reactions with nearly full active material utilization.

2. *Multiple plateau profile (curve 2):* two-step discharge indicating different reaction mechanisms or thermodynamic forces of the active material(s).

3. *Sloping discharge (curve 3):* indicative of changing reaction mechanisms, unstable reactants, increasing internal resistance, thermal effects, etc.

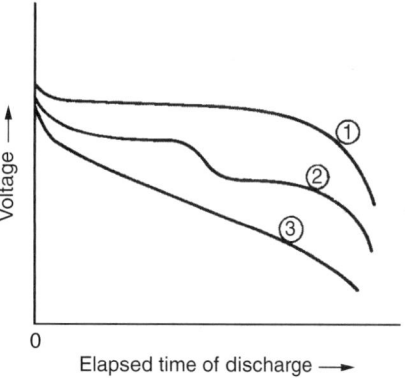

**FIGURE 5.4**   Battery discharge voltage profiles.

## 5.6   DISCHARGE CURRENT

As noted in Chap. 4 and Sec. 5.5, *IR* losses and polarization effects cause reduced voltage and lower capacity as current drain is increased. At extremely low current drains, the discharge voltage and capacity can approach theoretical values, but chemical deterioration during the prolonged discharge may become a factor and cause a reduction in capacity (Sec. 5.11).

At the other extreme, a cell that has been discharged at high current and has quickly reached cutoff voltage will typically become highly polarized, meaning that a large amount of capacity likely remains in the cell. Following an open circuit rest period if the voltage rises significantly, additional cell capacity might still be available. By subsequently discharging the cell at successively lower rates, the remaining capacity might be extracted. Figure 5.5 shows a procedure whereby cells are discharged initially at high rate to a specified end voltage, followed by a series of progressively lower rate discharge steps. Note that the figure depicts the use of an open circuit equilibration period between discharge steps to allow the voltage to rebound to the expected working voltage for the depleted cell (i.e., to the dashed lines).

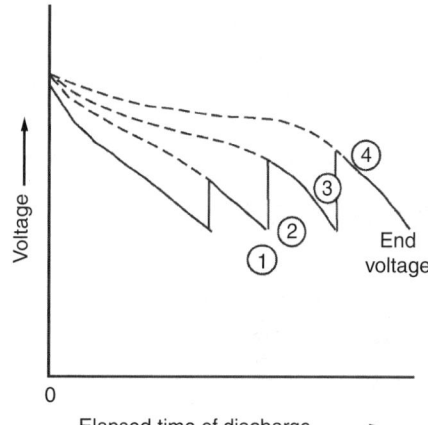

**FIGURE 5.5**   Voltage profile for a battery discharged in sequential steps going from high to low current.

### 5.6.1   C-Rate[a]

Discharge tests of cells of varied capacity and designs are difficult to compare to each other. The use of a fixed current (mA) or even a current density (mA/cm$^2$) does not adequately reveal a battery's true capabilities. An expression that has come into use for establishing a common basis for discharge (or charge) rate between dissimilar batteries is the *C*-rate, derived from the ratio of the cell capacity *C* (Ah) to current *I* (A):

$$x = C/I$$

where *C*-rate is a relative rate expressed in units of hours (*x*), typically written as a multiple or fraction of *C*. When *x* < 1, such as would be typical in high-rate discharge or charge, the *C*-rate is expressed as *yC*, where

---

[a]Note that a battery's nominal *C*-rate capacity may be specified in different ways by various battery manufacturers, who should be consulted for further details or clarifications. Also, the International Electrotechnical Commission (IEC) Subcommittee SC-21A has published a "Guide to the Designation of Current in Alkaline Secondary Cell and Battery Standards" (IEC 61434).

$y = 1/x$. When $x > 1$ (lower rates), the $C$-rate is written either as a fraction ($C/x$) or as a decimal number (i.e., $0.2C$ for $C$-rate = $C/5$).

The $C$-rate has units of hours but is best viewed as a discharge rate. Specifically, $C$-rate specifies the current required for battery of a given capacity to be discharged in a specific time period. This ratio allows batteries of any design or capacity rating to be compared on more equivalent terms. The current required to achieve a given $C$-rate is calculated by dividing capacity by a specified $C$-rate.

For example, for a battery rated at 5 Ah, the specified current for a 10-h discharge, expressed as the $0.1C$ or $C/10$ $C$-rate, is 0.5 A. Conversely, a 250 mAh battery, discharged at 50 mA, has a calculated $C$-rate as follows:

$$C\text{-rate} = \frac{0.05}{0.250} = \frac{C}{5} \text{ (or } 0.2C)$$

Alternatively, the $C$-rate for a battery discharge at $C/10$ that is rated at 5 Ah capacity based on a 2-h discharge (typically designated as $C_n$, where $n$ = discharge time) may be expressed as follows:

$$0.1C_2$$

In this example, the $C/10$ rate is equal to 0.5 A, or 500 mA.

***Constant Power.*** The constant power discharge mode ($E$-rate) is analogous to the $C$-rate, and can be used to express the discharge or charge rate in terms of power:

$$E\text{-rate} = \text{Energy (Wh)/power (W)}$$

For example, the power level at the $0.5E_5$ or $E_5/2$ rate for a battery rated at 1200 mWh, at the $0.2E$ or $E/5$ rate, is 600 mW.

## 5.7    DISCHARGE PROFILES/REGIMES

Battery operation occurs within a complex environment of internal and external forces with numerous feedback loop interactions that convolute the analysis. The sections below suggest different potential confounding factors to consider in battery design and application.

### 5.7.1    Equipment Voltage Regulation

Most electrical equipment operates best within a fairly narrow voltage range, which in turn impacts the potential capacity obtainable from the battery. Low cutoff voltage and wide voltage range results in the highest delivered battery capacity. Similarly, the upper voltage limit of the equipment should be compatible with battery voltage characteristics.

Figure 5.6 compares a battery with a flat discharge with one having a sloping discharge. Using a lower voltage limit of 85% of nominal (–15%), a flat discharge profile (curve 1) gives longer service. But if lower cutoff voltages are acceptable, other batteries with sloping discharge profiles could be used, potentially resulting in greater equipment run-time (curve 2).

In other cases, such as with multicell series-connected batteries that are over-discharged, a

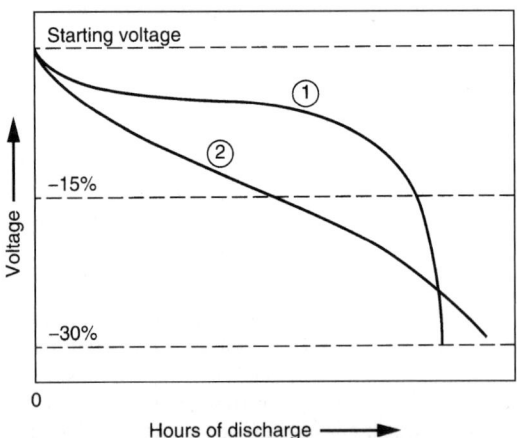

**FIGURE 5.6**   Comparison of batteries having either a flat ① or sloping ② discharge voltage profile.

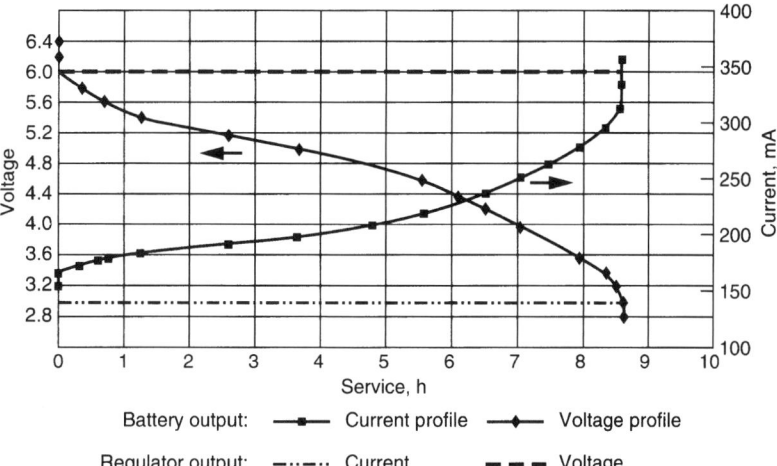

**FIGURE 5.7**   Voltage and current profiles for a battery and regulator. Battery output—1 W; regular output—840 mW.

low cut-off voltage may cause negative voltages in one or more cells, causing safety issues. With some cells, such as in a lithium-sulfur dioxide primary battery, this could result in venting or rupture.

In applications requiring a narrow voltage range, either cells with flat discharge profiles must be selected or else a voltage regulator will be required. Voltage regulators convert the fluctuating battery voltage to a constant output voltage, but with energy inefficiencies. Figure 5.7 shows voltage and current profiles of the battery and regulator at a constant battery output of 1 watt. At an 84% conversion efficiency, the output from the regulator is constant at a predetermined 6 V and 140 mA (constant power = 840 mW).

### 5.7.2   Discharge Mode

The discharge mode for a battery can have a significant effect on energy output for various reasons. Therefore, the mode of discharge used to test a battery should be similar to the intended battery application.

Specifically, while the energy (Wh) derived from a battery under different electrical load profiles may be constant, the extracted capacity (Ah) will vary dependent on the voltage levels during the discharge. The amount of energy delivered by a cell will require different amounts of battery capacity dependent on the voltage levels under the stipulated discharge regime. Three of the basic modes under which the battery may be discharged are as follows:

1. *Constant resistance.* The resistance of the load remains constant throughout the discharge (the current decreases during the discharge proportional to the decrease in the battery voltage).

2. *Constant current.* The current remains constant during the discharge.

3. *Constant power.* The current increases during the discharge as the battery voltage decreases, thus discharging the battery at constant power level (power $[W]$ = current $[I]$ × voltage $[V]$).
   The discharge profiles are illustrated under these three different conditions in Figs. 5.8 to 5.10.

   *Scenario 1: Uniform Initial Discharge Current*   In Fig. 5.8, at the start of the discharge, the current and power $(P)$ are the same for all three modes. For constant resistance discharge the current drops (Fig. 5.8$a$) proportional to the cell voltage decrease (Fig. 5.8$b$) according to Ohm's law:

$$I = V/R$$

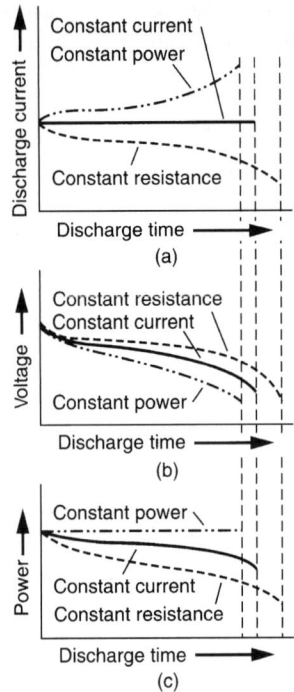

**FIGURE 5.8** Discharge profiles, using the same initial current and power at start of discharge, with constant resistance, constant current, and constant power discharge modes: (*a*) Current profiles during discharge. (*b*) Voltage profiles during discharge. (*c*) Power profiles during discharge.

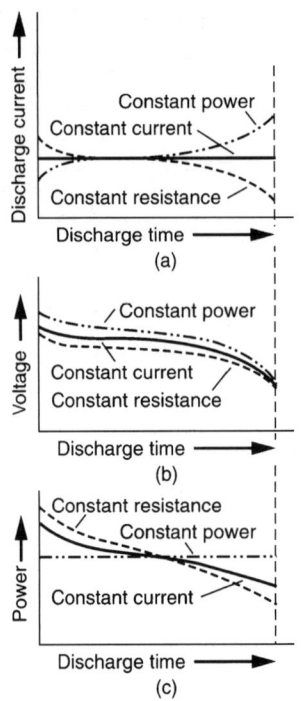

**FIGURE 5.9** Discharge profiles, maintaining the same discharge time, with constant resistance, constant current, and constant power discharge modes: (*a*) Current profiles during discharge. (*b*) Voltage profiles during discharge. (*c*) Power profiles during discharge.

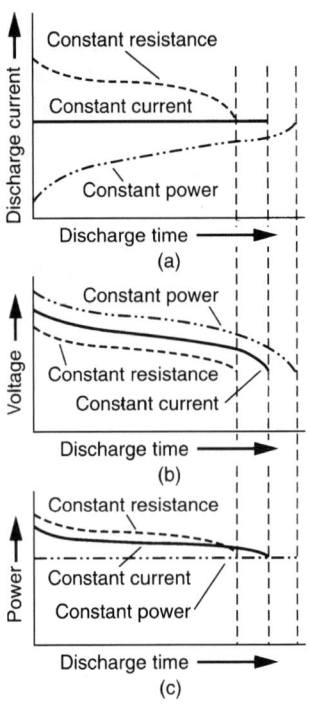

**FIGURE 5.10** Discharge profiles, providing the same output power at end of discharge, with constant resistance, constant current, and constant power discharge modes: (*a*) Current profile during discharge. (*b*) Voltage profile during discharge. (*c*) Power profile during discharge.

In the case of a constant-current discharge, the discharge time is reduced compared to constant-resistance due to higher average discharge. Finally, in the constant-power mode, current increases and voltage decreases even more dramatically than other modes according to the relationship.

$$I = P/V$$

Figure 5.8*c* is a plot of the power level for each mode of discharge.

*Scenario 2: Equal Discharge Time* Figure 5.9 shows relationships similar to scenario 1, but with discharge loads adjusted to provide the same hours of output (to a given end voltage) for all three modes of discharge. As expected, constant power draws less current initially and more current at end of discharge. Constance resistance discharge is the opposite.

*Scenario 3: Minimum End-of-Discharge Power* Many electrical circuits draw a set level of power. When the power draw drops below circuit requirements (typically at the cutoff voltage), the discharge terminates. In Fig. 5.10, the discharge loads are selected so that the final power output meets the minimum requirement for all three discharge modes. In the constant-resistance discharge mode, discharge current (Fig. 5.10*a*) mirrors battery voltage (Fig. 5.10*b*) with power deteriorating most rapidly, proportional to the square of the battery voltage (Fig. 5.10*c*). With this mode, the initial current and power draws are high, reducing capacity more rapidly.

For constant current circuitry, the current is held constant throughout the discharge at the minimum level required at end of life for acceptable equipment performance, thus offering lower average current drain than constant resistance loads.

Constant-power circuitry draws the lowest current at the beginning of the discharge with increases proportional to the decrease in battery voltage. Constant-power output is maintained at the minimal level required by the equipment throughout the entire discharge with the lowest average current drain and longest life of any mode.

The above generalizations will depend on the selected cutoff voltage, the useful voltage range for the specified electrochemistry, and the changes to internal resistance of the cell over the full discharge range. When voltage levels vary widely, the constant power discharge mode is most advantageous.

### 5.7.3    Battery Performance Case Study

To properly compare batteries, the testing mode should replicate the intended application as per Fig. 5.11.

In Fig. 5.11a, a typical AA-size primary cell is discharged using three different load conditions (constant resistance, constant current, and constant power) but with average discharge current manipulated to deliver the same discharge time to a fixed end voltage (1.0 V) similar to the testing shown in Fig. 5.9b, where a resistive load was used to approximate a constant-current or constant-power application. While the results are useful, the discharge current and power over the time interval (see Figs. 5.9a and c, respectively) show deviations for the different modes of discharge, meaning battery performance comparisons are somewhat convoluted.

Figure 5.11b shows results, using the same three discharge regimes, but for a cell with lower internal resistance and hence, a higher operating voltage. While the voltage level in Fig. 5.11b is higher than Fig. 5.11a for constant resistance discharge, the discharge capacities are about the same. However, under constant-current and constant-power discharge modes, the low resistance cell used in Fig. 5.11b delivers much greater capacity relative to constant resistance testing (i.e., the reverse of the results shown in Fig. 5.11a).

Two cells with similar capacity but having different internal characteristics show quite different performance results when tested under certain discharge modes. A cell with low internal resistance (Fig. 5.11b) shows improved capacity under constant-power discharge, while a constant-resistance discharge shows the least capacity, clearly illustrating a big difference from a high resistance cell (Fig. 5.11a), where all results are the same.

Differences in battery design and performance characteristics may contribute to flawed analyses when even a minor variation exists between the discharge testing mode and the actual application requirements. For example, Fig. 5.11c shows the discharge characteristics for a third type of cell with slightly higher capacity but also higher internal resistance than the cell shown in Fig. 5.11a. Despite minimal differences for the three modes of discharge down to the specified 1.0 V end voltage, the constant-power discharge shows a slight decrease in discharge time in Fig. 5.11c compared with Fig. 5.11a. Conversely, constant-current and constant-resistance discharge modes show a small but real capacity increase in Fig. 5.11c, despite the higher resistance of this third cell type.

Note that if these cells used lower cutoff voltage criteria, the run-time, at least for the cells tested in Figs. 5.11a and c, increases for the constant-resistance and constant-current discharge modes.

### 5.7.4    Intermittent Discharge

The battery discharge regimes detailed above were based on uninterrupted discharges under constant conditions. In actual practice batteries often have intermittent load profiles. A partially discharged battery may recover to some extent when placed on open circuit due to chemical and physical changes in the active materials. When placed back on load, a jagged discharge voltage profile results, as illustrated in Fig. 5.12, often providing increased battery output. Battery capacity often increases when the intermittent discharge occurs at higher current with longer recovery periods between discharges.

### 5.7.5    Varied Duty Cycle

Another common discharge regime for batteries includes applications where current varies, such as changing from "receive" to "transmit" in the operation of a radio transceiver. The service life of the battery is determined when the cutoff voltage is reached under the highest discharge load.

Another typical battery application requires a periodic high-rate pulse to be superimposed over a lower background current, such as backlighting for a liquid crystal diode (LCD) watch application, a smoke detector alarm signal, or computer disc engagement. A pulse discharge profile is plotted in Fig. 5.13. The voltage drop

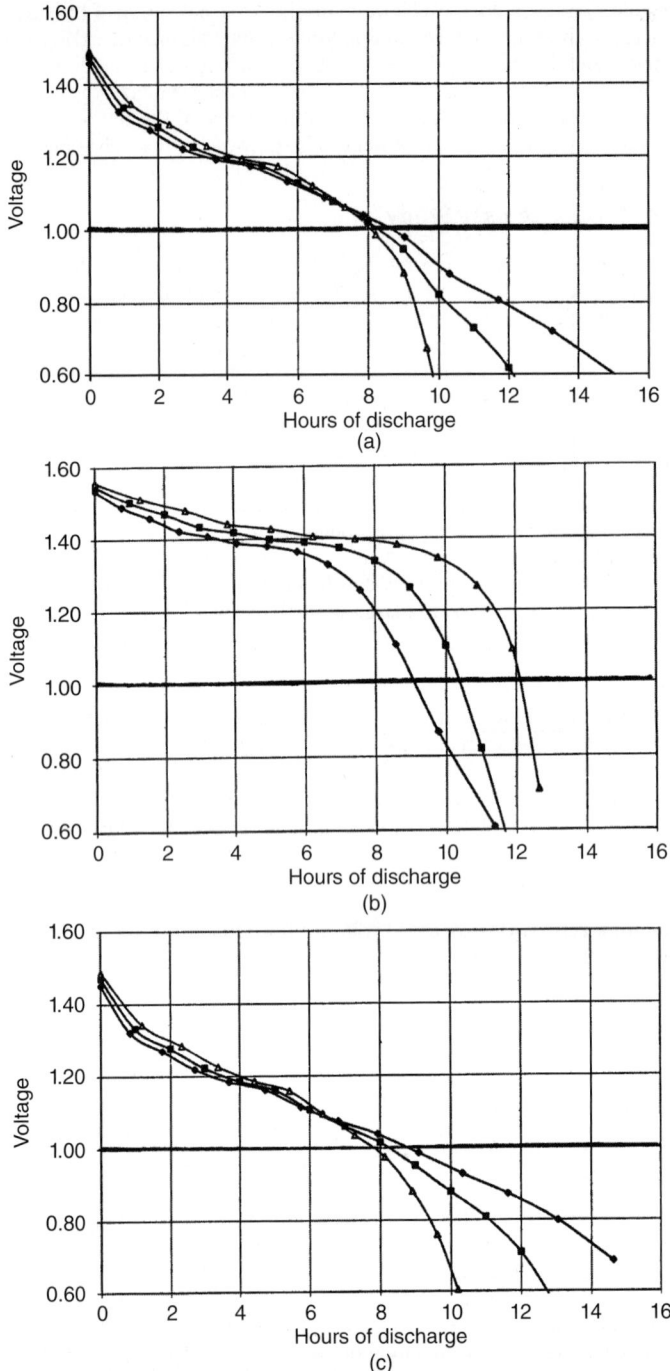

**FIGURE 5.11** Characteristics of a AA-size primary battery discharged under constant-resistance, constant-current, and constant-power conditions at 5.9 Ω, —•—; 200 mA, —▪—; and 235 mW, —▲—; where (*a*) is a typical cell, (*b*) is a low-impedance cell, and (*c*) is a high-capacity/high-resistance cell.

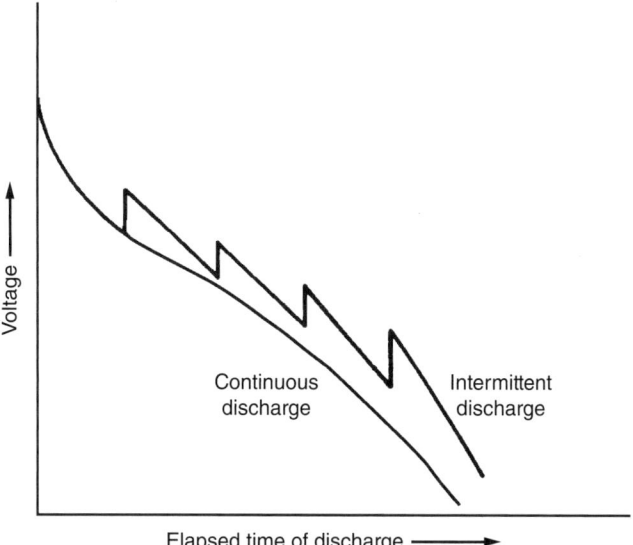

**FIGURE 5.12**    Effect of intermittent discharge on battery capacity.

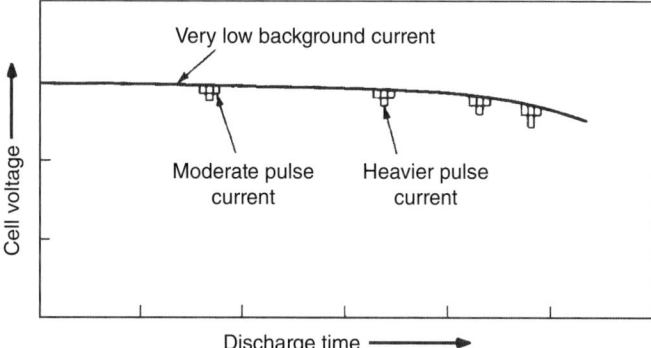

**FIGURE 5.13**    Typical discharge characteristics of a battery subjected to a periodic high-rate pulse.

during pulse load application will vary depending on battery internal resistance as well as cell and pack design. In Fig. 5.13, note that the voltage spread widens due to the increase in internal resistance as the battery is discharged.

The voltage profile of the pulse can vary significantly depending on the characteristics of the electrical circuitry and the battery response. Figure 5.14 shows the characteristics of 9 V primary batteries subjected to a 100-ms alarm pulse in a smoke detector. The curve in Fig. 5.14a shows the response of a zinc-carbon battery, the voltage dropping sharply initially and then recovering. Curves in Figs. 5.14b and 5.14c are voltage responses for two different zinc/alkaline/manganese dioxide batteries, where after an initial drop the voltage fades at varying rates under the pulse load.

The type of response shown in Fig. 5.14a occurs when chemical degradation products are present on an electrode surface. This layer is called a passivating film and often serves to protect the active material from further corrosion. However, until the surface film is chemically etched from the surface during subsequent discharge, the

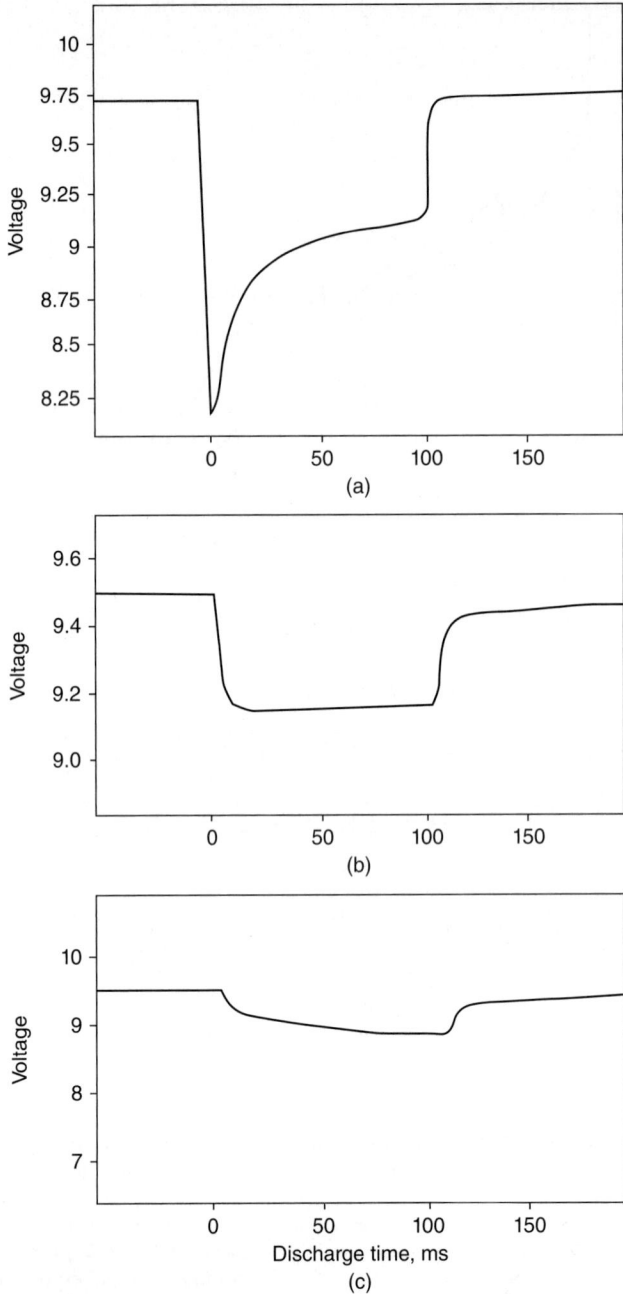

**FIGURE 5.14**   Discharge characteristics of a 9-V battery subjected to a 100-ms pulse (smoke detector pulse tests): (*a*) zinc-carbon battery, (*b*) and (*c*) zinc/alkaline/manganese dioxide battery.

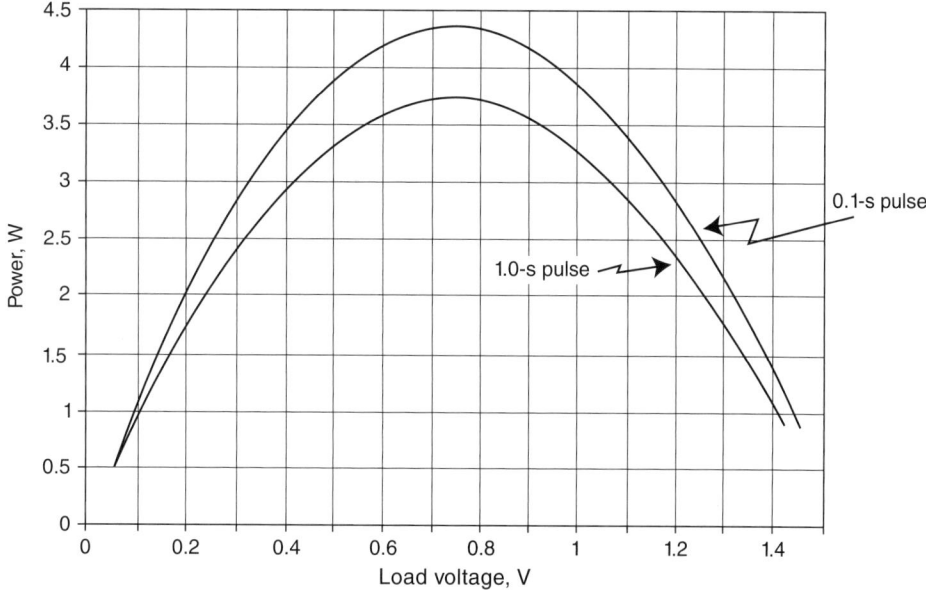

**FIGURE 5.15**    Peak power test (final output power at the end of 0.1- and 1.0-s constant voltage pulses) (undischarged zinc/alkaline/manganese dioxide AA-size battery. (*From Ref. 3.*)

voltage will drop significantly before recovering as fresh surfaces are exposed (see Sec. 5.11 on voltage delay). Passivation is a complex phenomenon but correlates directly to the degree of cell impedance (see Chap. 4).

The pulse capability of a cell is best measured by determining peak power. Peak power occurs when, ideally, a nonpassivated cell is pulsed at a current that yields a load voltage equal to one-half of the open circuit voltage (i.e., when the resistance of the external circuit is equal to the internal resistance of the battery). Peak power is found by plotting the output power against the load voltage using a series of progressively greater short-term pulses.[4] Figure 5.15 shows plots of the power output at the end of 0.1- and 1.0-s pulses over the full range of load voltages for a fresh zinc/alkaline/manganese dioxide AA cell.

## 5.8    TEMPERATURE EFFECTS

Battery temperature will greatly affect capacity retention, output rate, voltage, and capacity and recharge characteristics for secondary cells. Reduced chemical activity and increased internal resistance dominate at lower temperatures as shown in Fig. 5.16. Cells under constant current discharge, but at progressively decreasing temperatures from warmest ($T_4$, approximating normal room temperatures) to coldest ($T_1$), will have reduced capacity and progressively lower load voltage (i.e., greater slope of the discharge curve). Generally, for most commercial electrochemical cells, best performance falls in a range of 20 to 40°C. Moderately elevated temperatures often provide improved rate, voltage, and capacity. However, accelerated chemical activity may also degrade the cell even while under test (a phenomenon, known as *self-discharge*), causing a loss of capacity under load.

Figure 5.17 shows the effect of temperature and discharge rate on the battery capacity. In this figure, fresh cells stored at different temperatures ($T_1$ to $T_6$) were discharged using a wide range of load currents. Plots of the resulting measured Ahs show greater cell capacity losses at colder temperatures and with increasing discharge load. Note that voltage levels are also depressed under these conditions. However, anomalies may occur in some tests, such as shown with the highest temperature cell ($T_6$). Curve $T_6$ shows unexpected significant capacity loss at high temperature and low-rate discharge. However, this negative result, due to self-discharge, is negated and capacity again increases at higher discharge rates as a result of reduced cumulative corrosion effects as well as enhanced reactivity and self-heating of the cell.

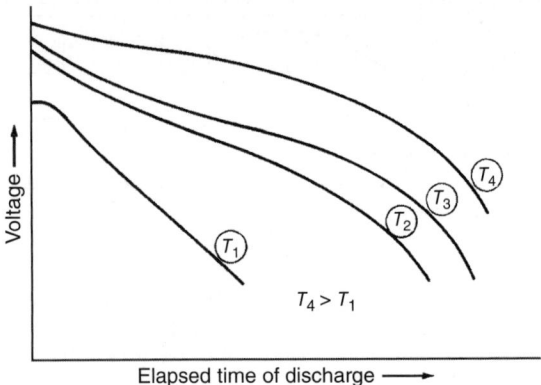

**FIGURE 5.16** Effect of temperature on battery capacity. $T_1$ to $T_4$—increasing temperatures.

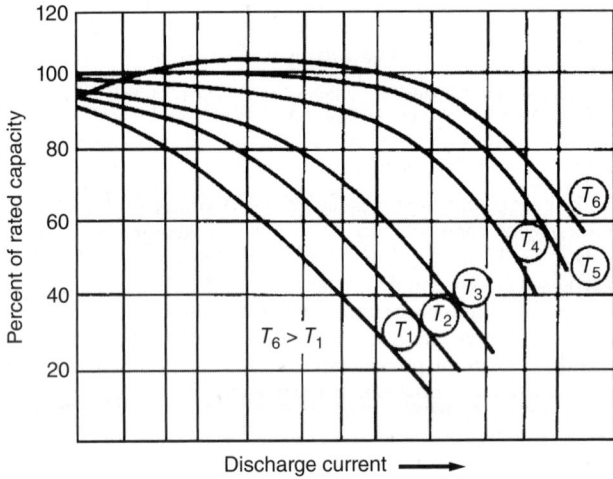

**FIGURE 5.17** Effect of discharge load on battery capacity at various temperatures. $T_1$ to $T_6$—increasing temperatures; $T_4$—normal room temperature.

## 5.9  SERVICE LIFE

Various graphical depictions are commonly used to summarize a battery's general performance parameters. Figure 5.18 compares battery run-time as a function of discharge current at different temperatures.

Ideally, these plots are logarithmically linear, but performance will be somewhat degraded due to internal resistance increases at high rate and self-discharge at the low rate. Peukert's equation is used to estimate the performance under various hypothetical conditions as follows:

$$I(\exp n) \times t = C$$

$$\text{or } n \log I + \log t = \log C$$

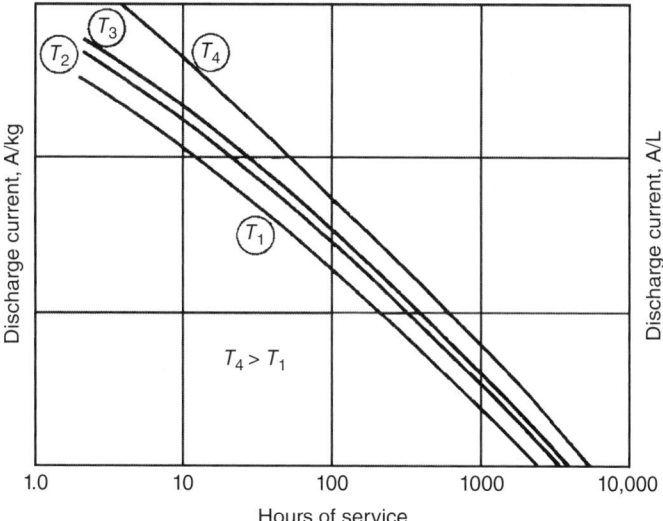

**FIGURE 5.18** Battery service life at various discharge loads and temperatures (log-log scale). $T_1$ to $T_4$—increasing temperature.

where $I$ = discharge rate, $t$ = discharge time, and $C$ = constant. The slope of the linear portion of the plot is used to determine the value of $n$. A case study example is presented in Chap. 13. Other references are also available detailing various mathematical relationships for describing battery performance, including any nonlinearity.[5]

The Ragone plot, another common graph used in battery analysis, correlates specific energy (or energy density) of a battery system with specific power (or power density). This graph is very effective in showing the impact of discharge load (or power) on the delivered energy. Chap. 23, detailing battery selection methodologies, shows Ragone plots of specific power (W/kg) versus specific energy (Wh/kg) for various primary and secondary batteries.

## 5.10 BATTERY CHARGING

Battery charging, discussed further in Parts 3 to 5, may be based on a variety of methodologies. For instance, a battery used as a standby power source is typically connected into the operating circuit in parallel with a second electrical energy source. Specified currents or set voltages may be used in such cases to supply both the equipment load requirements as well as to recharge and maintain the full capacity of the battery. In other cases, a battery charger may simply be used to quickly recharge full capacity at which time the battery is disconnected and placed on open circuit.

Generally, a primary battery, used within any piece of electrical equipment containing a charger, should never be charged and should also be electrically isolated from the charging circuit. For example, memory backup batteries are usually protected from charging by the use of diodes (or current limiting resistors). The life and safety of a secondary cell depend significantly on the design of the charging circuit. Proper designs, reliable components, and rigorous testing are needed for most chargers (cell phones, grid power systems, and electric vehicles, to name a few).

## 5.11 BATTERY STORAGE

Batteries deteriorate upon storage depending on cell design, the electrochemical system, temperature, contaminants, leakage currents, storage time, etc. The shelf or service life of a battery may be impacted by internal resistance increases, self-discharge, or reduced cell voltage. Chapter 23 presents storage data for several battery systems at various temperatures. Low-temperature storage often, but not always, extends the shelf life. However, stored batteries are warmed before discharge to obtain maximum performance.

The development of a passivating film on electrode surfaces during storage often dramatically improves the shelf life. However, upon subsequent discharge, the voltage may be severely depressed until the electrodes become depassivated (see Sec. 5.7.5). This effect, known as voltage delay, is detailed in Fig. 5.19. The extent of the voltage delay is dependent on and increases with greater storage time and temperature. The delay also worsens with higher discharge current and lower discharge temperature.

Self-discharge is not a fixed or easily characterized quantity in many cases. Factors such as the prior cell discharge/storage history, discharge rate and temperature, discharge product deposition/accumulation, the depth of discharge, etc., may all impact storage capacity losses. Even the partial destruction or reformation of the protective film, due to various test conditions, may hurt or help storage capacity retention. Ideally, cells will be tested under the intended application conditions and any deviations in specified usage or recommended storage parameters are known and accommodated.

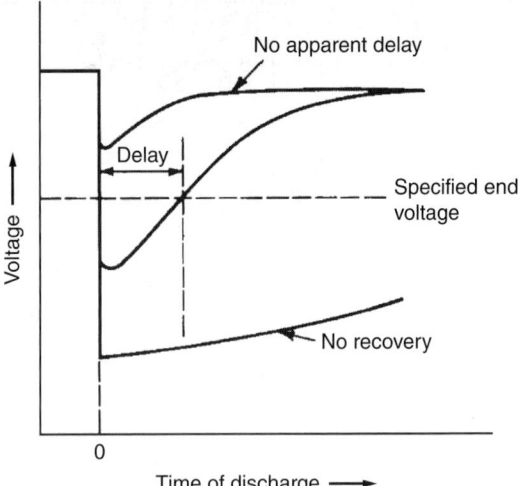

**FIGURE 5.19**  Voltage delay response after long-term storage.

# REFERENCES

1. M. Winter and B. Brodd, *Chem. Revs.* **104**:4245–4270 (2004).

2. D. Linden and T. B. Reddy, *Battery Power Products Technol.* **5**(2) (March/April 2008).

3. P. Ruetschi, "Alkaline Electrolyte—Lithium Miniature Primary Batteries," *J. Power Sources* **7**(2):165–180 (1982).

4. D. I. Pomerantz, "The Characterization of High Rate Batteries," *IEEE Trans. Electron.* **36**(4):954 (1990).

5. R. Selim and P. Bro, "Performance Domain Analysis of Primary Batteries," *Electrochem. Technol. J. Electrochem. Soc.* **118**(5):829 (1971).

# CHAPTER 6
# MATHEMATICAL MODELING OF BATTERIES

**Shriram Santhanagopalan, Ralph E. White**

## 6.1 INTRODUCTION

Mathematical modeling of batteries can be described as a process of developing an equation or a set of equations to describe the performance of a battery. For example, a simple aggregate model to predict the capacity of a cell as a function of the discharge current can be obtained by regressing coefficients on a single equation using experimental data collected from that cell. More complicated models can be developed based on equations used to describe cell components such as electrodes, separator, and electrolyte. Such a model can, for instance, account for the spatial distribution of current or potential drop across different parts of the cell. This model can then be used to explain the performance of the cell at different C-rates based on limitations to the charge-carrying ability of the individual cell components. Both the lumped model and the detailed model can be extended to capture the response of a battery with multiple electrode pairs by accounting for the electrical connections between the electrodes and external connections of the cells to the busbar in different series or parallel configurations, as needed for the voltage and capacity requirements.

The level of detail in a mathematical model depends on its intended use. For example, detailed three-dimensional models (three spatial coordinates and time) capturing the geometry of individual particles within multiple electrode pairs have been developed to study the thermal characteristics of localized failure in electrode pairs and cells. Some historical perspectives and various mathematical battery models (i.e., empirical and mechanistic) are discussed below, followed by detailed examples and case studies, including a porous electrode model, and thermal and degradation models.

### 6.1.1 Evolution of Battery Models

The earliest mathematical models for batteries were simply empirical relationships between measured parameters, such as the battery voltage, overall resistance, density of the electrolyte, pressure within the can or temperature of the cell, versus the remaining capacity under different operating conditions. These models are still used today, and perhaps the best-known example is Peukert's relationship[1] (see Chap. 5). This equation has been used to represent the discharge capacity as a function of discharge current for a lead-acid cell, for example, as shown in Fig. 6.1, which shows a comparison between the capacity predicted by Peukert's equation and the experimentally measured value for the cell capacity. This simple relationship has been used under a variety of different scenarios for many decades in the battery industry.

A second empirical example used to monitor the state-of-charge (SOC) of a battery is shown in Fig. 6.2, where the available lead-acid battery capacity is shown as a function of the specific gravity of the electrolyte during discharge at the C/2-rate. In this case, a simple correlation exists between the available capacity of the

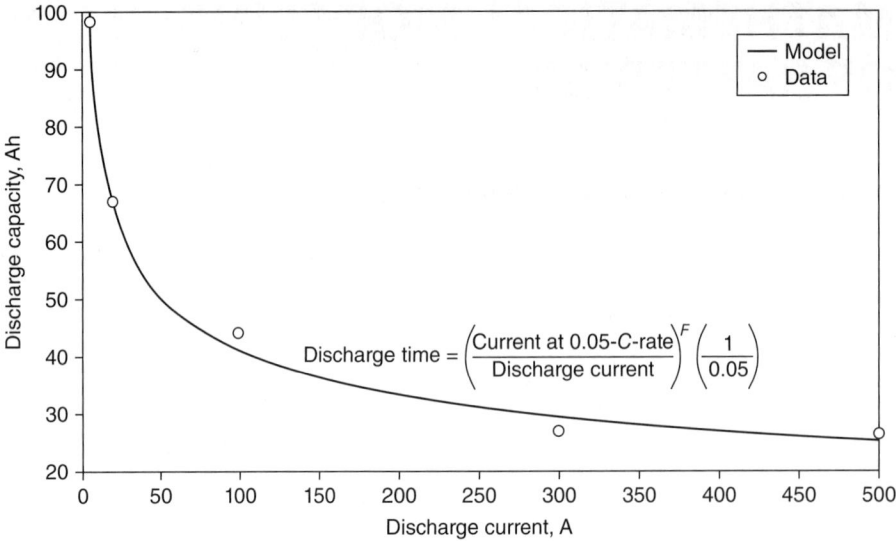

**FIGURE 6.1** Cell capacity versus discharge rate for a 100 Ah lead-acid battery following Peukert's equation with current at 0.05-C-rate equal to 4.98 A and the Peukert coefficient ($F$) set to 1.3.

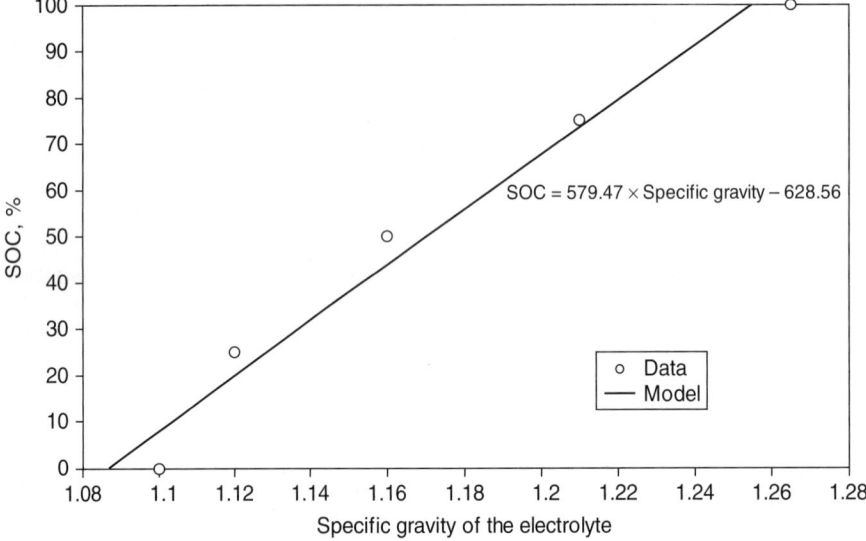

**FIGURE 6.2** Linear relationship between the specific gravity of the electrolyte and the state-of-charge (SOC) of a lead-acid battery.

cell (SOC, %) and electrolyte molarity. Details on this experimentally determined relationship between the electrolyte concentration and battery performance are discussed further in Sec. 6.4.2.

Due to the predictive success of Peukert's equation and other experimental analyses (i.e., causes for cell capacity losses), other modeling studies were initiated. One early step toward understanding limitations in battery performance was quantifying the loss in efficiency of the battery relative to theoretical expectations and then being able to attribute such losses to constraints across different parts of the cells. The limitations within the cell were categorized by root cause: electrodes, electrolyte, grids, and so on.

The use of AC impedance (see Chap. 4) as a diagnostic tool for electrochemical systems emerged in the 1970s[2,3] and with that evolved the concept of representing the battery as a circuit consisting of traditional electrical components, such as resistors and capacitors, as shown in Fig. 6.3.

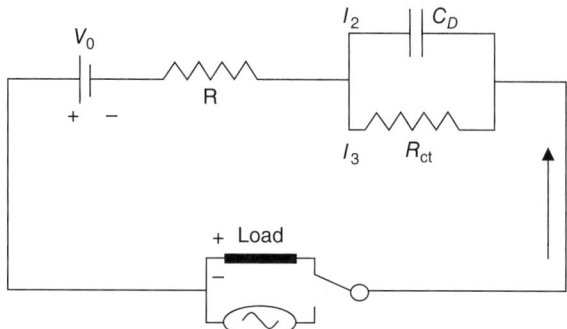

**FIGURE 6.3**    Equivalent circuit representation of a cell.

Figure 6.3 depicts a typical circuit diagram used for a single cell.[4] Battery performance is determined by specifying values for the circuit elements shown in Fig. 6.3. The voltage source $V_0$ represents the open circuit voltage (OCV) of the battery, which establishes the thermodynamic limitations on cell performance. Resistance $R$ refers to ohmic losses inside the battery that arise from current passing through the electrolyte, contact resistances, and the like. The two parameters $R_{ct}$ and $C_D$ together represent charge transport across the interface: $R_{ct}$ refers to the Faradaic part of the charge transfer resistance, and $C_D$ is a pseudocapacitance term often used to characterize mass transfer limitations. These parameters then characterize the ideal behavior of the battery as $V_0$ changes during discharge. The other parameters shown in Fig. 6.3 can be assigned different values, based on experimental data such as the voltage curves from various discharge rates of the battery. Initial deviations from the OCV are due to limitations in the transport of ions, ohmic effects, reaction kinetics, etc. Nonideal responses are reflected in the parameter values. For example, inefficiencies in the electrochemical reaction rates result in changes in the value of $R_{ct}$.

The circuit analog model of a battery fostered investigation of mechanistic approaches. Specifically, each change observed in cell performance was classified as a subprocess with a different time constant. Circuit analog models are still used in the battery industry because the relationship between the battery voltage and load current can often be expressed as a simplified analytical expression. The computational requirements for such models are minimal, making them suitable to test in hardware and to estimate the SOC rapidly. However, the circuit elements used in circuit-based models do not readily translate into physically meaningful parameters that greatly aid battery design.

Linking together of battery design parameters, such as electrode coating thickness to observed experimental behavior, has led to the formulation of physics-based models that employ universal laws (e.g., charge, mass, momentum, and energy balances) and to characterization of the behavior of individual components within a cell as functions of their material properties. These component models can then be integrated to form rigorous mechanistic models to describe the behavior of the cell. Composite electrode structures, spanning geometries of a few microns to several centimeters, have been used to develop mathematical models based on readily measurable physical properties, such as effective conductivities and diffusivities. Design parameters, such as electrode geometry, tab location, and integration of battery electronic circuitry with other components in a device, now typically rely on using sophisticated computer-aided design (CAD) tools and physics-based models.[5] Despite past difficulties, development of complex and mathematically rigorous models is now aided by the advent of user-friendly interfaces on commercial software such as MATLAB™, ANSYS Fluent™, Siemens Battery Design Studio, and COMSOL Multiphysics™.

Thus, battery models have significantly evolved from empirical relationships and rules of thumb. However, the principal objective remains the same: to predict battery output (in terms of energy or power) under the specified load conditions over the required time. The following sections describe the process of developing a mathematical model for a cell, the choice of model equations and parameters, and some examples of implementing such models for optimal design of batteries.

## 6.2   MATHEMATICAL MODELING

Developing a mathematical model for a battery involves identifying the physical processes that take place and how materials within each component respond during the operation of the battery. A systematic approach for modeling the response of cell components to the various processes involves the use of generalized laws that describe the behavior of the materials under different scenarios. The simplest example is the representation of current flow through a copper wire: When current flows through a battery, the busbars or tabs that connect the electrodes to the external load (or power source) experience a voltage drop. One physical process that occurs within the copper conductor is heating of the cable—especially at any welded joints—which increases with an increase in the current. Having established various physical processes, the next step is to identify general rules that quantify various observed phenomena. For example, current flow follows Ohm's law, which states that the potential $V$ across a metallic cable depends on the amount of current $I$ that flows and the metal's resistance to current flow $R$:

$$V = IR \tag{6.1}$$

Equation (6.1) shows that higher resistance increases the potential drop required to conduct a given amount of current, confirming the observation that corroded terminals (having lower electronic conductivity) reduce cell discharge voltage more than cells with unoxidized electrical contacts. A second phenomenon, heating of the welded connections, was first quantified by Joule using the following relationship:

$$\Delta H = I^2 R t \tag{6.2}$$

where $\Delta H$ is the heat generated, $I$ is current, $R$ is the resistance at the weld, and $t$ is the duration of the current flow. Equations (6.1) and (6.2) can be used to describe adequately some physical observations during the passage of current through a busbar and these equations constitute the mathematical model in this example. As a result weld materials can be chosen based on the calculated amount of heat that will be generated for a given operating condition. Alternatively, one can determine the maximum amount of current that can be passed safely without damaging the welds. As long as one can measure the conductivity of any given cable material experimentally and independently prove that each of the above laws holds true under the operating conditions of interest, the performance can be modeled with these equations. For example, by specifying the operating conditions (i.e., the values for $I$ and $t$, the discharge rate and the duration of discharge, respectively), $V$ and $\Delta H$ (voltage and heat generated) can be measured for use in modeling. Intrinsic material properties, such as the conductivity of metals and other similar parameters, on the other hand, do not vary with the cell design or operating condition. Thus, every mathematical model will include input variables, measured variables, and physical parameters.

Next, a mathematical model for a battery will be modified to include component models for other physical phenomena occurring during battery operation. The complexity of a mathematical model depends on the number of processes and the desired level of detail for the physical phenomena. For example, if heating of the weld-joints is minimal, the flow of current across the busbar can be modeled using Eq. (6.1) only. On the other hand, if the resistance of the cable changes with temperature, additional equations must be used that describe the change in the parameter, $R$, shown in Eq. (6.2). An efficient model must balance the complexity associated with describing the data set and the analytical sophistication of the results. The following sections describe complementary approaches used to develop such models. Finally, the various components are integrated to study the interaction among the different parts of the battery in response to the mutually interactive physical phenomena of interest. Based on the degree of comprehension the models provide, they are usually classified as empirical or mechanistic models.

### 6.2.1    Empirical Models

An empirical model can be developed by assuming the form of an expression that relates operating conditions, such as the rate of discharge or the load across the battery, with measured quantities, such as the temperature or voltage of the battery. Prior knowledge of the battery's behavior or trial and error is used to derive such expressions. However, the various limitations within the cell may also provide guidelines. For example, the equivalent circuit shown in Fig. 6.3 accounted for deviation of the cell voltage ($V$) from the OCV ($V_0$) as described in Sec. 6.1. One common objective for battery models is to predict the cell voltage as a function of the SOC of the battery as demonstrated below for the circuit shown in Fig. 6.3.

The voltage drop across each of the resistors $R$ and $R_{ct}$ follows Ohm's law [see Eq. (6.1)]. The rate of charge buildup in capacitor $C_D$ equals the current that flows through the capacitor. This is expressed mathematically as follows:

$$I_2 = \frac{dq}{dt} \tag{6.3}$$

Kirchoff's node-and-loop rules relate the currents that flow across the different branches of the circuit and the voltage across each branch. These rules state that the voltage across any branch is the sum of all voltage drops along the branch and that the sum of all currents that enter or exit a node on the circuit is zero. For example, the total current, $I$, branches out into $I_2$ and $I_3$ as shown in Fig. 6.3. Hence, according to Kirchoff's laws, we have the following equation:

$$I = I_2 + I_3 \tag{6.4}$$

The constraint on the voltages across each branch yields the following equations:

$$V = V_0 + IR + I_3 R_{ct} \tag{6.5}$$

$$V = V_0 + IR + \frac{q}{C_D} \tag{6.6}$$

Equations (6.3) through (6.6) above can be rearranged as follows to obtain the relationship between the change in the applied current with time, $dI/dt$, and the resultant voltage drop ($V - V_0$):

$$R\frac{dI}{dt} + \frac{1}{C_D}\left(1 + \frac{R}{R_{ct}}\right)I = \frac{dV}{dt} + \frac{1}{R_{ct}C_D}(V - V_0) \tag{6.7}$$

Equation (6.7) now includes the component models [Eqs. (6.3) through (6.6)] for each element in the cell model. For the case of constant current, the solution of Eq. (6.7) takes the following form[4]:

$$V = \frac{Q_0}{C_D}e^{-t/R_{ct}C} + V_0 + IR + IR_{ct}(1 - e^{-t/R_{ct}C_D}) \tag{6.8}$$

This model equation relates the change in cell voltage, $V$, to the input current, $I$. The parameter $Q_0$ refers to the total capacity of the battery. The change in cell capacity during charge or discharge is calculated by integrating the current passed as follows:

$$Q = Q_0 - \int_0^t I\,dt \tag{6.9}$$

The values for the circuit elements such as $V_0$, $C_D$, $R$, and $R_{ct}$ are adjusted to represent relevant experimental data. Figure 6.4 shows cell voltage versus capacity during charge and discharge at various rates. Some experimental data are also shown for comparison. The results for a wide range of $C$-rates are obtained by utilizing the circuit parameters extracted using data at one discharge (or charge) rate.

Similar results can be obtained for constant power loads by replacing the current ($I$) in Eq. (6.7) with $P/V$, where $P$ is the power drawn by the load. Figure 6.5 shows a comparison of modeling data, using the same set of equations described above, with experimental results for the parameter set shown in Table 6.1 for a lithium-ion cell.

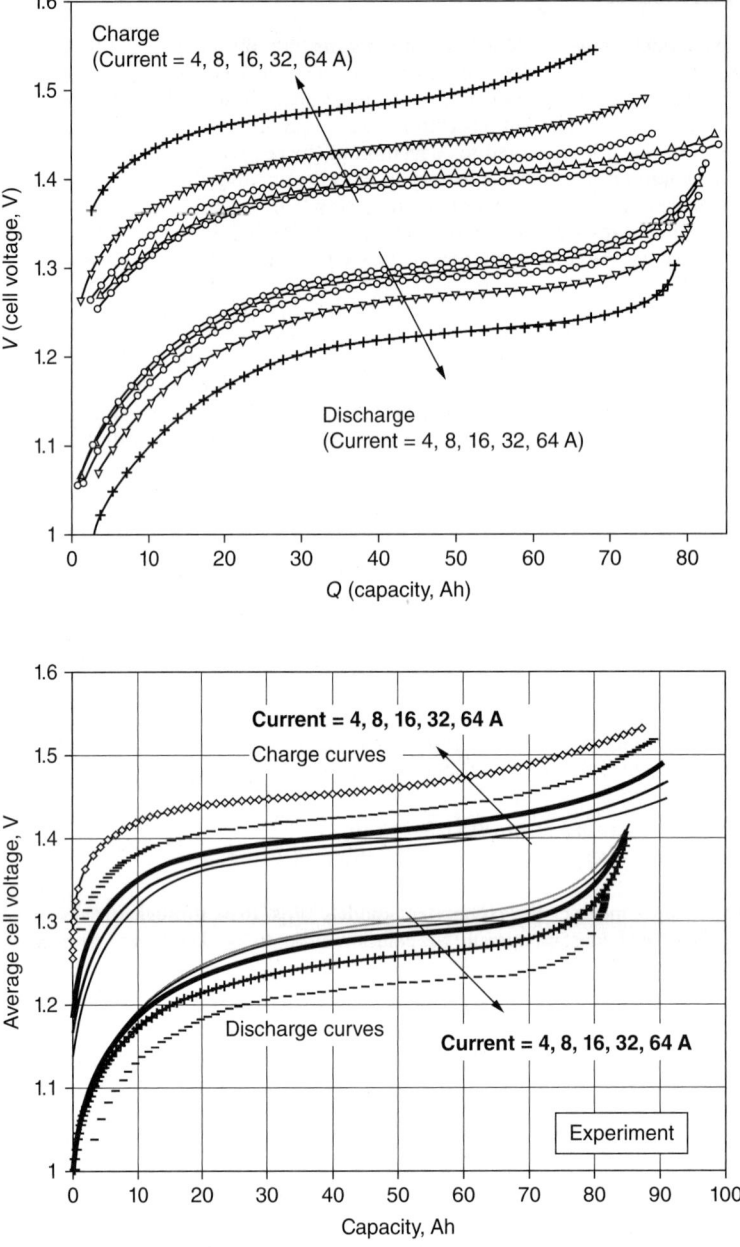

**FIGURE 6.4** Model predictions for a NiMH cell. Results from a model using the empirical equivalent circuit shown in Fig. 6.3 are shown in the upper plot and the corresponding experimental data on the lower plot.[4]

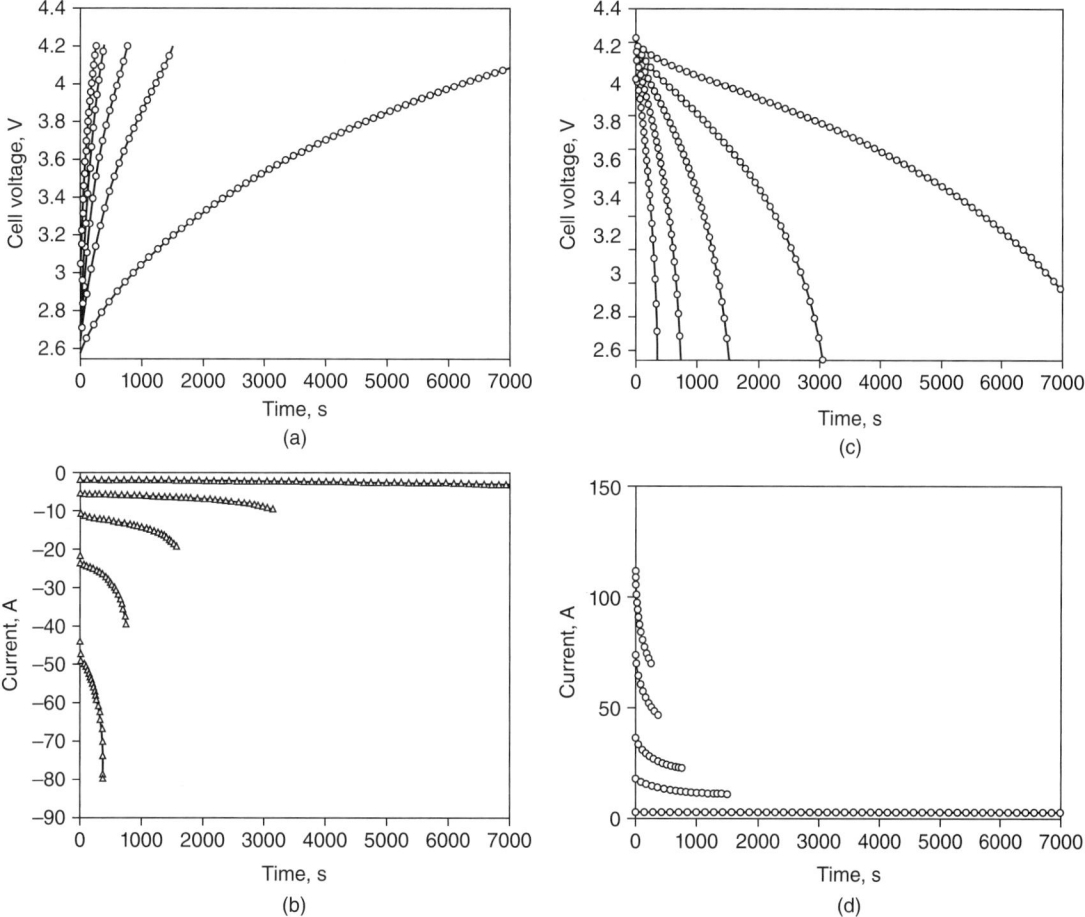

**FIGURE 6.5**  Model predictions versus experimental data[6] from the equivalent circuit model under constant power discharge for a lithium-ion cell at 0°C. (*a*) Cell voltage during charge at constant power; the power supplied was set to 10, 25, 50, 100, 200 W, respectively, for each curve shown in the figure. (*b*) The charge current corresponding to the cell voltages shown in (*a*); since the voltage rises rapidly at higher rates of charge, the current tapers at the end of charge in order to maintain constant power during the entire process. (*c*) Cell voltage during discharge at constant power; the power drawn from the cell was set to the same values as described in (*a*). (*d*) Cell current during discharge, corresponding to (*c*). The symbols represent the experimental data and the solid lines the model predictions.

**TABLE 6.1**  Parameters[6] Used in the Equivalent Circuit Model to Predict the Response of a Lithium-Ion Cell as Shown in Fig. 6.5

| Parameter | Discharge | Charge |
|---|---|---|
| $\tau$, s | 5 | 5 |
| $C_D$, F | 12,500 | 16,667 |
| $R$, m$\Omega$ | 1.637 | 1.637 |
| $R_{ct}$, m$\Omega$ | 0.4 | 0.3 |

## 6.2.2    Mechanistic Models

Mechanistic models relate the battery characteristics to physical properties of the constituent materials. Such properties are usually measurable in independent experiments. For example, a mechanistic model for the Ohm's law equation shown in Eq. (6.1) can be built by describing the resistance parameter $R$ in terms of physically measurable properties of the copper busbar, namely, the electronic conductivity of the metal ($\sigma_c$), the cross-sectional area ($A_c$), and the length of the busbar ($L$). Each of these properties is characteristic of a busbar. The resistance $R$ is related to these parameters as follows:

$$R = \frac{L}{\sigma_c A_c} \tag{6.10}$$

and hence Eq. (6.1) can be rewritten as:

$$V = I \frac{L}{\sigma_c A_c} \tag{6.11}$$

Note that Eq. (6.11) can be used for a cable of any given dimension, made up of any material whose conductivity is known. The use of Eq. (6.1), however, requires that we measure the resistance parameter $R$ every time the busbar is replaced.

We now proceed to developing mechanistic models to describe the other physical processes that take place within the battery. A few common electrochemical battery processes include movement of ions in the electrolyte, movement of electrons within the electrodes, and chemical and electrochemical reactions. See Chaps. 1 and 4 for background information on the basic equations governing each of these processes. In this section, we employ these concepts to build a mechanistic model for a battery.

***Charge Transport by Electrons.***    The total voltage $V$ across a cell can be approximated as the sum of the potential drops across electrodes and electrolyte, and other losses arising from contact resistances. In the following sections, subscript 1 will be used to denote the properties/variables in the electrodes and subscript 2 to represent the corresponding variables in the electrolyte. We already considered voltage drop due to the flow of electrons across metal cables in Eq. (6.11) above. The differential voltage drop across the electrodes ($\nabla\phi_{1,j}$) is also governed by Ohm's law:

$$\nabla\phi_{1,j} = -\frac{\mathbf{i}_1}{\sigma_j^{\text{eff}}}, \quad j = n \text{ or } p \tag{6.12}$$

where $\mathbf{i}_1$ is a vector called the current density (current per unit area) and $\sigma_j^{\text{eff}}$ is the effective electronic conductivity of the electrode material within electrode $j$ ($j = n$ for the negative electrode or the anode and $j = p$ for the positive electrode or the cathode). Usually a battery electrode is composed of several components such as solid solutions of different metals, or composed of active material, binders, and other components. The effective conductivity term is used to account for the presence of additional electrode components besides the active material and is calculated as the sum of the conductivities of individual components, scaled in proportion to the composition of the electrode:

$$\sigma_j^{\text{eff}} = \sum_k w_k \sigma_k \tag{6.13}$$

where $w_k$ is the proportion (e.g., mass fraction) of the individual components $k$ that constitute the electrode, and $\sigma_k$ refers to the electronic conductivity of the pure component $k$. Alternatively, $\sigma_j^{\text{eff}}$ can be measured directly after the electrode is assembled.

***Charge Transport by Ions.***    A unique feature of electrochemical devices is the transport of charge by ions. Once the charge moves past the electrodes and undergoes an electrochemical reaction, charge transport from one electrode to another is facilitated by ions. This transport of charge by the movement of ions is more complicated than the current-carrying mechanism involving movement of electrons. Usually, there are several species of ions present in the electrolyte. The total current density in the electrolyte ($i_2$) is the sum of the current densities carried by species $k$:

$$\mathbf{i}_2 = \sum_k i_k \tag{6.14}$$

The current density carried by species $k$, $i_k$, is proportional to its flux $N_k$[7]:

$$\mathbf{i}_k = F \sum_k \mathbf{N}_k \tag{6.15}$$

The proportionality factor in Eq. (6.15) is the Faraday's constant, which is the amount of charge carried by each mole of ions. The flux of ion $k$ is defined as the product of the number of $k$ ions per unit volume of the electrolyte (i.e., the concentration of species $k$) and the velocity of each ion:

$$\mathbf{N}_k = c_k \mathbf{v}_k \tag{6.16}$$

The concentration of the electrolyte is a readily measurable quantity; the velocity of an ion is proportional to the charge it carries ($z_k$) and the potential gradient in the solution phase ($\nabla \phi_2$), which is the electrical driving force for ion movement:

$$\mathbf{v}_k = -u_k F z_k \nabla \phi_2 \tag{6.17}$$

The proportionality constant $u_k$ in Eq. (6.17) represents the mobility of the ion and is obtained from equivalent conductance measurements. The negative sign indicates that the ions move from a region of higher potential to lower. Equations (6.14) to (6.17) can be rearranged to obtain[7]:

$$\mathbf{i}_2 = -\left( F^2 \sum_k c_k u_k z_k \right) \nabla \phi_2 \tag{6.18}$$

Equation (6.18) closely resembles Ohm's law, and the electrical conductivity for the electrolyte ($\kappa$) is now given by:

$$\kappa = \left( F^2 \sum_k c_k u_k z_k \right) \tag{6.19}$$

As demonstrated in Eq. (6.13), Eq. (6.19) relates the properties of the component ions to the conductivity of the electrolyte. Hence, knowing the composition of the electrolyte, one can model the movement of ions in the electrolyte. Alternatively, one can experimentally measure the electrical conductivity ($\kappa$) in Eq. (6.19).

In deriving Eq. (6.18), an implicit assumption that the concentration of the electrolyte was uniform precluded the effects of concentration gradients present within the cell. However, this assumption can be easily relaxed by incorporating flux terms arising from concentration differences using Fick's laws of diffusion. Equation (6.16) then becomes:

$$\mathbf{N}_k = c_k \mathbf{v}_k - D_k \nabla c_k \tag{6.20}$$

where $D_k$ is the diffusion coefficient of ion $k$. In the case of flow batteries (see Chap. 22B), there is a convective velocity in addition to $\mathbf{v}_k$ in Eq. (6.20). Thus, the modified flux is now given by:

$$\mathbf{N}_k = c_k (\mathbf{v}_k + \mathbf{v}) - D_k \nabla c_k \tag{6.21}$$

Here $\mathbf{v}$ is the velocity of electrolyte flow. Combining Eqs. (6.17) and (6.21) gives[7]:

$$\mathbf{N}_k = -z_k u_k F c_k \nabla \phi - D_k \nabla c_k + c_k \mathbf{v} \tag{6.22}$$

Equation (6.22) represents the case of dilute electrolytic solutions. More sophisticated models that consider the mutual interaction of ions within the electrolyte and the effects of temperature on the conductivity of the electrolyte are also available.[8]

***Driving Forces for Charge Transfer across the Interface.***    Storing charge in batteries requires conversion of chemical energy into electrical energy or vice versa. Faraday's law dictates the maximum amount of charge generated for a given amount of active material. When no net current flows across the plates of the battery, the driving force for charge transfer is referred to as the equilibrium potential ($E^0$) and is related to the free energy of the system by Faraday's law[9]:

$$E^0 = -\frac{\Delta G}{nF}$$ (6.23)

The negative sign implies that the free energy is reduced when the battery is discharged. In practice, the generation of electrical energy from chemicals depends on the temperature and the concentration of the chemical species taking part in the reactions generating the electrical energy. The OCV ($E$) under typical operating conditions is modeled as the equilibrium value $E^0$ corrected for temperature and concentration variations in the system, using the Nernst equation:

$$E = E^0 + \frac{RT}{nF}\ln\left(\frac{c_{\text{Oxd}}}{c_{\text{Red}}}\right)$$ (6.24)

where $c_{\text{Oxd}}$ is the concentration at the electrode surface of the species that releases electrons to the external circuit of the battery, and $c_{\text{Red}}$ is the surface concentration of the ions that flow through the electrolyte from one electrode plate to another, thus completing the electric circuit. More complicated models, relating the surface concentration of the reacting species to the OCV of the battery, also exist.[10]

An alternative to rigorous relationships between the OCV and the surface concentrations is the use of empirical expressions. Intercalation electrodes (e.g., lithium-ion cathodes), where the equilibrium potential $E^0$ is not constant, are such a case. Modeling the open circuit potential is based on measuring the voltage of the individual electrodes with respect to a standard reference at a very slow charge or discharge rate. Figure 6.6 shows examples of such measurements.

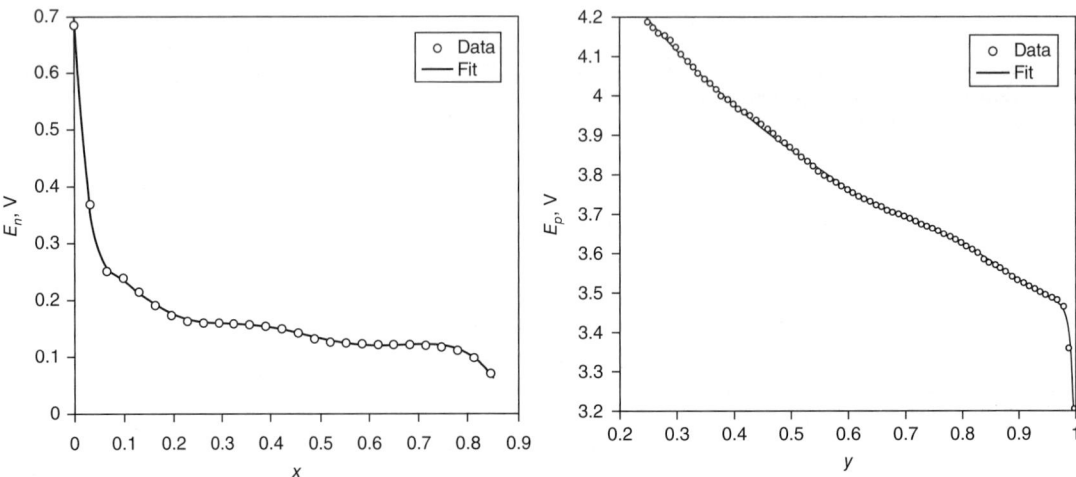

**FIGURE 6.6**    Open circuit voltage versus stoichiometry of lithium ions in an intercalation electrode: the curve on the left shows the experimental data from an anode composed of mesocarbon microbeads (MCMB) fit to an empirical expression, and the curve on the right shows similar results from a nickel-cobalt-oxide ($LiNiCoO_2$) cathode.[22]

***Rate of Charge Transfer.***    As with any chemical reaction, the efficiency of charge transfer also depends on the kinetic barriers across the electrode/electrolyte interface that the reaction must overcome. The rate of reaction correlates with the local overpotential at the reacting interface $j$ by the Butler-Volmer expression[7]:

$$i_j = i_{0,j}\left[\exp\left(\frac{\alpha_{a,j}n_jF\eta_{s,j}}{RT}\right) - \exp\left(\frac{-\alpha_{c,j}n_jF\eta_{s,j}}{RT}\right)\right] \tag{6.25}$$

where $i_{0,j}$ is the exchange current density normal to the reacting surface. Often this term includes concentration dependence of the reacting species at the interface and is written as follows:

$$i_{0,j} = i_{0,j}^{\text{ref}}f(c,c_s) = i_{0,j}^{\text{ref}}\left(\frac{c}{c^{\text{ref}}}\right)^{\gamma}\left(\frac{c_s}{c^{s,\text{ref}}}\right)^{\delta} \tag{6.26}$$

The parameter $i_{0,j}^{\text{ref}}$ is the electrochemical analog of the rate constant for a chemical reaction. The function $f$ relates the concentrations of the reacting species in the electrolyte and at the surface of the electrode ($c$ and $c_s$, respectively) to the exchange current density. The superscripts (ref) in Eq. (6.26) are the reference conditions, and the parameters $\gamma$ and $\delta$ correspond to the order of the reaction with respect to the participating species. Beyond simulating a typical chemical reaction rate equation, the Butler-Volmer expression for a charge-transfer reaction includes an exponential dependence of the current density on the local overpotential $\eta_{s,j}$, which is the difference between the potential at the electrode surface $\phi_{1,s}$ and that in the electrolyte at the interface $\phi_{2,s}$:

$$\eta_{s,j} = \phi_{1,s} - \phi_{2,s} \tag{6.27}$$

Alternatively, the overpotential term may include a reference potential that accounts for the potential difference across the interface at the open-circuit conditions, wherein the term $E$ [see Eq. (6.24)], corresponding to the electrode $j$, is subtracted from $\eta_{s,j}$:

$$\eta_j = \eta_{s,j} - E_j \tag{6.28}$$

If $\eta_j$ [i.e., Eq. (6.28)] is used instead of $\eta_{s,j}$ [Eq. (6.27)] in the Butler-Volmer expression [Eq. (6.25)], the concentration dependent term $f$ in Eq. (6.26) is modified accordingly to accommodate concentration terms from Eq. (6.24).[7] Similarly, if the reaction involves an intermediate step such as adsorption, the kinetic expressions for each step of the mechanism are stipulated with the final expression for the charge transfer reaction usually expressed in the form of Eq. (6.25).

***Distribution of Ions.***    Equation (6.24) relates the electron flow driving force to the concentration of the participating chemical species at the electrode-electrolyte interface. All the concentration terms ($c$ and $c_s$) are defined at the reacting interface. It is difficult to monitor the ion concentrations at the electrode surface. A material balance for the ions relates the concentration in the bulk of the solution to that at the electrode surface. These material balances specify the changes in ion concentrations that occur over time in accordance with the flux of the ions[11]:

$$\frac{\partial c_k}{\partial t} = -\nabla\cdot(\mathbf{N}_k) + R_k \tag{6.29}$$

The flux used in the material balance is consistent with the one used to determine the conductivity of the electrolyte [Eq. (6.22)]. The term $R_k$ refers to the production rate of species $k$. At the electrode-electrolyte interface, the ion concentration changes because of the electrochemical reaction, and hence Eq. (6.15) is used to relate the amount of ions involved in the reaction to the amount of ions present at the electrode-electrolyte boundary. For highly concentrated electrolytes, interactions among the ions must be considered. For example, the diffusion of one species of ions is impacted by all other ion types present within the electrolyte. Such complexities are usually handled by defining an *effective* property that considers such interactions. In this case, the following expression (see Sec. 8.4.6 of Ref. 11 for definition of the effective diffusion coefficients) is used for the ion flux:

$$\hat{\mathbf{N}} = c(\hat{\mathbf{v}} + \hat{\mathbf{v}}) + \hat{D}\nabla c \tag{6.30}$$

In Eq. (6.30), properties such as the diffusion coefficient $(\hat{D})$ are interpreted as effective properties. Note that the effective flux $(\hat{\mathbf{N}})$ is now a function of the electrolyte concentration $(c)$ and not the concentration of the individual ions $(c_k)$. The velocity term $\hat{\mathbf{v}}$ now relates to an effective field within the electrolyte and can be expressed in terms of the transport number $(t_+^0)$:

$$c\hat{\mathbf{v}} = (1 - t_+^0)\frac{\mathbf{i}_2}{F} \tag{6.31}$$

Expressions such as Eq. (6.30) use values for diffusivity or conductivity of the electrolyte measured using the actual mixture, instead of component diffusivities.

## 6.3    MODELING POROUS ELECTRODES

Porous battery electrodes are often utilized to improve the efficiency of the electrodes by providing access for the electrolyte to the active material within the electrode. By enhancing the accessibility of active materials to electrolyte ions, the charge transfer reaction benefits. At the same time, potential drop is minimized across the solution phase within the electrode. The material balance for transport of ions across a porous electrode closely follows Eq. (6.29) with the concentration terms now based on the fraction of the electrode volume occupied by the electrolyte. Hence, a porosity term $\varepsilon$, based on the effective properties similar to those discussed in Sec. 6.2, is used to simulate the transport limitations along the tortuous path through the electrodes. For example, the conductivity of the electrolyte within a porous electrode is corrected for the geometric effects as follows[12]:

$$\kappa_{eff} = \varepsilon^b \hat{\kappa} \tag{6.32}$$

The exponent $b$, called the tortuosity factor, is often an empirical term. Recent work using computer tomography of battery electrodes has resulted in direct estimates for the tortuosity factor.[13] In a porous electrode reactions are distributed throughout the volume of the electrode. Hence, the flux of the ions and reaction rates are now measured as quantities averaged across the volume of the electrode $(V)$. As a result, the material balance for the porous electrode becomes[16]

$$\frac{\partial(\varepsilon c)}{\partial t} = -\nabla \cdot (\bar{\mathbf{N}}) + \bar{R} \tag{6.33}$$

where $\bar{\mathbf{N}}$ is the volume averaged flux given by

$$\bar{\mathbf{N}} = \frac{1}{V}\int_V \hat{\mathbf{N}} \, dV \tag{6.34}$$

where $\hat{\mathbf{N}}$ is the effective flux [per Eq. (6.30)]. The volume averaged production rate $\bar{R}$ is calculated using a similar expression. The current densities for the charge transfer reactions in the Butler-Volmer expression are stated on a unit volume basis, which for a one-dimensional case is given by:

$$j = \frac{di}{dx} = i_{0,j}a\left[\exp\left(\frac{\alpha_{a,j}n_jF\eta_{s,j}}{RT}\right) - \exp\left(\frac{-\alpha_{c,j}n_jF\eta_{s,j}}{RT}\right)\right] \tag{6.35}$$

where $a$ is the available reaction surface per unit volume of the electrode.

Intercalation of the reaction ions into the electrode active material in a porous electrode has been modeled in several ways. The simplest treatment considers the phenomenon as diffusion of ions into a solid solution. Fick's law is used to represent this process. The electrode particles are usually represented using a

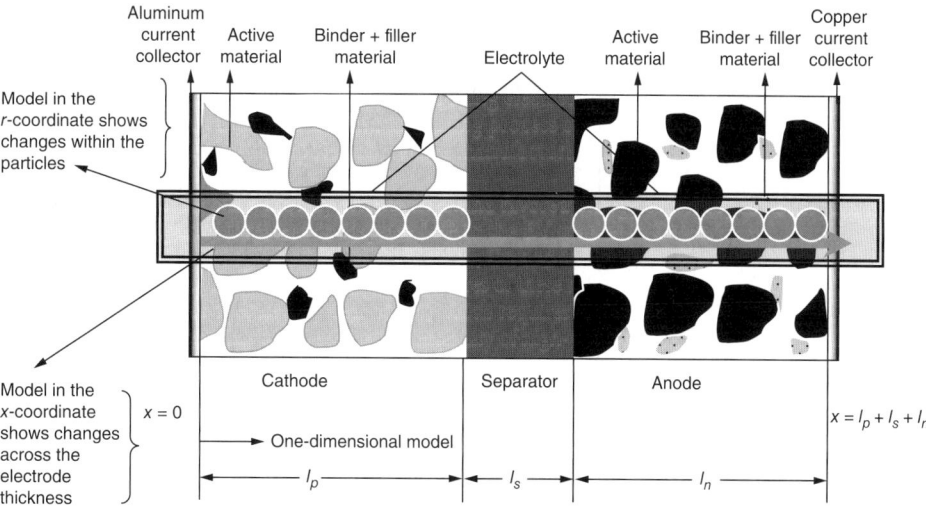

**FIGURE 6.7**  Schematic of a lithium-ion cell used to develop a one-dimensional model along the thickness of the electrodes.

regular geometry that has equivalent characteristics. For instance, in Fig. 6.7 the particles in the electrode are represented in the model as hypothetical spheres that have the same surface area to volume ratio as the actual active material particles on the porous electrode. Intercalation of ions within spherical particles is then governed by the diffusion equation:

$$\frac{\partial c_s}{\partial t} = D_s \left( \frac{\partial^2 c_s}{\partial r^2} + \frac{2}{r} \frac{\partial c_s}{\partial r} \right) \tag{6.36}$$

The subscript $s$ in Eq. (6.36) is used to refer to the solid particles. The concentration of the ions at the surface of the particles is mathematically connected to the electrolyte concentration at the interface through the Butler-Volmer equation [see Eq. (6.35)].

The equations outlined in this section constitute the mathematical framework for a generic mechanistic model of a battery. Figure 6.8 summarizes the utility of a mechanistic model in cell design. Several thought experiments can be conducted by altering the different design parameters, such as the particle size, as well as material properties, such as the conductivity. The model is used to identify critical factors that limit performance of the cell at high rates of charge or discharge.

## 6.4  BATTERY MODELS—CASE STUDIES

This section illustrates the use of the above equations for some common battery chemistries.

### 6.4.1  Kinetic Model of a Silver Vanadium Oxide Cell

The silver vanadium oxide (SVO) cell (see Chap. 28) is commonly used as a primary cell in medical devices. The cathode reaction can be written as follows[14]:

$$\text{Ag}_2^+\text{V}_4^{5+}\text{O}_{11} + (x+y)\text{Li}^+ + (x+y)e^- \rightarrow \text{Li}_{x+y}^+\text{Ag}_{2-x}^+\text{V}_{4-y}^{5+}\text{O}_{11} + x\text{Ag}^0 \tag{6.37}$$

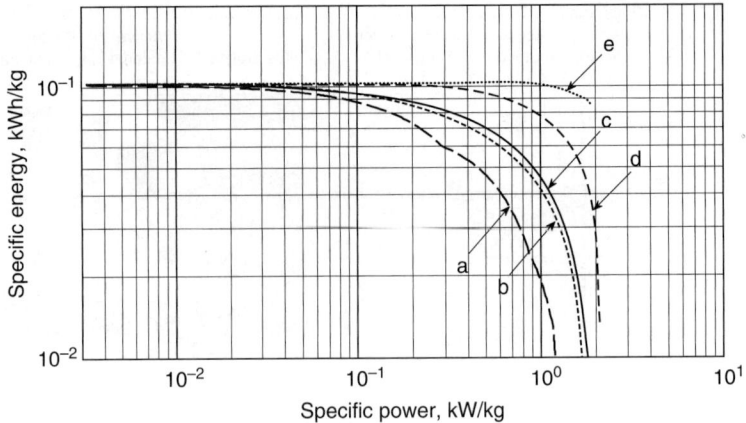

**FIGURE 6.8**  Simulated Ragone plots constructed using a mechanistic model. (*a*) Model predictions show that the original design cell delivers very low specific energies at high-power applications. (*b*) Increasing the electronic conductivity of the cathode matrix (e.g., by addition of conductive carbon) shows some improvement. (*c*) Further enhancement in electronic conductivity shows little change in cell performance. (*d*) Diffusion limitations are relaxed by increasing the solid-phase diffusion coefficient at the cathode (e.g., by doping). (*e*) Further limitations within the solid phase are eliminated by reducing the particle size to a few nanometers.

It is assumed that two electrochemical reactions take place at the cathode:

$$Ag_2^+V_4^{5+}O_{11} + xLi^+ + xe^- \rightarrow Li_x^+Ag_{2-x}^+V_4^{5+}O_{11} + xAg^0 \tag{6.38}$$

$$Ag_2^+V_4^{5+}O_{11} + yLi^+ + ye^- \rightarrow Li_y^+Ag_2^+V_y^{4+}V_{4-y}^{5+}O_{11} \tag{6.39}$$

The first reaction (6.38) corresponds to the reduction of silver, and the second (6.39) corresponds to the reduction of the vanadium ions. The stoichiometric parameter $x$ varies from 0 to 2, and $y$ varies from 0 to 4. The OCV as a function of composition is shown in Fig. 6.9.

The low operating current of these cells allows for a simple mechanistic model that excludes all transport limitations, such as diffusion in the electrolyte. The cell is classified as a kinetics-controlled regime with the material balance of Eq. (6.29) written by setting the flux equal to zero and the reaction rate equal to the charge transfer reaction:

$$\frac{\partial \theta_j}{\partial t} = \frac{\partial (c_j/c_{max})}{\partial t} = -\frac{aV}{n_j F c_{max}} i_j \tag{6.40}$$

where the subscript $j$ is set to $S$ for reaction (6.38) and to $V$ for (6.39). The term $aV$ refers to the reaction surface area across the entire volume of the electrode $V$ and is set equal to $2.0 \times 10^4 \, cm^2/cm^3$. The theoretical maximum for the concentration ($c_{max}$) is equal to 124.35 mol/cm³. The Butler-Volmer equation for each of the reactions (6.38) and (6.39) is written using the parameters in Table 6.2. The total current density ($i_2$) normal to the electrode surface is given as the sum of the individual reaction current densities (see Eq. [6.14]).

$$i_2 = i_S + i_V \tag{6.41}$$

Equations (6.25), (6.40), and (6.41) are used to relate the cell voltage to the applied current density. Figure 6.10 shows good agreement between model predictions and the experimental data for different current densities.

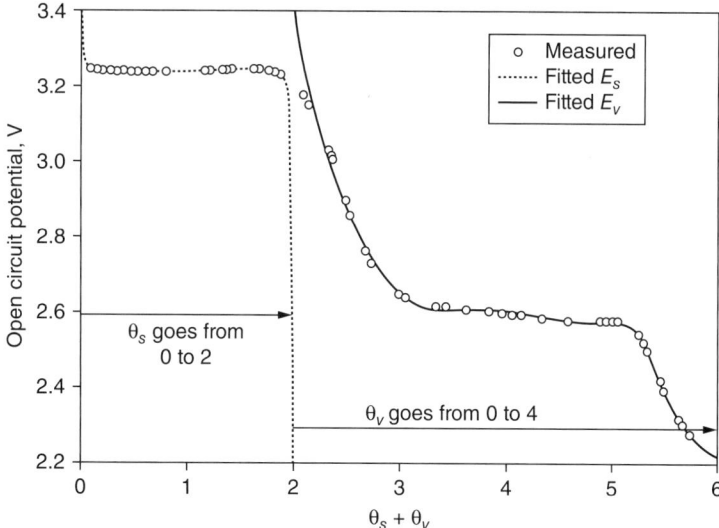

**FIGURE 6.9** Open circuit voltage (OCV) for a silver vanadium oxide cathode. Empirical expressions are used to relate the OCV of the electrode to the stoichiometric coefficients.[14,15]

**TABLE 6.2** Parameters for the Silver Vanadium Oxide Battery Model[14]

| Parameter | Silver reduction reaction | Vanadium reduction reaction |
|---|---|---|
| $i_{0,j}$, A/cm$^2$ | $10^{-10} (2 - \theta_s)^2$ | $10^{-8}$ |
| $\alpha_{a,j}$ | 0.5 | 0.5 |
| $\alpha_{c,j}$ | 0.5 | 0.5 |
| $n_j$ | 2 | 4 |
| $\eta_j$, V | $E - E_S$ | $E - E_v$ |

### 6.4.2 Lead-Acid Battery Model

A lead-acid battery model developed by Nguyen[17] specifies the following reactions:

$$PbO_2 + HSO_4^- + 3H^+ + 2e^- \rightarrow PbSO_4 + 2H_2O \quad \text{(positive electrode)} \tag{6.42}$$

$$Pb + HSO_4^- \rightarrow PbSO_4 + 2e^- + H^+ \quad \text{(negative electrode)} \tag{6.43}$$

At each electrode, the material balance is given by Eq. (6.33). The reaction term $\overline{R}_k$ is given by the Butler-Volmer equation for reaction at the positive electrode [Eq. (6.42)] and at the negative electrode [Eq. (6.43)]. In addition, the volume of the product formed in these reactions (i.e., PbSO$_4$) is much higher than the

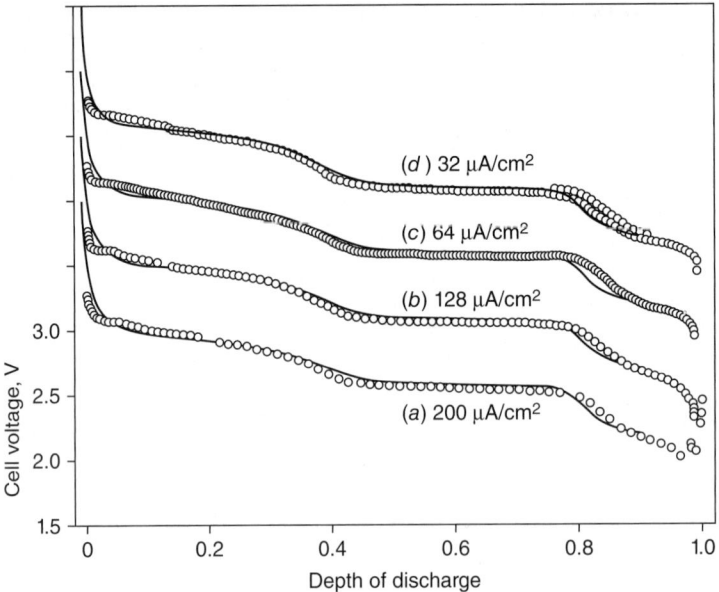

**FIGURE 6.10** Predicted cell voltage ($E$) for different current densities ($i_2$) using model Eq. (6.36). The curves for different current densities are offset at 0.5 V intervals for clarity.

reactants ($PbO_2$ or Pb particles). This induces a change in the porosity of the electrode during charge/discharge. This change in porosity is modeled by using Faraday's law as follows[18]:

$$\frac{\partial \varepsilon}{\partial t} = \frac{1}{n_j F}\left[\left(\frac{M}{\rho}\right)_{Product} - \left(\frac{M}{\rho}\right)_{Reactant}\right]\left(\frac{di_j}{dx}\right) \tag{6.44}$$

The total current is carried across the electrode, from the separator/electrode interface to the current-collector/electrode interface, by both the electrons within the electrode matrix and by the ions in the electrolyte that fill the pores within the electrode [see Eq. (6.14)]:

$$i = i_1 + i_2 \tag{6.45}$$

The current flow normal to the surface is given by Eq. (6.12) for the electrode matrix ($i_1$) and by Eq. (6.18) for electrolyte ($i_2$), after modifying the conductivity, as shown in Eq. (6.32). Next, ion transport is modeled for the influence of concentration gradients by using transport numbers. The resultant expression for the current density in the electrolyte phase is given by[19]:

$$i_2 = -\kappa_{eff}\left[\nabla\phi_2 + \frac{2RT}{F}(1-t_+^0)\frac{\nabla c}{c}\right] \tag{6.46}$$

The resultant profiles for the distribution of the porosity across the thickness of the electrode are shown in Fig. 6.11. The porosity at the electrode/separator interface decreases for both the anode and the cathode, owing to the precipitation of $PbSO_4$, having lower particle density than the starting Pb and $PbO_2$ active materials. The

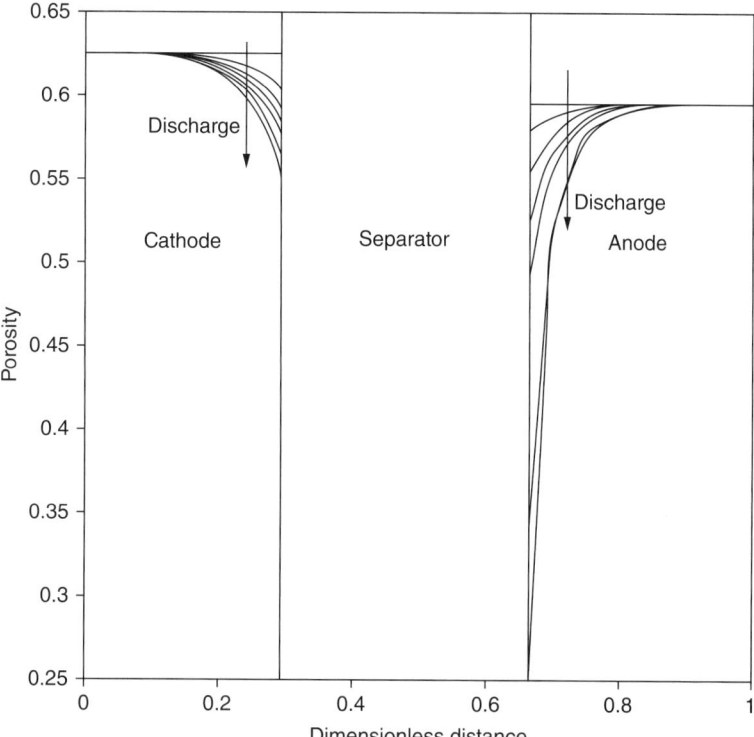

**FIGURE 6.11**   Porosity distribution across the thickness of the electrodes as a function of time during discharge in a lead-acid cell. The porosity at the electrode-separator interface decreases in both the anode and the cathode/separator interfaces due to the formation of PbSO$_4$.[18]

reduction in porosity is higher at the negative electrode since the difference in density between the reactant (metallic lead) and the product (PbSO$_4$) is higher. A direct effect from the porosity reduction at the electrode surface is that the entire volume of the electrode will then have more restricted access to the electrolyte. As a result, reactions are highly nonuniform, as shown in Fig. 6.12, and the reaction rates, as measured by the volumetric current densities adjacent to the current-collector surfaces, are close to zero, indicating poor utilization of the electrodes.

## 6.5   THERMAL MODELING OF BATTERIES

Many battery chemistries are impacted by temperature extremes. Abnormal temperature sometimes reduces performance, but in other cases may lead to concerns over safe operation of the battery. An energy balance equation is used to model heat transfer to the environment that results from heat generation within the battery due to Joule heating within the electrodes and electrolyte or due to chemical/electrochemical reactions. A general form of the energy balance equation is shown in Eq. (6.47)[11]:

$$\frac{\partial(\rho c_p T)}{\partial t} = \nabla \cdot (\lambda \nabla T) + h \tag{6.47}$$

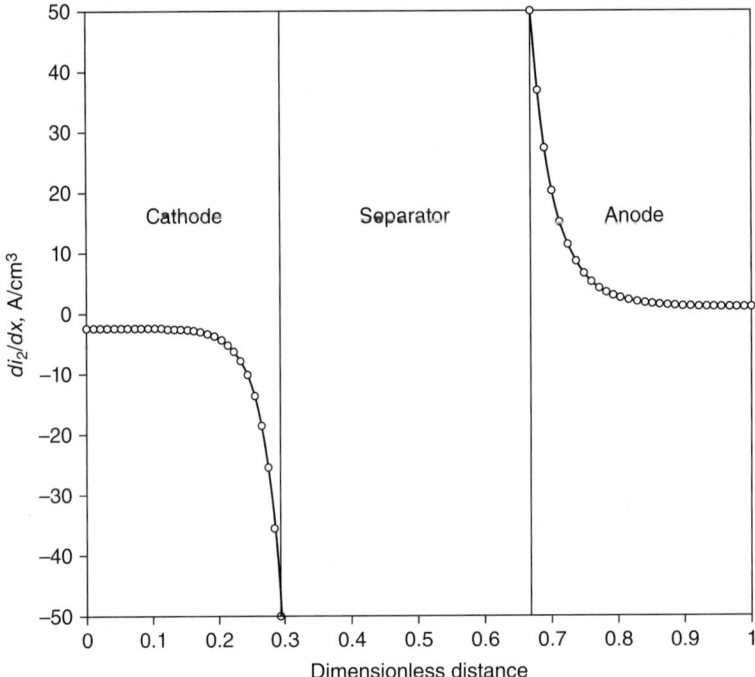

**FIGURE 6.12**    Reaction rate distribution across the thickness of a lead-acid cell [see Eq. (6.35)]. The blockage of the electrode surface by the formation of PbSO$_4$ creates a nonuniform distribution of the reaction across the thickness.[18]

The left side of Eq. (6.47) denotes the rate at which energy is consumed or generated per unit volume. The first term on the right side represents heat transfer by conduction following Fourier's law. The term $h$ refers to heat generated due to the reactions that take place during the operation of the battery. Typically, for an electrochemical reaction, this term is expressed as follows[20]:

$$h = \frac{\partial i_2}{\partial x}\left[\phi_1 - \phi_2 - \left(E_j - T\frac{\partial E_j}{\partial T}\right)\right] - i_1\frac{\partial \phi_1}{\partial x} - i_2\frac{\partial \phi_2}{\partial x} \tag{6.48}$$

The first component in this equation represents the heat generated from the charge-transfer reaction, the second term is the Joule heat generated due to the current flow across the solid matrix, and the last term is the corresponding value for current flow in the electrolyte. Additional complexities, such as heat transfer due to differential phase changes, heats of mixing, radiation effects, etc., can be treated by incorporating the amount of heat generated from such phenomena in Eq. (6.48). The term $T\dfrac{\partial E_j}{\partial T}$ corrects the OCV for changes in entropy with temperature. OCVs measured at different temperatures may be used to evaluate this term empirically. Changes in other properties such as the diffusivity or conductivity with temperature are often approximated by the Arrhenius equation:

$$\Phi = \Phi_{ref}\exp\left[-\frac{E_a}{R}\left(\frac{1}{T} - \frac{1}{T_{ref}}\right)\right] \tag{6.49}$$

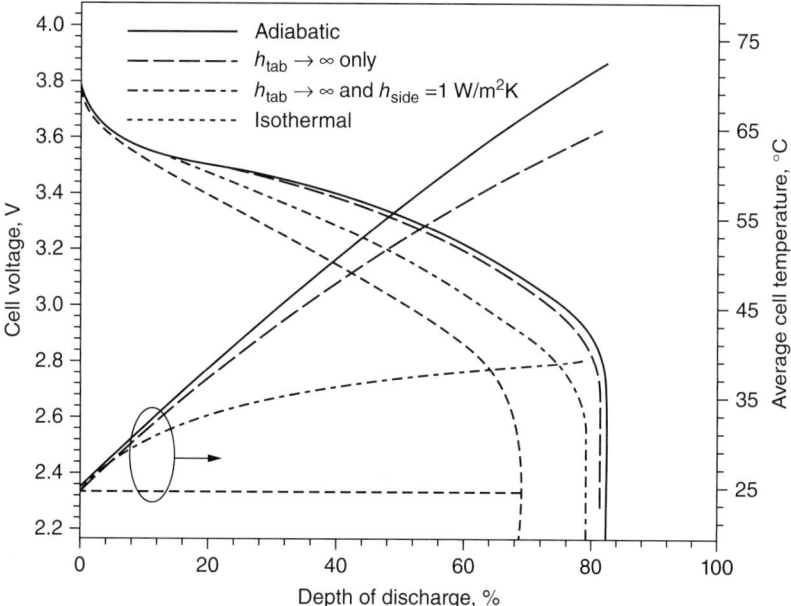

**FIGURE 6.13**  Comparison of the cell performance with various degrees of heat exchange with the environment.[20]

where $\Phi$ may represent $D_{eff}$, $\kappa_{eff}$, $D_s$, etc., $\Phi_{ref}$ is the corresponding property measured at the reference temperature ($T_{ref}$), and $E_a$ is the activation energy.

Figure 6.13 illustrates the change in battery performance with different degrees of convective cooling during a 3C-rate discharge of a lithium-ion cell. For the adiabatic case, increase in the cell temperature favors higher reaction rates and enhances transport within the electrolyte following Eq. (6.49) (see Sec. 5.8). The model predicts that without a cooling system in place, the difference in cell temperature from the ideal isothermal case can be as high as 45°C. The other two cases shown indicate the effect of using suitable packing material to implement rapid cooling of the substrate and the influence of additional convection along the side walls of the cell. Including a simple heat-transfer model for the walls thus provides significant insight into the design of an efficient cooling system, particularly for large format cells.

In the case of a nickel-metal hydride (NiMH) battery, the primary electrochemical reactions as well as reactions involving oxygen both contribute to heat generation within the cell. The primary reactions are as follows[21]:

$$\textit{Positive electrode:}\ \ NiOOH + H_2O \underset{\text{Charge}}{\overset{\text{Discharge}}{\rightleftarrows}} Ni(OH)_2 + H^-$$

$$(6.50)$$

$$\textit{Negative electrode:}\ \ MH + OH^- \underset{\text{Charge}}{\overset{\text{Discharge}}{\rightleftarrows}} H_2O + M + e^-$$

In addition, the inadequate utilization of one electrode at end of charge leads to evolution of oxygen, which undergoes subsequent reactions:

$$2OH^- \to \frac{1}{2}O_2 + H_2O + 2e^- \qquad \text{(positive)}$$

$$4MH + O_2 \to 4M + 2H_2O \qquad \text{(negative)}$$

$$\text{(6.51)}$$

Reactions of the following type involving phase changes within the metal hydride electrode leading to the formation of β-MH have also been proposed:

$$(x - y)H + MH_y \to MH_x \qquad \text{(6.52)}$$

Equation (6.52) is a chemical reaction, and hence the heat generated by this reaction is calculated as the product of the rate of the reaction and the enthalpy of reaction.

The heat generated from the above reactions is compared in Fig. 6.14 to the Joule heating terms [(see Eq. (6.48)]. The model simulates charging of a NiMH cell at the 1-$C$ rate. At the beginning of charge, the heat generated from the MH [Eq. (6.52)] is balanced by the endothermic primary reactions. Also, the oxygen evolution reaction does not take place to a significant extent; as a result, heat contributions from Eq. (6.51) are negligible. Toward the end of charge, the enthalpy changes in favor of heat absorption in reactions [Eq. (6.50)]. In addition, overcharge leads to significant evolution of $O_2$. As a result, there is a dramatic increase in the heat generating within the cell toward the end of charge.

## 6.6    *DEGRADATION MODELS*

Ideally, a mathematical model of a battery will provide some insight into the future performance of the battery. For example, based on the results shown in Fig. 6.13, a lithium-ion cell operating at elevated temperatures may be desirable for improved performance if only limited cycles are required; however, prolonged cycling under

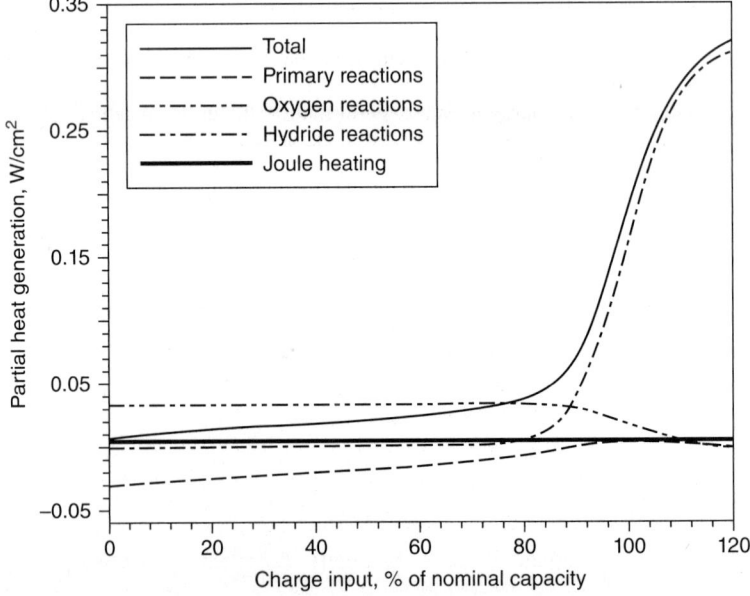

**FIGURE 6.14**    Contribution of different chemical reactions to heat generated within a NiMH cell.[21]

these hotter conditions leads to faster deterioration of the cell performance. In developing a life model for a battery, understanding the mechanism of degradation is a critical step in developing a model to predict life of a battery. For example, nickel-based lithium-ion cell electrodes will develop surface oxidation on the particles, resulting in an additional cathodic impedance Alternatively, cobalt-based cathodes develop increased impedance due to phase changes at higher voltages. Similarly, dissolution of manganese ions is a major factor contributing to capacity loss when cycling a $LiMn_2O_4$-based electrode.

Determining the values for the parameters used in the models is another major challenge in using physics-based models for life prediction. Physics-based models require considerably more parameters than empirical models. While many parameters are determined from the operating conditions and the fresh cell component designs, subsequent changes in certain parameters over several cycles are not easily determined. Additionally, many parameters will vary with operating conditions or the history of the battery.

The simplest life prediction models use linear extrapolation: plotting the capacity of the cell versus the cycle number and obtaining the slope and the intercept of a straight line by regression. For cells cycled under mild operating conditions (e.g., shallow depths of discharge [DOD]) that do not stress the cell prior to end of life, linear extrapolation has been found to provide a good degree of confidence in predicting the end of life of the cell. This technique simplifies the ability to extract the coefficients.

Depending on the range of operating conditions, more than one set of coefficients may be required for successful prediction of the cell performance. For example, if various cells are subject to cycling regimes with different DOD, the degradation rates will differ—and consequently, the coefficients in the empirical fits for each case are different. Some predictions made using empirical models are shown in Fig. 6.15. The accuracy of this method relies on the functional terms used in the expression. A complicated polynomial expression may provide a better prediction compared to a linear equation. Commercial software tools for nonlinear regression are also readily available. In practice, curve-fitting techniques are more typically used to interpolate between

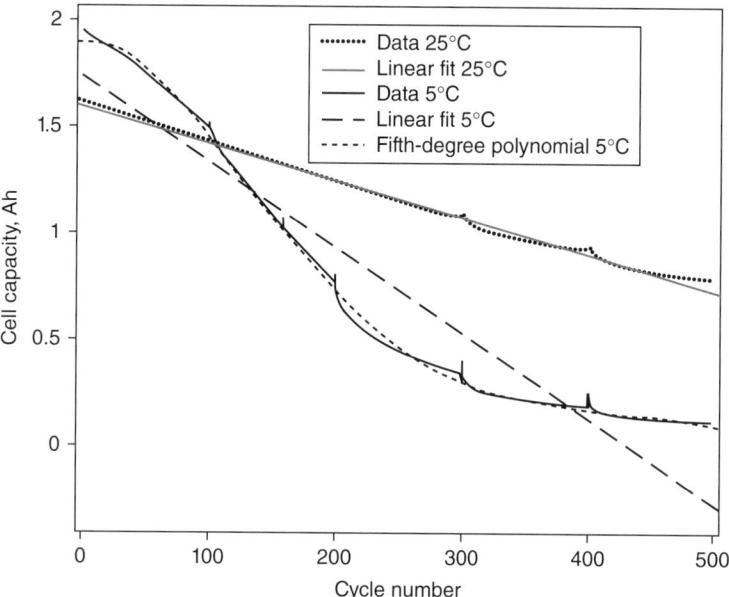

**FIGURE 6.15**  Cell capacity versus cycle number data for a lithium-ion cell fit to empirical expressions. The linear equation fits the data at milder conditions (cycling at 25°C) better than the data at a more rigorous condition (cycling at 5°C). Predictions made using the linear expression are closer to the experimental observation for the data at 25°C. For the data collected at 5°C, a more complicated expression (in this case, a fifth-degree polynomial) is required to represent the data more accurately.[25]

known operating scenarios rather than to project beyond the available experimental data. This shortcoming is typical of all empirical prediction techniques, yet it does not prevent curve-fitting from being a valued analytical tool.

An alternate approach is to use a mechanistic model similar to the one shown in Table 6.3 and adjust parameters, such as the diffusivity and the exchange current density, periodically as to obtain a good fit between the model predictions and experimentally observed performance.[21] Such an approach is usually referenced as semiempirical modeling. Figure 6.16 illustrates the parameter fluctuations observed for a lithium-ion cell that loses about 30% of its initial capacity during the first 800 cycles.

**TABLE 6.3**    Summary of Equations Used to Represent a Porous Intercalation Electrode. The Corresponding Equation Numbers from the Text Are Shown in the Last Column[20]

| Variable | Governing equation | Equation number in the text |
|---|---|---|
| Solid-phase potential ($\phi_{1,j}$) | $\mathbf{i}_1 = -\sigma_j^{\text{eff}} \nabla \phi_{1,j}$ | (6.12) |
| Solution-phase potential ($\phi_{2,j}$) | $\mathbf{i}_2 = -\kappa_{\text{eff}} \left[ \nabla \phi_2 + \frac{2RT}{F} \left(1 - t_+^0\right) \frac{\nabla c_2}{c_2} \right]$ | (6.46) |
| Solid-phase current density ($\mathbf{i}_1$) | $\mathbf{i}_{\text{tot}} = \mathbf{i}_1 + \mathbf{i}_2$ | (6.45) |
| Solution-phase current density ($i_2$) | $\dfrac{di_2}{dx} = i_{0,j} a \left[ \exp\left(\dfrac{\alpha_{a,j} n_j F \eta_{s,j}}{RT}\right) - \exp\left(\dfrac{-\alpha_{c,j} n_j F \eta_{s,j}}{RT}\right) \right]$ | (6.35) |
| Solid-phase concentration ($c^s$) | $\dfrac{\partial c_s}{\partial t} = D_s \left( \dfrac{\partial^2 c_s}{\partial r^2} + \dfrac{2}{r} \dfrac{\partial c_s}{\partial r} \right)$ | (6.36) |
| Solution-phase concentration ($c_2$) | $\varepsilon \dfrac{\partial c_2}{\partial t} = D_{\text{eff}} \dfrac{\partial^2 c_2}{\partial x^2} + \dfrac{(1 - t_+)}{F} \dfrac{\partial \mathbf{i}_2}{\partial x}$ | (6.29) |

A third approach uses a mechanistic model based on degradation processes. An electrochemical reaction that consumes part of the cell's deliverable capacity over several cycles may be proposed. For example, resistance increases can be modeled as an anode particle surface film formation process caused by reduction of the solvent during charge.[23] (See Chap. 17A on solid-electrolyte interface [SEI] layers.) Since passivation (Chaps. 4 and 5) is a charge transfer reaction involving reduction of Li$^+$ into salts, a Butler-Volmer type equation assumes the following form:

$$\frac{di_{\text{side}}}{dx} = -i_{0,\text{side}} a \exp\left( \frac{\alpha_{c,\text{side}} n_{\text{side}} F \eta_{\text{side}}}{RT} \right) \tag{6.53}$$

The formation of the film introduces an additional resistance at the surface of the anode particles and consequently a drop in the overpotential at the anode:

$$\eta_n = \phi_{1,s} - \phi_{2,s} - E_n - \frac{1}{a_n} \left( \frac{\partial i_2}{\partial x} \right) \left( \frac{\delta_{\text{film}}}{\kappa_{\text{film}}} \right) \tag{6.54}$$

The thickness of the film ($\delta_{\text{film}}$) is calculated using Faraday's law:

$$\frac{\partial \delta_{\text{film}}}{\partial t} = -\frac{1}{Fa_n} \left( \frac{\partial i_{\text{side}}}{\partial x} \right) \left( \frac{M_{\text{side}}}{\rho_{\text{side}}} \right) \tag{6.55}$$

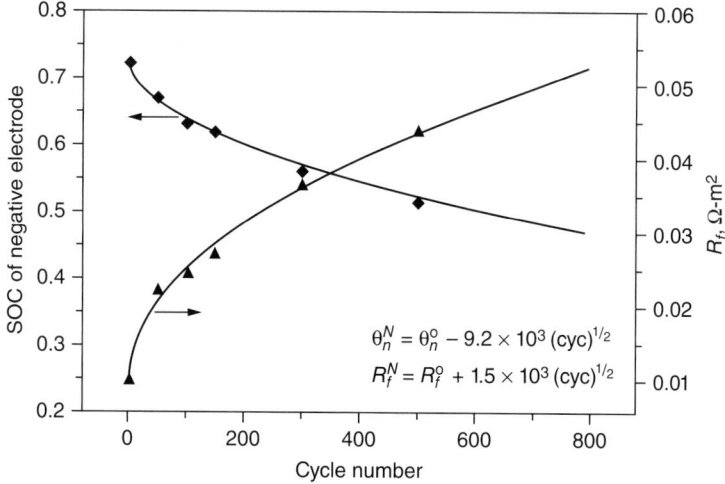

**FIGURE 6.16**   Change in parameters with cycling of a lithium-ion cell. Adjustable parameters in the semiempirical model are the state-of-change (SOC) of the negative electrode at the beginning of discharge ($\theta_n^N$) and the resistance of the film formed on the anode ($R_f^N$). The changes in these parameter values with cycle number were obtained by adjusting the experimental curves at various cycle numbers.[22]

where $M_{side}$ and $\rho_{side}$ represent the molecular weight of the side reaction product and the density of the film, respectively.

Figure 6.17 shows the thickness of film formed on the anode particle surface as predicted by the mechanistic model under different operating conditions. The model predicts a rapid growth of the film during the first few

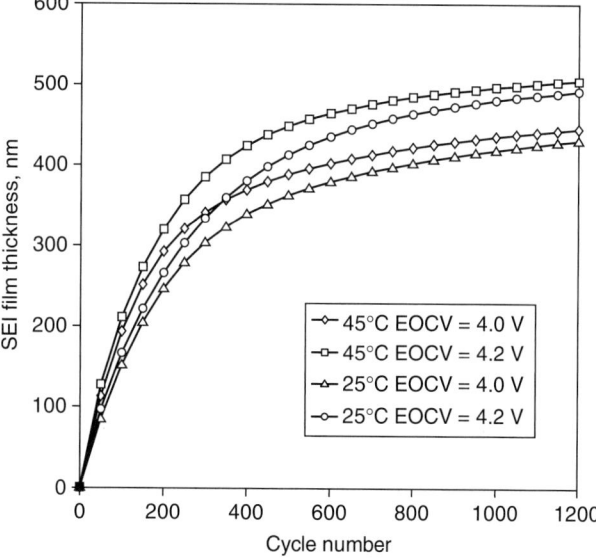

**FIGURE 6.17**   Solid-electrolyte-interface (SEI) film thickness as a function of the cycle number as predicted by a mechanistic model for a lithium-ion cell.

cycles that eventually levels off. For the higher end of charge voltage (EOCV), the anode is maintained at highly reducing potentials for longer periods of time, and as a result, the growth rate of the film is higher in a cell cycled with a higher EOCV. Under identical conditions, a cell cycled at a higher temperature has lower transport limitations, and hence the duration of a given cycle is longer than it would be at a lower temperature. This results in additional charging time since the model for the film growth depends on the amount of time taken to charge the cell [see Eq. (6.55)]. The cell cycled at 45°C shows higher thickness of the film during the first few cycles and hence more capacity fade. However, after about 300 cycles, the enhancements in charging time due to the temperature are offset by the additional resistance created by the growth of the film. As a result, the cell cycling at 25°C, but with a higher EOCV, loses more capacity.

These results indicate that if operating at a relatively higher temperature for about 300 cycles, the cell can be programmed to have a higher EOCV (and hence deliver higher capacity), whereas if the application demands a longer cycle life, a more conservative EOCV should be used. Similar conclusions can be drawn from empirical models once data under different operating conditions is available, whereas only mechanistic models can uniquely provide greater insights into the physical phenomena contributing to degradation. Once the rate constant for the film formation side reactions and the conductivity of this film are determined by independent experiments, the model may be used for cell design under a variety of different scenarios as long as the proposed degradation mechanisms are valid.

## 6.7    DETERMINING THE RIGHT MODEL

A good mathematical model should strike a balance between the limited details associated with the input parameters available to the end user and the amount of insight the model can provide to improve battery design. Limitations to the mechanistic models are attributed to the tedium involved in developing and solving the model equations as well as the large number of parameters required by such models. Often many of these parameters cannot be obtained from direct experimental measurements. On the other hand, circuit analog models provide limited insight into the physical phenomena that take place within the battery. For example, a drop in capacity at higher rates of discharge can be modeled as an increase in the pseudocapacitance parameter $C_D$ shown in Fig. 6.3. However, whether this change is caused by a limitation in the diffusion of ions within the electrode or by losses of electrolyte conductivity over time cannot be determined. Consequently, making improvements to the cell design based on circuit analog models is difficult; and in this example, it is not obvious whether an increase in the porosity of the electrode plate would resolve the issue or if the electrolyte formulation needs improvement.

Typically, fine-tuning of the parameters such as the conductivity of the electrolyte or the porosity of the electrodes is carried out at the cell design phase. Hence, a mechanistic model is invaluable at this design phase. In all cases, the assumptions behind a mathematical model must be carefully explored before employing the conclusions drawn from simulations.

## List of Symbols

| | |
|---|---|
| $a$ | Specific area ($m^2/m^3$) |
| $A$ | Area of the electrode ($m^2$) |
| $A_c$ | Cross-sectional area ($m^2$) |
| $b$ | Tortuosity factor |
| $c_k$ | Concentration of the ion $k$ ($mol/m^3$) |
| $c$ | Volume averaged concentration of the electrolyte ($mol/m^3$) |
| $c^s$ | Volume averaged concentration of the ion within the electrode ($mol/m^3$) |
| $c_p$ | Specific heat capacity (J/kg/K) |
| $C_D$ | Double-layer capacitance (F) |
| $D$ | Diffusion coefficient of the ion in the electrolyte ($m^2/s$) |
| $D_s$ | Diffusion coefficient of the ion in the electrode ($m^2/s$) |
| $E_a$ | Activation energy (J/mol) |

| $E^0$ | Equilibrium potential (V) of the electrode |
|---|---|
| $F$ | Faraday's constant (96,485 C/mol) |
| $G$ | Gibb's free energy (J/mol) |
| $h$ | Rate of heat generation per unit volume (W/m$^3$) |
| $\Delta H$ | Amount of heat generated (J) |
| $\mathbf{i}$ | Current density (A/m$^2$) |
| $i_{0,j}$ | Exchange current density (A/m$^2$) |
| $I$ | Current (A) |
| $j$ | Volumetric current density (A/m$^3$) |
| $L$ | Length (m) |
| $M$ | Molecular weight (kg/mol) |
| $n$ | Number of electrons transferred |
| $N$ | Flux (mol/m$^2$/s) |
| $\bar{N}$ | Volume averaged flux (mol/m$^3$/s) |
| $\hat{\mathbf{N}}$ | Effective flux of an ion (mol/m$^2$/s) |
| $q$ | Charge (C) |
| $Q$ | Capacity of the cell (C) |
| $Q_0$ | Initial capacity of the cell (C) |
| $R$ | Ohmic resistance within a cell (Ohm) |
| $R_{ct}$ | Charge transfer resistance (Ohm) |
| $\underline{R_k}$ | Rate of reaction involving ions of species $k$ (mol/m$^3$/s) |
| $\bar{R}$ | Volume averaged reaction rate (mol/m$^3$/s) |
| $R$ | Universal gas constant (8.314 J/mol/K) |
| $t$ | Time (s) |
| $t_+^0$ | Transport number |
| $T$ | Temperature (K) |
| $u$ | Mobility of an ion (cm$^2$ mol/J/s) |
| $\mathbf{v}$ | Convective velocity (m/s) |
| $V$ | Volume of the electrode (m$^3$) |
| $V$ | Cell voltage (V) |
| $V_0$ | Open circuit voltage (OCV) of the cell (V) |
| $x$ | Spatial variable (m) |
| $z$ | Charge carried by an ion |

## Greek

| $\alpha$ | Transfer coefficient |
|---|---|
| $\delta_{film}$ | Thickness of the SEI film (m) |
| $\varepsilon$ | Porosity |
| $\kappa$ | Ionic conductivity of the electrolyte (S/cm) |
| $\lambda$ | Thermal conductivity (W/K) |
| $\eta$ | Overpotential (V) |
| $\phi$ | Potential (V) |
| $\rho$ | Density (kg/m$^3$) |
| $\sigma$ | Electronic conductivity of the electrode (S/cm) |
| $\tau$ | Time constant (s$^{-1}$) |
| $\theta$ | Dimensionless concentration |

## Subscripts and Superscripts

| $c$ | Current collector |
|---|---|
| eff | Effective |
| $n$ | Negative electrode |

| | |
|---|---|
| $p$ | Positive electrode |
| ref | Reference condition |
| side | Side reaction |
| 0 | Initial or standard condition |
| 1 | Electrode matrix |
| 2 | Electrolyte |
| $s$ | Solid phase |
| $\hat{\Lambda}$ | Effective value of the parameter $\Lambda$ |

## Abbreviations

| | |
|---|---|
| CAD | Computer-aided design |
| DOD | Depth of discharge |
| MCMB | Mesocarbon microbeads |
| SEI | Solid-electrolyte interface |
| SOC | State-of-charge |

## *REFERENCES*

1. W. Peukert, Über die Abhängigkeit der Kapacität von der Entladestromstärcke bei Bleiakkumulatoren. *Elektrotechnische Zeitschrift* **20** (1897).

2. A. Lasia, *Electrochemical Impedance Spectroscopy and Its Applications*, Springer Science, New York, 2014.

3. R. De Levie, "Response of Porous and Rough Electrodes," in P. Delahay, C. W. Tobias (eds.), *Advances in Electrochemistry and Electrochemical Engineering*, Vol. 6, John Wiley & Sons, New York, 1971.

4. M. W. Verbrugge, R. S. Conell, "Electrochemical and Thermal Characterization of Battery Modules Commensurate with Electric Vehicle Integration," *J. Electrochem. Soc.* **149**(1):A45–A53 (2002).

5. A. Pesaran, Progress of the Computer-Aided Engineering of Electric Drive Vehicle Batteries (CAEBAT), Presented at the 2013 U.S. DOE Vehicle Technologies Office Annual Merit Review and Peer Evaluation Meeting, Arlington, VA, May 14, 2013. NREL Report No. PR-5400-58202.

6. M. Verbrugge, "Adaptive Characterization and Modeling of Electrochemical Energy Storage Devices for Hybrid Electric Vehicle Applications," in M. Schlesinger (ed.), *Modern Aspects of Electrochemistry*, Vol. 43, Springer-Verlag, New York, pp. 417–524, 2009.

7. J. Newman, K. E. Thomas-Alyea, *Electrochemical Systems*, 3rd ed., John Wiley & Sons, Hoboken, NJ, 2004.

8. C. Lin, R. E. White, H. J. Ploehn, "Modeling the Effects of Ion Association on Alternating Current Impedance of Solid Polymer Electrolytes," *J. Electrochem. Soc.* **149**(7):E242–E251 (2002).

9. D. McQuarrie, J. D. Simon, *Molecular Thermodynamics*, University Science Books, Sausalito, CA, 1999.

10. T. Ohzuku, A. Ueda, "Phenomenological Expression of Solid-State Redox Potentials of $LiCoO_2$, $LiCo_{1/2}Ni_{1/2}O_2$ and $LiNiO_2$ Insertion Electrodes," *J. Electrochem. Soc.* **144**(8):2780–2785 (1997).

11. Slattery, J. C., *Advanced Transport Phenomena*, Cambridge University Press, New York, 1999.

12. S. Whitaker, "Diffusion and Dispersion in Porous Media," *AIChE J.* **13**(3):420–427 (1967).

13. B. Tjaden, D. J. L. Brett, P. R. Shearing, "Tortuosity in Electrochemical Devices: A Review of Calculation Approaches," *International Materials Reviews* (2016), http://dx.doi.org/10.1080/09506608.2016.1249995.

14. P. M. Gomadam, D. R. Merritt, E. R. Scott, C. L. Schmidt, P. M. Skarstad, J. W. Weidner, "Modeling Li/$CF_x$-SVO Hybrid-Cathode Batteries," *J. Electrochem. Soc.* **154**(11):A1058–A1064 (2007).

15. A. M. Crespi, P. M. Skarstad, H. W. Zandbergen, "Characterization of Silver Vanadium Oxide Cathode Material by High-Resolution Electron Microscopy," *J. Power Sources* **54**(1):68–71 (1995).

16. J. Newman, W. Tiedemann, "Porous-Electrode Theory with Battery Applications," *AIChE J.* **21**(1):25–41 (1975).

17. T. V. Nguyen, *Modeling and Characterization of a Lead-Acid Cell*, Ph.D. Dissertation, Texas A & M University, College Station, TX, 1988.

18. T. V. Nguyen, R. E. White, H. Gu, "The Effects of Separator Design on the Discharge Performance of a Starved Lead-Acid Cell," *J. Electrochem. Soc.* **137**(10):2998–3004 (1990).

19. K. E. Thomas, R. M. Darling, J. Newman, "Mathematical Modeling of Lithium Batteries," in W. A. van Schalkwijk, B. Scrosati (eds.), *Advances in Lithium-Ion Batteries*, Kluwer Academic/Plenum Publishers, New York, pp. 345–392, 2002.

20. W. Gu., C. Y. Wang, "Thermal and Electrochemical Coupled Modeling of a Lithium-Ion Cell in Lithium Batteries," in *Proceedings of the Electrochemical Society*, Vol. 99–25(1), Plenum Publishers, Pennington, NJ, pp. 748–762, 2000.

21. C. Y. Wang, W. B. Gu, B. Y. Liaw, "Thermal-Electrochemical Modeling of Battery Systems," *J. Electrochem. Soc.* **147**(8):2910–2922 (2000).

22. P. Ramadass, B. Haran, R. White, B. N. Popov, "Mathematical Modeling of the Capacity Fade of Li-Ion Cells," *J. Power Sources* **123**(2):230–240 (2003).

23. P. Ramadass, B. Haran, P. M. Gomadam, R. White, B. N. Popov, "Development of First Principles Capacity Fade Model for Li-Ion Cells," *J. Electrochem. Soc.* **151**(2):A196–A203 (2004).

24. Q. Zhang, R. E. White, "Capacity Fade Analysis of a Lithium-Ion Cell," *J. Power Sources* **179**:793–298 (2008).

25. S. Santhanagopalan, J. Stockel, R. E. White, "Life Prediction for Lithium-Ion Batteries," in J. Garche, C. Dyer, P. Moseley, Z. Ogumi, D. Rand, B. Scrosati (eds.), *Encyclopedia of Electrochemical Power Sources*, Vol. 5, Elsevier Publications, Amsterdam, pp. 418–437, 2009.

# BATTERY PRODUCT
# OVERVIEW

# CHAPTER 7
# BATTERY SYSTEM DESIGN

**Daniel D. Friel**

## 7.1 INTRODUCTION

Correct design of the battery pack is important to assure optimum, reliable, and safe operation of electronic devices and equipment. Many battery problems may be prevented if proper precautions are taken. Specifically, battery designers must consider the design of the battery pack itself, the interface with any battery monitoring or protection devices or electronics, and integration of the battery pack into the battery-operated equipment.

This chapter is intended to illustrate the proper methods of constructing a battery pack from a collection of electrochemical cells (typically rechargeable nickel- or lithium-based cells). Design considerations for devices using individual, loose cells, often with primary alkaline or primary lithium, will briefly be discussed. Devices, ranging from portable and hand-held to transportable, will be discussed with a review of considerations for larger vehicle and stationary battery applications.

The performance of a collection of cells in a battery pack can be different from that of an individual cell depending on multiple factors. Specifications provided by the manufacturers should only be used as a guide. The final assembly must be thoroughly tested in order to determine the performance of multicell batteries in a series/parallel configuration in a battery pack.

## 7.2 BATTERY CONSTRUCTION

Assembly of a battery pack from a collection of electrochemical cells involves the following steps:

1. Cell-to-cell connections (both series and parallel connections)
2. Physical constraint or encapsulation of cells (should not inhibit vent activation)
3. External case design and materials (optional for some battery packs)
4. Terminals and contact materials

Figure 7.1 shows a typical battery pack design.

### 7.2.1 Cell-to-Cell Connections

The least preferred method of battery connection is the use of pressure contacts. Although this technique is used with some inexpensive consumer battery products, it can be the cause of battery failure where high reliability is desired. This type of connection is prone to corrosion at the contact points. In addition, under shock and vibration, intermittent loss of contact may result.

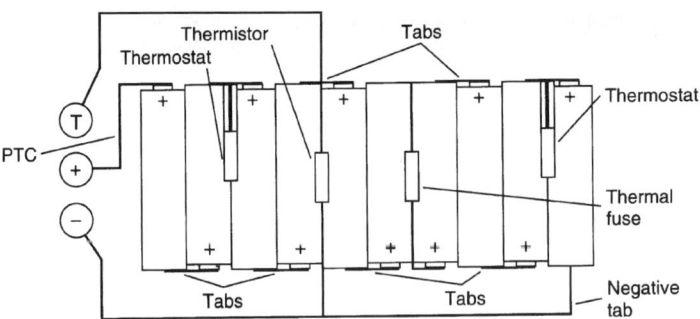

**FIGURE 7.1**   Example locations of protective devices, cells, and tabs in a battery pack.

Welding of conductive tabs (i.e., metal strips) between cells is the preferred method of intercell connection for most battery systems. At least two weld spots should be made at each connection joint. Care must be taken to ensure a proper weld without burning through the cell container. Excessive welding temperatures could also result in damage to the internal cell components. Weld testing is conducted by pulling the weld apart. The welded joint must maintain structural integrity while the base metal tears. For tabs, the weld diameter, as a rule of thumb, should be three to four times the tab thickness. For example, a 0.125-mm-thick tab should have a tear diameter of 0.375 to 0.5 mm. Figure 7.2 demonstrates examples of poor welds.

**FIGURE 7.2**   Examples of poor welds that can cause latent failures.[1]

The tab materials for most applications are nickel-based since the corrosion resistance of the nickel and its ease of welding result in reliable permanent connections. The electrical resistance of the tab material must be matched to the application to minimize voltage loss: Nickel-plated steel material has 50% higher resistance than that of pure nickel for an equivalent-size tab and is lower cost. But while this resistance may be of little significance in the circuit, the difference must be considered in the design.

For applications with high discharge currents, such as in power tools or electric vehicles, normal internal cell protection devices such as a positive temperature coefficient (PTC) overcurrent device may not be utilized. To protect against a cell short or other potentially damaging cell failure, particularly in large parallel combinations of cells, the tab interconnect can be constructed to have high resistance, causing internal melting due to heating when shorted, serving as a fuse element.

The connection tabs must also be kept clean and straight, but with perhaps a little flexibility to avoid mechanical stress on the welds. Tab edges should be prevented from cutting into the cells, particularly in high shock or vibration end applications.

## 7.2.2   Physical Constraint or Encapsulation of Cells

Some applications may require that the cells within the battery be rigidly fixed in position to meet shock, vibration, or other requirements. Encapsulation of the cells can be accomplished with plastic holders, epoxy, foams, or other suitable materials. (Electronic circuit boards contained within the battery pack may be similarly fixed. Conformal coating is recommended to protect the circuitry from cell leaking or venting events, which could cause shorting of the circuit components.)

The preferred method to keep the cells immobile is through careful case design *without* the use of encapsulation potting materials that are often time-consuming to apply and difficult to control. Plastic materials offer more options for optimal solutions.

In any case, care must be taken to prevent the encapsulation method and material from blocking the vent mechanisms of the cells. One technique is to orient the cell vents in the same direction and encapsulate the battery to a level below the vent. Encapsulation material must also not inhibit heat transfer paths to surrounding cells or other thermal management devices. New materials that improve heat transfer away from a cell, such as phase change materials (PCMs), may be employed. Figure 7.3 shows an example of dangerous use of encapsulation since the vent ends of the cells are completely obstructed.

**FIGURE 7.3**    Example of poor use of encapsulation.[1]

## 7.2.3   External Case Design and Materials

Careful design of the case housing should include the following:

1. Materials must be compatible with the cell chemistry chosen. For example, aluminum reacts with alkaline electrolytes and must be protected where cell venting may occur.

2. Materials should also be compatible with the end-use environment of the device. Some healthcare disinfectant cleaning chemicals can degrade and weaken certain plastic materials.

3. Flame-retardant materials may be required to comply with end-use requirements. Underwriters Laboratories (UL), the Canadian Standards Association (CSA), the United Nations (UN), and other agencies may require testing to ensure safety compliance.

4. Adequate battery venting must allow for the release of vented cell gases. In sealed batteries, this requires the use of a pressure relief valve or breather mechanisms. Case enclosures should also consider the expansion of cells during use. Lithium-ion polymer "bag" or "pouch" cells can expand during use and with age, thus increasing the size of the enclosure needed.

5. The design must provide for effective dissipation of heat to limit the temperature rise during use and especially during charge. (High temperatures should be avoided as they increase self-discharge, could

cause cell venting, and generally are detrimental to battery life. Temperature gradients in the battery pack can also lead to cell imbalance, which can degrade performance and safety.) The design may also be required to include enough cell-to-cell spacing to permit one cell to reach thermal runaway without igniting neighboring cells. This cell propagation prevention requirement is becoming more common in larger battery packs.[2]

Figure 7.4 compares the temperature profiles of groups of cells with and without a battery case.

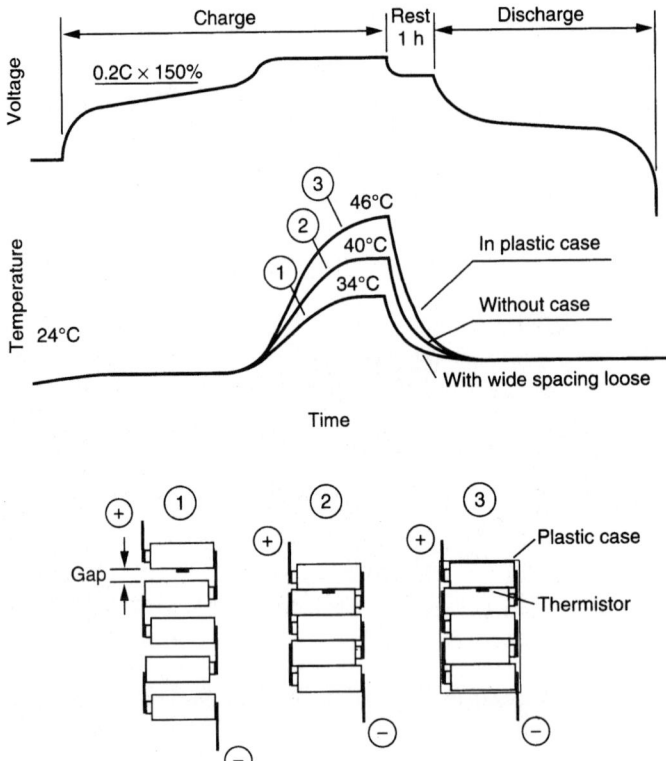

**FIGURE 7.4**   Temperature response characteristics during charge and discharge of the battery pack. (Note that the internal cell temperatures can be higher than the measured skin temperatures of the cell, further increasing the detrimental effects.)

### 7.2.4   Terminals and Contact Materials

Design of the interface between a snap-in battery and the device must consider a number of factors:

1. The terminal material selection must be compatible with the battery and the usage environment. Noncorrosive materials should be selected, such as solid nickel or other metals.

2. The normal force provided by the terminal contacts must be great enough to hold the battery in place, even when the device is dropped and to prevent electrical degradation and any resulting instability.

3. Contacts must be able to resist permanent set: This refers to the ability of the contact to resist permanent deformation after a number of battery insertions.

4. Terminal contact resistance should be minimized. Temperature rise at high currents due to the resistance of the contact material must be limited. Excessive temperature increase could lead to stress relaxation and loss in contact pressure, as well as to the growth of oxide films that raise contact resistance.

5. A common way to minimize contact resistance is to provide a wiping action of the device contact to the battery contact when the battery is inserted in place.

6. Coatings should be selected to satisfy requirements not met by the substrate material, such as conductivity, wear, and corrosion resistance. Gold is an optimal coating due to its ability to meet most of the requirements, but other materials may be used.

Figure 7.5 illustrates a typical battery connector with wiping action receptacles.

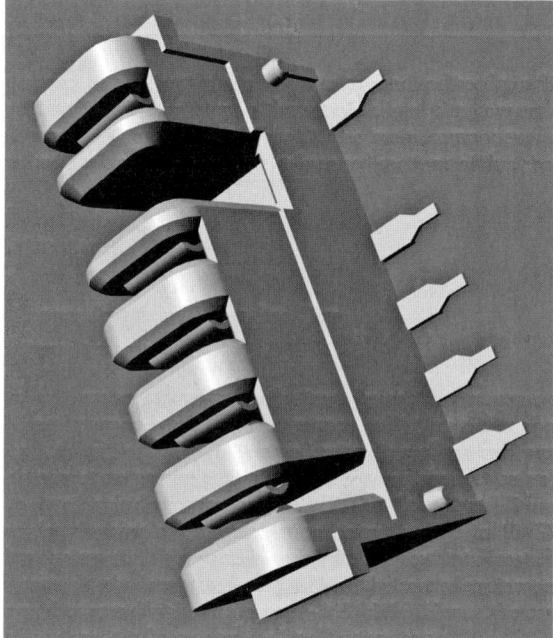

**FIGURE 7.5** Battery pack terminal connector design example (AMP connector).[13]

## 7.3  CHARGING/MONITORING/CONTROL

Since the introduction of rechargeable lithium cells in the mid-1990s, electronics originally utilized for safety and charge control have expanded considerably to now provide multiple "smart" functions for the battery, end-use device, and complete system (including the charger). These functions optimize the performance of the battery pack, control charge and discharge, enhance safety, and provide the user with information on the condition and "health" of the battery. The continued improvement in electronic miniaturization has enabled more analog measurement and digital processing power to be included in Li-ion rechargeable battery systems, either in the battery pack, the charger, or in the host battery-using equipment. Since their introduction, such smart battery electronics have become more sophisticated, more integrated, more precise, and less costly.[3]

Although non-Li-ion battery systems do not require the same safety electronics, most of these still utilize charge control electronics, often due to the unique charge termination techniques required for nickel-based rechargeable cells.

While generically referenced as a "smart battery," there are varying degrees of smart or intelligent battery systems. Specific examples include Smart Battery System (SBS) and System Management Bus (SMBus) products that conform to a set of specifications for interoperability.[4,5]

Batteries of all sizes can incorporate electronic controls for enhanced performance, safety, and reliability. Guidelines for the proper design and use of rechargeable battery packs have expanded since some high-profile accidents in previous years.[6] Guidelines suggest the use of electronics for maintaining the operation of the battery cells within safe limits for particular applications ranging from cellular phones,[7] tablets, laptop computers[8] to electric vehicles and energy storage systems. Mobile portable battery devices such as "self-balancing scooters" (i.e., hoverboards),[9] which even have their own Underwriters Laboratory (UL) guidelines,[10] add to the diversity of systems.

Electronics in a battery pack can range from basic protection functions that prevent or mitigate abusive conditions to sophisticated measurement, calculation, and communication processors that provide protection, monitoring, and communications to a host or end-user device. Although common in smart phones, tablets, laptop computers, and lawn and power tools, more advanced electronics are also now common in battery and hybrid electric vehicles, energy storage systems, and even electric golf carts.[11]

### 7.3.1   Features and Benefits

Some of the features and benefits of embedding advanced electronics into the battery pack, host device system, or charger are discussed next.

*Charge Control.*   Battery electronics can monitor the battery during charge and assist in controlling the charge rate and charge termination method using measurements such as time, maximum voltage, voltage delta (differential), temperature delta (differential), and rate of temperature change, to name a few. These parameters can be utilized to cut off the charge or to switch to a lower charge rate or another charge method.

Constant current-constant voltage hybrid charging can be controlled and options can be incorporated into the electronics for pulse charging, "reflex" charging (a brief periodic discharge pulse during charge), or other appropriate control features, such as precharging cells at a lower charge rate when at low capacity levels or low temperatures.

Finally, charge protection from overcurrent and overvoltage conditions can also be included and tailored for the particular end product requirements.

*Discharge Control.*   Discharge control is also provided to regulate such items as discharge rate, end-of-life cutoff voltage (to prevent over-discharge), and cell equalization (balance), and assist with thermal management.

Individual cells, as well as the entire battery pack, can be monitored for voltage or temperature during discharge and direct action taken to alter the discharge current or inform the host device to terminate or slow the discharge.

Total battery pack current can also be monitored to detect overcurrent and short-circuit conditions to prevent damaging the cells.

*Cell Balancing.*   Cell balancing can be used to improve the performance of multicell battery packs by maintaining all the series cell elements in voltage or capacity balance with one another. Maintaining cell balance therefore increases the usable capacity of the battery pack and improves cycle life.

Cell imbalance is often the cause of poor battery pack life in multicell series packs, typically those greater than four series cells. Even if the individual cells are well matched when the battery pack is assembled, cells will diverge with time due to slight differences in capacity, self-discharge rate, etc.

Temperature gradients across a large multicell battery pack will also alter the rate of divergence such that cell balancing or equalization is often required in applications with large arrays of cells, such as electric and hybrid vehicles.

Some chemistries are easier to rebalance using overcharge techniques, while most lithium rechargeable chemistries require alternate approaches such as bypass balancing or charge transfer balancing.[3]

*Communications.* Communicating the battery information to the end-user or host device can be accomplished by both simple and sophisticated methods, depending on the requirements. Basic measurement data, such as voltage, temperature, and current, can be relayed to the host device for use in charge or discharge control or to calculate battery state-of-charge (SOC), state-of-health (SOH), or state-of-function (SOF). If the battery pack contains on-board calculation capability, then more information can be calculated locally and communicated to the end-user or host device.

Information such as battery SOC estimates the remaining battery capacity by factoring in such variables as the discharge rate and time, temperature, self-discharge, cell impedance, past history, charge rate, and charge duration. The SOC, remaining capacity, or run-time can be displayed locally on the battery pack by a sequence of illuminated LEDs or an LCD display.

Detailed data can also be directly communicated to the host device via a standard communications link such as the Inter-Integrated Circuit (I2C) bus or the derivative SMBus. This communications option provides significantly more information than can be displayed via a local method and is often utilized by the end device for use in a more detailed graphic form, such as on a laptop computer's main screen. Alternatively, single-wire data communications (DQ-bus, HDQ, or others) or simple level-based analog threshold signals can also be used to communicate that the battery is operating outside of normal limits and that external action should be taken by the charger or the host end-use device. For larger systems more robust and longer distance communications links, such as CAN Bus (J1939), RS-485, or other protocols and physical layers, can be used.

*Historical Information.* Battery data are collected during the life of the battery and can be used to make changes to the operating algorithm to maintain optimum performance. Other information can also be collected beyond initial manufacturing information (date of manufacture, chemistry, configuration), including detailed battery history, cycle count, and other such data, which can provide a complete accounting of the battery's usage over time.

Data such as maximum temperature, time at temperature, time at voltage, and similar parameters can also be utilized to determine battery aging. This data can be retained even if a battery failure occurs so that warranty returns can be properly evaluated and analyzed.

*Customization.* The continued advancement of electronics technology allows the monitoring electronics to be custom tailored to the chemistry of the battery as well as the specific requirements of the cell manufacturer and host end device. This improves the accuracy, safety, and reliability of the battery pack as well as the performance of the device using the battery.

"Smart" battery electronics today range from simple protection circuits with limited communications to sophisticated all-in-one circuits that provide protection, fuel-gauging, balancing, charge control, history, and adaptive algorithms that compensate for the battery as it ages. As battery systems have increased in size, such smart electronics have also expanded to monitor and control larger arrays of series and parallel connected cells. Customized as well as off-the-shelf smart battery management electronics are readily available from semiconductor firms such as Analog Devices, Maxim Integrated Products, and Texas Instruments.[12]

### 7.3.2  Battery Electronics Functional Requirements

There are several main elements to consider in the design of batteries containing electronics. These are discussed next.

*Monitoring and Measurement.* There are multiple parameters that can be measured directly by battery pack electronics to provide the basic information about the battery components. These include the individual cell voltages and the complete top to bottom pack voltage, the pack current, individual or grouped cell temperatures, and time. Other parameters impacting the battery pack may also be measured such as the coolant temperature or air-intake temperature in a thermal control system, the external charger voltage prior to connection, the communications bus voltage, etc. Some parameters classified as measurements may be calculated from multiple inputs and also used or classified as measurements.

The difference between "measurement" and "monitor" is primarily in the resolution of the voltage, current, temperature or time parameter, and the frequency of doing so. A "measurement" of one of these analog signals often involves a conversion to a digital representation of the parameter's value. Alternatively, "monitoring" one of these parameters may not involve an analog to digital conversion, but simply an analog comparison of the parameter to a simple threshold limit. Fuel-gauging and similar advanced functions often require "measurements" that can be processed digitally while basic safety protection functions may only require simple "monitoring" to know when a parameter exceeds a limit. In summary, a "measurement" will give a precise value for a cell voltage, for example; but "monitoring" the cell voltage may only provide a value when the cell voltage falls outside a set window or range.

The accuracy of the measurements should fit the requirements of the chemistry and the intended application: Precise fuel-gauging and charge control often require accurate measurements while protection from abusive conditions may require less stringent monitoring. For high-reliability gas-gauging of remaining capacity and run-time, it is important that these measurements be made as accurately as possible to provide the best data for the control algorithms and predictive functions.

When precise measurements are not required, signals will only be generated when the monitored values (cell voltages, pack currents) exceed preset thresholds. Simple comparator-based monitoring is often sufficient when protecting the battery from operation outside desired limits.

Voltage measurements can be critical as the charge control and termination depend on the battery voltage, which, for some chemistries, should be accurate to 25 mV or better at the cell level. An inaccurate measurement could result in under- or overcharge, which could lead to short service life or damage to the battery. In the case of the lithium rechargeable cells, overcharge could be a safety hazard. Similarly, on discharge, terminating the discharge prematurely results in a shortened run-time, while over-discharge could result in damage to the cells.

Errors in the measurement of current not only affect the calculation of capacity and the SOC "gas gauge," but also influence the termination of charge and discharge, as the termination voltages may vary depending on the current, particularly in high-current devices such as power tools. Complicating this measurement is the fact that current is not a constant value during discharge; there are often multiple device power modes and high current pulses as short as milliseconds.

Combinations of certain measurements can provide additional information, such as a cell's DC resistance or impedance, for example, when voltage and current measurements are synchronous.

Temperature is another important parameter, as the performance of batteries is highly temperature dependent, and exposure to high temperatures can cause irreversible damage to the cells. Temperature gradients across the battery pack that cause the cells to reach different temperatures will create cell imbalance, which limits battery life.

The key points are to select monitoring electronics that match the application requirements. A nickel battery pack for a power tool may only require pack-level temperature and voltage monitoring, while the same application using rechargeable lithium may require individual cell voltage monitoring and limited temperature monitoring.

***Calculation Algorithms.***    Having calculation ability in the battery pack allows the system to adjust to the environment or usage conditions and permits safer operation and better performance. The level of calculation required can vary from a predefined logic state machine to a more versatile processor that can factor more variables when making adjustments. Calculation can also occur elsewhere, such as in the device itself, which is typical for most modern cell phones and tablets where the battery pack is not removable. (Previously, these devices had removable—and replaceable—battery packs.)

For example, consider the precharge function typically used with lithium rechargeable chemistries. When the cell voltage or temperature is below preset thresholds, a separate charge path that reduces the charge current is enabled until the temperature or cell voltage increases above the preset thresholds. Alternatively, the proper charge current could be calculated based on the readings and then communicated to the charging system. Similar level-based conditions are often used for charging based on temperature—both low and high. (Note that the location of the charge control device depends on the application—it may be in the battery for small battery devices but physically outside the unit, in the form of a large "contactor" relay, for large battery systems.)

More sophisticated calculation can be required for accurate gas-gauging to represent battery SOC, state-of-power (the battery's ability to provide high discharge currents), SOH (battery life information), or SOF. Although some chemistries can provide relatively useful gas-gauging information by monitoring the open-circuit voltage of the battery, most lithium rechargeable chemistries require more sophisticated approaches.

Calculations transform any measured data through the use of simple or complex algorithms, depending on the host device application's requirements and the cell chemistry. Prior knowledge of battery characteristics, such as capacity at various discharge loads and temperatures, charge acceptance, self-discharge, etc., are required to determine future battery performance. Early battery electronics used simple linear models for these parameters, which severely limited accuracy in predicting the battery's performance. As noted in the descriptions in various chapters in this Handbook, battery performance is often very nonlinear. Self-discharge, for example, is a complex relationship influenced by temperature, time, SOC, and other factors. Further, the performance of batteries using the same chemistry varies with design, size, manufacturer, age, etc. A good calculation engine and algorithm will account for these relationships and help assure safe, reliable operation.

Calculations can also be used to maximize the performance from the battery pack during actual use by considering calculated values, such as cell impedance along with voltage, current, and temperature measurements. Techniques that operate the cells to the edge of their performance envelope require not only precise measurements, but also well-known models of the cells' characteristics and performance under various usage conditions. Processors to perform these measurements and calculations in real time, while under heavy loads, can maximize the performance obtained from the cells in the battery pack. High-end power tool products and hybrid and electric vehicles often use such sophisticated calculations.

In the case of hybrid electric vehicles, the state-of-power calculation is valuable to the vehicle controller: Can the battery support a high-power load for enough time until the internal combustion engine can be restarted? If not, then the engine may not be stopped until the battery has reached a higher state-of-power capability.

As with monitoring and measuring, properly matching the calculation requirements with the battery chemistry's needs and the end-application requirements is critical to a high performance, low cost, reliable design.

***Communication Systems and Protocols.***   Communications can range from detailed measurement or calculated data over a sophisticated communications bus to a single line "go/no-go" signal that indicates that the battery pack is operating outside of preset limits.

Battery packs have for years used a single interface line to represent the temperature of the battery via the voltage on the line. The voltage is a representation of the pack temperature with a negative temperature coefficient (NTC) thermistor. The resistance of an NTC temperature sensing device located in the battery pack is monitored externally, often by the charger. Low resistances represent high temperatures and vice versa. Nickel-based chemistries often use this signaling approach to detect end-of-charge via a change in the rate of rise of the temperature. This same approach can still be utilized by chemistries that do not exhibit any temperature changes with full charge. For example, an early technique with simple lithium rechargeable battery packs is to simply mimic the temperature of a "hot" pack, which signals the charger to terminate charging.

When more information is to be conveyed between the battery, charger, and host device, a digital interface, such as the Inter-Integrated Circuit (I2C), Serial Peripheral Interface (SPI), DQ/HDQ, 1-Wire, Systems Management Bus (SMBus), or similar protocol, is often used. These are standardized data communications interfaces with low-power characteristics well suited for battery applications. The electrical and data protocols are defined and available in many prepackaged parts for use in battery packs, chargers, and end-equipment devices. Automotive and larger energy storage battery systems may utilize local interconnect network (LIN) or controller area network (CAN) bus interfaces for additional robustness.

Between the battery and charger, information that is communicated may include the required charge conditions, such as maximum charge current, maximum charge voltage, and perhaps maximum temperature to initiate charge separately from a maximum charge continuation temperature. As previously mentioned in the precharge example, other charge-gating information may also be communicated relating to the specific conditions of the battery pack prior to initiation of charge.

Information from the battery to the end-equipment host device can be utilized to maximize the run-time of the device while also preventing abusive discharge conditions. In laptop computers various power management techniques are employed by the computer systems as the battery's SOC decreases. High current loads, such as the spin-up current from a motor, can be delayed briefly while the notebook reduces loads elsewhere, perhaps momentarily dimming the screen backlight or powering down other subsections. Screen and backlight time-outs are also often altered based on the battery state. Advanced smart batteries can provide information to the end-user system so that such decisions can be easily determined.

Fuel-gauging that represents the run-time of the device in meaningful terms instead of SOC percentages can also be communicated. Similar application-specific information that brings more meaning to the end equipment and provides a more user-friendly experience is also possible. Smart batteries can provide information ranging from the time remaining during charge as well as discharge, the number of usage cycles, and the approximate remaining useful life.

The reliability of the communication link is critical both to accurately convey the data and to prevent unauthorized access. Modern battery systems include error detection and correction in the communication protocol at a minimum. Other links use encryption or challenge-response authentication to limit unauthorized battery packs to operate in the end equipment or charger.

### 7.3.3   The Smart Battery System (SBS)[4]

A formalized electronic battery management system was created by leading cell suppliers, laptop computer makers, and semiconductor manufacturers in 1995 to standardize the electrical interface between battery packs, chargers, and notebook computers. This SBS (also called SMBus System) has been widely adopted by notebook computer makers and other portable device manufacturers for many industrial and general-purpose battery systems. The physical form factor of the battery packs is not standardized (although some standard sizes exist, such as the DR202). The standardization is only for the communications interface.

The SMBus defines additional protocols and electrical requirements on top of the I2C specification developed by Philips Corp. These protocols include error-detecting mechanisms, minimum voltage levels, and similar timing and power requirements. Typical portable battery systems utilize SMBus V1.1, while fixed nonbattery systems may also use SMBus V2.0 for other devices, such as backlight controllers found in a typical notebook computer.

The SBS also includes specifications for the data content and transfer between a host device such as a notebook computer, the smart battery, and a smart charger. The Smart Battery Data Specification and the Smart Battery Charger Specification detail the interaction and data requirements for each device. SBS smart batteries provide up to 34 data values, both measured and calculated, that can be utilized by the host device or charger to enhance battery performance and system power management. Similarly, there are three levels of smart chargers that can be utilized in SBS platforms.

Most small, complex consumer battery devices utilize the SBS standards to enable multiple battery suppliers and chemistry interchangeability (typically within the lithium rechargeable options).

The goal of the smart battery interface is to provide adequate information for power management and charge control regardless of the particular battery's chemistry. The smart battery consists of a collection of cells or single-cell batteries and is equipped with specialized hardware that provides present state, calculated, and predicted information to the host. The electronics need not be inside the smart battery if the battery is not removable from the device.

Many semiconductor companies such as Maxim Integrated Products and Texas Instruments supply battery monitor, charger, or host controller products that comply to the various SBS standards to provide easy interoperability. Most have extended the data set beyond the 34 core values in order to provide advanced functionality and custom features.

## 7.4   ELECTRICAL MALFUNCTIONS/SAFETY

Batteries are sources of energy and when used properly will deliver their energy in a safe manner. But it is possible for a battery pack to vent, rupture, or even explode if it is abused. The design of the battery pack should include protective devices and other features that can prevent, or at least minimize, this problem.

Some of the most common causes of battery failure are as follows:

1. Short-circuiting of the battery
2. Excessive high-rate discharge or charge
3. Overcharge above or over-discharge below the recommended operating voltage of the cells (also includes voltage reversal, or the discharging of the cell below 0 V)
4. Imbalance between series cells (can lead to overcharge or over-discharge of cells)
5. Improper charge control

These conditions may cause heating and a subsequent increase in internal pressure within the cells, resulting in an activation of the vent device or a rupture or explosion of the battery. Internal cell shorts can also cause failures, although these are rare. These may occur due to impurities being accidentally introduced into the cells during manufacture. There are a number of means to minimize the possibilities of these occurrences. Additional failure mechanisms can occur due to the improper assembly of individual cells into the battery pack. For example, poor cell connector tab welds, lack of proper insulation between tabs, and improper case assembly can all lead to latent battery failures.

The use of high-quality individual cells does not guarantee a safe battery pack assembly. All factors must be considered carefully, including the mechanical assembly of the pack, any internal protection devices or electronics, contacts, monitoring components, and the pack casing.

### 7.4.1    Short-Circuiting of the Battery

When a battery is short-circuited through the external terminals, the chemical energy is converted into heat within the battery. In order to prevent short-circuiting, the positive and negative terminals of the battery should be physically isolated. Effective battery design will incorporate the following:

1. The battery terminals should be recessed within the external case.
2. If connectors are used, the battery should incorporate the female connection. The connector should also be polarized to only permit correct insertion (see Fig. 7.5). This figure illustrates a multipin female connector that receives bladed connections for battery power and other signals (if used). Such a connector is often molded into the battery case.[13]

Even with recessed terminals, it is also necessary to include some means of circuit interruption if a short circuit occurs. There are a number of devices that can perform this function, including

1. Fuses or circuit breakers.
2. Thermostats designed to open the battery circuit when the temperature or current reaches a predetermined upper limit.
3. PTC devices that, at normal currents and temperatures, have a very low value of resistance. When excessive currents pass through these devices or the battery temperature increases, the resistance increases by orders of magnitude, limiting the current. These devices are incorporated internally in some cells by the cell manufacturer. When using cells with internal protection, it is advisable to use an external PTC selected to accommodate both the current and the voltage levels of the battery application. Note that PTC devices will not prevent a short-circuit condition from discharging the cell or pack fully. A continually shorted cell or pack will still be discharged through the PTC, although at a slow rate.
4. Electronic protection methods that monitor for high-current (short-circuit) conditions and interrupt current using a power metal oxide semiconductor field-effect transistor (MOSFET) switch.

The above protection methods go beyond external mechanisms. Proper battery pack assembly is also critical to preventing internal short circuits within the battery pack.

### 7.4.2    Excessive High-Rate Discharge or Charge

As with short-circuit conditions, a high-rate charge or discharge can abuse the cells within a battery pack. Items (3) and (4) mentioned above (a PTC device in the cell or pack or electronic monitoring) are the most common protections.

With larger battery systems such as electric vehicles or even power tools, a PTC in the cell or battery pack is not suitable due to the high peak current demands of these applications. Instead, electronic monitoring is most suitable. When high currents are detected for a predetermined period of time (which may vary with temperature), then current interrupt devices may be activated either within the battery pack or elsewhere in the system. For a power or lawn tool, this activation may be inside the tool or the battery pack. For an electric vehicle, the current interrupt mechanism is likely to be inside the motor controller system.

### 7.4.3   Overcharge or Over-Discharge beyond Recommended Operating Voltages

Maintaining the correct operating voltages of series-connected cells in a battery pack often requires individual cell monitoring and is common in lithium rechargeable systems. Nickel-based rechargeable systems often do not require individual cell monitoring, but instead require that connected cell strings are limited to 10 cells in series. This limit permits the pack voltage to stay within a range such that an individual cell reaching 0 V would be detectable at the pack voltage level.

Discharge beyond the cell manufacturer's recommended operating range is not suggested, so precautions should be taken at the system level if such discharge is possible. If external conditions are likely to cause discharge below recommended limits, then protection within the battery pack may be required as is often utilized in lithium rechargeable chemistries.

When discharged in a series configuration, the capacity of the weakest cell in the series string of a multicell battery will be depleted before the others. If the discharge is continued, the voltage of the low-capacity cell will reach 0 V and then reverse. The heat generated may eventually cause pressure buildup in the cell and subsequent venting or rupture.

Some cells are designed to withstand over- or undervoltage conditions. The cells may be designed with internal protection, such as fuses or thermal cutoff devices, to interrupt the charge or discharge if an unsafe condition develops.

Individual series cell voltage monitoring in lithium systems insures cell voltages stay within operational limits by automatically interrupting the charge or discharge current via an electronic switch (MOSFETs) or mechanical switch (relay or contactor).

### 7.4.4   Imbalance between Series Cells

When multiple series stacks are paralleled within a battery pack, charging may occur when a defective or a low-capacity cell is present in one of the stacks. The remaining stacks of cells will charge the stack with the defective cell. *Therefore, series elements should be connected in parallel at the cell level* (as shown on the left in Fig. 7.6).

When larger battery systems utilize parallel connected strings of cells, then current limiting controls must be put in place to prevent large current flows between cell strings that are not at the same voltage (i.e., unbalanced strings).

This condition of cell unbalance could be exacerbated with rechargeable cells, as the individual cell capacities could change during cycling. To minimize this effect, rechargeable batteries should at least be constructed with "matched" cells, that is, cells having nearly identical capacities. Cells are sorted, within grades, after at least one cycle of charge and discharge.

Cell imbalance, however, can occur after the cells are assembled into a battery pack and utilized in the end application. Such cell imbalance can result from uneven thermal gradients across the battery that cause some cells to reach higher temperatures than others. This temperature gradient will cause a difference in cell self-discharge and degradation, potentially leading to cell imbalance. If imbalance occurs within the battery pack, then corrective action must be taken to prevent an accumulation of imbalance. Some chemistries permit a low-rate overcharge to correct imbalance, while other chemistries, such as lithium, require rebalancing by electronic methods.

### 7.4.5   Improper Charge Control

A combination effect of the previous fault conditions can occur with improper charge control of a battery pack. If the charger system does not properly monitor the battery pack, perhaps due to a legacy charger with a new battery, or vice versa, then damage can occur.

To mitigate this, the battery pack protection devices as previously mentioned must be included to consider the impact of improper charge control. Examples include charge time-out timers, electronic pack monitoring, temperature monitoring, or other methods to detect improper charge control.

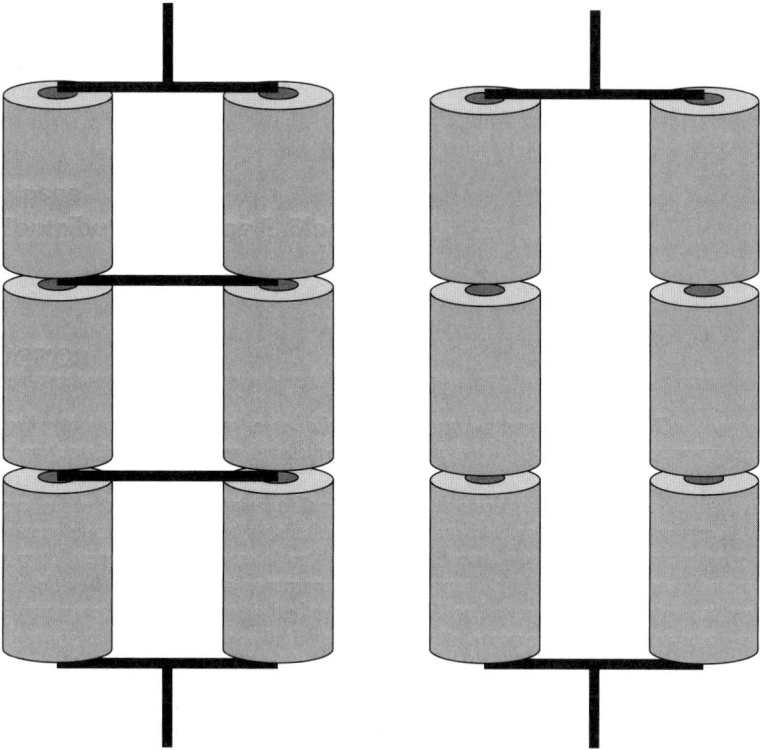

**FIGURE 7.6**    Preferred method (*left*) to connect parallel cell strings.

## 7.5    *MECHANICAL MALFUNCTIONS/SAFETY*

### 7.5.1    Cell Packaging Requirements

Recent battery pack requirements from regulatory agencies have pushed the mechanical pack design to include new restrictions on cell location within a battery pack so as to prevent a catastrophic thermal event from propagating to other cells. As battery pack sizes increase for new applications such as lawn power equipment, electric vehicles, and energy storage systems, the requirement for failure containment has grown.

At present, arrays of more than four cells should have cell separations greater than 2 mm and cell orientation that prevents venting of one cell from potentially igniting another cell.[2]

Similarly, the use of PCM or other heat-absorbing techniques inside the battery pack is recommended.

Shipping regulations, primarily driven by air-transport organizations, require mechanical malfunction testing to insure the ability of battery packs to safely survive drop or other damage that might be introduced during shipping and handling.

### 7.5.2    Design to Prevent Improper Insertion of Discrete Batteries[14]

When designing products using individual single-cell batteries, special care must be taken in the layout of the battery compartment. If provisions are not made to ensure the proper placement of the single-cell batteries, a situation may result in which some that are improperly inserted could be exposed to charge. This could lead to leakage, venting, rupture, or even explosion.

To minimize the possibility of physically reversing a battery, the proper battery orientation should be clearly marked on the device, with simple and clear instructions. Blind battery compartments, where the individual batteries are not visible, should be avoided. The best practice is to use oriented or polarized battery holders.

A suggested approach is to design the cell cavities for single cells so there are no strings of cells that could allow an incorrect reversal insertion of one cell. This does add cost to the device by requiring additional contacts, but it ensures that the circuit is correctly completed (by virtue of the physical connection of the cells by the device's circuit). Such a design is strongly suggested when a device can accept primary and rechargeable cells of a particular size, such as the AA or AAA sizes, which are commonly available in primary alkaline, rechargeable nickel, or primary lithium.

## 7.6   EXAMPLE BATTERY DESIGNS

Specific examples of design techniques for rechargeable battery packs in consumer devices, using lithium rechargeable cells, are detailed next.

### 7.6.1   Cell Balancing

Multicell rechargeable battery packs should be built using cells having matched capacities. In a series-connected multicell battery, the cell with the lowest capacity will determine the duration of the discharge, while the one with the highest capacity will control the capacity returned during the charge. If the cells are not balanced, the battery will not be charged to its designed capacity. To minimize the mismatch, the cells within a multicell battery should be selected from one production lot, and the cells selected for a given battery should have nearly identical capacities. Due to the need for current limits during charge of lithium ion cells, individual cell capacities cannot be balanced with a top-off or trickle charge.

Alternatively, techniques for minimizing imbalance should be employed as previously mentioned: reduce thermal gradients, limit differences in charge or discharge rates, etc. Modern electronic circuits available for lithium rechargeable batteries also include options to rebalance individual cells using resistive dissipative or active charge transfer methods.

Discharging secondary lithium cells beyond manufacturers' recommended limits should be minimized, particularly at low discharge rates. Brief excursions below minimum voltages at high discharge rates may be permitted. Furthermore, safeguards must be included to control charging so as to prevent damage to the battery due to abusive charging. Proper control of the charge and discharge process is critical to the ultimate life and safety of the battery.

The two major considerations to be addressed include:

1. Voltage and current control to prevent overcharge (overvoltage) and over-discharge (undervoltage). These controls can be located in the battery pack for redundancy or can be part of the device's system design, which includes the charger.

2. Temperature sensing and response to maintain the battery temperature within the range specified by the battery manufacturers.

### 7.6.2   Simple Protective Devices

The controls for voltage and current during charge for most lithium rechargeable chemistry batteries are contained in the charger.

Protective devices should be installed within the battery pack to stop the charge in the event of an unacceptable temperature rise or other abnormal condition. The thermal devices that can be used include the following:

1. *Thermistor.* This device is a calibrated resistor that varies inversely with temperature. The nominal resistance is set at 25°C and is in the kilohm range with 10K being the most common. By proper placement within the battery pack, a measurement of the temperature of the battery is available and $T_{max}$, $T_{min}$, and $\Delta T/\Delta t$ or other

such parameters can be established for charge control. Also, during discharge the battery temperature may be measured and used to modify the discharge loads (i.e., terminate discharge until battery temperature decreases in order to prevent thermal damage to the battery).

2. *Thermostat (temperature cutoff [TCO]).* This device operates at a fixed temperature and is used to cut off the charge (or discharge) when a preestablished internal battery temperature is reached. TCOs are usually resettable. They are connected in series within the cell stack.

3. *Thermal fuse.* This device is wired in series with the cell stack and will open the circuit when a predetermined temperature is reached. Thermal fuses are included as a protection against thermal runaway and are normally set to open at approximately 30–50°C above the maximum battery operating temperature. They do not reset and are often placed on or near the power control devices in a battery pack, such as power MOSFETs used to interrupt current flow in or out of the battery.

4. *PTC device.* This is a resettable device, connected in series with the cells, whose resistance increases rapidly when a preestablished temperature is reached, thereby reducing the current in the battery to a low and acceptable current level. It will respond to high circuit current beyond design limits (such as a short circuit) and acts like a resettable fuse. It will also respond to high temperatures surrounding the PTC device, in which case it operates like a TCO device. PTCs are also often embedded in the lithium rechargeable cells unless a high charge or discharge rate is required of the cell.

Figure 7.1 shows a schematic of a typical battery circuit, indicating the electrical location of these protective devices. The location of the thermal devices in the battery assembly is critical to ensure that they will respond properly, as the temperature may not be uniform throughout the battery pack. Examples of recommended locations in a battery pack are shown in Fig. 7.1. Other arrangements are possible, depending on the particular battery design and application.

It should be noted that for lithium-rechargeable chemistries, temperature is less useful as a charge control mechanism. Cell voltage is the best control mechanism for the lithium rechargeable chemistries.

### 7.6.3   Examples of Discharge and Charge Control

Electronic circuitry can be used to maximize battery service life by cutting off the discharge as close to the specified end or cut-off voltage as possible. Ending the discharge at an excessively high voltage will result in a loss of a significant amount of battery capacity. Ending at a very low voltage, and thus discharging the battery beyond its safe cut-off, could cause permanent damage to the battery (see Chap. 17A). Similarly, on charge, accurate control, as discussed previously, will enable a maximum charge under safe conditions without damage to the battery.

Modern battery packs and devices will utilize electronics in the device or battery to provide monitoring that ensures the operational limits of the battery pack are not exceeded. These electronic circuits can also provide enhanced safety and reliability, fuel-gauging, warranty data recording, and long-term battery health information. Some devices such as smart phones, music players, and portable speakers may use system-side monitoring and protection circuits, while larger devices such as laptop computers and power tools use monitoring circuits located inside the battery pack.

The ability to interrupt charge and discharge current to protect the battery may also reside in a separate location from where the monitoring occurs. Power tools, for example, may only interrupt charge current inside the charger, while a laptop computer may contain a battery pack that can interrupt both charge and discharge currents inside the battery pack itself.

Discharge and charge control are especially important for a lithium-rechargeable battery that is recharged at a relatively high, constant (typically) current to a given voltage with subsequent taper-current charging at a constant voltage to a given current cutoff. Exceeding the maximum voltage is a potential safety hazard and could cause irreversible damage to the battery. Charging to a lower voltage will reduce the capacity of the battery, although in some applications such as uninterruptible power supply (UPS) systems, charging to a lower voltage is preferred for cycle and calendar life. As the battery is continuously charged, the lower capacity is an acceptable side effect in this application since long battery life is a more critical need.

Other useful information can also be monitored and recorded by any charge or discharge control electronic circuitry either in the battery pack or on the device (system) side. This information can be used for warranty return analysis, life-time predictions, fuel-gauging, and similar advanced features.

For lithium-rechargeable battery packs, the voltage of each individual cell in the battery pack is monitored on a continuous basis. Due to safety concerns, secondary cell voltage monitors are often employed in the event that the primary monitor fails. These secondary monitors typically only look for an overvoltage condition and once detected, activate a permanent fuse in the charge current path. Depending on the specific lithium-ion battery chemistry that is used, the upper voltage limit on charge, as specified by the manufacturer, is usually limited between 4.1 and 4.5 V. On discharge, the cell voltage should not fall below 2.3 to 2.7 V. Newer lithium-ion and polymer formulations have significantly different over- and undervoltage limits. Phosphate-based lithium-ion voltages have maximum limits near 3.8 V and minimums below 2.0 V. Titanate-based chemistries are even lower, with maximum limits near 2.5 V and minimums near 1.8 V. In all cases, the tolerance for primary detection is typically ±25 to 50 mV for overvoltage conditions and ±50 to 100 mV for undervoltage conditions, although higher precision permits operating closer to the cell voltage limits recommended by the cell manufacturers.

As with any battery system, high temperature will cause irreversible damage. With lithium-ion and polymer cells, temperature can alter how the cells should be charged or discharged. Guidelines from industry organizations limit charge currents when temperatures exceed suggested thresholds, while cell manufacturers often have similar limits for discharge currents at low temperatures.[6] Internal battery temperature, for most applications, should be kept below 75°C. TCO, with a trip of 70°C and reset temperature in the range of 45 to 55°C, is routinely used. Temperatures in excess of 100°C could result in permanent cell damage. For this situation, permanent type fuses are used, typically set for 104°C with a tolerance of ±5°C. Temperature in lithium rechargeable chemistries is more difficult to detect compared to nickel-based chemistries, which exhibit a more linear trend. Internal cell temperatures are difficult to detect, and once a large rise occurs, it may be too late to take effective action. Thermal runaway can occur at temperatures as low as 130°C.

Normally, current limits are incorporated into the protection circuits located in the battery pack or device. These circuits monitor the current in or out of the battery cells via a very low value series-sense resistance placed in-line with the power path. These circuits must be operating continuously and respond quickly to open a power MOSFET or similar device to interrupt the current. Short-circuit protection on discharge as well as overcurrent protection on charge (from a faulty charger) are often employed in lithium battery packs. As a backup, a PTC device or fuse is placed in series with the battery pack. It is advisable to place the PTC between the pack assembly and the output of the battery. By placing it at this point, the PTC will not interfere with the operation of the upper or lower voltage detection of the electronic control circuit. However, for some high-rate devices, such as power tools and electric vehicles, a PTC is not utilized since short duration peak currents must be tolerated. In these devices, the electronic overcurrent monitors may have multiple detection thresholds that are able to respond to not only the magnitude of the current, but also the time duration of the current.

## 7.7   SUMMARY

As battery chemistry technology improves, the requirements for battery pack monitoring, measurement, control, communications, and assembly will change. New requirements for battery packs to insure safety in new use cases, such as on-board aircraft, in intrinsically safe environments, or large format home use will require new design guidelines. Just as the technology of the cells improves and changes, designs for battery packs will also continue to change.

But even with today's chemistries and formulations, the presence of larger cells and more cells connected together require new design options. Datacenter backup and electric vehicles reaching 500 V or more require very different design techniques to insure reliable and safe operation. Similarly, lower voltage but larger capacity battery designs will also require new methods to insure safety and performance.

## REFERENCES

1. R.S. Tichy, "The Dangers of Counterfeit Battery Packs," Electronic Design Magazine, October 23, 2008, www.Micro-Power .com (Micro-Power Electronics, now Integer).

2. J. Jeevarajan, C. Lopez, and J. Orieukwu, "Can Cell to Cell Thermal Runaway Propagation be Prevented in a Li-ion Battery Module?" https://ntrs.nasa.gov/search.jsp?R=20140012758. September 25, 2014.

3. Datasheets for smart battery monitors, bq20z95, bq76940, bq78PL114, and bq77PL900. Texas Instruments, www.ti.com.

4. Smart Battery Data Specification, Rev. 1.1, System Management Bus Specification, www.sbs-forum.org; Smart Battery System Implementers Forum, part of the System Management Interface Forum, Inc.

5. D. Friel, "How Smart Should a Battery Be," *Battery Power Products and Technology*, March 1999.

6. "A Guide to the Safe Use of Secondary Lithium Ion Batteries in Notebook-Type Personal Computers" and "Safe Use Manual for Lithium Ion Rechargeable Batteries in Notebook Computers," Japan Electronics and Information Technology Industries Association (JEITA), www.jeita.or.jp, and Battery Association of Japan (BAJ), www.baj.or.jp; April 2007.

7. IEEE ANSI STD. 1725(TM)-2006, "IEEE Standard for Rechargeable Batteries for Cellular Telephones," IEEE, 3 Park Avenue, New York, www.ieee.org.

8. IEEE Std. 1626(TM)-2008, "IEEE Standard for Rechargeable Batteries for Multi-Cell Mobile Computing Devices," IEEE, 3 Park Avenue, New York, www.ieee.org.ref.

9. UL Std 2272—Standard for Electrical Systems for Personal E-Mobility Devices, https://standardscatalog.ul.com/standards/en/standard_2272_1.

10. UL 2272—"Battery Systems for Use in Self Balancing Scooters," https://industries.ul.com/blog/the-new-ul-2272-standard-gets-a-handle-on-hoverboard-safety.

11. EZ-Go Elite Lithium-Battery Powered Electric Golf Carts, http://www.ezgo.txtsv.com/golf/elite-lithium-0.

12. Texas Instruments, www.ti.com/lsds/ti/power-management/battery-management-products-overview.page.

13. TE Connectivity, Battery Interconnection System Products Receptacle Assemblies, e.g., part numbers 1-1123688-7 or 1-1437118-0. http://www.te.com/usa-en/product-1-1123688-7.html (original source was AMP/Tyco).

14. Duracell Alkaline Technical Bulletin, www.Duracell.com/OEM.

# CHAPTER 8
# BATTERY STANDARDS

**Steven Wicelinski**

## 8.1　GENERAL

Battery standardization efforts began in 1912, when a committee of the American Electrochemical Society recommended standard methods of testing dry cells. In 1917, work began on the first nationally recognized specification to include cell sizes, battery arrangements, standard tests, and performance requirements, eventually leading to the first national publication in 1919, issued as an appendix to a circular on dry cells from the National Bureau of Standards. This group further evolved into the present American National Standards Institute (ANSI) Accredited Standards Committee C18 on Portable Cells and Batteries. Subsequently, other professional societies have developed battery-related standards, including battery standards issued by international, national, military, and federal organizations. Manufacturers' associations and trade organizations have published standards as well. Related application standards published by Underwriters Laboratories, the International Electrotechnical Commission (IEC), and other organizations that cover battery-operated equipment may also be of interest. In addition, organizations such as SAE International (formerly known as The Society of Automotive Engineers) publish testing methods, specifications, and standards relevant to battery components and systems.

One of the major goals of battery standardization is interchangeability of common consumer replaceable batteries in electronic devices across the globe. Battery interchangeability is achieved by specifying the preferred values for the physical aspects of the battery, such as dimensions, polarity, terminals, nomenclature, and markings. In addition, performance characteristics, such as service life or capacity, may be described and specified with test conditions for verification.

Most batteries, particularly primary cells, will be replaced at some time. Therefore, certain characteristics of the battery (size, shape, voltage, and terminals) must be specified as standard values to insure compatibility. Otherwise, interchangeability will be limited. These characteristics are absolute requirements in order to fit the appliance receptacle to make proper contact and to provide the proper voltage. In addition to the end user's need for replacement information, the original equipment manufacturer (OEM) appliance designer must have a reliable source of information about these parameters in order to design a battery compartment and circuits that will accommodate the tolerances on battery products available for purchase by the end user.

## 8.2　INTERNATIONAL STANDARDS

International standards are rapidly gaining in importance. This has been further accelerated by the creation of the European Common Market and the 1979 Agreement on Technical Barriers to Trade. The latter requires the use of international standards for world trade purposes.

The IEC is the designated organization responsible for standardization in the fields of electricity, electronics, and related technologies. Promoting international cooperation on all questions of electrotechnical standardization and related matters is its basic mission. This organization was founded in 1906 and consists of national committees that represent more than 80% of the world's population and 95% of the world's production and consumption of electricity. The International Standards Organization (ISO) is responsible for international standards in fields other than electrical. IEC and ISO are gradually adopting equivalent development and documentation procedures, while ever closer ties are being established between these two international organizations.

The ANSI is the sole U.S. representative of the IEC through the U.S. National Committee (USNC), which coordinates all IEC activities in the United States. ANSI also serves as the U.S. interface with emerging regional standards developing bodies such as the European Committee for Electrotechnical Standardization (CENELEC), Council for Harmonization of Electrotechnical Standardization of the Nations of the Americas (CANENA), Pan American Standards Commission (COPANT), African Organisation for Standardisation (ARSO), and other groups. ANSI does not develop standards, but rather facilitates standards development by establishing consensus among accredited, qualified groups. ANSI standards are published as U.S. National Standards.

To further IEC's overall mission, certain objectives have been established:

1. Efficiently meet the requirements of the global marketplace
2. Ensure maximum use of standards and conformity assessment schemes
3. Assess and improve the quality of products and services covered by standards
4. Establish conditions for interchangeability
5. Increase the efficiency of electrotechnical industrial processes
6. Contribute to the improvement of human health and safety
7. Work toward protection of the environment

The objectives of the international battery standards are to:

1. Define a standard of quality and provide guidance for its assessment
2. Ensure the electrical and physical interchangeability of products from different manufacturers
3. Limit the number of battery types
4. Provide guidance on matters of safety

The IEC sponsors the development and publication of standard documents. This development is carried out by working groups of experts from participating countries. These experts represent consumer, user, producer, academia, government, and trade and professional interests in the consensus development of these standards. The groups of experts in IEC working on battery standards are:

1. Technical Committee (TC) 21 and Sub-Committee (SC) 21A Secondary Batteries, and
2. Technical Committee (TC) 35 Primary Batteries (nonrechargeable)

The designation for the ANSI Committee on Portable Cells and Batteries is C18.

The following section details the IEC standards that pertain to primary and rechargeable/secondary batteries. Many countries utilize these standards either by simply adopting them in full as their national standards or by harmonizing their own national standards to the IEC standards. ANSI battery standards play an important role by harmonizing requirements with IEC standards when feasible.

## 8.3   *COMMON BATTERY STANDARDS*

Since the creation of the IEC TC21 Committee in the 1930s and the IEC TC35 Committee in the 1940s and along with the ANSI C18 Committee, several standard documents have been developed and published. Tables 8.1*a* to 8.1*d* list some of the widely known standards for batteries.

**TABLE 8.1a**    International Standards (IEC—International Electrotechnical Commission)

| Publication | Title | Electrochemical systems |
|---|---|---|
| IEC 60086-1 | Primary Batteries; Part 1, General, and Part 2, Specification | Zinc-carbon |
| IEC 60086-2 | Sheets | Zinc/air |
| | | Alkaline-manganese dioxide |
| | | Nickel oxyhydroxide |
| | | Silver oxide |
| | | Lithium/carbon monofluoride |
| | | Lithium/manganese dioxide |
| | | Lithium/thionyl chloride |
| IEC 60086-3 | Watch Batteries | |
| IEC 60095 | Lead-Acid Starter Batteries | Lead-acid |
| IEC 60254 | Lead-acid Traction Batteries | Lead-acid |
| IEC 61951-1 | Portable Sealed Rechargeable Single Cells; Part I: Nickel-Cadmium | Nickel-cadmium |
| IEC 61960 | Secondary Lithium Cells and Batteries for Portable Applications | Lithium-ion |
| IEC 60622 | Sealed Nickel-Cadmium Prismatic Rechargeable Single Cells | Nickel-cadmium |
| IEC 60623 | Vented Nickel-Cadmium Prismatic Rechargeable Cells | Nickel-cadmium |
| IEC 60952 | Aircraft Batteries | Nickel-cadmium |
| | | Lead-acid |
| IEC 60896 | Stationary Lead-Acid Batteries | Lead-acid |
| IEC 61056 | General Purpose Lead-Acid Batteries | Lead-acid |
| IEC 61427 | Secondary Cells and Batteries for Photovoltaic Energy Systems | |
| IEC 61951-2 | Portable Sealed Rechargeable Single cells; Part 2: Nickel-Metal Hydride | Nickel-metal hydride |
| IEC 61959 | Mechanical Tests for Sealed Portable Secondary Cells and Batteries | |
| IEC 61982 | Secondary Batteries (Except Li) for the Propulsion of Electric Vehicles | |
| IEC 62620 | Secondary Lithium Cells and Batteries for Use in Industrial Applications | Lithium-ion |
| IEC 62660 | Secondary Lithium-Ion Cells for the Propulsion of Electric Vehicles | Lithium-ion |

*Note:* See Table 8.10a for IEC Safety Standards.

**TABLE 8.1b**    National Standards (ANSI—American National Standards Institute)

| Publication | Title | Electrochemical systems |
|---|---|---|
| ANSI C18.1M, Part 1 | Standard for Portable Primary Cells and Batteries with Aqueous Electrolyte | Zinc-carbon<br>Alkaline-manganese dioxide<br>Silver oxide<br>Zinc/air |
| ANSI C18.2M, Part 1 | Standard for Portable Rechargeable Cells and Batteries | Nickel-cadmium<br>Nickel-metal hydride<br>Lithium-ion |
| ANSI C18.3M, Part 1 | Standard for Portable Lithium Primary Cells and Batteries | Lithium/carbon monofluoride<br>Lithium/manganese dioxide |
| ANSI C18.4M | Standard for Portable Cells and Batteries—Environmental | Primary |

*Note:* See Table 8.10a for ANSI Safety Standards.

**TABLE 8.1c**  U.S. Military Standards (MIL)

| Publication | Title | Electrochemical systems |
|---|---|---|
| MIL-B-18 | Batteries Non-Rechargeable | Zinc-carbon |
| MIL-B-8565 | Aircraft Batteries | Various |
| MIL-B-11188 | Vehicle Batteries | Lead-acid |
| MIL-B-49030 | Batteries, Dry, Alkaline (Nonrechargeable) | Alkaline-manganese dioxide |
| MIL-B-55252 | Batteries, Magnesium | Magnesium |
| MIL-B-49436 | Batteries, Rechargeable, Sealed Nickel-Cadmium | Nickel-cadmium |
| MIL-B-49450 | Vented Aircraft Batteries | Nickel-cadmium |
| MIL-B-49458 | Batteries, Non-Rechargeable | Lithium/manganese dioxide |
| MIL-B-49461 | Batteries, Non-Rechargeable | Lithium/thionyl chloride |
| MIL-B-55130 | Batteries, Rechargeable, Sealed Nickel-Cadmium | Nickel-cadmium |
| MIL-B-81757 | Aircraft Batteries | Nickel-cadmium |
| MIL-PRF-49471 | Batteries, Non-Rechargeable, High Performance | Various |

**TABLE 8.1d**  Manufacturers' and Professional Associations

| Publication | Title | Battery type covered |
|---|---|---|
| Society of Automotive Engineers | | |
|    SAE AS 8033 | Aircraft Batteries | Nickel-cadmium |
|    SAE J 537 | Storage Batteries | Lead-acid |
| Battery Council International | Battery Replacement Data Book | Lead-acid |

### 8.3.1  Cross-References of ANSI and IEC Battery Standards

Tables 8.2a and 8.2b list some of the more popular ANSI battery standards and cross-references to the international standard publications for primary and secondary batteries.

### 8.3.2  Listing of IEC Standard Round Batteries

The 13th edition of IEC 60086-2 for primary batteries lists over 100 types with dimensional, polarity, voltage, and electrochemical requirements. The fourth edition of IEC 61951-2 for rechargeable nickel-metal hydride cells (batteries) lists 25 sizes with diameter and height specified in chart form. Several rechargeable nickel-metal hydride and nickel-cadmium batteries are also packaged to be interchangeable with the popular sizes in the primary replacement market. These have physical shapes and sizes that are identical to primary batteries and have equivalent voltage outputs under load. These batteries carry, in addition to the rechargeable nomenclature, the equivalent primary battery size designations and therefore must comply with the dimensional requirements set forth for primary batteries. Table 8.3a lists the dimensions of select round primary batteries, and Table 8.3b lists some nickel-metal hydride rechargeable batteries that are interchangeable with the primary batteries.

In addition to the many designations in national and international standards, of which there may typically be both old and new versions, there are trade association designations. Cross-references to many of these may be found in sales literature and point-of-purchase information.

**TABLE 8.2a**  ANSI/IEC Cross-Reference for Primary Batteries

| ANSI | IEC | ANSI | IEC |
|------|------|--------|---------|
| 13A | LR20 | 1158SO | SR58 |
| 13AC | LR20 | 1160SO | SR55 |
| 13D | R20C | 1162SO | SR57 |
| 14A | LR14 | 1163SO | SR59 |
| 14AC | LR14 | 1164SO | SR59 |
| 14D | R14C | 1165SO | SR57 |
| 15A | LR6 | 1166A | LR44 |
| 15AC | LR6 | 1170SO | SR55 |
| 15D | R6C | 1175SO | SR60 |
| 15N | ZR6 | 1179SO | SR41 |
| 24A | LR03 | 1406SO | 4SR44 |
| 24AC | LR03 | 1412A | 4LR61 |
| 24D | R03 | 1414A | 4LR44 |
| 24N | ZR03 | 1604A | 6LR61 |
| 908A | 4LR25X | 1604AC | 6LR61 |
| 910A | LR1 | 1604D | 6F22 |
| 918A | 4LR25-2 | 1604F | 6F22 |
| 918D | 4R25-2 | 5000LC | CR2016 |
| 1107SO | SR44 | 5003LC | CR2025 |
| 1131SO | SR44 | 5004LC | CR2032 |
| 1133SO | SR43 | 5018LC | CR17345 |
| 1134SO | SR41 | 5024LC | CR-P2 |
| 1135SO | SR41 | 5032LC | 2CR5 |
| 1136SO | SR48 | 7000Z | PR48 |
| 1137SO | SR48 | 7002Z | PR41 |
| 1138SO | SR54 | 7003Z | PR44 |
| 1139SO | SR42 | 7005Z | PR70 |

**TABLE 8.2b**  ANSI/IEC Select Cross-References for Rechargeable Batteries

| ANSI | IEC |
|------|------|
| 1.2H1 | HR03 |
| 1.2H2 | HR6 |
| 1.2H3 | HR14 |
| 1.2H4 | HR20 |

**TABLE 8.3a**  Dimensions of Round Primary Batteries

| IEC designation | Diameter, mm | | Height, mm | |
|-----------------|---------|---------|---------|---------|
| | Maximum | Minimum | Maximum | Minimum |
| R03 | 10.5 | 9.8 | 44.5 | 43.5 |
| R1 | 12.0 | 10.9 | 30.2 | 29.1 |
| R6 | 14.5 | 13.7 | 50.5 | 49.5 |
| R14 | 26.2 | 24.9 | 50.0 | 48.6 |
| R20 | 34.2 | 32.3 | 61.5 | 59.5 |
| R41 | 7.9 | 7.55 | 3.6 | 3.3 |

*(Continued)*

**TABLE 8.3a**    Dimensions of Round Primary Batteries (*Continued*)

| IEC designation | Diameter, mm | | Height, mm | |
|---|---|---|---|---|
| | Maximum | Minimum | Maximum | Minimum |
| R42 | 11.6 | 11.25 | 3.6 | 3.3 |
| R43 | 11.6 | 11.25 | 4.2 | 3.8 |
| R44 | 11.6 | 11.25 | 5.4 | 5.0 |
| R48 | 7.9 | 7.55 | 5.4 | 5.0 |
| R54 | 11.6 | 11.25 | 3.05 | 2.75 |
| R55 | 11.6 | 11.25 | 2.1 | 1.85 |
| R56 | 11.6 | 11.25 | 2.6 | 2.3 |
| R57 | 9.5 | 9.15 | 2.7 | 2.4 |
| R58 | 7.9 | 7.55 | 2.1 | 1.85 |
| R59 | 7.9 | 7.55 | 2.6 | 2.3 |
| R60 | 6.8 | 6.5 | 2.15 | 1.9 |
| R62 | 5.8 | 5.55 | 1.65 | 1.45 |
| R63 | 5.8 | 5.55 | 2.15 | 1.9 |
| R64 | 5.8 | 5.55 | 2.7 | 2.4 |
| R65 | 6.8 | 6.6 | 1.65 | 1.45 |
| R66 | 6.8 | 6.6 | 2.6 | 2.4 |
| R67 | 7.9 | 7.65 | 1.65 | 1.45 |
| R68 | 9.5 | 9.25 | 1.65 | 1.45 |
| R69 | 9.5 | 9.25 | 2.1 | 1.85 |
| R1220 | 12.5 | 12.2 | 2.0 | 1.8 |
| R1620 | 16 | 15.7 | 2.0 | 1.8 |
| R2016 | 20 | 19.7 | 1.6 | 1.4 |
| R2025 | 20 | 19.7 | 2.5 | 2.2 |
| R2032 | 20 | 19.7 | 3.2 | 2.9 |
| R2320 | 23 | 22.6 | 2.0 | 1.8 |
| R2430 | 24.5 | 24.2 | 3.0 | 2.7 |
| R11108 | 11.6 | 11.4 | 10.8 | 10.4 |

**TABLE 8.3b**    Dimensions of Some Popular Nickel-Metal Hydride Rechargeable Batteries That Are Interchangeable with Primary Batteries[*]

| IEC designation[†] | Consumer designation | ANSI designation | Diameter, mm | | Height, mm | |
|---|---|---|---|---|---|---|
| | | | Maximum | Minimum | Maximum | Minimum |
| HR03 | AAA | 1.2H1 | 10.5 | 9.5 | 44.5 | (43.3) |
| HR6 | AA | 1.2H2 | 14.5 | 13.5 | 50.5 | (49.2) |
| HR14 | C | 1.2H3 | 26.2 | 24.9 | 50.0 | (48.5) |
| HR20 | D | 1.2H4 | 34.3 | 32.2 | 61.5 | (59.5) |

[*]Figures in parentheses are reference values.
[†]From IEC Standard 61951-2.

### 8.3.3    Standard SLI and Other Lead-Acid Batteries

SLI (starting-lighting-ignition) battery sizes have been standardized by both the automotive industry through SAE International and the battery industry through the Battery Council International (BCI).[1,2] The BCI nomenclature follows the standards adopted by its predecessor, the American Association of Battery Manufacturers (AABM). The latest standards are published annually by the BCI. Table 8.4 is a listing of the standard SLI and other lead-acid batteries abstracted from the BCI publication.[3]

**TABLE 8.4**  Standard SLI and Other Lead-Acid Batteries

| | BCI group numbers, dimensional specifications, and ratings | | | | | | | | |
|---|---|---|---|---|---|---|---|---|---|
| | Maximum overall dimensions | | | | | | | Performance ranges | |
| | Millimeters | | | Inches | | | | Cold cranking performance amps. at 0°F (−18°C) | Reserve capacity min at 80°F (27°C) |
| BCI group number | L | W | H | L | W | H | Assembly figure no. | | |
| Passenger car and light commercial batteries 12 V (six cells) | | | | | | | | | |
| 21 | 208 | 173 | 222 | $8^3/_{16}$ | $6^{13}/_{16}$ | $8^3/_4$ | 10 | 310–400 | 50–70 |
| 21R | 208 | 173 | 222 | $8^3/_{16}$ | $6^{13}/_{16}$ | $8^3/_4$ | 11 | 310–500 | 50–70 |
| 22F | 241 | 175 | 211 | $9^1/_2$ | $6^7/_8$ | $8^5/_{16}$ | 11F | 220–425 | 45–90 |
| 22HF | 241 | 175 | 229 | $9^1/_2$ | $6^7/_8$ | 9 | 11F | 400 | 69 |
| 22NF | 240 | 140 | 227 | $9^7/_{16}$ | $5^1/_2$ | $8^{15}/_{16}$ | 11F | 210–325 | 50–60 |
| 22R | 229 | 175 | 211 | 9 | $6^7/_8$ | $8^5/_{16}$ | 11 | 290–350 | 45–90 |
| 24 | 260 | 173 | 225 | $10^1/_4$ | $6^{13}/_{16}$ | $8^7/_8$ | 10 | 165–625 | 50–95 |
| 24F | 273 | 173 | 229 | $10^3/_4$ | $6^{13}/_{16}$ | 9 | 11F | 250–700 | 50–95 |
| 24H | 260 | 173 | 238 | $10^1/_4$ | $6^{13}/_{16}$ | $9^3/_8$ | 10 | 305–365 | 70–95 |
| 24R | 260 | 173 | 229 | $10^1/_4$ | $6^{13}/_{16}$ | 9 | 11 | 440–475 | 70–95 |
| 24T | 260 | 173 | 248 | $10^1/_4$ | $6^{13}/_{16}$ | $9^3/_4$ | 10 | 370–385 | 110 |
| 25 | 230 | 175 | 225 | $9^1/_{16}$ | $6^7/_8$ | $8^7/_8$ | 10 | 310–490 | 50–90 |
| 26 | 208 | 173 | 197 | $6^3/_{16}$ | $6^{13}/_{16}$ | $7^3/_4$ | 10 | 310–440 | 50–80 |
| 26R | 208 | 173 | 197 | $6^3/_{16}$ | $6^{13}/_{16}$ | $7^3/_4$ | 11 | 405–525 | 60–80 |
| 27 | 306 | 173 | 225 | $12^1/_{16}$ | $6^{13}/_{16}$ | $8^7/_8$ | 10 | 270–810 | 102–140 |
| 27F | 318 | 173 | 227 | $12^1/_2$ | $6^{13}/_{16}$ | $8^{15}/_{16}$ | 11F | 360–660 | 95–140 |
| 27H | 298 | 173 | 235 | $11^3/_4$ | $6^{13}/_{16}$ | $9^1/_4$ | 10 | 440 | 125 |
| 29NF | 330 | 140 | 227 | 13 | $5^1/_2$ | $8^{16}/_{16}$ | 11F | 330–350 | 95 |
| 27R | 306 | 173 | 225 | $12^1/_{16}$ | $6^{13}/_{16}$ | $8^7/_8$ | 11 | 270–700 | 102–140 |
| 33 | 338 | 173 | 238 | $13^5/_{16}$ | $6^{13}/_{16}$ | $9^3/_8$ | 11F | 1050 | 165 |
| 34 | 260 | 173 | 200 | $10^1/_4$ | $6^{13}/_{16}$ | $7^7/_8$ | 10 | 375–770 | 100–115 |
| 34R | 260 | 173 | 200 | $10^1/_4$ | $6^{13}/_{16}$ | $7^7/_8$ | 11 | 675 | 110 |
| 35 | 230 | 175 | 225 | $9^1/_{16}$ | $6^7/_8$ | $8^7/_8$ | 11 | 310–500 | 80–110 |
| 36R | 263 | 183 | 206 | $10^3/_6$ | $7^1/_4$ | $8^1/_8$ | 19 | 650 | 130 |
| 40R | 278 | 175 | 175 | $10^{15}/_{16}$ | $6^7/_8$ | $6^7/_8$ | 15 | 590–600 | 110–120 |
| 41 | 293 | 175 | 175 | $11^9/_{16}$ | $6^7/_8$ | $6^7/_8$ | 15 | 235–650 | 65–95 |
| 42 | 242 | 175 | 175 | $9^1/_2$ | $6^{13}/_{16}$ | $6^{13}/_{16}$ | 15 | 260–495 | 65–95 |
| 43 | 334 | 175 | 205 | $13^1/_8$ | $6^7/_8$ | $8^1/_{16}$ | 15 | 375 | 115 |
| 45 | 240 | 140 | 227 | $9^7/_{16}$ | $5^1/_2$ | $8^{15}/_{16}$ | 10F | 250–470 | 60–80 |
| 46 | 273 | 173 | 229 | $10^3/_4$ | $6^{13}/_{16}$ | 9 | 10F | 350–450 | 75–95 |
| 47 | 242 | 175 | 190 | $9^1/_2$ | $6^7/_8$ | $7^1/_2$ | 24(A, F)$^a$ | 370–550 | 75–85 |
| 48 | 278 | 175 | 190 | $12^1/_{16}$ | $6^7/_8$ | $7^9/_{16}$ | 24 | 450–695 | 85–95 |
| 49 | 353 | 175 | 190 | $13^7/_8$ | $6^7/_8$ | $7^9/_{16}$ | 24 | 600–900 | 140–150 |
| 50 | 343 | 127 | 254 | $13^1/_2$ | 5 | 10 | 10 | 400–600 | 85–100 |
| 51 | 238 | 129 | 223 | $9^3/_8$ | $5^1/_{16}$ | $8^{13}/_{16}$ | 10 | 405–435 | 70 |
| 51R | 238 | 129 | 223 | $9^3/_8$ | $5^1/_{16}$ | $8^{13}/_{16}$ | 11 | 405–435 | 70 |
| 52 | 186 | 147 | 210 | $7^5/_{16}$ | $5^{13}/_{16}$ | $8^1/_4$ | 10 | 405 | 70 |

*(Continued)*

**TABLE 8.4** Standard SLI and Other Lead-Acid Batteries (*Continued*)

| | BCI group numbers, dimensional specifications, and ratings | | | | | | | | |
| --- | --- | --- | --- | --- | --- | --- | --- | --- | --- |
| | Maximum overall dimensions | | | | | | | Performance ranges | |
| | Millimeters | | | Inches | | | | Cold cranking performance amps. at 0°F (−18°C) | Reserve capacity min at 80°F (27°C) |
| BCI group number | L | W | H | L | W | H | Assembly figure no. | | |
| Passenger car and light commercial batteries 12 V (six cells) | | | | | | | | | |
| 53 | 330 | 119 | 210 | 13 | $4^{11}/_{16}$ | $8^{1}/_{4}$ | 14 | 280 | 40 |
| 54 | 186 | 154 | 212 | $7^{5}/_{16}$ | $6^{1}/_{16}$ | $8^{3}/_{8}$ | 19 | 305–330 | 60 |
| 55 | 218 | 154 | 212 | $8^{5}/_{8}$ | $6^{1}/_{16}$ | $8^{3}/_{8}$ | 19 | 370–450 | 75 |
| 56 | 254 | 154 | 212 | 10 | $6^{1}/_{16}$ | $8^{3}/_{8}$ | 19 | 450–505 | 90 |
| 57 | 205 | 183 | 177 | $8^{1}/_{16}$ | $7^{3}/_{16}$ | $6^{15}/_{16}$ | 22 | 310 | 60 |
| 58 | 255 | 183 | 177 | $10^{1}/_{16}$ | $7^{3}/_{16}$ | $6^{15}/_{16}$ | 26 | 380–540 | 75 |
| 58R | 255 | 183 | 177 | $10^{1}/_{16}$ | $7^{3}/_{16}$ | $6^{15}/_{16}$ | 19 | 540–580 | 75 |
| 59 | 255 | 193 | 196 | $10^{1}/_{16}$ | $7^{5}/_{8}$ | $7^{3}/_{4}$ | 21 | 540–590 | 100 |
| 60 | 332 | 160 | 225 | $13^{1}/_{16}$ | $6^{5}/_{16}$ | $8^{7}/_{8}$ | 12 | 305–385 | 65–115 |
| 61 | 192 | 162 | 225 | $7^{9}/_{16}$ | $6^{3}/_{8}$ | $8^{7}/_{8}$ | 20 | 310 | 60 |
| 62 | 225 | 162 | 225 | $8^{7}/_{8}$ | $6^{3}/_{8}$ | $8^{7}/_{8}$ | 20 | 380 | 75 |
| 63 | 258 | 162 | 225 | $10^{3}/_{16}$ | $6^{3}/_{8}$ | $8^{7}/_{8}$ | 20 | 450 | 90 |
| 64 | 296 | 162 | 225 | $11^{11}/_{16}$ | $6^{3}/_{8}$ | $8^{7}/_{8}$ | 20 | 475–535 | 105–120 |
| 65 | 306 | 192 | 192 | $12^{1}/_{16}$ | $7^{1}/_{2}$ | $7^{9}/_{16}$ | 21 | 650–850 | 130–165 |
| 66 | 306 | 192 | 194 | $12^{1}/_{16}$ | $7^{9}/_{16}$ | $7^{5}/_{8}$ | 13 | 650–750 | 130–140 |
| 70 | 208 | 180 | 186 | $8^{3}/_{16}$ | $7^{1}/_{16}$ | $7^{5}/_{16}$ | 17 | 260–525 | 60–80 |
| 71 | 208 | 179 | 216 | $8^{3}/_{16}$ | $7^{1}/_{16}$ | $8^{1}/_{2}$ | 17 | 275–430 | 75–90 |
| 72 | 230 | 179 | 210 | $9^{1}/_{16}$ | $7^{1}/_{16}$ | $8^{1}/_{4}$ | 17 | 275–350 | 60–90 |
| 73 | 230 | 179 | 216 | $9^{1}/_{16}$ | $7^{1}/_{16}$ | $8^{1}/_{2}$ | 17 | 430–475 | 80–115 |
| 74 | 260 | 184 | 222 | $10^{1}/_{4}$ | $7^{1}/_{4}$ | $8^{3}/_{4}$ | 17 | 350–550 | 75–140 |
| 75 | 230 | 180 | $196^{b}$ | $9^{1}/_{16}$ | $7^{1}/_{16}$ | $7^{11}/_{16}{}^{b}$ | 17 | 430–690 | 90 |
| 76 | 334 | 179 | 216 | $13^{1}/_{8}$ | $7^{1}/_{16}$ | $8^{1}/_{2}$ | 17 | 750–1075 | 150–175 |
| 78 | 260 | 180 | 186 | $10^{1}/_{4}$ | $7^{1}/_{16}$ | $7^{5}/_{16}$ | 17 | 515–770 | 105–115 |
| 79 | 307 | 179 | 188 | $12^{1}/_{16}$ | $7^{1}/_{16}$ | $7^{3}/_{8}$ | 35 | 770–840 | 140 |
| 85 | 230 | 173 | 203 | $9^{1}/_{16}$ | $6^{13}/_{16}$ | 8 | 11 | 430–630 | 90 |
| 86 | 230 | 173 | 203 | $9^{1}/_{16}$ | $6^{13}/_{16}$ | 8 | 10 | 430–640 | 90 |
| 90 | 242 | 175 | 175 | $9^{1}/_{2}$ | $6^{7}/_{8}$ | $6^{7}/_{8}$ | 24 | 520–600 | 80 |
| 91 | 278 | 175 | 175 | 11 | $6^{7}/_{8}$ | $6^{7}/_{8}$ | 24 | 600 | 100 |
| 92 | 315 | 175 | 175 | $12^{1}/_{2}$ | $6^{7}/_{8}$ | $6^{7}/_{8}$ | 24 | 650 | 130 |
| 93 | 353 | 175 | 175 | $13^{7}/_{8}$ | $6^{7}/_{8}$ | $6^{7}/_{8}$ | 24 | 800 | 150 |
| 94R | 315 | 175 | 190 | $12^{3}/_{8}$ | $6^{7}/_{8}$ | $7^{1}/_{2}$ | 24 | 640–765 | 135 |
| 95R | 394 | 175 | 190 | $15^{9}/_{16}$ | $6^{7}/_{8}$ | $7^{1}/_{2}$ | 24 | 850–950 | 190 |
| 96R | 242 | 175 | 175 | $9^{9}/_{16}$ | $6^{13}/_{16}$ | $6^{7}/_{8}$ | 15 | 590 | 95 |
| 97R | 252 | 175 | 190 | $9^{15}/_{16}$ | $6^{7}/_{8}$ | $7^{1}/_{2}$ | 15 | 557 | 90 |
| 98R | 283 | 175 | 190 | $11^{3}/_{16}$ | $6^{7}/_{8}$ | $7^{1}/_{2}$ | 15 | 620 | 120 |
| 99 | 207 | 175 | 175 | $8^{3}/_{16}$ | $6^{7}/_{8}$ | $6^{7}/_{8}$ | 34 | 360 | 50 |
| 100 | 260 | 179 | 188 | $10^{1}/_{4}$ | 7 | $7^{5}/_{16}$ | 35 | 770 | 115 |
| 101 | 260 | 179 | 170 | $10^{1}/_{4}$ | 7 | $6^{11}/_{16}$ | 17 | 540 | 115 |

**TABLE 8.4**  Standard SLI and Other Lead-Acid Batteries (*Continued*)

| | BCI group numbers, dimensional specifications, and ratings | | | | | | | | |
|---|---|---|---|---|---|---|---|---|---|
| | Maximum overall dimensions | | | | | | | Performance ranges | |
| | Millimeters | | | Inches | | | | Cold cranking performance amps. at 0°F (−18°C) | Reserve capacity min at 80°F (27°C) |
| BCI group number | L | W | H | L | W | H | Assembly figure no. | | |
| Passenger car and light commercial batteries 6 V (three cells) | | | | | | | | | |
| 1 | 232 | 181 | 238 | $9^1/_8$ | $7^1/_8$ | $9^3/_8$ | 2 | 400–545 | 105–165 |
| 2 | 264 | 181 | 238 | $10^3/_8$ | $7^1/_8$ | $9^3/_8$ | 2 | 475–650 | 136–230 |
| 2E | 492 | 105 | 232 | $19^7/_{16}$ | $4^1/_8$ | $9^1/_8$ | 5 | 485 | 140 |
| 2N | 254 | 141 | 227 | 10 | $5^9/_{16}$ | $8^{15}/_{16}$ | 1 | 450 | 135 |
| 17HF[c,d] | 187 | 175 | 229 | $7^3/_8$ | $6^7/_8$ | 9 | 2B | — | — |
| Heavy-duty commercial batteries 12 V (six cells) | | | | | | | | | |
| 4D[e] | 527 | 222 | 250 | $20^3/_4$ | $8^3/_4$ | $9^7/_{16}$ | 8 | 490–1125 | 225–325 |
| 6D | 527 | 254 | 260 | $20^3/_4$ | 10 | $10^1/_4$ | 8 | 750 | 310 |
| 8D[e] | 527 | 283 | 250 | $20^3/_4$ | $11^1/_8$ | $9^7/_{16}$ | 8 | 850–1250 | 235–465 |
| 28 | 261 | 173 | 240 | $10^5/_{16}$ | $6^{13}/_{16}$ | $9^7/_{16}$ | 18 | 400–535 | 80–135 |
| 29H | 334 | 171 | 232 | $13^1/_8$ | $6^3/_4$ | $9^1/_8$ | 10 | 525–840 | 145 |
| 30H | 343 | 173 | 235 | $13^1/_2$ | $6^{13}/_{16}$ | $9^1/_4$ | 10 | 380–685 | 120–150 |
| 31A | 330 | 173 | 240 | 13 | $6^{13}/_{16}$ | $9^7/_{16}$ | 18 (A,T)[a] | 455–950 | 100–200 |
| Heavy-duty commercial batteries 6 V (three cells) | | | | | | | | | |
| 3 | 298 | 181 | 328 | $11^3/_4$ | $7^1/_8$ | $9^3/_8$ | 2 | 525–660 | 210–230 |
| 4 | 334 | 181 | 328 | $13^1/_8$ | $7^1/_8$ | $9^3/_8$ | 2 | 550–975 | 240–420 |
| 5D | 349 | 181 | 238 | $13^3/_4$ | $7^1/_8$ | $9^3/_8$ | 2 | 720–820 | 310–380 |
| 7D | 413 | 181 | 238 | $16^1/_4$ | $7^1/_8$ | $9^3/_8$ | 2 | 680–875 | 370–426 |
| Special tractor batteries 6 V (three cells) | | | | | | | | | |
| 3EH | 491 | 111 | 249 | $19^5/_{16}$ | $4^3/_8$ | $9^{13}/_{16}$ | 5 | 740–850 | 220–340 |
| 4EH | 491 | 127 | 249 | $19^5/_{16}$ | 5 | $9^{13}/_{16}$ | 5 | 850 | 340–420 |
| Special tractor batteries 12 V (six cells) | | | | | | | | | |
| 3EE | 491 | 111 | 225 | $19^5/_{16}$ | $4^3/_8$ | $8^7/_8$ | 9 | 260–360 | 85–105 |
| 3ET | 491 | 111 | 249 | $19^5/_{16}$ | $4^3/_8$ | $9^3/_{16}$ | 9 | 355–425 | 130–135 |
| 4DLT | 508 | 208 | 202 | 20 | $8^3/_{16}$ | $7^{15}/_{16}$ | 16L | 650–820 | 200–290 |
| 12T | 177 | 177 | 202 | $7^1/_{16}$ | $6^{15}/_{16}$ | $7^{15}/_{16}$ | 10 | 460 | 160 |
| 16TF | 421 | 181 | 283 | $16^9/_{16}$ | $7^1/_8$ | $11^1/_8$ | 10F | 600 | 240 |
| 17TF | 433 | 177 | 202 | $17^1/_{16}$ | $6^{15}/_{16}$ | $7^{15}/_{16}$ | 11L | 510 | 145 |
| General-utility batteries 12 V (six cells) | | | | | | | | | |
| U1 | 197 | 132 | 186 | $7^3/_4$ | $5^3/_{16}$ | $7^5/_{16}$ | 10(X)[a] | 120–375 | 23–40 |
| U1R | 197 | 132 | 186 | $7^3/_4$ | $5^3/_{16}$ | $7^5/_{16}$ | 11(X)[a] | 200–280 | 25–37 |
| U2 | 160 | 132 | 181 | $6^5/_{16}$ | $5^3/_{16}$ | $7^1/_8$ | 10(X)[a] | 120 | 17 |
| Electric golf car/utility batteries 6 V (three cells) | | | | | | | | | |
| GC2 | 264 | 183 | 290 | $10^3/_8$ | $7^3/_{16}$ | $11^7/_{16}$ | 2 | [g] | [g] |
| GC2H[f] | 264 | 183 | 295 | $10^3/_8$ | $7^3/_{16}$ | $11^5/_8$ | 2 | [g] | [g] |
| Electric golf car/utility batteries 8 V (four cells) | | | | | | | | | |
| GC8 | 264 | 183 | 290 | $10^3/_8$ | $7^3/_{16}$ | $11^7/_{16}$ | 31 | — | — |

(*Continued*)

**TABLE 8.4**    Standard SLI and Other Lead-Acid Batteries (*Continued*)

| | BCI group numbers, dimensional specifications, and ratings | | | | | | | | |
| | Maximum overall dimensions | | | | | | | Performance ranges | |
| | Millimeters | | | Inches | | | | Cold cranking performance amps. at 0°F (−18°C) | Reserve capacity min at 80°F (27°C) |
| BCI group number | L | W | H | L | W | H | Assembly figure no. | | |
| Commercial batteries (deep cycle) 12 V (six cells) | | | | | | | | | |
| 920 | 356 | 171 | 311 | 14 | $6^3/_4$ | $12^1/_2$ | 37 | — | — |
| 921 | 397 | 181 | 378 | $15^3/_4$ | $7^1/_8$ | $14^7/_8$ | 37 | — | — |
| Marine/commercial batteries 8 V (four cells) | | | | | | | | | |
| 981 | 527 | 191 | 273 | $20^3/_4$ | $7^1/_2$ | $10^3/_4$ | 8 | — | — |
| 982 | 546 | 191 | 267 | $21^1/_2$ | $7^1/_2$ | $10^1/_2$ | 8 | — | — |

[a]Letter in parentheses indicates terminal type.
[b]Maximum height dimension shown includes batteries with raised-quarter cover design. Flat-top design model height (minus quarter covers) reduced by approximately $^3/_8$ in (10 mm).
[c]Rod-end types—Extend top ledge with holes for hold-down bolts.
[d]Not in application section but still manufactured.
[e]Ratings for batteries recommended for motor coach and bus service are for double insulation. When double insulation is used in other types, deduct 15% from the rating values for cold cranking performance.
[f]Special-use battery not shown in application section.
[g]Capacity test in minutes at 75 A to 5.25 V at 80°F (27°C); cold cranking performance test not normally required for this battery.

## 8.4    IEC AND ANSI NOMENCLATURE SYSTEMS

It is unfortunate that the various standards identified in Tables 8.1 to 8.4 do not share the same nomenclature system. The independent nomenclature systems of various battery manufacturers worsen this situation even more. Various nomenclature systems are presented next.

### 8.4.1    Primary Batteries

The IEC nomenclature system for primary batteries, which became effective in 1992, is based on the electrochemical system and the shape and size of the battery. The letter designations for the electrochemical system and the type of cell remain the same as in the previous IEC system for primary batteries. The new numerical designations are based on a diameter/height number instead of the arbitrary size classification used previously. The first digits specify the diameter of the cell in millimeters and the second the height of the cell (millimeters times 10). An example is shown in Table 8.5a. The codes for the shape and electrochemical system are given in Tables 8.5b and 8.5c, respectively. For reference, the ANSI letter codes for the electrochemical systems are also listed in Table 8.5c. The ANSI nomenclature system does not use a code to designate shape.

**TABLE 8.5a**    IEC Nomenclature System for Primary Batteries, Example

| Nomenclature | Number of cells | System letter | Shape | Diameter, mm | Height, mm | Example |
|---|---|---|---|---|---|---|
| CR2025 | 1 | C | R | 20 | 2.5 | A unit round battery having dimensions shown and electro-chemical system letter C of Table 8.5c (Li /MnO$_2$) |

**TABLE 8.5***b*    IEC Nomenclature for Shape, Primary Batteries

| Letter designation | Shape |
|---|---|
| R | Round-cylindrical |
| P | Nonround |
| F | Flat (layer built) |
| S | Square (or rectangular) |

**TABLE 8.5***c*    Letter Codes Denoting Electrochemical System of Primary Batteries

| ANSI | IEC | Negative electrode | Electrolyte | Positive electrode | Nominal voltage (V) |
|---|---|---|---|---|---|
| * | — | Zinc | Ammonium chloride, zinc chloride | Manganese dioxide | 1.5 |
| | A | Zinc | Ammonium chloride, zinc chloride | Oxygen (air) | 1.4 |
| LB | B | Lithium | Organic | Carbon monofluoride | 3 |
| LC | C | Lithium | Organic | Manganese dioxide | 3 |
| | E | Lithium | Nonaqueous inorganic | Thionyl chloride | 3.6 |
| LF | F | Lithium | Organic | Iron sulfide | 1.5 |
| | G | Lithium | Organic | Copper dioxide | 1.5 |
| A† | L | Zinc | Alkali metal hydroxide | Manganese dioxide | 1.5 |
| Z‡ | P | Zinc | Alkali metal hydroxide | Oxygen (air) | 1.4 |
| SO§ | S | Zinc | Alkali metal hydroxide | Silver oxide | 1.55 |

| | | |
|---|---|---|
| * | No suffix | Carbon-zinc |
| | D | Carbon-zinc, heavy duty |
| | F | Carbon-zinc, general purpose |
| † | A | Alkaline |
| | AC | Alkaline industrial |
| ‡ | Z | Zinc/air |
| § | SO | Silver oxide |
| | N | Nickel oxyhydroxide |

Nomenclature for existing batteries was grandfathered. Examples of the nomenclature for some of these primary batteries are shown in Table 8.5*d*. Examples of the IEC nomenclature system for primary batteries are shown in Table 8.5*e*.

**TABLE 8.5***d*    IEC Nomenclature for Typical Primary Round, Flat, and Square Cells or Batteries*

| IEC designation | Nominal dimensions, mm | | | | | ANSI designation | Common designation |
|---|---|---|---|---|---|---|---|
| | Diameter | Height | Length | Width | Thickness | | |
| Round batteries | | | | | | | |
| R03 | 10.5 | 44.5 | | | | 24 | AAA |
| R1 | 12.0 | 30.2 | | | | — | N |
| R6 | 14.5 | 50.5 | | | | 15 | AA |
| R14 | 26.2 | 50.0 | | | | 14 | C |
| R20 | 34.2 | 61.5 | | | | 13 | D |
| Flat cells | | | | | | | |
| F22 | | | 24 | 13.5 | 6.0 | | |
| Square batteries | | | | | | | |
| S4 | | 125.0 | 57.0 | 57.0 | | | |

*Chart is not complete—only a sampling of sizes is shown. Dimensions are used for identification only: Complete dimensions can be found in the relevant specification sheets listed in IEC 60086-2.

**TABLE 8.5e**    Examples of IEC Nomenclature for Primary Batteries

| IEC nomenclature | Number of cells | System letter | Shape | Cell | C, P, S, X, Y | Parallel | Groups in parallel | Example |
|---|---|---|---|---|---|---|---|---|
| R20 | 1 | None | R | 20 | * | | | A unit round battery using basic R20 type cell and electrochemical system letter (none) in Table 8.5c |
| LR20 | 1 | L | R | 20 | * | | | Same as above, except using electrochemical system letter L in Table 8.5c |
| 6F22 | 6 | None | F | 22 | * | | | A six-series multicell battery using flat F22 cells and electrochemical system letter (none) in Table 8.5c |
| 4LR25-2 | 4 | L | R | 25 | * | | 2 | A multicell battery consisting of two parallel groups, each group having four cells in series of the R25 type and electrochemical system letter L in Table 8.5c |
| CR17345 | 1 | C | R | See Section 8.4.1 | | | | A unit round battery, with a diameter of 17 mm and height of 34.5 mm, and electrochemical system letter C in Table 8.5c |

*If required, letters C, P, or S will indicate different performance characteristics and letters X and Y different terminal arrangements.

## 8.4.2  Rechargeable Batteries

The documentation for standardization of rechargeable batteries is not as complete as the documentation for primary batteries. Most of the primary batteries are used in a variety of portable applications using user-replaceable batteries. Hence, primary batteries especially need standards to ensure interchangeability. Developing such standards have been active projects by both IEC and ANSI for many years.

Early rechargeable batteries were mainly larger batteries and were usually application-specific with multicell designs. The large majority of standardized rechargeable batteries were lead-acid, manufactured for automotive SLI use. Standards for these batteries were developed by SAE, BCI, and the Storage Battery Association of Japan. More recently, rechargeable batteries have been developed for portable applications, in many cases in the same cell and battery sizes as the primary batteries. Starting with the portable-sized nickel-cadmium batteries, IEC and ANSI have developed standards for the nickel-metal hydride and lithium-ion batteries. The currently available standards were listed above in Tables 8.1a and 8.1b.

Table 8.6a lists the letter codes that are being used by IEC and those adopted by ANSI for secondary or rechargeable batteries. The IEC nomenclature system for nickel-metal hydride batteries is shown in Table 8.6b. The first letter in the IEC nomenclature designates the electrochemical system; the second letter, the shape; the first number, the diameter; and the second number, the height. In addition, the letters L, M, and H may be used, arbitrarily, to classify the rate capability as low, medium, or high, respectively. The last part of the designation is reserved for two letters that indicate various tab terminal arrangements, such as CF—none, HH—terminal at positive end and positive sidewall, or HB—terminals at positive and negative ends, as shown in Tables 8.6a and 8.6b.

**TABLE 8.6*a*** Letter(s) Denoting Electrochemical System of Secondary Batteries

| ANSI | IEC* | Negative electrode | Electrolyte | Positive electrode | Nominal voltage, V |
|------|------|--------------------|-----------|--------------------|--------------------|
| H | H | Hydrogen absorbing alloy | Alkali metal hydroxide | Nickel oxide | 1.2 |
| K | K | Cadmium | Alkali metal hydroxide | Nickel oxide | 1.2 |
| P | PB | Lead | Sulfuric acid | Lead dioxide | 2 |
| I | IC | Carbon | Organic | Lithium cobalt oxide | 3.6 |
| I | IN | Carbon | Organic | Lithium nickel oxide | 3.6 |
| I | IM | Carbon | Organic | Lithium manganese oxide | 3.6 |

*For portable batteries.

**TABLE 8.6*b*** IEC Nomenclature System for Rechargeable Nickel-Metal Hydride Cells and Batteries

| Nomenclature* | System letter | Shape | Diameter, mm | Height, mm | Terminals | Example |
|---------------|---------------|-------|--------------|------------|-----------|---------|
| HR 15/51 (R6) | H | R | 14.5 | 50.5 | CF | A unit round battery of the H system having dimensions shown, with no connecting tabs |

*Nomenclature dimensions are shown rounded off. ( ) indicates interchangeable with a primary battery.
*Source:* IEC 61951-2.

## 8.5  *MISCELLANEOUS STANDARDIZATION CATEGORIES*

In order to ensure interchangeability of products from different manufacturers around the world, battery standards also address aspects of terminals, electrical performance, and markings.

### 8.5.1  Terminals

Terminals are another aspect of the shape characteristics for batteries. It is obvious that without standardization of terminals and the other shape variables, a battery may not be available to match the receptacle facilities provided in the appliance. Table 8.7 lists terminal arrangements for batteries.

**TABLE 8.7** Terminal Arrangements for Batteries

| | |
|---|---|
| Cap and base | Terminals that have the cylindrical side of the battery insulated from the terminal ends |
| Cap and case | Terminals in which the cylindrical side forms part of the positive end terminal |
| Screw types | Terminals that have a threaded rod and accept either an insulated or a metal nut |
| Flat contacts | Flat metal surfaces used for electrical contact |
| Springs | Terminals that are flat metal strips or spirally wound wire |
| Plug-in sockets | Terminals consisting of a stud (nonresilient) and a socket (resilient) |
| Wire | Single or multistranded wire leads |
| Spring clips | Metal clips that will accept a wire lead |
| Tabs | Metal flat tabs attached to battery terminals |

When applicable, the terminal arrangement is specified in the standard within the same nomenclature designators used for shape and size. The designators thus determine all interchangeable physical aspects of the batteries in addition to the voltage.

## 8.5.2   Electrical Performance

In terms of the requirement to provide fit and function in the end product, the actual appliance does not require specific values of electrical performance. The correct battery voltage, needed to protect the appliance from overvoltage, is assured by the battery designation. Batteries of the same voltage but having differences in capacity can be used interchangeably, but will operate for different service times. The minimum electrical performance of the battery is therefore cited and specified in the standards.

1. *Application tests.* This is the preferred method of testing the performance specified for primary batteries. Application tests are intended to simulate the actual use of a battery in a specific application. Table 8.8*a* illustrates typical application tests.

**TABLE 8.8*a***   Example of Application Tests for R20 Type Batteries

| Nomenclature | | | | R20P | R20S | LR20 |
|---|---|---|---|---|---|---|
| Electrochemical system | | | | Zinc-carbon (high power) | Zinc-carbon (standard) | Zinc/manganese dioxide |
| Nominal voltage | | | | 1.5 | 1.5 | 1.5 |
| Application | Load, $\Omega$ | Daily period | End point | Minimum average duration | | |
| Portable lighting | 2.2 | * | 0.9 | 220 min | 85 min | 750 min |
| Portable Stereo | 600 mA | 2 h | 0.9 | — | — | 11 h |
| Radio | 10 | 4 h | 0.9 | 33 h | 18 h | — |
| Toys | 2.2 | 1 h | 0.8 | 5.5 h | 2 h | 16 h |

*4 min/15 min, 8 h/day.

2. *Capacity (service output) tests.* A capacity test is generally used to determine the quantity of electric charge a battery can deliver under specified discharge conditions. This method is the one that has been generally used for rechargeable batteries. It is also used for primary batteries when an application test would be too complex to simulate realistically or too lengthy to be practical for routine testing. Table 8.8*b* lists some examples of capacity tests.

**TABLE 8.8*b***   Example of Capacity Tests

| Nomenclature | | | | SR54 |
|---|---|---|---|---|
| Electrochemical system | | | | S |
| Nominal voltage | | | | 1.55 |
| Application* | Load, k$\Omega$ | Daily period | End point | Minimum average duration |
| Capacity (rating) test | 15 | 24 h | 1.2 | 580 h |

*Application for this battery is watches. As an application test could take up to 2 years to test, a capacity test is specified.

Test conditions in the standard must consider and therefore specify the following:

Battery temperature

Discharge rate (or load resistance)

Discharge termination criteria (typically loaded voltage)

Discharge duty cycle

For rechargeable cells, charge rate, termination criteria, and other conditions of charge, humidity or other conditions of storage may also be required.

### 8.5.3 Markings

Markings on both primary and secondary (rechargeable) batteries may consist of some or all of the printed information given in Table 8.9 in addition to the nomenclature discussed.

**TABLE 8.9**   Marking Information for Batteries

| Marking information | Primary batteries | Primary small batteries | Rechargeable round batteries |
|---|---|---|---|
| Nomenclature | × | × | × |
| Date of manufacture or code polarity | × | ×× | × |
| Nominal voltage | × | × | × |
| Name of manufacturer/supplier | × | ×× | × |
| Charge rate/time | × | ×× | × |
| Rated capacity | | | × |
| | | | × |

×—on battery.
××—on battery or package.

## 8.6  REGULATORY AND SAFETY STANDARDS

With the increasing complexity and energy of batteries and the concern about safety, greater attention is being given to developing regulations and standards with the goal to promote safe operation in use and transport. Stand-alone safety documents on primary and rechargeable batteries have been published by IEC and ANSI. These safety standards typically specify tests (e.g., short-circuit, shock, vibration, thermal abuse, over-discharge, and crush) and requirements for the safe operation of batteries under intended use and reasonably foreseeable misuse. In addition, Underwriters Laboratories (UL) has published several battery safety standards aimed at the safe operation of UL-approved equipment.[4]

Table 8.10*a* provides a list of organizations working on safety standards and the safety standards they prepared that cover various primary and secondary battery systems.

**TABLE 8.10*a***   Safety Standards

| Publication | Title |
|---|---|
| American Standards Institute | |
| ANSI C18.1M, Part 2 | American National Standard for Portable Primary Cells and Batteries with Aqueous Electrolyte— Safety Standard |
| ANSI C18.2M, Part 2 | American National Standard for Portable Rechargeable Cells and Batteries—Safety Standard |
| ANSI C18.3M, Part 2 | American National Standard for Portable Lithium Primary Cells and Batteries—Safety Standard |
| International Electrotechnical Commission | |
| IEC 60086-4 | Primary Batteries—Part 4: Safety for Lithium Batteries |
| IEC 60086-5 | Primary Batteries—Part 5: Safety of Batteries with Aqueous Electrolyte |
| IEC 61982-4 | Secondary Batteries (except lithium) for Propulsion of Electric Road Vehicles—Safety Requirements of Nickel-Metal Hydride Cells and Modules |
| IEC 62281 | Safety of Primary and Secondary Lithium Batteries During Transport |
| IEC 62133 | Safety for Portable Sealed Secondary Cells and Batteries |
| IEC 62485 | Safety Requirements for Secondary Batteries and Battery Installations |
| IEC 62619 | Safety Requirements for Secondary Lithium Batteries for Use in Industrial Applications |
| IEC 62660 | Secondary Lithium-Ion Cells for Propulsion of Electric Road Vehicles—Safety Requirements |
| Underwriters Laboratories | |
| UL1642 | Standard for Lithium Batteries |
| UL2054 | Standard for Household and Commercial Batteries |
| UL 62133 | Standard for Safety for Secondary Cells and Batteries |

While the various groups involved in developing safety standards are dedicated to the principle of harmonization, there are still differences in the procedures, tests, and criteria between the various standards. It is recommended that users of these standards follow them on a judicious basis, and place their battery or application in the proper context.

Table 8.10*b* provides a list of organizations that have focused on the safe transport of various goods and the regulations they have published. These regulations include procedures for the transport of batteries, including lithium batteries.

**TABLE 8.10*b***   Transportation Recommendations and Regulations

| Organization | Title |
|---|---|
| Department of Transportation (DOT) | Code of Federal Regulations—Title 49 Transportation |
| Federal Aviation Administration (FAA) | TSO C042, Lithium Batteries (referencing RTCA Document DO-227 "Minimum Operational Performance Standards for Lithium Batteries") |
| International Air Transport Association (IATA) | Dangerous Goods Regulations |
| International Civil Aviation Organization (ICAO) | Technical Instructions for the Safe Transport of Dangerous Goods |
| United Nations (UN) | Recommendations on the Transportation of Dangerous Goods Manual of Tests and Criteria |

In the United States, this responsibility for regulating the transport of goods rests with the Department of Transportation (DOT) through its Research and Special Programs Administration (RSPA).[5] Regulations are published in the Code of Federal Regulations (CFR49), which include the requirements for the shipment and transport of batteries under all modes of transportation. Under the DOT, the Federal Aviation Administration (FAA) is responsible for the safe operation of aircraft and has also issued regulations covering the use of batteries in aircraft.[6,7] Similar organizations are part of the governments of most countries throughout the world.

Internationally, transport is regulated by such organizations as the International Civil Aviation Organization (ICAO),[8] the International Air Transport Association (IATA),[9] and the International Maritime Organization. Their regulations are guided by the United Nations (UN) through their Committee of Experts on the Transport of Dangerous Goods. Currently, the UN Committee of Experts has developed guidelines covering the transport of lithium primary and secondary batteries. These recommendations, which include testing (e.g., altitude simulation, thermal, vibration, shock, short-circuit, crush, over-charge, and forced discharge) and performance criteria,[10,11] are addressed to governments and international organizations concerned with regulating the transport of various lithium battery products.

As these standards, regulations, and guidelines can be changed on a periodic basis, the current edition of each document should be used.

*Note/Warning:*   *It is imperative that only the latest version of each standard be used. Due to the periodic revision of these standards, only the latest version can be relied upon to provide reliable specifications of battery dimensions, terminals, marking, design features, testing conditions for performance verification, mechanical tests, test sequences, safety, shipment, storage, use, and disposal.*

## REFERENCES

1. SAE International, 400 Commonwealth Drive, Warrendale, PA 15096, www.sae.org.

2. Battery Council International, 401 North Michigan Ave., Chicago, IL 60611, www.batterycouncil.org.

3. Battery Council International, *Battery Replacement Data Book.*

4. Underwriters Laboratories, Inc., 333 Pfingsten Road, Northbrook, IL 60062.

5. Department of Transportation, Office of Hazardous Materials Safety, Research and Special Programs Administration, 400 Seventh St., SW, Washington, DC 20590.

6. Department of Transportation, Federal Aviation Administration, 800 Independence Ave., SW, Washington, DC 20591.

7. RTCA, 1828 L St., NW, Suite 805, Washington, DC 20036, info@rtca.org.

8. International Civil Aviation Organization, 1000 Sherbrooke Street West, Montreal, Quebec, Canada.

9. International Air Transport Association, 2000 Peel St., Montreal, Quebec, Canada.

10. United Nations, *Recommendation on the Transport of Dangerous Goods,* New York, NY, and Geneva, Switzerland.

11. United Nations, *Manual of Tests and Criteria,* New York, NY, and Geneva, Switzerland.

# CHAPTER 9
# AN INTRODUCTION TO PRIMARY BATTERIES

**David Linden, Thomas B. Reddy**

## 9.1  GENERAL CHARACTERISTICS AND APPLICATIONS OF PRIMARY BATTERIES

The primary battery is a convenient source of power for portable electric and electronic devices, including portable lighting, remote controllers, portable radio, hearing aids, watches, toys, memory backup, and a wide variety of other applications, providing freedom from utility power. Major advantages of the primary battery are that it is convenient, simple, and easy to use; requires little, if any, maintenance; and can be sized and shaped to fit the application. Other general advantages are good shelf life, reasonable energy and power density, reliability, and acceptable cost.

However, now that secondary (i.e., rechargeable) cell designs have improved dramatically in recent years, the role of primary cells has evolved. Consumption of primary cells has remained strong due to the proliferation of small consumer electronic devices and demand from lesser developed countries, where electricity infrastructure for recharging is less pervasive. The trade-offs of costs, battery energy and power, service life, run-time, life-cycle cost, waste generation, recycling, etc., which dictate the decision process for selecting primary or secondary cells, are discussed further in the sections below and in Chap. 23 on applications engineering/battery selection for consumer electronic products.

Commercial primary batteries have existed for over 100 years (powering original telegraph and telephone systems), but until 1940, the zinc-carbon battery, known as the Leclanché cell, was the only one in wide use. Subsequently, significant advances were made, not only with the zinc-carbon system, but with new, superior types of primary batteries. Capacity has improved from less than 50 Wh/kg with the early zinc-carbon batteries to now more than 500 Wh/kg with several electrochemistries. Historically, this dramatic improvement gave advantage over secondary nickel-cadmium and lead-acid cells, which were also prevalent 100 years ago. The shelf life of primary batteries in 1940s was limited to about 1 year, when stored at moderate temperatures (0–35°C); the shelf life of present-day conventional batteries is at least 2 to 5 years. In the 1970s, primary lithium batteries were developed with 10-year storage life and storage temperature capability from 70°C down to −40°C. In addition, the power density has been improved many fold. Historical advances in primary battery performance were shown graphically in Chap. 1 and are further shown in Fig. 9.1.

The proliferation of electronic equipment from 1970 to 1990 stimulated the concurrent development of primary battery technology. Demand for improved portable power sources for consumer electronics (Sony Walkman, Nintendo Gameboy, personal digital assistants [PDAs], digital cameras, smoke detectors, etc.); space and military applications (smart munitions and Strategic Defense Initiative); and implantable medical devices (pacers and defibrillators) drove much of the effort.

During this period, the zinc/alkaline/manganese dioxide battery began to replace the zinc-carbon Leclanché battery as the leading primary battery, capturing the major share of the U.S. market. Environmental concerns over mercury, used in several batteries in the past, led to new technologies, including consumer lithium and

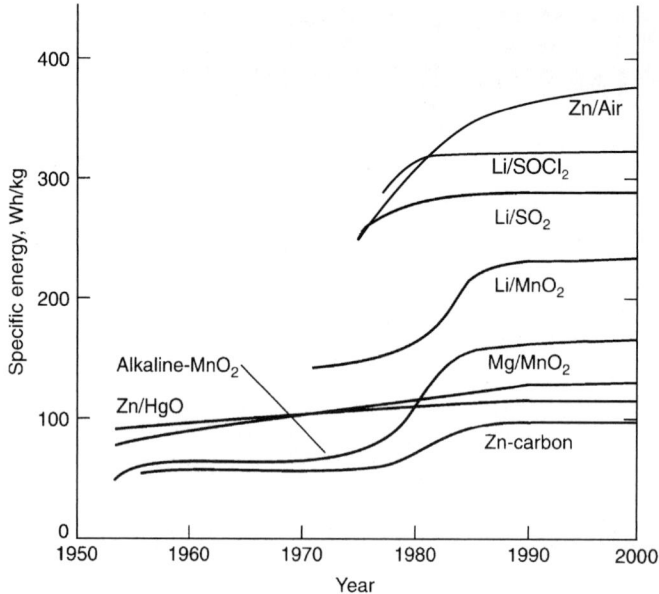

**FIGURE 9.1**  Advances in development of primary batteries in the 20th century. Continuous discharge at 20°C; 40–60 h rate; AA- or similar-size battery.

zinc/air primary cells. These cells quickly replaced zinc/mercuric oxide and cadmium/mercuric oxide batteries. The development and mass marketing of lithium primary batteries, using metallic lithium as the anode active material, was a major accomplishment of this era. The high specific energy of the lithium battery (double that of most conventional aqueous primary batteries), coupled with superior temperature range and shelf life, opened up a wide range of applications—from small coin and cylindrical batteries for memory backup and cameras to very large batteries, such as used for missile silo backup power.

While work continues to increase the energy density of primary cells, the advent of more advanced rechargeable cells and the reduction in power consumption of many electronic devices have, perhaps, slowed the pace of innovation. Nevertheless, advances have been made in other important performance characteristics, such as power density, shelf life, and safety.

Higher energy and power capabilities have resulted in a substantial reduction in battery size and weight. Without this evolution, portable two-way radios, PDAs, digital cameras, or other new portable high-power applications, typically powered in the past by secondary batteries or utility power, would not have been commercially successful. Primary batteries were more convenient and eliminated all recharging and maintenance. The extended shelf life of many primary batteries has similarly resulted in new uses, such as implanted medical devices and memory backup, leading to more reliable and enduring electronic equipment.

## 9.2   PRIMARY BATTERY TYPES AND CHARACTERISTICS

Although a number of anode-cathode combinations can be used as primary battery systems (see Chap. 1), only a relatively few have achieved practical success. Zinc has been by far the most popular anode material for primary batteries because of high electrochemical equivalence, compatibility with aqueous electrolytes, reasonably good shelf life, low cost, and good availability. Aluminum has high electrochemical potential, and high electrochemical equivalence and availability, but due to passivation, it has not been developed successfully into a practical active primary battery system. Magnesium also has attractive electrical properties (high energy density and good shelf life) and low cost, but it has found only limited use in military applications. Lithium, which has the highest gravimetric energy density and standard potential of all the metals, has continued to receive much of the focus. Using a number of different nonaqueous electrolytes with a variety of cathode materials has pushed the energy and power density envelope of primary systems. The following section summarizes various design and performance parameters for the major primary batteries of the last century.

## 9.2.1  Characteristics of Primary Batteries

Typical characteristics and applications of the different types of primary batteries are summarized in Table 9.1.

***Zinc-Carbon Battery (Chap. 11A).***    The Leclanché zinc-carbon dry cell battery had been the most widely used of all the primary batteries in the 20th century because of its low cost, relatively good performance, and ready availability. Cells and batteries of many sizes and characteristics were manufactured to meet the requirements of a wide variety of applications. From 1945 to 1965, significant improvements in capacity and shelf life were achieved through the use of new materials (such as chemical and electrolytic manganese dioxide and zinc chloride electrolyte) and through cell design modifications (such as the paper-lined cell). Despite low cost, the Leclanché battery has lost most of its market share in industrialized countries to newer primary batteries, particularly zinc/alkaline/manganese dioxide cells, with superior performance characteristics.

***Zinc/Alkaline/Manganese Dioxide Battery (Chap. 12A).***    In the two decades after 1990, the major portion of the primary battery market had shifted to the Zn/alkaline/$MnO_2$ battery. This system had become the battery of choice due to higher current, low temperature, and long shelf life capabilities. While more expensive on a unit basis than the Leclanché battery (typically double the price), alkaline cells are still likely to be slightly more costly even on an energy basis (<$0.02/Wh vs. >$0.02/Wh, respectively).[1] However, alkaline cells are often more cost-effective for

**TABLE 9.1**    Major Characteristics and Applications of Primary Batteries

| System | Characteristics | Applications |
|---|---|---|
| Zinc-carbon (Leclanché), Zinc/$MnO_2$ | Common, low-cost primary battery; available in a variety of sizes | Flashlight, portable radios, toys, novelties, instruments |
| Magnesium ($Mg/MnO_2$) | High-capacity primary battery; long shelf life | Formerly used for military receiver—transmitters and aircraft emergency transmitters (EPIRBs) |
| Mercury (Zn/HgO) | Highest capacity (by volume) of conventional types; flat discharge; good shelf life | Hearing aids, medical devices (pacemakers), photography, detectors, military equipment, but in limited use at present due to environmental hazard of mercury |
| Mercad (Cd/HgO) | Long shelf life; good low- and high-temperature performance; low energy density | Special applications requiring operation under extreme temperature conditions and long life; in limited use |
| Alkaline (Zn/alkaline/$MnO_2$) | Most popular general-purpose battery; good low-temperature and high-rate performance; low cost | Most popular primary battery; used in a variety of portable battery operated equipment |
| Silver/zinc (Zn/$Ag_2O$) | Highest capacity (by weight) of conventional types; flat discharge; good shelf life; costly | Hearing aids, photography, electric watches, missiles, underwater and space application (larger sizes) |
| Zinc/air (Zn/$O_2$) | Highest energy density; low cost; not independent of environmental conditions | Special applications, hearing aids, pagers, medical devices, military electronics |
| Lithium/soluble cathode | High energy density; long shelf life; good performance over wide temperature range | Wide range of applications requiring high energy density, long shelf life, e.g., from utility meters to military electronics applications |
| Lithium/solid cathode | High energy density; good rate capability and low-temperature performance; long shelf life; competitive cost | Replacement for conventional button and cylindrical cell applications, such as digital cameras |
| Lithium/solid electrolyte | Extremely long shelf life; low-power battery | Medical electronics |

applications requiring high-rate or low-temperature capability, where the alkaline battery lasts 2 to 10 times longer. In addition, because of the advantageous shelf life, the alkaline cell is often selected for applications in which the battery is used intermittently and exposed to uncontrolled storage conditions, but where it must perform dependably when required (such as in consumer flashlights and smoke alarms). More recent advances have been in designing batteries to provide improved high-rate performance. The global proliferation of alkaline cell manufacturing and heavy competition in the marketplace have also driven down the cost significantly.

***Zinc/Mercuric Oxide Battery (Chap. 12B).***   The zinc/mercuric oxide battery was another important zinc anode primary system in the past. This battery was developed during World War II for military communication applications because of its good shelf life and high volumetric energy density. In the postwar period, it was used in small button, flat, or cylindrical cell configurations as the power source in electronic watches, calculators, hearing aids, photographic equipment, and similar applications requiring a reliable long-life miniature power source. Because of the environmental problems associated with mercury contamination of the municipal waste stream, mercuric oxide cells have been replaced by zinc/air and lithium batteries, which have subsequently evolved to provide greatly superior performance for many applications.

***Cadmium/Mercuric Oxide Battery (Chap. 12B).***   The substitution of cadmium for the zinc anode results in a mercuric oxide cell with lower-voltage but higher stability, including a 10-year shelf life, as well as good high and low temperature performance. The energy of this battery is about 60% of the zinc/mercuric oxide battery, but because of the toxicity of both mercury and cadmium, this battery is no longer commercially available.

***Zinc/Silver Oxide Battery (Chap. 12C).***   The primary zinc/silver oxide battery is similar in design to the small zinc/mercuric oxide button cell, but it has a higher specific energy and performs better at low temperatures. These characteristics make this battery system desirable for use in hearing aids, specialized hand-held digital devices, and electronic watches. However, because of its high cost (due to use of silver) and the development of other battery systems, this primary battery is now limited to small button cells where the higher cost is justified and to larger cells for military applications, where cost is less of an issue.

***Zinc/Air Battery (Chap. 18A).***   The zinc/air battery system is noted for its high energy density, but it had been formerly used only in large, low-power batteries for signaling and navigational-aid applications. With the development of enhanced air electrodes, the high-rate capability of the system was improved and small button-type batteries are now used widely in hearing aids, electronics, and similar applications. These batteries have a very high energy density, as no active cathode material is needed. This electrochemistry is sensitive to extreme temperatures, humidity, and other environmental factors, such as carbonation and has poor activated shelf life and low power density as well. Nevertheless, because of their attractive energy density, zinc/air batteries continue to garner interest and are finding new applications for both primary cells and secondary cells (Chap. 18B).

***Magnesium Batteries (Chap. 11B).***   While magnesium has attractive electrochemical properties, commercial interest in magnesium primary batteries is minimal because of the generation of hydrogen gas during discharge and the relatively poor stability of a partially discharged cell. However, the high energy density and long shelf life of fresh undischarged magnesium dry cell batteries met the needs for military communications equipment in the past, where high-temperature storage was required. In addition various metal/air batteries (Chap. 18) and reserve cells (Chap. 29) still use magnesium anodes.

***Aluminum Batteries.***   Aluminum is another viable anode material with a high theoretical energy density, but problems such as polarization and parasitic corrosion have inhibited the development of a commercial product. The best promise for its use is as a "mechanically rechargeable" or reserve battery (see Chap. 11B and Chap. 29, respectively).

***Lithium Batteries (Chap. 13).***   Lithium anode batteries are a relatively recent development (only invented around 1970). These primary cells use pure lithium metal foil anodes with lithium-ion conductive, nonaqueous electrolytes and lithium reactive cathodes. They have the advantage of high energy density and specific energy, as well as operation over a very wide temperature range with long shelf life and are gradually replacing some conventional (i.e., alkaline) battery systems. However, except for sensor/monitor/instrumentation devices, watch, memory backup, military, and a few other niche applications, they have not yet captured the general purpose markets as was anticipated (presumably due to continued higher cost and more recently, due to safety concerns when transported in bulk).[2]

As with the zinc systems, numerous lithium primary batteries, ranging in capacity from <5 mAh to 10,000 Ah, with various designs and chemistries, have been deployed over the years with lithium metal anodes.

Lithium primary batteries generally fall into three categories:

1. Small, low-power solid, immobilized or gelled electrolytes and cathodes (such as used in cardiac pacers where long life and safety are required).

2. Solid composite cathodes with liquid electrolyte (such as the coin, button, or cylindrical cells used for small electronic devices and other applications where high energy, broad temperature ranges, and long shelf life are critical).

3. Liquid/soluble cathodes (such as used in certain batteries for military and industrial applications of <1 Ah to >100 Ah size in either flat-plate or wound-cell configurations, where weight, volume, and robustness are required despite higher costs).

The first category is rather unique as discussed in the next section.

***Solid Electrolyte Batteries.*** The solid-electrolyte batteries are different from the other battery systems in that ionic conductivity is required through a solidified material that does not conduct electrons. Any such leakage currents would allow direct short-circuiting to occur. Typical batteries use liquid electrolytes where the solute (an ionic species) is dissolved and dissociated in a liquid medium that minimizes any electron flow. Batteries using solid electrolytes are typically low-power (microwatt) devices, and are used for applications requiring long life, especially at high temperatures. The first solid-electrolyte batteries used a silver anode and silver iodide for the electrolyte, but lithium anodes and lithium iodide-based polymer electrolytes now predominate (Chap. 28). More recently new developments, mostly in secondary battery systems, have led to vastly improved polymer, gel, and solid-state electrolytes with reported ionic conductivity approaching that of liquid electrolytes (Chap. 22).

## 9.3   PRIMARY BATTERY PERFORMANCE COMPARISONS

### 9.3.1   General

A qualitative comparison of the various primary battery systems is given in Table 9.2. This listing illustrates the performance advantages of the lithium anode batteries. Nevertheless, due to low cost, availability, and adequate performance in many consumer applications, conventional primary cells still maintain a major share of the market.

The characteristics of the major primary battery systems are summarized in Table 9.3. Refer to Chap. 1 for the theoretical and practical electrical characteristics of various primary battery systems. Note that only about 25% to 35% of the theoretical capacity is attained under practical conditions as a result of design inefficiencies and nonoptimal discharge conditions.

**TABLE 9.2**  Comparison of Primary Batteries[*]

| System | Voltage | Specific energy (gravimetric) | Power density | Flat discharge profile | Low-temperature operation | High-temperature operation | Shelf life | Cost |
|---|---|---|---|---|---|---|---|---|
| Zinc/carbon | 5 | 4 | 4 | 4 | 5 | 6 | 8 | 1 |
| Zinc/alkaline/manganese dioxide | 5 | 3 | 2 | 3 | 4 | 4 | 7 | 2 |
| Magnesium/manganese dioxide | 3 | 3 | 2 | 2 | 4 | 3 | 4 | 3 |
| Zinc/mercuric oxide | 5 | 3 | 2 | 2 | 5 | 3 | 4 | 5 |
| Cadmium/mercuric oxide | 6 | 5 | 2 | 2 | 3 | 2 | 3 | 6 |
| Zinc/silver oxide | 4 | 3 | 2 | 2 | 4 | 3 | 5 | 6 |
| Zinc/air | 5 | 2 | 3 | 2 | 5 | 5 | — | 3 |
| Lithium/soluble cathode | 1 | 1 | 1 | 1 | 1 | 2 | 1 | 5 |
| Lithium/solid cathode | 1 | 1 | 1 | 2 | 2 | 3 | 2 | 3 |

[*]1 to 8—best to poorest.

**TABLE 9.3** Characteristics of Primary Batteries

| System | Zinc-carbon (Leclanché) | Zinc-carbon (zinc chloride) | Mg/MnO₂ | Zn/Alk./MnO₂ | Zn/HgO | Cd/HgO | Zn/Ag₂O | Zinc/air | Li/SO₂* | Li/SOCl₂* | Li/MnO₂* | Li/FeS₂* | Solid state |
|---|---|---|---|---|---|---|---|---|---|---|---|---|---|
| **Chemistry:** | | | | | | | | | | | | | |
| Anode | Zn | Zn | Mg | Zn | Zn | Cd | Zn | Zn | Li | Li | Li | Li | Li |
| Cathode | $MnO_2$ | $MnO_2$ | $MnO_2$ | $MnO_2$ | $HgO$ | $HgO$ | $Ag_2O$ or $AgO$ | $O_2$ (air) | $SO_2$ | $SOCl_2$ | $MnO_2$ | $FeS_2$ | $I_2(P2VP)$ |
| Electrolyte | $NH_4Cl$ and $ZnCl_2$ (aqueous solution) | $ZnCl_2$ (aqueous solution) | $MgBr_2$ or $Mg(ClO_4)$ (aqueous solution) | KOH (aqueous solution) | KOH or NaOH (aqueous solution) | KOH (aqueous solution) | KOH or NaOH (aqueous solution) | KOH (aqueous solution) | Organic solvent, salt solution | $SOCl_2$ w/$AlCl_4$ | Organic solvent, salt solution | Organic solvent, salt solution | Solid |
| **Cell voltage, V†:** | | | | | | | | | | | | | |
| Nominal | 1.5 | 1.5 | 1.6 | 1.5 | 1.35 | 0.9 | 1.5 | 1.5 | 3.0 | 3.6 | 3.0 | 1.5 | 2.8 |
| Open-circuit | 1.5–1.75 | 1.6 | 1.9–2.0 | 1.5–1.6 | 1.35 | 0.9 | 1.6 | 1.45 | 3.1 | 3.65 | 3.3 | 1.8 | 2.8 |
| Midpoint | 1.25–1.1 | 1.25–1.1 | 1.8–1.6 | 1.23 | 1.3–1.2 | 0.85–0.75 | 1.6–1.5 | 1.3–1.1 | 2.9–2.75 | 3.6–3.3 | 3.0–2.7 | 1.6–1.4 | 2.8–2.6 |
| End | 0.9 | 0.9 | 1.2 | 0.8 | 0.9 | 0.6 | 1.0 | 0.9 | 2.0 | 3.0 | 2.0 | 1.0 | 2.0 |
| Operating temperature, °C | −5 to 45 | −10 to 50 | −40 to 60 | −40 to 50 | 0 to 55 | −55 to 80 | 0 to 55 | 0 to 50 | −55 to 70 | −60 to 85 | −20 to 55 | −20 to 60 | 0 to 200 |
| **Energy density at 20°C‡:** | | | | | | | | | | | | | |
| Button size: Wh/kg | | | | 81 | 100 | 55 | 135 | 415 | | | 230 | | 220–280 |
| Button size: Wh/L | | | | 361 | 470 | 230 | 530 | 1350 | | | 545 | | 820–1030 |
| Cylindrical size: Wh/kg | 65 | 85 | 100 | 154 | 105 | | | Prismatic 500 | 260 | Bobbin 590, Spiral-wound 495 | Bobbin 270, Spiral-wound 261 | 310 | |
| Cylindrical size: Wh/L | 100 | 165 | 195 | 461 | 325 | | | Prismatic 1250 | 415 | Bobbin 1100, Spiral-wound 970 | Bobbin 620, Spiral-wound 546 | 560 | |
| Discharge profile (relative) | Sloping | Sloping | Moderate slope | Moderate slope | Flat | Flat | Flat | Flat | Very flat | Flat | Flat | Medium to high | Moderately flat (at low discharge rates) |
| Power density | Low | Low to moderate | Moderate | Moderate | Moderate | Moderate | Moderate | Low | High | Medium (but dependent on specific design) | Moderate | Medium to high | Very low |
| Self-discharge rate at 20°C, % loss per year‡ | 10 | 7 | 3 | 3 | 4 | 3 | 6 | 3 (if sealed) | 2 | 1–2 | 1–2 | 1–2 | <1 |

| | 1 | 2 | 3 | 4 | 5 | 6 | 7 | 8 | 9 | 10 | 11 | 12 | 13 |
|---|---|---|---|---|---|---|---|---|---|---|---|---|---|
| Advantages | Lowest cost; good for noncritical use under moderate conditions; variety of shapes and sizes; availability | Low cost; better performance than regular zinc-carbon | High capacity compared with zinc-carbon; good shelf life | High capacity; good low-temperature and high-rate performance; low cost | High volumetric energy density; flat discharge; stable voltage | Good performance at high and low temperatures; long shelf life | High energy density; good high-rate performance | High volumetric energy density; long shelf life (sealed) | High energy density; best low-temperature, high-rate performance; long shelf life | High energy density; long shelf life because of protective film | High energy density; good low-temperature, high-rate performance; cost-effective replacement for small conventional type cells | Replacement for Zu/alkaline/MnO$_2$ batteries for high-rate performance | Excellent shelf life (10–20 years); wide operating temperature range (to 200°C) |
| Limitations | Low energy density; poor low-temperature, high-rate performance | High gassing rate; performance lower than premium alkaline batteries | High gassing (H$_2$) on discharge; delayed voltage | Electrolyte leakage may occur | Expensive; moderate gravimetric energy density, poor-low-temperature performance | Expensive; low-energy density | Expensive, but cost-effective on button battery applications | Not independent of environment—flooding, drying out; limited power output | Pressurized system; potential safety problems; toxic components; shipment regulated | Voltage delay after storage | Available in small sizes; larger sizes being considered; shipment regulated | Higher cost than alkaline batteries | For very low discharge rates; poor low temperature performance |
| Status | High production, but losing market share | High production, but losing market share | NLA§ | High production; most popular primary battery | Phased out because of toxic mercury | In limited production being phased out because of toxic components except for some special applications | In production | Moderate production; key use in hearing aids | Moderate production, mainly military | Produced in wide range of sizes and designs; mainly for special applications | Increasing consumer production | Produced in AAA and AA sizes; 9 V batteries also available | In production for special applications |
| Major types available | Cylindrical single-cell bobbin and multicell batteries (see Chap. 11A) | Cylindrical single-cell bobbin and multicell batteries (see Chap. 11A) | Cylindrical single-cell bobbin and multicell batteries (see Chap. 11B) previously available | Button, cylindrical, and multicell batteries (see Chap. 12A) | NLA§ | NLA§ | Button batteries (see Chap. 12C) | (See Chap. 18) | Cylindrical batteries (see Chap. 13) | (See Chap. 13) | Button and small cylindrical batteries (see Chap. 13) | Cylindrical cells and 9 V (see Chap. 13) | (See Chap. 28.) |

*See Chap. 13 for other lithium primary batteries.

†Data presented are for 20°C, under favorable discharge condition. See details in Chap. 11 (aqueous cells), Chap. 12 (alkaline cells), and Chap. 13 (lithium cells).

‡Rate of self-discharge usually decreases with time of storage.

§No longer readily available commercially.

Also, as discussed in Chap. 5, the performance characteristics of single-cell batteries are typically superior to battery packs, and the capabilities are often based on approximations, with each system presented under favorable discharge conditions. Battery performance will depend on the cell and battery design and on the specific usage conditions.

### 9.3.2   Voltage and Discharge Profile

A comparison of the discharge curves for various primary batteries is presented in Fig. 9.2. The zinc anode batteries generally have a highly sloping discharge voltage profile between 1.5 and 0.9 V. The lithium anode batteries usually have higher voltages, some on the order of 3 V, with an end or cutoff voltage of about 2 V. The lithium/manganese dioxide and magnesium/manganese dioxide batteries have a moderate slope compared to flat discharge profiles of the other batteries.

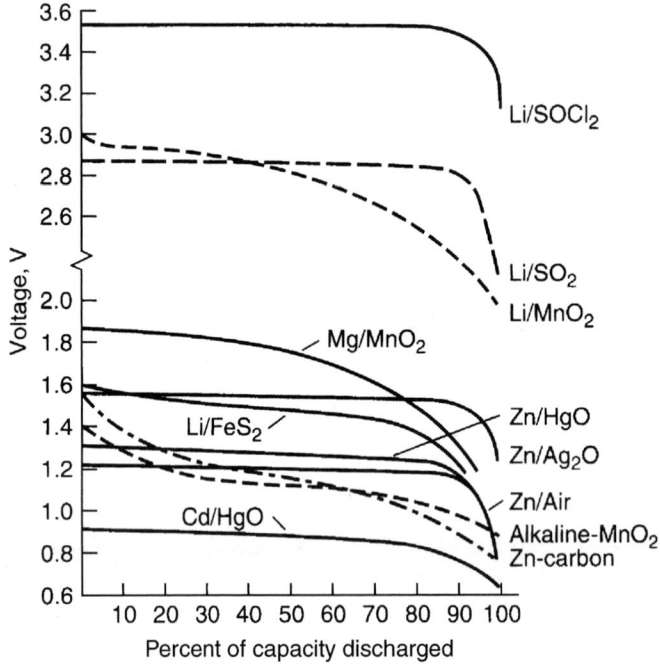

**FIGURE 9.2**   Discharge profiles of primary battery systems 30–100 h rate.

### 9.3.3   Power and Energy Metrics

Figure 9.3 presents a comparison of the specific power (i.e., gravimetric power density) of the different primary battery systems at various discharge rates at 20°C. This figure shows the hours of service each battery type (normalized to 1 kg battery weight) will deliver at various power levels to the commonly specified end voltage for each different battery type. The specific energy (gravimetric energy density in this case) can then be determined by

$$\text{Specific energy} = \text{specific power} \times \text{hours of service}$$

or

$$\text{Wh/kg} = \text{W/kg} \times \text{h} = \frac{A \times V \times h}{kg}$$

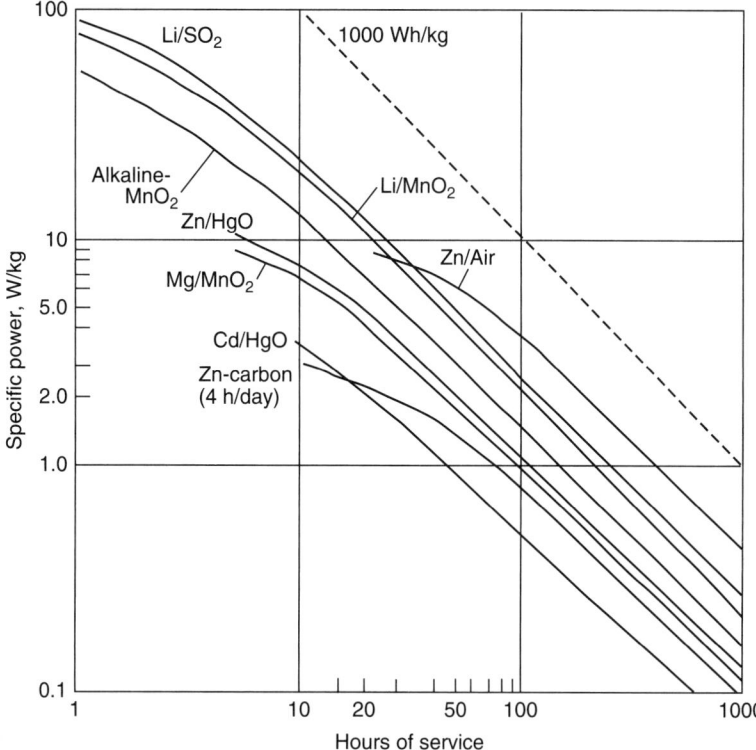

**FIGURE 9.3**   Comparison of typical performance of primary battery systems—specific power versus hours of service.

The conventional zinc-carbon battery has the lowest energy density of the primary batteries shown, with the exception, at low discharge rates, of the low-voltage cadmium/mercuric oxide electrochemical system. The zinc-carbon battery performs best at light discharge loads or with intermittent discharge, where rest or recovery periods help to improve the run-time compared with a continuous discharge (particularly at high discharge rates).

The reduced slopes of high current/high power portions of the performance curves show, in Fig. 9.3, the limited capabilities of battery systems at high rate. The dashed line, labeled 1000 Wh/kg, indicates the data points where the capacity or energy density of the battery would hypothetically remain constant at all discharge rates. However, as typical for most battery systems, the performance curves, even at low rate discharge, are less than that of the ideal curve (the 1000 Wh/kg line). Cell electrochemistries with high-rate discharge capability will have less severe flattening of the line's slope at higher discharge rates, however. As a further example of potential subtle performance variations, note that the zinc-carbon battery falls off sharply with an increasing discharge rate. However, heavy-duty zinc-carbon cells, using zinc chloride salts (see Chap. 11A), give better performance under high-rate discharge conditions. Meanwhile, the zinc/alkaline/manganese dioxide, the zinc/mercuric oxide, the zinc/silver oxide, and the magnesium/manganese dioxide batteries all have about the same specific energy and performance at 20°C. The zinc/air system has a higher specific energy at the low discharge rates, but falls off sharply at moderately high loads, indicating its low specific power. The lithium batteries are characterized by high specific energy, due in part to the higher cell voltage. The lithium/sulfur dioxide battery, in particular, is able to deliver high capacity at high discharge rates.

Volumetric energy density (Wh/L) is, at times, a more useful parameter than specific energy (i.e., gravimetric energy density, Wh/kg), particularly for button and small batteries, where the weight is insignificant. Lithium anode cells with 0.54 gm/cc lithium density have less benefit over denser batteries, such as the zinc/alkaline and zinc/mercuric oxide batteries when compared on a volumetric basis, as detailed in Chap. 1 and later in this chapter. Also, many chapters in Part 4 provide comparisons of the energy density (Wh/L) to specific energy (Wh/kg) as well as power density (W/L) to specific power (W/kg) for various battery types at varied discharge rates and temperatures, allowing estimates of battery run-time/service life.

### 9.3.4    Performance Comparisons of Representative Primary Batteries

Figure 9.4 compares the performance of a number of primary battery systems in a typical button-cell configuration (International Electrotechnical Commission [IEC] standard size 44). The data are based on the rated capacity at 20°C at about the $C/500$ rate (i.e., 500 h discharge). The performance of the different systems can be compared, but one should recognize that battery manufacturers may design and fabricate batteries, of the same size and with the same electrochemical system, with differing capacities and other characteristics, depending on the application requirements and the particular market segment the manufacturer is addressing.

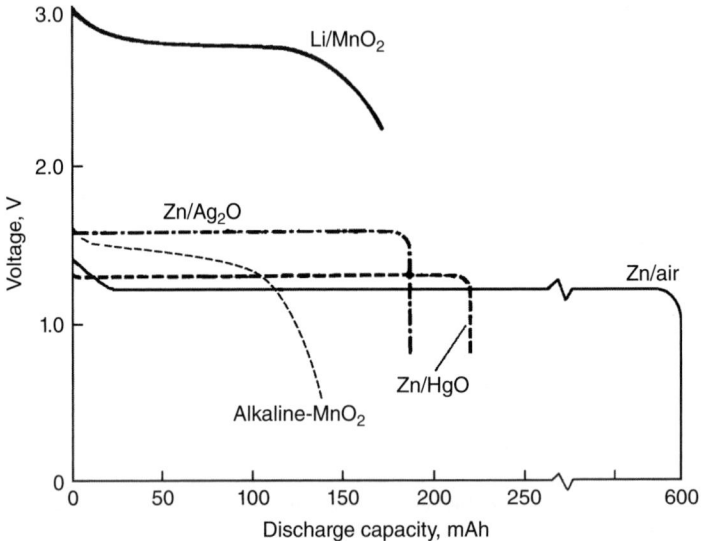

**FIGURE 9.4**   Typical discharge curves for primary battery systems, 11.6 mm diameter, 5.4 mm high, 20°C (Li/MnO$_2$ battery is 1/3 N-size).

Table 9.4 summarizes the typical performance obtained with the different primary battery systems for several cylindrical type batteries. The discharge curves for the AA-size batteries are shown in Fig. 9.5.

### 9.3.5    Effect of Discharge Load and Duty Cycle

The effect of the discharge load on the battery's capacity was shown in Fig. 9.3 and is again illustrated for several primary battery systems in Fig. 9.6. While the Leclanché zinc-carbon battery was noted to perform poorly with increasing discharge rates and the zinc/alkaline/manganese dioxide system to perform better with increasing discharge rates, neither system rivals the lithium battery for energy density or high-rate discharge. Specifically, for low-power applications, the energy derived from lithium cells is four times that of the zinc-carbon cell and three times greater for alkaline cells (zinc-manganese oxide). At the heavier loads, such as those required for toys, motors, and pulse discharges, lithium cells may provide 24 times the capacity and alkaline cells may

**TABLE 9.4**    Comparison of Cylindrical-Type Primary Batteries*

| Working voltage, V | Zinc-carbon (standard) | Zinc-carbon (heavy-duty ZnCl₂) | Zn/MnO₂† (alkaline) | Zn/HgO* | Mg/MnO₂* | Li/SO₂ | Li/SOCl₂ (bobbin type) | Li/MnO₂ (spiral wound) | Li/FeS₂ |
|---|---|---|---|---|---|---|---|---|---|
| | 1.2 | 1.2 | 1.2 | 1.25 | 1.75 | 2.8 | 3.3 | 2.8 | 1.5 |
| **D-size cells** | | | | | | | | | |
| Ah | 4.5 | 7.0 | 18.5 | 14 | 7 | 7.75 | 19 | 11.1 | |
| Wh | 5.4 | 8.4 | 22.8 | 17.5 | 12.2 | 21.7 | 64.6 | 30.0 | |
| Weight, g | 85 | 93 | 148 | 165 | 105 | 85 | 93 | 115 | |
| Wh/kg | 65 | 90 | 154 | 105 | 115 | 255 | 695 | 261 | |
| Wh/L | 100 | 160 | 407 | 325 | 225 | 397 | 1235 | 546 | |
| **N-size cells** | | | | | | | | | |
| Ah | 0.40 | | 1.00 | 0.8 | 0.5 | | | | |
| Wh | 0.48 | | 1.20 | 1.0 | 0.87 | | | | |
| Weight, g | 6.3 | | 9.0 | 12 | 5.0 | | | | |
| Wh/kg | 75 | | 133 | 85 | 170 | | | | |
| Wh/L | 145 | | 364 | 330 | 290 | | | | |
| **AA-size cells** | | | | | | | | | |
| Ah | 0.8 | 1.05 | 2.80 | 2.5 | | 0.95 | 2.4 | 1.4‡ | 3.1 |
| Wh | 0.96 | 1.25 | 3.39 | 3.1 | | 2.66 | 8.41 | 3.9 | 4.495 |
| Weight, g | 14.7 | 15 | 2.30 | 30 | | 15 | 18 | 17 | 14.5 |
| Wh/kg | 65 | 84 | 1.47 | 103 | | 177 | 467 | 235 | 310 |
| Wh/L | 125 | 162 | 4.18 | 400 | | 334 | 1007 | 525 | 562 |

*These batteries may no longer be available.
†Zn/MnO₂ (Alkaline) data to a 0.8-V cutoff.
‡2/3 A size.

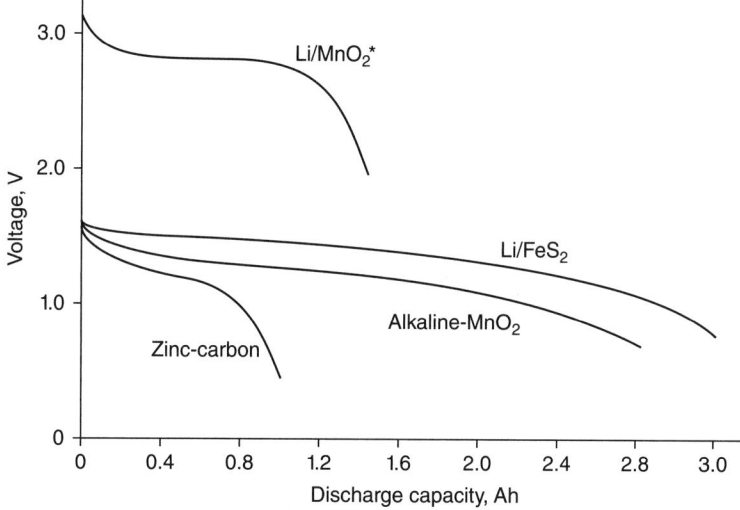

**FIGURE 9.5**    Typical discharge curves for primary battery systems. AA-size cells, approx. 20 mA discharge rate. *2/3 A-size battery.

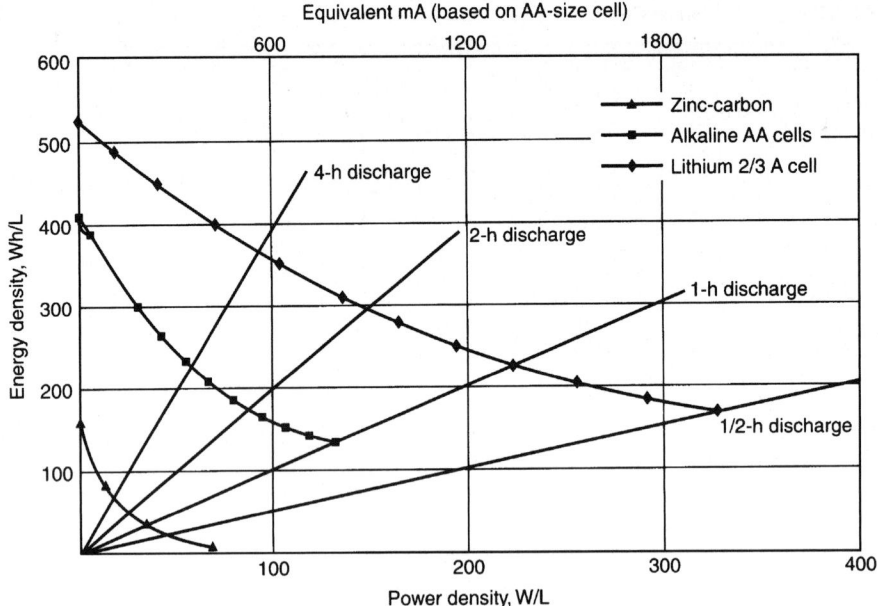

**FIGURE 9.6**   Comparison of primary battery systems under various continuous discharge loads at 20°C.

provide eight times the capacity of the zinc-carbon cells. Hence, premium batteries may at times be preferable, based on both performance and costs.

### 9.3.6   Effect of Temperature

The performance of the various primary batteries over a wide temperature range is illustrated in Fig. 9.7 on a specific energy basis (Wh/kg) and in Fig. 9.8 on an energy density basis (Wh/L). Lithium/soluble-cathode systems (Li/SOCl$_2$ and Li/SO$_2$) show the best performance throughout the entire temperature range, with the higher-rate Li/SO$_2$ system having the best capacity retention at the very low temperatures. The zinc/air system has a high energy density at normal temperatures, but only at light discharge loads. The lithium/solid-cathode systems, represented by the Li/MnO$_2$ system, show high performance over a wide temperature range, superior to the conventional zinc anode systems. Figure 9.8 shows an improvement in relative position of the denser, heavier battery systems when compared on a volumetric basis.

**Note:** As stated earlier, these data are generalized representations of each battery system under favorable discharge conditions. With the variability in performance due to manufacturer, design, size, discharge conditions, end voltage, and other factors, results may vary under different conditions of use. Refer to the appropriate chapters in Part 4 for more precise performance details.

### 9.3.7   Shelf Life of Primary Batteries

The shelf life characteristics of the major primary battery systems are plotted in Fig. 9.9. The data show the rate of loss as a percent capacity reduction per year, from 20 to 70°C. The relationship is approximately linear when capacity loss is plotted on a logarithmic scale against the log of the inverse temperature (i.e., log 1/T in Kelvin). The data assume that the rate of capacity loss remains constant throughout the storage period, which is not necessarily the case with most battery systems. For example, as shown in Chap. 13, the rate of loss for several

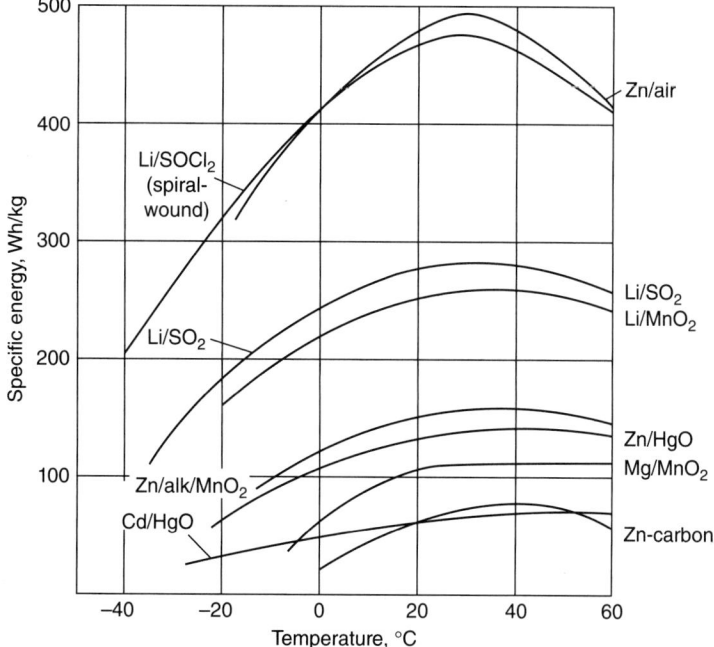

**FIGURE 9.7**    Specific energy of primary battery systems.

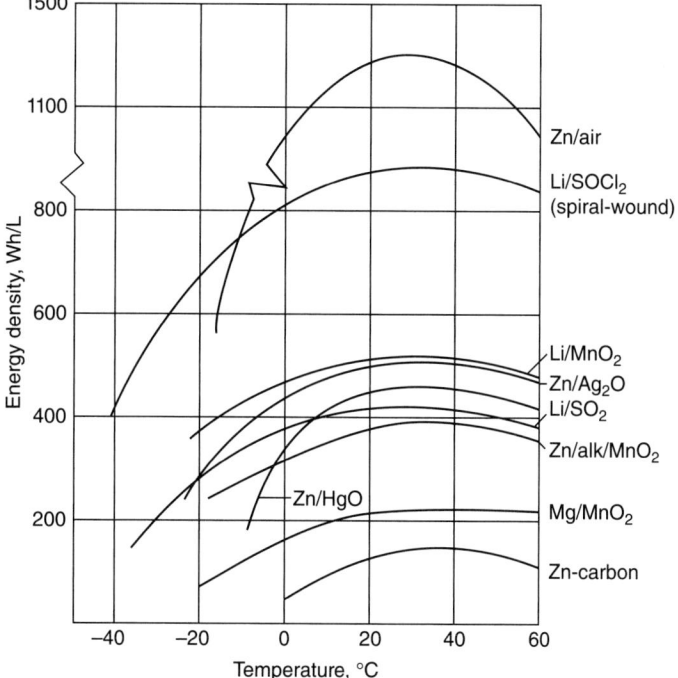

**FIGURE 9.8**    Volumetric energy density of primary battery systems.

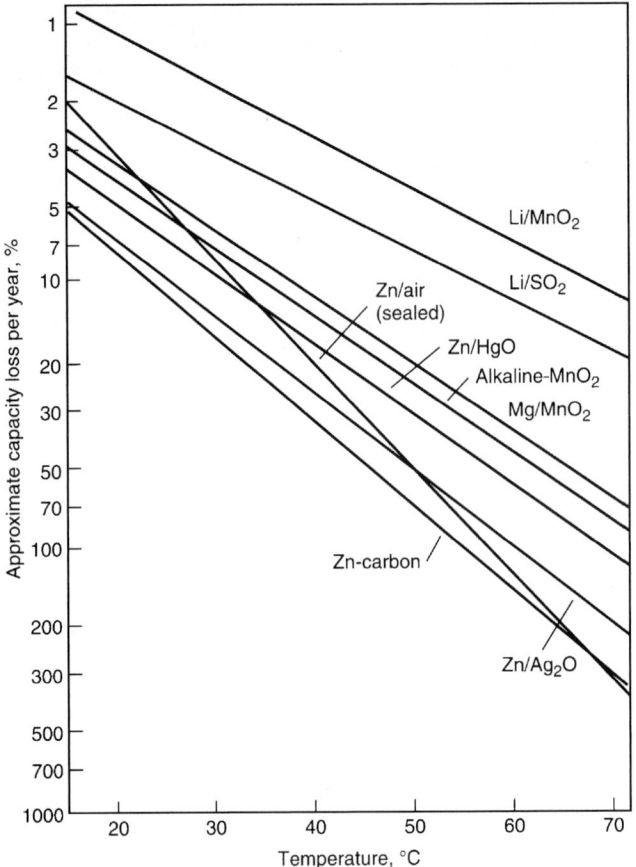

**FIGURE 9.9** Shelf life characteristics of primary battery systems.

types of lithium batteries tapers off as the storage period is extended. The data may also tend to overestimate the capability of each battery system, as there are many variations in battery design and formulation not always detailed in manufacturer ratings. The discharge conditions and battery size also have an influence on charge retention. The deliverable capacity is usually lowest under the more severe discharge conditions.

The ability to store batteries improves as the storage temperature is lowered. Cold storage of batteries, above the freezing point of the electrolyte, was used to extend their shelf life. As the shelf life of most batteries has been improved, manufacturers are no longer recommending cold storage, and suggest room temperature storage to be adequate, provided that excursions to high temperature are avoided.

### 9.3.8  Cost

Costing for various battery types has undergone a major paradigm shift over the last decade. The advent of new technologies, rapid commercialization strategies, large-scale operations, increasing raw material supplies, and aggressive global pricing have made and lost fortunes for companies and investors. Also, pricing does not always follow costs. Battery companies will always try to maximize margins and will often offer specialized packaging and electronics to provide added value. Economies of scale, the cost of energy supplies (i.e., electric utility power, solar energy, etc.), and swings in raw material supply and demand will further complicate the cost projections. Given continued fluctuations in currency exchange rates and trade balances as well as the pervasive role of governments/consortiums/trade associations to subsidize battery development and implementation, the true costs of any given cell technology is difficult to ascertain and will undoubtedly vary greatly. Nonetheless, as battery materials become more abundant and as cell design and construction improve, cheaper batteries will become available.

## 9.4   RECHARGING PRIMARY BATTERIES

Recharging primary batteries is a fundamentally unacceptable practice. Many modern cells not only employ extremely flammable, potentially toxic materials, many of the cell designs are tightly sealed and are not provided with an adequate mechanism to permit the release of gases that often form during charging. Buildup of internal pressure, including evolution of hydrogen gas due to electrolysis of water, has caused cells to leak, rupture, and explode, resulting in personal injury and death, damage to equipment and surroundings as well as exposure to toxins and other hazards. Even if not labeled with a cautionary notice advising against recharge, no primary battery should ever be recharged.

While some publications and internet postings may report that primary cells can be recharged, such actions are not advised under any circumstances. Unless done in a secure facility (i.e., a room or chamber with the ability to contain and handle fire, explosion, and toxins) and after a detailed safety review, the potential for harm and damage far outweighs the potential to achieve a few added cycles. Battery researchers typically conduct all charging experiments under carefully controlled conditions and usually at low charge rates but still assume that cells may fail with severe consequences. Primary batteries are not designed to be recharged, and charging should not be attempted with any primary battery.

Note that while some battery companies have marketed rechargeable zinc/alkaline/manganese dioxide cells (see Chap. 16B), these systems were carefully redesigned to allow recharge but only under specific conditions and after a full risk mitigation plan was instituted.

## REFERENCES

1. https://www.alibaba.com/product-detail/Carbon-zinc-dry-Cell-OEM-Welcome_60613089748.html?spm=a2700.7735675.30.55.qLOyFc&s=p $0.02, AA, 0.8 Ah dry cells and https://www.alibaba.com/product-detail/Smartlock-Battery-1-5V-AA Size_60620478904.html?spm=a2700.7735675.30.154.AAMY41&s=p, $0.075, AA alkaline, 2.4 Ah, extracted May 29, 2017.

2. http://www.thenational.ae/news/uae-news/lithium-battery-fire-risk-linked-to-dubai-plane-crash, extracted June 3, 2017.

# CHAPTER 10
# AN INTRODUCTION TO SECONDARY BATTERIES

**Thomas B. Reddy, Kirby W. Beard**

## 10.1 APPLICATIONS OF SECONDARY BATTERIES

Large secondary or rechargeable batteries are used for starting, lighting, and ignition (SLI) automotive applications (Chap. 14), industrial truck/materials handling equipment (Chap. 26B), military and government equipment (Chaps. 14, 15, and 25B), and emergency and standby power (Chaps. 14 and 17). Smaller, secondary batteries are used for power tools (Chap. 24A), lighting (Chap. 24B), toys, photography, communications, and various consumer electronic devices (Chap. 23—laptop computers, camcorders, etc.). More recently, secondary batteries have proliferated in electric and hybrid electric vehicles (Chap. 26A) and grid energy storage applications (Chap. 27). The applications of secondary batteries fall into two major categories:

1. *Energy storage:* Applications where batteries are charged by a primary energy source and deliver energy when main power is not available or is inadequate to handle the load requirement. Examples are automotive and aircraft systems, uninterruptible power supplies (UPSs) and standby power sources, and hybrid electric vehicle applications.

2. *Portable/mobile power:* Applications where the battery is discharged and recharged after regular use. These applications are often in competition with primary cells (consumer electronics) or are used where primary batteries would be unacceptable (electric vehicles, industrial trucks, and some stationary applications). The selection decisions are based on convenience, cost savings, or power drain.

## 10.2 GENERAL CLASSIFICATIONS

Batteries can be classified in many different ways. One of the most common groupings is by electrolyte type: aqueous or nonaqueous. While other types exist (solid state, molten salt, etc.), dividing rechargeable cells based on the use of water as the electrolyte solvent is a key distinction, not only due to the existence of water, but also due to the types of electrodes (and ionic species) present. The active materials used in a specific electrochemistry are typically not interchangeable with other systems.

A further refinement to these categories splits up aqueous systems by pH: acid, neutral, and base. Hence, the secondary battery chapters in this Handbook are organized as follows:

- *Acid electrolytes:* Chap. 14 (lead acid)
- *Alkaline and neutral electrolytes:* Chap. 15 (nickel-based cathodes) and Chap. 16 (metal oxide cathodes)
- *Nonaqueous:* Chap. 17A (lithium ion) and Chap. 17B (lithium metal anodes)

The nonaqueous electrolytes can be further classified as either organic or inorganic liquids. Most of the lithium-ion cells use organic solvents with lithium salts. Rechargeable inorganic systems have not been fully commercialized, but have used liquid cathodes that serve as both electrolyte and active material, including liquids such as sulfur dioxide and bromine.

The specialized electrochemical systems in Chap. 18B (metal-air alkaline), Chap. 19 (fuel cells) and Chap. 20 (capacitors) also function as secondary cells with further details provided in the listed chapters. These electrochemical couples use a wide range of electrolytes as follows:

- *Metal-air:* Almost all use alkaline electrolytes.
- *Fuel cells:* May use alkaline, molten salts/carbonates, inorganics, etc.
- *Capacitors:* Use both acid and organic electrolytes.

## 10.3  MARKET TRENDS

The worldwide secondary battery market was approximately $65 billion in 2015.[1] The lead-acid battery is the most popular, with the SLI battery accounting for a major share of the market. This share is declining gradually due to increasing applications for other types of batteries. In 2015, there were ~6 billion lithium-ion cells produced, with a market value of about $25 billion.[1] A major growth area has been the nonautomotive consumer applications for small secondary batteries. Lithium-ion batteries have emerged in the last decade to capture a 75% share of the market for small, sealed consumer batteries.

Additional sales trends for secondary batteries are shown in Tables 10.1 and 10.2 (courtesy of The Freedonia Group, December 2016). Table 10.1 shows the relative market position of secondary batteries over the past 15 years

**TABLE 10.1**    World Battery Demand by Product (million dollars)*

| Item | 2004 | 2009 | 2014 | 2019 | 2024 |
|---|---|---|---|---|---|
| World battery demand | **36,345** | **56,145** | **82,470** | **120,000** | **166,500** |
| Primary | 12,990 | 16,455 | 19,865 | 24,050 | 28,150 |
| Secondary | 23,355 | 39,690 | 62,605 | 95,950 | 138,350 |
| Price deflator (2013 = 100) | 76.2 | 99.1 | 98.9 | 92.9 | 77.0 |
| **World battery demand (million 2013 USD)** | **47,670** | **56,650** | **83,410** | **129,130** | **216,160** |

*Table IV-6, Battery Markets in the United States by Application and End User, published December 2016 by The Freedonia Group.

**TABLE 10.2**    World Secondary Battery Demand by Chemistry (million dollars)*

| Item | 2004 | 2009 | 2014 | 2019 | 2024 |
|---|---|---|---|---|---|
| World battery demand | 36,345 | 56,145 | 82,470 | 120,000 | 166,500 |
| Percentage of secondary | 64.3 | 70.7 | 75.9 | 80.0 | 83.1 |
| World secondary battery demand | **23,355** | **39,690** | **62,605** | **95,950** | **138,350** |
| Lead-acid | 15,564 | 27,415 | 41,540 | 59,800 | 80,050 |
| Lithium ion | 4206 | 7433 | 14,800 | 28,500 | 49,500 |
| Nickel based | 2436 | 3331 | 4330 | 5340 | 6105 |
| Other | 1149 | 1511 | 1935 | 2310 | 2695 |
| Secondary battery deflator (2013 = 100) | 73.9 | 98.8 | 99.0 | 90.0 | 72.1 |
| Secondary battery demand (million 2013 USD) | 31,610 | 40,170 | 63,230 | 106,640 | 191,890 |

*Table IV-12, Battery Markets in the United States by Application and End User, published December 2016 by The Freedonia Group.

with projections for the next 5 years. Clearly, while the battery market is growing significantly (i.e., quadrupling over a 20-year span), secondary cells are outpacing primary cells by threefold (i.e., primary cells double while secondary cells increase sixfold over this same time period). In addition, the market size for secondary cells will go from almost double the volumes of primaries to nearly five times greater by 2024.

Table 10.2 breaks the secondary battery market into the three main types, discussed in the next section (lead-acid, lithium ion, and Ni-based alkaline), plus a miscellaneous category. The secondary battery market will grow from approximately two-thirds of battery sales to almost 85% of all batteries between 2004 and 2024. Lead-acid batteries are shown to grow fivefold, but lithium ion will expand by an order of magnitude, owing to a rather low starting point. Nonetheless, though, lithium ion is starting to approach the size of the lead-acid business (going from ¼ to almost ⅔ the size of lead acid in 2 decades).

These data also confirm the prior estimates of the secondary battery market, growing from $60 billion a few years ago to over $100 billion in the next few years (~2020).

## 10.4  *GENERAL CHARACTERISTICS*

Conventional aqueous secondary batteries are characterized by high power density, stable discharge voltage, and good low-temperature performance. Energy density, specific energy, and charge retention are typically poorer than primary batteries. Lithium-ion technologies have higher energy densities and charge retention, but until recently they did not have the best power density or good temperature range due in large part to the use of aprotic solvents with lower conductivity than the aqueous electrolyte. The use of high surface area electrodes and advanced materials has helped remedy the situation (see Chap. 17). The specific energy and energy density of various secondary systems are as follows[2-4]:

- *Lead acid*: 30 to 50 Wh/kg, 70 Wh/L
- *Ni-cadmium (Ni-Cd)*: 45 to 80 Wh/kg, 100 Wh/L
- *Ni-metal hydride*: 60 to 120 Wh/kg, >250 Wh/L
- *Li-ion*: 120 to 300 Wh/kg, 400 to 650 Wh/L
- *Li-sulfur or similar systems:* 500 to 800 Wh/kg, 350 to 800 Wh/L
- *Metal (Zn, Li) air:* >1000 Wh/kg, 1500 Wh/L (may require anode replacement each cycle)

Since the invention of the first secondary battery, the lead-acid cell in 1859 by Planté, and the subsequent commercialization of the nickel-iron alkaline battery by Edison in 1908, a number of systems were developed, including Ni-Cd in 1909, silver-zinc in the 1940s, and lithium metal anode batteries in the 1980s (e.g., MoliCel®). Early applications included electric automobile, industrial haul vehicles, and stationary applications. The Ni-Fe cells were durable, but had high cost, high maintenance requirements, and lower specific energy.[5]

The pocket-plate Ni-Cd battery still exists today, but has been supplemented by sintered-plate designs, which led to its use in high-power/high-energy uses, such as aircraft engine starting. The invention of sealed Ni-Cd cells in the 1950s led to its dominance in the portable rechargeable market. Sealed Ni-Cd cells eventually gave up market share to sealed lead-acid and later Ni-metal hydride, but today lithium-ion batteries are far superior in specific energy and energy density.

Lithium-ion cells have a much shorter, but perhaps, more prolific history. Primary lithium batteries were known and used for decades. However, in less than one decade lithium secondary batteries were perfected and launched on a path to global success, exceeding all other major categories except the SLI batteries and even surpassing all the alkaline rechargeable cells combined. While superior energy density and high-rate capability were important to this progress, form factor and voltage may have been the most valuable attributes of lithium-ion cells. The sealed aqueous batteries could not be made in thin flat footprints that were needed for the expanding field of consumer electronics. Also, the ability to use a single lithium cell (or fewer cells) due to higher voltages (almost triple alkaline cells and double lead acids) greatly benefits the design of the control electronics. The future of lithium cells is wide-open and there is no turning back, especially since the costs per watt-hour have improved so dramatically. The typical characteristics and applications of secondary batteries are summarized in Table 10.3.

**TABLE 10.3** Major Characteristics and Applications of Secondary Batteries

| System | Characteristics | Applications |
|---|---|---|
| Lead acid | | |
| Automotive | Popular, low-cost secondary battery, low specific energy, high rate, and low temperature performance; maintenance-free designs | Automotive SLI, golf carts, lawn mowers, tractors, aircraft, marine, microhybrid vehicles |
| Traction (motive power) | Designed for deep 6–9 h discharge, cycling service | Industrial trucks, materials handling, electric and hybrid electric vehicles, special types for submarine power |
| Stationary | Designed for standby float service, long life, VRLA designs | Emergency power, utilities, telephone, UPS, load leveling, energy storage, emergency lighting |
| Portable | Sealed, maintenance-free, low cost, good float capability, moderate cycle life | Portable tools, small appliances and devices, portable electronic equipment |
| Nickel-cadmium | | |
| Industrial and FNC (Fiber Ni-Cd) | Good high rate, low-temperature capability, flat voltage, excellent cycle life | Aircraft batteries, industrial and emergency power applications, communication equipment |
| Portable | Sealed, maintenance-free, good high-rate low-temperature performance, good cycle life | Consumer electronics, portable tools, pagers, appliances, photographic equipment, standby power, memory backup |
| Nickel-metal hydride | Sealed, maintenance-free, higher capacity than nickel-cadmium batteries; high energy density and power | Consumer electronics and other portable applications; hybrid electric vehicles |
| Nickel-iron | Durable, rugged construction, long life, low specific energy | Materials handling, stationary applications, railroad cars |
| Nickel-zinc | High specific energy, extended cycle life, high-power capability | Bicycles, scooters, consumer electronics such as power tools |
| Silver-zinc | High specific energy, very good high-rate capability, low cycle life, high cost | Training targets, drones, submarines, other military equipment, launch vehicles and space power |
| Silver-cadmium | High specific energy, good charge retention, moderate cycle life, high cost | Portable equipment requiring a lightweight, high-capacity battery; space satellites |
| Nickel-hydrogen | Long cycle life under shallow discharge, long life | Primarily for aerospace applications such as LEO and GEO satellites |
| Ambient-temperature rechargeable "primary" types ($Zn/MnO_2$) | Low cost, good capacity retention, sealed and maintenance-free, limited cycle life and rate capability | Cylindrical cell applications, rechargeable replacement for zinc-carbon and alkaline primary batteries, consumer electronics (ambient-temperature systems) |
| Lithium-ion | High specific energy and energy density, long cycle life; high power capability | Portable and consumer electronic equipment, electric vehicles (EVs, HEVs, PIIEVs), space applications, electrical energy storage |

## 10.5 CHARACTERISTICS OF SPECIFIC SECONDARY BATTERIES

The transformation between electric energy and chemical energy in secondary cells should proceed nearly reversibly with high energy efficiency and minimal physical changes. Chemical reactions on both charge and discharge can deteriorate cell components, causing reduced shelf and cycle life, loss of energy, resistance increases, and poor performance at high or low temperatures. Ideally, batteries will also be recycled and safe to use. A limited number of rechargeable batteries with the required attributes are known and less than a dozen basic types have been generally successful, as summarized below.

### 10.5.1    Lead-Acid Batteries (Chap. 14)

Lead-acid batteries are very adaptable and often tailored to customer requirements. As of 2015, they have the largest market share of all secondary batteries at 35 to 45 billion USD (see Sec. 10.3). The charge-discharge processes are stable and reliable over a wide temperature range. Low costs with good cycle-life and very high recycle rate make this a popular choice. Lead toxicity is an issue, but exposure problems have been controlled.

Various designs are listed in Table 10.3 (and Chap. 14), ranging from small sealed cells with a capacity of 1 Ah to large 12,000 Ah cells. The automotive SLI battery has dominated this category for over a century. While designs have evolved slowly, some new technologies, such as the enhanced flooded battery (EFB), also known as the maintenance-free battery, and valve-regulated lead-acid (VRLA) designs, have impacted sales of the conventional, "flooded" battery. Lead alloys are used to reduced water loss (minimizing or eliminating the need to add water) and to reduce self-discharge rates, allowing for shipping and storage in an activated (wet, charged) state.

Stop-start and mild hybrid electric vehicles, and potentially, 48 V car systems have created the need for new battery designs. EFB batteries with carbon additives and modified separators have provided improved performance at comparable costs. Industrial lead-acid batteries are typically larger and more robust. Categories include:

1. Motive power/traction types for materials-handling trucks, tow tractors, mining vehicles, and personnel carriers.
2. Diesel locomotive engine starting.
3. Stationary service: telecommunications systems, electric utilities, emergency and standby power systems, UPS, and railroad signaling and car power systems. Most of these applications have moved to a VRLA construction.

Positive plate designs include:

1. Tubular and pasted flat plates (motive power, diesel engine cranking, and stationary applications)
2. Planté designs

Pure lead is used in Planté cells, mainly in stationary batteries, and either lead-antimony or lead-calcium grid alloys are used in flat plate cells. Other designs have found niche applications, such as the "round cell," used for telephone backup power, instead of the normal prismatic design.

The sealed lead acid cell, based on the use of an absorbed glass mat (AGM), an immobilized, "starved" electrolyte, and a resealable pressure valve (VRLA), was an important development in lead-acid battery technology in which oxygen was recombined (similar to the sealed Ni-Cd cell design). Oxygen produced on overcharge diffuses from the positive to the negative electrode, recombining with hydrogen and eliminating over 95% of the outgassing. The VRLA battery is now used in more than 70% of the telecommunication batteries and more than 80% of the UPS applications plus in various emergency lighting and consumer-type applications. These small sealed lead-acid batteries are constructed as either prismatic cells with parallel plates (1–30 Ah) or as cylindrical cells with spirally wound electrodes and 25 Ah capacity. These cells may also use gelled rather than absorbed electrolyte to prevent leakage. The grids may be lead-calcium-tin alloy(s) or pure lead.

Other applications include submarine service, reserve power in marine applications, and backup system in confined spaces where engine-generators cannot be used, such as indoors and in underground mining equipment. Applications in load leveling for utilities and solar photo-voltaic systems are being studied where low costs are needed.

Newer designs, such as the Ultrabattery™ and the EFBs that use negative plates (constructed partially or fully with carbon), alone or in tandem with standard negative plates, are being evaluated for electric vehicles and grid storage applications.

## 10.5.2 Alkaline Secondary Batteries

The other main category of conventional, aqueous secondary batteries is the alkaline electrolyte (KOH or NaOH solutions) system (see Table 10.3, and Chaps. 15 and 16). Alkaline electrolytes are typically less reactive than acid electrolytes with their respective active materials. Furthermore, unlike the lead-acid cell, the charge-discharge mechanism in the alkaline electrolyte does not consume or generate water, leaving the composition or concentration of the electrolyte unchanged during charge and discharge. The two main categories of cathodes in this group include nickel oxyhydroxide and the different metal oxides.

*Nickel-Iron Batteries (Chap. 15A).*   The nickel-iron battery was a key player up until the 1970s, when displaced by industrial lead-acid batteries in stationary and industrial transit uses. The main cell components are nickel-plated steel, making for a durable but low-energy battery with storage, temperature, and cost issues.

*Nickel-Cadmium Batteries (Chaps. 15B and 15C).*   Ni-Cd batteries have numerous vented cell designs and sizes with pocket- or sintered-plate constructions. The pocket-type cells are rugged and have long life and low maintenance beyond occasional topping with water. The sintered-plate construction is a more recent development, having higher energy density with better performance, but it is more expensive. Newer plate designs use nickel foam, nickel fiber, or plastic-bonded (pressed-plate) electrodes. Sealed cell designs have excess capacity in the negative electrode that allow oxygen recombination and are available in prismatic, button, and cylindrical configurations.

*Nickel-Zinc Batteries (Chap. 15E).*   The nickel-zinc is a cross between Ni-Cd and Ag-Zn cells with better energy density than Ni-Cd, but limited cycle life (due to dendrite formation). Given the potential superior cost:performance ratio, work has continued on this system.

*Hydrogen and Metal Hydride Electrode Batteries (Chaps. 15D and 15F).*   Hydrogen, whether stored as a cryogenic liquid, a gas, or absorbed in a molecular sponge, offers an acceptable negative active material when used with conventional nickel cathodes. Nickel-hydrogen batteries were used exclusively for aerospace programs in low earth orbit (LEO) and geosynchronous earth orbit (GEO) satellites (see Chap. 25B) until recently when Li-ion cells made inroads due largely to better economics. The sealed nickel-metal hydride battery, where the hydrogen is absorbed by a metal alloy, has better specific energy and energy density than Ni-Cd and was widely adapted for portable electronic applications, hybrid electric vehicles until, again, Li-ion cells took over in many areas.

*Silver Oxide Batteries (Chap. 16A).*   The silver-zinc battery is noted for high energy density, low internal resistance, and stable voltages and had been the leader when high energy density is required. The high cost of silver, even if recycled, limited this cell to military and space uses or other premium applications. Also, cycle life, activated life, and its low temperature performance were weaknesses.

*Miscellaneous (Chaps. 16 and 18).*   Batteries, based on the zinc/alkaline/manganese dioxide chemistry primary battery electrochemistry, were briefly sold as a secondary battery, but proved to be too limited in life/capacity. Some efforts still continue along with studies of iron oxide (Chap. 16C) and zinc-air (Chap. 18B) systems.

## 10.5.3 Lithium Secondary Batteries

*Lithium-Ion Batteries (Chap. 17A).*   Lithium-ion batteries now dominate the secondary consumer electronics market: laptop computers, cell phones, tablets, and e-books. Remarkably, sales volumes have increased over fivefold in the last 15 years, despite a drop in cell costs from around \$1500/kWh in 2004 to about \$200/kWh today (nearly a 90% cost reduction).[6] Production capacity was estimated to be 250 million/cells per month in 2010,[7] but this figure could vary erratically as cell sizes and types evolve (i.e., relatively more large prismatic or flat wound cells and fewer small cylindrical cells). An alternative measure to cell units can be expressed in total cell gigawatt-hour production, given the pending commission of dozens of Li-ion (and potentially fuel cell) factories (Chap. 2E, Ref. 3). Lithium-ion cells dominate due to high energy density and specific energy, long cycle life, acceptable environmental tolerance, and for having a reasonable track record for adequate safety, despite various misgivings. The adoption of sophisticated battery management circuitry (Chaps. 23 and 31) has also helped to improve performance and safety.

*Lithium-Metal Batteries (Chap. 17B).* The future of lithium cells will likely shift to lithium-metal anodes or hybrid systems with silicon, carbon, or other constituents. These systems will also likely be paired with new cathodes (sulfur) and potentially even new solid state, polymer, hybrid, or other advanced electrolyte systems.

## 10.6   COMPARISON OF PERFORMANCE CHARACTERISTICS FOR SECONDARY BATTERY SYSTEMS

### 10.6.1   General

The current choices for commercial secondary cells are mainly limited to lead acid, a few selected alkaline systems (mostly Ni-based cathodes), and lithium ion. Other systems are improving and a few will likely be widely adopted in the next 5 to 10 years. However, for now the main concern for most battery users is which of the three basic types is best for a particular application. This section details how these major secondary battery types compare based on key performance characteristics. Table 10.4 summarizes the choices and lists various theoretical and practical characteristics of secondary battery systems (see Chaps. 14 to 17 for the most current data). Typically, only about 25% to 35% of the theoretical capacity of a battery system is attained under practical conditions.

Note that the systems being presented in this section are likely tested under favorable discharge conditions with optimized designs. The performance of a battery system is highly dependent on the specific designs and conditions of the use. A qualitative comparison of the various secondary battery systems is presented in Table 10.5. Often various designs with the same electrochemistry will differ in performance.

### 10.6.2   Voltage and Discharge Profiles

The discharge curves of the conventional secondary battery systems, at the $C/5$ rate, are compared in Fig. 10.1. The lead-acid battery operates around 2 V and alkaline systems range from about 1.65 to 1.1 V. At higher discharge rates and lower temperatures, cell differences within a particular electrochemical system could be large.

Most cells except for the silver oxide systems and the rechargeable zinc/manganese dioxide battery have fairly flat, smooth discharge profiles.

The discharge curve for the graphite/lithiated cobalt oxide system shows much higher voltage and also a more sloping profile, typically, due to the lower conductivity of the nonaqueous electrolytes and less thermodynamically favorable reaction (due to intercalation mechanisms). A lithium-ion cell at 3.7 V can replace three Ni-Cd or NiMH cells.

### 10.6.3   Effect of Discharge Rate on Performance

Figure 10.2 compares cell run-time at different power levels (similar to a Ragone plot) except where hours of service are used to indicate specific energy (Wh/kg). A standardized 1 kg battery with a higher curve is indicative of superior capacity at higher discharge loads. The specific energy can be calculated by the following equation:

$$\text{specific energy} = \text{specific power} \times \text{hours of service}$$

or

$$Wh/kg = W/kg \times h = \frac{A \times V \times h}{kg}$$

**TABLE 10.4** Characteristics of the Major Secondary Battery Systems

| Common name | Lead-acid | | | | Nickel-cadmium | | | |
| --- | --- | --- | --- | --- | --- | --- | --- | --- |
| | SLI | Traction | Stationary | Portable | Vented pocket plate | Vented sintered plate | Sealed | FNC |
| **Chemistry** | | | | | | | | |
| Anode | Pb | Pb | Pb | Pb | Cd | Cd | Cd | Cd |
| Cathode | PbO$_2$ | PbO$_2$ | PbO$_2$ | PbO$_2$ | NiOOH | NiOOH | NiOOH | NiOOH |
| Electrolyte | H$_2$SO$_4$ (aqueous solution) | H$_2$SO$_4$ (aqueous solution) | H$_2$SO$_4$ (aqueous solution) | H$_2$SO$_4$ (aqueous solution) | KOH (aqueous solution) | KOH (aqueous solution) | KOH (aqueous solution) | KOH (aqueous solution) |
| **Cell voltage (typical), V** | | | | | | | | |
| Nominal | 2.0 | 2.0 | 2.0 | 2.0 | 1.2 | 1.2 | 1.2 | 1.2 |
| Open circuit | 2.1 | 2.1 | 2.1 | 2.1 | 1.29 | 1.29 | 1.29 | 1.35 |
| Operating | 2.0–1.8 | 2.0–1.8 | 2.0–1.8 | 2.0–1.8 | 1.25–1.00 | 1.25–1.00 | 1.25–1.00 | 1.25–0.85 |
| End | 1.75 (lower operating and end voltage during cranking operation) | 1.75 | 1.75 (except when on float service) | 1.75 (where cycled) | 1.0 | 1.0 | 1.0 | 1.00–0.65 |
| Operating temperatures, °C | −40 to 55 | −20 to 40 | −10 to 40[†] | −40 to 60 | −20 to 45 | −40 to 50 | −20 to 70 | −50 to 60 |
| **Specific energy and energy density (at 20°C)** | | | | | | | | |
| Wh/kg | 40 | 25 | 10–20 | 30 | 27 | 30–37 | 35 | 10–40 |
| Wh/L | 80 | 80 | 50–70 | 90 | 55 | 58–96 | 100 | 15–80 |
| Discharge profile (relative) | Flat | Flat | Flat | Flat | Flat | Very flat | Very flat | Flat |
| Power density | High | Moderately high | Moderately high | High | High | High | Moderate to high | Very high |
| Self-discharge rate (at 20°C), % loss per month[§] | 20–30 (Sb-Pb) 2–3 (maintenance-free) | 4–6 | — | 4–8 | 5 | 10 | 15–20 | 10–15 |
| Calendar life, years | 3–6 | 6 | 18–25 | 2–8 | 8–25 | 3–10 | 5–7 | 5–20 |
| Cycle life, cycles[†] | 200–700 | 1500 | — | 250–500 | 500–2000 | 500–2000 | 300–1000 | 500–10,000 cycles in LEO satellite tests to 35% DOD at 10°C |
| Advantages | Low cost, ready availability, good high rate, high- and low-temperature operation (good cranking service), good float service, new maintenance-free designs | Lowest cost of competitive systems (also see SLI) | Designed for "float" service; lowest cost of competitive systems (also see SLI) | Maintenance-free; long life on float service; low- and high-temperature performance; no "memory" effect; operates in any position | Very rugged, can withstand physical and electrical abuse; good charge retention, storage, and cycle life; lowest cost of alkaline batteries | Rugged, excellent storage, good specific energy, and high-rate and low-temperature performance | Sealed, no maintenance, good low-temperature and high-rate performance, long life cycle, operates in any position | Sealed, no maintenance, high power capability even at low temperature, long cycle life at low depth of discharge, fast charging |
| Limitations | Relatively low cycle life, limited energy density, poor charge retention and storability, hydrogen evolution | Low energy density, less rugged than competitive systems, hydrogen evolution | Hydrogen evolution | Cannot be stored in discharged condition, lower cycle life than sealed nickel-cadmium, difficult to manufacture in very small sizes | Low energy density | High cost, "memory" effect, thermal runaway problem | Sealed lead-acid battery better at high temperature and float service, "memory" effect | Lower energy density than sintered plate design |
| Major battery types available | Prismatic cells: 40–100 Ah at 20-h rate | Based on positive plate design; 45–200 Ah per positive plate | Based on positive plate design: 5–400 Ah per positive to 1440 Ah plate | Sealed cylindrical cells: 2.5–25 Ah; prismatic cells to 1440 Ah | Prismatic cells: 5–1200 Ah | Prismatic cells: 1.5–100 Ah | Button cells to 0.5 Ah; cylindrical cells to 12 Ah | Prismatic designs to 490 Ah |

[*]Based on C/LiCoO$_2$ lithium-ion battery (see Chap. 17A) (characteristics vary with battery system and design).
[†]Dependent on depth of discharge.
[‡]High rate Zn/AgO battery.
[§]Self-discharge rate usually decreases with increasing storage time.

| Nickel-iron (conventional) | Nickel-zinc | Zinc/silver oxide (silver-zinc) | Cadmium/silver oxide (silver-cadmium) | Nickel-hydrogen | Nickel-metal hydride | Rechargeable "primary" types, Zn/MnO$_2$ | Lithium ion systems* |
|---|---|---|---|---|---|---|---|
| Fe<br>NiOOH<br>KOH (aqueous solution) | Zn<br>NiOOH<br>KOH (aqueous solution) | Zn<br>AgO<br>KOH (aqueous solution) | Cd<br>AgO<br>KOH (aqueous solution) | H$_2$<br>NiOOH<br>KOH (aqueous solution) | MH<br>NiOOH<br>KOH (aqueous solution) | Zn<br>MnO$_2$<br>KOH (aqueous solution) | C<br>LiCoO$_2$<br>Organic solvent graphite anode |
| 1.2<br>1.37<br>1.25–1.05<br>1.0 | 1.65<br>1.73<br>1.6–1.4<br>1.2 | 1.5<br>1.86<br>1.7–1.3<br>1.0 | 1.1<br>1.41<br>1.4–1.0<br>0.7 | 1.4<br>1.32<br>1.3–1.15<br>1.0 | 1.2<br>1.4<br>1.25–1.10<br>1.0 | 1.5<br>1.5<br>1.3–1.0<br>0.9 | 4.0<br>4.1<br>3.7<br>3.0 |
| −10 to 45 | −20 to 50 | −20 to 60 | −25 to 70 | 0 to 50 | −20 to 65 | −20 to 40 | −20 to 50 |
| | Prismatic / Cylindrical | | | | HEV / Commercial | | |
| 30<br>55<br>Moderately flat | 60–100 / 70–110<br>110–200 / 200–360<br>Flat | 105‡<br>180<br>Double plateau at low rate | 70<br>120<br>Double plateau | 64 (CPV)<br>105 (CPV)<br>Moderately flat | 47 / 90–110<br>177 / 430<br>Flat | 100<br>286<br>Sloping | 203<br>570<br>Sloping |
| Moderate to low | High | Very high (for high-rate designs) | Moderate to high | Moderate | High | Moderate | Moderate (energy cells); high (power cells) |
| 20–40 | 20 | 5 | 5 | Very high except at low temp. | 15–30 | | 2 |
| 8–25 | — | 6–18 cells (wet) | 3 (vented)<br>4 (sealed) | — | 5–10 | 5–7 | |
| 2000–4000 | To 900 at 80% DOD | 10–50 (HR) | 300–800 | 1500–6000<br>40,000 at 40% DOD | 500–1000 (300,000 for HEV) | 15–25 | 1000+ |
| Very rugged, can withstand physical and electrical abuse; long life (cycling or stand) | High energy density; relatively low cost; good low-temperature performance; high-power capability | High energy density; high discharge rate; low self-discharge | High-energy density; low self-discharge; good cycle life | High energy density; long cycle life at low DOD; can tolerate overcharge | High energy density; sealed; good cycle life | Good shelf life; low cost | High specific energy and energy density; low self-discharge; long cycle life; good rate-capability |
| Low power and energy density; high self-discharge; hydrogen evolution; high cost and high maintenance cost | Subject to zinc dendrite shorting | High cost; low cycle life; decreased performances at low temperatures | High cost; decreased performance at low temperatures | High initial cost; self-discharge proportional to H$_2$ pressure and temperature | Memory effect; must be charged at moderate temperature | Limited cycle life; low drain applications; small sizes only | Requires the use of a battery management system; safety issues |
| Decreasing significance in developed countries | Prismatic and cylindrical (AA, sub-C, and D) types available for light vehicles and consumer applications such as power tools and digital cameras | Prismatic cells: <1 to 1000 Ah; special types to 5000 Ah | Prismatic cells: for space application | Aerospace applications (up to 100 Ah) | Button and cylindrical cells to 12 Ah; large prismatics to 250 Ah | AAA and AA cylindrical cells to 2.0 Ah multicell bundles | Cylindrical and prismatic cells available in many chemistries LiCoO$_2$/graphite typical for consumer applications |

**TABLE 10.5**  Comparison of Secondary Batteries*

| System | Energy density | Power density | Flat discharge profile | Low-temperature operation | Charge retention | Charge acceptance | Efficiency | Life | Mechanical properties | Cost |
|---|---|---|---|---|---|---|---|---|---|---|
| Lead acid | | | | | | | | | | |
| Pasted | 4 | 4 | 3 | 3 | 4 | 3 | 2 | 3 | 5 | 1 |
| Tubular | 4 | 5 | 4 | 3 | 3 | 3 | 2 | 2 | 3 | 2 |
| Planté | 5 | 5 | 4 | 3 | 3 | 3 | 2 | 2 | 4 | 2 |
| Sealed | 4 | 3 | 3 | 2 | 3 | 3 | 2 | 3 | 5 | 2 |
| Lithium metal | 1 | 3 | 3 | 2 | 1 | 3 | 3 | 4 | 3 | 4 |
| Lithium ion | 1 | 2 | 3 | 2 | 1 | 1 | 1 | 1 | 2 | 2 |
| Nickel-cadmium | | | | | | | | | | |
| Pocket | 5 | 3 | 2 | 1 | 2 | 1 | 4 | 2 | 1 | 3 |
| Sintered | 4 | 1 | 1 | 1 | 4 | 1 | 3 | 2 | 1 | 3 |
| Sealed | 4 | 1 | 2 | 1 | 4 | 2 | 3 | 3 | 2 | 2 |
| Nickel-iron | 5 | 5 | 4 | 5 | 5 | 2 | 5 | 1 | 1 | 3 |
| Nickel-metal hydride | 2 | 1 | 2 | 2 | 3 | 1 | 2 | 2 | 2 | 3 |
| Nickel-zinc | 2 | 1 | 2 | 3 | 4 | 3 | 3 | 4 | 3 | 3 |
| Silver-zinc | 1 | 1 | 4 | 3 | 1 | 3 | 2 | 5 | 2 | 4 |
| Silver-cadmium | 2 | 3 | 5 | 4 | 1 | 5 | 1 | 4 | 3 | 4 |
| Nickel-hydrogen | 2 | 3 | 3 | 4 | 5 | 3 | 5 | 2 | 3 | 5 |
| Silver-hydrogen | 2 | 3 | 4 | 4 | 5 | 3 | 5 | 2 | 3 | 5 |
| Zinc-manganese dioxide | 2 | 4 | 5 | 3 | 1 | 4 | 4 | 5 | 4 | 2 |

*Rating: 1 to 5, best to poorest.

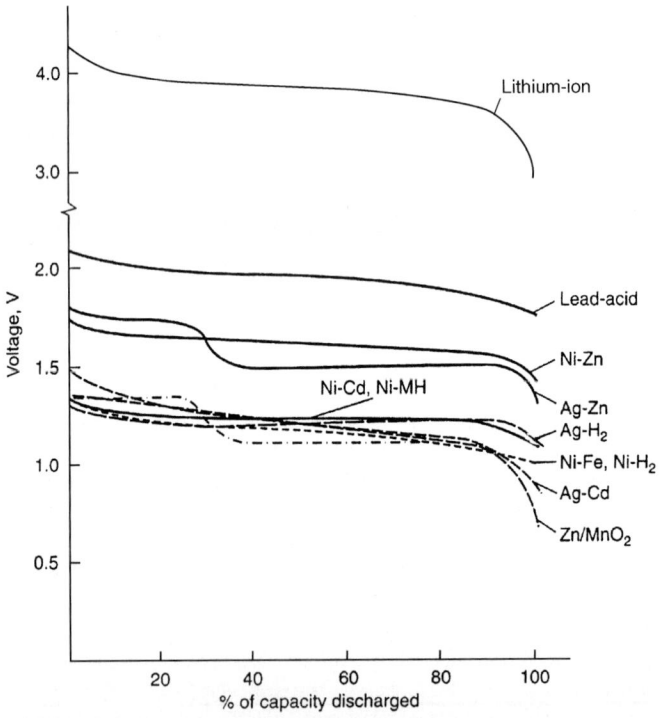

**FIGURE 10.1**  Discharge profiles of conventional secondary battery systems and rechargeable lithium-ion battery at approximately C/5 discharge rate.

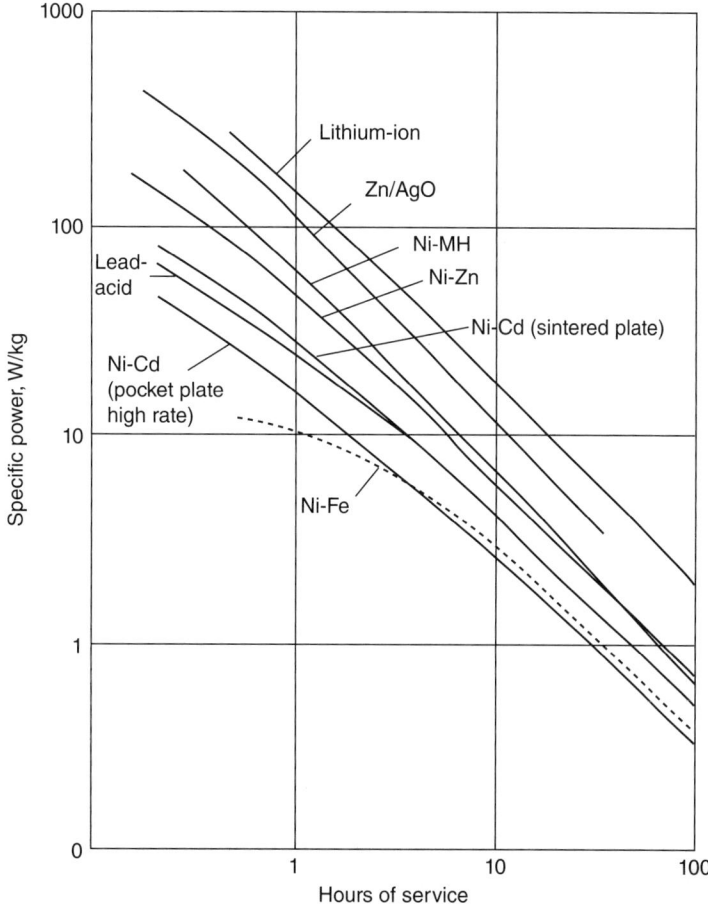

**FIGURE 10.2**  Comparison of performance of secondary battery systems at 20°C.

### 10.6.4   Effect of Temperature

Figure 10.3 shows the impact on delivered energy at various temperatures. Each battery system is plotted from −40 to +60°C at $C/5$ discharge rate. This lithium-ion system has the highest energy density over its range (to −20°C), while the alkaline cells, but especially the nickel-zinc and nickel-metal hydride batteries, have better low temperature performance than lead-acid batteries. Alternatively, lead-acid cells suffer less capacity losses at higher temperatures. These data were collected under favorable discharge conditions and would be different under other, less ideal conditions.

### 10.6.5   Charge Retention

Aqueous secondary batteries have rather poor storage properties compared to aqueous primary batteries. While these secondary batteries can be recharged or maintained on "float" charge, periodic testing and/or maintenance is often required. Most alkaline secondary batteries, especially the nickel cathode batteries, can be stored fully discharged without impacting capacity. Lead-acid batteries, however, form lead sulfate during storage and

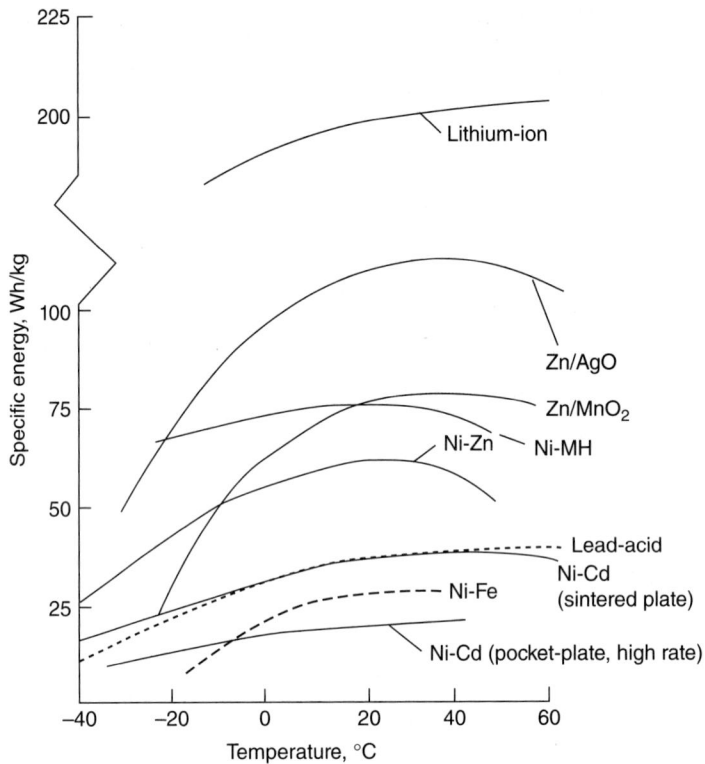

**FIGURE 10.3**   Effect of temperature on specific energy of secondary battery systems at approximately C/5 discharge rate.

will suffer capacity losses if not recharged regularly. Lithium-ion batteries will retain capacity over long periods if properly stored. But lithium cells may lose some capacity in storage that cannot be fully recovered (called permanent storage losses).

Figure 10.4 shows the charge retention for selected cell designs at varied temperatures. Typically, the rate of capacity losses decreases with increasing storage time.

Lithium-ion cells and cells with the metal oxide cathodes (i.e., AgO and $MnO_2$) have the best charge retention characteristics of the secondary battery systems. In general, these cells may be able to retain capacity for up to 1 year if high temperatures are avoided. Neither the conventional nickel cathode nor traditional lead-acid cells retain capacity for over a year if heated above room temperature.

Of course, charge retention is dependent on the electrode types, cell design, electrolyte concentration, and component purity or additives, as well as other factors. Older SLI batteries, using antimonial-lead grids, lasted about 6 months storage at room temperature, while modern maintenance-free batteries with lead-calcium lose only about 20% to 30% capacity per year.

### 10.6.6   Life

Battery cycle life and calendar life were listed in Table 10.4. However, specific performance is dependent on the battery design and the test conditions. For example, the depth of discharge (DOD), as illustrated in Fig. 10.5, as well as the charging regime strongly influence the battery's cycling capacity.[8]

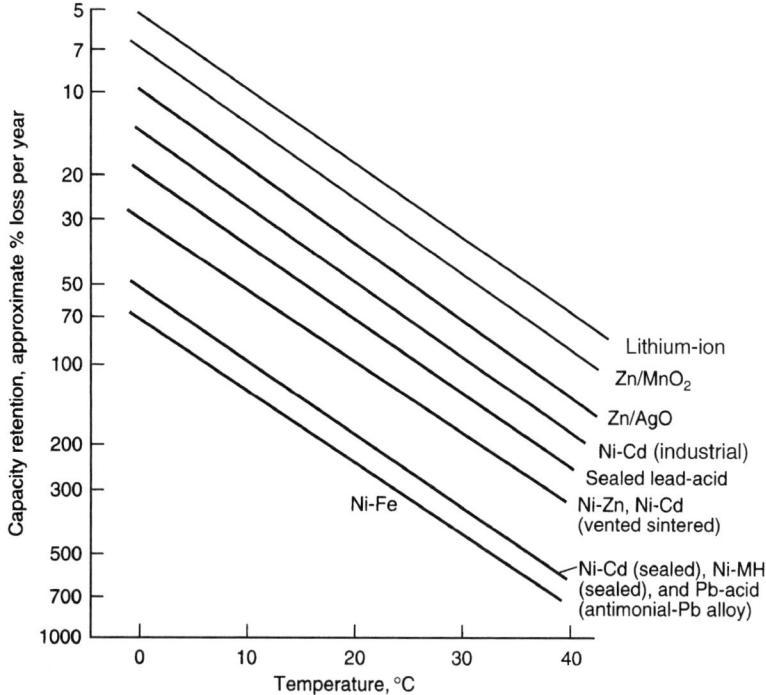

**FIGURE 10.4**    Capacity retention of secondary battery systems.

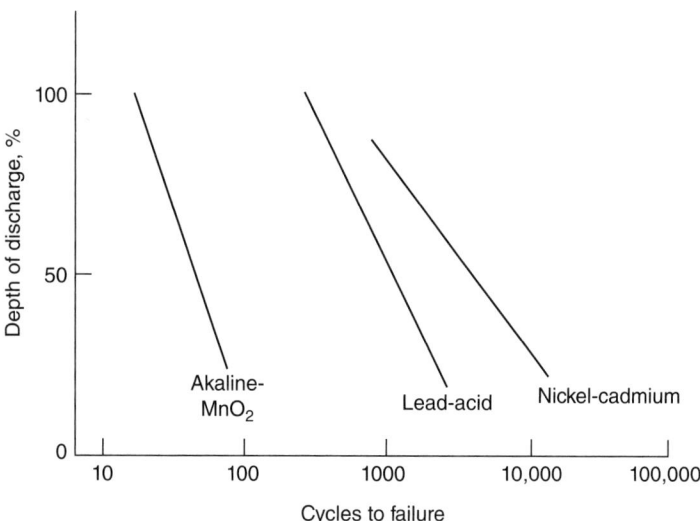

**FIGURE 10.5**    Effect of depth of discharge on cycle life of secondary battery systems.

Nickel-iron and the vented pocket-type Ni-Cd batteries, followed by Ni-H$_2$ batteries, are the best aqueous systems with regard to cycle life and total lifetime.

Traditional lead-acid batteries do not cycle well at high DOD, but improvements are being made. Flooded cells may last up to 2000 cycles, but 5000 cycles at 70% DOD have been achieved with VLRA (see Chap. 14). Also, float charging a cell at constant voltage, such as in UPS systems, can provide 20+ years of standby service. Much research has gone into improving the storage properties of anodes with high reducing potentials, such as zinc, and lithium, which suffer from side reactions (see Chaps. 15E and 17 for further details).

### 10.6.7   Charge Characteristics

Charging voltage profiles for various secondary aqueous systems with constant current charge are shown in Fig. 10.6. Constant current charge is a common, basic method of charging, but is usually supplemented by constant voltage or modified constant voltage methods, depending on battery type and the application needs. High constant voltage charge levels may cause thermal runaway, especially in nickel cathode cells, but if current limits and temperature compensation are used, problems can be avoided. The aqueous cells may allow some overcharged if some degree of water electrolysis can be tolerated. The metal oxide (AgO and MnO$_2$) and lithium cells, in particular, are most sensitive to charging; overcharging is very detrimental to battery life. Figure 10.7 shows typical constant current–constant voltage charging characteristics of an 18650 lithium-ion battery.

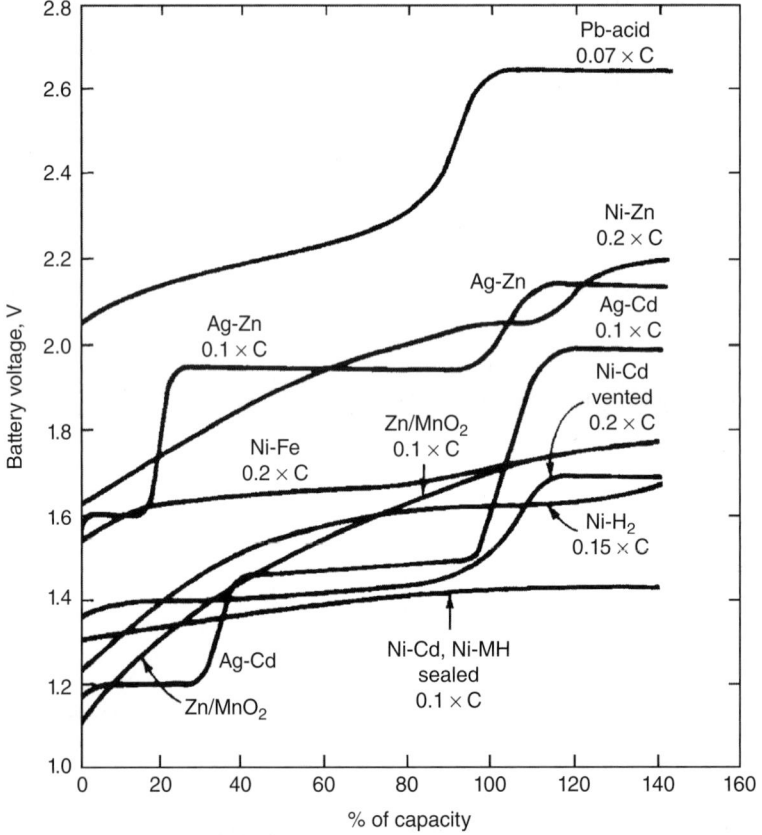

**FIGURE 10.6**   Typical charge characteristics of secondary battery systems, constant current charge at 20°C. (Adapted from S. U. Falk and A. J. Salkind, *Alkaline Storage Batteries,* Wiley, New York, 1969.)

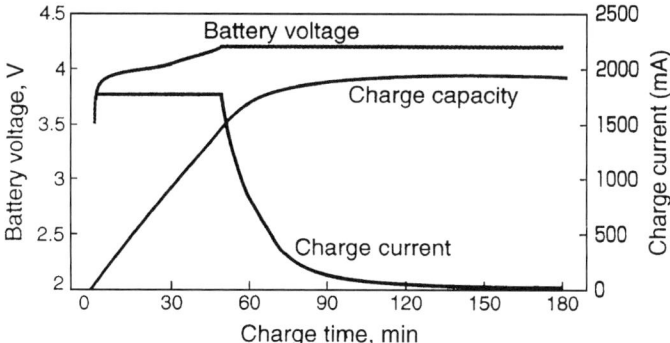

**FIGURE 10.7** Charging characteristics of a typical cylindrical 18650 lithium-ion battery at 20°C. Battery is charged at constant current to 4.2 V followed by a taper charge at this voltage to a current limit.

Table 10.6 summarizes typical charge conditions for different systems. However, manufacturers' recommendations should always be consulted to preserve battery capacity/life and insure safety.

"Fast" charge has become standard due to consumer demands (i.e., >90% recharge in <1 h is a typical goal). Depending on the situation, charge must be monitored to prevent excessive gas generation and high pressure or temperature. Venting not only dries the cell out, but is also a safety hazard due to the release of flammable gases (hydrogen and organic vapors) and the creation of internal dendritic shorts.

In general, new electronic circuitry is used in many systems (and is especially required in Li-ion cell packs) to prevent overcharging, to facilitate "fast" charging, to sense when a potentially deleterious or unsafe condition

**TABLE 10.6**  Charging Characteristics of Secondary Batteries

| System | Charging methods[*] | | Recommended constant current charge rate, $C$ (A) | Over-charge ability | Temperature range for charging, °C | Efficiencies[†] | |
|---|---|---|---|---|---|---|---|
| | Preferred | Not recommended | | | | Ah, % | Wh, % |
| Lithium ion | cc, cv | | 0.20 | None | 0 to 50 | 99 | 95 |
| Lead acid | | | | | | | |
| Pasted, planté | cc, cv | | 0.07 | Fair | −40 to 50 | 90 | 75 |
| Tubular | cc, cv | | 0.07 | Fair | −40 to 50 | 80 | 70 |
| Nickel-cadmium: | | | | | | | |
| Industrial vented | cc, cv | | 0.2 | Very good | −50 to 40 | 70 | 60 |
| Sintered vented | cc, cv | | 0.2 | Very good | −55 to 75 | 70–80 | 60–70 |
| Sealed | cc | cv | 0.1–0.3[‡] | Very good | 0 to 40 | 65–70 | 55–65 |
| Nickel-metal hydride | cc | cv | 0.1[‡] | Fair | 0 to 40 | 65–70 | 55–65 |
| Nickel-iron | cc | cv | 0.2 | Very good | 0 to 45 | 80 | 60 |
| Nickel-zinc | cc, cv | | 0.1–0.4 | Fair | −20 to 40 | 85 | 70 |
| Silver-zinc | cc | | 0.05–0.1 | Poor | 0 to 50 | 90 | 75 |
| Silver-cadmium | cc | | 0.01–0.2 | Fair | −40 to 50 | 90 | 70 |
| Zn/MnO$_2$ | cv | cc w/o v. limit | | Fair | 10 to 30 | | 55–65 |

[*]Constant current (cc) includes two-rate charging, and constant voltage (cv) includes modified constant voltage charging.
[†]All data are related to normal rates of charge and discharge and room temperature operation.
[‡]Fast charge procedures can be used with charge control.

***Source:*** Based on Falk and Salkind (S. U. Falk and A. J. Salkind, *Alkaline Storage Batteries,* Wiley, New York, 1969).

may arise, and to cut off the charge or reduce the charging rate to safe levels. Similarly, discharge controls are also being used to maintain cell balance and to prevent over-discharge. See Chaps. 23 and 31 for details on these battery management systems (BMSs).

### 10.6.8   Cost

The cost of a secondary battery is analyzed based on initial (upfront) cost as well as life cycle costs. The number of charge-discharge cycles delivered over a battery's lifetime can provide a measure of dollar per cycle or dollar per total kilowatt-hour values. The cost of charging, maintenance, and associated equipment may also have to be considered in this evaluation. In other applications, such as standby or back-up power, the most important factor may be the calendar life of the battery (rather than the cycle life), and the cost is evaluated on a dollar per operating year basis.

The lead-acid battery has long been the least costly of the secondary batteries, but lithium-ion cells (or maybe nickel-zinc or lead-carbon batteries) may potentially be more cost effective in time. Once cell materials and designs are better optimized and after economies of scale are realized, the cost metrics will undoubtedly change.

However, under special use conditions any battery could have an advantage. For example, nickel-hydrogen cells may still be attractive for certain space applications, such as for use on Mars where hydrogen fuel may be provided by electrolysis of water using solar electric power.

The cost of cylindrical lithium-ion batteries has been decreasing rapidly as production rates have increased and was at $0.20/Wh.[7] However, the U.S. Department of Energy's cost target for electric vehicle batteries has been set at $100/kWh at the cell level by 2025 (with a goal of $80/kWh).[6]

## ACKNOWLEDGMENT

Tables 10.1 and 10.2 are provided by The Freedonia Group.

## REFERENCES

1. C. Pillot, The rechargeable battery market and main trends 2014-2025, Presented at the Advanced Automotive Battery Conference, Mainz, 2016.

2. T. Placke, Strategies for mitigating active lithium losses in high-energy lithium-ion cells, 35th Annual International Battery Seminar and Exhibit, Ft. Lauderdale, FL, March 26–29, 2018.

3. N. Koratkar, Overcoming the fundamental problems in Li-S battery, 35th Annual International Battery Seminar and Exhibit, Ft. Lauderdale, FL, March 26–29, 2018.

4. F. Gittleson, Power and energy trade-offs in Li-air batteries, 35th Annual International Battery Seminar and Exhibit, Ft. Lauderdale, Fl., March 26–29, 2018.

5. A. J. Salkind, D. T. Ferrell, A. J. Hedges, "Secondary Batteries 1952–1977," *J. Electrochem. Soc.* **125**(8), August 1978.

6. B. Cunnigham, D. Howell, S. Boyd, T. Duong, P. Faguy, S. Gillard, W. James, U.S. Department of energy's electric vehicle battery research program and progress, 35th Annual International Battery Seminar and Exhibit, Ft. Lauderdale, FL, March 26–29, 2018.

7. H. Takeshita, *Proceedings of the 27th International Battery Seminar*, Ft. Lauderdale, FL, March 15–18, 2010.

8. L. H. Thaller, "Expected Cycle Life vs. Depth of Discharge Relationships of Well-Behaved Single Cells and Cell Strings," *J. Electrochem. Soc.* **130**(5), May 1983.

# ELECTROCHEMICAL CELL DESIGNS (PLATFORM TECHNOLOGIES)

# PRIMARY BATTERIES

# CHAPTER 11
# AQUEOUS PRIMARY CELLS

## 11.0  GENERAL OVERVIEW

The aqueous primary battery has existed for over 150 years. The first batteries, known as zinc-carbon cells, evolved into two types: Leclanché and zinc chloride systems. Both types remain among the most widely used of all primary batteries worldwide, although their use in the United States and Europe is declining due to higher power electronic device requirements. Conversely, the increased use of flashlights, portable radios, and other moderate- and light-drain applications has fostered the growth of zinc-carbon batteries in emerging countries. The cell has low cost, wide availability, and acceptable performance for a large number of applications.

These aqueous primary cells use a metal anode and a manganese dioxide cathode with a chloride salt aqueous solution electrolyte of nearly neutral pH. Carbon (acetylene black) is mixed with manganese dioxide to improve conductivity and retain electrolyte solution. As the cell is discharged, the metal foil is oxidized and the manganese dioxide is reduced. A simplified overall cell reaction for the zinc-carbon cell is

$$Zn + 2MnO_2 \rightarrow ZnO \cdot Mn_2O_3$$

In actual practice, the chemical processes that occur in the Leclanché cell are significantly more complicated, and controversy over the details of the electrode reactions continues.[1] Chemical "recuperation" reactions can operate simultaneously with the discharge reactions.[2]

The zinc-carbon battery industry continues to grow worldwide. Recent projections indicate that the overall global battery market is expected to grow from \$82.5 billion in 2014 to \$120 billion in 2019.[3] Primary cells will account for 90% of all batteries by unit volume and 20% of sales.[3-5] In 2014, zinc-carbon batteries were reported to be 6% of the total battery sales value (approx. 25% of primary batteries).[3] Table 11.1 shows a stable but flat position for zinc-carbon battery markets relative to the overall growth of primary battery markets (in particular).

Annual growth for the zinc-carbon global market from 1997 through 2012 was estimated at 5% per year. In the United States, zinc-carbon cells totaled \$370 million dollars in 1998 but were estimated to decline steadily at −2% to −5% per year. Asia, emerging countries, and Eastern European markets have driven the global demand for the inexpensive zinc-carbon battery system. The majority of primary batteries presently sold in Eastern and Central Europe have been zinc-carbon types.

New heavier-drain toys, lighting, and communication devices entering the consumer market have contributed to a decline in zinc-carbon cells and the growth of zinc-alkaline cells (see Chap. 12). Improved alkaline cells and the expanded use of rechargeable cells will continue to negatively impact the sales of zinc-carbon batteries, especially in the United States. A comparison of the characteristics of zinc-carbon cells is shown in Table 11.2.

Additionally, magnesium and aluminum have also been used as anode materials in primary aqueous batteries. High standard potential, low atomic weight, and multivalent states provide high gravimetric and volumetric electrochemical equivalence. Magnesium has been used commercially in a magnesium/manganese dioxide $(Mg/MnO_2)$ battery. The chemical reactions are generically similar to the Zn-carbon cells with Mg, rather than Zn, reacting electrochemically with $MnO_2$.

**TABLE 11.1**   World Zinc-Carbon Cell Demand (million dollars)*

| Actual/projected sales (based on 2016 sales data) | 2004 | 2009 | 2014 | 2019 | 2024 |
|---|---|---|---|---|---|
| Total world battery sales | 36,345 | 56,145 | 82,470 | 120,000 | 166,500 |
| World primary battery sales | 12,990 | 16,455 | 19,865 | 24,050 | 28,150 |
| (Primary batteries, % of total) | 35.7 | 29.3 | 24.1 | 20.0 | 16.9 |
| Zinc-carbon sales | 3352 | 4134 | 4879 | 5505 | 5920 |
| (Zinc-carbon, % of primary) | 25.8 | 25.1 | 24.6 | 22.9 | 21.0 |

*Table IV-7, World Primary Battery Demand by Chemistry, published December 2016 by The Freedonia Group.

**TABLE 11.2**   Major Advantages and Disadvantages of Leclanché and Zinc Chloride Batteries

| Standard Leclanché battery | | |
|---|---|---|
| Advantages | Disadvantages | General comments |
| Low cell cost | Low energy density | Good shelf life if refrigerated |
| Low cost per watt hour | Poor low-temperature service | For best capacity, the discharge should be intermittent |
| Large variety of shapes, sizes, voltages, and capacities | Poor leakage resistance under abusive conditions | Capacity decreases as the discharge drain increases |
| Various formulations | Low efficiency under high current drains | Steadily falling voltage is useful if early warning of end of life is important |
| Wide distribution and availability | Comparatively poor shelf life | |
| Long tradition of reliability | Voltage falls steadily with discharge | |

| Standard zinc chloride battery | | |
|---|---|---|
| Advantages | Disadvantages | General comments |
| Higher energy density | Requires excellent sealing system due to increased oxygen sensitivity | Steadily falling voltage with discharge |
| Better low-temperature service | | Good shock resistance |
| Good leak resistance | | Low to medium initial cost |
| High efficiency under heavy discharge loads | | |

This battery has twice the capacity of the zinc-carbon battery of equivalent size with an ability to retain this capacity during storage, even at elevated temperatures (Table 11.3). This excellent storability is due to a protective film that forms on the surface of the magnesium anode.

However, magnesium batteries have "voltage delay," parasitic magnesium corrosion, and poor storage life once the protective film has been removed (after partial discharge or during intermittent use), generating hydrogen and heat. While used successfully in military applications, such as radio transceivers and emergency or standby equipment, $Mg/MnO_2$ cells have not found wide commercial acceptance. Furthermore, lithium primary and lithium-ion rechargeable batteries are now used to meet the more stringent power needs of newer equipment.

**TABLE 11.3**   Major Advantages and Disadvantages of Magnesium Batteries

| Advantages | Disadvantages |
|---|---|
| Good capacity retention, even under high-temperature storage | Delayed action (voltage delay) |
| Twice the capacity of corresponding Leclanché batteries | Evolution of hydrogen during discharge |
| Higher battery voltage than zinc-carbon batteries | Heat generated during use |
| Competitive cost | Poor storage after partial discharge |

Aluminum has not been used successfully in an active primary battery despite its potential advantages. Aluminum, similar to magnesium, generates a protective film, which is detrimental to battery performance. As a result, battery voltage is considerably below theoretical and voltage delay can be significant for partially discharged batteries or for those that have been stored. Modifications to minimize the protective oxide film (i.e., using suitable electrolytes or amalgamation) are offset by accelerated corrosion and poor shelf life.

## Acknowledgments

Table 11.1 is provided by The Freedonia Group.

Content for this chapter has been taken from Chap. 9, Zinc-Carbon Batteries—Leclanché and Zinc Chloride Cell Systems, Brooke Schumm, Jr., and Chap. 10, Magnesium and Aluminum Batteries, Patrick J. Spellman, in *Linden's Handbook of Batteries*, 4th ed., T. B. Reddy and D. Linden, eds., McGraw-Hill, 2011.

## References

1. D. Glover, A. Kozawa, and B. Schumm, Jr. (eds.), *Handbook of Manganese Dioxides, Battery Grade*, International Battery Material Association (IBA, Inc.), IC Sample Office, 1989.

2. N.C. Cahoon, in N.C. Cahoon and G.W. Heise (eds.), *The Primary Battery*, Vol. 2, Chap. 1, Wiley, New York, 1976.

3. The Freedonia Group, Inc., https://www.freedoniagroup.com/World-Batteries.html, extracted from the World Wide Web on October 3, 2017 (exported tables generated/forwarded by The Freedonia Group on 10/4/2017 from IP address 207.89.36.85. © MarketResearch.com, Inc., 2000–2017).

4. https://www.upsbatterycenter.com/blog/global-battery-market-industry-report-review (extracted from the World Wide Web on October 3, 2017).

5. http://www.upsbatterycenter.com/blog/wp-content/uploads/2014/08/global1.jpg (extracted from the World Wide Web on October 3, 2017).

# SECTION A

# ZINC-CARBON BATTERIES—LECLANCHÉ AND ZINC CHLORIDE CELL SYSTEMS

**Brooke Schumm, Jr.**

## 11A.1   OVERVIEW OF ZINC-CARBON CELLS

The first aqueous primary cell was developed by a telegraph engineer, Georges-Lionel Leclanché, in 1866. The design resulted from the need to replace the corrosive ammonium chloride or mineral acid solutions used at the time so as to provide a more reliable power source for telegraph offices and railroad signaling. The cell was the first to use a single, low-corrosive (neutral pH) salt solution.

The cell remains relatively inactive until an external circuit is connected. It is inexpensive, safe, easily maintained, and provides excellent shelf (storage) life with adequate performance characteristics. The Leclanché cell consists of an amalgamated zinc bar serving as the negative electrode anode, a solution of ammonium chloride as the electrolyte, and an approximate one-to-one mixture of manganese dioxide and powdered carbon packed around a carbon rod as the positive electrode or cathode. The positive electrode was placed in a porous pot, which was, in turn, placed in a square glass jar along with the electrolyte and zinc bar. By 1876, Leclanché had evolved the design, removing the need for the porous pot by adding a resin (gum) binder to the manganese dioxide-carbon mix. In addition, the mix was compressed into a block by the use of hydraulic pressure at a temperature of 100°C. Leclanché's inventiveness brought together the major components of today's zinc-carbon battery and set the stage for conversion from the "wet" cell to the "dry" cell.

Carl Gassner is credited with constructing the first "dry" cell in 1888. It was similar to the Leclanché system except that ferric hydroxide and manganese dioxide were used as the cathode. The "dry" cell concept grew from the desire to make the cell unbreakable and spill-proof. The zinc anode foil was formed into a cup, replacing the typical glass jar. Next, an immobilized electrolyte was created by using a paste containing plaster of Paris and ammonium chloride. A cylindrical core of cathode mix (called a bobbin) was wrapped in a cloth and saturated with a zinc chloride-ammonium chloride electrolyte. This design reduced local chemical action and improved the shelf life. Gassner and others later replaced the plaster of Paris with wheat flour to form a gelled electrolyte. Such a battery was distributed as a portable lighting power source at the 1900 World's Fair in Paris. These advances were instrumental in establishing industrial production and commercialization of the "zinc-carbon dry cell" and led to the evolution of "dry-cell" portable power.

Zinc-carbon technology has continued to evolve. New manganese dioxide materials, produced by electrolytic and chemical techniques, afforded higher capacity and substantially higher activity than the natural manganese ores. The use of acetylene black carbon as a substitute for graphite provided a more conductive cathode structure plus higher absorption properties with enhanced processing characteristics, resulting in the production of an improved product at lower costs. A better understanding of the reaction mechanisms, improved separators, and venting/sealing systems also advanced the state of the art.

Since the 1960s, effort has been directed toward substantially improving performance in heavy-drain applications over that of the Leclanché cell. Starting in the 1980s the focus has been on environmental concerns, including the elimination of mercury, cadmium, and other heavy metals. During the past century, the discharge and storage life of the zinc-carbon battery has increased over 400% compared with the 1910 version.[1-7]

## 11A.2   ELECTROCHEMISTRY

The basic zinc-carbon cell ($Zn/NH_4Cl$, $ZnCl$, $H_2O/MnO_2$, C) reaction is

$$Zn + 2MnO_2 \rightarrow ZnO \cdot Mn_2O_3$$

However, a variety of intermediate reaction steps can and do exist. Furthermore, the chemistry reactions are complicated by the use of nonstoichiometric manganese oxide, more accurately represented as $MnO_x$, where $x$ is 1.9[+]. The efficiency of the chemical reaction may also depend on the electrolyte concentration, cell geometry, discharge rate, discharge temperature, depth of discharge, diffusion rates, and the type of $MnO_2$. A more comprehensive description of the cell reaction is as follows[4]:

1. *For cells with ammonium chloride as the primary electrolyte:*

   *Light discharge:* $Zn + 2MnO_2 + 2NH_4Cl \rightarrow 2MnOOH + Zn(NH_3)_2Cl_2$

   *Heavy discharge:* $Zn + 2MnO_2 + NH_4Cl + H_2O \rightarrow 2MnOOH + NH_3 + Zn(OH)Cl$

   *Prolonged discharge:* $Zn + 6MnOOH \rightarrow 2Mn_3O_4 + ZnO + 3H_2O$

2. *For cells with zinc chloride as the primary electrolyte[a]:*

   *Light or heavy discharge:* $Zn + 2MnO_2 + 2H_2O + ZnCl_2 \rightarrow 2MnOOH + 2Zn(OH)Cl$, or

   $4Zn + 8MnO_2 + 9H_2O + ZnCl_2 \rightarrow 8MnOOH + ZnCl_2 \cdot 4ZnO \cdot 5H_2O$

   *Prolonged discharge:* $Zn + 6MnOOH + 2Zn(OH)Cl \rightarrow 2Mn_3O_4 + ZnCl_2 \cdot 2ZnO \cdot 4H_2O$

The theoretical specific capacity is 224 Ah/kg, based on Zn and $MnO_2$ and the simplified cell reaction. A practical cell with added electrolyte, carbon black, and water will more typically have a specific capacity of 96 Ah/kg. A general-purpose large Leclanché cell, under certain discharge conditions, may approach this level. But most cells range from 75 Ah/kg on very light loads to 35 Ah/kg on heavy-duty, intermittent discharge conditions.

## 11A.3   CELL DESIGNS AND CONSTRUCTION

### 11A.3.1   Types of Cells and Batteries

During the last 150 years, the development of the zinc-carbon battery has been marked by gradual performance enhancements. While miniaturization of electronic components has reduced power demands, the decreases have been offset by the addition of new features requiring high power, such as small motors, halogen bulbs in lighting devices, etc. Therefore, zinc-carbon cells must now meet heavier discharge requirements. Traditional zinc-carbon Leclanché cells, which utilize a starch paste separator, are now being replaced by zinc chloride batteries constructed with thinner paper separators, which increases the proportion of cell volume available for active materials. However, developing countries still prefer pasted Leclanché products due to lower cost.

Zinc-carbon batteries, including both Leclanché and zinc chloride types, will likely be available during a transitional phase in both general-purpose and premium battery grades. Details for pasted and paper-lined constructions are described below.

*Leclanché Batteries*
   *General-Purpose Applications.*   Intermittent low-rate discharges; low cost.

---

[a]Note: 2MnOOH is sometimes written as $Mn_2O_3 \cdot H_2O$ and $Mn_3O_4$ as $MnO \cdot Mn_2O_3$. Electrochemical discharge of MnOOH versus zinc (prolonged discharge) does not provide a useful operating voltage for typical applications.

This most traditional battery, which is not too different from the one introduced in the late 19th century, uses ammonium chloride ($NH_4Cl$) as the main electrolyte component along with zinc chloride, a starch paste separator, and natural manganese dioxide ($MnO_2$; NMD) ore as the cathode. These batteries are the least expensive and are recommended for general-purpose use and when cost is more important than superior service or performance.

*Industrial Heavy-Duty Applications.*    Intermittent medium- to heavy-rate discharges or medium-rate continuous discharge; low to moderate cost.

The industrial heavy-duty battery has been mostly converted to a zinc chloride system. However, some types continue to include both ammonium chloride and zinc chloride ($ZnCl_2$) as well as synthetic electrolytic or chemical manganese dioxide (EMD or CMD) alone or in combination with natural ore as the cathode. The separator is typically a paste-coated paper liner type, but some designs may still use a starch paste.

### Zinc Chloride Batteries

*General-Purpose Applications.*    Low-rate discharges both intermittent and continuous; low cost.

This battery has replaced the Leclanché general-purpose battery in all developed countries. The electrolyte is zinc chloride; however, some manufacturers may add small amounts of ammonium chloride. NMD ore is still used as the cathode. These cells are competitive in cost to the Leclanché general-purpose batteries, but perform in some respects comparable to premium cells. The battery also exhibits a low leakage characteristic.

*Industrial Heavy-Duty Applications.*    Low to intermediate, continuous, and intermittent heavy-rate discharges; low to moderate cost.

This battery has generally replaced the industrial Leclanché heavy-duty battery. It is a true zinc chloride cell (except for some small amounts of ammonium chloride used by some companies) and possesses the heavy-duty characteristics of the premium zinc chloride type. NMD ore is used along with EMD as the cathode. These cells use paper separators coated with cross-linked or modified starches, which enhance their stability in the electrolyte. Such batteries are competitive in cost to the Leclanché heavy-duty industrial batteries and are recommended for heavy-duty applications where cost is an important consideration. This battery also exhibits a low leakage characteristic.

*Extra/Super Heavy-Duty (Premium) Applications.*    Medium and heavy continuous, and heavy intermittent discharges; higher cost than other zinc chloride types.

The extra/super heavy-duty type of battery is the premium grade of the zinc chloride line. This cell is composed mainly of an electrolyte of zinc chloride with perhaps a small amount of ammonium chloride, usually not exceeding 1% of the cathode weight. The cathode uses EMD, exclusively. These cells use paper separators coated with cross-linked or modified starches. Many manufacturers use proprietary separators for their zinc-carbon type batteries. This battery type is recommended when good performance is desired but at a higher cost. Both low-temperature and electrolyte leakage characteristics are benefited.

In general, the more advanced zinc-carbon batteries have a lower cost per minute of service. The price difference between classes is about 10% to 25%, but the performance difference can be from 30% to 100% in favor of the higher grades depending upon the application drain.

## 11A.3.2   Construction

The zinc-carbon battery is made in many sizes and a number of designs but in two basic constructions: cylindrical and flat. Similar chemical ingredients are used in both constructions.

*Cylindrical Configuration.*    In the common Leclanché cylindrical battery (Fig. 11A.1), the zinc can serves as both the cell container and the anode. The manganese dioxide is mixed with acetylene black, wet with electrolyte, and compressed under pressure to form a bobbin. A wax impregnated carbon rod is inserted into the bobbin. The rod serves as the current collector for the positive electrode. It also provides structural strength but is porous enough to permit the escape of gases, which typically accumulate in the cell, without allowing leakage of the electrolyte. The separator, which physically separates the two electrodes and provides the means for ion transfer through the electrolyte, can be a cereal grain paste wet with the electrolyte. Alternatively, a kraft paper impregnated with a starch or polymer coating is used to provide a thinner separator, lower internal resistance, and increased active materials volume. Single cells are covered with metal, cardboard, plastic, or paper jackets for aesthetic purposes and to help contain any electrolyte leakage.

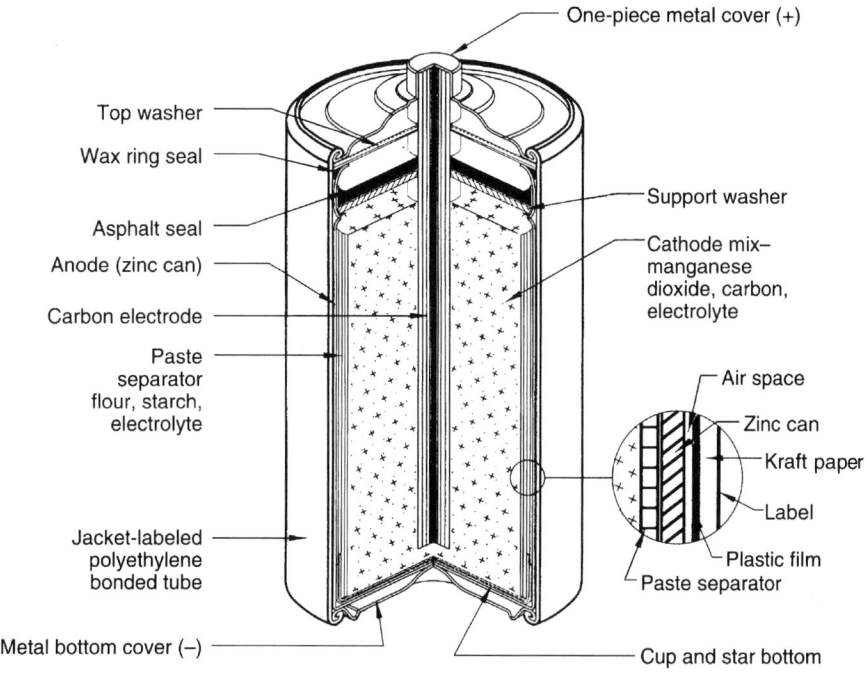

One-piece metal cover (+)

Top washer

Wax ring seal

Asphalt seal

Anode (zinc can)

Carbon electrode

Paste
separator
flour, starch,
electrolyte

Support washer

Cathode mix–
manganese
dioxide, carbon,
electrolyte

Air space

Zinc can

Kraft paper

Label

Plastic film

Paste separator

Jacket-labeled
polyethylene
bonded tube

Metal bottom cover (–)

Cup and star bottom

**FIGURE 11A.1**  Typical cutaway view of a cylindrical Leclanché battery ("Eveready") paste separator, asphalt seals.

Construction of the zinc chloride cylindrical battery differs from that of the Leclanché battery in that it usually possesses a resealable lid gasket that allows venting. Unlike the Leclanché cell, the carbon rod serving as the current collector is totally impregnated with wax to plug any vent paths. Venting is then restricted to the crimped lid sealant path only. This resealable vent prevents the cell from drying out and limits oxygen ingress into the cell during shelf storage, but still allows hydrogen gas evolved from corrosion of the zinc to safely vent. In general, the assembly and finishing processes resemble that of the original cylindrical batteries.

*Inside-Out Cylindrical Construction.*    Another cylindrical cell is the "inside-out" construction shown in Fig. 11A.2. This construction does not use the zinc anode as the container but instead has a molded impervious inert carbon container that also serves as a cathode current collector. The zinc anode, shaped as a rod with vanes and coated with a thin layer of separator paste, is placed in the cell's interior core and backfilled with a cathode mix. This version resulted in more efficient zinc utilization and less leakage, but it is now out of production.

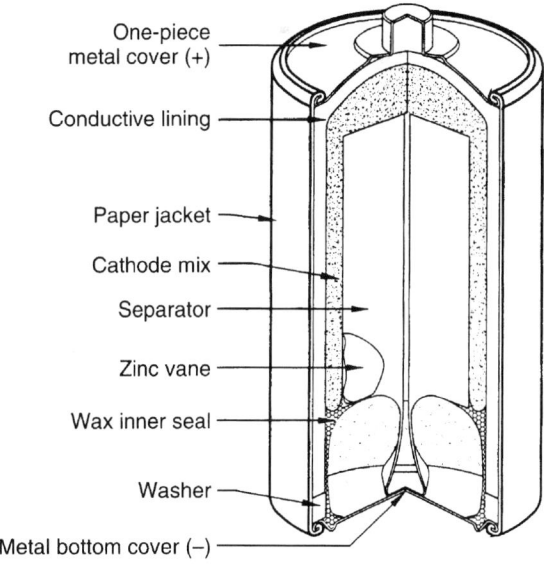

One-piece
metal cover (+)

Conductive lining

Paper jacket

Cathode mix

Separator

Zinc vane

Wax inner seal

Washer

Metal bottom cover (–)

**FIGURE 11A.2**  Typical cutaway view of a cylindrical Leclanché battery ("Eveready") inside-out construction.

***Flat Cell and Battery.***   The flat cell, typically used in 9-V transistor radio batteries, is illustrated in Fig. 11A.3. In this construction, a duplex (i.e., bipolar) electrode is formed by coating or laminating a conductive carbon-filled polymer (paint or film) to one side of a zinc plate to provide electrical contact directly between the zinc anode of one cell and the cathode of an adjacent cell. This layer functions as a cathode current collector but blocks the electrolyte from "shorting" adjacent cells in that the electrically conductive layer transfers electrons but not ions. Adhesive is used to bond and seal the painted zinc surface directly to a vinyl plastic band that encapsulates each cell. No expansion chamber or carbon rod is used as in the cylindrical cell. When using a laminated plastic film rather than paint to cover the back of the anode, conductive polyisobutylene will usually result in improved sealing to the vinyl bands, but at volumetric penalty.

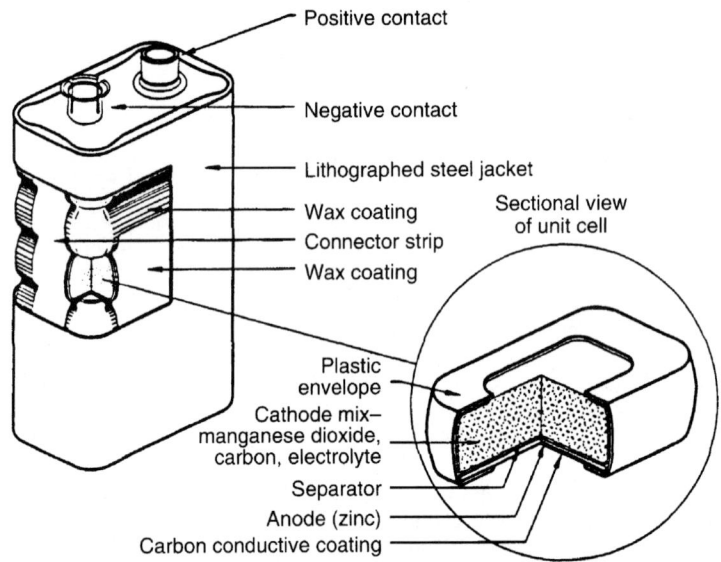

**FIGURE 11A.3**   Typical cutaway view of a Leclanché flat cell and battery (e.g., "Eveready" No. 216).

Flat cell designs minimize packaging and electrical component volumes and improved multicell battery pack volume efficiencies, thereby increasing the energy density. The volumetric energy density of an assembled battery using flat cells is nearly twice that of cylindrical cell assemblies.

Metal contact strips with snap-fit terminals are attached to the ends of the assembled battery to form the battery terminals. The orientation of the stack subassembly (cathode up or down) varies depending on each manufacturer's method of assembly. The entire assembly is usually encapsulated in wax or plastic. Some manufacturers also sleeve the assembly in shrink film after waxing for added protection.

***Special Designs.***   Certain special applications required unique designs, thus demonstrating the level of innovation achieved with this electrochemistry over its history. The Polaroid film pack battery, developed in the 1970s, was a multicell stack of thin components produced as continuous multilayer films on a web coating machine. This manufacturing concept was a predecessor to the techniques used today to produce lithium-ion cells. (See Sec. 11A.4.6 for further details on this battery application.)

## 11A.3.3   Cell Components

***Zinc.***   Battery-grade zinc is 99.99% pure. Typical alloys used to make cans formerly contained 0.3% cadmium and 0.6% lead, but modern lubrication and forming techniques have reduced these amounts. Currently, zinc cans have 0.03% to 0.06% cadmium and 0.2% to 0.4% lead, depending on the method used in the forming process. Lead, while insoluble in the zinc alloy, contributes to the forming qualities of the can and acts as a corrosion inhibitor by increasing the hydrogen overvoltage of the zinc in much the same manner as mercury, which had been used. Cadmium strengthens the zinc cans and increases their corrosion-resistance in ordinary dry-cell electrolytes. For cans made by a deep drawing process, less than 0.1% cadmium is required for optimal forming properties. Zinc cans are commonly made by three different processes:

1. Fabricated cans, which are now largely obsolete, are made by forming a rolled zinc sheet into a cylinder and soldering the seam together with a punched-out zinc disk for the bottom.

2. Deep-drawn zinc cans are made by shaping rolled zinc sheets into a can through a number of pressing/punching steps (a method historically used in the United States prior to the demise of U.S. zinc-carbon battery manufacturing).

3. Impact extrusion from a thick, flat zinc disk or calot is now the method of choice to form the cans. Used globally, this method reshapes the zinc by forcing it to flow under pressure from the calot shape into the can shape. Calots are either cast from molten zinc alloy or punched from a zinc sheet of the desired alloy.

Metallic impurities such as copper, nickel, iron, and cobalt cause corrosive reactions with the zinc in the battery electrolyte and must be avoided particularly in "zero" mercury constructions. In addition, iron in the alloy makes zinc harder and less workable. Tin, arsenic, antimony, magnesium, etc., make the zinc brittle and prone to perforation.[4,5]

U.S. federal environmental legislation now prohibits the land disposal of items containing cadmium and mercury when these components exceed specified leachable levels. Therefore, some states and municipalities banned land disposal of hazardous batteries, required collection programs, and prohibited sale of batteries containing added cadmium or mercury. Some European countries have similarly prohibited the sale and disposal of batteries containing these materials. For these reasons, levels of both of these heavy metals have been reduced to near zero. Manganese is a satisfactory substitute for cadmium and has been included in the alloy at levels similar to that of cadmium to provide stiffening. The handling properties of zinc alloyed with manganese or cadmium are equivalent; however, unlike with cadmium no corrosion resistance is imparted to the alloy when manganese is used.

***Bobbin.***   Bobbin is the term used for the wet powder mixture of $MnO_2$, carbon black, and electrolyte ($NH_4Cl$ and/or $ZnCl_2$, and water) shaped into the positive electrode, but it is also called the black mix, depolarizer, or cathode. The powdered carbon serves the dual purpose of retaining electrolyte and of adding electrical conductivity to $MnO_2$, which has inherently high electrical resistance. The cathode mixing and forming processes are also important since they determine the homogeneity of the cathode mix and the compaction characteristics associated with the different methods of manufacture. This becomes more critical in the case of the zinc chloride cell, where the cathode contains proportions of liquid that range between 60% and 75% by volume.

Of the various forming methods available, "mix extrusion" and "compaction-then-insertion" are the two most widely used methods. Additionally, a wide variety of techniques are used for mixing the cathode pastes. The most popular methods use cement-style barrel mixers, mash mixers, and rotary mill mixers. These machines offer the ability to manufacture large quantities of mix in relatively short times and minimize the shearing effect upon the carbon black, which reduces its ability to retain solution. The bobbin usually contains ratios of manganese dioxide to powdered carbon from 3:1 to as much as 11:1 by weight.

***Manganese Dioxide ($MnO_2$).***   The manganese dioxide used in dry cells are generally categorized as NMD, activated manganese dioxide (AMD), CMD, and EMD. EMD is a more expensive material that has a gamma-phase crystal structure. CMD has a delta-phase structure, and NMDs have the alpha, beta, and gamma phases of $MnO_2$. EMD, while more expensive, results in a higher cell capacity with improved rate capability and is used in heavy-duty or industrial applications. As shown in Figs. 11A.4*a* (Leclanché) and 11A.4*b* (zinc-chloride), polarization can be significantly reduced by the choice of natural ore. (As noted above, polarization is also reduced by adding EMD or CMD as a substitute for part of the natural ore.)

Naturally occurring ores (from Gabon, Africa; Brazil; Greece; and Mexico), which are high in battery-grade material (70–85% $MnO_2$), and synthetic forms (90–95% $MnO_2$) generally provide electrode potentials and capacities proportional to their manganese dioxide content. Manganese dioxide potentials are also affected by the pH of the electrolyte. Performance characteristics depend upon the crystalline state, hydration, and reactivity. Power capabilities depend upon the electrolyte, the separator characteristics, the internal resistance, and the overall construction of the cell.[2,4]

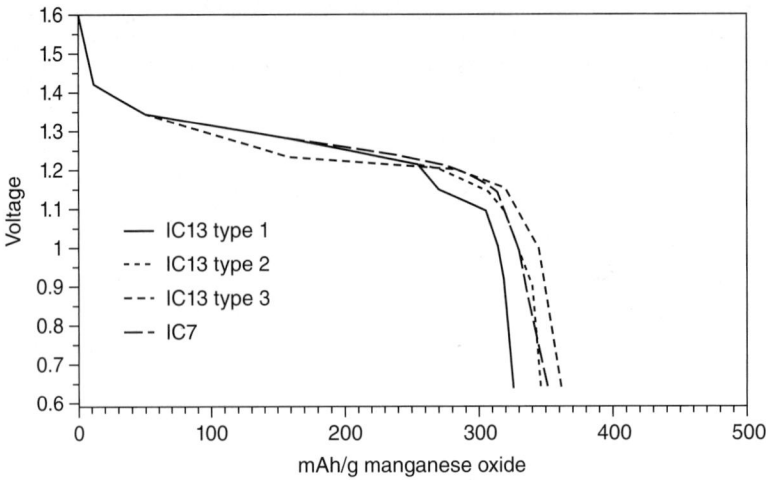

**FIGURE 11A.4a**    Ore sample performance, Leclanché 6.38% ore mix (13 mA/g ore).

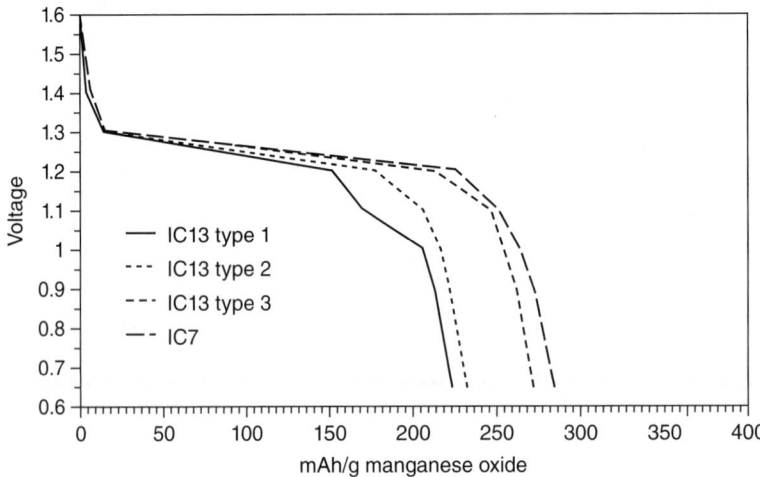

**FIGURE 11A.4b**    Ore sample performance, zinc chloride 6.71% ore mix (13 mA/g ore).

***Carbon Black.***    Because manganese dioxide is a poor electrical conductor, chemically inert carbon or carbon black is added to the cathode mix to improve conductivity. Specifically, the manganese dioxide particles are coated with carbon during the mixing process, which provides electrical conductivity to the particle surface and also serves the important functions of absorbing the electrolyte and providing compressibility and elasticity to the cathode mix during processing.

Graphite was once used as the principal conductive media and is still used to some extent. Acetylene black, by virtue of its properties, has displaced graphite in this role for both Leclanché and zinc chloride cells. Acetylene black also retains more electrolyte in the cathode mix. However, excessive shear during the mixing process may degrade carbon particles and reduce electrolyte retention. Zinc chloride cells, which contain much higher electrolyte levels, are most impacted. Cells containing acetylene black usually give superior intermittent service, a common use for zinc-carbon batteries, while graphite serves well for high pulse currents or for continuous drains.[4,7]

*Electrolyte.*   The Leclanché cell uses an aqueous ammonium chloride/zinc chloride mixture with the former predominating. Zinc chloride cells typically use only $ZnCl_2$, but they may contain a small amount of $NH_4Cl$ to ensure high-rate performance. Examples of typical electrolyte formulation for the zinc-carbon battery systems are listed in Table 11A.1.

**TABLE 11A.1**   Electrolyte Formulations[*]

| Constituent | Weight % |
|---|---|
| Electrolyte I | |
| $NH_4Cl$ | 26.0 |
| $ZnCl_2$ | 8.8 |
| $H_2O$ | 65.2 |
| Zinc-corrosion inhibitor | 0.25–1.0 |
| Electrolyte II | |
| $ZnCl_2$ | 15–40 |
| $H_2O$ | 60–85 |
| Zinc-corrosion inhibitor | 0.02–1.0 |

[*]Electrolyte I based on Kozawa and Powers.[1]
Electrolyte II based on Cahoon.[2]

Generally some zinc oxide is included in the electrolyte in order to prevent excess corrosion of the zinc.

*Corrosion Inhibitor.*   Mercuric or mercurous chloride, which forms an amalgam with the zinc, has been the traditional corrosion inhibitor. Cadmium and lead, which reside in the zinc alloy, also provide zinc anode corrosion protection. Other materials, such as potassium chromate or dichromate or organic compounds, will form oxide films on the zinc and protect via passivation. Surface-active organic compounds will also improve the wetting characteristic of the surface and stabilize the cell voltages. Inhibitors are not usually applied to the zinc can but instead are introduced into the cell via the electrolyte or as part of the coating on the paper separator.

Elimination of mercury and cadmium due to health and environmental restrictions affected the sealing and storage properties, especially for zinc chloride cells that exhibit more dissolution of zinc and more hydrogen gas evolution in the lower pH electrolytes. Materials considered to replace mercury include glycols, silicates, and metals such as gallium, indium, lead, tin, and bismuth, either alloyed into the zinc or added to the electrolyte as a soluble salt.

*Carbon Rod.*   The carbon rod, used in cylindrical cells, is inserted into the bobbin that serves as the cathode current collector and is typically made of compressed carbon, graphite, and binder, formed by extrusion and cured by baking. The rods have high electrical conductivity and a sufficient degree of porosity to perform as a vent mechanism without the need for a mechanical seal. The alternative, asphalt gaskets, allow hydrogen and carbon dioxide diffusion, resulting from heavy discharge or elevated temperature storage, through the lid seal area. Raw carbon rods are relatively porous but are impregnated with enough oils or waxes to slow the loss of water vapor or electrolyte leakage without overly restricting the gas diffusion. Ideally, the treated carbon will also not allow passage of oxygen into the cell, which could add to zinc corrosion during storage. Typically, diffusion mechanisms are variable and less reliable than the use of mechanical seals.[4,5]

Zinc chloride cells with plastic, mechanically resealable vents will utilize fully impregnated, nonporous carbon rods that prevent drying and limit oxygen ingress into the cell during shelf storage.

*Separator.*   The separator physically separates and electrically insulates the zinc (negative) from the bobbin (positive), but permits electrolyte flow and ionic conduction between electrodes. The two major separator types in use are gelled paste or paper coated with cereal grain pastes or other gelling agents, such as methycellulose.

Paste separators are dispensed into the zinc can, and the preformed bobbin (with the carbon rod) is inserted, forcing the paste to extrude up the can walls between the zinc and the bobbin where the paste subsequently sets or gels. Certain paste formulations require the use of two part formulations that are stored cold, mixed, and then immediately dispensed before gelling occurs. Other paste formulations need elevated temperatures (60–96°C) to gel. The gelling time and temperature also depend on the concentration of the electrolyte constituents. A typical paste electrolyte uses zinc chloride, ammonium chloride, water, and starch and/or flour as the gelling agent.

Coated-paper separator uses a special paper coated with flour/milled grains or other gelling agents on one or both sides. The paper, cut to the proper length and width, is shaped into a cylinder and, with the addition of a bottom paper, is inserted into the cell against the zinc wall. The cathode mix is then added to the can, followed by carbon rod insertion and tamping/compression to ensure separator contact. The compression releases some electrolyte from the cathode mix, wetting the paper liner to complete the operation. Preformed bobbins with rods are also used to simplify the cell assembly.

Since paste separator coatings are relatively thick compared with the typical paper liner, about 10% additional manganese dioxide can be accommodated in a paper-lined cell, resulting in an increase in capacity, due to reduced separator material usage per gram of cathode.[4,5,8]

**Seals (Sealants).**    The seal used to enclose the active ingredients can be asphalt pitch, wax and resin, or plastic (polyethylene or polypropylene). The seal must prevent the phenomenon of "air line" corrosion from oxygen ingress.[2,4]

Leclanché cells typically utilize thermoplastic materials and viscous liquids/semisolids for sealing, which are inexpensive and uncomplicated. A plastic washer is usually placed above the cathode bobbin, providing an air space between the seal and the top of the bobbin to allow for expansion. Melted asphalt pitch is then dispensed onto the washer and is heated until it flows and bonds to the zinc can. Unfortunately, this system displaces active materials and is easily ruptured by excessive generation of evolved gases, making it unsuitable for elevated temperature applications.

Premium Leclanché and almost all zinc chloride cells use injection-molded plastic seals, which allow for positive venting seals and greater reliability. Molded seals, which are mechanically swaged fit onto the zinc can, often include locking mechanisms and void spaces for various sealants or resealable vent designs. A shrink wrap or tape may be applied to the seal joint to further retard leakage through the seal or can perforations. Venting without moisture loss is critical since disruption of the separator/electrode interface may significantly reduce cell performance after storage. Use of molded seals in zinc chloride cell construction has resulted in good shelf storage characteristics.

**Jacket.**    The battery jacket can be made of various materials: metal, paper, plastic, polymer films, plain or asphalt-lined cardboard, foil, or other combinations. The jacket provides strength, protection, leakage control, electrical isolation, decoration, and manufacturer labeling. In many manufacturers' designs, the jacket is an integral part of the sealing system. It locks some seals in place, provides a vent path for the escape of gases, or acts as a reinforcing material to allow seals to flex under internal gas pressures. In the inside-out construction, the jacket serves as the container with a molded-in carbon-wax collector (Fig. 11A.2).

**Electrical Contacts.**    The top and bottom of most batteries are capped with tin-plated steel (or brass) terminals to aid conductivity, to prevent air exposure of the zinc, and in many designs to enhance the appearance of the cell. Some of the bottom covers are swaged onto the zinc can, others are locked into paper jackets or captured under the jacket crimp. Top covers are almost always pressed onto the carbon electrode with an interference fit. All designs help minimize the electrical contact resistance.

## 11A.4   PERFORMANCE CHARACTERISTICS

From the early 1900s through the 1990s, the portable power industry was based on electric and electronic devices (battery-operated telephones, electric doorbells, toys, lighting devices, and countless other applications), placing increasing performance demands on "dry battery" manufacturers. The advent of radio broadcasting and World War II military applications significantly increased that demand. Finally, 50 years ago, demands for an inexpensive battery to power flashlights, portable transistor radios, electric clocks, cameras, electronic toys, and other convenience applications maintained pressures to improve performance, as detailed below.

## 11A.4.1  Voltage

***Open-Circuit Voltage.***    The open-circuit voltage (OCV) of the zinc-carbon battery derives from the half-cell potentials of the active anode and cathode materials: zinc and manganese dioxide, respectively. As most zinc-carbon batteries use similar anode alloys, the OCV usually depends upon the type or mixture of manganese dioxide used in the cathode and the composition and pH of the electrolyte system. EMDs have greater purity than the natural NMDs, which contain a significant quantity of manganite (MnOOH), having lower voltage. Figure 11A.5 shows the OCV for fresh Leclanché and zinc chloride batteries containing various mixtures of natural and electrolytic manganese dioxide.

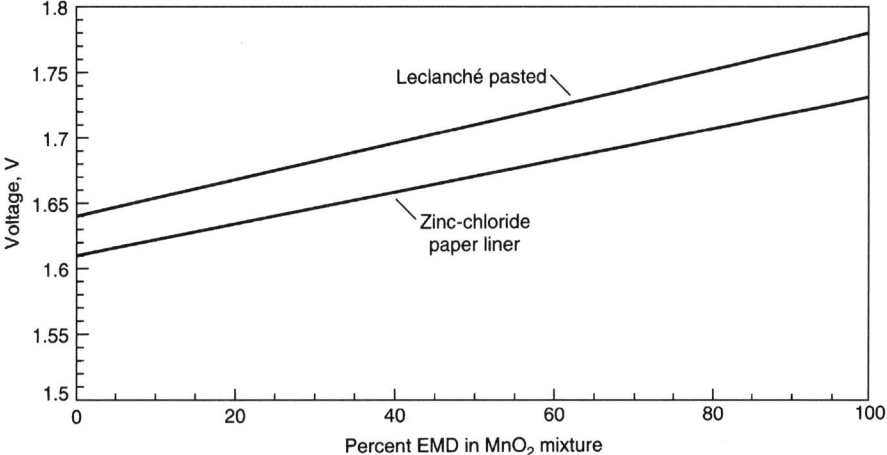

**FIGURE 11A.5**    Comparison of open-circuit voltage for batteries using mixtures of natural and electrolytic manganese dioxide.

***Closed-Circuit Voltage.***    The closed-circuit voltage (CCV), or working voltage, of the zinc-carbon battery is a function of the load, or current drain, the cell is required to deliver. Heavier current flow (due to lower resistance of the external circuit) depresses cell voltage, as illustrated in Table 11A.2 for both Leclanché and zinc chloride D-size cells.

The exact value of the CCV is determined mainly by the internal resistance of the battery but also by the ability to transport ions, solid reaction products, and water (see sections below). The physical geometry of the cell, the electrolyte solution volume, electrode porosity, and solute materials are critical characteristics that affect the diffusion coefficients of various species. Transport is enhanced by use of highly mobile ions, high

**TABLE 11A.2**    Initial Closed-Circuit Voltage of a Typical D-size Zinc-Carbon Battery as a Function of Load Resistance at 20°C

| Voltage, V | | Load resistance | Initial current, mA | |
|---|---|---|---|---|
| ZC* | LC* | Ω | ZC | LC |
| 1.61 | 1.56 | ∞ | 0 | 0 |
| 1.59 | 1.52 | 100 | 16 | 15 |
| 1.57 | 1.51 | 50 | 31 | 30 |
| 1.54 | 1.49 | 25 | 62 | 60 |
| 1.48 | 1.47 | 10 | 148 | 147 |
| 1.45 | 1.37 | 4 | 362 | 343 |
| 1.43 | 1.27 | 2 | 715 | 635 |

*ZC: Zinc chloride battery; LC: Leclanché battery.

liquid volumes, high electrode porosity, and high electrode-separator interfacial surface area. Transport characteristics are diminished by slow ionic transport, low electrolyte solution volumes, and precipitated reaction product, which block diffusion paths (see Chap. 4 for more details). Temperature, age, and depth of discharge greatly affect the internal resistance and transport factors as well.

As zinc-carbon batteries are discharged, the CCV and, to a lesser extent, the OCV drop in magnitude. The drop in OCV is attributable to the decrease in the active material manganese dioxide and the increase in the product of the reaction, manganite. Reduction of the CCV is the result of increased electrical resistance and a decrease in transport characteristic.

***End Voltage.*** The end voltage, or cutoff voltage (COV), is defined as a point along the discharge curve below which no usable energy can be drawn for the specified application. Typically, 0.9 V has been found to be the COV for a 1.5-V cell when used in a flashlight. Some radio applications can utilize the cell down to 0.75 V or lower, while other electronic devices may tolerate a drop to only 1.2 V. Lower end voltage will impact functionality, resulting in a dimming of flashlights and reduced volume and/or range for radios. Narrow voltage ranges with flat discharge profiles are preferred, unless the declining voltage is desired for end of battery life warning, as in a flashlight.

## 11A.4.2 Discharge Characteristics

Both Leclanché and zinc chloride batteries have performance trade-offs that are affected by a variety of factors (see Chap. 5). It is necessary to evaluate specifics about the application (discharge conditions, cost, weight, etc.) in order to make a proper selection of a battery. Many manufacturers provide data for this purpose.

Typical discharge curves for general-purpose D-size Leclanché and zinc chloride batteries, of equivalent capacity, discharged 2 h per day at 20°C, are shown in Fig. 11A.6. A sloping discharge voltage and a substantial reduction in voltage with increasing current, as a result of manganese dioxide depletion, are typical. The zinc chloride cells are less impacted at higher rates. On the 50-mA drain, both constructions provide nearly equivalent performance because zinc-carbon batteries are cathode limited.

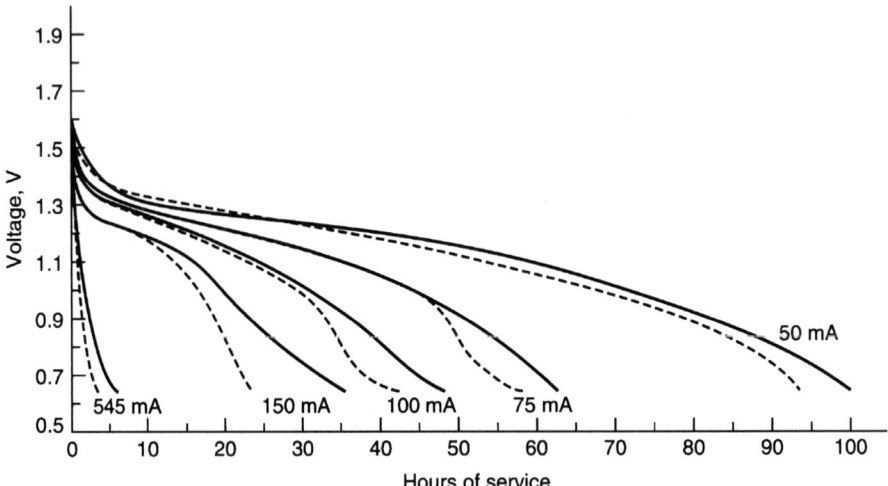

**FIGURE 11A.6** Typical discharge curves for general-purpose Leclanché and zinc chloride D-size batteries, discharged 2 h/day at 20°C. Solid line: Zinc chloride; broken line: Leclanché.

***Effect of Intermittent Discharge.*** Performance of zinc-carbon batteries varies depending upon the type of discharge. The performance of Leclanché batteries is significantly better when used under intermittent (i.e., discontinuous) discharge conditions because (1) a chemical recuperation reaction replaces a small portion of active ingredients during the rest periods and (2) transport phenomena redistribute reaction products.[2]

Zinc chloride batteries can support heavier drains and respond to intermittent discharges with longer discharge cycles. This system relies upon its improved transport mechanism to support heavier drains and to redistribute reaction products. Figure 11A.7 illustrates the general effects of intermittency and discharge rate on the capacity of a general-purpose D-size battery. On extremely low-current discharges, the benefit of intermittent rest and discharge is minimal for both systems because the reaction rate proceeds more slowly than the diffusion rate and results in a balanced condition even during discharge. Under conditions of extremely low rate of discharge, factors such as age will reduce the total delivered capacity. Therefore, most selected applications fall under intermittent moderate- (radio) to high-rate (flashlight) categories, where the delivered energy can be more than triple continuous discharge uses.

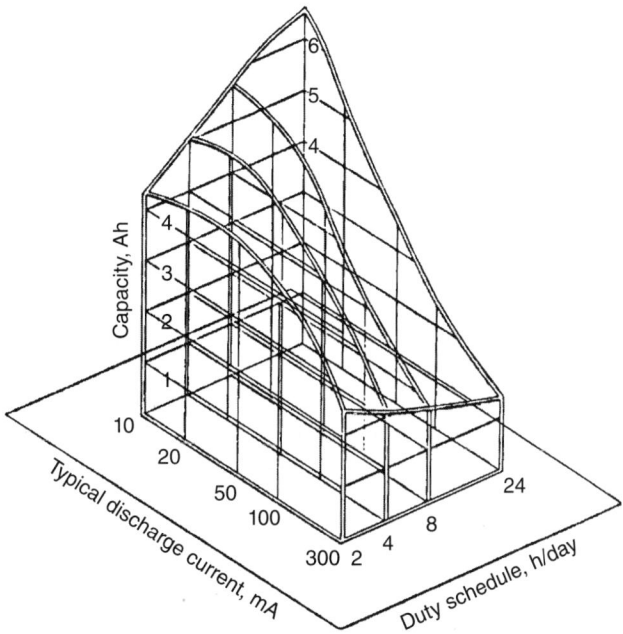

**FIGURE 11A.7**   Battery performance (capacity to 0.9 V) as a function of discharge, load, and duty schedule for a general-purpose D-size zinc-carbon battery at 20°C. (*Source: Eveready Battery Energy Data.*)[10]

Standard flashlight current drains with incandescent bulbs are 300 mA (3.9 Ω per cell) and 500 mA (2.2 Ω per cell), which correspond to two-cell flashlights using PR2 and PR6 lamps, respectively, or three-cell flashlights using PR3 and PR7 lamps, respectively. The beneficial effects of intermittent discharge are clearly shown in Fig. 11A.8, which compares Leclanché general-purpose D-size batteries on four different discharge regimens: continuous, light intermittent flashlight (LIF), heavy intermittent flashlight, and an hourly daily cassette tape player simulation test. Table 11A.3 lists the American National Standards Institute (ANSI) application tests currently being used to evaluate both cell systems.

***Comparative Discharge Curves: Size Effect.***   The performance of heavy-duty zinc chloride AAA-, AA-, C-, and D-size batteries is shown in Fig. 11A.9

**TABLE 11A.3**   Standard Application Tests Specified in ANSI Battery Specifications

| Typical use or test | Discharge schedule |
|---|---|
| Pulse test (PHOTO) | 15 s ON/min × 24 h/day |
| Portable lighting (GPI) | 5 min ON/day |
| Portable lighting (LIP) | 4 min ON/h × 8 h/day |
| Portable lighting (LANTERN) | 0.5 h ON/h × 8 h/day |
| Transistor radios | 4 h ON/day |
| Transistor radio (small 9 V) | 2 h ON/day |
| Personal tape recorder, cassette | 1 h ON/day |
| Toys and motors | 1 h ON/day |
| Pocket calculator | 0.5 h ON/day |
| Hearing aid | 12 h ON/day |
| Electronic | 24 h ON/day |

***Source:*** Based on ANSI C18.1M-2009.[9]

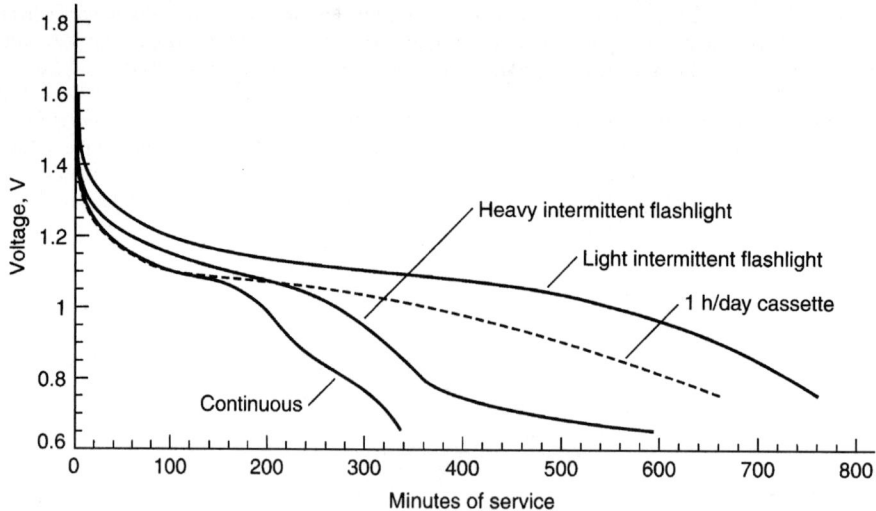

**FIGURE 11A.8**   A general-purpose D-size zinc-carbon battery discharged through 3.9 Ω at 20°C, under various discharge conditions.

(see Table 11A.7 for a list of cell sizes). Note that the AAA- through D-size batteries contain proportionally increasing amounts of active materials with larger electrode surface area, yielding higher voltage levels under identical loads.

This relatively high resistance (150 Ω or about 10 mA) shows discharge profiles for C- and D-size cells at 20°C having superior voltage levels and life compared with smaller AA and AAA cells. A 150-Ω load creates higher specific electrode current discharges (i.e., higher C-rate) in lower capacity cells. Also, while C- and D-size batteries would provide similar output under either intermittent or continuous load, the smaller cells would benefit from intermittent discharge usage. With a relatively low resistance of 10 Ω (about 150 mA), AAA- and AA-size batteries deliver about 30% less service when continuously discharged.

Leclanché batteries, using an intermittent discharge, would function better than zinc chloride cells due to improved dissipation of reaction products. Zinc chloride batteries have higher transport characteristic and do not benefit as greatly from intermittent use.

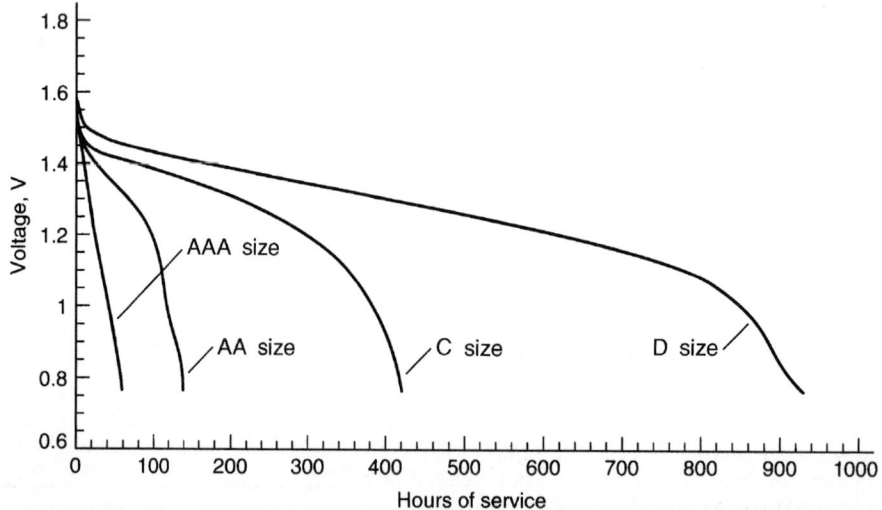

**FIGURE 11A.9**   Zinc-carbon batteries, continuous discharge through 150 Ω at 20°C.

The relative performance of both Leclanché and zinc chloride AAA, AA, C, and D batteries roughly follows a 1:2:8:16 proportion to the 0.9-V cutoff for lower rate discharges and a 1:2:12:24 proportion for high-rate drain, illustrating the advantage of the lower current density discharge for the larger batteries.

***Comparative Discharge Curves: Battery Grades.***   Figure 11A.10 compares both Leclanché and zinc chloride general-purpose, heavy-duty, and the extra/super heavy-duty D-size batteries, discharged continuously through a 2.2-$\Omega$ load at 20°C. A performance ratio of 1:1.3 was observed for Leclanché and zinc chloride GP batteries to the 0.9-V cutoff. The same ratio for the HD batteries was 1:1.5.

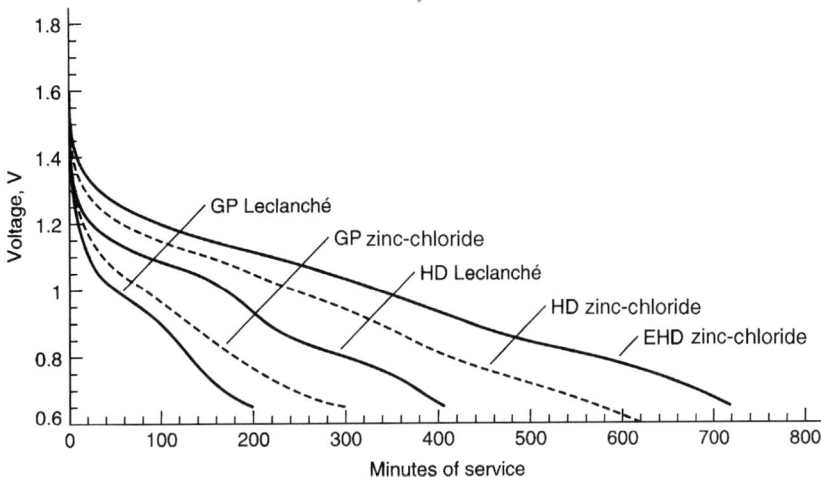

**FIGURE 11A.10**   Comparison of Leclanché and zinc chloride D-size batteries of various grades, continuously discharged through 2.2 $\Omega$ at 20°C. GP: General purpose; HD: Heavy duty; EHD: Extra heavy duty.

Figure 11A.11 shows a comparison of the same battery grades discharged intermittently through a 2.2-$\Omega$ load on the ANSI LIF test. On this regimen, the performance ratio to the 0.9-V cutoff is 1:1 for the GP batteries

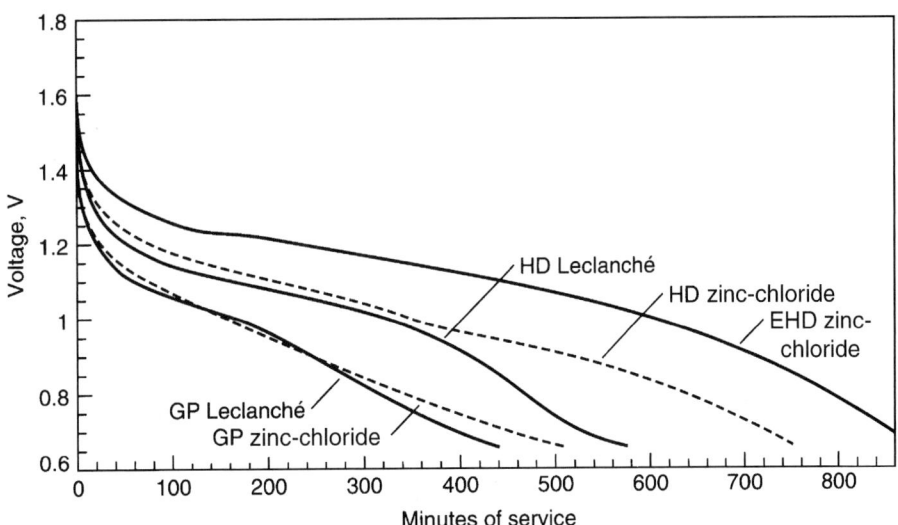

**FIGURE 11A.11**   Comparison of Leclanché and zinc chloride D-size cells of various grades discharged on the ANSI LIF test (4 min/h, 8 h/day) through 2.2 $\Omega$ at 20°C. GP: General purpose; HD: Heavy duty; EHD: Extra heavy duty.

and 1:1.3 for the HD batteries. Testing with intermittent discharge, which allows for a rest period for recovery, results in increased performance for all batteries and evidences a decreased difference in performance between the grades.

The same battery grades discharged continuously through a higher 3.9-$\Omega$ load gives ratios of 1:1.3 GP and 1:1.4 HD to the 0.9-V cutoff, showing less of a difference than the heavier 2.2-$\Omega$ discharge rate. With an intermittent 3.9-$\Omega$ load discharge for 1 hour a day, service life (i.e., total capacity yield) increases and differences are reduced among all cell types.

These battery grades are compared once again in Fig. 11A.12 on a moderate discharge through a 24-$\Omega$ resistor for 4 h continuously with 20 h of rest (e.g., ANSI transistor radio and electronic equipment battery test regime). At this more moderate discharge load, the performance ratios to a 0.9-V cutoff are even more tightly grouped.

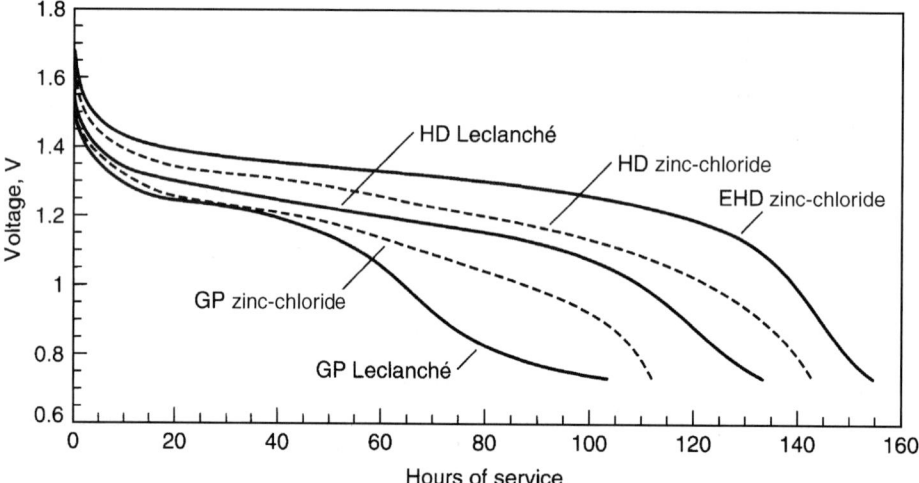

**FIGURE 11A.12**    Comparison of Leclanché and zinc chloride D-size batteries of various grades discharged through 24 $\Omega$ for 4h/day at 20°C. GP: General purpose; HD: Heavy duty; EHD: Extra heavy duty.

Differences in output under continuous discharge are wider among the different grades of batteries of the same size with the Leclanché cells showing lower output. Intermittent discharges tend to reduce the differences between systems and grades. Similarly, higher discharge currents tend to increase the performance difference.

Figures 11A.13a and b summarize the relationships for Leclanché and zinc chloride heavy-duty D-size batteries under continuous discharge to different voltage cutoffs. Leclanché and zinc chloride general-purpose D-size batteries show less capacity and more spread in the curves, since internal resistances are higher for both types.

Performance differences between identical batteries of the same grade but from different manufacturers may also be quite different (up to 25% when using the ANSI LIF test with a 2.2-$\Omega$ load at 20°C to 0.9-V cutoff).

### 11A.4.3    Internal Resistance

Internal resistance ($R_{in}$) is the opposition or resistance to the flow of electric and ionic currents within a cell or battery, as influenced by battery size and construction, temperature, cell age, and depth of discharge, and is the root cause of capacity differences due to discharge rate variations.

***Electronic Resistance.***    Electronic resistance includes the resistance of the current-carrying parts: metal covers, carbon rods, conductive cathode components, and so on. An approximation of the internal electronic resistance of a battery can be made by determining the OCV and the peak flash current *I*, using very low

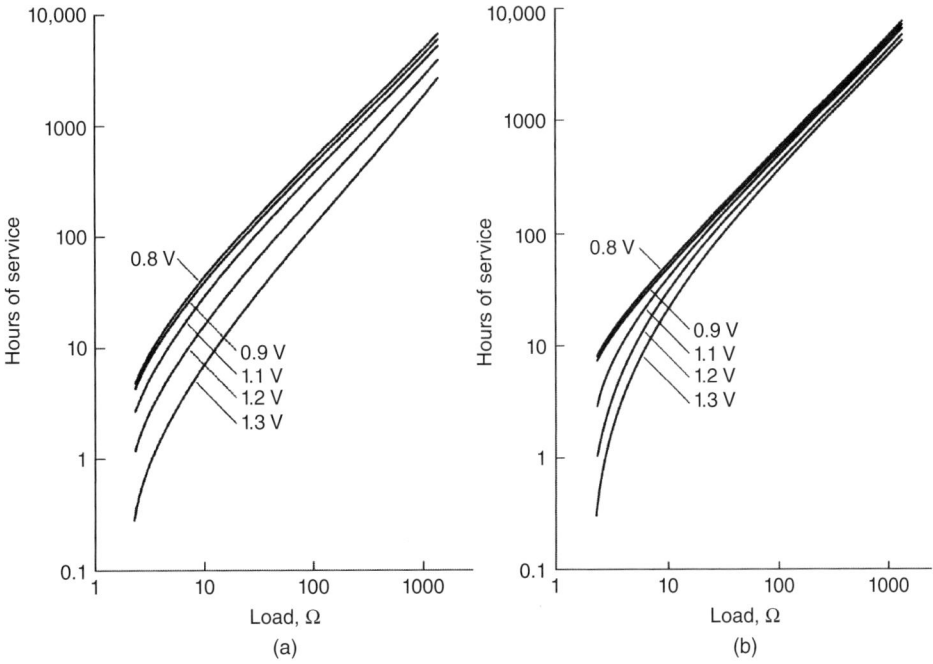

**FIGURE 11A.13**   Discharge load versus hours of service to different cutoff (end) voltages at 20°C, continuous discharge. (*a*) Heavy-duty D-size Leclanché battery. (*b*) Heavy-duty D-size zinc chloride battery.

resistance meters. The ammeter resistance should be low enough that the total circuit resistance does not exceed 0.01-$\Omega$ and is no more than 10% of the cell's internal resistance. The internal electronic resistance is roughly determined by Ohm's law:

$$R_{in} = OCV/I$$

where $R_{in}$ = internal resistance, $\Omega$
   OCV = open-circuit voltage
      $I$ = peak flash current, A

A more accurate method of calculation is made using the voltage-drop method. In this method, a small initial load is applied on the battery to stabilize the voltage. A load approximating the application load is then applied. The internal resistance is calculated by

$$R_{in} = (V_1 - V_2)R_L/V_2$$

where $R_{in}$ = internal resistance
      $V_1$ = initial stabilized CCV, V
      $V_2$ = CCV reading at the application load, VO
      $R_L$ = application load, $\Omega$

The application load time should be kept to a pulse of 5 to 50 ms to minimize effects due to polarization. These methods measure the voltage loss due to the electrical resistance component but do not take into account voltage losses due to polarization (ionic resistance).

*Ionic Resistance.* Ionic resistance encompasses factors resulting from the movement of ions within the cell. These include electrolyte conductivity, ionic mobility, electrode porosity, electrode surface area, secondary reactions, etc. The polarization effect is best illustrated by a trace of the pulse/time profile, as shown in Fig. 11A.14.

The total resistance ($R_T$) is expressed using Ohm's law by

$$R_T = dR = dV/dI$$

which also equals

$$(V_1 - V_2)/(I_1 - I_2)$$

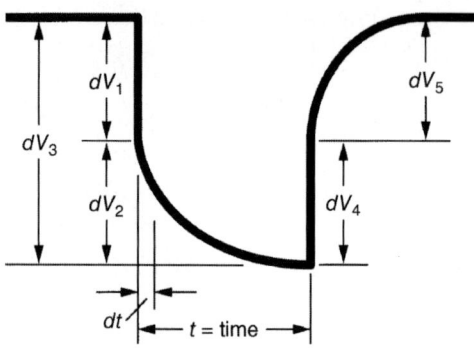

**FIGURE 11A.14** Voltage pulse/time profile illustration of curve shape and voltage components to calculate internal resistance.

where $V_1$ and $I_1$ = the voltage and current just prior to pulsing

$V_2$ and $I_2$ = the voltage and current just prior to the pulse load removal

$dV_3$ = total voltage drop shown

The internal resistance of the battery component is expressed as $dV_1$ in Fig. 11A.14 and the polarization effect component is the voltage drop $dV_2$. Since some energy was removed by the pulse, a more correct expression for the battery resistance is the voltage drop expressed by $dV_4$.

Measurement of the battery voltage drop ($dV_4$) is very difficult to capture; therefore, the pulse duration ($dt$) is minimized to reduce the polarization effect voltage drop ($dV_2$). The pulse duration is generally kept in the range of 5 to 50 ms. For accurate and repetitive results, it is recommended that duration times be kept constant by "read and hold" voltage measurements.

Since $dV_2$ is slightly greater than $dV_1$, one can see that the resistance due to polarization ($R_p$) is greater than the internal resistance of the battery ($R_{ir}$) by the formula

$$R_T = R_{ir} + R_p$$

Partial, light discharge or a light background load prior to the pulse and internal resistance measurements provides equilibration for consistent measurements.

Table 11A.4 shows the general relationship of flash current and internal resistance of the more popular cell sizes.

Zinc-carbon batteries dissipate the effects of polarization better on intermittent discharge as detailed above. Resting between discharges allows the zinc surface to "depolarize." One such effect is the dissipation of

**TABLE 11A.4** Flash Current and Internal Resistance for Various Battery Sizes

| Size | Typical maximum flash current, A | | Approximate internal resistance, $\Omega$ | |
|---|---|---|---|---|
| | LC* | ZC* | LC | ZC |
| N | 2.5 | ... | 0.6 | ... |
| AAA | 3 | 4 | 0.4 | 0.35 |
| AA | 4 | 5 | 0.30 | 0.28 |
| C | 5 | 7 | 0.39 | 0.23 |
| D | 6 | 9 | 0.27 | 0.18 |
| F | 9 | 11 | 0.17 | 0.13 |
| 9 V (battery) | 0.6 | 0.8 | 5 | 4.5 |

*LC: Leclanché; ZC: Zinc chloride.

**Source:** Eveready Battery Engineering Data.[10]

concentration polarization at the anode surface. This effect is more pronounced as heavier drains and longer duty schedules are applied. The internal resistance of the zinc chloride batteries is slightly lower than that of the Leclanché batteries. This results in a smaller voltage drop for a given battery size.

The internal resistance of zinc-carbon batteries increases with the depth of discharge. Some applications use this feature to establish low-battery alarms to predict end of battery life situations (such as in the smoke detector). Figure 11A.15 shows the relative battery internal resistance versus depth of discharge of a 9-V Leclanché battery.

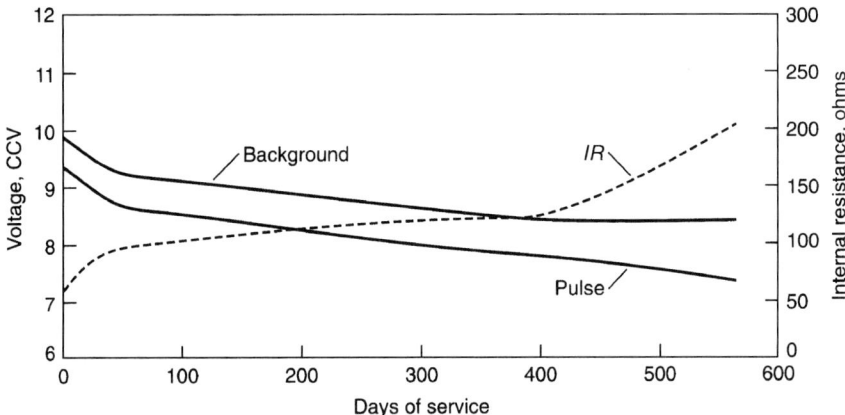

**FIGURE 11A.15** Comparison of voltage and internal resistance during discharge of a 9-V battery on smoke detector test. Background load = 620,000 $\Omega$ continuous; pulse load = 1500 $\Omega$ 10 ms every 40 s.

Another reason for increased internal resistance is the progressive plugging of the cathode with discharge reaction products. In the case of the Leclanché system, diamine-zinc chloride crystals form; in the case of the zinc chloride system, zinc oxychloride crystals form. Also the inherent conductivity of the manganese dioxide decreases.

### 11A.4.4 Effect of Temperature

Zinc-carbon batteries operate best in a temperature range of 20 to 30°C. The energy output of the battery increases with higher operating temperatures, but prolonged exposure to high temperatures (50°C and higher) will cause rapid deterioration. The capacity of the Leclanché battery falls off rapidly with decreasing temperatures, yielding no more than about 65% capacity at 0°C, and is essentially inoperative below −20°C. Zinc chloride batteries provide 80% capacity at 0°C compared to room temperature capacity. The effects are more pronounced at heavier current drains; a low current drain would tend to result in a higher capacity at lower temperatures than a higher current drain (except for a beneficial heating effect that may occur at the higher current drains).

At −20°C, typical zinc chloride electrolytes (25–30% zinc chloride by weight) form slurries. Below −25°C, ice formation is likely. Under these conditions, it is not surprising that performance is dramatically reduced. These data represent performance at flashlight-type current drains (300 mA for a D-size cell). A lower current drain would result in a higher capacity than shown. Additional characteristics of this D-size battery at various temperatures are listed in Table 11A.5.

Special low-temperature batteries were developed using low freezing-point electrolytes and a design that minimizes internal cell resistance, but they did not achieve popularity due to the superior overall performance of other types of primary batteries. For best device operation at low ambient temperatures, the Leclanché battery should be kept warm. A battery worn under the user's clothing, employing body heat to maintain a satisfactory operating temperature, was once used by the military to achieve reliable operation at low temperatures. Addition of other salts or gum karaya can boost low-temperature performance at the expense of high-temperature (>40°C) shelf life.

**TABLE 11A.5**  Temperature Effect on Internal Resistance

| Battery size | System* | Resistance, Ω | | | |
|---|---|---|---|---|---|
| | | −20°C | 0°C | 20°C | 45°C |
| | | Single cell batteries | | | |
| AAA | ZC | 10 | 0.7 | 0.6 | 0.5 |
| AA | LC | 5 | 0.8 | 0.5 | 0.4 |
| AA | ZC | 5 | 0.8 | 0.5 | 0.4 |
| C | LC | 2 | 0.8 | 0.5 | 0.4 |
| C | ZC | 3 | 0.5 | 0.4 | 0.3 |
| D | LC | 2 | 0.6 | 0.5 | 0.4 |
| D | ZC | 2 | 0.4 | 0.3 | 0.2 |
| | | Flat cell batteries | | | |
| 9 V | LC | 100 | 45.0 | 35.0 | 30.0 |
| 9 V | ZC | 100 | 45.0 | 35.0 | 30.0 |
| | | Lantern batteries | | | |
| 6 V | LC | 10 | 1.0 | 0.9 | 0.7 |
| 6 V | ZC | 10 | 1.0 | 0.8 | 0.7 |

*LC: Leclanché; ZC: Zinc chloride.
**Source:** Eveready Battery Engineering Data.[10]

## 11A.4.5   Battery Life

*Service Life.*    The service life of a general-purpose Leclanché battery is summarized in Fig. 11A.16, which plots capacity yields at various continuous discharge loads and temperatures normalized for unit weight (amperes per kilogram) and unit volume (amperes per liter). The approximate service life under specific discharge conditions

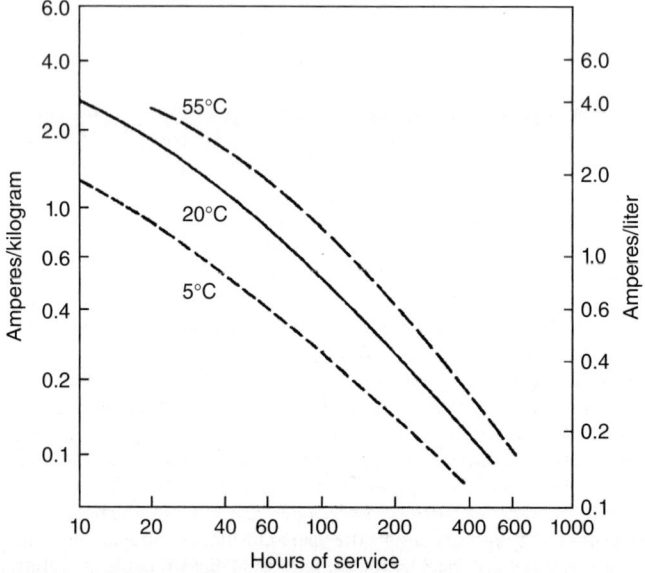

**FIGURE 11A.16**  Service hours for general-purpose zinc-carbon battery, discharged 2 h/day to 0.9 V at three temperatures.

**TABLE 11A.6**  Manufacturer's Data for AA-Size Zinc-Carbon Batteries

| Schedule | Drain @ 1.2 V, mA | Load Ω | Cutoff voltage, V | | | | | |
|---|---|---|---|---|---|---|---|---|
| | | | 1.3 | 1.2 | 1.1 | 1.0 | 0.9 | 0.8 |
| Typical service of Eveready No. 1015 general-purpose battery | | | | | | | | |
| | | | Hours | | | | | |
| 4 h/day | 28 mA | 43 | 2 | 5 | 12 | 20 | 24 | 27 |
| 1 h/day | 120 mA | 10 | 0.1 | 0.4 | 1.2 | 2.6 | 3.9 | 4.5 |
| 1 h/day | 308 mA | 3.9 | 0.09 | 0.2 | 0.4 | 0.7 | 0.9 | 1.0 |
| | | | Pulses | | | | | |
| 15 s/min/24 h/day (pulse) | 667 mA | 1.8 | 6 | 14 | 30 | 51 | 68 | 73 |
| Typical service of Eveready No. 1215 super heavy-duty battery | | | | | | | | |
| | | | Hours | | | | | |
| 4 h/day | 28 mA | 43 | 4 | 10 | 21 | 31 | 36 | 39 |
| 1 h/day | 120 mA | 10 | 0.2 | 0.4 | 2.5 | 5.2 | 6.4 | 7.0 |
| 1 h/day | 308 mA | 3.9 | 0.1 | 0.3 | 0.5 | 1.2 | 1.7 | 1.9 |
| | | | Pulses | | | | | |
| 15 s/min/24 h/day pulse) | 667 mA | 1.8 | 7 | 14 | 30 | 89 | 139 | 160 |

*Source:* Eveready Battery Engineering Data.[10]

can be estimated from these curves. Alternatively, these curves or similar ones for other cell types/sizes and discharge conditions can be used to estimate the size and weight of the battery required to meet a specific service requirement.

Manufacturers' catalogs should be consulted for specific performance data in view of the many cell formulations and discharge conditions. Table 11A.6 presents typical data from a manufacturer of two formulations of the AA-size battery.

*Shelf Life.*   Zinc-carbon batteries gradually lose capacity while idle. This deterioration is greater for partially discharged batteries than for fresh, unused cells and results from parasitic reactions such as zinc corrosion, chemical side reactions, and water loss. The rate of capacity loss is negatively affected by high temperatures. Hence, refrigerated storage will increase the shelf life. Figure 11A.17 shows the retention of capacity of a zinc-carbon battery after storage at 40, 20, and 0°C. The capacity retention of a zinc chloride battery is higher than that of the Leclanché cell because of the improved separators (coated paper separator types), sealing systems, and other materials used in that design.

Leclanché-type batteries, using the asphalt or pitch-type seals in conjunction with paste-type separators, have the poorest capacity retention. Zinc chloride batteries, using highly crosslinked starch-coated paper separators in conjunction with molded polypropylene or polyethylene seals, provide the best retention.

Batteries stored at −20°C are expected to retain approximately 80% to 90% of their initial capacity after 10 years. Since low temperatures retard deterioration, storage at low temperatures is an advantageous method for preserving battery capacity. A storage temperature of 0°C is ideal.

Freezing usually may not be harmful as long as there is no repeated broad temperature cycling. Use of case materials or seals with widely different coefficients of expansion may lead to cracking. When batteries are removed from cold storage, they should be allowed to reach room temperature in order to provide satisfactory performance. Moisture condensation during warm-up should be prevented, as this may cause electrical leakage or short-circuiting.

## 11A.4.6  Special Flat-Pack Zinc/Manganese Dioxide P-80

*Battery.*   The zinc-carbon system has been used in special designs to enhance particular performance characteristics or for new or unique applications. In the early 1970s, Polaroid introduced a new instant camera-film

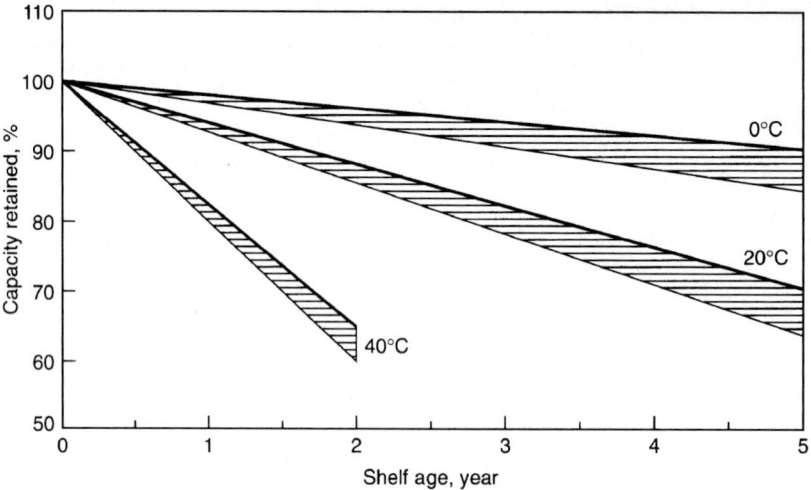

**FIGURE 11A.17**   Capacity retention after storage at 40, 20, and 0°C for paper-lined plastic seal zinc chloride batteries.

system, the SX-70. A major innovation in that system was the inclusion of a battery (model P-80) in the flat-film pack rather than in the camera. The film pack contained a battery designed to provide enough energy for the pictures in the pack. The concept was that the photographer would not have to be concerned about the freshness of the battery as it was changed with each change of film.[11]

***Battery Construction.***    The P-80 battery uses chemistry quite similar to Leclanché round cells, although the shape and construction are unique. Figure 11A.18 details one cell (1.5 V). The electrode area is approximately 5.1 cm × 5.1 cm. The anode, comprised of zinc powder and binder (without mercury), is coated on a conductive vinyl film or web.

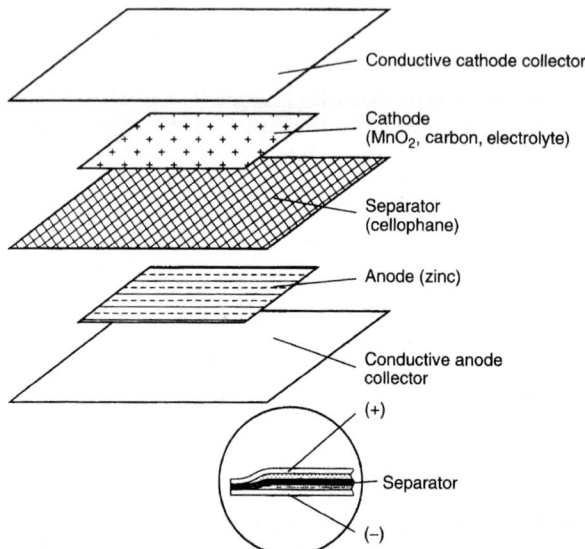

**FIGURE 11A.18**   Exploded view of a single cell from Polaroid P-80 battery pack.

The manganese dioxide is mixed in a thick paste that contains the electrolyte salts. The electrolyte is mainly zinc chloride with some ammonium chloride. The anode and the cathode are separated by a thin film of cellophane. The complete 6-V battery has four cells. The four identical cells are connected by vinyl frames to each other and the aluminum collector plates. A special conductive coating allows the aluminum to bond to the plastic materials.

***Battery Parameters.***    The key battery parameters of the flat battery are similar to those of the cylindrical one. The flat configuration provides low resistance by virtue of the geometry. The thin layers need to stay in intimate contact to maintain the low resistance, and gassing effects have to be minimized.

- *Open-circuit (no-load) voltage:* The OCV in this battery is dependent on the manganese dioxide activity and the system pH. The cathode slurry is adjusted to a constant pH to minimize battery-to-battery voltage variation. For example, the P-80 battery is adjusted so the voltage is 6.40 V at 56 days and 6.30 V after 12 months of shelf storage. The on-load voltage is sensitive to the synthetic manganese dioxide used.

- *Closed-circuit (on-load) voltage:* The CCV is used as an indicator of the battery's capability to deliver energy at high currents. In the case of the P-80 battery, a 1.63-A load is used, since that is one of the operating requirements for the camera. The CCV is measured at 55 ms to minimize polarization effects. The normal CCV is 5.58 V at 56 days and 5.35 V after 12 months of shelf storage. The on-load voltage is sensitive to the synthetic $MnO_2$ used.

- *Internal resistance and voltage drop ($\Delta V$):* The battery's internal resistance is measured by using the voltage drop ($\Delta V$) at a given load for a specified pulse period. The activity at the zinc surface is a major factor in establishing the $\Delta V$ and is dependent on both zinc particle size and amount of reaction product present. The polarization effect occurs when the load is held for an extended time period, as in the case of charging the camera's strobe circuit. Total resistance is, therefore, a summation of the two resistances—the internal resistance of the battery and the resistance due to polarization effects, the latter being very time sensitive. To minimize the effects of polarization resistance, the pulse period for $\Delta V$ measurements was minimized.

  The 56-day point is of interest, as that is the normal age when the battery is released for assembly into film packs. At that time, every battery is measured for electrical characteristics and defective ones are screened.

  The total internal resistance is expressed by the following:

$$R_t = R_i + R_p$$

  where $R_i$ = battery internal resistance

  $R_p$ = polarization resistance effect

  For the P-80 battery, $R_i$ was 0.50 $\Omega$ and $R_p$ was 0.12 $\Omega$.

- *Capacity:* The capacity simulator mimics the energy used to charge the camera strobe. The pulse consists of an OCV at rest, followed by a pulse at a 2-A load to result in a 50 wattsecond (50 Ws) pulse. The 50-Ws cycle test is maintained until the final CCV reaches the 3.7 cutoff voltage, where the number of cycles is determined.

  During the time while the 50-Ws load is maintained, the polarization drop occurs. The time to produce the 50 Ws increases with each cycle as the battery is "consumed." A 30-s rest between cycles is used. Initially, the voltage drop is fairly constant; however, near the end of the test, the resistance increases. The test is maintained to 3.7 V to indicate the cutoff point of the camera.

Figure 11A.19 illustrates the voltage profile at different discharge loads and relative capacity.

## 11A.4.7    Cell and Battery Types and Sizes

Zinc-carbon batteries are made in a number of different sizes with different formulations to meet a variety of applications. The single-cell and multicell batteries are classified by electrochemical system, either Leclanché or zinc chloride, and by grade: general-purpose, heavy-duty, extra heavy-duty, photoflash, and so on. These grades are assigned according to their output performance under specific discharge conditions.

Table 11A.7 lists the more popular battery sizes with typical performance at various loads under a 2-h per day intermittent discharge, except for the continuous toy battery test. The performance of these batteries under several intermittent discharge conditions is given in Table 11A.8.

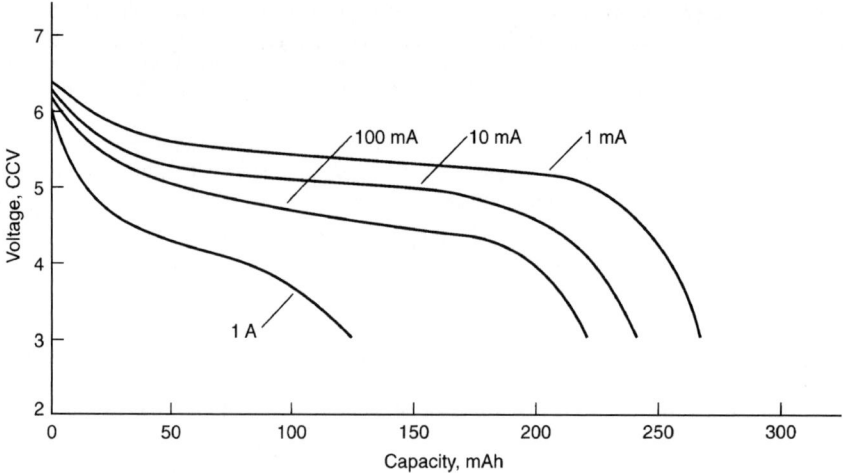

**FIGURE 11A.19** Polaroid P-80; battery voltage profile at various discharge loads.

**TABLE 11A.7** Characteristics of Zinc-Carbon Batteries

| | | | | Maximum dimensions, mm | | Typical service, 2 h/day[*] | | | |
| | | | | | | Leclanché | | Zinc chloride | |
| Size | IEC | ANSI, NEDA | Weight, g | Diameter | Height | Drain, mA | Service, h | Drain, mA | Service, h |
|---|---|---|---|---|---|---|---|---|---|
| N | R1 | 910 | 6.2 | 12 | 30.2 | 1 | 480 | | |
| | | | | | | 10 | 45 | | |
| | | | | | | 15 | 20 | | |
| AAA | R03 | 24 | 8.5 | 10.5 | 44.5 | 1 | — | 1 | 520 |
| | | | | | | 10 | — | 10 | 55 |
| | | | | | | 20 | — | 20 | 26 |
| AA | R6 | 15 | 15 | 14.5 | 50.5 | 1 | 950 | 1 | 1200 |
| | | | | | | 10 | 80 | 10 | 110 |
| | | | | | | 100 | 4 | 100 | 8 |
| | | | | | | 300 | 0.6 | 300 | 1 |
| C | R14 | 14 | 41 | 26.2 | 50 | 5 | 380 | 5 | 800 |
| | | | | | | 25 | 75 | 20 | 150 |
| | | | | | | 100 | 6 | 100 | 20 |
| | | | | | | 300 | 1.7 | 300 | 5.5 |
| D | R20 | 13 | 90 | 34.2 | 61.5 | 10 | 400 | 10 | 700 |
| | | | | | | 50 | 70 | 50 | 135 |
| | | | | | | 100 | 25 | 100 | 55 |
| | | | | | | 500 | 3 | 500 | 6 |
| F | R25 | 60 | 160 | 34[†] | 92[†] | 25 | 300 | 25 | 400 |
| | | | | | | 100 | 60 | 100 | 85 |
| | | | | | | 500 | 5.5 | 500 | 9 |
| G | R26 | — | 180 | 32[†] | 105[†] | — | | | |
| No. 6 | R40 | 905 | 900 | 67 | 170.7 | 5 | 8000 | | |
| | | | | | | 50 | 700 | | |
| | | | | | | 100 | 350 | | |
| | | | | | | 500 | 70 | | |

[*]Typical values of service to 0.9 V cutoff.
[†]Typical values.

**TABLE 11A.8**    ANSI Standards for Zinc-Carbon and Alkaline-Manganese Dioxide Batteries

| | | | | | Specifications requirements | |
|---|---|---|---|---|---|---|
| | | | | | Zinc-carbon batteries | Alkaline-manganese dioxide batteries |
| Size | Use | Ohms | Schedule | Cutoff voltage | Initial* | Initial* |
| N | | | | | 910D | 910A |
| | Portable lighting | 5.1 | 5 min/day | 0.9 | NA | 100 min |
| | Pager | (10.0 then 3000.0 | 5 s/h 3595 s/h) | 0.9 | NA | 888 h |
| AAA | | | | | 24D | 24A |
| | Pulse test | 3.6 | 15 s/min 24 h/day | 0.9 | 150 pulses | 450 pulses |
| | Portable lighting | 5.1 | 4 min/h 8 h/day | 0.9 | 48.0 min | 130.0 min |
| | Recorder | 10.0 | 1 h/day | 0.9 | 1.5 h | 5.5 h |
| | Radio | 75.0 | 4 h/day | 0.9 | 24.0 h | 48.0 h |
| AA | | | | | 15D | 15A |
| | Pulse test | 1.8 | 15 s/min 24 h/day | 0.9 | 100 pulses | 450 pulses |
| | Motor/toy | 3.9 | 1 h/day | 0.8 | 1.2 h | 5 h |
| | Recorder | 10.0 | 1 h/day | 0.9 | 5.0 h | 13.5 h |
| | Radio | 43.0 | 4 h/day | 0.9 | 27.0 h | 60 h |
| C | | | | | 14D | 14A |
| | Portable lighting | 3.9 | 4 min/h 8 h/day | 0.9 | 350 min | 830 min |
| | Toy | 3.9 | 1 h/day | 0.8 | 5.5 h | 14.5 h |
| | Recorder | 6.8 | 1 h/day | 0.9 | 10.0 h | 24.0 h |
| | Radio | 20.0 | 4 h/day | 0.9 | 30 h | 60.0 h |
| D | | | | | 13D | 13A |
| | Portable lighting | 1.5 | 4 min/15 min 8 h/day | 0.9 | 150 min | 540 min |
| | Portable lighting | 2.2 | 4 min/h 8 h/day | 0.9 | 120 min | 950 min |
| | Motor/toy | 2.2 | 1 h/day | 0.8 | 5.5 h | 17.5 h |
| | Recorder | 3.9 | 1 h/day | 0.9 | 10 h | 26.0 h |
| | Radio | 10.0 | 4 h/day | 0.9 | 33 h | 90.0 h |
| 9 V | | | | | 1604D | 1604A |
| | Calculator | 180 | 30 min/day | 4.8 | 380 min | 630 min |
| | Toy | 270 | 1 h/day | 5.4 | 7 h | 14 h |
| | Radio | 620 | 2 h/day | 5.4 | 23 h | 38 h |
| | Electronic | 1300 | 24 h/day | 6.0 | NA | NA |
| | Smoke detector | Currently under consideration. | | | | |
| 6 V | | | | | 908D | 908A |
| | Portable lighting | 3.9 | 4 min/h 8 h/day | 3.6 | 5 h | 21 h |
| | Portable lighting | 3.9 | 1 h/day | 3.6 | 50 h | 80 h |
| | Barricade | 6.8 | 1 h/day | 3.6 | 165 h | 300 h |

*Performance after 12 month storage:
  zinc-carbon batteries: 80% of initial requirement
  alkaline-manganese dioxide batteries: 90% of initial requirement
*Source:* ANSI C18.1M-2009.[9]

**TABLE 11A.9**    ANSI/NEDA Dimensions of Zinc-Carbon Batteries

| ANSI | IEC | Diameter, mm | | Overall height, mm | | Length, mm | | Width, mm | |
|---|---|---|---|---|---|---|---|---|---|
| | | Max | Min | Max | Min | Max | Min | Max | Min |
| 13C | R20S | 34.2 | 32.3 | 61.5 | 59.5 | | | | |
| 13CD | R20C | 34.2 | 32.3 | 61.5 | 59.5 | | | | |
| 13D | R20C | 34.2 | 32.3 | 61.5 | 59.5 | | | | |
| 13F | R20S | 34.2 | 32.3 | 61.5 | 59.5 | | | | |
| 14C | R14S | 26.2 | 24.9 | 50.0 | 48.5 | | | | |
| 14CD | R14C | 26.2 | 24.9 | 50.0 | 48.5 | | | | |
| 14D | R14C | 26.2 | 24.9 | 50.0 | 48.5 | | | | |
| 14F | R14S | 26.2 | 24.9 | 50.0 | 48.5 | | | | |
| 15C | R6S | 14.5 | 13.5 | 50.5 | 49.2 | | | | |
| 15CD | R6C | 14.5 | 13.5 | 50.5 | 49.2 | | | | |
| 15D | R6C | 14.5 | 13.5 | 50.5 | 49.2 | | | | |
| 15F | R6S | 14.5 | 13.5 | 50.5 | 49.2 | | | | |
| 24D | R03 | 10.5 | 9.5 | 44.5 | 43.3 | | | | |
| 903 | — | | | 163.5 | 158.8 | 185.7 | 181.0 | 103.2 | 100.0 |
| 904 | — | | | 163.5 | 158.8 | 217.9 | 214.7 | 103.2 | 100.0 |
| 908 | 4R25X | | | 115.0 | 107.0 | 68.2 | 65.0 | 68.2 | 65.0 |
| 908C | 4R25X | | | 115.0 | 107.0 | 68.2 | 65.0 | 68.2 | 65.0 |
| 908CD | 4R25X | | | 115.0 | 107.0 | 68.2 | 65.0 | 68.2 | 65.0 |
| 908D | 4R25X | | | 115.0 | 107.0 | 68.2 | 65.0 | 68.2 | 65.0 |
| 915 | 4R25Y | | | 112.0 | 107.0 | 68.2 | 65.0 | 68.2 | 65.0 |
| 915C | 4R25Y | | | 112.0 | 107.0 | 68.2 | 65.0 | 68.2 | 65.0 |
| 915D | 4R25Y | | | 112.0 | 107.0 | 68.2 | 65.0 | 68.2 | 65.0 |
| 918 | 4R25-2 | | | 127.0 | — | 136.5 | 132.5 | 73.0 | 69.0 |
| 918D | 4R25-2 | | | 127.0 | — | 136.5 | 132.5 | 73.0 | 69.0 |
| 926 | — | | | 125.4 | 122.2 | 136.5 | 132.5 | 73.0 | 69.0 |
| 1604 | 6F22 | | | 48.5 | 46.5 | 26.5 | 24.5 | 17.5 | 15.5 |
| 1604C | 6F22 | | | 48.5 | 46.5 | 26.5 | 24.5 | 17.5 | 15.5 |
| 1604CD | 6F22 | | | 48.5 | 46.5 | 26.5 | 24.5 | 17.5 | 15.5 |
| 1604D | 6F22 | | | 48.5 | 46.5 | 26.5 | 24.5 | 17.5 | 15.5 |

*Source:* ANSI C18.1M-2009.[9]

The AA-size battery predominates and is used in penlights, photoflash, and electronic applications. The smaller AAA size is used in remote control devices and other small electronic applications. The C- and D-size batteries are used mainly in flashlight applications, and the F-size is usually assembled into multicell batteries for lanterns and other applications requiring these large batteries. Flat cells are used in battery assemblies, in particular, the 9-V battery used in smoke detectors and electronic applications such as transistor radios.

Table 11A.9 lists some of the major multicell zinc-carbon batteries that are available commercially. The capabilities of these batteries can be estimated by using the International Electrotechnical Commission (IEC) designation to determine the cell compliment (e.g., NEDA 6, IEC 4R25 battery consists of four F-size cells connected in series). Table 11A.10 gives cross-references to the zinc-carbon batteries and manufacturers' designations. The most recent manufacturers' catalogs and web sites should be consulted for specific performance data to determine the suitability of their products for a particular application.

**TABLE 11A.10**   Cross-Reference of Zinc-Carbon Batteries

| ANSI | IEC | Duracell | Everyday | Rayovac | Panasonic | Toshiba | Varta | Military |
|------|-----|----------|----------|---------|-----------|---------|-------|----------|
| 13C | R20 | M13SHD | EV50 | GP-D | — | — | — | — |
| 13CD | R20 | M13SHD | EV150 | HD-D | UM1D | — | — | — |
| 13D | R20 | M13SHD | 1250 | 6D | UMIN | R20U | 3020 | — |
| 13F | R20 | — | 950 | 2D | UM1 | R20S | 2020 | BA-30/U |
| 14C | R14 | M14SHD | EV35 | GP-C | — | — | — | — |
| 14CD | R14 | M14SHD | EV135 | HD-C | UM2D | — | — | — |
| 14D | R14 | — | 1235 | 4C | UM2N | R14U | 3014 | — |
| 14F | R14 | — | 935 | 1C | UM2 | R14S | 2014 | BA-42/U |
| 15C | R6 | M15SHD | EV15 | GP-AA | — | — | — | — |
| 15CD | R6 | M15SHD | EV115 | HD-AA | UM3D | — | — | — |
| 15D | R6 | M15SHD | 1215 | 5AA | UM3N | R6U | 3006 | — |
| 15F | R6 | — | 1015 | 7AA | UM3 | R6S | 2006 | BA-58/U |
| 24D | R03 | — | 1212 | — | UM4N | — | — | — |
| 24F | R03 | — | — | — | — | — | — | — |
| 210 | 20F20 | — | 413 | — | — | — | — | BA-305/U |
| 215 | 15F20 | — | 412 | — | 15 | — | V72PX | BA-261/U |
| 220 | 10F15 | — | 504 | — | W10E | — | V74PX | BA-332/U |
| 221 | 15F15 | — | 505 | — | MV15E | — | — | — |
| 900 | R25-4 | — | 735 | 900 | — | — | — | — |
| 903 | 5R25-4 | — | 715 | 903 | — | — | — | BA-804/U |
| 904 | 6R25-4 | — | 716 | 904 | — | — | — | BA-207/U |
| 905 | R40 | — | EV6 | — | — | — | — | BA-23 |
| 906 | R40 | — | EV6 | — | — | — | — | BA-23 |
| 907 | 4R25-4 | — | 1461 | 641 | — | — | — | BA-44/U |
| 908 | 4R25 | M908 | 509 | 941 | 4F | — | — | BA-200/U |
| 908C | 4R25 | M908SHD | EV90 | GP-6V | — | — | 430 | — |
| 908CD | 4R25 | M908SHD | EV90HP | — | — | — | 431 | — |
| 908D | 4R25 | M908SHD | 1209 | 944 | — | — | 430 | — |
| 915 | 4R25 | M915 | 510S | 942 | — | — | — | BA-803/U |
| 915C | 4R25 | M915SHD | EV10S | — | — | — | — | — |
| 915D | 4R25 | M915SHD | — | 945 | — | — | — | — |
| 918 | 4R25-2 | — | 731 | 918 | — | — | — | — |
| 918C | 4R25-2 | — | EV31 | — | — | — | — | — |
| 918D | 4R25-2 | — | 1231 | 928 | — | — | — | — |
| 922 | — | — | 1463 | 922 | — | — | — | — |
| 926 | 8R25-2 | — | 732 | 926 | — | — | — | — |
| 1604 | 6F22 | — | 216 | 1604 | 006P | — | 2022 | BA-90/U |
| 1604C | 6F22 | M9VSHD | EV22 | GP-9V | — | — | — | — |
| 1604CD | 6F22 | M9VSHD | EV122 | HD-9V | 006PD | — | — | — |
| 1604D | 6F22 | M9VSHD | 1222 | D1604 | 006PN | 6F22U | 3022 | — |

*Source:* Manufacturers' catalogs.

# REFERENCES

1. D. Glover, A. Kozawa, and B. Schumm, Jr. (eds.), *Handbook of Manganese Dioxides, Battery Grade*, International Battery Material Association (IBA, Inc.), IC Sample Office, 1989.

2. N.C. Cahoon, in N.C. Cahoon and G.W. Heise (eds.), *The Primary Battery*, Vol. 2, Chap. 1, Wiley, New York, 1976.

3. S. Rubin, *The Evolution of Electric Batteries in Response to Industrial Needs*, Chap. 5, Dorrance, Philadelphia, 1978.

4. G. Vinal, *Primary Batteries*, Wiley, New York, 1950.

5. R. Huber, in K.V. Kordesh (ed.), *Batteries*, Vol. 1, Chap. 1, Decker, New York, 1974.

6. R.J. Brodd, A. Kozawa, and K.V. Kordesh, "Primary Batteries 1951–1976," *J. Electrochem. Soc.* **125:**271C–283C (1978).

7. B. Schumm, Jr., in *Modern Battery Technology,* C.D.S. Tuck (ed.), Ellis Horwood, Ltd., London, 1991, pp. 87–111.

8. C.L. Mantell, *Batteries and Energy Systems*, 2d ed., McGraw-Hill, New York, 1983.

9. "American National Standards Specification for Dry Cells and Batteries," ANSI C18.1M-2009, American National Standards Institute, Inc., 2009.

10. Eveready Battery Engineering Data: information is available at www.Energizer.com; technical information website. These data are frequently updated and current.

11. M. Dentch and A. Hillier, Polaroid Corp., *Progress in Batteries and Solar Cells*, Vol. 9 (1990).

# SECTION B

# MAGNESIUM AND ALUMINUM PRIMARY AQUEOUS BATTERIES

**Patrick J. Spellman**

## 11B.1  *TECHNOLOGY OVERVIEW*

Magnesium and aluminum cell designs were derived from the zinc-carbon cell technology and were largely the result of efforts sponsored by the U.S. military prior to the widespread use of lithium cells. The main advantages in using Mg or Al are that these elements have higher half-cell potentials relative to Zn and that the reactions involve multivalent ions, while Zn has only a single oxidation state.

## 11B.2  *CELL CHEMISTRY*

### 11B.2.1  Electrochemistry: Magnesium Anodes

The magnesium primary battery uses a magnesium alloy for the anode, manganese dioxide mixed with acetylene black to provide conductivity, an aqueous electrolyte consisting of magnesium perchlorate with barium and lithium chromate as corrosion inhibitors, and magnesium hydroxide as a buffering agent to improve storability (pH of about 8.5). The amount of water is critical, as water participates in the anode reaction and is consumed during the discharge.[1]

The discharge reactions of the magnesium/manganese dioxide battery are

$$
\begin{aligned}
\textit{Anode:} \quad & Mg + 2OH^- = Mg(OH)_2 + 2e \\
\textit{Cathode:} \quad & \underline{2MnO_2 + H_2O + 2e = Mn_2O_3 + 2OH^-} \\
\textit{Overall:} \quad & Mg + 2MnO_2 + H_2O = Mn_2O_3 + Mg(OH)_2
\end{aligned}
$$

The theoretical potential of the battery is greater than 2.8 V, but this voltage is not realized in practice. The observed values are decreased by about 1.1 V, giving an OCV of 1.9 to 2.0 V, still higher than for the zinc-carbon battery.

The rest potential of magnesium in neutral and alkaline electrolytes is a mixed potential, determined by the anodic oxidation of magnesium and the cathodic evolution of hydrogen. The kinetics of both of these reactions are strongly modified by the properties of the passive film, its history of formation, prior anodic (and to a limited extent cathodic) reactions, the electrolyte environment, and magnesium alloying additions. The key to a full appreciation of the magnesium electrode lies in an understanding of the predominant $Mg(OH)_2$ film,[2] the factors that govern its formation and dissolution, as well as the physical and chemical properties of the film.

The corrosion of magnesium under storage conditions is slight. A film of $Mg(OH)_2$ that forms on the magnesium provides good protection, and treatment with chromate inhibitors increases this protection. As a result of the formation of this tightly adherent and passivating oxide or hydroxide film on the electrode surface, magnesium is one of the most electropositive metals to find use in aqueous primary batteries. However, when the protective film is broken or removed during discharge, corrosion occurs with the generation of hydrogen

$$
Mg + 2H_2O \rightarrow Mg(OH)_2 + H_2
$$

During the anodic oxidation of magnesium, the rate of hydrogen evolution increases with increasing current density due to destruction of the passive film, which exposes more (cathodic) sites on the bared magnesium surface. This phenomenon has often been referenced as the "negative difference effect." An appreciable rate of anodic oxidation of magnesium can only take place on the bare metal surface. Magnesium salts generally exhibit low levels of anion conductivity, and one could theoretically invoke a mechanism wherein $OH^-$ ions migrate through the film to form reaction product $Mg(OH)_2$ at the magnesium-film interface. In practice, this does not occur at a sufficiently rapid rate and instead the film becomes disrupted, probably by mechanical mechanisms, as the result of anodic current flow.[3] A theoretical model for the breakdown of the passive film has been proposed.[4-7] This model involves, in successive order, metal dissolution at the metal-film interface, film dilatation, and film breakdown. This wasteful reaction is a problem, not only because of the need to vent the hydrogen from the battery and to prevent it from accumulating, but also because it uses water that is critical to the battery operation, produces heat, and reduces the efficiency of the anode.

The efficiency of the magnesium anode is about 60% to 70% during a typical continuous discharge and is influenced by such factors as the composition of the magnesium alloy, battery components, discharge rate, and temperature. On low drains and intermittent service, the anode efficiency can drop to 40% to 50% or less. The anode efficiency is also reduced with decreasing temperature.

Considerable heat is generated during the discharge of a magnesium battery, particularly at high discharge rates, due to the exothermic corrosion reaction (about 82 kcal per gram-mole of magnesium) and the losses resulting from the difference between the theoretical and operating voltage. Proper battery design must allow for the dissipation of this heat to prevent overheating and shortened life. On the other hand, this heat can be used to advantage at low ambient temperatures to maintain the battery at higher and more efficient operating temperatures.

A consequence of the passive film on these metals is the occurrence of a voltage delay (i.e., voltage depression when placed on load, as discussed in Chap. 5). Disruption of the protective film on the surface of the metal by the flow of current is needed to expose bare metal to the electrolyte (see Fig. 11B.1). When the current is interrupted, the passive film does indeed reform, but never to the original degree of passivity. Thus, both the magnesium and the aluminum batteries are at a significant disadvantage in very low or intermittent service applications.[3] Voltage delay, as shown in Fig. 11B.2, is usually less than 1 s, but can be longer (up to a minute or more) for discharges at low temperatures and after prolonged storage at high temperatures.

## 11B.2.2   Electrochemistry: Aluminum Anodes

The standard potential for aluminum in the anode reaction,

$$Al \rightarrow Al^{3+} + 3e$$

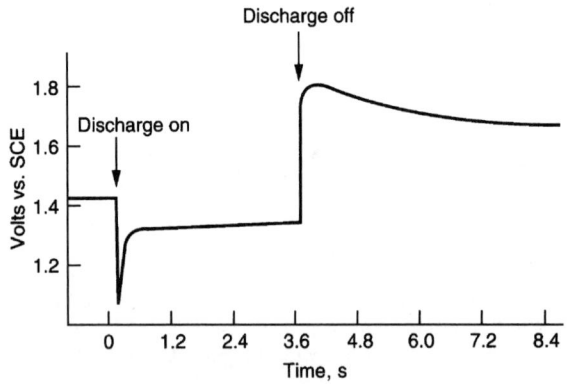

**FIGURE 11B.1**   Voltage profile of magnesium primary battery at 20°C.

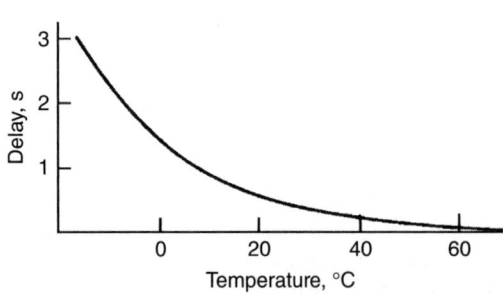

**FIGURE 11B.2**   Voltage delay versus temperature, $Mg/MnO_2$ battery.

is reported as −1.7 V. A battery with an aluminum anode should have a voltage about 0.9 V higher than the corresponding zinc battery. However, this voltage is not attained, and the voltage of an $Al/MnO_2$ battery is only about 0.1 to 0.2 V higher than that of a zinc battery. The $Al/MnO_2$ battery never progressed beyond the experimental stage because of the problems with the oxide film: prolonged voltage delay, excessive corrosion when depassivated, and the tendency for aluminum to corrode unevenly. The experimental batteries that were fabricated used a two-layer aluminum anode (to minimize premature failure due to can perforation), an electrolyte of aluminum or chromium chloride, and a manganese dioxide-acetylene black cathode similar to the conventional zinc/manganese dioxide battery. The reaction mechanism is

$$Al + 3MnO_2 + 3H_2O \rightarrow 3MnO \cdot OH + Al(OH)_3$$

## 11B.3   CELL CONSTRUCTION

Magnesium/manganese dioxide primary batteries are generally constructed in a cylindrical configuration in one of two ways.

### 11B.3.1   Standard Construction

The construction of the magnesium battery is similar to the cylindrical zinc-carbon battery. A cross section of a typical battery is shown in Fig. 11B.3. A magnesium alloy can, containing small amounts of aluminum and zinc, is used in place of the zinc can. The cathode consists of an extruded mix of manganese dioxide, acetylene black for conductivity and moisture retention, corrosion inhibitor, pH buffer, and aqueous electrolyte as detailed above. A carbon rod serves as the cathode current collector. The separator is an absorbent kraft paper similar to the paper-lined zinc battery structure. Sealing of the magnesium battery is critical, as it must both retain water during storage but provide a means for the escape of hydrogen gas generated by the corrosion reaction during the discharge. A mechanical vent is used, consisting of a top plastic seal washer with a small hole positioned under the retainer ring that is deformed under pressure, releasing the excess gas.[8]

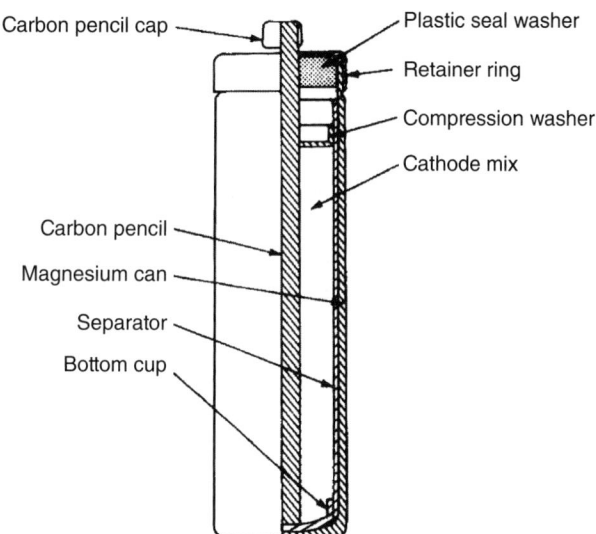

**FIGURE 11B.3**   Cylindrical construction of magnesium primary battery.

### 11B.3.2   Inside-Out Construction

The inside-out design, first presented in Sec. 11A.3.2, is based on a highly conductive, impervious carbon structure in the shape of a cup, which serves as the battery container, with a center rod molded into the center interior of the cup and attached to the cup bottom. A separator-encased magnesium tube is then positioned in the annular space. Cathode mix, consisting of manganese dioxide, carbon black, and inhibitors with aqueous magnesium bromide or perchlorate as the electrolyte, is packed in both the interior and exterior cavities filling the void between the anode tube and the center carbon rod and exterior carbon cup. This configuration doubles the number of electrode surface areas. The positive terminal is made by a metallic piece bonded during the forming process to the closed end of the carbon cup. The negative terminal to which the Mg anode tube is attached consists of a plastic insulating and sealing ring placed on the open end of the cup. The entire battery assembly is then enclosed in a crimped tin-plated steel jacket.[9-11]

## 11B.4   PERFORMANCE CHARACTERISTICS OF Mg/MnO$_2$ BATTERIES

Battery configuration has an important influence on the performance of the magnesium/manganese dioxide battery because of the heat generated during the discharge. As discussed in Sec. 11B.2.1, proper battery design must allow for the dissipation of this heat to prevent overheating, premature dry-out, and shortened performance—or for using this heat to improve performance at low ambient temperatures. In some low-temperature applications it is advantageous to insulate the battery against heat loss. Actual discharge tests will be required to obtain precise performance data under a variety of possible conditions and battery designs.

The battery and equipment design must also consider the hydrogen that is generated during discharge. The hydrogen must be vented and kept from reaching explosive levels.

### 11B.4.1   Discharge Performance

***Standard Construction.***    Typical discharge curves for the cylindrical magnesium/manganese dioxide primary battery are shown in Fig. 11B.4. The discharge profile is generally flatter than for the zinc-carbon batteries; the magnesium battery is also less sensitive to changes in the discharge rate. The average discharge voltage is on the order of 1.6 to 1.8 V, about 0.4 to 0.5 V above that of the zinc-carbon battery; the typical end voltage is 1.2 V. The performance characteristics of the cylindrical magnesium battery, type 1LM, on continuous and intermittent discharge are summarized in Figs. 11B.5 and 11B.6 and Table 11B.1. The batteries were discharged at 20°C. Figure 11B.5 provides a summary of the battery's performance under continuous load to 1.1 V end voltage.

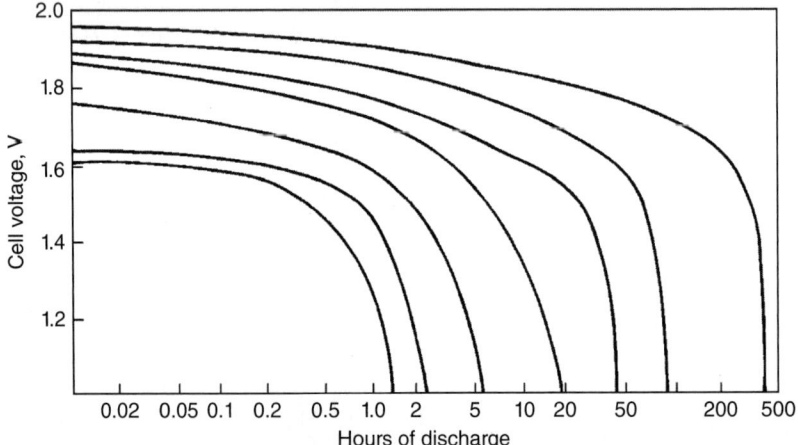

**FIGURE 11B.4**   Typical discharge curves of magnesium/manganese dioxide cylindrical battery at 20°C.

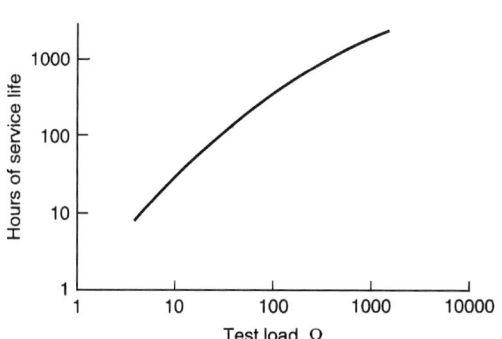

**FIGURE 11B.5**   1LM service life (hours to 1.1 V) versus test load at room temperature. (*Courtesy of Rayovac Corporation.*)

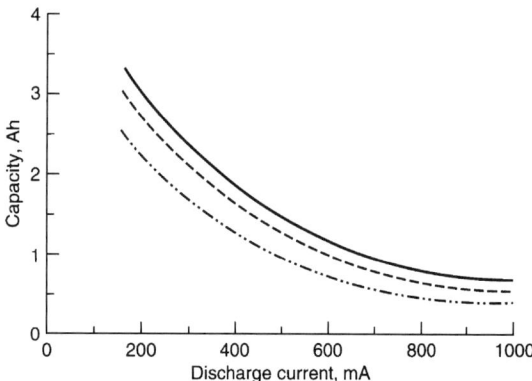

**FIGURE 11B.6**   1LM service life (ampere-hours) versus constant-current discharge. Dotted line: 1.4 V end voltage; dashed line: 1.2 V end voltage; solid line: 1.0 V end voltage. (*Courtesy of Rayovac Corporation.*)

Figure 11B.6 shows the relationship of discharge current to delivered ampere-hour capacity of the battery on continuous constant-current discharge to several end voltages. The intermittent discharge characteristics are illustrated in Table 11B.1. The sizable reduction in performance under low-rate or long-term discharge is attributed to the corrosion reaction between the discharging magnesium anode and the cell electrolyte. The reaction, which results in the evolution of hydrogen and the concomitant reduction of water, causes a loss of total cell efficiency. Loss of capacity on continuous low-rate (high-resistance) discharges is also observed.

The performance of the magnesium primary battery at low temperatures is also superior to that of the zinc-carbon battery, operating to temperatures of −20°C and below. Figure 11B.7 shows the performance of the magnesium battery at different temperatures based on the 20-h discharge rate. The low-temperature performance is influenced by the heat generated during discharge and is dependent on the discharge rate, battery size, battery configuration, and other such factors. Actual discharge tests should be performed if precise performance data are needed.

**TABLE 11B.1**   Performance, in hours, of 1LM Batteries on Continuous and Intermittent Discharge

| Type of discharge | End voltage | |
| --- | --- | --- |
| | 1.1 V | 0.8 V |
| 4 Ω, continuous | 8.9 | 9.9 |
| 4 Ω, LIFT* | 10.7 | 11.6 |
| 4 Ω, HIFT† | 11 | 12 |
| 4 Ω, 30 min/h, 8 h/day | 9.72 | 10.60 |
| 25 Ω constant resistance | | |
|    Continuous | 100 | 104 |
|    4 h/day | 84.2 | 88.4 |
| 500 Ω constant resistance | | |
|    Continuous | 1265 | 1312 |
|    4 h/day | 752 | 776 |

*Light industrial flashlight test, 4 min/h, 8 h/day.
†Heavy industrial flashlight test, 4 min/15 min, 8 h/day.
***Source:*** Rayovac Corporation.

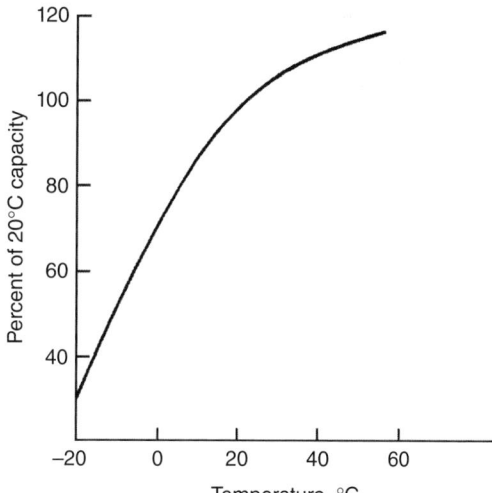

**FIGURE 11B.7**   Performance versus temperature of magnesium/manganese dioxide cylindrical battery.

On extended low-rate discharges, the magnesium battery may split open. This rupture is due to the formation of magnesium hydroxide, which occupies about one and one-half times the volume of the magnesium. It expands and presses against the cathode mix, which has hardened appreciably from the loss of water during the discharge. This opening of the can causes the voltage to rise about 0.1 V, also increasing capacity due to the air that can enter into the reaction.

The service life of the magnesium/manganese dioxide primary battery, normalized to unit weight (kilogram) and volume (liter), at various discharge rates and temperatures is summarized in Fig. 11B.8. The data are based on a rated performance of 60 Ah/kg and 120 Ah/L.

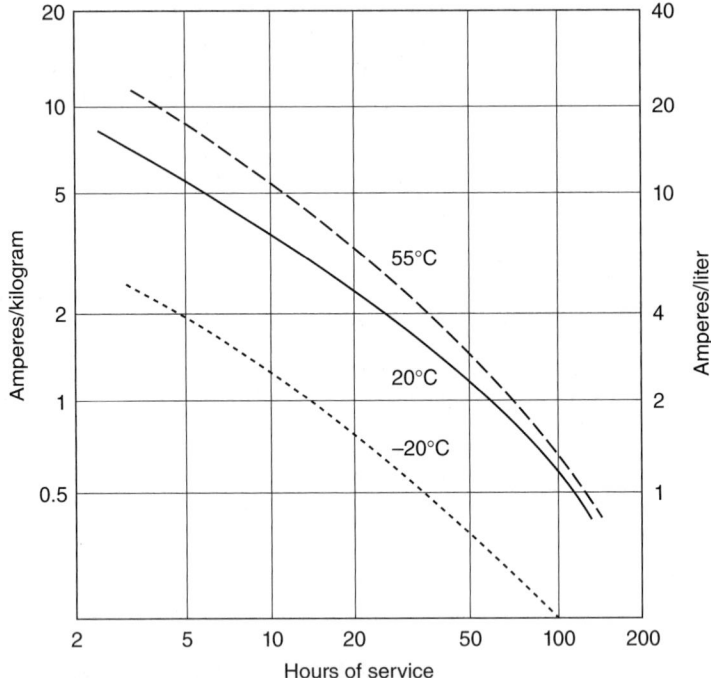

**FIGURE 11B.8** Service life of magnesium/manganese dioxide primary battery at various discharge rates and temperatures (to 1.2 V/cell end voltage).

***Inside-Out Construction.*** The discharge characteristics of the cylindrical inside-out magnesium primary batteries are shown in Fig. 11B.9 for various discharge rates and at 20°C and −20°C. This structure has better high-rate and low-temperature performance than the conventional structure. These batteries can be discharged at low rates at temperatures as low as −40°C. Discharge curves are characteristically flat. They also have good and reproducible low-drain, long-term discharge characteristics as they do not split under these discharge conditions. Discharges for a 2½-year duration are realized with a D-size battery at a 270-µA drain at 20°C.

### 11B.4.2  Shelf Life

The shelf life of the magnesium/manganese dioxide primary battery at various storage temperatures is compared with the shelf life of the zinc-carbon battery in Fig. 11B.10. The magnesium battery is noted for its excellent shelf life. The battery can be stored for periods of 5 years or longer at 20°C with a total capacity loss of 10% to 20% and at temperatures as high as 55°C with losses of about 20%/year.

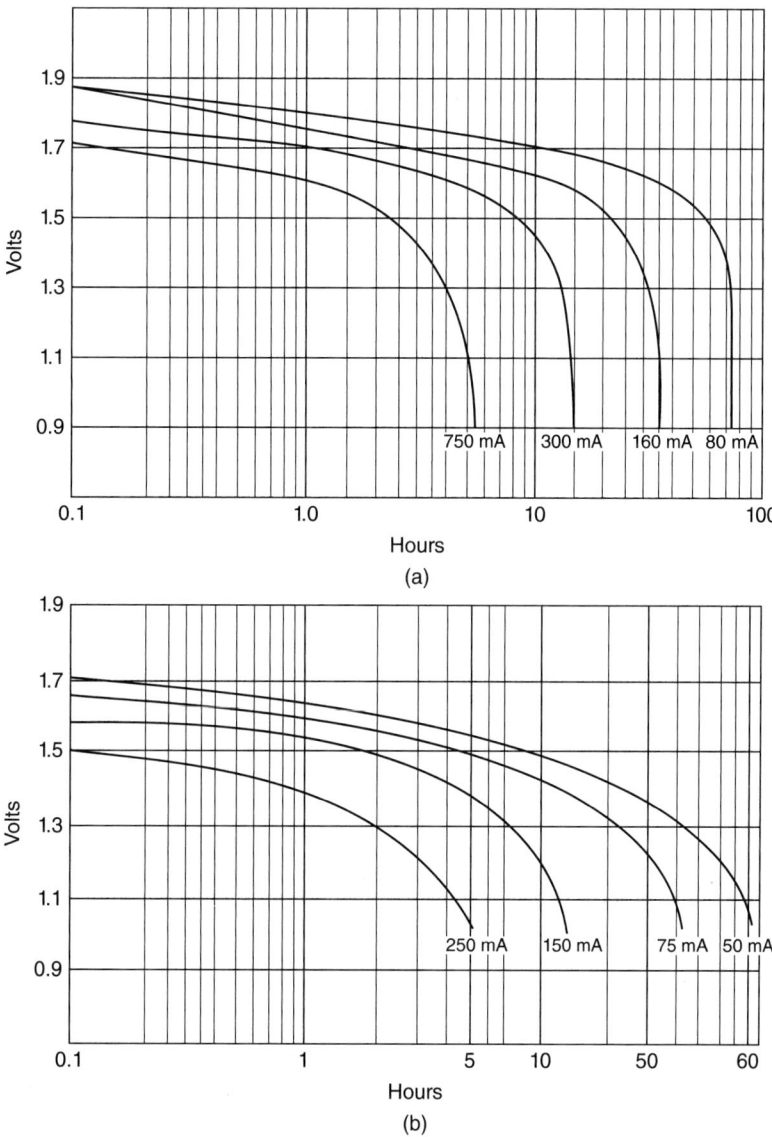

**FIGURE 11B.9**  Typical discharge curves of magnesium inside-out primary battery, D size. (a) 20°C. (b) –20°C. (*Courtesy of ACR Electronics, Inc.*)

### 11B.4.3  Types and Sizes of Mg/MnO$_2$ Batteries

The cylindrical magnesium/manganese dioxide batteries were manufactured in several of the popular standard ANSI sizes, as summarized in Table 11B.2. Most of the production of the conventional battery was used for military radio transceiver applications, and mainly in the 1LM size.[12,13] The batteries are no longer available commercially.

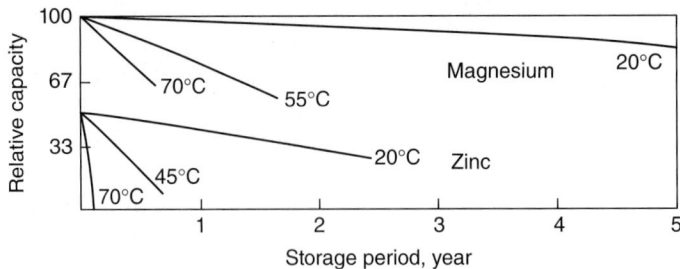

**FIGURE 11B.10**    Comparison of service versus storage of magnesium/manganese dioxide and zinc-carbon batteries.

**TABLE 11B.2**    Cylindrical Magnesium Primary Batteries

| Battery type | Diameter, mm | Height, mm | Weight, g | Capacity, Ah* | |
| | | | | Conventional structure† | Inside-out cell‡ |
|---|---|---|---|---|---|
| N | 11.0 | 31.0 | 5 | 0.5 | |
| B | 19.2 | 53.0 | 26.5 | 2.0 | |
| C | 25.4 | 49.7 | 45 | — | 3.0 |
| 1LM | 22.8 | 84.2 | 59 | 4.5 | |
| D | 33.6 | 60.5 | 105 | — | 7.0 |
| FD | 41.7 | 49.1 | 125 | — | 8.0 |
| No. 6 | 63.5 | 159.0 | 1000 | — | 65 |

*50-h discharge rate.
†Manufacturer: Rayovac Corp.
‡Manufacturer: ACR Electronics, Inc., Hollywood, FL (no longer manufactured).

## 11B.4.4    Other Types of Magnesium Batteries

Magnesium primary batteries have been developed in other structures and with other cathode materials, but these designs have not achieved commercial success. Flat cells, using a plastic-film pouch or envelope, were designed but were never produced commercially.

The use of organic depolarizers, such as *meta*-dinitrobenzene (*m*-DNB), in place of manganese dioxide was of interest because of the high capacity that could be realized with the complete reduction of *m*-DNB to *n*-phenylenediamine (2 Ah/g). The discharge of actual batteries, while having a flat voltage profile and a higher ampere-hour capacity than the manganese dioxide battery, had a low operating voltage of 1.1 to 1.2 V per cell. Cell energy was not significantly higher than for the magnesium/manganese dioxide batteries. The *m*-DNB battery was also inferior at low temperatures and high current drains. Commercial development of these batteries never materialized.

Two other versions of aqueous primary magnesium batteries include alkaline magnesium/air and reserve batteries. These cell designs are covered in Chaps. 18 and 29. Also, as an outgrowth of the primary battery efforts, magnesium rechargeable batteries were constructed but not commercialized.[14,15] The typical approach was to use organic electrolyte such as tetrahydrofuran (THF) with $Mg(butylAlCl_3)_2$ salt, and an intercalating cathode such as $Mg_xMo_6S_8$ ($x = 0–2$). This battery system has an operating voltage of 1.1 V, long cycle life with little capacity fade, and specific energy comparable to lead-acid and NiCd systems (~60 Wh/kg).[16,17] The rate capability would be suitable for low-rate applications such as load leveling or solar support.[18] Other studies have used $C_2H_5MgF$ in diethyl ether and a AgO cathode.[19]

## REFERENCES

1. J.L. Robinson, "Magnesium Cells," in N.C. Cahoon and G.W. Heise (eds.), *The Primary Battery,* Vol. 2, Wiley-Interscience, New York, 1976, Chap. 2.

2. G.R. Hoey and M. Cohen, "Corrosion of Anodically and Cathodically Polarized Magnesium in Aqueous Media," *J. Electrochem. Soc.* **105**:245–250 (1958).

3. J.E. Oxley, R.J. Ekern, K.L. Dittberner, P.J. Spellman, and D.M. Larsen, "Magnesium Dry Cells," in *Proc. 35th Power Sources Symp.,* IEEE, New York, 1992, pp. 18–21.

4. B.V. Ratnakumar and S. Sathyanarayana, "The Delayed Action of Magnesium Anodes in Primary Batteries. Part I: Experimental Studies," *J. Power Sources* **10**:219–241 (1983).

5. S. Sathyanarayana and B.V. Ratnakumar, "The Delayed Action of Magnesium Anodes in Primary Batteries. Part II: Theoretical Studies," *J. Power Sources* **10**:243–261 (1983).

6. S.R. Narayanan and S. Sathyanarayana, "Electrochemical Determination of the Anode Film Resistance and Double Layer Capacitance in Magnesium-Manganese Dioxide Cells," *J. Power Sources* **15**:27–43 (1985).

7. B.V. Ratnakumar, "Passive Films on Magnesium Anodes in Primary Batteries," *J. Appl. Electrochem.* **18**:268–279 (1988).

8. D.B. Wood, "Magnesium Batteries," in K.V. Kordesch (ed.), *Batteries,* Vol. 1: *Manganese Dioxide,* Marcel Dekker, New York, 1974, Chap. 4.

9. R.R. Balaguer and F.P. Schiro, "New Magnesium Dry Battery Structure," in *Proc. 20th Power Sources Symp.,* Atlantic City, NJ, 1966, p. 90.

10. R.R. Balaguer, "Low Temperature Battery (New Magnesium Anode Structure)," Report: ECOM-03369-F, 1966.

11. R.R. Balaguer, "Method of Forming a Battery Cup," U.S. Patent 3,405,013, 1968.

12. D.M. Larsen, K.L. Dittberner, R.J. Ekern, P.J. Spellman, and J.E. Oxley, "Magnesium Battery Characterization," in *Proc. 35th Power Sources Symp.,* IEEE, New York, 1992, p. 22.

13. L. Jarvis, "Low Cost, Improved Magnesium Battery," in *Proc. 35th Power Sources Symp.,* New York, 1992, p. 26.

14. P. Novak, R. Imhof, and O. Haas, "Magnesium Insertion Electrodes for Rechargeable Nonaqueous Batteries—A Competitive Alternative to Lithium?" *Electrochimica Acta* **45**:351–367 (1999).

15. D. Aurbach, Y. Gofer, Z. Lu, A. Schechter, O. Chusid, H. Gizbar, Y. Cohen, et al., "A Short Review on the Comparison between Li Battery Systems and Rechargeable Magnesium Battery Technology," *J. Power Sources* **97–98**:28–32 (2001).

16. D. Aurbach, Y. Gofer, A. Schechter, L. Zhohdghua, and C. Gizbar, "High Energy, Rechargeable Electrochemical Cells with Nonaqueous Electrolytes," U.S. Patent 6,316,141, November 13, 2001.

17. N. Amir, Y. Vestfrid, O. Chusid, Y. Gofer, and D. Aurbach, "Progress in Nonaqueous Magnesium Electrochemistry," *J. Power Sources* **174**:1234–1240 (2007).

18. D Aurbach, "Advances in R&D of Electrolyte Solutions for Recharging Batteries," Twenty-Sixth International Battery Seminar, Fort Lauderdale, FL, March 2009.

19. S. Ito, O. Yamamoto, T. Kanbara, and H. Matsuda, "Nonaqueous Electrolyte Secondary Battery with an Organic Magnesium Electrolyte Compound," U.S. Patent 6,713,213 B2, March 30, 2004.

# CHAPTER 12
# PRIMARY ALKALINE BATTERIES

## 12.0  GENERAL OVERVIEW

Primary aqueous alkaline systems, typically referenced as alkaline batteries due to the use of alkaline electrolytes, have been known for over 125 years, but it was not until World War II that a practical battery was developed in response to a requirement for a battery with a high capacity-to-volume ratio that would withstand storage under tropical conditions.[1,2]

Alkaline batteries all typically employ a metallic anode (typically zinc), caustic electrolyte solutions (typically potassium hydroxide) and cathodes of various oxides of manganese, mercury, silver, and other similar compounds. The initial alkaline cells were based on zinc anodes and mercuric oxide cathodes. Over the years, different variations were successfully commercialized as detailed in the three sections of this chapter. An overview of each of these systems follows.

### 12.0.1  Zinc/Manganese Dioxide

The most successful alkaline system to be developed was the zinc/manganese dioxide battery. This cell ($Zn/MnO_2$) has grown to be the dominant primary battery system used in portable devices requiring disposal batteries. As the portable device market has increased, so has the market share of this battery system. Alkaline batteries have become the primary battery of choice in the United States and most other developed nations. While this alkaline cell was commercially introduced in 1959, it wasn't until the 1980s that it became widely recognized as being superior to the carbon-zinc type primary battery, which used the same $Zn/MnO_2$ electrode materials but with a near neutral pH aqueous electrolyte (see Chap. 11). Eveready Battery Co. (now Energizer®) is credited with leading the development of the cylindrical alkaline cell, patented by Lew Urry, whose original battery is displayed in the Smithsonian National Museum of American History along with Edison's light bulb.

This superior performance is significantly manifested in the higher-drain devices such as electronic toys, audio devices, cameras, and remote controller. Some advantages are listed in Table 12.1 and the sales projections are shown in Table 12.2. However, these advantages are offset by higher cost, compared to the carbon-zinc battery.

### 12.0.2  Zinc/Mercuric Oxide

The Zn/HgO battery is noted for its high capacity per unit volume, constant voltage output, and good storage characteristics. The characteristics of this system were particularly advantageous in applications such as hearing aids, watches, cameras, some early pacemakers, and small electronic equipment, where it was widely used. The battery has also been used as a voltage reference source and in electrical instruments and electronic equipment, such as sonobuoys, emergency beacons, rescue transceivers, radio and surveillance sets, small scatterable mines, and early satellites. These applications, however, did not become widespread, except for military and special uses, because of the relatively higher cost of the mercuric oxide system.

**TABLE 12.1**   Major Advantages of Alkaline-Manganese Dioxide Battery Compared to Carbon-Zinc Battery

Higher energy density
Superior service performance at all drain rates, higher capacity
Superior cold-temperature performance
Lower internal resistance
Longer shelf life
Less leakage

**TABLE 12.2**   World Alkaline Cell Demand (million dollars)*

| Actual/projected sales (based on 2016 sales data) | 2004 | 2009 | 2014 | 2019 | 2024 |
|---|---|---|---|---|---|
| Total world battery sales | 36,345 | 56,145 | 82,470 | 120,000 | 166,500 |
| World primary battery sales | 12,990 | 16,455 | 19,865 | 24,050 | 28,150 |
| (Primary batteries, % of total) | 35.7 | 29.3 | 24.1 | 20.0 | 16.9 |
| Primary alkaline sales | 7132 | 8925 | 10,569 | 12,950 | 15,550 |
| (Alkaline cells, % of primary) | 54.9 | 54.2 | 53.2 | 53.8 | 55.2 |

*Table IV-7, World Primary Battery Demand by Chemistry, published December 2016 by The Freedonia Group.

Cadmium has been used to replace zinc, resulting in a very stable battery with excellent storage and performance at extreme temperatures due to the low solubility of cadmium in alkali solutions over a wide range of temperatures. However, cadmium is costly, and cell voltage is low, less than 1.0 V. Hence, cadmium/mercuric batteries were only used in special applications requiring the particular performance capabilities of the system, such as gas and oil well logging, telemetry from engines and other heat sources, alarm systems, remote equipment data-monitoring, surveillance buoys, weather stations, and emergency equipment.[3]

However, the sale of mercuric oxide batteries has ended due to environmental problems associated with mercury and cadmium. The 1996 Mercury-Containing and Rechargeable Battery Management Act (P.L. 104-142) prohibits the sale of mercuric oxide batteries in the United States unless manufacturers provide for recycling and disposal. Furthermore, these batteries have been removed from the International Electrotechnical Commission (IEC) and the American National Standards Institute (ANSI) standards. In many applications, they have been replaced by alkaline-manganese dioxide (Chap. 12A), zinc/silver oxide (Chap. 12C), zinc/air (Chap. 18), and lithium batteries (Chap. 14).

The major characteristics of these two mercuric oxide systems are summarized in Table 12.3.

### 12.0.3   Zinc/Silver Oxide

The zinc/silver oxide system (Zn/AgO) offers high capacity, a steady discharge voltage, and good storage capacity retention. The zinc/silver oxide battery will discharge at a flat, constant voltage typically between 1.5 and 1.6 V at both high and low discharge rates. The battery retains more than 95% of its initial capacity after 1 year of room-temperature storage and also has good low-temperature discharge capabilities, delivering about 70% of its nominal capacity at 0°C and 35% at −20°C. These features have enabled the zinc/silver oxide battery to be an

**TABLE 12.3**    Characteristics of the Zinc/Mercuric Oxide and Cadmium/Mercuric Oxide Batteries

| Advantages | Disadvantages |
|---|---|
| Zinc/mercuric oxide battery | |
| High energy-to-volume ratio, 450 Wh/L | Batteries were expensive; although widely used in miniature sizes, but only for special applications in the larger sizes |
| Long shelf life under adverse storage conditions | |
| Over a wide range of current drains, recuperative periods are not necessary to obtain a high capacity from the battery | After long periods of storage, cell electrolyte tends to seep out of seal, which is evidenced by white carbonate deposit at seal insulation |
| High electrochemical efficiency | Moderate energy-to-weight ratio |
| High resistance to impact, acceleration, and vibration | Poor low-temperature performance |
| Very stable open-circuit voltage, 1.35 V | Batteries for disposal are considered hazardous wastes under environmental regulations |
| Flat discharge curve over wide range of current drains | |
| Cadmium/mercuric oxide battery | |
| Long shelf life under adverse storage conditions | Batteries are more expensive than zinc/mercuric oxide batteries due to high cost of cadmium |
| Flat discharge curve over wide range of current drains | System has low output voltage (open-circuit voltage = 0.90 V) |
| Ability to operate efficiently over wide temperature range, even at extreme high and low temperatures | Moderate energy-to-volume ratio |
| Can be hermetically sealed because of inherently low gas evolution level | Low energy-to-weight ratio |
| | Batteries for disposal are considered hazardous wastes under environmental regulations |

important micro power source for electronic devices and equipment, such as watches, calculators, electronic thermometers, glucometers, cameras, and other applications that require small, thin, high-capacity, long-service-life batteries that discharge at a constant voltage. The commercial primary zinc/silver oxide batteries are manufactured mainly in the button cell configuration, with the sizes ranging from 5 to 250 mAh. There are a few applications for this battery system in larger sizes, such as for the military (see Chap. 29), but its use is limited by the high cost of silver.

Of the three oxidation states for silver oxide, the monovalent state or silver (I) oxide ($Ag_2O$) is most commonly used for commercial button cells. The divalent silver oxide or silver (II) oxide (AgO) not only has a higher theoretical capacity, but it also has the disadvantages of a dual voltage discharge and greater instability in alkaline solutions. The divalent silver oxide button cell was sold commercially as "Ditronic" or "Plumbate" batteries but discontinued years ago. The trivalent silver oxide or silver (III) oxide ($Ag_2O_3$) is very unstable and is not used in batteries.

The major advantages and disadvantages of the zinc/monovalent silver oxide battery are summarized in Table 12.4.

**TABLE 12.4**    Major Advantages and Disadvantages of Zinc/Silver Oxide Primary Batteries

| Advantages | Disadvantages |
|---|---|
| High energy density | Use limited to button and miniature cells because of high cost |
| Good voltage regulation, high-rate capability | |
| Flat discharge curve can be used as a reference voltage | |
| Comparatively good low-temperature performance | |
| Leakage and salting negligible | |
| Good shock and vibration resistance | |
| Good shelf life | |

## Acknowledgments

Table 12.2 is provided by The Freedonia Group.

Content for this chapter has been taken from Chap. 11, Alkaline-Manganese Dioxide Batteries, John C. Nardi and Ralph J. Brodd; Chap. 12, Mercuric Oxide Batteries, Nathan D. (Ned) Isaacs; and Chap. 13, Button Cell Batteries: Silver Oxide–Zinc and Zinc-Air Systems, Joseph Passaniti, Denis Carpenter, and Rodney McKenzie, in *Linden's Handbook of Batteries*, 4th ed., T. B. Reddy and D. Linden, eds., McGraw-Hill, 2011.

## References

1. C. L. Clarke, U.S. Patent 298,175 (1884).
2. S. Ruben, "Balanced Alkaline Dry Cells," *Proc. Electrochem. Soc. Gen. Meeting,* Boston, Oct. 1947.
3. B. Berguss, "Cadmium-Mercuric Oxide Alkaline Cell," *Proc. Electrochem. Soc. Meeting,* Chicago, Oct. 1965.

# SECTION A

# MANGANESE DIOXIDE

**John C. Nardi, Ralph J. Brodd**

## 12A.1   OVERVIEW

The alkaline-manganese dioxide or zinc-manganese dioxide battery ($Zn/KOH/MnO_2$) cell is produced in two different cell types: the cylindrical type and the miniature button style. In addition, multiple alkaline cells are coupled into multi-cell configurations such as the 9 V radio/smoke detector type batteries. Demand for the alkaline $MnO_2$ cell was about $10.5 billion worldwide in 2014 and is expected to increase to over $15.5 billion by 2024 due to the demand for battery-powered devices, especially for smaller, thinner, and lighter portable devices.[1]

Major U.S. battery companies, such as Energizer®, Duracell®, and Rayovac®, have additionally developed a range of products. The standard alkaline cell is designed for the widest range of device applications. Some companies provide an economy or value grade alkaline, designed for long life but at low to moderate drain rates such as radios, remote controls, and clocks. Lastly, premium cells are also available for delivering superior performance in high power devices, such as digital cameras, photoflash, and high power lighting.

Several new alkaline cell types, based on the nickel oxyhydroxide addition to the cathode formulation (the Panasonic® Oxyride battery and the Duracell Powerpix™ battery), were introduced in the past decade, claiming a doubling of life at power levels sufficient for the higher drain devices. In addition, competition from rechargeable batteries, such as the nickel-metal hydride, and the primary 1.5 V lithium chemistries advertised to last up to seven times longer than regular alkaline batteries, impacts traditional alkaline battery sales.

The battery manufacturers have continued to respond to the higher power and constant current drains required by the new portable devices being commercialized in recent years. Improvements in performance, aided by a combination of improved materials, design, and chemistry, have benefited the sales growth of alkaline cells.

When compared to the carbon-zinc system, the alkaline cell is built inside-out and upside-down, as shown in Fig. 12A.1. The positive electrode mix (EMD, graphite, and potassium hydroxide electrolyte) is molded into a steel can whose bottom is at the top of the pictured cutaway. A paper separator basket or two paper strips are inserted, and a potassium hydroxide gel containing powdered zinc is dispensed into the basket. The electrolyte also includes an inhibitor to mitigate zinc corrosion and ensure long shelf life. A negative collector assembly consisting of a brass nail and a plastic seal are inserted, making contact with the zinc gel. A flat cover is then placed over the open part of the can and crimped shut, becoming the negative end of the cell. The steel can bottom is also provided with a cover, sometimes having a center dimple, which forms the positive end of the finished cell.

Over the past 50 years many improvements have been made to this cell design. Initial advanced concepts (i.e., gelled zinc powder anode and vented plastic seal) were enhanced by use of butt-seam metal cans, which provided an increased internal volume. The addition of organic inhibitors to the anode was discovered to reduce the rate of gassing caused by impurities or contaminants in the zinc anode that had previously caused bulging and leakage problems. Another major development was the introduction of a plastic label and lower profile seal, which even further increased the cell's internal volume, allowing the addition of more active materials and thus a greater discharge capacity.

One of the most important developments of the alkaline cell in the 1980s was the gradual reduction in the amount of added mercury in the anode. Most early alkaline cells contained up to 6% mercury in the zinc anode, but with the development of cathode materials with lower impurities and better processing techniques, the level of added mercury was gradually reduced to zero. Worldwide concern over the environmental impact of the cell components after their disposal had necessitated the removal of mercury. Today, most countries have banned batteries that contain mercury. These improvements in materials and construction have allowed the alkaline-$MnO_2$ battery to gain as much as a 60% to 70% increase in specific energy output since its first introduction.

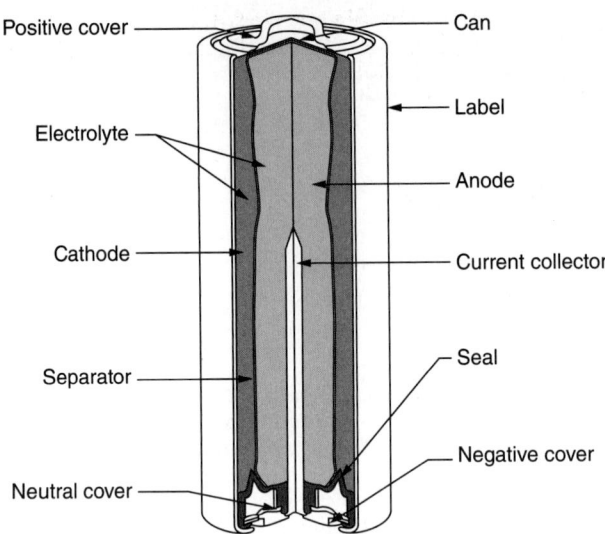

**FIGURE 12A.1**   Schematic view of typical alkaline cell construction. (*Courtesy of Energizer, Inc.*)

These improvements have allowed alkaline batteries to keep pace with the needs of consumers and their demand for smaller and higher energy devices. Continuing research by the major battery companies will likely provide further technological improvements that help maintain market share.

The miniature alkaline button and coin cells use the same zinc/alkaline electrolyte/manganese dioxide configuration as the cylindrical cells and compete with other types of miniature cells, such as silver oxide and zinc/air as well as lithium-based chemistries. The major use of these cells is to power watches, hearing aids, and specialty items. This cell consists of a shallow steel can that serves as the positive contact while containing the cathode mix and a copper-clad steel cover that serves as the negative contact while containing a potassium hydroxide gel with zinc powder. Table 12A.1 lists the advantages and disadvantages of the alkaline-manganese dioxide miniature battery compared to other lithium-based miniature systems.

**TABLE 12A.1**   Comparison of Advantages and Disadvantages of Miniature Alkaline-Manganese Dioxide Cell to Other Miniature Systems

| Advantages | Disadvantages |
| --- | --- |
| Lower cost | Sloping discharge |
| Lower internal resistance | Lower energy density |
| Good low-temperature performance | Shorter shelf life |
| Equivalent leakage resistance | |

## 12A.2   CELL ELECTROCHEMISTRY

The active components of the alkaline-manganese dioxide cell include powdered zinc, an aqueous KOH electrolyte, and electrolytically produced manganese dioxide (EMD). The EMD is used instead of either chemically produced $MnO_2$ or the natural ore due to higher manganese content, increased activity, and higher purity. The KOH electrolyte is a high-purity concentrated solution typically in the range of 35% to 52%, which provides high conductivity and reduced gassing rate for sealed alkaline cells subject to various uses and storage conditions.

The zinc powder anode provides a high surface area for high-rate capability, i.e., low local current density, and facilitates the homogeneous distribution of the solid and liquid phases in the anode compartment to minimize concentration polarization of the reactants and products.

During discharge, the manganese dioxide cathode first undergoes a one-electron reduction to the oxyhydroxide with expansion and distortion of its lattice in concentrated alkaline electrolytes.

$$MnO_2 + H_2O + e \rightarrow MnOOH + OH^- \tag{12A.1}$$

The MnOOH product forms a solid solution with the reactant, which produces the characteristic sloping discharge.[2] Of at least nine crystal structural forms of $MnO_2$ that exist, only the gamma form (known as the natural mineral nsutite) has adequate alkaline discharge characteristics, because its surface is not prone to blockage by the reaction product. Nsutite is an intergrowth of the beta or pyrolusite form and ramsdellite that is the structurally disordered form of manganese dioxide that is found in alkaline cells. It is composed of the $1 \times 1$ tunnel structure of pyrolusite and the $1 \times 2$ tunnel form of the ramsdellite, as depicted in Fig. 12A.2.[3] Table 12A.2 shows the different manganese oxide structures.[4]

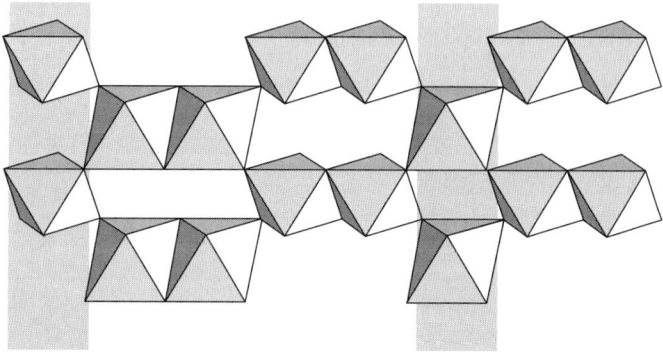

**FIGURE 12A.2**  Schematic representation of intergrowth between pyrolusite and ramsdellite lattices.

**TABLE 12A.2**  Different Manganese Oxide Structures

| Mineral | Space Group | Z | a(Å) | b(Å) | c(Å) | β, γ(°) | Reference |
|---|---|---|---|---|---|---|---|
| Pyrolusite (β) | P4₂/mnm | 2 | 4.398 | — | 2.873 | 90 | Baur, 1976 |
| Ramsdellite | Pbnm | 4 | 4.533 | 9.27 | 2.866 | 90 | Byström, 1949 |
| Nsutite (γ) | [intergrowth] | 4 | 4.45 | 9.305 | 2.85 | 90 | De Wolff, 1959 |
| Birnessite | P ml | 1 | 2.84 | — | 7.27 | 120 | Giovanoli et al., 1970 |
| ε-MnO₂ | P6₃/mmc | 1 | 2.80 | — | 4.45 | 120 | De Wolff et al., 1978 |
| Spinel (λ) | Fd3m | 16 | 8.029 | — | — | 90 | Mosbah et al., 1983 |
| Hollandite (α) | I2/m | 2 | 10.026 | 2.8782 | 9.729 | 91.03 | Post et al., 1982 |
| Psilomelane | C2/m | 2 | 13.929 | 2.8459 | 9.678 | 92.39 | Turner and Post, 1988 |
| Todorokite | P2/m | 8 | 9.764 | 2.8416 | 9.551 | 94.06 | Post and Bish, 1988 |
| Manganite (γ) | B2₁/d | 8 | 8.88 | 5.25 | 5.71 | 90 | Dachs, 1963, 1973 |
| Groutite (α) | Pbnm | 4 | 4.560 | 10.7 | 2.87 | 90 | Glasser and Ingram, 1968 |
| Groutellite | [Pbnm] | 4 | 4.7 | 9.531 | 2.864 | 90 | JCPDS 42-1316 |
| Feitkneichtite (β) | P ml | 1 | 3.32 | — | 4.71 | 120 | Feitnecht et al., 1962 |
| Pyrochroite | P ml | 1 | 3.322 | — | 4.734 | 120 | Christensen, 1965 |

The cathode expands about 17% in volume when forming the MnOOH reaction product. MnOOH can also undergo some undesirable chemical side-reactions, depending on the conditions and extent of the discharge. In the presence of the zincate ion, MnOOH, based on its equilibrium with soluble Mn(III), can form the complex compound hetaerolite, $ZnMn_2O_4$. Although electroactive, hetaerolite is not as easily discharged as MnOOH, and results in an increased cell impedance. In addition, the $MnOOH/MnO_2$ solid solution can undergo recrystallization into a less active form, resulting in a noticeable loss of cell voltage under certain very slow discharge conditions.[5]

Overall, at least the first half of the $MnO_2$ discharge reaction is a simple proton-electron insertion reaction with no structural change except for expansion and distortion of the lattice. Toward the end of the first electron discharge, however, the reaction proceeds through a soluble $Mn^{3+}$ species along with a variety of $Mn^{3+}$ and $Mn^{2+}$ intermediate products, depending on the discharge conditions.

During discharge at the lower voltages, the MnOOH can be further discharged, as depicted by the following equation:

$$3MnOOH + e \rightarrow Mn_3O_4 + OH^- + H_2O \tag{12A.2}$$

This reaction produces a flat discharge curve that only occurs under low-rate discharge conditions. No additional volume change occurs during this reaction in the cathode. This step only provides about one-third of the capacity of the first $MnO_2$ reaction. During deep discharge, a further reduction to $Mn(OH)_2$ is possible but seldom occurs.

The following reactions provide a more detailed alkaline cathode reaction scenario.

$$MnO_2 + xH_2O + xe \rightarrow MnOOH_x + xOH^- \qquad (0 < x < {\sim}0.6) \tag{12A.3}$$

For $x > {\sim}0.6$ other products result via the soluble species to yield MnOOH, $Mn(OH)_2$, $Mn_3O_4$, and $ZnMn_2O_4$.

During the early stage of the cell's discharge, the anode reaction in KOH produces the following soluble zinc ion reaction, which can be found in the separator and cathode:

$$Zn + 4OH^- \rightarrow Zn(OH)_4^{-2} + 2e \tag{12A.4}$$

At a certain point in the discharge, depending on the composition of the anode and the rate and depth of discharge, the electrolyte will become saturated with zincate that then causes the reaction product to change to the insoluble $Zn(OH)_2$. Eventually, the anode will become depleted of water and the zinc hydroxide dehydrates to ZnO by the following two reactions:

$$Zn + 2OH^- \rightarrow Zn(OH)_2 + 2e \tag{12A.5}$$

$$Zn(OH)_2 \rightarrow ZnO + H_2O \tag{12A.6}$$

These changes in the different zinc discharge products cannot be easily noted in the discharge curve since the standard reaction potentials for reactions (12A.5) and (12A.6) are very similar. However, under certain conditions, the formation of the oxide can be sufficiently high that any undischarged zinc is passivated. Such conditions would include high rate, low temperatures, and poor electrolyte conductivity. Typically, high surface area zinc is used to minimize any cell impedance increase by the anode.

The overall total one-electron reaction of the alkaline cell during a continuous discharge is as follows:

$$2MnO_2 + Zn + 2H_2O \rightarrow 2MnOOH + Zn(OH)_2 \tag{12A.7}$$

Since water is a reactant in this reaction, the amount of water in a cell is quite important, especially in high-rate discharge applications. Therefore, the total water management in a cell is an important variable that battery manufacturers must control in order to provide good performance over a wide range of discharge conditions. Some battery manufacturers have included additives to the cell, such as $TiO_2$ and $BaSO_4$, in order to better manage this important characteristic. Also, many different ZnO morphologies may exist that could affect the anode's performance.

However, at the low- or intermittent-drain rates, the total cell reaction for 1.33 electrons per mole is

$$3MnO_2 + 2Zn \rightarrow Mn_3O_4 + 2ZnO \tag{12A.8}$$

This reaction clearly indicates that under such conditions there is no water management concern.

The open-circuit voltage of an undischarged alkaline cell is typically between 1.55 and 1.65 V, depending on the purity and activity of the cathode components, the ZnO content of the anode, and the storage temperature of the cell.

Due to the natural corrosive nature of basic solutions, zinc metal can reduce water and form hydrogen gas. Such a reaction does occur in the alkaline cell and reduces the overall cell capacity (due to zinc corrosion) if allowed to become significant. Hydrogen gas evolution can occur during long-term storage of undischarged cells or after partial discharge. The amount of gas formed during the latter event depends on the discharge rate, delivered capacity, and storage temperature. This gas buildup can cause the cell to bulge and eventually leak. In addition, the hydrogen gas that forms can reduce the manganese dioxide, even further reducing the cell's available capacity.

While the hydrogen evolution rate of pure zinc is low, the inevitable presence of impurities (i.e., ppm level trace amounts) exacerbates rate of gassing by acting as cathodic sites on the zinc. This gassing can be reduced or minimized in several of the following ways: (1) alloying the zinc with known gassing inhibiting elements, (2) reducing the impurity levels of the cell components, (3) adding ZnO to the anode, and (4) adding inorganic or organic gassing inhibitors (e.g., polyethylene glycol) to the anode. Mercury is the best and most efficient inhibitor, but due to its toxic nature and the international drive for green chemistry has been banned for use in alkaline cells worldwide.

Alloying elements are incorporated into the zinc to both inhibit gassing and improve performance. The main group of such elements includes lead, bismuth, thallium, and indium. The levels of these elements have been empirically determined based on the performance required by the battery manufacturer.

As already mentioned, the impurity levels in the zinc should be as low as possible, with most of them coming from the natural ore or during processing. Contaminant levels are typically determined by the efficiency of the zinc powder purification process.

Another important aspect of the zinc reaction in alkaline solution is the ionic solution equilibrium. The zinc metal anode is in equilibrium with the zinc ions in the KOH solution. Therefore, on stand, the zinc electrode reaction is continually dissolving and redepositing in the anode compartment of the cell. Detailed scanning electron micrograph (SEM) studies have shown that both type I and type II zinc oxides are present in the discharged zinc particles in the porous anode. The morphology of these two different zinc types appears to depend on the drain rate, i.e., the current density, of the zinc during discharge in a cell and is caused by a solution-precipitation reaction. While most of the zinc particles are discharged, the core of undischarged zinc is covered with these two oxides. At low drain rates, the distribution of ZnO is uniformly distributed within the anode, while at high drain rates, the ZnO mainly forms near the separator.[6,7]

## 12A.3   CELL DESIGNS AND CONSTRUCTION

### 12A.3.1   Components and Materials

*Cathode Components.*    The typical composition of an alkaline cathode and its functions are listed in Table 12A.3. The cathode is basically a mixture of manganese dioxide and carbon (typically graphite) and binder (typically Portland cement or polymers), along with cell electrolyte. However, other materials, such as other binders, additional water, and/or more electrolyte solution, may also be present.

**TABLE 12A.3**    Composition of Typical Alkaline Cell Cathode

| Component | Percent of cathode (%) | Function |
|---|---|---|
| Manganese dioxide | 80–90 | Active material |
| Carbon | 2–10 | Electronic conductor |
| KOH | 7–10 | Ionic conductor |
| Binder | 0–1 | Maintain cathode integrity |

*Manganese Dioxide.*    The manganese dioxide is the positive electrode of the cell or the oxidizing component. It must be highly active and very pure as it basically determines the battery's OCV and shape of the discharge curve during use.

The production of high-quality EMD involves multiple prolonged steps. Natural manganese dioxide deposits are mined and then calcined to form MnO. The MnO is then dissolved in sulfuric acid to produce a manganese sulfate solution. The solution goes through an electrolyte purification step to remove most of the deleterious heavy metal impurities, e.g., Fe, Cu, Co, Ni, Mo, and Cr. This purified solution is then placed in an electrolysis cell and heated to near boiling. The typical electrolysis cell of most EMD manufacturers consists of a titanium anode and copper cathode. However, in the past, lead and graphite cathodes had also been employed. The anode reaction to deposit solid $MnO_2$ alkaline cell product is as follows:

$$Mn^2 + 2H_2O \rightarrow MnO_2 + 4H^+ + 2e \qquad (12A.9)$$

Hydrogen is formed at the cathode according to reaction (12A.10):

$$2H^+ + 2e \rightarrow H_2 \uparrow \qquad (12A.10)$$

Thus, the overall reaction for the plating of EMD is

$$MnSO_4 + 2H_2O \rightarrow MnO_2 + H_2SO_4 + H_2 \qquad (12A.11)$$

The important plating variables to produce an EMD suitable for use in alkaline cells require precise control of the bath's temperature, current density, and component concentrations. Once the EMD is removed from the anode, it is crushed, washed, ground, and dried. Each battery manufacturer has its own EMD specification, so one type does not fit all requirements.

The analysis of a typical EMD is shown in Table 12A.4. The low levels of impurities are essential in order to minimize the hydrogen gassing at the anode if these elements become soluble and diffuse to the anode. The other listed parameters are also important, and their listed ranges all go toward providing an EMD suitable for alkaline cell use.

**TABLE 12A.4**    Typical Analysis of Electrolytic Manganese Dioxide (EMD)

| Component | Typical value* | Component | Typical value* |
|---|---|---|---|
| $MnO_2$ content | >91% | Ti | <5 ppm |
| Mn | >60% | Cr | <7 ppm |
| Peroxidation | >95% | Ni | <4 ppm |
| $H_2O$, 120°C | <1.50% | Co | <2 ppm |
| $H_2O$, 120–400°C | >3.0% | Cu | <4 ppm |
| Real density | 4.45 g/cm³ | V | <2 ppm |
| K | <300 ppm | Mo | <1 ppm |
| Na | <4000 ppm | As | <1 ppm |
| Mg | <500 ppm | Sb | <1 ppm |
| Fe | <100 ppm | Pb | <100 ppm |
| C | 0.07% | $SO_4^{2-}$ | ≤0.85% |

*Based on analyses of typical alkaline-grade EMD.

Other important EMD characteristics include surface area and hardness. The surface area, dictated by the porosity and particle size distribution, determines the current density in the cathode and is especially important for high-rate discharge applications. EMD is typically a very hard material, and this hardness affects the milling of the EMD and tool wear of the equipment used to make the cathode mixes and molding equipment. Premature tool and mill wear can introduce iron impurities into the pure EMD and add cost to the overall battery manufacturing process.

*Carbon.*    EMD is a relatively poor conductor in its undischarged state and even worse when partially discharged. To overcome this problem, carbon, typically in the form of graphite, is added to the cathode mix to enhance its overall electronic conductivity. The graphite provides a conductive matrix so the electrons can be evenly distributed throughout the cathode, thus lowering the overall current density in the cathode. However, a balance between the amount of added carbon and the EMD level must be maintained. Additional carbon provides a more conductive cathode matrix, but it reduces the amount of active material in the cathode. Therefore, the ratio of carbon to EMD in the cathode needs to be optimized for the required applications of the battery. Over the years, many changes in the type of carbon added to the alkaline cell cathode have occurred. Natural graphite, synthetic graphite, acetylene black, and, most recently, expanded graphite have been used to improve the cathode conductivity. In all cases, this conductor must be pure to avoid addition of any more impurities to the cell. The expanded graphite allows less carbon to be used as its synthesis expands the graphite planes while maintaining its conductivity within the carbon planes.[8] This graphite has a higher liquid absorption value, and the particle size can be optimized for the required cathode formulation.

*Other Components.*    Potassium hydroxide and water are used to form the cathode electrolyte and are added during the mixing of the cathode ingredients to form a moist paste, making the cathode mix easier to handle and mold. Depending on the battery manufacturer, other ingredients, such as binders and additives, are used to produce a dense and stable cathode with good electronic and ionic conductivity. The battery must also perform efficiently under a variety of discharge conditions, including low and high continuous and intermittent discharge over a wide range of temperatures.

**Anode Components.**    The anode is composed of a mixture of ingredients that allow for good cell performance and provide for easy manufacturing. The typical composition of an alkaline anode is listed in Table 12A.5.

**TABLE 12A.5**    Typical Composition of Alkaline Cell Anode

| Component | Range (%) | Function |
|---|---|---|
| Zinc powder | 60–70 | Negative electrode material |
| Aqueous KOH (25–50%) | 25–35 | Ionic conductor |
| Gelling agent | 0.4–1.0 | Control viscosity |
| ZnO | 0–2 | Zinc plating; gassing suppressor |
| Surfactant/gassing inhibitor | 0–0.1 | Gassing suppressor; improves performance |

*Zinc Powder.*    Zinc is the electrochemically active component of the alkaline cell's negative electrode. High-purity zinc is required for use in alkaline cells and is commercially obtained either by a thermal distillation process (called thermal zinc) or by electrolytic deposition from an aqueous solution, i.e., electrolytic zinc. This zinc is converted to a powder by atomizing a thin stream of the molten metal with high-pressure air. Depending on the setup and requirements, the particle shape of this zinc can range from "potatoes" to "dog bones." Improvements in the process have allowed zinc manufacturers to better control the size and shape of the final zinc powder in order to better meet increasing demands for performance improvements and cost savings. Typical battery-grade zinc ranges in particle size from 20 to 500 μm with a log-normal distribution. Alloying elements are added to the pure zinc to better control the normal gassing that does occur in a basic electrolyte. Such metallic additives can include indium, lead, bismuth, and aluminum in varying ratios. Such additives have become very important since mercury addition to the anode has been banned. A typical analysis of battery-grade zinc is listed in Table 12A.6. While typical levels are shown, some battery-grade zincs have lower levels of impurities.

Recent research in developing a zinc powder for high-rate discharge applications has resulted in the patenting of a blended zinc powder.[9] This blended powder contains selected portions of two different particle size powder distributions. Advantageously, it allows battery manufacturers to maximize an alkaline cell's performance while minimizing the cost of the zinc.

*Anode Gel.*    The anode gel serves to suspend the zinc particles and to maintain them in contact with one another. Starch or cellulosic derivatives, polyacrylates, and/or ethylene maleic anhydride copolymers can be used as anode gelling agents. Common gelling agents can include sodium carboxymethyl cellulose or the sodium salt

**TABLE 12A.6**   Typical Impurity Analysis of Battery-Grade Zinc Powder

| Element | Typical level (ppm)* |
|---------|---------------------|
| Ag | 1.56 |
| Al | .14 |
| As | .01 |
| Ca | .20 |
| Cd | 4.2 |
| Co | .05 |
| Cu | 1.5 |
| Cr | .10 |
| Fe | 4.0 |
| Ni | .20 |
| Mg | .03 |
| Mo | .035 |
| Sb | .09 |
| Si | .20 |
| Sn | .10 |
| V | .001 |

*Based on analyses of typical alkaline-grade zinc powder.

of an acrylic acid copolymer. Typically, the selected gel is fully mixed with the zinc powder and any other additives prior to dispensing into the anode cavity of the cell. As with the other cell components, these materials must also be of high purity in order to minimize gassing. This is especially true of the carbonate, chloride, and iron levels. Depending on the cell's primary application, the volume fraction range of the gelling agent can vary. The lower limit is based on maintaining good electronic conductivity in the anode, while the upper limit is defined by limiting the accumulation of reaction products that could eventually passivate the undischarged zinc and hinder ionic diffusion within the anode. A prior patent has suggested the use of the cross-linked polymer polyvinylbenzyltri-methylammoniumhydroxide.[10] It is claimed that the use of this gelling agent allows better high-rate discharge performance.

*Anode Collector.*   The anode collector used in the alkaline cell has typically been a high-purity cartridge brass, but silicon bronzes have also been used. The collector in most current designs has a pin or nail shape, but strip type collectors have been used in the past. The collector is part of the collector assembly that also consists of the seal and cover. Once the collector is inserted into the anode gel, its surface becomes rapidly coated with zinc, thus acting more like a zinc electrode than brass. This provides good electronic contact with the zinc particles and suppresses gassing in the anode due to any impurities in the brass that could be gassing promoters. In order to provide a rapid zinc plating, the brass collector can undergo a special cleaning or surface coating. One such patented method involves the electroplating of the collector wire with indium that forms an indium-plated wire with a thickness of about 0.1 to 10 μm.[11] This coating reduces the amount of gassing that may occur in the cell, especially in mercury-free alkaline cells, prior to the collector becoming fully coated with zinc.

### Miscellaneous Components

*Separators.*   The separator insulates the cathode electronically from the anode. However, it must remain ionically conductive and be chemically stable in the concentrated alkaline electrolyte under both oxidizing and reducing conditions while also being strong, flexible, uniform, impurity-free, and very absorptive. There are many ways to produce such a material, but the more frequently used material is a nonwoven or felt-like material.

The typical separator material can consist of cellulose, vinyl polymers, polyolefins, or combinations. Depending on the battery manufacturer, the separator can consist of two inserted cross strips or a preformed "convolute" separator basket. Other types of separators that have not proven as successful include gelled, inorganic, and radiation-grafted separators. A cellophane separator has also been used typically if there is a concern about zinc dendrite growth through the separator. A prior patent claimed the use of a reinforced separator that can withstand the forces applied during manufacture and contain any fragmented electrode particles formed when the cell is dropped.[12]

*Containers, Seals, and Finish.*    The can or external container of the alkaline cell, unlike the carbon-zinc can, does not take part in the discharge reaction and merely functions as an inert container that provides an external contact for the positive electrode. The can is typically made of mild steel that is thick enough to maintain its shape during discharge as the cathode is known to expand and hydrogen gas can form during storage or discharge, creating internal pressure. Over the years, the can thickness has been reduced in order to provide more internal space for the active battery materials.[13] The can is formed by the deep drawing of a steel strip.

The can materials must also be of high purity as the can does contact the cathode. Depending on the cell construction, the interior contact can be the steel itself or it can be treated to improve the cathode-to-can contact. Such a treatment of the steel could be a nickel plating or coating with a conductive carbon paint. Such treatments are typically used only for high-rate applications.

The seal is typically a plastic material, such as nylon or polypropylene and is combined with several metal parts, including the brass collector and cover, so as to form the collector assembly. After crimping in place, closing off the open end of the cylindrical can, the seal will prevent leakage of the electrolyte from the cell and provide electrical insulation between the can (positive electrode) and the anode (negative electrode).

The remaining components of the alkaline cell include the label and metal disks, placed at either end of the cell to provide the negative and positive electrode contacts. The label on most current alkaline cell designs is a plastic label that is heat-shrunk onto the can (another recent design feature that allows larger can diameter and greater discharge capacity).

*Miniature Cell.*    The components that make up the miniature cell are essentially the same as those of the cylindrical cell, but scaled down. The can that contains the molded cathode pellet is typically made of mild steel plated on both sides with nickel. Some can designs are even made of a tri-clad metal material. The seal is a thin plastic gasket, and the separator is the typical nonwoven material. The anode cup, which contains the anode gel, is then pressed into the seal to complete the cell. The outside of the cell contains the manufacturer's logo and cell-size identification (see Fig. 12A.5 for further details).

## 12A.3.2   Cell Types

*Cylindrical Design.*    Figure 12A.1 showed a partial cutaway drawing of a typical alkaline battery that is representative of most alkaline batteries currently being produced. Figure 12A.3 shows one method for assembling the battery. The cylindrical steel can is the container for the cell and serves as the current collector for the cathode. The cathode mix can be added to the can in two ways, depending on the battery manufacturer's equipment setup. In one process, a prescribed amount of compacted mixture of manganese dioxide, carbon, and other additives are added to the can. The cathode is then directly molded into the can by inserting a ram down the center of the mix, compacting the cathode to the required density and providing good can-to-cathode contact. In the second method, and currently the most common one, the cathode mix is first formed into cylindrical rings outside of the cell, then three or four such rings are inserted into the can. Next, separator strips are inserted into the hollow cavity of the rings using either the two-strip method or a convolute separator. Once the separator is in place, the required amount of anode gel is added based on precise calculations, whereby the fill capacity of the anode is close to or slightly greater than that of the cathodes. This design ratio prevents excessive gassing if and when the cell is deeply discharged. Next, the collector assembly is added and the can top is crimped to provide a leak-proof seal. The brass collector pin now provides the external contact for the negative electrode. The cell will next have top and bottom covers added and a plastic label applied. The covers serve a twofold purpose. In addition to providing a label with the manufacturer's distinctive artwork, it also indicates the polarity of the cell. Proper battery polarity is critical for proper operation and becomes increasingly important for a device that uses multiple cells since one inserted backward could be charged by the others, causing

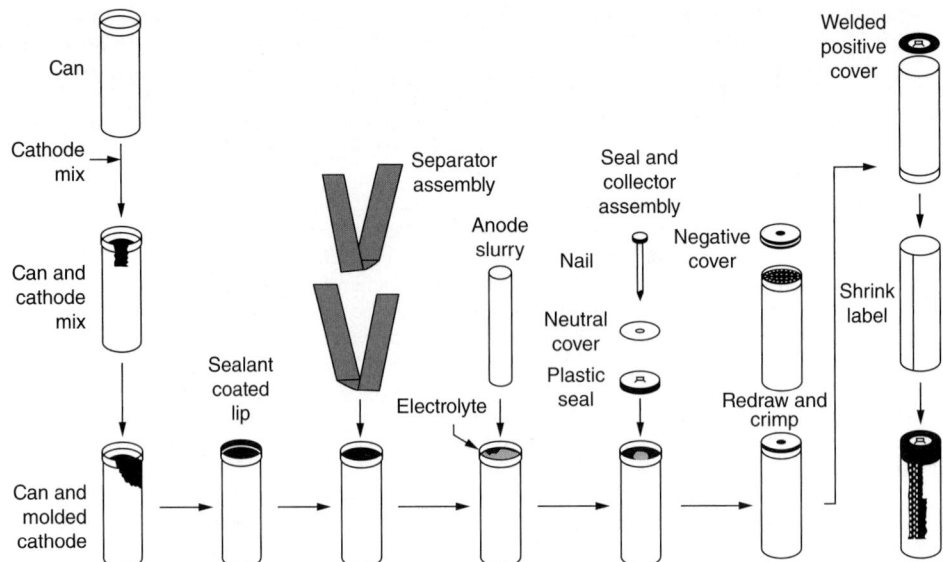

**FIGURE 12A.3**    Assembly process for cylindrical alkaline-manganese dioxide cell. (*Courtesy of Energizer, Inc.*)

the cell to prematurely leak. In order to prevent misorientation, some manufacturers have also added a reversal protection, which consists of an insulating ring on the positive end of the cell.

Figure 12A.4 shows the more typical process of alkaline cell cathode production. Molded cathode rings are inserted into the can, a ram compacts them against the can, and the cell is finished in the normal way.

In continuing attempts to increase the performance of the alkaline cell, some novel innovations regarding the anode and cathode have recently been patented. One involves the use of zinc ribbons, which allows for a significantly increased high-rate performance.[14] Another involves the formation of a cathode having lobes or a sinusoidal cathode design.[15] This increases the surface area of the cathode, thus decreasing the current density, which allows for a better high-rate performance. Another recent patent, regarding the manganese dioxide itself,

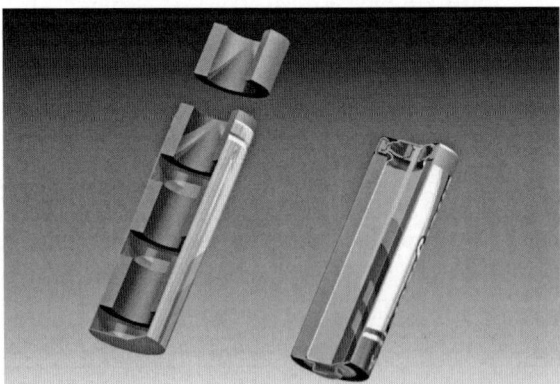

**FIGURE 12A.4**    Illustration showing insertion of four cathode rings in can and finished cell. (*Courtesy of Energizer, Inc.*)

discloses better performance using a particulate $MnO_2$ simultaneously having a micropore surface area of greater than 8.0 $m^2/g$ and BET surface area between 20 and 31 $m^2/g$.[16]

***Miniature Cell Configuration.*** A cross-sectional view of a miniature alkaline-manganese dioxide cell is shown in Fig. 12A.5. This button cell construction is similar to that of the cylindrical cell. There is a bottom cup that holds the cathode pellet, followed by round disks of separator paper, the plastic seal, and the top cover containing the anode mix. The top cover is pressed into the plastic seal, providing a leak-proof seal to prevent leakage.

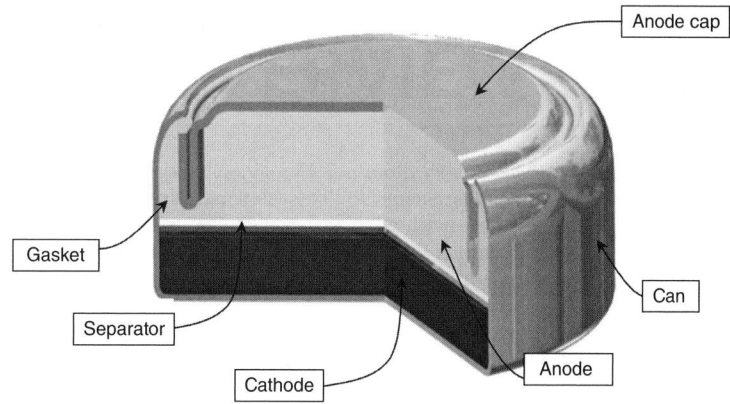

**FIGURE 12A.5** Cross-section illustration of miniature alkaline-manganese dioxide battery. (*Courtesy of Energizer, Inc.*)

The five common sizes of miniature alkaline cells and their dimensions are listed in Table 12A.7.

**TABLE 12A.7** Dimensions of Miniature Alkaline Cells

| | | Dimensions (mm) | | | | | |
|---|---|---|---|---|---|---|---|
| | | A/B | | M | N | Φ | |
| Designation | Voltage | Max. | Min. | Min. | Min. | Max. | Min. |
| LR41 | 1.5 | 3.6 | 3.3 | 3.0 | 3.8 | 7.9 | 7.55 |
| LR55 | 1.5 | 2.1 | 1.85 | 3.8 | 3.8 | | |
| LR54 | 1.5 | 3.05 | 2.75 | 3.8 | 3.8 | | |
| LR43 | 1.5 | 4.2 | 3.8 | 3.8 | 3.8 | | |
| LR44 | 1.5 | 5.4 | 5.0 | 3.8 | 3.8 | | |

A/B: cell height; *M*: diameter of flat negative contact; *N*: diameter of flat positive contact; Φ: diameter of the cell.
*Source: IEC Int'l. Standard Part 2: Primary Batteries*, 11th edition, 2006.

## 12A.4  PERFORMANCE CHARACTERISTICS

### 12A.4.1  Battery Types and Sizes

Alkaline-manganese dioxide cells are made in a wide variety of sizes, dictated by the many different portable devices and their discharge characteristics. Figure 12A.6 shows the cylindrical cell sizes currently using the alkaline-manganese dioxide chemistry. While most cans are quite common and easily recognized, there are several

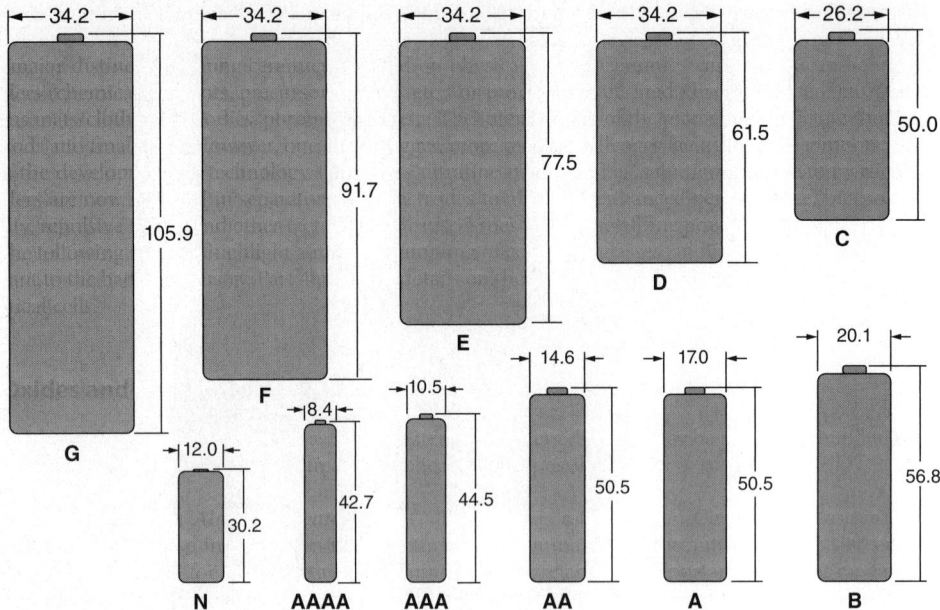

**FIGURE 12A.6**  Typical alkaline cell sizes and their dimensions in millimeters.

that are not so common, e.g., the A, B, F, and G cell sizes. As an example, the B size is commonly used in bicycle lights in Europe, while the F has been used in lantern batteries.

Recognizing the need for smaller, thinner, and lighter device applications, the AAAA (4A) premium alkaline battery has been commercially introduced. While initially having limited distribution, it has become more popular with increased sales of smaller, more portable electronic devices. The AAAA cell is even thinner than AAA cells with applications that include Bluetooth™ devices, flash audio players, remote controls, and noise-canceling headsets.

## 12A.4.2  Testing Standards

With the many different battery sizes and the significant number of battery manufacturing companies worldwide, it is important to establish standard tests and designation codes for cylindrical and miniature primary cells so they can be easily compared to actual devices that they would power. This has been accomplished by ANSI. The current testing standards are outlined in their current publication, *ANSI C18.1M, Part 1-2008, American National Standard for Portable Primary Cells and Batteries with Aqueous Electrolyte—General and Specifications.* For a brief history of the standardization of these cells, see Ref. 17. As the number of new devices being commercialized increases, this organization meets regularly to determine if current testing requirements need to be updated to reflect the latest device requirements. In addition to standard cell designations (D, C, AA, etc.), battery manufacturers also use their own markings that can be found on the battery and its packaging. More information about the individual manufacturer's battery specifications and performance is available on their individual websites.[18] The IEC and ANSI also have their own designations. Table 12A.8 lists some of the current tests that several of the more common cylindrical alkaline batteries undergo for advertising purposes, along with their different codes.

## 12A.4.3  Cell Leakage

Commercial alkaline cells are of a sealed construction using a compressible polymer grommet (seal) between the can and the top current collector. Leakage can occur by two different mechanisms: (1) nonelectrochemical leakage resulting from manufacturing defects and/or poor cell design that can occur at either electrode

**TABLE 12A.8**  Designations and Typical ANSI Tests for Cylindrical Alkaline Cells

| Size | IEC designation | ANSI designation | Test | Load | Cycle | End voltage (V) | Min. avg. duration |
|------|-----------------|------------------|------|------|-------|-----------------|--------------------|
| D | LR20 | 13A | Portable stereo | 600 ohms | 2 h on, 22 h off | 0.9 | 11 h |
|   |      |     | Portable lighting | 2.2 ohms | 4 min on, 56 min off; 8 h on, 16 h off cycle | 0.9 | 15.8 h |
|   |      |     | Toy | 2.2 ohms | 1 h on, 23 h off | 0.9 | 17.5 h |
|   |      |     | Radio | 10 ohms | 4 h on, 20 h off | 0.9 | 90 h |
| C | LR14 | 14A | Portable stereo | 400 mA | 2 h on, 22 h off | 0.9 | 8 h |
|   |      |     | Portable lighting | 3.9 ohms | 4 min on, 56 min off; 8 h on, 16 h off cycle | 0.9 | 13.8 h |
|   |      |     | Toy | 3.9 ohms | 1 h on, 23 h off | 0.8 | 14.5 |
|   |      |     | Radio | 20 ohms | 4 h on, 20 h off | 0.9 | 85 h |
| AA | LR6 | 15A | Digital camera | 1500 mW, 650 mW | 1st load for 2 s, then 2nd load for 28 s; 5 min on, 55 min off | 1.05 | 50 pulses |
|   |      |     | Toothbrush | 500 mA | 2 min on, 13 min off; 24 hours | 0.8 | 2.5 h |
|   |      |     | CD | 250 mA | 1 h on, 23 h off | 0.9 | 6 h |
|   |      |     | Toy | 3.9 ohms | 1 h on, 23 h off | 0.8 | 5 h |
|   |      |     | Remote control | 24 ohms | 15 s on, 45 s off; 8 h on, 16 h off | 1.0 | 33 h |
|   |      |     | Radio/clock | 43 ohms | 4 h on, 20 h off | 0.9 | 60 h |
| AAA | LR03 | 24A | Photoflash | 600 mA | 10 s on, 50 s off; 1 h on, 23 h off | 0.9 | 170 pulses |
|   |      |     | Portable lighting | 5.1 ohms | 4 min on, 56 min off; 8 h on, 16 h off cycle | 0.9 | 2.2 h |
|   |      |     | Digital audio | 100 mA | 1 h on, 23 h off | 0.9 | 7.5 h |
|   |      |     | Remote control | 24 ohms | 15 s on, 45 s off; 8 h on, 16 h off | 1.0 | 14.5 h |
| AAAA | LR8 | 25A | Lighting | 5.1 ohms | 5 min on, 23 h and 55 min off | 0.9 | 1.3 h |
|   |      |     | Laser pointer | 75 ohms | 1 h on, 23 h off | 1.1 | 22 h |
| N | LR1 | 910A | Portable lighting | 5.1 ohms | 5 min on, 23 h and 55 min off | 0.9 | 1.6 h |
|   |      |     | Pager | 10 then 3K ohms | 1st load for 5 s, then 2nd load for 3595 s; 24 h | 0.9 | 888 h |

(called either positive or negative leakage, depending on whether it occurs between the seal and can or at the collector, respectively), and (2) electrochemical-related leakage associated with leakage that occurs only at the negative electrode.

Most manufacturers use a surface active coating such as asphalt, polyimide, etc., in the seal area to smooth out surface imperfections in the cell parts to provide a better seal. The amount of seal compression should not exceed the elastic limit of the grommet. Details of the cell design, choice of grommet material, as well as manufacturing processes play a role in cell leakage. Factors that influence leakage include the interior surface of the can (e.g., nickel, stainless steel, or gold), the choice of polymer gasket material, the choice of the anode gelling agent, and possible damage to the seal area during cell manufacture.

Electrochemical-related leakage occurs only at the negative electrode. Here, leakage, or seepage of the electrolyte, occurs on the negative electrode of the cell and is evident by "salting" or the appearance of white crystals at the negative terminal only. As shown in Fig. 12A.7, oxygen ingress into the seal area is reduced to OH⁻ at the air-electrolyte interface, increasing the local concentration of OH⁻ in the seal area. The reaction of oxygen with the zinc anode has the same effect. The increased concentration between the seepage film compared to the concentration of the bulk cell electrolyte gives rise to an osmotic pressure difference. This osmotic force can reach several atmospheres and will force electrolyte into the seal area under high pressure. Grommet materials with a high water absorption, such as nylon, readily transmit water into the reaction zone in the seal area as opposed to a vinylidine chloride-acrylonitrile (Saran), which has a low water transmission rate. The choice of gelling agent in the zinc electrode may also play a role by improving the wetting, which facilitates movement of water into the reaction zone. Cells stored in dry nitrogen have lower leakage rates than do cells with polished can walls.

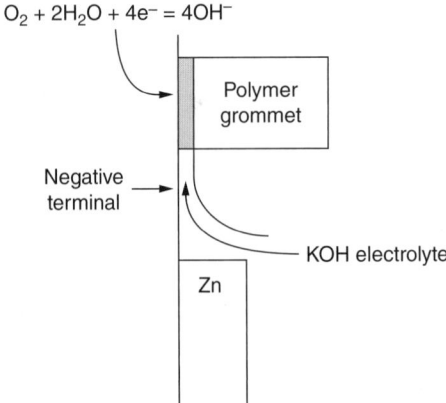

$$O_2 + 2H_2O + 4e^- = 4OH^-$$

Polymer grommet

Negative terminal

KOH electrolyte

Zn

**FIGURE 12A.7**  Illustration of possible leakage path in a cylindrical alkaline cell.

The moist environment between the can wall and the grommet forms a reaction zone for the reduction of oxygen to produce OH⁻ ions. The increase in the local hydroxide concentration causes osmotic pressure between the bulk electrolyte and the electrolyte in the reaction zone, which drives the cell electrolyte out through the seal area where it reacts with $CO_2$ in the atmosphere to produce the white precipitate, potassium carbonate, e.g., $K_2CO_3$.[19,20] This situation is particularly disconcerting as it provides an immediate visual clue that such a cell is defective.

### 12A.4.4   Evolta™ and Oxyride™ Batteries

Competition in the battery business has always been very keen. Most improvements by alkaline battery manufacturers have come in the form of improved active materials, an increased internal volume, and/or reduced internal resistance. These improvements in Ah capacity, longer storage life, and high-rate discharge have not been evident as the outside dimensions have not changed over the years. Significantly different alkaline-based batteries are seldom introduced. However, in 2004 when Panasonic first introduced their Oxyride battery to the Japanese market, they claimed an impressive improvement over the current alkaline-manganese dioxide battery for use in high-power applications. Panasonic literature states that this cell includes a finer grained graphite and manganese dioxide, allowing for a denser fill of material. The positive electrode also contains nickel oxyhydroxide, allowing the battery to maintain a higher operating voltage. These cells also utilize a "vacuum-pouring" technology during the production process, allowing more electrolyte to be packed into each battery for increased durability. This same chemistry, used in the Duracell PowerPix battery, provides enhanced power capabilities for digital devices but at more than double the cost of normal alkaline-manganese dioxide cells. These cells have a higher open-circuit voltage of 1.7 V compared to the alkaline-manganese dioxide cell's typical open-circuit voltage of 1.60 to 1.65 V. This higher voltage could cause problems, especially with devices containing an incandescent light bulb or no voltage regulator.

Panasonic launched their new AA Evolta product in Japan in April 2008 and in the United States in May 2008. This cell claimed to be the *Guinness Book of World Records'* "longest-lasting AA alkaline battery cell" along with having a reported shelf life of 10 years. The increased performance derives from the addition of titanium oxyhydroxide to the cathode, a thinner can, thinner seal, and increased collector length. The discharge performance of this cell can be seen by visiting Panasonic's website.[18]

## REFERENCES

1. The Freedonia Group, Inc., https://www.freedoniagroup.com/World-Batteries.html, extracted from the World Wide Web on October 3, 2017 (exported tables generated/forwarded by The Freedonia Group on 10/4/2017 from IP address 207.89.36.85. © MarketResearch.com, Inc., 2000–2017.) Battery Markets in the United States by Application and End User, published December 2016 by The Freedonia Group.

2. A. Kozawa and R.A. Powers, *J. Chem. Educ.* **49:**587 (1972).

3. R. Burns and V. Burns, *Manganese Dioxide Symposium,* Vol. 1, Cleveland, p. 306, 1975.

4. Y. Chabre and J. Pannetier, *Prog. Solid St. Chem.* **23:**12 (1995).

5. D. M. Holton, et al., in *Proc. 14th International Power Sources Symposium*, Brighton, England, Pergamon, NY, 1984.

6. Q. C. Horn and Y. Shao-Horn, *J. Electrochem. Soc.* **150**(5):A652 (2003).

7. R. W. Powers and M. Brieter, *J. Electrochem. Soc.* **116:**1652 (1952).

8. J. C. Nardi, U.S. Patent 6,828,064, December 7, 2004.

9. D. Fan, U.S. Patent 7,364,819, April 29, 2008.

10. C. Robert, U.S. Patent 6,916,577, July 14, 2005.

11. D. Mihara, U.S. Patent 5,622,612, April 22, 1997.

12. R. Janmey, U.S. Patent 6,828,061, December 7, 2004.

13. R. Ray, U.S. Patent 6,855,454, February 15, 2005.

14. N. C. Tang, U.S. Patent 6,221,527, April 24, 2001.

15. P. J. Slezak, U.S. Patent 6,869,727, March 22, 2005.

16. S. Davis, U.S. Patent 6,863,876, March 8, 2005.

17. ANSIC18 Committee Doc. 18/382/DOC/, November 21, 2002.

18. www.energizer.com; www.duracell.com; www.rayovac.com; www.sanyo.com; www.panasonic.com; www.varta.com.

19. M. N. Hull and H. I. James, *J. Electrochem. Soc.* **124:**332 (1977).

20. S. M. Davis and M. N. Hull, *J. Electrochem. Soc.* **125:**1918 (1978).

# SECTION B

# MERCURIC OXIDE

**Nathan D. (Ned) Isaacs**

## 12B.1   CELL OVERVIEW

Although no longer commerciality viable, the mercury cell has played a major role in the advancement of electronic technology.

## 12B.2   ELECTROCHEMISTRY

The generally accepted basic cell reaction for the zinc/mercuric oxide cell is

$$Zn + HgO \rightarrow ZnO + Hg$$

For the overall reaction, $\Delta G^0 = 259.7$ kJ. This gives a thermodynamic value for $E^0$ at 25°C of 1.35 V, which is in good agreement with the observed values of 1.34 to 1.36 V for the open-circuit voltage of commercial cells.[1] Based on this equation, 1 g of zinc provides 819 mAh and 1 g of mercuric oxide provides 247 mAh.

Some types of zinc/mercuric oxide cells exhibit open-circuit voltages between 1.40 and 1.55 V. These cells contain a small percentage of manganese dioxide in the cathode and are used where voltage stability is not of major importance for the application.

The basic cell reaction for the cadmium/mercuric oxide cell is

$$Cd + HgO + H_2O \rightarrow Cd(OH)_2 + Hg$$

For the overall reaction, $\Delta G^0 = -174.8$ kJ. This gives a thermodynamic value for $E^0$ at 25°C of 0.91 V, which is in good agreement with the observed values of 0.89 to 0.93 V. From the basic reaction it can be calculated that 1 g of cadmium should provide 477 mAh.

## 12B.3   CELL DESIGNS AND CONSTRUCTIONS

### 12B.3.1   Components

*Electrolyte.*   Two types of alkaline electrolyte can be used in the zinc/mercuric oxide cell: potassium hydroxide or sodium hydroxide. Both of these basic compounds are very soluble in water, allowing highly concentrated solutions to be used. Zinc oxide will also dissolve to a degree in the solution, thus helping suppress hydrogen generation.

Potassium hydroxide electrolytes generally contain between 30% and 45% w/w KOH and up to 7% w/w zinc oxide. For proper low temperature operation, the concentrations of both potassium hydroxide and the zinc oxide must be reduced, but some instability (hydrogen gas generation) will then likely result at higher temperatures.

Sodium hydroxide electrolytes do not function well at low temperatures or at heavier current drains. Typical sodium hydroxide electrolytes employ similar concentration ranges. Sodium hydroxide electrolytes are best suited for use in slow rate discharge applications because of the reduced tendency of this electrolyte to seep out of the cell seal over an extended period of time.

Generally, cadmium/mercuric oxide cells are best constructed with potassium-based alkaline electrolytes only. As cadmium is practically insoluble in all concentrations of aqueous potassium hydroxide solutions, the electrolyte allows optimization for low-temperature operation.

The freezing-point curve for caustic potash solutions is shown in Fig. 12B.1. The eutectic freezing point is below −60°C at 31% w/w KOH. Improvements in low-temperature performance are possible by the addition of a small percentage of cesium hydroxide to the electrolyte.

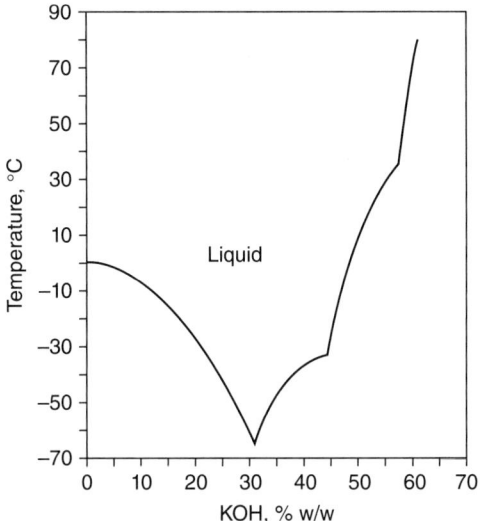

**FIGURE 12B.1**    Freezing-point curve for aqueous caustic potash solutions.

***Zinc Anode.***    Alkaline electrolytes act as ionic carriers in the cell reactions. The reaction at the zinc negative electrode may be written as

$$Zn + 4OH^- \rightarrow Zn(OH)_4^{2-} + 2e$$

$$Zn(OH)_4^{2-} \rightarrow ZnO + 2OH^- + H_2O$$

These reactions imply the dissolution of the zinc electrode, with the crystallization of zinc oxide from the electrolyte. The reaction at the anode can be simplified to

$$Zn + 2OH^- \rightarrow ZnO + H_2O + 2e$$

Dissolution of the zinc electrode in alkaline solutions on open circuit is minimized by adding zinc oxide in the electrolyte and amalgamating the zinc in the electrode. Mercury in the range of 5% to 15% w/w prevents zinc from dissolving. Impurity levels in the zinc must be minimized since minor cathodic inclusions in the electrode can increase the hydrogen generation reaction despite the precautions indicated.[2,3]

***Cadmium Anode.***    The reaction at the anode is

$$Cd + 2OH^- \rightarrow Cd(OH)_2 + 2e$$

Accordingly, water is depleted during discharge, necessitating the use of an adequate quantity of electrolyte in the cell and the desirability for high water ratios in the electrolyte. Cadmium has a high hydrogen overvoltage in the electrolyte, making the use of an amalgamation irrelevant since the electrode potential is some 400 mV less electropositive than zinc.

Conventional cadmium metal powders lack sufficient activity, requiring cadmium anodes produced by (1) electroforming the anode, (2) electroforming powder by a special process followed by pelleting, or (3) precipitating by a special process as a low-nickel alloy and pelleting. All of these processes have been used by different manufacturers to give cells with various performance parameters.[4]

*Mercuric Oxide Cathode.*    The overall cathode reaction may be written as

$$HgO + H_2O + 2e \rightarrow Hg + 2(OH)^-$$

Mercuric oxide is stable in alkaline electrolytes and has a very low solubility. Being nonconductive, graphite must be added to provide a conductive HgO cathode matrix. As the discharge proceeds, the ohmic resistance of the cathode falls and the graphite assists in the prevention of mass agglomeration of mercury droplets. Other additives that have been used to prevent agglomeration of the mercury are manganese dioxide (which also increases the cell voltage to 1.4–1.55 V), lower valence manganese oxides, and silver powder (which forms a solid-phase amalgam with the cathode product).

Graphite levels, ranging from 3% to 10%, and manganese dioxide from 2% to 30% are optimal. Silver powder, used only in special purpose cells because of cost considerations, may reach 20% of the cathode weight. Again, high-purity cathode materials are critical as trace impurities that are soluble in the electrolyte can migrate to the anode and initiate hydrogen evolution. An excess of mercuric oxide capacity of 5% to 10% is optimal to "balance" the cell and prevent hydrogen generation in the cathode at the end of discharge.

*Materials of Construction.*    Strong caustic alkali solutions are very aggressive toward cell components. External contacts require corrosion resistance, galvanic compatibility with the equipment interface, and to some degree, cosmetic appearance. Homogeneous metal parts, plated or clad metals, insulating parts (whether injection-, compression-, or transfer-molded polymers or rubbers) must be selected for chemical capability.

With the exception of the anode contact with a modified top/anode interface, the preferred construction materials are generally the same for both cadmium/mercuric oxide and zinc/mercuric oxide cells. Cellulose, cellulose derivatives, and low-melting-point polymers are not suitable. Nickel can be used on both anode and cathode parts.

## 12B.3.2  Cell Construction

The mercuric oxide batteries were manufactured in three basic structures—button, flat, and cylindrical configurations with several design variations within each configuration.

*Button Configuration.*    The button configuration of the zinc/mercuric oxide battery is shown in Fig. 12B.2. The top is copper or copper alloy on the inner face and nickel or stainless steel on the outer face. This part may also be gold plated, depending on the application. Within the top is a dispersed mass of amalgamated zinc powder (gelled anode), and the top is insulated from the can by a nylon grommet. The whole top-grommet-anode assembly presses down onto an absorbent film that contains most of the electrolyte, the remainder being dispersed in the anode and cathode. Below the absorbent is a permeable barrier, which prevents cathode material from migrating to the anode. The mercuric oxide/graphite cathode is consolidated into the can, and a sleeve support of nickel-plated steel prevents collapse of the cathode mass as the battery discharges. The can is nickel-plated steel, and the whole cell is compressed and crimped at the top edge of the can as shown in Fig. 12B.2.

The cadmium/mercuric oxide button battery uses a similar configuration.

*Flat-Pellet Configuration.*    A larger zinc/mercuric oxide battery design is shown in Fig. 12B.3. In these cells, the zinc powder is amalgamated and pressed into a pellet with sufficient porosity to allow electrolyte impregnation. A double top is used, with an integrally molded polymer grommet as a safeguard to relieve excessive gas pressures and maintain a leak-resistant structure. The outer top is of nickel-plated steel, and the inner top is nickel-plated steel but tin plated on its inner face. This cell also uses two nickel-plated steel cans with an adaptor tube between

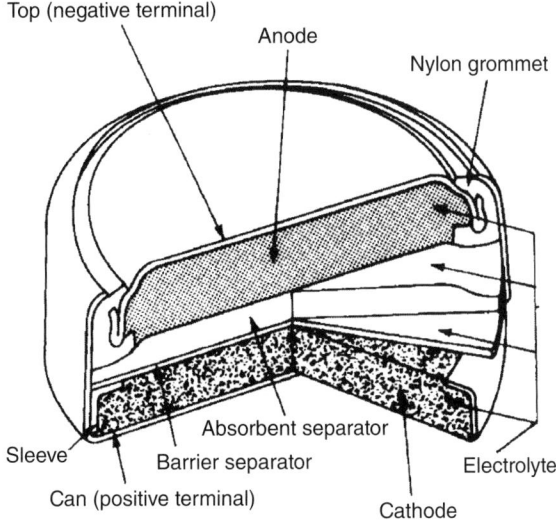

**FIGURE 12B.2**  Zinc/mercuric oxide battery—button configuration. (*Courtesy of Duracell, Inc.*)

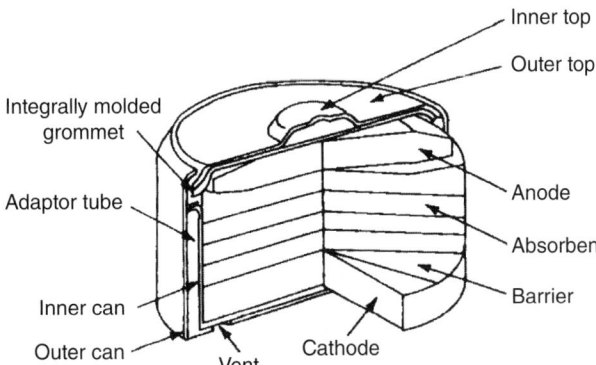

**FIGURE 12B.3**  Zinc/mercuric oxide battery—flat-pellet configuration. (*Courtesy of Duracell, Inc.*)

the two, the seal being achieved by pressing the top-grommet assembly against the inner can and crimping over the outer can. A vent hole is pierced into the outer can so that if gas is generated within the cell, it can escape between the inner and outer cans, any entrained electrolyte being absorbed by the paper adaptor tube.

***Cylindrical Configuration.***    Large cylindrical zinc/mercuric oxide batteries are constructed from annular pressings, as shown in Fig. 12B.4. The anode pellets are rigid and pressed against the cell top by the neoprene insulator slug. A number of variations of the cylindrical cell exist that use dispersed anodes, where contact with the anode is made either by a nail welded to the inner top or a spring extending from the base insulator to the top.

***Wound-Anode Configuration.***    Another design of the zinc/mercuric oxide battery that operates better at low temperatures is the wound-anode or jelly-roll structure shown in Fig. 12B.5. Structurally the cell is similar to the

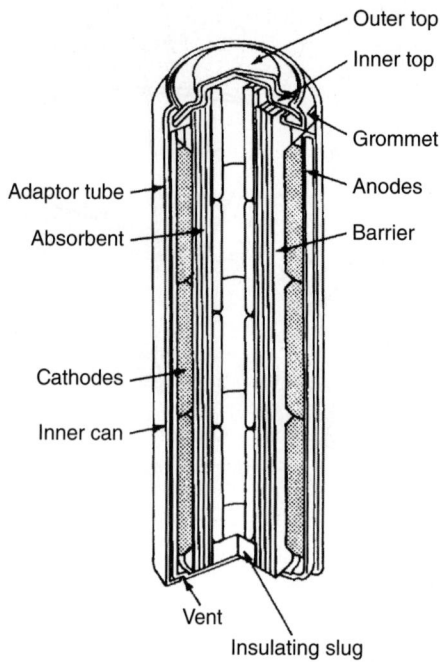

**FIGURE 12B.4**   Zinc/mercuric oxide battery—cylindrical configuration. (*Courtesy of Duracell, Inc.*)

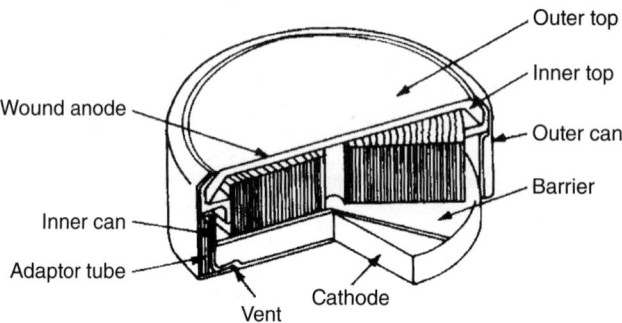

**FIGURE 12B.5**   Zinc/mercuric oxide battery—wound-anode configuration. (*Courtesy of Duracell, Inc.*)

flat cell shown in Fig. 12B.3, but the anode and absorbent have been replaced by a wound anode, which consists of a long strip of corrugated zinc interleaved with a strip of absorbent paper. The paper edge protrudes at one side and the zinc strip at the other. This provides a large surface area anode. The roll is held in a plastic sleeve and the zinc is amalgamated in situ. The paper swells in the electrolyte and forms a tight structure, which is compressed in the cell at the assembly stage with the zinc edge in contact with the top.

Electrolyte formulations can be adjusted for low-temperature operation or long storage life at elevated temperature, or as a compromise between the two. The performance is optimized by careful adjustment of the anode geometry.

*Low-Current-Drain Configurations.*    Batteries designed for operation at low current drain require modification of the structure to prevent internal electrical shorting due to formation of conductive pathways originating from both anode and cathode. After partial discharge, metallic mercury globules are particularly troublesome in this respect. The problem can be minimized by the use of silver powder in the cathode.

All potential short-circuit deposits need to be blocked to prevent premature self-discharge. In watch button cell designs, multiple barrier layers and a polymer insulator washer effectively seal off the anode from the cathode by compressing these layers against the support ring. These batteries are discharged at the 1- to 2-year rate.[5]

## 12B.4    PERFORMANCE CHARACTERISTICS

### 12B.4.1    Zinc/Mercuric Oxide Batteries

*Voltage.*    The open-circuit voltage of the zinc/mercuric oxide battery is 1.35 V. Its voltage stability under open-circuit or no-load conditions is excellent, and these batteries have been widely used for voltage reference purposes. The no-load voltage is nonlinear with respect to both time and temperature. A voltage-time curve is shown in Fig. 12B.6. The no-load voltage will remain within 1% of its initial value for several years. A voltage-temperature curve is shown in Fig. 12B.7. Temperature stability is even better than age stability. From −20 to +50°C, the total no-load voltage range is in the region of 2.5 mV.

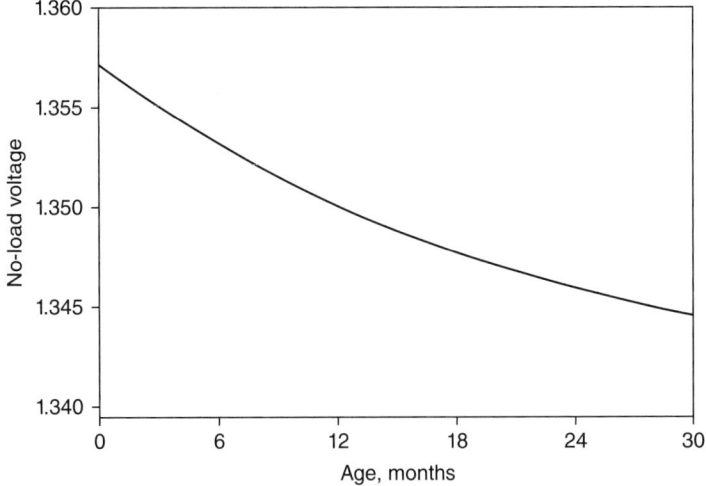

**FIGURE 12B.6**    No-load voltage versus time, zinc/mercuric oxide battery, 20°C.

*Discharge Performance.*    A flat discharge curve, characteristic of the zinc/mercuric oxide battery, is shown in Fig. 12B.8 for a pressed-powder anode battery at 20°C. The end-point voltage is generally considered to be 0.9 V, although at higher current drains the batteries may discharge usefully below this voltage. At low current drains, the discharge profile is very flat and the curve is almost "squared off."

The capacity or service of the zinc/mercuric oxide battery is about the same on either continuous or intermittent discharge regimes over the recommended current drain range, irrespective of the duty cycle.

Under overload conditions, however, a considerable shift in available capacity can be realized by the use of "rest" periods, which may increase service life considerably. Batteries designed for the low rate discharge do not have this issue unless a high-current-drain pulse is superimposed on a continuous low-drain base current.

*Effect of Temperature.*    The zinc/mercuric oxide battery is best suited for use at normal and elevated temperatures from 15 to 45°C. Discharging batteries at temperatures up to 70°C is also possible if the discharge period

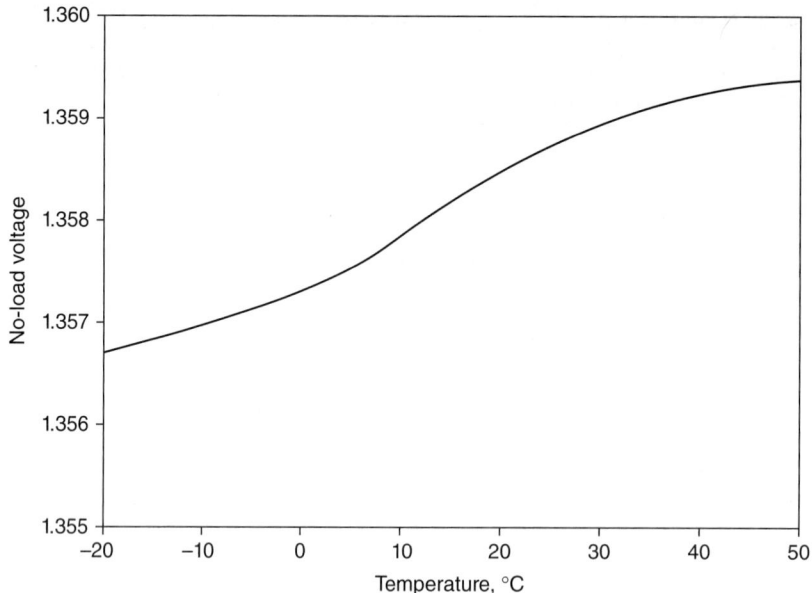

**FIGURE 12B.7**   No-load voltage versus temperature, zinc/mercuric oxide battery.

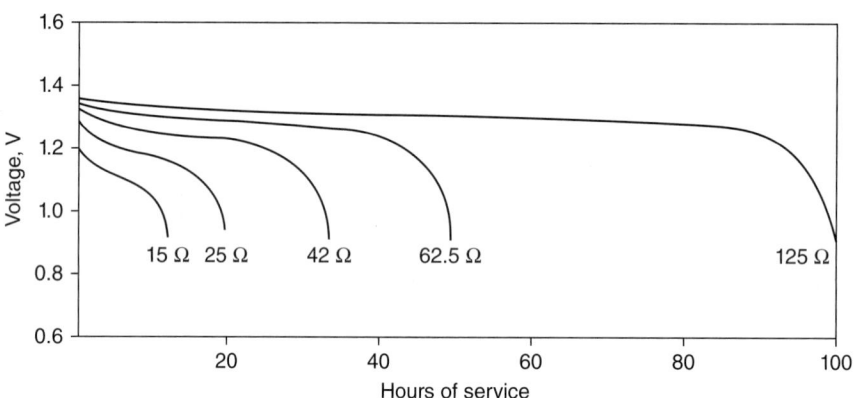

**FIGURE 12B.8**   Discharge curves, zinc/mercuric oxide battery, 1000 mAh size, 20°C.

is relatively short. The zinc/mercuric oxide battery generally does not perform well at low temperatures. Below 0°C, discharge efficiency is poor unless the current drain is low. Figure 12B.9 shows the effect of temperature on the performance of two types of zinc/mercuric oxide batteries at nominal discharge drains.

The wound-anode or "dispersed"-powder anode structures are better suited to high rates and low temperatures than the pressed-powder anode.

**Impedance.**   Impedance and voltage levels during discharge are shown in Fig. 12B.10 for a zinc/mercuric oxide button battery (measured at a frequency of 1 kHz because of their use in hearing aids).[6] However, the value obtained is frequency-dependent to some degree, particularly above 1 MHz, which must be considered.

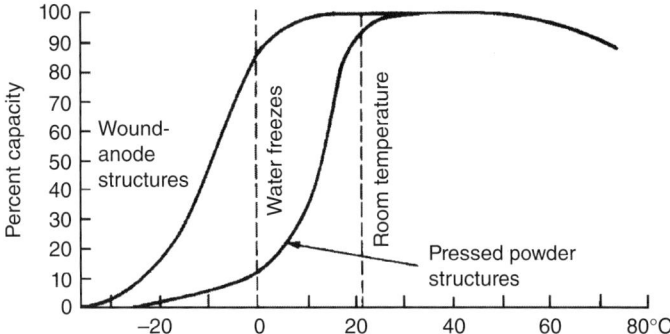

**FIGURE 12B.9**  Effect of temperature on performance of zinc/mercuric oxide batteries.

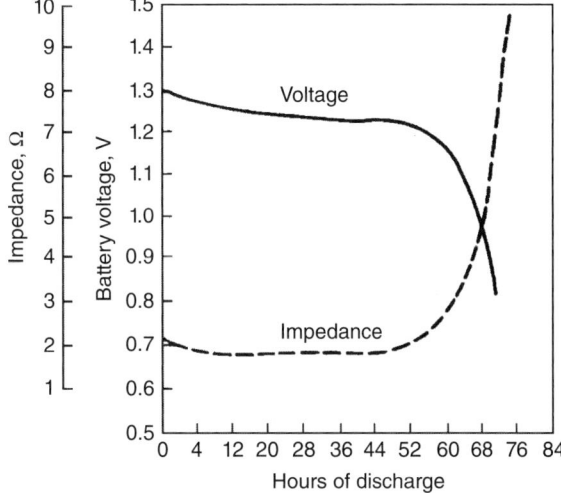

**FIGURE 12B.10**  Internal impedance, zinc/mercuric oxide battery, 350 mAh size, 20°C, 1 kHz, 250 Ω load.

***Storage.***  Zinc/mercuric oxide batteries have good storage characteristics. In general they will store for over 2 years at 20°C with a capacity loss of 10% to 20%, and 1 year at 45°C with about a 20% loss. Storage at lower temperatures, such as down to −20°C, will, as with other battery systems, increase storage life.

The storability will depend on the discharge load and also on the cell structure. Failure in storage is usually due to the breakdown of cellulosic compounds within the cell which, at first, results in a reduction of the limiting-current density at the anode. Further breakdown produces low-drain internal electrical paths and a real loss of capacity due to self-discharge. Eventually, complete self-discharge can occur, but below 20°C these processes take many years.

Long storage lives are within the capabilities of the mercuric oxide system. For example, a wound-anode cell with a noncellulosic barrier has a capacity loss around 15% over 6 years. With cells designed for long-term storage, dissolution of mercuric oxide from the cathode and its transfer to the anode become a significant factor in capacity loss.

***Service Life.***   The performance of the zinc/mercuric oxide cell at various temperatures and loads is summarized in Fig. 12B.11, based on an 800 mAh battery with a dispersed anode.

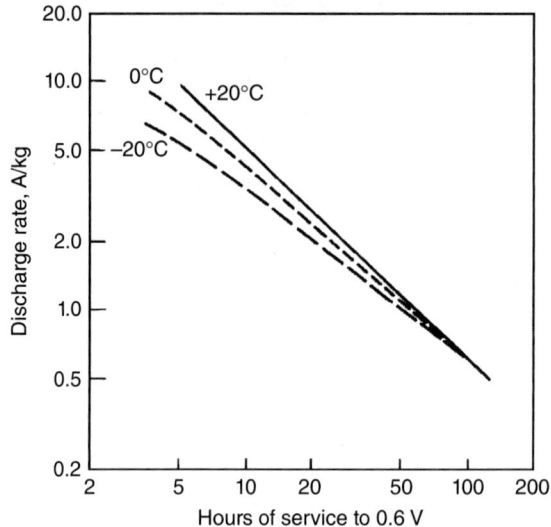

**FIGURE 12B.11**   Service life of typical zinc/mercuric oxide battery (dispersed anode) on a weight basis.

## 12B.4.2   Cadmium/Mercuric Oxide Batteries

***Discharge.***   An outstanding feature of the cadmium/mercuric oxide battery is its ability to operate over a wide temperature range. The usual operating range is from −55 to +80°C, but with the low gassing rate and high thermal stability, cell operating temperatures up to 180°C have been achieved with special designs.

Figure 12B.12 shows the discharge curves for a typical button battery at various temperatures. Excellent voltage stability and flat discharge curves are cell characteristics, but occurring at a low operating voltage (open-circuit voltage is only 0.9 V). Figure 12B.13 shows the effect of temperature on the capacity at various discharge loads. A high percentage of the 20°C capacity is available at the lower temperatures. The end-point voltage is usually taken as 0.6 V, although at higher current densities and lower temperatures more useful life can be obtained to lower end voltages.

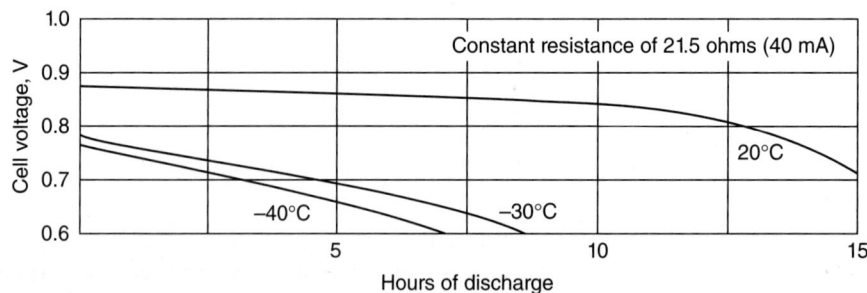

**FIGURE 12B.12**   Discharge curves—cadmium/mercuric oxide button battery (500 mAh size).

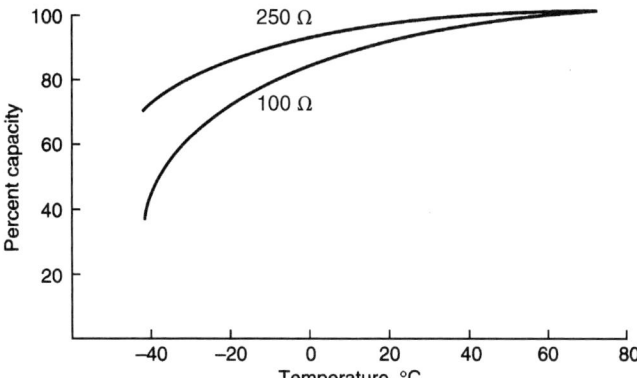

**FIGURE 12B.13** Effect of temperature on capacity of a cadmium/mercuric oxide button battery (3000 mAh size).

The performance of the cadmium/mercuric oxide battery is summarized in Fig. 12B.14, as derived from typical button cell data.

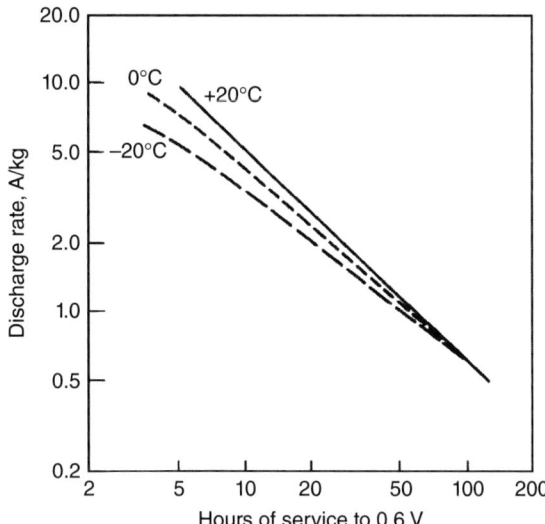

**FIGURE 12B.14** Service life of typical cadmium/mercuric oxide batteries on a weight basis.

***Storage.*** Storage life over the temperature range of −55 to +80°C is remarkably good, and if the barrier-absorbent system is designed to withstand elevated-temperature storage, the major self-discharge mechanism is by dissolution of the mercuric oxide and its transfer to the anode. A shelf life of 10 years at ambient temperatures with less than 20% capacity loss is within the capabilities of the system. Elevated-temperature storage is exceptionally good (approximately 15% loss per year at 80°C), and since neither electrode should generate hydrogen, the cells can be hermetically sealed with minimal risk of electrolyte leakage or cell distortion.[5]

## *REFERENCES*

1. P. Ruetschi, "The Electrochemical Reactions in Mercuric Oxide-Zinc Cell," in D. H. Collins (ed.), *Power Sources,* vol. 4, Oriel Press, Newcastle-upon-Tyne, England, 1973, p. 381.

2. D. P. Gregory, P. C. Jones, and D. P. Redfearn, "The Corrosion of Zinc Anodes in Aqueous Alkaline Electrolytes," *J. Electrochem. Soc.* **119**:1288 (1972).

3. T. P. Dirkse, "Passivation Studies on the Zinc Electrode," in D. H. Collins (ed.), *Power Sources,* vol. 3, Oriel Press, Newcastle-upon-Tyne, England, 1971, p. 485.

4. D. Weiss and G. Pearlman, "Characteristics of Prismatic and Button Mercuric Oxide-Cadmium Cells," *Proc. Electrochem. Soc. Meeting,* New York, October 1974.

5. P. Ruetschi, "Longest Life Alkaline Primary Cells," in J. Thompson (ed.), *Power Sources,* vol. 7, Academic, London, 1979, p. 533.

6. S. A. G. Karunathilaka, N. A. Hampson, T. P. Haas, R. Leek, and T. J. Sinclair. "The Impedance of the Alkaline Zinc-Mercuric Oxide Cell. I. Behaviour and Interpretation of Impedance Spectra." *J. Appl. Electrochem.* **11** (1981).

# SECTION C

# SILVER OXIDE

## Joseph Passaniti, Denis Carpenter, Rodney McKenzie

## 12C.1  CELL TECHNOLOGY

Although the use of silver provides for an expensive cell, superior performance capabilities can be realized.

## 12C.2  ELECTROCHEMISTRY

The most common type of zinc/silver oxide cell, the button cell, consists of three main electrochemical components: fine powdered zinc metal anodes, an aqueous alkaline electrolyte with dissolved zincates, and a compressed silver oxide cathode pellet. The active components are contained in an anode top lid and a cathode bottom can, separated by an ionic conductive barrier membrane and sealed with a nylon gasket.

The overall electrochemical reaction of the zinc/monovalent silver oxide cell is

$$Zn + Ag_2O \rightarrow 2Ag + ZnO \quad (1.59\,V)$$

The zinc/divalent silver oxide cell has a two-step electrochemical reaction

$$Step\,1:\ Zn + AgO \rightarrow Ag_2O + ZnO \quad (1.86\,V)$$

$$Step\,2:\ Zn + Ag_2O \rightarrow 2Ag + ZnO \quad (1.59\,V)$$

## 12C.3  CELL DESIGNS AND CONSTRUCTION

### 12C.3.1  Components

*Zinc Anode.*  Zinc is used for the negative electrode in aqueous alkaline batteries because of its high half-cell potential, low polarization, and high limiting current density (up to 40 mA/cm$^2$ in a cast electrode). Zinc has low equivalent weight, thus resulting in a high theoretical capacity of 820 mAh/g. The low polarization of zinc allows for a high discharge efficiency of 85% to 95% (the ratio of useful capacity to theoretical capacity).

The zinc powder is first created by air or gas atomization of molten zinc. As in other alkaline zinc anode cells, care is taken during all processes to avoid contamination of the zinc with other metals, especially iron, as the purity of the zinc is critical to the performance and leakage resistance of the finished cells.

Zinc metal is thermodynamically unstable in aqueous alkali. Pure zinc will very slowly reduce water to hydrogen gas and zinc oxide

$$Zn + H_2O \rightarrow ZnO + H_2$$

Commercial zinc often contains trace heavy metal impurities that act as catalytic sites that rapidly increase the rate of hydrogen generation. The generation of hydrogen within a tightly sealed battery may lead to cell distortion, leakage, or, if the pressure is sufficient, rupture. Zinc alloys containing copper, iron, antimony, arsenic, or tin are known to increase the zinc corrosion rate, while zinc alloyed with mercury, cadmium, aluminum, bismuth, or lead will reduce the corrosion rate.[1,2] Anode gassing can be reduced by the use of zinc alloys, organic inhibitors and mercury addition. While the amount of mercury per cell was minimized (typically about 3% of the anode weight), the battery industry has substituted the use of mercury in all batteries with organic inhibitors and the use of lower gassing zinc alloys.

The oxidation of zinc during discharge is a complex phenomenon. The reactions can be written simplistically as[3,4]

$$Zn + 2OH^- \rightarrow Zn(OH)_2 + 2e \qquad E^0 = +1.249 \text{ V}$$

$$Zn + 4OH^- \rightarrow ZnO_2^{-2} + 2H_2O + 2e \quad E^0 = +1.249 \text{ V}$$

As the electrolyte becomes saturated, zinc oxide will precipitate, releasing the bound water.

$$Zn(OH)_2 \rightarrow ZnO + H_2O$$

***Silver Oxide Cathode.***   Silver oxide is known to have three oxidation states: monovalent ($Ag_2O$), divalent ($AgO$), and trivalent ($Ag_2O_3$).[2] The trivalent silver oxide is very unstable and is not used for batteries. The divalent form had been used in button cells, generally mixed with other metal oxides. The monovalent silver oxide is the most stable and is the one primarily used for commercial button cell batteries.

The reaction product of the discharge of the monovalent silver oxide cathode is highly conductive silver metal.

$$Ag_2O + H_2O + 2e \rightarrow 2Ag + 2OH^- \qquad E^0 = +0.342 \text{ V}$$

Prior to discharge, however, the monovalent silver oxide is a very poor conductor of electricity. Without any additives, a monovalent silver oxide cell would initially exhibit a very high cell impedance and an unacceptably low closed-circuit voltage (CCV). To improve the initial CCV, the monovalent silver oxide is generally blended with 1% to 5% powdered graphite. As the cathode discharges, the silver metal produced maintains a low internal cell resistance and a high CCV. The theoretical capacity of the monovalent silver oxide is 231 mAh/g by weight or 1640 Ah/L by volume. The addition of graphite reduces the cathode capacity due to lower packing density and lower silver oxide content.

Unlike the other silver oxides, the monovalent silver oxide is stable to decomposition in alkaline solutions. Some decomposition to silver metal may occur due to the impurities from the graphite. Low-quality graphite, high amounts of graphite, and increased cell storage temperature result in greater silver oxide decomposition rates.[5]

Due to the high cost of silver bullion, some manufacturers may reduce the amount of silver in the cathode by the use of other cathode active additives. One common additive is manganese dioxide ($MnO_2$). With increasing amounts of $MnO_2$ added to the cathode, the voltage curve changes from a constant voltage throughout the discharge to a curve where the voltage gradually decreases as the cathode is depleted (Fig. 12C.1). This gradual drop in voltage has been considered an indicator of the state of cell depletion; the decreasing voltage indicates the cell is nearing its end of useful life.

Another additive that can serve a dual function is silver nickel oxide ($AgNiO_2$). Silver nickel oxide is produced by the reaction of nickel oxyhydroxide ($NiOOH$) with monovalent silver oxide in hot aqueous alkaline solution[6,7]

$$Ag_2O + 2NiOOH \rightarrow 2AgNiO_2 + H_2O$$

Silver nickel oxide has properties of high electrically conductive, like graphite, as well as good cathode reactivity, like $MnO_2$. The coulometric capacity of silver nickel oxide (263 mAh/g) is higher than $Ag_2O$ and discharges at 1.5 V against zinc (Fig. 12C.2). Silver nickel oxide can replace both the graphite and part of the monovalent silver oxide, reducing the cost of the cell.

Divalent silver oxide is unstable in alkaline solutions, decomposing to monovalent silver oxide and oxygen gas as follows[8]:

$$4AgO \rightarrow 2Ag_2O + O_2$$

Addition of lead, cadmium compounds or gold to the divalent silver oxide improves stability.[9–13]

***Anode-Cathode Reaction Products.***   The zinc/divalent silver oxide battery exhibits a two-step discharge curve. The first occurs at 1.8 V, corresponding to the reduction of divalent to monovalent silver oxide:

$$2AgO + H_2O + 2e \rightarrow Ag_2O + 2OH^- \qquad E^0 = +0.607 \text{ V}$$

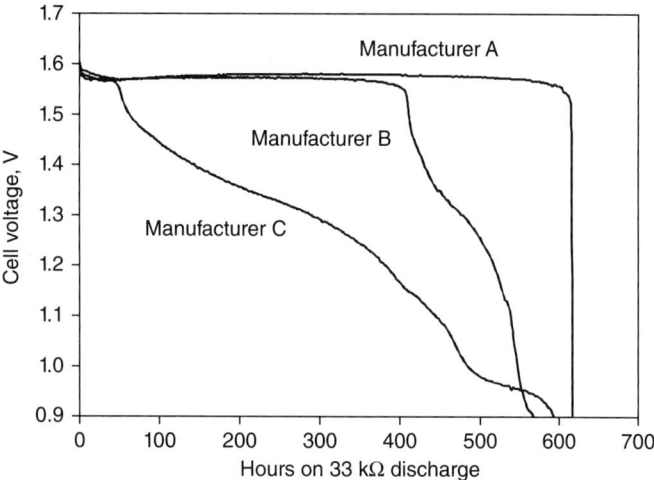

**FIGURE 12C.1**   Voltage profiles of zinc/silver oxide 377 cells from three different manufacturers displaying an increasing dilution of the silver oxide cathode with lower cost manganese dioxide. Discharge on 33 kΩ, 21°C. Data from a competitive audit. (*Courtesy of Rayovac*)

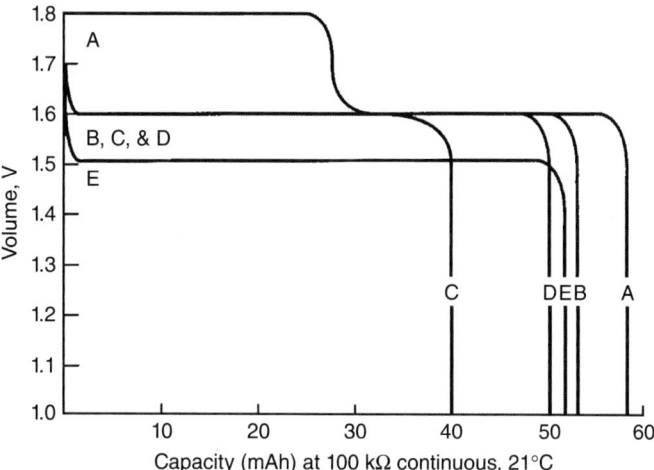

**FIGURE 12C.2**   Comparative performances of various zinc/silver oxide chemistries, type 392 button cell, 7.8 × 3.6 mm. (A) Zn/AgO; (B) Zn/"double-treatment" AgO; (C) Zn/Ag$_2$O; (D) Zn/AgO-silver plumbate; (E) Zn/AgNiO$_2$. Discharge on 100 kΩ, 21°C.

As the discharge continues, the voltage drops to 1.6 V, corresponding to the reduction of monovalent silver oxide to silver metal as shown.

$$Ag_2O + H_2O + 2e \rightarrow 2Ag + 2OH^- \quad E^0 + 0.342 \text{ V}$$

The overall electrochemical reaction of the Zn/AgO cell is

$$Zn + AgO \rightarrow Ag + ZnO$$

This two-step discharge is not desirable for many electronic applications where tight voltage regulation is required.

The elimination of the two-step discharge was resolved by several methods.[11,14-16] One commercial approach was to treat a compressed pellet of AgO with a mild reducing agent such as methanol. The treatment forms a thin outer layer of $Ag_2O$ around a core of AgO. The treated pellet is consolidated into a can and then reacted with a stronger reducing agent such as hydrazine, which reduces a thin layer of silver metal across the pellet's exposed surface. This cathode possesses a surface of silver metal and $Ag_2O$ with the $Ag_2O$ layer shielding the high potential AgO and the thin, conductive silver layer reducing the cell impedance. In use, only the monovalent silver oxide voltage is observed; yet, the cell delivers more capacity from the divalent silver oxide. Even with this double surface treatment, the cells delivered, with the same weight of silver, 20% to 40% more hours of service than the standard monovalent silver oxide cells.

Cells produced by this "double-treatment" process are termed Ditronic™ cells and have about a 30% capacity advantage over a conventional $Ag_2O$ cathode at the same operating voltage.

Control of the treatment process is critical to the shelf life of the cell. Reducing the outer surface of the divalent silver oxide pellet to either only monovalent silver oxide or to only silver metal does not have the advantages of the dual process. The same is true if either coating is not sufficiently thick (Table 12C.1).

**TABLE 12C.1**   Effect of the Coating Thickness for the "Double-Treatment" Method of Divalent Silver Oxide

| $Ag_2O$ thickness on pellet, mm | Ag thickness on consolidation, mm | Final cathode capacity, mAh/g | Voltage level with storage | | |
|---|---|---|---|---|---|
| | | | 1 month | 3 months | 6 months |
| 0.2 | 0.12 | 372 | 1.73 | 1.77 | 1.80 |
| 0.6 | 0.12 | 360 | 1.61 | 1.63 | 1.71 |
| 1.0 | 0.12 | 326 | 1.60 | 1.59 | 1.59 |
| 0.2 | 0.24 | 360 | 1.60 | 1.59 | 1.59 |
| 0.6 | 0.24 | 348 | 1.60 | 1.59 | 1.59 |
| 1.0 | 0.24 | 315 | 1.60 | 1.59 | 1.59 |

During storage, the cells eventually exhibited the phenomenon referred to as "voltage up" and "impedance up" (Fig. 12C.3). The divalent silver oxide slowly oxidizes the conductive silver layer back to the resistive monovalent silver oxide.

$$Ag + AgO \rightarrow Ag_2O$$

As the metallic silver layer is depleted, the cell demonstrates an increase in open-circuit voltage and impedance. The high impedance, due to high internal resistance, results in a low CCV and a nonfunctional cell.

A second approach to eliminate the two-step discharge was through the use of silver plumbate as a cathode additive material as referenced in Fig. 12C.2.[17] Silver plumbate cathode material was prepared by reacting an excess of divalent silver oxide as a coarse powder with lead sulfide (PbS) in a hot alkaline solution. The product of the reaction is a mixture of remaining divalent silver oxide (AgO), monovalent silver oxide ($Ag_2O$), and silver plumbate ($Ag_5Pb_2O_6$). The sulfur is oxidized to the sulfate and is washed from the reaction product as detailed below.

$$2PbS + 19AgO + 4NaOH \rightarrow Ag_5Pb_2O_6 + 7Ag_2O + 2Na_2SO_4 + 2H_2O$$

After reacting, the AgO particles retain a core of AgO but have an outer coating of monovalent silver oxide and silver plumbate. The silver plumbate compound is conductive, stable, and cathode active. The $Ag_2O$ serves to mask the AgO, while the conductive $Ag_5Pb_2O_6$ improves the cell impedance. Unlike silver metal, the

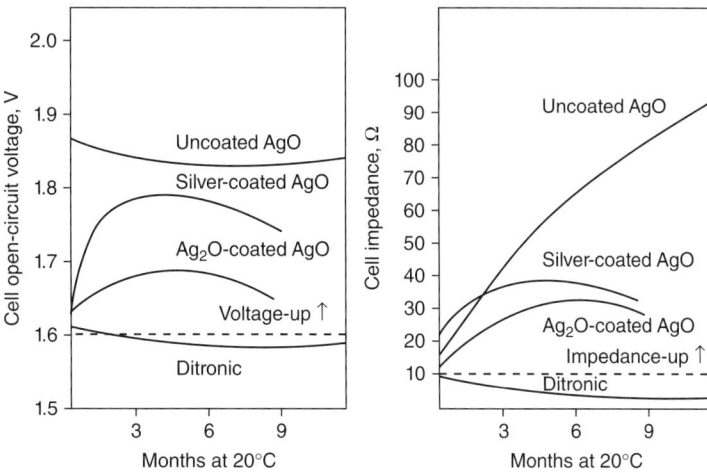

**FIGURE 12C.3** "Voltage-up" and "impedance-up" of Zn/AgO battery during 1-year storage at 21°C, type 392 button cell, 7.8 × 3.6 mm.

conductive silver plumbate is not oxidized by the AgO and the cathode impedance remains low throughout the cell life.

Cathode material prepared by the silver plumbate process will discharge through four reaction steps (Fig. 12C.4):

| Rxn I | $2AgO + H_2O + 2e$ | $\rightarrow Ag_2O + 2OH^-$ | $E^0 = +0.607$ V |
|-------|-------------------|----------------------------|------------------|
| Rxn II | $Ag_2O + H_2O + 2e$ | $\rightarrow 2Ag + 2OH^-$ | $E^0 = +0.342$ V |
| Rxn III | $Ag_5Pb_2O_6 + 4H_2O + 8e$ | $\rightarrow 5Ag + 2PbO + 8OH^-$ | $E^0 = +0.2$ V |
| Rxn IV | $PbO + H_2O + 2e$ | $\rightarrow Pb + 2OH^-$ | $E^0 = +0.580$ V |

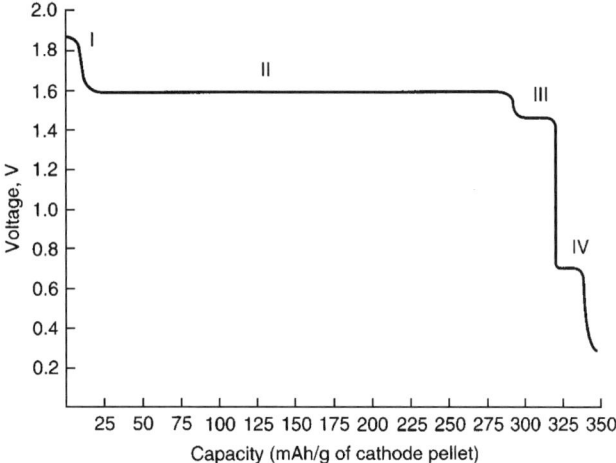

**FIGURE 12C.4** Cathode limited discharge of the zinc/AgO-silver plumbate system; 300 Ω continuous in flooded beaker cell at 21°C. Cathode pellet weight = 0.12 g. Roman numerals indicate the reaction steps.

The open-circuit voltages (OCVs) of the silver plumbate cells are found to be stable at about 1.75 V. However, once placed on discharge, the cell voltage quickly drops to the monovalent silver oxide operating voltage of about 1.6 V, eliminating the AgO plateau (Fig. 12C.2). As button cells are anode limited, the cells may be depleted in zinc capacity before the $Ag_5Pb_2O_6$ and PbO reduction reactions can be observed.

The silver plumbate approach has advantages over the dual-treatment process in that the treatment is simpler but retains a capacity advantage over monovalent silver oxide. The product from the reaction of divalent silver oxide with 8% lead sulfide has a coulometric capacity of 345 to 360 mAh/g.

The silver plumbate process has the disadvantage that the button cells do contain a small amount of lead, 1% to 4% of the cell weight. An alternate approach was developed to use bismuth sulfide in place of the lead sulfide in the material preparation reaction.[18] The reaction product retains the advantages of the silver plumbate material but without the toxicity of lead. Bismuth is not considered toxic and is used in medical and cosmetic applications, both externally and internally within the body.[19] The product of the reaction of bismuth sulfide with divalent silver oxide is believed to be silver bismuthate ($AgBiO_3$).

$$Bi_2S_3 + 28AgO + 6NaOH \rightarrow 2AgBiO_3 + 13Ag_2O + 3Na_2SO_4 + 3H_2O$$

Like the silver plumbate compound, the silver bismuth compound is conductive and cathode active. The monovalent silver oxide produced by the reaction coats the divalent silver oxide particles, while the conductive silver bismuthate reduces the cell impedance, allowing for a high cell CCV. The silver bismuthate will discharge against zinc in alkaline solutions at about 1.5 V. Therefore, in anode-limited button cells only the monovalent silver oxide voltage is observed.

Unlike monovalent silver oxide systems, additives such as graphite or manganese dioxide cannot be added to the divalent silver oxide. Graphite enhances the decomposition of AgO to $Ag_2O$ and oxygen. Manganese dioxide is readily oxidized by AgO to alkali-soluble manganate compounds.

Although the divalent silver oxide has a higher theoretical capacity (432 mAh/g by weight or 3200 Ah/L by volume) than the monovalent silver oxide, the use of the divalent form in button batteries was limited and is no longer marketed commercially.[2,8] This is due primarily to the difficulty in eliminating the two-step discharge and declining prices as the zinc/silver oxide button cells became a commodity.

*Alkaline Electrolyte.*    The electrolytes used for zinc/silver oxide cells are based upon 20% to 45% aqueous solutions of potassium hydroxide (KOH) or sodium hydroxide (NaOH). Zinc oxide (ZnO) is dissolved in the electrolyte as the zincate salt to help control zinc gassing. The zinc oxide concentration varies from a few percent to a saturated solution.

Potassium hydroxide (KOH) is preferred due to higher electrical conductivity, which allows cells to discharge over a wider range of current demands (Fig. 12C.5).[20,21] Sodium hydroxide (NaOH) is used mainly for long life cells not requiring a high-rate discharge (Fig. 12C.6). The sodium hydroxide exhibits less creep, and such cells are less apt to leak than the potassium hydroxide cells. Leakage is evidenced as frosting or salting around the seal. However, leakage issues with potassium hydroxide cells have been resolved by most manufacturers through improvements in seal technology.

Electrolyte gelling agents such as polyacrylic acid, potassium or sodium polyacrylate, sodium carboxymethyl cellulose, or various gums are generally blended into the zinc powder to improve electrolyte accessibility during discharge.

*Barriers and Separators.*    A physical barrier is required to keep the zinc anode and silver cathode apart in the tight volume constraints of a button cell. Failure of the barrier will result in internal cell shorting and cell failure. A silver oxide cell requires a barrier with the following properties:

1. Permeability to water and hydroxyl ions
2. Stability in strong alkaline solutions
3. Oxidation resistance to solid silver oxide or dissolved silver ions
4. Resistance to migration of dissolved silver ions from cathode to anode

Because of the slight solubility of silver (I) oxides in alkaline electrolyte, little work was done with zinc/silver oxide cells until 1941 when André suggested the use of a cellophane barrier.[22] Cellophane prevents migrating

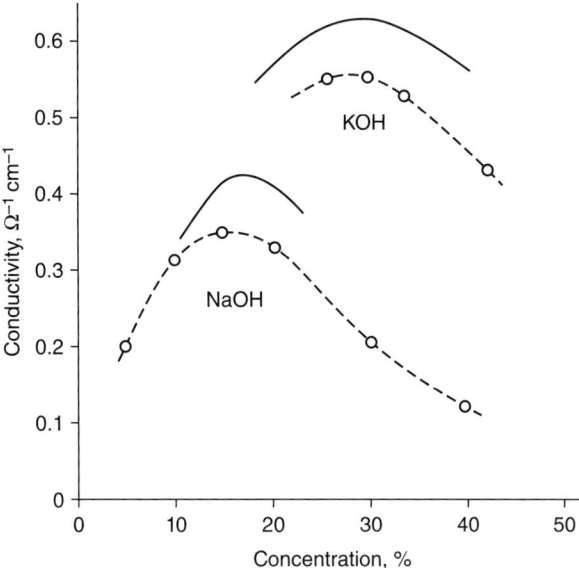

**FIGURE 12C.5**    Specific conductivity of alkaline hydroxide solutions. Solid line: 25°C, Ref. 5; dotted line: 15°C, Ref. 6.

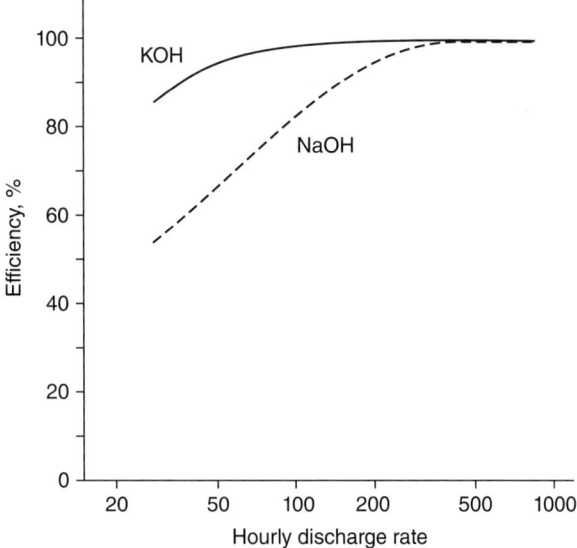

**FIGURE 12C.6**    Dependence of discharge efficiency on discharge rate of zinc/silver oxide button battery at 20°C.

silver ions from reaching the anode by reducing them to insoluble silver metal.[23,24] The cellophane is oxidized and destroyed in the process, making it less effective for long-life cells.

Many types of laminated membranes are presently available. A commonly used alternate barrier material is prepared from a radiation graft of methacrylic acid onto a polyethylene membrane.[23,24] The graft makes the film wettable and permeable to the electrolyte. Studies have shown that a lower resistance polyethylene barrier membrane is suitable for high-rate KOH cells, while higher resistance polyethylene is suitable for low-rate NaOH cells. Cellophane is used in conjunction with the grafted membrane as a sacrificial barrier. The lamination of cellophane to either side of the polyethylene membrane results in a synergistic action for stopping silver migration.[15]

A separator is commonly used in conjunction with a barrier membrane layer as added protection to the barrier. It is located between the barrier and anode cavity and is multifunctional both during cell manufacture and in performance. Separators in zinc/silver cells are typically fibrous woven or nonwoven polymers such as polyvinyl alcohol (PVA). The fibrous nature of the separator gives it stability and strength that protect the more fragile barrier layers from compression failure during cell closure or through penetration of zinc particles. The separator also acts to moderate the effects of dimensional stresses in the barrier layers developed during the lamination processes. These stresses are relieved as the barrier membranes wet.

### 12C.3.2   Cell Construction

Figure 12C.7 is a cross-sectional view of a typical zinc/silver oxide button type battery. Zinc/silver oxide button cells are designed to be anode limited; the cell has 5% to 10% more cathode capacity than anode capacity. If the cell were cathode limited, a zinc-nickel or zinc-iron couple could form between the anode and the cathode can, resulting in the generation of hydrogen gas.

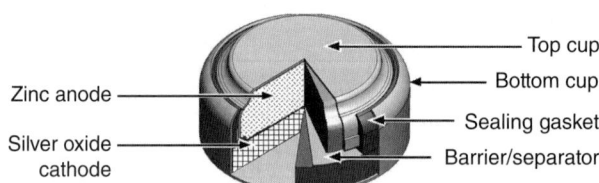

**FIGURE 12C.7**   Cutaway view of typical zinc/silver oxide button type battery.

The cathode material for zinc/silver oxide cells is monovalent silver oxide ($Ag_2O$) mixed with 1% to 5% graphite to improve the electrical conductivity. The $Ag_2O$ cathode material may also contain manganese dioxide ($MnO_2$) or silver nickel oxide ($AgNiO_2$) as cathode extenders. A small amount of polytetrafluoroethylene (Teflon™) may be added to the mix as a binder and to aid pelleting.

The anode is a high surface area, amalgamated, gelled zinc metal powder housed in a top cup, which serves as the external negative terminal for the cell. The top cup is pressed from a triclad metal sheet, which has an outer surface protective layer of nickel over a core of steel. The inner surface that is in direct contact with the zinc is high-purity copper or tin.

The cathode pellet is consolidated into the positive cup, which is formed from nickel-plated steel and serves as the positive terminal for the cell. To keep the anode and cathode separated, a barrier disk of cellophane or a grafted polymeric membrane is placed over the consolidated cathode. The entire system is wetted with potassium or sodium hydroxide electrolyte.

The gasket serves to seal the cell against electrolyte loss and to insulate the top and bottom cups from contact. The gasket material is made from an elastic, electrolyte-resistant plastic such as nylon. The seal may be improved by coating the gasket with a sealant such as polyamide or bitumen to prevent electrolyte leakage at the seal surfaces.

## 12C.4   PERFORMANCE CHARACTERISTICS

### 12C.4.1   Open-Circuit Voltage

The OCV of the $Zn/Ag_2O$ battery is about 1.60 V, but will vary slightly (1.595–1.605 V) with electrolyte concentration, concentration of zincate in the electrolyte, and temperature exposure.[25] The reaction of the silver oxide with carbon dioxide during battery manufacturing can raise the OCV to 1.65 V due to the formation of silver carbonate. The increase in voltage, however, is temporary and will drop to the operating voltage level of 1.58 V within seconds, for example, in a watch. The depth of discharge has little effect on the OCV of a monovalent silver battery; a partially used battery has the same OCV as a new one.

The OCV of the zinc/divalent silver oxide battery will vary from 1.58 to 1.86 V, depending on the ratios of Ag to $Ag_2O$ to AgO in the cathode. The OCV will decrease with greater $Ag_2O$ to AgO ratios and with the presence of silver metal in the cathode. With divalent silver oxide batteries the depth of discharge does have an effect on the OCV; a partially used battery will have more $Ag_2O$ and silver metal than a new one and may have a lower OCV.

### 12C.4.2   Discharge Characteristics

Figure 12C.8 exhibits the typical curves for a type 389, $11.6 \times 3.0$ mm monovalent silver oxide battery on constant resistance discharge. These are typical voltage curves; the discharge voltage profiles of other size batteries would be similar. The service life will vary depending upon the size of the battery and the applied load.

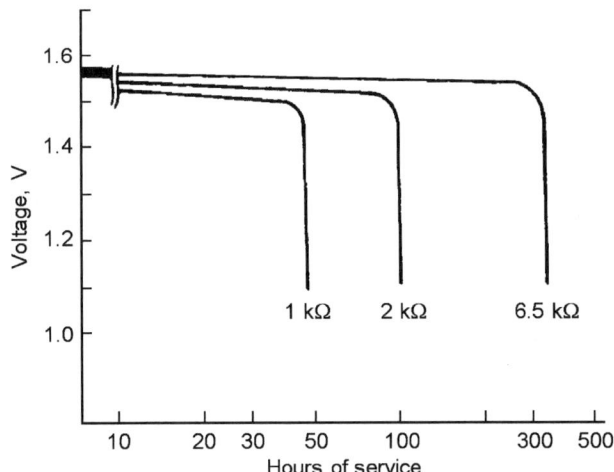

**FIGURE 12C.8**   Typical discharge curves of zinc/silver oxide battery at 20°C, type 389 button cell $11.6 \times 3.0$ mm.

The discharge characteristics of the various types of silver oxide batteries are also covered in Sec. 12C.3.1.

Figure 12C.9 shows the initial closed-circuit voltage of representative sizes of zinc/silver oxide batteries at various loads and temperatures. The zinc/silver oxide button battery is capable of operating over a wide temperature range. The battery can deliver more than 70% of its 20°C capacity at 0°C and 35% at −20°C at the more moderate loads. At heavier loads, the loss is greater. Higher temperatures tend to accelerate capacity deterioration, but temperatures as high as 60°C can be tolerated for several days with no serious effect.

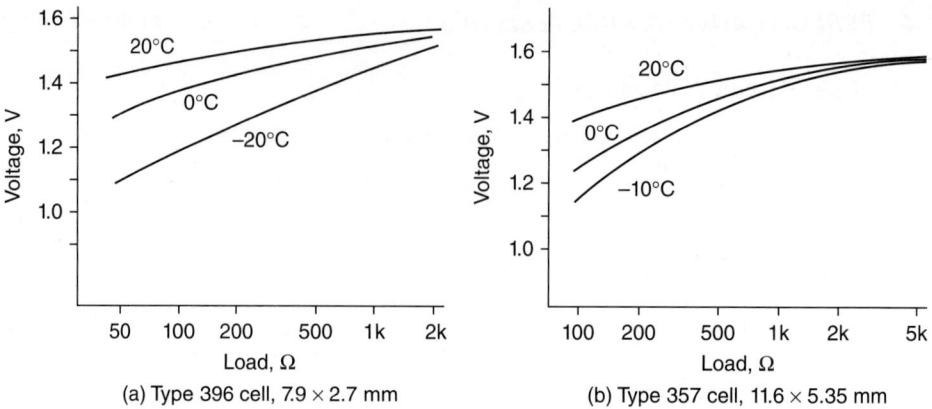

**FIGURE 12C.9**   Closed-circuit voltage of zinc/silver oxide batteries' various resistances. (*a*) Type 396 cell, 7.9 × 2.7 mm. (*b*) Type 357 cell, 11.6 × 5.35 mm.

Figure 12C.10 shows the pulse performance of treated divalent silver oxide batteries using sodium hydroxide and potassium hydroxide as electrolytes. Potassium hydroxide electrolyte yields a cell that will discharge at a higher operating voltage than the sodium hydroxide electrolyte.

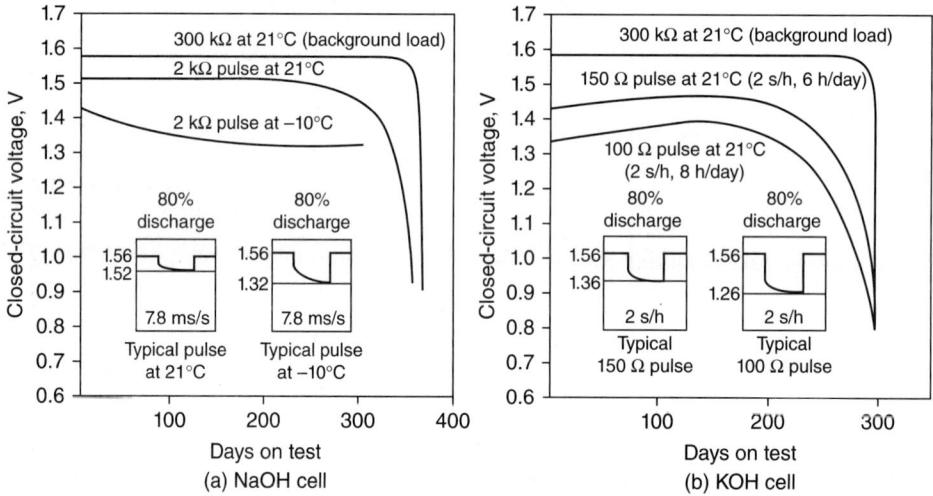

**FIGURE 12C.10**   Closed-circuit voltage curves for Zn-AgO cells with NaOH and KOH as electrolytes, type 392 cell, 7.8 × 3.6 mm. (*a*) NaOH cell, analog watch test. (*b*) KOH cell, LCD watch with backlight test.

The manufacturers of these two types of batteries do not distinguish them by service life tests. In fact, similar mAh output is obtained at loads lighter than the 500 h rate. The industry uses pulse CCV tests to differentiate the higher rate KOH version from the low-rate NaOH version.

The impedance of a zinc/silver oxide battery is influenced primarily by the conductive diluents in the cathode, the barrier resistivity, and the electrolyte type and concentration.

These factors are balanced by battery manufacturers to obtain the desired values required to meet the applications. As the cell is discharged, the impedance will decline as the resistive silver oxide is reduced to conductive metallic silver (Fig. 12C.11).

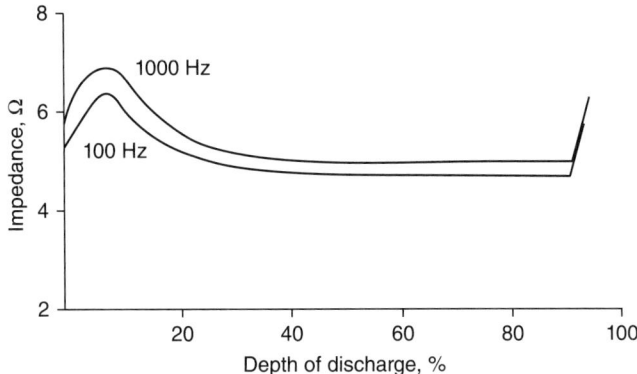

**FIGURE 12C.11**    Impedance of a Zn/Ag$_2$O battery at 100 and 1000 Hz during discharge, type 357 cell, 11.6 × 5.35 mm.

### 12C.4.3    Shelf Life

Major improvements in seal technology and in cell stability have extended the shelf life of watch batteries to more than 5 years. The effect of temperature and humidity on leakage of button batteries was reported by Hull.[26] Leakage was caused by mechanical means (improper seal, fibers in the seal, scratches) or electrochemical means (high oxygen content or high humidity). Batteries are now designed to operate watches for 5 years without leakage.

Stability of these batteries after high temperature storage or prolonged storage at room temperature is influenced by cathode stability and barrier selection. With the monovalent silver oxide cathode, gassing in aqueous potassium or sodium hydroxide at 74°C is not a problem.

With modified cathodes, however, such as divalent silver oxide, silver plumbate, or silver nickel oxide, gas suppression is necessary. CdS, HgS, SnS$_2$, and WS$_2$ were found to reduce oxygen evolution, while BaS, NiS, MnS, and CuS increased oxygen evolution from AgO.[11] Failure on storage of these zinc/silver oxide batteries is closely connected with barrier selection. Cellulosic membranes were used for many years in Zn/Ag$_2$O cells, but their use in Zn/AgO cells was unsuccessful because of massive silver diffusion. While solubility of AgO and Ag$_2$O was reported to be the same (4.4 × 10$^{-4}$ mol/L in 10N NaOH),[24] AgO decomposition to Ag$_2$O occurred spontaneously, resulting in more silver diffusion with Zn/AgO cells than with Zn/Ag$_2$O cells. The small amount of soluble silver reaching the zinc caused accelerated corrosion and hydrogen evolution. In addition, silver was plated within the barrier layer, forming electronic shorts that internally discharge the cell. Laminated Permion membranes have been used to stop silver migration to the zinc. Figure 12C.12 shows an Arrhenius plot of the storage characteristics of various low- and high-rate zinc/silver oxide systems. The data show that 10 years of storage at 21°C is possible.

### 12C.4.4    Service Life

Figure 12C.13 is a monograph that can be used to calculate the service life of the various sized batteries at various current drains at 20°C.

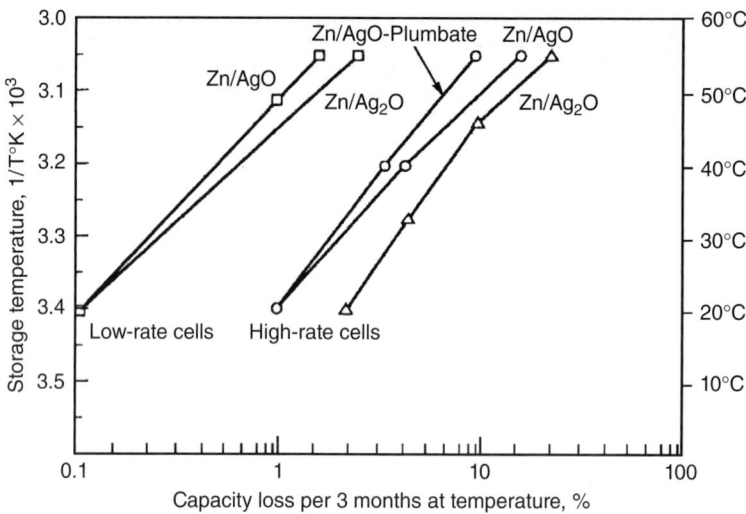

**FIGURE 12C.12** Arrhenius plot of various zinc/silver oxide chemistries. Type 357 cell, 11.6 × 5.35 mm, 6500 Ω continuous discharge to 0.9 V at 21°C. Projected 10% loss: □ > 10 years, ○ > 5 years, Δ > 3 years.

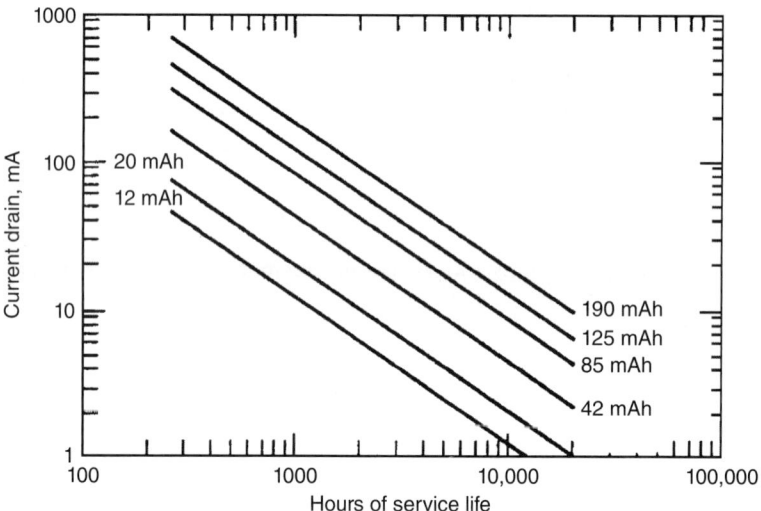

**FIGURE 12C.13** Service life of zinc/silver oxide batteries at 20°C.

## 12C.4.5 Cell Sizes and Types

The characteristics of commercially available zinc/monovalent silver oxide button batteries are summarized in Table 12C.2.

**TABLE 12C.2**   Characteristics of Commercially Available Zinc/Silver Oxide Batteries

| Rayovac model number | ANSI[*] | IEC[†] | Drain rate | Rated load, kΩ | Nominal capacity, mAh | Approximate volume, cm² | Maximum dimensions dia. × ht., mm | Approximate weight, g |
|---|---|---|---|---|---|---|---|---|
| 376 | 1196SO | SR626 | High | 47 | 26 | 0.09 | 6.8 × 2.6 | 0.4 |
| 361 | 1173SO | SR58 | High | 30 | 22 | 0.1 | 7.9 × 2.1 | 0.44 |
| 396 | 1163SO | SR59 | High | 45 | 35 | 0.13 | 7.9 × 2.7 | 0.56 |
| 392 | 1135SO | SR41 | High | 15 | 35 | 0.17 | 7.8 × 3.6 | 0.61 |
| 393 | 1137SO | SR48 | High | 15 | 90 | 0.26 | 7.8 × 5.4 | 1.04 |
| 370 | 1188SO | SR69 | High | 45 | 35 | 0.15 | 9.5 × 2.1 | 0.6 |
| 399 | 1165SO | SR57 | High | 20 | 53 | 0.19 | 9.5 × 2.7 | 0.79 |
| 391 | 1160SO | SR55 | High | 15 | 43 | 0.22 | 11.6 × 2.1 | 0.83 |
| 389 | 1138SO | SR54 | High | 15 | 85 | 0.32 | 11.6 × 3.0 | 1.21 |
| 386 | 1133SO | SF43 | High | 6.5 | 120 | 0.44 | 11.6 × 4.2 | 1.56 |
| 357 | 1131SO | SR44 | High | 6.5 | 190 | 0.57 | 11.6 × 5.35 | 2.22 |
| 337 | NA | SR416 | Low | 100 | 8.3 | 0.02 | 4.8 × 1.65 | 0.13 |
| 335 | 1193SO | SR512 | Low | 150 | 6 | 0.03 | 5.8 × 1.25 | 0.13 |
| 317 | 1185SO | NA | Low | 70 | 11 | 0.04 | 5.8 × 1.65 | 0.18 |
| 379 | 1191SO | NA | Low | 70 | 14 | 0.06 | 5.8 × 2.15 | 0.23 |
| 319 | 1186SO | NA | Low | 70 | 16 | 0.07 | 5.8 × 2.7 | 0.26 |
| 321 | 1174SO | SR65 | Low | 70 | 14 | 0.06 | 6.8 × 1.65 | 0.24 |
| 364 | 1175SO | SR60 | Low | 70 | 19 | 0.08 | 6.8 × 2.15 | 0.33 |
| 377 | 1176SO | SR66 | Low | 45 | 26 | 0.09 | 6.8 × 2.6 | 0.4 |
| 346 | 1164SO | SR721 | Low | 100 | 9.5 | 0.06 | 7.9 × 1.25 | 0.23 |
| 341 | 1192SO | SR714 | Low | 68 | 13 | 0.07 | 7.9 × 1.45 | 0.3 |
| 315 | 1187SO | SR67 | Low | 70 | 16 | 0.08 | 7.9 × 1.65 | 0.32 |
| 362 | 1158SO | SR58 | Low | 70 | 22 | 0.1 | 7.9 × 2.1 | 0.44 |
| 397 | 1164SO | SR59 | Low | 45 | 35 | 0.13 | 7.9 × 2.7 | 0.56 |
| 329 | NA | NA | Low | 20 | 36 | 0.15 | 7.9 × 3.1 | 0.6 |
| 384 | 1134SO | SR41 | Low | 15 | 35 | 0.17 | 7.8 × 3.6 | 0.61 |
| 373 | 1172SO | SR68 | Low | 45 | 24 | 0.12 | 9.5 × 1.65 | 0.44 |
| 371 | 1171SO | SR69 | Low | 45 | 35 | 0.15 | 9.5 × 2.1 | 0.61 |
| 395 | 1162SO | SR57 | Low | 20 | 53 | 0.19 | 9.5 × 2.7 | 0.81 |
| 394 | 1161SO | SR45 | Low | 15 | 64 | 0.26 | 9.5 × 3.6 | 0.96 |
| 366 | 1177SO | SR1116 | Low | 30 | 30 | 0.17 | 11.6 × 1.65 | 0.7 |
| 381 | 1170SO | SR55 | Low | 20 | 43 | 0.22 | 11.6 × 2.1 | 0.8 |
| 390 | 1159SO | SR54 | Low | 15 | 85 | 0.32 | 11.6 × 3.0 | 1.21 |
| 344 | 1139SO | SR42 | Low | 15 | 105 | 0.38 | 11.6 × 3.6 | 1.35 |
| 301 | 1132SO | SR43 | Low | 6.8 | 110 | 0.44 | 11.6 × 4.2 | 1.68 |

[*]ANSI: American National Standards Institute.
[†]IEC: International Electrotechnical Commission.
***Source:*** Rayovac Corporation.

# REFERENCES

1. F. Kober and H. West, "The Anodic Oxidation of Zinc in Alkaline Solutions," Extended Abstracts, The Electrochemical Society, Battery Division 12, 66–69 (1967).

2. A. Fleischer and J. Lander (eds.), *Zinc-Silver Oxide Batteries*, Wiley, New York, 1971.

3. W. M. Latimer, *Oxidation Potentials*, Prentice Hall, Englewood Cliffs, NJ, 1952.

4. D. R. Lide (editor-in-chief), *Handbook of Chemistry and Physics*, 73rd ed., CRC Press, Boca Raton, FL, 1992.

5. A. Shimizu and Y. Uetani, "The Institute of Electronics and Communication Engineers of Japan," Tech. Paper CPM79-55, 1979.

6. T. Nagaura and T. Aita, U.S. Patent 4,370,395 (1981).

7. T. Nagaura, "New Material AgNiO₂ for Miniature Alkaline Batteries," *Progress in Batteries and Solar Cells* **4:**105–107 (1982).

8. E. A. Megahed, "Small Batteries for Conventional and Specialized Applications," *The Power Electronics Show and Conference*, San Jose, CA, pp. 261–272 (1986).

9. B. C. Cahan, U.S. Patent 3,017,448 (1959).

10. P. Ruetschi, in *Zinc-Silver Oxide Batteries*, A. Fleischer and J. J. Lander, eds., Wiley, New York, p. 117 (1971).

11. E. A. Megahed and C. R. Buelow, U.S. Patent 4,078,127 (1978).

12. A. Tvarusko, *J. Electrochem. Soc.* **116:**1070A (1969).

13. S. M. Davis, U.S. Patent 3,853,623 (1974).

14. E. A. Megahed, C. R. Buelow, and P. J. Spellman, U.S. Patent 4,009,056 (1977).

15. E. A. Megahed and D. C. Davig, "Long Life Divalent Silver Oxide-Zinc Primary Cells for Electronic Applications," in *Power Sources*, Vol. 8, Academic, London, 1981.

16. E. A. Megahed and D. C. Davig, "Rayovac's Divalent Silver Oxide-Zinc Batteries," *Progress in Batteries and Solar Cells.* **4:**83–86 (1982).

17. E. A. Megahed and A. K. Fung, U.S. Patent 4,835,077 (1989).

18. J. L. Passaniti, E. A., Megahed, and N. Zreiba, U.S. Patent 5,389,469 (1994).

19. "Bismuth," in *Minerals, Facts, and Problems*, Bureau of Mines Bulletin 675, U.S. Department of the Interior (1985).

20. E. J. Rubin and R. Babaoian, "A Correlation of the Solution Properties and the Electrochemical Behavior of the Nickel Hydroxide Electrode in Binary Aqueous Alkali Hydroxides," *J. Electrochem. Soc.* **118:**428 (1971).

21. "Kagaku Benran," Maruzen, Tokyo, 1966.

22. H. André, *Bull. Soc. Franc. Elect.* **6:**1, 132 (1941).

23. V. D'Agostino, J. Lee, and G. Orban, "Grafted Membranes," in A. Fleischer and J. J. Lander (eds.), *Zinc-Silver Oxide Batteries*, Wiley, New York, 1971, pp. 271–281.

24. R. Thornton, "Diffusion of Soluble Silver-Oxide Species in Membrane Separators," General Electric Final Report, Schenectady, NY (1973).

25. S. Hills, "Thermal Coefficients of EMF of the Silver (I) and the Silver (II) Oxide-Zinc-45% Potassium Hydroxide Systems," *J. Electrochem. Soc.* **108:**810 (1961).

26. M. N. Hull and H. I. James, "Why Alkaline Cells Leak," *J. Electrochem. Soc.* **124:**332–339 (1977).

# CHAPTER 13
# LITHIUM PRIMARY BATTERIES

**Thomas B. Reddy**

## 13.0  *GENERAL CHARACTERISTICS OF LITHIUM PRIMARY CELLS*

Lithium metal is attractive as a battery anode material because of its light weight, high voltage, high electrochemical equivalence, and good conductivity. Because of these outstanding features, the use of lithium has dominated the development of high-performance primary batteries during recent decades.[1] (Chapter 17 covers lithium secondary batteries.)

Serious development of high-energy-density battery systems was started in the 1960s and concentrated on nonaqueous primary batteries using lithium as the anode. The lithium batteries were first used in the early 1970s in selected military applications, but their use was limited as suitable cell designs, formulations, and safety considerations had yet to be resolved. Lithium primary cells and batteries have since been designed, using a number of different chemistries, in a variety of sizes and configurations. Sizes range from less than 5 to 10,000 Ah; configurations range from small coin and cylindrical cells for memory backup and portable applications to large prismatic cells for standby power.

Lithium primary batteries are now being used in increasing quantities in a variety of applications, including cameras, memory backup circuits, security devices, calculators, watches, etc. However, lithium primary batteries have not attained the dominant market share that was anticipated because of their high initial cost, concerns with safety, the advances made with competitive systems, and the cost-effectiveness of the alkaline/manganese battery. Nevertheless, worldwide sales of lithium primary batteries doubled from 2004 to 2014, when they reached an estimated USD 2.94 billion.[2] Sales are expected to remain relatively flat over the next decade, especially as lithium ion rechargeable batteries continue to take over more applications previously served by primary cells.

### 13.0.1  Advantages of Lithium Cells

Primary cells using lithium anodes have many advantages over conventional batteries. The advantageous features include the following:

1. *High voltage.* Lithium batteries have voltages up to about 4 V, depending on the cathode material, compared with 1.5 V for most other primary battery systems. The higher voltage reduces the number of cells in a battery pack by a factor of at least 2.

2. *High specific energy and energy density.* The energy output of a lithium battery (up to 870 Wh/kg and 1180 Wh/L) is two to five times better than that of conventional zinc anode batteries.

3. *Operation over a wide temperature range.* Many of the lithium batteries will perform over a temperature range from about 70 to −40°C, with some capable of performance to 150°C or as low as −80°C.

4. *Good power density.* Some of the lithium batteries are designed with the capability to deliver their energy at high current and power levels.

5. *Flat discharge characteristics.* A flat discharge curve (constant voltage and resistance through most of the discharge) is typical for many lithium batteries.

6. *Superior shelf life.* Lithium batteries can be stored for long periods, even at elevated temperatures. Storage of up to 10 years at room temperature has been achieved and storage of 1 year at 70°C has also been demonstrated. Shelf lives over 20 years have been projected from reliability studies.

The performance advantages of several types of lithium batteries compared with conventional primary and secondary batteries are shown in Chap. 9 (An Introduction to Primary Batteries). The advantage of the lithium cell is shown graphically in Figs. 9.1 to 9.9, which compare the performance of the various primary cells. Only the zinc/air, zinc/mercuric oxide, and zinc/silver oxide cells, which are noted for their high energy density, approach the capability of the lithium systems at 20°C. The zinc/air cell, however, is very sensitive to atmospheric conditions; the others do not compare as favorably on a specific energy basis nor at lower temperatures.

## 13.0.2  Classification of Lithium Primary Cells

Lithium batteries use nonaqueous solvents for the electrolyte because of the reactivity of lithium in aqueous solutions. Organic solvents, such as acetonitrile, propylene carbonate (PC), and dimethoxyethane (DME), and inorganic solvents, such as thionyl chloride, are typically employed. A compatible solute is added to provide the necessary electrolyte conductivity. Solid- and molten-salt (ionic liquid) electrolytes are also used in some other primary and reserve lithium cells. Many different materials have been considered for the active cathode material; sulfur dioxide, thionyl chloride, manganese dioxide, iron disulfide, and carbon monofluoride are now in common use. The term "lithium battery," therefore, applies to many different types of chemistries, each using lithium as the anode but differing in cathode material, electrolyte, and chemistry as well as in design and other physical and mechanical features.

Lithium primary batteries can be classified into several categories, based on the type of electrolyte (or solvent) and cathode material used. These classifications, examples of materials that were considered or used, and the major characteristics of each are listed in Table 13.1 and discussed in further detail below.

**TABLE 13.1**    Classification of Lithium Primary Batteries[*]

| Cell classification | Typical electrolyte | Power capability | Size, Ah | Operating range, °C | Shelf life, years | Typical cathodes | Nominal cell voltage, V | Key characteristics |
|---|---|---|---|---|---|---|---|---|
| Soluble cathode (liquid or gas) | Organic or inorganic (with solute) | Moderate to high power, W | 0.5–10,000 | −80 to 70 | 5–20 | $SO_2$<br>$SOCl_2$<br>$SO_2Cl_2$ | 3.0<br>3.6<br>3.9 | High energy output, high power output, low-temperature operation, long shelf life |
| Solid cathode | Organic (with solute) | Low to moderate power, mW–W | 0.03–1200 | −40 to 50 | 5–8 | $V_2O_5$<br>$AgV_2O_{5.5}$<br>$MnO_2$<br>$CFx$<br>$CuS$<br>$FeS_2$<br>$FeS$ | 3.3<br>3.2<br>3.0<br>2.6<br>1.7<br>1.5<br>1.5 | High energy output for moderate power requirements, nonpressurized cells |
| Solid electrolyte (see Chap. 28) | Solid state | Very low power, μW | 0.003–2.4 | 0 to 100 | 10–25 | $PbI_2/PbS/$<br>$PbI_2(P2VP)$ | 1.9<br>2.8 | Excellent shelf life, solid state—no leakage, long-term microampere discharge |

[*]For reserve lithium batteries, see Chap. 29.

*Soluble-Cathode Cells.*　These cells use liquid or gaseous active cathode materials, such as sulfur dioxide ($SO_2$) or thionyl chloride ($SOCl_2$), that dissolve in the electrolyte or serve directly as the electrolyte solvent. Their operation depends on the formation of a passive layer on the lithium anode resulting from a reaction between the lithium and the cathode material. This prevents further chemical reaction (self-discharge) between anode and cathode or reduces it to a very low rate. These cells are manufactured in many different configurations and designs (such as high and low rate) and with a very wide range of capacities. Conductive carbon powders bonded to metal current collectors typically serve as the site at which the cathode reactions occur. Cells are generally fabricated in a cylindrical configuration in the smaller sizes, up to about 35 Ah, using a bobbin construction for the low-rate cells and a spirally wound (jelly-roll) structure for the high-rate designs. Prismatic containers, having flat parallel plates, are generally used for the larger cells up to 10,000 Ah in size. Flat or "pancake-shaped" configurations have also been designed. These soluble cathode lithium cells are used for low to high discharge rates. The high-rate designs, using large electrode surface areas, are noted for their high power density and are capable of delivering the highest current densities of any active primary cell.

*Solid-Cathode Cells.*　The second type of lithium anode primary cell uses solid rather than soluble gaseous or liquid materials for the cathode. With these solid-cathode materials, the cells have the advantage of not being pressurized or necessarily requiring a hermetic-type seal, but they do not have the high-rate capability of the soluble-cathode systems. They are designed, generally, for low- to medium-rate applications such as memory backup, security devices, portable electronic equipment, digital cameras, watches, calculators, and small lights. Button, flat, and cylindrical-shape cells are available in low-rate and the moderate-rate jelly-roll configurations. A number of different solid cathodes are being used in lithium primary cells, as listed in Table 13.1. The discharge of the solid cathode cells is not as flat as that of the soluble cathode cells, but at the lower discharge rates and ambient temperature, their capacity (energy density) may be higher than that of the lithium/sulfur dioxide cell.

*Solid-Electrolyte Cells.*　These cells are noted for their extremely long storage life, in excess of 20 years, but are capable of only low-rate discharge in the microampere range. They are used in applications such as memory backup, cardiac pacemakers, and similar equipment where current requirements are low, but long life is critical (see Chap. 28).

In Fig. 13.1, the size or capacity of these three types of lithium cells (up to the 30 Ah size) is plotted against the current levels at which they are typically discharged. The approximate weight of lithium in each of these cells is also shown.

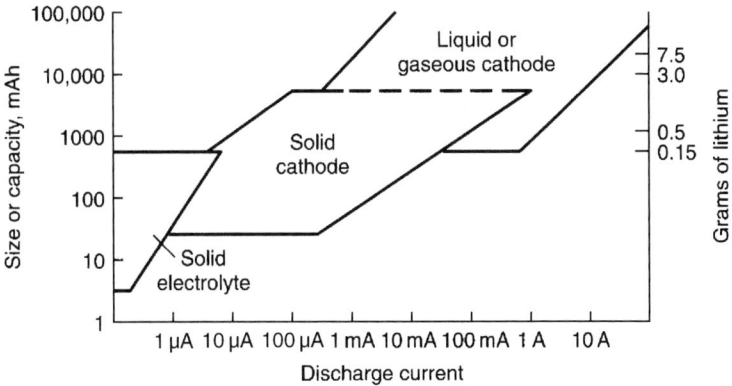

**FIGURE 13.1**　Classification of lithium primary cell types.

## 13.0.3   General Electrochemistry

***Lithium.***   The main requirements for electrode materials used for high-performance (high specific energy and energy density) batteries are a high electrochemical equivalence (high coulombic output for a given weight of material) and a high electrode potential. It is apparent from data in Apps. B and C as well as in Chap. 1, which list the characteristics of metals used as battery anodes, that lithium is an outstanding candidate. Its standard potential and electrochemical equivalence are the highest of the metals; it excels in theoretical gravimetric energy density; and, with its high potential, it is inferior only to aluminum and magnesium on a volumetric energy basis (watthours per liter). Aluminum, however, has not been used successfully as an anode except in reserve systems, and magnesium has a low practical operating voltage. Furthermore, lithium is preferred to the other alkali metals because of its better mechanical characteristics and lower reactivity. Calcium has been investigated as an anode, in place of lithium, because of its higher melting point (838°C compared with 180.5°C for lithium). To date, practical cells using calcium have not been produced.

Lithium is one of the alkali metals, and it is the lightest of all the metallic elements, with a density about half that of water. When first made or freshly cut, lithium has the luster and color of bright silver, but it tarnishes rapidly in moist air. It is soft and malleable, can be readily extruded into thin foils, and is a good conductor of electricity. Table 13.2 lists some of the physical properties of lithium.[3,4]

**TABLE 13.2**   Physical Properties of Lithium

| | |
|---|---|
| Melting point | 180.5°C |
| Boiling point | 1347°C |
| Density | 0.534 g/cm$^3$ (25°C) |
| Specific heat | 0.852 cal/g (25°C) |
| Specific resistance | $9.35 \times 10^6 \, \Omega \cdot cm$ (20°C) |
| Hardness | 0.6 (Mohs' scale) |

Lithium reacts vigorously with water, releasing hydrogen and forming lithium hydroxide

$$2Li + 2H_2O \rightarrow 2LiOH + H_2$$

This reaction is not as vigorous as that of sodium and water, probably due to the fairly low solubility and the adherence of LiOH to the metal surface under some conditions; however, the heat generated by this reaction may ignite the hydrogen that is formed and the lithium will then also burn. Because of this reactivity, however, lithium must be handled in a dry atmosphere and, in a battery, be used with nonaqueous electrolytes. (The lithium/air and lithium/water batteries are described in Chaps. 18 and 29, respectively.)

***Cathode Materials.***   A number of inorganic and organic materials have been examined for use as the cathode in primary lithium batteries.[1,5] The critical requirements for this material to achieve high performance are high battery voltage, high energy density, and compatibility with the electrolyte (i.e., being essentially nonreactive or insoluble in the electrolyte). Preferably the cathode material should be conductive, although there are few such materials available and solid-cathode materials are usually mixed with a conducting material, such as carbon, and applied to a conductive grid to provide the needed conductivity. If the cathode reaction products are a metal and a soluble salt (of the anode metal), this feature can improve cathode conductivity as the discharge proceeds. Other desirable properties of the cathode material are low cost, availability (noncritical material), and favorable physical properties, such as nontoxicity and nonflammability. Table 13.3 lists some of the cathode materials that have been studied for primary lithium batteries and gives their cell reaction mechanisms and the theoretical cell voltages and capacities.

***Electrolytes.***   The reactivity of lithium in aqueous solutions requires the use of nonaqueous electrolytes for lithium anode batteries.[5] Polar organic liquids are the most common electrolyte solvents for the active primary cells, except for the thionyl chloride (SOCl$_2$) and sulfuryl chloride (SO$_2$Cl$_2$) cells, where these

**TABLE 13.3** Cathode Materials Currently or Previously Used in Lithium Primary Batteries

| Cathode material | Molecular weight | Valence change | Density, $g/cm^3$ | Theoretical faradic capacity (cathode only) | | | Cell reaction mechanism (with lithium anode) | Theoretical cell | |
|---|---|---|---|---|---|---|---|---|---|
| | | | | Ah/g | $Ah/cm^3$ | g/Ah | | Voltage, V | Specific energy, Wh/kg |
| $SO_2$ | 64 | 1 | 1.37 | 0.419 | — | 2.39 | $2Li + 2SO_2 \rightarrow 2Li_2S_2O_4$ | 3.1 | 1170 |
| $SOCl_2$ | 119 | 2 | 1.63 | 0.450 | — | 2.22 | $4Li + 2SOCl_2 \rightarrow 4LiCl + S + SO_2$ | 3.65 | 1470 |
| $SO_2Cl_2$ | 135 | 2 | 1.66 | 0.397 | — | 2.52 | $2Li + SO_2Cl_2 \rightarrow 2LiCl + SO_2$ | 3.91 | 1405 |
| $Bi_2O_3$ | 466 | 6 | 8.5 | 0.35 | 2.97 | 2.86 | $6Li + Bi_2O_3 \rightarrow 3Li_2O + 2Bi$ | 2.0 | 640 |
| $Bi_2Pb_2O_5$ | 912 | 10 | 9.0 | 0.29 | 2.64 | 2.41 | $10Li + Bi_2Pb_2O_5 \rightarrow 5Li_2O + 2Bi + 2Pb$ | 2.0 | 544 |
| $(CF)_x$ | 31 | 1 | 2.7 | 0.86 | 2.32 | 1.16 | $xLi + (CF)_x \rightarrow xLiF + xC$ | 3.1 | 2180 |
| $CuCl_2$ | 134.5 | 2 | 3.1 | 0.40 | 1.22 | 2.50 | $2Li + CuCl_2 \rightarrow 2LiCl + Cu$ | 3.1 | 1125 |
| $CuF_2$ | 101.6 | 2 | 2.9 | 0.53 | 1.52 | 1.87 | $2Li + CuF_2 \rightarrow 2LiF + Cu$ | 3.54 | 1650 |
| $CuO$ | 79.6 | 2 | 6.4 | 0.67 | 4.26 | 1.49 | $2Li + CuO \rightarrow Li_2O + Cu$ | 2.24 | 1280 |
| $Cu_4O(PO_4)_2$ | 458.3 | 8 | — | 0.468 | — | 2.1 | $8Li + Cu_4O(PO_4)_2 \rightarrow Li_2O + 2Li_3PO_4 + Cu$ | 2.7 | — |
| $CuS$ | 95.6 | 2 | 4.6 | 0.56 | 2.57 | 1.79 | $2Li + CuS \rightarrow Li_2S + Cu$ | 2.15 | 1050 |
| $FeS$ | 87.9 | 2 | 4.8 | 0.61 | 2.95 | 1.64 | $2Li + FeS \rightarrow Li_2S + Fe$ | 1.75 | 920 |
| $FeS_2$ | 119.9 | 4 | 4.9 | 0.89 | 4.35 | 1.12 | $4Li + FeS_2 \rightarrow 2Li_2S + Fe$ | 1.8 | 1304 |
| $MnO_2$ | 86.9 | 1 | 5.0 | 0.31 | 1.54 | 3.22 | $Li + Mn^{IV}O_2 \rightarrow Mn^{III}O_2(Li^+)$ | 3.5 | 1005 |
| $MoO_3$ | 143 | 1 | 4.5 | 0.19 | 0.84 | 5.26 | $2Li + MoO_3 \rightarrow Li_2O + Mo_2O_5$ | 2.9 | 525 |
| $Ni_3S_2$ | 240 | 4 | — | 0.47 | — | 2.12 | $4Li + Ni_3S_2 \rightarrow 2Li_2S + 3Ni$ | 1.8 | 755 |
| $AgCl$ | 143.3 | 1 | 5.6 | 0.19 | 1.04 | 5.26 | $Li + AgCl \rightarrow LiCl + Ag$ | 3.267 | 583 |
| $Ag_2CrO_4$ | 331.8 | 2 | 5.6 | 0.16 | 0.90 | 6.25 | $2Li + Ag_2CrO_4 \rightarrow Li_2CrO_4 + 2Ag$ | 3.35 | 515 |
| $AgV_2O_{5.5}$* | 297.7 | 3.5 | — | 0.282 | — | — | $3.5Li + AgV_2O_{5.5} \rightarrow Li_{3.5}AgV_2O_{5.5}$ | 3.24 | 655 |
| $V_2O_5$ | 181.9 | 1 | 3.6 | 0.15 | 0.53 | 6.66 | $Li + V_2O_5 \rightarrow LiV_2O_5$ | 3.4 | 490 |

*Multiple-step discharge; see Ref. 11 (experimental values to +1.5 V cutoff).

inorganic compounds serve as both the solvent and the active cathode material. The important properties of the electrolyte are:

1. It must be aprotic, that is, have no reactive protons or hydrogen atoms, although hydrogen atoms may be in the molecule.

2. It must have low reactivity with lithium (or form a protective coating on the lithium surface to prevent further reaction) and the cathode.

3. It must be capable of forming an electrolyte of good ionic conductivity.

4. It should be liquid over a broad temperature range.

5. It should have favorable physical characteristics, such as low vapor pressure, stability, nontoxicity, and nonflammability.

A listing of the organic solvents commonly used in lithium batteries is given in Table 13.4. These solvents are typically employed in binary or ternary combination. These organic electrolytes, as well as thionyl chloride (mp −105°C, bp 78.8°C) and sulfuryl chloride (mp −54°C, bp 69.1°C), are liquid over a wide temperature range with low freezing points. This characteristic provides the potential for operation over a wide temperature range, particularly low temperatures.

**TABLE 13.4**    Properties of Organic Electrolyte Solvents for Lithium Primary Batteries

| Solvent | Structure | Boiling point at $10^5$ Pa, °C | Melting point, °C | Flash point, °C | Density at 25°C, g/cm³ | Specific conductivity with 1M LiClO$_4$, S/cm$^{-1}$ |
|---|---|---|---|---|---|---|
| Acetonitrile (AN) | $H_3C-C{\equiv}N$ | 81 | −45 | 5 | 0.78 | $3.6 \times 10^{-2}$ |
| γ-Butyrolactone (BL) | (ring structure) | 204 | −44 | 99 | 1.1 | $1.1 \times 10^{-2}$ |
| Dimethylsulfoxide (DMSO) | $H_3C-S(=O)-CH_3$ | 189 | 18.5 | 95 | 1.1 | $1.4 \times 10^{-2}$ |
| Dimethylsulfite (DMSI) | $O{=}S(OCH_3)_2$ | 126 | −141 | | 1.2 | |
| 1,2-Dimethoxyethane (DME) | $H_2C-O-CH_3$ / $H_2C-O-CH_3$ | 83 | −60 | 1 | 0.87 | |
| Dioxolane (1,3-D) | (ring structure) | 75 | −26 | 2 | 1.07 | |
| Methyl formate (MF) | $H-C(=O)-O-CH_3$ | 32 | −100 | −19 | 0.98 | $3.2 \times 10^{-2}$ |
| Propylene carbonate (PC) | $H_3C-CH-CH_2$ (carbonate ring) | 242 | −49 | 135 | 1.2 | $7.3 \times 10^{-3}$ |
| Tetrahydrofuran (THF) | $H_2C-CH_2-CH_2-CH_2$ (ring with O) | 65 | −109 | −15 | 0.89 | |

The Jet Propulsion Laboratory (Pasadena, CA) has evaluated several types of lithium primary batteries to determine their ability to operate planetary probes at temperatures of −80°C and below.[6] Individual cells were evaluated by discharge tests and electrochemical impedance spectroscopy. Of the five types considered (Li/SOCl$_2$, Li/SO$_2$, Li/MnO$_2$, Li-BCX, and Li-CF$x$), lithium-thionyl chloride and lithium-sulfur dioxide were found to provide the best performance at −80°C. Lowering the electrolyte salt to ca. 0.5 M was found to improve performance with these systems at very low temperatures. In the case of D-size Li/SOCl$_2$ batteries, lowering the electrolyte salt (LiAlCl$_4$) concentration from 1.5 to 0.5 M led to a 60% increase in capacity on a baseline load of 118 Ω with periodic 1-min pulses at 5.1 Ω at −85°C.

Lithium salts, such as LiClO$_4$, LiBr, LiCF$_3$SO$_3$, LiI, and LiAlCl$_4$, are the electrolyte solutes most commonly used to provide ionic conductivity. The solute must be able to form a stable electrolyte that does not react with the active electrode materials. It must be soluble in the organic solvent and dissociate to form a conductive electrolyte solution. Maximum conductivity with organic solvents at room temperature is normally obtained with a 1-M solute concentration, but generally the conductivity of these electrolytes is about one-tenth that of aqueous systems. To accommodate this lower conductivity, close electrode spacing and cells designed to minimize impedance and provide good power density are used.

***Cell Couples and Reaction Mechanisms.*** The overall discharge reaction mechanism for various lithium primary batteries was shown in Table 13.3, which also lists the theoretical cell voltage. The mechanism for the discharge of the lithium anode is the oxidation of lithium to form lithium ions (Li$^+$) with the release of an electron.

$$Li \rightarrow Li^+ + e$$

The electron moves through the external circuit to the cathode, where it reacts with the cathode material, which is reduced. At the same time, the Li$^+$ ion, which is small (0.06 nm in radius) and mobile in both liquid- and solid-state electrolytes, moves through the electrolyte to the cathode, where it reacts to form a lithium compound.

A more detailed description of the cell reaction mechanism for the different lithium primary batteries is given in the below sections for those battery systems.[1,7]

## 13.0.4 Basic Cell Designs and General Features

***Summary of Designs.*** A listing of the major lithium primary batteries now in production or advanced development and a summary of their constructional features, key electrical characteristics, and available sizes are presented in Table 13.5. The types of batteries, their sizes, and some characteristics are subject to change depending on design, standardization, and market development. Manufacturers' data should be obtained for specific characteristics. The performance characteristics of these systems, under theoretical conditions, were given in Table 13.3. Comparisons of the performance of the lithium batteries with comparably sized conventional primary batteries are covered in Chap. 9. Detailed characteristics of some of these batteries are covered in Chaps. 13A to 13G, and in Chap. 28.

*Soluble-Cathode Lithium Primary Batteries.* Two types of soluble-cathode lithium primary batteries are currently available (Table 13.1). One uses SO$_2$ as the active cathode dissolved in an organic electrolyte solvent. The second type uses an inorganic solvent, such as the oxychlorides SOCl$_2$ and SO$_2$Cl$_2$, which serves as both the active cathode and the electrolyte solvent. These materials form a passivating layer or protective film of reaction products on the lithium surface, which inhibits further reaction. Even though the active cathode material is in contact with the lithium anode, self-discharge is inhibited by the protective film, which proceeds at very low rates, and the shelf life of these batteries is excellent. This film, however, may cause a voltage delay to occur, that is, a time delay to break down the film and for the cell voltage to reach the operating level when the discharge load is applied. These lithium batteries have a high specific energy and, with proper design, such as the use of high surface area electrodes, are capable of delivering high specific energy at high specific power.

These cells generally require a hermetic-type seal. Sulfur dioxide is a gas at 20°C (bp −10°C) that is not fully absorbed or dissolved in the solvent such that the undischarged cell has an internal pressure of 3 to $4 \times 10^5$ Pa at 20°C. The oxychlorides are liquid at 20°C, but with boiling points of 78.8°C for SOCl$_2$ and 69.1°C for SO$_2$Cl$_2$, a moderate pressure can develop at high operating temperatures. In addition, as SO$_2$ is a discharge product in the oxychloride cells, the internal cell pressure increases as the cell is discharged.

**TABLE 13.5** Characteristics of Lithium Primary Batteries

Soluble cathode batteries

| System | Cathode | Electrolyte | | | Construction | Voltage, V | | Specific energy,[†] Wh/kg | Energy density,[†] Wh/L | Power density | Discharge profile | Available sizes |
| | | Solvent | Solute | Separator | | Nominal | Working[*] (20°C) | | | | | |
|---|---|---|---|---|---|---|---|---|---|---|---|---|
| Lithium/sulfur dioxide (Li/SO₂) | SO₂ with carbon and binder on Al screen | AN | LiBr | Microporous polypropylene | Spiral "jelly-roll" cylindrical construction; glass-to-metal seal | 3.0 | 2.9–2.7 | 260 | 415 | High | Very flat | Cylindrical batteries up to 34 Ah |
| Lithium/thionyl chloride (Li/SOCl₂) | SOCl₃ with carbon and binder on Ni or SS | SOCl₂ | LiAlCl₄ | Glass nonwoven | Wafer construction | 3.6 | 3.6–3.4 | 275 | 630 | Low | Flat | 0.4–1.7 Ah |
| Low rate | | | | | "Bobbin" in cylindrical construction | 3.6 | 3.5–3.3 | 590 | 1100 | Medium | Flat | Cylindrical batteries 1.2–35 Ah |
| High capacity | | | | | Prismatic with flat plates | 3.6 | 3.5–3.3 | 480 | 950 | Medium | Flat | 12–10,000 Ah |
| High rate | | | | | Spiral "jelly-roll" cylindrical construction or flat disk | 3.6 | 3.5–3.2 | 495 | 970 | Medium to high | Flat | Cylindrical: 1.2–14 Ah Flat disk: up to 2300 Ah |
| | | SOCl₂ with halogen additives | LiAlCl₄ | Glass mat | Spiral "jelly-roll" cylindrical construction | 3.9 | 3.8–3.3 | 485 | 1070 | Medium | Flat | 2–30 Ah |
| Lithium/sulfuryl chloride (Li/SO₂Cl₂) | SO₂Cl₂ with carbon and binder SS screen | SO₂Cl₂ (some with additives) | LiAlCl₄ | Glass | Spiral "jelly-roll" cylindrical construction; glass-to-metal seal | 3.95 | 3.5–3.1 | 480 | 1040 | Medium to high | Flat | 7–30 Ah |
| Lithium/carbon monofluoride | CFx with carbon and binder on nickel collector | PC + DME or BL | LiBF₄ or LiAsF₆ | Polypropylene | "Coin" construction; crimped seal Pin type | 3.0 | 2.7–2.5 | 215 | 550 | Low to medium Low | Moderately flat Humped | Coin batteries to 500 mAh Small cylinders 25–50 mAh |
| Li/CFx | | | | | Spiral "jelly-roll" cylindrical construction; crimped or glass-to-metal seal | | | 350 (commercial) 800 (military) | 560 | | | Cylindrical batteries to 5 Ah (commercial) and 1200 Ah (military) |
| | | | | | Rectangular with flat plates | | | 440 (biomedical) | 900 | | | Rectangular batteries to 40 Ah |

| Battery type | Cathode | Solvent | Salt | Separator | Construction | Nominal voltage* | Voltage range* | Energy density (Wh/kg)† | Energy density (Wh/L)† | Internal resistance | Discharge profile | Sizes |
|---|---|---|---|---|---|---|---|---|---|---|---|---|
| Lithium/copper oxide (Li/CuO) | CuO pressed in cell can | 1,3D | LiClO₄ | Nonwoven glass | "Bobbin" inside-out cylindrical construction | 1.5 | 1.5–1.4 | 280 | 650 | Low | High initial voltage drop, then moderately flat | Cylindrical batteries 500–3500 mAh |
| Lithium/iron disulfide (LiFeS₂) | FeS₂ with carbon and binders | 1–3D + DME | LiI | Microporous polyethylene | "Jelly-roll" cylindrical construction; crimped seal | 1.5 | 1.6–1.4 | 310 | 562 | Medium to high | High initial drop, then flat | AAA and AA sizes to 3.1 Ah |
| Lithium/manganese dioxide (Li/MnO₂) | MnO₂ with carbon and binder on supporting grid | PC + DME | Li salt | Polypropylene | "Coin construction" with flat electrodes | 3.0 | 3.0–2.7 | 230 | 545 | Low to medium | Moderately flat | Coin batteries 25–1000 mAh |
|  |  | Organic solvent | Li salt | Polypropylene | "Jelly-roll" cylindrical construction; crimped and hermetic seals | 3.0 | 2.8–2.5 | 261 | 546 | Medium to high | Moderately flat | 2/3 A Cylindrical batteries typical, larger cells available to 11 Ah |
|  |  | Organic solvent | Li salt | Polypropylene | "Bobbin" cylindrical construction | 3.0 | 3.0–2.8 | 270 | 620 | Low to medium | Moderately flat | Cylindrical batteries to 2.5 Ah |
| Lithium/silver vanadium oxide (Li/AgV₄O₁₁) | AgV₂O₅.₅ with graphite and carbon | PC, DME | LiAsF₆ | Microporous polypropylene | Rounded prismatic and D-shaped cross section | 3.2 | 3.2–1.5 | 270 | 780 | Low to medium | Multiple plateaus | Special sizes for implantable medical devices |

*Working voltages are typical for discharges at favorable loads.
†Energy densities are for 20°C under favorable discharge conditions. See details in appropriate sections.

The lithium/sulfur dioxide (Li/SO$_2$) battery is the most advanced of these lithium primary batteries. These batteries are typically manufactured in cylindrical configurations in capacities up to 34 Ah. They are noted for their high specific power (about the highest of the lithium primary batteries), high energy density, and good low-temperature performance. They are used in military and specialized industrial, space, and commercial applications where these performance characteristics are required.

The lithium/thionyl chloride (Li/SOCl$_2$) battery has one of the highest specific energies of all the practical battery systems. Figures 9.7 and 9.8 illustrate the advantages of the Li/SOCl$_2$ battery over a wide temperature range at moderate discharge rates. Figure 13.2 compares typical discharge profile of the Li/SOCl$_2$ cell with the Li/SO$_2$ cell. At 20°C, at moderate discharge rates, the Li/SOCl$_2$ cell has a higher working voltage and about a 50% advantage in service life. The Li/SO$_2$ cell, however, does have better performance at low temperatures and high discharge rates and a lower voltage delay after storage. Li/SOCl$_2$ cells have been fabricated in many sizes and designs ranging from small button and cylindrical cells with capacities below 1 Ah, to large prismatic cells with capacities as high as 10,000 Ah. Low-rate cells have been used successfully in many applications, especially as memory backup, for many years; high-rate cells are used in special applications.

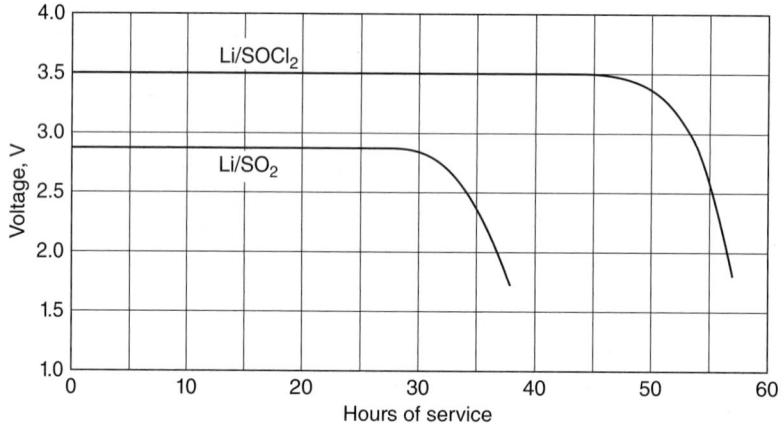

**FIGURE 13.2**  Comparison of performance of Li/SO$_2$ and Li/SOCl$_2$ C-size batteries; 100-mA discharge load at 20°C.

The lithium/sulfuryl chloride (Li/SO$_2$Cl$_2$) battery has potential advantages because of its higher voltage (3.9 open-circuit voltage) and resultant higher specific energy. Suitable cathode electrode formulations and cell designs have been investigated to achieve the full capability of this electrochemical system. Figure 13.3 shows a comparison of the cathode polarization for Li/SO$_2$Cl$_2$ and Li/SOCl$_2$ batteries. Halogen additives such as chlorine have been used as a means to improve performance. Halogen additives are also used in some cells.

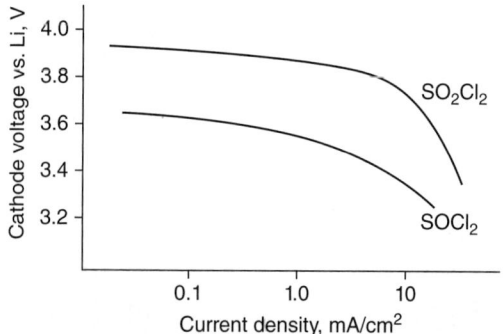

**FIGURE 13.3**  Comparison of cathode polarization curves, Li/SO$_2$Cl$_2$ versus Li/SOCl$_2$ batteries.[8]

Calcium has been investigated as an anode material in place of lithium in thionyl chloride cells. Safer operation was anticipated with calcium since its melting temperature of 838°C is not likely to be reached by any internally driven cell condition. While the discharge voltage is about 0.4 V lower than for the $Li/SOCl_2$ cell (open-circuit voltage is 3.25 V), the $Ca/SOCl_2$ cell has a flat discharge profile and about the same volumetric ampere-hour capacity. Shelf-life characteristics are also similar to those of the lithium anode cell.[9,10] However, calcium is significantly more difficult to process than lithium and passivation is a more significant factor. To date, no calcium-thionyl chloride batteries have been commercialized.

*Solid-Cathode Lithium Primary Cells.* The solid-cathode lithium batteries are generally used in low- to moderate-drain applications and are manufactured mainly in small flat or cylindrical sizes ranging in capacity from 25 mAh to about 11 Ah, depending on the particular electrochemical system. Larger batteries have been produced in cylindrical and prismatic configurations. A comparison of the performance of solid-cathode lithium batteries and conventional batteries is presented in Chap. 9.

The solid-cathode batteries have the advantage, compared with the soluble-cathode lithium primary batteries, of being nonpressurized and thus not requiring a hermetic-type seal. A mechanically crimped seal with a polymeric gasket is satisfactory for most applications. On light discharge loads, the energy density of some of the solid-cathode systems is comparable to that of the soluble-cathode systems, and in smaller battery sizes it may be greater. Their disadvantages, again compared with the soluble-cathode batteries, include lower rate capability, poorer low-temperature performance, and a more sloping discharge profile.

To maximize their high-rate performance and compensate for the lower conductivity of the organic electrolytes, designs are used for these lithium cells to increase electrode area, such as a larger-diameter coin cell instead of button cells, or the spirally wound jelly-roll construction for the cylindrical cells.

A number of different cathode materials have been used in the solid-cathode lithium cells. These were listed above in Tables 13.3 and 13.5, which presented some of the theoretical and practical performance data of these cells. The major features of the solid-cathode lithium cells are compared in Table 13.6. Many of the characteristics are similar, such as high specific energy and energy density and good shelf life. An important property is the 3 V cell voltage obtained with several of these cathodes. Some cathode materials have been used mainly in the coin or button cell designs, while others, such as the manganese dioxide cathode, have been used in coin, cylindrical, and prismatic cells, as well as in both high (spirally wound) and low (bobbin) rate designs.

**TABLE 13.6** Characteristics of Typical Lithium/Solid-Cathode Batteries

| Type of battery | Operating voltage, V | Characteristics |
|---|---|---|
| $Li/MnO_2$ | 3.0 | High specific energy and energy density; wide operating temperature range (−40 to +85°C); performance at relatively high discharge rates; minimal voltage delay; relatively low cost; available in flat (coin) and cylindrical batteries (high and low rates). |
| $Li/CFx$ | 2.8 | Highest theoretical specific energy, low- to moderate-rate capability; wide operating temperature range (−20 to 85°C); flat discharge profile; available in flat (coin), cylindrical and prismatic designs. |
| $Li/CuO$ | 1.5 | Highest theoretical volumetric coulombic capacity (Ah/L); long storage life; low- to moderate-rate capability; operating temperature range up to 125 to 150°C; no apparent voltage delay. Potential replacement for alkaline-manganese but not currently available. |
| $Li/FeS_2$ | 1.5 | Replacement for conventional zinc-carbon and alkaline-manganese dioxide batteries; higher power capability than conventional batteries and better low-temperature performance and storability. Currently available in AA and AAA sizes as a direct replacement for alkaline-manganese. Finding increasing use in digital cameras. |
| $Li/AgV_2O_{5.5}$ | 3.2 | High specific energy and energy density multiple-step discharge; good rate capability; used in implantable and other medical devices. See Chap. 28. |
| $Li/V_2O_5$ | 3.3 | High energy density; two-step discharge; used in reserve cells (see Chap. 29). |

Although a number of different solid-cathode lithium batteries have been developed and even manufactured, more recently the trend is toward reducing the number of different chemistries that are manufactured. The lithium/manganese dioxide ($Li/MnO_2$) battery was one of the first to be used commercially and is still the most popular system. It is relatively inexpensive, has excellent shelf life, good high-rate and low-temperature performance, and is available in coin and cylindrical cells. The lithium/carbon monofluoride ($Li/CFx$) battery is another of the early solid-cathode batteries and is attractive because of its high theoretical capacity and flat discharge characteristics. It is also manufactured in coin, cylindrical, and prismatic configurations. The higher cost of polycarbon monofluoride has affected the commercial potential for this system, but it is finding use in biomedical, military, and space applications where cost is not a factor. This system is also finding increased use in digital cameras.

The lithium/vanadium pentoxide ($Li/V_2O_5$) battery has a high volumetric energy density, but with a two-step discharge profile. Its main application has been in reserve batteries (Chap. 29). The lithium/silver vanadium oxide ($Li/AgV_2O_{5.5}$) battery is used in medical applications, such as defibrillators, which have pulse load requirements as this battery is capable of relatively high-rate discharge.[11] The other solid-cathode lithium batteries operate in the range of 1.5 V and were developed to replace conventional 1.5 V button or cylindrical cells. The lithium/copper oxide ($Li/CuO$) cell is noted for its high coulombic energy density and has the advantage of higher capacity or lighter weight when compared with conventional cylindrical cells. It is capable of performance at high temperatures and has a long shelf life under adverse conditions. It is not currently available commercially. The iron disulfide ($Li/FeS_2$) cell has similar advantages over the conventional cells, plus the advantage of high-rate performance. Once available in a button cell configuration, it is now being marketed commercially in high-rate cylindrical AA and AAA sizes as a replacement for alkaline-manganese dioxide batteries. The remaining solid-cathode systems listed in the tables above are not known to be commercially available.

Typical discharge curves for the major solid-cathode batteries are shown in Fig. 13.4. The discharge curves of the $Li/SO_2$ and $Li/SOCl_2$ batteries showing their flatter discharge profile are also plotted for comparison purposes.

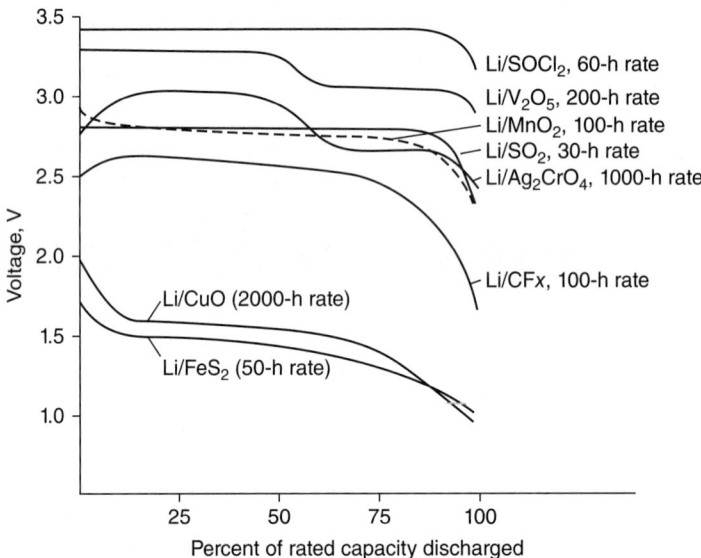

**FIGURE 13.4**   Typical discharge curves of lithium/solid-cathode batteries.

A comparison of the performance of several of the solid-cathode batteries in a low-rate button configuration and the higher-rate cylindrical configuration is presented in Chap. 9. In the button configuration the lithium batteries have an advantage in specific energy (Wh/kg) over many of the conventional batteries. This advantage may not be too important in these small battery sizes, but the lithium batteries have an advantage of lower cost, particularly when compared with the silver cells, and longer shelf life. In addition, the zinc/mercuric oxide battery, which once dominated the button battery market, and the cadmium/mercuric oxide battery are not currently available due to environmental factors.

In the larger cylindrical sizes (Table 9.4), the lithium cells have an advantage in both volumetric and gravimetric energy density. In some designs, this advantage is even more significant at higher discharge loads. Figure 9.3 shows another comparison of the performance of solid-cathode and soluble-cathode lithium cells with aqueous cells.

It is important when making these comparisons to identify the specific discharge conditions of the application since the comparative performance of each battery system can vary depending on the discharge conditions. For example, as shown in Fig. 13.5, the lithium/copper oxide battery, designed for optimum performance at low discharge rates, has a comparatively high energy output when discharged at these light discharge loads, but the output drops off considerably at high rates. The similarly sized high-rate spirally wound configuration for the lithium/manganese dioxide cell has a lower energy output at the low discharge rates, but can maintain this performance as the discharge rate is increased. The performance of each of the battery systems, under various discharge conditions, is presented in the sections discussing each specific system.

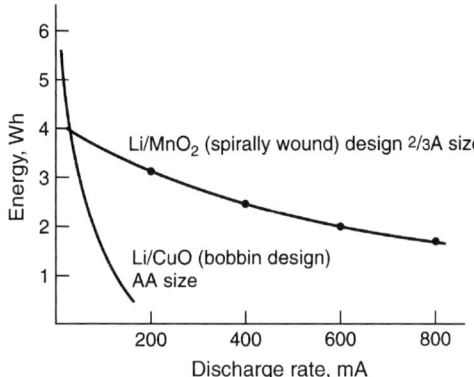

**FIGURE 13.5**   Comparison of Li/CuO and Li/MnO$_2$ batteries at 20°C. Batteries are equivalently sized.

The selection of a lithium versus a conventional cell thus becomes a trade-off between the lower initial cost of most of the conventional cells, the performance advantages of the lithium cells, and the key requirements of the specific application.

### 13.0.5   Safety and Handling of Lithium Batteries

*Factors Affecting Safety and Handling.*   Attention must be given to the design and use of lithium cells and batteries to ensure safe and reliable operation. As with most battery systems, precautions must be taken to avoid physical and electrical abuse because some batteries can be hazardous if not used properly. This is important in the case of lithium cells since some of the components are toxic or flammable,[12] and the relatively low melting point of lithium (180.5°C) indicates that cells must be prevented from reaching high internal temperatures.

Because of the variety of lithium cell chemistries, designs, sizes, and so on, the procedures for their use and handling are not the same for all cells and batteries and depend on a number of factors such as the following:

1. *Electrochemical system.* The characteristics of the specific chemicals and cell components influence operational safety.

2. *Size and capacity of cell and battery.* Safety is directly related to the size of the cell and the number of cells in a battery. Small cells and batteries, containing less material and, therefore, less total energy, are "safer" than larger cells of the same design and chemistry.

3. *Amount of lithium used.* The less lithium that is used, implying less energetic cells, the safer they should be.

4. *Cell design.* High-rate designs, capable of high discharge rates, versus low-power designs where discharge rate is limited, use of "balanced" cell chemistry, adequate intra- and intercell electrical connections, and other features affect cell performance and operating characteristics.

5. *Safety features.* The safety features incorporated in the cell and battery will obviously influence handling procedures. These features include cell-venting mechanisms to prevent excessive internal cell pressure, thermal cutoff devices to prevent excessive temperatures, electrical fuses, positive temperature coefficient (PTC) devices, and diode protection. Cells are hermetically or mechanically crimped-sealed, depending on the electrochemical system, to effectively contain cell contents if cell integrity is to be maintained.

6. *Cell and battery containers.* These should be designed so that cells and batteries will meet the mechanical and environmental conditions to which they will be exposed. High shock, vibration, extremes of temperature, or other adverse conditions may be encountered in use and handling, and the cell and battery integrity must be maintained. Container materials should also be chosen with regard to their flammability and the toxicity of combustion products in the event of fire. Container designs should also be optimized to dissipate the heat generated during discharge and to release pressure in the event of cell venting.

**Safety Considerations.**    The electrical and physical abuses that may arise during the use of lithium cells are listed in Table 13.7 together with some generalized comments on corrective action. The behavior of specific cells is covered in the other sections of this chapter. The manufacturer's data should be consulted for more details on the performance of individual cells. Material safety data sheets (MSDSs) should also be obtained.

**TABLE 13.7**    Considerations for Use and Handling of Lithium Primary Batteries

| Abusive condition | Corrective procedure |
|---|---|
| High-rate discharging or short-circuiting | Low-capacity or low-rate batteries may be self-limiting<br>Electrical fusing, thermal protection<br>Limit current drain; apply battery properly |
| Forced discharge (cell reversal) | Voltage cutoff<br>Use low-voltage batteries<br>Limit current drain<br>Special designs ("balanced" cell)<br>Use of diode in parallel across cells to bypass current |
| Charging | Prohibit charging<br>Diode protection to prevent or limit charging current |
| Overheating | Limit current drain<br>Fusing, thermal cutoff, PTC devices<br>Design battery properly<br>Do not incinerate |
| Physical abuse | Avoid opening, puncturing, or mutilating cells<br>Maintain battery integrity |

*High-Rate Discharges or Short-Circuiting.*    Low-capacity batteries, or those designed as low-rate batteries, may be self-limiting and not capable of high-rate discharge. The temperature rise will thus be minimal and there will be no safety problems. Larger or high-rate cells can develop high internal temperatures if short-circuited or operated at excessively high rates. These cells are generally equipped with safety vent mechanisms to avoid more serious hazards. Such cells or batteries should be fuse-protected (to limit the discharge current). Thermal fuses or thermal switches should also be used to limit the maximum temperature rise. PTC devices are used in cells and batteries to provide this protection.

*Forced Discharge or Voltage Reversal.*    Voltage reversal can occur in a multicell series-connected battery when the better performing cells can drive the poorer cell below 0 V, into reversal, as the battery is discharged toward 0 V. In some types of lithium cells, this forced discharge can result in cell venting or, in more extreme cases, cell rupture. Precautionary measures include the use of voltage cutoff circuits to prevent a battery from reaching a low voltage, the use of low-voltage batteries (since this phenomenon is unlikely to occur with a battery containing only a few cells in series), and limiting the current drain, since the effect of forced discharge is more pronounced on high-rate discharges. Special designs, such as the "balanced" $Li/SO_2$ cell (see Chap. 13A), also have been developed that are capable of withstanding this discharge condition. The use of a current collector in the anode maintains lithium integrity and may provide an internal shorting mechanism to limit the voltage in reversal. (See Chap. 7 for further battery pack design details.)

*Charging.*    Lithium primary batteries, as well as the other primary batteries, are not designed to be recharged. If they are, they may vent or explode. Batteries that are connected in parallel or that may be exposed to a charging source (as in battery-backup CMOS memory circuits) should be diode-protected to prevent charging (see Chap. 7).

*Overheating.*    As discussed, overheating should be avoided. This can be accomplished by limiting the current drain, using safety devices such as fusing and thermal cutoffs, and designing the battery to provide necessary heat dissipation.

*Incineration.*    Lithium cells are either hermetically or mechanically sealed. They should not be incinerated without proper protection because they may rupture or explode at high temperatures.

Currently, special procedures govern the transportation and shipment of lithium batteries, and procedures for the use, storage, and handling of lithium batteries have been recommended.[13–14] Disposal of some types of lithium cells also is regulated. The latest issue of these regulations should be consulted for the most recent procedures (see Chap. 8 for details). The U.S. Federal Aviation Agency has adapted technical standard order, TSO-C142-Lithium Batteries, governing the installation and use of lithium primary batteries on commercial aircraft.[15] U.S. DOT, IATA, ICAO, and other governmental agencies issue regulations governing the shipment of lithium batteries.

# SECTION A

# LITHIUM/SULFUR DIOXIDE

## 13A.1  LITHIUM/SULFUR DIOXIDE (Li/SO₂) BATTERIES

One of the more advanced lithium primary batteries, used mainly in military and in some industrial and space applications, is the lithium/sulfur dioxide ($Li/SO_2$) system. The battery has specific energy and energy density of up to 300 Wh/kg and 415 Wh/L, respectively, in large sizes. The $Li/SO_2$ battery is particularly noted for its capability to handle high current and high power requirements, excellent low-temperature performance, and long shelf life.

## 13A.2  CELL CHEMISTRY

The $Li/SO_2$ cell uses lithium as the anode and a porous carbon cathode electrode with sulfur dioxide as the active cathode material. The cell reaction mechanism is

$$2Li + 2SO_2 \rightarrow Li_2S_2O_4 \downarrow \text{(lithium dithionite)}$$

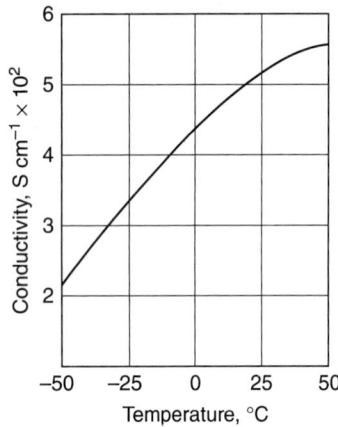

**FIGURE 13A.1**  Conductivity of acetonitrile/lithium bromide/sulfur dioxide electrolyte (70% $SO_2$).

As lithium reacts readily with water, a nonaqueous electrolyte consisting of sulfur dioxide and an organic solvent, typically acetonitrile, with dissolved lithium bromide is used. The specific conductivity of this electrolyte is relatively high and decreases only moderately with decreasing temperature (Fig. 13A.1), thus providing a basis for good high-rate and low-temperature performance. About 70% of the weight of the electrolyte/depolarizer is $SO_2$. The internal cell pressure, in an undischarged cell, due to the vapor pressure of the liquid $SO_2$, is $3-4 \times 10^5$ Pa at 20°C. The pressure at various temperatures is shown in Fig. 13A.2. The mechanical features of the cell are designed to contain this pressure safely without leaking and to vent the electrolyte if excessively high temperatures and the resulting high internal pressures are encountered.

During discharge, the $SO_2$ is used up and the cell pressure reduced somewhat. The discharge is generally terminated by the full use of available lithium in designs where the lithium is the limiting electrode or by blocking of the cathode by precipitation of the discharge product (cathode limited). Current designs are typically limited by the cathode so that some lithium remains at the end of discharge. The good shelf life of the $Li/SO_2$ cell results from the protective lithium dithionite film on the anode formed by the initial reaction of lithium and $SO_2$. It prevents further reaction and loss of capacity during storage.

Most $Li/SO_2$ cells are now fabricated in a balanced construction where the lithium:sulfur dioxide stoichiometric ratio is in the range of $Li:SO_2 = 0.9-1.05:1$. With the earlier designs, where the ratio was on the order of $Li:SO_2 = 1.5:1$, high temperatures, cell venting, or rupture and fires due to an exothermic reaction between residual lithium and acetonitrile, in the absence of $SO_2$, could occur on deep or forced discharge. Lithium cyanide, methane, and other organic products can also be generated through this reaction. In the balanced cell, the anode is protected by residual $SO_2$ and remains passivated. The conditions for the hazardous reaction are minimized since some protective $SO_2$ remains in the electrolyte.[16] A higher negative cell voltage, occurring when the

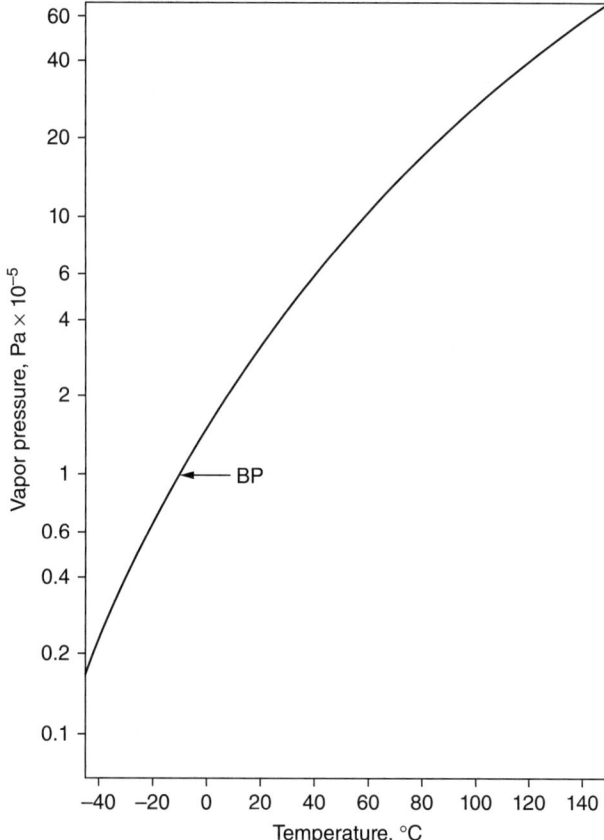

**FIGURE 13A.2**  Vapor pressure of sulfur dioxide at various temperatures.

balanced cell goes into reversal due to over-discharge, is also beneficial when diode protection is used to bypass the current through the cell and minimize the adverse effects of reversal.

The use of a current collector, typically an inlayed stripe of copper metal, also helps to maintain the integrity of the anode and leads to formation of a short-circuit mechanism since copper dissolution on cell reversal causes plated copper on the cathode to form an internal ohmic bridge.

## 13A.3   CONSTRUCTION

The Li/$SO_2$ cell is typically fabricated in a cylindrical structure, as shown in Fig. 13A.3. A jelly-roll construction is used, made by spirally winding rectangular strips of lithium foil, a microporous polypropylene separator, the cathode electrode (a Teflon-acetylene black mix pressed on an expanded aluminum screen), and a second separator layer. This design provides the high surface area and low cell resistance to obtain high-current and low-temperature performance. The roll is inserted in a nickel-plated steel can, with the positive cathode tab welded to the pin of a glass-to-metal (GTM) seal and the anode tab welded to the cell case, the top is welded in place, and the electrolyte/depolarizer is added. The safety vent releases when the internal cell pressure reaches excessive levels, typically 2.41 MPa (350 psi) caused by inadvertent abusive use, such as overheating

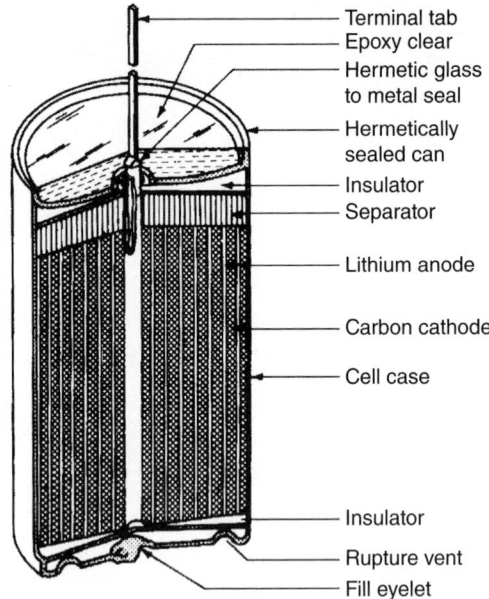

Terminal tab
Epoxy clear
Hermetic glass to metal seal
Hermetically sealed can
Insulator
Separator
Lithium anode
Carbon cathode
Cell case
Insulator
Rupture vent
Fill eyelet

**FIGURE 13A.3**   Lithium/sulfur dioxide batteries.

or short-circuiting, and prevents cell rupture or explosion. The vent activates at approximately 95°C, well above the upper temperature limit for operation and storage, safely relieving the excess pressure and preventing possible cell rupture. Additional construction details have been previously described.[16] It is important to employ a corrosion-resistant glass to prevent lithiation of the glass due to the potential difference between the cell case and the pin of the GTM seal.

## 13A.4   PERFORMANCE CHARACTERISTICS AND APPLICATIONS

### 13A.4.1   Cell Output

*Voltage.*   The open-circuit voltage of the Li/SO$_2$ battery is 2.95 V. The nominal voltage is usually specified as 3 V. The specific voltage on discharge is dependent on the discharge rate, discharge temperature, and state of charge; typical working voltages range between 2.7 and 2.9 V. See the sections below (Figs. 13A.4, 13A.5, and 13A.7) for some typical voltage profiles. The cutoff voltage, the voltage by which most of the battery capacity has been exhausted, is typically 2 V.

*Discharge.*   Typical discharge curves for the standard-rate Li/SO$_2$ battery at 20°C are given in Fig. 13A.4a. The high cell voltages and the flat discharge profile are characteristic of the Li/SO$_2$ battery. Another unique feature is the ability of the Li/SO$_2$ battery to be efficiently discharged over a wide range of current or power levels, from high-rate short-term or pulse loads to low-drain continuous discharges for periods of 5 years or longer. At least 90% of the battery's rated capacity may be expected on the long-term discharges. Figure 13A.4b shows the discharge curves for a high-rate D-size battery at four rates up to 3 A.

The Li/SO$_2$ battery is capable of higher-rate discharges on pulse loads. For example, a squat D cell designed in a high-rate construction can deliver pulse loads as high as 37.5 A, producing 59 W of power.[17] For high-rate designs, extended discharges, however, at rates above the 2 h rate may cause overheating. The actual heat rise

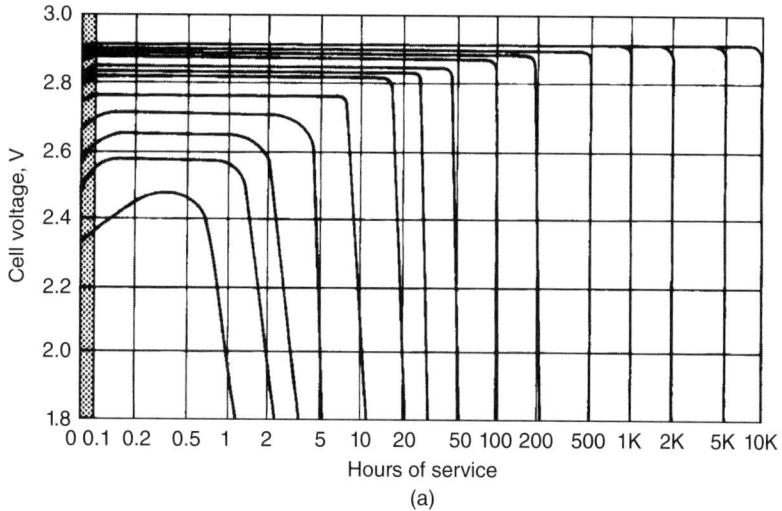

(a)

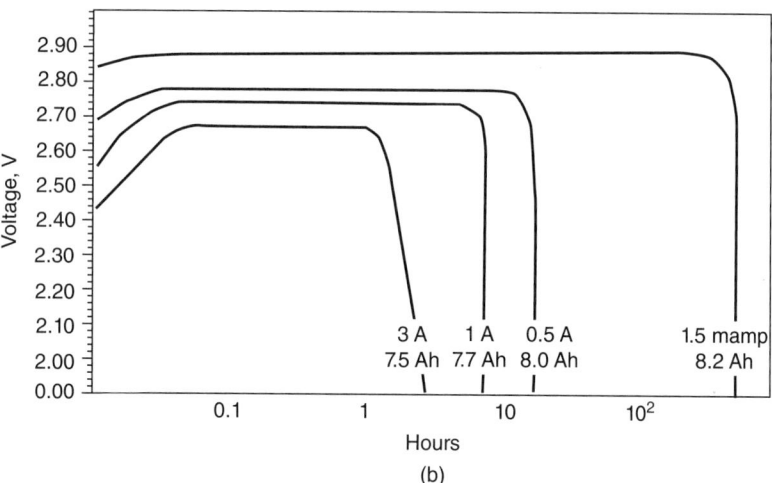

(b)

**FIGURE 13A.4**  (*a*) Typical discharge characteristics of standard-rate Li/SO$_2$ battery at various loads at 20°C. (*b*) Discharge characteristics of high-rate Li/SO$_2$ D-size battery at four rates at 23°C.

depends on the battery design, type of discharge, temperature, and voltage. As discussed in Sec. 13.0.5, the design and use of the battery should be controlled to avoid overheating.

A study has shown that the high-rate pulse output of the lithium/sulfur dioxide battery may be enhanced by a variety of design variables.[17] Multiple tabbing (one to three) of both anode and cathode, optimizing the composition of the cathode mix and reducing the aspect ratio (length/width) of the electrodes were all found to reduce polarization during high-rate, 10 s pulse discharge. D-size cells and thin D-size cells (1.1 in diameter × 2.20 in high) with anodes and cathodes containing two tabs using an optimized cathode mix were found capable of producing 99 and 97 W, respectively, under 50 A, 10 s pulses. Ultimately, a 5/4 C-size cell without multiple tabbing but using the optimized cathode mix was selected for reasons of volumetric efficiency to produce a 74-cell, 110 V battery capable of providing 5500 W, 10 s pulses for a U.S. Navy application.

A similar design optimization study has resulted in the production of a $Li/SO_2$ D cell with a room-temperature capacity of 9.1 Ah at 250 mA and 8.8 Ah at 2 A.[18] This compares to 7.75 Ah for the standard design and was achieved through an optimization study in which the aspect ratios of both anode and cathode were varied along with the use of three types of carbon in the cathode and the use of a central cathode tab. When deep discharged down to between 2.0 and 0.0 V, these cells were found to generate less heat than the standard cells. The high-capacity cells were used to construct U.S. Military BA-5590 batteries, which were tested to the requirements of MIL-PRF-49471. These batteries met the specification requirements for performance and safety.

### 13A.4.2    Factors Affecting Performance

*Effect of Temperature.*    The $Li/SO_2$ battery is noted for its ability to perform over a wide temperature range, from −40 to 55°C. Discharge curves for a standard-rate $Li/SO_2$ battery at various temperatures are shown in Fig. 13A.5. Significant, again, are the flat discharge curves over a wide temperature range, the good voltage regulation, and the high percentage of the 20°C performance available at the temperature extremes. As with all battery systems, the relative performance of the $Li/SO_2$ battery is dependent on the rate of discharge. In Fig. 13A.6, the discharge performance of a standard-rate cell is plotted as a function of load and battery temperature.

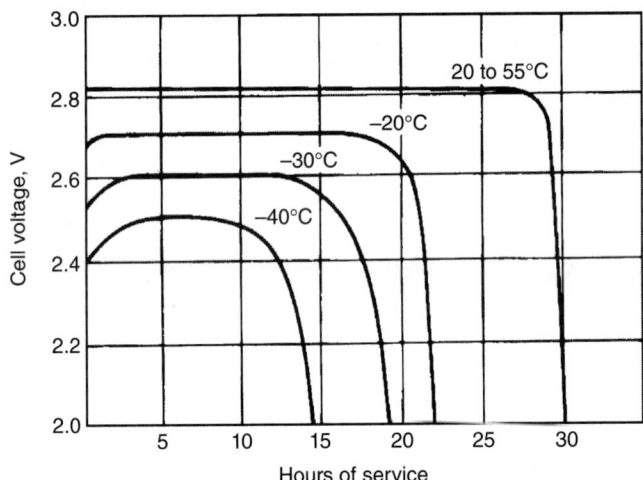

**FIGURE 13A.5**    Typical discharge characteristics of $Li/SO_2$ battery at various temperatures, C/30 discharge rate.

*Internal Resistance and Discharge Voltage.*    The $Li/SO_2$ battery has a relatively low internal resistance (about one-tenth that of conventional primary batteries) and good voltage regulation over a wide range of discharge loads and temperatures. The midpoint voltage for the discharge of a standard-rate $Li/SO_2$ battery (to an end voltage of 2 V) at various discharge rates and temperatures is plotted in Fig. 13A.7.

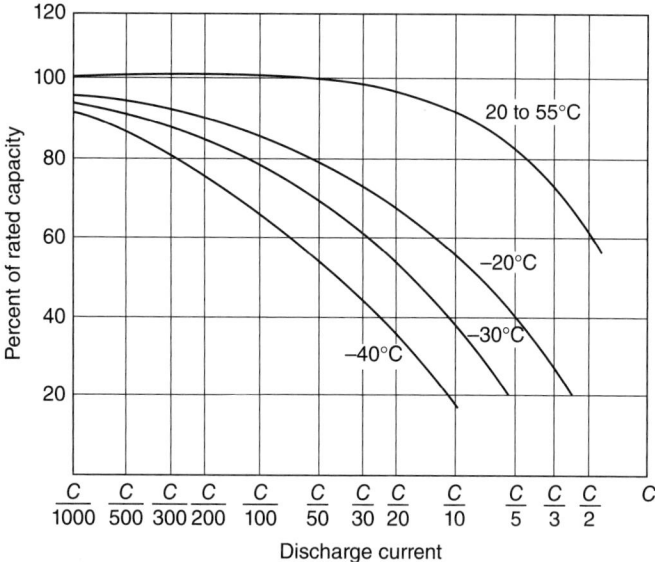

**FIGURE 13A.6**  Performance of Li/SO$_2$ batteries as a function of discharge temperature and load.

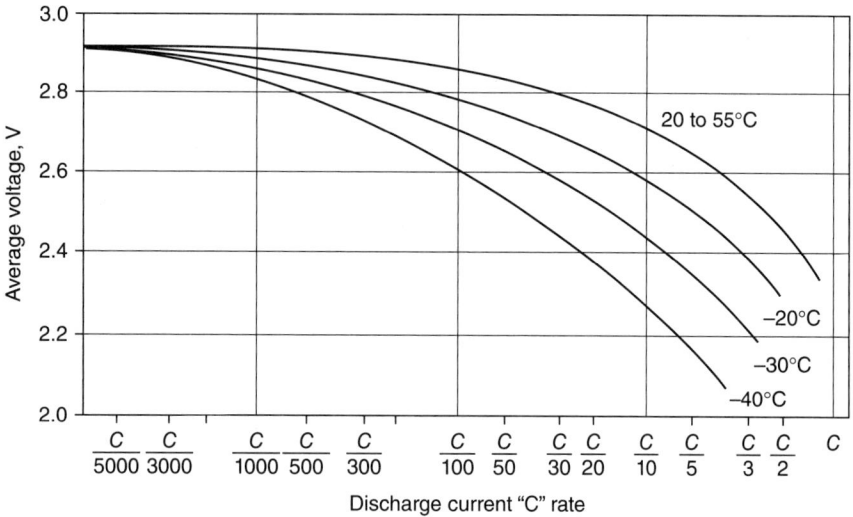

**FIGURE 13A.7**  Midpoint voltage of Li/SO$_2$ batteries during discharge.

***Service Life.***  The capacity or service life of the Li/SO$_2$ battery at various discharge rates is given in Fig. 13A.8. The data are normalized for a 1-kg or 1-L size battery and presented in terms of hours of service at various discharge rates. The linear shape of this curve is again indicative of the capability of the Li/SO$_2$ battery to be efficiently discharged at these extreme conditions. These data can be used in several ways to calculate the approximate performance of a given battery or to select a Li/SO$_2$ battery of suitable size for a particular application, recognizing that the specific energy of the larger-size batteries is higher than that of the smaller ones.

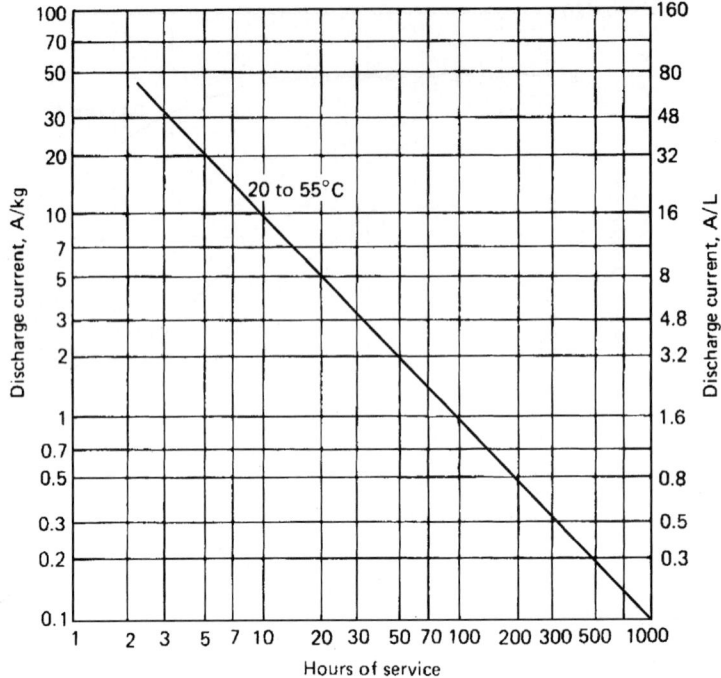

**FIGURE 13A.8**   Service life of high-rate Li/SO$_2$ batteries; 2.0 V end voltage.

The service life of a battery at a given current load can be estimated by dividing the current (in amperes) by the weight or volume of the battery. This value is located on the ordinate, and the service life, at a specific current and temperature, is read on the abscissa.

The weight or volume of a battery needed to deliver a required number of hours of service at a specified current load can be estimated by locating a point on the curve corresponding to the required service hours and discharge temperature. The battery weight or volume is calculated by dividing the value of the specified current (in amperes) by the value of amperes per kilogram or amperes per liter obtained from the ordinate.

**Shelf Life.**   The Li/SO$_2$ battery is noted for its excellent storage characteristics, even at temperatures as high as 70°C. Most primary batteries lose capacity while idle or on standby due to anode corrosion, side chemical reactions, or moisture loss. With the exception of the magnesium battery, most of the conventional primary batteries cannot withstand temperatures in excess of 50°C and should be refrigerated if stored for long periods. The Li/SO$_2$ battery, however, is hermetically sealed and protected during storage by the formation of a film on the anode surface. Capacity losses during stand are minimal. If cells are partially discharged and then stored, the self-discharge rate is accelerated.

Data on 2-year-old BA-5590 batteries consisting of 10 Li/SO$_2$ D-size cells discharged in series at 2 A at +21 and −30°C showed a 6.5% capacity loss at the higher temperature, but no loss at the lower temperature.[19] Fourteen-year storage data were also obtained on BA-5598 batteries consisting of five "squat" D-size cells in series. These batteries showed only an 8% capacity loss when discharged at room temperature at 2 A, but virtually no loss at cold temperature. In both cases, a lower operating voltage was observed after storage. Using multiple groups of batteries, stored for 4, 6, and 14 years under ambient conditions, the data shown in Fig. 13A.9 were obtained. The capacity loss for the first two years is approximately 3%/yr, but the rate of loss decreases significantly after that period. High-temperature storage of batteries was also carried out at +70 and +85°C, as shown in Fig. 13A.10. At 70°C, these batteries showed 92% capacity retention after 1 month and 77% capacity retention

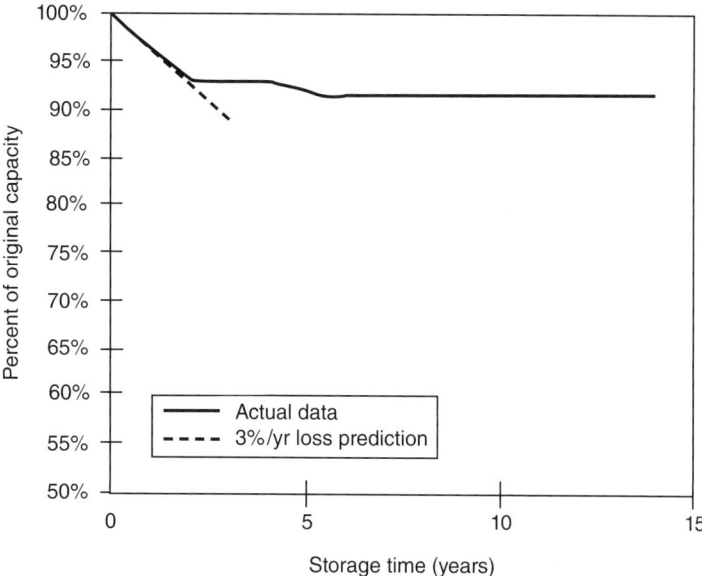

**FIGURE 13A.9**    Capacity retention of Li/SO$_2$ batteries after ambient storage and discharge at the 2 A rate.

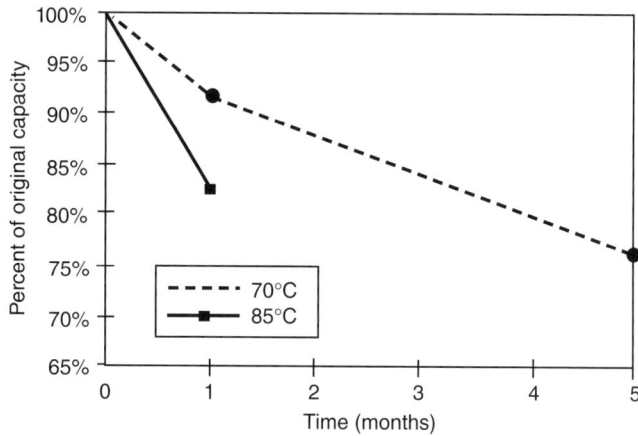

**FIGURE 13A.10**    Effect of storage time/temperature on capacity of Li/SO$_2$ batteries.

after 5 months. At 85°C, 82% capacity retention was observed after 1 month's storage. This study concluded that there was no obvious benefit to making long-term storage predictions based on accelerated aging tests at high temperature.

***Voltage Delay.***    After extended long-term storage at elevated temperatures, the Li/SO$_2$ battery may exhibit a delay in reaching its normal operating voltage when placed on discharge, especially at high current loads and low temperatures. This start-up or voltage delay is caused by the protective film formed on the lithium anode, the characteristic responsible for the excellent shelf life of the cell. The specific delay time for a battery depends on such factors as the history of the battery, the specific cell design and components, the storage time and temperature, discharge load, and temperature. Typically, the voltage delay is minimal or nonexistent for discharges at moderate to low rates at temperatures above $-20°C$. No delay is evident on discharge at 20°C, even after storage at 70°C for 1 year. On discharge at $-30°C$, the delay time is less than 200 ms after 8 weeks of storage at 70°C on discharges lower than the 40 h rate. At higher rates, the voltage delay increases with increasing storage temperature and time. At the 2 h discharge rate, for example, the maximum start-up time is about 80 s after 8 weeks of storage at 70°C; it is 7 s after 2 weeks of storage.[20] The start-up voltage delay can be eliminated by preconditioning with a short discharge at a higher rate to depassivate the anode until the operating voltage is reached since the delay will return only after another extended storage period.

### 13A.4.3 Cell and Battery Types and Sizes

Li/SO$_2$ batteries are manufactured in a number of cylindrical cell sizes, ranging in capacity to 34 Ah. Some of the cells are manufactured in standard ANSI (American National Standards Institute) cell sizes in dimensions of popular conventional zinc primary batteries. While these single batteries may be physically interchangeable, they are not electrically interchangeable because of the higher cell voltage of the lithium cell (3.0 V for lithium, 1.5 V for the conventional zinc cells). Table 13A.1 lists some of the sizes and rated capacities of Li/SO$_2$ batteries that are currently manufactured.

**TABLE 13A.1**   Typical Lithium/Sulfur Dioxide Cylindrical Cells

| Size | Open-circuit voltage, V | Nominal voltage, V | Nominal capacity, mAh (drain, mA) | Max. recommended continuous current, mA | Outside diameter max., mm | Height max., mm | Weight, g | Transport status |
|------|------|------|------|------|------|------|------|------|
| 1/2 AA | 3 | 2.8 | 450 (50) | 250 | 14.2 | 27.9 | 8 | Nonrestricted |
| AA | 3 | 2.8 | 950 (80) | 500 | 14.2 | 50.3 | 15 | Nonrestricted |
| 2/3 A | 3 | 2.8 | 800 (80) | 750 | 16.3 | 34.5 | 12 | Nonrestricted |
| "Long" A | 3 | 2.8 | 1700 (80) | 1500 | 16.3 | 57.7 | 18 | Nonrestricted |
| 1/3 C | 3 | 2.9 | 860 (80) | 1000 | 25.9 | 20.3 | 18 | Nonrestricted |
| 2/3 C | 3 | 2.8 | 2200 (650) | 2000 | 25.9 | 35.9 | 30 | Nonrestricted |
| C | 3 | 2.8 | 3200 (1000) | 2500 | 25.6 | 49.5 | 47 | Class 9 |
| C | 3 | 2.8 | 3750 (250) | 2500 | 25.9 | 50.4 | 40 | Class 9 |
| 5/4 C | 3 | 2.8 | 5000 (200) | 2500 | 25.6 | 60.2 | 58 | Class 9 |
| 5/4 C | 3 | 2.8 | 5000 (200) | 2500 | 25.9 | 59.3 | 53 | Class 9 |
| 2/3 "Thin" D | 3 | 2.8 | 3500 (120) | 2000 | 28.95 | 42.29 | 40 | Class 9 |
| "Thin" D | 3 | 2.8 | 5750 (200) | 2500 | 29.1 | 59.9 | 63 | Class 9 |
| D | 3 | 2.8 | 7750 (250) | 2500 | 34.5 | 59.8 | 85 | Class 9 |
| D | 3 | 2.8 | 7750 (250) | 2500 | 34.2 | 59.3 | 85 | Class 9 |
| D | 3 | 2.8 | 9200 (250) | 2500 | 34.2 | 59.3 | 85 | Class 9 |
| D | 3 | 2.8 | 7500 (250) | 4000 | 34.2 | 59.3 | 85 | Class 9 |
| "Fat" D | 3 | 2.8 | 8000 (270) | 2500 | 39.5 | 50.3 | 96 | Class 9 |
| F | 3 | 2.8 | 11,500 (1000) | 3000 | 31.9 | 100.3 | 125 | Class 9 |
| DD | 3 | 2.8 | 16,500 (500) | 3000 | 33.3 | 120.6 | 175 | Class 9 |
| "Long Fat DD" | 3 | 2.8 | 34,000 (1000) | 3000 | 41.7 | 141.0 | 300 | Class 9 |

*Source:* SAFT Batteries.
Cells leakproof up to +95°C. Most cells are UL recognized.
Operating temperature range: −60 to +70°C.

### 13A.4.4  Use and Handling of Li/SO$_2$ Cells and Batteries: Safety Considerations

The Li/SO$_2$ battery is designed as a high-performance system and is capable of delivering a high capacity at high discharge rates. The cell should not be physically or electrically abused, safety features should not be bypassed, and manufacturers' instructions should be followed.

Abusive conditions could adversely affect the performance of the Li/SO$_2$ battery and result in cell venting, rupture, explosion, or fire. Preventive measures are discussed in Sec. 13.0.5.

The Li/SO$_2$ battery is pressurized and contains materials that are toxic or flammable. Properly designed batteries are hermetically sealed so there will be no leakage or out-gassing, and they are equipped with safety vents that release if the batteries reach excessively high temperatures and pressures, thus preventing an explosive condition.

The Li/SO$_2$ batteries can deliver very high currents. Because high internal temperatures can develop from continuous high current drain and short-circuit, batteries must be protected by electrical fusing and thermal cutoffs. Charging of Li/SO$_2$ batteries may result in venting, rupture, or even explosion and should never be attempted. Cells or groups of cells connected in parallel should be diode-protected to prevent one group from charging another. The balanced Li/SO$_2$ cell is designed to handle forced discharges or cell reversal and will perform safely within the specified bounds, but design limits should not be exceeded in any application.

Proper battery design, using the Li/SO$_2$ cell, should follow these guidelines:

1. Use electrical fusing and/or current-limiting devices to prevent high currents or short-circuits.

2. Protect with diodes if cells are paralleled or connected to a possible charging source.

3. Minimize heat buildup by adequate heat dissipation and protect with thermal cutoff devices.

4. Do not inhibit cell vents in battery construction.

5. Do not use flammable materials in the construction of batteries.

6. Allow for release of vented gases.

7. Incorporate resistor with an activation switch to ensure complete depletion of active materials after normal discharge. This allows disposal as nonhazardous waste.

8. In certain cases, a diode is placed in parallel with the cell to limit the voltage excursion in reversal.

Currently special procedures govern the transportation, shipment, and disposal of Li/SO$_2$ batteries as well as other lithium batteries.[12-15] Procedures for the use, storage, and handling of these batteries also have been recommended. The latest issue of these regulations should be consulted for the most recent procedures.

### 13A.4.5  Applications

The desirable characteristics of the Li/SO$_2$ battery and its ability to deliver a high energy output and operate over a wide range of temperatures, discharge loads, and storage conditions have opened up applications for this primary battery that heretofore were beyond the capability of primary battery systems.

Major applications for the Li/SO$_2$ battery are in military equipment, such as night-vision devices, radio transceivers, and portable surveillance devices, taking advantage of its light weight and wide-temperature operation. Table 13A.2 lists the most common types of military Li/SO$_2$ and Li/MnO$_2$ batteries, their characteristics, and applications. These batteries are constructed to meet the requirements of MIL-PRF-49471 B (CR) and the applicable specification sheets for the particular battery type. Other military applications, such as sonobuoys and munitions, have long shelf-life requirements, and the active Li/SO$_2$ primary battery can replace reserve batteries used earlier. Some industrial applications have developed, particularly to replace secondary batteries and eliminate the need for recharging. Consumer applications have been limited to date because of restrictions in shipment and transportation and concern with its hazardous components.[21]

**TABLE 13A.2**   U.S. Military Lithium Nonrechargeable Batteries (MIL-PRF-49471[*])

| Type designation | Open-circuit voltage (series/parallel), V | Nominal voltage (series/parallel), V | Nominal energy,[†] Wh | Weight, g | Typical/applications |
|---|---|---|---|---|---|
| Lithium/sulfur dioxide batteries | | | | | |
| BA-5093/U | 27 | 23.4 | 77.2 | 635 | Respirators |
| BA-5557A/U | 30/15 | 16/13 | 54 | 410 | Digital message devices |
| BA-5588A/U | 15 | 13 | 35 | 290 | PRC-68 and PRC-126 radios, respirators |
| BA-5590A/U[‡] | 30/15 | 26/13 | 185 | 1021 | SINCGARS (single channel ground and airborne radio system) radios, chemical agent detectors, satellite radios, jammers, loudspeakers, range finders, countermeasures |
| BA-5590B/U[‡] | 30/15 | 26/13 | 185 | 1021 | Same as BA-5590A/U |
| BA-5598A/U | 15 | 13 | 87 | 650 | PRC-77 radios, direction finders, sensors |
| BA-5599A/U | 9 | 7.8 | 50 | 450 | Test sets, sensors |
| Lithium/manganese dioxide batteries | | | | | |
| BA-5312/U | 13.2 | 10.8 | 41 | 275 | PRC-112G survival radio |
| BA-5347/U | 6.6 | 5.4 | 40 | 290 | Thermal weapons sights, test sets |
| BA-5360/U | 9.9 | 8.1 | 65 | 320 | Digital communications devices |
| BA-5367/U | 3.3 | 2.7 | 3.25 | 20 | Night vision devices |
| BA-5368/U | 13.2 | 10.8 | 12 | 140 | PRC-90 survival radios |
| BA-5372/U | 6.6 | 5.4 | 2.3 | 20 | Memory hold function, encoding devices |
| BA-5380/U | 6.6 | 5.4 | 45 | 230 | Ground navigation sets, chemical agent monitors, respirators |
| BA-5388/U | 16.5 | 13.5 | 49 | 500 | PRC-68 and PRC-126 radios, respirators |
| BA-5390/U[‡] | 33/16.5 | 27/13.5 | 250 | 1350 | SINCGARS radios, chemical agent detectors, satellite radios, jammers, loudspeakers, range finders, countermeasures |
| BA-5390A/U[‡] | 33/16.5 | 27/13.5 | 250 | 1350 | Same as BA-5390/U |

*Notes:* Contributed by Mr. Patrick Lyman, U.S. Army Material Command.
[*]MIL-PRF-49471 will be replaced by MIL-PRF-32271 in DOD procurements.
[†]Nominal energy rating for temperature range of 25 ± 10°C (77 ± 18°F).
[‡]The BA-5590A/U and BA-5390A/U have built-in state-of-charge indicators (SOCIs); the BA-5590B/U and BA-5390/U do not.

# SECTION B

# LITHIUM/THIONYL CHLORIDE

## 13B.1   LITHIUM/THIONYL CHLORIDE (Li/SOCl₂) BATTERIES

The lithium/thionyl chloride (Li/SOCl$_2$) battery has one of the highest cell voltages (nominal voltage 3.6 V) and energy densities of any practical battery system. Specific energy and energy densities range up to about 590 Wh/kg and 1100 Wh/L, the highest values being achieved with the low-rate batteries. Figures 9.2, 9.7, 9.8, and 13.2 illustrate some of the advantageous characteristics of the Li/SOCl$_2$ cell.

Li/SOCl$_2$ batteries have been fabricated in a variety of sizes and designs, ranging from wafer or coin cells with capacities as low as 420 mAh, cylindrical cells in bobbin, and spirally wound electrode structures to large 10,000 Ah prismatic cells, plus a number of special sizes and configurations to meet particular requirements. The thionyl chloride system originally suffered from safety problems, especially on high-rate discharges and over-discharge, and a voltage delay that was most evident on low-temperature discharges after high-temperature storage.[22]

Low-rate batteries have been used commercially for many years for memory backup and other applications requiring a long operating life, such as toll tags and RF transponders. The large prismatic batteries have been used in military applications as an emergency backup power source. Medium- and high-rate batteries have also been developed as power sources for a variety of electric and electronic devices. Some of these batteries contain additives to the thionyl chloride and other oxyhalide electrolytes to enhance certain performance characteristics. These are covered in Chap. 13C.

## 13B.2   CELL CHEMISTRY

The Li/SOCl$_2$ cell consists of a lithium anode, a porous carbon cathode, and a nonaqueous SOCl$_2$:LiAlCl$_4$ electrolyte. Other electrolyte salts such as LiGaCl$_4$ have been employed for specialized applications. Thionyl chloride is both the electrolyte solvent and the active cathode material. There are considerable differences in electrolyte formulations and electrode characteristics. The proportions of anode (lithium foil), cathode (porous carbon substrate), and thionyl chloride will vary depending on the manufacturer and the desired performance characteristics. Significant controversy exists as to the relative safety of anode-limited versus cathode-limited designs.[23] Some cells have one or more electrolyte additives. Catalysts, metallic powders, or other substances have been used in the carbon cathode or in the electrolyte to enhance performance.

The generally accepted overall reaction mechanism is

$$4\mathrm{Li} + 2\mathrm{SOCl}_2 \rightarrow 4\mathrm{LiCl}\!\downarrow + \mathrm{S} + \mathrm{SO}_2$$

The sulfur and sulfur dioxide are initially soluble in the excess thionyl chloride electrolyte, and there is a moderate buildup of pressure due to the generation of sulfur dioxide during the discharge. The lithium chloride, however, is not soluble and precipitates within the porous carbon cathode as it is formed. Sulfur may precipitate in the cathode at the end of discharge. In most cell designs and discharge conditions, this blocking of the cathode is the factor that limits the cell's service or capacity. Formation of sulfur as a discharge product can also present a problem because of a possible reaction with lithium, which may result in a thermal runaway condition.

The lithium anode is protected by reacting with the thionyl chloride electrolyte during stand, forming a protective LiCl film on the anode as soon as it contacts the electrolyte. This passivating film, while contributing to the excellent shelf life of the cell, can cause a voltage delay at the start of a discharge, particularly on low-temperature discharges after long stands at elevated temperatures. The presence of trace qualities of moisture leads to the formation of HCl, which increases passivation, as does the presence of trace levels (ppm) of iron. Some products have special anode treatments or electrolyte additives to overcome or lower this voltage delay.

The low freezing point of thionyl chloride (below $-110°C$) and its relatively high boiling point ($78.8°C$) enable the cell to operate over a wide range of temperatures. The electrical conductivity of the electrolyte decreases only slightly with decreasing temperature. Some of the components of the $Li/SOCl_2$ systems are toxic and flammable; thus exposure to open or vented cells or cell components should be avoided.

## 13B.3   CELL DESIGNS/CONSTRUCTIONS

Thionyl chloride cells are constructed in a variety of configurations. The cell designs listed below are reviewed in Sec. 13B.4 along with related performance data.

- Bobbin-type cylindrical batteries
- Spirally wound cylindrical batteries
- Flat or disk-type $Li/SOCl_2$ cells
- Large prismatic $Li/SOCl_2$ cells

## 13B.4   PERFORMANCE CHARACTERISTICS AND APPLICATIONS

### 13B.4.1   Bobbin-Type Cylindrical Batteries

$Li/SOCl_2$ bobbin batteries are manufactured in a cylindrical configuration, most in sizes conforming to ANSI standards. These batteries are designed for low- to moderate-rate discharge and are not typically subjected to continuous discharge at rates higher than the $C/50$ rate. They have a high energy density. For example, the D-size cell delivers 19.0 Ah at 3.4 V, compared with 15 Ah at 1.5 V for the conventional zinc-alkaline cells (see Tables 9.4 and 13B.1).

**TABLE 13B.1**   Characteristics of Extended Life Wafer-Type and Cylindrical Bobbin-Type $Li/SOCl_2$ Cells

| Size | 1/2 AA | AA | C | 1/10 D | 1/6 D | D | DD |
|---|---|---|---|---|---|---|---|
| Rated capacity, Ah | 1.2 | 2.4 | 8.5 | .1 | 1.7 | 19 | 35 |
| Rated voltage, V | 3.6 | 3.6 | 3.6 | 3.6 | 3.6 | 3.6 | 3.6 |
| Dimensions, max | | | | | | | |
|   Diameter, mm | 14.5 | 14.5 | 26.2 | 32.9 | 32.9 | 32.9 | 32.9 |
|   Height, mm | 25.2 | 50.5 | 50 | 6.5 | 10.2 | 61.5 | 124.5 |
|   Volume, cm³ | 4.16 | 8.34 | 27.0 | 5.2 | 8.2 | 52.3 | 105.8 |
| Weight, g | 9.6 | 18 | 49.5 | 16.2 | 21 | 93 | 190 |
| Maximum current for   continuous use, mA | 20 | 60 | 75 | 10 | 10 | 100 | 450 |
| Specific energy, Wh/kg | 438 | 467 | 610 | 216 | 283 | 695 | 645 |
| Energy density, Wh/L | 1010 | 1007 | 1117 | 673 | 726 | 1235 | 1158 |

Operating temperature range: $-55$ to $+85°C$. UL Component Recognition: MH12193.
**Source:** Tadiran Lithium Batteries.

**Construction.**   Figure 13B.1 shows the constructional features of the cylindrical $Li/SOCl_2$ cell, which is built as a bobbin-type construction. The anode is made of lithium foil which is swaged against the inner wall of a stainless or nickel-plated steel can; the separator is made of nonwoven glass fibers. The cylindrical, highly porous cathode, which takes up most of the cell volume, is made of Teflon-bonded acetylene black. The cathode also incorporates a current collector, which is a metal cylinder in the case of the larger cells and a pin in the case of smaller cells that do not have an annular cavity.

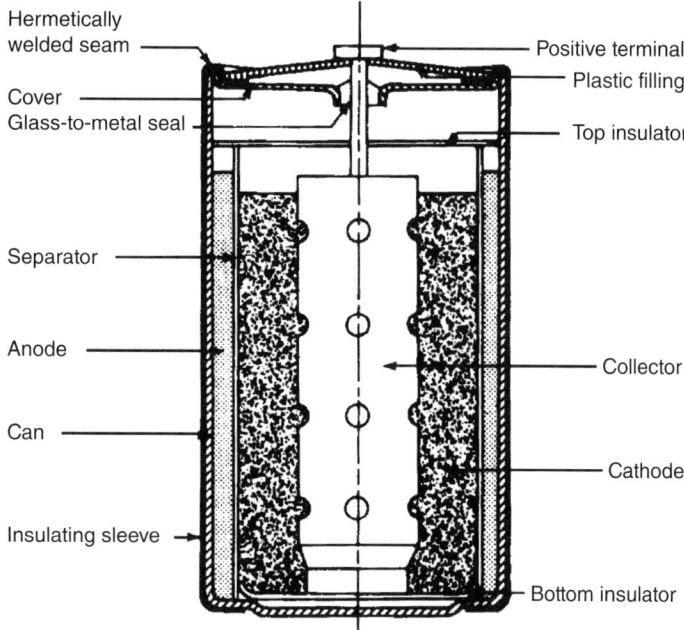

**FIGURE 13B.1**    Cross-section of bobbin-type Li/SOCl₂ battery.[24]

***Performance.***    The open-circuit voltage of the Li/SOCl₂ cell is 3.65 V; typical operating voltages range between 3.3 and 3.6 V with an end voltage of 3.0 V. Typical discharge curves for the Li/SOCl₂ battery are shown in Fig. 13B.2*a* for the D-size cell. The Li/SOCl₂ cell discharges are characterized by a flat profile with good performance over a wide range of temperatures and low- to moderate-rate discharges. Figure 13B.2*b* shows the operating voltage of the bobbin D-cell at various drain rates and temperatures. The relationship of capacity with current is given in Fig. 13B.3, showing the performance from −40 to 80°C. The Li/SOCl₂ cell is capable of performance at unusually high temperatures. At 145°C, as shown in Fig. 13B.4, the cells deliver most of their capacity at high rates and up to 70% at low discharge rates (20 days of discharge).[25] Li/SOCl₂ cells are used to build battery packs that are employed in oil exploration and most withstand temperatures to 150°C as well as high levels of shock and vibration.

Figure 13B.5 shows the behavior of AA cells on continuous low-rate discharge at 25°C. The discharge curve is very flat at these low-current drains, but capacity loss below the 2.4 Ah rating occurs below the 1000 h rate due to parasitic self-discharge.

The capacity or service life of the high-capacity bobbin-type Li/SOCl₂ cell, normalized for a 1-kg and 1-L size cell, at various discharge temperatures and loads, is summarized in Fig. 13B.6.

The long shelf life of the Li/SOCl₂ battery is due to the stability of the lithium anode in contact with the electrolyte, as a result of a protective LiCl film that forms on the lithium surface. The long shelf life can also be credited to the stability of other cell components. For example, the can and cover are cathodically protected by the lithium, and the carbon, stainless-steel collector, and glass separator are all inert in the electrolyte. Figure 13B.7 shows the loss of capacity after 3 years at 20°C, a loss of about 1% to 2% per year. Storage at 70°C results in a capacity loss of about 5% per year. Cells should preferably be stored in an upright position; storage on the side or upside-down may result in higher capacity loss.

After storage the Li/SOCl₂ battery may exhibit a delay in reaching its operating voltage because of the formation of the LiCl film on the lithium surface. The voltage delay becomes more pronounced with a heavier discharge load and lower discharge temperature. The voltage delay of the Li/SOCl₂ cells can be improved by an in situ coating of the lithium anode with an ionic conductor-solid-electrolyte interface. The improvement is

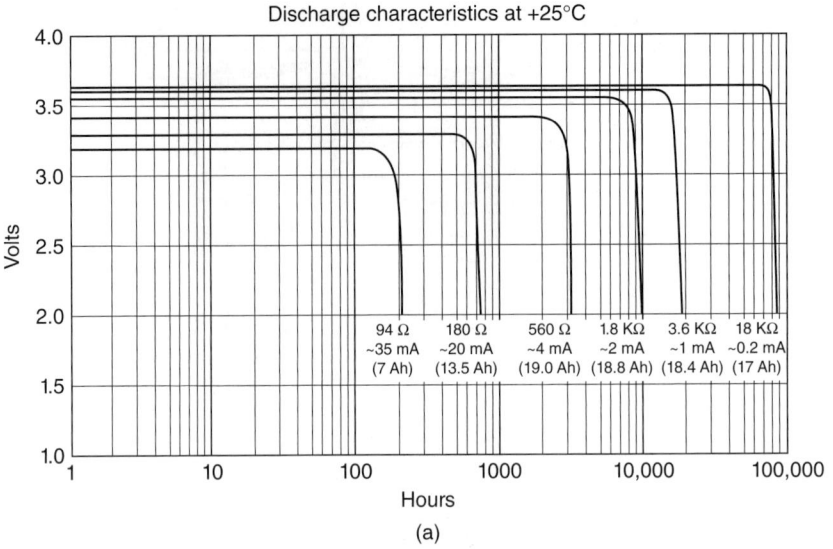

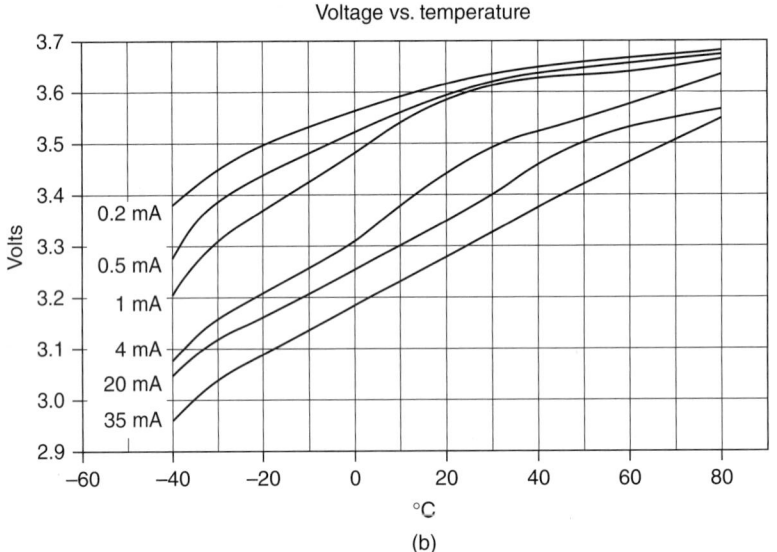

**FIGURE 13B.2** (*a*) Discharge characteristics of high-capacity Li/SOCl$_2$ cylindrical D-size bobbin battery at +25°C. (*b*) Operating (plateau) voltage of the same battery as a function of temperature at various drain rates.[24]

shown in Fig. 13B.8, which compares the minimum voltage and the load after 2 years of storage for both the standard construction and the coated one. It shows the dependence of the closed-circuit voltage on the discharge current of AA cells after 2 years of storage at 25°C. Once the discharge is started, the passivation film is dissipated gradually, the internal resistance returns to its normal value, and the plateau voltage is reached. The passivation film may be removed more rapidly by the application of high-current pulses for a short period or, alternatively, by short-circuiting the batteries momentarily several times until the cell is activated. The use of a pulse provides more reproducible results.

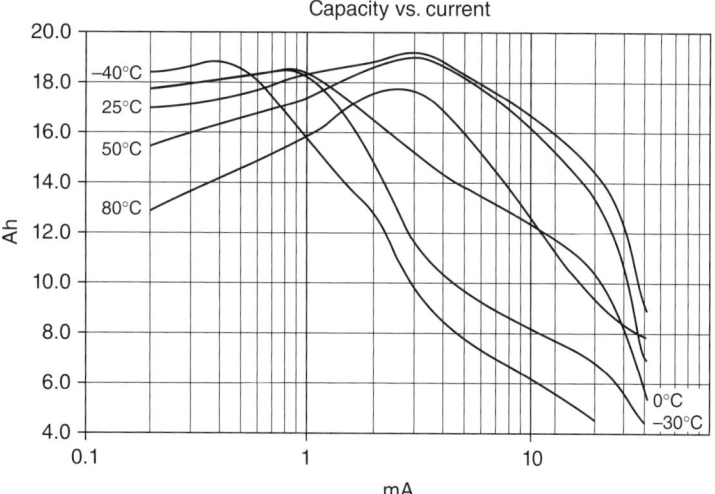

**FIGURE 13B.3**   Performance characteristics of high-capacity cylindrical bobbin D-size batteries as a function of drain rate at various temperatures.[24]

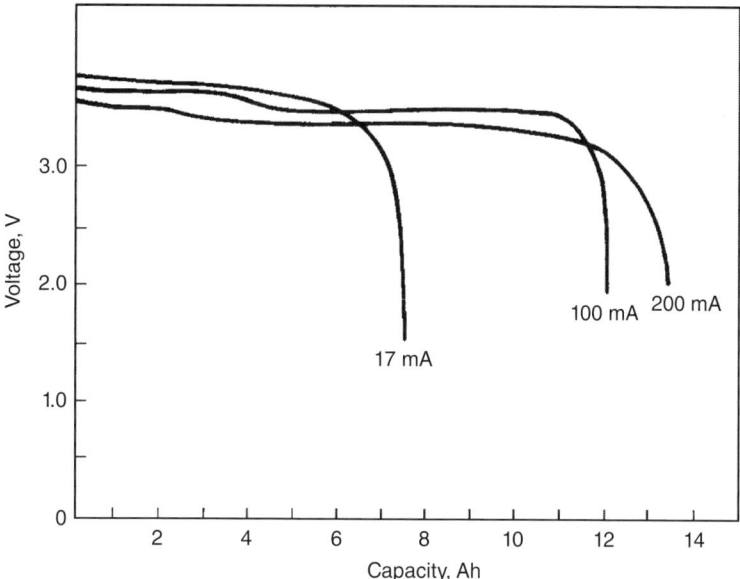

**FIGURE 13B.4**   Discharge characteristics of Li/SOCl$_2$ cylindrical D-size bobbin battery at 145°C.[25]

***Special Characteristics.***   The bobbin batteries are designed to limit the possibility of hazardous operation and to eliminate the need (in some designs) for a safety vent. This is achieved by minimizing the reactive surface area and increasing the heat dissipation, thus limiting the short-circuit current and a hazardous temperature rise, respectively. These cells also are cathode-limited, a feature that was found safer than anode-limited cells for this design.[26] The batteries have withstood short-circuits, forced discharge, and charging under certain conditions with no hazardous condition.[24,25,27] Batteries should not be disposed of in fire or subjected to long-term exposure at temperatures near 180°C because they may explode.

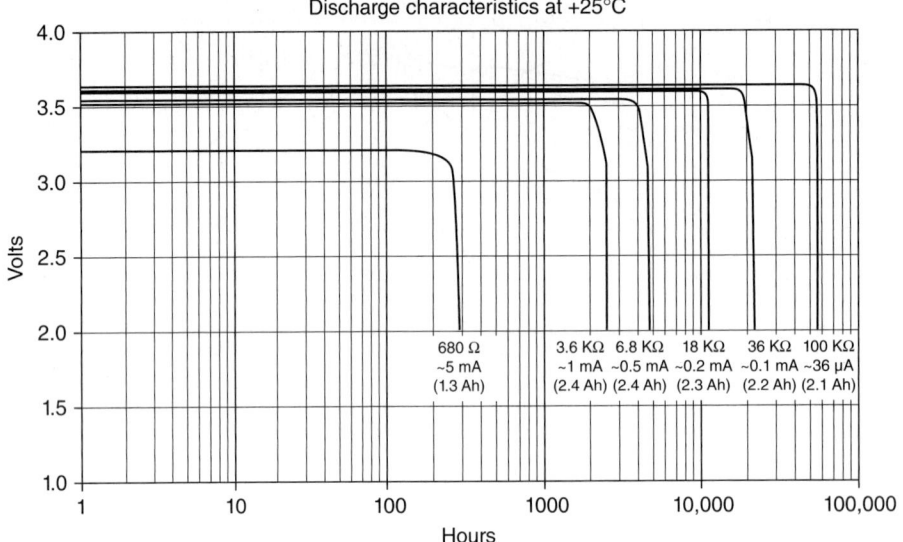

**FIGURE 13B.5** Discharge characteristics of high-capacity Li/SOCl$_2$ cylindrical AA-size bobbin battery on low-rate discharge at +25°C.[24]

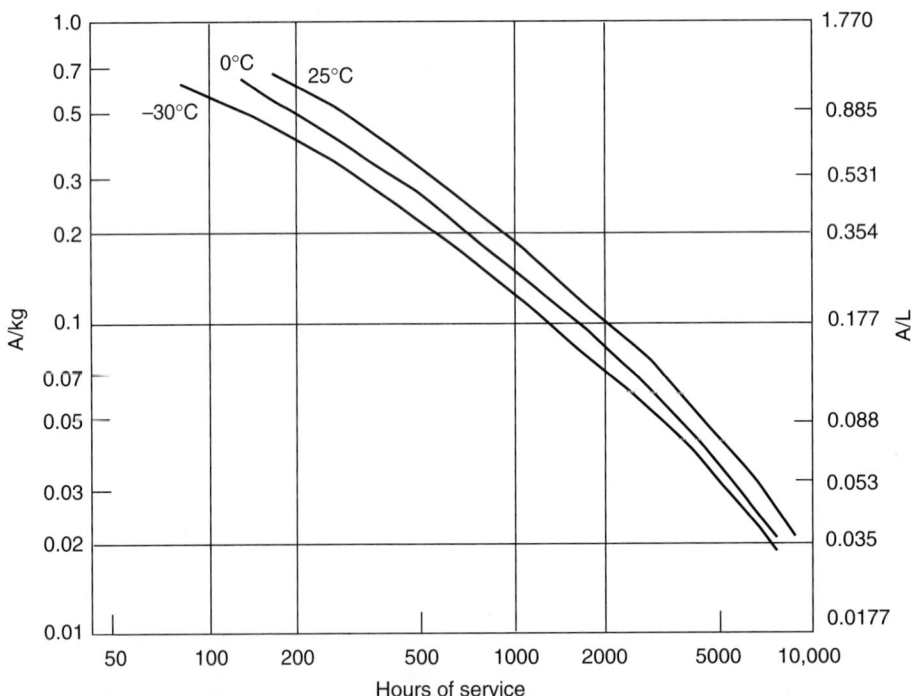

**FIGURE 13B.6** Service life of Li/SOCl$_2$ cylindrical high-capacity bobbin batteries to 2.0 V cutoff.

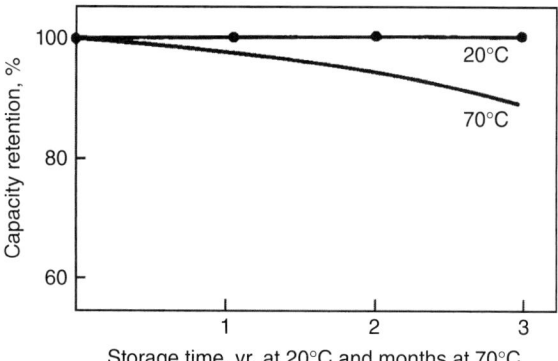

**FIGURE 13B.7**  Capacity retention of Li/SOCl$_2$ cylindrical bobbin battery.[24]

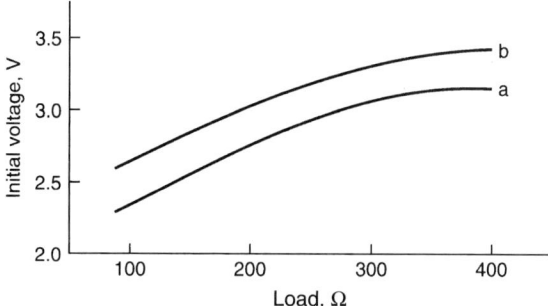

**FIGURE 13B.8**  Li/SOCl$_2$ cylindrical AA-size bobbin batteries—minimum voltage versus load after 2-year storage at 25°C; (*a*) standard construction; (*b*) with loading on lithium anode.

***Battery Sizes.***   The bobbin-type Li/SOCl$_2$ batteries are manufactured in the standard ANSI cell sizes as well as in special cell and battery configurations. Although some of these batteries may be physically interchangeable with conventional zinc batteries, they are not electrically interchangeable because of their higher voltages.

Table 13B.1 lists the properties of some of the typical bobbin-type batteries that are manufactured. These characteristics may vary with the manufacturer. Manufacturer's data should be consulted for specific data as well as for the characteristics of their other batteries.

## 13B.4.2   Spirally Wound Cylindrical Batteries

Medium to moderately high-power Li/SOCl$_2$ batteries that are designed with a spirally wound electrode structure are also available. These batteries were developed primarily to meet military specifications where high drains and low-temperature operation are required. They are also used in selected industrial applications where these features are also needed.

A typical construction is shown in Fig. 13B.9. The cell container is made of stainless steel, a corrosion-resistant GTM feedthrough is used for the positive terminal, and the cell cover is laser sealed or welded to provide a hermetic closure. Safety devices, such as a vent and a fuse or a PTC device, are incorporated in the cell to protect against buildup of internal pressure or external short-circuits.

The discharge curves for a D-size battery are plotted in Fig. 13B.10, showing the higher performance at moderate drains compared to the bobbin cell (see Fig. 13B.2).

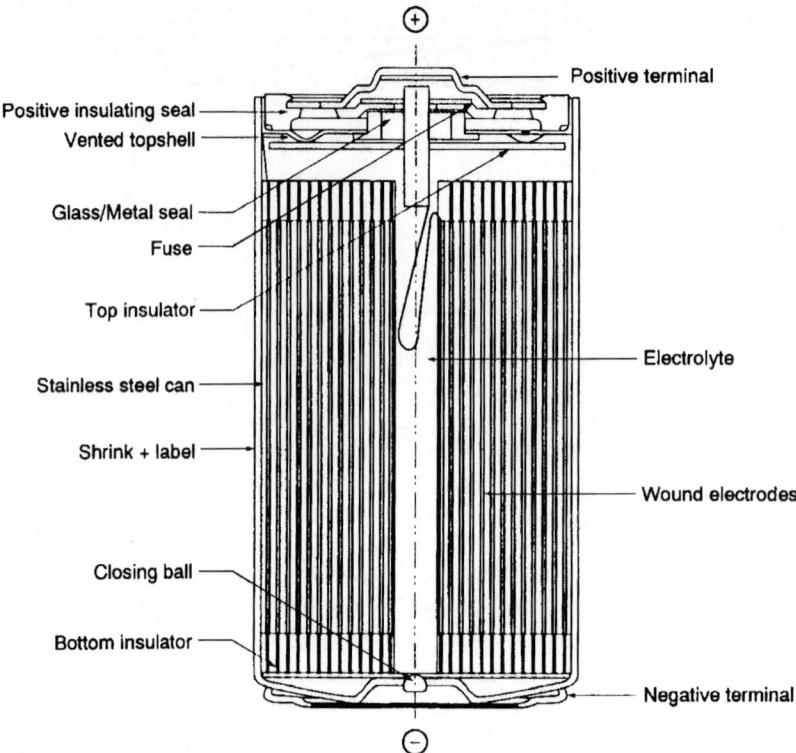

**FIGURE 13B.9** Cutaway view of lithium/thionyl chloride spirally wound electrode battery. (*Courtesy of SAFT Batteries.*)

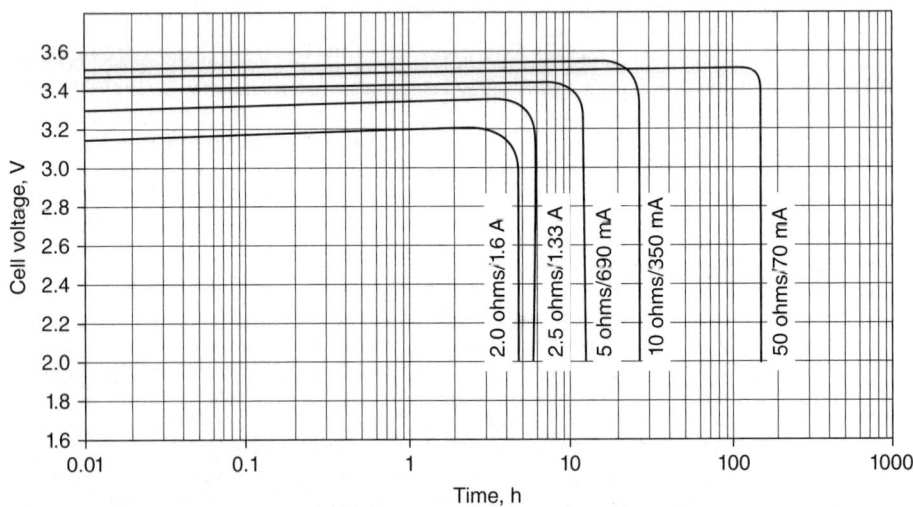

**FIGURE 13B.10** Discharge characteristics of spirally wound Li/SOCl₂ D-size battery, medium discharge rate at 20°C. (*Courtesy of SAFT Batteries.*)

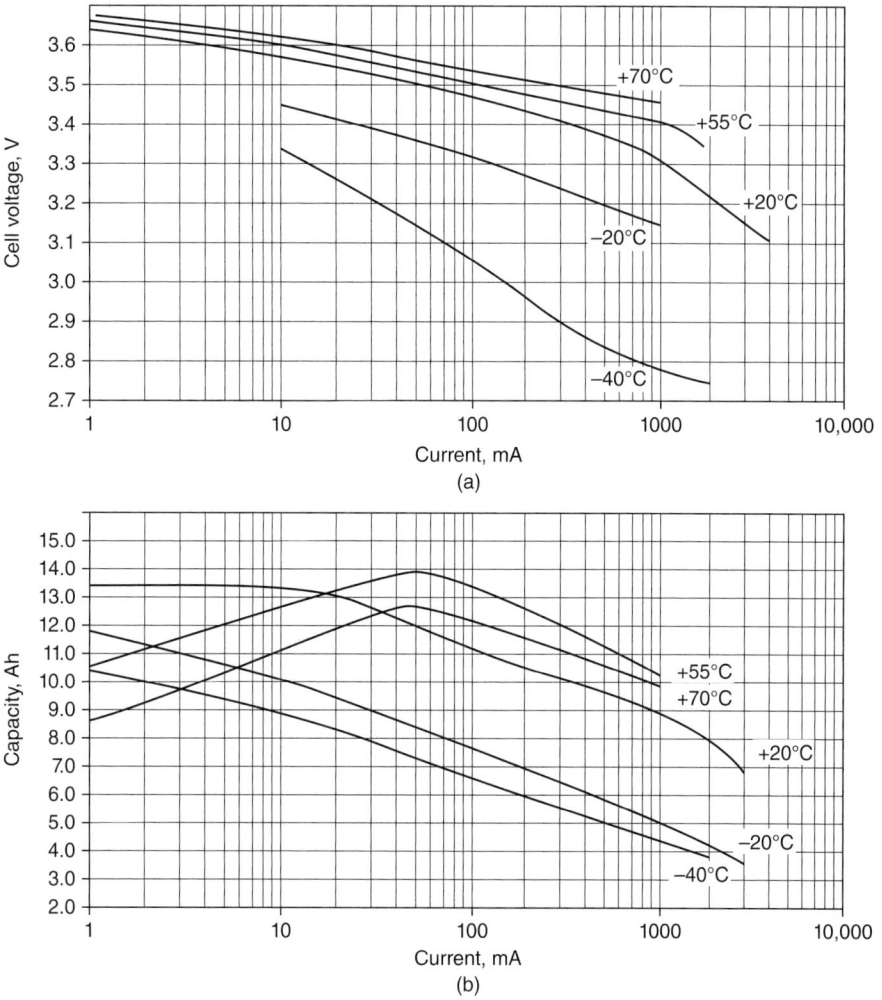

**FIGURE 13B.11**  Discharge characteristics of spirally wound Li/SOCl₂ D-size battery at various temperatures. (*a*) Voltage versus current. (*b*) Capacity versus current. (*Courtesy of SAFT Batteries.*)

Figure 13B.11 shows the performance characteristics of the D-size battery, providing the relationship of voltage and capacity with current drain at several temperatures.

Like the other Li/SOCl₂ batteries, these batteries have an excellent storage capability over a wide temperature range due to the buildup of a protective lithium chloride layer on the lithium. Capacity loss on storage at ambient conditions is stated to be less than 3% per year. These products are said to be resistant to passivation. Table 13B.2 lists the characteristics of typical cylindrical spirally wound Li/SOCl₂ batteries.

## 13B.4.3  Flat or Disk-Type Li/SOCl₂ Cells

The Li/SOCl₂ system was also designed in a flat or disk-shaped cell configuration with a moderate to high discharge rate capability. These batteries are hermetically sealed and incorporate a number of features to safely handle abusive conditions, such as short-circuit, reversal, and overheating, within design limits.

**TABLE 13B.2**   Characteristics of Typical Cylindrical Spirally Wound Li/SOCl₂ Batteries

| Size | 1/3 C | C | C (light) | D | D | D |
|---|---|---|---|---|---|---|
| Capacity at 20 Ah | 1.2 | 5.8 | 3.6 | 13.0 | 12.0 | 14.0 |
| Rated current, mA | 10 | 15 | 15 | 15 | 50 | 300 |
| Nominal voltage, V | 3.6 | 3.6 | 3.6 | 3.6 | 3.6 | 3.6 |
| Dimensions (max) | | | | | | |
|   Diameter, mm | 26.2 | 26.0 | 26.0 | 33.4 | 33.4 | 32.05 |
|   Height, mm | 18.6 | 50.4 | 50.4 | 61.6 | 61.6 | 61.7 |
| Maximum current for continuous use, A | 0.4 | 1.3 | 1.3 | 1.8 | 1.0 | Not specified |
| Weight,g | 24 | 51 | 51 | 100 | 100 | 104.5 |
| Operating temperature range, °C | −60/+85 | −60/+85 | −60/+85 | −65/+85 | −65/+120 | −40/+150 |
| Transport | Non-restricted | Class 9 | Non-restricted | Class 9 | Class 9 | Class 9 |

*Source:* SAFT Batteries.
Open circuit voltage = 3.67 V. Individual cells fitted with nonresettable 5 A fuse protection.

The battery consists of a single or multiple assembly of disk-shaped lithium anodes, separators, and carbon cathodes stacked in a sealed stainless-steel case containing a ceramic feedthrough for the anode and insulation between the positive and negative terminals of the cell.[28]

The batteries were originally manufactured in small and large diameter sizes by Altus Corp., and were last produced in large sizes only for U.S. Navy applications by HED Battery Corporation. The characteristics of these batteries are summarized in Table 13B.3. Discharge curves for large batteries are shown in Fig. 13B.12. Typically the cells have a high energy density, flat discharge profiles, and the capability of performance over the temperature range of −40 to 70°C. On storage they can retain 90% of the capacity after storage of 5 years at 20°C, 6 months at 45°C, or 1 month at 70°C.

**TABLE 13B.3**   Characteristics of Disk-Type Batteries Currently Available

| Nominal capacity (Ah) | Diameter (cm) | Height (cm) | Weight (kg) | Test current (A) | Average voltage (V) | Actual capacity (Ah) | Specific energy Wh/kg | Energy density (Wh/L) |
|---|---|---|---|---|---|---|---|---|
| 1200 | 20.32 | 12.7 | 7.63 | 20 | 3.34 | 1170 | 510 | 947 |
| 2400 | 40.64 | 5.84 | 15.1 | 8 | 3.42 | 2300 | 523 | 1043 |
| 2400 | 40.64 | 5.84 | 15.1 | 50 | 3.28 | 2000 | 434 | 871 |

*Source:* HED Battery Corp.

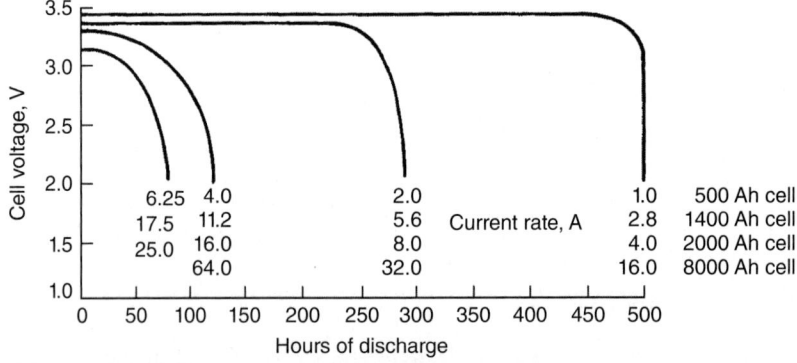

**FIGURE 13B.12**   Performance characteristics of disk-type Li/SOCl₂ cells. High-capacity cell; typical performance at 0 to 25°C range to 2.5 V cutoff.

The cell design includes the following features:

1. *Short-circuit protection.* Structure of interconnects fuses at high currents, providing an open circuit.
2. *Reverse-voltage chemical switch.* Upon cell reversal, it allows cell to endure 100% capacity reversal, up to 10 h rate, without venting or pressure increase.
3. *Antipassivation (precoat lithium anode).* Reduces voltage delay by retarding growth of LiCl film; large cells stored for 2 years reach operating voltage within 20 s.
4. *Self-venting.* Ceramic seal is designed to vent cell at predetermined pressures.[28]

These cells are used as multicell batteries in naval applications.

Studies of these designs have involved 1000 and 1200 Ah cells for application in a U.S. Navy Long-Range Mine Reconnaissance System (LMRS).[29,30] These are scaled-down versions of 2350 Ah cells, which had shown the ability to operate at the C/40 rate, providing a power density of 2.3 W/kg. Both 1000 and 1200 Ah cells were 20.3 cm in diameter with an annular cavity at the center of the disk. The former unit was 9.53 cm high, while the latter was 12.07 cm high. Both designs incorporate a ceramic-to-metal seal capable of carrying 60 A, and both were limited by the capacity of the carbon cathode with $Li/SOCl_2$ capacity ratio balanced. The 1000 Ah units were tested individually and as 4- and 12-cell batteries with 0.5 cm intercell insulators and compressed between 1.59 cm aluminum end plates by tie-rods. The 12-cell battery consisted of three stacks of four cells with a diameter of 45.3 cm designed to fit within the hull of LMRS. Test data are summarized in Table 13B.4. Based on the results of this testing, a 30-cell battery weighing about 205 kg would deliver 100 kWh at 100 V for operational power up to 5 kW. Subsequently, the cell capacity was increased to 1200 Ah by increasing the cell height.[30] These cells were subjected to a series of safety tests as defined by NAVSEA INST 9310.1B (June 13, 1992) and U.S. Navy Technical Manual S9310-AQ-SAF-010. The 1200 Ah units were subjected to intermittent and sustained short-circuits, forced discharge into voltage reversal, charging tolerance high-temperature discharge, and high-temperature exposure after low temperature (0°C) discharge. No cells produced venting, loss of material, or case breach of any kind during these tests, nor were there indications of internal shorts or potentially violent conditions. The pulsed and sustained soft-shorts produced significant heating and pressure, but these were within the capability of the battery to operate safely. At sustained currents in excess of 110 A, the cathode appears to clog rapidly, limiting capacity. The exothermic response obtained when the battery was quickly heated to 75°C after cold discharge at 40 A at 0°C is a result of accelerated anode repassivation. The subsequent 55°C short-circuit behavior confirms this hypothesis. There was an indication that this response would have led to a thermal runaway. The 40 A, 55°C discharge demonstrated that the battery could operate safely in the absence of cooling for an extended period of time in a simulated vehicle structure. The tolerance to a moderate charging voltage indicated a margin level in potential failures of diodes, and the forced reversal test demonstrated a moderate tolerance to these conditions. A fuse in the negative terminal assembly is being considered to withstand high-rate short-circuits. This test program demonstrated the feasibility of using a large lithium/thionyl chloride propulsion battery for LMRS and other similar undersea applications.

**TABLE 13B.4**  Performance Characteristics of 1000 Ah LMRS Lithium/Thionyl Chloride Cells and Batteries (Number of Cells Tested Indicated in Parenthesis after Each Test)

| Configuration | Rate | Ah | kWh | Wh/kg |
|---|---|---|---|---|
| Single (1) | C/22–C/67 | 931 | 3.12 | 108 |
| Single (5) | C/25–C/67 | 913 | 3.00 | 105 |
| Single (2) | C/40 | 927 | 3.09 | 111 |
| 4-cell | C/25–C/50 | 1053 | 3.58 | 125 |
| 4-cell | C/40 | 1075 | 3.67 | 126 |
| 4-cell | C/60 | 1004 | 3.41 | 119 |
| 12-cell | C/20–C/40 | 896 | 3.03 | 106 |
| 12-cell | C/20–C/40 | 1016 | 3.44 | 121 |

## 13B.4.4   Large Prismatic Li/SOCl₂ Cells

These large, high-capacity Li/SOCl₂ batteries were developed mainly as a standby power source for those military applications requiring a power source that is independent of AC line power and the need for recharging.[31-33] They generally were built in a prismatic configuration, as shown schematically in Fig. 13B.13. The lithium anodes and Teflon-bonded carbon electrodes were made as rectangular plates with a supporting grid structure, separated by nonwoven glass separators, and housed in a hermetically sealed stainless-steel container. The terminals were brought to the outside by GTM feedthrough or by a single feedthrough isolated from the positive steel case. The cells were filled through an electrolyte filling tube.

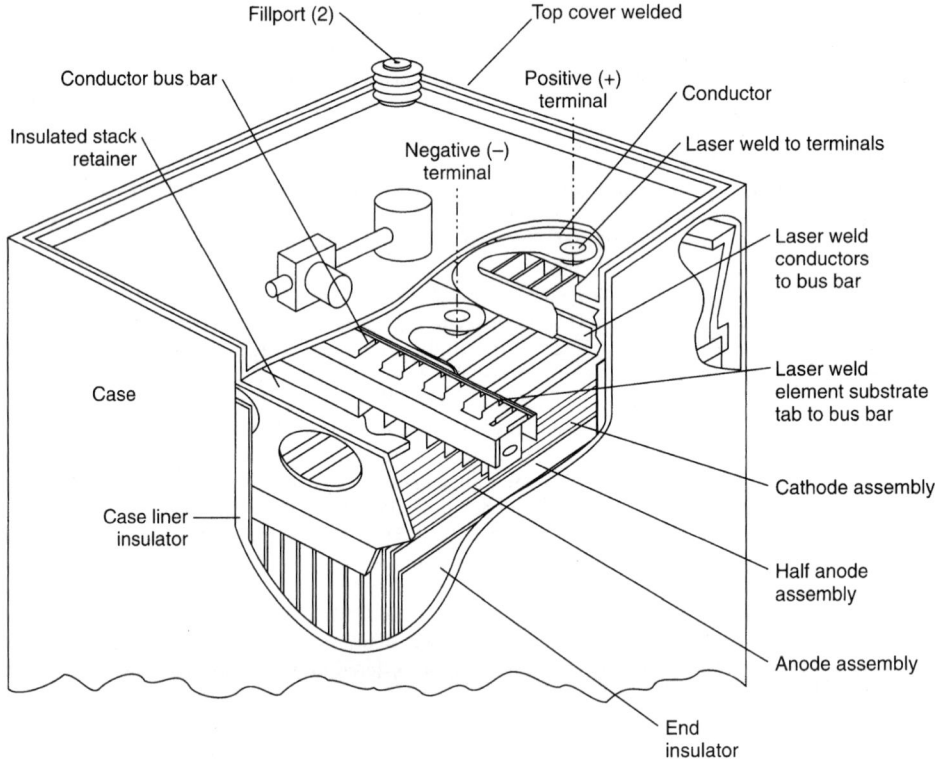

**FIGURE 13B.13**   Cutaway view of 10,000 Ah Li/SOCl₂ battery.[33]

The characteristics of several prismatic batteries are summarized in Table 13B.5. These cells had a very high energy density. They were generally discharged continuously at relatively low rates (200–300 h rate), but were capable of heavier discharge loads. A typical discharge curve is shown in Fig. 13B.14. The voltage profile was flat, and the cell operated just slightly above ambient temperature at this discharge load. During the course of the discharge there was a slight buildup of pressure, reaching a value of about $2 \times 10^5$ Pa at the end of the discharge. A higher rate pulse discharge is shown in Fig. 13B.15. The 2000 Ah cell was discharged continuously at a 5 A load, with 40 A pulses, 16 s in duration, superimposed once every day. A steady discharge voltage was obtained throughout most of the discharge, with only a slight reduction in voltage during the pulse. The batteries were capable of performance from −40 to 50°C; shelf-life losses were estimated at 1% per year.[33] These batteries have been decommissioned and are no longer in use but remain the largest lithium batteries ever built.

**TABLE 13B.5**  Characteristics of Large Prismatic Li/SOCl$_2$ Batteries

| Capacity, Ah | Height, mm | Length, mm | Width, mm | Weight, kg | Specific energy, Wh/kg | Energy density, Wh/L |
|---|---|---|---|---|---|---|
| 2000 | 448 | 316 | 53 | 15 | 460 | 910 |
| 10,000 | 448 | 316 | 255 | 71 | 480 | 950 |
| 16,500 | 387 | 387 | 387 | 113 | 495 | 970 |

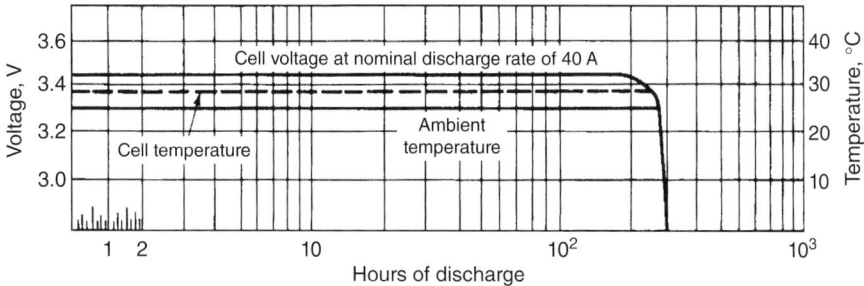

**FIGURE 13B.14**  Discharge curves for 10,000 Ah Li/SOCl$_2$ battery.

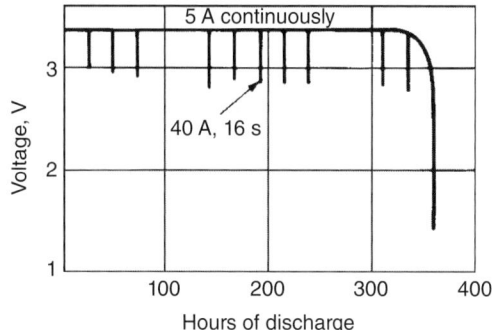

**FIGURE 13B.15**  Discharge of high-capacity 2000 Ah Li/SOCl$_2$ battery.

## 13B.4.5  Applications

The applications of the Li/SOCl$_2$ system take advantage of the high energy density and long shelf life of this battery system. The low-drain cylindrical batteries are used as a power source for CMOS memories, utility meters, and radio frequency identification (RFID) tags such as the EZ Pass Toll collection system, programmable logic controllers, and wireless security alarm system. Wide application in consumer-oriented applications is limited because of the relatively high cost and concern with the safety and handling of these types of lithium batteries.

The higher rate cylindrical and the larger prismatic Li/SOCl$_2$ batteries are used mainly in military applications where high specific energy is needed to fulfill important mission requirements. A significant application for the large 10,000 Ah batteries was as standby power source as nine-cell batteries for the Missile Extended System Power in the event of loss of commercial or other power. These batteries have been decommissioned.

A lithium/thionyl chloride battery was developed for use on the Mars Microprobe Mission, a secondary payload on the Mars 98 Lander Mission, which disappeared on entry into the Martian atmosphere in December 1999.[34] The Microprobe power source is a four-cell lithium/thionyl chloride battery with a second redundant battery in parallel. The eight 2 Ah cells are arranged in a single-layer configuration in the aft-body of the microprobe. The lithium primary cells (and battery configuration) have been designed to survive the maximum landing impact which may reach 80,000 G, and then be operational on the Martian surface to −80°C. Primary lithium-thionyl chloride batteries were selected for the microprobes based on high specific energy and low-temperature performance. A parallel plate design was selected as the best electrode configuration for surviving the impact without shorting during impact. A cross section of the final 2 Ah Mars cell design showing the parallel plate electrode arrangement is shown in Fig. 13B.16. For this cell, the cathodes are blanked from sheets of a Teflon-bonded carbon composition attached to nickel-disc current collectors and connected in parallel. The 10 full-disc anodes are also connected in parallel and are electrically isolated from the case and cover. The assembly fixture helps with component alignment and handling during stack assembly and during connection of the cathode and anode substrate tabs to the cover and the GTM seal anode terminal pin, respectively. The D-size diameter case is 2.22 cm in height. The cover was redesigned after initial tests to minimize the chance of a GTM seal fracture during impact. The Tefzel spacer, located between the cover and stack, helps in handling the stack during substrate tab connections and provides for the proper degree of cathode and separator compression once the cover is tungsten inert gas (TIG) welded to the case. The Mars cells were required to deliver 0.55 Ah of capacity at −80°C. A low-temperature thionyl chloride electrolyte consisting of a 0.5 M $LiGaCl_4$ in $SOCl_2$ was developed during the initial phase of the program. As the result of this effort, the battery was able to operate at −80°C on a 1 A discharge as shown in Fig. 13B.17. The battery provided over 0.70 Ah at this extremely low temperature. The ability to withstand the 80,000 G impact was demonstrated by Air Gun tests performed at −40°C into frozen desert sand followed by a simulated mission profile at −60°C in an environmental chamber. The battery supplied power for a drill that provided a core sample of subsurface soil for water analysis. In addition, the power requirement for the 20 min water experiment was increased from 2.5 to ~6 W, and increased power levels were required for telemetry at both −60 and −80°C. The total low temperature capacity and major tasks are listed in Table 13.6. The drill operation required an initial current of 1 A for 25 ms, after which the current was in the range of 75 to 85 mA for the duration of the task. The soil sample heating operation lasted for 20 min with power in excess of 6 W required. The high-rate transmission started out at 10.4 W (9.7 V), but the power level dropped off toward the end to 6.4 W (7.6 V) after 9 min. The cell delivered a total of 0.724 Ah of low temperature capacity. In the end, this program extended the state of the art for lithium/thionyl chloride battery technology by demonstrating its ability to withstand 80,000 G impact and then operate at temperatures down to −80°C.

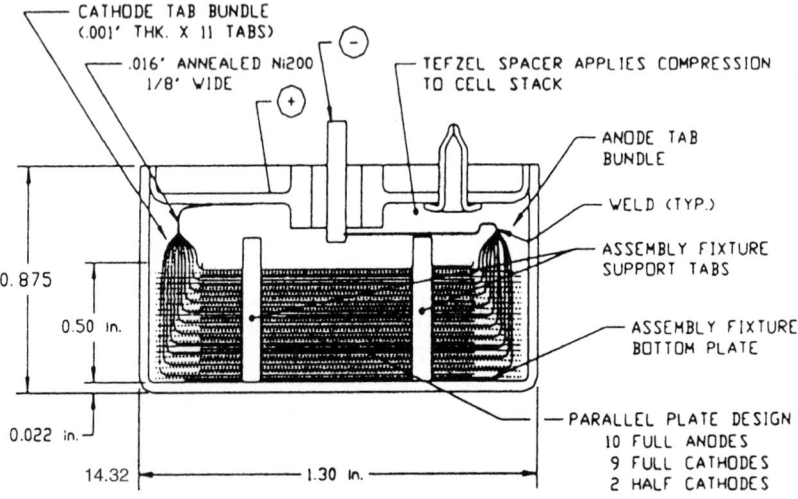

**FIGURE 13B.16** Vertical cross section of the final 2 Ah cell design. (*Courtesy of Yardney Technical Products, Inc.*)

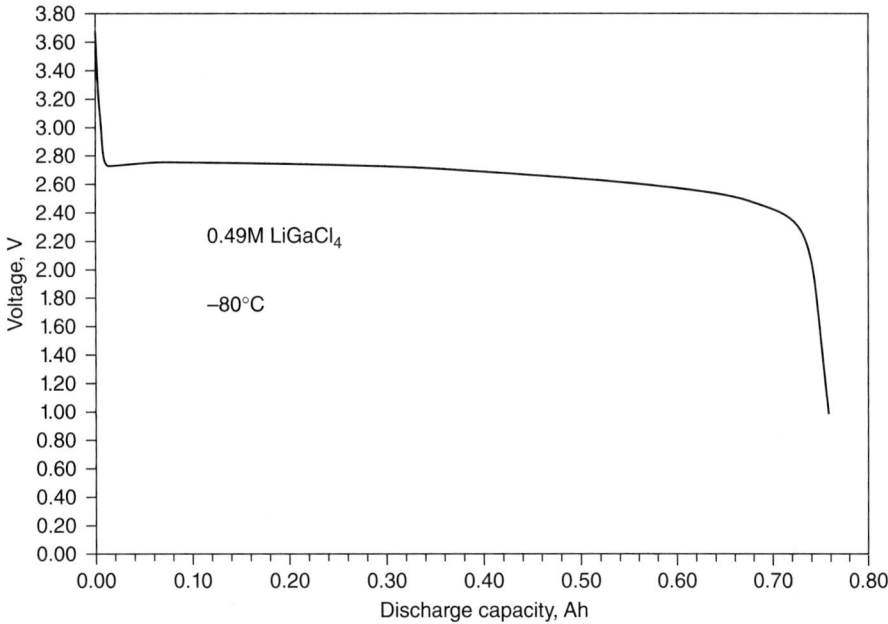

**FIGURE 13B.17**    Capacity in ampere-hours for a 1 A discharge at −80°C for Mars Microprobe battery. (*Courtesy of Yardney Technical Products, Inc.*)

**TABLE 13B.6**    Results of Air Gun Test on Mars Microprobe Battery

| Postimpact battery discharge, Ah | |
| --- | --- |
| Output on profile | 0.515 |
| Additional output at −80°C | 0.157 |
| Total output | 0.724 |
| Major tasks, V | |
| Calib. 9Ω, −60°C | 9.5 |
| Drill 136Ω, −60°C | 11.7 |
| H2O 16Ω, −60°C | 10.5 |
| High X-mit 9Ω, −60°C | 9.7 |
| X-mit 59Ω, −80°C | 7.6 |

# SECTION C

# LITHIUM/OXYCHLORIDE

## 13C.1    LITHIUM/OXYCHLORIDE BATTERIES

The lithium/sulfuryl chloride ($Li/SO_2Cl_2$) battery is the other oxychloride that has been used for primary lithium batteries. The $Li/SO_2Cl_2$ battery has three potential advantages over the $Li/SOCl_2$ battery:

1. A higher energy density as a result of a higher operating voltage (3.9 V open-circuit voltage) as shown in Fig. 13.3 and less solid product formation (which may block the cathode) during the discharge.

2. Inherently greater safety because sulfur, which is a possible cause of thermal runaway in the $Li/SOCl_2$ battery, is not formed during the discharge of the $Li/SO_2Cl_2$ battery.

3. A higher rate capability than the thionyl chloride battery as, during the discharge, more $SO_2$ is formed per mole of lithium, leading to a higher conductivity.

Nevertheless, the $Li/SO_2Cl_2$ system is not as widely used as the $Li/SOCl_2$ system because of several drawbacks:

1. Cell voltage sensitivity to temperature variations.

2. Higher self-discharge rate.

3. Lower rate capability at low temperatures.

Another type of lithium/oxychloride battery involves the use of halogen additives to both the $SOCl_2$ and $SO_2Cl_2$ electrolytes. These additives give an increase in the cell voltage (3.9 V for the $Li/BrCl$ in the $SOCl_2$ system; 3.95 V for the $Li/Cl_2$ in the $SO_2Cl_2$ system), energy density and specific energy (up to 1070 Wh/L and 485 Wh/kg), and safer operation under abusive conditions.

## 13C.2    LITHIUM/SULFURYL CHLORIDE (Li/SO$_2$Cl$_2$) CHEMISTRY

The $Li/SO_2Cl_2$ battery is similar to the thionyl chloride battery using a lithium anode, a carbon cathode, and the electrolyte/depolarizer of $LiAlCl_4$ in $SO_2Cl_2$. The discharge mechanism is

$$Anode: \qquad Li \rightarrow Li^+ + e$$

$$Cathode: \qquad SO_2Cl_2 + 2e \rightarrow 2Cl^- + SO_2$$

$$Overall: \qquad 2Li + SO_2Cl_2 \rightarrow 2LiCl \downarrow + SO_2$$

The open-circuit voltage is 3.909 V.

## 13C.3    CELL TYPES

### 13C.3.1    Basic Cell Designs

Cylindrical, spirally wound $Li/SO_2Cl_2$ cells were developed experimentally, but were never commercialized because of limitations with performance and storage. Bobbin-type cylindrical cells, using a sulfuryl chloride/$LiAlCl_4$ electrolyte and constructed similar to the design illustrated in Fig. 13B.1, also showed a variation of voltage with temperature and a decrease of the voltage during storage. This may be attributed to reaction of chlorine, which is present in the electrolyte and formed by the dissociation of sulfuryl chloride into $Cl_2$ and $SO_2$. This condition can be ameliorated by including additives in the electrolyte. Bobbin cells, made with the improved electrolyte, gave significantly higher capacities at moderate discharge currents, compared to the thionyl chloride cells.[35] This system has been employed for reserve lithium/sulfuryl chloride batteries as well[36] (see Chap. 29).

### 13C.3.2  Halogen-Additive Lithium/Oxychloride Cells

Another variation of the lithium/oxyhalide cell involves the use of halogen additives in both the $SOCl_2$ and the $SO_2Cl_2$ electrolytes to enhance the battery performance. These additives result in: (1) an increase in the cell voltage (3.9 V for BrCl in the $SOCl_2$ system [BCX], 3.95 V for $Cl_2$ in the $SO_2Cl_2$ system [CSC]), and (2) an increase in energy density and specific energy to about 1054 Wh/L and 486 Wh/kg for the CSC system.

The lithium/oxyhalide cells with halogen additives offer among the highest energy density of primary battery systems. They can operate over a wide temperature range, including high temperatures, and have excellent shelf lives. They are used in a number of special applications—oceanographic and space applications, memory backup, and other communication and electronic equipment.

These lithium/oxychloride batteries are available in hermetically sealed, spirally wound electrode cylindrical configurations, ranging from AA to DD size in capacities up to 30 Ah. These batteries are also available in the AA size containing 0.5 g of Li and in flat disk-shaped cells. Figure 13C.1 shows a partial cutaway view of a typical cell. Table 13C.1 lists several different lithium-oxychloride batteries manufactured and their key characteristics. Two types of halogen-additive lithium/oxychloride batteries have been developed as follows:

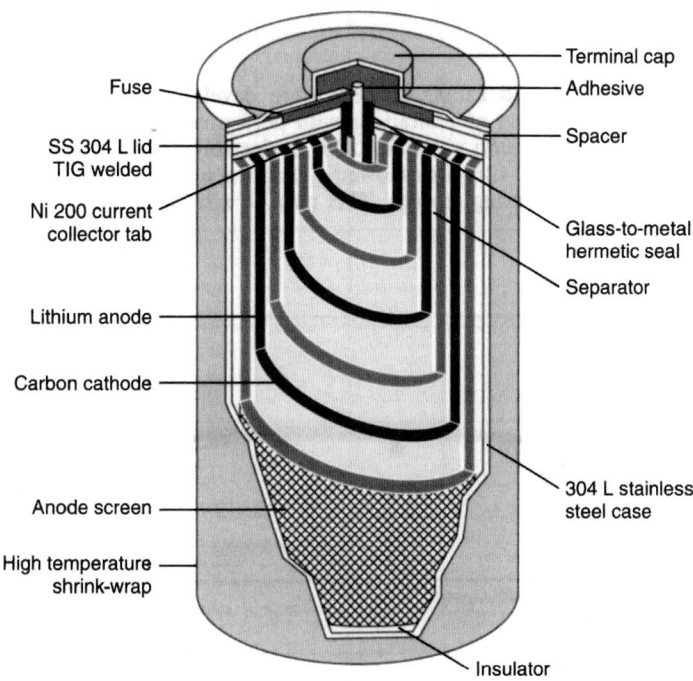

**FIGURE 13C.1**  Cross section of lithium/oxychloride cell. (*Courtesy of Electrochem Solutions Div. Greatbatch Ltd.*)

**TABLE 13C.1**   Typical Halogen Additive Oxychloride Batteries

| | BrCl in $SOCl_2$ | | | | $Cl_2$ in $SO_2Cl_2$ | | |
|---|---|---|---|---|---|---|---|
| | AA | C | D | DD | C | D | DD |
| Voltage, V | | | | | | | |
|   Open circuit | | 3.9 | | | | 3.9 | |
|   Average operating | | 3.4 | | | | 3.3–3.5 | |
| Rated capacity, 100 h rate, Ah | 2.0 | 7.0 | 15.0 | 30.0 | 7.0 | 15.0 | 30.0 |
| Dimensions | | | | | | | |
|   Diameter, mm | 13.7 | 25.6 | 33.5 | 33.5 | 25.6 | 33.5 | 33.5 |
|   Height, mm | 49.2 | 48.4 | 59.3 | 111.5 | 48.4 | 59.3 | 111.4 |
|   Volume, $cm^3$ | 7.25 | 24.9 | 52.3 | 98.3 | 24.9 | 52.3 | 98.2 |
| Weight, g | 16 | 55 | 115 | 216 | 52 | 116 | 213 |
| Maximum current capability, mA | 100 | 500 | 1000 | 3000 | 1000 | 2000 | 4000 |
| Specific energy/energy density | | | | | | | |
|   At 100 h rate | | | | | | | |
|     Wh/kg | 453 | 445 | 433 | 486 | 478 | 452 | 486 |
|     Wh/L | 965 | 984 | 975 | 1068 | 998 | 990 | 1054 |
| Operating temperature range, °C | | | −55 to +85 | | | −20 to +93 | |

Self-discharge rated at 3%/yr at 25°C for both types.
***Source:*** Electrochem Solutions Div., Greatbatch, Inc.

## 13C.4   *PERFORMANCE CHARACTERISTICS AND APPLICATIONS*

### 13C.4.1   Li/SOCl₂ System with BrCl Additive (BCX)

This battery has an open-circuit voltage of 3.9 V and an energy density of up to 1070 Wh/L at 20°C. The BrCl additive is used to enhance the performance. The cells are fabricated by winding the lithium anode, the carbon cathode, and two layers of a separator of nonwoven glass into a cylindrical roll and packaging them in a hermetically sealed can with a GTM feedthrough. The performance of the D-size battery at various temperatures and discharge rates is shown in Figs. 13C.2*a* to 2*c*. The discharge curves are relatively flat with a working voltage of

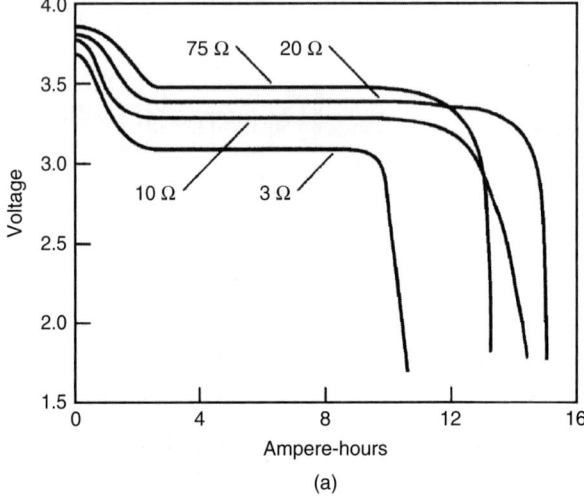

(a)

**FIGURE 13C.2**   Performance characteristics of Li/SOCl₂ with BrCl additive. D-size batteries. (*a*) Discharge characteristics at 20°C. (*Courtesy of Electrochem Solutions Div., Greatbatch, Inc.*) (*b*) Capacity as a function of discharge temperature (100% represents rated capacity at room temperature). (*c*) Loaded voltage as a function of temperature.

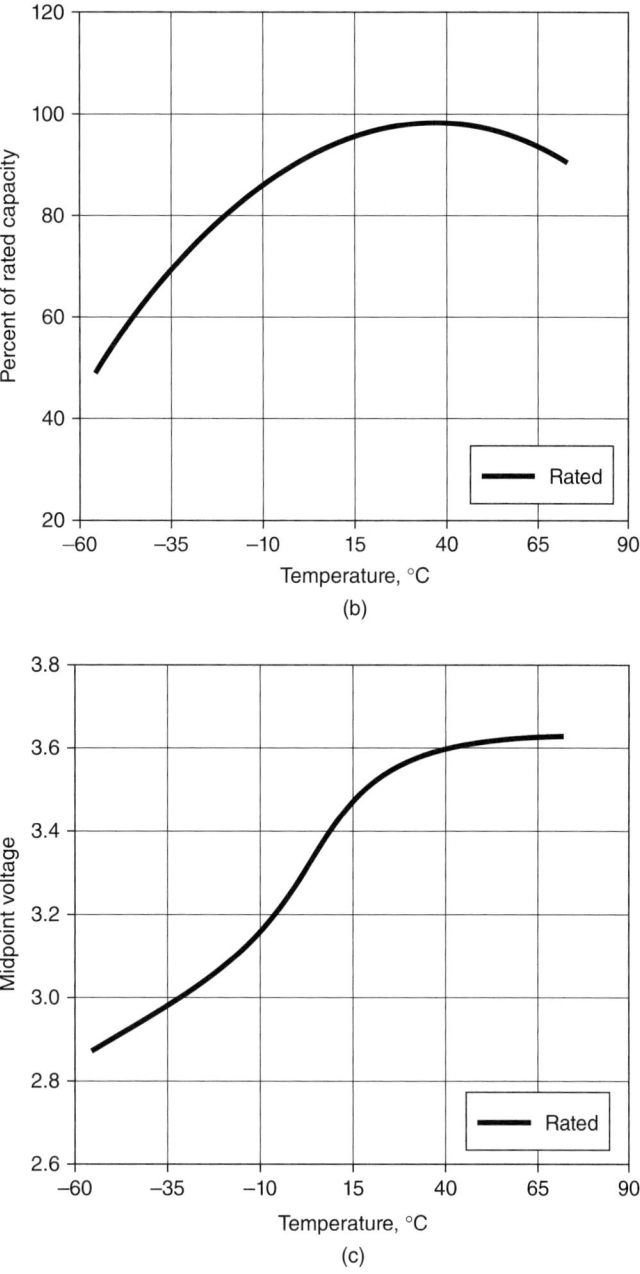

FIGURE 13C.2   (*Continued*)

about 3.5 V. The batteries are capable of performance over the temperature range of −55 to 72°C. The capacity loss is rated at 3%/yr at 25°C. Capacity loss on storage is higher than with lithium systems using thionyl chloride only.

The addition of BrCl to the depolarizer may also prevent the formation of sulfur as a discharge product, at least in the early stage of the discharge, and minimize the hazards of the $Li/SOCl_2$ battery attributable to sulfur or discharge intermediates. The cells show abuse resistance when subjected to the typical tests, such as short-circuit, forced discharge, and exposure to high temperatures.[37]

### 13C.4.2    Li/SO$_2$Cl$_2$ with Cl$_2$ Additive (CSC)

This battery has an open-circuit voltage of 3.9 V and an energy density of up to 1050 Wh/L. The additive is used to decrease the voltage-delay characteristic of the lithium/oxyhalide cells. The typical operating temperature of these cells is −20 to 93°C. The cylindrical cells are designed in the same structure as those shown in Fig. 13C.1.

Typical performance characteristics for this battery type are shown in Figs. 13C.3*a* and 3*b*. The cells show abuse resistance similar to the Li/BrCl in $SOCl_2$ cells when subjected to abuse tests. Capacity loss is also rated at 3%/yr at 25°C.

Another study has evaluated the effect of ambient temperature storage for up to 6 years.[38] The interrelation of voltage stability, capacity retention, self-discharge, and voltage delay has been delineated. This source should be consulted to obtain detailed data on this system.

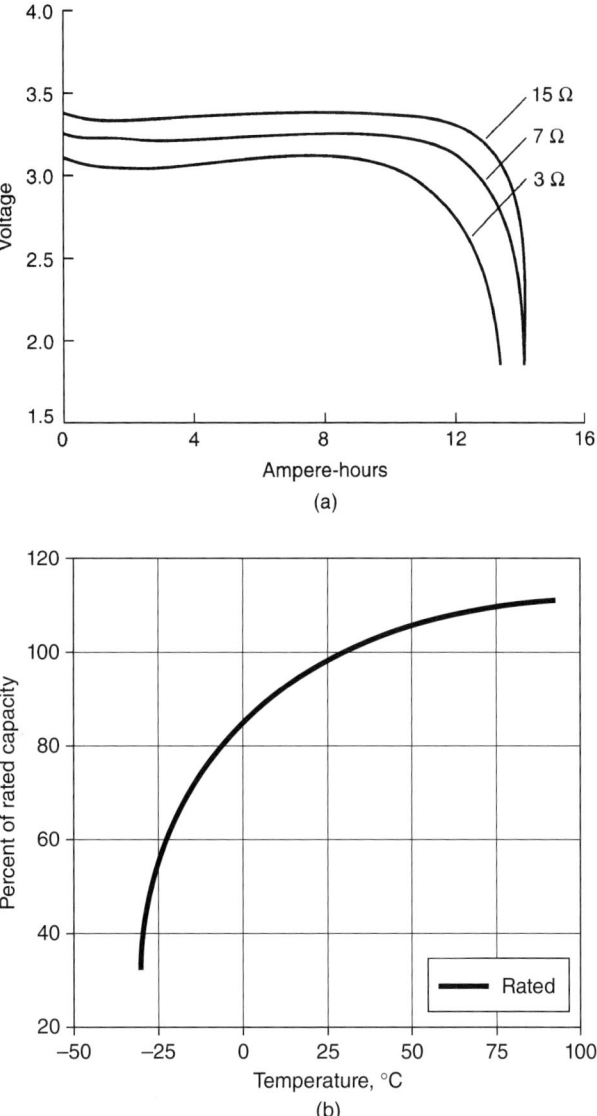

**FIGURE 13C.3**   Performance characteristics of $Li/SO_2Cl_2$ with $Cl_2^-$ additive in D-size batteries. (*a*) Discharge at 20°C. (*b*) Capacity versus discharge temperature; 100% capacity delivered at 20°C. (*Courtesy of Electrochem Power Solutions Div., Greatbatch Inc.*)

# SECTION D
# LITHIUM/MANGANESE DIOXIDE

## 13D.1 LITHIUM/MANGANESE DIOXIDE (Li/MnO$_2$) BATTERIES

The lithium/manganese dioxide (Li/MnO$_2$) battery was one of the first lithium/solid-cathode systems to be used commercially and is now the most widely used primary lithium battery. It is available in many configurations, including coin, bobbin, spirally wound cylindrical, and prismatic configurations in multicell batteries, and in designs for low, moderate, and moderately high-drain applications. The capacity of batteries available commercially ranges up to 11.1 Ah. Larger-sized batteries are available for special applications and have been introduced commercially. Its attractive properties include a high cell voltage (nominal voltage 3 V), specific energy about 280 Wh/kg and an energy density above 588 Wh/L (depending on design and application), good temperature performance, long shelf life, good storability (even at elevated temperatures), and low cost.

The Li/MnO$_2$ battery is used in a wide variety of applications such as long-term memory backup, safety and security devices, cameras, many consumer devices, and in military electronics. It has gained an excellent safety record during the period since it was introduced.

The performance of a Li/MnO$_2$ battery is compared with comparable mercury, silver oxide, and zinc batteries in Chap. 9, illustrating the higher energy output of the Li/MnO$_2$ battery.

## 13D.2 CELL CHEMISTRY

The Li/MnO$_2$ cell uses lithium for the anode, and an electrolyte containing lithium salts in a mixed organic solvent, such as PC and 1,2-dimethoxyethane, and a specially prepared heat-treated form of MnO$_2$ for the active cathode material.

The cell reactions for this system are

$$
\begin{array}{ll}
\textit{Anode:} & x\text{Li} \rightarrow \text{Li}^+ + \text{e} \\
\textit{Cathode:} & \text{Mn}^{\text{IV}}\text{O}_2 + x\text{Li}^+ + \text{e} \rightarrow \text{Li}_x\text{Mn}^{\text{III}}\text{O}_2 \\
\hline
\textit{Overall:} & x\text{Li} + \text{Mn}^{\text{IV}}\text{O}_2 \rightarrow \text{Mn}^{\text{III}}\text{O}_2
\end{array}
$$

Manganese dioxide, an intercalation compound, is reduced from the tetravalent to the trivalent state, producing Li$_x$MnO$_2$ as the Li$^+$ ion enters into the MnO$_2$ crystal lattice.[1,39]

The theoretical voltage of the total cell reaction is about 3.5 V, but an open-circuit voltage of a new cell is typically 3.3 V. Cells are typically predischarged to lower the open-circuit voltage to reduce corrosion.

## 13D.3 CONSTRUCTION

The Li/MnO$_2$ electrochemical system is manufactured in several different designs and configurations to meet the range of requirements for small, lightweight, portable power sources.

### 13D.3.1 Coin Cells

Figure 13D.1 shows a cutaway illustration of a typical coin cell. The manganese dioxide pellet faces the lithium anode disk and is separated by a nonwoven polypropylene separator impregnated with the electrolyte. The cell is crimped-sealed, with the can serving as the positive terminal and the cap as the negative terminal.

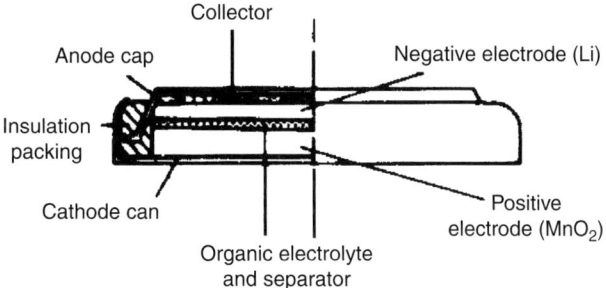

**FIGURE 13D.1** Cross-sectional view of Li/MnO$_2$ coin-type battery. (*Courtesy of Duracell, Inc.*)

## 13D.3.2 Bobbin-Type Cylindrical Cells

The bobbin-type cell is one of the two Li/MnO$_2$ cylindrical cells. The bobbin design maximizes the energy density due to the use of thick electrodes and the maximum amount of active materials, but at the expense of electrode surface area. This limits the rate capability of the cell and restricts its use to low-drain applications.

A cross section of a typical cell is shown in Fig. 13D.2. The cells contain a central lithium anode core surrounded by the manganese dioxide cathode, separated by a polypropylene separator impregnated with the electrolyte. The cell top contains a safety vent to relieve pressure in the event of mechanical or electrical abuse. Welded-sealed cells are manufactured in addition to the crimped-seal design. These cells, which have a 10-year life, are used for memory backup and other low-rate applications.

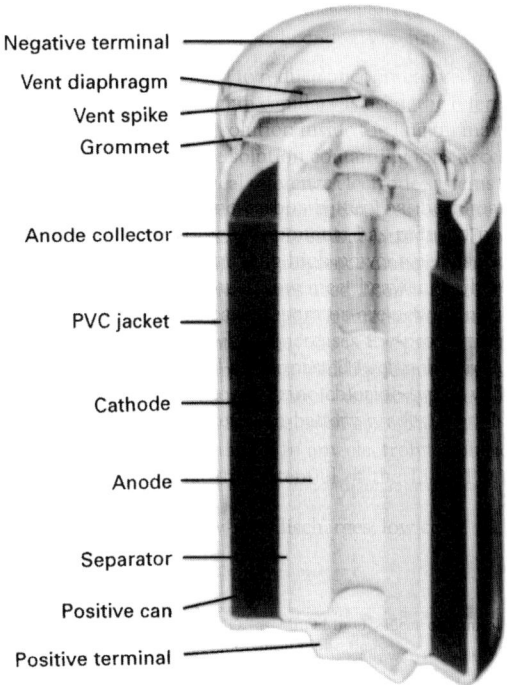

**FIGURE 13D.2** Cross-sectional view of Li/MnO$_2$ bobbin battery. (*Courtesy of Duracell, Inc.*)

### 13D.3.3   Spirally Wound Cylindrical Cells

The spirally wound cell, illustrated in Fig. 13D.3, is designed for high-current pulse applications as well as continuous moderate-rate operation. The lithium anode and the cathode (a thin, pasted electrode on a supporting grid structure) are wound together with a microporous separator interspaced between the two thin electrodes to form the jelly-roll construction. With this design, a high electrode surface area is achieved and the rate capability increased.

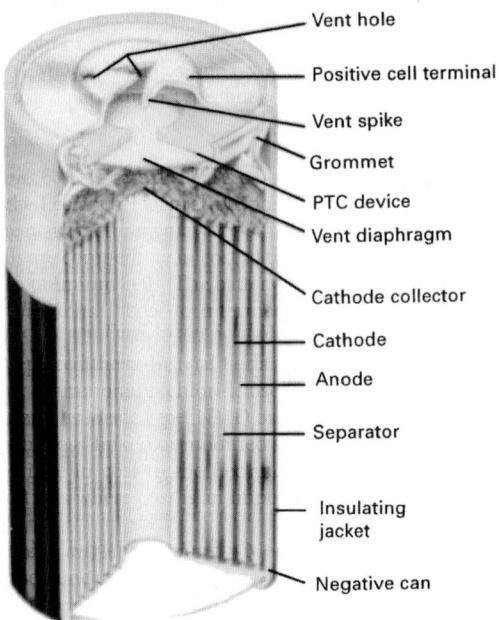

**FIGURE 13D.3**   Cross-sectional view of Li/MnO$_2$ spirally wound electrode battery. (*Courtesy of Duracell, Inc.*)

High-rate spirally wound cells contain a safety vent to relieve internal pressure in the event the cell is abused. Many of these cells also contain a resettable PTC device that limits the current and prevents the cell from overheating if short-circuited accidentally (see also Sec. 13D.4.4). Some manufacturers produce these cells with a peripheral laser-welded seal.

### 13D.3.4   Multicell 9 V Battery

The Li/MnO$_2$ system has also been designed in a 9 V battery with 1200 mAh capacity in the ANSI 1604 configuration as a replacement for the conventional alkaline zinc battery. The battery contains three prismatic cells, using an electrode design that utilizes the entire interior volume, as shown in Fig. 13D.4. An ultrasonically sealed plastic housing is used for the battery case.

### 13D.3.5   Foil (Pouch) Cell Designs

Other cell design concepts are being used to reduce the weight and cost of batteries by using lightweight cell packaging. One of these approaches is the use of heat-sealable thin metal foil/polymer film multilayer laminates in a prismatic (or alternatively, flat/oval wound) cell configuration in place of metal containers. The design of a cell with a capacity of about 16 Ah is illustrated in Fig. 13D.5. The cell contains 10 anode and 11 cathode plates in a parallel plate array.[40]

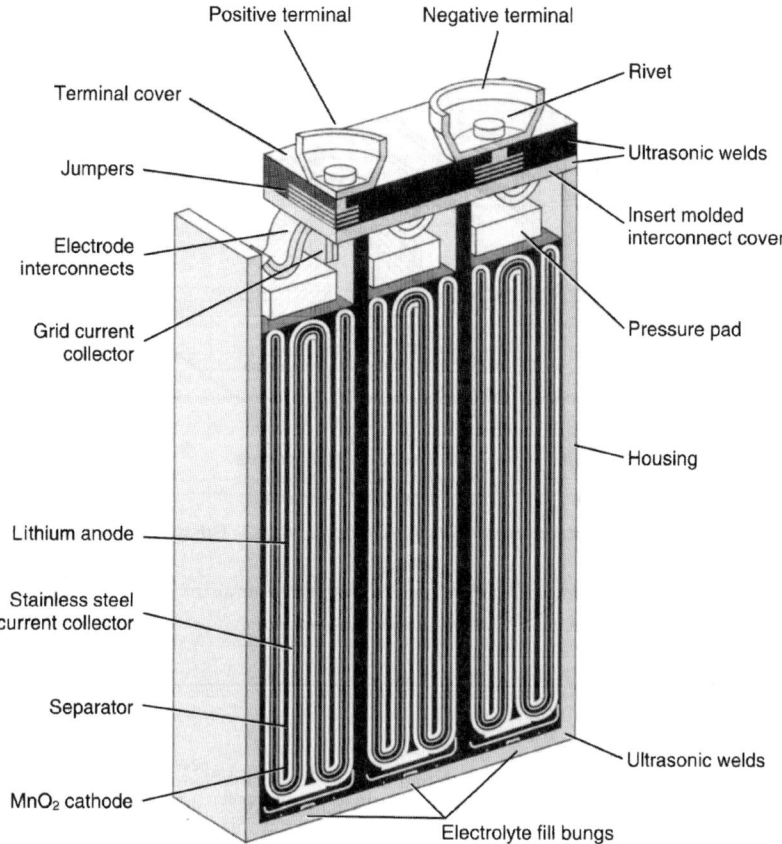

**FIGURE 13D.4**    Cross-sectional view of three-cell 9 V Li/MnO₂ battery. (*Courtesy of Ultralife Batteries, Inc.*)

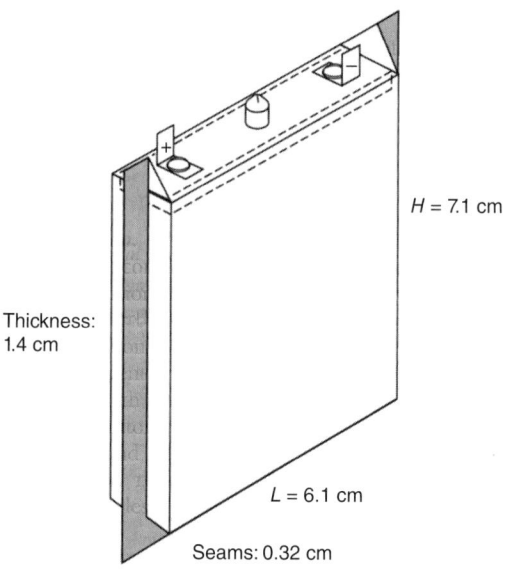

**FIGURE 13D.5**    Foil-cell design. (*From Ref. 40.*)

## 13D.4   PERFORMANCE CHARACTERISTICS AND APPLICATIONS

### 13D.4.1   Cell Output Performance

*Voltage.*   The open-circuit voltage of the Li/MnO$_2$ battery is typically 3.1 to 3.3 V after predischarge. The nominal voltage is 3.0 V. The operating voltage during discharge ranges from about 3.1 to 2.0 V and is dependent on the cell design, state of charge, and discharge conditions. The end or cutoff voltage, the voltage by which most of the capacity has been expended, is 2.0 V, except under high rate, low-temperature discharges, when a lower end voltage may be specified.

*Discharge Characteristics of Coin-Type Batteries.*   Typical discharge curves for the Li/MnO$_2$ coin cells are presented in Fig. 13D.6. The discharge profile is fairly flat at these low to moderate discharge rates throughout most of the discharge, with a gradual drop near the end of life. This gradual drop in voltage can serve as a state-of-charge indicator to show when the battery is approaching the end of its useful life.

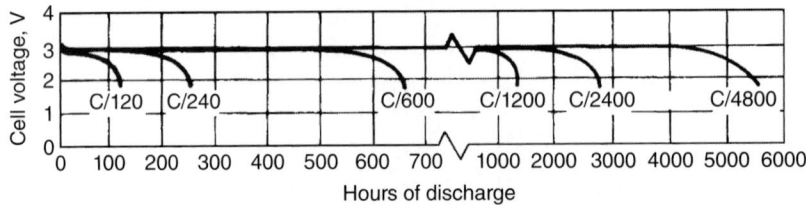

**FIGURE 13D.6**   Typical discharge curves of Li/MnO$_2$ coin-type batteries. (*Courtesy of Duracell, Inc.*)

Some applications (such as an LED watch with backlight) require a high pulse load superimposed on a low background current. The performance of a coin-type battery under these conditions is shown in Fig. 13D.7.

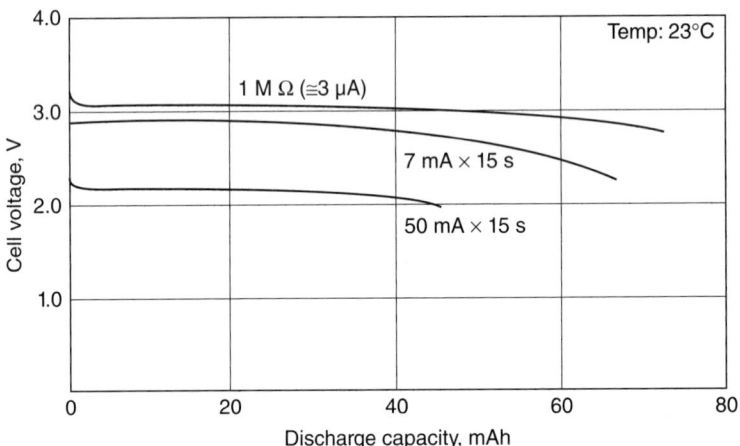

**FIGURE 13D.7**   Pulse characteristics of Li/MnO$_2$ coin-type battery (80 mAh size) at 23°C. Test conditions: continuous load—1 MΩ ≈ 3 μA; pulse load—7 mA × 15 s and 50 mA × 15 s. (*Courtesy of Sanyo Electric Co., Ltd.*)

The Li/MnO$_2$ coin-type battery is capable of performing over a wide temperature range, from about −20 to 70°C, as shown in Fig. 13D.8.

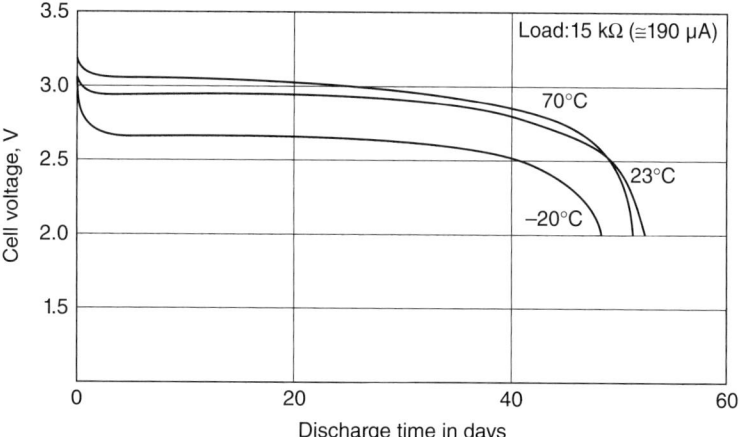

**FIGURE 13D.8**   Typical discharge performance of Li/MnO$_2$ coin-type battery (230 mAh) at various temperatures. (*Courtesy of Sanyo Electric Co., Ltd.*)

The discharge characteristics of the Li/MnO$_2$ battery are summarized in Fig. 13D.9, which shows the percent capacity delivered at various temperatures and discharge loads.

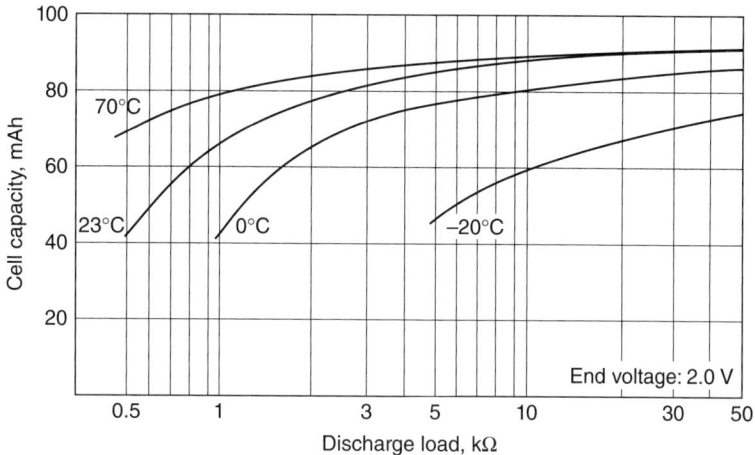

**FIGURE 13D.9**   Delivered capacity of Li/MnO$_2$ coin-type (80 mAh size) at various temperatures and loads. (*Courtesy of Sanyo Electric Co., Ltd.*)

***Discharge Characteristics of Cylindrical Bobbin Batteries.***   Typical discharge curves for the Li/MnO$_2$ cylindrical bobbin batteries are given in Fig. 13D.10. These bobbin electrode batteries are designed for use at low to moderate discharge rates, delivering higher capacities at these discharge rates than the spirally wound electrode batteries of the same size (see Table 13D.1). The discharge profile is fairly flat at these low rates throughout most of the discharge, with the typical gradual slope near the end of the discharge. The effect of a high pulse load superimposed on a low background current is shown in Fig. 13D.11.

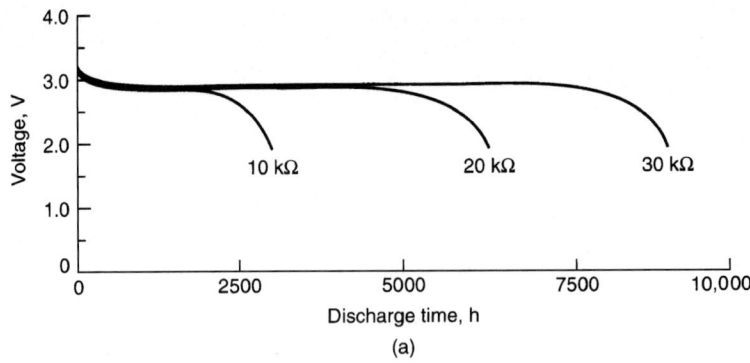

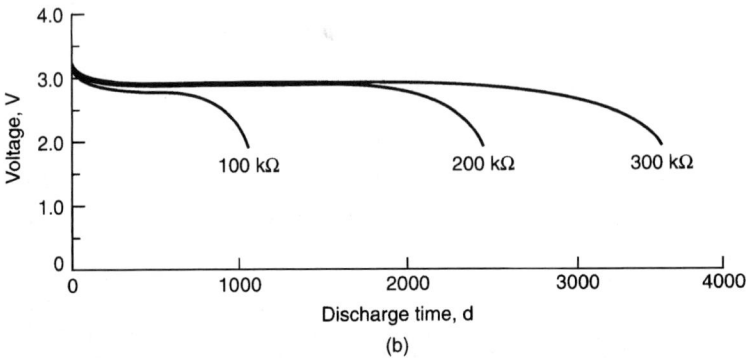

**FIGURE 13D.10**  Discharge characteristics of Li/MnO$_2$ cylindrical bobbin battery (850 mAh size) at 20°C. (*a*) Discharge time in hours. (*b*) Discharge time in days. (*Courtesy of Duracell, Inc.*)

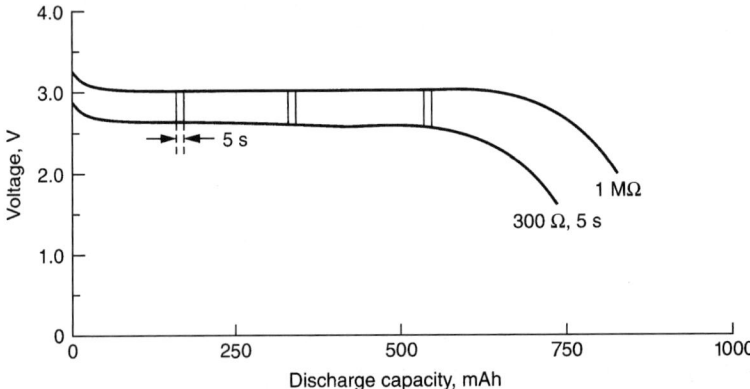

**FIGURE 13D.11**  Pulse discharge characteristics of Li/MnO$_2$ cylindrical bobbin cell (850 mAh size) at 20°C. Test conditions: continuous load—1 MΩ ≈ 2.9 µA; pulse load—300 Ω ≈ 10 mA; duration—5 s; pulses—3; time between pulses—3 h. (*Courtesy of Duracell, Inc.*)

The performance of the Li/MnO$_2$ cylindrical bobbin battery at temperatures from $-20$ to 60°C is shown in Fig. 13D.12. Operation of the coin-type and cylindrical bobbin electrode batteries at the lower temperatures is limited to the lower discharge rates.

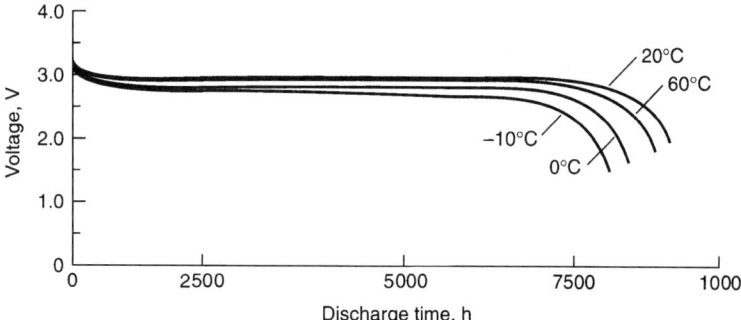

**FIGURE 13D.12**    Discharge performance of Li/MnO$_2$ cylindrical bobbin cell (850 mAh size) at various temperature; 30 kΩ discharge rate. (*Courtesy of Duracell, Inc.*)

***Discharge Characteristics of Cylindrical Spirally Wound Batteries.***    Typical discharge curves for Li/MnO$_2$ cylindrical spirally wound batteries at various constant-current discharge loads and temperatures are given in Fig. 13D.13. These batteries are designed for operation at fairly high rates and low temperatures. Their discharge profile is flat under most of these discharge conditions. The midpoint voltage when discharged at various loads and temperatures is plotted in Fig. 13D.14.

The characteristics of the batteries under constant power discharge are shown in Fig. 13D.15. These data are expressed in terms of *E*-rate, which is calculated in a manner similar to calculating the *C*-rate, but based on the rated watt-hour capacity. For example, the *E*/5 rate for a cell rated at 4 Wh is 800 mW.

The discharge characteristics of the cylindrical spirally wound Li/MnO$_2$ battery at various temperatures and loads are summarized in Fig. 13D.16. Figure 13D.16*a* shows the percent capacity delivered on constant-resistance loads and Fig. 13D.16*b* shows the percent capacity delivered on constant-current loads. The good performance of the Li/MnO$_2$ battery at the lower rate discharges is evident, and it still delivers a higher percentage of its capacity at relatively high discharge rates compared to conventional aqueous primary cells.

The discharge characteristics of a larger (D-size) spiral-wound Li/MnO$_2$ battery are shown in Fig. 13D.17. These three figures (13D.17*a–c*) show the discharge curves at three rates (250 mA, 2.0 A, and 3.0 A) at temperatures from $-40$ to $+72$°C. The discharge characteristics indicate the fall-off in performance at lower temperatures.

***Discharge Characteristics of Three-Cell 9 V Li/MnO$_2$ Battery.***    The performance of the 9 V, 1.2 Ah Li/MnO$_2$ battery is shown in Fig. 13D.18. Typically discharge curves on a 900-Ω load at temperatures from $-20$ to $+23$°C are shown in Fig. 13D.19*a*. Figure 13D.19*b* shows the realized capacity in ampere-hours for loads from 60 to 900 Ω at room temperature. The lithium battery has a higher voltage and delivers significantly more service than the comparable zinc-alkaline and carbon-zinc batteries as shown in Fig. 13D.18.

## 13D.4.2    Factors Affecting Performance

***Internal Resistance.***    The internal resistance of the Li/MnO$_2$ battery, as with most battery systems, is dependent on the cell size, design, electrode, separator, as well as the chemistry. Inherently, the conductivity of the organic solvent-based electrolytes is lower than that of the aqueous electrolytes, and the Li/MnO$_2$ system, therefore, has a higher impedance than conventional aqueous cells of the same size and construction. Designs that increase electrode area and decrease electrode spacing, such as coin-shaped flat cells and spirally wound jelly-roll configurations, are used to reduce the resistance. Further, the lithium cells will perform relatively more

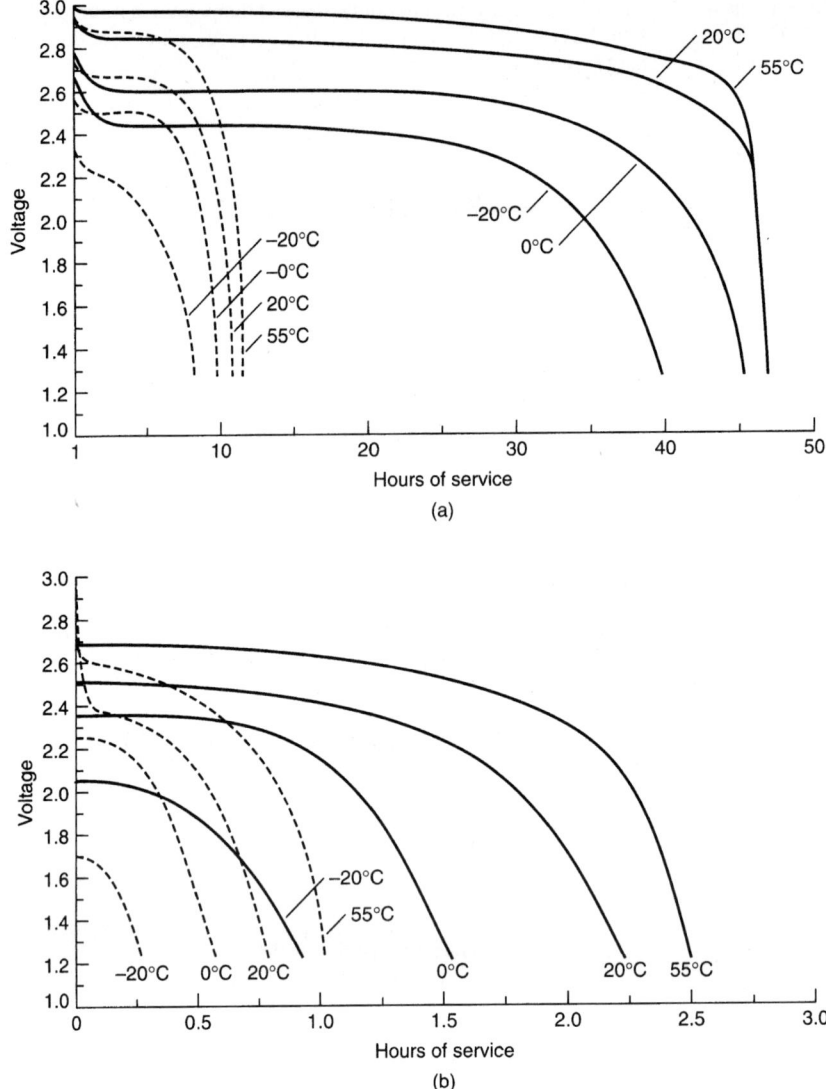

**FIGURE 13D.13**   Discharge characteristics of cylindrical (spirally wound electrode) Li/MnO$_2$ battery (CR123A-size). (*a*) Discharge at 30 and 125 mA. Broken line—125 mA; solid line—30 mA. (*b*) Discharge at 500 and 1000 mA. Broken line—1 A; solid line—0.5 A.

efficiently at the lower temperatures because the conductivity of the organic solvents is less sensitive to temperature changes than it is for the aqueous solvents.

Figure 13D.20 shows the change in internal resistance of a 280 mAh coin-type battery during a low-rate discharge at 20°C. Typically, the resistance is a mirror image of the voltage profile. It remains fairly constant for most of the discharge and increases at the end of life.

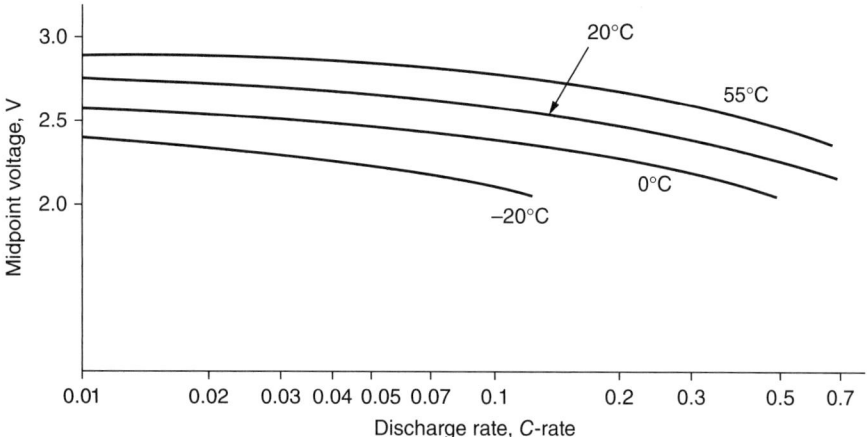

**FIGURE 13D.14**    Midpoint voltage of cylindrical (spirally wound) Li/MnO$_2$ batteries during discharge; 2 V end voltage.

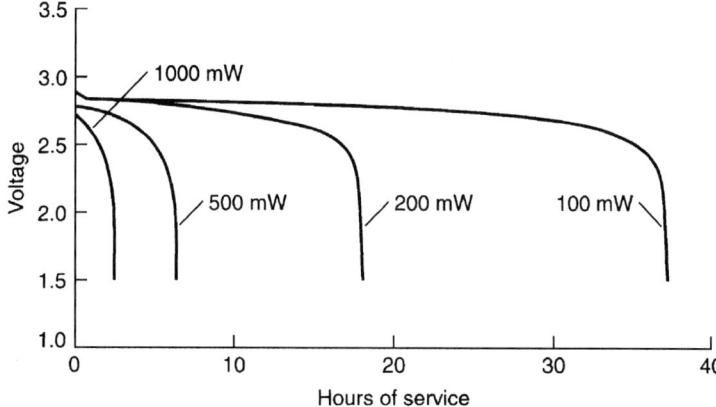

**FIGURE 13D.15**    Discharge characteristics of cylindrical (spirally wound electrode) Li/MnO$_2$ cells (CR123A-size) under constant-power mode at 20°C.

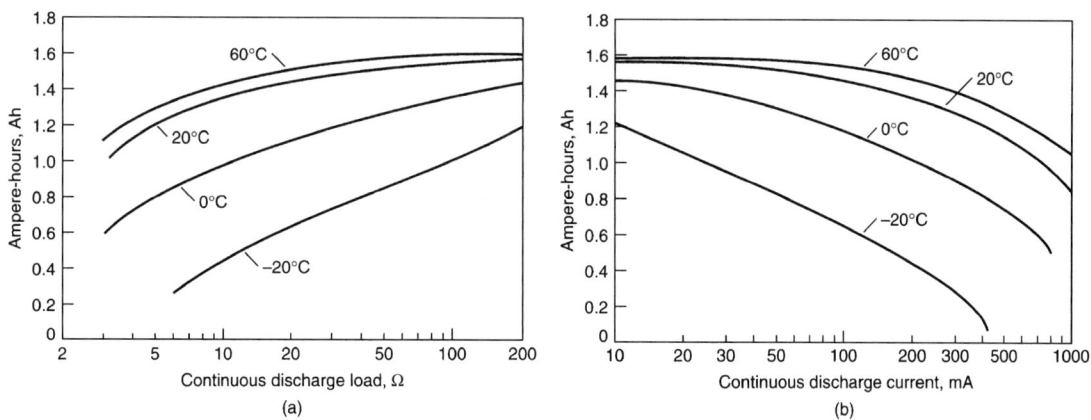

**FIGURE 13D.16**    Summary of discharge characteristics of cylindrical (spirally wound) Li/MnO$_2$ battery (CR123A-size); capacity versus discharge load to 2.0 V per cell. (*a*) Constant-resistance loads. (*b*) Constant-current loads.

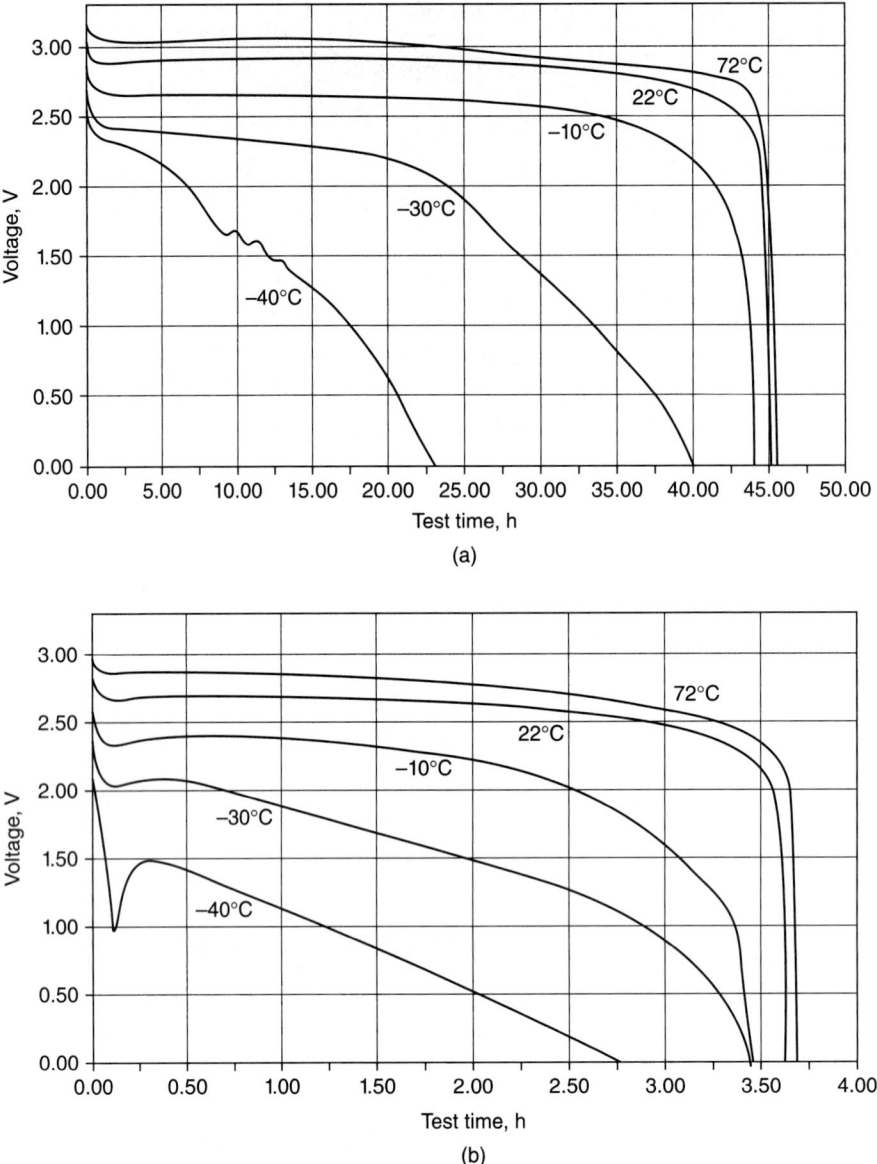

**FIGURE 13D.17**    Discharge characteristics of spiral-wound Li/MnO$_2$ D-size battery. (*a*) Discharge curve at 250 mA amp rate at 5 temperatures. (*b*) Discharge curves at 2.0 A and 5 temperatures. (*c*) Discharge curves at 3.0 A and 5 temperatures. Temperatures for all discharges are: +72, +22, −10, −30, and −40°C. (*Courtesy of Ultralife Batteries, Inc.*)

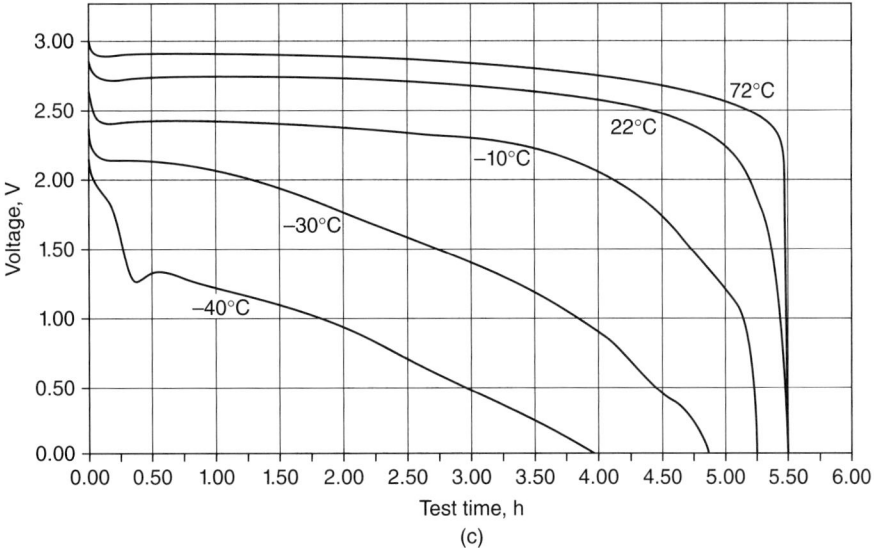

**FIGURE 13D.17**    (*Continued*)

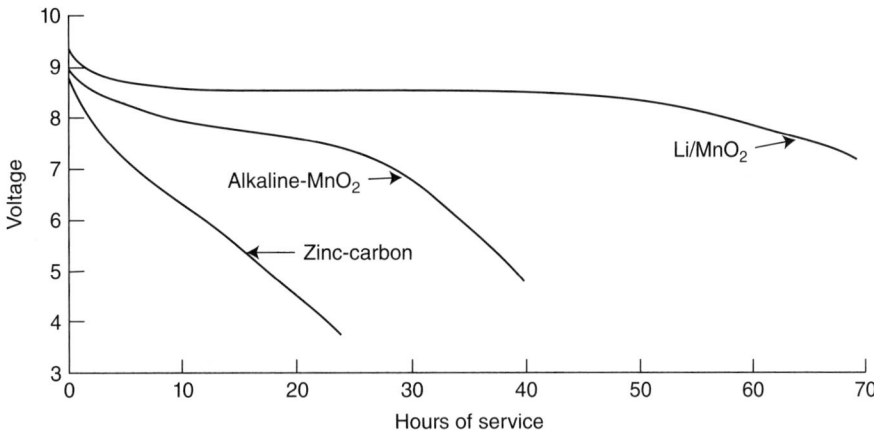

**FIGURE 13D.18**    Typical discharge curves for an ANSI 1604 battery, 9 V, 500 Ω discharge load at 20°C.

***Service Life.***    The capacity or service life of the different types of Li/MnO$_2$ cells, normalized for a 1 g and 1 cm$^3$ cell, at various discharge loads and temperatures, is summarized in Fig. 13D.21. These data can be used to approximate the performance of a given cell or to determine the size and weight of a cell for a particular application.

***Shelf Life.***    The storage characteristics of two Li/MnO$_2$ cells are shown in Fig. 13D.22. This system is very stable in all of the configurations, with a loss of capacity of less than 0.5%/yr for hermetic and laser-sealed cells. Coin cells show capacity loss of less than 1% annually at room temperature. The cells do not have any noticeable voltage delay at the start of most discharges, except at low temperature on high discharge rates.

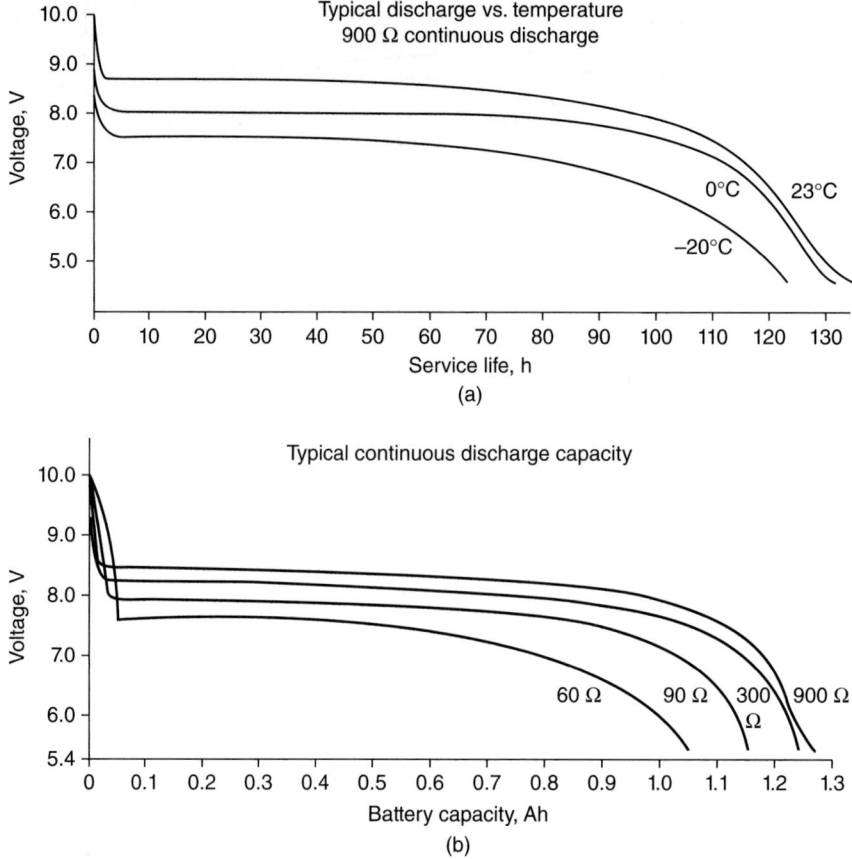

**FIGURE 13D.19**    Discharge characteristics of 9 V Li/MnO$_2$ battery. (*a*) Discharge versus temperature; 900 Ω continuous discharge. (*b*) Continuous discharge at room temperature on loads from 60 to 900 Ω. (*Courtesy of Ultralife Batteries, Inc.*)

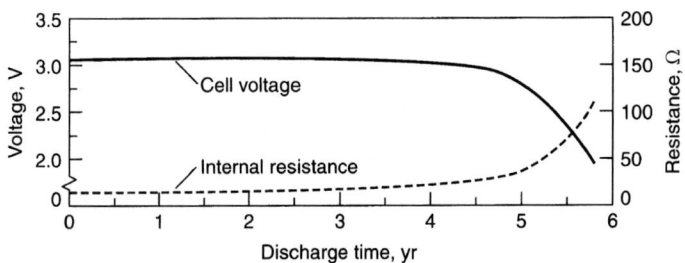

**FIGURE 13D.20**    Internal resistance of Li/MnO$_2$ coin-type battery (280 mAh size), 5-μA drain, at 20°C.

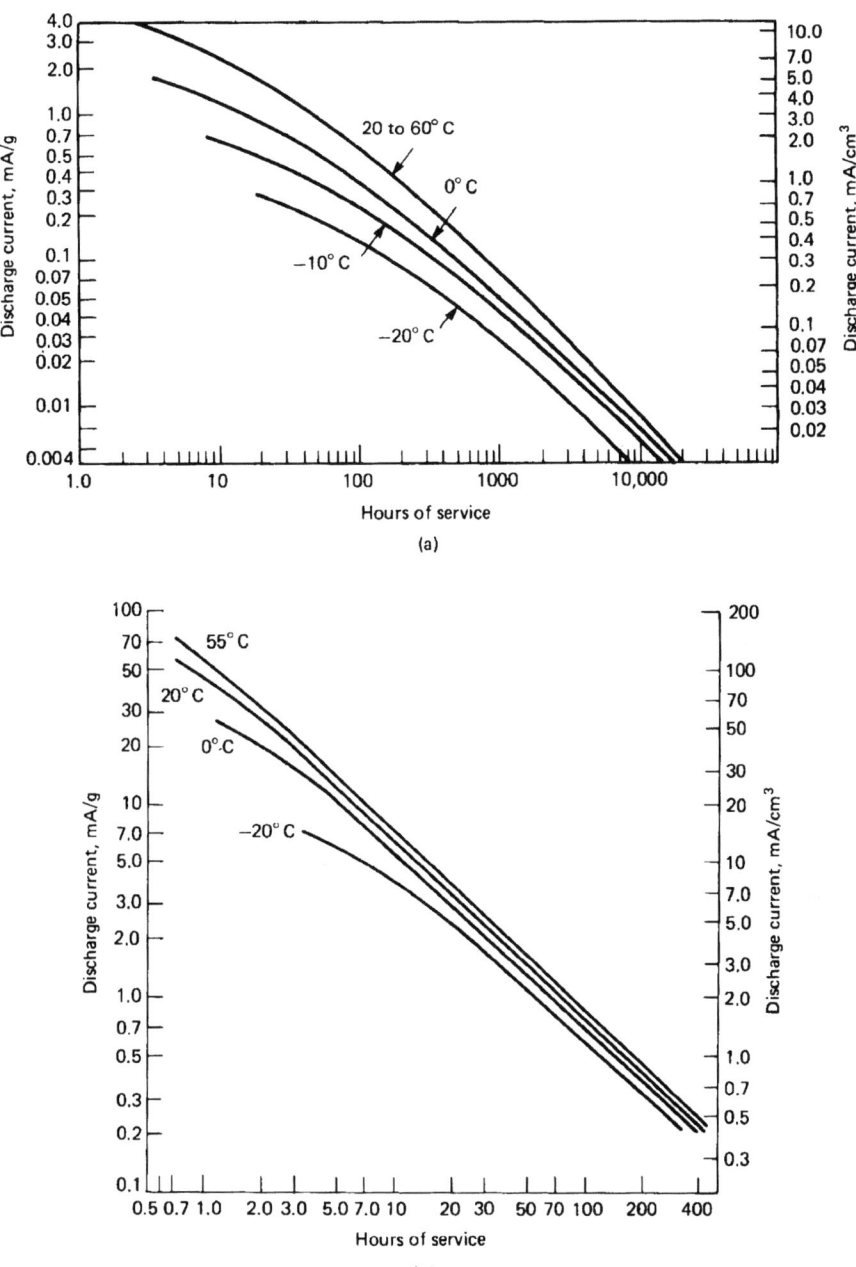

**FIGURE 13D.21**   Service life of Li/MnO₂ cells to 2 V end voltage. (*a*) Low-rate coin-type batteries. (*b*) Small cylindrical batteries.

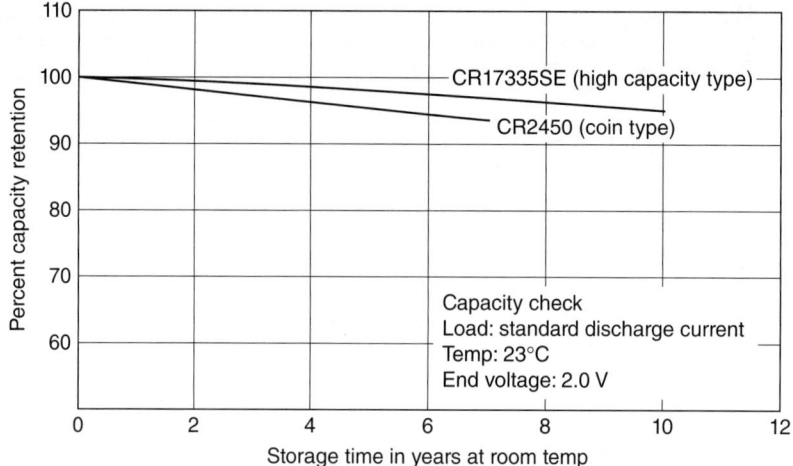

**FIGURE 13D.22**   Storage characteristics of a typical spiral-wound, laser-sealed cylindrical cell and a typical coin cell.

### 13D.4.3   Cell and Battery Sizes

The Li/MnO$_2$ cells are manufactured and commercially available in a number of flat and cylindrical batteries ranging in capacity from about 30 mAh to 11.1 Ah. The physical and electrical characteristics of some of these are summarized in Table 13D.1. In some instances interchangeability with other battery systems is provided by doubling the size of the battery to accommodate the 3 V output of the Li/MnO$_2$ cell compared with 1.5 V of the conventional primary batteries, for example, battery type CR-V3. Table 13D.2 shows the characteristics of two commercially available thin cells in foil-laminate packaging.

**TABLE 13D.1**   Typical Li/MnO$_2$ Batteries

| | (a) Low-rate coin cell batteries | | | | | |
|---|---|---|---|---|---|---|
| | Electrical characteristics (20°C) | | | Dimensions, mm | | |
| Model no. | Nominal voltage, V | Nominal capacity,[*] mAh | Continuous drain, mA | Diameter | Height | Weight, g |
| CR1025 | 3 | 30 | 0.1 | 10.0 | 2.5 | 0.7 |
| CR1216 | 3 | 25 | 0.1 | 12.5 | 1.6 | 0.7 |
| CR1220 | 3 | 35 | 0.1 | 12.5 | 2.0 | 1.2 |
| CR1612 | 3 | 41 | 0.1 | 16.0 | 1.2 | 0.8 |
| CR1616 | 3 | 55 | 0.1 | 16.0 | 1.6 | 1.2 |
| CR1620 | 3 | 75 | 0.1 | 16.0 | 2.0 | 1.3 |
| CR1632 | 3 | 140 | 0.1 | 16.0 | 3.2 | 1.8 |
| CR2012 | 3 | 55 | 0.1 | 20.0 | 1.2 | 1.4 |
| CR2016 | 3 | 90 | 0.1 | 20.0 | 1.6 | 1.6 |
| CR2025 | 3 | 165 | 0.2 | 20.0 | 2.5 | 2.3 |
| CR2032 | 3 | 225 | 0.2 | 20.0 | 3.2 | 2.9 |
| CR2330 | 3 | 265 | 0.2 | 23.0 | 3.0 | 3.8 |
| CR2354 | 3 | 560 | 0.2 | 23.0 | 5.4 | 5.8 |
| CR2412 | 3 | 100 | 0.2 | 24.5 | 1.2 | 2.0 |
| CR2450 | 3 | 620 | 0.2 | 24.5 | 5.0 | 6.3 |
| CR2477 | 3 | 1000 | 0.2 | 24.5 | 7.7 | 10.5 |
| CR3032 | 3 | 500 | 0.2 | 30.0 | 3.2 | 6.8 |

[*]Nominal capacity shown above is based on standard drain and cutoff voltage down to 2.0 V at 20°C.

**TABLE 13D.1** Typical Li/MnO₂ Batteries (*Continued*)

(*b*) Specialized high-power, cylindrical-type batteries (spiral structure, laser sealing)

| IEC type | Nominal voltage, V | Nominal capacity, mAh | Diameter | Height | Weight, g |
|---|---|---|---|---|---|
| CR17335 | 3 | 1600 | 17.0 | 33.5 | 17 |
| CR17335 | 3 | 1350 | 17.0 | 33.5 | 16 |
| CR17450 | 3 | 2400 | 17.0 | 45.0 | 23 |
| CR17450 | 3 | 2600 | 17.0 | 45.0 | 23 |

Operational temperature range: −40 to +85°C.

(*c*) Standard high-power, cylindrical batteries (spiral-wound, hermetic)

| Size | Nominal voltage, V | Nominal capacity, Ah | Diameter | Height | Weight, g | Continuous current, A |
|---|---|---|---|---|---|---|
| C | 3.0 | 4.8 | 25.8 | 50.0 | 61 | 2.0 |
| 5/4 C | 3.0 | 6.1 | 25.8 | 60.5 | 71 | 2.5 |
| D | 3.0 | 11.1 | 34.0 | 60.5 | 115 | 3.3 |

(*d*) Specialized low-power, cylindrical-type batteries

| Model | Nominal voltage, V | Nominal capacity, mAh | Diameter | Height | Weight, g |
|---|---|---|---|---|---|
| CR14250 | 3 | 850 | 14.5 | 25.0 | 9 |
| CR12600 | 3 | 1500 | 12.0 | 60.5 | 15 |
| CR17335 | 3 | 1800 | 17.0 | 33.5 | 17 |
| CR17450 | 3 | 2500 | 17.0 | 45.0 | 22 |

Operational temperature range: −40 to +85°C.

(*e*) Specialized cylindrical type primary lithium batteries (spiral structure, crimp-sealing)

| IEC type | Nominal voltage, V | Nominal capacity, mAh | Diameter | Height | Weight, g |
|---|---|---|---|---|---|
| CR-1/3 | 3 | 160 | 11.6 | 10.8 | 3.3 |
| 2CR-1/3N | 6 | 160 | 13.0 | 25.2 | 9.1 |
| CR2 | 3 | 850 | 15.6 | 27.0 | 11 |
| CR123A | 3 | 1400 | 17.0 | 34.5 | 17 |
| CR-V3 | 3 | 3300 | 28.6 (L) × 14.6 (W) × 52.2 (H) | | 38 |
| CR-P2 | 6 | 1400 | 34.8 (L) × 19.5 (W) × 35.8 (H) | | 37 |
| 2CR5 | 6 | 1400 | 34 (L) × 17 (W) × 45 (H) | | 40 |

Operational temperature range: −40 to +60°C.

**TABLE 13D.2** Characteristics of Commercially Available Li/MnO₂ Cells in Thin, Foil-Laminate Packaging

| Dimensions (thickness × width × length), mm | Average voltage, V | Nominal capacity to 1.5 V | Maximum discharge (mA-continuous) | Weight, g | Pulse capacity, mA |
|---|---|---|---|---|---|
| 5.00 × 44.45 × 54.61 | 3.0 | 1.5 Ah | 250 | 15.0 | Up to 500 |
| 2.16 × 32.16 × 40.36 | 3.0 | 400 mAh | 25 | 3.5 | Up to 130 |

Operating/temperature: 0 to 71°C.
Capacity of 1.5 Ah cell determined at 10 mA and of the 400 mAh cell at 6 mA.
*Source: Courtesy of Ultralife Batteries, Inc.*

Performance characteristics of these cells are summarized in Table 13D.3. When used in the BA-7847 battery, they provide capacities greater than 19.5 Ah on 250 mA discharge at room temperature. This corresponds to a specific energy of about 300 Wh/kg. These batteries also passed the applicable UN/IATA shipping tests. When tested on the military L-test (8 W for 2 min followed by 5 W to a 4.0 V cutoff), these batteries ran for 9.5 h at $-10°C$ but only 0.5 h at $-20°C$. Further improvement in low-temperature performance is being sought. These batteries must comply with the requirements of MIL-PRF-49471 for the particular battery type. A list of BA-type lithium/manganese dioxide batteries currently qualified by the U.S. Army is given in Table 13D.3.

**TABLE 13D.3**    Performance Data at Room Temperature for $Li/MnO_2$ Pouch Cells Designed for BA-7847 Batteries

| Discharge rate, mA | Capacity, Ah | Energy, Wh | Specific energy, Wh/kg | Energy density, Wh/L |
|---|---|---|---|---|
| 250 | 9.94 | 26.68 | 402 | 737 |
| 500 | 9.80 | 25.77 | 384 | 712 |
| 1000 | 9.27 | 23.91 | 356 | 661 |
| 2000 | 9.00 | 22.68 | 339 | 627 |

*Source: Courtesy of Ultralife Batteries, Inc.*

Battery packs are also being employed for Emergency Positioning Indicating Radio Beacons (EPIRBs) and pipeline test vehicles. Smaller batteries are also available commercially in foil-laminate packages for use in specialized applications such as toll collection transponders, RFID tags for shipping and inventory control, and smart security tags.

The specific conditions for the use and handling of $Li/MnO_2$ batteries are dependent on the size as well as the specific design features. Manufacturers' recommendations should be consulted.

# SECTION E

# LITHIUM/CARBON MONOFLUORIDE

## 13E.1    LITHIUM/CARBON MONOFLUORIDE (Li/CFx) BATTERIES

The lithium/carbon monofluoride Li/(CFx) battery was one of the first lithium/solid-cathode systems to be used commercially. It is attractive as its theoretical specific energy (about 2190 Wh/kg) is among the highest of the solid-cathode systems. Its open-circuit voltage is 3.2 V, with an operating voltage of about 2.5 to 2.7 V. Its practical specific energy and energy density ranges up to 250 Wh/kg and 635 Wh/L in smaller sizes and 820 Wh/kg and 1180 Wh/L in larger sizes. The system is used primarily at low to medium discharge rates.

## 13E.2    CELL CHEMISTRY

The active components of the cell are lithium for the anode and polycarbon monofluoride (CFx) for the cathode. The value of $x$ is typically 0.9 to 1.2. Carbon monofluoride is an interstitial compound formed by the reaction between carbon powder and fluorine gas. While electrochemically active, the material is chemically stable in the organic electrolyte and does not thermally decompose up to 400°C, resulting in a long storage life. Different electrolytes have been used: 1 M lithium tetrafluoroborate (LiBF$_4$) in δ-butyrolactone (GBL) for cylindrical cells and LiBF$_4$ in a mixture of GBL and DME or a mixture of PC and DME for coin cells.

The simplified discharge reactions of the cell are

$$
\begin{aligned}
\textit{Anode:} \quad & x\text{Li} \rightarrow x\text{Li}^+ + xe \\
\textit{Cathode:} \quad & \underline{\text{CF}x + xe \rightarrow x\text{C} + x\text{F}^-} \\
\textit{Overall:} \quad & x\text{Li} + \text{CF}x \rightarrow x\text{LiF} + x\text{C}
\end{aligned}
$$

The polycarbon monofluoride changes into carbon, which is more conductive as the discharge progresses, thereby increasing the cell's conductivity, improving the regulation of the discharge voltage and increasing the discharge efficiency. The crystalline LiF precipitates in the cathode structure.[1,43,44]

## 13E.3    CONSTRUCTION

The Li/CFx system is adaptable to a variety of sizes and configurations. Batteries are available in flat coin or button, cylindrical, and rectangular shapes, ranging in capacity from 0.020 to 25 Ah; larger sized batteries have been developed for specialized applications.

Figure 13E.1 shows the construction of a coin-type battery. The Li/CFx cells are typically constructed with an anode of lithium foil rolled onto a collector and a cathode of Teflon-bonded polycarbon monofluoride and acetylene black on a nickel collector. Nickel-plated steel or stainless steel is used for the case material. The coin cells are crimped-sealed using a polypropylene gasket.

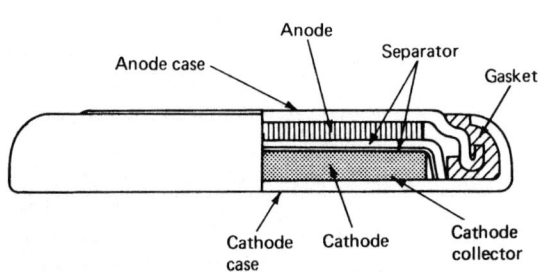

**FIGURE 13E.1**    Cross-sectional view of Li/CFx coin-type battery. (*Courtesy of Panasonic Corp. of North America.*)

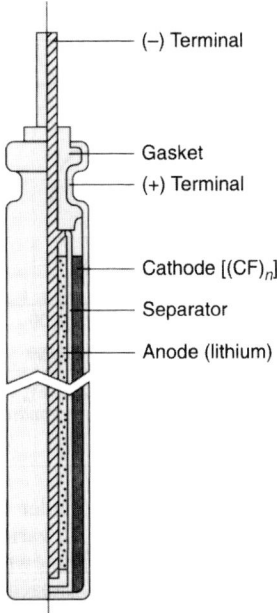

**FIGURE 13E.2**   Cross-sectional view of Li/CF$x$ pin-type battery. (*Courtesy of Panasonic Corp. of North America.*)

The pin-type batteries use an inside-out design with a cylindrical cathode and a central anode in an aluminum case, as shown in Fig. 13E.2.

The cylindrical batteries use a spirally wound (jelly-roll) electrode construction, and the batteries are either crimped or hermetically sealed. Their construction is similar to the cylindrical spiral-wound electrode design of the Li/MnO$_2$ battery shown in Fig. 13D.3. The larger cells are provided with low-pressure safety vents.

## 13E.4   PERFORMANCE CHARACTERISTICS AND APPLICATIONS

### 13E.4.1   Performance

***Coin-Type Batteries.***   Figure 13E.3 presents the discharge curves at 20°C for a typical Li/CF$x$ coin-type battery rated at 165 mAh. The voltage is constant throughout most of the discharge, and the coulombic utilization is close to 100% under low-rate discharge. Figure 13E.4 presents the discharge curves for the same battery at various discharge temperatures. The behavior of the battery on a pulse discharge at 20°C is shown in Fig. 13E.5.

The performance of the coin battery (165 mAh capacity) is summarized in Fig. 13E.6. Figure 13E.6*a* shows the average load voltage (plateau voltage during discharge) and Fig. 13E.6*b* shows the capacity for discharges at various loads and temperatures.

Figure 13E.7 summarizes the discharge performance data for Li/CF$x$ coin-type batteries normalized for a 1 g and 1 mL battery. These data can be used to approximate the size or performance of a battery for a particular application.

***Cylindrical Cells.***   The cylindrical batteries are designed to operate at higher discharge rates than the coin batteries. Figure 13E.8 presents the discharge curves on a 1 kΩ load at several temperatures. In some cases, an initial low voltage is observed with the Li/CF$x$ battery; that is, the voltage drops initially below the operating level on load and recovers gradually as the discharge progresses. This is attributed to the fact that CF$x$ is an insulator, but the resistance of the cathode decreases during the discharge as conductive carbon is produced.

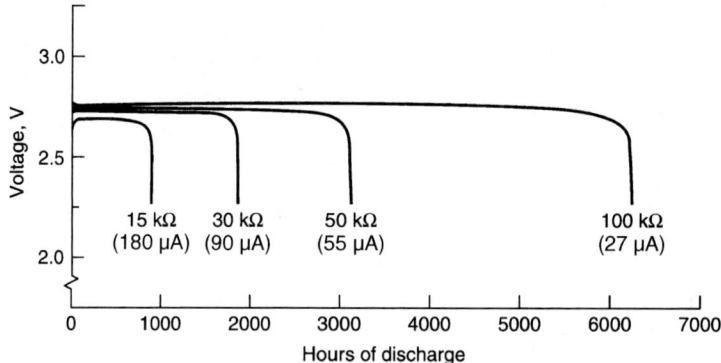

**FIGURE 13E.3**    Typical discharge curves of Li/CFx coin-type battery at 20°C; rated capacity 165 mAh. (*Courtesy of Panasonic Corp. of North America.*)

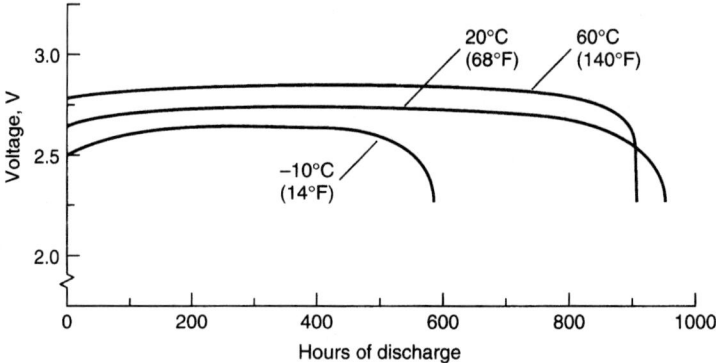

**FIGURE 13E.4**    Typical discharge curves of Li/CFx 165 mAh coin-type battery at various temperatures; 15-kΩ discharge load; 180 μA. (*Courtesy of Panasonic Corp. of North America.*)

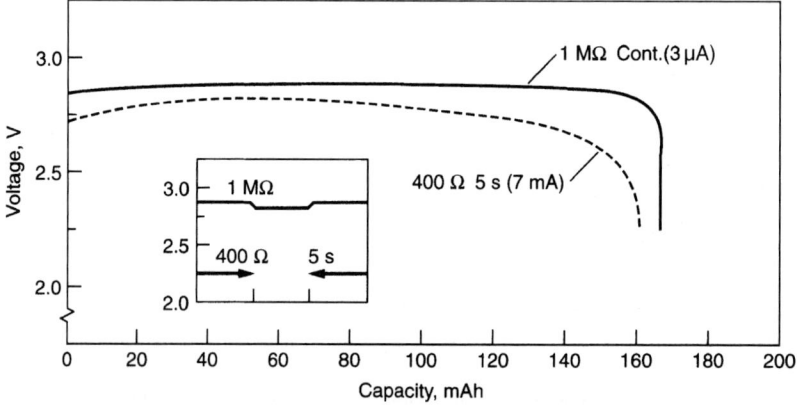

**FIGURE 13E.5**    Pulse discharge characteristics of Li/CFx coin-type (165-mAh size) at 20°C. (*Courtesy of Panasonic Corp. of North America.*)

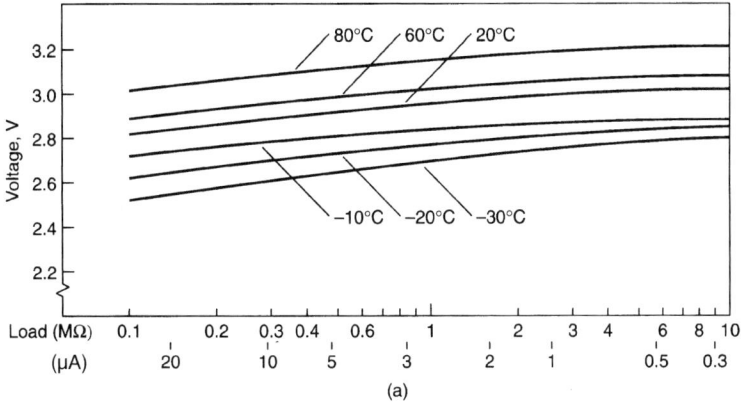

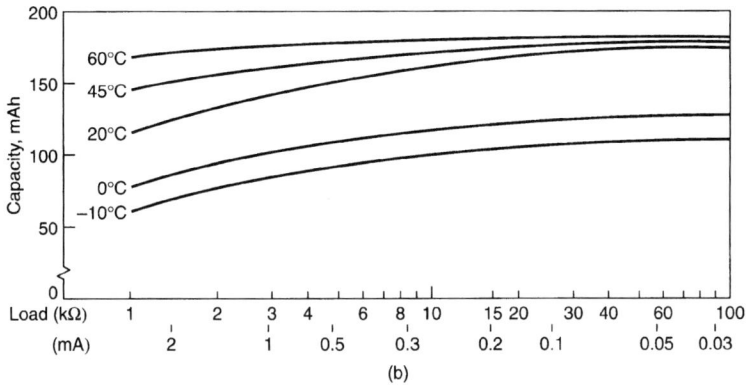

**FIGURE 13E.6** Discharge characteristics of Li/CFx coin-type battery (165 mAh size). (*a*) Operating voltage versus discharge load; voltage at 50% discharge. (*b*) Capacity versus discharge load; cutoff at 2.0 V. (*Courtesy of Panasonic Corp. of North America.*)

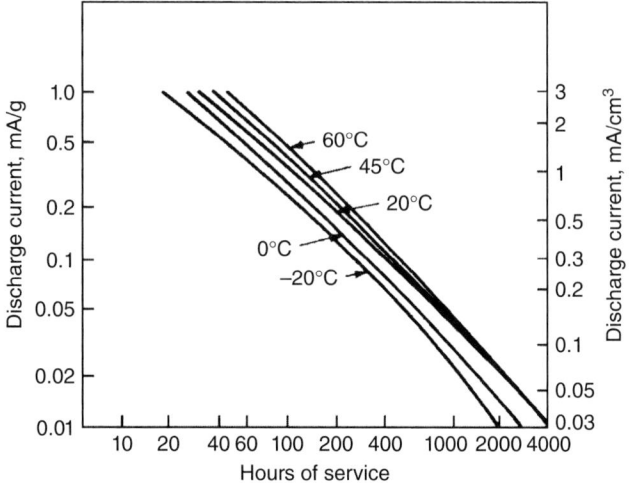

**FIGURE 13E.7**  Service life of Li/CFx coin-type batteries at various discharge rates and temperatures; 2.0 V end voltage.

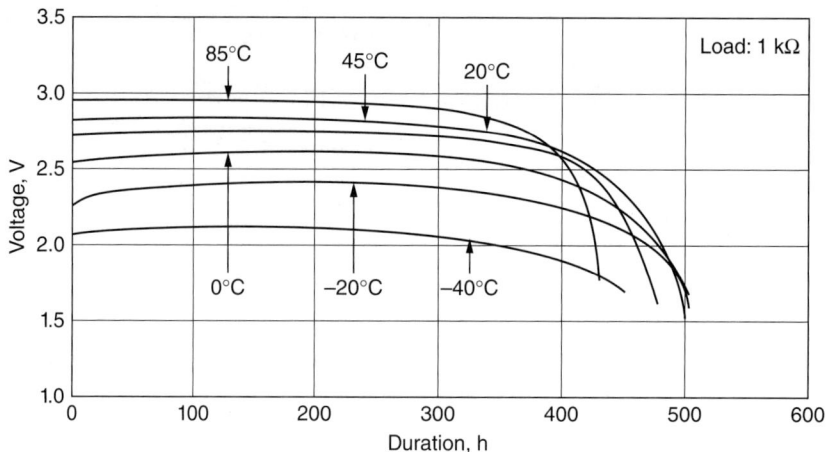

**FIGURE 13E.8**   Discharge curves for Li/CF$x$ 2/3 A-size cylindrical battery on a 1 kΩ load at temperatures from −40 to +85°C.

The average load voltages for 2/3 A-size cylindrical batteries at various temperatures and rates are given in Fig. 13E.9. The performance data are then summarized in Fig. 13E.10, which shows the effect of temperature and load on the service life of a battery normalized to unit weight (grams) and volume (cubic centimeters).

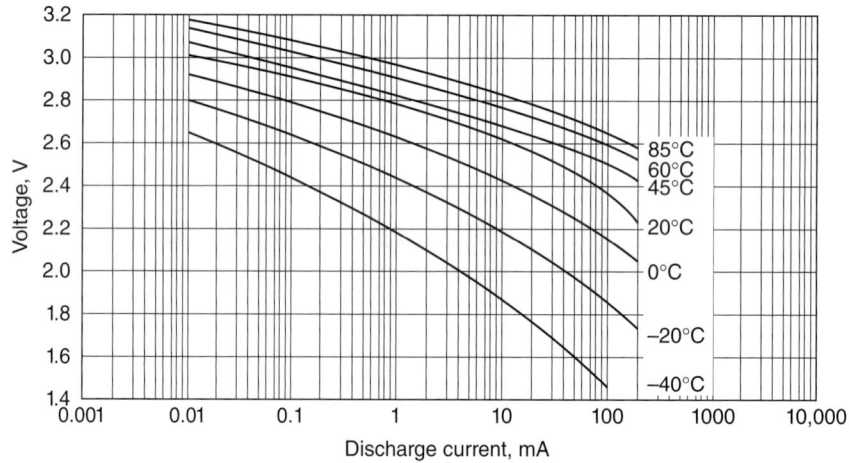

**FIGURE 13E.9**   Midpoint voltage for Li/CF$x$ 2/3 A-size cylindrical battery as a function of temperature and discharge current.

***Shelf Life.***   The Li/CF$x$ cells have extremely good storage characteristics. Tests over more than 10 years of storage show a self-discharge rate of about 0.5%/yr at 20°C for coin cells and 1.0%/yr for cylindrical batteries. These rates decrease on longer term storage.[45] This cell is also well suited for applications requiring low current drain over an extended period of time. This is illustrated in Fig. 13E.11, which shows the discharge characteristics of the 2/3 A-size cell at a 20-μA discharge rate over a period of 7 years. Voltage delay after storage is not apparent with these cells, except under severe discharge conditions.

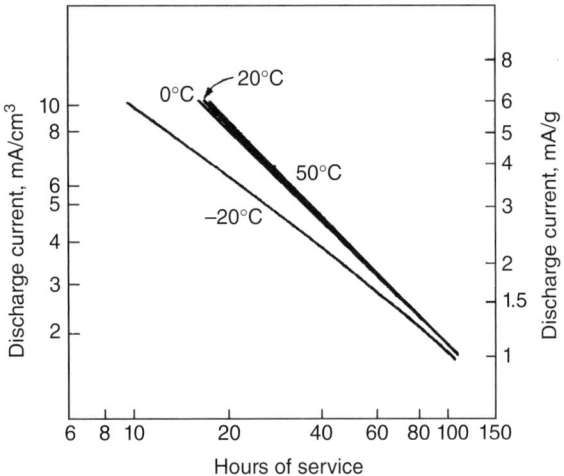

**FIGURE 13E.10**  Service life of Li/(CF)$n$ cylindrical battery at various discharge rates and temperatures; 1.8 V end voltage. (*Courtesy of Panasonic Corp. of North America.*)

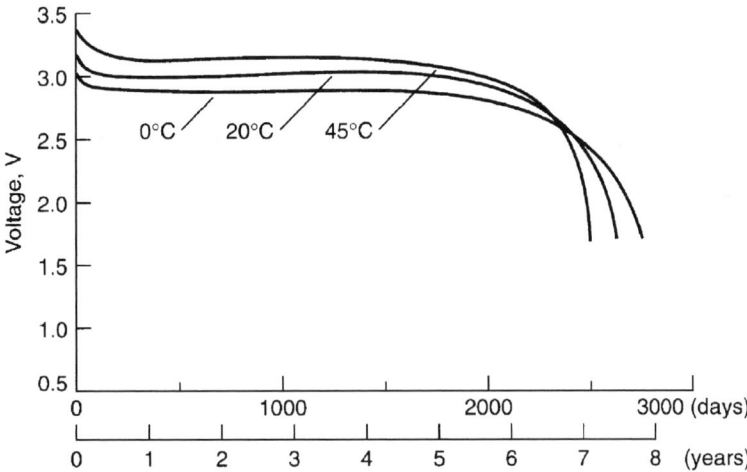

**FIGURE 13E.11**  Long-term discharge of Li/CF$x$ cylindrical battery. BR 2/3 A-size; 150-kΩ discharge. (*Courtesy of Panasonic Corp. of North America.*)

## 13E.4.2  Cell and Battery Types

The Li/(CF)$_n$ batteries are manufactured in a number of coin, cylindrical, and pin configurations. The major electrical and physical characteristics of some of these batteries are listed in Tables 13E.1a and b. Manufacturers' specifications should be consulted for the most recent listings of commercially available cells.

**TABLE 13E.1a**    Characteristics of Lithium/Carbon Monofluoride (Li/CF$x$) Batteries

| | Coin batteries, 3 V | | | | |
|---|---|---|---|---|---|
| | Standard load | | Dimensions and weight | | |
| Model no. | Nominal capacity,* mAh | Continuous drain, mA | Diameter, mm | Height, mm | Weight, g |
| BR1220 | 35 | 0.03 | 12.5 | 2.00 | 0.7 |
| BR1225 | 48 | 0.03 | 12.5 | 2.50 | 0.8 |
| BR1632 | 120 | 0.03 | 16.0 | 3.20 | 1.5 |
| BR2032 | 190 | 0.03 | 20.0 | 3.20 | 2.5 |
| BR2325 | 165 | 0.03 | 23.0 | 2.50 | 3.2 |
| BR2330 | 255 | 0.03 | 23.0 | 3.00 | 3.2 |
| BR3032 | 500 | 0.03 | 30.0 | 3.20 | 5.5 |

*Nominal capacity shown is based on standard drain and cutoff voltage down to 2.0 V at 20°C.

| | Pin type, 3 V | | | | |
|---|---|---|---|---|---|
| | | Dimensions (mm) | | | |
| Model no. | Nominal capacity, mAh | External diameter | Height | Basic battery weight, g | Continuous drain, mA |
| BR425 | 25 | 4.2 | 25.9 | 0.55 | 0.5 |
| BR435 | 50 | 4.2 | 35.9 | 0.85 | 1.0 |

| | Cylindrical type, 3 V | | | | | |
|---|---|---|---|---|---|---|
| | | Dimensions (mm) | | | | |
| Model no. | Nominal capacity,* mAh | External diameter | Height | Basic battery weight, g | Continuous drain, mA | Operating temp, 0°C |
| BR-C | 5000 | 26.0 | 50.5 | 42.0 | 5.0 | −40 ~ +85 |
| BR-A | 1800 | 17.0 | 45.5 | 18.0 | 2.5 | −40 ~ +85 |
| BR-1/2AA | 1000 | 14.5 | 25.5 | 8.0 | 2.5 | −40 ~ +100 |
| BR-2/3A | 1200 | 17.0 | 33.5 | 13.5 | 2.5 | −40 ~ +85 |
| BR-AG | 2200 | 17.0 | 45.5 | 18.0 | 2.5 | −40 ~ +85 |
| BR-2/3AG | 1450 | 17.0 | 33.5 | 13.5 | 2.5 | −40 ~ +85 |

*Nominal capacity is based on standard drain rate and cutoff voltage of 2.0 V at 20°C.

**TABLE 13E.1b**    Characteristics of Large Lithium/Carbon Monofluoride (Li/CF$x$) Batteries

| | Single-cell batteries | | | |
|---|---|---|---|---|
| | | Dimensions (cm) | | |
| Part number | Capacity, Ah | Diameter | Height | Weight, g |
| LCF-111 | 240 | 6.62 | 16.51 | 880 |
| LCF-112 | 35 | 3.02 | 13.84 | 170 |
| LCF-117 | 1200 | 11.43 | 26.67 | 3950 |
| LCF-119 | 400 | 11.43 | 9.53 | 1575 |
| LCF-122 | 18 | 3.37 | 6.06 | — |
| LCF-123 | 35 | 3.37 | 11.72 | — |
| LCF-313 | 40 | $6.45(L) \times 3.43(W) \times 7.09(H)$ | | 230 |

**TABLE 13E.1*b*** Characteristics of Large Lithium/Carbon Monofluoride (Li/CF*x*) Batteries (*Continued*)

| Multicell batteries | | | | | | | |
|---|---|---|---|---|---|---|---|
| | | | Dimensions, cm | | | | |
| Part number | Capacity, Ah | Nominal voltage | H | L | W | Weight, g | Comments |
| MAP-9036 | 23.5 | 39 | 17.1 | 20.3 | 14.0 | 4586 | Former shuttle range safety system |
| MAP-9046 | 3.74 (×2) | 30 (×2) | 15.9 | 17.3 | 7.6 | 3405 | Two independent voltage sections |
| MAP-9225 | 240 | 15 | 24.9 | 30.7 | 6.5 | 6000 | |
| MAP-9257 | 80 | 18 | 12.4 | 18.5 | 14.8 | — | |
| MAP-9319 | 240 | 21 | 42.9 | 29.7 | 9.7 | — | |
| MAP-9325 | 120/7.2 | 12/15 | 17.1 | 18.6 | 9.2 | — | Optional casing |
| MAP-9334 | 80 | 6 | 16.8 | 7.6 | 4.8 | — | Minuteman III GRP batteries |
| MAP-9381 | 70 | 39 | 31.3 | 20.0 | 9.7 | — | Integrated capacitor bank |
| MAP-9382 | 80/70 | 33/12 | 20.1 | 17.6 | 14.1 | — | Two independent voltage sections |
| MAP-9389 | 280 | 15 | 23.6 | 33.8 | 11 | | |
| MAP-9392 | 40 | 39 | 17.1 | 20.3 | 14.0 | | X-33 range safety system |

***Source:*** Eagle-Picher Technologies.

Larger sizes of cells and batteries have also been developed for military, governmental, and space applications, as given in Table 13E.1*b*.[46] Spiral-wound and prismatic cells are used to build the multicell batteries given in this table. The smaller cylindrical cells employ a 0.030 cm thick steel case, but larger units, such as the 1200 Ah cell, are reinforced with an epoxy-fiberglass cylinder to provide additional strength, with an increase in weight about half that of increasing the steel wall thickness. All these cells employ a Zeigler-type compression seal, a unique cutter vent mechanism, and two layers of separator. The first is a microporous layer to prevent particulate migration, and the second is a nonwoven polyphenylene sulfide material to provide high-strength, high-temperature stability and good electrolyte wicking action. These low-rate designs provide a specific energy of 600 Wh/kg and an energy density of 1000 Wh/L in the DD size and higher values for larger units. Capacity to a 2.0 V cutoff as a function of temperature at four rates from 0.04 to 1.00 A is shown in Fig. 13E.12 for the

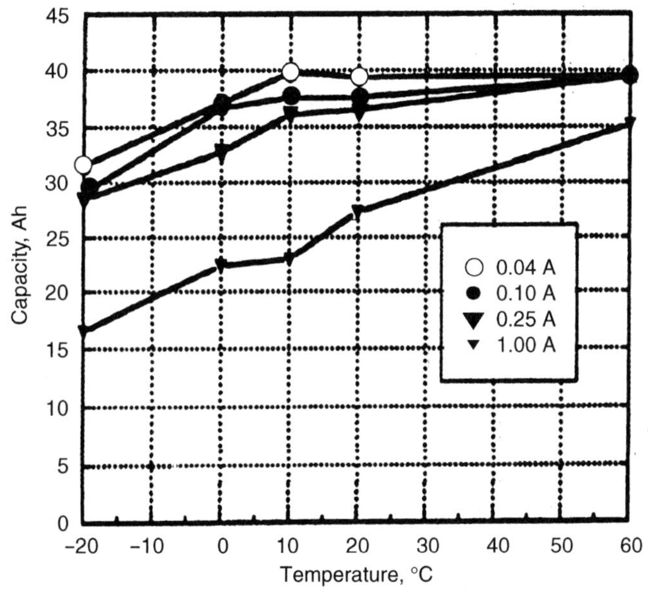


**FIGURE 13E.12** DD Li/CF*x* discharge performance as a function of rate and temperature.

DD design. The capacity of this battery is relatively independent of temperature at the three lower currents and above 10°C, but decreases at the higher rate and lower temperatures. Discharge curves for the 1200 Ah reinforced cylindrical battery at the 2000 h (ca. 500 mA) and the 1000 h rates (ca. 1.0 A) are shown in Fig. 13E.13. The trailing knee in these discharge curves has been attributed to electrolyte starvation at the end of the discharge. These batteries effectively demonstrate the ability of the lithium/carbon monofluoride system to provide very high specific energies and energy densities in these low-rate designs.

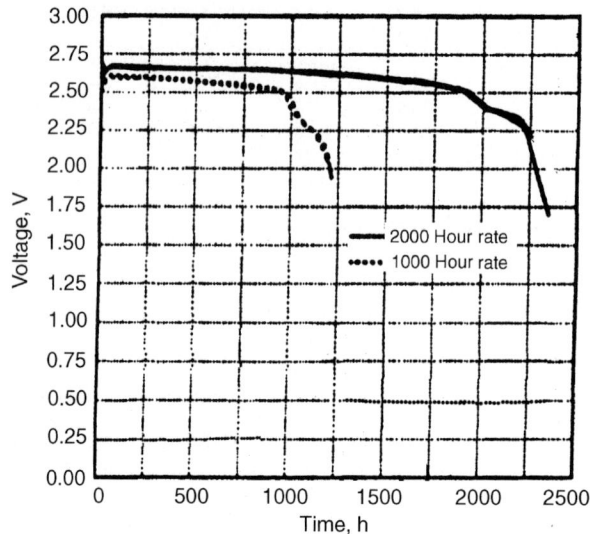

**FIGURE 13E.13**   Typical discharge curves for 1200 Ah Li/CF$x$ battery.

### 13E.4.3   Applications and Handling

The applications of the Li/CF$x$ battery are similar to those of the other lithium/solid-cathode batteries, again taking advantage of the high specific energy and energy density and long shelf life of these batteries. The Li/CF$x$ coin batteries are used as a power source for watches, portable calculators, memory applications, and electronic translators. The low-capacity miniature pin-type batteries have been used as an energy source for LEDs and for fishing lights and microphones. The cylindrical batteries can also be used in memory applications, but their higher drain capability also covers use in cameras, electrical locks, emergency signal lights, and utility meters. The very large cells are used for military and space applications.

Handling considerations for the Li/CF$x$ systems are similar to those for the other lithium/solid-cathode systems. The limited current capability of the coin and low-capacity batteries restricts temperature rise during short-circuit and reversal. These batteries can generally withstand this abusive use even though they are not provided with a safety vent mechanism. The larger batteries are provided with a venting device, but short-circuit, high discharge rates, and reversal should be avoided as these conditions could cause the cell to vent. Charging and incineration likewise should be avoided for all batteries. The manufacturer's recommendations should be obtained for handling specific battery types.

### 13E.4.4   Recent Advances in Lithium Carbon Monofluoride Technology

***Use of Mixtures of Carbon Monofluoride and Manganese Dioxide.***   The use of both CF$x$ and MnO$_2$ in the cathode of a lithium primary battery was first described in a U.S. patent issued in 1982.[47] This patent claims the use of mixtures of CF$x$ and MnO$_2$, but also claims a cell in which a layer of CF$x$ is disposed on top of a manganese dioxide layer. Little data are presented in this patent to support the claims.

The high cost of carbon monofluoride relative to manganese dioxide has limited its use in many applications. A more recent report[48] describes a study in which a mixture of CF$x$ with $x = 1$ and heat-treated MnO$_2$ was used to construct a lithium primary D-cell. The proportions of the mixture were stated to be a 50/50 blend, and the discharge curve shows two plateaus of approximately equal duration. The cells were stated to have a balanced design. The electrolyte was only described as an inorganic lithium salt in an organic electrolyte mixture. A copolymer film separator was also used. The D cells were discharged at 0.050, 0.250, and 2.0 A at 21°C and at 2.0 A at −30°C. Results are summarized in Table 13E.2. All three room-temperature discharge curves show two plateaus of approximately equal duration. The 2.0 A discharge shows running voltages of 2.64 and 2.41 V, which are ascribed to MnO$_2$ and CF$x$, respectively. On 0.250 A discharge at 21°C, these cells exhibit a specific energy of 380 Wh/kg and an energy density of 923 Wh/L. These parameters represent increases of 35% and 57% in specific energy and energy density compared to standard D cells using manganese dioxide only. On 2.0 A discharge at −30°C, the cells with hybrid cathodes provide a capacity of 12.0 Ah, which is 79% of that obtained at 21°C at the same rate. This corresponds to a specific energy of 227 Wh/kg and an energy density of 552 Wh/L, both figures being 67% of the values obtained at 21°C. These results indicate that the hybrid cathode cells provide superior performance to cells using manganese dioxide alone, particularly at low temperature.

**TABLE 13E.2**  Performance Data for High-Capacity Li/CF$x$-MnO$_2$ D Cells at Different Rates and Temperatures to a 2.0 V Cutoff

| Temperature, °C | Rate, A | Capacity, Ah | Specific energy, Wh/kg | Energy density, Wh/L |
|---|---|---|---|---|
| +21 | 0.050 | 16.6 | 407 | 990 |
| +21 | 0.250 | 16.2 | 380 | 923 |
| +21 | 2.0 | 15.2 | 338 | 823 |
| −30 | 2.0 | 12.0 | 227 | 552 |

Mixtures of carbon monofluoride and silver vanadium oxide have also been employed for biomedical applications (see Chap. 28).

***Subfluorinated and Semi-Ionic Carbon Fluoride Materials.***  Recent studies have shown that subfluorinated carbon fluorides (SFCFs) provide enhanced performance at low temperatures to −40°C. An initial study compared partially fluorinated natural graphites with $x$ values of 0.33, 0.48, 0.52, and 0.63 to a commercial CF$x$ with $x = 1.08$.[49] Structural studies showed that the SFCF material consisted of domains of fluorinated carbon intimately mixed with the graphite precursor particles.[50] The partially fluorinated CF$x$ materials exhibit higher power capability at room temperature and superior low-temperature performance compared to CF$_{1.08}$. Figure 13E.14 shows discharge curves at room temperature for a Li/CF$x$ coin cell with $x = 0.52$ at different discharges to the 2.5-C-rate. The cathode mix contained 80% CF$x$, 10% acetylene black, and 10% binder and was used with a 1.2 M LiBF$_4$ in PC/DME (7/3) electrolyte. A Ragone plot shows better high-rate power capability above 6.4 kW/kg for the SFCFs compared to the standard commercial cells.[49] Figure 13E.15 shows discharge curves for Li/CF$x$ coin cells with $x = 0.65$ at −40°C with and without a 3% predischarge at a C/40 rate. These cells used an electrolyte of 1 M LiBF$_4$ in PC/DME (20/80). Predischarged cells with CF$_{0.65}$ provided a specific capacity of 610 mAh/g to a 1.5 V cutoff compared to 200 mAh/g for control cells with CF$_{1.08}$ tested under the same conditions.

Another study to optimize the electrolyte composition for SFCFs found that 0.5 M LiBF$_4$ in PC/DME (20/80) provided improved low-temperature performance compared to higher salt concentrations and eliminated the need for the predischarge step prior to low-temperature discharge.[51]

The use of CF$x$ materials with semi-ionic character prepared in a two-step fluorination process and SFCF materials prepared from multiwalled carbon nanotubes (MWCNs) were also found to have superior properties in terms of rate capability and low-temperature performance compared to conventional CF$x$ materials.[52]

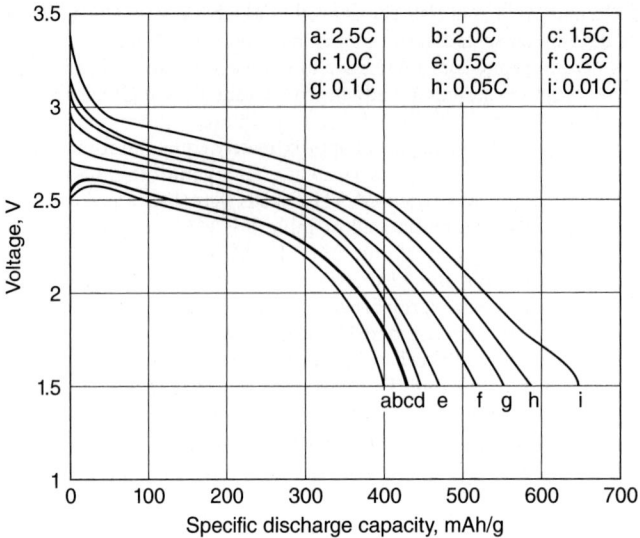

**FIGURE 13E.14** Discharge profiles for Li/CF$_{0.52}$ coin cells discharged at different rates.

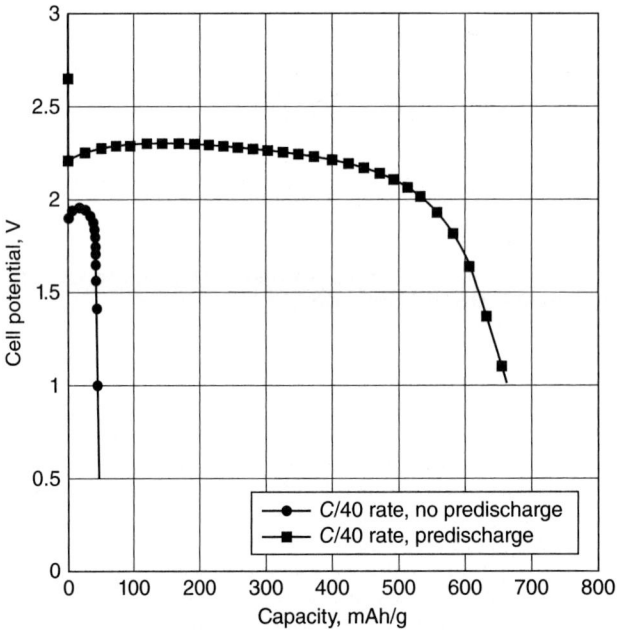

**FIGURE 13E.15** Discharge data from Li/CF$_{0.65}$ test cells at −40°C with and without a room-temperature predischarge of 3% of the total cell capacity.

# SECTION F
# LITHIUM/IRON DISULFIDE

## 13F.1 *LITHIUM/IRON DISULFIDE (Li/FeS₂) BATTERIES*

Iron sulfide, in both the monosulfide (FeS) and the disulfide (FeS₂) forms, has been considered for use in solid-cathode lithium batteries. Only the disulfide battery has been commercialized because of its performance advantage due to its higher sulfur content and higher voltage. The monosulfide electrode has the advantage of reduced corrosion, longer life, and a single voltage plateau compared to the disulfide electrode, which discharges in two steps.

These batteries have a nominal voltage of 1.5 V$^a$ and can therefore be used as replacements for aqueous batteries having a similar voltage. Button-type Li/FeS₂ batteries were manufactured as a replacement for zinc/silver oxide batteries but are no longer marketed. They had a higher impedance and a slightly lower power capability but were lower in cost and had better low-temperature performance and storability.

Li/FeS₂ batteries are now manufactured in a cylindrical configuration. These batteries have better high-drain low-temperature performance than the zinc/alkaline-manganese dioxide batteries. The capacity of these two systems on constant-current discharge at four rates is compared in Fig. 13F.1 for the AA-size cells.

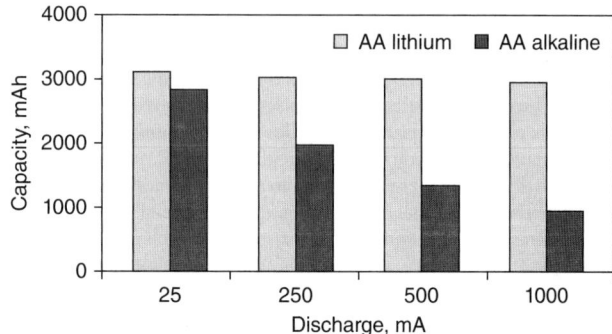

**FIGURE 13F.1** Comparison of capacity (mAh) of AA Li/FeS₂ and alkaline-manganese cells on constant-current discharge at four rates (25, 250, 500, and 1000 mA) at 21°C. (*Courtesy Energizer Battery Co.*)

## 13F.2 *CELL CHEMISTRY*

These cells employ a cathode of FeS₂ mixed with carbon and a mixed Teflon® organic binder coated on an aluminum foil, an anode of lithium alloyed with 0.5% aluminum, and a 20 μm high-porosity polyethylene separator.[53] The electrolyte is a 0.75 M solution of LiI in a 65:35 (V/V) mixture of 1,3 dioxolane and 1,2 dimethoxyethane, which is reported to increase in conductivity as the temperature decreases. The cell reactions at high rate and ambient temperature are

$$
\begin{aligned}
\textit{Anode:} &\quad 4Li \rightarrow 4Li^+ + 4e \\
\textit{Cathode:} &\quad FeS_2 + 4e \rightarrow Fe + 2S^{-2} \\
\hline
\textit{Overall:} &\quad 4Li + FeS_2 \rightarrow Fe + 2Li_2S
\end{aligned}
$$

---

$^a$ANSI Standard C18.3M, Part 1-2009.

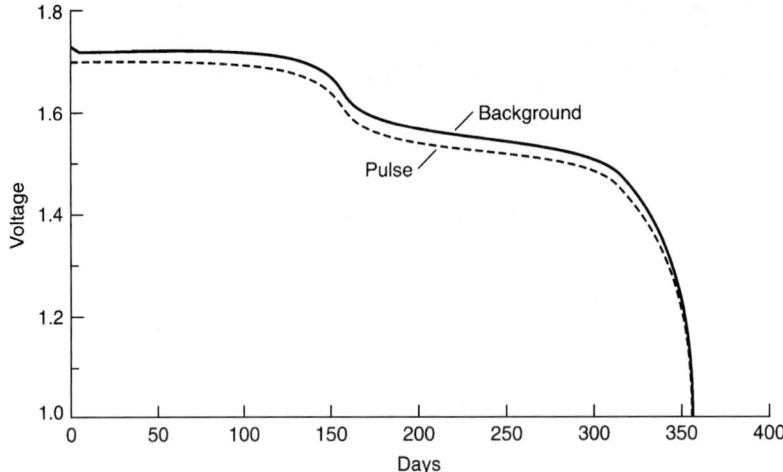

**FIGURE 13F.2** Stepped discharge curve of Li/FeS$_2$ AA-size batteries on light drain at 21°C. Five thousand ohm background with 25 ohm, 1 s/week pulse. (*Courtesy of Energizer Battery Co., Inc.*)

At low rate and/or high temperature, a two-step discharge process occurs as seen in Fig. 13F.2. The cell reactions are then given by:

$$2Li + FeS_2 \rightarrow Li_2FeS_2$$

$$2Li + Li_2FeS_2 \rightarrow 2Li_2S + Fe$$

## 13F.3 CONSTRUCTION

Li/FeS$_2$ batteries may be manufactured in a variety of designs, including the button and both bobbin and spiral-wound electrode cylindrical cells. A bobbin construction is most suitable for light-drain applications. The spiral-wound electrode construction is needed for the heavier drain applications, and it is this design that has been commercialized.

The construction of the spiral-wound cylindrical battery is shown in Fig. 13F.3. These batteries typically have several safety devices incorporated in their design to provide protection against such abusive conditions such as short-circuit, charging, forced discharge, and overheating. Two safety devices are shown in the figure—a pressure relief vent and a resettable thermal switch (a PTC device). The safety relief vent is designed to release excessive internal pressure to prevent violent rupture if the battery is heated or abused electrically.

The primary purpose of the PTC is to protect against external short-circuits, though it also offers protection under certain other electrical abuse conditions. It does so by limiting the current flow when the cell temperature reaches the PTC's designed activation temperature. When the PTC activates, its resistance increases sharply, with a corresponding reduction in the flow of current and, consequently, internal heat generation. When the battery (and the PTC) cools, the PTC resistance drops, allowing the battery to discharge again. The PTC will continue to operate in this manner for many cycles if an abusive condition continues or recurs. The PTC will not "reset" indefinitely, but when it ceases to do so, it will be in the high-resistance condition. The characteristics of PTCs (or any other current-limiting devices in the battery) may place some limitations on performance. These are discussed in more detail in Sec. 13F.4

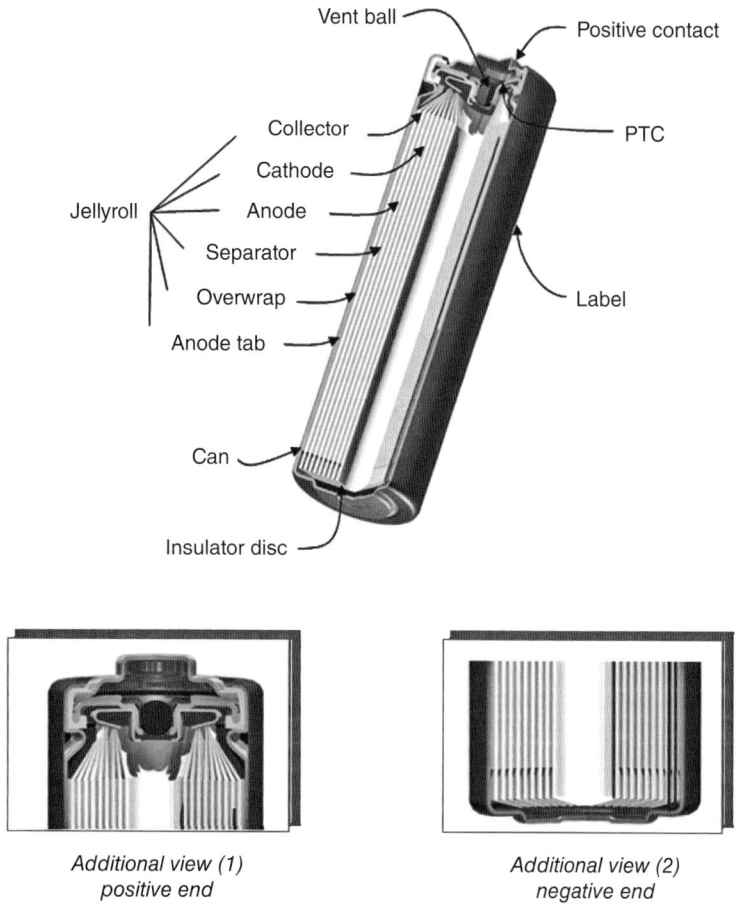

**FIGURE 13F.3** Partial cutaway view of spiral-wound Li/FeS$_2$ cell with additional details of positive (1) and negative (2) ends. (*Courtesy of Energizer Battery Co.*)

## 13F.4   PERFORMANCE CHARACTERISTICS AND APPLICATIONS

### 13F.4.1   Cell Output Performance

**Voltage.**   The nominal voltage of the Li/FeS$_2$ system is given as 1.5 V and the open-circuit voltage of undischarged cells is approximately 1.78 V. The voltage on load drops within milliseconds, as shown in Fig. 13F.4.

**Discharge.**   Li/FeS$_2$ batteries typically have a higher operating voltage and a flatter discharge profile than aqueous zinc/alkaline manganese dioxide 1.5 V batteries. This is illustrated in Fig. 13F.5, which compares the performance of these two battery systems at relatively light and heavy constant-current discharge rates. These characteristics of the Li/FeS$_2$ battery result in higher energy and power output, especially on heavier drains where the operating voltage differences are greatest.

Performance characteristics of AA Li/FeS$_2$ batteries under constant-current and constant-power discharge modes are shown in Figs. 13F.6 and 13F.7. The improvement in performance over recent times in an ANSI digital still camera (DSC) test is shown in Fig. 13F.8.

### 13D.4.4   Applications and Handling

The main applications of the $Li/MnO_2$ system currently range up to several ampere-hours in capacity, taking advantage of its higher energy density, better high-rate capability, and longer shelf life compared with the conventional primary batteries. The $Li/MnO_2$ batteries are used in memory applications, watches, calculators, cameras, and RFID tags. At the higher drain rates, motor drives, automatic cameras, toys, personal digital assistants (PDAs), digital cameras, and utility meters are excellent applications.

The low-capacity $Li/MnO_2$ batteries can generally be handled without hazard, but, as with the conventional primary battery systems, charging and incineration should be avoided as these conditions could cause a cell to explode.

The higher capacity cylindrical batteries are generally equipped with a venting mechanism to prevent explosion, but the batteries, nevertheless, should be protected to avoid short-circuits and cell reversal, as well as charging and incineration. Most of the high-rate batteries are also equipped with an internal resettable current and thermal protective system called a PTC device. When a cell is short-circuited or discharged above design limits and the cell temperature increases, the resistance of the PTC device quickly increases significantly. This limits the amount of current that can be drawn from the cell and keeps the internal temperature of the cell within safe limits. Figure 13D.23 shows the operation of the PTC device when a cell is short-circuited. After a short-circuit peak of about 10 A, the current is abruptly limited and maintained at the depressed level. When the short-circuit is removed, the cell reverts to its normal operating condition. The short delay of several minutes before the PTC operates permits the cell to deliver pulse currents at higher values than the maximum permitted under continuous drain.

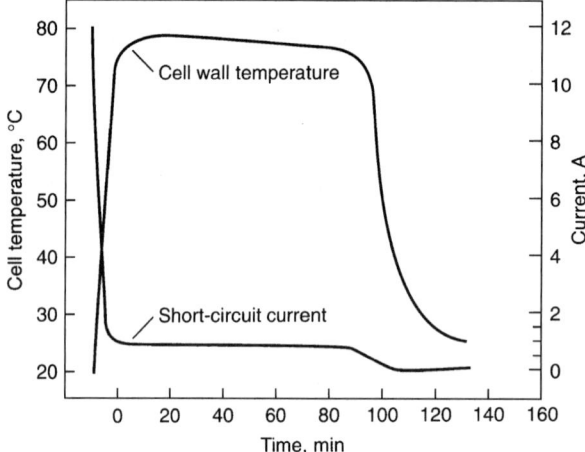

**FIGURE 13D.23**    Short circuit of Duracell XL™ CR123A battery.

Military applications of lithium/manganese dioxide batteries are increasing.[41] At room temperature, they provide higher energy density and slightly higher specific energy than the $Li/SO_2$ batteries commonly employed by the U.S. military. A recently developed $Li/MnO_2$ D-cell provides 14.0 Ah at the 250 mA rate and 13.0 Ah at the 2.0 A rate at room temperature. These cells were produced using a specially heat-treated manganese oxide that is more highly active. When employed in cathodes in a standard cell design with a $LiClO_4$-DME-PC electrolyte, this material provides a specific energy of 339 Wh/kg and an energy density of 742 Wh/L on 250 mA discharge. At −40°C, these cells also provide 3.39 Ah at 250 mA and 0.46 Ah at 2.0 A discharge rates, more than twice the capacity of standard $Li/MnO_2$ cells under these conditions. The higher capacity cell is being utilized in military batteries.

A lithium/manganese dioxide pouch cell in a foil-laminate package has also been developed for use in the BA-7847 battery, which powers the Thermal Weapons Sight and other U.S. military electronics.[42] These cells have dimensions of 8.25 mm × 61 mm × 72 mm and are employed in a 2p2s configuration within the prismatic BA-7847 case.

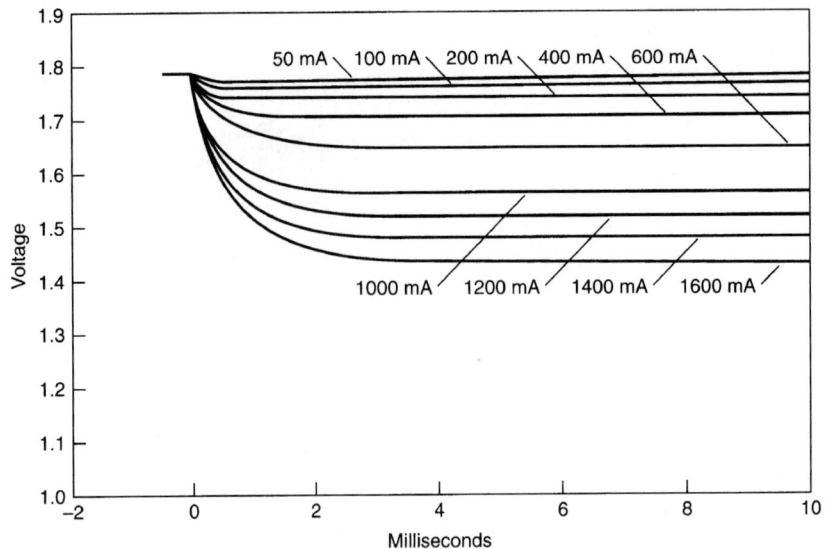

**FIGURE 13F.4**  On-load voltage of Li/FeS₂ AA-size cells. (*Courtesy of Energizer Battery Co., Inc.*)

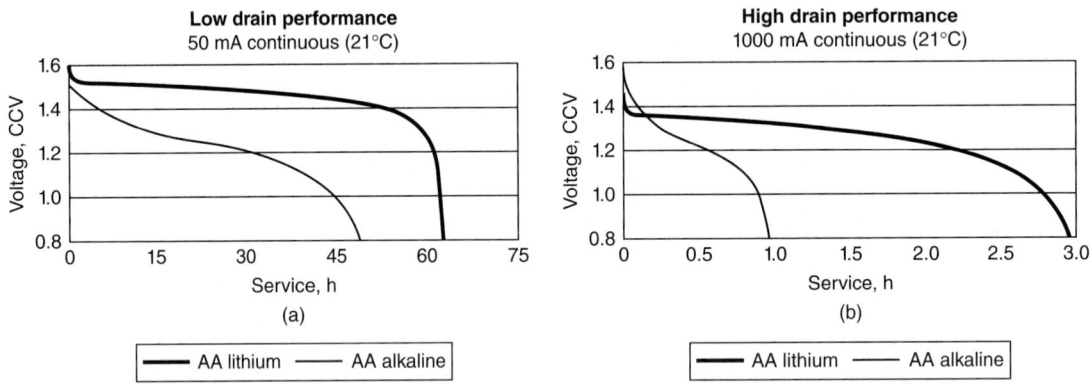

**FIGURE 13F.5**  Comparison of AA Li/FeS₂ and alkaline-manganese batteries at two drain rates, (*a*) 50 and (*b*) 1000 mA, at 21°C. (*Courtesy Energizer Battery Co.*)

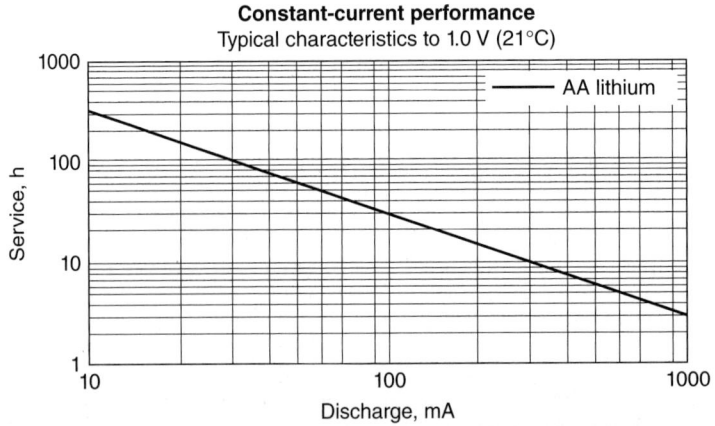

**FIGURE 13F.6**  Constant current performance shows hours of service as a function of discharge rate (mA) for AA Li/FeS₂ batteries. (*Courtesy Energizer Battery Co.*)

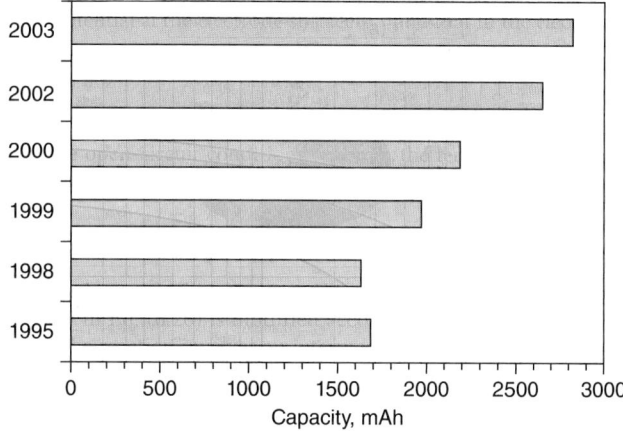

**FIGURE 13F.7** Constant power performance showing hours of service as a function of constant power discharge rate (mW) for AA Li/FeS$_2$ battery. (*Courtesy Energizer Battery Co.*)

**FIGURE 13F.8** Improvement in AA high-rate performance on ANSI DSC test: 1A continuous test to 1.0 V.

## 13F.4.2 Factors Affecting Performance

*Operating Temperature.* Li/FeS$_2$ batteries are also suitable for use over a broad temperature range, generally −40 to 60°C. Service life is improved at elevated temperatures. In some applications there may be further limits on the maximum discharge temperature due to current limiting, which are part of the cell or battery device designs. Service life is reduced as the discharge temperature is lowered below room temperature, though the performance of the Li/FeS$_2$ battery is affected much less by low temperature than are aqueous systems.

*Effects of Current-Limiting Devices.* Some current-limiting devices, such as thermal fuses and PTCs, are designed to respond to high temperatures. Both the ambient temperature and internal cell heating can affect these devices, so any of the following factors may play a role:

• Surrounding air temperature
• Thermal insulating properties of battery container

**TABLE 13F.1** Characteristics of Commercial Li/FeS$_2$ Batteries

| Model no. | Size | Max. diameter, mm | Max. height, mm | Typical weight, g | Typical volume, cc | Max. continuous current, A | Max. pulsed current (2 s on/8 s off), A | Shelf life at 21°C, yrs | Capacity at 21°C, mAh |
|---|---|---|---|---|---|---|---|---|---|
| L92 | AAA | 10.5 | 44.5 | 7.6 | 3.8 | 1.5 | 2.0 | 15 | 1200 |
| L91 | AA | 14.5 | 50.5 | 14.5 | 8.0 | 2.0 | 3.0 | 15 | 3000 |
| EA92 | AAA | 10.5 | 44.5 | 7.6 | 3.8 | 1.0 | 1.5 | 10 | 1200 |
| EA91 | AA | 14.5 | 50.5 | 14.5 | 8.0 | 1.5 | 2.0 | 10 | 3000 |

Storage temperature: −40 to 60°C; operating temperature: −40 to 60°C; typical IR drop: 90–150 mΩ; shelf-life rating to 90% of initial capacity for L92 and L91 models and to 80% of initial capacity for EA92 and EA91 models.
*Source:* Energizer Battery Co.

- Heat generated by equipment components during use
- Cumulative heating effects of multicell batteries
- Discharge rates and durations
- Frequency and duration of rest periods

It may be necessary to consult the manufacturer or conduct testing to determine limitations in specific applications.

*Impedance.* AC impedance is an electrical characteristic that is frequently used as an indicator of performance for aqueous batteries. The correlation is only poor at best with Li/FeS$_2$ batteries. There is a protective film that forms on the surface of the lithium anode. This film is an important factor in the excellent shelf life of the Li/FeS$_2$ cell. As the cell ages, this protective film increases with age. As the film increases, the impedance does as well. However, this film is easily disrupted when the battery is put on load, making impedance inappropriate as an indicator of expected Li/FeS$_2$ battery performance, especially after storage.

*Storage Temperature.* Storage at high temperature will reduce the service life of Li/FeS$_2$ batteries, as it will with other systems. However, because of the very low levels of impurities in the materials used and the high degree of seal effectiveness required in lithium batteries, service maintenance of Li/FeS$_2$ batteries after high-temperature storage is better than expected with aqueous systems. The typical storage temperature range of Li/FeS$_2$ batteries is −40 to 60°C. Accelerated storage tests at temperatures to 85°C have resulted in an estimated shelf life at room temperature between 26 and 40 years to 80% of initial capacity. They are rated for a 10-year shelf life for the EA models and a 15-year shelf life for the L models (see Table 13F.1).

## 13F.4.3  Cell Types and Applications

Table 13F.1 lists the characteristics of the cylindrical Li/FeS$_2$ batteries that are currently available commercially. These batteries have better high-drain and low-temperature performance than the conventional zinc cells and are intended to be used in applications that have a high current drain requirement, such as cameras, digital audio devices, CD players, portable lighting, toys, and games. In one particular camera test, two Li/FeS$_2$ AA cells provided approximately 1000 flashes compared to 400 for a high-rate AA alkaline battery.

Button-type Li/FeS$_2$ batteries are no longer manufactured. Multicell batteries using the Li/FeS system have not been manufactured for commercial use.

# SECTION G

# LITHIUM/COPPER OXIDE

## 13G.1   *LITHIUM/COPPER OXIDE (Li/CuO) CELLS*

The lithium/copper oxide (Li/CuO) system is characterized by a high specific energy and energy density (about 280 Wh/kg and 650 Wh/L) as copper oxide has one of the highest volumetric capacities of the practical cathode materials (4.16 Ah/cm$^3$). The battery has an open-circuit voltage of 2.25 V and an operating voltage of 1.2 to 1.5 V, which makes it interchangeable with some conventional batteries. The battery system also features a long shelf life with a low self-discharge rate and operation over a wide temperature range.

Li/CuO batteries have been designed in button and cylindrical configurations up to about 3.5 Ah in size, mainly for use in low- and medium-drain, long-term applications for electronic devices and memory backup. Higher-rate designs as well as hermetically sealed batteries with GTM seals have also been manufactured.

Figure 13G.1 compares the performance of a Li/CuO AA-size cylindrical battery with the zinc/alkaline/MnO$_2$ battery. The Li/CuO cell has a significant capacity advantage at low discharge rates, but loses this advantage at higher discharge rates.

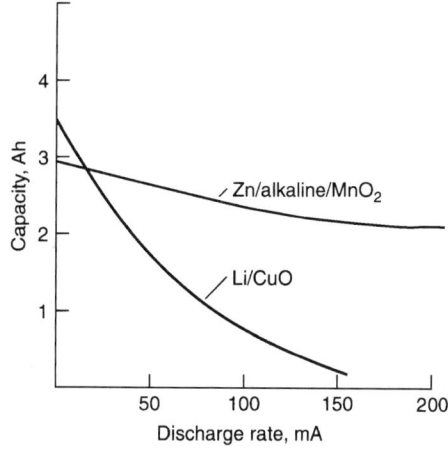

**FIGURE 13G.1**   Comparison of Li/CuO and Zn/alkaline/MnO$_2$ AA-size batteries at 20°C.

## 13G.2   *CELL CHEMISTRY*

The discharge reaction of the Li/CuO cell is

$$2Li + CuO \rightarrow Li_2O + Cu$$

The discharge proceeds stepwise, CuO $\rightarrow$ Cu$_2$O $\rightarrow$ Cu, but the detailed mechanism has not been clarified.[1,54] A double-plateau discharge has been observed at high-temperature (70°C) discharges at low rates, which blends into a single plateau under more normal discharge conditions.[55]

## 13G.3   CONSTRUCTION

The construction of the Li/CuO button-type battery shown in Fig. 13G.2*a* is similar to other conventional and lithium/solid-cathode cells. Copper oxide forms the positive electrode and lithium the negative. The electrolyte consists of lithium perchlorate in an organic solvent (dioxolane).

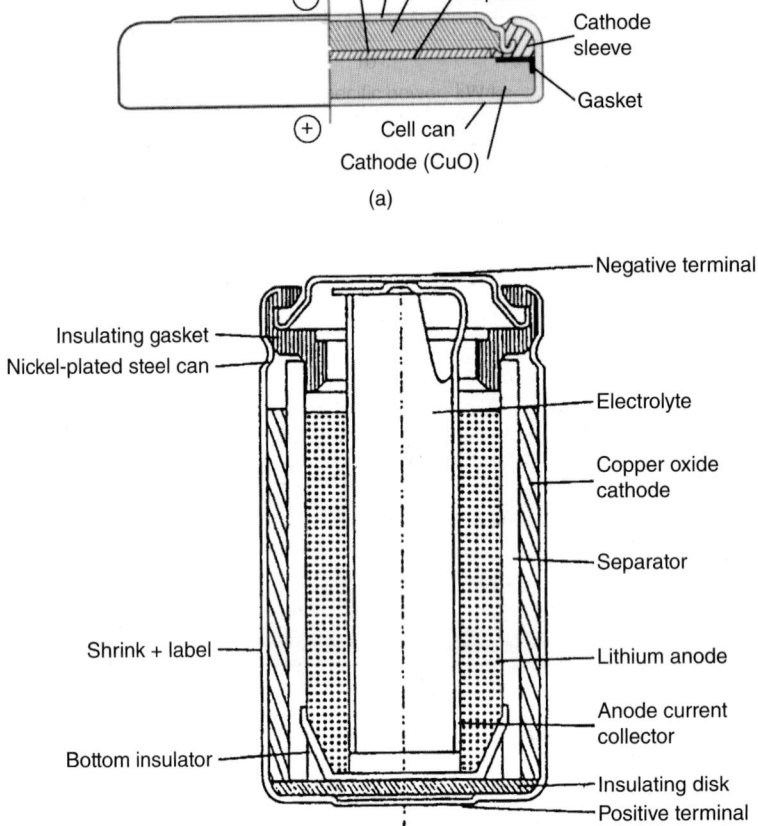

**FIGURE 13G.2**   Lithium/copper oxide batteries (*a*) Button configuration. (*Courtesy of Panasonic Corp. of North America.*) (*b*) Cylindrical battery, bobbin construction. (*Courtesy of SAFT America, Inc.*)

The cylindrical batteries (Fig. 13G.2*b*) use an inside-out bobbin construction. A cylinder of pure porous nonwoven glass is used as the separator, nickel-plated steel for the case, and a polypropylene gasket for the cell seal. The can is connected to the cylindrical copper oxide cathode and the top to the lithium anode.

## 13G.4  PERFORMANCE CHARACTERISTICS AND APPLICATIONS

### 13G.4.1  Performance

*Button Battery.*    The performance of the 60 mAh Li/CuO button cell under various discharge conditions and temperatures is shown in Fig. 13G.3.

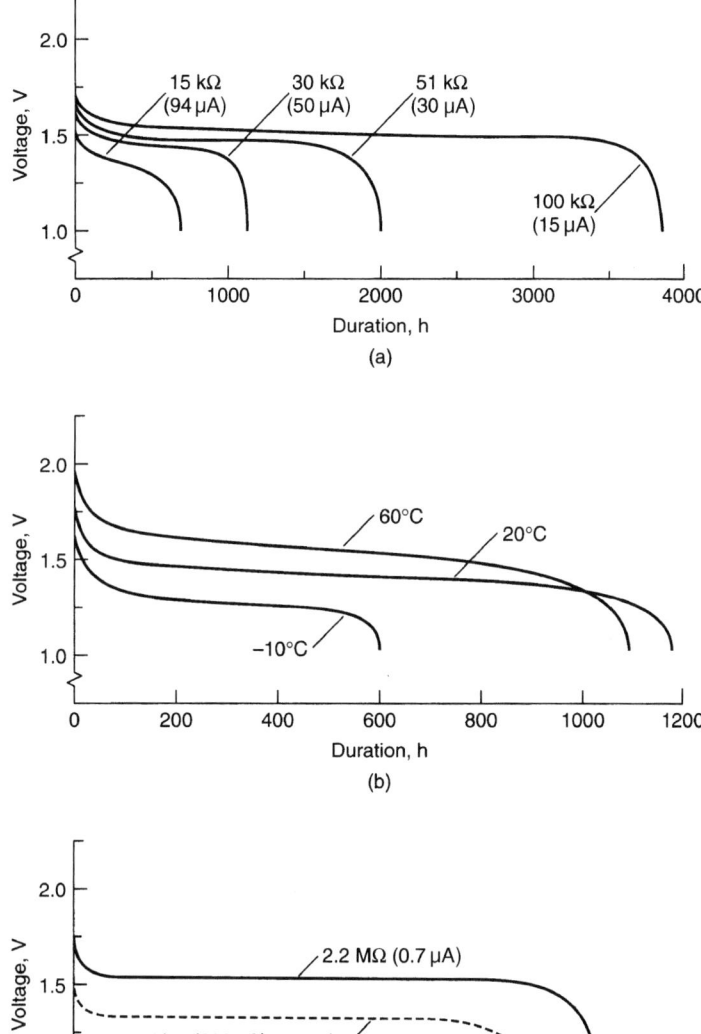

**FIGURE 13G.3**    Discharge characteristics of Li/CuO button-type battery, 60 mAh size. (*a*) Load characteristics. (*b*) Temperature characteristics. (*c*) Pulse discharge characteristics. (*Courtesy of Panasonic Corp. of North America.*)

***Cylindrical Bobbin Li/CuO Battery.***    Typical discharge curves for this system are shown in Fig. 13G.4. After a high initial load voltage, the discharge profile is flat at the relatively light loads. The bobbin construction does not lend itself to high-rate discharges, and the battery capacity is significantly lowered with increasing discharge rates. The Li/CuO cylindrical battery operates over a wide temperature range, typically from −20 to 70°C, although the battery can operate outside these limits but with changes in the discharge profile or load capability. Discharge curves at several different temperatures are shown in Fig. 13G.5. The performance of the battery at temperatures from −40 to 70°C and at various loads is summarized in Fig. 13G.6. The high capacity of the battery at the lighter loads falls off sharply with increasing load and decreasing temperatures.

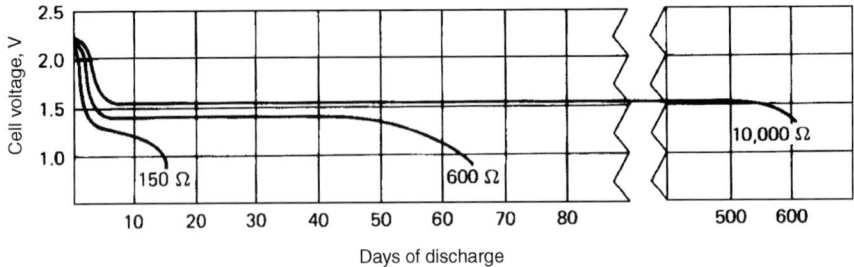

**FIGURE 13G.4**    Typical discharge curves for Li/CuO AA-size battery at 20°C. (*Courtesy of SAFT America, Inc.*)

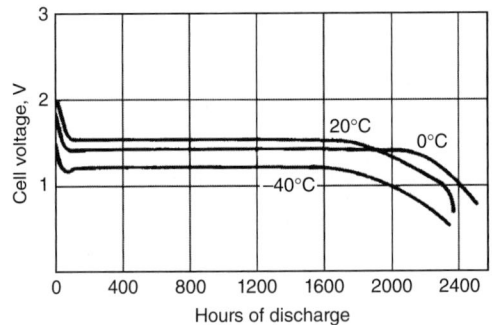

**FIGURE 13G.5**    Effect of temperature on Li/CuO AA-size battery, 1-kΩ load. (*Courtesy of SAFT America, Inc.*)

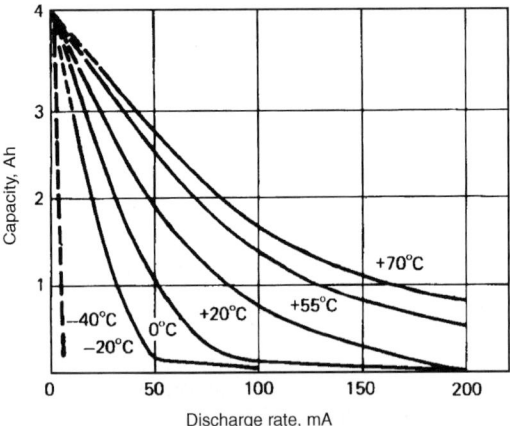

**FIGURE 13G.6**    Capacity of Li/CuO AA-size battery as a function of discharge load and temperature. (*Courtesy of SAFT America, Inc.*)

The long-term storage capability of these Li/CuO cells is illustrated in Fig. 13G.7. Figure 13G.7a shows that there is only a minimum loss of capacity after 10 years of storage at room temperature, less than 0.5% per year. Performance after storage at high temperatures is plotted in Fig. 13G.7b. The retention of residual capacity in partially discharged cells is said to be equivalent to that of fully charged cells.

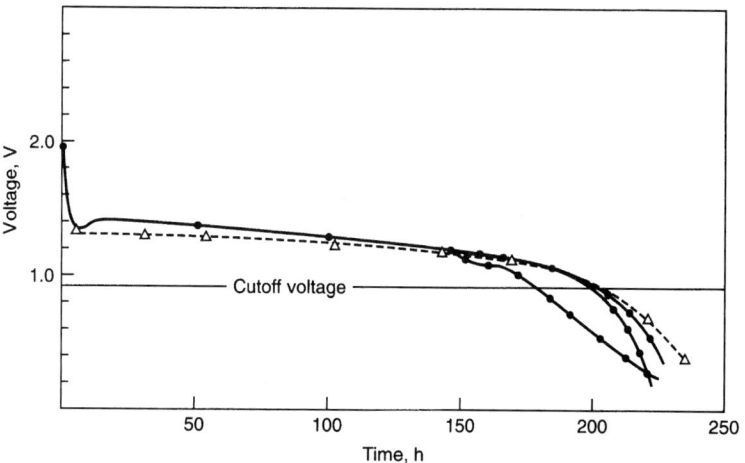

| | OCV (V) | Impedance (Ω) | Capacity (Ah) |
|---|---|---|---|
| Fresh (average)<br>-----△----- | 2.36 | 9 | 3.25 |
| 10 yr storage<br>————●———— | 2.33 | 10 | 3.11 |

(a)

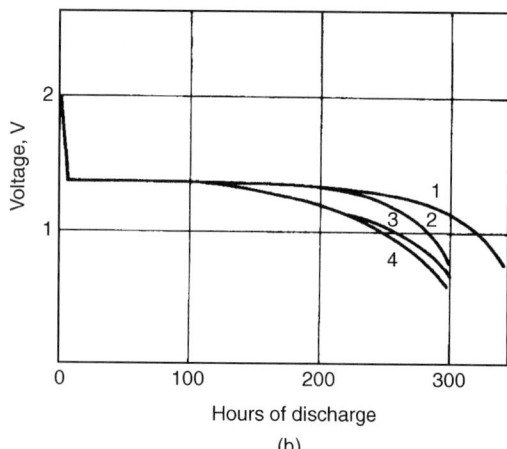

(b)

**FIGURE 13G.7**    Effect of storage on performance on Li/CuO cylindrical batteries. (a) Discharges at 20°C before and after 10-year storage at 20°C. (b) Discharges at 20°C after 70°C storage. Curve 1—No storage; curve 2—6 months' storage; curve 3—12 months' storage; curve 4—18 months' storage. (*Courtesy of SAFT America, Inc.*)

***High-Temperature Cells.***   Specially designed hermetically sealed batteries have been developed for use at the high temperatures encountered, for example, by the oil-well logging industry, which uses down-hole tools operating to 150°C, the maximum temperature at which the Li/CuO cells can operate.

***Spirally Wound Cells.***   Cylindrical batteries in C and D sizes have been designed with spirally wound electrodes to meet higher drain requirements. Figure 13G.8 shows the discharge performance of a Li/CuO D-size battery at various temperatures and at relatively low and high discharge rates.

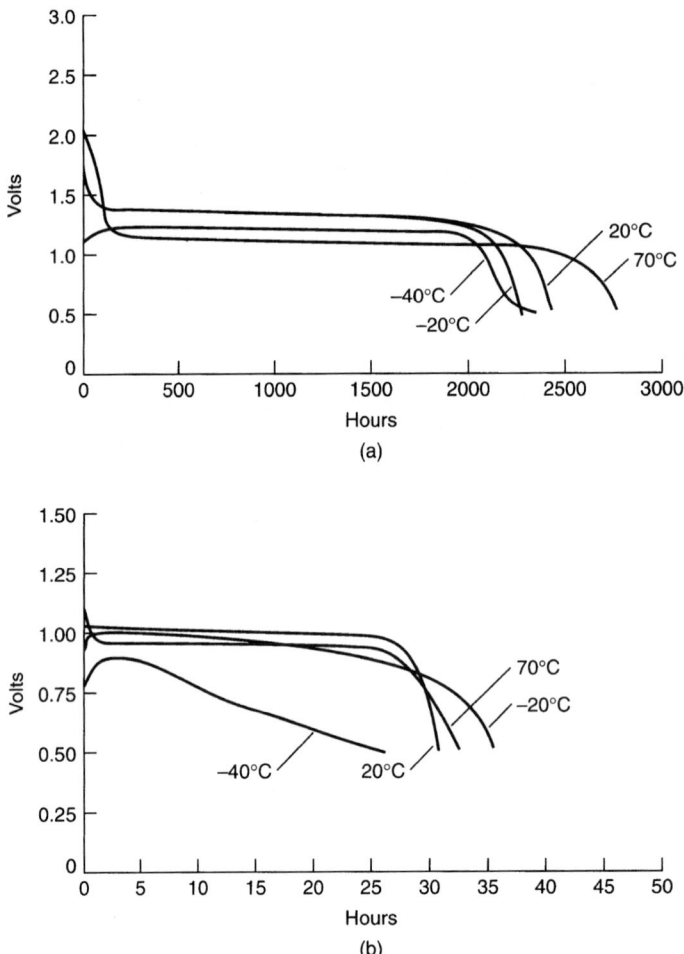

**FIGURE 13G.8**   Performance of high- and low-rate Li/CuO D-size batteries. (*a*) Discharge at 147 Ω. (*b*) Discharge at 1.5 Ω.

## 13G.4.2   Cell Types and Applications

The Li/CuO batteries that have been available in the button and small cylindrical (bobbin) configurations are listed in Table 13G.1. Under the low-drain conditions, these batteries have a significant capacity advantage over the conventional aqueous batteries. Combined with their excellent storability and operation over a wide temperature range, these batteries provide reliable power sources for applications such as memory backup, clocks, electric meters, and telemetry with high-temperature cells, in high-temperature environments. Specially designed units were also manufactured to meet higher drain applications. These batteries are no longer available commercially. Since this technology remains a potential competitor for alkaline-manganese at low-drain rates, information on its properties is retained in this edition.

**TABLE 13G.1**   Characteristics of Lithium/Copper Oxide Batteries

|  | Li/CuO | | |
|---|---|---|---|
|  | Button | 1/2 AA | AA |
| Nominal voltage, V | 1.5 | 1.5 | 1.5 |
| Dimensions (max) | | | |
|    Diameter, mm | 9.5 | 14.5 | 14.5 |
|    Height, mm | 2.7 | 26.0 | 50.5 |
|    Volume, cm$^3$ | 0.2 | 4.3 | 8.3 |
| Weight, g | 0.6 | 7.3 | 17.4 |
| Rated capacity, Ah[*] | 0.060 | 1.4 | 3.4 |
| Specific energy/energy density | | | |
|    Wh/kg | 150 | 285 | 290 |
|    Wh/L | 450 | 485 | 610 |
| Weight of lithium, g | — | 0.4 | 0.9 |
| Maximum current, mA | 0.3 | 20 | 40 |

[*]At approximately C/1000 rate.
*Source:* SAFT America, Inc. and Panasonic of North America Corp.

# SECTION H
# LITHIUM/SILVER VANADIUM OXIDE

## 13H.1   *LITHIUM/SILVER VANADIUM OXIDE BATTERIES*

The lithium/silver vanadium oxide system has been developed for use in biomedical applications, such as cardiac defibrillators, neurostimulators, and drug delivery devices. A description of this system and its applications is found in Chap. 28.

# SECTION I

# LITHIUM/WATER AND LITHIUM/AIR

## 13I.1   LITHIUM/WATER AND LITHIUM/AIR BATTERIES

These two technologies are described in Chaps. 18 and 29.

## ACKNOWLEDGMENT

Background information for sales volumes of primary cells is provided by The Freedonia Group as detailed in the following table.

World Lithium Primary Cell Demand (million dollars)[*]

| Actual/projected sales (based on 2016 sales data) | 2004 | 2009 | 2014 | 2019 | 2024 |
|---|---|---|---|---|---|
| Total world battery sales | 36,345 | 56,145 | 82,470 | 120,000 | 166,500 |
| World primary battery sales | 12,990 | 16,455 | 19,865 | 24,050 | 28,150 |
| (Primary batteries, % of total) | 35.7 | 29.3 | 24.1 | 20.0 | 16.9 |
| Primary lithium sales | 1551 | 2177 | 2943 | 3830 | 4700 |
| (Lithium, % of primary) | 11.9 | 13.2 | 14.8 | 15.9 | 16.7 |

[*]Table IV-7, World Primary Battery Demand by Chemistry, published December 2016 by The Freedonia Group.

## REFERENCES

1. J. P. Gabano, *Lithium Batteries*, Academic, London, 1983.

2. Battery Markets in the United States by Application and End User, published December 2016 by The Freedonia Group (from exported table generated/forwarded by The Freedonia Group on 10/4/2017 from IP address 207.89.36.85. © MarketResearch.com, Inc., 2000–2017).

3. Technical data, Foote Mineral Co., Exton, PA; Lithium Corp. of America, Gastonia, NC.

4. H. R. Grady, "Lithium Metal for the Battery Industry," *J. Power Sources* **5**:127 (1980), Elsevier Sequoia, Lausanne, Switzerland.

5. J. T. Nelson and C. F. Green, "Organic Electrolyte Battery Systems," U.S. Army Material Command Rep. HDL-TR-1588, Washington, DC, March 1972.

   J. O. Besenhard and G. Eichinger, "High Energy Density Lithium Cells, pt. I, Electrolytes, and Anodes," *J. Electroanal. Chem.* **68**:1 (1976), Elsevier Sequoia, Lausanne, Switzerland.

   G. Eichinger and J. O. Besenhard, "High Energy Density Lithium Cells, pt. II, Cathodes and Complete Cells," *J. Electroanal. Chem.* **72**:1 (1980), Elsevier Sequoia, Lausanne, Switzerland.

6. F. Deligiannis, B. V. Ratnakumar, H. Frank, E. Davies, and S. Surampudi, *Proc. 37th Power Sources Conf.*, pp. 373–377 (1996), Cherry Hill, NJ.

7. A. N. Dey, "Lithium Anode Film and Organic and Inorganic Electrolyte Batteries," in *Thin Solid Films*, Vol. 43, Elsevier Sequoia S. A., Lausanne, Switzerland, 1977, p. 131.

8. S. Gilman and W. Wade, "The Reduction of Sulfuryl Chloride at Teflon-Bonded Carbon Cathodes," *J. Electrochem. Soc.* **127**:1427 (1980).

9. A. Meitav and E. Peled, "Calcium-Ca(AlCl$_4$)$_2$-Thionyl Chloride Cell: Performance and Safety," *J. Electrochem. Soc.* **129**:3 (1982).

10. R. L. Higgins and J. S. Cloyd, "Development of the Calcium-Thionyl Chloride Systems," *Proc. 29th Power Sources Conf.*, Electrochemical Society, Pennington, NJ, June 1980.

    M. Binder, S. Gilman, and W. Wade, "Calcium-Sulfuryl Chloride Primary Cell," *J. Electrochem. Soc.* **129**:4 (1982).

11. E. S. Takeuchi and W. C. Thiebolt, "The Reduction of Silver Vanadium Oxide in Lithium/Silver Vanadium Oxide Cells," *J. Electrochem. Soc.* **135**:11 (1988).

    E. S. Takeuchi, "Lithium/Solid Cathode Cells for Medical Applications," *Proc. Int. Battery Seminar*, Boca Raton, FL, 1993.

    A. Crespi, "The Characterization of Silver Vanadium Cathode Material by High-Resolution Electron Microscopy," *Proc. 7th Int. Meet. Lithium Batteries*, Boston, MA, May 1994.

12. N. I. Sax, *Dangerous Properties of Industrial Materials*, Van Nostrand Reinhold, New York, 1984.

13. *Transportation*, Code of Federal Regulations CFR 49, U.S. Government Printing Office, Washington, DC; Exemption DOT-E-7052, Department of Transportation, Washington, DC: "Technical Instructions for the Safe Transport of Dangerous Goods by Air," International Civil Aviation Organization, DOC 9284-AN/905, Montreal, Quebec, Canada.

14. E. H. Reiss, "Considerations in the Use and Handling of Lithium-Sulfur Dioxide Batteries," *Proc. 29th Power Sources Conf.*, Electrochemical Society, Pennington, NJ, June 1980.

15. Technical Standard Order: TSO-C142, Lithium Batteries, U.S. Dept. of Transportation, Federal Aviation Administration, Washington, DC (2000).

16. T. B. Reddy, *Modern Battery Technology*, Sec. 5.2, C. D. S. Tuck, ed., Ellis Horwood, New York (1991).

17. M. Mathews, *Proc. 39th Power Sources Conf.*, pp. 77–80 (2000), Cherry Hill, NJ.

18. S. Charlton, R. Costa, and C. Negrete, *Proc. 41st Power Sources Conf.*, pp. 29–31 (2004), Philadelphia, PA.

19. M. Sink, *Proc. 38th Power Sources Conf.*, pp. 187–190 (1998), Cherry Hill, NJ.

20. H. Taylor, "The Storability of Li/SO$_2$ Cells," *Proc. 12th Intersociety Energy Conversion Engineering Conf.*, American Nuclear Society, LaGrange Park, IL, 1977.

21. D. Linden and B. McDonald, "The Lithium-Sulfur Dioxide Primary Battery—Its Characteristics, Performance and Applications," *J. Power Sources* **5**:35 (1980), Elsevier Sequoia, Lausanne, Switzerland.

22. R. C. McDonald et al., "Investigation of Lithium Thionyl Chloride Battery Safety Hazard," Tech. Rep. N60921-81-C0229, Naval Surface Weapons Center, Silver Spring, MD, January 1983.

23. S. C. Levy and P. Bro, *Battery Hazards and Accident Prevention*, Sec. 10.3.2, Plenum Publishing Corp., New York (1994).

24. Tadiran Batteries, Port Washington, NY.

25. M. Babai and U. Zak, "Safety Aspects of Low-Rate Li/SOCl$_2$ Batteries," *Proc. 29th Power Sources Conf.*, Electrochemical Society, Pennington, NJ, June 1980.

26. K. M. Abraham and R. M. Mank, "Some Safety Related Chemistry of Li/SOCl$_2$ Cells," *Proc. 29th Power Sources Conf.*, Electrochemical Society, Pennington, NJ, June 1980.

27. R. L. Zupancic, "Performance and Safety Characteristics of Small Cylindrical Li/SOCl$_2$ Cells," *Proc. 29th Power Sources Conf.*, Electrochemical Society, Pennington, NJ, June 1980.

28. J. F. McCartney, A. H. Willis, and W. J. Sturgeon, "Development of a 200 kWh Li/SOCl$_2$ Battery for Undersea Applications," *Proc. 29th Power Sources Conf.*, Electrochemical Society, Pennington, NJ, June 1980.

29. A. Zolla, J. Westernberger, and D. Noll, *Proc. 39th Power Sources Conf.*, pp. 64–68 (2000), Cherry Hill, NJ.

30. C. Winchester, J. Banner, A. Zolla, J. Westenberger, D. Drozd, and S. Drozd, *Proc. 39th Power Sources Conf.*, pp. 5–9 (2000), Cherry Hill, NJ.

31. K. F. Garoutte and D. L. Chua, "Safety Performance of Large Li/SOCl$_2$ Cells," *Proc. 29th Power Sources Conf.*, Electrochemical Society, Pennington, NJ, June 1980.

32. F. Goebel, R. C. McDonald, and N. Marincic, "Performance Characteristics of the Minuteman Lithium Power Source," *Proc. 29th Power Sources Conf.*, Electrochemical Society, Pennington, NJ, June 1980.

33. D. V. Wiberg, "Non-Destructive Test Techniques for Large Scale Li/Thionyl Chloride Cells" *Proc. Int. Battery Seminar*, Boca Raton, FL, 1993.

34. P. G. Russell, D. Carmen, C. Marsh, and T. B. Reddy, *Proc. 38th Power Sources Conf.*, pp. 207–210 (1998), Cherry Hill, NJ.

35. E. Elster, S. Luski, and H. Yamin, "Electrical Performance of Bobbin Type Li/SO$_2$Cl$_2$ Cells," *Proc. 11th Int. Seminar Batteries*, Boca Raton, FL, March 1994.

36. S. McKay, M. Peabody, and J. Brazzell, *Proc. 39th Power Sources Conf.*, pp. 73–76 (2000), Cherry Hill, NJ.

37. C. C. Liang, P. W. Krehl, and D. A. Danner, "Bromine Chloride as a Cathode Component in Lithium Inorganic Cells," *J. Appl. Electrochem.* (1981).

38. D. M. Spillman and E. S. Takeuchi, *Proc. 38th Power Sources Conf.*, pp. 199–202 (1998), Cherry Hill, NJ.

39. H. Ikeda, S. Narukawa, and S. Nakaido, "Characteristics of Cylindrical and Rectangular Li/MnO$_2$ Batteries," *Proc. 29th Power Sources Conf.*, Electrochemical Society, Pennington, NJ, 1980.

40. T. B. Reddy and P. Rodriguez, "Lithium/Manganese Dioxide Foil-Cell Battery Development," *Proc. 36th Power Sources Conf.*, Cherry Hill, NJ, 1994.

41. X. Wang, J. Bennetti, M. Mathews, and X. Zhang, *Proc. 42nd Power Sources Conf.*, pp. 69–72 (2006), Philadelphia, PA.

42. Z. Pi and X. Zhang, *Proc. 42nd Power Sources Conf.*, pp. 65–68 (2006), Philadelphia, PA.

43. A. Morita, T. Iijima, T. Fujii, and H. Ogawa, "Evaluation of Cathode Materials for the Lithium/Carbon Monofluoride Battery," *J. Power Sources* **5**:111 (1980), Elsevier Sequoia, Lausanne, Switzerland, 1980.

44. D. Eyre and C. D. S. Tuck, *Modern Battery Technology*, Sec. 5.3, C. D. S. Tuck (ed.), Ellis Horwood, New York, 1991.

45. R. L. Higgins and L. R. Erisman, "Applications of the Lithium/Carbon Monofluoride Battery," *Proc. 28th Power Sources Symp.*, Electrochemical Society, Pennington, NJ, June 1978.

46. T. R. Counts, *Proc. 38th Power Sources Conf.*, 143–146 (1998), Cherry Hill, NJ.

47. V. Z. Leger, U.S. Patent No. 4,327,166 (April 1982).

48. X. Wang, J. Mastroangelo, and X. Zhang, *Proc. 43rd Power Sources Conf.*, pp. 541–545 (2008), Philadelphia, PA.

49. P. Lam and R. Yazami, *J. Power Sources* **153**:354–359 (2006).

50. J. Whitacre et al., *J. Power Sources* **160**:517 (2006).

51. J. F. Whitacre et al., *Electrochem and Solid-State Letters* **10**:A166–A170 (2007).

52. R. Yazami, *25th International Florida Battery Seminar*, Ft. Lauderdale, FL, March 2008.

53. A. Webber, *Proc. 41st Power Sources Conf.*, pp. 25–28 (2004), Philadelphia, PA.

54. T. Iijima, Y. Toyoguchi, J. Nishimura, and H. Ogawa, "Button-Type Lithium Battery Using Copper Oxide as a Cathode," *J. Power Sources* **5**:1 (1980), Elsevier Sequoia, Lausanne, Switzerland.

55. J. Tuner et al., "Further Studies on the High Energy Density Li/CuO Organic Electrolyte System," *Proc. 29th Power Sources Conf.*, Electrochemical Society, Pennington, NJ, June 1980.

# SECONDARY BATTERIES

PART 4B

SECONDARY BATTERIES

# CHAPTER 14
# LEAD-ACID BATTERIES

**John Olson, Geoffrey J. May, Antonio L. Ferreira, George Zguris**
**(Emeritus Contributors: Kathryn R. Bullock, Alvin J. Salkind)**

*This chapter is dedicated to the memory of Professor Detchko Pavlov (1930–2017), who contributed significantly to the knowledge and technology of lead-acid batteries.*

## 14.1   *GENERAL OVERVIEW*

Lead-acid batteries encompass a group of varied electrochemical cell designs that all use sulfuric acid electrolyte and lead active material and grids. This oldest of the secondary batteries has been in use for over a century in applications such as telephone systems, power tools, automobiles, mobile communication devices, emergency lighting, photovoltaic (PV) and wind electrical grid systems, as well as a power source for mining and materials-handling equipment. The wide availability and variety of designs, sizes, and system voltages have arisen due to the high recyclability,[1] low price, safety, and durability/reliability of the lead-acid electrochemistry.

While the lead-acid battery is considered by many to be an antiquated technology, this electrochemistry still remains the most dominant secondary battery. Far from being stagnant, lead-acid continues to respond to new market demands with innovations and improvements. A lead-acid battery is almost always the least expensive storage battery for any application, while still providing good performance and life characteristics. The production and use of special lead-acid battery constructions (e.g., valve-regulated lead-acid [VRLA] battery and enhanced flooded battery [EFB] for new applications such as start-stop hybrid cars) have continued to grow, typically by replacing the market share of traditional flooded electrolyte lead-acid batteries. The lead-acid battery market is still the largest share of the global demand for secondary batteries, but it is slowly losing ground to lithium-ion batteries. The technology is sustainable with a fully developed recycling infrastructure, but that benefit is offset by the use of toxic lead metal, viewed by many as an environmental hazard. Yet with proper use and recycling, environmental contamination is negligible.

The overall advantages and disadvantages of the lead-acid battery, compared with other systems, are listed in Table 14.1.

The lead-acid battery is manufactured in a variety of sizes and designs, ranging from less than 1 to over 10,000 Ah. Table 14.2 lists many of the various types of lead-acid batteries that are available.

### 14.1.1   History

Practical lead-acid batteries began with the research and inventions of Raymond Gaston Planté in 1860, although batteries containing sulfuric acid or lead components were discussed earlier.[2] Table 14.3 lists the events in the technical development of the lead-acid battery. In Planté's fabrication method, two long strips of lead foil and intermediate layers of coarse cloth were spirally wound and immersed in a solution of about 10% sulfuric acid. The early Planté cells had little capacity, since the amount of stored energy depended on the corrosion of the lead foil to lead dioxide to form the positive active material (PAM), and similarly the negative electrode was

**TABLE 14.1**   Major Advantages and Disadvantages of Lead-Acid Batteries

| Advantages | Disadvantages |
|---|---|
| Popular low-cost secondary battery—capable of manufacture on a local basis, worldwide, from low to high rates of production | Moderate cycle life (300–1500 cycles)[*] |
| | Limited energy density—typically 30–40 Wh/kg |
| Available in large quantities and in a variety of sizes and designs—manufactured in sizes from smaller than 1 Ah to several thousand Ah | Long-term storage in a discharged condition can lead to irreversible polarization of electrodes (sulfation) |
| | Difficult to manufacture in very small sizes |
| Good high-rate performance—suitable for engine starting. Good low- and high-temperature performance | Hydrogen evolution in some designs can be an explosion hazard (flame arrestors are installed to prevent this hazard) |
| Electrically efficient—turnaround efficiency of over 85%, comparing discharge energy out with charge energy in | Stibene and arsine evolution in designs with antimony and arsenic in grid alloys can be a health hazard |
| High cell voltage—open-circuit voltage of >2.0 V is the highest of all aqueous-electrolyte battery systems | Thermal runaway in improperly designed batteries or charging equipment |
| Good float service | |
| Easy state-of-charge indication | |
| Good charge retention for intermittent charge applications (if grids are made with high-overvoltage alloys) | |
| Available in maintenance-free designs | |
| Low cost compared with other secondary batteries | |
| Cell components are easily recycled | |

[*]Up to 5000 deep cycles can be attained with newer designs.

**TABLE 14.2**   Types and Characteristics of Lead-Acid Batteries

| Type | Construction | Typical applications |
|---|---|---|
| SLI (starting, lighting, ignition) | Flat-pasted plates generally with maintenance-free or VRLA[*] construction | Automotive, marine, aircraft, diesel engines in vehicles and for stationary power |
| Traction | Flat-pasted plates; tubular and gauntlet plates, VRLA | Industrial trucks (materials handling) |
| Vehicular propulsion | Flat-pasted plates; tubular and gauntlet plates; also, composite construction, VRLA | Electric vehicles, golf carts, hybrid vehicles, mine cars, personnel carriers |
| Submarine | Tubular plates; flat-pasted plates | Submarines |
| Stationary (including energy-storage types such as charge retention, solar photovoltaic, load leveling) | Planté[†]; Manchester[†]; tubular and gauntlet plates; flat-pasted plates; circular conical plates, VRLA | Standby emergency power: telephone exchange, UPS[*], load leveling, signaling |
| Portable | Flat-pasted plates, VRLA; spirally wound electrodes; tubular plates | Consumer and instrument applications: portable tools, appliances, lighting, emergency lighting, radio, TV, alarm systems |

[*]VRLA: Valve-regulated lead-acid; UPS: Uninterruptible power supply.
[†]Now rarely used.

**TABLE 14.3**    Events in Technical Development of Lead-Acid Battery

| | | Precursor systems |
|---|---|---|
| 1836 | Daniell | Two-fluid cell; copper/copper sulfate/sulfuric acid/zinc |
| 1840 | Grove | Two-fluid cell; carbon/fuming nitric acid/sulfuric acid/zinc |
| 1854 | Sindesten | Polarized lead electrodes with external source |
| | | Lead-acid battery developments |
| 1860 | Planté | First practical lead-acid battery, corroded lead foils to form active material |
| 1881 | Faure | Pasted lead foils with lead oxide-sulfuric acid pastes for positive electrode, to increase capacity |
| 1881 | Sellon | Lead-antimony alloy grid |
| 1881 | Volckmar | Perforated lead plates to provide pockets for support of oxide |
| 1882 | Brush | Mechanically bonded lead oxide to lead plates |
| 1882 | Gladstone and Tribe | Double sulfate theory of reaction in lead-acid battery: $$PbO_2 + Pb + H_2SO_4 \rightleftharpoons 2PbSO_4 + 2H_2O$$ |
| 1883 | Tudor | Pasted mixture of lead oxides on grid pretreated by Planté method |
| 1886 | Lucas | Formed lead plates in solutions of chlorates and perchlorates |
| 1890 | Phillipart | Early tubular construction—individual rings |
| 1890 | Woodward | Early tubular construction |
| 1910 | Smith | Slotted rubber tube, Exide tubular construction |
| 1920 to present | | Materials and equipment research, especially expanders, oxides, and fabrication techniques |
| 1935 | Haring and Thomas | Lead-calcium alloy grid |
| 1935 | Hamer and Harned | Experimental proof of double sulfate theory of reaction |
| 1956–1960 | Bode and Voss Ruetschi and Cahan Burbank Feitknecht | Clarification of properties of two crystalline forms of $PbO_2$ (alpha and beta) |
| 1970s | McClelland and Devitt | Commercial spiral-wound sealed lead acid battery that work on the recombinant cycle |
| | | Expanded metal grid technology; composite plastic/metal grids; sealed and maintenance-free lead-acid batteries; glass fiber and improved separators; through-the-partition intercell connectors; heat-sealed plastic case-to-cover assemblies; high energy density batteries (above 40 Wh/kg); conical grid (round) cell for long-life float service in telecommunications facilities |
| 1980s | | Sealed valve-regulated batteries; quasi-bipolar engine starter batteries; improved low-temperature performance; world's largest battery installed (Chino, CA); 40 MWh lead-acid load leveling |
| 1990s | | Electric-vehicle interest re-emerges; bipolar battery designs for high-power use in uninterruptible power supplies, power tool market, and electronic backup; thin foil cells, small cells for consumer and current road applications |
| 2009 | | Development of lead-carbon batteries, extended life flooded batteries for microhybrids, improved high-rate partial state-of-charge (HRPSOC) VRLA, microhybrids for stop-start application, using bipolar battery |

formed by roughening of another foil (on cycling) to form a porous surface. Primary cells were used as the power sources for this formation. The capacity of Planté cells was increased on repeated cycling as corrosion of the substrate foils created more active material and increased the surface area. In the 1870s, magnetoelectric generators became available to Planté, and about this time the Siemens dynamo began to be installed in central electric plants. Lead-acid batteries found an early market to provide load leveling and to average out the demand peaks. They were charged at night, similar to the procedure now planned for modern load-leveling energy-storage systems.

Subsequent to Planté's first developments, numerous experiments were done on accelerating the formation process and coating lead foil with lead oxides on a lead plate pretreated using the Planté method. Attention then turned to other methods for retaining active material, and two main technological paths evolved.

1. Coating a lead oxide paste on cast or expanded grids, rather than foil, in which the active material developed structural strength and retention properties by a "cementation" process (interlocked crystalline lattice) through the grid and active mass. This is generally referred to as the flat-plate design.

2. The tubular electrode design, in which a central conducting wire or rod is surrounded by active material and the assembly is encased in an electrolyte porous insulated tube, which can be either square, round, or oval.

Simultaneous with the advances in developing and retaining active material was the work toward strengthening the grid by casting it from lead alloys such as lead-antimony (e.g., Sellon, 1881) or lead-calcium (e.g., Haring and Thomas, 1935).[3] The technical knowledge for an economical manufacture of reliable lead-acid batteries was in place by the end of the 19th century and subsequent growth of the industry was rapid. Improvements in design; manufacturing equipment and methods; recovery methods; active material utilization and production; and supporting structures and components such as separators, cases, and seals continue to improve the economic and performance characteristics of lead-acid batteries. Battery development continues today with the main focus toward the growing hybrid car market. Work sponsored by the Advanced Lead-Acid Battery Consortium (ALABC) on carbon and other additives in the active materials has improved the battery charge acceptance for improved high-rate, partial state-of-charge (HRPSoC) performance. Lead-carbon batteries represent the advanced performance capabilities of the modern battery.

### 14.1.2  Manufacturing Data and Battery Usage

The largest use of lead-acid batteries is in automotive vehicle applications for starting, lighting, and ignition (SLI) purposes. Similar usage is prevalent in aircraft, boats, and off-road and farm equipment vehicles. The increased use of electronics in most modern cars has also resulted in increased electrical capacity (Ah) requirements. The 12 V systems now commonly used are designed with capacities in the 40 to 100 Ah range, and weigh between 11 and 45 kg. The high-rate current capability, in the standard cold crank test, can be as high as 900 A or more. Data on the market growth of lead-acid batteries are given in Table 14.4. The automobile industry is moving toward the production of greener designs. Various platform technologies are described below.

**TABLE 14.4**  Market Growth of Lead-Acid Batteries in the United States[*]

|  | 1960 | 1980 | 1991 | 1999[*] | 2016 (Est) |
|---|---|---|---|---|---|
| SLI units (original equipment and replacement) | 34 | 62 | 76 | 100 | 125 |
| SLI sales, $ | 330 | 1675 | 2100 | 2700 | 5500 |
| Industrial, $ | 70 | 380 | 550 | 1015 | 1900 |
| Consumer, $ | 1 | 55 | 100 | 150 | 100 |
| Total, $ | 400 | 2110 | 2750 | 3965 | 7500 |

[*]All units in millions values are at manufacturers' pricing. Battery prices are affected by the price of lead.
Lead prices varied from $0.40 to over $4.00/kg between 1978 and 2016 (lead price Dec. 2016 = $2.5 kg).
*Source:* London Metal Exchange.

## 14.2  CHEMISTRY

### 14.2.1  General Characteristics

The lead-acid battery uses lead dioxide as the active material of the positive electrode and metallic lead, in a high-surface-area porous structure, as the negative active material (NAM). The physical and chemical properties of these materials are listed in Table 14.5.[4] Typically, a charged positive electrode contains both $\alpha$-PbO$_2$ (orthorhombic) and $\beta$-PbO$_2$ (tetragonal). The equilibrium potential of the $\alpha$-PbO$_2$ is more positive than that of $\beta$-PbO$_2$ by 0.01 V. The $\alpha$ form also has a larger, more compact crystal morphology, which is less active electrochemically and slightly lower in capacity per unit weight; it does, however, promote longer cycle life. Neither of the two forms is fully stoichiometric. Their composition can be represented by PbO$_x$, with $x$ varying between 1.85 and 2.05. The PAM, which is formed electrochemically from the cured plate, is a major factor influencing the performance and life of the lead-acid battery. In general, the negative (sponge lead) electrode controls cold-temperature performance (such as engine starting).

**TABLE 14.5**  Physical and Chemical Properties of Lead and Lead Oxides (PbO$_2$)

| Property | Lead | $\alpha$-PbO$_2$ | $\beta$-PbO$_2$ |
|---|---|---|---|
| Molecular weight, g/mol | 207.2 | 239.19 | 239.19 |
| Composition | | PbO:$_{1.94-2.03}$ | PbO:$_{1.87-2.03}$ |
| Crystalline form | Face-centered cubic | Rhombic (columbite) | Tetragonal (rutile) |
| Lattice parameters, nm | $a = 0.4949$ | $a = 0.4977$ $b = 0.5948$ $c = 0.5444$ | $a = 0.491-0.497$ $c = 0.337-0.340$ |
| X-ray density, g/cm$^3$ | 11.34 | 9.80 | ~9.80 |
| Practical density at 20°C (depends on purity), g/cm$^3$ | 11.34 | 9.1–9.4 | 9.1–9.4 |
| Heat capacity, cal/deg·mol | 6.80 | 14.87 | 14.87 |
| Specific heat, cal/g | 0.0306 | 0.062 | 0.062 |
| Electrical resistivity, at 20°C, $\mu\Omega$/cm | 20 | ~$100 \times 10^3$ | |
| Electrochemical potential in 4.4M H$_2$SO$_4$ at 31.8°C, V | 0.356 | ~1.709 | ~1.692 |
| Melting point, °C | 327.4 | | |

*Source:* Ref. 4.

The electrolyte is a sulfuric acid solution, of typically about 1.28 specific gravity or 37% acid by weight in a fully charged condition. As the cell discharges, both electrodes are converted to lead sulfate. The process reverses on charge:

*Negative electrode:*
$$Pb \underset{\text{charge}}{\overset{\text{discharge}}{\rightleftharpoons}} Pb^{2+} + 2e$$

$$Pb^{2+} + SO_4^{2-} \underset{\text{charge}}{\overset{\text{discharge}}{\rightleftharpoons}} PbSO_4$$

*Positive electrode:*
$$PbO_2 + 4H^+ + 2e \underset{\text{charge}}{\overset{\text{discharge}}{\rightleftharpoons}} Pb^{2+} + 2H_2O$$

$$Pb^2 + SO_4^{2-} \underset{\text{charge}}{\overset{\text{discharge}}{\rightleftharpoons}} PbSO_4$$

*Overall reaction:*
$$Pb + PbO_2\ 2H_2SO_4 \underset{\text{charge}}{\overset{\text{discharge}}{\rightleftharpoons}} 2PbSO_4 + 2H_2O$$

As shown, the basic electrode processes in the positive and negative electrodes involve a dissolution-precipitation mechanism and not a solid-state ion transport or film-formation mechanism.[4] As the sulfuric acid in the electrolyte is consumed during discharge, producing water, the electrolyte is an "active" material and in certain battery designs can be the capacity-limiting material. This capacity-limiting effect of the electrolyte is an important design consideration in VRLA batteries.

On charge, as the cell approaches full charge and the majority of the $PbSO_4$ has been converted to Pb or $PbO_2$, the cell voltage becomes greater than the gassing voltage (about 2.39 V per cell) and the overcharge reactions begin, resulting in the production of hydrogen and oxygen (gassing) and the resultant loss of water.

$$\text{Negative electrode:} \qquad 2H^+ + 2e \rightarrow H_2$$

$$\text{Positive electrode:} \qquad H_2O \rightarrow \frac{1}{2}O_2 + 2H^+ + 2e$$

$$\text{Overall reaction:} \qquad H_2O \rightarrow H_2 + \frac{1}{2}O_2$$

The general performance characteristics of the lead-acid cell, during charge and discharge, are shown in Fig. 14.1. As the cell is discharged, the voltage decreases due to depletion of material, internal resistance losses, and polarization. If the discharge current is constant, the voltage under load decreases smoothly to the cutoff voltage and the specific gravity decreases in proportion to the ampere-hours discharged.

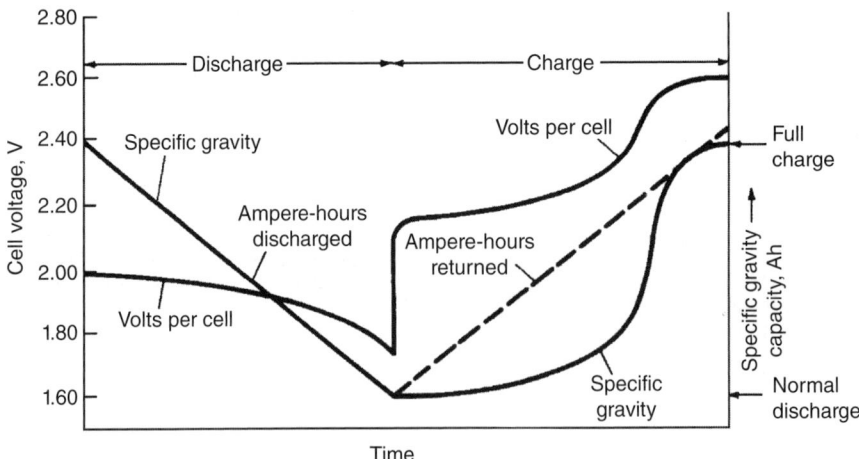

**FIGURE 14.1**   Typical voltage and specific gravity characteristics of lead-acid cell at constant-rate discharge and charge.

An analysis of the behavior of the positive and negative plates can be done by measuring the voltage between each electrode and a reference electrode, the "half-cell" voltage. Figure 14.2 illustrates this analysis, using a cadmium reference electrode.

***Specific Gravity.***   The selection of specific gravity (relative density) used for the electrolyte depends on the application and service requirements (see Sec. 14.3.10). The electrolyte concentration must be high enough for good ionic conductivity and to fulfill electrochemical requirements, but not so high as to cause separator deterioration or corrosion of other parts of the cell, which would shorten life and increase self-discharge. The electrolyte concentration is deliberately reduced in high-temperature climates. During discharge, the specific

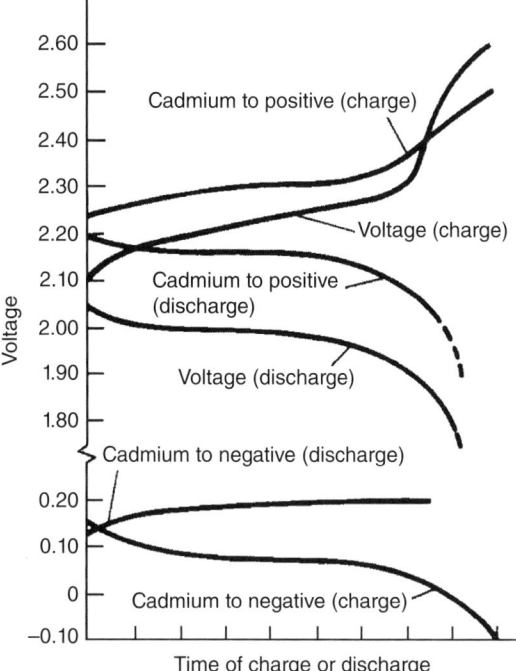

**FIGURE 14.2**    Typical charge-discharge curves of lead-acid cell.
(*From Ref. 5.*)

gravity decreases in proportion to the ampere-hours discharged (Table 14.6). The specific gravity is thus a means of checking the state of charge of the battery. On charge, the change in specific gravity should similarly be proportional to the ampere-hour charge accepted by the cell. However, there is a lag because complete mixing of the electrolyte does not occur until gassing commences near the end of the charge.

**TABLE 14.6**    Specific Gravity of Lead-Acid Battery Electrolytes at Different States of Charge for Various Designs*

| State of charge | Specific gravity | | | |
|---|---|---|---|---|
| | A | B | C | D |
| 100% (full charge) | 1.330 | 1.280 | 1.265 | 1.225 |
| 75% | 1.300 | 1.250 | 1.225 | 1.185 |
| 50% | 1.270 | 1.220 | 1.190 | 1.150 |
| 25% | 1.240 | 1.190 | 1.155 | 1.115 |
| Discharged | 1.210 | 1.160 | 1.120 | 1.0 |

Assumes flooded cell design.
*Specific gravity may range from 100 to 150 points between full charge and discharge depending on cell design: A—electric vehicle battery; B—traction battery; C—SLI battery; D—stationary battery.

## 14.2.2  Open-Circuit Voltage Characteristics

The open-circuit voltage for a lead-acid battery is a function of temperature and electrolyte concentration as expressed by the Nernst equation,

$$E = 2.047 + \frac{RT}{F} \ln\left(\frac{\alpha H_2 SO_4}{\alpha H_2 O}\right)$$

A graph of the open-circuit voltage versus electrolyte concentration at 25°C is given in Fig. 14.3. Above 1.10 specific gravity, the plot is fairly linear but shows strong deviations at lower concentrations. The open-circuit voltage is also affected by temperature. The temperature coefficient of the open-circuit voltage of the lead-acid battery is shown in Fig. 14.4. Where $dE/dT$ is positive, such as above 0.5 M $H_2SO_4$, the reversible potential of the system increases with increasing temperature. Below 0.5 M, the temperature coefficient is negative. Most lead-acid batteries operate above 2 M $H_2SO_4$ (1.120 specific gravity) and have a thermal coefficient of about +0.2 mV/°C.

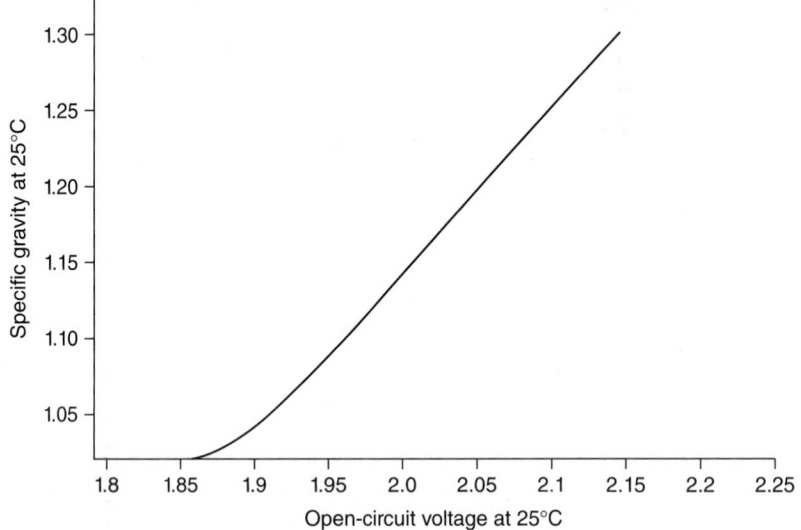

**FIGURE 14.3**   Open-circuit voltage of lead-acid cell as a function of electrolyte specific gravity.

## 14.2.3  Polarization and Resistive Losses

When a battery is being discharged, the voltage under load is lower than the open-circuit voltage. The thermodynamically stable state for batteries is the discharged state. Work (charging) must be done to cause the equilibria of the electrochemical reactions to go toward $PbO_2$ in the positive and Pb in the negative. Thus, the voltage of the power source for recharging the lead-acid battery must be higher than the Nernst voltage of the battery on open circuit.

These deviations from the open-circuit voltage during charge or discharge are due to resistive losses in the battery or polarization. These losses can be measured by use of an interrupted discharge or charge, where the *IR* losses can be estimated by Ohm's law ($\Delta E/\Delta I = R$) within a few seconds to a few minutes after the current is stopped. The effect of polarization can take several hours to measure in order for diffusion to allow the plate interiors to equilibrate. AC impedance spectroscopy techniques are also of value. Polarization of the two plates is measured by using a reference electrode. The standard reference of hydrogen on platinum is not practical for most measurements on lead-acid batteries, and several other sulfate-based reference electrode systems are used. A review of reference electrodes neglects several very practical sulfate electrodes.[6] Still, a commonly used electrode for battery maintenance is the cadmium "stick," but it is not especially stable (±20 mV/day).

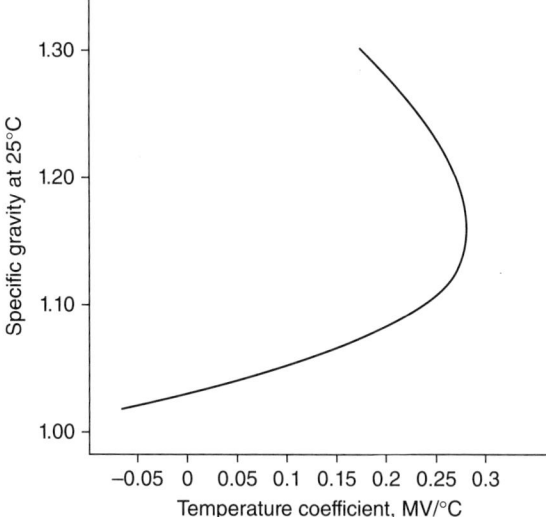

**FIGURE 14.4** Temperature coefficient of open-circuit voltage of lead-acid cell as a function of electrolyte specific gravity.

Mercury-mercurous sulfate reference electrodes are stable and are available from several vendors. A novel $Pb/H_2SO_4/PbO_2$ reference electrode has been patented.[7] This electrode measures the polarization on charge or discharge directly, without the need for a correction of different thermal coefficients of the electromotive force (EMF). The change in polarization between the start and the end of discharge is typically several hundred mV, and the cell capacity is limited by the plate group (positive or negative) that falls over the discharge knee first. When both plate groups in a cell change about equally, the capacity limitation is likely due to depletion of $H_2SO_4$ in the electrolyte. On charge, the polarization is a good measure that both positives and negatives have been recharged. Polarization voltages stabilize at some value when plates are recharged and are gassing freely.

## 14.2.4  Self-Discharge

The rate of self-discharge (loss of charge capacity when no external load is applied) of the lead-acid cell is moderate, but it can be reduced significantly by incorporating certain design features.

The rate of self-discharge depends on several factors. Water is thermodynamically unstable in the lead-acid battery, and on open circuit electrolysis can occur. Oxygen is evolved at the positive electrode and hydrogen at the negative, at a rate dependent on temperature and acid concentration (the gassing rate increases with increasing acid concentration) as follows:

$$PbO_2 + H_2SO_4 \rightarrow PbSO_4 + H_2O + \tfrac{1}{2}O_2$$

$$Pb + H_2SO_4 \rightarrow PbSO_4 + H_2$$

For most positives, the formation of $PbSO_4$ by self-discharge is slow, typically much less than 0.5%/day at 25°C. The self-discharge of the negative is generally more rapid, especially if the plate is contaminated with various catalytic metallic ions. For example, antimony lost from the positive grids by corrosion can diffuse to the negative, where it is deposited, accelerating hydrogen formation. New batteries with lead-antimony grids lose about 1% of charge per day at 25°C, but the charge loss increases by a factor of 2 to 5 as the battery ages. Batteries with nonantimonial lead grids lose less than 0.5% of charge per day regardless of age, as illustrated in Fig. 14.5a.[8] Maintenance-free and charge-retention-type batteries, where the self-discharge rate must be minimized, use

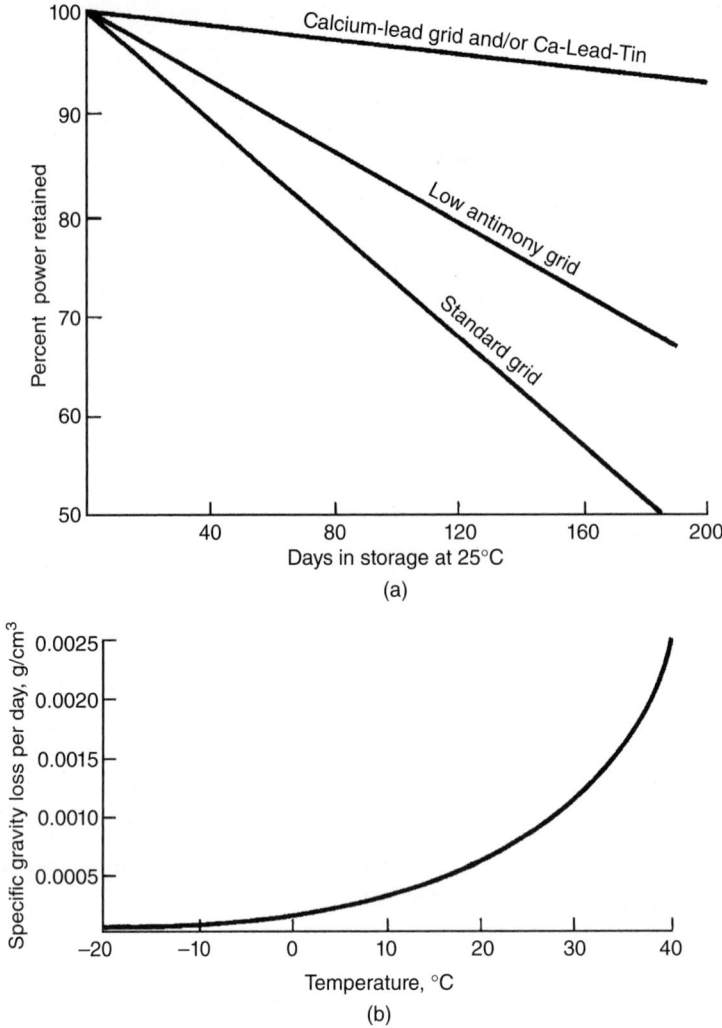

**FIGURE 14.5**    (a) Capacity retention during stand or storage at 25°C. (*From Ref. 8.*) (b) Loss of specific gravity per day with temperature of a new, fully charged lead-acid battery with 6% antimonial lead grids. (*From Ref. 9.*)

low-antimony or antimony-free alloy (such as calcium-lead) grids. However, because of other beneficial effects of antimony, its complete elimination may not be desirable, and low-antimony–lead alloys are a useful compromise.

Self-discharge is temperature-dependent, as shown in Fig. 14.5b.[9] The graph shows the fall in specific gravity per day of a new fully charged battery with 6% antimonial lead grids. Self-discharge can thus be minimized by storing batteries at temperatures between 5 and 15°C.

## 14.2.5    Characteristics and Properties of Sulfuric Acid

The major characteristics and properties of the sulfuric acid electrolyte, as they apply to the operation of the lead-acid battery, are listed in Table 14.7. The freezing points of sulfuric acid solutions at various concentrations

**TABLE 14.7** Properties of Sulfuric Acid Solutions*

| Specific gravity | | Temperature coeff. α | H₂SO₄ | | | Freezing point, °C | Electrochemical equivalent (per liter of acid), Ah |
|---|---|---|---|---|---|---|---|
| At 15°C | At 25°C | | Wt., % | Vol., % | Mol/L | | |
| 1.00 | 1.000 | — | 0 | 0 | 0 | 0 | 0 |
| 1.05 | 1.049 | 33 | 7.3 | 4.2 | 0.82 | −3.3 | 22 |
| 1.10 | 1.097 | 48 | 14.3 | 8.5 | 1.65 | −7.7 | 44 |
| 1.15 | 1.146 | 60 | 20.9 | 13.0 | 2.51 | −15 | 67 |
| 1.20 | 1.196 | 68 | 27.2 | 17.7 | 3.39 | −27 | 90 |
| 1.25 | 1.245 | 72 | 33.2 | 22.6 | 4.31 | −52 | 115 |
| 1.30 | 1.295 | 75 | 39.1 | 27.6 | 5.26 | −70 | 141 |
| 1.35 | 1.345 | 77 | 44.7 | 32.8 | 6.23 | −49 | 167 |
| 1.40 | 1.395 | 79 | 50.0 | 38.0 | 7.21 | −36 | |
| 1.45 | 1.445 | 82 | 55.0 | 43.3 | 8.2 | −29 | |
| 1.50 | 1.495 | 85 | 59.7 | 48.7 | 9.2 | −29 | |

*To calculate the specific gravity for any temperature, °C, SG $(t)$ = SG (15°C) + $\alpha \times 10^{-5}$ (15 − $t$).

are also plotted in Fig. 14.6. The freezing point of aqueous sulfuric acid solutions varies significantly with concentration. Batteries must therefore be designed so that the electrolyte concentration is above the value at which the electrolyte would freeze when exposed to the anticipated cold. Alternatively, the battery can be insulated or heated so that it remains above the electrolyte freezing temperature.

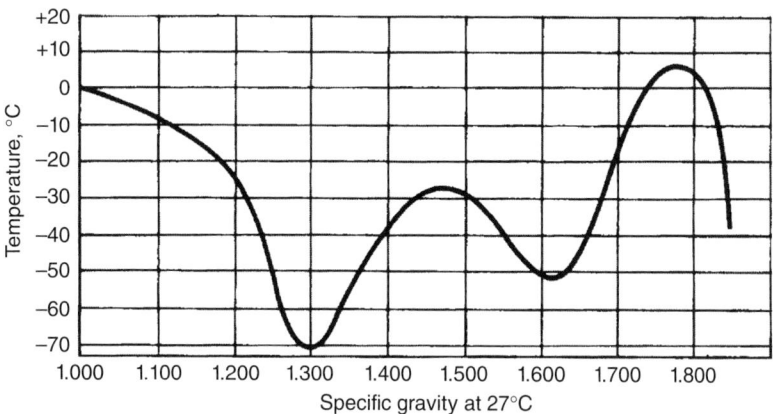

**FIGURE 14.6** Freezing points of sulfuric acid solutions at various specific gravities. The inflection points result from the different water to SO₃ hydration ratios.

Figure 14.7 shows the specific resistivity of sulfuric acid solutions at various specific gravities as a function of temperature from −40 to 40°C.

A comparison of specific gravity with freezing-point data will show that battery type A from Table 14.6 will freeze at −30°C when fully discharged, while battery type D will freeze at about −5°C, a factor which must be considered in the design of the battery and the battery housing. The acid concentration for most lead-acid batteries for use in temperate climates is usually between 1.26 and 1.28 specific gravity. Higher-concentration electrolytes tend to attack some separators and other components; lower concentrations tend to be insufficiently conductive in a partially charged cell and freeze at low temperatures. In high-temperature climates, a lower concentration is used, and in stationary cells with larger proportional electrolyte volumes and no high-rate demands, electrolytes with specific gravity as low as 1.21 are used.

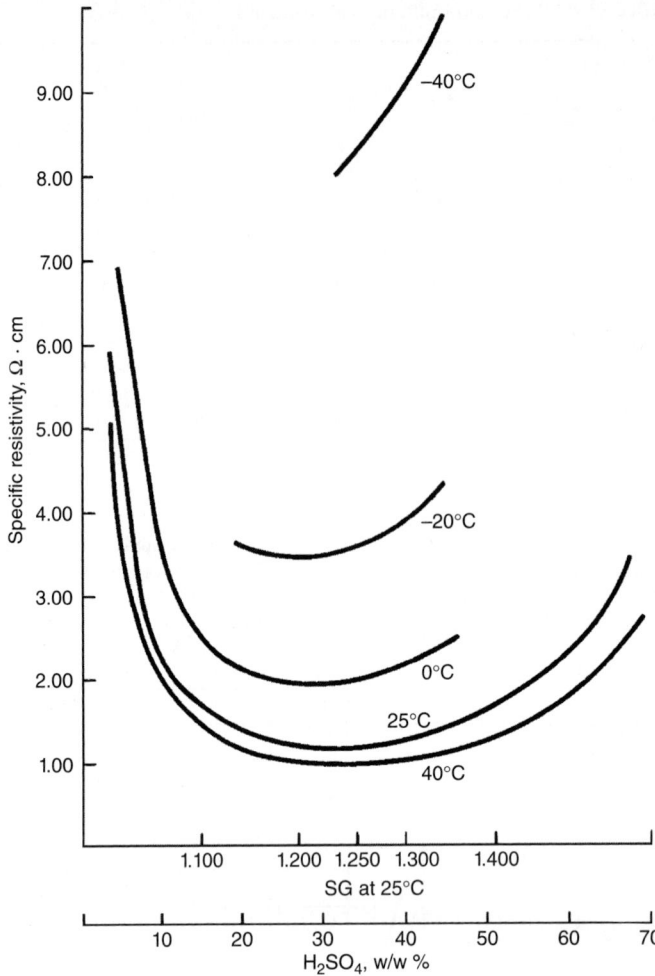

**FIGURE 14.7**    Specific resistivity of sulfuric acid solutions at various specific gravities and temperatures.

Figure 14.8 indicates the method of preparing sulfuric acid solutions of any specific gravity from concentrated sulfuric acid.[10]

## 14.2.6    VRLA Oxygen Recombination Cycle

Although the design and the construction of the VRLA battery are different than those of the flooded batteries, its chemistry is that of the traditional lead-acid battery. A unique aspect of the VRLA design is that the majority of the oxygen generated within the cells at normal overcharge rates is recombined at the negative plate (oxygen cycle). Grids made of high-purity lead, often alloyed with tin, are used in VRLA cells to reduce gassing reactions. In rectangular battery designs, more rigid grids made of lead-calcium-tin alloys are generally used. Lead-antimony alloys that have been used in traditional batteries must be avoided. High lead purity minimizes hydrogen evolution on overcharge and reduces the rate of self-discharge on stand. The oxygen cycle minimizes hydrogen formation due to preferential reduction of oxygen. VRLA batteries must nonetheless be operated in ventilated areas, because small quantities of hydrogen, carbon oxides, and oxygen are released through the pressure-release valve.

Oxygen will be reduced by lead at the negative plate in the presence of $H_2SO_4$ as quickly as it can diffuse to the lead surface

$$Pb + HSO_4^- + H^+ + \tfrac{1}{2}O_2 \underset{charge}{\overset{discharge}{\rightleftharpoons}} PbSO_4 + H_2O$$

In a flooded lead-acid battery, virtually all the $H_2$ and $O_2$ escape from the cell rather than recombine. In the VRLA battery, the closely spaced plates are separated by a nonwoven glass mat that is composed of fine glass strands in a porous structure. This design, called absorbed glass mat (AGM), is only partially filled with

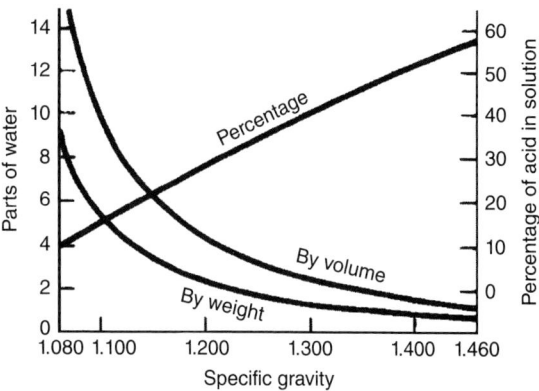

**FIGURE 14.8**   Preparation of sulfuric acid solutions of any specific gravity from concentrated sulfuric acid. (*From Ref. 10.*)

electrolyte, thus creating a "starved-electrolyte" condition that allows for gas transfer between the plates necessary to allow the oxygen recombination reaction. Additional discussion of VRLA chemical kinetics can be found in the literature.[11]

In the VRLA cell, a pressure-release valve maintains an internal pressure that aids recombination by retaining the gases long enough within the cell for diffusion to take place. The net result is that water is cycled electrochemically to take up the excess overcharge current beyond that used for conversion of active material, rather than being released from the cell as gases. Thus, the cell can be overcharged sufficiently to convert virtually all the active material without loss of significant water.

## 14.3  FLOODED BATTERIES—CONSTRUCTIONAL FEATURES, MATERIALS, AND MANUFACTURING METHODS

Lead-acid batteries consist of several major components, as shown in a cutaway view of an automotive SLI battery in Fig. 14.9. Batteries for other applications have analogous components, as illustrated and described in Secs. 14.8 to 14.10. The applications of the various cells and batteries dictate the design, size, quantities, and types of materials that are used.

The lead components of a typical lead-acid battery constitute more than 60% of its total weight. A breakout of the weights of the components of several types of lead-acid batteries is shown in Fig. 14.10.

The battery components are fabricated and processed as shown in the flow sheets of Figs. 14.11*a* and *b*. The major starting material is highly purified lead.[12] Lead is used for the production of alloys (for subsequent conversion to grids) and for the production of lead oxides (for subsequent conversion to paste and ultimately to the lead dioxide PAM [Fig. 14.11*a*] and the sponge lead NAM).

Automotive lead-acid batteries (SLI batteries) are produced mainly in high-volume plants with a great deal of automation. Many modern factories are capable of producing quantities on the order of 100,000 batteries per day. On average, an automated facility might require less than 500 employees, including all staffing levels. The automation has been prompted by environmental, reliability, and cost considerations.

### 14.3.1  Alloy Production

Pure lead is generally too soft to be used as a grid material. Exceptions that use pure lead plates are some special, very thick plate Planté or pasted-plate batteries, some spiral-wound batteries, some valve-regulated cells and batteries (see Sec. 14.3.2), and cylindrical cells (see Sec. 14.10.2).[13]

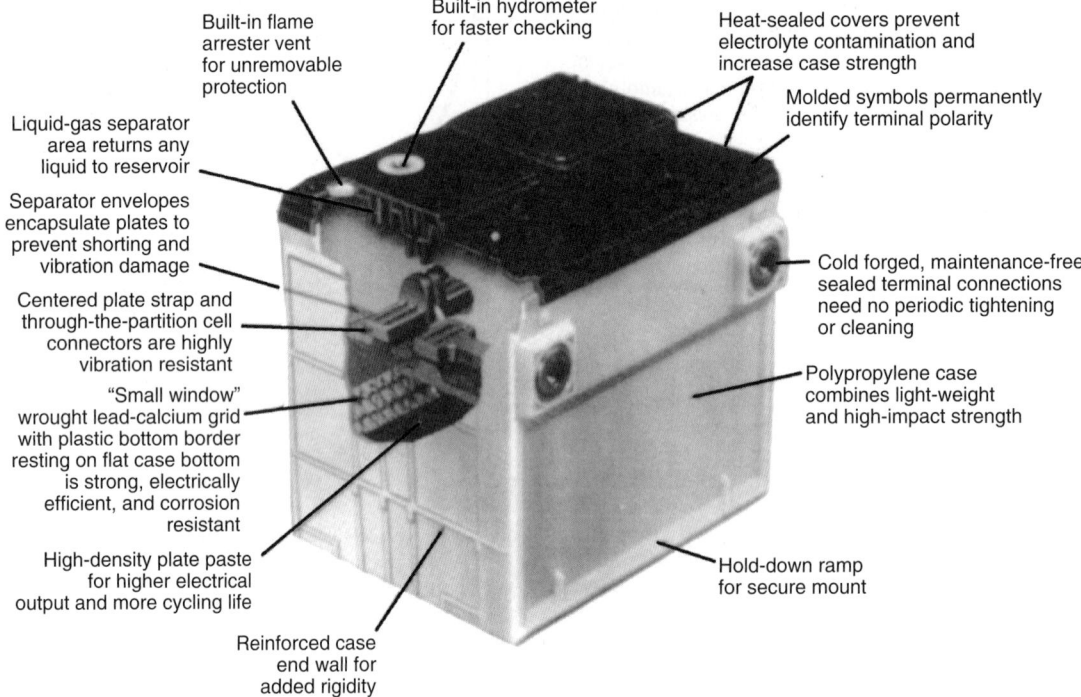

Built-in flame
arrester vent
for unremovable
protection

Built-in hydrometer
for faster checking

Heat-sealed covers prevent
electrolyte contamination and
increase case strength

Molded symbols permanently
identify terminal polarity

Liquid-gas separator
area returns any
liquid to reservoir

Separator envelopes
encapsulate plates to
prevent shorting and
vibration damage

Centered plate strap and
through-the-partition cell
connectors are highly
vibration resistant

"Small window"
wrought lead-calcium grid
with plastic bottom border
resting on flat case bottom
is strong, electrically
efficient, and corrosion
resistant

High-density plate paste
for higher electrical
output and more cycling life

Reinforced case
end wall for
added rigidity

Cold forged, maintenance-free
sealed terminal connections
need no periodic tightening
or cleaning

Polypropylene case
combines light-weight
and high-impact strength

Hold-down ramp
for secure mount

**FIGURE 14.9**  Typical maintenance-free lead-acid battery overview.

Lead was hardened in older designs, by the addition of antimony metal. The amount of antimony has varied between 5% and 12% by weight, generally dependent on the availability and cost of antimony. Typical modern alloys, especially for deep-cycling applications, contain 2% to 6% antimony. The trend in grid alloys is to go to even lower antimony contents, in the range of 1.5% to 2% antimony, in order to reduce the maintenance (water addition) that the battery will require. As the antimony content goes below 4%, the addition of small amounts of other elements is necessary to prevent grid fabrication defects and grid brittleness. These elements such as sulfur, copper, arsenic, selenium, tellurium, and various combinations of these elements act as grain refiners to decrease the lead grain size.[14–16]

Some of the alloying elements, not previously described as grain refiners, fall into two broad classes of elements that are beneficial or detrimental to grid production or battery performance. Beneficial elements include tin, which operates synergistically with antimony and arsenic to improve metal fluidity and castability. Silver, cobalt, and selenium are also thought to improve corrosion resistance. Detrimental elements include iron, which increases drossing; nickel, which affects battery operation; and manganese, which attacks paper separators. Cadmium has been used in ternary grid alloys to enhance process ability in antimony alloys and to minimize the detrimental effects of antimony. Cadmium, however, should not be used because of its toxicity and difficulty of removal during lead recovery (recycling) operations. Bismuth exists in many lead ore feedstocks and has no special detrimental effects.

A second class of lead alloys has been developed that uses calcium or other alkaline earth elements for stiffening. These alloys were developed originally for telephone service applications.[3,17] Antimony from the grids is dissolved during battery operation and migrates to the negative plates where it redeposits, which results in increased hydrogen evolution and water loss. For telephone applications, more stable battery operation and less frequent watering were desired. The composition of the alloy depends somewhat on the grid manufacturing process. Calcium is used in the range of 0.03% to 0.20%, but for corrosion resistance the preferred range is 0.03% to 0.06%. A variation has been to substitute strontium for calcium. Barium has been investigated but is generally felt to be detrimental to performance. Tin is used to enhance the mechanical and corrosion-resistant

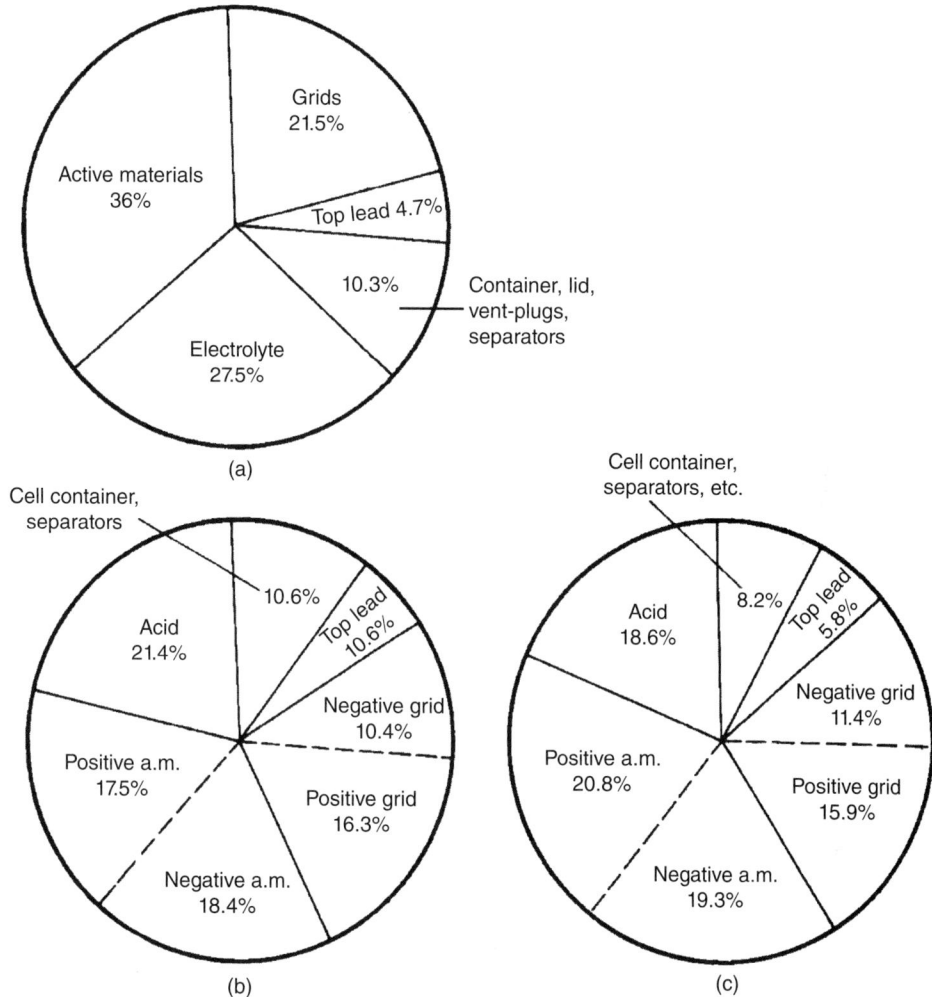

**FIGURE 14.10** Weight analysis of typical lead-acid batteries. (*a*) Starting, lighting, and ignition (SLI) battery. (*b*) Tubular industrial battery. (*c*) Flat-plate traction battery. (*From Ref. 9.*)

properties of Pb-Ca alloys and is usually used in the range of 0.25% to 2.0% by weight with higher levels of addition for positive grids. Aluminum is also added in small quantities to stabilize the drossing loss of the alkaline earth element (calcium or strontium) from the molten alloy. Grain refining is done by the alkaline earth metal, and no other elements (impurities) are desired. Silver may be added to improve the corrosion resistance of alloys used for positive grids. The properties of the alloys are summarized in Table 14.8.[14]

## 14.3.2 Grid Production

Two general classes of grid production methods describe virtually all modern production, but two other classes of production techniques might become more widespread in the future. These are listed in Table 14.9.

The purposes of the grid are to mechanically hold the active material and conduct electricity between the active material and the cell terminals. Additional mechanical support is sometimes gained by the construction

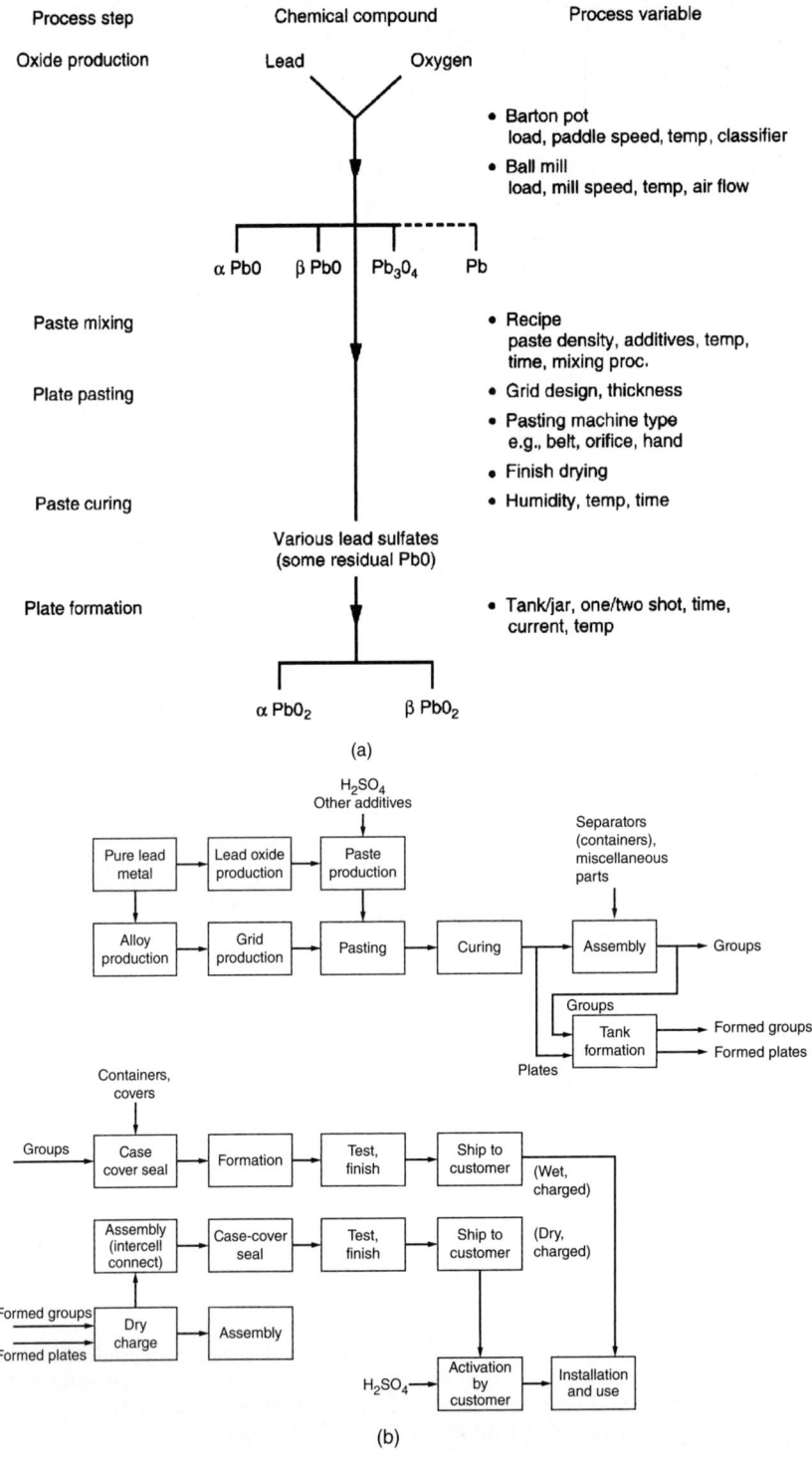

**FIGURE 14.11**   (*a*) Chemical compounds and process parameters in production of SLI batteries. (*b*) Production flow sheet for lead-acid batteries.

**TABLE 14.8**  Properties of Lead Alloys

### Alloys of the 1970s

| Property | Conventional antimony | Low antimony | Cast lead-calcium-tin 0.1Ca 0.3Sn | Cast lead-calcium-tin 1.1Ca 0.75Sn | Lead-strontium-tin-aluminum | Lead-cadmium-antimony | Wrought-lead-calcium-tin 0.065Ca 0.75Sn |
| --- | --- | --- | --- | --- | --- | --- | --- |
| Ultimate tensile strength, Pa × 10$^{-6}$ | 38–46 | 33–40 | 40–43 | 47–50 | 53 | 33–40 | 60 |
| Percent elongation | 20–25 | 10–15 | 25–35 | 20–30 | 15 | 25 | 10–15 |

| Property | Cast conventional antimony | Cast low-antimony | Cast lead-calcium | Cast lead-strontium | Cast lead-cadmium antimony | Wrought lead-calcium-tin (1st generation) |
| --- | --- | --- | --- | --- | --- | --- |
| Ease of grid manufacture | Good | Fair | Fair | Fair to good | Fair | Good |
| Mechanical | Good | Fair | Fair to good | Fair to good | Fair | Good |
| Corrosion | Fair | Fair | Good | Good | Fair | Good |
| Battery performance | Poor | Fair | Good | Good | Good | Fair to good |
| Economics | Good | Good | Fair | Poor | Fair | Fair to good |

### Alloys of the 1980s and 1990s

| | Cast alloys | | | | | Wrought alloys | | | Cast and wrought |
| --- | --- | --- | --- | --- | --- | --- | --- | --- | --- |
| Property | Lead-calcium-tin 0.1Ca 0.3Sn | Lead-calcium 0.1Ca | Lead-calcium-tin with aluminum | Lead-calcium with aluminum | Lead-calcium-tin 0.065Ca 0.3Sn | Lead-calcium-tin 0.065Ca 0.5Sn | Lead-calcium 0.075Ca | Low antimony 2.5–3.0% Sb | Lead 0.01–1.5Sn |
| Ultimate tensile strength, Pa × 10$^{-6}$ | 40–43 | 37–39 | 40–43 | 37–39 | 43–47 | 47 | 43 | 37–40 | Conductivity and corrosion-equivalent to pure lead |
| Percent elongation | 25–35 | 30–45 | 25–35 | 30–45 | 15 | 15 | 25 | 25–40 | |

| | Cast alloys | | Wrought alloys | |
| --- | --- | --- | --- | --- |
| Property | Low antimony | Lead-calcium | Wrought lead-calcium-tin (2nd generation) | Wrought low antimony |
| Ease of grid manufacture | Fair to good | Good (aluminum) | Good | Good |
| Mechanical | Fair | Fair to good | Good | Good |
| Corrosion | Fair | Good | Good | Fair to good |
| Battery performance | Fair to good | Good | Fair to good | Fair |
| Economics | Good | Good (lower tin) | Good | Good |

*Note:* Alloy constituents given in weight percent.

**TABLE 14.9**   Grid Production Methods

Book-mold cast
   Gravity cast
   Injection molded (die cast)
Mechanically worked (Planté, Manchester)
Continuous cast, drum cast
Continuous cast, wrought-expanded, cast-expanded
   Casting
   Working
   Expansion
      Progressive die expansion
      Precision expanded
      Rotary expanded
      Rotary expansion
      Diagonal/slit expansion
   Punching
Composite

method or by various wrappings on the outside of the plate. Metals other than lead alloys have been investigated to provide electrical conductivity, and some (copper, aluminum, silver) are more conductive than lead. These alternate conductors are not corrosion-resistant in the sulfuric acid electrolyte and are often more expensive than lead alloys. Titanium has been evaluated as a grid material; it is not corroded after special surface treatments but is very expensive. It has also been used to reinforce grids. Lead-plated copper grids are used in the negatives of some industrial and submarine batteries.

The grid design is generally a rectangular framework with a tab or lug for connection to the post strap. For cast grids, the framework features a heavy external frame and a lighter internal structure of horizontal and vertical bars. In some grid designs the frame tapers with the greater width near the lug; the internal bars may also be tapered. A more recent advance in grid design is the "radial" grid, with the vertical wires displaced along the frame, pointing directly toward the tab area in order to increase grid conductivity (Fig. 14.12). The radial design has been further refined to a composite of lead alloy radial conductor arrangement cast into a

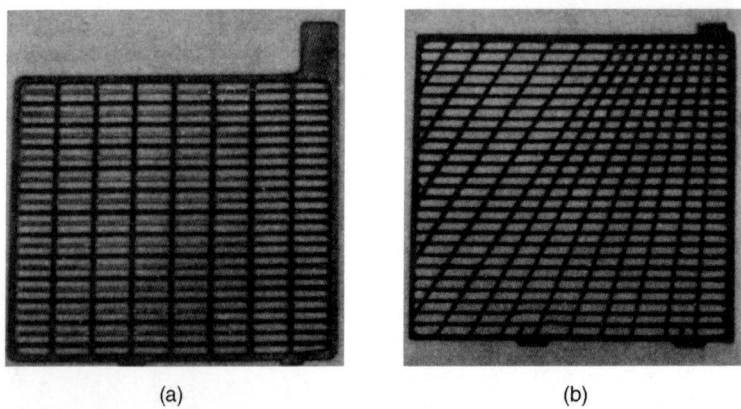

(a)                (b)

**FIGURE 14.12**   Examples of lead-acid battery cast grids. (*a*) Conventional cast flat grid. (*b*) Radial-design grid.

rectangular plastic frame. An example of this composite grid is shown in Fig. 14.13. The grids used in the cylindrical cell design (Fig. 14.14) incorporate both concentric and radial members.[18] This system has been in commercial production since 1970. An example of a balanced positive grid design is shown in Fig. 14.15.[19]

"Book-mold" casting historically accounted for most grid production. Permanent molds are made from steel (Meehanite®) blocks by machining grooves to form the grid frames and internal lattice structure. The molds are filled when closed with an amount of lead sufficient to form the grid and leave an excess gate or sprue, which is subsequently trimmed off by a cutting or stamping operation. The grid alloy is put into the mold from a ladle in a recirculation lead alloy stream, from a metering valve in a nonrecirculation lead stream or from a hand-filled ladle. A variation on book-mold casting is injection molding or die casting of battery grids,

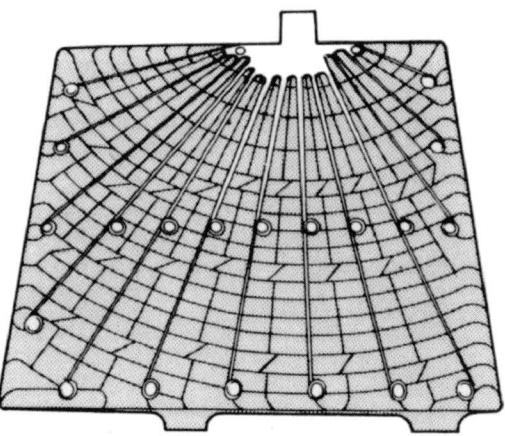

**FIGURE 14.13**   Composite grid, radial conductor. Grid combines diagonal conducting members with light robust plastic frame.

where the lead alloy is forced into a clamped mold by injection pressure. Depending on the alloy characteristics, injection molding can be capable of very high production rates (see Figs. 14.16 and 14.17).

The Planté-type method of grid manufacture involves mechanical treatment of a strip or slab of lead alloy. The traditional procedures were either to cut grooves into a thick lead plate, thereby increasing its surface area, or to crimp and roll up lead strips into rosettes that were inserted into round holes in a cast plate. More recent

**FIGURE 14.14**   Balanced positive design takes into account grid corrosion and growth and promotes the maintenance of contact of the grid with the active material, while maintaining the shape of the plate and its angle with the horizontal. This concept has also been carried into the prismatic grid structure. (*Courtesy of AT&T.*)

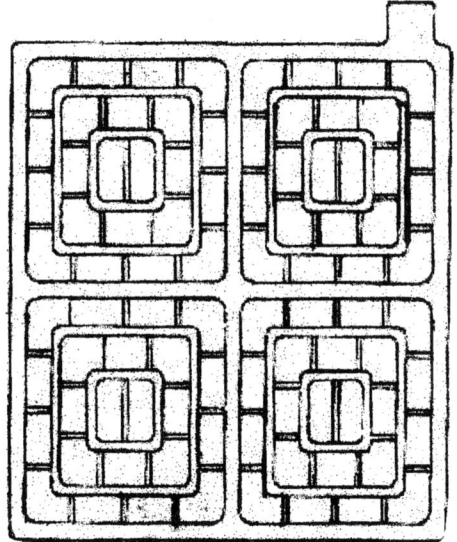

**FIGURE 14.15**   Balanced rectangular positive grid design. This design promotes active material contact and accounts for grid corrosion and growth in a prismatic cell. (*From Ref. 19.*)

**FIGURE 14.16** Wirtz automatic grid casting machine model 220C. (*Courtesy of Wirtz Manufacturing Co.*)

**FIGURE 14.17** Wirtz industrial grid casting machine model 450C. (*Courtesy of Wirtz Manufacturing Co.*)

processes use pressure die casting to produce a grid with thin lamellae suitable for subsequent processing. The resultant plates are formed electrolytically into positives in the classic Planté designs (Fig. 14.18).

Another grid production method is circumferential continuous casting onto a mold cut into the surface of a drum. Successful high-speed production of up to 150 grids per minute is possible. Continuous cast grids are not symmetrical about a central planar axis and need to be overpasted to hold the active material in place.

A fourth grid production method, expansion from wrought or cast lead alloy strip, has supplanted book-mold casting as the preferred method for the manufacture of SLI battery grids. The advantages of this method are lower grid weight, the capability to manufacture a wide variety of sizes with a minimum investment in tooling, a very high-rate production capability (up to 600 plates per minute), and very uniform grid sizes. Lead-calcium-tin alloys (Fig. 14.19) are preferred. A strip is produced from cast slabs by a variety of proprietary metal-working processes, and the thin strip is slit to the width specified by the battery manufacturer. The worked metal increases in strength as it decreases in thickness during this processing. More recently, grid production by punching from wrought strips has been developed and has many advantages in terms of grid design leading to improved battery performance.

The machinery to produce grids from wrought strips has been developed and put into production by several manufacturers. Four types of machinery are involved: progressive die expansion, precision expansion, rotary expansion, and diagonal slit expansion. Progressive die expansion has been the most extensively utilized of the four methods, but rotary expansion is important. Continuous drum casting of automotive grids is also used to produce negative Pb-Ca-Sn plates for automotive batteries.

Whatever grid production method is used, there may be the need for small cast parts for plate and cell interconnections and connection to external equipment. These parts have traditionally been cast in fixed molds, sometimes with mold inserts to allow a variety of similar parts to be made in each mold. Newer battery production methods often produce these various interconnections automatically in the course of battery assembly.

(a)

(b)

**FIGURE 14.18** Planté plates.

### 14.3.3 Lead Oxide Production

Lead is used to make the active materials as well as the grids. The lead must be highly refined (usually virgin or primary lead) to preclude contamination of the battery. It is described as corroding-grade lead in American Society for Testing and Materials (ASTM) specification B29.[12] Lead is oxidized to lead oxide by either of two processes: the

**FIGURE 14.19** Expanded wrought grid for lead-acid batteries.

Barton pot or the ball mill.[20] In the Barton pot process, a fine stream of molten lead is swept around inside a heated pot-shaped vessel, and oxygen from the air reacts with fine droplets or particles to produce an oxide coating around each droplet. Typical Barton pot "leady oxide" contains 15% to 30% free lead, which usually exists as the core of each fine spherically shaped particle. Barton pots are available in a variety of sizes up to 1000 kg/h output.

Ball milling describes a larger variety of processes. Lead pieces are put into a rotary mechanical mill, and the attrition of the pieces causes fine metallic flakes to form. These are oxidized by an airstream, which also serves to remove the leady oxide particles through collection in a baghouse. The feedstock for ball mills can range from small cast slugs weighing less than 30 g to full pigs of lead weighing approximately 30 kg. Typical ball mill oxides also contain 15% to 30% free lead in the shape of a flattened platelet core surrounded by an oxide coating.

Some battery positive plates use an additive of red lead ($Pb_3O_4$), which is more conductive than PbO, to facilitate the electrochemical formation of $PbO_2$. Red lead is produced from leady oxide by roasting this material in airflow until the desired conversion is complete. Such processing reduces the free lead content and generally increases the oxide particle size. Varieties of other oxides and lead-containing materials have been used to produce battery plates but are of only historical interest. Positive plates for the Lucent Technologies batteries (formerly Bell Laboratories) were initially made with tetrabasic lead sulfate ($4PbO·PbSO_4$), which is a precursor for $\alpha$-$PbO_2$. These plates now contain up to 25% red lead ($Pb_3O_4$) in order to facilitate the electrochemical formation process.

### 14.3.4   Paste Production

Lead oxide is converted to a plastic dough-like material so it can be affixed to the grids. Leady oxide is mixed with water and sulfuric acid in a mechanical mixer. Three types of mixers are commonly used: the change can or pony mixer, the muller, and a vertical muller. The pony mixer is the traditional unit. A preweighed amount of leady oxide is placed into the mixing tub, and this is wetted first with water and then with sulfuric acid solution. Dry paste additives, if any, are premixed into the leady oxide before water addition. These additives can be plastic modified microglass fibers to enhance the mechanical strength and electrical performance of the dried paste, expanders to maintain negative-plate porosity in operation, carbon (especially for extended-life flooded SLI), and various other proprietary additives that ease processing or are believed to improve battery performance.

Muller mixers are usually filled first with the water component, then with additives, followed by oxide, and finally with the acid. As mixing proceeds, the paste viscosity increases, then decreases, as measured by the amount of power consumed by the mixer motor. The paste becomes hot from the mechanical mixing and from the reaction of $H_2SO_4$ with the leady oxide. The paste temperature is controlled by cooling jackets on the mixer or by evaporation of water from the paste. The amounts of water and acid for a given amount of oxide will be different for the two mixer types and will also depend on the intended use of the plates: SLI plates are generally made at a low bulking agent—the more acid used, the lower the plate density will be. The total amount of additives, liquids, and the type of mixer used will affect the final paste consistency (viscosity). Paste mixing is controlled by the measurement of paste density using a cup with a hemispherical cavity and by the measurement of paste consistency with a penetrometer. In making paste for advance extended-life flooded SLI batteries where the carbon additive is added at high levels, the paste density will be lowered due to the low density of the carbon material. When using such additives, a reconsideration of the targeted paste density should be made.

Another option is to use a continuous paste mixer. This type of paste mixer can be used for all types of lead-acid batteries. The mixer has the ability to uniformly distribute fibers and additives (such as carbon) in a paste mix. Moreover, having a uniform distribution that avoids clumping of fibers also eliminates pasting problems that can cause costly line downtime. A diagram of the process is shown in Fig. 14.20.

**Typical process flow**

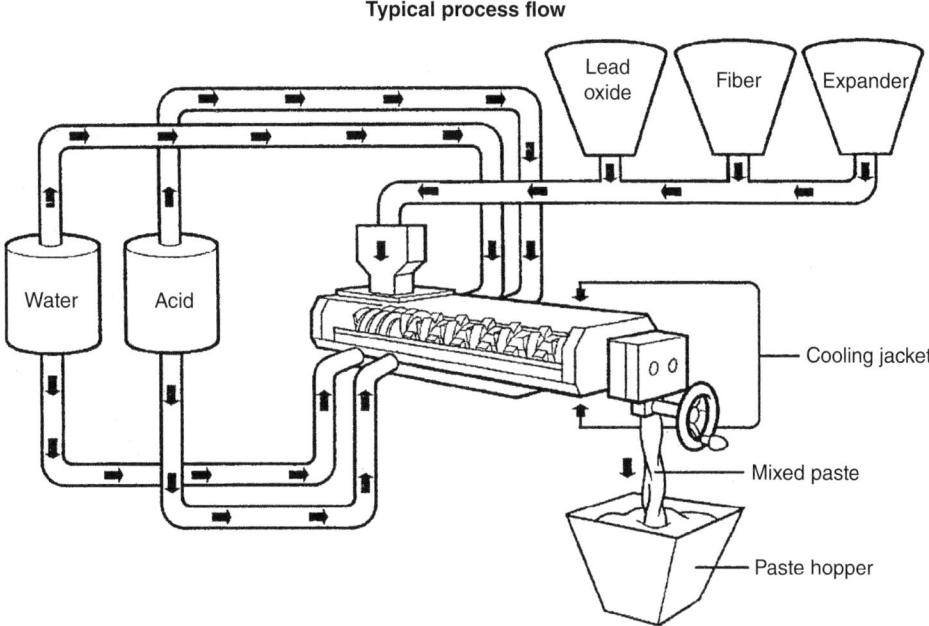

**FIGURE 14.20**   Typical process flow diagram for paste production.

### 14.3.5   Pasting

Pasting is the process by which the paste is actually integrated with the grid to produce a battery plate. This process is a form of extrusion, and the paste is pressed into the grid interstices. Two types of pasting machines are used: a fixed orifice paster that pushes the paste into both sides of the plate simultaneously and a belt paster in which the paste is pressed into the open side of a grid that is being conveyed past a paste hopper on a porous belt. The amount of paste applied to a plate by a belt paster is regulated by the spacing of the hopper above the grid on the belt and the type of trowel (roller or rubber squeegee) used at the hopper exit. Using identical paste and grids, a trowel roller machine packs the paste both thicker and more densely than a rubber squeegee machine. As plates are pasted on either belt pasting machine, water is forced out of the paste, into the belt, and ultimately to a sump on or near the machine. The sump material can be used in place of some of the liquids for subsequent batches of negative paste.

Grids are automatically or manually placed onto the belt before being moved under the paste hopper. Most smaller-sized plates are made as "doubles" joined at the feet (cast) or at the tab edge (wrought expanded), or as panels of varying number of plates. Typically, the belt is 35 to 50 cm wide and can handle such doubles. Industrial stationary or traction plates (being larger) are pasted by lengthwise feed into the machine or are hand-pasted. After pasting, plates are racked or stacked for curing. Stacked plates contain enough moisture to stick together, so before stacking, the plate surfaces are dried somewhat by a rapid passage through high-temperature driers or heated platens. Some carbon dioxide from the combustion process might be absorbed on the surface such that the surface is made "harder." The flash drying process may also help start the curing reactions. Thicker plates are usually placed with the long edge upward in racks rather than being stacked horizontally on pallets after flash drying. Wrought-expanded plates and some cast plates are cut into discrete plate portions by a slitter machine in the pasting line. Some manufacturers also have the plate lugs brushed clean of paste and surface oxidation on the same machinery.

In Europe, and less commonly in the United States, many of the heavy-duty battery positive plates are made in porous tubular sheaths (Fig. 14.21). The grid is cast or injection-molded of lead, with long-finned spines attached to a header bar and a connection lug (see Fig. 14.21*b*). The spines are placed in nonwoven multitube

(a)

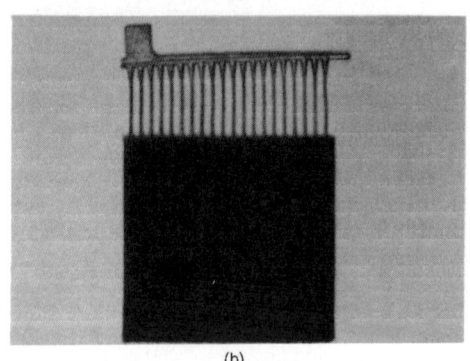

(b)

**FIGURE 14.21**   Tubular and gauntlet plates. (*a*) Tubular. (*b*) Gauntlet.

gauntlets and a very fluid paste is added with the gauntlet filtering out some of the fluid. To a lesser extent, woven plastic or glass sheaths are used, with the nonwoven tube being the more common. See Sec. 14.3.7 for typical property differences among different gauntlet types. A plastic cap then plugs the open sheath ends at the bottom of the plate.

### 14.3.6   Curing

The curing process is used to make the paste into a cohesive, porous mass and to help produce bonding between the paste and the grid. The curing process oxidizes residue-free lead, produces the physical and chemical structure of the plate to allow the plate to be handled efficiently through the battery-making process, and sets the stage for the future electrical performance of the battery. Several different curing processes are used for lead oxides, depending on paste formulation and the intended use of the battery.[21,22]

The typical cure for SLI plates is "hydroset" at low temperature and low humidity for 24 to 72 h. The temperature is preferably between 25 and 40°C; the water content is that contained in the flash-dried plates, typically 8% to 20% $H_2O$ by weight. The plates are usually covered by canvas, plastic, or other materials to help retain both temperature and moisture. Some manufacturers use enclosed rooms for the hydroset, and these rooms may be heated where required. As the plates cure, they reach a peak temperature, and then temperature and humidity decrease. Hydroset typically produces tribasic lead sulfate, which gives high energy density. A uniform cure in the plates is important for battery manufacture. Additives can impact the consistency of the plate curing, such as adding a microglass paste fiber. Figure 14.22 shows plates that have very uniform cure.[21] The microglass additive will also stop tetrabasic lead sulfate from forming, unless a seed crystal is used.

The type of cure that is obtained can be greatly influenced by additives. The addition of 0.5% to 2.0% of H&V-modified paste additive (PA-10-6™) prevents a standard tetrabasic core process from forming tetrabasic lead oxide.[23] Various seed crystals are being used in lead-acid batteries that provide a likelihood of a tetrabasic cure being formed. SureCure® is used as a seed crystal to promote rapid formation of tetrabasic lead sulfate during paste mixing and curing.[24,25] It is added at 1% to the paste batch with the other dry ingredients. Seed crystals are reported to offer benefits[22] such as:

- Shorter cure time
- Lower material and curing cost
- Better crystal structure that is more uniform
- Improved formation
- Reduced capital investment in curing chambers

**FIGURE 14.22**    Lead-acid plate batch after curing. Note radial grid to tab.

The use of curing ovens, where temperature and humidity can be precisely controlled, ensures that sufficient moisture is available to oxidize the remaining free lead in the paste. Peak temperatures in the range of 65 to 90°C are used. Another process to force completion of curing is to dip the partially cured plates into dilute sulfuric acid. This latter process ("pickling") is also used for cure of powder-filled tubular positive plates. Cured plates are stored until use. Shelf life is not critical, but the high cost of inventory will limit the storage time.

### 14.3.7    Assembly and Separator Materials

The simplest cell consists of one negative, one positive, and one separator in between wetted with electrolyte. Most practical SLI-flooded batteries contain about 7 to 30 plates. Electrode plates are usually enveloped in a microporous polyethylene (PE) separator, but most battery designs just envelope the positive plate. Individual or leaf separators are generally not used, having been replaced by the sealable microporous PE separator. In hot tropical climates, especially in areas with poor road conditions, a laminate synthetic pulp that is filled with silica particles and glass fibers is still in general use. This separator also has high mechanical strength that resists tearing and cracking, thus allowing automatic assembly. Envelope separators are also used in motive-power and standby batteries. The physical properties of these separators are shown in Table 14.10.[26–28]

In SLI batteries, a thinner back-web separator has been developed and is used to provide for decreased electrical resistivity to improve battery performance such as the cold cranking current. The back-web thicknesses used in SLI are in the range of 150 to 200 μm. An industrial application would use a thicker, back-web PE separator of 450 μm. An off-road, industrial, or heavy-duty SLI battery would have a PE separator that has a glass veil attached to the ribs of the separator. This glass veil is used to apply pressure and to hold the PAM in the plate from falling out due to vibration or shock. The glass veil is a chopped glass fiber mat in the 20 to 60 gsm range. The glass fibers are typically in the 11 to 18 μm range and the fibers are very long, 12 mm or longer. The mat will have a binder content of 15% to 25%. The percentage of the binder tends to decrease as the sheet gets thicker. To help reinforce the PAM, additives such as fine microglass fibers could also be added to the active material in the 1% to 2.5% range.[29]

To address the extra cycle requirements needed in microhybrid designs, the standard flooded battery has been redesigned to allow the battery to have enhanced cycling under partial state-of-charge conditions. These EFB designs use a combination of changes in the plates, alloy, electrolyte, and separator. These batteries will have a glass mat or synthetic nonwoven material against the positive plate to apply pressure to the positive plate. Having pressure on the active material has been shown to improve cycle life. The positive grid

**TABLE 14.10** Properties of Separator Systems and Materials

| Typical properties of industrial PE* separators | | |
| --- | --- | --- |
| Separator properties | Unit | Typical results |
| Back-web thickness | μm | 400–550 |
| Electrical resistance | mΩcm² | 210–270 |
| Porosity | % | 55–58 |
| Back-web oil | % | 15–17 |
| Total oil | % | 19–21 |
| Moisture resistance | % | 3 |

| Typical properties of SLI*-PE separators | | |
| --- | --- | --- |
| Separator properties | Unit | Typical results |
| Back-web thickness | μm | 60–200 |
| Electrical resistance | mΩcm² | 50–60 |
| Porosity | % | 50–60 |
| Total oil | % | 10–21 |
| Moisture resistance | % | 3–5 |
| Puncture | N | 5–13 |
| Elongation XMD | % | 200–500 |

| Typical properties of gauntlets | | | | |
| --- | --- | --- | --- | --- |
| Gauntlet properties | Unit | Standard | Reinforced | Woven |
| Electrical resistance | mΩcm² | 180 | 350 | 500 |
| Porosity | % | 74 | 60 | 40 |
| Acid absorbed | g/cm² | 0.12 | 0.10 | 0.05 |
| Acid retained | g/g | 2.7 | 2.0 | 0.8 |

*Note:* Test methods per Battery Council International, *Flooded Separators*, section 3B.
*PE: Polyethylene; SLI: Starting, lighting, ignition.

may be an expanded metal, punched, or other continuous strip plate. Instead of using a typical cellulose tissue paper for the pasting paper, companies are investigating alternate materials, such as spunbonded or carded synthetic webs, glass veils, or the conversion of the separator system to an all-glass system.[30] With the increased use of hybrid absorptive mats (HAGM) in VRLA, these very tough AGMs can have acceptable results in a flooded cell.

For large, industrial flooded cells, the separator construction used has remained basically unchanged. The plate is wrapped with thin, continuous, glass fibers (slyver) that have been laminated to a nonwoven glass mat. The glass fibers assist in allowing the gas bubbles generated during charging to rise to the top of the cell. The glass nonwoven mat is usually a chopped strand glass mat composed of glass fiber diameters in the 10 to 19 μm range, which is bonded together with about 16% to 24% acrylic binder. A plastic outer wrapper (with die cut holes) is used to wrap and seal in place the glass laminate against the plate. Then a plastic boot is usually placed on the bottom of the plate. When the plates are in the case, an industrial grade separator is used to separate the negative and positive plates. Experiments have demonstrated that polyester synthetic nonwovens could be used to replace the slyver-glass-plastic wrapper. However, this has never gained much market penetration due to the extremely long time to test a new separator concept for these cells.

In tubular constructions, gauntlets have moved toward the nonwoven type and away from the woven gauntlet. The nonwoven gauntlet has also been improved, moving away from a carded nonwoven material to a spunbound one that provides for greater fabric strength properties.[31,32]

Plates and separators are stacked manually or by a stacking machine. Stacked elements are staged on roller conveyors or carts as input to the inter-plate welding operations. Welding is done by two general methods: melting of the lugs in a mold with the lugs facing upward, or immersion of lugs facing downward into pools of molten lead alloy contained in a preheated mold. The first method is the traditional assembly method for lead-acid batteries. In this method, the plate lugs fit up through slots in a mold "comb," and the shape and the size of the group strap are delineated by the "dam" and "back iron" portions of the tooling. Some battery manufacturers use slotted "crowfoot" or "comb" posts to fit over the plate lugs to speed up the welding process. The second welding method is called the "cast-on strap" process and is generally used for SLI cells. Stacked elements are loaded into slots of the cast-on machine. A mold that has cutouts corresponding to the desired straps and posts is preheated and filled with the appropriate molten lead alloy, making sure not to join lead or lead-calcium alloys with antimony alloys. The mold and the stacked elements are moved until the plate lugs are immersed in the strap cutouts. External cooling solidifies the strap onto and around each lug, and the elements are moved to a point where they can be dropped into a battery case. Visual examination can differentiate between the two welding methods: fixture-welded plate straps are usually thicker and smoother than cast-on straps; usually, cast-on straps also show a convex meniscus of metal between adjacent plate lugs on the underside of the strap if the lug is properly cleaned of paste. A good weld is required between each plate lug and the strap so that high-rate discharge performance is maximized. The resultant assemblage of plates and separators is known as an element, and the welded subelements are known as groups. Electrical testing for short circuits is usually done on elements before further assembly.

Cast-on battery elements are either continuously connected or made in discrete one-cell modules. The first method requires that long intercell connections be used, which travel over the intercell partition and are seated in a slot in this partition; this is known as the loop-over-partition design. In the second cast-on method, tabs on the ends of the plate straps are positioned over holes that have been prepunched into the intercell partitions of a battery case. These tabs are welded together manually with a very small torch or automatically by a resistance welding machine. The latter also squeezes the tabs and the intercell partition to provide a leak proof seal.

Industrial traction cells and old-style SLI cells have been connected into batteries after the cell cases and covers are sealed together. Traction batteries are needed in thousands of different sizes for various applications, and the standard unit of construction is a cell, not a quantity of plates and separators. A heavy steel tray is fabricated and coated with an acid-resistant coating (urethane, epoxy, etc.). Traction cells are placed into the tray and shimmed as necessary, and intercell connections are welded on. Heavy flexible wires (made from welding cables) are welded to the end cells for connection to the external circuit.

### 14.3.8  Case-to-Cover Seal

Four different processes have been used to seal battery cases and covers together. Enclosed cells are necessary to minimize safety hazards related to the acidic electrolyte, to the potentially explosive gases produced on overcharge, and to electrical shock. Most SLI batteries and many modern traction cells are sealed with fusion of the case and cover. The fusion comes from preheating each on a platen, then forcing the two together mechanically, or from ultrasonic welding of the case and cover. Fusion-sealed batteries are virtually impossible to repair. At best the elements can be salvaged, but the cover and usually the case are discarded and replaced. A few SLI batteries are sealed using an epoxy cement that fills a groove in the cover; the battery is inserted and positioned so that the case and intercell partition lips fit into the epoxy filled groove. Heat is used to activate the catalyst to set the epoxy.

Historically, batteries used tar (asphalt) sealed cases and individual cell covers. The tar seal allows easy repair to the battery. Traditionally, all batteries were made this way before about 1960, but heat seals have been the standard for SLI batteries for many years.

Stationary batteries in plastic cases are sealed with epoxy or urethane glues, with solvent cement, or (for polyvinyl chloride [PVC] copolymer cases and covers) with a thermal seal. Terminals are cast or welded on. Heat sealing is also widely used. Some very large stationary and traction cells are made so that the coolant can be circulated through the terminals, and others are made with terminals with copper inserts for increased conductivity and mechanical strength.

### 14.3.9  Tank Formation

Formation is the electrical activation of the plates. Plates or assembled groups can be formed before assembly into the case. When SLI plates are formed, these are usually formed as "doubles," with two to five panels stacked together in a slotted plastic formation tank, spaced an inch or less from stacks of counter-electrode pasted panels in adjacent slots. The stacks are arranged so that all positive lugs protrude out of one side of the tank top and all negative lugs protrude out of the other side of the tank top. All lugs with the same polarity are connected by welding to a heavy lead bar, and the two bars are connected to a low-voltage, constant-current power supply. The tank is filled with electrolyte, and current is passed until the plates have been formed. The positives are converted to a deep brownish black and the negatives to a soft gray that shows a bright metallic streak when scratched. Industrial plates are usually formed singly. Sometimes these are also formed using dummy plates or grids. A variety of tank materials have been used but the most common are PVC, PE, or lead. The tanks are arranged so that the acid can be drained and refilled because formation increases the electrolyte concentration.

A variety of formation conditions are used, with variations in electrolyte density, charging rate (current), and temperature. The electrolyte is typically diluted, in the range of 1.050 to 1.150 specific gravity. The charging rate is usually fixed, but some manufacturers use a sequence of two or three different charging rates for different periods of time.

Tank-formed groups or plates are somewhat unstable (negatives will spontaneously oxidize in air), and therefore are "dry charged" before use (see Sec. 14.3.11).

Modern electronic chargers for formation operate in a constant-current mode with computer control of the current and time. Some formation schedules include charge at three or more different currents, which start at low currents, go to higher currents, and then revert to lower currents. Current adjustment during formation minimizes damage to the cells by high temperatures and reduces the need for cell cooling by water spray or forced air.

## 14.3.10   Case Formation

The more usual method of formation is to assemble the battery, fill it with electrolyte, seal covers and weld terminals, and then apply the formation charge. This method is used for SLI and most stationary and traction batteries. A variety of formation conditions are used, similar to those for tank formation. The two major formation processes are the two-shot formation process (used for stationary and traction batteries) and the one-shot formation process (used for most SLI batteries). In the two-shot formation, the electrolyte is dumped to remove the low-density initial electrolyte and refilled with more concentrated electrolyte, chosen so that the cell electrolyte will equilibrate at the desired density (Table 14.11). Typical values of the electrolyte specific gravity at full charge after formation are given in Table 14.12.

**TABLE 14.11**   Formation Processes (Flooded Types)

|  | One-shot | Two-shot |
|---|---|---|
| Typical application | SLI | All others, some SLI |
| Electrolyte concentrations, sp gr: |  |  |
|    Initial | 1.200 | 1.005–1.150 |
|    Final | 1.280 | 1.150–1.230 |
| Subsequent processing | None | Dump and refill with 1.280–1.330 sp gr electrolyte; continue charge for several hours |

**TABLE 14.12**   Specific Gravity of Electrolytes at Full Charge at 25°C

|  | Specific gravity | |
|---|---|---|
| Type of battery | Temperate climates | Tropical climates |
| SLI | 1.260–1.290 | 1.210–1.230 |
| Heavy duty | 1.260–1.290 | 1.210–1.240 |
| Golf cart | 1.260–1.290 | 1.240–1.260 |
| Golf cart (electric vehicle) | 1.275–1.325 | 1.240–1.275 |
| Traction | 1.275–1.325 | 1.240–1.275 |
| Stationary | 1.210–1.225 | 1.200–1.220 |
| Diesel starting (railroad) | 1.250 |  |
| Aircraft | 1.260–1.285 | 1.260–1.285 |

## 14.3.11   Dry Charge

The state of charge of wet batteries declines with long periods of storage, especially at warm temperatures. A loss of 1% to 3% of capacity each day is possible with SLI batteries that contain antimony-lead grids. The loss on stand can be much lower for maintenance-free batteries (0.1–3%/day). When lead-acid batteries must be stored for a long time, especially in high ambient temperatures, or when batteries are shipped for export, their performance can be stabilized by removal of the electrolyte by one of several methods.

When the electrolyte is removed, the battery is termed "dry-charged" (i.e., charged and dry) or "charged and moist." The first process is done before the battery elements are assembled inside the case and cover. The plates can be tank-formed, water-washed, and then dried in an inert gas before the element welding portion of assembly. Alternately, the welded element can be tank-formed, washed, and then dried in an inert gas. The latter process is simpler to carry out, but it is necessary that the separators can be rewetted easily after being washed and dried. The assembly (case, elements, cover) is completed and the battery is sealed. The battery can be stored in this dry-charged state for up to several years before reactivation and use.

Several processing innovations have been commercialized to convert wet-charged batteries into moist or semidry batteries. In one process, most of the electrolyte is removed by centrifugation. Another process uses an inorganic salt (sodium sulfate) in the electrolyte, which minimizes degradation during storage and assists in an eventual reactivation. A battery is formed, dumped, refilled with electrolyte that contains the additive, tested with a high-rate discharge, and then finally dumped. The high-rate electrical discharge (to simulate engine cranking) allows testing of an assembled "damp-dry" battery. These processes are not generally used now, as the lower rate of self-discharge of batteries with Pb-Ca-Sn alloys has extended the shelf life of wet batteries.

## 14.3.12   Testing and Finishing

Electrical tests are used to check the performance of batteries before they are sold and often before they are put into use. The type of test employed depends on the intended use of the battery. SLI batteries are tested by brief discharges at very high currents (200–1500 A) to simulate engine-cranking performance. Stationary and traction cells are discharged at a rate specified by the user, usually in the range of 1 to 10 h if stationary and 4 to 8 h for traction. The discharge for SLI batteries is usually done by dissipation through a fixed low-value resistor or by a brief, high-rate electrical discharge. Heavy-duty batteries are discharged through a resistor, a transistorized load, or an inverter.

The final manufacturing steps consist of improving the battery appearance by washing, drying, painting, installing vent plugs, and labeling. Product liability requirements in many countries mandate that the user be warned of the hazardous nature of the battery, especially that the electrolyte is corrosive and that gases are formed that can be explosive.

Traction batteries are physically sized to fit a myriad of different forklift trucks, and so the final assembly for a traction battery consists of inserting preformed and pretested cells into a sturdy metal box (tray), making intercell connections, making cable connections, and occasionally adding plastic (urethane) sealing material in the spaces between cell covers.

## 14.3.13   Shipping

Small batteries (SLI and golf cart types) are palletized several layers high for long-distance shipment. The batteries are cushioned between layers and are held laterally by banding or by a plastic sheet that is shrink- or stretch-wrapped around a full pallet. Pallets need to be very sturdy to withstand the battery weight and handling abuse.

There are strict regulations controlling the shipment of lead-acid batteries, which are more stringent for flooded rather than VRLA batteries, but if these regulations are followed, batteries may be shipped safely and economically. VRLA batteries may be shipped by air if packaged correctly.

## 14.3.14   Activation of Dry-Charged Batteries

When batteries have been dry-charged, they have to be reactivated before use. Activation consists of unpacking the battery, filling the cells with electrolyte (which sometimes is shipped with the battery in a separate package), charging the battery (if time is available), and testing the battery performance. When dry-charged batteries are activated, the materials that were used to seal the vent holes must be removed and discarded.

## 14.4   VALVE-REGULATED LEAD-ACID BATTERIES—GENERAL CHARACTERISTICS

The major difference between a VRLA battery and a conventional flooded battery or EFB is the separator component. The VRLA battery uses an AGM or gelled separator that allows the VRLA battery to operate with the oxygen recombination cycle. A VRLA battery has a one-way, pressure-relief valve designed to seal the cell unless its internal pressure exceeds a design maximum. The resealable valves are normally closed to prevent the entrance of oxygen from the outside air. VRLA batteries have two typical designs: a spiral-wound cell (jelly-roll construction) in a cylindrical container, and a flat-plate cell in a prismatic container. The cylindrical containers can maintain higher internal pressures without deformation and can be designed to have a higher release pressure than the prismatic cells. In some designs, an outer metal container is used to prevent deformation of the plastic cases at higher temperatures and internal cell pressures. The range of venting pressures includes a high of 25 to 40 psi for a metal-sheathed, spirally wound cell to 1 to 4 psi for a large prismatic battery.

The electrolyte is commonly immobilized in two ways:

- *Gelled electrolyte:* The liquid electrolyte is typically immobilized by mixing it with fumed or colloidal silica.
- *AGM:* In AGM batteries, the electrolyte is immobilized by absorption in highly porous nonwoven glass fiber separators between the positive and negative electrodes. AGM separators are made primarily from glass microfibers. The immobilization of the electrolyte allows batteries to operate in different orientations without spillage. In larger industrial applications, the batteries can be installed on their sides, permitting compact installations that use up to 40% less floor space and volume.

The VRLA battery has a lower amount of acid versus a flooded battery. As a result, VRLA designs are more prone to thermal runaway under abusive overcharge conditions that pose little hazard for flooded cells. This is particularly true when VRLA batteries are subjected to operations at elevated temperatures. Recently, high-temperature (over 40°C) effects and advantages and disadvantages have been discussed.[11,33] A detailed review of VRLA applications has been given in a series of articles.[34] To ensure long battery life under conditions of higher temperature operations or overheating due to oxygen recombination, precautions must be taken.

The major advantages and disadvantages of VRLA batteries are listed in Table 14.13.

**TABLE 14.13**   Major Advantages and Disadvantages of VRLA Batteries

| Advantages | Disadvantages |
|---|---|
| Maintenance-free | Should not be stored in discharged condition |
| Excellent life on float service | Relatively low energy density |
| High-rate capability | Thermal runaway can occur with incorrect charging or improper thermal management |
| No "memory" effect (compared to nickel-cadmium battery) | More sensitive to higher temperature environments than conventional lead-acid batteries unless designed for high temperature service |
| "State of charge" can usually be determined by measuring voltage |  |
| Relatively low cost |  |
| Available from small single-cell units (2 V) to large 12 V monoblocs |  |
| Certain designs can be installed on their side, simplifying maintenance |  |

### 14.4.1   VRLA Cell Construction—Cylindrical Cells

Many design and material changes were necessary to develop the VRLA battery.[35] A cross section of a VRLA cylindrical cell and a breakdown of the basic components contained in the cell are shown in Figs. 14.23 and 14.24. Both the positive and the negative grids are made from 99.99% pure lead with 0.6% tin added for deep-discharge recovery. The lead grid is relatively thin, 0.6 to 0.9 mm, to provide for a high plate surface area and resultant high discharge rates.

### 14.4.2   VRLA Cell Construction—Prismatic Cells

A cutaway view of a prismatic lead-acid cell is shown in Fig. 14.25*a*. An exploded view of a three-cell monobloc battery, using this prismatic cell, is given in Fig. 14.25*b*.

### 14.4.3   Performance Characteristics

*Open Circuit Voltage.*   The open-circuit voltage relationships to the state of charge and the concentration of electrolyte are analogous to those previously described for flooded lead-acid batteries.

*Discharge Characteristics.*   The discharge voltage profiles of typical VRLA single-cell batteries, at various temperatures ranging from −40 to 65°C for various discharge rates, are shown in Fig. 14.26 (see Table 14.14 for capacity data on various size cells). Figure 14.27 shows a set of discharge voltage curves for high rates of discharge at 25°C.

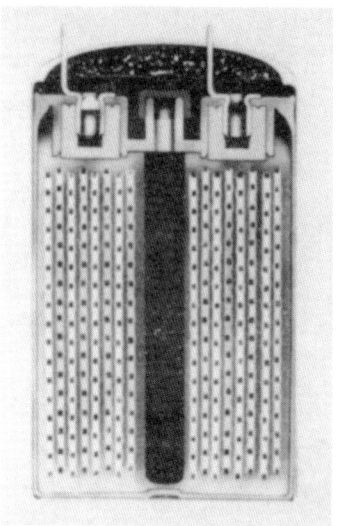

**FIGURE 14.23** Cross section of the VRLA cell. Components identified in Fig. 14.24. (*Courtesy of EnerSys Energy Products, Inc., formerly Hawker Energy Products, Inc.*)

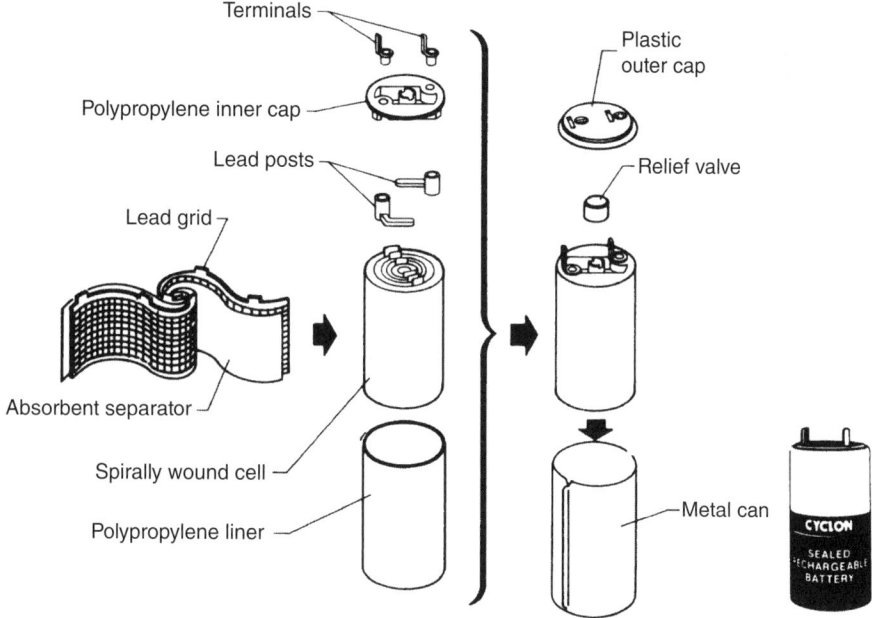

**FIGURE 14.24**   Components of the VRLA cell. (*Courtesy of EnerSys Energy Products, Inc., formerly Hawker Energy Products, Inc.*)

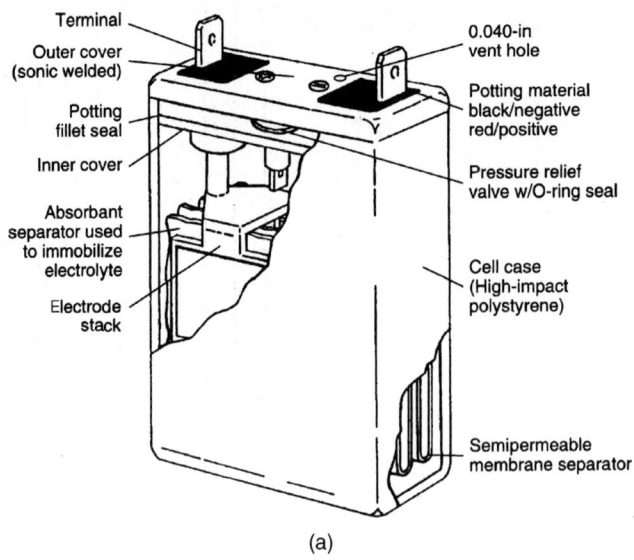

(a)

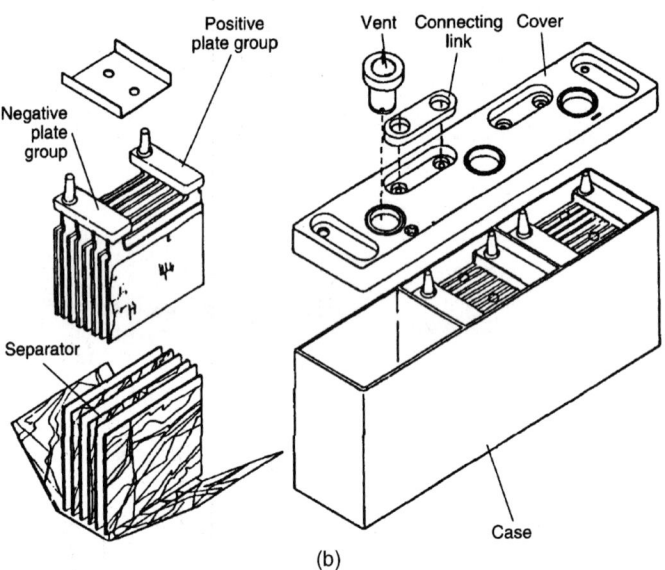

(b)

**FIGURE 14.25** Typical prismatic lead-acid cell and monobloc battery. (*a*) Cutaway view. (*Courtesy of Eagle-Picher Industries, Inc.*) (*b*) Exploded view. (*Courtesy of Johnson Controls, Inc.*)

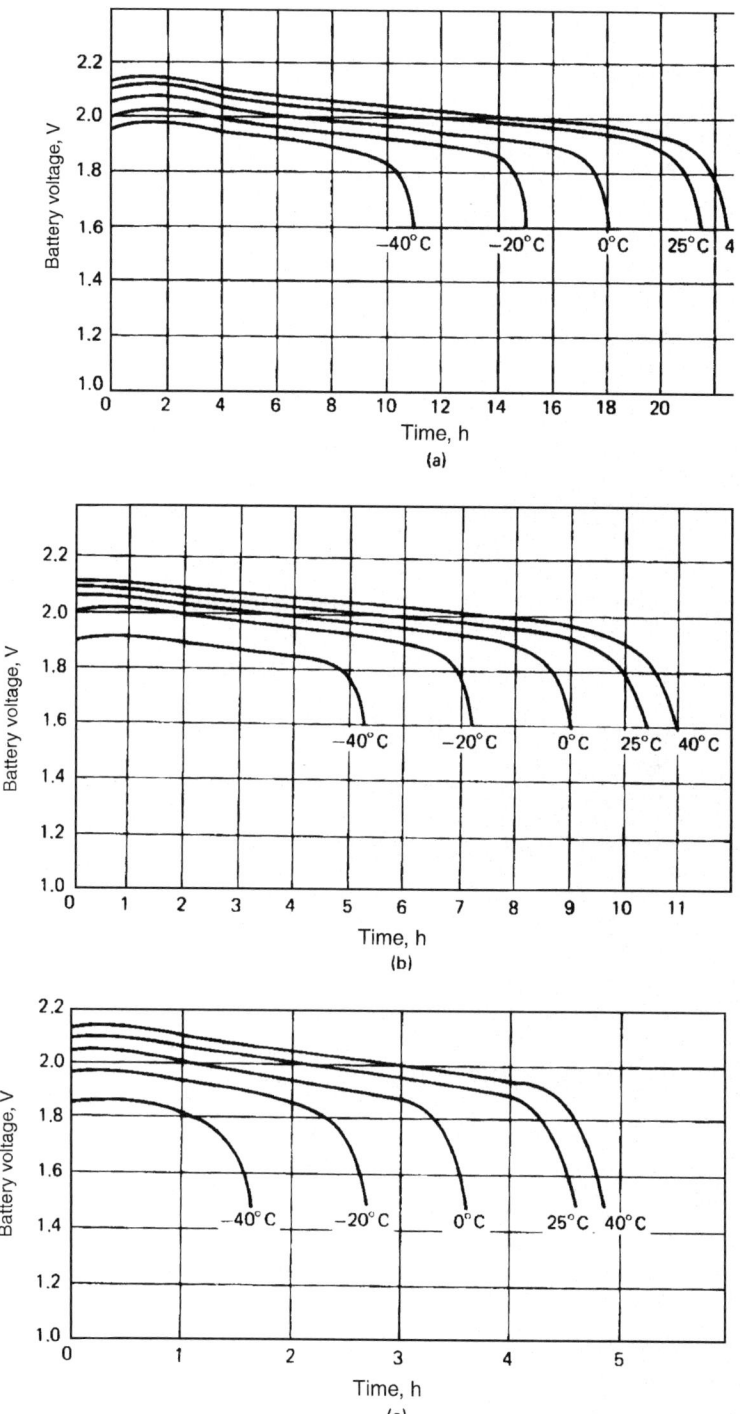

**FIGURE 14.26**  Discharge curves of cylindrical VRLA D and X single-cell batteries. Discharge rate (*a*) at *C*/20, (*b*) at *C*/10, (*c*) at *C*/5, (*d*) at *C*/12.5, (*e*) at 1*C* (see Table 14.14 for capacity data).

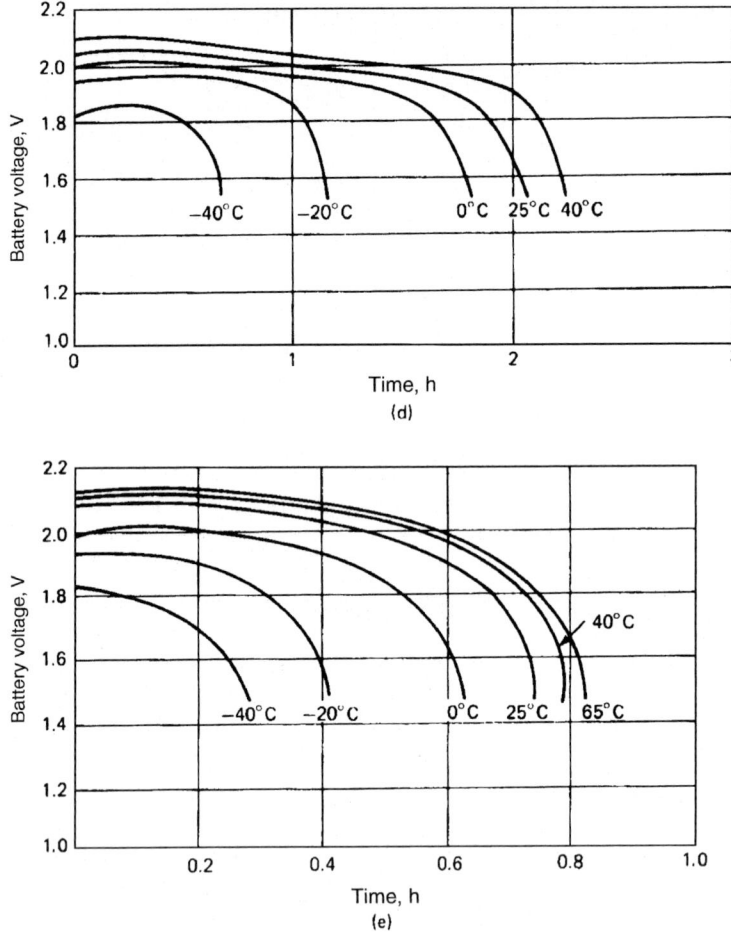

**FIGURE 14.26** (*Continued*)

***High–Rate Discharge.***  The capacity of the VRLA battery is increased greatly when an intermittent pulse discharge is used. The voltage-time curves for a battery at room temperature at the 10C discharge rate, both on continuous discharge and for a 16.7% duty cycle (10 s pulse, 50 s rest), are shown in Fig. 14.28 for 25 and −20°C.

This is an evidence for the phenomenon known as "concentration polarization." As a discharge current is drawn from the cell, the sulfuric acid in the electrolyte reacts with the active materials in the electrodes. This reaction reduces the concentration of the acid at the electrode-electrolyte interfaces. Consequently, the cell voltage drops. During the rest period, the acid in the bulk of the solution diffuses into the electrode pores to replace the acid that has been depleted. During a pulse discharge, the acid is not depleted as quickly, and the total cell capacity is increased.

The curves in Fig. 14.29 illustrate the maximum power that can be delivered as a function of the discharge rate at room temperature and at −20°C. The maximum power obtainable increases as the temperature increases.

**TABLE 14.14**   VRLA Cylindrical Batteries

| Model | Capacity, Ah C/10 | Capacity, Ah C/20 | Capacity, Ah 1C | Dimensions, mm Height | Dimensions, mm Diameter, width | Dimensions, mm Length | Weight (typical), g | Specific energy @C/20 Wh/kg | Specific energy @C/20 Maximum discharge, A |
|---|---|---|---|---|---|---|---|---|---|
| | | | | | Single cells | | | | |
| D | 2.5 | 2.7 | 1.8 | 67.3 | 34.3 | N/A | 180 | 30.0 | 40 |
| X | 5.0 | 5.4 | 3.2 | 80.3 | 49.5 | N/A | 390 | 27.6 | 40 |
| J | 12.5 | 13.0 | 9.0 | 135.7 | 51.8 | N/A | 840 | 30.8 | 60 |
| BC | 25.0 | 26.0 | 17.5 | 172.3 | 65.3 | N/A | 1580 | 32.9 | 250 |
| DT | 4.5 | 4.8 | 3.7 | 102.9 | 34.3 | N/A | 272 | 35.3 | 40 |
| E | 8.1 | 8.4 | 6.2 | 108.7 | 44.5 | N/A | 549 | 30.6 | 40 |
| | | | Monobloc batteries (preassembled batteries in common sizes) | | | | | | |
| D, 4 V | 2.5 | 2.7 | 1.8 | 70 | 45 | 78 | 360 | | 40 |
| D, 6 V | 2.5 | 2.7 | 1.8 | 70 | 45 | 113 | 540 | | 40 |
| X, 4 V | 5.0 | 5.4 | 3.2 | 77 | 54 | 96 | 740 | | 40 |
| X, 6 V | 5.0 | 5.4 | 3.2 | 77 | 54 | 139 | 1110 | | 40 |
| E, 4 V | 8.0 | 8.6 | 5.8 | 102 | 54 | 96 | 1110 | | 40 |
| E, 6 V | 8.0 | 8.6 | 5.8 | 102 | 54 | 139 | 1670 | | 40 |

*Source:* EnerSys Energy Products, Inc.

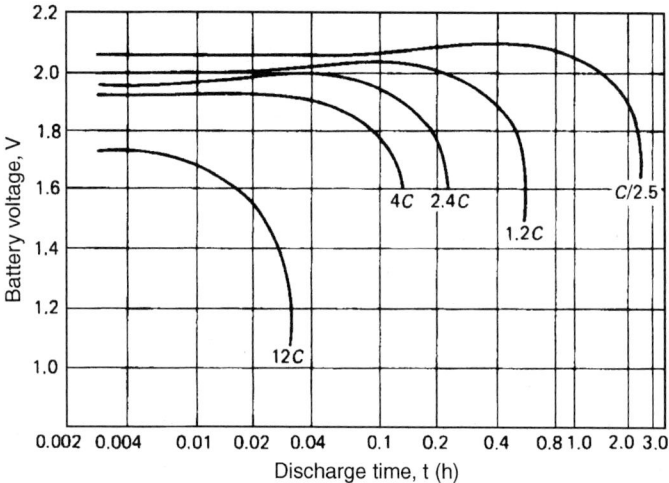

**FIGURE 14.27**   Discharge curves of VRLA cylindrical D and X batteries at 25°C, high discharge.

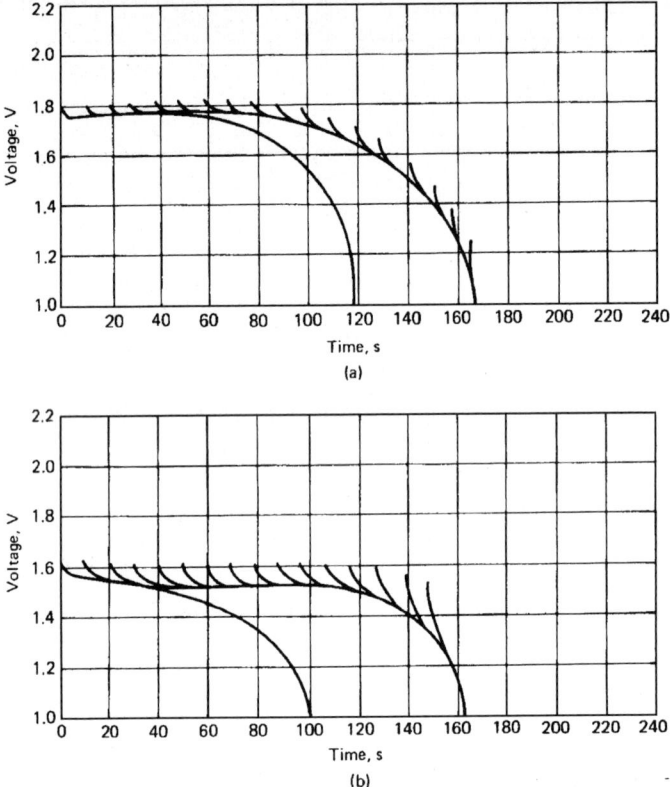

**FIGURE 14.28** High-rate pulse performance at 10C discharge rate for VRLA cylindrical units. (a) 25°C. (b) −20°C. The upper curves are pulsed, while the lower curves are continuous discharges.

***Discharge Level.*** Discharging the VRLA battery beyond the point at which 100% of the capacity has been removed can shorten the life of the battery or impair its ability to accept a charge.

The voltage point at which 100% of the usable capacity of the cell has been removed is a function of the discharge rate, as shown in the upper envelope of the curve of Fig. 14.30. The lower curve shows the minimum voltage level to which the battery may be discharged with no effect on recharging capability. For optimum life and charge capability, the cell should be disconnected from the load at the voltages within the gray area between the two curves.

Under overdischarge conditions, the sulfuric acid electrolyte can be depleted of the sulfate ion and become mainly water, which can create several problems. A lack of sulfate ions as charge conductors will increase the cell resistance and limit charge current. Longer charge times or alteration of the charge voltage may be required before normal charging may resume.

Another potential problem in a severe deep-discharge condition is that the lead sulfate present at the plate surfaces can go into solution in the aqueous electrolyte. Upon recharge, the water and the sulfate ion in the lead sulfate convert to sulfuric acid, leaving a precipitate of lead metal in the separator. This lead metal can result in dendritic short circuits between the plates and subsequently cell failure.

***Storage Characteristics.*** Most batteries lose their stored energy when allowed to stand on open circuit due to the fact that the active materials are in a thermodynamically unstable state.[36] The rate of self-discharge is dependent on the chemistry of the system and the temperature at which it is stored.

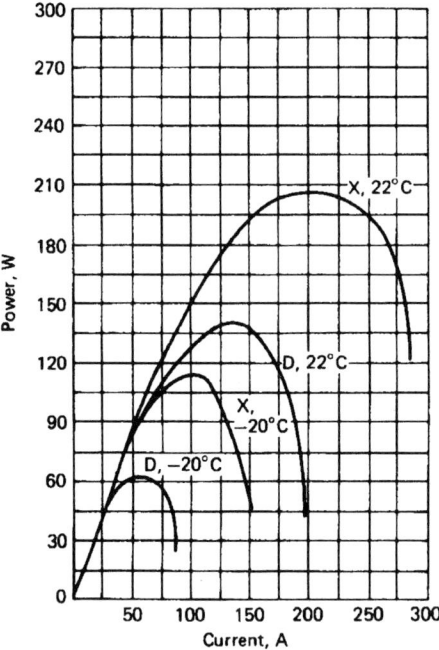

**FIGURE 14.29**   Instantaneous maximum peak power of VRLA batteries at 22 and −20°C.

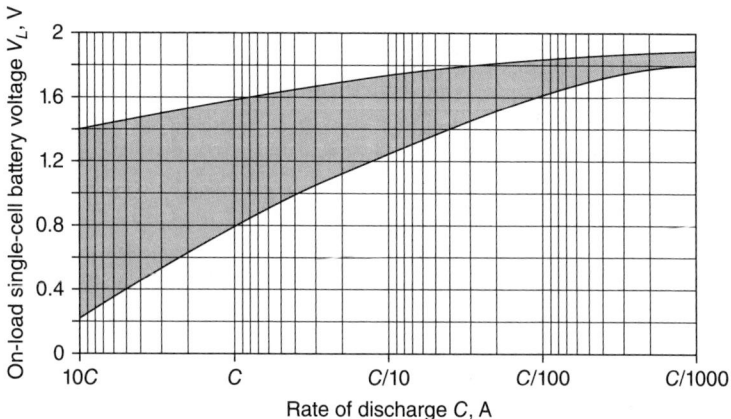

**FIGURE 14.30**   Acceptable voltage discharge levels of VRLA cells.

It is important to recognize that the self-discharge rate of the VRLA battery is nonlinear; thus, the rate of self-discharge changes as the state of charge of the cell changes. When the cell is in a high state of charge, that is, 80% or greater, the self-discharge is very rapid. The cell may discharge from 100% to 90% at room temperature in a matter of a week or two. Conversely, at the same temperature it may take 10 weeks or longer for the same cell to self-discharge from 20% state of charge down to 10% state of charge.

Figure 14.31 is a curve of the remaining usable capacity in a VRLA battery versus months of storage at various temperatures. This curve is convenient in determining the approximate remaining capacity after a given storage time at a particular temperature.

*Cycle Life.* The life of all the VRLA batteries is variable, depending on the type of use, environment, cycling, and charge to which the battery is subjected during its life. Figure 14.32 shows a curve of capacity versus cycles for a 2.5 Ah cell (2.35 Ah at 5 h rate) cycled at 1 cycle per day at a $C/5$ discharge rate to 1.6 V per cell and an 18 h charge at 2.5 V constant voltage. The cell takes from 20 to 25 cycles to achieve rated capacity, exceeds rated capacity, and then begins to fall off slowly. The initial increase in capacity is a function of forming the cell.

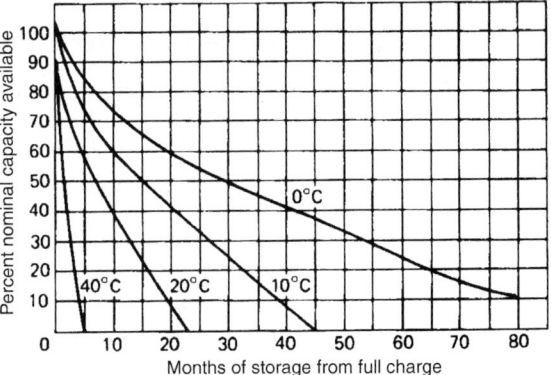

**FIGURE 14.31** Remaining usable capacity of VRLA batteries after storage.

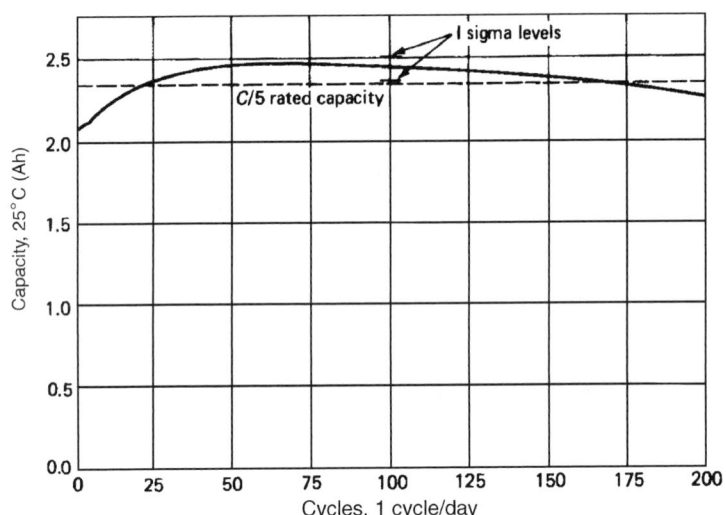

**FIGURE 14.32** Effect of cycling on cell capacity, VRLA cell, $C/5$ discharge rate. Rated at 2.35 Ah at $C/5$ rate.

Proper charging has an important effect on cycle life. Undercharging leads to failure to fully charge the battery and progressive loss of capacity. Overcharging of a battery leads to positive paste softening and excess water loss, again leading to shortened cycle life.

Figure 14.33 shows the general effect of the depth of discharge (DOD) on the cycle life; typically, at 100% DOD, 200 cycles are characteristic. It demonstrates that high cycle life can be achieved by slightly oversizing the battery for the application to reduce the DOD.

*Float Life.* The expected float life of the VRLA battery is greater than 8 years at room temperature, arrived at by using accelerated testing methods, specifically, at high temperatures.

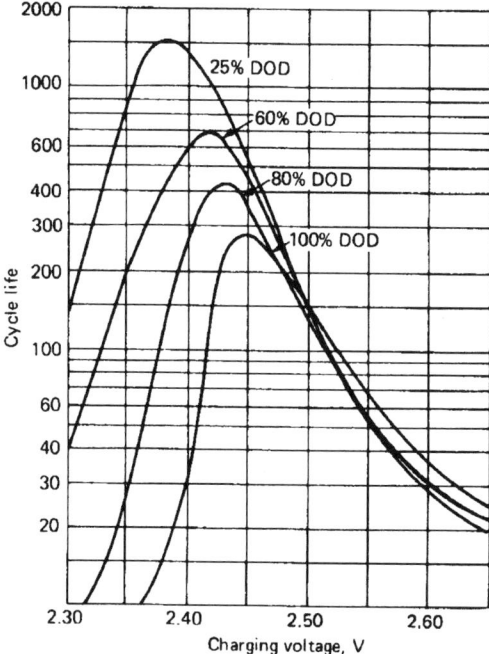
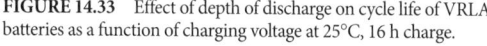

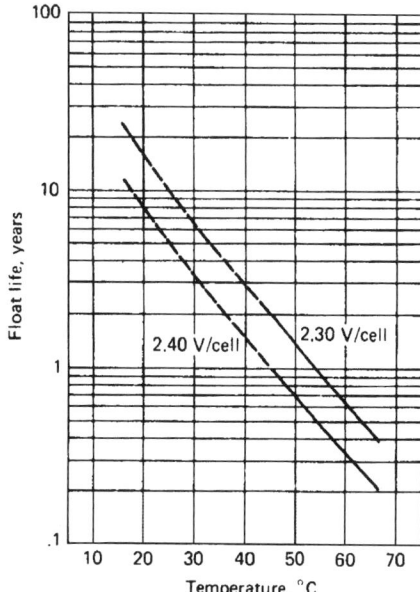

**FIGURE 14.33**  Effect of depth of discharge on cycle life of VRLA batteries as a function of charging voltage at 25°C, 16 h charge.

**FIGURE 14.34**  Float life of VRLA batteries.

The primary failure mode of the VRLA battery can be defined as growth of the positive plate. Because this growth is the result of chemical reactions within the cell, the rate of growth increases with increasing temperature. In Fig. 14.34, the float life is plotted against temperature. The solid lines represent data from float-life tests performed at two float voltages, 2.3 and 2.4 V per cell. This graph can be used to determine the expected float life at various temperatures. End of life is defined as the failure of the cell to deliver 80% of rated capacity.

### 14.4.4  Battery Types and Sizes

A listing of VRLA cylindrical batteries is given in Table 14.14. The performance of these units at various current drains at 25°C is given in Fig. 14.35. A number of multicell batteries are available that use these cells in various series/parallel configurations.

Tables 14.15 and 14.16 list some of the typical VRLA prismatic lead-acid batteries that are manufactured. Information on other manufacturers' products can be obtained by consulting their websites. The principal suppliers include C&D Technologies, East Penn, EnerSys, Exide Technologies, Exide Industries, NorthStar Battery, Johnson Controls, Maura, Hitachi, Furukawa, Narada, Leoch, ChinaShoto, Coslight, and GS Yuasa. Unlike some of the other types of lead-acid batteries, there is no standardized list of sizes. Hence, sizes, weights, and capacity ratings may vary from manufacturer to manufacturer.

The products described above, and similar products from other manufacturers, have been specifically engineered for use in indoor/outdoor cabinets, have a flame-retarded case and cover, and can be installed in any orientation other than inverted, which is not recommended.

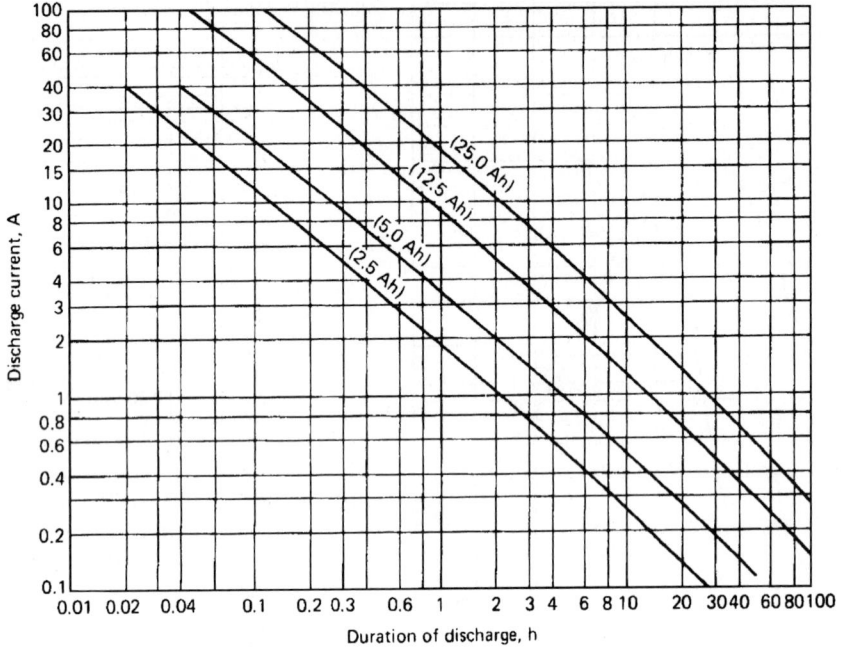

**FIGURE 14.35**   Discharge times of four types of VRLA cylindrical batteries at 25°C.

**TABLE 14.15**   Typical VRLA Batteries for Telecom Applications

|  | NSB40 | NSB70 | NSB75 | NSB90 | NSB125 |
|---|---|---|---|---|---|
| Height | 176 mm 6.93 in | 176 mm 6.93 in | 200 mm 7.87 in | 213 mm 8.39 in | 275 mm 10.81 in |
| Length | 197 mm 7.76 in | 331 mm 13.02 in | 261 mm 10.27 in | 341 mm 13.42 in | 345 mm 13.57 in |
| Width | 165 mm 6.50 in | 165 mm 6.50 in | 173 mm 6.80 in | 173 mm 6.80 in | 173 mm 6.80 in |
| Weight | 16.0 kg 35.3 lbs | 27.3 kg 60.0 lbs | 27.3 kg 60.0 lbs | 37.8 kg 83.1 lbs | 54.0 kg 119 lbs |
| Terminal | M6 × 1.25 | M6 × 1.25 | M6 × 1.25 | M6 × 1.25 | M6 × 1.25 |
| C/10 Cap | 40 Ah | 66 Ah | 69 Ah | 96 Ah | 129 Ah |
| Impedance (1 kHz) | 4.5 mΩ | 2.7 mΩ | 2.6 mΩ | 2.0 mΩ | 2.0 mΩ |
| Conductance @ 25°C (77°F) | 1052 S | 1589 S | 1398 S | 1806 S | 2103 S |
| Short-circuit current | 2000 A | 3200 A | 3200 A | 4300 A | 5000 A |

*(Courtesy of NorthStar Battery Company.)*

**TABLE 14.16**  Specifications for Typical Pure Lead-Tin VRLA Batteries

| Product (capacity) | Internal res. of fully charged battery, mΩ @ 25°C | Nominal short-circuit current for charged battery | Dimensions | | | Weight, lb. (kg) |
|---|---|---|---|---|---|---|
| | | | Length, in (mm) | Width, in (mm) | Height, in (mm) | |
| G13EP (13 Ah) | 8.5 | 1400 A | 6.910 (175.51) | 3.282 (83.36) | 5.113 (129.87) | 10.8 (4.9) |
| G13EPX (13 Ah) | 8.5 | 1400 A | 6.998 (177.75) | 3.368 (85.55) | 5.165 (131.19) | 12.0 (5.4) |
| G16EP (16 Ah) | 7.5 | 1600 A | 7.150 (181.61) | 3.005 (76.33) | 6.605 (167.77) | 35.5 (6.1) |
| G16EPX (16 Ah) | 7.5 | 1600 A | 7.267 (184.58) | 3.107 (78.92) | 6.666 (169.32) | 14.7 (6.7) |
| G26EP (26 Ah) | 5.0 | 2400 A | 6.565 (166.75) | 6.920 (175.77) | 4.957 (125.91) | 22.3 (10.1) |
| G26EPX (26 Ah) | 5.0 | 2400 A | 6.636 (168.55) | 7.049 (179.04) | 5.040 (128.02) | 23.8 (10.8) |
| G42EP (42 Ah) | 4.5 | 2600 A | 7.775 (197.49) | 6.525 (165.74) | 6.715 (170.56) | 32.9 (14.9) |
| G42EPX (42 Ah) | 4.5 | 2600 A | 7.866 (199.80) | 6.659 (169.14) | 6.803 (172.80) | 35.1 (15.9) |
| G70EP (70 Ah) | 3.5 | 3500 A | 13.020 (330.71) | 6.620 (168.15) | 6.930 (176.02) | 53.5 (24.3) |
| G70EPX (70 Ah) | 3.5 | 3500 A | 13.020 (330.71) | 6.620 (168.15) | 6.930 (176.02) | 56.0 (25.4) |

*Source:* EnerSys Energy Products, Inc.

## 14.5  LEAD-CARBON AND OTHER ADVANCED DESIGNS

Over the past decade, new applications have emerged in which VRLA batteries are frequently cycled but are seldom, if ever, fully charged or discharged. Batteries in HRPSoC applications satisfy many needs, including energy for braking and acceleration in hybrid electric vehicles (HEVs), fast-charge motive applications, energy storage in PV and wind generators, and grid energy storage. Most of these applications require a battery that is able to accept a high-rate charge and deliver a high-rate discharge.

In traditional applications, the lead-acid battery positive electrode has typically been the first to fail. In HRPSoC cycling, lead sulfate crystals grow progressively larger on the negative plates, increasing the battery's resistance and decreasing its power and charge acceptance leading to rapid failure.

### 14.5.1  Carbon-Enhanced Designs

A large research effort in the industry has led to new designs of battery, both flooded and VRLA, that are able to meet the requirements. The key to this has been the development of special carbons that improve the behavior of the negative plates.

There are a number of ways in which carbon can modify the performance of the negative plate of a lead-acid battery: (i) by capacitive effects, (ii) by extending the surface area on which the electrochemical charge and discharge processes take place, and (iii) by physical processes. Capacitive effects are favored by carbons that have large specific surface areas and are in good contact with the grid current collector and the spongy lead matrix of the active mass. The carbon does not, however, need to be intimately mixed with the sponge lead. For surface area effects, the carbon is also required to be conductive and in contact with the current collector, but since the carbon promotes bulk rather than surface processes, the surface area can be less than the capacitive process. Where carbon is used to take advantage of physical processes, it does not have to be conductive, but it does have to be very intimately mixed with the sponge lead and the particle size sufficient for its function not to be reduced over time. To the extent that these requirements are conflicting and because there are a huge variety of different carbon materials, a lot of research has been directed at optimizing combinations of different carbons but much basic work remains to be done to elucidate the fundamental mechanisms.[37]

### 14.5.2   Carbon Negative Current Collectors

Some or all of the metallic parts of the negative grid may potentially be replaced with carbon. Various concepts have been studied with rigid carbon foams, lead electroplated graphite foil, and flexible carbon felts. The rigid carbon foams had outstanding life and active mass utilization, but the brittle nature of these materials made manufacture problematic. Graphite foil, electroplated with lead, had a low level of utilization but high durability in HRPSoC cycling, suggesting that lead sulfate formation was inhibited. A more promising concept has been developed by ArcActive in New Zealand in which the lead grid has been replaced with a carbon felt activated by treatment with an electric arc under controlled conditions. The felt is impregnated with active material and attached to lead alloy current collectors. This construction shows excellent behavior in HRPSoC operation. To date it has been developed for automotive applications, but it has good potential for energy-storage applications in larger formats especially as the high-rate capability required for automotive service is not critical for most energy-storage duty cycles.

### 14.5.3   Carbon Negative Electrodes

It is possible to substitute the NAM entirely with carbon. In this case, the electrode has no faradaic energy storage but functions as a capacitor, and when carbon materials with suitable physical form and high specific surface area are used, the electrode becomes a supercapacitor. This electrode can be paired with a conventional lead dioxide positive plate to make an asymmetric supercapacitor. The energy density of this type of a device is low compared to a lead-acid battery, and it has a much more steeply sloping discharge voltage. But this cell offers a very long cycle life and can also be recharged rapidly. This concept has been developed by Axion Power in the United States as an energy-storage system.

### 14.5.4   Supercapacitor/Battery Hybrids

It is also possible to combine a carbon-based supercapacitor with a conventional negative electrode to produce a composite negative electrode. When this electrode is used with a standard positive plate, the hybrid construction offers substantially improved behavior in HRPSoC cycling. The two components of the negative electrode are connected together in parallel, and the capacitor part of the electrode acts as a buffer to share current with the negative plate and reduce the rate of charge and discharge, as shown in Fig. 14.36.[38]

This design is available commercially as the UltraBattery® and offers important advantages over both conventional lead-acid batteries and asymmetric lead-based supercapacitors. The advantages include (i) the avoidance of irreversible sulfation of the negative plate in HRPSoC cycling and the need for intermittent conditioning cycles where the battery is charged for an extended period, (ii) improved high-rate charge acceptance, (iii) better self-balancing of cells in series strings, and (iv) an energy density and voltage profile on discharge in line with a lead-acid battery. The UltraBattery has lasted over 100,000 miles in a Honda Insight® HEV in a road test at the GM Millbrook Proving Ground near London, in a test sponsored by the ALABC. The use of components from the capacitor and lead battery industries has made this design relatively easy to manufacture at a low cost. The UltraBattery has been licensed and a manufacturing process has been developed by Furukawa. East Penn Manufacturing Company has also licensed the design. Cost estimates are about 70% of the cost of the nickel

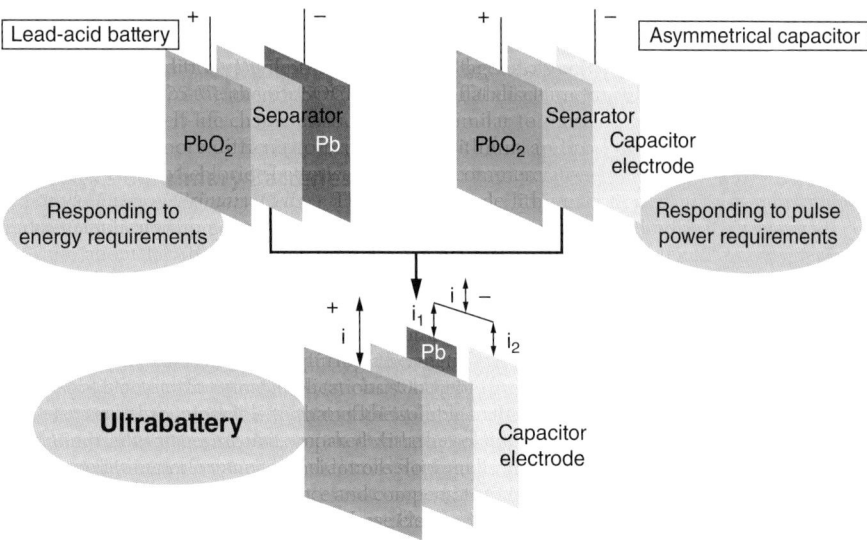

**FIGURE 14.36**  Diagram of the UltraBattery® design. (*From Ref. 38.*)

batteries used in HEVs. These new approaches to HEV propulsion may provide an opportunity within the next few years to develop a mass market in HEVs.

### 14.5.5 Bipolar Lead-Acid Batteries

Bipolar constructions have been researched over many years, and more recently a number of concepts are showing greater promise of technical and commercial success. In a bipolar battery, apart from the end-plates, the electrodes have one side operating as the positive and the other as the negative separated by a solid plate that is impervious to ion flow, electronically conductive, and corrosion resistant. For lead-acid batteries, selection of the bipolar plate is the key along with a reliable edge seal around the plate that isolates the electrodes on either side. The use of porous alumina impregnated with lead has been attempted without success. Conductive titanium suboxides (Ebonex) incorporated in resin fabricated into thin sheets has been extensively examined by Atraverda but have not been commercialized. Ebonex has reasonable electronic conductivity and is inert in a lead-acid cell environment; but as a membrane, the resistance is relatively high. Silicon is also a candidate and although it is a semiconductor, it can be made sufficiently conductive to operate as a membrane in a bipolar lead-acid battery. This concept is being developed by Gridtential in the United States. Lead sheet is an excellent membrane, provided that it is sufficiently corrosion resistant, and Advanced Battery Concepts have a design which uses a polymer support for the lead sheet. Battery performance data for this design show good results. A successful bipolar lead-acid design would offer an attractive energy-storage battery.

## 14.6  CHARGING AND CHARGING EQUIPMENT

### 14.6.1  General Considerations

In the charging process, DC electric power is used to reform the active chemicals of the battery system to their high-energy, charged state. In the case of the lead-acid battery, this involves the conversion of lead sulfate in the positive electrodes to lead oxide ($PbO_2$), the conversion of lead sulfate of the negative electrode to metallic lead (sponge lead), and the restoration of the electrolyte from a low-concentration sulfuric acid solution to the higher concentration of approximately 1.21 to 1.30 specific gravity. Since a change of phase from

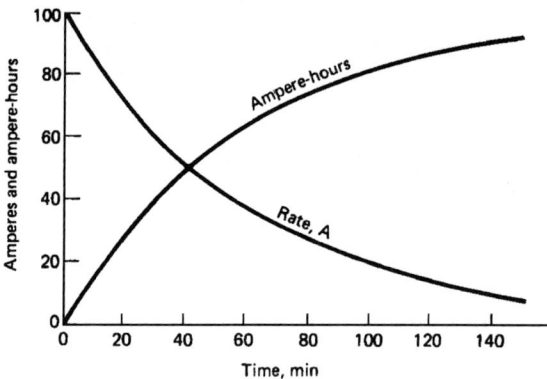

**FIGURE 14.37** Graphic illustration of ampere-hour law. (*From Ref. 10.*)

solid to solution is involved with the sulfate ion, charging lead-acid batteries has mass diffusion considerations and is quite temperature sensitive. During charge and discharge the solid materials that go into solution as ions are reprecipitated as different solid compounds. This causes a redistribution of the active material. The rearrangement will tend to make the active material contain a crystal structure with fewer defects, which can result in less electrochemical activity. Therefore, the lead-acid battery is not as reversible physically as it is chemically.[39] This physical deactivation can be minimized by proper charging and paste additives.

Charging of lead-acid batteries and other aqueous electrolyte batteries are complicated by the secondary reactions of water electrolysis. A lead-acid battery can generally be charged at any rate that does not produce excessive gassing, overcharging, or high temperatures. The battery can absorb a very high current during the early part of the charge, but there is a limit to the safe current as the battery becomes charged due to increased gassing. This is shown in Fig. 14.37, which is a graphic representation of the ampere-hour rule,

$$I = Ae^{-t}$$

where $I$ is the charging current, $A$ is the number of ampere-hours previously discharged from the battery, and $t$ equals time. Because there is considerable latitude, there are a number of charging regimes, and the selection of the appropriate method depends on a number of considerations, such as the type and design of the battery, service conditions, time available for charging, number of cells or batteries to be charged, and charging facilities. Figure 14.38 shows the relation of cell voltage to the state of charge and the charging current. The figure shows that a fully discharged battery can absorb high currents with the charging voltage remaining relatively low. However, as state of charge increases, the voltage increases to excessively high values if the charge is maintained at the high rate, leading to overcharge, gassing, and lower life. For fastest charge, the initial current is high but then the current should be reduced to reasonable values as the battery reaches full charge.

In automotive, marine, and other vehicle applications, the DC electric power is usually provided by an onboard generator or alternator driven by the prime engine. These devices have a voltage and current

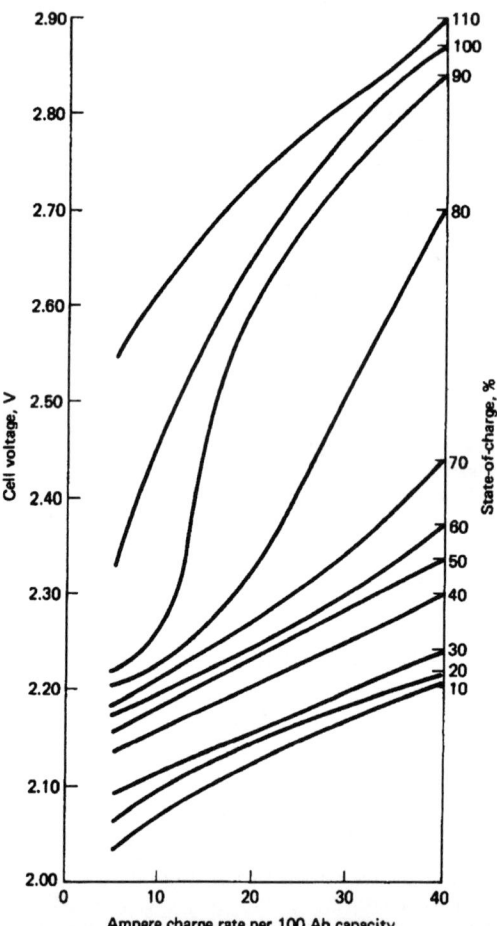

**FIGURE 14.38** Charging voltage of lead-acid battery at various states of charge. (*From Ref. 40.*)

limiter to prevent overcharging. The proper limit is dependent on the chemistry and physical construction of the cell or battery. For the traditional automotive batteries that used antimonial lead alloy as grid material, voltage limits in the range of 14.1 to 14.6 V for a nominal 12 V battery are typical. With current maintenance-free batteries, which use a calcium-lead alloy grid or other grid material with high hydrogen overvoltage, higher charging voltages, in the range of 14.5 to 15.0 V, can be used without the danger of overcharge. Batteries in automobile and similar applications today see what is close to cycling rather than float service, but the charging controls are such that very little gas is evolved on charge. This minimizes the requirement for watering, but makes accurate control of the charge necessary.

In many nonautomotive applications, charging is done separately from the system using the battery. The direct current necessary for charging is usually obtained by rectifying alternating current. These chargers include wall-hung units and mobile units as well as floor-mounted units. Newer charger designs have microprocessor controls, can sense battery condition, temperature, voltage, charge current, and so on, and are capable of changing charging rates during the charge. Most rectifiers produce some AC ripple with the direct current, which causes additional heating of the battery. This should be minimized, especially near the end of the charge when batteries tend to get hot. Pulse charging and the use of asymmetric alternating current have been proposed as a means to overcome this problem, but practical lead-acid batteries have such large capacitances that the pulses are smoothed out and the effects minimized.[38]

### 14.6.2    Charging VRLA Batteries

Charging a VRLA battery, like charging flooded lead-acid batteries, is a matter of replacing the energy depleted during discharge. Because of gassing and recombination inefficiency, it is necessary to return more than 100% of the energy removed during discharge.[41] The amount of energy necessary for recharge depends on how deeply the battery was discharged, the method of recharge, the recharge time, and the temperature.[42] In high-temperature environments, the charge current or voltage should be controlled by the battery temperature.[43] The overcharge required in the lead-acid battery is associated with the generation of gases and corrosion of the positive-grid materials. In conventional flooded lead-acid batteries, the gases generated are released from the system, resulting in a loss of water. Overcharging of flooded batteries is not a major concern since the lost water can be replaced (except in a sealed battery). The VRLA battery is more sensitive to overcharge since the lost water (although greatly reduced) cannot be replaced and dry-out of the VRLA battery is a life limiting factor. The loss of water in a VRLA battery can lead to a greater chance of thermal runaway.

### 14.6.3    Methods of Charging Lead-Acid Batteries

Proper recharging is important to obtain optimum life from any lead-acid battery under any conditions of use. Some of the rules for proper charging are given below and apply to all types of lead-acid batteries.

1. The charge current at the start of recharge can be any value that does not produce an average cell voltage in the battery string greater than the gassing voltage (about 2.4 V per cell).

2. During the recharge and until 100% of the previous discharge capacity has been returned, the current should be controlled to maintain a voltage lower than the gassing voltage. To minimize charge time, this voltage can be just below the gassing voltage.

3. When 100% of the discharged capacity has been returned under this voltage control, the charge rate will have normally decreased to the charge "finishing" rate. The charge should be finished at a current no higher than this rate, normally about 5 A/100 Ah of capacity (at the 20 h rate).

A number of methods for charging lead-acid batteries have evolved to meet these conditions. The most effective for the majority of applications are those based on constant voltage. These charging methods are commonly known as:

1. Constant voltage and modifications
2. Constant current and modifications
3. Taper charge

**4.** Float charging

**5.** Pulse charging

**6.** Trickle charging

**7.** Rapid charging

***Constant-Voltage Charging.***     In constant-voltage charging, the charge is limited to a voltage generally below the gassing voltage. Early in the charge, the voltage is below the limit and current flows at the maximum of the charger capability. When the voltage limit is reached, the current decays. Figure 14.39 shows the recharge times at various charge voltages for a VRLA cell discharged to 100% of capacity. The charger required to achieve these times at given voltages must be capable of at least the 2*C* rate. If the constant-voltage charger used has less than the 2*C* rate of charge capability, the charge times should be lengthened by the hourly rate at which the charger is limited; that is, if the charger is limited to the *C*/10 rate, then 10 h should be added to each of the charge voltage-time relationships. If the charger is limited to the *C*/5 rate, then 5 h should be added, and so on. There are no limitations on the maximum current imposed by the charging characteristics of the battery.

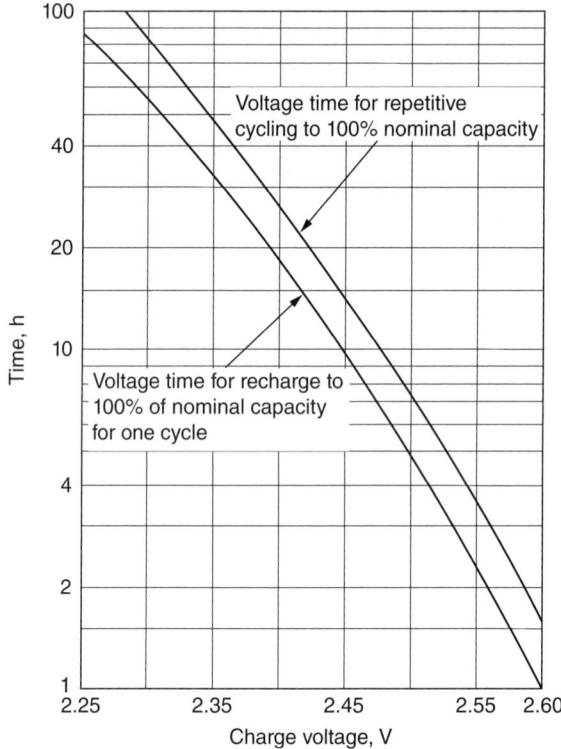

**FIGURE 14.39**   Charge voltage versus time on charge at 25°C for a VRLA battery.

Figure 14.40 shows a set of curves of charge current versus time for 2.5 Ah VRLA batteries charged by a constant voltage of 2.45 V with chargers limited to 2, 1, and 0.3 A currents. As shown, the only difference in these three charges is the length of time necessary to recharge the battery. The time for the current decay is lengthy, leading to modifications that shorten the total charge time.

In normal industrial applications, modified constant-voltage charging methods are used. Modified constant-voltage charging is used for on-the-road vehicles and utility, telephone, and uninterruptible power system

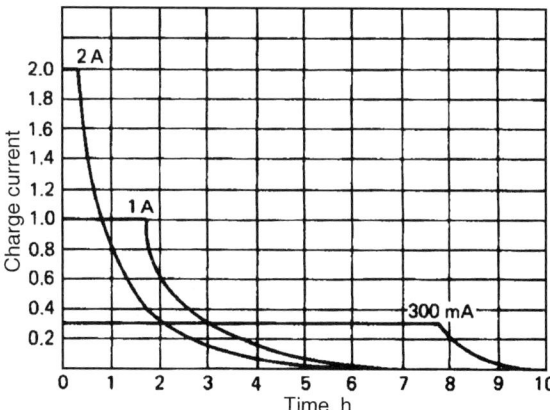

**FIGURE 14.40**  Charge current versus time at 2.45 V constant voltage with various current limits (2.5 Ah VRLA battery, C/10 rate).

applications where the charging circuit is tied to the battery. In this case, the charging circuit has a current limit, and this value is maintained until a predetermined voltage is reached. Then the voltage is maintained constant until the battery is called on to discharge. Decisions must be made regarding the current limit and the constant-voltage limit. This is influenced by the time interval when the battery is at the constant voltage and in a 100% state of charge. For this "float"-type operation with the battery always on charge, a low charge current is desirable to minimize overcharge, grid corrosion associated with overcharge, water loss by electrolysis of the electrolyte, and maintenance to replace this water. To achieve a full recharge with a low constant potential requires proper selection of the starting current, which is based on the manufacturer's specifications.

The modified constant-voltage charge, with constant-current start and finish rates, is common for deep-cycling batteries that are typically discharged at the 6 h rate to a depth of 80%; the recharge is normally completed in an 8 h period. The charger is set for the constant potential of 2.39 V per cell (the gassing voltage), and the starting current is limited to 16 to 20 A/100 Ah of rated capacity based on the 6 h discharge rate. This initial current is maintained constant until the average cell voltage in the battery reaches 2.39 V. The current decays at a constant voltage to the finishing rate of 4.5 to 5 A/100 Ah, which is then maintained to the end of the charge. Total charge time is controlled by a timer. The time of charge is selected to ensure a recharge input capacity of a predetermined percent of the ampere-hour output of the previous discharge, normally 110% to 120%, or 10% to 20% overcharge. The 8 h charging time can be reduced by increasing the initial current limit rate.

***Constant-Current Charging.***    Constant-current charging is accomplished by the application of a nonvarying constant-current source. This charge method is especially effective when several cells are charged in series, since it tends to reduce charge imbalances between batteries. Constant-current charging charges all cells equally because it is independent of the charging voltage of each cell in the battery. Constant-current recharging, at one or more current rates, is not widely used for lead-acid batteries. This is because of the need for current adjustment unless the charging current is kept at a low level throughout the charge (ampere-hour rule), which will result in long charge times of 12 h or longer. Typical charger and battery characteristics for the constant-current charge, for single- and two-step charging are shown in Fig. 14.41.

Constant-current charging at half the 20 h rate can be used to decrease the sulfation in batteries that have been overdischarged or undercharged. This treatment, however, may diminish battery life and should be used only with the advice of the battery manufacturer.

Constant current is also used to charge the VRLA battery. Figure 14.42 shows a family of curves of VRLA battery voltage versus percent capacity of previous discharges returned at different constant-current charging rates. As shown by these curves at different charge rates, the voltage increases sharply as the full charge state is approached. This increase in voltage is caused by the plates going into overcharge when most of the active material on the plates has been converted from lead sulfate to lead on the negative plate and to lead dioxide on the positive plate. The voltage increase will occur at lower states of charge when the cell is being charged at higher

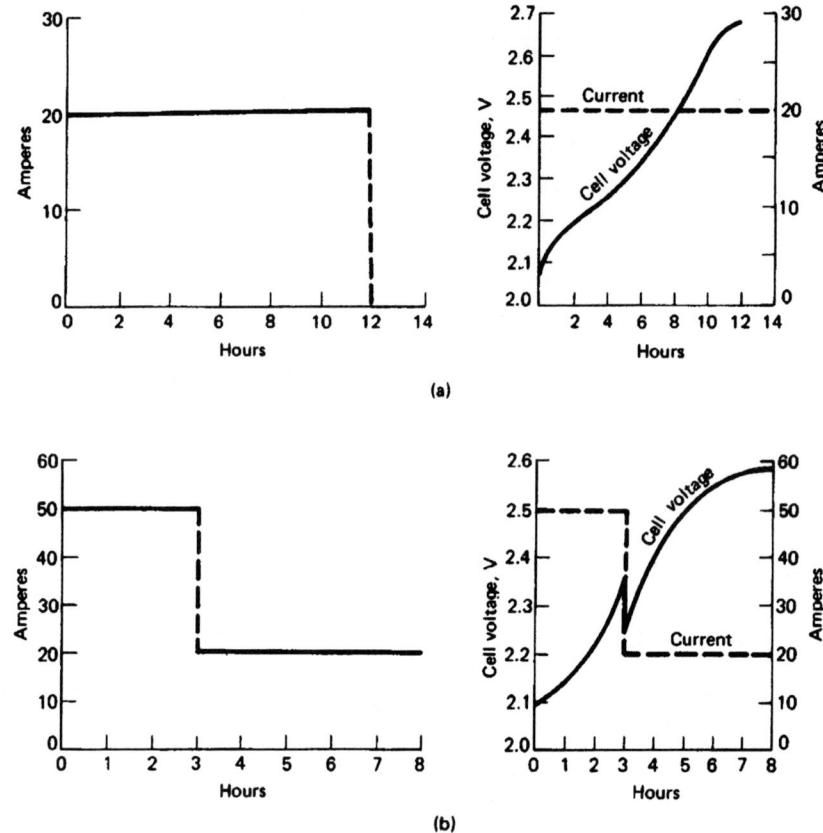

**FIGURE 14.41** Typical charger and battery characteristics for constant-current charging of lead-acid batteries. (*a*) Single-step constant-current charging. (*b*) Two-step constant-current charging. (*From Ref. 9.*)

rates because of increased gassing. The voltage curves in Fig. 14.42 are somewhat different from those for a conventional lead-acid battery due to the effect of the recombination of gases on overcharge within the system. The decrease in voltage during overcharge is due to the recombination of oxygen at the negative plate and does not occur in flooded batteries. The VRLA battery is capable of recombining the oxygen produced on overcharge up to the $C/3$ rate of constant-current charge. At higher rates, the recombination reaction is exceeded by the rate of gas generation and venting occurs.

Continued application at rates above $C/500$ after the battery is fully charged can be detrimental to life. At overnight charge rates ($C/10$–$C/20$), the large increase in voltage at the nearly fully charged state is a useful indicator for terminating or reducing the rates for a constant-current charger. If the rate is reduced to $C/500$, the battery can be left connected continuously and give 8 to 10 years of life at 25°C. Figure 14.43 is a plot of voltage versus time for a battery charging at the $C/15$ rate of constant current at 25°C. This curve shows that the battery is not fully charged at the time the voltage increase occurs and must receive additional charging.

*Taper Charging.* Taper charging is a variation of the modified constant-voltage method, using a low-cost fer-roresonant charger. The characteristics of taper charging are illustrated in Fig. 14.44. The initial rate is limited, but the taper of voltage and current is such that the 2.39 V per cell at 25°C is exceeded prior to the 100% return of the discharge ampere-hours. This method does result in gassing at the critical point of recharge, and the cell

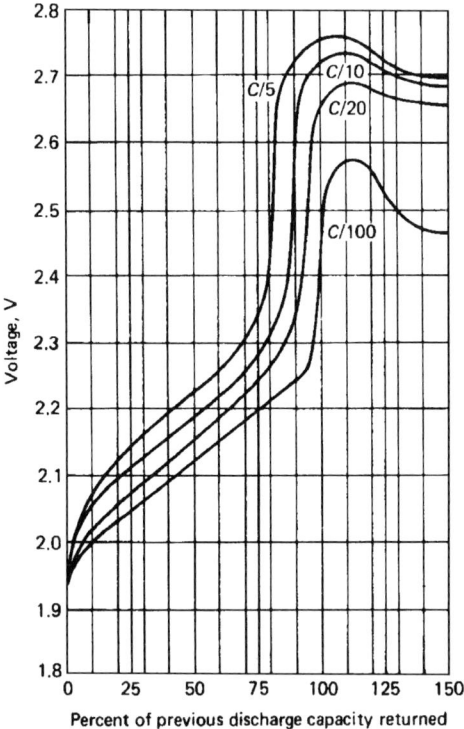

**FIGURE 14.42**    Voltage curves for VRLA batteries charged at various constant-current rates at 25°C.

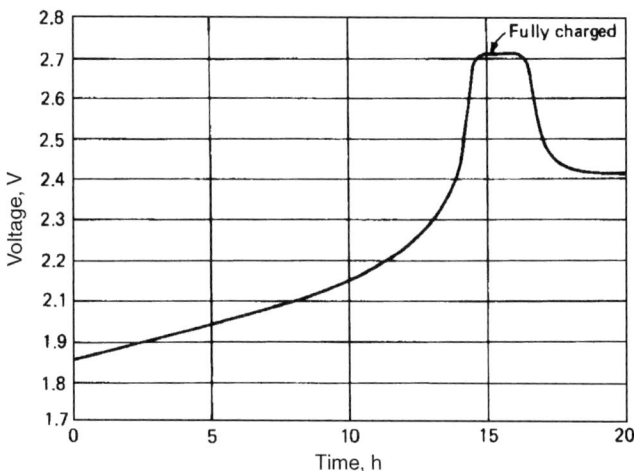

**FIGURE 14.43**    Constant-current charge at C/15 rate, 25°C for a VRLA battery.

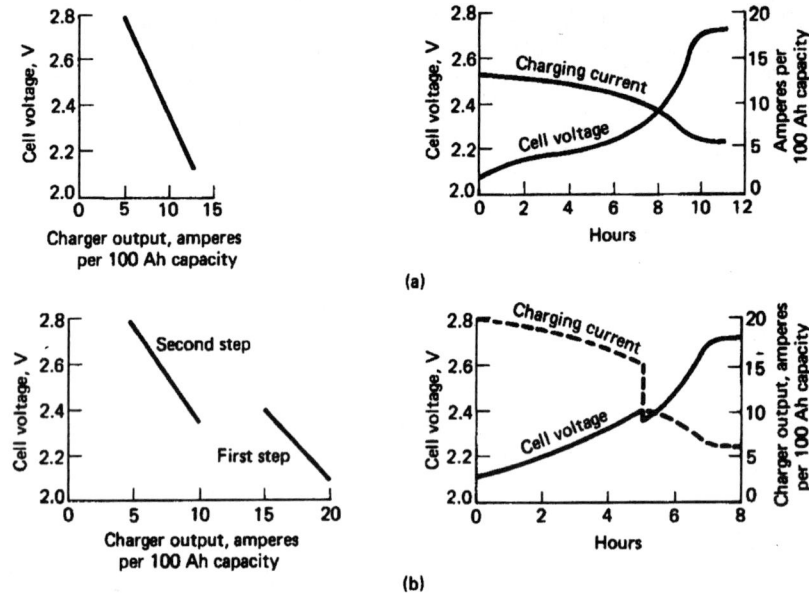

**FIGURE 14.44** Typical charger and battery characteristics for taper charging of lead-acid batteries. (*a*) Single-step taper charge. (*b*) Two-step taper charge. (*From Ref. 9.*)

temperature is increased. The degree of gassing and temperature rise is a variable depending on the charger design, and battery life can be degraded from excessive battery temperature and overcharge gassing. The gassing voltage decreases with increasing temperature; correction factors given in Table 14.17 provide the voltage correction factors at temperatures other than 25°C. This type of charging has been superseded by modern electronically controlled rectifiers.

**TABLE 14.17** Correction Factors for Cell Gassing Voltage

| Electrolyte temperature, °C | Cell gassing voltage, V | Correction factor, V |
|---|---|---|
| 50 | 2.300 | −0.090 |
| 40 | 2.330 | −0.060 |
| 30 | 2.365 | −0.025 |
| 25 | 2.390 | 0 |
| 20 | 2.415 | +0.025 |
| 10 | 2.470 | +0.080 |
| 0 | 2.540 | +0.150 |
| −10 | 2.650 | +0.260 |
| −20 | 2.970 | +0.508 |

The end of the charge is often controlled by a fixed voltage rather than a fixed current. Therefore, when a new battery has a high counter-EMF, this final charge rate is low and the battery often does not receive sufficient charge within the time period allotted to maintain the optimum charge state. During the latter part of life when the counter-EMF is lower, the charging rate is higher than the normal finishing rate, and so the battery receives excessive charge, which degrades life. Thus, the taper charge does degrade battery life, which must be justified by the use of the less expensive equipment.

The VRLA battery can withstand charge voltage variations, but some caution in using taper current chargers is recommended. The output characteristics are such that as the voltage of the battery increases during charge, the charging current decreases. This effect is achieved by use of the proper wire size and the turns ratio. Basically, the turns ratio from primary to secondary determines the output voltage at no load, and the wire size in the secondary determines the current at a given voltage. The transformer is essentially a constant-voltage transformer that depends entirely on the AC line voltage regulation for its output-voltage regulation. Because of this method of voltage regulation, any changes in input line voltage directly affect the output of the charger. Depending on the charger design, the output-to-input voltage change can be more than a direct ratio; for example, a 10% line-voltage change can produce a 13% output-voltage change.

When considering the cost advantage of using a half-wave rectifier versus a full-wave rectifier in a taper-current charger, it should be noted that the half-wave rectifier supplies a 50% higher peak-to-average-voltage ratio than the full-wave rectifier. Therefore, the total life of the battery for a given average charge voltage can be reduced for the half-wave type of charger because of the higher peak voltages. A DC ripple can lead in time to decreased performance through degradation of the active material and the grid. An AC ripple can be a more significant factor in premature battery failure, especially in float or uninterruptible power systems. The repeated charging and discharging of the battery shortens the battery life through heat generation and corrosion.

There are several charging parameters that must be met. The parameter of main concern is the recharge time to 100% nominal capacity for cycling applications.

***Float Charging.*** Float charging is a low-rate constant-potential charge also used to maintain the battery in a fully charged condition. This method is used mainly for stationary batteries that can be charged from a DC bus.

When the VRLA battery is to be float-charged as in a standby application, the constant-voltage charger should be maintained between 2.2 and 2.3 V for maximum life. Continuous charging at greater than 2.4 V per cell is not recommended because of accelerated grid corrosion. Figure 14.45 gives the approximate

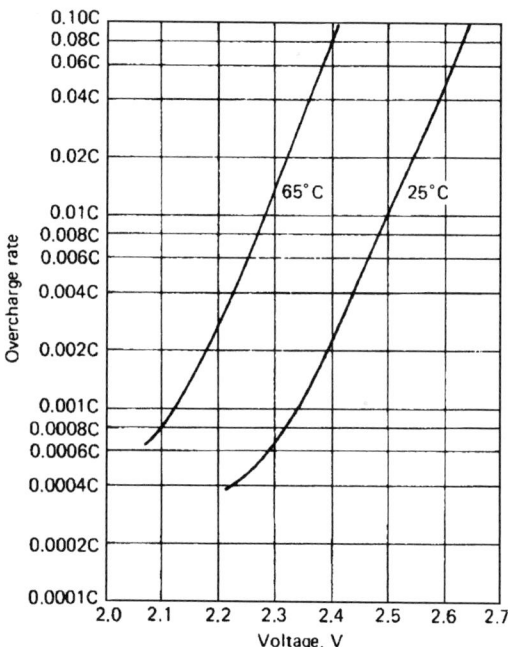

**FIGURE 14.45** Overcharge current and voltage for VRLA batteries.

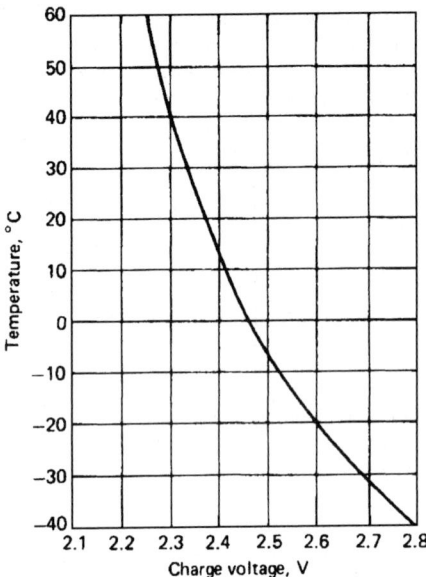

**FIGURE 14.46**   Recommended charge voltage at various temperatures (temperature compensation) for VRLA batteries.

values of voltage a battery will attain when float-charged at 25 and 65°C, or the charge rate a battery will accept if it has been charged for a sufficient period of time so that it is in a state of overcharge equilibrium. These curves can also be used to determine the approximate value of continuous constant current (trickle charge) that will maintain the proper float voltage. As an example, if a battery were trickle-charged at the $0.001C$ rate, its average voltage per cell on overcharge would be 2.35 V at 25°C. Conversely, if a cell were constant-voltage-charged at 2.35 V, its overcharge rate would be $0.001C$.

High temperatures accelerate the rate of the reactions that reduce the life of a battery. At increased temperatures, the voltage necessary for returning full capacity in a given time is reduced because of the increased reaction rates within the battery. To maximize life, a negative charging temperature coefficient of approximately −2.5 mV/°C per cell is used at temperatures significantly different from 25°C. Figure 14.46 shows the recommended charging voltage at various temperatures for a sealed battery float-charged at 2.35 V per cell at 25°C. It is obvious from this curve that at extremely low temperatures, a significantly greater temperature coefficient than −2.5 mV/°C is required to achieve full recharge of the cell. Figure 14.46 also shows the voltage compensation under cycling service. The voltage compensation keeps the charging current at about the same value that it would be at 25°C when the battery temperature is different. Temperature compensation of the charging voltage prevents thermal runaway of the batteries when they are used at high temperatures and ensures adequate charging if the battery temperature is low.

When trickle-charging, it may be necessary to increase the charge rate at higher temperatures to maintain the proper float voltage. From Fig. 14.45, it can be seen that for a battery trickle-charged at the $0.001C$ rate at 25°C, the float voltage would be 2.34 V. However, at the same rate at 65°C, the float voltage would be approximately 2.12 V, which is below the open-circuit voltage of the cell. At 65°C, the trickle-charge current would need to be increased to approximately $0.01C$ to maintain the proper float voltage.

*Pulse Charging.*   Pulse charging is also used for traction applications, particularly in Europe. In this case, the charger is periodically isolated from the battery terminals, and the open-circuit voltage of the battery is automatically measured (an impedance-free measurement of the battery voltage). If the open-circuit voltage is above a preset value, depending on a reference temperature, the charger does not deliver energy. When the open-circuit voltage decays below that limit, the charger delivers a DC pulse for a fixed time period. When the battery state of charge is very low, charging current is connected almost 100% of the time because the open-circuit voltage is below the present level or rapidly decays to it. The duration of the open-circuit and the charge pulses are chosen so that when the battery is fully charged, the time for the open-circuit voltage to decay is exactly the same as the pulse duration. When the charger controls sense this condition, the charger is automatically switched over to the finish rate current and short charging pulses are delivered periodically to the battery to maintain it at full charge. In many industrial applications, high-voltage batteries may be used and difficulty can be encountered in keeping the cells in a balanced condition. This is particularly true when the cells have long periods of standby use with different rates of self-decay. In these applications, the batteries are completely discharged and recharged periodically (usually semiannually) in what is called an equalizing charge, which brings the whole string of cells back to the complete charge state. On completion of this process, the liquid levels in the cells must be checked and water should be added to the depleted cells as required. With the newer types of maintenance-free cells, which are semisealed, such equalizing charges and differential watering of the cells may not be possible, and special precautions are taken in the charger design to keep the cells at an even state of charge.

***Trickle Charging.*** A trickle charge is a continuous constant-current charge at a low (about $C/100$) rate, which is used to maintain the battery in a fully charged condition, recharging it for losses due to self-discharge, as well as to restore the energy discharged during intermittent use of the battery. This method is typically used for SLI and similar type batteries when the battery is removed from the vehicle or its regular source of charging energy.

***Rapid Charging.*** In many applications, it is desirable to be able to rapidly recharge the battery within an hour or less. As is the case under any charging condition, it is important to control the charge to maintain the morphology of the electrode, to prevent a rise in the temperature, particularly to a point where deleterious side reactions (corrosion, conversion to nonconducting oxides, high solubility of materials, decomposition) take place, and to limit overcharge and gassing. As these conditions are more prone to occur during high-rate charging, charge control under these conditions is critical.

The availability of small, low-cost but sophisticated semiconductor chips has made effective methods of controlling the charging voltage-current-profile feasible. These devices can be used to either terminate the charge, limit the charge current, or switch between charge regimes when potentially damaging conditions arise during the charge.

The large surface area of the thin plates used in some VRLA batteries reduces the current density to a level far lower than normally seen in the fast charge of conventional lead-acid batteries, thereby enhancing the fast-charge capabilities.[44]

Figure 14.47 shows the charge rate or the current the VRLA battery can accept for a 1 h charge at three different voltages. The charger has a capability of delivering up to a $5C$ charge rate. The battery has a high charge acceptance at the beginning of the charge time; in fact, in the case of the 2.55 V per cell charge, the cell accepted the full current capability of the charger for the first 3 to 4 min. In the case of the 2.7 V per cell charge, there was a considerable amount of overcharging starting at 30 min, which caused internal heating and a consequent increase in charge current.

Figure 14.48 shows a set of curves of normalized charge efficiency versus time in minutes for the three different voltages. This efficiency figure was calculated by dividing the total ampere-hour capacity returned by the previous discharge capacity removed. On the 2.55 V charge, 100% of the capacity removed on the previous cycle was returned in 15 min. With the 2.7 V charge, a 60% overcharge was returned at the end of the 60 min charge.

Figure 14.49 shows the plots of the discharge time in minutes versus cycle number for the three charge voltages. Also, a set of reference batteries was charged at 2.5 V constant voltage for 16 h and discharged at the $1C$ rate. This reference curve is displayed to show the expected capacity at the $1C$ rate. It can be seen from these data that the 2.55 V per cell curve most closely approximates the reference line. The battery charged at 2.7 V per cell received too much overcharge and, therefore, the capacity degraded after 15 cycles. The battery

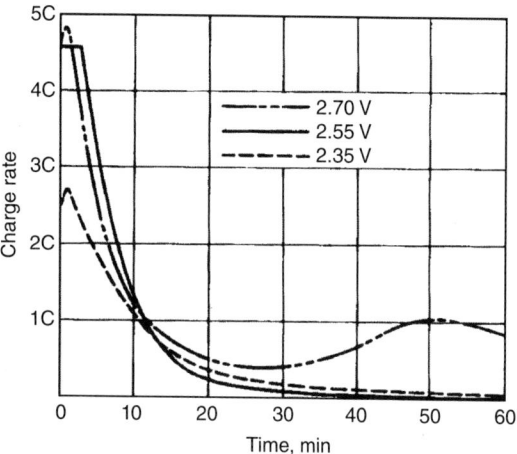

**FIGURE 14.47** Charge rate versus time for three charge voltages (VRLA battery).

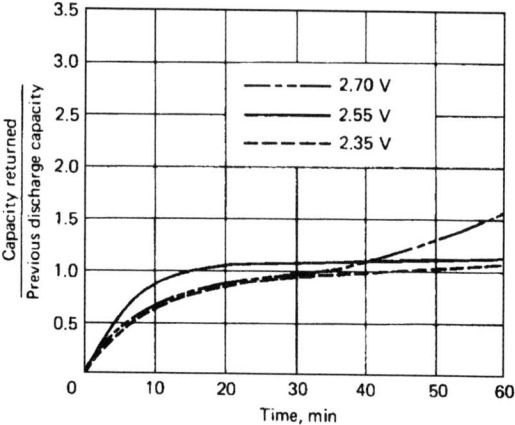

**FIGURE 14.48** Charge efficiency versus time for three charge voltages (VRLA battery).

charged at 2.35 V achieved a value of approximately 75% of the reference and continued to cycle at that level.

These data show that the thin-plate VRLA battery can be fast-charged to 100% of the rated capacity in less than 1 h. A constant-voltage charger set at 2.5 to 2.55 V per cell and capable of the 3 to 4C rate of charge is preferred. It should be noted, as discussed, that charging at 2.7 V per cell for prolonged periods will damage the battery.

### 14.6.4 Charge (Coulombic) Efficiency

Charge or coulombic efficiency is the ratio of the current that is actually used for electrochemical conversion of the active material to the total current supplied to the cell on recharge. The current that is not used for charging is consumed in parasitic reactions within the cell such as corrosion and gas production.

The charging efficiency is high for a VRLA battery and generally lower for a conventional flooded battery. Charge efficiency is a direct function of the state of charge. The charge efficiency of a battery is high until it approaches full charge, at which time the overcharge reactions begin and the charge efficiency decreases.

Figure 14.50 shows a curve of charge efficiency versus voltage at various constant voltages. Increasing voltage decreases the efficiency because of increased gassing. The efficiency shows a marked decrease below the open-circuit voltage, typically 2.15 to 2.18 V, because the charge voltage is not high enough to support the charging reaction.

Figure 14.51 is a curve of efficiency versus log rate at various constant-current charge rates. As can be seen from the curve, at rates up to C/10, the efficiency approaches 100%. At higher rates, there is a decrease in efficiency because as the cell approaches the fully charged

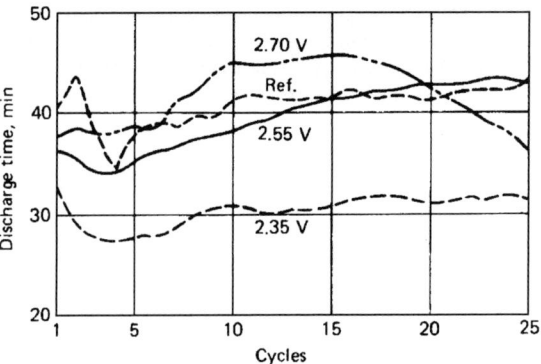

**FIGURE 14.49** Effect of cycling on discharge time for three VRLA battery charge voltages.

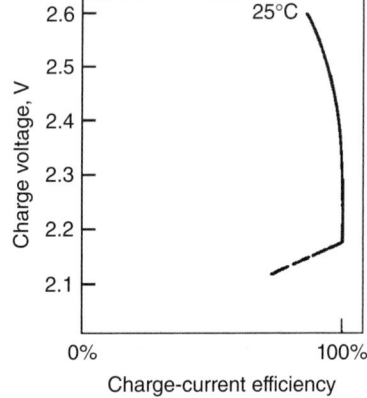

**FIGURE 14.50** Constant-voltage charge efficiency for a VRLA battery.

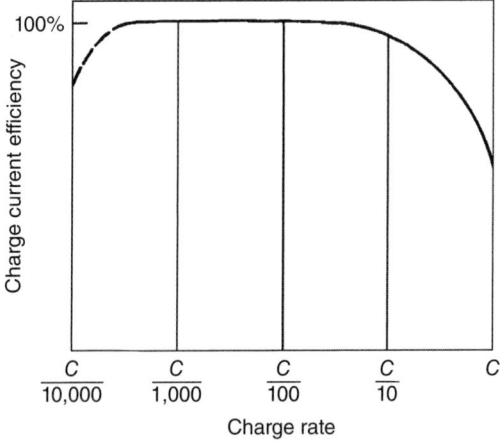

**FIGURE 14.51** Constant-current charge efficiency for a VRLA battery.

state, the surfaces of the plates become fully charged. This increases the charging reaction rates and results in increased voltages and gassing. At low charge rates, the efficiency drops because the charge current is equivalent to the parasitic currents and the battery voltage approaches the open-circuit value.

Figure 14.52 shows the charge acceptance during charge at various temperatures and charge rates.

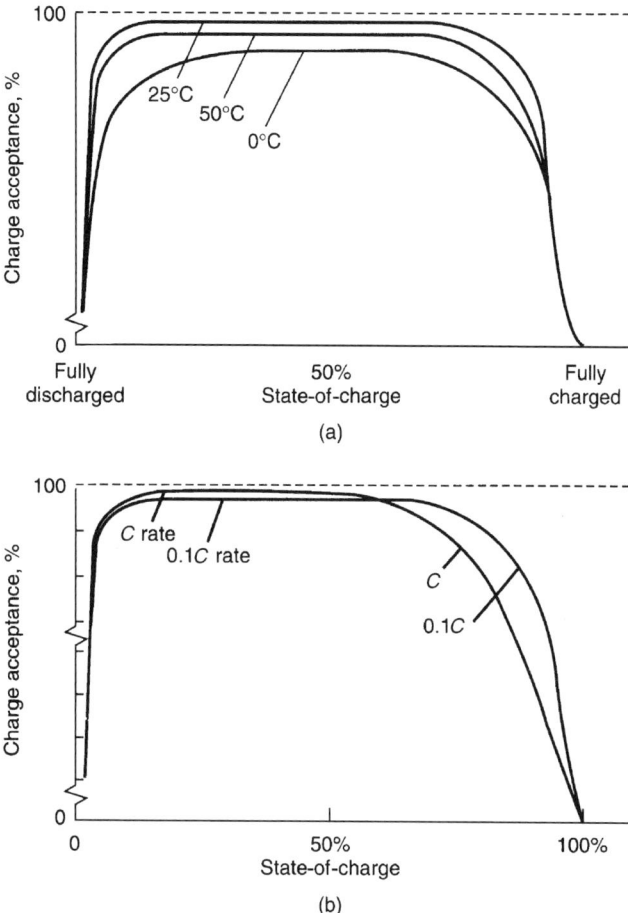

**FIGURE 14.52**   Charge acceptance of sealed lead-acid batteries. (*a*) At various temperatures. (*b*) At various charge rates.

## 14.7  *MAINTENANCE, SAFETY, AND OPERATIONAL FEATURES*

### 14.7.1  Maintenance

It is common for industrial flooded lead-acid batteries to function for periods of 10 years or longer. Proper maintenance can ensure this extended useful life. Five basic rules of proper maintenance are as follows:

1. Match the charger to the battery charging requirements.
2. Avoid overdischarging the battery or storage in a discharged condition.

**3.** Maintain the electrolyte at the proper level (add water as required).

**4.** Keep the battery clean.

**5.** Avoid overheating the battery.

In addition to these basic rules, as the battery is made of individual cells connected in series, the cells must be properly balanced periodically.

***Charging Practice.***    Poor charging practice is responsible for short battery life more than any other cause. Fortunately, the inherent physical and chemical characteristics of lead-acid batteries make control of charging quite simple. If the battery is supplied with DC energy at the proper charging voltage, the battery will draw only the amount of current that it can accept efficiently, and this current will reduce as the battery approaches full charge. Several types of devices can be used to ensure that the charge will terminate at the proper time. The specific gravity of the electrolyte should also be checked periodically for those batteries that have a removable vent and adjusted to the specified value (see Tables 14.7 and 14.12).

***Overdischarge.***    Overdischarging the battery should be avoided. The capacity of large batteries, such as those used in industrial trucks, is generally rated in ampere-hours at the 6 h discharge rate to a final voltage of 1.75 V per cell. These batteries can usually deliver more than rated capacity, but this should be done only in an emergency and not on a regular basis. Discharging cells below the specified voltage reduces the electrolyte to a low concentration, which has a deleterious effect on the pore structure of the battery. Battery life has been shown to be a direct function of the DOD, as illustrated in Fig. 14.53.[45]

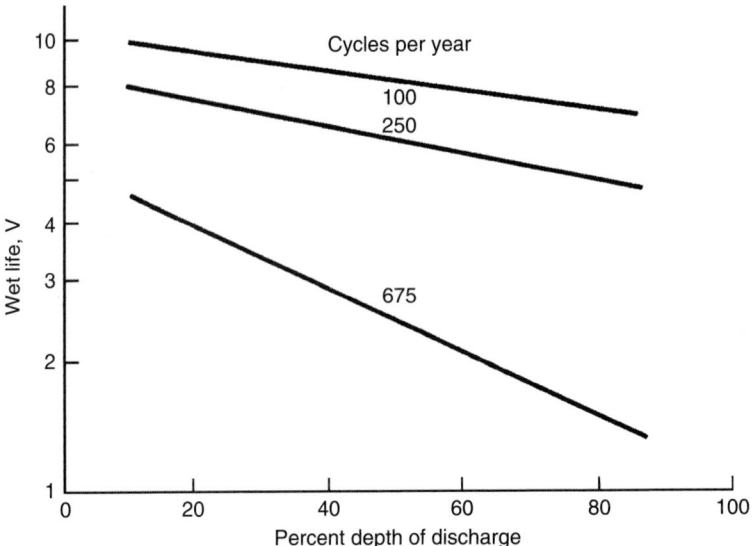

**FIGURE 14.53**    Effect of depth of discharge and number of cycles per year on wet life at 25°C. (*From Ref. 45.*)

***Electrolyte Level.***    During normal operation, water is lost from a battery as the result of evaporation and by electrolysis into hydrogen and oxygen, which escape into the atmosphere. Evaporation is a relatively small part of the loss, except in very hot, dry climates. With a fully charged battery, electrolysis consumes water at a rate of 0.336 cm³/Ah overcharge. A 500 Ah cell overcharged 10% can thus lose 16.8 cm³, or about 0.3% of its water in each cycle. It is important that the electrolyte be maintained at the proper level in the battery. The electrolyte not only serves as the conductor of electricity but is a major factor in the transfer of heat from the plates. If the electrolyte is below the plate level, then an area of the plate is not electrochemically active; this causes a concentration of heat in other parts of the cell. Periodic checking of water consumption can also serve as a rough check on charging efficiency and may warn when adjustment of the charger is required.

Since replacing water can be a major maintenance cost, water loss can be reduced by controlling the amount of overcharge and by using hydrogen and oxygen recombining devices in each cell where possible. Addition of water is best accomplished after recharge and before an equalization charge. Water is added at the end of the charge to reach the high acid level line. Gassing during charge will stir the water into the acid uniformly. In freezing weather, water should not be added without mixing, as it may freeze before gassing occurs. Water added must be either distilled or demineralized water. Automatic watering devices and reliability testing can further reduce maintenance labor costs. Overfilling must be avoided because the resultant overflow of acid electrolyte will cause tray corrosion, ground paths, and loss of cell capacity. A final check of specific gravity should be made after water has been added to ensure correct acid concentration at the end of charge. A helpful approximation is

$$\text{Specific gravity} = \text{Cell open circuit voltage} - 0.845$$

which permits electrical monitoring of specific gravity on an occasional basis (see also Fig. 14.3).

*Cleanliness.*    Keeping the battery clean will minimize corrosion of cell post connectors and steel trays and avoid expensive repairs. Batteries commonly pick up dry dirt, which can be readily blown off or brushed away. This dirt should be removed before moisture makes it a conductor of stray currents. One problem is that the top of the battery can become wet with electrolyte any time a cell is overfilled. The acid in this electrolyte does not evaporate and should be neutralized by washing the battery with a solution of baking soda and hot water, approximately 1 kg of baking soda to 4 L (liters) of water. After application of such a solution, the area should be rinsed thoroughly with water.

*High Temperature—Overheating.*    One of the most detrimental conditions for a battery is high temperature, particularly above 55°C, because the rates of corrosion, solubility of metal components, and self-discharge increase with increasing temperature. High operating temperature during cycle service requires higher charge input to restore discharge capacity and local action (self-discharge) losses. More of the charge input is consumed by the electrolysis reaction because of the reduction in the gassing voltage at the higher temperature (see Sec. 14.7.4). While a 10% overcharge per cycle maintains the state of charge at 25 to 35°C, 35% to 40% overcharge may be required to maintain the state of charge at the higher (60–70°C) operating temperatures. On float service, float currents increase at the higher temperatures, resulting in reduced life. Eleven days float at 75°C is equivalent in life to 365 days at 25°C.

Batteries intended for high-temperature applications should use a lower initial specific gravity electrolyte than those intended for use at normal temperatures (see Table 14.12). Other design features, such as the use of more expander in the negative plate, are also important to improve operation at high temperatures.

*Cell Balancing.*    During cycling, a high-voltage battery having many cells in a series string can become unbalanced with certain cells limiting charge and discharge. Limiting cells receive more overcharge than other cells in the string, have greater water consumption, and thus require more maintenance. The equalization charge has the function of balancing cells in the string at the end of charge. In an equalization charge, the normal recharge is extended for 3 to 6 h at the finishing rate of 5 A/100 Ah, 5 h rated capacity, allowing the battery voltage to rise uncontrolled. The equalization charge should be continued until cell voltages and specific gravities rise to a constant, acceptable value. Frequency of equalization charge is normally a function of the accumulative discharge output and will be specified by the manufacturer for each battery design and application.

## 14.7.2  Safety

Safety problems associated with lead-acid batteries include spills of sulfuric acid, potential explosions from the generation of hydrogen and oxygen, and the generation of toxic gases such as arsine and stibine. All these problems can be satisfactorily handled with proper precautions. Wearing face shields and plastic or rubber aprons and gloves when handling acid is recommended to avoid chemical burns from sulfuric acid. If acid gets into the eyes, or comes in contact with skin or clothing, it should be flushed immediately and thoroughly with clean water. Medical attention should be obtained when eyes are affected. A bicarbonate of soda solution (100 g/L of water) should be used to neutralize any acid that is accidentally spilled. After neutralization the area should be rinsed with clear water.

Precautions must be routinely practiced to prevent explosions from ignition of the flammable gas mixture of hydrogen and oxygen formed during overcharge of lead-acid batteries. The maximum rate of formation is 0.42 L of hydrogen and 0.21 L of oxygen per ampere-hour overcharge at standard temperature and pressure. The gas mixture is explosive when hydrogen in air exceeds 4% by volume. A standard practice is to set warning devices to ring alarms at 20% to 25% of this lower explosive limit (LEL). Low-cost hydrogen detectors are available commercially for this purpose.

With good air circulation around a battery, hydrogen accumulation is normally not a problem. However, if relatively large batteries are confined in small rooms, exhaust fans should be installed to vent the room constantly or to be turned on automatically when hydrogen accumulation exceeds 20% of the LEL. Battery boxes should also be vented to the atmosphere. Sparks or flame can ignite these hydrogen atmospheres above the LEL. To prevent ignition, electrical sources of arcs, sparks, or flame must be mounted in explosion-proof metal boxes. Battery cells can similarly be equipped with flame arrestors in the vents to prevent outside sparks from igniting explosive gases inside the cell cases. It is good practice to refrain from smoking, using open flames, or creating sparks in the vicinity of the battery. A considerable number of the reported explosions of batteries come from uncontrolled charging in non-automotive applications. Often batteries will be charged, off the vehicle, for long periods of time with an unregulated charger. In spite of the fact that the charge currents can be low, fair volumes of gas can accumulate. When the battery is then moved, this gas vents, and if a spark is present, explosions have been known to occur.

During charging of the VRLA battery some hydrogen gas and also some carbon dioxide gas are given off, but at much lower levels than the flooded battery. However, ventilation of the charging area is still required. The hydrogen outgassing is essential with each cycle to ensure continued internal chemical balance. Carbon dioxide is produced by oxidation of organic compounds in the cell.

Another consideration is the potential failure of the charger. If the charger malfunctions, causing higher-than-recommended charge rates, substantial volumes of hydrogen and oxygen will be vented from the VRLA battery. This mixture is explosive and should not be allowed to accumulate, so adequate ventilation is required. Therefore, the VRLA battery should never be operated in a gastight container. The batteries should never be totally encased in a potting compound since this prevents the proper operation of the venting mechanism and free release of gas.

Some types of batteries can release small quantities of the toxic gases stibine ($SbH_3$) and arsine ($AsH_3$). These batteries have positive or negative plates that contain small quantities of the metals antimony and arsenic in the grid alloy to harden the grid and to reduce the rate of corrosion of the grid during cycling. $AsH_3$ and $SbH_3$ are generally formed when the arsenic or antimony alloy material comes into contact with nascent hydrogen, usually during overcharge of the battery, which then combine to form these colorless and essentially odorless gases. They are extremely dangerous and can cause serious illness and death. The Occupational Safety and Health Administration (OSHA) 1978 concentration limits for $SbH_3$ and $AsH_3$ are 0.1 and 0.05 ppm, respectively, as a maximum allowable weighted average for any 8 h period. Ventilation of the battery area is very important. Indications are that ventilation designed to maintain hydrogen below 20% LEL (approximately 1% hydrogen) will also maintain $SbH_3$ and $AsH_3$ below their toxic limits.

The ordinary 12 V SLI automotive battery is a minor shock hazard. The hazard level increases with higher-voltage systems, and systems in the range of 84 to 360 V are being used for electric vehicles. Systems as high as 1000 V are in use for fixed-location energy-storage systems for load leveling. However, these batteries are capable of delivering high currents if externally short-circuited. The resultant heat can cause severe burns and is a potential fire hazard. Particular caution should be used when any person working near the open terminals of cells or batteries is wearing metal rings or watchbands. Inadvertently placing these metal articles or tools across the terminals could result in severe skin burns.

Batteries are electrically alive even in the discharged state, and the following precautions should be practiced:

1. Keep the top of the battery clean and dry to prevent ground short circuits and corrosion.
2. Do not lay metallic objects on the battery. Insulate all tools used in working on batteries.
3. Remove jewelry and any other electrical conductors before inspecting or servicing batteries.

4. When lifting batteries, use insulated lifting tools to avoid risks or short circuits between cell terminals by lifting chains or hooks.

5. Make sure that gases do not accumulate in the charging area.

### 14.7.3 Effect of Operating Parameters on Battery Life

Operating parameters which have a strong influence on battery life are DOD, number of cycles used each year, charging control, type of storage, and operating temperature. In some cases, the battery design features that increase life tend to decrease the initial capacity, power, and energy output. It is important, therefore, that the design features of the battery be selected to match the operating and life requirements of the application as listed below.

1. Increasing the number of cycles performed per year decreases the wet life (see Fig. 14.53).

2. Increasing the DOD decreases cycle life, as illustrated in Fig. 14.54[45] and Sec. 14.9.3 (Fig. 14.71).

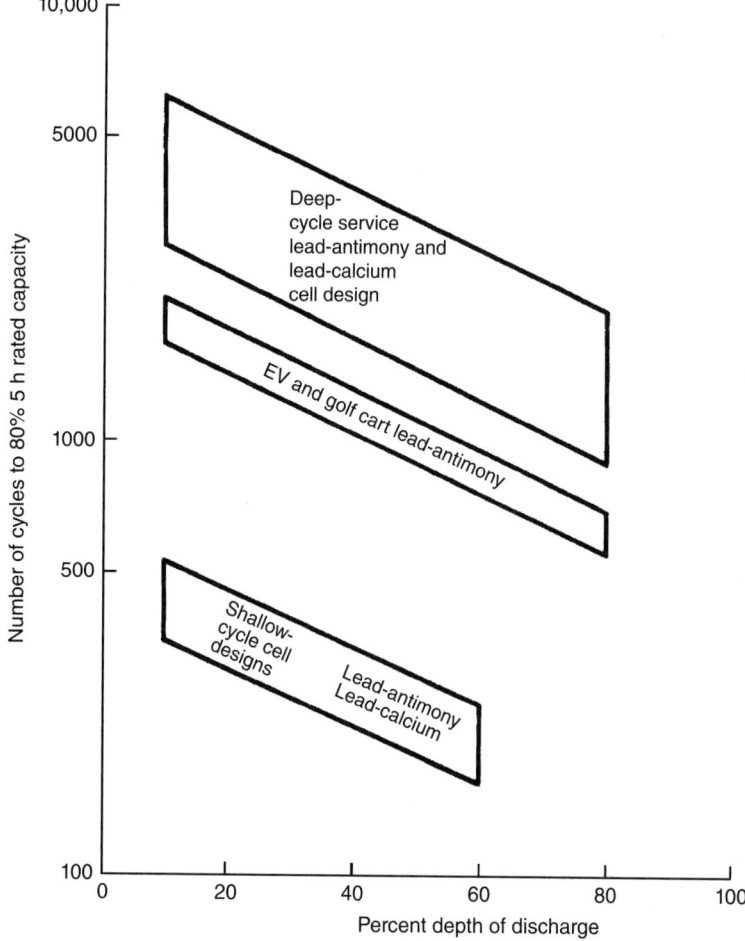

**FIGURE 14.54**   Effect of cell design and depth of discharge on cycle life of various types of lead-acid batteries at 25°C. (*From Ref. 45.*)

3. Excessive overcharging leads to increasing positive grid corrosion, active material shedding, and shorter wet life.

4. Storing wet cells in a discharged condition promotes sulfation and decreases capacity and life.

5. Proper charging operations with good equipment maintain the desired state of charge with a minimum of overcharge and leads to optimum battery life.

6. Stratification of the electrolyte in large flooded cells into levels of varying concentration can limit charge acceptance, discharge output, and life unless controlled during the charge process. During a recharge, sulfuric acid of higher concentration than the bulk electrolyte forms in the pores of the plates. This higher-density acid settles to the bottom of the cell, giving higher specific gravity acid near the bottom of the plates and lower specific gravity acid near the top of the plates. This stratification accumulates during the nongassing periods of charge. During the gassing periods of overcharge, partial stirring is accomplished by gas bubbles formed at and rising along the surfaces of the plates and in the separator system. During discharge, acid in the pores of the plates and near their surface is diluted; however, concentration gradients set up by longer charge periods are seldom compensated entirely, particularly if the discharge periods are shorter, as is usually the case. Diffusion processes to eliminate these concentration gradients are very slow, and stratification during repetitive cycling can become progressively greater. Two methods for stratification control are by deliberate gassing of the plates during overcharge at the finishing rate and by stirring of cell electrolyte by pumps (usually airlift pumps). The degree of success in eliminating stratification is a function of cell design, the design of the pump accessory system, and cell operating procedures.

### 14.7.4 Failure Modes

The failure modes of lead-acid batteries depend on the type of application and the particular battery design. Hence, the rationale for the manufacture of different batteries is that each type is designed to give optimum performance in a specific type of use. The more prevalent failure modes for the different types of lead-acid batteries are listed in Table 14.18.[46] Significantly, if a battery is properly maintained, most of the inherent failures are due to the degradation of the positive plate through either grid corrosion or paste shedding. These failures are irreversible, and when they occur, the battery must be replaced. Details of the failure modes of SLI batteries are given in Sec. 14.8.4.

**TABLE 14.18**    Failure Modes of Lead-Acid Batteries

| Battery type | Normal life | Normal failure mode |
| --- | --- | --- |
| SLI | Several years | Grid corrosion |
| SLI (maintenance-free) | Several years | Grid corrosion, dry out |
| Golf cart | 300–600 cycles | Positive shedding and grid corrosion, sulfation |
| Stationary (Industrial) | 6–25 years | Grid corrosion |
| Traction (Industrial) | Minimum 1500 cycles | Shedding, grid corrosion |

The failure mode and the time to failure can be modified by changes in the inner parameters (I), such as battery materials, processing, and design, or by the conditions of use, designated as the outer parameters (O). Some of these are listed in Table 14.19.[46]

## 14.8   SLI (AUTOMOTIVE) BATTERIES: CONSTRUCTION AND PERFORMANCE

### 14.8.1   General Characteristics

Traditionally, the most common use of the lead-acid battery is for SLI in automobiles and other vehicles with internal combustion engines. Almost all of these have used 12 V nominal electric systems. High cranking ability at low temperatures is still the major design factor. The design of lead-acid batteries is varied in order to

**TABLE 14.19** Modification of Lead-Acid Battery Failure Rate

| Failure mechanism | Rate of failure modification* |
|---|---|
| Shedding of positives | I: active mass structure, retention system, battery design |
| | O: number of cycles, depth of discharge, charge factor |
| Sulfation of negatives | I: active mass additives |
| | O: temperature, charge factor, maintenance |
| Positive grid corrosion (overall, localized, or positive grid growth) | I: grid alloy, manufacturing conditions, active mass |
| Separators | I: separator specification, electrolyte concentrations, battery design |
| | O: temperature, charge factor, maintenance |
| Case, cover, vents, external battery connections | I: battery materials and design |
| | O: maintenance, abuse |

*I—Inner parameters; O—outer parameters.

maximize the desired type of performance. Trade-offs exist for optimization among such parameters as power density, energy density, cycle life, "float-service" life, and cost.

High power density requires that the internal resistance of the battery be minimal. This affects grid design, the porosity, thickness, and type of separator, and the method of intercell connection. High power and energy densities also require that plates and separators be thin and very porous and that paste density be very low. High cycle life requires premium separators, high paste density, the presence of $\alpha$-$PbO_2$ or another bonding agent, modest DOD, and good maintenance. Low cost requires both minimum fixed and variable costs, high-speed automated processing, and no premium materials for the grid, paste, separator, and other cell and battery components.[2,10,15,47,48]

## 14.8.2 Automotive Applications

SLI batteries are also used in trucks, aircraft, industrial equipment, and motorcycles as well as in many other applications. They are used in off-road vehicles, such as snowmobiles, in boats to crank inboard and outboard engines, and in various farm and construction equipment. Military vehicles in the United States and NATO countries have standardized on a 24 V electric system that is provided by a series connection of two 12 V batteries.

SLI batteries today see more cycling-type service (compared with float service) because of the electrical load of the auxiliaries. Size and weight reduction as well as the battery geometry have become important. These factors have led to the redesign of the lead-acid battery for SLI applications. The most important changes are:

- Historically a change from high-Sb (4–5%) lead alloy grid to low-Sb (1–2%) and to Pb-Ca-Sn alloy grids, thus reducing hydrogen evolution
- Use of thinner electrodes
- Better separators with lower electrical resistance
- Plate tabs located in from corners, and grids redesigned for high conductivity
- EFB, maintenance-free construction
- VRLA construction

The term "SLI battery" has evolved into something of a misnomer. In addition to starting, lighting, and ignition, the automotive SLI battery may provide the power for many other functions. Although these features may not pose much of a burden individually, collectively they add up to a significant drain on the SLI battery. Some of today's automobiles require up to 2 to 3 kW of power or more. Table 14.20 shows some current or anticipated

**TABLE 14.20**   Features Requiring Power for Present and Future Automobile Designs (Exclusive of SLI Function)

| | |
|---|---|
| Alarms (may include flashing LEDs) | Communication devices |
| Computer | Audio-radio, tape or CD players |
| Electric suspension | Global positioning features (maps, routing, emergency location) |
| Automatic start-stop of engine | Electromagnetic valve trains |
| Air conditioning | Electric heating of catalytic converters |
| Electric heating of seats | Sensing (e.g., for airbag deployment) |
| Electric steering | Anti-lock braking |
| Power windows | Electrochromic mirrors |
| Rear seat entertainment center | Rear window de-icer/defogger |
| Electric door locks | Cigarette lighter (other functions) |
| Clock | Cruise control |
| Regenerative braking | |

features requiring power exclusive of the SLI functions. The typical SLI battery is not designed to handle the cycling demands prompted by some of these items, and further improvements are required. The use of higher voltage batteries has been considered to reduce electric current levels.

Simply scaling up the 12 V SLI lead-acid battery to higher voltages leads to a number of problems, not the least of which is the significant increase in the weight of the battery. Today's SLI batteries are maintenance free in that they don't require addition of water during their operating life, and the use of VRLA batteries is increasing.

The automotive industry, to comply with corporate average fuel economy (CAFE) standards, is moving to microhybrids (start-stop), mild hybrids, hybrids, and electric vehicles (EVs). These requirements for car companies have put new load and cycling requirements on the traditional SLI battery. Batteries under microhybrid requirements must operate under a start-stop requirement. This requires the battery to have good cycle life especially under HRPSoC. A traditional flooded battery construction cannot meet these requirements. The best technology to address these new load requirements is with an AGM VRLA. A new type of flooded lead-acid SLI is also being developed. This new technology is generally identified as enhanced flooded, or extended-life flooded batteries. These batteries are about 1.5 times the cost of standard maintenance-free batteries and use different alloys, much higher additive (i.e., carbon) loading in the active material, and a separator system that applies pressure against and coverage of the surface of the positive plate. A permanent type of pasting paper is used in place of a standard cellulose tissue. In addition, additives are being used in the electrolyte.

### 14.8.3   Construction

A traditional SLI battery, whose function is to start an internal combustion engine, discharges briefly but at a high current. Once the engine is running, a generator or alternator system recharges the battery and then maintains it on "float" at full charge or slight overcharge. In recent automobile designs, the parasitic electrical load of lights, motors, and electronics cause a gradual discharge of the battery when the engine is not in operation. Most automobile manufacturers are contemplating conversion to a microhybrid design with stop-start requirements. This factor, coupled with normal self-discharge, introduces a significant cycling component into the normal cranking/floating duty cycles. As a result, the traditional SLI battery does not provide the required cycle capabilities (see Secs. 14.7.4 and 14.8.4 for details).

The cranking ability of an SLI battery is directly proportional to the geometric area of plate surface, with the proportionality factor (i.e., current density) typically 0.155 to 0.186 cold-crank amperes (CCA) at $-17.8°C$ ($0°F$) per square centimeter of the positive-plate surface. Cranking performance is generally limited by the positive plate at higher temperatures ($>18°C$) and by the negative plate at lower temperatures ($<5°C$). The ratio of positive surface to negative surface is fixed by design. To maximize the cranking capacity, SLI battery designs emphasize grids with minimum electrical resistance (using a variety of radial and expanded grid designs), thin plates, and a higher concentration of electrolyte than motive-power or stationary batteries.

Usually an "outside-negative" ($n + 1$ negative plates interspersed with $2n$ separators and $n$ positive plates) design is used. However, in order to balance the cranking rating with the capacity requirement or electrical load, as well as to facilitate automatic assembly, SLI batteries with an even number of plates, or "outside-positive" designs, are widely produced in the United States.

The maintenance-free SLI battery has several characteristics that distinguish it from the flooded battery. The maintenance-free SLI battery requires no addition of water during its life, has significantly improved capacity retention during storage, and has minimal terminal corrosion. The construction of a typical maintenance-free SLI battery is illustrated in Fig. 14.55. This type of battery relies mainly on charge control to prevent electrolysis of water and dry-out as compared to the VRLA designs that rely on oxygen recombination.

The SLI maintenance-free battery has a large acid reservoir made possible by the use of smaller plates and placement of the elements directly on the bottom of the container, eliminating the sludge space. The positive plates are usually enveloped in a microporous PE separator that prevents active material from falling to the bottom of the container and creating a short circuit. An important feature of the maintenance-free battery is the use of nonantimonial (such as calcium-lead) or low-antimonial lead grids. The use of these grids reduces the overcharge current significantly, reducing the rate at which water is lost during overcharge, as well as improving the stand characteristics (see Sec. 14.6). The use of the expanded grid, produced from wrought lead-calcium strip, is also shown in Fig. 14.9. Most maintenance-free SLI batteries are built using lead-calcium-tin grids for the negative and low-antimony-lead grids for the positive electrode.

In another refinement of the SLI battery design, the plates are approximately one-fifth of the width of conventional SLI battery plates and are inserted parallel, rather than perpendicular, to the length of the battery case. This design reduces the internal impedance of the cells and gives very high CCA ratings, but was not commercialized.

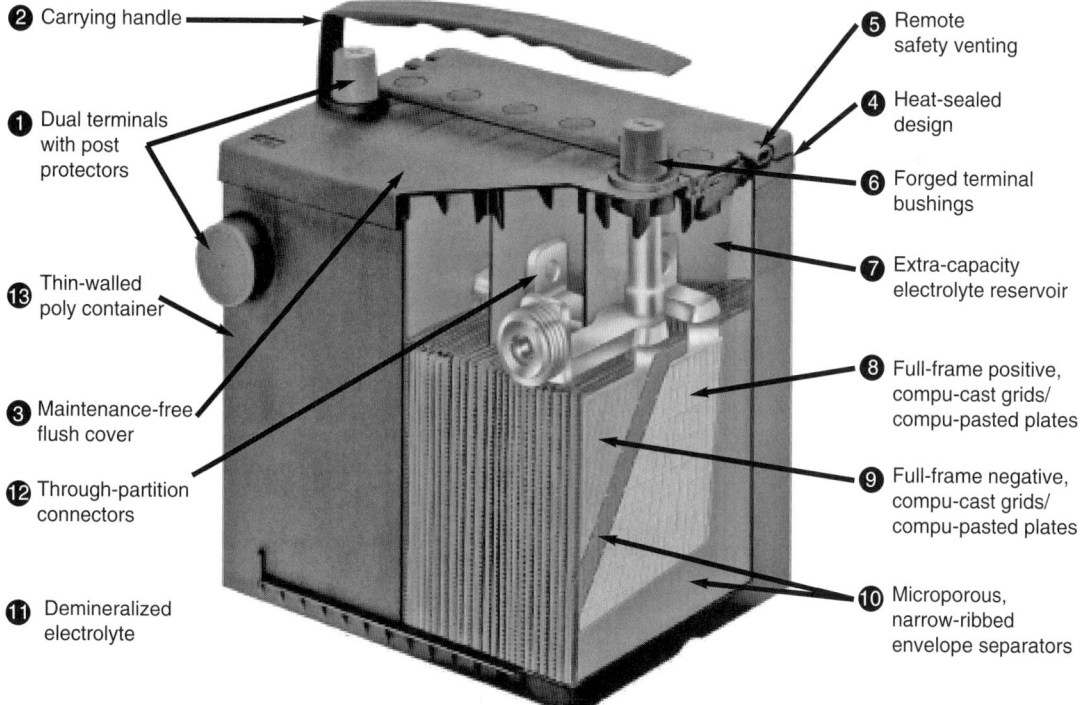

**FIGURE 14.55** Cutaway view of a lead-acid battery. (*Courtesy of East Penn Co.*)

Heavy-duty SLI batteries for trucks, buses, and construction equipment are designed similar to the passenger vehicle SLI batteries but use heavier and thicker plates with high-density paste, premium separators often with glass mats, anchor-bonding of the element to the bottom of the case, and other such features to enhance life. These modifications are necessary to provide maximum mechanical strength for these physically large batteries (from 285 mm to 530 mm case dimensions). Because the thick plates provide less cranking current than the thinner plates (since fewer can be included in a cell of a given size), series or series-parallel connections of batteries are used. Typically, the 12 V monoblocs are connected in series for cranking at 24 V and in parallel for running and recharging at 12 V. A few sizes of maintenance-free heavy-duty batteries have also been produced.

SLI-type batteries are also used in motorcycles and boats. Batteries for recreational marine use generally have thicker plates (to give more capacity) and higher-density paste. They have the same Battery Council International type designations as automotive batteries. See Sec. 8.3.3 for a listing of BCI battery types. Marine batteries are also manufactured in 4-cell 8 V monoblocs. Many of these special applications have been converted to a VRLA design.

Aircraft had previously employed SLI-type batteries with special spill-proof vent caps, which precluded loss of electrolyte during flight, but VRLA designs are now used in these applications.

### 14.8.4   Performance Characteristics

***Discharge Performance.***     Discharge curves, showing the discharge profile of SLI batteries at several constant-current discharge rates, are presented in Fig. 14.56. The typical final or end voltages at these discharge rates are

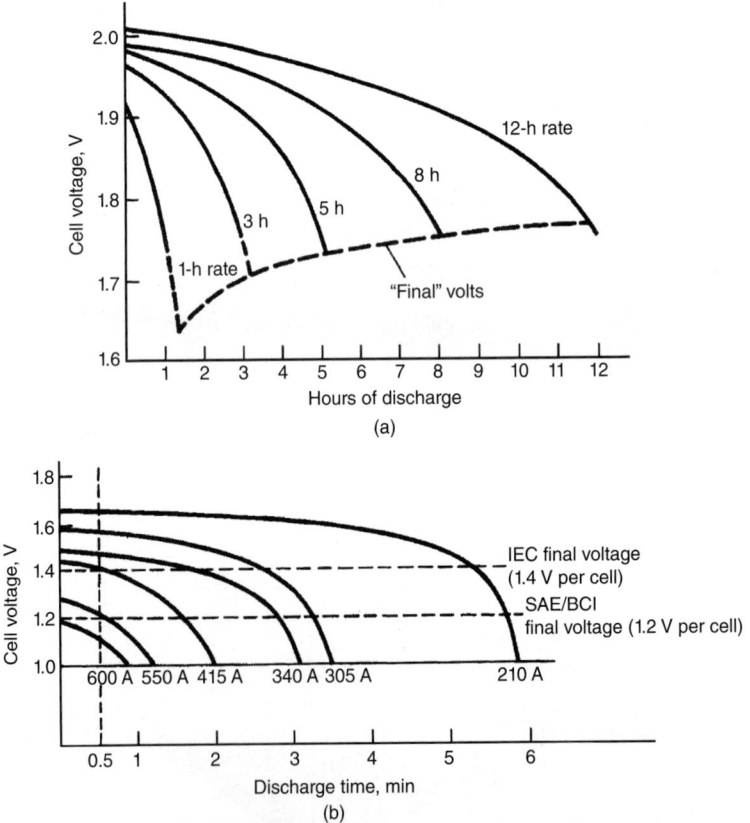

**FIGURE 14.56**   Discharge curves of lead-acid SLI batteries. (*a*) At various hourly rates and 25°C. (*b*) At various high rates and −17.8°C. Battery rated at 70 Ah, 20 h rate at 25°C.

also shown. Higher service capacity is obtained at the lower discharge rates. At higher discharge rates, the electrolyte in the pore structure of the plates becomes depleted and the electrolyte cannot diffuse rapidly enough to maintain the cell voltage. Intermittent discharge, which allows time for the electrolyte to recirculate, or forced circulation of the electrolyte will improve high-rate performance. In general, the lead-acid cell may be discharged without harm at any rate of current it will deliver, but the discharge should not be continued beyond the point where the cell approaches exhaustion or where the voltage falls below a useful value.

***Effect of Temperature on Performance.***    Figure 14.57a shows typical discharge curves for a lead-acid single-cell battery at several discharge temperatures. Figure 14.57b shows the discharge characteristics of a 12 V, 60 Ah battery when discharged at 340 A at temperatures from −30 to 25°C. Higher discharge voltages and capacities are obtained at the higher temperatures and lower discharge rates.

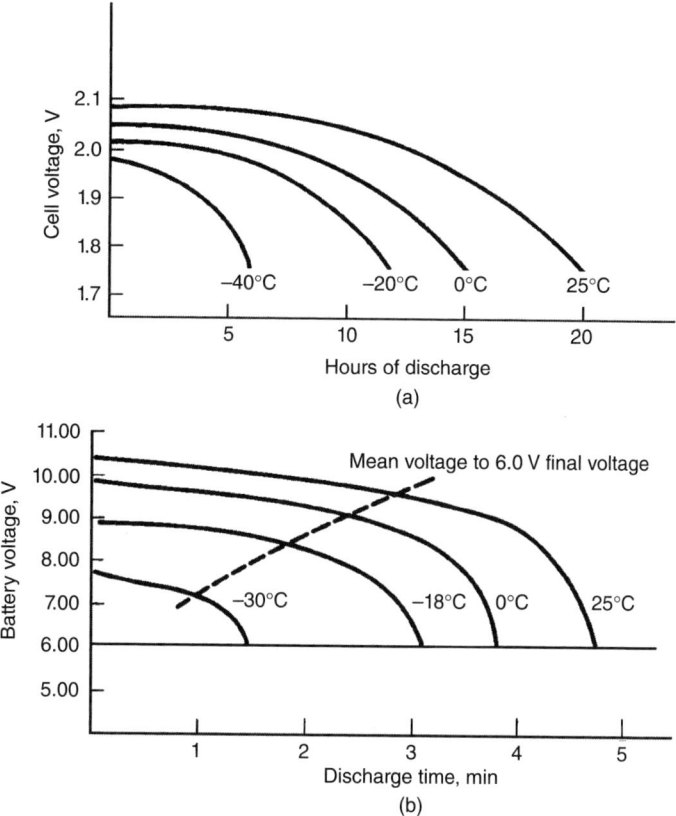

**FIGURE 14.57**    Discharge curves of lead-acid SLI batteries at various temperatures. (*a*) At C/20 rate. (*b*) At 340 A, 12 V battery, nominal capacity 60 Ah, 20 h rate at 25°C.

The effect of discharge rate and temperature on the capacity of the lead-acid battery is summarized in Fig. 14.58, which shows the percentage of the 20 h rate capacity delivered under different discharge conditions. Although the battery will operate over a wide temperature range, continuous operation at high temperatures may reduce life as a result of an increase in the rate of corrosion (see Sec. 14.7.1).

The performance of the lead-acid SLI-type cell at different temperatures and loads is given in another form in Fig. 14.59. The logarithm of the current drain is plotted against the logarithm of the service hours, in

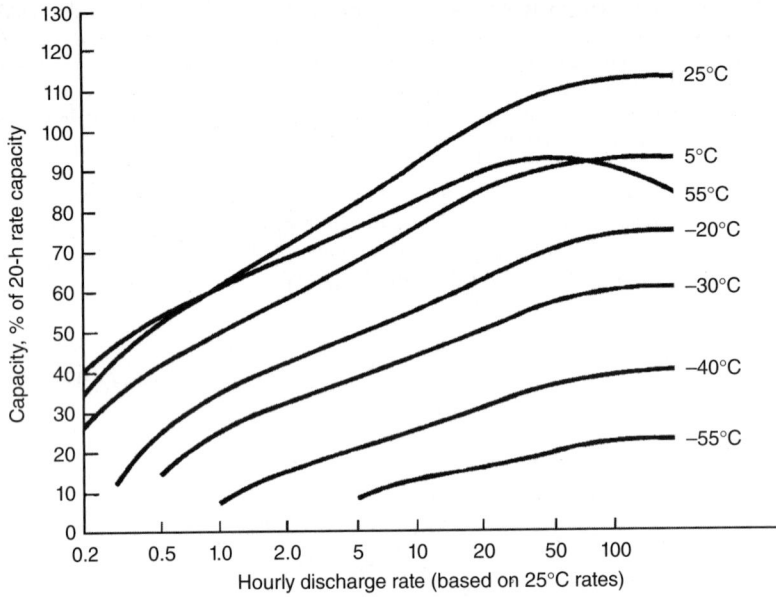

**FIGURE 14.58** Performance of lead-acid SLI batteries at various temperatures and discharge rates to 1.75 V per cell end voltage.

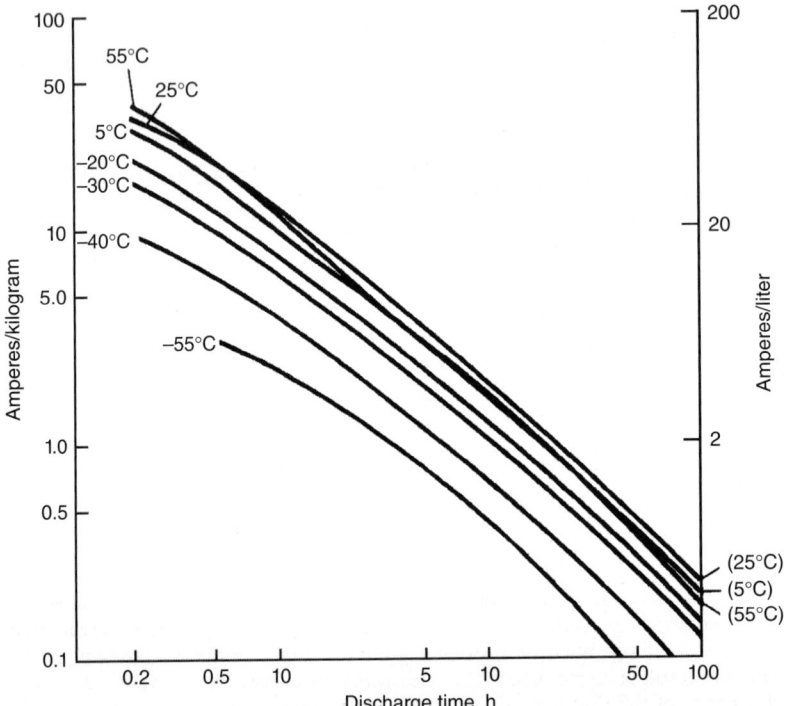

**FIGURE 14.59** Service life of lead-acid SLI battery to 1.75 V end voltage per cell.

accordance with Peukert's relationship. The linear relationship is maintained over a wide range, with divergences appearing on the high-rate and low-temperature discharges. In this figure, the data have been normalized to unit cell weight (kilograms) and unit cell volume (liters). Figure 14.59 can be used to approximate the performance of various size cells over the operating conditions shown or to determine the size and weight of a battery to meet a particular service requirement.

***Internal Resistance.*** The high current requirement for engine cranking demands that the batteries be designed with low resistance; for example, that conductors have large cross sections and minimal lengths; that separators have maximum porosity and minimum back-web thickness; and that the electrolyte be in the range of low resistance. The resistance can be evaluated by Ohm's law by determining the voltage difference at two levels of discharge current. The resistance of a lead-acid battery increases during a discharge almost linearly with the decrease of the specific gravity of the electrolyte. The difference in resistance between full charge and discharge is in the order of 40%. The effect of temperature on the resistance of the battery is shown in Fig. 14.60; the battery resistance increases by about 50% between 30 and −18°C.

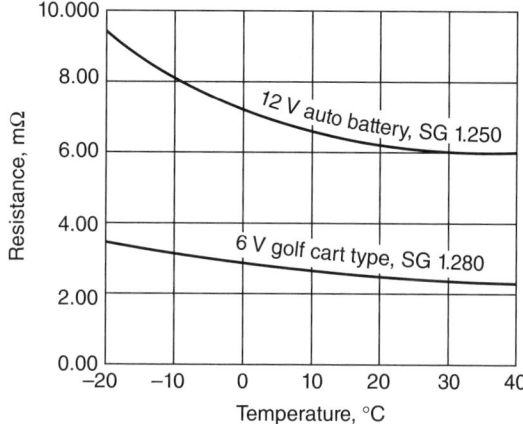

**FIGURE 14.60**    Comparison of lead-acid battery designs. Effect of temperature on battery resistance.

***Life and Failure Modes.*** The life of SLI batteries is affected by the design, the processing, and the operational environment of the battery. Because of the automated assembly methods used today, SLI batteries are fairly consistent in life under ideal operating conditions, but the wide variety of operating conditions tends to spread the failure distribution. Warranty coverage for a failed battery is often more dependent on marketing strategy than on the statistical expectations of the failure rate.

SLI battery design, materials, and operation have changed markedly in recent decades; life and failure mode distribution have also changed, as shown in Fig. 14.61*a*. Possible explanations for the shorter life in 1982 may be a reduction in battery size and more demanding cycling requirements. These averages include taxis, police cars, and other heavy-duty users, which account for the relatively low age for failed batteries. Figure 14.61*b* shows the failure modes for these batteries, which are described in more detail in Table 14.21. A higher incidence of short-circuited batteries used in warmer climates suggests that grid corrosion is still a major failure mode. The "worn-out" category includes low electrolyte level, and it should be noted that many maintenance-free SLI batteries are sealed so that water lost to evaporation and electrolysis cannot be replaced.

SLI batteries are not designed for deep-cycling service, and very short lives are generally obtained with such operation. The deep-cycling capability of SLI batteries is covered in Sec. 14.9.3.

***Standard Tests for Rating SLI Batteries.*** Several standard tests have been devised to evaluate and rate the performance of SLI batteries under conditions simulating the major requirements of their applications. The CCA test evaluates the capability of the battery to deliver power to crank an engine at cold temperatures. The CCA

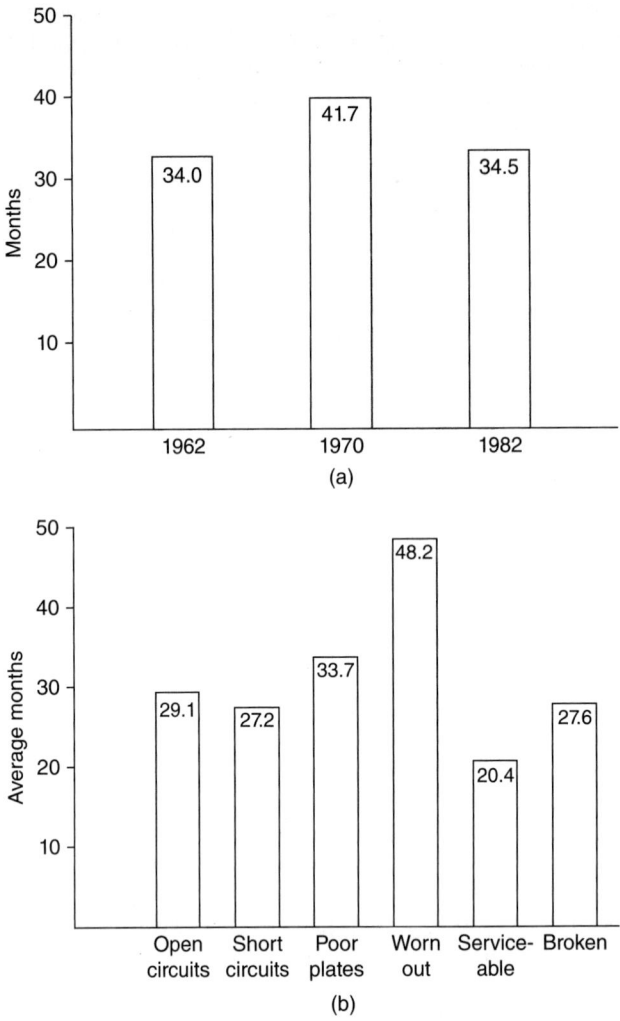

**FIGURE 14.61**   Failure modes of SLI batteries. (*a*) Average age of returned batteries. (*b*) Failure mode of returned batteries. (*From Ref. 1.*)

rating is the current that a fully charged battery can deliver at −17.8°C (0°F) for 30 s to a voltage of 1.2 V per cell. If the measured voltage is above or below this value at 30 s, the CCA value can be estimated by multiplying the discharge current by the correction factor shown in Fig. 14.62. Figure 14.56*b* illustrates the performance of a 70 Ah cell with a CCA rating of 550 A. The cranking ampere test (CA) is analogous to the CCA test but at a temperature of 0°C (32°F).

Reserve capacity is measured to test the battery's ability to provide power for lights, ignition, and the auxiliaries. The reserve capacity is defined as the number of minutes a fully charged battery can maintain a current of 25 A to 1.75 V per cell at 25°C.

Other SLI tests are included in the SAE battery test standard J537 on charge rate acceptance, overcharge life, and vibration resistance. A standard SLI life test is specified in SAE J240A. This test consists of a shallow

**TABLE 14.21**    Summary of Failure Modes of Lead-Acid SLI Batteries

| | |
|---|---|
| 1. Open circuits | 4. Worn out |
|    *a.* Terminal |    *a.* Worn out |
|    *b.* Cell to terminal |    *b.* Undercharge |
|    *c.* Cell to cell |    *c.* Low level (electrolyte) |
|    *d.* Broken straps |    *d.* Terminal corrosion |
|    *e.* Plates off |    *e.* Under formed |
| 2. Short circuits | 5. Serviceable |
|    *a.* Plate to strap |    *a.* Serviceable |
|    *b.* Plate to plate (plate fault) |    *b.* Discharged only |
|    *c.* Plate to plate (separator fault) | 6. Broken |
|    *d.* Plate to plate (sediment/moss) |    *a.* Broken container |
|    *e.* Vibration short circuit |    *b.* Broken cover |
| 3. Poor plates |    *c.* Damaged terminal |
|    *a.* Overcharge/overheat |    *d.* Internal damage |
|    *b.* Grid corrosion |    *e.* Other |
|    *c.* Paste adhesion | |
|    *d.* Paste sulfation | |
|    *e.* Paste under formed | |

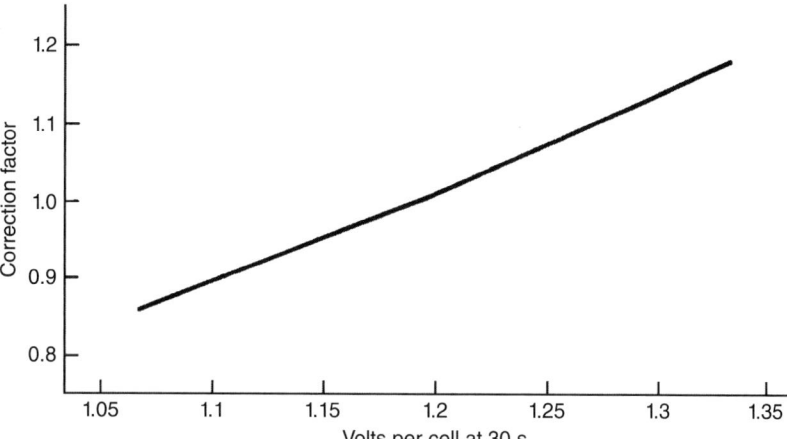

**FIGURE 14.62**    Correction factor for calculating cold-cranking ampere (CCA) rating.

discharge at 25 A followed by a brief charge at voltage and current limits for 10 min and a weekly stand followed by a high-rate discharge at the CCA rating performed at temperatures of 41°C or 75°C.

## 14.8.5    Cell and Battery Types and Sizes

SLI battery sizes have been standardized by both the automotive industry through the Society of Automotive Engineers (SAE), Warrendale, PA (now known as SAE International) and the battery industry through the Battery Council International (BCI), Chicago, IL.[1] The latest standards are published annually by the BCI.[1,49]

Internationally, standardization is handled by the International Electrotechnical Commission (IEC). More detailed information on these standards is found in Chap. 8 (Sec. 8.3.3).

### 14.8.6   Enhanced Flooded Batteries

EFBs have been developed for microhybrid cars, those that are equipped with start-stop technology. As the electrification of the vehicle increases further, the demands on the battery also increase and the EFBs may become insufficient. The use of VRLA batteries and ultimately lead-carbon batteries may be required.

In order to modify a normal flooded SLI battery into an EFB SLI battery the following changes are needed:

*Increase the Compression of the Plates.*   Typically, flooded SLI plates are inserted rather loosely in the container. EFB plates *must* be inserted with as high a compression as possible for the given assembly line conditions.

*Reinforce the Positive Plates.*   The PAM must be denser and the positive plate must be covered by a pasting paper that will reinforce it, preventing premature shedding.

*Reinforce the Negative Plates.*   The negative grid must be made thicker. The main reason is to provide a thicker lug; it is well known that the partial state of charge application leads to lug thinning of the negative plates.

*Carbon Additives.*   Another major change that is needed in the NAM is to render it more resistant to sulfate build up and make it more conductive. To this end, carbon additives are needed and their concentration exceeds what has been used in the NAM over the years. Carbon additives to NAM are specific for the partial state charge cycling demanded of EFBs. These carbon additives must allow for the dispersion of sulfate throughout the NAM and enable increased conductivity when the negative plates are heavily sulfated.

Recent test results (2016–2017) shown in Table 14.22 show that EFB technology can greatly enhance the performance of flooded automotive batteries in automotive partial state of charge cycling.

**TABLE 14.22**   Partial State of Charge Cycling Comparing EFB with Standard Flooded and VRLA-AGM Batteries

|  | Cycles at 50% DOD, 40 C | Cycles at 17.5% DOD, 27 C |
|---|---|---|
| OEM requirement for EFB | 270 | 1020 |
| Battery technology: |  |  |
|   Standard flooded | 130 | 580 |
|   EFB | 355 | 1650 |
|   VRLA-AGM | 605 | 1730 |

*Source:* Courtesy of Acumuladores Moura S.A. (Brazil).

## 14.9   DEEP-CYCLE, TRACTION, MOTIVE BATTERIES: CONSTRUCTION AND PERFORMANCE

Deep-cycle batteries are designed for low power, high capacity, and long life. The basic designs include flooded lead-acid, maintenance-free, and VRLA batteries. Applications for lead-acid batteries, other than the SLI and small sealed power units, fall into two categories, as shown in Table 14.23—those based on automotive-type constructions and those based on industrial-type constructions. Often several designs can be used for a single type of application. Stationary batteries, which may be either shallow or deep cycles, will be covered in more detail in the next section.

**TABLE 14.23**   Major Applications of Lead-Acid Batteries (Non-SLI Types)

| Automotive and small energy storage designs | | Industrial designs | | |
|---|---|---|---|---|
| Traction | Special | Stationary | Traction (motive power) | Special |
| Golf cart | Emergency lighting | Switch gear | Mine locomotives | Submarines |
| Off-road vehicles | Alarm signals | Emergency lighting | Industrial trucks | Ocean buoys |
| On-road vehicles | Photovoltaic | Telecommunication facilities | Large electric vehicles | |
| | Sealed cells (for tools, | Railway signals | | |
| | instruments, | Uninterrupted power supply | | |
| | electronic devices, | Photovoltaics | | |
| | etc.) | Load leveling and energy storage | | |

## 14.9.1   Marine Batteries

The marine battery market is another deep-cycle application and includes small and large leisure craft used for fishing, sailing, and travel as well as larger commercial vessels engaged in towing, passenger transportation, and workboat activities. In general, lead-acid battery systems are used (both flooded and VRLA) with system voltages ranging from 6 to 220 V where recharging is accomplished by an engine generator or alternator.

Marine service differs from automotive in several aspects. Lights, refrigeration, blowers, motors, radio, and other electrical equipment result in cycling service, often with a delay between discharge and charge. As a result, the batteries should have much greater capacity than would normally be specified for the same horsepower equipment in a shore-based application.

The key engineering features utilized by one prominent manufacturer of marine batteries[50] include the following:

- A special grid design with heavier vertical and horizontal members.
- High density active materials, both positive and negative.
- Positive plates double insulated with thick woven glass matting and then sealed in a microporous PE envelope.

In some designs, individual cells can be replaced while outer plastic cases provide high impact and environmental protection.

## 14.9.2   Construction

The prime requirement for deep-cycling batteries for traction applications is maximum cycle life, then, if possible, high energy density and low cost. In an electric forklift application, in fact, light weight may not be advantageous because the battery's weight usually is needed to counterbalance the payload. The life of these batteries is improved by the use of thick plates with high paste density, usually a high-temperature and high-humidity cure, low electrolyte density formation, premium separators, and one or more layers of glass fiber matting (to retain the active material in the positive plates). The major modes of failure are disintegration of the $PbO_2$ positive active mass and corrosion of the positive grids. The deep-cycling battery is usually designed to be capacity limited when new by the amount of electrolyte and not by the material in the plates. This serves to protect the plates and maximize their life. Both negatives and positives are degraded during use, but at end of life, battery capacity is generally limited by the positive plate.

A typical traction battery, using flat-pasted plates, is shown in Fig. 14.63. Cells are always made with an outside-negative design (e.g., $n$ positive plates, $n + 1$ negative plates). Deep-cycling traction batteries are built as

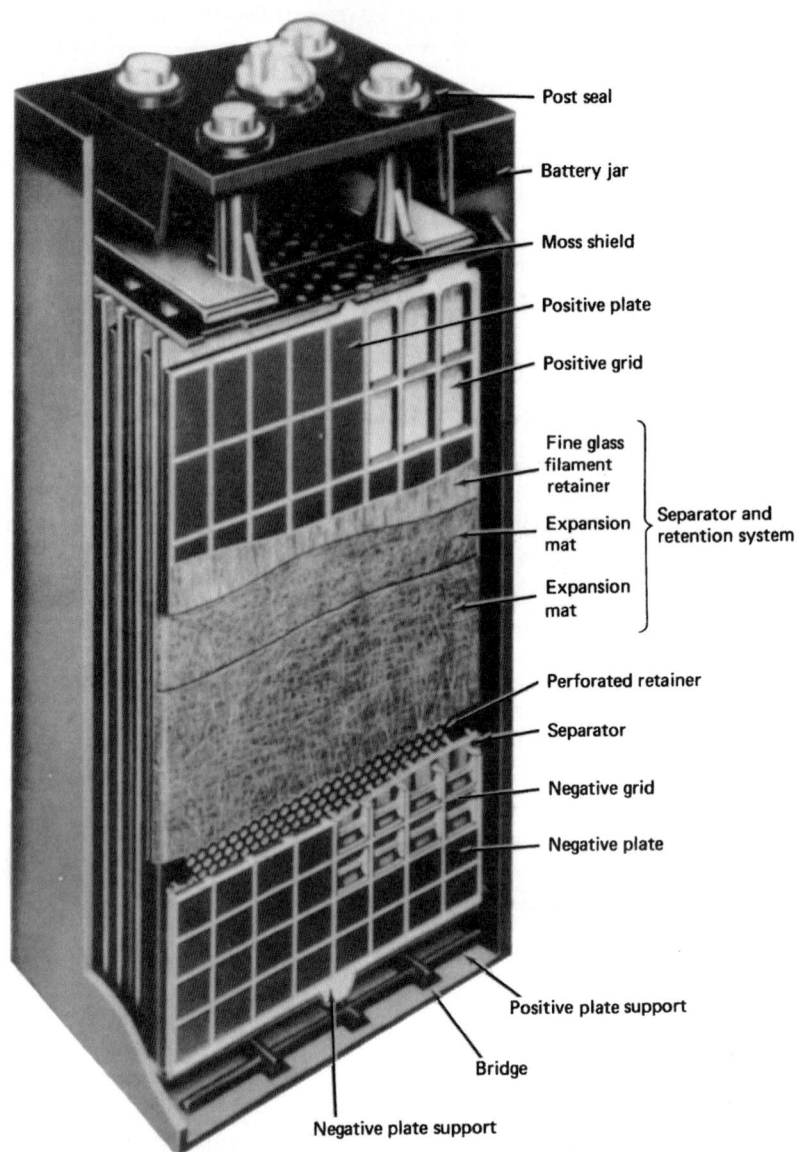

**FIGURE 14.63** Flat-pasted-plate lead-acid traction battery. (*Courtesy of C&D Technologies.*)

an assemblage of individual cells. If the battery's performance is limited by a catastrophic failure of one (or a few) cell(s), then those cells can be repaired or replaced in a cost-effective manner. Power requirements vary widely with the load, distance traveled, and lifting or climbing requirements. Battery sizes are determined by the forklift truck manufacturer and can be "calibrated" in the actual application by the use of an ampere-hour meter. A rough indication of the suitability of a traction battery for an application is the change in the specific gravity of the electrolyte during use. A larger battery size (or battery replacement or repair) is indicated when full operation cannot be achieved.

Although the flat-pasted (Faure) positive plate is typical for deep-cycling batteries in the United States, some cycling batteries in the United States and most cycling batteries in the rest of the world are built with tubular or gauntlet-type positives (Fig. 14.64). The tubular construction minimizes both grid corrosion and shedding, and long life is characteristic of these designs, but at a higher initial cost. Flat-pasted negative plates are used in conjunction with these positives and the cells are of the outside-negative design.

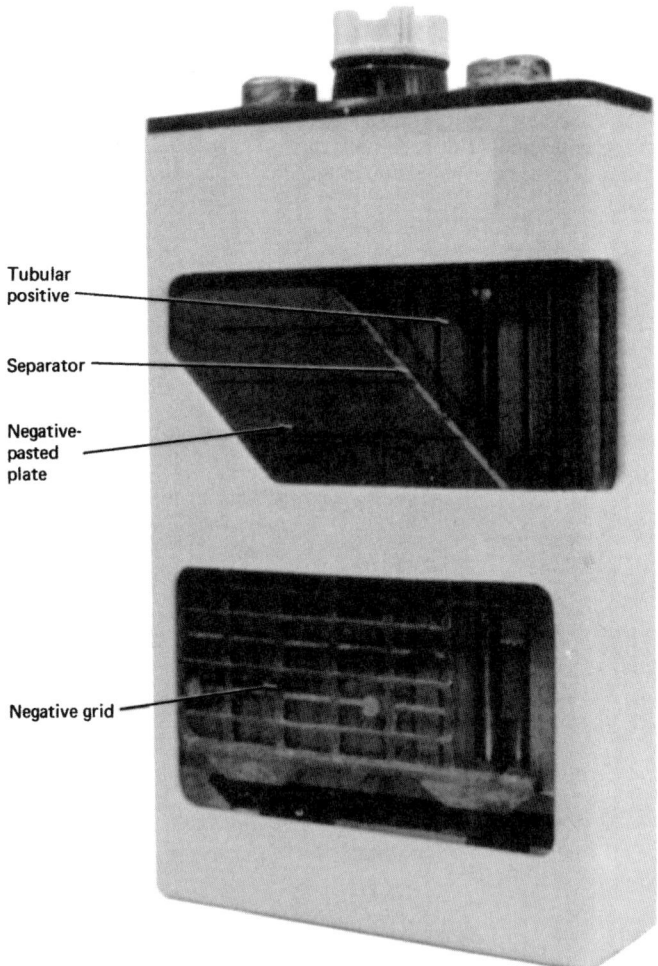

**FIGURE 14.64**   Lead-acid cell with tubular positive plates. (*Courtesy of Enersys, Inc.*)

Small traction batteries (such as for golf carts) are designed to be intermediate between full-sized traction batteries and SLI batteries. Traction design concepts sometimes utilized include high paste density, careful control of plate curing and formation, maximizing the content of the positive plates, glass matting against the positive, and tubular positives. SLI concepts sometimes utilized for golf cart and other electric vehicle batteries include thin cast radial grids, minimum separator resistance, through-the-partition intercell connection, and heat or epoxy-sealed plastic cases and covers. Cost is also an important factor.

Naval submarines of the diesel electric type require cycling batteries for propulsion. Nuclear submarines use similar batteries for standby and emergency service. These batteries are made with Pb-Ca-Sn grids because stibine and arsine produced on charge are unacceptable for personnel health in a closed environment. The plates are much larger than most traction cells—up to 600 cm wide and up to 1500 cm tall. Both flat-pasted and tubular positive plates are used.

### 14.9.3    Performance Characteristics

Batteries for traction and deep-cycle applications use cells with either pasted or tubular positive plates. In general, the performance of the two types of plates is similar, but the tubular or gauntlet plates show lower polarization losses because of the larger active surface area, better retention of the PAM, good compression of the active material against the spine, and reduced self-discharge. The self-discharge rates at room temperature for the two-plate structures, as measured by the drop in specific gravity, are shown in Fig. 14.65.

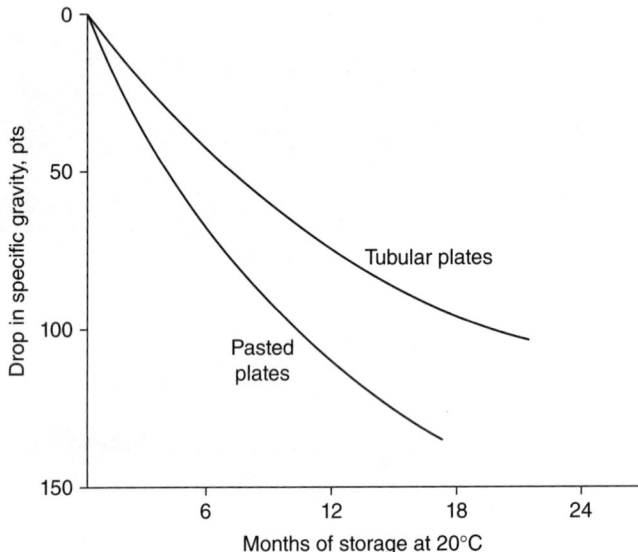

**FIGURE 14.65**    Retention of charge of pasted- versus tubular-plate traction batteries.

Typical discharge curves for the two types of traction cells are shown in Fig. 14.66. The relationship of discharge current to ampere-hour capacity, up to various end voltages, is shown in Fig. 14.67. These data are presented on the basis of the positive plate since cell design and performance data of traction batteries are generally based on the number and size of positive plates that are in the cell. As is typical with most batteries, the capacity decreases with increasing discharge load and increasing end voltage.

The same relationship and comparison of the performance of the pasted and tubular plates are plotted in Fig. 14.68. These data show the superiority of the tubular plate as the discharge rate is increased.

Figure 14.69 shows the increase in available service on intermittent discharge, carried out over different periods, as compared with a continuous discharge. The gain is more pronounced at the heavier discharge loads and when the intermittent discharge is spread out over a longer period, thus allowing more time for recovery.

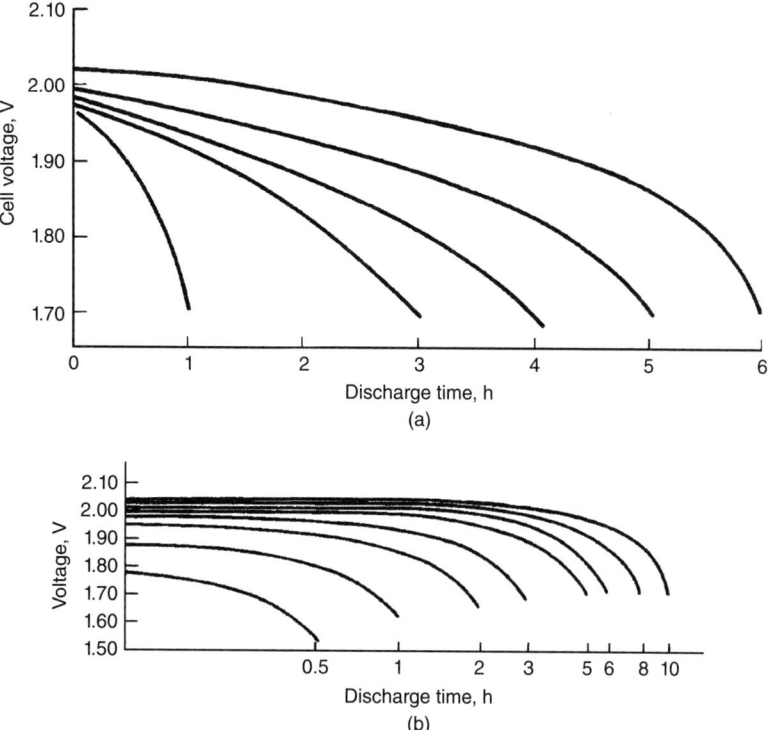

**FIGURE 14.66**   Discharge characteristics of traction batteries at 25°C. (*a*) Flat-pasted-plate batteries. (*b*) Tubular positive batteries.

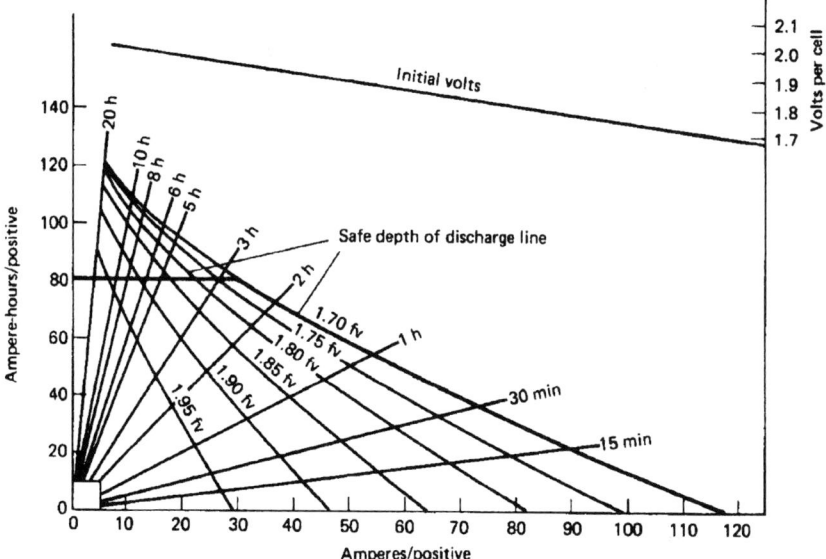

**FIGURE 14.67**   Performance characteristics of industrial flat-pasted plate traction battery to various final voltages (FV), at 25°C based on positive plate rated at 100 Ah at 6 h rate.

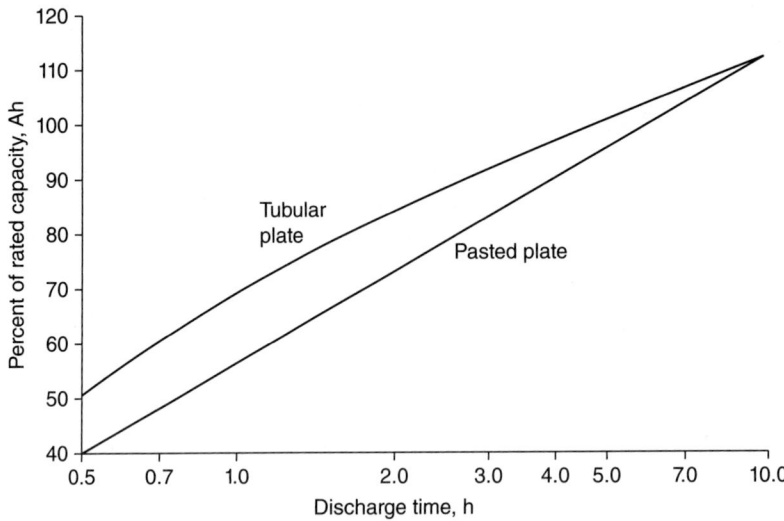

**FIGURE 14.68**   Effect of discharge rate on capacity of traction batteries at 25°C. Comparison of performance of flat-pasted-plate versus tubular-plate batteries.

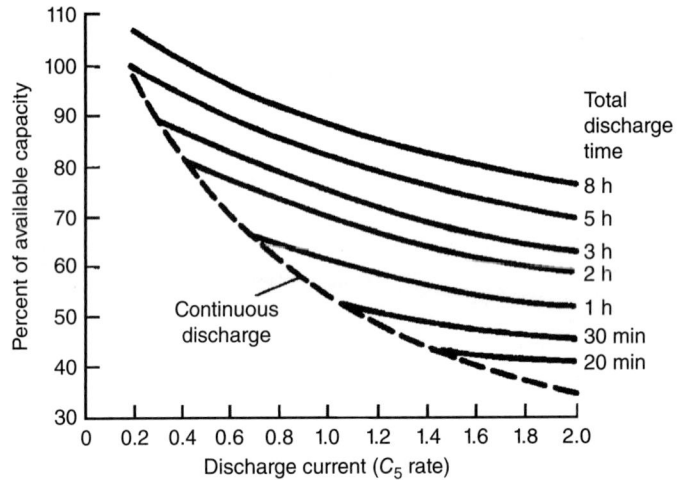

**FIGURE 14.69**   Capacity available on intermittent discharge of traction batteries at 25°C.

The effect of temperature on the discharge performance of traction-type batteries is illustrated in Fig. 14.70.

The cycle life characteristics of traction batteries are presented in Fig. 14.71. This figure shows the relationship of cycle life to DOD at the 6 h discharge rate, cycle life being defined as the number of cycles of 80% of rated capacity. It is quite evident that the deeper the cells are discharged, the shorter their useful life, and that 80% DOD should not be exceeded if full life expectancy is to be attained. Figure 14.67 shows the safe DOD for other discharge rates. At low rates, the discharge should be terminated at the higher end voltages as shown, until the 1.70 V line is intercepted; then, the discharges at the higher rates can be run to 1.7 V per cell final voltage. Typical cycle life expectancy is 1500 cycles (approximately 6 years).

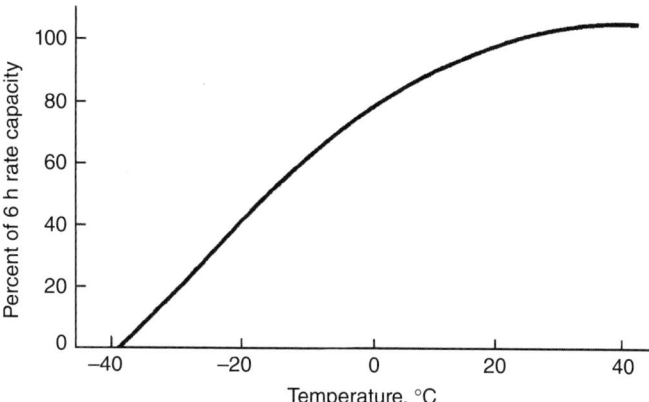

**FIGURE 14.70**   Effect of temperature on capacity of traction batteries, typical flat-plate design. (*From Ref. 40.*)

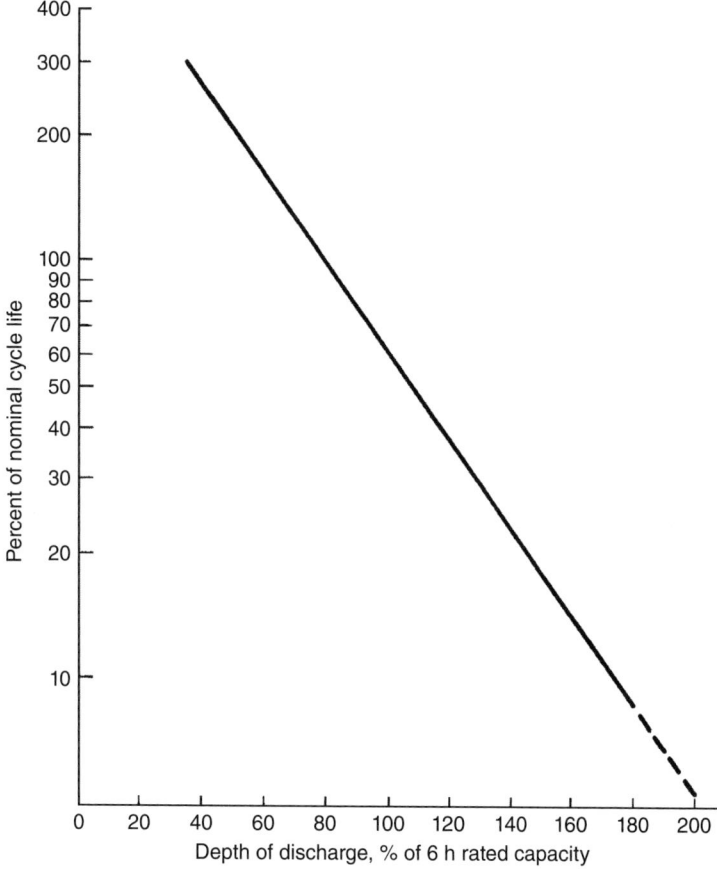

**FIGURE 14.71**   Cycle life versus depth of discharge of traction batteries.

The relationship of discharge current and service time for several small deep-cycling batteries is shown in Fig. 14.72. At very high discharge rates, Peukert's relationship does not hold as well as for the SLI types, and the performance deviates at shorter discharge times. Nevertheless, such deep-cycling batteries can be used for cranking service and may be preferred if the battery will be deeply or repeatedly discharged in operation. Conversely, an SLI battery generally makes a poor deep-cycling battery; SLI batteries are usually made with nonantimonial lead grids (U.S. practice) with lower density pastes. A comparison of the cycle life at a low discharge rate (25 A) of an SLI battery with a deep cycle design of the same physical size is shown in Fig. 14.73.

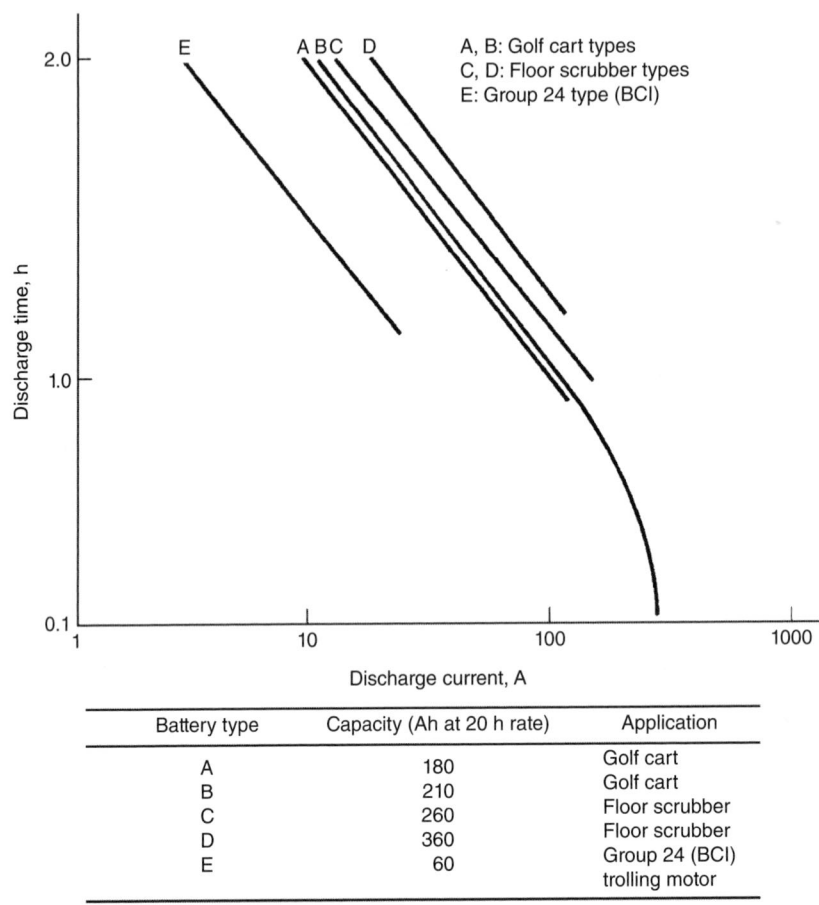

| Battery type | Capacity (Ah at 20 h rate) | Application |
|:---:|:---:|:---|
| A | 180 | Golf cart |
| B | 210 | Golf cart |
| C | 260 | Floor scrubber |
| D | 360 | Floor scrubber |
| E | 60 | Group 24 (BCI) trolling motor |

**FIGURE 14.72**   Performance of motive batteries.

### 14.9.4   Cell and Battery Types and Sizes

Traction or motive-power batteries are made in many different sizes, limited only by the battery compartment size and the required electrical service. The basic rating unit is the positive-plate capacity, given in amperehours at the 5 or 6 h rate. Table 14.24 lists the typical U.S. traction plate sizes using flat-pasted plates; between 5 and 33 plates are used to assemble traction cells, as also shown in the table. Ratings of the cell are the product of the capacity of a single positive plate multiplied by the number of positive plates. The cells, in turn,

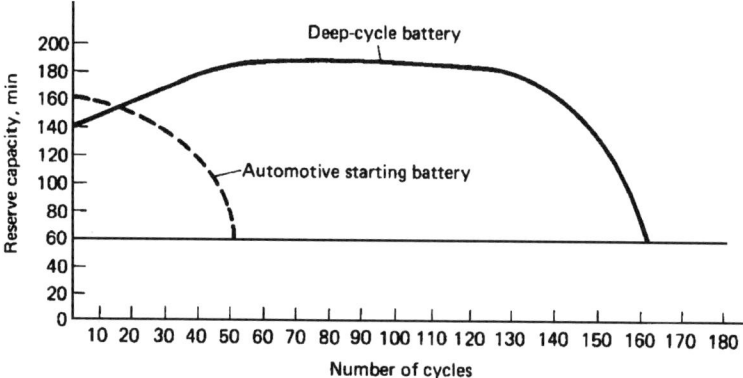

**FIGURE 14.73**    Cycle life characteristics at low discharge rate (25 A) for deep-cycle versus SLI-type batteries.

**TABLE 14.24**    Typical Traction Batteries (United States), Flat-Pasted Plates

| Positive-plate capacity, Ah at 6 h rate | Plate dimensions, mm | | | | | Cell size,[*][†] (positive plates per cell) |
| --- | --- | --- | --- | --- | --- | --- |
| | | Width | | Thickness | | |
| | Height | Positive | Negative | Positive | Negative | |
| 45 | 275 | 143 | 138 | 6.5 | 4.6 | 5–16 |
| 55 | 311 | 143 | 138 | 6.5 | 4.6 | 5–16 |
| 60 | 330 | 143 | 138 | 6.5 | 4.6 | 5–16 |
| 75 | 418 | 143 | 138 | 6.5 | 4.6 | 2–16 |
| 85 | 438 | 146 | 146 | 7.4 | 4.6 | 3–16 |
| 90 | 489 | 138 | 143 | 6.5 | 4.6 | 3–16 |
| 110 | 610 | 143 | 143 | 7.4 | 4.6 | 4–12 |
| 145 | 599 | 200 | 200 | 6.5 | 4.7 | 4–10,12,15 |
| 160 | 610 | 203 | 203 | 7.2 | 4.7 | 4–10,12,15 |

[*]All cells have $n$ positive plates and $n + 1$ outside negative plates.
[†]Typical cell characteristics: 6 positive, 85 Ah plates (510-Ah cell); weight: 45 kg; size: length, 127 mm; width, 159 mm; height, 616 mm.
   ***Source:*** C&D Technologies.

are assembled in a variety of battery layouts, with typical voltage in 6 V increments (e.g., 6, 12, 18, …, 96 V) resulting in almost 1000 battery sizes. Popular traction battery sizes are the 6-cell, 11-plates-per-cell, 75 Ah positive-plate (375 Ah cell) and the 6-cell, 13 plates-per-cell, 85 Ah positive-plate (510 Ah cell) batteries. Table 14.25 presents similar information on the tubular positive-plate batteries.

Several SLI group sizes have been used for deep-cycling applications, especially taller versions of otherwise SLI standard lengths and widths. Some of these are listed in Table 14.26.

***Fast-Charge Application.***    Lead-acid batteries in the motive-power application have one of the toughest jobs of all lead-acid batteries. In this application, the battery goes from a state of charge of 100% to less than 20%. The swing in electrolyte densities goes from the fully charged 1.280 to 1.150 g/cc. Traditionally, this happens over an 8-h work shift. Following the work shift, the traction battery is placed on a recharge that typically lasts for 12 h; following this charge period, there is a 4-h rest period before the next work shift. To maximize battery use, large

**TABLE 14.25**   Typical Traction Batteries, Tubular Plates

| Positive-plate capacity, Ah at 6 h rate | Dimensions, mm | | | | | Cell size,[†‡] (positive plates per cell) |
| | Height | Width | | Thickness | | |
| | | Positive | Negative | Positive | Negative | |
|---|---|---|---|---|---|---|
| 49 | 249 | 147 | 144 | 9.1 | * | 4–10 |
| 55 | 258 | 147 | 144 | 9.1 | * | 4–10 |
| 57 | 300 | 147 | 144 | 9.1 | * | 4–10 |
| 75 | 344 | 147 | 144 | 9.1 | * | 4–10 |
| 85 | 418 | 147 | 144 | 9.1 | * | 4–10 |
| 100 | 445 | 147 | 144 | 9.1 | * | 4–10 |
| 110 | 565 | 147 | 144 | 9.1 | * | 4–10 |
| 120 | 560 | 147 | 144 | 9.1 | * | 4–10 |
| 170 | 560 | 204 | 203 | 9.1 | * | 3–8 |

*Varies from 5 to 8 mm depending on manufacturer.
[†]All cells have $n$ positive and $n + 1$ outside negative plates. Negatives are flat-pasted plates.
[‡]Typical cell characteristics; 6 positive, 85 Ah plates (510 Ah cell); weight: 36 kg; size: length, 127 mm; width, 157 mm; height, 549 mm.
   *Source:* Enersys, Inc.

**TABLE 14.26**   Small Deep-Cycling Batteries

| BCI type | Volts | Dimensions, mm | | | Ratings | Typical operational current, A | Applications |
| | | L | W | H | | | |
|---|---|---|---|---|---|---|---|
| U1 | 12 | 197 | 132 | 186 | 30–45 Ah at 20 h | 25 ⎫ | Trolling motors |
| 24 | 12 | 260 | 173 | 225 | 75–90 Ah at 20 h | 25 ⎬ | Wheelchairs |
| 27 | 12 | 306 | 173 | 225 | 90–105 Ah at 20 h | 25 ⎭ | |
| GC2 | 6 | 264 | 183 | 260 | 75 min at 75 A | 75 (GC) ⎫ | Golf carts |
| (GC2H) | 6 | 264 | 183 | 260 | 95–90 min at 75 A | 300 (EV) | Electric vehicles |
| Not assigned | 6 | 264 | 183 | 260 | 100–100 min at 75 A | 300 (EV) ⎬ | |
| Not assigned | 12 | 261 | 181 | 279 | 105 Ah at 20 h | 150 (EV) ⎭ | |
| Not assigned | 6 | 295 | 178 | 276 | 200–230 Ah at 20 h | 50–75 ⎫ | Floor maintenance machinery |
| Not assigned | 12 | 241 | 166 | 239 | 50–70 Ah at 20 h | 50–75 ⎭ | |
| Not assigned | 12 | 518 | 276 | 445 | 350–400 Ah at 20 h | 30–50 | Mine cars |

   *Source:* BCI Technical Committee, Battery Council International.

operations use battery rooms, where vehicles change batteries with freshly charged ones replacing those that just ended the work shift. In such a traditional application, it is common to measure traction battery life in years, a good battery lasting anywhere from 4 to 5 years.

But the world of motive-power batteries is changing. The change is driven by economics and the relentless drive for greater operation efficiencies. Starting around 2000, manufacturers realized that they could no longer afford large battery room operations, thus was born the fast-charge application. More details on traction battery applications are presented in Chap. 26B.

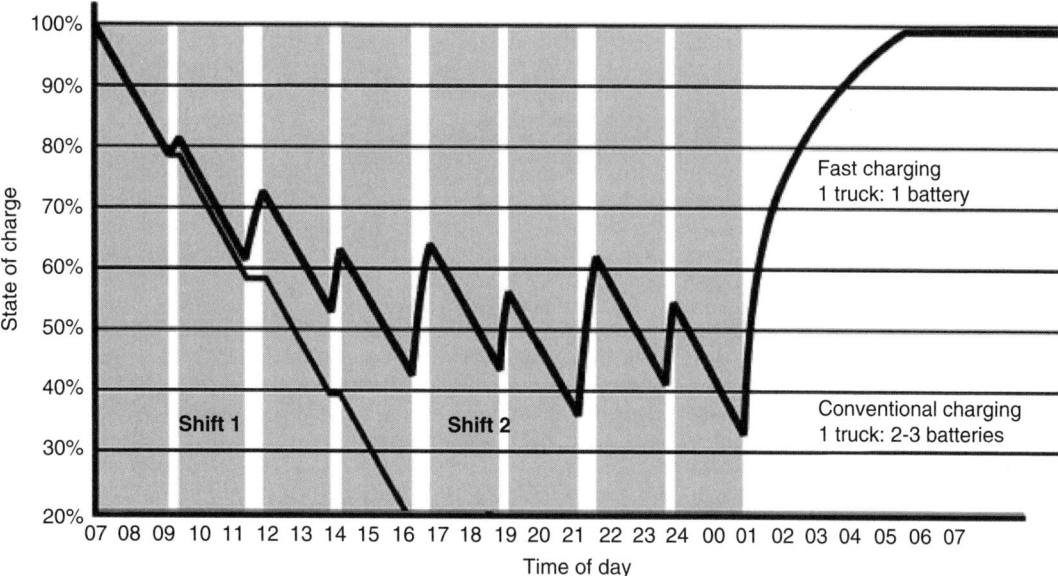

**FIGURE 14.74**  Conventional charging cycle versus fast charging.

In this new mode of using motive batteries, the battery stays in the vehicle. The recharging is done during the work shift, taking advantage of the idle time intervals available during the shift (i.e., coffee and lunch breaks). The key to a successful fast-charge operation is a good and powerful charger. Charger technology has evolved a lot in the last couple of decades and high-frequency chargers are now available that can put out large currents very efficiently. The motive-power battery now operates in a partial state of charge and accepts high recharge currents during short intervals. This is tough on the battery, and therefore the traditional life expectancy of 5 years no longer applies. Charging parameters and temperature control are necessary to ensure that overcharge and excessive temperatures are limited. Life is reflected in terms of capacity turnover of the number of cycles delivered. In this tougher application, the motive battery life time is somewhat reduced, but the efficiency of the operation is greatly enhanced, because there is no longer a dedicated battery room and the battery is more fully utilized. Figure 14.74 illustrates some of these changes.

## 14.10  STATIONARY AND ENERGY-STORAGE BATTERIES

### 14.10.1  General Considerations

Stationary or standby applications for batteries are characterized by a duty cycle where the battery spends most of the time on a constant-voltage floating charge sufficient to counter any tendency to self-discharge so that the battery may take over an external load without interruption if the electricity supply fails. In some applications, the battery may be charged intermittently to maintain it close to top-of-charge, but the duty cycle will remain similar. Battery life is generally determined by the corrosion of the positive grid at a constant potential that, in turn, is influenced by temperature, the grid alloy, grid thickness, and the metallurgical processing of the grid. The battery will be designed so that other failure modes are not life limiting. These types of batteries may be flooded or valve-regulated. For energy storage, the duty cycle necessarily includes extensive cycling that may be deep cycling as for traction batteries but often includes shallow cycling and periods when the battery may not be routinely returned to a fully charged condition. These applications require the battery to be modified to provide a full-service life, and although this market sector may be considered as an extension of the stationary sector as the batteries are permanently installed in a fixed location, it is quite distinct and will be described below.

The main application areas for standby batteries are telecommunications, uninterruptible power supplies (UPS), utilities, and general standby. Telecommunication networks are generally powered from a DC bus, usually at a nominal 48 V, supplied by a rectifier from a normal AC line with a battery floating on the bus. If the utility supply fails, the battery can supply the load for periods between 1 and 10 h with further backup from standby generators. This is often a statutory requirement in order to provide a secure service under emergency conditions. The battery may be a large installation in a central office or switching center, in a small exchange, or in smaller modules in a distributed form alongside individual equipment racks. Battery backup for cellular telephone networks tends to use shorter autonomy times for individual nodes, as the considerations for reliability are less stringent, but as the traffic is concentrated through back-haul, system integrity becomes more important.

UPS is mainly used for data processing, but it is also used for critical processes such as semiconductor device manufacture. The installations may be large and able to provide megawatts of power or limited to a few watts for individual computers. Autonomy times for UPS applications are usually short, of 5 to 30 min duration. Standby generation may take over the load, or the system may be configured such that data is protected before the standby supply is terminated. UPS installations not only provide backup power, but also improve power quality by using systems that rectify the incoming supply to provide a stable voltage for floating the battery. Subsequently, the resultant DC current is inverted back to an AC current but with less noise, spikes, sags, and other distortion to the waveform compared to the original AC power supply.

Utilities are major users of standby batteries for both generation and distribution. Batteries are used for all safety-critical systems in the power-generating plant, particularly nuclear power stations and switchgear tripping in the distribution network that must be reconfigured, disconnected, or reconnected irrespective of utility power.

General standby installations include emergency lighting, fire alarms, intruder alarms, and security systems. Many of these use smaller batteries with a capacity of less than 25 Ah that may be referred to as portable batteries.

Stationary lead/acid batteries may be divided into two basic types: flooded or VRLA. These, in turn, may be further categorized. Flooded batteries may have pasted plates, tubular plates, or Planté plates, or may be a special type of round cell. VRLA batteries may have gelled electrolyte or may use AGM separators. Batteries using gelled electrolyte may have tubular or pasted plates. Those using AGM separators all have pasted plates, but some use lead-calcium grid alloys and others use pure lead that may be alloyed with tin. Pure lead types may have flat plates or may use a wound construction. These cell types will be described below and their special characteristics highlighted.

### 14.10.2   Flooded Stationary Batteries

*General Considerations.*   Designs for flooded stationary batteries have evolved more slowly than for SLI and traction batteries, which reflects the longer service life of these batteries.[51,52] Heavy thick plates may be used, and there are older types of batteries, using Planté and pasted Faure plates, still in use. There are also round cells with a special type of construction designed for very long life (Sec. 14.3). The key standard applying to flooded stationary batteries is IEC 60896-11: 2002 Stationary lead-acid batteries—Part 11: Vented types—general requirements and methods of test.

*Pasted Plate.*   Pasted-plate cells using Pb-Ca-Sn grids are used in North America and in some other areas. The positive and negative plates are flat-pasted plates with a substantial thickness and are cast in Pb-Ca-Sn alloys, as seen in Fig. 14.75. The overall plate thickness for long-life cells is in the range

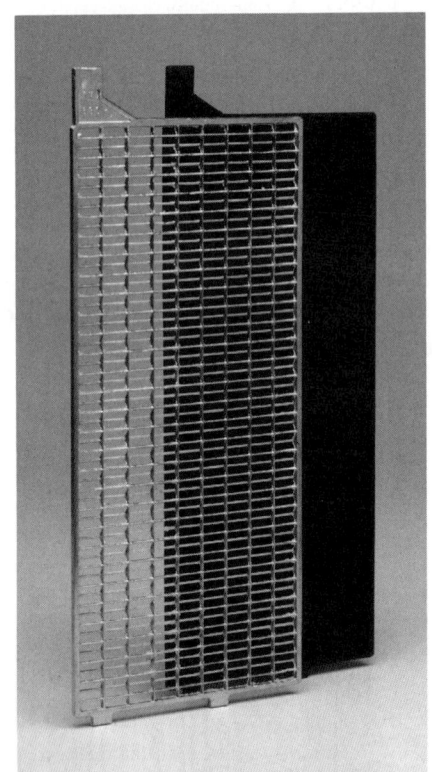

**FIGURE 14.75**   Pasted plates for stationary lead/acid cells; grid in foreground, pasted plate behind.

from 6 to 8 mm for the positive plates and 4 to 6 mm for the negative plates. The alloys used for the negative grid have sufficient Sn to ensure good casting behavior and use Ca as a hardening agent. The alloys used for the positive plate have higher Sn levels to improve the corrosion behavior and may have more moderate levels of Ca to avoid any tendency for intergranular attack. The positive plates are suspended from the top of the cell so that plate growth can occur in a benign manner in a downward direction. Hanging lugs supported by special features in the side of the container or nonconductive rods are used to hang the positive plates. A high paste density is used and special attention is paid to curing and drying to ensure a good conductive bond between the active material and the grids and freedom from cracking. The positive plates are wrapped in a glass retainer mat to ensure that shedding of the active material is minimized. The negative plates stand on ribs so that any sediment arising from shedding can accumulate harmlessly at the base of the cell. Cell designs normally have outer negative plates such that there are $n + 1$ negative and $n$ positives. The outer negatives may be thinner than the inner negatives for material efficiency. Microporous plastic separators are used. Cell containers are molded in transparent PVC, styrene acrylonitrile (SAN), acrylonitrile-butadiene-styrene (ABS), polycarbonate (PC), or PE so that the visual condition of the cells can be observed. Pillar seals are proprietary and need to have a high degree of integrity as well as the capability to avoid the buildup of stresses on the seal. Lead inserts molded into the lid and welded to the pillar or rubber grommets of various designs may be used. In some designs, the pillars can accommodate some movement to accommodate positive plate growth in service. Cells are fitted with vents that will trap acid mist generated by gassing and return acid to the cells. The vents should be flame retardant such that any accidental ignition of hydrogen external to the cell will not penetrate into the cell to cause an explosion. There is an excess of electrolyte such that maintenance watering is reduced to the minimum and the capacity is limited by the positive plate. The acid gravity tends to be moderate compared to other types of lead/acid batteries and the float voltage is adjusted accordingly. Cell capacities range from 100 to 4000 Ah. Service lives of up to 25 years are achieved in a well-controlled environment. This type of cell is used for telecommunications central office applications, nuclear power, national security installations, and for large critical UPS applications. Reliability is well established and because of that, they will continue to be specified.

Flooded pasted-plate cells may also be built with Pb-Sb grids or with pure Pb positive grids (Faure plates), but these are essentially of historic interest only. Pb-Sb grids, even with low Sb levels, result in higher water losses than Pb-Ca-Sn grids. The cyclic endurance of Pb-Sb grids is superior to Pb-Ca-Sn and for some applications they are preferred. Pure Pb grids offer an excellent corrosion life, but plate growth becomes the dominant failure mode and the cells need to be designed accordingly.

***Tubular Plate.*** Tubular plate cells have been historically more widely used in Europe, but they are now available in all parts of the world. They are often referred to as OPzS cells from Deutsches Institut für Normung (DIN) specifications for these types (DIN 40736-1, Lead-acid batteries—Part 1: Stationary vented cells with positive tubular plates in plastic containers [2015]). The positive plates have pressure die-cast Pb-Ca-Sn alloy spines, although some suppliers offer Pb-Sb spines as an alternative shown in Fig. 14.76. The spine diameter is normally 3 to 4 mm with the tube diameter 7 to 9 mm and the negative-plate thickness in the range from 5 to 6 mm. A multitubular fabric gauntlet is used as the retainer for the PAM. The negative plates are pasted-plate type, and generally microporous PE separators are used. Cell containers are usually molded in SAN so that the electrolyte level and the visual condition of the cell can be readily seen. Pillar seals vary in complexity between suppliers but offer freedom from corrosion and leakage over the service life of the battery. The specific gravity is higher than the pasted-plate cells. Sufficient excess of electrolyte is provided to extend the watering interval to

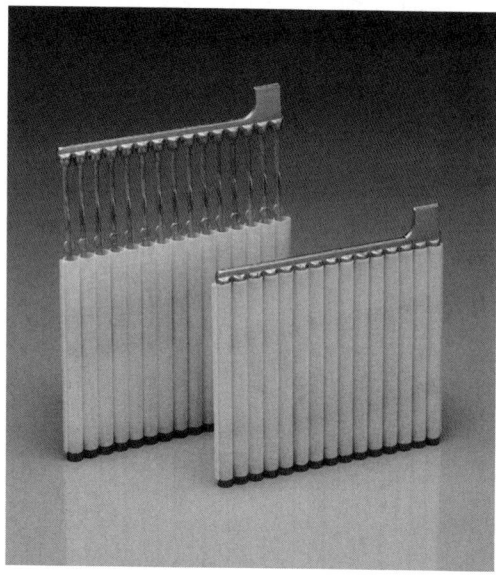

**FIGURE 14.76**  Tubular plates for stationary cells; grid with spines not fully inserted into the gauntlet at the rear, plate in foreground.

3 years and the overall service life is up to 15 years. Pb-Sb plates provide good cyclic endurance, and for areas where the reliability of the public utility is poor or for supporting solar power installations, tubular cells provide good service. They are used for telecommunications for standby power in main switching centers in fixed-line telephony but are being displaced by sealed VRLA tubular cells with gelled electrolyte.

*Planté Cells.*   Planté cells were used extensively in Europe for telecommunications and for utility applications, particularly for nuclear power. There is limited production for replacement, but they are no longer specified. These cells use a pure lead-positive grid with an extended surface consisting of fine lamellae, as shown in Fig. 14.77. This is 8 to 12 mm thick, but the negative plates are 4 to 6 mm thick, as there is a large amount of lead relative to active material used in the positive plates. The active material is formed by corroding the surface of the lead in situ. The lead is corroded under a controlled potential in a nonpassivating solution such as a mixture of sulfuric acid and potassium perchlorate, then the plates are discharged or reverse charged, extensively washed, and finally recharged in a normal sulfuric acid solution. The use of pure lead provides a long life in floating service and the manufacturing process ensures good conductivity across the grid/active material interface. The plates are hung from the top of the cell such that plate growth occurs harmlessly downward. Pasted plates with Pb-Sb alloy grids are used for the negative plates and microporous plastic separators. Cell cases and lids are molded in SAN to facilitate visual inspection. The pillar seals vary in design from simple rubber grommets to more elaborate arrangements with rubber boots and plastic fittings molded directly to the pillar. This type of battery has a very predictable behavior and a life of up to 25 years.

**FIGURE 14.77**   Planté plate for stationary lead/acid battery; cast plate to the rear, formed plate to the front.

*Round Cells.*   The round cells designed by AT&T Bell Laboratories (also referred to as Bell Cells) and supplied by Lineage Power were essentially flooded pasted-plate cells but had a number of features that provided for a service life of 40 years or more rather than the 15 to 20 years of life achieved with large flooded pasted-plate prismatic cells (see Fig. 14.78).[53,54] These have a circular pure lead positive grid shaped as a shallow cone with an angle of 10° and stacked horizontally with pure lead negative plates of a similar construction. The PAM is made with a mixture of chemically produced tetrabasic lead sulfate and red lead before formation which provides durability. The positives are welded together externally and the negatives are welded to a central conductor core, with an insulator at the outside edge. Microporous PE separators are used with a glass retainer mat. The cell containers and lids are molded in transparent PVC to permit visual inspection in service. The use of pure lead grids reduces plate growth and corrosion. The special shape is designed to counter the effect of plate growth and ensures that the grid and the active material remain in good contact throughout the life of the battery. Grid growth is caused by the formation of lead dioxide at the surface of the grid and distorting the grid, but in this case the shape change causes no adverse effects. This is achieved by maintaining the ratio of the cross-sectional area to surface area of the grid wires for the concentric wires.[13] The conversion of grid material into lead dioxide causes a small but measurable increase in capacity over time.[18] The same effect is seen in Planté cells. The acid gravity is moderately low and the float voltage is correspondingly low. This could cause a problem with recharge of the negative plate but is resolved by adding a small amount of Pt as a depolarizer to the negative expander to avoid increasing the cell voltage. The cells also have an elaborate pillar seal. A Pb-Sn alloy is used to prevent nodular corrosion and formation of an epoxy resin sheath around the pillar, which then has a rubber boot to seal it to the cover.[55,56] It is important for seal integrity that the area coated with epoxy remains above the electrolyte level. These cells have been extensively deployed in telecommunications central office applications in North America and in some other territories and following revisions to the design, particularly the pillar seal, are achieving long service lives.

Tests conducted in 1988 on cells that had been in service for 15 years showed that the capacity increase predicted from accelerated testing at the average service temperature of 22°C was 3.8% over this period or 0.25%/year and that plate growth was 0.4% over the period or 0.027%/year. The actual increases were greater than predicted (see Fig. 14.79a and b) but nonetheless sufficiently low to achieve a life of up to 40 years.

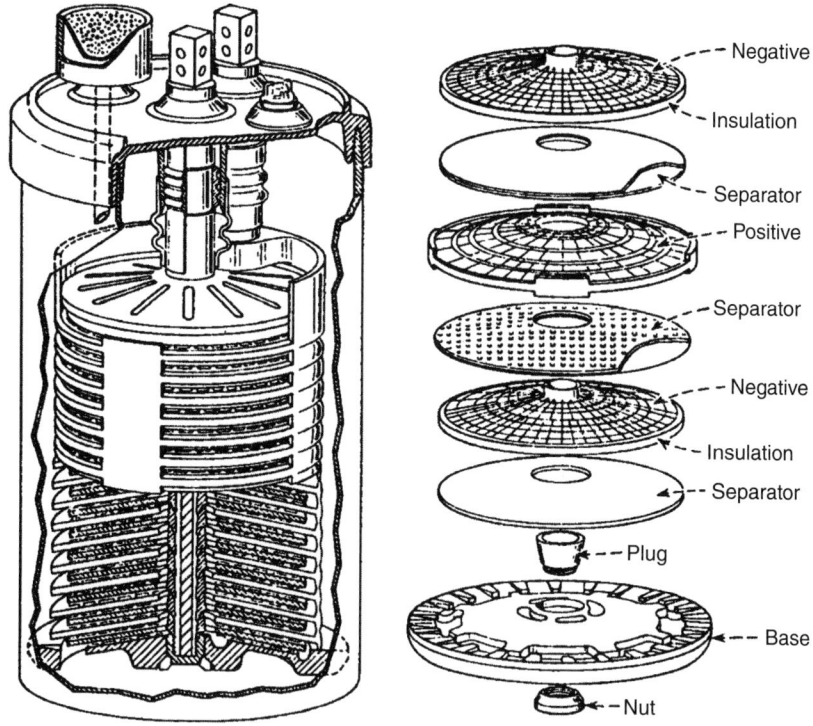

**FIGURE 14.78**    Cutaway and exploded views of the Bell System lead-acid battery. (*From Ref. 53.*)

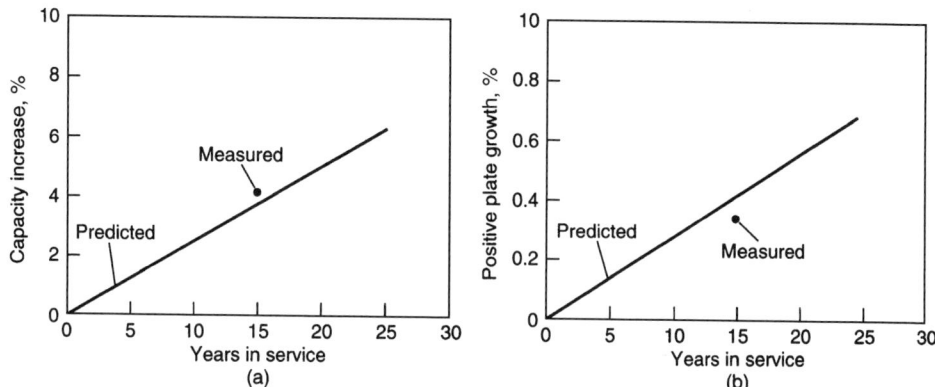

**FIGURE 14.79**    (*a*) Capacity after 15 years in service versus prediction from accelerated testing. (*From Ref. 55.*) (*b*) Positive-plate growth after 15 years in service versus prediction from accelerated testing.

***Performance Characteristics.***    Flooded batteries for stationary applications may use cells with flat-pasted, tubular, or Planté positive plates. Typical discharge curves for the flat-pasted-type stationary cell at various discharge rates at 25°C are shown in Fig. 14.80, and the effect of the discharge rate on the capacity of the cell is summarized in Fig. 14.81. Generally, the discharge rate for stationary cells is identified as the hourly rate (the current in amperes that the battery will deliver for a given time in hours and remain above a specified cutoff voltage) rather than the *C* rate used for other types of batteries.

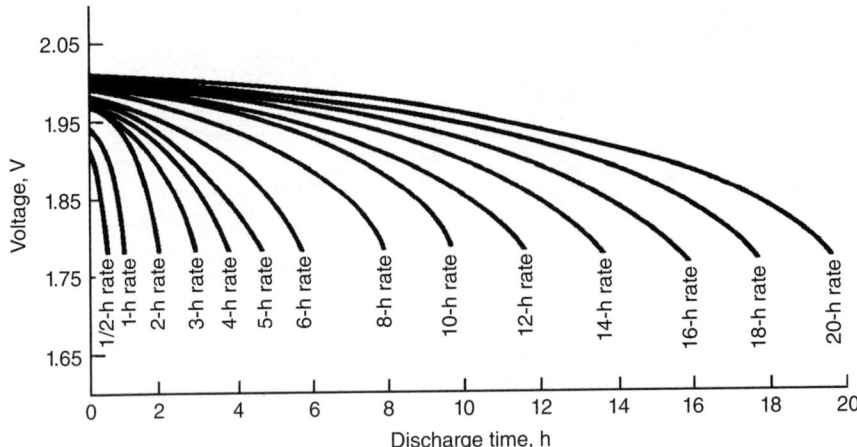

**FIGURE 14.80** Discharge curves of flat-pasted lead-acid stationary batteries (specific gravity 1.215) at various discharge rates at 25°C.

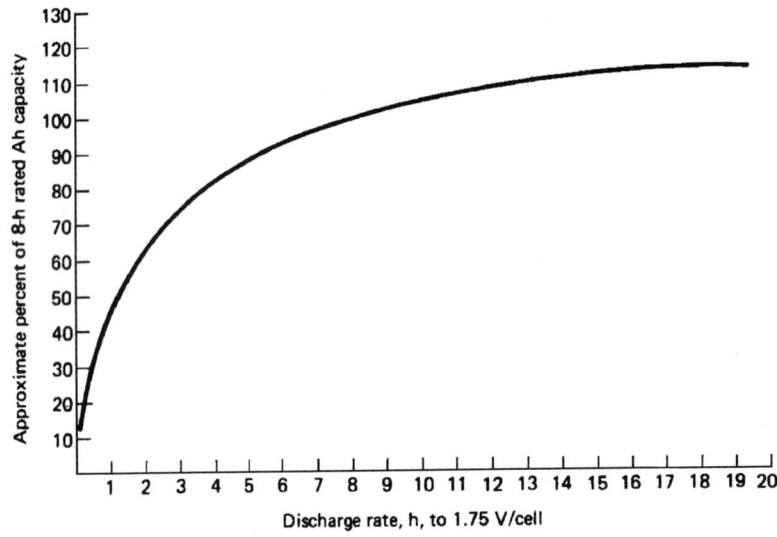

**FIGURE 14.81** Effect of discharge rate on cell capacity at 25°C for flat-pasted lead-acid stationary batteries (specific gravity 1.215) to 1.75 V end voltage.

Figures 14.82 to 14.84 are a series of curves showing the specific performance characteristics of the three types of stationary batteries at 25°C based on positive-plate design. An electrolyte with a specific gravity of 1.215 is used in all these batteries. The format used in these figures consists of two sections. The lower log-log section shows the capacity (expressed in discharge time) the particular positive plate will deliver at the specified current (expressed in amperes per positive plate) to various voltages including a final voltage. The upper semi-log section shows the cell voltage at various stages of the discharge at various discharge rates (also expressed in amperes

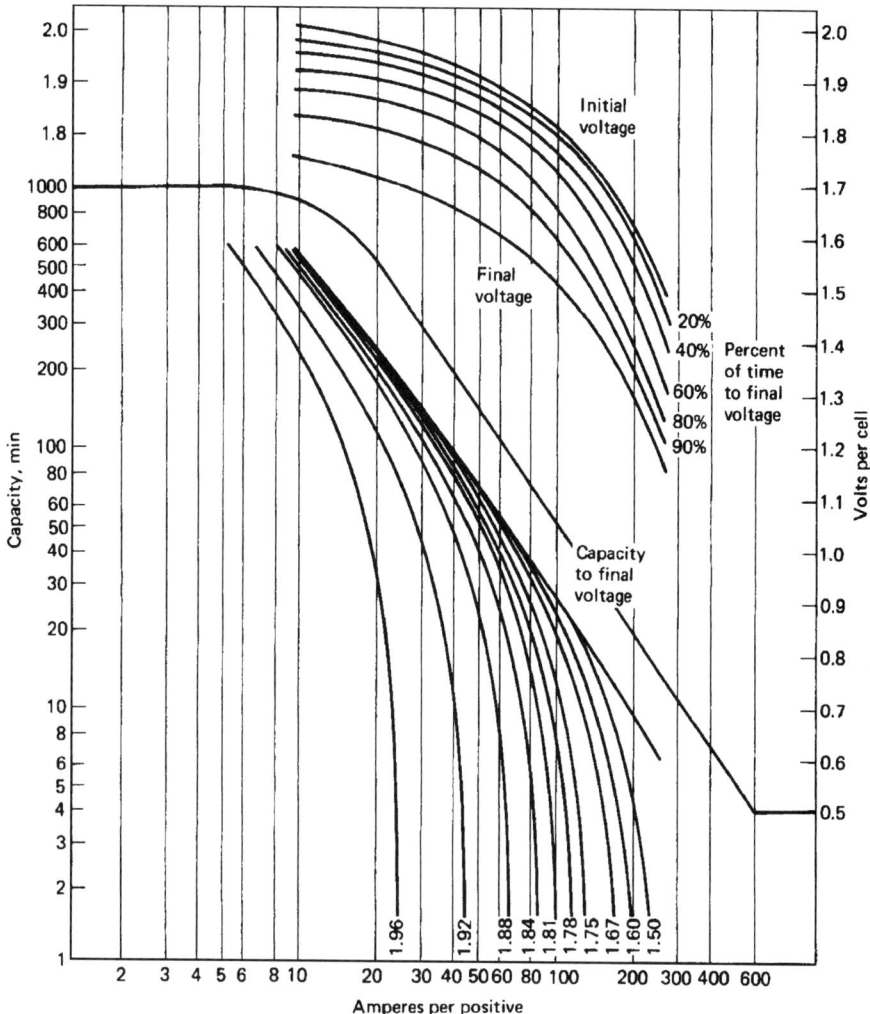

**FIGURE 14.82** Performance curves of lead-acid stationary batteries at 25°C (S-shaped curves, based on positive-plate performance). Antimony flat-pasted plate, 125 Ah at 8 h rate; 290 mm height, 239 mm width, 8.6 mm thickness. (*Courtesy of Enersys, Inc.*)

per positive plate). The final voltage is the voltage at which the cell voltage collapses rapidly if further discharged, and the cell may be damaged.

The energy density of flat-pasted positive-plate and tubular positive-plate batteries is similar. It is lower for the Planté positive-plate batteries. The high-rate performance of the flat-pasted positive cells is better because these plates can be made thinner than the tubular or Planté plates.

The optimal temperature for the use of stationary batteries ranges from 20 to 30°C, although temperatures from −40 to 50°C can be tolerated. The effect of temperature on the capacity of stationary batteries at different discharge loads is shown in Fig. 14.85. High-temperature operation, however, increases self-discharge, reduces cycle life, and causes other adverse effects.

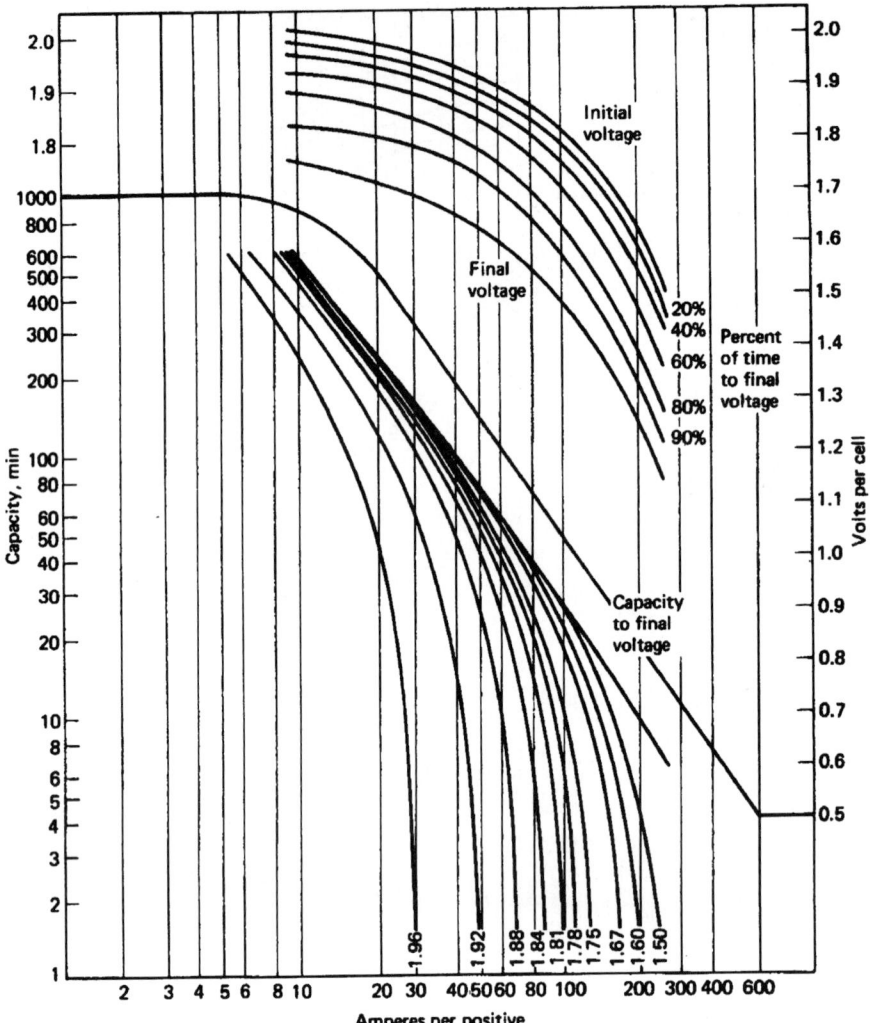

**FIGURE 14.83**   Calcium flat-pasted plate, 125 Ah at 8 h rate; 290 mm height, 239 mm width, 8.6 mm thickness. (*Courtesy of Enersys, Inc.*)

The rates of self-discharge of the various types of stationary batteries are compared in Fig. 14.86, which shows the relative float current at a specified float voltage. The float current under these conditions is a measure of self-discharge or local action. It is lowest for the Pb-Ca-Sn grid pasted positives and remains low throughout life. The float current is progressively higher for tubular Pb-Sb positives and pasted Pb-Sb positives at the beginning and throughout battery life. If the float current is not increased periodically, the antimonial cells will become progressively self-discharged and sulfated, but it will tend to increase the corrosion rate of the positive grids. In practice, the majority of flooded stationary cells now supplied use Pb-Ca-Sn grids in order to achieve long service lives and low rates of self-discharge.

For fully charged batteries, the self-discharge rate at 25°C for Pb-Ca-Sn positive-plate cells is about 1%/ month, 3% for Planté, and 5% to 15% for Pb-Sb positive cells depending on the Sb content. At higher temperatures, the self-discharge rate increases significantly, doubling for each 10°C rise in temperature.

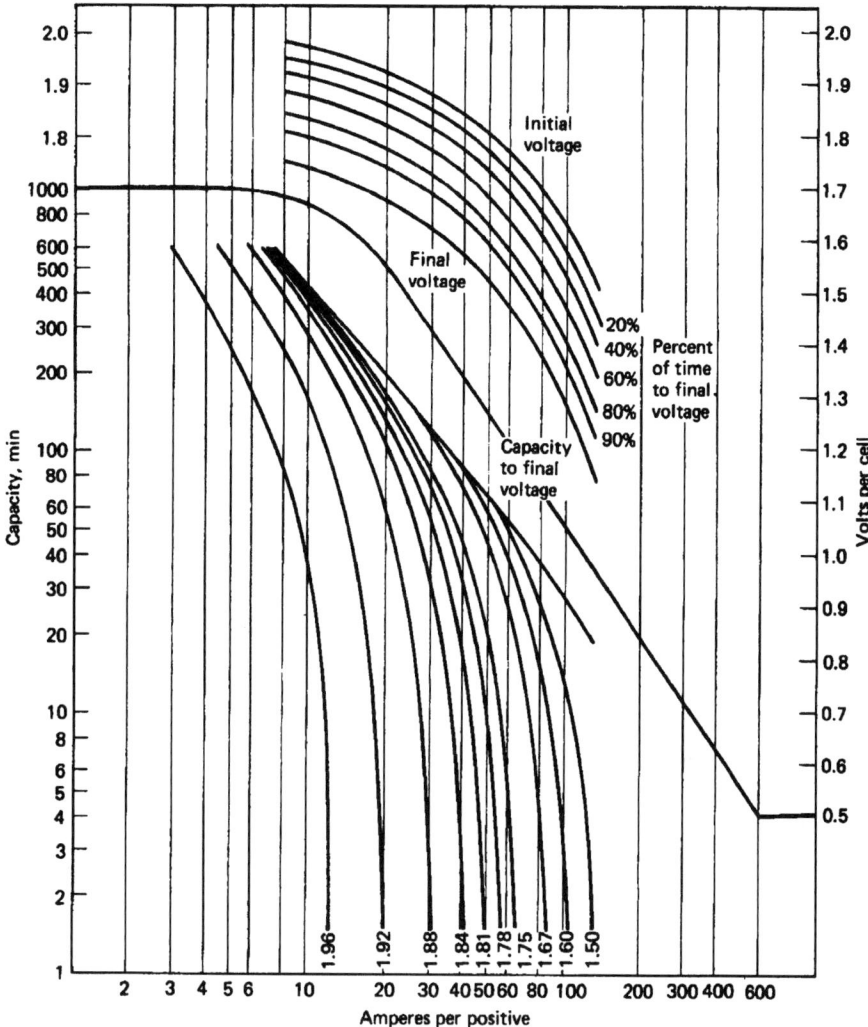

**FIGURE 14.84**   Ironclad tubular plate, 70 Ah at 8 h rate; 274 mm height, 203 mm width, 8.9 mm thickness. (*Courtesy of Enersys, Inc.*)

The float current for Pb-Ca and Pb-Sb positive-plate batteries is shown in Fig. 14.87 under float charge at voltages between 2.15 and 2.40 V per cell. It has been found that more than 50 mV positive and negative over-potential is necessary to prevent self-discharge so that 5 mA of float current per 100 Ah of battery capacity is required for Pb-Ca batteries. Pb-Sb batteries initially require at least 60 mA/100 Ah, but this increases to 0.6 A/100 Ah as the battery ages. The higher float current also increases the rate of water consumption, the evolution of hydrogen and the corrosion of the grids.

Various, and at times conflicting, claims about the life of stationary battery designs are made by the different manufacturers worldwide. Generally, flat-pasted plate Pb-Sb batteries have the shortest life (5–15 years), followed by flat-pasted plate Pb-Ca batteries (15–25 years), tubular batteries (20–25 years), and Planté batteries (25 years).

Life on float service has been found to be related to temperature following an Arrhenius relationship, as shown in Fig. 14.88. The growth rate constant $k$ is plotted for several different types of grid alloys. At 25°C the

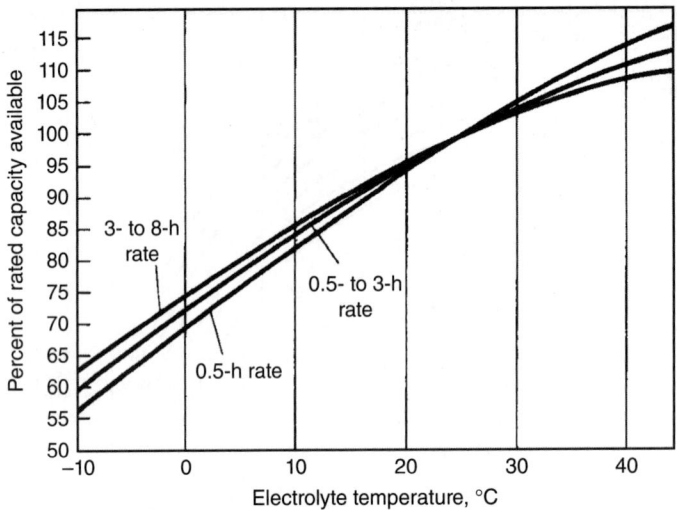

**FIGURE 14.85**   Performance of flat-pasted lead-acid stationary batteries at various temperatures and discharge rates. (*Courtesy of C&D Technologies.*)

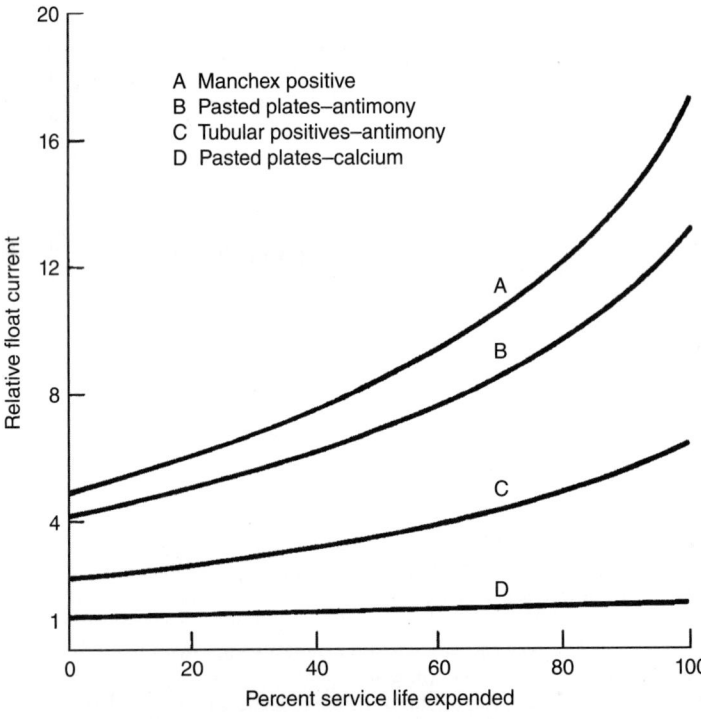

**FIGURE 14.86**   Relative self-discharge of lead-acid stationary batteries of different construction.

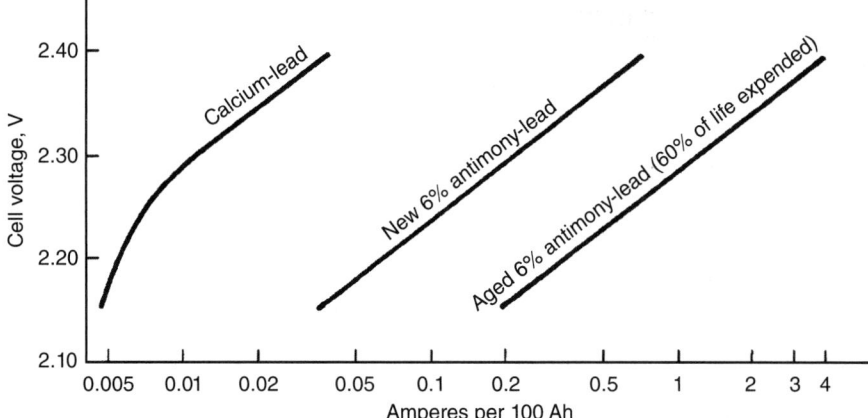

**FIGURE 14.87**  Float current characteristics of stationary batteries at 25°C, 100 Ah cells, fully charged, 1.210 specific gravity.

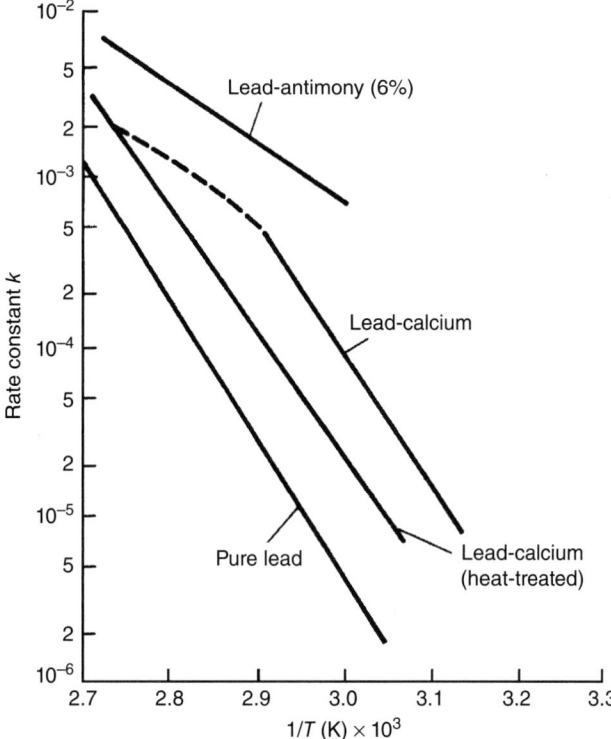

**FIGURE 14.88**  Corrosion rate constant log $k$ versus $1/T$ for different lead-alloy grids. (*From Ref. 53.*)

time to reach 4% growth, an upper limit before the battery's integrity is impaired, is calculated to be 14 years for Pb-Sb, 17 years for Pb-Ca, and 80 years for pure lead, but in practice the failure of other components in the cell may result in earlier failure.

***Cell and Battery Types and Sizes.***    Stationary batteries are offered in a very wide range of plate and cell sizes that have evolved over many years. For tubular cells in Europe, sizes are defined by the OPzS standard and for the pasted plate cells by the OGi standard. They are normally assembled on insulated metal racks from groups of cells or monoblocs in series with voltages from 12 to 400 V or more. Battery installations may have a number of series strings in parallel to achieve the specified capacity, and best practice is to limit the number of parallel strings to a maximum of six. Tables 14.27 and 14.28 provide a summary of typical specifications for pasted-plate and tubular flooded batteries.

**TABLE 14.27**    Typical Specifications for Pasted-Plate Flooded Stationary Cells

| Cell/battery type | EA.M | CA.M | DU | FTS-P | GC-M |
|---|---|---|---|---|---|
| Capacity range, Ah | 215–850 | 50–200 | 310–780 | 840–1810 | 875–3550 |
| Length range, mm | 130–257 | 178–310 | 241–424 | 282 | 315–422 |
| Width range, mm | 279 | 229 | 406 | 368 | 384 |
| Height range, mm | 475 | 375 | 683–732 | 577 | 686 |
| Weight range, kg | 27–85 | 26–60 | 121–280 | 99–145 | 171–321 |
| Grid alloy | Pb-Sb | Pb-Sb | Pb-Ca-Sn | Pb-Ca-Sn | Pb-Ca-Sn |
| Positive grid thickness, mm | 8.6 | 7.1 | 6.4 | 8.1 | 7.6 |
| Acid gravity | 1.215 | 1.215 | 1.215 | 1.215 | 1.215 |
| Container type | SAN | SAN | PC | PVC | PVC |
| Life expectancy, years | 20 | 20 | 20 | 20 | 20 |

*Note:* Cell/battery types CA.M are 6 V monoblocs, cell/battery types DU are 8 V monoblocs, and the 840 Ah FTS-P cell type is a 4 V monobloc. The capacity is at the 8 h rate at 25°C to 1.75 V per cell.
*Source:* EnerSys data.

**TABLE 14.28**    Typical Specifications for Tubular-Plate Flooded Stationary Cells

| Cell type | Number of terminals | Capacity, Ah | Length, mm | Width, mm | Height, mm | Weight, kg |
|---|---|---|---|---|---|---|
| 4OPzS200 | 2 | 217 | 103 | 206 | 403 | 17 |
| 6OPzS300 | 2 | 319 | 145 | 206 | 403 | 24 |
| 8OPzS420 | 2 | 466 | 145 | 206 | 520 | 32 |
| 6OPzS600 | 2 | 648 | 145 | 206 | 695 | 45 |
| 8OPzS800 | 4 | 856 | 210 | 191 | 695 | 61 |
| 10OPzS1000 | 4 | 1071 | 210 | 233 | 695 | 75 |
| 12OPzS1200 | 4 | 1293 | 210 | 275 | 695 | 88 |
| 12OPzS1500 | 4 | 1730 | 210 | 275 | 845 | 114 |
| 14OPzS1750 | 6 | 2092 | 214 | 399 | 820 | 144 |
| 16OPzS2000 | 6 | 2307 | 214 | 399 | 820 | 152 |
| 18OPzS2250 | 8 | 2669 | 212 | 487 | 820 | 184 |
| 20OPzS2500 | 8 | 2884 | 212 | 487 | 820 | 193 |
| 22OPzS2750 | 8 | 3238 | 212 | 576 | 820 | 225 |
| 24OPzS3000 | 8 | 3360 | 212 | 576 | 820 | 235 |

*Note:* These cells have Pb-Sb positive spines, SAN containers, an acid gravity of 1.240, and an expected life of 20 years on float at 2.23 V per cell at 20°C. The capacity is at the 8 h rate at 25°C to 1.75 V per cell.
*Source:* EnerSys data.

### 14.10.3    VRLA Batteries

For stationary applications, the great majority of batteries currently supplied are VRLA types mostly with AGM separators, but gelled electrolyte types are also important. The use of flooded types has declined, and while some of the types discussed above remain in service, there are applications in telecommunications, nuclear installations, and national security where they will continue to be specified.[57]

The key international standards for VRLA batteries both AGM and gel are IEC 60896-21: 2004 Stationary lead-acid batteries—Part 21: Valve-regulated types—Methods of test and IEC 60896-2: 2004—Requirements. These standards cover a range of service applications, and it is for the user to specify the level of compliance with certain tests. Manufacturers may indicate correctly that their products comply with these standards, but this does not provide an assurance of service life.

*VRLA Gelled Electrolyte Batteries.*  Gelled electrolyte cells may use either a flat, pasted-plate construction or tubular plates, as shown in Fig. 14.89. Tubular gel cells generally follow the DIN OPzV specification (DIN 40744 Lead-acid batteries—Stationary valve-regulated batteries with positive tubular plates and immobilized electrolyte [gel] [2015]). The capacity range is from 200 to 3000 Ah. They are also within the scope of IEC 60896-21/22: 2004. The positive plates have a tubular construction as for flooded tubular cells with spines in Pb-Ca-Sn alloys. Tin levels tend to be at the higher end of the range and calcium levels somewhat lower in order to improve the corrosion behavior and also the cyclic behavior of the product. Negative plates use flat cast grids and are processed conventionally. If pasted positive plates are used, the alloys are similar to tubular plates and processing is conventional. The separator system uses a microporous plastic separator with pores in the larger range of available materials. Materials with very small pores cannot be effectively filled with the gelled electrolyte and will not operate correctly. The electrolyte is gelled with finely divided silica. This has a thixotropic behavior and when mixed with a high shear mixer, will remain fluid for sufficient time to fill the cells. It then sets to a rigid gel, and at the beginning of life, the cell is fully saturated, but as the cell loses water, microscopic fissures develop in the gel and allow oxygen transport between the positive and negative plates. Cells dry out with the loss of a few percentage points of the electrolyte volume and the efficiency of recombination is such that they are very stable in this condition. Drying out is not a failure mode. Cell cases and lids are molded in flame retardant ABS. Pillar seals are proprietary and pressure relief valves may be simple Bunsen valves or may have more complex designs. The service life of this type of cell is 12 to 15 years at 20 to 25°C and they have a reasonably good cycle life. The rate capability is modest.

**FIGURE 14.89**  Tubular gel stationary cell. (*Courtesy of FIAMM Energy Technologies.*)

*VRLA AGM Batteries.*  The great majority of VRLA batteries, produced either as cells or monoblocs, use pasted plates with Pb-Ca-Sn grids and AGM separators (Fig. 14.90). Sizes range from 12 V, 1 Ah monoblocs to single cells with a capacity of 4000 Ah or more. Cell cases may be polypropylene (PP), ABS, PC/ABS, or PVC. The majority of types specified for telecommunication applications use flame retardant ABS cases, but PP cases are often used for batteries for UPS applications for economic reasons. The positive grid alloys have evolved over recent years from moderately high levels of calcium (0.07–0.08%) to lower levels (0.04–0.05%) with a corresponding increase in the tin level from 0.7% to 0.8% to ~1.2%. This improves the corrosion behavior of the alloy and also cycle life. One manufacturer has used Pb-Sb-Cd alloys in the past, but this has not been followed elsewhere in the industry. Grid design and thickness are important in achieving the required life on float. For a life of 10 to 12 years at 20 to 25°C, a grid thickness of 4 to 6 mm is required.

The AGM separator is a key element for VRLA batteries. The glass mat needs to fill the space between the electrodes under compression of about 40 kPa, absorb a maximum volume of electrolyte, and retain a small volume of connected gas porosity to permit oxygen transport between the plates for effective recombination. High porosity is required (90–95%) with a small pore size (5–8 μm). The material also needs to be fully inert in the cell environment. These requirements are met with glass microfibers (<1 μm) that are made into a separator by a paper-making process. The separator is normally formulated with larger-diameter fibers to reduce the cost of the material. Organic polymer fiber blends may also be added, which increases both the tensile strength and puncture strength. High levels of polymer fibers allow the AGM to be heat sealed into a pocket around the plate. Dimensional stability is important, as the material needs to retain pressure on the plates during charge and discharge. Higher levels of resilience are achieved with higher fine fiber contents, and the overall formulation is adjusted to give the correct balance of properties.

The details of the electrochemical design of the cell, separator specification, and compression and acid filling are all important in achieving reliable performance over life. For higher performance with shorter discharge times, thinner plates are used but at the expense of service life. Pillar seals use a variety of rubber sealing rings, mechanical compression, or thermosetting resins. The lid-to-case seal may also use a resin to bond the case and lid together or they may be heat sealed together with a hot-plate

**FIGURE 14.90** Valve-regulated lead/acid battery for standby applications with front access terminals. (*Courtesy of FIAMM Energy Technologies.*)

welder. Valves may be simple Bunsen valves or more complex arrangements and normally have a flame filter to prevent any external ignition of hydrogen from penetrating the cell. Venting pressures are low and the more important requirement is that air cannot enter the cell from outside.

VRLA cells with AGM separators may be built with pure lead or pure lead-tin grids instead of cast Pb-Ca-Sn grids. These are fabricated from continuously cast or wrought strips by punching the grid pattern into the strips. The strip is then pasted continuously and the plates are processed as for VRLA cells with Pb-Ca-Sn grids. The grids are typically 1.0 to 1.2 mm thick. Active material utilization is much better than in cells with thicker plates. The use of pure lead reduces the corrosion such that the life is equivalent to cells with Pb-Ca-Sn grids with much thicker plates. Alloying with a small amount of tin (0.4–0.6%) improves performance on cycling but slightly reduces the corrosion resistance. Other details of construction are similar to normal types of VRLA cells.

The gravimetric energy density of thin-plate pure lead cells is up to 50% better than conventional cells at moderate rates of discharge, and the relative improvement increases at higher rates. Somewhat greater improvements are achieved in the volumetric energy density.

Thicker plate VRLA cells are also manufactured with pure lead plates, which extends the corrosion life but leaves the rate performance unaltered. Service lives in excess of 25 years at 25°C and up to 10 years at 40°C are claimed. For higher temperature service, the case and lid material needs to be a polymer, such as PC/ABS, with a higher softening point that avoids any tendency of the container to distort over time.

***Performance Characteristics.***   The discharge performance of VRLA AGM batteries is mainly dependent on plate thickness and plate spacing. Batteries designed for higher rates of discharge, as required for UPS, will tend to have thinner plates, whereas batteries designed for telecommunications service, where the discharge rates are lower, will tend to have thicker plates. Battery service life is determined by the corrosion life of the positive grids, and thicker plates will have longer lives. But improved alloys have also resulted in longer lives in floating service. Batteries with pure lead or pure lead-tin grids may use thinner grids but without reduction in life due to the lower corrosion rate of these materials. The service temperature should normally be in the range from 20 to 25°C for maximum service life, but products can be adapted to have service lives of 10 years at 40°C with adjustments to the grid alloys, acid gravity, and the use of polymers for the battery case that will not soften or bulge at or slightly above the service temperature. PC, PC/ABS alloys, and polyphenylene oxide (PPO) have been found to be suitable.

Self-discharge is reasonably slow. Batteries with Pb-Ca-Sn alloys will retain 80% of nominal capacity when stored at 20°C for 12 months, or at 25°C for 8 months; but at 45°C the shelf life to 80% retained capacity is only 2 months. With pure lead grids, the shelf life is approximately doubled (24 months at 20°C). In normal temperate conditions, battery open-circuit voltage should be checked every 4 months and batteries should be recharged if the voltage falls below 2.10 V or according to the manufacturer's information.

The floating voltage depends on the specific gravity of the acid used, but the majority of VRLA AGM types should be floated at 2.27 to 2.28 V per cell at 20°C.

Batteries may be recharged on float, but if shorter recharge times are required, a boost charge at higher voltages (2.35–2.40 V per cell) may be used for a limited time. Charging at higher voltages for extended times will lead to electrolyte loss and positive grid corrosion that will shorten battery life. The float charge voltage should be compensated for temperatures above 25°C with small reductions in order not to limit service life.

Service lives of VRLA AGM batteries vary widely depending on the battery design and construction. Smaller general-purpose types have service lives of 3 to 5 years; larger batteries for telecommunications or utility applications have lives in the range 10 to 20 years; and batteries for UPS applications may be specified for service lives of 10 years or more, but are often designed for lives of 5 to 8 years.

Tubular gel cells are designed for lower-rate applications typical of telecommunications service, but other characteristics are similar. Float voltages are generally slightly lower at 2.23 to 2.25 V per cell. Service lives of up to 20 years may be achieved at 20°C.

***Cell and Battery Types and Sizes.*** VRLA AGM batteries are offered in a very wide range of sizes and capacities, and there have been few initiatives to specify standard dimensions. The exceptions are smaller types from 1 Ah to 65 Ah, front-access terminal batteries for telecommunications, and tubular OPzV cells.

The smaller types are offered in 4, 6, and 12 V monoblocs and originally followed dimensions specified by Japanese Standards, but they are now within the scope of IEC 61056 (IEC 61056-1: 2012, General purpose lead-acid batteries [valve-regulated types]—Part 1: General requirements, functional characteristics; IEC 61056-2: 2012, General purpose lead-acid batteries [valve-regulated types]—Part 2: Dimensions, terminals and marking). These sizes are widely used for security and fire alarms, for mobility aids, toys, emergency lighting, and other applications. They are offered in standard general purpose, longer float life, and longer cyclic life variants. There are also wound cylindrical cells and monoblocs available for the same applications but in different dimensions. Table 14.29 summarizes the characteristics of important types in 12 V monoblocs.

**TABLE 14.29** Typical Specifications of Small VRLA AGM Prismatic Cells for General-Purpose Applications

| Battery type | Capacity, Ah | Length, mm | Width, mm | Height, mm | Weight, kg |
|---|---|---|---|---|---|
| NP1.2-12 | 1.2 | 97 | 48 | 54 | 0.57 |
| NP2-12 | 2.0 | 150 | 20 | 89 | 0.70 |
| NP2.3-12 | 2.3 | 178 | 34 | 64 | 0.94 |
| NP3.2-12 | 3.2 | 134 | 67 | 64 | 1.20 |
| NP4-12 | 4.0 | 90 | 70 | 106 | 1.70 |
| NP7-12 | 7.0 | 151 | 65 | 98 | 2.65 |
| NP12-12 | 12.0 | 151 | 98 | 98 | 4.00 |
| NP18-12B | 17.2 | 181 | 76 | 167 | 6.20 |
| NP24-12 | 24.0 | 166 | 175 | 125 | 8.65 |
| NP38-12 | 38.0 | 197 | 165 | 170 | 13.8 |
| NP65-12 | 65.0 | 350 | 166 | 174 | 22.8 |

***Note:*** The capacity is given as the 20 h rate to 1.75 V per cell at 25°C.
***Source:*** Data from GS Yuasa.

Front access terminal VRLA AGM batteries (Fig. 14.90) are not standardized, but they are designed to fit standard equipment racks with four batteries side by side across the rack, which defines the battery width (105/110 or 125 mm). The battery length (395 or 561 mm) is defined by the depth of the rack. Table 14.30 shows a typical range. A tubular gel VRLA OPzV range is summarized in Table 14.31.

**TABLE 14.30** Typical Specifications for Front-Access VRLA AGM Telecommunications 12 V Monoblocs

| Battery type | Capacity, Ah | Length, mm | Width, mm | Height, mm | Weight, kg | Short circuit current, A | Internal resistance, mΩ |
|---|---|---|---|---|---|---|---|
| 12V30F | 31 | 280 | 97 | 159 | 11 | 1327 | 9.87 |
| 12V38F | 38 | 280 | 97 | 184 | 13 | 1500 | 8.53 |
| 12V62F | 62 | 280 | 97 | 264 | 19 | 2080 | 5.98 |
| 12V92F | 92 | 395 | 105 | 264 | 28 | 2410 | 5.19 |
| 12V100FC | 100 | 395 | 108 | 287 | 31 | 1930 | 6.46 |
| 12V101F | 101 | 510 | 110 | 235 | 34 | 2108 | 5.92 |
| 12V125F | 126 | 561 | 105 | 316 | 45 | 2355 | 5.30 |
| 12V155FS | 155 | 561 | 125 | 283 | 49 | 3325 | 3.80 |
| 12V170FS | 170 | 561 | 125 | 283 | 51 | 3360 | 3.75 |
| 12V190F | 190 | 561 | 125 | 316 | 57 | 3625 | 3.50 |

*Note:* The capacity is given as the 8 h rate to 1.75 V per cell at 25°C.
*Source:* EnerSys data.

**TABLE 14.31** Typical Specifications for Tubular Gel VRLA OPzV Cells

| Battery type | Capacity, Ah | Length, mm | Width, mm | Height, mm | Weight, kg | Short circuit current, A | Internal resistance, mΩ |
|---|---|---|---|---|---|---|---|
| 4OPzV200 | 224 | 105 | 208 | 399 | 17 | 2200 | 0.95 |
| 6OPzV300 | 337 | 147 | 208 | 399 | 25 | 3350 | 0.61 |
| 6OPzV420 | 499 | 147 | 208 | 515 | 35 | 3950 | 0.53 |
| 6OPzV600 | 748 | 147 | 208 | 690 | 49 | 4300 | 0.48 |
| 8OPzV800 | 998 | 212 | 193 | 690 | 66 | 4850 | 0.38 |
| 10OPzV1000 | 1248 | 212 | 235 | 690 | 80 | 6250 | 0.33 |
| 12OPzV1200 | 1497 | 212 | 277 | 690 | 95 | 7850 | 0.29 |
| 12OPzV1500 | 1643 | 212 | 277 | 759 | 106 | 9000 | 0.23 |
| 16OPzV2000 | 2190 | 216 | 400 | 816 | 149 | 10750 | 0.19 |
| 20OPzV2500 | 2738 | 214 | 489 | 816 | 190 | 13400 | 0.16 |
| 24OPzV3000 | 3286 | 214 | 578 | 816 | 238 | 16100 | 0.10 |

*Note:* The capacity is given as the 10 h rate to 1.80 V at 20°C.
*Source:* EnerSys data.

There are a wide variety of sizes and types of VRLA AGM batteries offered in 6 and 12 V monoblocs differentiated by design life and whether they are optimized for low- or high-rate discharge. Gel types are also offered in 6 and 12 V monoblocs with both tubular and pasted plates. Table 14.32 shows the performance of typical VRLA AGM monoblocs for UPS applications. These are optimized for discharge rates in the 5 to 15 min range, and the characteristics are given for constant-power discharges as the on-load voltage falls significantly at higher rates. Constant power is obtained by increasing the current as the discharge proceeds.

**TABLE 14.32** Typical Specifications for VRLA AGM 12 V Monoblocs Designed for UPS Applications

| Battery type | Capacity, Ah | Constant power rating at 25°C W per cell to 1.67 V per cell | | | | | | Weight, kg |
|---|---|---|---|---|---|---|---|---|
| | | 5 m | 10 m | 15 m | 20 m | 30 m | 60 m | |
| UPS12-300MR | 79 | 546 | 385 | 300 | 245 | 183 | 106 | 27 |
| UPS12-400MR | 103 | 716 | 506 | 400 | 328 | 244 | 139 | 34 |
| UPS12-540MR | 149 | 875 | 657 | 537 | 451 | 343 | 198 | 45 |

*Note:* The nominal capacity is given at the 20 h rate to 1.75 V per cell at 25°C.
*Source:* Reproduced with permission of C&D Technologies.

For larger VRLA AGM capacities single 2 V cells are used. These may either have rigid polymer cases, typically molded in ABS, or they may have PP cases that are not sufficiently rigid to contain the operating pressure. They are installed in steel racks that serve to restrain the cells and also support multicell modules. Larger VRLA batteries are usually installed horizontally so that any stratification of the electrolyte that may occur in taller cells is avoided. It is also important to ensure that in a horizontal installation the plates are oriented vertically.

## 14.11  GRID ENERGY-STORAGE APPLICATIONS

### 14.11.1  General Considerations

The use of energy storage in electricity networks has increased rapidly as more generating capacity using renewable energy sources has come into use. These sources, especially wind and solar PV generation, are intrinsically intermittent and the use of electricity has to be precisely balanced with input which can be, in part, achieved through the use of energy storage. The key requirement is not only to maintain the frequency within narrow limits, but also to maintain voltage stability, allow full use of renewable generation, and to reduce the overall costs of electricity. In conventional networks, the spinning reserve of generating equipment has sufficient inertia to maintain system frequency until more generating assets can be brought into use if demand increases. Correspondingly, if demand decreases, the frequency will increase until generating assets are disconnected. Energy storage by various means provides a further way of stabilizing the network. Large networks have installed pumped hydroelectric energy-storage schemes to augment spinning reserve, but the requirement for additional rapidly deployable energy-storage systems is continuing to increase. Electrochemical energy storage in batteries is attractive because they are compact, easy to install, economical, and provide virtually instant response in terms of both input from the battery and output from the network to the battery. The power conversion equipment, either inverting the DC battery input to AC for the network or rectifying the AC line to DC for output to the battery, is efficient, bidirectional, and has a very rapid response.

Lead-acid batteries have been used for energy-storage schemes for many years in utility installations and for smaller-scale domestic and commercial energy-storage schemes. The types used follow those developed for stationary and traction service, but there has been a lot of development to better adapt lead-acid batteries to the duty cycles required. Flooded types, both with pasted and tubular plates, may be used as well as VRLA types, either with pasted-plate AGM constructions or gel types with pasted or tubular plates.[58,59]

In pasted-plate cells, either flooded or VRLA, the PAM needs to have sufficient density and structure to provide durability on cycling. The charge and discharge processes are reconstructive transformations involving changes in volume and shape, and so a high-plate integrity is essential. This is achieved in the paste formulation and in the curing and drying stage after pasting, where conditions of higher temperatures and humidity favor the formation of tetrabasic lead sulfates that provide better cyclic performance. The correct conditions in curing also ensure that the active material is well bonded to the grids and that the interface between the active material and the grid is fully conductive. Red lead ($Pb_3O_4$) may be added to the paste formulation, which improves the efficiency of formation and the structure of the positive active mass. Phosphoric acid ($H_3PO_4$) may be added to the PAM. This increases the adhesion between the PAM and the grids and the cohesion of the active material.

Tubular cells may also be used for energy storage either with flooded or gelled electrolytes. The cycle life is extended because the construction provides excellent behavior (i.e., good retention of active material due to the robust tubes that do not allow shedding to occur).

The key standards applicable to battery energy storage are IEC 61427-1: 2013: Secondary cells and batteries for renewable energy storage and methods of test—Part 1: Photovoltaic off-grid applications and IEC 61427-2: 2015: Secondary cells and batteries for renewable energy storage and methods of test—Part 2: On-grid applications. More details on grid storage battery applications are presented in Chap. 27.

### 14.11.2 Lead Batteries in Energy-Storage Service

Table 14.33 shows the cycle life at different DODs for a VRLA battery with carbon additives designed for energy-storage applications. This type of battery has optimum performance between 25° and 30°C and 90% of the rated cycle life between 20° and 40°C.

**TABLE 14.33** Cyclic Performance at Different Depths of Discharge for a Carbon-Enhanced VRLA Battery Designed for Energy Storage Applications

| Battery type | Capacity, Ah | Nominal voltage, V | Cycle life at | | | Weight, kg |
| --- | --- | --- | --- | --- | --- | --- |
| | | | 50% DOD | 60% DOD | 70% DOD | |
| SLR-1000 | 1000 | 2 | 5500 | 5250 | 5000 | 67 |

*Source:* Data from GS Yuasa.

Lead-acid batteries have a long record in energy storage in both large and small installations. They have demonstrated reliable operation over many years in installations as large as 10 MW, 40 MWh for peak shaving, load levelling, load following, spinning reserve, transmission line support, frequency regulation, voltage control, VAR control, and black starts. Figure 14.91 shows a large UltraBattery installation used for frequency regulation, and Fig. 14.92 shows a large VRLA battery used for network support.

**FIGURE 14.91** Lead-based UltraBattery® used for frequency regulation services. (*Courtesy of East Penn Manufacturing Co./Ecoult.*)

**FIGURE 14.92**    A 2.7 MWh lead-acid VRLA battery used for utility energy storage. The batteries are rack mounted in the air-conditioned enclosure to the rear and the power conversion equipment is in the container to the left. (*Courtesy of NorthStar Battery.*)

Safety needs to be considered for all energy-storage installations. Lead batteries provide a safe system with an aqueous electrolyte and active materials that are not flammable. In a fire, the battery cases will burn but the risk of this is low, especially if flame retardant materials are specified. Li-ion batteries have a much higher energy density, highly reactive component materials, and a flammable electrolyte. Safety engineering needs to be of a very high standard to ensure that the risk of thermal runaway, fire, and explosion is managed. Other battery systems also have safety issues that need to be controlled.

An issue with all battery technologies is sustainability. There are strict regulations regarding collection and recycling of all types of batteries and mandated efficiency targets irrespective of the broader societal needs to ensure that all goods form part of a circular economy. For lead batteries, there is an established recycling infrastructure in place that operates economically in full compliance with all environmental regulations. For Li-ion and other chemistries used for battery energy storage, recycling processes do not recover significant value and will need to be substantially improved to meet current and future requirements.

### 14.11.3    Competitive Position of Lead-Acid Batteries versus Lithium-Ion Batteries

The competitive position between lead batteries and other types of batteries indicates that lead batteries are competitive in technical performance in static installations with Li-ion batteries. Table 14.34 provides a summary of the key parameters. Lead-acid batteries cover a range of different types of batteries which may be either flooded and require maintenance watering, or valve-regulated and only require inspection. For many energy-storage applications with intermittent charging input and output requirements, especially with solar PV input, batteries are not routinely returned to a fully charged condition and where the battery is required to absorb power as well as deliver power to the network, partial state-of-charge operation becomes the normal mode. There have been substantial improvements in lead-acid batteries in this area especially with the use of carbon additives to the negative plate, but this continues to be an area of active development and further improvements in performance should be achieved. There are also other types of novel lead batteries under development, particularly including batteries with a hybrid construction of supercapacitor elements combined with a conventional negative plate. These offer further improvements in shallow-cycle performance.

**TABLE 14.34**    Comparison of Technical and Other Features of Lead-Acid and Li-Ion Batteries for Energy Storage Service

| System | Lead-acid | Li-ion |
|---|---|---|
| Energy density | 35–40 Wh/kg* <br> 80–90 Wh/L | 150–180 Wh/kg <br> 300–350 Wh/L |
| Power density | 250 W/kg (1) <br> 500 W/L | 1250 W/kg <br> 1250 W/L |
| High temperature performance | to 40°C | to 50°C |
| Low temperature performance | to −30°C | to −20°C |
| Charge acceptance | Good | Better |
| Cycle life | 1500–5000 | 1000–5000 |
| Overall service life | 15 years | 10–15 years |
| Reliability | Proven | Needs to be assessed for longer times |
| Sustainability | Excellent | Recovery methods uneconomical |
| Safety | Excellent | Issues to be resolved |
| Cost (battery system only) | $150–$200/kWh | $600–$800/kWh |

*Bipolar lead-acid batteries are being developed which have energy densities in the range from 55 to 60 Wh/kg (120–130 Wh/L) and power densities of up to 1100 W/kg (2000 W/L).

Lead batteries have a long history of use in utility energy storage, and their capabilities and limitations have been carefully researched. Their reliability is well established and they can be adapted for a wide range of duty cycles within this sector, which will continue to ensure that they provide a good solution that is competitive to other approaches.

## ACKNOWLEDGMENTS

Major content for this chapter was provided in whole or in part by Alvin J. Salkind and George Zguris, Chap. 16, Lead-Acid Batteries, and Kathryn R. Bullock and Alvin J. Salkind, Chap. 17, Valve-Regulated Lead-Acid Batteries, *Linden's Handbook of Batteries*, 4th ed., T. B. Reddy and D. Linden, eds., McGraw-Hill, 2011.

## REFERENCES

1. Battery Council International, www.batterycouncil.org.

2. H. Bode, *Lead-Acid Batteries*, Wiley, New York, 1977.

3. H.E. Haring and U.B. Thomas, *Trans. Electrochem. Soc.* **68**:293 (1935).

4. P. Ruetschi, "Review of the Lead-Acid Battery Science and Technology," *J. Power Sources*, **2**:3 (1977/1978).

5. C. Mantell, *Batteries and Energy Systems*, 2nd ed., McGraw-Hill, New York, 1983.

6. D.J.G. Ives and G.J. Janz, *Reference Electrodes*, Academic, New York, 1961.

7. E.A. Willihnganz, U.S. Patent 3,657,639.

8. A. Sabatino, *Maintenance-Free Batteries, Heavy Duty Equipment Maintenance*, Irving-Cloud Publishing, Lincolnwood, IL, 1976.

9. Special Issue on Lead-Acid Batteries, *J. Power Sources* **2**(1):1–120 (1977/1978).

10. G.W. Vinal, *Storage Batteries*, 4th ed., Wiley, New York, 1955.

11. D. Berndt, "Valve-Regulated Lead-Acid Battery" and "Lead-Acid Batteries," *Conference on Oxygen Cycle in Lead and Batteries*, 7th ELBC, Dublin, Ireland, Sept. 2000.

12. ASTM Specification B29, "Pig Lead Specifications," American Society for Testing and Materials, Philadelphia, 1959.

13. A.G. Cannone, D.O. Feder, and R.V. Biagetti, "Lead-Acid Battery: Positive Grid Design Principles," *Bell Sys. Tech. J.* **49**:1279–1304 (1970).

14. A.T. Balcerzak, *Alloys for the 1980s*, St. Joe Lead Co., Clayton, MO, 1980.

15. *Grid Metal Manual*, Independent Battery Manufacturers Association (IBMA), Key Largo, FL, 1973.

16. N.E. Hehner, *Storage Battery Manufacturing Manual*, Independent Battery Manufacturers Association (IBMA), Key Largo, FL, 1976.

17. U.B. Thomas, F.T. Foster, and H.E. Haring, "Corrosion and Growth of Lead-Calcium Alloy Storage Battery Grids as a Function of Calcium Content," *Trans. Electrochem. Soc.* **92**:313–325 (1947).

18. A.G. Cannone, U.S. Patent 3,556,853, Jan. 19, 1971.

19. A.G. Cannone, U.S. Patent 4,980,252, Dec. 25, 1980.

20. N.E. Hehner and E. Ritchie, *Lead Oxides*, Independent Battery Manufacturers Association (IBMA), Key Largo, FL.

21. T. Ferreira et al., "Stronger, Cleaner Plates Make Better Batteries," *116th Convention of Battery Council International*, May 4, 2004.

22. D. Boden, "Sure Cure™ Technology, and Applications," *120th Convention of Battery Council International,* Tampa, FL, April 2008.

23. Trademark of Hollingsworth & Vose Company, East Walpole, MA.

24. Trademark of the Hammond Group, IN.

25. U.S. Patent 7,118,830, "Battery Paste Additive and Method for Producing Battery Plates," D. Boden, October 10, 2006.

26. Data from Daramic technical data sheets at www.daramic.com/products/daramic_products.cfm.

27. Data from Entek International data sheet at www.entek-international.com/Products/RhinoHide.html.

28. Data from Amer-sil website at www.amer-sil.com/Frames/Prod-AmerTube.htm.

29. T. Ferreira, "Development of an Inorganic Additive to Active Materials of Lead-Acid Batteries," *Long Beach Battery Conference*, 2002.

30. Yonezu, T. Masaharu, and T. Katsuhiro, "Pasted Type Lead Acid Battery." U.S. Patent 4,336,314, June 22, 1982.

31. V. Toniazzo, European Patent Application E.P 1,720,210 A1(200), "Non-Woven Gauntlet for Batteries."

32. V. Toniazzo, "New Generation of Non-Woven Gauntlets for Tubular Positive Plate," *J. Power Sources* **158**(2):1062–1068 (2006).

33. D. Pavlov, *Conference on Oxygen Cycle in Lead-Acid Batteries*, 7th ELBC, Dublin, Ireland, September, 2000.

34. P. Moseley, "Improving the Valve Regulated Lead-Acid Battery," *Proc. 1999 IBMA Conf., Battery Man*, pp. 16–29, Feb. 2000.

35. D.H. McClelland and John L. Devitt, U.S. Patent 3,862,861.

36. K.R. Bullock and D.H. McClelland, "The Kinetics of the Self-Discharge Reaction in a Sealed Lead-Acid Cell," *J. Electrochem. Soc.* **123**:327–331 (1976).

37. P.T. Moseley, D.A.J. Rand, and K. Peters, "Enhancing the Performance of Lead-Acid Batteries with Carbon—In Pursuit of an Understanding," *J. Power Sources* **295**:268–274 (2015).

38. www.furukawadenchi.co.jp/English/rd/nt_ultra.htm.

39. E. Ritchie, International Lead-Zinc Research Organization Project LE-82-84, Final Rep., New York, Dec. 1971.

40. *Gould Battery Handbook*, Gould Inc., Mendota Heights, MN, 1973.

41. R.F. Nelson, E.D. Sexton, J.B. Olson, M. Keyser, and A. Pesaran, "Search for an Optimized Cyclic Charging Algorithm for Valve-Regulated Lead–Acid Batteries," *J. Power Sources* **88**(1):44–52 (2000).

42. K.R. Bullock, D. Fent, and P. Ng, *Proc. 17th International Telecommunications Energy Conference*, pp. 8–13 (1995), IEEE 0-7803-2750-0/95.

43. R.O. Hammel, "Charging Sealed Lead Acid Batteries," *Proc. 27th Annual Power Sources Symp*, 1976.

44. R.O. Hammel, "Fast Charging Sealed Lead Acid Batteries," extended abstracts, pp. 34–36, *Electrochem. Soc. Meeting*, Las Vegas, NV, 1976.

45. "Handbook of Secondary Storage Batteries and Charge Regulators in Photovoltaic Systems." Exide Management and Technology Co., Rep. 1-7135, Sandia National Laboratories, Albuquerque, NM, Aug. 1981.

46. G.E. Mayer, "Critical Review of Battery Cycle Life Testing Methods," *Proc. 5th Int. Electric Vehicle Symp.*, Philadelphia, Oct. 1978.

47. M. Barak, *Electrochemical Power Sources*, Peter Peregrinus, Stevanage, U.K., 1980.

48. *Battery Service Manual*, 9th ed., Battery Council International, Chicago, 1982.

49. *Battery Replacement Data Book*, Battery Council International, 2000.

50. Product Literature, Rolls Battery Engineering, Nova Scotia, Canada.

51. G.J. May, Secondary Batteries—Lead-Acid Systems: Performance, in *Encyclopaedia of Electrochemical Power Sources*, J. Garche, C. Dyer, P.T. Moseley, Z. Ogumi, D.A.J. Rand, and B. Scrosati (Eds), 5, 2009, pp. 693–704.

52. G.J. May, Secondary Batteries—Lead-Acid Systems: Stationary Batteries, in *Encyclopaedia of Electrochemical Power Sources*, J. Garche, C. Dyer, P.T. Moseley, Z. Ogumi, D.A.J. Rand, and B. Scrosati (Eds), 4, 2009, pp. 859–874.

53. D.E. Koontz, D.O. Feder, L.D. Babusci, and H.J. Luer, "Lead-Acid Battery: Reserve Batteries for Bell System Use: Design of the New Cell," *Bell Sys. Tech. J.* **49**(7):1253–1278 (1970).

54. R.V. Biagetti and H.J. Luer, "A Cylindrical, Pure Lead, Lead-Acid Cell for Float Service," *J. Power Sources* **4**:309–319 (1979).

55. R.V. Biagetti, "The AT&T Lineage 2000 Round Cell Revisited: Lessons Learned; Significant Design Changes; Actual Field Performance v. Expectations," *INTELECT—Int. Telecommunications Energy Conf.*, Kyoto, Japan, Nov. 5–8, 1991.

56. A.G. Cannone, U.S. Patent 4,605,605, Aug. 12, 1986.

57. R. Wagner, Valve-Regulated Lead-Acid Batteries for Telecommunications and UPS Applications, in *Valve-Regulated Lead-Acid Batteries*, D.A.J. Rand, P.T. Moseley, and J. Garche (Eds.), Elsevier, 2004, pp. 435–465.

58. G.J. May, A. Davidson, and B. Monahov, "Lead Batteries for Utility Energy Storage: A Review," *J. Energy Storage* **15**:145–157 (2017).

59. D.A.J. Rand and P.T. Moseley, Energy Storage with Lead-Acid Batteries, in *Electrochemical Energy Storage for Renewable Sources and Grid Balancing*, P.T. Moseley and J. Garche (Eds), Elsevier, 2015, pp. 201–222.

# CHAPTER 15
# SECONDARY NICKEL CATHODE CELLS

## 15.0 OVERVIEW OF NICKEL-BASED CATHODES

### 15.0.1 Introduction

The use of nickel as a cathode material in secondary batteries began over a century ago with the advent of the nickel-iron cell by Junger in Europe and by Edison in the United States.[1] Nickel compounds have electrochemical properties as follows:

- ~1/2 V standard reduction potential (allows for coupling with various anodes to yield cells that operate within a range of 1–2 V)
- One electron valence change (limits specific energy compared to other metal oxides)
- 0.3 Ah/gm
- 2.16 Ah/cm$^3$

The cost, availability, and stability of the nickel compounds are key to the commercial success. More importantly, though, the ability to produce thin, high surface area, rugged electrodes in a relatively simple, low-cost manufacturing process has led to widespread use in batteries.

Nickel-based cathodes have been used commercially with five different anodes in six different main cell types as follows:

1. Iron
2. Cadmium
   a. Vented cells
   b. Sealed cells
3. Metal hydride
4. Zinc
5. Hydrogen

While other electrochemical couples are feasible, these five couples are the most widely known and most developed cell types.

The nickel-iron battery was adopted by Edison and others for use in the first DC electric power grids as well as for various motive power applications. Nickel-cadmium cells came into prominence for industrial, aerospace, and consumer applications, but were replaced more recently in part by the metal hydride system (less hazardous than cadmium). Nickel-zinc cells have offered promise and have had a measure of commercial success due to lower costs and having a less hazardous anode material. Meanwhile, nickel-hydrogen cells have had decades of success in the most demanding of applications, including most of the original space flights and satellite launches.

## 15.0.2   Nickel Cathode Chemistry

The cathode reactions result in the transfer of oxygen from one electrode to the other. The exact details of the reaction can be very complex; however, the electrolyte (alkaline solutions composed of potassium hydroxide, KOH, potentially with other hydroxide such as LiOH and NaOH) are presumed to undergo no permanent change in the overall reaction, unlike many other systems. The active cathode material is based on nickel oxyhydroxide, which is reduced to nickel hydroxide on discharge as follows:

$$NiOOH + H_2O + e^- \rightarrow Ni(OH)_2 + OH^- \qquad E^0 = 0.52 \text{ V}$$

The reactions are reversed on charge, but the nickel cathodes also experience a competing charging reaction as follows:

$$2OH^- \rightarrow H_2O + \tfrac{1}{2}O_2 + 2e^-$$

Depending on the particular anode and the specific cell design, hydrogen will typically be generated at the anode, but may also be suppressed by recombination (i.e., sealed cells) or other mechanisms.

Chapter 15D provides additional details on the nickel hydroxides used in many types of secondary nickel cathode cells. Supplemental materials, including both intentional additives (Co and Zn) and inadvertent impurities (sulfates and nitrates), varied physical forms (i.e., particle shape and crystallinity), and alternative processing steps (annealing and heat treating) may function to further alter the capabilities of the nickel cathodes.

## 15.0.3   Components and Cell Designs

As with other types of electrochemical cells, the cathodes may be designed and the manufacturing processes may be configured in a variety of ways. Flat plate and cylindrical cells with polymer and metal containers are common. Electrode plates are made by pasting, dough molding, sintering, and powder filling of pocket plate designs. Separators have evolved from asbestos and ceramic fibrous mats to more modern microporous plastic and fluoropolymer sheets/mats/films. The separators must act not only to prevent electrode contact, but also serve as electrolyte reservoirs necessary to compensate for outgassing and electrolyte side reactions. With certain anodes (i.e., zinc), the separator must also help prevent dendritic shorting. Additional aspects of the nickel cathode batteries are discussed in the sections below (Chaps. 15A to 15F).

# SECTION A

# NICKEL-IRON

**Gary A. Bayles**
**(Emeritus Contributors: Ralph J. Brodd, John F. Jackovitz)**

## 15A.1  INTRODUCTION

The nickel-iron rechargeable battery was introduced at the turn of the 20th century by Junger in Europe and by Edison in the United States.[1] Even today the batteries are produced in a fashion similar to the original construction. New constructions have been developed, which give better high-rate performance and have lower manufacturing costs. Today the nickel-iron battery is the most common rechargeable aqueous system using iron electrodes. The characteristics of the nickel-iron battery are compared to other iron battery systems in Tables 15A.1 and 15A.2.

As designed by Edison, the nickel-iron battery was and is almost indestructible. It has a very rugged physical structure and can withstand electrical abuse such as overcharge, over-discharge, discharged stand for extended periods, and short-circuiting. The battery is best applied where high cycle life at repeated deep discharges is required (such as traction applications) and as a standby power source with a 10- to 20-year life. Its limitations are low power density, poor low-temperature performance, poor charge retention, and gas evolution on stand. The cost of the nickel-iron battery lies between the lower cost lead-acid and the higher cost nickel-cadmium battery in most applications, with the exception of limited use applications in electric vehicles and mobile industrial equipment.

## 15A.2  CELL CHEMISTRY

### 15A.2.1  Electrochemistry

The active materials of the nickel-iron battery are metallic iron for the negative electrode, nickel oxide for the positive, and a potassium hydroxide solution with lithium hydroxide for the electrolyte. The nickel-iron battery is unique in many respects. The overall electrode reactions result in the transfer of oxygen from one electrode to the other. The exact details of the reaction can be very complex and include many species of transitory existence.[2-4] The electrolyte apparently plays no part in the overall reaction, as noted in the following reactions:

$$\text{Fe} + 2\text{NiOOH} + 2\text{H}_2\text{O} \xrightleftharpoons[\text{charge}]{\text{discharge}} 2\text{Ni(OH)}_2 + \text{Fe(OH)}_2 \quad \text{(first plateau)}$$

$$3\text{Fe(OH)}_2 + 2\text{NiOOH} \xrightleftharpoons[\text{charge}]{\text{discharge}} 2\text{Ni(OH)}_2 + \text{Fe}_3\text{O}_4 + 2\text{H}_2\text{O} \quad \text{(second plateau)}$$

The overall reaction is

$$3\text{Fe} + 8\text{NiOOH} + 4\text{H}_2\text{O} \xrightleftharpoons[\text{charge}]{\text{discharge}} 8\text{Ni(OH)}_2 + \text{Fe}_3\text{O}_4$$

**TABLE 15A.1** Comparison of Nickel Iron with other Iron Electrode Battery Systems

| System | Historical uses | Advantages | Disadvantages |
|---|---|---|---|
| Iron/nickel oxide (tubular) | Renewable energy storage, material handling vehicles, underground mining vehicles, miners' lamps, railway cars and signal systems, emergency lighting | Physically almost indestructible Not damaged by discharged stand Long life, cycling, or stand Withstands electrical abuse: overcharge, over-discharge, short-circuiting | High self-discharge Hydrogen evolution on charge and discharge Low power density Lower energy density than competitive systems Poor low-temperature performance Damaged by high temperatures Higher cost than lead-acid Low cell voltage |
| Iron/air | Motive power | Good energy density Uses readily available materials Low self-discharge | Low efficiency Hydrogen evolution on charge Poor low-temperature performance Low cell voltage |
| Iron/silver oxide | Electronics | High energy density High cycle life | High cost Hydrogen evolution on charge |

**TABLE 15A.2** System Characteristics of Nickel Iron Compared with Other Iron Electrode Battery Systems

| System | Nominal voltage, V | | Specific energy, Wh/kg | Energy density, Wh/L | Specific power, W/kg | Cycle life, 100% DOD |
|---|---|---|---|---|---|---|
| | Open-circuit | Discharge | | | | |
| Iron/nickel oxide | | | | | | |
|   Tubular | 1.4 | 1.2 | 30 | 60 | 25 | 4000 |
|   Developmental* | 1.4 | 1.2 | 55 | 110 | 110 | >1200 |
| Iron/air | 1.2 | 0.75 | 80 | | 60 | 1000 |
| Iron/silver oxide | 1.48 | 1.1 | 105 | 160 | — | >300 |

*Based on the Westinghouse nickel-iron battery.

The electrolyte remains essentially invariant during charge and discharge. It is not possible to use the specific gravity of the electrolyte to determine the state-of-charge as with the lead-acid battery. However, the individual electrode reactions do involve an intimate reaction with the electrolyte.

A typical charge-discharge curve of an iron electrode is shown in Fig. 15A.1.[5] The two plateaus on charge correspond to the formation of the stable +2 and +3 valent states of the iron reaction products.

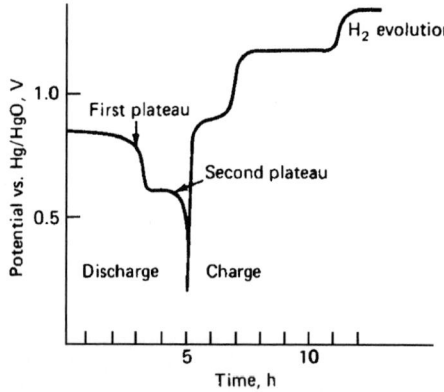

**FIGURE 15A.1** Discharge-charge curve of an iron electrode. (*From Ref. 5.*)

The reaction of the iron electrode can be written as

$$Fe + nOH^- \rightarrow Fe(OH)_n^{2-n} + 2e \quad \text{(first plateau)}$$

and

$$Fe(OH)_n^{2-n} \rightarrow Fe(OH)_2 + nOH^-$$

$$Fe(OH)_2 + OH^- \rightarrow Fe(OH)_3 + e \quad \text{(second plateau)}$$

Then

$$2Fe(OH)_3 + Fe(OH)_2 \rightarrow Fe_3O_4 + 4H_2O$$

Iron dissolves initially as the +2 species in alkaline media. The divalent iron complexes with the electrolyte to form $Fe(OH)_n^{2-n}$, a complex of low solubility. The tendency to supersaturate plays an important role in the operation of the electrode and accounts for many important aspects of the electrode performance characteristics. Continued charge forms the +3 valent iron which, in turn, interacts with +2 valent iron to form $Fe_3O_4$.

The superior life-cycling characteristics of the iron electrode result from the low solubility of the reaction intermediates and oxidized species. The supersaturation on discharge results in the oxidized material forming small crystallites near the reaction site. On charge, the low solubility also slows the crystal growth of the iron, thereby helping to ensure formation of the original active high surface area structure. The low solubility also accounts for poor high-rate and low-temperature performance as the discharged (oxidized) species precipitate at or near the reaction site and block the active surface.

## 15A.2.2  Additives/Enhancements

The performance characteristics are substantially improved, however, in the advanced nickel-iron batteries by the use of a superior electrode grid structure, such as fiber metal, which provides intimate contact with the iron active material throughout the volume of porous structure.

Addition of sulfide to the iron electrode radically changes the electrocrystallization kinetics. It increases the supersaturation and makes reaction more reversible. Sulfide also absorbs on the surface to block crystallization sites and raises the hydrogen evolution reaction on charge. The self-discharge rate of the iron electrode is correspondingly reduced by the sulfide as well, with studies showing that PbS outperforms FeS in this regard.[6] Lithium salt additions seem to make the electrode perform more reversibly, perhaps by enhancing the solubility of the reaction intermediates.

Nickel electrode reactions[7,8] are generally thought to be solid-state-type reactions wherein a proton is injected or rejected from the lattice reversibly on discharge and on charge, respectively.

$$\beta\text{-Ni(OH)}_2 \xleftarrow[\text{in KOH}]{\text{transformation}} \alpha\text{-Ni(OH)}_2$$

$$\begin{array}{ccc} \text{reduction} & \text{oxidation} & \text{oxidation} & \text{reduction} \\ \text{discharge} \updownarrow & \text{(charge)} & \text{(charge)} & \updownarrow \text{(discharge)} \end{array}$$

$$\beta\text{-NiOOH} \xrightarrow[\text{in KOH}]{\text{overcharge}} \gamma\text{-NiOOH}$$

The oxidation (charge) voltage for the $\alpha$ and $\beta$ materials is more positive than the discharge voltage by 60 and 100 mV, respectively. The $\beta\text{-Ni(OH)}_2$ is the usual electrode material. It is converted on charge to $\beta\text{-NiOOH}$ with about the same molar volume. On overcharge, the $\gamma$ structure can form. This form also incorporates water and potassium (and lithium) into the structure. Its molar volume is about 1.5 times the $\beta$ form. This is thought to be responsible in large part for the volume expansion (swelling) that occurs on charging the battery. The $\alpha$ form then results on discharge of the $\gamma$ form. Its molar volume is about 1.8 times the $\beta$ form, and the electrode can swell further on discharge. On discharge stand in concentrated electrolyte, the $\alpha$ form converts to the $\beta$ form. Cobalt additions (2–5%) improve the charge acceptance (reversibility) of the nickel electrode.

## 15A.3    CELL COMPONENTS AND CONSTRUCTION

### 15A.3.1    Conventional Pocket Plate Cell Construction

The construction of a tubular or pocket plate nickel-iron cell is shown in Fig. 15A.2. The active materials are filled in nickel-plate perforated steel tubes or pockets. The tubes are fastened into plates of desired dimensions and assembled into cells by interleaving the positive and negative plates. The container is fabricated from nickel-plated sheet steel. The cells may be assembled into batteries in molded nylon cases or mounted into wooden traps. The steel cases may be coated with plastic or rubber for insulation or spaced by insulating buttons.

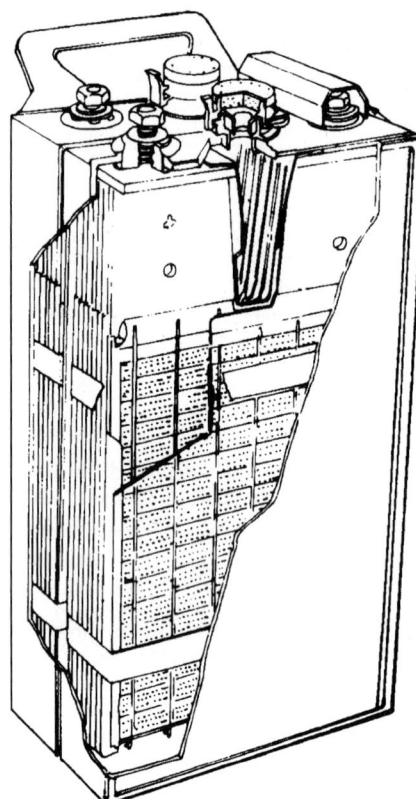

**FIGURE 15A.2**    Cross section of typical nickel-iron battery. (*Courtesy of SAFT America, Inc.*)

The manufacturing process has remained relatively unchanged for over 50 years. The processes are designed to produce materials of highest purity and with special particle characteristics for good electrochemical performance.

### 15A.3.2    Pocket Plate Negative Electrode

To produce the anode active material, pure iron is dissolved in sulfuric acid. The $FeSO_4$ is recrystallized, dried, and roasted (815–915°C) to $Fe_2O_3$. The material is washed free of sulfate, dried, and partially reduced in hydrogen. The resulting material ($Fe_3O_4$ and Fe) is partially oxidized, dried, ground, and blended. Small amounts of additives, such as sulfur, FeS, and HgO, are blended in to increase battery life by acting as depassivators, reducing gas evolution, or improving conductivity.

To make the anode current collector, steel strips or ribbon are perforated and nickel-plated. After drying and annealing, the strip is formed into a pocket, about 13 mm wide and 7.6 mm long. One end is left open and filled with the iron active material. A machine automatically introduces the active material and tamps it into the pockets. After filling, the negative pockets are crimped and pressed into openings in a nickel-plated steel frame.

### 15A.3.3  Tubular Positive Electrode

The positive active material consists of nickel hydroxide in alternate layers with nickel flake. High-purity nickel powder or shot is dissolved in sulfuric acid. The hydrogen evolved is used in making the iron active material. The acidity of the resulting solution is adjusted to pH 3 to 4 and filtered to remove ferric iron and other insoluble materials. If needed, the solution may be further purified to remove traces of ferrous iron and copper. Cobalt sulfate may be added in the proportion of 1.5% to improve nickel electrode performance. The nickel sulfate solution is sprayed into hot 25% to 50% NaOH solution. The resulting slurry is filtered, washed, dried, crushed, and screened to yield particles that pass 20-mesh but not 200-mesh screens.

Special nickel flake ($1.6 \times 0.01$ mm) is produced by electrodepositing alternate layers of nickel and copper on stainless steel. The electroplate is stripped and cut into squares. The copper is dissolved out in hot sulfuric acid, and the resulting nickel flakes are washed free of copper and dried at low temperature to prevent nickel oxide formation. With a modified process,[9] the flakes of proper shape and size can be produced as a single layer, eliminating the need for deposition of alternate copper layers. As in the negative electrode, the positive-electrode process starts with perforated steel ribbon that is nickel-plated and annealed. The ribbon is wound into tubes with an interlocked seam. Two types, right- and left-wound tubes, are produced, typically of 6.3 mm diameter. The tubes are filled with alternate layers of nickel hydroxide and nickel flakes. Each layer is tamped (144 kg/cm$^2$) to ensure good contact. There are 32 layers of flake per centimeter. To prevent the seam from opening during the rigors of charge and discharge, rings are placed around the tubes at uniform intervals of about 1 cm. The tubes are enclosed, and the pinched ends are locked into the nickel-plated steel grid frame. The "rights" and "lefts" are alternated so that any tendency to distort on the part of one tube is counteracted by the next one. The positive electrode can also be made in the pocket plate construction, as described above under "Pocket Plate Negative Electrode."

### 15A.3.4  Conventional Pocket Plate Cell Assembly

The configuration and size of the tubes and pockets determine the capacity for each plate. The plates are then assembled into electrodes to meet the capacity requirements of each cell.

Each plate group is assembled by bolting a terminal pole and a selected number of plates, depending on capacity, to a steel rod that passes through the grid at the top of the plates. Groups of positive and negative plates are intermeshed to form the element. A cell usually contains one or more negative than positive plates. The cells are made positive limiting for best cycle life.

The positive and negative plates are separated by hard rubber or plastic pins called "hair pins" or "hook pins," which fit into spaces formed by the tubular positive and flat negative electrodes.

### 15A.3.5  Advanced Metal Plaque Batteries

The desire to capitalize on the attractive features of the nickel-iron couple, such as ruggedness and long life, in applications requiring high-rate performance and low manufacturing costs led Westinghouse to pursue the development of advanced nickel-iron batteries in the 1980s with performance characteristics suitable for electric automobiles and other mobile traction applications. The capability of these batteries was intended to provide an electric vehicle with a range of at least 150 km between charges, acceleration rapid enough to merge into highway traffic, and a cycling life equivalent to 10 or more years of on-the-road service. Both the positive and the negative plates of the Westinghouse nickel-iron battery used fiber metal plaques as the substrate. The techniques for making plaques, impregnation and activation, stacking, and assembly are all amenable to high-volume production methods similar to those used in lead-acid battery manufacture.

Two methods of active nickel impregnation were developed and used in demonstration batteries. An electroprecipitation process (EPP), developed in the mid-1960s, demonstrated good performance, ruggedness, and long cycle life. The EPP deposits nickel hydroxide electrochemically into the porous substrate. Efficient use of the nickel material is achieved, with active material utilization of 0.14 Ah/g of total electrode. An alternate nickel electrode manufacturing process was also developed, which entails the preparation of a nickel hydroxide paste that is then loaded into the fiber metal substrate by roll pasting methods. Pasted nickel electrodes demonstrated performance equivalent to EPP electrodes (0.14 Ah/g of total electrode) while demonstrating a less expensive manufacturing process. The iron electrodes were also produced by a pasting process. Iron oxide, $Fe_2O_3$, was paste-loaded into the fiber metal electrode substrate and then furnace-reduced in a hydrogen atmosphere. These electrodes demonstrated 0.26 Ah/g of total electrode, or better, at $C/3$ discharge rates. Nonwoven polypropylene sheets are used as separators between electrodes.

### 15A.3.6  Electrolyte

The electrolyte is a 25% to 30% KOH solution with up to 50 g/L of LiOH added. The composition of the replacement electrolyte to compensate for losses due to spray from the vent cap is about 23% caustic with about 25 g/L LiOH. Occasionally, the electrolyte is replaced completely to rejuvenate the cell performance. The renewal electrolyte is about 30% KOH with 15 g/L LiOH.

Lithium additions to the electrolyte are important but not completely understood. Recent work on the mechanism of the lithium interaction suggests that $Li^+$ is reduced within the iron oxide lattice to produce intermediate $Li_xFe_yO_z$ intercalation compounds, which are subsequently reduced to metallic iron and lithium hydroxide.[10] Lithium hydroxide improves cell capacity and prevents capacity loss on cycling and also seems to facilitate nickel electrode kinetics. It expands the working plateau on charge and delays oxygen evolution. Some evidence exists for the formation of $Ni^{4+}$, which improves electrode capacity. Lithium also decreases the carbonate content in the electrolyte since $Li_2CO_3$ is not very soluble. It also decreases the tendency for swelling of the positive active material but increases the resistivity of the cell electrolyte.

Shortly after initiation of charge, hydrogen evolution begins on the iron electrode. The considerable hydrogen evolution on charge presumably helps counteract iron passivation in alkaline solution. Mercury additions also have a similar effect, but only in the early formation cycles.

## 15A.4  PRODUCT PERFORMANCE AND APPLICATIONS

### 15A.4.1  Performance

*Voltage.*  A typical discharge-charge curve of a commercial iron/nickel oxide battery is shown in Fig. 15A.3.[11] The battery's open-circuit voltage is 1.4 V; its nominal voltage is 1.2 V. On charge, at rates most commonly used, the maximum voltage is 1.7 to 1.8 V.

*Capacity.*  The capacity of the nickel-iron battery is limited by the capacity of the positive electrode and, hence, is determined by the length and number of positive tubes in each plate. The diameter of the tubes generally is held constant by each manufacturer. The 5 h discharge rate is commonly used as the reference for rating its capacity.

The conventional nickel-iron battery has moderate power and energy density and is designed primarily for moderate to low discharge rates. It is not recommended for high-rate applications such as engine starting. The high internal resistance of the battery lowers the terminal voltage significantly when high rates are required. The relationship between capacity and rate of discharge is shown in Fig. 15A.4.

If a battery is discharged at a high rate and then at a lower rate, the sum of the capacities delivered at the high and low rates nearly equals the capacity that would have been obtained at the single discharge rate. This is illustrated in Fig. 15A.5.

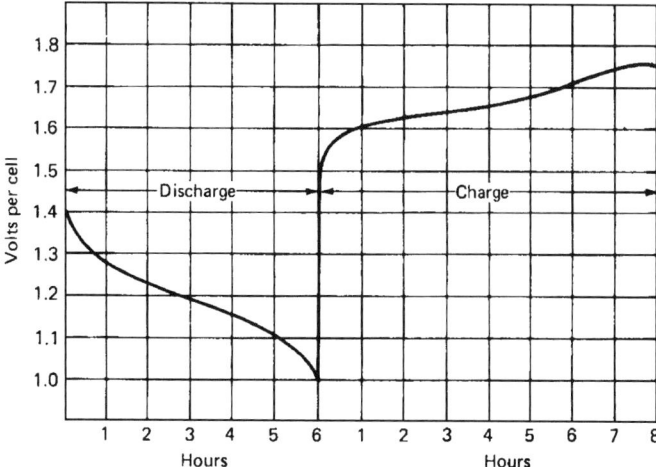

**FIGURE 15A.3**  Typical voltage characteristics during constant-rate discharge and recharge. (*From Ref. 11.*)

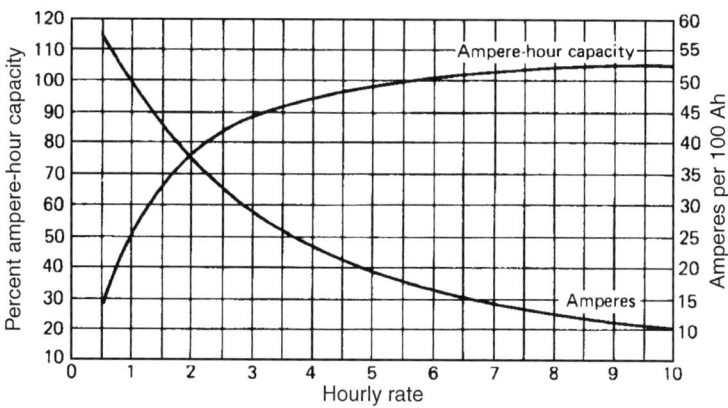

**FIGURE 15A.4**  Curves of capacity versus discharge rate at 25°C; end voltage 1.0 V per cell. (*From Ref. 11.*)

***Discharge Characteristics.***    The nickel-iron battery may be discharged at any current rate it will deliver, but the discharge should not be continued beyond the point where the battery nears exhaustion. It is best adapted to low or moderate rates of discharge (1 to 8 h rate). Figure 15A.6 shows the discharge curves at different rates of discharge at 25°C.

***Effect of Temperature.***    Figure 15A.7 shows the effects of temperature on the discharge. The capacity at 25°C is normally taken as the standard reference value. The decrease in performance is generally attributed to passivity of the iron electrode and decreased solubility of the reaction intermediate. At low temperature, increased resistivity and viscosity of the electrolyte along with slower nickel electrode kinetics contribute to the fall-off of capacity. Care must be exercised to keep the temperature from exceeding about 50°C as the self-discharge of the nickel positive electrode is accelerated. Also, the increased solubility of iron at high temperature can adversely affect operation of the nickel electrode by incorporating soluble iron into the nickel hydroxide crystal lattice. The battery is seldom used below −15°C.

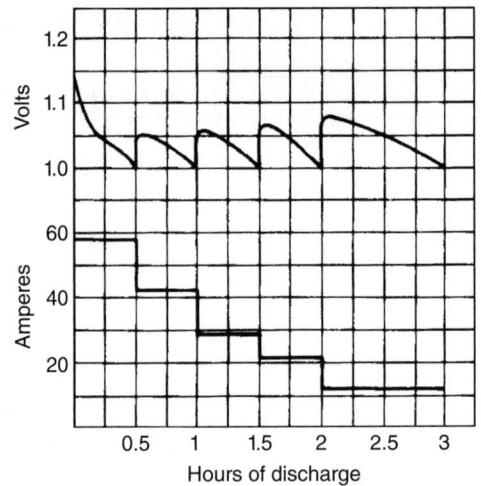

**FIGURE 15A.5** Effect of decreasing rate on battery voltage of nickel-iron cell.

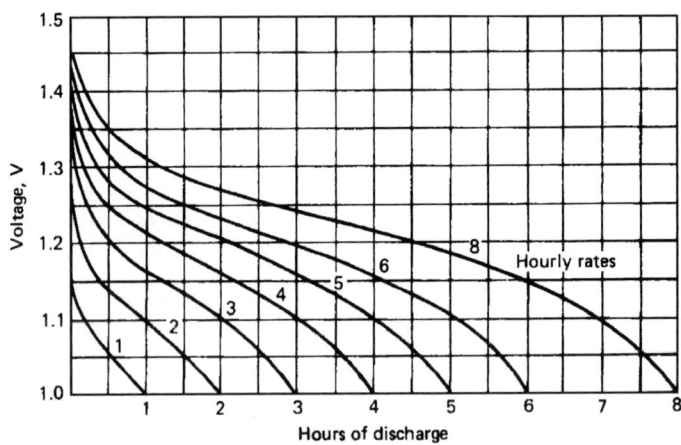

**FIGURE 15A.6** Time-voltage discharge curves of nickel-iron battery; end voltage 1.0 V per cell. (*From Ref. 11.*)

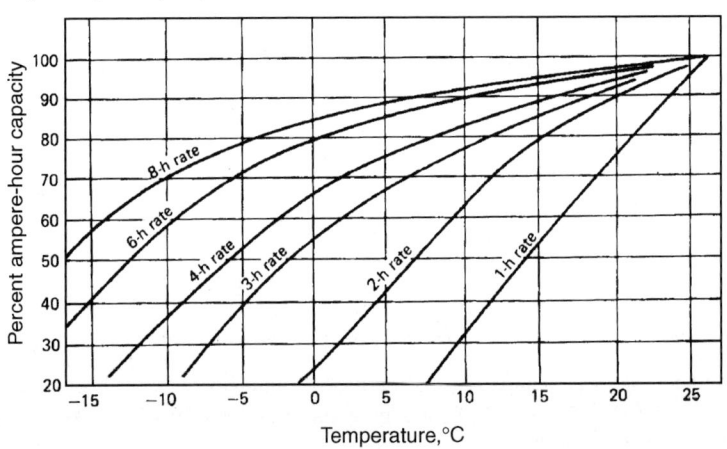

**FIGURE 15A.7** Effect of temperature on capacity at various rates. (*From Ref. 11.*)

***Hours of Service.***    The hours of service on discharge that a typical nickel-iron battery, normalized to unit weight (kilograms) and volume (liters), will deliver at various discharge rates and temperatures are summarized in Fig. 15A.8.

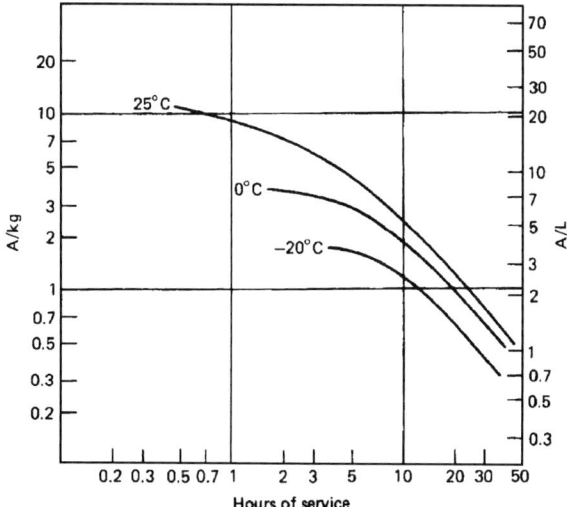

**FIGURE 15A.8**    Hours of service of nickel-iron battery at various discharge rates and temperatures; end voltage 1.0 V cell.

***Self-Discharge.***    The self-discharge rate, charge retention, or stand characteristic of the nickel-iron battery is poor. At 25°C a cell can lose 15% of its capacity in the first 10 days and 20% to 40% in a month. At lower temperatures, the self-discharge rate is lower. For example, at 0°C the losses are less than one-half of those experienced at 25°C.

***Internal Resistance.***    To a rough approximation, the internal resistance $R_i$ can be estimated for tubular Ni-Fe from the equation

$$R_i \times C = 0.4$$

where  $R_i$ = internal resistance, $\Omega$
       $C$ = battery capacity, Ah

For example, $R_i = 0.004\ \Omega$ for a 100 Ah battery. The value of $R_i$ remains constant through the first half of the discharge, then increases about 50% during the latter half of the discharge.

***Life.***    The main advantages of the tubular-type nickel-iron battery are its extremely long life and rugged construction. Battery life varies with the type of service but ranges from 8 years for heavy duty to 25 years or more for standby or float service. With moderate care, 2000 cycles can be expected; with good care, for example, by limiting temperatures to below 35°C, cycle life of 3000 to 4000 cycles has been achieved.

The battery is less damaged by repeated deep discharge than any other battery system. In practice, an operator will drive a battery-operated vehicle until it stalls, at which point the battery voltage is a fraction of a volt per cell (some cells may be in reverse). This has a minimal effect on the nickel-iron battery in comparison with other systems.

***Charging.***    Charging of the batteries can be accomplished by a variety of schemes. As long as the charging current does not produce either excessive gassing (spray out of the vent cap), or temperature rise (above 45°C), any current can be used. Excessive gassing will require more frequent addition of water. If the cell voltage is limited

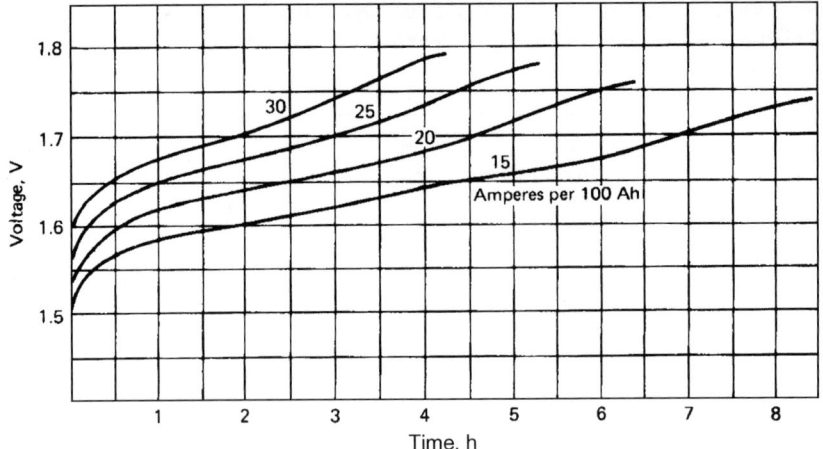

**FIGURE 15A.9**    Typical charging voltage for nickel-iron battery at various rates. (*From Ref. 11.*)

to 1.7 V, these conditions should not be a consideration. Typical charging curves are given in Fig. 15A.9. The ampere-hour (Ah) input should return 25% to 40% excess of the previous discharge to ensure complete charging. The suggested charge rate is normally between 15 and 20 A per 100 Ah of battery capacity. This rate would return the capacity in the 6 to 8 h time frame. The effect of temperature on charging is shown in Fig. 15A.10.

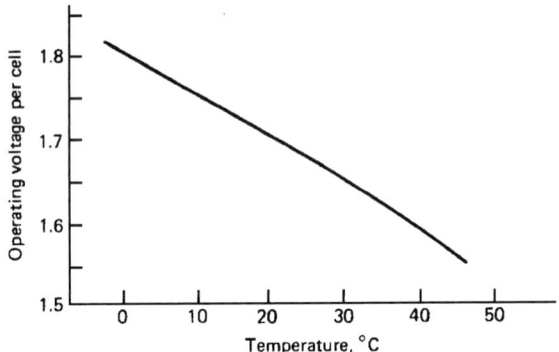

**FIGURE 15A.10**    Variation of relay operating voltage with temperature. (*From Ref. 11.*)

Constant current and modified constant potential (taper), shown in Fig. 15A.11, are common recharging techniques. The charging circuit should contain a current-limiting device to avoid thermal runaway on charge, which results in dangerous conditions and can severely damage the battery. When the battery nears full charge, the gassing reactions produce heat and the temperature rises, lowering the internal resistance and the cell EMF. Accordingly, the charge current increases under constant-voltage charge. This increased current further increases the temperature, and a vicious cycle is started. A modified constant-voltage charging with current limiting is therefore required.

Recharging each night after use (cycle charging) is the normal procedure. The batteries can be trickle-charged to maintain them at full capacity for emergency use. A trickle charge rate of 0.004 to 0.006 A/Ah of battery capacity overcomes the internal self-discharge and maintains the battery at full charge.

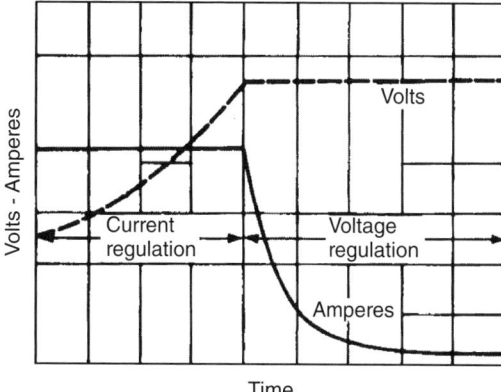

**FIGURE 15A.11**   Effects of "regulators" with voltage and current regulation. (*From Ref. 11.*)

***Performance Data.***   The performance characteristics of the advanced nickel-iron batteries, as typified by the Westinghouse system, are summarized in Table 15A.3. Figure 15A.12 shows a typical discharge curve for a 90-cell electric-vehicle battery at the $C/3$ rate. The battery capacity and energy as a function of discharge rate are shown in Fig. 15A.13. Cell power and voltage characteristics, as a function of discharge rate and state-of-discharge, are presented in Fig. 15A.14. The variation in battery capability with temperature, based on tests on five-cell modules, is shown in Fig. 15A.15.[12-14] The Eagle-Picher Company also developed an advanced nickel-iron battery using sintered-nickel electrodes similar to those described in Chap. 15B. The iron electrode is similar to the Swedish National Development Corporation iron electrode[5] which used a sintered iron-mesh anode construction, and a pore-forming material that could be leached out after treatment. The resulting pressed matrix was treated in $H_2$ at 650°C, providing an active material utilization approaching 65%. The performance of this battery is similar to that given in Figs. 15A.12 to 15A.15.[15]

**TABLE 15A.3**   Advanced Nickel-Iron Battery Performance Characteristics[*]
Demonstrated as of December 1991

| | |
|---|---|
| Capacity,[†] Ah | 210 |
| Specific energy,[†] Wh/kg | 55 |
| Energy density,[†] Wh/L | 110 |
| Specific power,[‡] W/kg | 100 |
| Cycle life[§] | >900 |
| Urban range, km | |
|    with regenerative braking | 154 |
|    without regenerative braking | 125 |
| Projected production cost, $/kWh (1990 $) | 200–250 |

[*]Based on the Westinghouse nickel-iron battery.
[†]At the $C/3$ rate.
[‡]30 s average at 50% state of charge.
[§]Cycle to 100% depth of discharge; life to 75% of rated energy.

## 15A.4.2   Applications and Summary

***Sizes of Nickel-Iron Batteries.***   Nickel-iron batteries have been available in sizes ranging from about 5 to 1250 Ah. Table 15A.4 lists the physical and electrical characteristics of typical nickel-iron batteries that were available from Varta in the 1980s.

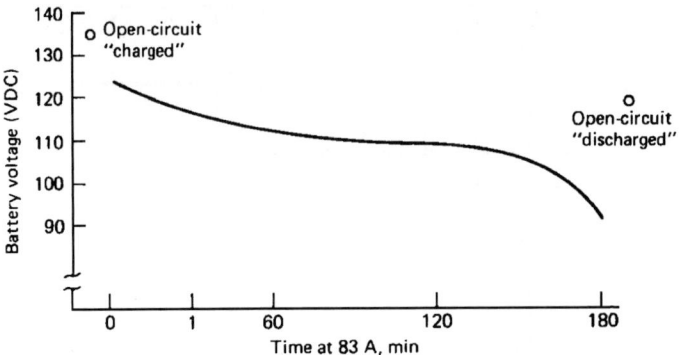

**FIGURE 15A.12**    Battery voltage at $C/3$ (83 A) discharge rate.

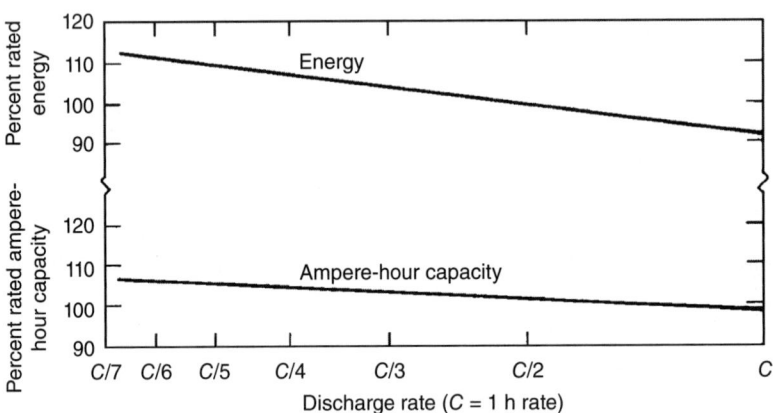

**FIGURE 15A.13**    Capacity as a function of discharge rate. (*Courtesy of Westinghouse Electric Corp.*)

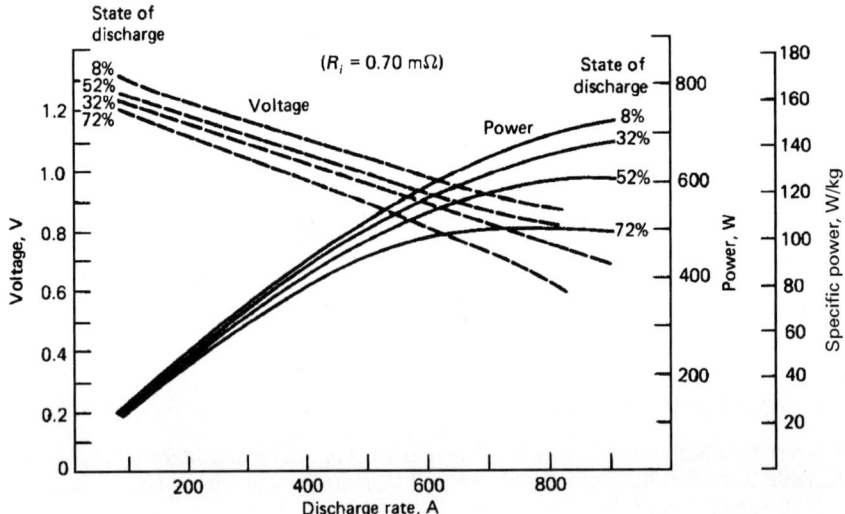

**FIGURE 15A.14**    Power characteristics of 210 Ah nickel-iron battery. (*Courtesy of Westinghouse Electric Corp.*)

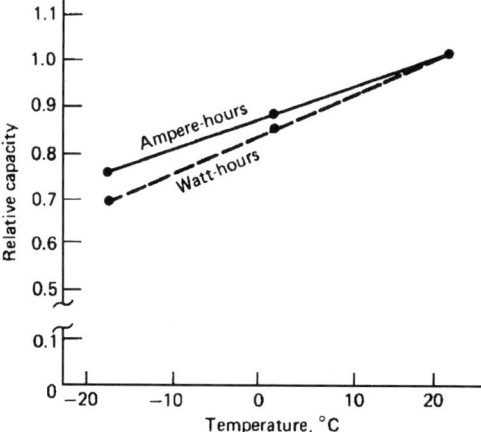

**FIGURE 15A.15** Effect of temperature on capacity and energy of nickel-iron battery (*C*/3 rate).

**TABLE 15A.4** Typical Nickel-Iron Batteries from Varta

| Nominal capacity, Ah | 169 | 225 | 280 | 337 | 395 | 450 | 560 | 675 |
|---|---|---|---|---|---|---|---|---|
| Nominal current, A: 5 h discharge | 34 | 45 | 56 | 67 | 79 | 90 | 112 | 135 |
| Cell weight, filled, kg | 8.8 | 10.8 | 12.9 | 15.3 | 17.4 | 19.5 | 24.3 | 28.6 |
| Installed weight, kg | 9.8 | 12.0 | 14.3 | 16.9 | 19.3 | 21.7 | 26.5 | 31.2 |
| Electrolyte (1.17 kg/L), kg | 1.8 | 2.2 | 2.6 | 3.0 | 3.4 | 3.8 | 4.9 | 5.9 |
| Cell dimensions, mm* | | | | | | | | |
| Length | 52 | 66 | 82 | 96 | 111 | 125 | 156 | 186 |
| Width | 130 | 130 | 130 | 130 | 130 | 130 | 135 | 135 |
| Height | 534 | 534 | 534 | 534 | 534 | 534 | 534 | 534 |
| Battery dimensions, mm† | | | | | | | | |
| Length: | | | | | | | | |
| Two cells | — | — | — | 265 | 295 | 321 | 343 | 343 |
| Three cells | — | — | — | 376 | 421 | 460 | | |
| Four cells | 284 | 367 | 431 | 487 | | | | |
| Five cells | 346 | 448 | 545 | | | | | |
| Six cells | 408 | 546 | | | | | | |
| Width | 161 | 161 | 161 | 161 | 161 | 161 | 197 | 228 |
| Height | 568 | 573 | 582 | 582 | 582 | 582 | 590 | 590 |

*See drawing (*a*).
†See drawing (*b*).
***Source***: Varta Batteries AG, Hanover, Germany.

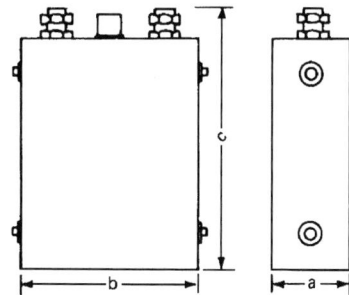

(*a*) Cell showing dimensions used in the above table.

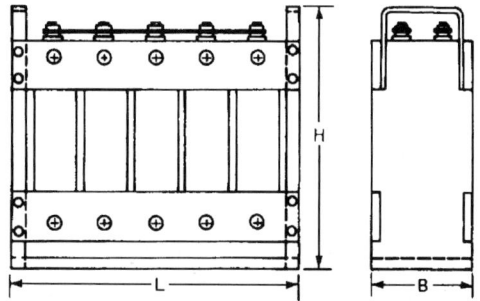

(*b*) Multicell battery showing dimensions used in the above table. Tolerances are 5, 3, and 3 mm for dimensions L, B, and H, respectively.

In recent years, nickel-iron has given way to the lead-acid and nickel-cadmium batteries in many markets, and are no longer manufactured by many of the original manufacturers. At present (2017), nickel-iron batteries are advertised by suppliers in the United States (Iron Edison, Zapp Works, Encell), Australia (IronCore, UNISUN), and China (Henan Xintaihang Power Source Co., Ltd, Zhuhai Ciyi Battery Co., Ltd., Sichuan Changhong Battery Co., Ltd.) among others.

In the United States, Iron Edison works closely with an overseas manufacturer on the design of their nickel-iron cell. Each cell uses a mechanically perforated pocket plate technology, and the steel structure of the plates means they cannot be weakened by repeated cycling, increasing longevity. These batteries use an electrolyte solution that acts as a metal preservative for the plates, protecting against degradation. The battery is flooded, requiring distilled water roughly every 6 weeks, and off-gases hydrogen, requiring a battery enclosure with active ventilation. Each cell is rated at 1.2 V DC. The Ah ratings available from Iron Edison range from 100 to 1000 Ah in 100 Ah increments.[16] Examples of Iron Edison cells and batteries are shown in Figs. 15A.16 and 15A.17.

**FIGURE 15A.16**   Commercial nickel-iron cells from Iron Edison. (*Courtesy of Iron Edison, Ref. 16.*)

***Special Handling and Use of Nickel-Iron Batteries.***   The battery should be operated in a well-ventilated area to prevent the accumulation of hydrogen. Under certain circumstances, hydrogen can be ignited by a spark to cause an explosion with a resulting fire. In multicell batteries, the usual precautions in dealing with high voltages should be taken.

If the battery is to be out of service for more than a month, it should be stored in the discharged condition. It should be discharged and short-circuited, then left in that condition for the storage period. Filling caps must be kept closed. If this procedure is not followed, several cycles are required to restore the capacity upon reactivation.

***Recent Advances in Iron Anode Materials.***   The rugged and economical nature of this couple continues to sustain a level of research interest, particularly in the area of large-scale renewable energy storage.[17-19] Efforts to develop advanced iron electrodes, improved rate capability, improved efficiency, and even to create a sealed maintenance-free design[20] are areas of continued development in that regard.

The increased use of carbon in battery electrodes, coupled with advances in the development of nanomaterials in the last 20 years, has presented new opportunities in iron electrode research. The ability to disperse iron in

**FIGURE 15A.17**    A nickel-iron battery from Iron Edison. (*Courtesy of Iron Edison, Ref. 16.*)

carbon nanostructures offers the potential for better active material utilization and cyclic efficiency. Iron active material utilizations as high as 510 mAh/g have been reported,[21] which is a significant improvement over the commonly achieved levels of 350 mAh/g, and a respectable approach toward the 962 mAh/g theoretical limit. The challenge in this work has been to maintain the nanoscale character, since the iron particle size has been shown to increase with repeated cycling, accompanied by a decrease in the surface area, which in turn produces a linear decline in the active material utilization of about 30 mAh/g for every m²/g of surface area lost.

Several methods of fabricating iron electrodes with nanoscale properties have been studied. A slow addition of a ferric chloride solution to a chilled sodium borohydride solution produces iron particles with sizes in the range of 30 to 70 nm.[21] Another approach[22] uses iron carbide particles in the 20 μm range as a starting material, which after cycling, ultimately produces particles with a size of 100 nm or less, presumably through repeated dissolution and redeposition of Fe and $Fe(OH)_2$. Yet, another approach[23] uses carbon substrates with nanoscale features upon which iron is deposited. Carbon nanofibers, nanotubes, or platelets having particle sizes in the 100 nm range are impregnated with an aqueous iron nitrate solution and then dried and calcined to produce an iron-carbon composite with finely dispersed $Fe_2O_3$. In an attempt to keep the iron particles from growing during cycling (which degrades capacity), the particles have been preferentially deposited within the nanotubes, rather than on the surface.[24] This approach has shown only limited success, since the penetration of the electrolyte into the nanotubes then becomes the limiting factor, and this limitation is only partly overcome by treating the nanotubes to create pores in the walls of the tubes. Iron carbon nanostructured electrodes have shown promise in increasing discharge rate capability and utilization.[25-27]

In both sintered porous iron electrodes and iron-carbon composite electrodes, adding sulfide compounds to the electrolyte or the electrode material appears to have the beneficial effects of increasing hydrogen over-potential, improving electrode capacity, and enhancing charge efficiency.[18,28-31]

## *REFERENCES*

1. S. U. Falk and A. J. Salkind, *Alkaline Storage Batteries*, Wiley, New York, 1969.

2. A. J. Salkind, C. J. Venuto, and S. U. Falk, "The Reaction at the Iron Alkaline Electrode," *J. Electrochem. Soc.* **111**:493 (1964).

3. R. Bonnaterre, R. Doisneau, M. C. Petit, and J. P. Stervinou, in J. H. Thompson (ed.), *Power Sources*, vol. 7, Academic, London, 1979, p. 249.

4. L. Ojefors, "SEM Studies of Discharge Products from Alkaline Iron Electrodes," *J. Electrochem. Soc.* **123**:1691 (1976).

5. B. Anderson and L. Ojefors, in J. H. Thompson (ed.), *Power Sources*, vol. 7, Academic, London, 1979, p. 329.

6. C. A. C. Souza, I. A. Carlos, M. Lopes, G. A. Finazzi, and M. R. H. de Almeida, "Self-Discharge of Fe-Ni Alkaline Batteries," *J. Power Sources* **132**:288–290 (2004).

7. J. L. Weininger, in R. G. Gunther and S. Gross (eds.), *The Nickel Electrode*, vol. 82–84, Electrochemical Society, Pennington, NJ, 1982, pp. 1–19.

8. D. Tuomi, "The Forming Process in Nickel Positive Electrodes," *J. Electrochem. Soc.* **123**:1691 (1976).

9. INCO ElectroEnergy Corp. (formerly ESB, Inc.), Philadelphia.

10. U. Casellato, N. Comisso, and G. Mengoli, "Effect of Li Ions on Reduction of Fe Oxides in Aqueous Alkaline Medium," *Electrochimica Acta* **51**:5669–5681 (2006).

11. "Nickel Iron Industrial Storage Batteries," Exide Industrial Marketing Divisions of ESB, Inc., 1966.

12. F. E. Hill, R. Rosey, and R. E. Vaill, "Performance Characteristics of Iron Nickel Batteries," *Proc. 28th Power Sources Symp.*, Electrochemical Society, Pennington, NJ, 1978, p. 149.

13. R. Rosey and B. E. Tabor, "Westinghouse Nickel-Iron Battery Design and Performance," EV Expo 80, EVC #8030, May 1980.

14. W. Feduska and R. Rosey, "An Advanced Technology Iron-Nickel Battery for Electric Vehicle Propulsion," *Proc. 15th IECEC*, Seattle, Aug. 1980, p. 1192.

15. R. Hudson and E. Broglio, "Development of the Nickel-Iron Battery System for Electric Vehicle Propulsion," *Proc. 29th Power Sources Conf.*, Electrochemical Society, Pennington, NJ, 1980.

16. N. Renteria, Corporate Communication, Iron Edison Corporation, November 11, 2017.

17. C. Yang, A. K. Manohar, and S. R. Narayanan, "A High-Performance Sintered Iron Electrode for Rechargeable Alkaline Batteries to Enable Large-Scale Energy Storage," *J. Electrochem. Soc.* **164**(2): A418–A429 (2017).

18. A. K. Manohar, S. Malkhandi, B. Yang, C. Yang, G. K. Surya Prakash, and S. R. Narayanan. "A High-Performance Rechargeable Iron Electrode for Large-Scale Battery-Based Energy Storage," *J. Electrochem. Soc.* **159**(8):A1209–A1214 (2012).

19. A. H. Abdalla, C. I. Oseghale, J. O. Gil Posada, and P. J. Hall, "Rechargeable Nickel–Iron Batteries for Largescale Energy Storage," *IET Renew. Power Gen.* **10**(10):1529–1534 (2016).

20. B. Hariprakash, S. K. Martha, M. S. Hegde, and A. K. Shukla, "A Sealed, Starved-Electrolyte Nickel-Iron Battery," *J. Appl. Electrochem.* **35**:27–32 (2005).

21. K. C. Huang and K. S. Chou, "Microstructure Changes to Iron Nanoparticles during Discharge/Charge Cycles," *Electrochemistry Communications* **9**:1907–1912 (2007).

22. K. Ujimine and A. Tsutsumi, "Electrochemical Characteristics of Iron Carbide as an Active Material in Alkaline Batteries," *J. Power Sources* **160**:1431–1435 (2006).

23. B. T. Hang, T. Watanabe, M. Egashira, S. Okadab, J. Yamaki, S. Hata, S.-H. Yoon et al., "The Electrochemical Properties of $Fe_2O_3$-Loaded Carbon Electrodes for Iron-Air Battery Anodes." *J. Power Sources* **150**:261–271 (2005).

24. B. T. Hang, H. Hayashi, S. H. Yoon, S. Okada, and J. Yamaki, "$Fe_2O_3$-Filled Carbon Nano-Tubes as a Negative Electrode for an Fe-Air Battery," *J. Power Sources* **178**:393–401 (2008).

25. H. Wang, Y. Liang, M. Gong, Y. Li, W. Chang, T. Mefford, J. Zhou et al., "An Ultrafast Nickel-Iron Battery from Strongly Coupled Inorganic Nanoparticle/Nanocarbon Hybrid Materials," *Nature Communications* **3**:917 (2012).

26. A. S. Rajan, S. Sampath, and A. K. Shukla, "An In Situ Carbon-Grafted Alkaline Iron Electrode for Iron-Based Accumulators," *Energy & Environmental Sci.* **7**:1110–1116 (2014).

27. W. Jiang, F. Liang, J. Wang, L. Su, Y. Wu, and L. Wang, "Enhanced Electrochemical Performances of $FeO_x$-Graphene Nanocomposites as Anode Materials for Alkaline Nickel-Iron Batteries," *RSC Adv.* **4**:15394–15399 (2014).

28. B. T. Hang, T. Watanabe, M. Egashira, I. Watanabe, S. Okada, and J. Yamaki, "The Effect of Additives on the Electrochemical Properties of Fe/C Composite for Fe/Air Battery Anode," *J. Power Sources* **155**(2):461–469 (2006).

29. J. O. Gil Posada and P. J. Hall, "Post-Hoc Comparisons among Iron Electrode Formulations Based on Bismuth, Bismuth Sulphide, Iron Sulphide, and Potassium Sulphide under Strong Alkaline Conditions," *J. Power Sources* **268**:810–815 (2014).

30. J. O. Gil Posada, and P. J. Hall, "Towards the Development of Safe and Commercially Viable Nickel–Iron Batteries: Improvements to Coulombic Efficiency at High Iron Sulphide Electrode Formulations," *J. Appl. Electrochem.* **46**(4):451–458 (2016).

31. B. Yang, S. Malkhandi, A. K. Manohar, G. K. Surya Prakash, S. R. Narayanan, "Organo-sulfur Molecules Enable Iron-based Battery Electrodes to Meet the Challenges of Large-Scale Electrical Energy Storage," *Energ. Environ. Sci.* **7**:2753–2763 (2014).

# SECTION B

# VENTED NICKEL CADMIUM

**R. David Lucero**
**(Emeritus Contributor: John K. Erbacher)**

## 15B.1    INTRODUCTION TO VENTED CELLS

### 15B.1.1    Pocket Plate

The vented pocket-plate battery is the oldest and most mature of the various designs of nickel-cadmium batteries available. It is a very reliable, sturdy, long-life battery that can be operated effectively at relatively high discharge rates and over a wide temperature range. It has very good charge retention properties, and it can be stored for long periods of time in any condition without deterioration. The pocket-plate battery can stand both severe mechanical abuse and electrical maltreatment such as overcharging, reversal, and short-circuiting. Little maintenance is needed on this battery. The cost is lower than for any other kind of alkaline storage battery; still, it is higher than that of a lead-acid battery on a per watt-hour basis. The major advantages and disadvantages of this type of battery are listed in Table 15B.1.

**TABLE 15B.1**    Major Advantages and Disadvantages of Industrial and Aerospace Nickel-Cadmium Batteries

| Advantages | Disadvantages |
|---|---|
| Long cycle life | Low energy density |
| Rugged, can withstand electrical and physical abuse | Higher cost than lead-acid batteries |
| Reliable, no sudden death | Contains cadmium |
| Good charge retention | Caustic alkaline electrolyte |
| Excellent long-term storage | Memory effect |
| Low maintenance | Temperature-controlled charging system required to |
| Flat discharge profile |     extend life |

The pocket-plate battery is manufactured in a wide capacity range, 5 to more than 1200 Ah, and it is used in a number of applications. Most of these are of an industrial nature, such as railroad service, switchgear operation, telecommunications, uninterruptible power supplies (UPSs), and emergency lighting. The pocket-plate battery was also used in military and space applications.

Pocket-plate batteries are available in three-plate thicknesses to suit the variety of applications. The high-rate designs use thin plates for maximum exposed plate surface per volume of active material. They are used for the highest rate discharge. The low-rate designs use thick plates to obtain maximum volume of active material per exposed plate surface. These types are used for long-term discharge. The medium-rate designs use plates of medium thickness and are suited for applications between, or combinations of, high-rate and long-term discharge.

### 15B.1.2    Sintered Plate

The sintered plate, which can be constructed in a thinner form than the pocket plate, was developed during the 1940s and has a lower internal resistance and gives superior high-rate and low-temperature performance compared to the pocket plate. It is used in high-power applications, such as engine starting, and in low-temperature environments, leading to the design of smaller batteries for portable equipment and subsequently to the sealed, maintenance-free nickel-cadmium battery covered in Chap. 15C.

**TABLE 15B.2**    Major Advantages and Disadvantages of Vented Sintered-Plate Nickel-Cadmium Batteries

| Advantages | Disadvantages |
|---|---|
| Flat discharge profile | Higher cost |
| Higher energy density (50% greater than pocket plate) | Memory effect (voltage depression) |
| Superior high-rate and low-temperature performance | Temperature-controlled charging system required to enhance life |
| Excellent long-term storage | |
| Good capacity retention; capacity can be restored by recharge | |

The sintered-plate nickel-cadmium battery is a mature development of the nickel-cadmium system, having a higher energy density, up to 50% greater than its predecessor, the pocket-type construction. The sintered plate can be constructed in a much thinner form than the pocket plate, and the cell has a much lower internal resistance and gives superior high-rate and low-temperature performance. A flat discharge curve is characteristic of the battery, and its performance is less sensitive than other battery systems to changes in discharge load and temperature. The sintered-plate battery has most of the favorable characteristics of the pocket-type battery, although it is generally more expensive. It is electrically and mechanically rugged, is very reliable, requires little maintenance, can be stored for long periods of time in a charged or uncharged condition, and has good charge retention. Batteries losing capacity through self-discharge can be restored to full service with a normal charge. The major advantages and disadvantages of this battery type are given in Table 15B.2.

For these reasons, vented sintered-plate nickel-cadmium batteries are used in applications requiring high-power discharge service such as aircraft turbine engine and diesel engine starting as well as other mobile and military equipment. The battery provides outstanding performance where high peak power and fast recharging are required. In many applications, the vented sintered-plate battery is used because it leads to a reduction in size, weight, and maintenance as compared to other battery systems. This is particularly true in systems subject to low-temperature operation. The rise in terminal voltage of the vented cell at the end of charge also provides a useful characteristic for controlling the charge.

### 15B.1.3    Advanced Plate Designs

The sintered-plate battery is expensive, complex to manufacture, and uses a large amount of nickel, making the technology impractical for thicker electrodes or for cells larger than 100 Ah. The pocket-plate battery is too heavy for many applications. Efforts to develop a high surface area, conductive-plate structure that would be light, easy to manufacture, and inexpensive led to the development of a new electrode structure, the fiber-structured electrode (i.e., the Fiber Nickel Cadmium Battery—FNC) developed by Deutsche Automobilgesellschaft GmbH (DAUG).

The fiber plates are manufactured from either a mat of pure nickel fibers or, more commonly, nickel-plated plastic fibers. To make the plastic fiber conductive, a thin layer of nickel is applied by electroless plating and thereafter a sufficiently thick layer of nickel for good conductivity is applied by electroplating. The plastic is then burned off, leaving a mat of hollow nickel fibers. The nickel-fiber plaque is welded to a nickel-plated steel tab. Figure 15B.1a shows the structure of a nickel-fiber plaque before impregnation, and Fig. 15B.1b shows an unformed positive electrode.

This fiber electrode technology, while originally developed for EV applications, was first used for industrial low- and medium-rate vented cells. It is now being used in all types of nickel-cadmium as well as nickel-metal hydride batteries, including high-rate batteries for engine-cranking and sealed cells with oxygen recombination (see Chap. 15C).

A more recent design with significantly improved performance characteristics is the plastic-bonded or pressed-plate electrode. These electrodes are used in industrial batteries and are a spin-off from those used in aircraft and sealed portable consumer batteries. In the plastic-bonded plate, which is mainly used in the cadmium electrode, the active material, cadmium oxide, is mixed with a plastic powder, normally PTFE, and a solvent to produce a paste. The paste is isotropic, and the materials are manufactured at the final density for the active material. As a result, dust problems are eliminated during manufacturing. The paste is extruded, rolled, or pasted onto a center current collector normally made of nickel-plated perforated steel. The plate structure is welded to nickel-plated steel tabs.

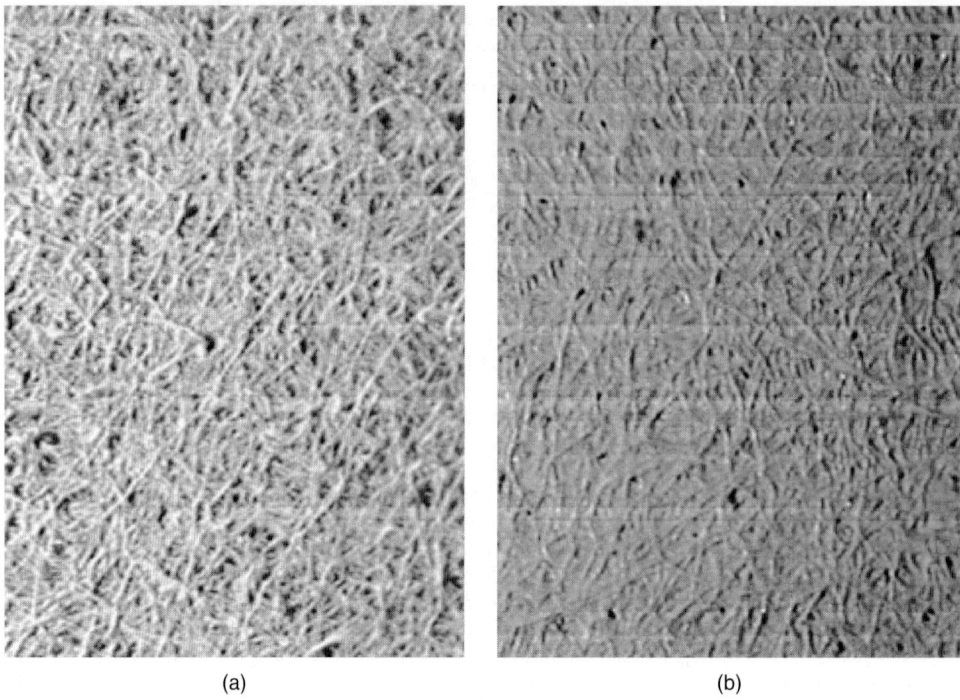

(a)                                    (b)

**FIGURE 15B.1**    (*a*) Nickel-fiber electrode structure before impregnation. (*Courtesy of Acme Electric Corp.*) (*b*) Unformed, pasted nickel positive electrode. (*Courtesy of Acme Electric Corp.*)

## 15B.2   CHEMISTRY

The basic electrochemistry is the same for the vented pocket-plate, sintered-plate, fiber and plastic-bonded plate types, as well as for other variations of the nickel-cadmium system. The reactions of charge and discharge can be illustrated by the following simplified equation:

$$2NiOOH + 2H_2O + Cd \underset{charge}{\overset{discharge}{\rightleftharpoons}} 2Ni(OH)_2 + Cd(OH)_2$$

On discharge, trivalent nickel oxyhydroxide is reduced to divalent nickel hydroxide with consumption of water. Metallic cadmium is oxidized to form cadmium hydroxide. On charge, the OPPOSITE reactions take place. The electromotive force (EMF) is 1.29 V.

The potassium hydroxide electrolyte is not significantly changed with regard to density or composition during charge and discharge, in contrast to the sulfuric acid in lead-acid batteries. The electrolyte density is generally approximately 1.2 g/mL. Lithium hydroxide is often added to the electrolyte for improved cycle life and high-temperature operation.

The electrochemistry of the charge and discharge of the positive electrode is quite complex and not well understood,[1] especially the role that cobalt plays in the active material.[2] For simplicity, consider the role of nickel hydroxide in the charge-discharge reaction.

During charge, the nickel hydroxide in the positive electrode is oxidized to nickel oxyhydroxide (NiOOH) and higher valence states of nickel. Potassium and water are also incorporated into the active material as potassium hydroxide according to the following equation[3]:

$$Ni(OH)_2 + xK^+ + (1+x)OH^- \rightleftharpoons NiOOH \cdot xKOH \cdot (H_2O) + e$$

The fraction of potassium that is bonded into the NiOOH lattice is represented by $x$. The value is small (much less than 1.0) and varies according to manufacturing process.

The cadmium hydroxide in the negative electrode is reduced to metallic cadmium during charge

$$Cd(OH)_2 + 2e \rightleftharpoons Cd + 2OH^-$$

The overall charge-discharge reaction is thus

$$2Ni(OH)_2 + 2xKOH + Cd(OH)_2 \underset{\text{discharge}}{\overset{\text{charge}}{\rightleftharpoons}} 2NiOOH \cdot xKOH \cdot (H_2O) + Cd$$

According to the above equation, it may seem that the change in the electrolyte concentration might offer a means of state-of-charge determination by measuring the specific gravity of the electrolyte. Unfortunately, the complication of potassium in the active material, accumulation of carbonates, along with the large volume of electrolyte makes this type of measurement unreliable and impractical.

The positive electrode does not accept charge, and converts nickel hydroxide to nickel oxyhydroxide, at the thermodynamically reversible potential.[3] In fact, with a low enough charge rate, gassing according to the equation below occurs:

$$4OH^- \rightarrow 2H_2O + O_2 + 4e$$

If the rate is increased appreciably, this will result in an oxygen overvoltage sufficiently high to allow the preferred conversion of the nickel hydroxide to the nickel oxyhydroxide instead of oxygen gassing. However, when about 80% conversion of nickel hydroxide to nickel oxyhydroxide is achieved, the competing oxygen-generating reaction occurs gradually and remains until 100% state-of-charge is achieved, and then the only reaction occurring is oxygen evolution.

The negative electrode accepts charge until it is essentially 100% charged, at which time the favored reaction is hydrogen gassing as shown in the following equation:

$$2H_2O + 2e \rightarrow H_2 + 2OH^-$$

The hydrogen gassing reaction, with a $C/10$ charge rate, occurs at a cell voltage close to 1600 mV as shown in Fig. 15B.2.

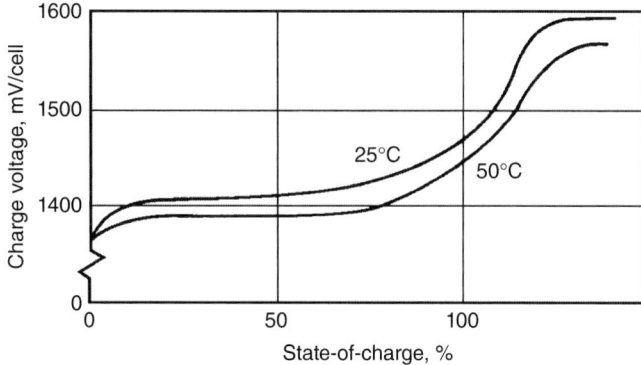

**FIGURE 15B.2**  Constant-current charge voltage of vented sintered-plate nickel-cadmium cell; $C/10$ charge rate.

The hydrogen overvoltage on the cadmium electrode is quite high, about 110 mV at the $C/10$ rate. Consequently there is a sharp rise in voltage as the negative electrode goes into overcharge. This rise in voltage is used in various charging schemes to control or terminate charging.

During overcharge, all the current is used to electrolyze water to hydrogen and oxygen, as shown in the overall reaction

$$2H_2O \rightarrow 2H_2 + O_2$$

This overcharge reaction consumes water and thereby decreases the level of electrolyte in the cell. The water loss can be limited by controlling the amount of overcharge so as to maximize the interval between needed water replenishments.

Cells are constructed with 50% excess capacity in the negative electrodes and thus are positive limited.

## 15B.3  CELL COMPONENTS AND CONSTRUCTION

### 15B.3.1  Pocket Plate Cells

A cutaway view of a modern pocket-plate cell is shown in Fig. 15B.3. The active material for the positive electrodes consists of nickel hydroxide mixed with graphite for conductivity and additives such as barium or cobalt compounds for improved life and capacity. The active material for the negative electrodes is prepared from cadmium hydroxide or cadmium oxide mixed with iron or iron compounds and sometimes also with nickel. The iron and nickel materials are added to stabilize the cadmium, prevent crystal growth and agglomeration, and improve conductivity. Typical active material compositions are shown in Table 15B.3.

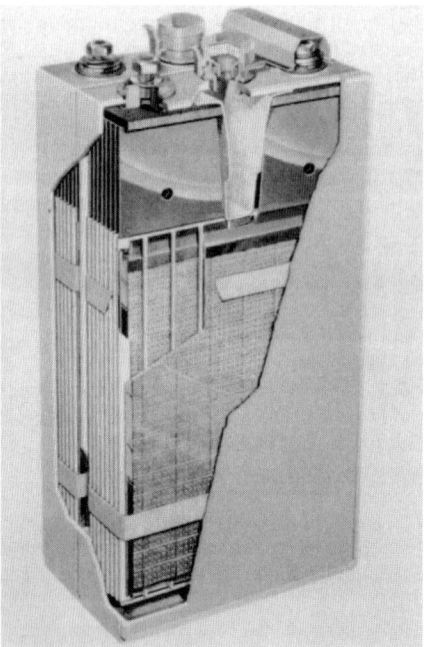

**FIGURE 15B.3**    Pocket plate cell.

The positive and negative electrodes of pocket-plate nickel-cadmium batteries are made using the same basic design to hold the active materials. The pocket plates are built up of flat pockets of perforated steel strips holding the active materials. The thin steel strips are perforated by hardened steel needles or by a technique

**TABLE 15B.3**  Typical Composition of Active Materials for Pocket-Plate Cells in the Discharged State

| Positive active material | | Negative active material | |
|---|---|---|---|
| Substance | Weight% | Substance | Weight% |
| Nickel (II) hydroxide | 80 | Cadmium hydroxide | 78 |
| Cobalt (II) hydroxide | 2 | Iron | 18 |
| Graphite | 18 | Nickel | 1 |
| | | Graphite | 3 |

using profiled roller dies. The specific hole area is between 15% and 30%. The strips are nickel-plated to prevent "iron poisoning" of the positive active material.

The active mass is either pressed into briquettes, which are fed into the pre-shaped perforated strip, or fed into the pre-shaped strip as a powder. The upper and lower steel strips are folded together by rollers. A number of these folded strips are arranged to interlock with each other to form long electrode sheets, which are then cut to electrode blanks. Electrodes are made from these blanks by providing them with steel frames for mechanical stability and for current takeoff.

The electrodes are made with different thicknesses (1.5 to 5 mm) to provide cells for high-, medium-, and low-rate discharge rates. The negative plate is always thinner (30% to 40%) than the positive.

The electrodes are bolted or welded to electrode groups. Plate groups of opposite polarity are intermeshed and electrically separated from each other by plastic pins and plate edge insulators. Sometimes separators or perforated plastic sheets or plastic ladders are used between the electrodes. The distance between plates of different polarity in an element may vary from less than 1 mm for high-rate cells to 3 mm for low-rate cells.

The elements are inserted into cell containers of plastic or stainless steel. Plastic containers are made from polystyrene, polypropylene, or flame-retarded plastics. Important advantages of plastic containers over steel containers are that they allow visual control of the electrolyte level and require no protection against corrosion. Also, they have lower weight and can be more closely packed in the battery. The main drawbacks are that they are more sensitive to high temperatures and require somewhat more space than steel containers.

## 15B.3.2  Sintered Plate Cells

Vented sintered-plate nickel-cadmium cells, in the discharged state, consist of flat positive nickel hydroxide and negative cadmium hydroxide plates, separated by materials that act as a gas barrier and electrical separator. The electrolyte, normally a 31% potassium hydroxide solution, completely covers the plates and separators: for this reason vented cells are referred to as "flooded cells."

In the sintered-plate design, the active materials are held within the pores of a sintered-nickel structure. Nickel hydroxide with 3% to 10% cobalt hydroxide is the active material of the positive plate, while cadmium hydroxide is the active material of the negative plate.

Vented cells are designed so that both electrodes reach full charge at about the same time. The positive electrode, as noted, will begin to evolve oxygen before it is fully charged. If this gas is allowed to reach the negative electrode due to failure of the gas barrier, it will recombine and generate heat. This will not only prevent the negative from reaching a full state-of-charge, but it will also result in reduced voltage due to depolarization of the cadmium electrode. To maintain the fullest capability, adequate precautions must be taken to prevent oxygen recombination at the negative plate. This is accomplished by providing a gas barrier between the positive and negative plates and by flooding the plates with excess electrolyte.

Figure 15B.4 shows details of a typical vented sintered-plate nickel-cadmium cell.

***Plate Types and Processing.***  A variety of plate formulations are used in vented sintered-plate nickel-cadmium cells produced by different manufacturers. The plates differ according to the nature of the substrate, method of sintering, impregnation process, formation, and termination techniques. The predominate plate-fabrication process used for vented sintered plates over the years has been described by Fleischer.[4] There are several reviews on electrode fabrication processes that have been used in flooded vented cells.[3,5,6]

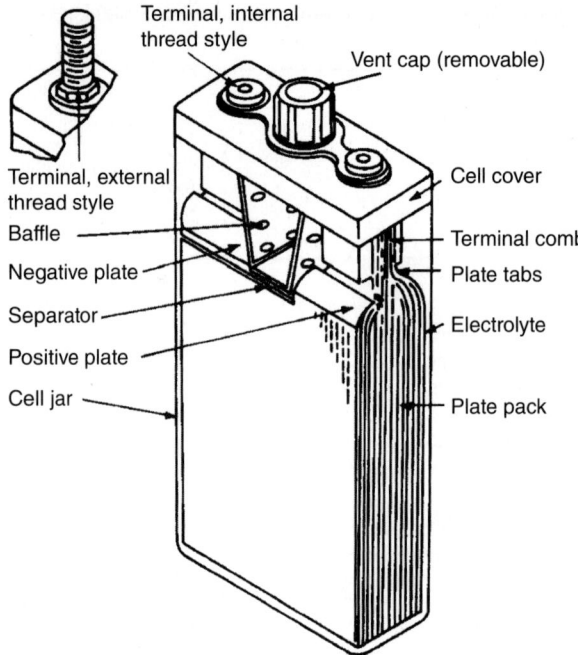

**FIGURE 15B.4**   Cross section of vented sintered-plate nickel-cadmium cell.

*Substrate.*   The substrate serves as a mechanical support for the sintered structure and as a current collector for the electrochemical reactions that occur throughout the porous sintered portion of the plate. It also provides mechanical strength and continuity during the manufacturing processes. Two types of substrate are typically used: (1) perforated nickel-plated steel or pure nickel strip in continuous lengths, and (2) woven screens of nickel or nickel-clad steel wire. A common perforated type may be 0.1 mm thick with 2 mm holes and a void area of about 40%. A typical screen may use 0.18 mm diameter wire with 1.0 mm openings.

*Plaque.*   The sintered structure before impregnation is generally referred to as "plaque." It usually has a porosity of 80% to 85% and ranges in thickness from 0.40 to 1.0 mm. Two generic sintering processes are used: (1) the slurry coating process and (2) the dry-powder process. Both processes employ special low-density battery grades of carbonyl nickel powder.

In the slurry coating process, the nickel powder is suspended in a viscous, aqueous solution containing a low percentage of a thixotropic agent. The nickel-plated strip with the desired perforated pattern is pulled through the suspension. The thickness is controlled by passing it through doctor blades, while wiping the edges free of slurry. The continuous strip is then dried before sintering in a reducing atmosphere at about 1000°C.

The dry-powder processes generally employ wire screen precut to the so-called master plaque dimension. The screens are placed in molds with loose powder on each side. They are then typically sintered in a belt furnace in a reducing atmosphere at 800 to 1000°C.

*Impregnation.*   A review of various impregnation processes, used to load the porous sintered structure of the positive with nickel hydroxide and of the negative with cadmium hydroxide, has been given by Pickett.[6] The plaque is impregnated with a concentrated solution of nitrate that is then converted to hydroxide by chemical precipitation[4] or electrochemical precipitation.[7,8,9] The most widely used process for vented cells involves a chemical precipitation, and with minor variations, follows, in principle, the process described in 1948.[4]

The plaque is impregnated with a concentrated solution of the nitrate, briefly rinsed, and the nitrate salts are precipitated as hydroxide with caustic. Following the addition of caustic, the plaque is made cathodic. This is called polarization. The polarization cycle usually consists of a high current charge ($C$ rate or higher) for approximately 1 h. The plaque is then rinsed and this sequence of steps is repeated a number of times so as to fill about 40% to 60% of the sintered pore volume (or until a targeted weight gain is achieved).

*Plate Formation.*    Following impregnation, the plates are mechanically brushed and electrochemically cleaned and formed by charging and discharging the electrode. In the master-plaque process, they are formed against inert counter electrodes (typically stainless steel or nickel) and can be performed in a loose pack or tight pack configuration. Formation is essential for properly converting hydroxide into the pores of the sinter structure, as well as the reduction of nitrates in the plates. Typical formation cycles for chemical plates consist of high-current cycling. This regime or time may vary for plaque type and capacity. In the case of the continuous strip process, the formation is done on a machine similar in appearance to a continuous strip electroplating machine. Plates blanked from the continuous strip have a clean, wiped area at the top that serves as attachment points for nickel or nickel-plated steel current-collector tabs. In the case of the master-plaque process, a coined or densified area is provided for attachment of these collector tabs.

**Separator.**    The separator system is a thin, multilayered combination. It consists of a cloth that electrically separates the positive and negative plates and an ion-permeable plastic membrane that serves as the gas barrier.

Electrical and mechanical separation of the plates is typically provided by either woven or felted nylon material. This material is relatively porous in order to provide a good ionic conduction path through the electrolyte with a microporous polypropylene separator.

The microporous polypropylene membrane, typically Celgard® (Celgard 3400, manufactured by Celgard LLC, Charlotte, NC, 28273),[10] is utilized as the gas barrier while at the same time it offers minimum ionic resistance. This thin gas barrier, which becomes relatively soft when wetted, is frequently placed between two layers of the cloth separator and receives significant mechanical support from them. Substantial improvements have been made to the toughness of the plastic membrane gas barrier.

**Plate-Pack Cell Assembly.**    Plate packs are assembled by alternately stacking positive and negative plates with the separator-gas barrier system interleaved between them. The cell terminals are bolted, riveted, or welded to the current-collector plate tabs. In the case of cells with many plates, the tabs from the outermost plates may need to be bent quite significantly inward to reach the cell terminals. Spacers at the terminals are sometimes used in these situations to keep the angle of the tabs at a minimum.

**Electrolyte.**    Potassium hydroxide electrolyte is used in a concentration of approximately 31% at full charge (specific gravity 1.30). Performance of the cell, particularly at low temperatures, is significantly dependent on this concentration.

Electrolyte purity can also have significant effects on cell performance. The level of potassium carbonate in the cell relates directly to the cell's performance. Increasing carbonate concentration changes the characteristics of the electrolyte, reducing the high-rate charge and discharge capability of the cell. Fresh electrolyte contains very low levels of carbonate. However, organic components in the cell are slowly oxidized in the presence of the electrolyte and oxygen, forming small amounts of carbonate. The carbonates accumulate as the cell ages and eventually reduce cell performance. Carbonate levels at the time of cell activation are on the order of 80 to 90 g/l due to reaction of the electrolyte with residues from the impregnation process. High-quality cells are designed with components that do not degrade in KOH. In addition, at least one manufacturer flushes new cells repeatedly with fresh electrolyte, lowering final carbonate levels to 6 to 8 g/l.

**Cell Container.**    The plate pack is placed into the cell container with the cell terminals extending through the cover. The cell container is usually made of a low-moisture-absorbent nylon and consists of the cell jar and matching cover that are permanently joined together at assembly by solvent sealing, thermal fusion, or ultrasonic bonding. The container is designed to provide a sealed enclosure for the cell, thus preventing electrolyte leakage or contamination, as well as providing physical support for the cell components. The terminal seal is generally provided by means of O-rings with Belleville washers and retaining clips.

***Vent Cap and Check Valve.*** The vent cap serves as a removable cap to provide the access required for replenishment of water to the electrolyte and also to function as a check valve to release gases generated when water is consumed during overcharge. The check valve prevents atmospheric contamination of the electrolyte. It consists of a nylon body with a hollow center post through which a cross-hole is drilled and around which an elastomeric sleeve is placed. This functions as a Bunsen valve to allow gas to escape from the cell but not to enter. Typical sleeves used for this application have developed significantly over time, and ethylene-propylene rubber seems to have the best characteristics for vented cells. Neoprene vent sleeves were previously used, but the neoprene is attacked by potassium hydroxide and can soften, swell, and split. It also frequently erodes at the interface between the vent cap and sleeve until the neoprene no longer seals. Before erosion occurs, a sleeve surface can soften due to electrolyte at the interface between the sleeve and vent cap, dry during a subsequent storage, and literally glue itself to the vent. When this occurs, the pressure will build up in the cell during charge until the sleeve breaks free or ruptures or the cell explodes.[10]

### 15B.3.3  Vented Cells with Advanced Plate Type

A plastic-bonded plate cell in a plastic container is shown in Fig. 15B.5.

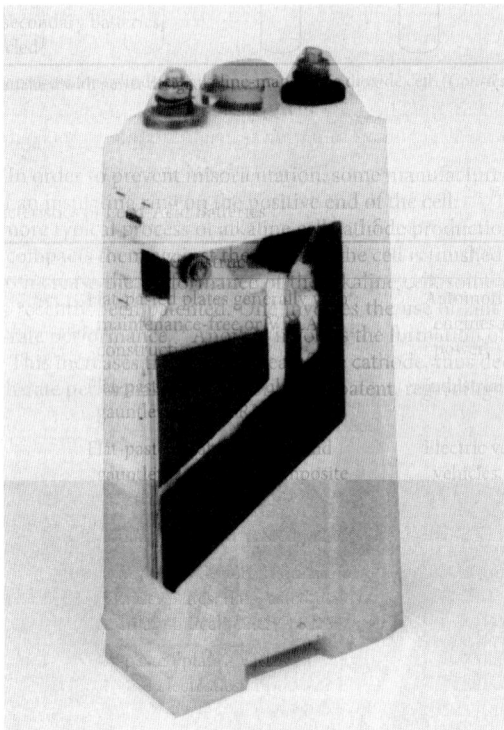

**FIGURE 15B.5**  Plastic-bonded plate cell.

Figure 15B.6 shows a partial cutaway view of a fiber nickel-cadmium (FNC) cell. The case and cover are polypropylene and are welded together. The electrode assembly shows a negative electrode, a corrugated separator, and a positive electrode. O-ring seals are employed in the bushings for the terminals to ensure gas retention, and a vent valve is seen in the case cover between the terminals. A catalytic gas-recombination plug in the vent valve is employed for some applications. The terminals are nickel-plated copper, and a 1.19 kg/l KOH electrolyte is typically employed.

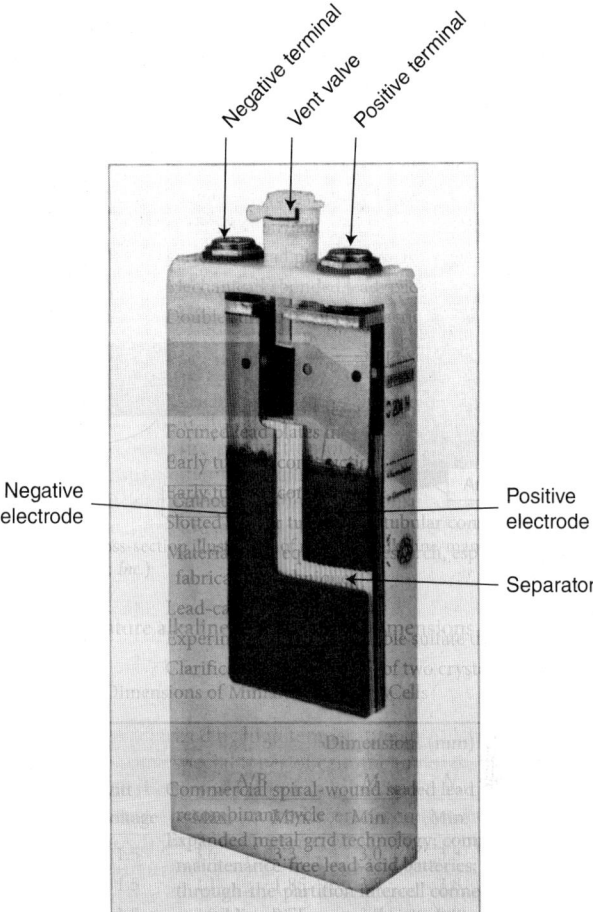

**FIGURE 15B.6**  Partial cutaway view of fiber nickel-cadmium (FNC) cell. (*Source: Hoppecke Batteries.*)

### 15B.3.4  Battery Modules

Regardless of plate or cell type, cells may also be assembled into batteries in different ways. Often 2 to 10 cells are mounted in a separate battery unit, several of which may be used to form the complete battery. A typical battery is shown in Fig. 15B.7. Cells in plastic containers are also assembled into batteries by putting the single cells close together on a rack or a stand and connecting them with intercell connectors. Cells in steel containers can be assembled in a similar way; however, here the cells must be spaced from one another and insulated from the rack.

Aerospace batteries are assembled with 19 to 21 cells per battery in a configuration similar to that shown above in Fig. 15B.8. In many cases, voltage monitoring at the half or quarter battery configuration is used to monitor cell balance and battery state-of-charge.

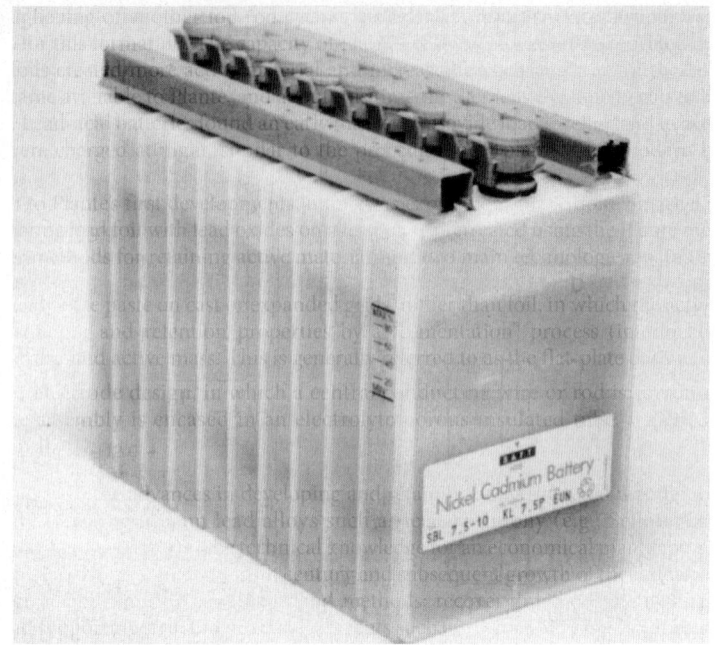

**FIGURE 15B.7**    Ten-cell welded polypropylene unit. (*Courtesy of SAFT America, Inc.*)

**FIGURE 15B.8**    Typical vented Ni-Cd aviation battery and top view of cell assembly. (*Courtesy of SAFT America, Inc.*)

## 15B.4   PERFORMANCE AND APPLICATIONS

### 15B.4.1   General Performance Characteristics

*Pocket Plate Cells*

*Energy Density and Specific Energy.*    Typical specific energy and energy density values for pocket-plate, single-cell batteries are 20 Wh/kg and 40 Wh/L, with the best values for commercially available units reaching 27 Wh/kg and 55 Wh/L. The corresponding values for complete pocket-plate batteries are 19 Wh/kg and 32 Wh/L and 27 Wh/kg and 44 Wh/L, respectively. These data are based on the nominal capacity and the average discharge voltage at the 5 h rate. The specific energy and energy density of larger fiber plate batteries approach 40 Wh/kg and 80 Wh/L. Batteries with plastic-bonded plates approach 56 Wh/kg and 110 Wh/L. This compares to a specific energy of 30 to 37 Wh/kg and an energy density of 58 to 96 Wh/L for sintered-plate designs (see "General Applications" in Sec. 15B.4.2).

*Discharge Properties.*    The nominal voltage of a nickel-cadmium battery is 1.2 V. Although discharge rate and temperature are of importance for the discharge characteristics of all electrochemical systems, these parameters have a much smaller effect on the nickel-cadmium battery than on, for instance, the lead-acid battery. Thus, pocket-plate nickel-cadmium batteries can be effectively discharged at high discharge rates without losing much of the rated capacity. They can also be operated over a wide temperature range.

Typical discharge curves at room temperature for pocket-plate and plastic-bonded plate batteries at various constant discharge rates are shown in Fig. 15B.9. Even at a discharge current as high as $5C$ (where $C$ is the numerical value of the capacity in Ah), a high-rate pocket-plate battery can deliver 60% of the rated capacity and a plastic-bonded battery as much as 80%. Battery capacities as a function of discharge rate and cutoff voltage are given in Fig. 15B.10.

Pocket plate nickel-cadmium batteries can be used at temperatures down to $-20°C$ with the standard electrolyte. Cells filled with a more concentrated electrolyte can be used down to $-50°C$. Figure 15B.11 shows the effect of temperature on the relative performance of a nickel-cadmium medium-rate battery with standard electrolyte.

Batteries can also be used at elevated temperatures. Although occasional operation at very high temperatures is not detrimental, 45 to 50°C is generally considered as the maximum permissible temperature for extended periods of operation. Recent tests on aviation batteries exposed to the higher temperature regime of Southwest Asia have led to changes in the high-temperature limits for operation and maintenance of these batteries, which can be operated at as high as 70°C.

Figure 15B.12 shows typical so-called starter curves for high-rate batteries. The batteries can deliver as much current as $20C$ A during 1 s loads to a final voltage of 0.6 V.

Occasional over-discharge or reversal of nickel-cadmium batteries is not detrimental, nor is complete freezing of the cells. After warming up, the cells will function normally again.

*Internal Resistance.*    Nickel-cadmium batteries generally have a low internal resistance. Typical DC resistance values are 0.4, 1, and 2 mΩ, respectively, for a charged 100 Ah high-, medium-, and low-rate pocket-plate single-cell battery. The internal resistance is largely inversely proportional to the battery size in a given series. Decreasing temperature and decreasing state-of-charge of a battery will result in an increase of the internal resistance. The internal resistance of fiber-plate batteries is 0.3 mΩ for a high-rate design and 0.9 mΩ for a low-rate design. Plastic-bonded plate batteries have an internal resistance as low as 0.15 mΩ.

*Charge Retention.*    Charge retention characteristics of vented pocket-plate batteries at 25°C are shown in Fig. 15B.13. Charge retention is temperature-dependent, the capacity loss at 45°C being about three times higher than at 25°C. There is virtually no self-discharge at temperatures lower than $-20°C$. Charge retention for fiber and plastic-bonded plate batteries has similar characteristics; their charge retention corresponds to that shown in Fig. 15B.13 for high-rate batteries.

*Life.*    The life of a battery can be given either as the number of charge and discharge cycles that can be delivered or as the total lifetime in years. Under normal conditions, a nickel-cadmium battery can reach more than 2000 cycles. The total lifetime may vary between 8 and 25 years or more, depending on the design and application and on the operating conditions. Batteries for diesel engine cranking normally last about 15 years, batteries for train lighting have normal lives of 10 to 15 years, stationary standby batteries have lives of 15 to 25 years, and aircraft batteries have lives of 3 to 5 years.

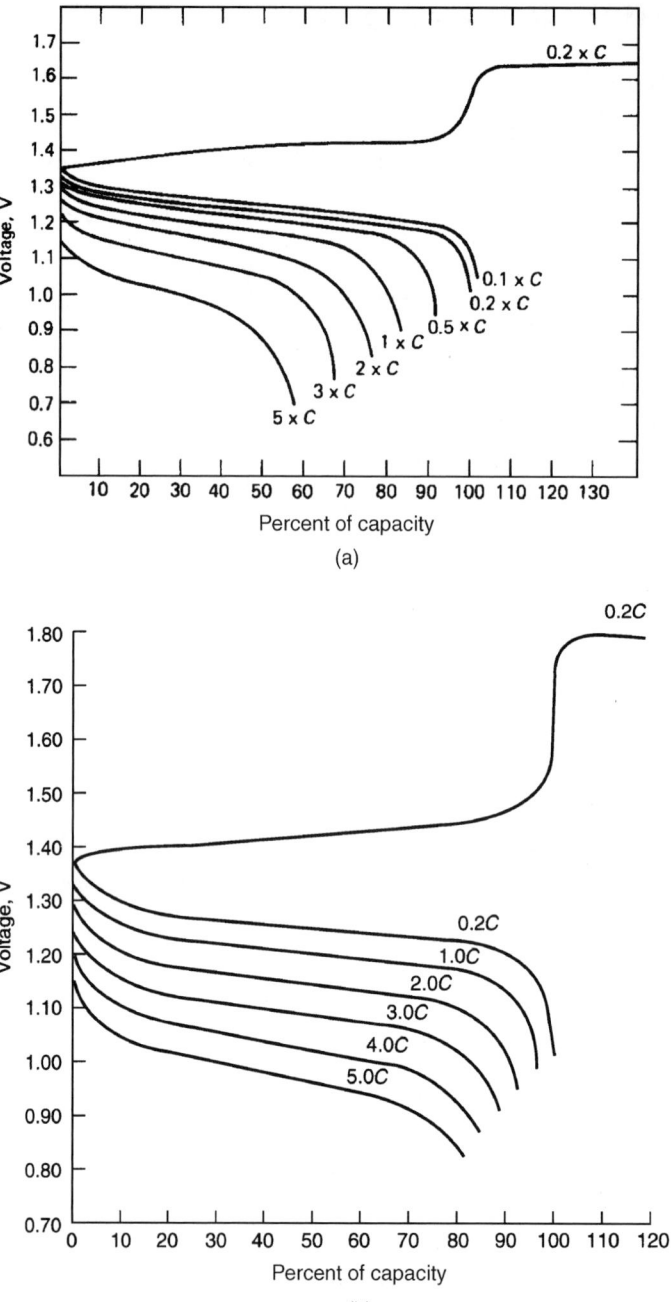

**FIGURE 15B.9**   Charge and discharge characteristics of nickel-cadmium batteries at 25°C. (*a*) Pocket-plate battery, high rate. (*b*) Plastic-bonded plate battery, high rate.

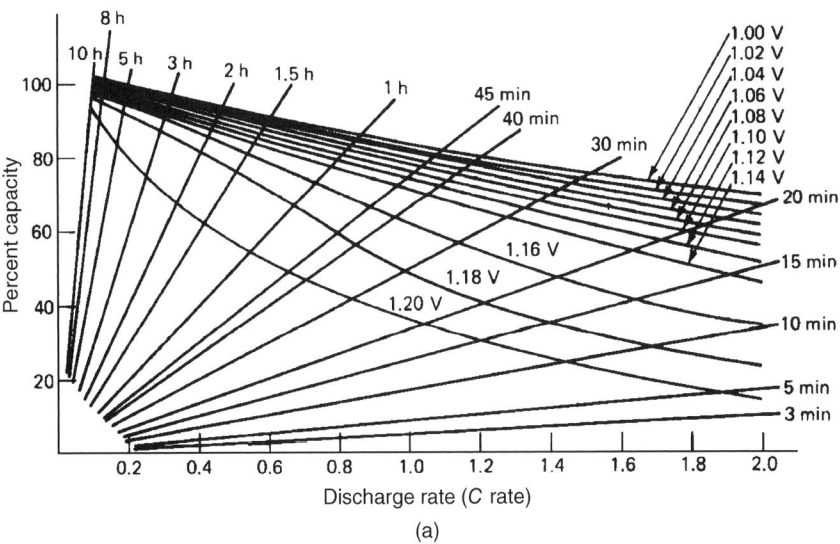

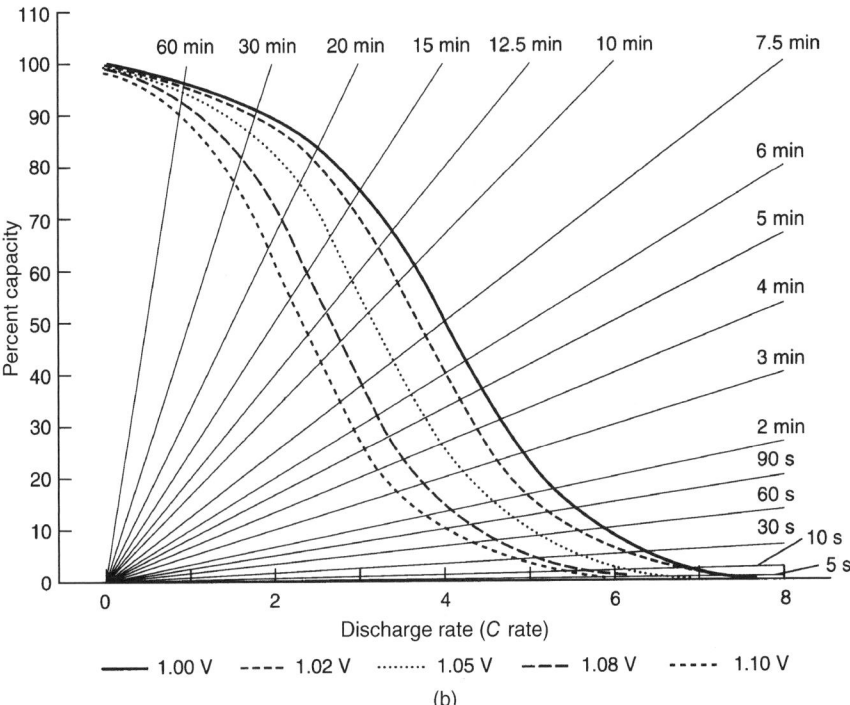

**FIGURE 15B.10** Discharge characteristics of nickel-cadmium batteries at 25°C; capacity as a function of discharge rate and cutoff voltage. (*a*) Pocket-plate battery, high rate. (*b*) Plastic-bonded plate battery, high rate.

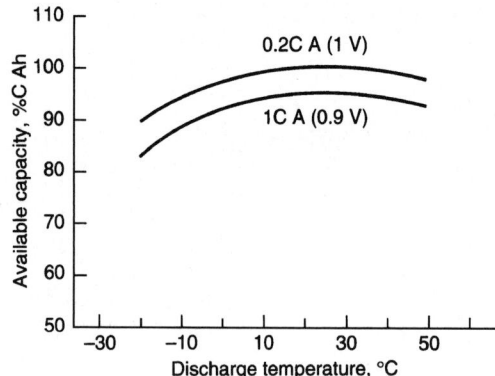

**FIGURE 15B.11**   Typical available capacity at different temperatures for nickel-cadmium medium-rate batteries with standard electrolyte, fully charged at 25°C.

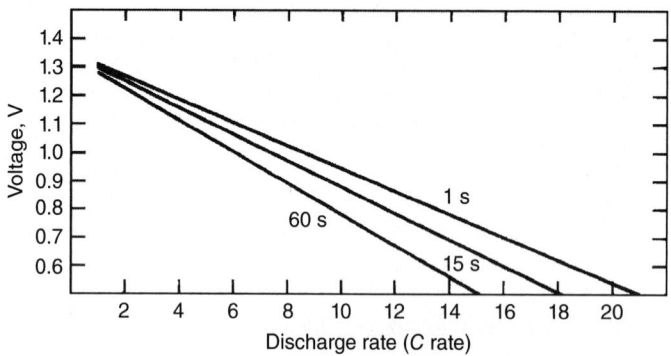

**FIGURE 15B.12**   Voltage-current curves for high-rate pocket-plate batteries at 25°C.

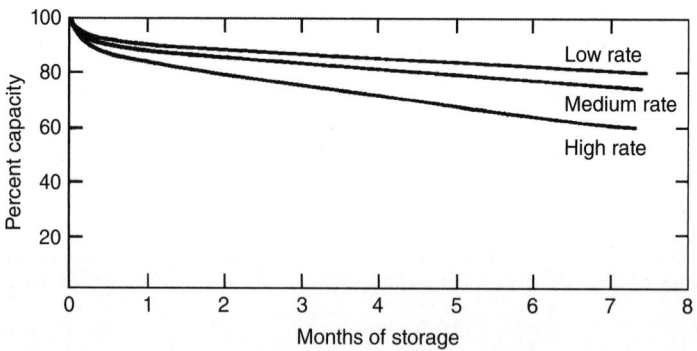

**FIGURE 15B.13**   Charge retention of pocket-plate batteries at 25°C.

Factors that affect battery life are the operating temperature, the discharge depth, and the charging regime. Low or moderate operating temperatures should always be preferred. Batteries operating at elevated temperatures or in cycling applications should be filled with electrolyte to which lithium hydroxide has been added.

The factors behind the excellent reliability and very long life of the nickel-cadmium batteries are the mechanically strong design, the absence of corrosive attack of the electrolyte on the electrodes and other components in the cell, and, furthermore, the ability of the battery to withstand electrical abuse, such as reversal or overcharging, and to stand long-time storage in any state-of-charge.

*Mechanical and Thermal Stability.*    Nickel-cadmium cells and batteries are mechanically very robust and can withstand severe mechanical abuse and rough handling in general. The electrode groups are carefully bolted or, in more recent designs such as FNC, welded together. The cell containers are made of steel or high-impact plastics.

The electrolyte does not attack any of the components in the cell, and, accordingly, there is no risk of decreased strength during the lifetime of the battery. Cases of so-called sudden death due to corroded lugs or terminals cannot occur.

The thermal resistance of the nickel-cadmium batteries is also very good. These batteries can withstand temperatures up to 85°C or more without mechanical damage. Cells in polypropylene or steel containers are the best in this respect. Saline or corrosive environments present no problems for cells in plastic containers.

*Memory Effect.*    The memory effect—the tendency of a battery to adjust its electrical properties to a certain duty cycle to which it has been subjected for an extended period of time—has been a problem with nickel-cadmium batteries in some applications. Pocket, fiber, and plastic-bonded plate cells do not show this tendency. See the section below labeled "Safety and Maintenance (Vented, Sintered Plate Cells)" and Chap. 15C for a description of the memory effect with sintered-plate nickel-cadmium batteries.

*Charging Characteristics.*    Pocket-plate nickel-cadmium batteries may be charged at constant current, constant voltage, or modified constant voltage. Constant-current charge characteristics are shown above in Fig. 15B.9. Charging is normally carried out at the 5 h rate for 7 h for a fully discharged battery. Overcharging is not detrimental but should be avoided as it leads to increased gassing and decomposition of water. Charging can be carried out in the temperature range of −50 to 45°C. However, at the extreme temperatures, the charging efficiency is lower.

Constant-voltage charging characteristics with current limitations are shown in Fig. 15B.14. The current is often limited to 0.1 to 0.4C A, and charging is normally carried out in the voltage range of 1.50 to 1.65 V per cell. The charging time may vary from 5 to more than 25 h, depending on the current limitation value and cell type.

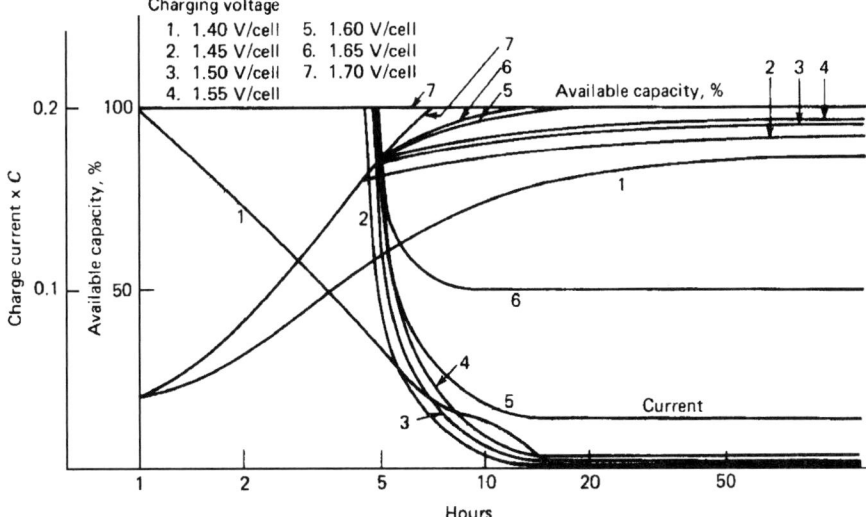

**FIGURE 15B.14**   Constant-voltage charging with current limitation 0.2C of medium-rate pocket plate batteries at 25°C.

In some applications, such as emergency and standby, it is necessary to keep the battery in a high state-of-charge. A convenient way is to connect the battery in parallel with the ordinary current source and the load, and to float the battery at 1.40 to 1.45 V per cell. The floating may be combined with a supplementary charge at fixed intervals or after each discharge.

The ampere-hour efficiency of the pocket plate battery is 72% when going from the discharged to the fully charged state. The watt-hour efficiency is approximately 60%. The best plastic-bonded plate batteries have an ampere-hour efficiency of 85% and a watt-hour efficiency of 73%.

### Sintered Plate Cells Performance Characteristics

*Discharge Properties.* The discharge curves for a typical vented sintered-plate nickel-cadmium battery at various constant-discharge loads are shown in Fig. 15B.15. The discharge curves for a typical battery at various temperatures are shown in Fig. 15B.16. The curves for this battery are characterized by a flat voltage profile, even at relatively high discharge rates and low temperatures. Voltages at various constant-current discharge loads and states of discharge are given in Fig. 15B.17.

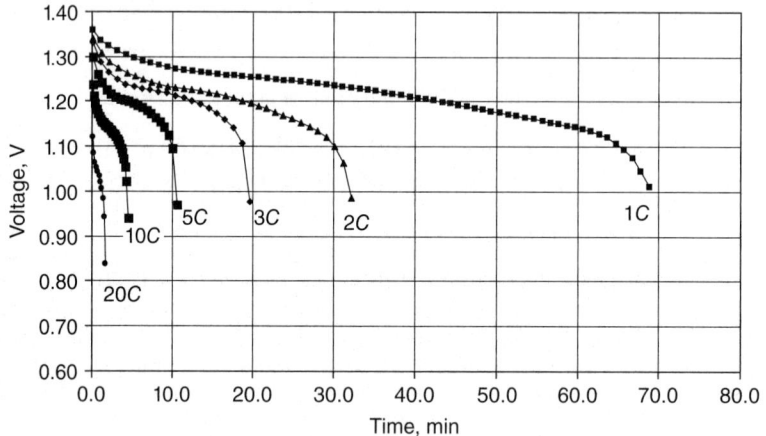

**FIGURE 15B.15**   Typical discharge curves at various *C* rates, 25°C.

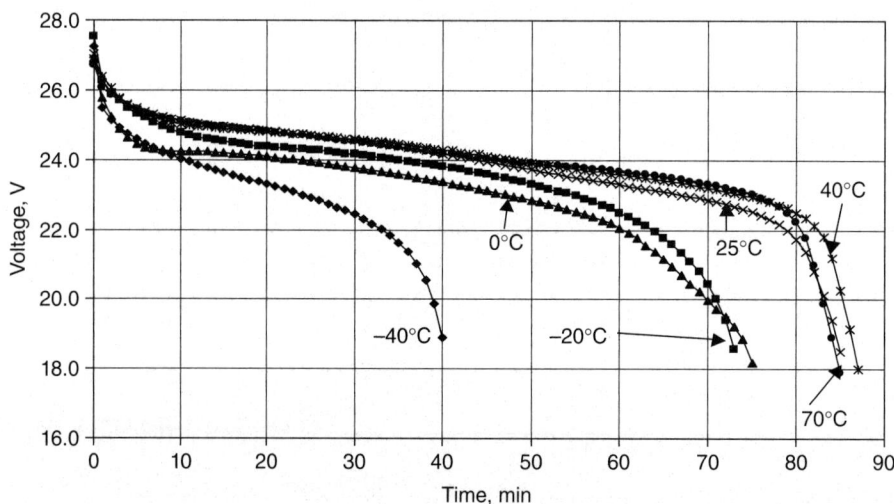

**FIGURE 15B.16**   Typical discharge curves at various temperatures, 1*C* rate, 20-cell battery.

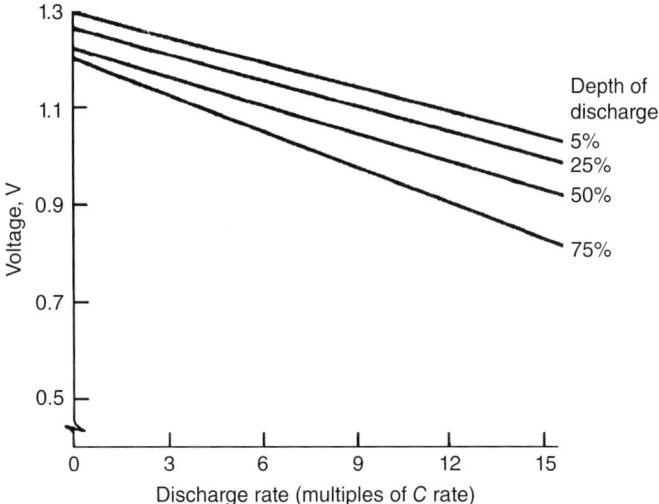

**FIGURE 15B.17**   Voltage as a function of discharge load and at various states of charge at 25°C.

The battery, because of its low internal resistance, is capable of delivering pulse currents as high as the 20 to 40C rate. For this reason, it can be used successfully for very high-power applications, such as engine starting.

*Factors Affecting Capacity.*   The total capacity that the fully charged sintered-plate vented battery is capable of delivering is dependent on both discharge rate and temperature, although the sintered-plate battery is less sensitive to these variables than most other battery systems. The relationships of capacity to discharge load and temperature are shown in Figs. 15B.18 and 15B.19, respectively.

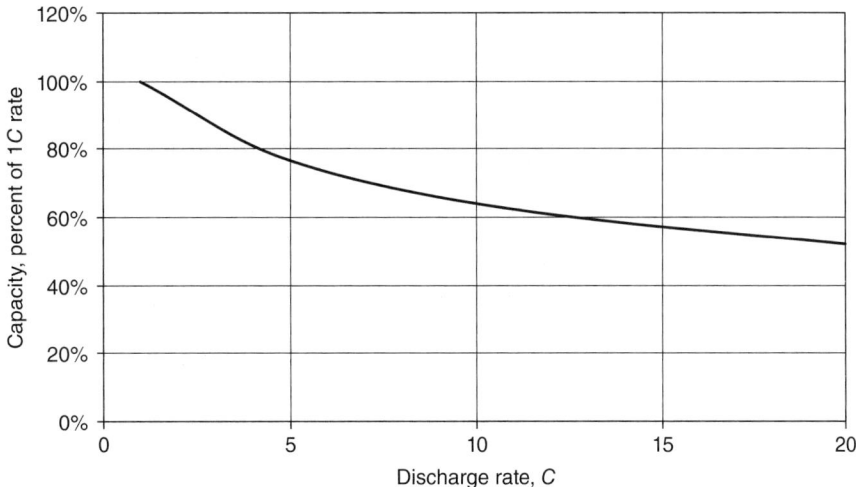

**FIGURE 15B.18**   Capacity derating as a function of discharge rate at 25°C.

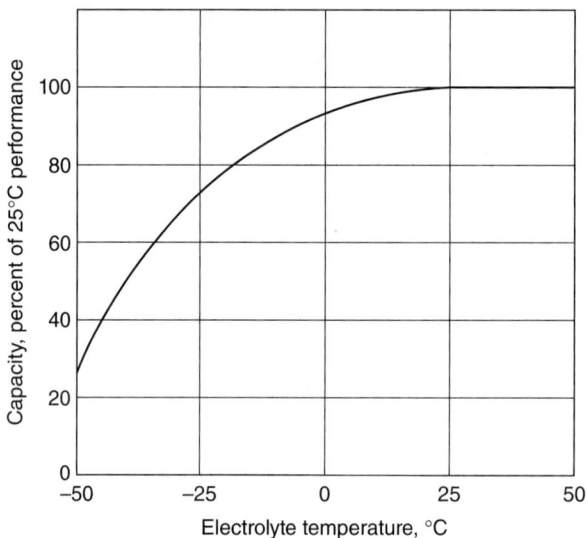

**FIGURE 15B.19**   Capacity derating as a function of discharge temperature at 1$C$ rate discharge.

Low-temperature performance is enhanced by the use of eutectic 31% KOH (1.30 specific gravity) electrolyte, which freezes at −66°C. Higher or lower concentrations will freeze at higher temperatures; for example, 26% KOH freezes at −42°C. As shown in Fig. 15B.19, more than 60% of the 25°C capacity is available at −35°C, with the temperature having an increasingly significant effect as it is lowered toward −50°C. At high discharge rates, heat that is generated may cause the battery to warm up, giving improved performance on immediate or subsequent discharges than would be expected under ambient conditions.

Vented sintered-plate batteries can also be discharged at elevated temperatures. Strict control is required, however, when charging at high temperature. As with most chemically based devices, exposure to high temperatures for extended periods of time will shorten the life of the battery.

The combined effects of increased discharge rate and low temperature may be approximated by multiplying the two derating factors.

*Variable-Load Engine-Start Power.*   The most common and demanding use of the vented sintered-plate nickel-cadmium battery is as the power source for starting turbine engines onboard aircraft. The discharge in this application occurs at relatively high rates for periods of 15 to 45 s. Typically, the load resistance when the start is initiated, particularly at low temperature in a marginal-start situation, is of the same order of magnitude as the effective internal resistance of the battery $R_e$. The apparent load resistance increases as the engine rotor gradually comes up to speed. This results in a typical discharge current, which slowly decreases from some high initial value, while the battery voltage recovers from an initial drop of perhaps 50% or more, back toward 1.2 V per cell, the effective zero load voltage. A representative graph of the battery-starter voltage and current, expressed as a function of time, is shown in Fig. 15B.20.

A common and useful measure of battery performance is the maximum power current. This property is generally defined as the load current at which the battery voltage would be $0.6N \times V$, or one-half of the effective open-circuit voltage (1.2 V/cell) and where $N$ is the number of cells in the battery. The instantaneous maximum power current decreases with decreasing state-of-charge due to rising internal resistance. Its value versus state-of-charge tends to behave exponentially, as shown in Fig. 15B.21. An approximation of $I_{mp}$ may also be measured by performing a "constant-potential" discharge at $0.6N \times V$ for 15 to 120 s. A typical discharge is shown in Fig. 15B.22.

The maximum power delivery $P_{mp}$ and the effective internal resistance $R_e$ are related to the value of $I_{mp}$ as follows:

$$P_{mp} = 0.6N\, I_{mp}$$

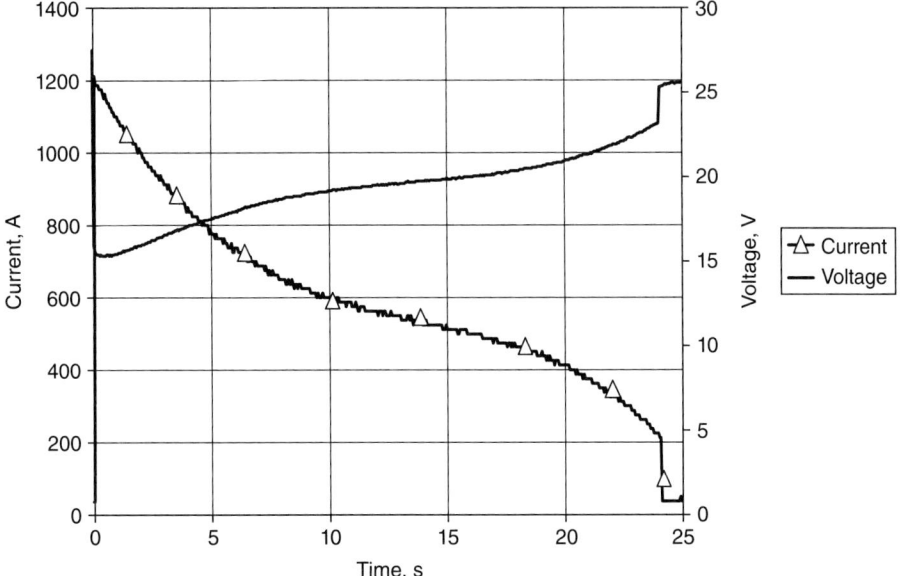

**FIGURE 15B.20** Battery voltage and current as a function of time for a typical turbine engine start (20-cell battery).

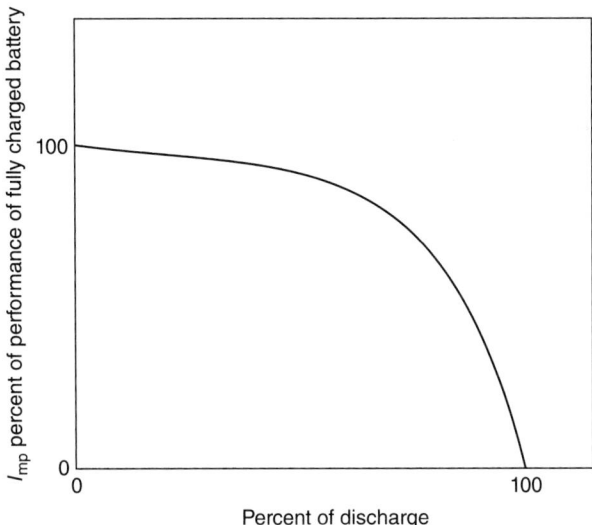

**FIGURE 15B.21** Maximum power current derating as a function of state-of-charge at 25°C.

and

$$R_e = \frac{0.6N}{I_{mp}}$$

*Factors Affecting Maximum Power Current.* The value of $I_{mp}$, which a battery is capable of delivering, is maximum at full charge and at 25°C electrolyte temperature. Derating effects due to state-of-charge and

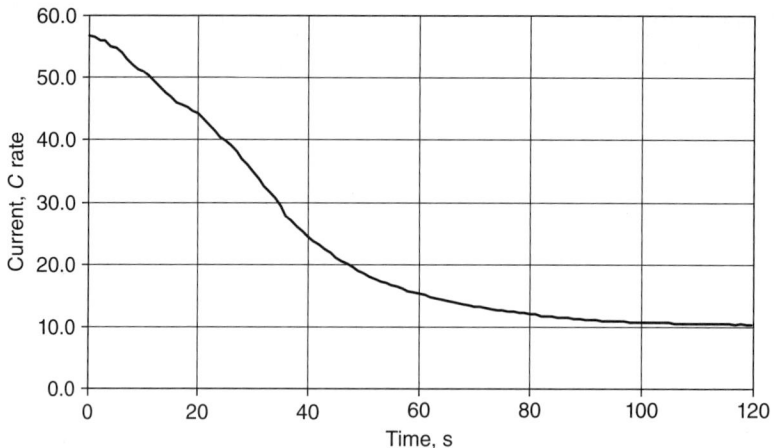

**FIGURE 15B.22**    Representative 0.6 V constant potential discharge at 25°C.

electrolyte temperature factors are shown in Figs. 15B.21 above and Fig. 15B.23 below, respectively. It will be noted that both relationships are nonlinear in that the effects on maximum power delivery, per unit of change, increase with decreasing state-of-charge and with decreasing temperature. As with capacity, the approximate effect of combined low electrolyte temperature and decreased state-of-charge may be determined by multiplying the individual factors. It should be noted, however, that high-rate discharge at low temperature may increase the battery temperature. This self-heating must be accounted for when determining the combined derating factors for a subsequent discharge. A negligible effect on $I_{mp}$ occurs with increases in electrolyte temperature above 25°C.

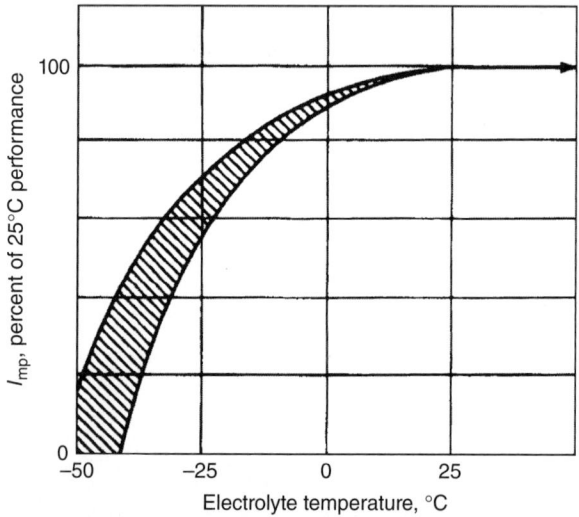

**FIGURE 15B.23**    Maximum power current derating as a function of battery temperature (fully charged).

*Energy/Power Density.*   Typical average values for the energy and power densities of the vented sintered-plate nickel-cadmium battery at 25°C are shown in Table 15B.4.

**TABLE 15B.4**   Energy and Power Characteristics of Vented Sintered-Plate Nickel-Cadmium Battery (Single Cell Basis)

| | |
|---|---|
| Specific capacity (single cell, C rate) | 25–31 Ah/kg |
| Capacity density | 48–80 Ah/L |
| Specific energy (C rate) | 30–37 Wh/kg |
| Energy density | 58–96 Wh/L |
| Specific power (at maximum power) | 330–460 W/kg |
| Power density | 730–1250 W/L |

*Service Life.*   The service life (discharge time) of the vented sintered-plate nickel-cadmium cell, normalized to unit weight (kilogram) and volume (liter) at various discharge rates at 25°C, is approximated in Figs. 15B.24 and 15B.25.

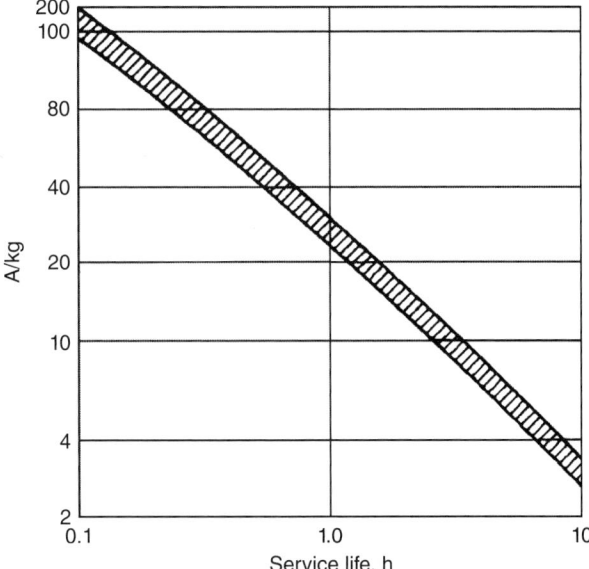

**FIGURE 15B.24**   Service life of typical vented sintered-plate nickel-cadmium battery (gravimetric) at 25°C.

*Charge Retention.*   Charge retention or capacity retention refers to the amount of dischargeable capacity remaining in a battery following prolonged storage under open-circuit conditions. Two mechanisms are responsible for the loss of charge, namely, self-discharge and electrical leakage between cells.

Self-discharge rates are an intrinsic property of cells. Typically, experimental results for the capacity retained as a function of open-circuit storage time best fit a semilogarithmic relationship such as that shown in Fig. 15B.26. The self-discharge rate of a cell is affected by impurity levels and the electrochemical stability of the electrodes.

The effect of temperature is shown in Fig. 15B.27, where the exponential time constant (tc), the time to retention of 36.8% of initial capacity, is plotted against temperature. Storage temperature is the most important factor affecting the self-discharge rate.

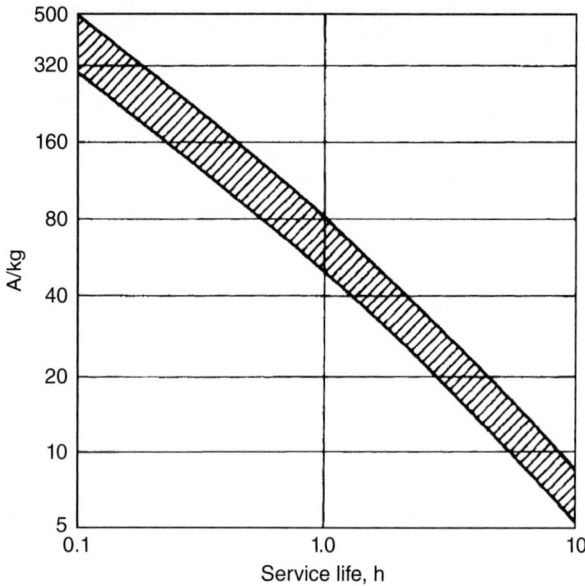

**FIGURE 15B.25** Service life of typical vented-sintered-plate nickel-cadmium battery (volumetric) at 25°C.

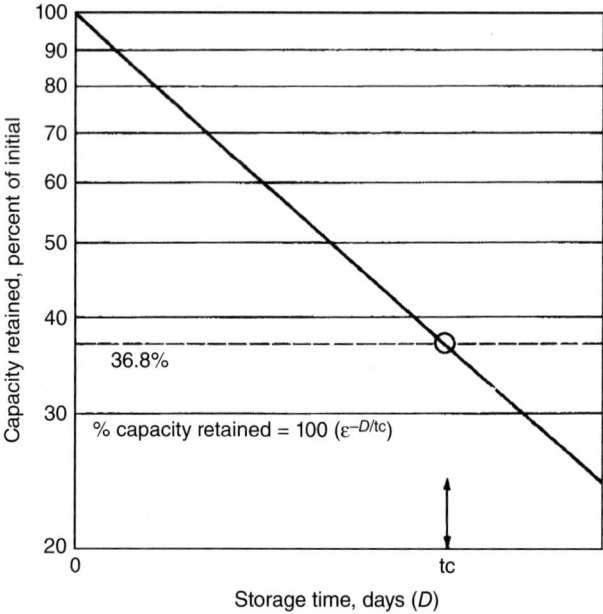

**FIGURE 15B.26** Capacity retention as a function of storage time.

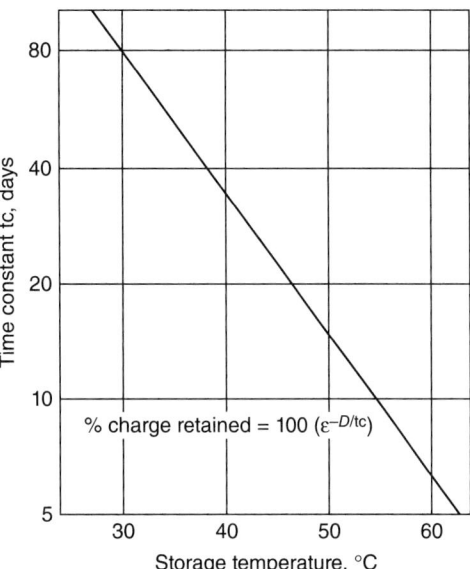

**FIGURE 15B.27**    Charge retention time constant as a function of storage temperature.

The second mechanism, the loss of charge due to electrical leakage, is influenced by the history of the battery's use and maintenance. Charge retention usually improves with cycling of the battery, and this will be true unless this cycling history has been abusive. The maintenance factor influencing charge retention is primarily battery cleanliness. Battery charge can leak from the terminals of one cell to the terminals of other cells across the cell tops if they are wet with potassium hydroxide. Loss of charge from this cause is relatively unpredictable, but it can usually be prevented by good housekeeping practices. Although surface leakage may affect only a portion of the cells in the battery, it is important since the capacity of the battery is limited to that of the lowest capacity cell. Additionally it unbalances the cells in the timing of the onset of overcharge voltage response.

It should be noted that loss of charge through these mechanisms is not permanent since the battery capability can be completely restored through comprehensive maintenance practices and recharging.

*Storage.*    The sintered-plate cell can be stored in any state-of-charge and over a very broad temperature range (−60 to 60°C) for an unlimited period. The battery should be clean and dry before placing it in storage. Intercell hardware may have a light coating of petroleum jelly to prevent corrosion. It should be fully discharged and shorted prior to storage periods greater than 30 days. Fully discharged batteries that have been stored longer than 30 days should be charged by a "slow charge" method. The "slow charge" method typically consists of incremental charge rates to voltage cutoff (VCO; i.e., $C$ rate to 1.57 V, $C/2$ rate to 1.6 V), with a final charge ($C/10$ or lower) for 2 h. It is the best practice to store the cells shorted and upright with proper electrolyte level at a temperature between 0 and 30°C. The preferable storage method is to allow the battery to discharge through a resistor until the battery voltage is close to zero. Because a vented Ni-Cd still has considerable power available even at very low states-of-charge, failure to completely discharge the battery prior to applying a shorting device can create a hazardous situation.

*Life.*    The life of the battery is strongly influenced by factors such as the design, the care with which it is maintained and reconditioned, as well as the way it is used; hence it is difficult to predict battery life. Best life performance is obtained with operation at normal temperatures, temperature-controlled charging, and minimum reconditioning. Some design features that improve the life expectancy of a battery are: (1) the use of modern separator materials and gas barriers, (2) the elimination of materials that degrade in KOH (e.g., O-rings), (3) the reduction of electrolyte impurity levels in manufacturing (by electrolyte flushing and replacement), and (4) the use of pure nickel components versus nickel-plated steel.

***Sintered Plate Cells: Charging Characteristics.***     The functional design of the vented cell battery differs from that of the sealed cell battery primarily by the inclusion of a gas barrier between the positive and negative electrodes. This gas barrier has one principal function, which is to prevent, as discussed in Chap. 15C, the cross-plate migration and recombination of generated gases within the cell. Preventing this recombination allows both positive and negative plates to return to full charge. This results in an overvoltage during onset of overcharge, which is used as the feedback signal to control the charging device. Because the gas is driven out of the cell, however, the vented cell consumes water, which must be replenished.

Charging of the vented sintered-plate nickel-cadmium battery following its discharge in cyclic use has four significant objectives. These may be stated as follows:

1. Restore the charge used during discharge as quickly as possible.
2. Maintain the fully charged capacity as high as possible during the use intervals between removals for maintenance.
3. Minimize the amount of water usage during overcharge.
4. Minimize the damaging effects of overcharge.

Fulfillment of the first objective is the principal reason for the design and use of vented cells, since the gas barrier provides the voltage signal, which may be utilized in several different ways to terminate the fast recharge. The charge may thus be accomplished at the desired high rate, without compromising the battery, by continuing that rate in overcharge. Objective **2** must inherently be balanced against objectives **3** and **4** in the design and control of the charging method. Generally, a continued good capacity between reconditioning is enhanced by providing more overcharge, while more overcharge inherently utilizes more water and, if sufficiently high, may result in damage to the battery. A compromise must therefore be struck. Usually about 101% to 105% of the ampere-hours removed on discharge are replaced on the subsequent charge.

Charging techniques that are used in onboard systems utilize the signal provided by the overvoltage of the vented cell in overcharge. This significant voltage rise is present at all charge rates, and its sharpness actually improves as the cell is cycled in high-rate discharge and recharge.

*Constant-Potential Recharging.*     Constant-potential (CP) charging, the oldest of the methods still in use, is typically utilized in general aviation aircraft. Similar to an automobile battery charging system, CP charging utilizes a regulated voltage output from the aircraft DC generator, which is mechanically coupled to the engine. The voltage is typically regulated at 1.40 to 1.50 V per cell. Figure 15B.28 illustrates the form of the charge current as a function of the state-of-charge of the vented cell battery during CP recharging. Although the initial current could be quite high if limited only by the voltage response of the battery, it most frequently is limited by the capability of the source as shown. As the battery approaches full charge, however, the "back EMF" of the battery

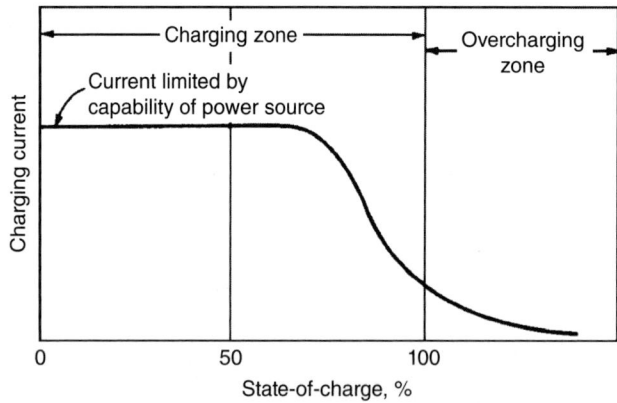

**FIGURE 15B.28**   Constant-potential charge current.

reduces the current to that required by the battery to provide an overvoltage equal to the regulated voltage of the charging source. CP charging requires very careful consideration of the selection of charge voltage and its proper maintenance in order to achieve the balance between objectives **2** and **3**, stated below in the section on temperature compensation. This is particularly difficult to achieve when the battery temperature experiences significant variation, since overvoltage is dependent on battery temperature. This balance may be made essentially independent of battery temperature effects by means of temperature compensation of the CP voltage.

*Constant-Current, Voltage-Controlled Recharging.*    A number of commercially available chargers based in general on constant-current charging with VCO control are utilized in modern aircraft. One of the simplest and most effective of these chargers applies an approximately *C*-rate constant-current charge to the battery and then terminates it when the battery voltage reaches a predetermined VCO value such as 1.50 V per cell. The control also reinitiates the constant-charge current whenever the open-circuit battery voltage falls to a predetermined lower level, such as 1.36 V per cell. The net result is that the charger will recharge the capacity used during an engine start, typically 10% of the battery's total, in approximately 6 min and then cut the charger off due to the sharp rise in voltage as the cells go into overcharge. Shortly thereafter, as the voltage falls below the turn-on voltage, the charge is reinitiated for a short period of time until the battery voltage again reaches the cutoff voltage. This simple on-off action continues at decreasing frequency and decreasing lengths of on-time, thereby maintaining the battery in a float condition at a completely full state-of-charge.

The battery voltage reduction, due to a discharge, inherently initiates the recharging of the battery without additional controls, thus automatically providing the recharge signal function regardless of the discharge rate or the reason. Adjustment of the cutoff and turn-on voltages as a function of battery temperature matches the mode of charging to the temperature characteristics of the vented sintered-plate nickel-cadmium battery, thereby maintaining the desired balance of objectives. Both cutoff and turn-on voltages are compensated at the same rate, thereby maintaining a constant differential between turn-on and cutoff. A diagram of the function of this simple basic charge control scheme is shown in Fig. 15B.29.

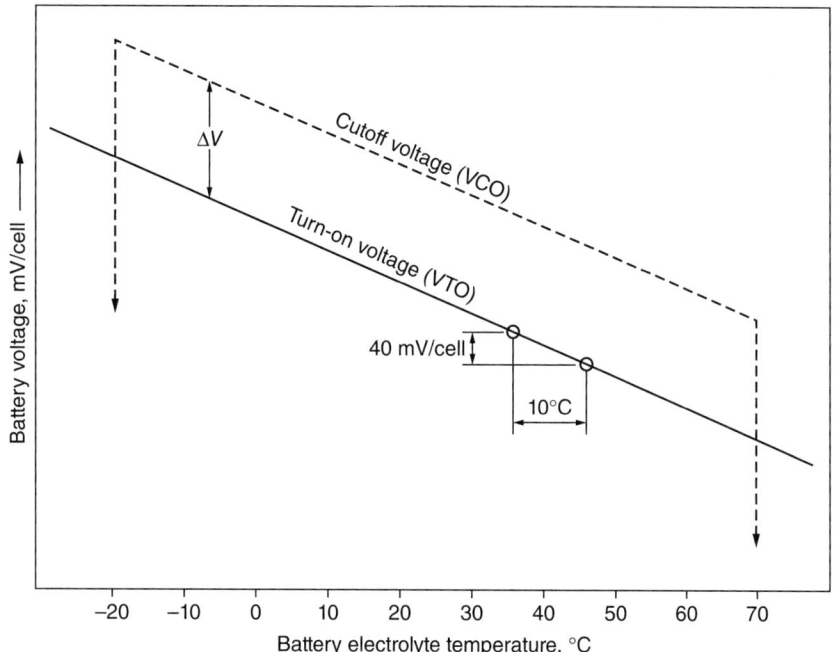

**FIGURE 15B.29**    Charger control voltages as a function of battery temperature, *C*-rate charge (nominal values).

Several other proprietary chargers, based in part on the simple techniques outlined here, may also be found in commercial use. Many of these chargers also provide auxiliary functions such as upper and lower battery temperature charge discontinuation, detection of malfunctioning cells in the battery by detecting half-battery voltage imbalance, and signaling the user in the event of either of these conditions.

*Other Charging Methods.*     The charging methods outlined in the preceding sections are those used in order to achieve fast recharging of a battery that has been discharged in normal use. Periodic maintenance of the vented sintered-plate nickel-cadmium battery, however, requires a full and complete discharge of each cell followed by a thorough recharge well into overcharge. This places both positive and negative plates in each cell of the battery in full and complete overcharge. The battery may then be returned to service with all plates of all cells in the same full charged condition, thus enabling the battery to work from the top-down.

The simplest maintenance charge method, requiring the least complex equipment to ensure this fully balanced, fully overcharged condition, is the low-rate charge. This technique utilizes a constant-current, approximately $C/10$ charge-overcharge current without voltage feedback control. At this low rate, the charge may be continued safely into overcharge without compromising the physical integrity of the components of the cell. This charge current should be maintained until at least twice the rated capacity of the battery has been replaced. Since this will inherently result in water usage, water level replenishment is best performed on a fully charged battery just prior to placing the battery back into service at the conclusion of this maintenance charging routine.

Batteries in standby service can be maintained in a fully charged condition by a float or trickle charge similar to pocket-plate batteries. The float voltage for vented sintered-plate batteries is 1.36 to 1.38 V per cell.

*Temperature Compensation of Charge Voltage.*     In both the CP and the constant-current VCO charging methods, it has been pointed out that the selection of voltage is a compromise between the minimization of water usage and the maintenance of a high state-of-charge. The inherent change of overcharge voltage as a function of battery temperature increases the difficulty of this compromise significantly. This voltage effect is shown as the Tafel curves in Fig. 15B.30. The relationship between the Tafel curves at various temperatures indicates a temperature coefficient of $-4$ mV/°C at constant-current conditions. In other words, overcharge voltage, at constant current, decreases by approximately 4 mV per cell for each 1°C rise in cell temperature.

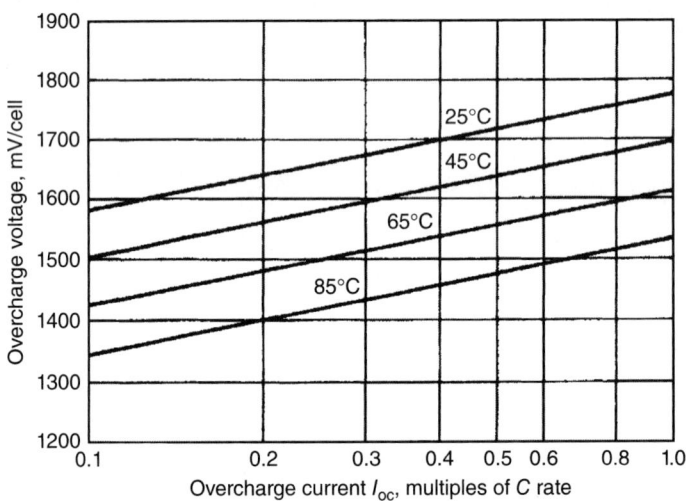

**FIGURE 15B.30**   Overcharge voltage as a function of current and temperature (Tafel curves).

As shown by the slope of these Tafel curves, overcharge voltage is also a linear function of the logarithm of overcharge current. That slope for vented sintered-plate nickel-cadmium batteries is approximately 200 mV per cell per decade of change in overcharge current. Thus with CP charging without temperature compensation, the overcharge current, and therefore both water usage and the overcharge damaging effects, will increase approximately 60% for each 10°C increase in electrolyte temperature.

A convenient technique for avoiding the detrimental effects of an increasing electrolyte temperature is to compensate the CP voltage, or the constant-current cutoff/turn-on voltages, for this change in battery temperature at the rate of $-4$ mV/°C per cell. This may be accurately accomplished through the use of thermistors or other temperature-sensitive electric devices installed in the battery case to signal cell temperature. It is important that cell temperature and not ambient temperature be sensed by this device. There may be significant differences between the two. This function for the constant-current charging system is also shown in Fig. 15B.29. The selection and use of the proper value of temperature compensation permits the battery charging system to function as though the battery were maintained at a constant temperature. Care must be exercised in the design and manufacture of these devices, since they must operate in an environment of potassium hydroxide, which is electrically conductive and also has a propensity to wet and creep on most surfaces. High-grade potting processes must therefore be used to insulate and protect all auxiliary electronic components and wiring placed inside the battery case.

### Safety and Maintenance (Vented, Sintered Plate Cells)

*Electrical Reconditioning.*    The periodic maintenance of the vented sintered-plate nickel-cadmium battery has six specific objectives:

1. Assess the timeliness of the preselected maintenance period schedule.
2. Restore the electrical performance, both capacity and power.
3. Detect and isolate cell failures and facilitate their replacement.
4. Physically clean the battery.
5. Replenish the water in the electrolyte.
6. Maintain the charging-system voltage calibration.

A relatively simple electrical procedure fulfills the first objective, namely a single discharge, initially at a relatively high rate to simulate engine start and second at a relatively low rate, approximating that of the 1 h value. This split-rate discharge of the battery as removed from the aircraft serves as a measure of its performance readiness while it was in the aircraft. The 15 s high-rate portion, at approximately the $I_{mp}$ rate of discharge, while measuring the voltage with the capacity removed, determines the relative engine start power capability. The capacity removed during the low-rate portion of the discharge at approximately the $C$ rate, when added to the ampere-hour capacity removed during the high rate test, determines the status of the available emergency energy capacity. The battery prior to performing this discharge should be in the same fully charged condition that would typically be encountered when it is in the normal installation. Comparison of this power delivery, and the total available capacity as removed from the aircraft, with the requirements of the application will allow the user to determine whether the maintenance schedule interval may be increased or whether it needs to be decreased.

The restoration of the electrical characteristics of the battery, known as electrical reconditioning or deep cycle maintenance, objective **2**, may be accomplished in two additional simple steps. The first is a thorough and complete discharge of each cell in the battery in order to discharge all the active material. The second step is the complete recharge of each plate of each cell into full gassing overcharge. The first step consists of a $C$-rate discharge to approximately 0.7 V. Once all cells have reached voltage, then resistors short cells to 0.010 V per cell or 16 h, whichever occurs first. The second step is accomplished by charging the battery with constant current at a value of one-tenth its ampere-hour rating ($C/10$) for at least 20 h. Since the capacity for some of the cells in the battery may be as much as 30% to 40% greater than the rated value, the total charge of $2C$ Ah is sufficient to ensure that both plates of all cells reach the full overcharge required. Adjust electrolyte levels accordingly. The battery should have deep cycle maintenance performed every 1000 flight hours or 500 starts, whichever occurs first or when any abnormal operation of the battery is observed.

There are other procedures used and recommended by the manufacturers of specialized proprietary equipment to recondition cells in a shorter period of time. Periodic maintenance in a qualified service center is necessary for optimum performance of the battery. The efficacy of each should be verified, and the added expense and complexity entailed in the use of these proprietary reconditioning devices justified in specific applications. Care must always be exercised in their use, however, to avoid sustained high-rate overcharge, which may damage the gas barrier material.

Evaluation of cell-to-battery case leakage current, part of objective **3**, when the battery is first received for maintenance will determine the electrical need for cell cleanup as well as the presence of cracked or leaking cell cases. This procedure may be conveniently carried out by the simple expedient of completing a circuit from each cell terminal to the battery case with a fused ammeter. A significant amount of leakage current through the ammeter to the case from anywhere in the cell electric circuit indicates the presence of a conductive path, through potassium hydroxide, on the external surfaces of the cell cases. Such a conductive path may result from spewing of the electrolyte during overcharge, which may indicate either overfilling or the existence of a cracked or leaky cell case. Isolation of the exact cause can be accomplished by determining the leakage nodal point in the cell string by repeating the measurements after physical cleanup of the battery, objective **4**.

Detection of the failure of the gas barrier, a very important part of objective **3**, may be reliably and conveniently accomplished by extending the $C/10$ charge to 24 h. This overcharge will indicate accurately the failure of the gas barrier by either or both of two principal measurements near the end of that overcharge. First, the overcharge gassing rate is extremely sensitive to gas barrier condition and gas recombination. When measured with a simple ball flowmeter, the 24 h gas rate will be less than 80% of normal if the barrier has failed significantly in the cell. The normal value is 11 mL/min for each ampere of the $C/10$ overcharge rate. The second indicator of gas barrier failure will be a 24 h overcharge voltage of less than 1.5 V if the cell is being charged at a 23°C ambient. Some downward adjustment of this voltage criterion may be made at the rate of 4 mV/°C if the battery is being charged at a higher ambient temperature.

*Mechanical Maintenance.*    The replenishment of water in the electrolyte to return the electrolyte to the level recommended, objective **5**, is the most important routine mechanical procedure employed during battery reconditioning. It is best accomplished near the end of the 24 h of the $C/10$ rate charge by replenishing with deionized water until the electrolyte reaches the recommended level for a battery in overcharge. A record of the amount of water usage in each cell should be maintained and compared with the battery manufacturer's statement of reserve electrolyte level in each cell. If the total water usage between maintenance fillings, after deducting the amount used during the maintenance procedure, exceeds the reserve available in that cell design, the maintenance interval must be shortened to prevent plate dry-out during use and resultant cell failure. Note that the 24 h $C/10$ reconditioning procedure will itself use approximately 0.4 mL of water for each ampere-hour of rated capacity during the 24 h reconditioning period. For example, 12 mL of water would be used during the reconditioning period for a 30 Ah rated cell, and this must be subtracted from the amount added to determine the amount actually used in service. It should also be noted that a cell with a damaged gas barrier may use less water.

That point in the maintenance procedure, following the thorough short-circuiting of each cell, may be utilized to perform physical maintenance. Cells may only be replaced while in the discharged state. Cleanup generally consists of a thorough rinsing with clear water followed by warm-air drying of the battery. This will dissolve and remove any accumulation of potassium hydroxide and carbonates from the outside of the cell jars. Vent caps should also be washed in warm deionized water, warm water forced through the vent, and then dried with warm air. This is safely accomplished only with the cells in the completely discharged state. Replacement of any cells not found defective until the conclusion of the $C/10$ overcharge requires discharging the cells a second time.

Other typical hardware problems include:

- Loose terminal nuts—indicated by burns or arcing on intercell links.
- Terminal seal failure—various heavy deposits around cell terminal; remove all hardware and the O-ring.
- Vent failure—various heavy deposits on or around the vent valve; the valve may have been installed improperly or the vent sleeve or O-ring has failed.
- Also inspect the vent sleeve to ensure it is not torn or broken. Power delivery of the battery may be enhanced by removing intercell link hardware, buffing contact surfaces, then replacing and re-torquing all connectors.

*System Inspection Criteria.*    Reinstallation of the reconditioned and fully charged battery into the aircraft system presents the opportunity for performing the system voltage calibration check in fulfillment of objective **6**. The only measurement required for this on a CP charging system is to record the value of the battery voltage after reactivating the system and following stabilization of the voltage. This stable value is the regulated float voltage to which the battery will be subjected during extended overcharge in use. This voltage measurement should be made at the battery with the engine running at a sufficiently high speed to produce a representative and stable value.

The battery voltage measurement on a constant-current VCO systems should be made just as the system reaches cutoff voltage. The regulated voltage on either of the two systems should then be adjusted to the manufacturer's recommended value if necessary. These adjustments must consider the effects of any automatic temperature compensation of voltage present in the system.

*Reliability—Failure Modes.*    The sintered plate construction is very robust and capable of operating in both high and low temperature extremes. The cell is capable of withstanding substantial abuse and still performing as intended. It can be discharged into reversal and given a substantial amount of overcharge, as long as the temperature is controlled. The gas barrier that prevents recombination of the oxygen on the cadmium electrode aids control of the temperature during charge. In the past, cellophane was mainly used as the gas barrier. It had a tendency to hydrolyze in the electrolyte and eventually decompose to carbonate and derivatives of the cellophane structure. In recent times, the cellophane has been replaced with Celgard 3400 or other similar materials.[10] These materials are polyethylene and polypropylene based, and do not degrade in the 30% KOH electrolyte. Should the cell's gas barrier fail, continuing in operation for enough cycles will result in the battery losing capacity and maximum power capability. Continued temperature increase in the cell can result in fusing, or melting, of the nylon separator and result in an internal short circuit of the cell.

Although the cellophane replacement used for the gas barrier alleviates the above failure mechanism, it introduces another problem if the cell is not manufactured and maintained properly. Without proper additives in the electrolyte, which find their way into the cadmium electrode during cycling, the cell can lose capacity in the negative electrode. The role of supplying an oxidized cellophane expander for the cadmium electrode needs to be replaced by cellulose derivatives and other additives to maintain the negative capacity.[11]

Several other failure modes that account for a small portion of cell and battery failures include the following:

- Internally short-circuited cells can result from cut-through of the electrical separator by burrs and other plate irregularities, aggravated by cell interplate pressures and vibration.

- Cracked and leaking cell cases may result from abusive handling of the cells during cell replacement procedures and maintenance, or from defective manufacturing or sealing procedures.

- Burned terminal contacts may result from faulty cleaning and buffing procedures during maintenance, insufficient link assembly torqueing, terminal screw failure, or conductive articles being dropped on the internal connectors of a charged battery.

*Memory Effect.*    In addition to these permanent failures, there is a reversible effect that may result in a gradual reduction of both power and capacity with cycling. This effect, sometimes referred to as "memory effect," "fading," or "voltage depression," results from charging following repetitive shallow discharges where some portion of the active materials in the cell is not used or discharged, such as in a typical engine-start use. This effect is noticed when the previously undischarged material is eventually discharged. The terminal voltage during the latter part of that full discharge may be lower by approximately 120 mV (hence, "voltage depression"). The total capacity is not reduced, however, if the discharge is continued to the lower voltages, as, for example, to the "knee" of the curve.

This effect is completely reversible by a maintenance cycle consisting of a thorough discharge followed by a full and complete charge-overcharge as described above in the section on "Electrical Reconditioning."

*Factors Influencing Gas Barrier Failure.*    Gas barrier failure is generally acknowledged to be caused or aggravated by excessive overcharge current, excessive overcharge temperatures, and discharge at high rates with low electrolyte levels. Gas barrier failure may not be detectable during reconditioning with other than the low-rate constant-current procedures. Barrier failures may actually occur during poorly structured maintenance and then manifest themselves at a later time after reinstallation in the aircraft. This possibility emphasizes the importance of an accurate assessment of the condition of the gas barrier at the end of the reconditioning period just prior to reinstallation. This assessment is accurately made by the measurement of overcharge gas flow following the extended $C/10$ overcharge, as described under the electrical reconditioning section.

One indication of the significant importance of the two factors of (1) temperature compensation of charger voltage and (2) effective maintenance practices is the existence of large-scale field data that document real-time failure differences of up to 100:1. These life differences exist between well-maintained batteries in temperature-compensated systems, on the one hand, and identical battery designs in uncompensated CP systems with frequent and poorly managed maintenance procedures, on the other. Recent improvements in gas barrier materials have significantly reduced these failures.

*Thermal Runaway.*    The loss of the gas barrier in one or more cells of a vented nickel-cadmium battery can lead to thermal runaway. Loss of this function allows oxygen, generated in overcharge, to reach the negative plate and recombine on it. This generates heat. The temperature increase that follows causes the internal cell voltage to decrease. Charge current then increases exponentially to increase cell voltage to match the charger voltage.

Thermal runaway occurs with the use of a voltage-regulated (CP) charge source on a battery containing cells with a failed gas barrier. Thermal runaway begins when the failed cells approach overcharge following recharge. The (over)charge current may reach a minimum and then gradually increase. Voltage inequities may exist at this point unless all cells are experiencing similar recombination (gas barrier damage). Oxygen recombination heats, and begins to increase the temperature of the failed cell or cells and thus their neighboring cells unless the battery is effectively air cooled. The resulting temperature increase, however, proceeds slowly due to the large thermal mass involved. It may take 2 to 4 h of (near) consecutive overcharging for a cell to reach boiling temperature.

If the boiling phase continues long enough, or is repeated, and the failed cell becomes dry, large inequities in cell voltage will appear. The voltage across the cell that has boiled dry will increase, thereby decreasing the charge current and the voltage across the cells that are still wet with electrolyte. The next event probably will be internal short-circuiting of the dried-out cell due to very high temperatures and voltage at the last remaining damp spots with consequent burning of the electrical separator insulation. The (over)charge current then increases sharply due to cell loss, and the process repeats itself with the next cell to go dry.

Because of extensive heating and boil-away times, thermal runaway may go undetected for many flight hours following the onset of gas barrier failure if the use of the system is not consecutive or continuous. This can confuse the perceived connection between the cause of the gas barrier damage and the resultant thermal runaway.

*Potential Hazards.*    Potential hazards that may be present during the use and maintenance of the vented sintered-plate nickel-cadmium battery fall into five general categories, as described in this section.

GAS FIRE AND/OR EXPLOSION.    Since all functional vented batteries generate a stoichiometric mixture of hydrogen and oxygen gases during overcharge and expel them normally from the cell into the battery container, a potential for explosion of these gases is always present. Two conditions are necessary for such an explosion, however, and both are recognized and accounted for in the design of a typical system. The first condition is the accumulation of a sufficient quantity of this gas mixture. This condition is minimized in all system designs by the incorporation of adequate battery case ventilation. Some designs rely on supplying a modest quantity of air to purging tubes on the battery from overboard vents in the aircraft. Others incorporate natural convection of the gases from a louvered battery case into a ventilated compartment. Air-cooled designs inherently accomplish the required ventilation due to the large volume of air used.

Unusual circumstances, however, may defeat any of these gas-purging techniques. It should also be remembered that batteries generate a significant amount of explosive gas during the maintenance procedure, and therefore maintenance should always be performed in a well-ventilated shop.

The second condition necessary for explosion of the gases is the presence of a source of ignition. Although normally there are no ignition sources inside the battery case, several abnormal possibilities do exist. One is the internal short-circuiting of a relatively dry cell in overcharge, resulting in an explosion inside the cell with a subsequent ejection of flames into the battery case. A second and more likely source of ignition may exist at an improperly maintained cell terminal due to the high temperatures generated during high-rate discharge. A third source of ignition may occur at the site of stray leakage currents.

Although the coincidence of both a sufficient amount of gas accumulation and a source of ignition is quite rare, it can happen and has happened. Because of this possibility, many batteries are also designed to be physically capable of managing a hydrogen or oxygen explosion and containing the effects entirely within the battery case. Typically these batteries will also be electrically functional for at least one *C*-rate discharge following such an explosion.

ARCING AND BURNING.    This potential hazard concerns excessive leakage currents through electrolyte paths outside the cells. Such currents can occur either between cells that are physically adjacent but with a wide voltage separation in the cell circuit or, more probably, from a cell to a grounded metallic case. Arcing is more likely to occur, however, in the circuit of an inappropriately protected auxiliary device located inside the battery case in the environment containing potassium hydroxide. Some examples of these devices are battery heaters, thermal detectors, and voltage sensors. The proper design of these auxiliary appliances must recognize the conductivity of KOH and the ability of that electrolyte to creep along wires and even into mechanically "sealed" insulation. Devices of this type should be tested with high dielectric voltages while submerged in a water-detergent mix before they are installed in the battery case environment.

The result of a sustained leakage current through relatively localized KOH conducting paths may be the ignition of the explosive environment by arcing, as discussed previously. The result might also be the carbonization of adjacent insulating materials and the subsequent burning of those materials within the battery case.

ELECTRICAL POWER.   One of the essential functional capabilities required of the vented nickel-cadmium battery is the ability to deliver high-power rates for engine starting. This very capability, however, presents a potential risk in the form of hot spots on improperly torqued cell terminals during high-rate discharge. It is also a potential hazard to the unwary maintenance technician operating with metallic tools or other objects, such as jewelry, in a careless manner in the vicinity of charged batteries. Since the short-circuit current of these cells (batteries) may exceed 1000 to 4000 A, it should be obvious that the exposed conductors of a charged battery should be treated with respectful caution. Very severe burns may occur if, for example, a ring should accidentally make contact between two adjacent cell terminals. Although one of the most obvious, this hazard is one of the most frequently encountered. Insulated cell hardware does provide a partial solution; however, caution and respect for the available power must always be exercised.

CORROSIVE KOH.   Because of the corrosive nature of the KOH used as an electrolyte, all material employed in the construction of the battery and its accessory appliances must be KOH immune. Materials such as nylon, polypropylene, nickel-plated steel, steel, and stainless steel are therefore used. The potential hazard of KOH corrosiveness, however, is primarily encountered during maintenance. The use of safety glasses and safety face shields should be mandatory while performing maintenance on these batteries. A very small amount of KOH in the eye, for example, without prompt, continued, and adequate flushing followed by medical treatment can result in the loss of eyesight. KOH is also corrosive to the skin, and the affected area should be thoroughly washed and rinsed with water, thereby minimizing the detrimental effect.

ELECTRIC SHOCK.   Most vented sintered-plate nickel-cadmium batteries are arrayed in groups of 10 to 30 cells, presenting maximum voltages of 15 to 45 V. There are applications, however, in which batteries are connected electrically in series strings of 90 to 200 cells or more. It should be apparent that the voltages presented by this number of cells in series may be lethal to anyone exposed to them. Personnel should also be cautioned, because of the high probability of electrolyte being present between the cell circuit and the conductive external surface of the battery, to exercise care by disconnecting series-connected batteries prior to personal exposure. Significant shock currents may be carried by relatively small amounts of KOH.

### 15B.4.2  Products and Applications

*Typical Products.*   Table 15B.5 contains data regarding prominent manufacturers of industrial nickel-cadmium batteries. Table 15B.6 lists the market segments and applications for these batteries.

**TABLE 15B.5**   Major Manufacturers of Industrial and Aerospace Nickel-Cadmium Batteries (Does Not Include Sintered-Plate Designs)[*]

| | Product range | | | |
|---|---|---|---|---|
| Manufacturer/country | Trademark | Pocket plate | Fiber plate | Plastic-bonded plate |
| Acme Electric, U.S. | Acme | | X | |
| Alcad Ltd., U.K. | Alcad | X | | |
| HBL Power Systems, India | HBL | X | | |
| Hoppecke Batterien, Germany | Hoppecke | | X | |
| Japan Storage Battery, U.S. | GS | X | | X |
| Marathon Battery, U.S. | Marathon | X | | X |
| SAFT, S.A., France | SAFT | X | | X |
| Tudor S.A., Spain | Tudor | X | X | |
| Varta, Germany | Varta | X | | |
| Yuasa, Japan | Yuasa | X | | |

[*]See "Typical Vented Sintered-Plate Nickel-Cadmium Single Cells" and "Battery Assemblies with Sintered Plate Cells" in Sec. 15B.4.

**TABLE 15B.6**  Market Segments and Applications for Vented Industrial Nickel-Cadmium Batteries

| | Pocket plate | | | | Fiber plate | | | Plastic-bonded plate | |
|---|---|---|---|---|---|---|---|---|---|
| Cell range* | H | M | L | XX | H or X | M | L | H | M |
| Capacity, Ah | 10–1000 | 10–1250 | 10–1450 | 23–47 | 10–220 | 20–450 | 20–490 | 11–190 | 20–200 |
| Applications | UPS, starting, switchgear | UPS, switchgear, auxiliary power, emergency power | Lighting, alarms, signaling, communications, standby power | Aircraft | UPS, satellites, starting, switchgear, traction, power stations and substations | UPS, switchgear, auxiliary power, emergency power | Lighting, UPS, alarms, signaling, telecommunications, standby power | UPS, starting, switchgear, traction, aircraft | Lighting, auxiliary power, traction |
| Railroad | X | X | X | | X | X | X | X | X |
| Mass transit | X | X | X | | X | X | X | X | X |
| Industry | X | X | X | | X | X | X | X | |
| Buildings | X | X | X | | X | X | X | X | |
| Hospitals | X | X | X | | X | X | X | X | |
| Oil and gas | X | X | X | | X | X | X | X | |
| Airports | X | X | X | | X | X | X | X | |
| Marine | X | X | X | | X | X | X | X | |
| Military | X | X | X | | X | X | X | X | X |
| Telecommunications | X | X | X | | X | X | X | X | X |
| Photovoltaics | | | X | | | | X | | |
| AGV/hybrid vehicles | | | | | X | | | X | |

*H or X: high-rate; M: medium-rate; L: low-rate; XX: ultra-high-rate.

While these companies may produce a variety of designs, certain applications (such as aerospace batteries) tend to be restricted to sealed cell designs (see Chap. 15C).

***General Applications.*** Because of their favorable electrical properties, excellent reliability, low maintenance, rugged design, and long life, nickel-cadmium batteries are used in a large variety of applications, as indicated in Table 15B.6. Most of these are of an industrial nature, but this type of battery is also used in many commercial, military, and space applications.

The nickel-cadmium battery was originally developed for traction applications, and since the early years of the 20th century, it has been used extensively in railroad applications. Today the nickel-cadmium battery is the system of choice in a variety of railroad and mass-transit installations around the world. Approximately 40% of all industrial nickel-cadmium batteries produced are used in train lighting and air-conditioning for rail cars, emergency and standby systems such as emergency brakes, door openers, and lighting in mass-transit and subway cars, diesel-engine cranking in locomotives and commuter cars, railroad signaling, communication along tracks, as well as standby power in rail stations and traffic control systems. The pocket-plate battery has traditionally dominated this market segment, but in recent years, with demands for higher energy per unit weight and volume, plastic-bonded and fiber-plate batteries have penetrated this market, particularly for high-speed trains, mass-transit cars, subway cars, and light rail vehicles. Where ruggedness and long durability are the main requirements, the pocket-plate battery still maintains its position.

In stationary applications where reliability is a must, nickel-cadmium batteries are used in standby and emergency installations where life and great economic values would be endangered by a power failure. Examples of such installations are emergency power in hospital operating theaters, standby power for all vital functions on offshore oil rigs, UPSs for large computer systems in banks and insurance companies, standby power in process industries, and emergency lighting and landing systems in airports.

The nickel-cadmium battery is also used in power-generating stations and power distribution networks where power supplies must not break down. The batteries are used in switch-gear applications and for control and monitoring functions. In centralized emergency lighting systems in hospitals, public buildings, sports arenas, and schools, nickel-cadmium batteries are often specified in building codes and by consultants in many industrialized countries.

In case of failure of the primary power supply, diesel generators or gas turbines are installed to take over the power supply. For reliable and fast-acting startup of these engines, nickel-cadmium batteries have proven to be the best emergency power source.

In portable applications where batteries are exposed to temperature extremes or rough handling, nickel-cadmium batteries are used for signal lamps, hand lamps, search lights, and portable instruments. Vented spill proof batteries are used in large devices, whereas sealed nickel-cadmium batteries dominate the smaller ones.

The industrial battery market is dominated by the lead-acid battery, and the nickel-cadmium battery is a niche-market product. The reason for this is the higher capital cost for nickel-cadmium batteries compared to lead-acid batteries. Where only energy is required, the lead-acid battery is the least expensive, as its cost per watt-hour is lower than that for nickel-cadmium batteries. However, in cost per watt or life cycle cost, nickel-cadmium batteries can compete with lead-acid batteries due to much better high-rate performance and longer life combined with low maintenance costs. A typical example is locomotive diesel-engine cranking, where a nickel-cadmium battery with only one-third of the ampere-hour capacity and a life four times that of the lead-acid battery can do the job. In applications with short-duration discharges—standby and emergency equipment are usually used for less than a half-hour—the rated capacity of a battery is of little importance. The size of the battery is chiefly determined by the power need. The nickel-cadmium battery is unmatched in industrial applications when reliability and durability are considered in a life-cycle cost calculation.

FNC batteries of the ultra-high-rate (XX) and high-rate (X) design are employed primarily in aircraft, military, and space applications. Because of the variety of applications for nickel-cadmium batteries, it is important to select the best technology for the application. The characteristics are somewhat different for the three technologies available today for industrial use.

The pocket-plate battery has the lowest cost of the three technologies and is known for high reliability and fail-safe operation. However, the energy and power density limit its use in some areas. The fiber-plate battery has lower internal resistance than the pocket-plate battery and is also available in ultra-high-, high-, medium-, and low-rate cells. Where very high energy and power density are required, the plastic-bonded plate may be the choice. The plastic-bonded and fiber-plate batteries are the only technologies possible for use in automated guided vehicles (AGVs). They have also cost and performance advantages in some traditional pocket-plate battery applications such as engine cranking, switchgear, and UPSs where only very short duration discharge is required.

***Typical Vented Sintered-Plate Nickel-Cadmium Single Cells.***    A listing of several typical vented sintered-plate nickel-cadmium cells is given in Table 15B.7. The 10, 20, and 36 Ah sizes are those typically employed in aircraft batteries. Other cells are available in sizes up to about 350 Ah. The larger cells are generally constructed in steel containers rather than the plastic containers now used for the aircraft-size cells.

**TABLE 15B.7**    Typical Vented Sintered-Plate Nickel-Cadmium Cell Properties

| Rated capacity, Ah | Height, cm | Width, cm | Thickness, cm | Weight, g |
|---|---|---|---|---|
| 1.5 | 10.16 | 2.92 | 1.70 | 86 |
| 2 | 8.74 | 3.81 | 1.83 | 95 |
| 5.5 | 10.31 | 5.51 | 2.39 | 236 |
| 5.5 | 10.36 | 5.51 | 2.39 | 272 |
| 7 | 18.85 | 6.65 | 1.29 | 299 |
| 6 | 11.60 | 5.89 | 2.69 | 354 |
| 13 | 11.93 | 7.95 | 3.02 | 486 |
| 10 | 14.48 | 5.89 | 2.69 | 445 |
| 12 | 14.38 | 5.86 | 2.69 | 422 |
| 20 | 17.42 | 7.95 | 3.53 | 1067 |
| 28 | 17.27 | 7.95 | 3.53 | 1149 |
| 23 | 20.57 | 8.08 | 2.72 | 903 |
| 36 | 17.42 | 7.95 | 5.08 | 1562 |
| 40 | 23.31 | 7.95 | 3.53 | 1453 |
| 42 | 23.31 | 7.95 | 3.53 | 1453 |
| 100 | 24.48 | 12.7 | 3.83 | 2860 |

*Source:* Courtesy of EaglePicher Technologies, LLC.

### Battery Assemblies with Sintered Plate Cells

*Typical Battery Designs.*    A typical battery configuration of vented sintered-plate nickel-cadmium cells is the conventional aircraft battery. An example of this use is shown in Fig. 15B.31 and in detail in Fig. 15B.32. This arrangement generally consists of a completely enclosing battery case and cover made of either stainless steel or steel with a KOH-resistant finish of epoxy or paint. The cover is typically secured with over-center-type latches. The battery case is provided with gas-purging vents or with freely convective gas-diffusion openings for dilution. The cells are encased in nylon-molded cell cases with terminals extending through a nylon cover sealed to the case. The cells are electrically connected in series with nickel-plated copper links from cell terminal to cell terminal and from the first and last cell to the battery termination and disconnect device. This battery termination extends through the battery case wall and is present on the outside surface of the battery case as a recessed double-male, polarized, high-current receptacle. Functional requirements of aircraft batteries are specified in SAE standard AS 8033A.

*Air Cooling/Heating.*    The major battery manufacturers produce some battery designs with provision for forced-air cooling. These designs generally take the form of plenum spaces below and above the cells, with cooling passages between the cells connecting the two plenums. The construction provides a means for the external connection of a high-volume, low-pressure air source. Supplying 23°C air will not only effectively cool an overheating battery, it will also rapidly warm a cold battery. The heat transfer coefficient from the battery core is improved up to 10 times by this technique. The thermal time constant for a battery with this feature may be as short as 10% to 20% of that of a standard non-air-cooled battery.

*Temperature Sensors.*    Sensors may be provided inside the battery case that sense either typical or average cell temperatures. They are equipped with a provision for external electrical connections. These devices may be of the on-off type, such as thermostats, or they may be the continuous type, such as thermistor assemblies.

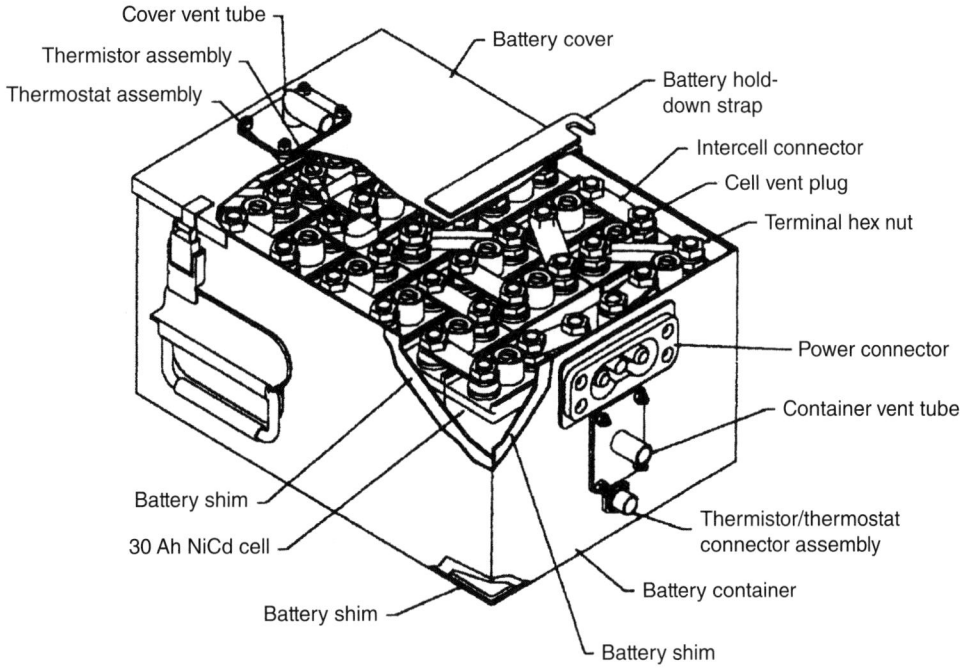

**FIGURE 15B.31**  Vented sintered-plate nickel-cadmium battery. (*Courtesy of EaglePicher Technologies, LLC.*)

Continuous types have the capability of providing continuous modulation of, for example, the regulated charging voltage of a CP charging source or of the cutoff/turn-on voltages of a constant-current VCO charging system.

*Battery Cases.*  Although the corrosive effect of KOH on bare steel is minimal and cosmetic in effect, additional KOH resistance in the case material is desirable. In addition to stainless steel and steel with a protective finish, some special applications use KOH-resistant plastic materials. It must be emphasized, however, that battery cases may require withstanding significant rough treatment in shock and vibration without losing their KOH-containment capability.

*Battery Electrical Termination.*  Aircraft batteries are normally terminated on the front case surface by a connector of the type shown in Fig. 15B.31. Special applications, however, may utilize direct cable connection to the first and last cell terminals, as well as various other special configurations capable of handling the high-current rates available in the event of a short circuit of the external battery circuit.

*Battery Heaters.*  Alternative to the airflow heating, heater blankets are sometimes employed on the inside or outside of the battery case. These blankets may be energized with any available electric energy source. Primarily, this source will be either the DC bus of the same voltage as the battery or an aircraft AC bus of higher potential. Heaters may also be energized from an auxiliary ground supply. Heater blankets have the inherent poor thermal time constant of non-air-cooled batteries.

*Extensions of Vented Sintered-Plate Nickel-Cadmium Designs.*  To avoid costly maintenance procedures associated with vented nickel-cadmium battery designs and to improve general battery reliability, the battery industry, in conjunction with the U.S. government, has incorporated lessons learned from vented sintered-plate nickel-cadmium batteries. There are two alternate versions of the standard nickel-cadmium battery that are being used in both commercial and military aircraft service, the vented low-maintenance nickel-cadmium battery and the maintenance-free nickel-cadmium aircraft battery.

The vented low-maintenance nickel-cadmium battery was derived from the standard vented sintered-plate nickel-cadmium cell design with the exception of a few internal components. As in the standard configuration,

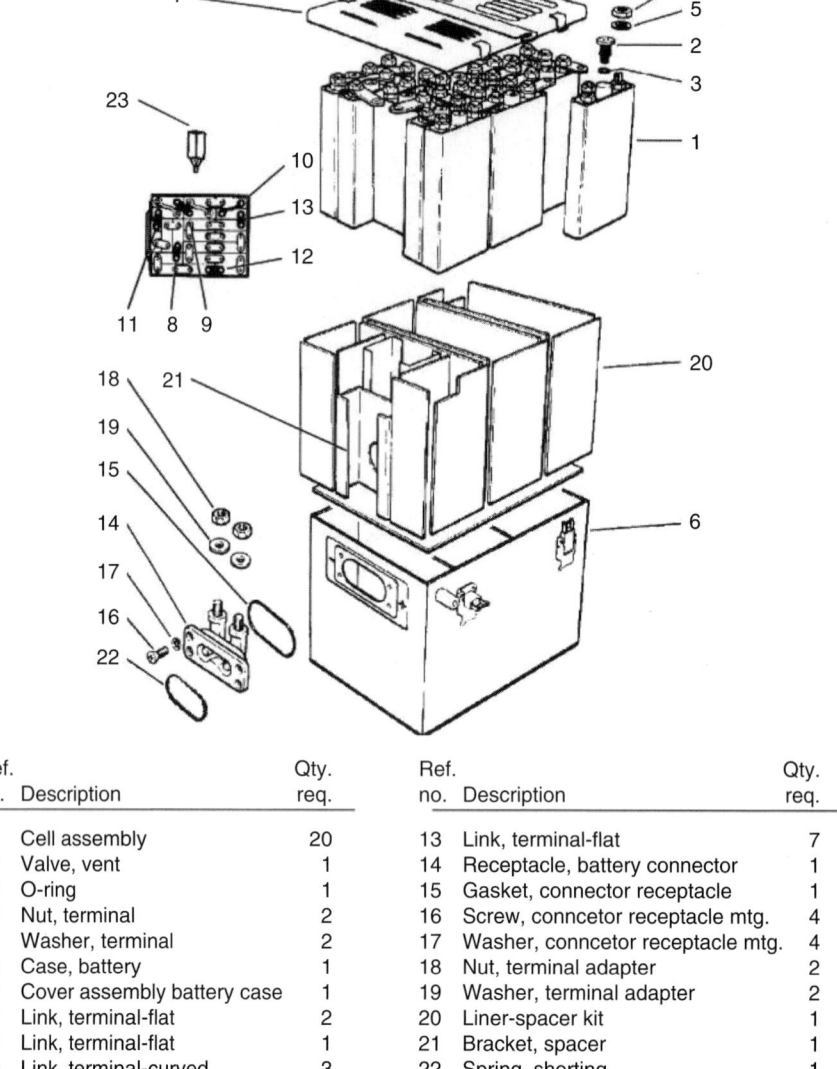

| Ref. no. | Description | Qty. req. | Ref. no. | Description | Qty. req. |
|---|---|---|---|---|---|
| 1 | Cell assembly | 20 | 13 | Link, terminal-flat | 7 |
| 2 | Valve, vent | 1 | 14 | Receptacle, battery connector | 1 |
| 3 | O-ring | 1 | 15 | Gasket, connector receptacle | 1 |
| 4 | Nut, terminal | 2 | 16 | Screw, conncetor receptacle mtg. | 4 |
| 5 | Washer, terminal | 2 | 17 | Washer, conncetor receptacle mtg. | 4 |
| 6 | Case, battery | 1 | 18 | Nut, terminal adapter | 2 |
| 7 | Cover assembly battery case | 1 | 19 | Washer, terminal adapter | 2 |
| 8 | Link, terminal-flat | 2 | 20 | Liner-spacer kit | 1 |
| 9 | Link, terminal-flat | 1 | 21 | Bracket, spacer | 1 |
| 10 | Link, terminal-curved | 3 | 22 | Spring, shorting | 1 |
| 11 | Link, terminal-flat | 1 | 23 | Wrench, vent | 1 |
| 12 | Link, terminal-flat | 7 | | | |

**FIGURE 15B.32**   Typical assembly of vented sintered-plate nickel-cadmium aircraft battery.

the positive electrodes are sintered nickel and chemically impregnated. The standard negative electrode has been replaced with a pasted cadmium oxide electrode, and the electrolyte composition has been modified to increase ionic condition. This modified cell configuration allows for an increase in cell voltage and, on the aircraft bus, reduces the degree of cell overcharge the battery is exposed to and therefore reduces the cell gassing. This results in significantly lower water consumption and subsequently leads to longer maintenance intervals for this battery. The maintenance period can be extended in application from 2 to 3 months to 12 months or longer with the same battery performance and reliability.

The maintenance-free nickel-cadmium aircraft battery uses the same geometric cell shape as the vented nickel-cadmium design, but allows gases to recombine inside the cell. This cell design uses all sintered electrodes but a modified electrolyte composition and has a number of features found in aerospace cell technology. It is not strictly a sealed design since the cell will vent at a high pressure, then reseal itself well above the ambient pressure. To avoid excessive overcharge and associated thermal runaway, the cell is charged and controlled by its own integrated charger and associated electronics. This battery design has been supplied to several military aircraft programs by EaglePicher Technologies, LLC.[12] Sealed Ni-Cd batteries are detailed in Chap. 15C.

## ACKNOWLEDGMENT

Major content for this chapter was provided in whole or in part by John K. Erbacher, Chap. 19, Industrial and Aerospace Nickel-Cadmium Batteries, *Linden's Handbook of Batteries*, 4th ed., T. B. Reddy and D. Linden, eds., McGraw-Hill, 2011.

## REFERENCES

1. J. McBreen, *The Nickel Oxide Electrode*, Modern Aspects of Electrochemistry, No. 21, Ralph E. White, J. O'M. Bockris, and B. E. Conway (eds.), Plenum Press, New York, p. 29, 1990.

2. D. F. Pickett and J. T. Maloy, *J. Electrochem. Soc.* **12:**1026 (1978).

3. S. U. Falk and A. J. Salkind, *Alkaline Storage Batteries*, Wiley, New York, 1969.

4. A. Fleischer, *J. Electrochem. Soc.* **94:**289 (1948).

5. G. Halpert, *J. Power Sources* **12:**117 (1984).

6. D. F. Pickett, in *Proceedings of the Symposium on Porous Electrodes, Theory and Practice*, H. C. Maru, T. Katan, and M. G. Klein (eds.), The Electrochemical Society, Pennington, NJ, 1982, p. 12.

7. M. B. Pell and R. W. Blossom, U.S. Patent 3,507,699 (1970).

8. R. L. Beauchamp, U.S. Patent 3,573,101 (1971); U.S. Patent 3,653,967 (1972).

9. D. F. Pickett, U.S. Patent 3,827,911 (1974); U.S. Patent 3,873,368 (1975).

10. Mil-B-81757, Performance Specification, Batteries and Cells, Storage, Nickel Cadmium, Aircraft General Specification, Crane Division, NSWC, July 1, 1984.

11. J. J. Lander, personal communication.

12. T. M. Kulin, 33rd Intersociety Engineering Conference on Energy Conversion, IECEC-98-145, August 2–6, 1998, Colorado Springs, CO.

## BIBLIOGRAPHY

### General

M. Barak (ed.), *Electrochemical Power Sources*, Peter Peregrinus, London, 1980.

P. Brunamonti, *Life Cycling at Elevated Temperatures Battery Types M81757/8-5 and M81757/15: Marathon Power Tech. Co., EDD 99–127*, Nov. 30, 1999, Crane Div., Naval Surface Warfare Center, Crane, IN 47522-5001.

S. U. Falk, and A. J. Salkind, *Alkaline Storage Batteries*, Wiley, New York, 1969.

H.-D. Jacksch, *Batterie Lexikon*, Pflaum Verlag, Munich, pp. 348–394, 1993.

R. Kinzelbach, *Stahlakkumulatoren*, Varta, Hannover, Germany, 1968.

Y. Miyake, and A. Kozawa, *Rechargeable Batteries in Japan*, JEC Press, Cleveland, OH, 1977.

B. Newman, *Life Cycling at Elevated Temperatures Battery Types M81757/15: SAFT America, Inc., EDD 99–122*, Nov. 17, 1999, Crane Div., Naval Surface Warfare Center, Crane, IN 47522-5001.

## Plastic-Bonded Electrode Technology

B. McRae and D. Nary, *Proceedings of the 38th Power Sources Conference,* pp. 123–126, 1998.

## FNC Technology

M. Anderman, C. Baker, and F. Cohen, *Proceedings of the 32nd Intersociety Energy Conversion Conference,* Vol. 1, p. 97465, 1997.

C. Baker, *Proceedings of the SAE Power Systems Conference,* Williamsbury, VA, 1997. See *Advanced Battery Technology,* April 1997.

C. Baker and M. Barekatien, *Proceedings of the SAE Power Systems Conference,* San Diego, CA, 2000.

FNC Vented Nickel-Cadmium Batteries, Hoppecke Batterien.

# SECTION C
# SEALED NICKEL CADMIUM

## R. David Lucero, Joseph A. Carcone
## (Emeritus Contributor: John K. Erbacher)

## 15C.1   INTRODUCTION/OVERVIEW OF SEALED CELLS

### 15C.1.1   Development of Sealed Cell Technology

Starting in the 1970s, work began on sealed Ni-Cd (SNC) aviation batteries but was not very successful due to lack of quality control in materials and chemistry. However, advances in space nickel-cadmium technologies indicated that quality control of materials, manufacturing, and assembly were needed to develop a low-maintenance, long-life battery for aviation applications. Programs to improve these aspects were initiated at Wright-Patterson AFB by 1980 and resulted in the development of the advanced, maintenance-free battery system (AMFABS), which was tested on several USAF aircraft in the early 1990s and eventually placed in service on the B-52 aircraft. This success led to use of an SNC battery with low maintenance on other military aircraft, and eventually it was adopted for commercial aircraft by the Boeing Aircraft Company. Unfortunately, the SNC technology with its more sophisticated charging requirement and increased quality control was more costly in the acquisition phase of aircraft, so implementation was slower than anticipated. Simultaneously other commercial battery companies developed a low-cost version of this technology, such as the Micro-Maintenance and Ultra Low Maintenance battery concepts from Marathon and SAFT, respectively, along with the fiber nickel-cadmium (FNC electrode) battery. Details of the SNC technology are sufficiently different from industrial nickel-cadmium technology to be considered a derivative of vented and sintered-plate nickel-cadmium batteries.

### 15C.1.2   Specialized Design Features

Sealed nickel-cadmium batteries incorporate specific battery design features to prevent a buildup of pressure in the battery caused by gassing during overcharge. As a result, batteries can be sealed and require no servicing or maintenance other than recharging. These unique characteristics for a secondary battery have created wide acceptance for use in a variety of applications, ranging from lightweight portable power (photography, toys, housewares) to high-rate, high-capacity power (electronic devices such as phones, computers, camcorders, power tools) and standby power (emergency lighting, alarm, memory backup). Some nickel-cadmium batteries are now incorporating smart battery control circuitry to give state-of-charge indication and to control overcharge and over-discharge.

The major advantages and disadvantages of the sealed nickel-cadmium battery are summarized in Table 15C.1. The important characteristics are as follows:

1. *Maintenance-free operation:* The batteries are sealed, contain no free electrolyte, and require no servicing or maintenance other than recharging.

2. *High-rate charging:* Sealed nickel-cadmium batteries are capable of recharge at high rates within 30 min under controlled conditions. Many batteries can be charged in 3 to 5 h without special controls, and all can be recharged within 14 h.

3. *High-rate discharge:* Low internal resistance and constant-discharge voltage make the nickel-cadmium battery especially suited for high-rate discharge or pulse-current applications.

**TABLE 15C.1**   Major Advantages and Disadvantages of Sealed Nickel-Cadmium Batteries

| Advantages | Disadvantages |
|---|---|
| Batteries are sealed; no maintenance required | Voltage depression or memory effect in certain |
| Long cycle life | applications |
| Good low-temperature and high-rate | Higher cost than sealed lead-acid battery |
| performance capability | Poor charge retention |
| Long shelf life in any state of charge | Sealed lead-acid battery better at high temperature |
| Rapid recharge capability | and float service |
| Safety vent system | Environmental concern with the use of cadmium |
| Low internal impedance | Lower capacity than other competitive batteries |

4. *Wide temperature range:* Sealed nickel-cadmium batteries can operate over the range from about −20 to +70°C and are particularly noted for their low-temperature performance.

5. *Long service life:* Over 500 cycles of discharge or up to 5 to 7 years of standby power are common for sealed nickel-cadmium batteries.

## 15C.2   CHEMISTRY

### 15C.2.1   Baseline Reactions

The active materials of the sealed nickel-cadmium battery are the same as for other types of nickel-cadmium batteries, namely, in the charged state, cadmium for the negative electrode, nickel oxyhydroxide for the positive, and a solution of potassium hydroxide for the electrolyte. In the discharged state, nickel hydroxide is the active material of the positive electrode, and cadmium hydroxide that of the negative.

During charge, nickel hydroxide, $Ni(OH)_2$, is converted to a higher valence oxide

$$Ni(OH)_2 + OH^- \rightarrow NiOOH + H_2O + e^-$$

At the negative electrode, cadmium hydroxide, $Cd(OH)_2$, is reduced to cadmium

$$Cd(OH)_2 + 2e^- \rightarrow Cd + 2OH^-$$

The overall discharge/charge reaction is

$$Cd + 2NiOOH + 2H_2O \underset{charge}{\overset{discharge}{\rightleftarrows}} Cd(OH)_2 + 2Ni(OH)_2$$

During operation, the active materials undergo changes in their oxidation states but little change in their physical states. Similarly, there is little if any change in the electrolyte concentration. The active materials of both electrodes, in both charged and discharged states, are relatively insoluble in the alkaline electrolyte, remain as solids, and do not dissolve while undergoing changes in their oxidation states. Because of these and other properties, nickel-cadmium batteries are characterized by long life in both cyclic and standby operations and by a relatively flat voltage profile over a wide range of discharge currents.

The operation of the sealed battery is based on the use of a negative electrode having a higher effective capacity than the positive. During charge, the positive plate reaches full charge before the negative and begins to evolve oxygen. The oxygen migrates to the negative electrode, where it reacts with and oxidizes or discharges the cadmium to produce cadmium hydroxide.

$$Cd + 1/2O_2 + H_2O \rightarrow Cd(OH)_2$$

A separator permeable to oxygen is used so that oxygen can pass through the separator to the negative electrode. Also, a limited amount of electrolyte is used (starved electrolyte system) as this facilitates the transfer of oxygen. This process is illustrated in Fig. 15C.1.

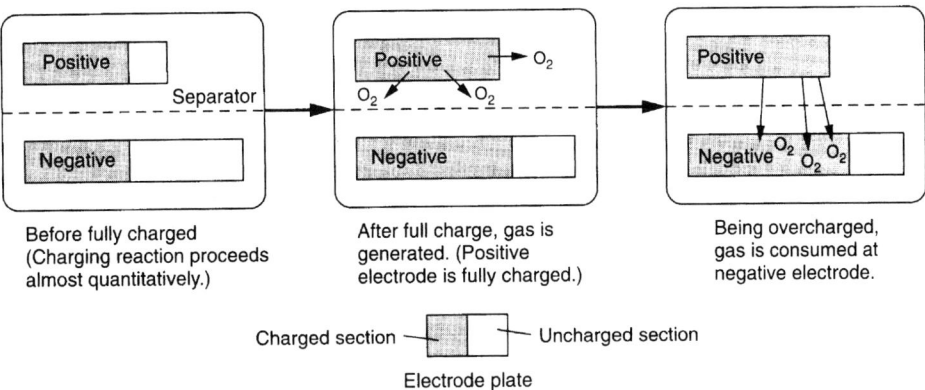

**FIGURE 15C.1**    Oxygen recombination process. (*Courtesy of Sanyo Mobile Energy Co.*)

At a steady state, the recombination reaction rate during overcharge must be no lower than the rate of oxygen generation to prevent buildup of pressure. The internal pressure is related to charge current, the reactivity of the negative electrode, the electrolyte level, and the temperature. Solid cadmium, gaseous oxygen, and liquid water must coexist in mutual contact for the recombination reaction to occur. If, for example, the electrolyte level is too high (the electrodes are in a flooded state), the oxygen is prevented from contacting the electrode, and the reaction rate at a given temperature and pressure is substantially lowered.

A safety venting mechanism is used in the battery design to prevent rupture in case of excessive pressure buildup due to a malfunction, high charge rate, or abuse.

### 15C.2.2    Sealed versus Vented Designs

The charging of aqueous-nickel batteries always occurs in competition with water electrolysis. Toward the end of the charging cycle, oxygen is typically evolved at the positive electrode, and hydrogen may be evolved at the negative electrode. The way that the cell deals with these evolved gases will determine whether the cell can be sealed. In sealed cells, the gases are recombined internally. In an open cell, the gases are allowed to vent, hence the name "vented cell."

The reactions that produce the gases are called the overcharge reactions and differ for sealed or vented cell. Overcharge reactions:

Vented (open) cell:

| | |
|---|---|
| *Positive:* | $4OH^- \rightarrow 2H_2O + O_2 + 4e^-$ |
| *Negative:* | $4H_2O + 4e^- \rightarrow 2H_2 + 4OH^-$ |
| *Net:* | $2H_2O \rightarrow 2H_2 + O_2$ |

The net result is the electrolysis of water to yield hydrogen and oxygen gas.

*Positive:* $\qquad 4OH^- \rightarrow 2H_2O + O_2 + 4e^-$

*Negative:* $\qquad 2Cd(OH)_2 + 4e^- \rightarrow 2Cd + 4OH^-$

*Net electrochemical:* $\qquad 2Cd(OH)_2 \rightarrow 2Cd + 2H_2O + O_2$

*Chemical recombination on negative:* $\qquad 2Cd + O_2 + 2H_2O \rightarrow 2Cd(OH)_2$

The result in a sealed cell is that electricity is converted into heat without any net chemical change in the cell. The overcharge reaction is exothermic, particularly the chemical recombination reaction in the sealed cell.

The overcharge reaction on the positive plate starts before the cell is fully charged, so that some oxygen evolution on charging is unavoidable. At higher temperatures, the oxygen evolution starts at a lower voltage. This results in lower charging efficiencies at higher temperatures and, in the case of the vented cell, an increased need for the addition of water. Additionally, this leaves the positive plate undercharged while the negative plate continues toward a full charged state. The resulting plate imbalance reduces battery capacity. To regain the lost capacity, vented batteries require deep discharge conditioning with each cell being clipped out (shorted).

## 15C.3    COMPONENTS/DESIGN/CONSTRUCTION

### 15C.3.1    General Cell Types

Sealed nickel-cadmium cells and batteries are available in several constructions. The most common types are the cylindrical-shaped batteries. Smaller button batteries and rectangular batteries are also manufactured.

***Cylindrical Batteries.***    The cylindrical battery is the most widely used type because the cylindrical design lends itself readily to mass production and because excellent mechanical and electrical characteristics are achieved with this design. Figure 15C.2 shows a cross section of the cylindrical battery.

The positive electrode is a highly porous sintered, foam, or fibered nickel structure into which the active material is introduced by embedding or impregnation with a molten nickel salt, followed by the precipitation of nickel hydroxide by immersion or electrochemical deposition in an alkaline solution. The negative electrodes are made by several methods: using a sintered-nickel substrate similar to the positive, by pasting or pressing the negative cadmium active material onto a substrate, or by a continuous electrodeposition process.

After processing, the positive and negative electrodes are cut to size and wound together in a jelly-roll fashion with a separator material between them. The separator material, usually unwoven nylon or polypropylene, is highly absorbent to the potassium hydroxide electrolyte and permeable to oxygen. The roll is inserted into a rugged nickel-plated steel can, and the electrical connections are made. The negative electrode is welded or press connected to the can and the positive is usually welded to the top cover. The very small amount of the electrolyte, enough for efficient operation, is absorbed by the separator. There is no free liquid electrolyte. The cover assembly incorporates a fail-safe one-time or resealable vent mechanism to prevent rupture in case of excessive pressure buildup, which could result from extreme overcharge or discharge rates.

***Button Batteries.***    Nickel-cadmium button cells and batteries usually have electrodes made from "pressed" plates. The active materials are compressed in molds into disks or plates, and the electrodes are assembled in a sandwich configuration, as shown in Fig. 15C.3.

In some cases, the electrodes are backed with expanded metal or screen to enhance electrical conductivity and mechanical strength. The button battery does not have a resealable fail-safe device, but its construction allows the battery to expand, either breaking the electrical continuity or opening the seal to relieve excess pressure caused by an abnormal circumstance. Button batteries are very suitable for low-current, low overcharge rate applications.

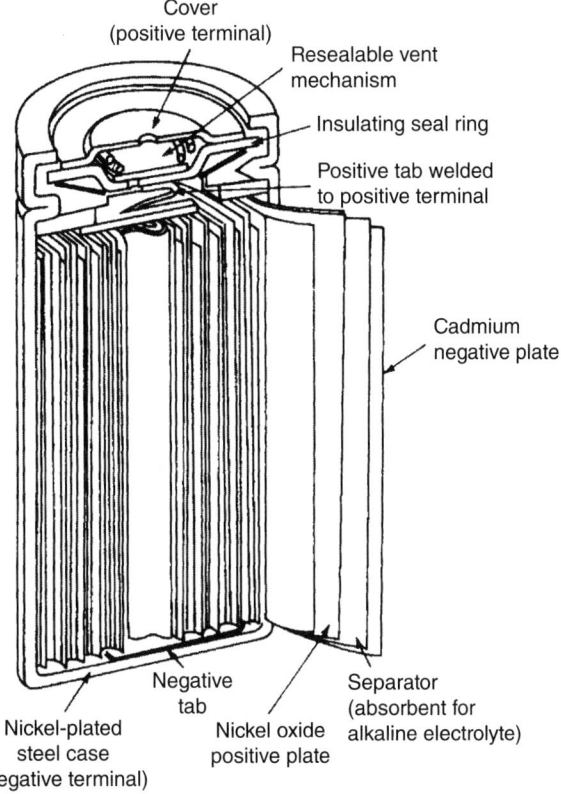

**FIGURE 15C.2**   Construction of a sealed nickel-cadmium cylindrical battery.

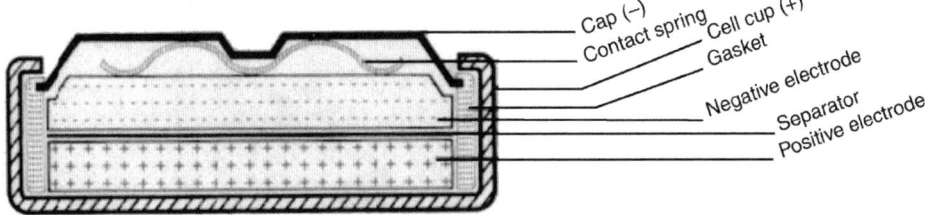

**FIGURE 15C.3**   Nickel-cadmium battery, button configuration.

***Rectangular Batteries.***   The flat or rectangular batteries are designed to meet the needs of lightweight and compact equipment. The rectangular shape permits more efficient battery assembly, eliminating the voids that occur with the assembly of cylindrical batteries. The volumetric energy density of batteries can be increased by a factor of about 20%.

Figure 15C.4 shows the structure of the rectangular battery. The plates are manufactured as described above for cylindrical cells, but the finished electrodes are cut to predetermined dimensions and placed in the metal casing. These are then hermetically sealed in place to the cover plate. All sides of the casing and cover-plate assembly are laser-welded together to prevent electrolyte leakage. A resealable safety venting system is built into these batteries similar to the ones used in the cylindrical designs.

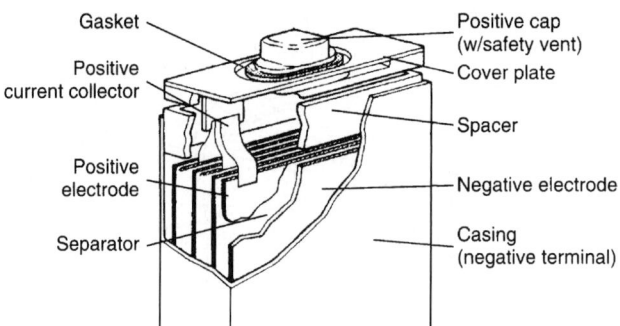

**FIGURE 15C.4**   Construction of a sealed rectangular nickel-cadmium battery. (*Courtesy of Sanyo Mobile Energy Co.*)

The rectangular battery is housed in a nickel-plated steel can using a construction similar to the vented battery, but incorporating the features needed for sealed-battery operation. This construction is particularly suited for high discharge rates because of the large electrode area. Figure 15C.5 is an illustration of a sealed rectangular battery, in this case using fiber-structured electrodes (see Chap. 15B and Sec. 15C.3.2 below).

## 15C.3.2   FNC Electrode Technology

The performance of the sealed cell greatly depends on the electrode technology. An ideal electrode will feature the following characteristics:

- Provide a high surface area conductive matrix to contact the active material
- Provide sufficient porosity for high active material loading and an open structure for good electrolyte penetration
- Have sufficient electrical conductivity to carry the current to the tab with minimal voltage drop, yet will still be light
- Fully contain the active material
- Able to accommodate the dimensional changes during battery charge and discharge without fatigue cracking
- Tolerate mechanical shock and vibration
- Chemically and thermally inert to the battery environment; will not introduce any undesirable impurities into the cell
- Utilize a simple process for the loading of the active material
- Strong enough to tolerate the cell manufacturing processes
- Versatile enough to allow manufacturing of various sizes, thickness, conductivity, and porosity
- Economical

It is in this area of plate design that the FNC technology (a fiber structure electrode) has made a significant step forward in comparison to older technologies. The core of the FNC technology is the three-dimensional nickel-plated fiber matrix. The nickel coating is optimized to the expected current density of the battery. Thus, there is no excess nickel. Electrodes of thickness ranging from 0.5 to 10 mm targeted at ultra-high (XX), high (X and H), medium (M), and low (L) rate designs are fabricated in a common process. The nickel fibers are very compact with one cubic centimeter of electrode volume nominally containing 300 m of conducting filament. This current collecting matrix is 90% porous, allowing excellent utilization of the active material. The result is improved low-temperature performance, a lower charge coefficient, and significantly higher power capability.

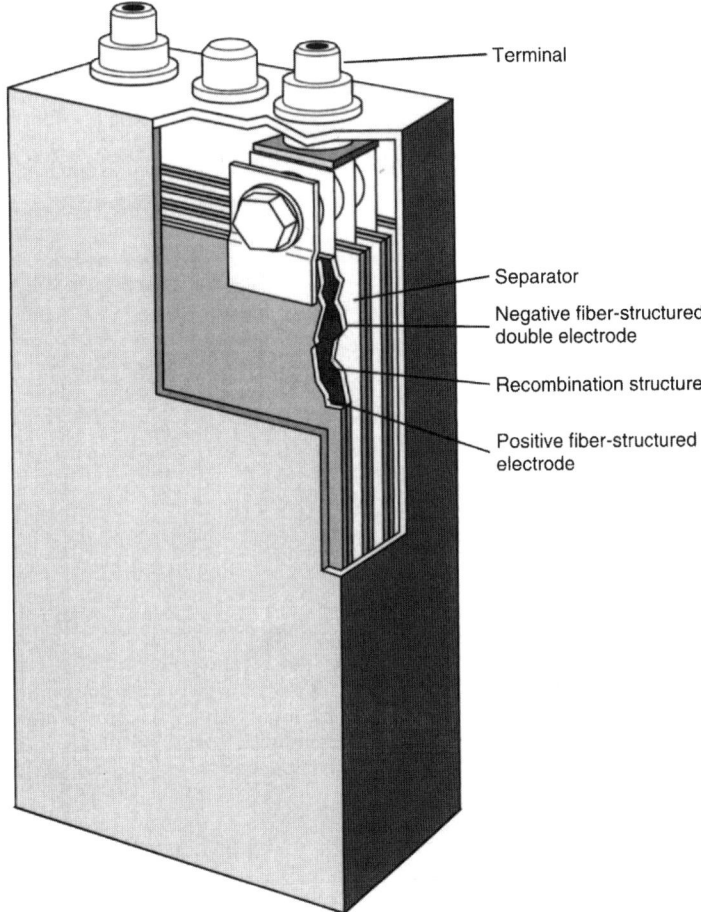

**FIGURE 15C.5**   Sealed rectangular nickel-cadmium battery using fiber-structured elec-trodes. (*Courtesy of Hoppecke Batteries.*)

Since the structure is highly porous, high loading of active materials as well as excellent penetration of electrolyte is achieved. No conductive diluents such as graphite or iron are required. Yet, due to the very high surface area of the fibers, the contact between the current carrying fiber matrix and the paste is very good. Because of this, resistive losses are low, resulting in improved efficiency. The paste is loaded into the electrodes mechanically, and no impurities are introduced in the process. Active material (nickel hydroxide in the positive plate and cadmium hydroxide in the negative) is mechanically embedded directly into the fiber plate. The pure active material contributes to longer life, lower self-discharge, and a more consistent and reliable product (see FNC photomicrographs in Chap. 15B for visual details). Consequently, a high surface area plate capable of high-current loads and very long life has been realized.

The processes and cell design associated with FNC technology have resulted in improved battery performance. Improved charging efficiency has reduced the gassing on overcharge, and with it, the frequency of water topping for the vented cells. The design of the FNC plate allows elastic expansion and contraction during charge and discharge, eliminating one of the main causes of nickel-cadmium plate degradation. This plate flexibility also provides increased shock and vibration tolerances. The flexibility of the electrode structure eliminates mechanical cracks and associated plate degradation, resulting in increased battery life.

### 15C.3.3    Sealed FNC Battery Operating Characteristics

With the development of sealed FNC technology, the first maintenance-free high-rate prismatic nickel-cadmium battery was introduced. In the sealed cell, an unfilled nickel-coated fiber plate is placed between two cadmium-filled negative plates. This effectively results in a split negative with an unfilled central region. This unactivated section serves as a catalytic site for rapid oxygen reduction. The main oxygen pathway to the recombination site is through the plate pores, which, in the FNC plate construction, are relatively large. This provides the oxygen with easy access to the large recombination reaction surfaces, as shown in Fig. 15C.6.

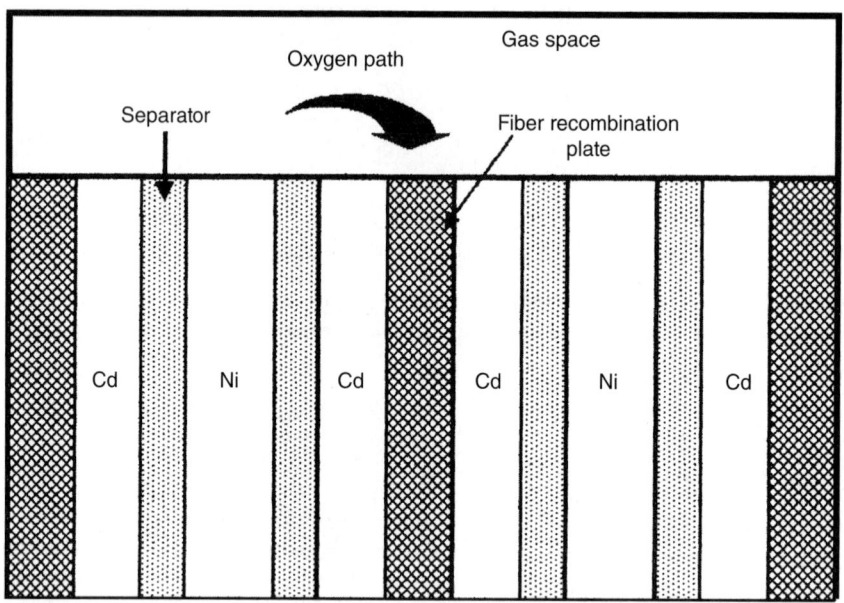

**FIGURE 15C.6**    Electrode structure of fiber nickel-cadmium cell.

Because rapid oxygen recombination eliminates the pressure buildup normally associated with sealed nickel-cadmium cells, high charging rates can be sustained even in the overcharge mode. Also, it is possible to use conventional nylon cell construction to produce prismatic sealed cells rather than the cylindrical design required for high pressure cells. The sealed FNC prismatic cell case is made of either polyamide (nylon) or stainless steel. The negative pressure within the cell (approximately 0.1 bar absolute) allows for pressure change due to oxygen generation during overcharge without causing the cell walls to expand.

All nickel-cadmium batteries must be overcharged to achieve a 100% state-of-charge. During the overcharge stage, excessively charged portions evolve oxygen and hydrogen. In vented batteries, these gases, along with water vapor, are vented outside the cell. The lost liquid must be replaced. The sealed FNC battery completely eliminates the loss of any gases from within the cell. Oxygen generated is rapidly recombined on the negative electrode. Hydrogen evolution is avoided by excess discharged cadmium on the negative electrode. This recombination process also keeps the plates in balance, eliminating the capacity loss that would otherwise result. Electrolyte spillage and corrosion is completely eliminated.

In the event of cell reversal or a failure to control charging voltage, hydrogen gas will be produced. A recombination plate of Pt/Pd-catalyzed plaque located within the cell provides for hydrogen recombination. The oxygen source for hydrogen recombination is provided by the self-discharge reaction on the positive electrode or the overcharge reaction on the following charge.

A safety valve at the top of the cell is provided to allow excessive pressure to escape should the battery be abused to the point that the electrolyte boils. This condition might be caused by a severe overcharge, where adequate heat dissipation is not provided. Under high temperature abuse (+100°C and above), the safety valve will open at approximately 45 psia over pressure, allowing water vapor to escape. Electrolyte will not be expelled, even with the cell in an inverted position. When the cell is allowed to cool, the valve will reseal and the negative pressure cell will return to a normal operating condition. A reduction in cell capacity may be anticipated due to the loss of water from within the cell.

Positive and negative plates are connected to their respective terminal posts by nickel tabs. The tabs are attached directly to the fiber plates by a patented welding process and then fastened directly to nickel-plated, solid-copper terminal posts. The electrical path of each cell type is designed for maximum electrical performance.

Plate stacking is the same as previously discussed. Single positive plates are separated from the negative cadmium electrode by an electrolyte wet separator. The cadmium electrode is in three parts: two fiber frameworks carrying the negative active material, and an unfilled fiber recombination electrode placed between them. The large recombination surface area is sufficient to handle a 2C rate charge on a fully charged battery. With the unfilled recombination plate being the primary gas path for oxygen recombination, a small pore size separator can be used. The separator is designed to be completely filled with electrolyte, thus contributing to improved high-rate performance. Additionally, the recombination plate acts as a reservoir for electrolyte, allowing for volumes in excess of 4 mL/Ah. This prevents stack dry-out as a possibility for premature cell failure.

Sealed FNC batteries are fail-safe. Even if subjected to extreme overcharge to the point at which the electrolyte boils, the battery will not go into thermal runaway. Instead, the hot cells dry out, with the loss of water vapor causing the battery impedance to increase. As the impedance increases, the current will decrease. After a time, the battery will no longer accept the charge current and it will cool down.

### 15C.3.4 Manufacturing Flexibility

The power capability of a battery will affect its potential applications. A high-power battery will be capable of delivering most of its capacity in a few minutes. To maximize battery power, it is necessary to minimize cell resistance. To that end, high-power cells are designed with high surface areas, thin electrodes, and high metallic contents. However, the above measures will increase battery weight, volume, and cost. Given these trade-offs, it is desirable to optimize the cell design for the application.

Practical manufacturing constraints inhibited the development of high-capacity low-rate sintered-plate batteries. In contrast, the FNC technology covers a wide range of power capabilities. The thickness of the fiber electrode and the amount of conductive metallic nickel on it are varied within an order of magnitude. This results in the capability to produce high-capacity, low-weight, low-cost batteries or ultra-high-power, higher weight, and higher cost cells using the same processes and equipment. For the user, this means that the characteristics of sinter foils and the various types of pocket- or foam-plate batteries no longer have to be considered separately. The FNC system has the same properties and basic characteristics over the entire range of applications.

## 15C.4    PERFORMANCE AND APPLICATIONS

### 15C.4.1 Performance of FNC Sealed Aircraft Batteries

The high current performance capability of the sealed FNC battery design is exemplified by a short-circuit test performed on a model KCF XX47 battery, which produced currents approaching 4000 A (see Fig. 15C.7). The KCF XX47 unit was designed to meet the requirements for large auxiliary power unit (APU) and direct engine starting. The constant voltage discharge of 12 V demonstrates the extraordinary high power capability of the battery (see Fig. 15C.8). The KCF XX47 battery's high-power performance is also demonstrated with start curves for a large, wide-body aircraft APU. The first two discharge sequences represent unsuccessful start attempts with the third showing a successful start. The minimum voltage required for this particular specification is 13 V, while the FNC battery provides over 16 V (see Fig. 15C.9).

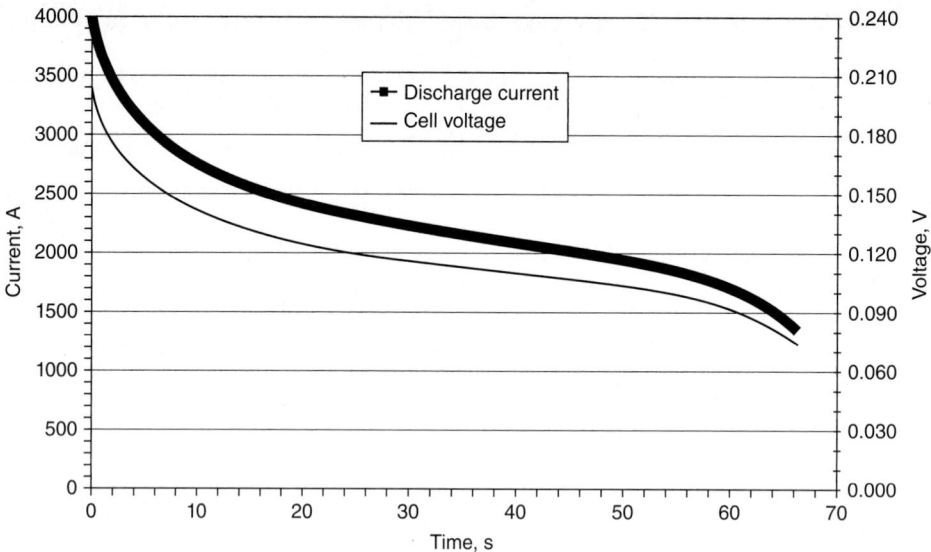

**FIGURE 15C.7**    Short-circuit current. Model KCF XX47 FNC cell (47 Ah).

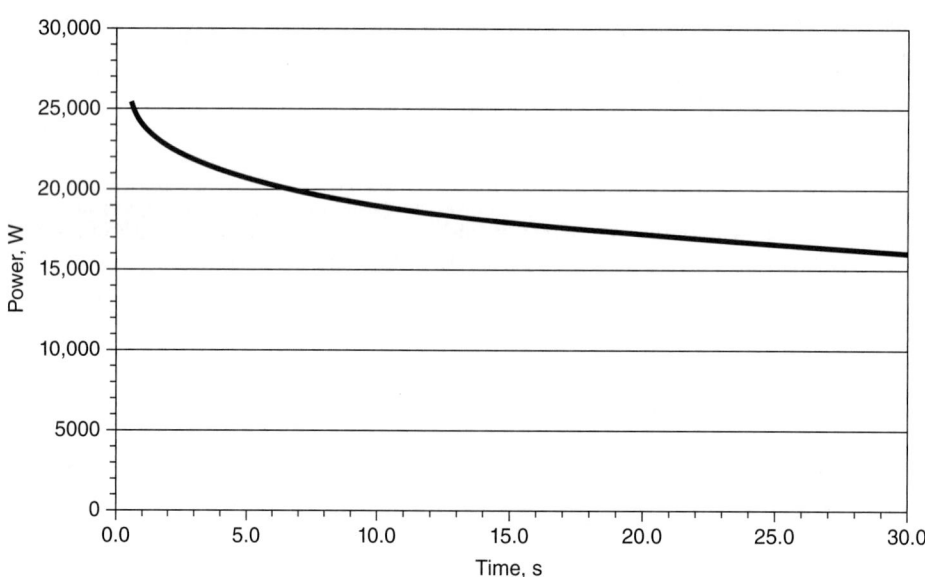

**FIGURE 15C.8**    Constant voltage (12.0 V) discharge. Room temperature. Model XX47 FNC battery (47 Ah rated).

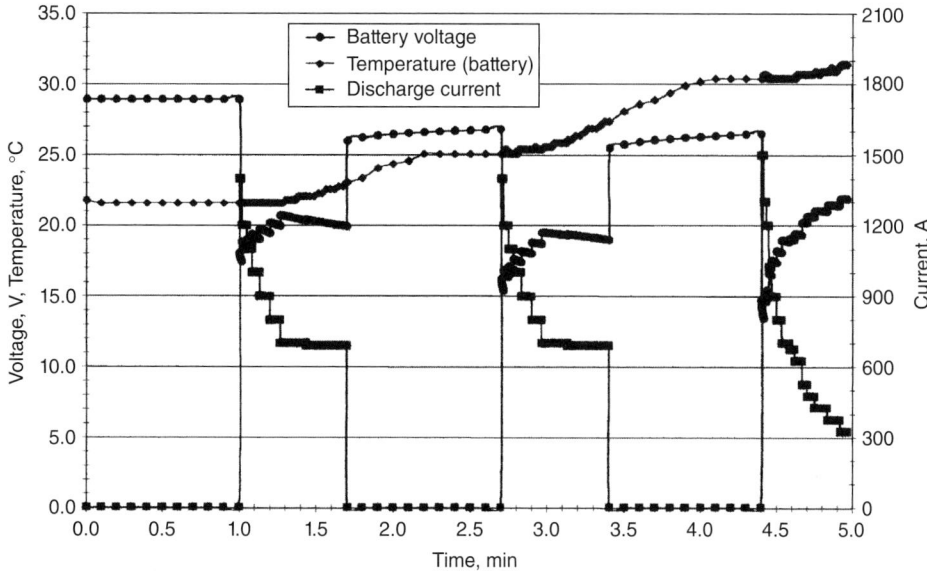

**FIGURE 15C.9**   APU starts: two unsuccessful, one successful. Model XX47 FNC battery (47 Ah rated).

Cold-temperature performance available with the sealed FNC cells is also impressive. Figure 15C.10 shows the capacity of a 28 V, 47 Ah battery for four different discharge rates with the battery soaked at a temperature of −18°C.

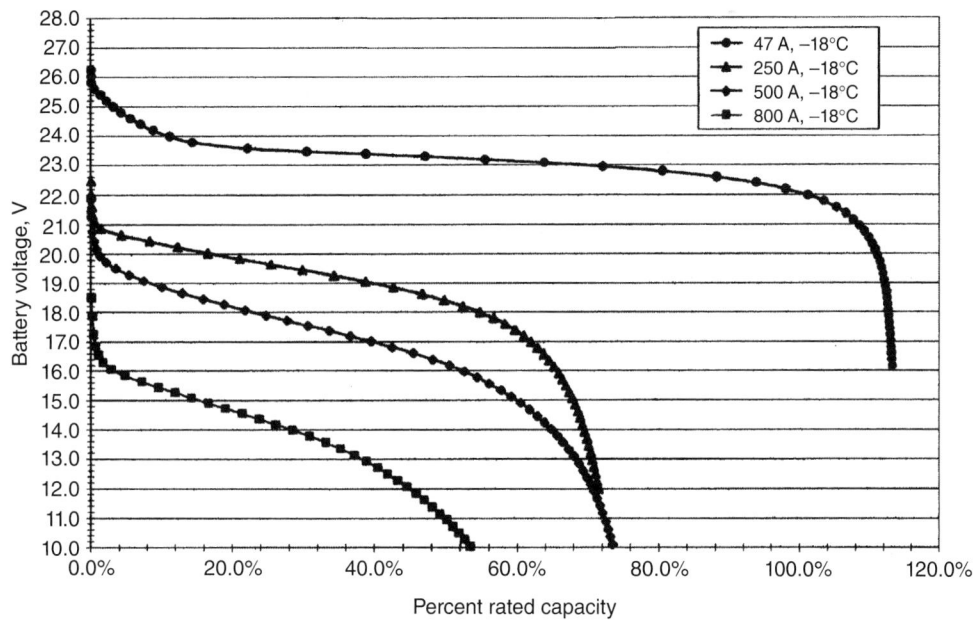

**FIGURE 15C.10**   Constant current discharge. Battery charged at 25°C. FNC battery discharged at −18°C. Model XX47 FNC battery (47 Ah rated).

Sealed FNC batteries have shown outstanding cycle life at both low and high rates of discharge. Cycle life data for low earth orbit (LEO) cycle testing has demonstrated a cycle life in excess of 10,000 cycles for 35% DOD, 10°C, $C/2$ cycling. By maintaining the stable end of discharge voltage, the low recharge coefficient (approximately 3%) demonstrates the superior charge efficiency of the sealed FNC cell. (See Chap. 25B for details on satellite applications.)

Deep discharge cycling is not necessary, eliminating the necessity for battery removal from the aircraft during maintenance. Capacity checks, if desired, can be accomplished with the battery installed by using a portable discharge/charger unit because the sealed FNC design does not exhibit the memory effect typically found in other NiCds.

Charging characteristics of the sealed FNC cells are simple, yet different from vented NiCds. Because of the recombination that takes place during overcharge, the normal $dV/dt$ behavior is not always observed. In addition, heat is generated during the overcharge from the recombination reaction, providing a reliable parameter for charge control. Changing from main mode to topping mode and charge termination is determined by battery temperature rise ($\Delta T$).

The preferred charge is at constant current with a voltage clip (maximum voltage) of 1.55 V per cell. This voltage is also sufficient to charge the battery at temperatures as low as −40°C. For many applications, this means that a heater blanket is not required.

A completely charged FNC battery has sufficient recombination to continue to accommodate a $2C$ overcharge rate.

### 15C.4.2 General Performance Characteristics of Typical Sealed Cylindrical and Button Cells

***Overall Capabilities.*** A typical charge-discharge cycle for the sealed nickel-cadmium cylindrical battery, at 20°C, is shown in Fig. 15C.11. The voltage increases slowly but steadily at the $C/10$ charge rate to a steady-state condition, decays slightly during the 1 h rest, and is relatively flat during the 1 h discharge to 1.0 V. The voltage recovers rapidly over the next hour, while at rest, to near 1.2 V.

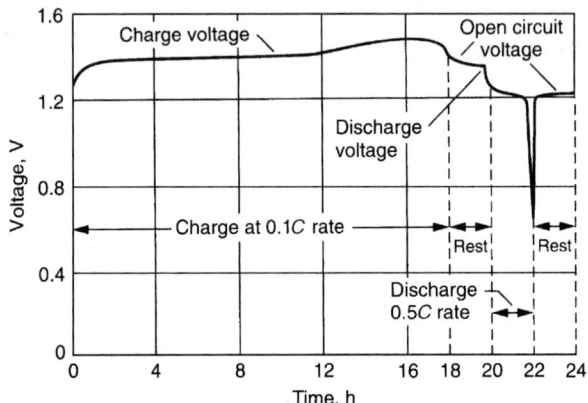

**FIGURE 15C.11** Voltage profile of nickel-cadmium battery in a typical charge/discharge cycle.

***Discharge Characteristics.*** Typical discharge curves for the cylindrical battery at 20°C at various discharge loads are shown in Fig. 15C.12. The flat voltage profile, after the initial voltage drop, is characteristic.

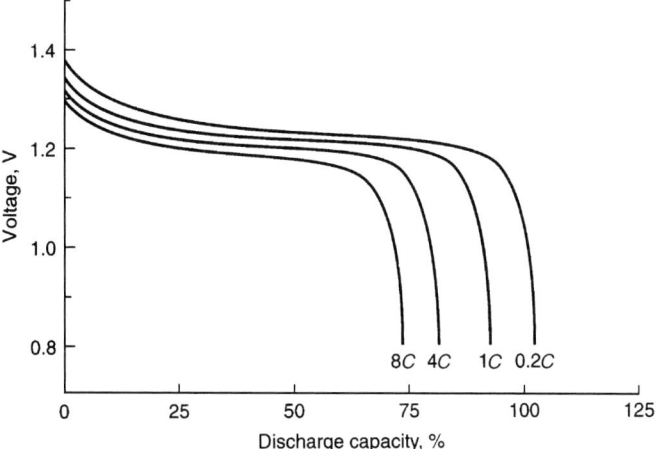

**FIGURE 15C.12**    Constant current discharge curves for sealed nickel-cadmium batteries at 20°C, charge 0.1C, 16 h. (*Courtesy of Sanyo Mobile Energy Co.*)

The capacity that can be obtained from a battery is dependent on the rate of discharge, the voltage at which discharge is terminated, the discharge temperature, and the previous history of the battery. Figure 15C.13 shows the percentage of rated capacity delivered during discharges at various rates and temperatures. The midpoint voltage during discharge decreases as the discharge rate increases (see Fig. 15C.16). If the battery were allowed to discharge to a lower cutoff voltage, a greater percentage of the $C/5$ rate capacity will be obtained. However, batteries should not be discharged to below the specified cutoff voltage as the battery or individual cells may be damaged.

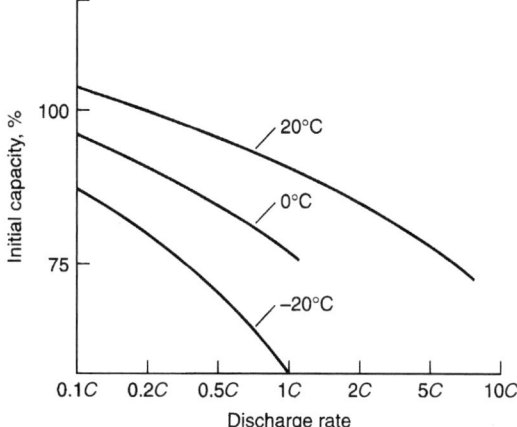

**FIGURE 15C.13**    Percent of $C/5$ rate capacity versus discharge rate to 1.0 V cutoff for typical sealed nickel-cadmium battery.

***Effect of Temperature.***    The sealed nickel-cadmium battery is capable of good performance over a wide temperature range. Best operation is between −20 and +30°C, although usable performance can be obtained beyond this range. The low-temperature performance, particularly at high rates, is generally better than that of the lead-acid battery but usually inferior to that of the vented sintered-plate battery. The reduction in performance at low temperatures is due to an increase in the internal resistance. At high temperatures, the loss can be due to a depressed operating voltage or to self-discharge.

Figure 15C.14 shows some typical discharge curves of the sealed battery at various temperatures at the 0.2$C$ and 8$C$ rates; Fig. 15C.15 shows typical discharge curves at −20°C. A flat discharge profile is still characteristic, but at a lower operating voltage than at room temperature. Figure 15C.16 shows the effect of temperature on the midpoint voltage. Ambient temperatures significantly above or below 20 to 25°C have a depressing effect on the average discharge voltage.

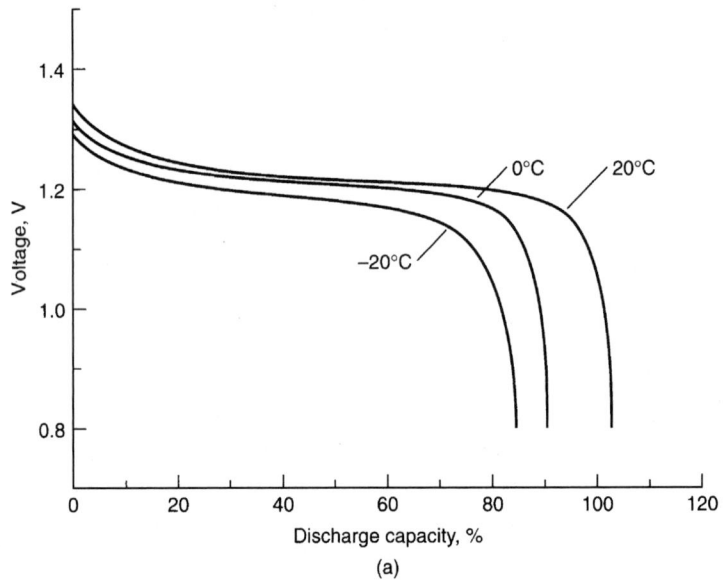

(a)

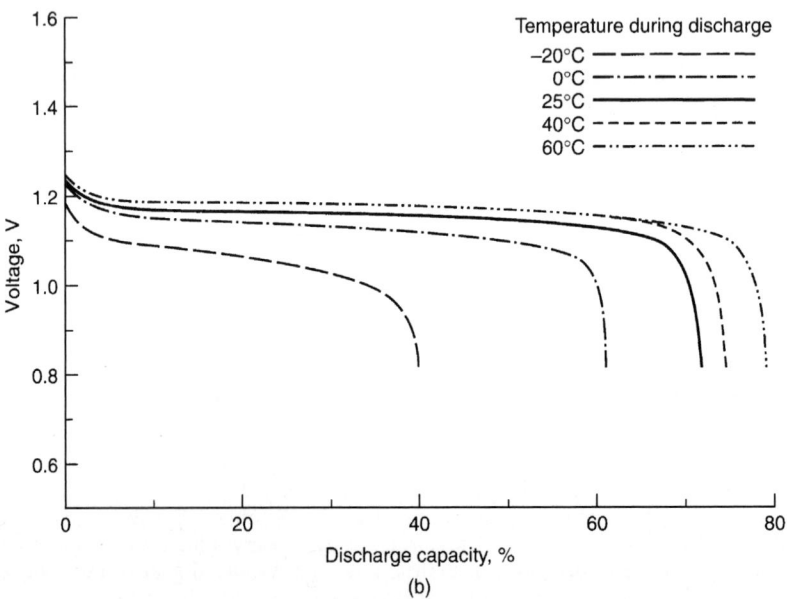

(b)

**FIGURE 15C.14**   Constant current discharge curves of sealed nickel-cadmium batteries at various temperatures: (*a*) 0.2*C* discharge rate; (*b*) 8*C* discharge rate.

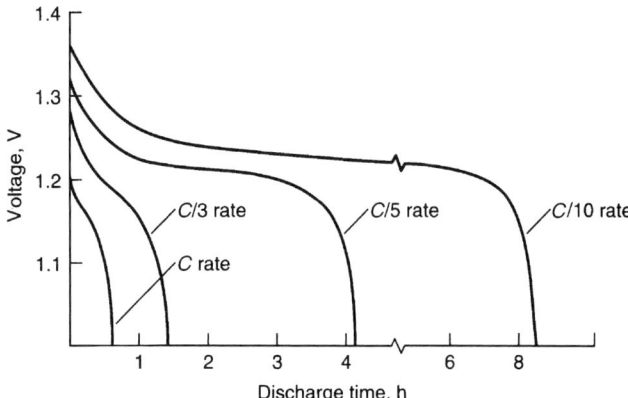

**FIGURE 15C.15**   Constant current discharge curves of sealed nickel-cadmium batteries at −20°C.

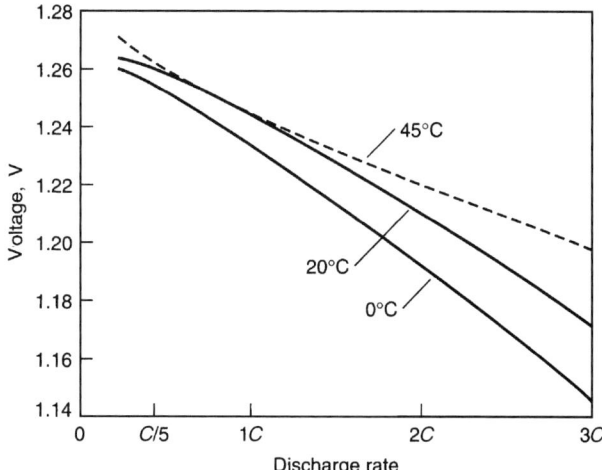

**FIGURE 15C.16**   Midpoint discharge voltage versus rate at various temperatures for sealed nickel-cadmium batteries. 1.0 V cutoff.

The effect of temperature and discharge rate on the capacity of a sealed nickel-cadmium battery is shown in Fig. 15C.17. These data are typical of standard batteries. Manufacturers should be contacted to obtain performance characteristics of specific batteries.

***Internal Impedance.***   The internal impedance of a battery is dependent on several factors, including ohmic resistance (due to conductivity, the structure of the current collector, the electrode plates, separator, electrolyte, or other features of the battery design), resistance due to activation and concentration polarization, and capacitive reactance. In most cases, the effects of capacitive reactance can be ignored. Polarization effects are dependent in a complicated way on current, temperature, and time; they decrease with increasing temperature. The effect may be negligible for pulses of short duration, that is, less than a few milliseconds.

The nickel-cadmium battery is noted for its low internal resistance due to the use of thin and large surface area plates with good electrical conductivity, a thin separator with good electrolyte retention, and an electrolyte having a high ionic conductivity. During discharge, the activation and concentration polarization effects

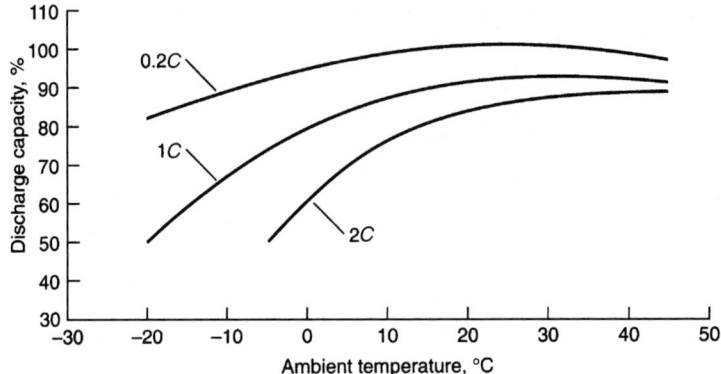

**FIGURE 15C.17**   Percentage of rated capacity versus temperature at different discharge rates for typical sealed nickel-cadmium batteries. 1.0 V cutoff.

are negligible, at least at low and moderate rates, and the internal resistance of the battery and the discharge voltage remain relatively constant from the state of full charge to the point where almost 90% of the capacity has been discharged. At that point, the resistance increases due to the conversion of active materials in the electrode plates, which tends to lower electrical conductivity. Figure 15C.18 shows the change in internal resistance with the depth of discharge for two batteries of different size and capacity. Figure 15C.19 illustrates the effect of temperature. The internal resistance increases as the temperature drops because the conductivity of the electrolyte and other components is lower at the lower temperatures.

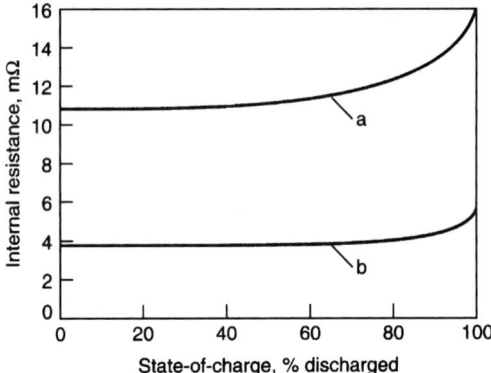

**FIGURE 15C.18**   Resistance versus state-of-charge at 20°C, discharged at 0.2C rate, for sealed nickel-cadmium batteries. (*a*) AA size battery. (*b*) Sub-C-size battery. (Typical for sintered plate electrode type batteries.)

With use over time, a nickel-cadmium battery gradually loses capacity, resulting in a gradual increase in internal resistance. This is caused by gradual deterioration of the separator and electrodes and by loss of liquid through the seals, which changes the electrolyte concentration and level. The net effect is an increase in internal impedance.

***Service Life.***   The service life of a sealed nickel-cadmium battery, normalized to unit weight (kilograms) and size (liters), at various discharge rates and temperatures is summarized in Fig. 15C.20. The curves are based on a capacity, at the C/5 discharge rate at 20°C, of 30 Ah/kg and 85 Ah/L, reflecting the performance of standard-type sealed cylindrical batteries. Manufacturers should be contacted for performance characteristics of specific batteries.

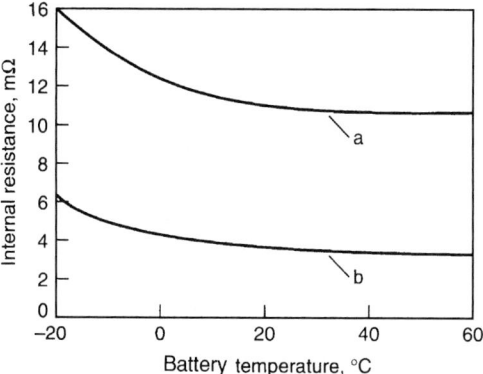

**FIGURE 15C.19** Resistance versus temperature for fully charged sealed nickel-cadmium batteries. (*a*) AA size battery. (*b*) Sub-C-size battery. (Typical for sintered plate electrode type batteries.)

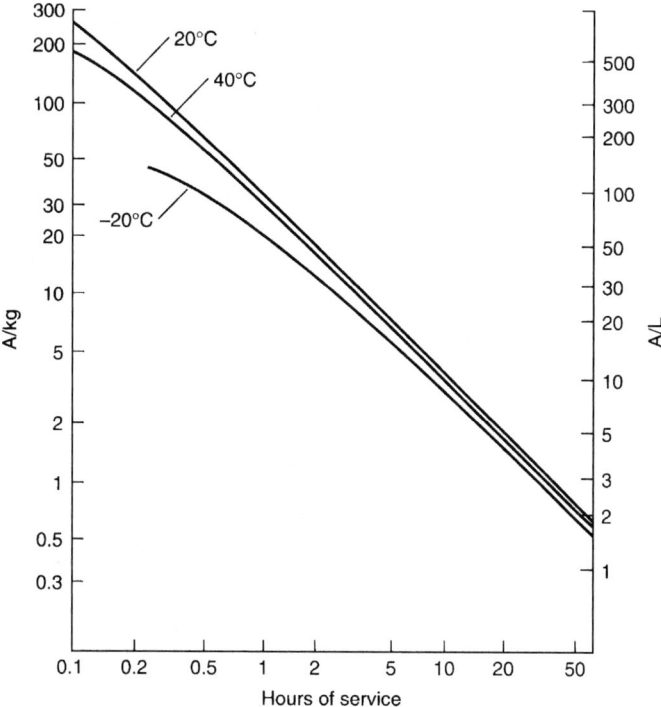

**FIGURE 15C.20** Service life of sealed nickel-cadmium battery at various constant-current discharge rates and temperatures; end voltage at 1.0 V.

***Reversal of Voltage Polarity.***    When three or more batteries are series-connected, the lowest-capacity battery can be driven into voltage reversal by the others. Thus, the larger the number of batteries in series, the greater the possibility of voltage reversal occurring. During reversal, hydrogen may evolve from the positive electrode and oxygen from the negative. Figure 15C.21 shows the complete discharge curve of a battery, including polarity reversal. Section 1 of the figure is the normal period of discharge with active materials remaining on both electrodes. Section 2 represents the period when the discharge has been extended and all of the active material on the positive electrode has been discharged and hydrogen gas is generated at this electrode. Active material still remains on the negative electrode and its normal discharge reaction continues. The battery voltage varies with the discharge current, but stays at about −0.2 to −0.4 V. In Section 3, the negative active material has been discharged and oxygen gas is generated at this electrode.

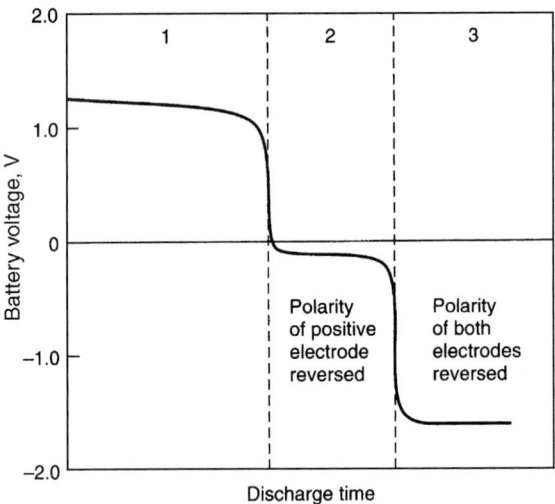

**FIGURE 15C.21**    Discharge of sealed nickel-cadmium battery showing polarity reversal.

Continued discharging during polarity reversal will lead to high battery pressure and opening of the safety vent. This then results in a loss of gas and electrolyte and breakdown of the capacity balance of the positive and negative electrodes.

Some battery designs provide a limited amount of built-in protection against deep reversal by adding a small amount of cadmium hydroxide to the positive electrode. The term used for the material added to the positive electrode for reversal protection is "antipolar mass" (APM). When the positive electrode is completely discharged, the cadmium hydroxide in that electrode is converted to cadmium, which, combining with the oxygen generated from the negative electrode, depolarizes the positive, preventing hydrogen generation for a time. This reaction occurs at about −0.2 V. This reaction can sustain for only a limited time, after which hydrogen is evolved from the positive electrode. Because hydrogen combines with the battery materials to only a limited extent, repetitive battery reversal will gradually increase a battery's internal pressure, ultimately causing the battery to vent.

Discharging to the point of reversal should be avoided. In order to prevent voltage reversal in any of the cells of a multicell battery, particularly with a series string of more than four cells, the battery should not be discharged to a voltage below 0.8 V per cell. In applications of multicell batteries where it is likely that the battery will be frequently discharged below 1.0 V per cell, a voltage-limiting device is recommended to avoid cell reversal.

***Types of Discharge.***    As discussed in Chap. 5, a battery may be discharged under different modes (such as constant resistance, constant current, or constant power), depending on the characteristics of the equipment load. The type of discharge mode selected has a significant impact on the service life delivered by a battery in a specified application. The voltage profiles of a nickel-cadmium battery discharged under the three different modes are plotted in Fig. 15C.22. The data are based on a discharge of a 650 mAh battery so that, at the end of the discharge (1.0 V per cell), the power output is the same for all modes of discharge. In this example, the power output is 130 mW. To discharge at 130 mW at 1.0 V, the constant-current discharge is 130 mA ($C$/5 rate) and the constant-resistance discharge is 7.7 Ω. As shown, the longest service life is obtained under the constant-power mode as the average current is the lowest under this mode of discharge.

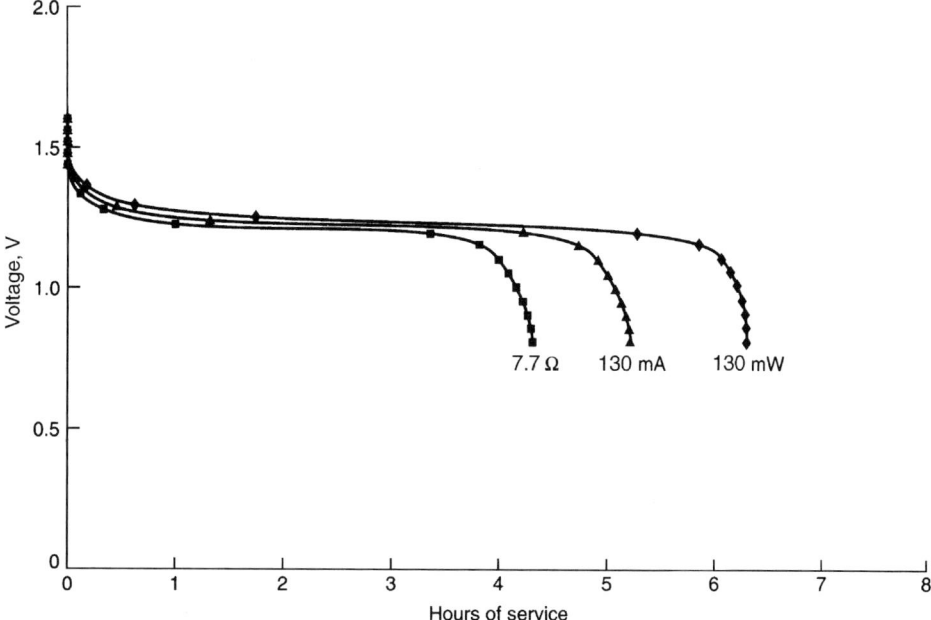

**FIGURE 15C.22**   Discharge curves of AA size (650 mAh) nickel-cadmium battery—constant power ( ) versus constant current ( ) versus constant resistance ( ).

***Constant-Power Discharge.***   The discharge characteristics of the nickel-cadmium battery under the constant-power mode, at several different power levels, are shown in Fig. 15C.23. These are similar to the data presented in Fig. 15C.12 for constant-current discharges, except that the performance is presented in hours of service instead of percent discharge capacity. The power levels are shown based on the $E$ rate. The $E$ rate is calculated in a manner similar to calculating the $C$ rate, but based on the rated watt-hour capacity. For example, for the $E$/5 power level, the power for a battery rated at 780 mWh is 156 mW.

***Shelf Life (Capacity or Charge Retention).***   Nickel-cadmium batteries lose capacity during storage. The rate of this self-discharge is a function of storage temperature and battery design. Figure 15C.24 can serve as a guide for the shelf life (capacity or charge retention) at several temperatures for typical standard type nickel-cadmium sealed batteries. Specifically designed batteries may have considerably different charge retention characteristics. For example, the button batteries designed for memory-backup application have significantly better charge retention characteristics than the standard lower resistance higher discharge rate cylindrical batteries.

Sealed nickel-cadmium batteries can be stored in a charged or a discharged condition. Except for extended storage at high temperatures, they can be restored to full capacity after storage by recharging (two or three charge-discharge cycles). Figure 15C.25 illustrates the capacity recovery after prolonged storage at several temperatures. The recovery time may be longer after high-temperature storage.

***Cycle Life.***   The cycle life is usually measured to the point when the battery will not deliver more than a given percentage (usually 60% to 80%) of its rated capacity. Sealed nickel-cadmium batteries have long cycle lives. Under controlled conditions over 500 cycles can be expected on a full discharge, as illustrated in Fig. 15C.26.

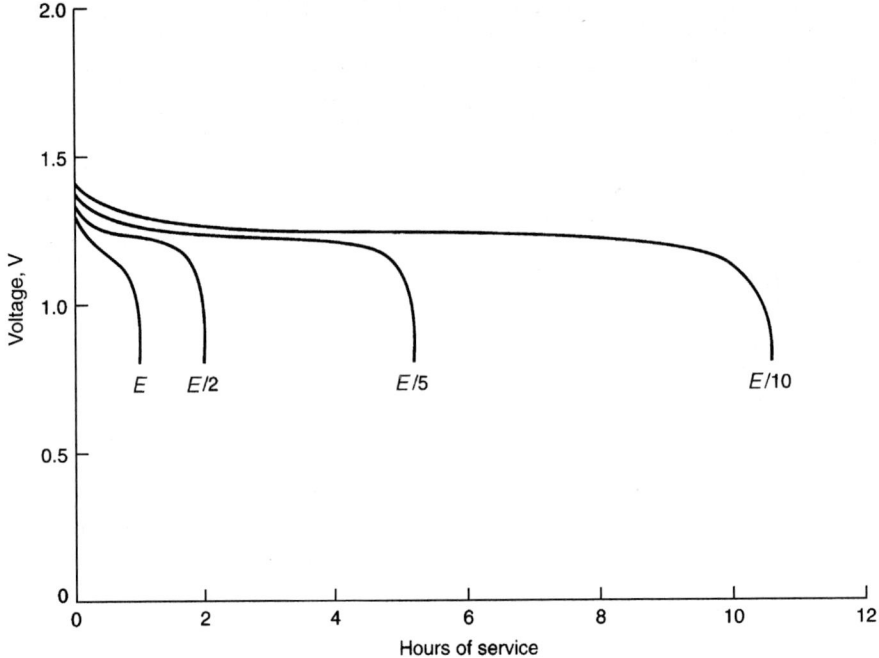

**FIGURE 15C.23**  Constant power discharge curves at various $E$ rates for sealed nickel-cadmium batteries at 20°C.

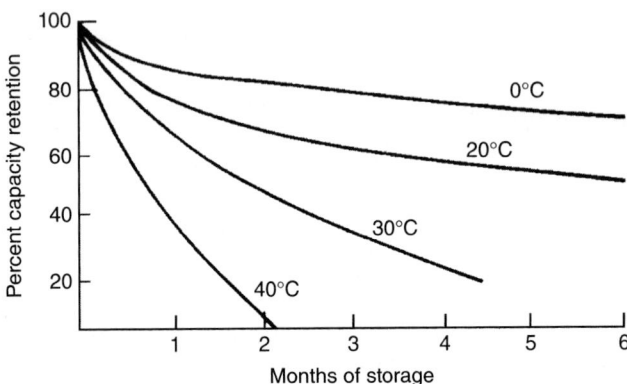

**FIGURE 15C.24**  Capacity retention (shelf life) of standard type sealed nickel-cadmium batteries.

On shallow discharges considerably higher cycle life can be obtained, as shown in Fig. 15C.27. Cycle life is also very dependent on the many conditions to which the battery has been exposed, including charge, overcharge, and discharge rates, frequency of cycling, the temperatures to which the battery has been exposed, and battery age, as well as battery design and battery components. Specially designed batteries, such as those using alkali-resistant materials, are also manufactured, which have longer life, particularly at the higher temperatures.

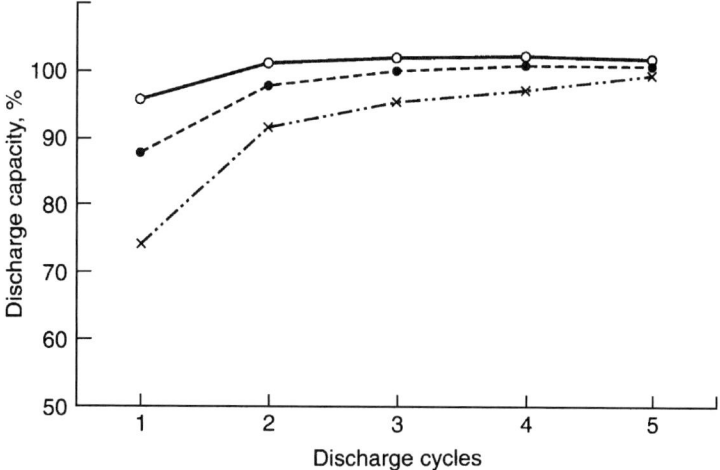

**FIGURE 15C.25**  Capacity recovery of standard type sealed nickel-cadmium batteries, discharge at 0.2C rate after 2-year storage. Storage temperatures: at 20°C (○), 35°C (●), and 45°C (×).

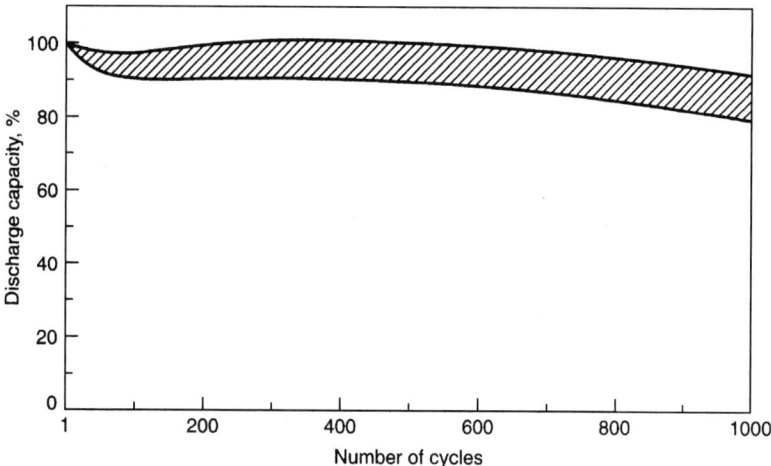

**FIGURE 15C.26**  Cycle life of sealed nickel-cadmium batteries at 20°C. Cycle conditions: charge—0.1C × 11 h; discharge—0.7C × 1 h. Capacity-measuring conditions: charge—0.1C × 16 h; discharge—0.2C; end voltage—1 V.

*Life Expectancy and Failure Mechanisms.*    The useful life of a nickel-cadmium battery can be measured either in terms of the number of cycles before failure or in units of time. It is virtually impossible to know all the detailed information necessary to make any kind of accurate prediction of battery life in a given application. The best that can be provided is an estimate based on laboratory test data and field experience or extrapolation of accelerated test data.

Basically, failure of a battery occurs when it ceases to operate the device, for whatever reason, at the pre-scribed performance level, despite the possibility that the battery may still be useful in another application with less demanding requirements.

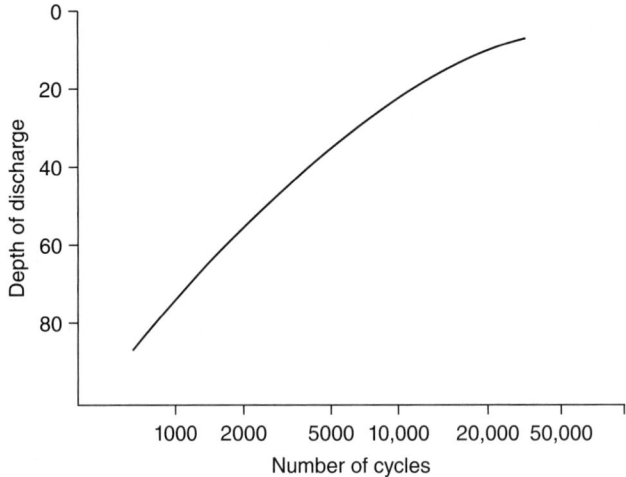

**FIGURE 15C.27**    Cycle life of sealed nickel-cadmium batteries at shallow discharge.

Failure of a nickel-cadmium battery can be classified into two general categories: reversible and irreversible failure. When a battery fails to meet the specified performance requirements but can, by appropriate reconditioning, be brought back to an acceptable condition, it is considered to have suffered a reversible failure. Permanent or irreversible failure occurs when the battery cannot be returned to an acceptable performance level by reconditioning or any other means.

*Reversible Failures*
VOLTAGE DEPRESSION (MEMORY EFFECT).    A sealed nickel-cadmium battery may suffer a reversible loss of capacity when it is cycled repetitively on shallow discharges (discharge terminated before its full capacity is delivered) and recharged. For example, as shown in Fig. 15C.28, if a battery is cycled repetitively, but only

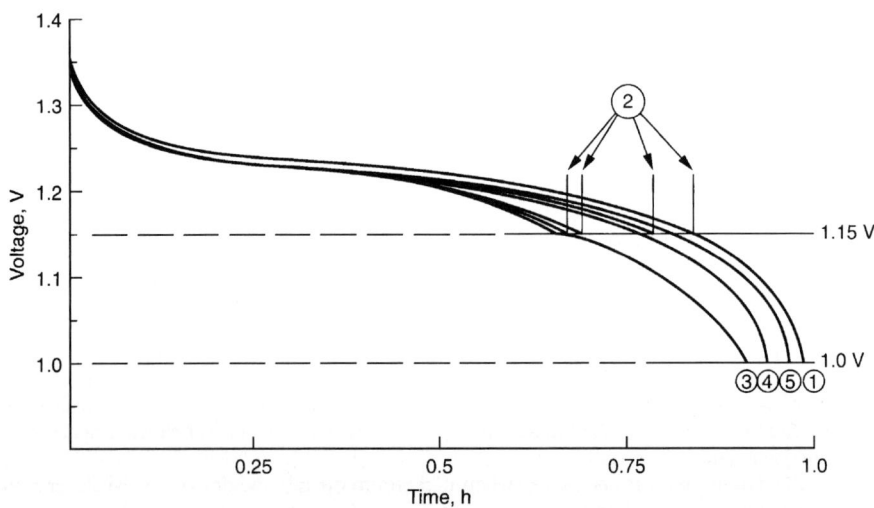

**FIGURE 15C.28**    Voltage depression and subsequent recovery.

partially discharged and then recharged, the voltage and delivered capacity will gradually decrease with cycling (curves 2 representing repetitive cycling). If the battery is then fully discharged (curve 3), the discharge voltage is depressed compared to the original full discharge (curve 1). The discharge profile may show two steps and the battery may not deliver the full capacity to the original cutoff voltage. This condition is known as "voltage depression." It also is referred to as "memory effect," as the battery appears to "remember" the lower capacity of the shallow discharge. Operation at higher temperatures accelerates this type of loss.

The battery can be restored to full capacity with a few full discharge-charge reconditioning cycles. The discharge characteristics on the reconditioning cycles are illustrated in Fig. 15C.28 (curves 4 and 5).

The voltage drop occurs because only a portion of the active materials is discharged and recharged during shallow or partial discharging. The active materials that have not been cycled, particularly the cadmium electrode, may undergo a change in physical characteristics and an increase in resistance. The effect has also been ascribed to structural changes in the nickel electrode.[1] Subsequent cycling restores the active materials to their original state.

The extent of voltage depression depends on the depth of discharge and can be avoided or minimized by the selection of an appropriate end voltage. Too high an end voltage, such as 1.16 V per cell, terminates the discharge prematurely. (A high end voltage should be used only if an extended cycle life is desired and the lower capacity can be tolerated, as in some satellite applications.) A small voltage depression may be observed if the discharge is terminated between 1.16 and 1.10 V per cell. The extent of the depression is dependent on the depth of discharge, which is also rate-dependent. Discharging to an end voltage below 1.1 V per cell should not result in a subsequent voltage depression. Discharging to too low of an end voltage, however, should be avoided.

This condition varies with the design and formulation of the electrode and may not be evident with all sealed nickel-cadmium batteries. Modern nickel-cadmium batteries use electrode structures and formation processes that reduce the susceptibility to voltage depression, and most users may never experience low performance due to memory effect. However, the use of the term "memory effect" persists, since it is often used to explain low battery capacity that is attributable to other problems, such as ineffective charging, overcharge, battery aging, or exposure to high temperatures.

OVERCHARGING. A similar reversible failure can occur with long-term overcharging, particularly at elevated temperatures. Figure 15C.29 shows the voltage "step" near the end of discharge that can be induced by long-term overcharging. The capacity is still available, but at a lower voltage than when it was freshly cycled. Again, this is a reversible failure; a few charge and discharge cycles will restore normal voltage and expected capacity.

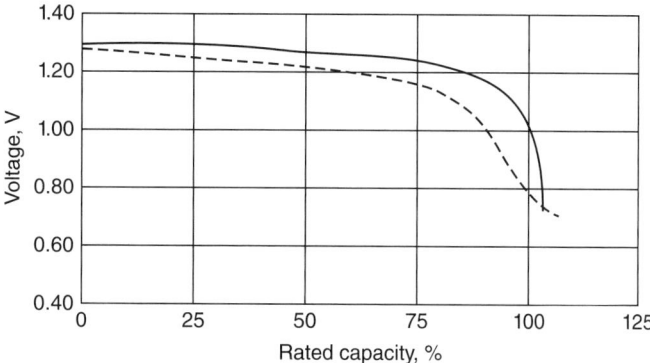

**FIGURE 15C.29** Typical discharge voltage profile of sealed nickel-cadmium batteries after long-term overcharge (dotted line) versus 16 h charge, both at C/10 rate.

IRREVERSIBLE FAILURES. Permanent failure in nickel-cadmium batteries results from essentially two causes: short-circuiting and loss of electrolyte. An internal short circuit may be of relatively high resistance and will be evidenced by an abnormally low on-charge voltage and by a drop of voltage as the battery's energy is dissipated through the internal short circuit. A short circuit may also be of such a low resistance that virtually all the charge current travels through it or the battery electrodes are totally shorted internally.

Any loss of electrolyte will cause degradation in capacity. Charging at high rates, repeated voltage reversal, and direct short-circuiting are ways that can cause loss of electrolyte through the pressure relief device. Electrolyte can also be lost over a long period of time through the battery seals, and capacity is lost in proportion to the reduction in electrolyte volume. Capacity degradation caused by electrolyte losses is more pronounced at high discharge rates.

High temperature degrades battery performance and life. A nickel-cadmium battery gives optimum performance and life at temperatures between 18 and 30°C. Higher temperatures reduce life by promoting separator deterioration and increasing the probability of short-circuiting. Higher temperatures also cause more rapid evaporation of moisture through the seals. These effects are all long term, but the higher the temperature, the more rapid the deterioration. Table 15C.2 lists the recommended temperature limits for sealed nickel-cadmium batteries.

**TABLE 15C.2**  Operating and Storage Limits for Sealed Nickel-Cadmium Batteries

|  | Temperature, °C | |
| --- | --- | --- |
| Type | Storage | Operating |
| Button | −40 to 50 | −20 to 50 |
| Standard cylindrical | −40 to 50 | −40 to 70 |
| Premium cylindrical | −40 to 70 | −40 to 70 |

### 15C.4.3  General Charging Characteristics

Sealed nickel-cadmium batteries are usually charged by means of the constant-current method. The $0.1C$ rate can be used and the battery is charged for 12 to 16 h (140%). At this rate, the battery can withstand overcharging without harm, although most sealed nickel-cadmium batteries can be safely charged at the $C/100$ to $C/3$ rate. At higher charge rates, care must be taken not to overcharge the battery excessively or develop high battery temperatures and pressures.

The voltage profile of a sealed nickel-cadmium battery during charge at the $C/10$ and $C/3$ rates is shown in Fig. 15C.30. A sharp rise in voltage to a peak near the end of the charge is evident.

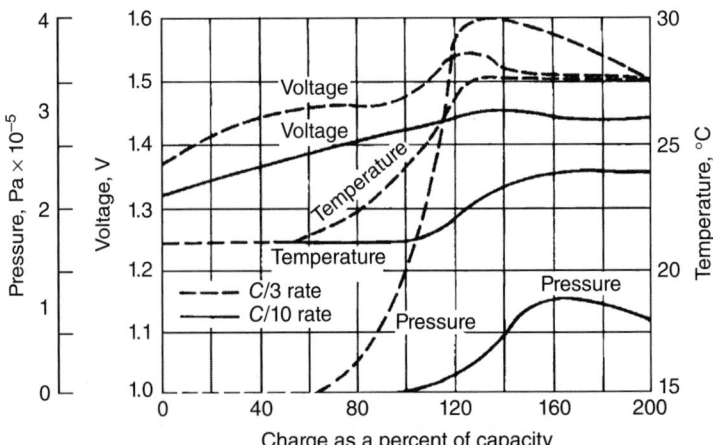

**FIGURE 15C.30**  Typical pressure, temperature, and voltage relationships of sealed nickel-cadmium battery during charge at constant current.

The voltage profile of sealed nickel-cadmium batteries is different from the one for a vented one, as illustrated in Fig. 15C.31. The end-of-charge voltage for the sealed battery is lower. The negative plate does not reach a state-of-charge as high as it does in the vented construction because of the oxygen recombination reaction.

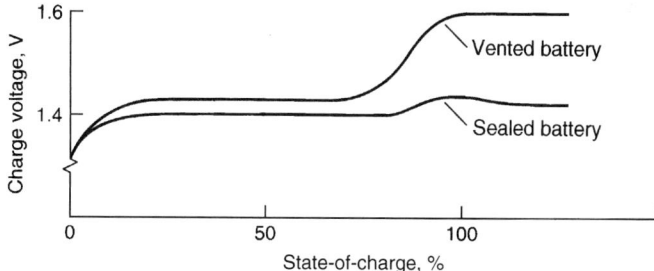

**FIGURE 15C.31**   Charge versus voltage for sealed and vented nickel-cadmium batteries at 25°C; 0.1C charge rate.

Constant-potential charging is not recommended for sealed nickel-cadmium batteries as it can lead to thermal runaway. It can, however, be used if precautions are taken to limit the current toward the end of charge.

***The Charge Process.***   The charge process is summarized in Fig. 15C.32. Figure 15C.32a plots the charge efficiency against total input energy, where

$$\text{charge efficiency} = \frac{\text{discharge energy (on subsequent discharge)}}{\text{charge input energy}}$$

At the start of the discharge (area 1), the charge energy is consumed by the conversion of the active materials into a chargeable form, and the charge efficiency is low. In area 2, charging is most efficient and almost all of the input energy is used to convert the discharged active material into the charged state. As the battery approaches the full charge state (area 3), most of the energy goes to the generation of oxygen, and the charge efficiency is low.

Figure 15C.32b presents the relationship of charge efficiency to charge rate. It shows that the charge efficiency, as well as the output capacity, is lower at a lower charge rate.

The charge efficiency also depends on the ambient temperature during charge. This relationship is shown in Fig. 15C.32c. There is a decrease in capacity in the high-temperature range due to a fall in potential for oxygen gas generation at the positive electrode.

The principle of the sealed battery is based on the ability of the negative electrode to recombine this oxygen gas and prevent the buildup of internal gas pressure. The capacity for this recombination is limited. Hence the maximum charge rate that can be tolerated is the one that keeps the rate of oxygen generation below the gas recombination rate so that the internal gas pressure does not build up excessively.

Overcharging at rates beyond the ability of oxygen recombination or heat dissipation can result in failure. "Fast" charging methods can be used successfully, but a means must be provided for monitoring and terminating the charge before excessive overcharging occurs. Temperature rise, voltage, or pressure can be monitored and used effectively as a cutoff.

***Voltage, Temperature, and Pressure Relationships.***   Figure 15C.30 also shows the relationship of voltage, temperature, and pressure of a typical sealed battery during charging at the C/10 and C/3 rates. The voltage increases gradually during the charge until the battery is about 75% to 80% charged. The voltage then rises more sharply due to the generation of oxygen at the positive electrode. The temperature remains relatively constant during the early part of the charge, as the charge reaction is endothermic. It then rises as the battery reaches the overcharge state due to the heat generated by the oxygen recombination reaction. Similarly, the internal pressure remains low until the battery goes into the overcharge condition, when most of the current is used to produce oxygen and the pressure rises. Finally, as the battery reaches full charge, the voltage drops because of a decrease in the battery's internal resistance due to the increase in the battery temperature. This drop in voltage can be used effectively in a control circuit to terminate the charge.

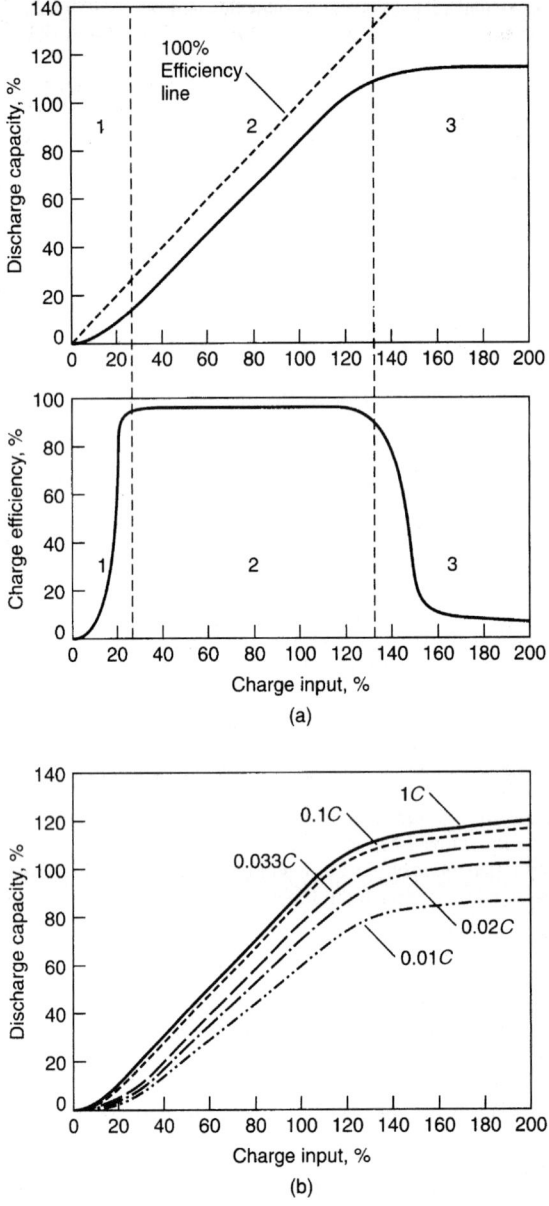

**FIGURE 15C.32** Charge process of sealed nickel-cadmium batteries. (*a*) Charge efficiency at 20°C. Charge: $0.1C \times 16$ h; discharge: 0.2C; end voltage: 1V. (*b*) Charge efficiency versus charge rate at 20°C. (*c*) Charge efficiency versus ambient temperature during charge at 0.1C rate.

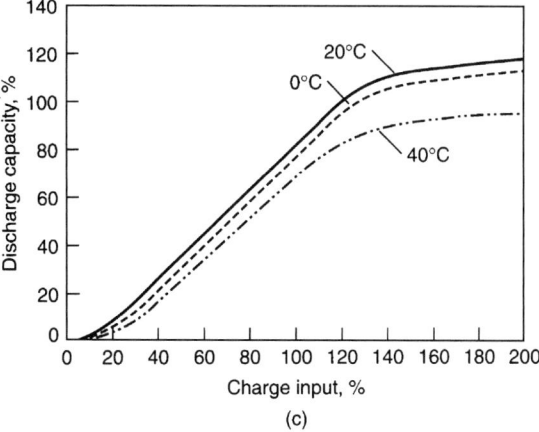

**FIGURE 15C.32**    (*Continued*)

As shown, when the battery is overcharged at acceptable rates, the pressure and temperature tend to stabilize. These steady-state conditions are governed by such factors as ambient temperature, overcharge rate, heat transfer characteristics of the cell and battery, cell design and components such as the separator, recombination capability of the negative electrode, and the resistance of the cell and battery. Charging at higher rates, such as the $C/3$ rate compared to the $C/10$ rate, results in higher temperatures and internal pressures. At higher charge rates, temperature and pressure will rise more significantly, particularly if the oxygen recombination rate is exceeded. Because of the possibility of venting and other deleterious effects on battery performance due to these high temperatures and pressures, it is necessary to terminate the charge before these conditions are reached.

*Voltage Characteristics during Charge.*    The voltage profile of a sealed nickel-cadmium battery during charge at various charge rates at 20°C is shown in Fig. 15C.33. The charge voltage also depends on temperature, as shown in Fig. 15C.34. The voltage and voltage peak decrease with a rise in temperature. Charging at temperatures between 0 and 30°C is best for sealed batteries. At lower temperatures, the voltage increases, recombination of oxygen is slower, and the internal gas pressure tends to increase. Charging rates must be reduced. Above 40°C, the charging efficiency is low, and higher temperatures cause battery deterioration.

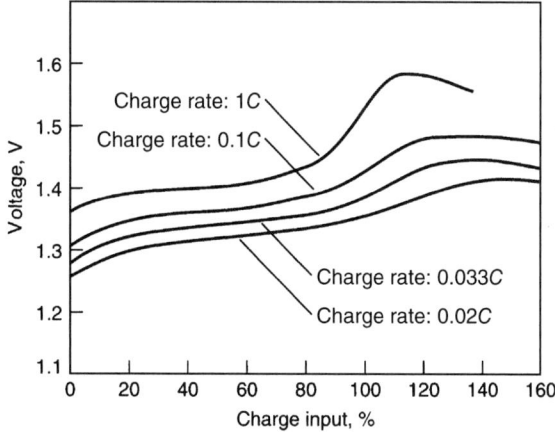

**FIGURE 15C.33**    Voltage profile during charge at various charge rates at 20°C.

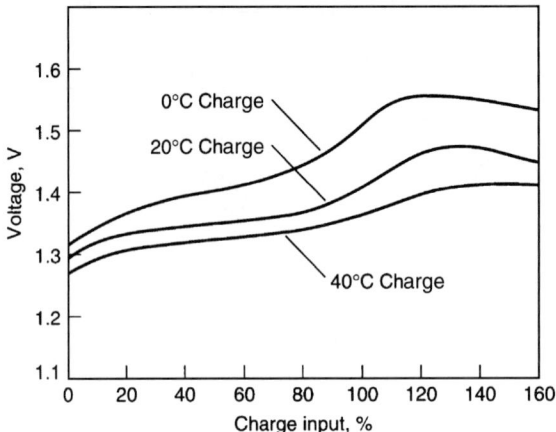

**FIGURE 15C.34**   Voltage profile during charge at $0.1C$ rate at various temperatures.

*Charge Methods.*   There are a number of different methods for charging sealed nickel-cadmium batteries. The standard method is a quasi-constant current charge at a relatively low rate. "Fast" charge methods are becoming more popular in order to reduce the time required for charging. A control circuit is needed when charging at these high rates to cut off the charge or reduce the charge current at the completion of charge. Figure 15C.35 shows the voltage and current profiles during charge for each of the charge control methods.

*Standard Method.*   This is the simplest method, which uses a relatively inexpensive circuit, controlling the charge current by inserting a resistance between the DC power supply and the battery (Fig. 15C.35a). The battery is charged at a constant current at a low ($C/10$) rate so that the generation of oxygen is below the recombination rate. This rate also limits the temperature rise. Excessive overcharge should be avoided. The battery should be charged to about 140% to 150% charge input.

*Timer Control.*   For moderate charge rates, a timer can be used to cut off the charge or reduce the charge current to the trickle charge level (Fig. 15C.35b). This is a relatively inexpensive control device and suitable for applications where the battery is usually fully discharged before charging. It is not suitable for applications where the battery is frequently charged without prior deep discharging as this could result in excessive overcharge. A thermal cutoff control should be used when charging at rates higher than the $C/5$ rate or without deep discharging to prevent the battery from reaching high temperatures.

*Temperature Detection.*   This control system uses a sensor to detect the temperature rise of the battery and terminate the charge (Fig. 15C.35c). A thermostat or thermistor is used as the detection device, and the detecting temperature is usually set at 45°C. It is important that the sensor be located so that it can accurately determine the battery's temperature. Charging in high ambient temperatures can result in an insufficient charge, while low ambient temperatures may result in overcharge. The cycle life with this method may be shorter than with the $-\Delta V$ method or peak voltage methods as the battery could be subjected to more overcharge.

*Negative Delta V ($-\Delta V$).*   This is one of the preferred charge control systems for sealed nickel-cadmium batteries (Fig. 15C.35d). The drop in voltage of the battery is detected after the battery voltage has reached its peak during charge. The signal can be used to terminate the charge or reduce the charge current to a trickle charge. The method provides a complete charge regardless of ambient temperature or residual capacity from the previous charge. A value of 10 to 20 mV per cell is usually used for the control. The method is not suitable for charging below the $0.5C$ as the $-\Delta V$ value is too low to be detectable.

*Trickle and Float Charge.*   Trickle-charge systems are used in two different situations: (1) in a standby power application where the battery is on continual charge to maintain it in a state of full charge (compensating for self-discharge) until it is connected to the load when the prime power fails, and (2) as a supplementary charge after the termination of rapid charging (Fig. 15C.35e). Charging is at the 0.02 to $0.05C$ rate, depending on the frequency and depth of discharge. A periodic discharge every 6 months followed by a charge is advisable to ensure optimum performance.

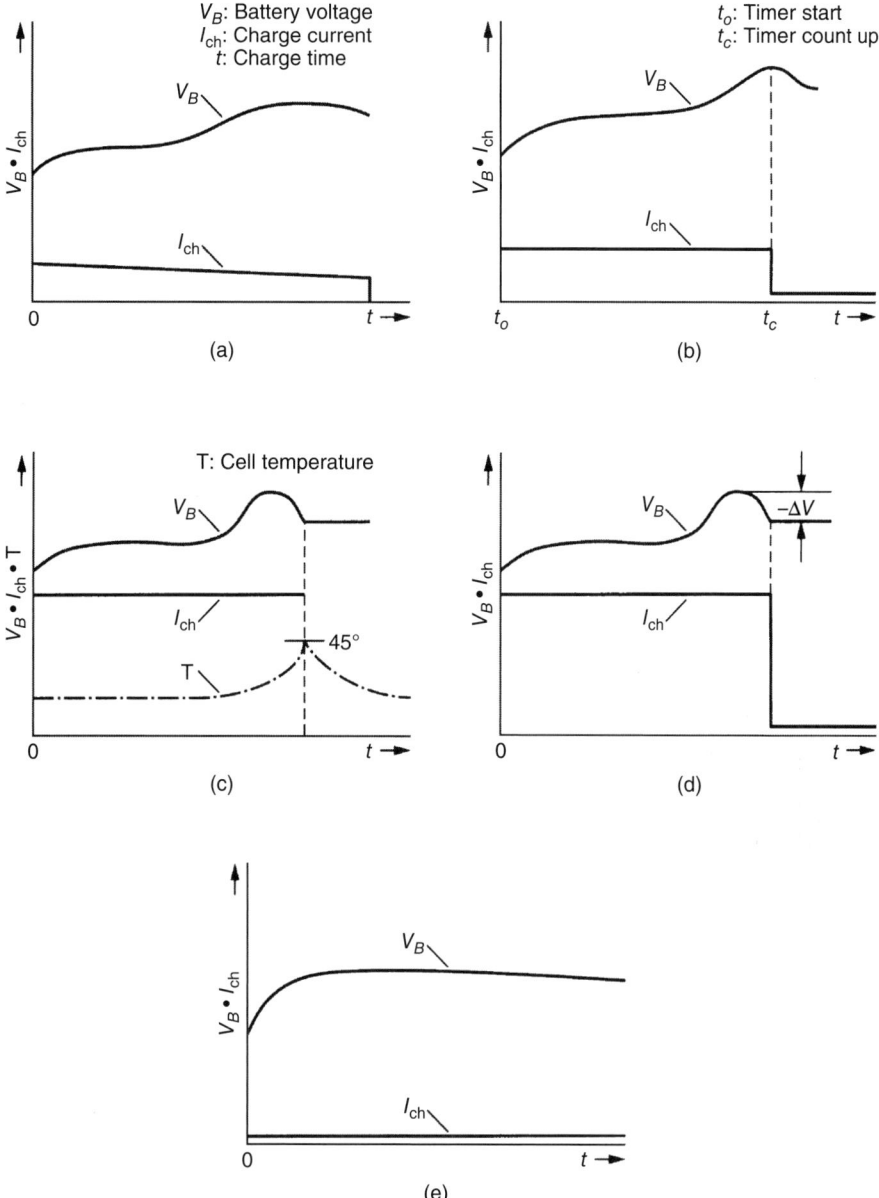

**FIGURE 15C.35** Methods for charging sealed nickel-cadmium batteries and charge control. (*a*) Semiconstant current ($V_B$: battery voltage; $I_{ch}$: charge current; *t*: time). (*b*) Timer control ($t_0$: timer start; $t_c$: timer count up). (*c*) Temperature detection (*T*: battery temperature). (*d*) $-\Delta V$ detection. (*e*) Trickle charge. (*Courtesy of Panasonic Industrial Co., Division of Matsushita Electric Corp. of America.*)

## 15C.4.4 Special-Purpose Batteries

Special-purpose sealed nickel-cadmium batteries are manufactured with specifically designed characteristics, overcoming some of the limitations of standard batteries to meet the requirements for certain applications. Manufacturers' recommendations should be followed because of the specific performance characteristics of these batteries.

***High-Capacity Batteries.*** These batteries incorporate design features, such as nickel foam substrate positive plates, pasted negative electrodes, thin-walled battery containers, and increased amounts of active material. These changes result in a 20% to 40% increase in capacity. These batteries are also designed with improved oxygen recombination capability and are capable of being charged at the 0.2$C$ rate or less without control. They are capable of fast 1 h charging using $-\Delta V$ charge control. Figure 15C.36 compares the discharge characteristics of the high-capacity battery with those of a standard battery. Figure 15C.37 shows the relationship of battery capacity and discharge current for the two designs.

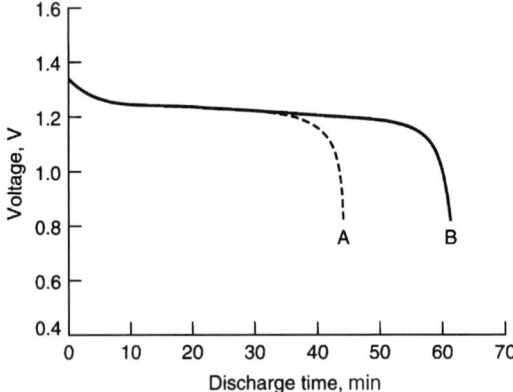

**FIGURE 15C.36** Comparison of discharge characteristics of sub-$C$ size standard battery (A) versus high-capacity battery (B) on discharge at 20°C. Discharge at $C$ rate, charge at 0.1$C$ rate for 16 h.

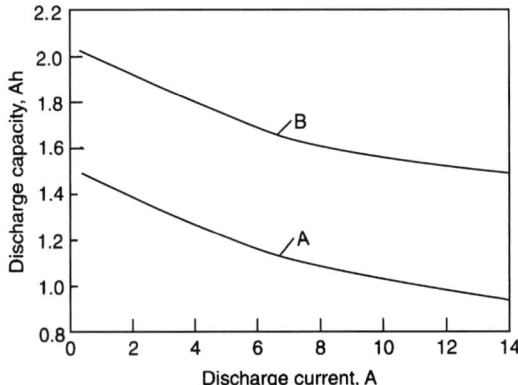

**FIGURE 15C.37** Comparison of performance of standard battery (A) versus high-capacity battery (B) (sub-$C$ size) at 20°C.

***Fast-Charge Batteries.*** These batteries have electrode structures and electrolyte distribution designs to enhance oxygen recombination. They can be charged at the fast 1 h rate with charge control (such as temperature sensing and $-\Delta V$ techniques) and at the $C/3$ rate without charge control because of their ability to withstand this level of overcharge. They are also capable of performance at high discharge rates, though this is achieved at the expense of a slightly reduced battery capacity. These batteries have improved internal heat conductivity, which results in a faster increase in surface temperature. This feature can be used advantageously in a temperature-sensing fast-charge system. Figure 15C.38 shows the charge characteristics of a fast-charge battery compared to a standard one. The internal gas pressure of the standard battery increases quickly during charging, whereas that of a fast-charge battery stabilizes.

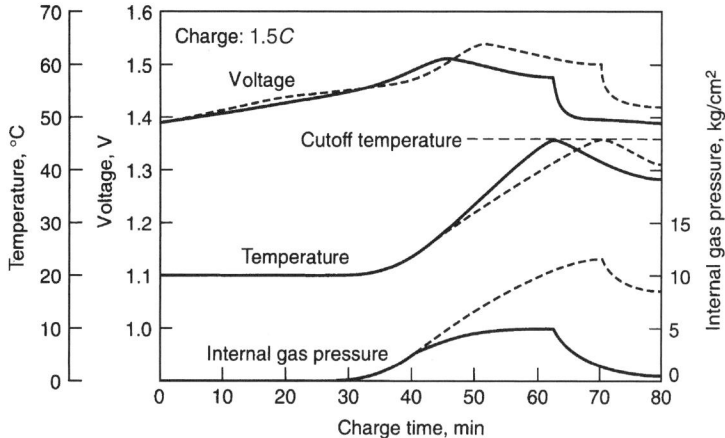

**FIGURE 15C.38**  Comparison of charge characteristics of fast-charge battery (solid line) versus standard battery (broken line).

***High-Temperature Batteries.***   These batteries are designed to operate at high temperatures without the service life deterioration and charging inefficiencies experienced with conventional designs. Figure 15C.39 compares the performance of the high-temperature battery with the standard battery as a function of ambient temperature during charge. This type of battery is capable of charge-discharge cycling at temperatures as high as 35 to 45°C and is particularly designed for trickle charging ($C/20$ to $C/50$ rate) at these high temperatures. The charge voltage of these batteries is slightly higher than that of the standard battery due to the designed-in control of the oxygen-generating potential.

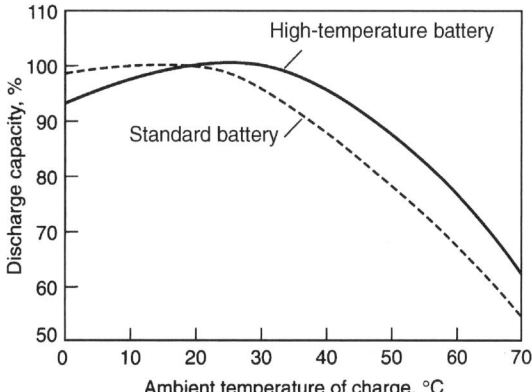

**FIGURE 15C.39**  Comparison of performance of high-temperature battery versus standard battery. Charge—$C/30$ rate; discharge—$1C$ rate at 20°C.

***Heat-Resistant Batteries.*** These batteries are designed for fast charging at high temperatures. For example, charging at the 0.3C rate is possible even at temperatures as high as 45 to 70°C. Their performance characteristics are similar to those of the standard battery. However, they have a superior service life when used at high temperatures because of the use of specially selected materials with minimum deterioration at high temperatures. Figure 15C.40 compares the service life for standard and heat-resistant batteries throughout the temperature range.

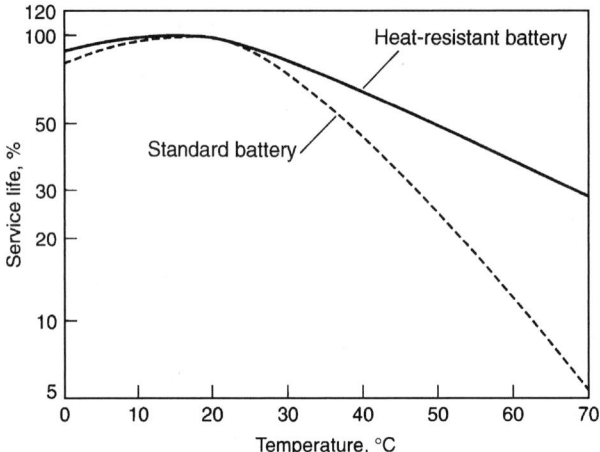

**FIGURE 15C.40** Comparison of performance of heat-resistant battery versus standard battery.

***Memory-Backup Batteries.*** These batteries are used to provide battery backup for volatile semiconductor memory devices. The key requirements for this type of battery are long life (up to 10 years in certain applications), low self-discharge, and good performance at low discharge rates. Figure 15C.41 shows the storage characteristics of the memory-backup battery. The low-rate discharge characteristics of the battery are plotted in Fig. 15C.42. As the backup battery is designed for low-rate use, its internal resistance is higher than that of the standard battery and its high-rate discharge characteristics are not as good.

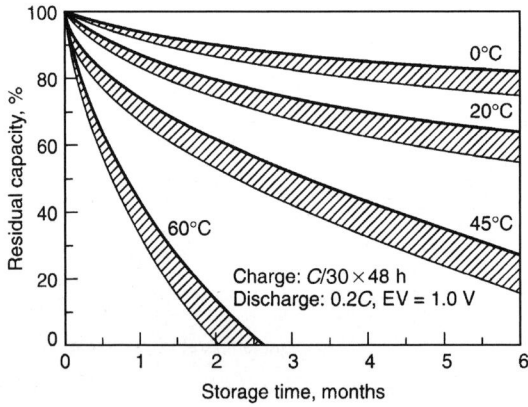

**FIGURE 15C.41** Storage characteristics of memory-backup batteries.

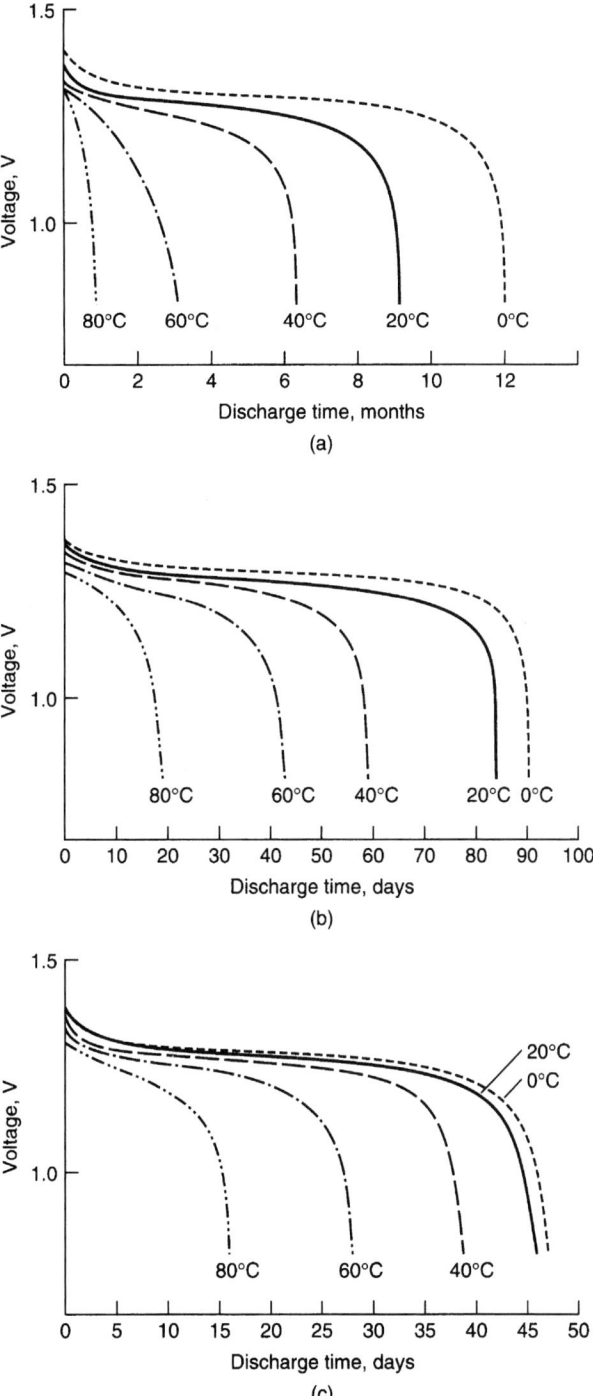

**FIGURE 15C.42**   Performance of memory-backup batteries. Charge—$C/30$ for 48 h at 20°C. Discharge rate: (a) $C/10,000$, (b) $C/2000$, (c) $C/1000$.

***Rectangular Batteries.*** The constructional features of the flat rectangular battery are described in Sec. 15C.3.1. The advantage of the rectangular battery is that it permits more efficient battery design, eliminating the voids that occur with the assembly of cylindrical batteries. The volumetric energy density of these batteries can be about 20% higher than a battery using a cylindrical design.

Most of the performance characteristics are similar to those of the standard cylindrical battery, except that it also incorporates some of the features of the high-capacity battery. Gas recombination has been improved to permit charging at the $0.2C$ rate or less and 1 h charging with charge control, preferably with $-\Delta V$ sensing. This is illustrated in Fig. 15C.43. The voltage profile on discharge is flat, as with the cylindrical battery, as shown in Fig. 15C.44. However, because the resistance of the rectangular battery is higher, performance at rates greater than $4C$ is not as good as with the cylindrical battery. Storage characteristics and cycle life are similar to those of the cylindrical battery.

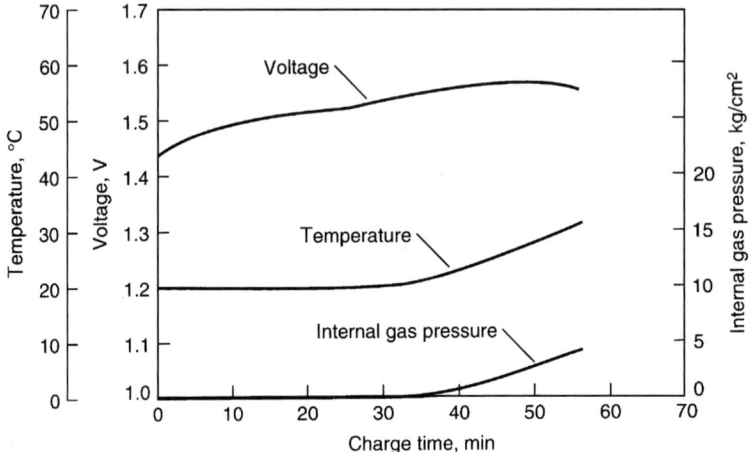

**FIGURE 15C.43** Charge characteristics of rectangular batteries at 20°C. Charge—$1.5C$ rate; $DV = 10$ mV.

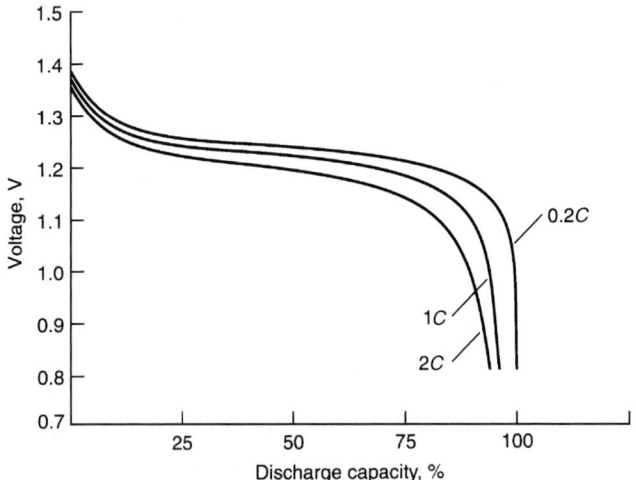

**FIGURE 15C.44** Discharge characteristics of rectangular batteries at 20°C. Charge—$0.1C$ rate for 16 h.

## 15C.4.5    Battery Types and Sizes

Table 15C.3 lists some of the types of sealed nickel-cadmium single-cell batteries that are manufactured and some of their physical and electrical specifications. Multicell batteries are also manufactured, using these cells, in a variety of output voltages and configurations.

**TABLE 15C.3**    Specifications of Typical Sealed Nickel-Cadmium Single-Cell Batteries

| Battery size | Typical capacity at 0.2C rate (mAh) | Dimensions, max., mm | | |
|---|---|---|---|---|
| | | Diameter | Height | Weight, g |
| | | Cylindrical batteries | | |
| **Standard batteries:** Charging: 0–45°C. Discharge: −20 to +60°C | | | | |
| F | 7000 | 33.2 | 91.0 | 224 |
| M | 12,000 | 43.1 | 91.0 | 395 |
| **Extended service life batteries**: Charging—standard: 0–45°C; quick charge: 10–45°C. Discharge: −20 to +60°C | | | | |
| AA | 600 | 14.3 | 50.3 | 22 |
| AA | 700 | 14.3 | 50.3 | 23 |
| AA | 600 | 14.3 | 48.9 | 22 |
| AA | 700 | 14.3 | 48.9 | 23 |
| SC | 1200 | 22.9 | 43.0 | 52 |
| **Fast charge batteries:** Charging—standard: 0–45°C; quick charge: 10–45°C; fast 1 h: 5–45°C. Discharge: −20 to +60°C | | | | |
| 4/5 SC | 1200 | 22.9 | 34.0 | 43 |
| SC | 1300 | 22.9 | 43.0 | 51 |
| SC | 1700 | 22.9 | 43.0 | 55 |
| C | 3000 | 26.0 | 50.0 | 86 |
| **High-temperature batteries:** Charging—standard: 0–70°C. Discharge: −20 to +70°C | | | | |
| AA | 600 | 14.3 | 48.9 | 23 |
| SC | 1600 | 22.9 | 43.0 | 49 |
| C | 2900 | 26.0 | 50.5 | 78 |
| F | 7000 | 33.2 | 91.0 | 224 |
| M | 10,000 | 43.1 | 91.0 | 395 |
| **Heart-resistant batteries:** Charging—standard: 0–70°C; quick: 10–70°C. Discharge: −20 to +70°C | | | | |
| AA | 600 | 14.3 | 50.2 | 22 |
| SC | 1200 | 22.9 | 43.0 | 52 |

Figure 15C.45 is a guide to determining the approximate battery size required for a given performance requirement or application. These data are based on the performance of a standard battery at 20 to 25°C. Allowance must be factored into the estimate to determine the performance under other discharge conditions.

Manufacturers' data should be consulted for specific details on dimensions, ratings, and performance characteristics as they may be different from those shown.

## 15C.4.6    Battery Sizes and Availability

Currently, nickel-cadmium cells and batteries are available from suppliers worldwide. However, manufacturing of sealed cells is predominantly conducted in Asia. In selecting a cell or battery for an application, the user may refer to the power data chart shown in Fig. 15C.45 to determine ampere-hour capacity requirement.

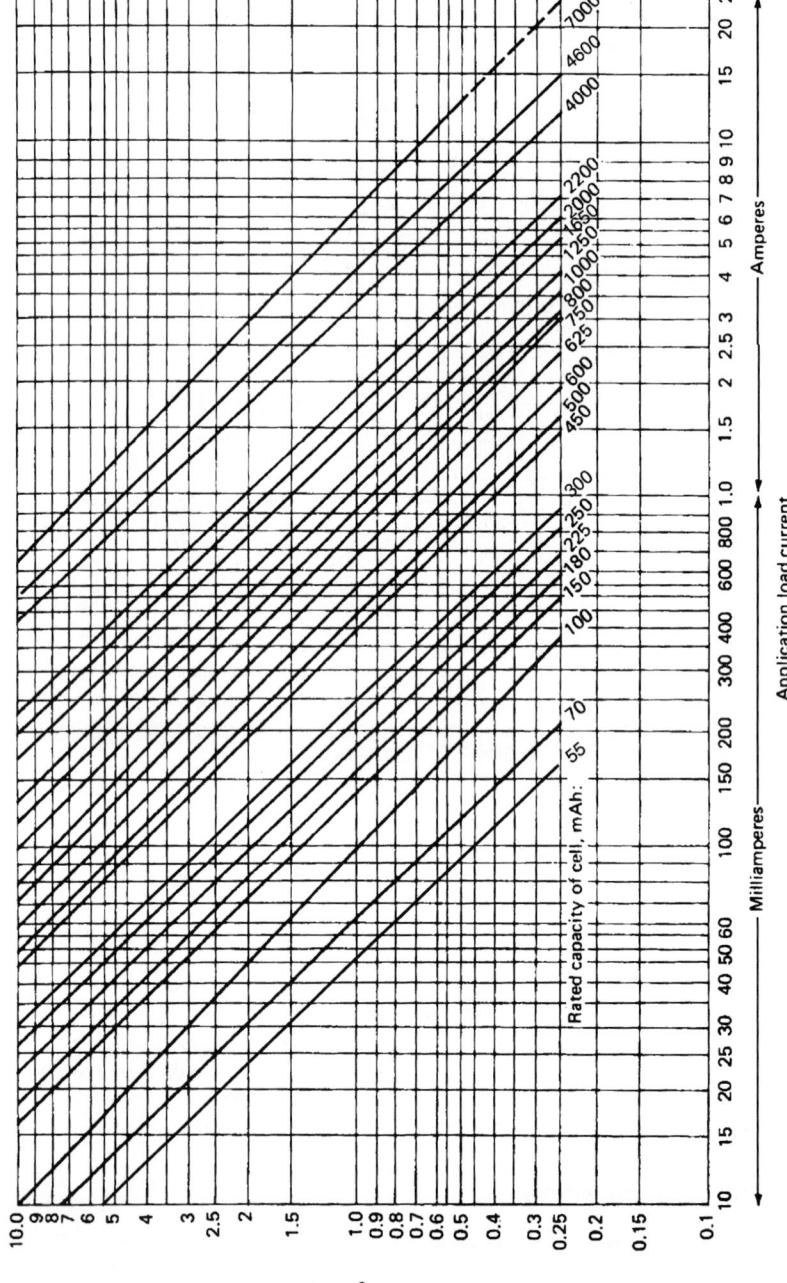

**FIGURE 15C.45** Selector guide for sealed nickel-cadmium cylindrical batteries. Guide can be used to determine approximate required battery size, given the load and desired run (service) time. Data based on fully charged battery and 20°C operating temperature.

The manufacturer's data sheet should also be consulted to determine suitability to meet requirements. Ultimately, a cell manufacturer or capable value-added battery assembler is necessary to facilitate finalizing production designs. Figure 15C.46 shows a 28 V, 47 Ah aircraft battery system, using the Acme FNC cell Model XX47, with a dedicated charger.

**FIGURE 15C.46**    FNC airborne battery system. Battery (*left*); charger (*right*). 28 V, 47 Ah battery. (*Courtesy of Acme Electric, USA.*)

## ACKNOWLEDGMENT

Major content for this chapter was provided in whole or in part by John K. Erbacher, Chap. 19, Industrial and Aerospace Nickel-Cadmium Batteries, *Linden's Handbook of Batteries*, 4th ed., T. B. Reddy and D. Linden, eds., McGraw-Hill, 2011.

## REFERENCE

1. Y. Sato, K. Ito, T. Arakawa, and K. Kobaya Kawa, "Possible Causes of the Memory Effect Observed in Nickel-Cadmium Secondary Batteries," *J. Electrochem. Soc.* **143**:L225, 1996.

## BIBLIOGRAPHY

### General

M. Barak (ed.), *Electrochemical Power Sources*, Peter Peregrinus, London, 1980.

P. Brunamonti, *Life Cycling at Elevated Temperatures Battery Types M81757/8-5 and M81757/15: Marathon Power Tech. Co., EDD 99–127*, Nov. 30, 1999, Crane Div., Naval Surface Warfare Center, Crane, IN 47522-5001.

S. U. Falk and A. J. Salkind, *Alkaline Storage Batteries*, Wiley, New York, 1969.

H.-D. Jacksch, *Batterie Lexikon*, Pflaum Verlag, Munich, pp. 348–394, 1993.

R. Kinzelbach, *Stahlakkumulatoren*, Varta, Hannover, Germany, 1968.

Y. Miyake and A. Kozawa, *Rechargeable Batteries in Japan*, JEC Press, Cleveland, OH, 1977.

B. Newman, *Life Cycling at Elevated Temperatures Battery Types M81757/15: SAFT America, Inc., EDD 99–122*, Nov. 17, 1999, Crane Div., Naval Surface Warfare Center, Crane, IN 47522-5001.

## Plastic-Bonded Electrode Technology

B. McRae and D. Nary, *Proceedings of the 38th Power Sources Conference*, pp. 123–126, 1998.

## FNC Technology

M. Anderman, C. Baker, and F. Cohen, *Proceedings of the 32nd Intersociety Energy Conversion Conference*, Vol. 1, p. 97465, 1997.

C. Baker, *Proceedings of the SAE Power Systems Conference*, Williamsburg, VA, 1997. See *Advanced Battery Technology*, April 1997.

C. Baker and M. Barekatien, *Proceedings of the SAE Power Systems Conference*, San Diego, CA, 2000.

FNC Vented Nickel-Cadmium Batteries, Hoppecke Batterien.

# SECTION D

# NICKEL-METAL HYDRIDE

**Michael Fetcenko, John Koch, Michael Zelinsky, Kwo Young**

## 15D.1    INTRODUCTION

### 15D.1.1    Overview

Nickel-metal hydride (NiMH) batteries have been a commercially important rechargeable battery system since their commercial introduction in 1989. The key driving forces for the growth of NiMH have been the rapid growth of hybrid electric vehicles (HEVs) and the development of NiMH batteries as direct replacements for alkaline primary batteries.

Rechargeable NiMH continues to have a strong presence in commercial HEVs, providing overall excellent performance, reliability, and cost. NiMH enables HEV applications through a combination of desirable performance attributes such as high energy and power, an excellent range of operating temperatures, and low cost. Further, NiMH has demonstrated excellent safety, abuse resistance, and cycle life that have translated into superior field reliability for advanced vehicular applications and overall performance in consumer applications.

While many consumer appliances that were once dominated by NiMH and NiCd, such as laptop computers and cell phones, have transitioned to lighter weight Li-ion batteries, continued improvement of NiMH technology has fostered expansion into new markets which, until recently, were the exclusive domain of alkaline primary cells. The introduction of customer-replaceable "pre-charged" NiMH rechargeable batteries with equal or better performance than alkaline, at little to no price premium, has achieved wide market acceptance. The key to this NiMH technology advancement is charge retention, which has significantly improved to a remarkable 85% for 1 year. Consumers have also benefited from improvements in NiMH cycle life and over-discharge resistance.

Recent advancements in temperature range, high temperature charge acceptance, and deep discharge cycle life are creating opportunities for NiMH batteries in stationary and industrial applications where incumbent batteries chemistries cannot keep up with evolving customer demands.

### 15D.1.2    General Characteristics

Table 15D.1 summarizes the key advantages and disadvantages of the sealed NiMH battery. In addition to the essential characteristics of low cost, reliability, cycle life, and operating temperature, the following features of NiMH[1,2] have helped establish the technology's preeminence:

- Flexible cell sizes from 0.06 to 250 Ah
- Safe operation at high voltage 750+ V
- Excellent volumetric energy and power, flexible packaging
- Straightforward application to series and series/parallel strings
- Choice of cylindrical or prismatic cells
- Safety in charge and discharge, including tolerance to abusive overcharge and over-discharge
- Maintenance free
- Excellent thermal properties ($-30$ to $70°C$)
- Capability to utilize high energy pulses

**TABLE 15D.1**    Major Advantages and Disadvantages of Sealed Nickel-Metal Hydride Batteries

| Advantages | Disadvantages |
|---|---|
| Higher energy density compared to lead acid and nickel cadmium | Higher cost than lead acid |
| High volumetric energy density comparable to Li-ion | Lower gravimetric energy density and specific power compared to Li-ion |
| Good elevated temperature and high rate capability | Decreased performance at low temperature |
| Good charge retention | Limited commercial availability of large-format batteries |
| Long cycle life on deep discharge | Limited supplier choice for automotive applications |
| Partial state of charge operation | |
| Rapid recharge capability | |
| Long shelf life | |
| Sealed maintenance-free design | |
| Safe and recyclable | |

- Simple and inexpensive charging and electronic control circuits
- Environmentally acceptable and recyclable materials.

## 15D.2   BATTERY CHEMISTRY

During discharge, nickel oxyhydroxide is reduced to nickel hydroxide,

$$\text{NiOOH} + \text{H}_2\text{O} + \text{e}^- \rightarrow \text{Ni(OH)}_2 + \text{OH}^- \qquad E'' = 0.52 \text{ V}$$

and the metal hydride (MH) is oxidized to the metal alloy (M) as follows:

$$\text{MH} + \text{OH}^- \rightarrow \text{M} + \text{H}_2\text{O} + \text{e}^- \qquad E^0 = -0.83 \text{ V}$$

The overall reaction on discharge is

$$\text{MH} + \text{NiOOH} \rightarrow \text{M} + \text{Ni(OH)}_2 \qquad E^0 = 1.35 \text{ V}$$

The process is reversed during charge.

The sealed NiMH cell uses an oxygen-recombination mechanism to prevent the buildup of pressure that may result from the generation of gases toward the end of the charge and in overcharge. This is shown schematically in Fig. 15D.1. During charge, the positive electrode reaches full charge before the negative and begins to evolve oxygen.

$$2\text{OH}^- \rightarrow \text{H}_2\text{O} + \tfrac{1}{2}\text{O}_2 + 2\text{e}^-$$

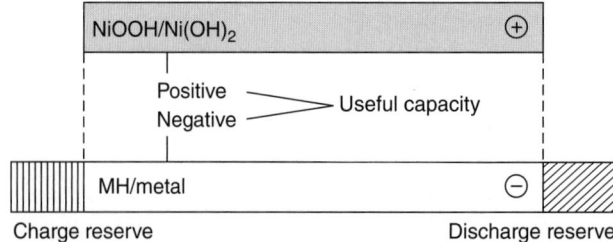

**FIGURE 15D.1**    Schematic representation of electrodes of sealed nickel-metal hydride cell divided into useful capacity, charge reserve, and discharge reserve.

The oxygen gas diffuses through the separator to the negative electrode facilitated by the starved electrolyte design and the use of an appropriate separator.

At the negative electrode, the oxygen reacts with the hydrogen electrode to produce water. This results in stabilizing the internal pressure of the battery.

$$4MH + O_2 \rightarrow 4M + 2H_2O$$

Furthermore, the negative electrode does not become fully charged, which prevents the generation of hydrogen. This is true for the initial stages of cycling where the only gas found inside the cell is oxygen. However, continued cycling of the cell results in hydrogen gas evolution and a proportional increase in hydrogen is observed. The reason for this behavior is the result of oxygen recombination on the surface of the metal hydride (MH) negative electrode, an exothermic reaction that causes localized heating of the MH alloy. This heating in turn affects the equilibrium pressure of the alloy resulting in the release of hydrogen and increased pressure during charging. The charge current must then be controlled at the end of charge and during overcharge to limit the generation of oxygen to below the rate of recombination to prevent the buildup of gases and pressure.

A term in the design of NiMH batteries is the negative to positive ratio (N/P). It is based on the use of a negative electrode (the MH electrode) that has a higher effective capacity than the positive (or nickel oxyhydroxide electrode). Overall, as shown in Fig. 15D.1, having an excess of MH capacity allows for gas recombination reactions to occur during overcharge (oxygen recombination) and over-discharge (hydrogen recombination, see "Polarity Reversal during Over-Discharge" in Sec. 15D.4). In addition, excess MH electrode capacity is provided to inhibit oxidation and corrosion of the MH alloy. N/P ratios vary by cell design and by manufacturer but are typically within the range of 1.3 to 2.0. The lower N/P values are used to maximize energy while higher values are used in power and cycle life designs. The useful capacity of the battery is thus determined by the positive electrode.

## 15D.3    CELL COMPONENTS AND CONSTRUCTIONS

### 15D.3.1    Metal Hydride Alloy

NiMH batteries are an unusual battery technology in that the MH active material is an engineered alloy made up of many different elements with the MH alloy formulas, typically based on binary systems with the two parts labeled A and B, varying to a significant degree.[3,4] The negative electrode contains either an $AB_5$ (LaCePrNdNiCoMnAl), an $A_2B_7$ (LaPrNdMgNiAlZr), or an $AB_2$ (VTiZrNiCrCoMnAlSn) disordered type MH active material.[5-8] $AB_5$ type alloys are more common, despite significantly lower hydrogen storage capacity as compared to $A_2B_7$ (320 vs. 380 mAh/g) and $AB_2$ (440 mAh/g). The advantages of the $AB_5$ alloys include low raw-material cost, easy activation and formation, flexibility in electrode processing, and high discharge-rate capability. On the other hand, there has been significant ongoing development to improve the properties of $A_2B_7$ and $AB_2$ materials to take advantage of their inherently higher energy, which is especially important for cost reduction.

Acceptable electrochemical utilization of MH materials as anodes in NiMH batteries requires that those materials meet a demanding list of performance attributes including hydrogen storage capacity, suitable metal-to-hydrogen bond strength, acceptable catalytic activity, discharge kinetics, and acceptable oxidation/corrosion resistance. Multielement, multiphase, disordered alloys of the $LaNi_5$, LaMgNi, and VTiZrNiCr types are attractive development candidates for atomic engineering due to a broad range of elemental addition and substitution possibilities, availability of alternative crystallographic phases that form the matrix for chemical modification, and a tolerance for nonstoichiometric formulas. Through the introduction of modifier elements, easy activation and formation has been achieved. Special processing steps (such as alloy melting and size reduction) suitable for these metallurgically challenging materials have been developed.

The MH active materials may be adjusted to obtain added capacity, power, and/or cycle life.

***AB_5 Metal Hydride Alloys.***    For the $AB_5$ system, typical formulas include

- $La_{5.7}Ce_{8.0}Pr_{0.8}Nd_{2.3}Ni_{59.2}Co_{12.2}Mn_{6.8}Al_{5.2}$ (atomic percent a/o)
- $La_{10.5}Ce_{4.3}Pr_{0.5}Nd_{1.3}Ni_{60.1}Co_{12.7}Mn_{5.9}Al_{4.7}$

While the capacity of the various $AB_5$ alloys is typically around 290 to 320 mAh/g, other overall performance attributes can be greatly influenced by changing the ratio of A to B elements in the alloy. It is common for the ratio of La/Ce to be reversed to emphasize cycle life and power. The total amount of Co, Mn, and Al significantly affects the ease of activation and formation, but increased cobalt has higher cost implications. After production of the $AB_5$ alloy ingot, it is common to further refine the microstructure of the material by a post anneal treatment at higher temperature for several hours. The annealing treatment can have a significant effect on capacity, discharge rate, and cycle life by adjusting crystallite size and grain boundaries, as well as by eliminating unwanted crystal structures that may be formed during ingot melting and casting. After annealing the ingot is first crushed and then ground to the desired final particle size range. Additional, special processing methods such as melt spinning and other rapid solidification techniques can also promote higher cycle life, although these processing methods are more expensive and there may be other trade-offs, such as discharge-rate capability.[9]

Commercial $AB_5$ alloys have a predominantly $CaCu_5$ crystal structure. However, within that structure, there are a range of lattice constants brought about by compositional disorder[10] within the material that are important to catalysis, storage capacity, and stability to the alkaline environment and embrittlement. These materials also precipitate metallic NiCo inclusions embedded in the surface oxide that is important to high-rate discharge.[11,12] Due to the high price of Pr and Nd, a version of $AB_5$ alloy free of these two elements has been developed and used widely in consumer batteries.[13,14]

***$A_2B_7$ Metal Hydride Alloys.***    $A_2B_7$ alloys also provide for compositional and processing choices. Popular $A_2B_7$ alloy formulas (a/o) include

- $La_{4.8}Ce_{0.4}Pr_{9.1}Nd_{5.4}Mg_{1.7}Ni_{68.8}Co_{3.0}Mn_{0.2}Al_{5.5}Zr_{0.2}$
- $Nd_{18.7}Mg_{2.5}Ni_{74.7}Co_{0.1}Al_{3.6}Zr_{0.2}$

$A_2B_7$ alloy capacity may range from 335 to 400 mAh/g. Ce is often excluded from the alloy in an effort to reduce the tendency to form $AB_5$ crystal structures. The addition of Co helps reduce lattice expansion, while Al forms a protective dense surface oxide and together both improve the cycle ability of the alloy. The addition of Mn has been shown to play a role in regulating the amount of the various crystal structures present, while the addition of Zr has been shown to improve the high-rate discharge capability of the alloys; however, Mn also deteriorates the self-discharge performance by forming microshorts in the separator.

Alloy preparation using conventional methods results in a combination of $PuNi_3$ and $CaCu_5$ crystal structures within the alloy, but Mg is present only in the $PuNi_3$ crystal structure. After production of the $A_2B_7$ alloy ingot, further refinement of the material microstructure is required using a post annealing treatment at an elevated temperature for several hours. The annealing treatment has been shown to be critical in reducing or eliminating the amount of $AB_5$ or other unwanted crystal structures that may be formed during ingot melting and casting. After annealing, the ingot is first crushed and then ground to the desired final particle size range. Additional, special processing methods such as melt spinning and other rapid solidification techniques may be required to eliminate unwanted $AB_5$ crystal structures depending on the alloy formulation.

***$AB_2$ Metal Hydride Alloys.***    $AB_2$ alloys also provide for compositional and processing choices. Popular $AB_2$ alloy formulas (a/o) include

- $V_{18}Ti_{15}Zr_{18}Ni_{29}Cr_5Co_7Mn_8$
- $V_5Ti_9Zr_{26.7}Ni_{38}Cr_5Mn_{16}Sn_{0.3}$
- $V_5Ti_9Zr_{26.2}Ni_{38}Cr_{3.5}Co_{1.5}Mn_{15.6}Al_{0.4}Sn_{0.8}$

$AB_2$ alloy capacity may range from 385 to 450 mAh/g. Higher vanadium-content alloys may suffer from higher rates of self-discharge due to the solubility of vanadium oxide and its consequent ability to form a special type of undesirable redox shuttle. The concentrations of Co, Mn, Al, and Sn are important for easy activation, formation, and long cycle life. The ratio of hexagonal C14 to cubic C15 crystal structure is important for improving capacity or power.

Alloy preparation is done using conventional melting and casting methods to produce an ingot. After casting, the ingot is processed using a hydride/de-hydride sequence to reduce the ingot into smaller pieces. These smaller pieces are then ground to the desired final particle size range. After grinding, oxygen is slowly introduced to allow for the growth of a protective oxide layer.

For all MH alloys, the metal/electrolyte surface oxide interface is a crucial factor in discharge rate capability and cycle life stability. Original $LaNi_5$ and $TiNi$[15] alloys extensively studied in the 1970s and 1980s for NiMH battery applications were never commercialized due to poor discharge rate and cycle life capability.[16,17] Lack of catalytic activity at the surface oxide limits discharge and lack of sufficient oxidation/corrosion resistance is a critical obstacle to long cycle life. The complicated chemical formulas and microstructures of present disordered $AB_5$, $A_2B_7$, and $AB_2$ alloys extend to the surface oxide. For the surface oxide, key factors include thickness, microporosity, and catalytic activity. Of importance to a high discharge rate was the discovery that ultrafine metallic nickel particles having a size on the order of 50 to 70 Å or less, and dispersed within the oxide are excellent for catalyzing the reaction of hydrogen and hydroxyl ions.[18]

The other critical design factor within the surface oxide is to achieve a balance between surface oxide passivation and corrosion. Porosity with the oxide is important to allow ionic access to the metallic catalysts and therefore promote high-rate discharge. While passivation of the surface oxide is problematic for high-rate discharge and cycle life, unrestrained corrosion is equally destructive. Oxidation and corrosion of the anode metals consumes electrolyte, changes the state-of-charge balance, and creates corrosion products that are sometimes soluble and capable of poisoning the positive electrode. Establishing a balance between passivation and corrosion for stability is a primary function of the compositional and structural disorder.

## 15D.3.2    Nickel Hydroxide

Positive electrodes for use in NiMH batteries, whether cylindrical or prismatic, can be of the sintered or pasted type, which are also common to NiCd batteries (see Chaps. 15B and C). The nickel hydroxide for use in NiMH batteries is fundamentally the same as that used in NiCd and NiFe (Chap. 15A), and from a simple viewpoint, the basic compound is the same as that used by T. Edison and W. Junger 100 years ago. However, today's high-performance nickel hydroxide is more advanced and continues to improve in capacity and utilization, power and discharge rate capability, cycle life, high temperature charging efficiency, and cost.

*Spherical Nickel Hydroxide.*    As mentioned previously, one type of nickel hydroxide is by far the most common—a high-density spherical type used in pasted electrodes that became commercial around 1990.[19,20] High-density spherical nickel hydroxide is made in a precipitation reactor where metal salts such as nickel sulfate are reacted with a caustic such as NaOH in the presence of ammonia. The nickel source may have additives such as cobalt and zinc to enhance performance. The important physical parameters within this type of nickel hydroxide are as follows:

- *Chemical formula:* The nickel hydroxide active material itself is most commonly a NiCoZn triprecipitate, a common composition is $Ni_{94}Co_3Zn_3$. However, the amounts of cobalt and zinc which are usually about 1% to 5% each can be adjusted for conductivity, oxygen overvoltage, and microstructure refinement with some design trade-offs in terms of active material capacity and cost. Other more complicated multielemental precipitates such as NiCoZnCaMg offer higher capacity, cycle life, and elevated temperature performance, but cannot be manufactured by conventional precipitation processes.

- *Tap density:* Usually around 2.2 g/cc tap density is a measure of the dry nickel hydroxide powder packing efficiency and influences the amount of active material that can be loaded into the pores of the nickel foam current collector.

- *Particle size:* An average particle size of about 10 μm.

- *Surface area:* Measured by the BET (Brunauer–Emmett–Teller) method, surface area refers not to the geometric area of each nickel hydroxide sphere, but rather to the total surface area of each particle that contributes to the charge-discharge reactions and can thus affect utilization and high-rate discharge capability. Typical BET surface area for high-density spherical nickel hydroxide is about 10 to 20 $m^2$/g.

- *Crystallinity:* Each nickel hydroxide sphere has an extremely high surface area corresponding to the nickel hydroxide crystallites themselves. Crystallinity is measured by x-ray diffraction, where the full width at half maximum (FWHM) of a reflection such as the (101) plane may yield a typical crystallite size of about 110 Å.

A variety of other factors contribute to performance, such as impurities from processing like residual sulfates, nitrates, sodium sulfate, and others.

***Sintered Nickel Hydroxide Electrodes.***    Sintered electrodes have the best rate and power capability,[21] but sacrifice capacity on a weight and volume basis and are more expensive to manufacture. Sintered electrodes involve an expensive and complicated sequence of manufacturing steps, and consequently it is typical that only companies with an existing capital investment for this kind of process would manufacture sintered electrodes. Sintered positive plates begin with the pasting of filamentary nickel[22] onto a substrate such as perforated foil. The nickel fibers are then "sintered" under an elevated temperature annealing furnace in a nitrogen/hydrogen atmosphere, and the binders from the pasting process are burned away to leave a conductive skeleton of nickel having a typical average pore size of about 30 μm.

Nickel hydroxide is next precipitated into the pores of the sinter skeleton using either a chemical or electrochemical impregnation process. The impregnated electrode is then formed or preactivated in an electrochemical charge/discharge process. Important design variables in the manufacture of sintered nickel hydroxide electrodes include

- Strength and pore diameter of the filamentary nickel skeleton
- Chemical composition of the nickel hydroxide active material
- Active material loading
- Amount of harmful impurities (e.g., nitrate, carbonate)

One aspect of sintered versus pasted nickel hydroxide technology is that sintered electrodes require the battery manufacturer to make a significant capital investment in facilities and equipment, and to have a great deal of internal expertise in processing. Pasted electrodes conversely place a great deal of emphasis on the expertise of the suppliers for both the nickel foam substrate and for the high-density spherical nickel hydroxide. Recent development in the pasted electrode has brought about exceptional improvements in power and high-rate discharge capability to a level comparable to that of the sintered electrode.

***Pasted Nickel Hydroxide Electrodes.***    The more common pasted nickel hydroxide positive electrode is typically produced by mechanically pasting high-density spherical nickel hydroxide into the pores of a foam metal substrate (Fig. 15D.2). The foam metal substrate is typically produced by coating polyurethane foam with a layer of nickel either by electroplating or by chemical vapor deposition, followed by a heat treatment process to remove the base polyurethane. The pore size may be decreased from about 400 to 200 μm for better conductivity. The density of the foam may also be adjusted to promote conductivity and power versus capacity and utilization.

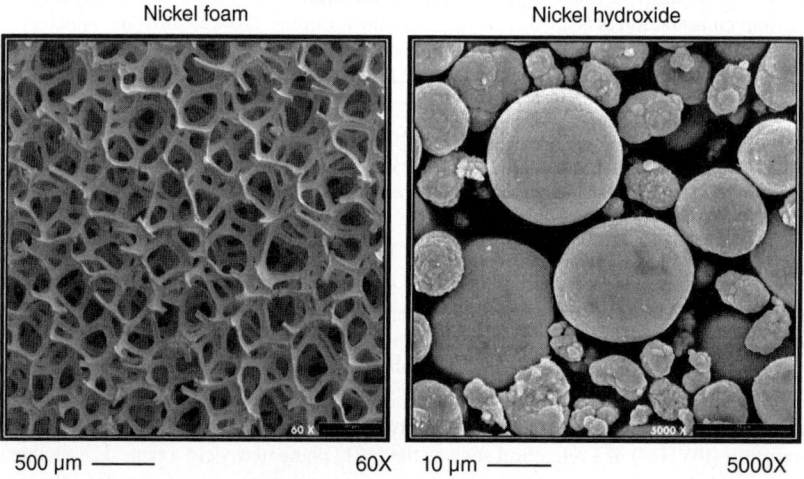

**FIGURE 15D.2**   Scanning electron microscope (SEM) micrographs of positive electrode nickel foam substrate and high-density spherical nickel hydroxide.

The foam[23] is then physically loaded with nickel hydroxide having an average diameter of about 10 μm in a paste containing conductive cobalt oxides, which form a conductive network to compensate for the large distance from the nickel hydroxide to the metal current collector and for the fact that the nickel hydroxide itself is relatively low in conductivity. A cross-sectional comparison of sintered and pasted positive electrodes is presented in Fig. 15D.3 under backscattered electron imaging where the bright areas indicate metallic nickel current collection.

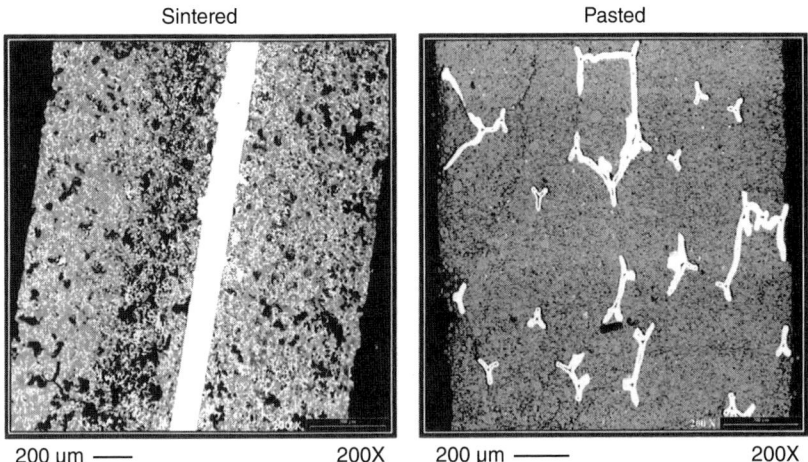

Sintered                        Pasted

200 μm ——        200X    200 μm ——        200X

**FIGURE 15D.3**    SEM micrographs under BEI imaging where bright areas indicate nickel metal current collection illustrating difference in active material distance to current collector for pasted and sintered positive electrodes.

The nickel hydroxide active material and electrode formula can be specially formulated for specific applications. For operation at temperatures above 35°C, some manufacturers may use other additives to the paste formula to inhibit premature oxygen evolution on charge (see High Temperature Nickel Hydroxide below).[24] In addition, paste formula modifications may adjust the type and quantity of conductive network additives such as cobalt metal and cobalt monoxide.[25-27] Usually, the paste additives are finely divided cobalt metal and cobalt monoxide that will dissolve and reprecipitate on surface of the nickel hydroxide active material. However, the coating may not be uniform and complete in coverage.

For ultra-high-power discharge, it is possible to add metallic nickel fibers to the paste formula to enhance conductivity. Addition of metallic nickel fibers lowers the amount of active material resulting in reduced capacity and specific energy.

***High Temperature Nickel Hydroxide.***    To combat the premature oxygen evolution mechanism resulting from temperatures above ambient, manufacturers often introduce oxygen evolution suppressants such as $Ca(OH)_2$, $CaF_2$, or $Y_2O_3$. As shown in Fig. 15D.4, introduction of these additives can reduce capacity loss under 65°C charging from about 50% to below 20%. NiMH manufacturers must carefully select the oxygen suppressant type, amount, and location to avoid deleterious effects such as power loss and cycle life reduction due to the nonconductive nature of many of these types of materials.

Another method is to modify the formula of the nickel hydroxide itself. The most common NiMH positive active material is NiCoZn. To improve high temperature performance, multielement formulas such as NiCoZnCaMg have also been developed.

***Core-shell β-α Nickel Hydroxide.***    A high-capacity (350 mAh/g) β-α $Ni(OH)_2$ with a long cycle life was developed recently.[28] The high capacity originates from the multielectron transfer nature of the α-$Ni(OH)_2$. The unique core-shell structure prevents the electrode failure due to the lattice expansion from β to α phase.

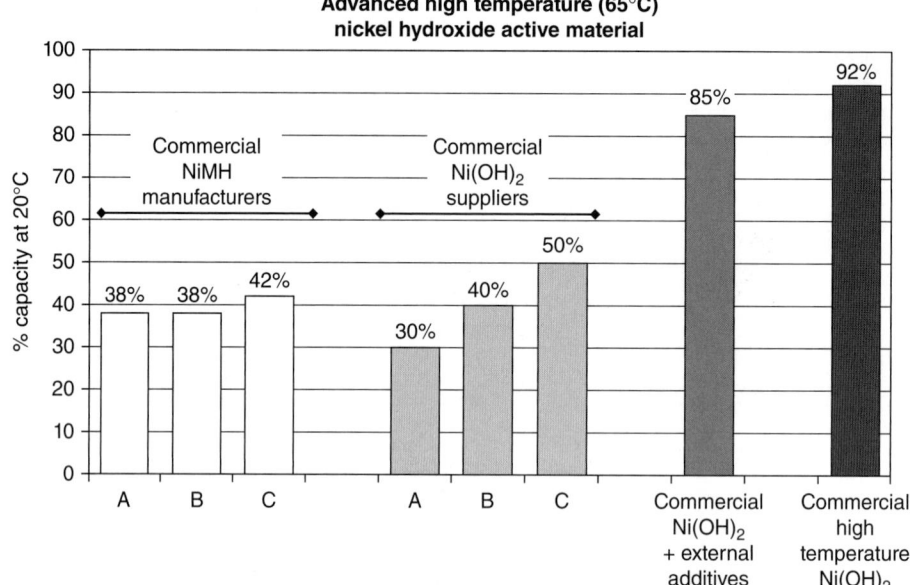

**FIGURE 15D.4**    Temperature performance of cylindrical *C*-size batteries using commercial nickel hydroxide.

## 15D.3.3  Electrolyte

The electrolyte in NiMH batteries of all types is routinely a mixture of about 30% potassium hydroxide in water, providing high conductivity over a wide temperature range. It is most common for the electrolyte to have a lithium hydroxide additive at a concentration of about 17 g/L to promote improved charging efficiency at the nickel hydroxide electrode by suppressing oxygen evolution, the competing reaction to charge acceptance.

An important feature of the electrolyte is related to fill fraction. Essentially all NiMH batteries utilize a sealed, starved electrolyte design. Similar to sealed NiCd cells, the electrodes are nearly saturated with electrolyte, while the separator is only partially saturated to allow for rapid gas transport and recombination.

Special electrolytes are also used in NiMH batteries to enhance high temperature operation. Instead of binary KOH/LiOH electrolytes, it is also possible to substitute a portion of the KOH with NaOH. The ternary KOH/NaOH/LiOH electrolyte is at a similar high concentration of about 7 M, but the contribution of NaOH promotes high-temperature charging efficiency. However, it is typical for this electrolyte to decrease cycle life through increased corrosion of the MH active materials.

Recently, ionic liquid was introduced to NiMH battery as an electrolyte material to reduce or eliminate the corrosive nature of alkaline solution.[29] Use of ionic liquid allows the use of high-capacity Si-anode[30] and expansion of voltage window. 1-Ethyl-3-methylimidazolium acetate, for example, shows a conductivity of 2.5 mS/cm at 25°C and an electrochemical window of 3.2 V.[31]

## 15D.3.4  Separator

The primary function of the separator is to prevent electrical contact between the positive and negative electrodes, while holding electrolyte necessary for ionic transport. Original separators for NiMH batteries were standard NiCd and Ni-H$_2$ separator materials. However, NiMH batteries proved to be more sensitive to self-discharge, especially when conventional nylon separators were used.[32] The presence of oxygen and hydrogen gas causes the polyamide materials in the nylon separator to decompose. The corrosion products from this decomposition allowed for poisoning of the nickel hydroxide, promoting premature oxygen evolution and forming compounds capable of redox shuttle between the two electrodes that further increases the rate of self-discharge.

In addition, the separator plays a crucial role relative to cycle life. In the starved electrolyte design, it is a common design principle to essentially saturate the electrodes with electrolyte at the assembly stage. The separator is designed to have a high electrolyte fill fraction in order to hold as much electrolyte as possible but not be overfilled so as to inhibit gas recombination. To the battery manufacturer, this means that during the first few charge/discharge cycles (formation), when the electrodes have not yet absorbed their full amount of intended electrolyte, charging must be initiated carefully to avoid venting.

The electrolyte design concept relates to capillary theory that the electrolyte will migrate to the smallest pores. In the NiMH battery, this translates to the nickel hydroxide positive electrode having the smallest pores, followed by the MH negative electrode, and finally the separator. At cell assembly, it is common for the separator to be about 90% filled, and then reduced to about 70% during the cell formation process of the first few charge/discharge cycles as both the positive and negative electrodes expand and contract opening interior regions for electrolyte absorption. This process continues to some degree over many hundreds of charge/discharge cycles, where NiMH cell failure is common when the separator fill fraction has been reduced to about 10% to 15% of its original level. As a result, separators that can absorb larger quantities of electrolyte at cell assembly, have small pores able to compete for electrolyte, and retain surface wettability are highly desirable. Inspection of the separator in failed batteries shows even these types of separators undergo some degradation and loss of electrolyte absorption ability, although not nearly to the same degree as earlier generation separators.

Consequently, there was a need for a more stable NiMH separator material to reduce self-discharge while still retaining electrolyte crucial for maintaining cycle life. In NiMH batteries, there is now widespread use of what is termed "permanently wettable polypropylene." In fact, the polypropylene is a composite of polypropylene and polyethylene, where the base composite fibers require special surface treatments to make them wettable to the electrolyte. Currently there are two major types of surface treatments:

*Acrylic Acid.*    This process involves the grafting of a chemical such as acrylic acid to the base fibers to impart wettability and is accomplished using a variety of techniques such as ultraviolet or cobalt radiation.[33]

*Sulfonation.*    This process imparts wettability to the polypropylene by exposing the base fiber material to fuming sulfuric acid. The separator surface is designed to be made hydrophilic to the electrolyte after completion of the treatment. Use of sulfonated separator materials is a key component in enabling "pre-charged" NiMH battery technology.

## 15D.3.5    Cell Construction Types

Sealed NiMH cells and batteries are constructed in cylindrical, button, and prismatic (both large and small) configurations, similar to those used for sealed nickel-cadmium batteries.

The electrodes are designed with highly porous structures having a large surface area to provide a low internal resistance and a capability for high-rate performance. The positive electrode in the cylindrical NiMH cell is a highly porous sintered, or foam-nickel substrate into which the nickel compounds are impregnated, or pasted, and converted into the active material by electrodeposition. Foams have generally replaced sintered plaque electrodes. Expanded metals and perforated sheets are lower cost, but they have poor high-rate capability. Sintered structures are much more expensive. The negative electrode, similarly, is a highly porous structure using a perforated nickel foil or grid onto which the plastic-bonded active hydrogen storage alloy is coated. The electrodes are separated with a synthetic nonwoven material, which serves as an insulator between the two electrodes and as a medium for absorbing the electrolyte.

*Cylindrical Configurations.*    The assembly of the cylindrical unit is shown in Fig. 15D.5. The electrodes are spirally wound and the assembly is inserted into a cylindrical nickel-plated steel can. The electrolyte is added and contained within the pores of the electrodes and separator.

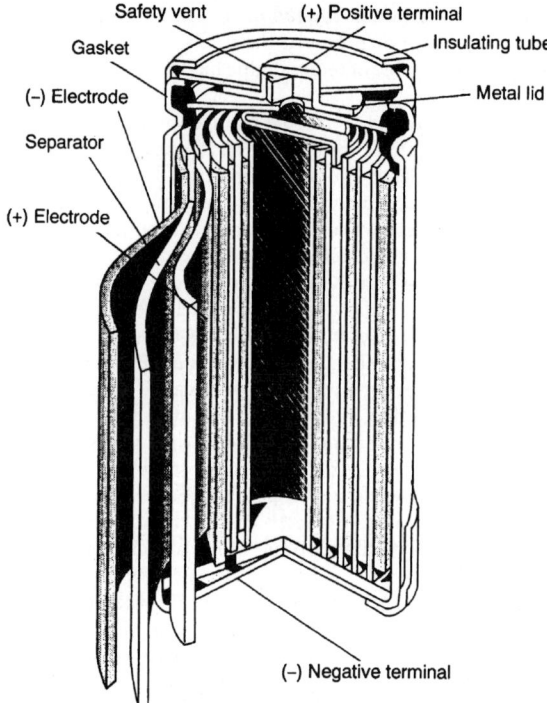

**FIGURE 15D.5**  Construction of a sealed cylindrical nickel-metal hydride battery.

The cell is sealed by crimping the top assembly to the can. The top assembly consists of a lid, which includes a re-sealable safety vent, a terminal cap, and a plastic gasket. The can serves as the negative terminal and the lid as the positive terminal, both insulated from each other by the gasket. The vent provides additional safety by releasing any excessive pressure that may build up if the battery is subjected to abuse.

***Button Configuration.***    The button configuration is illustrated in Fig. 15D.6. It is similar in construction to the nickel-cadmium button cell, except that the cadmium is replaced by the hydrogen storage alloy.

A...  Cap
B...  Gasket
C...  Negative electrode
D...  Separator
E...  Positive electrode
F...  Cup (Positive terminal)
G...  Positive terminal

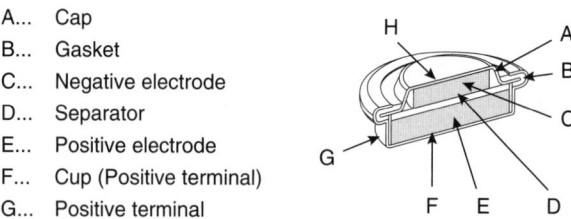

**FIGURE 15D.6**    Construction of a sealed nickel-metal hydride button cell.

### Prismatic Configuration

*Small Prismatic Configuration.*    The thin prismatic batteries are designed to meet the needs of compact equipment. The rectangular shape permits more efficient battery assembly, eliminating the voids that occur within the assembly of cylindrical cells. The volumetric energy density of the battery can be increased by a factor of about 20%. The prismatic cells also offer more flexibility in the design of batteries, since the battery footprint is not controlled by the diameter of the cylindrical cell.

Figure 15D.7 shows the structure of a prismatic battery. The electrodes are manufactured in a comparable manner as the electrodes for the cylindrical cell, except that the finished electrodes are flat and rectangular. Multiple flat electrodes are then stacked, alternating the positive and negative electrodes, interspaced by separator sheets. Positive electrode tabs are gathered and welded to the cover plate of the top lid assembly. The electrode stack is then placed in the nickel-plated steel can and the electrolyte is added. The cell is sealed by crimping the top assembly to the can. The top assembly is a lid that incorporates a resealable safety vent, a terminal cap, and a plastic gasket, similar to the one used on the cylindrical cell. An insulating heat shrink tube is placed over the metal can (jacket). The bottom of the metal can serves as the negative terminal and the top lid as the positive terminal. The gasket insulates the terminals from each other.

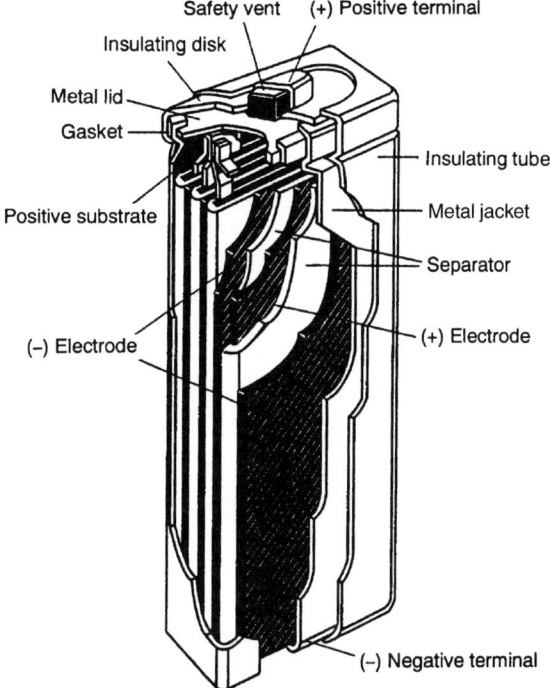

**FIGURE  15D.7**   Construction sealed thin prismatic nickel-metal hydride cell.

*Large Prismatic Configuration.*   The large prismatic configuration is similar in design to that of the small prismatic with the primary exceptions that positive and negative electrodes are welded to individual terminals rather than the lid and can, and the lid is welded to the can rather than relying on a crimp seal. The metal can (jacket) can be powder coated or taped with an insulating material to prevent cell-to-cell shorting. Two terminals protruding through the can lid serve as connections to the negative and positive electrodes with a gasket insulating the terminals from the can. Figure 15D.8 shows the structure of a large prismatic battery.

### Mono-Block Construction

*HEV Multicell Configuration.*   NiMH battery chemistry is especially well suited for HEV mono-block construction due to its high tolerance to overcharge and over-discharge, which pose problems for other battery chemistries. The mono-block design reduces cost by having a common pressure vessel construction and far fewer cell parts; a single vent assembly and shared hardware can be used in multicell modules. Further attributes of the mono-block construction include reduced volume, since interior cell walls can be shared between cells, and flexible choices of liquid or air cooling. An air cooled 7.2 V, 6.5 Ah plastic mono-block HEV NiMH battery is shown in Fig. 15D.9.

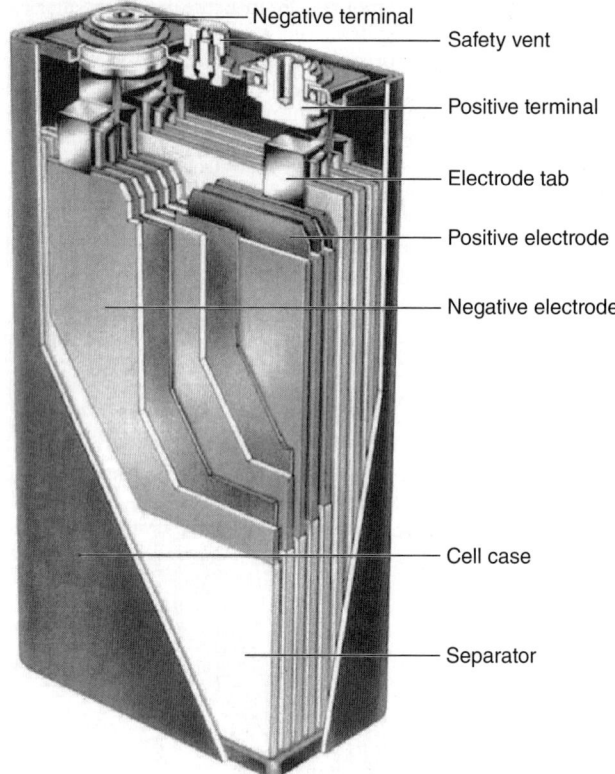

**FIGURE 15D.8**    Cutaway view of a prismatic NiMH cell.

**FIGURE 15D.9**    HEV plastic case prismatic module used in Toyota Prius.

Issues of mono-block construction include selection of the plastic casing material to avoid gas permeation and the need for individual cells to have well-matched capacity and impedance to avoid cell-to-cell imbalance. Further, mono-block construction must recognize that electrodes within each cell expand and contract during charge and discharge and that the ensuing swelling of the electrode stack must be compensated for in the mechanical design and loading of the mono-block.

*9-Volt Multicell Configuration.*    Another example of a mono-block design is the 9-V NiMH consumer batteries. The construction of a typical 9-V multicell battery is shown in Fig. 15D.10.

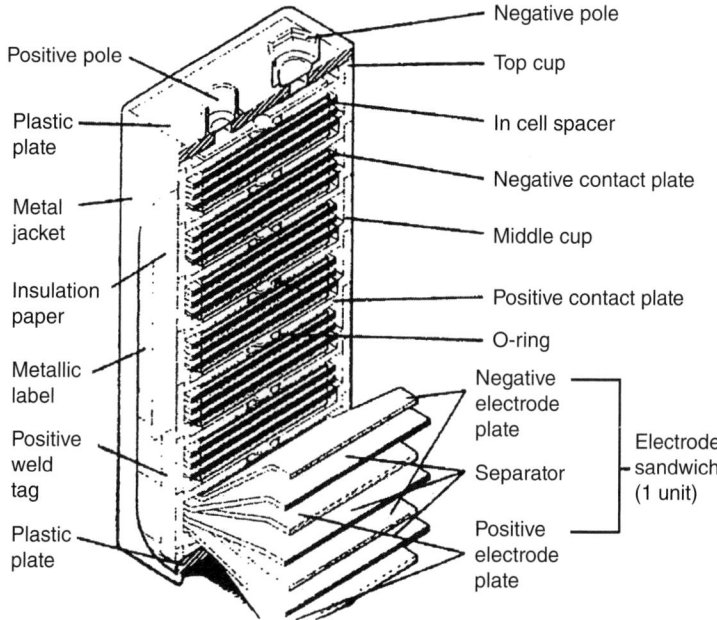

**FIGURE 15D.10**    Cutaway view of a sealed 9-V NiMH cell.

***Pouch Design.***    A new type of NiMH, copying from the Li-ion pouch cell, was introduced to increase the gravimetric energy density and the pack density (Fig. 15D.11). By using an MH alloy with a low hydrogen equilibrium pressure, the pouch cell was designed for electric vehicle (EV) application with a target energy density of 145 Wh/kg.

***Bipolar Construction.***    In recent years, several companies have begun offering NiMH battery modules using a bipolar construction design.[34,35] In the bipolar design, cells are stacked on top of one another such that the entire surface area between the cells becomes the pathway to transfer current from one cell to the next. This large cross-sectional area promotes uniform current flow across the cell, minimizing resistance and energy loss that result from normal intercell connections in traditional battery designs. As a result, the bipolar construction is well suited for high-rate applications. Figure 15D.12 illustrates the bipolar construction design of the Gigacell battery module manufactured by Kawasaki Heavy Industries.

The module shown above was designed specifically for very high charge currents associated with railroad applications and contains added features like cooling fans and heat sinks that are not used in simpler bipolar designs.[36] Additional benefits of the bipolar design include minimizing the uneven heat generation that may occur in traditional battery modules. Uniform cell heating leads to uniform cell aging, which ensures longer battery life.

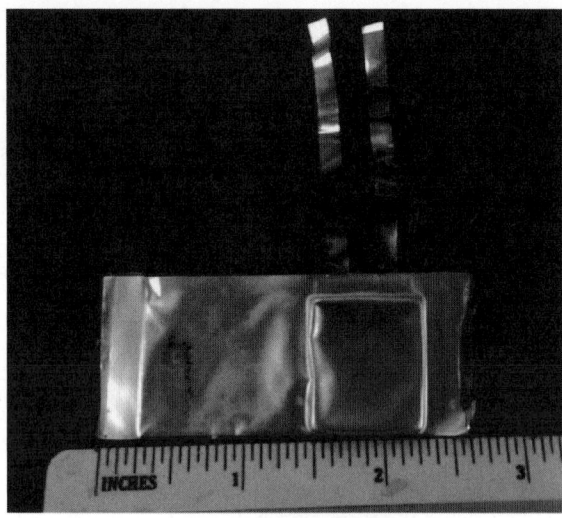

**FIGURE 15D.11**   A NiMH battery in a pouch format.

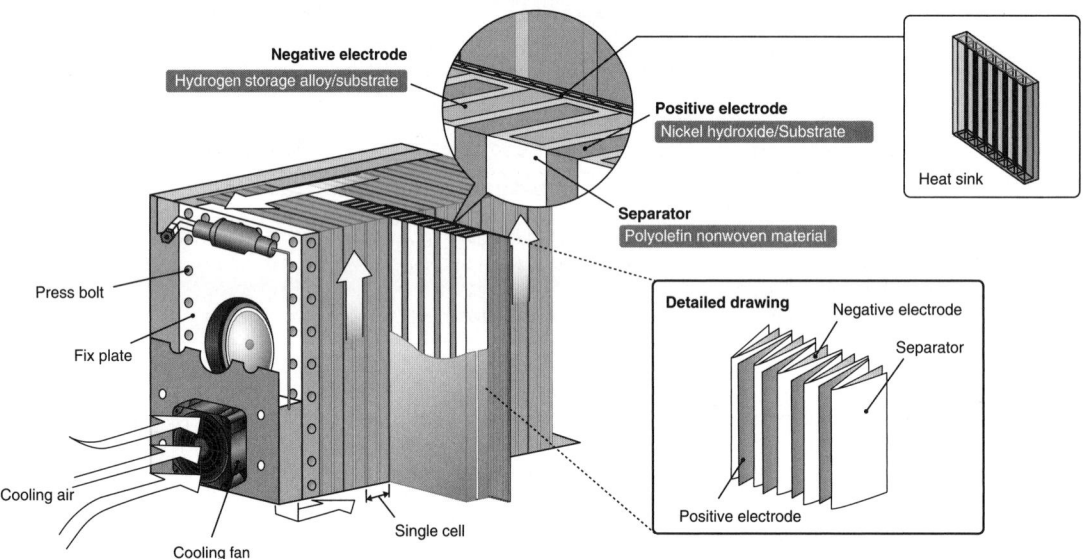

**FIGURE 15D.12**   Bipolar module construction design.

## 15D.4   *BATTERY APPLICATIONS AND PERFORMANCE*

### 15D.4.1   Cell and Battery Design Issues

*Cylindrical versus Prismatic Configuration.*   NiMH batteries are versatile in that both cylindrical and prismatic constructions can be utilized. Each type of construction has advantages and disadvantages, and a particular end use can predetermine the most suitable configuration. For NiMH applications below about 10 Ah, cylindrical construction dominates due to the low cost and highest speed of manufacture. Above 20 Ah, cylindrical construction is extremely difficult and the prismatic configuration dominates. In the 10 to 20 Ah cell size range, manufacturers are offering both cylindrical and prismatic designs although prismatic designs are most common.

Cylindrical cells for industrial and propulsion applications are similar to high-volume production consumer batteries in that the well-known "jelly roll" construction is used. However, most small portable cylindrical cells require only low to moderate discharge rate capability and electrode terminal connections are usually quite simple. Conversely, since industrial and propulsion NiMH applications require high to ultra-high discharge rate capability and low internal resistance, multiple tab current collection is used. This type of construction is termed edge welding and requires that each electrode have a current collecting strip on one of its sides. The current collecting strips (both for the positive and negative electrodes) are disposed on opposite sides of the jellyroll. After coil winding of the jellyroll, the edge current collector is welded in multiple locations to each electrode. Other aspects of cell assembly are virtually identical to small consumer batteries. The net result of the enhanced current collection is a reduction in specific energy due to added weight, but an increase in specific power due to decreased cell AC impedance (usually around 8 to 12 m$\Omega$ for small portable batteries and around 1 to 2 m$\Omega$ for industrial cylindrical cells). For industrial applications such as HEV, motorcycle, and bicycle applications, the most popular cylindrical cell sizes are standard C and D sizes, although a multitude of height changes within that diameter are also used. Some work on larger size cells such as F size has also been reported.[37]

Prismatic construction is conventional in that electrode stacks of alternating positive and negative electrodes with intermediate separators are used (as shown in "Prismatic Configuration" in Sec. 15D.3). The main design alternatives involve the thickness and number of each of the electrodes and the aspect ratio (relative proportions of cell height to width to thickness). Key design variations include the ratio of active materials to inactive components (such as the cell can and terminal, and current collectors). In all cases, the cell designer has the objective of emphasizing one or more properties of performance characteristics such as energy vs. power, while maintaining a minimum threshold of other performance such as cycle life. One example is that for EV NiMH prismatic cells,[38] a specific power of about 200 W/kg is acceptable for most vehicles and consequently, relatively thick positive and negative electrodes can be used to increase the ratio of active material to inactive cell components, thereby providing specific energy in the 63 to 80 Wh/kg range. Alternately, HEV prismatic NiMH batteries typically must deliver a specific power greater than 1300 W/kg and therefore electrode thickness is reduced as compared to the electrode thickness used for EV applications. NiMH batteries for HEV use have a typical specific energy ranging from about 42 to 68 Wh/kg.

***Metal versus Plastic Cell Cases.***    NiMH cylindrical cells exclusively use metal cell cases. One important reason for this is that the can itself is electrically connected to the MH negative electrode and serves as the negative terminal. Another important reason is that many applications require fast charge in which gas recombination can cause considerable internal pressure and only the strength of a metal container would suffice without significant volumetric penalties. Finally, the metal crimp seal to the cover plate assembly with a polysulfone gasket is inexpensive, fast, and reliable.

Both metallic and plastic cell cases are common for the prismatic NiMH batteries used in automotive applications such as EVs. Unlike cylindrical cells where the can itself is the negative terminal, prismatic designs have both a positive and negative terminal at the top cover plate. Key decision criteria for selecting metal cell cases include excellent thermal conductivity, inexpensive prototyping costs for changing cell dimensions, and small volumetric penalties.

The primary advantage of plastic cell cases is cost and electrical isolation in the $\geq$320 V battery packs common to today's EVs, where even high resistance leakage currents must be considered. Further design considerations for plastic cell cases include development costs for permanent molds, gas permeation, thermal conductivity, and sufficient plastic thickness for gas pressure containment without can wall bulging. NiMH battery modules using plastic cell cases are shown in Fig. 15D.13.

***Energy versus Power Design Trade-offs.***    Similar to other rechargeable battery technologies, NiMH batteries can be designed to emphasize energy, power, or some combination of the two. The application itself may dictate the choice, but perhaps not for obvious reasons. In some applications, a certain threshold power (such as 200 W/kg) is required for adequate performance, and once the power requirement is met, competing designs usually tout specific energy in the range of 62 to 80 Wh/kg. The motivation for the higher energy is longer runtime. The typical figure of merit for an energy-based battery cost is $/kWh, where the goal of $150/kWh has proven to be one of the most challenging development targets. NiMH cost is mainly controlled by the amount, cost, and

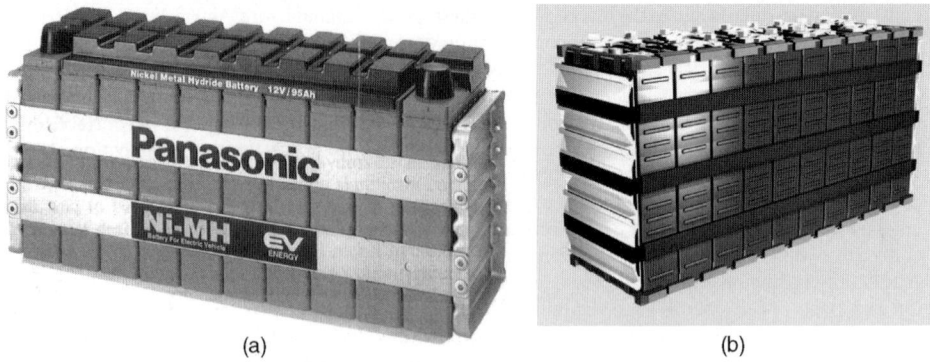

(a)                                                                (b)

**FIGURE 15D.13**   NiMH battery modules using plastic cases (*a*) and metal cases (*b*).

utilization of raw materials rather than processing labor, assembly, packaging, etc. Consequently, an important development activity for many NiMH manufacturers is to obtain higher utilization of less expensive active materials on a mAh/g basis and thereby reduce manufacturing costs.

For power applications, energy is of far less importance since the battery has a vastly different function. For HEVs, the main purpose of the battery is to accept and utilize the energy from regenerative braking and assist in acceleration at startup. The battery is commonly exposed to very high current pulses during both charge and discharge, but energy extraction is usually limited to relatively small depths of discharge. The crucial requirement for the NiMH Power battery is therefore specific power greater than 1000 W/kg. NiMH batteries available today are in the 1000–1700 W/kg specific power range, with development reports approaching 2000 W/kg.

***Cell, Module, and Pack Design.***   A typical 5.0 Ah C size NiMH portable cell has an AC impedance on the order of 8 to 15 m$\Omega$, and a $\Delta V/\Delta I$ resistance of 15 to 30 m$\Omega$. In contrast, a 100 Ah NiMH prismatic cell has an impedance on the order of 0.4 m$\Omega$, and a $\Delta V/\Delta I$ resistance of 0.9 m$\Omega$. Despite the low resistance, heating is still a concern due to the extremely high current pulses. Even the $I^2R$ heating effects at these high currents are small in comparison to heating due to overcharge. Consequently, an essential aspect of NiMH energy and power battery design involves thermal management.

Proper thermal management begins at the MH negative electrode, understanding that overcharge heat is generated at the surface of the hydride electrode where oxygen is recombined. Heat must migrate from the negative electrode to the cell can and therefore good thermal conductivity within the electrode and electrode stack bundle is important. Cells are usually packaged into 12 V modules, with cells bound together in a back-to-back arrangement. An important design feature is that the cells at each end will have a much higher exposed surface area available for convective cooling. This raises concerns of thermal imbalance within a module and a resultant state-of-charge imbalance that will lead to premature failure. Therefore, proper module design must include endplate and heat sink considerations.

Modules are packaged in a variety of configurations. Within the various packaging arrangements, important design considerations are distance between modules, air or water flow channels, and battery tray heat sink characteristics to equalize cooling from module to module. The merits of each approach are highlighted in Table 15D.2.

**TABLE 15D.2**   Summary on the Merits for Thermal Management of HEV, PHEV, and EV Batteries

|  | Air cooling | Water cooling |
| --- | --- | --- |
| Advantages | • Light weight<br>• Simple | • More effective heat transfer<br>• Average fluid temperature more consistent<br>• Integrated with vehicle cooling |
| Disadvantages | • Complicated distribution of air within pack<br>• Incoming air must be free of dirt and water from road<br>• Variable ambient temperature | • Increased weight<br>• Elaborate module design<br>• Higher average fluid temperature |

### 15D.4.2   Consumer Batteries—Precharged NiMH

NiMH technology continues to expand into new markets. Consumer products, which were previously the exclusive domain of alkaline primary cell, are now able to use rechargeable NiMH batteries with "ready to use" capability, providing the consumer a choice with equal or better performance than alkaline and little to no price premium. Hence, "precharged" NiMH batteries are achieving wide market acceptance.

The key to this NiMH technology advancement has been the improvement in charge retention to a remarkable 85% for 1 year (comparable to alkaline). Much work has gone into reducing self-discharge mechanisms resulting in two distinct classes of NiMH technology: conventional and advanced "precharged."

*Conventional NiMH.*   The rate of self-discharge is dependent on storage temperature and time—the higher the temperature, the greater the rate of self-discharge. This is illustrated in Fig. 15D.14, which shows charge retention for conventional sealed cylindrical NiMH batteries following storage at different temperatures for varying periods of time. The comparison is for a battery discharged at the rated discharge load (approximately the $1C$ rate) at the stated temperature. Note that conventional NiMH technology has a charge loss in the range of 20% to 30% for 30 days.

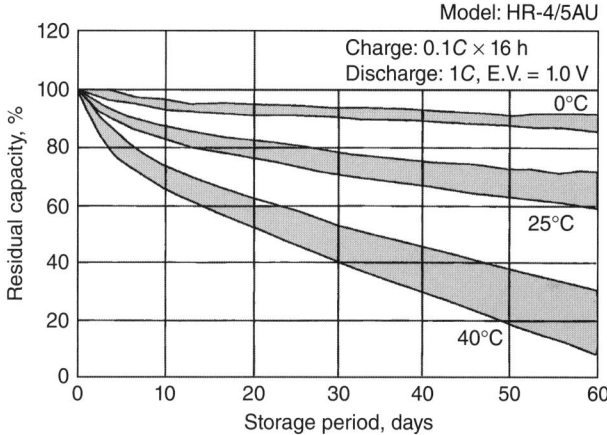

**FIGURE 15D.14**   Charge retention characteristics of conventional NiMH cylindrical batteries at various temperatures.

*Advanced "Precharged" NiMH.*   This class of NiMH technology is often referred to as "ready to use." These batteries are sold in the precharged state similar to alkaline primary cells and can be stored for up to 5 years with less than 15% loss in capacity. In order to achieve this low level of self-discharge, manufacturers typically use sulfonated separators, cobalt encapsulated $Ni(OH)_2$ with little added cobalt to the positive paste, and low corrosion $AB_5$ or $A_2B_7$ MH alloys. This is illustrated in Fig. 15D.15.

Storage of the NiMH cell, in a charged condition, has no permanent effect on its capacity. Any capacity loss due to self-discharge is reversible and batteries can recover to full capacity by simply recharging. However, storage at elevated temperatures, similar to operation at elevated temperatures, may deteriorate seals, MH alloys, and separators, leading to permanent damage, such as loss of capacity, cycle life, and overall battery lifetime. Hence, it is recommended the temperature range for storage of NiMH cells be maintained at 20 to 30°C.

### 15D.4.3   Discharge Performance

#### Discharge Voltage Profiles

*Cylindrical Batteries.*   Typical discharge curves for a cylindrical sealed NiMH battery under various constant current loads and temperatures are shown in Fig. 15D.16. (The data are based on the rate performance at

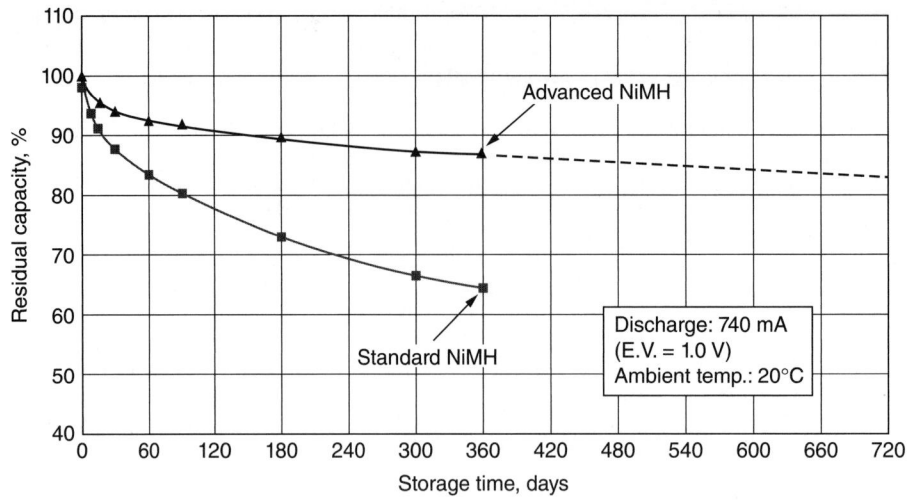

**FIGURE 15D.15** Charge retention characteristics of advanced versus conventional NiMH cylindrical batteries.

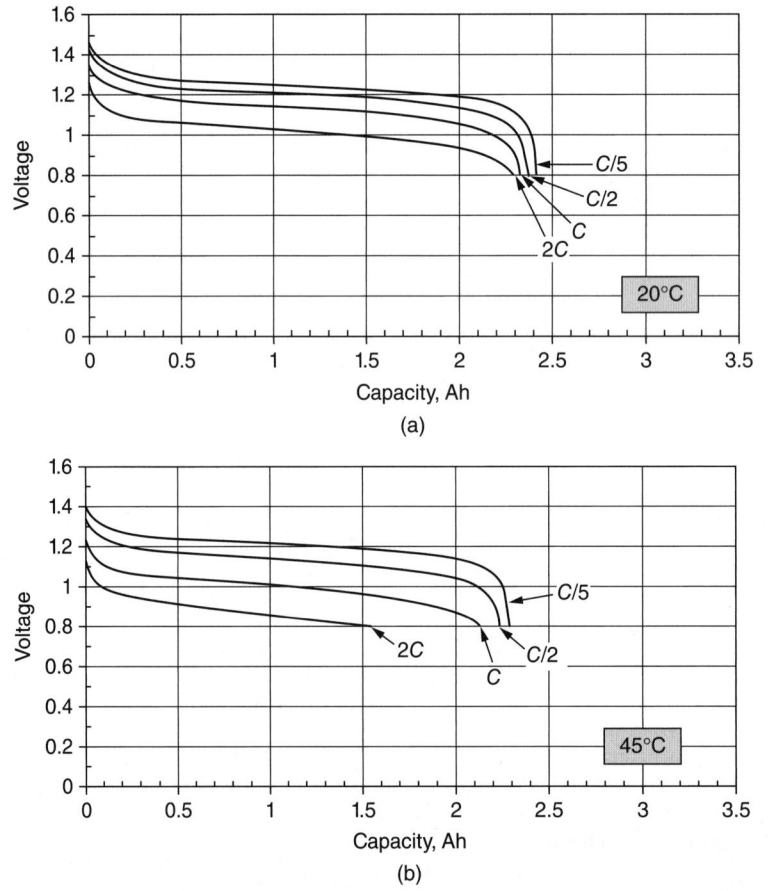

**FIGURE 15D.16** Discharge characteristics of NiMH cylindrical batteries. (*a*) Discharged at 20°C. (*b*) Discharged at 45°C. (*c*) Discharged at 0°C.

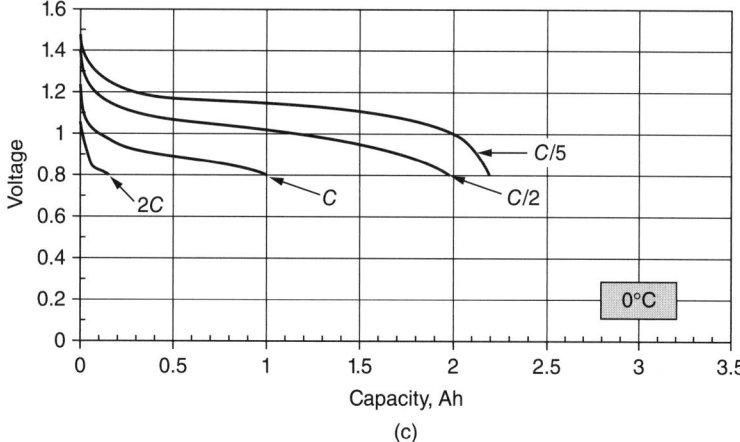

(c)

**FIGURE 15D.16**    (*Continued*)

20°C at the 0.2*C* discharge rate to 1.0 V.) A flat discharge profile is characteristic. The discharge voltage, as expected, is dependent on the discharge current and discharge temperature. Typically, with higher currents or lower temperatures, one will observe a lower operating voltage in the battery. This is due to a higher IR drop resulting from increasing current or increasing resistance from the lower temperatures. However, because of the relatively low resistance of NiMH batteries (as well as nickel cadmium batteries), this drop in voltage is less than experienced with other types of portable primary and rechargeable batteries.

*Button Batteries.*    Typical discharge curves for button-type sealed NiMH batteries at room and other temperatures are shown in Fig. 15D.17.

*Prismatic Batteries.*    Typical discharge curves for the prismatic sealed NiMH batteries at room and other temperatures are shown in Fig. 15D.18.

*9 V Battery.*    Typical discharge curves for the 9 V sealed NiMH batteries at room and other temperatures are shown in Fig. 15D.19.

***Specific Energy.***    The specific energy of NiMH batteries can vary anywhere from 42 to 110 Wh/kg depending on particular application requirements. For consumer applications where run-time is of paramount importance, NiMH batteries do not need high power capability or ultra-long cycle life. On the other hand, for extremely high-power charge and discharge, extra current collection, high N/P ratios, and other cell design and construction decisions can additively reduce specific energy. For the most common small portable NiMH batteries, specific energy is usually about 90 to 110 Wh/kg; for large prismatic batteries, usually about 65 to 75 Wh/kg; and for HEV batteries and other high-power applications, about 45 to 60 Wh/kg.

While gravimetric energy usually receives the attention for advanced battery technologies, in many cases volumetric energy density (in watt-hours per liter) is more important. NiMH has high energy density, having achieved 430 Wh/L.

***Specific Power.***    The power capability is a main strength of NiMH batteries relative to other advanced battery chemistries. For many years, it was widely believed that NiMH chemistry would never achieve a sufficiently high-rate discharge capability to replace NiCd. In fact, NiMH chemistry is now quickly replacing NiCd in these applications due to NiMH's higher energy and environmental concerns.

Voltage profiles up to 10*C* discharge rate for high power cylindrical NiMH cells are presented in Fig. 15D.20, attaining a specific power of 865 W/kg (Fig. 15D.21). NiMH HEV module power shown in Fig. 15D.22 is in excess of 1300 W/kg under both charge and discharge conditions.

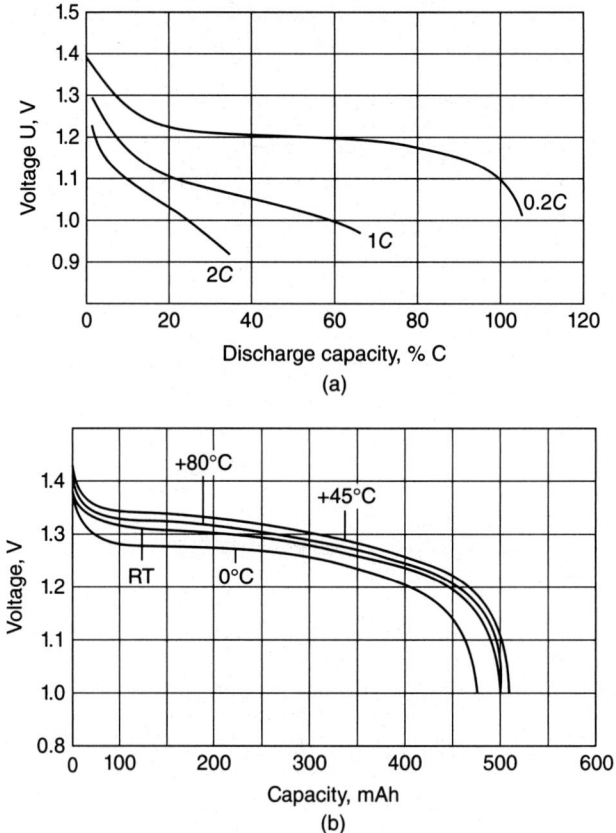

**FIGURE 15D.17**   Discharge characteristics of NiMH button batteries.
(*a*) Discharged at 20°C. (*b*) Discharged at different temperatures and at a 0.2*C* rate.

***Effect of Discharge Rate and Temperature on Capacity.***   The capacity of a battery is dependent on several factors including discharge current, temperature, and the cutoff or end voltage. The capacity can be increased by continuing the discharge to the lower end voltages, particularly at the higher current drain rates and lower temperatures, where the voltage drops off more rapidly than at the lighter drain rates. However, it should be noted there is a risk to the battery by discharging to too low a cutoff voltage as the cells may be permanently damaged (see "Polarity Reversal during Over-Discharge" below). Hence, the cutoff or end voltage for the NiMH battery is typically 1.0 V per cell.

The relationship between the capacity (expressed as a percentage of the capacity at 20°C on discharge at 0.2*C* rate) of the sealed NiMH battery and the discharge temperature and current (expressed in *C* rate) is summarized in Fig. 15D.23.

Typically, the best performance for NiMH batteries is obtained in the temperature range 0 to 40°C. Discharge performance characteristics of the battery are affected moderately at higher temperatures, while the effects at lower temperatures are more pronounced. Similarly, the effect of a change in temperature becomes more significant at higher discharge rates.

Reduced discharge performance over the life of the battery is mainly due to an increase in internal resistance resulting from the polarization of the MH electrode due to the generation of water during discharge. Capacity dependence as a function of temperature is strongly influenced by the selection of active MH material, active nickel hydroxide material, and the formulation of the electrolyte.

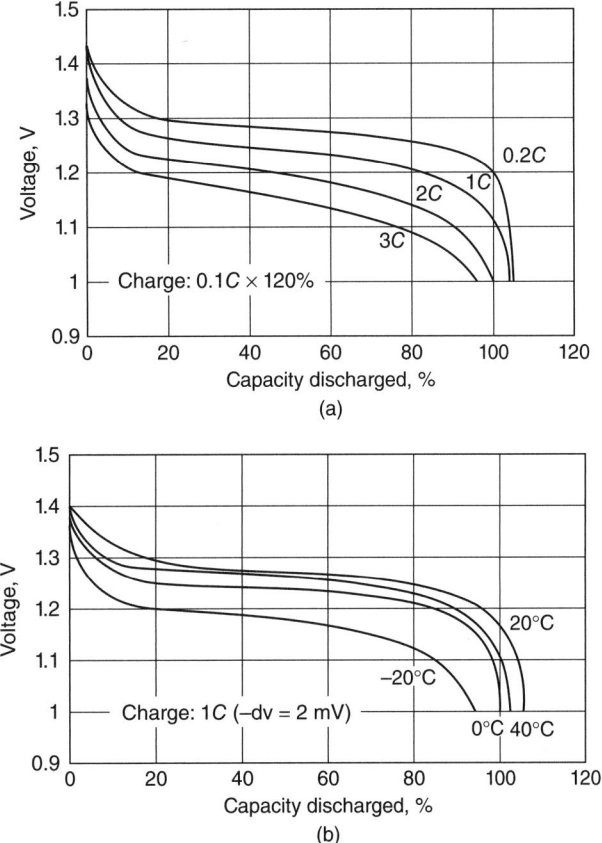

**FIGURE 15D.18**  Discharge characteristics of NiMH prismatic batteries. (*a*) Discharged at 20°C. (*b*) Discharged at different temperatures and at a 0.2*C* rate.

Elevated temperature performance is typically governed by the active positive material. As discussed earlier in Sec. 15D.3.2, elevated temperature charge efficiency can be greatly influenced by material and/or additive choices. The problem with elevated temperature charge acceptance is the oxygen evolution characteristics of the nickel hydroxide positive electrode. Normally, at room temperature, the nickel hydroxide electrode has almost complete charge acceptance until about 80% charge input when the competing reaction of oxygen evolution begins. At full charge, continued charging causes 100% oxygen evolution at the nickel hydroxide electrode, and the oxygen migration to the MH electrode forms the well-known oxygen recombination overcharge mechanism. The issue at elevated temperatures is that oxygen evolution occurs at much earlier states of charge, and total charge acceptance is reduced, as seen in Fig. 15D.24.

Another temperature-related property of NiMH chemistry is the storage effect on life (discussed in "Polarity Reversal during Over-Discharge" below). Prolonged exposure, either in storage or on back-up power usage, to temperatures above 45°C can reduce battery life due to degradation of the separator, oxidation, and corrosion of the MH alloy, and disruption of the cobalt conductive network in the positive electrode. Each of these mechanisms is highly dependent on the manufacturer's choices of the active electrode materials.

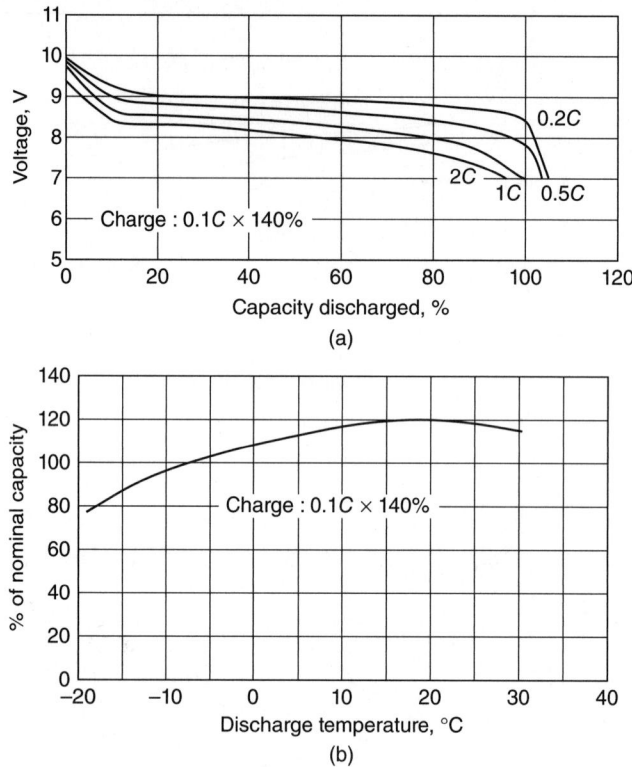

**FIGURE 15D.19**    Discharge characteristics of NiMH 9 V batteries. (*a*) Discharged at 20°C. (*b*) Discharged at different temperatures and at a 0.2*C* rate.

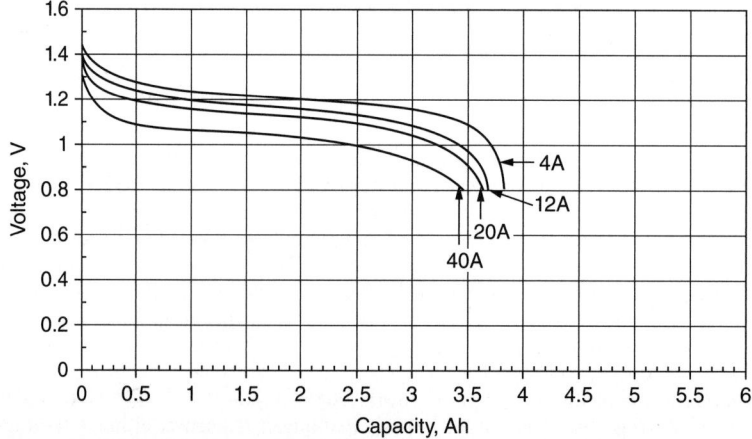

**FIGURE 15D.20**    Voltage-capacity profiles for high power NiMH cylindrical 3.5 Ah *C*-size batteries at different rates on continuous discharge.

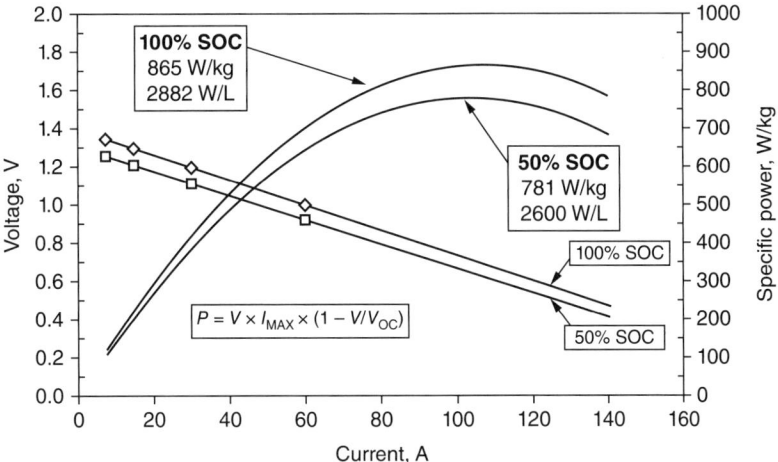

**FIGURE 15D.21**    Specific power for ultrahigh power NiMH cylindrical 3.5 Ah *C*-size batteries.

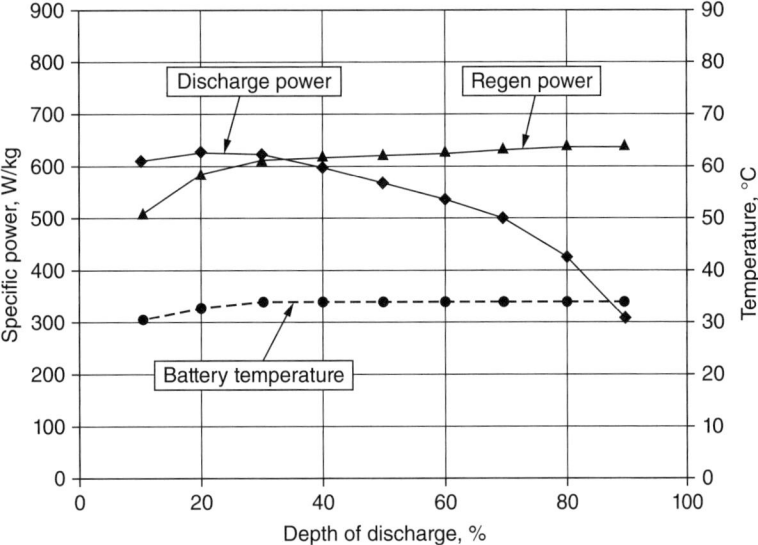

**FIGURE 15D.22**    Specific power of a 12 V, 20 Ah NiMH HEV battery module as a function of depth of discharge.

### 15D.4.4    Cycle/Life Performance

*Charge Retention.*    Charge retention is an area where NiMH manufacturers compete and the end user must be careful to compare all performance properties for a given design. It is understood that the state-of-charge of a NiMH battery decreases during storage due to self-discharge and is highly dependent on temperature as well as the separator material, active positive and negative electrode materials employed. The specific formula of the MH active material, separator material, and the quality of the nickel hydroxide active material all play important roles in either decreasing or increasing the self-discharge rate. Because there is such a large selection of materials

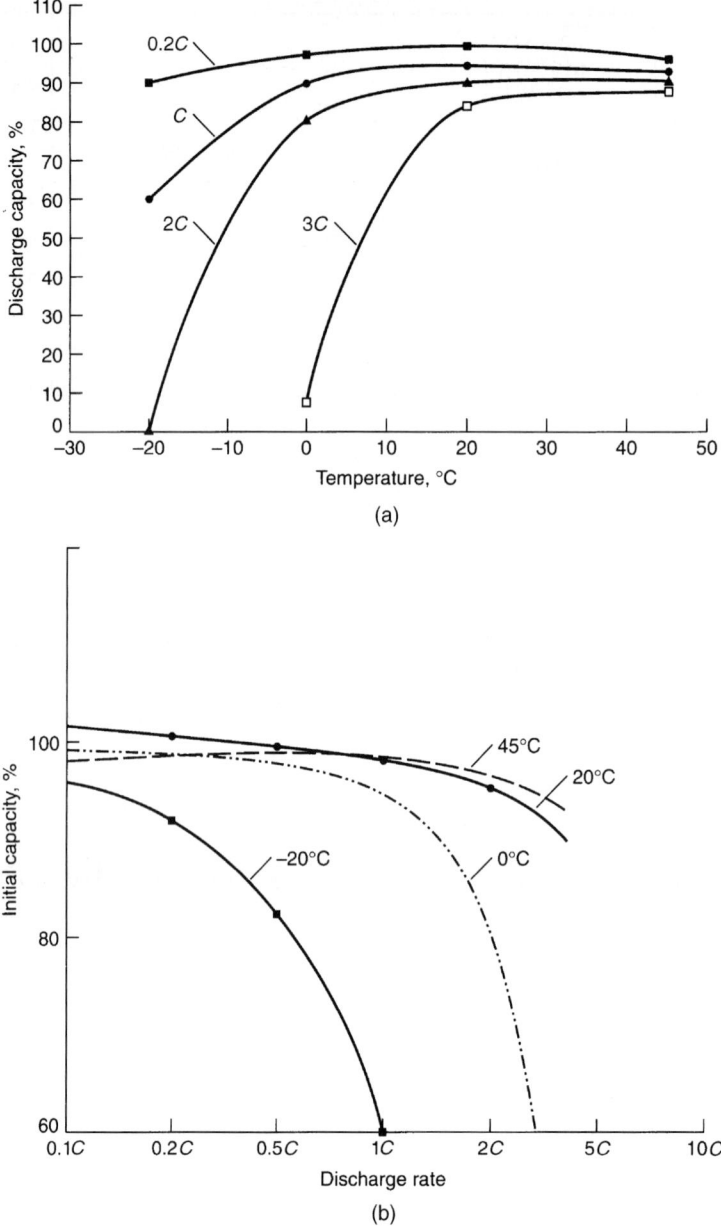

**FIGURE 15D.23** (*a*) Discharge capacity versus ambient temperature for sealed cylindrical NiMH batteries at various discharge rates; end voltage 1.0 V/cell. (*b*) Discharge capacity (% of 0.2*C* rate) versus discharge rate (*C* rate) for NiMH cylindrical batteries at various temperatures; end voltage 1.0 V/cell.

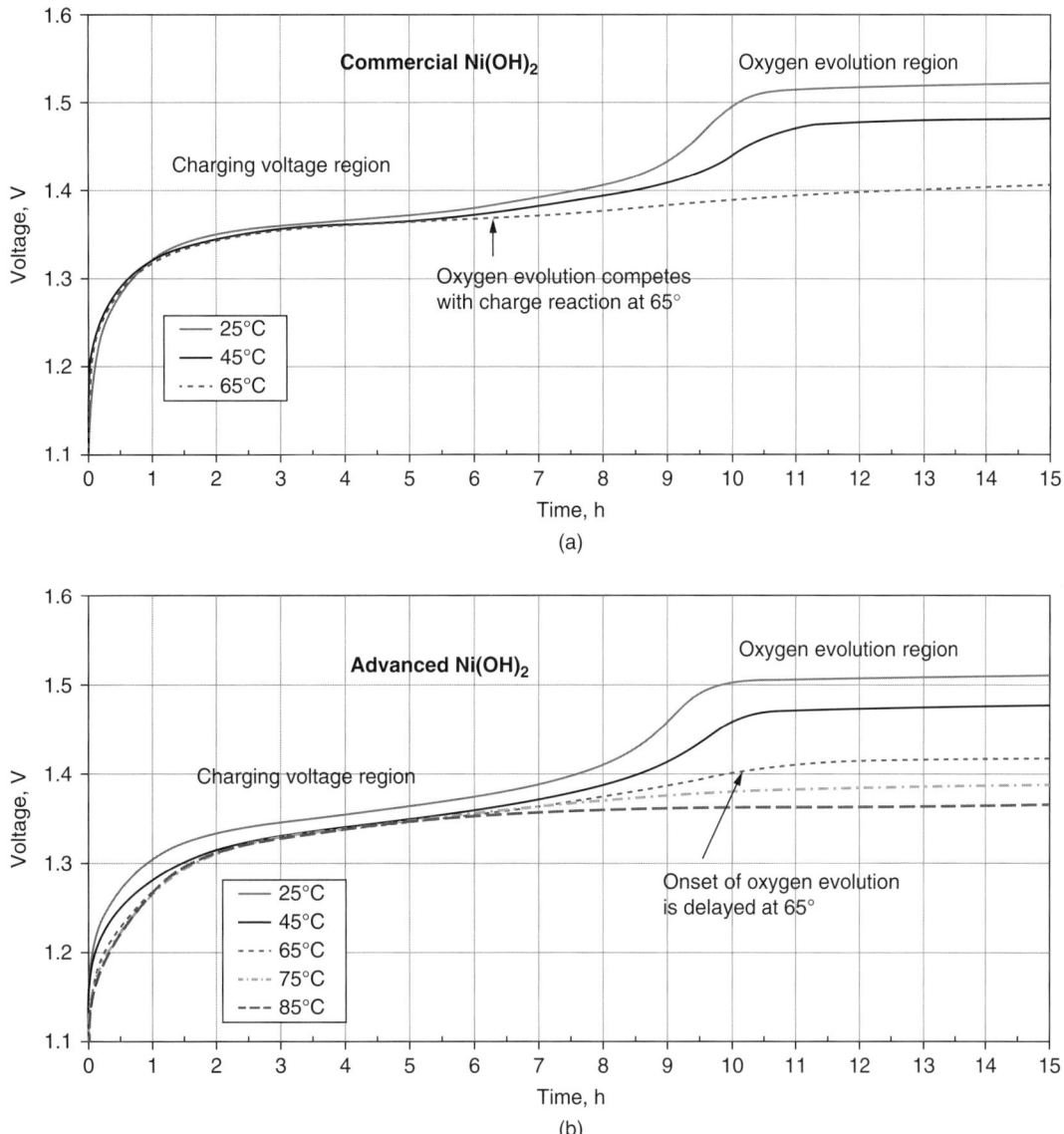

**FIGURE 15D.24** Charge characteristics of commercial nickel hydroxide as a function of temperature. (*a*) Commercial nickel hydroxide. (*b*) Advanced nickel hydroxide.

for use in NiMH batteries, charge retention performance can vary widely from one product to the next. However, choice of materials to optimize charge retention may have detrimental effects on other battery performance characteristics.

For example, advanced separator materials can reduce self-discharge from about 30% loss per month at room temperature to about 15% loss over the same time period. However, these same materials may also reduce cycle life 15% to 50%. The chemical mechanisms for how the separator reduces self-discharge are complicated, involving chemical grafting agents that can bind and thereby inactivate chemical species that may promote self-discharge. However, these separator treatments may have deleterious effects on electrolyte absorption and the ability of the separator to retain electrolyte during cycling. Separator dry-out is a very common mechanism of battery failure.

The choice of the MH alloy can provide similar design trade-offs. Higher capacity $AB_2$ alloys also have higher rates of self-discharge compared to lower capacity $AB_5$ or $A_2B_7$ alloys discussed previously. The mechanisms for the effect of the MH alloy on self-discharge are twofold. First, corrosion products from the hydride alloy may migrate to the nickel hydroxide positive electrode and promote oxygen evolution during storage. Second, other corrosion products such as vanadium with its multivalent oxides may form redox shuttle mechanisms similar to nitrate ions. Likewise, the quality of the nickel hydroxide material, the use of encapsulated nickel hydroxides, as well as residual impurities such as nitrates and carbonates can influence the above-cited self-discharge mechanisms. However, ultralow impurity levels and added encapsulation may inadvertently add to processing costs.

The positive electrode plays the most significant role in reducing self-discharge. Losses associated with self-discharge leave the cobalt conductive network open to degradation. As the cell gradually loses its charge, the cobalt conductive network can be reduced to $Co^{2+}$ or Co metal allowing the cobalt to go into solution and migrate elsewhere in the cell. Currently, there are two competing methods to combat this issue. One method, when long-term storage over several years is required, is to increase the levels of cobalt additives to combat breakdown of the conductive network and isolation of the active material. Another, increasingly more popular way to avoid cobalt loss and uniformity issues, is to use active materials that are "cobalt coated" or "cobalt encapsulated." Although more expensive, nickel hydroxide coated with cobalt hydroxide by the active material manufacturer has proven the cobalt to be in a more stable form. Further, it has been reported to offer increased utilization, high-rate discharge performance,[39] and significantly improved charge retention.

*Cycle Life.*    Cycle life for industrial-sized NiMH batteries has both similarities and differences to that of small portable NiMH batteries. Cycle life for small portable NiMH batteries can vary from manufacturer to manufacturer, but usually falls in the range of 500 to 1000 full cycles (100% DOD under 2-h charge/discharge). Design and chemistry factors affecting cycle life that are common to both large and small NiMH batteries include

1. *MH electrode:*
   - Active alloy formula (oxidation/corrosion properties)
   - Alloy processing effect on microstructure (particle disintegration)
   - Electrode construction (swelling in *x-y-z* direction, stability of conduction pathways)

2. *Nickel hydroxide electrode:*
   - Active material formula (swelling and poisoning resistance)
   - Conductive network stability (amount and type of cobalt oxides)
   - Substrate (pore size, strength, and resistance to fracture)

3. *Cell design:*
   - N/P ratio (amount of excess negative electrode capacity to influence cell pressure, MH corrosion, disintegration)
   - MH discharge reserve (over-discharge protection)
   - Separator (stability to corrosion, electrolyte absorption and retention, thickness, and resistance to short circuit)
   - Electrolyte (composition, amount, and fill fraction)
   - Vent pressure (weight loss, charge imbalance)
   - Electrode stack design (compression, electrode thickness, aspect ratio of height to width)

Factors affecting cycle life in industrial-sized NiMH batteries that differ from those factors affecting cycle life in portable NiMH batteries include

- Significantly higher battery voltages (42–320 V vs. 12 V) increases risk of abusive overcharge and over-discharge due to capacity or state-of-charge mismatch.

- Overall higher energy (0.1 kWh vs. 33 kWh) increases heat generation and therefore the criticality of thermal management (which can be further influenced by battery pack enclosure heat transfer limitations).

- Higher operating temperature. Air- and water-cooled vehicle batteries usually operate at a temperature of 35°C or higher, whereas operating temperatures for portable batteries may experience transient elevated temperatures, but on average are at or near room temperature.

- Series/parallel strings. In portable batteries, there are a large number of cell sizes available, ranging from 100 mAh button cells to 7 to 12 Ah D and F cells. There is a much smaller availability of cell sizes for power and prismatic NiMH batteries.

- End of life definition for portable batteries is usually based on capacity loss.

- Qualification and operation testing. The emphasis on power greatly influences testing and methodology. In portable battery cycle life testing, it is most common to use 1- or 2-h constant current charge and discharge, usually to 100% DOD each cycle. For EV cycle life testing, discharge is usually a variable current/time profile to simulate driving conditions—the so-called "DST" drive profile (see Chap. 26). The significance of pulsed discharge cycle life testing is that high current pulses dominate the test. On the other hand, most EV cycle life tests are done at 80% DOD and typical NiMH module cycle life is from 600 to 1200 cycles. HEV testing emphasizes power capability even more, and de-emphasizes depth of discharge further still. Typical HEV mode cycle life testing is under a high current pulse profile with a 2% to 10% state-of-charge swing. Typical NiMH cycle life under these conditions is over 300,000 cycles, which corresponds to nearly 150,000 miles in a vehicle. As can be seen in Fig. 15D.25, the difference in the depth of discharge (EV vs. HEV cycling) can play a significant role in cycle life.

Failure modes for NiMH batteries included short circuit due to mechanical penetration through the separator. The frequency of such events is usually small based on sound cell and electrode engineering and if manufacturing quality control is sufficient. Another failure mode may be due to abusive overcharge where excessive venting results in insufficient electrolyte within the separator. Abusive overcharge may result from charge imbalances caused by thermal differences from one part of the large battery to another. The problem may be compounded by the sophistication of the charger, where voltage and temperature sensing may not be on an individual module or cell basis. Another failure mode stems from over-discharge, where a cell or module within a high-energy battery is discharged below the minimum recommended cell voltage of 1.0 V. Over-discharge is likely caused by state-of-charge imbalance within a high-voltage string due to thermal gradients within a battery. Another source

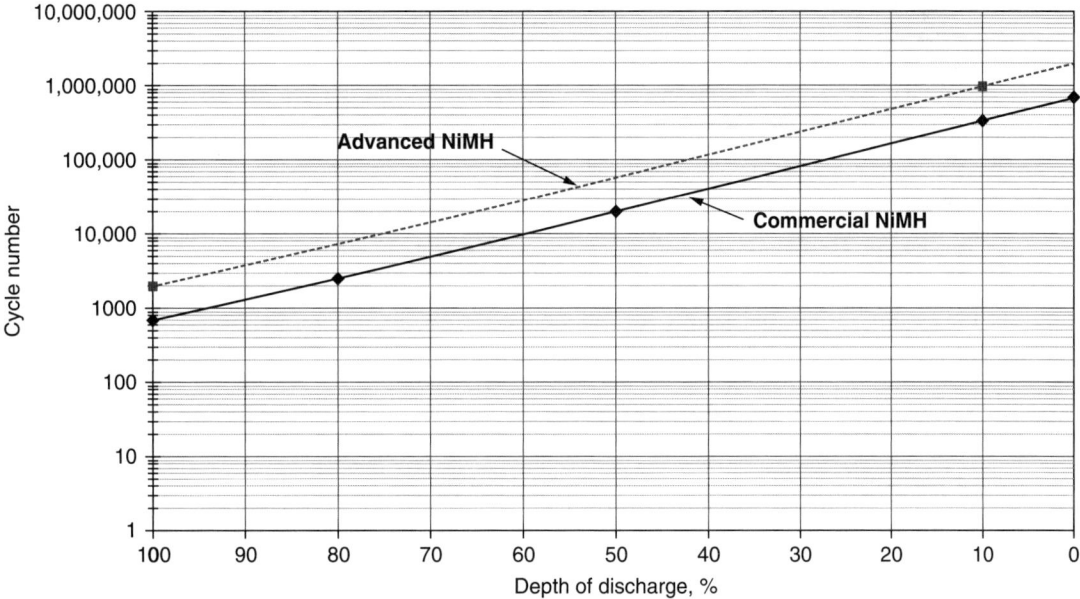

**FIGURE 15D.25** Relationship between DOD and cycle life for NiMH batteries.

of abusive overcharge and over-discharge is the "weak cell or weak module" concept. This involves the statistical predictability within a large number of cells as to the decay rate of capacity, power, and resistance as a function of cycle.

A common feature of the above-cited failure modes is the importance of maintaining state-of-charge balance within a battery pack that may contain several hundred individual cells. The method used to maintain equalized state-of-charge within an industrial NiMH battery is in effect to routinely bring all the cells to the same state-of-charge. This method of using overcharge to equalize state-of-charge is corrupted if the cell temperature within a battery pack is extreme or if cell-to-cell temperature gradients are too large. One of the biggest factors in replicating the excellent cycle life of small cylindrical NiMH batteries to large format batteries in pack applications is proper thermal management, as discussed in "Vehicle" in Sec. 15D.4.7.

If premature failure due to short circuit and abusive overcharge/over-discharge is prevented, the principal failure mode in large format/pack NiMH batteries is increasing internal resistance with cycling. The user of energy-specific NiMH battery will observe that run-time will diminish after long-term use. To the user of a power-specific NiMH battery, failure due to increasing internal resistance and resultant power loss will be observed as an inability of the battery to utilize high-energy pulses due to excessive heating caused by the high charging currents being supplied.

This primary failure mode of increasing internal resistance and power loss during cycling is caused by the same mechanisms as observed in small portable NiMH batteries: namely, separator dry-out as a result of electrolyte redistribution due to swelling of the MH and nickel hydroxide electrodes; consumption of electrolyte due to oxidation of the separator, MH active material and positive electrode materials; and loss of electrolyte through venting.[40] These mechanisms may be exaggerated for large NiMH batteries due to their prismatic construction. Cylindrical cells have one positive electrode, one negative electrode, and one separator for each electrode. NiMH prismatic batteries may have 20 positives, 21 negatives, and a corresponding number of separator sheets. The cylindrical can is more effective in pressure containment than a rectangular container, both in terms of gas pressure and the mechanical forces applied to the can from the electrode stack. Therefore, another critical factor for large NiMH batteries is the management of compressive forces within a module. Typically, restraining bands are used to secure a 10- or 11-cell module that has an endplate to equalize lateral forces on side wall of the can. Failure to adequately equalize compression within each cell in a module and within the internal cell stack itself will lead to premature failure due to unequal electrolyte distribution.

***Shelf Life.***    Over perhaps 6 months to a year of storage, traditional NiMH batteries may completely self-discharge. Further storage, under open circuit, may cause the cell voltage to gradually decline to 0 to 0.4 V, which can cause a breakdown of the cobalt conductive network in the positive electrode and/or increased surface oxidation of the MH active material.[41] The length of time the battery is stored under this low voltage condition, the temperature of low voltage storage, and cell design influence the ease and degree of recovery of the battery. For example, a few cycles of low-rate charge and discharge may be needed to recover cell capacity and power. If the degree of low voltage degradation is severe, the battery may not be recoverable.

Design factors that must be considered for good shelf life are the oxidation and corrosion resistance of the MH alloy, the amount of precharge on the MH electrode, the formula of the nickel hydroxide active material, and the quality of the cobalt conductive network in the positive electrode.

Users of industrial NiMH batteries may choose to leave the battery on low-current trickle charge if the battery will not be used for extended periods. Alternately, the battery may receive a periodic top-off charge designed to compensate for normal self-discharge capacity losses.

## 15D.4.5    Electro-Analytical Properties and Performance

***Coulombic/Voltaic Efficiency and Internal Resistance.***    The NiMH battery has a low internal resistance because of the use of thin plates with large surface areas and low resistance and an electrolyte having a high conductivity. Figure 15D.26 shows the change in internal resistance with the depth of discharge. The resistance remains relatively constant during most of the discharge cycle. Toward the end of the discharge cycle, the resistance also increases due to conversion of the active materials. The internal resistance also increases as the temperature drops because the resistance of the electrolyte and other components is higher at lower temperatures. The resistance of the NiMH battery increases with use and cycling.

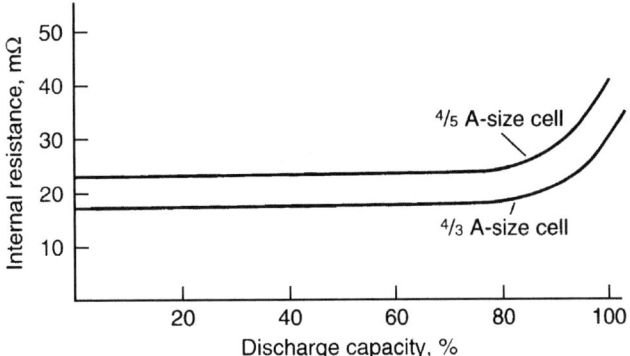

**FIGURE 15D.26**    Internal resistance versus discharge capacity of NiMH cylindrical batteries.

A key strength of NiMH technology is high efficiency due to low internal resistance. As shown in Fig. 15D.27, over 90% voltaic efficiency was observed for 60 Ah prismatic NiMH cells at 100 A; over 75% efficiency at 300 A.

Voltaic efficiency is largely determined by the linear resistance components in the cell, the electronic and ionic resistances that can be lowered by further engineering changes. Coulombic efficiency was determined to be 99% at 50% state-of-charge under an aggressive simulated HEV driving cycle that is a typical operating point for charge-sustaining HEVs. Under the EPA combined city-highway FTP driving schedule, energy efficiency is about 93% to 95%.

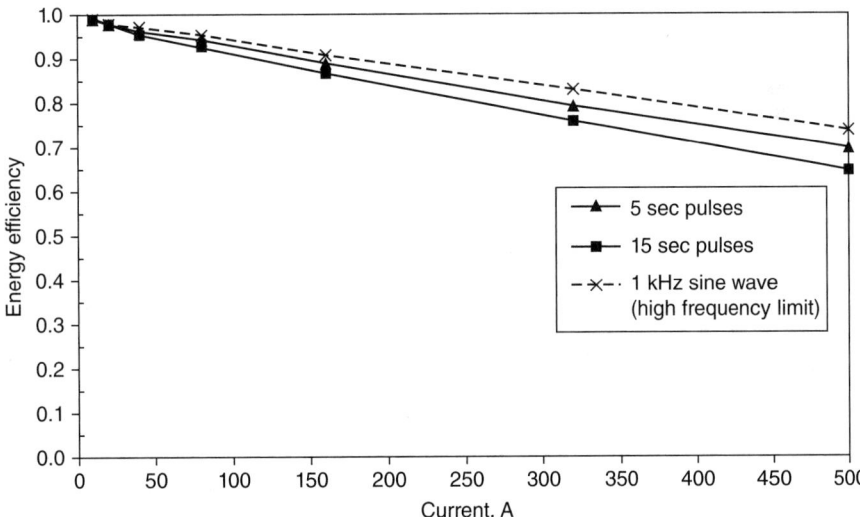

**FIGURE 15D.27**    Energy efficiency of a 60 Ah NiMH HEV battery module as a function of discharge rate.

***Polarity Reversal during Over-Discharge.***    When a multicell series-connected battery is discharged, the lowest capacity cell will reach full discharge before others. If the discharge is continued, this lower capacity cell can be driven into an over-discharge condition through 0 V and its polarity (voltage) reversed. This is illustrated in Fig. 15D.28.

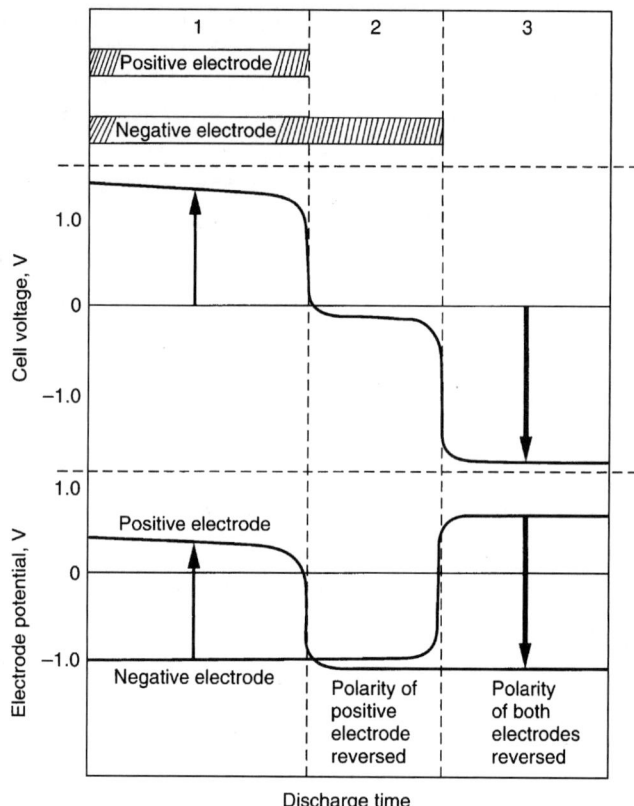

**FIGURE 15D.28**   Charge characteristics of commercial nickel hydroxide as a function of temperature.

Phase 1 of the figure is the normal phase of the discharge with active material remaining on both the positive and the negative electrodes.

During phase 2, the active material on the positive electrode has been fully discharged and continued discharge causes the generation of hydrogen gas to start. Some of this gas may be absorbed by the hydrogen storage metal alloy in the negative electrode and the remainder builds up in the cell. Active material, however, still remains on the negative electrode and the discharge continues. The cell voltage is dependent on the discharge current, but remains within −0.2 to about −0.4 V.

In phase 3, the active materials on both electrodes have been depleted and oxygen is produced at the negative electrode. Prolonged over-discharge leads to gassing, higher internal cell pressure, venting, and deterioration of the cell.

The greater the number of cells in series (in a multicell battery), the greater the possibility a reversal of polarity will occur. To minimize the effect, whenever three or more cells are connected in series, the cell selection method employed should group cell capacities within a narrow range of ±5%. The process of selecting cells of similar capacity is called "matching." Further, a cutoff voltage of 1.0 V per cell or higher should be used for discharge rates up to 1$C$ rate to prevent the possibility that any cell goes into reversal. Higher cutoff voltages should be used for batteries containing more than 10 cells in series and for discharge rates exceeding 1$C$.

***Type of Discharge.***   As discussed in Chap. 5, a battery may be discharged under different modes (such as constant resistance, constant current, or constant power), depending on the characteristics of the load. The type of discharge mode selected has a significant impact on the service life delivered by a battery in a specified application. The discharge profiles of a NiMH battery under the three different modes are plotted in Fig. 15D.29. Figure 15D.29*a* shows the voltage profile, Fig. 15D.29*b* shows the current profile, and Fig. 15D.29*c* shows the power profile during

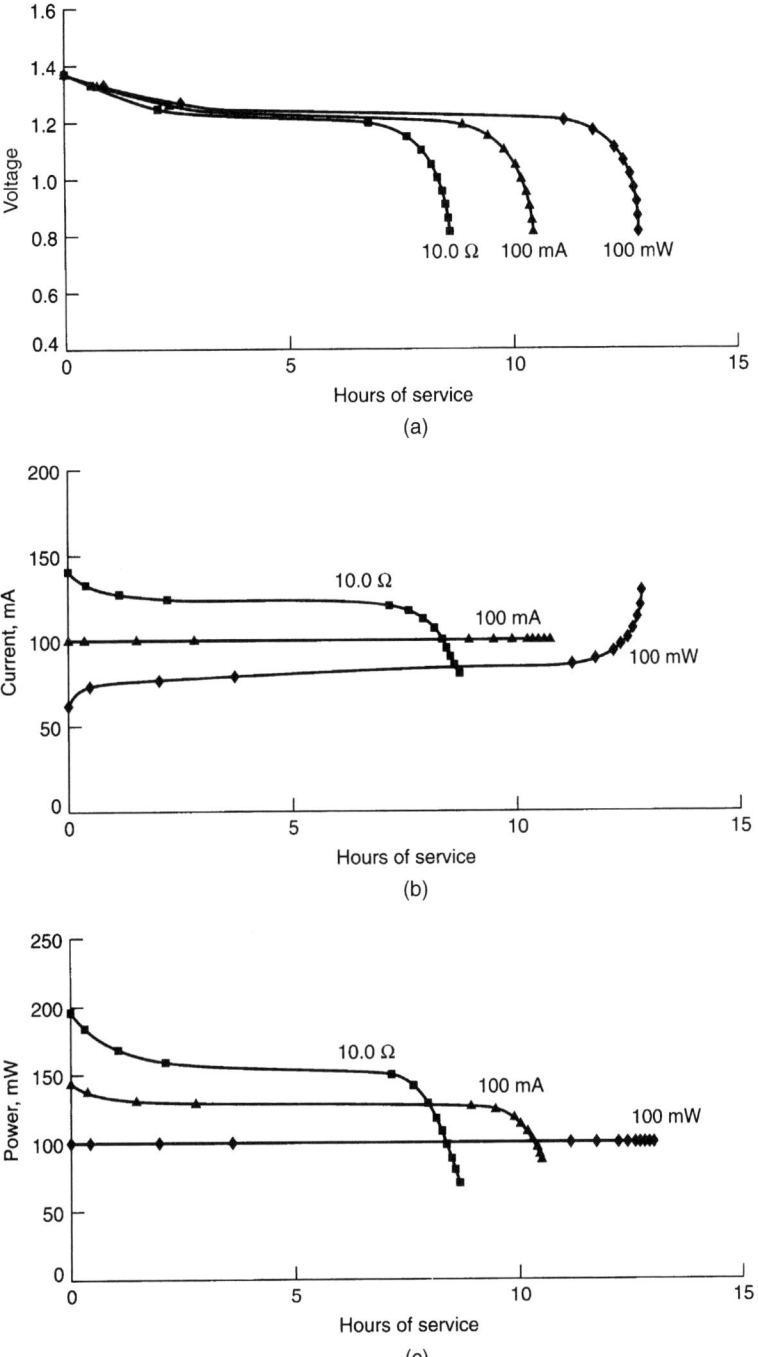

**FIGURE 15D.29** Discharge of NiMH cylindrical batteries—constant power versus constant current versus constant resistance: (*a*) voltage profile; (*b*) current profile; (*c*) power profile. All based on a battery rated at 1000 mAh.

discharge of the battery. As an example, the data are based on the discharge of a 1000 mAh battery such that, at the end of the discharge (1.0 V per cell), the power output is the same for all modes of discharge. In this example, the power output at the 1.0 V cutoff is 100 mW. To discharge at the 100 mW at 1.0 V, the constant current discharge is 100 mA ($C/10$ rate) and the constant resistance discharge is 10 $\Omega$. As shown, the longest service life is obtained under the constant power mode as the average current is the lowest under this mode of discharge.

***Constant Power Discharge Characteristics.*** The discharge characteristics of the NiMH battery under constant power mode, at several different power levels, are shown in Fig. 15D.30. These are similar to the data presented in Fig. 15D.29 for the constant current discharges. The power levels are shown based on $E$ rate. The $E$ rate is calculated in a manner similar to calculating the $C$ rate, but based on rated watt-hour capacity rather than ampere-hour capacity. For example, for the $E/2$ power level, the power for a battery rated at 1200 mWh (rated at 1000 mAh at the $C/5$ rate at 1.2 V) is 600 mW.

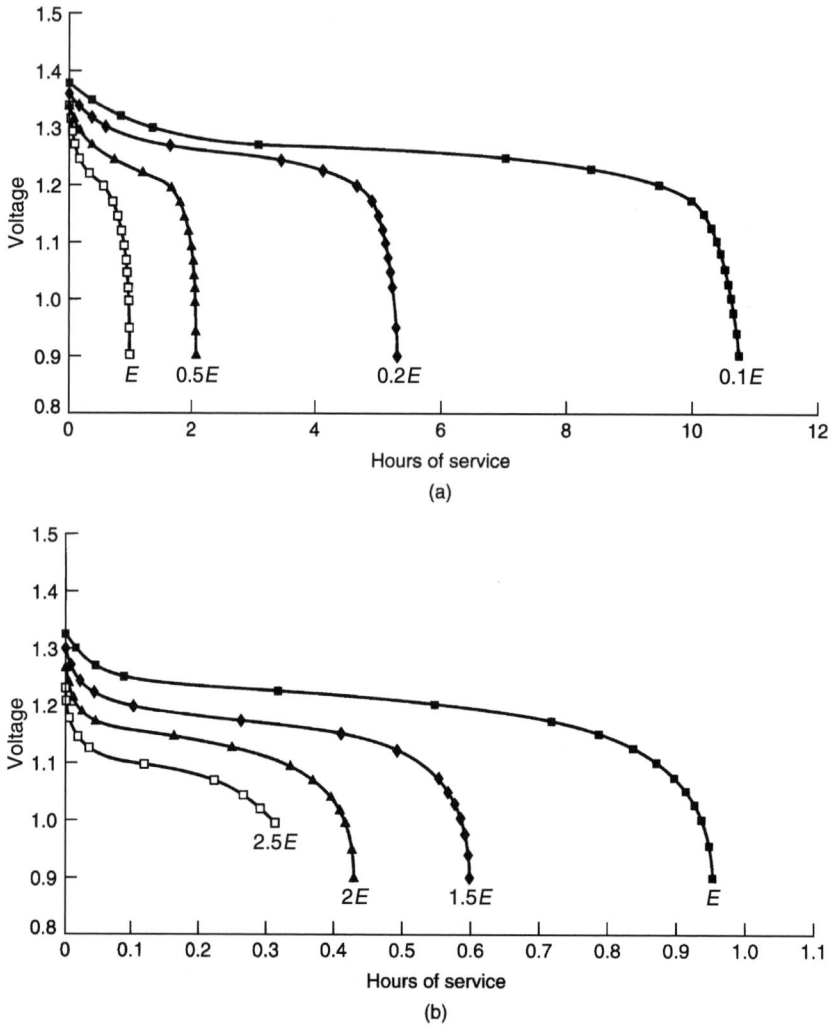

**FIGURE 15D.30** Constant power discharge curves for NiMH cylindrical batteries at 20°C. (*a*) 0.1$E$ to 1$E$ discharge rate; (*b*) 1$E$ to 2.5$E$ discharge rate.

## 15D.4.6 Charging Characteristics

***General Principles.***    Recharging is the process of replacing energy that has been discharged from the battery. Subsequent performance of a battery (as well as its overall life) is dependent on effective charging. The main criteria for effective charging are to:

- Recharge the battery to its full capacity
- Limit the extent of overcharge
- Avoid elevated temperatures and excessive temperature fluctuations

Recharging characteristics of the NiMH battery are generally similar to those of the sealed nickel cadmium battery. However, there are some distinct differences, particularly on the requirements for charge control, as the NiMH battery is more sensitive to overcharge. Caution should be exercised before interchangeably using the same battery charger.

The most common charging method for the sealed NiMH battery is a constant current charge, but with the current limited to avoid an excessive rise of temperature or exceeding the rate of the oxygen recombination reaction.

The voltage profiles of NiMH and nickel cadmium batteries during charge at a moderate constant current rate are compared in Figs. 15D.31*a* and *b*. The voltages of both battery systems rise as the battery accepts the charge. During the first phase of the charge, the temperature of both batteries increases slightly due to

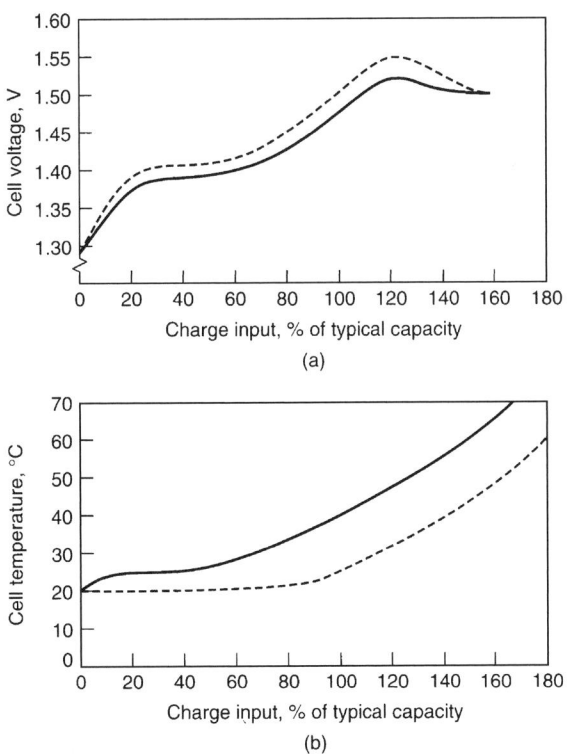

**FIGURE 15D.31**  Comparison of typical charge characteristics of Ni-Cad and NiMH batteries: (*a*) voltage characteristics; (*b*) temperature characteristics. Solid line: NiMH; broken line: Ni-Cad.

Joule heating which stems from the internal resistance of the batteries. Both battery chemistries are associated with endothermic charging processes ($Q_r = T\Delta S$ values are 27 and 40 KJ for NiCd and NiMH batteries, respectively, where $Q_r$ is the thermodynamic heat effect of the process related to the entropy change $\Delta S$). An increase in the temperature of the battery indicates that the heating resulting from the Joule effect is larger than the cooling effect due to reversible heat ($T\Delta S$) remaining relatively constant because its charge reaction is endothermic. As the batteries approach 75% to 80% recharge, the voltage rises more sharply due to the start of the oxygen evolution reaction at the positive electrode. The temperature increases sharply at this stage due to the high Joule heating associated with the oxygen evolution and recombination at the negative electrode. The increase in cell temperature causes the voltage to drop as the battery reaches full charge and goes into overcharge due to the endothermic reversible heat effect ($T\Delta S$) of the charging process.

The voltage profile of the NiMH battery does not show a peak as prominent as that of the NiCd battery. The shallower drop in a NiMH battery may be due to a somewhat higher exothermic recombination reaction ($\Delta H = -572$ KJ/mole $O_2$ for NiMH vs. $\Delta H = -550$ KJ/mole $O_2$ for NiCd) that compensates for the reversible endothermic effect of the charging reactions. Both the voltage drop after peaking ($-\Delta V$) and the temperature rise can be used to terminate the charge. However, while similar charge techniques can be used for both types of batteries, the conditions to terminate the charge may differ because of the different behavior of the two battery systems during charge.

The voltage of the sealed NiMH battery during charge depends on a number of conditions, including charge current and temperature. Figures 15D.32*a* and *b* show the voltage profile of the NiMH battery at different charge rates and temperatures. The voltage rises with an increase in charge current due to higher IR and overpotential during the electrode reaction. The voltage decreases with increasing temperature as the internal resistance and the overpotential during electrode reaction decreases. The voltage peak is not as evident at low charge rates and at higher temperatures.

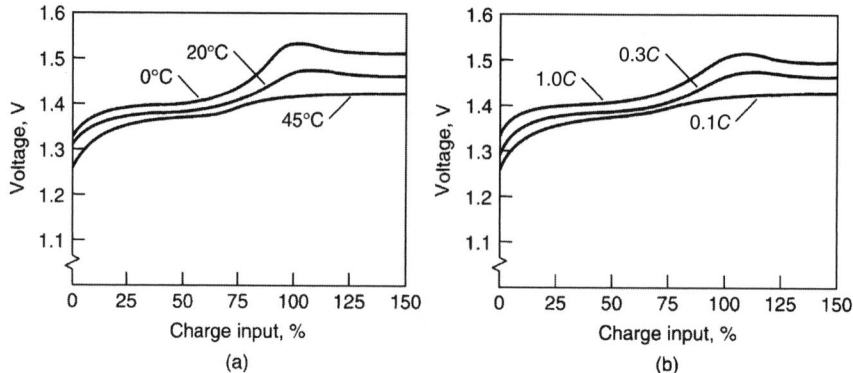

**FIGURE 15D.32**    Cell voltage versus charge input for NiMH cylindrical batteries. (*a*) At various temperatures (charge rate 0.3*C*). (*b*) At various charge rates at 20°C.

The increase in battery temperature during charge at various charge rates is shown in Fig. 15D.33. The internal cell pressure increases similarly. This rise in temperature and pressure at the higher charge rates emphasizes the need for proper charge control and effective charge termination when "fast charging" to avoid venting and other deleterious effects.

The charge efficiency is also dependent on temperature. It decreases at the higher temperatures due to increasing evolution of oxygen at the positive electrode. At lower temperatures, oxygen recombination is slowed and a rise in internal cell pressure may occur depending on the charge rate. Figure 15D.34 shows the available discharge capacity following charging at various temperatures and several charge rates. As shown, the battery capacity is reduced after elevated temperature charging. The extent of this effect is also dependent on the conditions of the discharge following the charge, as well as on other charge conditions.

Proper recharging is critical not only to obtain maximum capacity on subsequent discharges, but also to avoid elevated temperatures, overcharge, and other conditions that could adversely affect battery life.

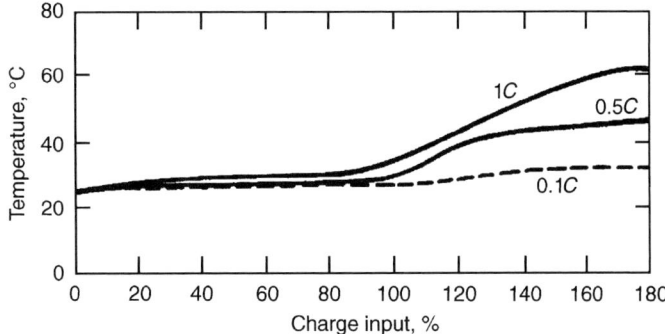

**FIGURE 15D.33**   Battery temperature during charge of NiMH cylindrical batteries.

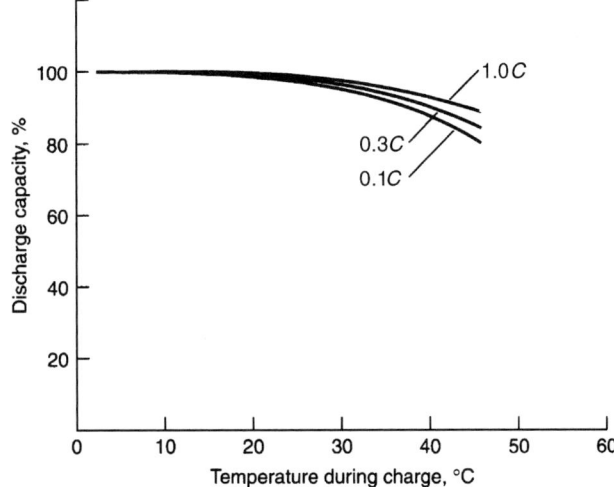

**FIGURE 15D.34**   Charge efficiency versus charge temperature at various charge rates for NiMH cylindrical batteries. Discharge at 0.2C rate to 1.0 V.

***Techniques for Charge Control.***   Characteristics of the NiMH battery define the need for charge control to terminate the charge thereby preventing the battery from being overcharged or exposed to elevated temperatures. The advantage of employing proper charge control to maximize the life of the battery is illustrated in Fig. 15D.35. The highest capacity levels are achieved with the 150% charge input, but at the expense of cycle life. The longest cycle life is attained with the 120% charge input, but with lower capacity due to insufficient charge input. Thermal cutoff (TCO) charge control may reduce cycle life because the battery is usually allowed to reach higher temperatures during charge. This method, however, is useful as a backup control in the event that the maximum temperature is exceeded during charge.

Some popular methods for charge control are summarized hereafter. The characteristics of these methods are illustrated in Fig. 15D.36. In many cases several methods are used, during a single charge, particularly to control high-rate charging.

*Timed Charge.*   Under this charge control method, charge is terminated after the battery has been charged for a predetermined amount of time. This method should only be used for charging at low rates to avoid excessive overcharge because the state-of-charge of the battery, prior to charging, cannot always be determined. This procedure is also used as "topping" charge to other charge termination methods to ensure a complete recharge.

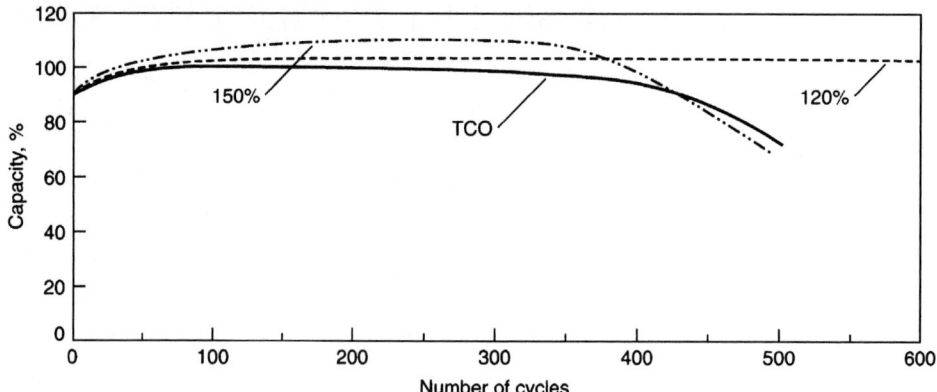

**FIGURE 15D.35**    Effect of charge control on cycle life of NiMH cylindrical batteries. 1C charge rate, discharge at 1C to 1.0 V. TCO: charge termination at 40°C; 120%: charge termination at 120% charge input; 150%: charge termination at 150% charge input.

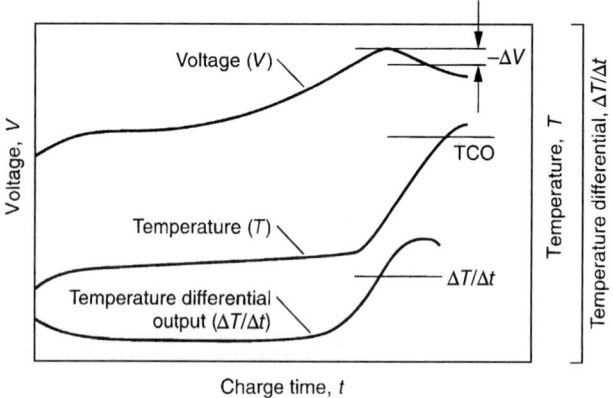

**FIGURE 15D.36**    Comparison of charge termination methods: TCO, $\Delta T/\Delta t$, and $-\Delta V$.

*Voltage Drop (–ΔV).*    With this technique, widely used with sealed nickel cadmium batteries, the voltage during charge is monitored and charge terminated when the voltage begins to decrease. This approach can be used with the NiMH battery, but as noted in "Discharge Voltage Profiles" in Sec. 15D.4, the peak voltage with the NiMH cell is not as prominent and may be absent in charge currents below the 0.3C rate, particularly at elevated temperatures. The voltage signal must be sensitive enough to terminate charge when the voltage drops, but not so sensitive that it will terminate charge prematurely due to noise or other normal voltage fluctuations. A 10 mV per cell drop is generally used for the NiMH battery.

*Voltage Plateau (0 ΔV).*    As the sealed NiMH battery does not always show an adequate voltage drop, an alternate method is to terminate charge when the voltage peaks and the slope is zero rather than waiting for the voltage to drop. The risk of overcharge is reduced as compared to the −ΔV method. A topping charge can follow to ensure a full recharge.

*Temperature Cutoff (T).*    Another technique for charge control is to monitor the temperature rise of the battery and terminate charge when the battery has reached a temperature that indicates the beginning of overcharge. It is difficult to determine this point precisely as it is influenced by ambient temperature, cell and battery design, charge rate, and other factors. For example, a cold battery may be overcharged before reaching the cutoff temperature while a warm battery may be undercharged. Usually this method is used in conjunction with other charge control techniques and mainly to terminate the charge in the event the battery reaches excessive temperatures before the other charge controls activate.

*Delta Temperature Cutoff (ΔT).* This technique measures the battery temperature rise above the ambient temperature during charging and terminates charge when this rise exceeds a predetermined value. In this way, the influence of ambient temperature is minimized. The cutoff value is dependent on several factors, including cell size, configuration and number of cells in the battery, and the heat capacity of the battery. Therefore, the cutoff value must be determined for each type of battery.

*Rate of Temperature Increase (ΔT/Δt).* In this method, the change in temperature with time is monitored and charge is terminated when a predetermined incremental temperature rise is reached. The influence of ambient temperature is virtually eliminated. A $\Delta T/\Delta t$ cutoff is a preferred charge control method for NiMH batteries because it provides long cycle life.

Note: Details on the design of batteries using protective devices and a description of the thermal protective devices that can be used for charge control are discussed in Chap. 7.

**Charging Methods.** NiMH batteries are extremely flexible and accept diverse charge methods. The principal factor in designing a charging algorithm is to prevent excessive overcharge, especially at high rates, to avoid heat buildup and electrolyte venting losses. Several methods of sensing overcharge are common, including time, absolute temperature, $\Delta T$, $\Delta T/\Delta t$, $-\Delta V$, and pressure rise. In all cases, the oxygen evolution/recombination mechanism to create heat is the basis for each charge termination approach. The method employed to terminate charge will depend on the charging rate (slow to fast charging) and should best limit the amount of heat generation in the NiMH battery in order to prevent damage.

*Low Rate Charge (12–15 h).* A convenient method to fully charge sealed NiMH batteries is to charge at a constant current at about $0.1C$ rate with time limited charge termination. At this current level, the generation of gas will not exceed the oxygen recombination rate. The charge should be terminated after 150% capacity input (approximately 15 h for a fully discharged battery). Excessive overcharge should be avoided as this can damage the battery. The temperature range for this charge method is 5 to 45°C, with the best performance obtained between 15 and 30°C.

*Quick Charge (4–5 h).* NiMH batteries can be recharged efficiently and safely at higher rates. Charge control is required to terminate the charge when the rate of oxygen recombination is exceeded or the battery temperature rises excessively. The fully discharged battery can be charged at a $0.3C$ rate for a charge time equivalent to a 150% charge input (approximately 4.5–5 h). In addition to the timer control, a TCO device should be used as a backup control to terminate charge at about 55 to 60°C to avoid exposing the battery to excessively elevated temperatures. This charging method may be used in an ambient temperature range of 10 to 45°C.

As a further precaution, the decrease in voltage ($-\Delta V$) should also be sensed to ensure that charge is terminated early enough to minimize overcharge. This is particularly advisable if the battery being charged was not fully discharged. A "topping" charge at the $0.1C$ rate, as described, may then be used to assure 100% recharge.

Charging sealed NiMH batteries at rates between 0.1 and $0.3C$ generally is not recommended. At these rates, the voltage and temperature charge profiles may not exhibit characteristics suitable for voltage-based cutoff control and the batteries may otherwise be exposed to harmful overcharge.

*Fast Charge (1 h).* Another method of charging NiMH batteries in an even shorter time is to charge at constant current at the 0.5 to $1C$ rates. At these high charge rates, it is essential that the charge be terminated early during overcharge. Timer control is inadequate as the time needed for charge cannot be predicted. A partially charged battery could easily be overcharged, while a fully discharged one could be undercharged, depending on how the timer control is set.

With fast charging, the decrease in voltage $-\Delta V$ and the increase in temperature $\Delta T$ can be used to terminate the charge. For better results, termination of fast charge can be controlled by sensing the rate of temperature increase $\Delta T/\Delta t$ with a TCO backup.

Figure 15D.37 shows the advantage of using a $\Delta T/\Delta t$ method compared to $-\Delta V$ in terminating a fast charge. The $\Delta T/\Delta t$ method can sense the start of the overcharge earlier than the $-\Delta V$ method. The battery is exposed to less overcharge and overheating, resulting in less loss of cycle life. A temperature increase of 1°C/min should be used for $\Delta T/\Delta t$; a temperature of 60°C is recommended for the TCO.

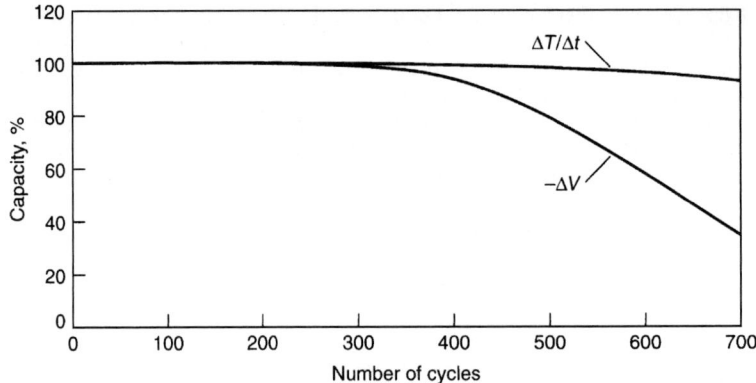

**FIGURE 15D.37**   Cycle life and capacity as a function of charge termination method for NiMH cylindrical batteries. Charge 1C, rest 30 min, discharge 1C to 1.0 V, rest for 2 h.

In the case of multicell batteries of three cells or more, $-\Delta V$ termination with TCO backup may be adequate. The $-\Delta V$ value usually is 10 to 15 mV per cell and 60°C for the TCO.

While the infrastructure to fast charge EV batteries is not fully in place, a strength of NiMH battery technology is the ability to accept fast charge when it is required and the proper power is available. As an example, a typical EV cell capacity may be ~100 Ah. Therefore, if 15 min charging is desired, currents of ~400 Amps (at 360+ V) must be available.

NiMH batteries can accept from 60% to 80% charge within 15 min at which point current must be reduced. Temperature rise due to internal resistance heating is small on the order of 15°C for a 33-kWh battery, whereas heating due to oxygen recombination is very large. Again, the crucial aspect of fast charge is proper sensing of the onset of overcharge.

*Trickle Charge.*   Many applications require the use of batteries that are maintained in a fully charged condition. This is done by trickle charging—charging at a rate that will replace the capacity loss due to self-discharge. A trickle charge at a current of between 0.03 and 0.05C rates is recommended. The preferred temperature range for trickle charging is between 10 and 35°C. Trickle charge may be used following any of the previously discussed charging methods.

*Three-Step Charge Procedure.*   A three-step procedure provides a means of rapidly charging a sealed NiMH battery to full charge without excessive overcharge or exposure to elevated temperatures.

1. The first step is a charge at a 1C rate terminated by using the $\Delta T/\Delta t$ method or the $-\Delta V$ method.

2. This is followed by a 0.1C topping charge, terminated by a timer after ½ to 1 h of charge.

3. The third step is a maintenance charge of indefinite duration at a current of between the 0.05 and 0.02C rates. The battery should also be protected with a TCO device to terminate the charge so that the temperature does not exceed 60°C.

**Charge Algorithms.**   Charge algorithms refer to the programming used to charge the NiMH battery pack. Conceptually, the battery can accept charge input at extremely high rates until the battery is about 80% fully charged, at which point the charge current must be reduced. In overcharge, the charge current should not exceed the 10-h charge rate, and generally, the total charge input should be below about 110% to 120%.

An inherent virtue of NiMH batteries is that simple charging methodology can be used, with inexpensive equipment, and that it is not necessary to monitor the voltage of each cell in the high voltage string.

One method used to reliably charge large format NiMH batteries utilizes constant current steps to a temperature-compensated voltage limit. Practically, this means that charging is done at high current until a predetermined voltage is reached which is an indication of a certain state-of-charge. At this voltage, the current is reduced or stepped down, until another predetermined voltage is reached. An important aspect of the charge algorithm is that the prescribed voltage set points are variable, based on temperature and current.

## 15D.4.7 Applications

*Consumer.* Since their market introduction, NiMH batteries have played a significant role in the evolving world of consumer electronics. While many early applications (digital cameras, cellphones, laptops) have become obsolete or transitioned to lighter weight Li-ion batteries, new devices and uses have emerged. Today's NiMH batteries power household appliances ranging from electric toothbrushes to robotic vacuum cleaners. However, the fastest growing use for consumer NiMH batteries is replacement of disposable alkaline batteries. Consumers have realized that a single package of rechargeable NiMH batteries can replace hundreds, if not thousands, of single-use alkaline cells.[42] Switching to NiMH eases the strain on the family budget as well as the local landfill.

*Vehicle.* Ten years after its introduction into consumer electronics, NiMH became the enabling technology for HEVs. As discussed earlier, nearly 12 million HEVs have been placed in service worldwide with NiMH batteries providing safe, reliable, life-of-vehicle performance. While battery weight reduces the attractiveness of NiMH for plug-in hybrids and battery EVs, the major supplier of HEVs believes NiMH offers distinct advantages for the HEV platform.[43] In addition to HEVs, start-stop vehicles are now being offered with NiMH batteries.[44]

*Stationary/Industrial.* Contrary to conventional wisdom, the operating characteristics of NiMH battery technology are well suited for use in stationary and industrial applications. Most of these applications have evolved around lead-acid batteries and lead continues to dominate the industrial market today. However, changes in usage, system configuration, and/or added functionality have, in many cases, exceeded the capabilities of lead acid and an advanced battery solution is sorely needed.

*Stationary/Backup Power (telecom).* The U.S. telephone network is a perfect example. Initially the network was powered from central offices where dedicated battery rooms and personnel could maintain batteries at ideal conditions. As the network evolved into a more distributed power system, batteries were moved away from the central office to backup remote power nodes housed in outdoor enclosures. The Institute of Electrical and Electronics Engineers (IEEE) recognizes the need for routine maintenance of valve-regulated lead-acid (VRLA) batteries to optimize performance and operating life,[45] but maintenance of remote battery installations is often neglected leading to premature battery failure. Leveraging the history of NiMH in HEVs, several companies are now introducing NiMH batteries specifically designed for telecom backup.[46]

*High-Temperature Applications.* In many cases, the outdoor telecom enclosures described above are simply metal cabinets that bake in the sun. Internal temperatures can exceed 60°C/140°F, leading to drastically short-ened battery life. Over the past few years, much work has been done to improve the heat tolerance of NiMH. Most of the effort has concentrated on cathode materials and electrolyte additives to improve charge acceptance at elevated temperature. Industrial grade NiMH batteries are now available with the ability to achieve full charge up to 75°C/167°F. Laboratory and field testing of these NiMH batteries in telecom equipment has demonstrated the ability to operate at these elevated temperatures for long durations. Capacity retention better than 95% after 14 months at 55°C/131°F have been reported.[47] Long life at elevated temperature is important to a wide range of applications beyond telecom: remote outdoor equipment, industrial process, and emergency lighting to name a few. High heat tolerance can also benefit applications with significant cooling loads like data centers and uninterruptible power supplies (UPSs).

*High-Cycle Applications.* In many regions of the world, electricity supply is not as stable as it is in the United States. In these areas, a backup battery must not only display long operating life, but must also have high cycle life. As shown earlier in Fig. 15D.25, NiMH is an excellent battery technology for cycling, capable of delivering thousands of discharge cycles to depths of 80% or more. Deep cycle capability is a very important consideration for photovoltaic storage and other emerging grid energy storage applications where battery capacity can be exercised multiple times per day (see Chap. 27). As this market continues to evolve, NiMH batteries are certain to play an important role. In the meantime, NiMH batteries are demonstrating valuable return on investment in railroad applications where regenerative braking energy from trains approaching a station is captured for later use in passenger comfort loads, line voltage regulation, and emergency power needs. Numerous installations of wayside railroad NiMH battery systems are providing significant energy savings and peak power reduction.[48] The typical wayside railroad storage system experiences approximately 4000 charge/discharge cycles daily, similar to the conditions experienced in large solar or wind farming applications.

*High-Rate Applications.*   The material handling equipment market is a prime example of an application where changes in usage demand a better battery. Similar to stationary backup power, this market has been dominated by lead-acid batteries. While these batteries are sufficient to handle most standard operations, slow charge rates and runtime limitations create an opportunity for NiMH in heavy use operations. Distribution centers and manufacturing operations that operate around the clock typically require three batteries for every vehicle in their forklift fleet. This reduces productivity, adds expense, and wastes floor space to charge, store, and swap-out batteries. There is a growing trend in the industry toward fast opportunity charging where every forklift uses a single battery that is charged as much as possible during breaks and other idle periods in an attempt to reduce the number of battery change outs. Because of heating issues, lead-acid batteries are limited to charge rates of 0.4C or less, whereas NiMH batteries have no problem charging at rates of 2C or higher. Numerous programs are currently underway to develop large NiMH batteries for use in forklifts and other material handling equipment (see Chap. 26B).

## 15D.4.8   Next Generation NiMH

Since the initial introduction of portable NiMH in 1987, great strides have been made in terms of increasing specific energy (from 52 up to 80–110 Wh/kg), specific power (from 180 up to 850–2000 W/kg), cycle life (from 300 up to 1000+), charge retention (from 70% loss in 1 month to less than 15% in 1 year), and cell sizes (from 1 to 4 Ah to a wide range from 30 mAh to 250 Ah).

Intensive development of NiMH battery technology continues in many diverse worldwide locations to reduce NiMH battery cost and further improve performance. Activities include development of new active materials and electrolytes, substitution of expensive raw materials, new processing techniques, novel battery designs, and more. As these developments progress, NiMH technology is poised to expand into markets and applications dominated by other battery chemistries. Higher storage density and lower impedance cell designs will enable high power operation at lower temperatures, opening new opportunities in vehicle and industrial applications. Optimization of monoblock and pouch construction designs will lead to cost-effective, large-format NiMH battery solutions for high cycle life stationary power, grid energy storage, and other applications where safety and reliability are paramount.

## *REFERENCES*

1. S. R. Ovshinsky, S. K. Dhar, M. A. Fetcenko, K. Young, B. Reichman, C. Fierro, J. Koch et al., *17th International Seminar & Exhibit on Primary and Secondary Batteries,* Ft. Lauderdale, Florida, March 6–9, 2000.

2. R. C. Stempel, S. R. Ovshinsky, P. R. Gifford, and D. A. Corrigan, *IEEE Spectrum,* Vol. 35,(11), November 1998.

3. S. R. Ovshinsky, *Materials Research Society*, MRS Fall Meeting, Boston, MA, November 1998.

4. K. Sapru, B. Reichman, A. Reger, and S. R. Ovshinsky, U.S. Patent 4,623,597, 1986.

5. S. R. Ovshinsky, M. Fetcenko, and J. Ross, *Science* **260:**176 (1993).

6. S. R. Ovshinsky, "Disordered Materials: Science and Technology," in D. Adler, B. Schwartz, and M. Silver (eds.), Institute for Amorphous Studies Series, Plenum Publishing Corporation, New York, 1991.

7. J. R. van Beek, H. C. Donkersloot, and J. J. G. Willems, *Proceedings of the 14th International Power Sources Symposium,* 1984.

8. R. Kirchheim, F. Sommer, and G. Schluckebier, *Acta Metall.* **30:**1059 (1982).

9. R. Mishima, H. Miyamura, T. Sakai, N. Kuriyama, H. Ishikawa, and I. Uehara, *J. Alloy. Compd.* **192:**176 (1993).

10. T. Weizhong and S. Guangfei, *J. Alloy Compd.* **203:**195 (1994).

11. P. H. L. Notten and P. Hokkeling, *J. Electrochem. Soc.* **138:**1877 (1991).

12. P. H. L. Notten, J. L. C. Daams, and R. E. F. Einerhand, *Ber. Bunsenges. Phys. Chem.* **96**(5) (1992).

13. W. Zhou, Q. Wang, D. Zhu, C. Wu, L. Huang, Z. Ma, Z. Tang et al. *Int. J. Hydrogen Energy* **41:**14852 (2016).

14. H. Li, Y. Fei, Y. Wang, L. Chen, and H. Jiang, *J. Rare Earths,* **33:**633 (2015).

15. K. Beccu, U.S. Patent 3,669,745, 1972.

16. M. H. J. van Rijswick, *Proceedings of the International Symposium on Hydrides for Energy Storage*, Pergamon, Oxford, 1978, p. 261.

17. M. A. Gutjahr, H. Buchner, K. D. Beccu, and H. Saufferer, *Power Sources 4*, in D. H. Collins (ed.), Oriel, Newcastle upon Tyne, United Kingdom, 1973, p. 79.

18. M. A. Fetcenko, S. R. Ovshinsky, B. Chao, and B. Reichman, U.S. Patent 5,536,591, 1996.

19. M. Oshitani, H. Yufu, K. Takashima, S. Tsuji, and Y. Matsumaru, *J. Electrochem. Soc.* **136:**1590 (1989).

20. M. Oshitani and H. Yufu, U.S. Patent 4,844,999, 1989.

21. V. Puglisi, *17th International Seminar & Exhibit on Primary and Secondary Batteries*, Ft. Lauderdale, FL, March 6–9, 2000.

22. G. Halpert, *Proceedings of the Symposium on Nickel Hydroxide Electrodes*, Electrochemical Society, Hollywood, FL, October 1989 (Electrochemical Society, Pennington, NJ, 1990), pp. 3–17.

23. V. Ettel, J. Ambrose, K. Cushnie, J. A. E. Bell, V. Paserin, and P. J. Kalal, U.S. Patent 5,700,363, 1997.

24. K. Ohta, H. Matsuda, M. Ikoma, N. Morishita, and Y. Toyoguchi, U.S. Patent 5,571,636, 1996.

25. I. Matsumoto, H. Ogawa, T. Iwaki, and M. Ikeyama, *16th International Power Sources Symposium*, 1988.

26. S. Takagi and T. Minohara, *Society of Automotive Engineers*, 2000-01-1060, March 2000.

27. K. Watanabe, M. Koseki, and N. Kumagai, *J. Power Sources* **58:**23 (1996).

28. K. Young, L. Wang, S. Yan, X. Liao, T. Meng, H. Shen, and W. C. Mays, *Batteries,* **3:**6 (2017).

29. T. Meng, K. Young, D. F. Wong, and J. Nei, *Batteries* **3:**4 (2017).

30. T. Meng, K. Young, D. Beglau, S. Yan, P. Zeng, and M. M. Cheng, *J. Power Sources* **302:**13 (2016).

31. ILCO Chemikalien GmbH. Ionic Liquid. Available online: http://www.ilco-chemie.de/downloads/Ionic%20Liquid.pdf.

32. M. A. Fetcenko, S. Venkatesan, and S. Ovshinsky, in *Proceedings of the Symposium on Hydrogen Storage Materials, Batteries, and Electrochemistry*, Electrochemical Society, Pennington, NJ, 1992, p. 141.

33. J. Cook, "Separator—Hidden Talent," Electric & Hybrid Vehicle Technology, 1999.

34. http://global.kawasaki.com/en/energy/solutions/battery_energy/about_gigacell/index.html.

35. http://www.fdk.com/whatsnew-e/release20170215-e.html.

36. http://www.nilar.com/design-technology/.

37. F. J. Kruger, *15th International Seminar on Primary and Secondary Batteries*, Ft. Lauderdale, FL, March 1998.

38. D. A. Corrigan, S. Venkatesan, P. Gifford, A. Holland, M. A. Fetcenko, S. K. Dhar, and S. R. Ovshinsky, *Proceedings of the 14th International Electric Vehicle Symposium*, Orlando, FL, 1997.

39. I. Kanagawa, *15th International Seminar on Primary and Secondary Batteries,* Ft. Lauderdale, Florida, March 1998.

40. M. A. Fetcenko, S. Venkatesan, K. C. Hong, and B. Reichman, *Proceedings of the 16th International Power Sources Symposium*, International Power Sources Committee, Surrey, United Kingdom, 1988, p. 411.

41. D. Singh, T. Wu, M. Wendling, P. Bendale, J. Ware, D. Ritter, and L. Zhang, *Mater. Res. Soc. Proc.* **496:**25–36 (1998).

42. http://www.gpbatteries.com/int_en/powerbank/usb/u411.

43. http://www.carscoops.com/2015/11/this-is-why-toyota-offers-two-different.html.

44. http://news.panasonic.com/global/press/data/2014/02/en140213-3/en140213-3.html.

45. 1188-2005—IEEE Recommended Practice for Maintenance, Testing, and Replacement of Valve-Regulated Lead-Acid (VRLA) Batteries for Stationary Applications.

46. K. Borders, "It's The Small Things That Matter," OSP Magazine, October 2015.

47. M. Zelinsky, "Market Development of NiMH Batteries For Stationary Applications," Battcon 2016.

48. WMATA Energy Storage Demonstration Project, Federal Transit Administration Final Report, June 2015.

# SECTION E

# NICKEL-ZINC

**Adam Weisenstein, Eivind Listerud, Allen Charkey**

## 15E.1   OVERVIEW

### 15E.1.1   General Characteristics

The nickel-zinc (zinc/nickel-oxide) battery is an alkaline rechargeable system with a relatively high energy density, excellent power capability, and high abuse tolerance. It is composed of a positive nickel electrode and a negative zinc electrode that are similar to individual electrodes used in other batteries such as nickel-cadmium, nickel-iron, nickel-metal hydride, and silver-zinc. Currently, the nickel-zinc system is capable of delivering from 60 to 110 Wh/kg and 90 to 250 Wh/L, depending on the specific design. The batteries can be cycled more than 500 times at 100% depth of discharge (DOD) and can achieve several thousand cycles at low DOD. Additional advantages of the nickel-zinc battery include fast recharge capability, good low temperature performance, a sealed maintenance-free design, and an environmentally acceptable chemistry. It is made with abundant low-cost materials that are readily recyclable. It has lower cost than lithium ion and does not require an extensive battery management system. Nickel-zinc is a lightweight and low-cost alternative to lead-acid, nickel-metal hydride, nickel-cadmium, and lithium-ion technologies. It is appropriate for use in applications such as heavy trucking, deep cycle marine, stop-start (microhybrid), and other transportation industries. Moreover, the fast recharge capability and the good cycle life make the nickel-zinc chemistry suitable for various stationary storage applications, like remote telecom, data center back-up, and grid storage.

### 15E.1.2   Background

The goal of the initial development of the nickel-zinc battery was to combine the long cycle life associated with the nickel electrode of nickel-cadmium batteries with the high specific energy of the zinc electrode, typically used in silver-zinc batteries. Historically, the nickel-zinc battery dates back at least to 1899 in a patent by Michalowski.[1] Further work with nickel-zinc battery-powered railcars was performed by Drumm in Ireland in the 1930s.[2] There was a serious effort in the 1960s to develop nickel-zinc as a longer life replacement for the silver-zinc battery in military applications.[3,4] Considerable effort was again focused on nickel-zinc in the 1970s in response to the increased interest in electric vehicles that resulted from an energy crisis and increasing gasoline prices. System development was hampered for many years by the limited cycle life associated with the zinc electrode.[5] This cycle life limitation was primarily due to the solubility of zinc in the alkaline electrolyte. This issue has been overcome by recent developments in stabilizing the zinc in the electrolyte and thereby reducing its solubility and increasing the cycle life of the battery. In the early 2000s, Evercel introduced a variety of large prismatic cells and batteries that achieved up to 600 cycles.[6] In addition, they introduced a gas recombination catalyst that controlled internal pressure and limited gas escape to scenarios with significant overcharge.[7] This recombination assembly allowed for a sealed, maintenance-free design. Subsequently, in 2008, PowerGenix introduced a sealed cylindrical nickel-zinc battery.[8] With ongoing advancements in manufacturing techniques and material development, there have been further improvements in the energy and power densities of the nickel-zinc system in recent years. This progress has attracted the interest of those looking to replace lead-acid batteries and those looking for a lower cost and safer alternative to lithium ion.

## 15E.2   CHEMISTRY

The nickel-zinc battery system uses a nickel/nickel oxide electrode (also known as the nickel-hydroxide/nickel oxyhydroxide electrode) as the positive and a zinc/zinc oxide electrode as the negative (see Sec. 15.0 for further details on cathodes, electrolyte, etc., common to the nickel cathode batteries). When the battery is discharged, nickel (III) oxyhydroxide is reduced to nickel (II) hydroxide, and metallic zinc (0) is oxidized to zinc (II) oxide/hydroxide. The reactions presented below are for illustration as the electrochemistry of zinc in alkaline solutions is considerably more complicated.[9] The theoretical open-circuit voltage of the overall electrochemical reaction is 1.73 V. When the battery is overcharged, oxygen is produced at the nickel electrode and hydrogen is produced at the zinc electrode. These gases may then be recombined to form water. In addition, oxygen produced at the nickel electrode during overcharge may recombine with metallic zinc directly at the zinc electrode. If the battery is over-discharged, hydrogen is formed at the nickel electrode and oxygen may be produced at the zinc electrode. In practical batteries, these reactions are influenced by the balance of active materials present and the active material utilization of the two electrodes. The simplified, representative electrochemical reactions are as follows:

**Discharge**

| | | |
|---|---|---|
| *Positive electrode:* | $2NiOOH + 2H_2O + 2e^- \rightarrow 2Ni(OH)_2 + 2OH^-$ | $E^0 = 0.49$ V |
| *Negative electrode:* | $Zn + 2OH^- \rightarrow Zn(OH)_2 + 2e^-$ | $E^0 = 1.24$ V |
| *Overall reaction:* | $2NiOOH + 2H_2O + Zn \rightarrow 2Ni(OH)_2 + Zn(OH)_2$ | $E^0 = 1.73$ V |

**Charge**

*Overall reaction:*   $2Ni(OH)_2 + Zn(OH)_2 \rightarrow 2NiOOH + 2H_2O + Zn$

**Overcharge**

*Positive electrode:*   $2OH^- \rightarrow \frac{1}{2}O_2 + H_2O + 2e^-$

*Negative electrode:*   $2H_2O + 2e^- \rightarrow H_2 + 2OH^-$

$Zn + \frac{1}{2}O_2 \rightarrow ZnO$   (Recombination of oxygen from the positive electrode)

*Overall reaction:*   $H_2O \rightarrow H_2 + \frac{1}{2}O_2$

## 15E.3   CELL DESIGN AND CONSTRUCTION

### 15E.3.1   Cell Components

*Positive Electrode.*   The electrochemically active material in the positive electrode is nickel hydroxide and is very similar to the nickel electrodes found in nickel-metal hydride and nickel-cadmium chemistries. The commonly utilized β-polymorph of nickel hydroxide has a theoretical specific capacity of 289 mAh/g; however, studies have been performed on the stabilization of the α-polymorph, which has exhibited a specific capacity up to 331 mAh/g.[10–12]

The most common method of positive electrode fabrication is a pasting or slurry coating process, where nickel hydroxide, conductive additives, and binders are combined with a solvent and deposited on a current collector. Traditional conductive additives include cobalt in either the hydroxide, metallic, or oxide form; nickel metal; natural or synthetic graphite; and activated carbons. Typically, polytetrafluoroethylene (PTFE) has been used as the primary binder along with gelling agents and plasticizers to form a workable and robust slurry. The current collector can be composed of metal foam, expanded metal, or metal foil that is typically nickel metal or nickel-plated steel. An image of a slurry-coated foam nickel positive electrode is shown in Fig. 15E.1.

Sintered and roll-bonded nickel electrodes have been utilized in nickel-zinc cells in the past with successful results. However, sintered nickel electrodes contain a relatively low amount of active materials and are expensive to manufacture, and roll-bonded electrodes use organic solvents and require a higher percentage of conductive additives.[8] Due to these drawbacks, the aqueous pasting or slurry coating technique is the most widely adopted for the manufacturing of nickel positive electrodes.

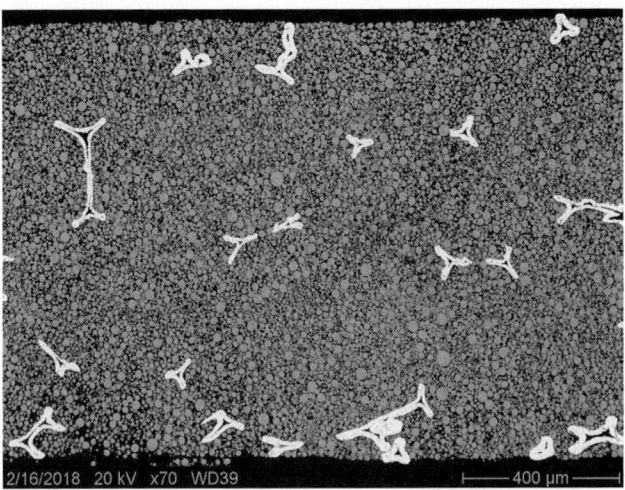

**FIGURE 15E.1**   The cross section of a nickel positive electrode made with slurry coated on foam. The bright areas indicate the nickel current collector. (*Courtesy of ZAF Energy Systems.*)

***Negative Electrode.***   The negative electrode is mainly made of zinc, in the starting form of either zinc metal or zinc oxide, blended with zinc stabilization and hydrogen suppression additives. Typically, most of the electrochemically active material is zinc oxide, which has a theoretical specific capacity of 659 mAh/g. The most common method of negative electrode fabrication is a pasting or slurry coating process, where zinc oxide, bulk additives, and binders are combined with a solvent and deposited on a current collector.

Zinc is partially soluble in the alkaline electrolyte and dissolves to form zincate anions (e.g., $Zn(OH)_4^{2-}$). This dissolution process can lead to shape change and loss in negative electrode capacity during discharge and dendritic growth during charge. Consequently, zinc stabilization additives are needed to increase the cycle life of the cell.[13,14] Traditionally, calcium has been added to the negative electrode to form calcium zincate during cycling. This compound has very low solubility in most alkaline electrolytes. Several other oxides, hydroxides, and stearates have also been utilized as successful zinc shape change inhibitors.[8] The zinc electrode tends to oxidize and evolve hydrogen, creating potential issues in the cell. Historically, hydrogen suppression additives such as lead, cadmium, and mercury were added to the negative electrode to combat this problem; however, due to environmental concerns, they have been eliminated from most electrodes.[8] A wide variety of oxide and hydroxide substitutes, such as indium, thallium, bismuth, and tin, have also been utilized.[9] Other additives include conductive aids and electrolyte wicking materials. Typically, PTFE has been used as the primary binder along with gelling agents and plasticizers to form a workable and robust slurry. Current collectors can be foam, expanded metal, and foil and are typically plated copper metal. A negative electrode made with a slurry, coated on copper foam, is shown in Fig. 15E.2.

***Separator.***   Nickel-zinc cells typically require a multicomponent separator system to act as both a zinc barrier and as an electrolyte reservoir. Typically, the zinc barrier is a microporous and tortuous polymer layer designed for high mechanical strength and dendritic penetration resistance. Multiple layers of the polymer separator are used to extend life by increasing the tortuosity and, therefore, the zincate travel path to the positive electrode. The multiple layers can also provide higher mechanical strength and resistance to dendritic penetration. The electrolyte reservoir is typically a nonwoven material capable of retaining large amounts of electrolyte. The number of layers and thickness of the reservoir is dependent on the capacity and cycle life requirements of the application.

***Electrolyte.***   Nickel-zinc electrolytes are composed of 20 to 35 weight% aqueous potassium hydroxide (KOH) with low amounts of lithium hydroxide. The ideal concentration of KOH is 28 weight% for conductivity and 32 weight% for freezing point.[15] However, the solubility of the zinc negative electrode is increased as the

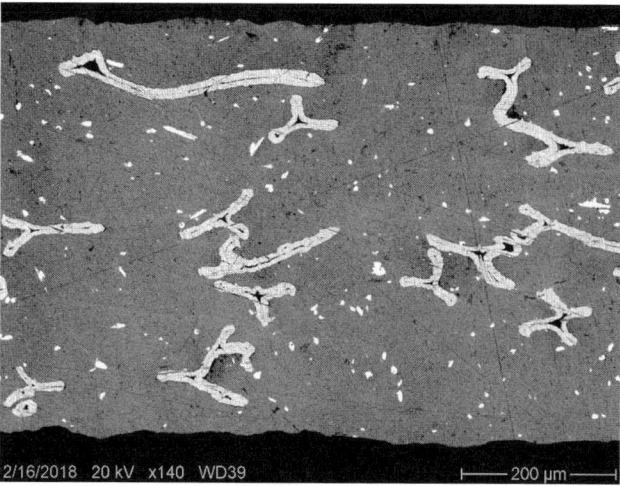

**FIGURE 15E.2** The cross section of a negative electrode made with slurry coated on copper foam. The bright areas indicate the copper current collector and other additives. (*Courtesy of ZAF Energy Systems.*)

concentration of KOH is increased, which limits the KOH% utilized in higher cycle life cells. To combat zinc shape change and migration, many additives have been added to the KOH electrolyte in the form of fluorides, borates, phosphates, and carbonates.[16–18]

## 15E.3.2  Cell and Battery Construction

Nickel-zinc cells and batteries have traditionally followed the same format as nickel-cadmium and nickel-metal hydride batteries in both prismatic and cylindrical configurations. Prismatic cell designs have been developed, ranging from 2 to 200 Ah, and the technology is readily scalable in prismatic, cylindrical, and monobloc designs.[9] An example of a sealed prismatic 12 V, 165 Ah (*C*/3 nominal rate) Group 31 battery and cell is shown in Fig. 15E.3, with a detailed layout of the cell construction shown in Fig. 15E.4.

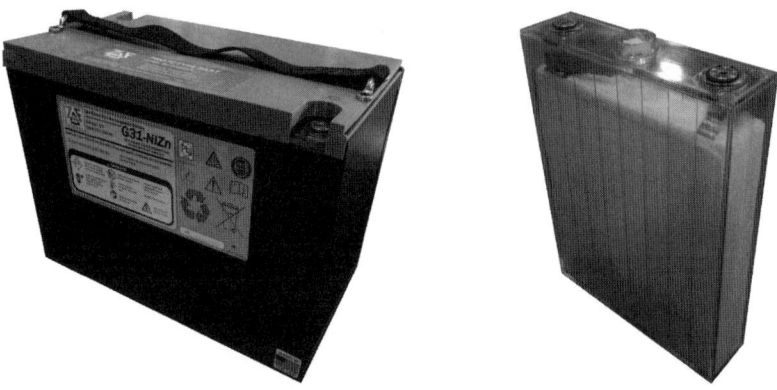

**FIGURE 15E.3** A G31 battery made with prismatic sealed 165 Ah Ni-Zn cells. (*Courtesy of ZAF Energy Systems.*)

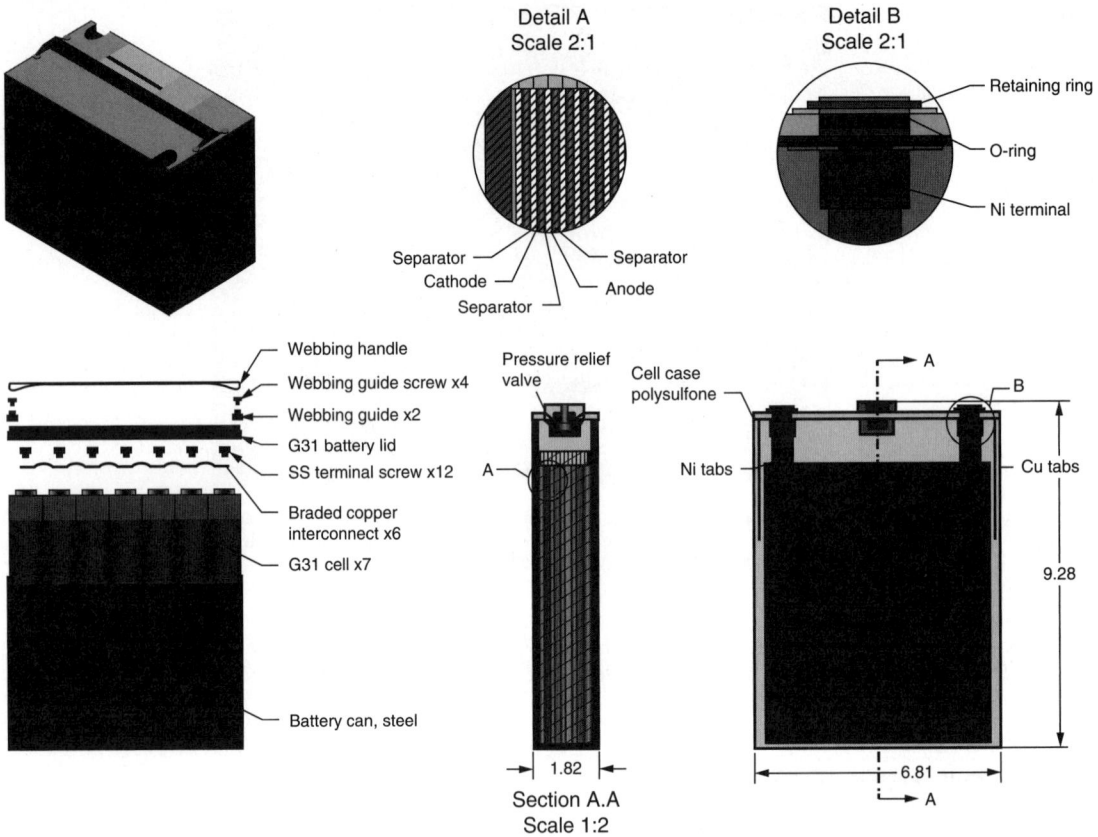

**FIGURE 15E.4** Prismatic cell construction for a Group 31 battery design. (*Courtesy of ZAF Energy Systems.*)

The typical prismatic cell design has alternating negative and positive electrodes in a 2:1 to 3:1 negative to positive capacity ratio. Cells designed with the higher negative to positive ratio will typically demonstrate longer cycle life; however, power and/or capacity will be reduced. A separator system will be between the alternating electrodes, which can include sets of the microporous polymer layers and electrolyte reservoir layers. Metallic conductors, commonly nickel or copper, are used for the electrode tabs as well as the cell and battery terminals.[9] The terminal to case sealing is typically accomplished through radial or face sealing O-rings or epoxy.

Most sealed cell designs are either starved or semi-starved with electrolyte. The electrolyte is typically vacuum filled into the cell due to production time constraints, as well as to fill electrode pores and electrolyte reservoirs as much as possible to combat the dry-out failure mode. Similar to valve-regulated lead-acid or Ni-Cd/NiMH (Chaps. 15C and 15D) batteries, the sealed prismatic design includes a resealable pressure relief valve that prevents excessive pressure within the cell due to overcharge or other catastrophic events. This pressure relief valve typically has an opening pressure of 2 to 20 psi and is sealed to the cell case via O-rings or epoxy. Some nickel-zinc cell designs use an auxiliary catalytic gas recombination electrode to facilitate gas management and help prevent dry-out by recombining the oxygen and hydrogen gases, created during charge, back into water. The cell case is typically molded from a commercial grade, alkali compatible, and mechanically robust resin of a thermoplastic, such as polysulfone. The cell case is sealed by ultrasonic welding or epoxy adhesive, or a combination of both. For a 12 V battery design, seven cells are connected in series via metal interconnects and placed into an enclosure. The battery enclosure is designed to constrain the cells, which expand and contract during cycling, and help protect the cells from impact and the environment.

### 15E.3.3    Design Considerations

Overall the nickel-zinc battery has a theoretical specific energy of 334 Wh/kg, which makes it a very attractive system for a number of applications. Practical batteries operate at a loaded discharge voltage of 1.55 to 1.65 V and may deliver up to 90 Wh/kg in larger prismatic formats, depending on the specific design. In recent years, PowerGenix and ZAF Energy Systems have demonstrated a specific energy density of 110 Wh/kg, using small cylindrical[7] and pouch cell formats, respectively. This is only about 32% of the theoretical specific energy, which indicates that there is room for system improvement. Nickel-zinc cells can be designed for power, capacity, and long cycle life configurations to accommodate the requirements of a particular application.

The power design, compared to the capacity design, will typically have thinner electrodes and a higher number of positive and negative electrodes in the cell stack. The positive electrodes in the power design may contain more conductive additives and a higher current collector to active material ratio. Additionally, the tabs will typically have a larger cross-sectional and electrode contact area for conduction purposes. The reduction in active material due to greater amounts of additives and larger current collectors and tabbing material will result in a loss of capacity. Long cycle life configurations will typically maximize the negative to positive capacity ratio and approach a 3:1 ratio in their design. A more robust separator system, including more than two layers of separator material, and an electrolyte with lower KOH molarity have also shown to increase the cycle life of nickel-zinc batteries.

## 15E.4    PERFORMANCE CHARACTERISTICS AND APPLICATIONS

### 15E.4.1    Failure Mechanisms

***Zinc Shape Change, Zinc Passivation, and Dendritic Shorting.***    Historically, using zinc as the negative electrode has presented challenges in obtaining specific energies close to the nickel-zinc system's theoretical value and achieving long cycle life. The main reason behind this is that the discharge product of the zinc electrode, i.e., zinc oxide, hydroxide, or zincate, can partially dissolve in alkaline electrolyte, creating various issues. The effect of the zinc dissolving in the electrolyte can manifest itself in several ways.

As zinc dissolves in the electrolyte during cycling, zinc species may move away from their original locations. During charge, zinc will deposit back on the electrode; however, deposition can occur in different locations, changing the electrode surface topology and creating zinc structures in the zinc electrode. This redistribution of zinc is often referred to as "shape change." The zinc may also cluster away from the electrode creating isolated zinc active material that will compromise power and utilization. During recharging, some of the dissolved zinc may also plate onto the electrode high points/protrusions rather than distribute in a uniform manner. Repeated charge cycles will lead to further growth of fernlike zinc dendrite structures to the point where they eventually penetrate the separator and create a short in the cell. This dendritic shorting is dictated by current density distribution, surface shape, and electrolyte molarity.

***Dry-Out.***    A common failure mode encountered in long-term cycling of nickel-zinc cells is the loss of electrolyte or dry-out.[19] In sealed cells, dry-out can be caused by the production of oxygen and hydrogen from the decomposition of water at the positive and the negative electrodes during overcharge, respectively. The equivalent water electrolysis is typically encountered at the end of charge, where charge voltage can play a large part in the rate of this reaction. Water loss can also occur by self-discharge of the negative and positive electrodes. Negative electrode additives, such as oxides and hydroxides of indium, thallium, bismuth, lead, mercury, and tin, have been utilized as hydrogen suppressants.[9] Recombinant devices have been used in sealed nickel-zinc cells to combat the dry-out failure mode. These devices include catalysts that will recombine the oxygen and hydrogen back into water that subsequently is incorporated back into the electrolyte.

***Positive Electrode Degradation.***    The positive electrode can also degrade over time causing a decrease in utilization and power. The two main culprits of degradation are carbon corrosion and gamma nickel oxyhydroxide formation. Carbon is typically used in the positive electrode to increase conductivity and high-power performance; however, the carbon will oxidize over the life of the cell causing an increase in resistance, mechanical degradation, and potassium carbonate species. The rate at which oxidation occurs is dictated by the carbon surface area, cycling voltages, and molarity of the electrolyte. The gamma polymorph formation in the positive electrode can cause utilization loss in the cell and is most commonly caused by cell overcharge.

## 15E.4.2   Performance and Life Capabilities

The general characteristics of typical nickel-zinc cells and batteries are shown in Table 15E.1.[8,9] The ranges provided for nickel-zinc specifications encompass power, capacity, and cycle life designs; however, extremely one-dimensional designs can fall outside of the ranges provided. Nickel-zinc cells and batteries have a very broad operating range from −30 to 75°C and a relatively high nominal voltage at 1.65 V, when compared to nickel-metal hydride and nickel-cadmium, which are nominally 1.2 V. Large prismatic nickel-zinc cells and batteries can deliver 60 to 90 Wh/kg and 90 to 170 Wh/L, depending on specific design characteristics. Nickel-zinc batteries also have high power discharge and charge capabilities, along with relatively low self-discharge rates. As with all batteries, cycle life specifications are highly variable and dependent upon the design, discharge and charge conditions, as well as the defined capacity level at which the test is terminated. That being said, nickel-zinc systems have demonstrated >500 cycles at 100% DOD and >8000 cycles at 10% DOD.

**TABLE 15E.1**   Typical Specification Overview of Large Prismatic Ni-Zn Cells

| Parameter | Value/type |
|---|---|
| Positive electrode electrochemistry | $Ni(OH)_2/NiOOH$ |
| Negative electrode electrochemistry | $ZnO/Zn$ |
| Electrolyte (% potassium hydroxide) | 20–35% |
| Separator | Microporous + wicking |
| Nominal cell voltage | 1.65 |
| Operating temperature range, °C | −30 to +75 |
| Theoretical specific energy, Wh/kg | 334 |
| Specific energy, Wh/kg | 60–90 |
| Energy density, Wh/L | 90–170 |
| Specific power, W/kg | 290 (for 10 min to 1 V) |
| Power density, W/L | 550 (for 10 min to 1 V) |
| Charge retention, % loss/month | <10% |
| Cycle life (cycles at 100% DOD) | >500 |

*Discharge Characteristics.*   During discharge, the lower cutoff voltage for a nickel-zinc system is typically between 1.0 and 1.3 V per cell. At high discharge rates and/or low temperatures, the voltage is allowed to drop to 1.0 V or lower; however, there is a risk of gas generation if the voltage is allowed to drop significantly below 1 V. For nominal discharge rates and optimum cycle life, the voltage cutoff is typically 1.25 V. Utilization as a function of $C$-rate is shown in Fig. 15E.5. In this test, a single 147 Ah ($C/3$) ZAF Energy Systems G31 size cell was discharged at room temperature at five different $C$ rates, from $C/20$ to $6C$. The cell was a prismatic design with a lightweight plastic cell case and a resealable pressure relief valve. The decrease in utilization from $C/20$ to $6C$ was only 13%, indicating that this nickel-zinc hybrid cell design is able to achieve a balance between power and energy and can be utilized for many low- or high-rate discharge applications.

The specific energy as a function of $C$-rate is shown in Fig. 15E.6. In this test, the nickel-zinc G31 sized cell performed at >70 Wh/kg at low $C$ rates and demonstrated a linear decline in specific energy of 2.87 Wh/kg for each unit increase in the $C$-rate, beginning at $1C$. The specific energy is highly dependent on the nickel-zinc cell and electrode design and can vary between 60 and 90 Wh/kg for large prismatic formats. The results presented were collected from a 147 Ah G31 cell with a hybrid design, intended to meet demands for both power and capacity. The voltage versus time for a constant power discharge is shown in Fig. 15E.7. The cell was able to deliver 552 and 342 s at constant power discharges of 1000 and 1200 W, respectively.

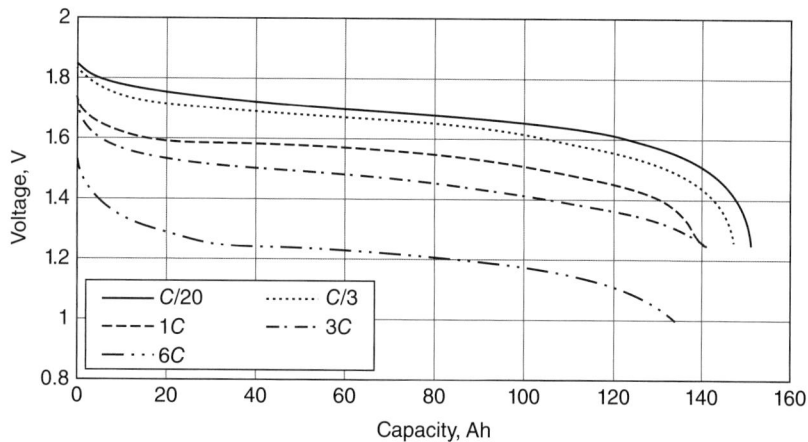

**FIGURE 15E.5**  Voltage profiles of a prismatic 147 Ah cell at various *C* rates. (*Courtesy of ZAF Energy Systems.*)

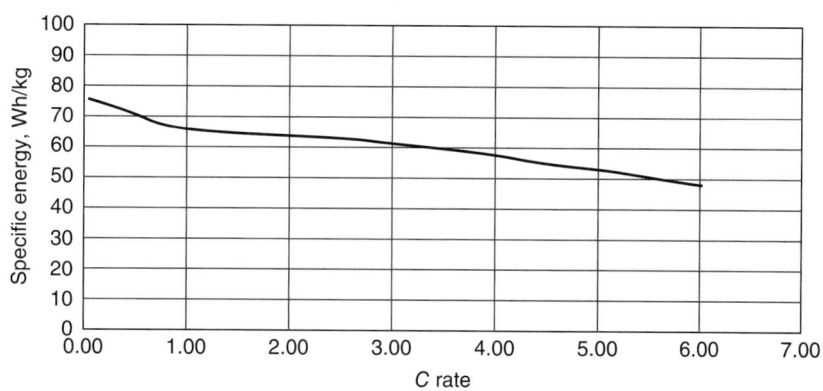

**FIGURE 15E.6**  Specific energy of a 147Ah prismatic cell as a function of *C* rates. (*Courtesy of ZAF Energy Systems.*)

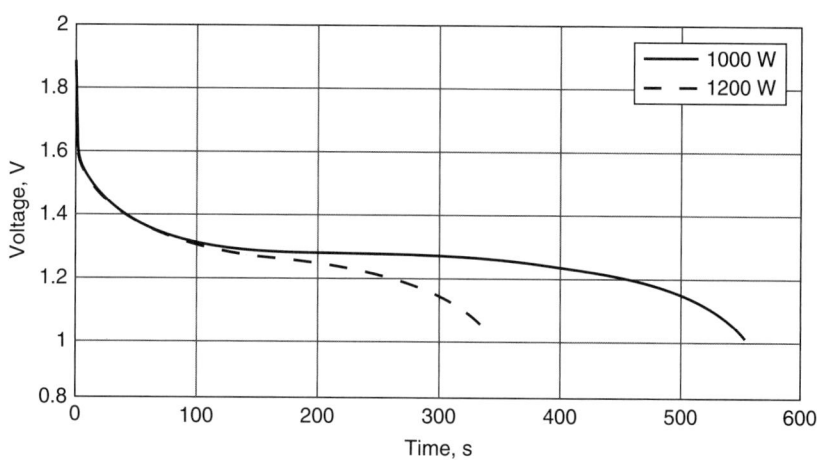

**FIGURE 15E.7**  Voltage profiles of a 147 Ah prismatic cell during constant power discharges at ambient conditions. (*Courtesy of ZAF Energy Systems.*)

***Temperature Considerations.***   Utilization as a function of temperature is shown in Fig. 15E.8. In this testing, a single G31 sized cell was discharged at a *C*/3 rate at multiple temperatures ranging from −30 to 40°C. The utilization of the cell was ≥96% at test temperatures of 0°C and higher, with minimal change in polarization over the discharge curve up to 80% DOD. After 80% DOD, the 0°C curve began to deviate and required a lower end of discharge voltage to achieve a similar capacity. The cell was able to achieve 71% and 62% utilization at −20 and −30°C, respectively; however, the midpoint voltage dropped to 1.56 and 1.43 V, respectively.

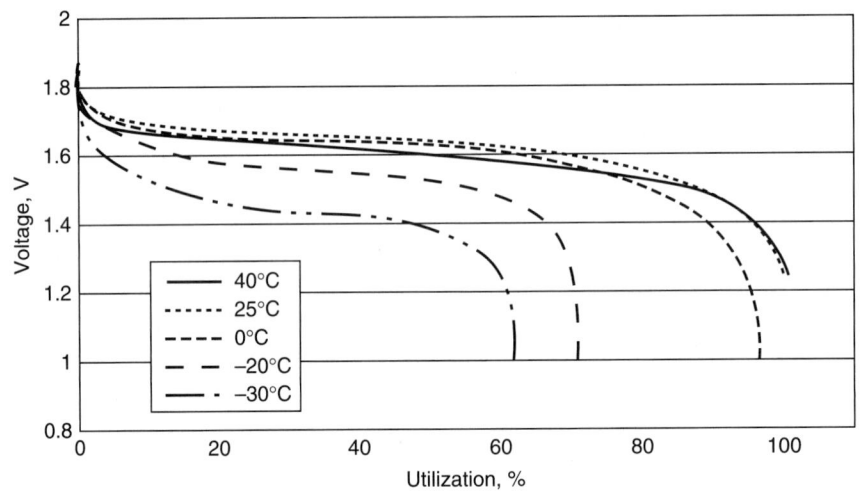

**FIGURE 15E.8**   Voltage profiles of a prismatic 147 Ah cell at various temperatures during *C*/3 discharges. (*Courtesy of ZAF Energy Systems.*)

The nickel-zinc chemistry has shown the ability to discharge at very high rates in low temperature applications. In this testing, a single G31 sized cell was discharged at 700, 800, and 900 A rates at −18°C, and 400, 500, 600, and 700 A at −30°C as shown in Figs. 15E.9 and 15E.10, respectively. During −18°C cold cranking tests, the G31 cell was able to discharge for 91.4, 71.4, and 49.9 s at 700, 800, and 900 A rates, respectively, before reaching the cutoff voltage. While in the −30°C cold cranking tests, it was able to discharge for 142.6, 97.6, 53.6, and 37.7 s at 400, 500, 600, and 700 A rates, respectively. In the −30°C cold cranking tests, the voltage fluctuates during the first 20 s of testing with the variability increasing with higher currents. This behavior is attributed to resistive cell heating. These results indicate a greater than 2/3 power retention from −18 to −30°C.

***Charge Characteristics.***   The constant current, constant voltage (CCCV) schedule is commonly used when charging nickel-zinc cells. A constant current is applied until the voltage reaches a specified level and the voltage is then held constant while the current is allowed to taper. The charge voltage is dependent on both electrode and cell design, and it is important to follow the manufacturer's recommendation for optimum cycle life. Typical voltage levels for the CCCV method range from 1.9 to 2.0 V.[8,9] This voltage is temperature sensitive and must be adjusted according to manufacturer's specifications when charging at conditions other than room temperature. The charging current also influences the upper voltage level,[9] which is critical when considering charge times to ≥80% state-of-charge (SOC). An example of SOC versus charge is shown in Fig. 15E.11 for a nickel-zinc cell. In this test, a constant current to constant voltage charge was performed at multiple currents and compensated voltages. For the *C*/3 and *C*/2 charge rates, a 1.93 V/cell voltage cutoff was used as the upper limit. This cutoff was increased to 1.95 V/cell for the 1*C* rate and further increased to 2.0 V for the 2*C* rate. The constant voltage portion of the charging schedule stayed consistent at 1.93 V/cell for all charge rates. The nickel-zinc chemistry has historically shown the ability for high charge acceptance. As shown in Table 15E.2, 80% SOC can be achieved in a little over an hour from a 0% SOC with a fast charge schedule (2*C*). With a slower charge schedule of *C*/3, optimized for a longer battery life, an 80% SOC can be achieved in less than 2.5 h.

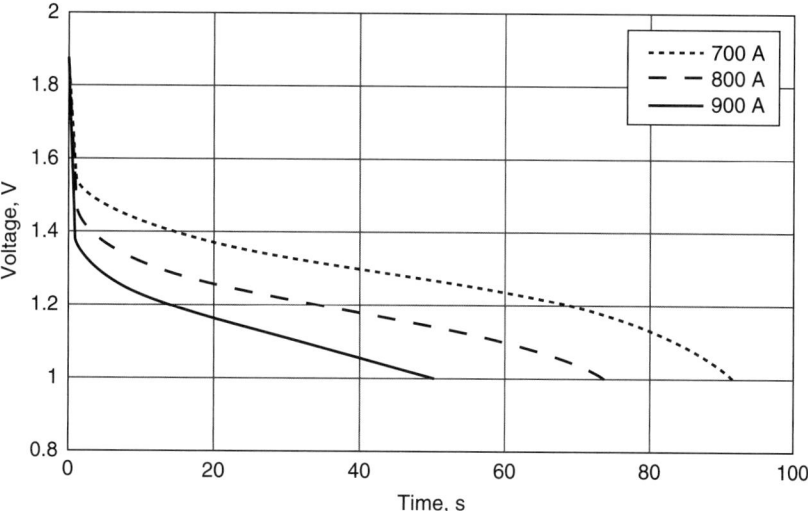

**FIGURE 15E.9**  Voltage profiles of a 147 Ah Ni-Zn cell at −18°C for rates of 700, 800, and 900 A. (*Courtesy of ZAF Energy Systems.*)

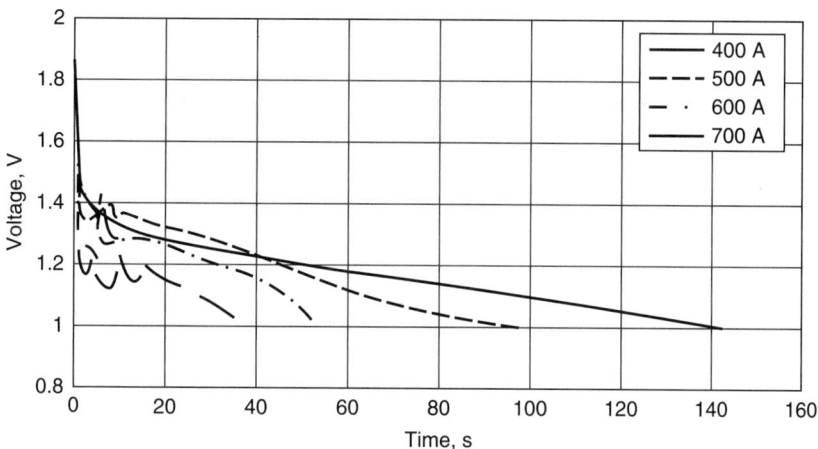

**FIGURE 15E.10**  Voltage profiles of a 147 Ah Ni-Zn cell at −30°C for rates of 400, 500, and 600 A. (*Courtesy of ZAF Energy Systems.*)

*Self-Discharge.*    Much progress has been made in the last few years on improving the self-discharge rate of the nickel-zinc battery. Historically, the typical discharge rate could be as high as 1% of the ampere-hour capacity per day.[8,9] Recent improvements in separator technology and electrode design have lowered the daily capacity loss to less than 0.25%. Self-discharge versus time for a 147 Ah nickel-zinc battery is shown in Fig. 15E.12. In this test, a charged nickel-zinc G31 size cell was monitored for voltage and capacity drop over a 60-day period. The nickel-zinc cells exhibited the largest drop in both capacity and voltage within the first 15 days of their shelf life. After 15 days, a linear degradation of 0.2 Ah/day is continued for the rest of the 60-day testing period. At this rate a total loss of 42 Ah over 6 months is predicted.

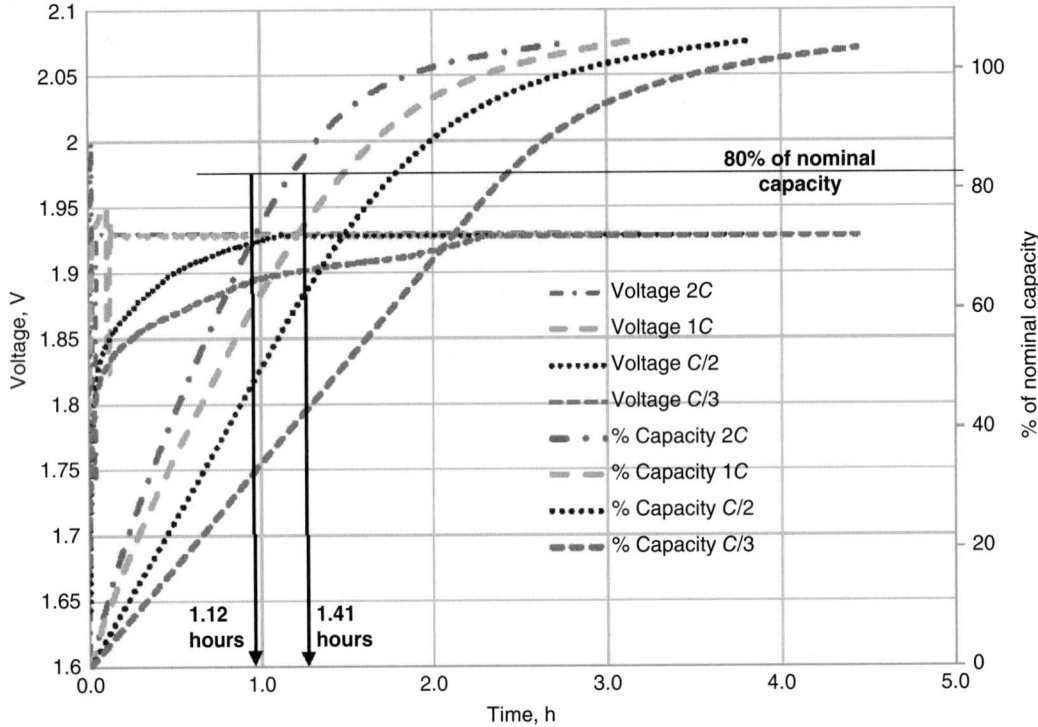

**FIGURE 15E.11**    Charge profiles and times versus various states-of-charge for multiple charge rates of a Ni-Zn cell. (*Courtesy of ZAF Energy Systems.*)

**TABLE 15E.2**    Charge Rate versus Time to 80% and 100% SOC
(Courtesy of ZAF Energy Systems)

| Charge rate | Time to 80% nominal capacity, h | Time to full charge, h |
|-------------|---------------------------------|------------------------|
| C/3 | 2.35 | 4.44 |
| C/2 | 1.70 | 3.83 |
| 1C | 1.41 | 3.18 |
| 2C | 1.14 | 2.76 |
| 3C | 1.12 | 2.75 |

***Life Cycle Performance.***    As shown in Figs. 15E.13 and 15E.14, nickel-zinc batteries have demonstrated the ability to deliver >500 cycles at very high DODs and >8000 cycles at low DODs, making this technology a viable solution for multiple applications. In the 80% DOD cycling test,[20] a nickel-zinc 30 Ah battery was cycled at a C/3 and C/5 discharge rates and achieved 600 cycles to 80% of nameplate capacity. In Fig. 15E.14, a 100 Ah prismatic cell was cycled at a 1C rate for charge and discharge, starting at a 50% SOC and cycling at 10% DOD until a decline in capacity was observed after about 8500 cycles.

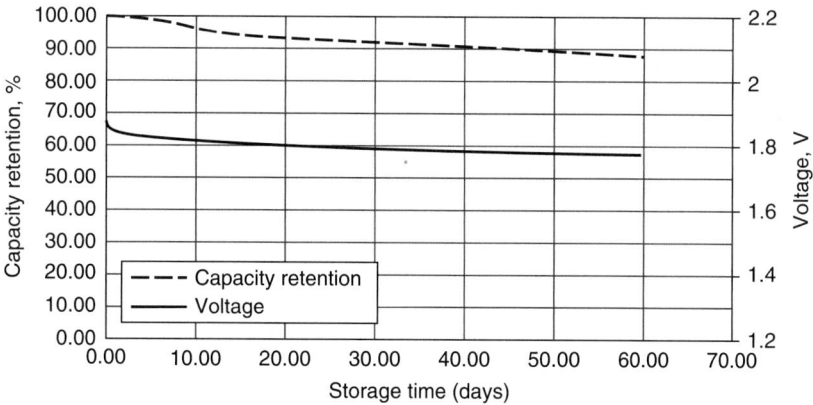

**FIGURE 15E.12**    Self-discharge data of a Ni-Zn cell. (*Courtesy of ZAF Energy Systems.*)

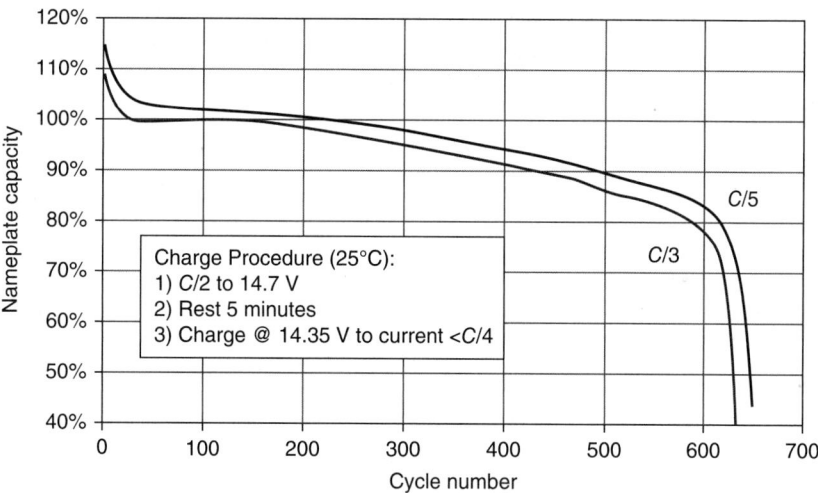

**FIGURE 15E.13**    Life cycle data of an 80% DOD test on a 30 Ah 12 V Ni-Zn battery for C/3 and C/5 discharges at 25°C.

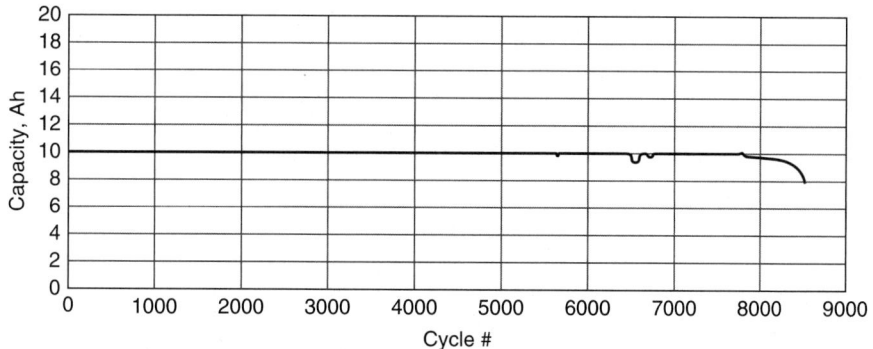

**FIGURE 15E.14**    Ten percent DOD cycle test of ZAF nickel-zinc 100 Ah cell at a 1C-rate charge and discharge. (*Courtesy of ZAF Energy Systems.*)

## 15E.4.3    Applications Engineering

*Battery System Considerations and Applications.*    Nickel-zinc provides a low-cost option for a long cycle-life alkaline rechargeable system. Given its high energy density and deep cycling capability, nickel-zinc becomes cost-competitive with lead acid for several applications when accounting for cost per cycled ampere-hour. The energy and power densities make nickel-zinc fit nicely in between the performance and application space of lithium ion and lead acids, with overlap in both directions, depending on the specific application.

*Temperature and Environmental Considerations.*    In terms of general use, it is helpful to note temperature and environmental considerations. All batteries are limited by the operational and nonoperational environments to which they are exposed. Alkaline batteries are limited in extreme low temperature applications primarily by the freezing point of the electrolyte, which varies from approximately $-25°C$ for 20 weight% KOH to $-60°C$ for 32 weight% KOH solutions. The nickel-zinc electrolyte is typically in the range of 20 to 35 weight% KOH. The presence of zinc anions in the electrolyte will also effectively lower the normal freezing point of the pure electrolyte. A lower KOH concentration of electrolyte is used in order to reduce zinc solubility and extend cycle life. It is preferable to maintain the battery above the freezing point of the electrolyte or permanent damage to the battery may occur. In some cases, nickel-zinc batteries can be optimized for extreme cold weather applications by the use of electrolyte additives that enhance conductivity at cold temperatures and depress the freezing point.[9]

Most nickel-zinc batteries are designed to operate in a temperature range from $-30$ to $60°C$. However, batteries can be optimized for operation below $-30°C$. At warmer temperatures the problem becomes one of charge efficiency. The charge efficiency drops at temperatures above $40°C$, resulting in reduced battery capacity.

*Battery Assembly and Packaging Considerations.*    Another notable aspect of the nickel-zinc battery system is that it is readily made in larger cell formats, such as those with 150 Ah or greater capacity. The safety aspect of nickel-zinc, as opposed to lithium-based chemistries, allows for these larger capacity formats that have advantages over smaller formats. Larger cells typically have greater energy and power densities than smaller cells since the ratio of active to inactive materials with respect to cell size is greater. At the battery system level, more interconnects are required for smaller cells in a parallel configuration to match a single large format cell in capacity. The impedance of the interconnections will add up and can add significant losses to the system, especially in high power applications. Nickel-zinc batteries have better packaging efficiency, regardless of cell format, being that it has become necessary to have intercell space and materials in Li-Ion packs to keep thermal runaway from propagating to neighboring cells.[21] This thermal management system is not necessary in a nickel-zinc system and volume can be utilized more efficiently.

*Deep Cycle Applications.*    There are several deep cycle applications that are suitable for the nickel-zinc system. One application is providing power to auxiliary power units (APUs) in heavy trucks. The APU supplies the hotel load for the driver and powers the heating and cooling systems in the truck cabin. This has become even more critical since several states and cities[22] have instituted anti-idle requirements for environmental reasons in recent years. This process requires deep DOD, typically on the order of a 10-h cycle. Weight and volumetric efficiency are also critical and with its higher energy density and good deep cycle capability, nickel-zinc batteries offer both weight and cost savings over traditionally used lead acids. Another motive power application that can benefit from the increased energy density and deep cycle capability of nickel-zinc encompasses the industrial equipment area, including forklifts, airport tugs, and hospital carts (see Chap. 26B). In addition, the greater charge acceptance of the nickel-zinc battery allows for faster turnaround time of equipment and maximized operating hours with multiple shifts.

Several marine deep cycle applications are particularly well suited for nickel-zinc chemistry. These include supplying power to the electric propulsion systems on larger ferries, running trolling motors, and powering hotel loads for smaller recreational boats. These applications are usually constrained in terms of both available volumetric footprint and maximum weight. Nickel-zinc is ideal for these applications since safety is a priority given the limited avenues of escape on a boat out at sea. Military applications that require deep cycle capability include torpedoes, swimmer delivery vehicles, and other submersibles.

Other land-based deep cycle applications for nickel-zinc batteries include electric bicycles and scooters (see Chap. 25A), wheelchairs, golf carts, and similar uses. In general, nickel-zinc batteries have good deep-cycle capability and are capable of exceeding 500 cycles at 100% DOD when operated in accordance with the manufacturer's specifications. It is important to follow the manufacturer's recommendation for charging when operating the battery at high rate of discharge as cell imbalance becomes an even more critical issue. Cells also tend to diverge in performance more rapidly as the DOD is increased.

***Stationary Storage Applications.*** Stationary power applications, whether for UPS, telecommunications, emergency lightning, or grid storage (see Chap. 27), often demand a combination of high-rate capability, good charge acceptance, long life, high temperature performance, low maintenance and cost, and safe use. Furthermore, energy density can also be critical in areas where real estate is expensive and/or physical access is challenging. One advantage of the nickel-zinc systems is that it possesses all of these characteristics to some degree and is therefore a very versatile and flexible storage technology that is suitable for the variety of needs required in stationary storage.

Grid storage, in terms of the adoption and full utilization of renewables like solar and wind, and stand-alone microgrids are two applications that are well suited for the versatility of the nickel-zinc system. The various electric power applications for energy storage, such as rapid reserve, load leveling, and transmission system stability, typically require different strengths in their respective battery chemistries in order to be cost-effective. However, there is no single chemistry that covers all power applications when cost and safety are considered. Therefore, most grid storage systems to date are made with a single chemistry[23] and only target a limited number of electric power applications. Recently, hybrid energy storage systems have been proposed[24] that will combine, for instance, lead-acid and lithium-ion, in order to leverage both their advantages and increase the utilization of the storage system and increase the cost benefit. However, this hybrid system will increase complexity in terms of both construction and management. The nickel-zinc system, being typically functional in the performance region between lead-acid and lithium-ion, in terms of both cost and electrical performance, offers a simpler option with the potential to cover more electric power applications. The environmental and safety aspects of the nickel-zinc chemistry also possess advantages over these hybrid systems.

***Standby/Float Charge Applications.*** Float charging may be used in various standby applications, such as emergency lighting, emergency power back-up systems, and uninterruptible power supplies (UPSs). During a float charge, the float voltage should be set so that the charge current will balance and offset the self-discharge rate resulting in no net overcharge of the battery. This float voltage range may vary depending on the battery design and the environmental conditions of the application. The battery manufacturer's recommendation should be followed to obtain maximum performance and life.

***Motive Storage Applications.*** One relatively new application for energy storage in the automotive field is the start-stop system. This development has been driven by the demand for lower emissions and higher fuel standards. It allows cars to shut off the engine to reduce idling time when the car comes to a stop in traffic. The battery will then repeatedly restart the engine in addition to powering all electronics, heating, and cooling while the engine is off. This process reduces overall fuel consumption and exhaust emissions. Typical starting-lighting-ignition (SLI) applications have been dominated by lead-acid batteries for decades, but these new start-stop systems require more demanding power and cycling profiles than those typical in SLI applications. Thus, there are ongoing efforts in industry to evaluate new battery technologies for start-stop applications. Nickel-zinc, new lead-acid technologies, and lithium ion chemistries are all being evaluated for this application. With its combination of power capability, cycle life, and low cost in a relatively small footprint, nickel-zinc offers a viable alternative to the other technologies.

***Safety.*** The nickel-zinc battery does not require any special handling or storage requirements beyond those of other typical commercial batteries. However, any battery may have some inherent safety and handling issues, such as electrical hazards or corrosive electrolyte spillage. A safety data sheet (SDS) and user manual should be obtained from the battery manufacturer.

***Cell Venting Hazards.*** Nickel-zinc batteries typically incorporate a safety vent to prevent overpressure conditions in the battery case, which might result in the rupture of the cell can or lid. During normal operation, the battery operates in a sealed manner and the vent is present as a safety feature only. The nickel-zinc battery typically uses a resealable valve so that venting does not lead to immediate battery failure. However, repeated venting will lead to dry-out and increased cell impedance. The low-pressure valve (typically less than 20 psi) also serves to regulate internal pressure, which assists in gas recombination within the battery. ZAF Energy Systems' nickel-zinc cells incorporate an internal gas recombination device that works as a catalyst for water formation from hydrogen and oxygen that may have been generated during overcharge. This recombination device helps control the internal pressure and limits venting of cells to scenarios with significant overcharge or over-discharge. At the system level, it is strongly recommended that natural ventilation or forced air flow exist to prevent hydrogen buildup and potential explosions. These potential hazards to personnel and equipment must be recognized and taken into consideration during operation of the battery.

***Environmental Impact and Recycling of Nickel-Zinc Batteries.***    The materials and components used in most nickel-zinc systems are readily recyclable. A typical ZAF Energy Systems' nickel-zinc cell will have an 85% material recovery yield. The material distribution is shown in Fig. 15E.15. The high material recovery rate also presents the potential for further cost savings in a battery production setting. The nickel-zinc system is also environmentally friendly, and all major components are made from natural abundant materials; therefore, supply shortages and geographic availability does not present a potential sustainability and cost issue. Another significant distinction between nickel-zinc and other chemistries, such as lead acid, lithium ion, nickel-metal hydride, and nickel-cadmium, is that none of the components in the nickel-zinc systems are flammable or explosive and all are RoHS (restriction of hazardous substances) compliant.

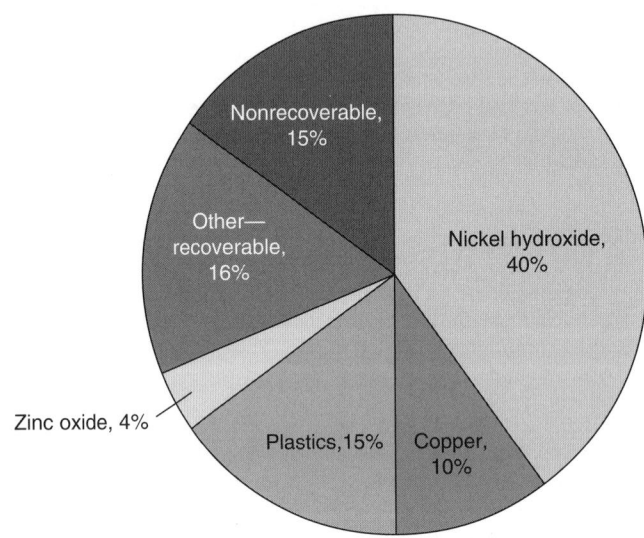

**FIGURE 15E.15**    Material and component distribution in a ZAF nickel-zinc cell. Eighty-five percent of the material is recyclable. (*Courtesy of ZAF Energy Systems.*)

## *ACKNOWLEDGMENTS*

The authors would like to acknowledge the significant contributions of Dr. Melissa McIntyre and Josh Rehanek of ZAF Energy Systems in preparing and editing this chapter.

## *REFERENCES*

1. T. de Michalowski, British Patent 15,370 (1899).

2. J. J. Drumm, U.S. Patent 1,955,155 (1934).

3. C. A. Ropp, Jr., U.S. Patent 3,558,358 (1971).

4. F. P. Kober and A. Charkey, *Nickel-Zinc: A Practical High Energy Density Battery*, Power Sources 3, D. H. Collins (ed.), Oriel Press Ltd., 1971.

5. F. R. McLarnon and E. J. Cairns, *J. Electrochem. Soc.* **138:**645–656 (1991).

6. D. Coates and A. Charkey, "Nickel-Zinc Batteries for Commercial Applications," *Intersociety Energy Conversion Engineering Conference (IECEC)*, Vancouver, BC, Canada, August 1999.

7. I. Levy and A. Charkey, U.S. Patent 4,810,598.

8. J. Phillips and S. Mohanta, "Nickel-Zinc Batteries," in T. B. Reddy (ed.), *Linden's Handbook of Batteries*, 4th ed., , McGraw Hill, New York, 2011.

9. D. Coates and A. Charkey, "Nickel-Zinc Batteries," D. Linden and T. B. Reddy (eds.), *Linden's Handbook of Batteries*, 3rd ed., McGraw Hill, New York, 2001.

10. H. Y. Wu, Y. L. Xie, and Z. A. Hu, *Int. J. Electrochem. Sci.* **8:**1839–1848 (2013).

11. J. Yao, Y. Li, Y. Zhu, and H. Wang, *J. Power Sources* **224:**236–240 (2013).

12. Y. Li, J. Yao, Y. Zhu, Z. Zou, and H. Wang, *J. Power Sources* **203:**177–183 (2012).

13. A. Charkey, U.S. Patent 5,658,694.

14. A. Charkey and D. Coates, U.S. Patent 5,863,676.

15. *Electrochemical Power Sources: Primary and Secondary Batteries*, IEE Energy Series 1, M. Barak (ed.), p. 330.

16. M. Eisenberg, U.S. Patent 4,224,391 (1980).

17. T. C. Alder, F. R. McLarnon, and E. J. Cairns, U.S. Patent 5,302,475 (1994).

18. J. Phillips and S. Mohanta, U.S. Patent 7,816,030 (2010).

19. A. Charkey, F. Cao, and G. Bowling, "Failure Analysis of the Nickel-Zinc Battery," *39th Power Sources Conference*, June 2000.

20. Based on testing of Evercel 30 Ah 12 V battery in February of 2000 at independent test lab (given through private communication).

21. E. Darcy, "Challenges with Achieving >180 Wh/kg Li-Ion Battery Modules that don't Propagate Thermal Runaway or Emit Flames/Sparks," The Battery Conference, Seoul, South-Korea, October 2015.

22. *Compilation of State, County, and Local Anti-Idling Regulations*, EPA420-B-06-004, April 2006.

23. A. A. Akhil, J. D. Boyes, P. C. Butler, and D. H. Doughty, "Batteries for Electrical Energy Storage Applications," T. B. Reddy (ed.), *Linden's Handbook of Batteries*, 4th ed., McGraw Hill, New York, 2011.

24. R. Moore, R. Nowlin, V. Vu, M. Parrot, J. Dermott, G. Miller, K. E. Ames et al., U.S. Patent 8,638,061 (2014).

## SECTION F

# NICKEL-HYDROGEN

**Jack N. Brill**

## 15F.1   GENERAL CHARACTERISTICS

A sealed nickel-hydrogen (Ni-H$_2$) secondary battery is a hybrid, combining battery and fuel-cell technologies.[1] The nickel oxide positive electrode comes from the nickel-cadmium cell, and the hydrogen negative electrode comes from the hydrogen-oxygen fuel cell. Major advantages and disadvantages are listed in Table 15F.1.

**TABLE 15F.1**   Major Advantages and Disadvantages of the Nickel-Hydrogen Battery

| Advantages | Disadvantages |
|---|---|
| High specific energy (60 Wh/kg) | High initial cost |
| Long cycle life (40,000 cycles at 40% DOD for LEO applications) | Self-discharge proportional to hydrogen pressure |
| Long lifetime in orbit (over 15 years for GEO applications) | Low volumetric energy density: |
| Cell can tolerate moderate overcharge and reversal | 50–90 Wh/L (IPV cell) |
| Hydrogen pressure gives an indication of state of charge | 20–40 Wh/L (battery) |

Salient features of this hybrid Ni-H$_2$ battery are a long cycle life that exceeds any other maintenance-free secondary battery system, high specific energy (gravimetric energy density) compared to other aqueous batteries, high power density (pulse or peak power capability), and a tolerance to overcharge and reversal. It is these features that make the Ni-H$_2$ battery system the energy storage subsystem currently employed in many aerospace applications, such as geosynchronous earth-orbit (GEO) commercial communications satellites, and low earth-orbit (LEO) satellites, such as the Hubble space telescope (HST).

Application of the Ni-H$_2$ battery has mainly been directed toward the aerospace field. Recently, however, programs have been started for terrestrial applications, such as long-life stand-alone photovoltaic systems.

## 15F.2   CHEMISTRY

The electrochemical reactions of the Ni-H$_2$ cell for normal operation, overcharge, and reversal are

*Normal operation:*

*Nickel electrode:*    $\text{NiOOH} + \text{H}_2\text{O} + \text{e} \underset{\text{charge}}{\overset{\text{discharge}}{\rightleftharpoons}} \text{Ni(OH)}_2 + \text{OH}^-$

*Hydrogen electrode:*    $\frac{1}{2}\,\text{H}_2 + \text{OH}^- \underset{\text{charge}}{\overset{\text{discharge}}{\rightleftharpoons}} \text{H}_2\text{O} + \text{e}$

*Net reaction:*    $\frac{1}{2}\,\text{H}_2 + \text{NiOOH} \underset{\text{charge}}{\overset{\text{discharge}}{\rightleftharpoons}} \text{Ni(OH)}_2$

*Overcharge:*

| | |
|---|---|
| *Nickel electrode:* | $2OH^- \rightarrow 2e + \frac{1}{2}O_2 + H_2O$ |
| *Hydrogen electrode:* | $\frac{1}{2}O_2 + H_2O + 2e \rightarrow 2OH^-$ |

*Reversal:*

*Hydrogen (negative) precharge:*

| | |
|---|---|
| *Nickel electrode:* | $H_2O + e \rightarrow OH + \frac{1}{2}H_2$ |
| *Hydrogen electrode:* | $\frac{1}{2}H_2 + OH^- \rightarrow H_2O + e$ |

*Positive precharge:*

| | |
|---|---|
| *Nickel electrode:* | $2NiOOH + 2H_2O + 2e^- \rightarrow 2Ni(OH)_2 + 2(OH)^-$ |
| *Hydrogen electrode:* | $2(OH)^- \rightarrow 2e^- + \frac{1}{2}O_2 + H_2O$ |
| *Net reaction:* | $2NiOOH + 2H_2O \rightarrow 2Ni(OH_2) + \frac{1}{2}O_2$ |

## 15F.2.1   Normal Operation

Electrochemically, the half-cell reactions at the positive nickel oxide electrode are similar to those occurring in the nickel-cadmium system. At the negative electrode, hydrogen gas is oxidized to water during discharge and is reformed, during charge, from the water by electrolysis. The net reaction shows hydrogen reduction of nickel oxyhydroxide to nickelous hydroxide on discharge with no net change in KOH concentration or in the amount of water within the cell.

## 15F.2.2   Overcharge

During overcharge, oxygen is generated at the positive electrode. An equivalent amount of oxygen is recombined electrochemically at the catalytic platinum electrode. Again, there is no change in KOH concentration or the amount of water in the cell with continuous overcharge. The oxygen recombination rate at the negative platinum electrode is very rapid, sustaining moderate rates of continuous overcharge, provided there is adequate heat transfer away from the cell to avoid thermal runaway. This is one of the operational advantages of the Ni-$H_2$ cell.

## 15F.2.3   Reversal

Two types of precharge are used with the nickel-hydrogen system. With hydrogen (negative) precharge during cell reversal, hydrogen is generated at the positive electrode and consumed at the negative electrode at the same rate. Therefore, the cell can be operated continuously in the cell reversal mode without pressure buildup or net change in electrolyte concentration. This is a unique feature of the system. If nickel (positive) precharge is used, oxygen gas is generated during reversal until the nickel positive material is consumed. The oxygen is consumed again during charge. Once the positive material is consumed, hydrogen is generated and consumed at the same rates. It is possible that the hydrogen-oxygen mixture could be in the combustible range for this mode of operation and that a rapid recombination could occur causing damage to the cell stack.

## 15F.2.4   Self-Discharge

The electrode stack is surrounded by hydrogen under pressure. A salient feature is that the hydrogen reacts electrochemically but not chemically to reduce the nickel oxyhydroxide. Actually, the nickel oxyhydroxide is reduced chemically, but at such an extremely low rate that performance for aerospace applications is not affected.

## *15F.3  BATTERY COMPONENTS AND DESIGNS*

Ni-H$_2$ cell stacks are assembled in three distinct configurations. These include the COMSAT back-to-back, the Air Force recirculating, and the hybrid, Mantech back-to-back designs. This section describes the electrode-stack components used for the fabrication of aerospace Ni-H$_2$ cells in these configurations. Figure 15F.1 shows the truncated disk electrode-stack components used in the COMSAT design. Figure 15F.2 shows the circular components (Figs. 15F.2*a* and *b*) for the Air Force recirculating and hybrid, Mantech back-to-back designs.

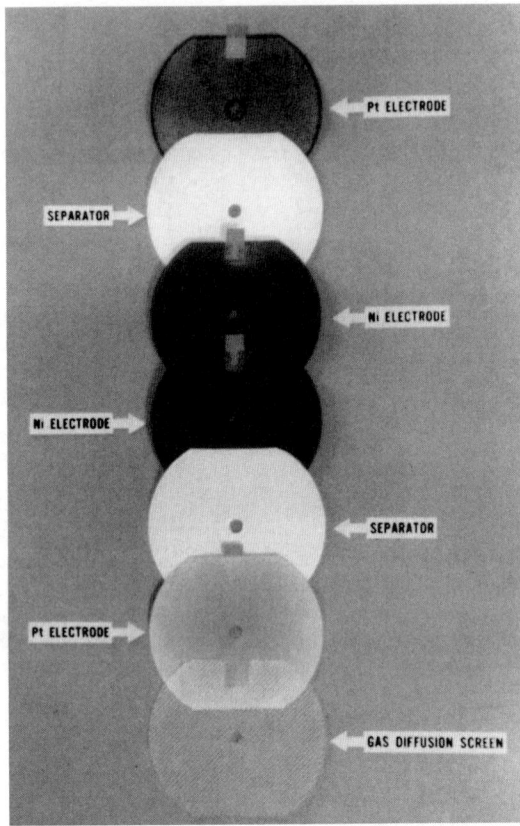

**FIGURE 15F.1**   COMSAT bus-bar configuration electrode-stack components.

### 15F.3.1   Cell Components

*Positive Electrodes (Sintered).*    The sintered positive electrode consists of a sintered porous nickel plaque that is impregnated with nickel hydroxide active material. The porous sintered plaque serves to retain the active nickel hydroxide material within its pores and conduct the electric current to and from the active material. Essential features of the sintered plaque are high porosity, large surface area, and electrical conductivity in combination with good mechanical strength.[2]

Active material is impregnated into the sintered plaque by an electrochemical impregnation process. There are two electrochemical impregnation processes used—aqueous and alcoholic. The aqueous impregnation process (Bell Laboratories process)[3] uses an aqueous-based nickel nitrate solution for the impregnation bath.

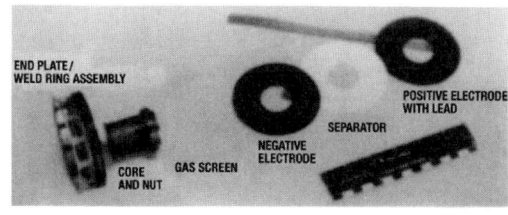

(a)

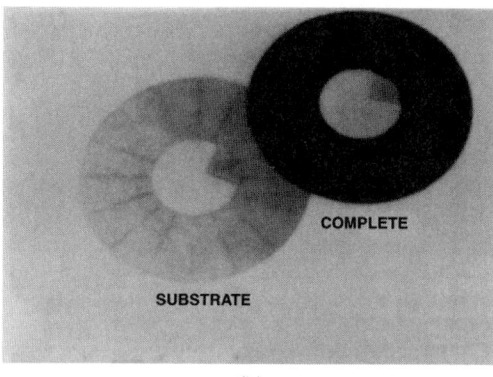

(b)

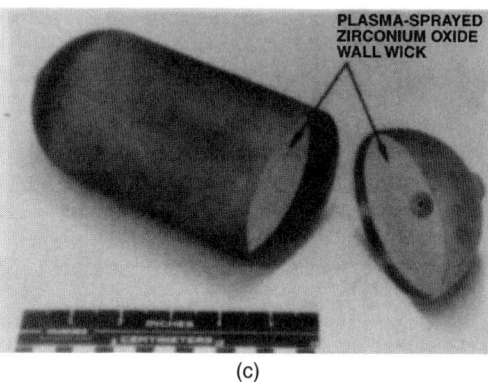

(c)

**FIGURE 15F.2** Air Force pineapple-slice configuration. (*a*) Stack components. (*b*) Negative electrode. (*c*) Pressure-vessel cylinder and dome.

The alcoholic impregnation process (Air Force process)[4] uses an alcohol-based nickel nitrate solution for the impregnation bath. Both processes provide the following advantages:

1. *Loading of active material:* Electrochemical impregnation gives very uniform loading of the active material within the pores of the nickel sinter.
2. *Loading level:* The loading level of active material can be accurately controlled by the electrochemical impregnation process. Typical loading values are $1.67 \pm 0.1$ g/cm$^3$ void volume for GEO applications and $1.65 \pm 0.1$ g/cm$^3$ void volume for LEO applications.

***Hydrogen Electrode.*** The hydrogen electrodes consist of a Teflon-bonded platinum black catalyst supported on a photo-etched nickel substrate with Teflon bonding. The sintered Teflon-bonded platinum electrodes were originally developed at Tyco Laboratories for the fuel-cell industry.[5] For Ni-H$_2$ cells, a hydrophobic Teflon backing was added to these platinum electrodes to stop water or electrolyte loss from the back side of the negative platinum electrode during charge and overcharge while readily allowing diffusion of hydrogen and oxygen gas. The use of Gortex® as the microporous Teflon membrane resulted from a development contract with HAC (Hughes Aircraft Corporation).[6] The platinum content is normally specified as $7.0 \pm 1.0$ mg/cm$^2$. The physical properties of this hydrogen electrode provide the right interface for the electrochemical reactions to occur without flooding or drying out the electrode at the separator interface.

***Separator Materials.*** Two types of separator materials have seen use in aerospace Ni-H$_2$ cells: (1) asbestos (fuel cell grade asbestos paper) and (2) Zircar (untreated knit ZYK-15 Zircar cloth). The predominant use in cells manufactured today is Zircar.

Fuel cell grade asbestos is a nonwoven fabric with a thickness of 0.25 to 0.38 mm. The asbestos fibers are made into a long roll of nonwoven cloth by a papermaking process. As an added precaution, the asbestos can be reconstituted in a blender and then reformed into a cloth to avoid any nonuniformity in the original structure that would allow oxygen to bubble through. The fuel cell grade asbestos has a high bubble pressure for oxygen gas; a pressure difference of more than $1.7 \times 10^5$ Pa is required across the separator cloth (250 μm thick) to force oxygen bubbles through the material. During overcharge, oxygen gas is forced off the backside of the positive electrode. The oxygen cannot channel or bubble through the separator to cause rapid recombination at the negative electrode. Zircar fibrous ceramic separators are available in textile product forms (Zircar Products, Inc.). These textiles are composed of zirconia fibers stabilized with yttria. These materials offer the extreme temperature and chemical resistance of the ceramic zirconia. They are constructed of essentially continuous individual filaments fabricated in flexible textile forms. Even with the fibrous structure, the inherently brittle nature of the ceramic material zirconia makes these separators fragile and susceptible to breaking. They must be handled with care. Untreated knit ZYK-15 Zircar cloth material is the tensile form used for Ni-H$_2$ cells.[7] Either one of two 250 to 380 μm thick layers of this separator material can be used. The second ZYK-15 layer is normally used as a backup to prevent oxygen channeling in the event of assembly damage to the first layer. The knit Zircar cloth has a very low oxygen bubble through pressure, and during charge and overcharge, oxygen gas readily permeates through the separator to recombine at the hydrogen platinum electrode to form water.

Both the asbestos and the Zircar separators serve the following functions:

1. They act as separators between positive and negative electrodes.
2. They serve as reservoirs for KOH electrolyte and remain stable in the electrolyte, allowing long-term storage and cycling.
3. They serve as media for charge and discharge current through the separator via ionic conduction of hydroxyl ions in the electrolyte.

***Gas Screen.*** A polypropylene gas diffusion screen is placed behind the hydrogen electrode to allow hydrogen gas and oxygen gas to diffuse to the back side of the negative electrode with the Teflon backing.

## 15F.3.2  Ni-H$_2$ Cell Construction

Sealed Ni-H$_2$ cells contain hydrogen gas under pressure within a cylindrical pressure vessel (see Fig. 15F.2c). They are referred to as individual pressure-vessel (IPV) cells because each individual cell is contained within its own pressure vessel. IPV cells are assembled using either single or dual electrode stack configurations inside the pressure vessel. An extension of IPV is the two-cell (2.5 V) CPV design made by connecting the dual electrode stacks in series within a single pressure vessel. IPV designs include cells having diameters of 6.35, 8.89, 11.43, and 13.97 cm.

Descriptions of the various cell designs follow. These designs represent the first generation of Ni-H$_2$ cell technology, which was developed in the 1970s and utilized in the 1980s along with the baseline designs currently in use.

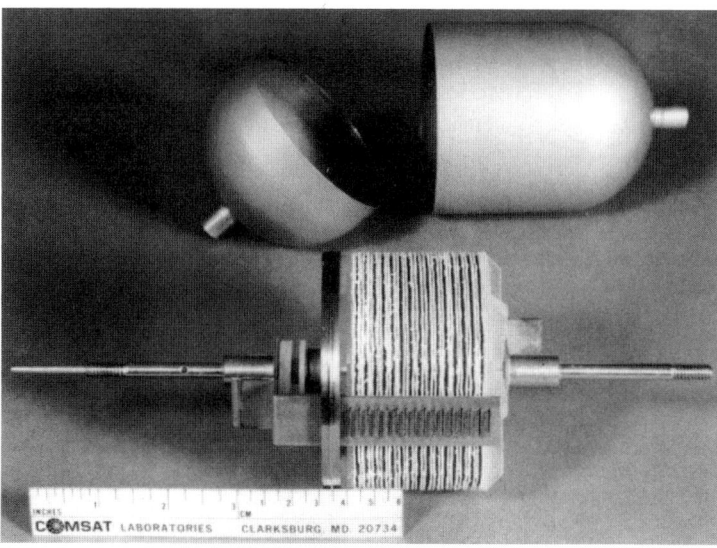

**FIGURE 15F.3** COMSAT NTS-2 Ni-H$_2$ cell components.

***COMSAT Ni-H$_2$ Cell.*** Components of the COMSAT NTS-2 Ni-H$_2$ cell are shown in Fig. 15F.3 with the electrode-stack assembly and weld ring positioned in front of the pressure shells.[8,9] These cells were built by Eagle-Picher Technologies (EPT) under an INTELSAT/COMSAT licensing agreement. The U.S. Navy's Navigation Technology Satellite 2 (NTS-2), launched on June 23, 1977, was the first flight demonstration of the Ni-H$_2$ battery.

*Electrode Stack.* Figure 15F.1 shows the basic arrangement of the electrode-stack components for the COMSAT back-to-back design. Two positive nickel oxide electrodes are positioned back-to-back. A separator is placed on each side of the positive electrodes. The negative platinum electrodes are placed with the platinum black surface next to the separator material. A plastic diffusion screen is placed on the back side of each negative electrode to facilitate gas diffusion to the back side of this electrode. These components constitute one module of the electrode stack. This arrangement is repeated until the number of modules is reached to provide the required capacity. A complete stack can be seen in Fig. 15F.3. The bus bars for the positive and negative electrodes are located along the outside of the electrode stack.

During charge and overcharge, oxygen gas that evolves at the nickel electrodes is forced out between the back-to-back positive electrodes. The oxygen diffuses into the gas space between the electrode stack and the pressure vessel wall into the region of the gas diffusion screens on the back of the negative electrode, and through the porous backing of the negative electrode, where it combines with hydrogen to form water. The partial pressure of oxygen is dependent on this diffusion process. The limiting step is the oxygen diffusion in the gas-phase pores of the Teflon-bonded electrode, not in the Teflon backing.[6] The fraction of oxygen gas should be less than 0.5% in the surrounding hydrogen gas when the cell is continuously overcharged at a C/2 rate.

*Pressure Vessel.* The pressure vessel (dome and cylinder), terminal bosses, and weld rings are all fabricated from Inconel alloy 718. The weld ring is manufactured by one of two methods. The first is manufactured using an investment casting process and is then machined to final dimensions. The second is manufactured by machining to final dimensions from an extruded or wrought Inconel metal. The outside diameter of the weld ring is machined as a T section to position the pressure shells on the weld ring and provide a backup support for an electron beam girth weld. The Inconel 718 pressure vessel shells are manufactured to a near uniform thickness using either a hydroforming or drawn process and then cut to length. The thickness is determined by the operating pressure and cycle requirements for the particular use. The pressure shells are "age hardened" using a standard heat treatment process. The terminal bosses for the compression seals are machined from Inconel 718 material and electron beam welded into the domes of the pressure vessel shells. Nylon plastic is injection-molded into these barrels. The Ziegler compression seal[10] is made by crimping the bosses. Cell designs commonly operate under maximum operating pressures between $4.1 \times 10^6$ Pa and $8.3 \times 10^6$ Pa. Depending on the particular use, the vessels are designed to provide a safety factor between 2:1 to over 4:1.

***Air Force Ni-H₂ Cell.***    Typical components for the Air Force Ni-H₂ cell are shown in Fig. 15F.2, including the electrode stack components, the negative electrodes with the chem-milled substrate, and the pressure vessel cylinder and dome with the plasma sprayed zirconium oxide wall wick for the electrolyte. These components are typically assembled in one of two configurations. Separators used in these designs are asbestos and Zircar alone or in combination with each other.

The first is commonly referred to as the recirculating electrode stack design. It is made of a number of modules composed of a gas screen, hydrogen electrode (negative), separator(s), and nickel electrode. The capacity of the cell is determined by the number of modules used to assemble the stack. A single gas screen and hydrogen electrode are used as the final module to maintain a hydrogen electrode for recombination on both sides of the positive throughout the stack. A wall wick is used on the interior of the pressure vessel to return electrolyte to the stack, hence the name recirculating design.

In this design, using asbestos or a combination of asbestos and Zircar separators, the oxygen generated on overcharge diffuses off the back of the nickel electrode. The diffusion path is very short; the oxygen gas simply travels through the gas screen to recombine at the next hydrogen electrode. Oxygen comes off the back side of the positive electrode of one module and recombines to form water at the next module. During overcharge, this transfer of water to the next module occurs throughout the electrode stack. The last module in the stack is simply a negative electrode and separator reservoir. The water formed at this electrode-separator combination goes to the wall wick and recirculated back to balance the electrolyte throughout the stack. With this recirculating design, the oxygen concentration is kept very low (below 0.2% in the surrounding hydrogen gas) during continuous overcharge at the *C* rate.

The second configuration is the Air Force Mantech back-to-back design. It is made in the same manner as the COMSAT design. A separator is placed on each side of the positive electrodes. The negative platinum electrodes are placed with the platinum black surface next to the separator material. A plastic diffusion screen is placed on the back side of each negative electrode to facilitate gas diffusion to the back side of this electrode. These components constitute one module of the electrode stack. This arrangement is repeated until the number of modules is reached to provide the required capacity. This design also uses the wall wick for return of electrolyte to the stack.

Designs using Zircar alone normally extend the separator to the pressure vessel wall. The Zircar contains large enough pores for oxygen gas to permeate through the pores of this separator to the negative electrode, where it recombines to form water. Oxygen can, of course, also emerge off the back side of the nickel electrode and diffuse through the gas screens as before. However, most of the oxygen permeates through the separator and there is little or no recirculation of water in cells with Zircar separators. For this design, the concentration of oxygen in the surrounding hydrogen gas is negligible.

The electrode-stack components are shaped like a pineapple slice (see Fig. 15F.2*b*) with provisions for the tab in the center hole. These electrode-stack components are assembled onto a polysulfone central core (see Fig. 15F.2*a*). The electrode tabs are brought out through this central core. The positive and negative tabs can be in opposite directions or in the same direction depending on the terminal configuration. This center core serves to align the electrode-stack components, provide a conduit for the positive and negative tabs, and insulate the positive and negative tabs from each other and from the electrode-stack components.

The cell capacity in a particular diameter is limited by the ability to manufacture a pressure vessel of sufficient length. In designing cells of larger capacity without increasing the diameter of the pressure vessel, two approaches are used. The first involves the use of a dual stack design to increase the capacities of cells for the individual cell diameters.[15] This is accomplished by using two stacks assembled on a single core as described above. The two stacks are separated by end plates and a weld ring and are connected electrically in parallel to attain a 1.25 V cell. The second approach utilizes a three-piece pressure vessel assembly.[16] A single electrode-stack is made with weld rings at each end. A cylinder is placed over the stack and joined at each end with the weld rings and two dome assemblies.

Heat transfer is better with the pineapple-slice configurations than with the COMSAT back-to-back configuration. Heat is transferred uniformly from the entire circumference of the pineapple-slice electrodes, whereas sections are removed from the circumference of the COMSAT back-to-back electrodes.

*Pressure Vessel.*    The pressure vessels used for the Air Force designs are essentially the same as those for the COMSAT designs. Certain designs utilize chemical milling to remove material from lower stress areas for weight reduction. Typically the weld areas are not reduced to compensate for the strength reduction of the age-hardened, Inconel 718 material in the heat-affected zone of the weld areas. The chemical milling is done prior to heat treating (age hardening) of the pressure vessel. The operating pressures and design margins are similar to those discussed for the COMSAT designs.

Two terminal designs have been used. One involves the terminal design using the compression seal described for the COMSAT designs. The second utilizes a hydraulic seal design. When this is used, the seal area is hydroformed as an integral part of the dome and cylinder. The terminal seals are hydraulic cold-flow Teflon seals.[14]

*Electrolyte Management.*    There are three mechanisms for the loss of electrolyte from the electrode stack: (1) by entrainment in the hydrogen and oxygen gases evolved during charge and overcharge, (2) by weeping of the negative electrode, and (3) by electrolyte displacement, that is, the electrolyte being pressed out of the positive electrodes in the cell stack by oxygen gas evolved during charge and overcharge.

Electrolyte loss by both entrainment and weeping of the negative electrode was determined to be negligible for negative electrodes with Gortex backing for both back-to-back and recirculating electrode-stack configurations. The major electrolyte loss mechanism is by displacement. When electrolyte is added to the cell, the void volume of the positive electrode is completely saturated with electrolyte. During activation, approximately 25% of the electrolyte in the positive electrodes is displaced by oxygen gas during charge and overcharge of the cell.[6] It was found that electrolyte loss by displacement occurred during initial cycling (activation) but eventually decreased to zero, leaving enough electrolyte to operate the cell efficiently.[11,12]

*Water Loss.*    Water loss from the electrode stack can result from evaporation and condensation of water vapor from the cell stack to the pressure-vessel wall when a large enough temperature difference exists between stack and wall (approximately 10°C difference). The plasma-sprayed zirconium oxide wall wick[13] shown in Fig. 15F.2c provides a return path for any water loss from the cell stack independent of the mechanism.

### 15F.3.3    Theoretical Specific Energy and Energy Density of Ni-H$_2$ Cells

Figures 15F.4 and 15F.5 depict the specific energy and energy density that could be projected for the different nickel-hydrogen cell designs. The actual values may vary depending on manufacturer.

1. In general, the specific energy increases as the capacity increases.
2. The choice and number of separators affect the weight (quantity of electrolyte) and thus the specific energy of the cell.
3. The energy density is primarily a function of the pressure range or free volume in the cell. Cells operating at higher maximum operating pressures have higher energy densities. This increase is a result of the reduction in weight of the pressure vessel at the higher pressures.

### 15F.3.4    Ni-H$_2$ Battery Design

Various Ni-H$_2$ battery designs have evolved through the years. They are tailored to the specific application and interface with the particular satellite. Mechanical and thermal requirements are the primary drivers for the configuration and interface of each battery. The nickel-hydrogen system is sensitive to temperature and performs best between −10 and +10°C. Thus, thermal control of the battery is important to minimize size and weight.

Several features have been integrated into the battery designs, which enhance the performance and reliability. These include pressure monitoring of cells through strain gauge or transducers, strain gauge voltage amplification circuits, individual cell voltage monitors, temperature monitoring, redundant individual cell heaters, and individual cell diode bypass protection. Bypass diodes on each cell protect the battery against a failure from an open-circuited cell. Protection in the charge direction is provided by three silicon diodes in series, while protection in the discharge direction is provided by one Schottky barrier diode. The diodes are mounted on heat sinks on the thermal sleeves near the base of the cells or on a separate panel attached to the battery base plate.

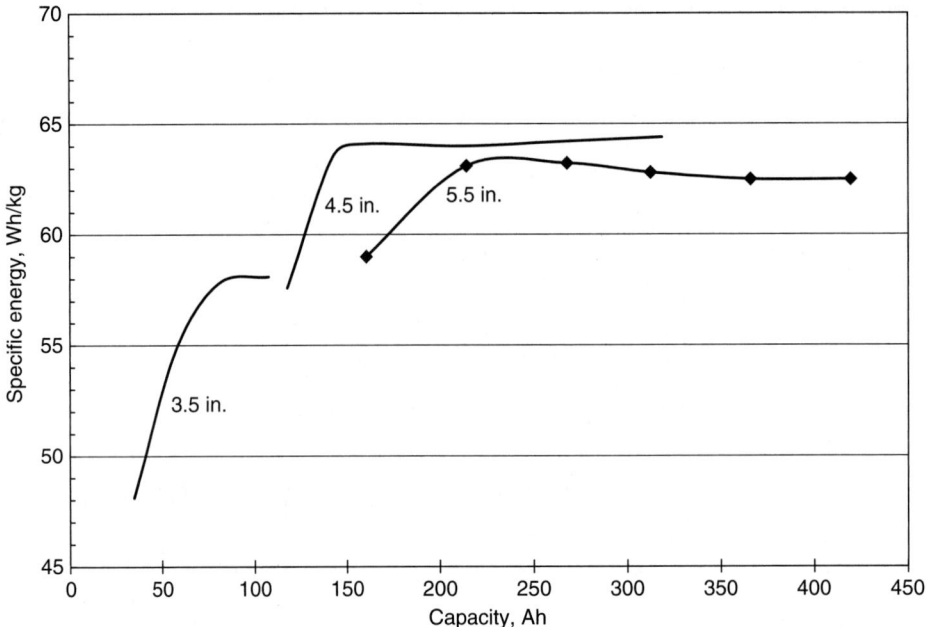

**FIGURE 15F.4**    Specific energy of Ni-H$_2$ cells.

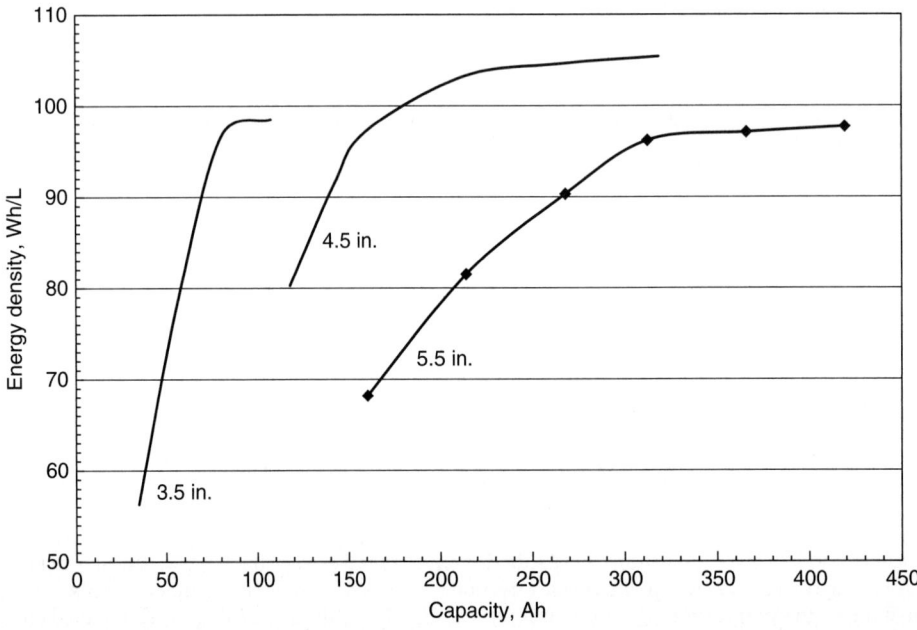

**FIGURE 15F.5**    Energy density of Ni-H$_2$ cells.

***Battery Configurations.***    Several different battery configurations can be seen in Figs. 15F.6 through 15F.10. One of the earliest Ni-H$_2$ battery designs was flown on the INTELSAT V program. Two 27-cell, 30 Ah Ni-H$_2$ batteries were used to provide the electric energy during launch, transfer orbit, and solar eclipses.[17]

**FIGURE 15F.6**    DMPS 100 Ah Ni-H$_2$ battery assembly.

The first battery using the dual electrode stack configuration was made by EPT, LLC, for the EUTELSAT II Program. The first of two flight sets was delivered in February 1990 and launched in August 1990.[18] This battery is shown in Fig. 15F.7.

The photovoltaic power subsystem for the International Space Station uses Ni-H$_2$ batteries for energy storage to support eclipse and contingency operations. These batteries are designed for LEO operation with a 6.5-year design life expectancy and are configured as orbital replacement units (ORUs), permitting replacement of worn-out batteries over the anticipated 30-year station life.[19] The baseline energy storage system design contains two batteries of 76 cells of 81 Ah capacity, packaged as 38-cell assemblies (Figs. 15F.8 to 15F.10), or approximately 184.7 kWh of stored energy. The initial batteries were placed in service on the space station in November 2000. Six batteries were placed in service in July 2009 during the visit of STS 127 to the ISS, and another six batteries were replaced in May 2010 during the servicing mission of STS 132.

### Design Features

*Mechanical Design.*    Typically, each battery will have a thermal sleeve around each cell. The cells are mechanically restrained by clamping them in a precision-machined sleeve. These sleeves can be made of either a metal such as aluminum or a composite made in a manner to provide electrical isolation, high thermal conductivity, and strength. The sleeve is isolated electrically from the cell by a blanket, such as CHO-THERM, which allows thermal transfer, wrapped around the cylindrical portion of the cell between the cell and sleeve. The space between the sleeves, blanket, and cell is normally filled with a material such as an RTV 566 to provide better thermal transfer as well as to bond the interfaces mechanically. The sleeves are then attached mechanically either to a base plate that is the interface to the satellite structure or to an interface such as extruded heat pipe assemblies that are a part of the satellite structure. The exposed surfaces of the cells are protected by a coating of Solithane or a combination of paint on the cell pressure vessel and Solithane. The desired battery voltage defines the number of cells used for the assembly.

**FIGURE 15F.7**    EUTELSAT II 58 Ah Ni-H₂ battery.

**FIGURE 15F.8**    International Space Station 81 Ah Ni-H₂ 38-cell assembly.

*Thermal Design.*    Each battery is designed to operate thermally within a specified set of limits dictated by the mission requirements and the satellite interface. These limits are normally in the range between $-10°C$ and $+15°C$ during the periods of battery operation. During periods of inactivity, such as the equinox periods of the geosynchronous missions, the temperature range can be reduced since the thermal output of the battery is less. Heat dissipated from within the cells is conducted radially to the pressure vessel wall, through the insulating

**FIGURE 15F.9**    TRW 81 Ah Ni-H$_2$ battery for a flight program. (*Photograph provided courtesy of TRW.*)

**FIGURE 15F.10**    MIDEX 23 Ah CPV Ni-H$_2$ battery assembly.

blanket/RTV 566 to the thermal sleeves, and down the thermal sleeve to the base plate or mounting interface. Depending on the interface, the heat is transferred through the mounting to second surface mirrors or to thermal heat pipes for dissipation. The charge control and heater assemblies must be operated in conjunction with the passive or active thermal dissipation to regulate operating temperature. The battery surfaces may be anodized to optimize thermal emissivity.

*Weight and Energy Density.*    Most battery designs are optimized through stress and thermal analysis. Figures 15F.11 and 15F.12 depict expected specific energies and energy densities for 22-cell, 28 V batteries. These, of course, may vary depending on the manufacturer and the intended use.

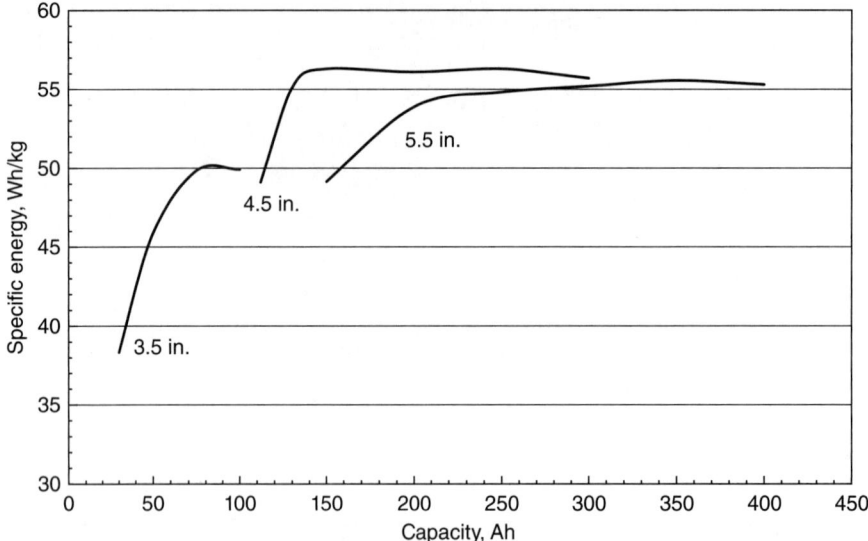

**FIGURE 15F.11**    Specific energy for 28 V Ni-H$_2$ battery.

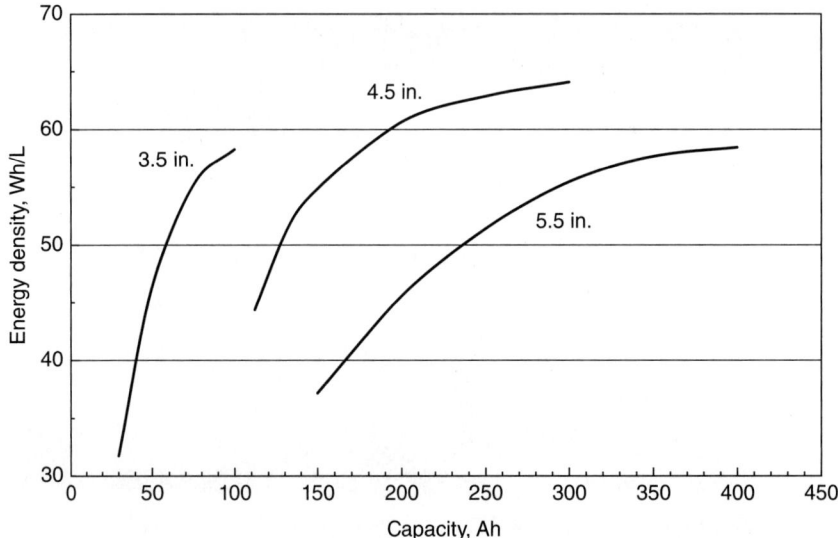

**FIGURE 15F.12**    Energy density of 28 V Ni-H$_2$ battery.

## 15F.4    APPLICATIONS AND PERFORMANCE

### 15F.4.1    General Applications

Nickel-hydrogen batteries are an excellent choice for any demanding application. Aerospace applications of Ni-H$_2$ batteries include various orbiting spacecraft, which can be divided into two categories: LEO and GEO.[a] Ni-H$_2$ batteries have also been developed and evaluated for terrestrial applications.[20,21] The advantages offered by Ni-H$_2$ batteries, including long calendar and cycle life, low maintenance, and high reliability, make them very attractive for terrestrial applications such as stand-alone photovoltaic systems or standby power for emergency or remote site use. The major drawback to the wider use of the Ni-H$_2$ battery is its high initial cost. The following are two examples of terrestrial applications addressing Ni-H$_2$ batteries.

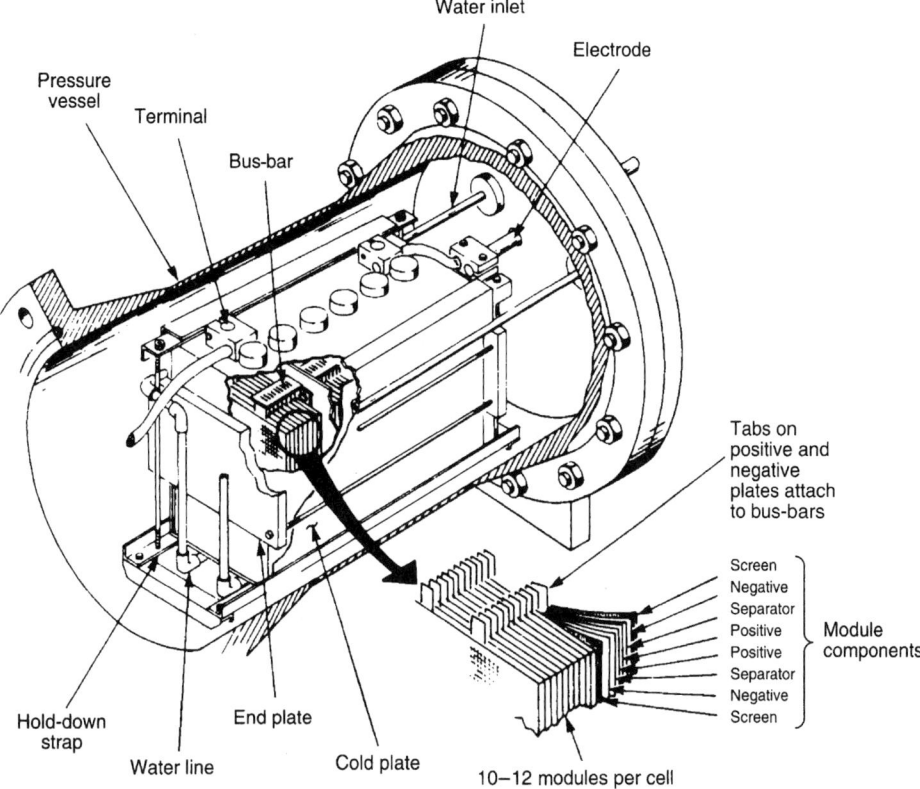

**FIGURE 15F.13**    6 V 100 Ah terrestrial Ni-H$_2$ battery. (*Courtesy of COMSAT Laboratories.*)

Starting in 1983, Sandia National Laboratories sponsored a cost-sharing program with COMSAT Laboratories and Johnson Controls, Inc. for the design and development of a sealed Ni-H$_2$ battery for deep-discharge terrestrial applications that would be cost-competitive with lead-acid batteries in a system designed for a 20-year life. The main thrust of this program was to reduce the cost of the aerospace technology without comprising the desirable features of the Ni-H$_2$ system. Figure 15F.13 shows a five-cell, 6 V 100 Ah Ni-H$_2$ battery assembled in a pressure vessel. Assemblies have been tested by Sandia National Laboratories in combination with

---

[a]See Chap. 25B for more details on these applications.[22,23]

photovoltaic arrays. The conclusion was expressed that the cells "continue to perform well, reinforcing the projection of a 20-year life, matching that of photovoltaic panels."[21]

EPT, LLC, has developed another design for use as standby power for a remote site in a terrestrial application. Again the primary effort in the design was to create a more cost-effective, reliable battery for terrestrial use. This design utilizes a combination of the dependent pressure vessel (DPV) and two-cell common pressure vessel (CPV) technologies. Five 2-cell DCPV units are assembled, creating a 12 V battery. The nominal capacity of the battery is 44 Ah at 10°C. This battery is shown in Fig. 15F.14.

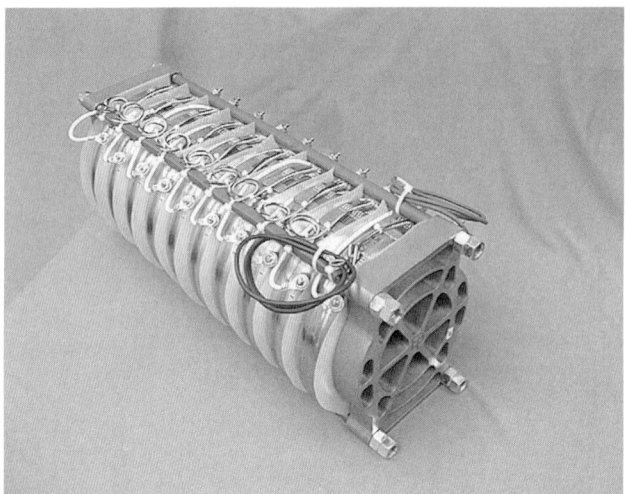

**FIGURE 15F.14**    DPV battery assembly.

## 15F.4.2    General Performance Characteristics

***Voltage Performance.***    Electrochemically impregnated nickel oxide electrodes are used in Ni-H$_2$ cells because of their excellent cyclic performance capabilities.[9,24] The capacity of these electrodes increases as the temperature decreases. The measured capacity of an electrochemically impregnated electrode is about 20% greater at 10°C than at 20°C. The capacity of the NTS-2 35 Ah cell at different temperatures is presented in Fig. 15F.15 to show the variation in capacity with temperature. These NTS-2 cells were discharged at the C/1.67 rate. Note that the mid-discharge voltage is between 1.2 and 1.25 V.

The high-rate discharge of an INTELSAT V cell at 200 A (12 min rate) is shown in Fig. 15F.16. The discharge profile is almost flat at 0.6 V; the potential drop of approximately 600 mV is due to the 3 m$\Omega$ terminal impedance of the cell (3 m$\Omega$ × 200 A = 600 mV). As previously seen in Fig. 15F.15, the mid-discharge voltage is between 1.2 and 1.25 V for a cell discharged at the C/1.67 rate. The aerospace IPV Ni-H$_2$ cells are not optimized for high-rate discharge but are optimized for maximum specific energy at discharge rates of between C/2 and C/1.5. At higher rates, the usable energy drops off because of the $I^2R$ losses in the terminals. For example, the INTELSAT V cell is capable of delivering about 50 Wh/kg up to the C rate (1 h rate) of discharge at 0°C. Above the C rate (30 A rate), however, the specific energy starts to drop off, as shown in Fig. 15F.17.

A salient feature of the Ni-H$_2$ cell is that the pressure is a direct indication of the state-of-charge of a cell (Fig. 15F.18). On charge, the hydrogen pressure increases linearly until the nickel oxide electrode approaches the fully charged condition. During overcharge, oxygen evolved at the positive electrode recombines at the negative electrode, and the pressure stabilizes. On discharge, the hydrogen pressure decreases linearly until the nickel oxide electrode is fully discharged. If the cell is reversed by over-discharging, hydrogen generated at the positive electrode is consumed at the negative electrode, and again the pressure is constant.

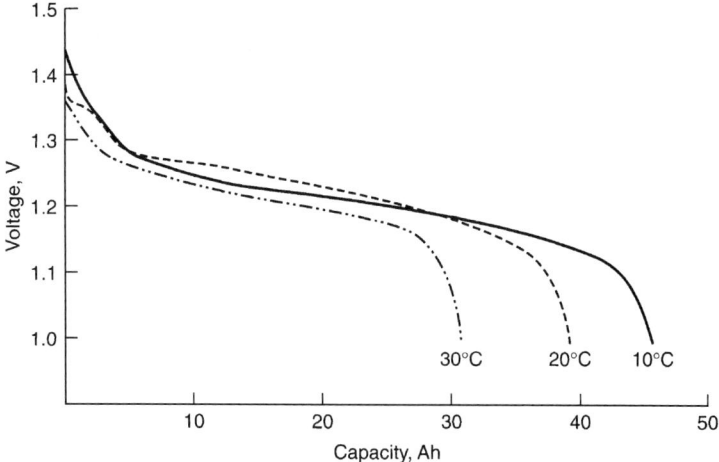

**FIGURE 15F.15**  Capacity of NTS-2 35 Ah cell at different temperatures. Discharge rate, C/1.67.

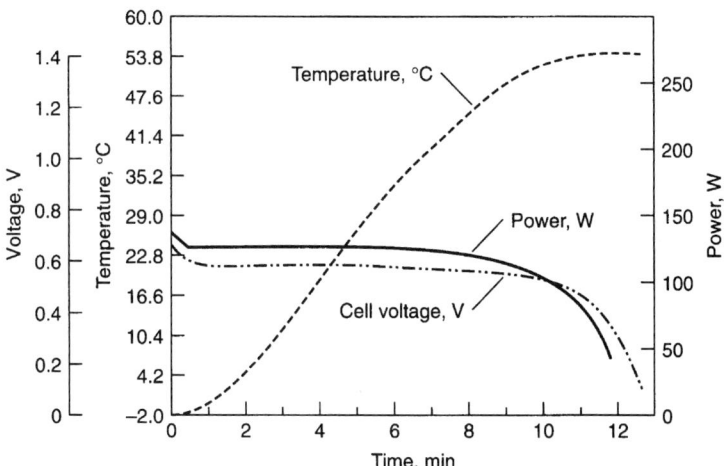

**FIGURE 15F.16**  Discharge of INTELSAT V 30 Ah cell at 200 A rate or 6.7 C-rate discharge.

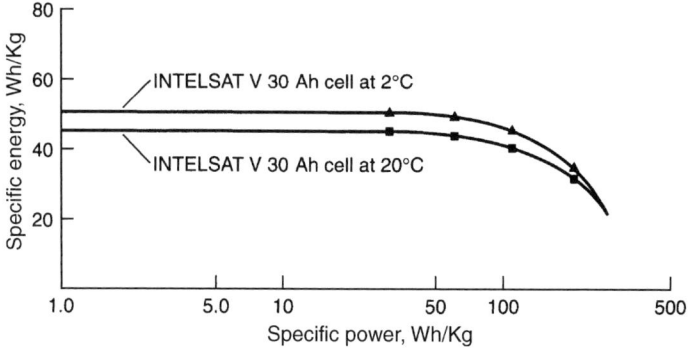

**FIGURE 15F.17**  Specific energy versus specific power of INTELSAT V Ni-H$_2$ cell.

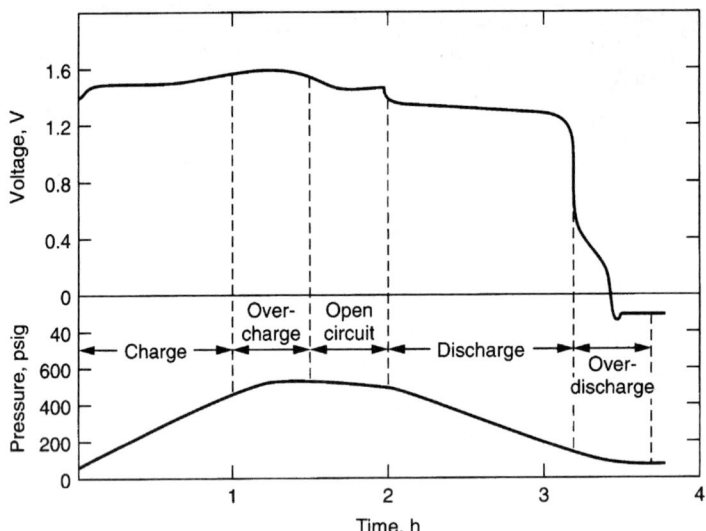

**FIGURE 15F.18**    Pressure and voltage characteristics of NTS-2 Ni-H$_2$ cell at 23°C.

The effects on cell voltage and capacity of discharge at different rates for a 90 Ah Hubble Space Telescope cell can be seen in Fig. 15F.19. The data were gathered from cells stabilized at 10°C. As the current increases, the capacity to 1.00 V decreases. Changes in the cell voltage and capacity can be attributed to the impedance of the cell (0.9 m$\Omega$).

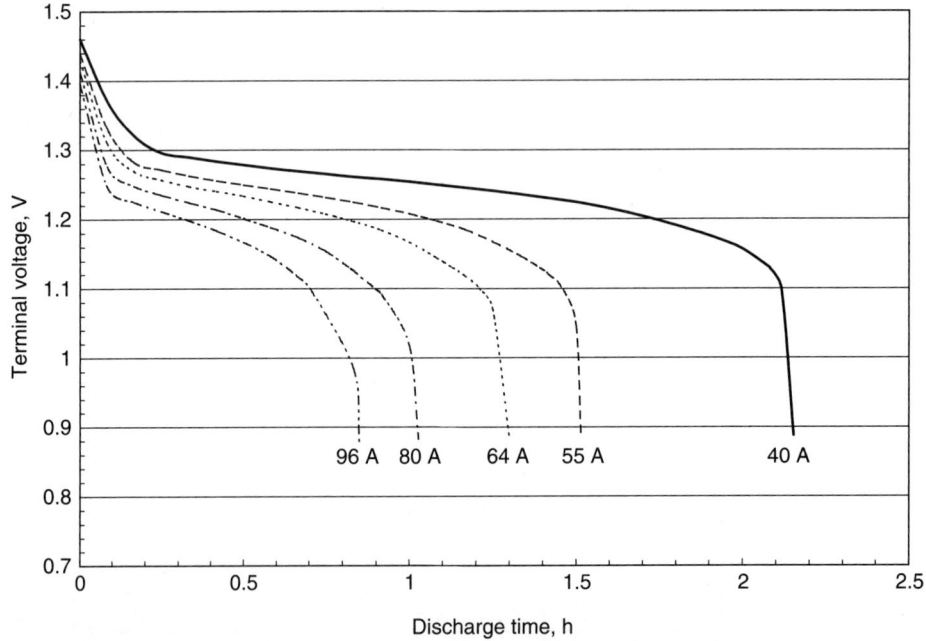

**FIGURE 15F.19**    Discharge of Hubble Space Telescope cell at different rates.

***Self-Discharge Characteristics of Ni-H₂ Cells.*** The self-discharge rate as a function of temperature was determined experimentally for Air Force 50 Ah cells used in the INTELSAT VI program.[25] Figure 15F.20 gives the data for the self-discharge of these 50-Ah cells at 10, 20, and 30°C. Figure 15F.21 shows the Arrhenius plot for these three temperatures. The slope of the straight-line regression fit to these three data points indicates an activation energy of 13.6 kcal/mol.[25]

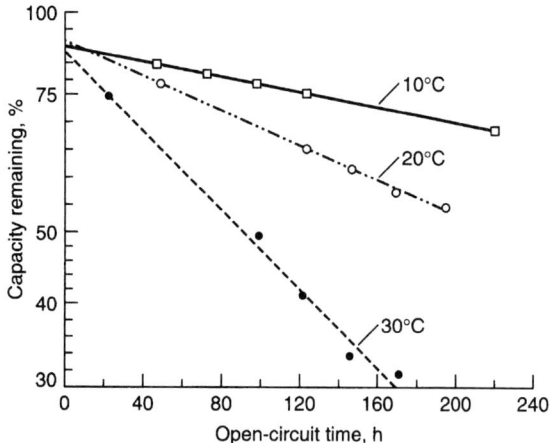

**FIGURE 15F.20** Self-discharge rates versus temperature for 50 Ah Ni-H₂ cell. (*Courtesy of COMSAT Laboratories.*)

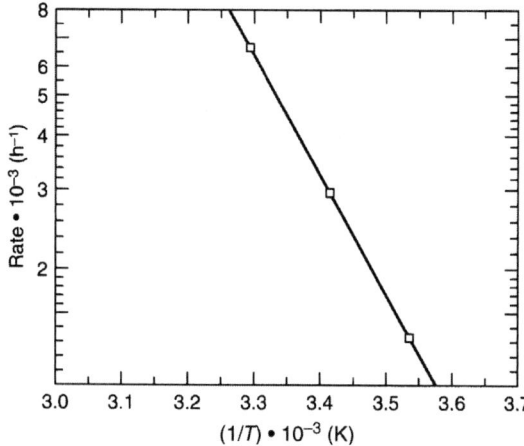

**FIGURE 15F.21** Arrhenius plot, self-discharge rates versus temperature. (*Courtesy of COMSAT Laboratories.*)

***Capacity as a Function of Electrolyte Concentration.*** The effects of electrolyte concentration on capacity were determined experimentally. Air Force positive electrodes and the Air Force 50 Ah Ni-H₂ cells from the INTELSAT VI program were used for this investigation. The Air Force positive electrodes were impregnated with active material by the alcoholic electrochemical impregnation process. The plaque was manufactured using the dry-powder process.

For the Air Force standard 50 Ah Ni-H$_2$ cell, the electrolyte concentration was determined to be 26% KOH in the charged state and 31% KOH in the discharged state.[25] Cells were activated with three different levels of electrolyte concentration: 25, 31, and 38 wt% concentration of KOH. The electrolyte concentration in these cells was determined by analyses in both the charged and the discharged conditions. Table 15F.2 presents cell capacity, electrolyte concentration, and average discharge voltage.

**TABLE 15F.2**   Cell Capacity and Voltage versus Electrolyte Concentration at 10°C

| | Electrolyte concentration | | |
|---|---|---|---|
| Parameter | 38% | 31% | 25% |
| Cell capacity, Ah | 64 | 56 | 43 |
| Number of positive plates | 40 | 40 | 40 |
| Capacity per plate, Ah | 1.60 | 1.40 | 1.08 |
| Electrolyte concentration: | | | |
|   Charged, wt% KOH | 32* | 26 | 21* |
|   Discharged, wt% KOH | 38 | 31 | 25 |
| Average discharge voltage, V | 1.247 | 1.268 | 1.290 |

*Estimated.

## 15F.4.3   PERFORMANCE DATA IN GEO APPLICATIONS

The Ni-H$_2$ batteries on the INTELSAT V satellites completed 9 years in orbit in 1990.

***Voltage Performance in Orbit.***   In-orbit performance of the INTELSAT V batteries was judged by the minimum end-of-discharge voltage observed during an eclipse season. The minimum battery voltage requirement is 28.6 V, or 1.10 V per cell average, which allows for one cell to fail via a short-circuit. The actual minimum battery voltages and the corresponding load currents and depths of discharge for each of the 14 batteries on the F-6 through F-15 satellites are presented in Table 15F.3 for the Fall 1990 eclipse season.[26] Also presented are the minimum cell voltages within each battery and the corresponding average cell voltage per battery. After up to 7 years in orbit, the minimum end-of-discharge battery voltages on the longest eclipse day ranged from 31.2 to 32.4 V for all 14 batteries. The cells within the batteries were well matched at their minimum end-of-discharge voltages during the Fall 1990 eclipse season (maximum deviation is ±20 mV between cells within the same battery). The one exception is cell 22 in battery 1 on the F-6 spacecraft. This cell was 40 mV below the average cell voltage. These battery voltages are well above the minimum voltage requirement.

**TABLE 15F.3**   Battery Loads and Minimum Voltages for Fall 1990 Eclipse Season

| | DOD, % | | Current, A | | Voltage, V | | Cell voltage, V | | | |
|---|---|---|---|---|---|---|---|---|---|---|
| | Battery 1 | Battery 2 | Battery 1 | Battery 2 | Battery 1 | Battery 2 | Batt. 1 av. | Batt. 1 min. | Batt. 2 av. | Batt. 2 min. |
| F-6 | 55.8 | 53.1 | 14.2 | 13.5 | 32.0 | 32.4 | 1.20 | 1.16 | 1.20 | 1.19 |
| F-8 | 54.0 | 54.4 | 13.7 | 13.8 | 32.0 | 32.0 | 1.20 | 1.18 | 1.20 | 1.18 |
| F-10 | 56.9 | 55.7 | 14.4 | 14.3 | 31.8 | 32.0 | 1.19 | 1.18 | 1.20 | 1.18 |
| F-11 | 55.3 | 60.0 | 14.1 | 15.4 | 32.0 | 32.0 | 1.20 | 1.18 | 1.20 | 1.19 |
| F-12 | 53.5 | 58.0 | 13.6 | 14.8 | 32.0 | 31.8 | 1.20 | 1.18 | 1.18 | 1.18 |
| F-13 | 67.0 | 59.0 | 16.9 | 15.0 | 31.2 | 31.8 | 1.17 | 1.15 | 1.19 | 1.17 |
| F-15 | 67.0 | 62.3 | 16.9 | 15.8 | 31.2 | 31.8 | 1.17 | 1.16 | 1.18 | 1.16 |

***Pressure Data.***   INTELSAT V batteries are reconditioned prior to each eclipse season. The reconditioning capacity and pressure data for INTELSAT V F-6 battery 2 are presented in Table 15F.4.[26] The pressure data were measured during the reconditioning discharge. The EOC pressure and the EOD pressure are the pressure at the start and the pressure at the end of the reconditioning discharge, respectively. The pressure constant is $\Delta P$ per measured capacity.

**TABLE 15F.4**   Reconditioning Capacity and Pressure Data for INTELSAT V F-6 Battery 2

| Eclipse season | Measured capacity, Ah | Max EOC pressure, lb/in.$^2$ | Min EOD pressure, lb/in.$^2$ | $\Delta P$, lb/in.$^2$ | Pressure constant $\Delta P$/measured capacity, lb/in.$^2$/Ah[*] |
|---|---|---|---|---|---|
| F83 | 38.1 | | No pressure data in database | | |
| S84 | 35.4 | | No pressure data in database | | |
| F84 | 37.7 | 516.39 | 13.87 | 502.62 | 13.33 |
| S85 | 37.6 | 518.49 | 17.90 | 500.59 | 13.31 |
| F85 | 37.5 | 515.14 | 17.23 | 497.9 | 13.27 |
| S86 | 37.9 | 519.34 | 15.32 | 504.02 | 13.29 |
| F86 | 37.6 | 519.73 | 22.03 | 497.70 | 13.23 |
| S87 | 38.3 | 514.34 | 13.87 | 505.47 | 13.19 |
| F87 | 37.2 | 519.73 | 22.03 | 497.7 | 13.37 |
| S88 | 38.3 | 525.78 | 16.20 | 509.58 | 13.30 |
| F88 | 37.8 | 521.86 | 17.90 | 503.96 | 13.33 |
| S89 | 36.9 | 526.91 | 18.67 | 508.24 | 13.77 |
| F89 | 40.2 | 534.22 | −0.57 | 534.79 | 13.30 |
| S90 | 38.6 | 551.73 | 19.22 | 532.51 | 13.79 |
| F90 | 36.0 | 530.87 | 38.04 | 492.83 | 13.68 |
| S91 | 39.5 | 546.52 | 17.23 | 529.29 | 13.39 |
| F91 | 39.0 | 545.30 | 17.90 | 527.40 | 13.52 |
| | | | | | Average 13.37 |

[*]1 lb/in.$^2$ = 6895 pa.

The data in Table 15F.4 show the following:

1. The strain-gauge bridge circuit provides useful pressure data.
2. No change occurred in the reconditioning EOD pressure with time for these INTELSAT V battery cells. The EOD pressure in Fall 1991 was almost the same as the EOD pressure at the beginning of life in orbit.
3. The significance of these data is that no oxidation or corrosion has occurred within these cells. Any oxidation of the cell components would result in a pressure increase at the end of the reconditioning discharge.

## 15F.4.4   Performance Data in LEO Applications

The HST was launched on April 24, 1990, with six 88 Ah Ni-$H_2$ batteries as the primary energy storage subsystem. This was the first reported nonexperimental mission to use Ni-$H_2$ batteries in an LEO application.[27] The batteries are being charged to a temperature-compensated voltage limit as described in Chap. 25B. The batteries are discharged to 7% to 10% of depth of discharge. As reported at the 1991 IECEC, "To date (April 1991) the performance of the batteries has been flawless."[27] Orbital data had shown an expected normal slow loss in useful capacity, which could eventually limit support of the HST. During servicing mission STS 125 in May 2009, after 18 years of service (13 years beyond design orbital life), the six batteries were replaced to extend the life of the HST.

## 15F.4.5    Advanced Batteries

***Advanced Designs for IPV Ni-H$_2$ Cells.***    A number of advanced concepts for IPV Ni-H$_2$ cells are being used to improve cycle life at deep DOD[28] mitigating failure modes commonly found in Ni-H$_2$ cells. They include (1) the use of alternative methods for recombination (i.e., the catalyzed wall wick), (2) the use of serrated-edge separators to facilitate the movement of gas within the stack while maintaining physical contact with the cell wall, (3) the incorporation of Belleville washers to yield an expandable stack capable of accommodating some of the nickel electrode expansion that is known to occur with cycling, and (4) the use of lower KOH concentrations to improve cycle life.

Cells utilizing the catalyzed wall wick are now in use. This concept offers an improved thermal design with recombination taking place on the pressure vessel wall. The heat from recombination is removed immediately through the pressure vessel wall to the cell thermal sleeve. This design also mitigates damage from internal stack popping since the recombination site is outside the cell stack.

The use of serrated edges for the separator is usually employed in designs utilizing asbestos. The irregular edge allows the unrestricted passage of oxygen along the edge of the cell stack while maintaining contact with the wall wick of the pressure vessel for recovery of electrolyte and removal of heat from the stack.

The Belleville washer acts as a spring, allowing further compression to accommodate any plate expansion during cycling.

The effects of KOH concentration on the cycle life of Ni-H$_2$ were investigated.[29] A breakthrough in the LEO cycle life of IPVs cells was reported and cell cycle life was improved by greater than a factor of 10 when KOH concentration was reduced from 31% to 26% in the fully discharged state. The lower concentrations, while enhancing the cycle life, result in a slightly higher mid-discharge operating voltage with a lower available capacity to the 1.00 V per cell limit.

***Advanced Battery Design Concepts.***    The CPV Ni-H$_2$ battery and the bipolar Ni-H$_2$ battery are two advanced battery design concepts investigated to improve the gravimetric and volumetric energy densities as compared to the IPV cell and battery Ni-H$_2$ technology.

*Common Pressure Vessel.*    Conceptually a CPV Ni-H$_2$ battery consists of a number of individual cells connected together in series and contained within one CPV.[30] For the IPV cell, each individual Ni-H$_2$ cell is contained within its own pressure vessel. Potential advantages for the CPV Ni-H$_2$ batteries include a significant increase in volumetric energy density (a decrease in volume), a decrease in manufacturing cost, a reduction in the complexity associated with the wiring and interconnection of IPV cells, an increase in specific energy, a reduction in the internal impedance of the battery, and improved heat transfer between the electrode stack and the pressure-vessel wall.

Several dual-cell CPV designs have been developed and tested. This design utilizes the dual-stack configuration used for IPV cells. For the CPV cell, the two stacks are connected in series, as shown in Fig. 15F.22. This dual cell CPV battery offers a 30% reduction in volume and a 7% to 14% reduction in mass compared to an equivalent battery with IPV cells.[31]

Batteries utilizing these cells have been used in several flight programs including LEO and interplanetary missions. Two batteries used on the Mars Global Surveyor and Mars Polar Lander flight programs can be seen in Figs. 15F.23 and 15F.24. These are 28 V batteries having capacities of 23 and 16 Ah, respectively.

A lightweight CPV Ni-H$_2$ battery was designed and developed jointly by COMSAT and Johnson Controls, Inc.[32] A prototype 10 in. diameter 26-cell, 24 Ah CPV battery was fabricated and tested to demonstrate the feasibility of this lightweight design for LEO applications. This battery used two 13-cell half stacks connected in series within the single CPV to provide a nominal 32 V battery. The 10 in. aerospace design used a semicircular cell with a double-tap design to enhance current distribution. The components of the prototype CPV battery are shown in Fig. 15F.25 with the fixed heat-fin cavity and lightweight pressure vessel.

Johnson Controls developed a new 5 in. diameter 9.6 Ah CPV battery with loose heat fins (Fig. 15F.26). The loose heat fin design was designed to overcome the problems encountered with the insertion of the cells into the heat fin cavity for the 10 in. CPV cell described.[33]

A 5 in. diameter, 28 V, 15 Ah CPV battery with the loose fin design was flown on the Clementine Program. This battery was manufactured by Johnson Controls under contract with the Naval Research Laboratory. This flight was launched and flown successfully in January 1994.

**FIGURE 15F.22**    EPT CPV design (2.5 V). (*Courtesy of Power Sub-systems Group, Eagle-Picher Technologies, LLC.*)

**FIGURE 15F.23**    Mars Global Surveyor 23 Ah CPV battery assembly.

**FIGURE 15F.24**    Mars Polar Lander 16 Ah CPV battery assembly.

**FIGURE 15F.25**    COMSAT/JCI CPV Ni-H$_2$ battery (10 in. diameter).

The advantages of the CPV Ni-H$_2$ battery make it a candidate for use in large multikilo-watthour LEO energy storage applications, such as the International Space Station, or constellation systems, such as Iridium®. It also appeals to the other end of the spectrum—the small 100 to 400 Wh applications that need low-cost lightweight batteries.[34]

EPT, LLC, supplied 10 in. diameter 28 V, 50 and 60 Ah CPV batteries for the Iridium program. Over 80 satellites using these CPV batteries have been launched. A 28 V, 60 Ah CPV battery manufactured for the Iridium program can be seen in Fig. 15F.27. This design offered impedances less than 25 m$\Omega$, a specific energy of 55 Wh/kg, and an energy density of 68 Wh/L.

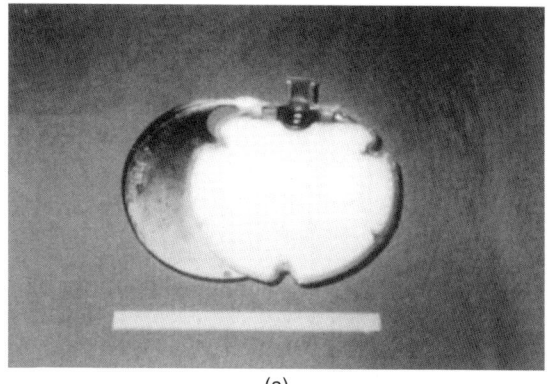

(a)

(b)

**FIGURE 15F.26**   JCI CPV Ni-H$_2$ battery (5 in. diameter). (*a*) Circular cell and loose heat fin. (*b*) 10-cell stack.

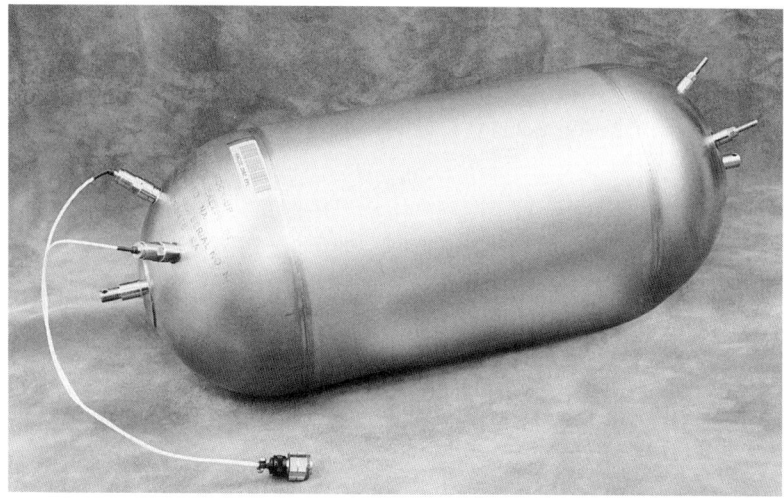

**FIGURE 15F.27**   IRIDIUM® 60 Ah CPV battery assembly.

*Bipolar Ni-H$_2$ Batteries.*   Studies have shown that the bipolar batteries promise savings in weight and volume as compared to IPV batteries.[35] The research has been directed toward large energy storage requirements for LEO applications such as the International Space Station program. Several bipolar Ni-H$_2$ batteries were designed, fabricated, and tested at NASA Lewis Research Center. The second one, assembled in 1983, was a 10-cell 6.5 Ah bipolar Ni-H$_2$ battery. Useful data were generated from tests of this 10-cell bipolar battery and results should aid the development work needed to improve performance.[35]

## REFERENCES

1. J. Dunlop, J. Giner, G. van Ommering, and J. Stockel, "Nickel-Hydrogen Cell," U.S. Patent 3,867,199, 1975.

2. S. U. Falk and A. J. Salkind, *Alkaline Storage Batteries,* Wiley, New York, 1969, sec. 2.5.

3. R. L. Beauchamp, "Positive Electrodes for Use in Nickel Cadmium Cells and the Method for Producing Same and Products Utilizing Same," U.S. Patent 3,653,967, April 4, 1972.

4. D. F. Pickett, H. H. Rogers, L. A. Tinker, C. Bleser, J. M. Hill, and J. Meador, "Establishment of Parameters for Production of Long Life Nickel Oxide Electrodes for Nickel-Hydrogen Cells," *Proc. 15th IECEC,* Seattle, WA, 1980, p. 1918.

5. L. W. Niedrach and H. R. Alford, *J. Electrochem. Soc.* **112:**117–124 (1965).

6. G. Holleck, "Failure Mechanisms in Nickel-Hydrogen Cells," *Proc. 1976 Goddard Space Flight Center Battery Workshop,* pp. 279–315.

7. E. Adler, S. Stadnick, and H. Rogers, "Nickel-Hydrogen Battery Advanced Development Program Status Report," *Proc. 15th IECEC,* Seattle, WA, 1980, p. 189.

8. G. van Ommering and J. F. Stockel, "Characteristics of Nickel-Hydrogen Flight Cells," *Proc. 27th Power Sources Conf.,* June 1976.

9. J. Dunlop, J. Stockel, and G. van Ommering, "Sealed Metal Oxide-Hydrogen Secondary Cells," in D. H. Collins (ed.), *Proceedings Of the 9th International Symposium on Power Sources,* 1974, Vol. 5, Academic, New York, 1975, pp. 315–329.

10. E. McHenry and P. Hubbauer, "Hermetic Compression Seals for Alkaline Batteries," *J. Electrochem. Soc.* **119:**564–568 (May 1972).

11. H. H. Rogers, S. J. Krause, and E. Levy, Jr., "Design of Long Life Nickel-Hydrogen Cells," *Proc. 28th Power Sources Conf.,* June 1978.

12. G. L. Holleck, M. J. Turchan, and D. DeBiccari, "Improvement and Cycle Testing of Ni/H2 Cells," *Proc. 28th Power Sources Symp.,* June 1978, pp. 139–141.

13. H. H. Rogers, U.S. Patent 4,177,325, December 4, 1979.

14. S. J. Stadnick, U.S. Patent 4,224,388, September 23, 1980.

15. L. Miller, J. Brill, and G. Dodson, "Multi-Mission Ni-H2 Battery Cells for the 1990s," *Proc. 24th IECEC,* Washington, DC, 1989, p. 1387.

16. T. M. Yang, C. W. Koehler, and A. Z. Applewhite, "An 83-Ah Ni-H2 Battery for Geosynchronous Satellite Applications," *Proc. 24th IECEC,* Washington, DC, 1989, p. 1375.

17. G. van Ommering, C. W. Koehler, and D. C. Briggs, "Nickel-Hydrogen Batteries for INTELSAT V," *Proc. 15th IECEC,* Seattle, WA, 1980, p. 1885.

18. P. Duff, "EUTELSAT II Nickel-Hydrogen Storage Battery System Design and Performance," *Proc. 25th IECEC,* Reno, NV, 1990, Vol. 6, p. 79.

19. R. J. Hass, A. K. Chawathe, and G. van Ommering, "Space Station Battery System Design and Development," *Proc. 23d IECEC,* 1988, Vol. 3, pp. 577–582.

20. D. Bush, "Evaluation of Terrestrial Nickel/Hydrogen Cells and Batteries," SAND88-0435, May 1988.

21. D. Bush, "Terrestrial Nickel/Hydrogen Battery Evaluation," SAND90-0390, July 1990.

22. D. E. Nawrocki, J. D. Armantrout, D. J. Standlee, and R. C. Baker, "The Hubble Space Telescope Nickel-Hydrogen Battery Design," *Proc. 25th IECEC,* Vol. 3, Reno, NV, 1990, pp. 1–6.

23. J. E. Lowery, J. R. Lanier, Jr., C. I. Hall, and T. H. Whitt, "Ongoing Nickel-Hydrogen Energy Storage Device Testing at George C. Marshall Space Flight Center," *Proc. 25th IECEC,* Reno, NV, 1990, pp. 28–32.

24. M. P. Bernhardt and D. W. Mauer, "Results of a Study on Rate of Thickening of Nickel Electrodes," *Proc. 29th Power Sources Conf.,* Electrochemical Society, Pennington, NJ, 1980.

25. J. F. Stockel, "Self-Discharge Performance and Effects of Electrolyte Concentration on Capacity of Nickel-Hydrogen (Ni/H$_2$) Cells," *Proc. 20th IECEC,* Vol. 1, 1986, p. 1171.

26. J. D. Dunlop, A. Dunnet, and A. Cooper, "Performance of INTELSAT V Ni-H2 Batteries in Orbit (1983–1991)," *Proc. 27th IECEC,* 1992.

27. J. C. Brewer, T. H. Whitt, and J. R. Lanier, Jr., "Hubble Space Telescope Nickel-Hydrogen Batteries Testing and Flight Performance," *Proc. 26th IECEC,* 1991.

28. J. J. Smithrick, M. A. Manzo, and O. Gonzalez-Sanabria, "Advanced Designs for IPV Nickel-Hydrogen Cells," *Proc. 19th IECEC,* San Francisco, CA, 1984, p. 631.

29. H. S. Lim and S. A. Verzwyvelt, "KOH Concentration Effects on the Cycle Life of Nickel-Hydrogen Cells," *Proc. 20th IECEC,* Miami Beach, FL, 1985, p. 1165.

30. D. Warnock, U.S. Patent 2,975,210, 1976.

31. T. Harvey and L. Miller, private communication on EPI handout.

32. M. Earl, J. Dunlop, R. Beauchamp, J. Sindorf, and K. Jones, "Design and Development of an Aerospace CPV Ni-H$_2$ Battery," *Proc. 24th IECEC,* Vol. 3, 1989, pp. 1395–1400.

33. J. Zagrodnik and K. Jones, "Development of Common Pressure Vessel Nickel-Hydrogen Batteries," *Proc. 25th IECEC,* Reno, NV, 1990.

34. J. Dunlop and R. Beauchamp, "Making Space Nickel-Hydrogen Batteries Lighter and Less Expensive," *AIAA/DARPA Meeting on Lightweight Satellite Systems,* Monterey, CA, August 1987, NTIS N88-13530.

35. R. L. Cataldo, "Life Cycle Test Results of a Bipolar Nickel-Hydrogen Battery," *Proc. 20th IECEC,* Vol. 1, 1985, pp. 1346–1351.

## *BIBLIOGRAPHY*

*NASA Handbook for Nickel-Hydrogen Batteries,* NASA Reference Publ. 1314, September 1993.

# CHAPTER 16
# SECONDARY ALKALINE, METAL OXIDE CATHODE CELLS

## 16.0  OVERVIEW

Secondary alkaline batteries with various electrode couples have existed for nearly two centuries. Chapter 15 covered cells that used a nickel-based cathode [$NiOOH/Ni(OH)_2$] with a variety of anodes. However, as revealed in Part 4A, a variety of other cathodes may potentially be used in alkaline cells, including a number of secondary alkaline battery systems. This chapter details three types of alkaline rechargeable cells with metal oxide cathodes:

   Chapter 16A: Silver Oxide

   Chapter 16B: Manganese Oxide

   Chapter 16C: Iron Oxide

Most of these batteries use zinc anodes, but other metals have also been used with some success, including cadmium and iron.

Due to the far-ranging nature of the electrochemical variations, cell designs have been adapted to be compatible with each specific system. In general, though, the cells use potassium hydroxide electrolytes and either a cellulose-based or alkali-resistant polymer (i.e., nylon) separator (typically a combination of both). Cells may be constructed in flat-plate, cylindrically wound, button/coin, or other cell configurations. As with most alkaline systems, maximum voltage is limited to less than 2 V (typically around 1.5 V) on open circuit due to electrolysis of water at higher levels. The three sections in this chapter provide the details on metal oxide cathode, alkaline, secondary batteries.

# SECTION A

# SILVER OXIDE CATHODES

**Alexander P. Karpinski**

## 16A.1 GENERAL CHARACTERISTICS

The rechargeable silver oxide batteries are noted for their high specific energy and power per unit weight and volume. The high cost of the silver electrode, however, has limited their use to applications where these qualities are prime requirements, such as in lightweight medical and electronic equipment, submersibles, torpedoes, and space applications. The characteristics of the silver oxide secondary batteries are summarized in Table 16A.1.

**TABLE 16A.1** Advantages and Disadvantages of Silver Oxide Secondary Cells

| Advantages | Disadvantages |
|---|---|
| Silver-zinc (zinc/silver oxide)* | |
| Highest energy and power per unit weight and volume | High cost |
| High discharge rate capability | Relatively low cycle life |
| Moderate charge rate capability | Sensitivity to overcharge |
| Good charge retention | |
| Flat discharge voltage curve | |
| Low maintenance | |
| Safety | |
| Silver-cadmium (cadmium/silver oxide)* | |
| High energy and power per unit weight and volume (approx. 60% of silver-zinc) | High cost |
| Cycle life (up to 250 cycles vented, 100 cycles sealed) | Decreased performance at low temperatures |
| Flat discharge voltage curve | |
| Low maintenance | |
| Nonmagnetic construction possible | |
| Safety | |
| Silver-iron (iron/silver oxide)* | |
| High energy and power capability | High cost |
| Good capacity maintenance | Water and gas management requirements |
| Overcharge capability | Not yet proven in field use |

*The corrected designation for these battery systems is shown in parentheses. However, the initial designation (silver-zinc, etc.) is more popular and generally used.

The first recorded use of a "silver battery" was by Volta with his now historic silver-zinc pile battery, which he introduced to the world in 1800.[1] This battery dominated the scene in the early 19th century, and during the next 100 years many experiments were carried out with cells containing silver and zinc electrodes. All these cells, however, were of the primary (nonrechargeable) type.

The first scientist to report a workable secondary silver battery was Jungner in Sweden in the late 1880s.[2] Although he experimented in the early stages with iron/silver oxide and copper/silver oxide batteries (which reportedly delivered as much as 40 Wh/kg), he settled on the cadmium/silver oxide battery for his experiments with electric car propulsion. The short cycle life and high cost of these batteries, however, made them commercially unattractive. During the next 40 years, other scientists (including Edison) experimented with various electrode formulations and separators, but without much practical success. It was the French professor Henri André who provided the key to the practical rechargeable zinc/silver oxide (silver-zinc) battery in 1941.[3,4] He described the use of a semipermeable membrane, cellophane, as a separator that would retard the migration of the soluble silver oxide to the negative plate and also impede the formation of zinc "trees," or dendrites, from the negative to the positive plate, the two major causes of cell short circuits.

In the 1950s, interest was revived in the silver-cadmium battery using the then newly available silver-zinc and nickel-cadmium technologies. This provided improved cycle life over the silver-zinc system. These batteries were first commercialized by Yardney International Corporation. Later, Westinghouse Corporation reported the commercial application of a silver-iron battery in which they sought to "eliminate the zinc plate problems with a trouble-free iron plate, ease the separator materials and life problem, and shift the deep discharge capacity stability to that limited by the silver plate."[5] The goal now, as for the past two centuries, is to provide the high energy content and power capability of the silver electrode in an improved-life, lower-cost, commercially viable secondary battery.

*Zinc/silver oxide batteries* provide the highest energy per unit weight and volume of any commercially available aqueous secondary batteries. They can operate efficiently at extremely high discharge rates, and they exhibit good charge acceptance at moderate rates and low self-discharge. The disadvantages are low cycle life (normally ranging, depending on design and use, from 10 to 150 deep cycles and up to 4000 partial depth of discharge [DOD] cycles), decreased performance at low temperatures, sensitivity to overcharge, and high cost. Rates as high as 20 times the nominal capacity (20C rate) can be obtained from specially designed silver-zinc batteries because of their low internal impedances. These high rates, however, must often be limited in duration to avoid a potentially damaging temperature rise within the cells.

*Cadmium/silver oxide batteries* have been viewed as a compromise between the high energy density but short life of the silver-zinc system and the long cycle life but low energy density of the nickel-cadmium system. Their energy density is roughly two to three times higher than that of nickel-cadmium, nickel-iron, or lead-acid batteries, with a relatively long cycle life, especially during shallow cycling. Charge retention is excellent. In addition, the ability to fabricate the cells without the use of magnetic materials has made them the battery of choice for several scientific satellite programs. The major disadvantage of the silver-cadmium system is cost; the cost per unit energy is even higher than for the silver-zinc battery.

*Iron/silver oxide batteries* provide high energy and power capability with long service life under deep-discharge use. They are capable of withstanding overcharge and overdischarge without damage and can provide good capacity maintenance with cycling. Disadvantages are, once again, cost and also the need for gas and water management in overcharge applications. Their nominal load voltage of 1.1 V is comparable to that of the silver-cadmium system, but lower than the 1.5 V level for silver-zinc batteries. Sufficient data have not been published for these batteries to date to permit complete characterization of their properties.

All three systems also offer the advantages of long dry shelf life and of providing a flat discharge voltage during the major portion of their discharge. This latter characteristic is related to the fact that as the silver oxide is reduced to metallic silver during discharge, the conductivity of the silver electrode increases and serves to counteract polarization effects. It is noteworthy that other couples successfully used the silver oxide electrode. These include metal hydride/silver oxide cells, hydrogen/silver oxide cells, and aluminum/silver oxide pile batteries, with the latter successfully used in torpedo applications.

## 16A.2 CHEMISTRY

### 16A.2.1 Positive-Electrode Reactions

The charge and discharge processes of the silver electrode in alkaline systems are of special interest because they are characterized by two discrete steps that manifest themselves as two plateaus in the charge and discharge curves. The reaction occurring at the silver electrode at the higher (peroxide) voltage plateau is shown as

$$2AgO + H_2O + 2e \underset{\text{charge}}{\overset{\text{discharge}}{\rightleftharpoons}} Ag_2O + 2OH^-$$

and at the lower, (monoxide) voltage plateau as

$$Ag_2O + H_2O + 2e \underset{\text{charge}}{\overset{\text{discharge}}{\rightleftharpoons}} 2Ag + 2OH^-$$

As shown, these reactions are reversible.

### 16A.2.2    Overall Cell Reactions

The overall electrochemical cell reactions, all of which use aqueous solutions of potassium hydroxide (KOH) for electrolyte, can be summarized as follows:

*Silver-zinc:*

$$AgO + Zn + H_2O \underset{\text{charge}}{\overset{\text{discharge}}{\rightleftharpoons}} Zn(OH)_2 + Ag$$

*Silver-cadmium:*

$$AgO + Cd + H_2O \underset{\text{charge}}{\overset{\text{discharge}}{\rightleftharpoons}} Cd(OH)_2 + Ag$$

*Silver-iron:*

$$4AgO + 3Fe + 4H_2O \underset{\text{charge}}{\overset{\text{discharge}}{\rightleftharpoons}} Fe_3O_4 \cdot 4H_2O + 4Ag$$

*Silver-metal hydride:*

$$AgO + 2MH \underset{\text{charge}}{\overset{\text{discharge}}{\rightleftharpoons}} Ag + 2M + H_2O$$

*Silver-hydrogen:*

$$AgO + H_2 \underset{\text{charge}}{\overset{\text{discharge}}{\rightleftharpoons}} Ag + H_2O$$

*Silver-aluminum:*

$$3AgO + 2Al \underset{\text{charge}}{\overset{\text{discharge}}{\rightleftharpoons}} 3Ag + Al_2O_3$$

These are simplified equations since there is still no general agreement on the detailed mechanisms of these reactions or on the exact form of all the reaction products. More details on zinc and cadmium anode reactions are provided in Chap. 15 as well as in Chap. 12.

The cell reactions for silver iron cells are

$$Fe + 2AgO + H_2O \underset{\text{charge}}{\overset{\text{discharge}}{\rightleftharpoons}} Fe(OH)_2 + Ag_2O \quad \text{(first plateau)}$$

$$Fe + Ag_2O + H_2O \underset{\text{charge}}{\overset{\text{discharge}}{\rightleftharpoons}} Fe(OH_2) + 2Ag \quad \text{(second plateau)}$$

$$3Fe(OH)_2 + Ag_2O \underset{\text{charge}}{\overset{\text{discharge}}{\rightleftharpoons}} Fe_3O_4 + 3H_2O + 2Ag \quad \text{(third plateau)}$$

In practice, only the first and second discharge plateaus are used. The overall reaction is

$$2Fe + 2AgO + 2H_2O \underset{\text{charge}}{\overset{\text{discharge}}{\rightleftharpoons}} 2Fe(OH)_2 + 2Ag \quad E^0 = 1.34 \text{ V}$$

## 16A.3    CELL CONSTRUCTION AND COMPONENTS

Secondary silver cells have been produced in prismatic, spirally wound cylindrical, and button-shaped configurations. The most common shape is prismatic. The construction of a typical prismatic cell is shown in Fig. 16A.1. This cell contains flat electrodes that are wrapped with multiple layers of separators to provide mechanical separation and inhibit migration of the silver to the zinc plate and the growth of zinc dendrites

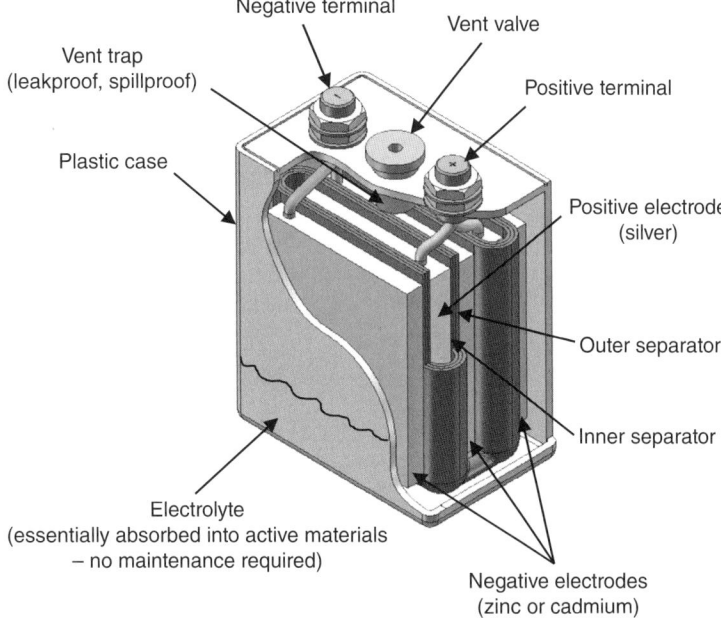

**FIGURE 16A.1**    Cutaway view of a typical prismatic zinc/silver oxide or cadmium/silver oxide secondary cell.

toward the positive plate. The plate groups are intermeshed, and the pack is placed in a tightly fitting case (Fig. 16A.2). Because of the relatively short shelf life of the activated silver cells, they are usually supplied by the manufacturers in the dry charged or dry unformed condition with filling kits and instructions. The cells are filled with electrolyte and activated just prior to use. They may also be supplied in the filled and ready-to-use condition if required by the user.

**FIGURE 16A.2**    Cell stack being assembled into a cell case; model LR-190, 210 Ah silver-zinc battery. (*Courtesy of Yardney Technical Products, Inc.*)

The mechanical strength of these cells is usually excellent. The electrodes are generally strong and are fitted tightly into the containers. The cell containers are made of high-impact plastics. Specific designs of these cells, when properly packaged, have met the high-shock, vibration, and acceleration requirements of launch vehicles, missiles, and torpedoes with no degradation.

### 16A.3.1 Silver Oxide Cathodes

The most common fabrication technique for silver electrodes is by sintering silver powder onto a supporting silver grid. The electrodes are manufactured either in molds (as individual plates or as master plates that are later cut to size) or by continuous rolling techniques. They are then sintered in a furnace at approximately 700°C.

Alternate techniques include dry processing and pressing, as well as slurry pasting of AgO or $Ag_2O$ onto a grid. If pasted, the plates are often sintered, converting the silver oxide into metallic silver and burning off the organic additives. The grid may be a woven or expanded metal form of silver, or silver-plated copper.

After being cut to size and having wires or tabs hot-forged onto an appropriately coined (compressed) area to carry current to the cell terminals, the electrodes are either electroformed (charged in tanks against inert counter electrodes) before assembly into cells or assembled into the cells in the metallic state and later charged in the cell to yield silver oxide.

Grid material; density and thickness; electrical lead type and size; and final electrode size, thickness, and density are all design variables that depend on the intended application for the cells. The silver powder particle size may be varied, with the finer powders approaching the theoretical silver utilization of 2.0 g/Ah. The use of very fine powder, however, results in an initial voltage dip (typically less than 120 ms) at medium (i.e., C-rate) to high discharge rates.

### 16A.3.2 Anodes

***Zinc Anodes.*** Zinc electrodes are most widely made by dry pressing, by a slurry or paste method, or by electrodeposition. In the dry-pressing method, a mixture of metallic zinc or zinc oxide, binder, and additives is compressed around a metal grid; this is normally done in a mold. The grid usually has the current-carrying leads prewelded in place. As the unformed powder electrodes have little strength, one component of the separator system, the negative interseparator, is usually assembled around the electrode as part of the fabricating operation. Rolling techniques have also been developed to permit continuous fabrication of dry-powder electrodes.[6]

In the paste or slurry method, a mixture of zinc oxide, binder, and additives is combined with water and applied continuously to a carrier paper or directly to an appropriate metal grid. Again, the negative interseparator is usually integral to provide needed physical strength. After drying, multiple layers of these pasted slabs may be pressed together about a pretabbed grid to form the final electrode. These plates may be assembled unformed into the cell, or they may be electroformed in a tank against inert counter electrodes.

Electrodeposited negative electrodes are manufactured by plating zinc in tanks onto metallic grids. The plates must then be amalgamated and pressed or rolled to the desired thickness and density, followed by drying.

The zinc electrode is acknowledged as the life-limiting component in both the silver-zinc and the nickel-zinc systems. Accordingly, much work has been done in the area of additives for these electrodes, both to reduce hydrogen evolution and to improve cycle life. The common additive to reduce hydrogen evolution has traditionally been mercury (1% to 4% of the total mix), but this is being replaced, for personnel safety and environmental reasons, by small amounts or mixtures of the oxides of lead, cadmium, indium, thallium, gallium,[7-11] and bismuth.[12,13] Many other (proprietary) additives have been introduced into the zinc electrode by various manufacturers in attempts to increase life.

Zinc electrodes also suffer capacity loss, which results from "shape change," or the migration of materials from the sides and top to the center and bottom of the electrode.

Several approaches have been taken to improve the stability of the zinc electrode, including (1) an excess of zinc to compensate for losses during cycling; (2) oversized electrodes on the basis that shape change starts on the electrode edges where current densities are higher; (3) use of binders such as polytetrafluoroethylene (PTFE), potassium titanate, neoprene latex, or other polymers to hold the active materials together; and (4) electrolyte additives.[12-14]

As is the case for the silver electrodes, the grid material, additives, and final electrode size, thickness, and density are all design variables that depend on the final application.

***Cadmium Anodes.***    Most silver-cadmium cells contain cadmium electrodes that are manufactured by pressed-power or pasting techniques. Although other methods have been used, such as impregnating nickel plaques with cadmium salts, as is done for nickel-cadmium cells, the most common method in silver-cadmium cells is to press or paste a mixture of cadmium oxide with a binder onto a silver or nickel grid. These processes are similar to those used for the pressed and pasted zinc electrodes.

***Iron Anodes.***    The iron electrodes used here are generally manufactured by powder-metallurgy techniques. To produce the anode active material, pure iron is dissolved in sulfuric acid. The $FeSO_4$ is recrystallized, dried, and roasted (815 to 915°C) to $Fe_2O_3$. The material is washed free of sulfate, dried, and partially reduced in hydrogen. The resulting material ($Fe_3O_4$ and Fe) is partially oxidized, dried, ground, and blended. Small amounts of additives, such as sulfur, FeS, and HgO, are blended in to increase battery life by acting as depassivators, reducing gas evolution, or improving conductivity.

To make the anode current collector, steel strips or ribbons are perforated and nickel-plated. After drying and annealing, the strip is formed into a pocket, about 13 mm wide and 7.6 mm long. One end is left open and filled with the iron active material. A machine automatically introduces the active material and tamps it into the pockets. After filling, the negative pockets are crimped and pressed into openings in a nickel-plated steel frame.

## 16A.3.3   Separators

***Zinc and Cadmium Anode Separators.***    The separators in the silver cells must meet the following major requirements:

1. Provide a physical barrier between positive and negative electrodes

2. Have minimum resistance to the flow of electrolyte and ions

3. Prevent migration of particles and dissolved silver compounds between positive and negative electrodes

4. Be stable in the electrolyte and cell operating environment

In general, secondary silver-zinc and silver-cadmium cells require up to three different separators, as shown in Fig. 16A.1. The inner separator, or positive interseparator, serves both as an electrolyte reservoir and as a barrier to minimize oxidation of the main separator by the highly oxidative silver electrode. This separator is usually made of an inert fiber such as nylon or polypropylene, usually with an added wetting agent.

The outer separator, or negative interseparator, also serves as an electrolyte reservoir and can also, ideally, stabilize the zinc electrode and retard zinc penetration of the main separator, and dendritic growth. Much work has been done to develop improved inorganic positive or negative electrode interseparators utilizing materials such as asbestos and potassium titanate fibers. Improvements in life have been reported as a result of this work.[7-11,15] However, many of these separators are no longer commercially available because of health hazard considerations.

The main, or ion exchange, separator remains the key to the wet life of the secondary silver cell. It was André's[3] use of cellophane as a main separator that first made the secondary silver cells feasible. The cellulosics (cellophane, treated cellophane, and fibrous sausage casing) are usually employed in multiple layers as the main separators for these cells. Again, much work has been done in recent years to develop improved separators utilizing such materials as radiation-grafted polyethylene,[16] inorganic separators,[10,11,15,17,24,26] and other synthetic polymer membranes. Improved cell life has been reported through the use of these new membranes either alone or in combination with cellulosics. Some of these have yet to be applied extensively to commercial silver cells, however, because of drawbacks sometimes involving high impedance, limited availability, and high cost.

***Iron Anode Separators.***    Multilayer microporous polyethylene, nonwoven felt polypropylene, and cellophanes are some of the materials typically used as separators in this system. It is important to note that the choice of separator has very little to do with the iron electrode, which is extremely stable in KOH and does not react with the separator system. Rather, the separator system must be selected to retard the migration of silver to the iron electrode and to withstand the oxidative effects of the silver electrode itself. The particular separator system chosen will therefore determine the cycle life, the shelf life, and the power capabilities. Consequently, the separator system is usually application-specific. Performance data for typical separator systems used in iron anode cells is shown in Sec. 16A.4.4.

### 16A.3.4   Cell Cases

The cell cases must be chemically resistant to attack by the corrosive concentrated potassium hydroxide electrolyte and to oxidizing effects of the silver electrodes. They must also be strong enough to contain any internal pressure generated in the cells and to maintain structural integrity throughout the anticipated range of environmental conditions that will be experienced by the cells.

The majority of secondary silver cells are assembled in plastic cases. The plastic most commonly used is an styrene-acrylonitrile copolymer (SAN). This material is relatively transparent and can be sealed easily by solvent cement or epoxy. However, its relatively low softening temperature (80°C) precludes it from use in some applications. A wide variety of other plastics have been used for cell cases. Table 16A.2 lists some of these materials and gives their characteristics. Metal cases have been used for some sealed cell applications; however, these

**TABLE 16A.2**   Properties of Cell Case and Cover Materials

| Material | Methyl methacrylate/ acrylonitrile-butadiene-styrene (MABS) | Acrylonitrile-butadiene-styrene (ABS) | | Polysulfone | Modified polyphenyl-ene oxide (PPO) | | Nylon[*] |
|---|---|---|---|---|---|---|---|
| Trade name | Terlux 2802 HD | MG37EP | Lustran 448 | Bakelite polysulfone P-1700 | Noryl 731 | Noryl SE1X | Zytel 151 or 151L |
| Specific gravity | 1.08 | 1.05 | 1.05 | 1.24 | 1.04–1.09 | 1.06–1.10 | 1.05–1.07 |
| Minimum tensile strength (psi) | 6960 | 4900 | 6100 | 10,200 | 8000 | 9800 | 8850 |
| Min. impact strength izod, notched (ft-lb/in.) | 1.31 @ 73°F 0.37 @ −22°F | 6.50 @ 73°F | 6.2 @ 73°F 1.20 @ −22°F | 6.50 @ 73°F | 3.0, min | 3.9, min | 1.29 |
| Conductivity (Btu/h/ft²/°F/in.) | | 1.3–2.3 | | 1.8 | — | — | 1.5 |
| Flexural modulus (psi) | | 355,000 | 348,000 | 380,000 | 351,000 | 363,000 | 247,000 |
| Specific heat (Btu/lb/°F)[†] | | 0.30–0.40 | | 0.31 | — | — | 0.3–0.4 |
| Heat defl. temp. | | | | | | | |
| °F @ 66 psi load | 273 | 210 | up to 252 | 358 | 274 | 262 | 275 |
| °F @ 264 psi load | 194 | 185 | 221, min | 345 | 240, min | 244, min | 131 |
| Use temp. (°F)[‡] | 167 | 140–210 | 190–230 | 320 | | | 180–250 |
| Hardness (Rockwell) | | R75–105 | R109 | R120 | R119 | — | R110 |
| Transparency | Yes | No | No | Yes | No | No | Translucent |
| Annealing temp (°F) | 185 ± 5 | 180 ± 5 | | 333 ± 5 | 214 ± 5 | | — |
| Manufacturer | BASF | SABIC | BAYER | AMOCO Performance Products | SABIC | | E.I. DuPont |

[*]Data are from ASTM D4066.
[†]Or cal/g/°C.
[‡]Maximum continuous, at no load.

**TABLE 16A.3**   Physical and Electrical Characteristics of KOH Solutions

| % KOH | Specific gravity at 15.6°C | Conductivity at 18°C, $\Omega^{-1} \cdot cm^{-1}$ | Specific heat at 18°C, cal/g·°C | Freezing point, °C | Boiling point, °C | | Vapor pressure, mm Hg | | Viscosity, cP | |
|---|---|---|---|---|---|---|---|---|---|---|
| | | | | | at 760 mm Hg | at 100 mm Hg | at 20°C | at 80°C | at 20°C | at 40°C |
| 0 | 1.0000 | | 0.999 | 0 | 100 | 52 | 17.5 | 355 | 1.00 | 0.66 |
| 5 | 1.0452 | 0.170 | 0.928 | −3 | 101 | 52.5 | 17.0 | 342 | 1.10 | 0.74 |
| 10 | 1.0918 | 0.310 | 0.861 | −8 | 102 | 53 | 16.1 | 327 | 1.23 | 0.83 |
| 15 | 1.1396 | 0.420 | 0.801 | −14 | 104 | 54 | 15.1 | 306 | 1.40 | 0.95 |
| 20 | 1.1884 | 0.500 | 0.768 | −23 | 106 | 56 | 13.8 | 280 | 1.63 | 1.10 |
| 25 | 1.2387 | 0.545 | 0.742 | −36 | 109 | 59 | 11.9 | 250 | 1.95 | 1.31 |
| 30 | 1.2905 | 0.505 | 0.723 | −58 | 113 | 62 | 10.1 | 215 | 2.42 | 1.61 |
| 35 | 1.3440 | 0.450 | 0.707 | −48 | 118 | 66 | 8.2 | 178 | 3.09 | 1.99 |
| 38 | 1.3769 | 0.415 | 0.699 | −40 | 122 | 69 | 7.0 | 156 | 3.70 | 2.35 |
| 40 | 1.3991 | 0.395 | 0.694 | −36 | 124 | 71 | 6.2 | 140 | 4.16 | 2.59 |
| 45 | 1.4558 | 0.340 | 0.678 | −31 | 134 | 80 | 4.5 | 106 | 5.84 | 3.49 |
| 50 | 1.5143 | 0.285 | 0.660 | +6 | 145 | 89 | 2.6 | 70 | 8.67 | 4.85 |

present problems in sealing and in electrically isolating the electrodes from the cases and are not used widely, except for button cells.

### 16A.3.5   Electrolyte and Other Components

The electrolyte used in secondary silver cells is generally an aqueous solution (35–45% concentration) of potassium hydroxide (KOH). Lower concentrations of the electrolyte provide lower resistivity and thus a higher voltage output under load as well as a lower freezing point. Concentrations below 45% KOH, however, are more corrosive to the cellulosic separators typically used in silver-based batteries and are not used for extended wet-life applications. Table 16A.3 depicts the critical parameters of KOH solutions. Various additives such as zinc oxide, lithium hydroxide, potassium fluoride, potassium borate, tin, and lead have been used to reduce the solubility of the zinc electrode.[14]

Since potassium hydroxide readily combines with carbon dioxide in the air to form potassium carbonate, thus reducing conductivity, cell vents are usually covered with a vent cap or a low-pressure relief valve.

Cell terminals are typically made of steel or brass and are almost always silver- or nickel-plated to improve corrosion resistance.

## *16A.4   PERFORMANCE CHARACTERISTICS AND BATTERY APPLICATIONS*

### 16A.4.1   Performance and Design Trade-Offs

The secondary silver batteries provide high energy capability combined with minimum weight and volume. The advantages and disadvantages of the various systems have been described earlier in this chapter. The performance of the batteries for specific applications will depend on the internal designs of the cells. It is rare that one can select an "off-the-shelf" battery that will meet all the requirements of a specific application.

Starting with the basic parameters, the cell design involves a series of compromises to obtain the most favorable combination of voltage, electrical capacity, and cycle life characteristics within the allowable battery weight and volume.

Assuming, for example, a nominal 1.5 V per silver-zinc battery at low current densities (0.01 to 0.03 A/cm²) and lower voltages at higher currents, the designer selects the number of cells for the application. The problem is increased if high current pulse loads are required and the battery must provide voltage above the minimum allowable at the high rate, while not exceeding the maximum allowable voltage at initial low rates. The size of the cell is then chosen by dividing the allowable volume by the calculated required number of cells.

The voltage, current, electrical capacity, and cycle life requirements must then be reviewed in conjunction with the allowable weight and the environmental conditions that the battery must be able to withstand. Each of these will be a factor in determining the choice of separator material for the cell. The stability and number of layers of separator must be sufficient to provide the desired wet life under these conditions while having a resistance low enough to prevent undue voltage drop at the high current load. Each of these requirements is also a factor in choosing the number of electrodes within the cell. As the number of electrodes (and thus the active electrode area) is increased, the current density during any discharge (amperes per square centimeter) is decreased, raising the output voltage. It should be noted that a cell design optimized for high discharge rates will, by nature of the design, have a reduced capacity under low discharge rates. This is a result of having many electrodes each having to be wrapped with the required number of layers of separators. Given a fixed internal volume, it follows that in such a high-rate (HR) cell, less space is available for active electrode material.

The cell must also be designed to contain enough active silver and zinc to supply the required electrical capacity for the desired number of cycles. Theoretically, 2.01 g of silver and 1.22 g of zinc are required in the cell for each ampere-hour of electrical capacity desired. Since these values are the theoretical capability of the pure material, and since some of the active materials will go into solution with each charge-discharge cycle, the designer must work with higher values—on the order of 3.5 g of silver and 3.0 g of zinc per nominal ampere-hour for long cycle-life cells. Other design variables, such as silver powder particle size, will also ultimately affect cell performance.

Because of these considerations, the performance curves shown in the following sections must be viewed as general characteristics of the systems and not necessarily of specific batteries for a particular application.

## 16A.4.2  Discharge Performance for Zinc/Silver Oxide Batteries

The open-circuit voltage of a fully charged zinc/silver oxide cell is 1.84 to 1.86 V. The discharge is characterized by two discrete steps, the first corresponding to the divalent oxide and the second to the monovalent oxide, as shown in Fig. 16A.3. The flat portion of the curves is referred to as "plateau voltage." This voltage is rate-dependent; at high rates the voltage steps may be obscured.

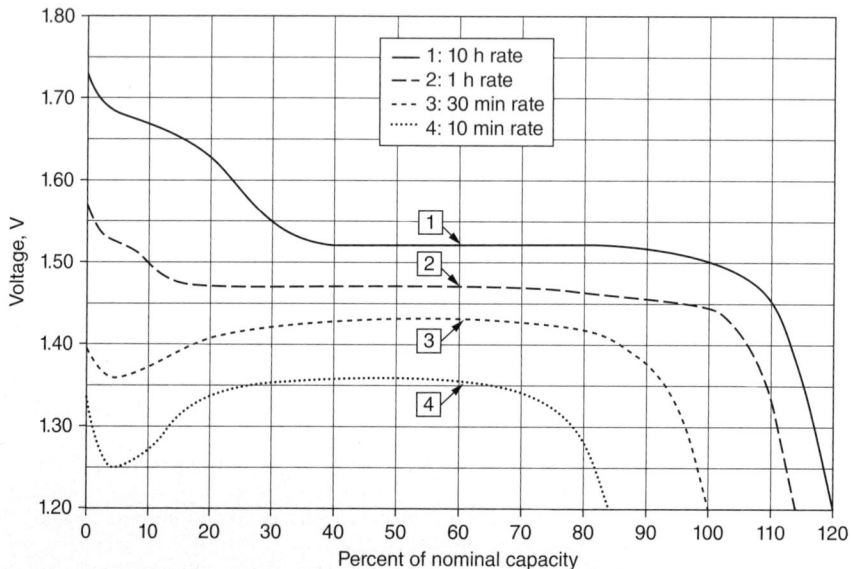

**FIGURE 16A.3**  Typical discharge curves of a silver-zinc battery at various rates, at 20°C.

The performance of the cells at various discharge rates and temperatures can be seen in Figs. 16A.4 to 16A.6, which show the effects on plateau voltage, capacity, and specific energy, respectively. The HR capability of the zinc/silver oxide battery is a complex process that can be characterized as the result of the electrical conductivity of the silver grid and the conductivity of the positive electrode as it is discharged, as well as the thin multiplate

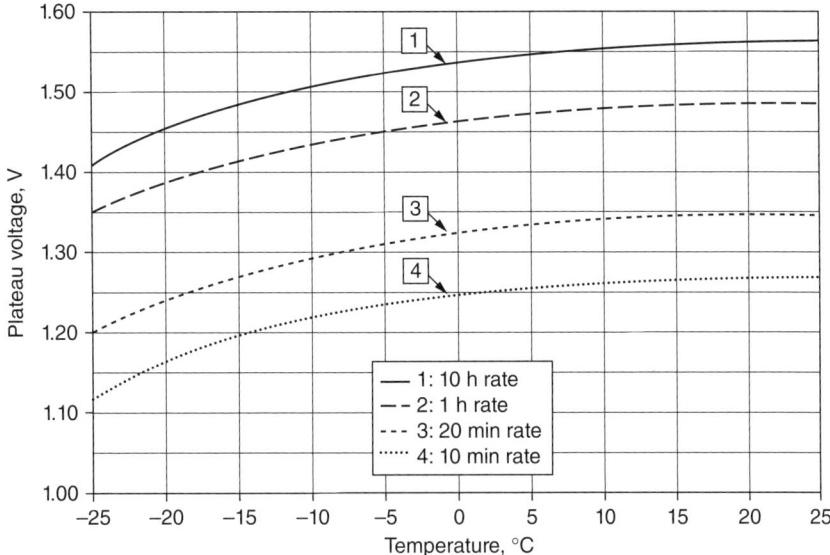

**FIGURE 16A.4** Typical effect of temperature on plateau voltage for high-rate silver-zinc cells.

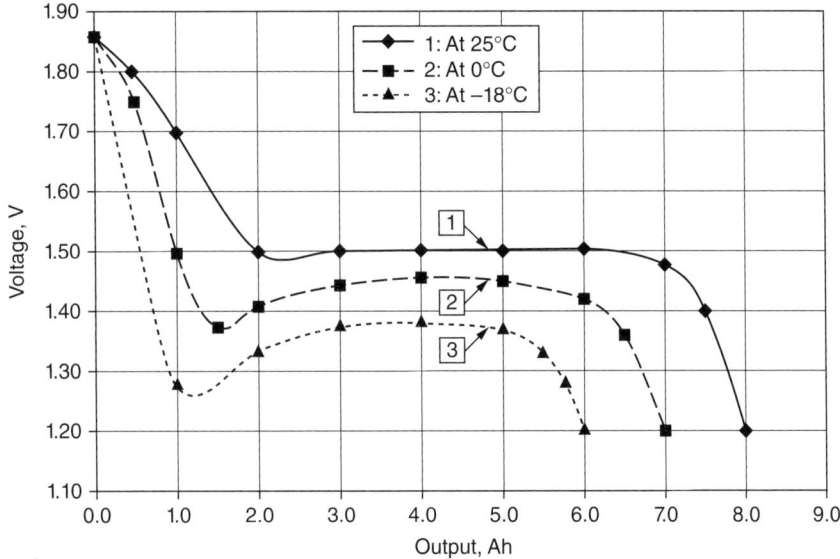

**FIGURE 16A.5** A typical discharge curve of an HR5 silver-zinc cell at various temperatures, with no insulation.

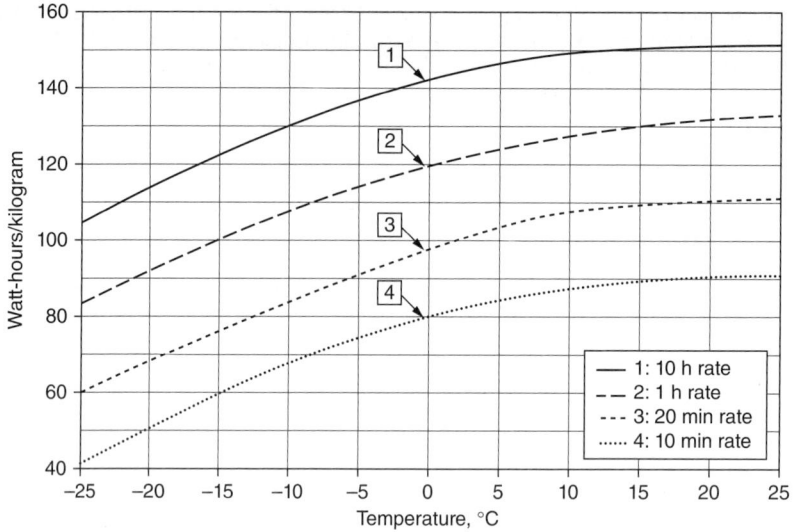

**FIGURE 16A.6**    Typical effect of temperature on specific energy of silver-zinc cells.

design of the cell. The performance of the battery falls off with decreasing temperature, particularly below −20°C. Allowing the battery to warm up or retaining the heat generated during the discharge can improve the performance at low ambient temperatures, over the range of values shown in Fig. 16A.5.

The performance characteristics of the zinc/silver oxide battery are summarized in Figs. 16A.7 and 16A.8, which can be used to determine the capacity, service life, and voltage under a variety of discharge conditions.

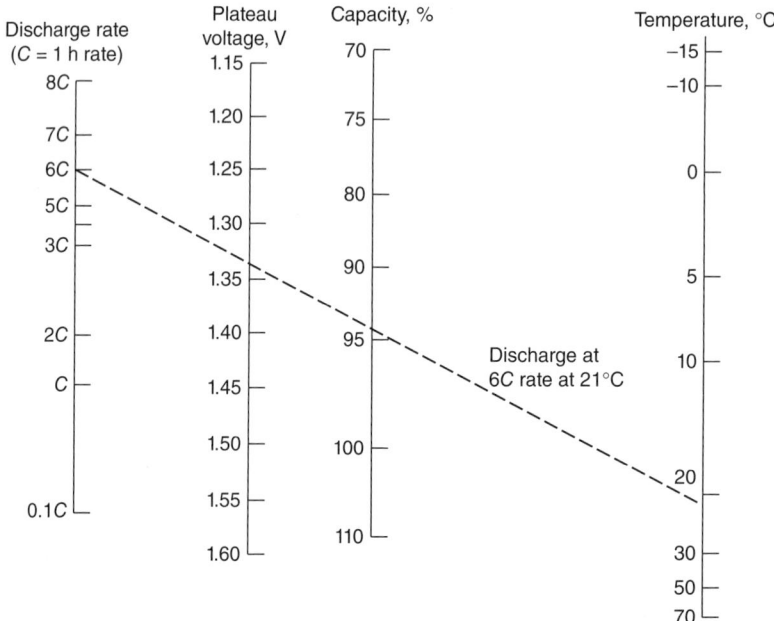

**FIGURE 16A.7**    Performance characteristics of silver-zinc batteries under various conditions. (To find the capacity and the plateau voltage of a silver-zinc battery, draw a straight line between the discharge rate and the ambient temperature at which the battery is stabilized.)

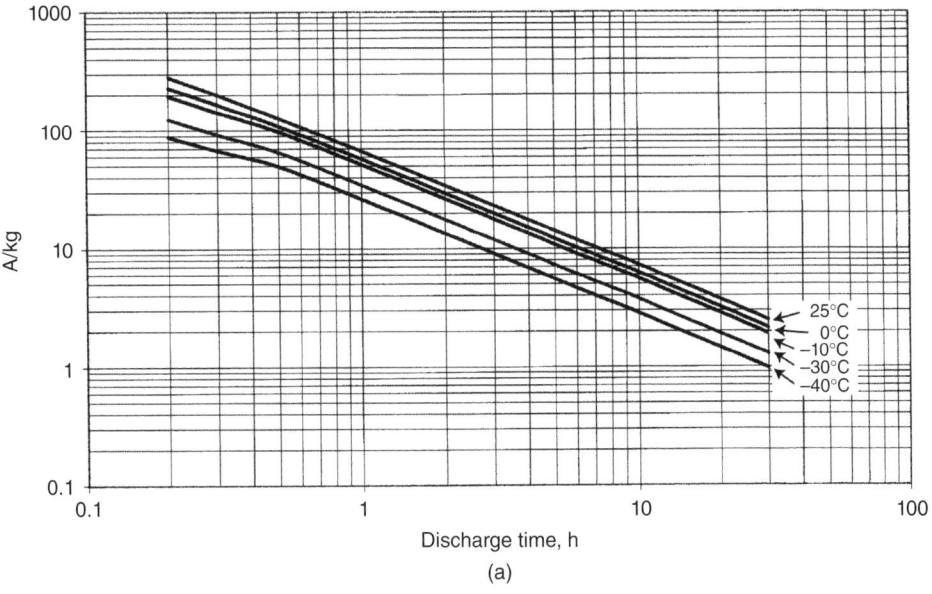

(a)

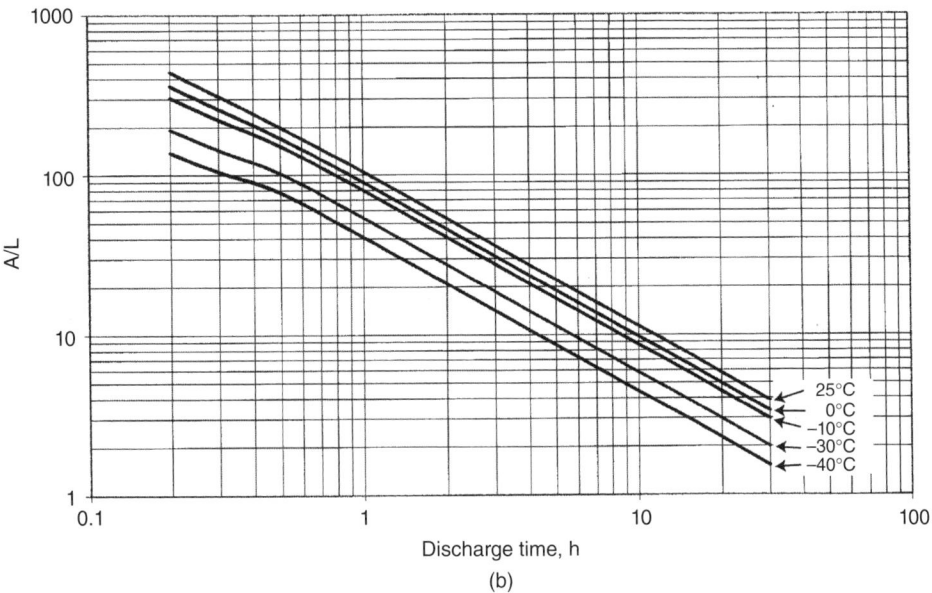

(b)

**FIGURE 16A.8**   (*a*) Service life of silver-zinc batteries at various discharge rates and temperatures (Amperes/kg vs. discharge time in hours). (*b*) Service life of silver-zinc batteries at various rates and temperature (Amperes/liter vs. discharge time in hours).

These figures present typical performance data. Performance differences can occur for each specific design and even for each battery, depending on cycling history, state-of-charge, storage time, temperature, and other conditions of use.

Figures 16A.3 to 16A.8 are specifically for HR designs. For many applications, trade-offs can be made to provide longer life at the expense of somewhat lower energy density. Alternative low-rate (LR) designs contain additional layers of separator, meaning, of necessity, fewer electrodes with higher impedance and lower capacity within a given volume. Typically, the LR battery cannot be discharged at higher than the 1 h rate and will provide about 3% to 5% lower average voltage and capacity than its HR counterpart at the 1 h rate. However, the LR batteries do provide substantial wet shelf life and cycle life advantages (see Table 16A.4).

**TABLE 16A.4** Nominal Life Characteristics of Secondary Silver Batteries[*]

|  | High rate (HR) Ag-Zn | Low rate (LR) Ag-Zn | Ag-Cd | Ag-Fe |
|---|---|---|---|---|
| Wet shelf life | 6–18 months | 1–2.5 years | 2–3 years | 2–4 years |
| Cycle life[†] | 10–50 cycles | 50–150 cycles | 150–1000 cycles | 100–300 cycles |

[*]These characteristics are nominal and vary with operating conditions and design of individual models.
[†]Cycle life characteristics are for deep (80–100% of full capacity) discharge cycles. Cycle life improves considerably with partial discharges.

### 16A.4.3    Discharge Performance for Cadmium/Silver Oxide Batteries

The open-circuit voltage of the cadmium/silver oxide battery is 1.38 to 1.42 V. Typical discharge curves at 20°C are given in Fig. 16A.9, showing the two-step discharge typical of the silver oxide electrode. The discharge characteristics are similar to those of the zinc/silver oxide battery except for the lower operating voltage; ampere-hour capacities per gram of silver are about the same.

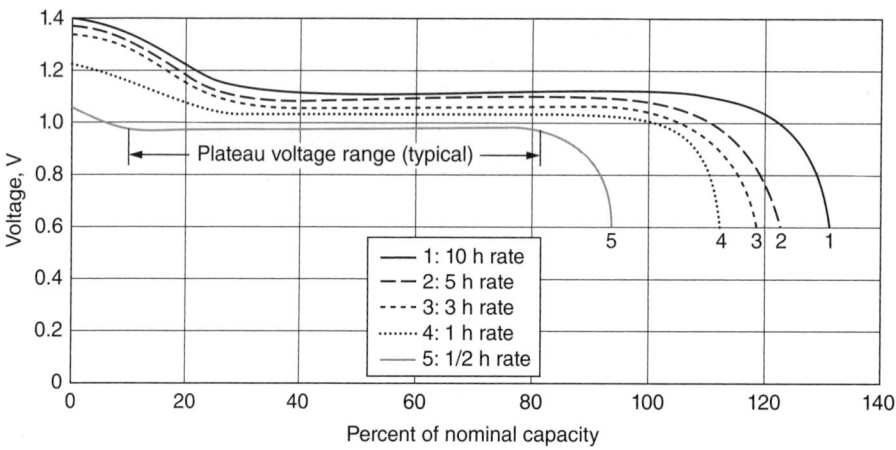

**FIGURE 16A.9**   Typical discharge curves of a silver-cadmium battery at various rates at 20°C.

The capacity and the discharge voltage of the battery are temperature-dependent, again similar to the zinc/silver oxide battery. The effects of temperature and discharge rate on voltage and specific energy are shown in Figs. 16A.10 and 16A.11, respectively. The recommended operational temperature range is −25 to 70°C, with the optimum performance obtained between 10 and 55°C. With external heating, the temperature range can be lowered to −60°C.

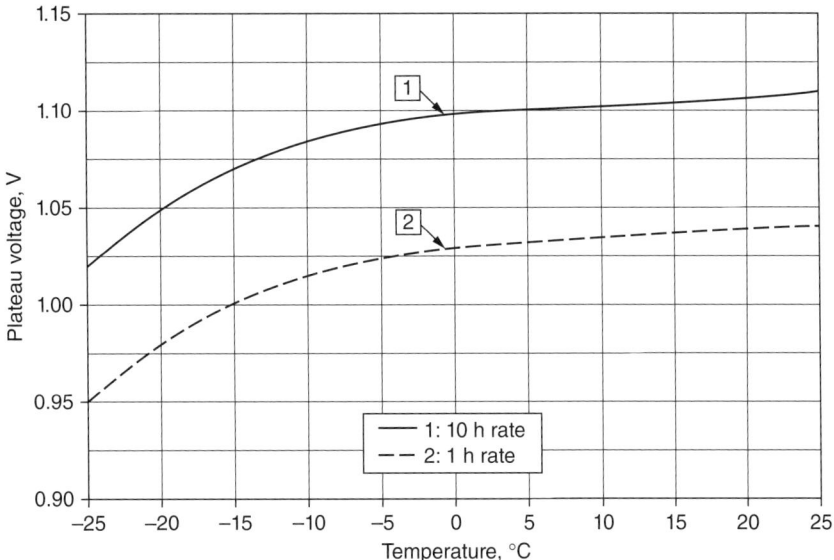

**FIGURE 16A.10**    Typical effect of temperature on plateau voltage of silver-cadmium cells (discharged without heaters).

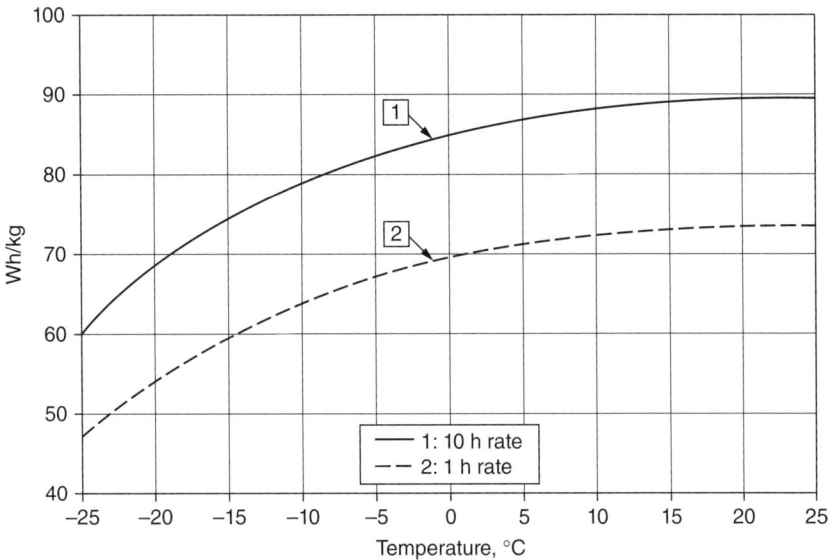

**FIGURE 16A.11**    Typical effect of temperature on specific energy of silver-cadmium cells.

The performance characteristics of the cadmium/silver oxide battery are summarized in Figs. 16A.12 and 16A.13*a* and *b*, which can be used to determine the capacity, service life, and voltage levels under a variety of discharge conditions.

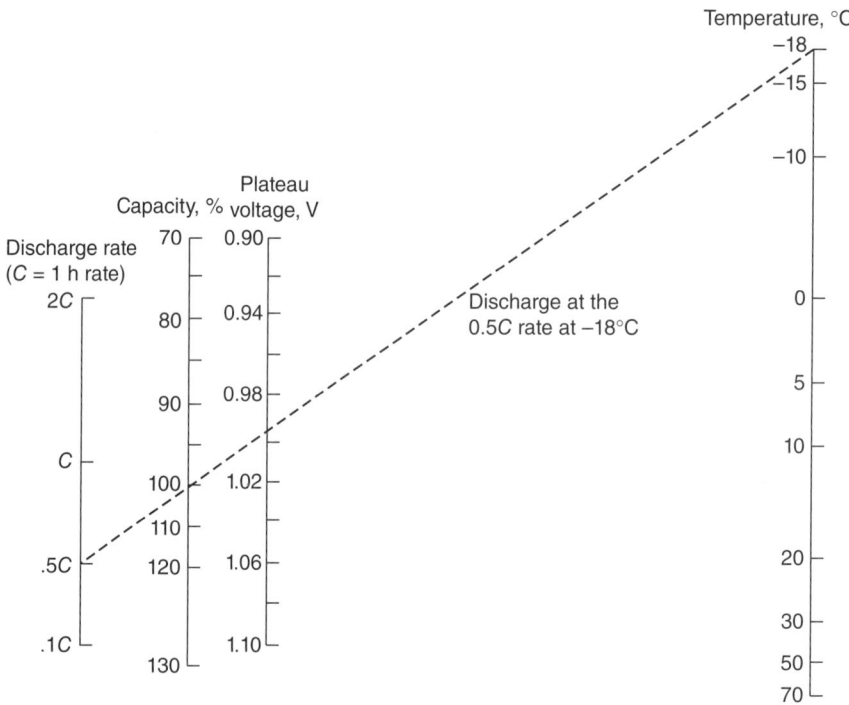

**FIGURE 16A.12**     Performance characteristics of silver-cadmium batteries under various conditions. (To find the capacity and the plateau voltage of a silver-cadmium battery, draw a straight line between the discharge rate and the ambient temperature at which the cell is stabilized.)

## 16A.4.4   Charge/Discharge Characteristics of Iron/Silver Oxide Batteries

Typical charge-discharge curves for a silver-iron battery of the type shown in "Silver-Iron Batteries" (see Sec. 16A.4.11) are given in Fig. 16A.14. The electrolyte is KOH of 1.31 specific gravity with 15 g/L LiOH added. The batteries can withstand several complete reversals without appreciable adverse effect on the capacity. Cycle life testing of typical separator systems, comparing cells with zinc and iron anodes, at 100% DOD and 10% overcharge is shown in Fig. 16A.15.

## 16A.4.5   Impedance

The impedance of the silver oxide cells is normally low but can vary considerably with many factors, including the separator system, current density, state-of-charge, stand time, cell age, temperature, and, importantly, cell size. In a study of the effect of storage time on the impedance of silver-zinc cells, initial values of 5 to 11 m$\Omega$ for partially charged cells were reported,[18] with the values rising to as much as 3 $\Omega$ following 8 months' storage at 21°C and 9 to 15 $\Omega$ following 8 months at 38°C. Cells stored in the fully discharged condition retained their low impedance (ranging from 2 to 10 m$\Omega$) throughout the entire test period. The high-impedance cells returned to normal low values, however, within several seconds of the start of discharge.

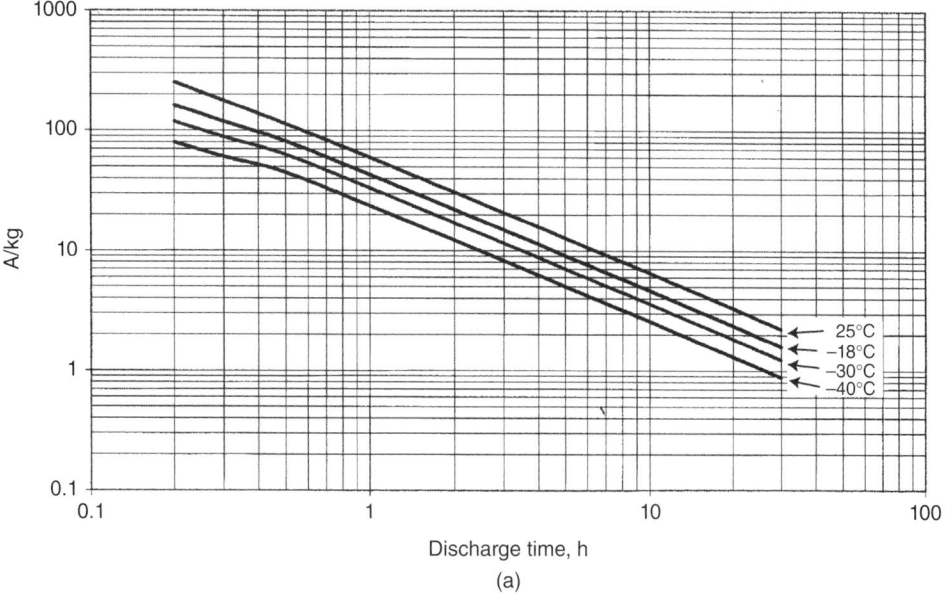

(a)

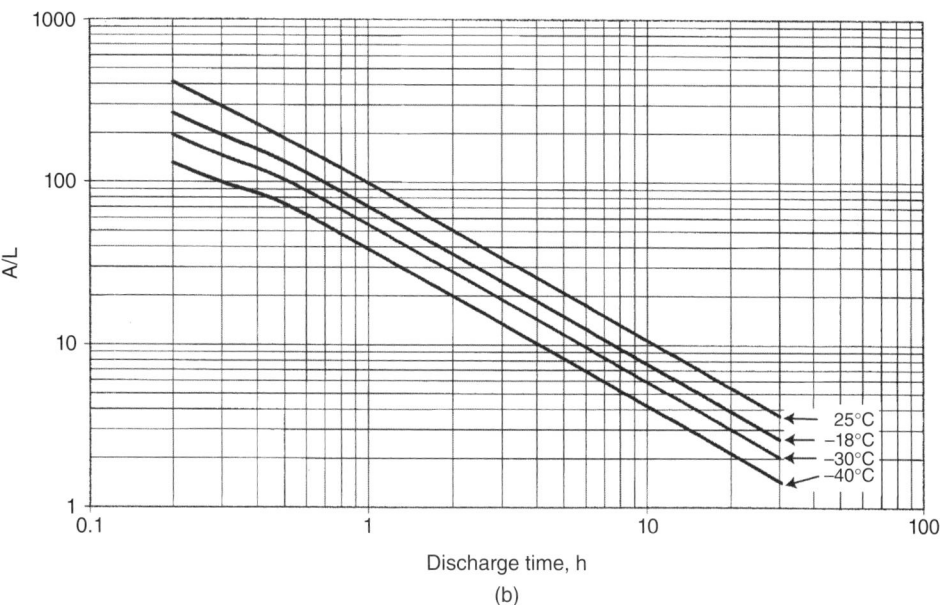

(b)

**FIGURE 16A.13**   (*a*) Service life of silver-cadmium batteries at various discharge rates and temperatures (Amperes/kg vs. discharge time in hours). (*b*) Service life of silver-cadmium batteries at various discharge rates and temperatures (Amperes/ liter vs. discharge time in hours).

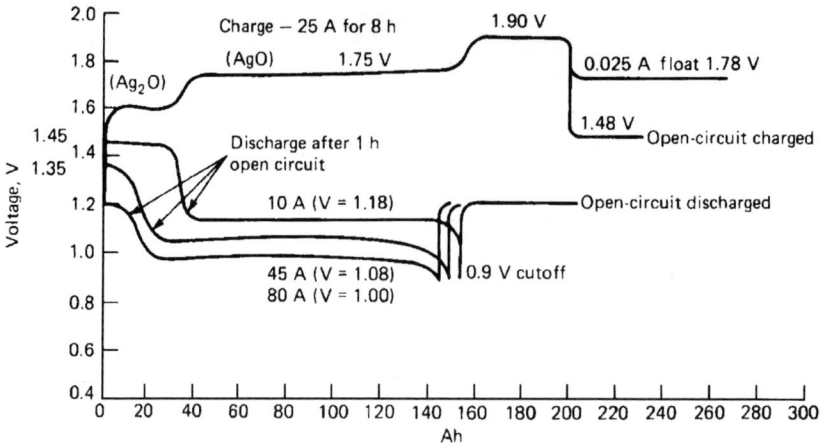

**FIGURE 16A.14**    Charge-discharge characteristics of a nominal 140 Ah iron/silver oxide battery. (*From Ref. 29.*)

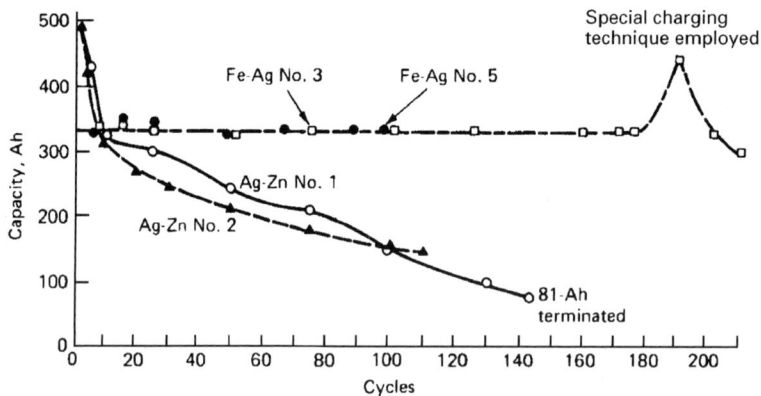

**FIGURE 16A.15**    Cyclic life performance of zinc/silver oxide and iron/silver oxide prototype batteries. (*From Ref. 29.*)

The AC impedance of the silver oxide batteries is highly dependent on the frequency of the load, with the impedance rising sharply above 5 kHz. The impedance of a 6-cell, 350 Ah silver-zinc battery, discharged to approximately 50% DOD, is shown in Fig. 16A.16, and the corresponding phase angle is shown in Fig. 16A.17.[19]

### 16A.4.6    Charge Retention

The charge retention of the activated and charged silver oxide cells is better than that of most secondary batteries, with retention of more than 85% of charge after 3 months' storage at 20°C.

As with other chemical reactions, the rate of loss of charge is dependent on the storage temperature (see Fig. 16A.18). Storage at −20 to 0°C is recommended for maximizing charge retention. In the dry and charged condition, properly sealed and stored cells will retain their charge for over 10 years. Here again, low-temperature storage is highly recommended.

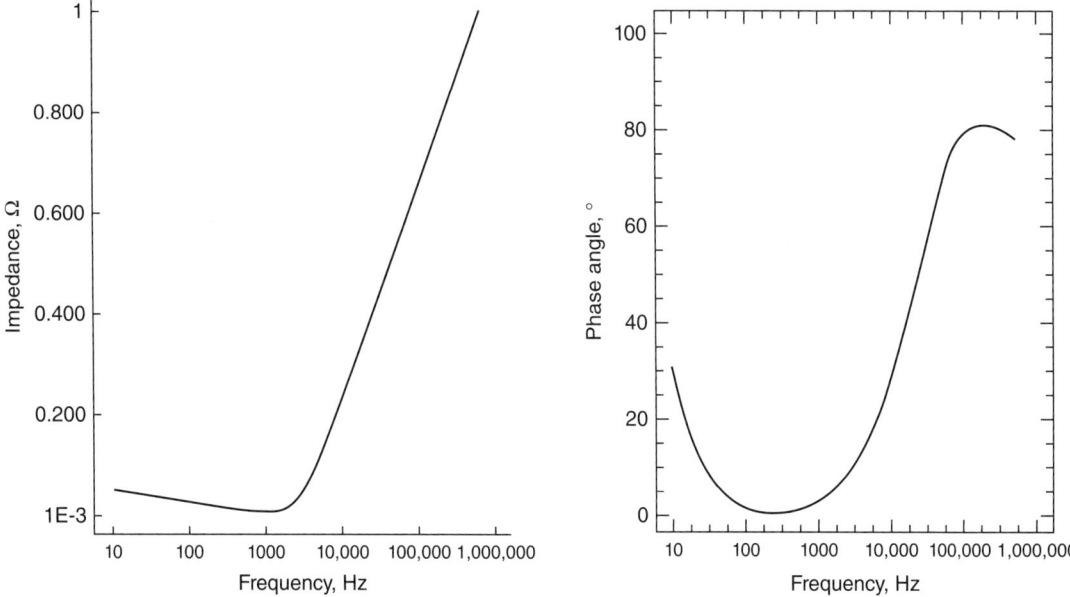

**FIGURE 16A.16**   Impedance magnitude versus frequency.

**FIGURE 16A.17**   Impedance phase angle versus frequency.

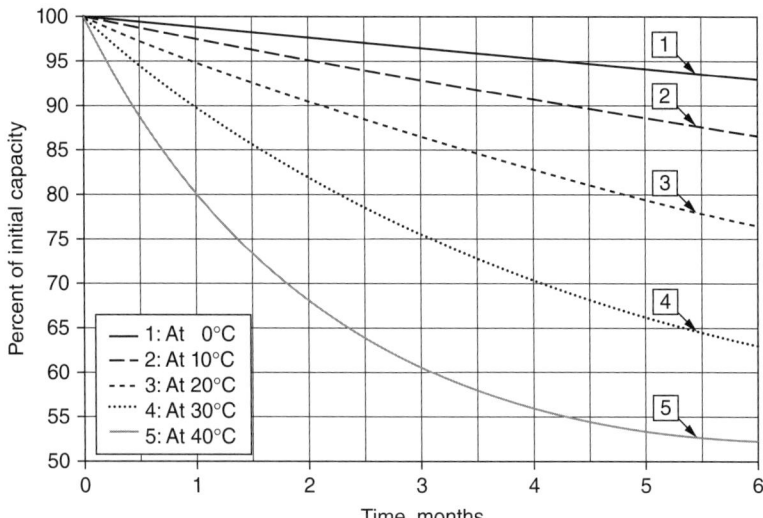

**FIGURE 16A.18**   Capacity retention of silver-zinc and silver-cadmium cells at various temperatures.

## 16A.4.7   Cycle Life and Wet Life

The separator system and the solubility of the active materials play critical roles in determining the wet and cycle lives of the silver-based cells. The separator must have a low electrolytic resistance for discharges at high rates, yet it must have high resistance to chemical oxidation by the silver species as well as low permeability to colloidal silver, zinc, cadmium, or iron.

Since cadmium and iron are relatively insoluble in concentrated alkaline electrolytes, the life of the silver-cadmium and silver-iron cells is therefore limited by the rate of silver migration through the various layers of the separator system. Failure (internal short circuit) occurs when a metallic bridge is established through the separator between positive and negative electrodes. Multiple layers of separator are used to extend the life capability, however, at the expense of higher internal resistance.

The life of the silver-zinc battery is further hindered by the high solubility of zinc in alkaline electrolytes. The problem manifests itself in two failure mechanisms: shape change and growth of dendrites. The migration of zinc from the electrode tops and edges where it becomes depleted, to the center and bottom where it densifies, resulting in capacity loss. Dendrites, sharp, needlelike structures of the metal, are formed during overcharge. They may perforate the separators and cause internal short circuits. The capacity degradation in silver-zinc batteries due to shape change is illustrated in Fig. 16A.19. The decline in capacity of silver-cadmium batteries is at a much slower rate, as depicted in Fig. 16A.20.

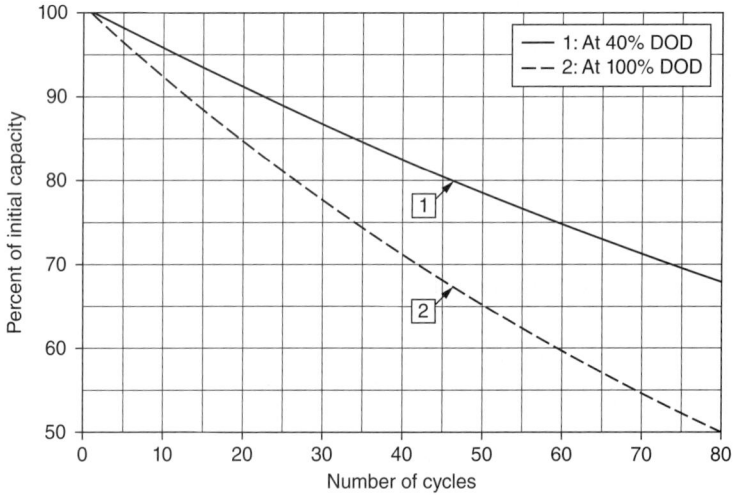

**FIGURE 16A.19**    Typical capacity degradation of silver-zinc cells discharged at low rate.

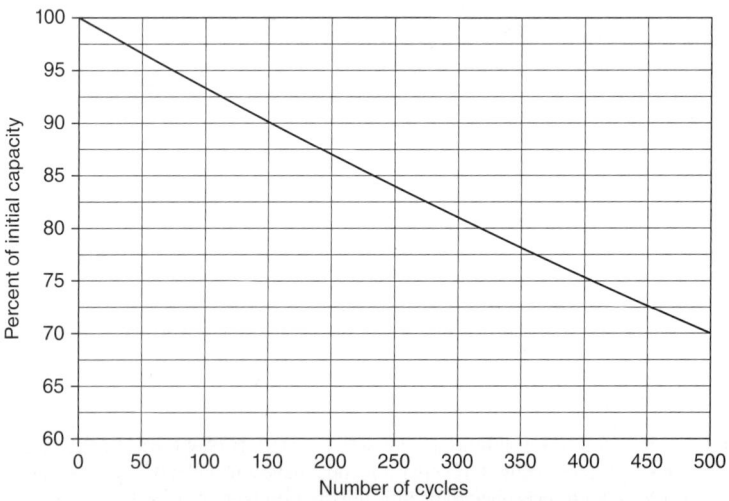

**FIGURE 16A.20**    Typical capacity degradation of sealed silver-cadmium cells discharged at low rate to 100% DOD.

Aside from normal capacity loss due to wet activated storage or cycling, dry-charged zinc-silver oxide cells may exhibit a one-time deviation in capacity (typically less than 20% of the initial capacity) during the second cycle discharge, called "second cycle syndrome." This deviation, which manifests itself as the inability to accept a full charge during the second cycle and has never been fully explained, is well known to the user community. It may be addressed in two ways:

1. If the reduced capacity is acceptable for the application, do nothing; the capacity returns to normal during the next cycle.

2. If not acceptable, the capacity can be increased by a partial discharge followed by vacuuming and recharging, preferably at a lower rate. Note that overcharging will not improve the capacity and may harm the cell.

The nominal life ratings for the silver-zinc, silver-cadmium, and silver-iron batteries are given in Table 16A.4. The life of the silver oxide cells will also vary greatly with operating and storage conditions. High rates of discharge to 100% DOD and high-temperature exposure (for more than 30 days) will significantly reduce the wet life and cycle life of the batteries. Cold-temperature storage (at less than $-10°C$), when not in use, on the other hand, will greatly increase the life of the cells. The cycle and wet lives will also increase with a decreasing DOD.

In a study to evaluate the capabilities of silver-cadmium batteries for satellite applications, an extensive test program was run on 3-cell, 3 Ah silver-cadmium batteries at various DOD.[20] These results are summarized in Table 16A.5, showing the increase in cycle life with decreasing DOD.[20] Another study on 250 Ah silver-zinc cells, cycled at less than 1% DOD, with 14 full-capacity cycles, resulted in a cycle life of 7280 cycles over a 38-month period.[21]

**TABLE 16A.5**   Cycle Life versus Depth of Discharge for 3 V, 3 Ah Sealed Cadmium/Silver Oxide Batteries

| Depth of discharge, % | Cycle life at first cell failure |
|---|---|
| 65 | 1800 |
| 50 | 3979 |
| 50 | >5400 (375 days) |
| 35 | >5400 (375 days) |
| 25 | >5400 (375 days) |

*Source:* Ref. 20.

### 16A.4.8  Charging Characteristics

*Efficiency.*    The *ampere-hour* efficiency (ampere-hour output/ampere-hour input) of the silver-zinc and silver-cadmium systems under normal operating conditions is high—greater than 99%—because practically no side reactions occur when charging at normal rates. The *watt-hour* efficiency (watt-hour output/watt-hour input) is about 70% under normal conditions because of the difference between charge and discharge voltages.

*Zinc/Silver Oxide Batteries.*    The manufacturers of these batteries recommend constant-current charging at the 10 to 20 h rate for most applications. However, constant-potential and pulsed charging techniques have also been applied.

A typical charge curve at constant current is shown in Fig. 16A.21. The two plateaus reflect the two levels of oxidation of the silver electrode: the first from silver to monovalent silver oxide ($Ag_2O$), which occurs at a potential of approximately 1.6 V; the second from the monovalent to the divalent silver oxide (AgO), which occurs at approximately 1.9 V. It should be noted that during this transition from the monovalent to the divalent state-of-charge, a momentary spike in the charge voltage, of up to 2.0 V, may occur prior to stabilizing at the 1.90 to 1.95 V plateau. To ensure a full charge, the charging system must be designed to ignore this temporary rise in voltage.

Charging is normally terminated when the voltage during charge rises to 2.0 V. Above 2.1 V the cell begins to generate oxygen at the silver electrode and/or hydrogen at the zinc electrode, decomposing water from the electrolyte. Overcharge is also detrimental to cell life in that it promotes the growth of zinc dendrites and subsequent short circuits.

The importance of proper charging to the life of these batteries cannot be overemphasized.

*Cadmium/Silver Oxide Batteries.*    Except for lower voltages on each of the plateaus (1.2 V on the lower level, 1.5 V on the upper level), the charging characteristics of the silver-cadmium battery are similar to those of silver-zinc. A typical charge curve is shown in Fig. 16A.22.

As with silver-zinc, silver-cadmium batteries are usually charged at constant current at the 10 to 20 h rates. The recommended cutoff voltage during charge is normally 1.6 V per cell.

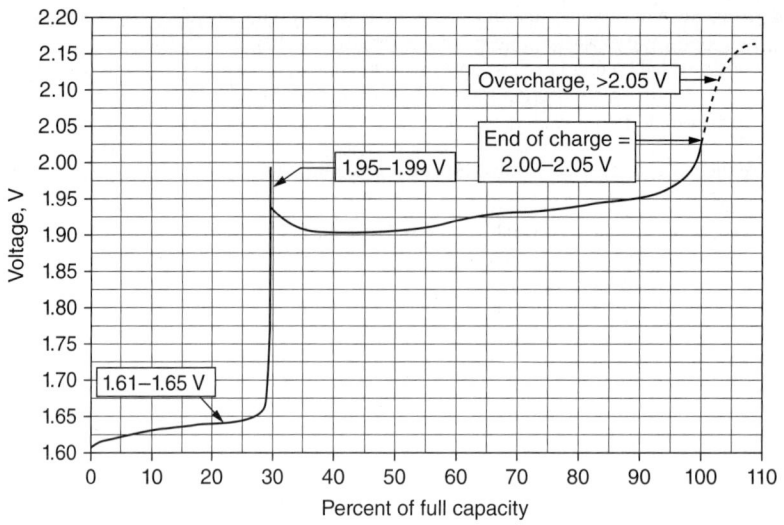

**FIGURE 16A.21**    A typical charge curve of a silver-zinc cell at the 10 to 20 h rate.

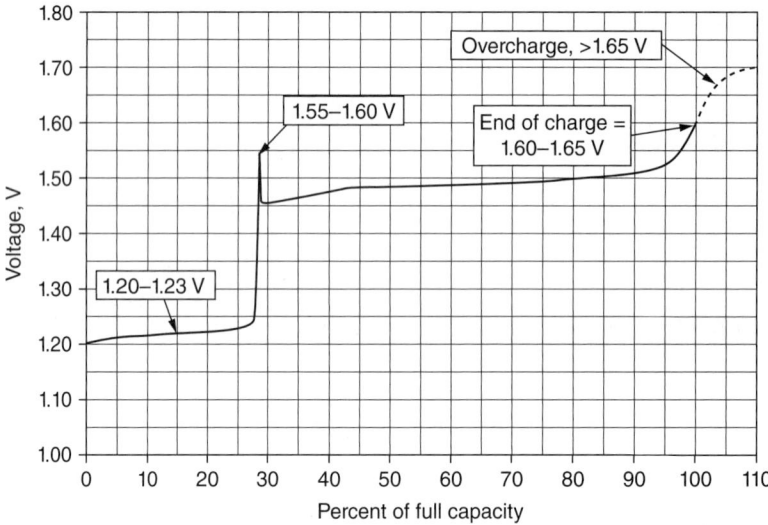

**FIGURE 16A.22**    A typical charge curve of a silver-cadmium cell at the 10 to 20 h rate, at room temperature.

The silver-cadmium battery, however, is less sensitive to overcharge than the silver-zinc battery. Other charge methods can be and have been adapted to specific applications.

### 16A.4.9    Cell Types and Sizes

Tables 16A.6 to 16A.8 present examples of former products from two major silver battery manufacturers—Yardney Technical Products, Inc. and Eagle-Picher Technologies, Inc. They are intended as a guide only, since the design parameters can be varied to meet specific customer requirements. Many applications for the

**TABLE 16A.6**   Nominal Characteristics of Typical Vented Zinc/Silver Oxide Batteries

| Cell type | Capacity, Ah | Cell dimensions, mm (including terminals) | | | Cell weight, g (including electrolyte) | Maximum continuous rate, A |
|---|---|---|---|---|---|---|
| | | Length | Width | Height | | |
| High-rate types | | | | | | |
| HR-02 | 0.2 | 5.6 | 16.0 | 49.3 | 6.5 | 2.0 |
| HR-05 | 1.3 | 13.7 | 27.4 | 39.6 | 21.3 | 4.0 |
| HR-1 | 2.0 | 13.7 | 27.4 | 51.3 | 31.2 | 6.0 |
| HR-2 | 4.5 | 15.0 | 43.7 | 63.5 | 68.0 | 20.0 |
| PMV-2(4.5)* | 5.0 | 15.2 | 43.7 | 64.3 | 72.6 | 100 |
| HR-5 | 8.5 | 20.3 | 52.8 | 73.7 | 127.6 | 60.0 |
| PM-15* | 21.8 | 20.3 | 58.9 | 125.5 | 295 | 200 |
| HR-21 | 35.9 | 20.6 | 58.4 | 191.5 | 439 | 160 |
| PM-30* | 44.0 | 25.4 | 77.7 | 166.4 | 607 | 400 |
| HR-40 | 46.0 | 25.1 | 82.6 | 180.3 | 646 | 200 |
| HR-105 | 121 | 35.2 | 96.9 | 137.4 | 950 | 120 |
| HR-140 | 190 | 72.4 | 82.5 | 183.4 | 1721 | 600 |
| PML-170* | 221 | 35.3 | 97.0 | 184.4 | 1520 | 120 |
| HR-190 | 238 | 39.4 | 152.6 | 165.4 | 2217 | 400 |
| PML-2500* | 2750 | 107.2 | 107.2 | 479.0 | 18,150 | 600 |
| MR-200 | 250 | 53.5 | 101.6 | 206 | 2156 | 200 |
| Low-rate types | | | | | | |
| LR-1 | 2.1 | 13.7 | 27.4 | 51.3 | 30.1 | 4.5 |
| LR-4 | 7.5 | 15.0 | 43.7 | 85.3 | 99.2 | 16.0 |
| LR-8 | 10.0 | 16.3 | 29.9 | 120.1 | 116.3 | 16.0 |
| LR-12 | 16.0 | 19.1 | 47.2 | 100.1 | 163.0 | 20.0 |
| LR-40 | 64.0 | 25.1 | 82.6 | 180.3 | 638 | 64.0 |
| LR-70 | 100 | 36.1 | 92.5 | 155.4 | 1055 | 160 |
| LR-90 | 155 | 54.9 | 82.9 | 179.3 | 1588 | 150 |
| LR-190 | 220 | 39.1 | 151.6 | 162.6 | 2048 | 200 |
| LR-350 | 560 | 107.4 | 107.4 | 222.3 | 5615 | 350 |
| LR-360 | 570 | 69.9 | 147.3 | 162.6 | 4391 | 300 |
| LR-660 | 840 | 79.2 | 161.3 | 177.8 | 6183 | 180 |
| Special deep submersible types | | | | | | |
| LR625 | 692 | 161 | 80 | 187 | 5,470 | 125 |
| LR-700(DS)† | 1060 | 107 | 107 | 486 | 11,200 | 900 |
| LR-750(DS)† | 1075 | 142 | 97 | 513 | 12,500 | 750 |
| LR-850 | 1200 | 119 | 114 | 479 | 13,200 | 800 |
| LR-875 | 1050 | 160 | 79.6 | 183 | 7,000 | 125 |
| LR-1000(DS) | 1072 | 137 | 137 | 513 | 18,500 | 1250 |

*Primary, manually activated.
†Pressure compensated
*Source:* Yardney Technical Products, Inc.

**TABLE 16A.7**    Nominal Characteristics of Typical Vented Zinc/Silver Oxide Cells

| | | High rate | | | | | | Low rate | | | | | | |
|---|---|---|---|---|---|---|---|---|---|---|---|---|---|---|
| | | Nominal capacity Ah rates | | | | | | Nominal capacity Ah rates | | | | Physical dimensions, mm | | |
| Cell type | Rated capacity, Ah | 15 min | 30 min | 60 min | Weight, g | Cell type | Rated capacity, Ah | 4 h | 10 h | 20 h | Weight, g | L | W | H |
| SZHR | | | | | | SZLR | | | | | | | | |
| 0.8 | 0.8 | 0.7 | 0.7 | 0.8 | 22.7 | 0.8 | 0.8 | 0.8 | 0.8 | 0.8 | 22.7 | 10.9 | 26.9 | 51.6 |
| 1.5 | 1.5 | 1.4 | 1.5 | 1.5 | 39.7 | 1.5 | 1.5 | 1.5 | 1.5 | 1.5 | 42.6 | 12.4 | 30.7 | 57.2 |
| 2.8 | 2.8 | 2.6 | 2.7 | 2.8 | 53.9 | 3.0 | 3.0 | 3.0 | 3.0 | 3.0 | 56.7 | 14.2 | 35.1 | 63.2 |
| 5.0 | 5.0 | 4.8 | 5.0 | 5.0 | 76.6 | 5.3 | 5.3 | 5.3 | 5.3 | 5.3 | 85.1 | 16.3 | 40.1 | 70.9 |
| 6.5 | 6.5 | 6.2 | 6.4 | 6.5 | 119.1 | 7.5 | 7.5 | 7.4 | 7.5 | 7.5 | 124.8 | 14.9 | 43.7 | 90.2 |
| 10.5 | 10.5 | 10.0 | 10.3 | 10.5 | 170.2 | 11.5 | 11.5 | 11.5 | 11.4 | 11.5 | 184.4 | 20.1 | 49.5 | 84.8 |
| 15 | 15 | 12 | 14 | 15 | 210.0 | 16.5 | 16.5 | 15.5 | 16.5 | 16.5 | 215.6 | 21.3 | 41.1 | 120.7 |
| 26 | 26 | 20 | 24 | 26 | 312.1 | 30 | 30 | 28.0 | 30.0 | 30.0 | 326.3 | 25.4 | 62.7 | 103.9 |
| 48 | 48 | * | 40 | 48 | 595.9 | 51 | 51 | 48 | 51 | 51 | 624.2 | 18.5 | 89.9 | 167.9 |
| 65 | 65 | * | 50 | 65 | 737.8 | 70 | 70 | 65 | 70 | 70 | 780.3 | 26.9 | 83.1 | 155.4 |
| 100 | 100 | * | 80 | 100 | 1107 | 115 | 115 | 100 | 110 | 115 | 1220 | 37.3 | 92.7 | 150.9 |
| 140 | 140 | * | * | 140 | 1944 | 160 | 160 | * | 150 | 160 | 2049 | 74.17 | 75.7 | 161.8 |

*Not applicable at this rate.
*Source:* Eagle-Picher Technologies.

**TABLE 16A.8**    Nominal Characteristics of Typical Cadmium-Silver Oxide Batteries

| Cell type | Capacity, Ah | Cell dimensions, mm (including terminals) | | | Cell weight, g (including electrolyte) | Maximum continuous rate, A |
|---|---|---|---|---|---|---|
| | | Length | Width | Height | | |
| YS-1 | 1.5 | 13.7 | 27.4 | 51.3 | 31.2 | 5.0 |
| YS-3 | 4.2 | 15.2 | 43.7 | 72.6 | 82.2 | 12.0 |
| YS-5 | 7.8 | 19.1 | 51.1 | 73.9 | 130.5 | 25.0 |
| YS-5 (sealed) | 6.8 | 20.1 | 52.8 | 73.9 | 141.8 | 15.0 |
| YS-10 | 14.5 | 18.8 | 58.9 | 122.2 | 246.7 | 30.0 |
| YS-16 (sealed) | 21.0 | 20.6 | 58.4 | 146.1 | 348.8 | 50.0 |
| YS-20 | 32 | 43.9 | 52.1 | 108.7 | 450.9 | 40.0 |
| YS-40 | 54 | 25.1 | 82.6 | 179.8 | 745.9 | 100 |
| YS-100 | 132 | 70.6 | 87.4 | 122.2 | 1503 | 150 |
| YS-150 | 240 | 45.2 | 106.4 | 272.0 | 2978 | 150 |
| YS-300 | 420 | 45.2 | 106.4 | 444.5 | 5190 | 150 |

*Source:* Yardney Technical Product, Inc.

high-energy silver batteries, in fact, require special designs, which often dictate new cell cases and cover designs, and tooling. These then become the "available" models for future applications.

## 16A.4.10    Special Features and Handling

Silver cells are capable of providing extremely high currents if short-circuited. Accordingly, provisions must be made to insulate all tools used with the batteries and to protect the cells against grounding in their application.

The electrolyte is a caustic solution of potassium hydroxide. Precautions such as the use of gloves and safety goggles are required when handling the electrolyte. In most applications, addition of electrolyte or water is not required. However, the manufacturer's recommendations for periodic maintenance and electrolyte checks should be followed closely.

Proper ventilation of these as well as other vented batteries, although a lesser problem compared to other battery systems, is required to avoid the accumulation of flammable hydrogen, especially during charge. For larger installations, forced air or fans may also be required to prevent undesirable temperature buildups. When close voltage regulation is required at cold temperatures, thermostatically controlled heaters are often used with the batteries.

Because of the sheer size and power of the batteries used, and because of critical personnel safety requirements (e.g., the U.S. Navy's NR-1 submersible had a 240 V, 850 Ah silver-zinc backup battery installed under the deck that vented into the operator's quarters), a whole new body of engineering technology has been developed for the application of these batteries for underwater power. Some of the special features developed include removal of all mercury, provision for fire walls inside cells, and provision for pressure compensation for those batteries that are external to the vessels' pressure hulls.[8] Electronic systems have also been developed to permit continuous scanning of individual cell voltages to maximize battery life. Special applications such as these can be successfully developed only if the battery designers, manufacturers, and users work closely together during the design of the product.

## 16A.4.11  Applications

*Silver-Zinc Batteries.*    Because of the fluctuating cost of silver, the major applications for these batteries have historically been, and continue to be, governmental. However, because of their high power and energy density, these batteries have found many varied uses where space and weight limitations are critical. In addition, the cost of silver for most applications can be offset by reclaiming the metal after the battery completes its useful life.

One of the original applications for the silver-zinc battery was for use in torpedoes.[22] Much of the original development work was sponsored by the U.S. Navy. Later, development expanded to other underwater applications—including mines, buoys, special test vehicles, swimmer aids, and manual submersibles such as the deep submergence rescue vehicles (DSRV) and advanced seal delivery system (ASDS), such as the unmanned underwater vehicles (UUVs) and NR-1 exploratory underwater vehicles—and various antisubmarine warfare (ASW) applications. The MK40 ADMATT target torpedo battery is illustrated in Fig. 16A.23. The ADMATT Propulsion Battery has two configurations, a $60 \times HR300DC/58 \times HR300DC$ medium performance battery and a $120 \times HR215DC/116 \times HR215DC$ high-performance battery. The discharge rate for the nominal 147 V medium-performance battery is 325 A, while the nominal 290 V high-performance battery is 650 A at room

**FIGURE 16A.23**  An ADMATT medium-performance propulsion battery. (*Courtesy of Yardney Technical Products, Inc.*)

temperature conditions; operating time is 35 min for the medium-performance battery and 6 min for the high-performance battery. The DSRV battery illustrated in Fig. 16A.24 is a pressure-compensated design using mineral oil to fill the cell and battery void during ocean descent. It is a 115 V, 700 Ah rated battery and the pressure compensation allows it to be mounted outside of the pressurized vessel.

Silver-zinc batteries have found wide use in numerous space applications, including launch-vehicle guidance and control, telemetry and flight termination power; Apollo lunar spacecraft; lunar and Mars rovers; lunar drill power; space shuttle payload launch; and getaway special batteries, as well as power for the life-support

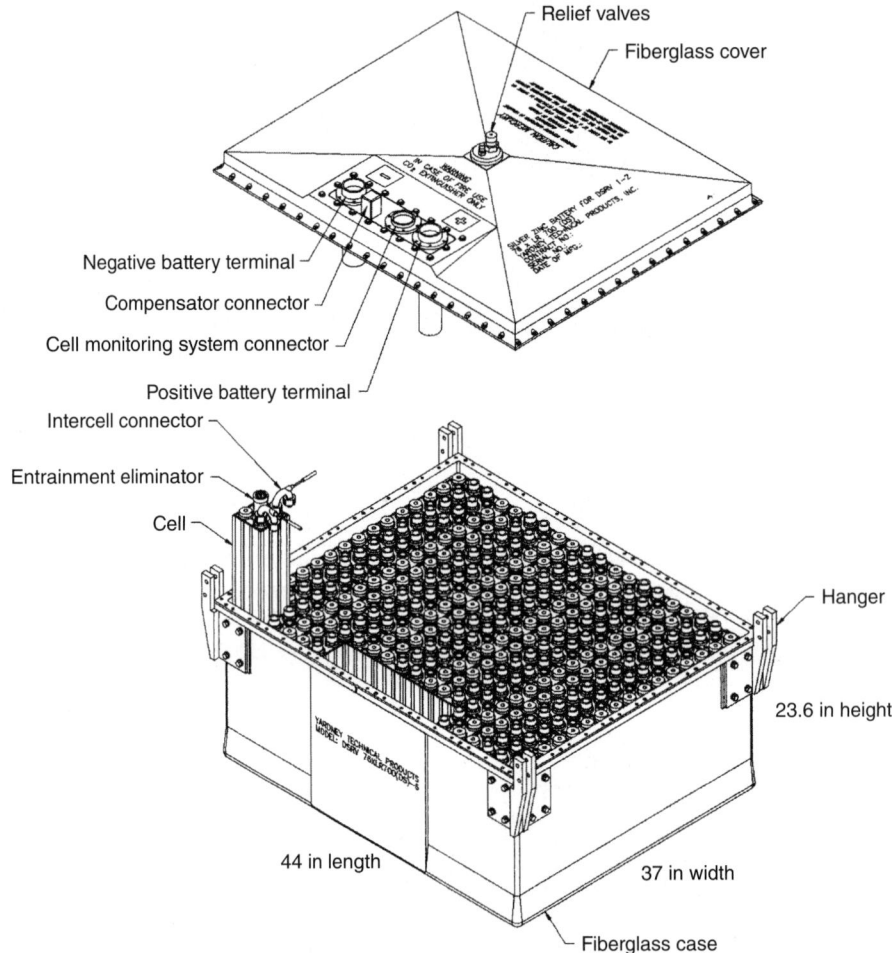

**FIGURE 16A.24** A DSRV pressure-compensated battery. Model 76 × LR700(DS). (*Courtesy of Yardney Technical Products, Inc.*)

equipment used by the U.S. astronauts during extra-vehicular activities (EVA). Figure 16A.25 shows a typical aerospace battery (Model 20 × HR2DC) consisting of twenty 2 Ah silver-zinc cells housed in a cast aluminum case, equipped with a pressure-relief valve, a pressurizing valve, and a battery connector.

***Silver-Cadmium Batteries.*** Silver-cadmium batteries have been used in a number of space applications requiring nonmagnetic properties. One such battery provided the main power for the *Giotto Halley Comet* intercept spacecraft. Another group of batteries provides backup power for the solar cells (for the periods when the earth

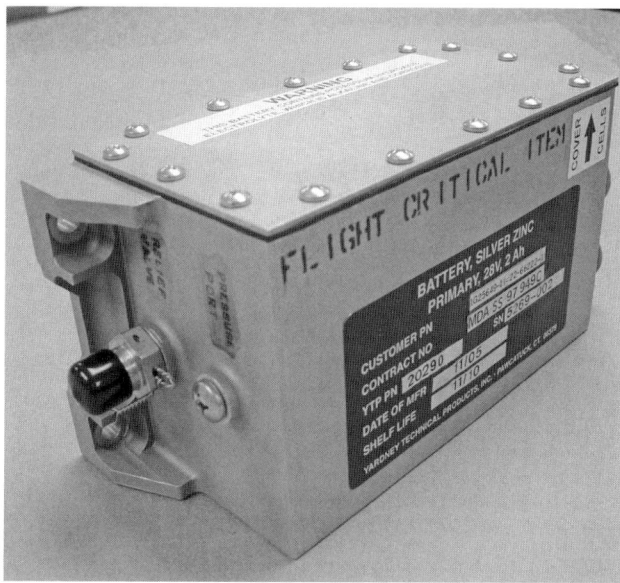

**FIGURE 16A.25**    A silver-zinc aerospace battery, 28 V, 2 Ah. (*Courtesy of Yardney Technical Products, Inc.*)

eclipses the sun) on board the Cluster II scientific spacecraft launched by the European Space Agency in the year 2000. Although only required to operate for 2 years, those batteries were still operating as of May 2009.

Ground applications include communications equipment, portable television cameras, portable lights, camera drives, medical equipment, vehicle motive power, and similar uses requiring high-energy-density rechargeable batteries.

***Silver-Iron Batteries.***    The silver-iron battery has been limited in use because of its high cost. Its theoretical energy density is essentially equal to that of the more popular silver-zinc system. The silver-iron battery has good cycle life compared with the silver-zinc and provides a battery of high reliability, long life, and better durability where high specific energy content is essential.[27-32] Figure 16A.26 shows a 3.5 kWh battery designed for telecommunication use, whereas Fig. 16A.27 shows a 9.5 kWh battery designed for use in a submersible.

*Temperature Effects.*    As with other alkaline battery systems, silver-iron performance can depend on the operating temperature. Cells designed for long-life and LR applications normally have a higher internal resistance than those designed for high rate and shorter life. The two designs therefore behave differently when the discharge temperature is decreased, as shown in Figs. 16A.28 and 16A.29.

*Battery Design Variations.*    Tests on experimental designs other than the monopolar,

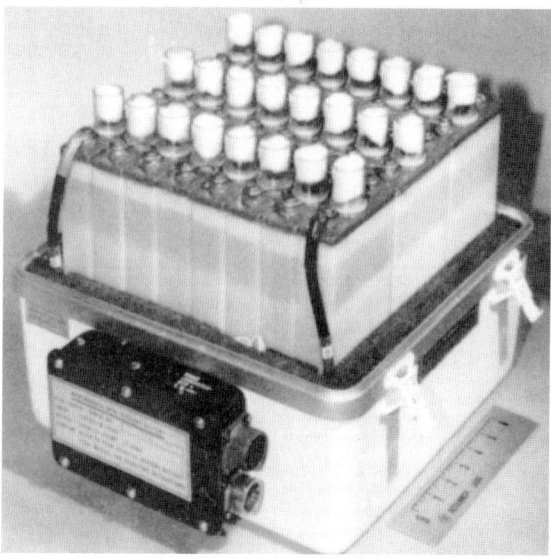

**FIGURE 16A.26**    A 3.5 kWh telecommunications iron/silver oxide battery. (*Courtesy of Westinghouse Electric Corp.*)

**FIGURE 16A.27**    A 9.5 kWh iron/silver oxide battery for a submersible vehicle. (*Courtesy of Westinghouse Electric Corp.*)

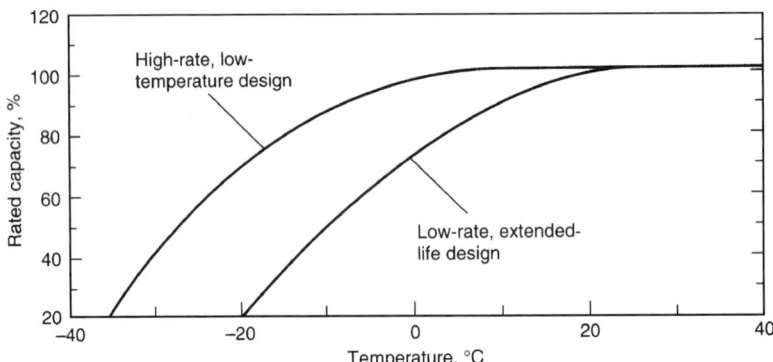

**FIGURE 16A.28**    Effect of temperature on discharge capacity for different battery designs, $C/10$ discharge rate. (*Courtesy of Westinghouse Electric Corp.*)

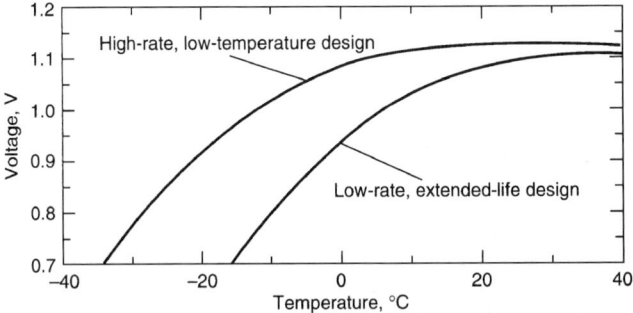

**FIGURE 16A.29**    Effect of temperature on discharge voltage for different battery designs; $C/10$ discharge rate. (*Courtesy of Westinghouse Electric Corp.*)

prismatic cells discussed above have included bipolar and cylindrically wound cells. Results of voltage polarization tests on those designs are compared to the prismatic design in Fig. 16A.30. The voltage characteristics of the jelly-roll design would be expected to improve considerably as assembly is refined and automated. Designs such as these would be suited for use in smaller portable systems such as communication devices that may require high power and energy density.

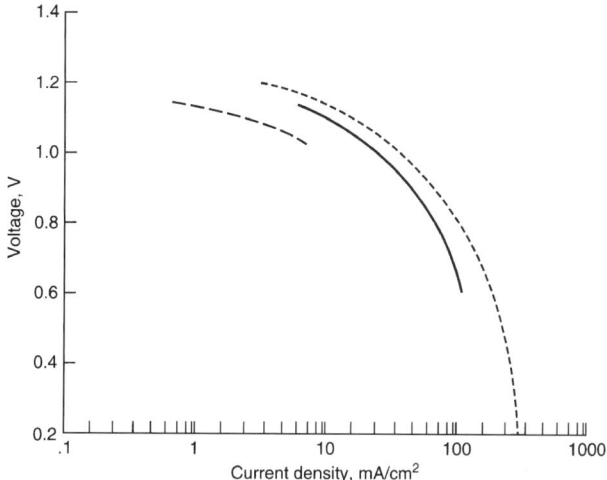

**FIGURE 16A.30**   Voltage polarization characteristics for a silver-iron system in experimental tests on three conventional types of cell designs. - - -, bipolar; ____, prismatic; __ __ __ , jelly roll. (*Courtesy of Westinghouse Electric Corp.*)

### 16A.4.12   Summary

Silver-zinc batteries offer a safer alternative to lithium-ion batteries used in notebook computers, cell phones, and consumer electronics. The user should keep in mind that no one type of battery is suitable for all applications. Optimum performance of a battery in an application can usually be achieved only by meeting the critical needs of the application and subordinating others. The best approach for battery selection is to work with the battery manufacturers during the early stages of equipment design, rather than attempting to retrofit the battery to the remaining space in the electronics compartment.

## *REFERENCES*

1. A. Volta, "On the Electricity Excited by the Mere Contact of Conducting Substances of Different Kinds," *Phil. Trans. R. Soc. London* **90**:403–431 (1800).

2. S.U. Falk and A.J. Salkind, *Alkaline Storage Batteries*, Wiley, New York, 1967.

3. H. André, *Bull. Soc. Fr. Electrochem.* (6th ser.) **1**:132 (1941).

4. H. André, U.S. Patent 2,317,711 (1943).

5. J.T. Brown, "Iron-Silver Battery—A New High Energy Density Power Source," Westinghouse Corp., Rep. 77-5E6-SILEL-RI, 1977.

6. "Design & Cost Study, Zinc/Nickel Oxide Battery for Electric Vehicle Propulsion," Yardney Electric Corp., Final Rep., Contract 31-109-38-3543, Oct. 1976.

7. R. Serenyi, "Recent Developments in Silver-Zinc Batteries," Yardney Electric Corp., Internal Rep. 2449–79, Oct. 1979.

8. G.W. Work and P.A. Karpinski, "Energy Systems for Underwater Use," *Marine Tech. Expo. Int. Conf*, New Orleans, LA, Oct. 1979.

9. A. Himy, "Substitutes for Mercury in Alkaline Zinc Batteries," *Proc. 28th Annual Power Sources Symp.*, 1978, pp. 167–169.

10. R. Serenyi and P. Karpinski, "Final Report on Silver-Zinc Battery Development," Yardney Electric Corp., Contract N00140-76-C-6726, Nov. 1978.

11. "Medium Rate Rechargeable Silver-Zinc 850 Ah Cell," Eagle-Picher Industries, Final Rep., USN Conract N00140-76-C-6729, Mar. 1978.

12. R. Serenyi, U.S. Patent 5,773,176.

13. *Proceedings of the 5th Workshop for Battery Exploratory Development*, Burlington, VT, July 1997, pp. 153–157.

14. R. Einerhand, W. Visscher, J. de Goeij, and E. Barendrecht, "Zinc Electrode Shape Change," *J. Electrochem. Soc.* **138**:7–17 (1991).

15. K. Choi, D. Bennion, and J. Newman, "Engineering Analysis of Shape Change in Zinc Secondary Electrodes," *J. Electrochem. Soc.* **123**:1616–1627 (1976).

16. K. Bass, P.J. Mitchell, G.D. Wilcox, and J. Smith, "Methods for the Reduction of Shape Change and Dendritic Growth in Zinc-Based Secondary Cells," *J. Power Sources* **35**:333–351 (1991).

17. A. Charkey, "Long Life Zinc-Silver Oxide Cells," *Proc. 26th Ann. Power Sources Symp.*, 1976, pp. 87–89.

18. V. D'Agostino, J. Lee, and G. Orban, "Grafted Membranes," in A. Fleischer and J. Lander (eds.), *Zinc-Silver Oxide Batteries*, Wiley, New York, 1971, Chap. 19, pp. 271–281.

19. C.P. Donnel, "Evaluation of Inorganic/Organic Separators," Yardney Electric Corp., Contract NAS3-18530, Oct. 1976.

20. A.P. Karpinski, B. Makovetski, S.J. Russell, J.R. Serenyi, and D.C. Williams, "Silver-Zinc: Status of Technology and Applications," *J. Power Sources* **80**:53–60 (1999).

21. *Proceedings of the 38th Power Sources Conference*, Cherry Hill, NJ, June 1998, pp. 175–178.

22. H.A. Frank, W.L. Long, and A.A. Uchiyama, "Impedance of Silver Oxide-Zinc Cells," *J. Electrochem. Soc.* **123**(1):1–9 (1976).

23. J.C. Brewer, R. Doreswamy, and L.G. Jackson, "Life Testing of Secondary Silver-Zinc Cells for the Orbital Maneuvering Vehicle," *Proc. 25th IECEC*, Reno, NV, Aug. 1990.

24. "Evaluation of Silver-Cadmium Batteries for Satellite Applications," Boeing Co., Test D2-90023, Feb. 1962.

25. A.P. Karpinski and J.A. Patten, "Performance Characteristics of Silver-Zinc Cells for Orbiting Spacecraft," *Proc. 25th IECEC*, Reno, NV, Aug. 1990.

26. A. Fleischer and J. Lander (eds.), *Zinc-Silver Oxide Batteries*, Wiley, New York, 1971.

27. O. Lindstrom, in D.H. Collins (ed.), *Power Sources*, vol. 5, Academic, London, 1975, p. 283.

28. *The Silver Institute Letter*, vol. 7, no. 3 (1977).

29. J.T. Brown, Extended Abstract No. 28, Battery Div., the Electrochemical Society, Las Vegas, NV, pp. 76–77 (1977).

30. E. Buzzelli, "Silver-Iron Battery Performance Characteristics," *Proc. 28th Power Sources Symp.*, Electrochemical Society, Pennington, NJ, 1978, p. 160.

31. G.A. Bayles, E.S. Buzzelli, and J.S. Lauer, "Progress in the Development of a Silver-Iron Communications Battery," *Proc. 34th Int. Power Sources Symp.*, Cherry Hill, NJ, June 1990.

32. G.A. Bayles, J.S. Lauer, E.S. Buzzelli, and J.F. Jackovitz, "Silver-Iron Batteries for Submersible Applications," *Proc. 3rd Annual Underwater Vehicle Conf.*, Baltimore, June 1989.

# SECTION B

# MANGANESE OXIDE CATHODES

**Josef Daniel-Ivad**
**(Emeritus Contributor: Karl Kordesch)**

## 16B.1   *GENERAL OVERVIEW*

The rechargeable alkaline manganese dioxide-zinc battery is an outgrowth of the primary battery. Zinc is used as the negative active material (the anode during discharge), manganese dioxide for the positive active material (the cathode during discharge), and a potassium hydroxide solution for the electrolyte.

The major advantages and disadvantages of the rechargeable alkaline manganese dioxide battery are listed in Table 16B.1.

**TABLE 16B.1**   Major Advantages and Disadvantages of Rechargeable Alkaline Manganese Dioxide Zinc Batteries

| Advantages | Disadvantage |
|---|---|
| Low initial cost (and possible lower operating cost than other rechargeable batteries) | Useful capacity about two-thirds of primary battery, but higher than most rechargeable batteries |
| Manufactured in a fully charged state | Limited cycle life |
| Good retention of capacity (compared to other rechargeable batteries) | Available energy decreases rapidly with cycling and DOD Higher internal resistance than NiCd and NiMH |
| Completely sealed and maintenance-free | |
| No "memory effect" problem | |
| No toxic materials, green certified | |

## 16B.2   *CELL ELECTROCHEMISTRY*

The discharge mechanism of electrolytic manganese dioxide (EMD), which is essentially $\gamma$-$MnO_2$, has been studied extensively. It is generally assumed that the first electron discharge step proceeds in a homogeneous reaction by the movement of protons and electrons into the lattice, resulting in a gradually decreasing value of $x$ in $MnO_x$ from $x = 2$ to $x = 1.5$. The reaction is a conversion of one solid structure ($MnO_2$) into another ($MnOOH$), with manganese (formally) in the trivalent state[1]

$$MnO_2 + H_2O + e \rightarrow MnOOH + OH$$

Soluble manganese species begin to appear if the discharge is continued, especially when the lower voltage second-electron range is approached.[2]

When EMD is recharged, the process is reversed. The number of discharge and charge cycles obtainable depends on the DOD, indicating that irreversible electrode processes are occurring. The relationship between cycle number and DOD was found to be essentially logarithmic. Coulometric studies indicate that the recharge efficiency reaches nearly 100% after a few cycles. The reason for the initial losses have been researched by various groups and new approaches to solving the issue have been proposed, but no commercially viable solution has been introduced to date.[3–8]

The cathode-dominated cycle characteristic of the rechargeable batteries and the logarithmic relationship of loss of capacity with cycling are shown in Fig. 16B.1. A zinc capacity of about 2 Ah is usually provided in an

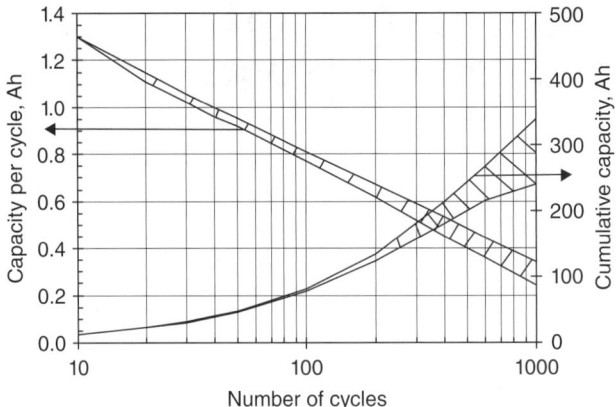

**FIGURE 16B.1**   Performance of rechargeable alkaline manganese dioxide zinc battery on cycling at 20°C, recharging after each discharge. (*Courtesy of Blizzard Technologies Inc.*)

AA-size battery to maintain a high first-cycle discharge capacity. This still prevents the $MnO_2$ from discharging beyond the one-electron capacity. When the discharge is limited to predetermined capacity at each cycle until the cutoff voltage of 0.9 V is reached, the estimated cycle life number and cumulative capacity can be determined from Fig. 16B.1.

Charging of the manganese dioxide-zinc battery must be controlled to limit the charge voltage to 1.75 V for standard charging of 16 h, to 1.65 V for prolonged charging of up to 1 week, and to 1.60 V for float charging. Charging to higher voltages produces a hexavalent manganate and oxygen gas. The soluble manganate disproportionates into tetravalent $MnO_2$ and a nonrechargeable divalent manganese species, which results in a loss in cycle capacity. The oxygen gas reacts with zinc to produce ZnO.

Attempts have been made to replace the EMD with a specially prepared bismuth-doped form of manganese dioxide (BMD) to harness a deep two-electron $MnO_2 \rightarrow Mn(OH)_2$ reversible discharge. However, no successful commercial implementation of such an approach is available.[9]

## 16B.3   CELL DESIGN AND COMPONENTS

The construction of the cylindrical rechargeable alkaline cell is shown in Fig. 16B.2. The construction is similar to the primary cell, using an inside-out design. The positive electrode consists of three or four cathodic rings, which are formed under high pressure to a slightly oversized diameter and then inserted into the steel can. The cathodic mix formulation uses electrolytic $MnO_2$, up to 10% graphite, and additives for better rechargeability. The negative electrode, consisting of a powder zinc mass containing gelled KOH, is in the center. A nail, located in the center of the gel, serves as the negative current collector. The cell is crimped-sealed and contains a vent mechanism.

The following features distinguish the rechargeable cell. Cathode additives such as barium, strontium, and calcium compounds are utilized to increase the cathode capacity and improve cycling.[10] The cathode also contains a catalyst, such as silver on acetylene black or carbon, to recombine any hydrogen that may form in the cell. The limiting zinc powder anode contains KOH and a gelling agent. The amount of zinc determines the DOD and thereby the capacity of the cell. Mercury is not added to the anode, and special zinc alloys and/or organic inhibitors in combination with a special anode preparation process are used to reduce the zinc corrosion and control dendritic zinc growth on recharge.[11] Zinc oxide is dissolved into the KOH to ensure that on charge (or overcharge) only oxygen, not hydrogen, can be formed by electrolysis. The separator, which is multilayered, contains regenerated cellulose with a high oxidation resistance to caustics. It also prevents internal short-circuiting due to zinc dendrite formation.[12,13]

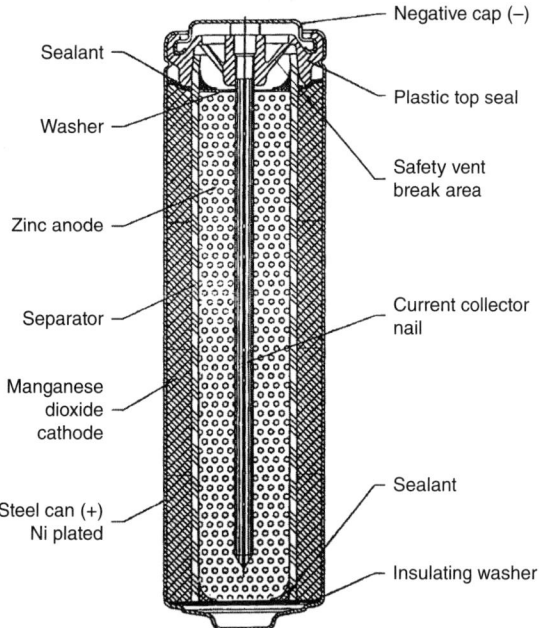

**FIGURE 16B.2**  An AA-size rechargeable alkaline manganese dioxide zinc battery. (*Courtesy of Blizzard Technologies Inc.*)

## 16B.4    BATTERY PERFORMANCE AND APPLICATIONS

### 16B.4.1    First Cycle Discharge

The rechargeable batteries are manufactured and shipped in a charged state, as are the primary cells. Because of good shelf life, these cells can retain most of this capacity (depending on storage prior to use) and do not need to be recharged before first use. The discharge characteristics of the rechargeable manganese dioxide-zinc batteries are similar to those of the primary batteries. However, due to the zinc-limited design of the rechargeable cell, its terminal voltage on a medium- or high-load discharge drops rapidly once 0.8 V is reached. Figure 16B.3 shows the discharge curves of fresh AA-size rechargeable batteries on the first cycle of discharge at several constant-current discharge loads.

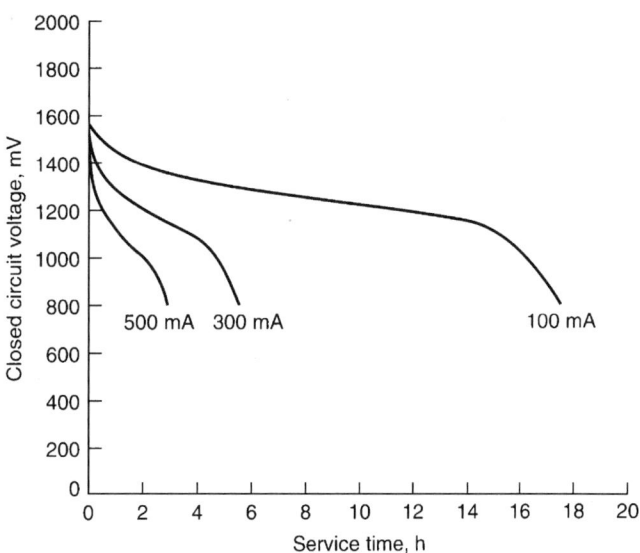

**FIGURE 16B.3**  First-cycle discharge characteristics of AA-size rechargeable alkaline manganese dioxide zinc batteries discharged continuously at different constant-current loads at 22°C. (*Courtesy of Blizzard Technologies Inc.*)

### 16B.4.2    Cycling

The rechargeable battery has its highest capacity on the first cycle, and that value at 20°C is about 70% to 80% of the capacity of the primary cell. On subsequent charge-discharge cycles, if the cells are completely discharged before being recharged, the capacity reduces gradually after each cycle. The shape of the discharge curve changes slightly during cycling, but the voltage level drops with cycling, as shown in Fig. 16B.4.[14]

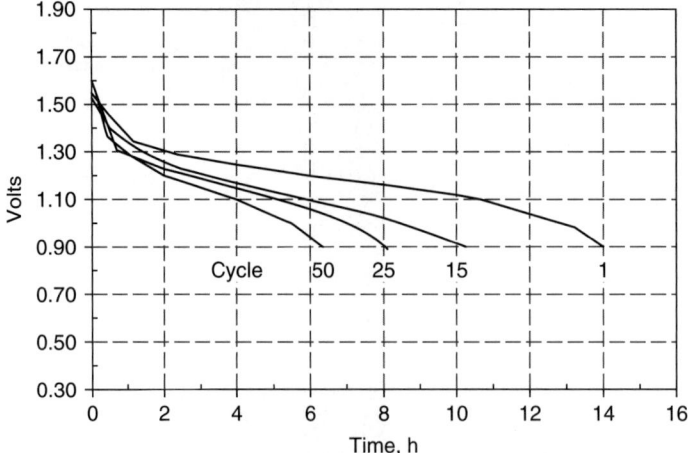

**FIGURE 16B.4**    Continuous-discharge characteristic of rechargeable alkaline manganese dioxide zinc AA-size batteries after cycling through a 10 Ω load at 20°C.

The number of useful cycles and the cycle capacity increase when the batteries are only partially discharged and recharged after use. Figure 16B.5 shows the increased cycle life obtained on a discharge of an AA-size battery on a 10 Ω load, applied daily for 4 h (about a 25% DOD), then recharged at a constant voltage of 1.65 V. More than 300 cycles can be obtained with the terminal voltage above 0.9 V.[14]

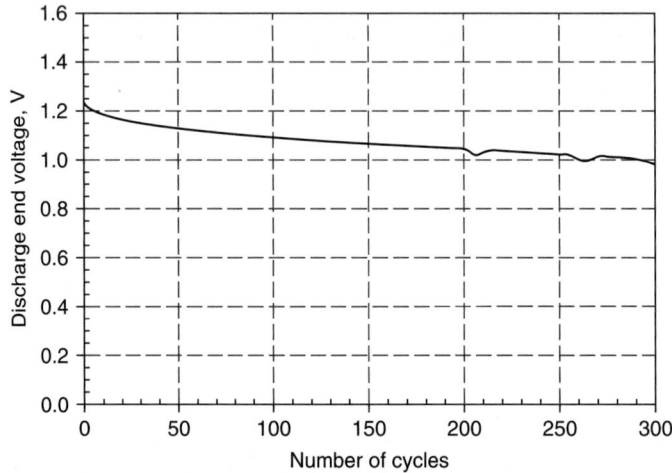

**FIGURE 16B.5**    Discharge characteristics of rechargeable alkaline manganese dioxide zinc AA-size batteries after cycling at 20°C; cells are discharged 4 h/day at 10 Ω, then recharged.

The cycle life, when discharged to lower DOD, is shown in Fig. 16B.6. This figure shows the results of repeated discharge of the rechargeable AA-size battery to approximately 15%, 20%, and 25% DOD, followed by recharge. The cycle life increases with reduced DOD and the voltage drop decreases upon lowering the DOD. When the DOD is very shallow, several thousand cycles have been demonstrated in lab testing, which is illustrated in Fig. 16B.7, showing the cycling performance for the shallow discharge–frequent recharge mode, achieving up to 5000 cycles without reaching the cutoff criteria.[14]

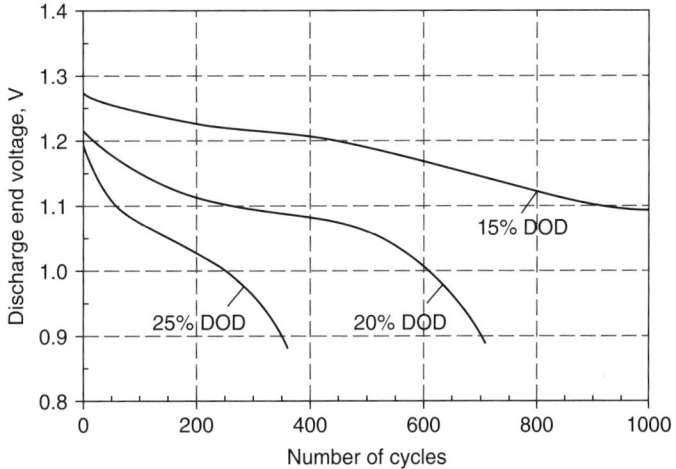

**FIGURE 16B.6**    The cycle life on a 10 Ω load at three levels of depth of discharge, full recharge after each shallow discharge.

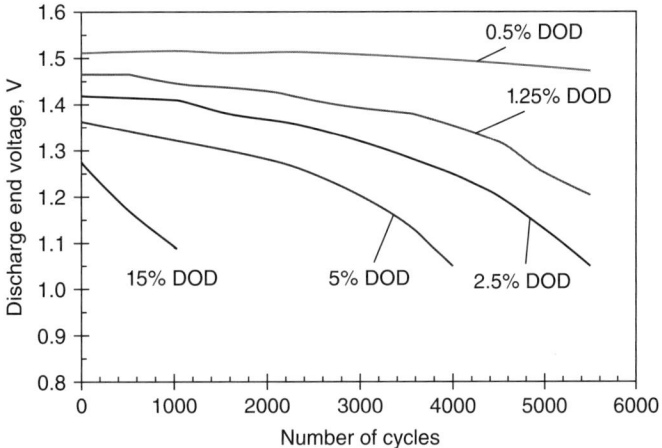

**FIGURE 16B.7**    Performance of a rechargeable alkaline manganese dioxide zinc battery on very shallow discharge cycling at 20°C.

### 16B.4.3  Effect of Temperature

The performance of the rechargeable alkaline manganese dioxide-zinc batteries at various temperatures is shown in Fig. 16B.8. At low temperatures down to −30°C, the batteries function, but performance is decreased. The decrease is more severe for moderate and higher rates. At higher temperatures up to 50°C, LR performance is unchanged, but performance at moderate and higher rates is improved. The performance of the AA-size cell at 45 and 65°C is shown in Fig. 16B.9. It should also be noted that the capacity and high current drain capability are higher at the higher temperatures due to better diffusion and higher $MnO_2$ utilization.[15,16]

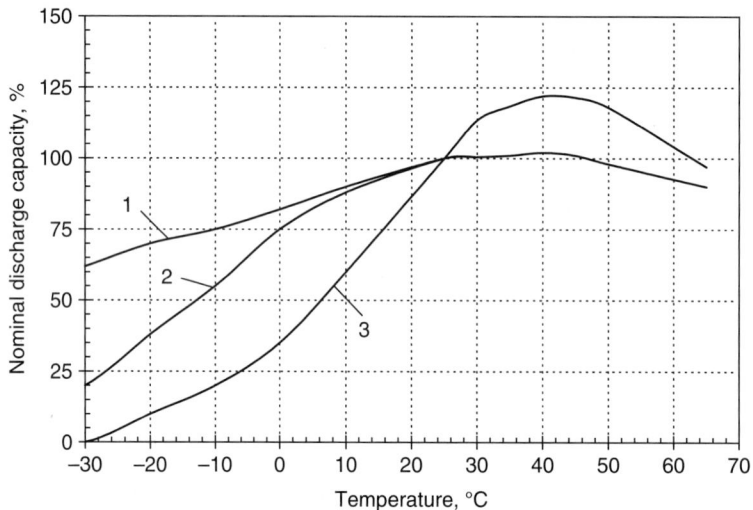

**FIGURE 16B.8**    Effect of temperature on capacity of rechargeable alkaline manganese dioxide zinc batteries for different rates of discharge, 1 = very low rate: 1–5 mA; 2 = low rate: 50–100 mA; 3 = moderate rate: 250–300 mA. (*Courtesy of Blizzard Technologies Inc.*)

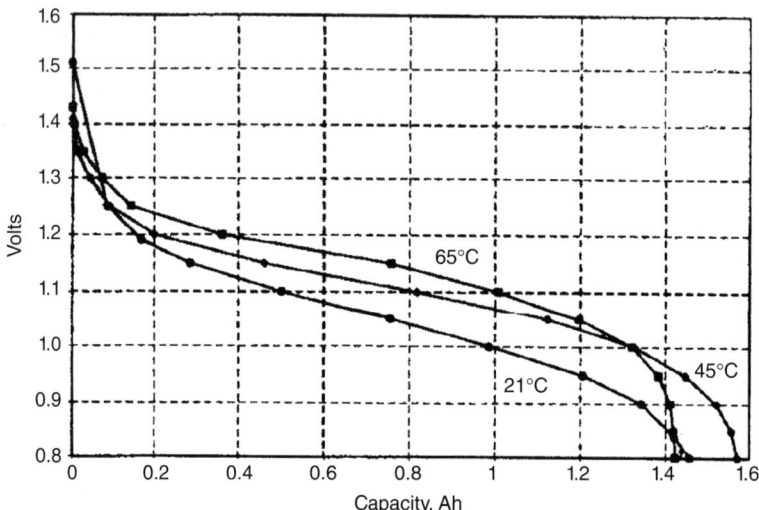

**FIGURE 16B.9**    Discharge of rechargeable AA-size manganese dioxide zinc batteries at different temperatures at a 3.9 Ω load. (*From Ref. 15.*)

## 16B.4.4  Shelf Life

The shelf life of fresh, unused (charged) rechargeable alkaline manganese dioxide-zinc batteries is about the same as that of the primary batteries (20% to 25% loss after 10 years when stored at room temperature). The data for open-circuit voltage during high-temperature storage are shown in Fig. 16B.10. Over a storage period of 12 weeks at 71°C, which represents more than 7 years at 21°C, the open-circuit voltage drops only 6%, demonstrating that no self-discharge reaction of significance takes place.

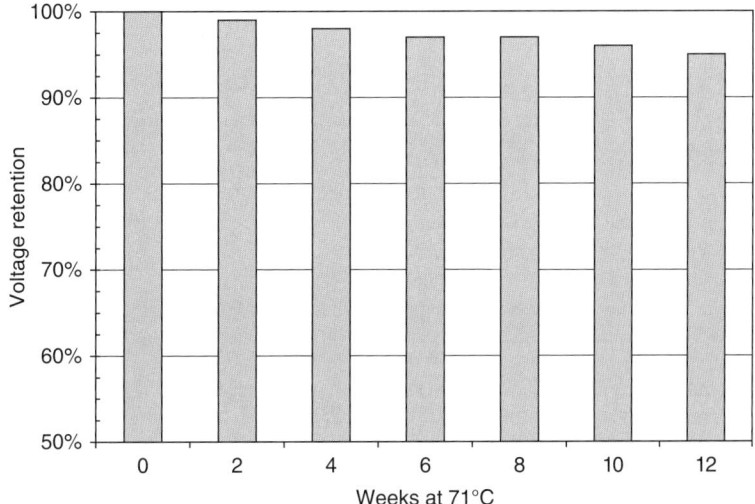

**FIGURE 16B.10**    Open-circuit voltage stability of rechargeable alkaline manganese dioxide zinc AA-size batteries at 71°C. (*Courtesy of Blizzard Technologies Inc.*)

The shelf life of cycled cells depends on whether they are stored in a charged or a discharged condition. Batteries stored in a charged state after cycling show about the same losses as a fresh, uncycled battery. Storage of batteries in a discharged condition, particularly at elevated temperatures (65°C), may be detrimental to the anode performance on subsequent cycles. However, under normal usage, batteries can be recharged close to the capacity level of the previous cycle.

## 16B.4.5  Charge Methods

In the charging process for the rechargeable alkaline manganese dioxide-zinc cell, the discharged positive active material, manganese oxyhydroxide (MnOOH), is oxidized to manganese dioxide ($MnO_2$) and zinc oxide (ZnO) in the negative is reduced to metallic zinc. Manganese dioxide can be further oxidized to higher oxides ($Mn^{+6}$ compounds), which are soluble, resulting in loss of rechargeability. Therefore, proper recharging is important to obtain optimum life. Charging over 1.75 V per cell for over a few days or over 1.70 V per cell for many weeks can damage the battery. Batteries should not be charged after 105% of the ampere-hours removed have been replaced. Batteries can be float-charged for extended periods at 1.60 V per cell.[16]

***Constant-Potential Charging.***    Constant-potential charging is the preferred method. This is equivalent to a taper current charge method. The voltage on charge should not exceed 1.65 to 1.68 V. Figure 16B.11 shows the voltage and current profiles during the charge.[14]

***Constant-Current Charging.***    Uncontrolled constant-current charging over an extended period of time leads to electrolysis of the electrolyte, which causes a buildup of internal gas pressure and results in the rupturing of

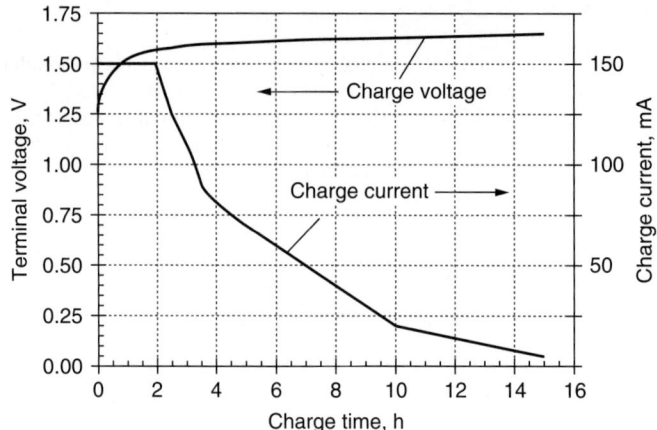

**FIGURE 16B.11**    Constant-potential charging of rechargeable alkaline manganese dioxide zinc AA-size batteries at 20°C.

the safety vent, allowing the release of the gases. Constant-current charging is feasible if the charge voltage is limited to 1.65 V per cell (resistance-free) and a shutoff control is incorporated in the charge circuit.

***Pulse Charging.***    Pulse charging can be used to permit quick charging of the rechargeable zinc-alkaline batteries. During a pulse pause, the circuit measures the cell voltage. Because no charge current flows through the battery during the pause, the true electrochemical voltage, without any ohmic resistance, is measured. This voltage is often called the "resistance-free" voltage (RFV). The pulse-charger circuit regulates the time period when the charge is on by comparing the actual resistance-free cell voltage to the preset cutoff voltage. As long as the RFV is lower than the cutoff voltage, charge current passes into the battery. If the RFV is equal to or higher than the charge cutoff voltage, the charge current is cut off. The charge voltage can be much higher than the specified charge cutoff voltage as long as the RFV does not exceed the charge cutoff voltage. This causes the initial charge current to be much higher, making fast charging possible. Pulse charging has also been shown to increase the cycle life of the battery due to improved replating of zinc.[17] Figure 16B.12 shows the current profile for pulse charging of one to four AA-size cells in parallel.[14]

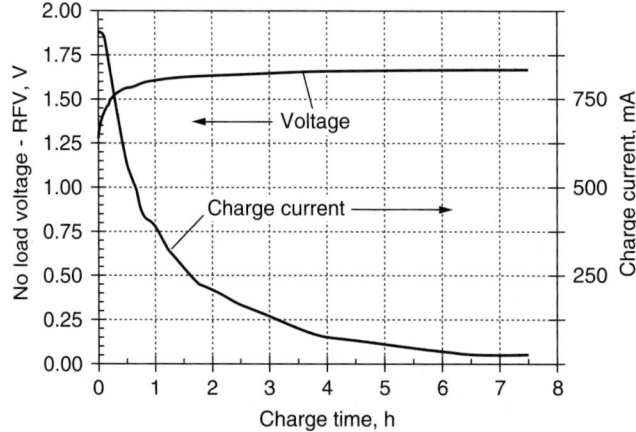

**FIGURE 16B.12**    Pulse charging of rechargeable alkaline manganese dioxide zinc AA-size batteries. Voltage limit set at a resistance-free voltage (RFV) of 1.65 V, initial charge current limit at 1 A.

## 16B.4.6 Types of Cells and Batteries

The characteristics of commercially available rechargeable alkaline manganese dioxide-zinc batteries are listed in Table 16B.2. Table 16B.3 shows the electrical performance of rechargeable alkaline manganese dioxide-zinc batteries on international standard tests for primary alkaline batteries, according to International Electrotechnical Commission (IEC) 60086-2 for AA size. Table 16B.4 shows these results for AAA size.

**TABLE 16B.2** Typical Rechargeable Alkaline Manganese Dioxide Zinc Batteries

| Cell type | Dimensions, mm | | Weight, G | Rated capacity, Ah* (initial discharge) |
| | Height | Diameter | | |
| --- | --- | --- | --- | --- |
| AAA | 44 | 10 | 11 | 0.90 at 75 Ω |
| AA | 50 | 14 | 22 | 2.00 at 43 Ω |
| C (built from 3 AAA) | 50 | 26 | 58 | 2.40 at 10 Ω |
| D (built from 3 AA) | 60 | 34 | 104 | 6.00 at 10 Ω |

*Based on discharge, through specified resistance, to 0.9 V per cell.
***Source:*** Blizzard Technologies Inc.

**TABLE 16B.3** Performance of Rechargeable Alkaline Manganese Dioxide Zinc AA-Size Batteries on International Standard Tests for Primary Alkaline Batteries According to Standard IEC 60086-2

| Application test | Load | Duty cycle | End voltage | IEC minimum average duration | Typical service duration |
| --- | --- | --- | --- | --- | --- |
| Radio | 43 Ω | 4 h/d | 0.9 V | 60 h | 75 h |
| Toy | 3.9 Ω | 1 h/d | 0.8 V | 4.0 h | 6.0 h |
| Cassette | 10 Ω | 1 h/d | 0.9 V | 11.5 h | 16 h |
| CD/MD/Games | 250 mA | 1 h/d | 0.9 V | 4.5 h | 6.4 h |
| Photoflash | 1000 mA | 10 s/m, 1 h/d | 0.9 V | 200 pulses | 315 pulses |
| Remote | 24 Ω | 15 s/min, 8 h/d | 1.0 V | 31 h | 40 h |

**TABLE 16B.4** Performance of Rechargeable Alkaline Manganese Dioxide Zinc AAA-Size Batteries on International Standard Tests for Primary Alkaline Batteries According to Standard IEC 60086-2

| Application test | Load | Duty cycle | End voltage | IEC minimum average duration | Typical service duration |
| --- | --- | --- | --- | --- | --- |
| Radio | 75 Ω | 4 h/d | 0.9 V | 44 h | 65 h |
| Cassette | 10 Ω | 1 h/d | 0.9 V | 5 h | 6.8 h |
| Lighting | 5.1 Ω | 4 min/h, 8 h/d | 0.9 V | 130 min | 190 min |
| Photoflash | 600 mA | 10 s/m, 1 h/d | 0.9 V | 140 pulses | 250 pulses |
| Remote | 24 Ω | 15 s/min, 8 h/d | 1.0 V | 14.5 h | 17 h |

## 16B.4.7 Applications

Practically all applications, where single-use alkaline cells are presently being used, can be operated with rechargeable alkaline manganese dioxide-zinc batteries. High usage of primary cells occurs in applications such as some toys and games, personal audio players, flashlights, remote controls, wireless mice/keyboards, and others.

Further, more and more applications have become wireless or require backup power, and hence, the need for batteries. In particular applications, where the DOD before recharge is low and when only backup or emergency power is required, rechargeable alkaline manganese dioxide-zinc batteries are the best economic choice. These types of applications usually have built-in batteries and the selection of the most suitable battery system is made by the original equipment manufacturer; examples include solar decorative lighting, solar-charged clocks and smoke detectors, sensors, thermostats, data monitor/transfer devices, wireless data communication, wireless mice and keyboards, and global positioning satellite tracking devices. Integrated solar charge capability allows for a true wireless and self-sufficient solution.

## ACKNOWLEDGMENTS

Major content for this chapter was provided in whole or in part by Josef Daniel-Ivad and Karl Kordesch, Chap. 28, Rechargeable Zinc/Alkaline/Manganese Dioxide Batteries, *Linden's Handbook of Batteries*, 4th ed., T. B. Reddy and D. Linden, eds., McGraw-Hill, 2011.

## REFERENCES

1. A. Kozawa, "Electrochemistry of Manganese Oxide," in K. Kordesch (ed.), *Batteries*, vol. 1, Dekker, New York, 1974, Chap. 3.

2. S.W. Donne, G.A. Lawrance, and D.A.J. Swinkels, "Redox Process at the Manganese Dioxide Electrode", *J. Electrochem. Soc.* **144**:2949–2967 (1997).

3. L. Bai, D.Y. Qu, B.E. Conway, Y.H. Zhou, G. Chowdhury, and W.A. Adams, "Role of Dissolution of Mn(III) Species in Discharge and Recharge of Chemically Modified $MnO_2$ Battery Cathode Materials," *J. Electrochem. Soc.* **140**:884–889 (1993).

4. D. Im and A. Manthiram, "Role of Bismuth and Factors Influencing the Formation of $Mn_3O_4$ in Rechargeable Alkaline Batteries Based on Bismuth-Containing Manganese Oxides," *J. Electrochem. Soc.* **150**(1):A68–A73 (2003).

5. M.R. Bailey and S.W. Donne, "The Effect of Barium Hydroxide on the Rechargeable Performance of Alkaline $\gamma$-$MnO_2$," *J. Electrochem. Soc.* **159**(7):A999–A1004 (2012).

6. M.R. Bailey and S.W. Donne, "Electrode Additives and the Rechargeability of the Alkaline Manganese Dioxide Cathode," *J. Electrochem. Soc.* **161**(3):A403–A409 (2014).

7. B. Hertzberg, L. Sviridov, E.A. Stach, T. Gupta, and D. Steingart, "A Manganese-Doped Barium Carbonate Cathode for Alkaline Batteries Batteries and Energy Storage," *J. Electrochem. Soc.* **161**(6):A835–A840 (2014).

8. G.G. Yadav, J.W. Gallaway, D.E. Turney, M. Nyce, J. Huang, X. Wei, and S. Banerjee, "Regenerable Cu-intercalated $MnO_2$ layered cathode for highly cyclable energy dense batteries," *Nature Comm.* 14424 (2017).

9. D. Qu , D. Diehl, B.E. Conway, W.G. Pell, and S.Y. Qian, "Development of high-capacity primary alkaline manganese dioxide/zinc cells consisting of Bi-doping of $MnO_2$," *J. Applied Electrochem.* **35**:1111–1120 (2005).

10. J. Daniel-Ivad, "Rechargeable Alkaline Cell Having Reduced Capacity Fade and Improved Cycle Life," U.S. Patent No. 7,754,386 (2010).

11. J. Daniel-Ivad, R.J. Book, and E. Daniel-Ivad, "Method of Manufacturing Anode Compositions for Use in Rechargeable Electrochemical Cells," U.S. Patent No. 7,008,723 (2006).

12. K. Kordesch, L. Binder, J. Gsellmann, W. Taucher, and Ch. Faistauer, "Rechargeable Alkaline Zinc-Manganese Dioxide Batteries," *36th Power Sources Conference*, Palisades Institute for Research Services, Inc., New York, 1994.

13. T. Messing, R. Jacus, and S. Megahed, "Improved Components for Rechargeable Alkaline Manganese-Zinc Batteries," *36th Power Sources Conference*, Palisades Institute for Research Services, Inc., New York, 1994.

14. J. Daniel-Ivad, Zinc-Manganese, in J. Garche, C. Dyer, P. Moseley, Z. Ogumi, D. Rand, and B. Scrosati (eds.) *Encyclopedia of Electrochemical Power Sources*, vol 5, Amsterdam: Elsevier; 2009, pp. 497–512.

15. J. Daniel-Ivad, K. Kordesch, and E. Daniel-Ivad, "Performance Improvements of Low-Cost RAM¨ Batteries," *Proc. 38th Power Sources Conf.*, Cherry Hill, NJ, pp. 155–158, 1998.

16. J. Daniel-Ivad, K. Kordesch, and E. Daniel-Ivad, "High-Rate Performance Improvements of Rechargeable Alkaline (RAM¨) Batteries," *Proc. Vol. 98-15 Aqueous Batteries of the 194th Electrochem. Soc. Meeting*, Boston, Nov. 1–6, 1998.

17. K.V. Kordesch, "Charging Methods for Batteries Using the Resistance-Free Voltage as Endpoint Indication," *J. Electrochem. Soc.* **119**:1053–1055 (1972).

# SECTION C

# IRON OXIDE CATHODES

**Gary A. Bayles**

## 16C.1 OVERVIEW

Iron-based electrodes have long been used as anodes in battery systems. In addition to the cell designs presented in Chap. 16A, iron anodes were introduced in nickel-iron rechargeable batteries at the turn of the century by Junger in Europe and Edison in the United States (see Chap. 15A). Also, iron-based cathodes such as FeS and $FeS_2$ have been used for many years in lithium primary batteries (see Chap. 13).

More recently, as detailed below, researchers have explored the use of iron compounds as the cathode (or positive active material) in alkaline cells. In the late 1990s, an iron oxide material having a high valence state, referred to as "super iron," was first reported for use as a cathode active material.[1]

## 16C.2 ELECTROCHEMISTRY

Iron normally exists as a metal or in the valence states of Fe(II) and Fe(III). The new "super-iron" cathode material is an Fe(VI)-containing compound that has a high specific capacity due to a three-electron change in its reduction reaction, as follows:

$$FeO_4^{2-} + 3H_2O + 3e \rightarrow FeOOH + 5OH^- \qquad E^0 = \sim 0.9 \text{ V}$$

The theoretical capacity of several of these Fe(VI) compounds is listed in Table 16C.1. These values compare well with more conventional cathode materials (see Chap. 1). The list of Fe(VI) salts includes $Cs_2FeO_4$, $Rb_2FeO_4$, $K_xNa_{(2-x)}FeO_4$, and $SrFeO_4$ as well as a transition metal Fe(VI) salt ($Ag_2FeO_4$).[2]

**TABLE 16C.1**   Theoretical Capacities of Fe(VI) Compounds

| Material | Molecular weight, g | Valence change | Electrochemical equivalence | |
|---|---|---|---|---|
| | | | mAh/g | g/Ah |
| $Li_2FeO_4$ | 133.7 | 3 | 601 | 1.66 |
| $Na_2FeO_4$ | 165.8 | 3 | 485 | 2.06 |
| $K_2FeO_4$ | 198.1 | 3 | 406 | 2.46 |
| $BaFeO_4$ | 257.2 | 3 | 313 | 3.19 |

## 16C.3 CELL DESIGN AND COMPONENTS

Iron oxide-zinc test cells evaluated to date have included both button and conventional cylindrical cells and are constructed similar to other alkaline rechargeable systems.

## 16C.4   BATTERY PERFOMANCE AND APPLICATIONS

The characteristics of Fe(VI) compounds have not been studied extensively in the past, mainly because of the perception that these materials are highly unstable. While $Li_2FeO_4$ and $Na_2FeO_4$ are soluble in alkaline hydroxide, $BaFeO_4$ and $K_2FeO_4$ show evidence of low alkaline solubility and high stability, as shown in Fig. 16C.1. Further, their stability is greater in the more concentrated alkaline solutions that are used as battery electrolytes. These data have been extrapolated to suggest that, over a 10-year period, there will be less than a 10% loss of Fe(VI) in concentrated potassium hydroxide solutions using highly purified materials.

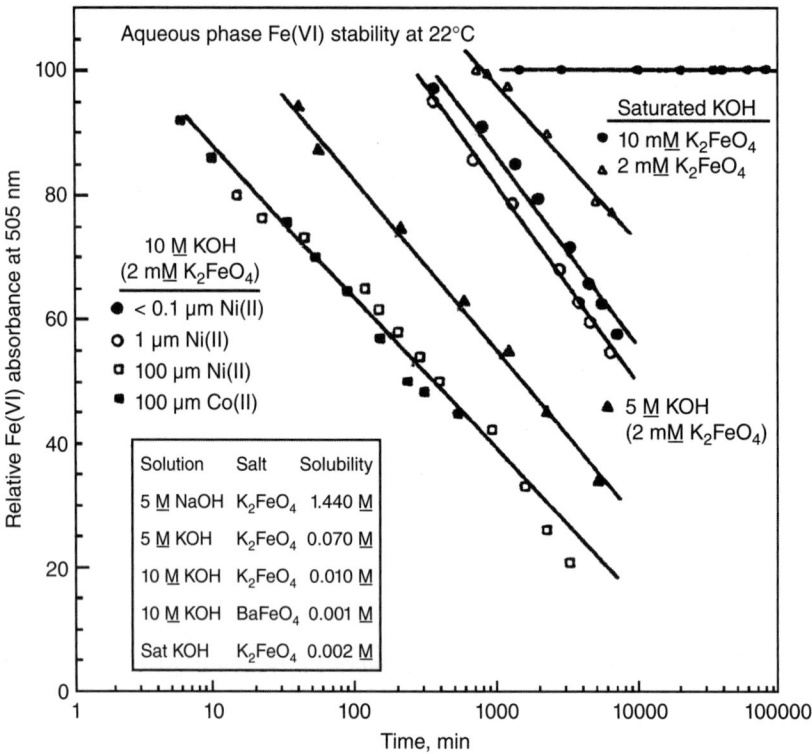

**FIGURE 16C.1**   Stability of Fe(VI) in alkaline electrolyte with various concentrations of OH⁻, $K_2FeO_4$ salts, and Co(II) and Ni(II) impurities. (*From Ref. 1.*)

Electrochemically, the $FeO_4^{2-}$ species have a high reduction potential, on the order of 0.9 V. Against an anode of zinc, the open-circuit potential was found to be 1.75 and 1.85 V for the $K_2FeO_4$ and $BaFeO_4$ cells, respectively. The proposed discharge reaction mechanism is as follows:

$$MFe(VI)O_4 + \tfrac{3}{2}Zn \rightarrow \tfrac{1}{2}Fe(III)_2O_3 + \tfrac{1}{2}ZnO + MZnO_2$$

where $M = K_2$ or Ba

The theoretical capacities and specific energy for the two batteries are given in Table 16C.2. These values can be compared with those of other electrochemical couples (see Chap. 1). The values for the zinc/iron oxide cells are higher than most of the other batteries with the exception of the lithium- and air-breathing systems.

The discharge characteristics and specific energy of experimental primary alkaline button cells, with zinc anodes and Fe(VI) compound cathodes, were measured and compared to those with $MnO_2$ cathodes. These

**TABLE 16C.2**  Theoretical Capacity and Specific Energy for MFeO₄ Batteries

| Couple | Open circuit voltage, V | Theoretical specific capacity, g/Ah | Theoretical specific capacity, Ah/g | Theoretical specific energy, Wh/kg |
|---|---|---|---|---|
| Zn/K₂FeO₄ | 1.75 | 3.68 | 0.271 | 475 |
| Zn/BaFeO₄ | 1.85 | 4.41 | 0.226 | 419 |

data are plotted in Fig. 16C.2 and illustrate the higher-energy output of the cells fabricated with the Fe(VI) cathodes. Similar results were obtained with cells fabricated in the conventional cylindrical construction.

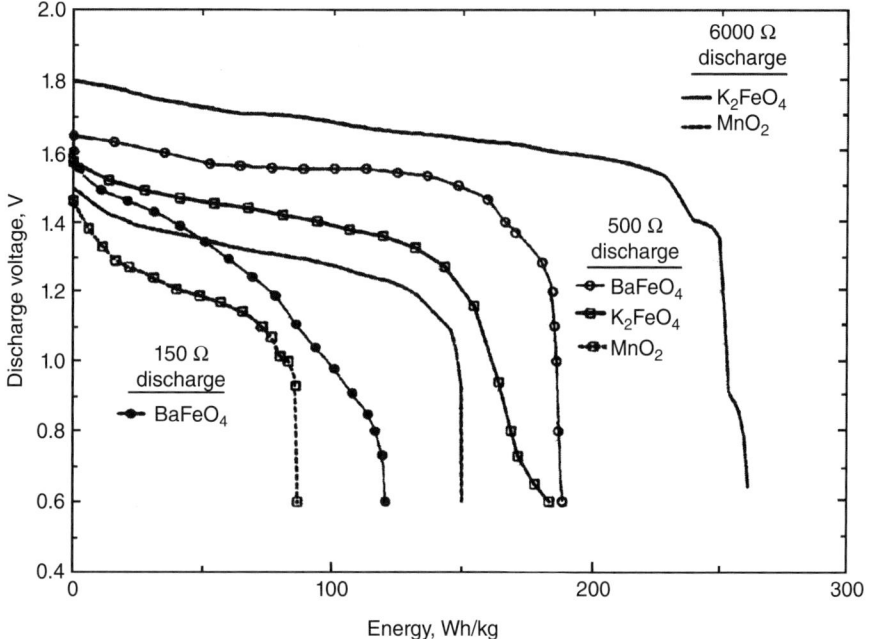

**FIGURE 16C.2**  Capacity of several experimental button cells using Fe(VI) compounds as cathodes and Zn as anodes, compared to conventional Zn/MnO₂ button cells. (*From Ref. 1.*)

The Fe(VI) compounds were also shown to be rechargeable. A button cell, using a metal hydride anode and a capacity-limited $K_2FeO_4$ cathode, was cycled for several cycles to a 75% DOD and for more than 400 cycles to a 30% DOD. The open-circuit voltage of the cell was 1.3 V and the midpoint voltage was 1.1 V, similar to the voltage characteristics of the nickel-metal hydride cell. Efforts to increase the film thickness and prevent the buildup of a passivating Fe(III) layer have been the focus of continuing studies, with the objective of improving practical capacity and long-term rechargeability of this couple.[3]

The Fe(VI) compounds are promising candidates for cathode materials for both primary and rechargeable alkaline batteries. The reported results demonstrate their higher specific energy compared to other cathode materials now used in alkaline batteries. Primary alkaline cells containing super-iron cathodes and metal boride anodes (e.g., $VB_2$, $TiB_2$, $TaB_2$) have also been investigated[4] and appear to have the potential to exceed capacity of the ubiquitous Zn-MnO₂ cell. More recent work has explored the use of the super-iron material as a lower-cost replacement for cathodes in nonaqueous lithium-ion couples, also with intriguing results.[3,5]

## *REFERENCES*

1. S. Licht, B. Wang, and S. Ghosh, *Science* **128:**1039–1042 (1999).
2. X. Yu and S. Licht, *J. Power Sources* **171:**966–980 (2007).
3. X. Yu and S. Licht, *J. Power Sources* **171:**1010–1022 (2007).
4. S. Licht, X. Yu, Y. Wang, and H. Wu, *J. Electrochem. Soc.* **155**(4):A297–A303 (2008).
5. S. Licht, *Energies* **3:**960–972 (2010).

# CHAPTER 17
# LITHIUM SECONDARY CELLS

## 17.0 OVERVIEW OF LITHIUM-BASED CELLS

### 17.0.1 Introduction

Lithium-based electrochemical systems have several key elements:

- The use of the most electropositive (highest reducing potential) element
- The smallest positive ion (other than hydrogen protons)
- Lowest density metal
- High reactivity with water (precludes the use of aqueous electrolytes)[a]

These characteristics allow the use of lithium metals, alloys, and compounds in a wide range of battery systems. However, the development of rechargeable cells, in particular, has not been simple, requiring billions of investment, especially over the last several decades. Cell component and production costs, safety, handling, etc. can create issues, but the industry has successfully dealt with most of these concerns for over 50 years. The market for lithium cells has expanded from a few niche applications to the most dominant battery technology of the 21st century. The high bar that has been established will be hard to surpass but provides motivation for more intense exploration of other non-lithium systems.

### 17.0.2 History and Background

The first lithium batteries were primary cells developed in the 1960s (see Chap. 13). Shortly thereafter, the recognition of reversible lithium reactions with various compounds, such as the chalcogenides (i.e., Ti and Mo sulfides) and transition metal oxides (i.e., $Li_xMnO_2$ and other intercalation materials), suggested possible technical approaches for a secondary lithium battery. In 1976 Exxon (now Exxon-Mobil) filed a patent for a functional secondary lithium battery based on Li metal and $TiS_2$.[b] The discovery of lithium ion conductive polymers (see Chap. 17B, Armand, 1978) suggested the potential for other, safer concepts (i.e., to prevent lithium dendrite shorting and thermal runaway).

Subsequently, with the development of improved electrolyte options and the invention of various lithium transition metal oxides, such as $LiCoO_2$ (see Chap. 17A, Goodenough), a number of new secondary lithium metal anode cells were introduced. Efforts to commercialize polymer and gelled polymer systems were fairly extensive: Mead-Hope JV (later Valence Technology and Lithium Technology Corp.), Hydro-Quebec/Yuasa/3M (later Avestor), EIC, etc. However, the trade-off of energy and rate plus nagging safety concerns prevented successful commercialization.

Several companies continued to pursue alternatives to Exxon's battery (such as Moli Energy in Canada). However, by the late 1980s the search resumed in earnest for other options. In 1990, an initial patent was filed

---

[a]Certain water-activated cell designs are feasible (Chaps. 18C and 29), and Li-ion cells with aqueous electrolytes have been proposed (U.S. Patent Application 20100248078A1, K. W. Beard, 2009).
[b]US Patent 4091191A, L. H. Gaines.

for a practical lithium-ion cell with a lithium titanate (LTO) anode (rather than lithium metal).[c] Theoretical energy density of up to 300 to 400 Wh/kg (about half that of a Li/LiCoO$_2$ couple) and over 1000 full discharge cycles at 2.5 to 3 V nominal were claimed.

However, the biggest breakthrough came a year later when Sony announced the commercial production of Li-ion cells based on a carbon anode. While these first cells only functioned at ~90 Wh/kg, improvements over the last 25 years have tripled the energy density (Chap. 17A).

Other efforts to develop lithium metal anode cells and new lithium ion components did not abate in the 1990s. Tadiran (Israel) and Bellcore/Telcordia (formerly Bell Laboratories) looked at the use of safety additives and gelled polymer laminates, respectively, to help improve their designs. Ambitious projects, such as the development, build, and test of a 24 V, 65 Ah, Li/LiCoO$_2$ battery module with liquid electrolyte, were undertaken (see Chap. 25B) even at a time when proper chargers (see Chaps. 23 and 31, Battery Management Systems) and separators did not exist for such demanding applications.

### 17.0.3   Future

Eventually (by the 2000s), mostly all mainstream activity shifted to carbon-based anodes with modified LiCoO$_2$ cathodes, such as NMC, LMO, NCA, etc. (see Chaps. 2A and 17A). However, as the performance increases of these new Li-ion cells taper off and as potential improvements to lithium metal anode systems are identified, work will continue to shift to systems such as Li-S (Chap. 17B), solid state (Chap. 22C), and hybrid (Chap. 22A) electrolytes and optional anodes (LTO and Si).

---

[c] US Patent 5284721A, K. W. Beard.

# SECTION A

# LITHIUM-ION BATTERIES

## Jeff Dahn, Grant M. Ehrlich

## 17A.1   *GENERAL CHARACTERISTICS*

Lithium-ion (Li-ion) batteries employ lithium storage compounds as the positive and negative electrode materials. As a battery is cycled, lithium ions ($Li^+$) are exchanged between the positive and negative electrodes. Li-ion batteries have been referred to as rocking chair batteries because the lithium ions "rock" back and forth between the positive and negative electrodes as the cell is charged and discharged. The positive electrode material has either a layered structure, such as in lithium cobalt oxide $LiCoO_2$ (LCO), lithium nickel manganese cobalt oxide $LiNi_{1-x-y}Mn_xCo_yO_2$ (NMC or NCM), and lithium nickel cobalt aluminum oxide $LiNi_{0.8}Co_{0.15}Al_{0.05}O_2$ (NCA), or a tunneled structure, such as in lithium manganese oxide $LiMn_2O_4$ (LMO) spinel or lithium iron phosphate $LiFePO_4$ (LFP). The positive electrode material is coated on an aluminum current collector. The negative electrode material is typically a graphitic carbon, also a layered material, coated on a copper current collector, or a composite of carbon and a metalloid such as silicon (Si) that forms an alloy with lithium. In the charge-discharge process, lithium ions are inserted or extracted from interstitial space between atomic layers of the layered or tunneled materials, or alloyed and dealloyed if a metal or metalloid such as Si is included.

The first batteries to be marketed by Sony in 1991 used LCO as the positive electrode material. LCO provides good specific capacity and capacity density, 274 mAh/g and 1363 $mAh/cm^3$, respectively. It also provides long cycle life, low self-discharge, is easily prepared, and is relatively insensitive to process variation and moisture. However, due in part to the cost of cobalt, the industry has transitioned to oxides with reduced cobalt content such as NMC and NCA, manganese-based materials such as LMO spinel, and phosphates such as LFP. Also, to avoid the cost of cobalt, the shift away from LCO has been driven in part by the need for materials having improved safety properties. In addition to lower cost, the alternatives to LCO offer other advantages, including higher rate capability, improved thermal stability, long cycle life, and improved capacity.

The first lithium-ion batteries to be commercialized by Sony employed cells with coke negative electrode materials. As improved graphites became available, the industry shifted to graphitic carbon negative electrode materials because graphitic carbon offers greater specific capacity than coke, approaching the theoretical capacity of carbon of one Li atom per six carbon atoms, equivalent to 372 mAh/g. More recently, silicon and $Li_{4/3}Ti_{5/3}O_4$ spinel (LTO) have been introduced. Silicon is used to provide improved energy density and LTO is used to provide improved thermal stability, greater cycle life, and improved rate capability, thereby broadening the applicability of the technology.

Dramatic improvements in the performance of lithium ion batteries and cost reduction have enabled similarly significant growth in the market for Li-ion batteries. From their commercial introduction in 1991, Li-ion sales grew at a rate of over 15% per year to over $22 billion in 2016, representing approximately 5.6 billion cells providing about 80 GWh of storage if all cells were charged one time. In 2017, nearly all personal portable electronic devices (cell phones, tablets, personal computers) used Li-ion power. Since their introduction to the power tool market in 2004, the 2017 market share for Li-ion batteries has grown to over 60%. Li-ion batteries are now used for grid-energy storage and frequency regulation, such as in the 20 MW, 170,000 cell system commissioned in 2014 by NextEra Energy in DeKalb, Illinois.[1] Larger systems are being installed, such as a 100 MW, 400 MWh battery being installed for Southern California Edison.[2] Between 2010 and 2016, Li-ion battery pack cost decreased from about $1000/kWh to about $250/kWh, and dramatic cost reductions continue. On a cell basis, the cost of Li-ion energy decreased from $2.60/Wh in 2000 to $0.15/Wh in 2016, a portion of which is attributable to an increase in energy density of about 45%. Li-ion battery cost decreased 14% per year between 2000 and 2015,[3] and average cylindrical cell cost decreased about 50% between 2005 and 2017 to about $0.20/Wh.[4]

The growth of the lithium-ion battery market is illustrated in Figs. 17A.1 and 17A.2, which provide the historical and predicted size of the lithium-ion market, based on dollar value and energy, respectively. As shown in Figs. 17A.1 and 17A.2, while historically the lithium-ion market has been dominated by electronic devices, electric vehicles (EVs) are expected to play a major role in the next decade. In total, the Li-ion battery market is predicted to grow from 80 GWh in 2016 to 210 GWh by 2025, representing a 13% annual growth rate. Much of the growth during the next decade is forecast to occur in the EV market, where 20% to 30% annual growth is forecast.[5]

**FIGURE 17A.1** Li-ion market, 2000–2025 (Christophe Pillot, Avicenne Energy).

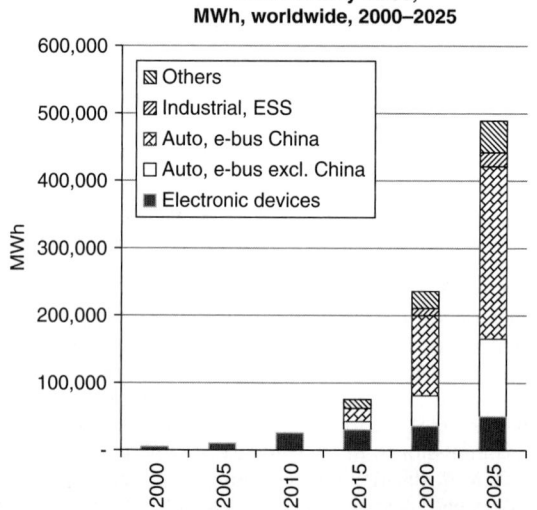

**FIGURE 17A.2** Li-ion market, 2000–2025 (Christophe Pillot, Avicenne Energy).

### 17A.1.1 Advantages and Disadvantages

Major advantages and disadvantages of Li-ion cells, relative to other types of cells, are summarized in Table 17A.1. The high specific energy (275 Wh/kg), energy density (730 Wh/L), and specific power (600–3000 W/kg) of commercial cells make them attractive for weight- or volume-sensitive applications. Li-ion batteries offer a low self-discharge rate (1–5% per month, depending on the state of charge and temperature), long cycle life (500–10,000 cycles), and a broad temperature range of operation (commercially available cells may be charged at 0–45°C and discharged at −40–65°C), enabling their use in a wide variety of applications. A wide array of sizes and shapes is now available from a variety of manufacturers. Single cells typically operate between 2.5 and 4.3 V (some smart phones use cells charged to 4.4 V), which is approximately three times that of NiCd or NiMH, and thus fewer cells are required for a battery pack of a given voltage than if a cell using an aqueous electrolyte were used. Li-ion batteries also offer high-rate capability. Commercially available cells intended for power tool applications provide about 88% of their capacity when discharged at 10$C$, and charging at 10$C$ provides 73% of the 0.5$C$ capacity. It is common for manufacturers to offer "energy cells," designed for optimal specific energy and energy density, and "power cells," designed for optimal specific power and power density while still retaining energy density greater than competitive technologies. The combination of these qualities within a cost-effective, hermetic package has enabled diverse application of the technology.

**TABLE 17A.1**   Advantages and Disadvantages of Li-Ion Batteries

| Advantages | Disadvantages |
|---|---|
| Sealed, maintenance-free cells | Moderate initial cost |
| Long cycle life | Degrade at high temperature |
| Broad temperature range of operation | Need for protective circuitry |
| Long shelf life | Capacity loss and potential for thermal |
| Low self-discharge rate | runaway when overcharged |
| Rapid charge capability | Possible venting and possible thermal |
| High-rate and high-power discharge | runaway when crushed |
| High coulombic and energy efficiency | May fail if rapidly charged at low |
| High specific energy and energy density | temperatures (<0°C) |
| No memory effect | |
| Cylindrical, prismatic, and polymer cells offer design flexibility | |

A disadvantage of some Li-ion batteries is that they degrade when discharged below 2 V and may vent when overcharged because they do not inherently have a chemical mechanism to manage overcharge, unlike most aqueous cell chemistries. Therefore, Li-ion batteries typically employ management circuitry and mechanical disconnect devices to provide protection from overdischarge, overcharge, or overtemperature conditions. For "0 volt" applications, i.e., applications where discharge to 0 V must be tolerated, configurations that can be discharged to 0 V are commercially available. Other disadvantages of Li-ion products are that they may permanently lose capacity at elevated temperatures (65°C), albeit at a lower rate than most NiCd or NiMH products, and may become unsafe if rapidly charged at low temperatures (<0°C).

### 17A.1.2 Designation and Markings

The International Electrotechnical Commission (IEC) publication 61960 provides standard designation, marking, electrical testing, and safety testing of Li-ion cells and batteries. Included in IEC 61960 are standard procedures for electrical tests, including charge and discharge at standard and high rate at 20°C and −20°C, charge retention, charge recovery after long-term storage, endurance cycles, and measurement of battery internal resistance.

The IEC designation and marking system provided in IEC 61960 for Li-ion cells utilizes five numbers in the case of cylindrical cells and six numbers in the case of prismatic cells. For cylindrical cells, the first two digits designate the diameter in millimeters and the next three digits designate the length in tenths of millimeters.

For example, an 18650 cell is 18 mm in diameter and 65.0 mm in length. For prismatic cells, the first two digits designate the thickness in tenths of millimeters, the next two designate the width in millimeters, and the last two designate the length of the cell in millimeters. For example, a 564656P prismatic cell is 5.6 mm thick, 46 mm wide, and 56 mm long. If a dimension is smaller than 1 mm, it can be written as t$N$, where $N$ is in tenths of a millimeter.

For batteries, the designation and marking system is expanded to $N_1A_1A_2A_3N_2/N_3/N_4-N_5$, where $N_1$ denotes the number of series connected cells; $A_1$ indicates the basis of negative electrode phase, where I is carbon, L is lithium metal or a lithium alloy, T is titanium, and X is for other materials; $A_2$ designates the positive electrode phase, and could be C, F, Fp, N, M, Mp, T, V, or X for cobalt, iron, iron phosphate, nickel, manganese, manganese phosphate, titanium, vanadium, and others, respectively; $A_3$ designates the shape of the cell, either R for cylinder and P for prismatic; $N_2$ is the maximum diameter (cylindrical) or thickness (prismatic) in millimeters; $N_3$ denotes the maximum width in millimeters for prismatic shapes; $N_4$ is the maximum overall height in millimeters; and $N_5$ denotes the number of parallel connected cells if two or more are used (omitted if the value is 1). The IEC designation scheme is shown schematically in Fig. 17A.3.

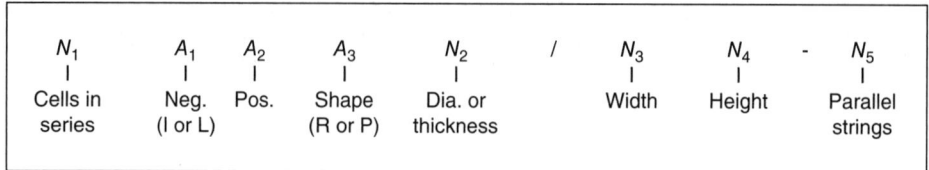

**FIGURE 17A.3**    IEC 61960 battery designation scheme.

Based on the publications of various manufacturers, it is clear that these nomenclatures are not yet fully standardized. This is presumably due to the large numbers of cell chemistries, which can include graphite, Si, SiO, or LTO negative electrodes and LCO, spinel, LFP, NMC, or NCA positive electrodes. Blends of electrode materials are also often used. The latest standards should be obtained for detailed information on nomenclature, performance, and safety guidelines. Also, ANSI standard C18.2M "Standard for Portable Rechargeable Cells and Batteries" includes standards for portable Li-ion batteries.

## 17A.2    CHEMISTRY

The electrochemically active electrode materials in Li-ion batteries are a lithium metal oxide or a lithium metal phosphate for the positive electrode material and typically a lithiated graphite for the negative electrode material. It is now common to find a silicon-containing material along with graphite in the negative electrode of high energy density Li-ion cells, as is further discussed in Sec. 17A.2.3. These materials are adhered to a metal foil current collector with a binder, typically polyvinylidene fluoride (PVDF), carboxymethylcellulose, and/or styrene-butadiene rubber, and a conductive diluent, typically carbon black or graphite. In some designs, thin layers (having a thickness of approximately 2 μm) of ceramic particles, such as $Al_2O_3$, are coated on one electrode surface to increase stability by reducing the likelihood of internal shorting.[6] The positive and negative electrodes are electrically isolated by a microporous polyethylene (PE) or a polypropylene (PP) separator film. Some separator films have ceramic particles coated on their surface(s) or impregnated within them (see Sec. 17A.2.6). Since the commercialization of Li-ion batteries by Sony in 1991, a broad array of variants have been introduced, including cylindrical, prismatic, and pouch-type or "polymer" Li-ion cells (see Fig. 17A.4). The active cell chemistries in cylindrical, prismatic, and pouch-type cells are the same, although many choices exist for the positive and negative electrode materials.

### 17A.2.1    Intercalation Processes

The active materials in conventional Li-ion cells operate by reversibly incorporating lithium in an intercalation process, a topotactic reaction where lithium ions (guests) are reversibly removed or inserted into a host without a significant structural change to the host. The positive material in a Li-ion cell is a metal oxide, and the metal oxide has either a layered or a tunneled structure. Graphitic carbon negative materials have a layered structure

**FIGURE 17A.4**  A polymer or "pouch"-type Li-ion cell. Polymer cells are available in a broad array of sizes and have a layered plastic-aluminum laminate packaging like that shown here.

similar to graphite. Thus, the metal oxide, graphite, and other materials act as hosts, reversibly incorporating the lithium ion guests to form sandwich-like structures.[7] Silicon-containing negative electrode materials are also used. Silicon-containing negative electrode materials operate by an alloying/dealloying process where significant structural changes to the host do occur. Silicon-containing materials are further discussed in Sec. 17A.2.2.

Intercalation materials, originally discovered by the Chinese 2700 years ago,[8] have been the subject of contemporary chemical research for only the last 80 years. Intercalation compounds are commonly found as electrode materials in a wide range of battery chemistries, and thus are not limited to Li-ion. For example, the $Ni(OH)_2$, found in NiCd and NiMH batteries, is an intercalation compound and deintercalates H, forming $NiOOH$ during charge of these cells. Sodium-ion batteries, under development today, use intercalation compounds for both electrodes. The intercalation of a variety of electron donors, including lithium, and electron acceptors, such as halogens, into graphite has been studied.[9] The field of graphite intercalation compounds is especially rich,[10] both in the diversity of the chemistry and the depth of study.

Alkali metal intercalation of graphite and related carbons, in particular $Li_xC_6$ ($0 \leq x \leq 1$), is well studied.[11] When a Li-ion cell is charged, the active positive electrode material is oxidized and the active negative electrode material is reduced. In this process, lithium ions are deintercalated from the positive material and intercalated into the negative material. The reactions that occur when a cell is charged and discharged are shown in Scheme 1 (Fig. 17A.5a), in which $LiMO_2$ represents the metal oxide positive material, such as $LiCoO_2$, and C represents the carbonaceous negative material, such as graphite. In Scheme 1, $x$ and $y$ are selected based on the molar capacities of the electrode materials for lithium. Normally, $x$ is about 0.65 and $y$ is about 0.16; therefore, $x/y$ is about 4. The reverse occurs on discharge. Because metallic lithium is not present in the cell, Li-ion batteries are chemically less reactive, safer, and offer longer cycle life than is possible with rechargeable lithium batteries that would employ lithium metal as the negative electrode material. The charge-discharge process in a Li-ion cell is further illustrated graphically in Fig. 17A.5b. In this figure, the layered active materials are shown on metallic current collectors.

## 17A.2.2  Positive Electrode Materials

Positive electrode materials in commercially available Li-ion batteries utilize a lithiated metal oxide or lithiated metal phosphate as the active material. The first Li-ion products marketed by Sony used LCO. Goodenough and Mizushima developed this material, as described in a series of patents.[12] Recently, cells have been developed that utilize less costly materials, such as LMO, NMC, LFP, and materials with higher specific capacity, such as NCA and NCA grades with greater Ni content and less Co and Al. $LiNi_{1/3}Mn_{1/3}Co_{1/3}O_2$, $LiNi_{0.5}Mn_{0.3}Co_{0.2}O_2$, and

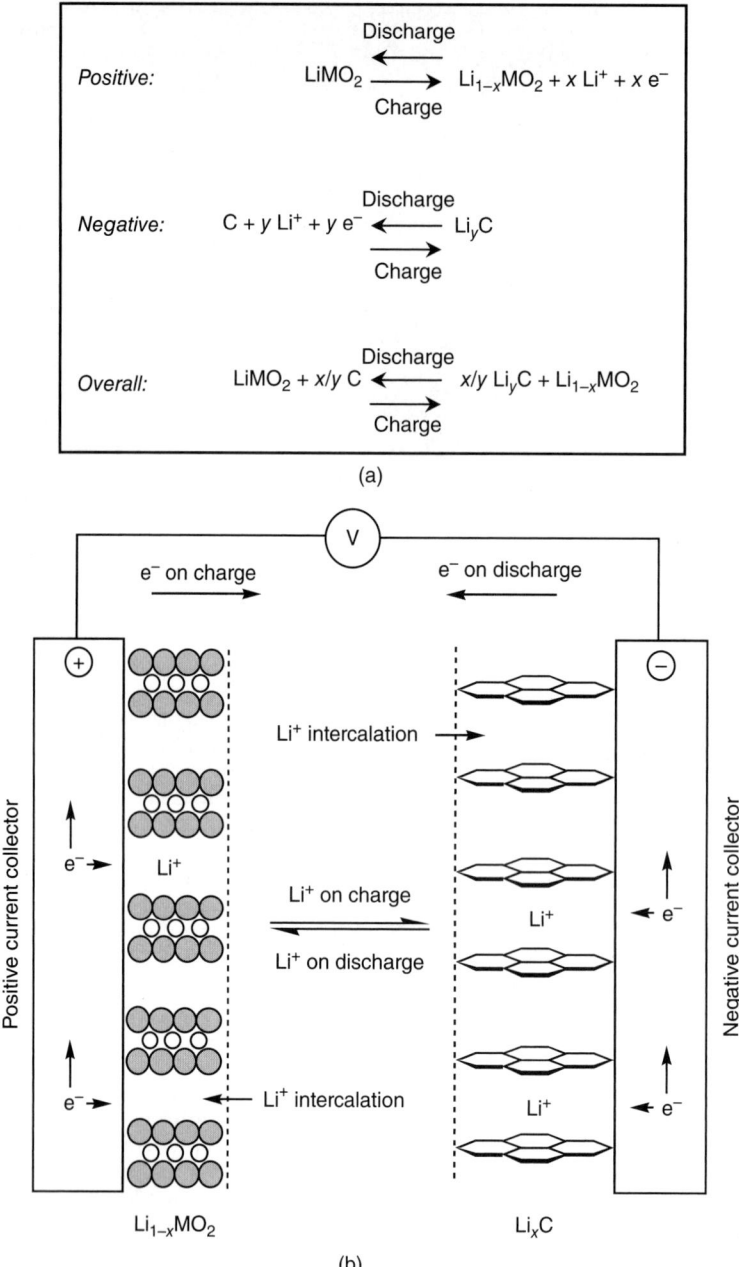

**FIGURE 17A.5**    (*a*) Scheme 1. Li-ion cell reactions. (*b*) Electrode processes in a Li-ion cell.

$LiNi_{0.6}Mn_{0.2}Co_{0.2}O_2$ are some of the most commonly commercialized types of NMC materials. Figure 17A.6 shows the relative amounts of various positive electrode materials used in 2016 and projected to be used in 2025. The market share of LMO is decreasing due to the difficulty in making energy dense cells with long lifetime with this material, and therefore, LMO will only be briefly discussed in this chapter.

**2016 positive active material: >180,000 tons**

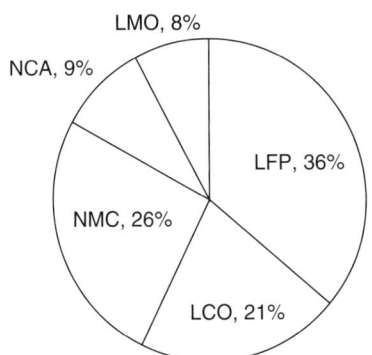

**2025 positive active material forecast: >400,000 tons**

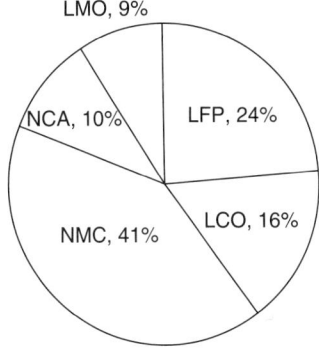

**FIGURE 17A.6**    Worldwide use of positive active material in 2016 and 2025 forecast.[4]

Viable positive electrode materials must satisfy a number of requirements, which are summarized in Table 17A.2. These factors guide materials selection and development. In commercial cells the positive electrode material is the source of all the active lithium ions in a lithium-ion cell. Therefore, to provide high capacity, these materials must incorporate a large amount of lithium as synthesized. Furthermore, the materials must reversibly exchange that lithium with minimal structural change to permit long cycle life, high coulombic efficiency, and high energy efficiency. To achieve high cell voltage and high energy density, the lithium intercalation reaction must occur at a high potential relative to lithium. When a cell is charged or discharged, an electron is removed from or returned to the positive material, respectively. In order that the charge and discharge processes occur at a high rate, the electronic conductivity and lithium-ion mobility in the material must be high. $LiFePO_4$ is an exception to this rule. In $LiFePO_4$, adequate lithium ion transport is achieved by use of electrode particles having a nanometer particle size. The positive electrode material must also be compatible with the other materials in the cell. In particular, the positive electrode material must not be soluble in the electrolyte. In addition, the positive electrode material must have an acceptable cost. To minimize cost, preparation from inexpensive materials in a low-cost process is preferred.

**TABLE 17A.2**    Requirements for Li-Ion Positive Electrode Materials

High free energy of reaction with lithium (high voltage vs. lithium metal potential)
Can incorporate large quantities of lithium
Reversibly incorporates lithium with minimal structural change
High lithium-ion diffusivity
Good electronic conductivity or can be made conductive with coatings
Insoluble in the electrolyte
Prepared from inexpensive materials
Low-cost synthesis
Packing density provides high energy density

***Characteristics of Positive Electrode Materials.***    A variety of positive electrode materials are commercially available. These materials can have one of three structure types: an ordered rock salt-type structure, a spinel-type structure, or an olivine-type structure. The ordered rock-salt structure is a layered structure in which the lithium atoms, transition metal atoms, and oxygen atoms occupy octahedral sites in alternate layers. Exemplary layered materials include LCO, NMC, and NCA. The ideal layered structure of $LiCoO_2$ is shown in Fig. 17A.7.

The spinel and olivine structures are both three-dimensional "framework" structures. The term *spinel* formally refers to the mineral $MgAl_2O_4$, although the term is used for materials, such as $LiMn_2O_4$, that have an equivalent structure. Likewise, the term *olivine* formally refers to the mineral $(MgFe)_2SiO_4$, although the term is

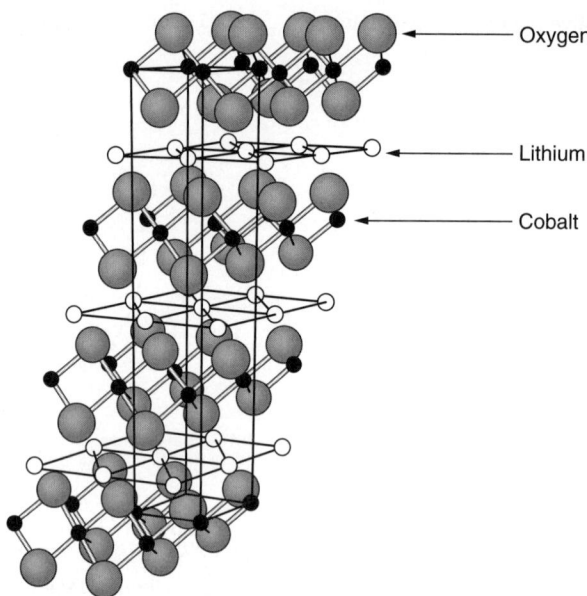

Oxygen

Lithium

Cobalt

**FIGURE 17A.7** Structure of $LiCoO_2$ highlighting the layers of mobile Li atoms. The repeating unit cell is indicated.

used for $LiFePO_4$ and $LiMnPO_4$ because they have an equivalent structure. The three-dimensional framework or tunnel structure of $LiMn_2O_4$ (spinel), which is based on $\lambda$-$MnO_2$, is illustrated in Fig. 17A.8. In spinel $LiMn_2O_4$, lithium fills one-eighth of the tetrahedral sites within the $\lambda$-$MnO_2$ structure and Mn-centered oxygen octahedra fill one-half of the octahedral sites.

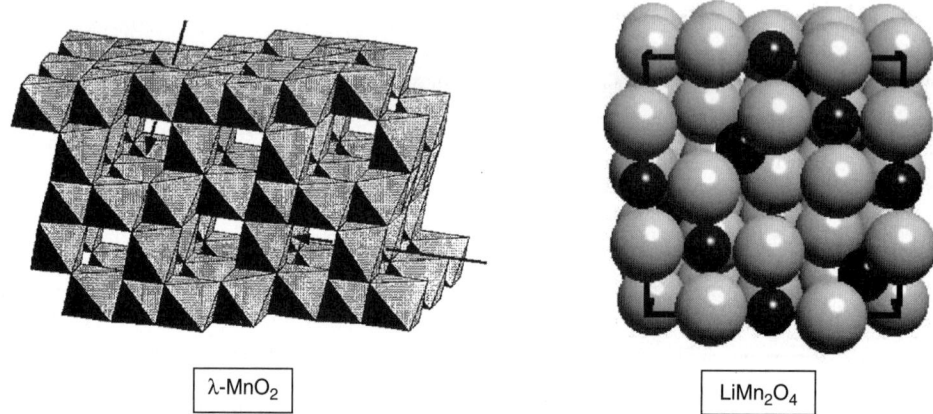

$\lambda$-$MnO_2$

$LiMn_2O_4$

**FIGURE 17A.8** Structure of $\lambda$-$MnO_2$ and $LiMn_2O_4$ spinel, showing the manganese-centered oxygen octahedra of $\lambda$-$MnO_2$. In the model of $LiMn_2O_4$, oxygen is gray and lithium is black. (*Courtesy CISR and Michael Tucker.*)

Materials having the olivine structure have a three-dimensional framework structure based on $PO_4$ tetrahedra and $FeO_6$ octahedra, as shown in Fig. 17A.9. In $LiFePO_4$, the Li atoms move along one-dimensional tunnels. Defects and imperfections in these tunnels can lead to poor rate capability.

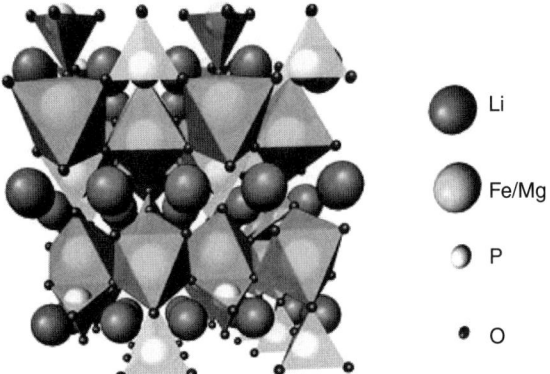

**FIGURE 17A.9**  The structure of $LiFe_{1-x}Mg_xPO_4$, which has an olivine-type structure, showing the $FeO_6$ octahedra and $PO_4$ tetrahedra.

A truly amazing accomplishment is that John Goodenough's research group is responsible for the first clear demonstrations that each of these three structure types could be used as positive electrode materials. Layered materials are described in "$Li_xCoO_2$ $(0 \leq x \leq 1.0)$: A New Cathode Material for Batteries of High Energy Density," which was published in the Materials Research Bulletin in 1980.[13] Spinel materials are described in "Lithium Insertion in Manganese Spinels," which was also published in the Materials Research Bulletin.[14] Olivine materials are described in "Phospho-olivines as Positive-Electrode Materials for Rechargeable Lithium Batteries."[15] These three papers had been cited 2532, 1507, and 7373 times, respectively, as of January 26, 2018. John Goodenough[16] has been recognized by numerous major awards for these achievements, including the Japan Prize in 2000, the Enrico Fermi Award in 2009, the National Medal of Science in 2013, the Draper Award in 2014, the Welch Award in Chemistry in 2017, and the Franklin Medal in 2018, which is fitting given that his research group is responsible for all major classes of lithium-ion positive electrode materials. The Royal Society of Chemistry now grants an award in his honor. At the time of this writing the only major award to elude Prof. Goodenough is the Nobel Prize. Nobel Prizes are normally awarded for inventions or discoveries that change the world. There can be no argument that the positive electrode materials discovered by Prof. Goodenough have done so.

The voltage and capacity characteristics of common positive electrode materials are summarized in Table 17A.3. The most commonly used positive electrode material, $LiCoO_2$, offers good capacity, 155 mAh/g when charged to 4.3 V versus Li/Li$^+$, and high average voltage, 3.9 V versus Li. NMC materials have the same structure and provide performance substantially equivalent to that of $LiCoO_2$, and have the benefits of lower raw materials cost and improved thermal stability during abuse.[17] $LiNi_{0.8}Co_{0.15}Al_{0.05}O_2$ offers higher capacity, up to 200 mAh/g, albeit at a voltage which is about 0.2 V lower than that of $LiCoO_2$ or $LiMn_2O_4$. Other NCA grades provide even greater specific capacity, up to 220 mAh/g. Recently, special coatings have been used to increase the performance of layered positive electrode materials by making them more tolerant to charging to higher voltages. Some commercially available cells employ coated positive electrode materials having performance that exceeds that listed in Table 17A.3. $LiMn_2O_4$ is also used commercially, particularly in applications that are cost sensitive or require exceptional stability upon abuse. $LiMn_2O_4$ has lower capacity, about 120 mAh/g, slightly higher voltage, 4.0 V versus lithium, but has greater capacity loss on storage or cycling, especially at elevated temperature, relative to cells that use $LiCoO_2$, NMC, LFP, or NCA. $LiFePO_4$ has a specific capacity of about 165 mAh/g and an average voltage of 3.45 V versus lithium. $LiFePO_4$ is virtually unreactive with electrolytes in the charged or discharged state up to at least 350°C. A disadvantage of $LiFePO_4$ is that it has both low capacity density and a low packing efficiency (tap density), making it difficult to produce $LiFePO_4$ cells having high energy density.

There are several new positive electrode materials that are being commercialized. As an example, $Li[Li_{1/3-2x/3}Ni_xMn_{2/3-x/3}]O_2$ $(0 \leq x \leq 0.5)$[18] and related materials containing Co, provide specific capacity, up to 300 mAh/g. However, these materials suffer from low crystallographic density, low tap density, and poor rate capability at this time, as detailed in the excellent review by Hy et al.[19] Also being developed are "5 V" materials, such as $LiNi_{0.5}Mn_{1.5}O_4$, which take the spinel structure like LMO. Finding suitable stable electrolytes for $LiNi_{0.5}Mn_{1.5}O_4$ has been problematic.

**TABLE 17A.3**   Characteristics of Some Positive Electrode Materials

| Material | Specific capacity, mAh/g | Midpoint V vs. Li at C/20 | Comments |
|---|---|---|---|
| $LiCoO_2$ | 155–185 | 3.9–3.95 | Co is expensive. Dominates smart phones. Capacity depends on upper voltage cutoff. |
| $LiNi_{1-x-y}Mn_xCo_yO_2$ (NMC) | 140–190 | ~3.8–3.9 | Safer and less expensive than $LiCoO_2$. Capacity depends on upper voltage cutoff. |
| $LiNi_{0.8}Co_{0.15}Al_{0.05}O_2$ | 200 | 3.73 | About as safe as $LiCoO_2$, high capacity. |
| Advanced NCA | 220 | 3.70 | Less Co for lower cost. Very high capacity. |
| $LiMn_2O_4$ | 120 | 4.05 | Inexpensive, safer than $LiCoO_2$, poor high temperature stability. Losing market share. |
| $LiFePO_4$ | 165 | 3.45 | Synthesis in inert gas leads to process cost. Very safe. Low volumetric energy. |
| $Li[Li_{1/9}Ni_{1/3}Mn_{5/9}]O_2$ | 275 | 3.8 | High specific capacity, R&D scale, low rate capability. Related materials contain Co. |
| $LiNi_{0.5}Mn_{1.5}O_4$ | 130 | 4.6 | Requires an electrolyte that is stable at a high voltage. Not commercialized. |

***Physical Properties of Positive Electrode Materials.***   The particle-size distribution, particle shape, specific surface area, and tap density of positive electrode materials all play important roles in determining the properties of Li-ion batteries that incorporate them. Because the particle size determines the solid-state diffusion path length, the particle-size distribution controls aspects of the rate capability. Also, the particle-size distribution and specific surface area control the properties of slurries used for coating electrodes. For lithium transition metal oxides, the specific surface area also controls the reactivity of the charged material with the electrolyte at elevated temperature. This is because interactions between the particles of positive electrode material and the electrolyte occur at the surfaces of the particles, thus the surface area should be minimized to reduce reactivity and improve safety. Finally, the tap density is a useful indicator of the ultimate density of a calendered electrode because a material having a high tap density normally provides a high density electrode.

Positive electrode materials are supplied in three common morphologies: predominantly single-crystal particles (LCO, LMO, and some new NMC grades, shown in Fig. 17A.10a); spherical polycrystalline particles (NMC and NCA, shown in Fig. 17A.10b), and nanomaterials (LFP, shown in Fig. 17A.10c).

The particles of LCO, LMO, and some new grades of NMC[20] typically have very smooth surfaces, as shown in Fig. 17A.10. The smooth surfaces result because each particle is a single crystal or perhaps made up of a few crystal grains. In contrast, the particles of NMC and NCA are normally spherical and made up of thousands of small primary particles, as shown in Fig. 17A.10b, hence they are polycrystalline materials. Since LCO, LMO, NMC, and NCA can support rapid Li-ion diffusion, the powder particles of these materials can be on the 3 to 20 µm scale. Figure 17A.10c shows a scanning electron microscope (SEM) image of $LiFePO_4$ from Hunan Reshine New Material Company. The $LiFePO_4$ has a very small particle size to accommodate the slow rate of Li diffusion in this material.

In order to understand the particle shapes and sizes of $LiCoO_2$, NMC, NCA, $LiFePO_4$, and $LiMn_2O_4$, it is helpful to understand how these materials are synthesized. $LiCoO_2$ is easy to make by mixing cobalt oxide or cobalt carbonate with lithium carbonate or lithium hydroxide and sintering the mixture at 700 to 1000°C in air. Various methods of producing $LiCoO_2$ are known.[21] Normally, $LiCoO_2$ is made with a Li:Co stoichiometric ratio of 1:1 or

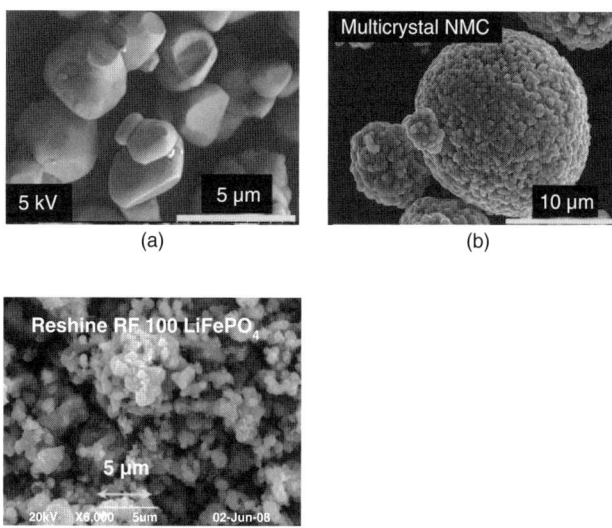

(a)                                      (b)

(c)

**FIGURE 17A.10** Common morphologies of positive electrode materials found in Li-ion cells: (*a*) Single crystal, (*b*) Spherical polycrystalline, and (*c*) nano materials. (*a and b Courtesy of Jing Li, Dalhousie University, and c courtesy of Reshine.*)

with a slight excess (0–5%) of lithium. $LiCoO_2$ sinters very quickly and particles comprising large single crystals with smooth surfaces (as shown in Fig. 17A.10*a*) grow in times as short as 1 h at the synthesis temperature.

NMC and NCA materials rely on a uniform and homogeneous distribution of cations in the transition metal layers of the structure (see Fig. 17A.7). The most common way to ensure this is to use a mixed transition metal hydroxide or carbonate precursor that has the cations perfectly mixed on the atomic scale. Such precursors are normally prepared by coprecipitation of mixed metal sulfate solutions in a base or in a carbonate. The precursors made by this route form spherical agglomerates of many primary particles, resulting in a rough surface morphology. The spherical particle size and tap density of the precursors can be controlled by the pH, temperature, and ammonia concentration during precipitation.[22] Oxides are produced by sintering the mixed transition metal hydroxide precursors with lithium carbonate or lithium hydroxide monohydrate. The spherical shape of the hydroxide precursors are maintained in the resulting NMC and NCA oxides, as shown in Fig. 17A.11.

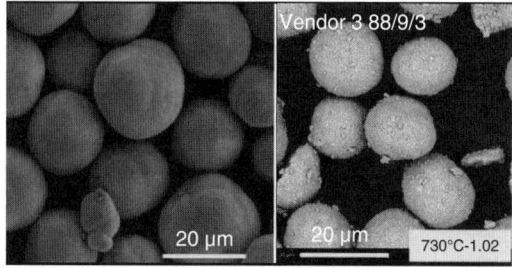

**FIGURE 17A.11** Left panel shows a typical lithium nickel cobalt aluminum oxide (NCA) precursor, in this case $Ni_{0.88}Co_{0.09}Al_{0.03}(OH)_2$. Right panel shows NCA, in this case $LiNi_{0.88}Co_{0.09}Al_{0.03}O_2$ prepared from the precursor on the left by heating with $LiOH \cdot H_2O$ at 730°C. (*SEM photos courtesy of Ning Zhang and Hongyang Li, Dalhousie University.*)

$LiFePO_4$, unlike the layered oxides and the spinel materials, must be synthesized in an inert atmosphere to prevent oxidation of the desired divalent Fe to trivalent Fe. The requirement for an inert atmosphere increases process complexity, and hence the cost of the resulting material. In addition, the primary $LiFePO_4$ particles have a particle size of less than about 500 nm to provide sufficient rate capability. Inexpensive $LiFePO_4$ is normally made from an $Fe^{3+}$ precursor, which must be reduced during synthesis to $Fe^{2+}$. This reduction can be accomplished by an elegant method using carbothermal reduction[23] because $LiFePO_4$ is stable at elevated temperatures in contact with carbon. The same is not true for the layered and spinel oxides, which are reduced by carbon at elevated temperature. The carbothermal reduction synthesis conveniently results in a thin layer of carbon formed on the $LiFePO_4$ particle surface. Ravet et al. showed that a thin carbon coating on the surfaces of the $LiFePO_4$ particles enhances electronic conductivity and improves rate capability.[24] The commercial $LiFePO_4$ materials shown in Fig. 17A.10c are apparently produced by pyrolysis of an organic precursor in a similar process.

***Li-Ion Electrode Capacity Ratio and Cell Voltage.***     Figure 17A.12 shows potential versus specific capacity of a $Li/LiCoO_2$ cell and a Li/graphite cell in which the specific capacity axis of the Li/graphite cell has been scaled by dividing by 2. Normally, researchers test each of the positive and negative Li-ion electrode versus metallic lithium to determine specific capacity, differential capacity, and charge-discharge cycle life. (Research coin cell systems are available from NRC Canada.) If such preliminary results are encouraging, full Li-ion cells with electrodes having active materials in appropriate ratios are constructed.

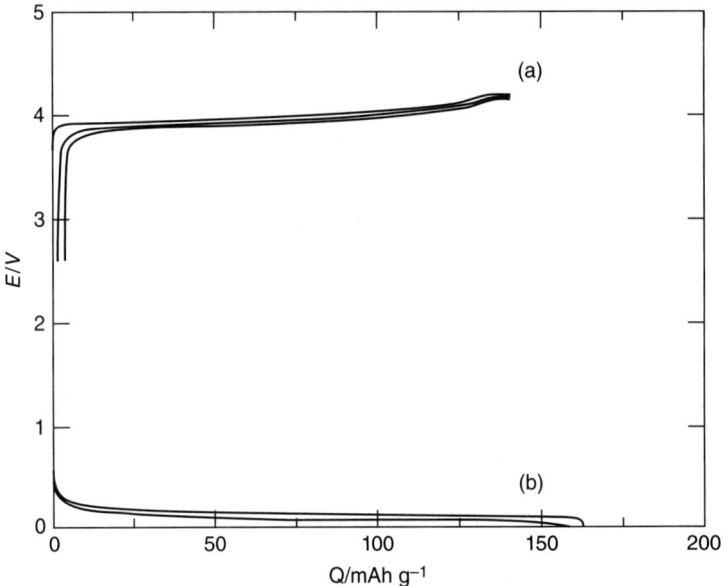

**FIGURE 17A.12**   Potential versus specific capacity for (*a*) $Li/LiCoO_2$ and (*b*) Li/graphite cells. The specific capacity axis for the Li/graphite cell has been divided by 2.[25] (*Reproduced with permission from the Journal of Power Sources, 174, 449–456 [2007].*)

The capacity of the positive and negative electrodes must be matched. For example, in Fig. 17A.12, the specific capacity of graphite is about 350 mAh/g and the specific capacity of $LiCoO_2$ (to 4.2 V) is about 140 mAh/g; thus this graphite would be properly included in a Li-ion cell using this $LiCoO_2$ at a mass per unit electrode area of about 50% of the $LiCoO_2$ mass per unit electrode area. In addition, about 10% excess graphite is typically added to the negative electrode to avoid lithium plating at a full state of charge. The voltage of the Li-ion cell is the difference between the curves in Fig. 17A.12 at a selected state of charge. $LiCoO_2$ can deliver more capacity if it is charged to higher potentials, e.g., to about 155 mAh/g at 4.3 V, but may need to be coated and/or doped with other elements to allow excursions to high voltages without detrimental impacts to cycle life or thermal stability.

*Electrochemical Properties of Positive Electrode Materials.*    Figure 17A.13 shows potential versus specific capacity for (a) $LiNi_{0.5}Mn_{1.5}O_4$ spinel, (b) $Li[Li_{0.1}Al_{0.1}Mn_{1.8}]O_4$ (LAMO, an aluminum-stabilized spinel), (c) $LiNi_{1/3}Mn_{1/3}Co_{1/3}O_2$ (NMC 111), (d) $LiFePO_4$, and (e) $Li_{4/3}Ti_{5/3}O_4$ spinel (a potential negative electrode material also written as $Li_4Ti_5O_{12}$). Figure 17A.13 shows that $LiNi_{0.5}Mn_{1.5}O_4$ has the highest potential and $LiFePO_4$ has the lowest potential among these positive electrode materials. LAMO has the smallest specific capacity. NMC 111 provides a very high specific capacity, 200 mAh/g if it is charged to 4.6 V, as shown in Fig. 17A.13. However, due to a compromise between capacity and cycle life, NMC materials are typically charged to only about 4.3 V where their capacities are similar to $LiCoO_2$.

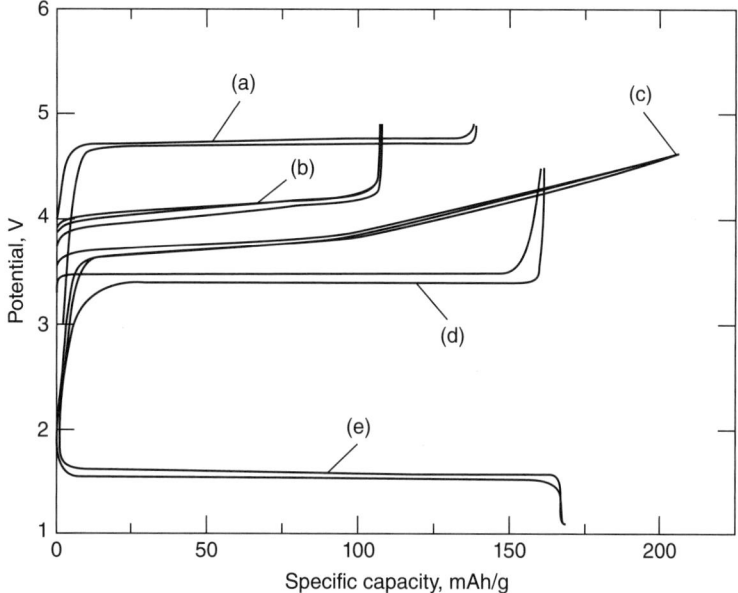

**FIGURE 17A.13**  Potential versus specific capacity for (*a*) $LiNi_{0.5}Mn_{1.5}O_4$ spinel, (*b*) $Li[Li_{0.1}Al_{0.1}Mn_{1.8}]O_4$ (LAMO), (*c*) $LiNi_{1/3}Mn_{1/3}Co_{1/3}O_2$ (NMC 111), (*d*) $LiFePO_4$, and (*e*) $Li_{4/3}Ti_{5/3}O_4$ spinel.

Figure 17A.14 shows the voltage (vs. Li/Li$^+$) of modern LCO, NMC, and NCA positive electrode materials. NMC532, NMC622, NMC811, and NCA801505 can reversibly deliver up to 200 mAh/g if charged to 4.5 V, 4.4 V, 4.3 V, and 4.3 V, respectively. NCA materials with higher Ni content such as NCA920503 can deliver 220 mA/g reversible capacity when charged to 4.3 V. Manufacturers make selections between the various positive electrode materials based on many requirements including cycle life (favors NMC), energy density (favors LCO due to high density, NMC charged to high voltage, or NCA), or cost (favors NMC or NCA). Particle sizes of the various materials can be adjusted by varying synthesis conditions to prepare materials for power cells (smaller particles) or energy cells (larger particles, generally).

There are a large number of common NMC grades including NMC111, NMC442, NMC532, NMC622, and NMC811. The relative advantages and disadvantages of these grades have been discussed in an excellent review article by Noh et al.[26] Figure 17A.15 summarizes the various NMC grades, provided one considers them all to be charged to the same upper cutoff potential of 4.3 V. Materials with less cobalt content are attractive commercially because of cobalt's cost; however, as shown in Fig. 17A.15, materials with greater nickel content tend to be less stable.

Many of these electrode materials display spectacular physics and chemistry associated with metal-insulator transitions, stacking rearrangements, and ordering of the intercalated lithium within the available sites. Perhaps the most famous is the feature that appears in the differential capacity of $Li/Li_xCoO_2$ cells at $x = \frac{1}{2}$ which is caused by the ordering of lithium atoms along rows in the structure.[27]

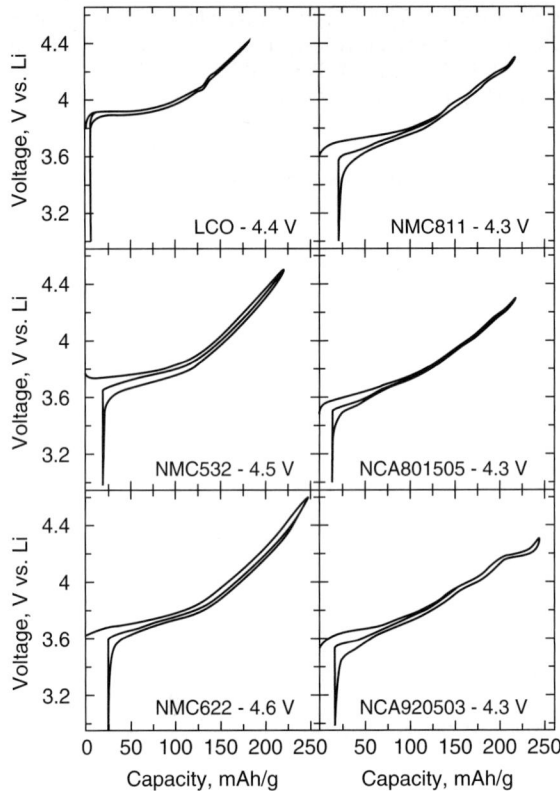

**FIGURE 17A.14**   Voltage versus specific capacity for modern positive electrode materials used in Li-ion cells. (*Courtesy of Jeremy Peters, Alex Louli, Ning Zhang and Jing Li, Dalhousie University.*)

Figure 17A.16 shows differential capacity versus potential for a Li/LiCoO$_2$ cell. The local minimum near 4.14 V ($x = \frac{1}{2}$ in Li$_x$CoO$_2$) corresponds to a structure in which Li is ordered along every second row of available sites in the Li layers shown in Fig. 17A.7. The peaks on either side of the minimum correspond to order-disorder phase transitions. Disruptions to the Co layer, by the substitution of Ni or by the inclusion of an excess of Li, destroys the order-disorder transition, even with a few percent of substituent atoms, and the feature in the differential capacity plot ($dQ/dV$ vs. $V$) disappears. Some commercial LiCoO$_2$ grades do not show this order-disorder feature.

Considerable effort has been expended by academic researchers and battery manufacturers to understand capacity loss mechanisms in Li-ion positive electrodes. Figure 17A.17 summarizes those mechanisms thought to be the most influential as they apply to the active and inactive phases in the positive electrode material. Electrolyte additives and positive electrode material surface coatings are thought to be effective ways to mitigate these issues.

## 17A.2.3   Negative Electrode Materials

***Historical Overview.***   Since the early 1970s, intercalation compounds have been used as electrode materials for secondary lithium batteries. Secondary lithium battery development effort throughout the 1970s and early

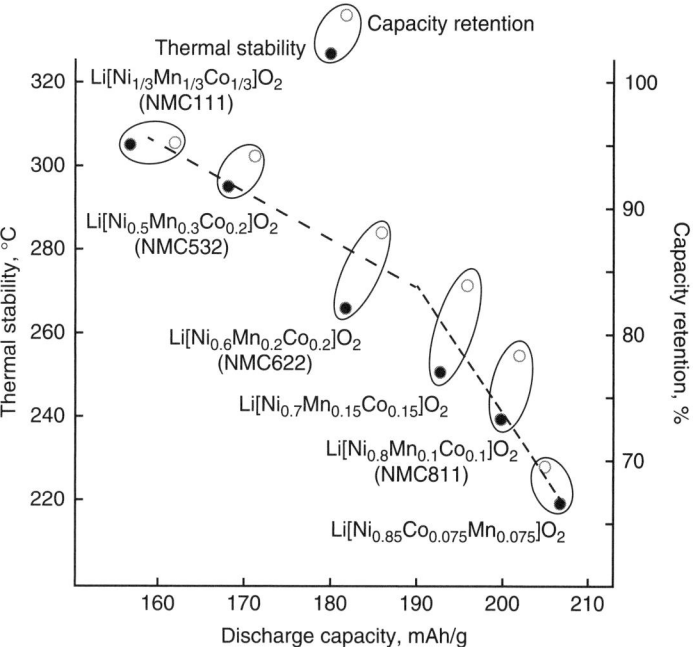

**FIGURE 17A.15** Thermal stability and capacity retention versus discharge capacity of NMC grades. (*Based on Noh et al.[26]*)

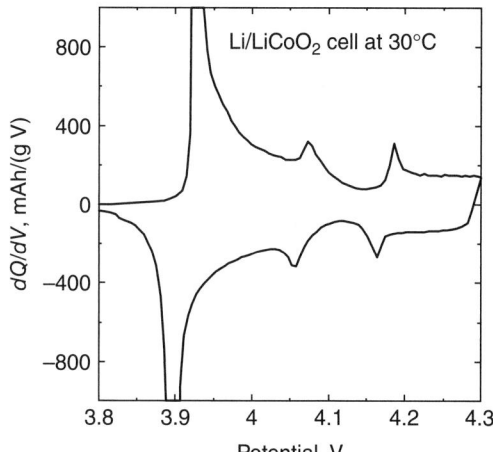

**FIGURE 17A.16** Differential capacity (*dQ/dV*) versus cell potential for a Li/LiCoO$_2$ cell charged and discharged at a 0.1*C* rate at 30°C. (*Courtesy of Wenbin Luo.*)

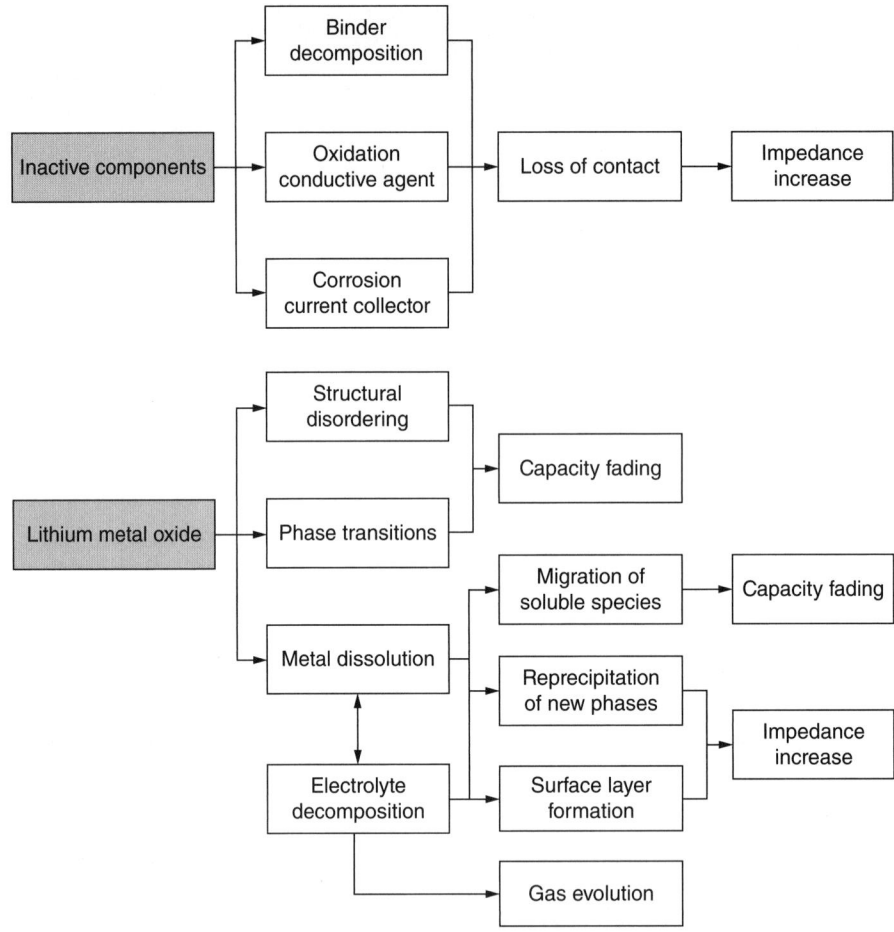

**FIGURE 17A.17**   Capacity loss mechanisms for Li-ion positive electrode materials.[28] (*Reproduced with permission from the J. Power Sources, 127, 58–64 [2004].*)

1980s focused on the use of lithium metal as the negative electrode because of the high specific capacity of the metal. Cells with impressive performance were developed and some were commercialized; however, safety issues with lithium metal batteries[29] caused the industry to concentrate on using lithium intercalation into carbon at the negative electrode instead of lithium metal.[30] The safety issues with lithium metal have been attributed to the changing morphology of lithium as a cell is cycled. Also, because reactivity is a function of surface area, the safety properties of negative electrodes may be correlated to their surface area. Thus, while the properties of lithium metal negative electrodes change with use, carbon electrodes offer stable surface morphology, resulting in consistent safety properties over their useful life.[30] By utilizing low-surface-area carbons, electrodes with acceptable self-heating rates may be fabricated.

The first Li-ion batteries marketed by Sony utilized petroleum coke at the negative electrode. Coke-based materials offer good capacity, 180 mAh/g, and are stable in the presence of propylene carbonate (PC), in contrast to graphitic materials, which can cointercalate PC and exfoliate unless stabilizing additives are used. The disorder in coke materials is thought to pin the graphene layers, inhibiting reaction or exfoliation in the presence of PC.[30] In the mid-1990s, most Li-ion cells utilized electrodes employing graphitic spheres, in particular mesocarbon microbeads (MCMBs). MCMBs offer higher specific capacity, 300 to 350 mAh/g, and low surface area, thus providing low irreversible capacity (IRC) and good safety properties. Recently, a wider variety of carbon types have been used in negative electrodes. Many commercial cells utilize artificial (i.e., synthetic) graphite or natural graphite, which are available at very low cost, and are usually highly graphitized to provide the highest specific capacity and improve packing efficiency. Figure 17A.18 shows the total mass used in the Li-ion battery industry and market share of artificial and natural graphite in 2015 and a projection for 2025.

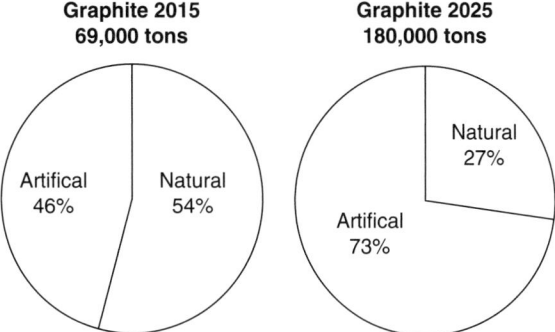

**FIGURE 17A.18**    Graphite used in Li-ion cells in 2015 and projections for 2025. (*Courtsey of Avicenne Energy, March 2017.*)

***Types of Carbon.***    Many types of carbon materials are commercially available. The structure of the carbon greatly influences its electrochemical properties, including lithium intercalation capacity and potential. The basic building block for carbon materials is graphene, which is a planar sheet of carbon atoms arranged in a hexagonal array, as shown in Fig. 17A.19. These sheets are stacked in a registered fashion to form graphite. In Bernal graphite, the most common type, ABABAB stacking occurs, resulting in hexagonal or 2H graphite. In a less common polymorph, ABCABC stacking occurs that is termed rhombohedral or 3R graphite.[7]

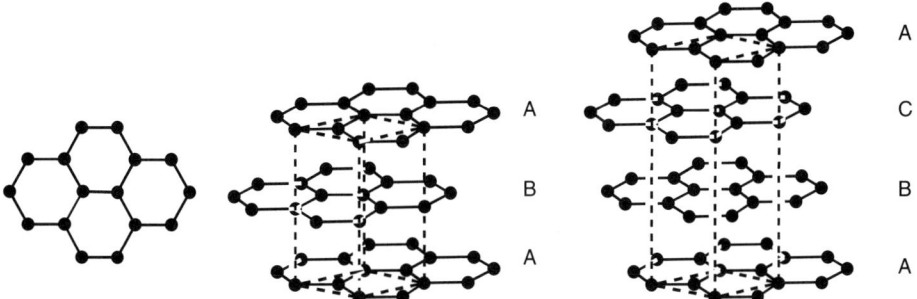

**FIGURE 17A.19**    The hexagonal structure of graphene, i.e., a layer of carbon, and the structures of hexagonal (2H) and rhombohedral (3R) graphite.

Most real materials contain disorder, including the 2H and 3R stacking orders as well as random stacking; thus a more precise way to identify a graphite is to specify the relative fractions of 2H, 3R, and random stacking. Various forms of carbon have been developed with a range of stacking disorders and different morphologies. Stacking disorders include those where the graphitic planes are parallel but shifted or rotated, termed turbostratic disorder,[31] and those in which the planes are not parallel, termed unorganized carbon.[30] Particle morphologies include flat plates, which are present in natural graphite, carbon fibers, and spheres.

Carbon materials can be considered as different aggregations of a basic structural unit (BSU) consisting of two or three parallel planes with a diameter of 2 nm.[32] The BSUs may be oriented randomly, resulting in carbon black, or oriented to a plane, axis, or point, resulting in a planar graphite, a whisker, or a spherule. The types of carbon may alternatively be organized based on the type of precursor material, as illustrated in Fig. 17A.20, because the precursor material and the processing parameters determine the nature of carbon produced. Materials that can be graphitized by treatment at high temperature (2000–3000°C) are termed soft carbons. Upon graphitization, the turbostratic disorder is removed progressively with increasing temperature, and strain in the material is relieved.[33] Hard carbons, such as those prepared from phenolic resin, cannot be readily graphitized, even when treated at 3000°C. Coke-type materials are prepared at about 1000°C, typically from an aromatic petroleum precursor.

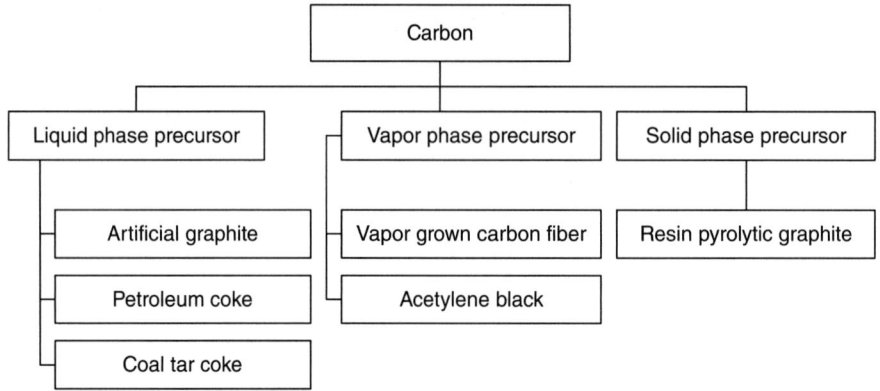

**FIGURE 17A.20**    Carbons classified by the precursor phase.

***Staging and Electrochemical Intercalation into Carbon.***    When lithium is intercalated into graphite, the ABAB structure transforms to an AAAA structure and distinct voltage plateaus are observed as distinct phases (stages) are formed.[10] Staging in graphite is illustrated in Fig. 17A.21, which shows the voltage of a Li/graphite cell over one cycle at a low rate for a highly ordered graphite. A classical model of lithium staging is illustrated in Fig. 17A.22. As shown, lithium forms "islands" within graphite instead of distributing homogeneously.

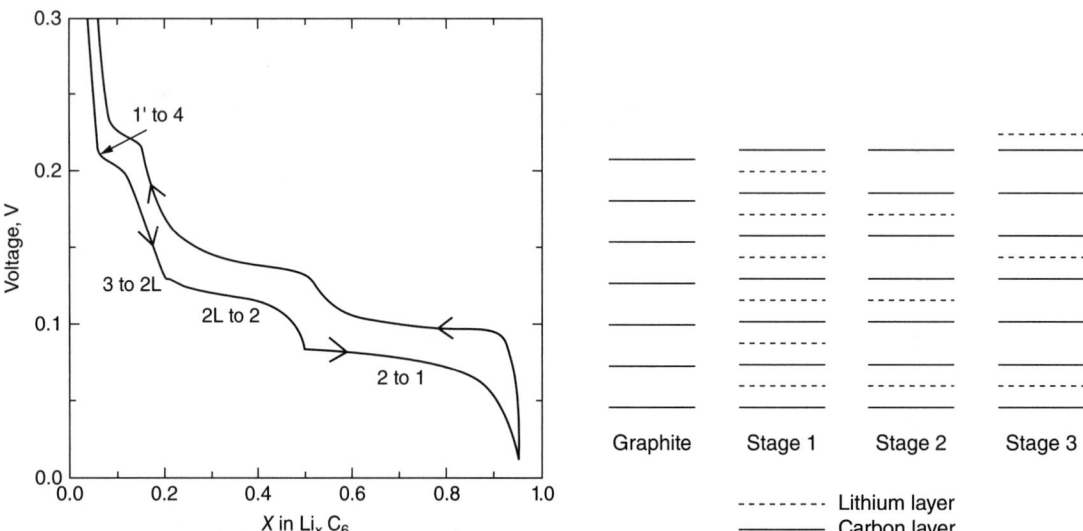

**FIGURE 17A.21**    The voltage of a Li/graphite cell illustrating lithium staging upon intercalation into graphite.[33] (*Reproduced with permission from Phys. Rev. B, **51**, 734 [1995].*)

**FIGURE 17A.22**    Schematic diagram of lithium staging in graphite.[33] (*Reproduced with permission from Phys. Rev. B, **51**, 734 [1995].*)

The most lithium-rich stage, $LiC_6$, is termed stage 1 and is formed at the lowest voltage, as indicated in Fig. 17A.21. As lithium is removed from the graphite, higher stages are formed, as indicated in Figs. 17A.21 and 17A.22.

The specific capacity of a graphitic carbon is determined by the fraction of adjacent graphene layers that are in turbostratic misalignment. It has been found that lithium cannot be inserted between parallel graphene sheets that are turbostratically misaligned. The specific capacity, $Q$, of graphite is then simply calculated as

$$Q = 372(1 - P) \text{ mAh/g}$$

where $P$ is the fraction of adjacent misaligned layers. Alternatively, the capacity ($x_{max}$) in $Li_xC_6$ can be written as

$$x_{max} = 1 - P$$

The fraction of layers in turbostratic misalignment can be determined by x-ray diffraction. A software package developed by Hang Shi can be used to determine $P$ by comparing calculated patterns to measured patterns, as illustrated in Fig. 17A.23.[33,34] As $P$ decreases, the diffraction peaks become sharper and more peaks appear.

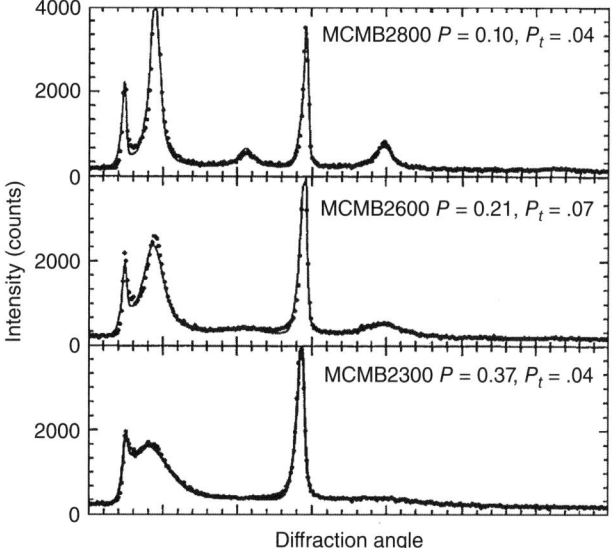

**FIGURE 17A.23**  X-ray diffraction patterns of three of the mesocarbon microbeads (MCMB) materials listed in Table 17A.4. The solid lines are fitted curves used to extract the value of $P$. $P_t$ is the probability of finding layers having rhombohedral (ABC) stacking instead of hexagonal (ABAB) stacking.[33,34] (*Reproduced with permission from Phys. Rev. B, **51**, 734 [1995].*)

Figure 17A.24 shows how the potential-capacity profile of lithiated graphite depends on the heat-treatment temperature and on $P$.[33,34] Note that in Fig. 17A.24 the curves have been sequentially offset by 0.1 V for clarity. As $P$ approaches zero, the capacity increases. This increase in capacity is understood to result from relief of turbostratic misalignment as the heat-treatment temperature is increased.

Table 17A.4 provides structural parameters of the graphitic carbons described in Figs. 17A.23, 17A.24, and 17A.25.

Figure 17A.25 shows the capacity of the graphites listed in Table 17A.4 as well as other graphites plotted versus the disorder probability $P$. Figure 17A.25 illustrates the linear correlation between the fraction of turbostratically misaligned layers and the reduction in capacity. Evidently, in order to obtain the highest specific capacity, it is important to use a graphitic carbon with the highest degree of order. In addition, particle size, specific surface area, tap density, impurity content, and surface treatments are other significant parameters that

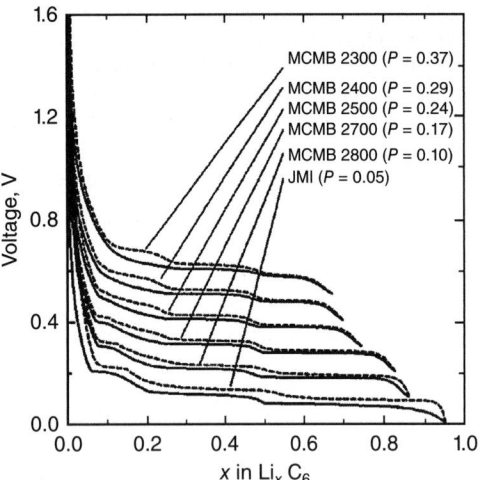

**FIGURE 17A.24**    Voltage versus capacity for the second discharge and charge cycle of lithiated artificial graphites that were initially heat-treated at different temperatures. In the figure the curves have been offset by 0.1 V for clarity. [33,34] (*Reproduced with permission from Phys. Rev. B*, **51**, *734 [1995].*)

**TABLE 17A.4**    Structural Parameters of the Graphitic Carbons Described by Figs. 17A.23, 17A.24, and 17A.25

| Material | Heat-treatment temperature (°C) | $d_{(002)}$ (Å) | P |
|---|---|---|---|
| JMI | N/A | 3.356 | 0.05 |
| MCMB2800 | 2800 | 3.352 | 0.10 |
| MCMB2700 | 2700 | 3.357 | 0.17 |
| MCMB2600 | 2600 | 3.358 | 0.21 |
| MCMB2500 | 2500 | 3.359 | 0.24 |
| MCMB2400 | 2400 | 3.363 | 0.29 |
| MCMB2300 | 2300 | 3.369 | 0.37 |

influence the lithium intercalation performance of graphite. As further described below, a small specific surface area is desirable.

During the first electrochemical reaction of lithium with graphite, some of the lithium transferred to the graphite reacts with the electrolyte to form a passivation layer at the electrode-electrolyte interface, which is commonly referred to as the solid electrolyte interphase (SEI). The SEI contains lithium that is no longer electrochemically active; thus the formation of the SEI results in IRC. The capacity difference between the charge and discharge curves in Fig. 17A.26 results from IRC. This initial capacity loss is an undesirable property of all current materials and occurs largely on the first cycle. However, once formed, the SEI layer protects the graphite surface from further reaction with the electrolyte.

***Properties of Commercial Graphite for Li-Ion Battery Negative Electrodes.***    The most significant selection factors for a graphite powder to be used in a negative electrode material for a Li-ion battery include the fraction of the layers in turbostratic misalignment, the specific surface area, the content of impurities, the particle size, the particle shape, and the tap density.

The fraction of layers in turbostratic misalignment is characterized by the disorder probability P, which represents the probability of turbostratic misalignment in adjacent parallel layers. Because P is correlated to the heat-treatment temperature of a synthetic graphite and to the d(002) spacing (the spacing between graphene sheets, see Table 17A.4), the heat-treatment temperature and the lattice parameters can often be used to characterize the graphite. As noted above, the reversible specific capacity of graphite is directly correlated to

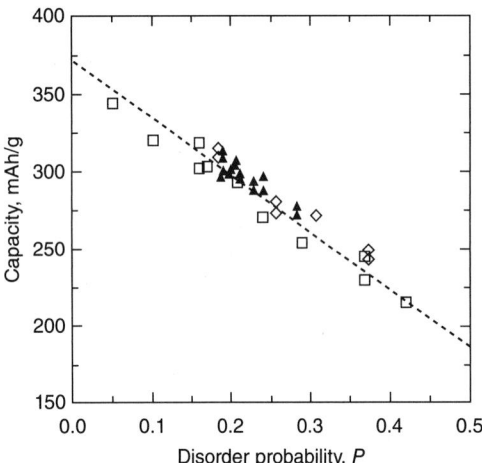

**FIGURE 17A.25**    Reversible specific capacity of Li/graphite cells versus disorder probability, i.e., the fraction of P of adjacent graphene layers in turbostratic misalignment. [33] (*Reproduced with permission from Phys. Rev. B*, **51**, *734 [1995].*)

P. P can be reduced by increasing heating time and/or temperature; however, this contributes to increased product cost.

The reactions that form the SEI and contribute to the IRC are proportional to the specific surface area. Fong et al. showed that a linear relationship exists between specific surface area and IRC.[35] In addition, the reactivity of the negative electrode with electrolyte also scales with the surface area,[36] suggesting that surface area should be minimized to avoid IRC and improve stability. However, to maintain reasonable rate capability, the specific surface area cannot be too small because this would require use of undesirably large particles. Graphite normally has a flaky particle morphology; however, the use of flake graphite results in electrodes with extremely high tortuosity after electrode compression, which leads to poor rate capability. Mesophase graphite and spherical or "potato-shaped" graphite, which provide improved tortuosity, are commonly found in Li-ion cells because the spherical geometry minimizes surface area and also minimizes Li-diffusion path lengths in the material. Impurities normally found in graphite are not detrimental to cell performance, but contribute dead weight to the electrode. Natural graphite is purified to reduce impurity content.

Graphite particle-size distribution and tap density are not always independent of the specific surface area and control the ability to make well-controlled electrode coatings and electrodes having high density after compression. High-density electrodes are important for maximizing Li-ion cell energy density.

Recent advances in graphitic materials for Li-ion batteries include the application of carbon coatings on the surface of graphite to reduce surface area and inhibit exfoliation. The carbon coatings reduce IRC and improve the thermal stability of the charged electrode materials in electrolyte. Graphite producers Toyo Tanso (Nozaki et al.,[37] Japan) and Carbonix Inc. (Park et al.,[38] Korea) have discussed the application of

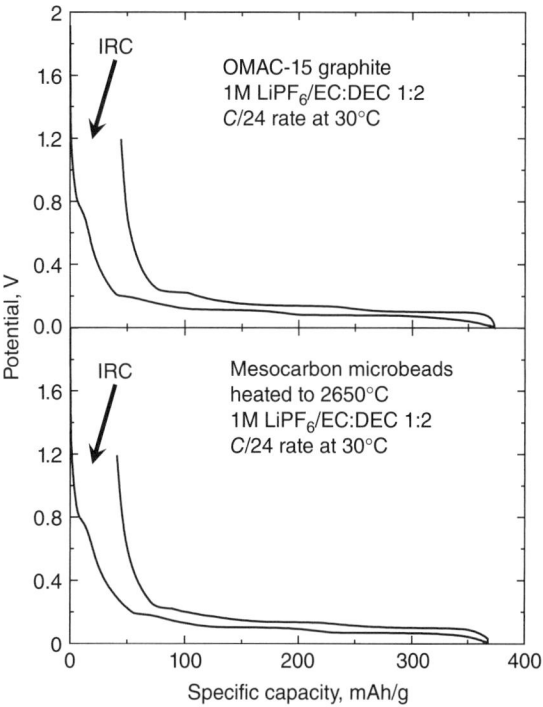

**FIGURE 17A.26** First lithiation and delithiation of graphite electrodes in Li/graphite cells illustrating the irreversible capacity (IRC) attributed to formation of the solid electrolyte interphase (SEI) on the graphite surface. (*Courtesy of Aaron Smith.*)

coatings on natural graphite. Figure 17A.27 shows SEM images of uncoated and coated natural graphite.

As shown in Fig. 17A.27, the uncoated natural graphite has a rough surface, whereas the coated natural graphite has a smooth surface. Also shown is that the coating does not change the overall particle size or shape. The coating decreases the Brunauer-Emmett-Teller (BET) surface area from about 5.7 $m^2$/g to about 0.6 $m^2$/g,

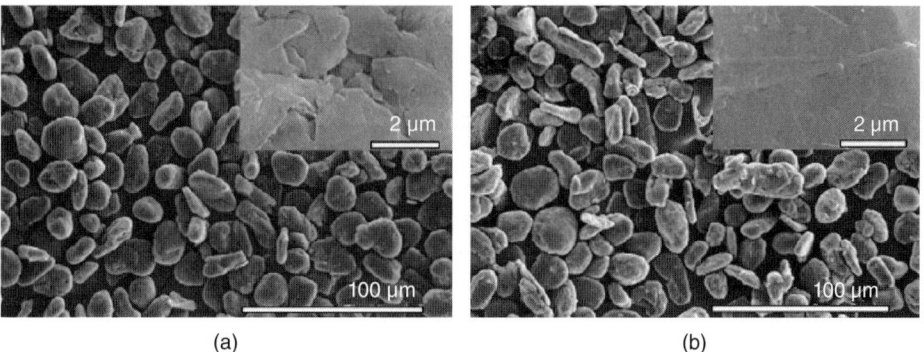

**FIGURE 17A.27** SEM images of (*a*) uncoated and (*b*) coated natural graphite.[38] (*Reproduced with permission from J. Power Sources, **190**, 553–557 [2009].*)

and reduces the IRC from about 55 mAh/g to about 25 mAh/g.[38] Also, as shown in Fig. 17A.28, differential scanning calorimetry (DSC) analysis of lithiated coated natural graphite ($LiC_6$) shows that the coated materials evolve heat more slowly, attributable to their smaller surface area, which slows the reaction rate between the $LiC_6$ and the electrolyte.[38]

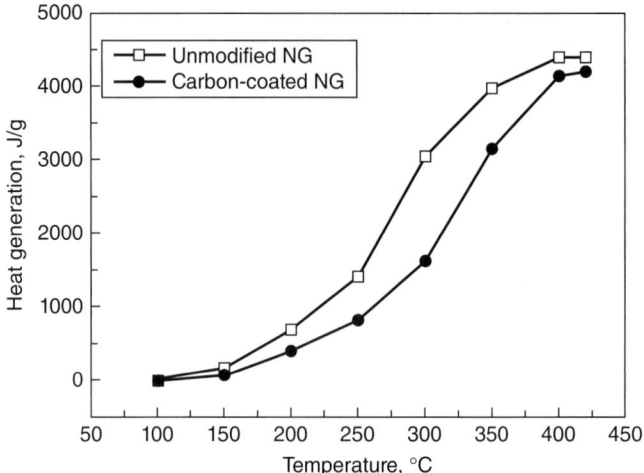

**FIGURE 17A.28** Heat evolution from lithiated uncoated and carbon coated natural graphite (NG) electrodes in the presence of electrolyte.[38] (*Reproduced with permission from J. Power Sources, **190**, 553–557 [2009].*)

Table 17A.5 summarizes the properties of commercially available graphite from two major Chinese manufacturers. First-cycle efficiencies as high as 94% are now common for graphite marketed for use in Li-ion batteries. Many of these materials provide capacity near the theoretical capacity of graphite, which is 372 mAh/g, indicating that $P$ in these materials is near zero.

### Lithium Titanate Negative Electrode Material, $Li_{4/3}Ti_{5/3}O_4$

*Overview.*    As described by Zhu et al.,[39] there are a number of lithium-titanium oxides that are suitable for lithium-ion negative electrodes. Of those materials, only lithium titanate, $Li_4Ti_5O_{12}$, commonly written as $Li_{4/3}Ti_{5/3}O_4$ in view of its spinel structure, has been commercialized in Li-ion cells. $Li_{4/3}Ti_{5/3}O_4$ (LTO) is an alternative to graphite and provides extremely long cycle life and better safety characteristics, albeit with reduced voltage and energy density. The improved stability provided by LTO is attributed to its reduced reactivity with electrolyte relative $LiC_6$ because the potential of lithium titanate is about 1.55 V versus Li, compared to about 0.1 V for $LiC_6$. However, the 1.55 V terminal voltage penalty of Li-ion cells using LTO and LTO's low packing

**TABLE 17A.5** Properties of Some Graphitic Carbons for Li-Ion Negative Electrodes

| Manufacturer | Grade | Type | Tap density, g/mL | BET surface area, m²/g | Particle size ($d_{50}$), μm | Reversible capacity, mAh/g | First-cycle efficiency, % |
|---|---|---|---|---|---|---|---|
| BTR New Energy Materials Inc. | 918 | Natural | 1.0–1.2 | 3–4.5 | 16–20 | >360 | >94 |
| BTR New Energy Materials Inc. | S360M | Artificial | 0.95–1.15 | 2.2–3.2 | 18.5–21.5 | >360 | >94 |
| BTR New Energy Materials Inc. | MSG18 | Natural | 1.0–1.1 | <1.7–2.7 | 16–19 | >355 | >92 |
| Shanshan Tech. | GF-1S2 | Natural | 1.04 | 3.3 | 11 | 363 | 92 |
| Shanshan Tech. | CMS-G15 | Natural | 1.42 | 1.2 | 16 | 324 | 94 |
| Shanshan Tech. (China) | CAG-3MT | Natural | 1.11 | 1.5 | 11 | 336 | 94 |

*Source*: www.btrchina.com and www.shanshantech.com.

efficiency compared to graphite result in lower energy density and specific energy than when using graphite. Nonetheless, such cells find application in stationary applications where extremely long cycle life is required, such as grid-energy storage and in implantable medical devices.

Some of the first studies of LTO were by the Murphy,[40] Dahn,[41] and Ohzuku groups.[42] LTO has a spinel structure like that of $LiMn_2O_4$, but with some Li occupying $16d$ metal sites. LTO is the end member of the solid solution series $Li_{1+x}Ti_{2-x}O_4$ with $x = 1/3$.[41]

The electrochemical behavior of LTO is shown by curve (e) in Fig. 17A.13. The charge-discharge curve contains a plateau that represents the coexistence between two phases, the $Li_{4/3}Ti_{5/3}O_4$ starting material and the $Li_{7/3}Ti_{5/3}O_4$ fully lithiated end phase. Ohzuku et al.[42] showed that these two phases have exactly the same lattice constant; thus the intercalation-deintercalation reaction proceeds topotatically, i.e., without any volume change. Accordingly, Ohzuku calls this material a "zero-strain" insertion electrode material and suggests that the zero-strain properties of LTO are the reason why LTO provides excellent charge-discharge cycle performance.

*Characteristics of Commercially Available $Li_{4/3}Ti_{5/3}O_4$.* LTO is now commercially available from a number of suppliers, including Süd-Chemie and BTR New Energy Materials Inc. Commercially available materials provide capacity of about 160 mAh/g, which is very close to the theoretical limit of 170 mAh/g. Table 17A.6 compares the properties of several LTO grades from BTR New Materials Technology (www.btrchina.com). Both high and low surface area materials are available, suitable for high-rate and longer lifetime applications, respectively. LTO materials are exceptionally stable and can provide thousands of cycles.

**TABLE 17A.6**  Properties of $Li_{4/3}Ti_{5/3}O_4$ from BTR New Materials Technology Inc.

| Manufacturer | Grade | $d(50)$, μm | Surface area, m²/g | Tap density, g/mL | Specific capacity @ 1C-rate, mAh/g | Irreversible capacity, % |
|---|---|---|---|---|---|---|
| BTR New Energy Materials Inc. (China) | LTO-1 | 1 | <16 | >0.9 | >160 | <6 |
| BTR New Energy Materials Inc. (China) | LTO-2 | 1 | <6 | >1.0 | >160 | <7 |
| BTR New Energy Materials Inc. (China) | LTO-S | 0.9 | <4 | >0.9 | >160 | <8 |

### Si-Containing Negative Electrode Materials

*Overview.* During the last two decades there has been an enormous amount of work on Si- and Sn-containing negative electrode materials for Li-ion batteries because Si and Sn offer significant volumetric and gravimetric energy advantages over graphite. However, at this point Sn-based negative electrodes have been abandoned by most researchers and manufacturers due to the high cost of Sn on a molar basis compared to Si. By contrast, Si-containing materials are now appearing in billions of Li-ion cells used in vehicles, phones, and other products. A thorough review by Obrovac and Chevrier[43] provides extensive information about various Si-alloys and other alloy materials that can be used as negative electrode materials in Li-ion cells.

Figure 17A.29, which shows the specific and volumetric capacities of various Li-metal alloys compared to graphite, illustrates why the switch to alloy-based negative electrode materials is attractive. However, all the alloys undergo large volume changes (up to 280%) when they react with lithium, which can cause mechanical failure, e.g., pulverization, as the material expands and contracts when cycled.

The electrochemical reaction between lithium and silicon has been extensively studied.[43] Beginning with crystalline silicon, during the first lithiation, amorphous $Li_xSi$ with $x$ near 3.5 is formed. If further lithium is added to the silicon, then $Li_{15}Si_4$ ($x = 3.75$) forms. When lithium is removed from the silicon, amorphous silicon is formed. Figure 17A.30 illustrates the reaction mechanism.

Al Magrabi et al.[44] carefully measured the voltage-capacity curve of silicon, beginning with sputtered amorphous Si (a-Si). As shown in Fig. 17A.31, Si provides exceptional reversible capacity, about 3580 mAh/g, which corresponds to $Li_{15}Si_4$. Also, the sputtered a-Si film has minimal IRC, nominally 2%. Also, as shown in Fig. 17A.31, the voltage during charge and discharge is offset, resulting in a hysteresis, which is a characteristic of Si.

The hysteresis between charge and discharge remains even if Si is lithiated and delithiated incredibly slowly. This is because the original Si-Si bonds in crystalline or amorphous Si are broken during the formation of $Li_xSi$ and only Li-Si and Li-Li bonds exist in $Li_{15}Si_4$. It is this bond breaking and reforming that leads to the inherent hysteresis. Intercalation compounds, such as graphite and LCO, do not show inherent hysteresis because the

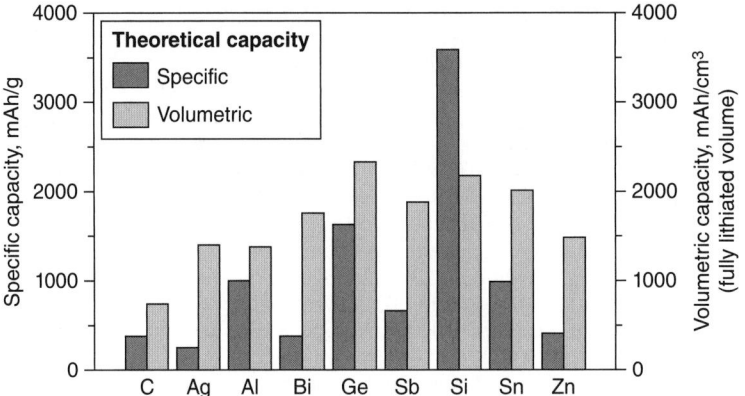

**FIGURE 17A.29** Specific and volumetric capacities of Li alloys compared to LiC$_6$. The volumetric capacities were determined based on the fully lithiated volume.

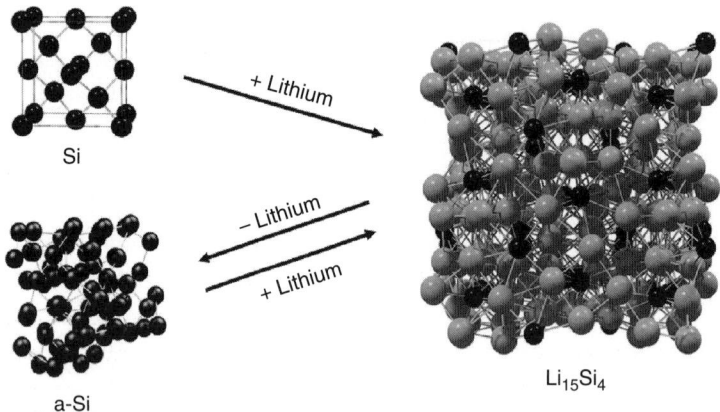

**FIGURE 17A.30** Reaction mechanism of silicon with lithium. After crystalline silicon (Si) is lithiated, subsequent cycles are between amorphous Si (a-Si) and Li$_{15}$Si$_4$. (*Courtesy of Mark Obrovac.*)

bonds in the host structure are not significantly altered during lithium intercalation. All Si-containing materials, including carbon encapsulated Si, nano-Si, Si alloys, and SiO, show this hysteresis between charge and discharge. While significant challenges exist in making commercial electrodes for Li-ion cells that have large silicon content,[43] SiO is particularly promising.

*SiO$_x$.* Silicon suboxide (SiO$_x$,where $x \sim 1$) is presently being used in Li-ion cells when blended with graphite in the negative electrode. Si-O is composed of nanosilicon grains in a matrix of SiO$_2$. According to patent applications by Shin-Etsu Corporation, SiO$_x$ can be made by vapor deposition from mixtures of silicon and SiO$_2$ powders heated to high temperatures. SiO has a theoretical reversible specific capacity near 1710 mAh/g. According to Al-Magrabi et al., the reactions of lithium with SiO are as follows[45]:

$$1\,\text{Li} + \text{SiO} \rightarrow 1/4\text{Li}_4\text{SiO}_4 + 3/4\,\text{Si (irreversible)} \qquad (17\text{A}.1)$$

followed by

$$45/16\,\text{Li} + 1/4\text{Li}_4\text{SiO}_4 + 3/4\text{Si} \leftrightarrow 1/4\text{Li}_4\text{SiO}_4 + 3/4\,\text{Li}_{3.75}\text{Si (reversible)} \qquad (17\text{A}.2)$$

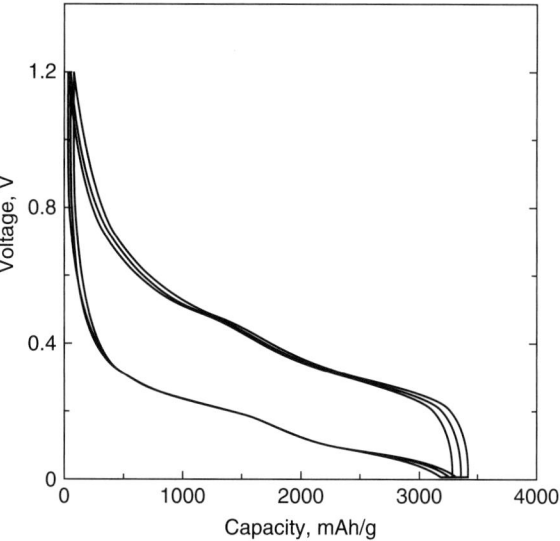

**FIGURE 17A.31** Voltage-specific capacity curve for the first three charge-discharge cycles of an a-Si sputtered film electrode illustrating the hysteresis characteristic of Si.[44] The reversible capacity is 3580 mAh/g and the irreversible capacity is 2%.

Reaction (17A.2) corresponds to Fig. 17A.30 in the presence of a $Li_4SiO_4$ matrix, which presumably helps to improve charge-discharge capacity retention. The $Li_4SiO_4$ is a product of Reaction (17A.1) and is unchanged in Reaction (17A.2).

Although $SiO_x$ has similar volumetric and specific capacity as the Si-alloy materials, it suffers from appreciable IRC loss (~30%) due to the formation of lithium silicates during initial lithiation, as shown by Reaction (17A.1).[46] Figure 17A.32 shows the specific capacity of Li/SiO cells.[47]

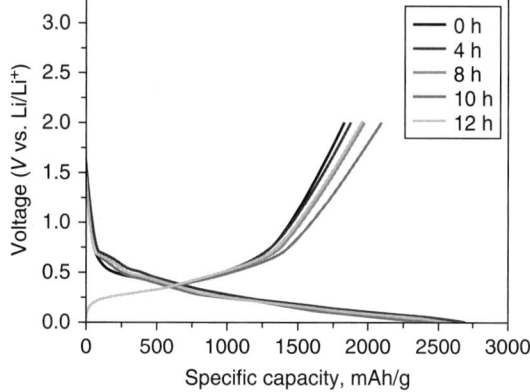

**FIGURE 17A.32** Voltage versus specific capacity for the first lithiation and delithiation of Li/SiO cells.[47]

The initial IRC can be compensated by reacting SiO with lithium at elevated temperature, thus causing Reaction (17A.1) to proceed outside the cell. Yom et al.[48] heated various SiO:Li molar ratios ranging from 7:1 to 12:1 in an effort to find the minimum IRC. After heat treatment, the materials were shown to be stable in both air and water. As shown in Fig. 17A.33, the initial coulombic efficiency was less than theoretical, suggesting that other irreversible processes, such as SEI formation or the formation of the Li passivation layer, are present.

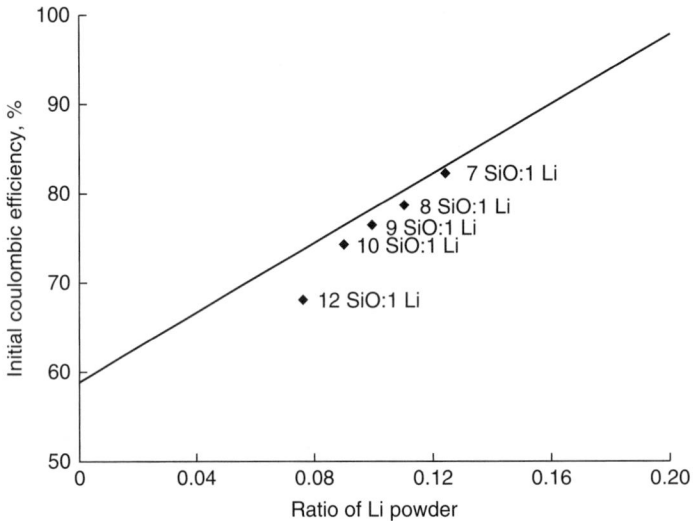

**FIGURE 17A.33** Initial coulombic efficiency for SiO reacted with various amounts of Li.[48] The line represents theoretical initial coulombic efficiency based on the content of Li powder. The material with the lowest irreversible capacity, 7 SiO:1 Li, had first-cycle capacities of 1220 mAh/g on discharge and 1001 mAh/g on charge, and thus an irreversible capacity of 17.9%. (*Reproduced with permission from J. Power Sources, **311**, 159–166 [2016].*)

Compared to other silicon-containing materials, SiO provides minimal particle pulverization and cracking during charge-discharge cycling. As shown in Fig. 17A.34, Chae et al.[46] have demonstrated that SiO can sustain over 400 full lithiation and delithiation cycles without cracking or fracturing, despite a reversible volume change of 117%.[43] This is an important feature of SiO that many other Si-containing materials do not provide.

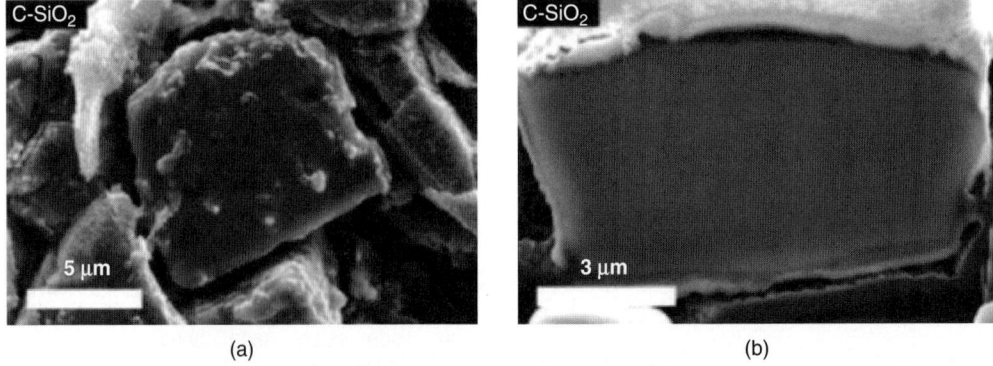

(a)                                                          (b)

**FIGURE 17A.34** SEM images of an SiO particle in the delithiated state and after 400 charge discharge cycles. In (*a*), an image of the electrode surface, a large gap between the particle and its neighbors is present due to the dramatic contraction during delithiation. The cross-section image of (*b*) shows no cracking or fracturing.[46]

Figure 17A.34 shows that graphite and SiO can be combined to provide effective electrodes containing between 5% and 25% by weight of SiO. The graphite helps maintain electrical contact to the SiO particles as they expand and contract, as shown in Fig. 17A.34a, and the particles themselves maintain their integrity, as shown in Fig. 17A.34b. A commercially available SiO-graphite composite, grade BSO-650 available from BTR New Materials Technology (www.btrchina.com), has about 79% by weight graphite and about 21% by weight SiO, and has a reversible capacity of 650 mAh/g and an IRC of about 11%. At this time, the charge-discharge cycle life of cells with graphite-SiO (or other graphite-Si material) composite negative electrodes is not as good as those with pure graphite negative electrodes. However, there are many applications, in particular EVs, where graphite-SiO (or another graphite-Si material) provides acceptable cycle life and improved energy density.

Amorphous Si is the active species in SiO after Reaction 17A.1 above (Li + SiO → 1/4Li$_4$SiO$_4$ + 3/4 Si) occurs during the first lithiation of SiO. Amorphous Si has a voltage-capacity profile, as shown in Fig. 17A.31, and has voltage hysteresis between lithiation and delithiation. Since graphite does not show this hysteresis, it means that during lithiation the graphite and the silicon voltage-capacity curves overlap in voltage and they lithiate more or less simultaneously, provided they were both fully delithiated. During delithiation, the voltage-capacity relations of Si and graphite do not overlap and graphite delithiates first, provided they were both fully lithiated.

Figure 17A.35 shows how a composite graphite-SiO negative electrode operates in a full Li-ion cell. The left panels show the charge of the Li-ion cell and demonstrates that the lithiation of the negative electrode components happens in the same voltage range. The right panels show the discharge of the Li-ion cell and demonstrate that graphite (lower voltage) will delithiate first, while the Li$_x$Si component of the SiO remains fully lithiated. This occurs due to the voltage hysteresis in the Si component. Provided that the Li-ion cell is only partially discharged, the Li$_x$Si component remains fully lithiated and volume change does not occur. Only during a deep discharge is the lithium removed from the Li$_x$Si.

Figure 17A.36 shows how this feature is used beneficially in an EV. Smart et al. describe the typical driving distances of EVs, shown in the top panel of Fig. 17A.36.[49] The bottom panel of this figure shows that only the graphite capacity of the composite electrode will be accessed for the majority of drives, and only when long trips are made will the silicon capacity be accessed. Therefore, in a typical EV, the silicon capacity will be rarely used

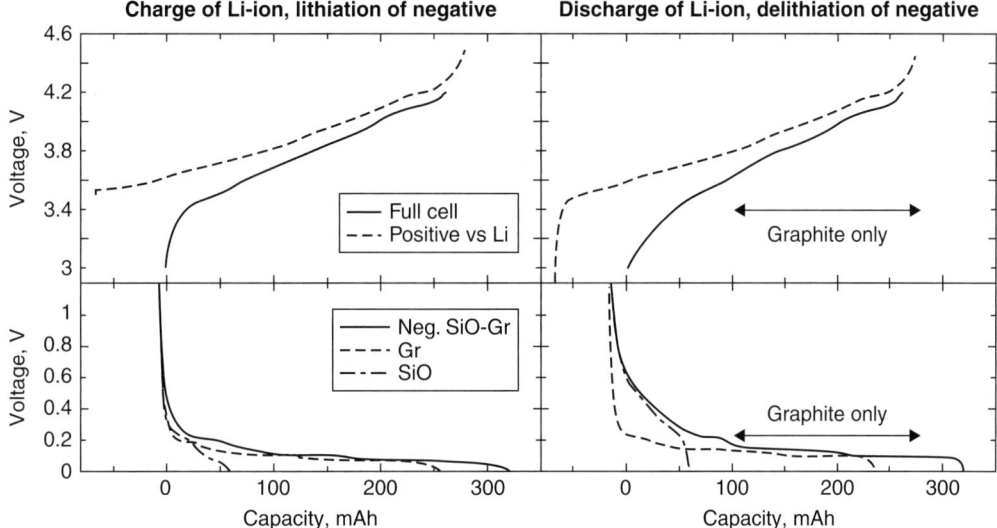

**FIGURE 17A.35**  Graphs of voltage versus capacity showing how a graphite-SiO composite electrode operates in a Li-ion cell. If the cell is fully charged and then partially discharged, only the graphite portion of the electrode is active, as indicated in the right panels. (*Courtesy of Alex Louli, Dalhousie University.*)

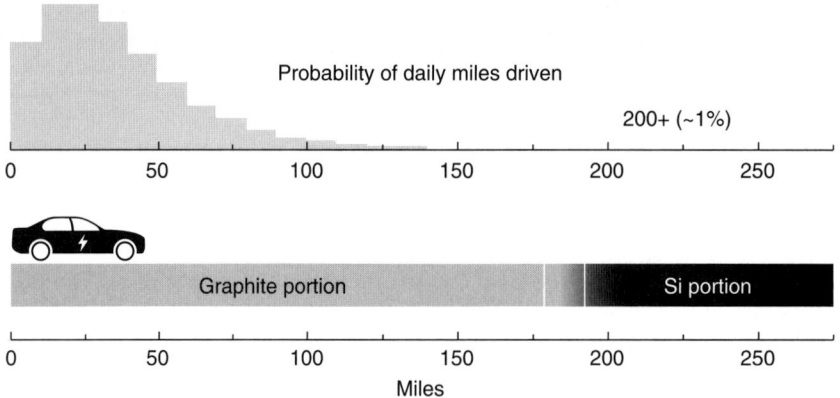

**FIGURE 17A.36**    The Si portion of a Li-ion cell with a graphite-Si composite negative electrode is only used during deep discharges (long trips) in an electric vehicle as shown in this figure. (*Courtesy of Vincent Chevrier.*)

and will function as a "range extender" and will only play a minor role in impacting cell lifetime. Li-ion cells used in some EVs incorporate SiO or other Si-containing materials to increase energy density, and thus increase driving range.

In the coming years, the fraction of Si-containing materials used in the negative electrodes of Li-ion cells is expected to increase. Hundred percent silicon negative electrodes are in development, such as the silicon nanowire materials produced by Amprius Technology.

### 17A.2.4  Nonaqueous Lithium Electrolytes

The vast majority of Li-ion batteries use liquid electrolytes, and only liquid electrolytes will be discussed here. Liquid electrolytes are solutions of a lithium salt in one or more organic solvents, typically carbonates. In a typical Li-ion battery employing a liquid electrolyte, the electrolyte is almost completely absorbed into the electrode and separator materials.

Most commercially used Li-ion electrolytes include $LiPF_6$ as the salt because $LiPF_6$ solutions offer high ionic conductivity, $10^{-2}$ S/cm, have a high lithium-ion transference number (~0.35), and provide acceptable safety properties. As reviewed below, many other salts have attracted industrial interest, notably $LiBF_4$, $LiN(CF_3SO_2)_2$, lithium bisoxalato borate (LiBOB), lithium bissulfonyl imide ($LiN(FSO_2)_2$,LiFSI), and $LiPO_2F_2$. Electrolytes in current use are formulated almost exclusively with carbonate solvents although ester solvents such as methyl acetate (MA) and methyl propionate (MP) are used in Li-ion cells designed for high-power, fast-charge, or low-temperature operations. Carbonates are aprotic, polar, and have a high dielectric constant; thus they can solvate lithium salts to high concentration (≥1 M). They also are compatible with cell electrode materials over a broad range of potential. Current formulations utilize ethylene carbonate (EC), dimethyl carbonate (DMC), ethyl methyl carbonate (EMC), and diethyl carbonate (DEC), and in some cases PC and fluoroethylene carbonate (FEC). The choice of solvents for a Li-ion electrolyte is influenced by any low-temperature, high-power, or fast-charge requirements of the application. Low-temperature, high-power, and fast-charge electrolytes utilize low-viscosity solutions with low freezing points.

*Salts.*    Salts commonly used in Li-ion cells are listed in Table 17A.7. $LiPF_6$ is the most common, because $LiPF_6$ solutions provide high conductivity and good safety properties. However, $LiPF_6$ is hygroscopic and the reaction with water yields hydrofluoric acid (HF), and therefore $LiPF_6$ must be handled in a dry environment. Organic salts have also been developed. They are more stable to water than $LiPF_6$, and thus easier to handle. In particular, lithium bistrifluoromethanesulfonimide ($LiN(CF_3SO_2)_2$) has received significant attention as an additive in carbonate-based electrolytes to provide improved high-temperature cell performance and reduced gas formation. New electrolyte salts introduced in the last decade include LiFSI and $LiPO_2F_2$. LiFSI provides increased electrolyte

**TABLE 17A.7**   Salts for Li-Ion Electrolytes

| Name | Formula | Mol. Weight (g/mol) | Typical impurities | Comments |
|---|---|---|---|---|
| Lithium hexafluorophosphate | $LiPF_6$ | 151.9 | $H_2O$, HF | Most common salt |
| Lithium tetrafluoroborate | $LiBF_4$ | 93.74 | $H_2O$, HF | Less hygroscopic than $LiPF_6$ |
| Lithium bisoxalatoborate | $LiB(C_2O_4)_2$ | 193.7 | $H_2O$ | Helps SEI formation |
| Lithium bistrifluoromethane sulfonimide | $LiN(CF_3SO_2)_2$ | 286.9 | $H_2O$ | Reduces gassing and improves high-temperature cycle life |
| Lithium difluorophosphate | $LiPO_2F_2$ | 107.9 | $H_2O$, HF | Additive to improve lifetime |
| Lithium bissulfonyl imide | $LiN(FSO_2)_2$ | 187 | $H_2O$ | Improves conductivity and life |

conductivity when mixed with $LiPF_6$, and improved cell lifetime in certain cases.[50] $LiPO_2F_2$ is also used as an additive to improve lifetime.[51] The solubility of $LiPO_2F_2$ in carbonate solvents is limited to about 0.1 M.

*Solvents.*   A wide variety of solvents, including carbonates, ethers, and acetates, have been evaluated for non-aqueous electrolytes. The industry has focused on the carbonates as they offer excellent stability, good safety properties, and compatibility, i.e., they are unreactive after SEI formation, with electrode materials. Fluorinated solvents such as FEC are used as cosolvents (over 10% of the electrolyte solution) in some commercial cells with Si-containing electrode materials and in cells designed for high voltage (>4.3 V) charge.[52] Markevic et al. have shown that FEC-based electrolyte solutions bring significant benefits over the most commonly used EC-based electrolyte solutions for different advanced lithium-ion battery systems, including those with Si anodes and high-voltage $LiCoPO_4$ and $LiNi_{0.5}Mn_{1.5}O_4$ cathodes. The benefits of FEC derive from the composition and properties of the SEI that is formed on the surface of anodes and cathodes as a result of the electrochemical reduction or oxidation of FEC. Esters such as MA and MP are used in cells designed for high-rate and low-temperature application. Neat carbonate solvents typically have an intrinsic solution conductivity of $<10^{-7}$ S/cm, dielectric constant $\geq 3$, and solvate lithium salts to a high concentration. Table 17A.8 presents the structure and properties of some commonly used solvents.

Electrolyte formulations in current Li-ion cells typically utilize three to five solvents and further include one or more additives. Additives are typically included in amounts of about 2% or less. Formulations with multiple solvents can provide better cell performance, higher conductivity, and a broader temperature range than is possible with a single-solvent electrolyte. For example, and as will be further explained below, EC is associated with low IRC and low capacity fade when used in conjunction with graphitic negative electrodes. EC is found in many commercial electrolyte formulations, but is a solid at room temperature. Multiple solvent formulations

**TABLE 17A.8**   Properties of Selected Solvents

| Characteristic | EC* | PC* | DMC* | EMC* | DEC* | FEC* | MA* | MP* |
|---|---|---|---|---|---|---|---|---|
| Structure | | | | | | | | |
| BP, °C | 248[53] | 242[53] | 90 | 109 | 126 | 212 | 57 | 80 |
| MP, °C | 36.4[53] | −48[53] | 4.6 | −55 | −74[54] | 20 | −98 | −88 |
| Density, g/mL | 1.32 | 1.20[55] | 1.07 | 1.0 | 0.97 | | 0.932 | 0.915 |
| Viscosity, cP | 1.86 (40°C) | 2.5 | 0.59 | 0.65 | 0.75 | | 0.38 | 0.43 |
| Dielectric constant | 89.6 (40°C) | 64.9[55] | 3.12 | 2.9 | 2.82 | | 6.7 | 6.2 |
| Donor number | 16.4 | 15 | 8.7[66] | 6.5[66] | 8[66] | | | |
| Mol. Wt. | 88.1 | 102.1 | 90.1 | 104.1 | 118.1 | 106.5 | 74.08 | 88.11 |

*EC = ethylene carbonate, PC = propylene carbonate, DMC = dimethyl carbonate, EMC = ethyl methyl carbonate, DEC = diethyl carbonate, FEC = fluoroethylene carbonate, MA = methyl acetate, MP = methyl propionate.

often include EC, thereby incorporating its desirable properties, while using other solvents to lower the freezing point and viscosity of the electrolyte. Further properties of electrolyte solvents and strategies for electrolyte formulation are included in several notable review articles.[56]

***Conductivity of Electrolytes.*** The conductivity of 1 M LiPF$_6$ solutions in solvents commonly used in Li-ion electrolytes is provided in Table 17A.9 and plotted in Fig. 17A.37 at temperatures ranging from –40 to 80°C. In general, these solutions offer high conductivity, 10$^{-2}$ S/cm, and a few solvents, such as PC and EMC, offer good conductivity at low temperature and a high boiling point.

**TABLE 17A.9**   Conductivity, in mS/cm, of 1 M LiPF$_6$ Solutions in Various Solvents

| Solvent[*] | –40.0°C | –20.0°C | 0.0°C | 20.0°C | 40.0°C | 60.0°C | 80.0°C |
|---|---|---|---|---|---|---|---|
| DEC | — | 1.4 | 2.1 | 2.9 | 3.6 | 4.3 | 4.9 |
| EMC | 1.1 | 2.2 | 3.2 | 4.3 | 5.2 | 6.2 | 7.1 |
| PC | 0.2 | 1.1 | 2.8 | 5.2 | 8.4 | 12.2 | 16.3 |
| DMC | — | 1.4 | 4.7 | 6.5 | 7.9 | 9.1 | 10.0 |
| EC | — | — | — | 6.9 | 10.6 | 15.5 | 20.6 |
| MA | 8.3 | 12.0 | 14.9 | 17.1 | 18.7 | 20.0 | — |
| MF | 15.8 | 20.8 | 25.0 | 28.3 | — | — | — |

[*]DEC = diethyl carbonate, EMC = ethyl methyl carbonate, PC = propylene carbonate, DMC = dimethyl carbonate, EC = ethylene carbonate, MA = methyl acetate, MF = methyl formate.
*Source*: Courtesy of Merck KGaA, Darmstadt, Germany.

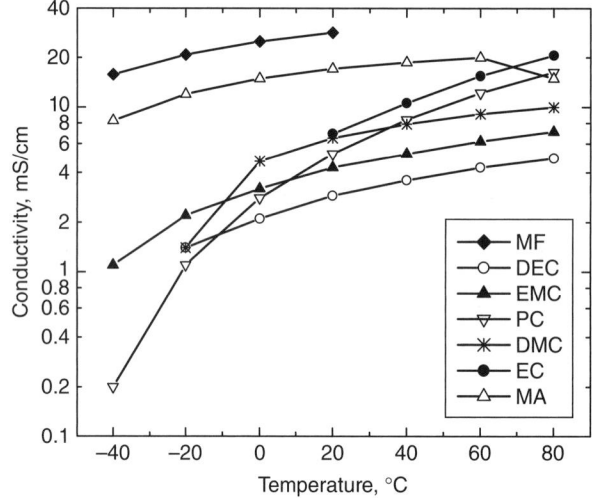

**FIGURE 17A.37**   Conductivity of 1 M LiPF$_6$ solutions in various solvents. (*Courtesy of Merck KGaA, Darmstadt, Germany.*)

The conductivity of binary 1:1 mixtures of EC with common Li-ion electrolyte solvents over a range of salt concentrations and temperatures is given in Table 17A.10. For many solvent pairs, conductivity is highest with 1 M LiPF$_6$, and these formulations are liquid from –40 to 80°C. The conductivity of 1 M LiPF$_6$ binary mixtures with EC is plotted in Fig. 17A.38. As shown, the EC:MA mixture offers the highest conductivity, although such levels of MA are often associated with high capacity fade.[57] Recent work has shown that esters such as MP can provide improved lifetime.[58] Also, Li et al. have shown that Li-ion cells with the MA cosolvent can have long lifetime and support high-rate charge.[59] Other mixtures, including EC:DEC, EC:DMC, and EC:EMC, offer suitable conductivity and lower capacity fade. In particular, EC:EMC mixtures offer 0.9 mS/cm conductivity at –40°C and low capacity fade.

**TABLE 17A.10**    Conductivity, in mS/cm, of LiPF$_6$ Solutions in Binary Mixtures, 1:1 by Weight

| Solvents* | Concentration | −40°C | −20°C | 0°C | 20°C | 40°C | 60°C | 80°C |
|---|---|---|---|---|---|---|---|---|
| | 0.25 M | — | — | 1.7 (C) | 4.2 | 5.8 | 7.3 | 8.8 |
| | 0.50 M | — | 2.5 (C) | 3.0 | 6.4 | 8.7 | 11.1 | 13.6 |
| EC:DEC | 1.00 M | 0.7 | 2.2 | 4.2 | 7.0 | 10.3 | 13.9 | 17.5 |
| | 1.25 M | 0.4 | 1.7 | 3.6 | 6.4 | 9.7 | 13.5 | 17.4 |
| | 1.50 M | — | — | — | 5.6 | — | — | — |
| | 1.75 M | — | — | — | 4.8 (S) | — | — | — |
| | 0.25 M | — | — | 4.2 | 5.8 | 7.8 | 9.7 | 11.5 |
| | 0.50 M | — | — | 6.5 | 9.3 | 12.8 | 16.0 | 19.1 |
| | 0.75 M | — | 3.8 | 6.9 | 10.3 | 14.0 | 17.9 | 21.6 |
| | 1.00 M | — | 3.7 | 7.0 | 15.0 | 19.5 | 24.0 | 24.0 |
| EC:DMC | 1.25 M | 0.7 | 2.7 | 5.6 | 9.3 | 13.7 | 18.4 | 23.3 |
| | 1.50 M | — | 2.2 | 5.4 | 9.3 | 14.1 | 19.2 | 24.7 |
| | 1.75 M | — | — | — | 7.5 | — | — | — |
| | 2.00 M | — | — | — | 6.7 | — | — | — |
| | 2.25 M | — | — | — | 0.9 (S) | — | — | — |
| | 0.25 M | — | — | 3.7 | 5.3 | 7.2 | 9.1 | 10.9 |
| | 0.50 M | — | 3.0 | 5.1 | 7.5 | 10.2 | 12.8 | 15.4 |
| EC:EMC | 1.00 M | 0.9 | 2.7 | 5.3 | 8.5 | 12.2 | 16.3 | 20.3 |
| | 1.25 M | 0.6 | 2.3 | 4.7 | 8.0 | 12.0 | 16.2 | 20.6 |
| | 3.50 M | — | — | — | 0.9 (S) | — | — | — |
| | 0.25 M | 2.4 (C) | 4.6 | 6.3 | 8.3 | 10.4 | 12.4 | — |
| | 0.50 M | 3.1 (C) | 6.7 | 9.8 | 13.1 | 16.0 | 19.3 | — |
| EC:MA | 1.00 M | 3.8 | 7.8 | 12.2 | 17.1 | 22.3 | 27.3 | — |
| | 1.25 M | — | 7.1 | 11.8 | 17.2 | 22.7 | 28.4 | — |
| | 3.0 M | — | 0.5 | 2.1 | 5.2 | — | 15.4 | 21.8 |
| | 3.5 M | — | — | — | 3.4 (S) | — | — | — |
| EC:PC | 1.00 M | C | 1.5 | 3.6 | 6.3 | 9.5 | 12.9 | |

*C = partially crystallized, DEC = diethyl carbonate, EMC = ethyl methyl carbonate, DMC = dimethyl carbonate, EC = ethylene carbonate, MA = methyl acetate, PC = propylene carbonate, S = saturated.
*Source*: Courtesy of Merck KGaA, Darmstadt, Germany.

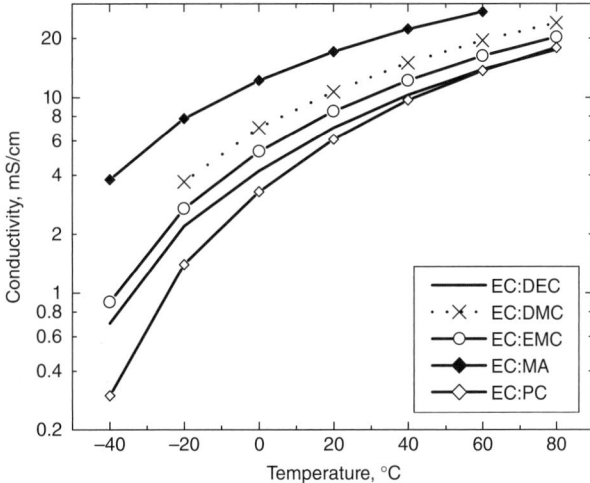

**FIGURE 17A.38**    The conductivity of 1 M LiPF$_6$ in binary solvent mixtures, 1:1 by weight. (*Courtesy of Merck KGaA, Darmstadt, Germany.*)

**TABLE 17A.11**    Conductivity, in mS/cm, of 1 M LiPF$_6$ in Ternary Mixtures

| Solvent[*] | Wt. ratio | −40°C | −20°C | 0°C | 20°C | 40°C | 60°C | 80°C |
|---|---|---|---|---|---|---|---|---|
| EC:PC:DMC | 20:20:60 | — | — | 6.9 | 10.6 | 14.5 | 18.4 | 22.2 |
| EC:PC:EA | 15:25:60 | 3 | 6.2 | 9.8 | 13.7 | 17.8 | 21.6 | 25.1 |
| EC:PC:EMC | 15:25:60 | 1 | 2.8 | 5.3 | 8.1 | 11.5 | 14.6 | 17.8 |
| EC:PC:MA | 15:25:60 | 4.1 | 8.1 | 12.9 | 17.8 | 22.8 | 27.6 | boils |
| EC:PC:MPC | 15:25:60 | 0.5 | 1.4 | 3.3 | 5.6 | 8.2 | 10.9 | 13.9 |
| EC:DMC:EMC | 15:25:60 | 1.4 | 3.2 | 5.3 | 7.6 | 10 | 12.1 | 14.1 |
| EC:DMC:MPC | 15:25:60 | 0.7 | 1.8 | 3.4 | 5.3 | 7.2 | 9 | 10.9 |

[*]DEC = diethyl carbonate, EMC = ethyl methyl carbonate, PC = propylene carbonate, DMC = dimethyl carbonate, EC = ethylene carbonate, MA = methyl acetate, MPC = methyl propyl carbonate, EA = ethyl acetate.
*Source*: Courtesy of Merck KGaA, Darmstadt, Germany.

The conductivity of 1 M LiPF$_6$ in selected ternary solvent mixtures is provided in Table 17A.11 and plotted in Fig. 17A.39. These mixtures contain 33% EC, as is common in Li-ion electrolytes, and provide high conductivity and a broad temperature range, illustrating the utility of multiple component mixtures. For example, four of these mixtures provide at least 1 mS/cm at −40°C, of which three are liquid at 80°C. Quaternary solvent mixtures have also been developed to provide electrolytes with better low-temperature performance. The conductivity of 1 M LiPF$_6$ in various quaternary mixtures is shown in Fig. 17A.40. As illustrated, these solutions offer over 1 mS/cm at −40°C and up to 0.6 mS/cm at −60°C.

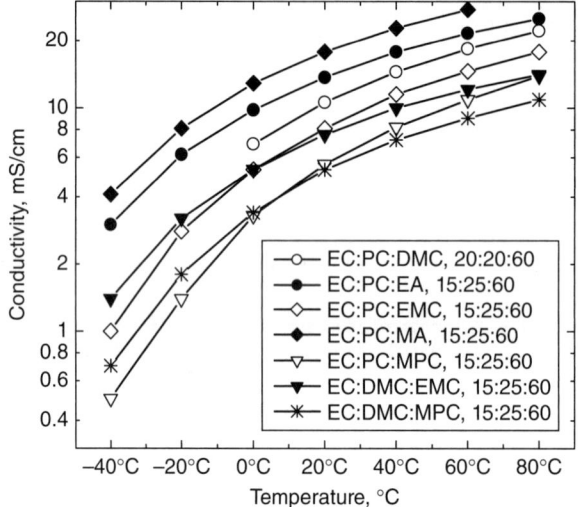

**FIGURE 17A.39**    The conductivity of 1 M LiPF$_6$ in ternary solvent mixtures. (*Courtesy of Merck KGaA, Darmstadt, Germany.*)

A recent development in electrolyte design is the "Advanced Electrolyte Model (AEM)" developed by Kevin Gering of Idaho National Laboratory.[60] The commercially available AEM software[61] allows key properties of most common Li-ion battery electrolytes to be calculated accurately, reducing the need for physical measurements. The results from the AEM have been validated by third parties. Figures 17A.41 and 17A.42 show the measured conductivities and viscosities of LiPF$_6$:EC:DMC:MA solutions and AEM predictions, respectively.[62] The AEM output is perfectly suited for physics-based models of Li-ion cells.

***Electrolyte Formulation, Irreversible Capacity, and the SEI.***    For lithium-ion cells to function, the electrolyte must be stable at both the anodic and cathodic potentials found in Li-ion cells, which are about 0 V to about

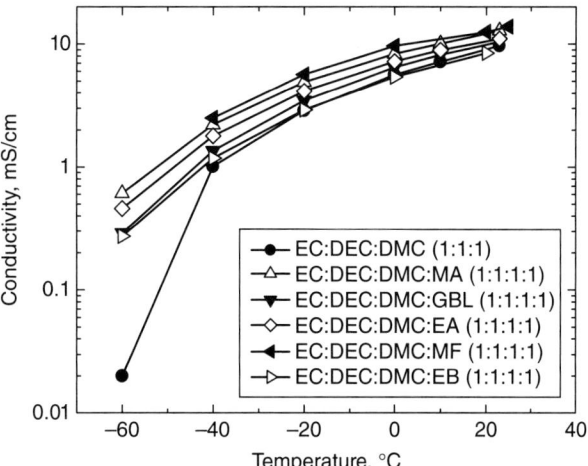

**FIGURE 17A.40**   The conductivity of 1 M $LiPF_6$ in quaternary solvent mixtures. (*Courtesy of Merck KGaA, Darmstadt, Germany.*)

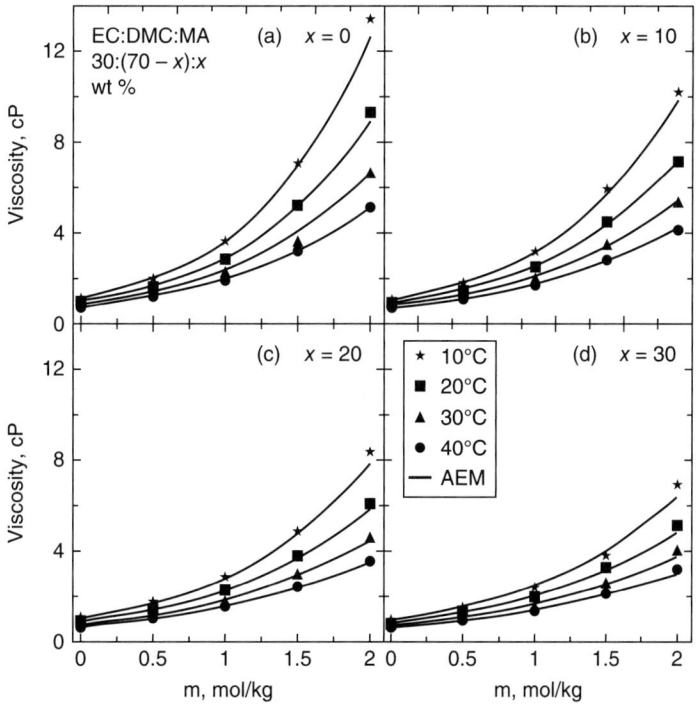

**FIGURE 17A.41**   Viscosity as a function of $LiPF_6$ concentration for electrolytes with solvent composition EC:DMC:MA 30:(70 − x):x, with (*a*) x = 0, (*b*) x = 10, (*c*) x = 20, (*d*) x = 30 for temperatures between 10 and 40°C. Symbols represent the result of physical measurements. Solid lines are predictions from the AEM.

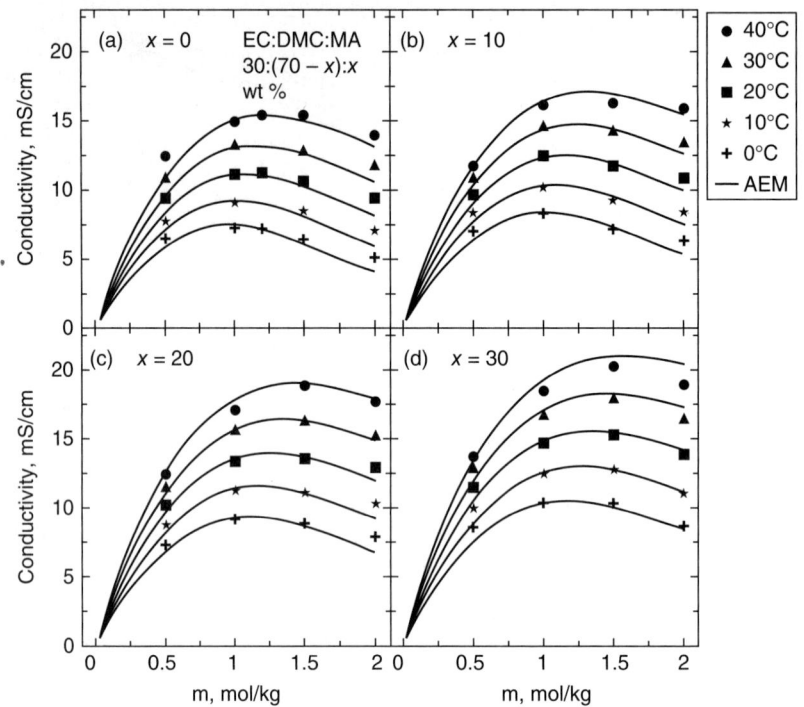

**FIGURE 17A.42** Conductivity as a function of $LiPF_6$ concentration for electrolytes with the solvent composition EC:EMC:MA $30:(70 - x):x$, with $0 \leq x \leq 30$ for temperatures between 0 and 40°C. Symbols represent the result of physical measurements. Solid lines are predictions from the AEM.

4.4 V, or even 4.5 V, versus Li. No practical solvents are thermodynamically stable with lithium or $Li_xC_6$ near 0 V versus Li, but many solvents undergo a limited reaction to form a passivation film, termed the SEI, on the electrode surface, and thus are kinetically stable. The resulting SEI spatially separates the electrolyte from the electrode particles, yet is ionically conductive, and thus allows passage of lithium ions. The SEI provides extrinsic stability to the system, allowing the fabrication of cells that are stable for years without significant degradation.[57]

When the SEI is formed, lithium is incorporated into the passivation film. This process is irreversible and is observed as a loss of capacity, primarily on a cell's first cycle. The amount of IRC is dependent on the electrolyte formulation and the electrode materials, in particular the type of carbon (and/or silicon-containing material) used in the negative electrode. Because the reaction occurs at the surface of a particle, materials with low specific surface area typically offer lower IRC.

A poorly formed SEI layer can result in capacity fade and/or poor capacity retention due to the consumption of lithium by continuous reaction with the solvent at the electrolyte surface. Such cells also can suffer from high impedance, due to electrolyte depletion, and thick deposits of electrolyte decomposition products. Under normal circumstances, even a stable SEI layer can take many cycles to form completely. In order to establish a stable SEI layer on the negative electrode surface, commercial cells typically undergo a regime of charging, floating, and discharging, sometimes at elevated temperatures. Minimizing the time of this cell "formation" step is essential in high-volume manufacturing.

Cells with electrolyte formulations that contain alkyl carbonates, in particular EC, have been shown to offer lower capacity fade, low IRC, and high capacity.[63] In EC-containing electrolytes, the passivation film formed on the surface of Li-ion electrodes is formed with a minimum amount of lithium. This SEI has been shown to consist primarily of $Li_2(OCO_2(CH_2)_2OCO_2)_2$,[64] and related reaction products, such as $Li_2CO_3$ and $LiOCH_3$,[65] which are derived from the electrolyte solvent and a lithiated species such as $Li_xC_6$ or $Li_xSi$. Many solvents, such as acetonitrile (AN) and ethyl acetate (EA) do not form stable passivation layers on graphite if 1 M $LiPF_6$ electrolyte salt is used. However, Yamada et al. have shown that highly concentrated solutions of $LiPF_6$ in AN (e.g., 4 M solutions) do form stable passivating layers.[66] Petibon showed a similar phenomenon for electrolytes based on EA.[67] Graphite can be cycled in other solvents that do not form a stable passivation film if an additive, such as a crown ether[68] or $CO_2$,[69] is added to the electrolyte. The use of additives can significantly change the chemistry of the SEI layer and enhance its ability to protect electrode surfaces from reaction with the electrolytes. The use of additives is discussed in detail in the next section.

## 17A.2.5  Electrolyte Additives

Undesired reactions in Li-ion batteries, which lead to capacity loss, can occur at either the positive or the negative electrode. For example, as discussed in the previous section, the formation and repair of the SEI consumes lithium at the negative electrode, resulting in IRC loss.[70] Electrolyte oxidation at the positive electrode,[71] and the presence of impurities such as water or HF, can also lead to capacity loss. These reactions can also result in the production of gas, which is particularly serious in prismatic or pouch cell designs, where cell bulging and the loss of stack pressure can result. The use of high-purity electrolytes, electrode coatings,[72] and special electrode materials[73] can be used to improve cycle life and reduce gassing. In particular, the use of various electrolyte additives in the form of organic molecules, salts, inorganic compounds, or gases[74] can significantly enhance performance by improving SEI stability and scavenging HF or water.[75] Additives can also significantly improve cell safety characteristics, as will be discussed below. Xu describes many aspects of Li-ion electrolytes and additives.[76]

Additives are typically present in electrolyte formulations at less than about 5% by weight, often 1% to 2% by weight. After cell formation by the manufacturer, many additives are completely consumed in the formation of the SEI layer and are difficult or impossible to detect in commercially available cells. Petibon et al. tracked the consumption of vinylene carbonate (VC) during the formation procedure and showed that if 1% VC is initially incorporated in the electrolyte, it is undetectable after formation.[77] Because additives are difficult to detect in commercial products and thus can be retained as a trade secret, the identity and amount of additives used is highly proprietary among the commercial cell makers. Nevertheless, a small number of additives are known to be in common use. VC was found to be highly beneficial, and this additive is used today in many commercial Li-ion cells. Aurbach et al.[78] report that VC is a reactive additive that reacts on both the anode and the cathode surfaces. The influence of this additive on the behavior of Li-graphite anodes is very positive since it improves the cyclability of the Li-graphite anode, especially at elevated temperatures, and reduces the IRC. Spectroscopic studies indicate that VC polymerizes on the lithiated graphite surfaces, forming polyalkyl Li-carbonate species that suppress both solvent and salt anion reduction. Broussely et al.[70] demonstrate the effectiveness of VC as an additive to promote the formation of a stable SEI that consumes little lithium after many cycles.

FEC is also a common additive found in many Li-ion cells. Merkevich et al.[52] show that FEC is important not only as an additive but also as a primary solvent in Li-ion cells with advanced cathodes and anodes.

Patoux et al.[79] describe the use of a number of additives, including 1,3-propane sultone, to reduce the self-discharge of cells with the 4.7 V positive electrode $LiNi_{0.5}Mn_{1.5}O_4$. In an excellent review, Zhang[80] describes hundreds of additives that can be used to stabilize the SEI on graphite, protect the positive electrode, remove HF, and mitigate the effects of overcharge. El-Ouatani et al.[81] describe the impact of VC in $LiCoO_2$/graphite, $LiFePO_4$/graphite, and $LiCoO_2/Li_{4/3}Ti_{5/3}O_4$ cells.

Extensive systematic studies of electrolyte additives and their impact on the lifetime of Li-ion cells are available. Some major studies involving dozens of additives have been described by Wang et al.[82] and Burns et al.[83] Wang et al. demonstrate the utility of additives such as ethylene sulfate, methylene methane disulfonate, and propene sultone in NMC-based Li-ion cells.[84] Also, Yang et al. demonstrate the utility of $LiPO_2F_2$ as an electrolyte additive.[51]

In addition to electrolyte additives that ultimately modify surface films, coatings on electrode particles can directly modify the surfaces in contact with the electrolyte. Commercial LCO, NMC, and NCA positive electrode materials are now routinely supplied with coatings.[85] A common coating is $Al_2O_3$, which has been shown to be beneficial in many literature studies. Arumugam et al. show that $Al_2O_3$ coatings can be applied by wet chemical, dry chemical, and atomic layer deposition (ALD).[86] $Al_2O_3$ coatings prepared by ALD have been carefully characterized and Li-ion cell performance has been shown to be enhanced.[87] Patoux et al.[73] describe the impact of a number of coatings on the capacity retention of the 4.7 V spinel $LiNi_{0.5}Mn_{1.5}O_4$.

Biphenyl is an additive that has been used in commercial cells as an overcharge protection additive. When exposed to high potential in an overcharged cell, biphenyl polymerizes, significantly decreasing the ionic conductivity of the electrolyte and effectively shutting down the cell. Other additives known to be in common use include FEC, succinonitrile, and lithium bistrifluoromethanesulfonimide. Xu et al. provide a review of electrolyte additives.[88]

There can be no doubt that electrolyte additives and electrode material coatings are beneficial and that Li-ion battery makers use them. However, because many additives are consumed during cell formation and because manufacturers do not disclose the additives they use, it is very difficult to determine what additives or coatings are employed within a particular Li-ion cell.

## 17A.2.6    Separator Materials

The separator in a battery electrically isolates the positive and negative electrodes, while allowing ion transport between them. Most commercially available liquid electrolyte Li-ion cells use a thin (10–25 µm) microporous polyolefin separator because polyolefins provide excellent mechanical properties and chemical stability at an acceptable cost. Nonwoven polyolefin materials have also been developed and are now being used.[89]

Requirements for Li-ion separators include:

- High machine direction strength to permit automated winding.
- High yield strength to prevent shrinkage, e.g., at elevated temperature.
- Resistant to puncture by electrode materials.
- Effective pore size of less than 1 µm to prevent penetration of the separator by electrode materials.
- Easily wetted by electrolyte to facilitate acceptance of the electrolyte during cell assembly.
- Compatible and stable in contact with electrolyte and electrode materials.

Microporous polyolefin separators in current use include single-layer PE or PP materials, and trilayer laminates of PE and PP. Also available are surfactant coated materials designed to offer improved electrolyte wetting. Figures 17A.43 and 17A.44 show top and cross-section views of PE, PP, and PP/PE/PP trilayer separators.

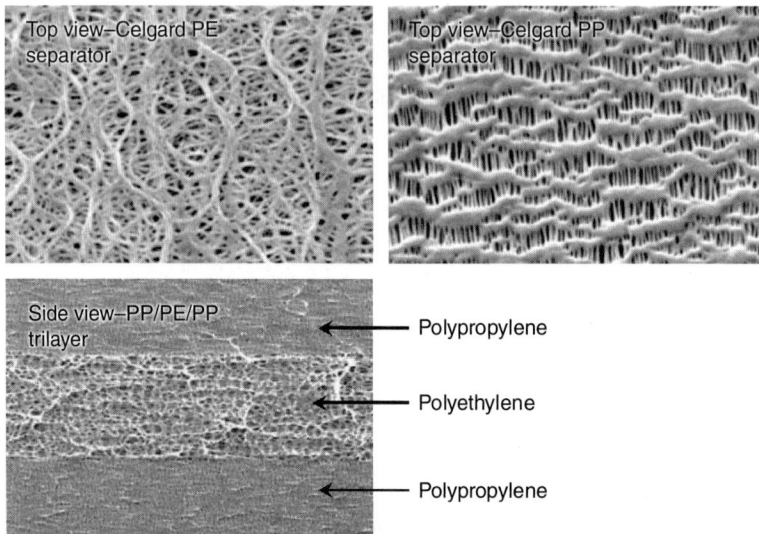

**FIGURE 17A.43**    SEM micrographs showing top views of polyethylene and polypropylene separator (top left and right, respectively), and a cross section view of a Celgard PP/PE/PP trilayer separator (bottom). (*Courtesy Celgard.*)

**Single layer PE**   **Single layer PP**   **Trilayer PP/PE/PP**

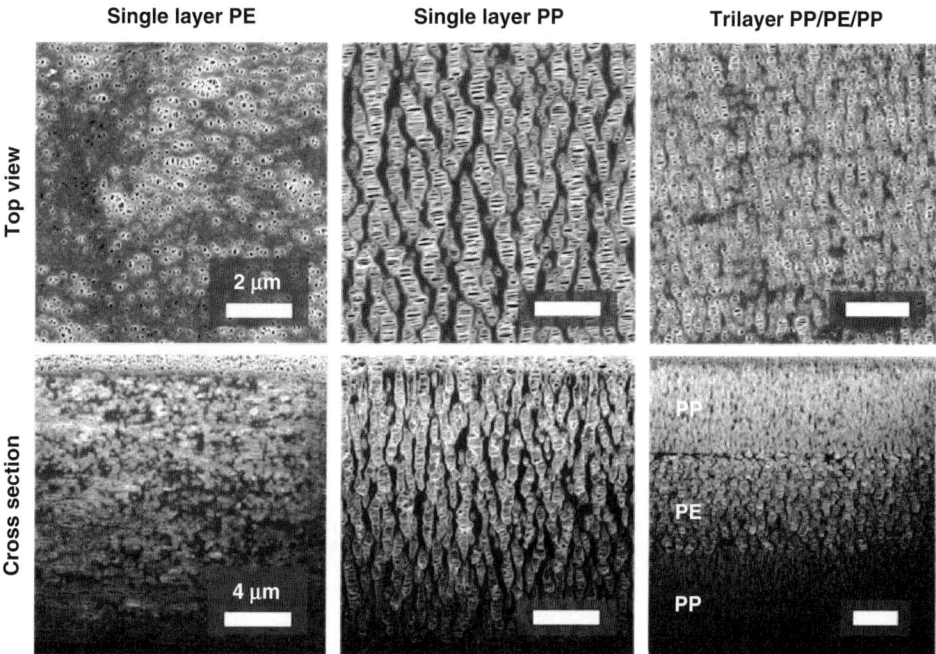

**FIGURE 17A.44**   Top view and cross-section SEM images of (left) single layer polyethylene (Celgard® K1640), (middle) single layer polypropylene (Celgard® PP1615), and (right) trilayer PP/PE/PP (Celgard® 2325) separators. (*Courtesy Marie Francine Lagadec and Vanessa Wood, ETH Zürich.*)

As shown in these figures, the pore structure of the PE, PP, and PP/PE/PP separators is distinct in both the in-plane and through-plane directions. These materials are fabricated by either a dry, extrusion-type process with uniaxial stretching, or by a wet, solvent-based process with biaxial stretching.[90] These processes have been the industry standard for more than four decades. While PE separators can be fabricated using both types of processes, PP is almost exclusively dry processed. Nonwoven polyolefin membranes, which are produced by bonding randomly oriented fibers using dry or wet chemical and mechanical methods,[91] are also used.[89]

The properties of commercial separators, including pore dimensions, porosity, and air permeability (e.g., Gurley number), have been reported[92] and are available on the manufacturers' Web sites. Commercial materials offer a pore size of 0.03 to 0.1 μm, and 30% to 55% porosity. Three-dimensional analysis has shown that commercially available materials have different microstructures; for example, different tortuosity and different contents of dead-end pores.[93] Figure 17A.45 provides an example of three-dimensional renderings of a PE and a PP separator.

The low melting point of polyolefins enables their use as a thermal fuse to stop ion transport and provide "shutdown" properties. If the temperature of a cell locally approaches the melting point of the polymer, 135°C for PE and 155°C for PP, porosity is lost when the polymer melts and the pores close, stopping the transport of Li$^+$ ions across the separator.[95] Preferably, after "shutdown," the separator is dimensionally stable and does not shrink. For improved dimensional stability after shutdown, trilayer materials (e.g., PP/PE/PP) include PP layers that are designed to maintain the integrity of the film, while the lower melting point of the PE layer is intended to shut down the cell if an overtemperature condition occurs. Such multicomponent separators have improved Li-ion battery safety. Ultimately, what is needed is a separator that has a shutdown component and a second component that does not melt at any temperature.[96] Ceramic-coated polyolefin separators have been designed for this purpose and are widely used in large Li-ion cells (>10 Ah).

Ceramic-coated separators are available from many manufacturers and generally have a thin (several micrometers thick) coated layer of a refractory metal oxide (e.g., $Al_2O_3$ or AlOOH) on one or both surfaces of the separator, as shown in Fig. 17A.46. These layers provide improved wetting properties and hinder or prevent dimensional changes of the separator when a Li-ion cell is thermally, electrically, or mechanically abused. There are numerous

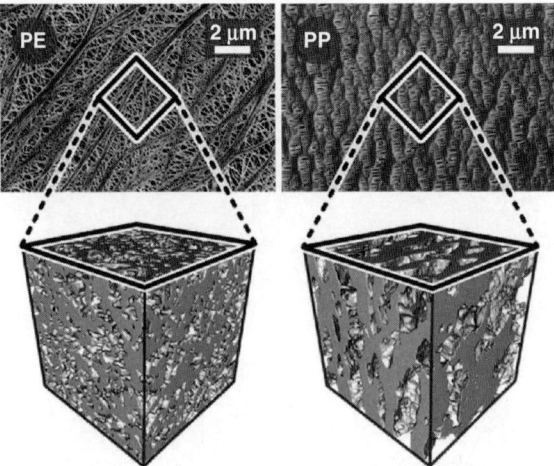

**FIGURE 17A.45**   SEM images and 3D renderings of (*left*) monolayer PE (Targray PE16A) and (*right*) monolayer PP (Celgard® PP1615) separators. The 3D renderings are based on open-source focused ion beam SEM tomography.[94] (*Courtesy Marie Francine Lagadec and Vanessa Wood, ETH Zürich.*)

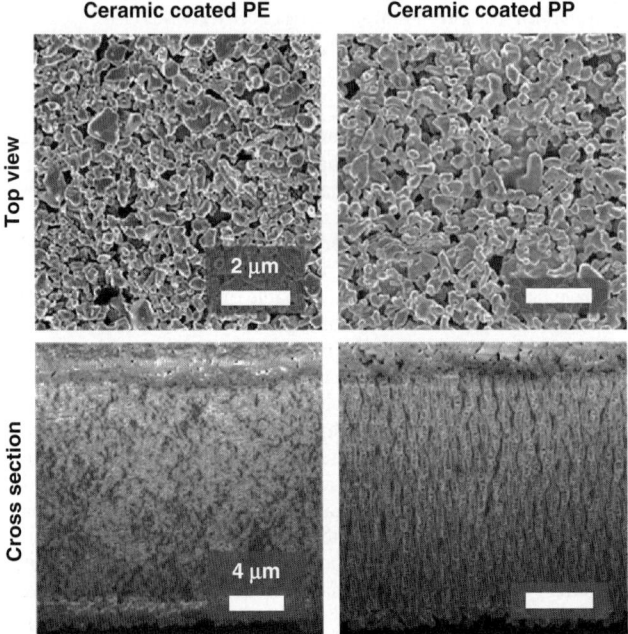

**FIGURE 17A.46**   Top and cross-sectional views of ceramic coated polyethylene and polypropylene separators having total thicknesses of 20 μm and 16 μm, respectively. (*Courtesy Marie Francine Lagadec and Vanessa Wood, ETH Zürich.*)

strategies for incorporating the ceramic particles on or within the separator. For example, some separators have the ceramic mixed within the polymer layers and still provide shutdown and dimensional stability during melting. Weber et al. have published reviews of separator materials including lists of major suppliers.[97]

An additional advantage of ceramic-coated separators is that the ceramic coating can be placed adjacent to the positive electrode in high-voltage Li-ion cells to prevent electric contact between the positive electrode and the polymeric separator. This limits electrochemical oxidation and damage to the polyolefin.[98]

## 17A.3   CONSTRUCTION

Li-ion batteries are presently being mass produced by more than 100 manufacturers worldwide. Cylindrical cells have a wound configuration. For prismatic cells, two configurations are practiced commercially: stacked, true-prismatic configurations and flat-mandrel-wound pseudoprismatic or "rolled electrode" configurations. Prismatic cells may have a laminate construction, to provide "pouch" or "polymer" cells.

Because Li-ion cells are fabricated in the discharged state, they must be charged before use. At least the first charge cycle is completed by the manufacturer, a process called "formation." During formation, lithium ions from the positive electrode material transport to the negative electrode, some of which are consumed to form a surface layer known as the SEI. For niche applications, cells may be cycled a few times and the cell capacity and voltage profile can be monitored for quality control. One method of testing includes discharging cells to about 30% capacity and then storing them for several weeks while the open-circuit voltage is monitored. This procedure allows cells having a high self-discharge rate to be identified. Self-discharge rates are normally less than 2%.

An important parameter in lithium ion cell design is the capacity ratio of the positive and negative electrodes, more specifically the ratio of the first-cycle charge capacity over the first-cycle discharge capacity of the positive electrode and the same ratio for the negative electrode. Ideally, the positive electrode's first-cycle capacity ratio is designed to match the negative electrode's first-cycle capacity ratio. If the positive electrode ratio exceeds the negative electrode ratio, lithium metal electroplating can occur, which can result in undesirable capacity fading and safety problems. If the negative electrode ratio exceeds the positive electrode ratio, the cell's reversible capacity is limited by the amount of lithium provided by the positive electrode and reduced by IRC associated with the negative electrode.

### 17A.3.1   Construction of Wound Cylindrical Li-Ion Cells

The construction of a wound cylindrical Li-ion cell is illustrated in Fig. 17A.47. The construction includes positive and negative electrodes separated by a 12 to 20 μm thick microporous separator, as described in Sec. 17A.2.6.

Positive electrodes commonly include a 10 to 20 μm Al foil current collector coated with the active material (on both sides) to a total thickness of typically 100 to 150 μm. Negative electrodes commonly include an 8 to 10 μm Cu foil current collector coated with a graphite-type active material to a total thickness of 100 to 150 μm. For applications where Cu is not suitable, e.g., 0 V applications, a different negative electrode current collector material, such as titanium or stainless steel, may be used.

Electrode active material coatings in energy cells (designed for high energy density) are typically about 70 to 100 μm thick on each side of the foil and densified, e.g., by calendaring. In energy cells, a single tab at the end of the wind is used to connect the current collectors to their respective terminals. The coating thickness is limited by the ability of the coating to be handled during the manufacturing process. This is especially true for wound cells, where coatings with excessive thickness can fracture when wound.

In cells designed for power applications, the coatings can be intentionally porous and thinner to accommodate the relatively low conductivity of nonaqueous electrolytes (about 10 mS/cm), and relatively slow $Li^+$ diffusion kinetics in the positive and negative electrode materials (about $10^{-9}$ $cm^2$/s). Typical coating thicknesses for power cells are about 50 μm on each side of the foil. Multiple tabs spaced along the current collector are a common feature in power cells.

The can, often used as the negative terminal, is typically nickel-plated steel. Alternatively, when the can is used as the positive terminal, the can is typically aluminum. Most commercially available cells utilize a header

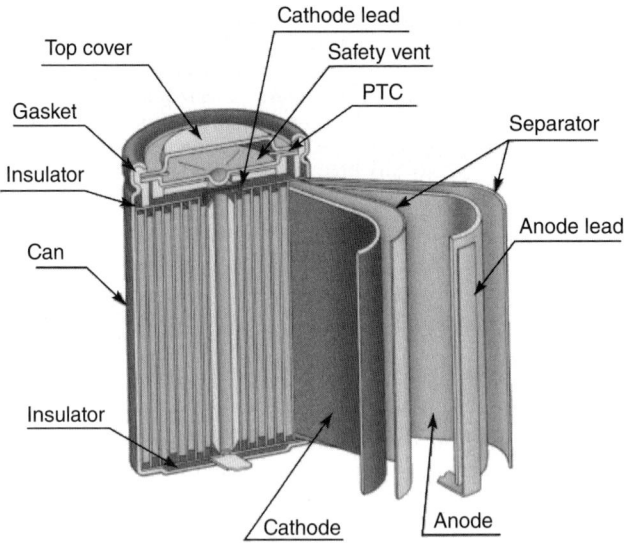

**FIGURE 17A.47** Schematic diagram of a wound cylindrical Li-ion cell. (*Courtesy of the University of South Carolina; Reproduced with permission from J. Power Sources.*)

that incorporates one or more disconnect devices, which are activated by temperature, pressure, or both, such as a positive temperature coefficient (PTC) device, a current interrupt device (CID), or a safety vent (see Chaps. 23 and 32B). A PTC device provides high electrical resistance at high temperatures, causing the material to melt and break the circuit. A CID is activated when pressure inside the cell causes a deformation that mechanically opens the circuit. A design incorporating a header having a PTC device and a scored burst disk is illustrated in Fig. 17A.48. The header-can seal is formed through a crimp. Some manufacturers, such as Panasonic, utilize a double crimp header design that incorporates an internally crimped seal within an outer crimp formed with an upper portion of the can. A vent, if incorporated into the header, may be designed to vent at about 2.5 to 3 megapascals (MPa), and a burst disk, if present, may be designed to vent at about 6 MPa. Most commercially available cells include both a vent and a burst disk.

Bottom vents or burst disks are also common. Figure 17A.49*a* shows a bottom view of a Sony US18650VC7 cell and an illustration of the bottom vent before and after rupture. The venting region in the Sony cell is engraved and the inside of the can is scored. In this cell, the bottom vent operates at about 3.5 MPa. Figure 17A.49*b*

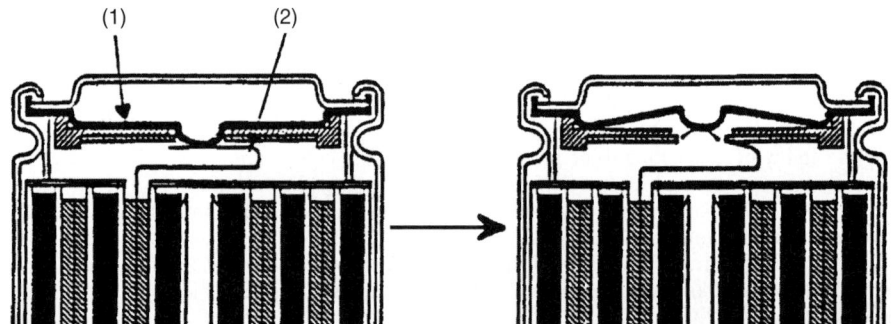

**FIGURE 17A.48** Detail of the construction of a cell header with a breaker and vent mechanism which can be activated by an abnormal rise in internal pressure, (1) Aluminum burst disk, (2) Aluminum lead. (*Courtesy of Sony Corp.*)

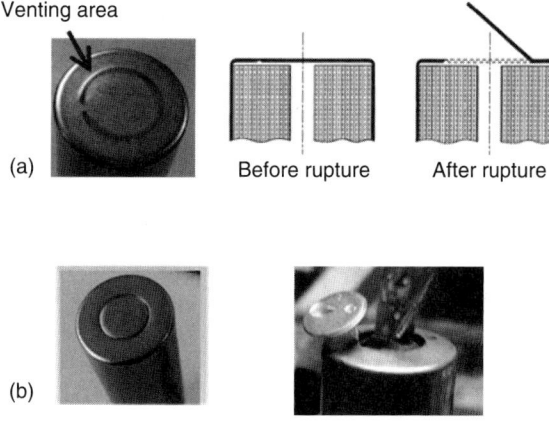

Venting area

(a)          Before rupture       After rupture

(b)

**FIGURE 17A.49**   (*a*) Bottom of a Sony US18650VC7 cell and illustration of the bottom vent before and after rupture, and (*b*) bottom of an LG cell showing an alternative design before and after operation. (*Courtesy Eric Darcy, NASA-JSC.*)

shows an alternative design in which the bottom of the can, having a nominal thickness of 0.3 mm, is scored through 90% of its thickness to leave a 30 μm thick score that defines the burst disk. In operation, this alternative design separates completely, operating at about 2 MPa.

## 17A.3.2   Construction of Wound Prismatic Li-Ion Cells

The construction of a wound prismatic cell is similar to the cylindrical configuration, the most notable exception being that a flat mandrel is used instead of a cylindrical mandrel. A schematic illustration of a wound prismatic cell is shown in Fig. 17A.50. A spring plate can be included between the can and the electrode winding to provide stack pressure to the electrodes.

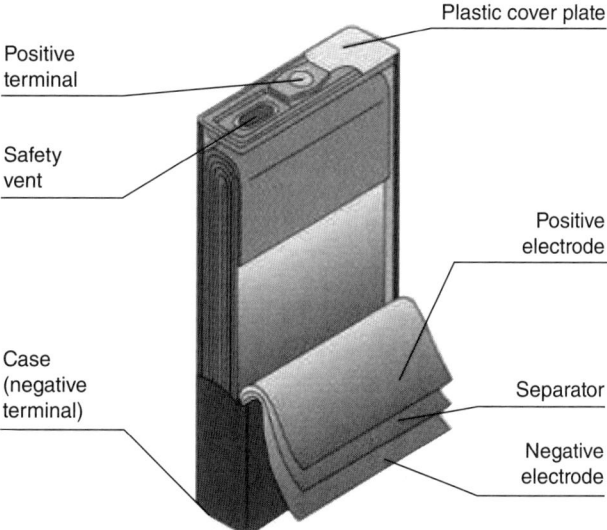

Positive terminal

Plastic cover plate

Safety vent

Positive electrode

Case (negative terminal)

Separator

Negative electrode

**FIGURE 17A.50**   Schematic illustration of a wound prismatic cell. (*Courtesy of Japan Storage Battery Co., Ltd.*)

### 17A.3.3    Construction of Stacked Prismatic Li-Ion Batteries

The construction of a stacked prismatic cell is illustrated in Fig. 17A.51. As in a wound cell, a microporous separator separates the positive and negative electrodes. Typically, each plate in the cell has a tab, and the tabs are bundled and welded to their respective terminals or to the cell case. Cell cases of aluminum, nickel-plated steel, or 304 L stainless steel have been used. As shown, the cover typically incorporates one or two terminals, a fill port, and a rupture disk. The terminal may be a glass-to-metal seal, although for low-cost applications, compression-type seals have been used. The terminal may incorporate devices similar to those found in the header of cylindrical products to provide pressure, temperature, and overcurrent interrupt in a single component. The case-to-cover seal is typically formed either by tungsten inert gas (TIG) or laser welding.

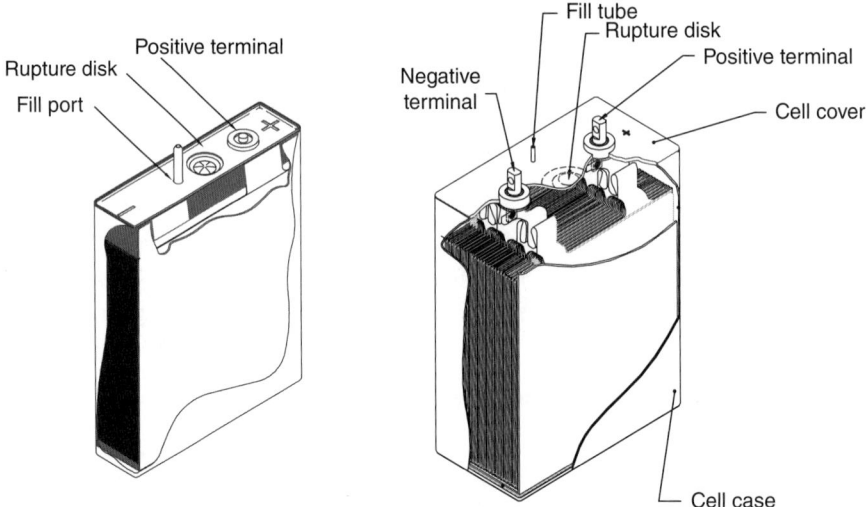

**FIGURE 17A.51**    Schematic view showing the header and electrodes of a 7 Ah case negative and a 7 Ah case negative and a 40 Ah case neutral flat plate prismatic Li-ion cell. (*Courtesy of Yardney Technical Products, Inc.*)

Prismatic batteries are attractive in retrofit and volume-sensitive applications because the dimensions of the battery can be selected to efficiently use the available space. For example, prismatic lithium-ion batteries have been used to replace NiCd and lead-acid batteries in aircraft and marine applications. Prismatic lithium-ion batteries have also been used in extraterrestrial applications, such as the 28 V, 10 Ah prismatic batteries developed for Mars Exploration Rover (MER), and the 2.6 kWh prismatic batteries developed for the Mars Science Laboratory (MSL) Curiosity. MER Opportunity landed on Mars on January 25, 2004 and remains active in 2018, exceeding its operating plan by over 13 years (in Earth time). MSL Curiosity landed on August 6, 2012 and also remains active in 2018.

### 17A.3.4    Construction of Polymer Li-Ion Cells

Polymer Li-ion cells, also called "pouch" or "LiPo" cells, use a flexible aluminized laminate film as the cell case material. The laminate construction allows for a thin and lightweight configuration suitable for high-power applications, but because of the lack of rigidity of the laminate envelope, the cells are susceptible to swelling. The laminate packaging is heat-sealable. Commercially available laminates include PP/Al foil/PP and polyethylene terephthalate (PET)/nylon/Al/PP laminates. The aluminum is treated for corrosion resistance, e.g., by chromate treatment or by passivating the surface with a ceramic coating.

Figure 17A.52 illustrates the construction of a polymer Li-ion cell using a flat-mandrel pseudoprismatic rolled electrode. The rolled electrode has flat tabs attached to each of the positive and negative electrodes, and the tabs protrude from the packaging after the laminate case is vacuum sealed. The packaging can be sealed using heat or

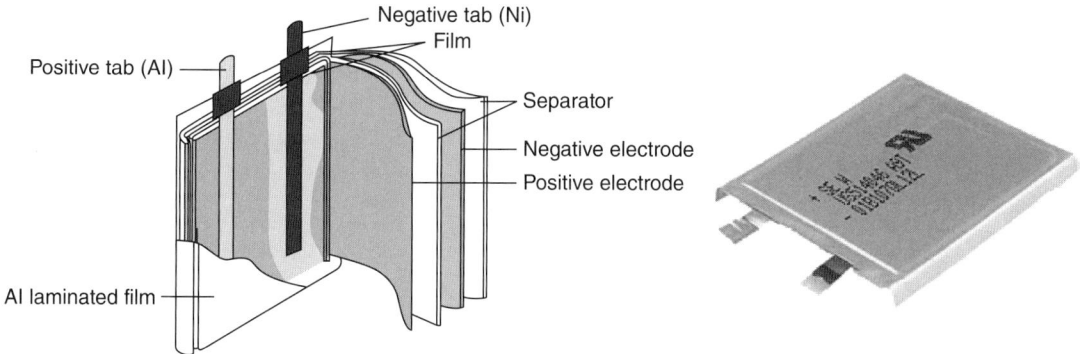

**FIGURE 17A.52**    Schematic diagram and photo of a polymer Li-ion cell. (*Schematic Courtesy BYD.*)

ultrasonic welding. The electrolyte can be added to an evacuated case, and after the cell is sealed by thermal or ultrasonic welding, exterior air pressure then applies about 95 kPa of pressure to the electrode stack.

A flow diagram illustrating the construction of a lithium-ion polymer cell is provided in Fig. 17A.53. The top row describes the electrode coating, slitting, and winding steps to produce a rolled electrode. The middle row shows the steps of cutting the aluminized film, forming the pocket for the rolled electrode, and then leaving excess film that is used to create an extra lobe to collect excess electrolyte or gas that is generated on the first charge cycle. The electrode roll is inserted, and the sides of the pouch sealed. After the electrolyte is added, the oversize pouch is vacuum sealed. After charging, the cells are compressed between flat plates and the extra lobe is removed. Some manufactures compress the cells at an elevated temperature for a few hours.

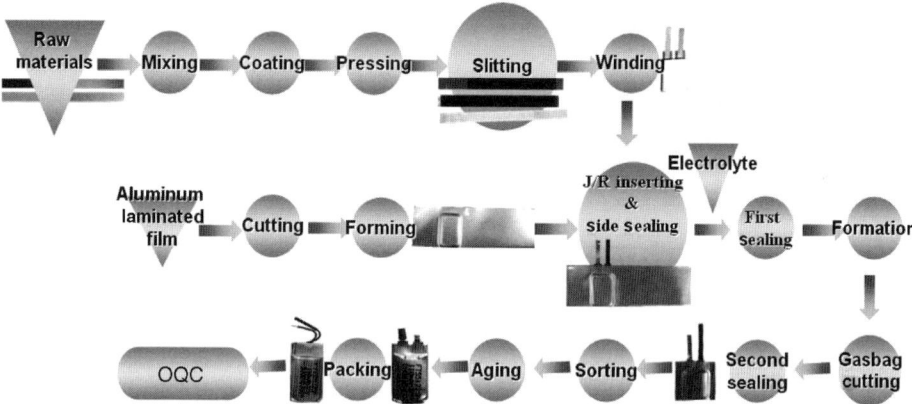

**FIGURE 17A.53**    Li-ion polymer battery construction process. (*Courtesy Hangzhou Future Power Technology Ltd.*)

## 17A.4   *CHARACTERISTICS AND PERFORMANCE*

The general performance characteristics of Li-ion batteries are outlined in Table 17A.12. As indicated in the table, the Li-ion chemistry provides an operating voltage of about 2.5 to 4.2 V, which is approximately three times that of NiCd or NiMH, and thus fewer cells are required for a battery of a given voltage. Li-ion batteries also provide high specific energy and energy density. Small (18650) cells having a specific energy of 275 Wh/kg and energy density of 735 Wh/L are commercially available. Also, Li-ion batteries designed for power applications offer

**TABLE 17A.12**    General Performance Characteristics of Li-Ion Cells (Cylindrical, Prismatic, and "Polymer") Using Common Cell Chemistries

| Characteristic | LiCoO$_2$/graphite NMC/ graphite NCA/graphite Energy cells | NMC/graphite LMO/graphite Power cells | LiFePO$_4$/graphite Power cells | LMO/Li$_{4/3}$ Ti$_{5/3}$O$_4$ |
|---|---|---|---|---|
| Voltage range, V | 2.5–4.2 typical 2.5–4.4 for some cells | 2.5–4.2 | 2.5–3.6 | 2.8–1.5 |
| Avg. voltage | 3.7-3.85 | 3.7 | 3.3 | 2.3 |
| Specific energy, Wh/kg | 175–275 cylindrical 140–240 polymer | 100–175 | 60–110 | 70 |
| Energy density, Wh/L | 400–735 cylindrical 370–600 polymer | 250–460 | 125–250 | 120 |
| Continuous rate capability | ~3$C$ | Over 30$C$ | 10 to 125$C$ | 10$C$ |
| Pulse-rate capability, $C$ | 5 | Over 100 | Up to 250 | 20 |
| Cycle life at 100% DOD | 500+ | 500+ | 1000+ | 4000+ |
| Calendar life, year | >5 | >5 | >5 | >5 |
| Self-discharge rate, %/month | 1–5 %/mo | 1–5 %/mo | 1–5 %/mo | 1–5 %/mo |
| Charge temperature, °C | 0 to ~45 | 0 to ~45 | 0 to ~45 | −20 to ~45 |
| Discharge temperature, °C | −20 to +60 | −30 to +60 | −30 to +60 | −30 to +60 |
| Memory effect | None | None | None | None |
| Power density, W/L (pulse) | ~2000 | ~10,000 | ~10,000 | ~2000 |
| Specific power, W/kg (pulse) | ~1000 | ~4000 | ~4000 | ~1100 |

NMC = LiNi$_{1/3}$Mn$_{1/3}$Co$_{1/3}$O$_2$, LiNi$_{0.5}$Mn$_{0.3}$Co$_{0.2}$O$_2$ or LiNi$_{0.42}$Mn$_{0.42}$Co$_{0.16}$, etc.
NCA = LiNi$_{0.8}$Co$_{0.15}$Al$_{0.05}$O$_2$, etc.
LMO = Li$_{1+x}$Mn$_{2-x}$O$_4$, etc.

high-rate capability. Many small (e.g., 18650) wound cells provide 20$C$ discharge capability, and more specialized power cells can provide 100$C$ discharge. Also, specialized prismatic and polymer cells can provide 60$C$ continuous and 200$C$ pulse discharge. Further, Li-ion cells provide a low self-discharge rate (about 2% per month at 20°C), years of calendar life, no memory effect, and a broad temperature range of operation. Li-ion batteries can be charged from 0 to 45°C and discharged from −40 to 60°C. This combination of properties within a variety of cost-effective and hermetic packages has enabled the diverse applicability of this technology.

Lithium-ion batteries continue to improve at a rapid pace. Consider, for example, that the "original" 1991 Sony 18650 cell provided 700 mAh and 90 Wh/kg, nominally 20% of the capacity and 33% of the energy of Sony's current US18650VC7 cell.

## 17A.4.1    Characteristics of Li-Ion Cells

As illustrated in Table 17A.13, Li-ion cells are available in cylindrical and prismatic configurations. Manufacturers' Web sites can be consulted for the latest specifications, and Table 17A.13 was current as of January 1, 2018. Although the cylindrical cells in this table are often characterized as "energy cells" or "power cells," many manufacturers offer a range of cells. The most common sizes for cylindrical cells are the 18650 and 26650 sizes, although other sizes (e.g., 14500) are available. Numerous Chinese battery manufacturers offer a huge range of polymer Li-ion battery sizes; for example, Harbin Coslight Power and BYD each offer over 100 Li-ion polymer cell sizes.

Cylindrical cells having the highest specific energy and energy density are now being charged to voltages as high as 4.4 V instead of the typical 4.2 V for LiCoO$_2$, NMC, and NCA positive electrodes. Each of these electrode materials delivers more specific capacity when charged to 4.4 V, and even more if charged to 4.6 V. Manufacturers have determined how to obtain acceptable cycle life for LCO cells charged to 4.4 V through a combination of electrolyte additives,[99] electrode material treatments, and coatings. Increased voltage is the easiest way to increase the energy density of a Li-ion battery; therefore, it is likely that further increases in upper charging voltage will occur over the next years. This will lead to energy increases in power cells as well.

***Impedance Characteristics.***    Power is frequently a factor when determining how many or what kind of cells are required for a given application. The power capability of a battery is determined by its cell voltage and its impedance.

**TABLE 17A.13** Specifications of Selected Commercially Available Li-Ion Cells

| Manufacturer | Cell | Chemistry | Capacity, Ah | Max current, A | Mass, g | Specific energy, Wh/kg | Energy density, Wh/L | DC IR, mΩ | AC IR, mΩ |
|---|---|---|---|---|---|---|---|---|---|
| **Energy cells** | | | | | | | | | |
| LG Chem | INR18650 MJ1 | NMC | 3.41 | 10A | 47 | 266 | 720 | 33 | |
| LG Chem | ICP103450A1 | | 2.0 | 4A | 41 | | | | |
| Samsung | INR18650-35E | | 3.5 | | 46 | 276 | 733 | 35 | |
| Sony | US18650VC7 | | 3.5 | | 47.4 | 269 | 735 | 31 | |
| Panasonic | NCR18650GA | | 3.34 | | 47 | 259 | 704 | 38 | |
| Kokam | SLPB065070180 | NMC | 12 | 48A | 170 | 257 | 480 | | 2.8 |
| **Power cells** | | | | | | | | | |
| LG Chem | INR18650HG2 | | 2.9 | | | 228 | 620 | | |
| Moli Energy | INR18650A | NCA | 2.5 | 20A | 45 | 200 | 530 | 20 | 12 |
| Moli Energy | IHR18650C | NMC | 2.0 | 20A | 45 | 160 | 425 | 20 | <15 |
| Kokam | SLPB8043128H | NMC | 3.2 | 128A | 80 | 141 | 263 | | 5 |
| Kokam | SLPB130255255N | LTO | 65 | 600A | 1760 | 77 | 149 | | 0.4 |

The impedance of batteries is complex; it can vary with the time of discharge (or charge), current, state of charge, and temperature. Thus, when sizing a battery for a desired application, it is important to specify the operating temperature, length of time the power will be needed, and minimum (or maximum) voltage cutoff.

Batteries typically show an Arrhenius dependence of impedance on temperature, i.e., a plot of log $Z$ (real) versus $1/T$ absolute is linear. There may be a discontinuity at the freezing point of the electrolyte, which is typically between –20 and –40°C, although lithium-ion cells with freezing points below –20°C have been demonstrated for specialty applications. At the freezing point of the electrolyte, the impedance dramatically increases.

The impedance of batteries increases with increasing time of passing current. The increased impedance arises at short times from the double-layer capacitance at the surface of the electrodes, and at long times from diffusion of lithium ions within the electrolyte and from lithium within the intercalation materials that serve as the positive and negative electrodes. In some chemistries, impedance strongly changes with state of charge, and thus never reaches a constant value. In other chemistries, the impedance may "saturate" at some relatively constant value after current has been passed for a duration of seconds to minutes, with the time to saturation depending on the design of the battery. The dependence of impedance on time is often characterized using AC impedance, often represented in a Nyquist plot, which is a graph of imaginary resistance versus real resistance. A schematic Nyquist plot is shown in Fig. 17A.54.

The shape of the profile in the Nyquist plot contains three features that relate to three components of impedance in batteries: the intercept with the real axis is the ohmic component that arises from resistance to electron and ion migration, the semicircle that arises from charge-transfer at the interfacial surface between the electrolyte and the electrode materials, and the tail at low frequencies that arises from diffusion of lithium ions in the electrolyte and from lithium within the negative and positive electrode materials. These three components of impedance have different dependencies on current. Ohmic impedance is independent of current. Charge-transfer impedance can **decrease** with increasing current at high current (a phenomenon known as Tafel kinetics). Diffusion impedance **increases** with increasing current. Also, the components of impedance all have different temperature dependencies, and thus the dependence of impedance on current may be different at different temperatures. Typically, the dependence of impedance on current is small above 0°C, where ohmic components may dominate cell impedance, and impedance may vary substantially with current at low temperature. Also, the direction of the dependence will depend on whether charge transfer or diffusion dominates the impedance of the particular cell design.

## 17A.4.2 Performance of Commercial Li-Ion Cells

As shown in Tables 17.12 and 17.13, there are numerous electrode material choices and cell construction formats available for Li-ion cells. To a first approximation, there is not a large difference between the performance

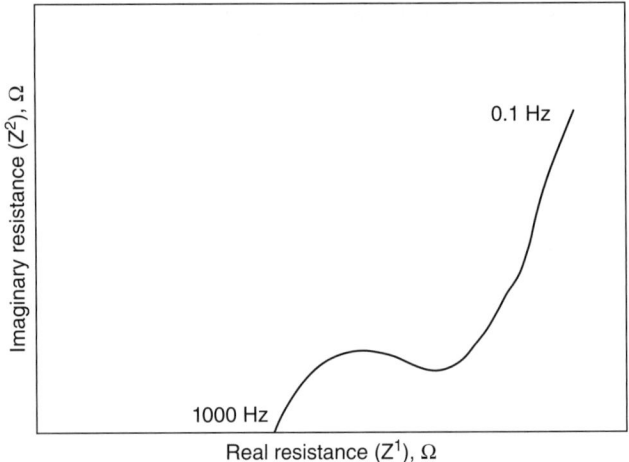

**FIGURE 17A.54** A schematic Nyquist plot for a lithium-ion battery. Frequency (inverse of time) decreases from left to right. (*Courtesy of Karen Thomas Alyea.*)

characteristics of a particular chemistry in cylindrical, prismatic, or polymer Li-ion formats. Therefore, this section will not provide examples of every possible cell configuration and instead, selected examples, which normally focus on 18650-size cells, will be provided. The following chemistries will be discussed: $LiCoO_2/$ graphite, $LiCoO_2$/hard carbon, NCA/carbon, NMC/carbon, NCO/MCMB, and $LiFePO_4$/carbon. Chemistries which have fallen out of favor in commercial cells, such as $LiMn_2O_4$, are reviewed in past editions of this book.

***LiCoO₂/Graphite.*** Figure 17A.55 shows discharge curves for an E-One Moli ICR-18650M cell. As shown, this cell provided 2.85Ah, or about 225 Wh/kg, and a voltage of about 3.8 V at 50% depth of discharge (DOD) at a $C/10$ (0.25A) rate at 20°C. At a $1C$ rate the capacity was similar, 2.77 Ah (−3%), although the energy delivered was reduced to 205 Wh/kg (−9%) because of the reduction in the cell voltage, which was 2.55 V at 50% DOD

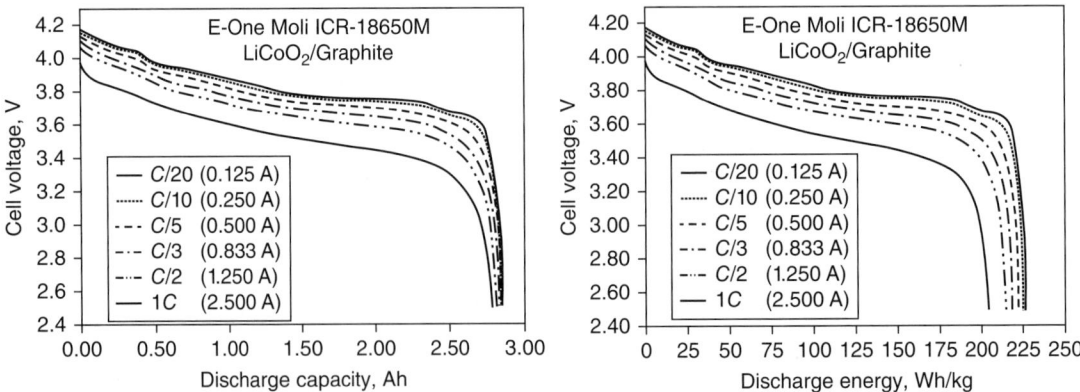

**FIGURE 17A.55** Cell voltage versus discharge capacity (Ah) and discharge energy (Wh/kg) for an E-One Moli ICR-18650M cell at 20°C. The cell was charged at 0.25 A ($C/10$) to 4.2 V and then held at 4.2 V until the current tapered to 0.025 A ($C/100$). The cell was then discharged at $C/20$, $C/10$, $C/5$, $C/2$, and at 1C to 2.5 V. (*Courtesy Marshall Smart, NASA/JPL-Caltech.*)

at 1*C*. The relatively flat shape of the discharge curve is characteristic of the LiCoO$_2$ positive electrode material and graphite negative electrode material used in this cell. Note the cell voltage of 3.7 V at 95% DOD.

The temperature performance of the E-One Moli ICR-18650M cell is shown in Fig. 17A.56, which shows the results of *C*/5 (0.5 A) discharge at temperatures ranging from −20°C to +30°C. As shown, the discharge capacity is little changed between 30 and 0°C (4% change), and the delivered capacity at −20°C was 2.5 Ah, or 87% of the capacity at 20°C. Also, the cell voltage was little changed between 30 and 0°C. At −10°C the voltage at 50% DOD was 3.6 V, and at −20°C the voltage at 50% DOD was 3.4 V, i.e., 0.35 V less than at 20°C.

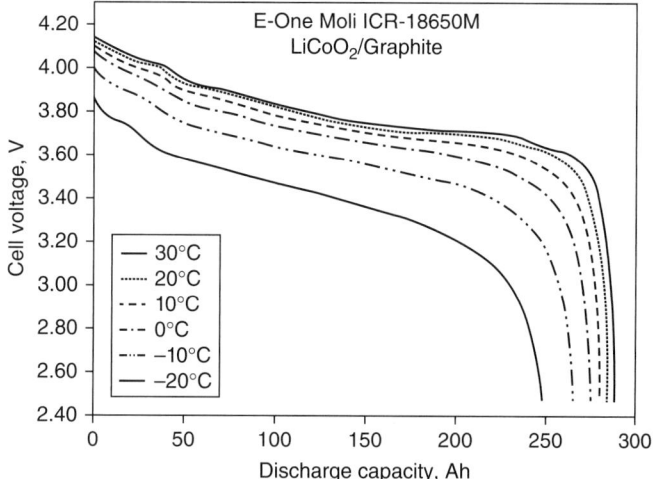

**FIGURE 17A.56**    Cell voltage versus discharge capacity (Ah) for an E-One Moli ICR-18650M cell at −20°C, −10°C, 0°C, 10°C, 20°C, and 30°C. The cell was charged at 0.25 A (*C*/10) to 4.2 V and then held at 4.2 V until the current tapered to 0.025 A (*C*/100). The cell was then discharged at *C*/5 (0.500 A) to 2.5 V. (*Courtesy Marshall Smart, NASA/JPL-Caltech.*)

***LiCoO$_2$/Hard Carbon.***    Figure 17A.57 shows discharge curves for a Sony HCM-18650 cell at 20°C. This cell provided 1.45 Ah, or about 130 Wh/kg, when discharged at a *C*/5 rate at 20°C, i.e., about half the specific energy of the LiCoO$_2$/graphite cell, and a voltage of about 3.9 V at 50% DOD at *C*/5. At a 1*C* rate the capacity was 1.3 Ah (−10% relative to *C*/10), and the energy density was 110 Wh/kg (−15%) because of the reduction in the cell voltage, which was 3.6 V at 50% DOD at 1*C*. The sloped shape of the discharge curve is characteristic of the hard carbon negative electrode material used in this cell. Note the cell voltage of 2.6 V at 95% DOD, compared to the 3.7 V provided by the LiCoO$_2$/graphite cell at 95% DOD.

The temperature performance of the Sony HCM-18650 cell is shown in Fig. 17A.58, which shows the results of *C*/5 (0.3 A) discharge at temperatures ranging from −20 to +25°C. As shown, the discharge capacity decreased 10% between 20 and 0°C, and the delivered capacity at −20°C was 1.15 Ah, or 80% of the capacity at 20°C. At −10°C the voltage at 50% DOD was 3.7 V, i.e., 0.2 V less than at 20°C.

Despite the fact that the LCO/hard carbon combination provides significantly less energy density than alternatives, e.g., 58% of the capacity provided by LCO/graphite, the slope and greater linearity of the voltage curve of the LCO/hard carbon cell is useful for applications where determination of the remaining capacity is based on the cell voltage.

***LiNi$_{1-x-y}$Co$_x$Al$_y$O$_2$ (NCA)/Carbon.***    Figure 17A.59 shows discharge curves for a Panasonic NCR-18650B cell. As shown, this cell provided 3.25 Ah, or about 257 Wh/kg, and a voltage of about 3.7 V at 50% DOD at a *C*/10 (0.28 A) rate. At a 1*C* rate the capacity was similar, 2.9 Ah (−10%), and the energy delivered was 224 Wh/kg (−3%).

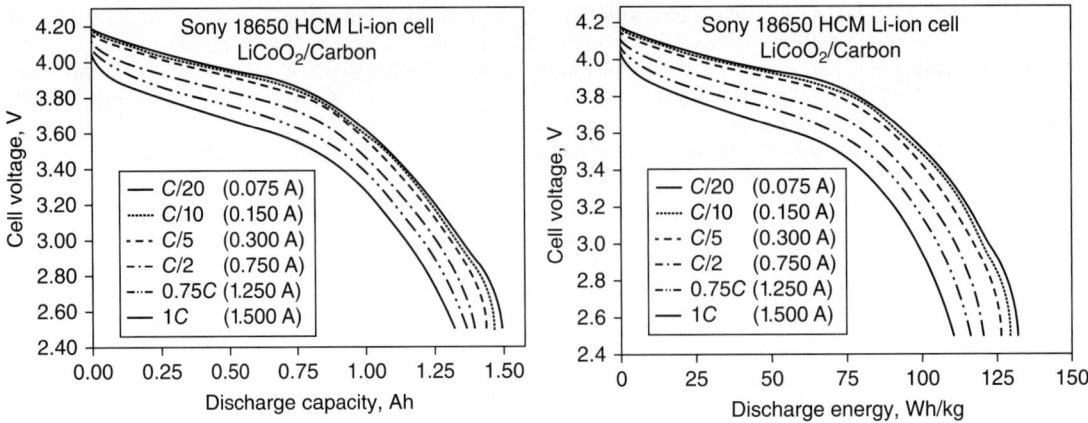

**FIGURE 17A.57**   Cell voltage versus discharge capacity (Ah) and discharge energy (Wh/kg) for a Sony HCM18650 cell at 20°C. The cell was charged at 0.15 A (C/10) to 4.2 V and then held at 4.2 V until the current tapered to 0.015 A (C/100). The cell was then discharged at C/20, C/10, C/5, C/2, 0.75C, and at 1C to 2.5 V. (*Courtesy Marshall Smart, NASA/JPL-Caltech.*)

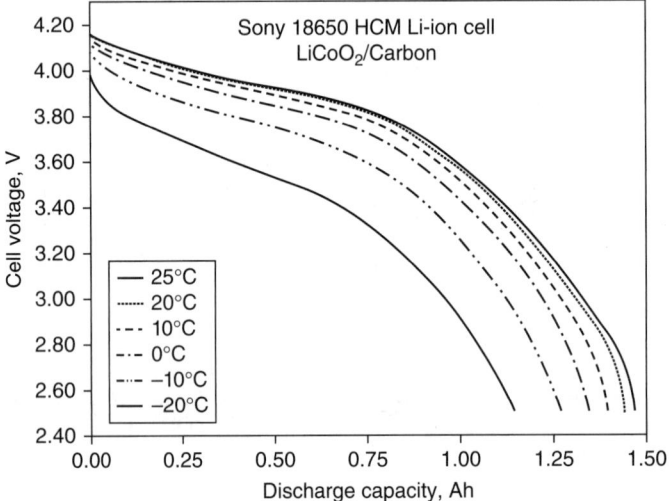

**FIGURE 17A.58**   Cell voltage versus discharge capacity (Ah) for a Sony HCM18650 cell at −20°C, −10°C, 0°C, 10°C, 20°C, and 25°C. The cell was charged at 0.15 A (C/10) to 4.2 V and then held at 4.2 V until the current tapered to 0.015 A (C/100). The cell was then discharged at C/5 (0.300 A) to 2.5 V. (*Courtesy Marshall Smart, NASA/JPL-Caltech.*)

The sloped shape of the discharge curve is characteristic of the NCA positive electrode material and graphite negative electrode material used in this cell. The cell voltage was 3.65 V at 95% DOD when discharged at a C/10 rate.

The temperature performance of the Panasonic NCR-18650B cell is shown in Fig. 17A.60, which shows the results of C/5 (0.56 A) discharge at temperatures ranging from −20 to +30°C. As shown, the discharge capacity decreased 13% between 20 and 0°C; however, the delivered capacity at −20°C was 1.2 Ah, indicating that the electrolyte was formulated for warmer temperatures. At 20°C, the voltage at 50% DOD was 3.7 V, whereas at 0°C the voltage at 50% DOD was 3.5 V.

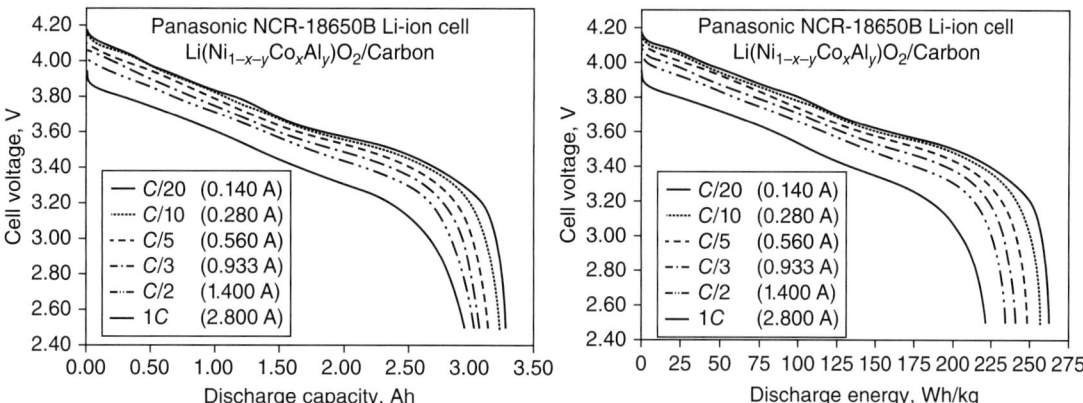

**FIGURE 17A.59**   Cell voltage versus discharge capacity (Ah) and discharge energy (Wh/kg) for a Panasonic NCR-18650B cell at 20°C. The cell was charged at 0.28 A (C/10) to 4.2 V and then held at 4.2 V until the current tapered to 0.028 A (C/100). The cell was then discharged at C/20, C/10, C/5, C/3, C/2, and at 1C to 2.5 V. (*Courtesy Marshall Smart, NASA/JPL-Caltech.*)

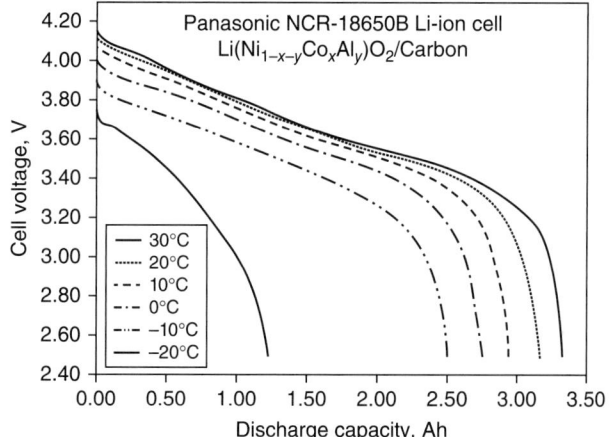

**FIGURE 17A.60**   Cell voltage versus discharge capacity (Ah) for a Panasonic NCR-18650B cell at −20°C, −10°C, 0°C, 10°C, 20°C, and 30°C. The cell was charged at 0.28 A (C/10) to 4.2 V and then held at 4.2 V until the current tapered to 0.028 A (C/100). The cell was then discharged at C/5 (0.560 A) to 2.5 V. (*Courtesy Marshall Smart, NASA/JPL-Caltech.*)

Illustrating the trades that are inherent in cell design, shown in Figs. 17A.61 and 17A.62, are discharge curves illustrating the rate capability and temperature capability of an E-One Moli INR-18650A designed for power. The cell has a center positive electrode tab and dual negative electrode tabs to provide a DC impedance (10 A/1 s) of 20 mΩ and an AC impedance of 12 mΩ at 1 kHz. The cell used an NCA positive electrode material and a graphite negative electrode material. At low rates and at ambient temperatures, the cell provided less capacity than the Panasonic NCR-18650B cell; however, at −20°C the Moli INR-18650A cell provided about double the capacity of the Panasonic NCR-18650B cell. Also, the Moli cell has 20 A discharge capability, and thus can provide power not available from the Panasonic NCR-18650B cell. The voltage increase in the discharge curves of Fig. 17A.62 at −20°C and −30°C is due to self-heating. The cycle life of the Moli INR-18650A cell is shown in Fig. 17A.63. When charged to 4.1 V or below, over 2500 cycles are available. Charging to 4.2 V provided a 10% increase in capacity; however, the higher charging voltage reduced the cycle life to about 1500 cycles.

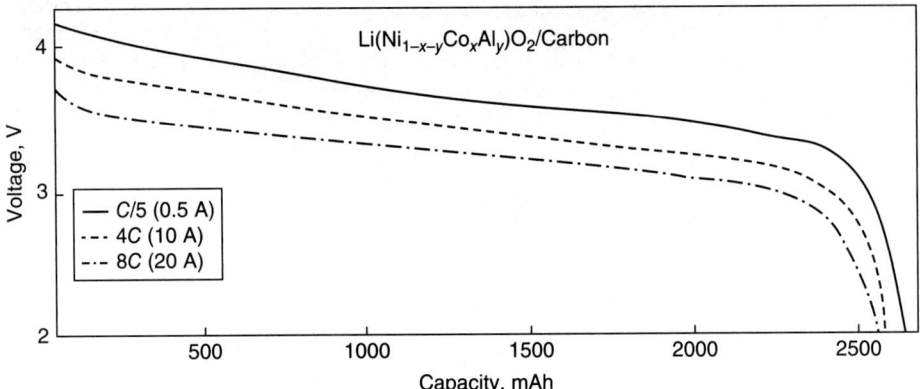

**FIGURE 17A.61**   Discharge capacity of an E-One Moli INR-18650A cell discharged at 23°C. The cell was charged at 2.5 A to 4.2 V with a taper to 50 mA and then discharged to 2 V.

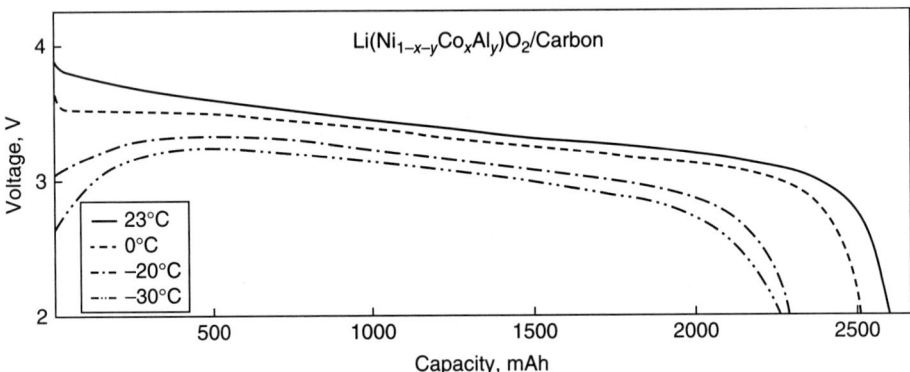

**FIGURE 17A.62**   Discharge capacity of an E-One Moli INR-18650A cell discharged at 4C (10 A). The cell was charged at 2.5 A to 4.2 V with a taper to 50 mA and then discharged at 10 A to 2 V.

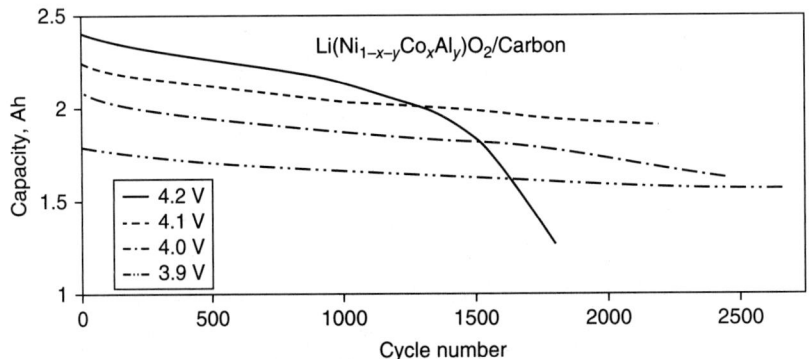

**FIGURE 17A.63**   Cycle life of an E-One Moli INR-18650A cell when charged to 3.9 V, 4.0 V, 4.1 V, or 4.2 V at 23°C. The cell was charged at 2 A to the cutoff voltage and then tapered to 50 mA, and discharged at 2 A to 3 V.

NCA has also been used in high-power prismatic cells. Figure 17A.64 shows discharge curves for a Yardney 5 Ah NCA/graphite cell discharged at 25°C at rates up to 48$C$ (240 A). At 0.5$C$ the cell delivered 6 Ah (100%), at 10$C$ 5 Ah (83%), and at 48$C$ 4.5 Ah (75%). Performance at 15°C was similar, at 0.5$C$ the cell delivered 5.9 Ah, at 10$C$ 4.9 Ah, and at 48$C$ 4.5 Ah. At 45°C, at 0.5$C$ the cell delivered 6 Ah, at 10$C$ 5.3 Ah, and at 48$C$ 4.8 Ah.

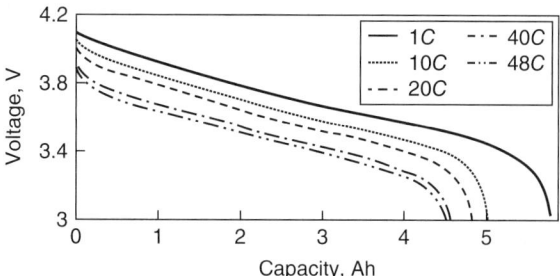

**FIGURE 17A.64**  Discharge curves for a Yardney 5 Ah NCA/graphite cell discharged at 25°C at rates from 1$C$ to 48$C$ (240 A). At 0.5$C$ the cell delivered 6 Ah, at 10$C$ the cell delivered 5 Ah, and at 48$C$ the cell delivered 4.5 Ah.

NCA has also been used in high-rate polymer cells. As shown in Figs. 17A.65 and 17A.66, polymer NCA/graphite cells can provide exceptional rate capability.

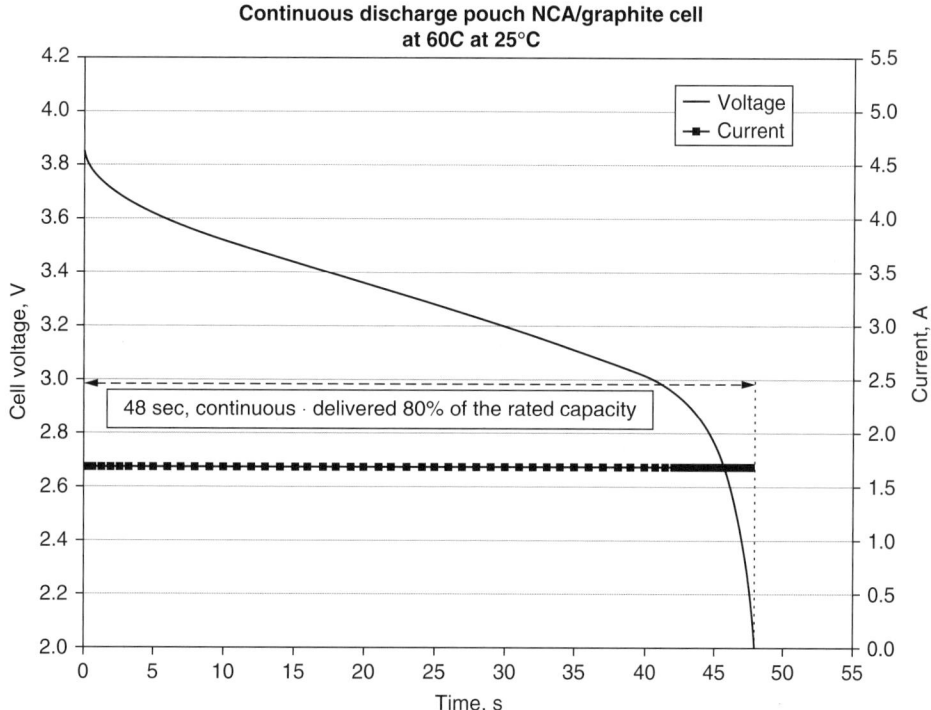

**FIGURE 17A.65**  60$C$ discharge of a Yardney polymer NCA/graphite high-rate cell. The cell delivered 80% of the rated capacity in 48 sec. with a voltage at 50% depth of discharge (DOD) of 3.3 V.

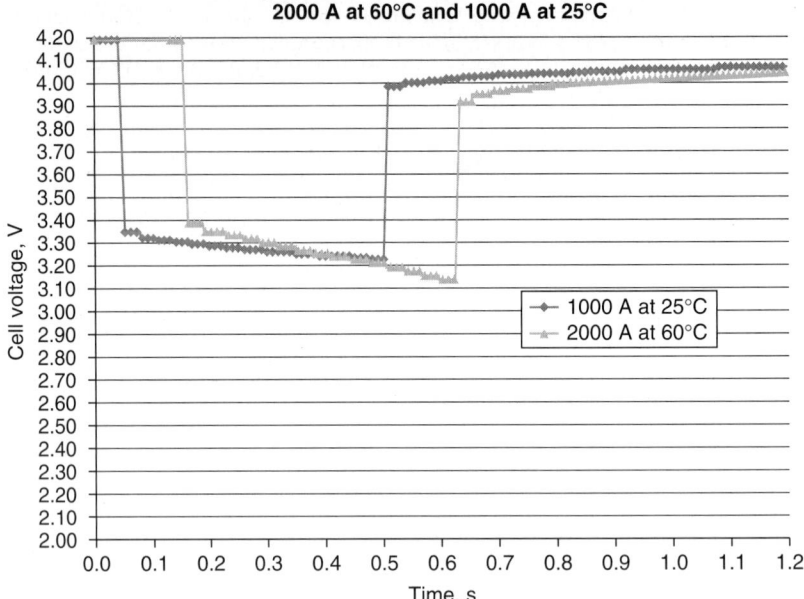

**FIGURE 17A.66**   Pulse discharge of a Yardney polymer NCA/graphite high-rate cell. The cell voltage was no lower than 3.2 V at 200C (1000 A) at 25°C.

Panasonic is a supplier of cells for Tesla vehicles that use an NCA positive electrode material. Figure 17A.67 illustrates that these cells are providing extended life in the field, based on analysis of 409 vehicles.[100] Over the first 50,000 miles, the batteries lost about 5% capacity on average. Also, the data suggests that the batteries will provide over 150,000 miles before reaching 90% capacity, and will reach 80% capacity after 278,000 miles or 14 years, on average.

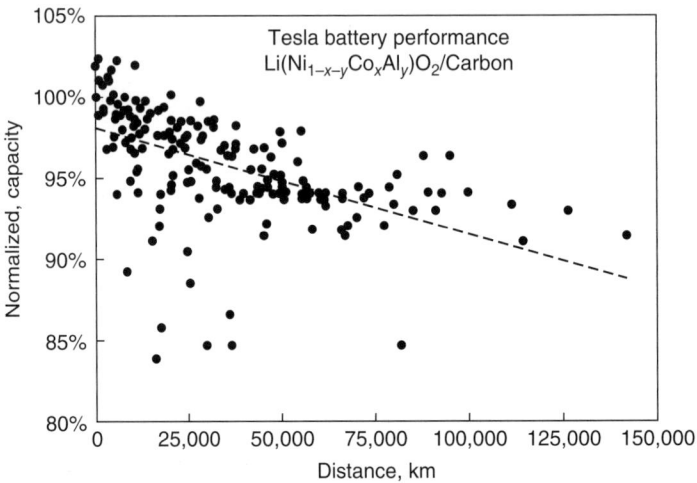

**FIGURE 17A.67**   Normalized remaining capacity versus distance for 409 Tesla vehicles. (*Source: https://docs.google.com/spreadsheets/d/t024bMoRiDPIDialGnuKPsg/edit#gid= 154312675, last accessed February 27, 2018, graph courtesy Stephen Glazier, Dalhousie University.*)

***LiNi$_{1-x-y}$Mn$_x$Co$_y$O$_2$ (NMC)/Carbon.*** NMC is being used in cells designed for high energy and in cells designed for high power. Cells comprising NMC can provide lower cost and better inherent safety than LiCoO$_2$. Figures 17A.68 and 17A.69 show discharge curves illustrating the rate capability and temperature capability of an E-One Moli IHR-18650C cell designed for power. The cell has a center positive electrode tab and dual negative electrode tabs to provide a DC impedance (10 A/1 s) of 20 mΩ and an AC impedance of less than 15 mΩ at 1 kHz. The cell used an NMC positive electrode material and a graphite negative electrode material. The cell provided less capacity than the Moli INR-18650A cell, which used an NCA positive electrode material. The cycle life of the Moli IHR-18650C cell is shown in Fig. 17A.70. When charged to 4.1 V or below over 4000 cycles are available. Charging to 4.2 V provided an 8% increase in capacity; however, the higher charging voltage reduced the cycle life to about 1500 cycles.

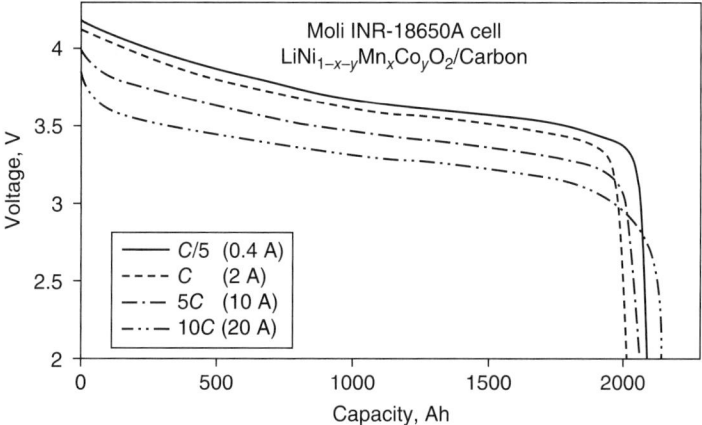

**FIGURE 17A.68** Discharge capacity of an E-One Moli IHR-18650C cell discharged at 23°C. The cell was charged at 2 A to 4.2 V with a taper to 50 mA, and then discharged to 2 V.

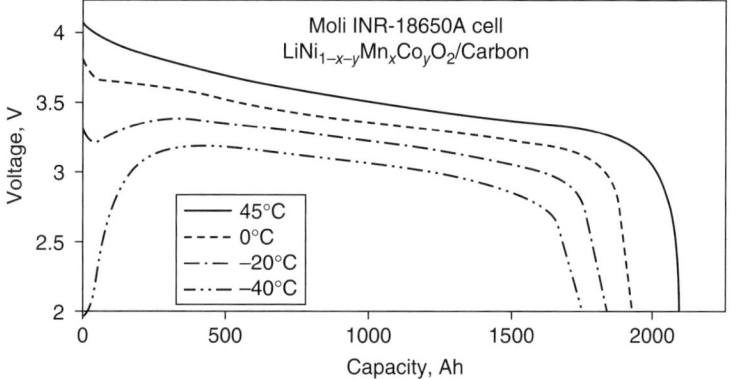

**FIGURE 17A.69** Discharge capacity of an E-One Moli IHR-18650C cell discharged at 5C (10 A). The cell was charged at 23°C at 2 A to 4.2 V with a taper to 50 mA, and then discharged at 10 A to 2 V.

***LiNi$_{1-x}$Co$_x$O$_2$ (NCO)/MCMB.*** Figure 17A.71 shows discharge curves for a Yardney NCP-43-3 cell developed for the Mars Science Laboratory after charging to 4.10 V and discharging to 2.50 V. As shown, this cell provided 48.4 Ah, or about 145 Wh/kg, and a voltage of about 3.65 V at 50% DOD at a C/4 (10.75 A) rate. The sloped shape of the discharge curve is characteristic of the NCO positive electrode material and the MCMB negative electrode material used in this cell.

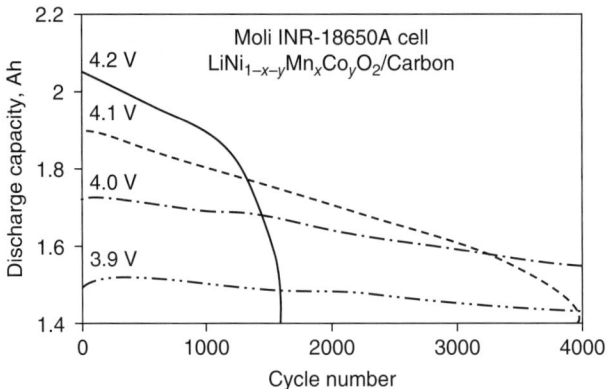

**FIGURE 17A.70**   Cycle life of an E-One Moli IHR-18650C cell when charged to 3.9 V, 4.0 V, 4.1 V, or 4.2 V at 23°C. The cell was charged at 2 A to the cutoff voltage and then tapered to 50 mA, and discharged at 2 A to 3 V.

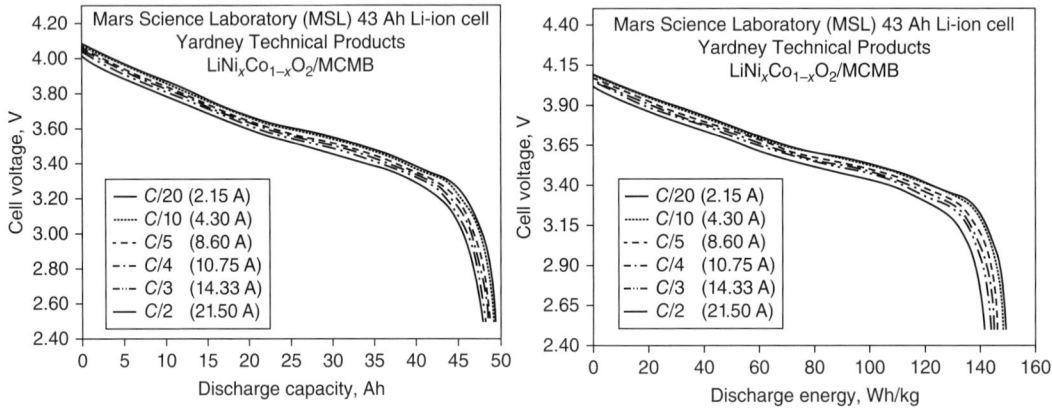

**FIGURE 17A.71**   Cell voltage versus discharge capacity (Ah) and discharge energy (Wh/kg) for a Yardney prismatic cell at 20°C. The cell was charged at 4.3 A ($C$/10) to 4.1 V and then held at 4.1 V until the current tapered to 0.43 A ($C$/100). The cell was then discharged at $C$/20, $C$/10, $C$/5, $C$/4, $C$/3, and at $C$/2 to 2.5 V. (*Courtesy Marshall Smart, NASA/JPL-Caltech.*)

The temperature performance of the Yardney prismatic cell is shown in Fig. 17A.72, which shows the results of $C$/5 (5.8 A) discharge at temperatures ranging from −30 to +20°C. As shown, the discharge capacity decreased 6% between 20 and 0°C. At −30°C the cell delivered 35 Ah and a voltage of 3.3 V at 50% DOD.

***LiFePO₄/Carbon.***   Lithium iron phosphate materials are known for providing high charge and discharge rate capability (up to 100$C$ or 3000 W/kg), albeit with less energy density due in part to the lower 3.3 V working voltage, and greater self-discharge than other chemistries. Figure 17A.73 shows discharge curves for an A123 IFpR-26650 cell. This cell provided 2.25 Ah, or about 105 Wh/kg, and a voltage of about 3.25 V at 50% DOD at a $C$ (1.1 A) rate at 23°C. The relatively flat shape of the discharge curve is characteristic of the $LiFePO_4$ positive electrode material and graphite negative electrode material used in this cell. Note that the cell voltage decreases minimally during discharge, about 0.2 V between 5% and 95% DOD.

Results from high rate discharge of the A123 IFpR-26650 cell at 20°C are shown in Fig. 17A.74. The capacity of the cell was effectively unchanged at rates up to 6.8$C$, the highest rate evaluated, and the cell voltage changed minimally, remaining near 3 V at 3.8$C$, illustrating the exceptional rate capability of this cell.

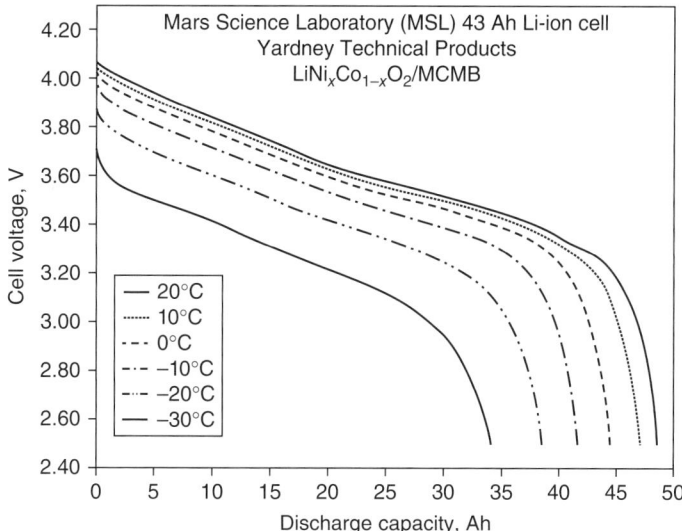

**FIGURE 17A.72**   Cell voltage versus discharge capacity (Ah) for a Yardney prismatic cell at −30°C, −20°C, −10°C, 0°C, 10°C, and 20°C. The cell was charged at 4.3 A (*C*/10) to 4.1 V and then held at 4.1 V until the current tapered to 0.430 A (*C*/100). The cell was then discharged at *C*/5 (5.80 A) to 2.5 V. (*Courtesy Marshall Smart, NASA/JPL-Caltech.*)

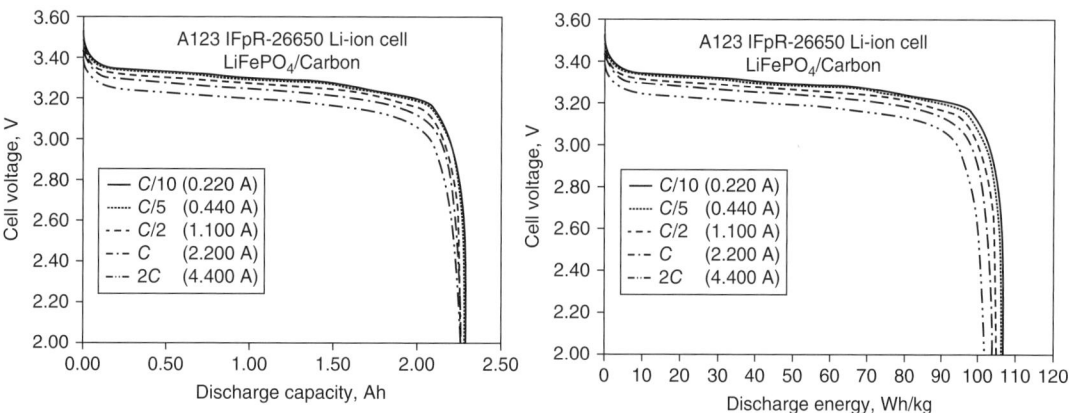

**FIGURE 17A.73**   Cell voltage versus discharge capacity (Ah) and discharge energy (Wh/kg) for an A123 IFpR-26650 cell at 23°C. The cell was charged at 0.22 A (*C*/10) to 3.6 V and then held at 3.6 V until the current tapered to 0.11 A (*C*/20). The cell was then discharged at *C*/10, *C*/5, *C*/5, *C*/2, and at 1*C* to 2.5 V. (*Courtesy Marshall Smart, NASA/JPL-Caltech.*)

*Self-Discharge.*   Self-discharge is a phenomenon in which chemical reactions internal to a cell reduce the stored charge of the battery in the absence of an external connection between the electrodes. Because of self-discharge, a charged cell will provide less charge after storage than when initially charged.

  Self-discharge can occur due to reversible and irreversible lithium-consuming reactions.[101] The reversible reactions reduce the stored charge and have little effect on the cell capacity, whereas irreversible lithium-consuming reactions reduce both the stored charge and the capacity of the cell. Yazami and Reynier suggest that

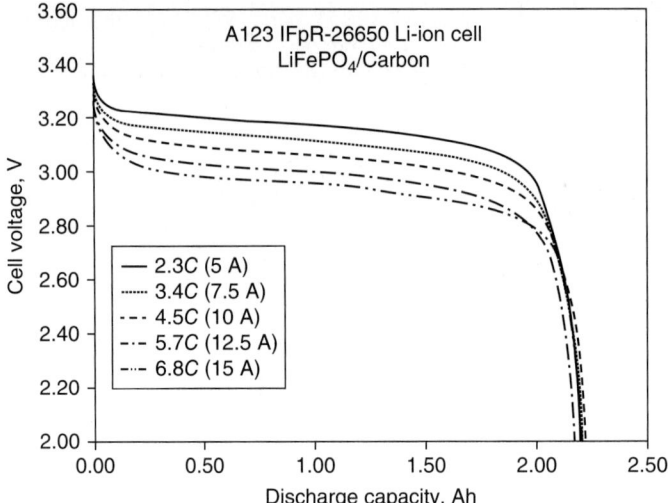

**FIGURE 17A.74**   Cell voltage versus discharge capacity (Ah) for an A123 IFpR-26650 cell discharged at 5 A (2.3$C$), 7.5 A (3.4$C$), 10 A (4.5$C$), 12.5 A (5.7$C$), and 15 A (6.8$C$) at 23°C. The cell was charged at a $C$/5 rate (0.44 A) to 3.6 V. (*Courtesy Marshall Smart, NASA/JPL-Caltech.*)

self-discharge in lithium-ion cells involves the formation of an ion-solvent complex that can either dissociate, resulting in self-discharge, or react to form an SEI layer, resulting in a reversible capacity loss.[102] Commercial cells have self-discharge rates of about 1% to 4% per month, depending on the temperature and state of charge. Self-discharge rates are nonlinear. Models which fit empirical data well are available.[103]

As shown in Fig. 17A.75 for NMC/graphite polymer cells,[104] self-discharge is strongly temperature dependent and the self-discharge rate decays exponentially. Also, as shown in Fig. 17A.76, the state of charge is a significant factor in the self-discharge rate, which is significantly less at less than 100% state of charge.

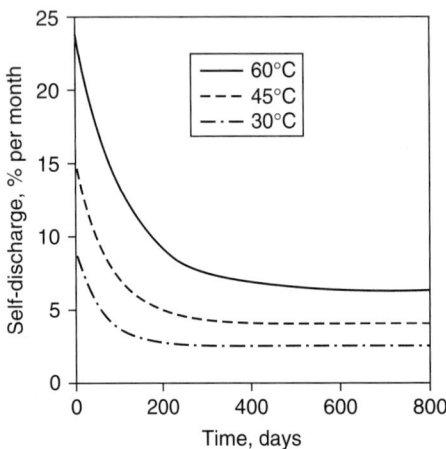

**FIGURE 17A.75**   Self-discharge of Kokam NMC/graphite 12 Ah SLPB 70205130P cells after charging to 100% state of charge at 30°C, 45°C, and 65°C. (*Data from Redondo-Iglesias.[104]*)

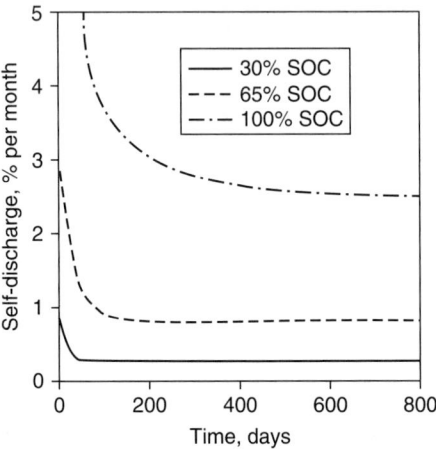

**FIGURE 17A.76**   Self-discharge of Kokam NMC/graphite 12 Ah SLPB 70205130P cells after charging to 30%, 65%, and 100% state of charge at 30°C. (*Data from Redondo-Iglesias.[104]*)

## 17A.5   SAFETY PROPERTIES

### 17A.5.1   Temperature Dependence of the Reactions between Charged Electrode Materials and Electrolyte

There have been numerous safety events documented involving lithium-ion batteries in portable electronic devices. Many of these have led to recalls involving significant numbers of batteries and the devices themselves.[105] In addition, new products like e-cigarettes[106] and hoverboards[107] have also had numerous safety problems involving Li-ion cells. There have been several serious incidents involving Li-ion cells on passenger aircraft,[108] freight,[109] and in EVs.[110] Nevertheless, the safety record of Li-ion batteries is extremely good and is continually improving due to the worldwide efforts of Li-ion battery manufacturers, original equipment manufacturers (OEMs), equipment designers, and regulatory agencies. If one assumes about 700 reported safety incidents per year from a global production of 7 billion cells, this translates to a safety incident rate of only 1 per 10 million cells.

All safety issues with Li-ion batteries ultimately stem from the fact that the charged positive and negative electrode materials react with the cell electrolyte at elevated temperature. Figure 17A.77 shows accelerating rate calorimeter (ARC) data for the reaction of various lithiated carbon negative electrode samples with 1 M $LiPF_6$-containing electrolyte.

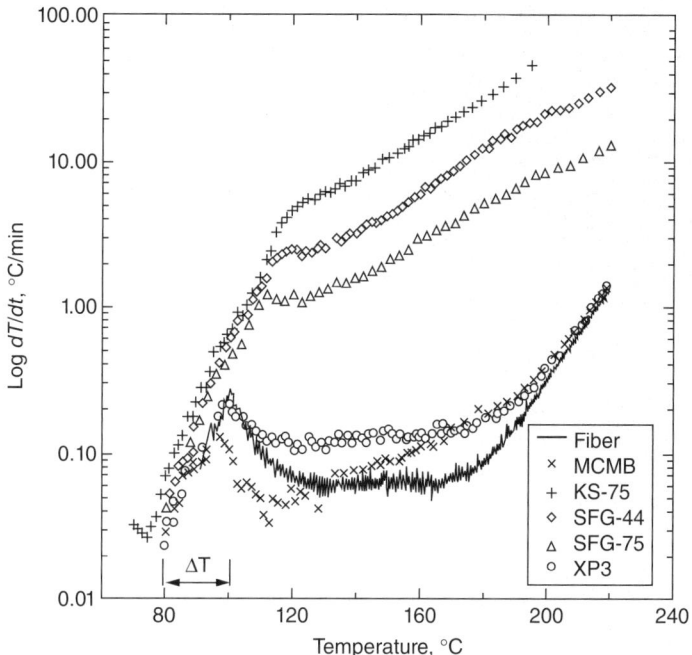

**FIGURE 17A.77**   Self-heating rate profiles of lithium-intercalated carbons in $LiPF_6$ EC:DEC (33:67) electrolyte. (*Reproduced with permission from J. Electrochem. Soc., 146, 3596 [1999].*[36])

Figure 17A.77 shows that all of these materials begin reacting with the electrolyte near 80°C at a low rate. (The reactivity of the electrolyte observed in this figure near 80°C is fundamentally the reason why Li-ion batteries begin to lose capacity when cycled above 60°C—reactions between the lithiated graphite and electrolyte do not just "turn on," but are exponentially activated by temperature.) As the temperature increases, the reaction rate increases strongly for high-surface area carbons such as the KS and SFG materials. The MCMB and fiber samples have a lower specific surface area. For these materials, the reaction rate reaches a plateau after about 100°C and increases dramatically above 200°C.

A common misconception is that lithium in $Li_{7/3}Ti_{5/3}O_4$ (charged lithium titanate) is not reactive with the electrolyte at elevated temperature, and therefore Li-ion cells with LTO negative electrodes would show no negative electrode reactivity at elevated temperature. Jiang et al.[111] showed that the heat of reaction between lithiated negative electrode materials and the electrolyte decreases linearly with the potential of the negative electrode material versus lithium, reaching zero near 3.0 V. This means that an LTO negative electrode at 1.5 V storing the same amount of lithium as a graphite electrode at 0.1 V would evolve about half of the total heat during a reaction with electrolyte.

There has been substantial work to characterize the reactions between lithiated negative electrode materials and the electrolyte. In short, it is noted that detectable negative electrode reactivity begins at a relatively low temperature, around 80°C.

The reactions between a number of charged positive electrode materials and the electrolyte have been characterized by MacNeil et al.[112] DSC was used to study the reactions between charged $LiCoO_2$, $LiNiO_2$, $LiNi_{0.8}Co_{0.2}O_2$, $Li_{1+x}Mn_{2-x}O_4$, $LiNi_{0.7}Co_{0.2}Ti_{0.05}Mg_{0.05}O_2$, $LiNi_{3/8}Co_{1/4}Mn_{3/8}O_2$ (an NMC composition), or $LiFePO_4$ and the electrolyte under identical conditions. The paper concludes: "In our opinion, the cathode materials can be ranked from most safe to least safe in the following order: $LiFePO_4$, $LiNi_{3/8}Co_{1/4}Mn_{3/8}O_2$, $Li_{1+x}Mn_{2-x}O_4$, $LiCoO_2$, $LiNi_{0.7}Co_{0.2}Ti_{0.05}Mg_{0.05}O_2$, $LiNi_{0.8}Co_{0.2}O_2$, [and] $LiNiO_2$." Although many of the materials studied are not currently used in Li-ion batteries, the paper identified $LiFePO_4$, NMC, and $LiMn_2O_4$ spinel as potentially the safest materials, in agreement with current knowledge. Wang et al.[113] characterized the reactivity of $LiCoO_2$, $LiNi_{0.8}Co_{0.15}Al_{0.05}O_2$ (NCA), $LiNi_{1/3}Mn_{1/3}Co_{1/3}O_2$ (NMC), and $LiNi_{0.42}Mn_{0.42}Co_{0.16}O_2$ (NMC) samples with electrolyte. Figure 17A.78 shows ARC results for the reactions of these materials with electrolyte. The NMC samples begin reacting strongly with the electrolyte at the highest temperatures. Wang et al. conclude that although the $LiCoO_2$ sample had the smallest specific surface area of all samples studied, it was the most reactive of all the samples below 180°C. $LiCoO_2$ and NCA reached self-heating rates of 10°C/min at approximately the same temperature. These results suggest that switching from $LiCoO_2$ to state-of-the-art NCA should not lead to significant safety compromises, if any. The NMC samples had the lowest self-heating rates of all the samples, at least below 250°C, suggesting that Li-ion cells with the best safety properties can be made using NMC.

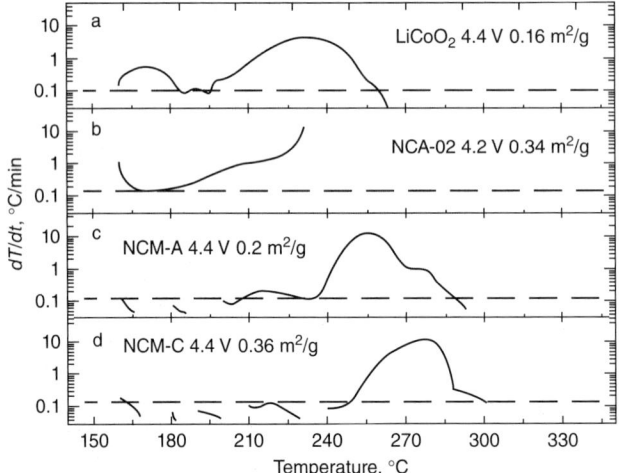

**FIGURE 17A.78**   Self-heating rate versus temperature for 56 mg of the charged positive electrode materials $LiCoO_2$, NCA, and two types of NMC with 18 mg of 1 M $LiPF_6$ EC:DEC. The horizontal dashed line indicates a self-heating rate of 0.12°C/min. The ARC was forced to a starting temperature of 160°C at 5°C/min before exotherm searching began.[113] (*Reproduced with permission from Electrochemistry Communications **9**, 2534 [2007].[113]*)

In recent papers by Noh et al.[26] and by Ma et al.[114] the safety properties of various NMC grades are compared under the same conditions. Both author groups reach the same conclusion. Thermal instability of the charged NMC positive electrodes decreases with an increase in charging voltage and with an increase in the Ni content. Li-ion cells with NMC622 positive electrodes (60% Ni) are now in wide use around the world, suggesting cells with NMC622 can be made very safe. It is unclear at this point if large cells with NMC811 can be made to pass all required safety tests.

Although the above data are compelling, calorimetry data are measured at relatively low temperatures and at low heating rates and are best used to understand abuse tolerance under similar conditions, such as hot-box or oven exposure testing. Under instances of extreme thermal abuse, e.g., nail penetration, crush, or internal short, other considerations, such as the total heat released during thermal abuse and the rate of heat release, must be considered when selecting materials for cell safety. Thus, charged positive electrode materials react with electrolyte, and these reactions can begin at temperatures as low as $130°C$ for charged $LiCoO_2$. Because both charged negative electrode materials and charged positive electrode materials react with electrolyte at temperatures as low as $130°C$ (for $LiCoO_2$ cells, at least), the design of Li-ion cells and their safety testing must ensure that the cells are safe at these temperatures, even upon prolonged exposure. Under conditions of electrical or mechanical abuse, e.g., under crush or internal short, where portions of cells may reach much higher temperatures, cells will remain safe if the rate of heat generation is less than the rate of heat dissipation. These factors are discussed in detail in an excellent review by the SAFT research group.[115]

## 17A.5.2 Regulatory Criteria for the Safety and Design of Li-Ion Batteries

It is in the interest of the consumer, the OEM, and the battery producers that lithium-ion batteries be as safe as possible. Li-ion battery safety is given the top priority by cell manufacturers and OEMs. To this end, a number of standards and certifications have been developed to help ensure the safety of Li-ion cells. Most OEMs will not use Li-ion batteries in their products unless they comply with one or more of these standards. In this section, the following standards are discussed:

1. Underwriters Laboratories (UL)1642, "Standard for Li-ion Batteries," 2012, and
2. International Electrotechnical Commission (IEC) 62133, "Secondary Cells and Batteries Containing Alkaline or Other Non-acid Electrolytes," Feb, 2017.

These standards have been selected because they include the essential safety tests that Li-ion batteries must pass or comply with to obtain these and other certifications. Other Li-ion battery standards exist; for example, the Battery Association of Japan (BAJ) and the United Nations have numerous standards. A UN committee of experts recommends tests for shipping by air that are then adopted by the International Air Transport Association (IATA).

Table 17A.14 lists and defines the safety tests and required outcomes to obtain UL1642 certification. These tests require that both fresh cells (fully charged) and cells that have been cycled to 25% of the specified cycle life (or cycled for 90 days, whichever is shorter) and then are fully charged meet the test requirements. The tests cover the three main areas of electrical, mechanical, and environmental abuse. Virtually all Li-ion cells that reach consumers for power tool and portable electronics applications pass these tests.

The IEC 62133 standard is similar to the UL1642 standard. However, there are some very important differences. First, the IEC standard includes a free-fall test where charged batteries and devices containing the batteries must be dropped three times from a 1 m height. No fire or explosion can occur. The IEC standard contains an additional overcharge test. Cells are charged at $2C$ continuously with a thermocouple attached to the cell casing. The test continues until the temperature reaches steady-state conditions or returns to room temperature. No fire or explosion can occur.

The most significant difference between the IEC and UL standards is the inclusion of a forced internal short-circuit test in the latter standard. The forced internal short-circuit test is extremely important because many of the safety events in the field are thought to have been caused by internal shorts. The proposed test procedure in the IEC guidelines is detailed and essentially consists of placing a small "L"-shaped Ni particle in between the negative electrode and the separator of a charged cell, followed by crushing the cell under a specified pressure.

**TABLE 17A.14**    Safety Tests for Lithium-Ion Batteries as Outlined by UL1642, 5th Edition, March 2012

| Test | Description | Number of cells | Required outcome |
|---|---|---|---|
| **Electrical tests** | | | |
| Short circuit (23°C) | <0.08 Ω to 0.2 V, monitor until T returns to 33°C. | 5 fresh charged, 5 cycled charged | No explosion, no fire |
| Short circuit (55°C) | <0.08 Ω to 0.2 V, monitor until T returns to 65°C. | 5 fresh charged, 5 cycled charged | No explosion, no fire |
| Abnormal charge | Charge at 3 times manufacturers' recommended rate for 7 h. | 5 fresh charged, 5 cycled charged | No explosion, no fire |
| Forced discharge | One discharged cell in series with the number of series-connected cells (charged) used in the device. Discharge the series assembly through a resistance < 0.08 Ω to 0.2 V. Monitor until T returns to 10°C above ambient. | 5 fresh charged, 5 cycled charged | No explosion, no fire |
| **Mechanical tests** | | | |
| Flat plate crush | Between flat surfaces to 13 kN. Prismatic cells to be tested in both directions. | 5 fresh charged, 5 cycled charged | No explosion, no fire |
| Impact test | 15.8 mm diameter bar placed across cell or battery. 9.1 kg weight dropped onto the bar from a height of 61 cm. Prismatic cells to be tested in both directions. | 5 fresh charged, 5 cycled charged | No explosion, no fire |
| Shock test | 3 axis, minimum 75 g, peak 125 to 175 g. | 5 fresh charged, 5 cycled charged | No explosion, no fire, no leaking, no venting |
| Vibration test | 0.8 mm amplitude, 10 to 55 Hz at a rate of 1 Hz/min and back again. | 5 fresh charged, 5 cycled charged | No explosion, no fire, no leaking, no venting |
| **Environmental tests** | | | |
| Heating test | Heat to 130°C at 5°C/min, and hold at 130°C for 10 min. Return to room temperature and examine. | 5 fresh charged, 5 cycled charged | No explosion, no fire |
| Temperature cycling test | 70°C: 4 h, –40°C: 4 h, room T: 4 h; repeat cycle 10 times. | 5 fresh charged, 5 cycled charged | No explosion, no fire, no leaking, no venting |
| Altitude test | 11.6 kPa for 6 h. | 5 fresh charged, 5 cycled charged | No explosion, no fire, no leaking, no venting |
| **Projectile test** | Cells are incinerated. | 5 fresh charged | Cell parts cannot penetrate the wire screen used in the test |

*Note*: The projectile test is one in which the cells are heated on a screen over a burner. When they explode or vent and burn, the cells must not puncture the screen on which they rest by projectiles produced by the event.

Figure 17A.79 shows the shape of the Ni particle, and Fig. 17A.80 shows insertion of the Ni particle into the jelly-roll of a prismatic cell according to IEC 61233. After the jelly-roll is reassembled, it is pressed between flat plates to create an internal short. Once the short is initiated, the cell is monitored and cannot catch fire to pass the test. The BAJ uses a similar test in which a nickel particle is placed in a cell's jelly-roll electrode configuration during cell assembly to simulate an internal short circuit.

There are many other standards that OEMs can require Li-ion cells to pass before they will purchase and use them. For example, the Institute of Electrical and Electronics Engineers (IEEE) standards 1625 and 1725 define the design, assembly, and safety response of Li-ion cells for computing and cellular phone applications, respectively. In order to obtain these certifications, cells must meet the UL1642 or IEC 62133 safety standards *and* meet a significant number of design and quality assurance guidelines. Many of these guidelines are designed to prevent the occurrence of cell defects that might ultimately cause an internal short circuit during the use of the cell. The IEEE 1625 standards also include protocols for the safe operation of multicell battery packs.

**Unit: mm**

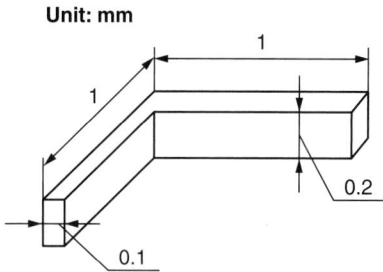

**FIGURE 17A.79** Shape of the Ni particle placed between the positive electrode and the separator in the IEC internal short test.[116]

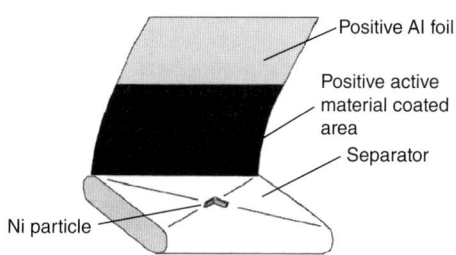

**FIGURE 17A.80** Ni particle insertion between the positive electrode and the separator for the IEC internal short test.[116]

## 17A.6 CONCLUSIONS AND FUTURE TRENDS

Initially used in small consumer electronics, e.g., cameras, cell phones, and computers, Li-ion batteries are now in virtually every type of electronic device, including larger consumer applications (such as lawn mowers, e-bikes, and vacuum cleaners), are found in virtually all EVs, and are dominant in industrial applications from backup power to grid-energy storage.

The acceptance of the technology has been driven by its unique ability to offer a high level of performance in many aspects, including energy density, specific energy, specific power, cycle life, calendar life, storage life, and temperature range, in a safe, low-cost product. Commercially available Li-ion cells providing specific energy near 300 Wh/L and energy density near 800 Wh/L are now common. The cost of mass-produced Li-ion cells is now around $125 Wh/kg and still decreasing.[4] It is conceivable that 50 kWh Li-ion battery pack for an EV will have a price near US $5000 in the near future. At this price, the payback time for an EV compared to a fossil fuel-powered car will only be a few years given the significantly lower cost of "fuel" and maintenance for the EV. Many auto makers have seen "the writing on the wall" and have stated that they will eliminate fossil fuel-powered cars from their fleets within a few decades consistent with statements by some governments that fossil fuel-powered cars will be banned in the foreseeable future.[117]

Low-cost and long-lived Li-ion cells also mean that solar-battery and wind-battery energy generation and storage facilities have become cost effective and are preferred in a number of places, such as islands in the Pacific and the Caribbean.[118] In fact, renewable energy generation and storage is growing at a very rapid pace. Germany, for example, has stated that its electricity will come entirely from renewables by 2050.[119]

Demand for cost reduction and improved safety will be everpresent as lithium-ion batteries continue their penetration into the broader battery market and applications historically serviced by other technologies. Because little room for improvement remains in the mechanical aspects of battery design, and, as illustrated in Fig. 17A.81, because the electrochemically active materials are the source of at least half of the cost of current

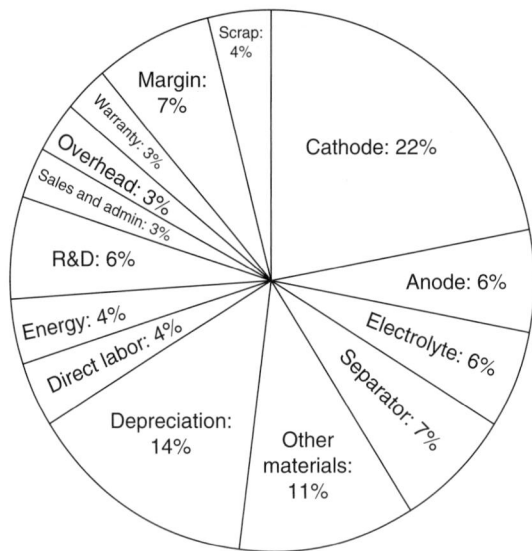

**FIGURE 17A.81**   Lithium-ion cost breakdown.[4]

batteries, there will be continued focus on the development of improved battery materials, in particular positive, negative, separator, and electrolyte materials. Improved positive electrode materials that offer greater energy, improved power, improved lifetime, less cost, and improved safety properties are in development, as are negative electrode materials, such as silicon-based materials, that offer further improvement in specific energy, energy density, rate capability, and longevity. Electrolyte additives and electrode material coatings will be employed to increase cycle and calendar life.

Depending on the cell selected, the positive electrode is responsible for about 22% of the cost of current commercial cells. As noted previously, the cost and availability of cobalt is an ongoing concern for the industry, providing ample motivation to develop materials with reduced cobalt content. Use of greater amounts of nickel and less cobalt provides reduced cost, albeit with reduced stability. Currently available nickel-based materials, such as NMC811 ($LiNi_{0.8}Mn_{0.1}Co_{0.1}O_2$), thus provide promising electrochemical performance. However, more stable variants will be needed for continued market penetration. Accordingly, there will be continued focus on improving the stability of low-cobalt materials, such as NCA- and NMC-type materials.

Increasing lithium and manganese content is a general route to improved capacity[120]; however, materials with increased lithium and manganese content have not been adopted commercially due to performance shortcomings including IRC,[121] capacity fade,[122] increased resistance, and transition metal dissolution.[123]

Materials with increased lithium content and reduced cobalt content that have received attention include the so-called two-component "layered-layered" materials of the general formula $xLi_2MnO_3 \cdot (1 - x)LiMO_2$ (M = Mn, Ni, Co) and the "layered-spinel" materials of the general formula $xLi_2MnO_3 \cdot (1 - x)LiM_2O_4$. Some aspects of these materials such as capacity in excess of 250 mAh/g are encouraging.[120] Initially, these materials were believed to have integrated structures in which a $Li_2MnO_3$ (layered) component is structurally integrated with either a layered $LiMO_2$ component or a spinel $LiM_2O_4$ component. More recent studies conclude that in fact these materials are single phase with randomly stacked domains that correspond to variants of the monoclinic structure.[124] Shukla et al. explain that the confusion with respect to the structure of these materials stems from the fact that at least three possible structures are consistent with the results of traditional methods of structural characterization, e.g., x-ray and neutron diffraction.[124] They further explain that the ambiguity can be resolved by analysis of discrete grains within primary particles to avoid the effects of grain orientation and grain boundaries.[125] Shukla et al. conclude that NMC materials with excess lithium have a bulk structure consisting of a single monoclinic phase with a spinel phase on the surface of some grain facets.[125] The spinel phase has greater nickel and cobalt content relative to the bulk composition. Also, discrete regions with only trigonal $LiMO_2$ or monoclinic

$Li_2MnO_3$ were never observed.[125] These results imply that past assumptions that surfaces are representative of bulk composition were likely incorrect and that more accurate understanding of actual bulk compositions will yield further insight.

There is also concern that there may not be enough lithium (not to mention cobalt) available to satisfy market requirements. Both automotive and grid-energy storage markets are growing very rapidly, prompting renewed search by mining companies for new resources. In addition, there is more investment in recycling lithium-ion batteries to capture the valuable lithium and transition metals in spent batteries and to avoid environmental costs.

Conversely, technically related battery technology having improved sustainability, such as Na-ion batteries,[126] alternatives such as using a sulfur-based thermochemical cycle to store heat (baseload concentrating solar power [CSP]),[127] Prussian Blue batteries,[128] aqueous zinc batteries,[129] and zinc hybrid cathode batteries,[130] etc., may provide a platform for the next generation of energy storage.

## ACKNOWLEDGMENTS

The authors acknowledge the assistance of numerous experts in preparing this chapter. Special thanks to Marshall Smart (JPL), Frank Puglia (EaglePitcher), Mark Obrovac (Dalhousie University), Paul Craig (E-One Moli Energy Canada), Vincent Chevrier (3M), Alex Louli (Dalhousie University), Eric Logan (Dalhousie University), Jing Li (Dalhousie University), Ning Zhang (Dalhousie University), Karen Thomas-Alyea (Samsung), Marie Francine Lagadec and Vanessa Claire Wood (ETH Zürich), and Michelle Fetzner (Cantor Colburn).

## REFERENCES

1. https://www.energystorageexchange.org/AESDB/projects/2017, accessed March 1, 2018.

2. Russell Gold, "Sun + Batteries = Peak Power," *Wall Street Journal*, February 13, 2018, at B1. See also https://cleantechnica.com/2018/01/24/100-mw-400-mwh-fluence-energy-storage-project-long-beach-worlds-largest-li-ion-battery-storage-project, accessed March 1, 2018.

3. B. Nykvist and M. Nilsson, "Rapidly Falling Costs of Battery Packs for Electric Vehicles," *Nat. Climate Change* **5**, 329–332 (2015).

4. C. Pillot, "The Rechargeable Battery Market and Main Trends 2016–2025," International Battery Seminar and Exhibit (March 20, 2017).

5. Goldman Sachs Global Lithium Market Outlook, 2016.

6. www.ubergizmo.com/15/archives/2009/12/panasonic_lithium_ion_battery_enters_production.html. Accessed February 22, 2018.

7. M. S. Whittingham and M. B. Dines, *Surv. Prog. Chem.* **9**, 55 (1980).

8. A. Weiss, *Angew. Chem.*, **75**, 755–761 (1963).

9. A. Herold, in *Intercalated Materials,* F. Levy (ed), D. Reidel Publishing, Dordrecht, the Netherlands, 1979, p. 323; D. M. Adams, *Inorganic Solids,* Wiley, New York, 1974.

10. H. Selig and L. B. Ebert, *Adv. Inorg. Chem. Radiochem.* **23**, 281 (1980).

11. J. R. Dahn, A. K. Sleigh, H. Shi, B. M. Way, W. J. Weydanz, J. N. Reimers, and Q. Zhong et al., in *Lithium Batteries—New Materials, Developments and Perspectives*, G. Pistoia (ed.) 1994, pp. 1–97. T. Zheng, J. N. Reimers, and J. R. Dahn, *Phys. Rev. B.* **51**, 734–741 (1995).

12. U.S. Patent 4,357,215. U.S. Patent 4,302,518.

13. K. Mizushima, P. C. Jones, P. J. Wiseman, and J. B. Goodenough, *Mater. Res. Bull.* **15**, 783–789 (1980).

14. M. M. Thackeray, W. I. F. David, P. G. Bruce, and J. B. Goodenough, *Mater. Res. Bull.* **18**, 461–472 (1983).

15. A. K. Padhi, K. S. Nanjundaswamy, and J. B. Goodenough, *J. Electrochem. Soc.* **144**, 1188 (1997).

16. Prof. Goodenough is a member of the National Academy of Engineering, the National Academy of Sciences, the Royal Society, French Academy of Sciences, the Real Academia de Ciencias Exactas, Físicas y Naturales of Spain, and has authored more than 550 articles, 85 book chapters and reviews, and five books, including two seminal works, Magnetism and the Chemical Bond (1963), and Les oxydes des metaux de transition (1973).

17. Z. Lu, D. D. MacNeil, and J. R. Dahn, "Layered Li[Ni$_x$Co$_{1-2x}$Mn$_x$]O$_2$ Cathode Materials for Lithium Ion Batteries," *Electrochem. Solid-State Lett.* **4**, A200–A203 (2001).

18. Z. Lu, D. D. MacNeil, and J. R. Dahn, "New Layered Cathode Materials Li[Ni$_x$Li$_{(1/3-2x/3)}$Mn$_{(2/3-x/3)}$]O$_2$ for Lithium Ion Batteries," *Electrochem. Solid-State Lett.* **4**, A191–A194 (2001).

19. S. Hy, H. Liu, M. Zhang, D. Qian, B.-J. Hwang, and Y. S. Meng, *Energy Environ. Sci.* **9**, 1931 (2016).

20. J. Li, A. R. Cameron, H. Li, S. Glazier, D. Xiong, M. Chatzidakis, and J. Allen et al., *J. Electrochem. Soc.* **164**, 1534–1544 (2017).

21. M. Yoshio, H. Tanaka, K. Tominaga, and H. Noguchi, *J. Power Sources* **40**, 347–353 (1992); K. Mizushima, P. C. Jones, P. J. Wiseman, and J. B. Goodenough, *Mater. Res. Bull.* **15**, 783–789 (1980); W. D. Johnson, R. R. Heikes, and D. Sestrich, *Phys. Chem. Solids* **7**, 1–13 (1958); E. Jeong, M. Won, and Y. Shim, *J. Power Sources* **70**, 70–77 (1998); E. Zhecheva, R. Stoyanova, M. Gorova, R. Alcantra, J. Moales, and J. L. Tirado, *Chem. Mater.* **8**, 1429–1440 (1996); B. Garcia, J. Farcy, J. P. Pereira-Ramos, J. Perichon, and N. Baffier, *J. Power Sources* **54**, 373–377 (1995); P. N. Kumta, D. Gallet, A. Waghray, G. E. Blomgren, and M. P. Setter, *J. Power Sources* **72**, 91–98 (1998); Y. Chiang, Y. Jang, H. Wang, B. Huang, D. Sadoway, and P. Ye, *J. Electrochem. Soc.* **145**, 887 (1998); T. J. Boyle, D. Ingersoll, T. M. Alam, C. J. Tafoya, M. A. Rodriguez, K. Vanheusden, and D. H. Doughty, *Chem. Mater.* **10**, 2270–2276 (1998); G. G. Amatucci, J.-M. Tarascon, D. Larcher, and L. C. Klein, *Solid State Ionics* **84**, 169–180 (1996); D. Larcher, M. R. Palacin, G. G. Amatucci, and J.-M. Tarascon, *J. Electrochem. Soc.* **144**, 408 (1997); M. Antaya, J. R. Dahn, J. S. Preston, E. Rossen, and J. N. Reimers, *J. Electrochem. Soc.* **140**, 575 (1993); M. Antaya, K. Cearns, J. S. Preston, J. N. Reimers, and J. R. Dahn, *J. Appl. Phys.* **75**, 2799 (1994); P. Frajnaud, R. Nagarajan, D. M. Schleich, and D. Vujic, *J. Power Sources* **54**, 362–366 (1995); E. Antolini, *J. Eur. Ceram. Soc.* **18**(10), 1405–1411 (1998).

22. A. van Bommel and J. R. Dahn, *J. Electrochem. Soc.*, **156**, A362–A366 (2009); A. van Bommel and J. R. Dahn, *Chem. Mater.* **21**, 1500–1503 (2009).

23. J. Barker, M. Y. Saidi, and J. L. Swoyer, *Electrochem. Solid-State Lett.* **6**, A53–A55 (2003).

24. N. Ravet, S. Besner, M. Simoneau, A. Vallee, M. Armand, and J. F. Magnan, U.S. Patent No. 6,855,273 (2005).

25. T. Ohzuku and R. J. Brodd, *J. Power Sources* **174**, 449–456 (2007).

26. H.-J. Noh, S. Youn, C. S. Yoon, and Y.-K. Sun, *J. Power Sources* **233**, 121–130 (2013).

27. J. N. Reimers and J. R. Dahn, *J. Electrochem. Soc.* **139**, 2091 (1992); A. Van der Ven, M. K. Aydinol, G. Ceder, G. Kresse, and J. Hafner, *Phys. Rev. B.* **58**, 2975 (1998).

28. M. Wohlfahrt-Mehrens, C. Vogler, and J. Garche, *J. Power Sources* **127**, 58–64 (2004).

29. "Cellular Phone Recall May Cause Setback for Moli," *Toronto Globe and Mail*, August 15, 1989 (Toronto, Canada). *Adv. Batt. Technology* **25**(10), 4 (1989).

30. J. R. Dahn, A. K. Sleigh, H. Shi, B. M. Way, W. J. Weydanz, J. N. Reimers, and Q. Zong et al., in *Lithium Batteries—New Materials, Developments and Perspectives*, G. Pistoia (ed.) 1994, pp. 1–97.

31. R. E. Franklin, *Proc. Roy. Soc.* (London) **A209**, 196 (1951).

32. M. Inagaki, *Solid State Ionics* **86–88**, 833–839 (1996).

33. T. Zheng, J. N. Reimers, and J. R. Dahn, *Phys. Rev. B.* **51**, 734 (1995).

34. H. Shi, J. N. Reimers, and J. R. Dahn, *J. Appl. Crystallogr.* **26**, 827–836 (1993).

35. R. Fong, U. von Sacken, and J. R. Dahn, *J. Electrochem. Soc.* **137**, 2009–2013 (1990).

36. D. D. MacNeil, D. Larcher, and J. R. Dahn, *J. Electrochem. Soc.* **146**, 3596–3602 (1999).

37. H. Nozaki, K. Nagaoka, K. Hoshi, N. Ohta, and M. Inagaki, *J. Power Sources* **194**, 486–493 (2009); Y.-S. Park, H. J. Bang, S.-M. Oh, Y.-K. Sun, and S.-M. Lee, *J. Power Sources* **190**, 553–557 (2009).

38. Y.-S. Park, H. J. Bang, S.-M. Oh, Y.-K. Sun, and S.-M. Lee, *J. Power Sources* **190**, 553–557 (2009).

39. G.-N. Zhu, Y.-G. Wang, and Y.-Y. Xia, *Energy Environ. Sci.* **5**, 6652 (2012).

40. D. W. Murphy, R. J. Cava, S. M. Zahurak, and A. Santaro, *Solid State Ionics* **9–10**, 413 (1983).

41. K. M. Colbow, R. R. Haering, and J. R. Dahn, *J. Power Sources* **26**, 397–402 (1989).

42. T. Ohzuku, A. Ueda, and N. Yamamoto, *J. Electrochem. Soc.* **142**, 1431 (1995).

43. M. N. Obrovac and V. L. Chevrier, *Chem. Rev.* **114**(23), 11444–11502 (2014).

44. M. A. Al-Maghrabi, N. van der Bosch, R. J. Sanderson, D. A Stevens, R. A. Dunlap, and J. R. Dahn, *Electrochem. Solid-State Lett.* **14** (4), A42–A44 (2011).

45. M. A.Al-Maghrabi, J. Suzuki, R. J. Sanderson, V. L. Chevrier, R. A. Dunlap, and J. R. Dahn, *J. Electrochem. Soc.* **160**, A1587 (2013).

46. S. Chae, N. Kim, J. Ma, J. Cho, and M. Ko, *Adv. Energy Mater.* **7**, 1700071 (2017).

47. T. Huang, Y. Yang, K. Pu, J. Zhang, M. Gao, H. Pan, and Y. Liu, *RSC Adv.* **7**, 2273–2280 (2017).

48. J. H. Yom, S. W. Hwang, S. M. Cho, and W. Y. Yoon, *J. Power Sources* **311**, 159–166 (2016).

49. J. Smart, W. Powell, and S. Schey, "Extended Range Electric Vehicle Driving and Charging Behavior Observed Early in the EV Project," SAE Technical Paper 2013-01-1441, 2013, available at https://doi.org/10.4271/2013-01-1441, accessed February 21, 2018.

50. D. Y. Wang, A. Xiao, L. Wells, and J. R. Dahn, *J. Electrochem. Soc.* **162**(1), A169–A175 (2015).

51. G. Yang, J. Shi, C. Shen, S. Wang, L. Xia, and H. Hu, *RSC Adv.* **7**, 26052 (2017).

52. E. Markevich, G. Salitra, and D. Aurbach, *ACS Energy Lett.* **2**, 1337–1345 (2017).

53. *Handbook of Organic Solvents*, D. R. Lide (ed.), CRC Press, Boca Raton, FL (1995).

54. M. S. Ding, K. Xu, S. Zhang, and T. R. Jow, *J. Electrochem. Soc.* **148**(4), A299–A304 (2001).

55. *Electrolyte Data Collection, Part 1d, Conductivities, Transference Numbers and Limiting Ionic Conductivities of Solutions of Aprotic, Protophobic Solvents II. Carbonates,* J. Barthel and R. Neueder (eds.), Chemistry Data Series, vol. XII, DECHEMA, Frankfurt (2000).

56. A. B. McEwen, H. L. Ngo, K. LeCompte, and J. L. Goldman, *J. Electrochem. Soc.* **146**, 1687–1695 (1999); A. B. McEwen, S. F. McDevitt, and V. R. Koch, *J. Electrochem. Soc.* **144**, L84 (1997); K. Xu, "Nonaqueous Liquid Electrolytes for Lithium-Based Rechargeable Batteries," *Chem. Rev.* **104**, 4303–4418 (2004); K. Xu, "Electrolytes and Interphases in Li-Ion Batteries and Beyond," *Chem. Rev.* **114**, 11503–11618 (2014).

57. S. T. Mayer, H. C. Yoon, C. Bragg, and J. H. Lee, "Low Temperature Ethylene Carbonate Based Electrolyte for Lithium-Ion Batteries," Polystor Corporation, Dublin, CA, 1997.

58. X. Ma, R. S. Arumugam, L. Ma, E. Logan, E. Tonita, J. Xia, and R. Petibon et al., *J. Electrochem. Soc.* **164** (14), A3556–A3562 (2017).

59. J. Li, X. Ma, H. Li, W. Stone, S. Glazier, E. Logan, E. M. Tonita, K. L. Gering, and J. R. Dahn, *J. Electrochem. Society* **165** (5), A1027–A1037 (2018).

60. K. L. Gering, *Electrochim. Acta* **225**, 175 (2017); K. L. Gering, *Electrochim. Acta* **51**, 3125 (2006).

61. Contact Kevin Gering at the Idaho National Laboratory, email: kevin.Gering@inl.gov.

62. E. R. Logan, E. M. Tonita, K. L. Gering, J. Li, X. Ma, L. Y. Beaulieu, and J. R. Dahn, *J. Electrochem. Soc.* **165** (2), A21–A30 (2018).

63. D. Guyomard and J. M. Tarascon, *J. Electrochem. Soc.* **54**, 92 (1995); T. Zheng, Y. Liu, E. W. Fuller, U. von Sacken, and J. R. Dahn, *J. Electrochem. Soc.* **142**, 2581 (1995); D. Aurbach, B. Markovsky, A. Schechter, Y. Ein-Eli, and H. Cohen, *J. Electrochem. Soc.* **143**, 3809 (1996).

64. D. Aurbach, Y. Ein-Eli, B. Markovsky, A. Zaban, S. Luski, Y. Carmeli, and H. Yamin, *J. Electrochem. Soc.* **142**, 2882 (1995).

65. H. Yoshida, T. Fukunaga, T. Hazama, M. Terasaki, M. Mizutani, and M. Yamachi, *J. Power Sources* **68**, 311–315 (1997).

66. Y. Yamada, M. Yaegashi, T. Abe, and A. Yamada, "A Superconcentrated Ether Electrolyte for Fast-charging Li-ion Batteries," *Chem. Commun.* **49**, 11194–11196 (2013).

67. R. Petibon, C. P. Aiken, L. Ma, D. Xiong, and J. R. Dahn, "The Use of Ethyl Acetate as a Sole Solvent in Highly Concentrated Electrolyte for Li-ion Batteries," *Electrochim. Acta.* **154**, 287–293 (2015).

68. Z. X. Shu, R. S. McMillian, and J. J. Murray, *J. Electrochem. Soc.* **140**, 922 (1993).

69. D. Aurbach, Y. Ein-Eli, B. Markovsky, A. Zaban, S. Luski, Y. Carmeli, and H. Yamin, *J. Electrochem. Soc.* **142**, 2882 (1995); O. Chusid, Y. Ein-Eli, M. Babai, Y. Carmeli, and D. Aurbach, *J. Power Sources* **43–44**, 47 (1993).

70. M. Broussely, Ph. Biensan, F. Bonhomme, Ph. Blanchard, S. Herreyre, K. Nechev, and R.J. Staniewicz, *J. Power Sources* **146**, 90–96 (2006).

71. K. Xu, "Nonaqueous Liquid Electrolytes for Lithium-Based Rechargeable Batteries," *Chem. Rev.* **104**, 4303–4418 (2004).

72. K.-S. Lee, S.-T. Myung, K. Amine, H. Yashiro, and Y.-K. Sun, *J. Mater. Chem.* **19**, 1995 (2009); Y.-K. Sun, S.-T. Myung, C. S. Yoon, and D.-W. Kim, *Electrochem. Solid-State Lett.* **12**, A163 (2009); Y.-K. Sun, S.-W. Cho, S.-W. Lee, C. S. Yoon, and K. Amine, *J. Electrochem. Soc.* **154**, A168 (2007); Y.-K. Sun, S.-T. Myung, B.-C. Park, J. Prakash, I. Belharouk, and K. Amine, *Nat. Mater.* **8**, 320 (2009); G. Li, Z. Yang, and W. Yang, *J. Power Sources* **183**, 741 (2008); Z. H. Chen and J. R. Dahn, *Electrochim. Acta* **49**, 1079 (2004).

73. S. Patoux, F. Le Cras, C. Bourbon, and S. Jouanneau, U.S. Patent Application Publication No. 2008/0107968 A1 (2008).

74. K. Abe, Y. Ushigoe, H. Yoshitake, and M. Yoshio, *J. Power Sources* **153**, 328 (2006); Y. Li, R. Zhang, J. Liu, and C. Yang, *J. Power Sources* **189**, 685 (2009); S. Patoux, L. Daniel, C. Bourbon, H. Lignier, C. Pagano, F. Le Cras, and S. Jouanneau, *J. Power Sources* **189**, 344 (2009); S. S. Zhang, *J. Power Sources* **162**, 1379 (2006); L. El-Ouatani, R. Dedryvere, C. Siret, P. Biensan, and D. Gonbeau, *J. Electrochem. Soc.* **156**, A468 (2009); K. Abe, K. Miyoshi, T. Hattori, Y. Ushigoe, and H. Yoshitake, *J. Power Sources* **184**, 449 (2008); G. H. Wrodnigg, J. O. Besenhard, and M. Winter, *J. Electrochem. Soc.* **146**, 470 (1999).

75. H. Yamane, T. Inoue, M. Fujita, and M. Sano, *J. Power Sources* **99**, 60 (2001).

76. K. Xu, "Nonaqueous Liquid Electrolytes for Lithium-Based Rechargeable Batteries," *Chem. Rev.* **104**, 4303–4418 (2004); K. Xu, "Electrolytes and Interphases in Li-Ion Batteries and Beyond," *Chem. Rev.* **114**, 11503–11618 (2014).

77. R. Petibon, J. Xia, J. C. Burns, J. R. Dahn, "Study of the Consumption of Vinylene Carbonate in $Li[Ni_{0.33}Mn_{0.33}Co_{0.33}]O_2/$ Graphite Pouch Cells," *J. Electrochem. Soc.* **161**, A1618–A1624 (2014).

78. D. Aurbach, K. Gamolsky, B. Markovsky, Y. Gofer, M. Schmidt, and U. Heider, *Electrochim. Acta* **47**, 1423 (2002).

79. S. Patoux, L. Daniel, C. Bourbon, H. Lignier, C. Pagano, F. Le Cras, and S. Jouanneau, *J. Power Sources* **189**, 344 (2009).

80. S. S. Zhang, *J. Power Sources* **162**, 1379 (2006).

81. L. El-Ouatani, R. Dedryvere, C. Siret, P. Biensan, and D. Gonbeau, *J. Electrochem. Soc.* **156**, A468 (2009).

82. D. Y. Wang, N. N. Sinha, R. Petibon, J. C. Burns, and J. R. Dahn, "A Systematic Study of Well-Known Electrolyte Additives in $LiCoO_2$/Graphite Pouch Cells," *J. Power Sources* **251**, 311–318 (2014); D. Y. Wang, J. Xia, L. Ma, K. J. Nelson, J. E. Harlow, D. Xiong, and L. E. Downie et al., "A Systematic Study of Electrolyte Additives in $Li[Ni_{1/3}Mn_{1/3}Co_{1/3}]O_2$ (NMC)/Graphite Pouch Cells," *J. Electrochem. Soc.* **161**, A1818–A1827 (2014).

83. J. C. Burns, A. Kassam, N. N. Sinha, L. E. Downie, L. Solnickova, B. M. Way, and J. R. Dahn, "Predicting and Extending the Lifetime of Li-Ion Batteries," *J. Electrochem. Soc.* **160**, A1451–A1456 (2013).

84. D. Y. Wang, J. Xia, L. Ma, K. J. Nelson, J. E. Harlow, and D. Xiong et al., "A Systematic Study of Electrolyte Additives in $Li[Ni_{1/3}Mn_{1/3}Co_{1/3}]O_2$ (NMC)/Graphite Pouch Cells," *J. Electrochem. Soc.* **161**, A1818–A1827 (2014).

85. X. XIA, J. Paulsen, J. Kim, and S.-Y. Han, Patent Application WO2016116862 A1 (2016).

86. R. S. Arumugam, L. Ma, J. Li, X. Xia, J. M. Paulsen, and J. R. Dahn, *J. Electrochem. Soc.* **163**, A2531–A2538 (2016).

87. D. Mohanty, K. Dahlberg, D. M. King, L. A. David, A. S. Sefat, D. L. Wood, and C. Daniel et al., *Scientific Reports*, 6:26532 (2016), DOI: 10.1038/srep26532.

88. K. Xu, "Nonaqueous Liquid Electrolytes for Lithium-Based Rechargeable Batteries," *Chem. Rev.* **104**, 4303–4418 (2004).

89. C. J. Weber, S. Geiger, S. Falusi, and M. Roth, "Material Review of Li ion Battery Separators," *AIP Conf. Proc.* **1597**, 66–81 (2014).

90. H. S. Bierenbaum, R. B. Isaacson, M. L. Druin, and S. G. Plovan, *Ind. Eng. Chem. Prod. Res. Dev.* **13**, 2 (1974).

91. H. Lee, M. Yanilmaz, O. Toprakci, K. Fu, and X. Zhang, *Energy Environ. Sci.* **7**, 3857–3886 (2014).

92. G. Venugopal, J. Moore, J. Howard, and S. Pendalwar, *J. Power Sources* **77**, 34–41 (1999).

93. M. F. Lagadec, M. Ebner, R. Zahn, and V. Wood, *J. Electrochem. Soc.* **163**, A992–A994 (2016); D. P. Finegan, S. J. Cooper, B. Tjaden, O. O. Taiwo, J. Gelb, G. Hinds, and D. J. L. Brett, *J. Power Sources* **333**, 184–192 (2016); M. F. Lagadec, M. Ebner, and V. Wood, Microstructure of Targray PE16A Lithium-Ion Battery Separator. doi: 10.5905/ethz-1007-32 (2016); M. F. Lagadec and V. Wood, Microstructure of Celgard® PP1615 Lithium-Ion Battery Separator. doi: 10.3929/ethz-b-000265085 (2018).

94. M. F. Lagadec, M. Ebner, and V. Wood, Microstructure of Targray PE16A Lithium-Ion Battery Separator. doi:http://doi.org/10.5905/ethz-1007-32 (2016); M. F. Lagadec, R. Zahn, and V. Wood, Microstructure of Celgard® PP1615 Lithium-Ion Battery Separator, submitted for publication (2018).

95. R. P. Quirk and M. A. A. Alsamarraie, in *Polymer Handbook*, J. Brandrup and E. H. Immergut (eds.), Wiley, New York, 1989.

96. C. J. Orendorff, *Electrochem. Soc. Interface* **21**, 61–65 (2012).

97. C. J. Weber, S. Geiger, S. Falusi, and M. Roth, "Material review of Li ion battery separators," *AIP Conf. Proc.* **1597**, 66–81 (2014); also see T. Nestler, R. Schmid, W. Münchgesang, V. Bazhenov, J. Schilm, T. Leisegang, and D. C. Meyer, "Ceramic Based Separators for Secondary Batteries," *AIP Conf. Proc.* **1597**, 155–184 (2014).

98. Y. Obana and H. Akashi, "Improvement of Cycle Performance for 4.4V Class 18650 Size Li-ion Battery," in 209th ECS Meeting, The Electrochemical Society, Pennington, NJ, Abstract #105, (2006).

99. K. J. Nelson, J. E. Harlow, and J. R. Dahn, *J. Electrochem. Soc.* **165** (3), A456–A462 (2018).

100. https://docs.google.com/spreadsheets/d/t024bMoRiDPIDialGnuKPsg/edit#gid=154312675, last accessed February 27, 2018.

101. A. H. Zimmerman, "Self-discharge Losses in Lithium-ion Cells," IEEE AESS Systems Magazine, February 2004.

102. R. Yazami and Y. F. Reynier, "Mechanism of Self-discharge in Graphite–lithium Anode," *Electrochim. Acta* **47**, 1217–1223 (2002). See also C. Wang, X. Zhang, A. J. Appleby, X. Chen, and F. E. Little, "Self-discharge of Secondary Lithium-ion Graphite Anodes," *J. Power Sources* **112**, 98–104 (2002).

103. M. Valentin de Hoog, J. Brazil, M. Thomas, and D. Mareels Iven, "Modeling Reversible Self-Discharge in Series-Connected Li-ion Battery Cells," IEEE TENCON Spring 2013—Conference Proceedings, April, 2013. 10.1109/TENCONSpring.2013.6584489.

104. E. Redondo-Iglesias, P. Venet, and S. Pelissier, "Global Model for Self-discharge and Capacity Fade in Lithium-ion Batteries Based on the Generalized Eyring Relationship," IEEE Transactions on Vehicular Technology, Institute of Electrical and Electronics Engineers, September 25, 2017, doi: 10.1109/TVT.2017.2751218.

105. See, for example, the HP laptop recall (http://www8.hp.com/us/en/hp-information/recalls.html), the Lenovo ThinkPad recall (https://support.lenovo.com/us/en/solutions/hf004122), and the Note 7 recall (http://www.samsung.com/us/note7recall/).

106. https://www.theguardian.com/society/video/2016/nov/04/no-smoke-without-fire-e-cigarette-explodes-in-mans-pocket-video.

107. https://www.npr.org/sections/thetwo-way/2016/07/06/484988211/half-a-million-hoverboards-recalled-over-risk-of-fire-explosions.

108. NTSB Incident Report No. NTSB/AIR-14/01, PB2014-108867. Available at https://www.ntsb.gov/investigations/AccidentReports/Reports/AIR1401.pdf.

109. James Graham, "FAA Plans to Impose Its Largest-ever Hazmat Fine," Air Cargo Week, December 14, 2017.

110. Tycho de Feijter, "Visiting the Scene of the May 1 Beijing Electric Bus Charging Station Fire," CarNewsChina.com, August 9, 2017.

111. J. Jiang and J. R. Dahn, *J. Electrochem. Soc.* **153**, A310 (2006).

112. D. D. MacNeil, Z. Lu, Z. Chen, and J. R. Dahn, *J. Power Sources* **108**, 8 (2002).

113. Y. Wang, J. Jiang, and J. R. Dahn, *Electrochem. Commun.* **9**, 2534 (2007).

114. L. Ma, M. Nie, J. Xia, and J. R. Dahn, "A Systematic Study on the Reactivity of Different Grades of Charged Li[Ni$_x$Mn$_y$Co$_z$]O$_2$ with Electrolyte at Elevated Temperatures using Accelerating Rate Calorimetry," *J. Power Sources* **327**, 145–150 (2016).

115. Jim Mc Dowell, Philippe Biensan, and Michel Broussely, "Industrial Lithium Ion Battery Safety—What Are the Tradeoffs?" IEEE document # 978-1-4244-1628-8/07 (2007).

116. IEC 61233.

117. Steven Castle, "Britain to Ban New Diesel and Gas Cars by 2040," N. Y. Times, July 26, 2017. J. Ewing, "France Plans to End Sales of Gas and Diesel Cars by 2040," N. Y. Times, July 6, 2017. See also "Volkswagen plans electric option for all models by 2030," BBC News, September 11, 2017.

118. D. B. Gray, "Tesla Switches on World's Biggest Lithium Ion Battery," Reuters, December 1, 2017. Jordan Golson, "Tesla built a huge solar energy plant on the island of Kauai," The Verge, March 8, 2017.

119. R. Kunzig, "Germany Could Be a Model for How We'll Get Power in the Future," National Geographic, November, 2015.

120. M. M. Thackeray, C. S. Johnson, J. T. Vaughey, N. Lia, and S. A. Hackney, *J. Mater. Chem.* **15**, 2257–2267 (2005).

121. Z. Lu and J. R. Dahn, "Understanding the Anomalous Capacity of Li/Li[Ni$_x$Li$_{(1/3-2x/3)}$Mn$_{(2/3-x/3)}$]O$_2$ Cells Using *In Situ* X-Ray Diffraction and Electrochemical Studies," *J. Electrochem. Soc.* **149** (7), A815–A822 (2002).

122. J. R. Croy, S.-H. Kang, M. Balasubramanian, and M. M. Thackeray, "Li$_2$MnO$_3$-based Composite Cathodes for Lithium Batteries: A Novel Synthesis Approach and New Structures," *Electrochem. Commun.* **13**, 1063–1066 (2011).

123. S. Kang and M. M. Thackeray, "Enhancing the Rate Capability of High Capacity xLi$_2$MnO$_3$ · (1 − x)LiMO$_2$ (M = Mn, Ni, Co) Electrodes by Li–Ni–PO$_4$ Treatment," *Electrochem. Commun.* **11**, 748–751 (2009).

124. A. K. Shukla, Q. M. Ramasse, C. Ophus, H. Duncan, F. Hage, and G. Chen, "Unravelling Structural Ambiguities in Lithium- and Manganese-rich Transition Metal Oxides," *Nat. Comm.*, DOI: 10.1038/ncomms9711 (2015).

125. A. K. Shukla, Q. M. Ramasse, C. Ophus, D. M. Kepaptsoglou, F. S. Hage, C. Gammer, C. Bowling, P. A. H. Gallegos, and S. Venkatachalam, "Effect of Composition on the Structure of Lithium- and Manganese-rich Transition Metal Oxides," *Energy Environ. Sci.* 2018, DOI: 10.1039/C7EE02443F (2018).

126. J.-Y. Hwang, S.-T. Myung, and Y.-K. Sun, "Sodium-ion Batteries: Present and Future," *Chem. Soc. Rev.* **46**, 3529–3614 (2017). Kei Kubota and Shinichi Komaba, "Review—Practical Issues and Future Perspective for Na-Ion Batteries," *J. Electrochem. Soc.* **162** (14), A2538–A2550 (2015).

127. Wong, Bunsen, "Sulfur Based Thermochemical Heat Storage for Baseload Concentrated Solar Power Generation," No. DE-EE0003588 (2014). doi: 10.2172/1165341.

128. U.S. Patent Publication No. 2014/0220392; U.S. Patent No. 9,123,966; U.S. Patent No. 9,893,382.

129. D. Kundu, B. D. Adams, V. Duffort, S. H. Vajargah, and L. F. Nazar, "A High-capacity and Long-life Aqueous Rechargeable Zinc Battery Using a Metal Oxide Intercalation Cathode," *Nature Energy*, vol. 1, Article No.: 16119 (2016), doi:10.1038/nenergy.2016.119.

130. PCT Publication No. WO/2012/012558.

# SECTION B
# LITHIUM METAL ANODE

## Daniel H. Doughty

## 17B.1 OVERVIEW

### 17B.1.1 General Characteristics

Lithium metal (Li)-based batteries were the initial focus for Li batteries up until the late 1980s, but safety concerns and difficulties in achieving satisfactory performance resulted in an overwhelming shift in focus to Li-ion batteries after the commercialization of the Sony Li-ion battery in 1991. Although there are currently only a few manufacturers of Li metal-based batteries, a strong revival of interest has occurred over the past decade or so as part of the focus on "Beyond Li-ion Batteries," which has led to tremendous research efforts worldwide. The principal motivator for this is that Li rechargeable batteries—operating at or near room temperature—offer a very desirable advantage relative to Li-ion (graphite negative electrode) batteries due to their potentially much higher gravimetric and volumetric energy density (Fig. 17B.1).[1] For many other battery characteristics, however, Li-ion batteries have better performance metrics. In addition to increasing energy content, Li metal rechargeable battery development activities have focused on other performance aspects such as achieving a longer cycle life, wider operational temperature limits, and improved safety.

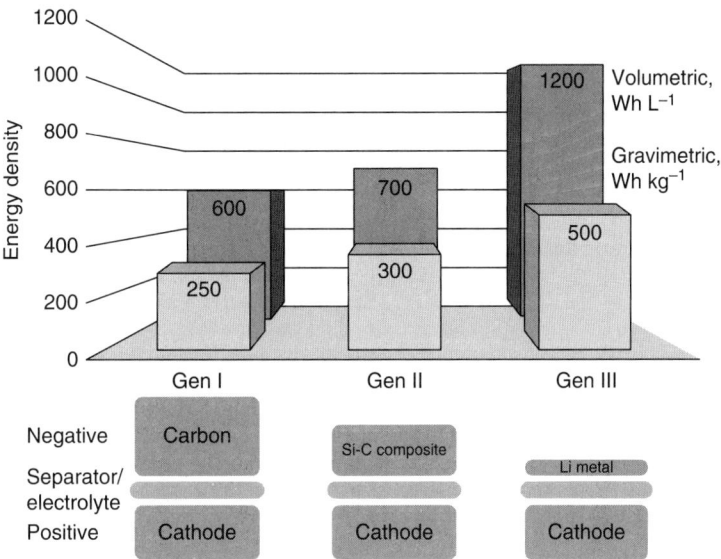

**FIGURE 17B.1** Evolution of Li-based batteries with different anodes (Gen I: graphite, Gen II: Si-graphite composite, and Gen III: thin Li metal). (*From Ref. 1.*)

Commercial sources for Li metal rechargeable batteries provide cells with capacities ranging from <0.1 mAh to several ampere-hours, with many more manufacturers supplying small cells in the mAh size. Bolloré Blue Solutions, however, produces much larger cells (up to 75 Ah). Li metal rechargeable batteries are often categorized based upon the type of electrolyte used: liquid (including ionic liquids [ILs]), solid polymer, or solid inorganic electrolytes. Some cell designs may include combinations of these three electrolyte categories. The positive electrode is generally either an oxygen (air) or a sulfur cathode or one of the metal oxide cathodes typically used in Li-ion batteries. Since the cathode materials used in Li-ion batteries are discussed in Chap. 17A, they will not be reviewed here.

## 17B.1.2   History

In the late 1970s, Armand proposed ionically conducting polymer electrolytes for use in solid-state battery designs,[2,3] and considerable development efforts subsequently ensued. The unique aspect of these batteries is that the electrolyte is a solid flexible film comprised of a polymeric "solvent" matrix with a dissolved lithium salt (LiX). Thin-film solid-polymer electrolyte batteries offered the possibility of improved safety, as well as simplicity of design/manufacturing. Poly(ethylene oxide) (PEO) was the first material utilized as a true (dry) solid polymer electrolyte (SPE),[4] and this remains the polymer of choice to date for such applications. Despite the large amount of research devoted to polymer electrolytes, Avestor (previously Argotech in Canada) batteries were the only ones to reach commercialization. The Avestor battery, originally designed as an electric vehicle (EV) battery module, was redesigned for telecommunication applications and was an outcome of a development program sponsored by the U.S. Advanced Battery Consortium (USABC) and U.S. Department of Energy (DoE). Although Avestor closed operations in 2006, their assets were purchased by Bathium (now Blue Solutions Canada Inc.), and the Bolloré company, Blue Solutions in France, commercialized Li metal polymer (LMP) batteries for the Bolloré BlueCar EV.[5]

Li metal rechargeable batteries with liquid electrolytes and solid cathodes were developed in the early 1980s at EIC (Norwood, MA, USA). These consisted of a Li metal anode, a variety of cathodes (e.g., vanadium oxide, $TiS_2$, and $MoO_xS_{(3-x)}$)[6-10] and an electrolyte solution containing tetrahydrofuran (THF), 2Me-THF, methyl furan (a stabilizer), and $LiAsF_6$. These battery systems were the first practical Li metal-organic electrolyte rechargeable cells that could achieve more than 100 charge/discharge cycles. They operated from –10 to 50°C and had a reasonable stability up to 70°C. They were found to be not safe enough for commercial uses, but could be used for space and military applications.

Another Li metal rechargeable battery that was developed and sold commercially was the molybdenum disulfide ($Li/MoS_2$) battery by Moli Energy (Vancouver, Canada). The $Li/MoS_2$ system was introduced in the mid-1980s in a cylindrical AA size.[11,12] The cell used thin Li metal anodes (125 μm), with a stoichiometric excess of Li of about three times the initial discharge capacity of the cell. The $MoS_2$ cathode slurry was coated onto a thin Al foil (150 μm). The operating voltage was 2.2 to 1.4 V. A spirally wound construction was used. The electrolyte was 1 M $LiAsF_6$ dissolved in a 50/50 mixture of propylene carbonate (PC) and ethylene carbonate (EC). These batteries were withdrawn from the market after several safety incidents occurred in the late 1980s,[13] and in 1990 the company was acquired by a Japanese consortium led by NEC (including Mitsui and Yuasa).

A rechargeable Li metal battery system that became a commercial success during the mid-1990s was developed and produced by Tadiran Batteries Ltd. (Israel). The battery was AA sized and comprised of a $Li_xMnO_2$ (0.3 < x < 1) cathode, an electrolyte solution containing 1,3-dioxolane (DOL), $LiAsF_6$, and tributylamine (TBA; a stabilizer) in trace amounts. The operating voltage was between 3.4 and 2.0 V. This battery system had several attractive features including an energy density >140 Wh/kg, a wide temperature range of operation (from –30 to 60°C), an excellent shelf life, a reasonable cycle life (more than 300 cycles at 100% depth of discharge [DOD]), and an internal safety mechanism.[14] This safety mechanism, which made this battery safe for use and of commercial value, was based upon a shutdown of the battery in abuse cases when heating occurred above 130°C, such as from short-circuiting and overcharge, via a rapid polymerization of the solvent.[15] This battery had a limited cycle life, however, unless the charging rates used were very low (i.e., <C/9, corresponding to current densities less than 0.5 mA/cm²) to avoid the extensive formation of mossy Li causing a premature end of cell life.[16] Because Li-ion cells became available with similar energy and longer cycle life, Tadiran discontinued marketing this cell.

The Li-air ($Li-O_2$) rechargeable battery system has an exceedingly high theoretical energy of 5200 Wh/kg (including oxygen). Working cells with high-temperature ceramic electrolytes (operating at 650°C) were reported in 1987,[17] and room-temperature organic electrolyte cells were first demonstrated by workers at EIC in 1996.[18,19] Open-circuit voltages approached 3.0 V and the working voltage was between 2.8 and 2.0 V. The cell used a poly(acrylonitrile) (PAN) polymer electrolyte membrane to which PC and salts were added and a cobalt compound catalyzed the air electrode. Metal-air secondary batteries are covered in Chap. 18B and 18C.

The lithium-sulfur (Li-S) rechargeable battery, which has a theoretical specific energy of 2500 Wh/kg, the highest of any sealed rechargeable system, was first reported in 1979 by Rauh and coworkers[20] at EIC. They observed good electrochemical reversibility at low current and elevated temperature (50°C). It was recognized that the solubility of polysulfide reaction products from sulfur reduction ($Li_2Sx$ polysulfides) strongly influences the performance of the cathode. The final discharge product, $Li_2S$, was insoluble and electronically insulating, which was a challenge for this system. The operating voltage was 2.3 to 1.7 V.

Thin-film, solid-state batteries are a specialized type of Li metal rechargeable battery developed for low-current semiconductor and microelectronic applications. These microbatteries—which employ a metallic Li anode, solid electrolyte, and a transition metal oxide cathode material—can be fabricated by high-volume manufacturing techniques on silicon wafers and are viable as on-chip or onboard power sources for microelectronics. The first Li battery of this type was enabled by the fabrication of a lithium phosphorus oxynitride (LiPON) glassy electrolyte by Bates.[21] LiPON exhibits a single Li-ion conducting phase between –26 and 140°C, with an average conductivity of $2.3 \times 10^{-6}$ S/cm at 25°C. LiPON is mechanically stable and has a 5 V stability window. Despite being rigid, it does not crack—since the anode and cathode undergo limited volume changes during cycling due to the low amount of Li plated/stripped—as evidenced by the long cycle life of these thin-film batteries. $Li/LiCoO_2$ batteries have been cycled over 40,000 times at 25°C to ~96% DOD with total capacity losses of less than 5%.

There is now strong renewed commercial interest in developing Li metal renewable cells with Li-ion cathodes and large companies (e.g., BASF, Toyota, BMW, Bosch, and Dyson) have acquired smaller companies and increased their investments in this area, although Bosch and Dyson have since ceased work on Li metal battery production related to the companies (SEEO and Sakti3) that they acquired. Safety and stability, however, still remain as impediments for commercialization. Additional details and information on other Li metal rechargeable batteries that were developed but not commercialized in the 1990s and earlier can be found in previous editions of this Handbook and will not be discussed here.

## 17B.2 CHEMISTRY

The objective of Li metal rechargeable battery development is to produce batteries that have high energy density, high power density, good cycle life, and high charge retention—provided reliably and safely. The judicious selection of cell components and designs is thus necessarily a compromise to achieve an optimum balance. Many of the characteristics and criteria for selection are similar to those for the primary Li batteries (Chap. 13) and Li-ion rechargeable batteries (Chap. 17A). The design process, however, is even more complex for Li metal rechargeable batteries since the reactions and volume changes that occur—especially during recharge (Li plating)—affect all of the cell characteristics and the performance on subsequent cycling.

### 17B.2.1 Negative Electrode

The search for high energy density primary and rechargeable batteries has inevitably led to the use of Li, as the electrochemical characteristics of this metal are uniquely favorable. Li, as the lightest and one of the most electropositive metals (dependent upon the electrolyte used), has a high specific capacity of 3862 mAh/g[22] (as compared to 372 mAh/g for the lithiated graphite $LiC_6$ used as the anode in Li-ion batteries). The volumetric comparison is also favorable for Li metal (2061 Ah/L as compared to 837 Ah/L for $LiC_6$), but the low density of Li metal (0.534 g/cm$^3$) means that the advantage is not as great. Li is also more easily handled (less reactive to $H_2O$ and contaminants) than the other alkali metals.

It is noteworthy that experimental studies have shown that Li metal tends to plate from solutions as either kinked whiskers/filaments or instead as particle-like nodules (Fig. 17B.2)[23,24] rather than as dendritic structures, but the term "dendrite" will be retained in this chapter due to its common usage. Some of the images and illustrations of Li morphology in publications are inaccurate depictions of the actual experimentally observed

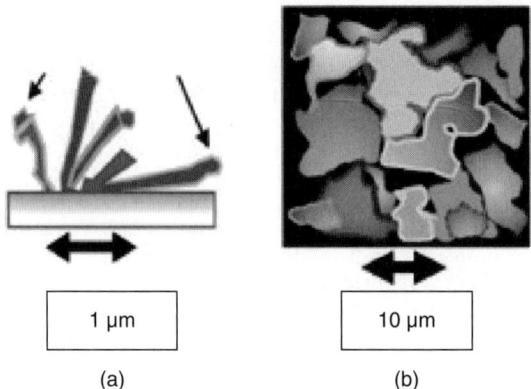

| 1 µm | 10 µm |
|:---:|:---:|
| (a) | (b) |

**FIGURE 17B.2**   Li deposition (*a*) 1 M EC/DMC (1/1 w/w)-LiPF$_6$ and (*b*) 1 M DOL-LiAsF$_6$ (with TBA additive). (*From Refs. 23 and 24.*)

structures. Although such images or related illustrations have a visual appeal, they can lead to misconceptions about the actual morphology of Li deposits.[25,26]

One important consideration for Li that differs from Li-ion batteries is that the low density of Li metal results in a significant volume change when cycling a Li metal rechargeable battery.[27–29] Figure 17B.3*a* and 17B.3*b* contrasts the difference in volume during cycling for a Li$_y$MnO$_2$/Li$_x$MnO$_2$ cell (in which Li$^+$ cations are deintercalated from one electrode and intercalated into another) with that for a comparable Li/Li$_x$MnO$_2$ cell in which Li$^+$ cations are deintercalated from the cathode and plated on the anode as Li metal during recharge.[27,29] This substantial volume change induces significant mechanical variations within a cell during cycling and is an important factor for cycling performance, especially for cells with rigid electrolytes.

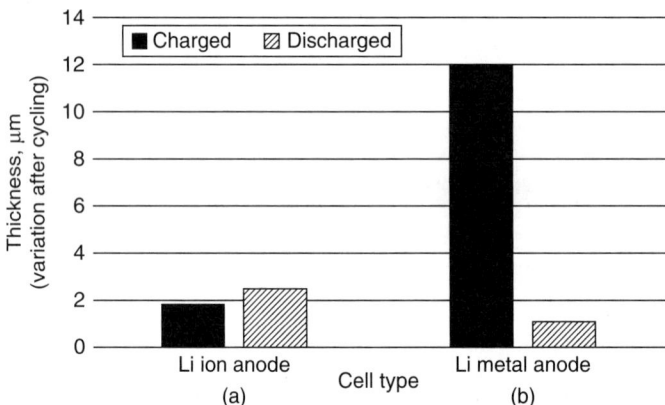

**FIGURE 17B.3**   Thickness variations of (*a*) Li$_y$MnO$_2$ (anode)/Li$_x$MnO$_2$ (cathode) at 130 psi stack pressure and (*b*) Li (anode)/Li$_x$MnO$_2$ (cathode) (200 psi) for flat-plate cells with 1 M PC/EC (85/15 v/v)-LiAsF$_6$ at different states of charge. (*From Ref. 27.*)

In contrast to Li primary batteries, commercial success has been a challenge for all but very small capacity Li metal rechargeable cells due to persistent safety and durability (cycle life) problems. Dendrite formation (i.e., filaments) and the resulting short-circuiting when the Li extends across the electrolyte and makes physical contact with the cathode (thus permitting a direct, uncontrolled reaction) is a primary failure mechanism (a low impedance failure) in Li metal rechargeable batteries.[30] Short-circuits can lead to thermal runaway (characterized by uncontrolled cell self-heating leading to fire or an explosion). Such deposits were first hypothesized to occur in Li/organic systems in 1974[31] and were first directly observed in 1980.[32] Dendrites were linked to cell failure in 1988.[33] Note that dendritic Li is problematic for another reason as well. The thermal stability of Li metal foil in many organic electrolytes is sufficient early in cell life, with minimal exothermic reactions occurring up to temperatures near the melting point of Li (180°C). But after extended cycling, the surface area of the Li increases significantly with a corresponding increase in the reactivity, which degrades the thermal stability of the system with the result that cells become increasingly sensitive to thermal abuse as they are cycled.

There is also a second principal mechanism for cell failure that often receives much less attention. In a Li-ion cell, the electrolyte is tailored to react during the first few cycles with the graphite surface to generate a solid electrolyte interphase (SEI) surface layer that largely passivates the electrode surface from further reactions in subsequent cycling. Some of the active material is consumed, but the optimization of the initial cycling conditions and selection of additives and other electrolyte components minimizes this and ensures the creation of a relatively stable and highly conductive (to Li$^+$) SEI. None of this occurs for a Li metal negative electrode. New Li surfaces are created during plating (and to some extent during stripping) in each cycle. These fresh Li surfaces react with the electrolyte to generate a new SEI. The larger the surface area of the metal, the greater the extent of the reactions. This then leads to the formation of "mossy Li" via the consumption of the Li as SEI reaction products and the entrapment of Li metal that is electrically isolated from the electrode (i.e., dead Li; see Fig. 17B.4).[34] Thus, a dendritic Li plating morphology (filaments) generally results in rapid Li degradation. The formation of mossy Li, depletion of electrolyte solvent ("drying out" of the cell), and possible continuous gas formation from these reactions increase the cell resistance, as well as the volume of the cell, often resulting in a (high impedance) cell failure.[23,30,35-37]

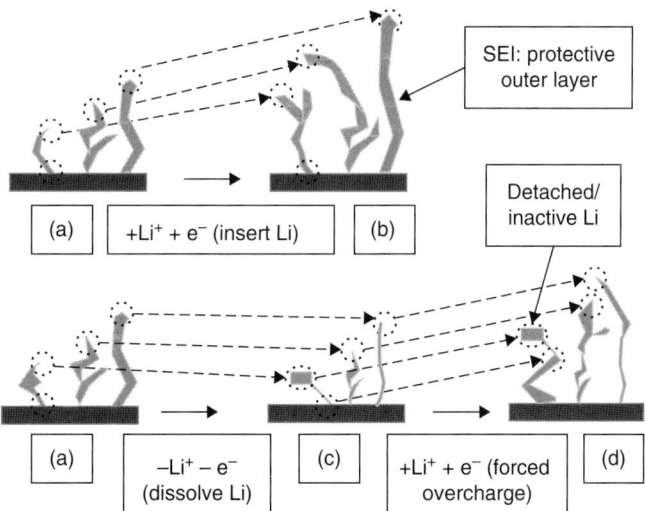

**FIGURE 17B.4** Growth mechanism of "mossy Li" including (*a*) an SEI layer as deposited, and (*b*) after further electrodeposition. Dashed line circles highlight the increased distance between these features over time due to Li deposition. The overall increased heights indicate the expansion of the total structure. After a dissolution step (*c*) the tips of the structure still contain Li metal ("dead Li") that is electrically isolated from the substrate but is still being held in place by the former SEI shell. After an additional electrodeposition step (*d*), the tops rise upward from new lithium moss growing underneath. (*From Ref. 34.*)

Dendrite formation was analytically modeled by Monroe and Newman[38] and later by Ferrese and Newman[39] for a system with a Li anode and a polymer electrolyte or separator. Their models predicted that the introduction of a highly rigid electrolyte between the two electrodes (the elastic modulus of the electrolyte exceeding ~1 GPa, compared to PEO having an elastic modulus of <1 MPa) would transmit pressure, thereby generating a smooth Li interface. Applied stack pressure in cells has been shown to stabilize the Li surface by suppressing or eliminating dendrite formation, and thus reducing mossy Li formation (Fig. 17B.4), resulting in improved cycling coulombic efficiency (CE) and increased cell cycle life.[27–29,40–45] Other factors that influence the Li metal morphology include the charge conditions utilized—during both plating and stripping,[46–50] as well as the electrode surface structure.[44,51,52] While these latter factors may be of limited utility for practical cells, they do provide critical insight during research studies about the determinants that govern the Li negative electrode performance. Another important consideration for research studies is the current collector used with the Li negative electrode. Depending upon the metal used, Li can, in some cases, alloy with the current collector, and this is relevant for some of the research results reported in the scientific literature for "Li metal" plating. At ambient temperature, Li does not alloy with Cu—nor does it alloy with Ni, Ti, W, Mo, Ta, or Fe (and thus stainless steel), but Li does alloy with Al, Ag, Au, Pd, and Pt.[46]

A recent paper has explored in depth the methodologies available to accurately determine the CE for Li plating/stripping.[53] In addition to cycling CE, however, another crucial consideration for practical Li metal rechargeable batteries that is often overlooked in research studies is the effect of rest/storage time. Chemical reactions between the Li metal and electrolyte may occur if the Li surface is not fully passivated during plating. This has been referred to as "Li corrosion," "Li self-discharge rate," or "Li encapsulation."[54–57] These ongoing chemical reactions at the negative electrode during storage can result in the continuous loss of electroactive Li metal with time (Fig. 17B.5).[56] Storage time-based Li corrosion measurements, perhaps using the CE determination methodology,[53] should be widely adopted for Li metal rechargeable battery research studies and cell evaluations to aid in determining the merit of utilizing new materials/compositions for practical cells.

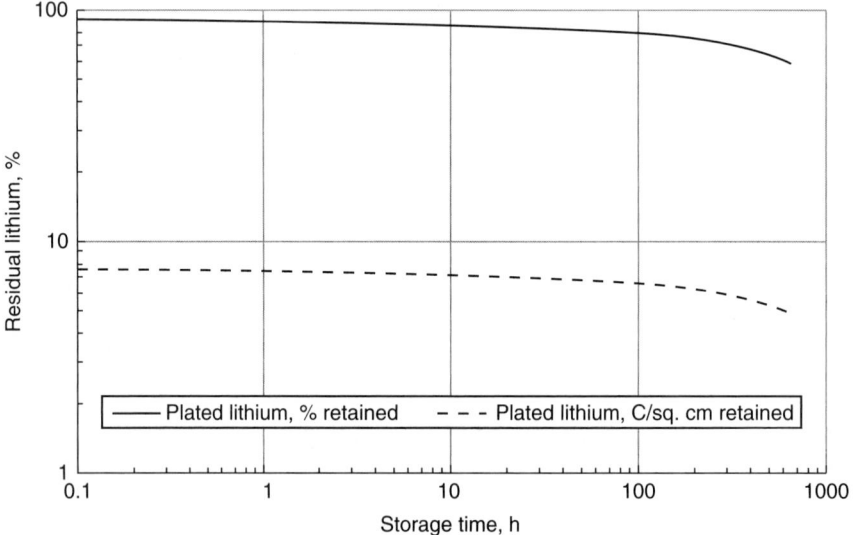

**FIGURE 17B.5**    Corrosion of Li plated onto a Ni current collector in a 1 M methyl acetate (MA)-LiAsF$_6$ electrolyte. (*From Ref. 56.*)

Eliminating the generation of high surface area Li over the life of the cell will substantially improve cycle life and safety. Many recent reviews have been published summarizing the efforts utilized to change the Li plating morphology and reactivity via improved electrolyte formulations, as well as the use of additives, protective layers, and solid electrolytes.[1,46,58–71] The failure to control the growth of the surface area of the Li anode remains a critical problem, limiting the commercialization of Li metal cells. More rapid plating of Li is often found to result in increased amounts of degradation reactions (Fig. 17B.6).[72] Finally, there is also often a trade-off between anode DOD and cycle life (measured by cumulative capacity attained during all cycles). The general trend observed is the less the Li plating per cycle ($\mu$m plated or mAh/cm$^2$ per cycle), the greater the cumulative capacity and longer the cycle life attained. This is likely attributable to both the degradation (loss of Li, drying out of solvent, generation of mossy Li) and electrode volume changes (expansion/contract and mossy Li buildup) noted above.[73] This is another important consideration when evaluating published research results and the merits of new innovations for practical cells.

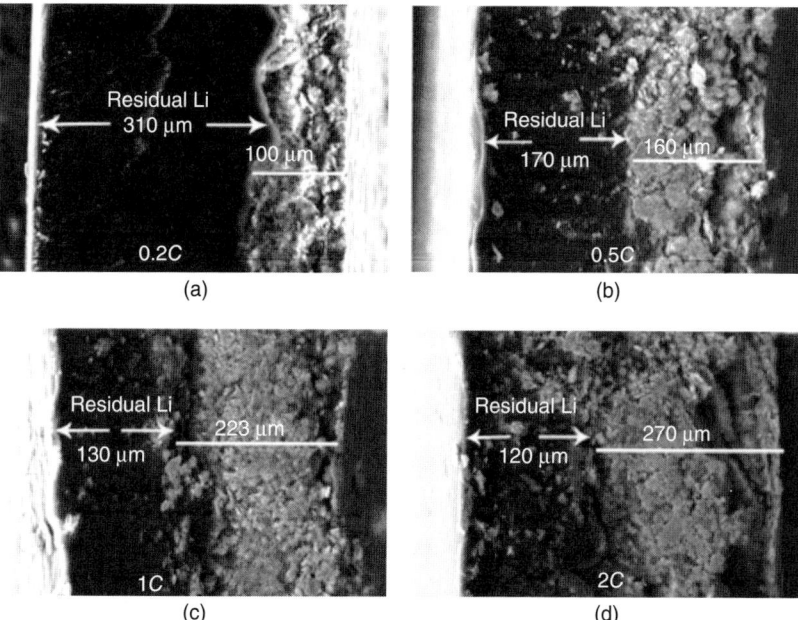

**FIGURE 17B.6**    Li anodes extracted from Li/NMC cells after 100 cycles cycled at different rates (charge/discharge): (*a*) 0.2C/1C, (*b*) 0.5C/0.5C, (*c*) 1C/1C, and (*d*) 2C/2C (the mossy Li layer previously in contact with the separator is on the right with the thickness indicated). (*From Ref. 72.*)

## 17B.2.2    Positive Electrodes

A wide variety of materials exist that may be selected for the positive electrodes of Li metal rechargeable batteries. Cathode materials that are used in Li-ion batteries are also suitable for Li metal rechargeable batteries. Li-ion battery cathode materials will not be discussed here except where specific information relates to Li metal rechargeable batteries. In addition, other cathodes are utilized. Significant work has been devoted to enabling Li metal rechargeable batteries with an oxygen (or air) cathode, but the open-air structure of the positive electrode and complex, often irreversible, chemical reactions associated with this technology continue to be challenging. For sealed battery systems, however, sulfur is particularly well matched with Li metal as both electrode active materials have very high capacities.

*Oxygen (Air) Cathode.*    The exceptional theoretical value for Li-air batteries resulted in them being widely marketed by researchers as capable of achieving an energy density 10 times that of Li-ion batteries. Tremendous research efforts were thus devoted to Li-O$_2$ rechargeable batteries over the past eight or so years, but—despite this—many of the same challenges remain that represent seemingly insurmountable hurdles for widespread commercialization of these batteries. These include the low round-trip energy efficiency (due to sluggish reaction kinetics and the large cathode overpotential necessary for recharge), low practical capacity, and poor cycle life.[74–79] For the Li metal negative electrode, the reactions with O$_2$, CO$_2$, H$_2$O, and other contaminants from the cathode are particularly vexing.[80,81] This has led many research groups to redirect their efforts toward other battery systems, such as Li-S batteries. The Li-air battery system is discussed in Chap. 18.

*Sulfur Cathode.*    The reduction of elemental sulfur (S$_8$) to sulfide ions (S$^{2-}$ as Li$_2$S) has a theoretical capacity of 1675 mAh/g, which is about an order of magnitude higher than that of the lithiated transition metal oxides used in Li-ion rechargeable batteries. Two approaches have been explored: incorporating either elemental sulfur or lithium polysulfides as solids into the C-S cathode or adding the active material to the electrolyte as a liquid (sulfur in the form of dissolved $Li_2S_x$ polysulfides) (i.e., a catholyte). For the latter, early work[20] showed that solvents with high basicity could dissolve large amounts of lithium polysulfides. In dimethyl sulfoxide (DMSO) or ethers such as THF,[82] the sulfur solubility as $Li_2S_x$ can exceed 10 M. Spectroscopic and electrochemical studies of polysulfides in nonaqueous solutions suggest that their dynamic equilibrium, redox chemistry, and kinetics are strongly affected by the solvent and its concentration.[20,83–87]

Solid C-S cathodes are typically prepared by coating a slurry that contains elemental sulfur, various forms of carbon (since sulfur and $Li_2S$ are insulators, an electronically conducting matrix is required), and a binder. The sulfur is often impregnated within the carbon in some manner. A variety of positive electrode current collectors may be used, but Al foil is common. Sulfur is reduced in a stepwise fashion and a series of lithiated polysulfides are formed. The first ambient temperature discharge of a Li-S cell consists of two plateaus, as shown in Fig. 17B.7. Regions 1 to 4 identify areas where different polysulfide species predominate. It is important to note that the sulfur species are in equilibrium with each other, so a mixture of species is expected to exist in the cathode (or electrolyte) at all times.

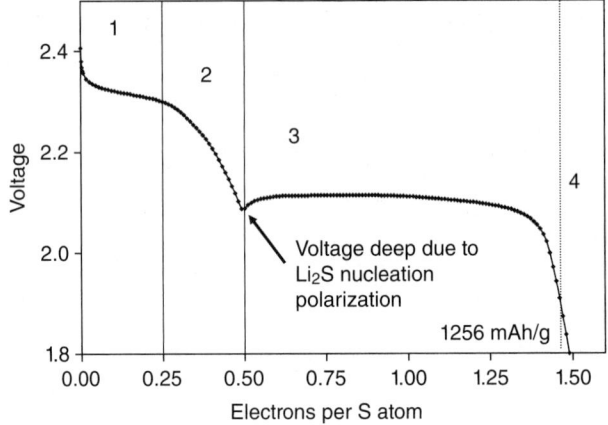

**FIGURE 17B.7**   First discharge curve of a Li-S cell. Sulfur utilization up to 1256 mAh/g sulfur was achieved at low rate (C/30). (*Courtesy of SION Power Corp., Tucson AZ.*)

In region 1, sulfur is reduced to form $Li_2S_8$ according to the reaction

$$S_8 + 2e^- + 2Li^+ \leftrightarrow Li_2S_8$$

The reduction continues with the formation of $Li_2S_6$ in region 2. Region 3 contains the lower-order polysulfides, $Li_2S_4$, $Li_2S_2$, and $Li_2S$. Although $Li_2S$ tends to be highly insoluble, the other polysulfides are soluble to varying extents in different solvents such as DOL, 1,2-dimethoxyethane (DME), and longer glymes and the other solvents and ILs that are commonly used in Li-S cells.[84–87] Note that the anion(s) in the electrolyte also influence(s) the polysulfide solubility.[85,86] Full reduction to $Li_2S$ is achieved only at low discharge rates because of the high polarization (region 4) that is caused by cathode porosity (i.e., electrolyte access) being blocked by the solid $Li_2S_2$ and $Li_2S$ reaction products (including sulfur redistribution at the electrolyte-cathode interface) and electronic isolation of the reactants due to the insulating nature of these reaction products. Polysulfide solubility and the accessibility of the reactants (i.e., $e^-$ and $Li^+$) are primary factors that limit full sulfur utilization.

The Li-S electrochemical performance is strongly influenced by the polysulfide shuttle mechanism[88] (Fig. 17B.8), which has been described and modeled.[89,90] This shuttle reaction affects many cell properties including self-discharge, charge-discharge efficiency, and the charge profile. The recharge of a Li-S cell produces, to some extent, high-order polysulfides (e.g., $Li_2S_8$) rather than elemental sulfur during the latter stages of the charge.[88] These longer polysulfides dissolve into the electrolyte and diffuse to the Li negative electrode where they react directly with the Li in a parasitic reaction to create lower-order (shorter) polysulfides. The dissolved polysulfides may also react during discharge at the cathode electrolyte-carbon interface to generate an

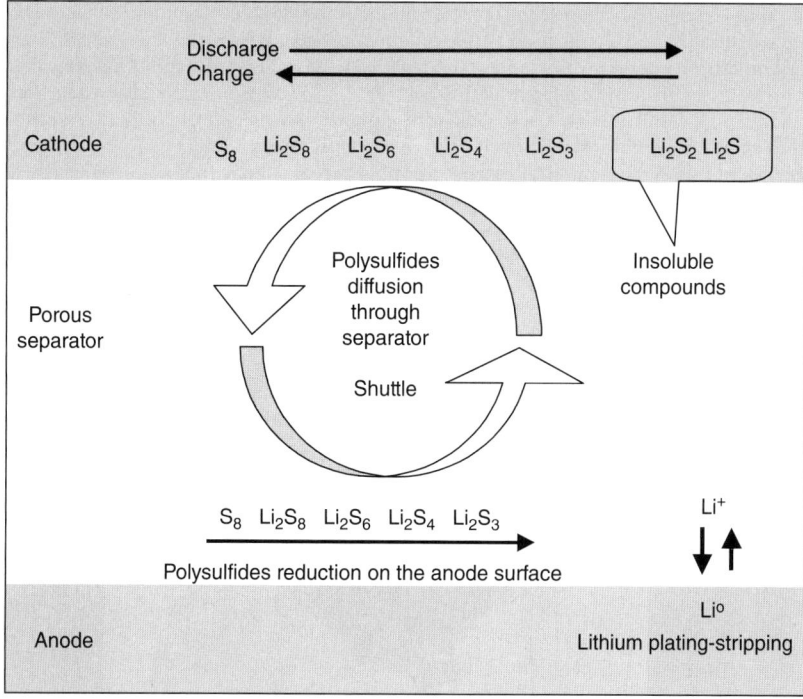

**FIGURE 17B.8**    Shuttle mechanism for Li-S batteries. (*From Ref. 88.*)

insulating layer of insoluble shorter polysulfides (i.e., sulfur redistribution) that further restricts cell reactions.[90,91] Storage time-based Li corrosion measurements may therefore be particularly important for evaluating Li-S batteries.

### 17B.2.3    Electrolytes

The choice of electrolyte for rechargeable Li batteries is critical. Ideally, the liquid electrolyte should have the following characteristics for Li-ion batteries:

- High ionic conductivity from −40 to 70°C ($\sim$10$^{-3}$ S/cm at 20°C) to minimize internal resistance
- Li$^+$ cation transference number approaching unity (to limit concentration polarization)
- Wide electrochemical voltage window (0–5 V), although this upper limit requirement is cathode dependent
- Thermal stability (up to 70°C or higher)
- Chemical and electrochemical compatibility with the cathode, separator, cell body, and other cell components
- Low volatility and flammability

In addition, the electrolyte must have the following characteristics:

- It should enable the plating of Li metal in a nodular, rather than dendritic (i.e., filaments) morphology.
- It should have a high chemical stability or passivation ability with Li metal to prevent a continuous side reaction on storage (Li corrosion).

***Liquid Electrolytes.***    When selecting aprotic liquid electrolyte solvents, ethers (such as DOL, DME, and other glymes) are the most prevalently used because of their low reactivity with Li. In contrast, carbonates such as EC, PC, and dimethyl carbonate (DMC) are more reactive with Li. Solvent stability can be gauged by consideration of the highest occupied molecular orbital (HOMO) and lowest unoccupied molecular orbital (LUMO) from molecular orbital (MO) theory.[92] A solvent with a higher (less negative) HOMO is a stronger electron-donor (and thus more susceptible to oxidation at the cathode), whereas a lower LUMO indicates that the solvent is a stronger electron-acceptor (and thus more susceptible to reduction at the Li anode). Figure 17B.9 shows that the solvent's MO energy values can be grouped together based upon the solvent functional groups and structure.[92] The HOMO/LUMO data suggest that the linear and cyclic carbonates are expected to be fairly stable to oxidation, but have poor reduction stability (i.e., stability to Li). In contrast, ethers are very stable to reduction by Li, but they are more reactive at the cathode at elevated potentials (generally >4 V vs. Li/Li$^+$). Thus, the use of ether solvents is often precluded for use with many battery cathodes due to this low oxidative stability, although this is not the case for sulfur due to its lower reaction potentials (Fig. 17B.7).

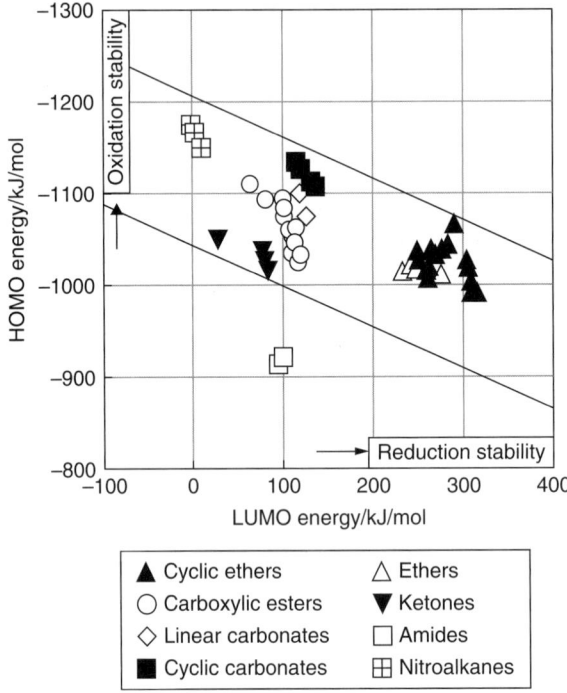

**FIGURE 17B.9**   Molecular orbital energy of solvents. (*From Ref. 92.*)

Note that this electrochemical stability is strongly influenced by the solvent interactions. When a solvent forms a coordinate bond to a Li$^+$ (i.e., when it donates electron density via an electron lone-pair to the electron-deficient cation), it is more difficult to oxidize the solvent (remove an electron),[93-95] but the coordinated solvent is also easier to reduce. The same factors hold true for uncoordinated and coordinated anions. This would suggest that highly concentrated electrolytes may enable ether usage with higher-voltage Li-ion intercalation battery cathodes, but this is generally not the case as the stability enhancement is a kinetic rather than thermodynamic effect. Degradation on the cathode occurs more slowly than for dilute (e.g., 1 M) electrolytes when such cells are held at high potential (i.e., fully charged), but it still transpires.[96]

Additives have been one central focus to improve Li plating from liquid electrolytes, but these have generally been ineffective due to their rapid depletion on cycling and other factors. For example, the addition of small amounts of Cs+ ions (i.e., 0.05 M CsPF$_6$ added to 1 M PC-LiPF$_6$) was reported to prevent dendrite formation due to a proposed self-healing electrostatic shielding (SHES) mechanism.[97] Although the initial Li plated from such electrolytes produces a Li deposition with a mirrorlike smooth surface due to Li columnar aligned-nanorod formation, this morphology results in very poor Li cycling due to a low CE (e.g., 76%) and the resulting rapid accumulation of an interphasial layer from mossy Li.[98,99] A remarkably similar plating morphology can also be achieved through the addition of small amounts of $H_2O$ to a 1 M PC-$LiPF_6$ electrolyte.[100] The most successful electrolyte additive to improve the performance of the Li anode has been $LiNO_3$.[101–109]

ILs are another class of "liquid solvents" that possess important attributes that make them attractive for use in electrochemical applications, and Li metal rechargeable batteries have been one such application. These attributes for some ILs include a wide electrochemical stability window, high thermal stability, and low safety hazards. The limitations to be overcome include high viscosity (especially with incorporated lithium salts), resulting in poor wetting of battery electrodes and separators, and the relatively low ionic conductivity of IL-LiX electrolytes at room temperature, not to mention purity difficulties and high cost. The principal justification for using ILs is their nonvolatile—and thus low flammability—property. Uniform, nondendritic Li metal plating has been obtained when cycling using IL-based electrolytes, but so have dendritic morphologies.[110–114] The variations may be due to differences in the SEI coating that has different characteristics from that with aprotic solvents, frequently being composed principally of anion degradation products (but also some components from the organic cations).[115–117] Notably, there is very little information available about the CE[53] for Li cycled in ILs, so it is difficult to assess the efficacy of such electrolytes for Li plating/stripping relative to those based on aprotic solvents.

***Solid Polymer Electrolytes.***    An alternative to the liquid electrolytes is an SPE formed by incorporating lithium salts into polymer matrices and casting into thin films.[118] These films can function both as the electrolyte and the separator. SPE electrolytes have a lower ionic conductivity relative to the liquid electrolytes, but the PEO-based electrolytes have a low reactivity with Li, which enhances the safety of the battery. The solid polymers offer the design advantages of a "nonliquid" battery, allowing flexibility to manufacture thin batteries in a variety of configurations. Initially, high-molecular-weight polymers such as PEO and lithium salts such as $LiClO_4$ and LiTFSI were used.[119,120] These PEO-lithium salt electrolytes have conductivities on the order of $10^{-5}$ to $10^{-8}$ S/cm at 25°C, which is two or more orders of magnitude lower than that of most organic liquid electrolytes. To improve the $Li^+$ transport properties of the SPE, gelling agents such as molecular solvents or ILs can be added to PEO-LiX and other SPEs.

Gelled electrolytes are another form of SPE prepared by trapping liquid solutions of lithium salts in aprotic organic solvents (for example, $LiClO_4$ in PC/EC) into a solid polymer matrix such as poly(vinylidene fluoride) (PVDF)[121] or PAN.[122,123] Such electrolytes are made by filling the polymer's porosity with the liquid electrolyte solutions via an immobilization procedure such as crosslinking, gelling, or casting. Crosslinking may be carried out by ultraviolet, electron-beam, or gamma-ray irradiation. Conductivities exceeding $10^{-3}$ S/cm at 25°C have been obtained. The stability of the gelled electrolytes with Li often is comparable to that of the corresponding liquid electrolyte with Li.

Single-ion conducting SPEs (ionomers) designed for use with Li metal anodes have also been prepared such as graft copolymers and comb-branched polyepoxide ethers.[124,125] The flexibility of tailoring the mechanical properties of graft copolymers, while maintaining a $Li^+$ transference number of unity,[126] is an attractive characteristic of these SPEs. Moreover, depending upon the polymers utilized, the voltage stability window can be extended to 4.5 V versus Li/Li$^+$ (more than 0.5 V higher than for the PEO-based polymer electrolytes). But a low conductivity requiring cell operation at elevated temperatures (60–80°C)[127] and difficult synthesis procedures remain key limitations for this approach.

***Solid Inorganic Electrolytes.***    Solid inorganic electrolytes can either be crystalline or amorphous (glassy) materials. Some solid inorganic electrolytes have been prepared by thin-film deposition techniques, but for other materials the powders are difficult to fabricate into commercially attractive large area, thin solid electrolytes for electrochemical cells.

The LiPON thin-film glassy solid electrolyte invented at Oak Ridge National Laboratory in the early 1990s is the most widely used solid electrolyte for thin-film batteries.[128] The key insight by Bates was that addition of nitrogen to the glass structure might enhance the chemical and thermal stability of a lithium glass, as it does for sodium phosphate and sodium silicate glasses. The LiPON electrolyte is deposited by radio frequency (RF) magnetron sputtering from a ceramic target of $Li_3PO_4$ using a nitrogen process gas to form the plasma.[129,130]

The films are amorphous and free of any columnar microstructure or boundaries. With a nitrogen/oxygen ratio as small as 0.1, the ionic conductivity is $10^{-6}$ S/cm, which is about 40-fold higher than for glassy films of nitrogen-free $Li_3PO_4$. More importantly, the cationic transport number of $Li^+$ is unity, the electrochemical stability extends to 5.5 V versus $Li/Li^+$ and LiPON is stable at both elevated temperatures and in contact with Li.[21] The conductivity, although three orders of magnitude lower than for many liquid electrolytes, is sufficient because 1 µm-thick films are adequate to create a pinhole-free barrier over most thin-film electrodes. Furthermore, the electronic resistivity of LiPON is very high, greater than $10^{14}$ Ω cm. Thus, LiPON meets the requirements for thin-film batteries, but it is unsuitable for larger batteries where a thicker electrolyte is required.

A wide variety of inorganic electrolytes are known with varying conductivity and performance (Fig. 17B.10).[131–135] Some sulfide-based inorganic electrolytes—such as $Li_7P_3S_{11}$, $Li_{10}GeP_2S_{12}$ (LGPS), and $Li_{9.54}Si_{1.74}P_{1.44}S_{11.7}Cl_{0.3}$—are reported to have an exceptional conductivity of $>10^{-2}$ S/cm at 25°C.[131,135–139] Most sulfides, however, have poor stability when in contact with Li, as well as with cathodes,[140–142] which has restricted their usage. In contrast, oxide materials such as garnetlike $Li_5La_3M_2O_{12}$ (M = Ta, Nb, Ba, or Zr) have garnered significant attention since they are reported to have excellent stability with both Li and common cathode materials, but their relatively low conductivity (~$10^{-5}$ S/cm) has also limited their applicability.[131,135,143] Other solid inorganic electrolytes include $Li_3N$, hydrides (e.g., $LiBH_4$, $LiBH_4$-LiX, $LiNH_2$, etc.), halides (e.g., LiI, $Li_3OCl$, etc.), and borates or phosphates (e.g., $Li_2B_4O_7$, $Li_3PO_4$, etc.).[135] The hydrides are stable with Li and have a low grain boundary resistance, but they are sensitive to moisture and incompatible with many cathodes. Similar characteristics are found for the halides. All of these materials tend to have a relatively low ambient temperature conductivity ($10^{-4}$–$10^{-8}$ S/cm).

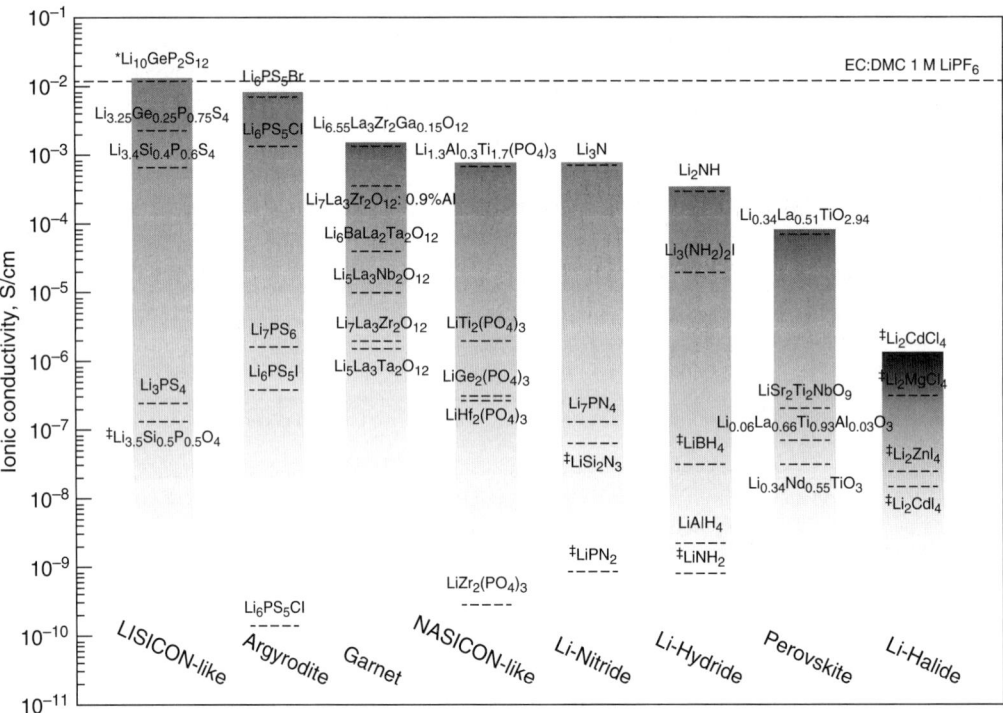

**FIGURE 17B.10**   Conductivity of inorganic conductors at room temperature. The conductivity of 1 M EC/DMC-$LiPF_6$ is shown for comparison as a dashed gray line.[132]

Note that first-principles calculations predict that most of the inorganic electrolytes have a limited electrochemical stability (Fig. 17B.11).[144] It was suggested that the favorable stability noted experimentally in some cases arises not from an inherent thermodynamic stability, but rather from sluggish reaction kinetics and interphases (due to reactions) that form between the inorganic electrolytes and electrodes.[144]

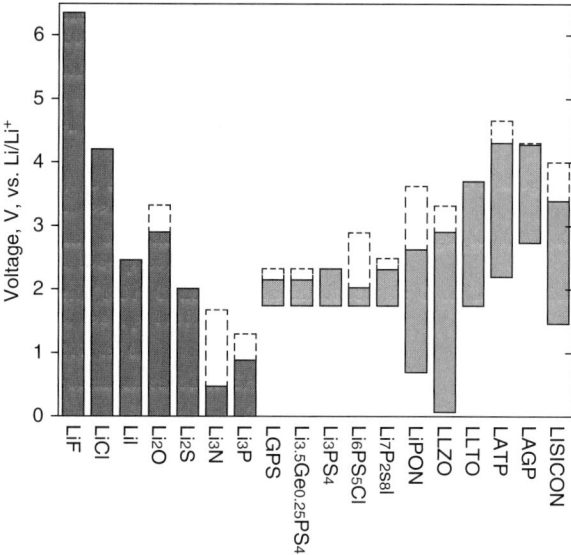

**FIGURE 17B.11**  Electrochemical stability widow (from first-principles calculations) (solid color bar) of inorganic electrolytes and other materials. The oxidation potential to fully delithiate the material is marked by a dashed line. (*From Ref. 144.*)

## 17B.3   CELL DESIGNS AND COMPARISONS

As noted above, the different types of ambient-temperature Li metal rechargeable batteries can be classified into three design categories:

- Liquid electrolyte cells
- SPE cells
- Solid inorganic electrolyte cells

The components, chemical reactions, and performance characteristics of typical examples of the three types are summarized and compared below.

### 17B.3.1   Liquid Electrolyte Cells

Li metal rechargeable batteries with liquid electrolyte are of high interest because of their advantage in gravimetric and volumetric energy density relative to Li-ion rechargeable batteries. The electrolytes employed, however, differ markedly from those for Li-ion batteries in terms of the solvent and salt used, as well as the salt concentration and additives.

***Liquid Electrolyte Intercalation Cathode Cells.***    As noted above, intercalation cathode materials that are useful in Li-ion rechargeable batteries may also be used in Li metal rechargeable batteries (see Chaps. 3A and 17A). The principal barrier to the development of practical commercial cells is controlling the surface morphology of the Li anode on cycling. Unfortunately, optimization of the electrolyte for use with the Li negative electrode often results in unfavorable characteristics for the use of the same electrolyte with high-voltage metal oxide cathodes. For conventional aprotic solvent-based electrolytes, the Li plating/stripping CE is strongly dependent upon the solvent (Fig. 17B.12), salt (Fig. 17B.13), and additives (e.g., VC, FEC, etc.) used.[145]

**FIGURE 17B.12**    SEM images of Li deposition morphologies from 1 M solvent-LiPF$_6$ electrolytes with (*a*) PC, (*b*) EC, (*c*) DMC, and (*d*) EMC (ethyl methyl carbonate) (CE noted as %). (*From Ref. 145.*)

**FIGURE 17B.13**    SEM images of Li deposition morphologies from 1 M PC-LiX electrolytes with (*a*) LiBOB, (*b*) LiPF$_6$, (*c*) LiAsF$_6$, (*d*) LiTFSI, (*e*) LiI, (*f*) LiDFOB, (*g*) LiBF$_4$, (*h*) LiCF$_3$SO$_3$, and (*i*) LiClO$_4$ (CE noted as %). (*From Ref. 145.*)

In general, however, the CE remains <96% for such electrolytes, which results in rapid degradation of the Li anode. In recent years, however, a new focal theme for Li metal plating has gained traction focused upon highly concentrated electrolytes (which have little or no uncoordinated solvent). An early report in 2008 indicated that increasing the salt concentration in PC-LiTFSI electrolytes suppressed dendritic Li growth during plating and improved the Li cyclability.[146] But the CE was still low and only a limited number of cycles (50) were achieved. In 2015, Qian and coworkers demonstrated that by replacing the carbonate (PC) solvent with an ether (DME), as well as substituting $LiN(SO_2F)_2$ (or LiFSI) for LiTFSI and using a high salt concentration (i.e., ≥4 M), it was possible to plate nodular (nondendritic) Li.[147] A Li/Li symmetric cell could be cycled for more than 6000 cycles at a 10 $mA/cm^2$ rate, whereas a Cu/Li cell (plating on a Cu current collector) was cycled for more than 1000 cycles at 4 $mA/cm^2$ with an average CE of 98.4%. It was further demonstrated that "anode-free" $Cu/LiFePO_4$ cells (in which all of the Li is initially in the cathode) could be cycled with an exceptionally high CE of >99.8%.[148] This paper noted, however, that a CE >99.9% is required to retain more than 80% of a cell's initial capacity for 1000 or more cycles (for 100% Li utilization each cycle). While such a CE is likely unachievable, practical cells can be prepared with these electrolytes with a much reduced excess amount of Li initially supplied to the cell. Another group has also reported favorable Li metal cycling for highly concentrated electrolytes, such as 2 M DME-LiFSI + 2 M DME-LiFTFSI ($LiN(SO_2F)(SO_2CF_3)$) for $Li/LiFePO_4$ cells.[149] Note, however, that this nodular morphology with excellent Li cycling ability was also achieved for the 1 M $DOL-LiAsF_6$ electrolyte (with a TBA stabilizer) noted earlier,[23] as well as with an electrolyte with 1,4-dioxane (DX), i.e., 1 M DX/DME (1/2)-LiFSI.[150]

Although IL-based electrolytes are often reported to have an exceptional oxidative stability (>5 V vs. $Li/Li^+$), that does not necessarily equate to a comparable stability when used with intercalation cathodes at high voltage (e.g., 4.6–4.8 V vs. $Li/Li^+$),[151] although relatively stable cycling can be achieved when cells are cycled with a lower upper voltage (e.g., 4.2 V vs. $Li/Li^+$) and at moderate rates (e.g., C/8).[152] Such electrolytes tend to have poor wettability with some battery components, thus necessitating the use of a separator optimized for such an application.[153] Efforts have been made to formulate electrolytes consisting of mixtures of an IL, LiX, and one or more aprotic solvents to overcome these wettability and ion transport limitations.

***Liquid Electrolyte Li-S Cells.*** Li-S rechargeable cells are the highest-energy Li metal, sealed rechargeable batteries available today. Sulfur is inexpensive and nontoxic. Li-S has the advantage of combining the highest capacity anode (Li metal) and cathode (sulfur) resulting in an exceedingly high theoretical specific energy (2500 Wh/kg). However, only a fraction of this is achievable for practical cells. Commercial cells are available with a specific energy of 360 Wh/kg, but the volumetric energy density (Wh/L) and cycle life of Li-S cells tend to be relatively low. Detailed evaluations of Li-S cells indicate that for these to be competitive with Li-ion cells the following characteristics are required: the sulfur load must be ≥6 $mg/cm^2$, the sulfur fraction ≥70%, the sulfur utilization ≥80%, and the E/S (electrolyte/sulfur) weight ratio 3/1 or better (lower).[154–156] Most cell testing reported in the scientific literature does not use comparable testing conditions, and the favorable results for new Li-S material evaluations when testing with thin cathodes having a low sulfur loading and an excess of electrolyte are generally not retained for the required cell conditions noted above.[157–159] Thus, a good portion of the scientific "advances" for Li-S batteries are unusable for practical cells. An additional complication for evaluations arises from the applied measurement protocols, as it has been shown that the rate capability evaluations of Li-S cells are sensitive not only to the applied current, but also to the cycling prehistory of the cell.[160]

Electrolyte optimization for Li-S batteries has been challenging.[161–163] Although the sulfur-based cathode has a relatively low operating voltage (thus not requiring materials with an exceptionally high oxidative stability), the radical sulfur reaction chemistry, poor Li anode cycling, dissolution of the $Li_2S_x$ polysulfides in the electrolyte, and sulfur redistribution have stymied efforts to achieve long cycle life. Bryantsev and coworkers published two studies examining the stability of solvents for Li-air batteries, but these studies likely have strong relevance for Li-S batteries as well due to similarities in oxygen and sulfur radical reactions.[164,165] For liquid electrolytes, efforts to reduce/eliminate polysulfide dissolution have included the use of IL-based electrolytes,[85,166] as well as highly concentrated electrolyte formulations—with and without fluorinated solvents (e.g., hydrofluoroether [HFE])—such as acetonitrile (AN)-LiTFSI-HFE (2:1:1 mol ratio), tetraglyme (G4)-LiTFSI-HFE (1:1:-4 mol ratio), and 0.5 M diglyme (G2)-LiTFSI + 3 M $G2-LiNO_3$.[167–169]

For commercial cells, Sion Power Corporation is the lead developer of high-energy Li-S batteries for Unmanned Aerial Vehicle (UAV) applications. Current Sion Li-S cells have a capacity of 2.8 Ah and a specific energy of 350 Wh/kg. Additional information about Sion Li-S cells may be found in the previous edition of this Handbook.

## 17B.3.2   Polymer Electrolyte Cells

SPE Li batteries contain all solid-state components: Li as the anode material, a thin SPE membrane as a solid electrolyte/separator, and typically either an intercalation cathode (with a transition metal chalcogenide, oxide, or phosphate) or a sulfur-based cathode. These features offer the potential for (a) improved safety because of the reduced activity of Li with the solid electrolyte, (b) flexibility in design as the cell can be fabricated in various sizes and shapes, and (c) high energy density. The cathode (usually containing some of the SPE) and the SPE membrane are coated onto a current collector to form a thin sheet, called the cathode laminate. The Li metal foil is then applied to the cathode laminate to form a layered structure with the SPE separating the Li from the cathode. These cells often use thin components with high surface areas to minimize the internal resistance and compensate for the lower conductivity of the SPE. The thickness depends on the specific cell design and required capacity. A thicker laminate delivers a higher capacity per unit area of electrode, but with lower efficiency at the higher current drains. LMP batteries have two major advantages with respect to standard batteries that use liquid electrolyte and inert porous polymeric separators:

- The electrode and electrolyte layers are laminated (usually by heating and pressing the layers), thus allowing more varied battery shapes without loss of contact.

- Even if a low-molecular-weight (liquid) plasticizer is added to obtain high conductivity at ambient and sub-ambient temperatures, there is no free liquid present in the battery, thus preventing leakage problems.

***Polymer Electrolyte Intercalation Cathode Cells.***   Avestor, Inc. was engaged in the development, manufacture, and commercialization of a family of SPE batteries with Li metal anodes that were being developed for stationary telecommunications market and EV applications. The LMP cell was a laminate of four thin materials:

- A metallic Li foil anode—the ultrathin Li foil (less than 50 μm thick) acted as both a Li source and a current collector.

- An SPE obtained by dissolving a lithium salt (LiTFSI) in a solvating (PEO-like) copolymer.

- A metallic oxide cathode based on a reversible intercalation compound of vanadium oxide ($V_3O_8$), blended with a lithium salt and polymer to form a plastic composite.

- An Al foil current collector.

The solid, dry, Li-ion conducting PEO polymer membrane served both as the electrolyte and as the separator between the anode and cathode foils. The result was a solid-state electrochemical cell, having neither liquid nor gel components.[170] The operating cell voltage was 3.1 to 2.0 V. The minimum operating temperature was 60°C, and the battery module contained a thermal management control system. The modules were equipped with cell equalization and balancing to maintain all cells at a uniform float voltage of 3.1 V. Safety systems, such as charge control and disconnect switching, were integrated to protect against operation under excessive, abnormal conditions. The module also had a mechanical pressure subsystem to ensure the stability of the interface between the anode and the electrolyte. The mechanical pressure system (50–100 psi) could only maintain the uniformity of the Li foil surface if the rate of plating was low. Consequently, the charging current was limited to a maximum of C/8 to prevent more rapid battery deterioration resulting in a shorter cell life.[170] The primary Avestor product was an 80 Ah SE48S80 battery that operated at 48 V. In August 2006, the company had produced and shipped its 20,000th battery. At that time, Avestor said that it had signed multimillion dollar, multiyear contracts with major telecommunications service providers in North America. By October 2006, however, the company closed. In March 2007, Avestor, Inc. was acquired by Bathium (now Blue Solutions Canada), which subsequently developed and commercialized LMP batteries to power the EU EV program BlueCar. The Blue Solutions LMP batteries[5] contain no heavy metals/toxic liquids and are completely recyclable. The anode, cathode, and SPE electrolyte thin films (a few dozen μm thick and varying in width dependent on application) are produced by extrusion and then assembled into batteries. A LiFePO$_4$ cathode active material is used. Optimal conductivity for the SPE is between 70 and 80°C. A prismatic design is used to achieve high capacity (75 Ah). The module for the series-connected cells is designed to be flexible to allow for variable applications. A pack is then created from several modules. For the BlueCar vehicle, the 35 kWh pack contains a series of six 5.8 kWh modules (each containing 20 cells). The pack is light and compact (100 Wh/kg and 100 Wh/L). The battery normally operates at 70°C, but it is estimated that the temperature can be lowered to 60°C for onboard applications and 50°C for stationary applications. Under normal use, the LMP battery has a service life exceeding 3000 cycles (after losing 20% of its rated capacity when used in mobile drive applications, it can be reallocated for use in stationary applications having lower power requirements). In 2017, however, it was announced that the Bolloré Blue Solutions SA LMP batteries (Table 17B.1)[5,171] were not cost competitive with other EV batteries, and thus the focus for these LMP batteries would be shifted to buses, services, and stationary storage, rather than for EV automobiles.[172]

The Monroe and Newman publications[38,173,174] spawned a search for a new class of SPEs that could provide the high modulus SPE (~1 GPa) that the calculations predicted will suppress or eliminate dendrite formation. Because segmental motion of polymer chains is very important for ionic mobility, high conductivity and high modulus are almost mutually exclusive goals in homopolymer SPE electrolytes. For example, PEO has an elastic modulus of less than 1 MPa, whereas glassy polymers such as polystyrene have a very high modulus (~3 GPa), but do not conduct $Li^+$ ions. Niitani and coworkers in 2005[175,176] and then Balsara and coworkers in 2007[177] (with many follow-on publications) synthesized "SEO"-type polystyrene/PEO block copolymers to obtain high-modulus SPEs that meet this requirement for Li metal batteries. In these nanostructured electrolytes, the PEO phase with dissolved salt is the ionic conducting part of the copolymer, while the polystyrene provides the high elastic modulus.

Block copolymers self-assemble into well-defined structures such as lamellae or cylinders with domain spacings on the order of tens of nanometers, with specifics depending on the molecular weight and volume fraction of each block and the conductivity dependent to some extent upon the molecular weight of the PEO segments. It is necessary to operate the cells at elevated temperature (e.g., 90°C) due to the low conductivity of the SEO SPE.[178] Although it was possible to increase the time for Li dendrite penetration of the SEO-based SPE (relative to comparable PEO electrolytes), Li penetration still occurred (Fig. 17B.14).[178]

It was found that the cycling of Li/Li symmetric cells resulted in a low-impedance cell failure due to dendrite (or Li globule) shorting,[179,180] while testing in Li/LiFePO₄ cells instead resulted in a high-impedance cell failure due to delamination of the electrode from the rigid SPE electrolyte.[181] A startup company, Seeo Inc.[182] founded in 2007 attempted to commercialize batteries based upon these SEO SPEs, but was unsuccessful in doing so and was acquired by Bosch in 2015.

Many other reports of Li metal batteries with SPEs have been published.[133,183,184] One approach with notable success

**TABLE 17B.1**  Bolloré Blue Solutions SA LMP Battery Characteristics

| | |
|---|---|
| Energy | 35 kWh |
| Peak power | 45 kW (30 s) |
| Nominal voltage | 410 V |
| Min/max voltage | 300 V/450 V |
| Capacity (at $C/4$) | 75 Ah |
| Energy density | 100 Wh/kg |
| Volume density | 100 Wh/L |
| Mass | 300 kg |
| Volume | 300 L |

*Source:* From Ref. 5.

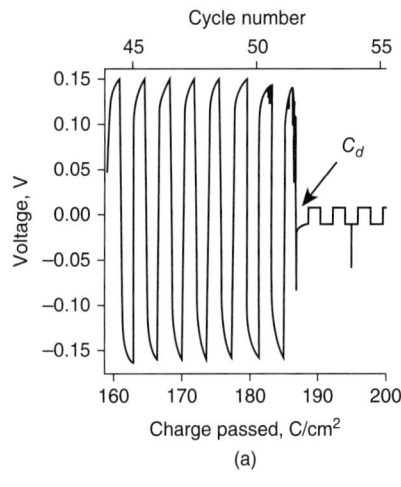

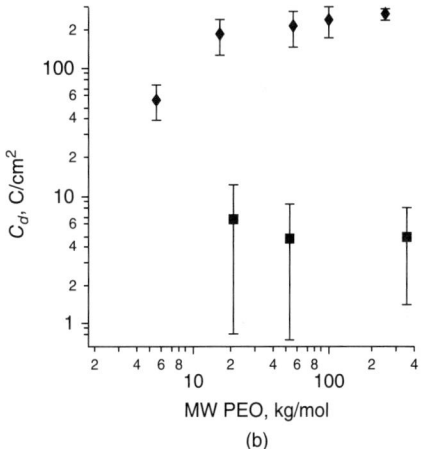

**FIGURE 17B.14**  (*a*) Typical Li/Li cell cycling for a SEO electrolyte (short circuit indicated by $C_d$) and (*b*) cycling data showing $C_d$ as a function of PEO molecular weight for SEO electrolytes (diamonds) and PEO electrolytes (squares). (*From Ref. 178.*)

has been the addition of ILs to PEO-based electrolytes to improve their Li$^+$ transport properties.[185–187] The operating temperature of the cells with PEO-LiX-IL SPEs could be reduced (relative to comparable cells with PEO-LiX SPEs) from 60 to 40°C. Such cells can be cycled with low capacity fading using Li metal anodes for hundreds of cycles (Fig. 17B.15),[188] but the relatively low conductivity of the electrolytes and high impedance of the Li-electrolyte interface restricts the rate performance of such cells, while the inclusion of PEO limits the use of these electrolytes to lower voltage (<4.0 V vs. Li/Li) cathodes such as LiFePO$_4$ and V$_2$O$_5$.[188–192]

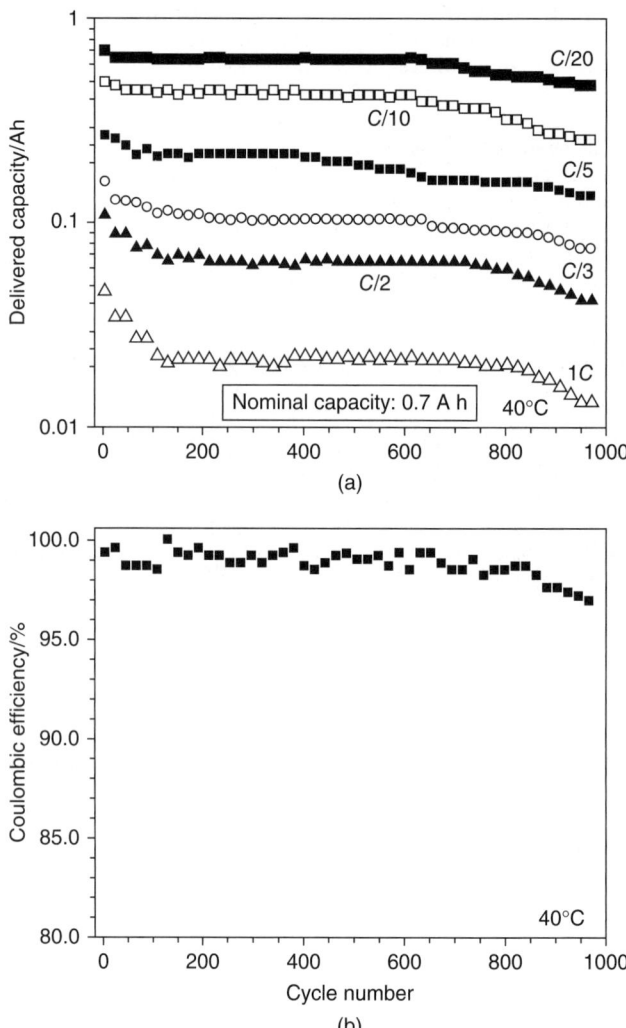

**FIGURE 17B.15**    Discharge capacity (*a*) and CE (*b*) evolution of a Li/LiFePO$_4$ LMP battery 0.7 Ah prototype with a PEO-IL-LiX SPE electrolyte at 40°C and different current rates. (*From Ref. 188.*)

The PEO, however, can be substituted with other polymers. For example, an SPE with an IL-LiX mixture in a PVDF polymer performed reasonably well (up to 4.3 V vs. Li/Li$^+$) in a Li/LiMn$_2$O$_4$ cell when cycled at 25°C at low rates (e.g., C/10).[193] Note that Li dendrite growth through the SPE is still possible for such electrolytes, but the growth rate was suppressed by the inclusion of the IL (Fig. 17B.16).[194]

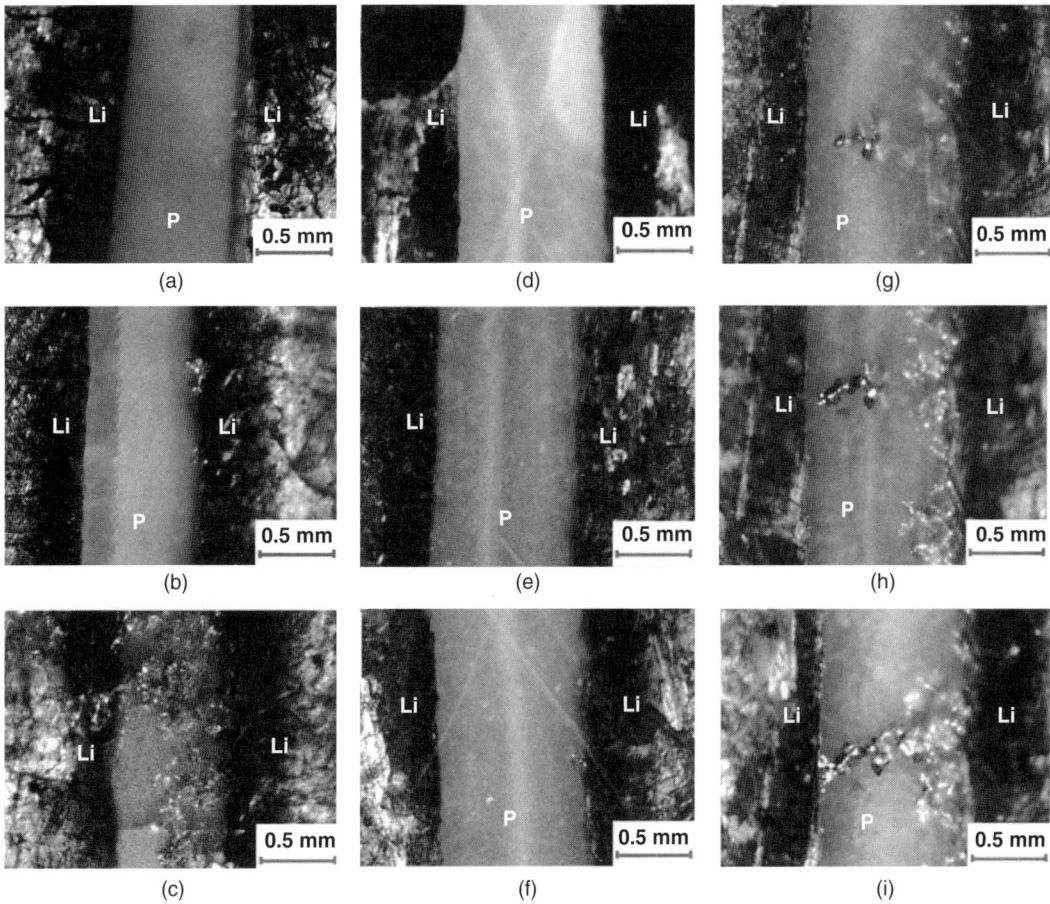

**FIGURE 17B.16**  Li dendrite growth in SPEs for Li/Li symmetric cells cycled at 0.5 mA/cm² and 60°C: Li/P(EO)$_{18}$-LiTFSI/Li cell at $t = $ (*a*) 0, (*b*) 15, and (*c*) 20 h and a Li/P(EO)$_{18}$-LiTFSI-IL/Li cell at $t = $ (*d*) 0, (*e*) 30, (*f*) 35, (*g*) 45, (*h*) 65, and (*i*) 75 hr. (*From Ref. 194.*)

*Polymer Electrolyte Li-S Cells.*    Sulfur and polysulfides can vulcanize unsaturated polymers, as well as replace fluorine in fluorinated polymers (to form thiols).[162] Thus, many of the efforts to use an SPE for Li-S batteries have focused on stable polymers such as PEO, but the low conductivity of PEO-LiX electrolytes remains problematic. One example of an alternative, which can serve as both a separator and a binder (in the cathode), is a poly(DDA)TFSI-IL-LiTFSI electrolyte, where DDA stands for diallyldimethylammonium.[195] This electrolyte, however, still did not prevent polysulfide dissolution in the SPE or reactions with the Li anode.

## 17B.3.3    Inorganic Electrolyte Cells

Efforts to develop larger-format full cells with new inorganic solid electrolytes continue with significant work ongoing worldwide due to the potential benefits of such cell construction, including high safety (nonflammability), a Li⁺ transference number near unity, physical blocking of polysulfide dissolution/migration, high energy density, and tailorable battery configurations. Challenges for this, however, persist. Although it is frequently stated that—for batteries with Li metal anodes—Li dendrites can be prevented by the use of solid inorganic electrolytes, this is often not true due to dendrite growth through grain boundaries and/or voids/cracks in the solid electrolyte.[196-198] It is often accepted that such batteries are dendrite-free, but there is little experimental proof to validate this belief. Thus, there is a need to investigate the Li dendrite formation behavior for Li metal-solid electrolyte interfaces when various cycling conditions are utilized. In addition, although very thin, ceramic or glassy inorganic electrolyte may be made flexible, this does not necessarily prevent defect generation and slow crack propagation (e.g., a copper wire is flexible, but can be broken by simply bending it repeatedly). Thus, mechanical abuse or stresses originating from active material volume changes within the electrodes and at the electrode-electrolyte interfaces upon extensive battery cycling may lead to the failure of ceramic materials due to cracking. Very thin materials are also much more susceptible to pinholes and other defects that may result in Li dendritic short-circuiting, especially for larger format production of thin ceramic membranes.

As noted above, sulfide-based inorganic solid electrolytes tend to have a very high ionic conductivity, but such electrolytes are inherently unstable in contact with Li. In contrast, oxide-based inorganic electrolytes have a lower conductivity, but are much more stable in contact with Li. Thus, there is a trade-off between conductivity and stability. The sulfide-based electrolytes can be processed without high-temperature methods due to their softness/ductility, but the oxide-based electrolytes are rigid and require high-temperature processing such as sintering. But all inorganic solid electrolytes, including the sulfides, have a high interfacial resistance with the electrodes due to poor contact and resistive surface layers formed by processing chemical transformation and/or by degradation due to reactions with the electrodes.[144] In particular, extensive difficulties are associated with the high-temperature thermal processing of many solid inorganic electrolytes required for their densification from powders.[199] Of the many oxide conductors synthesized, garnet LLZO ($Li_7La_3Zr_2O_{12}$) has received the most focus because of its high $Li^+$ conductivity ($>10^{-4}$ S/cm at 25°C), stable electrochemical properties (low reactivity with Li), and low grain boundary resistance. Garnet, however, is unstable in contact with moisture and decomposes during the high-temperature processing required (above 1100°C) to the lower conductivity tetragonal phase or $La_2Zr_2O_7$ (from volatilization of lithium) resulting in resistive surface layers for the electrolyte.[199] Glassy ceramic solid inorganic electrolytes such as LAGP ($Li_{1.5}Al_{0.5}Ge_{1.5}(PO_4)_3$) and LATP ($Li_{1.4}Al_{0.4}Ti_{1.6}(PO_4)_3$) have a high conductivity ($>10^{-4}$ S/cm) and can be sintered at a lower temperature of 800°C. These glassy oxide electrolytes are also stable against moisture, but they decompose when in contact with Li and have a high reactivity with electrode materials during the heat treatment. Chemical/electrochemical stability, high contact resistance, dendrite formation, mechanical stability during cycling, and processing difficulties continue to inhibit the commercialization of larger (than thin film) batteries with solid inorganic electrolytes.

***Inorganic Electrolyte Intercalation Cathode Cells.*** A number of small format, thin-film, Li metal rechargeable batteries have been commercialized for portable applications such as a power source for electronic devices, memory backup, and other types of auxiliary power sources. Thin-film, all-solid-state batteries are based on the technology developed by Bates et al.[200] and fabricated by a sequential series of physical vapor deposition processes. The key similarity is the use of a LiPON glassy electrolyte that is applied onto a variety of cathode thin films. The main problems are the low ionic conductivity of solid electrolytes and a large charge-transfer resistance at the electrode-solid electrolyte interface.

For a typical thin-film, solid-state Li cell, each of the component layers is 0.1 μm to several micrometers thick.[201] Ideally, the substrate is also a component of the device; otherwise even with very thin substrates, the weight and volume of the battery are largely determined by the inactive support. Thin-film batteries have been developed using a variety of supports, including silicon, quartz, mica, alumina, polymers, soda-lime glass, and metal foils. Many of the batteries are not only thin but also quite flexible.[202] These cells are fabricated by sequential layer deposition of the cell components using RF magnetron sputtering, except for the evaporation of Li and for the metallic current collector components, which are deposited by direct current (DC) magnetron sputtering. The deposition conditions for $LiCoO_2$[200] and the LiPON electrolyte are reported in the literature.[203] Positive current collectors of Au or Pt (0.1–0.3 μm), over a layer of cobalt (0.01–0.05 μm, to improve adhesion), have been used. Cells using either $LiCoO_2$ or $LiMn_2O_4$ positive electrode materials have been fabricated. The positive electrode layer of laboratory test cells is typically 0.05 to 5 μm thick and 0.04 to 25 $cm^2$ in area, depending on the capacity required by the application. Negative current collectors of Cu, Ti, or TiN (0.1–0.3 μm) are typical. To enhance the hermeticity of the cell, protective overcoated layers of LiPON (1 μm) or parylene[204] (6 μm) and Ti or Al (0.1 μm) have been used.

The capacity of these LiPON cells depends upon cathode choice, coating thickness, degree of crystallization of the cathode, and other processing conditions. Typical results are provided in the previous edition of this Handbook. Thick $LiCoO_2$ cathodes have the highest capacity. The capacity is typically rated by active cell area, but studies have shown[205] that cells can achieve 100 Wh/kg at 1 kW/kg and 100 Wh/L at 1 kW/L (excluding substrate weight and volume). The rate capability of thin-film batteries using LiPON films strongly depends on the cathode,[206] and $LiCoO_2$ is attractive for high-power applications. Additionally, the charge transfer at the electrode-electrolyte interface can influence the charge and discharge rate.[207] The use of alternate electrode materials such as $Li_4Ti_5O_{12}$, a material with zero strain insertion on lithiation,[208] has been shown to provide more rapid kinetics. Commercialization has proceeded at a rapid pace from the early developments at Oak Ridge National Laboratory with at least six commercial endeavors. Cells are available from these companies (e.g., Infinite Power Solutions and Front Edge Technology Inc.).[209,210]

***Inorganic Electrolyte Li-S Cells.***    All-solid-state Li-S cells have been prepared with inorganic electrolytes, in part motivated by a desire for improved safety, an increased energy density, and to prevent polysulfide dissolution and crossover to the anode.[211-214] Difficulties with such a cell configuration include the volume change of the anode, as well as that of the C-S cathode. This combined with chemical incompatibilities often results in a high interfacial contact resistance. Li dendrite penetration through the inorganic electrolyte is also a challenge, as noted above.[215]

### 17B.3.4    Hybrid Electrolyte Cells

It has been exceptionally difficult to meet the many requirements associated with stable cycling of both the Li negative electrode and the positive electrode (using different intercalation cathodes or the Li-S cathode) with a single electrolyte. Thus, some effort has been devoted to using hybrid electrolytes, which are designed to capitalize upon the favorable characteristics of different electrolytes (see Chap. 22A).

Hybrid SPE-inorganic electrolytes that combine garnet inorganic electrolyte (e.g., LLZO) particles with an SPE (e.g., PEO-LiX) are one example.[216-218] Such electrolytes do not require high-temperature sintering, are nonflammable, do not undergo mechanical degradation, and retain a low contact resistance with the electrodes. They may be used with all-solid-state Li cells with intercalation cathodes (e.g., $LiFePO_4$)[219-221] (but are still limited to lower voltage cathodes due to the limited oxidative stability of PEO), as well as for Li-S cells.[222] An interesting variant of this is an electrolyte with PEO-LLZO, but no added LiX salt.[223] Li cells can be prepared with this electrolyte and $LiFePO_4$ or $LiFe_{0.15}Mn_{0.85}PO_4$ cathodes with reasonable cycling stability. Note that all of these hybrid electrolytes require cell cycling at elevated temperature (e.g., 60°C) due to the low conductivity of the hybrid electrolytes.

## 17B.4    APPLICATIONS AND PERFORMANCE

Sion Power® is extending its new *Licerion*® rechargeable lithium metal technology to include Li-ion intercalation cathodes. At the core of Licerion technology is a protected metallic Li thin-film anode with multiple levels of physical and chemical protection to enhance the safety and life of Li metal anodes. The physical protection is comprised of ceramic-polymer composite membranes (Fig. 17B.17), which is combined with a specialized cell design and electrolyte system to provide smooth, dendrite-free lithium stripping and redeposition.

In collaboration with BASF, Sion Power has demonstrated the potential of this technology, and commercial production of Licerion batteries will commence in 2018 at Sion Power's facility in Tucson, Arizona. Individual Licerion cells, with dimensions of 10 cm × 10 cm × 1 cm, have a capacity of 25 Ah and offer the highest combination of energy density and specific energy available (exceeding 1000 Wh/L and 500 Wh/kg). The 100% DOD cycle life of Licerion cells reached nearly 350 cycles at 500 Wh/kg in 2017 (Fig. 17B.18). Sion Power's Licerion technology targets a diverse range of battery applications—from EVs to stratospheric drones to urban air transportation.

SolidEnergy uses ultrathin Li negative electrodes with an NMC positive electrode. The exceptionally high gravimetric and volumetric energy densities achieved for 3 Ah cells of 450 Wh/kg and 1200 Wh/L (Fig. 17B.19) greatly reduces the size of the cells relative to comparable Li-ion batteries (Fig. 17B.20).[224] A conventional separator is used with a concentrated electrolyte in concert with the NMC cathode, but a protective coating (solid electrolyte) is used to stabilize the Li anode (Fig. 17B.21).[224] The initial Hermes™ cells (High Energy Rechargeable Metal cells for Space) have been designed for applications including HALE (High Altitude Long Endurance) vehicles, VTOL (Vertical Take-Off and Landing) flying transportation, and drones.[224]

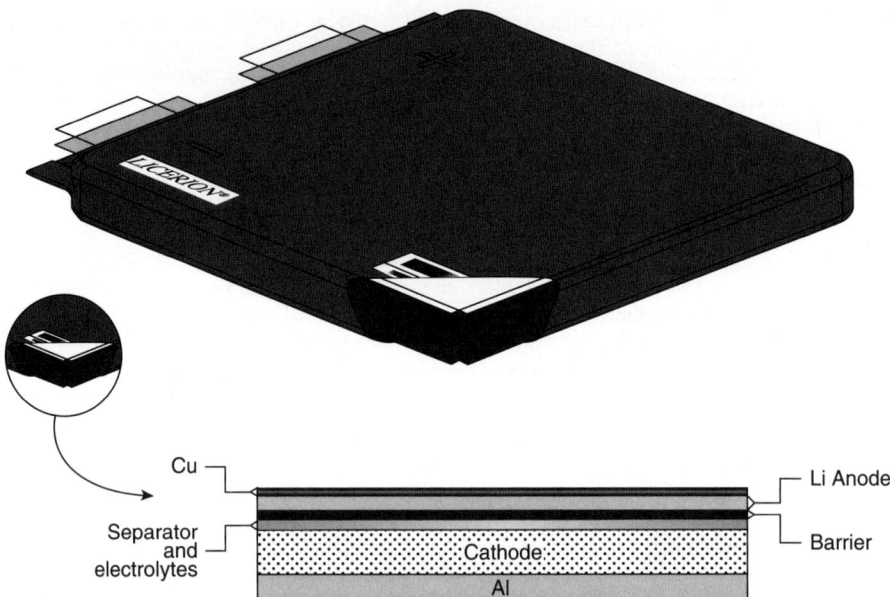

**FIGURE 17B.17**   Cutaway drawing of Licerion Cell. External dimensions are 10 cm × 10 cm × 1 cm and cell capacity is 25 Ah. (*Courtesy of Sion Power Corp., Tucson AZ.*)

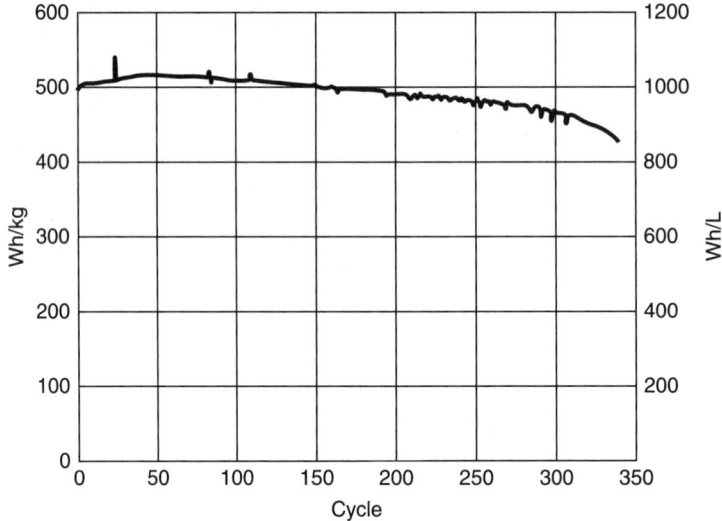

**FIGURE 17B.18**   Cycling behavior of Licerion® cell at *C*/4 discharge and *C*/12 charge rates (100% DOD cycle to 80% of rated capacity). Specific energy and energy density projected to 10 cm × 10 cm × 1 cm cell design using same active materials balance as laboratory 0.4 Ah cells and accounting for weight and volume of all large cell components. (*Courtesy of Sion Power Corp., Tucson AZ.*)

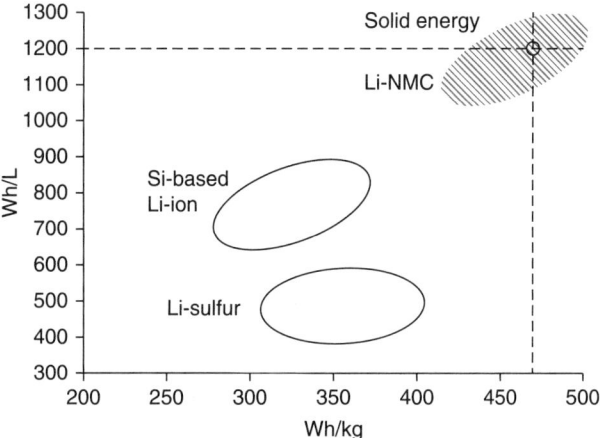

**FIGURE 17B.19**   Comparison of the SolidEnergy gravimetric and volumetric energy densities with other battery technologies. (*From Ref. 223.*)

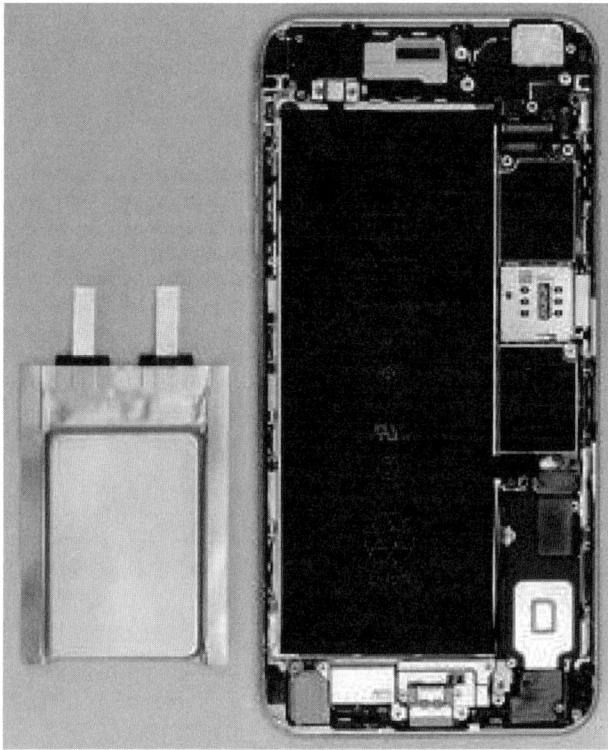

**FIGURE 17B.20**   Half the size and half the weight. A SolidEnergy Li metal 3 Ah cell (left) compared to an iPhone 6 Plus 3 Ah Li-ion cell. (*From Ref. 223.*)

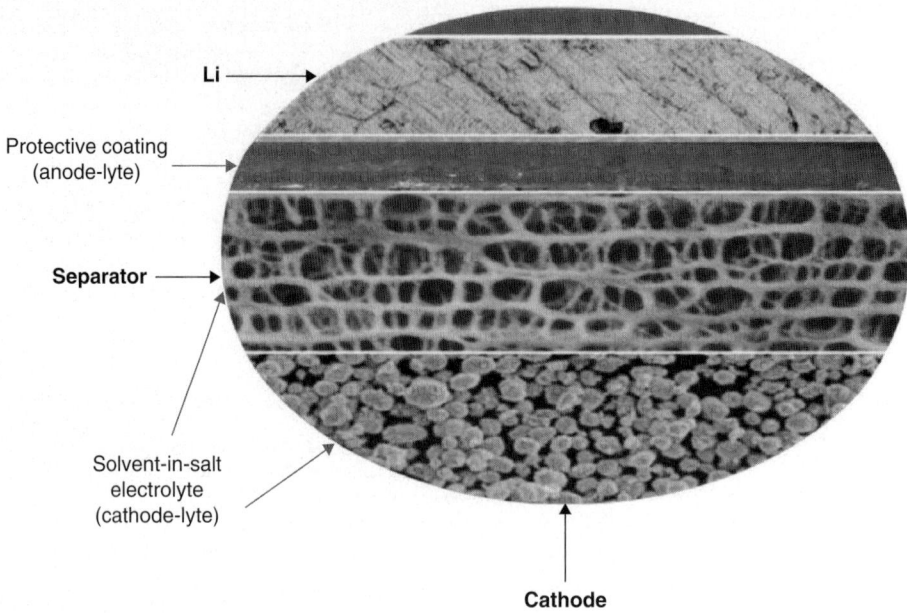

**FIGURE 17B.21** SolidEnergy's semisolid Li metal design using a hybrid electrolyte. (*From Ref. 223.*)

## 17B.5   CONCLUSION

The development of very high-energy storage devices will enable solutions to many of society's pressing needs. Li metal rechargeable batteries are commercially available in small sizes with good-to-excellent performance. Limitations in temperature stability, cycle life, and safety are being addressed. Larger cell sizes (>3 Ah) are still in development or in early commercialization stages. The future will likely see a range of high-capacity Li metal rechargeable batteries become commercialized with performance metrics that meet market demands.

## ACKNOWLEDGMENT

Thanks to Yuriy V. Mikhaylik, Sion Power, for providing information and figures for this chapter.

## REFERENCES

1. A. Varzi, R. Raccichini, S. Passerini, and B. Scrosati, *J. Mater. Chem. A* **4**, 17251 (2016).

2. M. B. Armand, J. M. Chabagno, and M. Duclot, "Extended Abstracts," *2nd Int. Meeting on Solid Electrolytes*, St. Andrews, Scotland, Sept. 1978.

3. M. B. Armand, J. M. Chabagno, and M. Duclot, in *Fast Ion Transfer in Solids*, P. Vashishta (ed.), p. 131, North Holland, New York, 1979.

4. M. Gauthier, D. Fauteux, G. Vassort, A. Bélanger, M. Duval, P. Ricoux, and J.-M. Chabagno et al., *J. Electrochem. Soc.* **132**, 1333 (1985).

5. Blue Solutions Registration Document 2016. https://www.blue-solutions.com/wp-content/uploads/2017/01/0612_BLUE_1701280_DR_2016_GB_MEL.pdf.

6. K. M. Abraham, J. L. Goldman, and M. D. Dempsey, *J. Electrochem. Soc.* **128**, 2493 (1981).

7. M. W. Rupich, L. Pitts, and K. M. Abraham, *J. Electrochem. Soc.* **129**, 1857 (1982).

8. K. M. Abraham, J. S. Foos, and J. L. Goldman, *J. Electrochem.* Soc. **131**, 2197 (1984).

9. G. L. Holleck and T. Nguyen, U.S. Patent 4,911,996 (1990).

10. K. M. Abraham, D. M. Pasquariello, and E. B. Willstaedt, *J. Electrochem. Soc.* **136**, 576 (1989).

11. D. Fouchard, in *Proc. 33rd Power Sources Symp.*, The Electrochemical Society, Pennington, NJ, 1988.

12. J. A. R. Stilb, *J. Power Sources* **26**, 233 (1989).

13. L. Dominey, in *Non-Aqueous Electrochemistry*, D. Aurbach (ed.), Chap. 8, pp. 437–460, Marcel Dekker, New York, 1999. Also see "Cellular Phone Recall May Cause Setback for Moli," *Toronto Globe and Mail*, August 15, 1989 and *Adv. Batt. Technology* **25**, 4 (1989).

14. P. Dan, E. Mengeritsky, Y. Geronov, D. Aurbach, and I. Weissman, *J. Power Sources* **54**, 143 (1995).

15. D. Aurbach, I. Weissman, A. Zaban, Y. Ein-Eli, E. Mengeritsky, and P. Dan, *J. Electrochem. Soc.* **143**, 2110 (1996).

16. D. Aurbach, E. Zinigrad, H. Teller, Y. Cohen, G. Salitra, H. Yamin, and P. Dan et al., *J. Electrochem. Soc.* **149**, A1267 (2002).

17. K. W. Semkow and A. F. Sammells, *J. Electrochem. Soc.* **134**, 2084 (1987).

18. K. M. Abraham and Z. Jiang, *J. Electrochem. Soc.* **143**, 1 (1996).

19. K. M. Abraham, *ECS Trans.* **3**, 67 (2008).

20. R. D. Rauh, K. M. Abraham, G. F. Pearson, J. K. Surprenant, and S. B. Brummer, *J. Electrochem. Soc.* **126**, 523 (1979).

21. X. Yu, J. B. Bates, G. E. Jellison, Jr., and F. X. Hart, *J. Electrochem. Soc.* **144**, 524 (1997).

22. D. Linden, *Handbook of Batteries,* 2nd ed., McGraw-Hill, Inc., New York, 1995, p. 36.9.

23. E. Zinigrad, E. Levi, H. Teller, G. Salitra, D. Aurbach, and P. Dan, *J. Electrochem. Soc.* **151**, A111 (2004).

24. J. Steiger, D. Kramer, and R. Mönig, *J. Power Sources* **261**, 112 (2014).

25. R. R. Chianelli, *J. Crystal Growth* **34**, 239 (1976).

26. https://areweanycloser.wordpress.com/2013/06/21/dendritic-lithium-and-battery-fires/.

27. D. P. Wilkinson, H. Blom, K. Brandt, and D. Wainwright, *J. Power Sources* **36**, 517 (1991).

28. D. Wainwright and R. Shimizu, *J. Power Sources* **34**, 31 (1991).

29. D. P. Wilkinson and D. Wainwright, *J. Electroanal. Chem.* **355**, 193 (1993).

30. B. Wu, J. Lochala, T. Taverne, and J. Xiao, *Nano Energy* **40**, 34 (2017).

31. R. Selim and P. Bro, *J. Electrochem. Soc.* **121**, 1457 (1974).

32. I. Epelboin, *J. Electrochem. Soc.* **127**, 2100 (1980).

33. I. Yoshimatsu, T. Hirai, and J. I. Yamaki, *J. Electrochem. Soc.* **135**, 2422 (1988).

34. J. Steiger, D. Kramer, and R. Mönig, *Electrochim. Acta* **136**, 529 (2014).

35. X.-B. Cheng, C. Yan, J.-Q. Huang, P. Li, L. Zhu, L. Zhao, and Y. Zhang et al., *Energy Storage Mater.* **6**, 18 (2017).

36. K. N. Wood, M. Noked, and N. P. Dasgupta, *ACS Energy Lett.* **2**, 664 (2017).

37. K.-H. Chen, K. N. Wood, E. Kazyak, W. S. LePage, A. L. Davis, A. J. Sanchez, and N. P. Dasgupta, *J. Mater. Chem. A* **5**, 11671 (22017).

38. C. Monroe and J. Newman, *J. Electrochem. Soc.* **150**, A1377 (2003).

39. A. Ferrese and J. Newman, *J. Electrochem. Soc.* **161**, A1350 (2014).

40. M. Gauthier, A. Belanger, and A. Vallee, U.S. Patent 6,007,935 (1999).

41. T. Hirai, I. Yoshimatsu, and J. Yamaki, *J. Electrochem. Soc.* **141**, 611 (1994).

42. T. Hirai, I. Yoshimatsu, and J. Yamaki, *J. Electrochem. Soc.* **141**, 2300 (1994).

43. E. Eweka, J. R. Owens, and A. Ritchie, *J. Power Sources* **65**, 247 (1997).

44. L. Gireaud, S. Grugeon, S. Laruelle, B. Yrieix, and J.-M. Tarascon, *Electrochem. Commun.* **8**, 1639 (2006).

45. H. J. Chang, N. M. Trease, A. J. Ilott, D. Zeng, L.-S. Du, A. Jerschow, and C. P. Grey, *J. Phys. Chem. C* **119**, 16443 (2015).

46. J.-G. Zhang, W. Xu, and W. A. Henderson, *Lithium Metal Anodes and Rechargeable Lithium Metal Batteries*, Springer International Publishing, 2017, Switzerland.

47. A. Aryanfar, D. J. Brooks, A. J. Colussi, and M. R. Hoffmann, *Phys. Chem. Chem. Phys.* **16**, 24965 (2014).

48. H. Yang, E. O. Fey, B. D. Trimm, N. Dimitrov, and M. S. Whittingham, *J. Power Sources* **272**, 900 (2014).

49. J. Zheng, P. Yan, D. Mei, M. H. Engelhard, S. S. Cartmell, B. J. Polzin, and C. Wang et al., *Adv. Energy Mater.* **6**, 1502151 (2016).

50. Q. Li, S. Tan, L. Li, Y. Lu, and Y. He, *Sci. Adv.* **3**, e1701246 (2017).

51. M.-H. Ryou, Y. M. Lee, Y. Lee, M. Winter, and P. Bieker, *Adv. Funct. Mater.* **25**, 834 (2015).

52. J. Park, J. Jeong, Y. Lee, M. Oh, M.-H. Ryou, and Y. M. Lee, *Adv. Mater. Interfaces* **3**, 1600140 (2016).

53. B. D. Adams, J. Zheng, X. Ren, W. Xu, and J.-G. Zhang, *Adv. Energy Mater.* 1702097 (2017).

54. R. D. Rauh and S. B. Brummer, *Electrochim. Acta* **22**, 75 (1977).

55. R. D. Rauh and S. B. Brummer, *Electrochim. Acta* **22**, 85 (1977).

56. F. W. Dampier and S. B. Brummer, *Electrochim. Acta* **22**, 1339 (1977).

57. J. O. Besenhard, *J. Electroanal. Chem.* **78**, 189 (1977).

58. Z. Li, J. Huang, B. Y. Liaw, V. Metzler, and J. Zhang, *J. Power Sources* **254**, 168 (2014).

59. K. Zhang, G.-H. Lee, M. Park, W. Li, and Y.-M. Kang, *Adv. Energy Mater.* **6**, 1600811 (2016).

60. X.-B. Cheng, R. Zhang, C.-Z. Zhao, F. Wei, J.-G. Zhang, and Q. Zhang, *Adv. Sci.* **3**, 1500213 (2016).

61. A. Mauger, M. Armand, C. M. Julien, and K. Zaghib, *J. Power Sources* **353**, 333 (2017).

62. C. Sun, J. Liu, Y. Gong, D. P. Wilkinson, and J. Zhang, *Nano Energy* **33**, 363 (2017).

63. J. Lang, L. Qi, Y. Luo, and H. Wu, *Energy Storage Mater.* **7**, 115 (2017).

64. T. Tao, S. Lu, S. Fan, W. Lei, S. Huang, and Y. Chen, *Adv. Mater.* 1700542 (2017).

65. X.-B. Cheng, R. Zhang, C.-Z. Zhao, and Q. Zhang, *Chem. Rev.* **117**, 10403 (2017).

66. Y. Guo, H. Li, and T. Zhai, *Adv. Mater.* **29**, 1700007 (2017).

67. S. F. Liu, X. L. Wang, D. Xie, X. H. Xia, C. D. Gu, J. B. Wu, and J. P. Tu, *J. Alloys Compd.* **730**, 135 (2018).

68. X.-B. Cheng, J.-Q. Huang, and Q. Zhang, *J. Electrochem. Soc.* **165**, A6058 (2018).

69. X.-L. Xu, S.-J. Wang, H. Wang, B. Xu, C. Hu, Y. Jin, and J.-B. Liu et al., *J. Energy Storage* **13**, 387 (2017).

70. R. Cao, W. Xu, D. Lv, J. Xiao, and J.-G. Zhang, *Adv. Energy Mater.* **5**, 1402273 (2015).

71. X.-Q. Zhang, X.-B. Cheng, and Q. Zhang, *Adv. Mater. Interfaces* 1701097 (2017).

72. D. Lu, Y. Shao, T. Lozano, W. D. Bennett, G. L. Graff, B. Polzin, and J. Zhang et al., *Adv. Energy Mater.* **5**, 1400993 (2015).

73. S. Jiao, J. Zheng, Q. Li, X. Li, M. H. Engelhard, R. Cao, and J.-G. Zhang et al., *Joule* **2**, 1 (2018).

74. K. G. Gallagher, S. Goebel, T. Greszler, M. Mathias, W. Oelerich, D. Eroglu, and V. Srinivasan, *Energy Environ. Sci.* **7**, 1555 (2014).

75. N. Imanishi and O. Yamamoto, *Mater. Today* **17**, 24 (2014).

76. J. Lu, K. C. Lau, Y. K. Sun, L. A. Curtiss, and K. Amine, *J. Electrochem. Soc.* **162**, A2439 (2015).

77. L. Grande, E. Paillard, J. Hassoun, J.-B. Park, Y.-J. Lee, Y.-K. Sun, and S. Passerini et al., *Adv. Mater.* **27**, 784 (2015).

78. X. Zhang, X.-G. Wang, Z. Xie, and Z. Zhou, *Green Energy Environ.* **1**, 4 (2016).

79. D. Aurbach, B. D. McCloskey, L. F. Nazar, and P. G. Bruce, *Nat. Energy* **1**, 16128 (2016).

80. H. Song, H. Deng, C. Li, N. Feng, P. He, and H. Zhou, *Small Methods* **1**, 1700135 (2017).

81. D. Geng, N. Ding, T. S. A. Hor, S. W. Chien, Z. Liu, D. Wuu, and X. Sun, *Adv. Energy Mater.* **6**, 1502164 (2016).

82. R. D. Rauh, F. S. Shuker, J. M. Marston, and S. B. Brummer, *J. Inorg. Nucl. Chem.* **39**, 1761 (1977).

83. M. Wild, L. O'Neill, T. Zhang, R. Purkayastha, G. Minton, M. Marinescu, and G. J. Offer, *Energy Environ. Sci.* **8**, 3477 (2015).

84. M. Hagen, P. Schiffels, M. Hammer, S. Dörfler, J. Tübke, M. J. Hoffmann, and H. Althues et al., *J. Electrochem. Soc.* **160**, A1205 (2013).

85. J.-W. Park, K. Ueno, N. Tachikawa, K. Dokko, and M. Watanabe, *J. Phys. Chem. C* **117**, 20531 (2013).

86. K. Ueno, J.-W. Park, A. Yamazaki, T. Mandai, N. Tachikawa, K. Dokko, and M. Watanabe, *J. Phys. Chem. C* **117**, 20509 (2013).

87. C. Zhang, A. Yamazaki, J. Murai, J.-W. Park, T. Mandai, K. Ueno, and K. Dokko et al., *J. Phys. Chem. C* **118**, 17362 (2014).

88. J. R. Akridge, Y. V. Mikhaylik, and N. White, *Solid State Ionics* **175**, 243 (2004).

89. Y. V. Mikhaylik and J. R. Akridge, *J. Electrochem. Soc.* **151**, A1969 (2004) and references therein.

90. R. Xu, J. Lu, and K. Amine, *Adv. Energy Mater.* **5**, 1500408 (2015).

91. X. Yu, H. Pan, Y. Zhou, P. Northrup, J. Xiao, S. Bak, and M. Liu et al., *Adv. Energy Mater.* **5**, 1500072 (2015).

92. X. Wang, E. Yasukawa, and S. Mori, *J. Electrochem. Soc.* **146**, 3992 (1999).

93. T. M. Pappenfus, W. A. Henderson, B. B. Owens, K. R. Mann, and W. H. Smyrl, *J. Electrochem. Soc.* **151**, A209 (2004).

94. K. Yoshida, M. Nakamura, Y. Kazue, N. Tachikawa, S. Tsuzuki, S. Seki, and K. Dokko et al., *J. Am. Chem. Soc.* **133**, 13121 (2011).

95. C. Zhang, K. Ueno, A. Yamazaki, K. Yoshida, H. Moon, T. Mandai, and Y. Umebayashi et al., *J. Phys. Chem. B* **118**, 5144 (2014).

96. S. Seki, N. Serizawa, K. Takei, K. Dokko, and M. Watanabe, *J. Power Sources* **243**, 323 (2013).

97. F. Ding, W. Xu, G. L. Graff, J. Zhang, M. L. Sushko, X. Chen, and Y. Shao et al., *J. Am. Chem. Soc.* **135**, 4450 (2013).

98. F. Ding, W. Xu, X. Chen, J. Zhang, Y. Shao, M. H. Engelhard, and Y. Zhang et al., *J. Phys. Chem. C* **118**, 4043 (2014).

99. Y. Zhang, J. Qian, W. Xu, S. M. Russell, X. Chen, E. Nasybulin, and P. Bhattacharya et al., *Nano Lett.* **14**, 6889 (2014).

100. J. Qian, W. Xu, P. Bhattacharya, M. Engelhard, W. A. Henderson, Y. Zhang, and J.-G. Zhang, *Nano Energy* **15**, 135 (2015).

101. Y. V. Mikhaylik, U.S. Patent 7,352,680 (2008).

102. D. Aurbach, E. Pollak, R. Elazari, G. Salitra, C. S. Kelley, and J. Affinito, *J. Electrochem. Soc.* **156**, A694 (2009).

103. X. Liang, Z. Wen, Y. Liu, M. Wu, J. Jin, H. Zhang, and X. Wu, *J. Power Sources* **196**, 9839 (2011).

104. S. S. Zhang and J. A. Read, *J. Power Sources* **200**, 77 (2012).

105. S. S. Zhang, *Electrochim. Acta* **70**, 344 (2012).

106. S. Xiong, K. Xie, Y. Diao, and X. Hong, *Electrochim. Acta* **83**, 78 (2012).

107. S. Xiong, K. Xie, Y. Diao, and X. Hong, *J. Power Sources* **246**, 840 (2014).

108. L. Carbone, M. Gobet, J. Peng, M. Devany, B. Scrosati, S. Greenbaum, and J. Hassoun, *J. Power Sources* **299**, 460 (2015).

109. W. Li, H. Yao, K. Yan, G. Zheng, Z. Liang, Y.-M. Chiang, and Y. Cui, *Nat. Commun.* **6**, 7436 (2015).

110. T. Nishida, K. Nishikawa, M. Rosso, and Y. Fukunaka, *Electrochim. Acta* **100**, 333 (2013).

111. H. Sano, H. Sakaebe, H. Senoh, and H. Matsumoto, *J. Electrochem. Soc.* **161**, A1236 (2014).

112. L. Grande, J. von Zamory, S. L. Koch, J. Kalhoff, E. Paillard, and S. Passerini, *ACS Appl. Mater. Interfaces* **7**, 5950 (2015).

113. H. Sano, M. Kitta, and H. Matsumoto, *J. Electrochem. Soc.* **163**, D3076 (2016).

114. G. M. A. Girard, M. Hilder, D. Nucciarone, K. Whitbread, S. Zavorine, M. Moser, and M. Forsyth et al., *J. Phys. Chem. C* **121**, 21087 (2017).

115. P. C. Howlett, D. R. MacFarlane, and A. F. Hollenkamp, *Electrochem. Solid-State Lett.* **7**, A97 (2004).

116. P. C. Howlett, N. Brack, A. F. Hollenkamp, M. Forsyth, and D. R. MacFarlane, *J. Electrochem. Soc.* **153**, A595 (2006).

117. S. Xiong, K. Xie, E. Blomberg, P. Jacobsson, and A. Matic, *J. Power Sources* **252**, 150 (2014).

118. A. Arya and A. L. Sharma, *Ionics* **23**, 497 (2017).

119. M. B. Armand, J. M. Chubagno, and M. Duclot, in *Fast Ion Transport in Solids*, P. Vashista, J. M. Mundy, G. K. Sherroy (eds.), North-Holland, Amsterdam, 1979.

120. M. B. Armand, *Solid State Ionics* **9 & 10**, 745 (1979).

121. A. S. Gozdz, C. N. Schmutz, J.-M. Tarascon, and P. C. Warren, U.S. Patent 5,456,000 (1995).

122. K. M. Abraham, in *Applications of Electroactive Polymers*, B. Scrosati (ed.), Chapman and Hall, London, 1993.

123. D. H. Shen, G. Nagasubramanian, C. K. Huang, S. Surampudi, and G. Halpert, in *Proc. 36th Power Sources Conf.*, pp. 261–263, Cherry Hill, NJ, 1994.

124. P. E. Trapa, Y.-Y. Won, S. C. Mui, E. A. Olivetti, B. Huang, D. R. Sadoway, and A. M. Mayes et al., *J. Electrochem. Soc.* **152**, A1 (2005).

125. X.-G. Sun and J. B. Kerr, *Macromolecules* **39**, 362 (2006).

126. P. E. Trapa, M. H. Acar, D. R. Sadoway, and A. M. Mayes, *J. Electrochem. Soc.* **152**, A2281 (2005).

127. R. Bouchet, S. Maria, R. Meziane, A. Aboulaich, L. Lienafa, J.-P. Bonnet, and T. N. T. Phan et al., *Nat. Mater.* **12**, 452 (2013).

128. N. J. Dudney, *Electrochem. Soc. Interface* **17**, 44 (2008).

129. J. B. Bates, N. J. Dudney, G. R. Gruzalski, R. A. Zuhr, A. Choudhury, C. F. Luck, and J. D. Robertson, *Solid State Ionics* **53–56**, 647 (1992).

130. J. B. Bates, N. J. Dudney, G. R. Gruzalski, R. A. Zuhr, A. Choudhury, C. F. Luck, and J. D. Robertson, *J. Power Sources* **43–44**, 103 (1993).

131. Y.-Z. Sun, J.-Q. Huang, C.-Z. Zhao, and Q. Zhang, *Sci. China Chem.* **60**, 1508 (2017).

132. J. C. Bachman, S. Muy, A. Grimaud, H.-H. Chang, N. Pour, S. F. Lux, and O. Paschos et al., *Chem. Rev.* **116**, 140 (2016).

133. C. Jiang, H. Li, and C. Wang, *Sci. Bull.* **62**, 1473 (2017).

134. B. Zhang, R. Tan, L. Yang, J. Zheng, K. Zhang, S. Mo, and Z. Lin et al., *Energy Storage Mater.* **10**, 139 (2018).

135. A. Manthiram, X. Yu, and S. Wang, *Nat. Rev.* **2**, 16103 (2017).

136. N. Kamaya, K. Homma, Y. Yamakawa, M. Hirayama, R. Kanno, M. Yonemura, and T. Kamiyama et al., *Nat. Mater.* **10**, 682 (2011).

137. G. Oh, M. Hirayama, O. Kwon, K. Suzuki, and R. Kanno, *Chem. Mater.* **28**, 2634 (2016).

138. Y. Seino, T. Ota, K. Takada, A. Hayashi, and M. Tatsumisago, *Energy Environ. Sci.* **7**, 627 (2014).

139. Y. Kato, S. Hori, T. Saito, K. Suzuki, M. Hirayama, A. Mitsui, and M. Yonemura et al., *Nat. Energy* **1**, 16030 (2016).

140. S. Wenzel, D. A. Weber, T. Leichtweiss, M. R. Busche, J. Sann, and J. Janek, *Solid State Ionics* **286**, 24 (2016).

141. S. Wenzel, S. Randau, T. Leichtweiß, D. A. Weber, J. Sann, W. G. Zeier, and J. Janek, *Chem. Mater.* **28**, 2400 (2016).

142. S. Wenzel, S. J. Sedlmaier, C. Dietrich, W. G. Zeier, and J. Janek, *Solid State Ionics* in-press (2018). https://doi.org/10.1016/j.ssi.2017.07.005.

143. V. Thangadurai, H. Kaack, and W. J. F. Weppner, *J. Am. Ceram. Soc.* **86**, 437 (2003).

144. Y. Zhu, X. He, and Y. Mo, *ACS Appl. Mater. Interfaces* **7**, 23685 (2015).

145. F. Ding, W. Xu, X. Chen, J. Zhang, M. H. Engelhard, Y. Zhang, and B. R. Johnson et al., *J. Electrochem. Soc.* **160**, A1894 (2013).

146. S.-K. Jeong, H.-Y. Seo, D.-K. Kim, H.-K. Han, J.-G. Kim, Y. B. Lee, and Y. Iriyama et al., *Electrochem. Commun.* **10**, 635 (2008).

147. J. Qian, W. A. Henderson, W. Xu, P. Bhattacharya, M. Engelhard, O. Borodin, and J.-G. Zhang, *Nat. Commun.* **6**, 6362 (2015).

148. J. Qian, B. D. Adams, J. Zheng, W. Xu, W. A. Henderson, J. Wang, and M. E. Bowden et al., *Adv. Funct. Mater.* **26**, 7094 (2016).

149. Q. Ma, Z. Fang, P. Liu, J. Ma, X. Qi, W. Feng, and J. Nie et al., *Chem. Electro. Chem* **3**, 531 (2016).

150. R. Miao, J. Yang, Z. Xu, J. Wang, Y. Nuli, and L. Sun, *Sci. Rep.* **6**, 21771 (2016).

151. S. Seki, Y. Ohno, H. Miyashiro, Y. Kobayashi, A. Usami, Y. Mita, and N. Terada et al., *J. Electrochem. Soc.* **155**, A421 (2008).

152. S. Seki, Y. Kobayashi, H. Miyashiro, Y. Ohno, Y. Mita, A. Usami, and N. Terada et al., *Electrochem. Solid-State Lett.* **8**, A577 (2005).

153. M. Kirchhöfer, J. von Zamory, E. Paillard, and S. Passerini, *Int. J. Mol. Sci.* **15**, 14868 (2014).

154. D. Eroglu, K. R. Zavadil, and K. G. Gallagher, *J. Electrochem. Soc.* **162**, A982 (2015).

155. M. Hagen, D. Hanselmann, K. Ahlbrecht, R. Maça, D. Gerber, and J. Tübke, *Adv. Energy Mater.* 1401986 (2015).

156. W. Xue, L. Miao, L. Qie, C. Wang, S. Li, J. Wang, and J. Li, *Curr. Op. Electrochem.* **6**, 92 (2017).

157. J. Brückner, S. Thieme, H. T. Grossmann, S. Dörfler, H. Althues, and S. Kaskel, *J. Power Sources* **268**, 82 (2014).

158. D. Lv, J. Zheng, Q. Li, X. Xie, S. Ferrara, Z. Nie, and L. B. Mehdi et al., *Adv. Energy Mater.* **5**, 1402290 (2015).

159. S. Urbonaite, T. Poux, and P. Novák, *Adv. Energy Mater.* **5**, 1500118 (2015).

160. T. Poux, P. Novák, and S. Trabesinger, *J. Electrochem. Soc.* **163**, A1139 (2016).

161. J. Scheers, S. Fantini, and P. Johansson, *J. Power Sources* **255**, 204 (2014).

162. S. Zhang, K. Ueno, K. Dokko, and M. Watanabe, *Adv. Energy Mater.* **5**, 1500117 (2015).

163. Q. Pang, X. Liang, C. Y. Kwok, and L. F. Nazar, *Nat. Energy* **1**, 16132 (2016).

164. V. S. Bryantsev, V. Giordani, W. Walker, M. Blanco, S. Zecevic, K. Sasaki, and J. Uddin et al., *J. Phys. Chem. A.* **115**, 12399 (2011).

165. V. S. Bryantsev and F. Faglioni, *J. Phys. Chem. A.* **116**, 7128 (2012).

166. A. Rosenman, E. Markevich, G. Salitra, D. Aurbach, A. Garsuch, and F. F. Chesneau, *Adv. Energy Mater.* **5**, 1500212 (2015).

167. M. Cuisinier, P.-E. Cabelguen, B. D. Adams, A. Garsuch, M. Balasubramanian, and L. F. Nazar, *Energy Environ. Sci.* **7**, 2697 (2014).

168. K. Dokko, N. Tachikawa, K. Yamauchi, M. Tsuchiya, A. Yamazaki, E. Takashima, and J.-W. Park et al., *J. Electrochem. Soc.* **160**, A1304 (2013).

169. B. D. Adams, E. V. Carino, J. G. Connell, K. S. Han, R. Cao, J. Chen, and J. Zheng et al., *Nano Energy* **40**, 607 (2017).

170. V. Dorval, C. St-Pierre, and A. Vallee, *Proc. of 2004 BATCON Conf.*, p. 19-1; available at www.battcon.com/PapersFinal2004/ValleePaper2004.pdf.

171. https://www.blue-solutions.com/en/blue-solutions/technology/lmp-batteries/.

172. https://www.blue-solutions.com/en/blue-solutions/technology/batteries-lmp/.

173. C. Monroe and J. Newman, *J. Electrochem. Soc.* **151**, A880 (2004).

174. C. Monroe and J. Newman, *J. Electrochem. Soc.* **152**, A396 (2005).

175. T. Niitani, M. Shimada, K. Kawamura, and K. Kanamura, *J. Power Sources* **146**, 386 (2005).

176. T. Niitani, M. Shimada, K. Kawamura, K. Dokko, Y.-H. Rho, and K. Kanamura, *Electrochem. Solid-State Lett.* **8**, A385 (2005).

177. M. Singh, O. Odusanya, G. M. Wilmes, H. B. Eitouni, E. D. Gomez, A. J. Patel, and V. L. Chen et al., *Macromolecules* **40**, 4578 (2007).

178. G. M. Stone, S. A. Mullin, A. A. Teran, D. T. Hallinan Jr., A. M. Minor, A. Hexemer, and N. P. Balsara, *J. Electrochem. Soc.* **159**, A222 (2012).

179. D. T. Hallinan, S. A. Mullin, G. M. Stone, and N. P. Balsara, *J. Electrochem. Soc.* **160**, A464 (2013).

180. K. J. Harry, X. Liao, D. Y. Parkinson, A. M. Minor, and N. P. Balsara, *J. Electrochem. Soc.* **162**, A2699 (2015).

181. D. Devaux, K. J. Harry, D. Y. Parkinson, R. Yuan, D. T. Hallinan, A. A. MacDowell, and N. P. Balsara, *J. Electrochem. Soc.* **162**, A1301 (2015).

182. www.seeo.com.

183. Q. Zhang, K. Liu, F. Ding, and X. Liu, *Nano Research* **10**, 4139 (2017).

184. C. Yang, K. Fu, Y. Zhang, E. Hitz, and L. Hu, *Adv. Mater.* **29**, 1701169 (2017).

185. J.-H. Shin, W. A. Henderson, and S. Passerini, *Electrochem. Solid-State Lett.* **8**, A125 (2005).

186. J.-H. Shin, W. A. Henderson, and S. Passerini, *J. Electrochem. Soc.* **152**, A978 (2005).

187. J.-H. Shin, W. A. Henderson, S. Scaccia, P. P. Prosini, and S. Passerini, *J. Power Sources* **156**, 560 (2006).

188. G.-T. Kim, S. S. Jeong, M.-Z. Xue, A. Balducci, M. Winter, S. Passerini, and F. Alessandrini et al., *J. Power Sources* **199**, 239 (2012).

189. M. Wetjen, G.-T. Kim, M. Joost, M. Winter, and S. Passerini, *Electrochim. Acta* **87**, 779 (2013).

190. I. Osada, J. von Zamory, E. Paillard, and S. Passerini, *J. Power Sources* **271**, 334 (2014).

191. H. de Vries, S. Jeong, and S. Passerini, *RSC Adv.* **5**, 13598 (2015).

192. I. Osada, H. de Vries, B. Scrosati, and S. Passerini, *Angew. Chem. Int. Ed.* **55**, 500 (2016).

193. A. Swiderska-Mocek and D. Naparstek, *Solid State Ionics* **267**, 32 (2014).

194. S. Liu, N. Imanishi, T. Zhang, A. Hirano, Y. Takeda, O. Yamamoto, and J. Yang, *J. Electrochem. Soc.* **157**, A1092 (2010).

195. M. Baloch, A. Vizintin, R. K. Chellappan, J. Moskon, D. Shanmukaraj, R. Dedryvère, and T. Rojo et al., *J. Electrochem. Soc.* **163**, A2390 (2016).

196. R. Sudo, Y. Nakata, K. Ishiguro, M. Matsui, A. Hirano, Y. Takeda, and O. Yamamoto et al., *Solid State Ionics* **262**, 151 (2014).

197. Y. Ren, Y. Shen, Y. Lin, and C.-W. Nan, *Electrochem. Commun.* **57**, 27 (2015).

198. Y. Suzuki, K. Kami, K. Watanabe, A. Watanabe, N. Saito, T. Ohnishi, and K. Takada et al., *Solid State Ionics* **278**, 172 (2015).

199. S.-W. Baek, J.-M. Lee, T. Y. Kim, M.-S. Song, and Y. Park, *J. Power Sources* **249**, 197 (2014).

200. J. B. Bates, N. J. Dudney, B. J. Neudecker, F. X. Hart, H. P. Jun, and S. A. Hackney, *J. Electrochem. Soc.* **147**, 59 (2000).

201. N. J. Dudney, *Electrochem. Soc. Interface* **17**, 44 (2008).

202. N. J. Dudney, "Thin Film Batteries for Energy Harvesting," in *Energy Harvesting Technologies*, S. Priya and D. J. Inman (eds.), pp. 349–357, Springer Publisher, Dec. 2008.

203. B. J. Neudecker, R. A. Zhur, and J. B. Bates, *J. Power Sources* **81–82**, 27 (1999).

204. www.vp-scientific.com/parylene_properties.htm.

205. N. J. Dudney, *Mat. Sci. Eng. B.* **116**, 245 (2005).

206. N. J. Dudney and Y. I. Jang, *J. Power Sources* **119**, 300 (2003).

207. Y. Origami, D. Shimizu, T. Abe, M. Sodom, and Z. Ogumi, *ECS Trans.* **16**, 45 (2009).

208. T. Ohzuku, A. Ueda, and N. Yamamoto, *J. Electrochem. Soc.* **142**, 1431 (1995).

209. www.infinitepowersolutions.com.

210. www.frontedgetechnology.com.

211. T. Yamada, S. Ito, R. Omoda, T. Watanabe, Y. Aihara, M. Agostini, and U. Ulissi et al., *J. Electrochem. Soc.* **162**, A646 (2015).

212. R.-C. Xu, X.-H. Xia, X.-L. Wang, Y. Xia, and J.-P. Tu, *J. Mater. Chem. A* **5**, 2829 (2017).

213. R.-C. Xu, X.-H. Xia, S.-H. Li, S.-Z. Zhang, X.-L. Wang, and J.-P. Tu, *J. Mater. Chem. A* **5**, 6310 (2017).

214. X. Huang, C. Liu, Y. Lu, T. Xiu, J. Jin, M. E. Badding, and Z. Wen, *J. Power Sources*, in-press (2018). https://doi.org/10.1016/j.jpowsour.2017.11.074.

215. X. Yu and A. Manthiram, *Acc. Chem. Res.* **50**, 2653 (2017).

216. J.-H. Choi, C.-H. Lee, J.-H. Yu, C.-H. Doh, and S.-M. Lee, *J. Power Sources* **274**, 458 (2015).

217. F. Langer, I. Bardenhagen, J. Glenneberg, and R. Kun, *Solid State Ionics* **291**, 8 (2016).

218. K. Fu, Y. Gong, J. Dai, A. Gong, X. Han, Y. Yao, and C. Wang et al., *Proc. Natl. Acad. Sci.* **113**, 7094 (2016).

219. S. H.-S. Cheng, K.-Q. He, Y. Liu, J.-W. Zha, M. Kamruzzaman, R. L.-W. Ma, and Z.-M. Dang et al., *Electrochim. Acta* **253**, 430 (2017).

220. F. Chen, D. Yang, W. Zha, B. Zhu, Y. Zhang, J. Li, and Y. Gu et al., *Electrochim. Acta* **258**, 1106 (2017).

221. R.-J. Chen, Y.-B. Zhang, T. Liu, B.-Q. Xu, Y.-H. Lin, C.-W. Nan, and Y. Shen, *ACS Appl. Mater. Interfaces* **9**, 9654 (2017).

222. X. Tao, Y. Liu, W. Liu, G. Zhou, J. Zhao, D. Lin, and C. Zu et al., *Nano Lett.* **17**, 2967 (2017).

223. J. Zhang, N. Zhao, M. Zhang, Y. Li, P. K. Chu, X. Guo, and Z. Di et al., *Nano Energy* **28**, 447 (2016).

224. http://www.solidenergysystems.com/.

# MISCELLANEOUS AND SPECIALTY BATTERIES

# CHAPTER 18
# METAL/AIR BATTERIES

## 18.0  GENERAL CHARACTERISTICS

The electrochemical coupling of a reactive anode to an air electrode provides a battery with an inexhaustible cathode reactant and, in some cases, very high specific energy and energy density. The capacity limit of such systems is determined by the ampere-hour capacity of the anode and the techniques for handling and storage of the reaction product. As a result of this performance potential, a significant effort has gone into metal/air battery development.[1,2] The major advantages and disadvantages of the metal/air battery system are summarized in Table 18.1.

Primary, reserve, electrically rechargeable, and mechanically rechargeable metal/air battery configurations have been explored and developed. In the mechanically rechargeable designs (that is, replacing the discharged metal electrode), the battery essentially functions as a primary battery and can use relatively simple "unifunctional" air electrodes that need to operate only in a discharge mode. Conventional electrical recharging of metal/air batteries requires either a third electrode (to sustain oxygen evolution on charge) or a "bifunctional" electrode (a single electrode capable of both oxygen reduction and evolution).

Table 18.2 lists the metals that have been considered for use in metal/air batteries with several of their electrical characteristics. Of the potential metal/air battery candidates, zinc has received the most attention because it is the most electropositive metal that is relatively stable in aqueous and alkaline electrolytes without significant corrosion, provided the appropriate inhibitors are used.

Zinc has been used for many years in commercial primary zinc/air batteries. Initially, the products were large batteries using alkaline electrolytes for such applications as railroad signaling, remote communications, and ocean navigational units requiring long-term, low-rate discharge. As thin electrodes were developed, the technology was applied to small (button-type), high-capacity primary cells used in hearing aids, pagers, and similar applications.

Zinc is also attractive for electrically rechargeable metal/air systems because of its relative stability in alkaline electrolytes and also because it is the most active metal that can be electrodeposited from an aqueous electrolyte. The development of a practical rechargeable zinc/air battery with an extended cycle life would provide a promising high-capacity power source for many portable applications (computers, communications equipment) as well as, in larger sizes, for electric vehicles. Problems of dendrite formation, nonuniform zinc dissolution and deposition, limited solubility of the reaction product, and unsatisfactory air electrode performance have slowed progress toward the development of a commercial rechargeable battery. However, there is a continued search for a practical system because of the potential of the zinc/air battery.

The lithium/air battery is attractive because lithium has the highest theoretical voltage and electrochemical equivalence (3860 Ah/kg of lithium) of any metal anode considered for a practical battery system. In the cell discharge reaction, lithium metal, atmospheric oxygen, and water are consumed, and LiOH is generated. The cell can operate at high coulombic efficiencies because of the formation of a protective film on the metal that retards rapid corrosion after formation. On open-circuit and low-drain discharge, the self-discharge of the lithium metal is rapid due to the parasitic corrosion reaction. This reaction degrades the anode's coulombic capacity and must be controlled if the full potential of the lithium anode is to be realized. This self-discharge also necessitates the removal of the electrolyte during stand.

**TABLE 18.1**  Major Advantages and Disadvantages of Metal/Air Batteries

| Advantages | Disadvantages |
| --- | --- |
| High energy density | Dependent on environmental conditions: |
| Flat discharge voltage |   Drying out limits shelf life once opened to air |
| Long shelf life (dry storage) |   Flooding limits power output |
| No ecological problems | Limited power output |
| Low cost (on metal use basis) | Limited operating temperature range |
| Capacity independent of load and temperature | $H_2$ from anode corrosion |
|   when within operating range | Carbonation of alkali electrolyte |

**TABLE 18.2**  Characteristics of Metal/Air Cells

| Metal anode | Electrochemical equivalent of metal, Ah/g | Theoretical cell voltage,* V | Valence change | Theoretical specific energy (of metal), kWh/kg | Practical operation voltage, V |
| --- | --- | --- | --- | --- | --- |
| Li | 3.86 | 3.4 | 1 | 13.0 | 2.4 |
| Ca | 1.34 | 3.4 | 2 | 4.6 | 2.0 |
| Mg | 2.20 | 3.1 | 2 | 6.8 | 1.2–1.4 |
| Al | 2.98 | 2.7 | 3 | 8.1 | 1.1–1.4 |
| Zn | 0.82 | 1.6 | 2 | 1.3 | 1.0–1.2 |
| Fe | 0.96 | 1.3 | 2 | 1.2 | 1.0 |

*Cell voltage with oxygen cathode.

The principal advantage of the lithium/air battery is its higher cell voltage, which translates into higher power and specific energy. However, in view of their availability, cost, and safety advantages, the development of metal/air batteries has previously concentrated on zinc and aluminum.

Other metals have also been investigated as electrode materials for metal/air batteries. Calcium, magnesium, and aluminum have attractive energy densities. Lithium/air,[3,4] calcium/air, and magnesium/air batteries[5,6] have been studied, but cost and problems such as anode polarization or instability, parasitic corrosion, non-uniform dissolution, safety, and practical handling have so far inhibited the development of commercial products. The voltage and the specific energy of the iron/air battery are relatively low. Development on this battery, therefore, has concentrated on an electrically rechargeable system, as the iron electrode is long-lived and more adapted to recharging.

Aluminum is attractive for use because of its geological abundance (third most abundant element in the earth's crust), its potentially low cost, and its relative ease of handling.[7–9] However, the aluminum/air battery has a charging potential that is too high to be electrically recharged in an aqueous system (water is preferentially electrolyzed). Therefore, the effort has been directed to reserve designs and mechanically rechargeable batteries. Figure 18.1 summarizes the work on the various types and designs of metal/air batteries.

### 18.0.1  Chemistry

The metal/air batteries being developed use neutral or alkaline electrolytes. The oxygen-reduction half-cell reaction during discharge may be written as

$$O_2 + 2H_2O + 4e \rightleftharpoons 4OH^- \quad E^0 = +0.401 \text{ V}$$

The theoretical cell voltages, the equivalent weights of the metals, and the theoretical specific energies obtained when this oxygen electrode is coupled with various metal anodes are given in Table 18.2. Polarization effects at

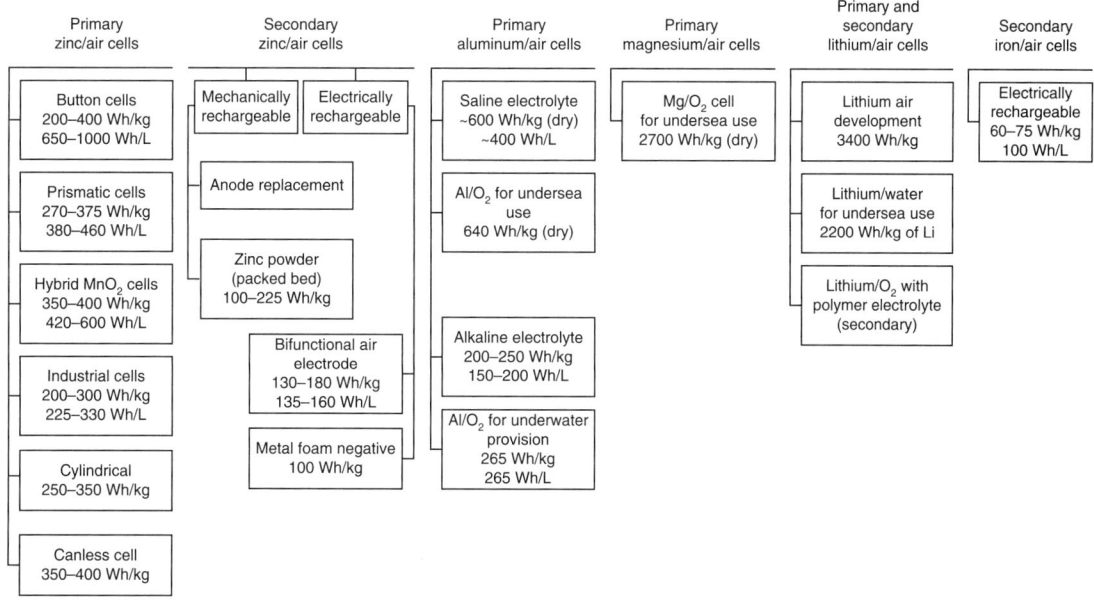

**FIGURE 18.1**   Metal/air batteries.

both electrodes degrade these voltages to those shown in the table at practical operating discharge rates. Note that the theoretical specific energy of metal/air batteries is based only on the negative electrode (anode or fuel electrode during discharge) as this is the only reactant that has to be carried in the battery. The other reactant, oxygen, is introduced into the battery from ambient air during discharge.

The discharge reaction at the negative or metal electrode (anode during discharge) is dependent on the specific metal used, the electrolyte, and other factors in the cell chemistry. The discharge reaction at the negative electrode can be generalized as

$$M \rightarrow M^{n+} + ne$$

The generalized overall discharge reaction may be written as

$$4M + nO_2 + 2nH_2O \rightarrow 4M(OH)_n$$

where M is the metal and the value of $n$ depends on the valence change for the oxidation of the metal, as listed in Table 18.2.

Most metals are thermodynamically unstable in an aqueous electrolyte and react with the electrolyte to corrode or oxidize the metal and generate hydrogen as follows:

$$M + nH_2O \rightarrow M(OH)_n + n/2H_2$$

This parasitic corrosion reaction, or self-discharge, degrades the coulombic efficiency of the anode and must be controlled to minimize this loss of capacity.

Other factors that can affect the performance of the metal/air battery are the following:

***Polarization.***    The voltage of a metal/air battery drops off more sharply with increasing current than that of other types of batteries because of diffusion and other limitations in the oxygen or air cathode. This means that these air systems are more suited for low- to moderate-power applications than to high-power ones.

*Electrolyte Carbonation.* As the cell is open to air, carbon dioxide can be absorbed. This can result in the crystallization of carbonate in the porous air electrode, which may impede air access and cause mechanical damage and a decreasing electrode performance. Potassium carbonate is also less conductive than the KOH electrolyte normally employed in metal/air batteries.

*Water Transpiration.* Again, as the cell is open to air, water vapor can be transferred if a vapor partial pressure difference exists between the electrolyte and the surrounding environment. Excessive water loss increases the concentration of the electrolyte and leads to drying out and premature failure. Gain of water can lead to dilution of the electrolyte. This gain can cause flooding of the air electrode pores and electrode polarization due to the inability of the air to reach the reaction sites.

*Efficiency.* The oxygen electrode at moderate temperatures displays a significant irreversibility during both charge and discharge. As a result there is generally about a 0.2 V difference between the actual charging voltage and the reversible potential, with the same situation on discharge. For example, a zinc/air battery generally discharges at a voltage of about 1.2 V, while the charging voltage is about 1.6 V or higher. This results in a loss of overall energy efficiency even before any other factors are considered.

*Charging.* Oxidation of catalysts and electrode supports during charging can be a problem for those systems that are recharged electrically, such as zinc/air and iron/air. Approaches to solving this problem generally involve either the use of oxidation-resistant substrates and catalysts, the use of a third electrode for charging, or charging the negative (metal) electrode material external to the cell.

## 18.0.2 Air Electrode

Successful operation of metal/air batteries depends on an effective air electrode. As a result of the interest in gaseous fuel cells and metal/air batteries over the past 40 years, a significant effort has been aimed at improved high-rate, thin air electrodes, including the development of better catalysts, longer lived physical structures, and lower cost fabrication methods for such gas diffusion electrodes.

An alternative approach is to use a low-cost air cathode with more modest performance, but this requires a greater cathode area in each cell. Figure 18.2 shows a type of electrode that is produced by a continuous process using low-cost materials.[10–12] This electrode is composed of two active layers bonded to each side of a current-collecting screen, with a microporous Teflon layer bonded to the air side of the electrode. The active layers are fabricated by passing a nonwoven web of carbon fibers (see Fig. 18.2b) through a slurry containing the catalyst, a dispersing agent, and a binder in a continuous process, with a drying and compacting step built into the process. The active layers, the screen, and the Teflon layer are then bonded in the continuous process. These electrodes are used in the aluminum air reserve standby batteries (see "Aluminum/Air Cells in Alkaline Electrolytes" in Sec. 18B.4.1).

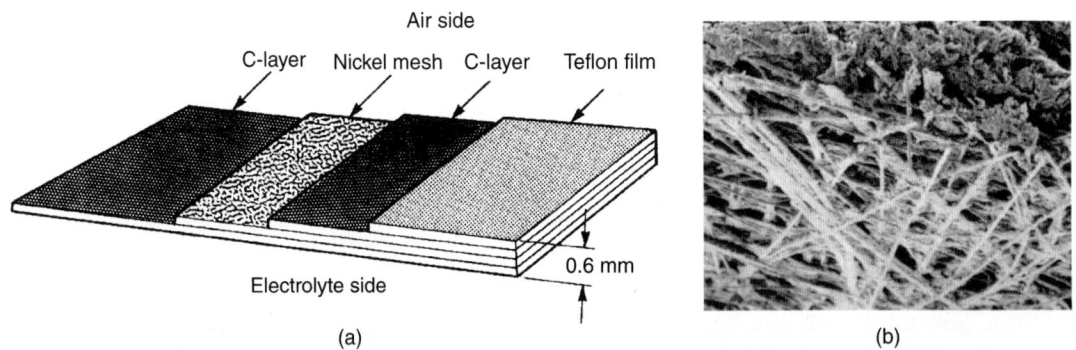

**FIGURE 18.2** (a) Laminated air cathode. (b) Carbon fiber substrate. (*Courtesy of Alupower, Inc.*)

## Acknowledgments

Content for this chapter was taken in part from Chap. 13, Button Cell Batteries: Silver Oxide–Zinc and Zinc-Air Systems, Joseph Passaniti, Denis Carpenter, and Rodney McKenzie; Chap. 18, Iron Electrode Batteries, Gary A. Bayles; and Chap. 33, Metal/Air Batteries, Terrill B. Atwater and Arthur Dobley, *Linden's Handbook of Batteries*, 4th ed., T. B. Reddy and D. Linden, eds., McGraw-Hill, 2011.

## References

1. D. A. J. Rand, "Battery Systems for Electric Vehicles: State of Art Review," *J. Power Sources* **4:**101 (1979).

2. K. F. Blurton and A. F. Sammells, "Metal/Air Batteries: Their Status and Potential—A Review," *J. Power Sources* **4:**263 (1979).

3. H. F. Bauman and G. B. Adams, "Lithium-Water-Air Battery for Automotive Propulsion," Lockheed Palo Alto Research Laboratory, Final Rep., COO/1262-1, October 1977.

4. W. P. Moyer and E. L. Littauer, "Development of a Lithium-Water-Air Primary Battery," *Proc. IECEC,* Seattle, WA, August 1980.

5. W. N. Carson and C. E. Kent, "The Magnesium-Air Cell," in D. H. Collins (ed.), *Power Sources,* 1966.

6. R. P. Hamlen, E. C. Jerabek, J. C. Ruzzo, and E. G. Siwek, "Anodes for Refuelable Magnesium-Air Batteries," *J. Electrochem. Soc.* **116:**1588 (1969).

7. J. F. Cooper, "Estimates of the Cost and Energy Consumption of Aluminum-Air Electric Vehicles," *ECS Fall Meeting,* Hollywood, FL, October 1980; Lawrence Livermore, UCRL-84445, June 1980; update UCRL-94445 rev. 1, August 1981.

8. R. P. Hamlen, G. M. Scamans, W. B. O'Callaghan, J. H. Stannard, and N. P. Fitzpatrick, "Progress in Metal-Air Battery Systems," *International Conference on New Materials for Automotive Applications,* October 10–11, 1990.

9. A. S. Homa and E. J. Rudd, "The Development of Aluminum-Air Batteries for Electric Vehicles," *Proceedings of the 24th IECEC,* vol. 3, 1989, pp. 1331–1334.

10. W. H. Hoge, "Air Cathodes and Materials Therefore," U.S. Patent 4,885,217, 1989.

11. W. H. Hoge, "Electrochemical Cathode and Materials Therefore," U.S. Patent 4,906,535, 1990.

12. W. H. Hoge, R. P. Hamlen, J. H. Stannard, N. P. Fitzpatrick, and W. B. O'Callaghan, "Progress in Metal-Air Systems," *Electrochem. Soc.,* Seattle, WA, October 14–19, 1990.

# SECTION A

# ZINC/ALKALINE

**Arthur Dobley, Terrill B. Atwater**
**(Emeritus Contributors: Joseph Passaniti, Denis Carpenter, Rodney McKenzie)**

## 18A.1  GENERAL

Zinc/air batteries have been commercially available in primary button-type batteries for nearly 100 years. Since the late 1990s, 5 to 30 Ah prismatic batteries as well as larger primary industrial-type batteries have been available. Electrically rechargeable batteries are being considered for both portable and electric-vehicle applications, but the control of the replating of zinc during recharging and the development of an efficient high-rate bifunctional air electrode remain a challenge. In some designs, a third oxygen-evolving electrode is used for recharging, or recharging is done external to the cell to avoid the need for the bifunctional air electrode. Another approach to avoid the difficulties with electrical recharging is the "mechanically rechargeable" battery, where the spent zinc electrode and/or the discharged products are removed and physically replaced.

The developmental progression of the primary zinc/air system lends itself to be described in four generations.

First generation (GEN 1) zinc/air batteries were introduced in the 1930s, resemble automotive SLI batteries in construction, and are used in remote applications such as buoys and railroad signaling (see Sec. 18A.4.3). They are designed for low-rate (<1 A), multiyear service, and have moderate specific energy.

The second generation (GEN 2) zinc/air is the button cell, commercialized in the 1970s for hearing aids. With specific energies of more than 400 Wh/kg, the button cell is typically limited to about 10 mW in power and has a service life of 1 month.

Third generation (GEN 3) zinc/air cells employ molded plastic cell housings, sealed and joined with epoxy adhesives (see Sec. 18A.4.2). Batteries using 30 Ah GEN 3 zinc/air cells were first available in 2003 in a 12/24 V, 750 Wh battery. The battery is designed for moderate power (up to 50 W) and a service life of several months. It powers tactical radios for up to a week or more for typical duty cycles.

The fourth generation (GEN 4) began development in the late 1990s and early 2000s (see Sec. 18A.4.2 for more detailed information on GEN 4 batteries). Figure 18.1 contains a summary of the different types of zinc/air batteries.

## 18A.2  CHEMISTRY

The overall cell reaction for a zinc/air battery on discharge in an alkaline electrolyte may be represented as

$$Zn + \tfrac{1}{2}O_2 + H_2O + 2(OH)^- \rightarrow Zn(OH)_4^{2-} \quad E^0 = 1.62 \text{ V}$$

The initial discharge reaction at the zinc electrode can be simplified to:

$$Zn + 4(OH)^- \leftrightarrow Zn(OH)_4^{2-} + 2e$$

This reaction occurs as a result of the solubility of the zincate anion in the electrolyte and proceeds until the zincate level reaches the saturation point. There is no well-defined solubility limit, since the degree of supersaturation is time-dependent. After partial discharge, the solubility exceeds the equilibrium solubility level, with subsequent precipitation of zinc oxide, as follows:

$$Zn(OH)_4^{2-} \rightarrow ZnO + H_2O + 2(OH)^-$$

The overall cell reaction then becomes

$$Zn + \tfrac{1}{2}O_2 \leftrightarrow ZnO$$

This transient solubility is one of the main reasons for the difficulty in making a successful rechargeable zinc/air battery. The location of the precipitation of the reaction product cannot be controlled, so that on a subsequent recharge, the amount of zinc deposited on different parts of the electrode area of the cell can vary.

## 18A.3   BATTERY DESIGN AND CONSTRUCTION

### 18A.3.1   Button and Coin Cell Primary Zinc/Air Batteries

Button and coin cells are the most prevalent form for zinc/air batteries to take; therefore, it is this configuration that will be used as exemplary examples of the zinc/air electrochemistry and cell design considerations. Development of other formats, such as cylindrical and prismatic forms, has been somewhat limited due to economics, but also by the frustrations of air management and the poor performance of zinc/air when use is intermittent. Moisture sensitivity in large cells becomes a greater problem when rate capability demands a largely open cathode to air interface. Unless the cathode can be isolated from the environment during the idle periods, the cells will suffer the adverse effects of water gain or loss, as well as the degradation caused by the accumulation of carbonates at the air-electrolyte interface.

*Cell Design.*   Similar in many ways to their predecessors, the zinc/mercury oxide and zinc/silver oxide button cells, the zinc/air button cell is designed with an eye to ultimate electrical capacity within a defined volume. As illustrated in Fig. 18A.1, the zinc/air button cell can pack more zinc in the same volume because the system does not require a thick metal oxide cathode to complete the cell reaction. Each cell size has an analogue size in the two older chemistries, with the exception of the size 10 cell (IEC PR70), which was developed by Rayovac in the mid-1980s. It was introduced to the hearing instrument industry as an enabling technology for in the ear (ITE) and completely in the canal (CIC) devices.

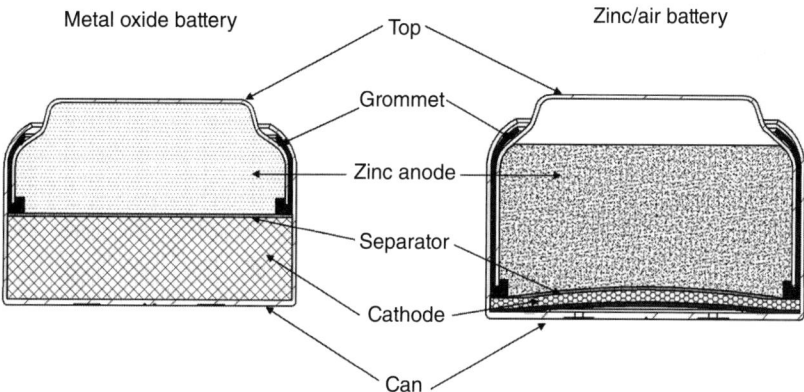

**FIGURE 18A.1**   Cross-sections of metal oxide and zinc/air button batteries.

As seen in Fig. 18A.2, there are no spare parts in the construction of the zinc/air cell. Each part performs multiple functions and must do so while occupying a minimum of volume of its own. The resulting assembly contains the maximum amount of zinc that can be discharged by the user, delivering the longest number of hours or days of battery performance. The negative top of the cell is an impervious container for the zinc, acts as the

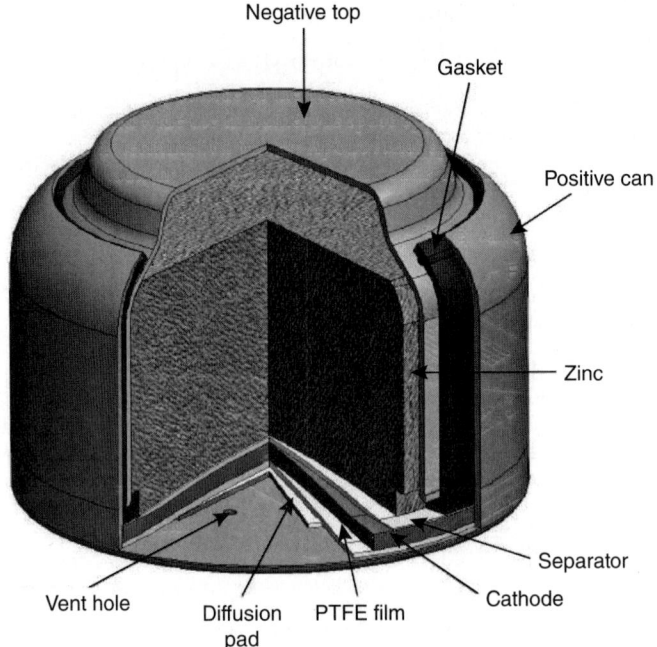

**FIGURE 18A.2**   A typical zinc/air button battery (components not to scale).

current collector, and resists the pressure of closure that seals it against the gasket. The top also makes the electrical contact that brings the electronic current out of the cell to an external device or circuit. The positive can acts as the outer container of the cell, connecting electrically to the cathode disk, and applies the compressive forces of closure against the gasket that stands between the can and the top. Vent holes in the can admit air to the cathode, and the outer surface provides electrical contact to the external circuit. Finally, the gasket between the can and the top provides a seal against electrolyte leakage and prevents electrical shorting between the top and the can. The gasket is usually made of a hard polymeric material, often an engineering plastic such as polyamide, and may be coated with additional sealant material to enhance its resistance to leakage.

Within the cell container, the cathode rests flat (see Fig. 18A.3), occupying the inside bottom of the can. It is a multilayer structure consisting of a gas diffusion membrane.[1,2] The first layer is a hydrophobic, microporous membrane that is permeable to gases such as oxygen and water vapor, but too fine in pore size to allow liquid to pass from inside the cell to the outside. In most commercial zinc/air batteries, this layer is made up of expanded PTFE film.

The next layer is the conductive mix of carbon and catalysts that are wetted with electrolyte. This creates the conditions for oxygen to be held at activated sites and thus made available for the electrochemical reaction that produces hydroxide ions in the electrolyte, and water is consumed from the electrolyte. To support this reaction, there needs to be a source of electrons. They come in from the external circuit, driven by the potential of the zinc in the anode. The electronic path for the current is through the current collector, usually a wire mesh screen or expanded metal layer impressed into the cathode active material. In the case of most button or coin cells, this is a disk blanked from a larger sheet which fits snugly into the inside diameter of the cathode can.

Air enters the cathode through vent holes that are punched into the flat bottom of the can. The access of air is controlled by a combination of size and number of the vent holes, as well as the degree to which the porosity of the PTFE sheet adjacent to the cathode material has been occluded or compressed during manufacture of that subassembly. Depending on the intended use of the cell, and the electrical current needed to meet that use, the parameters of vent size, their number, and the cathode rate capability might all be increased or decreased to suit the application.

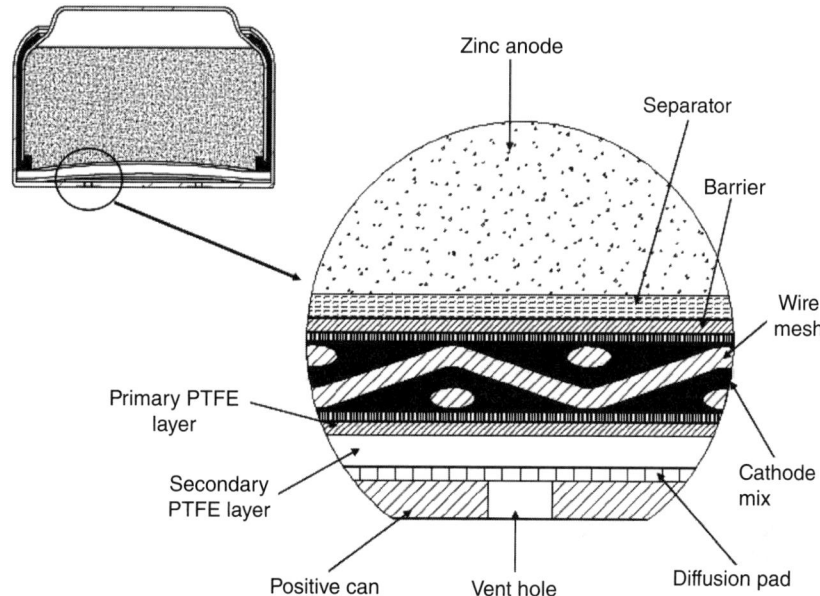

**FIGURE 18A.3**   Key constructional features of the zinc/air cathode.

On the anode side of the cathode structure are the separator/barrier layers, wettable to the electrolyte, but capable of preventing zinc or zinc oxide from directly contacting the cathode. If this were to happen, a direct electron path would be established, and the cell would self-discharge. The barrier is most commonly a microporous layer of polymer film that does not break down in the caustic environment of the cell and remains a good ionic conductor throughout discharge. The separator is most commonly a nonwoven cellulosic that is highly absorbent to electrolyte and helps to prevent zinc dendrite shorting. If the separator/barrier system interferes with the conductivity of the cell, then the rate capability, the capacity of the cell, or both will be adversely affected.

The negative top of the button cell contains all of the zinc and the majority of the electrolyte. Only a small portion of the electrolyte wets into the cathode, stopping at the hydrophobic layer. In order to allow for the growth of the anode, it is not possible to initially fill the entire top. Zinc metal has a density of approximately 7.14 g/cc, whereas the density of ZnO is about 5.47 g/cc. Zinc going to zinc oxide gains in mass by a factor of 1.25, with a corresponding volume increase due to the decrease in density of 1.63 g/cc. Space for that growth must be accounted for in the initial fill of anode material into the cell. Under equilibrium conditions, the electrolyte mass does not change during discharge, and occupies essentially the same volume at the end of discharge as it did when first assembled.

High-purity zinc is the active constituent of the anode. It is normally a distribution of finely divided, atomized particles. In some instances, it is alloyed to reduce the tendency for catalytic corrosion of the metal. Historically, as in virtually all button cells, it was also amalgamated with a small amount of mercury typically less than 25 mg/cell to reduce hydrogen overpotential, largely due to the segregation of trace metals to the grain boundaries of the zinc during the solidification that takes place as the zinc is atomized. Mercury would preferentially concentrate at the grain boundaries. Insoluble in the zinc oxide, the mercury would further concentrate into the remaining zinc, and eventually be released as beads of liquid metal suspended in the discharged anode. After the introduction of zero mercury zinc/air cells in 2011, the use of mercury in most commercial products has been effectively eliminated.

The electrolyte used in zinc/air cells is normally a caustic solution of KOH at a concentration of about 30%. The electrolyte is highly conductive and wets through the cathode structure readily, providing excellent ionic access to both zinc and cathode active sites. At this concentration, the electrolyte is at equilibrium with water vapor at roughly 50% relative humidity (RH) for ambient temperatures. Performance of the zinc/air system will be compromised if the rate of ingress or egress of water is significant. Gain of water will dilute the electrolyte slightly, but more significant is the increase in the volume of the anode and the reduction of the space that was initially left for the discharge product. The gain in water does not have a large effect on rate capability, but capacity of the battery will be cut short as the internal materials push the cathode against the inside bottom of the cell and reduce the effective cathode active area. When the environmental conditions promote drying out of the cell, there is a more immediate detrimental effect on the battery's electrical performance. Good ionic contact within the cell may be affected as gaps develop at the electrode interface and sufficient water is not available in the cathode to support the destruction of peroxide and subsequent hydroxyl formation. Dry conditions in the anode promote early cementation of the partially discharged anode, and there is usually an increase in the battery's internal impedance.

Both the chemical and mechanical design of zinc/air cells need to balance four main electrical performance characteristics. These performance characteristics include (1) open-circuit voltage (OCV), (2) closed circuit or operating voltage, (3) internal cell impedance, and (4) limiting current.

OCV represents the potential for current to flow and is usually 1.4 to 1.5 V for commercial zinc/air cells. Closed circuit is the actual voltage that the cell can support when a defined load has been applied that is demanding a current to flow through a test circuit or through an actual device under test. The power being delivered is the product of voltage and current delivered by the cell under a given discharge condition. This current is partially controlled by the internal cell impedance and the limiting current and is generally in the range of a few milliwatts for a zinc/air button cell.

The internal impedance of the battery is usually measured with an LCR meter and accounts for the AC components of resistance, capacitance, and inductance. For fresh zinc/air button cells, the impedance rises as one goes to low frequency levels. The impedance drops as cells are discharged and remains low until the end of discharge as the last of the available zinc is converted to zinc oxide.

The most dynamic and perhaps interesting electrical property of the zinc/air system is the limiting current, so named because it occurs when an increasing external load is applied, the cell has reached a point where there is no corresponding increase in current, and the voltage of the cell drops precipitously. Below the limiting current value, the cell is not internally rate limited; it is the external circuit that is limiting the flow of electrons. At the limiting current, there is a rate-limiting process that has been maximized. At the beginning of discharge, the rate-limiting process of the fresh cell is usually oxygen diffusion within the cathode, PTFE diffusion rate, or the vent hole configuration of the can. As discharge continues, the anode active electrode area (metallic zinc surface) decreases and becomes the rate-limiting process.

*Cell Sizes.*    Zinc/air button and coin cells are available in a variety of sizes. Capacities range from 35 to 1000 mAh. Table 18A.1 lists the physical and electrical characteristics of currently available zinc/air button and coin cells. The button cells are primarily used for hearing aid applications while the coin cell has been used in pager applications. With the advances in cochlear implants, a special higher power PR44 (675 Cochlear) has been designed for that application.

Zinc/air batteries for hearing aid applications have been improved and refined over the years to meet stringent device and user requirements. With the advent of digital aids in the late 1990s, the battery design has had to adapt to higher current and pulse requirements. Today's zinc/air hearing aid batteries offer up to twice the capacity of the original designs of the late 1970s.

These improvements have been achieved by optimizing the internal anode volume without exceeding the standard external cell dimensions.[3,4] The zinc content is maximized without compromising the internal free volume needed for expansion as zinc metal is converted to zinc oxide. If the free internal volume is not adequate, the cell could have excessive expansion, leakage, or a premature end of life failure.

**TABLE 18A.1**    Characteristics of Zinc/Air Button and Coin Batteries

| Generic type | IEC no. | ANSI no. | Max. height, mm | Max. diameter, mm | Average weight, g | Standard drain, mA | High power drain, mA | Rated capacity, mA |
|---|---|---|---|---|---|---|---|---|
| 5 | — | — | 2.15 | 5.8 | 0.2 | 0.4 | — | 33 |
| 10 | PR70 | 7005ZD | 3.6 | 5.8 | 0.3 | 0.7 | 1.0 | 75–105 |
| 312 | PR41 | 7002ZD | 3.6 | 7.9 | 0.5 | 1.2 | 2.0 | 145–180 |
| 13 | PR48 | 7000ZD | 5.4 | 7.9 | 0.8 | 2.0 | 3.0 | 265–310 |
| 675 | PR44 | 7003ZD | 5.4 | 11.6 | 1.8 | 5.0 | 8.0 | 600–650 |
| 675 Cochlear | — | — | 5.4 | 11.6 | 1.8 | 10 | 20 | 550–570 |
| 2330 | — | — | 3.0 | 23.2 | 4 | 4.0 | — | 950 |

*Source:* Rayovac, Duracell, Energizer, Power One, Zeni, Panasonic.

## 18A.3.2    Portable Primary Zinc/Air Batteries

Primary zinc/air button-type batteries, as described above, constitute the majority of applications of the zinc-air electrochemistry. The button cell configuration is an effective way to package the zinc/air system in small sizes, but scaling up to larger sizes tends to lead to performance and leakage problems. However, these can be overcome with prismatic cell designs described in Sec. 18A.4.2.

A typical prismatic cell uses a metal or plastic tray, which holds the zinc anode/electrolyte blend while the separator and cathode are bonded onto the rim of the tray. The anode/electrolyte blend is similar to the anode blend used in zinc/alkaline primary cells, containing zinc powder in a gelled aqueous potassium hydroxide electrolyte. The cathode is a thin gas diffusion electrode comprising two layers, an active layer and a barrier layer. The active layer of the cathode, which interfaces with the electrolyte, uses a high surface area carbon and a metal oxide catalyst bonded with Teflon. The high surface area carbon is required for oxygen reduction and the metal oxide catalyst ($MnO_2$) for peroxide decomposition. The barrier layer, which interfaces with the air, consists of carbon bonded together with Teflon. A high concentration of Teflon prevents electrolyte from weeping from the cell. Prismatic zinc/air cells have been designed with moderately high rate and high capacity.

## 18A.3.3    Industrial Primary Zinc/Air Batteries

Large primary zinc/air batteries have been used for many years to provide low-rate, long-life power and are available in various versions: preactivated, water-activated, and gelled electrolyte (see Sec. 18A.4.3 for details). Until recently, the zinc contained a few percent of mercury to minimize self-discharge after activation. The newer batteries use additives and alloys to eliminate the mercury and minimize hydrogen generation and corrosion.

Preactivated industrial-type zinc/air cells, such as the Edison Carbonaire cell, are manufactured in a 1100 Ah size, contained in a tinted transparent acrylic plastic housing that allows the electrolyte level to be visually checked and the condition of the zinc plates and precipitates to be monitored.

Water-activated cells and batteries are supplied sealed and activated by removing the seals and adding the appropriate amount of water to the included potassium hydroxide powder. Periodic inspection and addition of water are the only required maintenance steps. These cells are manufactured in a 1100 Ah size and are available in multicell batteries, with the cells connected in series or parallel.

The gelled electrolyte cell eliminates the possibility of leakage during operation. The zinc electrode is composed of zinc powder mixed with a gelling agent and the electrolyte. The discharge reaction product is zinc oxide rather than calcium zincate. The Gelaire battery is manufactured by Saft America, Inc. in a nominal 1200 Ah size and is available in multicell batteries with the cells connected in series or parallel.

## 18A.3.4    Primary Hybrid-Air/Manganese Dioxide Batteries

Another primary zinc/air battery design uses a hybrid cathode that contains a significant amount of manganese dioxide (see Sec. 18A.4.4 for details). During low-rate operation, the battery functions as a zinc/air system. At high rates, as the oxygen may be depleted, the discharge function at the cathode is taken over by the manganese dioxide. This zinc/air battery has capacity similar to zinc/air at low discharge rates but still functions well at high pulse currents similar to a manganese dioxide battery.

## 18A.3.5    Electrically Rechargeable Zinc/Air Batteries

Electrically rechargeable zinc/air batteries use a bifunctional oxygen electrode so that both the charge process and the discharge process take place within the battery structure. Section 18A.4.5 provides further details on the bifunctional cathodes.

## 18A.3.6    Mechanically Rechargeable Zinc/Air Batteries

Mechanically rechargeable (or refuelable) batteries are designed with a means to remove and replace the discharged anodes or discharge products. The discharged anode or discharge products can be recharged or reclaimed external to the cell. This avoids the need for a bifunctional air cathode and the shape change problems resulting from the charge-discharge cycling of an in situ zinc electrode (refer to Sec. 18A.4.6).

## 18A.4  PERFORMANCE AND APPLICATIONS

### 18A.4.1  Primary Button and Coin Cell Capabilities

*Voltage.*  The nominal OCV for a zinc/air battery is 1.4 V. This value can vary from manufacturer to manufacturer because of differences in anode and cathode chemistry. Typically, the open-circuit value can range from 1.4 to 1.5 V. The initial closed-circuit voltage at 20°C ranges from 1.15 to 1.35 V depending on discharge load. The discharge is relatively flat, with the typical end voltages falling between 0.9 and 1.1 V.

In order to ensure freshness and long-term shelf life, the zinc/air battery's air holes are covered with a tape tab. The tab is designed to mute air ingress to the point that it lowers the OCV. This lower tabbed voltage helps to determine if the tape tab is properly attached to the battery.

If the battery tape tab does not properly adhere, allowing excess air access, the OCV will be above 1.40 V. This would be the same OCV as if the zinc/air battery had been left untabbed for a couple of hours. If this condition occurs when the battery is in storage, the battery may dry out and not function when used.

It is important that the tape tab does not lower the voltage of the battery too much. A tabbed cell OCV of less than 1.0 V may not rise fast enough when untabbed to properly start the device the battery is powering.

Figure 18A.4 illustrates the time it takes to achieve a functional voltage based on the initial tabbed voltage. The lower the tabbed voltage, the longer it will take to reach the functional voltage. Rise time can be influenced by changes in the air cathode chemistry, cell limiting current, or air hole design.

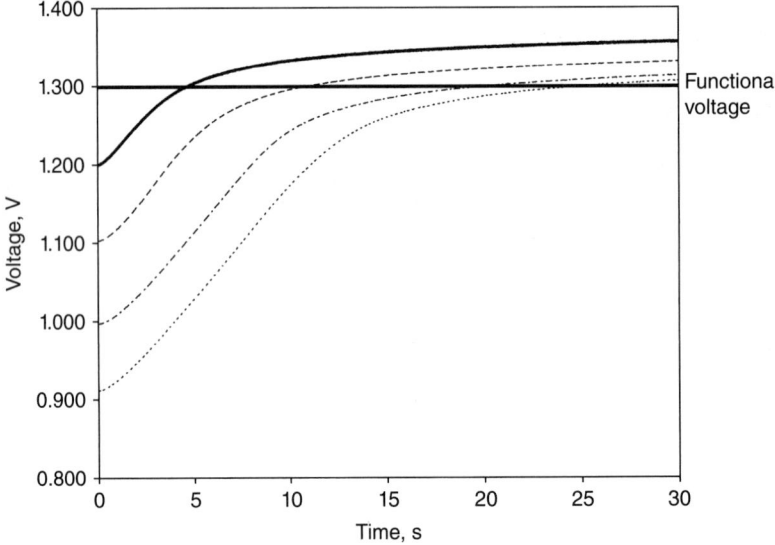

**FIGURE 18A.4**  Typical rise time for zinc/air batteries of selected tabbed OCV.

*Energy Density.*  Zinc/air batteries have the highest volumetric energy density of any other primary button or coin cell chemistry system. The common hearing aid batteries range from 1300 Wh/L in the PR70 (size 10) to 1400 Wh/L in the PR44 (size 675).

*Cell Internal Impedance.*  The effects of discharge level and signal frequency on the internal impedance characteristics of a PR48 (size 13) button battery are presented in Fig. 18A.5. Fresh cells have the highest impedance at low frequencies. This low frequency impedance decreases with depth of discharge.

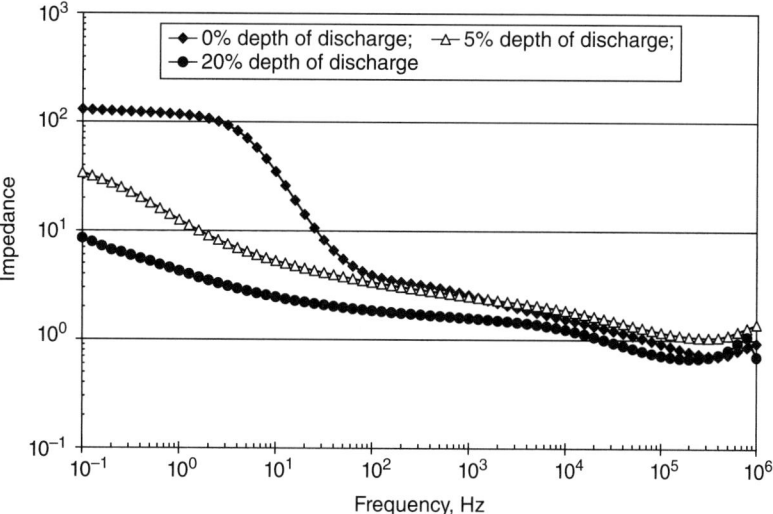

**FIGURE 18A.5**  Impedance versus frequency and depth of discharge for a PR48 (13) zinc/air battery.

***Discharge Characteristics.***  A set of discharge curves, typical of zinc/air button batteries at 20°C, is presented in Fig. 18A.6. At low to moderate current levels, the discharge curves are relatively flat. As the discharge current goes up, the discharge voltage goes down. When the discharge rate approaches the cell's limiting current, the delivered capacity is decreased and cell polarization increases.

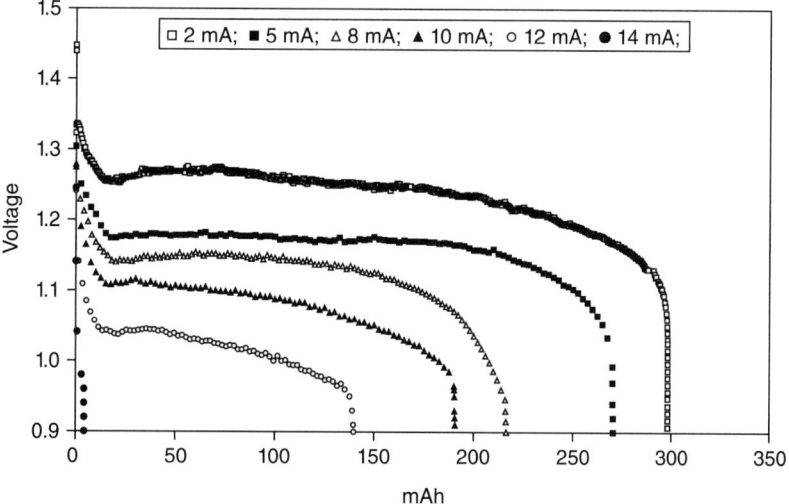

**FIGURE 18A.6**  Discharge curves for a PR48 (13) zinc/air battery at six different rates of discharge.

***Voltage-Current Performance.***  The catalytic activity of the cathode, along with the amount of oxygen access to the cathode, generally defines the voltage-current profile of a zinc/air battery. Oxygen access is defined by the degree of air access to the battery's cathode. Improved oxygen access to the cathode increases the cell's power output. Oxygen access can be improved by increasing the number or size of the air access holes in the battery case. If the number and size of the air holes are kept constant, the power capability of the battery can be increased by raising the limiting current of the cathode. To minimize the detrimental effects of water vapor transport, air access needs to be properly balanced.[5,6] In a low humidity environment, a high rate of water vapor loss accelerates dry out, lowering the battery's overall capacity. In a high humidity environment, a high rate of water vapor gain takes up the free volume designed for the zinc anode discharge expansion, reducing the battery's overall capacity along with causing possible swelling or leakage. Understanding the maximum current requirements of the application using the zinc/air battery is important to minimizing the detrimental effects of water vapor transport.

Figure 18A.7 illustrates the effect of increasing the total air hole circumference on the cathode limiting current for a PR48 (size 13) zinc/air battery. Limiting current is the maximum current a zinc/air battery or cathode can deliver for a given set of conditions. The limiting current is measured by potentiostatically setting the battery or cathode to a voltage of 0.9 V and then measuring the current output after 1 to 5 min from the initiation of the test.

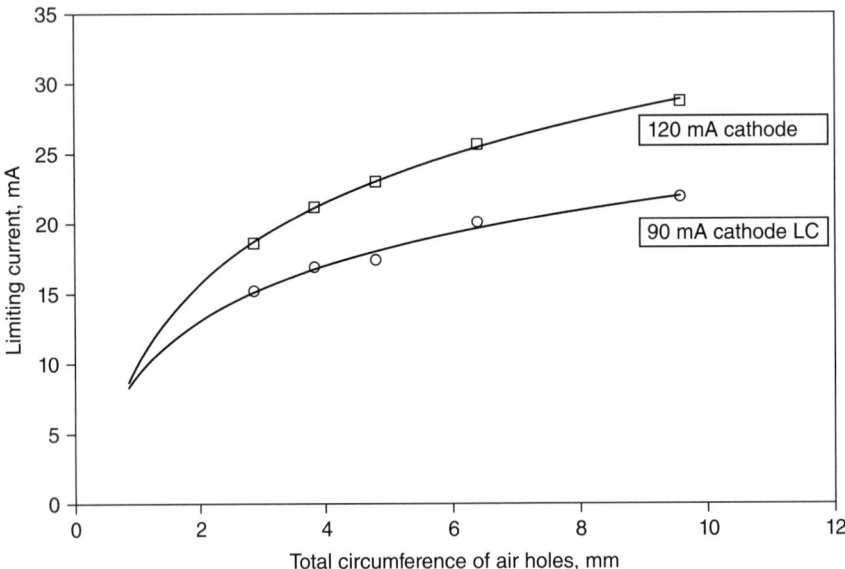

**FIGURE 18A.7**    Battery limiting current as a function of air hole circumference and cathode limiting current in a PR48 (13) battery.

Since the limiting current is a measure of air access to the cathode, this measurement can also be used to determine water vapor transport rates (see Fig. 18A.8). Since internal electrolyte evaporation is the primary source of cell weight loss, the water vapor transport rate is directly related to cell limiting current in a low-humidity environment.

In Fig. 18A.9, a cell limiting current versus water vapor transport relationship also exists in high-humidity conditions. A high cell limiting current increases the water vapor transport rate, increasing the cell weight gain.

The maximum power output of the zinc/air battery can be improved by increasing the catalytic activity of the cathode. Catalytic additives are commonly mixed into the carbon mix component of the cathode.[7,8] Various forms of $MnO_x$ are typically used as catalysts for the zinc/air cathode. Figure 18A.10 compares the Tafel plots of a carbon cathode with and without a manganese oxide catalyst. The use of a manganese oxide catalyzed cathode instead of the carbon cathode in a cell will produce a higher operating voltage.

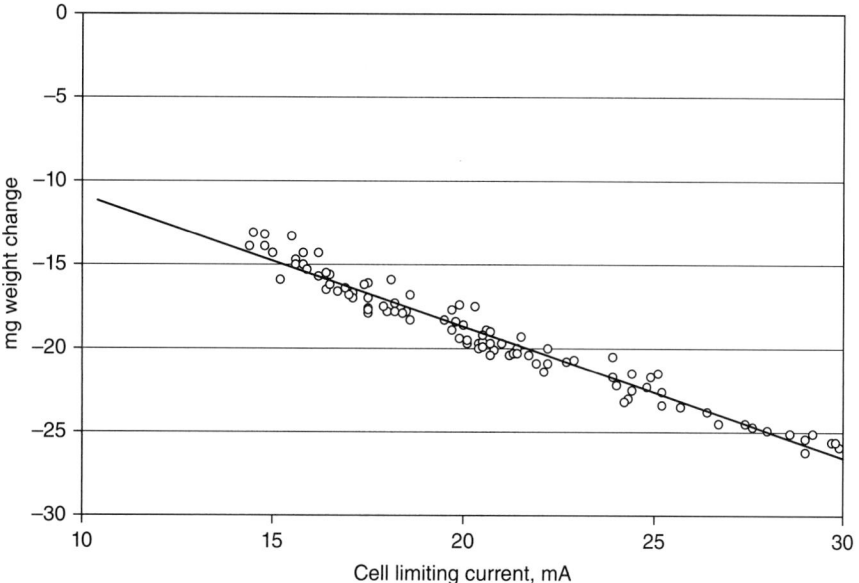

**FIGURE 18A.8**    PR48 (13) weight loss versus cell limiting current after 6 days at 20°C, 20% RH.

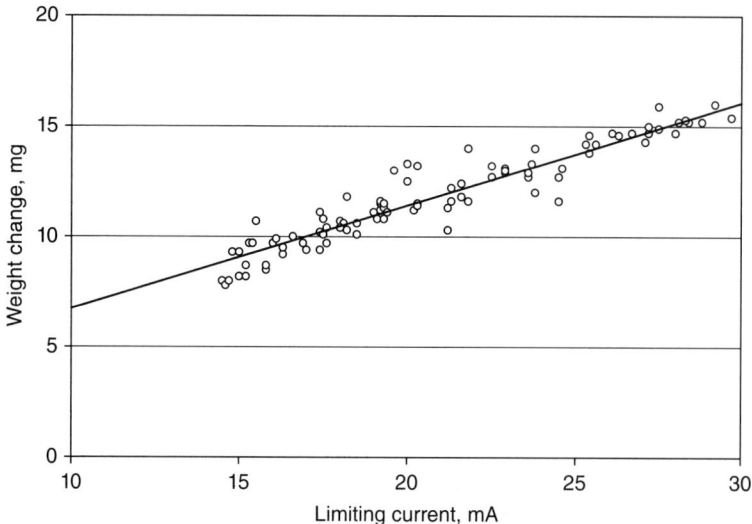

**FIGURE 18A.9**    PR48 (size 13) weight gain versus cell limiting current after 6 days at 20°C, 80% RH.

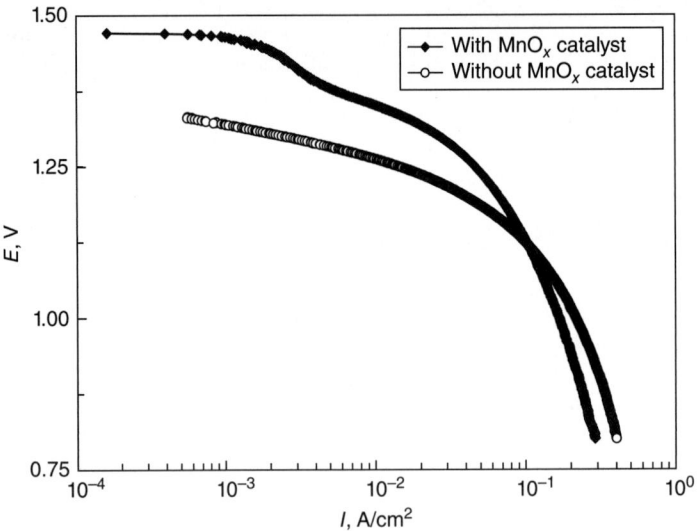

**FIGURE 18A.10**   Voltage as a function of current for an activated carbon cathode with and without a manganese oxide catalyst.

The average voltage-current profiles for various sized zinc/air button batteries are shown in Fig. 18A.11. The average voltage of the battery falls as the current increases, until the battery becomes oxygen starved. Once the battery is oxygen starved, it cannot support the load. Increasing the diameter of the battery will increase its constant current capability.

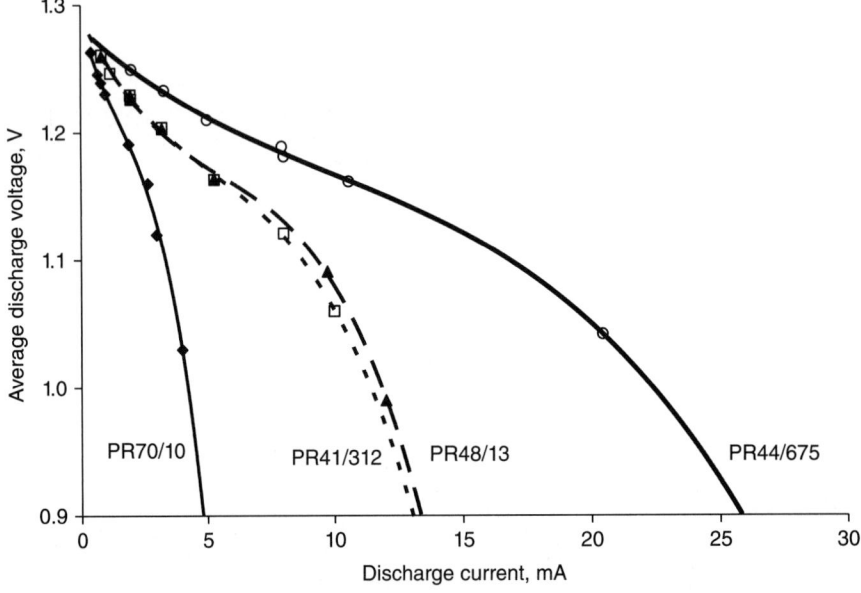

**FIGURE 18A.11**   Average voltage-current profiles for various zinc/air button batteries at 20°C.

***Pulse Load Performance.***    Zinc/air batteries can handle pulse currents much higher than the limiting current ($I_L$) of the battery. The maximum pulse current level for a given battery depends on the nature of the pulse. This capability results from the reservoir of oxygen that builds up within the cell when the current load is below the cell's limiting current.

As long as the average current ($I_{ave}$) of the pulse load does not exceed the cell limiting current ($I_L$), the zinc/air battery is able to sustain the pulse load. Figure 18A.12 illustrates the resulting voltage profile of a series of PR41 (312) batteries with an $I_L$ of 12 mA that had been subjected to an ever-increasing series of 1 s, 15 mA pulses over a 5 mA background current drain. In this illustration, the duty cycle of the 15 mA pulse was increased from 10% to 50%. As the duty cycle increases, the average current is increased, reducing the overall running voltage of the cell. Since the average current of the pulse regime never exceeded the battery's limiting current, the battery was able to sustain the pulse regime. If the pulse regime's average current exceeds the battery's limiting current, the battery will become oxygen starved, and the cell's voltage level will collapse.

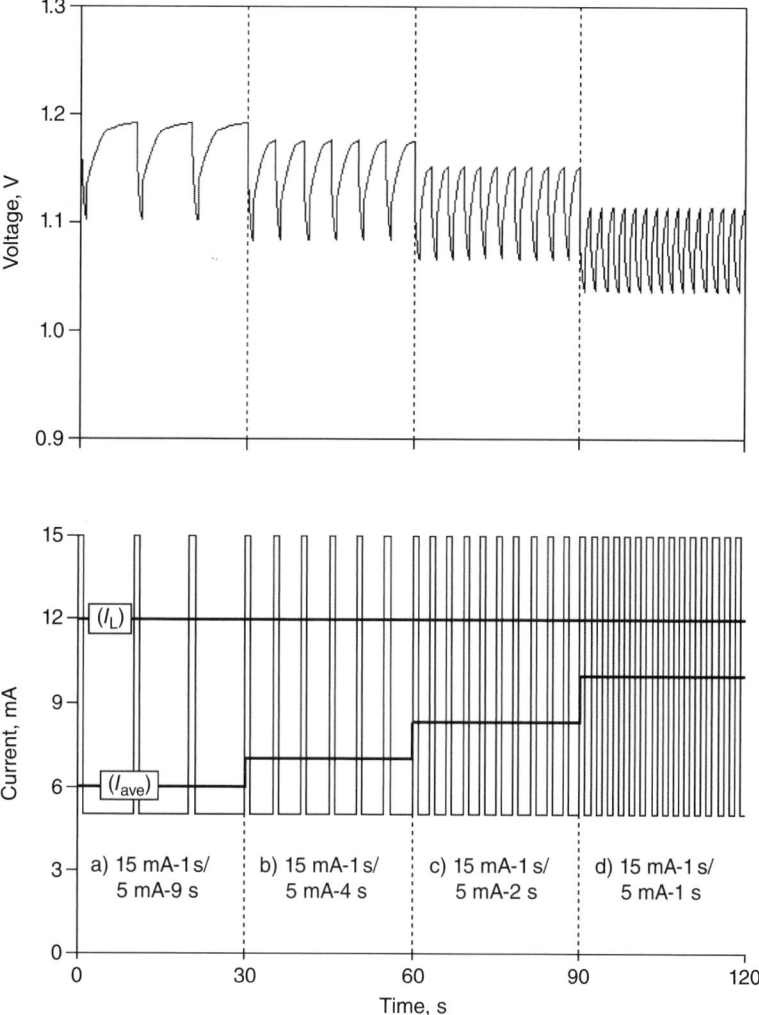

**FIGURE 18A.12**    Pulse loads profiles of a PR41 (312) zinc/air battery after 20 mAh of discharge under the pulsing conditions.

Although the battery may not be oxygen starved, the voltage level achieved during the pulse regime must be considered. In our example, if the device using the batteries does not work below 1.1 V, increasing the pulse duty cycle will cause premature failure in the device. If, however, the device works down to 0.9 V, increasing the duty cycle will have no effect on the function of the device. Very high, short-duration pulses can be achieved by the zinc/air cell as long as the average current of the pulse and the background current do not exceed the battery's limiting current and the voltage level of the pulse does not go lower than the functional voltage level of the device.

**Effect of Temperature.**  The effect of temperature at various discharge rates is illustrated in Fig. 18A.13. The decrease in voltage levels as the temperature decreases is primarily due to electrolyte effects. Adding potassium hydroxide to water lowers the freezing point, and at concentrations typical for zinc/air cells, the electrolyte will freeze at lower than −40°C. Also, as the temperature lowers, the conductivity of the electrolyte is decreased, slowing the discharge reaction. An application running at low discharge rate can function at a lower temperature than an application requiring a higher rate of discharge.

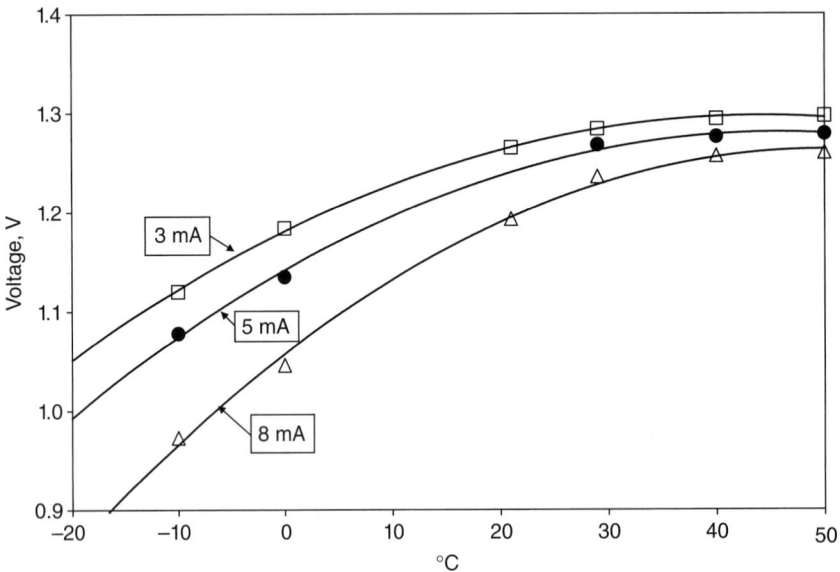

**FIGURE 18A.13**   Discharge voltage level of a PR44 (675) zinc/air battery as a function of discharge current and temperature.

**Effect of Altitude.**  The cathode of a zinc/air cell relies on oxygen from the external environment being available at a given partial pressure (21% of 760 mm at sea level or 160 mm). In low current demand applications, the rate of oxygen diffusion is limited by the actual oxygen used in discharge, but the situation changes when the demand is oxygen limited. At increased elevation, the barometric pressure goes down, as does the partial pressure of each gas present, including oxygen. As a result, the limiting current of a cell decreases as it operates at higher altitudes where the concentration of oxygen is lower than it is at sea level. The effect has been noted by active hearing aid users who have hiked to greater elevations. Airliners are pressurized to a "pressure altitude" around 8000 ft. This is enough to decrease limiting current by about 25%. The relative partial pressure of oxygen at various points on interest, expressed as a percent of the oxygen available at sea level (760 mm barometric pressure) is displayed in Table 18A.2.

**TABLE 18A.2**  Partial Pressure of Oxygen at Altitude

| Elevation, ft | Barometric pressure, mm Hg | Partial pressure of oxygen, mm Hg | Percent of sea level pressure | Points on earth |
|---|---|---|---|---|
| −1500 | 802 | 168.0 | 105.5 | Dead Sea, Israel-Jordan (−1317 ft) |
| −500 | 774 | 162.5 | 101.8 | Death Valley, CA (−282 ft) |
| 0 (sea level) | 760 | 160.0 | 100.0 | London, England |
| 500 | 746 | 156.7 | 98.2 | Montmartre, Paris, France (423 ft) |
| 1000 | 733 | 153.9 | 96.4 | Vaalserberg, The Netherlands (1053 ft) |
| 2000 | 707 | 148.4 | 93.0 | High Willhays, Cumbria, U.K. (2037 ft) |
| 5000 | 633 | 132.8 | 83.2 | Denver, CO (5280 ft) |
| 10,000 | 523 | 109.8 | 77.2 | Cascade Mountain, Canadian Rockies (9836 ft) |
| 20,000 | 349 | 73.4 | 46.0 | Denali (Mt. McKinley) (20,320 ft) |
| 30,000 | 226 | 47.48 | 29.8 | Mt. Everest (29,028 ft) |

Tests of cells in a chamber that had been evacuated to simulate the pressure at altitude declined in limiting current (i.e., the current that can be made to flow if the closed circuit voltage [CCV] is held to 0.9 V) and in current density (i.e., the current that can be made to flow if the CCV is held to 1.1 V). The current delivered exceeded slightly the predicted levels of the theoretical model. An example of a PR48 (size 13) cell is given in Fig. 18A.14.

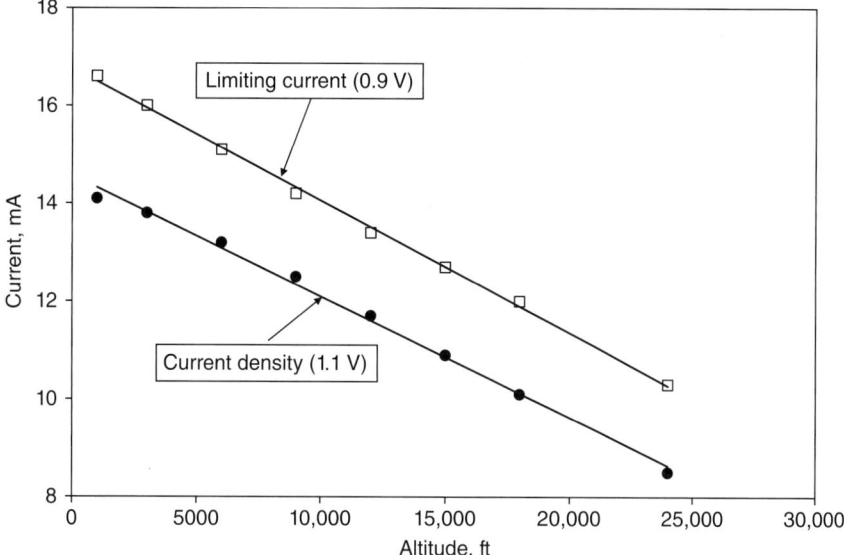

**FIGURE 18A.14**  Change in limiting current (0.9 V) and current density (1.1 V) for a PR48 (13) size cell operated at a pressure equivalent of altitude.

***Storage Life.***  Four principal mechanisms affect the capacity of zinc/air batteries during storage and operating service. One mechanism, self-discharge of the zinc (corrosion), is an internal reaction; the other three are caused by gas transfer. The gas transfer mechanisms are direct oxidation of the zinc anode, carbonation of the electrolyte, and electrolyte water gain or loss.

During storage, the air access holes of the zinc/air battery are sealed to prevent gas transfer decay. Only enough oxygen is allowed into the cell to give a sealed OCV of greater than 1 V. Oxygen, one of the cell's reactants, is severely restricted from entering the cell during storage. Limiting air access gives zinc/air batteries excellent shelf-life performance.

The primary mechanism affecting the shelf life of a zinc/air battery is the self-discharge reaction. Zinc is thermodynamically unstable in alkaline electrolyte and reacts to form zinc oxide and hydrogen gas. To control this reaction, additives are used in the anode. Mercury historically was one of the additives used to control this self-discharge reaction. Environmental concerns have forced mercury's removal from the cell anode chemistry, resulting in new additives to control the self-discharge of the zinc/air system.

Capacity retention results of PR41 (size 312) and PR48 (size 13) cells are presented in Fig. 18A.15. Under low rate conditions, the batteries lose about 3% a year, while increasing the discharge rate two to three times increases the rate of loss to 7% to 8% a year. Improving self-discharge storage retention can result in trade-offs with other cell performance parameters such as the discharge voltage level.

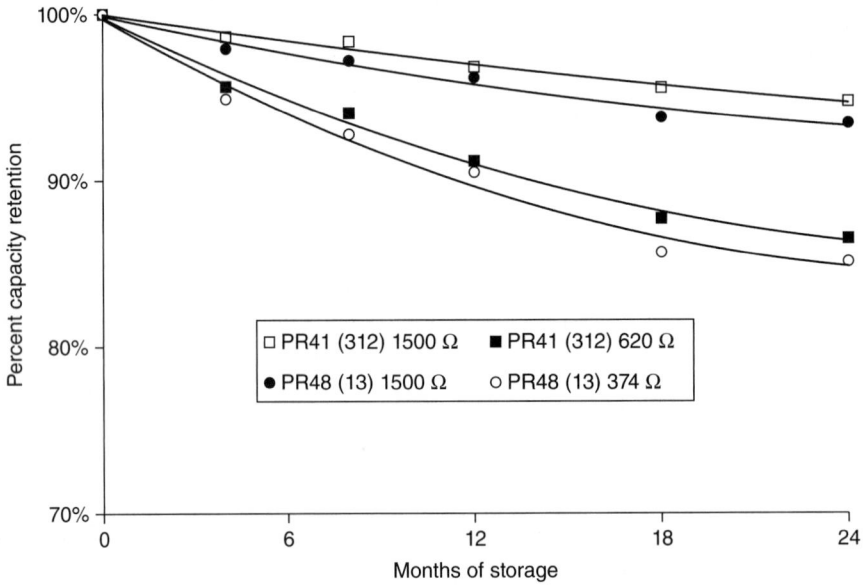

**FIGURE 18A.15**   Capacity retention of PR41 (312) and PR48 (13) batteries at different rates of discharge.

Elevated temperature will increase the rate of the self-discharge reaction and is used as an analytical tool to accelerate the effects of additives on the self-discharge performance of zinc/air batteries.

***Factors Affecting Service Life.***   The zinc/air system is open to the atmosphere, even when the tab is in place restricting gas transmission into the cell. The most immediate effect that the external environment can have on the cell is that of relative humidity. Other effects that are generally of less concern, but are well documented, include carbonation of electrolyte, direct oxidation, and the impact that high-altitude use can have on rate capability. In the use profile of a hearing aid, the most common use for zinc/air batteries, these effects are difficult, if not impossible, to notice.

*Carbonation of Electrolyte.*   While alkaline electrolytes have significant solubility for carbon dioxide, most zinc/air cells are used within weeks of being opened to the external environment when the tab is removed. As a result, carbonation will not be a factor in the use of the product. Extremely light drain or intermittent duty that would extend the use of the product beyond a month can challenge the zinc/air cell's service life, first in response to relative humidity, and then to possible carbonate crystals forming in the electrolyte by the gas diffusion membrane of the cathode. Crystallization can produce a pathway for direct electrolyte leakage if this occurs.

*Direct Oxidation.*   Direct oxidation is not a significant factor in the consumption of the zinc in the zinc/air cell as long as the tab is kept in place and oxygen access to the cell is properly restricted. Any alkaline zinc cell needs oxygen to discharge the zinc and release the electrons that will flow back to the positive terminal through the external circuit. Normal discharge occurs when the hydroxyl ions interact with the zinc, producing zinc hydroxide species and eventually ZnO. There is a significant solubility in the electrolyte for zinc, zinc oxide, and the zincates (hydroxyl species). Oxygen is also soluble in the aqueous KOH electrolyte, leading to a secondary means to oxidize the zinc that is present in metallic form. The source of this oxygen is from the gas-liquid interface found by the gas diffusion layer of the cathode.

*Effect of Water Vapor Transfer on Service Life.*   The primary cause of service life reduction in a zinc/air cell is water vapor transfer. As illustrated in Fig. 18A.16, water vapor transfer occurs when a partial pressure difference exists between the vapor pressure of the internal cell electrolyte and that of the outside environment. The internal cell vapor pressure is determined by the cell's electrolyte at a given temperature. If the external humidity is lower than the cell's internal relative humidity (dry day), the cell will lose water. If the external humidity is higher than the cell's internal relative humidity (humid day), the cell will gain water. Excessive water loss causes the electrolyte to concentrate, increasing cell impedance and promoting carbonation. Eventually the loss of water will cause the cell to dry out to the point where direct oxidation can occur. Excessive water gain dilutes the cell electrolyte, reducing conductivity. The addition of water vapor to the cell can flood the cathode and fill up the anode free-space cavity that is designed for zinc oxide expansion, eventually causing loss of rate capability, battery swelling, or leakage.

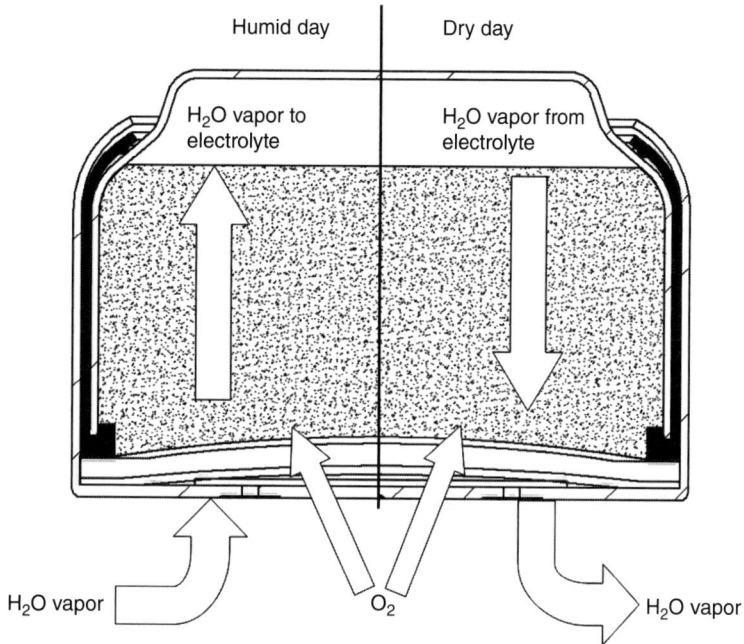

**FIGURE 18A.16**   Water vapor transfer mechanism in a zinc/air cell.

Figure 18A.17 illustrates the relationship between KOH concentration and relative humidity at room temperature. Based on the cell design requirements, the desired electrolyte concentration can range from 25% to 40% by weight. At a given temperature, lowering the electrolyte concentration will raise the internal cell

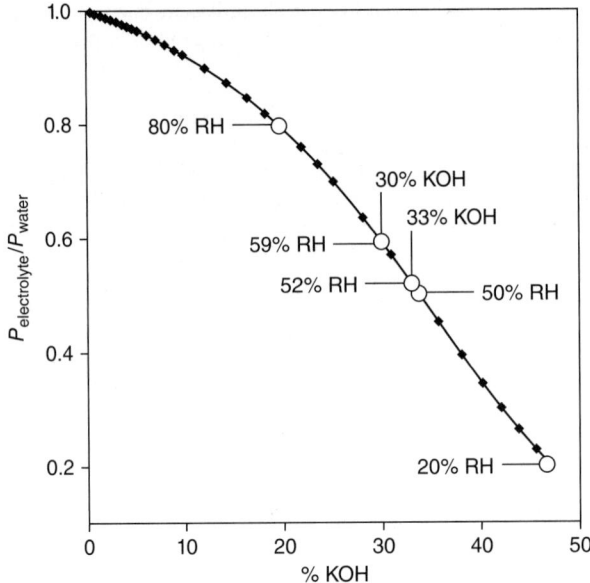

**FIGURE 18A.17** Closed container relative humidity as a function of KOH concentration.

relative humidity. The lowering of the electrolyte concentration will slow the rate of water vapor transport in a high-humidity environment but increase the rate in a low-humidity environment. The opposite effect occurs with vapor transport if the concentration of electrolyte is raised.

A thorough understanding of the intended application is required to properly design a zinc/air battery. Knowing the rate requirements and functional voltages of the application along with environmental conditions of use can determine the trade-offs that can be made to optimize cell performance.

The effect of water vapor transport is demonstrated in the evaluation described below. Table 18A.3 compares the average limiting currents and open-stand weight changes of three commercially available PR41 (size 312) zinc/air batteries, typically used in hearing aids. The batteries, with their seal tabs removed, were initially weighed and placed in three different 20°C relative-humidity environments. After 7 days untabbed, the batteries were weighed again and the change in battery weight was determined. It is assumed that cell weight change is due to the water vapor exchange with the environment, as illustrated previously in Fig. 18A.16.

**TABLE 18A.3** Average Cell Weight Change after 7 Days at 20°C on Open Stand in Various Relative Humidity Conditions for Three Commercially Available Size PR41 (312) Zinc/Air Batteries

| Design | Cell limiting current, mA | Cell weight change, mg | | | |
| | | 20% RH | 50% RH | 80% RH | Total range |
| --- | --- | --- | --- | --- | --- |
| A | 7.5 | −7.3 | 0 | 5.8 | 13.1 |
| B | 10.4 | −10.4 | −1.9 | 6.7 | 17.1 |
| C | 13.9 | −11.7 | −1.4 | 9.7 | 23.6 |

With the lowest average limiting current, design A had the lowest total range of cell weight change after 7 days. Design C's average limiting current was 85% higher and had 80% more total weight change than design A.

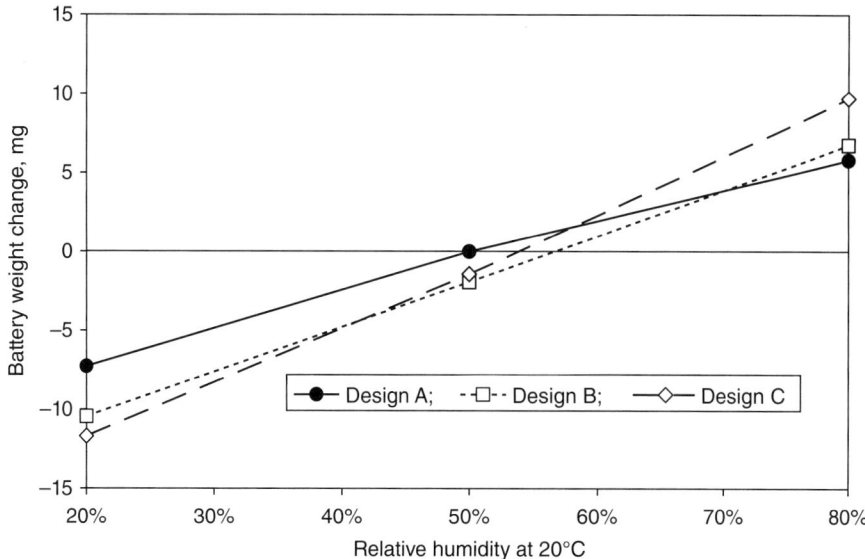

**FIGURE 18A.18**   Change in battery weight of three different manufactures of PR41 (312) zinc/air batteries after 7 days open stand at 20%, 50%, and 80% relative humidities.

Figure 18A.18 compares the average 7-day cell weight changes of the three designs in three different humidity environments. The graph illustrates how the designs are targeted for relative humidity environment. Design A crosses the no-weight-change line at 50% relative humidity, while design B crosses the no-weight-change line at between 55% and 60% relative humidity. If design A and design B had the same limiting current, design A would have less weight change in the low-humidity environment, but more weight change in the high-humidity environment.

Initially, the three designs were submitted to two test loads at various relative humidities. The first evaluation was on the lower drain 1500 $\Omega$, 12 h/day test. The typical duration for this test at 50% relative humidity is about 16 to 18 days. Figure 18A.19 compares the performance of the three designs from this test in 20% and 50% relative humidity environments. To normalize the different capacities, the results of each test are plotted as a percentage of each design's 50% RH test based on a 1500 $\Omega$, 12 h/day load at 20°C.

The second evaluation was at the moderate drain 620 $\Omega$, 16 h/day test. The typical duration for this test at 50% relative humidity is about 5 to 6 days. Figure 18A.20 compares the performance of the three designs on this test in 20%, 50%, and 80% relative humidity. To normalize the different capacities of the designs, the results of each capacity test are plotted as a percentage of each design's average 620 $\Omega$, 16 h/day result at 50% relative humidity.

For the short duration moderate 620 $\Omega$ test, none of the designs experienced a significant loss of performance in high- and low-humidity conditions. However, when the typical test time frame was changed to 16 to 18 days, a significant loss to performance occurred at 20% relative humidity. Design B with the middle limiting current range and higher internal battery relative humidity had a 15% performance loss in the low-humidity test environment.

Figure 18A.21 summarizes the percent of the initial performance the various designs obtained on the moderate drain 620 $\Omega$ test after 7 days open stand. For instance, a sample of design C was left open for 7 days in a 20°C, 20% relative humidity environment. In that time, the cells lost an average of 11.7 mg water vapor and then when tested on the 620 $\Omega$ test in the same 20% relative humidity environment performed at 65% of the typical initial, 20% RH results. Design A, with the lowest limiting current, had the best retention after open stand, since it restricts water vapor transport up to 85% better than the other designs.

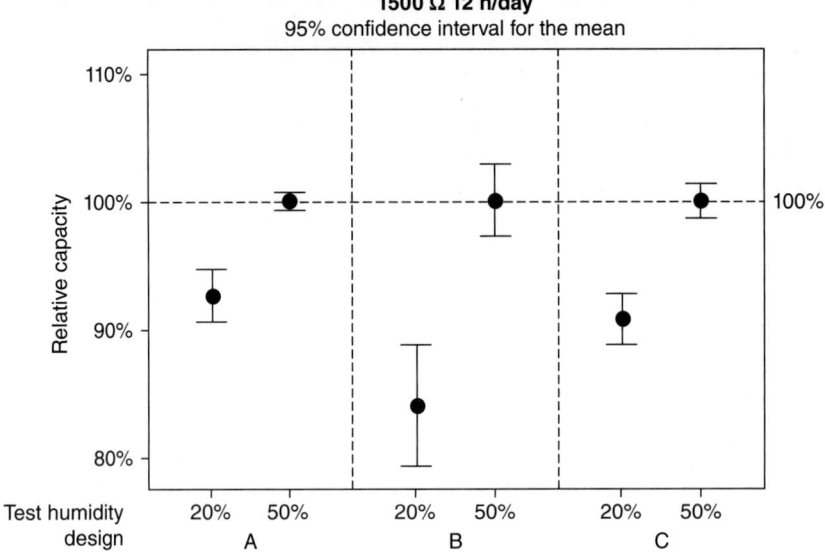

**FIGURE 18A.19**   The initial 20% relative humidity capacity performance as a percentage of each design's 50% relative humidity result on the 1500 Ω, 12 h/day test at 20°C.

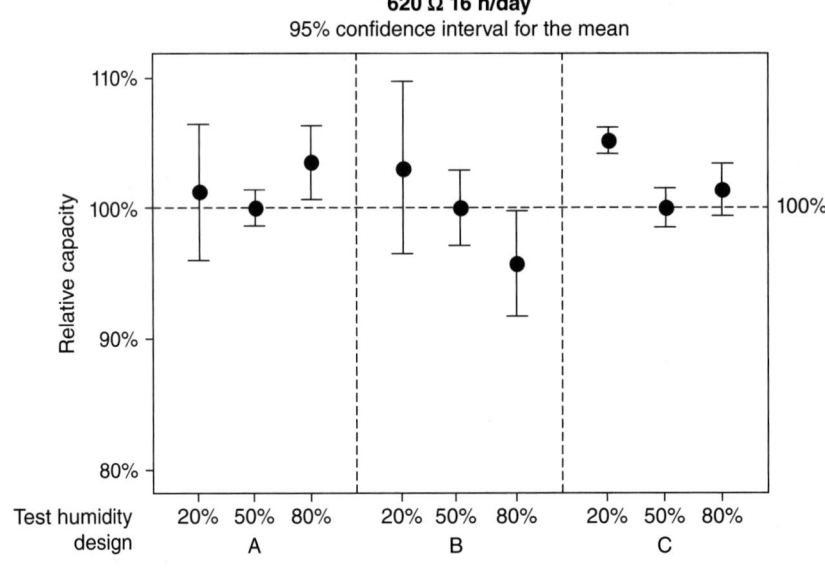

**FIGURE 18A.20**   The initial 20% and 80% relative humidity capacity performance as a percentage of each design's 50% relative humidity performance on the 620 Ω, 16 h/day test at 20°C.

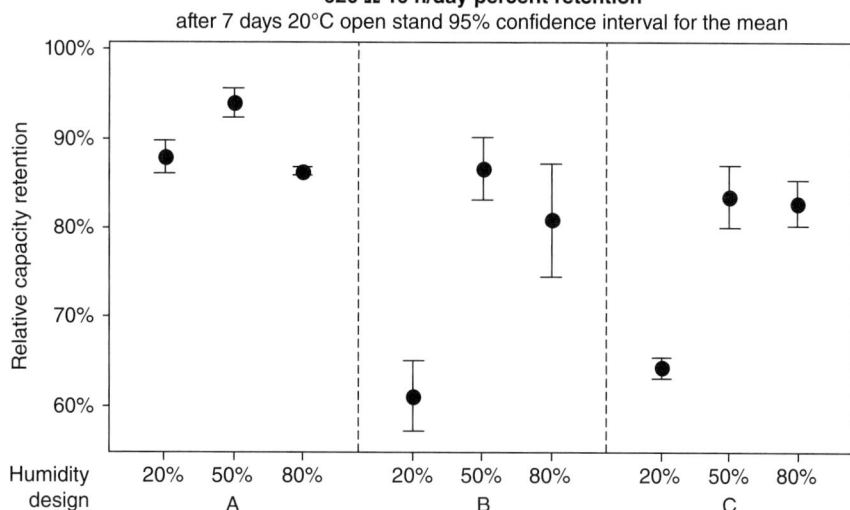

**FIGURE 18A.21**    The percentage retention of the initial capacity on the 620 Ω, 16 h/day test after 7 days open stand in 20%, 50%, and 80% relative humidity open stand at 20°C.

## 18A.4.2    Portable Primary Zinc/Air Performance

Figure 18A.22 shows the basic schematic of a prismatic zinc/air cell. The thickness of the cell determines the anode capacity of the cell, and the cross-sectional surface area determines the maximum rate capability.[10,11]

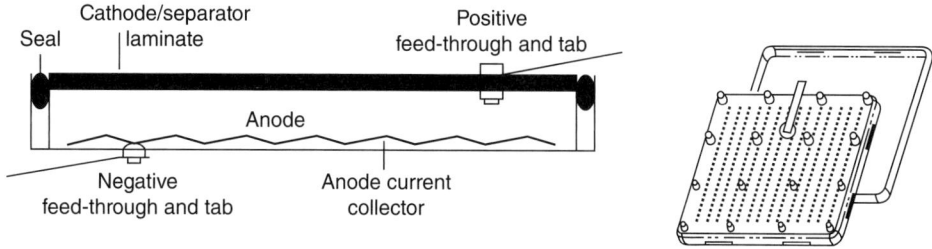

**FIGURE 18A.22**    Design of a prismatic primary zinc/air cell. (*Courtesy of Electric Fuel Corp.*)

In addition to prismatic cell designs, cylindrical zinc/air cells (see Fig. 18A.23) have been designed.[12–14]

The high specific energy, low cost, and safety of the zinc/air primary battery make it an attractive choice for many portable electronics applications. It is particularly advantageous for applications where the battery energy is consumed within a range of 1 to 14 days, since the high specific energy and energy density of the zinc/air system can be realized and the impact of environmental interactions (dry-out, flooding, and carbonation) is low. Typical cell discharge curves at 25°C are shown in Fig. 18A.24. The cell voltage is relatively flat throughout most of the discharge, with little capacity remaining beyond 0.9 V per cell. A summary of discharge characteristics of representative state-of-the-art prismatic zinc/air cells is given in Table 18A.4.

Three approaches are being taken to the design of prismatic zinc/air cells for portable batteries (GEN 3). The first is a metal case prismatic cell. This design is essentially an adaptation of button cell technology.

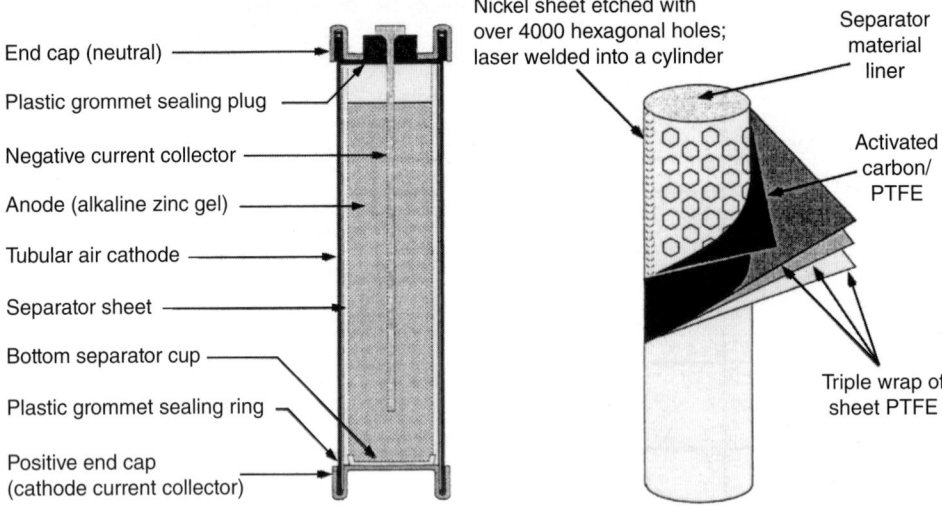

**FIGURE 18A.23**   Design of a cylindrical primary zinc/air cell. (*Courtesy of Rayovac Corp.*)

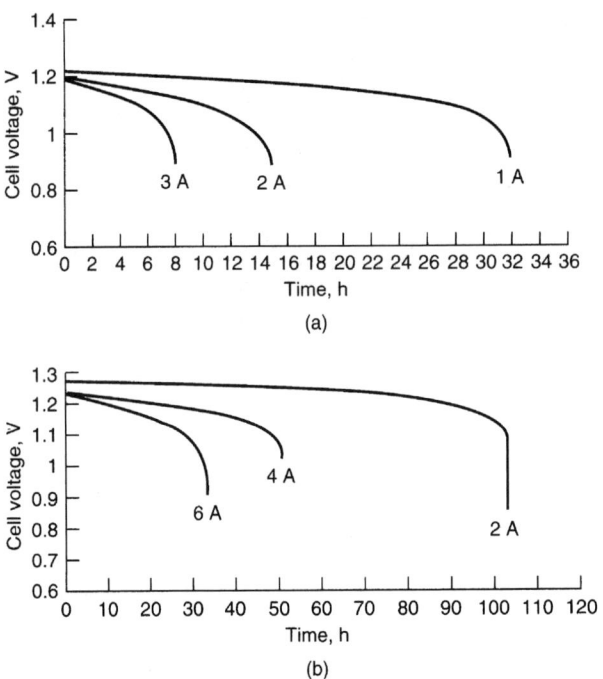

**FIGURE 18A.24**   Discharge curves for prismatic primary zinc/air cells at 25°C. (*a*) High-rate cell. (*b*) High-capacity cell. (*Courtesy of Matsi, Inc.*)[15]

**TABLE 18A.4**  Specifications of Prismatic Zinc/Air Cells

| Variable | Cellular phone cell | Auxiliary power cell |
|---|---|---|
| Facial dimensions, cm (length × width) | 4.6 × 2.7 | 7.6 × 7.6 |
| Height, cm | 0.43 | 0.6 |
| Weight, g | 15 | 87 |
| Capacity, Ah | 3.6 | 30 |
| Specific energy, Wh/kg | 300 | 500 |
| Energy density, Wh/L | 800 | 1250 |

In this design, a cathode subassembly, contained in a nickel-plated steel can, is crimp-sealed onto an anode subassembly, contained in a copper-lined nickel-plated stainless steel can. A molded plastic insulator seal separates the anode and cathode assemblies. This design has performed well for smaller sizes (5 Ah or less).

The second design uses plastic for the case of the prismatic zinc/air cell. This design employs adhesive technology to bond the cell anode and cathode subassemblies. The plastic cell design is preferred for large capacity cell sizes (>5 Ah) due to technological limitations imposed on the metal cell design. In particular, leak-tight crimp seals become a challenge as cell dimensions increase due to the need for close dimensional tolerances. The key challenges for the plastic cell include the development of the proper designs and materials for the cathode and cell seals and for the current feedthroughs. The latter is required for the plastic cell but not the metal cell, in which the cans serve as terminals for electrical contact. Figures 18A.25 and 18A.26 show a cell and battery designed for auxiliary power and for remote applications.

**FIGURE 18A.25**  Zinc/air cell for auxiliary power and for remote applications. (*Courtesy of Electric Fuel Corp.*)

The third prismatic zinc/air cell design consists of the fourth-generation (GEN 4) zinc/air systems, whose development began in the late 1990s and early 2000s. These cells are designed to use the air cathode as the cell housing, folded in half around the zinc electrode and its edges sealed for leaktight integrity. Figure 18A.27 shows a GEN 4 zinc/air cell. The "canless" design of the GEN 4 zinc/air electrochemical cell allows for increased power density and increased specific power. These increases come at the expense of increased parasitic reactions.

**FIGURE 18A.26**    Zinc/air (BA-8180) battery for auxiliary power and for remote applications. (*Courtesy of Electric Fuel Corp.*)

**FIGURE 18A.27**    GEN 4 zinc/air cell. (*Courtesy of Electric Fuel Corp.*)

Prismatic cells are designed so they can be stacked as multicell batteries for use in various portable electronic equipment. Stacking of the cells requires a provision, such as a spacer, to permit air access to the cathode and a fan to provide forced flow of air. The thickness of the spacer is dependent on the dimensions of the cell and the required current density. If the spacer is too thin, the cell can become oxygen starved, while if too thick, it increases the battery weight and volume unnecessarily. Figure 18A.28 shows a typical zinc/air cell stack. An alternative approach to dealing with oxygen diffusion is by providing a positive pressure of air by designing a fan and air channels into the battery design (see Fig. 18A.29).

Cylindrical zinc/air cells have been designed primarily in the AA cell size. These cells allow for a direct replacement of zinc alkaline manganese dioxide cells. The zinc/air technology uses a very thin cathode, allowing for the bulk of the cell to contain the anode/electrolyte mixture. The relatively high surface area of AA cells provides high power discharge rates. Batteries constructed from arrays of these cells do not provide for forced flow of air, but it has been shown that thermal gradients within the battery pack do provide convective flow.

**FIGURE 18A.28**  Typical zinc/air cell stack. Representative stack consists of GEN 4 zinc/air cells. (*Courtesy of Electric Fuel Corp.*)

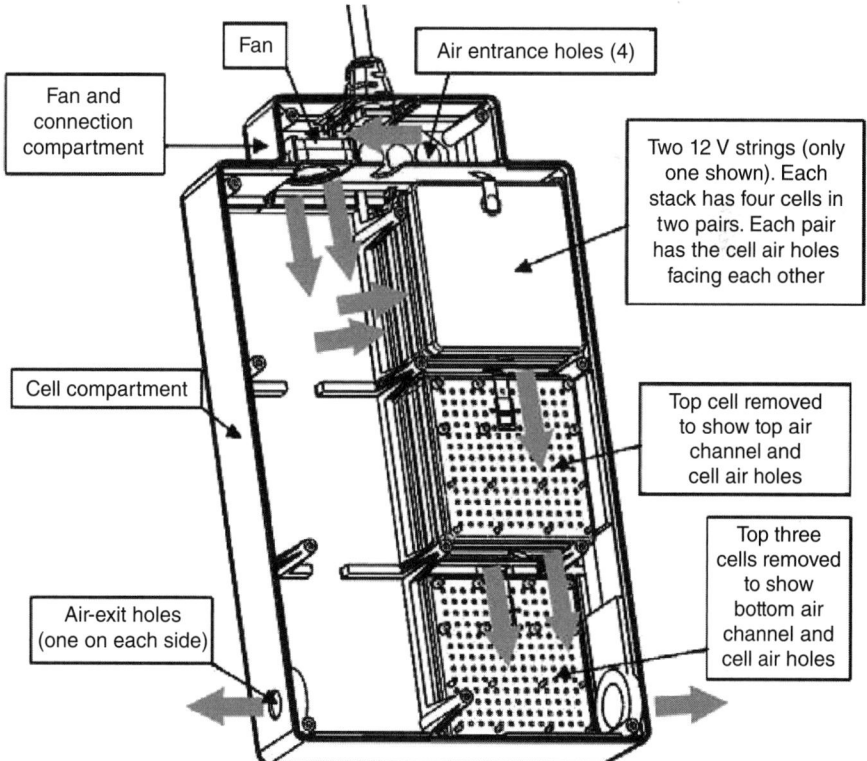

**FIGURE 18A.29**  BA-8180 battery design. (*Courtesy of Electric Fuel Corp.*)

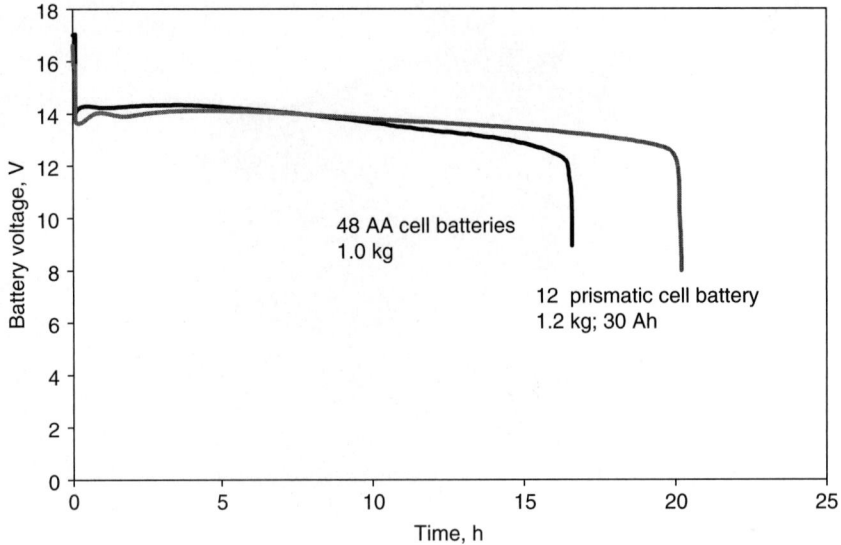

**FIGURE 18A.30**    Discharge profile for 12 V zinc/air batteries discharged at 18 W continuous. (*Data from U.S. Army tests.*)

Figure 18A.30 shows a typical discharge curve for two 12 V zinc/air battery configurations: (1) twelve 30 Ah prismatic zinc/air cells in series, and (2) 48 AA zinc/air cells consisting of four parallel strings of 12 cells in series.

The advancement of the zinc/air electrochemical cell can be seen in Fig. 18A.31. The figure presents the performance characteristics of each generation in the form of specific energy versus specific power. At more than 400 Wh/kg, the button cell (GEN 2) has the highest specific energy, while the power-optimized GEN 4 cell

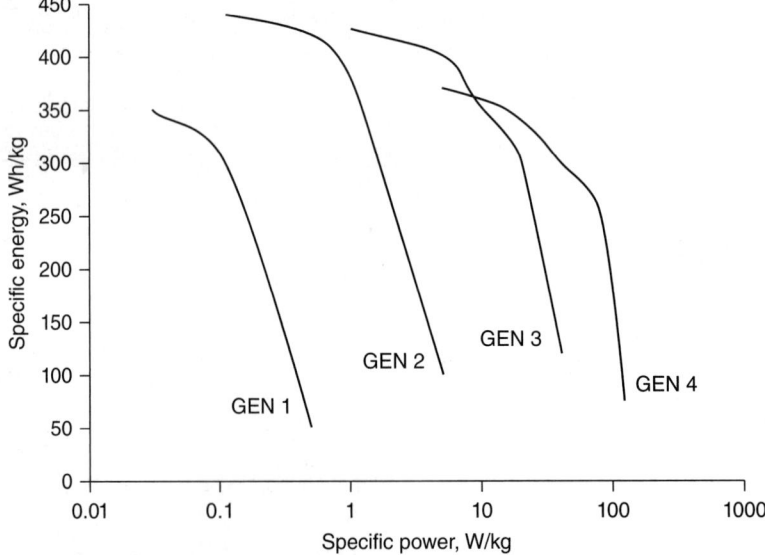

**FIGURE 18A.31**    Discharge characteristics for zinc/air cells designed for portable electronic equipment. (*Data from U.S. Army tests.*)

has the highest specific power, exceeding 100 W/kg. However, the GEN 4 cell operates at a reduced specific energy when compared to the GEN 3 cell. The figure shows that, at low rates (<10 W/kg), the specific energy of GEN 3 approaches that of the button cell, but falls off rapidly as the rate increases. This is because the anode is very thick (more than 5 mm, at least twice those of typical GEN 4 cells). For reference, a GEN 3 cell delivering 30 Ah is used in the BA-8180 to power radiotelephones over a period of several days to a week at low cost.

### 18A.4.3  Industrial Primary Zinc/Air Capabilities

Large primary zinc/air batteries have been used to provide power for applications such as railroad signaling, seismic telemetry, navigational buoys, and remote communications. They are available in either water-activated (containing dry potassium hydroxide) or preactivated versions.[16] Preactivated versions are also available with a gelled electrolyte to minimize the possibility of leakage.

***Preactivated and Water-Activated Types.*** A preactivated Edison Carbonaire cell, available in two- and three-cell configurations, is illustrated in Fig. 18A.32. The construction features are shown in Fig. 18A.33 identifying the wax-impregnated carbon cathode block, the solid zinc anodes, and the lime-filled reservoir, which absorbs carbon dioxide and removes soluble zinc compounds from solution and precipitates them as calcium zincate. The transparent cases reveal the condition of the lime bed, which turns darker as it is converted to zincate.

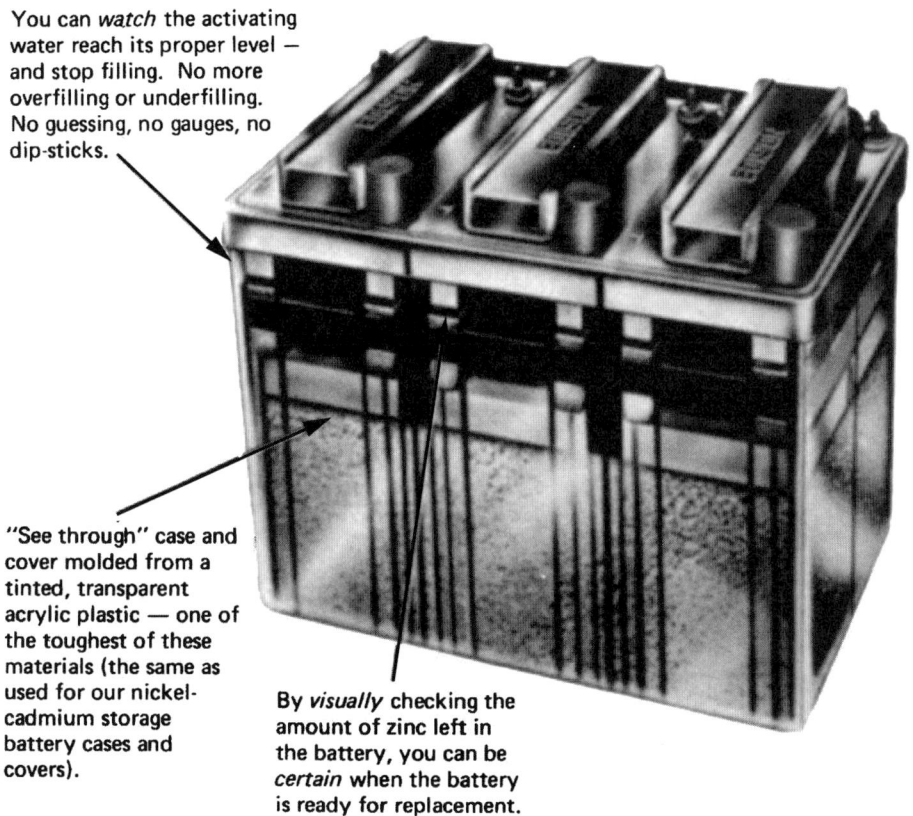

You can *watch* the activating water reach its proper level — and stop filling. No more overfilling or underfilling. No guessing, no gauges, no dip-sticks.

"See through" case and cover molded from a tinted, transparent acrylic plastic — one of the toughest of these materials (the same as used for our nickel-cadmium storage battery cases and covers).

By *visually* checking the amount of zinc left in the battery, you can be *certain* when the battery is ready for replacement.

**FIGURE 18A.32**  Edison Carbonaire zinc/air battery. (*Courtesy of SAFT America, Inc.*)

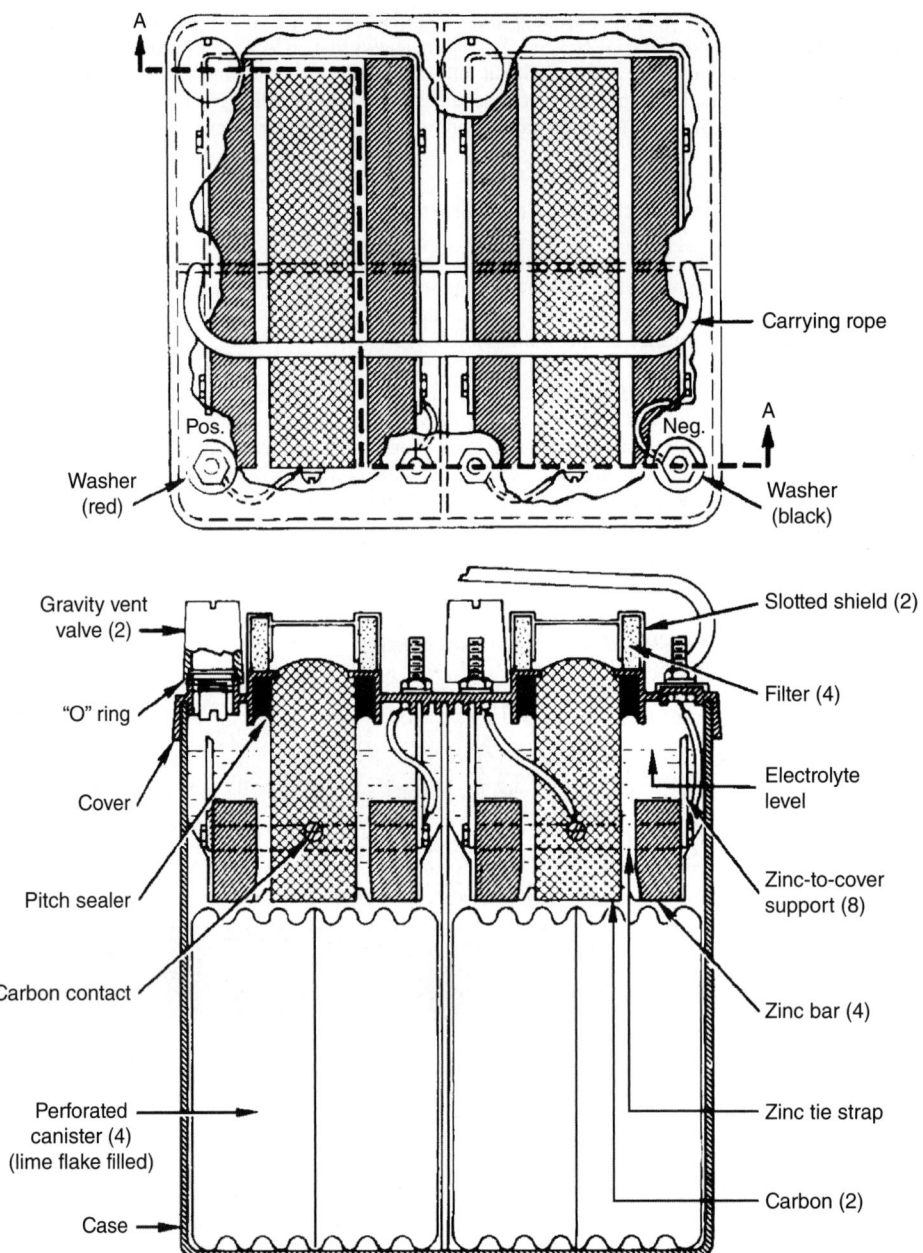

**FIGURE 18A.33**    Top and side views of type ST-2 Carbonaire zinc/air battery. (*Courtesy of SAFT America, Inc.*)

Water-activated cells require the addition of water to the dry caustic (potassium hydroxide) electrolyte and the lime flake. The maximum continuous discharge rate for this battery at 25°C is 0.75 A. A preactivated 1100 Ah three-cell battery weighs about 2 kg, giving an energy density of about 180 Wh/kg.

***Gelled-Electrolyte Types.***    An alternative version uses a gelled electrolyte T that is added during manufacture. Figure 18A.34 shows a cross section of the Gelaire battery.

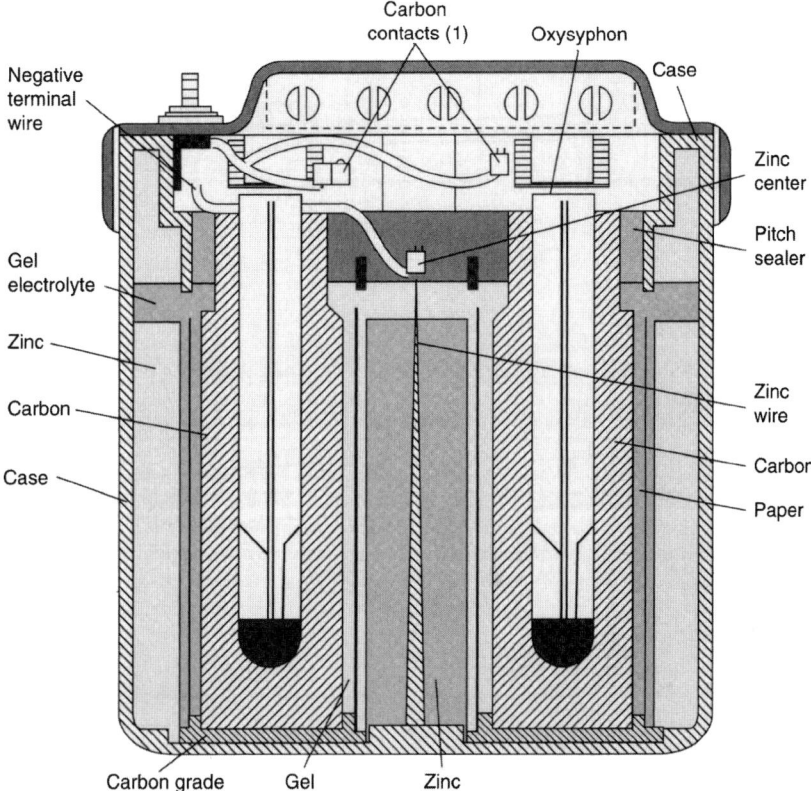

**FIGURE 18A.34**    Cross section of Gelaire cell. (*Courtesy of SAFT America, Inc.*)

## 18A.4.4    Primary Hybrid-Air/Manganese Dioxide Batteries

Another approach to primary zinc/air batteries is to use a hybrid cathode that contains excess manganese dioxide.[17] The battery functions as a zinc/air system at low rates and as a normal alkaline cell at high rates, whereby the manganese dioxide serves as the oxidizer. This means that such a battery should essentially have the capacity of a zinc/air battery when discharged at low rates, but it should have the pulse current capability of a manganese dioxide battery. After the high-current pulse, the manganese dioxide is partially regenerated by air oxidation so that the pulse current capability is restored.

Figure 18A.35 is a side view of a flat-pack cell. The specific energy of this battery is about 350 Wh/kg. Single and multicell batteries are available in capacities of 40 to 4800 Ah.

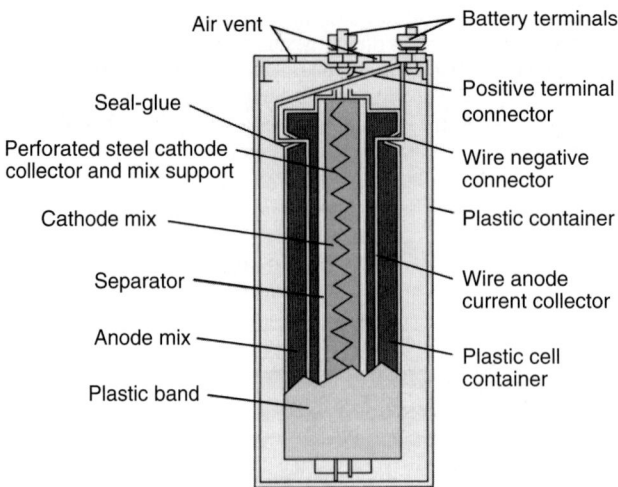

**FIGURE 18A.35** Side view of hybrid zinc/air-manganese dioxide cell. (*Courtesy of Celair Corp.*)

### 18A.4.5 Electrically Rechargeable Zinc/Air Battery Capabilities

A bifunctional oxygen electrode enables a zinc/air battery to be electrically recharged (i.e., charge and discharge occur directly within the cell).

The basic reactions of an electrically rechargeable zinc/air cell using a bifunctional oxygen electrode are shown in Fig. 18A.36. Advances in electrically rechargeable zinc/air cells have concentrated on the bifunctional air electrode.[18-21] Electrodes based on La, Sr, Mn, and Ni perovskites have demonstrated good cycle life. Figure 18A.37 shows the gains in cycle life for the bifunctional air cathode achieved going from Phase I to Phase II of a research and development program.

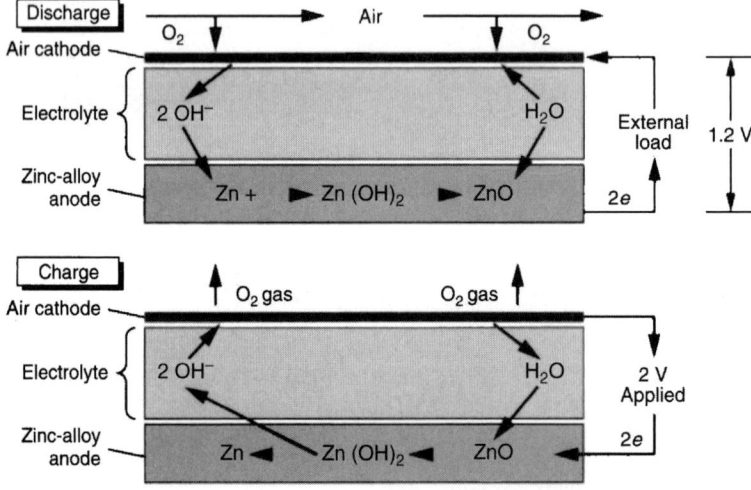

**FIGURE 18A.36** Basic operation of electrically rechargeable zinc/air cell. (*Courtesy of AER Energy Resources, Inc.*)

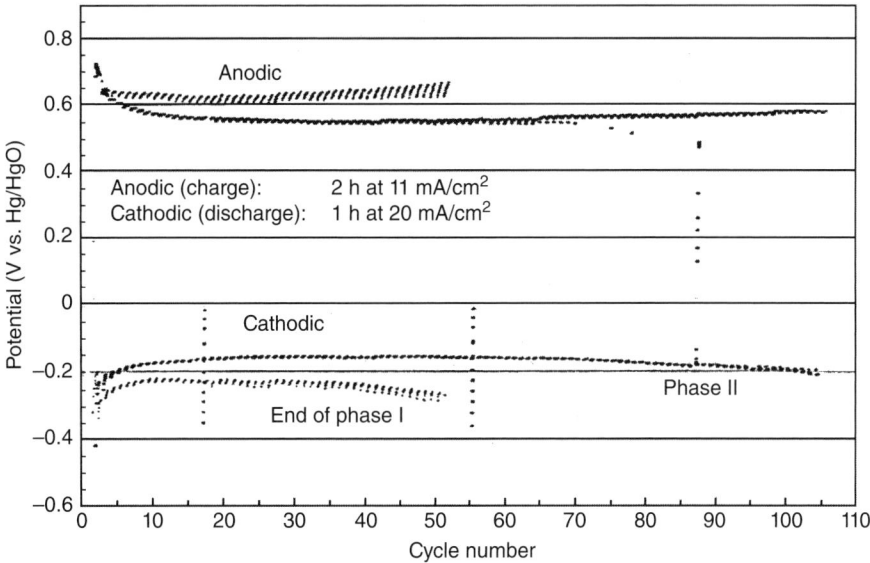

**FIGURE 18A.37**  Advances in bifunctional air electrodes. LSNC perovskite plus Shawinigan black carbon. Area = 25 cm². 8 M KOH at room temperature. (*Courtesy of Alupower, Inc.*)

An electrically rechargeable zinc/air cell having a bifunctional oxygen electrode, designed for use in computers and other electronic communication equipment, is shown in Fig. 18A.38. The cell is a prismatic or thin rectangular design. A high-porosity zinc structure, which maintains its integrity and morphology during cycling, is used. The air electrode is a corrosion-resistant carbon structure, containing a large number of small pores and a catalyst. The structure is permeable to oxygen, hydrophobic, and supported by a low-impedance current collector. The flat zinc negative and the air electrode plates face each other, separated by a high-porosity separator with low electrochemical resistance and the ability to absorb and retain the potassium hydroxide electrolyte. The cell case is injection-molded polypropylene with openings to permit the inflow of oxygen during discharge and release of the oxygen generated during charge.

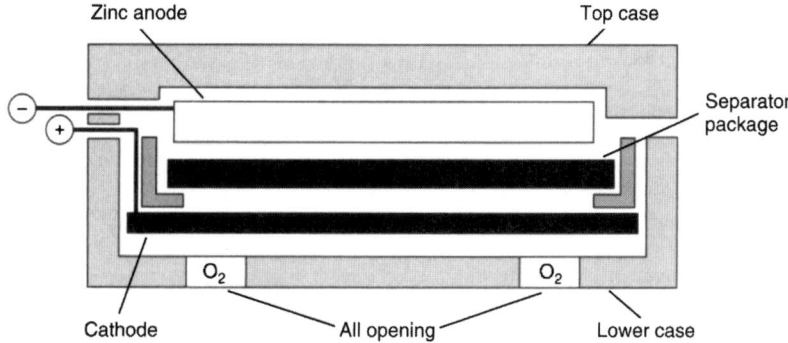

**FIGURE 18A.38**  Cross section of electrically rechargeable zinc/air cell. (*Courtesy of AER Energy Resources, Inc.*)

A critical factor in the design of the cell and battery is the means of controlling the flow of air into and out of the cell, which must be matched to the needs of the application. Excessive amounts of air could result in drying out the cell; too little air (oxygen-starved) will result in a drop-off of performance. The stoichiometric quantity of air required is 18.1 $cm^3$/min per ampere of continuous current. An air manager is used to control the flow of air by opening air access to the cathode during discharge and sealing the battery from the air when it is not in use to minimize self-discharge. A fan, powered by the battery, also is used to assist the airflow.

***Electrically Rechargeable Systems for Electric Vehicles.*** A similar rechargeable zinc/air cell, operating at room temperature, was being developed for use in electric vehicles. The cell uses a planar bipolar configuration. The negative electrode consists of zinc particles in a paste form, similar to the electrode used in alkaline-manganese dioxide primary cells. The bifunctional air electrode consists of a membrane of carbon and plastic with appropriate catalysts. The electrolyte is potassium hydroxide with gelling agents and fibrous absorbing materials. A typical cell is rated at 100 Ah with an average operating voltage of 1.2 V.

Specific energies up to 180 Wh/kg at the 5 to 10 h discharge rates and a battery life of about 1500 h have been achieved. Technical limitations are limited power density and a relatively short separator life. The air must be managed to remove carbon dioxide, and to provide humidity and thermal management. Table 18A.5 provides some of the characteristics of this battery, which is no longer under development.[22,23]

**TABLE 18A.5**   Characteristics of Zinc/Air Traction Battery

| | |
|---|---|
| Physical characteristics | |
|     Cell size | $33 \times 35 \times 0.75$ cm |
|     Cell weight | 1.0 kg (typical) |
| Cell voltage | |
|     Open circuit | 1.5 V |
|     Average | 1.2 V |
|     High load | 1.0 V |
|     Charging | 1.9 V |
| Configurations | |
|     General purpose | 120 Wh/kg, 120 W/kg peak power |
|     High energy | 180 Wh/kg at 10 W/kg |
|     High power | 200 W/kg peak at 100 Wh/kg |

***Source:*** Dreisbach Electromotive, Inc. (DEMI).

## 18A.4.6   Mechanically Refueled Systems—Anode Replacement

Mechanically replaceable zinc/air batteries were seriously considered for powering portable military electronic equipment in the late 1960s because of their high specific energy and ease of recharging. This battery contained a number of bicells connected in series to provide the desired voltage. Each bicell, as illustrated in Fig. 18A.39, consisted of two air cathodes connected in parallel and supported by a plastic frame that together formed an envelope for the zinc anode. The anode, which was a highly porous zinc structure enclosed in an absorbent separator, was inserted between the cathodes. The electrolyte, KOH, was contained in a dry form in the zinc anode and only water was needed to activate the cell. "Recharging" was accomplished by removing the spent anode, washing the cell, and replacing the anode with a fresh one. These batteries were never deployed because of their short activated life, poor intermittent operation, and the development of new high-performance primary lithium batteries that were superior in rate capability and ease of handling in the field.[24,25]

***Mechanically Refueled Electric Vehicles.*** A design similar to the portable mechanically rechargeable zinc/air battery has been considered for electric vehicle applications. The battery would be refueled "robotically" at a fleet servicing location or at a public service station by removing and replacing the spent anode cassettes. The discharged fuel would be electrochemically regenerated, using a modified zinc electrowinning process, at central facilities that serve regional distribution networks.[26]

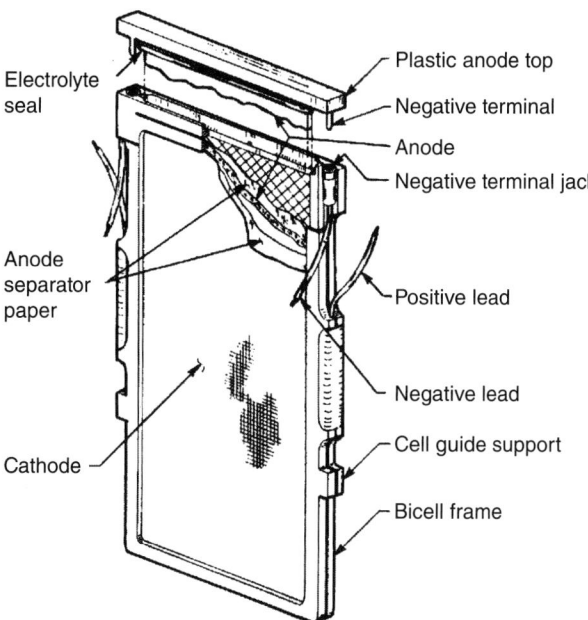

Electrolyte seal

Plastic anode top

Negative terminal

Anode

Negative terminal jack

Anode separator paper

Positive lead

Negative lead

Cathode

Cell guide support

Bicell frame

**FIGURE 18A.39**   Zinc/air bicell.

This zinc/air battery consisted of modular cell stacks, each containing a series of individual bicells. Each bicell consists of an anode cassette containing a zinc-based electrolyte slurry, contained between air cathodes, and a separator system. The slurry is maintained in a static bed without circulation. In addition, the battery contains subsystems for air provision and heat management and is adapted for fast mechanical replacement of the cassettes.

The technology has been evaluated in a full-size 264 V, 110 kWh battery weighing 650 kg in a van that was converted to electric drive. The battery delivered 230 Wh/kg and 230 Wh/L with a power density of 100 W/kg.

Another approach to powering electric vehicles with mechanically rechargeable zinc/air batteries is a hybrid configuration where the zinc/air battery is combined with a rechargeable battery, such as a high-power lead-acid battery.[27] With this approach, the performance of each battery can be optimized, using the high specific energy zinc/air battery as the energy source with a high specific power rechargeable battery to handle the peak power requirements. The power battery can also be sized to handle the anticipated peak load and duty cycle. In operation, during periods of light load, the zinc/air battery handles the load and charges the rechargeable battery through a voltage regulator. The load is shared by both batteries during peak load conditions. When fully discharged, the zinc/air battery is recharged by removing and replacing the zinc oxide discharge product, which can be regenerated externally and efficiently in designated facilities. The advantage of this hybrid design is illustrated in Fig. 18A.40, a Ragone plot comparing the performance of the hybrid with the performance of the individual batteries. In this example, the hybrid lead-acid battery is one specifically designed for high-rate performance.

***Mechanically Refueled System—Zinc Powder Replacement.***[28-30]   Figure 18A.41 is a sketch of an 80 cm² laboratory cell using a packed bed of zinc powder, which can be replaced when depleted. Natural convection is utilized for electrolyte circulation. During operation, electrolyte flows downward through the zinc bed and upward around the back of the current collector, which is either graphite or copper. Figure 18A.42 shows the voltage profile on constant-current discharge for each of these current collectors.

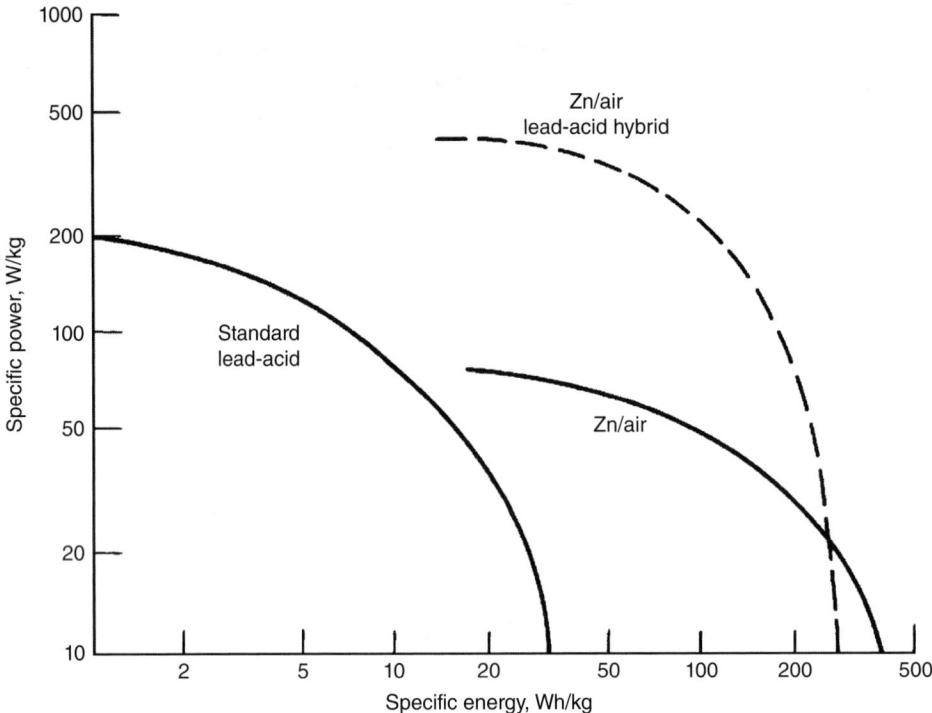

**FIGURE 18A.40** Comparison of Zn/air lead-acid hybrid battery with individual lead-acid and Zn/air batteries. Lead-acid battery uses special high-rate design.

The cell was designed so that the zinc bed and electrolyte could be pumped out at the end of discharge and replaced with a fresh charge of zinc and electrolyte to simulate operation in an electric vehicle. The cell was discharged at 2 A for 4 h. Most of the electrolyte and the residual particles were then sucked out of the anode side of the cell through a tube passing through a hole in the top of the cell and connected to a water jet aspirator. Without rinsing, fresh particles and electrolyte were placed in the cell through the hole and a second discharge was carried out. Following this, about 90% of the particles were removed less carefully, and the cell was refilled and discharged for a third time. The data in Fig. 18A.43 shows that the three discharges were essentially the same.

Based on these experiments, a conceptual design was made for a 55 kW (peak power) electric-vehicle battery. Projected specific energy of the battery was 110 Wh/kg at 97 W/kg under a modified Simplified Federal Urban Driving Schedule (SFUDS). These values were increased to 228 Wh/kg at 100 W/kg when the battery was designed for optimum capacity, and to 100 Wh/kg at 150 W/kg when designed for optimum power output, based on the results of discharge experiments at 45°C.

Efficient regeneration of the zinc particles is required to provide for a practical, efficient system. It is projected that for the practical application of this system, the spent electrolyte and residual particles would be removed at local service centers and the vehicle would be quickly refueled by the addition of regenerated zinc powder and electrolyte. The system under development[30] involves stopping the discharge of the battery described when the voltage falls below a practical value rather than when the voltage becomes zero. Under these conditions, no precipitation has occurred and the electrolyte is clear. The processing of products removed from the cell is then simply one of redeposition of zinc onto the particles.

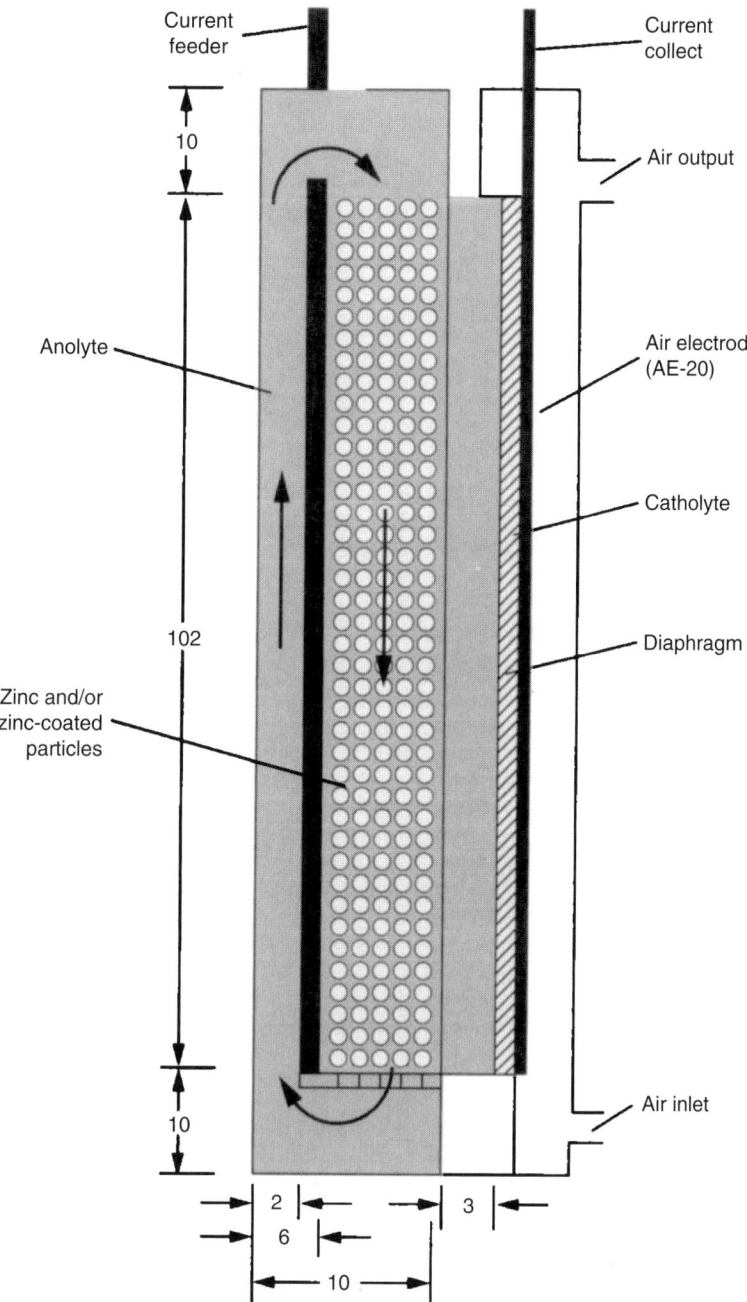

**FIGURE 18A.41**    Schematic of mechanically refueled 80 cm$^2$ laboratory zinc/air cell. (*Courtesy of Lawrence Berkeley Laboratory.*)

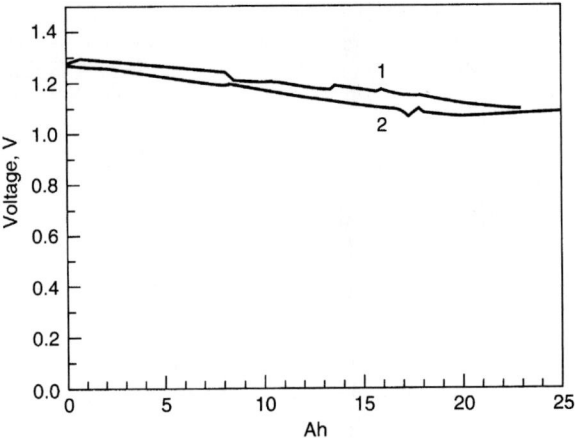

**FIGURE 18A.42**   Constant-current discharges of mechanically refueled zinc/air battery, graphite versus copper feeders. Anolyte/catholyte—45% KOH; anode—30-mesh zinc; cathode—AE-20 air electrode; $I = 2A$; $A = 78$ cm$^2$. Curve 1—1.5-mm copper current feeder; curve 2—4.0-mm graphite current feeder. (*Courtesy of Lawrence Berkeley Laboratory.*)

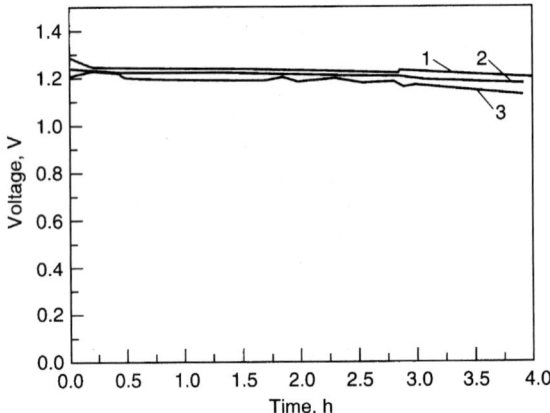

**FIGURE 18A.43**   Voltage versus time during subsequent mechanical recharging for mechanically refueled zinc/air battery. Anolyte/catholyte—45% KOH; anode—20-mesh zinc particles; cathode—AE-20 air electrode; $I = 2$ A; $A = 76$ cm$^2$. Curve 1—first run; curve 2—100% of anolyte/particles suctioned out, cell refilled with fresh ones, no rinsing; curve 3–90% of anolyte/particles suctioned out, cell refilled with fresh ones, no rinsing. (*Courtesy of Lawrence Berkeley Laboratory.*)

## *REFERENCES*

1. G. W. Elmore and H. A. Tanner, U.S. Patent 3,419,900.

2. A. M. Moos, U.S. Patent 3,267,909.

3. J. Oltman, B. Dopp, and J. Burns, U.S. Patent 5,567,538.

4. J. Oltman, U.S. Patent 6,245,452 B1.

5. A. Ohta, A. Hanafusa, H. Yoshizawa, and Z. Ogumi, "Design of Air Holes on Button Type Zinc-Air Batteries. I. New Evaluation Method of Both Water Vapor and Oxygen Permeabilities," *Denki Kagaku (Electrochemistry)*, **65**(5) (1997).

6. A. Ohta, H. Yoshizawa, A. Hanafusa, and Z. Ogumi, "Design of Air Holes on Button Type Zinc-Air Batteries. II. Simulation of Gas Flow Though Air Holes," *Denki Kagaku (Electrochemistry)*, **66**(4) (1998).

7. J. Passanti and R. Dopp, U.S. Patent 5,308,711.

8. A. Ohta, Y. Morita et al., "Manganese Oxide as a Catalyst for Zinc-Air Cells," *Proc. Battery Material Symp.*, 1985.

9. Energizer Zinc Air Prismatic Handbook, https://www.energizer.com/, Winter 2009.

10. T. Atwater, R. Putt, D. Bouland, and B. Bragg, "High-Energy Density Primary Zinc/Air Battery Characterization," *Proceedings of the 36th Power Sources Conference,* Cherry Hill, NJ, 1994.

11. R. Putt, N. Naimer, B. Koretz, and T. Atwater, "Advanced Zinc-Air Primary Batteries," *Proceedings of the 6th Workshop for Battery Exploratory Development,* Williamsburg, VA, 1999.

12. J. Passanitti, "Development of a High Rate Primary Zinc-Air Cylindrical Cell," *Proceedings 5th Workshop for Battery Exploratory Development,* Burlington, VT, 1997.

13. J. Passanitti, "Development of a High Rate Primary Zinc-Air Cylindrical Cell," *Proceedings of the 38th Power Sources Conference,* Cherry Hill, NJ, 1998.

14. J. Passanitti and T. Haberski, "Development of a High Rate Primary Zinc-Air Battery," *Proceedings of the 6th Workshop for Battery Exploratory Development,* Williamsburg, VA, 1999.

15. R. A. Putt and G. W. Merry, "Zinc-Air Primary Batteries," *Proceedings of the 35th Power Sources Symposium*, IEEE, 1992.

16. Sales literature, SAFT, Greenville, NC.

17. Celair Corp., Lawrenceville, GA.

18. A. Karpinski, "Advanced Development Program for a Lightweight Rechargeable AA Zinc-Air Battery," *Proceedings of the 5th Workshop for Battery Exploratory Development,* Burlington, VT, 1997.

19. A. Karpinski, B. Makovetski, and W. Halliop, "Progress on the Development of a Lightweight Rechargeable Zinc-Air Battery," *Proceedings of the 6th Workshop for Battery Exploratory Development,* Williamsburg, VA, 1999.

20. A. Karpinski and W. Halliop, "Development of Electrically Rechargeable Zinc/Air Batteries," *Proc. 38th Power Sources Conf.,* Cherry Hill, NJ, 1998.

21. AER Energy Resources, Inc., Atlanta, GA.

22. L. G. Danczyk, R. L. Scheffler, and R. S. Hobbs, "A High Performance Zinc-Air Powered Electric Vehicle," *SAE Future Transportation Technology Conference and Exposition*, Portland, OR, August 5–7, 1991, paper 911633.

23. M. C. Cheiky, L. G. Danczyk, and M. C. Wehrey, "Second Generation Zinc-Air Powered Electric Minivans," *SAE International Congress and Exposition*, Detroit, MI, February 24–28, 1992, paper 920448.

24. S. M. Chodosh, M. G. Rosansky, and B. E. Jagid, "Metal-Air Primary Batteries, Replaceable Zinc Anode Radio Battery," *Proceedings of the 21st Annual Power Sources Conference,* Electrochemical Society, Pennington, NJ, 1967.

25. D. Linden and H. R. Knapp, "Metal-Air Primary Batteries, Metal-Air Standard Family," *Proceedings of the 21st Annual Power Sources Conference,* Electrochemical Society, Pennington, NJ, 1967.

26. Electric Fuel, Ltd., Jerusalem, Israel.

27. R. A. Putt, "Zinc-Air Batteries for Electric Vehicles," *Zinc/Air Battery Workshop,* Albuquerque, NM, December 1993.

28. H. B. Sierra Alcazar, P. D. Nguyen, G. E. Mason, and A. A. Pinoli, "The Secondary Slurry-Zinc/Air Battery," LBL Rep. 27466, July 1989.

29. G. Savaskan, T. Huh, and J. W. Evans, "Further Studies of a Zinc-Air Cell Intended for Electric Vehicle Applications, Part I: Discharge," *J. Appl. Electrochem.* (Aug. 1991).

30. T. Huh, G. Savaskan, and J. W. Evans, "Further Studies of a Zinc-Air Cell Intended for Electric Vehicle Applications, Part II: Regeneration of Zinc Particles and Electrolyte by Fluidized Bed Electrodeposition," *J. Appl. Electrochem.* (Aug. 1991).

# SECTION B

# AQUEOUS ELECTROLYTES

**Arthur Dobley, Terrill B. Atwater**
**(Emeritus Contributor: Gary A. Bayles)**

## 18B.1   OVERVIEW

Metal/air batteries have an advantage over conventional systems as only one reactant (the anode material) needs to be contained within the battery. Certain metal anode materials, as discussed in this section, are also capable of functioning in aqueous saline solutions, unlike zinc/air batteries (Sec. 18A) that require more chemically reactive alkaline electrolytes.

### 18B.1.1   Aluminum/Air Batteries

Aluminum has long attracted attention as a potential battery anode because of its high theoretical ampere-hour capacity, voltage, and specific energy. While these values are reduced in a practical battery because of the inability to operate aluminum and the air electrodes at their thermodynamic potentials and because water is consumed in the discharge reaction, the practical energy density still exceeds that of most battery systems. The inherent hydrogen generation of the aluminum anode in aqueous electrolytes is such that the batteries are designed as reserve systems with the electrolyte added just before use, or as "mechanically" rechargeable batteries with the aluminum anode replaced after each discharge. Electrically rechargeable aluminum/air batteries are not feasible using aqueous electrolytes.

### 18B.1.2   Magnesium/Air Batteries

Magnesium/air batteries have not been successfully commercialized, but an effort has been directed toward undersea applications using the dissolved oxygen in seawater as the reactant. The battery uses a magnesium alloy anode and a catalytic membrane cathode, and it is activated by the seawater. The main advantage of this system is that, with the exception of the magnesium, all of the reactants are supplied by the seawater. The battery can have a specific energy of about 700 Wh/kg.

The concentration of oxygen in seawater is only 0.3 mol/m$^3$, corresponding to 28 Ah/t of seawater. Therefore, the cathode must have an open structure to ensure that there is sufficient contact with the seawater. Further, as seawater is highly conductive, it is not feasible to use more than one cell. A DC-to-DC converter is used to increase the low cell voltage to the required voltage range.

### 18B.1.3   Iron/Air Batteries

The electrically rechargeable iron/air cell has a lower specific energy than the mechanically rechargeable cells, but it has the advantage of potentially lower life-cycle costs. Unlike zinc, the iron electrodes do not suffer a severe redistribution of active materials or gross shape change upon prolonged electrical cycling. The iron/air cell is another candidate as a motive power source, especially for electric vehicles.

## 18B.2   CHEMISTRY

### 18B.2.1   Aluminum/Air Batteries

The discharge reactions for the aluminum/air cell are

$$
\begin{array}{ll}
\textit{Anode:} & \text{Al} \rightarrow \text{Al}^{+3} + 3\text{e} \\
\textit{Cathode:} & \underline{\text{O}_2 + 2\text{H}_2\text{O} + 4\text{e} \rightarrow 4\text{OH}^-} \\
\textit{Overall:} & 4\text{Al} + 3\text{O}_2 + 6\text{H}_2\text{O} \rightarrow 4\text{Al(OH)}_3
\end{array}
$$

The parasitic hydrogen-generating reaction is

$$\text{Al} + 3\text{H}_2\text{O} \rightarrow \text{Al(OH)}_3 + \tfrac{3}{2}\text{H}_2$$

Aluminum can be discharged in neutral (saline) solutions as well as in caustic solutions. The neutral electrolytes are attractive because of the relatively low open-circuit corrosion rates and the reduced hazards of these solutions compared with concentrated caustic. Saline systems were under development for relatively low-power applications, such as ocean buoys and portable battery applications, with specific energies of a "dry" battery as high as 800 Wh/kg. Seawater batteries for underwater vehicle propulsion and other applications, using oxygen present in the ocean, rather than air, or operating as corrosion cells, also are of interest because of the potentially high-energy output.

*Aluminum/Air Cells in Alkaline Electrolytes.*   Alkaline systems have an advantage over saline systems because the alkaline electrolyte has a higher conductivity and a higher solubility for the reaction product, aluminum hydroxide. Thus, the alkaline aluminum/air battery is a candidate for high-power applications such as standby batteries, propulsion power for unmanned underwater vehicles, and has been proposed for electric vehicle propulsion. The specific energy can be as high as 400 Wh/kg. Aluminum/air batteries (as well as zinc/air batteries), because of their high-energy densities, can also be used as power sources for recharging lower energy rechargeable batteries in remote areas where line power is not available.

*Aluminum/Air Cells in Neutral Electrolytes.*   Aluminum/air cells using neutral electrolytes have been developed for portable equipment, stationary power sources, and marine applications. Aluminum alloys are now available for saline cells with low polarization voltages, which can operate with coulombic efficiencies in the range of 50% to 80%. Alloying elements are required to enhance the disruption of the anodic surface film when current is drawn. Interestingly, in neutral electrolytes the corrosion reaction, resulting in the direct evolution of hydrogen, occurs at a rate linearly proportional to the current density, starting from near zero at zero current.[1]

Cathodes, such as those described earlier, are satisfactory. However, there are some extra limitations that apply in a saline solution. Nickel is not a suitable substrate where extensive periods on open circuit are involved. Under these conditions, the potential of the active material in contact with the screen is high enough to oxidize the screen. One way to minimize this problem is to continue to draw a very low current during no-load periods, which is sufficient to keep the cathode potential from rising to its open-circuit value.

A suitable neutral electrolyte is a 12 wt% solution of sodium chloride, which is near the maximum conductivity. Current densities are limited to 30 to 50 mA/cm$^2$ as a result of the limitation imposed by the conductivity of the electrolyte. Such batteries may also be operated in seawater, with obvious limitations in current capability as a result of the lower conductivity of seawater.

Electrolyte management is required because of the behavior of the reaction product, aluminum hydroxide. It has a transient high solubility in the electrolyte and tends to become gel-like when it first precipitates. In an unstirred system, the electrolyte starts to become "unpourable" when the total charge produced exceeds 0.1 Ah/cm$^3$. Up to this point, the electrolyte and the reaction product can be poured out of a cell and more saline solution added to continue the discharge until all of the aluminum is consumed. If the discharge is continued without draining the electrolyte, it will proceed satisfactorily until the total discharge reaches approximately 0.2 Ah/cm$^3$. At this point, the cell contents are nearly solid.

Approaches to minimizing the amount of electrolyte required have been studied.[2] In one approach, the electrolyte was stirred in a reciprocating manner, which minimized gel formation and produced a finely divided product that was dispersed in the electrolyte. A total electrolyte capacity of 0.42 Ah/cm$^3$ was achieved using reciprocated 20% potassium chloride electrolyte. A similar result was achieved by injecting a pulsed air stream at the bottom of each cell. This has the additional advantage that it sweeps the hydrogen out of each cell in a concentration below the flammability limit. An electrolyte utilization of 0.2 Ah/cm$^3$ was achieved in a system from which the electrolyte could be easily drained.

### 18B.2.2   Magnesium/Air Batteries

The discharge reaction mechanisms of the magnesium/air battery are

$$\text{Anode:} \quad Mg \rightarrow Mg^{+2} + 2e$$
$$\text{Cathode:} \quad O_2 + 2H_2O + 4e \rightarrow 4OH^-$$
$$\text{Overall:} \quad Mg + \tfrac{1}{2}O_2 + H_2O \rightarrow Mg(OH)_2$$

The theoretical voltage of this reaction is 3.1 V, but in practice, the open-circuit voltage is about 1.6 V.

Magnesium anodes tend to react directly with the electrolyte with the formation of magnesium hydroxide and the generation of hydrogen.

$$Mg + 2H_2O \rightarrow Mg(OH)_2 + H_2$$

This reaction stops in alkaline electrolytes because of the formation of an insoluble film of magnesium hydroxide on the electrode surface, which prevents further reaction. Acid tends to dissolve the film. An important consequence of the film on magnesium electrodes is that there is a delayed response to an increase in the load because of the need to disrupt the film to create new bare surfaces for reaction. "Pure" magnesium anodes usually do not give good cell performance, and several magnesium alloys have been developed for use as anodes tailored to provide the desired characteristics.

### 18B.2.3   Iron/Air Batteries

The cell reactions are as follows:

$$O_2 + 2Fe + 2H_2O \xrightleftharpoons[\text{charge}]{\text{discharge}} 2Fe(OH)_2 \quad \text{(first plateau)}$$

$$3Fe(OH)_2 + \frac{1}{2}O_2 \xrightleftharpoons[\text{charge}]{\text{discharge}} Fe_3O_4 + 3H_2O \quad \text{(second plateau)}$$

Iron dissolves initially as the +2 species in alkaline media. The divalent iron complexes with the electrolyte to form $Fe(OH)_n^{2-n}$, a complex of low solubility. The tendency to supersaturate plays an important role in the operation of the electrode and accounts for many important aspects of the electrode performance characteristics. Continued charge forms the +3 valent iron which, in turn, interacts with +2 valent iron to form $Fe_3O_4$.

The superior life-cycling characteristics of the iron electrode result from the low solubility of the reaction intermediates and oxidized species. The supersaturation on discharge results in the oxidized material forming small crystallites near the reaction site. On charge, the low solubility also slows the crystal growth of the iron, thereby helping to ensure formation of the original active high surface area structure. The low solubility also accounts for poor high-rate and low-temperature performance as the discharged (oxidized) species precipitate at or near the reaction site and block the active surface. The performance characteristics are substantially improved, however, in advanced nickel-iron batteries by the use of a superior electrode grid structure, such as fiber-metal, which provides intimate contact with the iron active material throughout the volume of the porous structure.

The oxygen electrode reactions follow the kinetic path with peroxide as an intermediate. The oxygen electrode reactions in simple form are as follows:

$$O_2 + 2H_2O + 2e \rightarrow H_2O_2 + 2OH^-$$

$$H_2O_2 + 2e \rightarrow 2OH^-$$

The single most important life-limiting factor in this battery system is the stability of the air electrode, which loses its ability to function reversibly as it undergoes repeated charges and discharges. The oxygen and peroxide evolved on charge and discharge may attack the substrate, alter the activity of the catalyst, and delaminate the wetproofing film. Separate air (oxygen) electrodes and circuits can be employed in the charge and discharge modes; however, considerations of system weight and volume favor the use of a bifunctional electrode, that is, a single electrode capable of sustaining either oxygen reduction or evolution. These electrodes must be stable over the potential range of both reactions, a fact that poses constraints on material stability and electrode design.

## 18B.3  DESIGN AND CONSTRUCTION

### 18B.3.1  Aluminum/Air Batteries

A number of batteries using saline electrolytes have been designed. In general, they are built as reserve batteries and activated by adding the electrolyte to the battery. A saltwater battery, illustrated in Fig. 18B.1, was designed for field recharging of nickel-cadmium and lead-acid storage batteries.[3]

**FIGURE 18B.1**   Aluminum/air field recharger. 600 Wh, 6 V. (*Courtesy of Alupower, Inc.*)

### 18B.3.2  Magnesium/Air Batteries

Figure 18B.2 illustrates a cell design for an undersea mission intended to deliver 3 to 4 W for 1 year or longer at a total weight of 32 kg. In this design, the oxygen-reduction cathode is positioned on the circumference of a cylinder with a total cathode area of 3 m². The anode is a 19 kg cylinder of magnesium, located internal to the

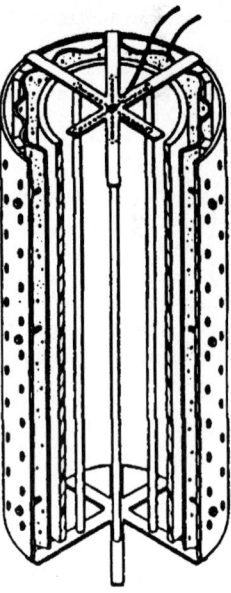

**FIGURE 18B.2**   Schematic of concentric cylinder configuration of seawater cell. The output cylinder is composed of porous fiberglass coated with an antifoulant. The corrugated structure is the air cathode, which is exterior to the magnesium anode. The entire structure is open to the seawater electrolyte. (*Courtesy of Westinghouse Corp.*)

air cathode. The weight of the cathode is about 1.8 kg, and the remainder of the weight is for support structure and other necessary hardware. The single-cell battery has a long shelf life in its dry, unactivated state. It is immediately active on deployment when it is immersed in the seawater electrolyte.

### 18B.3.3   Iron/Air Batteries

Relatively little work has been done on iron/air batteries in recent years. Section 18B.4.3 details the design and performance of several historical development efforts.

## 18B.4   *PERFORMANCE AND APPLICATIONS*

### 18B.4.1   Aluminum/Air Batteries

*Aluminum/Air Cells in Saline Electrolytes.*   Figure 18B.3 shows the charge and discharge characteristics for a 2 Ah, 24 V sealed nickel-cadmium battery being charged within 4 h. The aluminum/air battery can recharge this nickel-cadmium battery about seven times before the aluminum is depleted. The specific energy of a dry battery, with enough metal for the anode and salt for the electrolyte to provide for a complete discharge, is about 600 Wh/kg.

*Ocean Power Supplies.*   Batteries based on the use of oxygen dissolved in seawater have an advantage over others as all reactants, except for the anode material, are supplied by the seawater. In these batteries, a cathode, which is open to the ocean, is spaced around an anode so that the reaction products can fall out into the ocean.[4] Relatively large surface areas are required as there is not much oxygen in seawater. In addition, because of the conductivity of the ocean, there can be no series arrangement of cells. Higher voltages are obtained by the use of a DC-to-DC converter.

Many instruments and devices used in the ocean have to operate over long periods of time, and aluminum is a candidate for the anode for missions requiring months or years of service.

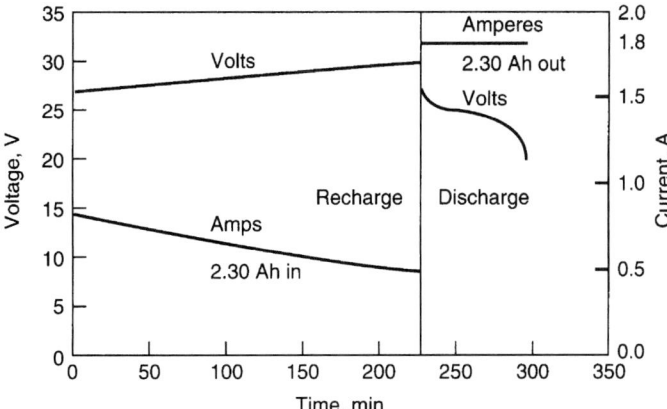

**FIGURE 18B.3**   Charge and discharge of a nickel-cadmium battery, aluminum/air field recharger. (*Courtesy of Alupower, Inc.*)

Figure 18B.4 shows a flat-plate aluminum/dissolved oxygen battery.[5] The battery, which is about 1.5 m high, has a dry specific energy of 500 Wh/kg and can operate at power densities of up to 1 W/m². This battery can be installed beneath a buoy, as shown in the illustration, and used with a DC-to-DC converter to charge a lead-acid battery.

*Aluminum/Air Cells in Alkaline Electrolytes.*   The concept of operating aluminum/air batteries at high energy and power densities was described in the early 1970s, but successful commercialization was impeded because of technological limitations, including the high open-circuit corrosion rate of aluminum alloys in alkaline electrolyte, the nonavailability of thin, large-dimension air cathodes, and the difficulty of handling and removing the cell reaction products (precipitated aluminum hydroxide) to prevent cell clogging.

Significant advances have been made in reducing the corrosion of aluminum alloys in caustic electrolytes.[6,7] An aluminum anode, containing magnesium and tin, has approximately two orders of magnitude reduction in corrosion current at open-circuit and operates at greater than 98% coulombic efficiency over a wide range of current densities. Even at open-circuit, the alloy can remain in a caustic electrolyte with a relatively low rate of self-discharge. Prior to this development, the amount of hydrogen and heat evolved during open-circuit stand was usually so high as to require that the electrolyte be drained from the alloy during this period to prevent it from boiling.

Techniques to handle the aluminum so that it can be continuously fed as chips or pellets into the electrochemical system have also been developed.[8] One design uses aluminum particles having diameters of 1 to 5 mm.[9] The electrode is a pocket whose walls are composed of cadmium-plated expanded steel screen. The electrode is fed by a system that keeps the cell maintained with an aluminum particulate at an optimum level. Figure 18B.5 shows the performance of a cell operating at 50°C using 8N KOH-containing stannate. The battery, with an electrode area of 360 cm², was able to deliver a current of 56 A at 1.35 V for 110 h, with the automatic addition of aluminum every 20 min.

Management of the electrolyte (i.e., removal of reaction products) is required as the conductivity of the electrolyte decreases with increasing aluminate concentration. As shown in Fig. 18B.6, the voltage decreases if the aluminate is not removed. Several techniques have been developed to remove the reaction products and are discussed later in this section.

*Applications of Alkaline Aluminum/Air Batteries.*   The alkaline aluminum/air batteries being developed cover a wide range of applications from emergency power supplies to field-portable batteries for remote power applications and underwater vehicles. Most of these are designed as reserve batteries, which are activated before use, or "mechanically" recharged by replacing the exhausted aluminum anodes.

RESERVE POWER UNITS, STANDBY BATTERY.[10,11]   This is a reserve battery, used with conventional lead-acid batteries, to provide a standby power supply with extended service life. The aluminum/air battery is about one-tenth the weight and one-seventh the volume of a lead-acid battery containing the same energy.

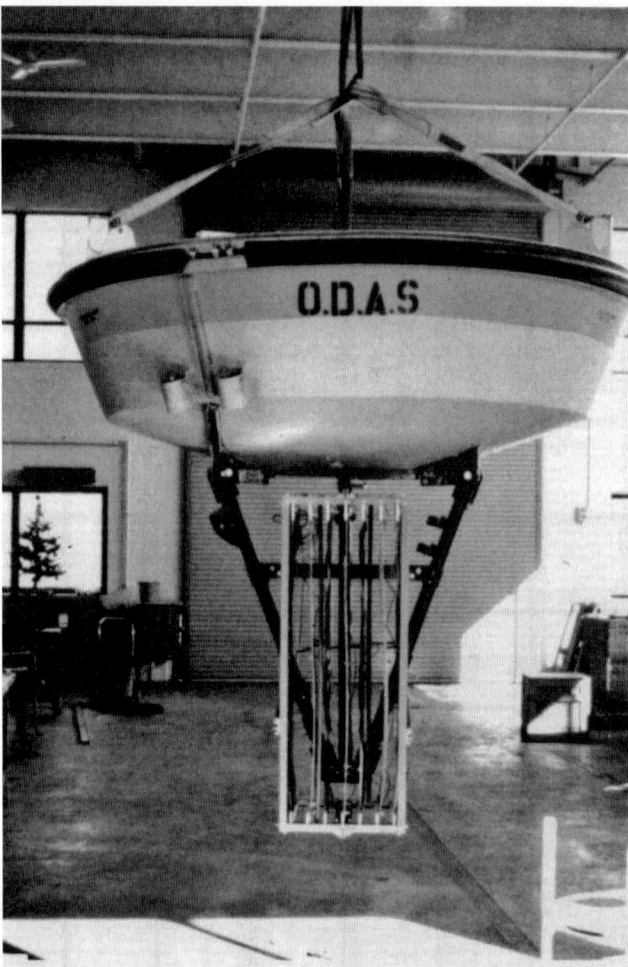

**FIGURE 18B.4**  Aluminum dissolved oxygen flat-plate battery (attached to Woods Hole Oceanographic Institution buoy.) (*Courtesy of Alupower, Inc.*)

The basic elements of the power supply design are shown in Fig. 18B.7. The aluminum/air battery consists of an upper cell stack, a lower electrolyte reservoir (which is isolated from the stack during periods of nonuse), and auxiliary systems to pump and cool the electrolyte and circulate air through the battery. The electrolyte is an 8 M solution of potassium hydroxide containing a stannate additive. During operation, the electrolyte becomes increasingly saturated and then supersaturated with potassium aluminate. Eventually the conductivity of the electrolyte decreases to the point where the battery is unable to sustain the load. At that point, it has reached the end of its capacity based on total electrolyte volume (VLD). The electrolyte can be changed at this point, and the discharge continued to the point where the aluminum in the anode is exhausted (ALD). Figure 18B.8 shows the performance for a nominal 1200 W battery discharged in the two modes. Operation in the electrolyte-capacity limited mode will yield a total discharge time of 36 h, while operation to anode exhaustion, incorporating one electrolyte change, will result in a total discharge time of 48 h. The overall energy density and specific energy are greater than 150 Wh/L and 250 Wh/kg.

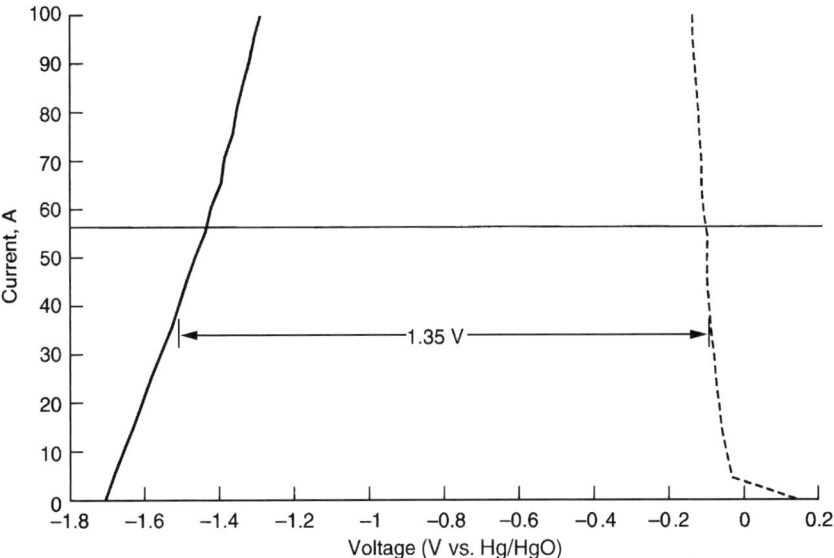

**FIGURE 18B.5**    Polarization curves of particulate-feed aluminum/air battery. - - -, positive electrode; ——, negative electrode. (*Courtesy of Sorapec.*)

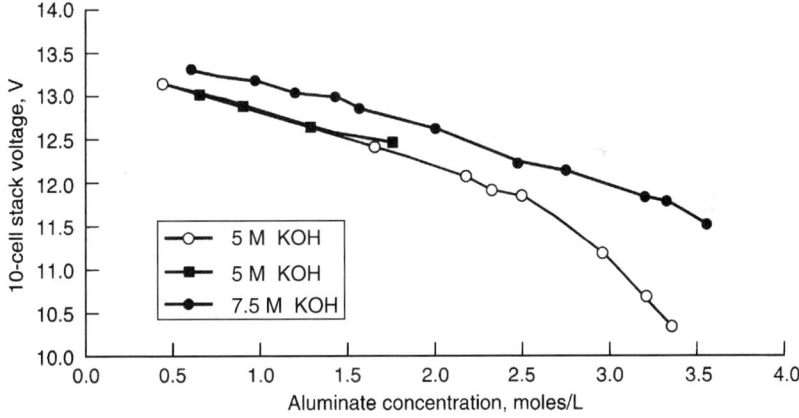

**FIGURE 18B.6**    Voltage versus aluminate concentration. (*Courtesy of Eltech Systems.*)

The control system for this power unit is arranged so that in the event of a power outage, the lead-acid batteries provide the backup for the first 1 to 3 h. As the voltage of the lead-acid battery begins to fall, the aluminum/air battery is activated by a controller that initiates the pumping of electrolyte from a reservoir through the aluminum/air cell stack. Once the aluminum/air battery reaches full power (about 15 min from activation), it provides full power to the load and recharges the depleted lead-acid batteries. The electrical characteristics of the unit are shown in Fig. 18B.9. The aluminum/air battery has limited restart capability but can be refurbished by replacing the cell stack and the electrolyte.

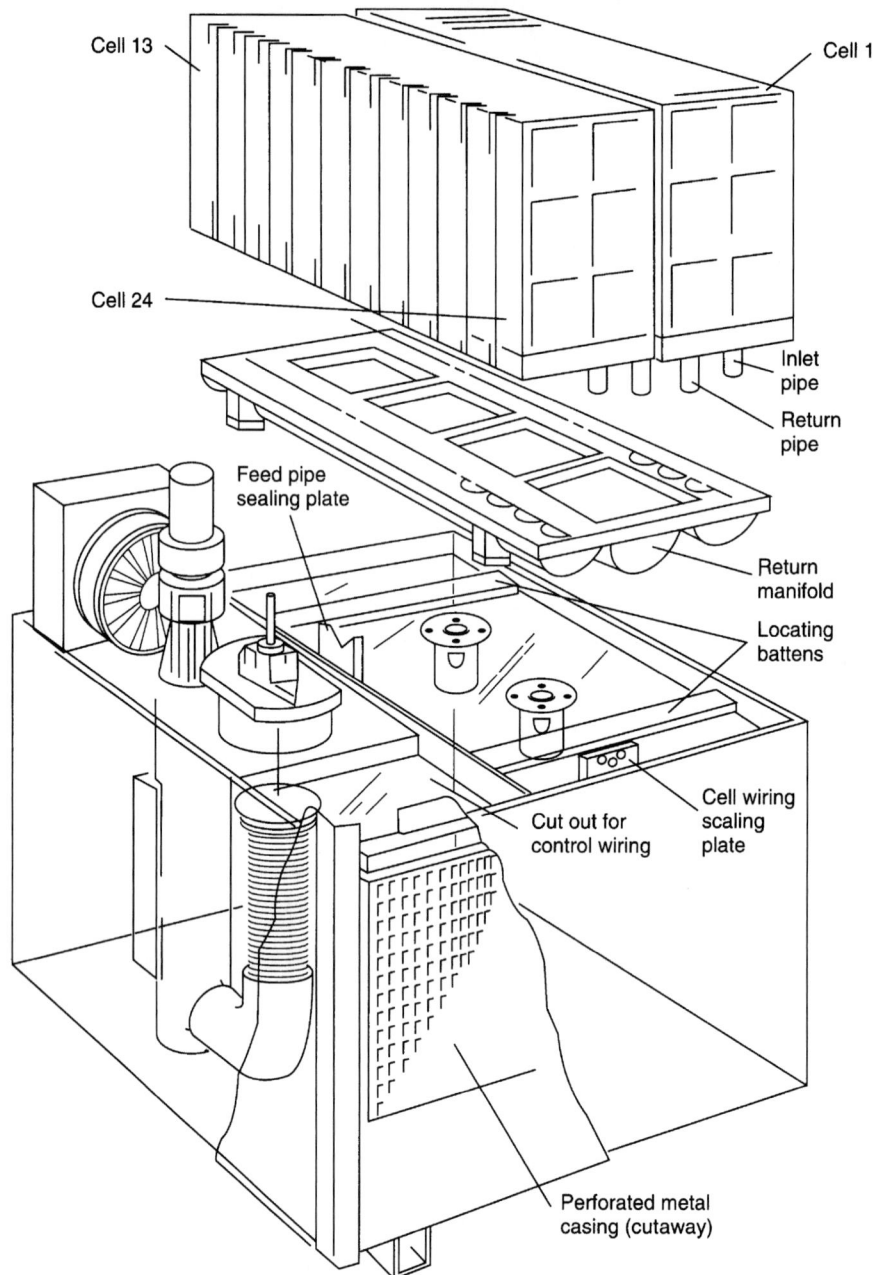

**FIGURE 18B.7**   Aluminum/air reserve power unit, standby battery. (*Courtesy of Alupower, Inc.*)

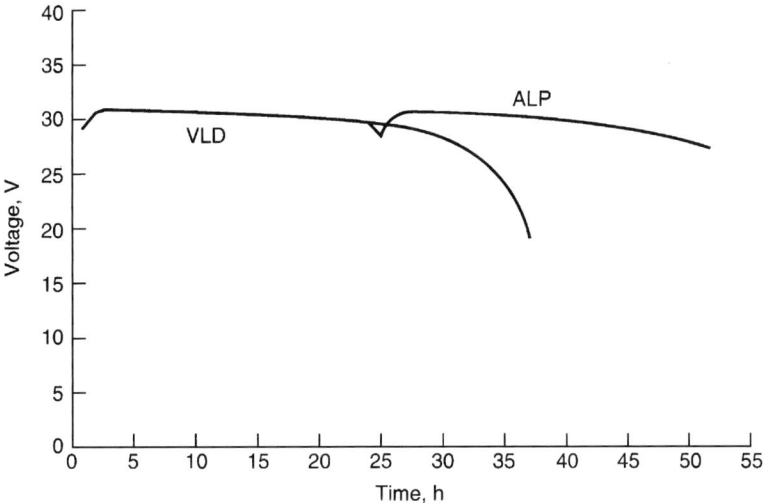

**FIGURE 18B.8** Comparison of volume (VLD) versus anode (ALD) limited discharges of aluminum/air reserve power unit. (*Courtesy of Alupower, Inc.*)

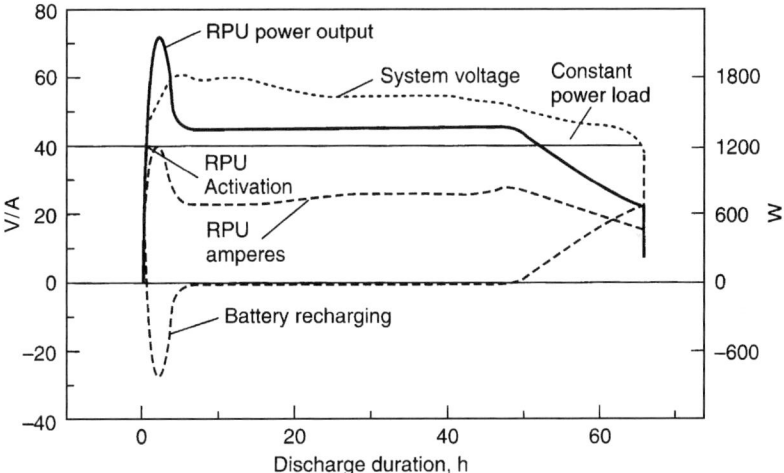

**FIGURE 18B.9** Discharge profile of 6 kW aluminum/air reserve power unit. (Direct connection to power system.) (*Courtesy of Alupower, Inc.*)

BATTLEFIELD POWER UNIT. This power source, called the Special Operations Forces Aluminum Air (SOFAL) battery,[12] was developed as a reserve system to support specialized military communications equipment in covert field operations. The SOFAL weighs approximately 7.3 kg after activation and powers 12 and 24 VDC (volt [direct current]) equipment with pulse currents up to 10 A, continuous drains up to 4 A, and has a design capacity of 120 Ah. To minimize weight, this battery is carried to the field dry and can be activated with any source of water.

The SOFAL unit consists of 16 series-connected cells (Fig. 18B.10*a*) with intercell connections provided by a printed circuit board. The cell stack, which weighs 3.5 kg dry, is shown in Fig. 18B.10*b*. Activation of the system is accomplished with 2.5 L of water through a manifold system to each cell where it dissolves a cast block of

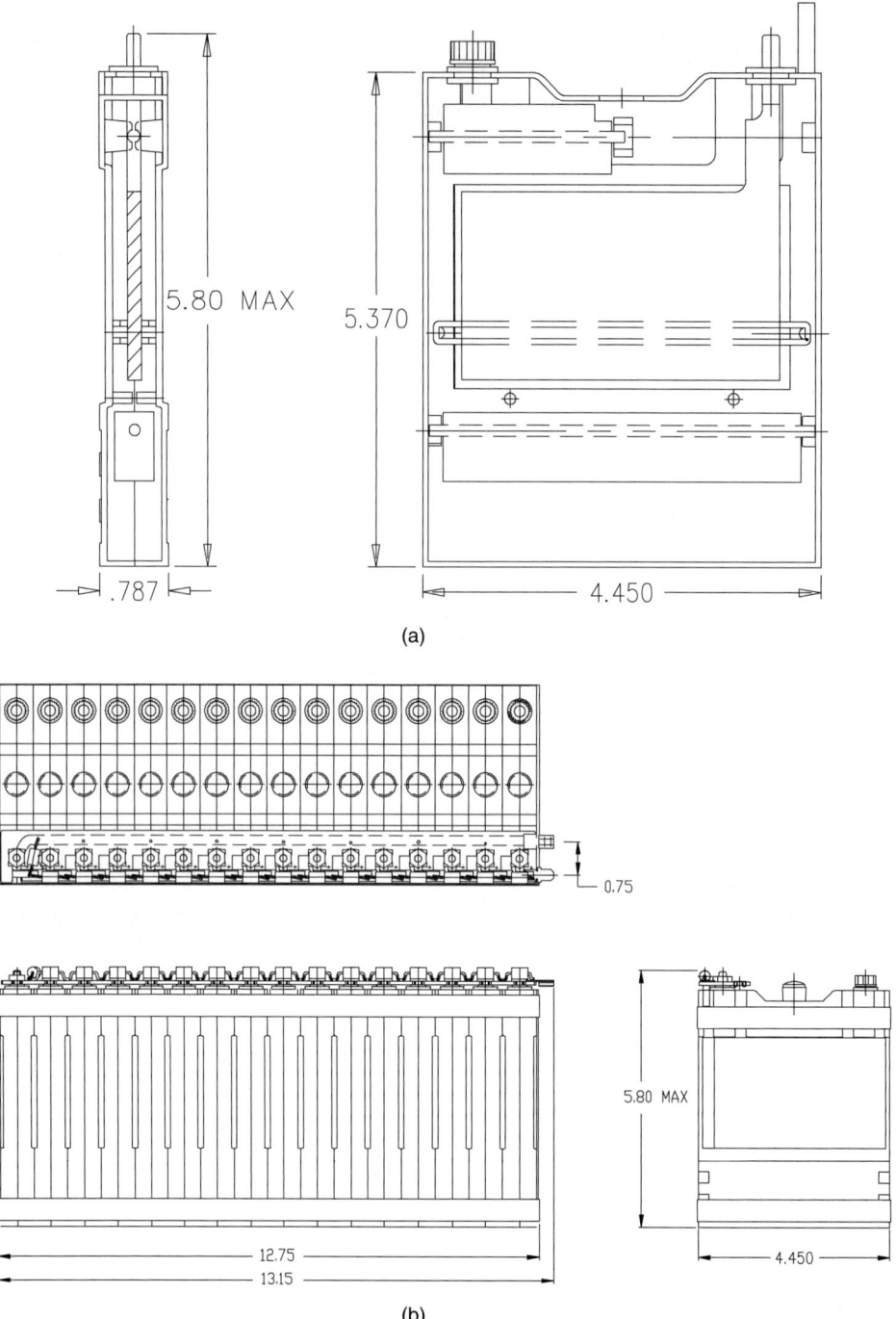

(a)

(b)

**FIGURE 18B.10**    SOFAL battery. (*a*) Cell design configuration, (*b*) battery block design configuration, (*c*) full-scale design layout.

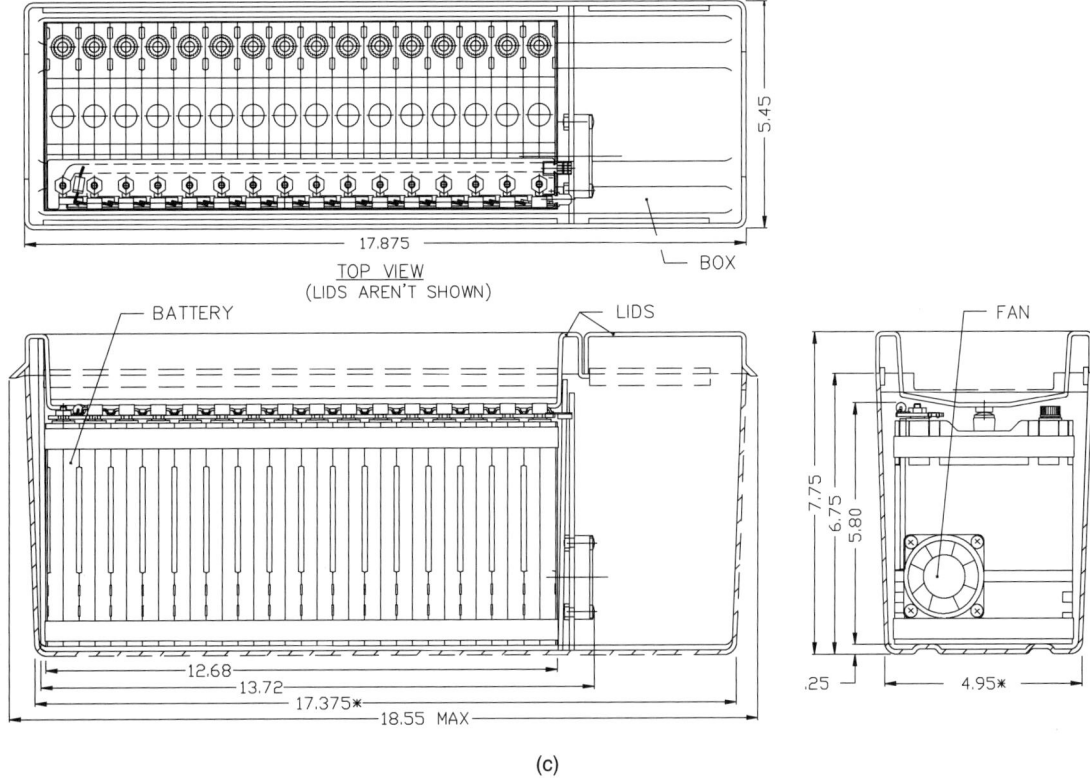

17.875
TOP VIEW
(LIDS AREN'T SHOWN)

BOX

5.45

BATTERY

LIDS

FAN

12.68
13.72
17.375*
18.55 MAX

7.75
6.75
5.80

.25

4.95*

(c)

**FIGURE 18B.10**    (*Continued*)

stannated potassium hydroxide giving a 30% (w/w) KOH solution. After activation, each cell has an open-circuit voltage of 1.7 or 27.2 V for the cell stack. Electrochemistry of the battery is similar to that described earlier. Dissolution of the KOH and corrosion of the aluminum provide heat to operate the system even at low temperature. The unit has been designed to access 1.6 L/min of air, which provides for low-power operation. If ambient air flow is insufficient, a small fan, activated by the battery, will provide the required airflow and will dissipate excess heat during high-power use. The SOFAL unit provides up to 2 weeks of service after activation.

The electronics module package (EMP) shown in Fig. 18B.10c contains the electronic components, provides the mounting for an internal rechargeable battery, and houses the fan. The electronics package contains the power management circuitry to keep the internal secondary battery fully charged, provides both 24 V and a regulated 12 V output, and can be used to power electronics directly or recharge external batteries. Figure 18B.11 shows the discharge profile of the SOFAL battery on a 2 A continuous load on 24 V operation. Figure 18B.12 shows the discharge curves for two cells from the SOFAL battery. Cell no. 1 was activated with 8 M KOH, while cell no. 2 employed KOH pellets in the cell and was activated with water. Both cells were discharged at 0.5 A, and provided 135 Ah capacity, but cell no. 1 operated at slightly higher voltage, particularly at the end of discharge. Following discharge, the cell stack can be replaced to provide a new battlefield power unit.

UNDERWATER PROPULSION.    Another area of application for alkaline aluminum batteries is in self-contained extended-duration power supplies for underwater vehicles such as unmanned vehicles for submarine and mine surveillance, long-range torpedoes, swimmer delivery vehicles, and submarine auxiliary power.[13,14] In these applications, the oxygen can be carried in pressurized or cryogenic containers, or it can be obtained from the decomposition of hydrogen peroxide or from oxygen candles. The aluminum/oxygen system can produce

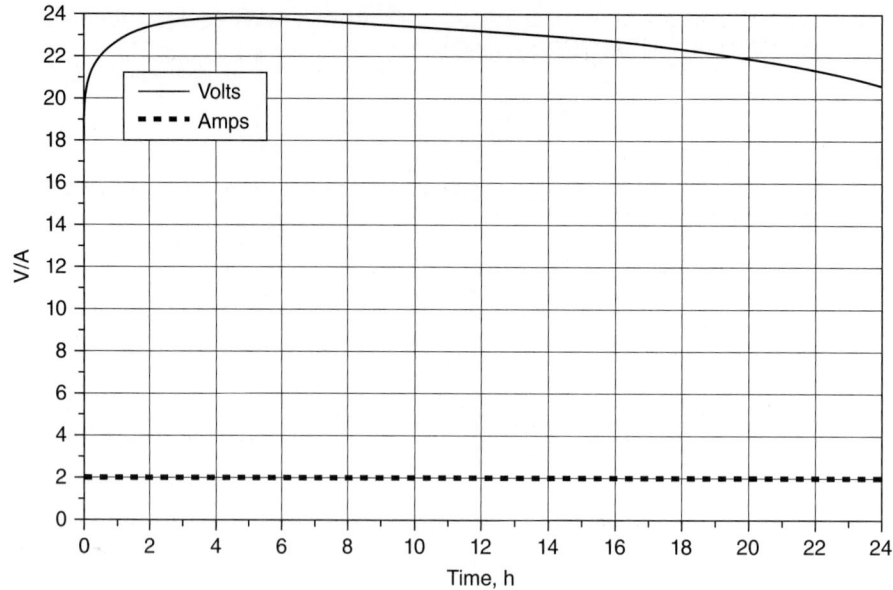

**FIGURE 18B.11**     Sixteen cell SOFAL battery. Constant current 2.0 A discharge.

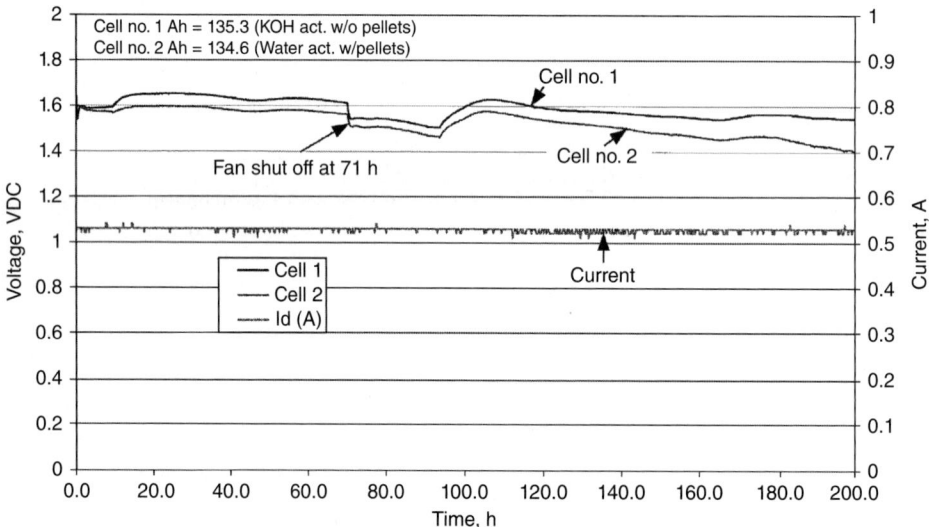

**FIGURE 18B.12**     SOFAL test cells on 0.5 A resistive discharge.

almost twice as much energy per kilogram of oxygen as a hydrogen/oxygen fuel cell, as the operating voltage of 1.2 to 1.4 V is almost twice that of the fuel cell. One type of aluminum/oxygen battery for the propulsion of underwater vehicles is shown in Fig. 18B.13, and its characteristics are listed in Table 18B.1.

**TABLE 18B.1**    Characteristics of Aluminum/Oxygen Battery

| | |
|---|---|
| Performance | |
|    Power | 2.5 kW |
|    Capacity | 100 kWh |
|    Voltage | 120 V nominal |
|    Endurance | 40 h at full power |
| Fuel | 25 kg aluminum anodes |
| Oxidant | 22 kg oxygen at 4000 lb/in$^2$ |
| Buoyancy | Neutral, including aluminum hull section |
| Time to refuel | 3 h |
| Dimensions | |
|    Mass | 360 kg |
|    Battery diameter | 470 mm |
|    Hull diameter | 533 mm |
|    System length | 2235 mm |
| Performance | |
|    Energy density | 265 Wh/L |
|    Specific energy | 265 Wh/kg |

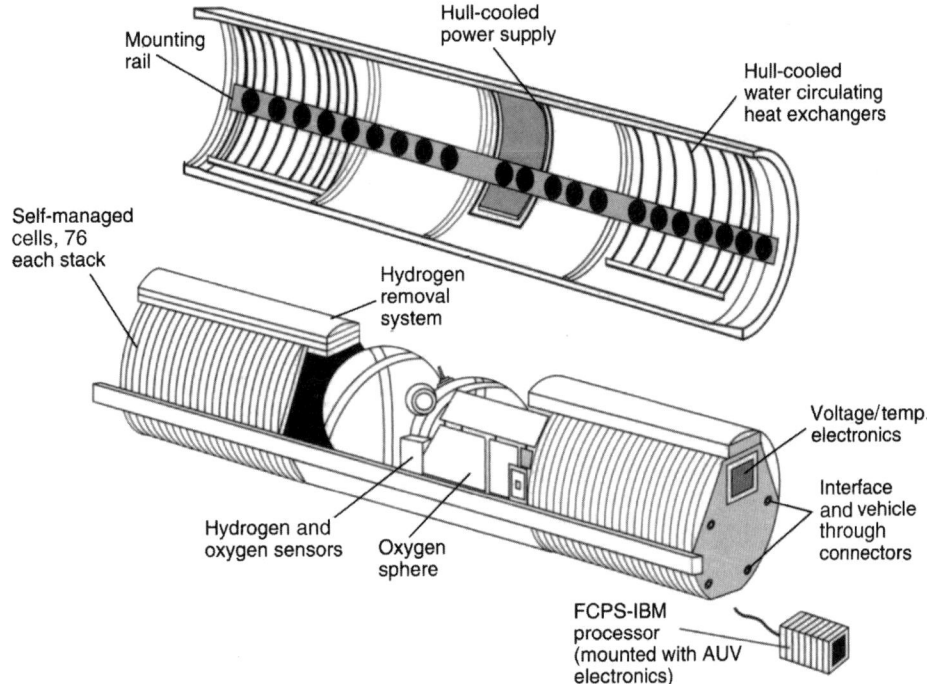

**FIGURE 18B.13**    Aluminum/oxygen power system. (*Courtesy of Alupower, Inc.*)

This battery uses a "self-managing" electrolyte system, where the required electrolyte circulation and precipitation take place within the cell chamber, without the requirement for external pumps. This has the advantage of allowing the design of battery systems without electrolyte circulation pumps. There are no shunt currents between cells as each cell is independent, and there are no electrolyte paths between cells. In addition, the cells can be conformal with the system they are designed to power. Figure 18B.14 shows the design of a system using 19-in diameter (48.25 cm) cells. Each cell is about 0.5 in. (1.25 cm) thick. Thermal and concentration gradients and the resulting convection currents within the cell precipitate the reaction product to the bottom. With this type of system, it is possible to utilize about 0.8 Ah/cm$^3$ of electrolyte. Figure 18B.15 gives a discharge curve for a cell that is exactly one-half the cell shown in Fig. 18B.14. The cell was divided down the center to provide for redundancy. The discharge was carried out at a constant power level of 18 W and a current density of about 50 mA/cm$^2$. The figure shows that the voltage remains relatively constant, between 1.4 and 1.5 V, for most of the run.

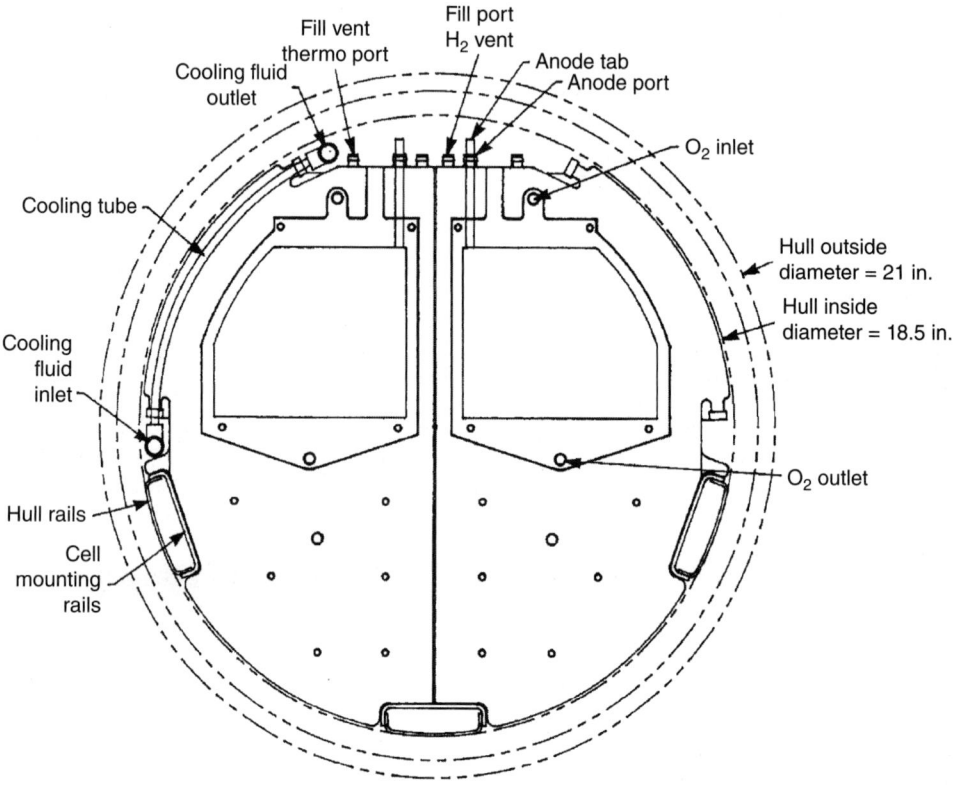

**FIGURE 18B.14**    Self-managing cell system. (*Courtesy of Alupower.*)

To maximize the capacity, the amounts of aluminum and electrolyte are matched so that at the end of discharge the aluminum is consumed and the electrolyte is completely filled with reaction product. Under this condition, the module is either discarded or rebuilt after use. Alternatively, a higher concentration of electrolyte can be used and discharge stopped before the onset of precipitation. In this mode of operation, the amount of aluminum incorporated into the cell can be sufficient for several runs with only the electrolyte being replaced between runs.

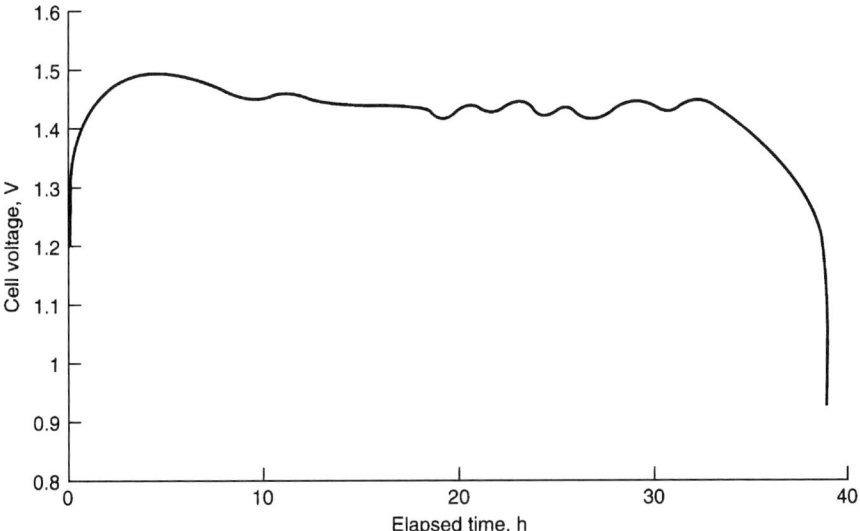

**FIGURE 18B.15** Discharge curve for self-managing cell. (*Courtesy of Alupower.*)

Another requirement of the underwater power system is the hydrogen-removal system, which is needed to safely remove the hydrogen that is generated by corrosion of the anode. Catalytic recombination is especially attractive for applications where space and energy efficiency are needed.[15] The unit shown above in Fig. 18B.13 uses a hydrogen-removal system, but the amount of hydrogen generated is not excessive since a low-corrosion aluminum alloy is used.

Another approach to removing the reaction product is a filter/precipitator system,[15] as shown in Fig. 18B.16. The aluminate concentration is controlled by pumping the electrolyte out of the cell stack and through the filter/precipitator. The filter promotes the growth of the aluminum trioxide and regeneration of KOH as follows:

$$KAl(OH)_4 \rightarrow KOH + Al(OH)_3$$

The crystal cake gradually increases in thickness with a subsequent increase in the pressure drop across the filter. When the pressure drop reaches a predetermined level, the cake is pulsed off the filter by backflushing (flow reversal) and collected in the bottom of the precipitate tank.

### 18B.4.2   Magnesium/Air Batteries

Figure 18B.17 shows the discharge of a test battery (such as in Fig. 18B.2) with periodic voltage spikes when the load is increased.[16]

### 18B.4.3   Iron/Air Batteries[17]

Most work on iron/air batteries has been discontinued in favor of zinc/air. Prior work included improvements such as sulfide addition to the iron electrode, which radically changes the electrocrystallization kinetics. It increases the supersaturation and makes reaction more reversible. Sulfide also absorbs on the surface to block crystallization sites and raises the hydrogen evolution reaction on charge. The self-discharge rate of the iron electrode is correspondingly reduced by the sulfide as well, with studies showing that PbS outperforms FeS in this regard.[18] Lithium salt additions seem to make the electrode perform more reversibly, perhaps by enhancing the solubility of the reaction intermediates. Most recent research focused on the iron anode.[19–21]

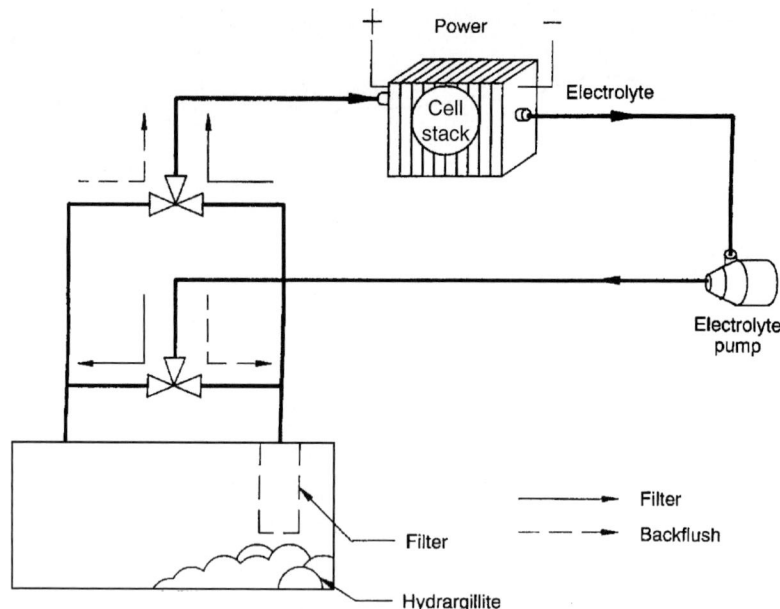

**FIGURE 18B.16**    Conceptual design of filter/precipitator system when integrated with aluminum/oxygen battery. (*Courtesy of Eltech Systems.*)

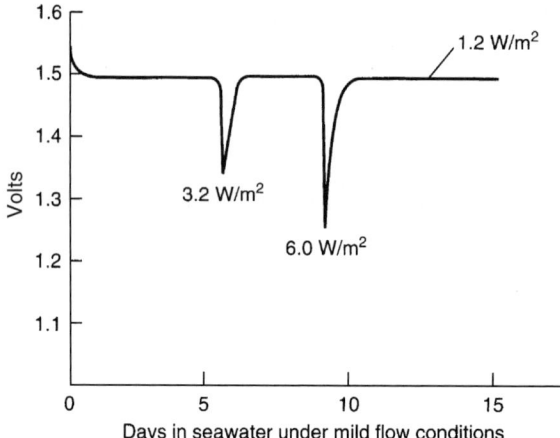

**FIGURE 18B.17**    Discharge profile of seawater power source at 80 μA, 20°C. Current spikes represent charging of a small silver-iron battery. Neutral pH was maintained by periodic addition of hydrochloric acid. (*Courtesy of Westinghouse Corp.*)

The Swedish National Development Corporation's iron/air cell used the sintered iron-mesh anode construction.[22-24] A pore-forming material could be included to control the development of the optimum electrode structure. The resulting pressed matrix was treated in $H_2$ at 650°C. The pore-forming material could be leached out after treatment. The active material utilization approached 65%. The air electrode was a porous-nickel double-layer

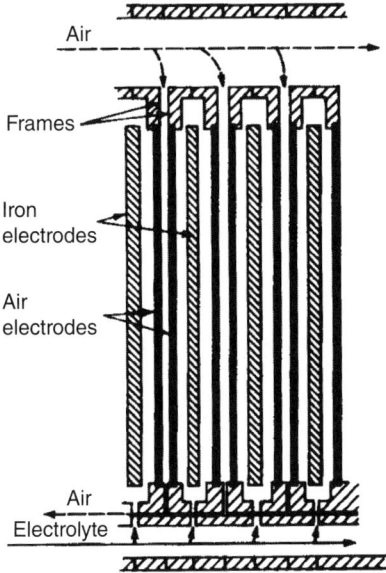

**FIGURE 18B.18** Cross section of Swedish National Development Corporation's iron/air battery pile.

structure (0.6 mm thick) composed of sintered nickel of coarse and fine porosities. The coarse layer on the electrolyte side was catalyzed with silver and impregnated with hydrophobic agents. The electrodes were welded into a polymer frame and formed into cells, as shown in Fig. 18B.18. There were two air electrodes for each iron electrode. Air was forced past the electrode at about two times the stoichiometric requirement during operation. A schematic and a photo of a 30 kWh battery are shown in Figs. 18B.19 and 18B.20, respectively.

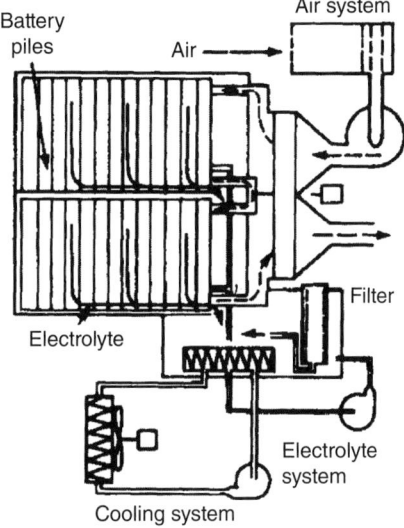

**FIGURE 18B.19** Schematic cross section of Swedish National Development Corporation's iron/air battery, including auxiliary system.

**FIGURE 18B.20**   Swedish National Development Corporation's 30 kWh iron/air battery system. (*Courtesy of Swedish National Development Corp.*)

Electrolyte circulation was used to control heat balance and remove gases generated during operation. Carbon dioxide was scrubbed from the incoming air using NaOH. The air was then humidified to minimize electrolyte loss. Overall the auxiliary systems require less than 10% of the system output.

Typical charge-discharge curves for an average battery in the iron/air battery are shown in Fig. 18B.21. The marked difference in charge and discharge voltages accounts largely for the low overall system efficiency. Figure 18B.22 shows the power-producing characteristics. The system is capable of over 1000 cycles, limited by the gradual deterioration of the air electrode.

The Westinghouse iron/air battery used a construction similar to that described for the Swedish National Development Corporation's system.[25] The sintered-iron electrode was somewhat similar to that described before. The iron electrode for this iron/air battery had a high active iron content and lower cycle life compared with the electrode described previously. Particles of iron powder were sintered to form a structure without the steel fiber substrate. Electrodes with this construction demonstrated up to 0.44 Ah/g. The air electrode was bifunctional and used a Teflon-bonded carbon-based structure with complex silver catalysts (silver content was less than 2 mg/cm$^2$) supported on a silver-plated nickel screen. The Westinghouse system used a horizontal flow concept to improve performance and control gas and heat. Good life was demonstrated for over 300 cycles with

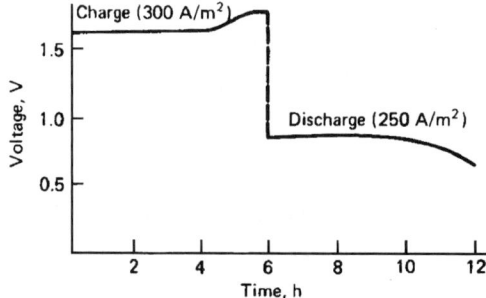

**FIGURE 18B.21**   Charge-discharge voltages for battery in Swedish National Development Corporation's iron/air battery.

an air electrode of potentially very low cost. A summary of the projected characteristics of the 40 kWh battery is given in Table 18B.2.

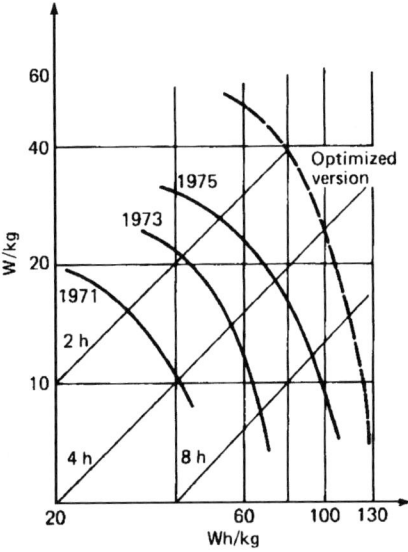

**FIGURE 18B.22**    Performance of Swedish National Development Corporation's iron/air batteries.

**TABLE 18B.2**    Characteristics of Iron/Air Electric-Vehicle Battery

| | |
|---|---|
| Electric vehicle | |
|    Weight | 900 kg curb weight |
|    Range | 240 km |
| Battery | |
|    Energy | 40 kWh |
|    Power | 10 kW continuous power |
|    Weight | 530 kg |
|    Volume | 0.04 m³ |
|    Cost | $150/kWh |

The Siemens cell was similar except that the air electrode was fabricated with two layers: a hydrophilic layer of porous nickel on the electrolyte side for oxygen evolution and a hydrophobic layer (carbon black bonded with Teflon® [PTFE] and catalyzed with silver) on the air side for oxygen reduction. The dual porosity helped to shield the silver catalyst from oxidation. As many as 200 cycles were achieved.[26]

## *REFERENCES*

1. A. R. Despic, "The Use of Aluminum in Energy Conversion and Storage," *First European East-West Workshop on Energy Conversion and Storage*, Sintra, Portugal, March 1990.

2. N. P. Fitzpatrick and D. S. Strong, "An Aluminum-Air Battery Hybrid System," *Elec. Vehicle Develop.* **8**:79–81 (July 1989).

3. T. Dougerty, A. Karpinski, J. Stannard, W. Halliop, V. Alminauskas, and J. Billingsley, "Aluminum-Air Battery for Communications Equipment," *Proceedings Of the 37th Power Sources Conference,* Cherry Hill, NJ, 1996.

4. C. L. Opitz, "Salt Water Galvanic Cell with Steel Wool Cathode," U.S. Patent 3,401,063, 1968.

5. D. S. Hosom, R. A. Weller, A. A. Hinton, and B. M. L. Rao, "Seawater Battery for Long-Lived Upper Ocean Systems," *IEEE Ocean Proceedings,* vol. 3, October 1–3, 1991.

6. J. A. Hunter, G. M. Scamans, and J. Sykes, "Anode Development for High Energy Density Aluminium Batteries," *Power Sources 13* (Bournemouth, England, April 1991).

7. R. P. Hamlen, W. H. Hoge, J. A. Hunter, and W. B. O'Callaghan, "Applications of Aluminum-Air Batteries," *IEEE Aerosp. Electron. Mag.* **6**:11–14 (October 1991).

8. S. Zaromb, C. N. Cochran, and R. M. Mazgaj, "Aluminum-Consuming Fluidized Bed Anodes," *J. Electrochem. Soc.* **137**:1851–1856 (June 1990).

9. G. Bronoel, A. Millott, R. Rouget, and N. Tassin, "Aluminum Battery with Automatic Feeding of Aluminium," *Power Sources 13* (Bournemouth, England, April 1991); also French Patents 88.15703, 1988; 90.07031, 1990; 90.14797, 1990.

10. W. B. O'Callaghan, N. Fitzpatrick, and K. Peters, "The Aluminum-Air Reserve Battery—A Power Supply for Prolonged Emergencies," *Proceedings of the 11th International Telecommunications Energy Conference,* Florence, Italy, October 15–18, 1989.

11. J. A. O'Conner, "A New Dual Reserve Power System for Small Telephone Exchanges," *Proceedings of the 11th International Telecommunications Energy Conference,* Florence, Italy, October 15–18, 1989.

12. A. P. Karpinski, J. Billingsley, J. H. Stannard, and W. Halliop, *Proceedings of the 33rd IECEC,* 1998.

13. K. Collins et al., "An Aluminum-Oxygen Fuel Cell Power System for Underwater Vehicles," Applied Remote Technology, San Diego, 1992.

14. D. W. Gibbons and E. J. Rudd, "The Development of Aluminum/Air Batteries for Propulsion Applications," *Proceedings of the 28th IECEC,* 1993.

15. D. W. Gibbons and K. J. Gregg, "Closed Cycle Aluminum/Oxygen Fuel Cell with Increased Mission Duration," *Proceedings of the 35th Power Sources Symposium,* IEEE, 1992.

16. J. S. Lauer, J. F. Jackovitz, and E. S. Buzzelli, "Seawater Activated Power Source for Long-Term Missions," *Proceedings of the 35th Power Sources Symposium,* IEEE, 1992.

17. S. U. Falk and A. J. Salkind, *Alkaline Storage Batteries,* Wiley, New York, 1969.

18. C. A. C. Souza, I. A. Carlos, M. Lopes, G. A. Finazzi, and M. R. H de Almeida, "Self-Discharge of Fe-Ni Alkaline Batteries," *J. Power Sources* **132**:288–290 (2004).

19. B. T. Hang, T. Watanabe, M. Egashira, S. Okadab, J. Yamaki, S. Hata, S-H. Yoon et al., "The Electrochemical Properties of $Fe_2O_3$-Loaded Carbon Electrodes for Iron-Air Battery Anodes," *J. Power Sources* **150**:261–271 (2005).

20. B. T. Hang, H. Hayashi, S. H. Yoon, S. Okada, and J. Yamaki, "$Fe_2O_3$-Filled Carbon Nano-tubes as a Negative Electrode for an Fe-Air Battery," *J. Power Sources* **178**:393–401 (2008).

21. B. T. Hang, T. Watanabe, M. Egashira, I. Watanabe, S. Okada, and J. Yamaki, "The Effect of Additives on the Electrochemical Properties of Fe/C Composite for Fe/Air Battery Anode," *J. Power Sources* **155**:461–469 (2006).

22. B. Anderson and L. Ojefors, in J. H. Thompson (ed.), *Power Sources,* vol. 7, Academic, London, 1979, p. 329.

23. L. Carlsson and L. Ojefors, "Bifunctional Air Electrode for Metal-Air Batteries," *J. Electrochem. Soc.* **127**:525 (1980).

24. L. Ojefors and L. Carlson, "An Iron-Air Vehicle Battery," *J. Power Sources* **2**:287 (1977/1978).

25. J. F. Jackovitz and C. T. Liu, *Extended Abstracts: 9th Battery and Electrochemical Contractors' Conf.,* USDOE, Alexandria, VA, November 12–16, 1989, pp. 319–324.

26. H. Cnoblock, D. Groppel, D. Kahl, W. Nippe, and G. Siemsen, in D. H. Collins (ed.), *Power Sources,* vol. 5, Academic, London, 1975, p. 261.

# SECTION C

# LITHIUM ANODE

## Arthur Dobley, Terrill B. Atwater

## 18C.1  BACKGROUND

Lithium/air batteries consist of lithium (Li) anodes electrochemically coupled to atmospheric oxygen through an air cathode. Lithium/air batteries are also referred to as lithium/oxygen batteries. For lithium/air batteries, oxygen gas introduced into the battery through an air cathode is essentially an unlimited cathode reactant source. Theoretically, with oxygen as a cathode reactant, the capacity of the battery is limited by the Li anode. The theoretical specific energy of the Li/oxygen cell is 13.0 kWh/kg, the highest for a metal/air battery. In addition to this very high specific energy, the Li/air battery offers a flat discharge voltage profile, environmental friendliness, and long storage life.

Disadvantages, which depend on the cell design, include reliance on the environment, drying out, low discharge rates, and safety concerns. In the past, lithium/air batteries solely used alkali aqueous electrolytes.[1] These alkali aqueous cells suffered from parasitic corrosion of the lithium metal anode and cell failures. Cell designs utilizing a nonaqueous electrolyte and/or a protected lithium anode alleviate the parasitic corrosion reactions of the Li anode that plagued past lithium/air batteries based on aqueous alkaline electrolytes. The nonaqueous electrolyte-based cell design also overcomes safety concerns of the Li/air system. Most recent advancements have focused on nonaqueous systems for both primary and secondary lithium/air chemistries.

## 18C.2  ELECTROCHEMISTRY

Lithium/air batteries consist of lithium (Li) anodes electrochemically coupled to atmospheric oxygen through an air cathode. Oxygen gas ($O_2$) introduced into the battery through the air cathode is essentially an unlimited cathode reactant source. Because of this, the air cathode is the most important component of the system. In the nonaqueous lithium/air system, the lithium metal reacts with $O_2$ to produce electricity according to the following reactions:

Discharge:

$$4Li \rightarrow 4Li^+ + 4e^- \quad \text{(lithium electrode) (anode)}$$

$$O_2 + 4e^- \rightarrow 2O^{2-} \quad \text{(gas electrode) (cathode)}$$

$$4Li + O_2 \rightarrow 2Li_2O \quad \text{(cell)} \quad E^0 = 2.91 \text{ V}$$

$$2Li + O_2 \rightarrow Li_2O_2 \quad \text{(cell)} \quad E^0 = 3.10 \text{ V}$$

The system and cells are capable of being recharged. Electricity is applied to the cell to convert the lithium oxide (stored in the cathode) back to lithium metal and oxygen gas. The reactions involved in recharging the cells are as follows:

Charge:

$$4Li^+ + 4e^- \rightarrow 4Li \quad \text{(lithium electrode) [cathode (reduction)]}$$

$$2O^{2-} \rightarrow O_2 + 4e^- \quad \text{(gas electrode) [anode (oxidation)]}$$

$$2Li_2O \rightarrow 4Li + O_2 \quad \text{(cell)}$$

$$Li_2O_2 \rightarrow 2Li + O_2 \quad \text{(cell)}$$

Theoretically, with unlimited oxygen, the capacity of the battery is limited by the Li anode. The theoretical specific energy of the Li/oxygen cell, as shown with the above reactions, is 13 kWh/kg (excluding oxygen). In addition, the Li/air battery offers lightweight components, rechargeability, and potentially low cost. This calculation does not include the need to store the reaction product(s) within the cell.

## 18C.3    CELL DESIGN AND COMPONENTS

### 18C.3.1    Anodes

The anodes for lithium/air batteries are typically lithium metal on a current collector. This type of electrode construction is simple and works well for many applications. A more sophisticated design incorporates a protective layer. This protective layer is often a ceramic or glass lithium-ion conductor. The first patented ceramic and glass protective layer for lithium metal anodes opened a new field within lithium/air cells (Fig. 18C.1). Examples of other work that followed included the use of LiSICON,[2] LiPON,[3] LATP,[3] and LiGC as solid electrolytes.[4] Many protected lithium anodes are constructed with a special interlayer between the lithium metal and the ion conductor.

One of the first protected lithium electrodes used LMP (a solid ion conducting layer), a special interlayer, and lithium metal (see Figs. 18C.2 and 18C.3).[5] This protected lithium anode is stable in both aqueous and nonaqueous systems and has been used successfully in a variety of cell types. The protected lithium anode was built into a lithium/air cell with aqueous electrolyte and successfully discharged at three different current densities (Fig. 18C.4). Many of these protected lithium anodes can be used for the lithium/air chemistry and also lithium/water and even lithium-ion systems.

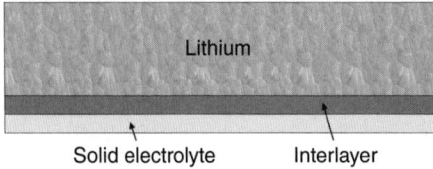

**FIGURE 18C.1**    Cross-sectional illustration of protected lithium electrode: lithium/interlayer/solid electrolyte: U.S. Patents 7,282,295, 7,282,296, 7,282,302, 7,390,591, and 7,491,458. (*Courtesy of PolyPlus Battery Company.*)

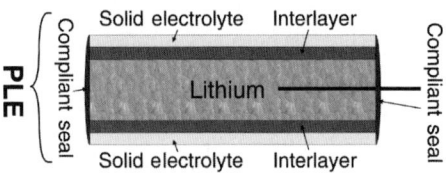

**FIGURE 18C.2**    Cross-sectional view of protected lithium electrode having a compliant seal; U.S. Patent 8,404,388. (*Courtesy of PolyPlus Battery Company.*)

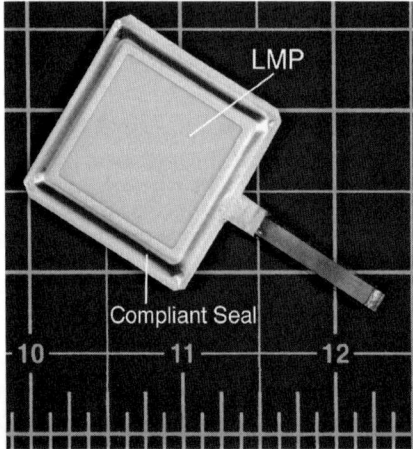

**FIGURE 18C.3**    Fully functional protected lithium electrode; lithium electrode is stable to a broad range of protic and aprotic solvents including water; 2400 Wh/kg with a 2.8 V cathode, patent pending. (*Courtesy of PolyPlus Battery Company.*)

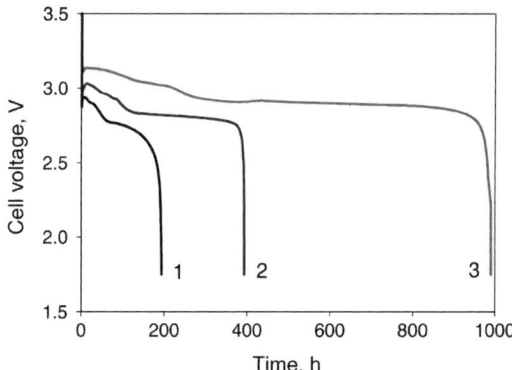

**FIGURE 18C.4**   The protected lithium anode was built into the lithium/air cells and discharged in aqueous electrolyte at three current densities: (1) 1.0 mA/cm², (2) 0.5 mA/cm², and (3) 0.2 mA/cm². The specific energy of these cells is about 800 Wh/kg (includes all components except the external battery case). (*Courtesy of PolyPlus Battery Company.*)

## 18C.3.2   Electrolytes and Separators

The majority of research on electrolytes for lithium/air batteries has advanced the use of nonaqueous electrolytes. This was first done with lithium salts in organic carbonates and conductive polymer membranes.[6] Many electrolytes used for Li-ion chemistry were transitioned to lithium/air chemistry. Salts and organic carbonates commonly used in electrolytes are listed in Table 18C.1. Polymer electrolytes and solid-state electrolytes have also been used. Each presents its own challenges and the difficulty of matching it with the air cathode. Electrolyte work has also been done on aqueous systems[5,7,8] and ionic liquids.[9,10] Separators used for lithium/air cells are typically polyolefins such as Setela® and Celgard®. Glass fibers and solid ion conducting membranes are also used.

**TABLE 18C.1**   The Most Commonly Used Salts and Solvents for Lithium-Air Cells

| Salts | Solvents |
|---|---|
| $LiPF_6$ | Propylene carbonate (PC) |
| $LiBF_4$ | 1,2-Dimethoxyethane (DME) |
| $LiCF_3SO_3$ | Ethylene carbonate (EC) |
| $LiN(SO_2CF_3)_2$ | Diethyl carbonate (DEC) |
| $LiClO_4$ | Dimethylcarbonate (DMC) |

## 18C.3.3   Cathodes

The limiting factor in nearly all metal/air batteries is the air cathode (also known as an oxygen cathode).[6] The slow rate of the chemical reaction is the diffusion of oxygen into the cell through the air cathode. The performance of the lithium/air battery has also been limited by the air cathode.

A typical cathode preparation entails depositing carbon, binder, and catalyst on a metal current collector. This process can be done by coating, impregnating, or pressing. Additionally, a substrate may be used to increase surface area, or an atmospheric membrane could be placed on top of the cathode to protect it from the environment. This produces an air cathode suitable for laboratory testing or actual field use.

A commercially available air cathode that is carbon based has a double-sided electrode.[11-13] This electrode consists of two carbon layers sandwiched around a current collector and then covered with a polytetrafluoroethylene (PTFE) film. The carbon layers contain the high surface carbon and the metal catalysts. Catalysts incorporated into the carbon electrode enhance the oxygen reduction kinetics and increase the specific capacity of the cathode. The importance of catalysts can be seen in Fig. 18C.5. Several cathodes were constructed with different metal catalysts such as silver, platinum, and ruthenium. Oxides such as manganese, cobalt, and a manganese-cobalt mixture also provide catalytic activity, as seen in Fig. 18C.5. The PTFE film acts as an atmospheric water barrier. Preventing water from entering the battery increases safety and performance.

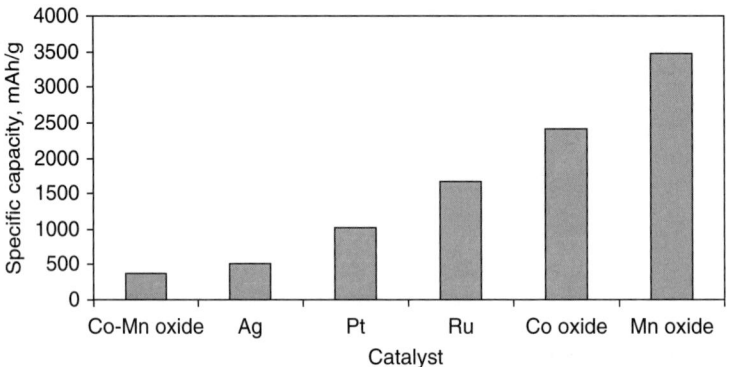

**FIGURE 18C.5**   The specific capacities of lithium/air pouch cells utilizing various catalysts. Specific capacity is based on grams of carbon in the air cathode, which is customary. The discharge current was 1.0 mA and corresponds to 0.1 mA/cm². (*Courtesy of Yardney Technical Products, Inc.*)

## 18C.4   APPLICATIONS AND PERFORMANCE

### 18C.4.1   Initial Test Cells

The first nonaqueous lithium/air battery using polymer electrolytes was reported by Abraham and Jiang.[6] It comprised a lithium-ion conductive polymer electrolyte membrane between a thin Li metal foil anode and a thin carbon composite on which oxygen, the electro-active cathode material, is reduced during discharge. Figure 18C.6 shows the cell structure, which is encapsulated in a metallized plastic envelope with pores on the cathode surface, which are covered by tape prior to activation. The cathode is composed of 20 wt% acetylene black (or graphite powder) and 80 wt% polymer electrolyte catalyzed with cobalt phthalocyanine, in some cases, and pressed on a Ni or Al screen current collector. The electrolyte was composed of 12% polyacrylonitrile (PAN), 40% ethylene carbonate, 40% propylene carbonate, and 8% $LiPF_6$, all by weight, formed into a film 75 to 100 μm thick. The lithium electrode was 50 μm thick.

Figure 18C.7 shows an intermittent discharge curve for the Li/PAN-based electrolyte/$O_2$ battery at a current density of 0.1 mA/cm, using an acetylene black cathode in an atmosphere of flowing oxygen. The capacity was found to be proportional to the carbon weight. This battery exhibited an open circuit voltage (OCV) of 2.85 V prior to discharge. The OCV remained steady during intermittent discharge, indicating a two-phase equilibrium at the electrode surface. Raman spectra of the reaction product absorbed on the electrode surface showed it was $Li_2O_2$ and that the following reaction was occurring during discharge:

$$2Li + O_2 \rightarrow Li_2O_2 \qquad E^0 = 3.10 \text{ V}$$

It was also determined that the absorbed lithium peroxide on a catalyzed electrode could be reoxidized to oxygen. Figure 18C.8 shows the first discharge and recharge, followed by the second discharge of one battery.

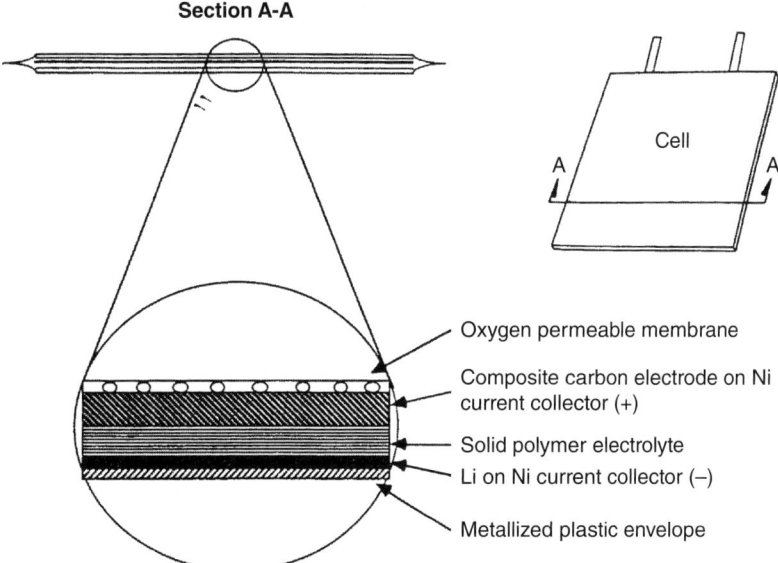

**FIGURE 18C.6**  Lithium/oxygen battery with solid polymer electrolyte in metallized plastic envelope.

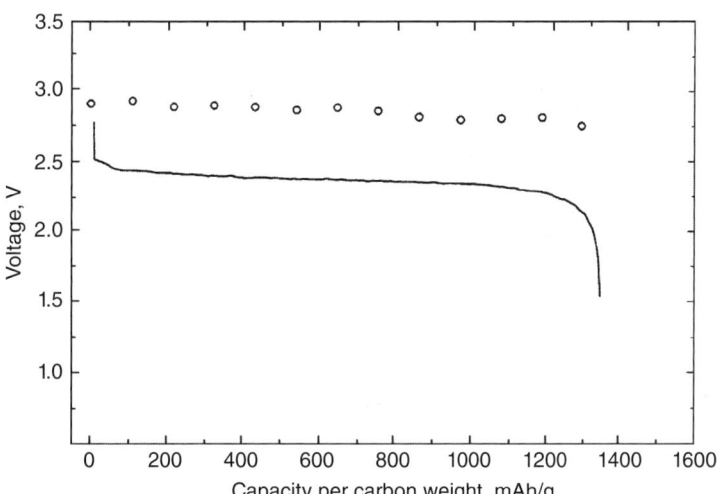

**FIGURE 18C.7**  The intermittent discharge curve and the open-circuit voltages of a Li/PAN-based polymer electrolyte/oxygen battery at a current density of 0.1 mA/cm² at room temperature in an atmosphere of oxygen. The cathode contained acetylene black carbon. The cell was discharged in 1.5 h increments with an open-circuit stand of about 15 min between discharges. Open circles: OCV; solid line: load voltage.

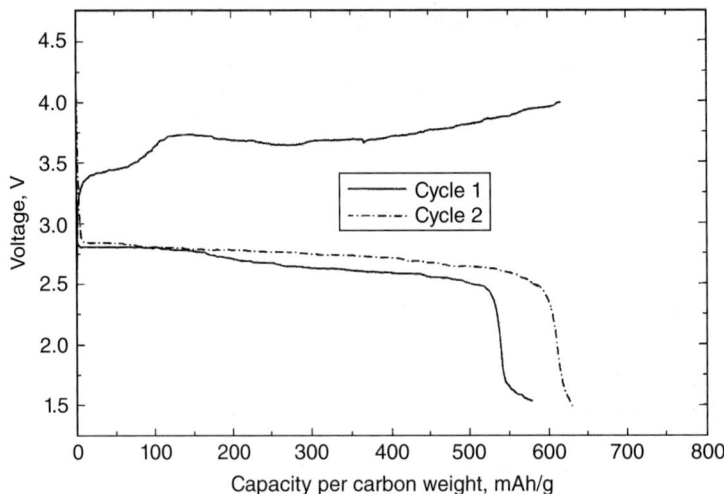

**FIGURE 18C.8** Cycling data for a Li/PAN-based polymer electrolyte/oxygen battery at room temperature in an atmosphere of oxygen. The cathode contained 20 w/o catalyzed Chevron carbon black and 80 w/o polymer electrolyte. The battery was discharged at 0.1 mA/cm² and charged at 0.05 mA/cm².

Although this system is of considerable technological interest, its active life was limited by the diffusion of oxygen through the PAN electrolyte, where it reacted chemically with lithium. This started the interest in the nonaqueous lithium/air system and in other cell designs.

## 18C.4.2 POUCH CELL TESTS

Lithium/air cells have been built using several types of cell construction techniques, such as using Swagelok® fittings,[14] pouch cell,[15] coin cell,[16] and plastic cases.[12,15] A common construction is the pouch cell design due to its ease of fabrication and design adaptability. These cells are suitable for testing in different environments. The cell construction involves layering the cell components and sealing them in a plastic case resembling a packet. The lithium metal anode, separator, electrolyte, and carbon air cathodes are sealed inside the metallized plastic packaging material. A pouch cell used for testing is shown in Fig. 18C.9.

The anode is composed of a lithium metal foil, pressed into a nickel current collector with a nickel tab. The area of the anode is slightly larger than the 10 cm². The separator used was Setela, which is a microporous polyolefin film. The liquid electrolyte used was 1 M LiPF$_6$ in 1:1:1 EC/DEC/DMC. (EC is ethylene carbonate, DEC is diethyl carbonate, and DMC is dimethyl carbonate.) The air cathode is a carbon composite made by combining carbon, a metal catalyst, and a binder, deposited on a metal current collector. The binder used was PTFE. The air cathode structure is a layered composite with an increased capacity. A thin PTFE film between the air cathode and the atmosphere provides hydrophobicity to the cathode to repel atmospheric water and creates channels for oxygen diffusion.

The cell was run in an oxygen atmosphere, at room temperature, and about 1 atmosphere of pressure. The lithium/air pouch cell had a capacity of 91 mAh with a relatively flat discharge profile (Fig. 18C.10). Discharging further to 1.5 V resulted in 100 mAh total capacity. The corresponding energy yield is 246 mWh. With 0.028 g of carbon impregnated into the air cathode current collector, the specific capacity is 3137 mAh/g of carbon. The goal for this lithium/air cell was 3000 Wh/kg for a full cell.[a]

---

[a]The value of 3000 Wh/kg is the weight of the cell components and its packaging.

**FIGURE 18C.9**   A picture of a lithium/air pouch. The "window" toward the top is the air cathode that allows oxygen into the cell. The cell is about 3 in. on each side. (*Courtesy of Yardney Technical Products, Inc.*)

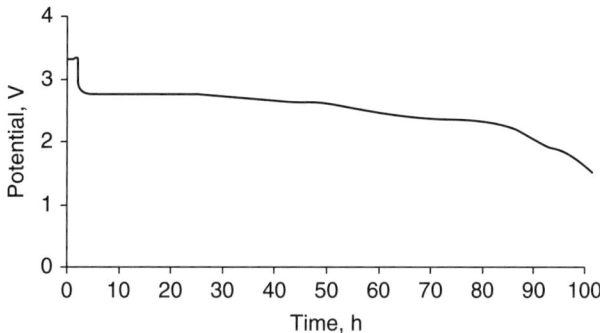

**FIGURE 18C.10**   Discharge profile of Mn oxide catalyzed cathode in a lithium/air pouch cell at 0.1 mA/cm$^2$ shown with an initial open-circuit rest of 2 h. (*Courtesy of Yardney Technical Products, Inc.*)

### 18C.4.3   Rechargeability

Rechargeable lithium/air cells have recently attracted much attention for use in the electric vehicle market. To date, a rechargeable cell with long life and high efficiency has been elusive.[17] Most of the work has focused on Li metal anodes, nonaqueous electrolytes, and air cathodes with and without catalysts. The electrochemistry of the actual cell in use may vary due to the specific materials used in the electrodes and electrolyte. Many challenges still exist for the rechargeable lithium/air battery to replace existing batteries used in the electric vehicle market.

## *REFERENCES*

1. E. L. Littauer and K. C. Tsai, "Anodic Behavior of Lithium in Aqueous Electrolytes, ii. Mechanical Passivation," *J. Electrochem. Soc.* **123:**964 (1976); "Corrosion of Lithium in Aqueous Electrolytes," ibid. **124:**850 (1977); "Anodic Behavior of Lithium in Aqueous Electrolytes, iii. Influence of Flow Velocity, Contact Pressure and Concentration," ibid. **125:**845 (1978).

2. D. L. Foster, J. R. Read, M. Shichtman, S. Balagopal, J. Watkins, and J. Gordon, "High Energy Lithium-Air Batteries for Soldier Power," http://oai.dtic.mil/oai/oai?verb=getRecord&metadataPrefix=html&identifier=ADA481576. Accessed October 2009. Paper from unspecified conference.

3. N. Imanishi, S. Hasegawa, T. Zhang, A. Hirano, Y. Takeda, and O. Yamamoto, "Lithium Anode for Lithium-Air Secondary Batteries," *J. Power Sources* **185:**1392 (2008).

4. I. Kowalczk, J. Read, and M. Salomon, "Li-Air Batteries: A Classic Example of Limitations Owing to Solubilities," *Pure Appl. Chem.* **79**(5):851 (2007).

5. S. J. Visco, E. Nimon, B. Katz, L. D. Jonghe, and M.-Y. Chu, "The Development of High Energy Density Lithium/Air and Lithium/Water Batteries with No Self-Discharge," *210th Meeting of the Electrochemical Society*, Cancun, Mexico, 2006.

6. K. M. Abraham and Z. Jiang, *J. Electrochem. Soc.* **143:**1 (1996).

7. T. Zhang, N. Imanishi, S. Hasegawa, A. Hirano, J. Xie, Y. Takeda, O. Yamamoto et al., "Water-Stable Lithium Anode with the Three-Layer Construction for Aqueous Lithium-Air Secondary Batteries," *Electrochem. Solid State Lett.* **12**(7):A132 (2009).

8. M. B. Marx and J. A. Read, "Performance of Carbon/Polyetraflouroethylene (PTFE) Air Cathodes from pH 0 to 14 for Li-Air Batteries," Army Research Laboratory Summary Report ARL-TR-4334 (2007).

9. H. Ye, J. Huang, J. J. Xu, A. Khalfan, and S. G. Greenbaum, "Li Ion Conducting Polymer Gel Electrolytes Based on Ionic Liquid/PVDF-HFP Blends," *J. Electrochem. Soc.* **154**(11):A1048 (2007).

10. T. Kuboki, T. Okuyama, T. Ohsaki, and N. Takami, "Lithium/Air Batteries Using Hydrophobic Room Temperature Ionic Liquid Electrolyte," *J. Power Sources* **146:**766 (2005).

11. A. Dobley, R. Rodriguez, and K. M. Abraham, "High Capacity Cathodes for Lithium-Air Batteries," *Electrochemical Society Conference*, Honolulu, HI, Oct. 2004.

12. A. Dobley, C. Morein, and R. Roark, "Lithium Air Cells with High Capacity Cathodes," *Electrochemical Society 210th Meeting Proceedings*, Cancun, Mexico, 2006.

13. A. Dobley, J. DiCarlo, and K. M. Abraham, "Non-aqueous Lithium-Air Batteries with an Advanced Cathode Structure," *41st Power Sources Conference,* Philadelphia, PA, June 2004.

14. S. D. Beattie, D. M. Manolescu, and S. L. Blair, "High-Capacity Lithium-Air Cathodes," *J. Electrochem. Soc.* **156**(1):A44 (2009).

15. A. Dobley, C. Morein, and R. Roark, "Design Options for Emerging Lithium-Air Technology," *212th Electrochemical Society Conference*, Washington, DC, October 2007.

16. J. Ostroha, "Lithium-Air System Development," *11th Electrochemical Power Sources R&D Symposium*, Baltimore, MD, July 2009.

17. A. C. Luntz and B. D. McCloskey, "Nonaqueous Li-Air Batteries: A Status Report," *Chem. Rev.* **114:**11721–11750 (2014).

# CHAPTER 19
# FUEL CELLS

## H. Frank Gibbard, Zhigang Qi
## (Emeritus Contributors: David Linden, Arthur Kaufman)

## 19.1  GENERAL OVERVIEW

### 19.1.1  Introduction and Background

Fuel cells have been of interest for over 170 years as a potentially more efficient and less polluting means for converting hydrogen and carbonaceous or fossil fuels to electricity compared to conventional heat engines. As these fuels react in an electrochemical cell, the term "fuel cell" has become popular to describe these devices. A significant application of the fuel cell has been the use of the hydrogen/oxygen fuel cell by NASA, using cryogenic fuels, in space vehicles for over 50 years, including space shuttles. Terrestrial applications such as portable power, back-up power, utility power, and motive power have been ongoing for some time, and significant progress has been achieved. By October 2017, about 200,000 CHP (combined heat and power [CHP]) systems, 20,000 fuel cell forklifts, 10,000 fuel cell backup power systems, and 20,000 DMFC (direct methanol fuel cell) systems have been deployed worldwide.

Interest in small fuel cells has arisen for dispersed or onsite electric generators, remote devices, and other such applications in the sub-kilowatt power range, replacing small engine generators and larger-sized batteries. Great progress has been made with fuel cell systems in the sizes above 50 W, especially for extended long-term service. However, the development of yet lower power-level portable fuel cells (which can be "recharged," e.g., by replacing a small container of fuel) that are competitive in size and performance with batteries remains a challenge.

A fuel cell is a galvanic device that continuously converts the chemical energy of a fuel (and an oxidant) to electrical energy directly.[1–3] A simple fuel cell is illustrated in Fig. 19.1. Two catalyzed electrodes are immersed in an electrolyte ($H_2SO_4$ acid in this illustration) and separated by a gas barrier. The fuel, in this case hydrogen, is bubbled across the surface of one electrode while the oxidant, in this case oxygen from ambient air, is bubbled across the other electrode. When the electrodes are electrically connected through an external load, the following events occur:

1. The hydrogen dissociates on the catalytic surface of the fuel electrode, forming hydrogen ions and electrons.

2. The hydrogen ions migrate through the electrolyte (and a gas barrier) to the catalytic surface of the oxygen electrode.

3. Simultaneously, the electrons move through the external circuit, doing useful work, to the same catalytic surface.

4. The oxygen, hydrogen ions, and electrons combine on the oxygen electrode's catalytic surface to form water.

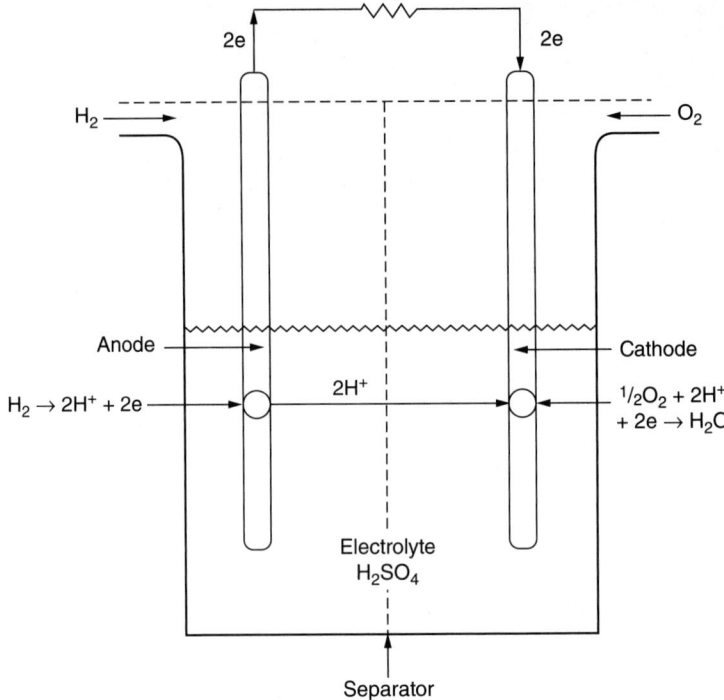

**FIGURE 19.1**   Schematic illustration of a fuel cell.

The net reaction, $H_2 + 0.5O_2 = H_2O$, is that hydrogen and oxygen produce water and electrical energy (and some heat). At 25°C, the Gibbs free energy of the reaction $\Delta G°$ is −237.1 kJ/mol, and thus the thermodynamic reversible voltage $E°$ is 1.23 V for a hydrogen/oxygen fuel cell ($E° = -\Delta G°/2F$, where $F$ is the Faraday constant, 96,485 C/mol).

Like batteries, fuel cells convert energy electrochemically and are not subject to the Carnot cycle limitation of thermal engines, thus offering the potential for highly efficient conversion. The essential difference between a fuel cell and a battery is the manner of supplying the source of chemical energy. In a fuel cell, the fuel and the oxidant are supplied continuously from an external source when power is desired. The fuel cell can produce electrical energy as long as the reactants are fed to the electrodes. In a battery, the fuel and oxidant are an integral part of the device. The battery will cease to produce electrical energy when the limiting reactant is consumed. The battery must then be replaced or recharged. The electrode materials and the catalysts in the fuel cell undergo no chemical or physical changes during the cell reaction, but only serve, respectively, for the exchange of electrons between the reactants and the electrodes, and for fostering the electroreduction or electrooxidation reactions. In contrast, the active materials in batteries are contained in the electrodes, and chemical changes within the electrodes generally cause deformation during cycling, leading to deterioration and ultimately to failure of the battery.

The important components of a single cell are shown in Fig. 19.2 and listed below along with their functions.

1. The *anode* (fuel electrode) provides a common interface for the fuel and electrolyte, catalyzes the fuel oxidation reaction, and conducts electrons from the reaction site to the external circuit (or to a current collector that, in turn, conducts the electrons to the external circuit).

2. The *cathode* (oxidant electrode) provides a common interface for the oxidant and the electrolyte, catalyzes the oxidant reduction reaction, and conducts electrons from the external circuit to the oxidant electrode reaction site.

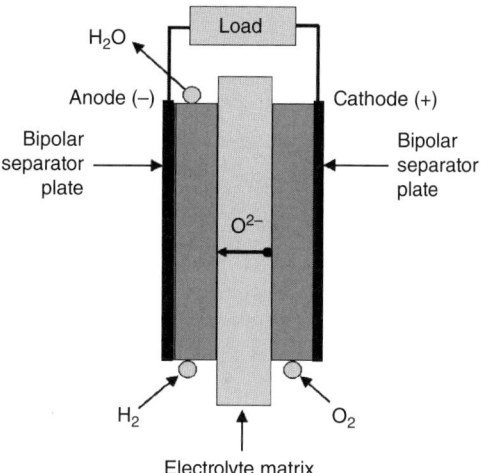

**FIGURE 19.2**   Schematic illustration of a SOFC.

3.  The *electrolyte* transports the ionic species involved in the reactions while preventing the conduction of electrons. In addition, in practical cells, the role of gas separation is usually provided by the electrolyte system. In systems that employ concentrated aqueous electrolytes, such as phosphoric acid in the phosphoric acid fuel cell (PAFC), or potassium hydroxide in the alkaline fuel cell, this is accomplished by retaining the electrolyte in the pores of a solid matrix disposed between the electrodes. Currently, the most popular electrolyte in use for portable ambient temperature fuel cells is a polymer electrolyte membrane such as Nafion®.

### 19.1.2   Classification

In one aspect, fuel cells can be classified into two categories[4]:

1.  Direct fuel systems, where fuels, such as hydrogen, methanol, formic acid, and hydrazine, can react directly in the fuel cell.
2.  Indirect fuel systems in which the fuel, such as natural gas, propane, or other fossil fuels, is first converted by reforming to a hydrogen-rich gas mixture that is subsequently oxidized on the fuel cell anode.

## *19.2   ELECTROCHEMISTRY OF VARIOUS FUEL CELL TECHNOLOGIES*

### 19.2.1   Types of Fuel Cells

Fuel cells are typically classified into five categories based on the type of electrolyte used.[3,5]

***Solid Oxide Fuel Cell (SOFC).***   These cells use a solid oxygen-ion-conducting metal oxide electrolyte. As shown in Fig. 19.2, the ionic species, $O^{2-}$, is formed at the cathode and migrates through the electrolyte to the anode. Electrodes are made of nonprecious metals. They operate at about 700~1000°C, with an electrical efficiency of up to 60% and are slow to start up, but once running, provide high-grade waste heat, which can be used in turbines for further electricity generation, and additionally to heat buildings using the remaining heat.

They may find application in small portable devices and in large industrial applications. The anode, cathode, and overall reactions are as follows:

$$
\begin{aligned}
\textit{Anode:} &\quad H_2 + O^{2-} = H_2O + 2e^- \\
\textit{Cathode:} &\quad 0.5O_2 + 2e^- = O^{2-} \\
\textit{Overall:} &\quad H_2 + 0.5O_2 = H_2O
\end{aligned}
$$

**Molten Carbonate Fuel Cell (MCFC).**    These cells use a mixed alkali-carbonate molten-salt electrolyte and operate at about 650°C. The ionic species, $CO_3^{2-}$, is formed at the cathode and migrates through the electrolyte to the anode. Their electrodes are nonprecious-metal based. They are used in continuous operating systems from 100 kW to MW. The anode, cathode, and overall reactions are as follows:

$$
\begin{aligned}
\textit{Anode:} &\quad H_2 + CO_3^{2-} = H_2O + CO_2 + 2e^- \\
\textit{Cathode:} &\quad 0.5O_2 + CO_2 + 2e^- = CO_3^{2-} \\
\textit{Overall:} &\quad H_2 + 0.5O_2 + CO_2 \,(\text{cathode}) = H_2O + CO_2 \,(\text{anode})
\end{aligned}
$$

**Phosphoric Acid Fuel Cell (PAFC).**    This type has been the most commonly used fuel cell for stationary commercial sites such as hospitals, hotels, and office buildings. The electrolyte is concentrated phosphoric acid imbedded in a solid matrix such as SiC or polybenzimidazole (PBI) membrane. The ionic species $H^+$ is formed at the anode and migrates through the electrolyte to the cathode. Their electrodes are typically resin-bonded, carbon-supported, platinum-based catalyst layers on wet-proofed carbon-fiber substrates. The fuel cell operates at about 160~200°C and is highly efficient in CHP generation with up to 85% total efficiency (40% electricity and 45% heat). The anode, cathode, and overall reactions are as follows:

$$
\begin{aligned}
\textit{Anode:} &\quad H_2 = 2H^+ + 2e^- \\
\textit{Cathode:} &\quad 0.5O_2 + 2H^+ + 2e^- = H_2O \\
\textit{Overall:} &\quad H_2 + 0.5O_2 = H_2O
\end{aligned}
$$

**Alkaline Fuel Cell (AFC).**    These were used by NASA on the manned space missions, and operate well at above 70°C. They use circulated aqueous potassium hydroxide solution as the alkaline electrolyte. The ionic species $OH^-$ is formed at the cathode and migrates through the electrolyte to the anode. Their electrodes are typically nonprecious-metal mesh structures, although precious metals may be used at the anodes. Disadvantages of this system are that the electrolyte absorbs $CO_2$ from the atmosphere (carbonates) and that the fuels and oxidants must not contain or produce carbon dioxide. Recent research focuses on developing hydroxide exchange solid membranes to minimize the impact of $CO_2$. The anode, cathode, and overall reactions are as follows:

$$
\begin{aligned}
\textit{Anode:} &\quad H_2 + 2OH^- = 2H_2O + 2e^- \\
\textit{Cathode:} &\quad 0.5O_2 + 2e^- + H_2O = 2OH^- \\
\textit{Overall:} &\quad H_2 + 0.5O_2 = H_2O
\end{aligned}
$$

**Proton Exchange Membrane Fuel Cell (PEMFC).**    These cells (Fig. 19.3) use a perfluorinated sulfonic acid ion-exchange membrane, e.g., DuPont's Nafion, at temperatures generally ranging from somewhat above ambient to about 80°C. The ionic species is $H^+$, which is formed at the anode and migrates through the electrolyte to the cathode. The electrodes are typically ionomer-bonded, carbon-supported, platinum-based catalyst layers on wet-proofed carbon fiber substrates. PEMFCs are especially notable for their rapid startup and are being developed for use in transportation, small stationary and portable applications. The reactions at the anode, the cathode, and overall are the same as those in PAFCs.

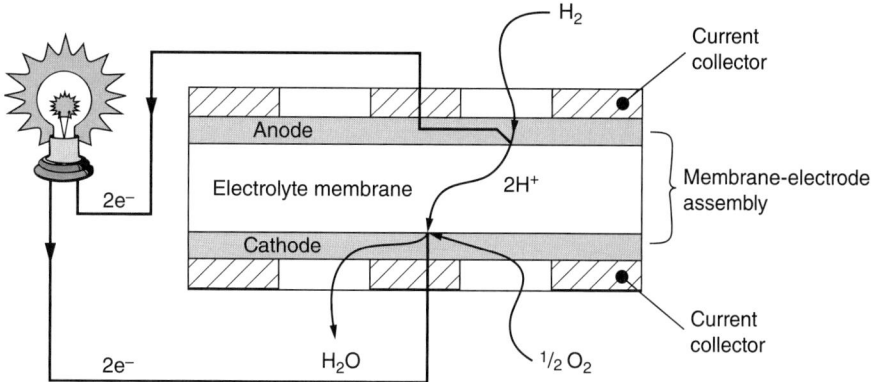

**FIGURE 19.3**    Schematic illustration of a PEMFC.

DMFCs belong to the PEMFC category because DMFCs typically also use proton-exchange membranes (Fig. 19.4). The ionic species $H^+$ is formed at the anode and migrates through the electrolyte to the cathode. DMFCs offer the major advantage of utilizing a logistically attractive liquid fuel without the burden of external processing, and have received a great deal of attention and are in low-volume commercial production for sub-kilowatt applications. The anode, cathode, and overall reactions are as follows:

$$Anode: \quad CH_3OH + H_2O = CO_2 + 6H^+ + 6e^-$$

$$Cathode: \quad 1.5O_2 + 6H^+ + 6e^- = 3H_2O$$

$$Overall: \quad CH_3OH + 1.5O_2 = CO_2 + 2H_2O$$

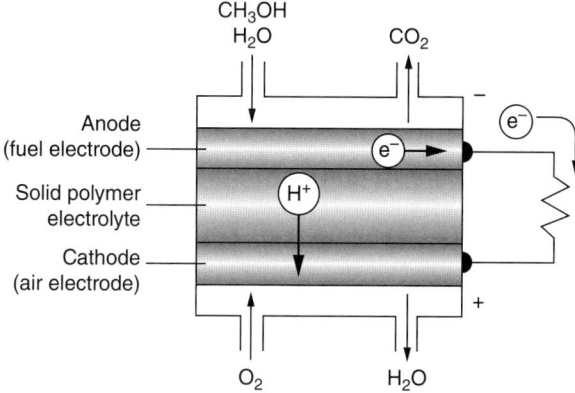

**FIGURE 19.4**    Schematic illustration of a DMFC.

## 19.2.2    Electrochemical Characteristics

Ideally, a single $H_2/O_2$ fuel cell could produce 1.23 V at standard conditions, as shown by the dotted line in Fig. 19.5. In practice, the useful voltage outputs are less than the ideal and decrease with increasing discharge rate (current density) as shown by the curve in Fig. 19.5. An equation describing this curve can be written as:

$$E = E^\circ - \eta_{c,act} - \eta_{a,act} - \eta_{c,conc} - \eta_{a,conc} - iR_{int}$$

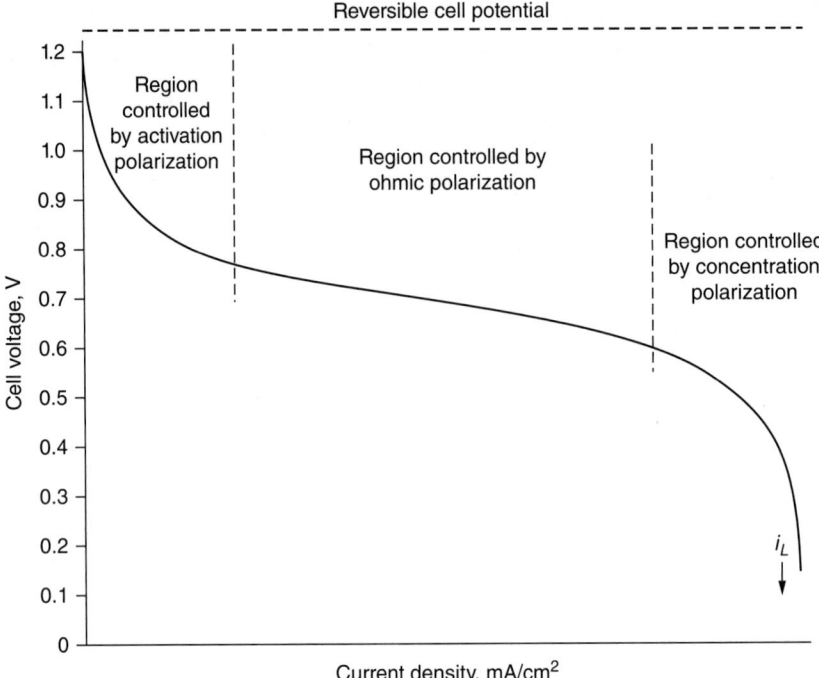

**FIGURE 19.5**    Fuel cell polarization curve.

where  $\eta_{c,act}$ and $\eta_{a,act}$ = activation overpotentials at the cathode and anode, V
$\eta_{c,conc}$ and $\eta_{a,conc}$ = concentration overpotentials at the cathode and anode, V
$i$ = current density, A/cm$^2$
$R_{int}$ = the internal resistance of the cell, $\Omega$ cm$^2$
$E°$ = reversible cell potential, V

The losses or reductions in voltage from the ideal are referred to as "polarization" or "overpotential," including

1. Activation polarization that represents energy losses associated with the electrode reactions. Most chemical reactions involve an energy barrier that must be overcome for the reactions to proceed. For electrochemical reactions, the activation energy lost in overcoming this barrier takes the form

$$\eta_{act} = a + b \ln i$$

where $a$, $b$ = constants, and $\eta_{act}$ = activation polarization = $\eta_{a,act} + \eta_{c,act}$, V.

2. Ohmic polarization that represents the summation of all the ohmic losses within the cell, including electronic impedances through electrodes, gas-diffusion media, bipolar plates, current collectors, and contacts, and ionic impedance through the electrolyte and the electrodes. These losses follow Ohm's law

$$\eta_{ohm} = iR_{int}$$

3. Concentration polarization that represents the energy losses associated with mass transport effects. For instance, the performance of an electrode reaction may be inhibited by the inability for reactants to diffuse to or products to diffuse away from the reaction site. In fact, a limiting current density $i_L$ will be reached when the reactant arriving at the catalyst surface is completely consumed, and then the current is completely limited by the reactant diffusion processes. Concentration polarization can be represented by

$$\eta_{conc} = (RT/nF) \ln (1 - i/i_L)$$

where $\eta_{conc}$ = concentration polarization = $\eta_{c,conc} + \eta_{a,conc}$
  $R$ = gas constant = 8.314, J mol$^{-1}$ K$^{-1}$
  $T$ = temperature, K
  $n$ = number of electrons in cell reaction
  F = Faraday constant = 96,485 C/mol
  $i_L$ = limiting current, A/cm$^2$

These polarization effects in conjunction with the electrical efficiency considerations usually result in a single-cell operating voltage between 0.6 and 0.8 V. Fuel cell performance can be increased by increasing cell temperature and reactant partial pressure. However, for small or portable fuel cells, operation near ambient conditions is usually a requirement, particularly when the fuel cell is to be used as a replacement for batteries.

## 19.3  FUEL CELL SYSTEM DESIGN AND CONSTRUCTION

### 19.3.1  Cell Stack Design

Since the voltage generated in the discharge of a single cell is typically in the 0.6 to 0.8 V range (about 0.4 V for DMFCs), multiple cells are connected in series to obtain practical voltage levels. These are most commonly arrayed as a bipolar stack, as illustrated for a PEMFC in Fig. 19.6. The electrochemically active cell elements shown schematically in Figs. 19.2 to 19.4 are commonly referred to as membrane-electrode assemblies (MEAs). The MEAs are interleaved with bipolar plates that have multiple functions: (1) conduction of electronic current from cell to cell, (2) dispersion of the active hydrogen and oxygen (air) gases through flow channels on the two opposing surfaces, (3) prevention of mixing of these gases, and (4) providing a means for removing heat that is generated during operation of the fuel cell. Since the cells are connected in series, the voltage of the stack can be raised by increasing the number of cells.

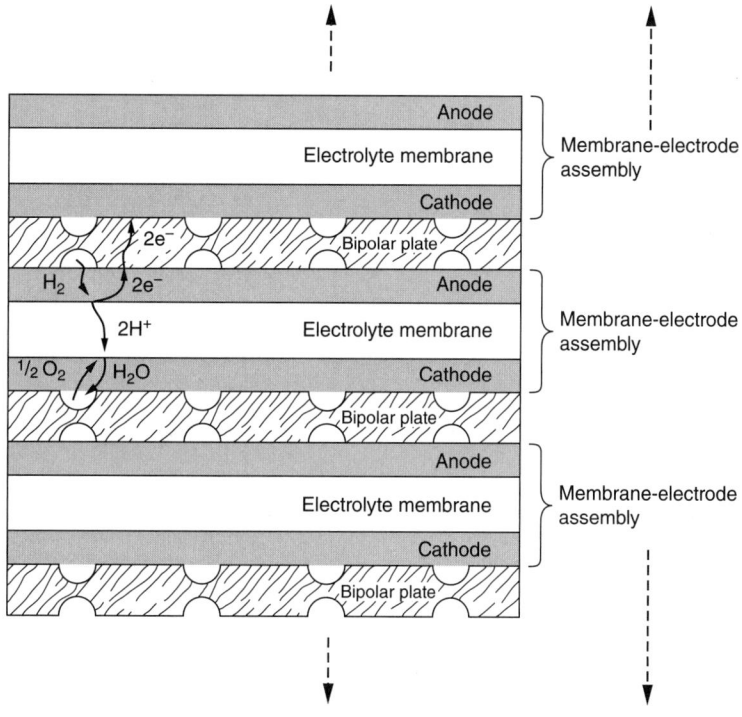

**FIGURE 19.6**   Illustration of a PEMFC stack.

An alternative to the bipolar stacking arrangement is the so-called planar monopolar approach, whereby the individual MEAs are nominally arranged in a side-by-side configuration such that the electronic current is edge-collected at one electrode and passed on to the opposite electrode of the adjacent MEA.[6] The voltage of the array is increased by increasing the number of MEAs in the line, and possibly by using multiple series-connected arrays. The current is increased by increasing the cell active area and/or by connecting arrays in parallel. This approach eliminates the need for bipolar plates, thereby providing the opportunity for significant volume (especially height) and weight reduction. In order to be effective, the design must minimize the tendencies toward nonuniform current distribution and substantial ionic and electronic losses associated with in-plane conduction within the electrolyte-membranes and the electrodes (and interconnects), respectively.

In addition to the planar single-cell geometry described to this point, the adaptation of SOFC technology for small power sources has focused in part on the use of cells in the form of small-diameter tubes. Each tube in an array has an anode, cathode, electrolyte, and provisions for fuel and air delivery. Cells are arranged in series-parallel combination, so as to obtain the desired voltage-current characteristics. This design is analogous to the commonly used method of constructing batteries with cylindrical cells.

## 19.3.2 Total Fuel Cell System Design

As a power generator, a practical fuel cell system is a power plant consisting of the eight subsystems, schematically shown in Fig. 19.7 and described below.

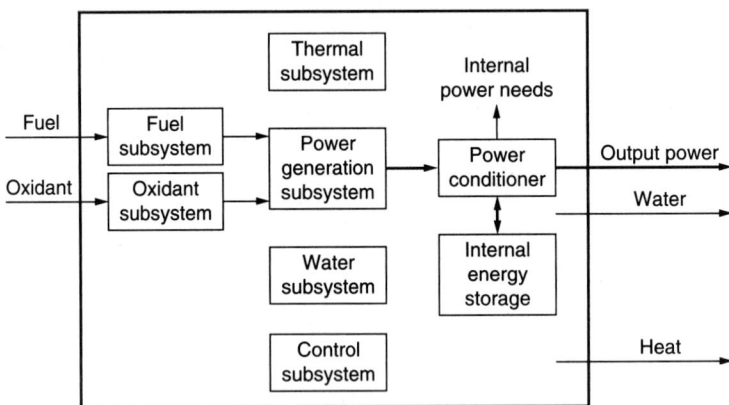

**FIGURE 19.7**  Schematic illustration of a fuel cell system.

### Subsystems.

*Power Generation Subsystem.*    Power generation subsystem consists of one or more fuel cell stacks, usually connected in series electrically to produce a power output ranging from a few to several hundred volts. This section converts the fuel and the oxidant into DC power.

*Fuel Subsystem.*    Fuel subsystem manages the fuel supply to the power generation subsystem. This subsystem can range from simple flow controls to a much more complex fuel-processing facility. Fuel cell systems utilizing virtually pure hydrogen typically provide this fuel to the anodes at a modest gauge pressure (generally in the 10 to 100 kPa range) and dead-ended, allowing the anodes to consume the amount of hydrogen required to sustain the electrochemical reaction at any given rate. A momentary "purge" of the exit port is implemented to allow accumulated impurities and water to be discharged from the anode compartment at selected time intervals.

If the power source uses processed fuel to generate a hydrogen-rich gas, the $H_2$ is utilized at the anodes to the extent practical to balance fuel consumption, processor temperature, and cell performance. Depending on specific system characteristics, this extent could be in the 70% to 90% range. The unused fuel is combusted in the fuel processor section to generate heat needed therein.

PEMFCs fueled directly by methanol (DMFCs) are typically fed by a solution of methanol in water at the anodes (although the methanol is generally stored in neat form and enriches a circulating methanol-water solution as needed). Carbon dioxide formed at the anodes must be separated from the circulating stream. With a proper design, neat methanol can also be sent to the fuel cell anode without any prior dilution with water as achieved by former MTI Micro Fuel Cells, Inc. Air is not humidified either in the MTI design. The water needs for methanol reaction and for humidifying the MEA solely rely on the water produced by the fuel cell itself. Such a DMFC system is significantly simplified, and the power and energy densities of the MTI design are much higher than those of traditional DMFC systems.[3]

*Oxidant Subsystem.*    Oxidant subsystem manages the oxidant (typically air) supply to the power generation subsystem. Depending on size, type, and applications of the fuel cell power plant, it can be a fan, a blower, or a compressor. Diffused-air (e.g., air-breathing) units are generally limited in their applicability because of air supply rate issues and the impact on size and weight; also, the requisite openness of the air compartments in such devices tends to render them vulnerable to atmospheric conditions. Hence, diffused-air fuel cells are usually practical only in certain particularly low-power applications, up to perhaps 25 W, where extreme simplicity of operation is essential.[7]

Forced-air fuel cells are practical over the entire output range of power sources. The reactant air for this type of system is generally delivered to the fuel cell cathodes at whatever pressure is necessary to overcome the pressure drop through that section and associated plumbing. This is typically a small fraction of atmospheric pressure, in the range of 1 to 20 kPa, depending on stack design characteristics. The air-moving devices are usually small fans, pumps (such as rotary-vane, diaphragm, or, less commonly, piezoelectric types), blowers, or compressors. The cathodes' utilization rate of oxygen in the reactant air will vary in accordance with operating conditions, especially the ambient temperature for air-cooled fuel cells because the air also acts for cooling purposes for air-cooled fuel cells. The exit air is generally discharged to the atmosphere; however, DMFC systems often recover a portion of the water in this stream to offset water consumption at the anodes.

*Water Management Subsystem.*    Water management subsystem maintains a water balance of the fuel cell system. A key design issue for PEMFCs is the management of water with respect to the fuel cell stack. Today's technology primarily utilizes the trifluoromethanesulfonic-acid-based electrolyte membrane that requires a certain level of water content in order to conduct protons efficiently, with water molecules effectively serving as carriers for the protons in their migration across the membrane. Accordingly, the system design must provide for a reasonably high relative humidity in the reactant passages that are in fluid communication with the membrane.

*Thermal Management Subsystem.*    Thermal management subsystem controls the stack and some heat generation parts in a fuel cell power plant in a suitable temperature range. It may also supply heat to needed components or devices in the fuel cell power plant. The thermal management requirements for conventional PEMFCs are intimately associated with water management. As discussed above, the level of hydration of the electrolyte membrane must be maintained in order to prevent dry out and thus loss of proton conduction in the membrane and the electrodes. The temperatures within the stack must accordingly be restricted. Cooling of the individual cells of the stack is carried out to ensure that temperatures are moderate and rather uniform among cells throughout the stack.

Liquid cooling in larger PEMFCs is very effective because of the relatively high thermal conductivity and high volumetric heat capacity of liquids. However, since small fuel cell applications benefit from simple systems, the preferred approach for small fuel cells is usually air cooling. This may be based on the use of forced air delivery through channels spaced appropriately throughout the stack of cells or, alternatively, across external surfaces of the stack along with conduction of heat from interior to exterior sections (enhanced by favorable geometric profiles). The latter approach could employ finned extensions of the bipolar plates to expand the cells' external surface area. Nevertheless, the potential benefits of liquid cooling are significant, and embodiments emphasizing minimized complexity have been implemented in small power sources.[8] These are typically based on a circulating water loop with external air cooling of the water stream. The high operating temperature and the configuration of the cell arrays in SOFC systems allow a simple air-cooled system.

The thermal management burden is also eased in the case of DMFC systems since these can use the circulating methanol-water stream to manage fuel cell temperature.

*Power Conditioner.*    Power conditioner converts the output from the power generation section to the type of power and quality required by the application. This subsystem could be a DC-DC converter or a DC-DC converter plus a DC-AC inverter.

*Internal Energy Storage Device.*    This device starts up the fuel cell system by providing the power needed for fuel cell components such as the control boards, monitors, sensors, and actuators when the fuel cell system is in an idling state. It also supplies additional power to the load when the power output of the stack is inadequate, while it can be charged by the stack via the power conditioner when its energy capacity becomes low. It is typically a battery pack or a supercapacitor.

*Control Subsystem.*   Control subsystem manages the operating parameters of the fuel cell power plant, e.g., temperatures, mass flows, and power conditioning, and coordinates the operations of the other subsystems. The flow rates of the reactants must be controlled as a function of load to prevent starvation of reactant and to ensure that neither excessive water buildup (low flow rate) nor cell dry out (high flow rate) is encountered. Avoiding such imbalances in flow requires measurement of fuel-cell current and corresponding speed adjustment in the air-moving device. Too high of a fuel cell temperature must be avoided to prevent operation in a dry-out condition. This requires that the stack cooling fan or external heat-rejection fan for liquid cooling be either turned on or ramped to a higher speed in response to a high-temperature signal from the temperature sensor. Since systems fueled by virtually pure hydrogen generally run dead-ended, a timer or coulometer is utilized to impose a brief open-close cycle to a solenoid valve in the exit line to purge the anode compartment of accumulated impurities and water on a regular basis. Other control means are provided on an application-specific basis, as appropriate.

### 19.3.3   Fuel Supply Systems

The anode active materials used in fuel cells are generally gaseous or liquid fuels such as hydrogen, methanol, or formic acid, which are fed into the anode side of the fuel cell and are oxidized to produce electrons. Hydrocarbons such as natural gas can also be used as the raw fuel, which needs to be processed into a hydrogen-rich gas mixture prior to being oxidized at the fuel cell anode. Oxygen, most often from air, is the predominant oxidant and is fed into the cathode.

The viability of small fuel cells in their various applications is heavily dependent on the fuel that is utilized. The predominant fuels implemented in PEM-based small fuel cells to date are hydrogen and methanol. Hydrogen fuel may be obtained directly from storage or indirectly via processing of another chemical or fuel entity. DMFCs generally store the fuel as neat methanol. Small SOFC systems, operating at much higher temperatures, are far more tolerant of fuel by-products and typically utilize and process compressed/liquefied hydrocarbon gases.[9]

***Compressed Hydrogen Storage.***   The simplest form of fuel storage and utilization for fuel cell systems is compressed hydrogen. Since volume and weight are often a priority, this requires the use of lightweight, high-pressure canisters and pressure regulators, and the associated cost factors must be taken into account. Figure 19.8 shows a 50 W system from the late 1990s that was designed for 1 kWh over a period of 20 h. It should be noted

**FIGURE 19.8**   50 W fuel cell with compressed hydrogen gas.

that small (1.5 L, 1.3 kg) commercially available cylinders operating at 5000 psi can store hydrogen with approximately 1750 Wh of energy content (lower heating value).[10] Canisters capable of carrying up to 10% wt. hydrogen are available for unmanned aerial vehicles (UAVs). Larger lightweight cylinders with operating pressure capability greater than 10,000 psi have been developed for use in automotive fuel cell systems.[11]

***Indirect Hydrogen Storage.*** Storage of hydrogen by indirect means for small fuel cell systems includes hydrogen generation from metal hydrides and chemical hydrides.

*Metal Hydrides.* The storage of hydrogen in the form of hydrides of metal-alloy powders is often an attractive and convenient energy storage mode for small fuel cell systems. This is attributable to their simplicity of operation and compactness with the volumetric hydrogen density close to liquid hydrogen. The energy densities of these materials can range up to 500 to 1000 Wh/L. This reflects a hydrogen loading approaching 2% by weight, which is characteristic of the maximum obtainable in alloys (typically $AB_2$ type, where, for instance, A is Zr or a mixture of Zr and Ti, and B is a mixture of transition metals) that have useful hydrogen pressures at ambient temperatures. Many metal hydrides are rechargeable by exposing them to hydrogen gas at a pressure greater than the equilibrium hydrogen pressure at a given temperature.

Magnesium-based alloys have been formulated to obtain hydrogen loadings on the order of 5% by weight, but they require discharge temperatures in the neighborhood of 300°C, which necessitates combustion means with a percentage of the hydrogen being sacrificed to generate heat and makes the system bulkier and more complex.

*Chemical Hydrides.* Primary hydride systems of various types have seen used in small fuel cell systems. These are irreversible (throwaway) chemical systems that generate hydrogen on demand. The active reactant is generally an alkali or alkaline earth metal hydride (sometimes in complex forms). This chemical typically reacts with water to form hydrogen and a metal oxide (or mixed metal oxide). Analogous chemical hydride systems are also being explored.

The potential advantages of chemical hydride systems include high specific energy, as the hydrogen yield by weight can be a far higher percentage of the reactant weight in comparison with reversible metal hydrides. On the other hand, volumetric energy densities for these systems are not as attractive because of the relatively low densities of the reactants. The hydrogen-generating reaction and theoretical hydrogen yield are shown in Table 19.1 for representative chemical hydrides. The mass of water required to generate the hydrogen is included in the calculations of percent hydrogen for each chemical hydride. The specific energy advantage can be enhanced if product water from the fuel cell can be recovered for use in the chemical hydride reaction, in which case the system can be refueled simply via a stored reserve of reactant powder or granules.

**TABLE 19.1** Representative Chemical Hydride Reactions and Theoretical Hydrogen Yields

| Reaction | Theoretical hydrogen yield, % | Theoretical specific energy, Wh/kg[*] |
|---|---|---|
| $LiH + H_2O \rightarrow LiOH + H_2$ | 7.8 | 2540 |
| $CaH_2 + 2H_2O \rightarrow Ca(OH)_2 + 2H_2$ | 5.2 | 1700 |
| $NaBH_4 + 2H_2O \rightarrow NaBO_2 + 4H_2$ | 10.9 | 3590 |

[*]Theoretical, including weight of hydride and water, in $H_2$/air fuel cell based on 1.23 V per cell.

The challenges associated with chemical hydrides include the requirement that the respective reactants be brought into contact such that the rate of reaction just meets the fuel cell's hydrogen needs. Also, since the reaction products are disposed, as opposed to being regenerated, the cost of the replenishing reactant chemicals must be taken into account in evaluating the economics of system operation. The most likely use of chemical hydrides in small fuel cells is for high-performance military systems such as UAVs.

***Fuel Processing Systems.*** The applicability of small fuel cells could clearly be greatly enhanced if compact systems using conventional fuels are implemented. In most cases, this requires a fuel processor to convert the fuel into a hydrogen-rich gas that would be delivered to the fuel cell. Much of the challenge in such an approach relates to attaining a sufficiently compact and low-cost fuel processor.

A variety of common fuels and chemicals can be considered as candidate fuels for small fuel cells. These include the following.

*Ammonia.*   Ammonia is commonly used in industry and agriculture in the form of a liquid stored at its own modest vapor pressure. Liquid ammonia offers high specific energy and volumetric energy density based on a relatively simple thermocatalytic hydrogen-generating dissociation reaction:

$$NH_3 \rightleftharpoons 3/2H_2 + 1/2N_2 \qquad \Delta H° = 45.9 \text{ kJ/mol}$$

Thus, ammonia can yield hydrogen at about 17% of its own weight. This corresponds to about 3 kWh/kg and almost 2 kWh/L; however, a fraction of the hydrogen formed must be consumed in generating heat to sustain the endothermic dissociation reaction.

Liquefied petroleum gas (LPG) is generally preferred as a fuel over ammonia because of its greater distribution infrastructure, even higher energy density, and lower cost per unit of energy content, and because of ammonia's reputation as a toxic chemical. Nevertheless, ammonia could play a role in selected small fuel cell applications as a result of its far simpler hydrogen-generating process.

*Methanol.*   Methanol is a widely available chemical that is relatively easy to handle and store as a liquid at atmospheric pressure. It can also be converted to a hydrogen-rich gas by a process that is the simplest among those for carbon-containing fuels. The endothermic reaction

$$CH_3OH + H_2O \rightarrow CO_2 + 3H_2 \qquad \Delta H° = 49.3 \text{ kJ/mol}$$

is carried out at modest temperature (about 250°C) and methane cannot be formed when conventional methanol steam-reforming catalysts are used at such a temperature. Here again, whereas LPG is favored from the points of view of energy content and cost, methanol is an attractive fuel for small fuel cells because its processing burden is greatly eased. Small PEMFC systems using processed methanol are being provided by several companies to a variety of customers.

*Ethanol.*   Ethanol is similar to methanol in its handling and storage. Its reaction with steam can be expressed as

$$C_2H_5OH + 3H_2O \rightarrow 2CO_2 + 6H_2 \qquad \Delta H° = 173.3 \text{ kJ/mol}$$

Unlike methanol, it is considered nontoxic. Its availability and cost as a fuel are irregular. However, ethanol processing requires the rupture of a carbon-carbon bond and needs to be carried out at temperatures around 350°C. Downstream processing is required, at least for low-temperature PEMFCs.

*Liquefied Petroleum Gases.*   LPG (principally propane in the United States, but sometimes principally butane, as in Japan) is often the preferred fuel for dispersed fuel cell systems. It has substantially higher specific energy than both ammonia and methanol along with lower cost per unit energy. Indeed, in the absence of pipeline natural gas, it is the fuel of choice for stationary fuel cells. However, small fuel cells generally have a different set of requirement criteria. In a relative sense, fuel cost is less important, and compactness, simplicity, and hardware cost are more important. This drawback should not be overgeneralized; for instance, missions with very long durations could be an exception. Similar to natural gas and transportation-type fuels, LPG requires high temperatures to process. Also, this type of fuel is attractive for SOFCs because its processing integrates well with high-temperature, by-product-tolerant fuel cells.

*Natural Gas.*   Stationary fuel cell systems that have access to a natural gas pipeline would typically be fueled by this methane-rich gas. In addition to eliminating the storage burden, pipeline natural gas provides a lower cost per unit energy than LPG, and methane yields more hydrogen per unit weight than propane in a fuel cell's fuel processor. Here again, however, these factors are unlikely to be the prevailing issues in the case of small fuel cell systems. In cases where onboard storage of fuel is required (for portability or mobility), LPG would be preferred based on logistics and compactness.

*Transportation-Type Fuels.*   Aviation-type fuels (such as JP-8) are preferred for military applications since these are in general use as readily available "logistic fuels," and are the safest because of their very low vapor pressure. Diesel fuel (widely used in heavy vehicles, to a lesser degree in automobiles, and often as a general-purpose fuel in underdeveloped regions) is less desirable for small fuel cells because of its higher molecular weight and typically higher sulfur content. Low-sulfur (15 ppm) diesel fuel is now employed in the United States. Indeed, these factors pose a difficult challenge to the processing of aviation fuel as well. This is attributable to the difficulty of vaporizing and breaking down its large molecules without carbon formation and/or catalyst poisoning by sulfur at some point in the process. The compactness and low hardware cost required in most small fuel cell applications exacerbate these issues, but aviation fuel use remains a target for small SOFC systems. Gasoline is more readily processed than aviation fuel because of its lower molecular weight, and concomitant easier vaporization and reactivity, and its relative absence of sulfur. However, gasoline is not a preferred fuel for small fuel cell systems because of handling and safety issues, related to its relatively high vapor pressure and flammability.

***Methodologies for Fuel Processing.*** The generation of hydrogen from carbon-containing fuels (with the exception of methanol) requires a high-temperature (usually catalytic) process.[3] The fuel is reacted (1) with steam catalytically (steam reforming, or SR), (2) with substoichiometric oxygen from air (partial oxidation, homogeneous or catalytic, POX), or (3) with steam and oxygen catalytically (autothermal, ATR). SR is an endothermic reaction carried out typically at 700°C or higher. POX is an exothermic process, usually carried out at higher temperatures (perhaps as high as 900 to 1000°C). The ATR process is almost thermally neutral and typically operates at or somewhat above SR temperatures. Representative reactions for these processes using methane as the fuel can be expressed as follows:

$$\text{SR: } CH_4 + H_2O \rightarrow CO + 3H_2 \qquad \Delta H° = 205.9 \text{ kJ/mol}$$
$$\text{POX: } CH_4 + 0.5O_2 \rightarrow CO + 2H_2 \qquad \Delta H° = -247 \text{ kJ/mol}$$
$$\text{ATR: } CH_4 + 0.23O_2 + 0.54H_2O \rightarrow CO + 2.54H_2 \qquad \Delta H° = 0 \text{ kJ/mol}$$

In simplistic terms, the SR process generally provides the highest hydrogen yield and consequently the highest efficiency. Because of the endothermic reaction, its thermal management tends to be the most complex and bulky. Conversely, POX is typically the least efficient but has the simplest configurations. The ATR process tends to be between the other two in both aspects.

The selection of a preferred fuel processor type is decided based on the application requirements. For example, a conventional stationary fuel cell system operating continuously on natural gas might be best suited to the SR processor to minimize fuel cost, while a small, mobile system requiring rapid startup might be best served via a POX or ATR system.

***Process Gas Upgrading.*** The high-temperature processes described above yield a reformate gas that is high in carbon monoxide content (usually greater than 10%). Low-temperature fuel cells, like PEMFC, generally require further processing to maximize the hydrogen yield and minimize the fuel cell anode-catalyst inhibiting effects of CO. Thus, the reformate gas is then passed through a catalytic water-gas shift-reactor at lower temperature (sometimes in two stages, the second at lower temperature) where the following reaction takes place:

$$CO + H_2O \rightarrow CO_2 + H_2 \qquad \Delta H° = -41.2 \text{ kJ/mol}$$

Here again, as in the case of methanol steam reforming, the formation of methane under these conditions is prevented via the specificity of the shift-reactor catalyst. For PEMFCs, further reduction in CO concentration is usually necessary (from about 2% down to less than 20 ppm). This is often carried out by way of catalytic preferential oxidation of CO in the presence of hydrogen with the addition of air at a flow rate that is a small multiple (~2) of the stoichiometric rate required for complete oxidation of the CO.

From the above discussion, it is evident that steam or water vapor is an essential player in the fuel processor, whether it is in the reforming reaction itself or in the subsequent shift-conversion stage. The source of this water must come from storage, make-up, or condensation and recovery from the fuel cell system. In any event, the design and logistics for water management must be provided for the system, and the selected mode must best reflect the requirements for the specific application. Among potential military users of portable fuel cell systems, there is a strong preference for systems that do not require the carrying of excess water onboard the system.

The complexity of the overall high-temperature fuel processing system for carbonaceous fuels indicates the challenge associated with adapting conventional fuels for use in small fuel cell systems. Considerable effort is required in designing and optimizing the system to achieve the requisite miniaturization and low cost in these applications. Even in the best case it appears that only the SOFC system, which can tolerate partially processed fuel (CO is a fuel for SOFCs), would be a potentially viable small power source for fuels requiring high-temperature processing.

## 19.4   *FUEL CELL PRODUCTS, APPLICATIONS, AND PERFORMANCE*

### 19.4.1   Overview

Small fuel cells that provide power at ratings around 1000 W can be implemented in numerous applications as shown in Table 19.2. These existing and candidate applications may include power systems that are fuel-cell only, fuel-cell/battery hybrids, fuel-cell/capacitor hybrids, and fuel-cell/solar/battery hybrids, depending on the nature of the system's requirements.

The interest in small fuel cells stems primarily from their potential to replace batteries with systems having higher specific or volumetric energy density, and small engine generators with more portable, efficient, and environmentally friendly conversion systems. However, because of the unique requirements of portable devices, there are limitations in fuel cell technology that present challenges for its deployment in the sizes below 20 W as replacements for batteries.

**TABLE 19.2**   Small Fuel Cell Applications

| |
|---|
| Remote power, including battery charging |
| Portable power, including soldier-wearable power |
| Mobile power, including vehicle-based auxiliary power |
| Unmanned vehicle and robotic power |
| Cellular phone power |
| Power for portable digital devices |
| Backup power |
| General-purpose power |

The energy-storage and power-generating elements of a fuel cell system are separate entities. In a battery, with the exception of redox flow batteries, the energy-storage and power-generating elements are the same. Hence, the fuel cell system could be designed to relate optimally to its operating mode—the fuel cell power section to satisfy the power requirements, or to the fuel storage section to satisfy the energy requirements. This decoupling of energy and power can be particularly advantageous in applications where the energy requirement is great and the power requirement is minimal—i.e., in applications of long duration. In such applications, the fuel cell section, with its auxiliaries, becomes relatively insignificant in size and weight within the overall system; and the system's energy density approaches that of the fuel storage subsystem alone. The mission duration beyond which fuel cells would tend to be favored over batteries, by providing a smaller and/or lighter system, depends on the specific application requirements.

Certain applications are well suited to a fuel-cell/battery hybrid system by nature of their duty cycle. Those that have high peak-to-average load ratios and relatively short-duration peaks are generally attractive candidates. Such a system allows the fuel cell to be rated near the average power, while a relatively small battery provides excess power on demand and is recharged by the fuel cell during normal load operation. Hybrid systems thus exploit the strengths of both batteries and fuel cells—the wide dynamic range of the former and the high energy content per unit weight or volume of the latter.

Solar/battery power systems can also be combined advantageously with fuel cells in various applications. Use of the fuel cell can often allow solar power to be exploited without the need for excessive battery size and weight associated with prolonged or unpredictable lack of availability of solar energy. Small fuel cells are also expected to become an alternative to small engine-generator sets in some applications. Fuel cell systems are expected to demonstrate advantages over engine-generators in the areas of life, reliability, efficiency, noise, and emissions. While larger engine-generators tend to have size and weight advantages over comparable fuel cell systems operating on the same fuel, these advantages are expected to diminish as systems are scaled down in power rating. There are prospects of fuel cells competing effectively in these categories in the low power area. A key competitive challenge will be in the area of cost.

Hydrogen and other energy-rich fuels have a higher energy density than the active materials normally used in batteries. Table 19.3 lists the theoretical specific energy and energy density of several of these materials, which are significantly higher than those of batteries and are practical for use in portable fuel cells. Of these, hydrogen stands out, not only because of its high specific energy, but because it can be directly converted to electrical energy in a fuel cell operating at ambient temperatures. Natural gas, propane, gasoline, and other fossil fuels cannot be considered as they cannot be converted directly. Incorporating a fuel processing unit is typically not feasible for a small portable device for battery replacement. Methanol is the only liquid fuel that, at this time, shows promise for direct conversion at reasonable temperatures.

The necessity for containing and supplying hydrogen to the fuel cell, in a practical and safe method, substantially reduces its practical specific energy. Several methods are being used, including compressed gas cylinders, reversible storage in hydrides, and chemical methods for generating hydrogen that require specific methods for generating and controlling the generation of hydrogen. Table 19.3 also lists the theoretical values of the various

**TABLE 19.3**  Characteristics of Fuels for Use in Portable Fuel Cells

|  | Theoretical[*] | | Current state of art[†] | |
| --- | --- | --- | --- | --- |
|  | Wh/kg | Wh/L | Wh/kg | Wh/L |
| Hydrogen |  |  |  |  |
|    Hydrogen (gas) | 32,705 |  |  |  |
|    Hydrogen (liquid) cryogenic | 32,705 | 2310 |  |  |
| Pressurized $H_2$ containers |  |  |  |  |
|    70 MPa | 3925 |  |  |  |
| Metal hydride |  |  |  |  |
|    MH (2% $H_2$) | 655 |  | 164 | 426 |
|    MH (7% $H_2$) | 2290 | 3400 |  |  |
| Chemical hydrides |  |  |  |  |
|    LiH + $H_2O$ | 2539 |  |  |  |
|    $NaBH_4$ + $2H_2O$ | 3590 |  | 592[‡] |  |
|    30% $NaBH_4$ solution | 2375 | 2080 |  |  |
| Methanol (MeOH) |  |  |  |  |
|    100% MeOH | 6088 | 4810 | 289–805[§] | 141–385[§] |
|    MeOH-$H_2O$ solution (equimolar) | ~3900 | ~3350 |  |  |

[*]Based on 1.23 V for $H_2/O_2$ fuel cell.
[†]Based on actual watt-hour output of a fuel cell running on the specified $H_2$ source.
[‡]Includes container/packaging and required water.
[§]Depends on power and run time—see Fig. 19.9.

methods for supplying hydrogen and, as available, the status of current technology. However, it is not appropriate to compare only the fuel supply of the fuel cell (omitting the fuel cell stack and other fuel cell components) to a complete battery system. A more reasonable method of comparison is discussed below and illustrated in Fig. 19.9.

Figure 19.9 compares the performance of several primary and secondary batteries with that of a fuel cell, showing the total mass of each system designed, in this example, to deliver 20 W, for different time duration of

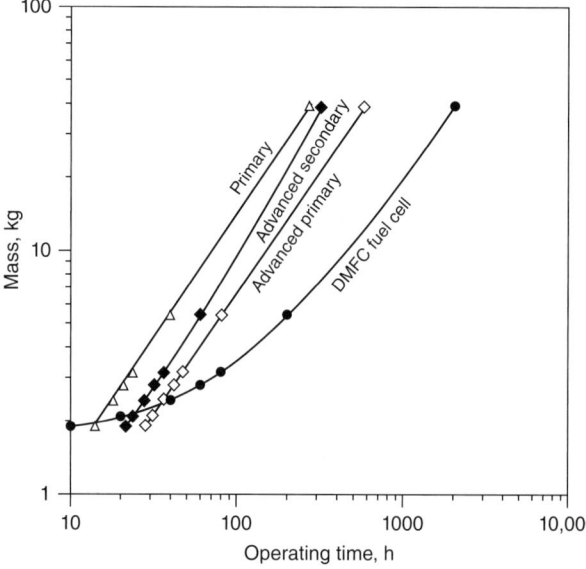

**FIGURE 19.9**  Comparison of electrochemical systems—mass versus service time (based on 20 W output and the following specific energies: primary, 145 Wh/kg; advanced primary, 300 Wh/kg; advanced secondary, 225 Wh/kg).

operational service. The secondary battery systems deliver their rated capacity even at the highest discharge rates shown in the figure; hence, their performance is characterized by a straight sloping line. The slope is equivalent, as shown, to their specific energy, and the mass of the battery is reduced almost proportionally to the reduction in service time. At the low operational times, the curve for the fuel cell levels off, reflecting the inactive mass of the system, i.e., the fuel cell stack and the "balance of plant" (BOP) components required to operate the system. At the longer service times, the mass of the power unit becomes insignificant, and the system mass asymptotically approaches the specific energy of the fuel and its containment system.

This figure graphically illustrates the respective advantages of batteries and fuel cells. The battery shows its advantage on the relatively short-term applications as the fuel cell is penalized by the mass of the power unit. On longer-term applications, the fuel cell benefits from the higher specific energy of the replacement fuel compared to that of nearly all battery systems. A similar relationship exists if the comparison is made on a volumetric basis.

This figure points out the direction that fuel cell development must take to compete successfully with battery systems for relatively short (say, less than 10 h) mission times in the sub-kilowatt range. The weight and size of the power unit must be reduced substantially, as the emphasis in the design of portable equipment is toward lower size and weight even at the sacrifice of service time. The lack of inexpensive, highly efficient, miniaturized components such as gas and liquid pumps of the size and capacity needed for small fuel cells is noteworthy. Such components, most often found in specialized uses such as medical devices, may become more widely available as the need for them is driven by the introduction of small fuel cells for large-market applications.

An interesting consideration is a possible trade-off in the design of the fuel cell component and the fuel source. In the case of the DMFC, for instance, water is required for the reaction of methanol. The discharge product of the fuel cell, water, can be used if the water management or recovery is incorporated in the fuel cell—a one-time cost of increased size, weight, and complexity of the fuel cell. Alternatively, water can be added to the fuel source at the expense of a recurring lower specific energy of the diluted fuel source.

Table 19.4 lists the year 2020 hydrogen storage system targets by the Department of Energy (DOE) of the United States for 2.5~150 W portable power applications.[12] The targets are based on the lower heating value of hydrogen without considering the fuel cell efficiency. Targets are for a complete hydrogen storage and delivery system, including tank, valves, regulators, piping, mounting brackets, insulation, cooling/heating, and other BOP components. The capacities refer to what could be delivered to the fuel cell power plant during normal use. It is required that all targets be met from the start of service to the end of service life.

**TABLE 19.4**    U.S. DOE 2020 Hydrogen Storage System Targets for 2.5~150 W Portable Power Applications

| Storage parameter | Unit | Single use | Rechargeable |
|---|---|---|---|
| Hydrogen capacity | g $H_2$ | >1~50 | >1~50 |
| System gravimetric capacity | kWh/kg system | 1.3 | 1.0 |
| | kg $H_2$/kg system | 0.04 | 0.03 |
| System volumetric capacity | kWh/L system | 1.7 | 1.3 |
| | kg $H_2$/L system | 0.05 | 0.04 |
| Storage system cost | $/Wh net | 0.1 | 0.5 |
| | $/g $H_2$ stored | 3.3 | 17 |
| Durability/operability | | | |
| External operating temperature range | °C | −40~60 | −40~60 |
| Min./max. $H_2$ delivery temperature to fuel cell | °C | 10/85 | 10/85 |
| Min. $H_2$ delivery pressure from storage system | bar (abs) | 1.5 | 1.5 |
| Max. $H_2$ delivery pressure from storage system | bar (abs) | 3 | 3 |
| Max. external surface tem. of storage container | °C | ≤40 | ≤40 |
| $H_2$ discharge rates | | | |
| Minimum full flow rate | (g $H_2$/s)/kW | 0.02 | 0.02 |
| Start time to full flow at 20°C | S | 5 | 5 |
| Start time to full flow at −20°C | S | 10 | 10 |
| Transient response time between 10% and 90% | S | 2 | 2 |
| Fuel purity | % $H_2$ | Meets applicable standards | Meets applicable standards |

## 19.4.2    Fuel Cell Operating Requirements

Small fuel cells can be exploited most effectively if they can stand by and operate at ambient temperatures (and can therefore start rapidly), operate on ambient air, respond rapidly to load changes, and have a nonmigrating (solid) electrolyte and a reasonably high power density. The fuel cell type that best suits these criteria is the PEMFC. The PEMFC can stand by under freezing conditions, and can generally operate under these conditions as well, taking advantage of self-generated heat. However, external means, such as power from a battery or an electric grid, are sometimes required to execute a subfreezing start-up or to prevent freezing.

PEMFCs using either hydrogen or methanol are indeed the type that has received the predominant share of development and implementation in the small fuel cell arena. Nevertheless, despite their high operating temperature, innovation in SOFCs as relevant to small power sources has opened the door to customer examination of hardware utilizing this technology.

In order to minimize the weight and size, a fuel cell system should be designed to be as simple as possible. Air-cooled stacks using fans to supply air without any external humidification become the best choice. Ambient (nonhumidified) reactant air is highly preferred in small fuel cells in order to achieve the simplicity and compactness that are generally sought in these power sources. The use of ambient air, however, requires design measures to prevent the membrane from drying out. The flow rate of air must be limited to reduce the drying effect, and the cell design often must be tailored to take advantage of the product water. The threat of drying clearly becomes far more acute as the ambient temperature increases and/or relative humidity decreases, and as the current density of the stack is increased, thereby increasing the fuel cell's temperature in relation to ambient.

The water management burden is not limited to preventing membrane dry-out. The need to operate at relatively high oxygen utilization rates (relatively low air flow rates) increases the tendency to form water droplets within the cell from the formation of product water at the cathode. This can lead to accumulation of water on the surface of or within the electrode substrate or in the air distribution channels of the cathode flow-field. Such events could result in serious performance losses from the ensuing restriction of air access to impacted regions of the cathode electrocatalysts. Consequently, the cell design approach must also serve to prevent such accumulation of water droplets.

DMFC systems, which use electrolyte membranes of the same type as those used in conventional PEMFCs, do not have the same water management issues. Dry-out conditions do not exist since the feed to the anodes is a methanol-water liquid mixture, provided that sufficient excess water generated at the cathode is returned to the anode feed. The cell design, however, must take into account the potential impact of water accumulation on the cathode side. SOFC systems, operating at very high temperatures, of course do not have water management issues.

Figures 19.10 and 19.11, respectively, show air-cooled stacks with rated power output of 220 and 1800 W developed by Pearl Hydrogen Technology Co. Ltd.[13] Their characteristics based on the surrounding temperature range 15~35°C and RH% 30~90% are listed in Tables 19.5 and 19.6, respectively.

**FIGURE 19.10**   Pearl Hydrogen's 220 W PEMFC Stack. (*Permission from Pearl Hydrogen.*)

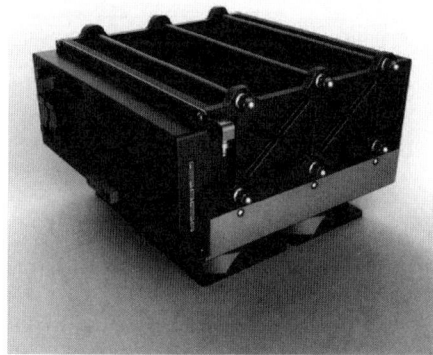

**FIGURE 19.11**   Pearl Hydrogen's 1800 W PEMFC Stack. (*Permission from Pearl Hydrogen.*)

**TABLE 19.5**   Characteristics of Pearl Hydrogen's 220 W PEMFC Stack

| Parameter | | Based on Pearl's PASH technology |
|---|---|---|
| Performance | Rated power | 220 W |
| | Rated voltage | 24 V |
| | Rated current | 9.2 A |
| | DC Voltage range | 20~38 V |
| | Electrical efficiency | ≥50% |
| Fuel | Pure hydrogen | ≥99.95% |
| | Pressure | 0.4~0.5 bar |
| | Hydrogen consumption | 3L/min (@ rated power) |
| | **Air** | |
| Oxidant/coolant | Pressure | Ambient |
| Physical characteristics | Mass | 1800 g |
| | Length × width × height | 208 mm × 138 mm × 98 mm |
| Operation condition | Environment temperature | −10~40°C |
| | Relative humidity | 10~95% |

**TABLE 19.6**   Characteristics of Pearl Hydrogen's 1800 W PEMFC Stack

| Parameter | | Based on Pearl's PASH technology |
|---|---|---|
| Performance | Rated power | 1800 W |
| | Rated voltage | 36 V |
| | Rated current | 50 A |
| | DC voltage range | 30~58 V |
| | Electrical efficiency | ≥50% |
| Fuel | Pure hydrogen | ≥99.95% |
| | Pressure | 0.5~0.6 bar |
| | Hydrogen consumption | 21.6 L/min (@ rated power) |
| | **Air** | |
| Oxidant/coolant | Pressure | Ambient |
| Physical characteristics | Mass | 4800 g |
| | Length × width × height | 288 × 178 × 216 mm |
| Operation condition | Environment temperature | −10~40°C |
| | Relative humidity | 10~95% |

### 19.4.3   Prototype and Commercial Fuel Cell Systems

PEMFCs, DMFCs, and SOFCs have been represented in various forms in commercial and precommercial hardware focusing on small fuel cell applications. Examples of these types are presented below.

### Proton Exchange Membrane (PEMFCs).

*Direct-Hydrogen-Fueled PEMFCs.*   Compressed hydrogen has been implemented as a fuel in various small fuel cell systems. The fuel cell system (from the late 1990s) shown in Fig. 19.8 is indicative of compressed hydrogen being competitive in short-duration missions (20 h). The specific energy approaches 200 Wh/kg (1000 Wh at 50 W at a total system weight of 5.22 kg) with decades-old fuel cell and hydrogen-cylinder technologies. Compressed hydrogen is now a candidate for use in a variety of systems. An example is a sixth-generation power supply system for UAVs by EnergyOr Technologies, Inc.[8] utilizing compressed hydrogen fuel. This system is illustrated in Fig. 19.12, and its technical specifications are presented in Table 19.7.

**FIGURE 19.12**  EnergyOr EO-310-XLE Power Supply System—fueled by pressurized hydrogen. (*Permission from EnergyOr Technologies, Inc.*)

**TABLE 19.7**  Technical Specifications EO-310-XLE Power Supply System (Permission from EnergyOr Technologies, Inc.)

| | | |
|---|---|---|
| System performance | Rated net output power | 310 W |
| | Max. continuous net output power | 450 W |
| | Peak net output power (take-off) | 1000 W |
| | DC output voltage range | 32–45 V |
| | System efficiency at 310 W | 54% |
| | Design lifetime | Up to 3000 h |
| | Net energy available at 310 W | 1790 Wh |
| Environment | Ambient temperature (max.) | 40°C |
| | Flight altitude | 1000 m |
| Physical | Total system mass (including $H_2$ delivery system, $H_2$ fuel and battery) | 3.95 kg |
| | Dimensions/volume | Fully configurable depending on UAV airframe |

A rotatory wing UAV powered by Pearl Hydrogen's 1800 W PEMFC stack using compressed hydrogen during a high-altitude test is shown in Fig. 19.13.[13] This UAV can navigate for about 4 h with one tank of hydrogen.

***Direct Methanol Fuel Cell (DMFC).***   DMFC systems have been pursued vigorously by SFC Energy AG.[14] Applications in a variety of commercial systems under the EFOY trade name are being addressed. Reference 14 details the specifications of systems operating at 45 W and 110 W nominal power.

Dalian Institute of Chemical Physics (DICP)[15] of the Chinese Academy of Sciences (CAS) has delivered a series of DMFC systems with rated power ranging from 25 to 200 W. Without fuel the DICP-50W system weighs 5.2 kg (Fig. 19.14). The fuel, neat methanol, is stored in three cartridges, each weighing 0.8 kg and producing 1200 Wh of energy; three cartridges provide 72 h of operation with a system-specific energy of 474 Wh/kg.

**FIGURE 19.13** A rotatory wing UAV powered by Pearl Hydrogen's 1800 W PEMFC system. (*Permission from Pearl Hydrogen.*)

**FIGURE 19.14** DICP-50W DMFC system. (*Permission from DICP.*)

Figure 19.15 shows a DICP's DMFC/battery hybrid system that has a nominal output power 650 W and maximum output power 1250 W. The system weighs 9 kg (without fuel) and the neat methanol weighs 7 kg, delivering 11.3 kWh of energy for 17.5 h of operation at rated power, yielding a system-specific energy of 700 Wh/kg.

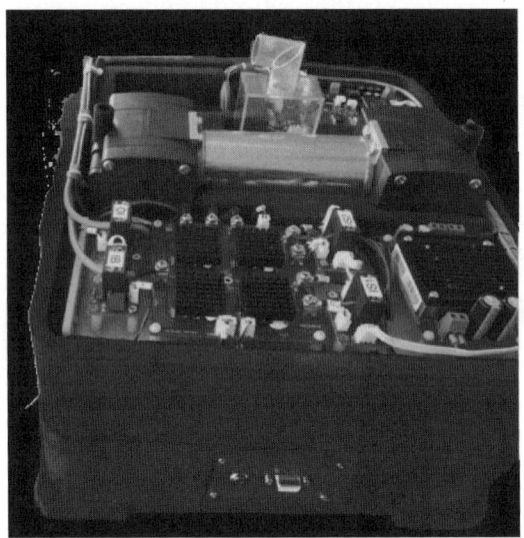

**FIGURE 19.15** DICP's 650 W DMFC system. (*Permission from DICP.*)

***Solid Oxide Fuel Cells (SOFCs).*** At first glance, an SOFC, traditionally operating at around 1000°C, does not appear to be a promising candidate for use in small fuel cells. The ratio of surface area to volume for small systems would appear to require too much insulation to prevent excessive heat loss and consequent loss of fuel energy to maintain the system at its operating temperature. Nevertheless, systems ranging from 25 to 250 W have been developed using a "microtubular" design based on cells only a few millimeters in diameter and using an oxygen-ion conducting ceramic electrolyte capable of operating in the temperature range of 600 to 800°C. Such cells can be brought to operating temperature quickly and can withstand the associated thermal shock.

The SOFC is well suited to the use of hydrocarbon fuels such as propane or butane, as only minimal fuel processing is required at the high operating temperature of the system.

Adaptive Materials, Inc. has developed SOFC hardware for small fuel cell applications, including a unit designed for unmanned ground and aerial vehicles.[9] Figure 19.16 depicts a 25 W SOFC with a mass of 1.5 kg (without fuel), a volume of 2.0 L, and the remarkable specific energy of 661 Wh/kg, including the mass of the fuel and its container.

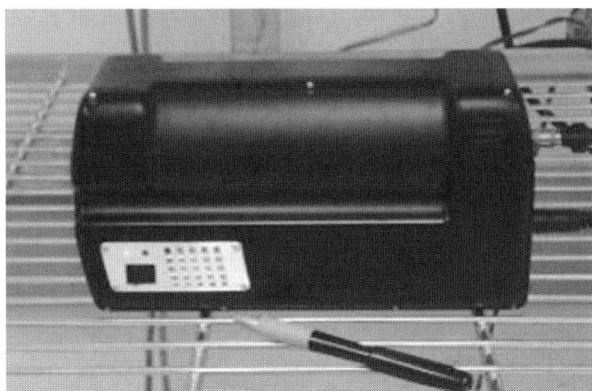

**FIGURE 19.16**    A 25 W portable SOFC. (*Source: Adaptive Materials, Inc.*)

As with many other fuel cell systems under development, the initial market for SOFC appears to be for military systems such as UAVs, field battery chargers, small robotic vehicles, and electronics and sensors. This market, though it requires ruggedization for harsh conditions, can also tolerate a relatively high initial price and a lifetime in the hundreds or low thousands of hours. Success in the military arena may also open the larger markets for industrial and consumer applications. The worldwide availability of propane and butane in canisters from 0.5 to 10 kg would provide a ready fuel infrastructure for consumer applications that is totally lacking for hydrogen gas.

### 19.4.4    Summary of Commercial Systems

Fuel cells are safe, user and environmental friendly, and can be quiet and relatively simple in design, making them superior to the engine-generators and batteries for many applications. However, the cost of the fuel cell is higher because it has not attained large-scale commercial production status, and the lack of hydrogen supply infrastructure is a hurdle. The pace of commercial and precommercial activity in small fuel cells has been fostered by ongoing technology improvements and greater recognition by potential customers of the prospective operational advantages and viability of fuel cell systems in applications of interest.

Small fuel cells currently being evaluated represent a variety of fuel cell technologies. These are predominantly based on low temperature PEM, including PEMFCs and DMFCs, but SOFC systems are participating, and further pursuit of such systems is anticipated. This is particularly likely where use of common fuels would have a significant advantage.

The period ahead is expected to provide further assessment of fuel cell systems in small fuel cell applications and concomitant commercial interest. The degree of commercial success of these systems will be determined in an ongoing manner based on factors such as cost, reliability, fuel logistics, and the size of markets advantageously served within these parameters.

### 19.4.5    Development Trends and Market Factors

By October 2017, organizations or governments in several countries and regions proposed timetables to ban the sales of polluting vehicles (Table 19.8). China is in process of making such a timetable, and it is expected that countries such as Japan, Korea, and other European countries will propose their own timetable sooner or later.

**TABLE 19.8**   Timetables to Ban on Polluting Vehicles

| Country | Starting year | Details |
|---------|---------------|---------|
| Norway | 2025 | Ban on oil-based vehicles |
| Netherlands | 2025 | Ban on oil-based vehicles |
| California (United States) | 2030 | Ban on traditional oil-based vehicles |
| Germany | After 2030 | Ban on traditional IEC-based vehicles |
| India | 2030 | Total ban on oil-based vehicles |
| France | 2040 | Total ban on oil-based vehicles |
| United Kingdom | 2040 | Total ban on traditional diesel/gasoline-based vehicles |

Whether or not such proposals will be implemented by the governments are not certain. Nevertheless, fuel cell-powered vehicles will increase significantly, among other electric vehicles such as battery-powered vehicles and plug-in hybrid vehicles.

A fuel cell passenger car can run for 500 km or more with each refueling that takes about 3~5 min, but lack of hydrogen infrastructure is a hurdle to large number deployment of fuel cell vehicles. In order to change this situation, several major countries or regions have laid out plans to build hydrogen stations, as summarized in Table 19.9. There are about 220 hydrogen refueling stations worldwide in 2016/2017, led by Japan (78), and followed by California of the United States (52), Germany (50), United Kingdom (15), Denmark (11), Korea (10), and China (4). Outside California, 12 stations are planned in the northeast United States. By 2020, a total of more than 550 stations, led by Japan with 160 stations, and Korea and China each reaching 100 stations is projected. By 2030, more than 5000 stations are expected, with China, Japan, and the United Kingdom each reaching 1000 stations, followed by 600 stations in France and 500 stations in Korea; the numbers in Germany and the United States are also expected to be in the 500~1000 range each.

**TABLE 19.9**   Hydrogen Refueling Station Status and Projection

| Year | Countries and region | | | | | | | | |
|------|-----------------|---------|--------|---------|------------|-------|-------|-------|-------|
|      | United Kingdom | Germany | France | Denmark | California | Japan | Korea | China | Total |
| 2016/2017 | 15 | 50 | | 11 | 52 | 78 | 10 | 4 | ~220 |
| 2018 | | 100 | | | 67 | 90 | | | ~300 |
| 2020 | 65 | | | | 87 | 160 | 100 | 100 | ~550 |
| 2025 | 330 | 400 | | | 100 | 320 | 210 | | ~1500 |
| 2030 | 1150 | >500 | 600 | | >500 | 900 | 500 | 1000 | ~5000 |

Fuel cell technology is complicated and interdisciplinary, comprising chemistry, materials science, fluid dynamics, thermodynamics, electronics, electricity, software, controls, machinery, and so on. It has proceeded on a bumpy road. The first practical use of fuel cells occurred in 1960s for the manned space programs. The PAFC started small-scale commercialization in 1970s. In the mid-1980s, thanks to the use of solid polymer electrolyte, PEMFC gained quick development, and a peak was reached in around year 2000. A very low point was reached when the U.S. Three Big Car companies faced bankruptcy during the financial crisis in 2008/2009, and the former Secretary of DOE, Dr. John Chu, announced that the U.S. government would not support fuel cell programs. His announcement was a major blow to the fuel cell and related companies in the United States and some other countries. However, Japan, Korea, and many European countries did not follow the United States in slowing down their fuel cell development, and they have now surpassed the United States to become the leading players.

The development of fuel cells is impacted by many factors. Externally, these factors include fossil fuel supplies; environmental issues such as global warming and air, water, and soil pollution; attitude and determination of governments; mature incumbent technologies; and politics. Internally, these factors include performance, durability, lifetime, cost, supply chain, and hydrogen infrastructure.

After about a half century's development, the fuel cell industry is still small, fragile, and unprofitable. Many earlier players disappeared primarily due to lack of continuous cash infusion. However, through so many years' efforts, the fuel cell has crept to the early commercialization stage. In 2013, 2014, and 2016, Hyundai, Toyota, and Honda introduced small series produced fuel cell vehicles to the markets. The lifetime of their FCVs comes close to meeting the 5000 h DOE target, and the cost has dropped to about 5~10% of the earlier prototypes, selling at around USD 50,000. In August 2017, Ballard Power Systems announced that a fuel cell bus powered by its FCveloCity®-HD6 fuel cells had achieved more than 25,000 h of revenue service, and several other fuel cell buses neared the 25,000-h DOE target. Some PEMFC and PAFC systems for stationary applications were reported to surpass the 60,000 operating hours DOE target several years ago. Apparently, the lifetime of fuel cells is becoming comparable to that of traditional technologies, which is a key to its commercialization.

The most successful fuel cell program is Japan's Ene-Farm program. Through this program, about 200,000 sub-kilowatt micro-CHP units have been installed. Those CHP systems produce 700~750 W electrical power, slightly over 1000 W heat, with a combined heat and electrical efficiency as high as 95% and unit cost around USD 10,000. The Japanese government plans cumulatively to install 1.4 and 5.3 million such units by 2020 and 2030, respectively.

PEMFCs are quite successful in powering materials handling devices such as forklifts. There are about 20,000 fuel cell forklifts deployed worldwide, primarily in the United States, but making inroads in Europe and Japan.

Deployment of fuel cell vehicles based on PEMFC technology is currently the most intense pursuit, especially in Japan, Korea, and China. A few thousand fuel cell passenger cars, a few hundred fuel cell buses, and a few hundred fuel cell trucks are operating worldwide. Toyota's production plan for FCV Mirai is 700, 2000, and 3000 units in 2015, 2016, and 2017, respectively, and reaches 30,000 in 2020. China focuses more on fuel cell buses and trucks, as those vehicles have more space for fuel cell system installation, and hydrogen refueling stations can be built at the bus or truck terminals. For each passenger car, light truck, or bus, and mid-to-large size truck or bus installing a 30-kW fuel cell system, the Chinese Central Government provides 200,000, 300,000, and 500,000 RMB subsidies, respectively, and more than 40 local governments provide matching subsidies as well.

Deployment of DMFCs is dominated by SFC Energy. Its EFOY Comfort provides 40~105 W electrical power, aiming at leisure markets for caravans and campers, while its EFOY Pro series provides 45~500 W power for industrial and security applications.

Table 19.10 summarizes the fuel cell shipments and megawatts for portable, stationary, and transportation applications from 2011 to 2016.[16] The power range for portable, stationary, and transportation applications is 1 W~20 kW, 0.5 kW~400 kW, and 1 kW~100 kW, respectively. The shipment numbers in Table 19.10 are rounded to the nearest 100 and the megawatts to the nearest 0.1 MW. The shipment of portable fuel cells peaked in 2014 and shrank significantly in 2015 and further in 2016, indicating that the portable market has been slowing down. The shipment in stationary systems shows a steady but minor increase, implying that the stationary

**TABLE 19.10**   Shipments and Megawatts by Application

| Application | 2011 | 2012 | 2013 | 2014 | 2015 | 2016 | Total |
|---|---|---|---|---|---|---|---|
| Portable | | | | | | | |
| Shipments | 6900 | 18,900 | 13,000 | 21,200 | 8700 | 4000 | 72,700 |
| Megawatts | 0.4 | 0.5 | 0.3 | 0.4 | 0.9 | 0.3 | 2.8 |
| Stationary | | | | | | | |
| Shipments | 16,100 | 24,100 | 51,800 | 39,500 | 47,000 | 54,800 | 233,300 |
| Megawatts | 81.4 | 124.9 | 186.9 | 147.8 | 183.6 | 200.8 | 925.4 |
| Transportation | | | | | | | |
| Shipments | 1600 | 2700 | 2000 | 2900 | 5200 | 6400 | 20,800 |
| Megawatts | 27.6 | 41.3 | 28.1 | 37.2 | 113.6 | 277.5 | 525.3 |
| **Total** | | | | | | | |
| **Shipments** | **24,600** | **45,700** | **66,800** | **63,600** | **60,900** | **65,200** | **326,800** |
| **Megawatts** | **109.4** | **166.7** | **215.3** | **184.4** | **298.1** | **478.6** | **1453.5** |

**TABLE 19.11**   Shipments and Megawatts by Fuel Cell Type

| Type | 2011 | 2012 | 2013 | 2014 | 2015 | 2016 | Total |
|---|---|---|---|---|---|---|---|
| PEMFC | | | | | | | |
| Shipments | 20,400 | 40,400 | 58,700 | 58,400 | 53,500 | 46,900 | 278,300 |
| Megawatts | 49.2 | 68.3 | 68.0 | 72.7 | 151.8 | 311.2 | 721.2 |
| DMFC | | | | | | | |
| Shipments | 3600 | 3000 | 2600 | 2500 | 2100 | 2200 | 16,000 |
| Megawatts | 0.4 | 0.3 | 0.2 | 0.2 | 0.2 | 0.2 | 1.5 |
| SOFC | | | | | | | |
| Shipments | 600 | 2300 | 5500 | 2700 | 5200 | 16000 | 32,300 |
| Megawatts | 10.6 | 26.9 | 47.0 | 38.2 | 53.3 | 53.7 | 229.7 |
| PAFC | | | | | | | |
| Shipments | — | — | — | — | 100 | 100 | 200 |
| Megawatts | 4.6 | 9.2 | 7.9 | 3.8 | 24.0 | 46.6 | 96.1 |
| MCFC | | | | | | | |
| Shipments | | | | | | | 100 |
| Megawatts | 44.5 | 62.0 | 91.9 | 70.5 | 68.6 | 66.9 | 404.4 |
| AFC | | | | | | | |
| Shipments | | | | | | | |
| Megawatts | 0.1 | 0.0 | 0.3 | 0.0 | 0.2 | 0.0 | 0.6 |
| **Total** | | | | | | | |
| **Shipments** | **24,600** | **45,700** | **66,800** | **63,600** | **60,900** | **65,200** | **326,800** |
| **Megawatts** | **109.4** | **166.7** | **215.3** | **185.4** | **298.1** | **478.6** | **1453.5** |

market is flattening out. The shipments for transportation applications increased sharply in 2015 and 2016, which reflects that the market is in the quick expanding stage.

Table 19.11 summarizes the fuel cell shipments and megawatts by fuel cell types from 2011 to 2016.[16] For PEMFC, shipments were flat, but the megawatts increased significantly in 2015 and further in 2016, reflecting the rapid increase in fuel cell cars and buses that typically use fuel cells rated higher than 50 kW, while the number of lower power units used for telecom backup and portable devices decreased. For DMFC, both the shipments and the megawatts show a general decline trend, indicting a slowing market, in accordance with the results for portable application shown in Table 19.11. For SOFC, the shipments increased significantly in 2015 and further in 2016, but the megawatts were nearly the same in 2015 and 2016, which implies that the market is expanding nicely for lower power units. For PAFC, the shipments were similar in 2015 and 2016, but the megawatts nearly doubled, indicating that more, larger units were installed in 2016. For MCFC, the shipments declined in 2015 and 2016 from 2014 while the megawatts were flat, indicating a more challenging business environment for MCFCs. For AFC, there were basically no shipments and megawatts, meaning that it is far from commercialization.

Overall, in recent years, PEMFCs and SOFCs expanded quickly, PAFCs and MCFCs remained flat, DMFCs shrank, and AFCs were still dormant. PEMFCs are mainly used for cars, buses, trucks, forklifts, and micro-CHP, while SOFCs are mainly used for micro-CHP.

## ACKNOWLEDGMENTS

Major content for this chapter was provided in whole or in part by David Linden and H. Frank Gibbard, Chap. 37 (Introduction to Fuel Cells), and Arthur Kaufman and H. Frank Gibbard, Chap. 38 (Small Fuel Cells), *Linden's Handbook of Batteries*, 4th ed., T. B. Reddy and D. Linden, eds., McGraw-Hill, 2011.

# REFERENCES

1. K. Kordesch and G. Simader, *Fuel Cells and Their Applications*, VCH Publishers, NY, 1996.

2. B. V. Tilak, R. S. Yeo, and S. Srinivasan, "Electrochemical Energy Conversion—Principles," in J. O'M. Bockris, B. E. Conway, E. Yeager, and R. E. White (eds.), *Comprehensive Treatise of Electrochemistry*, Vol. 3, Plenum Press, New York, 1981, pp. 39–122.

3. Z. Qi, *Proton Exchange Membrane Fuel Cells*, CRC Press, Boca Raton, FL, London, New York, 2014.

4. S. R. Narayan and T. I. Valdez, "High-Energy Portable Fuel Cell Power Sources," *Interface*, The Electrochemical Society, Winter, 2008, pp. 40–44.

5. S. Srinivasan, *Fuel Cells: From Fundamentals to Applications*, Springer, New York, 2006.

6. S. Calabrese Barton, T. Patterson, E. Wang, T. F. Fuller, and A. C. West, "Mixed-Reactant, Strip-Cell Direct Methanol Fuel Cells," *J. Power Sources* **96:**329–336 (2001).

7. M. Daugherty, D. Haberman, N. Stetson, S. Ibrahim, O. Lokken, D. Dunn, M. Cherniack et al., *Proceedings of the Conference on Portable Fuel Cells*, Lucerne, Switzerland, June 21–24, 1999, pp. 69–78.

8. www.energyor.com.

9. M. de Jong, J. J. Kowal, E. Ferry, J. Cristiani, and M. Dominick. "CERDEC Fuel Cell Team: Military Transitions for Soldier Fuel Cells," Fuel Cell Seminar, Phoenix AZ, October 27–30, 2008. (Available in the public domain.)

10. www.luxfercylinders.com.

11. Quantum Technologies, www.qtww.com.

12. https://energy.gov/eere/fuelcells/.

13. www.pearlhydrogen.com.

14. www.sfc.com/en/.

15. www.dicp.ac.cn.

16. www.FuelCellIndustryReview.com.

# CHAPTER 20
# ELECTROCHEMICAL CAPACITORS

**Andrew F. Burke**

## 20.1   INTRODUCTION

### 20.1.1   Comparisons of Electrochemical Capacitors and Batteries

Electrical energy storage is required in many applications including telecommunication devices such as cell phones and pagers, standby power systems, and electric/hybrid vehicles. The requirements for the energy-storage device in a particular application are given in terms of energy stored (Wh) and maximum power (W) as well as size and weight, initial cost, and life. A storage device to be suitable for a particular application must meet all the requirements. As power requirements for many applications become more demanding, it is often reasonable to consider separating the energy and power requirements by providing for the peak power by using a pulse power device (capacitor) that is charged periodically from a primary energy-storage unit (battery). For applications in which significant energy is needed in pulse form, traditional capacitors as used in electronic circuits cannot store enough energy in the volume and weight available. For these applications, the development of high energy density capacitors (supercapacitors, ultracapacitors, or electrochemical capacitors) has been undertaken by various groups around the world. This chapter considers in detail why such capacitors are being developed, how they function, and the present status and projected development of electrochemical capacitor technology.

The most common electrical energy-storage device is the battery. Batteries have been the technology of choice for most applications, because they can store large amounts of energy in a relatively small volume and weight and provide suitable levels of power for many applications. Shelf and cycle life have been a problem with most types of batteries, but people have learned to tolerate this shortcoming due to the lack of an alternative. In recent times, the power requirements in a number of applications have increased markedly and have exceeded the capability of batteries of standard design. This has led to the design of special high-power, pulse batteries often with the sacrifice of energy density and cycle life. Electrochemical capacitors are being developed as an alternative to pulse batteries. To be an attractive alternative, capacitors must have much higher power and much longer shelf and cycle life than batteries. In this instance, the word "much" means at least one order of magnitude higher. Electrochemical capacitors have much lower energy density than batteries and their low energy density is in most cases the factor that determines the feasibility of their use in a particular high-power application.

For capacitors, the trade-off between the energy density and the $RC$ (i.e., resistance-capacitance) time constant of the device is an important design consideration. In general, for a particular set of materials, a sacrifice in energy density is required to get a large reduction in the time constant, and thus a large increase in power capability. The characteristics of a number of electrochemical capacitors and pulse batteries are given in Tables 20.1 and 20.2.

**TABLE 20.1**    Comparisons of the Energy and Power Characteristics of Electrochemical Capacitors and Batteries

| Device technology | Nominal cell voltage | Specific energy, Wh/kg | Matched impedance, kW/kg | 90% Efficiency, kW/kg |
|---|---|---|---|---|
| Carbon/carbon double-layer capacitors | 2.7 | 5 | 10–25 | 2.5–10 |
| Lithiated graphite/carbon supercapacitors | 3.8 | 12 | 10–25 | 2.5–5 |
| Lithium-ion batteries | | | | |
|    Iron phosphate | 3.25 | 90–150 | 2–4 | 0.70–1.4 |
|    Lithium titanate | 2.4 | 35–100 | 2–6 | 0.7–2.5 |
|    NiCoMnO$_2$ | 3.7 | 100–200 | 1–4 | 0.5–2.0 |
| Ni metal hydride HEV | 1.2 | 46 | 1.1 | 0.40 |
| Lead-acid HEV | 2.0 | 26 | 0.4 | 0.15 |
| Zinc-air | 1.3 | 450 | 0.6–1.2 | 0.20–0.40 |

**TABLE 20.2**    Characteristics of Large Electrochemical Capacitors

| Device | $V_{rated}$ | C, F | R, mΩ* | RC, s | Wh/kg[†] | W/kg (95%)[‡] | W/kg matched impedance | Weight, kg | Vol., L |
|---|---|---|---|---|---|---|---|---|---|
| Maxwell | 2.7 | 2885 | 0.375 | 1.1 | 4.2 | 994 | 8836 | 0.55 | 0.414 |
| Maxwell | 2.7 | 605 | 0.90 | 0.55 | 2.35 | 1139 | 9597 | 0.20 | 0.211 |
| Ioxus | 2.85 | 3095 | 0.33 | 1.0 | 5.0 | 1355 | 12,065 | 0.51 | 0.41 |
| Ioxus | 2.85 | 1348 | 0.56 | 0.85 | 3.7 | 1372 | 12,292 | 0.295 | 0.228 |
| Skeleton Technologies | 2.85 | 350 | 1.2 | 0.42 | 4.0 | 2714 | 24,200 | 0.07 | 0.037 |
| Skeleton Technologies | 2.85 | 3450 | 0.13 | 0.45 | 5.4 | 3353 | 29,809 | 0.52 | 0.39 |
| Skeleton Technologies | 3.4 | 3200 | 0.48 | 1.5 | 8.9 | 1730 | 15,400 | 0.40 | 0.096 |
| Yunasko[§] | 2.7 | 510 | 0.9 | 0.46 | 5.0 | 2919 | 25,962 | 0.078 | 0.055 |
| Yunasko[§] | 2.75 | 480 | 0.25 | 0.12 | 4.45 | 10,241 | 91,115 | 0.060 | 0.044 |
| Yunasko[§] | 2.75 | 1275 | 0.11 | 0.13 | 4.55 | 8791 | 78,125 | 0.22 | 0.15 |
| Yunasko[§] | 2.7 | 7200 | 1.4 | 10 | 26 | 1230 | 10,947 | 0.119 | 0.065 |
| Yunasko[§] | 2.7 | 3200 | 1.5 | 7.8 | 30 | 3395 | 30,200 | 0.068 | 0.038 |
| Ness Maxwell | 3.0 | 3650 | 0.27 | 0.98 | 6.5 | 1875 | 16,666 | 0.50 | 0.394 |
| Ness | 2.7 | 1800 | 0.55 | 1.0 | 3.6 | 975 | 8674 | 0.38 | 0.277 |
| Ness | 2.7 | 3640 | 0.30 | 1.1 | 4.2 | 928 | 8010 | 0.65 | 0.514 |
| Ness (cyl.) | 2.7 | 3160 | 0.4 | 1.3 | 4.4 | 982 | 8728 | 0.522 | 0.379 |
| DAE-China | 2.7 | 1660 | 0.6 | 1.0 | 6.1 | 1734 | 15,420 | 0.197 | |
| DAE-China | 2.7 | 440 | 2.3 | 1.0 | 5.5 | 1536 | 13,662 | 0.058 | |
| JSR Micro | 3.8 | 1100 | 1.15 | 1.21 | 10 | 2450 | 21,880 | 0.144 | 0.077 |
| | | 2300 | 0.77 | 1.6 | 7.6 | 1366 | 12,200 | 0.387 | 0.214 |
| | | 3225 | 1.0 | 3.2 | 11 | 1167 | 10,374 | 0.348 | 0.213 |
| DAE-China | 3.8 | 850 | 4.7 | 3.5 | 12.4 | 993 | 8828 | 0.087 | |

*Steady-state resistance, including pore resistance.
[†]Energy density at 400 W/kg constant power, $V_{rated} - \frac{1}{2}V_{rated}$.
[‡]Power based on $P = 9/16 \times (1 - EF) \times V^2/R$, EF = efficiency of discharge.
[§]All devices except those with § are packaged in metal/plastic containers; those with a § are laminated-pouch packaged.

Two approaches to the calculation of the peak power density are indicated in the first table. The first and more standard approach is to determine the power at the so-called matched impedance ($P_{mi}$) condition at which one-half the energy of the discharge is in the form of electricity and one-half is heat. The maximum power at this point is given by

$$P_{mi} = V_{oc}^2/4\ R_b$$

where $V_{oc}$ is the open-circuit voltage of the device and $R_b$ is its resistance. The discharge efficiency at this point is 50%. For many applications in which a significant fraction of the energy is stored in the energy-storage device before it is used by the system, the efficiency of the charge/discharge cycle is important to the system efficiency. In these cases, the use of the energy-storage device should be limited to conditions that result in high efficiency for both charge and discharge. The discharge/charge power for a battery as a function of efficiency ($P_{ef}$) is given by

$$P_{ef} = (EF)\,(1 - EF)\,\left(V_{oc}^2 / R_b\right)$$

where EF is the efficiency of the high-power pulse. For EF = 0.95, $P_{ef}/P_{mi} = 0.19$. Hence, in applications in which efficiency is a primary concern, the useable power of the battery is much less than the peak power ($P_{mi}$) often quoted by the manufacturer of the battery.

In the case of electrochemical capacitors, the peak power for a pulse ($P_{uc}$) at a voltage midway between $V_o$ and $V_o/2$, where $V_o$ is the rated voltage of the device, is given by

$$P_{uc} = (9/16)\,(1 - EF)\,\left(V_o^2 / R_{uc}\right)$$

where $R_{uc}$ is the resistance of the capacitor. Peak power values are shown in Table 20.1 for both matched impedance and high efficiency discharges of the batteries and capacitors. It is apparent that in nearly all cases the power from the electrochemical capacitors is much higher than that from the batteries. Note that it is not correct to compare the high efficiency power density for the capacitors with the matched impedance power density for the batteries as is often done. The power capability of both types of devices is primarily dependent on their resistance, and knowledge of the resistance is key to determining the peak useable power capability. Hence, measurement of the resistance of a device in the pulsed mode of operation is critical to an evaluation of its high-power capability.

In addition to high-power capability, the other reason for considering electrochemical capacitors for a particular application is their long shelf and cycle life. This is especially true of capacitors using activated carbon electrodes. Most secondary (rechargeable) batteries if left on the shelf unused for many months will degrade markedly and be essentially useless after this time due to self-discharge and internal corrosion effects. Electrochemical capacitors will self-discharge over a period of time to a low voltage, but they will retain their capacitance and thus will be capable of recharge to their original condition. Experience has shown that capacitors can be left unused for several years and they remain in nearly their original condition. Electrochemical capacitors can be deep cycled at high rates (discharge times of seconds) for 500,000 to 1,000,000 cycles at room temperature with a relatively small change in characteristics (10–20% degradation in capacitance and resistance). This is not possible with batteries even if the depth of discharge is kept small (10–20%). The life of the electrochemical capacitors is significantly less at high temperatures (>50°C).

Hence, relative to batteries, the advantages of electrochemical capacitors as pulse power devices are high power density, high efficiency, and long shelf and cycle life. The primary disadvantage of capacitors is their relatively low energy density (Wh/kg and Wh/L) compared to batteries limiting their use to applications in which relatively small quantities of energy are required before the capacitor can be recharged. Electrochemical capacitors can, however, be recharged in very short times (seconds or fraction of seconds) compared to batteries if a source of energy is available at the high power levels required.

### 20.1.2 Energy Storage in Electrochemical Capacitors

The most common electrical energy-storage devices are capacitors and batteries. Capacitors store energy by charge separation. The simplest capacitors store the energy in a thin layer of dielectric material supported by metal plates that act as the terminals for the device. The energy stored in a capacitor is given by $\frac{1}{2}CV^2$, where $C$ is its capacitance (in farads, F) and $V$ is the voltage between the terminal plates. The maximum voltage of the capacitor is dependent on the breakdown characteristics of the dielectric material. The electrical charge, $Q$ (coulombs), stored in the capacitor is given by $CV$. The capacitance of the dielectric capacitor depends on the dielectric constant ($k$) and the thickness ($\delta_{th}$) of the dielectric material and its geometric area ($A$).

$$C = kA/\delta_{th}$$

In a battery, energy is stored in chemical form as the active material in its electrodes. Energy is released in electrical form by connecting a load across the terminal of the battery, permitting the electrode materials to react electrochemically with the required ions that are transferred through the electrolyte in which the electrodes are immersed. The useable energy stored in the battery is given as $VQ$, where $V$ is the voltage of the cell and $Q$ is the electrical charge (or $I_t$, current flow over time) transferred to the load during the chemical reaction. The voltage is dependent on the active materials (chemical couple) of the battery and is close to the open-circuit voltage ($V_{oc}$) for these materials.

An electrochemical capacitor, sometimes referred to as an ultracapacitor or supercapacitor, is an electrical energy-storage device that is constructed much like a battery (see Fig. 20.1) in that it has two electrodes immersed in an electrolyte with a separator between the electrodes. The electrodes are fabricated from high surface area, porous materials having pores of diameter in the nanometer (nm) range. The surface area of the electrode materials used in an electrochemical capacitor is much greater than that used in battery electrodes being 500 to 2000 $m^2/g$. Charge is stored in the micropores at or near the interface between the solid electrode material and the electrolyte. The charge and energy stored are given by the same expressions as cited previously for the simple dielectric capacitor. However, calculation of the capacitance of the electrochemical capacitor is much more difficult as it depends on the complex phenomena occurring in the micropores of the electrodes.[1-5] It is convenient to discuss the mechanisms for energy storage in electrochemical capacitors in terms of double-layer and psuedocapacitance processes separately. The physics and chemistry of these processes as they apply to electrochemical capacitors are explained in great detail in Refs. 1 to 5. In the following sections, the mechanisms are discussed briefly in terms of how they relate to the properties of the electrode materials and electrolyte.

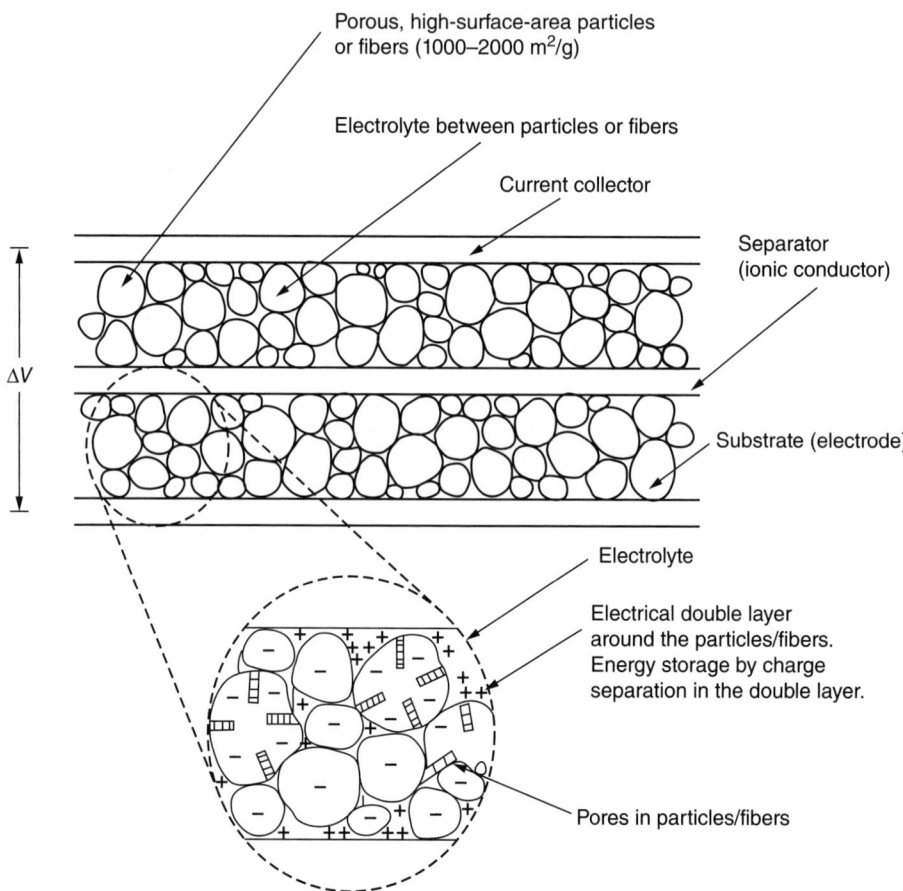

**FIGURE 20.1**   Schematic of an electrochemical capacitor.

***Electric Double-Layer Capacitors (EDLC).***    Energy is stored in the electric double-layer capacitor (EDLC) as charge separation in the double layer formed at the interface between the solid electrode material and the liquid electrolyte in the micropores of the electrodes. The constant-current, charge-discharge characteristic of the EDLC is shown in Fig. 20.2. The simple linear voltage versus time response is characteristic of devices using activated carbon in both electrodes.

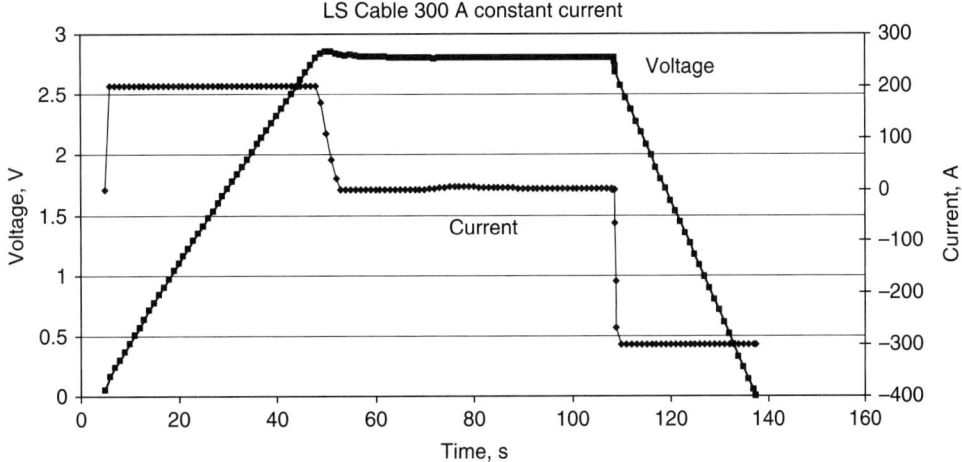

**FIGURE 20.2**    Charge-discharge response of an EDLC device.

A schematic of an electrochemical capacitor, as shown in Fig. 20.1, functions as follows: The ions displaced in forming the double layers in the pores are transferred between the electrodes by diffusion through the electrolyte. The energy and charge stored in an EDLC are $\frac{1}{2}CV^2$ and $CV$, respectively. The capacitance is dependent primarily on the characteristics of the electrode material (surface area and pore size distribution). The specific capacitance of an electrode material can be written as

$$C/g = (F/cm^2)_{act} \times (cm^2/g)_{act}$$

where the surface area is the active area in the pores on which the double layer is formed. In simplest terms, the capacitance per unit of active area is given by

$$(F/cm^2)_{act} = (k/\delta_{dleff})$$

where $\delta_{dleff}$ is the thickness of the double layer.

As discussed in Refs. 6 to 10, determination of the effective $(F/cm^2)_{act}$ is complex and not well understood. The thickness of the double layer is very small (a fraction of a nanometer in liquid electrolytes), resulting in a value for a theoretical specific capacitance of 15 to 30 $\mu F/cm^2$. For a surface area of 1000 $m^2/g$, this would result in a potential capacitance of 150 to 300 F/g of electrode material. As indicated in Table 20.3, the measured specific capacitances of carbon materials being used in electrochemical capacitors are in most cases less than these high values being in the range of 75 to 175 F/g for aqueous electrolytes and 40 to 120 F/g using organic electrolytes. These lower values of F/g result because for most carbon materials only a fraction of the surface area in pores can be accessed by the ions in the electrolyte, and measured values of $(F/cm^2)_{act}$ for porous carbons are less than the theoretical values being in the range of 8 to 20 $\mu F/cm^2$. This is especially true for the organic electrolytes for which the size of ions is much larger than in an aqueous electrolyte. Porous carbons for use in capacitors should have a large fraction of their pore volume in pores of diameter 0.5 to 5 nm. Materials with small pores (<0.5 nm) show a large falloff in capacitance at discharge currents greater than 100 mA/cm$^2$ especially using organic electrolytes. Materials with the larger pore diameters can be discharged at current densities greater than 500 mA/cm$^2$ with a minimal decrease in capacitance.

**TABLE 20.3**  Double-Layer Specific Capacitance for Various Carbon Electrode Materials

| Material | Density, g/cm³ | Electrolyte | Specific gravimetric capacitance, F/g | Specific volumetric capacitance, F/cm³ |
|---|---|---|---|---|
| Carbon cloth | 0.35 | KOH | 200 | 70 |
| | | Organic | 100 | 35 |
| Activated carbon | 0.7 | KOH | 160 | 112 |
| | | Organic | 100 | 70 |
| Aerogel carbon (activated) | 0.2 | KOH | 140 | |
| | | Organic (PC) | 80 | 16 |
| Particulate carbon from TiC | 0.5 | KOH | 220 | 110 |
| | | Organic | 120 | 60 |
| Advanced graphitic carbon | 0.7 | Organic | 180 | 126 |

The cell voltage of the electrochemical capacitor is dependent primarily on the electrolyte used. For aqueous electrolytes, the maximum cell voltage is 1 to 1.6 V and for organic electrolytes, the cell voltage is 2.5 to 3.0 V.[11]

***Electrochemical Capacitors Utilizing Pseudocapacitance.***    For an ideal double-layer capacitor, the charge is transferred into the double layer and there are no faradaic reactions between the solid material and the electrolyte. In that case, the capacitance ($dQ/dV$) is a constant and independent of voltage. For devices that utilize pseudocapacitance, most of the charge is transferred at the surface or in the bulk near the surface of the solid electrode material. Hence, in this case, the interaction between the solid material and the electrolyte involves faradaic reactions, which in most instances can be described as charge transfer reactions. The charge transferred in these reactions is voltage dependent resulting in the pseudocapacitance ($C = dQ/dV$) being voltage dependent. Three types of electrochemical processes have been utilized in the development of electrochemical capacitors using pseudocapacitance. These are (1) surface adsorption of ions from the electrolyte, (2) redox reactions involving ions from the electrolyte, and (3) the doping and undoping of an active conducting polymer material in the electrode. The first two processes are primarily surface mechanisms and are hence highly dependent on the surface area of the electrode material. The third process, involving the conducting polymer material, is more of a bulk process and the specific capacitance of the material is much less dependent on its surface area, although relatively high surface area with micropores is required to distribute the ions to and from the electrodes in a cell. In all cases, the electrodes must have high electronic conductivity to distribute and collect the electron current. An understanding of the charge transfer mechanism can be inferred from $C(V)$, which is often determined using cyclic voltammetry.

For assessing the characteristics of devices, it is convenient to use the average capacitance ($C_{av}$) calculated from

$$C_{av} = Q_{tot}/V_{tot}$$

where the $Q_{tot}$ and $V_{tot}$ are the total charge and voltage change for a charge or discharge of the device between specified voltages. This permits a determination of the specific capacitance ($C_{av}/g$) of the material for the electrolyte of interest. As shown in Table 20.4, the specific capacitance of carbons using pseudocapacitance reactions is much higher than that of carbon materials with double-layer charge storage. It is thus expected that the energy density of devices developed using the pseudocapacitance materials will be higher.

**TABLE 20.4**  Pseudo-Capacitance for Various Electrode Materials and Electrolytes

| Electrode material | Faradaic process | Electrolyte | Voltage range | Specific gravimetric capacitance, F/g | Specific volumetric capacitance, F/cm³ |
|---|---|---|---|---|---|
| Activated carbon | Double layer | EC/DMC/LiBF₄ | 2.7–1.35 | 80–110 | 48–66 |
| Activated carbon | Redox | Li2SO₄ + KI | 1.5–0.75 | 240 | 145 |
| Li titanate oxide/carbon nanotubes | Intercalation Li⁺ | EC/DMC/LiBF₄ | 3.0–1.0 | 225 | 50 |
| Graphene/polypyrrole composite | Intercalation H⁺ | Aqueous/1 M H₂SO₄ | 1.5–0 | 270 | NA |

*Redox-Capacitance in Aqueous Electrolyte Cells.*    Most of the research on the development of electrochemical capacitors using aqueous electrolytes has used sulfuric acid or potassium hydroxide as the electrolyte. In these cases, the maximum voltage of symmetric cells (both electrodes activated carbon) is 1 V or less. This makes it difficult to attain energy densities for the cell greater than 2 Wh/kg for a complete cell (5–7 Wh/kg based only on the weight of the active material in the cell). General discussions of electrochemical cell research using aqueous electrolytes are given in Ref. 3. A most promising approach[12–14] is to use a near-neutral aqueous electrolyte ($Li_2SO_4$) with a redox salt (KI) permitting a maximum voltage of 1.5 to 1.6 V. Electrode and cell tests have shown an electrode specific capacitance of 240 F/g, 145 F/cm³. This resulted in an energy density of 14 Wh/kg for the cell based on the active material of the electrodes.

*Hybrid Advanced Capacitors.*    Electrochemical capacitors can be fabricated with one electrode being of a double-layer (carbon) material and the other electrode being of a pseudocapacitance material (see Fig. 20.3). Such devices are often referred to as *hybrid capacitors*. Detailed discussions of research on hybrid electrochemical capacitors are given in Refs. 15 to 17. As discussed in those references, devices have been assembled and tested using pre-lithiated graphite and metal oxides (for example, nickel oxide, lead oxide, lithium titanate oxide [LTO], manganese oxide) as the pseudocapacitance material in one of the electrodes. As shown in Figs. 20.4a and b, their charge/discharge characteristics (*V* vs. *t*) are significantly different (nonlinear) than double-layer devices.

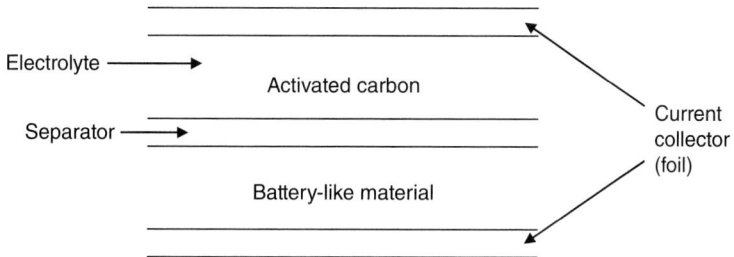

**FIGURE 20.3**    Schematic of a hybrid electrochemical capacitor.

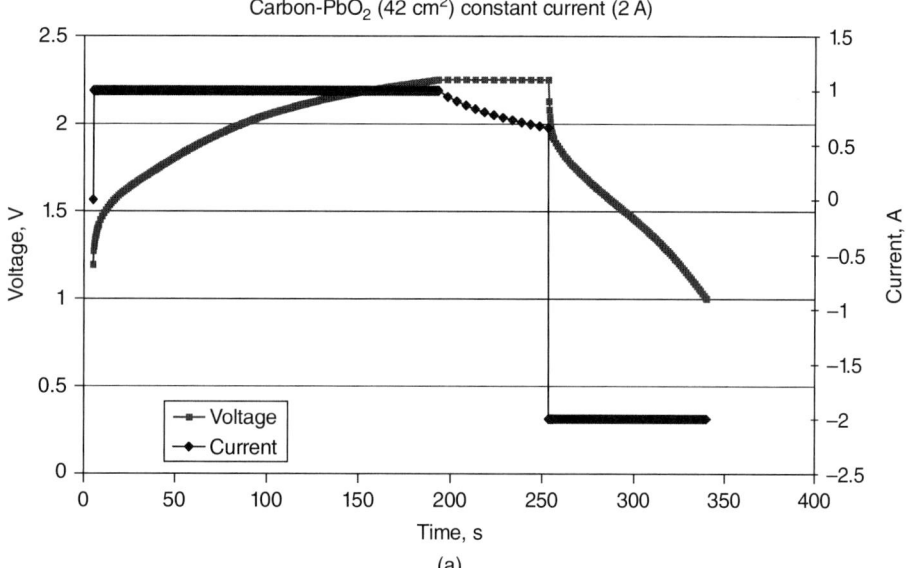

(a)

**FIGURE 20.4**    (*a*) Voltage versus current trace for a carbon-$PbO_2$ hybrid capacitor. (*b*) Voltage versus current trace for the Yunasko mixed metal oxides hybrid capacitor.

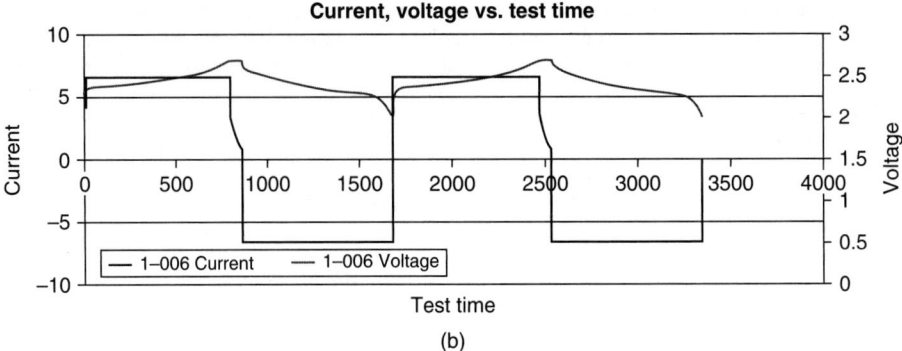

**FIGURE 20.4**    (*Continued*)

The energy densities of the hybrid devices are significantly higher than that of the double-layer cells using the same carbon and electrolyte. This is due to several factors. First, the charge stored in the carbon per unit mass is much greater than in the symmetric device because the useful voltage range of the carbon in the hybrid device is nearly twice that in the symmetric device. Second, the minimum (shoulder) voltage of the device is set by open-circuit voltage of the faradaic electrode chemistry. As shown in Sec. 20.4.5, these factors could result in increased energy densities as high as 20 to 50 Wh/kg. Research on hybrid electrochemical capacitors can be found in Refs. 15 to 17. Test data on hybrid devices are given in Sec. 20.4.1.

## 20.2   CHEMISTRY

In electrochemical capacitors, the role of chemistry in their design and operation is quite different than in battery cells, in which chemical reactions play the key role in the energy-storage process. In capacitors, as discussed in the previous section, charge transfer into the double layer of carbon is the critical process in determining the performance of the device. As discussed in the next section, the material properties of the electrode materials are given in terms of their ability to store charge, which is key to the overall design of the devices. The role of the electrolyte is to permit the transfer of the charge/ions between the electrodes.

## 20.3   MATERIAL PROPERTIES AND CELL DESIGN

### 20.3.1   Activated Carbon Electrodes

The electrodes in electrochemical capacitors are in general thin coatings applied to a metallic current collector. The active material is mixed with a binder to form a slurry that can be applied at a controlled thickness, rolled, and dried to form a porous electrode.[18,19] The thickness of the electrode is generally in the range of 100 to 200 $\mu$ and it has a porosity of 65% to 80%. In order to achieve a low resistance, the contact resistance between the active material coating and the current collector must be very small. This requires special attention to preparing the surface of the current collector before applying the electrode coating.[20,21]

As noted previously, a key electrode material property is its specific capacitance (F/g, F/cm$^3$). The capacitance of an electrode of known thickness and surface area dimensions ($\delta$th, $A_x$) can be calculated directly with good accuracy from its specific capacitance and density $\rho_c$:

$$C = F/g\,(\rho_c)\,\delta th\,A_x$$

As indicated in Table 20.3, the specific capacitance of activated carbon can vary over a wide range (80–220 F/g), depending on how it is processed and on the electrolyte used in the cell. The density of the carbon layer ($\rho_c$) can vary from 0.3 to 0.8 g/cm$^3$. The specific capacitance of the carbon depends on its surface area (m$^2$/g), pore size distribution, and intrinsic surface double-layer capacitance ($\mu$F/cm$^2$). It is well recognized that the pore size distribution can have a large effect on the specific capacitance of the carbon and that it is essential to properly match the solvated and desolvated size of the electrolyte ions and the pore size of the carbon. The physical process by which the ions diffuse in and out of the pores is not well understood making the correlation of specific capacitance and carbon surface area uncertain.[6–10] Research on optimizing carbons for double-layer capacitors will be critical to achieving large improvements in the energy density of carbon/carbon devices.

The specific capacitance of the electrode materials can also vary significantly with current density (A/cm$^2$). It is common for researchers to show the specific capacitance of their active electrode material as F/g versus A/g$_{actmat}$. F/g versus A/cm$^2$ can be determined from the A/g data if the g/cm$^2$ of the loading of the active material on the electrode is given. Hence, in evaluating materials for electrochemical capacitors, the specific capacitance of the active material should be measured for current densities up to at least 300 mA/cm$^2$ using relatively thin electrodes (<200 $\mu$). The most direct method for evaluating active materials is to form thin electrodes from the material and to perform constant current tests of small cells assembled from the electrodes (see Table 20.5 taken from Ref. 22).

**TABLE 20.5**  Effect of Current Density on the Specific Capacitance, F/g

| I, A | Current density, mA/cm$^2$ | C, F | R, $\Omega$ | R$_{surface}$, $\Omega$cm$^2$ | Specific capacitance, F/g, dry electrode |
|------|------|------|------|------|------|
| 0.2  | 66   | 5.72 | 0.123 | 0.37 | 163 |
| 0.3  | 100  | 5.58 | 0.151 | 0.45 | 159 |
| 0.5  | 167  | 5.3  | 0.120 | 0.36 | 151 |
| 0.75 | 250  | 4.96 | 0.144 | 0.43 | 142 |
| 1.0  | 333  | 4.80 | 0.164 | 0.49 | 153 |

*Note:* Lab cells, area 3 cm$^2$, sulfuric acid electrolyte, electrode thickness 200 $\mu$.[22]

## 20.3.2  Advanced Carbon Material Electrodes

The energy density of an electrochemical capacitor depends on both the specific capacitance (F/g) of the electrode materials and the maximum voltage that can be used in the cell. Most of the commercial devices currently being marketed utilize electrodes formed using particulate, activated carbons, but research is continuing to develop higher specific capacitance carbons. Other forms of carbon being studied are aerogel[23,24] and nanotube[25,26] carbons and carbons from polymers.[27–29] One of the advantages of these carbons is that the pore size and structure of the carbon can be closely controlled in the manufacturing process. To date (2017), none of these advanced carbons have shown more attractive characteristics for use in electrochemical capacitors than activated carbon powders (see Tables 20.3 and 20.4).

The use of carbon nanotubes to form the electrodes is continuing. This can be done by growing a bed of nanotubes normal to the current collector surface using a gaseous hydrocarbon feedstock from catalytic sites.[30] Considerable progress has been made in the preparation of this type of carbon material suitable for electrochemical capacitors. The key issues concerning the attractiveness of the carbon nanotubes are their low intrinsic specific capacitance (20–80 F/g) and low material density (less than 0.3 g/cm$^3$). In the case of aerogel carbons, the specific capacitance has been 80 to 140 F/g for organic and aqueous electrolytes, respectively, but the density has been low. However, as discussed in the next section, the structure of these advanced carbons has made them attractive for embedding nanoparticles of pseudocapacitive materials in the structure. In this way, composite materials with very high specific capacitance can be formed.

## 20.3.3  Advanced Composite Material Electrodes

Research[31–35] is being done to develop composite materials for electrodes with specific capacitance much higher than activated carbon. These materials in most cases involve the fabrication of nanosize particles of metal oxides and graphene that can be mixed with activated carbons or embedded (decorated) on nanostructured carbons such as aerogels or nanotubes. As indicated in Table 20.4, the specific capacitance of these composite materials can be 200 to 300 F/g. Using organic electrolytes, the voltages of cells using these materials can be 3 to 4 V. The charge storage mechanisms in the advanced materials are more complex than simple double-layer formation and involve either surface charge transfer or intercalation of the electrolyte ions into the nanoparticles in the porous carbon structure. These composite materials can be used in hybrid capacitors that utilize double-layer or pseudocapacitive material in one electrode and a battery-like, faradaic material in the other electrode[36] to achieve energy densities of 30 to 35 Wh/kg.

One of the key issues with these composite, pseudocapacitive materials is their stability in cycling and the resultant effects on the cycle life of devices assembled using these materials. It is unlikely that cycle life approaching that of EDLC devices is possible with the composite, pseudocapacitive materials.

### 20.3.4 Hybrid Capacitors Using a Lithiated Faradaic Electrode

The commercial hybrid electrochemical capacitors use a lithiated graphite negative electrode, an activated carbon positive electrode, and an organic electrolyte. Considerable research[31,32] has been done on cells using LTO in the negative electrode. The main limitations using LTO are that the rated voltage of the cell is only 2.5 to 3 V and the charge capacity of LTO is only about 150 mAh/g. In addition, the electronic conductivity of LTO is also very low resulting in the need to attach the LTO to a nanoscale carbon structure to achieve high device power capability.

In the case of lithiated graphite as the negative electrode, the primary difficulty is how to get the lithium into the graphite in a controlled manner. However, the rated cell voltage can be 3.5 to 4 V and the charge capacity is about 350 mAh/g. The high voltage and charge capacity increase the potential for high energy density in cells. Commercial cells with an energy density of 10 to 13 Wh/kg, 15 to 20 Wh/L (see Table 20.2) have been developed. Most cells presently use either a sacrificial lithium metal foil or stabilized lithium metal powder as the source of the lithium intercalated into the graphite during the first charge. Research[37] is continuing to find better approaches to lithiate graphite or to develop negative electrode materials that do not need lithiation.

### 20.3.5 Current Collector Materials

In most cases, the capacitive devices are assembled by applying a thin layer/film of active material on to a high conductivity, current collector. The key issues related to the current collector are a near-zero contact resistance with the thin film of carbon and the long-term stability of the material (metal or conducting plastic) and the coating in the environment of the cell (voltage and electrolyte). These issues are particularly important for electrochemical capacitors because of their very low resistance, cycle life of hundreds of thousands of cycles, and 10 to 15 years of calendar life. Of special interest is research dealing with cleaning and coating metal foils[20,21] and conducting plastic sheets,[38] which can be used for bipolar capacitors and batteries.

The density and electrical conductivity of the current collector material can have a large effect on both the energy density and resistance of the assembled cell. The characteristics of a number of metals used for the current collector in electrochemical capacitors are shown in Table 20.6. The current collector can contribute significantly to the weight of the cell and its resistance, especially if a material other than aluminum is used in the current collector. The weight of the current collector is minimized by making it very thin, but that can result in an unacceptable contribution to the resistance of the cell.

**TABLE 20.6**    Characteristics of Current Collector Materials

| Material | Density, g/cm$^3$ | Conductivity, $\Omega$cm |
|---|---|---|
| Aluminum | 2.7 | $2.65 \times 10^{-6}$ |
| Copper | 9.0 | $1.68 \times 10^{-6}$ |
| Stainless steel | 7.9 | $74 \times 10^{-6}$ |

## 20.3.6 Electrolytes

The capacitance of an electrochemical capacitor is dependent primarily on the specific capacitance (F/g) of the electrode material, but the cell voltage and resistance are primarily dependent on the electrolyte[11] used in the device. Three types of electrolytes have been used in electrochemical capacitors: aqueous (sulfuric acid and KOH), organic (propylene carbonate and acetonitrile), and ionic liquids. Recent research[12–14] has been done using near-neutral, aqueous electrolytes that have relatively high ionic conductivity but are much less corrosive than sulfuric acid and KOH. Salts are added to the solvents to provide the ions that move in and out of the double layers formed in the micropores of the carbon. The characteristics of the various electrolytes are given in Table 20.7. The ionic conductivity of the electrolyte depends on both the solvent and the salt used in it. The specific capacitance of the carbon material depends on the salt added and the diameter of the salt molecules relative to the pore size distribution of the electrode material.[12–14] In the case of cells utilizing redox reactions, the salts added to the solvent are key to achieving high specific capacitance in the cells.[3] Detailed discussions of various combinations of electrolyte solvents and salts are given in Ref. 7.

**TABLE 20.7**  Properties of Various Electrolytes

| Solvent | Salt | Density, g/cm$^3$ | Resistivity, $\Omega$cm | Cell voltage |
|---|---|---|---|---|
| **Aqueous (water)** | | | | |
| KOH | | 1.29 | 1.9 | 1.0 |
| Sulfuric acid | | 1.2 | 1.35 | 1.0 |
| Near-neutral | $Li_2SO_4$ + KI | 1.03 | 10.6 | 1.5 |
| **Organic** | | | | |
| Propylene carbonate | TEABF4 | 1.2 | 66 | 2.5–3.0 |
| Acetonitrile | TEABF4 | 0.78 | 18 | 2.5–3.0 |
| **Ionic liquid** | | | 12 | |
| Imidazolium | [EtMeIm]$^+$ [BF$_4$]$^-$ | 1.3 | 72 (25°C) | 2.2–4.3 |
| Pyrrolidinium | [nBuMePyrr]$^+$ [N(CF$_3$SO$_2$)$_2$]$^-$ | | 455 (25°C) | 5.5–2.5 |

There are large differences in the ionic resistivity and cell voltage (useable electrochemical window) of the electrolytes. These differences lead to corresponding large differences in the performance of devices using the various electrolytes. Since the energy density is proportional to the square of the cell voltage, increasing the cell voltage is a key objective of electrochemical capacitor research. Using activated carbon, the cell voltage is 2.5 to 3.0 V/cell using organic electrolytes and 1 to 1.5 V/cell using aqueous electrolytes. The cell voltage is also dependent to a limited extent on the carbon used in the device. Cell voltages up to 3.5 V/cell have been reported with structured graphitic carbons.[39,40]

The differences in the ionic resistivity of the electrolytes have a large effect on resistance and consequently the power capability of a device. The resistivity of propylene carbonate is about a factor of three higher than that of acetonitrile. For this reason, electrochemical capacitors with the best performance (highest energy density and power capability) use acetonitrile as the electrolyte. Cells using acetonitrile show less reduction in performance at low temperatures (T < −20°C) than those using propylene carbonate. However, cells using propylene carbonate can operate at higher temperatures than those using acetonitrile due to its higher vapor pressure at those temperatures. The boiling point of propylene carbonate is 240°C compared to 82°C for acetonitrile.

There is a continuing controversy[41] concerning the safety of acetonitrile especially in vehicles because of its toxicity and flammability. There has been much research to develop a low resistivity, nontoxic solvent to replace acetonitrile. Some progress has been made by mixing organic solvents,[7] but to date acetonitrile continues to be the solvent of choice in devices of very high power.

Research[7,42–44] is being done to develop electrochemical capacitors using ionic liquid electrolytes. These electrolytes are attractive for several reasons. First, they can be thermally stable for temperatures as high as

300°C with near-zero vapor pressure and are nonflammable with very low toxicity. Further, the useable electrochemical window is large leading to cell voltages as high as 4 to 5 V in double-layer capacitors. The major difficulties with the ionic liquids are their high ionic resistivity at or near room temperature and the high cost. The resistivity of an ionic liquid is strongly temperature sensitive and requires a temperature of about 125°C to have a resistivity comparable to that of acetonitrile. Blending ionic liquids with acetonitrile[43,44] can greatly reduce the flammability of the electrolyte with only minor changes in room-temperature conductivity and the voltage window. However, acetonitrile-free blends, which are both nonflammable and nontoxic, show large increases in resistivity and a 0.5 to 1.0 V reduction in the voltage window.[7,44]

## 20.4  PERFORMANCE CHARACTERISTICS OF DEVICES

### 20.4.1  Small Carbon/Carbon Devices (Capacitance <10 F)

Most of the small electrochemical capacitor units sold are modules of two or more cells in series having a voltage of 4 to 8 V. The capacitance of the units is <1.0 F with time constants of 1 to 10 ms. Devices using both aqueous and organic electrolytes are commercially available. As expected, the energy density of the devices using organic electrolytes is higher than those using aqueous electrolytes. The small units are available as coin cells (thin cylindrical disks) and thin, prismatic cells (credit card-like). Most of the small devices are used in consumer electronics such as pagers, cell phones, and computers in conjunction with batteries either in power-assist or battery backup modes. The price of the small devices is relatively low being about 50 cents/unit, but still much higher than the traditional ceramic capacitors. Requirements for these small units are usually given in terms of capacitance, $RC$ time constant, and volume or thickness (for the prismatic cells). Energy density is usually of secondary importance with the ability to provide periodic, multimillisecond pulses being of primary concern. In order to provide the short pulses, the time constant of the device should be <50 ms.

As indicated in Table 20.8, small devices with time constants in the 10 to 200 ms range have been developed by several manufacturers.[45–48] Testing of the small units[49] has indicated that they can be pulse charged and discharged with pulse widths as short as 1/50th of the $RC$ time constant without a significant reduction in the effective capacitance of the device. Hence, these devices respond as near-ideal devices (constant capacitance and resistance) as long as the charge and discharge periods of pulsing are 5 to 10 times longer than the $RC$ time constant of the device. As indicated in Table 20.8, the small devices have very high-power capability (greater than 10 kW/L). These devices can provide pulse currents of several amperes at high efficiency

**TABLE 20.8**    Physical and Performance Characteristics of Small Double-Layer Devices

| V | Capac., F | $R$, mΩ | Weight, g | Volume, cm³ | $RC$, ms | Spec. energy, Wh/L | Spec. energy, Wh/kg | Matched impedance,* kW/L* | 95% efficiency* |
|---|---|---|---|---|---|---|---|---|---|
| 2.4 | 0.18 | 45 | 0.6 | 0.44 | 8.1 | 0.25 | 0.18 | 73 | 8.2 |
| 2.4 | 0.3 | 34 | 0.9 | 0.6 | 10.2 | 0.3 | 0.2 | 70.5 | 7.9 |
| 2.4 | 0.65 | 18 | — | 0.93 | 11.7 | 0.42 | — | 86 | 9.7 |
| 2.4 | 1.1 | 26 | — | 0.8 | 28.6 | 0.825 | — | 69 | 7.8 |
| 2.4 | 2.3 | 28 | 1.2 | 1.02 | 64 | 1.35 | 1.15 | 50.5 | 5.7 |
| 2.4 | 4.0 | 22 | 1.5 | 1.4 | 88 | 1.7 | 1.6 | 47 | 5.3 |
| 2.7[†] | 10.5 | 25 | 2.5 | 1.5 | 262 | 4.8 | 2.9 | 29.2 | 3.3 |
| 2.7[†] | 15 | 30 | 4.15 | 2.83 | 438 | 3.6 | 2.5 | 14.6 | 1.65 |
| 7[‡] | 0.047 | 120 | — | 2.2 | 5.6 | 0.11 | — | 46.4 | 5.2 |
| 4.2[‡] | 0.022 | 200 | — | 1.1 | 4.4 | 0.04 | — | 20.0 | 2.5 |

*Pulse matched impedance power: $V_0^2/4R$; 95% eff. power: $9/16\,(1 - \mathrm{EF})\,V_0^2/4R$, $\mathrm{EF} = 0.95$.
[†]Cylindrical devices, all others are flat credit card-type devices.
[‡]Multicell devices using aqueous electrolyte, all other single-cell devices use an organic electrolyte.

(greater than 90%). However, the energy density (Wh/kg) of the small devices is relatively low being in the range of 0.1 to 1.0 Wh/kg depending on size and electrolyte being used. As noted previously, energy density is of secondary importance with short pulse capability with high power being the critical requirement.

## 20.4.2    Large Carbon/Carbon Devices (Capacitance >100 F)

Activated carbon/carbon electrochemical capacitor devices (single cells and modules) are commercially available from a number of companies—Maxwell, Panasonic, NessCap, Nippon, Chem-Con, etc. All of these companies market large devices with capacitance of 1000 to 5000 F. The carbon/carbon technology is the most suitable for vehicle applications because of its high power and long cycle life. The performance of cells from the various manufacturers is given in Refs. 45 to 48 and summarized in Table 20.1. The reported energy densities (Wh/kg) correspond to the useable energy from the devices based on constant power discharge tests from $V_0$ to $\frac{1}{2}V_0$. Peak power densities are given for both matched impedance and 95% efficiency pulses. For most applications with electrochemical capacitors (ultracapacitors), the high efficiency power density is the appropriate measure of the power capability of the device. The energy density for most of the available devices is between 4 and 5 Wh/kg, but the 95% power density varies over a wide range from 900 to 10,000 W/kg. In recent years, the energy density of the devices has been gradually increased for the carbon/carbon (double-layer) technology to 5.5 to 6.5 Wh/kg and the cell voltage has increased from 2.7 to 3.0 V using acetonitrile as the electrolyte. Both the energy density and power capability are lower for cells using propylene carbonate as the electrolyte.

For vehicle applications, the cells are connected in series to form higher-voltage modules. The module voltage utilized depends on the application and varies from 16 to about 60 V. The characteristics of modules from several companies are summarized in Table 20.9. The electrical characteristics (capacitance and resistance) of the module follow directly from the cell characteristics. Note, however, that the weight and volume of the modules are significantly greater than the cells alone with packaging factors of 0.5 to 0.6. All the modules being marketed utilize balancing circuits for each cell to prevent overvoltage of the cell and to minimize cell-to-cell variability during cycling. For this reason, it is best to base the energy storage and power capacity of the modules on the cell weight and volume, but to include the packaging factors in determining the weight and volume of the capacitor unit to be installed in a vehicle. More detailed discussions of module characteristics and cell balancing are given in "Module and Lifetime Considerations" in Sec. 20.4.7.

## 20.4.3    Performance of Cells Using Advanced Materials

Most of the devices listed in Table 20.9 are commercially available in 2018 and represent the state of the art for devices that utilize activated carbon in their electrodes. Several companies are in the process of developing electrochemical capacitors with higher energy density, but only JSR Micro is marketing cells and modules in 2018 having an energy density greater than 10 Wh/kg for the cells. The JSR Micro technology utilizes pre-lithiated graphite in the negative electrode and activated carbon in the positive electrode. This hybrid cell operates between 3.8 and 2.2 V. Several other manufacturers are developing hybrid cells, but are not yet marketing

**TABLE 20.9**   Summary of the Characteristics of Supercapacitor Modules

| Module | Weight/volume, kg/L | Voltage | Energy (Wh)/specific energy, Wh/kg | Power (kW) at 90% efficiency | Packaging weight factor | Packaging volume factor |
|---|---|---|---|---|---|---|
| Ness (166F) | 16/13.7 | 48 | 41/2.56 | 17.5 | 0.52 | 0.468 |
| Maxwell (145 F) | 13.5/13.4 | 48 | 36/2.7 | 14.5 | 0.627 | 0.484 |
| DAE-China (175F) | 12.7/10.1 | 48 | 40.3/3.17 | 17.5 | 0.567 | 0.565 |
| Ness (500F) | 5/4.3 | 16 | 13.6/2.26 | 6.0 | 0.52 | 0.56 |
| Maxwell (430F) | 5.0/4.85 | 16 | 11.8/2.36 | 4.8 | 0.564 | 0.445 |
| DAE-China (514 F) | 5.4/4.2 | 16 | 13.1/2.43 | 5.2 | 0.436 | 0.561 |
| Yunasko (205F) | 3.2/2.6 | 16 | 5.44/1.7 | 27.2 | 0.413 | 0.35 |

**TABLE 20.10**   Performance of Advanced Electrochemical Capacitor Devices of Various Technologies

| Technology type | Developer | Status | $V_{rated}$ | C, F | RC, s | Wh/kg* | W/kg 95%[†] |
|---|---|---|---|---|---|---|---|
| Carbide-based carbon with graphene | Skeleton Technologies | Prototype | 3.4 | 3200 | 1.5 | 8.9 | 1730 |
| Activated carbon/organic electrolyte | Ness/Maxwell | Comm. product | 3.0 | 3650 | 0.98 | 6.5 | 1875 |
| Activated carbon/organic electrolyte | Skeleton Technologies | Comm. product | 2.85 | 3450 | 0.26 | 5.4 | 5891 |
| Pre-lithiated graphite/activated carbon | JSR Micro | Comm. product | 3.8 | 1100 | 1.2 | 10 | 2450 |
|  |  |  |  | 1366 | 1.6 | 7.6 | 1366 |
| Pre-lithiated graphite/activated carbon | DAE-China | Prototype | 3.8 | 850 | 3.5 | 12.4 | 993 |
| Activated carbon mixed with metal oxides in both electrodes | Yunasko | Prototype | 2.7 | 3200 | 7.8 | 30–35 | 3395 |
| Hybrid carbon/$PbO_2$/ sulfuric acid | UCDavis | Lab[‡] | 2.2 | 13 | 2.8 | 9.7 | 1300 |
| $Li_4Ti_5O_{12}$/CNF/ activated carbon/organic electrolyte | Tokyo Univ. Agric./ Nippon Chemi-Con | Lab[‡] | 3.0 | NA | NA | 30 (estim.) | 3000 (estim.) |

*Useable energy density.
[†]Power density: $P = 9/16 \times (1 - EFF) \times V^2/R$, EFF = 0.95.
[‡]Unpackaged, weight of all active materials in the device.

them in 2018. The performance characteristics of advanced cells of various technologies are shown in Table 20.10, based primarily on testing of the cells at UC Davis.[50,51] The electrode technology and status (product, prototype, lab) of the advanced cells are noted. All the devices cited have significantly higher energy density than the commercially available carbon/carbon cells in Table 20.2. Most of the advanced devices are hybrid cells using activated carbon in one of the electrodes. The cycle life of most of the advanced cells is much less than the carbon/carbon devices, but the use of carbon in one electrode can lead to a relatively long cycle life of at least 100 K cycles, which is much higher than lithium batteries for complete (deep) discharges. The JSR Micro cells are claimed[52] to have a cycle life of one million cycles, which is the same as the carbon/carbon cells. The primary problem with the pre-lithiated hybrid cells has been gassing of some cells even at room temperature.

The power capabilities shown for the advanced cells shown in Table 20.10 are in most cases higher than for most of the commercially available carbon/carbon cells shown in Table 20.2.

### 20.4.4   Electrochemical Capacitor Modeling

This section is concerned with various approaches to modeling electrochemical capacitors. Some of the modeling is semiempirical in character ("Equivalent Circuit and AC Impedance" in Sec. 20.4.4) and other approaches are more mathematical, but all the approaches are dependent on knowledge of the properties of the materials used in the electrodes of the devices being modeled.

***Equivalent Circuit and AC Impedance.***   As discussed in previous sections, an electrochemical capacitor is assembled with porous electrodes consisting of a microporous material such as activated carbon. The capacitance and thus the electrical energy storage take place in the double layers formed in the micropores of the carbon. It is convenient to model these complex processes in terms of equivalent electrical circuits, as indicated in Fig. 20.5. As shown in the figure, the circuit consists of multiple *RC* elements connected in a ladder that accounts for the transport of the ions into the double layers along the length (depth) of the micropores.

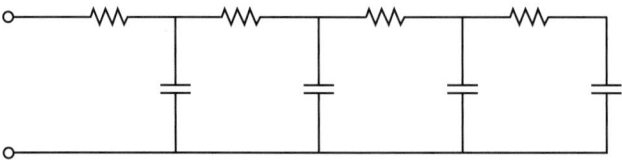

**FIGURE 20.5**   Equivalent circuit—multiple *RC* elements in series.

Experience[53] has shown that for applications in which significant changes in power demand occur over a few time constants, the single $RC$ element model predicts with reasonable accuracy the response of electrochemical devices. For other applications involving more rapid power changes, an equivalent circuit consisting of multiple $RC$ elements is needed. If the device is modeled by a simple $RC$ circuit, the values for $R$ and $C$ can be determined from DC constant current tests (see "Application of the Test Procedures to Carbon/Carbon Double-Layer Capacitors" in Sec. 20.4.6). However, if the device is modeled using the ladder circuit, the multiple $R$ and $C$ values can be determined from AC impedance testing[54,55] in which an AC voltage is applied to a device, and its impedance is measured as a function of frequency, $\omega$.

***Mathematical Modeling.***    Electrochemical capacitors can be modeled starting from the governing equations if simplifying assumptions are made concerning the properties of the carbon materials used in the electrodes and the processes involved with the transport of the ions in the electrolyte between the electrodes. Detailed mathematical models of the electrochemical capacitors are presented in Refs. 56 to 58. These models are one-dimensional ($x$—normal to the electrodes in the direction of current flow) and time dependent ($t$). Analytical solutions can be derived for DC constant current and constant power discharges and AC current responses to sinusoidal voltage perturbations of the capacitors if $R$ and $C$ are assumed constant. The mathematical solutions are helpful in understanding the design trade-offs for devices and the interpretation of test data especially for carbon/carbon double-layer devices. As shown in Fig. 20.1, the capacitor is a composite of solid, porous carbon, and a liquid electrolyte. The electrical current (electrons) is taken to and from the device through a metal current collector and is distributed in the electrode by the carbon particles. The positive and negative ions that form the double-layer capacitance in the micropores of the carbon move through the electrolyte by diffusion due to its ionic conductivity. The separator allows the diffusion of the ions between the positive/negative electrodes, but blocks the conductivity of the electrons that are then forced into the external circuit. The equations needed to describe these processes in the carbon and electrolyte can be relatively simple if the details of ion transport in and out of the pores are neglected and included only in terms of the specific capacitance of the carbon material. The equations to be solved are Ohm's law for the electron and ionic currents and the conservation of charge at all points in the electrode. The equations to be solved are the following:

*Ohm's law:*

$$i_1 = -\sigma \, \partial \, \Phi_1/\partial x$$
$$i_2 = -\kappa \, \partial \, \Phi_2/\partial x$$

where $i_1$ is the electron current density in the carbon and $i_2$ is the ionic current density in the electrolyte, and $\Phi_1$ and $\Phi_2$ are the potential in the carbon and electrolyte.

*Conservation of charge:*

$$I(t) = i_1 + i_2$$

where $I(t)$ is the applied current to the cell.

$$\partial \, i_1/\partial x = -\partial \, i_2/\partial x = a \, i_n$$

where $a$ is the area of the carbon per cm$^3$ and $i_n$ is the interface current A/cm$^2$ in the pores of the carbon.

$$i_n = -C \, \partial(\Phi_1 - \Phi_2)/\partial t$$

where $C$ is the specific capacitance (F/cm$^2$) of the carbon.

The above equations can be combined and written as

$$(\sigma \, \kappa/\sigma + \kappa) \, [\partial^2 \, \Phi_1/\partial x^2 - \partial^2 \, \Phi_2/\partial x^2] = -aC\partial(\Phi_1 - \Phi_2)/\partial t$$

and in nondimensional form

$$\partial^2\eta/\partial^2\xi = \partial\eta/\partial\tau \tag{20.1}$$

where $\eta = \Phi_1 - \Phi_2/V_0$, $\xi = x/\delta$, $\tau = aC\,\delta^2\,(\sigma + \kappa)/\sigma\,\kappa$, $\delta$ is the thickness of the electrode, $\sigma$ is the conductivity of the carbon, $\kappa$ is the conductivity of the electrolyte, and $\eta$ is the overpotential (voltage in the double layer) at the surface of the carbon. All the equations are written for 1 cm$^2$ of the electrode area. All currents are in A/cm$^2$.

For charging the electrode, the boundary conditions are the following:

$x = 0$ (surface of the metal current collector), $\qquad i_2 = I(t)$, $i_1 = 0$, $\eta = 0$

$x = \delta$ (surface of the separator), $\qquad i_2 = 0$, $i_1 = I(t)$

The equations shown apply to a single electrode. It is assumed that both electrodes in the cell are identical so that the energy stored and the resistance of the electrodes are identical and the solutions for the electrodes can be combined to form the total solution for the cell.

Methods for solving equations of the form of Eq. (20.1) are well known from nonsteady, one-dimensional heat transfer analysis.[59] Solutions for both constant DC currents and AC sinusoidal currents for carbon/carbon electrochemical capacitors are given in Refs. 56 to 58.

*Constant Current Discharges.* For the case of a constant current discharge of a cell, the solution to Eq. (20.1) can be written as

$$V = V_0 - IR_0 - 2It/\delta\,aC - I\Sigma_n[4(\kappa + (-1)^n \sigma)^2\,\delta]/(\sigma\kappa(\sigma + \kappa)n^2\Pi^2)\{1 - \exp(-n^2\Pi^2 t/\tau)\} \tag{20.2}$$

where $R_0$ ($\Omega$-cm$^2$) $= 2\delta/(\sigma + \kappa) + \delta_{sep}/\kappa_{sep} + R_{contact}$

The resistance ($\Omega$-cm$^2$) of the cell is time dependent and is given by

$$R = R_0 + \Sigma_n[4(\kappa + (-1)^n \sigma)^2\,\delta]/(\sigma\kappa(\sigma + \kappa)n^2\,\Pi^2)\{1 - \exp(-n^2\Pi^2 t/\tau)\}$$

The resistance at $t = 0$ is $R_0$. The solution for the cell performance given in Eq. (20.2) represents the cell as a simple *RC* circuit with a time-dependent resistance.

In most cases, the conductivity of the carbon is much greater than that of the electrolyte so the cell resistance becomes

$$R = 2\delta/\sigma + \delta_{sep}/\kappa_{sep} + R_{contact} + 4\delta/\kappa\Sigma_n(1/\Pi^2 n^2)\{1 - \exp(-n^2\Pi^2 t/\tau)\} \tag{20.3}$$
$$\tau = aC\delta^2/\kappa$$

The capacitance and resistance of the cell are $C_{cell} = \frac{1}{2}A_x\delta aC$ and $R_{cell} = 2\delta/\kappa A_x$. Hence to first approximation,

$$\tau = (RC)_{cell}$$

For large values of $t/\tau$, the final summation becomes $\Pi^2/6$ and the steady-state value of $R$ is

$$R_{ss} = 2\delta/\sigma + \delta_{sep}/\kappa_{sep} + R_{contact} + \frac{2}{3}\delta/\kappa$$

The exponent in the Eq. (20.3) is $n^2\Pi^2 t/(RC)_{cell}$. Even for $t = (RC)_{cell}$ and $n = 1$, the contribution of the terms in the summation is small. Hence, the simple solution, which neglects the effects of ion transport in the micropores of the carbon, indicates that the cell resistance is essentially the steady-state value except for discharge times equal to a small fraction of the *RC* time constant of the cell. However, as discussed in "Application of the Test Procedures to Carbon/Carbon Double-Layer Capacitors" in Sec. 20.4.6, test data indicate that for constant current discharges, the cell resistance does not approach the steady-state value until times of 1 to 2 *RC* time constants are reached, indicating the ion transport in the micropores is important for small times.

The value of $\kappa$, the ionic conductivity of the electrolyte, used in the calculations should include the effects of the porosity of the carbon.[60] Hence,

$$\kappa = \kappa_0 \, \varepsilon^{1.5}$$

where $\kappa_0$ is the bulk conductivity of the electrolyte and $\varepsilon$ is the porosity of the carbon.

*Constant Power Discharges.*    The previous analysis was concerned with constant current discharges. If the capacitance $C$ and resistance $R$ of a cell are assumed to be constant, it is straightforward to derive an expression for the voltage for a constant-power discharge. The governing equation for the discharge is

$$V_0 - V = IR + {}_v\!\int dq/C, \; dq = I \, dt \tag{20.4}$$

where $V_0$ is the voltage immediately after the initiation of the discharge.

For discharge at constant power $P$, the current is given by

$$I = P/V$$

and Eq. (20.4) becomes

$$1 - V/V_0 = PR/V_0^2/V/V_0 + P/V_0^2/C \left\{\int dt/V/V_0\right\} \tag{20.5}$$

Defining $z = V/V_0$, $K_1 = PR/V_0^2$, $K_2 = P/CV_0^2$, Eq. (20.5) becomes

$$1 - z = K_1/z + K_2 \int dt/z \tag{20.6}$$

where $z_0' = 1 - (IR)_0/V_0 = 1 - PR/V_0^2 = 1 - K_1$, $K_1/K_2 = RC$

Equation (20.4) can be differentiated and then integrated in closed form to obtain

$$K_1[\ln z - \ln z_0'] - \tfrac{1}{2}(z^2 - z_0'^2) = K_2 \, t \tag{20.7}$$

Inputting the defined variables, Eq. (20.5) becomes

$$t/RC = [\ln V/V_0 - \ln(1 - K_1)] - (\tfrac{1}{2})/K_1[(V/V_0)^2 - (1 - K_1)^2] \tag{20.8}$$

Equation (20.6) can be rewritten as

$$t = \tfrac{1}{2}CV_0^2/P[(1 - K_1)^2 - (V/V_0)^2] + RC \ln[V/V_0/(1 - K_1)]$$

$K_1 = PR/V_0^2 = I_0 R/V_0$ is an indicator of the efficiency EF of the discharge. $EF_0 = 1 - K_1$.

The energy density of the constant power discharge is then

$$\text{Wh/kg} = tP/\text{weight of cell in kg}$$

Consider the following example:

$$C = 2900 \text{ F}, \, R = 0.375 \text{ m}\Omega, \, \text{weight} = 0.55 \text{ kg}, \, RC = 1.09 \text{ s}, \, P = 500 \text{ W}, \, 909 \text{ W/kg}$$

$$t_{\text{calculated}} = 14.1 \text{ s}, \, 3.6 \text{ Wh/kg}, \, V_{\text{final}}/V_0 = \tfrac{1}{2}; \text{ measured } t = 15.1 \text{ s}, \, 3.8 \text{ Wh/kg}$$

## 20.4.5   Design Analysis for Electrochemical Capacitors

As an example of the design analysis, the case of a hybrid capacitor is considered (see Fig. 20.3). The same approach can be used for double-layer capacitors and devices using metal oxides with carbon in both electrodes.

All hybrid capacitor designs being considered utilize carbon in at least one of the electrodes. The device is designed such that the charge transferred between the electrodes is set by the charge capacity and voltage change of the carbon electrode. The battery-like electrode is designed such that its operating range (depth of discharge in the capacitor) in charge/discharge is relatively small (10–20%) so that its expected life can be very long compared to what would be expected in a battery. The cell voltage of the hybrid device is determined from the sum of the operating range of the carbon electrode and standard potential of the battery-like electrode in the electrolyte of interest. The performance calculated using the method outlined below is the ideal performance of the device assuming that all interfacial resistances are negligible and the specific capacitance of the carbon is not rate dependent. In addition, it is assumed that the resistance of the device is primarily determined by the macroresistance of the porous electrodes—neglecting the micropore resistance of the carbon. Calculating the pore resistance is difficult and uncertain.[61] For high-performance supercapacitors, test data for cell resistance indicate that the difference between the steady-state resistance, which includes the resistance of the micropores, and resistance at $t = 0$, which includes only the resistance of the macropores, is relatively small (see "Application of the Test Procedures to Carbon/Carbon Double-Layer Capacitors" in Sec. 20.4.6). Hence, neglecting the resistance of micropores should not greatly underestimate the resistance of the cell.

**Inputs.**    The physical dimensions and material properties of all the components used in the device must be known. The key inputs for the carbon electrode are its thickness and the properties of the carbon-specific capacitance (F/g), density (g/cm³), and porosity (%). For the battery-like electrode, the key properties are the charge capacity (amp-seconds per gram [As/g]), density (g/cm³), and porosity (%) of the electrode material. The current collectors are described in terms of their material (lead, copper, nickel, or aluminum) and thickness per side coated with electrode material. The electrolyte is specified by composition (sulfuric acid, KOH, acetonitrile plus salts), density (g/cm³), and specific resistivity ($\Omega$-cm). The separator is specified in terms of its thickness and porosity. The area of the cell cross section can be specified or the calculations done on a 1 cm² basis.

**Step 1:** Calculate the weight of the carbon electrode ($W_{carb}$) and, from that, the capacitance ($C_{carb}$) of the electrode. Based on the assumed voltage swing (delta $V_{carb}$) of the electrode and its capacitance, the charge transferred to the electrode during charge/discharge is given by

$$\text{Chg} = (C_{carb})\,(\Delta V_{carb}),\ C_{carb} = W_{carb}\,(F/g)_{carb}$$

**Step 2:** A key constraint on the cell design is that the charge transferred to the battery-like electrode must be equal to that transferred to the carbon electrode. The battery-like electrode will be sized such that to accommodate that charge, the change in the depth of discharge of this electrode should be only 10% to 20%. The change in state-of-charge is symbolized as $\Delta(SOC_{bl})$. The required weight ($W_{bl}$) of the battery-like electrode is given by

$$W_{bl} = \text{Chg}/[\Delta(SOC_{bl})\,(As/g)_{bl}]$$

The thickness ($th_{bl}$) of the battery-like electrode is then given by

$$(th_{bl}) = W_{bl}/(dens_{bl} \times A_{cell})$$

**Step 3:** Next calculate the weight of the electrolyte in the electrodes and the separator by accounting for the electrolyte in the macropores of these components. The electrolyte weight in each layer is given by

$$W_{elypor} = th \times A_{cell} \times porosity \times (dens)_{ely}$$

The electrolyte weight is the sum of the weights of the electrolyte in each of the component layers of the device.
**Step 4:** The weight of the current collectors is simply

$$W_{curcl} = th_{curcl} \times A_{cell} \times (dens)_{curcl}$$

**Step 5:** Next calculate the weight of the cell. This is done by simply adding up the weights of the components.

$$W_{cell} = W_{curcl} + W_{carb} + W_{bl} + W_{ely}$$

**Step 6:** The energy stored in the cell is the energy stored in the carbon as the voltage changes from $V_{max}$ to $[V_{max} - \Delta(V_{carb})]$.

$$E_{carb} = \tfrac{1}{2}(C_{carb})\,[2V_{max}\Delta(V_{carb}) - \Delta(V_{carb})]^2$$

Additional energy is stored in the cell when the voltage of the cell approaches the standard potential of negative electrode, but that energy is not included in the estimate of the energy density of the cell.

The energy density of the cell is

$$(Wh/kg)_{cell} = E_{carb}/W_{cell}$$

This energy density does not include the packaging weight, but does include the weights of all the active components of the cell.

**Step 7:** The resistance of the cell is calculated by relating the bulk resistivity of the electrolyte ($R_{ely}$) to that in the porous electrodes by the relationship

$$R_{elypor} = (R_{ely})\,(\text{porosity})^{-1.5}$$

The specific resistance ($\Omega$-cm$^2$) of each electrode layer is given by

$$\Omega\text{-cm}^2 = R_{elypor} \times \text{th}/2$$

The specific resistance of the cell is then the sum of the specific resistances of the carbon and battery-like electrodes and the separator. This approach neglects the pore resistance of the electrodes, electrical resistance of the carbon and battery-like electrode materials, and the interface resistance between the layers of the cells. The cell resistance is then given by

$$R_{cell} = (\Omega\text{-cm}^2)_{cell}/A_{cell}$$

This calculated resistance should be considered that of an ideal cell and the resistance of an actual cell will certainly be higher.

The resistance $R_{cc}$ due to the current collectors depends on the path lengths $L_{cc}$ for the current flow to the collection tabs and the conductivity of the current collector material (see Table 20.6 for the properties of current collector materials).

$$R_{cc} = R_{ccmat}/(wd_{el} \times th_{cc}/L_{cc}),$$

where $R_{ccmat}$ = resistivity of the cc material and $wd_{el}$ = width of electrode, and $th_{cc}$ = thickness of the current collector

**Step 8:** The power characteristics of the cell can be calculated from the cell voltage and resistance using the relationship

$$P_{max} = (9/16)\,(1 - EF)\,(V_{cell}^2/R_{cell})$$

where EF is the efficiency of the discharge.

The rated voltage of the cell $V_{cell}$ is given by the maximum allowable voltage $V_{max}$ for the electrolyte/carbon interface and the minimum voltage $V_{cell,min}$ is either $V_{stpot,neg}$ or $V_{max} - \Delta(V_{carb})$. The cells operate between $V_{max}$ and $V_{cell,min}$.

The power density for the device design being analyzed is

$$(W/kg)_{max} = P_{max}/W_{cell}$$
$$P_{max} = 9/16\,(1 - EF)\,V_{max}^2/R_{cell},\ EF = 0.95$$

This method of analysis has been applied to estimate the performance (energy density and power capability) of electrochemical capacitors using various combinations of materials. The results are shown in Table 20.11. As indicated in the table, the general approach outlined can be applied to electrochemical capacitors other than the hybrid-type using appropriate material properties and voltages for each capacitor design.

**TABLE 20.11**   Calculated Energy Density and Power Density Characteristics of Various Electrochemical Capacitor Technologies

| Type | Capacitance, F/g or mAh/g | V | Specific energy, Wh/kg[*][†] | Specific energy, Wh/L[*][†] | Specific resistance $\Omega cm^{2\ddagger}$ | Time const., RC, s | Power, kW/kg 95% eff. |
|---|---|---|---|---|---|---|---|
| Act. Carbon/Act. Carbon/ sulfuric acid | 150 F/g | 1.0–0.5 | 1.7 | 2.2 | 0.17 | 0.29 | 1.2 |
| Act. Carbon/Act. Carbon/ acetonitrile | 100 | 2.7–1.35 | 5.7 | 7.6 | 0.78 | 0.18 | 6.4 |
| Act. Carbon/Act. Carbon/acetonitrile | 120 | 3.0–1.5 | 6.8 | 9.1 | 0.78 | 0.22 | 6.4 |
| Act. Carbon/PbO₂/sulfuric acid | 150 F/g, 220 mAh/g | 2.25–1.0 | 16 | 39 | 0.12 | 0.36 | 8.9 |
| Act. Carbon/Act. Carbon/aqueous NaSO₄ + KI | 240 F/g (redox) | 1.5–0.75 | 3.9 | 6.2 | 3.1 | 3.5 | 0.26 |
| Act. Carbon/NiOOH/KOH | 150 F/g, 290 mAh/g | 1.6–0.6 | 14 | 31 | 0.16 | 0.71 | 4.0 |
| Li₄Ti₅O₁₂-Act. Carbon/Act. Carbon/Acetonitrile | 160 mAh/g 50% overcap, 120 F/g | 3.0–1.5 | 20  1.4 g/cm³ | 28 | 3.4 | 3.7 | 1.1 |
| Pre-lithiated graphite/Act. Carbon | 370 mAh/g, 75% overcap. Gr/pre-lithiated Li₂DHBN [x], 120 F/g | 4.0–2.0 | 38  1.6 g/cm³ | 61 | 3.0 | 3.2 | 3.1 |

[*]Useable energy—energy available between the voltage limits indicated.
[†]Unpackaged, all other weights and volumes included.
[‡]Does not include interface or micropore resistances, includes all others.

## 20.4.6   Testing Electrochemical Capacitors

There have been many studies of materials for electrochemical capacitors and testing of small laboratory and prototype devices as well as a wide range of larger commercial products. Much of the testing of materials and small laboratory devices has involved the application of cyclic voltammetry and impedance spectroscopy test approaches.[3,62,63] These approaches in most cases utilize small currents and limited voltage ranges and/or AC frequencies and are intended primarily to determine the electrochemical characteristics of the materials and electrodes used in the capacitors. Testing of the larger prototype and commercial devices is usually done using DC test procedures similar to those used to test batteries. This section discusses DC test procedures and how they can be used to characterize/evaluate electrochemical capacitors intended for various industrial and vehicle applications.

***Summary of Test Procedures.***   There are similarities and differences in the test procedures for electrochemical capacitors and high-power batteries. A detailed summary of test procedures for electrochemical capacitors is given in Refs. 3 and 64. It is customary to perform constant-current and constant-power tests of both types of devices. From the constant-current tests, the charge capacity (capacitance in farads and/or Ah) and resistance of the devices are determined. From the constant-power tests, the energy-storage characteristics (Wh/kg vs. W/kg—the Ragone curve) are determined. The current and power levels used in the testing are selected such that the charge and discharge times are compatible with the capabilities of the devices being tested. In the case of the capacitors, the test discharge times are usually in the range of 5 to 60 s and for the batteries several minutes to a significant fraction of an hour even for high-power batteries. The differences in the recharge times for the devices are also large. For example, the capacitors can be fully charged in 5 to 10 s without difficulty, but the high-power batteries require a minimum of 10 to 20 min for a complete charge even when the initial charge current is set at a maximum value. In addition to the constant-current and constant-power tests, the capacitors and batteries are tested using charge/discharge pulses of 5 to 15 s. For these tests, the current and power levels for the capacitors and high-power batteries are comparable (on a normalized basis). Test cycling tests,[3,64] consisting of a sequence of charge and discharge pulses (power density for a specified time) meant to simulate how the devices would be used in particular applications, are conducted to evaluate both capacitors and batteries. The tests to be performed on capacitors and batteries are summarized in Tables 20.12 and 20.13.

**TABLE 20.12**    Performance Characteristics of Electrochemical Capacitors

---

1 Energy density (Wh/kg vs. W/kg)
2 Cell voltage (V) and capacitance (F)
3 Series and parallel resistance ($\Omega$ and $\Omega cm^2$)
4 Power density (W/kg) for a charge/discharge at 95% efficiency
5 Temperature dependence of resistance and capacitance especially at low temperatures (20°C)
6 Cycle life for full discharge
7 Self-discharge at various voltages and temperatures
8 Calendar life (hours) at fixed voltage and high temperature (40–60°C)

---

**TABLE 20.13**    Testing of electrochemical capacitors

---

1 Constant current charge/discharge
  • Capacitance and resistance for discharge times between 5 and 60 s
2 Pulse tests to determine resistance
3 Constant power charge/discharge
  • Determine the Ragone curve for power densities between 100 and
    at least 1000 W/kg for the voltage between $V_{rated}$ and $\frac{1}{2} V_{rated}$
  • Test at increasing W/kg until discharge time is less than 5 s.
    The charging is often done at constant current with a charge time
    of at least 30 s
4 Sequential charge/discharge step cycling
  • Testing done using the PSFUDS (Pulsed Simple FUDS) test cycle
    with the maximum power step being 500 W/kg and higher
  • From the data, the round-trip efficiency for charge/discharge is
    determined
5 Tests modules with at least 15–20 cells in series

---

***Application of the Test Procedures to Carbon/Carbon Double-Layer Capacitors.***    In this section, the various test procedures discussed in the previous section are applied to carbon/carbon capacitors to determine their capacitance, resistance, energy density, and power capability. These devices use activated carbon in both electrodes and in nearly all cases, an organic electrolyte. The dominant energy-storage mechanism in these devices is charge separation (double-layer capacitance).

*Capacitance.*    The capacitance of a device can be determined directly from constant current discharge data. By definition,

$$C = I/dV/dt \text{ or } C = I\,(t_2 - t_1)/(V_1 - V_2), \; V = V(t)$$

Since the voltage trace is not exactly linear, the value of $C$ (calculated) depends to some extent on the values of $V_1$ and $V_2$ used. The voltage ranges that have been used in most cases are $V_0$ to $V_0/2$ and $V_0$ to 0. In the case of $V_0$, it is important to include the *IR* drop in the determination of the effective $V_1$ value. The results in Table 20.14 indicate that the determination of the capacitance of devices is relatively insensitive to this test procedure.

**TABLE 20.14**    Effect of Voltage Range and Test Current on the Measured Capacitance

| Device/developer | $V_0$ to 0 V | | $V_0$ to 1.35 V | |
|---|---|---|---|---|
| 3000F/Maxwell | 100 A: 2880 F | 200 A: 2893 F | 100 A: 3160 F | 200 A: 3223 F |
| 3000F/Nesscap | 50 A: 3190 F | 200 A: 3149 F | 50 A: 3214 F | 200 A: 3238 F |
| | 3.8–2.2 V | | 3.8–2.6 V | |
| 2000F/JSR Micro | 80 A: 1897 F | 200 A: 1817 F | 80 A: 1941 F | 200 A: 1938 F |

*Resistance.*   The resistance of a capacitor or a battery can be determined using one of several methods.

- *IR* drop at the initiation of a constant current discharge
- Current pulses (5–30 s) at specified states-of-charge (SOC)
- Voltage recovery at the interruption of a discharge or charge current
- Measurement of the AC impedance at 1 kHz

The method used most routinely involves analysis of the *IR* drop and the voltage variation at the initiation of a constant current discharge. Determination of the resistance of the capacitor is complicated by the fact that the voltage decreases due to both the resistance and capacitance of the device. In addition, due to the porous character of the electrode, the resistance of the capacitor varies with time until the current distribution in the electrodes is fully established. This problem has been analyzed mathematically in "Mathematical Modeling" in Sec. 20.4.4. The results of the analysis indicate that the initial value of the resistance $R_0$ can be as low as half the steady-state value of $R_{ss}$. A good estimate of the steady-state resistance can be obtained by extrapolating the linear portion of the voltage-time trace back to $t = 0$, as shown in Fig. 20.6 and utilizing that *IR* drop value to calculate $R_{ss}$. For many applications of electrochemical capacitors, it is the steady-state resistance that is most relevant for the calculation of power capability/electrical losses/heating and not the $R_0$ value that is smaller. It is important to define what resistance value is being reported. The effect of pore resistance on the cell resistance is indicated by the difference between $R_0$ and $R_{ss}$.

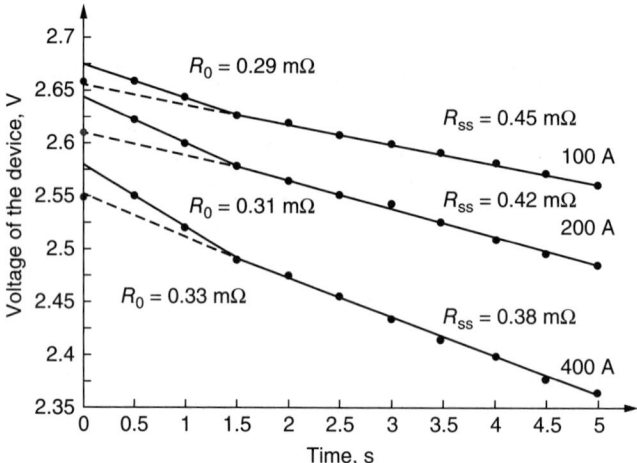

**FIGURE 20.6**   Method for determining the steady-state resistance by extrapolating the voltage trace to $t = 0$ (Nesscap/2.7 V/3000 F).

The resistance of a capacitor or a battery can also be inferred from the recovery of the voltage at the end of a current pulse when the current is removed ($I = 0$). Some researchers prefer this method rather than that involving the initiation of the pulse because the current is zero and the effect of the capacitance of the device on the voltage is not present. However, there is the effect of the charge redistribution in the device with time at $I = 0$ and the effect on the voltage is not well understood. As a result, there is an uncertainty as to the time after the setting of $I = 0$ at which the voltage should be read and $R$ calculated from $\Delta V/I$. This effect is illustrated in Fig. 20.7. These results indicate that the current initiation and interruption methods yield nearly the same value of resistance for both $R_0$ and $R_{ss}$. The voltage recovery time is relatively short—being approximately equal to the *RC* time constant of the device tested.

A comparison of the resistance measured using the pulse initiation and rebound methods is also shown in Table 20.15 for the Maxwell 2800F cell. The comparisons indicate that the two methods yield essentially the same value for the resistance of the cell.

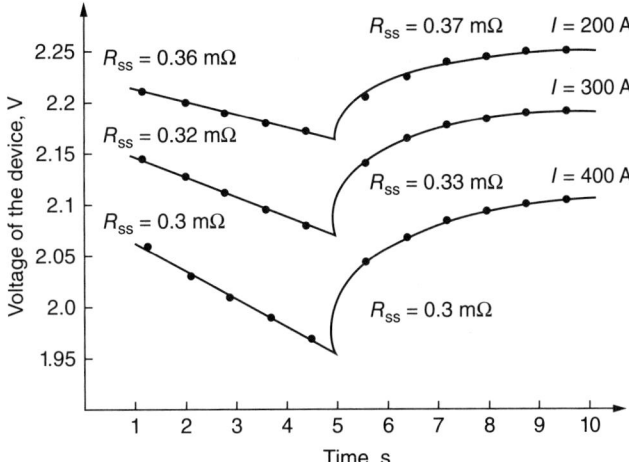

**FIGURE 20.7**   Resistance determination using voltage recovery after current interruption (Ioxus 2.85 V 3000 F at 60% SOC).

**TABLE 20.15**   Resistance of Maxwell 2800 F Capacitor Using the Current Initiation and Interrupt Methods

| Current, A | Voltage at interrupt | Current initiation at 2.7 V $R$, mΩ | Current interrupt $R$, mΩ |
|---|---|---|---|
| 200 | 2.3 | 0.36 | 0.4 |
| | 1.3 | | 0.4 |
| 300 | 2.3 | 0.37 | 0.37 |
| | 1.3 | | 0.37 |
| 400 | 2.3 | 0.4 | 0.35 |
| | 1.3 | | 0.375 |

For electrochemical capacitors, it is common for manufacturers to list the resistance measured with an AC impedance meter at 1 kHz. This value of resistance is always significantly lower than the DC value often by about a factor of two. The power capability of devices should not be calculated using the AC impedance meter value of the resistance.

*Energy Density.*   The total energy stored in a carbon/carbon capacitor can be calculated from the relationship $E = \frac{1}{2}CV_0^2$. If the voltage of the capacitor is restricted to the range $V_0$ to $V_0/2$, only 75% of the stored energy can be used. Hence, the useable energy density is given by

$$\text{Wh/kg} = \tfrac{3}{8}CV_0^2/(\text{device weight})$$

This simple relationship is often used to calculate the energy density from constant current test data. However, the most reliable approach to determining the energy stored in a device is to measure the watt-hours stored for a range of constant-power densities (W/kg). In general, tests should be made for power densities between 100 and 1000 W/kg or even higher for high-power devices. The Ragone plot data (Wh/kg vs. W/kg) for a typical commercially available 3500 F capacitor are shown in Table 20.16. Note that the energy density gradually decreases with W/kg. This is the case for all electrochemical capacitors. The value of energy density quoted by the device manufacturer is often calculated from the energy corresponding to $\frac{1}{2}CV_0^2$ using the rated voltage and specified capacitance. This value is too high as it is not the useable energy density and also it corresponds to a low-power density of 100 W/kg or lower. As shown in Tables 20.16 and 20.17, the effective capacitance $C_{\text{eff}}$ of a device decreases significantly with W/kg and often agrees closely with the value claimed by the manufacturer only for relatively low-power densities. Hence, combining the useable energy factor (0.75) and the

**TABLE 20.16**    Test Data for the NessCap 3500 F, 3 V Device

Device characteristics:
Packaged weight 500 g
Packaged volume 394 cm³

| Constant Current Discharge Data | | | | |
|---|---|---|---|---|
| Current, A | Time, s | Capacitance, F | Steady-state $R_{ss}$, m$\Omega$ | $R_0$, m$\Omega$ |
| 60 | 90.8 | 3632 | | |
| 85 | 64.2 | 3638 | | |
| 130 | 42.2 | 3657 | | |
| 200 | 27.0 | 3673 | | |
| 300 | 17.8 | 3657 | | |
| 400 | 12.7 | 3629 | 0.28 | 0.188 |
| 500 | 9.9 | 3600 | 0.27 | 0.22 |

Discharge 3.05 V to 1.50 V.
Resistance calculated from extrapolation of the voltage to $t = 0$.
Capacitance calculated from $C = I \times t_{\text{disch}}/$delta from $V_t = 0$.

| Constant Power Discharge Data | | | | | | |
|---|---|---|---|---|---|---|
| Power, W | W/kg | Time, s | Wh | Wh/kg | Wh/L | $C_{\text{eff}}$ |
| 100 | 200 | 119 | 3.31 | 6.6 | 8.4 | 3532 |
| 200 | 400 | 58.6 | 3.26 | 6.5 | 8.3 | 3478 |
| 400 | 800 | 28.8 | 3.23 | 6.4 | 8.1 | 3446 |
| 500 | 1000 | 22.7 | 3.15 | 6.3 | 8.0 | 3361 |
| 600 | 1200 | 18.7 | 3.12 | 6.2 | 7.9 | 3329 |
| 700 | 1400 | 15.9 | 3.09 | 6.2 | 7.8 | 3298 |
| 800 | 1600 | 13.9 | 3.09 | 6.2 | 7.8 | 3298 |

Pulse power at 95% efficiency:
$P = 9/16\,(1 - \text{eff})\,V_R^2/R_{ss}$, $R_{ss} = 0.27$ m$\Omega$, $(\text{W/kg})_{95\%} = 1875$, $(\text{W/L})_{95\%} = 2379$.
Matched impedance power: $P = V_R^2/4R_{ss}$, $(\text{W/kg}) = 16{,}666$.

effective capacitance reduction factor (0.9) from Tables 20.16 and 20.17, the simple calculation of energy density from $\frac{1}{2}CV_0^2$ can overestimate the energy density of a device by at least ⅓.

*Power Capability.*    There is confusion and unreliable information in the literature concerning the power capability of ultracapacitors and batteries.[64,65] This confusion stems to a large extent from the persistent use of the simple formula $P = V_0^2/4R$ to calculate the maximum power capability of electrochemical devices. This formula grossly overestimates the maximum power as it corresponds to operation of the device at the matched impedance point at which one-half of the discharge energy is electricity and one-half is heat. The corresponding efficiency is 50%, which makes that operating condition unuseable for nearly all applications. It is more reasonable to express the power capability of devices in terms of the pulse efficiency (EF). This can be done using the following relationships for ultracapacitors and batteries:

$$\text{Ultracapacitors: } P = (9/16)\,(1 - \text{EF})\,(V_0^2/R)$$
$$\text{Batteries: } \qquad P = (\text{EF})\,(1 - \text{EF})\,(V_{\text{oc}}^2/R)$$

These relationships are for pulse power and not constant-power discharges. In the case of the ultracapacitor, the power pulse is occurring at a voltage of ¾$V_0$ and is intended to remove only a relatively small fraction of the energy stored in the device. The battery relationship can be applied at any SOC by using the $V_{\text{oc}}$ and $R$ for that SOC. Note that the power from both the matched impedance and efficiency EF relationships is proportional to $V^2/R$. The key parameters in determining the power capability are thus $R$ and $V_0$. High-power devices necessarily must have low resistance. Hence, once the resistance of a device is known, its power capability follows directly. It is unfortunate that device manufacturers often do not provide information concerning the resistance

**TABLE 20.17**   Test Data for the Skeleton 2.85 V Device

Device characteristics:
Packaged weight 524 g
Packaged volume 390 cm$^3$

| Constant Current Discharge Data | | | | |
|---|---|---|---|---|
| Current, A | Time, s | Capacitance, F | Steady-state $R_{ss}$, mΩ | $R_0$, mΩ |
| 60 | 84.1 | 3541 | | |
| 85 | 58.6 | 3495 | | |
| 130 | 38.2 | 3473 | | |
| 200 | 24.4 | 3461 | | |
| 300 | 16.3 | 3469 | 0.14 | 0.067 |
| 400 | 12.0 | 3357 | 0.125 | 0.0875 |
| 500 | 9.6 | 3357 | 0.13 | 0.074 |

Discharge 2.85 V to 1.425 V.
Resistance calculated from extrapolation of the voltage to $t = 0$.
Capacitance calculated from $C = I \times t_{disch}$/delta from $V_t = 0$.

| Constant Power Discharge Data | | | | | |
|---|---|---|---|---|---|
| Power, W | W/kg | Time, s | Wh | Wh/kg | Wh/L |
| 100 | 191 | 104.6 | 2.91 | 5.55 | 7.45 |
| 200 | 382 | 51.7 | 2.87 | 5.48 | 7.36 |
| 400 | 763 | 25.4 | 2.82 | 5.39 | 7.24 |
| 500 | 954 | 20.4 | 2.83 | 5.40 | 7.26 |
| 600 | 1145 | 16.8 | 2.80 | 5.34 | 7.18 |
| 700 | 1336 | 14.4 | 2.80 | 5.34 | 7.18 |
| 800 | 1527 | 12.5 | 2.78 | 5.31 | 7.13 |

Pulse power at 95% efficiency:
   $P = 9/16\,(1 - \text{eff})\,V_R^2/R_{ss}$, $R_{ss} = 0.13$ mΩ, (W/kg)$_{95\%}$ = 3353, (W/L)$_{95\%}$ = 4505.
Matched impedance power: $P = V_R^2/4R_{ss}$, (W/kg) = 29,809.

of their devices. This makes careful measurement of the resistance, as discussed in the previous section, very important. For simple power pulses using capacitors, the ratio of the matched impedance to the efficiency power is $(4/9)/(1 - \text{EF})$. In the case of batteries, the ratio is $(\frac{1}{4})/[\text{EF}(1 - \text{EF})]$. The ratios as functions of EF are given in Table 20.18. For ultracapacitors, the efficiency specified by the United States Advanced Battery Consortium (USABC)[66,67] is 95% that results in the useable maximum power being only about 1/10 of the matched impedance power ($V^2/4R$). Hence, for capacitors using the $V^2/4R$ formula to estimate the useable maximum power does not yield a realistic value for most applications, especially vehicle applications. Note that in Table 20.9 both the matched impedance and EF = 95% power densities are presented for the various devices.

**TABLE 20.18**   Ratio of the Efficiency and the Matched Impedance Powers

| Efficiency EF | Ultracapacitor | Batteries |
|---|---|---|
| 0.5 | 1.1 | 1.0 |
| 0.6 | 0.9 | 0.96 |
| 0.7 | 0.68 | 0.84 |
| 0.8 | 0.45 | 0.64 |
| 0.9 | 0.22 | 0.36 |
| 0.95 | 0.11 | 0.19 |

*Pulse Cycle Testing.*    Since in many applications ultracapacitors experience highly transient operation, pulse cycle testing should be included in evaluating their performance capabilities. The pulse cycles are simply a sequence of discharge and charge pulses of specified currents (A) or powers (W) of specified time duration (seconds). The pulsed simple federal urban driving schedule (PSFUDS), which was first defined in Ref. 68, has been used extensively to test ultracapacitors and high-power batteries. The test cycle, specified in terms of W/kg-time steps, is given in Table 20.19 and shown graphically in Fig. 20.8. It can be utilized to test devices of all sizes and performance capabilities by adjusting the W/kg and time duration of the maximum power steps. The test data of most interest in using the PSFUDS cycle is the round-trip efficiency. Typical data using this cycle are also shown in Table 20.20.

**TABLE 20.19**    Time-Power Steps for the PSFUDS Test Cycle (with Diagram)

| Step no. | Time step duration, s | Charge C/discharge D | $P/P_{max}$ $P_{max} = 500$ W/kg |
|---|---|---|---|
| 1 | 8 | D | 0.20 |
| 2 | 12 | D | 0.40 |
| 3 | 12 | D | 0.10 |
| 4 | 50 | C | 0.10 |
| 5 | 12 | D | 0.20 |
| 6 | 12 | D | 1.0 |
| 7 | 8 | D | 0.40 |
| 8 | 50 | C | 0.30 |
| 9 | 12 | D | 0.20 |
| 10 | 12 | D | 0.40 |
| 11 | 18 | D | 0.10 |
| 12 | 50 | C | 0.20 |
| 13 | 8 | D | 0.20 |
| 14 | 12 | D | 1.0 |
| 15 | 12 | D | 0.10 |
| 16 | 50 | C | 0.30 |
| 17 | 8 | D | 0.20 |
| 18 | 12 | D | 1.0 |
| 19 | 38 | C | 0.25 |
| 20 | 12 | D | 0.40 |
| 21 | 12 | D | 0.20 |
| 22 | ≥50 | Charge to $V_0$ | 0.30 |

**Testing of Hybrid, Pseudocapacitive Devices.**    Most of the electrochemical capacitors that have been available for testing are of the carbon/carbon type that use activated carbon in both electrodes and double-layer capacitance for energy storage. The testing of devices that use intercalation carbon or other battery-like (pseudocapacitive) materials in at least one electrode is considered in this section. These devices are often referred to as hybrid electrochemical capacitors. Most of the hybrid capacitors presently available use pre-lithiated graphite in the negative electrode and activated carbon in the positive electrode. Some testing of hybrid capacitors has been done and differences between testing carbon/carbon and hybrid capacitor devices are becoming apparent. These differences are discussed in this section with emphasis on how they affect test procedures and data interpretation.

*Capacitance.*    As is the case for carbon/carbon cells, the capacitance can be determined from constant current discharge data. However, as shown in Figs. 20.9a and b, the character of voltage versus time traces for the hybrid capacitors can be quite different than for carbon/carbon cells.

As seen in Figs. 20.9a and b, the key differences are the nonlinearity of the hybrid device traces, especially in charging, and the well-defined voltage below which the capacitance of the device is small. As would be expected, the character of the V versus t trace must be considered in testing a particular hybrid capacitor device. In the case of the hybrid carbon device (Fig. 20.9a), the voltage should be restricted to be between the rated voltage (3.8 V) and that of the shoulder (2 V).

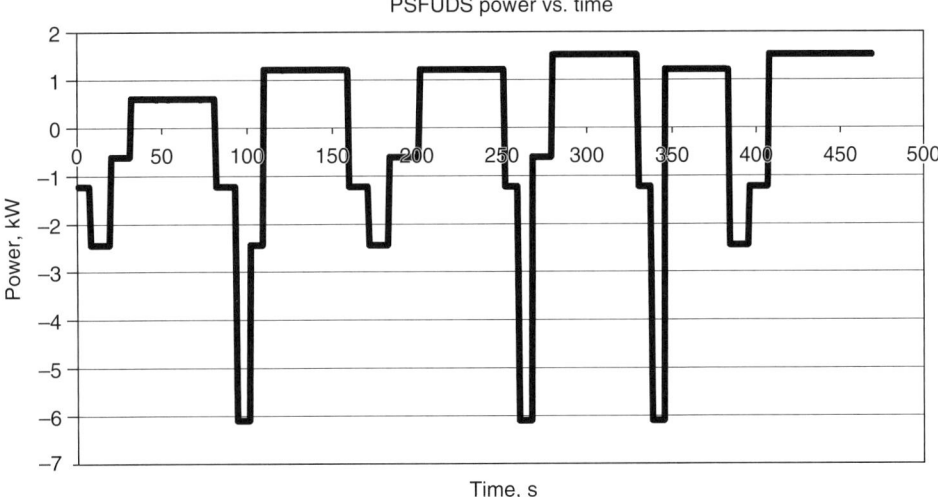

**FIGURE 20.8**  PSFUDS test results for a 45 V Ness ultracapacitor module.

**TABLE 20.20**  Round-trip Efficiencies for the Ness 45 V Module on the PSFUDS Cycle

| Cycle* | Energy in, Wh | Energy out, Wh | Efficiency, % |
|--------|---------------|----------------|---------------|
| 1 | 102.84 | 97.94 | 95.2 |
| 2 | 101.92 | 97.94 | 96.1 |
| 3 | 101.67 | 97.94 | 96.3 |

*PSFUDS power profile based on maximum power of 500 W/kg and the weight of the cells alone.

Test data for constant-current and constant-power discharges of several hybrid capacitors are given in Table 20.21.

It is evident from Table 20.21 (note the data for the Yunasko 5000 F cell) that the selection of voltage limits makes a greater difference in the calculation of the capacitance for hybrid capacitors than for carbon/carbon cells. The best approach is to use the complete range between the rated and shoulder voltages to calculate the capacitance, but correct the initial voltage ($V_1$) for the $IR$ drop as is done for the carbon/carbon devices. For hybrid capacitors, it is necessary to look closely at the $V$ versus $t$ trace before adapting a particular method for the calculation of capacitance.

*Resistance.*    The same methods can be used for determining the steady-state resistance $R_{ss}$ of the hybrid carbon capacitors as was used for the carbon/carbon devices. The $V$ versus $t$ traces for constant current discharges of the hybrid carbon devices become linear within a few seconds and the $IR$ drop can be determined by extrapolating back to $t = 0$. Hence, $R_{ss} = (\Delta V)_{t=0}/I$. When testing any new hybrid device, one should check the linearity of $V$ versus $t$ trace near the initiation of discharge to determine whether the simple method of linear extrapolation is applicable. Pulse tests with the JSR Micro devices yielded resistance values that are in good agreement with those obtained using the linear extrapolation method.

*Energy Density.*    In the simplest form, the energy stored in a hybrid capacitor can be expressed as

$$E_{stored} = \tfrac{1}{2}C_{eff}\left(V_{rated}^2 - V_{min}^2\right)$$

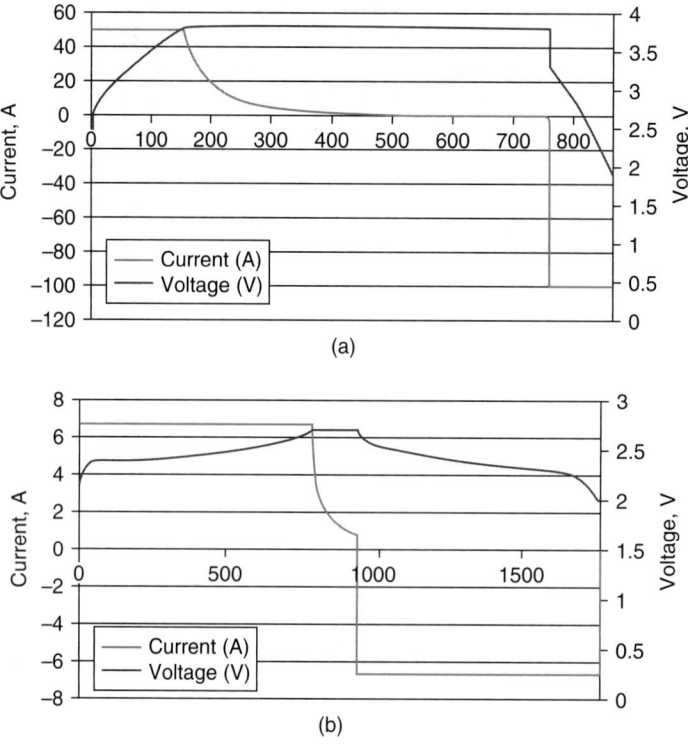

**FIGURE 20.9**   Voltage versus time traces for the constant current discharges of hybrid capacitors. (*a*) Pre-lithiated graphite/activated carbon (DAE-China-900 F device). (*b*) Activated carbon mixed with metal oxides in both electrodes (Yunasko-5000 F device).

**TABLE 20.21**   Test Data for Several Hybrid Supercapacitors: (a) JSR Energy, (b) DAE, (c) Yunasko

| (a) Characteristics of the JSR Energy 1100 F Cell[*] | | | |
|---|---|---|---|
| Current, A | Discharge time, s | $C$, F | Steady-state $R$, m$\Omega$[*] |
| 20 | 86.4 | 1096 | — |
| 40 | 41.9 | 1078 | — |
| 60 | 27.2 | 1067 | — |
| 75 | 21.4 | 1063 | 1.2 |
| 100 | 15.7 | 1057 | 1.15 |
| 150 | 10.1 | 1056 | 1.1 |

| Power, W | W/kg | Discharge time, s | Wh | Wh/kg[*] | Wh/L[*] | $C_{eff}$ |
|---|---|---|---|---|---|---|
| 50 | 347 | 106.7 | 1.47 | 10.2 | 14.1 | 1105 |
| 83 | 576 | 61.9 | 1.43 | 9.9 | 12.1 | 1075 |
| 122 | 847 | 40.1 | 1.36 | 9.4 | 11.5 | 1023 |
| 180 | 1250 | 26.2 | 1.31 | 9.1 | 11.1 | 985 |
| 240 | 1667 | 19.1 | 1.27 | 8.8 | 10.7 | 955 |

[*]Laminated pouch cell weight 144 g, 118 cm³, 1.22 g/cm³.
Constant discharge 3.8 V – 2.2 V.
Peak pulse power at 95% efficiency, $R = 1.15$ m$\Omega$:
  $P = 9/16 \times 0.05 \times (3.8)^2/0.00115 = 353$ W, 2452 W/kg.

**TABLE 20.21**  Test Data for Several Hybrid Supercapacitors: (a) JSR Energy, (b) DAE, (c) Yunasko (*Continued*)

| (b) Characteristics of the DAE 3.8 V Cell* | | | | |
|---|---|---|---|---|
| Current, A | Discharge time, s | Ah | As | C, F |
| 15 | 104.6 | 0.44 | 1569 | 867 |
| 30 | 47.3 | 0.39 | 1419 | 802 |
| 50 | 26.2 | 0.36 | 1310 | 749 |
| 75 | 15.8 | 0.33 | 1185 | 765 |
| 100 | 10.5 | 0.29 | 1050 | 778 |
| 150 | 6.2 | 0.26 | 930 | 705 |

| Power, W | Discharge time, s | Wh | Ws | Wh/kg | $C_{eff}$ |
|---|---|---|---|---|---|
| 40 | 80.3 | 1.08 | 3877 | 12.38 | 733 |
| 80 | 44.3 | 0.98 | 3544 | 11.32 | 655 |
| 120 | 28 | 0.93 | 3360 | 10.73 | 622 |
| 175 | 18.6 | 0.90 | 3255 | 10.39 | 603 |
| 250 | 11.8 | 0.82 | 2950 | 9.42 | 549 |
| 350 | 7 | 0.68 | 2450 | 7.82 | 456 |

*Constant discharge 3.8 V − 1.9 V; *Laminated pouch cell weight 87 g.

(c) Performance Characteristics of Yunasko/2.7 V/5000 F Hybrid Device*

| Constant current | $V_r$ to 2 V | | | $V_r$ to 1.35 V | | | |
|---|---|---|---|---|---|---|---|
| Current, A | Time, s | Ah | R, mΩ | Time, s | Ah | C, F | R, mΩ |
| 50 | 83.7 | 1.16 | — | 105 | 1.46 | 4256 | — |
| 100 | 36.1 | 1 | 1.58 | 44.9 | 1.25 | 4170 | 1.59 |
| 150 | 25.1 | 1.05 | 1.59 | 29.5 | 1.23 | 4060 | 1.58 |
| 200 | 7.1 | 0.39 | 1.58 | 21.1 | 1.17 | 3901 | 1.58 |
| 250 | 4.1 | 0.28 | 1.56 | 15.2 | 1.06 | 3830 | 1.57 |

| Power, W | W/kg | Time, s | Wh | Wh/kg | Time, s | Wh | Wh/kg |
|---|---|---|---|---|---|---|---|
| 55 | 743 | 164 | 2.5 | 33.8 | 172 | 2.63 | 35.5 |
| 150 | 2094 | 58.1 | 2.5 | 33.8 | 62.8 | 2.7 | 36.5 |
| 300 | 4095 | 16.6 | 1.4 | 18.9 | 28.3 | 2.38 | 32.2 |
| 350 | 4730 | 11.9 | 1.16 | 15.7 | 22.4 | 2.18 | 29.5 |
| 400 | 5405 | 8.3 | 0.92 | 12.4 | 17.3 | 1.92 | 25.9 |
| 500 | 6756 | 4.3 | 0.6 | 8.1 | 10.8 | 1.5 | 20.3 |

| Current pulse, (10 s) | Discharge R, mΩ | Rebound R, mΩ |
|---|---|---|
| 150 | 1.59 | 1.58 |
| 200 | 1.57 | 1.56 |
| 250 | 1.56 | 1.56 |

*Laminated pouch cell weight 74 g, 38 cm$^3$, 1.95 g/cm$^3$.
Peak pulse power at 95% efficiency, $R = 1.15$ mΩ:
$P = 0.95 \times 0.05 \times (2.7)^2/0.0015 = 231$ W.
$(W/kg)_{95\%} = 3120$, $(W/L)_{95\%} = 6078$.

assuming that capacitance $C_{eff}$ is a constant. In the case of the carbon/carbon devices, $V_{min} = \frac{1}{2}V_{rated}$. In the case of the hybrid capacitor, $V_{min}$ is the minimum voltage at which the device stores significant charge. $C_{eff}$ has been calculated from the test data for the hybrid capacitors. It is clear from Table 20.16 by comparing the calculated $C_{eff}$ values with the measured capacitances in the constant current tests that the simple $\frac{1}{2}CV^2$ relationship will not yield an accurate assessment of the Wh stored energy of hybrid capacitors as it did in the case of carbon/carbon devices. Hence, the energy density of hybrid capacitors should be obtained from testing them over a range of power densities. The simple $\frac{1}{2}CV^2$ relationship will overestimate the energy stored in the hybrid capacitors in most cases. As is the case for carbon/carbon devices, energy density decreases with increasing power density due to the effect of resistance on the operating voltage range of the device.

*Power Capability and Pulse Cycle Tests.*    Pulse testing of hybrid capacitors to obtain the resistance and the round-trip efficiency on the PSFUDS is essentially the same as for carbon/carbon devices. The power capability of hybrid capacitors can be calculated using the same relationships used for carbon/carbon devices when $V_{rated}$ and the pulse resistance $R$ are known.

## 20.4.7    Cost and System Considerations for Capacitors and Batteries

*Costs of Materials and Devices.*    Reducing the present high cost/price of electrochemical capacitors/ultracapacitors is a key issue in achieving high market penetration in the future, especially for midsize and large devices. There are many applications for which ultracapacitors are presently excluded and not seriously considered because they remain too expensive even though their selling price has decreased significantly in recent years. The cost of manufacture of any product is closely tied to production volume with the cost decreasing rapidly with increased volume up to relatively high production rates. Sales of capacitors in many millions of units per year are necessary to reduce the unit costs to levels at which the large markets can develop. Semiautomated production facilities presently exist at a number of companies for ultracapacitors of all sizes. It is common to speak of the price of devices in terms of cents per farad (cents/F) or $/Wh stored. It is easier to interpret the price information on the cents/F basis as it does not concern the cell voltage or the fraction of energy stored that can be used in a particular application. For example, for a 10 F device, if the price is quoted as 10 cents/F, the device cost would be $1. Similarly, a 2500 F device would cost $25 at 1 cent/F. In 2018, the price of 3000 F carbon/carbon 2.7 V devices in large orders from a high-volume manufacturer is 0.5 to 1.0 cents/F with large volume producers projecting prices of 0.25 to 0.5 cents/F in the near future. The price of hybrid capacitors is about 1 to 2 cents/F partly because of their lower volumes of production. Improving the energy density of the hybrid devices should also reduce their price on a $/Wh basis.

The cost to manufacture a carbon/carbon device depends on the material and production costs. At the present time, material costs are relatively high. The cost of carbon suitable for use in ultracapacitors can be as high as $100/kg with the average price being in the $30 to $50/kg range. The cost of the electrolyte solvent is also high in the range of $5 to $10/L for both propylene carbonate and acetonitrile. The ionic salts that dissociate in the solvent into the positive and negative ions, which move into and out of the double layer in the microporous carbon to store the energy, are also expensive (i.e., $50–$100/kg). Since the analysis of ultracapacitors is straightforward, material costs can be calculated[69,70] with good accuracy. The result of a typical costing exercise is shown in Table 20.22. Note the strong dependency of the cents/F and $/Wh unit costs for the device on the unit material costs. Presently, the price of ultracapacitors is relatively high because both the material and manufacturing costs are high. With more automated production and reduced material costs, as indicated previously, the price of supercapacitors in high volume can be in the range of 1 to 2 cents/F for small devices and 0.25 to 0.5 cents/F for large devices similar to those needed for vehicle applications.

**TABLE 20.22**    Material Costs for a 2.7 V, 3500 F Capacitor*

| Carbon | | | Electrolyte | Salt | Total device | | | | |
|---|---|---|---|---|---|---|---|---|---|
| F/g | Usage, g/device | $/kg | $/L (ACN) | $/kg | Matl. $ | $/kg | $/Wh | $/kW | Ct./F |
| 75 | 187 | 50 | 10 | 125 | 17.0 | 29 | 6.4 | 29 | 0.48 |
| 120 | 117 | 100 | 10 | 125 | 15.5 | 26 | 6 | 26 | 0.44 |
| 75 | 187 | 5 | 2 | 50 | 3.6 | 6.0 | 1.3 | 6 | 0.10 |
| 120 | 117 | 10 | 2 | 50 | 2.5 | 4.2 | 0.93 | 4.2 | 0.070 |

*4.5 Wh/kg, 1000 W/kg—95% eff.

Because of their relatively low energy density, ultracapacitors cannot compete with batteries in terms of $/Wh, but they can compete in terms of $/kW and $/unit for a particular application. Both energy-storage technologies must provide the same power and cycle life and sufficient energy (Wh) for the application. The weight of the battery is usually set by the system power requirement and cycle life and not the minimum energy-storage requirement. Satisfying only the minimum energy-storage requirement would result in a smaller, lighter battery than is needed to meet the other requirements. On the other hand, the weight and volume of the ultracapacitor are determined by the minimum energy-storage requirement. The power and cycle life requirements are usually easily satisfied. Hence, the ultracapacitor unit can be a more optimum solution for many applications and its weight can be less than that of the battery even though its energy density is less than one-tenth that of the battery.

Consider the example of a charge sustaining hybrid car such as the Prius. If the energy stored in the capacitor unit is 125 Wh and that in the battery unit is 1500 Wh, the unit costs of the capacitors and battery would be related by

$$(\$/Wh)_{cap} = 0.012(\$/kWh)_{bat}$$

The corresponding capacitor costs in terms of cents/F and $/kWh are given by

$$(cents/F)_{cap} = 0.125 \times 10^{-3} \times (\$/kWh)_{bat} \times V_{cap}^2$$

$$(\$/kWh)_{cap} = 9.6 \times 10^4 (cents/F)_{cap}/V_{cap}^2$$

The capacitor costs (cents/F) required for the capacitor and battery unit costs to be equal for a range of battery unit costs ($/kWh) are shown in Table 20.23.

**TABLE 20.23**    Relationships between Capacitor and Battery Costs

| Battery cost, $/kWh | Battery cost,* $/kW | Ultracap cost, cents/F $V_{cap} = 2.6$ | Ultracap cost, cents/F $V_{cap} = 3.0$ | Ultracap cost,[†] $/kWh $V_{cap} = 3.0$ | Ultracap cost, $/kW $V_{cap} = 3.0$ |
|---|---|---|---|---|---|
| 200 | 20 | 0.17 | 0.23 | 2453 | 4.9 |
| 300 | 30 | 0.25 | 0.34 | 3626 | 7.3 |
| 400 | 40 | 0.34 | 0.45 | 4800 | 9.6 |
| 500 | 50 | 0.42 | 0.56 | 5973 | 11.9 |

*Battery: 100 Wh/kg, 1000 W/kg.
[†]Capacitor: 5 Wh/kg, 2500 W/kg, cell characteristics.

The results shown in Table 20.23 indicate that for the charge sustaining hybrid application, ultracapacitor costs of 0.25 to 0.5 cents/F are competitive with lithium battery costs in the range of $200 to $500/kWh. Note also that the $/kW costs of the capacitor cells are about one-fourth those of the batteries.

***Combinations of Capacitors and Batteries.***    Combining ultracapacitors and batteries can significantly reduce the stress on the batteries, particularly in plug-in hybrid electric vehicle (PHEV) applications in which high-power electric motors are used and the batteries are subject to high-current pulses in both vehicle acceleration and regenerative braking. There has been some laboratory testing of combinations of capacitors and batteries,[71–73] but little work directly in vehicles. In general, if batteries are available that can meet both the energy and power requirements of the vehicle design, the vehicle designers choose to use batteries alone even though they realize batteries plus ultracapacitors would have some advantages. As a consequence, the battery selected is sized by the power requirement and not the energy requirement resulting in a larger and more expensive battery than would have been the case if supercapacitors had been combined with the battery.[71–73]

In other words, designers will not select a battery/capacitor combination unless there are clear, large advantages to do so. This may be the case when one considers the use of advanced lithium batteries (Wh/kg >200) in PHEVs and electric vehicles (EVs). These vehicles use high-power electric motors often greater than 100 kW and require high-power capability batteries to meet the peak power of the electric motors, as shown in Table 20.24.

**TABLE 20.24**    Battery Sizing and Power Density for a PHEV with Various Ranges and Electric Motor Power

| Battery range and size | | | 200 Wh/kg cells | | | 300 Wh/kg cells | | |
| --- | --- | --- | --- | --- | --- | --- | --- | --- |
| Range miles | kWh needed* | kWh stored† | kg† | 100 kW, kW/kg | 150 kW, kW/kg | kg† | 100 kW, kW/kg | 150 kW, kW/kg |
| 10 | 2.52 | 3.6 | 18 | 5.6 | 8.3 | 12 | 7.8 | 9.17 |
| 15 | 3.78 | 5.4 | 27 | 3.7 | 5.6 | 18 | 5.4 | 8.1 |
| 20 | 5.04 | 7.2 | 36 | 2.8 | 4.2 | 24 | 4.1 | 7.4 |
| 30 | 7.56 | 10.8 | 54 | 1.9 | 2.9 | 36 | 2.8 | 4.2 |
| 40 | 10.1 | 14.4 | 72 | 1.4 | 2.1 | 48 | 2.1 | 3.2 |

*Vehicle energy usage from the battery: 250 Wh/mi.
†Useable state-of-charge for batteries—70%; weights shown are for cells only.

In these cases, the motor peak power can be met using a supercapacitor (see Table 20.17, Skeleton 3200F) in combination with the battery independent of the electric range of the vehicle. This would permit use of a battery designed to have near-maximum energy density and cycle life and lower cost ($/kWh) than a power battery, which would have significantly lower energy density and higher cost. It seems unlikely that the high energy density (>300 Wh/kg) batteries will have the high-power capability they would need as indicated in Table 20.24.

### Module and Lifetime Considerations

*Module Characteristics and Design.*    The cell voltage of electrochemical capacitors is relatively low so that in most applications the cells are combined into modules, which are placed in series to achieve the operating system voltage of 200 to 600 V. Cooling and voltage-temperature management of the cells are done on a module basis. Cooling[74,75] of capacitors is less difficult than batteries because of the lower resistance (higher efficiency) of the capacitors. Several capacitor developers are marketing modules having voltages in the range of 16 to 48 V (see Table 20.9). As would be expected, the weight and volume of the modules are significantly greater than those of the cells alone resulting in packing factors of 0.5 to 0.7 for weight and 0.4 to 0.6 for volume. Hence, the system energy densities of the electrochemical capacitors are significantly less than the cell values for each of the capacitor technologies.

*Cycle Life Considerations for Packs.*    One of the advantages of electrochemical capacitors relative to batteries is their long cycle life, which may be as long as one million ($10^6$) deep-discharge cycles for cells at room temperature and less than rated cell voltage. Unfortunately, estimating the lifetime (years) of a pack of capacitor cells in a particular application is much more complicated than simply cycling cells at room temperature. One primary reason for this difficulty is that in most applications (vehicles in particular) many cells are connected in series to attain the required system voltage. In addition, the temperature varies across the pack even with cooling and the cells spend some time at voltages approaching their maximum rated voltage even with cell-balancing circuitry. As discussed in the following paragraphs, these factors significantly reduce the pack lifetime from that expected based on single-cell testing.

The estimation of the cell and pack lifetime is considered in detail analytically in Refs. 3 and 76. The analysis is based on the assumption that the cell lifetime statistics can be expressed in terms of a Weibull distribution.

$$F(t) = 1 - \exp[-(t/\alpha)^\beta]$$

where $F$ is the fraction of cell failures, $t$ is the test time, $\alpha$ is the characteristic lifetime of a cell, and $\beta$ is the shape factor.

Cell lifetime testing must be done using test conditions that result in the same aging mechanisms (cell failures) as in the application of interest. For electrochemical capacitors, the testing could be either cycling at specified power levels, voltage ranges, and temperatures or floating at specified temperatures and voltages. For vehicle applications, it is likely that lifetime data[76] from the cell float tests are the most appropriate. Such data can be curve fit to obtain values for $\alpha$ and $\beta$ for the single cells. Both parameters can vary over wide ranges depending on the cell voltage and temperature of the tests. In the case of activated carbon cells, the characteristic lifetime ($\alpha$)

can vary from >10,000 h at room temperature and 2.3 V to <500 h at 60°C and 2.8 V. The shape factor (β) can vary between about 4 for room temperature and 2.3 V to about 15 for 60°C and 2.8 V. A low value of β means that cell failures occur gradually over time, while a high value of β indicates that nearly all the cells fail together over a short period of time.

The lifetime characteristics of a pack of electrochemical capacitors depend strongly on the number ($N$) of the cells connected in series. Assuming the pack failure statistics are also Weibull, the shape factor of the pack is the same as that of the cells in the pack, and each cell failure is independent of other cells, the pack failure function $F_{pack}$ can be written as

$$F_{pack} = 1 - \exp[-(t/\alpha_c)^\beta]^N$$

$$R_{pack} = \exp[-(t/\alpha_c)^\beta]^N$$

where $R_{pack}$ is the fraction of cells not failed.

Defining a characteristic time for the pack ($\alpha_p$) and equating

$$[(t/\alpha_c)^\beta]^N = (t/\alpha_p)^\beta$$

one finds

$$\alpha_p = \alpha_c / N^{1/\beta}$$

Hence, the characteristic time of the pack is much less than that of the cells. For example, for a pack with 200 cells in series, the characteristic time is reduced by a factor of 3.76 if β is 4 and by a factor of 1.42 if β is 15. For most packs, the characteristic time would likely be ⅓ to ½ that of the cells.

The next factors to consider in the estimation of the lifetime of a pack are the effects of nonuniformities in voltage and temperature of the cells. Even with cooling and cell-balancing circuits, there will be nonuniformities in the pack. This will especially be the case for applications in which the pack provides dynamic, high power as in vehicles. The effects of nonuniformities are considered analytically in Refs. 76 and 77. The analysis is based on the following assumptions concerning the effect of temperature and voltage on the failure of single cells: in the case of temperature, a 10°C decrease in temperature doubles the cell life; in the case of voltage, a 0.1 V decrease in voltage doubles the cell life. The analytical forms of the temperature ($T$ K) and voltage ($V$) effects on the cell characteristic lifetime τ are

$$\tau = a \exp(b/T), \tau/\tau_0 = \exp[-6155(T - T_0)/T_0^2]$$

$$\tau = A \exp(-BV), \tau/\tau_0 = \exp[-6.93(V - V_0)]$$

These relationships project a reduction in characteristic time of about $1/\sqrt{2}$ for variations of 5°C in temperature and 0.05 V in voltage. This corresponds to a maximum temperature difference of 10°C with an average difference of 5°C for an average temperature of 30°C. Similarly, the maximum voltage difference is 0.1 V with an average difference of 0.05 V.

Applying these relationships to a particular application requires detailed knowledge of the application and specification of the cell-operating conditions and tolerable cell failure rates. Consider the following example of a capacitor pack in a hybrid passenger car. The pack failure requirements are 98% reliability for 5 years and 80% reliability for 12 years. The corresponding mileage values are 50,000 mi and 120,000 mi. If the average speed is 25 mph, the operating time requirements are 2000 h and 4800 h for the pack. The pack voltage is 300 V with 125 cells in series. The question to be answered is what cell characteristic time (hours) is needed for the cells if the shape factor of their failure distribution is 10. For a Weibull distribution, the relationship between the proportion ($P$) of failures, time to failure ($t_F$), and distribution characteristics ($\alpha_{pack}$, β) is

$$t_F = \alpha_{pack}[-\ln(1 - P)]^{1/\beta}$$

For the times to failure (2000 and 4800 h), the corresponding $\alpha_{pack}$ values are 2954 and 5581 h. Taking the maximum value of 5581, the cell $\alpha_c$ is 9041 h for $N = 125$ and β = 10. Assuming that the average temperature variation is 5°C and the average voltage variation is 0.05 V/cell in the pack from the base values of 30°C and 2.5 V/cell, the cell statistics on a float-type test at the base values of temperature and voltage should exhibit a $\alpha_c$ of 18,082 and $a\beta$ =10. This corresponds to a float time of about 2 years at 30°C and 2.5 V/cell.

*Cell-Balancing Considerations*

BACKGROUND.    The ultracapacitor unit in many applications will have relatively high voltage (60–500 V) and provide high power in both discharge and charge modes. As in the case of a battery pack, the capacitor unit will consist of many cells (20–200) connected in series. If each of the cells were identical, having exactly the same capacitance, series resistance, and parallel resistance, the voltage of all the cells would be the same at all times and be equal to the average cell voltage (*V*/number of cells). There would not be a concern that the voltage of some of the cells would exceed, at times, the maximum specified by the cell manufacturer. This maximum voltage is often referred to as the working (continuous) voltage limit of the cell, and experiencing voltages above that limit for times longer than a few seconds can significantly reduce the lifetime of the cell. This is not a safety issue per se as the cells can withstand considerably higher voltages without the pressure-relief vent being activated. Limiting the maximum cell voltage in the unit is, then, primarily an issue of cell life, the maximum useable energy of the unit, and system efficiency in high-power cycling. Controlling the variability of the voltage between the cells is referred to as cell balancing. The objectives of the cell balancing are to minimize the differences in the cell voltages and to restrict the maximum voltage of any cell to less than the continuous working voltage. It is desirable that the cell-balancing system provides a means of monitoring the voltages of the cells in order to ensure that they are maintaining their voltages in the proper range. Monitoring the voltage and temperature of the cells is needed to ensure long cycle life (10 years or longer).

The complexity of the balancing approach required and the absolute need for cell balancing depends to a large extent on the magnitude of the differences of the characteristics of the cells (capacitance and resistances). These differences are dependent on the uniformity of the materials used in the cells and quality control in the manufacture of the cells. The specification of ±15% to 20% variability for capacitance and resistance often seen on spec sheets is grossly greater than tolerable for high-voltage strings of capacitors. Experience (available test data[78]) has shown that the variability in the capacitance can be quite small for relatively large cells with the variation between the maximum and minimum being 1% to 1.5% and the standard deviation of the distribution of the capacitance being about 0.5%. Experience has shown larger percentage variations in the resistance of low resistance (fraction of a mΩ) cells, but the standard deviation is close to the accuracy of the resistance measurement (0.01 mΩ). Experience for the self-discharge voltage of a batch of cells after 1 h has shown a maximum variation of less than 5 mV and a standard deviation of less than 1 mV. Hence, present manufacturing practice for capacitor cells seems to be reasonably good and improving so that the prospects for use of relatively simple approaches to cell balancing seem to be good.

Relating the variability in cell characteristics to the cell-to-cell voltage variability for complex discharge/charge cycles is not a simple matter.[5,79,80] Of primary interest are the magnitude of the maximum cell variations, especially those in the direction of high voltage, and whether the magnitudes of the voltage variations tend to increase for long-term cycling without cell balancing and/or equalization. Available data[81,82] seem to indicate that the magnitude of cell-to-cell voltage variations do not increase with cycling, but rather seem to stabilize even without cell balancing. This seems to be the case even as the cells age.

Variations in cell capacitance will lead to the largest cell-to-cell voltage variations, but these variations will not tend to increase with extended charge/discharge cycling because the capacitance differences have self-compensating effects for charge and discharge pulses. Variations in cell resistances (series and parallel) lead to smaller cell-to-cell voltage variations, and these variations also tend to be self-compensating for cycles that consist of sequential charge/discharge pulses. Variations in self-discharge (parallel resistance) can lead to significant differences in cell voltages during extended rest periods without cell balancing. This could lead to large differences in cell voltages when charge/discharge cycling is resumed. For vehicle applications involving charge/discharge cycling, the effect of variations in cell characteristics on cell-to-cell voltage variations can be determined from cycle testing of the capacitor unit for relatively short times. The cycling time should be long enough to reach thermal equilibrium.

APPROACHES TO CELL BALANCING.    All the capacitor manufacturers are developing cell-balancing circuits for use with long series strings of their cells. Modules are supplied with balancing circuits installed. There are a number of approaches to cell balancing. These approaches (see Fig. 20.10) range in complexity from placing a simple resistor in parallel with each cell to connecting an active circuit with a power source to charge or discharge each cell separately as needed to reduce cell-to-cell voltage variations. The simple approaches are likely

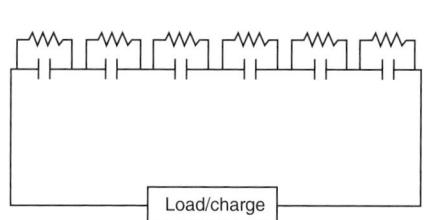

With passive balancing, a resistive ladder is connected to each node in the string of series-connected capacitors.

(a)

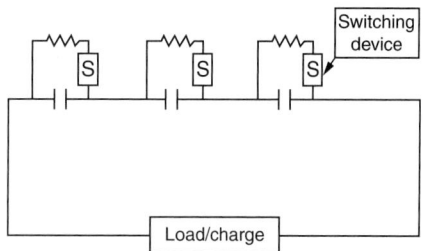

For active balancing, a switching device is placed in series with each balancing resistor. The switches are controlled by voltage-detection circuits that only turn a switch "on" when a capacitor's voltage approaches its continuous-working-voltage rating.

(b)

**Active voltage-balancing electronics**

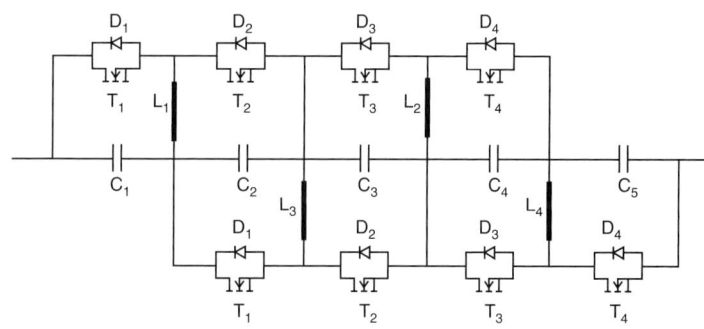

**Active charge equalization device based on back-boost topology**

(c)

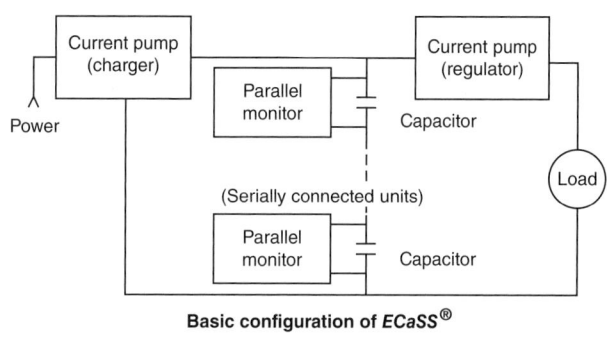

**Basic configuration of *ECaSS*®**

(d)

**FIGURE 20.10**    Cell-balancing circuits.

to be adequate for most applications if the variability of the cell characteristics is relatively small (a few percent); the more complex approaches will be needed if the cell variability is relatively large either due to poor manufacturing quality control, large temperature gradients in the capacitor unit, and/or the effects of aging.

It should be recognized at the outset that the currents involved with the balancing of the cells are small, being 1 A or less, so that in most applications they are much smaller than the pulse currents in/out of the cells. This means that during the high-power portions of the capacitor charge/discharge, the effects of the cell balancing on the cell voltages is small. Cell balancing has the largest effect during periods of low-power demand or rest. Hence, regardless of the cell-balancing approach used, the variability of the characteristics of the cells must be relatively small if cell-to-cell voltage variations are to be tolerable.

As discussed in "Module and Lifetime Considerations" in Sec. 20.4.7, the lifetime (time to significant failures) in the cells decreases by about a factor of 2 for a 0.1 V increase in maximum cell voltage (if that voltage increase is experienced a significant fraction of the time). In general, the lifetime of the carbon/carbon cells decreases markedly as the maximum set voltage for the cell-balancing circuit is increased beyond about 2.6 V/cell. Hence, it appears that the cell voltages should be limited to an average of about 2.6 V/cell with a cell-to-cell variation of less than 0.1 V/cell. This should result in long float and cycle life times of greater than 500,000 cycles and 15,000 h on float at near-room temperature (25–30°C).

## *REFERENCES*

1. Conway, B.E., *Electrochemical Capacitors: Scientific Fundamentals and Technological Applications,* Kluwer Academic/Plenum, 1999.

2. Chandrasekhar, P., *Conducting Polymers, Fundamentals and Applications,* Kluwer Academic Publishers, 1999.

3. Beguin, F., and Frackowiak, E., *Supercapacitors-Materials, Systems, and Applications,* Wily-VCH, 2013.

4. Nishino, A., and Naoi, K., *Technologies and Materials for Large Supercapacitors,* CMC International, 2010.

5. Yu, A., Chabot, V., and Zhang, J., *Electrochemical Supercapacitors for Energy Storage and Delivery,* CRC Press, 2013.

6. Simon, P., and Gogotsi, Y., Capacitive energy storage in nano-structured carbon-electrolyte system, *Acc. Chem. Res.* **46:** 1094–1103 (2013).

7. Kiyohara, K., Sugino, T., and Asaka, K., Electrolytes in porous electrodes: Effects of the pore size and dielectric constant of the medium, *J. Chem. Phys.* **132:**144705 (2010).

8. Interface, Electrochemical Capacitors Powering the 21st Century, publication of the Electrochemical Society, vol. 17, No. 1, Spring 2008.

9. Taberna, P.L., Portet, C., and Simon, P., The role of the interfaces on supercapacitor performance, *Proceedings of the 2nd European Symposium on Supercapacitors and Applications*, Luasanne, Switzerland, November 2006.

10. Simon, P., and Burke, A.F., Nanostructure carbon: Double-layer capacitance and more, Electrochemical Society, Interface Magazine, April 2008.

11. Zhong, C., Deng, Y., Hu, W., Sun, D., Han, X., Qiao, J., and Zhang, J., *Electrolytes for Electrochemical Supercapacitors,* CRC Press, 2016.

12. Fic, K., Lota, G., Meller, M., and Frackowiak, E., Novel insights into neutral medium as electrolyte for high-voltage supercapacitors, *Energy Environ. Sci.* **5:**5842–5850 (2012).

13. Abbas, Q., Babuchowska, P., Frackowiak, E., and Beguin, F., Sustainable AC/AC hybrid electrochemical capacitors in aqueous electrolyte approaching the performance of organic systems, *J. Power Sources* **326:**652–659 (2016).

14. Menzel, J., Fic, K., and Frackowiak, E., Hybrid aqueous capacitors with improved energy/power density performance, *Prog.Nat. Sci–Mat. Int.* **25:**642–649 (2015).

15. Rong, C., Chen, S., Han, J., Zhang, K., Wang, D., Mi, X., and Wei, X., Hybrid supercapacitors integrated rice based activated carbon with $LiMn_2O_4$, *J. Renew. Sustain. Energy* **7:**023104 (2015).

16. Yoo, H.D., Han, S.D., Bayliss, R.D., Andrew, A., Gewirth, A.A., Bostjan Genorio, B., Rajput, N.N., Kristin, A., and Persson, K.A., et al., "Rocking-chair"-type metal hybrid supercapacitors, *ACS Appl. Mater. Interfaces* **8:**30853–30862 (2016).

17. Lazzari, M., Sovavi, F., and Mastraggostini, M., Dynamic pulse power and energy of ionic-liquid supercapacitor for HEV applications, *J. Electrochem. Soc.* **156**(8):A661–A666 (2009).

18. Pandolfo, A.G., and Hollenkamp, A.F., Carbon properties and their role in supercapacitors, *J. Power Sources* **157:**11–27 (2006).

19. Choi, J.H., Lee, C., Cho, S., Moon, G.D., Kim, B.S., Chang, H., and Jang, H.D., High capacitance and energy density supercapacitor based on biomass-derived activated carbons with reduced graphene oxide binder, *Carbon* **132**:16–24 (2018).

20. Taberna, P.L., Portet, C., and Simon, P., Electrode surface treatment and electrochemical spectroscopy study on carbon/carbon supercapacitors, *Appl. Phys. A* **82**:639–646 (2006).

21. Portet, C., Taberna, P.L., Simom, P., and Laberty-Robert, C., Modification of AL current collector surface for sol-gel deposit for carbon-carbon supercapacitor applications, *Electrochim. Acta* **49**:905–912 (2004).

22. Burke, A.F., Kershaw, T., and Miller, M., Development of Advanced Electrochemical Capacitors using Carbon and Lead-Oxide Electrodes for Hybrid Vehicle Applications, UC Davis Institute of Transportation Studies report, UCD-ITS-RR-03-2, June 2003 (paper available on www.its.ucdavis.edu).

23. Fang, B., Wei, K., Maruyama, K., and Kumagai, M., High capacity supercapacitors based on modified carbon aerogel, *J. Appl. Electrochem.* **35**:229–233 (2005).

24. Yang, C., Chen, C., Pan, Y., Li, S., Wang, F., Li, J., and Li, N., et al., Flexible highly specific capacitance aerogel electrodes based on cellulose nanofibers, carbon nanotubes and polyaniline, *Electrochim. Acta* **182**:264–271 (2015).

25. Signorelli, R., Ku, D., Kassakian, J., and Schindall, J., Fabrication and Electrochemical Testing of First Generation Carbon-Nanotube Based Ultracapacitor Cells, *Proceedings of 17th International Seminar on Double-layer Capacitors and Hybrid Energy Storage Devices*, pp. 70–78, Deerfield Beach, December 2007.

26. Lu, W., and Dai, L., Carbon Nanotube Supercapacitors, INTECH, www.intertechopen.com/books/carbon-nanotubes.

27. Basnayaka, P.A., Ram, M.K., Stefanakos, L., and Kumar, A., Graphene/polypyrrole nanocomposite as electrochemical supercapacitor electrode: Electrochemical impedance studies, *Graphene* **2**:81–87 (2013).

28. Zheng, X., Yu, H., Xing, R., Ge, X., Sun, H., Li, R., and Zhang, Q., Multi-growth site graphene/polyaniline composites with highly enhanced specific capacitance and rate capability for supercapacitor application, *Electrochim. Acta* **260**:504–513 (2018).

29. Zhang, J., and Zhao, X.S., Conducting polymers directly coated on reduced graphene oxide sheets as high performance supercapacitor electrodes, *J. Phys. Chem.* **116**:5420–5426 (2012).

30. Ruch, P., Kotz, W., and Wokaun, A., Electrochemical characterization of single-wall carbon nanotubes for electrochemical double-layer capacitors using non-aqueous electrolyte, *Electrochem. Acta* **54**:4451–4458 (2009).

31. Naoi, K., Naoi, W., Aoyagi, S., Miyamoto, J., and Kamino, T., New generation "nanohybrid supercapacitor," *Acc. Chem. Res.* **46**(5):1075–1083 (2013).

32. Iwama, E., Simon, P., and Naoi, K., Ultracentrifugation: An effective novel route to ultrafast nanomaterials for hybrid supercapacitors, *Curr. Opin. Electrochem.* **6**:120–126 (2017).

33. Wu, C., Yang, S., Cai, J., Zhang, Q., Zhu, Y., and Zhang, K., Activated microporous carbon derived from Almond Shells for high energy density asymmetric supercapacitors, *ACS Appl. Mater. Interfaces*, **8**:15288–15296 (2016).

34. Campagnol, N., Romero-Vara, R., DEleu, W., Stappers, L., Binnemans, K., DeVos, D.E., and Fransaer, J., A hybrid supercapacitor based on porous carbon and the Metal-Organic Framework MIL-100(Fe), *Chem. Electro Chem.* **1**:1182–1188 (2014).

35. Salanne, M., Rotenber, B., Naoi, K., Kaneko, K., Taberna, P.L., Grey, C.P., and Dunn, B., et al., Efficient storage mechanisms for building better supercapacitors, *Nat. Energy* **1**:16070 (2016).

36. Chernukhin, S., Tretyakov, D., and Maletin, Y., Hybrid Electrochemical Energy Storage Device, US Pat. Appl., 2014/0085773 A1, publ. March 27, 2014.

37. Jezowski, P., Crosnier, O., Deunf, E., Poizot, P., Beguin, F., and Brousse, T., Safe and recyclable lithium-ion capacitors using a sacrificial organic lithium salt, *Nat. Mater.* **17**:167–173 (2018).

38. Yun, J.H., Han, G.B., Lee, Y.M., Lee, Y.G., Kim, M., Park, J.K., and Cho, K.Y., Low resistance flexible current collector for lithium—ion battery, *Electrochem. Solid-State Lett.* **14**(8):A116–A119 (2011).

39. Fujino, T., Lee, B., Oyama, S., and Noguchi, M., Characterization of Advanced Mesophase Carbons Using a Novel Mass Production Method, *Proceedings of the 15th International Seminar on Double-layer Capacitors and Hybrid Energy Storage Devices*, Deerfield Beach, Florida, December 2005.

40. Okamura, M., et al., The Nanogate Capacitor: A Potential Replacement for Batteries, *Proceedings of the 22nd International Battery Seminar and Exhibit*, Florida, pp. 14–17, March 2005.

41. Furukawa, T., The Reliability, Performance, and Safety of DLCAP, *Proceedings of the 2nd International Symposium on Large Ultracapacitor Technology and Applications*, Baltimore, Maryland, May 2006.

42. Eftekhari, A., Supercapacitors utilizing ionic liquids, *Energy Storage Mater.* **9**:47–69 (2017).

43. Demarconnay, L., Calvo, E.G., Timperman, L., Anouti, M., Lemordant, D., Raymundo-Piñero, E., and Arenillas, A., et al., Optimizing the performance of supercapacitors based on carbon electrodes and protic ionic liquids as electrolytes, *Electrochim. Acta* **108**:361–368 (2013).

44. Lin, Z., Taberna, P.-L., and Simon, P., Graphene-based supercapacitors using eutectic ionic liquid mixture electrolytes, *Electrochim. Acta* **206**:446–451 (2016).

45. Web site: www.cap-xx.com/products.

46. Web site: www.nec-tokin.com/english/products/supercapacitors.

47. Web site: www.lscable.com/products.

48. Web site: www.tjdoublewin.com/products.

49. Burke, A.F., Ultracapacitor Technology: Present and Future, *Proceedings of the Advanced Capacitor World Summit 2003*, Washington, D.C., August 11–13, 2003.

50. Burke, A.F., and Miller, M., Electrochemical Capacitors as Energy Storage in Hybrid-Electric Vehicles: Present Status and Future Prospects, EVS-24, Stavanger, Norway, May 2009 (paper on the CD of the meeting).

51. Burke, A.F., and Park, J., Tests of state-of-the art supercapacitors using aqueous and organic electrolytes, presented at AABC Europe, Mainz, Germany, January 2018.

52. JM Energy, www.jmenergy.co.jp/en.

53. Dougal, R.A., Gao, L., and Liu, S., Ultracapacitor model with automatic order selection and capacity scaling for dynamic system simulation, *J. Power Sources* **126**(1–2):250–257 (2004).

54. Barsoukov, B., and MacDonald, J.R., *Impedance Spectroscopy-Theory, Experiment, and Applications,* Wiley-Interscience, 2005.

55. Farma, R., Deraman, M., Awitdrus, Talib, I.A., Omar, R., Manjunatha, J.G., and Ishak, M.M., et al., Physical and electrochemical properties of supercapacitor electrodes derived from carbon nanotubes and biomass carbon, *Int. J. Electrochem. Sci.* **8**:257–273 (2013).

56. Farahmandi, C.J., Analytical Solution to an Impedance Model for Electrochemical Capacitors, Advanced Capacitor World Summit 2007, San Diego, California, June 2007, also Electrochemical Society Proceedings PV96-25, 1996.

57. Srinivasan, V., and Weidner, J.W., Mathematical modeling of electrochemical capacitors, *J. Electrochem. Soc.* **146**: 1650–1658 (1999).

58. Dunn, D., and Newman, J., Predictions of specific energies and specific powers of double-layer capacitors using a simplified model, *J. Electrochem. Soc.* **147**(3):820–830 (2000).

59. Carslaw, H.S., and Jaeger, J.C., *Conduction of Heat in Solids,* Oxford Press, 1947.

60. Newman, J.S., *Electrochemical Systems,* Prentice-Hall Publishers, 1991.

61. Griffin, J.M., Forse, A.C., Tsai, W.-Y., Taberna, P.-L., Simon, P., and Grey, C.P., *In situ* NMR and electrochemical quartz crystal microbalance techniques reveal the structure of the electrical double layer in supercapacitors, *Nat. Mater.* **14**:812–819 (2015).

62. DeLevie, R., *Electrochemical Response of Porous and Rough Electrodes, Advances in Electrochemistry and Electrochemical Engineering,* Paul Delahay (ed.), vol. 6, Interscience Publishers, 1967.

63. Carlen, M., Christen, T., and Ohler, C., Energy-Power Relations for Supercaps from Impedance Spectroscopy Data, *Proceedings of the 9th International Seminar on Double-Layer Capacitors and Similar Energy Storage Devices*, Deerfield Beach, Florida, December 1999.

64. Burke, A.F., and Miller, M., The power capability of ultracapacitors and lithium batteries for electric and hybrid vehicle applications, *J. Power Sources* **196**(1):514–522 (2011).

65. Zhao, J., Gao, Y., and Burke, A.F., Performance testing of supercapacitors: Important issues and uncertainties, *J. Power Sources* **363**:1–14 (2017).

66. FreedomCar Ultracapacitor Test Manual, Idaho National Engineering Laboratory Report DOE/NE-ID-11173, September 21, 2004.

67. Battery Test Manual for Plug-in Hybrid Electric Vehicles, U.S. Department of Energy, INL/EXT-07-12536, March 2008.

68. Miller, J.R., and Burke, A.F., Electric Vehicle Capacitor Test Procedures Manual, Idaho National Engineering Laboratory Report DOE/ID-10491, October 1994.

69. Anderman, M., Could Ultracapacitors Become the Preferred Energy Storage Device for Future Vehicles? *Proceedings of the 5th International Advanced Automotive Battery Conference*, Honolulu, Hawaii, June 15–17, 2005.

70. Burke, A.F., and Miller, M., Ultracapacitor Update: Cell and Module Performance and Cost Projections, *Proceedings of the15th International Seminar on Double-layer Capacitors and Hybrid Energy Storage Devices*, Deerfield Beach, Florida, Dec 5–7, 2005.

71. Angerer, C., Krapf, S., Wassiliadis, N., and Lienkamp, M., Reduction of aging-effects by supporting a conventional battery pack with ultracapacitors, 2017 Twelfth International Conference on Ecological Vehicles and Renewable Energies (EVER).

72. Zhao, C., Yin, H., and Ma, C., Quantitative evaluation of LiFePO$_4$ battery cycle life improvement using ultracapacitors, *IEEE Trans. Power Electron.* **31**(6):3989–3993 (2016).

73. Burke, A.F., and Zhao, H., Considerations in the use of supercapacitors in combination with batteries in vehicles, on the CD for EVS30, Stuttgart, Germany, October 2017.

74. Parvini, Y., Siegel, J.B., Stepanopoulou, A.G., and Vahidi, A., Supercapacitor electrical and thermal modeling, identification, and validation for a wide range of temperature and power applications, *IEEE Trans. Ind. Electron.* **63**(3):1574–1585 (2016).

75. Wang, K., Zhang, L., Ji, B., and Yuan, J., Thermal analysis on the stackable supercapacitor, *Energy* **59**:440–444 (2013).

76. Miller, J.R., Butler, S.M., and Goltser, I., Electrochemical Capacitor Life Predictions Using Accelerated Test Methods, *Proceedings of the 42nd Power Sources Conference*, paper 24.6, p 581, Philadelphia, Pa., June 2006.

77. Miller, J.R., and Butler, S.B., Capacitor System Life Reduction caused by Cell Temperature Variation, *Proceedings of the Advanced Capacitor World Summit*, San Diego, California, July 2006.

78. Burke, A.F., Characterization of a 25 Wh Ultracapacitor Module for High-Power, Mild Hybrid Applications, *Proceedings of the Large Capacitor Technology and Applications Symposium*, Honolulu, Hawaii, June 13–14, 2005.

79. Jung, D.Y., Shield Ultracapacitor Strings from Overvoltage Yet Maintain Efficiency, Electronic Design, May 27, 2002.

80. Kim, Y., Ultracapacitor Technology Powers Electronic Circuits, Power Electronics Technology, October 2003.

81. Kotz, R., Sauter, J.C., Ruch, P., Dietrich, P., Büchi, F.N., Magne, P.A., and Varenne, P., Voltage Balancing of a 250 V Supercapacitor Module for a Hybrid Fuel Cell Vehicle, *Proceedings of the 16th International Seminar on Double-layer Capacitors and Hybrid Energy Storage Devices*, Deerfield Beach, Florida, December 2007.

82. Burke, A.F., and Miller, M., Cell Balancing Considerations for Long Series Strings of Ultracapacitors in Vehicle Applications, *Proceedings of the Advanced Capacitor World Summit*, San Diego, California, July 11–13, 2005.

# CHAPTER 21
# THERMAL BATTERIES

**Paul F. Schisselbauer, Nicholas Shuster, Chase B. Whitman, Monica V. Stoka
(Emeritus Contributor: Charles M. Lamb)**

## 21.1  GENERAL CHARACTERISTICS

The importance of thermal batteries has grown over the years with each technical advance in performance capability. Currently, state-of-art thermal batteries offer the following features:

1. Highest power density of any practical battery systems.
2. Storage capability of more than 25 years.
3. Operating ability over extreme temperature ranges.
4. Ability to be produced on automated manufacturing equipment, ensuring consistency in performance and minimum cost.

These factors have led to their use in the most demanding strategic and tactical applications.

Thermal batteries are primary reserve batteries that employ inorganic salt electrolytes. These electrolytes are relatively nonconductive solids at ambient temperatures. Integral to the thermal battery are pyrotechnic materials scaled to supply sufficient thermal energy to melt the electrolyte. The molten electrolyte is highly conductive, and high currents may then be drawn from the cells.

The activated life of a thermal battery depends on several factors involving cell chemistry and construction. Once activated, and as long as the electrolyte remains molten, thermal batteries may supply current, discharging the active materials to the point of functional exhaustion. On the other hand, even with excess active materials present, the batteries will eventually cease functioning due to the loss of internal heat and subsequent resolidification of the electrolyte. Hence, two of the primary factors behind thermal battery active life include:

1. Compositions and masses of the active cell-stack materials (i.e., anodes and cathodes)
2. Other construction details, including the overall battery shape and the types and amounts of thermal insulation

Depending on the battery design, which is ultimately determined by the specific requirements of the application, the activated thermal battery may supply electric power for only a few seconds or may function for over an hour.

Initiation of a thermal battery is normally provided by an energy impulse from an external source to a built-in initiator. The initiator, an electric igniter or a percussion primer, ignites the cell-stack pyrotechnics. Rise time, the time interval between the initiation impulse and that time at which the battery can sustain a current at voltage, varies as a function of battery size, design, and chemistry. Rise times of several hundred milliseconds are not uncommon for large units. Small batteries have been designed to reliably achieve operating conditions within 10 to 20 ms.

The shelf life of an unactivated thermal battery is typically 10 to 25 years, depending upon the design. Once activated and discharged, though, they are not reusable or rechargeable. Current developments in extending the activated life capabilities of thermal batteries have widened their suitability and application potential in new military as well as industrial/civilian systems.

Thermal batteries were first developed in Germany in the 1940s, and were used primarily for weapons applications.[1-3] Batteries containing multiple cells and integral pyrotechnic materials have been produced since 1947.[4] Because of their high reliability and long shelf life, thermal batteries are ideally suited for military ordnance purposes. Consequently, they have been widely used in missiles, bombs, mines, decoys, jammers, torpedoes, space exploration systems, emergency escape systems, and similar applications. Figure 21.1 illustrates typical thermal battery configurations.

**FIGURE 21.1**    Typical thermal batteries. (*Courtesy of EnerSys Advanced Systems.*)

Some of the advantages of thermal batteries include the following:

1. Very long shelf life (up to 25 years) in a ready state without degradation in performance.
2. Almost instant activation; fast start designs can provide useful power in hundredths of a second.
3. Peak-power densities, exceeding 16 W/cm$^2$.
4. Very high demonstrated reliability and ruggedness following long-term storage over a wide temperature range and severe dynamic environments.
5. No required maintenance; they can be permanently installed in equipment.
6. Generally, negligible self-discharge; an unactivated battery can support almost no current.
7. Wide operating temperature range.
8. No outgassing; the batteries are hermetically sealed.
9. Custom designs for specific voltage, start time, current, and physical configuration requirements.

The disadvantages of thermal batteries include the following:

1. Generally short activated lives (typically less than 10 min), but they can be designed to operate for more than 2 h.
2. Low to moderate energy densities and specific energies.
3. Surface temperatures that typically reach 230°C or higher.
4. Voltage output that is nonlinear and decreases with life.
5. Onetime use; once activated, they cannot be turned off or reused (recharged).

Additional details on applications for thermal batteries are discussed in Chap. 29.

## 21.2   CELL CHEMISTRY

A wide variety of cell chemistries have been developed and used in thermal batteries. The chemistries include the following:

- Lithium/iron disulfide (Li/FeS$_2$)
- Lithium/cobalt disulfide (Li/CoS$_2$)
- Calcium/calcium chromate (Ca/CaCrO$_4$)
- Calcium/potassium dichromate (Ca/K$_2$Cr$_2$O$_7$)
- Calcium/lead chromate (Ca/PbCrO$_4$)
- Calcium/vanadium pentoxide (Ca/V$_2$O$_5$)
- Calcium/tungsten trioxide (Ca/WO$_3$) and others

At this time, the most widely used chemistries are Li/FeS$_2$ and Li/CoS$_2$. These chemistries offer greater benefits than the other thermal battery systems. However, there are some applications where one of the other less used chemistries could offer special advantages. As an example, the requirement for a very fast activation time with a relatively short activated life could be provided by the Ca/LiCl-KCl/K$_2$Cr$_2$O$_7$ system or the Ca/LiCl-KCl/PbCrO$_4$ system. For the purpose of this Handbook, our discussion will be limited to the chemistries most widely used. To learn more about the older legacy systems, the reader is referred to industry publications and reports.

### 21.2.1   Lithium/Iron Disulfide

There are three common lithium anode configurations: Li(Si) alloy, LiAl alloy, and Li metal in metal matrix, Li(m), where the matrix is usually iron powder. With the difference that the alloy anodes remain solid and the lithium in the Li(Fe) mix is molten in an activated cell, all three anodes participate in the cell reaction similarly. All may be used with the same FeS$_2$ cathode and the same electrolytes. These electrolytes may be the basic LiCl-KCl eutectic electrolyte, the LiCl-LiBr-LiF electrolyte for best ionic conductivity, or a lower-melting-point electrolyte such as LiBr-KBr-LiF for extended activated life. Since the FeS$_2$ is a good electronic conductor, the electrolyte layer is necessary in order to prevent direct anode-to-cathode contact and cell short-circuiting. When molten, the electrolyte between the anode and the cathode is held in place by capillary action through the use of a chemically compatible (inert) binder material. MgO is the preferred material for this application.[5]

The Li/FeS$_2$ electrochemical system has become the preferred system because it does not contain any parasitic chemical reactions. The extent of self-discharge depends on the type of electrolyte used and the cell temperature.[6] The predominant discharge path for cathodes is

$$3Li + 2FeS_2 \rightarrow Li_3Fe_2S_4 \ (2.1 \ V)$$

$$Li_3Fe_2S_4 + Li \rightarrow 2Li_2FeS_2 \ (1.9 \ V)$$

$$Li_2FeS_2 + 2Li \rightarrow Fe + 2Li_2S \ (1.6 \ V)$$

Most batteries are designed to use only the first and sometimes the second cathode transition to avoid changes in cell voltage.

The transitions that occur at the anode depend on the alloy used. For LiAl the reaction is

$$\beta\text{-LiAl} \ (ca. \ 20 \ wt\% \ Li) \rightarrow \alpha\text{-Al (solid solution)}$$

Below approximately 18.4 wt% lithium (lower limit for all β-LiAl) and above 10 wt% lithium (upper limit for α-Al), the alloy is two-phase α, β-LiAl. This fixes the alloy voltage on a plateau. This plateau is about 300 mV less than the voltage afforded by pure lithium metal.

The composition transitions for Li(Si) are

$$Li_{22}Si_5 \rightarrow Li_{13}Si_4 \rightarrow Li_7Si_3 \rightarrow Li_{12}Si_7$$

An anode voltage plateau is defined for compositions falling between each adjacent pair of alloys. That is, the first plateau occurs between $Li_{22}Si_5$ and $Li_{13}Si_4$. The 44 wt% Li(Si) composition falls here, and begins its discharge approximately 150 mV less than that of pure lithium.

The use of $FeS_2$ as a cathode material can cause a large voltage transient or spike of 0.2 V or more per cell, which is evident immediately after activation and lasts from milliseconds to a few seconds. This phenomenon is related to the impact of temperature, the amounts of electroactive impurities in the raw material (iron oxides and sulfates), elemental sulfur from $FeS_2$ decomposition, and the activity of lithium not being fixed in the cathode. In applications where the voltage has to be well regulated, this spike is not acceptable. The voltage transient can be virtually eliminated by the addition of small amounts of $Li_2O$ or $Li_2S$ (typically 0.16 mol Li per mol $FeS_2$) to the catholyte ($FeS_2$ and electrolyte blend), a method known as multiphase lithiation.[7] The spike can also be reduced (but not eliminated) by thoroughly washing or vacuum treating the $FeS_2$ to remove acid-soluble impurities and elemental sulfur.

The $Li/FeS_2$ electrochemical system has a number of important advantages over other systems, including $Ca/CaCrO_4$. These advantages include:

- Tolerance of a wide range of discharge conditions, from open-circuit to high current densities
- High current capabilities, 3 to 5 times that of $Ca/CaCrO_4$
- Highly predictable performance
- Simplicity of construction
- Tolerance to processing variations
- Stability in extreme dynamic environments

As a result of these advantages, this system has become the predominant choice for a wide range of high-reliability military and space applications.

## 21.2.2 Lithium/Cobalt Disulfide

As a cathode versus lithium in molten salt electrolyte cells, cobalt disulfide exhibits a slightly lower voltage than iron disulfide. Cobalt disulfide has a greater thermal stability with respect to loss of sulfur, however. The decomposition reactions for cobalt disulfide at elevated temperatures are

$$3CoS_2 \rightarrow Co_3S_4 + S_2(g)$$

$$3Co_3S_4 \rightarrow Co_9S_8 + 2S_2(g)$$

and for iron disulfide at elevated temperatures

$$2FeS_2 \rightarrow 2FeS + S_2(g)$$

As a rough indicator of the relative stabilities, $FeS_2$ will have a sulfur vapor pressure ($p_{S2}$) of 1 atm in equilibrium with it at 700°C, whereas $p_{S2} = 1$ atm for $CoS_2$ at 800°C. It is, therefore, no surprise that the substitution of $CoS_2$ for $FeS_2$ can yield a cell that is more stable at high temperature, and is therefore useful in batteries with activated lives of over 1 h.[8] In an active battery, the decomposition of $FeS_2$ to FeS and elemental sulfur becomes significant above approximately 550°C. The free sulfur can combine directly with the Li anode in a highly exothermic reaction. Not only would this reduce available anode capacity, but the extra heat can cause even more thermal decomposition of the cathode. $CoS_2$, which is stable up to 650°C, allows higher initial stack temperatures to be sustained without excessive degradation of the cathode.[9] It has also been demonstrated that cells with $CoS_2$ cathodes have a lower internal resistance later in activated life than do $FeS_2$ cathodes.

### 21.2.3 Calcium/Calcium Chromate

The reactions that take place in a $Ca/CaCrO_4$ thermal cell during activation must be in critical balance for the cell to function properly. Upon activation (application of heat), the calcium anode reacts with lithium ions in the LiCl-KCl eutectic electrolyte to form liquid beads of Ca-Li alloy. This alloy becomes the operational anode in the subsequent electrochemical reaction. The anodic half-cell reaction is

$$CaLi_x \rightarrow CaLi_{x-y} + yLi^+ + ye^-$$

The Ca-Li alloy beads also react with dissolved $CaCrO_4$, forming a coating of $Ca_5(CrO_4)_3Cl$.[10,11] This Cr(V) compound is the same species that is formed in the cathodic half-cell reaction:

$$3CrO^{-2} + 5Ca^{+2} + Cl^- + 3e^- \rightarrow Ca_5(CrO_4)_3Cl$$

This product acts as a separator or mass transport barrier between the cathode and the anode to limit electrochemical self-discharge. If the integrity of this separator is breached, the battery can experience a "thermal runaway" condition, whereby the active electrochemical components are chemically consumed with accompanying generation of large amounts of excess heat. At the same time, if battery conditions are such that alloy formation exceeds usage, the excess alloy can cause periodic shorting, the "alloy noise" sometimes seen in cold-stored batteries.

The balance between chemical and electrochemical reactions in this system is dependent on the source of materials (particularly $CaCrO_4$), processing variations, density of compression-formed pellets, operating temperature of the cell, rate of current drain, and other variables.[12] Consequently, this system has been gradually phased out in favor of the more stable and predictable lithium/iron disulfide cell chemistry, which also has a higher energy density.

## 21.3   SYSTEM DESIGN AND CONSTRUCTION

A number of electrochemical systems have been used in thermal batteries. As materials and techniques have improved the state-of-the-art (SOA), older designs have gradually disappeared. Battery designs with older technologies still exist, however, and continue to be manufactured, especially in cases where redesign and requalification is economically unacceptable. Table 21.1 lists some of the more common types of electrochemical systems that have been used over the years.

Thermal battery unit cells consist of an alkali or alkaline earth metal anode, a fusible salt electrolyte typically comprised of a mixture of alkali-halide salts, and more recently a metal disulfide ($MS_2$) cathode. A pyrotechnic

**TABLE 21.1**  Types of Thermal Batteries

| Electrochemical system: anode/electrolyte/cathode | Operating cell voltage | Characteristics and/or applications |
|---|---|---|
| $Ca/LiCl-KCl/K_2Cr_2O_7$ | 2.8–3.3 | Very fast activation times; short lives; used in "pulse" applications |
| $Ca/LiCl-KCl/WO_3$ | 2.4–2.6 | Medium-short lives; low electrical noise; not severe physical environments |
| $Ca/LiCl-KCl/CaCrO_4$ | 2.2–2.6 | Medium lives; severe dynamic environments |
| $Mg/LiCl-KCl/V_2O_5$ | 2.2–2.7 | Medium-short lives; severe physical environments |
| $Ca/LiCl-KCl/PbCrO_4$ | 2.0–2.7 | Fast activation; short lives |
| $Ca/LiBr-KBr/K_2Cr_2O_4$ | 2.0–2.5 | Short lives; used in high-voltage, low-current applications |
| $Li(alloy)/LiCl-LiBr-LiF/FeS_2$ | 1.6–2.1 | Short to medium lives, high-current capacity; severe physical environments |
| $Li(metal)/LiCl-KCl/FeS_2$ | 1.6–2.2 | Long lives, high-current capacity; severe physical environments |
| $Li(alloy)/LiBr-KBr-LiF/CoS_2$ | 1.6–2.1 | Long lives (past 1 h), high-current capacity; severe physical environments |

heat source (heat pellet) is inserted between each cell in a series cell-stack configuration and serves to activate the cell by raising its temperature well above the melting point of the electrolyte salt.

## 21.3.1   Cell Components

*Anode Materials.*   Until the 1980s, most thermal battery designs employed a calcium metal anode with a calcium foil generally attached to an iron, stainless steel, or nickel foil current collector or backing. Since first being introduced in the mid-1970s, lithium has become the most widely used anode material in thermal batteries, essentially replacing the calcium technologies. There are two major configurations of lithium anodes: lithium alloy and lithium metal. The most commonly used alloys are lithium-aluminum (LiAl), with about 20 wt% lithium and lithium-silicon Li(Si) containing about 44 wt% lithium. Lithium-boron alloy has also been evaluated; however, because of processing difficulties and cost, this alloy has not found widespread use.

LiAl and Li(Si) alloys are processed into powders which, after blending with various other constituents, are cold-pressed into anode wafers or pellets that generally range in thickness from 0.3 to 1.5 mm. Within a multicell stack, the alloy pellet is backed with an iron, stainless steel, or nickel current collector that serves to physically separate the anode from its adjacent pyrotechnic pellet thereby protecting it from thermal damage during battery activation. Lithium alloy anodes are intended to remain in the solid state during operation and therefore must be maintained below melt or partial melt temperatures. A 44 wt% Li(Si) alloy will partially melt at 709°C, while α, β-LiAl will exhibit partial melting at 600°C. If these temperatures are exceeded, the molten anode will contact the cathode material, resulting in a highly exothermic reaction, cell shorting, and thermal runaway.

Lithium metal anodes function in activated cells at temperatures above the melting temperature of lithium, 181°C. To prevent the molten lithium from flowing out of the anodes and short-circuiting the battery, it is combined with a high-surface-area binder of metal powder or metal foam. The binder holds the lithium in place by surface tension.

Lithium metal anodes are prepared by combining the binder material with molten lithium, followed by pressing the solidified mixture into thin foil, typically 0.07 to 0.65 mm thick. The foil is then cut into cell-sized parts. The anode foil parts are enclosed in iron-foil cups, which provide added protection against the migration of any free lithium (which can result in cell shorting) and also serve as electron collectors (electrical connections). Such anodes can function at cell temperatures greater than 700°C without significant loss of performance.[13] Each thermal battery designer or manufacturer has developed a number of anode configurations, from which the most suitable may be selected, depending upon specific battery performance requirements.

*Electrolytes.*   Historically, most thermal battery designs have used a molten eutectic mixture of lithium chloride and potassium chloride as the electrolyte (45:55 LiCl-KCl by weight, MP = 352°C). Halide mixtures containing lithium have been preferred because of their high conductivities and general overall compatibility with the anodes and cathodes. Compared to most lower-melting oxygen-containing salts, the halide mixtures are less susceptible to gas generation via thermal decomposition or other side reactions. More recent electrolyte variations, containing bromides, have been developed for thermal batteries to achieve a lower melting point (and thus extend the operating life) or to reduce the internal resistance (and raise the current capability) of the batteries. These include LiBr-KBr-LiF (MP = 320°C), LiCl-LiBr-KBr (MP = 321°C), and the all-Li$^+$ electrolyte LiCl-LiBr-LiF (MP = 430°C).[14] Electrolytes employing mixed cations (e.g., Li$^+$ and K$^+$, instead of all-Li$^+$) are subject to the establishment of Li$^+$ concentration gradients during discharge. These concentration gradients can give rise to localized freezing out of salts, especially during high current density operation.[15]

At battery-operating temperatures, the viscosity of molten salt electrolytes is very low (ca. 1 cp). In order to immobilize the molten electrolytes, binders are added to the salts during compounding. Earlier blends, originally developed for Ca/CaCrO$_4$ systems and the original LiAl/FeS$_2$ batteries, employed clays, such as kaolin, and fumed silica as effective binders for the salts. These siliceous materials will react with Li(Si) and lithium metal anodes; however, high-surface-area MgO is sufficiently inert for the more reactive anodes and is presently the binder of choice in most systems.

*Cathode Materials.*   Historically, a wide variety of cathode materials have been used for thermal batteries, to include calcium chromate (K$_2$CrO$_4$), potassium chromate (K$_2$CrO$_4$), potassium dichromate (K$_2$Cr$_2$O$_7$), and various metal oxides (V$_2$O$_5$ and WO$_3$). Current SOA thermal batteries employ primarily metal disulfide (MS$_2$) cathodes, including FeS$_2$, CoS$_2$, and NiS$_2$. The criteria for suitable cathodes include high voltage against a suitable anode, compatibility with molten alkali-halide electrolytes, and thermal stability to approximately 600°C.

Two of the more commonly used cathodes—$FeS_2$ and $CoS_2$—are thermally stable to approximately 550°C and 650°C, respectively.

***Pyrotechnic Heat Sources.*** The two principal heat sources that have been used in thermal batteries are heat paper and heat pellets. Heat paper is a paperlike composition of zirconium and barium chromate powders supported in an inorganic fiber mat. Heat pellets are pressed tablets or pellets consisting of a mixture of iron powder and potassium perchlorate.

The $Zr$-$BaCrO_4$ heat paper is manufactured from pyrotechnic-grade zirconium powder and $BaCrO_4$, both with particle sizes below 10 μ. Inorganic fibers, such as ceramic and glass fibers, are used as a structure for the mat.[16] The mixture, together with water, is formed into a paper, either as individual sheets by use of a mold or continuously through a paper-making process. The resultant sheets are cut into parts and dried. Once dried, the material must be handled very carefully since it is very susceptible to ignition by static charge and friction. Heat paper has a burning rate of 10 to 300 cm/s and a usual heat content of about 1675 J/g (400 cal/g). Heat paper combusts to an inorganic ash with electrical resistivity. If inserted between cells, it must be used in combination with highly conductive intercell connectors. In some battery designs, combusted heat paper serves as an electrical insulator between cells. In these applications it may have an additional layer of ceramic fibers only, known as base sheet, to enhance its dielectric properties. In most modern pellet-type batteries, heat paper is used only as an ignition or fuse train, if at all. In this application, the heat paper fuse, which is ignited by the initiator, in turn ignites the heat pellets, which are the primary heat source in these batteries.

Heat pellets are manufactured by cold-pressing a dry blend of fine iron powder (1–10 μ) and potassium perchlorate. The heat content of $Fe$-$KClO_4$ pellets ranges from 920 J/g for 88 wt% iron to 1420 J/g for 82 wt% iron. The iron content is well in excess of stoichiometric requirements. Once combusted the excess iron sinters into a solid pellet that provides sufficient electronic conductivity as to eliminate the need for separate intercell connectors. By retaining its physical shape, the heat pellet is very stable under dynamic environments (such as shock, vibration, and spin). This contributes greatly to the general ruggedness of battery designs that incorporate heat pellets. By virtue of their considerable thermal mass, heat pellets advantageously serve as heat reservoirs, retaining considerable heat after combustion and extending the useful life of the battery, particularly under minimum temperature conditions.

Burn rates are generally slower than those of heat paper, and the energy required to ignite them is greater. For that reason, the heat pellet is less susceptible to inadvertent ignition during battery manufacture. Heat pellets (and especially unpelletized heat powder) must nevertheless be handled with extreme care and protected from potential ignition sources.

### 21.3.2 Cell Construction

A number of factors, including the cell chemistry, the operating environments, and the preferences of the designer, determine the choice of cell design. Basically, all cell designs fall into three categories: cup cells, open cells, and pelletized cells. To meet specific performance requirements, some designs may incorporate aspects of more than one cell category. Figure 21.2 illustrates the relative thickness ranges of the different cell designs.

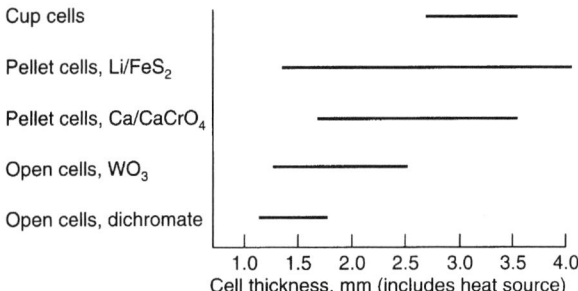

**FIGURE 21.2** Thicknesses of thermal battery cells.

*Cup Cells.*    The typical cup cell features a two-layer anode (calcium or magnesium) having an active anode material on both sides of a central current collector. On either side of the anode is an electrolyte pad made of glass tape impregnated with an eutectic electrolyte. Next to each electrolyte are depolarizer pads consisting of cathode material ($CaCrO_4$ or $WO_3$) in an inorganic fiber matrix (paper). The cell is enclosed in a nickel-foil cup and cover that are tightly crimped (Fig. 21.3*a*). Some designs also incorporate inorganic fiber mat gaskets and nickel eyelets to help prevent the molten electrolyte from leaking out of the activated cell. $Zr/BaCrO_4$ heat paper pads located on either side provide heat to the cup cell.

Cup cells have the advantage of large reactive surfaces (they are two-sided or bipolar), and may contain relatively large amounts of reactive materials. Their disadvantages are that they are difficult to seal against electrolyte leakage and they have low heat capacity. The $Ca/CaCrO_4$ cell chemistry is also prone to "alloying" (producing excess molten Ca-Li alloy), which can short-circuit the cells. In order to obtain required short activation times, cup cells typically have to be premelted or prefused prior to assembly into cell stacks. Intercell electrical connections are accomplished by spot-welding the cell output leads between each cell, which presents a potential reliability problem.

Currently, cup cells have limited application and are found primarily in older battery designs.

*Open Cells.*    The open-cell design is similar in construction to the cup cell, except that it is not enclosed in a cup (Fig. 21.3*b*). Elimination of the cup is possible because the amount of electrolyte is reduced to the extent that practically all of it is bound to the glass fiber cloth matrix by surface tension. Some designs use homogeneous electrolyte-depolarizer pads; others have discrete parts. Open-cell designs typically incorporate a

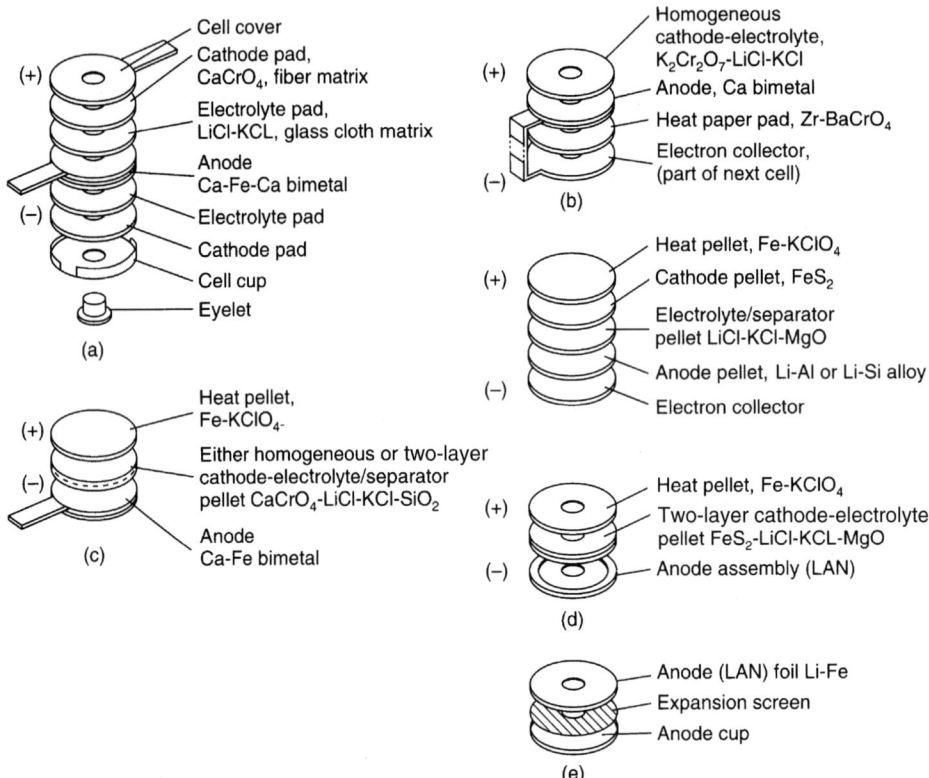

**FIGURE 21.3**   Variations in cell configurations. (*a*) Cup cell. (*b*) Open cell. (*c*) $Ca/CaCrO_4$ pellet cell. (*d*) Li alloy/ $FeS_2$ and $Li/FeS_2$ pellet cell. (*e*) Li metal/$FeS_2$, (LAN) anode assembly.

combination anode-electron collector, usually in the shape of a "dumbbell." This combination part has anode material vacuum-deposited on one end (which serves as anode in one cell), while the other end is an electron collector in the next series-connected cell. A narrow bridge connects the two ends of the dumbbell. The bridge serves as an intercell connector, eliminating the need for spot welds. $Zr/BaCrO_4$ heat paper pads heat the open cells, which are assembled between the folded dumbbells.

The open-cell design is used in relatively short-life applications and in pulse batteries. Their parts can be made very thin to promote very rapid heat transfer and obtain short activation times.

***Pellet Cells.***    In pellet cells, the anode, separator, cathode, and heat source are in pellet (wafer) form. For pellet production, the cell component chemicals are processed into powders, and the powders are uniaxially compacted into wafers. The pressure applied during compaction of cell components is critical to resultant pellet densities. If densities of the cathode, separator, and heat pellet vary extensively, poor battery performance can be expected. Heat pellet density variations can impact ignition sensitivity and burn rates, while variation to cathode and separator pellet densities can inhibit reactivity. Common powder constituents comprising each pellet type are as follows:

1. *Anode pellet:* contains lithium and silicon.
2. *Separator pellet:* contains various mixtures of salt and binder powders, also known as the electrolyte. Some historical alkali halide salt mixtures include LiCl-KCl eutectic, LiBr-KBr-LiF, or LiCl-LiBr-KBr. A commonly used inert binder is magnesium oxide (MgO). In an activated thermal battery, the MgO holds the molten salts in place by capillary action, or surface tension, or both.
3. *Cathode pellet:* contains metal disulfide active material, such as $FeS_2$ or $CoS_2$, blended with various salt chemistries such as the ones mentioned in the separator pellet.
4. *Heat pellet:* contains a blend of iron powder and potassium perchlorate ($KClO_4$). During activation, the $KClO_4$ reacts with the iron to produce heat. The heat transfers into the remaining iron and adjacent pellets. The excess iron serves as a thermal reservoir and conductive path for the duration of the run time.

Pellet cell variations include (1) the use of a two-layer pellet with discrete electrolyte and cathode layers formed into one part, and (2) the use of a homogeneous pellet that has the electrolyte and cathode powders blended together (see Fig. 21.3c). Typical Li/Metal $S_2$ cells are illustrated in Figs. 21.3d and 21.3e. Further details on pellet cell construction are displayed in Fig. 21.4 and Table 21.2.

### 21.3.3   Battery Design

***Cell-Stack Designs.***    All thermal batteries are designed to satisfy a specific set of performance requirements, each of which includes output voltage, current drain, and activated life. In designing a thermal battery, the output voltage determines the number of cells that must be connected in series. Since each cell produces a fixed maximum voltage (from 1.6 to 3.3 V on open circuit, depending on the cell chemistry used), the battery output will be in multiples of discrete cell voltages. Batteries containing over 180 series-connected cells with an overall output voltage near 400 V have been successfully manufactured. Typical batteries contain 14 to 80 cells, and have an output voltage of 28 to 140 V. Figure 21.5 illustrates two different cell-stack configurations, one with cup cells and the other with pellet-type cells.

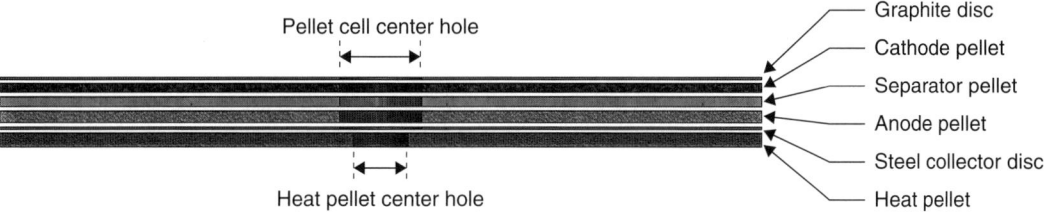

**FIGURE 21.4**    Pellet cell construction and components. (*Courtesy of EnerSys Advanced Systems.*)

**TABLE 21.2**   Typical Thermal Battery Pellet Cell Construction

| Component | Chemical composition | Typical ratios, w% | Density, g/cm² | Forming force, tons |
|---|---|---|---|---|
| 1. Graphite disc | Graphite (hexagonal) | 100 | — | — |
| 2. Cathode pellet | FeS₂/electrolyte/MgO | 64/16/20 | 3.00 | 200 |
| 3. Separator pellet | Electrolyte/MgO | 65/35 | 1.75 | 100 |
| 4. Anode pellet | Lithium/silicon | 44/56 | 1.00 | 100 |
| 5. Steel collector disc | 304 Stainless steel | — | 7.75 | — |
| 6. Heat pellet | Fe/KClO₄ | 88/12 | 3.40 | 60 |

*Notes:*
1. Graphite disc—cut to match the dimensions of the cathode and typically formed into the pellet during compaction. This disc serves as a thermal buffer to protect the cathode from decomposition during activation.
2. A pelletized cathode—$FeS_2$ or $CoS_2$ and electrolyte with an MgO binder.
3. A pelletized electrolyte powder blend—consisting of a salt mixture and an MgO binder.
4. A pelletized anode powder—consisting of lithium and silicone.
5. Steel collector disc—laser cut, stamped, or water jet-cut stainless steel sheet designed to provide a thermal buffer between the heat pellet and anode pellet.
6. A pelletized heat powder blend—of pyrotechnic-grade iron powder and $KClO_4$.

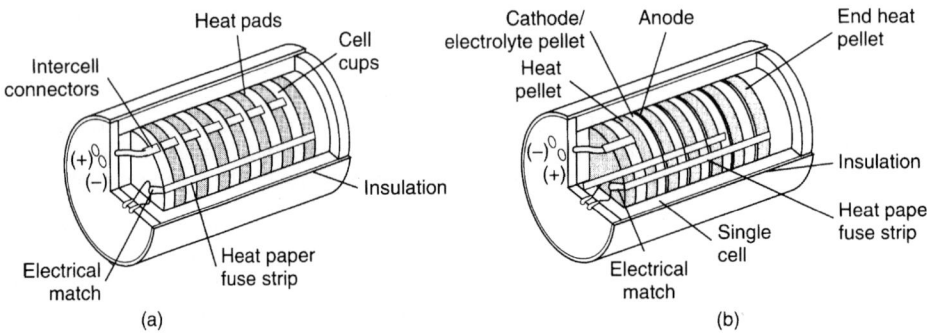

**FIGURE 21.5**   Typical thermal battery assemblies. (*a*) Cup cells. (*b*) Pellet cells.

The current-carrying capacity of each cell is determined by the reactive surface area of the cell, which is directly related to the cell size (diameter). As with cell voltages, the maximum useful current densities (amperes per unit area) differ greatly among cell chemistries (see Tables 21.3 and 21.4). The effective cell area, and hence the current-carrying capacity of a battery, can be adjusted by electrically connecting any number of cells in parallel.

Thermal batteries can be designed to provide multiple output voltages by electrically connecting the required number of cells in series. The multiple-voltage outputs can be drawn either from cells that are common to more than

**TABLE 21.3**   Attainable Current Density of Different Cell Designs

| | Current density, mA/cm² | | |
|---|---|---|---|
| Cell design | 10 s rate | 100 s rate | 1000 s rate |
| Cup cell | 620 | 35 | |
| Open cell/dichromate | 54 | | |
| Pellet cell/two-layer Ca/CaCrO₄ | 790 | 46 | |
| Pellet cell/DEB Ca/CaCrO₄ | 930 | 122 | |
| Pellet cell/Li/FeS₂ | >2500 | 610 | 150 |

**TABLE 21.4**  Typical Power and Energy Densities of Li/FeS$_2$ Thermal Batteries

| Battery volume, cm$^3$ | Power density, W/cm$^2$ | Energy density, Wh/L | Activated life, s |
|---|---|---|---|
| 20 | 11.25 | 46.87 | 15 |
| 29 | 1.44 | 34.20 | 85 |
| 70 | 2.59 | 35.97 | 50 |
| 108 | 0.65 | 32.41 | 180 |
| 170 | 1.98 | 109.80 | 200 |
| 171 | 10.64 | 118.26 | 40 |
| 183 | 2.29 | 63.75 | 100 |
| 306 | 0.51 | 39.65 | 280 |
| 311 | 2.25 | 75.03 | 700 |
| 552 | 0.15 | 67.63 | 1600 |
| 1176 | 0.40 | 101.19 | 900 |
| 1312 | 0.17 | 85.37 | 1800 |
| 3120 | 1.11 | 83.3 | 270 |

one output or from isolated cells whose output is not shared. An electrically isolated group of cells must be used for circuits that cannot tolerate "crosstalk" from other circuits in a system. It is also possible to combine cell-stack sections with different cell chemistries in the same battery. Such combinations yield the specific performance characteristics of both chemistries from a common unit. An example of this is a battery that combines a cell stack with a chemistry that has a very short start time with a different cell stack that can provide a high current over a long activated life. Where such combinations are used, the outputs from the different cell-stack sections are often diode-isolated to prevent one section from charging the other. Some thermal battery designs combine two or more discrete batteries into an assembly that may have a number of different, mutually isolated voltage outputs with widely varying current capabilities.

Cells comprising a cell stack are typically held in place by the closing compression applied when the battery cover is secured by welding it to the battery case. Some battery designs incorporate an inner case to maintain compression on the cell stack while the outer case and cover combination provides hermetic enclosure for the unit. Figure 21.6 depicts a battery design that employs an inner cell-stack case.

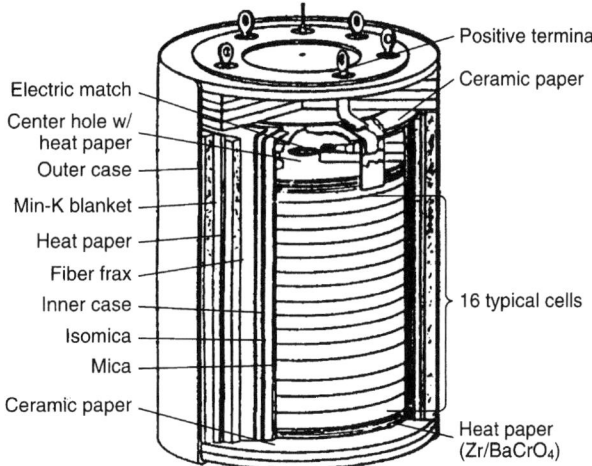

**FIGURE 21.6**  A typical thermal battery assembly with an inner case. (*Courtesy of SAFT Batteries.*)

***Methods of Activation.***    Thermal batteries are activated by applying an external signal to an initiation device that is incorporated in the battery. There are four typical methods of activation: electric signal to an electric igniter; mechanical impulse to a percussion primer; mechanical shock to an inertial activator; and optical energy (laser) signal to a pyrotechnic material.

Electric igniters typically contain one or more bridge wires and a heat-sensitive pyrotechnic material. Upon application of an electric current, the bridge wire ignites the pyrotechnic, which in turn ignites the heat source in the thermal battery. Igniters generally fall into two categories: squibs and electric matches. A typical squib is enclosed in a sealed metal or ceramic enclosure and contains one or two bridge wires. The most commonly used types require a minimum activation current of 3.5 A and have a maximum no-fire limit of 1 A or 1 W (whichever is greater). Electric matches do not have a sealed enclosure and typically contain only one bridge wire. They require an activation current of 500 mA to 5 A and should not be subjected to a no-fire test current of more than 20 mA. Squibs are 4 to 10 times more expensive than electric matches, but they are required for applications that may encounter environments with electromagnetic radiation.

Percussion primers are pyrotechnic devices that are activated by impact from a mechanical striking device. Typically, a primer is activated by an impact of 2016 to 2880 g·cm applied with a 0.6 to 1.1 mm spherical radius firing pin. Primers are installed in primer holders that are integral parts of the battery enclosure.

Inertial or setback activators are devices that are activated by a large-magnitude shock or rapid acceleration, as is generated upon firing of a mortar or artillery round. They are designed to react to a predetermined combination of g force and its duration. Inertial activators are typically firmly mounted inside the battery structure in order to withstand severe dynamic environments.

Optical energy (laser) activation of thermal batteries is accomplished by firing a laser beam through an optical window installed in the outer enclosure of the battery and igniting a suitable pyrotechnic material inside the unit. This method has found utility in applications where severe electromagnetic interference would be disruptive to an electrical firing method.

Thermal batteries can be equipped with more than one activation device. The multiple activators can be of the same type or of any combination required by the application.


***Insulation Materials.***    Thermal batteries are designed to maintain hermeticity throughout their service lives, even though their internal temperatures reach or even exceed 600°C. The thermal insulation used to retard heat loss from the cell stack and minimize peak surface temperatures must be anhydrous and must have high thermal stability. Ceramic fibers, glass fibers, certain high-temperature polymers, and their combinations have been used as thermal insulators. Older battery designs may still have asbestos insulation, which was widely used before the 1980s.

Electrical insulation materials for conductors, terminals, initiators, and other electrically conductive components are typically made of mica, glass or ceramic fiber cloths, and high-temperature resistant polymers.

Thermal insulation is located around the periphery and at both ends of the cell stack. Some designs also incorporate high-temperature epoxy potting materials as insulation and structural support for the initiators and electric conductors on the terminal end (header) of the batteries. Long-life batteries (+20 min) incorporate high-efficiency thermal insulation materials such as Min-K® (Morgan Advanced Materials) or Microtherm® (Promat Inc.). Extended-life batteries (1 h and longer) may incorporate vacuum blankets and/or double cases with vacuum space between them to retard heat loss. Figure 21.7 shows a typical arrangement of thermal insulation and an encapsulated header assembly with an initiator (squib).

An effective method for extending activated battery life and reducing the effects of heat on thermally sensitive components located near the battery is to use an external thermal blanket. Provided that it is protected from contamination, an external thermal blanket can be more effective than internal insulation, primarily because the hot gases that are generated inside the battery during activation cannot penetrate it. External insulation, mounting methods, and the surrounding environment have a significant effect on the heat loss from the battery and all of these must be taken into consideration in the design of thermal batteries.

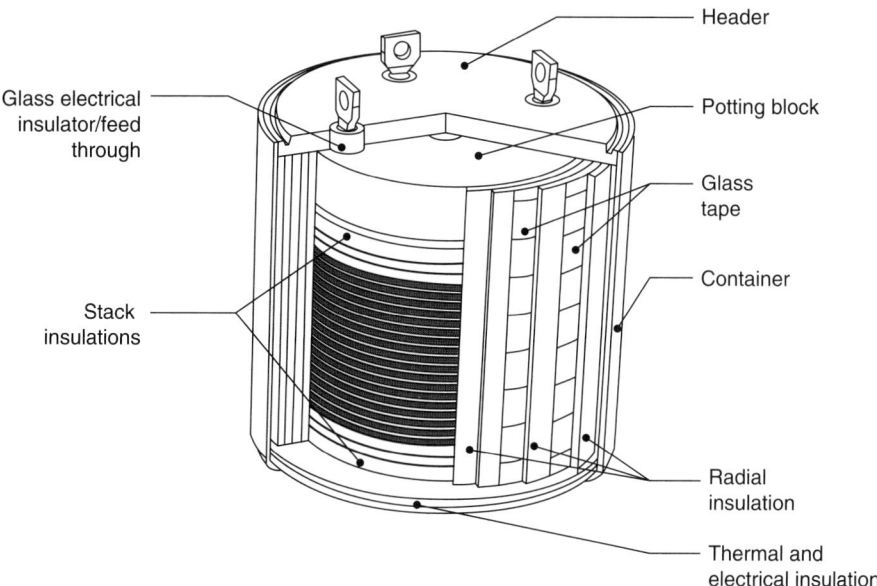

**FIGURE 21.7**    Typical thermal battery insulation layers. (*Courtesy of EnerSys Advanced Systems.*)

## 21.4   PERFORMANCE CHARACTERISTICS AND APPLICATION SPECIFICATIONS

Thermal batteries are custom-designed to satisfy a specific set of performance requirements. These include not only output voltage, current, activated life, and voltage rise time (start), but also storage and activated-life environments, mounting requirements, surface temperature exposure, activation method and energy, and other needs. For this reason, it is very important that the user and the system designer have a close technical interface with the battery designer during the design and development phases of the battery.

### 21.4.1   Activation Time

The activation time (rise time) is the time interval from the application of energy to the initiation device until the battery output voltage reaches the minimum specified limit. The activation time is affected by the operating temperature, applied load, and cell chemistry. Lowering the operating temperature or increasing the load typically increases the activation time. Typical pellet cell thermal batteries have activation times from 0.30 to 2.0 s, depending largely on the cell diameter. Figure 21.8 illustrates the effects of temperature on battery rise time and activated life.

### 21.4.2   Voltage Regulation

Thermal battery output voltages are not linear. After reaching a peak level, typically within 1 s after activation, the voltage starts to decay until it eventually drops below the minimum useful level. Voltage regulation is the range between the specified minimum and maximum limits. Typically, the minimum voltage limit is 75% of the

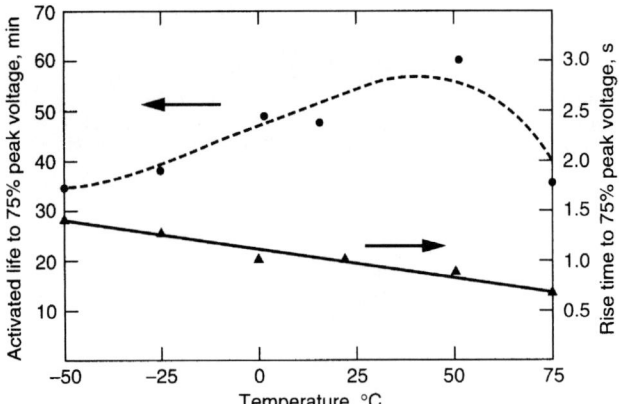

**FIGURE 21.8**    Activated life and rise time of a Li/FeS$_2$ battery.[17]

peak voltage. The battery output profile (consisting of the rise time, peak voltage, and rate of decay) depends on the cell chemistry, and is strongly affected by the operating temperature and applied load. Figure 21.9 illustrates the effects of discharge load on a typical battery output profile.

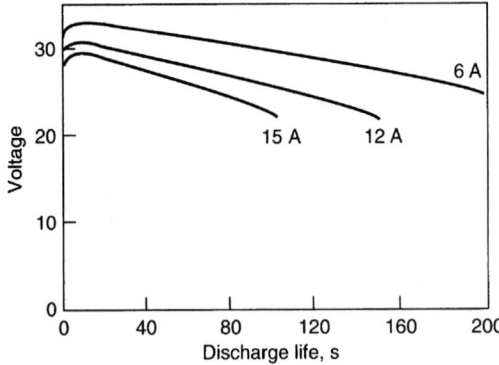

**FIGURE 21.9**    Discharge voltage curves of typical Li/FeS$_2$ thermal battery at three different current drain rates.

### 21.4.3   Activated Life

The activated (operating) life is typically specified as the time from the initial application of the activation energy until the battery voltage drops below the minimum specified limit. Activated life is affected by the cell chemistry used, the operating temperature environment, and the current drain. Typically, thermal batteries are thermally balanced (total cell mass versus caloric input) to have the longest activated lives between the high and low operating temperature limits, or near-room ambient. Lives will get shorter near each temperature limit. At the low limit, the electrolyte will start resolidifying sooner, whereas at the high limit the thermal degradation of FeS$_2$ occurs at a faster rate, depleting active materials.

### 21.4.4   Interface Considerations

The following performance and design characteristics must be noted when designing a system that interfaces with a thermal battery:

1. An unactivated battery has a very high internal resistance (M$\Omega$). Once activated, an individual cell's resistance is between 0.003 and 0.02 $\Omega$, depending on the cell design. The internal resistance of the battery is equal to the sum of the resistances of all series-connected cells.

2. Some cell chemistries, such as Li/FeS$_2$, are tolerant of back-charging from an external power source. Others must not be subjected to back-charging at all.

3. Electric initiators contain bridge wires that may not burn through during activation and, if not disconnected, may act as a parasitic load on the external ignition circuit.

4. Leakage paths that can adversely load the battery may develop in an activated battery between electrically live components and the battery case or activator circuits. System requirements such as case grounding, cell-stack common output, and activator circuit grounding must be specified so that special insulation provisions can be incorporated into the battery design.

5. The surface temperature of an activated battery may reach 400°C. The type of battery mounting, the heat transfer properties of the mounting, the effects of high temperature on the surrounding components, and the proximity of combustible materials must be considered. The battery surface temperature can usually be reduced significantly by incorporating added (or more efficient) thermal insulation. This can increase cost and battery volume. Figures 21.10 and 21.11 illustrate typical surface temperatures of thermal batteries.

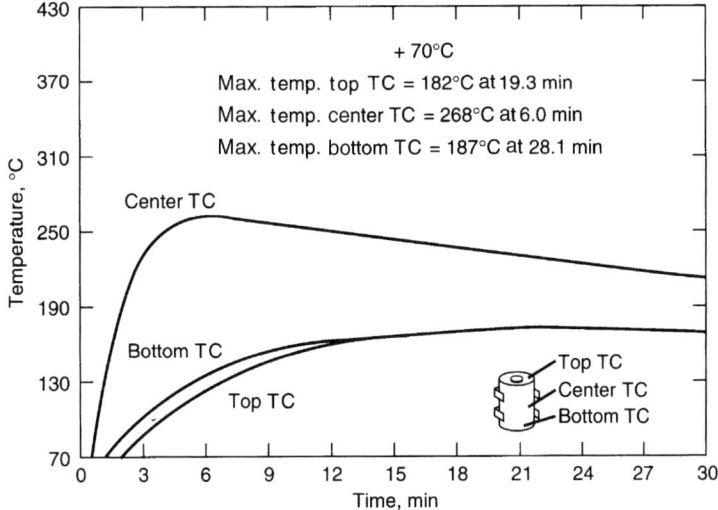

**FIGURE 21.10**   Surface temperature profiles for a long-life thermal battery.[18]

### 21.4.5   Testing and Surveillance

The safety and reliability of thermal batteries has been a matter of continuing study since they were first developed. To identify defective units, most designs are 100% tested for hermeticity, polarity, electrical insulation resistance, and activation circuit resistance (if applicable) as part of the manufacturing process. Most units are also radiographed. Prior to commencement of production, a sample group of 10 to as many as 500 batteries is

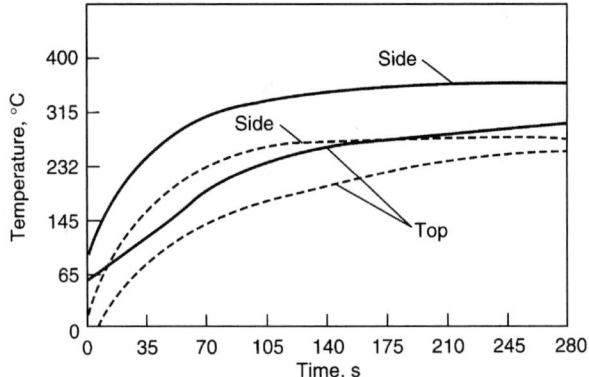

**FIGURE 21.11**   Surface temperature profiles for a medium-life thermal battery. Solid line—tested at 71°C; broken line—tested at −53°C.

subjected to qualification tests. This series of tests includes the most severe environmental and discharge conditions to which the particular battery design will be exposed in actual field use. Almost all thermal batteries are fabricated in homogeneous groups or lots, and samples from each lot are discharged to demonstrate compliance with the performance requirements. Usually the samples are discharged at maximum specified loads, often with concurrently imposed environmental forces. By using such test programs, reliability values greater than 99% and safety values greater than 99.9% have been demonstrated.

Lithium thermal batteries designed for use in U.S. Navy systems are subject to safety tests per Navy technical manual S9310-AQ-SAF, "Batteries, Navy Lithium Safety Program Responsibilities and Procedures." These tests are designed to ensure that the battery design is safe not only in proper storage and use, but also when subjected to inadvertent misuse and conditions caused by accidents, such as back-charging, short circuits, and fires.

### 21.4.6   New Developments

The primary aim of new research and development in the thermal battery area has been to increase the energy density and specific energy of the practical unit. Two possible approaches to this goal are (1) to decrease the total volume and mass of the battery, and (2) to increase the voltage or the current-carrying capacity per cell volume and mass.

In the area of decreasing battery mass, investigations have been conducted by substituting lighter-weight materials for the currently used stainless steel battery housings. Titanium, aluminum, composites, and other materials have been suggested and tried with varying degrees of success. Titanium cases and headers have been successful but suffer from higher cost.

Efforts to deposit active material films by plasma-spraying powders onto stainless steel substrates have yielded promising results. This has been accomplished both for $FeS_2$ cathodes[19] and for Li(Si) anodes.[20] This technology could potentially reduce the mass and volume of a thermal battery by virtue of the higher active material densities afforded. Several organizations have investigated tape casting[21,22] or conventional spraying[23] to achieve thinner components. Nanomaterials are currently being investigated as a potential method to improve thermal battery performance. The use of nano powders can offer higher voltages and increased energy density.[24] Development efforts to produce cells with higher voltage have demonstrated the potential of employing molten nitrate electrolytes with lithium anodes.[25] These systems have the added benefit of lowering the battery operating temperature by more than 200°C.

Efforts have also been directed toward increasing the activated life of batteries past 2 h up to 4 h. This has required the development of more efficient thermal insulation, such as the use of double-walled vacuum enclosures (cases) and multilayered insulating blankets, as well as lower melting point electrolyte compositions.

The long-time use of Fe-KClO$_4$ heat pellets as combination heat/cathodes has attracted increased interest with recent claims of improvement.[26] This technology, in which the iron oxide produced by the heat source is also the cathode, was employed by Catalyst Research Corporation as early as 1981.[27]

## ACKNOWLEDGMENT

Major content for this chapter was provided in whole or in part by Charles M. Lamb, Chap. 36, Thermal Batteries, *Linden's Handbook of Batteries*, 4th ed., T. B. Reddy and D. Linden, eds., McGraw-Hill, 2011.

## REFERENCES

1. G. O. Erb, "Theory and Practice of Thermal Cells," *Publication BIOS/Gp 2/HEC 182 Part II*, Halstead Exploiting Centre, June 6, 1945.

2. O. G. Bennett et al., U.S. Patent 3,575,714, Apr. 20, 1971.

3. B. H. van Domelen, and R. D. Wehrle, "A Review of Thermal Battery Technology," *Intersoc. Energy Convers. Conf.*, 1974.

4. F. Tepper, "A Survey of Thermal Battery Designs and Their Performance Characteristics," *Intersoc. Energy Convers. Conf.*, 1974.

5. Z. Tomczuk, T. Tani, N. C. Otto, M. F. Roche, and D. R. Vissers, *J. Electrochem. Soc.* **129**(5):925–932 (1992).

6. R. A. Guidotti, R. M. Reinhardt, and J. A. Smaga, "Self-Discharge Study of Li-Alloy/FeS$_2$ Thermal Cells," *Proc. 34th Int. Power Sources Symp.,* 1990, pp. 132–135.

7. R. A. Guidotti, "Methods of Achieving the Equilibrium Number of Phases in Mixtures Suitable for Use in Battery Electrodes, e.g., for Lithiating FeS$_2$," U.S. Patent 4,731,307, Mar. 15, 1988.

8. R. A. Guidotti and F. W. Reinhardt, "The Relative Performance of FeS$_2$ and CoS$_2$ in Long-Life Thermal-Battery Applications," *Proc. 9th Int. Symp. Molten Salts*, 1994.

9. R. A. Guidotti and F. W. Reinhardt, "Characterization of the Li(Si)/CoS$_2$ Couple for a High-Voltage, High-Power Thermal Battery," *SAND2000-0396*, 2000.

10. R. A. Guidotti, and F. W. Reinhardt, "Anodic Reactions in the Ca/CaCrO$_4$ Thermal Battery," *SAND83-2271*, 1985.

11. R. A. Guidotti, and W. N. Cathey, "Characterization of Cathodic Reaction Products in the Ca/CaCrO$_4$ Thermal Battery," *SAND84-1098*, 1985.

12. R. A. Guidotti, F. W. Reinhardt, D. R. Tallant, and K. L. Higgins, "Dissolution of CaCrO$_4$ in Molten LiCl-KCl Eutectic," *SAND83-2272*, 1984.

13. G. C. Bowser, D. E. Harney, and F. Tepper, "A High Energy Density Molten Anode Thermal Battery," *Power Sources* **6** (1976).

14. R. A. Guidotti, and F. W. Reinhardt, "Evaluation of Alternate Electrolytes for Use in Li(Si)/FeS$_2$ Thermal Batteries," *Proc. 33rd Power Sources Symp.,* 1988, pp. 369–376.

15. L. Redey, J. A. Smaga, J. E. Battles, and R. Guidotti, "Investigation of Primary Li-Si/FeS$_2$ Cells," *ANL-87-6*, Argonne National Laboratory, Argonne, IL, June 1987.

16. W. H. Collins, U.S. Patent 4,053,337, Oct. 11, 1977.

17. R. K. Quinn, and A. R. Baldwin, "Performance Data for Lithium-Silicon/Iron Disulfide Long Life Primary Thermal Battery," *Proc. 29th Power Sources Symp.*, 1980.

18. H. K. Street, "Characteristics and Development Report of the MC3573 Thermal Battery," *SAND82-0695*, 1983.

19. H. Ye et al., "Novel Design and Fabrication of Thermal Battery Cathodes Using Thermal Spray," Spring Meeting of the Materials Research Society, San Francisco, CA, April 5–9, 1999.

20. C. J. Crowley et al., "Development of Fabricating Processes for Plasma-Sprayed Li-Si Anodes," *Proc. 40th Power Sources Conference,* 2002, pp. 303–306.

21. J. K. Pugh, A. Lang, E. Dayalan, and D. Harney, "Tape Cast Technology as Applied to Thermal Batteries," *Proc. 43rd Power Sources Conference,* 2008, pp. 369–372.

22. J. Edington, G. Swift, and C. Lamb, "Development of Thin Components for Thermal Batteries," *Proc. 43rd Power Sources Conference,* 2008, pp. 177–180.

23. S. B. Preston, Z. Johnson, and R. Guidotti, "Development of Coating Process for Production of Low-Cost Thermal Batteries," *Proc. 43rd Power Sources Conference,* 2008, pp. 373–376.

24. R. Carpenter, and G. Di Benedetto, "Enhanced Nanostructuring Approach for Thermal Battery Cathode Materials," *Proceedings of the 47th Power Sources Conference*, 2016.

25. M. H. Miles, "Lithium Batteries Using Molten Nitrate Electrolytes," *Proc. 14th Annual Battery Conf.,* Long Beach, 1999.

26. D. R. Dekel, and D. Laser, U.S. Patent Appl. 2007/0292748.

27. C. S. Winchester, NSWC Carderock, personal communication.

## BIBLIOGRAPHY

Askew, B. A., and R. Holland, "A High Rate Primary Lithium-Sulfur Battery," *Power Sources* **4** (1972).

Baird, M. D., A. J. Clark, C. R. Feltham, and L. H. Pearce, "Recent Advances in High Temperature Primary Lithium Batteries," *Power Sources* **7** (1978).

Birt, D., C. Feltham, G. Hazzard, and L. Pearce, "The Electrochemical Characteristics of Iron Sulfide and Immobilized Salt Electrolytes," *Power Sources* **7** (1978).

Bowser, G. C., and J. R. Moser, U.S. Patent 3,891,460, June 24, 1975.

Bowser, G. C., and J. R. Moser, U.S. Patent 3,930,888, Jan. 1976.

Bush, D. M., and D. A. Nissen, "Thermal Cells and Batteries Using the $Mg/FeS_2$ and $LiAl/FeS_2$ Systems," *Proc. 28th Power Sources Symp.*, 1978.

Collins, W. H., U.S. Patent 1,482,738, Aug. 10, 1977.

De Gruson, J. A., "Improved Thermal Battery Performance," *AFAPL-TR-79-2042*, Eagle-Picher Industries, 1979.

Delnick, F. M., R. A. Guidotti, and D. K. McCarthy, "Chromium (V) Compounds as Cathode Materials in Electrochemical Power Sources," U.S. Patent 4,508,796, Apr. 2, 1985.

Guidotti, R. A., and F. W. Reinhardt, "Lithiation of $FeS_2$ for Use in Thermal Batteries," *Proc. 2nd Annual Battery Conf. on Applications and Advances,* 1987, paper 87DS-3.

Guidotti, R. A., F. M. Reinhardt, and W. F. Hammeter, "Screening Study of Lithiated Catholyte Mix for a Long-Life $Li(Si)/FeS_2$ Thermal Battery," *SAND 85-1737*, 1988.

Hansen, M., *Constitution of Binary Alloys*, McGraw-Hill, New York, 1958.

Harney, D. E., U.S. Patent 4,221,849, Sept. 9, 1980.

Kuper, W. E., "A Brief History of Thermal Batteries," *Proc. 36th Power Sources Conf.,* Cherry Hill, N.J., June 1994.

Quinn, R. K., A. R. Baldwin, J. R. Armijo, P. G. Neiswander, and D. E. Zurawski, "Development of a Lithium Alloy/Iron Disulfide 60-Minute Primary Thermal Battery," *SAND79-0814*, 1979.

Schneider, A. A. et al., U.S. Patent 4,119,796, Oct. 10, 1978.

Searcey, J. Q., and J. R. Armijo, "Improvements in $Li(Si)/FeS_2$ Thermal Battery Technology," *SAND82-0565*, 1982.

Szwarc, R., "Study of Li-β Alloy in LiCl-KCI Eutectic Thermal Cells Utilizing Chromate and Iron Disulfide Depolarizer," Gepp-TM-426, General Electric Co., Neut. Dev. Dept., 1979.

# CHAPTER 22
# EMERGING TECHNOLOGIES

## 22.1 OVERVIEW

The chapters included in Part 4 of the Handbook embody the mainstream battery technologies, practiced either historically or currently. Some notable exceptions not detailed in this book include radioisotope, biological, and "cold fusion" electrochemical devices. Perhaps, someday these or other fringe ideas[a] will prove practical and become commonplace. However, for now the field of electrochemistry is teeming with new ideas that are within reach of becoming commercial products. In many cases, these new inventions rely on a component improvement, but the more successful innovations are most likely to involve unique combinations of materials, components, and system architectures. Concepts such as "Fourth Generation R&D"[b] and "Dominant Designs and Platform Technologies"[c] illustrate pathways for achieving both technical and financial success. The sections in this chapter provide a flavor for the progress being made in the field of electrochemistry that may one day shape the entire progression of energy-storage technology.

## 22.2 CHAPTER TOPICS

Three areas of major current scientific endeavor, not fully covered in prior chapters, are discussed in the following chapter sections and are summarized as noted below:

Chapter 22A: Hybrid Electrolytes—the use of electrolytes that include two or more phases or types of ion-conductive materials, often including a solid-state electrolyte (see Chap. 22C) with either a liquid or gelled electrolyte.

Chapter 22B: Redox Flow Batteries—an electrochemical cell design that combines elements of fuel cells with traditional batteries.

Chapter 22C: Solid-State Electrolytes—ion-conductive materials that are based only on solids, typically polymer, glass, or ceramic, that do not melt, dissolve, or flow under normal use conditions.

While other innovations are being made on a regular basis, such as silicon-based lithium ion cell anodes and carbon-based lead-acid cell negative electrodes, the topics above were chosen so as to provide a few illustrative examples of the current trends for potential major advancements. Dozens of other equally promising research

---

[a]U.S. Patent Number 7,579,117, Electrochemical Cell Energy Device Based on Novel Electrolyte. Abstract: A novel electrolyte system technology, based on a supercritical fluid solvent using any of a variety of conventional dissolved species with organic salts, hydrates and aqueous-based systems being preferred, that is useful in a variety of electrochemical applications, including batteries, capacitors, fuel cells, sensors, fusion reactors and other similar types of electrolytic cells. Filed: August 22, 2006; Date of Patent: August 25, 2009; Inventor: Kirby W. Beard.

[b]*Fourth Generation R&D: Managing Knowledge, Technology, and Innovation,* William L. Miller and Langdon Morris, John Wiley & Sons, Inc., ©1999.

[c]*Strategic Management of Technology and Innovation,* 4th ed., Robert A. Burgelman, Clayton M. Christensen, Steven C. Wheelwright, McGraw-Hill/Irwin, ©2004.

areas could easily have been included. However, there are many other concepts that probably should not be the focus of intensive research. While there may be no bad ideas, some technologies or business ventures should be confined to skunk works or university basement laboratories before exposure to critical review. Missteps and mistakes are a necessary part of any new breakthrough.

## 22.3   CAUTIONARY ADVICE

The criteria for selecting any subject matter for scientific investigation and subsequent commercialization rely on both technical and economic factors. As noted in Chap. 32E, funding for project work is always a great challenge. The time lines for developing and implementing radical new battery technologies are nearly prohibitive from an economic justification point of view.

Therefore, researchers, business executives, and investors/stakeholders are highly obligated to ensure the most consequential use of resources. As detailed in various chapters of this Handbook (i.e., especially, as noted in the Preface), the rationing of scarce resources is particularly important in the battery industry. Major funding should typically be applied to only the most fundamentally sound concepts. Picking winning and losing technologies is not easy, and striking a proper balance between overfunding of less consequential technologies and underfunding of great ideas is the typical challenge.

Most significantly, electrochemical reactions are difficult to fully specify or adequately control. While the innovations detailed in this chapter appear promising, there is still considerable uncertainty over potential commercial rewards for any new concept. The history of battery R&D is one of many successes but also numerous failures.

# SECTION A

# HYBRID ELECTROLYTES

**Rose E. Ruther, Nancy J. Dudney, Kang Xu**

## 22A.1   *OVERVIEW*

Solid-state batteries promise improved safety, longer cycle life, greater operational temperature range, and higher energy density compared to their counterparts with organic liquid electrolytes, especially if the dendrite issue of the $Li^0$ electrode (i.e., lithium metal foil) could be eliminated.[1-3] To meet these targets, a solid electrolyte should have high ionic conductivity, robust mechanical properties, electrochemical stability over a wide voltage window, and intimate contacts with active electrode materials at both interfaces. The solid electrolyte should also be compatible with low-cost processing methods, by which large-area thin sheets or laminated electrodes free of pinholes could be mass-produced. To date, no single electrolyte has emerged that meets all these requirements. The combination of two or more electrolytes to form a "hybrid" from different categories (liquid, solid, polymeric, etc.) provides a possible path forward to achieve practical solid-state batteries that can be manufactured at large scale. This section provides an overview of the challenges and advantages of these hybrid electrolytes for solid-state or hybrid batteries.

## 22A.2   *SOLID-STATE AND HYBRID CATHODES*

A prime motivation for developing solid electrolytes is the possibility to allow lithium (or sodium) metal anodes, which deliver extremely high capacities (3860 mAh/$g_{Li}$ and 1166 mAh/$g_{Na}$) at low potential. State-of-the-art cathodes such as $LiNi_{1-x-y}Co_xAl_yO_2$ (NCA) and $LiNi_{1-x-y}Co_xMn_yO_2$ (NCM) offer only a fraction of the capacity of alkali metal anodes ($\leq$200 mAh/g). Therefore, matching the high capacity of Li or Na metal requires thick cathodes with high areal loadings. Processing thick solid-state cathodes is one of the major challenges to achieve solid-state batteries with high energy density.[4] The ideal solid-state cathode would have high active material loading (>90 wt% and >2 mAh/cm$^2$) with high electronic and ionic conductivity to be competitive with commercial cathodes that use liquid electrolytes. Only a few all-solid-state cathodes have been reported with reasonable loadings (70 wt% active material and >10 mg/cm$^2$ or <1.4 mAh/cm$^2$).[5,6] Achieving facile ion transport in all-solid-state cathodes is also nontrivial since it requires intimate contact between the ceramic electrolyte and the ceramic cathode material with minimal interfacial resistance. At room temperature the interfacial conductivity between solid electrolytes and solid-state cathodes is typically between $10^{-5}$ and $10^{-3}$ S/cm$^2$ for flat bilayers formed by physical vapor deposition (Fig. 22A.1).[7-9] This conductivity is one to three orders of magnitude lower than across a liquid electrolyte interface.[10] However, this is not the case for all solid electrolyte-to-cathode interfaces. For a number of promising thin film batteries with a Lipon electrolyte,[11,12] the overall area-specific internal resistance of the cell, which includes the bulk electrolyte with both Li anode and cathode interfaces, is only 100 to 300 $\Omega\cdot$cm$^2$ at about 50% state of charge. Because the bulk electrolyte and interface resistances are comparable in magnitude, extraction of the cathode interface contribution is very uncertain and would require a complex battery architecture and use of a carefully designed reference electrode for an accurate determination. A conservative estimate is included in Fig. 22A.1 for comparison.

Powder-based electrode processing and long-term electrochemical cycling place other constraints on all-solid-state cathodes. Oxide solid electrolytes such as the garnet $Li_7La_3Zr_2O_{12}$ (LLZO) or NASICON-type

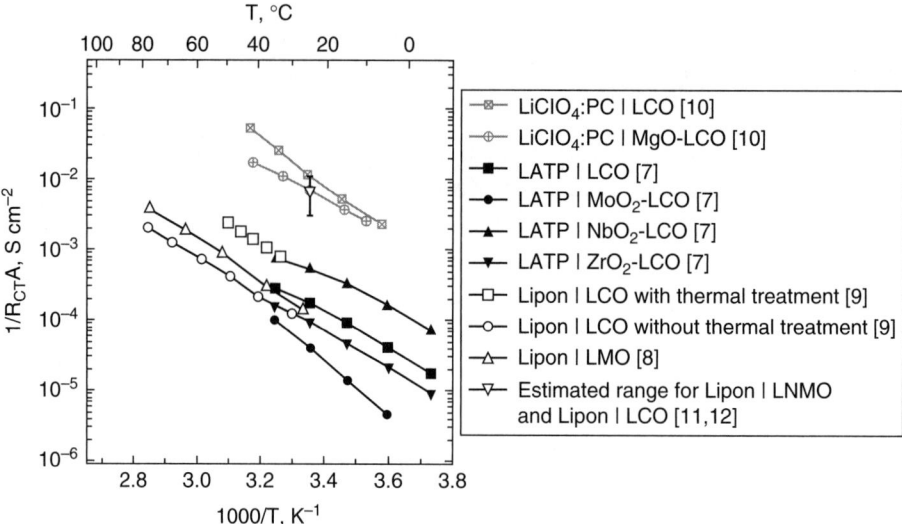

**FIGURE 22A.1**    Temperature dependence of the interfacial resistance for cathode interfaces to liquid (two lines at the top) and solid (remaining lower data points) electrolytes. The data were collected from the available literature on flat, nonporous bilayers.

$Li_{1.5}Al_{0.5}Ge_{1.5}(PO_4)_3$ (LAGP) are brittle and require high co-sintering temperatures (600–700°C) to facilitate good contact between the electrolyte and cathode material. High-temperature processing often results in decomposition reactions and formation of resistive interfacial layers.[13–15] Sulfur-containing solid electrolytes such as $xLi_2S$-$yP_2S_5$ (LPS) are soft and deformable, which enables the formation of dense cathode composites with good contacts at low temperatures. However, sulfides are thermodynamically unstable against high voltage (>3.0 V) cathodes and readily react to form resistive interfacial layers.[16–19] One practical workaround to improve interfacial stability is to coat the cathode with another material such as $LiNbO_3$ or $Li_2SiO_3$ to prevent interdiffusion.[16–18] However, even if a good interface is formed initially, the repeated volume expansion and contraction that accompanies delithiation and lithiation of active materials can result in cracking and delamination at the cathode/electrolyte interface.[20,21] Counterintuitively, models developed by Bucci et al. predict that more compliant solid electrolytes (like sulfides) are more prone to microcracking and fracture than stiffer oxide electrolytes.[22]

A hybrid architecture that combines certain "soft" components of noncrystalline and organic nature (liquid or polymer electrolytes) in the cathode with a ceramic electrolyte would overcome many of these challenges arising from the difficulty in forming good interfaces in all-solid-state cathodes. The inclusion of liquid electrolyte may negate some of the advantages brought by all-solid-state concept such as reductions in packaging.[4] However, hybrid cells could still improve safety over their all-liquid counterparts if they reduce the total volume of the flammable liquid. Results from calorimetry and nail penetration tests showed that solid electrolyte membranes improved safety over porous polypropylene separators even if a conventional cathode with an organic electrolyte was used.[23]

Cathodes for hybrid solid-state batteries have incorporated conventional organic electrolytes,[23–26] polymer electrolytes,[21,27–29] plastic crystal electrolytes,[30] and ionic liquids.[31–34] These secondary electrolytes could be part of a cathode composite that includes a solid electrolyte phase,[32,33] or it could be the only electrolyte in a cathode that interfaces with a solid electrolyte separator.[23–31] In either form, the addition of the secondary electrolyte to the cathode was intended to improve contact, reduce interfacial impedance, and improve performance. For example, the addition of a small amount of ionic liquid to solid electrolyte/cathode composites significantly improved capacity and capacity retention in hybrid cells compared to their all-solid-state counterparts.[32,33] In some cases, even the addition of a simple polymer or gel interlayer between a cathode and a solid electrolyte separator improved adhesion and wetting.[21,25] Adding a gel interlayer between a garnet solid electrolyte and a lithium iron phosphate (LFP) cathode decreased the interfacial resistance by over two orders of magnitude.[25] Despite the promise of this approach, only a few hybrid cathodes have been reported with sufficient thickness and active material loading to enable high-energy cells.[27]

## 22A.3  MULTILAYER INTERFACES TO LITHIUM METAL ANODES

Creating a stable, low-impedance interface to lithium metal also poses significant challenges. Very few electrolytes are thermodynamically stable in contact with lithium metal.[18,19] Garnet-type LLZO and Lipon ($Li_xPO_yN_z$), which are among the few solid electrolytes compatible with lithium metal anodes, may be kinetically stable or form thin, ionically conductive passivation layers at the interface with lithium. For electrolytes that are stable against lithium metal, the interfacial resistance between lithium metal and solid electrolytes is highly sensitive to synthesis and processing conditions. For example, the area-specific impedance at the interface between lithium metal and garnet solid electrolytes varied by almost two orders of magnitude depending on the preparation (Table 22A.1).[26,28,35,36] Low-impedance interfaces between garnet solid electrolytes and lithium metal have been successfully demonstrated by controlling the grain size,[35,36] reducing the amount of lithium carbonate at the surface,[28,35] and coating the solid electrolyte to improve surface wetting.[26]

**TABLE 22A.1**  Lithium Metal-Electrolyte Interfacial Resistance

| Interface | Temperature | ASI, $\Omega \cdot cm^2$ | Reference(s) |
|---|---|---|---|
| LLCZN \| Li | RT | 1710 | 26 |
| LLCZN-$Al_2O_3$ \| Li | RT | 34 | |
| LLZO (small grain) \| Li | 25°C | 37 | 35, 36 |
| LLZO (large grain) \| Li | 25°C | 130 | |
| LLZO (large grain, exposed to air) \| Li | 25°C | 880 | |
| LLZT \| Li | 25°C | 1260 | 28 |
| LLZT-LiF \| Li | 25°C | 345 | |
| Lipon \| Li | RT | 21 | 9 |

*Note:* Comparison of garnet solid electrolytes produced by different synthesis and treatment methods to Lipon.

Most solid electrolytes such as the sulfides and NASICON-type are not stable against lithium metal and form resistive interfacial layers.[18,19,37–40] Nonetheless, some of these solid electrolytes have been successfully integrated into cells with lithium metal anodes through the addition of a secondary electrolyte phase to act as a buffer. The interlayer between lithium metal and the solid electrolyte could be a solid electrolyte such as Lipon,[41,42] polymer,[28,43–45] organic liquid,[34,46,47] or gel electrolyte.[25] Much of the interest in these multilayer, protected lithium metal anodes has been targeted toward Li-sulfur[46,47] and Li-air[34,43,48] batteries with aqueous and nonaqueous catholytes. However, this approach may offer advantages in solid-state batteries by improving the contact to lithium metal.[25,28] Manthiram and coworkers introduced a porous polymer separator filled with a liquid electrolyte between lithium metal and a LISICON solid electrolyte. The liquid/polymer interphase wetted the lithium metal much better than the solid electrolyte and improved cycling more effectively than increased mechanical pressure.[46,47]

## 22A.4  CERAMIC-POLYMER COMPOSITES FOR SOLID ELECTROLYTE SEPARATORS

Most solid-state batteries reported in the literature use very thick solid electrolyte membranes to separate the anode and the cathode, with typical thicknesses ranging from 200 µm up to 2 mm.[5,49–55] This is one to two orders of magnitude thicker than the porous polymer membranes commonly used in lithium-ion batteries with liquid electrolytes, which drastically reduces cell energy density. Processing solid electrolytes into thin, nonporous sheets is very challenging. Thin solid electrolyte films have been successfully deposited by atomic layer deposition[56,57] and sputtering,[58] and efforts are underway to scale these techniques to high-throughput battery manufacturing.[59]

Another attractive approach that is more compatible with current manufacturing methods is to form composite separators that combine polymer and ceramic electrolytes. The goal is for the composite to simultaneously have sufficient stiffness to block lithium dendrites, flexibility to accommodate volume changes during cycling, and high lithium-ion conductivity.[60-63]

Studies on composite polymer electrolytes (CPEs) have yielded mixed results. In many cases, the addition of the solid electrolyte phase did not significantly increase the ionic conductivity compared to the neat polymer electrolyte or the polymer blended with a nonconducting ceramic filler such as $Al_2O_3$.[64-66] Polymer electrolytes such as those based on poly(ethylene oxide) typically have low conductivity (~$10^{-7}$–$10^{-6}$ S/cm) at room temperature. Improvements in conductivity have been reported for CPEs with solid electrolyte nanoparticles[67] or nanowires.[68,69] In these studies, the fraction of solid electrolyte was low (≤15 wt%), and the increased conductivity was attributed to enhanced lithium-ion transport in the interphase surrounding the particles or nanowires.

Ideally, lithium-ion transport in CPEs would occur through the bulk of the ceramic electrolytes, which tend to have higher conductivity than polymers. However, the interfacial resistance between polymer and solid electrolytes can be very large, effectively blocking this path for ion transport (Fig. 22A.2).[70,71] The reported interfacial resistance between solid and polymer electrolytes is highly variable, spanning over five orders of magnitude. Tenhaeff and coworkers demonstrated dramatic reductions in interfacial impedance through simple changes in processing and fabrication of solid electrolyte/polymer electrolyte bilayers.[63] In some cases, the impedance between the ceramic and the polymer was negligible, even lower than the values reported for solid electrolyte interfaces to liquid electrolytes.[72-74] Further work is needed to understand the fundamental factors that influence ion transport across electrolyte interfaces. Most importantly, efforts are urgently needed in establishing reliable and reproducible methodologies for accurately measuring the ion transport properties, which have not been well resolved even in the more mature realm of liquid electrolytes. The difficulty can be expressed by a paradoxical statement that an ideal interface is "invisible," and hence, nonmeasurable. The field would also benefit from more measurements of the mechanical properties of CPEs, which are critical to enable lithium metal anodes.[60,61,64]

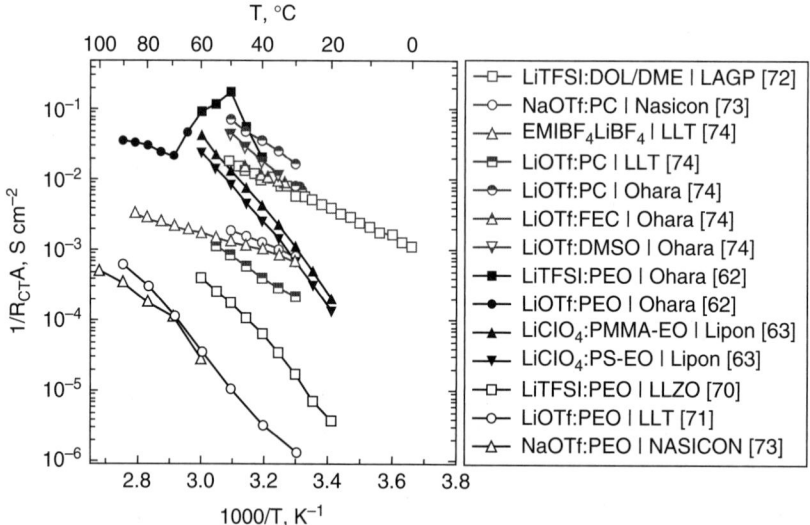

**FIGURE 22A.2** Temperature dependence of the interfacial resistance for solid electrolyte interfaces to liquid (gray) and polymer (black) electrolytes. The data were collected from the available literature on flat, nonporous bilayers.

## 22A.5  OUTLOOK

Electrolyte properties often limit battery performance. Only a limited number of candidate electrolytes have been discovered for solid-state batteries, and none are without shortcomings. Selection of a single solid electrolyte forces a trade-off between conductivity, processability, electrochemical window, interfacial resistance, and cycling stability. Judicious selection of two or more electrolytes can overcome many of these limitations, but also presents new challenges. In particular, charge transfer across additional electrolyte interfaces must be optimized. The formation of stable interfaces with minimal impedance is highly sensitive to both the choice of materials and the processing conditions. More fundamental studies are needed to unpack all the variables that impact charge transfer. Intelligent design of electrolyte interfaces is crucial to enable high-energy, high-power, and low-cost solid-state and hybrid batteries.

## REFERENCES

1. Sun, C. W., Liu, J., Gong, Y. D., Wilkinson, D. P., and Zhang, J. J., Recent Advances in All-Solid-State Rechargeable Lithium Batteries. *Nano Energy* **33**:363–386 (2017).

2. Takada, K., Progress and Prospective of Solid-State Lithium Batteries. *Acta Mater.* **61**(3):759–770 (2013).

3. Kim, J. G., Son, B., Mukherjee, S., Schuppert, N., Bates, A., Kwon, O., and Choi, M. J., et al., A Review of Lithium and Non-Lithium Based Solid State Batteries. *J. Power Sources* **282**:299–322 (2015).

4. Kerman, K., Luntz, A., Viswanathan, V., Chiang, Y. M., and Chen, Z. B., Review-Practical Challenges Hindering the Development of Solid State Li Ion Batteries. *J. Electrochem. Soc.* **164**(7):A1731–A1744 (2017).

5. Ohtomo, T., Hayashi, A., Tatsumisago, M., Tsuchida, Y., Hama, S., and Kawamoto, K., All-Solid-State Lithium Secondary Batteries Using the $75Li_2S \cdot 25P_2S_5$ Glass and the $70Li_2S \cdot 30P_2S_5$ Glass-Ceramic as Solid Electrolytes. *J. Power Sources* **233**:231–235 (2013).

6. Seino, Y., Ota, T., and Takada, K., High Rate Capabilities of All-Solid-State Lithium Secondary Batteries Using $Li_4Ti_5O_{12}$-Coated $LiNi_{0.8}Co_{0.15}Al_{0.05}O_2$ and a Sulfide-Based Solid Electrolyte. *J. Power Sources* **196**(15):6488–6492 (2011).

7. Okumura, T., Nakatsutsumi, T., Ina, T., Orikasa, Y., Arai, H., Fukutsuka, T., and Iriyama, Y., et al., Depth-Resolved X-Ray Absorption Spectroscopic Study on Nanoscale Observation of the Electrode-Solid Electrolyte Interface for All Solid State Lithium Ion Batteries. *J. Mater. Chem.* **21**(27):10051–10060 (2011).

8. Iriyama, Y., Nishimoto, K., Yada, C., Abe, T., Ogumi, Z., and Kikuchi, K., Charge-Transfer Reaction at the Lithium Phosphorus Oxynitride Glass Electrolyte/Lithium Manganese Oxide Thin-Film Interface and Its Stability on Cycling. *J. Electrochem. Soc.* **153**(5):A821–A825 (2006).

9. Iriyama, Y., Kako, T., Yada, C., Abe, T., and Ogumi, Z., Charge Transfer Reaction at the Lithium Phosphorus Oxynitride Glass Electrolyte/Lithium Cobalt Oxide Thin Film Interface. *Solid State Ionics* **176**(31–34):2371–2376 (2005).

10. Iriyama, Y., Kurita, H., Yamada, I., Abe, T., and Ogumi, Z., Efffects of Surface Modification by MgO on Interfacial Reactions of Lithium Cobalt Oxide Thin Film Electrode. *J. Power Sources* **137**(1):111–116 (2004).

11. Li, J. C., Ma, C., Chi, M. F., Liang, C. D., and Dudney, N. J., Solid Electrolyte: The Key for High-Voltage Lithium Batteries. *Adv. Energy Mater.* **5**(4):6 (2015).

12. Dudney, N. J., and Jang, Y. I., Analysis of Thin-Film Lithium Batteries with Cathodes of 50 nm to 4 micron Thick $LiCoO_2$. *J. Power Sources* **119**:300–304 (2003).

13. Kim, K. H., Iriyama, Y., Yamamoto, K., Kumazaki, S., Asaka, T., Tanabe, K., and Fisher, C. A., et al., Characterization of the Interface between $LiCoO_2$ and $Li_7La_3Zr_2O_{12}$ in an All-Solid-State Rechargeable Lithium Battery. *J. Power Sources* **196**(2):764–767 (2011).

14. Miara, L., Windmuller, A., Tsai, C. L., Richards, W. D., Ma, Q. L., Uhlenbruck, S., and Guillon, O., et al., About the Compatibility between High Voltage Spinel Cathode Materials and Solid Oxide Electrolytes as a Function of Temperature. *ACS Appl. Mater. Interfaces* **8**(40):26842–26850 (2016).

15. Iriyama, Y., Kako, T., Yada, C., Abe, T., and Ogumi, Z., Reduction of Charge Transfer Resistance at the Lithium Phosphorus Oxynitride/Lithium Cobalt Oxide Interface by Thermal Treatment. *J. Power Sources* **146**(1–2):745–748 (2005).

16. Haruyama, J., Sodeyama, K., and Tateyama, Y., Cation Mixing Properties Toward Co Diffusion at the $LiCoO_2$ Cathode/Sulfide Electrolyte Interface in a Solid-State Battery. *ACS Appl. Mater. Interfaces* **9**(1):286–292 (2017).

17. Sakuda, A., Hayashi, A., and Tatsumisago, M., Interfacial Observation between $LiCoO_2$ Electrode and $Li_2S$-$P_2S_5$ Solid Electrolytes of All-Solid-State Lithium Secondary Batteries Using Transmission Electron Microscopy. *Chem. Mater.* **22**(3):949–956 (2010).

18. Zhu, Y. Z., He, X. F., and Mo, Y. F., First Principles Study on Electrochemical and Chemical Stability of Solid Electrolyte-Electrode Interfaces in All-Solid-State Li-Ion Batteries. *J. Mater. Chem. A* **4**(9):3253–3266 (2016).

19. Richards, W. D., Miara, L. J., Wang, Y., Kim, J. C., and Ceder, G., Interface Stability in Solid-State Batteries. *Chem. Mater.* **28**(1):266–273 (2016).

20. Bucci, G., Swamy, T., Chiang, Y. M., and Carter, W. C., Random Walk Analysis of the Effect of Mechanical Degradation on All-Solid-State Battery Power. *J. Electrochem. Soc.* **64**(12):A2660–A2664 (2017).

21. Knutz, B., and Skaarup, S., Discharge of Solid-State $Li_3N+TiS_2$ Composite Electrodes. *Solid State Ionics* **18-9**:783–787 (1986).

22. Bucci, G., Swamy, T., Chiang, Y. M., and Carter, W. C., Modeling of Internal Mechanical Failure of All-Solid-State Batteries During Electrochemical Cycling, and Implications for Battery Design. *J. Mater. Chem. A* **5**(36):19422–19430 (2017).

23. Jung, Y. C., Kim, S. K., Kim, M. S., Lee, J. H., Han, M. S., Kim, D. H., and Shin, W. C., et al., Ceramic Separators Based on $Li^+$-Conducting Inorganic Electrolyte for High-Performance Lithium-Ion Batteries with Enhanced Safety. *J. Power Sources* **293**:675–683 (2015).

24. Fu, K. K., Gong, Y. H., Liu, B. Y., Zhu, Y. Z., Xu, S. M., Yao, Y. G., and Luo, W., et al., Toward Garnet Electrolyte-Based Li Metal Batteries: An Ultrathin, Highly Effective, Artificial Solid-State Electrolyte/Metallic Li Interface. *Sci. Adv.* **3**(4):11 (2017).

25. Liu, B. Y., Gong, Y. H., Fu, K., Han, X. G., Yao, Y. G., Pastel, G., and Yang, C. P., et al., Garnet Solid Electrolyte Protected Li-Metal Batteries. *ACS Appl. Mater. Interfaces* **9**(22):18809–18815 (2017).

26. Han, X. G., Gong, Y. H., Fu, K., He, X. F., Hitz, G. T., Dai, J. Q., and Pearse, A., et al., Negating Interfacial Impedance in Garnet-Based Solid-State Li Metal Batteries. *Nat. Mater.* **16**(5):572–579 (2017).

27. Chen, R. J., Zhang, Y. B., Liu, T., Xu, B. Q., Lin, Y. H., Nan, C. W., and Shen, Y., Addressing the Interface Issues in All-Solid-State Bulk-Type Lithium Ion Battery via an All-Composite Approach. *ACS Appl. Mater. Interfaces* **9**(11):9654–9661 (2017).

28. Li, Y. T., Xu, B. Y., Xu, H. H., Duan, H. N., Lu, X. J., Xin, S., and Zhou, W. D., et al., Hybrid Polymer/Garnet Electrolyte with a Small Interfacial Resistance for Lithium-Ion Batteries. *Angew. Chem.-Int. Edit.* **56**(3):753–756 (2017).

29. Zhou, W. D., Wang, S. F., Li, Y. T., Xin, S., Manthiram, A., and Goodenough, J. B., Plating a Dendrite-Free Lithium Anode with a Polymer/Ceramic/Polymer Sandwich Electrolyte. *J. Am. Chem. Soc.* **138**(30):9385–9388 (2016).

30. Gao, H. C., Xue, L. G., Xin, S., Park, K., and Goodenough, J. B., A Plastic-Crystal Electrolyte Interphase for All-Solid-State Sodium Batteries. *Angew. Chem.-Int. Edit.* **56**(20):5541–5545 (2017).

31. Liu, L. L., Qi, X. G., Ma, Q., Rong, X. H., Hu, Y. S., Zhou, Z. B., and Li, H., et al., Toothpaste-Like Electrode: A Novel Approach to Optimize the Interface for Solid-State Sodium-Ion Batteries with Ultralong Cycle Life. *ACS Appl. Mater. Interfaces* **8**(48):32631–32636 (2016).

32. Oh, D. Y., Nam, Y. J., Park, K. H., Jung, S. H., Cho, S. J., Kim, Y. K., and Lee, Y. G., et al., Excellent Compatibility of Solvate Ionic Liquids with Sulfide Solid Electrolytes: Toward Favorable Ionic Contacts in Bulk-Type All-Solid-State Lithium-Ion Batteries. *Adv. Energy Mater.* **5**(22):7 (2015).

33. Zhang, Z. Z., Zhang, Q. H., Shi, J. A., Chu, Y. S., Yu, X. Q., Xu, K. Q., and Ge, M. Y., et al., A Self-Forming Composite Electrolyte for Solid-State Sodium Battery with Ultralong Cycle Life. *Adv. Energy Mater.* **7**(4):11 (2017).

34. Zhang, T., and Zhou, H. S., A Reversible Long-Life Lithium-Air Battery in Ambient Air. *Nat. Commun.* **4**:7 (2013).

35. Cheng, L., Wu, C. H., Jarry, A., Chen, W., Ye, Y. F., Zhu, J. F., and Kostecki, R., et al., Interrelationships Among Grain Size, Surface Composition, Air Stability, and Interfacial Resistance of Al-Substituted $Li_7La_3Zr_2O_{12}$ Solid Electrolytes. *ACS Appl. Mater. Interfaces* **7**(32):17649–17655 (2015).

36. Cheng, L., Chen, W., Kunz, M., Persson, K., Tamura, N., Chen, G. Y., and Doeff, M., Effect of Surface Microstructure on Electrochemical Performance of Garnet Solid Electrolytes. *ACS Appl. Mater. Interfaces* **7**(3):2073–2081 (2015).

37. Wenzel, S., Randau, S., Leichtweiss, T., Weber, D. A., Sann, J., Zeier, W. G., and Janek, J., Direct Observation of the Interfacial Instability of the Fast Ionic Conductor $Li_{10}GeP_2S_{12}$ at the Lithium Metal Anode. *Chem. Mater.* **28**(7):2400–2407 (2016).

38. Wenzel, S., Weber, D. A., Leichtweiss, T., Busche, M. R., Sann, J., and Janek, J., Interphase Formation and Degradation of Charge Transfer Kinetics between a Lithium Metal Anode and Highly Crystalline $Li_7P_3S_{11}$ Solid Electrolyte. *Solid State Ionics* **286**:24–33 (2016).

39. Whiteley, J. M., Woo, J. H., Hu, E. Y., Nam, K. W., and Lee, S. H., Empowering the Lithium Metal Battery Through a Silicon-Based Superionic Conductor. *J. Electrochem. Soc.* **161**(12):A1812–A1817 (2014).

40. Hartmann, P., Leichtweiss, T., Busche, M. R., Schneider, M., Reich, M., Sann, J., and Adelhelm, P., et al., Degradation of NASICON-Type Materials in Contact with Lithium Metal: Formation of Mixed Conducting Interphases (MCI) on Solid Electrolytes. *J. Phys. Chem. C* **117**(41):21064–21074 (2013).

41. Visco, S. J., Nimon, Y. S., Katz, B. D., and De Jonghe, L. C., Active Metal Fuel Cells. US Patent 8,709,679. Issued April 29, 2014.

42. Visco, S. J., Katz, B. D., Nimon, Y. S., and De Jonghe, L. C., Protected Active Metal Electrode and Battery Cell Structures with Non-Aqueous Interlayer Architecture. US Patent 9,123,941. Issued September 1, 2015.

43. Zhang, T., Imanishi, N., Shimonishi, Y., Hirano, A., Takeda, Y., Yamamoto, O., and Sammes, N., A Novel High Energy Density Rechargeable Lithium/Air Battery. *Chem. Commun.* **46**(10):1661–1663 (2010).

44. Kubanska, A., Castro, L., Tortet, L., Dolle, M., and Bouchet, R., Effect of Composite Electrode Thickness on the Electrochemical Performances of All-Solid-State Li-Ion Batteries. *J. Electroceram.* **38**(2–4):189–196 (2017).

45. Chinnam, P. R., and Wunder, S. L., Engineered Interfaces in Hybrid Ceramic—Polymer Electrolytes for Use in All-Solid-State Li Batteries. *ACS Energy Lett.* **2**(1):134–138 (2017).

46. Yu, X. W., Bi, Z. H., Zhao, F., and Manthiram, A., Hybrid Lithium-Sulfur Batteries with a Solid Electrolyte Membrane and Lithium Polysulfide Catholyte. *ACS Appl. Mater. Interfaces* **7**(30):16625–16631 (2015).

47. Yu, X. W., Bi, Z. H., Zhao, F., and Manthiram, A., Polysulfide-Shuttle Control in Lithium-Sulfur Batteries with a Chemically/Electrochemically Compatible NaSICON-Type Solid Electrolyte. *Adv. Energy Mater.* **6**(24):8 (2016).

48. Visco, S. J., Nimon, V. Y., Petrov, A., Pridatko, K., Goncharenko, N., Nimon, E., and De Jonghe, L., et al. Aqueous and Nonaqueous Lithium-Air Batteries Enabled by Water-Stable Lithium Metal Electrodes. *J. Solid State Electrochem.* **18**(5):1443–1456 (2014).

49. Ito, S., Fujiki, S., Yamada, T., Aihara, Y., Park, Y., Kim, T. Y., and Baek, S. W., et al., A Rocking Chair Type All-Solid-State Lithium Ion Battery Adopting $Li_2O-ZrO_2$ Coated $LiNi_{0.8}Co_{0.15}Al_{0.05}O_2$ and a Sulfide Based Electrolyte. *J. Power Sources* **248**:943–950 (2014).

50. Kato, Y., Hori, S., Saito, T., Suzuki, K., Hirayama, M., Mitsui, A., and Yonemura, M., et al., High-Power All-Solid-State Batteries Using Sulfide Superionic Conductors. *Nat. Energy* **1**:16030 (2016).

51. Yubuchi, S., Ito, Y., Matsuyama, T., Hayashi, A., and Tatsumisago, M., 5 V Class $LiNi_{0.5}Mn_{1.5}O_4$ Positive Electrode Coated with $Li_3PO_4$ Thin Film for All-Solid-State Batteries Using Sulfide Solid Electrolyte. *Solid State Ionics* **285**:79–82 (2016).

52. Iwamoto, K., Aotani, N., Takada, K., and Kondo, S., Rechargeable Solid-State Battery with Lithium Conductive Glass, $Li_3PO_4-Li_2S-SiS_2$. *Solid State Ionics* **70**:658–661 (1994).

53. Ohta, S., Komagata, S., Seki, J., Saeki, T., Morishita, S., and Asaoka, T., All-Solid-State Lithium Ion Battery Using Garnet-Type Oxide and $Li_3BO_3$ Solid Electrolytes Fabricated by Screen-Printing. *J. Power Sources* **238**:53–56 (2013).

54. Ohta, S., Kobayashi, T., Seki, J., and Asaoka, T., Electrochemical Performance of an All-Solid-State Lithium Ion Battery with Garnet-Type Oxide Electrolyte. *J. Power Sources* **202**:332–335 (2012).

55. van den Broek, J., Afyon, S., and Rupp, J. L. M., Interface-Engineered All-Solid-State Li-Ion Batteries Based on Garnet-Type Fast $Li^+$ Conductors. *Adv. Energy Mater.* **6**(19):11 (2016).

56. Kozen, A. C., Pearse, A. J., Lin, C. F., Noked, M., and Rubloff, G. W., Atomic Layer Deposition of the Solid Electrolyte LiPON. *Chem. Mater.* **27**(15):5324–5331 (2015).

57. Nisula, M., Shindo, Y., Koga, H., and Karppinen, M., Atomic Layer Deposition of Lithium Phosphorus Oxynitride. *Chem. Mater.* **27**(20):6987–6993 (2015).

58. Bates, J. B., Dudney, N. J., Gruzalski, G. R., Zuhr, R. A., Choudhury, A., Luck, C. F., and Robertson, J. D., Fabrication and Characterization of Amorphous Lithium Electrolyte Thin-Films and Rechargeable Thin-Film Batteries. *J. Power Sources* **43**(1–3):103–110 (1993).

59. Yersak, A. S., Sharma, K., Wallas, J. M., Dameron, A. A., Li, X. M., Yang, Y. G., and Hurst, K. E., et al., Spatial Atomic Layer Deposition for Coating Flexible Porous Li-Ion Battery Electrodes. *J. Vac. Sci. Technol. A* **36**(1):11 (2018).

60. Kalnaus, S., Tenhaeff, W. E., Sakamoto, J., Sabau, A. S., Daniel, C., and Dudney, N. J., Analysis of Composite Electrolytes with Sintered Reinforcement Structure for Energy Storage Applications. *J. Power Sources* **241**:178–185 (2013).

61. Kalnaus, S., Sabau, A. S., Tenhaeff, W. E., Dudney, N. J., and Daniel, C., Design of Composite Polymer Electrolytes for Li Ion Batteries Based on Mechanical Stability Criteria. *J. Power Sources* **201**:280–287 (2012).

62. Tenhaeff, W. E., Perry, K. A., and Dudney, N. J., Impedance Characterization of Li Ion Transport at the Interface between Laminated Ceramic and Polymeric Electrolytes. *J. Electrochem. Soc.* **159**(12):A2118–A2123 (2012).

63. Tenhaeff, W. E., Yu, X., Hong, K., Perry, K. A., and Dudney, N. J., Ionic Transport Across Interfaces of Solid Glass and Polymer Electrolytes for Lithium Ion Batteries. *J. Electrochem. Soc.* **158**(10):A1143–A1149 (2011).

64. Leo, C. J., Rao, G. V. S., and Chowdari, B. V. R., Studies on Plasticized PEO-Lithium Triflate-Ceramic Filler Composite Electrolyte System. *Solid State Ionics* **148**(1–2):159–171 (2002).

65. Nairn, K. M., Best, A. S., Newman, P. J., MacFarlane, D. R., and Forsyth, M., Ceramic-Polymer Interface in Composite Electrolytes of Lithium Aluminium Titanium Phosphate and Polyetherurethane Polymer Electrolyte. *Solid State Ionics* **121**(1–4):115–119 (1999).

66. Wang, Y. J., and Pan, Y., $L_{1.3}Al_{0.3}Ti_{1.7}(PO_4)_3$ Filler Effect on $(PEO)LiClO_4$ Solid Polymer Electrolyte. *J. Poly. Sci. B* **43**(6):743–751 (2005).

67. Wang, W. M., Yi, E. Y., Fici, A. J., Laine, R. M., and Kieffer, J., Lithium Ion Conducting Poly(ethylene oxide)-Based Solid Electrolytes Containing Active or Passive Ceramic Nanoparticles. *J. Phys. Chem. C* **121**(5):2563–2573 (2017).

68. Liu, W., Liu, N., Sun, J., Hsu, P. C., Li, Y. Z., Lee, H. W., and Cui, Y., Ionic Conductivity Enhancement of Polymer Electrolytes with Ceramic Nanowire Fillers. *Nano Lett.* **15**(4):2740–2745 (2015).

69. Yang, T., Zheng, J., Cheng, Q., Hu, Y. Y., and Chan, C. K., Composite Polymer Electrolytes with $Li_7La_3Zr_2O_{12}$ Garnet-Type Nanowires as Ceramic Fillers: Mechanism of Conductivity Enhancement and Role of Doping and Morphology. *ACS Appl. Mater. Interfaces* **9**(26):21773–21780 (2017).

70. Dudney, N. J., *Composite Electrolyte to Stabilize Metallic Lithium Anodes*; Project ID: ES182; Vehicle Technologies Program Annual Merit Review and Peer Evaluation Meeting, 2013.

71. Abe, T., Ohtsuka, M., Sagane, F., Iriyama, Y., and Ogumi, Z., Lithium Ion Transfer at the Interface between Lithium-Ion-Conductive Solid Crystalline Electrolyte and Polymer Electrolyte. *J. Electrochem. Soc.* **151**(11):A1950–A1953 (2004).

72. Busche, M. R., Drossel, T., Leichtweiss, T., Weber, D. A., Falk, M., Schneider, M., and Reich, M. L., et al., Dynamic Formation of a Solid-Liquid Electrolyte Interphase and Its Consequences for Hybrid-Battery Concepts. *Nat. Chem.* **8**(5):426–434 (2016).

73. Sagane, F., Abe, T., Iriyama, Y., and Ogumi, Z., $Li^+$ and $Na^+$ Transfer through Interfaces between Inorganic Solid Electrolytes and Polymer or Liquid Electrolytes. *J. Power Sources* **146**(1–2):749–752 (2005).

74. Abe, T., Sagane, F., Ohtsuka, M., Iriyama, Y., and Ogumi, Z., Lithium-Ion Transfer at the Interface between Lithium-Ion Conductive Ceramic Electrolyte and Liquid Electrolyte—A Key to Enhancing the Rate Capability of Lithium-Ion Batteries. *J. Electrochem. Soc.* **152**(11):A2151–A2154 (2005).

# SECTION B

# REDOX FLOW BATTERIES

## H. Frank Gibbard

## *22B.1   OVERVIEW*

### 22B.1.1   Introduction and Background

Redox flow batteries (RFB) are rechargeable batteries in which the electrochemically active materials are dissolved in a solution and circulated through or past electrodes whose only function is electron transfer to or from the active materials as the battery is charged or discharged. The advantages of such batteries include the following:

- The possibility of very long life owing to the absence of strains in solid materials due to changes in volume during cycling.
- The flexibility of design as a result of the separation of the attributes of power (depending on the cell construction and method of operation) and energy (determined by the volume and composition of the negative and positive electrolytes, herein called "negalyte" and "posilyte," respectively).

These advantages were recognized in the early work of Kangro,[1,2] who proposed several RFB chemistries based on different oxidation states of inorganic ions.

The principles of operation of an RFB, illustrated in Fig. 22B.1, are very similar to those of a fuel cell, and indeed, some authors have described the RFB as a "regenerative fuel cell." An important distinction between the RFB and a fuel cell is, however, that both the positive and negative materials are stored in tanks in the RFB, whereas the fuel cell generally utilizes oxygen from the atmosphere. Other authors have included in the RFB category systems in which one component may be a gas, such as the hydrogen/bromine flow battery, or may be a solid, such as the zinc/bromine system. In this section, the latter type of system is designated as a hybrid flow battery, and the designation of the RFB system is limited to the case where all components are present in liquid phases. Other cases are described in the extensive review by Soloveichik.[3]

In the RFB shown schematically in Fig. 22B.1, the positive and negative active materials contained in the storage tanks are circulated through the cell by means of pumps for the liquid electrolytes. The reactive species pass through porous electrodes, where they exchange electrons with the inert electrodes plates and undergo redox reactions to form different soluble ions, after which the spent fluid is pumped back to the storage tanks. Ionic current passes through the cell via the separator, typically an ion-exchange membrane or, in some cases, a microporous or nanoporous sheet.[4]

### 22B.1.2   Features and Benefits

The following desirable characteristics are targeted in the development of RFBs:

- High cell voltage
- High solubility of active materials
- High energy density of electrolytes

- High efficiency
  - Rapid electrode kinetics
  - Low internal resistance
  - Minimal side reactions
- High power density of cell electrodes
- Wide operating temperature range
- Low electrolyte toxicity
- Low electrolyte corrosivity
- Minimal impact of inadvertent mixing of negalyte and posilyte
- Long calendar and cycle lives (i.e., 20 years and 10,000 deep cycles)
- Low costs for electrode stack materials, electrolytes, and balance of plant

Figure 22B.1 depicts the operation of a single RFB cell. Multicell batteries are constructed in series-bipolar arrays of cells, in much the same way that proton exchange membrane (PEM) fuel cells are designed.

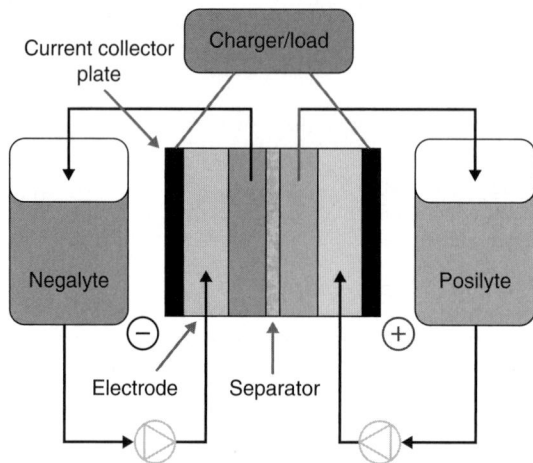

**FIGURE 22B.1**     Redox flow battery cell schematic.

## 22B.2    TYPES OF REDOX FLOW BATTERIES

### 22B.2.1    Iron-Chromium

In the 1970s an extensive study of redox couples that might be developed for large-scale energy-storage applications was carried out at NASA.[4] The most promising system identified in this work was the iron-chromium system, with the following reactions:

*Positive electrode:* $Fe^{3+} + e^- \rightarrow Fe^{2+}$               $E^0 = +0.77$ V

*Negative electrode:* $Cr^{2+} \rightarrow Cr^{3+} + e^-$               $E^0 = -0.41$ V

*Net cell reaction for discharge:* $Fe^{3+} + Cr^{2+} \rightarrow Fe^{2+} + Cr^{3+}$               $E^0 = +1.18$ V

The Fe/Cr system was scaled to the 20-kW power level as a part of the New Energy and Industrial Technology Development Organization (NEDO) Moonlight Project in Japan in the early 1980s, and a 60-kW system was tested by Sumitomo Electric.[5] The system was found to have several technical problems, including the loss in capacity when iron and chromium ions diffused across the cation-exchange separator and were not found to be separable into their original storage tanks. This was ameliorated by deliberately mixing the ions in each tank, at the concentrations required to achieve the desired electrical capacity. A more intractable problem was the low redox potential of the charged negative electrode, which resulted in the evolution of hydrogen gas. Eventually, the Japanese developers of Fe/Cr batteries abandoned work on this system in favor of the all-vanadium system described later in this chapter.

Nevertheless, two American companies—EnerVault and Deeya—were founded in 2008 and 2009 with the goal of commercializing Fe/Cr. Both companies ultimately filed for bankruptcy, Deeya after changing its name to Imergy and "pivoting" from Fe/Cr to the all-vanadium RFB. EnerVault carried its development work to the scale of 250 kW and an electrical capacity of 1 MWh, but ultimately failed through a combination of technical and business problems.

### 22B.2.2 All-Vanadium

The all-vanadium RFB is by far the most advanced toward the goal of large-scale commercialization. It was invented by Maria Skyllas-Kazacos in 1988.[6] The electrochemistry of this system is a "tour de force" in that it relies on only one element to provide four stable or metastable oxidation states to provide all the ionic species that are required to supply the reduced and oxidized species for both the posilyte and negalyte of an RFB. The electrochemical reactions for this system for discharge are as follows:

$$\text{Positive electrode: } VO_2^+ + 2H^+ + e^- \rightleftharpoons VO^{2+} + H_2O \qquad E^0 = +0.991 \text{ V}$$

$$\text{Negative electrode: } V^{2+} \rightleftharpoons V^{3+} + e^- \qquad E^0 = -0.255 \text{ V}$$

$$\text{Net cell reaction: } VO_2^+ + 2H^+ + V^{2+} \rightleftharpoons VO^{2+} + H_2O + V^{3+} \qquad E^0 = +1.246 \text{ V}$$

In terms of the desirable attribute aspects of the RFB listed in the Overview, the all-vanadium system has the following characteristics:

- Moderately high cell voltage
- Vanadium solubility of 1.6 to 2.5 M
- Modest energy density of combined electrolytes of 15 to 25 Wh/L
- DC round-trip cycling efficiency in the low 80s
- Electrolyte stability over a wide range of temperature ($-20$ to $+70°C$)
- Moderately high toxicity and corrosivity
- No irreversible change upon mixing of anolyte and catholyte
- The potential for very long cycle life and calendar life
- For large systems with long discharge duration, cost competitiveness with all other stationary energy-storage batteries including lithium-ion

These characteristics, particularly the potential for very long life and the ability to return the system to its original condition after deliberate mixing of the positive and negative electrolytes, have led to the status of the all-vanadium system as the most advanced RFB for large-scale commercialization.

The technology of the original all-vanadium system, although developed to the multi-kilowatt and megawatt scale in many projects over the past two decades, had several drawbacks that have recently been addressed by researchers at several academic and U.S. Government laboratories. Work at Pacific Northwest National Laboratory[7] (PNNL) describes the use of supporting electrolyte containing chloride ion in an all-vanadium RFB. The original electrolyte described by Skyllas-Kazacos used a supporting electrolyte composed of sulfuric acid. Batteries made with this electrolyte had two significant limitations as noted below.

First, the upper temperature of operation for the charged electrolyte was approximate 35°C, above which vanadium (+5) precipitates from the posilyte. Formation of such solids in the electrodes or flow channels can render the RFB inoperable. This problem requires active cooling of the electrolyte; and such refrigeration equipment is expensive and decreases system reliability as well as electrical efficiency.

Second, the upper limit for the concentration of vanadium ions in the all-sulfate electrolyte is only about 1.6 M. This dictates the use of larger storage tanks and greater pumping capabilities, leading to greater system costs and lower electrical efficiency. With an upper vanadium concentration of 2.5 M and an upper temperature range of 55°C, the PNNL mixed-acid electrolyte substantially increases the functionality of the all-vanadium RFB. The PNNL technology has been licensed to a limited number of companies for use in their RFBs. These are UniEnergy Technologies (UET), a spin-off of PNNL; Imergy; and WattJoule Corp.

The electrolyte contributes a substantial fraction of the cost of the DC all-vanadium RFB; the other major contributor is the bipolar cell stack in which chemical energy is converted to electrical, and vice versa.

Until recently, a drawback for RFBs was the low operating current density (typically 60–120 mA/cm²) and consequential low power density of 70 to 150 mW/cm² at the electrodes of such stacks. A quantum improvement in the useful current density resulted from research at University of Tennessee (Knoxville) and Oak Ridge National Laboratory,[8] leading to a cell design called "zero-gap architecture." This technology was licensed exclusively to WattJoule Corp., which has achieved nominal operating current density and power density of 350 mA/cm² and 420 mW/cm², with a maximum specific power of more than 1500 mW/cm². The increase in power density enables a decrease in the amount of materials needed to manufacture the cell stack, yielding a considerably lower system cost.

The readiness of the all-vanadium system to meet grid-scale energy-storage needs is proven by the installations of a 10 MW, 60 MWh system in Hokkaido by Sumitomo and of one of the largest energy-storage systems in the world, a 200 MW/800 MWh system to be manufactured by Rongke Power in Dalian, China.

### 22B.2.3  Vanadium-Bromine

Although the all-vanadium system has a substantial lead over other RFBs in its commercialization status, the question, naturally, concerns what system should be developed, next, that will have important advantages over vanadium. Since the price and source of supply of vanadium are somewhat uncertain, a system that uses less of this material, or none at all, is of interest. Other advantages useful for an advanced system, as identified in the above discussion of strengths and weaknesses of the all-vanadium system, led M. Skyllas-Kazacos and her coworkers to propose the development of the vanadium-bromine system, which they designated Gen 2, after Gen 1, the all-vanadium RFB.[9] The operation of the system is illustrated in Fig. 22B.2.

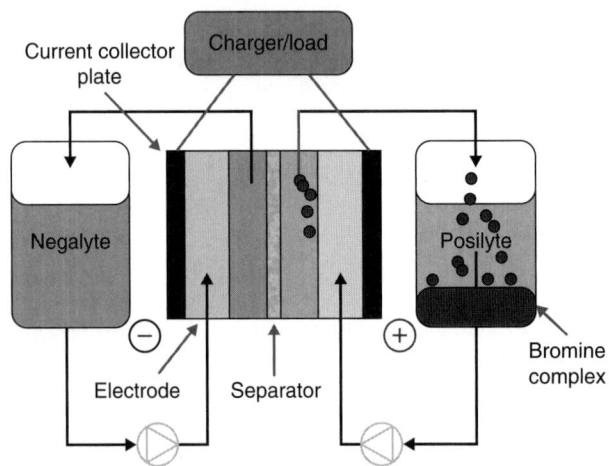

**FIGURE 22B.2**    Vanadium-bromine redox flow battery cell schematic.

The active materials are dissolved in electrolyte solutions designated as the anolyte and the catholyte and are stored in tanks, external to the electrochemical cell. The cell electrodes serve only to transfer electrons to and from the active species and do not undergo chemical or physical changes during the charging and discharging processes. During charging, the following reactions occur at the electrodes:

*Negative electrode reaction:* $V^{3+} + 2e^- \rightarrow 2V^{2+}$

*Positive electrode reaction:* $2Br^- + QBr_n \rightarrow QBr_{n+2} + 2e^-$

*Net cell reaction:* $V^{3+} + 2Br^- + QBr_n \rightarrow 2V^{2+} + QBr_{n+2}$

The V-Br system is interesting in that it resembles the all-vanadium RFB in a significant way: if the active materials cross the separator and mix the contents of the posilyte and negalyte tanks, the system can be easily returned to its original state by deliberately moving some of the aqueous phase from one tank to the other. This is a consequence of the fact that although one *element* does not provide all the oxidation states of the active materials, one *substance*, vanadium bromide, does.

Because elemental bromine has a rather high vapor pressure, when it is produced at the positive electrode it will be present in an aqueous phase as the element $Br_2$ and as polybromides such as $Br_3^-$ and $Br_5^-$. The bromine in these complex ions will also be in equilibrium with elemental bromine in the gas phase above the posilyte. This is evident when the vanadium-bromine battery is cycled: a red-brown cloud may be seen in the vapor space over the posilyte, which appears on charge and disappears during discharge. For environmental reasons, and because the high bromine activity in the posilyte increases the driving force for the diffusion of bromine-containing species across the separator and into the negalyte compartment, it is highly desirable to reduce the thermodynamic activity of bromine in the posilyte. Fortunately, because of the extensive work carried out to perfect the zinc-bromine battery, it was found necessary to find means of decreasing bromine activity in aqueous electrolytes. This has led to the development of complexing agents such as quaternary ammonium bromide salts, with the designation QBr, which react with dissolved bromine according to the reaction

$$QBr + nBr_2 \rightarrow QBr_{2n+1}$$

In the operation of the cell in Fig. 22B.2, bromine produced at the positive electrode reacts with the complexing agent QBr to form a nonaqueous polybromide oil ($QBr_{2n+1}$, where $n = 1, 2, 3$, or 4) that is substantially insoluble in the aqueous phase and is carried in droplets to the catholyte tank, where it settles to the bottom due to its greater density than that of the aqueous phase. A recent article on the zinc-bromine battery provides many references to the current state of the art in bromine complexation.[10]

Early-stage experiments at WattJoule Corp. have given promising results for the performance of V-Br single cells, including the following conditions and results:

- 25 $cm^2$ active area of electrodes
- Operating temperature of 45°C
- Proprietary bromine-complexing agent
- Liquid electrolyte energy density of ≥50 Wh/kg
- Operating electrode current density of ≥200 mA/$cm^2$
- Maximum power density of ≥1000 mW/$cm^2$
- Round-trip DC electrical efficiency of 80%

This performance, particularly an operating current density of five to ten times higher than that reported by previous investigators, gives some promise for the future development of the V-Br system to be a practical large-scale energy-storage system.

## 22B.3    ALTERNATIVE REDOX FLOW BATTERY SYSTEMS

### 22B.3.1    Past Work

In addition to the Fe-Cr, all-vanadium, and V-Br systems, many other candidate RFBs based on inorganic redox couples have been proposed, investigated, and in a few cases, scaled up to multi-kilowatt stage of development. For example, Table 1 of Ref. 11 lists 10 RFB chemistries that have been evaluated. Additionally, many other close relatives to the systems referenced in this article as RFBs have been tested, some with the expenditure of millions of dollars, without producing commercially useful energy-storage systems.[12] These include systems in which one component is a gas, or in which one component is plated as a metal. In the author's opinion, an approach more likely to succeed in the advanced development of a viable redox flow system is to not utilize precious resources in finding new combinations of well-known inorganic redox couples. Rather, it may be more productive to concentrate research on new redox couples that are likely to meet the criteria described in the first part of this chapter. In particular, couples with low toxicity and low corrosivity are likely to be adopted by companies, possessing the resources needed to implement RFBs, based on less complications for large-scale manufacture and wide commercial use.

### 22B.3.2    Future Work

If this proposed course of action immediately above is to be adopted, then it seems logical to harness the unlimited flexibility in the structure and properties of organic and organometallic chemistry. Many researchers are currently pursuing this path,[13,14] following the popular opinion that "green chemistry" is most likely to be acceptable in the light of environmental problems that are now becoming obvious. The challenges in developing such organic materials are substantial, including the demonstration of stability required to enable long calendar and cycle life, low cost, and the desirability of preventing capacity loss through mixing of the posilyte and negalyte solutions. The solution of these problems will have the greatest probability of success if organic and organometallic chemists, materials scientists, and battery scientists and engineers cooperate early in the discovery and development of new RFB chemistries.

## REFERENCES

1. W. Kangro, "Process for Storage of Electrical Energy," German Patent, June, 1949.

2. W. Kangro and H. Pieper, "On the Problem of Storing Electrical Energy in Liquids," *Electrochim. Acta* **7**:435–448 (1962).

3. G.L. Soloveichik, "Flow Batteries: Current Status and Trends," *Chem. Rev.* **115**:11533–11558, 2015.

4. N.H. Hagedorn, "NASA Redox Storage System Development Project. Final Report," DOE/NASA/12726-24 NASA TM-83677, October, 1984.

5. T. Shigematsu, "Redox Battery for Energy Storage," SEI Technical Review—Number 73–October 2011.

6. M. Skyllas-Kazacos, M. Rychick, and M. Robins, U.S. Patent 4,786,567, "All-Vanadium Redox Battery" 1988.

7. L. Li, S. Kim, Z. Yang, W. Wang, J. Zhang, B. Chen, Z. Nie, and G. Xia, U.S. Patent 9,819,039, "Redox Flow Batteries Based on Supporting Solutions Containing Chloride," 2017.

8. M. Mench, T. Zawodzinski, and C.N. Sun, "High Power High Efficiency Flow Type Battery," U.S. Patent Application 2015/0072261, March 2015.

9. M. Skyllas-Kazacos, "G1 and G2 Vanadium Redox Batteries for Renewable Energy Storage," http://www.eurosolar.org/new/pdfs_neu/electric/IRES2006_Skyllas-Kazacos.pdf.

10. B.G. McMillan, M. Spicer, A. Wark, and L. Berlouis, "Complexing Additives to Reduce the Immiscible Phase Formed in the Hybrid ZnBr Flow Battery," *J. Electrochem. Soc.* **164**(13):A3342–A3348 (2017).

11. L.F. Arenas, C. Ponce de Leon, and F. Walsh, "Engineering Aspects of the Design, Construction and Performance of Modular Redox Flow Batteries for Energy Storage," *J. Energy Storage* **11:**119–153 (2017).

12. P. Leung, X. Li, C. Ponce de Leon, C.T. John Low, L. Berlouis, and F.C. Walsh, "Progress in Redox Flow Batteries, Remaining Challenges and Their Applications," *RSC Adv.* **2:**10125–10156 (2012).

13. J.D. Milshtein, A.P. Kaur, M.D. Casselman, J.A. Kowalski, S. Modekrutti, P.L. Zhang, and N.H. Attanayake, et al., "High Current Density, Long Duration Cycling of Soluble Organic Active Species for Non-aqueous Redox Flow Batteries," *Energy Environ. Sci.* **9:**3531 (2016).

14. R.G. Gordon, A.A. Guzik, and M.J. Aziz, "Aqueous Flow Batteries Using Organics and Organometallics," International Flow Battery Forum, Manchester, U.K., 2017-06-28.

# SECTION C

# SOLID-STATE ELECTROLYTES (CERAMIC, GLASS, POLYMER)

**Ron Turi**

## 22C.1 INTRODUCTION

In battery applications, a solid-state electrolyte (SSE) material refers to a medium that conducts ions through a crystal lattice, amorphous glass, or a nonflowing polymer matrix. Some SSE materials conduct electric current that is a benefit in battery electrodes, while SSE materials in separator layers between electrodes must be electronic insulators. SSE materials often conduct a specific mobile ion and are designed for specific battery chemistries. For example, beta-alumina glasses that conduct $Na^+$ are used in commercial Na-S and Na-NiCl$_2$ grid energy-storage batteries that operate at temperatures above the 300°C melting point of Na metal. However, most of the current development of SSE materials for the battery industry focuses on lithium-ion transport in order to address the unmet needs of next-generation batteries for the emerging market for electric vehicle batteries. This review focuses on the development of lithium-ion SSE materials in recent years.

### 22C.1.1 Benefits

SSE materials generally provide improved electrochemical stability at battery electrodes, reduced flammability, and elevated thermal stability compared to liquid-based electrolyte solutions. For lithium-ion batteries, substituting SSE for liquid electrolyte solutions stands to improve or eliminate failure modes associated with safety hazards and with performance degradation in lithium-ion liquid battery chemistries. Eliminating the liquid solvents of an electrolyte solution also eliminates or at least reduces the first-cycle capacity losses associated with the formation of the solid electrolyte interphase (SEI) passivation layer which consumes 8% to 10% of the available lithium content in the cell. Just as important, the greater electrochemical stability of the SSE often enables lithium cells to operate at a higher voltage and to use lithium metal anodes, which in turn enable significant increases in materials utilization and specific capacity.[1] Taken together, these increases effectively lower the cell materials unit cost to widely anticipated levels under US $100/kWh and prospectively raise specific energy well above the 250 Wh/kg level, as claimed in best commercial practice for lithium-ion cells. These improvements in battery point metrics along with the reduced risk of safety failures propel SSE development across all battery end uses, but have the greatest impact on large format applications such as electric vehicle batteries.

Also, having an SSE enables a simple and viable means of manufacturing bipolar electrodes for bipolar cells. The challenge in building bipolar cells with liquid electrolyte is containing the liquid between negative and positive electrode pairs. Any leak of electrolyte between pairs of electrodes causes the cell to short circuit. With SSE, the risk of electrolyte leakage is eliminated. Battery voltage in electric vehicle applications has trended to higher voltage, which enables vehicles to operate using lower electric currents and reduces resistive losses in conductors in wiring harnesses and in motors. However, as voltages rise to enable greater energy efficiency for electric vehicles, the capacity of the cells decreases and the number of cells in series increases. This trend suggests that smaller cells that are inherently connected would be more viable in emerging electric vehicle platforms. Instead of welding hundreds of cells together in series, bipolar cells are built in interconnected series with an inherent electrical junction. Also, a reduced cell capacity is easier to achieve in the area of a single electrode—an ideal situation for bipolar cells since these have only one electrode surface. A successful lithium SSE can unlock bipolar battery designs to contribute to the viability of high-voltage batteries, e.g., for electric vehicles.

### 22C.1.2   Challenges

Viable SSE materials for lithium battery chemistries seek to meet or exceed the ionic conductivity of liquid electrolyte solutions, which provide $10^{-3}$ to $10^{-2}$ mS/cm at room temperature. While this target is achievable on the laboratory scale and in prototype cells, efforts to scale up SSE materials for industry production usually encounter engineering issues that limit immediate commercial practice. Ceramic-based materials, for example, tend to be hard and brittle, which leads to inadequate mechanical durability (i.e., cracking) and to high interfacial resistance at electrode surfaces due to poor contact integrity compared to liquid electrolyte solutions. Also, many ceramic and glass-based SSE materials face complications related to synthesizing the compounds due to the use of reactive precursors or the inability of nascent process technology to limit the formation of grain boundaries. Also, not all of the promising types of SSE materials are electrochemically stable against lithium anodes or highly electropositive cathode materials. For example, SSE ion conductors with reducible cations, such as $Ti^{4+}$ in NAISICON-like and perovskite materials, are susceptible to reduction at lithium or graphitic anodes.

SSE polymers work well at temperatures above the glass transition point ($T_g$), but ionic conductivity often decreases to impractical levels at lower temperatures as the polymer morphology becomes a rigid glass or crystalline solid. Polyethylene oxide (PEO)-based SSE materials with dissolved lithium salts are available and typically operate above 60°C. Other polymer media that entrain liquid electrolyte solutions—such as the Bellcore-Telcordia Technology—are discussed in Chap. 22A.

### 22C.1.3   Technical Solutions

Present commercialization programs address the various challenges of implementing SSE materials by developing composites with different types of SSE materials. For example, some work focuses on the development of polymer-ceramic composite SSE materials aimed at improving SSE flexibility and conformity to electrode material surfaces while maintaining the general advantages of SSE over liquid electrolyte systems. Also, a class of nonglassy polymer SSE materials is being developed using design principles similar to those that guide ceramic and rigid glass SSE materials.

As more SSE materials are discovered, the structure of the materials and the corresponding measurements of ionic transport lead to improved modeling and deeper understanding of design factors that govern solid-state ionic transport. Various modifications of the Arrhenius model incorporate Marcus theory and multiple body problem concepts to explain deviations from linear empirical data fitting within a given transport regime.

## 22C.2   SOLID-STATE ION TRANSPORT MECHANISMS IN RIGID CRYSTALLINE AND GLASSY SOLIDS

Because the atoms or ions in rigid crystalline and glassy solids remain in fixed positions, the transport of ions often requires structural irregularities that result in vacant sites. Mobile ions with sufficient activation energy hop from vacancy to vacancy and form an ionic current when driven by an electric field.

One mode of ion transport through crystalline solids is classified as movement through an *intrinsic defect* mechanism in which vacant sites form by the dislocation of lattice atoms to interstitial positions in the crystal—known as Frenkel defects. In a salt crystal, intrinsic defect vacancies form as a single ion moves from its regular, ordered lattice position to an interstitial location in the lattice. Ions that are smaller or have lower charge density are more likely to move from the regular ionic lattice into interstitial positions, e.g., $Ag^+$ in AgI and $F^-$ in CaF2. Frenkel developed an Arrhenius-type model to account for the dependency of ionic conductivity on temperature in salt crystals, attributing the energy term as an estimate of the heat of formation of defects and the energy required for ion movement.[2] The Arrhenius model is well known in the following form,

$$\sigma = (\sigma_0/T)\, e^{(E_A/k_B T)}$$

where the ionic conductivity $\sigma$ depends on absolute temperature $T$ and $\sigma_0$ is a reference ionic conductivity value related to the frequency of events that can overcome an energy barrier $E_A$ relative to the internal energy at that temperature by the Boltzmann number, $k_B$. In the Frenkel model, the concentration of defects increases with temperature and effectively increases the system energy to overcome the activation energy required for ions to move through the crystal lattice.

Schottky expanded the concept of intrinsic defects in ionic conduction to include vacancy pairs that are formed by the displacement of a cation-anion pair. The Arrhenius-type analysis also applies to Schottky defects where temperature increases both vacancy and dislocation concentrations to more effectively overcome the activation energy. For ionic compounds with $Li^+$ and other alkaline metal ions, calculated values of the heat of formation suggest that Frenkel defects require less energy to form than Schottky defects. In both cases, it is clear that high temperatures are required to form intrinsic defects that enable ionic conduction.

Another mode of transport through solids involves an *extrinsic defect* mechanism in which dopant atoms are added to disrupt the regular structure of the crystal lattice. Many battery electrodes function as ionic conductors with external defects. For example, solid-state conduction of $Li^+$ occurs successfully within conventional graphite anode materials and within transition metal oxides and phosphate cathode materials—all operating at ambient temperatures in commercial lithium-ion cells. As a lithium-ion cell discharges, it enters and disrupts the crystal lattice of a cathode such as $NiO_2$ that accommodates the $Li^+$ with adjustments in the lattice structure and a reduction of the nickel oxidation state. The smaller $Li^+$ transport in metal oxides can be viewed as transport by an external defect mechanism that follows an Arrhenius-type temperature relationship.

The activation energy for ion mobility through extrinsic defects is lower than the amount of heat that is required to form intrinsic defects. As a result, ion transport by extrinsic defects prevails at lower temperatures, while intrinsic defects dominate ion transport at higher temperatures. Table 22C.1 details the factors necessary to consider in developing SSEs. Development of SSE materials revolves around understanding the effect of materials composition and structure on ionic transport, defect formation, and concentration.

**TABLE 22C.1**    Critical Factors Affecting SSE Capabilities

| Factor | Variable | Mechanism | Effect |
|---|---|---|---|
| Intrinsic defect | Concentration | Frenkel or Schottky | Formation energy |
| Extrinsic defect | Dopant and concentration | Substitutions of ions—aliovalent, oversized | Disordering lattice structure, enlarging physical pathways |
| Mobile ion | Concentration | Size, charge | Mobility |
| Characteristics of the unit cell | Molar volume | Physical pathways and bottlenecks, molecular geometry | Barriers to vacancy hopping |
| Composition of the unit cell | Ligand in kernel | Mixed ligands | Managing mobile ion distribution, disorder and energy of attraction |
| | Cations in kernel | Disordering ligand bonds, enabling valence changes | Within the kernel—increasing charge mobility, enabling defect formation and mobility |
| Electronic conductivity | Cations with empty LUMO | Unfilled d-orbital transition metal ions | Dielectric material for electronic insulation |

As shown in Fig. 22C.1, Bachman et al.[3] provide a succinct overview of the selection of mobile ion, ligand, and kernel elements by mapping each category on a periodic table.

The use of these elements in a number of crystalline materials and the conductivity ranges of each family—normalized to room temperature values—is summarized by Bachman et al.[3] in Fig. 22C.2. Note that some values for room-temperature ionic conductivity are extrapolated from data at higher temperatures using an Arrhenius-type correlation.

Legend:
- Diffusive species
- Ligand
- Cation forming the polyhedra skeleton

Periodic table:

| 1 H | | | | | | | | | | | | | | | | | 2 He |
|---|---|---|---|---|---|---|---|---|---|---|---|---|---|---|---|---|---|
| 3 Li | 4 Be | | | | | | | | | | | 5 B | 6 C | 7 N | 8 O | 9 F | 10 Ne |
| 11 Na | 12 Mg | | | | | | | | | | | 13 Al | 14 Si | 15 P | 16 S | 17 Cl | 18 Ar |
| 19 K | 20 Ca | 21 Sc | 22 Ti | 23 V | 24 Cr | 25 Mn | 26 Fe | 27 Co | 28 Ni | 29 Cu | 30 Zn | 31 Ga | 32 Ge | 33 As | 34 Se | 35 Br | 36 Kr |
| 37 Rb | 38 Sr | 39 Y | 40 Zr | 41 Nb | 42 Mo | 43 Tc | 44 Ru | 45 Rh | 46 Pd | 47 Ag | 48 Cd | 49 In | 50 Sn | 51 Sb | 52 Te | 53 I | 54 Xe |
| 55 Cs | 56 Ba | 57 La | 72 Hf | 73 Ta | 74 W | 75 Re | 76 Os | 77 Ir | 78 Pt | 79 Au | 80 Hg | 81 Tl | 82 Pb | 83 Bi | 84 Po | 85 At | 86 Rn |

| 58 Ce | 59 Pr | 60 Nd | 61 Pm | 62 Sm | 63 Eu | 64 Gd | 65 Tb | 66 Dy | 67 Ho | 68 Er | 69 Tm | 70 Yb | 71 Lu |
|---|---|---|---|---|---|---|---|---|---|---|---|---|---|

**FIGURE 22C.1** SSE material categories. (*Reprinted with permission from Chemical Reviews: J.C. Bachman, S. Muy, A. Grimaud, et al., Inorganic Solid-State Electrolytes for Lithium Batteries: Mechanisms and Properties Governing Ion Conduction. Copyright 2016 American Chemical Society.*)

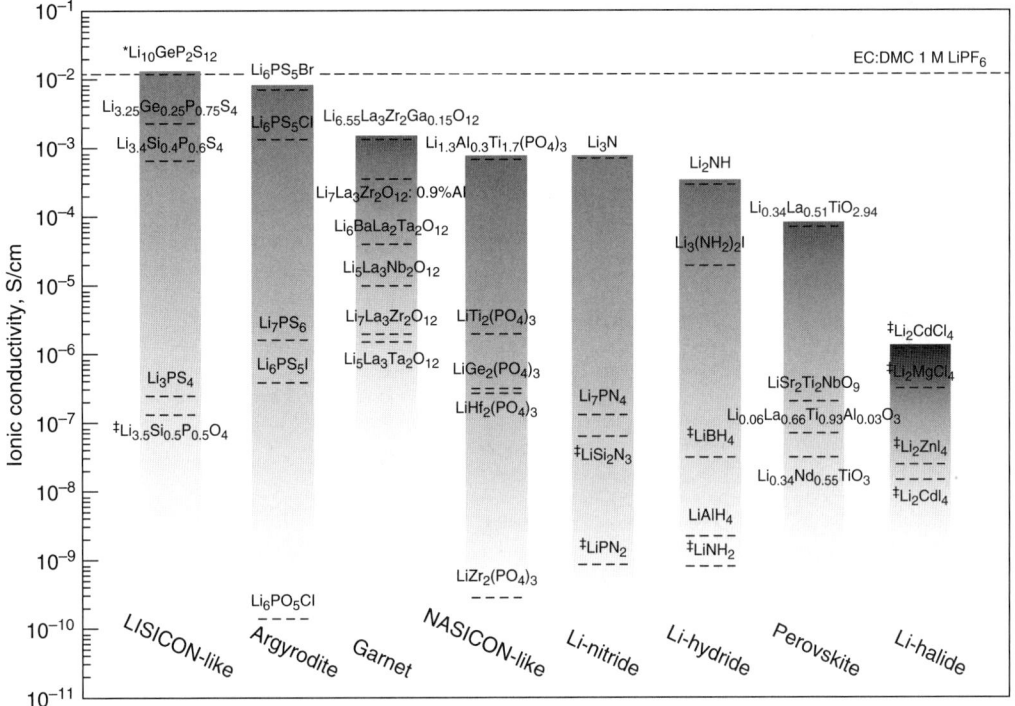

**FIGURE 22C.2** Conductivity ranges for various SSEs. (See footnotes in referenced source.) (*Reprinted with permission from Chemical Reviews: J.C. Bachman, S. Muy, A. Grimaud, et al., Inorganic Solid-State Electrolytes for Lithium Batteries: Mechanisms and Properties Governing Ion Conduction. Copyright 2016 American Chemical Society.*)

## 22C.3   LITHIUM-ION SOLID-STATE ELECTROLYTES—CERAMIC AND RIGID GLASS

The major families of lithium-ion solid-state ionic conductors as identified by Bachman et al.[3] include LISICON-like, argyrodite, Garnet, NAISICON-like, Li-nitride, Li-hydride, perovskite, and Li-halide. Details on these are summarized in the sections below.

### 22C.3.1   LISICON-Like

The lithium-ion SSE ionic material with the highest room-temperature ionic conductivity reported to date[4] is $Li_{10}GeP_2S_{12}$. As shown in Fig. 22C.3, the ionic conductivity reported for this material is comparable to liquid electrolyte solutions not only at room temperature, but also at lower temperatures where no other ionic conductor is within an order of magnitude to the $Li_{10}GeP_2S_{12}$ material. As a result, the focus for this material shifts to solving the issues associated with sulfur and phosphorous precursors and electrochemical stability with respect to lithium in the anode.

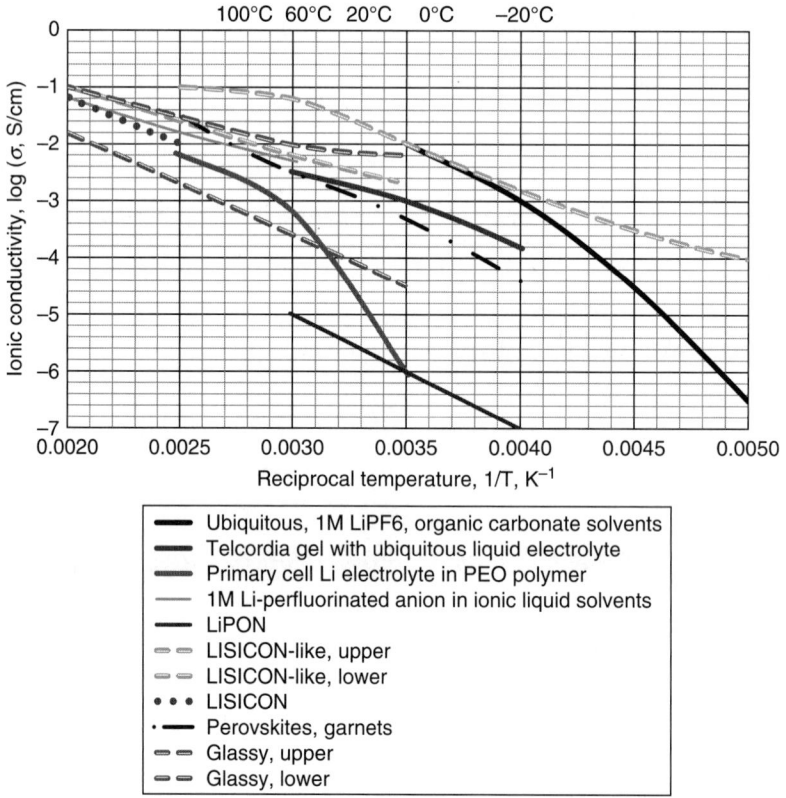

**FIGURE 22C.3**   Notional map of ionic conductivity for emerging classes of solid-state lithium-ion conductors.

### 22C.3.2   Argyrodite

In recent years, the more promising advances revolve around LISICON-like and argyrodite families of ceramic crystalline and glassy SSE materials that are characterized by an abundance of ligands with decentralized, accessible lone pair electrons—namely sulfur and phosphorous. These generate multiple states for $Li^+$ occupancy that create a dynamic equilibrium with vacant sites. Prior-generation SSE materials contained significant amounts of oxygen and nitrogen as the ligand functions, e.g., LiPON materials. However, the larger ligand atoms result in a greater dispersion of lone pair electrons, and a mixture of aliovalent atoms further increases the nonuniform distribution of lone pair electrons to create disorder in the tetrahedral crystal structure that may contribute to lowering the activation energy for ion transport. Argyrodite materials usually include halogens, which further increase lone pairs in the unit cell of the crystal and further increase disorder in the ligand mixture—all aimed at lowering activation energy for ion transport.

### 22C.3.3 Garnet

Garnet SSE materials have excellent electrochemical stability windows in cells, spanning lithium metal to >5 V cathode materials,[5] although poor interfacial contact with electrode materials is a major impediment to use. Nevertheless, the high inherent ionic conductivity of Garnet SSE materials drives a significant amount of research. As a class, Garnet SSE materials have a high lithium-ion loading and yet also have vacant sites to create a dynamic equilibrium within the unit cell.

Lithium-ion concentration and vacancy concentrations are important in determining the rate of ion transport in these materials. While defect formation is regarded as a temperature-dependent process, the lithium-ion concentration depends on the chemical formula and the volume of the unit cell. Although the amount of lithium in the Garnet family chemical formulas appears high (with lithium-ion residing in multiple tetrahedral and octahedral sites in the unit cell), adjusting ionic conductivity for the volume of the unit cell reveals that most of the lithium ions in the Garnet do not contribute to ionic conductivity. In contrast, the LISICON-like, argyrodite, and perovskite materials appear to utilize the lithium-ion content to a greater extent.[3]

Thangadurai et al.[6] present solid-state Li NMR data to explain that lithium ions located in octahedral sites in Garnet crystals are mobile, while lithium ions in tetrahedral site are not. The figure from Thangadurai et al. (Fig. 22C.4) depicts the three possible lithium-ion occupancy states for opposite, adjacent Garnet tetrahedrons. The octahedral sites form between the tetrahedral crystals and interconnect to form a conduction channel for the lithium ions. The lithium ions can travel in the channel or enter a tetrahedral site that is effectively a dead end. The high concentration of lithium ions in a unit cell enables tetrahedral sites to fill, in part, with sufficient lithium ions remaining to populate the octahedral channel. At the same time, lithium ions may be detained in the tetrahedral site by the lone pairs of ligand atoms as the cation in the unit cell—usually $La^+$—can accept and

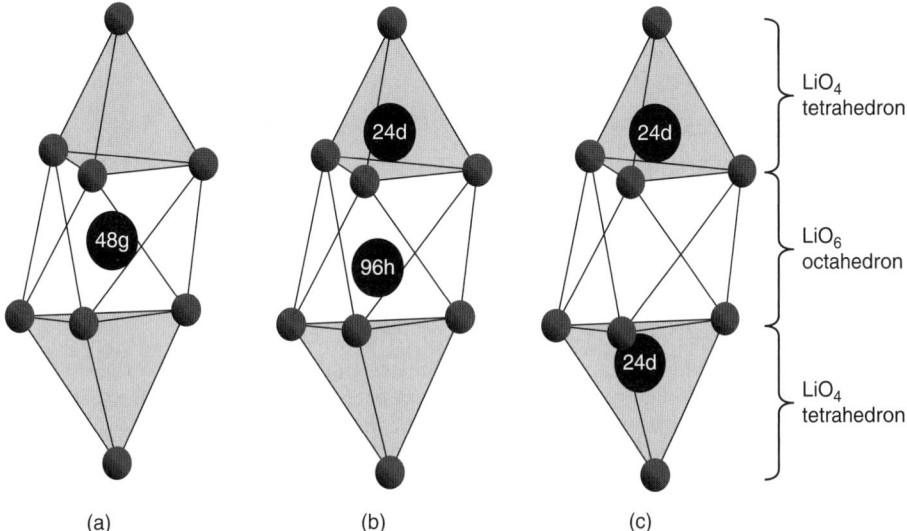

(a)                                  (b)                                  (c)

**FIGURE 22C.4** Lithium-ion occupancy states for Garnet tetrahedrons. (*Reprinted under license from the Royal Society of Chemistry: V. Thangadurai, S. Narayanan, D. Pinzaru, Garnet-Type Solid-State Fast Li Ion Conductors for Li Batteries: Critical Review. Copyright 2014 Royal Society of Chemistry.*)

then donate ligand electrons to stabilize the ligand-$Li^+$ interaction. $La^+$ in the Garnet structure has minimal electrons in d orbitals, which makes it a good choice for a nonconductor SSE material.

Thangadurai et al. studied over 50 Garnet formulations to examine the details of $Li^+$ to $Li^+$ repulsion in Garnet structures and also analyzed how this facilitates $Li^+$ mobility. Based on this analysis, SSE Garnet materials have structures that force nonuniform distribution of lithium ions in the available sites, which, along with the mutual repulsion of mobile ions, may facilitate $Li^+$ mobility within a unit cell under dynamic equilibrium.

It is interesting to note that throughout their exhaustive compilation of Garnet structures, Thangadurai et al. report activation energy values that fall between 0.3 and 0.6 eV, with a few exceptions. This suggests that even with all of the variations in chemical formula and synthesis procedures for the reported Garnet materials, transport mechanism is impacted by a factor of two or less. Arrhenius graphs of this data show relatively small changes in the slope while the absolute values of the ionic conductivity do vary by orders of magnitude among the Garnet materials. A cursory analysis suggests that substitution of Zr for La improves ionic conductivity, which confirms the trend reported by Bachman et al.[3] Also, Peng et al.[7] found that the Garnet SSE with combinations of Ge and Nb improves ionic conductivity, perhaps from similar disruptions to the crystal structure that widen paths for ion transport or introduce dynamic interactions for the lithium ions in the structures.

### 22C.3.4   Other Crystalline or Glassy SSE Materials

Although hydride-based SSE materials do not exhibit high room-temperature ionic conductivity, Suzuki et al.[8] demonstrate cells using a mixture of $LiBH_4$ and $LiNH_2$ that operate well at elevated temperatures. These SSE materials are low cost and may fulfill fitness for use in applications that require high-temperature operation, e.g., data centers. Also, these hydride SSE materials may enable new battery chemistries with other low-cost components.[9,10]

NAISICON-like materials often contain $Ti^{4+}$ cations that react with lithium in anodes. However, a composite material using protective PVDF-HFP polymer layers can protect the SSE from degradation.[11]

LiPON is a glassy SSE material that works well in very small cells that are often built through sputtering, layer by layer onto a silicon base and packaged as an electronic component for circuit boards. Although larger format cells are not commercial, the importance of LiPON in the small format electronics backup power end uses does drive studies to understand the mechanism that governs $Li^+$ conduction.[12] New processes and precursor materials for LiPON could improve larger-scale commercialization and expand its use as these may improve interfacial contact.[13]

Other materials combine structures of more than one class of materials to produce glassy hybrids such as LISICON with Garnet-like crystals.[14]

## 22C.4   LITHIUM-ION SOLID-STATE ELECTROLYTES—POLYMER MATERIALS

For polymer SSE materials above the glass transition temperature $T_g$, the Vogel-Tammann-Fulcher (VTF) equation provides an empirical relationship that accounts for the rapid change in transport properties approaching $T_g$. This relationship has similar form to the Arrhenius-type equation and is given by,

$$\sigma = (\sigma_0 / T^x)\, e^{(E_A / k_B(T - T_g))}$$

where the ionic conductivity $\sigma$ depends on absolute temperature $T$ and $\sigma_0$ is a reference ionic conductivity value related to the frequency of events that can overcome an energy barrier $E_A$ relative to the internal energy at that temperature by the Boltzmann number $k_B$, and $x$ is determined experimentally. This equation usually fits ionic conductivity from 10 to 100 K above the $T_g$. This form of the VTF equation allows for comparison of activation energy and reference ion conductivity values with nonpolymer materials.

Until recent years, PEO-based polymers dominated the development of polymer lithium-ion SSE materials. "Stars" are typically formed by cross-linking multiple cross-linking agents at a single vertex along the PEO chain to result in a network of chain segments that were flexible, but restrained from flowing. Dissolving a salt such as lithium triflate into the cross-linked polymer network provides ionic conductivity, since the two lone pair electrons of the ether oxygens in the polymer solvate and ionize the salt. Ionic conductivity at room temperature is poor, but improves at cell temperatures above 60°C since electrolyte ionization and the mobility of polymer segments increase. PEO remains a polymer of interest for new battery chemistries[15] and some that combine PEO with ionic liquids.[16]

Polymers beyond PEO and blends of polymers[17,18] have expanded the original concepts with some promising results. A general review of polymers covered by Agrawal and Pandey[19] in 2008 showed the shortfall of PEO-based materials and the promise of composites with nanoscale ceramic particles.

In recent years, a startup company, Ionic Materials, developed a new class of polymer electrolytes that is based on other ether-rich polymers and a set of proprietary additives. Initial reports indicate promising results. Other polymers have sulfonic acid moieties that result in materials that resemble those used in PEM fuel cell applications, only aiming at lower operating temperatures.[20]

## 22C.5   COMMERCIALIZATION AND ENGINEERING SOLUTIONS

Kerman et al.[21] and Sun et al.[22] review many of the practical issues that remain to be solved for commercial SSE implementation for lithium batteries. Because the electrochemical stability, interfacial contact, and synthesis development vary from material to material, and research is progressing each year, the best source of detailed information is found in the reference articles. However, some of the approaches to solving these issues are outlined in the following sections.

### 22C.5.1   Composite SSE Materials

The combination of crystalline or glassy SSE materials with SSE polymers follows a conventional method in preparing battery components, similar to the manufacturing of coated electrodes for conventional lithium-ion cells with liquid electrolyte systems. However, the SSE composite must not form pores between particles or layers and there must be a sufficient amount of polymer to fill any gaps in the composite.

Layered materials are simple composites with a continuous layer of crystalline or glassy SSE materials coated with an SSE polymer to enable intimate contact with electrode surfaces. In the case of Garnet SSE materials, this approach can overcome issues with lithium anode contact and dendrite growth.[23]

A direct approach is to prepare slurries of crystalline or glassy SSE materials using SSE polymers as binders. Tao et al.[24] report a successful use of PEO as a binder for a Garnet SSE material in a thin separator layer and incorporated into carbon foam with cathode material for the positive electrode—using both layer and binder composite forms in the same cell construction.

It is interesting to note that Liang et al.[25] and Zhang et al.[26] achieved good ionic conductivity in cells with composite SSE materials prepared using a Garnet SSE and a PVDF-HFP (polyvinylidene fluoride-hexafluoropropylene) polymer, which is a family of binders for conventional type lithium-ion cells as used in the Telcordia "polymer" technology with liquid electrolyte solutions. Such straightforward approaches avoid the use of SSE polymers that often do not bind particles as well as film-forming polymers used in liquid lithium-ion cells.

### 22C.5.2   Interfacial Issues

Because argyrodite SSE materials contain sulfur and phosphorous—and often halogens—lithium in the anode electrode can react with these species to degrade the SSE or at least degrade the anode-SSE interface. This occurs with lithium metal foil anodes as well as graphitic anodes that charge at voltages close to lithium metal voltage as some results appear to reflect.[27] Some studies of argyrodite SSE materials use $Li_4Ti_5O_{12}$ anode materials that operate ~1.5 V from lithium metal voltage,[28] although this lowers the cell potential by the same voltage. However, the sulfur content appears to be compatible with cathodes, especially for positive materials that contain sulfides.[29–31]

At the anode interface, lithium metal contact in SSE Garnet materials is poor and results in lithium dendrite growth into and through Garnet SSE materials. This issue can be reduced by adding an Al dopant to the surface of Garnet materials,[4,32] although this technique introduces complications in SSE materials synthesis. Another approach is to synthesize dense Garnet SSE, although the resulting material does not conform to interfaces with the active material. As a solution, Fu et al.[33] demonstrated a two-layer SSE Garnet with a porous layer that mechanically stabilizes a second layer that is denser and is less susceptible to lithium dendrite penetration. In addition, prewetting the Garnet SSE materials with molten lithium metal can improve subsequent wetting and contribute to reducing dendrite issues.[34] The addition of cation dopants can further improve the ability of these Garnet SSE materials to conform to electrode interfaces.[35]

Since Ge is one of the species added to SSE materials, its use by Luo et al.[36] as an atomic layer deposition (ALD) layer on Garnet SSE materials addresses the anode interface issues concerning lithium dendrite growth and lithium wetting. Direct deposition of electrode materials onto Garnet SSE surfaces enables detailed microscopic studies of these interfaces,[37] utilizing PEO-based polymer layers. Another solution to assure good interfacial contact is to deposit electrode materials onto SSE materials.[38]

## 22C.5.3   SSE Synthesis Process

The preparation of polymer SSE materials often involves polymerizing or cross-linking the polymer and dissolving electrolyte salt directly or in solution into a polymer film or melt. The preparation of crystalline and glassy SSE materials is more complex. For example, sulfur and phosphorous precursors make the synthesis of argyrodite materials complex and difficult to control, leading to the development of various processing techniques and procedures.[27]

While complex synthesis processes and the lithium reactivity of various SSEs can be discouraging factors, the room-temperature ionic conductivity of sulfide-rich SSE remains attractive. Part of the success in achieving high ionic conductivity has to do with the ability to compress many of these sulfide-bearing SSE materials to eliminate grain boundaries and the associated barriers to ionic flow.[29] Other studies on crystalline and glassy materials demonstrate that applying compressive forces reduces ionic conductivity.[3]

Because crystalline and glassy SSE materials are often sintered at elevated temperatures, the cooling steps can influence ionic conductivity by affecting the extent of grain boundary formation and even voids in the material.[39] Another approach to minimize the effect of cooling rate on ionic conductivity is to add aliovalent cations to the SSE material in order to lower the sintering temperature and time.[40]

## *REFERENCES*

1. A. Manthiram, X. Yu, and S. Wang, "Lithium battery chemistries enabled by solid-state electrolytes," *Nat. Rev. Mater.* **2**(4):16103 (2017).

2. K. Funke, "Solid state ionics: From Michael Faraday to green energy—the European dimension," *Sci. Technol. Adv. Mater.* **14**(4) (2013).

3. J. C. Bachman, S. Muy, A. Grimaud, H.-H. Chang, N. Pour, S. F. Lux, and O. Paschos, et al., "Inorganic Solid-State Electrolytes for Lithium Batteries: Mechanisms and Properties Governing Ion Conduction," https://pubs.acs.org/doi/abs/10.1021/acs.chemrev.5b00563. Accessed March 21, 2018.

4. N. Kamaya, K. Homma, Y. Yamakawa, M. Hirayama, R. Kanno, M. Yonemura, and T. Kamiyama, et al., "A lithium superionic conductor," *Nat. Mater.* **10**(9):682–686 (2011).

5. X. Han, Y. Gong, K. K. Fu, X. He, G. T. Hitz, J. Dai, and A. Pearse, et al., "Negating interfacial impedance in garnet-based solid-state Li metal batteries," *Nat. Mater.* **16**(5):572 (2017).

6. V. Thangadurai, S. Narayanan, and D. Pinzaru, "Garnet-type solid-state fast Li ion conductors for Li batteries: Critical review," *Chem. Soc. Rev.* **43**(13):4714 (2014).

7. H. Peng, L. Feng, L. Li, Y. Zhang, and Y. Zou, "Effect of Ge substitution for Nb on Li ion conductivity of Li5La3Nb2O12 solid state electrolyte," *Electrochim. Acta* **251**:482–487 (2017).

8. S. Suzuki, J. Kawaji, K. Yoshida, A. Unemoto, and S. Orimo, "Development of complex hydride-based all-solid-state lithium ion battery applying low melting point electrolyte," *J. Power Sources* **359**:97–103 (2017).

9. J. A. Weeks, S. C. Tinkey, P. A. Ward, R. Lascola, R. Zidan, and J. A. Teprovich, "Investigation of the reversible lithiation of an oxide free aluminum anode by a LiBH4 solid state electrolyte," *Inorganics* **5**(4):83 (2017).

10. Y. Yan, R.-S. Kühnel, A. Remhof, L. Duchêne, E. C. Reyes, D. Rentsch, and Z. Łodziana, et al., "A lithium amide-borohydride solid-state electrolyte with lithium-ion conductivities comparable to liquid electrolytes," *Adv. Energy Mater.* **7**(19):1700294 (2017).

11. Y. Xia, X. Wang, X. Xia, R. Xu, S. Zhang, J. Wu, and Y. Liang, et al., "A newly designed composite gel polymer electrolyte based on poly (vinylidene fluoride-hexafluoropropylene) (PVDF-HFP) for enhanced solid-state lithium-sulfur batteries," *Chem.-A Eur. J.* **23**(60):15203–15209 (2017).

12. Y. Aizawa, K. Yamamoto, T. Sato, H. Murata, R. Yoshida, Craig A. J. Fisher, and Takehisa Kato, et al., "In situ electron holography of electric potentials inside a solid-state electrolyte: Effect of electric-field leakage," *Ultramicroscopy* **178**:20–26 (2017).

13. A. J. Pearse, T. E. Schmitt, E. J. Fuller, F. El-Gabaly, C.-Fu Lin, K. Gerasopoulos, and A. C. Kozen, et al., "Nanoscale solid state batteries enabled by thermal atomic layer deposition of a lithium polyphosphazene solid state electrolyte," *Chem. Mater.* **29**(8):3740–3753 (2017).

14. P. Lu, F. Ding, Z. Xu, J. Liu, X. Liu, and Q. Xu, "Study on (100-x)(70Li2S-30P2S5)-xLi2ZrO3 glass-ceramic electrolyte for all-solid-state lithium-ion batteries," *J. Power Sources* **356**:163–171 (2017).

15. W. Li, L. Chen, Y. Sun, C. Wang, Y. Wang, and Y. Xia, "All-solid-state secondary lithium battery using solid polymer electrolyte and anthraquinone cathode," *Solid State Ionics* **300**:114–119 (2017).

16. O. Sheng, C. Jin, J. Luo, H. Yuan, C. Fang, H. Huang, and Y. Gan, et al., "Ionic conductivity promotion of polymer electrolyte with ionic liquid grafted oxides for all-solid-state lithium–sulfur batteries," *J. Mater. Chem. A* **5**(25):12934–12942 (2017).

17. H. Duan, Y.-X. Yinab, X.-X. Zeng, J.-Y. Li, J.-L. Shi, Y. Shi, and R. Wen, et al., "In-situ plasticized polymer electrolyte with double-network for flexible solid-state lithium-metal batteries," *Energy Storage Mater.* **10**:85–91 (2018).

18. B. Jinisha, K. M. Anilkumar, M. Manoj, V. S. Pradeep, and S. Jayalekshmi, "Development of a novel type of solid polymer electrolyte for solid state lithium battery applications based on lithium enriched poly (ethylene oxide)(PEO)/poly (vinyl pyrrolidone)(PVP) blend polymer," *Electrochim. Acta* **235**:210–222 (2017).

19. R. C. Agrawal and G. P. Pandey, "Solid polymer electrolytes: materials designing and all-solid-state battery applications: an overview," *J. Phys. D-Appl. Phys.* **41**(22):223001 (2008).

20. X. Judez, H. Zhang, C. Li, J. A. González-Marcos, Z. Zhou, M. Armand, and L. M. Rodriguez-Martinez, "Lithium bis (fluorosulfonyl) imide/poly (ethylene oxide) polymer electrolyte for all solid-state Li–S cell," *J. Phys. Chem. Lett.* **8**(9):1956–1960 (2017).

21. K. Kerman, A. Luntz, V. Viswanathan, Y.-M. Chiang, and Z. Chen, "Practical challenges hindering the development of solid state Li ion batteries," *J. Electrochem. Soc.* **164**(7):A1731–A1744 (2017).

22. C. Sun, J. Liu, Y. Gong, D. P. Wilkinson, and J. Zhang, "Recent advances in all-solid-state rechargeable lithium batteries," *Nano Energy* **33**:363–386 (2017).

23. Y. Li, B. Xu, H. Xu, H. Duan, X. Lü, S. Xin, and W. Zhou, et al., "Hybrid polymer/Garnet electrolyte with a small interfacial resistance for lithium-ion batteries," *Angew. Chem. Int. Ed.* **56**(3):753–756 (2017).

24. X. Tao, Y. Liu, W. Liu, G. Zhou, J. Zhao, D. Lin, and C. Zu, et al., "Solid-state lithium–sulfur batteries operated at 37°C with composites of nanostructured Li7La3Zr2O12/carbon foam and polymer," *Nano Lett.* **17**(5):2967–2972 (2017).

25. Y. F. Liang, S. J. Deng, Y. Xia, X. Wang, X. H. Zia, J. B. Wu, and C. D. Gu, et al., "A superior composite gel polymer electrolyte of Li 7 La 3 Zr 2 O 12-poly (vinylidene fluoride-hexafluoropropylene) (PVDF-HFP) for rechargeable solid-state lithium ion batteries," *Mater. Res. Bull.* **102**:412–417 (2018).

26. W. Zhang, J. Nie, F. Li, Z. L. Wang, and C. Sun, "A durable and safe solid-state lithium battery with a hybrid electrolyte membrane," *Nano Energy* **45**:413–419 (2018).

27. S. Chida, A. Miura, N. C. Rosero-Navarro, M. Higuchi, N. H. H. Phuc, H. Muto, and A. Matsuda, et al., "Liquid-phase synthesis of Li 6 PS 5 Br using ultrasonication and application to cathode composite electrodes in all-solid-state batteries," *Ceram. Int.* **44**(1):742–746 (2018).

28. J. Auvergniot, A. Cassel, D. Foix, V. Viallet, V. Seznec, and R. Dedryvère, "Redox activity of argyrodite Li6PS5Cl electrolyte in all-solid-state Li-ion battery: An XPS study," *Solid State Ionics* **300**:78–85 (2017).

29. M. Tatsumisago, M. Nagao, and A. Hayashi, "Recent development of sulfide solid electrolytes and interfacial modification for all-solid-state rechargeable lithium batteries," *J. Asian Ceram. Soc.* **1**(1):17–25 (2013).

30. R. C. Xu, X. L. Wang, S. Z. Zhang, Y. Xia, X. H. Xia, J. B. Wu, and J. P. Tu, "Rational coating of Li 7 P 3 S 11 solid electrolyte on MoS 2 electrode for all-solid-state lithium ion batteries," *J. Power Sources* **374**:107–112 (2018).

31. R. Xu, X. Xia, X. Wang, Y. Xia, and J. Tu, "Tailored Li 2 S–P 2 S 5 glass-ceramic electrolyte by MoS 2 doping, possessing high ionic conductivity for all-solid-state lithium-sulfur batteries," *J. Mater. Chem. A* **5**(6):2829–2834 (2017).

32. Y. Arinicheva, H. Zhenga, C. L. Tsaic, J. Nonemachere, J. Malzbendere, D. Fattakhova-Rohlfinga, and O. Guillon, et al., "Intrinsic improvement of LLZO solid-state electrolyte to suppress Li dendrite growth," *ECS Meeting Abstracts*, 480, 2018.

33. K. K. Fu, Y. Gong, G. T. Hitz, D. W. McOwen, Y. Li, S. Xu, and Y. Wen, et al., "Three-dimensional bilayer garnet solid electrolyte based high energy density lithium metal–sulfur batteries," *Energ. Environ. Sci.* **10**(7):1568–1575 (2017).

34. K. K. Fu, Y. Gong, B. Liu, Y. Zhu, S. Xu, Y. Yao, and W. Luo, et al., "Toward garnet electrolyte–based Li metal batteries: An ultrathin, highly effective, artificial solid-state electrolyte/metallic Li interface," *Sci. Adv.* **3**(4):e1601659 (2017).

35. Z. Fu, Y. Gonga, L. Zhang, E. Grittona, G. L. Godbeya, Y. Rena, and D. W. McOwen, et al., "Mechanical properties of Li7La2. 75Ca0. 25Zr1. 75Nb0. 25O12 garnet electrolyte—A preliminary study of a porous layer support all-solid state battery," *ECS Meeting Abstracts* 550–550 (2017).

36. W. Luo, Y. Gong, Y. Zhu, Y. Li, Y. Yao, Y. Zhang, and K. K. Fu, et al., "Reducing interfacial resistance between garnet-structured solid-state electrolyte and Li-metal anode by a germanium layer," *Adv. Mater.* **29**(22):1606042 (2017).

37. C. Wang, Y. Yang, X. Liu, H. Zhong, H. Xu, Z. Xu, and H. Shao, et al., "Suppression of lithium dendrite formation by using LAGP-PEO (LiTFSI) composite solid electrolyte and lithium metal anode modified by PEO (LiTFSI) in all-solid-state lithium batteries," *ACS Appl. Mater. Interfaces* **9**(15):13694–13702 (2017).

38. R. Wang, J. S. Dauberta, M. Ning, Y. Yang, Y. Liua, and G. N. Parsons, et al., "Development of a Li-ion electrochemical platform for in-situ tem of solid state electrode/electrolyte interfaces," *ECS Meeting Abstracts* 103 (2017).

39. H. Peng, Y. Zhang, L. Li, and L. Feng, "Effect of quenching method on Li ion conductivity of Li5La3Bi2O12 solid state electrolyte," *Solid State Ionics* **304**:71–74 (2017).

40. X. Yang, D. Kong, Z. Chen, Y. Sun, and Y. Liu, "Low-temperature fabrication for transparency Mg doping Li 7 La 3 Zr 2 O 12 solid state electrolyte," *J. Mater. Sci.-Mater. El.* **29**(2):1523–1529 (2018).

# BATTERY APPLICATIONS

# CHAPTER 23
# BATTERY SELECTION FOR CONSUMER ELECTRONICS

**John A. Wozniak**

## 23.1  INTRODUCTION

A proliferation of consumer electronic applications in recent years has pushed battery technology to the forefront of concerns when designing a new product. The traditional 3C products (Computers, Communication, and Consumer Electronics) have been augmented by applications such as portable cleaning machines, personal care products, and other nontraditional applications. Run-time, talk time, standby time, and shelf life are all important buzzwords that contribute directly to how well a product sells in the marketplace, and the values used for these variables depend on the choice of battery. Recent changes in regulatory requirements for lithium batteries can also influence decision making. This chapter will cover the needs of typical and emerging consumer electronic devices, home applications, common battery chemistries, and key selection criteria.

There have been many advances in battery technology in recent years: new chemistries, higher energy densities, higher power densities, new form factors, and new coatings for improved reliability. Yet in the end, there is still no single perfect battery that performs optimally under all electrical and environmental conditions. This ideal battery would have unlimited energy and power capabilities, operate well under all environmental conditions, be inexpensive, have unlimited shelf life, and be completely safe and consumer-proof. One may find two or three of these characteristics that a single battery may approach, but other characteristics will suffer.

The component materials of batteries exhibit a wide range of electrochemical properties, and the continued push to achieve higher energy and power densities requires caution on the part of the designer. As consumer electronic applications seek smaller and more powerful batteries, there runs a parallel risk of keeping this power under control. Selection of the proper battery for a specific application is a study in trade-offs. In considering these trade-offs, one must also keep in mind the education of the customer in the proper use and care of the battery in their device.

Many factors must be considered when selecting the battery that best meets the needs of a specific application. The characteristics of available batteries must be compared with the needs of the electronic device. It is critical to consider these needs early in the development of the device since the battery will have a direct impact on the size and weight of the device. Addressing the various trade-offs from the beginning of development is the most effective means of ensuring reasonable compromises resulting in a good product design.

Key considerations include the following:

- *Type of battery:* Primary (single use) or secondary (rechargeable)
- *Voltage:* Nominal or operating voltage, profile of the discharge curve, maximum and minimum permissible voltages
- *Physical size:* Weight, shape, size, and terminal requirements
- *Capacity:* Required Ah or Wh to achieve run, talk, or standby times

- *Load current and profile:* Constant power, constant current, constant impedance, or other; value of load current or profile; constant, variable, or pulsed load and duty cycle requirements
- *Temperature requirements:* Operating and storage temperature ranges
- *Shelf life:* State-of-charge during storage; storage time as a function of temperature, humidity, and other environmental factors; active/standby/sleep modes
- *Charging (if rechargeable):* Float or charge cycling; cycle life requirements, simplicity, and availability of charging source, charging efficiency
- *Safety and reliability:* Permissible variability and failure rates; use of potentially hazardous or toxic materials; operation under severe, hazardous, or abusive conditions; failure mode (out gassing, leakage, swelling)
- *Cost:* Initial cost; operating cost or life-cycle cost; use of exotic or critical materials with potentially volatile pricing; cost of charging circuit or charger, if rechargeable
- *Regulatory requirements:* Country of origin and country of delivery concerns; special shipping concerns; recycling requirements and labeling
- *Environmental conditions:* Shock and vibration, acceleration or other mechanical demands and forces; atmospheric conditions (pressure, humidity, altitude, etc.)

### 23.1.1  Typical Portable Applications

The demand for portability in the consumer electronics world has driven the demand for a wide range of electrochemical battery technologies. The wide range of power and environmental requirements for these devices requires an equivalent wide range of battery technologies to best match these needs. Portable consumer electronics is a rapidly expanding area as an increasing number of portable devices are being introduced that are designed to operate solely with batteries or, in some cases such as notebook computers, to operate with either batteries or AC line power.

Consumer electronics typically include electronic devices that are intended for everyday use. Communication, entertainment, and office applications dominate this product class and the Consumer Technology Association (CTA) estimated 2017 U.S. retail revenue to be over $290 billion.[1] Devices classified as consumer electronics include personal computers, cell phones, DVD/CD/video players, MP3 players, Bluetooth headsets, GPS navigational systems, televisions, digital cameras, camcorders, electronic toys, smart home devices, e-cigarettes, and calculators. Even simple devices such as laser pointers and hearing aids are considered in this category. The trend in the industry is convergent devices where a single device provides multiple functions. One example is how PDAs (personal digital assistants) have been virtually eliminated from the marketplace through the incorporation of office organizational functions, such as digital business cards and scheduling, into cell phones. Table 23.1 lists typical current drains of common portable electronic devices. This serves to illustrate

**TABLE 23.1**   Current Drain in Battery-Operated Portable Electronics

| Device | Current drain, mA |
|---|---|
| CD players | 100–350 |
| Cell phones (talking) | 300–600 |
| Digital camera | 500–1200 |
| Camcorders | 500–1000 |
| Notebook computer | 200–3000 |
| Memory backup | Microamperes |
| Radio | 20–50 |
| Radio-controlled toys | 600–1500 |
| TV (portable) | 300–700 |
| Travel shaver | 300–500 |
| Remote controls | 10–50 |
| Watches (LED) | 10–40 |

the wide range of requirements from microamperes to several amperes. Although the values are specified in milliamperes, many devices present a variable load to the battery, while others require more of a constant power drain that results in higher currents as the battery voltage declines.

Other portable devices in the home include such items as personal grooming appliances, cleaning apparatus, and uninterruptible power supplies. These devices have a wide range of power needs that can be completely divergent from the typical 3C applications, with power density being more important than energy density. These devices often require 10 to 20 A or more from a single-battery cell. Battery selection for these types of devices will be covered in Chap. 24. Battery solutions for consumer electronics can be divided into two main groups: primary single-use batteries and secondary rechargeable batteries. Primary batteries are typically used in low to medium power applications, whereas secondary batteries are used in nearly all other applications. The following sections will cover typical applications for both primary and secondary batteries and conclude with some specific examples of the trade-offs to consider when making a final choice. Table 23.2 lists some common applications and their suitability for primary and secondary battery solutions. There is significant overlap and, in the end, comparison of key characteristics of the battery choices to the device requirements will result in an acceptable solution.

**TABLE 23.2**  Application of Batteries: Primary versus Secondary

| Application | Primary or secondary battery |
|---|---|
| Portable tools | Secondary |
| Hand vacuum | ↑ |
| Notebook computer | |
| Cordless telephone | |
| Camcorders | |
| Video player | |
| Portable shaver | |
| Cell phone | |
| Audio players | |
| Digital camera | |
| Toys | |
| Hearing aid | |
| Remote control | |
| Watches | |
| Smoke detector | ↓ |
| Memory backup | Primary |

## 23.2   PRIMARY BATTERIES

Primary batteries are appropriate when power demands are relatively low. Devices such as garage door openers and remote controls that are only active for extremely short periods of time as well as very few times per day are often well suited for primary batteries. Watches and digital clocks are other examples where continuous, extremely low power demands may allow a primary battery to be used for at least a year without replacement. Backup memory power is also a typical application where a device such as an Internet-enabled phone or notebook computer may have a secondary battery for power, but may also incorporate a primary battery to keep volatile memory alive (e.g., RTC, or real-time clock).

Primary batteries come in a variety of shapes and sizes from coin cells to large cylindrical and prismatic cells (lantern batteries). Along with these diverse form factors, there is a wide range of chemistries, including carbon-zinc, alkaline, lithium metal, and others.

Primary batteries are usually less expensive to produce, with a smaller physical size than secondary batteries capable of supporting the same power demands. However, with more global awareness of the effects of electronic waste on the environment, many devices that once were powered by primary batteries (digital cameras, MP3 players) are now designed to accept secondary batteries instead of, or in addition to, primary batteries.

The most common primary battery chemistries used in consumer electronics are zinc/air, alkaline ($Zn/MnO_2$), lithium/manganese dioxide ($Li/MnO_2$), lithium/iron disulfide ($Li/FeS_2$), and lithium/sulfur dioxide ($Li/SO_2$). Lithium sulfur dioxide batteries are largely used in military electronics, but serve as a reference system for consumer applications. There are several other lithium primary chemistries, as covered in Chaps. 13 and 18C, but either cost or safety concerns preclude their use in consumer electronic applications.

Specific details regarding zinc/air batteries can be found in Chap. 18A. These batteries have a high energy density and stable voltage curve in addition to being environmentally friendly. However, they are extremely sensitive to relative humidity and the oxygen level in the air. They also have limited shelf life when exposed to air and do not perform well with intermittent loads. A principal application is hearing aids where compact size and constant low power demands are required.

Alkaline primary batteries are arguably the most ubiquitous cells available. Ease of replacement and low initial cost make them attractive for inexpensive consumer products such as toys, clocks, radios, and remote controls. Their relative insensitivity to discharge rate and duty cycle, a wide operating temperature range, and variety of shapes and sizes make them suitable for most consumer electronic applications. The key drawback for alkaline battery use is the need for regular replacement in medium- to high-drain rate applications such as audio and video players. They also exhibit a sloping discharge curve that can limit their use where constant voltage is required.

Lithium-manganese dioxide batteries are commonly used in button or coin cell form for watches, garage door openers, and memory backup. Good shelf life, pulse capability, and energy density make these batteries the most commonly used of the lithium primary cells. Reduced volatility compared to $Li/SO_2$ and relatively low cost make them attractive for many consumer electronic applications. They are, however, subject to the same regulatory restrictions as other lithium primary chemistries.

Lithium iron disulfide batteries are manufactured in AAA and AA sizes for consumers. They have a lower voltage (1.5 V), but similar energy density as $Li/MnO_2$. Performance characteristics are also similar, but the lower operating voltage makes them useful as a higher power replacement for alkaline batteries. This makes them preferable to alkaline cells in higher pulse power applications such as digital cameras with photo flash capability. They are also available in two grades, "ultimate" and "advanced." The former is an energy product while the latter provides higher power.

Positive characteristics of lithium sulfur dioxide batteries are high energy density, good low-temperature performance, superior rate capability, and long shelf life. Relatively high cost and safety and environmental concerns have limited their use in consumer applications to higher priced products like security systems and some telecommunications systems. They are more appropriate for military, aerospace, and some biomedical applications, although they are included in subsequent figures for reference purposes.

## 23.2.1  Comparing Primary Battery Characteristics

Characteristics of conventional primary batteries based on theoretical limits are summarized in Chaps. 1 and 9. Various tables also list characteristics based on the actual performance of a practical battery under near optimal conditions for each specific type. It is important to note the following:

- The actual capacity available from a battery is *significantly* less than the theoretical capacity of the active materials.
- The actual capacity is also less than the theoretical capacity of a practical battery because an actual battery has weight and volume attributed to nonenergy-producing materials used in construction of the battery in addition to the active materials.
- The capacity of a battery can vary greatly from the values listed in Chaps. 1 and 9. The values are based on optimum conditions for that battery, and the real-world conditions for use are rarely optimal. Measurements should be made under actual real-world conditions for the product before a final judgment is made.

The following figures compare some key characteristics of the above-described primary batteries. More detailed characteristics can be found in the appropriate chapter of this Handbook. Those data, rather than the generalized data in this section, should be used to evaluate the specific performance of each battery.

Figure 23.1 compares power on a volumetric basis with output duration, while Fig. 23.2 is a Ragone plot presenting power and energy data on a gravimetric basis. Lithium batteries have a considerable advantage in weight over their alkaline counterparts.

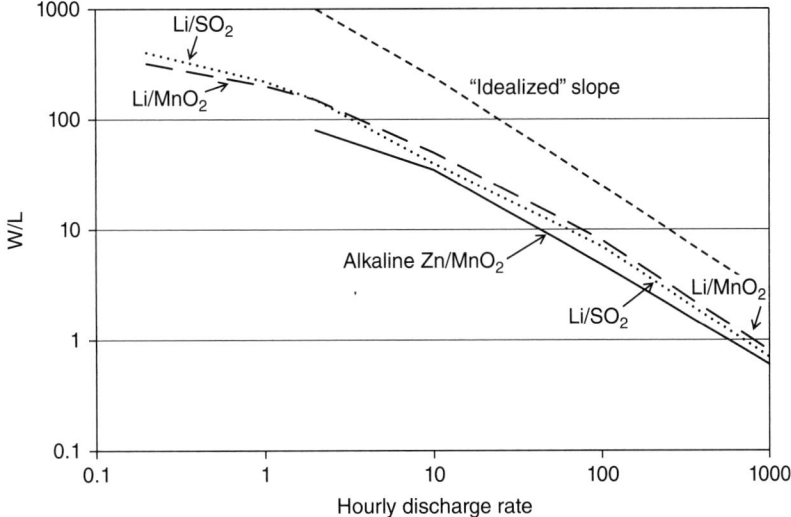

**FIGURE 23.1**  Performance characteristics of primary batteries on a volumetric basis, 20°C.

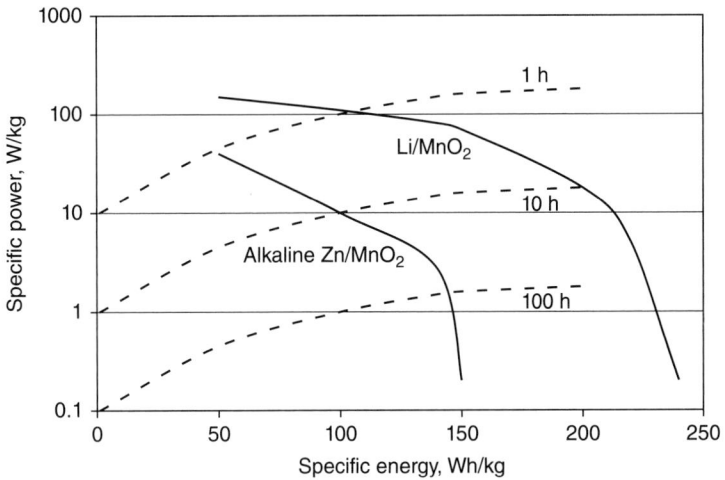

**FIGURE 23.2**  Performance capabilities of primary batteries—specific energy versus specific power.

The comparative performances, as a function of temperature, of various primary batteries are shown on a gravimetric basis (Wh/kg) in Fig. 9.7 (Chap. 9) and on a volumetric basis (Wh/L) in Fig. 9.8 (Chap. 9). In general, primary battery performance drops off rather quickly at low temperatures. Again, lithium batteries have better low-temperature characteristics than the zinc-based chemistries.

## 23.3   SECONDARY BATTERIES

Rechargeable batteries for consumer electronics have become an $800 billion per year industry that is continuing to grow. This includes not only the batteries themselves, but the raw materials, components, and packaging that go into manufacturing a rechargeable battery pack. The environmental "friendliness" of reusing the battery for months or years is a key selling point in the marketplace. Although these batteries are somewhat bulkier than primary batteries of equivalent power, consumers demand the convenience of not having to regularly replace the battery. Of course, a key factor in designing a secondary battery into an application is the charger. Decisions need to be made early in development as to whether the charger is embedded in the device or a stand-alone option, or both. There may be a higher initial cost of ownership, but the *lifetime* cost of ownership is typically much lower.

Cell phones, notebook computers, music and video players and recorders, cordless phones, and navigational devices are all heavily dependent on secondary batteries for power. These are devices that are used many hours per day, and the device is expected to last for years. Secondary batteries can be used in nearly every consumer electronics application. The eco-friendly consumer will even use secondary batteries in simple devices such as remote controls and electronic toys that have commonly used primary batteries in the past.

Common secondary battery chemistries include nickel-cadmium (NiCd), nickel-metal hydride (NiMH), and a variety of lithium-ion chemistries. These lithium-ion batteries include lithium cobalt oxide ($Li/CoO_2$), lithium manganese oxide spinel ($Li/Mn_2O_4$), lithium iron phosphate ($Li/FePO_4$), and a variety of blended oxides $Li/(Ni-Co-Mn)O_2$ where manganese, nickel, and/or aluminum may replace some of the cobalt in $Li/CoO_2$. Nearly all of the lithium-ion chemistries have carbon/graphite anodes, so the key differences lie in the cathode materials.

Nickel-cadmium is quickly being replaced by NiMH as the environmental concerns for cadmium have become more prevalent. It is rare to find new products using NiCd batteries with the exception of some developing countries where cost is the primary concern. NiCd batteries are typically lower cost, but regulatory restrictions and cost of disposal can make them unattractive in many countries. One key advantage that NiCd maintains over NiMH is high-temperature performance. Both are good for high-drain applications. NiCd batteries are typically used in low-cost consumer devices such as cordless phones and cheap power and gardening tools. NiMH batteries can be retrofitted into many NiCd applications with proper design to get more capacity in the same space. NiMH batteries can still be found in some low-cost cameras and audio players and recorders, but these applications are typically where lithium-ion batteries are a better fit. Both NiCd and NiMH batteries come in a wide variety of form factors ranging from button cells to F-size cylindrical.

With a high operating voltage, high energy density, good cycle life, and reasonable operating temperature range, lithium-ion has become the workhorse of the portable consumer electronics industry. There are few electronic applications that cannot take advantage of the improvements lithium ion has to offer over other secondary batteries. Cost has become the key factor that could exclude lithium-ion from some low-cost devices. The cost of using lithium-ion batteries comes not only from the cells, but from the need for protection devices and more complex chargers. Notebook computers, cell phones, video and audio recorders and players as well as GPS devices have all become more portable and ubiquitous due to lithium-ion battery technology. Lithium-ion cells are typically manufactured in cylindrical, prismatic, or "pouch" forms.

It is important to note that the term "polymer cells" is often misused to describe "pouch" cells. These are lithium-ion cells that are built into a relatively thin sealed pouch rather than a metallic can. Many of these cells use a polymerized gel electrolyte, thus the term "polymer." However, there are many that use liquid electrolyte just like their cylindrical or prismatic counterparts. These are typically referred to as "starved liquid" electrolyte cells because they have little free electrolyte in the cell. The pouch form factor allows for thinner cells and a wide variety of *x-y* dimensional variability. They tend to have better cycle life than those cells in metallic cans, because the pouch form allows the electrodes to swell some. This swelling, however, is also a design challenge when using pouch cells in battery packs where fixed dimensions are a must.

Chapter 17A covers details of the various lithium-ion chemistries, but the next section will compare some of the key characteristics of the secondary battery chemistries.

### 23.3.1   Comparing Secondary Battery Characteristics

Characteristics of conventional secondary batteries based on theoretical limits are summarized in Chaps. 1 and 10. Various tables also list characteristics based on the actual performance of a practical battery under near optimal conditions for each specific type. The same factors must be noted as in Sec. 23.2 for primary batteries.

**TABLE 23.3**  General Secondary Battery Comparison for Consumer Applications

| Characteristic | NiCd | NiMH | Lithium-ion |
|---|---|---|---|
| Cell voltage | 1.2 | 1.2 | 3.6–3.7 |
| Cycle life at 80% DOD | 1000+ | 500+ | 500–800 |
| Temperature range | −40 to 70°C | −40 to 50°C | −20 to 60°C |
| Memory effect | Yes | Yes | No |
| High-rate discharge | 10C+ | Up to 5C | >5C (typical) |
| Fast charge time | <1 h | 2 h | 1 h |
| Capacity after 1 year storage at 25°C, % | <30 | <20 | >80 |
| Energy density at 10 h discharge rate, Wh/kg | 60 | 95 | 250 |

Table 23.3 compares some key characteristics of the most common secondary battery types. These data are typical of what was commercially available at the time of publication. Exotic versions of these chemistries are not included. For example, there are now some lithium-ion chemistries that offer fast charge at the expense of slightly reduced energy density. Figure 23.3 is a Ragone plot of these common secondary battery chemistries. One can see the huge weight savings that can be achieved by using lithium-ion instead of the nickel chemistries. This weight savings is often critical in how well a consumer electronic product sells in the marketplace.

Further comparisons of some of the key differences between the different lithium-ion chemistries that are available in the market today are presented in Table 23.4. Li/FePO$_4$ has a lower operating voltage, but generally higher power capabilities than the other lithium-ion chemistries. Li/FePO$_4$ also requires less safety circuitry and exhibits a greater resistance to thermal runaway.

The next section will discuss specific criteria for battery selection and begin with the first big question: primary or secondary?

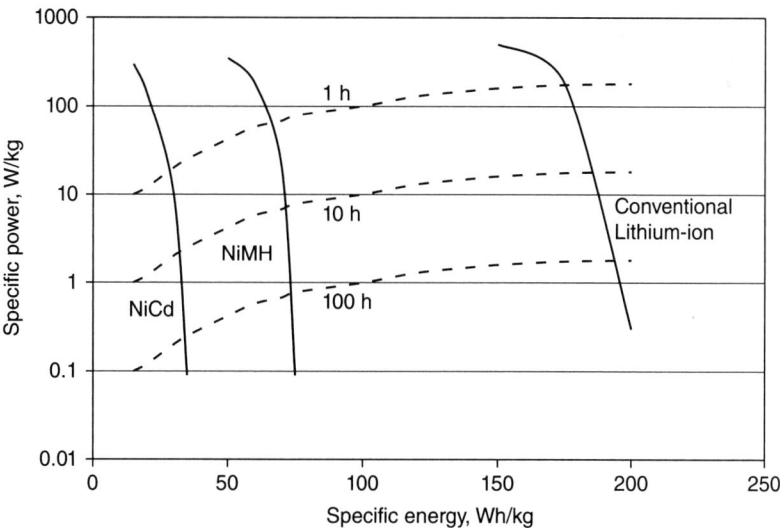

**FIGURE 23.3**  Performance capabilities of secondary batteries—energy versus specific power.

**TABLE 23.4**   General Rechargeable Lithium-Ion Battery Comparisons

| Characteristic | Li/CoO$_2$ | Li/Mn$_2$O$_4$ | Li/(Ni-Co-Mn)O$_2$ | Li/FePO$_4$ |
|---|---|---|---|---|
| Cell voltage | 3.6+ | 3.6 | 3.5–3.8 | 3.2 |
| Cycle life | 400–500 | 400–500 | 400–800 | 1000+ |
| Temperature range, °C | −20 to 60 | −20 to 60 | −20 to 50 | −20 to 60 |
| Discharge rate | Up to 2$C$ | Up to 5$C$ | Up to 1.5$C$ | 10$C$+ |
| Charge rate, h | 2 | 1–2 | 2–3 | <1 |
| Energy density at $C/5$ discharge, Wh/kg | 200 | 150 | 230 | 120 |

## 23.4   PERFORMANCE CRITERIA FOR BATTERY SELECTION

### 23.4.1   Primary versus Secondary

Is a single use or rechargeable battery more appropriate? This appears to be a simple question, but the answer is more complex. Primary batteries are typically used in low to medium power applications, whereas secondary batteries can be used in nearly all consumer electronic applications. The exception is the Li/FeS$_2$ primary, which offers high power capability at increased cost compared to alkaline manganese. Primary batteries for consumer electronics offer the convenience of simple replacement rather than recharge, but the long-term cost of operation can quickly escalate in medium- to high-power devices. This is the principal trade-off when considering which battery is appropriate for a device. The development of standard consumer sizes (AAA, AA, C) maintenance-free rechargeable batteries has allowed electronic designers of low and moderately priced devices to forego this question completely. Since NiCd and NiMH batteries can provide a voltage similar to primary alkaline batteries, the question becomes irrelevant. The only thing a consumer must do is choose between the low initial cost of primary batteries and the higher initial cost of rechargeable solutions, along with a suitable charger. This means that end-user preferences must also be considered. In the end, other factors may come into play that will sway the decision.

### 23.4.2   Voltage Concerns

A critical element in selecting a battery is the operating voltage range required for a specific application. Figure 23.4 is a voltage comparison chart. Many electronic components have a minimum voltage requirement near 3.0 V. The higher operating voltage of the lithium chemistries, both primary and secondary, can simplify

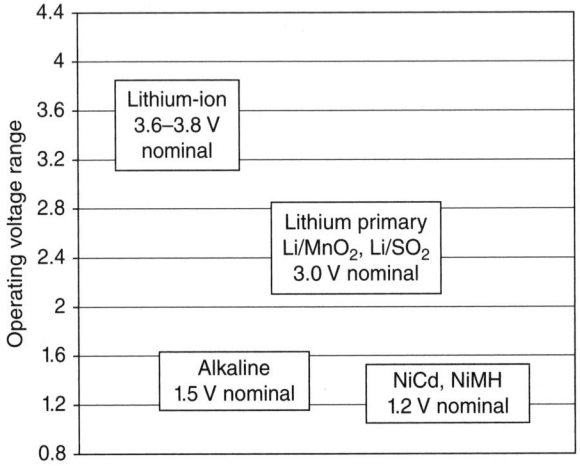

**FIGURE 23.4**   Voltage comparison chart for primary and secondary cells.

battery design by requiring only a single cell, instead of two or three in series. One must also note that in order to utilize the full capacity of NiCd or NiMH in place of alkaline batteries, the device must be capable of operating down closer to 1.0 V instead of 1.2 V for single-cell applications. Basically, the nominal operating voltage for the rechargeable chemistries is at the bottom end of the voltage range for the primary battery. In addition to the *range* of voltage, the actual discharge profile of that voltage is important if the entire functional capacity of the battery is to be used. Figure 23.5 illustrates the flatter discharge curves of conventional secondary batteries (S) compared to most primary cells (P).

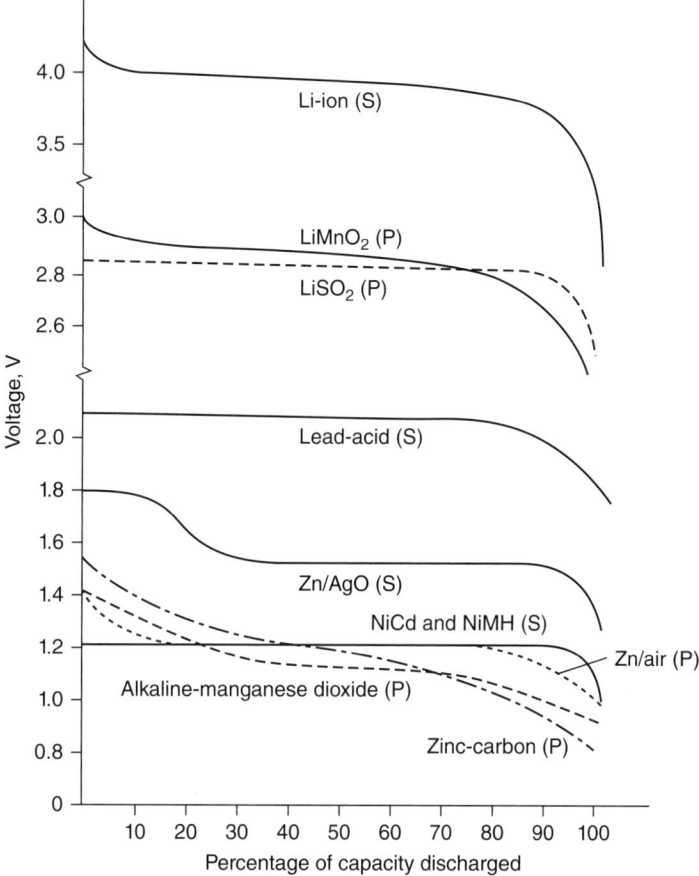

**FIGURE 23.5**   Discharge profiles of primary (P) and secondary (S) batteries.

### 23.4.3   Physical Size

Battery sizes will be classified into four general cell groups for this section: button/coin cells, cylindrical cells, prismatic cells, and pouch cells.

Coin cell part numbers usually describe the cell dimension and chemistry. For example, a BR2032 coin cell is decoded as follows:

- BR = chemistry/manufacturer designation
- 20 = 20 mm diameter
- 32 = thickness/10 = 3.2 mm

Some coin cells come socketed, some with tabs for welding or soldering, and some with wires attached. Check specifications from the manufacturer.

Cylindrical cells tend to follow ANSI guidelines as shown in Table 23.5, but are often available in fractions of the standard (4/3A, 1/2AA, 2/3D). The standard lithium-ion cell is actually a "short fat A" cell or 18650 (18 mm diameter by 65 mm long). The original standard NiMH cells were 17670 cells (17 mm diameter by 67 mm long). Settling on a lithium-ion cell size that would not fit into the same space as NiMH was a conscious decision within the industry to prevent accidental overcharging of lithium-ion cells.

**TABLE 23.5**    Standard Cylindrical Cell Sizes (mm)

| Full size | Diameter | Length |
|---|---|---|
| N | 12 | 30 |
| AAA | 10.5 | 44.5 |
| AA | 14.5 | 50 |
| A | 17 | 50 |
| Af (fat A) | 18 | 67 |
| SC (sub-C) | 23 | 43 |
| C | 25.8 | 50 |
| D | 33 | 61 |

Prismatic cells also come in a wide variety of sizes, and part numbers often designate the actual size in millimeters. For example, an ICP103450 prismatic cell can be decoded as follows:

- ICP = chemistry/manufacturer designation
- 10 = 10 mm nominal thickness
- 34 = 34 mm width
- 50 = 50 mm length

It is important to note that the *nominal* thickness is *not* what should be designed into a product. Prismatic cells will swell over cycle life and at higher temperatures. It is good engineering to allow 10% above the nominal thickness for swelling over the life of the product. Ask the manufacturer for a *maximum* thickness specification and the factors that influence it. This is especially important if cells are to be stacked on top of one another and welded together in the battery pack. Figure 23.6 illustrates how vertical strapping between cells must allow for this swelling. The risk of assembling a battery as shown in the "bad design" is that the nickel strapping can actually pull away from the cell when swelling occurs. The end result can be as simple as a broken connection and a nonfunctional battery pack, or as dangerous as creating short circuits within the battery pack. It is also important to note that most prismatic cells greater than 10 mm in thickness have a built-in safety device such as a thermal fuse or bimetallic switch. Smaller size prismatic cells do not typically incorporate this feature.

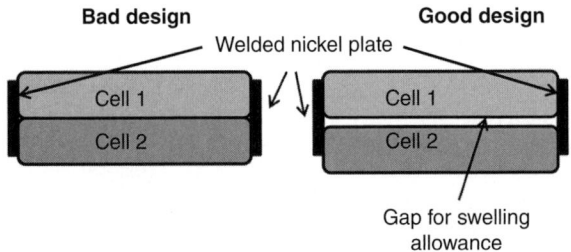

**FIGURE 23.6**    Proper stacking assemblies of prismatic cells.

Pouch batteries also come in a wide variety of sizes. They can be manufactured less than 2 mm thin and typically up to 7 or 8 mm thick. The maximum thickness depends on the manufacturing technique. Pouch batteries can use a wound, folded, or stacked electrode design. This is covered in Chap. 17A. Again, when designing into a product it is important to allow for the maximum thickness after swelling. Early pouch batteries were susceptible to swelling because of insufficient drying, resulting in moisture in the cell. As manufacturing has improved, the main reason for swelling beyond the maximum specification is damage to the cell that allows environmental moisture to intrude. Always check the manufacturer's specification for maximum swelling and the conditions under which this takes place.

### 23.4.4   Capacity

The rated capacity of a battery in milliampere-hours or milliwatt-hours is a significant factor in determining the number, size, and type of cell needed to meet run-time, talk time, or standby time expectations for a particular device. This capacity is directly related to the energy density of the materials used in the cell. Table 23.6 is a comparison of primary and secondary battery chemistries in AA size (14.5 mm × 50 mm). The capacity rating that a manufacturer gives a cell is based on a specific drain rate at room temperature (20 to 25°C). This drain rate is typically $C/5$ for secondary cells and $C/100$ or more for primary cells. It is important to note that very little development work has been done in the AA-size lithium-ion batteries. As a result, the capacity in this size does not reflect the increased energy density over other secondary chemistries.

**TABLE 23.6**   Capacity Comparisons for AA-Size Cells

| Chemistry | Voltage | Capacity (mAh) | Drain rate | Energy, mWh |
|---|---|---|---|---|
| Primary | | | | |
| Alkaline | 1.2 | 2850 | 25 mA | 3420 |
| Li/FeS$_2$ | 1.5 | 3000 | 500 mA | 4500 |
| Li/SO$_2$ | 3.0 | 2450 | 2 mA | 7350 |
| Li/MnO$_2$ | 3.0 | 2000 | 10 mA | 6000 |
| Secondary | | | | |
| NiCd | 1.2 | 700 | 140 mA | 840 |
| NiMH | 1.2 | 2100 | 420 mA | 2520 |
| Lithium-ion | 3.6 | 800 | 160 mA | 2880 |

### 23.4.5   Load Current and Profile

Table 23.6 is based on specific drain rates. Battery chemistries perform differently under different load conditions. This means that a battery that may appear to have better capacity than another according to a specification sheet may actually perform more poorly if the load in the application is much higher.

Figure 23.7 compares the performance of several AA-size primary and rechargeable batteries. The actual capacity drawn from the batteries is shown over a range of discharge currents. Typically, the primary batteries perform better under low-drain conditions, but they lose this advantage as the drain rate increases. Again, it is not completely fair to plot AA lithium-ion, since little work has been done in this size. The current generation of 18650 lithium-ion cells provides 3200 mAh of capacity and has twice the volume of AA cells. Thus, an optimized AA lithium-ion cell should provide a capacity of 1600 mAh.

### 23.4.6   Temperature Requirements

Operating temperature of a portable device is another key factor in selecting a battery. Some consumer electronic applications are designed to be used indoors and have much more conservative thermal demands. Others must be used outside, sometimes in extreme conditions, and this can be problematic for a battery. It was shown in Figs. 9.7 and 9.8 (Chap. 9) how temperature affects the gravimetric and volumetric energy

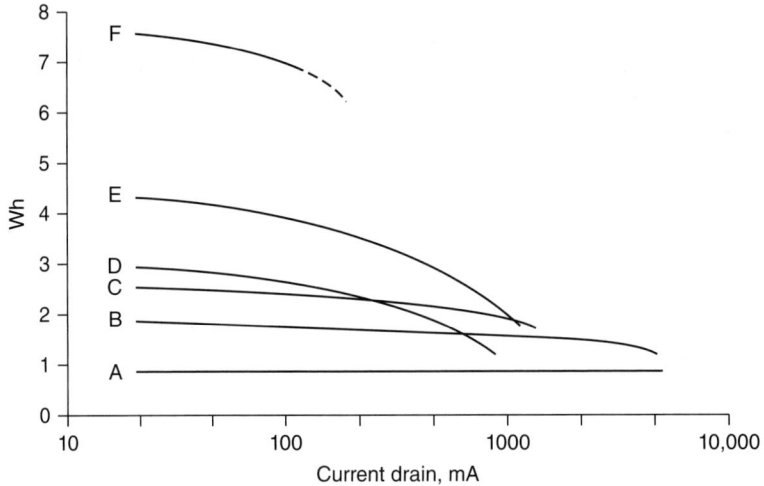

**FIGURE 23.7** Performance of AA-size (or equivalent) batteries at various current drains at 20°C. A: NiCd; B: NiMH; C: lithium-ion; D: zinc-alkaline; E: Li/MnO$_2$ (2/3A size); F: zinc/air (button type). A–C are secondary batteries. D–F are primary batteries.

density of primary cells. The effect of temperature on secondary batteries is much less severe, but the overall operating range of secondary cells is more restricted. This is shown in Fig. 23.8.

An operating characteristic that must also be considered is the effect of temperature on charging the secondary batteries. Both NiCd and NiMH batteries can use the cell temperature to determine the fully charged status of the battery. Operating in a high-ambient environment can mask this termination or cause charging to terminate early.

Lithium-ion cells present a different problem. Typically, they can only be charged at the maximum rate within a relatively narrow temperature window (~20–45°C). Below or above this temperature, much lower currents and/or voltages must be used. This can result in extremely long charge times in certain thermal environments. In fact, the self-heating during high-rate discharge can elevate the battery temperature of conventional lithium-ion batteries to the point where it must cool down before charging can be initiated. This is evident in the operating specifications of the cell where the "operational temperature" may be specified as −20 to 60°C, but the normal *charging* temperature range may be only −10 to 45°C. At temperatures below 5 or 10°C, a trickle charge may be required to prevent lithium plating. At temperatures above 45°C, the voltage may be limited as well.

### 23.4.7 Shelf Life

How long will a battery sit on a shelf or in a storeroom before it is put into use? The self-discharge rate, or charge retention, of a particular battery determines this limitation. Primary batteries have self-discharge rates that are typically an order of magnitude or more *lower* than secondary batteries. This allows them to have a shelf life measured in years, rather than months for secondary batteries. An exception are newer NiMH batteries, which boast capacity retention of 70% to 85% after 1 year at room temperature compared to 20% to 30% for conventional NiMH. Temperature also has a profound effect on shelf life, with a general rule being that the rate doubles for every 10°C increase in temperature. This means that the shelf life of *all* batteries can be maximized by keeping them cool, as was seen in previous chapters (Chap. 9 for primary cells and Chap. 10 for secondary cells) and as shown in Fig. 23.9.

### 23.4.8 Charging

The charging of secondary batteries varies considerably from simple float charging over days to rapid charging in less than an hour. Consumer electronic applications typically require something in between, although there

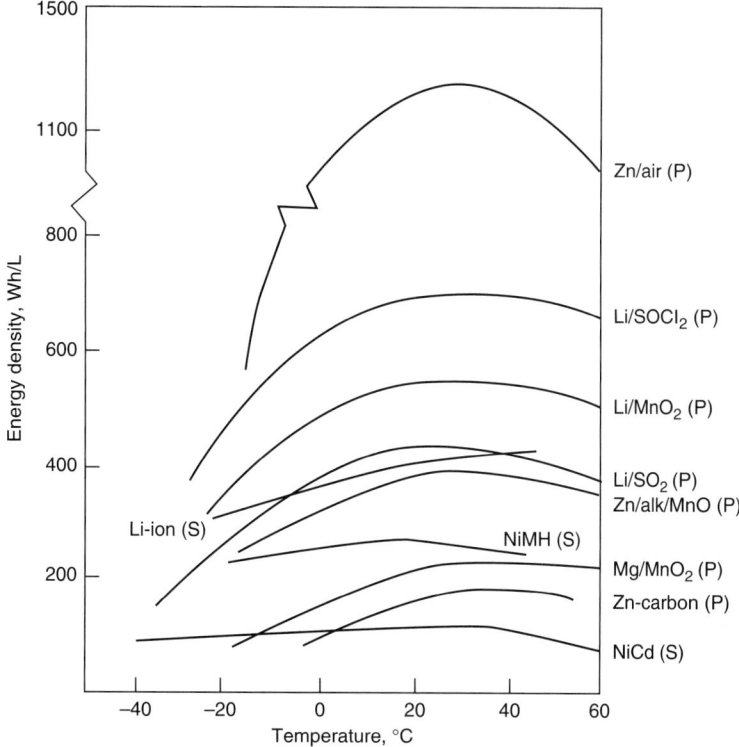

**FIGURE 23.8**    Effect of temperature on primary (P) and secondary (S) batteries.

are some rapid charge applications. The *rate* of charge can have a direct impact on cycle life of the battery. Higher charge rates typically mean lower charge efficiency and elevated temperatures, resulting in accelerated capacity loss with lithium-ion batteries. With NiCd and NiMH, high charge rates can cause premature termination of charge and a resulting loss of capacity. However, there are secondary cells that are specifically designed for high-power applications that can tolerate higher rates without loss of performance. The trade-off is that these cells typically have much lower energy densities.

Figure 23.10 illustrates the effect of charge rate on the capacity of typical lithium-ion cells charged with a conventional constant current-constant voltage (CC-CV) method.

This effect of charge rate on cycle life can be exacerbated by increased or decreased temperature, depending on what additives may be used in the electrolyte. It is critical to follow the cell manufacturers' recommendations for charging as closely as possible.

## 23.5   *SAFETY AND REGULATORY CONCERNS*

Batteries are energy storage devices that contain relatively large amounts of energy in a small space. If mishandled, abused, or used outside of certain environmental specifications, they can release this energy in a short period of time. Rechargeable batteries pose an additional risk in that they are connected to an essentially unlimited power source when charging. Using rechargeable batteries requires multiple levels of safety to protect against misuse by the consumer.

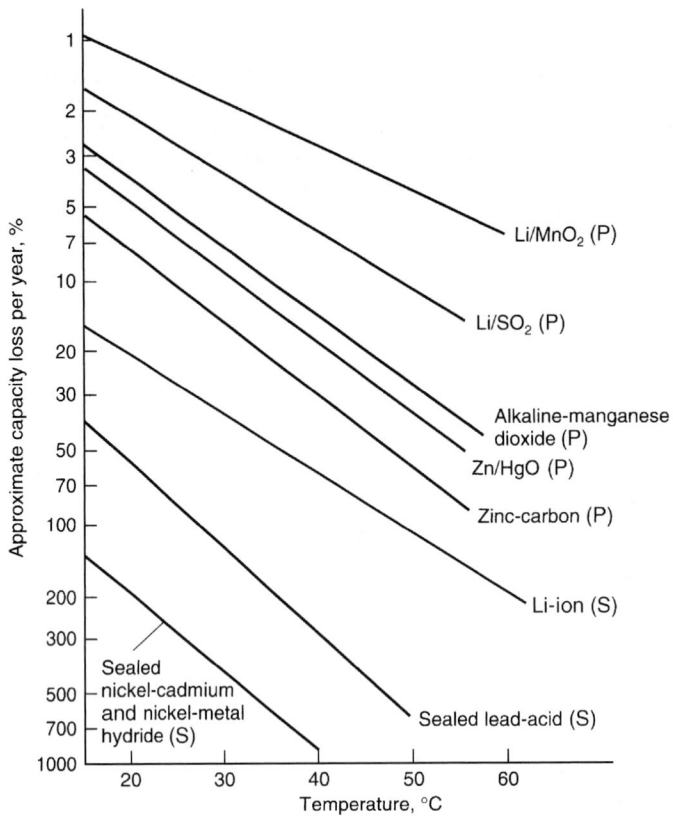

**FIGURE 23.9** Shelf life (charge retention) characteristic of a variety of batteries—primary (P) and secondary (S).

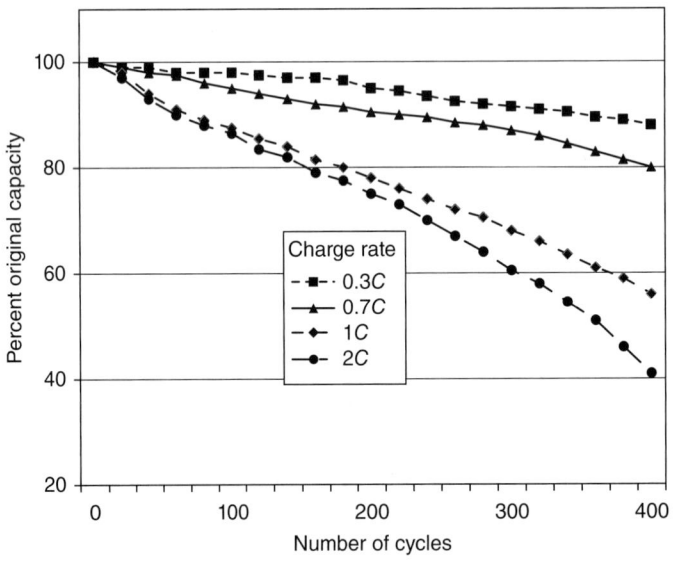

**FIGURE 23.10** Effect of charge rate on typical lithium-ion battery capacity during cycling.

The first critical layer is the charger. The charger should have protection to avoid excessive current and to disable charging when certain environmental conditions exist (e.g., high temperature).

The second layer is the battery pack itself. Good battery pack design protects terminals from short circuit and protects the cells from damage if a removed battery is dropped. The electronics in a battery pack protect against common conditions that can cause harm to the cells in the battery pack. Typical protection devices include PTCs (positive temperature coefficient current limiter), thermostats, chemical fuses, and dedicated safety integrated circuits. These protect against conditions such as overvoltage, undervoltage, over temperature, high current, and more (see Chaps. 7 and 31 for further details).

The final layer of protection is the cell itself. "Self-quenching" separator material has been used to improve cell safety in the event of overheating. In addition, a reliable vent mechanism to relieve excessive internal pressures is essential in batteries built into metal cans. If a thermal event were to occur without these vents, pressure inside the cells could rise to a level that results in splitting of the can and ejection of contents. Some cells also have internal current interrupt devices or thermal fuses to permanently disable the cell.

It is good practice to perform a failure modes and effects analysis (FMEA) prior to designing the battery (also see Chap. 32B). This type of study examines potential failure modes and ranks them based upon the severity of the effect of the failure. Potential catastrophic failures can be mitigated through independent, redundant safety mechanisms that address the potential fault conditions. An FMEA should ideally be done on the cells themselves, the battery pack, and the entire battery pack/host device combination. This concept is addressed in IEEE 1625 and 1725 documentation.

Regulatory requirements must also be considered, particularly for a product that will be shipped between countries. There are transportation regulations that define packaging, labeling, and documentation, and these vary from country to country and are constantly changing. Certain markings are also required on the batteries from government agencies. Again, these vary from country to country. Additionally, there are environmental regulations that govern recycling or disposal of the batteries. Planning for end of life is essential in the consumer electronics industry (see Chap. 8).

## 23.6   ECONOMIC/PRODUCIBLE CRITERIA

Cost effectiveness of primary and secondary battery solutions versus life cycle analysis can be used to evaluate which is more appropriate for a particular application. Table 23.7 is an example of such an analysis, comparing zinc-alkaline primary batteries to rechargeable NiMH batteries for a typical low-power, portable electronic device application. A low usage rate makes the primary battery solution much more cost effective in addition to

**TABLE 23.7**   Cost Effectiveness of Primary (Zinc/Alkaline/$MnO_2$) versus Secondary (Nickel-Metal Hydride) Batteries

Assumptions
Nominal voltage: 3 V
Drain rate: 150 mA
2 AA Zn/$MnO_2$ batteries (2500 mAh): $0.64
2 AA NiMH batteries (2300 mAh): $6.20
NiMH charger: $4.00

| Usage (h/day) | Charge NiMH (days) | Change Zn/$MnO_2$ (days) | NiMH battery payback (days) |
|---|---|---|---|
| 0.5 | 30.7 | 33.3 | 530 |
| 1.0 | 15.3 | 16.7 | 266 |
| 2.0 | 7.7 | 8.3 | 133 |
| 4.0 | 3.8 | 4.2 | 66 |
| 6.0 | 2.6 | 2.8 | 44 |
| 8.0 | 1.9 | 2.1 | 33 |

offering the convenience of not having to be recharged periodically. The payback computation is based on little degradation in capacity of the NiMH batteries over the time in question. Another factor to keep in mind is that NiMH capacity is based on discharge down to 1.0 V, so the device must be able to operate from 3 V down to 2 V with the 2 AA cells. Otherwise, the effective capacity of the NiMH solution is reduced and the payback time increased.

Another example is shown in Table 23.8, where power demands are higher. A 1-A load is typical of some digital video cameras and motorized remote-controlled toys. In this example, a high powered primary lithium battery is compared to the rechargeable NiMH battery. Conventional alkaline primary cells simply do not perform well at these higher loads. It is clear that even with moderate to low usage, the payback time for the rechargeable solution is rather short. This is due principally to the higher cost of the more advanced lithium primary batteries.

**TABLE 23.8**    Cost Effectiveness of Primary (Lithium Iron Disulfide) versus Secondary (Nickel-Metal Hydride) Batteries

Assumptions
Nominal voltage: 3 V
Drain rate: 1000 mA
2 AA Li/FeS$_2$ batteries (3000 mAh): $4.10
2 AA NiMH batteries (2300 mAh): $6.20
NiMH charger: $4.00

| Usage (h/week) | Charge NiMH (days) | Change Li/FeS$_2$ (days) | NiMH battery payback (days) |
|---|---|---|---|
| 1.0 | 14 | 21 | 53 |
| 3.0 | 4.7 | 7 | 17.5 |
| 7.0 | 2 | 3 | 7.5 |
| 14.0 | 1 | 1.5 | 4 |
| 21.0 | 0.7 | 1 | 2.5 |

If a decision is made to use a rechargeable battery solution, the cost of a lithium-ion solution should be considered where weight and size are critical. In general, cylindrical lithium-ion batteries have a lower dollar per watt-hour cost than prismatic or pouch batteries. This is due to the high level of automation and the huge volumes that the key manufacturers are producing. A typical lower capacity (2200 mAh) 18650 cell can cost as little as $0.20/Wh, with premium cells (3200 mAh) commanding as much as twice that amount. Cost varies widely between vendors, with Japanese and Korean sources being more expensive than Chinese manufacturers. Of course, one must also consider the cost of quality. Vendors should be thoroughly audited, and it is wise to select a battery with multiple sourcing to ensure good supply and competitive pricing.

Lithium-ion pouch batteries can cost 1.2 to 1.5 times more than an equivalent capacity cylindrical cell. The manufacturing technique for pouch batteries is slower and less cost effective. This, combined with somewhat lower factory yields, has kept lithium-ion pouch batteries more expensive than other form factors.

Lithium-ion batteries require more sophisticated charging circuits and protection circuits compared to NiCd or NiMH. A complete protection circuit for a single-cell lithium-ion application can cost $1.25, an expense not needed for NiCd or NiMH. In addition, a lithium-ion charger can cost twice that of a NiCd/NiMH charger. Although "universal" lithium-ion chargers are available, it is typically good practice to design a charger for the specific application. This usually ends up being more cost effective as the flexibility in voltage and current that a "universal" charger incorporates is eliminated.

To estimate the total cost of ownership, one must consider much more than the cost of the battery itself. Protection circuitry, charge circuitry, regulatory certification, disposal cost and responsibility, and shipping costs all need to be factored into the total cost. One additional consideration is any special assembly techniques that may add cost to the battery in the factory.

## 23.7 OVERALL SELECTION CRITERIA

Selecting a battery solution for a consumer electronics application involves narrowing down the available options, then weighing critical criteria to come up with the best solution. Keep in mind, there may not be one best solution. Consider the following five things to help reduce the available number of options: function/application, performance, cost, integrity, and safety. Questions to be asked when considering these factors will be covered next.

### 23.7.1 Narrowing Choices

A particular application may make it immediately clear whether recharge ability is required. Once this initial question is answered, many other application-dependent answers should be gathered.

- Primary or secondary?
- Will a charger be required in the device or as a stand-alone option?
- Resting state-of-charge can be an important factor. Will the device be required to power up at the time of purchase, or will charging be necessary?
- How will the battery be packaged to be compatible with the application?
- What overall size is acceptable, and how will the battery connect to the device?
- Will the battery be user-replaceable or embedded in the device?
- What type of protective circuitry (if any) will be required?

Once a basic size and shape and packaging are decided upon, the performance characteristics must be considered. This is where many of the factors discussed in Sec. 23.4 need to be sorted through to eliminate incompatible chemistries. The following performance characteristics should be included in this process. Further comparisons and trade-offs will be discussed later in this section.

- Energy density
- Voltage range, depth of discharge, shape of curve
- Discharge rates, continuous, pulse
- Temperature, both operating and storage requirements
- Reliability, environmental factors
- Self-discharge, shelf life
- Recharge ability, cycle life, capacity fade, charge rate, trickle charge, protection

Naturally, cost is a key consideration for consumer products, but some devices are less sensitive than others to the cost of the battery. The battery solution may be 10% of the total device cost of a notebook computer, but as much as 50% of the cost for an MP3 player. Cost-associated factors include the following:

- Initial cost of battery
  - Compared to other components
  - Budgetary constraints
- Operating cost of battery—cost per cycle (primary vs. secondary)
- Cost implications for host device; protective devices, charger, special assembly
- Disposal costs and responsibility

Integrity is a broad-ranging topic that is often not considered thoroughly enough. This incorporates topics relating to design, marketing, and manufacturing. The following key points should be considered:

- Reputation: cell vendor, battery pack assembler
- Sourcing
  - Do volumes require multiple sources?
  - Is there a competing alternative technology available (fuel cells, etc.)?
- Perceptions
  - How will the customer perceive the product?
  - How will the product compete in the marketplace?
- Government/regulatory
  - Safety
  - Transportation: strict guidelines are in place for lithium-based products
  - Recycling requirements

The final narrowing factor to be considered, but certainly not the least important, is safety. Safety concerns are not simply whether or not special circuits are required. They also encompass *how* and *where* the device and battery will be used. Anticipating what a consumer will actually do with a device is critical. Will it be left in a hot car? Will it be dropped a lot? Will it be exposed to high humidity or submerged? These are conditions of misuse that must be considered during design. There are also abusive conditions that are difficult to design for, but risk can be mitigated by considering potential abuse. These questions lead to the following safety considerations:

- Mechanical misuse/abuse: what level of shock/vibration tolerance is needed?
- Thermal misuse/abuse: temperature extremes during operation, storage, or charging
- Electrical hazards: battery pack or cell shorting risks
- Disposal/environmental
- Toxicity: special dangers and precautions

### 23.7.2   Performance Criteria Trade-Offs

All performance characteristics cannot be equally important for a specific application. Capacity becomes a critical parameter in consumer electronic applications where run-time of the device is used as a marketing tool. What can be sacrificed for higher battery capacity? Higher capacity may be achieved by simply using a larger battery. Are increased size and/or weight acceptable for the device in question? Other factors that can be reduced to increase capacity are things like cycle life, charge rate, and high- or low-temperature performance. In *all* cases, safety should *not* be sacrificed.

It is often useful to role-play and think like a consumer for a particular portable device. Table 23.9 lists some common applications along with characteristics to be considered. For each application, the relative importance or

**TABLE 23.9**   Relative Importance of Characteristics for Specific Applications

| Application characteristic | Notebook computer | Cell phone | MP3 player | Memory backup |
|---|---|---|---|---|
| Min. operating voltage | 6V | 3 V | 3 V | 3 V |
| Max. current drain | 3–4 A | 800 mA | 60 mA | 100 µA |
| Run-time | 2–3 h | 2–4 h talk, 24 h+ standby | 6–8 h | 24 h |
| Weight | Somewhat important | Very important | Very important | Not important |
| Size | Very important | Very important | Very important | Not important |
| Cost | Somewhat important | Somewhat important | Very important | Not important |

value of the characteristic is given. It is interesting to note that for a low current drain application, where the battery may be quite small from a capacity standpoint, the size, weight, and cost are of little importance because the battery is such a minor part of the device.

This can be taken even further within a particular application, such as notebook computers. These devices come in a wide range of sizes and target many different types of users. These applications range from 17-in displays that function as desktop computer replacements to 10-in screens that function as web browsers (netbooks). Table 23.10 illustrates the relative importance of some key battery characteristics for several classes of portable computers.

**TABLE 23.10**    Battery Criteria for Portable Computers

|  | Display size | Cost | Power | Run-time | Weight/size |
|---|---|---|---|---|---|
| Portable workstation | 15–17 in |  | +++ |  |  |
| Desktop replacement | 15–20 in | ++ | ++ | + |  |
| Mainstream | 14–16 in | +++ | + | ++ | + |
| Thin/light | 13–14 in |  | ++ | +++ | ++ |
| Ultraportable | 10–12 in |  |  | +++ | +++ |
| Netbook | <10 in | +++ |  | ++ | +++ |

Battery selection strategies can be summarized as follows:

- Workstations typically have high-powered graphics and CPU/chipsets. They also sell for a premium price and are rarely unplugged from AC power, thus power is king.

- Cost-sensitive programs (mainstream, netbook, desktop replacement) should use the lowest capacity that meets the minimum run-time requirements.

- More mobile devices (thin/light, ultraportable, netbooks) are extremely weight and size sensitive. Performance can also be important, so multiple batteries may be required to reach multiple market segments.

- Netbooks also need to be low cost. This may require a sacrifice in size or weight. The size question can also become an industrial design question. The thinness of a device may be a marketing point. If thinness is more important than cost, perhaps a pouch battery solution is needed. These are all examples of the trade-offs to consider when selecting a battery.

## 23.8    *RISK REDUCTION CRITERIA*

After considering device requirements, examining different battery choices, analyzing battery characteristics, and weighing relative importance of a variety of criteria, a battery solution may then be designed. But has anything been overlooked? This final section lists some common pitfalls to avoid.

- Do not neglect shelf life. Some batteries, such as lithium-ion, may require periodic recharging to avoid irrecoverable capacity loss. If a product runs the risk of sitting in a store or inventory pipeline for several months, consider this parameter. Also consider *where* the batteries will be stored, such as a hot warehouse, and implement environmental controls if the chemistry requires it.

- Leave enough room for the battery.
  - All batteries have tolerances in their size specifications. Designing a battery to use 18.1 mm diameter cells may eliminate several potential sources with maximum diameters of 18.3 or 18.5 mm.
  - Prismatic and pouch batteries also have swelling specifications for which an allowance must be made.

- Do not disregard temperature. Batteries have a limited operating temperature range. Using a battery beyond the specified range will result in decreased performance and possible loss of capacity over time. This can also result in safety-related issues.

- Consider choices for sourcing. Many batteries come in standard sizes from several manufacturers. If volume deliveries are critical, qualifying multiple sources may be necessary. Once locked in to a unique size, you are at the mercy of the supplier when it comes to pricing and delivery.

- Do not overlook the device power profile. Consider the entire discharge voltage profile of the battery under the expected load to be certain that run-time requirements can be met.

- Whenever possible, use actual performance data to make decisions. Battery specifications are never exact, and cells are typically rated with nominal and minimum capacities. Be certain to know the specifications.

- Do not neglect cycle life. Different battery chemistries age differently. Trade-offs between cycle life and cell capacity are commonly needed. *Always* expect shorter run-times as the battery ages.

- Fully understand the effects of charging on capacity and cycle life. A particular device may need faster charging, but this usually results in more rapid degradation of capacity and shortened cycle life.

- Do not discount parasitic drain. Some devices have a volatile memory that continues to draw power even when the device is off. Many "smart" batteries require communication with the host device, resulting in some parasitic power drain on the battery itself. Some safety circuits have small drain rates. Be careful!

- Consider unusual battery behavior. Battery voltage drops as the load increases or when the temperature is low. A sudden surge in the power demand, especially in low temperature, can result in a voltage dip below a device's cutoff voltage.

A final thought: *Design the battery into the system as early as possible!*

## REFERENCE

1. CTA: Press Releases, www.cta.tech.

# CHAPTER 24
# PERSONAL POWER EQUIPMENT

## 24.0   OVERVIEW

Batteries have been used to power various types of personal, portable electrical equipment for over a century. Designed for applications that enhance personal comfort and provide added utility, these devices include various consumer electronics, as detailed in Chap. 23, but also cover a wide range of equipment intended to replace gasoline-powered engines (ICE's [internal combustion engines]), manual labor, corded (plug-in/AC power) appliances, or mechanically powered (i.e., spring wound) systems. The battery requirements for this type of power equipment are similar in many ways to those required for electronics except that the range of use conditions is broader and more complex. Environmental conditions (temperature, water/moisture, etc.), abuse factors (impact, crush, chemicals, etc.), and energy/power demands are much more extreme than required for a typical cell phone or tablet.

The origin of battery-powered portable devices can probably be best traced to 1915 with the approval of the Edison Flameless Electric Miners' Cap Lamp manufactured by Mine Safety Appliances.[a] In the prior decade, thousands of miners were killed in underground explosions and fires caused by head lanterns that used flammable fuels. Figure 24.1 shows the modern miner's helmet with attached head lamp.[b]

This application grew out of dire necessity. However, today, the assortment of power-assisted products, intended for use by individuals, is only limited by the imagination of product designers and the financial backing of the manufacturers of battery-powered equipment. Items such as battery-powered work tools, flashlights, and toys have been well exploited in the past. But now there are many new twists and other new devices finding their way to market. Such uses include electronic fishing lures, lost key finders, personal heating, ventilation and air conditioning (HVAC) equipment (fans/heaters/coolers/air purifiers), electronic cigarettes, drones, food slicer/dicer/choppers, and thousands of other practical and novelty items.

However, the biggest category of small portable battery-powered equipment is, perhaps, the lawn, home, and garden power tools. A review of several consumer/residential applications is presented in Chap. 25A, Power Tools, and Chap. 25B, Rechargeable Flashlights. Additionally, though, various subsequent chapters are included that survey other more expensive, complex battery applications. Chapter 25, for example, details advanced portable/transportable/mobile equipment (i.e., autonomous robots, aerospace/space craft, underwater vehicles, special purpose electromechanical equipment, industrial tools, etc.), which require more sophisticated and often larger/more powerful battery systems than presented in this chapter. And finally, Chaps. 26 through 29 are dedicated to specialized uses ranging from electric vehicles to implantable medical devices.

---

[a]http://americanhistory.si.edu/collections/object-groups/mining-lights-and-hats (extracted March 13, 2018).
[b]Miner helmets with head lamps, No. 10 shaft, 7000 ft. depth (~2100 m), Resolution Copper mine at Superior, AZ, United States. (Photograph courtesy of Kirby W. Beard.)

**FIGURE 24.1**   Miner's helmet with battery head lamp.

# SECTION A

# POWER TOOLS

## Lisa Michelle King, Rouse Roby Bailey

## 24A.1  INTRODUCTION

The history of cordless power tools began with the world's first cordless electric drill introduced in 1961 by Black & Decker. The drill was powered by four integral ½ D nickel-cadmium cells and produced 36 W of output power. A year later, outdoor lawn and garden equipment (a hedge trimmer with a removable battery pack) and four cordless industrial drills were introduced.

During the 1960s and 1970s, Ni-Cd chemistry established itself as the power source of choice for power tools. The concept of removable battery packs, which could be shared across a variety of products, was also introduced. This accelerated the adoption of cordless power tools in the marketplace. As Ni-Cd technology advanced, pack operating voltages climbed and finally settled in at 18 V in the mid-1990s and became the de facto industry standard for handheld professional cordless tools. This voltage struck a balance between power and ergonomics.

Delivered power was largely limited by the impedance of the Ni-Cd cells. At 18 V, system maximum output power was limited to 400 to 500 W and was largely self-limiting, not requiring any special electronic control. The introduction of power Li-ion cells in the mid-2000s changed the landscape dramatically. The electronic content of cordless tools proliferated with Li-ion because the lower impedance and increased power density also added to safety concerns of the system for overcurrent, undervoltage, and overtemperature control. Output power has doubled with the efficiencies gained through electronic control, while ergonomics has also greatly improved due to the increase in energy and power density in Li-ion (see Fig. 24A.1).

Since the early days, the breadth of cordless power tool products available to consumers has exploded and has grown into a $15B global market by 2016. Steady growth of this market is 5% annually, mainly due to the need for power tools in construction and development taking place around the world. Ergonomic and power improvements in lithium-ion over aqueous chemistries, such as nickel-cadmium or lead acid, have also improved the portability of outdoor and home products that have been traditionally petroleum- or AC-powered.

In this chapter, we will discuss the families of cordless portable consumer tools that include power tools, garden tools, and vacuum cleaners. Each category has a diverse set of consumers: from professional industrial and construction trade users, who depend on these tools daily for their livelihood, to the enthusiast who prefers high-quality tools for moderate regular hobby use, to the homeowner who infrequently performs light-duty tasks.

## 24A.2  POWER CELL DESIGN

Handheld consumer tool applications that require continuous current drains between 3 $C$- and 15 $C$-rate are considered "power" applications. As of 2015, many of these products are based on lithium-ion chemistry due to the higher specific energy and power, and longer shelf-life, versus nickel-based alkaline rechargeable products. Other requirements for batteries in this consumer category include wide temperature operating range, high charge rate, inherent safety design features, good shelf life and storage characteristics, and moderate cycle life. Pack ergonomics play an important role in the portability of the tools.

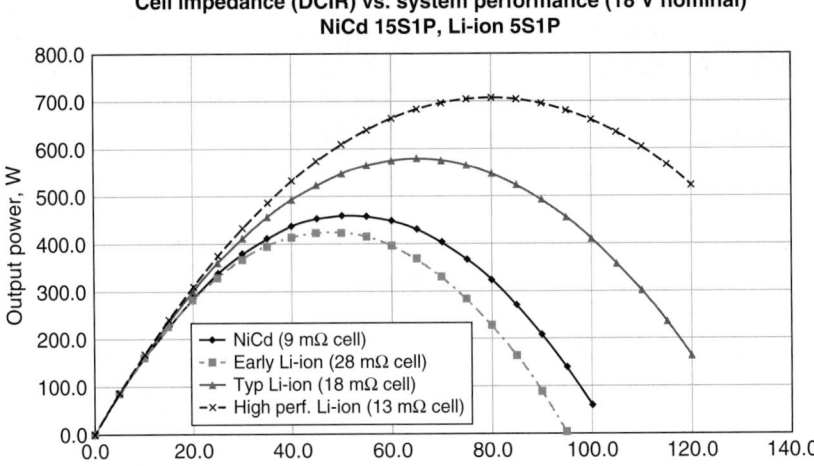

**FIGURE 24A.1**    Improvement of 18 V pack performance: nickel-cadmium versus lithium-ion incarnations.

When portable power tools were introduced into the market, the common battery chemistry was nickel-cadmium. Common nickel-cadmium pack configurations were 12 and 18 V, containing 10 and 15 cells, respectively. Typical cell capacities varied from 1.1 to 2.4 Ah, with the heavier duty products having more capacity or run time, and with it, more weight. Nickel-cadmium tools make up about one-quarter of the total global tool batteries today, decreasing every year.

Lithium-ion cells, when introduced in the late 2000s, provided modest gains in pack volumetric energy density and specific energy. The desire from the tool industry was not to change the power ratings for the applications very much, so the 12 and 18 V preferences remained. Cell capacities that started around 1.1 Ah with the first lithium-iron phosphate packs have now risen to as high as 4 Ah with the introduction of higher nickel-content lithium-metal oxide power cell chemistries such as NCM (nickel cobalt manganese) and NCA (nickel cobalt aluminum).

To date, cylindrically wound cell designs have been adopted for power applications because of the ability to use long, continuous, high surface area electrodes, which yields a low internal resistance and a low cost of mass production. In the future, it is expected that more flexible formats, such as pouch cells, may be introduced in this category due to the push for better ergonomics, higher power, and higher energy content. Until the cost per watt-hour becomes competitive with cylindrical, however, the focus will remain on cylindrical.

Nickel-cadmium power cell designs consist of either paste-paste or sinter-paste electrodes, utilizing stamped foil and expanded metal grid. Connection to the bared edge of the positive electrode is made via many resistance welds to a stamped current collector, which sits atop the jellyroll and is welded directly to the cap. The negative connection to the can is made by a single resistance weld in the bottom. The separators are nonwoven polyolefin type, providing a tortuous path to hinder growth of dendrites. Cells are fitted with a resealing safety vent to release pressure in the event of an abuse condition.

The lithium-ion power cell's general construction is similar to high-energy designs such as those used for personal computing/information technology (PC/IT). However, the power cell's electrode coatings, foils, and separator are much thinner with higher porosity, and the energy ratio of negative-positive plates is lower, which optimizes mass transfer and maximizes capacity. Different types of additives are used to improve conductivity, and multiple tabs are attached to the power cell's positive and negative electrodes, further lowering cell internal resistance.

The most significant difference between the lithium PC/IT-type designs and the power cell designs is the safety devices internal to the cell. Both types of cells contain shutdown separators and safety vents, but whereas power cells are capable of currents an order of magnitude greater than energy cells, power cells also incorporate fusible positive tabs and current interrupt devices (CIDs). Positive temperature coefficient (PTC) devices are not used in power cells: the potential abusive high current and temperature conditions are typically close enough to the thermal runaway activation temperature to warrant permanent disabling of the cell by activation of the CID or fuse. Internal protective tapes on the electrode tabs and gasket seal polymers may also be of a higher melting point, compared to those in PC/IT-type cells, to ensure that the protections remain intact during high-current use.

## 24A.3    PACK AND SYSTEM DESIGN AND SAFETY

### 24A.3.1    Packs

Removable packs manufactured for consumer power applications are common in their primary design elements: thermoplastic nonconductive housing, secure latching, terminals for connecting to external system tools and chargers, metal interconnections between cells, battery management system (BMS) board, thermistor, and fusing.

Pack housings are commonly made from filled, chemical-resistant, fire-resistant, and impact-resistant polymer blends. Materials must have good thermal cycling characteristics, since power tool packs can get quite hot in operation as well as be used in cold climate. Housings may or may not incorporate ventilation.

Many different materials are used for the interconnections, depending on product price point, the designed resistance, and chemical compatibility requirements. Constraint of the cells and interconnection weldments are designed for shock and vibration considerations. Pack fusing physically prevents extreme overcurrent conditions, such as short-circuit, and can be achieved by either fusible link, or electrical fuse, or both.

The BMS for a power tool pack will have the key safety monitoring software and hardware to alert the system tools or chargers to unsafe conditions such as overcurrent, overvoltage or undervoltage, and over- and under-temperature. Packs contain thermistors to provide signal to the BMS for temperature control. Boards are conformal coated to protect electronics from contamination such as dust and moisture.

### 24A.3.2    Tools

System tools interact with the battery by receiving signals from the BMS. The tools contain their own controllers and software with application-dependent overcurrent limits and standard overtemperature and undervoltage limits, all of which can shut down tool operation. Typical cutoff currents for power tools are in the 80 to 100 A range, and in lithium-ion systems, undervoltage limits are typically 2.5 V per cell.

### 24A.3.3    Chargers

Li-ion system chargers will have a primary level overvoltage protection chip, which responds to high pack voltage, preventing overcharge, the most severe safety concern for lithium-ion. A charger may also contain an undervoltage protection chip, blocking charge of low voltage, over-discharged packs, which could also be a safety concern due to the potential for extensive copper plating in cells.

A pack BMS will ultimately have an onboard secondary level, cell-by-cell, overvoltage protection chip, which prevents the pack from further charging in the event the charger primary overvoltage control is inoperable.

Chargers will also read the battery thermistor to prevent charging if pack temperature is too low (lithium plating) or if it is too high (gas generation, thermal runaway). Typical power tool chargers limit the minimum temperature of charging to 0°C and many step the current down in the range from 0 to 10°C. The maximum temperature at which most chargers will operate is 60°C. However, many charger designs incorporate a "hot pack delay" whereby the charger goes into standby until pack cools down to a lower temperature set point, at which time it resumes charging, typically 50°C.

## 24A.4    APPLICATIONS

Within the power tool category are many specialized applications, which can be divided into four broader categories: drilling and fastening, cutting and grinding, garden equipment, and vacuum cleaners. Each category demands similar average loads to perform applications, though durations, duty cycles, and special considerations may vary (see Fig. 24A.2 and Table 24A.1).

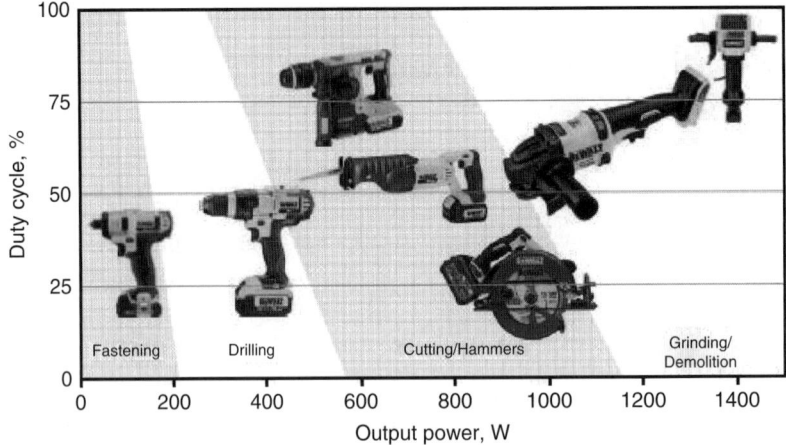

**FIGURE 24A.2**    Average power requirements for cordless drilling, fastening, cutting, and grinding.

**TABLE 24A.1**    Current Loads in Cordless Personal Power Equipment

| Equipment type | Typical current drain | | Duty cycle |
| | Average, A | Peak, A | |
| --- | --- | --- | --- |
| Circular saw | 30 | 60 | Short |
| Miter saw | 30 | 60 | Short |
| Grinder | 20 | 60 | Variable |
| Drill | 10 | 25 | Short |
| String trimmer | 7 | 10 | Variable |
| Push mower | 15 | 25 | Continuous |
| Hand vacuum | 8 | 20 | Short |
| Upright vacuum | 15 | 30 | Continuous |

### 24A.4.1    Drilling and Fastening

The drilling and fastening category includes tools such as screw drivers, impact drivers, drills, impact wrenches, and hammers, as shown in Fig. 24A.3. Most common screw driver tool voltages are 4 and 8 V, and many contain integral or nonremovable batteries. Removable packs for this category of tools are typically in the 12 to 18 V range, with some of the larger hammers for concrete drilling running on 36 or 54 V platforms. Packs kitted with drill drivers are nominally in the range of 1.3 to 2.0 Ah; these small pack sizes can give a user enough run time to apply tens of thousands of fasteners. Currents average 10 A for most applications, with higher current draws in higher torque situations such as drilling hard materials like concrete and knotty wood or for applications requiring large holes such as a plumbing and electrical rough-in work.

**FIGURE 24A.3**   Examples of impact driver, compact drill, and hammer drill.

Ergonomics are critical to this category, with the market trending toward more compact units. Challenging usage considerations for these handheld tools are drop shock and application vibration when in a hammer mode. End users in the automotive repair and plumbing industry may also have concerns around petrochemical and water compatibility of the tool and pack housing. A unique safety concern in drilling is locked rotor, where a bit locks up in the base material and the motor current spikes; tool electronic controls may limit the duration of such excursions by cutting power to protect the battery and the motor.

### 24A.4.2   Cutting and Grinding

End users are increasingly choosing heavier duty DC cordless tools over corded tools for several reasons. Many worksites lack mains AC power, and there is a necessity for mobility onsite. Jobsite safety is improved by removing AC extension cords and the associated trip and fall hazards. Some cutting tools have traditionally run on petroleum and should not be used indoors due to combustion fume dangers. Use of brushless motors has improved performance and motor efficiency through use of electronic control. Higher end packs are now designed with higher energy and higher power levels to address these larger applications.

Saws are the most widely used tool in the cutting category, and portable circular saws are the most common tool among them. Other cordless portable saws available in the market include reciprocating, table, band, chop, cutoff, and miter saws (Fig. 24A.4). The various handheld saws can be used to make short duration, odd and

**FIGURE 24A.4**   Cordless compound miter saw.

random cuts. Materials are varied but the most common ones are typical building materials such as wood, concrete, metal, or plastic.

Grinders are very similar in operation to one another; variations are based on the size of the abrasive wheel. The most common use for grinders is to even out surfaces such as in car body repair work or in welding to flatten a metal weld seam, creating a seamless appearance. Another odd use for grinders is as a pipe cutter.

The use of the grinding tool is very particular to the end user's behavior. If the tool is being used over a larger surface, such as polishing, a light force will be applied to the tool, and the current drawn from the battery will be smaller, in the range of 20 A. When grinding seams, however, an end user may apply their body weight to find the optimal force, or "sweet spot," that quickly removes material.

Common pack voltages in this category tend to be on the higher end, at 18 to 54 V, to deliver more power. Application currents average about 20 to 30 A for longer cuts, with short peaks in the range of 60 to 80 A. Applications such as weld seam grinding can run at very high continuous currents such as 50 to 60 A. Packs kitted with these types of tools can range on the higher end from 3 to 9 Ah. These applications run at a 30 $C$-rate for short periods and 15 $C$-rate for longer periods, which can cause cells to get very hot, affecting cycle life and causing nuisance failures such as opening of the CID or venting. One special consideration for protection of the battery pack concerns prevention of grinding contamination through the use of ventilation screens in packs and debris deflection guards to protect the pack from the debris field. Another consideration for wet cutting applications, such as concrete, is design of a water-tight compartment for the pack.

### 24A.4.3   Garden Equipment

Over the last several years, there has been tremendous growth in the battery-powered outdoor equipment category. The most popular tools in this category are, by far, string and hedge trimmers. Blowers and push mowers have had slower adoption but are expected to pick up over the next several years as more people convert from AC to lithium batteries, with battery power levels reaching those of AC and petroleum-powered tools. Other common tools include chain saws, pruners, and tillers, as shown in Fig. 24A.5. The most common complaint among homeowners using battery-powered equipment is lack of run time to complete their lawn and garden maintenance in one charge.

Many of the tools that were made for use with nickel-cadmium have now converted to Li-ion for ergonomic and ease-of-maintenance reasons. Common pack voltages are 18 V for trimmers, 36 V for larger trimmers and chainsaws, and 54 V for mowers. Currents for many of the outdoor tools average 15 A, some with a peak of 35 A. Pack capacities are in the range of 1.5 to 7.5 Ah. Special considerations for garden equipment include local ambient temperatures, adequate run time (energy) for upkeep of large properties, and contamination protection with packs exposed to the elements.

### 24A.4.4   Vacuum Cleaners

Due to the busy lifestyles of the modern family, a very recent trend in the home products category is the switch from AC uprights and canisters to cordless stick, upright, and robot units. Specialized applications, which are specific to the home, include pet hair removal (requires specially designed beater bar and high suction), HEPA (high efficiency particulate air) filtration for allergens (requires high power to create suction through fine filtration), robotics (requires high battery capacity due to constant run mode and continuous communications with a home base), and wet cleaning (requires high suction power for dispensing and vacuuming cleanser).

Where a handheld vacuum, Figure 24A.6, might get a mid-power cell (about 35 m$\Omega$ DC), the higher power requirement tasks, such as uprights, need high power cells such as those used in professional power tools (about 20 m$\Omega$ DC). Handheld and stick vacuums operate with currents in the range of 8 to 12 A, occasionally with 20 A peak at start up. Some designs incorporate a momentary switch that prevents lock on. Upright vacuum currents range from 15 to 20 A average, 30 A peak at start up. Many vacuum cleaner products have an integral, nonremovable battery as compared to other power tools. Robotic vacuums have an average continuous current of 2 A.

**FIGURE 24A.5**   Common cordless garden handheld tools: tiller, string trimmer, hedge trimmer, chain saw, pruner, and blower.

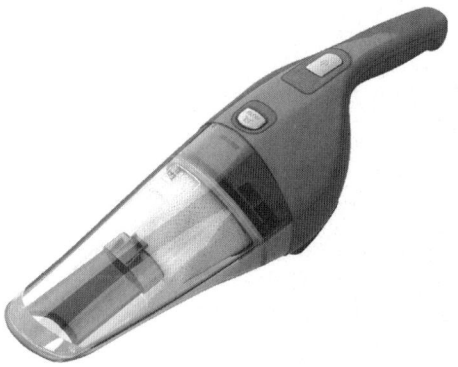

**FIGURE 24A.6**   Handheld vacuum cleaner.

# SECTION B

# RECHARGEABLE FLASHLIGHTS

**Kirby W. Beard**

## 24B.1   INTRODUCTION

The handheld flashlight is an iconic example of the utility of having battery-powered devices for personal convenience. The first patent for a basic handheld battery-powered flashlight followed the invention of the electric light bulb in 1879 (U.S. Patent Application 223898T, Thomas A. Edison) and the sale of the first commercial battery cell (National Carbon Company, 1896). U.S. Patent No. 737,107, the Electric-Circuit Closer, August 26, 1903, by Conrad Huber detailed a design that held cylindrical cells within a tube, typical even today in many flashlights (see Fig. 24B.1).[1,2] The flashlight body provided for a convenient grip and included an on/off switch and a parabolic reflector attached to one end.

The design, type, and scale of portable flashlights proliferated over the next 50 years to include both small, AA cell "penlights" and "lantern" batteries with headlight style bulbs and large 6 V canister type batteries.[3] Fifty years later, many more choices exist: LED strobe lights powered by a single, small lithium ion cell; button cell laser pointers; detachable 18 V lighting accessories included in various power tool kits; etc.

However, beyond having benefit over candles and oil lamps, the advent of handheld electrical lighting marked a milestone for the entire field of advanced consumer electronics, leading to the viability of portable computers, wireless communication devices, portal power tools, etc. While flashlights with primary cells were acceptable for occasional use, the cost, performance, and logistics of using primary cells for heavy or daily service were not acceptable. The more advanced lighting applications (e.g., portable spot lights, commercial photography, video cameras, etc.) required better battery options. The first breakthroughs in powering these more challenging lighting products, having repeated regular, heavy duty cycles, came with the introduction of secondary nickel-cadmium (see Chap. 15) and silver oxide cells (see Chap. 16) by the 1950s.

At that time, heavy duty power equipment relied on the use of removable or auxiliary battery packs that weighed at least 1 kg but often well over 10 kg. It was a major task to carry and recharge the portable battery packs needed for even fairly simple remote power needs. The large early rechargeable batteries were even less practical for powering smaller personal electronic devices. The battery was often far larger than the device being powered. The future of consumer electronics would have been hindered without the advancements that occurred in both batteries and electronics. As an example of this disconnect, the first portable cell phone (Motorola DynaTAC 8000X) weighed nearly 1 kg, had 1 h of talk time, and cost nearly $4000 in 1983.[4] Advancements in both batteries and electronics were needed to support the growth of consumer electronic devices. As shown in Fig. 24B.2, within a decade more modern cell phone designs evolved to allow a more practical cell phone.

Overall, then, the evolution of portable personal power devices has depended on the parallel development of both battery power sources and electronic circuitry. Progress on both fronts has been rapid over the last half century with a proliferation of small, long operational duration devices, such as the Samsung Gear and Apple watches (i.e., a Motorola DynaTAC shrunken down to fit on the wrist and last all day). This progress stems from the work of millions of people developing, manufacturing, and marketing thousands of products. However, what is less well known is the role rechargeable portable lighting devices played in the early development stages of the small secondary battery industry. Using the rechargeable flashlight as a developmental model, the following sections described one phase of this technological progression.

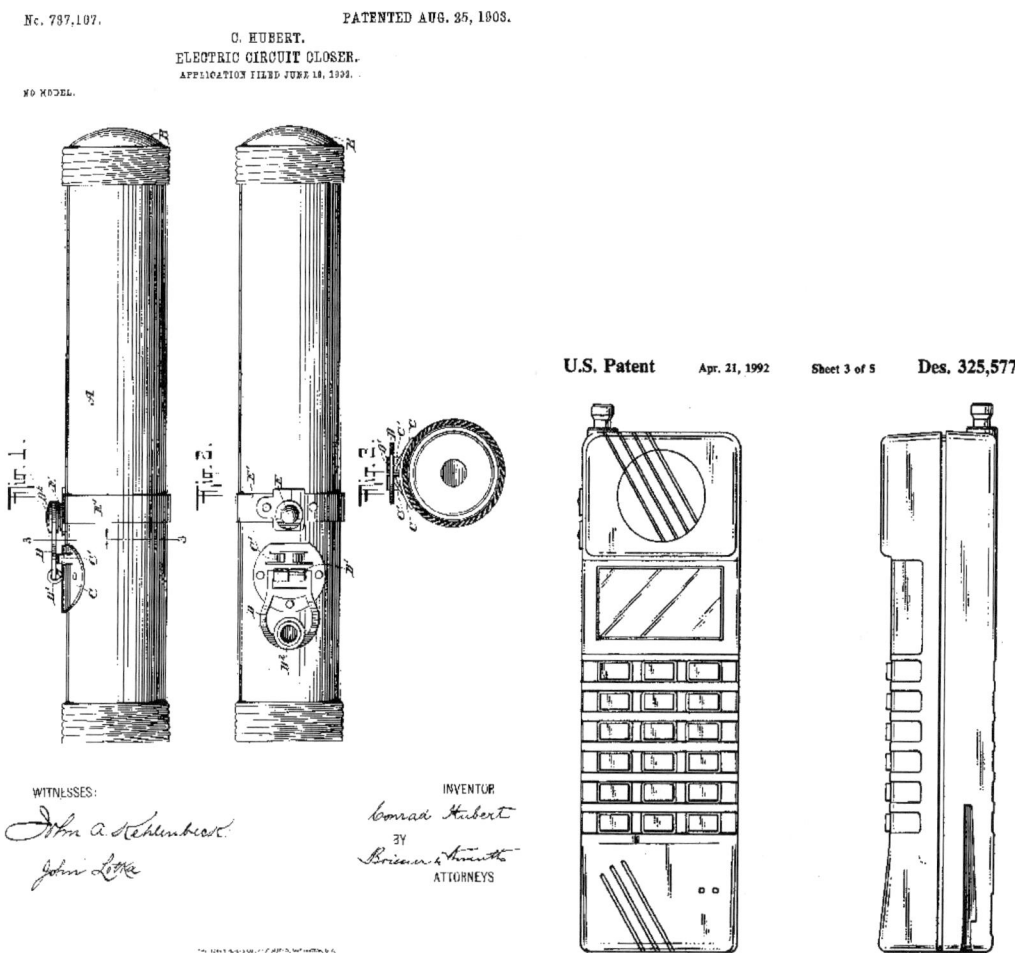

**FIGURE 24B.1** First patented handheld flashlight (U.S. Patent 737,107, Conrad Huber, 1903).

**FIGURE 24B.2** U.S. Patent Des. 325, 577 (Nokia cell phone, 1992).

## 24B.2  THE GATES ENERGY PRODUCT/SEARS RECHARGEABLE FLASHLIGHT

While there are numerous key milestones in the market expansion of personal power products, a rechargeable flashlight produced by Gates Energy Products in 1980 is emblematic of the type of innovation demonstrated in the battery industry. This product was sold exclusively by Sears (the leading retailer of that era) and is considered by today's standards as rather simplistic. The flashlight was, nonetheless, a noteworthy advancement in product design that resulted from a consumer-centric focus. Specifically, the flashlight was based on using a standard Ray-O-Vac® two-cell cylindrical tube design (i.e., a housing to fit two D-size primary cells).[5]

This flashlight was and still is a common style that uses two cells, placed end to end in the cylindrical housing, with a screw-on reflector lamp top piece and bottom cap. The top can be removed to replace the light bulb and the bottom to replace the primary cells.

Gates Energy Products, working with Sear's buyers, designers and engineers, substituted the two primary cells with two equivalent sized components: (1) a single 2 V Gates sealed lead acid (VRLA) D size cell, and (2) a cylindrical charger designed to the same dimensional tolerances as a typical D cell. The Gates cell was the first high voltage, nonvented lead-acid cell on the market. By eliminating the venting of acid fumes, through the use of absorbed glass mat separator and a pressure relief valve, the Gates cell could be used in proximity to electronic circuits without fear of corrosion or leakage (see Chap. 14).

The flashlight used a modified light bulb, applicable to a 2 V load (rather than a 3 V bulb, as required with two carbon or alkaline cells in series). The single Gates cell was permanently sealed into the casing just beneath the reflector compartment and was not replaceable—perhaps, a first of this particular design for a consumer electronic device with a rechargeable cell. The cell had positive and negative terminals on one end at the top of the flashlight to which wire leads were attached. One lead ran to the switch circuit and then both leads were connected to an AC adapter. This charger was a special custom cylindrical design developed specifically for this application. Despite a number of challenges in tailoring the size of the charger to fit in the space normally occupied by a second D cell, the result was a flashlight identical to the standard two-cell flashlight.

This arrangement of the cell and charger provided a design whereby the user simply removed the bottom screw cap, pulled out the charger (with an attached ~0.5 m cord), and plugged the charger blades into a standard 120 V, 60 Hz wall outlet. The single Gates D cell provided 2.5 Ah of capacity, enough to give up to several hours of light even at low temperatures (see Chap. 14). Recharge time for the Gates cell was typically in the 4 to 6 h range to near full capacity, but occasional extended charging was needed to eliminate sulfation. The cell kept charge for nearly 6 months and could provide several hundred full depth of discharge cycles and thousands of shorter duration. Additionally, lead-acid cells could be readily recycled.

## 24B.3  SUMMARY

While the Sears rechargeable flashlight lacked the high level of sophistication present in today's electronic devices, there are a number of parallels with more modern equipment. Issues of run time, recharge time, cycle life, storage losses, temperature capabilities, etc., were all issues in the 1980s for this flashlight, just as they are today with cell phones and other consumer electronics. The design decision to place the charger on-board the Sears flashlight was considered necessary at the time. As a first of its kind electronic device, it was assumed users would not desire having to transport and locate their charger each time the light grew dim.

Other small consumer devices from that era would rarely have a charger built in to the device due to size constraints. The added weight of a typical AC adapter was assumed to be excessive. However, the small size and the novel electronics of the Gates charger changed the parameters. In contrast it would be rare today to find on-board chargers on most portable consumer electronic devices. While there are a variety of wall mounted flashlights (i.e., plugged into and hung from a wall outlet), most of these only serve as a temporary source of lighting in a power outage. The brightness and duration of the light is not typically adequate for utility applications.

The Sears/Gates flashlight serves as a benchmark in the design of portable electronic devices. However, the preferences for an on-board charger and fixed batteries have evolved. In particular, the added weight and volume of an onboard power converter is now considered by most consumers as less desirable than the inconvenience of having detachable cords and chargers that must be stored, carried, located, attached, etc.

As in the past, when developments in both light bulbs and batteries were needed to enable the invention of the portable flashlight, a combination of several new technologies is leading to the creation of the next-generation consumer electronic devices. For example, the use of lower power LED light sources, on-board solar arrays, and wireless charging are trends that will impact future product designs. The key for the battery industry will be to maintain close associations with various stakeholders to achieve necessary breakthroughs. Gates relied on leadership from the customer (Sears), cooperation with a key component supplier (the manufacturer of the Ray-O-Vac plastic flashlight housing), and an advanced electronics equipment supplier (the customized AC adapter) to be successful.

Finally, though, Gates supported a fairly generous warranty program, even with a new battery technology that was largely untested in the consumer mass market. Gates would often replace either the cell or the charger if customers were having issues. Clearly, the future goal for battery manufacturers, OEM electronic suppliers, retailors, and other stakeholders will be to work cooperatively to provide consumer devices with nearly "invisible" power sources: batteries that have more than adequate run time under almost all conditions of use and abuse; recharge in a nearly automatic fashion (e.g., opportunity charging); and do not need replacement batteries before the device becomes obsolete or worn. Finally, the ability to easily reclaim batteries, housing materials (plastics), and electrical circuitry is another goal product designers should keep in mind.

## *REFERENCES*

1. http://www.ideafinder.com/history/inventions/flashlight.html (extracted from the world wide web, March 31, 2018).

2. http://www.flashlightmuseum.com/Eveready-Flashlight-2-Cell-Celluloid-Light-7-Long-with-Blank-Endcap-Similar-to-Eveready-7-3-Cell-2C-1899 (extracted from the world wide web, March 31, 2018).

3. http://www.flashlightmuseum.com (extracted from the world wide web, March 31, 2018).

4. https://www.pcworld.com/article/131450/in_pictures_a_history_of_cell_phones.html (extracted from the world wide web, March 31, 2018).

5. http://www.flashlightmuseum.com/Rayovac-Flashlight-2-Cell-Yellow-Plastic-Light-with-Black-Trim-2D-1975.

# CHAPTER 25
# ADVANCED BATTERY SYSTEMS

## 25.0   OVERVIEW

The two prior chapters provide a review of the two largest battery application categories: consumer electronics (Chap. 23: cell phone, computer, etc.) and small portable personal devices (Chap. 24: power tools, flashlights, toys, etc.). The subsequent chapters provide some details and highlights for a number of other important, specialized battery applications: battery powered transport (Chap. 26: electric vehicles, tow vehicles, and electric aircraft), electric grid storage (Chap. 27), implantable medical devices (Chap. 28), and special reserve cells (Chap. 29). While these chapters cover the main battery-powered product categories, the one final major category of batteries, covered in this current chapter, are the advanced power system applications.

These battery systems include primary and secondary applications in both advanced consumer products and the defense/military/aerospace sector. The first category includes items such as electric motorcycles and all-terrain vehicles (where battery systems may cost over $10,000), and the latter category includes aircraft, spacecraft, advanced autonomous robots, large drones, marine systems (ships, underwater vehicles, remote/tethered equipment, etc.), and other highly engineered, complex, and expensive deployments of batteries and electronics. The battery successes in this field are notable: discovery of the Thresher and Titanic by Ballard[a,b] and the 10+ year mission of the Mars Rover.[c] The failures of advanced battery systems are also well known: the power system failure on the Apollo 13 moon mission[d] and the recent reported battery failure and loss at sea of Argentina's submarine, the San Juan.[e]

Some of these battery projects involve several decades of dedicated research and development. Most of these applications are multimillion dollar efforts requiring a team of people to complete. The system hardware is often complex and uses cutting edge technology that necessitate numerous battery reconfigurations and retests before full deployment. And perhaps of greatest importance, these efforts often serve as the catalyst for developing new electrochemical systems and/or helping to fully qualify known battery technologies.

In Chap. 25A, the latest developments in light electric vehicles (hoverboards, e-bikes, electric motorcycles, etc.) are discussed. Chapter 25B details a variety of aerospace battery applications. Additional extensive information on other advanced battery systems is available from government, aerospace/defense contractor, and private company sources.

---

[a]https://news.nationalgeographic.com/2017/11/titanic-nuclear-submarine-scorpion-thresher-ballard/ (web: May 29, 2018).
[b]http://www.whoi.edu/oceanus/feature/engineers-honored-for-pioneering-undersea-robot (web: May 29, 2018).
[c]https://mars.nasa.gov/resources/21503/new-day-for-longest-working-mars-rover/ (web: May 29, 2018).
[d]https://www.space.com/17250-apollo-13-facts.html (web: May 29, 2018).
[e]http://www.latimes.com/world/mexico-americas/la-fg-argentina-missing-submarine-20171122-story.html (web: May 29, 2018).

# SECTION A

# LIGHTWEIGHT ELECTRIC VEHICLES (RIDING THE WAVE OF TECHNOLOGY)

**Rob Sweney**

## 25A.1    INTRODUCTION: EVs YOU CAN RIDE

Lightweight electric vehicles (LEVs) are characterized with a single verb: you "ride" them. These electric "rideables" with 1 to 4 wheels were previously limited to children's toys, powered wheelchairs, niche commuter tools, garage-built novelties, or overpriced playthings. Early generations were hampered by available battery technology, low production volumes, and unimpressive speed, range, and affordability. Their batteries were expensive and lacked enabling energy and power density. In the past decade, however, a convergence of performance and cost improvements in batteries, processors, sensors, electric motors, and manufacturing[f] has vaulted many LEVs into mainstream adoption as everyday transportation products. E-bicycles are the best-selling electric vehicle in China[1] and are a similarly common sight on European commuter bike paths. E-kickscooters, e-skateboards, e-wheels, and hoverboards pack enough range and power in a hand-carried package that they are now used by urban commuters and serious hobbyists alike. Some LEV campaigns have made international headlines, such as the 2015 hoverboard fad, which ended with a proliferation of uncertified/unregulated, patent-infringing[2] products and multiple safety incidents involving battery fires.[3] E-scooters are besting gasoline vehicle economics in Taipei and Berlin, proving to be the first financially viable application for battery-swap stations.[4] Lastly, e-motorcycles, from the dirt motocross course to city streets and racetracks, are now competing head-to-head with their internal combustion engine (ICE) counterparts. The rapid success of many disparate LEVs can be traced back to their convergence on an enabling battery solution: commodity lithium ion cells, primarily the 18650 format (Sec. 25A.4.2 ). Tables 25A.1a and 25A.1b summarize a range of characteristics of various LEVs.

## 25A.2    LEVs: AS SIMILAR AS THEY ARE DIFFERENT

At first glance the products in this chapter span an incredible range of shapes and sizes, and they might not appear to have a unifying set of battery performance characteristics. Still, these vehicles have many commonalities that lead to similar battery pack solutions that are distinct from the battery designs found in full size battery electric vehicles (BEVs). LEV battery solutions enable nontraditional architectures such as a single wheel (e-wheels) or two side-by-side wheels (hoverboards[g]), low vehicle masses of only 8 kg (e-skateboards); yet they also can pack more range and power than a battery electric car (e-motorcycles). Table 25A.2 groups LEVs into two broad categories based on the size of their battery packs and compares LEV product characteristics to the automotive sector.

---

[f]While dramatic improvements in batteries play an enabling role in this class of products, even greater advancements in price and performance of electronics, including electrically commutated brushless motors, have been key.

[g]The hoverboard is a special subcategory of LEVs that is "self-balancing" for riders. It has two wheels with independent electric motors driving each wheel (see Sec. 25A.5.3 for more on hoverboards).

**TABLE 25A.1a**  Lightweight Electric Vehicles Are as Similar as They Are Different

| Category | E-kickscooter E-wheel Hoverboard | E-skateboard | E-wheelchair | E-bicycle (e-bike) | E-scooter |
|---|---|---|---|---|---|
| Example product(s) | Swagger 5 by Swagtron[a] | Boosted Plus by Boosted Boards[b] | Innuovo Electric Wheelchair[c] | Cortland by Faraday Bicycles[d] | Gogoro 2 Smartscooter by Gogoro[e] |
| Photograph | | | Not pictured | | |
| Human assisted? | No | No | No | Yes | No |
| MSRP, $[f] | $599 | $1399 | $2199 | $3499 | $2700 |
| Vehicle mass, kg | 11.8 | 7.7 | 22.6 | 18 | 122 |
| Tire diam.,[g] cm | 21.6 | 8.5 | 23 | 66 | 48 |
| Range, mi | 18.6 | 12–14 | 12.5 | 15–20 | 60–65 |
| Top speed, mi/h | 18 | 22 | 3 | 20 | 56 |
| Battery mass, kg | 1.5 | 2.3 | 2.8 | 1.45 | 18 (2 × 9) |
| Battery energy, Wh | 222 | 199 | 288 | 306 | 2500 |
| Batt. peak power, W | 740 | ~2100 | ~500 | ~600 | ~7500 |
| Contin. E-rate,[h] W/Wh | ~1.1 | 1–2 | 0.25 | 1–1.5 | 0.5–1 |
| Peak E-rate, W/Wh | 3.33 | 11 | 1.75 | 2 | 3 |
| Rated consumpt, Wh/mi | 12–15 | 14–18 | 23–25 | 15–20 | 40 |

[a]Swagtron.
[b]John Ulmen, Boosted Boards.
[c]https://www.amazon.com/2018-NEW-Approval-Electric-Wheelchair/dp/B073V11B1Q/ (retrieved April 29, 2018).
[d]Faraday Bicycles.
[e]Gogoro, Inc.
[f]Swagtron, Innuovo Electric Wheelchair on amazon.com (retrieved April 29, 2018), Boosted Boards on buy.boostedboards.com (retrieved April 29, 2018), Faraday Cortland on faradaybikes.com (retrieved April 29, 2018), Gogoro 2 Smartscooter (79800 New Taiwan Dollars exchange rate to U.S. dollar on April 29, 2018) on store.gogoro.com (retrieved April 29, 2018).
[g]Driven tire(s).
[h]E-rate is a nondimensional discharge rate of a battery cell, calculated by dividing discharge power by the nominal battery energy. E-rate is closely related to C-rate. Both are measured units of 1/hours. C-rate in A/Ah multiplied by average discharge voltage and divided by the nominal voltage at beginning of life equals E-rate in W/Wh.

Some clear commonalities in product feature set emerge from this study. LEVs are primarily designed to transport a single person, and they do so far more efficiently than a car (Wh/mi basis). This enables significantly less battery energy for a given travel range and results in higher E-rates and larger daily depth-of-discharge swings. LEVs now offer 15 to 60+ mile range per charge, a threshold of utility that covers the majority of daily travel requirements for individuals, especially city dwellers (Sec. 25A.3), despite low cargo capacity, limited multipassenger capability, and compromised all-weather comfort. As a "discretionary purchase," the relative sales volumes of any single LEV category are orders of magnitude lower than an automotive category, except for e-bikes in China (Sec. 25A.4). The price-sensitive markets for LEVs and the small, low-cost battery packs lead to lower warranty expectations, less stringent crash and safety regulations, lower lifetime requirements, and modular, swappable, and/or upgradeable battery packs. They employ off-the-shelf battery cells, commodity

**TABLE 25A.1b** Lightweight Electric Vehicles Are as Similar as They Are Different (*Continued*)

| Category | E-motorcycle, off-road/mx | E-motorcycle, street | E-motorcycle, racing | Electric utility vehicle (UTV, ATV) | Battery electric car (BEV) |
|---|---|---|---|---|---|
| Example product(s) | Redshift by Alta Motorcycles[a] | Zero SR ZF14.4 by Zero Motorcycles[b] | Ego by Energica Motor Company[c] | Polaris Ranger EV and Ranger EV Li-ion[d] | 2017 Bolt EV by Chevrolet[e] |
| Photograph | | | | | |
| Human assisted? | No | No | No | No | No |
| MSRP, $[f] | $10,900 | $16,495 | $26,460 | $11,299 Pb acid $22,999 Li-ion | $37,495 |
| Vehicle mass, kg | 116 | 188 | 258 | 602.4 Pb acid 584 Li-ion | 1624 |
| Tire diam., cm | 63 | 63 | 63 | 64 | 65 |
| Range, mi | 50–55 | 90–135 | 90–125 | 20–40 Pb acid 30–50 Li-ion | 238 |
| Top speed, mi/h | 70 | 102 | 150 | 25 | 92 |
| Battery mass, kg | 32 | ~80–90 | 90 | 312 Pb acid n/d Li-ion | 435 |
| Battery energy, Wh | 5770 | 12,600 | 11,700 | 11,700 Pb acid 12,400 Li-ion | 60,000 |
| Batt peak power, W | ~41,000 | ~62,000 | ~122,000 | ~30,000 | ~166,000 |
| Contin. $E$-rate, W/Wh | 1–2 | 0.5 | ~3 | ~1 | 0.25 |
| Peak $E$-rate, W/Wh | 7 | 5 | 10.5 | 3 | 2.75 |
| Rated consumpt., Wh/mi | 105–115 | 90–115 | 95–125 (375 on racetrack) | 200–400 | 252 |

[a]Rob Sweney, Alta Motors.
[b]Zero Motorcycles.
[c]Energica Motor Company.
[d]Polaris.
[e]General Motors LLC.
[f]Alta Redshift from Alta Motors, Zero SR ZF14.4 from Zero Motorcycles, Energica Ego from Energica Motor Company, Chevrolet Bolt EV from Chevrolet USA.

battery management electronics, and/or battery modules supplied by a third party.[h] As LEVs grow in popularity, manufacturers are leaning on custom module and pack solutions to enable a unique cosmetic identity and differentiated vehicle performance, yet they still design around off-the-shelf cells (see all examples listed in Table 25A.1). In contrast, BEV battery packs (and often battery cells) are considered core intellectual property, and are designed and manufactured in-house or in a joint venture with the leading battery cell suppliers. See Tables 25A.3 and 25A.4 for examples of common LEV battery packs.

---

[h]For example, many e-bicycle manufacturers purchase finished, quick-connection battery packs from Bosch, Panasonic, Samsung SDI, and other component suppliers.[27]

**TABLE 25.A.2** Selected Product Features for LEV Categories versus Battery Electric Cars

| Category | LEV vs. BEV $\downarrow$ = lower, $\uparrow$ = higher | E-kickscooters and other small battery LEVs (this chapter) | E-motorcycles and other large battery LEVs (this chapter) | Battery electric cars (BEVs) |
|---|---|---|---|---|
| Rated consumption with single rider, Wh/mi | $\downarrow$ | 20–40 | 90–170 | 250–350 |
| Battery pack energy, kWh | $\downarrow$ | 0.2–2.5 | 6–12 | 20–100 |
| Vehicle mass, kg | $\downarrow$ | 8–20 | 116–188 | 1200–2500 |
| Continuous $E$-rate, W/Wh | $\uparrow$ | 1–2 | 0.5–3 | 0.2–0.4 |
| Peak $E$-rate, W/Wh | $\uparrow$ | 2–10 | 5–10 | 3–5 |
| Typical consumption, Wh/mi | $\downarrow$ | 15–40 | 100–200 | 250–350 |
| Typical daily depth of discharge, % | $\uparrow$ | 20–80 | 20–40 | ≤20 |
| Product MSRP, $ | $\downarrow$ | $500–$3500 | $10,000–$26,000+ | $28,000–$80,000+ |
| Annual sales of a model | $\downarrow$ | 1000–5000, 10,000+ for e-bikes | 500–1000 | 10,000–40,000+ |
| Annual batt. production throughput/model, MWh | $\downarrow$ | 0.5–3, 3–10 for e-bikes | 1–6 | 300–3000 |
| Crash and safety regulations[*] | $\downarrow$ | ≤60 VDC systems | ≤60 VDC or high-voltage systems, no or low regulation | High-voltage systems, stringent regulations |
| Battery warranty | $\downarrow$ | 0.5–2 years | 1–5 years | 5–8+ years |
| Battery pack replacement considerations | $\downarrow$ | $100–$300, customer swap or 1 h service, minimal special tooling | $800–$4000, customer swap or 1–2 h service with specialized tooling | $5500–$16,000, 4–8 h service with specialized tooling[†,‡] |

[*]United Nations Economic Council for Europe (UNECE) R.100 Revision 2, United States Federal Motor Vehicle Safety Standards (FMVSS) 49 CFR 571.305.[43]

[†]https://insideevs.com/heres-how-much-a-chevrolet-bolt-replacement-battery-costs/ (retrieved May 5, 2018).

[‡]https://cleantechnica.com/2017/10/04/nissan-leaf-replacement-battery-will-cost-5499/ (retrieved May 5, 2018).

## 25A.3  TICKET TO RIDE: POWER, ENERGY, AND UTILITY

One unique feature of LEV batteries is their high continuous and peak discharge rates (see Table 25A.2) compared to BEVs. LEVs require greater depth of discharge per mile when compared to passenger cars because LEVs suffer from comparatively high aerodynamic drag and tire rolling resistance and they have smaller battery sizes.

### 25A.3.1  Bluff Bodies

Vehicle aerodynamic drag is dominated by form or pressure drag, which results from the creation of pockets of low pressure air in the wake behind the moving vehicle. A nondimensionalized coefficient of drag $C_D$ multiplied by a frontal area intersecting the air particles during forward motion, $A_{\text{frontal}}$ is used to quantify the relative pressure difference that opposes forward motion. The equation to calculate aerodynamic drag force multiplies this *drag area* by the *dynamic pressure* of air, $\frac{1}{2} \cdot \rho_{\text{ambient}} \cdot v^2$ (i.e., kinetic energy per unit volume of displaced air particles): $F_{\text{aerodynamic\_drag}} = \frac{1}{2} \cdot \rho_{\text{ambient}} \cdot v^2 \cdot C_D \cdot A_{\text{frontal}}$.[5]

**TABLE 25A.3** Examples of Common Battery Cell or Module Building Blocks Used in LEV and Power Tool Battery Packs. See Sec. 25.A.4.2 for Further Discussion of the 18650 Cell Format

| Example battery cell | Sealed lead acid (SLA) NP 12–12[*] | Nickel cadmium (NiCd) sub-C | Nickel metal hydride (NiMH) | Lithium ion 18650, 1C, "high capacity" | Lithium ion 18650, 10 A, "high rate" | Lithium ion 18650, 30 A, "high power" |
|---|---|---|---|---|---|---|
| Photograph | | Not pictured[†] | Not pictured[†] | | | |
| Nominal voltage, V | 12 V | 1.2 | 1.2 | 3.65 | 3.6 | 3.6 |
| Nominal capacity, Ah | 12 Ah | 2.5 | 3.75 | 3.5 | 3.35 | 3.0 |
| Nominal energy, Wh | 144 Wh | 3 | 4.5 | 12.7 | 12.2 | 11 |
| Mass, g | 4050 | 55 | 58 | 48 | 47 | 46 |
| Dimensions, mm | L151 × W98 × H97.5 | Ø22 × H42.5 | Ø18.1 × H67.0 | Ø18.3 × H65.0 | Ø18.3 × H65.1 | Ø18.3 × H65.0 |
| Max rated continuous E-rate, W/Wh | 2–3 | 8 | 7 | 1–2 | 2.5–3 | 5–6 |
| % of nominal energy delivered at 1 E-rate continuous discharge | 60 | 95 | 95 | 92 | 92 | 96 |
| Energy density at 1 E-rate continuous discharge, Wh/L, Wh/kg | 60 Wh/L 21 Wh/kg | 176 52 | 261 78 | 676 240 | 654 240 | 620 230 |
| Common parallel-series configuration, e.g., in LEV battery packs | 1-parallel, 4-series for 48 V e-bicycle | 1-parallel, 14-series for 18 V power tool | Not common | 4-parallel, 7-series for 25 V power wheelch. | 2- to 3-parallel, 10- to 14-series for 36–48 V e-bicycle | 2- to 3-parallel, 7- to 12-series for 25–43 V e-skateboard |

[*]Yuasa Battery, Inc.
[†]See Chap. 15.

**TABLE 25.A.4** Example Lead Acid and Lithium Ion Battery Packs for E-Bicycles

| Example battery pack | 48 V sealed lead acid pack for e-bike | 36 V lithium ion pack for e-bike |
|---|---|---|
| Cells in parallel, and series | 1-parallel × 4-series | 4-parallel × 10-series |
| Nominal voltage, V | 48 | 36 |
| Nominal capacity, Ah | 12 | 12–14 |
| Nominal energy, Wh | 576 | 432–504 |
| Energy at 1 E-rate disch, Wh | ~346 | ~400–470 |
| Mass, kg | 16.3 | 2.5–3.0 |
| Volume, L | 5.8 | 1.9–3.0 |
| Energy density at 1 E-rate continuous discharge, Wh/L, Wh/kg | 60 21 | 150–250 120–190 |
| Packing factor of cells in pack | ~100% by volume and weight | 25–50% by volume 50–80% by weight |

An LEV rider presents a relatively large frontal area and a "flat plate" shape (see Fig. 25A.1[5]). The result is that LEVs have 50% to 100% of an automobile's drag, $C_D \cdot A_{frontal}$, despite only 5% to 15% the ridden weight. Drag is thus 300% to 2000% per unit mass of a passenger car. LEV rolling resistance compares unfavorably to passenger cars as well, owing to poor tire quality or high traction designs. Table 25A.5 reveals that the $C_{rrl}$ of LEV tires is typically 200% that of a passenger car, with the exception of e-bicycles.

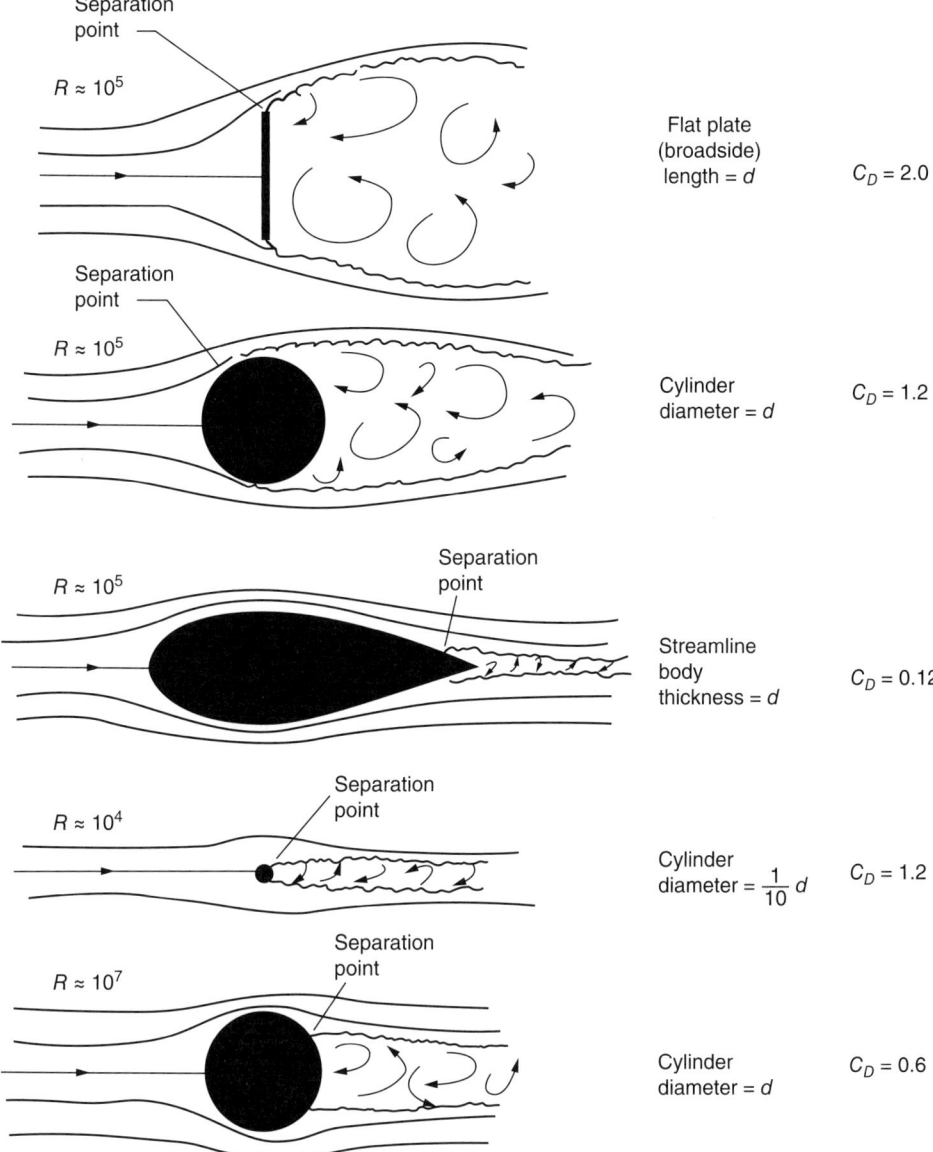

**FIGURE 25A.1**  Drag coefficients of various shapes.

**TABLE 25A.5**   Derivation of Power Requirements and Energy Consumption of LEVs (Governing Equations in Footnotes[*])

|  | E-skateboard | E-bicycle | E-motorcycle | BEV |
|---|---|---|---|---|
| Vehicle curb mass, kg | 8 | 18 | 250 | 1600 |
| Rider mass, kg | 80 | 80 | 80 | 80 |
| Rider mass percentage of total mass, % | 91 | 82 | 24 | 5 |
| Aerodynamic flat plate drag, $C_D \cdot A_{frontal}$, m$^2$ | $1.0 \times 0.4$ | $1.15 \times 0.63$ | $0.7 \times 0.65$ | $0.3 \times 2.33$ |
|  | $= 0.4$ | $= 0.71$ | $= 0.46$ | $= 0.7$ |
| Drag per ridden mass, $\dfrac{C_D \cdot A_{frontal}}{100 \text{ kg}} \left[\dfrac{\text{m}^2}{100 \text{ kg}}\right]$ | 0.45 | 0.74 | 0.14 | 0.04 |
| Rolling resistance coefficients, $C_{rr1} \mid C_{rr2}$ | 0.018 \| 0.0 | 0.007 \| 0 | 0.02 \| 0 | 0.010 \| 0 |
| Total efficiency, $\eta_{total}$ | 0.85 | 0.8 | 0.8 | 0.8 |
| **(a) Level ground** |  |  |  |  |
| Example cruise speed, mi/h | 10 | 20 | 60 | 60 |
| Calculated cruise battery power, W | 107 | 463[†] | 8753 | 15,637 |
| % of power, bluff body aerodynamic drag | 24 | 84 | 75 | 65 |
| Cruise battery energy consumption, Wh/mi | 11 | 23[†] | 146 | 261 |
| Cruise batt. energy per total mass, Wh/mi · 100 kg | 12 | 24 | 44 | 16 |
| **(b) Hill climb** |  |  |  |  |
| Example 25% gradient hill climb speed, mi/h | 5 | 5 | 20 | 20 |
| Hill climb battery power, W | 563 | 662[†] | 9190 | 45,474 |
| Hill climb battery energy consumption, Wh/mi | 113 | 132[†] | 460 | 2274 |
| % of power contributed by road gradient | 98 | 98 | 95 | 98 |

[*]Forces in Newtons.
Coefficients of rolling resistance, $C_{rr1}$ and $C_{rr2}$ | Air density $\rho_{ambient} = 1.2$ kg/m$^3$.
Coefficient of drag, $C_D$ | Frontal area, $A_{frontal}$ in m$^2$.
Velocity, v in m/s | Gradient angle, $\theta_{gradient}$ in radians = tan$^{-1}$ (% gradient).
Force of weight due to gravity, $F_{weight}$ = mass · $g$ where accel due to gravity $g = 9.8$ m/s$^2$

$F_{roadloads} = F_{rolling} + F_{aerodynamic\_drag} + F_{gradient}$, where
$F_{aerodynamic\_drag} = 1/2 \cdot \rho_{ambient} \cdot v^2 \cdot C_D \cdot A_{frontal}$
$F_{gradient} = F_{gravity} \cdot \sin\theta_{gradient}$
$F_{normal} = F_{gravity} \cdot \cos\theta_{gradient}$
$F_{rolling} = F_{normal} C_{rr1} + C_{rr2} \cdot v$
$P_{roadloads} = F_{roadloads} \cdot v$
$P_{battery} = P_{roadloads}/\eta_{tire\_to\_batt}$
$\eta_{tire\_to\_batt} = \eta_{tire} \cdot \eta_{driveline} \cdot \eta_{motor} \cdot \eta_{inverter}$

[†]E-bicycle calculations include rider contribution, so the battery power and energy consumption would be reduced by any rider pedal contribution.

## 25A.3.2   Peaks and Valleys

Table 25A.5 details the power and energy requirements for two constant speed scenarios: (a) level ground and (b) hill climb.

The e-skateboard is limited by rolling resistance drag during level ground and constant speed due to a low aerodynamic frontal area, low speed, and a high rolling resistance coefficient (its rubber or polyurethane wheels are designed for traction rather than efficiency). Conversely, the e-bicycle and e-motorcycle have efficient tires but cruise at faster speeds that penalize their large aerodynamic frontal areas and poor drag coefficients (i.e., bluff body drag dominant). The BEV has balanced rolling and aero resistances (excellent $C_{rr1}$ but high mass, excellent $C_D$ but large frontal area). The road gradient power dominates total power consumption in the 25% hill climb, which can be common in some cities and rare in others.[6]

### 25A.3.3    Real World Test Ride

Figure 25A.2 presents two real-world discharge profiles of different LEV products.[7,8]

The $E$-rate (root mean square)[i] and state-of-charge are plotted over time for a Boosted Boards e-skateboard hill climb and descent in Fig. 25A.2*a*. The $E$-rate versus depth-of-discharge[j] is plotted for an Alta Redshift e-motorcycle ridden on a motocross racing circuit in Fig. 25A.2*b*. These real-world profiles exemplify the high continuous and peak discharge ratings of LEVs.

### 25A.3.4    How Far Can I Go?

LEV designs are a trade-off between range (battery energy, mass, volume, and cost) and vehicle size and costs. Lithium ion battery packs are typically a poor trade-off, especially for the smaller LEV categories. Tables 25A.2 and 25A.5 exhibit that LEVs consume half the energy of BEVs at <1/6 the weight.[9] Smaller LEVs are designed to be charged indoors; this is enabled either by limiting the weight of the entire vehicle or by limiting the weight of the battery and making it removable. The relevant weight limit is generally 8 to 12 kg, similar to a piece of carry-on luggage.

Figure 25A.3 plots the cumulative probability of travel more than a certain average distance in a single day.[10] In Japan and European nations, an LEV with an 8 mi (13 km) range would cover 25% of typical single day travel outright, assuming amenable weather and route/road availability. In reality, an LEV can cover significantly more travel for urban populations, assuming limited cargo and multimodal travel (i.e., "first/last mile" LEV trips on metro transit). The contemporary small LEVs in Table 25A.1 already offer 10 to 20 mile (16–33 km) range with 3 kg battery packs.

### 25A.3.5    TAP/SWIPE/SWAP and Go: Micro Mobility on Demand

Urban LEV-sharing networks are rapidly popping up across the globe, allowing customers to rent vehicles on a pay-per-minute and/or pay-per-mile basis within a geographic area. This can result in 2 to 5+ rides per vehicle per day with significantly higher daily mileage. In some cities, the services have eclipsed public transit in total daily ridership.[11–13]

Starting in the mid-2000s[k,12] using traditional bicycles, cities have partnered to install networks of docking/locking stations, customer payment points, and network information hubs.[14] Users rent and return the vehicles to any station with available capacity.[l] In 2014, LEVs entered the segment when an e-bicycle share network, BiciMAD, was rolled out in Madrid, with many other cities following suit.[15] LEV sharing introduces additional operational challenges due to charging or battery swapping. So far, many cities have opted to integrate bike shares into their existing (unpowered) dock station network, thus relying on modular battery packs that system operators swap for remote recharge. A few cities have constructed or retrofitted these dock positions to include battery chargers and electrical interconnects that seamlessly mate to the e-bicycle for recharging.

Since 2012, a slew of smartphone-activated "hubless" or "dockless" LEV networks forego the docking station model and use near-field sensors, GPS antennas, and cellular radios (which have dropped in cost from $200 to sub-$20) so that vehicles can be located, checked out, parked, and locked anywhere.[18] Hubless e-scooters were introduced between 2012 and 2016 in Taiwan, Europe, and North America.[16] Modern e-scooters have sufficient range per charge for multiple rides or even multiple days of service. Riders receive discounts or credits for

---

[i]The root mean square (rms) metric is $x_{\mathrm{rms}} = \sqrt{1/n \cdot (x_1^2 + x_2^2 + \cdots + x_n^2)}$ for discharge rate measurements $\{x_1, x_2, \ldots, x_n\}$. The rms metric is traditionally used to estimate resistive (joule) heating from current flow through wires $P_{\mathrm{heat}} \propto I^2 \cdot R$ for current $I$ and resistance $R$. Calculating the rms of $E$-rate and $C$-rate is a useful approximation of battery heating.

[j]Depth-of-discharge is the reverse of state-of-charge: DOD = 1 − SOC.

[k]Paris introduced Vélib, the first successful bike-share scheme, in 2007. There were many precursors worldwide that succumbed to theft, poor utilization, and other complications.[11]

[l]Although the docking stations themselves do not have to be wirelessly connected, customers typically rely on smartphone apps to find which stations have bikes available for rent and which stations have docks available for drop-off.

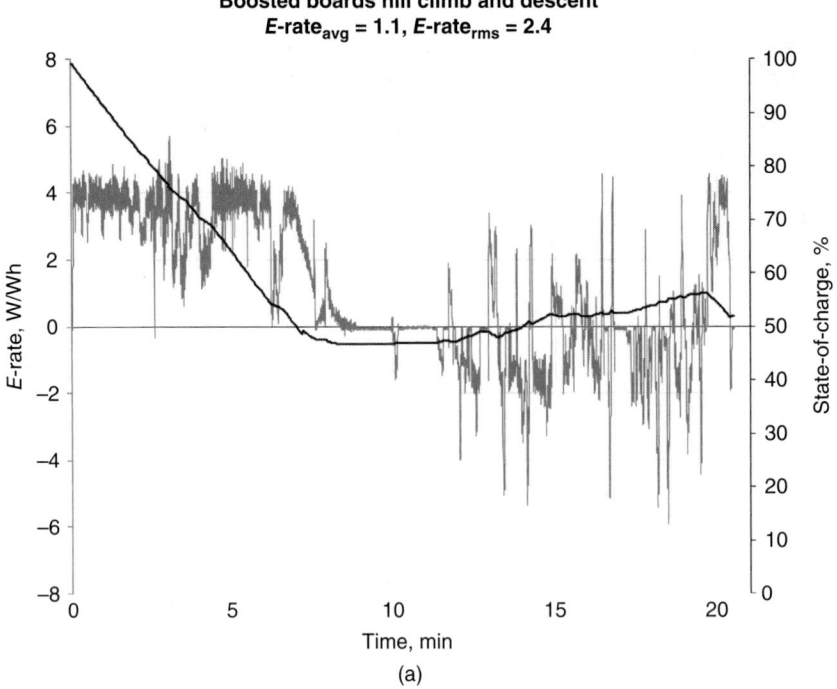

(a)

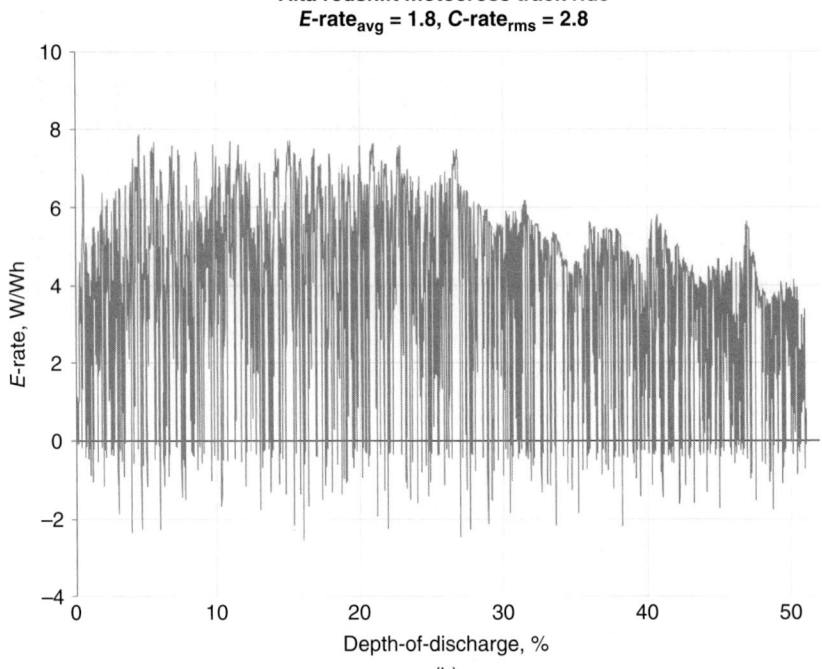

(b)

**FIGURE 25A.2** Real world discharge profiles for two LEV platforms. *E*-rate provided for (*a*), *C*-rate and *E*-rate provided for (*b*).[i] (*Data provided by [a] John Ulmen, Boosted Boards, and [b] Rob Sweney, Alta Motors.*)

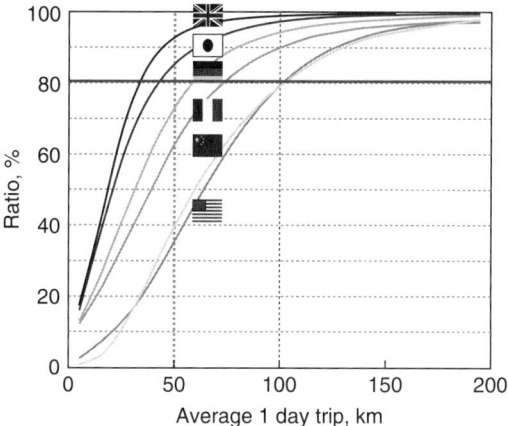

**FIGURE 25A.3**   Cumulative probability of individual average daily travel distance by country. (*2018 Nissan©.*)

**FIGURE 25A.4**   Image of shared electric scooters from Scoot Networks. Vehicles do not need to be parked at a hub if they have adequate charge remaining.

returning the scooter for recharge to a charging dock or parking spot owned by the private system operator (e.g., Fig. 25A.4).[17] Dockless e-bike-share networks were started in 2012 and were soon followed by dockless e-kickscooter networks. Customers can park and ride these small LEVs at will, which offers great travel flexibility but often results in disorganized or hazardous parking and sidewalk usage.[29] See Fig. 25A.5.

E-scooter use has been well regulated, but (as of 2018) e-bicycles and e-kickscooters are less regimented regarding sidewalk usage, speed limits, and helmet requirements. Startups often deploy rollout strategies to beat competitors to market and deal later with local regulators (who may impound vehicles, ban new products, issue stopgap ordinances, or rush to update codes).[18] Despite these micromobility wars,[19] LEVs are often the fastest

**FIGURE 25A.5** Dockless e-bicycles and e-kickscooters are both a blessing and a curse for urban residents. (*Image courtesy of Lime.*)

way to get from point A to B in a crowded city, so sharing networks can offer sustainable and urban-friendly alternatives to automobiles and affordable transportation solutions for communities underserved by public transit.[20]

Many of the private system operators have not standardized their solutions to battery charging. Shared e-bicycles are not easily transported to charging hubs and rely on swappable battery packs. Shared e-kickscooters use fixed battery packs, and rely on independent contractors to collect, charge, and redistribute the e-kickscooters on a per vehicle, per night basis.[19,21] In the long term, standard battery packs could improve operational efficiency for these networks, with the possibility of customer-owned and customer-swapped battery packs. As batteries continue to improve in energy density and cost performance, such battery packs could even be carried in handbags. Having already achieved 180 Wh/kg, a 1 kg LEV battery pack can provide adequate range (e.g., 5–6 miles) and power (e.g., 500+ W) for personal daily LEV usage (see Sec. 25A.4.1).

User-facing battery swap stations, implemented, for example, by Gogoro for their Smartscooter, allow owners to exchange standardized battery packs at nearby kiosks in only a few seconds. Gogoro sells their scooters, but offers battery swapping as a subscription service. GoStations are deployed every kilometer or less in Taipei.[22] In other cities, Gogoro has e-scooter on-demand services that are coupled with battery swap networks.[23]

Figure 25A.6 shows a subscriber installing two new 9 kg battery packs at GoStation battery swap kiosk. This new model for LEV use is also shaping the design of the battery systems. Vehicle sharing is intended to maximize the vehicle's utilization rate, meaning batteries will be cycled more frequently and more completely than with a single user. The direct cost of the battery falls on the service provider, necessitating more rational optimization of usage and total lifetime cost. In contrast, an individual LEV owner may use an oversized and overpriced battery pack, needlessly based on worst case assumptions. A business model that enables companies to maximize near-term returns but upgrades to new battery packs when available is preferable. Swapping of batteries, therefore, allows seamless implementation of the next generation of batteries. Battery pack production over the vehicle lifetime will benefit from: (1) battery manufacturing economies of scale, (2) electricity grid services revenue streams for the centralized store of batteries that are "idle" and connected to the electrical grid (e.g., demand-response), and (3) second use of end-of-life battery packs and battery recycling.

**FIGURE 25A.6**    A Gogoro subscriber installs two new battery packs from a GoStation battery swap kiosk.

## 25A.4    *IT MAKES THE WHEELS GO ROUND*

Presently, lithium ion batteries command near 100% market share as the energy storage component of many LEV categories sold in the Japan, the United States, and Europe, with strong growth in China.[24] Category exceptions to this trend,[25] including electric wheelchairs (lead-acid),[26] are those product segments and geographies that have extreme price sensitivity—primarily electric two wheelers in mainland China—yet, even there, lithium ion will attain majority market share within a decade.[24] Additionally, the number of electric two wheelers in operation globally, which was already substantial at 140 million units in 2010, has doubled in 8 years.[25] Likewise, e-kickscooters, e-skateboards, and other nascent LEV product segments have experienced explosive year-over-year sales growth.[27]

### 25A.4.1    A Virtuous Cycle

A positive feedback loop evolved between the use of lithium ion batteries and the market expansion of LEVs in China, where in the early 2000s sealed lead-acid powered pedal-assist e-bicycles proliferated.[m,n,28-31] The lead-acid batteries in these products (see Table 25A.4) were heavy but affordable, typically costing $20 to $50 for an e-bicycle that retailed for $300 to $350 equivalent (including batteries). As LEV prices dropped, rate-capable lithium ion costs also dropped and their performance improved,[o] making lithium ion-powered LEVs a compelling option to the mass market, especially e-bicycle consumers in Japanese and European cities. Figure 25A.7 depicts the rapid improvement in price and performance of lithium ion cells as compared to the incumbent technology of sealed lead-acid batteries.[32-36]

---

[m]Sealed lead-acid batteries have been in use for over a century, but their usage for e-bicycles and other LEVs required the advent of lightweight, affordable motor control inverters and traction motors. Motor controllers rely on silicon digital signal processors and silicon power MOSFETs, both of which experienced drastic improvements in affordability and performance in the 1990s due to the computer market boom and the introduction of TrenchFET technology.[32] Long life, power dense traction motors rely on rare earth magnets (primarily, neodymium iron boron or NdFeB), which also entered mass production in the 1990s, largely due to the rapid market growth of magnetic hard disk drives used in computers.[27]

[n]NiCd cells, first introduced into portable power tools in the 1960s,[32] were later used in niche LEV offerings in the 1990s. One popular example was the Peugeot Scoot'Elec, an e-scooter with a ~25 mi range that was powered by an 18 V, 100 Ah, 1.8 kWh NiCd battery pack outputting ~3500 W peak power (~2 $E$-rate).[32]

[o]As has been discussed elsewhere in this Handbook, high-rate lithium ion cells used in LEVs benefited at first from investments and production scale of 18650 batteries for the consumer electronics industry, and in later years from self-sustaining progress in LEVs, power tools, and battery electric vehicles.

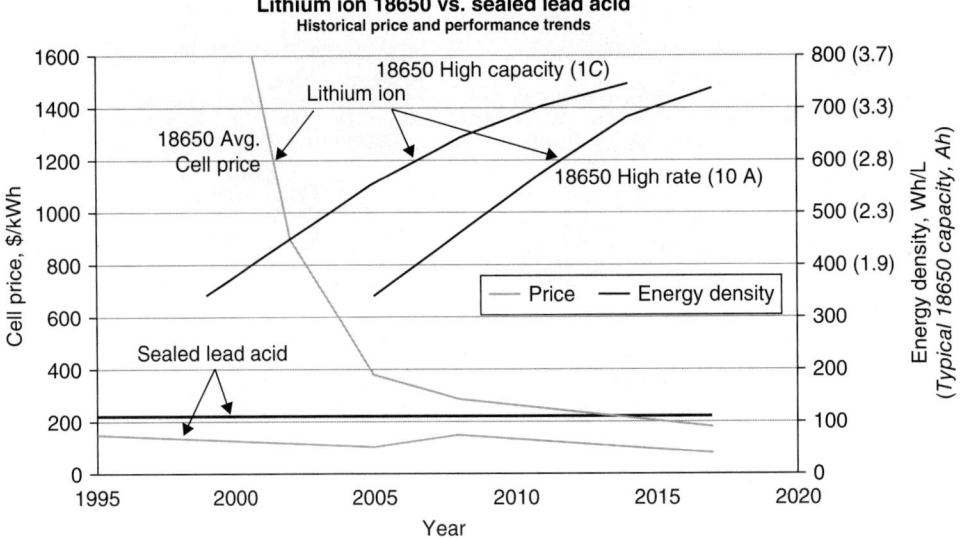

**FIGURE 25A.7**    Lithium ion 18650 versus sealed lead acid historical price and performance trends.

Between 2005 and 2015, high-rate lithium ion cells doubled in volumetric energy density while their prices were cut two to four times. The demand for commodity lithium ion LEV batteries fostered a new industry, drove manufacturing investment, and encouraged innovation. LEV lithium ion battery packs now offer three to five times more usable energy per volume, five to eight times more usable energy per weight, but cost two to four times more per usable energy than their lead acid counterparts. For example, a 300 to 500 Wh e-bicycle lithium ion battery costs approximately $100 to $200. This cost premium is less burdensome for LEV products in Western markets (see Table 25A.2). Modern LEVs have improved trade-offs in affordability, performance, and utility due to their lithium ion batteries. The future will hold more of the same: demand will rise, and the virtuous cycle will repeat. Figures 25A.8 through 25A.10 detail some market statistics.[37]

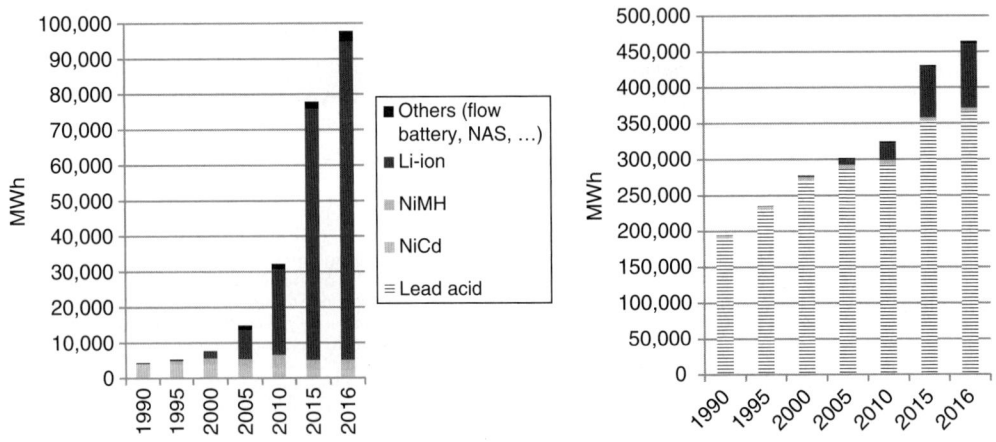

**FIGURE 25A.8**    Global market size in MWh for rechargeable energy storage. Lithium ion has been largest growth segment at a 25% CAGR from 2010 to 2016 and exceeding 90 GWh in 2016 versus 360 GWh for lead acid.

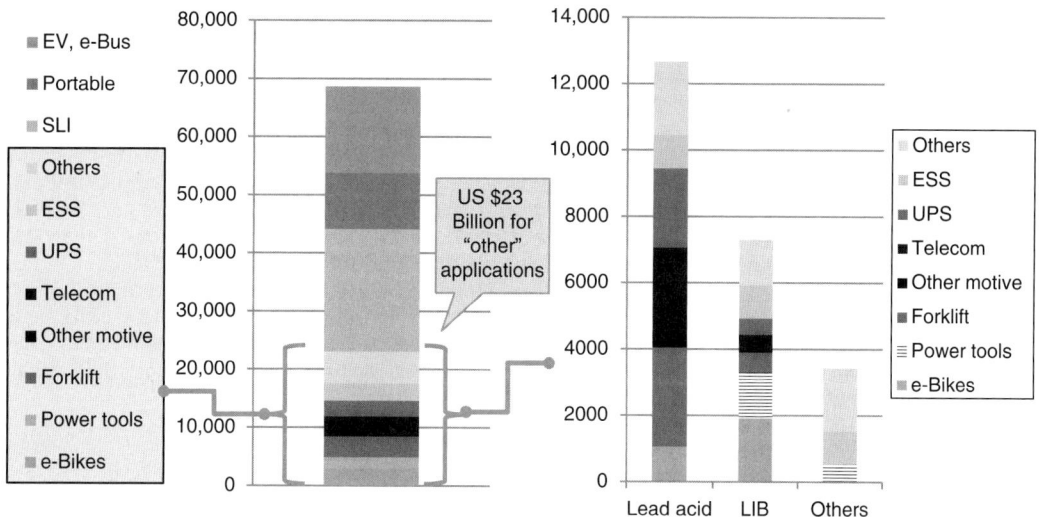

**FIGURE 25A.9** Global market size in U.S. dollars for rechargeable energy storage at pack-level (cells, cells assembly, battery management system, and connectors). Lithium ion market size in dollar terms exceeds lead acid for e-bicycles.

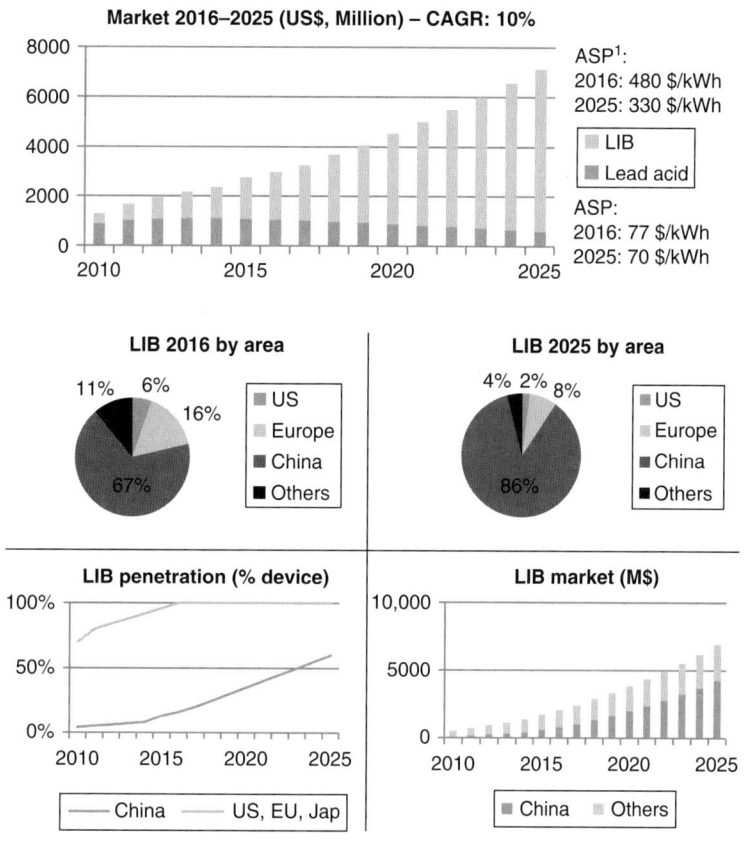

**FIGURE 25A.10** Global and regional market statistics for batteries used specifically for e-bicycles. Hundred percent lithium ion market share for Japan, United States, and European Union, and growing market penetration for lithium ion in China.

### 25A.4.2    No Two Batteries Are the Same

While designs vary considerably between LEV types and manufacturers, most LEV battery packs utilize "high-rate" battery chemistries, typically rated for 3 to 10 C-rate. Most use commodity 18650 cells as shown in Table 25A.3. Flat laminated pouch designs are growing in popularity (e.g., Zero Motorcycles 29 Ah cells[38]). 18650s were first introduced for consumer electronics products, primarily laptops, and have since proliferated to use in power tools, LEVs, electronic cigarettes, and battery electric cars. Tesla Motors selected the 18650 form factor for their first battery electric car model due to existing manufacturing scale (i.e., Panasonic's existing laptop battery business).[39] Tesla continues using the 18650 cell, but has introduced a battery pack based on a 21 mm diameter and 70 mm tall cell (the 2170) with ~146% more volume. Increased active material per unit packaging better allocates the overhead and reduces battery piece part count, while the cylindrical format and similar jellyroll height leverage existing manufacturing processes to ensure low production costs. LEV manufacturers have started introducing battery packs with 2170 cells in 2018.

## 25A.5    WEAR YOUR HELMET: BATTERY MANAGEMENT AND SAFETY IN LEVs

With regards to safety, LEVs bridge two worlds. One world is that of battery-powered consumer electronics, where basic electrical safety and radio frequency emissions are regulated, but products are generally regarded to be harmless.[p,40–42] The other world is that of passenger vehicles, where human error, design flaws, or malfunctions can easily cause injury or death. Where a particular LEV falls on this spectrum is a function of its speed, weight, battery capacity, recharging method, and maximum operating voltage. Regulatory bodies are beginning to introduce new type approvals for LEV vehicles just as many new categories are coming to market.[q,43] Figure 25A.11 exhibits typical battery packaging.

**FIGURE 25A.11**    Cutaway view of LEV battery pack architecture, including BMS circuit board. Image courtesy of Segway (and Ninebot by Segway).

---

[p]Battery-powered consumer electronics have been the cause of numerous product recalls. Sony recalled of over 9.6 million laptop batteries between 2004 and 2006.[30,31] More recently, in 2016 to 2017, the Samsung Galaxy Note 7 was the subject of a high profile recall of 4.3 million smartphones due to battery fires and was globally banned from aircraft, costing the company $5 billion.[33]

[q]Europe introduced a regulatory framework for new type approvals of electric two-wheeler products for calendar year 2017 that may be mirrored in other LEV classes: e-bicycles are regulated under category L1e-A, "powered cycles," or L1e-B, "two-wheel mopeds." For L1e-A, the motor has to be cut off at a speed of 25 km/h and the maximum continuous motor power should not exceed 1000 W. The L1e-B category is subject to a maximum design vehicle speed of 45 km/h and a maximum continuous rated power of 4000 W.[34]

## 25A.5.1    Cell Failures and Propagation

Any multicell battery pack carries some risk of "propagation," wherein the extreme heat from an exothermic failure of one or more cells within the pack causes adjacent cells to overheat in turn (and often, to combust). Exothermic failure called thermal runaway[r,44] can be initiated in multiple ways (see Fig. 25A.12[45]), the most common being:

- *external abuse*, such as a vehicle crash that impinges on one or more battery cells
- *cell manufacturing defect*, such as a metal particle embedded into the cell during fabrication that develops into a short circuit[s]
- *battery management error*, such as a failure in the charge electronics and battery monitoring system that leads to overcharging of a cell[t]

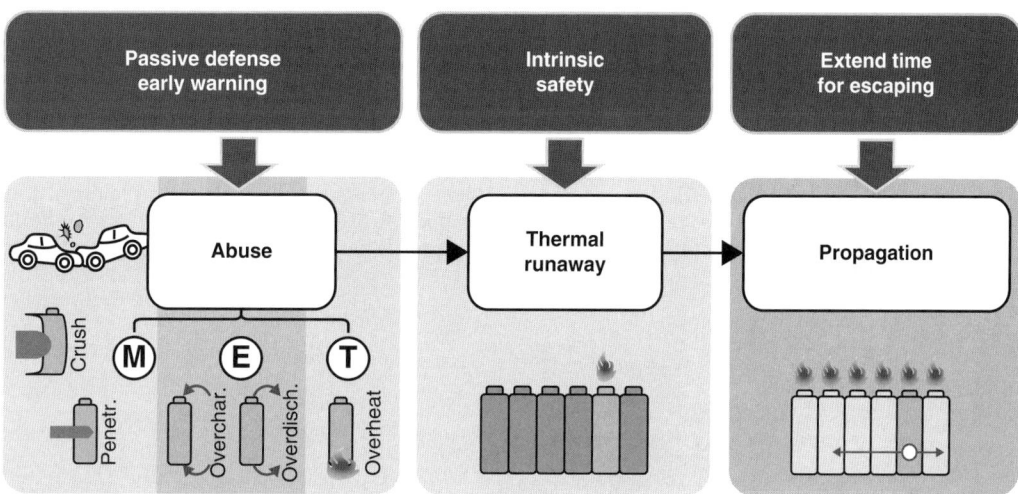

**FIGURE 25A.12**    Three tiers of safety for battery packs.

LEVs are typically designed for higher $E$ rates and have intrinsic safety advantages due to their high current capabilities, better heat transfer, and broader temperature stability. For smaller LEVs (e.g., ≤40 cells, 18650-format, 200–500 Wh), the total combustible energy of the pack is generally considered a serious but limited safety hazard; thus, manufacturers may prioritize lower cost and higher packing density ahead of abuse tolerance and propagation resistance or prevention.[u] The battery pack from an e-scooter or e-motorcycle, on the other hand, may contain 10 to 20+ times more combustible energy—enough that any event poses a serious safety hazard. Hence, propagation resistance is an important selling point.[46] Because LEVs have limited space and mass allocation for noncell battery pack components, propagation-resistant battery pack technology in LEVs has resulted in innovations not utilized by the automotive electric vehicle segment.[47]

---

[r]Lithium ion cells, when heated, can reach decomposition temperatures, characterized by a cascade of events that release increasing amounts of heat, fire, and hot-vented gases.[36,38]

[s]An "internal short" can cause either slow or instantaneous failures, depending on the resistance of the short, ranging from benign self-discharge to thermal runaway.[36,38]

[t]Cell overcharging can result from defective electrical circuits or a battery failure despite the use of cell-level safety devices.[36,38]

[u]Propagation resistance refers to a battery pack design that can contain the heat and flame from cell failures and control hot vent gas discharge, thereby delaying time to propagation or safely eliminating propagation altogether.

### 25A.5.2    Bruised Elbows, Scratched Knees

LEVs typically allow a rider easy escape after an accident or a noncrash event where thermal runaway may occur. However, the reduced risk of injury from battery-related events is replaced by the need for personal protective equipment (helmet, padding, etc.). Hence, crash safety is mostly decided by the rider and is not a standard aspect of vehicle design. Also, the perception of these vehicles as being discretionary or hobby purchases lowers expectations of LEV product liability relative to automobiles.

### 25A.5.3    Garage-Safe

Battery fires from LEVs can pose hazards to their owners beyond crash events, especially when they occur during battery charging. The hoverboard fad in the United States, described as "one of the fastest global boom-to-bust cycles in modern business,"[47] experienced a number of tragedies.

By the end of 2015, 40,000 units were shipping every day to the United States alone. Unlicensed imported products led to catastrophe: numerous hoverboard-related smoking, sparking, and fire incidents caused injuries, damaged millions of dollars of property, and claimed lives when homes burned to the ground.[48] By February 2016, in addition to a slew of patent infringement lawsuits, major retailers, airlines, universities, and transit authorities banned hoverboards. Underwriters Laboratories (UL) announced safety standards that few products on the market had met, and the U.S. Consumer Product Safety Commission notified manufacturers that boards noncompliant with the new UL standards would be subject to seizures and recalls.[v] Sales plummeted, and hundreds of factories were forced to close their production lines. As of 2018, the hoverboard market is healthy once again, with leading manufacturers selling patent-protected or licensed products with UL-certified battery packs (see Fig. 25A.13[49]).

**FIGURE 25A.13**   Hoverboards were first introduced in the United States at CES, Las Vegas in January 2015. As of 2018, newer generation hoverboards like the Swagtron T1 sport UL-certified battery packs that are safer than earlier models. (*Credit: Swagtron.*)

---

[v]Then, in July 2016, the CPSC recalled over 500,000 hoverboards from different manufacturers.[40]

### 25A.5.4 High Voltage

Larger LEV batteries often stack more cells in series to attain higher operating voltages,[w] and this can introduce additional cost and complexity into their design and manufacturing. Most regulatory bodies consider 60 V DC to be the dividing line between "low voltage" and "high voltage" for an electric vehicle.[50] The presence of high voltage on the vehicle introduces a greater risk of electric shock (if users are exposed to live circuitry) and drives much more careful attention to safety topics such as preventing access to live metal parts, protection from water, redundant safety measures, and monitoring to ensure that safety systems are working properly. Many countries also mandate specific high voltage safety training for personnel building and servicing high-voltage vehicles. For smaller vehicles, the additional complexity and components of high-voltage systems are not worth the benefit.

### 25A.5.5 Battery Management Systems

LEV battery packs, as with most other multicell battery applications, require a battery management system (BMS). Chapter 31 provides details on charging and BMS controls. The BMS will monitor individual cell voltages and temperatures within the pack, working to maintain safe operation and maximize usable energy and battery life. The BMS may perform higher functions such as tracking state-of-charge and controlling access to the battery pack terminals using relays or transistors.

The low cost, low complexity design of LEV electronic systems often means that the processor(s) running BMS monitoring and control algorithms may serve double duty by performing vehicle-level control functions. In some implementations, the BMS is even powered directly off of the cells within the pack, rather than from a separate (usually electrically isolated) supply, as is more common in electric cars. Due to this, and the relatively small capacity of the battery system, the LEV BMS benefits from careful design of off-state power consumption (including special "hibernate" power states) in order to prevent over-discharge of the battery pack during long periods of nonoperation. LEV owner manuals often recommend that owners plug their vehicles in during long-term storage to maintain charge. The higher $E$-rate and daily depth of discharge of LEVs relative to BEVs means that pack voltage will change more rapidly during use and cells will go through complete charge/discharge cycles at a faster pace. This can affect sensor selection, balancing circuit design, and other BMS electronics.

## 25A.6  RIDING INTO THE FUTURE

LEVs are quickly becoming an integral part of the new urban landscape. They can cut through citywide gridlock while offering personal point-to-point transit and easy parking. Their relatively low cost and small footprint also make them the ideal medium for urban transportation as a service. Beyond the city, electrically propelled motorcycles and utility vehicles are following more traditional business models but offering novel features. LEVs are easy to use, have low maintenance requirements, are quiet, and, in some classes, they can outpace fossil-fuel burning incumbents on the racetrack. Advances in battery technology are directly responsible for making these products and markets possible.

---

[w]In the price competitive LEV market, battery pack voltage flexibility may be constrained by the ratings of motors and inverters available from the existing low-cost supply base. If not, even a 500 Wh battery pack with 40 lithium ion cells offers numerous interconnection choices to the LEV battery pack designer. Raising the voltage (i.e., more cells in series, fewer in parallel, e.g., going from 5-parallel × 8-series to 4-parallel × 10-series) will reduce the current required to provide a given power output (recall: $P_{DC} = V_{DC} \cdot I_{DC}$). Then, reducing the current allows for smaller, lighter, and more flexible electrical conductors or reduces resistive losses in equal-size conductors. Conductor sizing is only one piece of the puzzle. LEV designers must also optimize around the power transistors (often Silicon MOSFETs or IGBTs) in the motor control inverter. These power transistors switch the DC battery power at high frequency into AC waveforms to power and control the traction motor(s), and they have voltage ratings and thermal and packaging constraints, among many other variables. Traction motors themselves must achieve long operating life under the voltage stress of those AC waveforms, which impacts electrical conductor insulation and thermal management choices.

The unique aspects of these vehicles have also led engineers to battery solutions that are distinguished from those found in larger EVs. Pack configurations trend toward small capacities that leverage "high-rate" battery chemistries that can match the dual requirements of hill climb and high energy consumption per mile. The LEV also yields different decisions to be made with regards to battery voltage, cooling method, design life, and safety.

LEVs sit at a unique intersection between a consumer electronics device and a large electric vehicle. Short model cycle times and affordable full vehicle prices mean that the LEV segment is a breeding ground for new business models in transportation and new vehicle categories. LEVs are poised to transform everyday mobility in cities as we know it. One day soon, small LEVs, with their lightweight and tightly integrated battery packs, may become so ubiquitous and so indistinguishable from their unpowered cousins that the lines between a bicycle and an e-bicycle may blur. Batteries will be the protagonist in that future.

## REFERENCES

1. Per total units sales volumes. Christophe Pillot, Avicenne Energy.

2. https://qz.com/641471/the-us-has-banned-all-hoverboard-imports-that-arent-from-segway/ (retrieved 2018/05/07).

3. https://mashable.com/2016/04/16/rise-and-fall-hoverboard/#kloXZfuK9Squ (retrieved 2018/05/07).

4. https://qz.com/1084282/the-future-of-transportation-may-be-about-sharing-batteries-not-vehicles (retrieved 2018/04/29).

5. Talay, Theodore A. "Introduction to the aerodynamics of flight [NASA SP-367]." Langley Research Center. Langley, VA, 1975.

6. https://www.citylab.com/transportation/2014/02/10-truly-hellish-hills-american-cyclists/8511/ (retrieved 2018/05/05).

7. John Ulmen, Boosted Boards.

8. Rob Sweney, Alta Motors.

9. Gogoro, Inc.

10. Shigetoshi Tokuoka, NISSAN MOTOR CO., LTD. "Development of the Nissan LEAF." (2011 Hybrid and Electric Vehicle Symposium.) 560-2, Okatsukoku, Atsugi-shi, Kanagawa 243-0192, Japan.

11. https://www.seattlebikeblog.com/2017/12/15/bike-share-pilots-daily-ridership-blows-past-prontos-lifetime-totals-rivals-both-streetcars-combined/ (retrieved 2018/05/10).

12. https://medium.com/transit-app/docked-vs-dockless-bikes-five-months-in-a86ac801f4c7 (retrieved 2018/05/10).

13. https://www.citylab.com/transportation/2018/03/scoot-scoot/555746/ (retrieved 2018/05/08).

14. https://www.economist.com/news/christmas-specials/21732701-two-wheeled-journey-anarchist-provocation-high-stakes-capitalism-how (retrieved 2018/05/10).

15. https://gizmodo.com/why-dont-more-cities-have-e-bike-shares-1595348781 (retrieved 2018/05/10).

16. https://wagner.nyu.edu/rudincenter/2016/12/scooter-share-primer (retrieved 2018/05/11).

17. Image of shared electric scooters from Scoot Networks.

18. http://www.chicagotribune.com/bluesky/techandculture/sns-tns-bc-tech-culture-bike-sharing-20180409-htmlstory.html (retrieved 2018/05/11).

19. https://www.forbes.com/sites/jimmcpherson/2018/04/03/the-micro-mobility-wars-have-begun/#7067ba3b90c3 (retrieved 2018/05/10).

20. https://www.recode.net/2018/4/29/17286194/scooters-bird-limebike-spin-san-francisco-dockless (retrieved 2018/05/10).

21. https://therideshareguy.com/i-signed-up-to-be-a-bird-electric-scooter-charger-heres-what-its-like-2/ (retrieved 2018/05/10).

22. Gogoro, Inc.

23. https://www.wired.com/story/gogoro-electric-scooters-japan/ (retrieved 2018/05/10).

24. Christophe Pillot, Avicenne Energy.

25. Frost and Sullivan.

26. The Freedonia Group: "A more expensive battery inflates the cost of the wheelchair and makes it more problematic for consumers to afford and insurance companies to subsidize....[Sales growth] will be limited by more restrictive Medicare reimbursement protocols and reduced reimbursement rates, which will make it more difficult for seniors to afford battery-powered mobility vehicles like scooters."

27. https://www.cnet.com/news/electric-scooters-bikes-dockless-ride-share-bird-lime-jump-spin-scoot/ (retrieved 2018/05/15).

28. Jacek Korec and Chris Bull, Power Stage Group, Texas Instruments. "History of FET Technology and the Move to NexFET™." May, 2009. http://www.ti.com/lit/ml/slpa007/slpa007.pdf (retrieved 2018/05/09).

29. https://www.ecmag.com/section/your-business/100-years-innovation-history-electric-drill (retrieved 2018/04/29).

30. https://en.wikipedia.org/wiki/Peugeot_Scoot%27Elec (retrieved 2018/05/09).

31. https://www.electricbike.com/e-bike-patents-from-the-1800s/ (retrieved 2018/05/09).

32. Christophe Pillot, Avicenne Energy.

33. Matteson, Schuyler & Williams, Eric. (2015). Residual learning rates in lead-acid batteries: Effects on emerging technologies. Energy Policy. 85. 10.1016/j.enpol.2015.05.014.

34. Takeshita, IIT. AABC 2012 Conference.

35. Takeshita, IIT. Battery Japan 2013 BJ-3 Conference.

36. Straubel, JB, Tesla. Silicon Valley/SEEDZ Energy Storage Symposium on May 21, 2014.

37. Christophe Pillot, Avicenne Energy.

38. Farasis Energy: Pouch, NMC, 160 × 230 × 6 mm, 3.65 V, 29 Ah. http://www.farasis.com/solutions/cells/ (retrieved 2018/05/09).

39. Eberhard, Martin. "A Bit About Batteries". November 30, 2006. https://www.tesla.com/blog/bit-about-batteries (retrieved 2018/05/09).

40. http://www.kyria.co.uk/blog-the-25th-anniversary-of-the-lithium-ion-battery/ (retrieved 2018/05/09).

41. https://www.nytimes.com/2006/08/15/technology/15battery.html (retrieved 2018/05/08).

42. https://www.cnet.com/news/samsung-galaxy-note-7-return-exchange-faq/ (retrieved 2018/05/08).

43. Commission Delegated Regulation (EU) 2018/295 of 15 December 2017 amending Delegated Regulation (EU) No 44/2014, as regards vehicle construction and general requirements, and Delegated Regulation (EU) No 134/2014, as regards environmental and propulsion unit performance requirements for the approval of two- or three-wheel vehicles and quadricycles: http://eur-lex.europa.eu/eli/reg_del/2018/295/oj (retrieved 2018/05/09).

44. Battery Safety Council Forum 3. "Lithium-ion Cell Internal Shorting: 1. Early Detection 2. Simulation." Washington, DC. January 12, 2017. http://www.prba.org/wp-content/uploads/17-Battery-Safety-Council-January-2017-Barnett.pdf.

45. Reprinted from Energy Storage Materials, Vol 10, Xuning Feng, Minggao Ouyang, Xiang Liu, Languang Lu, Yong Xia, and Xiangming He, "Thermal runaway mechanism of lithium ion battery for electric vehicles: A review," 246–267 (2018), with permission from Elsevier.

46. https://chargedevs.com/features/alta-motors-says-its-electric-dirt-bike-has-world-class-energy-density/ (retrieved 2018/05/08).

47. http://fortune.com/hoverboard-industry/ (retrieved 2018/05/07).

48. https://www.forbes.com/sites/dianahembree/2017/06/30/exploding-hoverboards-top-consumer-watchdog-blacklist-for-summer-toys/#5b2b161b60d2 (retrieved 2018/05/07).

49. Swagtron.

50. United Nations Economic Commission for Europe (UN ECE) R.100 Revision 2 http://www.unece.org/fileadmin/DAM/trans/main/wp29/wp29regs/2013/R100r2e.pdf (retrieved 2018/05/08); United States Federal Motor Vehicle Safety Standard (FMVSS) No. 305 (https://www.federalregister.gov/documents/2017/09/27/2017-20350/federal-motor-vehicle-safety-standards-electric-powered-vehicles-electrolyte-spillage-and-electrical [retrieved 2018/05/08]).

# SECTION B

# AEROSPACE BATTERY APPLICATIONS

**Kirby W. Beard**
**(Emeritus Contributor: Jack N. Brill)**

## 25B.1   INTRODUCTION

Batteries designed for use in advanced aircraft, satellites, missiles, and other high altitude or deep space environments represent the ultimate battery challenge. Nothing in designing batteries for these applications is simple, quick, direct, or cheap. The typical requirements include extreme temperatures (both hot and cold), severe shock and vibrations, prolonged periods in storage, and the most stringent requirements for energy, power, and cycle life on both a weight and volume basis. Costs are not often a top priority, but the extensive resources and time periods required to develop, test, and deploy aerospace batteries may be a determinant in whether a project is enacted.

Aerospace applications can be generally categorized as shown in Fig. 25B.1. This chart does not include general aviation, commercial aircraft, drones, or other such more common aircraft applications.

There are, of course, many variants to this chart, but the table clearly shows the wide variety of the potential battery technologies in various nonterrestrial applications. This section is not intended to be a compendium of all such applications, but instead provides anecdotal details on a variety of aerospace battery development programs. These examples include various battery technologies and both primary and secondary batteries. In addition it should be noted that battery requirements are often not well delineated in these applications due to both the evolving nature of the mission requirements and secrecy concerns. For example, battery requirements for the Strategic Defense Initiative, enacted during the Reagan administration (often called the Star Wars program), did not always include complete specifications. The specific energy for one particular missile battery was left as "To Be Determined." The entire multimillion dollar battery program was executed without ever knowing what battery capacity was actually required. In other cases, the power needs of the spacecraft are not finalized until new equipment technologies are perfected or mission profiles are finalized.

The sections below provide a few examples of different types of aerospace battery applications. Many other case studies are available and should be consulted for further insights. For example, Sony recently presented data[1] on the use of an original 18650 cell design (hard carbon anode/$LiCoO_2$ cathode cells) in the Rosetta space craft and lander: launched in 2004 with comet landing in 2014 (10 years, 3 billion miles). Some other prominent events include the Apollo 13 moon mission (i.e., failure of the on-board power system) and the Mars Rover (i.e., the extended mission success).

## 25B.2   MILITARY AIRCRAFT BATTERIES

### 25B.2.1   First Lead-Acid, Military Aircraft Battery: F/A-18A

In the late 1970s the U.S. Navy and McDonnel Douglas were developing a new aircraft for carrier service—the F/A-18A jet fighter. Traditionally, sealed nickel-cadmium (NiCd) batteries were used on all military aircraft due to its proven reliability over a long history in demanding environments. However, several issues with the

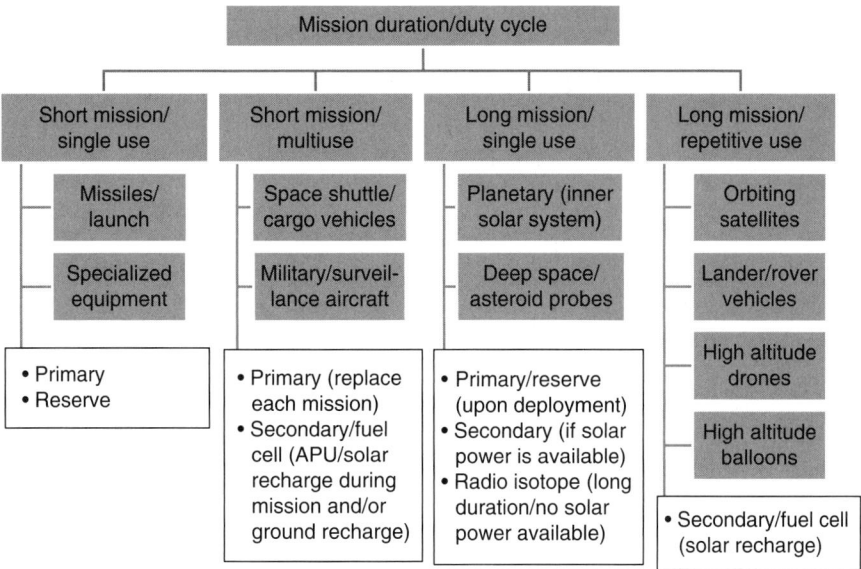

**FIGURE 25B.1**    Aerospace battery applications.

performance of NiCd batteries resulted in the U.S. Navy deciding to search for alternative technologies. Specifically, at this point in time, Gates Energy Products was in the process of commercializing the first sealed valve regulated lead-acid (VLRA), absorbed glass mat (AGM) battery. The first commercial cell was a 2.5 Ah, 2 V, D-size cell. The Navy requirement, however, dictated the use of a larger cell, a 12.5 Ah size for a 12-cell, 24 V module.

The function of the battery was to provide ground power prior to or after the mission, start-up power for the on-board auxiliary power unit (APU) and emergency power in case of APU and/or engine failure during flight. Hence, while energy and weight requirements were important, the APU start-up required excellent cold-cranking power over a range of temperature conditions and various states-of-charge. Of course cycle life and maintenance needs were other key considerations. Overall, the lead-acid batteries provided some weight and maintenance benefits over NiCd, but it was the cold temperature, high power performance that was the deciding factor in the conversion to the VLRA, AGM technology.

However, the lead-acid battery system was, nonetheless, not without issues. Specifically, to meet the operating temperature requirements, the battery housing was outfitted with internal silicone rubber heating blankets that were woven between the 12 cylindrical Gates cells, which were in a 4 × 3 array. The battery could be heated by auxiliary ground power units, but was also designed so that in remote locations the battery itself could provide enough current flow to the heaters to reach a point within a limited timeframe where under cold conditions APU start would be successful.

In order to survive shock and vibration testing, the bottoms of the cells were potted in a few centimeters of a thermally conductive silver-filled epoxy and a thermoformed plastic header was bolted down on the tops of the cells with a precise level of compression. The initial prototypes suffered several different failures. In one vibration test, before the cells were properly restrained, the contents of the cells were totally pulverized. The lead grids, fiberglass mats, and acid were reduced to a slurry that could be poured out of the cell.

Hence, a key finding from this work was that while cell design (a cylindrically wound construction with welded tabs/lugs at the top) is important, the cell by itself was not able to survive the vibration test. Additional engineering fixes were required to support and restrain the cells. A specialized welded aluminum battery housing to further restrain the cells from moving was also needed to achieve success. Similarly, even the electrochemistry needed a boost from the use of battery heating blankets.

### 25B.2.2   All-Electric Military Aircraft

By the early 1990s at about the time lithium metal anode secondary cells were first being commercialized, one aerospace company decided to evaluate the concept of an all-electric military aircraft. The power system would forego the traditional use of an APU and rely on battery power to directly start the jet engines (a severe cold-cranking test). More importantly, an all-electric aircraft would have also allowed the elimination of all hydraulic power systems on the plane. This concept is now being used (i.e., fly-by-wire), but not with lithium metal anode batteries. The on-board alternator would provide electrical power in flight and serve to recharge the battery. In an emergency (alternator failure) the battery would provide enough power to land the plane.

The initial proposed battery design used lithium metal anodes and $LiCoO_2$ cathodes with a traditional electrolyte solution of that era ($LiAsF_6$/$LiBF_4$/methyl formate). To achieve high-rate capabilities for the engine start, an innovative anode substrate was proposed among other component modifications. The proposed specifications for the required battery pack are shown in Table 25B.1 with some comparison data on the sealed lead acid cell used on the F/A-18A aircraft.

**TABLE 25B.1**   Advanced Military Aircraft Batteries

| Single cells | Units | All-electric aircraft (Li/CoO$_2$) | F/A-18A aircraft[*] (sealed lead acid) |
|---|---|---|---|
| Voltage | V | 4 (nom.), 3.75 at low rate/mid-point | 2 (nom.) |
| Capacity | Ah | 65 | 12.5 est. |
| Weight | kg | 1.75 | 0.84 est. |
| Volume | L | 0.65 | 0.3 est. |
| Specific energy | Wh/kg | 140 | 30 est. |
| Energy density | Wh/L | 375 | 80 est. |
| Peak current | A | 425 up to 25% DOD | 100 est. |
| Specific power | W/kg | 450 | 120 est. |
| Power density | W/L | 1230 | 333 est. |

| Battery pack | Units | All-electric aircraft (Li/CoO$_2$) | F/A-18A aircraft[†] (sealed lead acid) |
|---|---|---|---|
| Nom. volt | V | 24 (nom.), 22.5 at low rate/mid-point | 24 (nom.) |
| Capacity | Ah | 65 | 7.5 (U.S.), 10 (EUR) |
| Weight | kg | 16 est. | 10.5 |
| Volume | L | 7.1 | |
| Specific energy | Wh/kg | 100 | 17–23 |
| Energy density | Wh/L | 225 | |
| Specific power | W/kg | 300 | |
| Power density | W/L | 675 | |

[*]Estimates from preliminary cell specifications, Gates Energy Products, Battery Application Manual, 1982.

[†]Enersys/Hawker Technical Data Sheet for MIL-B- 8565 F-18 Hornet Battery (web: 5/24/2018, file:///C:/Users/Administrator/Downloads/Enersys_Hawker_Maintenance_Free_Sealed_Lead_Acid_Range_June-2012_Version1_EN%20v4%20SINGLE%20PAGES%20low%20res%20(4).pdf).

## 25B.3   DEEP SPACE MISSIONS (JUPITER GALILEO PROBE)

In the late 1980s, the Jet Propulsion Laboratory undertook a mission to investigate the moons (53 known moons), rings, and the atmosphere of Jupiter. A spacecraft would be placed into orbit around Jupiter and a probe would then be sent into the planet. Prior studies of the atmosphere were limited to remote sensing, but direct measurements of composition, temperatures/pressures, etc. were desired. It was known that the heat and pressure would eventually destroy the probe, but the goal was to relay instrument readings from the probe back to the parent spacecraft and then to Earth for as long as possible.

The spacecraft was initially intended to be launched directly to Jupiter to 1982, but various delays put the launch off for seven more years. Problems with the Centaur rocket required a rethinking of the mission. During the delays, the program evolved to include the use of planetary gravitational assist (i.e., a sling-shot maneuver) to accomplish the mission with a less powerful launch vehicle. Ultimately, the spacecraft, designated the Galileo, was carried into space on the Atlantis space shuttle on October 18, 1989. Galileo was first sent around Venus, with a return flight to Earth followed by an accelerated ejection from earth orbit to Jupiter. The complexity of linking Venus, Earth, and Jupiter together in this trajectory (called VEEGA for Venus, Earth, Earth Gravity Assist), dictated a limited time frame for executing the launch. The development and build of the spacecraft, the probe equipment, and the probe batteries, therefore, needed to be synchronized to this launch window.

A lithium sulfur dioxide ($LiSO_2$) electrochemistry was chosen for the probe due to a combination of energy density, temperature capabilities, and storage characteristics. Among the many trade-offs that were considered in designing this spacecraft were the compromises between payload and vehicle weight. The long mission to Jupiter limited the amount of instrumentation and battery packs that could be carried. Even the largest rocket engines of the day would not support a very extensive load of equipment. The value of the mission would be lessened if the probe's weight was severely curtailed.

A reputable aerospace battery manufacturer (Alliant Techsystems) and an established cell design were chosen in order to shorten delivery schedules and eliminate risk. The focus could then shift to meeting delivery schedules and confirming performance under proposed flight conditions. Multiple identical battery modules were constructed to allow for testing of these residuals in the laboratory over the multiyear duration of the flight. Any changes in battery response could then be known prior to deployment of the probe into the Jupiter atmosphere. The battery specifications[4] are shown in Table 25B.2.

**TABLE 25B.2**    Galileo Probe Lithium Sulfur Dioxide Battery Specifications

| Specification | Unit | Value |
|---|---|---|
| Number of modules | Each | 3 |
| Battery pack design | Series/parallel arrays | 13 cells in series, 3 modules in parallel |
| Current loads (during descent) | A | 0.44 initial to 8.6 w/squibs to 4.5 end |
| Voltage profile | V | >37 OCV to 27 EOL |
| Module length | cm | 35.56 |
| Module height | cm | 8.89 |
| Module width | cm | 7.11 |
| Total weight | kg | 18.7 |
| Battery weight factor | % of total probe weight | 6.0 |
| Leakage current | ηA (max.) | 150 |

In addition, a number of electronic controls were implemented to provide redundancy and isolate faults. Separate connectors were used for testing of the cells and for use under load. Other demands placed on the battery included power to ignite thermal battery squibs (eight cell taps with up to 6.3 A loads for parachutes, etc.) and to provide 155 days of power to the clock during the coast phase.

The Galileo probe was released on July 12, 1995, after a 5-year flight, and entered the Jovian atmosphere on December 7. The 339 kg, 1.3 m diameter, 0.86 m high probe had an entry speed of 47.8 km/s with external heat shield temperature reaching 14,000 K and wind speeds of 724 km/h. The 21 Ah $LiSO_2$ battery with a mission requirement of 16.3 Ah was intended to last from 48 min (dependent on capacity) to 75 min (dependent on pressure failure). Ultimately, the probe descended 153 km and transmitted data for 57.6 min.[2,3]

The probe battery module supported the functioning of various scientific instruments and was successfully able to transmit the following atmospheric data to the Galileo spacecraft over the course of the descent:

- Temperature
- Pressure and deceleration
- Composition
- Vapor/condensate clouds
- Particulates
- Electromagnetic emissions from lightning and high energy particles

Clearly, battery technology that can last 6 years over 4 billion kilometers in deep space cold storage after surviving the rigors of launch and various gravitation assist maneuvers and then perform to specification during a rough reentry into Jupiter's hot, dense, electromagnetic atmosphere is quite remarkable. The scientific value of these data cannot be underestimated. Even as recent as 2018, the analyses continue to reveal new findings.[4] The follow-up mission report concluded as follows[5]:

1. Battery performance met mission requirements
2. Ground based and flight tests prior to probe deployment predicted acceptable actual performance would be achieved
3. The battery selection was successful due to
    a. Good energy density
    b. Proven manufacturing capabilities
    c. Good extended storage
    d. Minimized voltage delay upon activation

## 25B.4   TERRESTRIAL SATELLITE APPLICATIONS: LEO AND GEO[6]

### 25B.4.1   Overview

LEO (low earth orbit) and GEO (geosynchronous earth orbit) satellites represent two different types of battery applications. LEO batteries require 3 to 6 years at approximately 6000 cycles per year (18,000–36,000 total cycles). The GEO batteries must last 15 to 20 years, but are cycled about 100 times per year (1500–2000 total cycles). Ni-$H_2$ batteries were used extensively for these applications, but recently lithium batteries have been deployed.

### 25B.4.2   GEO Batteries

*Battery Requirements.*    Communications satellites operate continuously without interruption, including during eclipses, which occur each day during the equinox seasons. When the satellite is obscured by the earth's shadow, a solar array power outage will occur for up to 72 min during the midseason (about March 21 or September 23). Hence, over this 45-day period, the batteries supply power to the spacecraft during the darkness and then must be recharged during the sunlight portion of each eclipse day. During the summer and winter solstice periods (approximately 138 days each) between eclipse seasons, the batteries are not used and only need trickle or float charge.

*Charge Control.*    At peak eclipse a typical Ni-$H_2$ battery is discharged up to 70% of its beginning-of-life rated capacity. After 15 years of operation, the battery still must meet the same initial load requirements. In some cases batteries are designed to operate at 100% DOD to reduce weight, leaving no reserve capacity. GEO satellite batteries are optimally recharged at a fixed charge/discharge ratio, returning 105% to 115% of the capacity at a high charge rate, before switching to trickle charge for the remainder of the 24 h eclipse day.

### 25B.4.3    LEO Batteries

*Battery Requirements.*    LEO satellites are closer to earth and complete an orbit on average every 96 min at a typical 555 km altitude, 28.3° inclination. During the 15 orbits that occur each day, the sunlight and eclipse periods will vary with each orbit. For example, during December 1991, the eclipse durations vary from a maximum of 35.58 min on December 1 to a minimum of 26.97 min on December 30.

*Charge Control.*    The battery is charged by solar arrays during the sunlight period and provides spacecraft power during the eclipse period. Such deep duty cycles generate heat during charge that must be dissipated. The large variation in the depth of discharge (due to variation in eclipse duration) requires an adaptable, sophisticated charger. For Ni-$H_2$ batteries charging at temperatures between 0 and 10°C will provide nearly 100% ampere-hour charge efficiency and ~85% watt-hour efficiency. Lithium ion cells have different responses, but still require careful control of temperatures and charge rates.

In one application[7] a temperature-compensated voltage-limit charging method was used.[8,9] The battery voltage limits and charge rates varied as a function of temperature over the life (i.e., going from high rate to a $C/100$ trickle charge and with a 1.513 V limit per cell at 0°C for the 22-cell battery. The battery is only charged at this voltage limit to about 73 Ah (83% of its rated capacity); however, thermal stability is maintained with an overall battery watt-hour efficiency of 80% to 85% with these control limits. Higher voltage charge limits would decrease the coulombic efficiency and the overall energy efficiency, and increase the heat generation, possibly exceeding the allowable temperature limits. Cell pressure may also serve as an indication of the state-of-charge of the battery and can be used for partial state-of-charge conditions.

### 25B.4.4    GEO Performance Data

A summary of the Ni-$H_2$ batteries on the INTELSAT V satellites after 9 years in orbit is detailed below.

*Voltage Performance in Orbit.*    INTELSAT V batteries are judged by the minimum end-of-discharge voltage observed during an eclipse season, which has a minimum requirement of 28.6 V, or 1.10 V per cell average and allows for one cell to fail via a short circuit. The actual battery voltages, load currents, and depths of discharge for each of the 14 batteries on the satellites are presented in Table 25B.3 for the Fall 1990 eclipse season.[10] Minimum cell voltages within each battery and the corresponding average cell voltage are also shown. The minimum end-of-discharge battery voltages on the longest eclipse day ranged from 31.2 to 32.4 V after 7 years. The cells within the batteries had uniform end-of-discharge voltages during the Fall 1990 eclipse season.

**TABLE 25B.3**    Battery Loads and Minimum Voltages for Fall 1990 Eclipse Season

|  | DOD, % | | Current, A | | Voltage, V | | Cell voltage, V | | | |
|---|---|---|---|---|---|---|---|---|---|---|
|  | Battery 1 | Battery 2 | Battery 1 | Battery 2 | Battery 1 | Battery 2 | Batt. 1 avg. | Batt. 1 min | Batt. 2 avg. | Batt. 2 min |
| F-6 | 55.8 | 53.1 | 14.2 | 13.5 | 32.0 | 32.4 | 1.20 | 1.16 | 1.20 | 1.19 |
| F-8 | 54.0 | 54.4 | 13.7 | 13.8 | 32.0 | 32.0 | 1.20 | 1.18 | 1.20 | 1.18 |
| F-10 | 56.9 | 55.7 | 14.4 | 14.3 | 31.8 | 32.0 | 1.19 | 1.18 | 1.20 | 1.18 |
| F-11 | 55.3 | 60.0 | 14.1 | 15.4 | 32.0 | 32.0 | 1.20 | 1.18 | 1.20 | 1.19 |
| F-12 | 53.5 | 58.0 | 13.6 | 14.8 | 32.0 | 31.8 | 1.20 | 1.18 | 1.18 | 1.18 |
| F-13 | 67.0 | 59.0 | 16.9 | 15.0 | 31.2 | 31.8 | 1.17 | 1.15 | 1.19 | 1.17 |
| F-15 | 67.0 | 62.3 | 16.9 | 15.8 | 31.2 | 31.8 | 1.17 | 1.16 | 1.18 | 1.16 |

The one exception is cell 22 in battery 1 on the F-6 spacecraft. This cell was 40 mV below the average cell voltage but still well above the minimum voltage requirement.

***Pressure Data.*** Prior to each eclipse season the batteries are reconditioned as per the data for INTELSAT V F-6 battery 2 as presented in Table 25B.4.[10] The pressure data during the reconditioning discharge are listed at the start (i.e., end of charge) and at the end (i.e., end of discharge) of the reconditioning cycle.
The data reveal the following:

1. The strain-gauge bridge circuit provides useful pressure data.
2. The reconditioning EOD pressure did not change with time.
3. Results indicate that no oxidation or corrosion has occurred.

**TABLE 25B.4** Reconditioning Capacity and Pressure Data for INTELSAT V F-6 Battery No. 2

| Eclipse season | Measured capacity, Ah | Max. EOC pressure, lb/in$^2$ | Min. EOD pressure, lb/in$^2$ | $\Delta P$, lb/in$^2$ | Pressure constant $\Delta P$/measured capacity, lb/in$^2$/Ah* |
|---|---|---|---|---|---|
| F83 | 38.1 | No pressure data in database | | | |
| S84 | 35.4 | No pressure data in database | | | |
| F84 | 37.7 | 516.39 | 13.87 | 502.62 | 13.33 |
| S85 | 37.6 | 518.49 | 17.90 | 500.59 | 13.31 |
| F85 | 37.5 | 515.14 | 17.23 | 497.9 | 13.27 |
| S86 | 37.9 | 519.34 | 15.32 | 504.02 | 13.29 |
| F86 | 37.6 | 519.73 | 22.03 | 497.70 | 13.23 |
| S87 | 38.3 | 514.34 | 13.87 | 505.47 | 13.19 |
| F87 | 37.2 | 519.73 | 22.03 | 497.7 | 13.37 |
| S88 | 38.3 | 525.78 | 16.20 | 509.58 | 13.30 |
| F88 | 37.8 | 521.86 | 17.90 | 503.96 | 13.33 |
| S89 | 36.9 | 526.91 | 18.67 | 508.24 | 13.77 |
| F89 | 40.2 | 534.22 | −0.57 | 534.79 | 13.30 |
| S90 | 38.6 | 551.73 | 19.22 | 532.51 | 13.79 |
| F90 | 36.0 | 530.87 | 38.04 | 492.83 | 13.68 |
| S91 | 39.5 | 546.52 | 17.23 | 529.29 | 13.39 |
| F91 | 39.0 | 545.30 | 17.90 | 527.40 | 13.52 |
| | | | | | Average 13.37 |

*1 lb/in$^2$ = 6895 pa.

## 25B.4.5 LEO Performance Data

The Hubble Space Telescope (HST) was launched on April 24, 1990, with six 88 Ah Ni-H$_2$ batteries and was the first reported regular mission to use Ni-H$_2$ batteries in a LEO application.[11] The batteries are charged to a temperature-compensated voltage limit as described above and discharged to a 7% to 10% of depth. As reported at the 1991 IECEC, "To date (April 1991) the performance of the batteries has been flawless."[11] Orbital data had shown an expected normal slow loss in useful capacity, which could eventually limit support of the HST. During servicing mission STS 125 in May 2009, after 18 years of service (13 years beyond design orbital life), the 6 batteries were replaced to extend the life of the HST.

## 25B.5  SUMMARY

Aircraft, missile, satellite, and other aerospace applications represent one of the most challenging environments for batteries. While costs will always, eventually, be a major factor, in many cases the batteries represent an enabling technology for the proposed missions. Therefore, new, innovative, and often high-risk options will be considered despite the cost, complexity, or the extended development cycles. The potential for technology transfer from these programs is often mentioned as a reason to justify the efforts.

## ACKNOWLEDGMENT

Major content for this chapter was provided in whole or in part by Jack N. Brill, Chap. 24, Nickel-Hydrogen Batteries, *Linden's Handbook of Batteries,* 4th ed., T. B. Reddy and D. Linden, eds., McGraw-Hill, 2011.

## REFERENCES

1. Y. Nishi, Past, present and future of LIB. Can new technologies open up new horizons? 35th Annual International Battery Seminar and Exhibit, March 26–29, 2018, Ft Lauderdale, Florida.
2. https://solarsystem.nasa.gov/missions/galileo/in-depth/ (extracted from the World Wide Web, May 7, 2018).
3. https://nssdc.gsfc.nasa.gov/nmc/spacecraftDisplay.do?id=1989-084E (extracted from the World Wide Web, May 7, 2018).
4. https://www.sciencedaily.com/releases/2018/04/180430131826.htm (extracted from the World Wide Web, May 7, 2018).
5. https://ntrs.nasa.gov/archive/nasa/casi.ntrs.nasa.gov/19970013722.pdf.
6. Content from *Linden's Handbook of Batteries*, Chapter 24, Nickel-Hydrogen Batteries, 4th edition, Jack N. Brill.
7. NASA's Marshall Space Flight Center (MSFC) and Lockheed Missile and Space Company (LMSC), 88 Ah Ni-$H_2$ batteries that replaced the Ni-Cd batteries on board the Hubble Space Telescope (HST) satellite as per NASA's Marshall Space Flight Center (MSFC) and Lockheed Missile and Space Company (LMSC).
8. D. E. Nawrocki, J. D. Armantrout, et al., "The Hubble Space Telescope Nickel-Hydrogen Battery Design," *Proc. 25th IECEC,* Reno, NV, 1990, Vol. 3, pp. 1–6.
9. J. E. Lowery, J. R. Lanier Jr., C. I. Hall, and T. H. Whitt, "Ongoing Nickel-Hydrogen Energy Storage Device Testing at George C. Marshall Space Flight Center," *Proc. 25th IECEC,* Reno, NV, 1990, pp. 28–32.
10. J. D. Dunlop, A. Dunnet, and A. Cooper, "Performance of INTELSAT V Ni-$H_2$ Batteries in Orbit (1983–1991)," *Proc. 27th IECEC,* 1992.
11. J. C. Brewer, T. H. Whitt, and J. R. Lanier, Jr., "Hubble Space Telescope Nickel-Hydrogen Batteries Testing and Flight Performance," *Proc. 26th IECEC,* 1991.

# CHAPTER 26
# BATTERY-POWERED TRANSPORTATION

## 26.0  GENERAL OVERVIEW

The transport of people, equipment, and materials/supplies has been a key human need since ancient times when humans, animals, and nature (wind, water currents, and gravity) were the primary sources of transportation power. Modern civilizations rely on various types of engines (steam, combustion, electric, etc.) to haul these loads. During the last century electric vehicles were commonplace in certain realms: forklifts, tow vehicles, pallet movers, and many other utilitarian applications. But in just the last few decades, an exponentially greater effort has arisen to develop and commercialize battery-powered transport vehicles across a wide spectrum of applications. Electric road vehicles (automobiles) are the most prominent example, but other efforts are nonetheless well underway to provide battery-powered delivery systems for land, air, and sea.

In evaluating this market a number of metrics are appropriate for comparing battery systems as well as competitive means of propulsion. Table 26.1 summarizes some key measures useful both alone and in combination to making application engineering decisions for transportation battery systems.

Summaries for the three exemplary market niches covered in this chapter are discussed below.

### 26.0.1   Electric and Hybrid Vehicles

Section A details the latest activities and state-of-the-art solutions for electric automobiles. Clearly, great progress has been made and there is no chance of turning back from electrification of the world's automotive fleets used for transporting people and items. Of course, the use of grid electric power for rail lines and other closed loop systems will continue to grow, but the transition to battery-powered vehicles for independent local use is a major new phenomenon.

### 26.0.2   Traction and Motive Power Vehicles

Traction and motive power vehicles are a subset of the transportation vehicle market that is mostly focused on industrial and special purpose utility applications. Section B covers the fundamentals for this large, important, and growing application area. These vehicles typically replace human power, wired electric grid power systems, and/or combustion engines for moving personnel, equipment, supplies, etc. typically within bounded areas such as industrial factories, airports, mines, business complexes, etc. Equipment that is better classified as "recreational" or "personal use," such as golf carts, small drones, and electric scooters, are not typically included in this category but are discussed in other chapters.

### 26.0.3   Electric Aircraft

To demonstrate the far-reaching potential for battery-powered transportation, Sec. C details progress on the efforts to certify crewed, battery-powered electric aircraft. While gliders and drones with battery-powered propellers have existed for years, recent new endeavors are intended to compete directly from take-off to landing

**TABLE 26.1**    Application Engineering Metrics for Battery-Powered Transportation Systems

| Category | Optimization objectives | Main units of measure |
|---|---|---|
| Battery physicals | Size and weight | kg, L, kg/L, form factors |
| Battery energy | Discharge output | Wh (Ah), Wh/kg, Wh/L |
| Battery power | Discharge rate | W (A), W/kg, W/L |
| Battery duration | Run time | h, km (mi), duty cycles |
| Battery life | Service life | year, km (mi), recharge cycles |
| Battery charging | Charge acceptance | A, W, h |
| Voltage regulation | Minimum and maximum charge/discharge | V (OCV and load) |
| Storage capability | Capacity retention | %Loss/month or year |
| Temperature capability | Charge, discharge, storage | °C |
| Battery resistance | Ohmic loss/voltage drop | $\Omega$, $\Delta V$ (joule heating) |
| Vehicle capability | Operational parameters | km/kWh, h/kWh, km/h |
| Purchase costs | Battery, battery system, and total vehicle | \$'s and \$'s per various unit measures listed above |
| Operating costs | Cycling, maintenance, and total life cycle | \$'s and \$'s per various unit measures listed above |
| Safety | Various | User specific/intangibles |
| Reliability | Various | User specific/intangibles |
| Sustainability | Various | User specific/intangibles |

with small, fixed wing aircraft using aviation fuel alone. Similar to electric vehicles, battery-electric aircraft have potential to offer reduced operating and maintenance costs. Other possibilities, such as hybrid systems with biofuel engines and solar arrays, are also viable options.

## 26.0.4   Summary

Battery-powered transportation is an old concept with many new modifications. Chapter 26 provides examples of three different application areas. Additional details on other similar types of battery-powered vehicles and equipment are discussed in other chapters. For example, Chap. 25A details a variety of smaller transportation devices intended for use by an individual person (i.e., electric scooters/platforms, bikes, and motorcycles). Similarly, considerable efforts have been ongoing for decades to develop and implement numerous other battery-powered transport units. Main categories include autonomous mobile robots, surface and undersea marine craft, such as small electric trolling motors for recreational fishing, large tugboats, and coal haulers (see Chap. 1).[a] The use of batteries in military torpedoes/unmanned underwater vehicles is detailed in Chap. 29, also.

Other examples, such as the Mars Rovers[b] and the Alvin[c] (the first deep sea human-occupied vehicle [HOV] submersible), demonstrate dramatic successes in the field of battery-powered transport equipment used in the most challenging environments. Hybrid and full battery power systems are also now filtering into numerous other industries, including trucking, public transportation (buses), railways, package delivery drones, etc. The three sections that follow offer guidance on the keys to successfully implementing battery technology into transportation applications. Detailed analyses of the applications (energy and power requirements over the full duty cycle under the full range of environmental and service conditions) and a thorough, in-depth understanding of the relevant battery technologies are at the center of these efforts.

---

[a]https://cleantechnica.com/2017/12/02/china-launches-worlds-first-electric-cargo-ship-will-use-haul-coal/ (published December 2, 2018; extracted from world wide web on January 22, 2018).

[b]https://www.researchgate.net/publication/238794735_LiIon_Rechargeable_Batteries_on_Mars_Exploration_Rovers (published December 9, 2014; extracted from world wide web on January 22, 2018).

[c]http://www.whoi.edu/main/hov-alvin (published 2017; extracted from world wide web on January 22, 2018).

# SECTION A

# ELECTRIC AND HYBRID VEHICLES

**Dennis A. Corrigan, Alvaro Masias**

## *26A.0 OVERVIEW*

### 26A.0.1 Introduction

At the beginning of the 21st century, a new generation of electric drive passenger vehicles, including electric and hybrid vehicles, was introduced after many decades of total domination by conventional internal combustion engine (ICE) vehicles. These alternative propulsion vehicles are viewed around the world with great promise to provide major societal value:

- To eliminate or reduce exhaust emissions from automobiles especially in urban areas of high air pollution.
- To provide strategic flexibility in national energy policy by reducing dependence on foreign oil for transportation.
- To reduce carbon dioxide greenhouse gas emissions, addressing concerns over global climate change.

Electric vehicles (EVs), also termed battery electric vehicles (BEVs), utilize rechargeable batteries to power electric motors for propulsion.[1-6] BEVs are principally recharged from external power sources, usually the electrical grid. The range of EVs has been extended in a recent innovation that incorporates a small ICE-powered generator that can utilize gasoline fuel. These vehicles, known as plug-in hybrid electric vehicles (PHEVs), can operate in all-electric mode like an EV or in the hybrid mode for extended trips.[3-5] Together, EVs and PHEVs comprise a category known as plug-in electric vehicles (PEVs), which utilize grid electricity, providing maximum benefit to society's objectives for environmental and energy security needs.

Hybrid electric vehicles (HEVs) utilize electric motors powered by batteries in conjunction with ICEs, but they do not utilize grid electricity and do not need to be plugged in.[1-6] While they are not zero-emission vehicles, HEVs can also substantially improve the energy efficiency over conventional ICE vehicles and make significant contributions to society's objectives for environmental and energy security needs. The commercialization of HEVs has progressed more rapidly because they are economically more feasible without incentives.

### 26A.0.2 History of Electric Vehicles

EVs date all the way back to 1837, shortly after the invention of the primary battery around 1800 and the electric motor in 1832. After the invention of rechargeable lead-acid batteries in 1860, they became a practical device, even before the development of the ICE. EVs were highly competitive with vehicles powered by steam engines and ICEs for the rest of the 19th century. In the golden age of EVs (1900–1912), over 30,000 EVs were introduced into the United States and several times that worldwide. They were powered by lead-acid batteries and the Edison nickel-iron battery. EVs were especially popular in cities, to overcome the pollution issues (manure) of the day by replacing horse-drawn vehicles. EVs also did not share a principal disadvantage of gas-powered cars: the difficulty and dangers of hand cranking ICEs. An early EV is shown in Fig. 26A.1.

Ironically, the development of another battery-powered device, the electric starter motor, allowed the gas-powered car to prevail. Once the disadvantage of the hand cranking issue was thus dealt with, the advantages of gas-powered cars, especially longer range and lower cost, allowed them to fully dominate the automotive industry for the rest of the 20th century. EVs cost several-fold as much as gas-powered vehicles due to the high cost

**FIGURE 26A.1**   Nickel-iron battery inventor Thomas Edison with 1912 Detroit Electric EV. (*Courtesy of the National Museum of American History.*)

of batteries. Throughout the rest of the 20th century, EVs were relegated to niche applications such as milk trucks in England, golf carts, and forklift applications. However, automotive companies continued periodic cycles of R&D into EVs in response to environmental issues and episodes of high fuel costs such as the Organization of Petroleum Exporting Countries (OPEC) oil embargo of the 1970s.

In 1990, as one approach to address serious air pollution in urban areas, the California Air Resources Board (CARB) enacted a zero-emission vehicle (ZEV) mandate that initially required automotive original equipment manufacturers (OEMs) to make available for sale EVs equivalent to 2% and 10% of their fleets in 1998 and 2003, respectively. In response, U.S. automakers vigorously developed a variety of EVs. In partnership with the U.S. Department of Energy (DOE), they formed the United States Advanced Battery Consortium (USABC) to develop advanced batteries with higher energy densities than commercially available lead-acid batteries. EVs developed in response to the mandate were mostly powered by lead-acid and nickel-metal hydride batteries although nickel-cadmium, sodium-sulfur, lithium-ion batteries were also evaluated. Particularly noteworthy was the General Motors EV1 (see Fig. 26A.2), a purpose-built EV famous due to its high performance, acceleration from 0 to 60 mi/h in 7 s, and a range approaching 200 mi (320 km) when equipped with nickel-metal hydride (NiMH) batteries.[7] Other car companies developed limited production EVs as well, including the Chrysler EPIC minivan and Ford Ecostar (as shown in Fig. 26A.2); and the Toyota RAV4 electric and the Honda EV Plus.

The mandate was not fully implemented, and EVs developed in response to the ZEV mandate were only offered in limited quantity because they could not be manufactured and sold with a profit. Disadvantages to the customer were limited range, "slow refueling" (i.e., charging), and most importantly the high cost of these EVs.

**FIGURE 26A.2**   Electric vehicles developed in response to the California zero-emission vehicle (ZEV) mandate include (left to right) GM EV1, Chrysler EPIC Minivan, and Ford Ecostar.

In particular, the battery cost was the key issue. High battery costs resulted in limited range EVs costing more than double that of comparable gas-powered vehicles.

The California mandate also created great conflict and controversy with disputes between stakeholders including government, the auto industry, the oil industry, electric utilities, EV proponents, and EV antagonists. The spirit of these times has been captured in the 2006 documentary "Who Killed the Electric Car?" that views the demise of the California mandate as a great conspiracy.[8] In reality, many socioeconomic factors played a role, but the biggest issue preventing commercialization was the high cost of batteries. However, the California mandate (which still exists in a form modified many times) continues to be a force promoting continued technological progress in BEVs.

Despite the roll back of the ZEV mandate, societal motivations for EVs remained strong, especially environmental concerns with increased carbon dioxide emissions and global warming, and also strategic issues related to the limited supply of petroleum and the geopolitical distribution of existing supply. While the enthusiasm for commercialization of BEVs stalled, the government focus largely shifted to the development and demonstration of fuel cell EVs. This interest was spurred by hope that new proton exchange membrane (PEM) fuel cell technology could enable long-range EVs that could be rapidly recharged by hydrogen. The U.S. government took a leadership role in the development and demonstration of fuel cell vehicles in the FreedomCAR initiative founded and promoted by the Bush administration in 2001. However, in the first decade of the 21st century, although technical progress in fuel cell vehicles was promising, commercialization was seriously hampered by significant cost and infrastructure issues (e.g., hydrogen production, distribution, and storage).

Meanwhile, international interest in BEVs continued, particularly in developing countries such as China and India. These large, densely populated countries have much more serious issues than the United States in terms of petroleum supply and air pollution. The number of gas-powered vehicles has been severely restricted in numerous major cities. Additionally, vehicles used in these overcrowded cities do not have the same acceleration and range requirements that have driven auto development in the industrialized world. Even with range limitations, EVs may reach much wider acceptance in the more populous developing countries.

In India, a small EV with modest performance called the REVA ("one who moves" in Sanskrit) was developed that first went on sale in 2001. Several thousand vehicles were produced. REVA was subsequently acquired by the large Indian automotive company Mahindra. In China, BYD (build your dreams) Auto was founded in 2003 and began production of EVs in 2008. Today, it is among the world's largest manufacturers of EVs.

In 2003, Tesla Motors was founded as a new U.S. automotive company exclusively focused on BEVs. In 2008, Tesla introduced the high-powered Tesla Roadster electric sports car with a 0 to 60 mi/h acceleration time of less than 4 s. Powered by a series-parallel arrangement of 6831 lithium-ion 18650 laptop computer battery cells, this vehicle was capable of a range of over 200 mi (320 km). Tesla has followed with other models with wider appeal including the Model S luxury midsize EV and Model X electric sport utility vehicle (SUV).

In 2007, General Motors (GM) announced their intention to develop a high-performance PHEV known as the Chevy Volt. It utilized a 16 kWh lithium-ion battery pack to enable an all-electric range of 40 mi (65 km). GM described the series HEV design as an extended-range electric vehicle (E-REV) and indeed it was expected to operate mostly as an EV since the majority of trips in the United States are within the 40 mi (65 km) range. The Volt was introduced in late 2010, and over 100,000 vehicles were sold by 2015. Other automotive companies such as Ford, Toyota, and BYD have subsequently introduced PHEVs including the Ford C-Max, Toyota Prius Prime, and BYD Qin. Frequently, smaller battery packs were used for reduced cost, but with a trade-off resulting in a lower all-electric range and blended ICE operation with electric drive during most trips.

Starting in 2009 with the Obama administration, the U.S. government renewed focus on BEV R&D with the U.S. DRIVE partnership, a successor to FreedomCAR, with a goal of 1 million PEVs in commercial use by 2015. Coupled with renewable electricity produced by solar and wind energy, a transition to EVs could theoretically eliminate vehicular emissions of greenhouse gases (GHGs), solving a critical environmental problem. Additionally, transitions to PEVs could alleviate strategic energy supply issues with respect to petroleum imports. While the 1 million PEV goal was not attained in 2015, over 600,000 PEVs were sold in the United States by 2017 (see Fig. 26A.3).[9]

With sales of over 160,000 PEVs in 2016, the U.S. market share reached 0.9%. A variety of new models were available including the Nissan Leaf, Chevy Volt PHEV, and Chevy Bolt EV shown in Fig. 26A.4. In the United States, political sentiments to promote PEVs by regulation and incentives may be waning. However, strong incentives and regulations in China and Europe have accelerated PEV commercialization globally. In 2016, 200,000 PEVs were sold in Europe. China is now the world leader in PEV commercialization with over 350,000 vehicles sold in 2016 reaching a market share of 1.5% in the world's largest automotive market.[10]

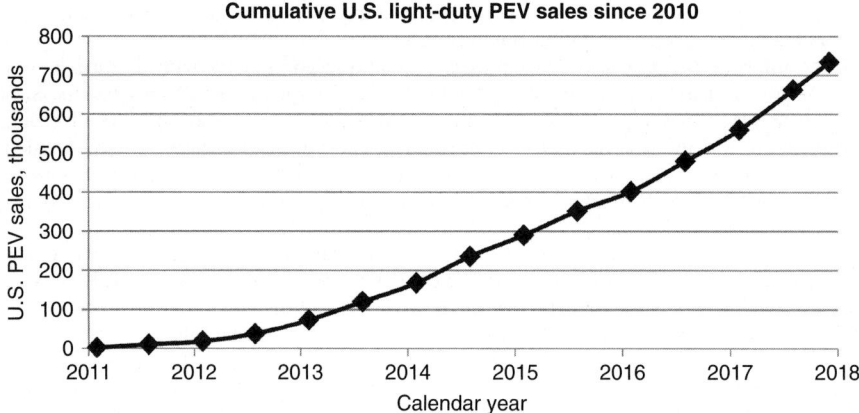

**FIGURE 26A.3**   Cumulative total of plug-in electric vehicles (EVs and PHEVs) sold in the United States. (*Data from Argonne National Lab.*)[9]

**FIGURE 26A.4**   PEVs commercialized in the 21st century including Nissan Leaf EV, Chevrolet Volt PHEV, and Chevrolet Bolt EV. (*Courtesy of Nissan and GM.*)

The development and commercialization of PEVs has been symbiotic with the more widespread commercialization of HEVs that occurred about a decade earlier.

### 26A.0.3   History of Hybrid Electric Vehicles

HEVs date back to the end of the 19th century. Ferdinand Porsche developed a gasoline-electric hybrid vehicle in 1899. Several HEVs were exhibited at the Paris Auto Show in 1906. However, though engineering concepts were developed, HEVs did not attain early commercial success. ICE power plants were hybridized to use electric motor-generator sets in some special applications such as submarines and diesel locomotives. It is also true that, in some sense, the conventional automobile can be considered as an HEV since it incorporates electric power in the starter which is powered by a battery that is recharged by the alternator or a generator.

HEVs remained in the R&D realm of the automotive industry throughout most of the 20th century. Serious development of hybrid vehicles accelerated around 1970 in response to various strategic oil supply issues. More intensive development was initiated with the Partnership for a New Generation of Vehicles (PNGV) initiated in 1993 by the DOE. The PNGV program was created by the Clinton administration as a partnership between the U.S. OEMs (Chrysler, Ford, and GM) and government agencies with the goal of producing highly fuel-efficient automobiles in a decade. Successor organizations, including FreedomCAR and U.S. Drive, have worked together with the DOE and USABC on the development of batteries for HEVs and the establishment of performance and cost targets that they have mutually adopted.

By 2000, the U.S. OEMs participating in the PNGV program had created three 5-passenger concept vehicles that achieved greater than 72 mi/gal (3.27 L/100 km): the GM Precept, the Ford Prodigy, and the Chrysler ESX-3. Development of these prototype vehicles provided the necessary technology for the subsequent commercialization of hybrid vehicles. In 2001, the DOE PNGV program was incorporated into a new program called FreedomCAR that emphasized development of hydrogen fuel cell vehicles, but continued development of HEV technology as well.

Meanwhile, the first mass-produced hybrid car, the Toyota Prius, was introduced into the Japanese domestic market in 1997. The Prius was a distinctively styled sedan that utilized an electric drivetrain powered by a 288 V nickel-metal hydride battery pack consisting of small but high-power cylindrical cells. The electric motor delivered 21 kW of traction power from a high-power NiMH battery pack with less than 2 kWh of energy. The electric drivetrain was coupled to a high-efficiency 43 kW Atkinson cycle engine through a planetary gear train in a series-parallel architecture. This full hybrid vehicle achieved a fuel economy in excess of 41 mi/gal. It utilized regenerative braking energy to enable electric launch as well as the motor-assist function.

The first HEV sold into the U.S. market was the vividly aerodynamic and sporty Honda Insight, a three-seater hybrid vehicle introduced in 1999. This mild hybrid vehicle utilized a small 144 V nickel-metal hydride battery pack and an electric drivetrain that delivered 10 kW of traction power. The Insight utilized a parallel drive architecture that coupled the electric drivetrain to the 52 kW ICE drivetrain directly through the crankshaft. This vehicle utilized regenerative braking energy for power assist during acceleration, but it had insufficient electric drive to provide electric launch.

The Insight equipped with a manual transmission had a fuel economy of 49 mi/gal in the city and 61 mi/gal on the highway. This vehicle also met ultralow emission vehicle (ULEV) emission standards. Honda also introduced a version of the Insight that utilized a continuously variable transmission (CVT) that achieved comparable fuel economy and met the more stringent super ultralow emission vehicle (SULEV) emission standards. This vehicle convincingly demonstrated the dual advantages of increased fuel economy and reduced emissions.

The Toyota Prius with upgraded technology was successfully introduced into the United States in 2001. It utilized a prismatic 274 V NiMH battery pack produced by Panasonic EV Energy (PEVE), a joint venture battery manufacturer formed by Toyota and Panasonic. A new model in 2004 achieved improved fuel economy with a yet smaller 202 V NiMH battery. This vehicle also utilized a boost power converter to provide 500 V power to the electric drivetrain. A later Prius (third generation) introduced in 2009 provided enhanced performance with an improved fuel economy of about 50 mi/gal. The Toyota Prius was by far the most popular HEV of the first decade of the 21st century with over 1.5 million units sold globally by 2009. Since then Toyota has continued to dominate the HEV field particularly with a family of Prius vehicles augmented by the smaller Prius c subcompact vehicle and the larger Prius v, as shown in Fig. 26A.5. Additionally, Toyota has commercialized a variety of other hybrid vehicles including luxury vehicles under the brand name Lexus.

**FIGURE 26A.5**  Toyota family of hybrid vehicles including, clockwise from top left, Prius Prime, Prius, Prius v, and Prius c. (*Courtesy of Toyota.*)

Early in the new century, the Prius and the Insight were joined in the U.S. market by a variety of other hybrid vehicles. In 2003, Honda introduced the parallel mild hybrid version of the Honda Civic. GM introduced multiple HEV models including stop-start and mild hybrid models such as the Chevy Silverado Truck HEV and the Saturn VUE HEV (now discontinued). The Ford Escape was introduced as the first hybrid SUV in 2004, using a power-split transmission architecture enabling full hybridization. Ford sold about 20,000 units per year of this vehicle advertised as the most fuel-efficient SUV in America. Excellent durability has been demonstrated with this robust vehicle that exceeded 200,000 mi (320,000 km) in a fleet of New York taxi cabs. Ford built on this success with the introduction of the Ford Fusion Hybrid, billed as the most efficient midsize sedan in America. Other popular HEV models available in 2017 include the Ford C-MAX hybrid, Lincoln MKZ hybrid, Chevrolet Malibu hybrid, Buick LaCrosse eAssist, Honda Accord hybrid, and a dozen HEV models in all from Toyota and Lexus.

The principal advantage of HEVs is fuel economy. Hybrid vehicles provide a fuel economy increase of about 50% for city driving, where substantial regenerative braking energy can be recovered in stop-and-go driving. More modest savings are achieved in highway driving, yielding around a 10% improvement. Consequently, the city fuel economy is comparable if not better than the highway fuel economy. HEVs feature small gas tanks and an extended urban range in comparison to conventional ICE vehicles. Aided by high fuel prices, the U.S. HEV market share increased to around 3%, but seems to have stagnated somewhat[11] since then as fuel prices have declined, as shown in Fig. 26A.6. With higher fuel prices, HEV market shares are higher in Europe and especially in Japan[10] where it now exceeds 20%.

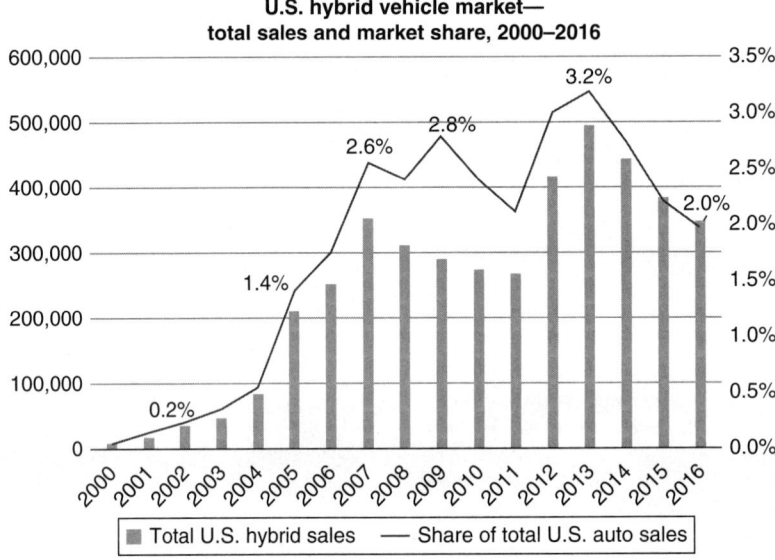

**FIGURE 26A.6**    US HEV sales and market share of light duty vehicles sold. (*Courtesy of PIRA Energy Group—data from S&P Global Platt Analytics, Ward's Auto.*)

Stop-start vehicles (SSVs), a class of HEVs with the lowest degree of power train electrification, have the capability to turn the engine off during coasting, braking, and stops with frequent engine starts during acceleration. Origins of this technology may trace to laws in Switzerland toward the end of the 20th century that required drivers to manually turn off the car ignition at stoplights to reduce air pollution and save fuel. The technology and hardware to perform this automatically is a small fraction of the cost of implementing full hybrid vehicle functions, but also yields a modest fuel economy benefit on the order of 5% to 10%. However, due to SSV affordability, the market penetration of SSVs might be substantially larger than that of the more efficient power-assist HEVs, so a larger overall fuel economy improvement on a fleet level could be achieved.

The commercialization of SSVs started over a decade ago in Europe first, motivated by regulation and aided by the ease of implementation in the manual transmission vehicles that are widespread there. SSVs are now in common use in Europe[10] with over 60% of the market share in 2016. It is more difficult to implement in automatic transmission vehicles predominant in North America. However, in 2012, the Obama administration increased Corporate Average Fuel Economy (CAFE) standards that call for a light vehicle corporate fuel economy of 54.5 mi/gal in the United States by 2025. Subsequently, SSVs are being introduced at a rapid rate. In 2015, SSV comprised a market share of 7.5%; in 2016, it was 15%. This is several times the market share of other HEVs and EVs together making this the biggest market movement toward hybrid and electric vehicles.[12]

## 26A.1    ELECTRIC AND HYBRID VEHICLE ENGINEERING

### 26A.1.1    Vehicle Propulsion Fundamentals

All automobiles including electric and hybrid vehicles require power to accelerate and sustain speed.[1-6] Force must be applied to accelerate the vehicle mass and to overcome rolling resistance and aerodynamic drag:

$$F = ma + mgC_{rr} + \tfrac{1}{2}\rho C_D A v^2$$

where $F$ = force required at the wheels of the vehicle
   $m$ = mass of the vehicle
   $a$ = acceleration of the vehicle
   $g$ = acceleration of free fall due to gravity
   $C_{rr}$ = coefficient of rolling resistance between tires and road surface
   $\rho$ = density of the ambient air
   $C_D$ = coefficient of drag of the vehicle in the direction of travel
   $A$ = cross-sectional area of the vehicle
   $v$ = speed in the direction of travel

Strictly speaking, there are two additional terms. There is a small force required related to rotational acceleration of rotating parts. There is also a force related to propelling the vehicle up a grade (negative for a descending grade) and this force can be very large for steep grades.

The power required to propel the vehicle depends on this force and the vehicle velocity according to:

$$P = Fv = mav + mgC_{rr}v + \tfrac{1}{2}\rho C_D A v^3$$

In a uniform acceleration, the power demand increases nearly linearly with time peaking at the end of the acceleration period. However, this power demand is somewhat more than directly proportional to velocity, due to higher velocity dependence of the aerodynamic drag force term.

For an EV, the power capability must be sufficient to meet acceleration requirements typically specified in terms of acceleration time from 0 to 60 mi/h. Acceleration from 0 to 60 mi/h within the 10 s typical for a 1400 kg midsize passenger car requires about 80 to 120 kW. Most of the power is required for acceleration, but around 10% is required to overcome aerodynamic drag and 3% to 5% can be required to overcome the rolling resistance.

Electric and hybrid vehicles can also brake using the electric propulsion motor in such a way as to store the braking energy in the battery.[1-6] This is known as regenerative (or regen) braking since it provides charge input to regenerate stored energy in the battery. The braking force is provided through the transfer of kinetic energy of vehicle motion into electrical energy that is then stored as chemical energy in the battery. In this case, the braking force required is less than the force needed to decelerate the vehicle mass, since the rolling resistance and aerodynamic drag forces also act to slow down the vehicle:

$$F = ma - mg \times C_{rr} - \tfrac{1}{2}\rho C_D A v^2$$

The power required to brake the vehicle is the product of this force and the vehicle speed:

$$P = Fv = mav - mg \times C_{rr}v - \tfrac{1}{2}\rho C_D A v^3$$

Taking into account again the approximate proportionality of power to velocity, the braking power required declines about linearly with time in a uniform deceleration. Thus, for braking, the peak regen braking power requirement occurs at the beginning of the braking event. Braking is partially assisted by the aerodynamic drag and rolling resistance forces. Thus, moderate decelerations comparable to acceleration performance require moderately less power than acceleration. However, deceleration during braking may be up to several times quicker than acceleration, thus the maximum power required for braking may be several times higher than that required for acceleration. Mechanical braking is used to supplement regenerative braking in EVs providing a margin to ensure safe braking.[3,4]

High-power batteries are now capable of providing excellent power performance for a wide range of EVs, including high-speed sports cars. Energy, the capability to provide power over time, is a related but separate key requirement to provide practical range for EVs.[1-6] It is the energy requirement that is challenging for batteries. Conventional ICE-powered vehicles typically provide a range of 300 to 400 mi (500–650 km) powered by gasoline with a theoretical specific energy of 13,000 Wh/kg. This is nearly 400 times more than the 35 Wh/kg specific energy typical of lead-acid batteries. A superficial (incorrect) conclusion would be that an EV version of a conventional gas-powered car with a 300 mi (500 km) range would have a range of less than 1 mi (1.6 km).

While still challenging, practical ranges can be obtained with BEVs due to several factors including the high efficiencies of the electric drivetrain. First, ICEs are heat engines that are subject to Carnot inefficiencies. The actual operating efficiency of an ICE can be as low as 20% or less in actual operation. The efficiency of battery discharge can exceed 90%. Even accounting for losses in the electric motor and power electronics, an EV drivetrain efficiency in excess of 80% is feasible. Thus, there can be a 4:1 efficiency advantage from electric drive in using onboard energy. Additionally, EVs can capture regenerative braking energy by the operation of motors in reverse where they function as generators. Especially in stop-and-go city driving, this can provide for substantial energy efficiency benefits. Finally, electric motors and power electronics do not weigh as much as automotive engine systems and transmissions. Thus, the battery weight in an EV can be several times the weight of the gasoline tank in a conventional gas-powered vehicle. By utilizing advanced batteries, a practical vehicle range of over 100 mi (160 km) is feasible even with lead-acid batteries.

The energy consumption during driving depends strongly on the size, weight, and type of vehicle as well as the way the vehicle is driven. Representative energy consumption results are usually obtained through measurements during dynamometer testing of vehicles subjected to standard driving schedules, such as those established by the U.S. Environmental Protection Agency (EPA) for fuel economy standards. Standard driving schedules include the urban dynamometer driving schedule (UDDS) that represents city driving and the highway fuel economy test (HWFET) that represents highway driving. The velocity profile over time for these driving schedules is shown in Fig. 26A.7. The energy requirement can be determined from the average power draw under these conditions.

Typical energy consumption results for a midsize EV are 225 Wh/mi (140 Wh/km) for city driving and 275 Wh/mi (170 Wh/km) for highway driving, according to the UDDS and HWFET schedules, respectively.[13] Under the US06 schedule, commonly used to represent the more aggressive highway driving now typical in the United States, the energy consumption is typically higher at about 400 Wh/mi (250 Wh/km). In contrast to results with conventional gas-powered vehicles, EVs are more efficient under city driving conditions where regenerative braking can provide substantial energy efficiencies in stop-and-go driving. The energy input from regenerative braking is subtracted from the energy required to propel the vehicle to determine the net energy consumption. Using the average of these driving cycle results yields 300 Wh/mi (190 Wh/km) as a typical energy consumption, so that a 100 mi (160 km) range would require a battery pack of 30 kWh. A 40 kWh battery pack would yield a range of 133 mi (215 km). However, we are used to conventional vehicles with a range of 300 to 400 mi (500–650 km). This would require a battery pack energy of 90 to 120 kWh. With commercially available batteries, such a battery pack would be very large and weigh 900 to 1200 kg even when using 100 Wh/kg technology. Some lowering of these energy requirements can be achieved by improved vehicle design for higher efficiency. However, the weight, size, and cost of these large battery packs are the key challenges for the EV industry.

Another challenge for BEVs relates to recharge of the battery at the end of a trip. Rapid recharge comparable with gasoline refueling of a conventional vehicle at the gas station is not feasible due to practical charge power limitations. Recharge of a 50 kWh battery pack in 3 min would require 1 MW of power capability. A more practical solution is charging at home or at work locations with full charge expected in 8 to 10 h. Even this would require installation of 220 V electrical power to provide around 6 kW of power needed for overnight charging.

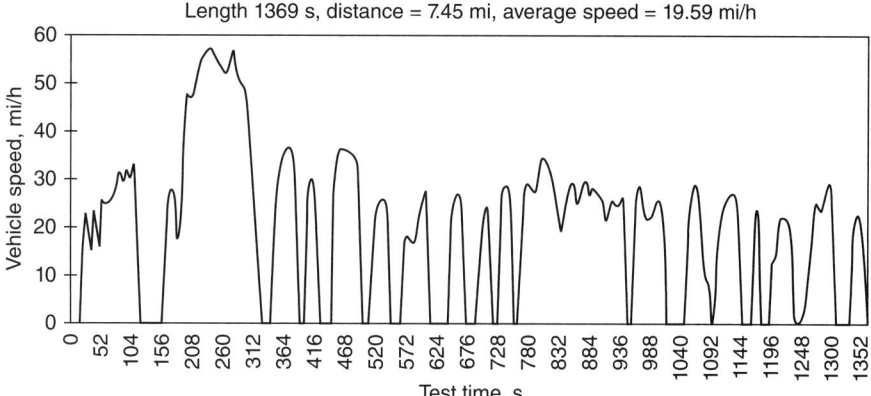

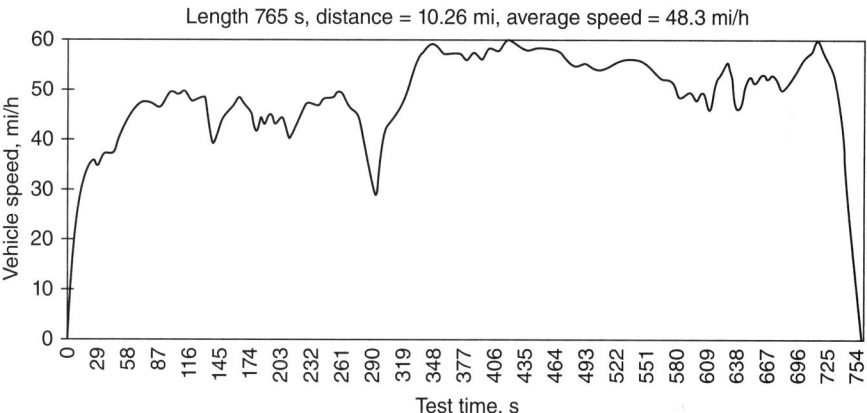

**FIGURE 26A.7**    EPA standard UDDS driving schedule representing city driving (top) and HWFET schedule representing highway driving (bottom). (*Courtesy of US EPA.*)

In BEVs, propulsion batteries must supply power and energy for propulsion, as summarized above. However, they must also supply power and energy for the vehicle electrical system to run lights, controls, and accessories as well as the heating and cooling climate control. In HEVs, the power and energy of the ICE drivetrain contributes to propulsion. However, the sizing of the battery must also consider electrical accessories including electric power steering and air conditioning as well as engine starting.

### 26A.1.2    Hybrid and Electric Vehicle Drivetrains

Hybrid and electric vehicles span a wide range of electrification from the conventional ICE vehicles with no electric propulsion to full BEVs with no ICE. Drivetrain designs are shown schematically in Fig. 26A.8. The most highly electrified vehicles are PEVs, comprised of BEVs and PHEVs. The EV drivetrain shown in Fig. 26A.8*a* provides welcome simplification over the conventional ICE drivetrains with complicated transmissions. In many cases, EV electric motors are coupled directly to the driveshaft without a need for a multispeed transmission.

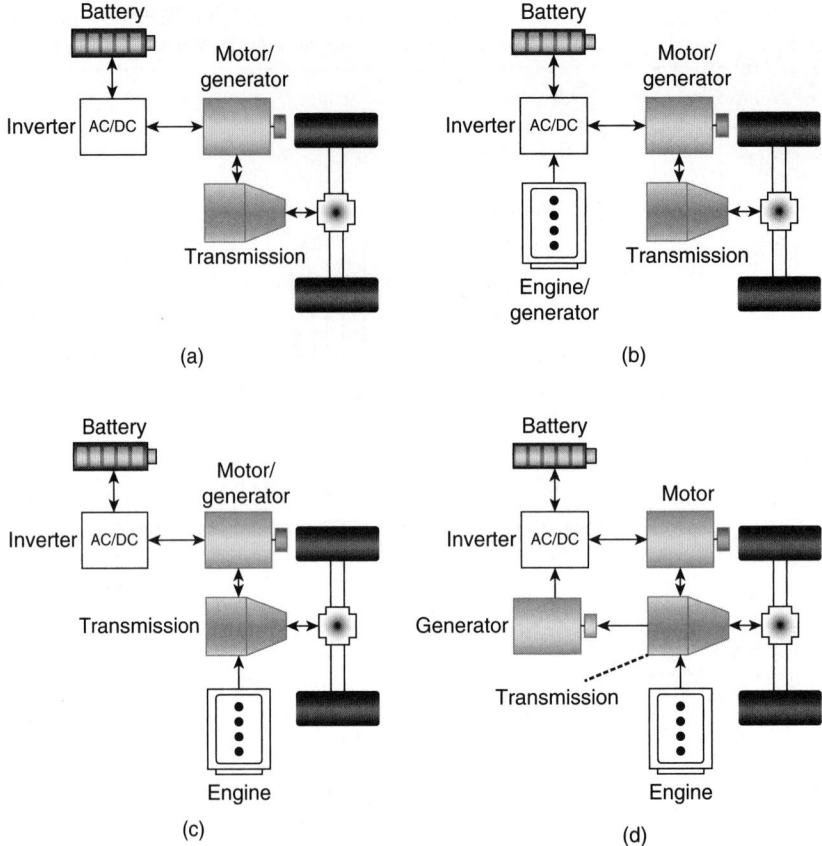

**FIGURE 26A.8**   Schematic diagram for various electrified power trains, including: (*a*) EV, (*b*) Series hybrid, (*c*) Parallel hybrid, and (*d*) Series-parallel hybrid.

In PHEVs, the basic design is the series power flow design shown in Fig. 26A.8*b*. The battery provides all vehicle power while the engine recharges the battery extending its range. Power-assist HEVs, which do not have plug-in capability, typically utilize a parallel or series-parallel architecture. Mild HEVs such as belt-alternator-starter (BAS) designs utilize the simple and efficient parallel design of Fig. 26A.8*c*. Full power-assist hybrids such as the Toyota Prius utilize the series-parallel architecture of Fig. 26A.8*d*. This offers an excellent combination of efficiency and operational versatility albeit with the highest complexity and cost.

## 26A.1.3   Electrified Vehicle Features

Additional capabilities over those in conventional ICE-powered vehicles are provided by a series of features that are added with increasing electrification going from stop-start microhybrid vehicles to mild and strong hybrid vehicles to PHEVs to EVs.

***Stop-Start.***   Stop-start capability is the ability to turn the engine off during coasting, braking, and stops. This eliminates fuel consumption during engine idling. When the engine is turned off, the auxiliary functions such as air-conditioning and electronics are powered by the battery. This feature is the lowest level of electrification requiring less than 10 kW of power for engine starting.

***Regenerative Braking.*** Regenerative braking is the capability to recover kinetic energy during braking through a generator or an electric motor operating backwards as a generator that charges the battery. Regenerative braking is used in tandem with mechanical braking to ensure safe braking. The electrical power and energy required depends on the degree of electrification ranging from around 10 kW to over 100 kW.

***Motor Assist.*** Motor assist is the ability to provide electric motor torque to propel the vehicle generally with higher efficiencies than an ICE-drivetrain can provide. In hybrid vehicles, motor assist provides 10 to 80 kW supplementing torque provided by the ICE drivetrain during blended operation.

***EV Drive.*** EV drive is the ability to operate in all-electric mode provided for plug-in vehicles. All propulsion and braking torque is provided by the battery and the electric motor. Blended operation utilizing the ICE power is not required for EVs or for PHEVs except for cases where the all-electric range has been exceeded requiring charge-sustaining operation. High power levels around 100 kW or higher are required for EV drive, since full power operation must be possible from the EV drivetrain.

### 26A.1.4 Types of Electric and Hybrid Vehicles

Battery requirements vary substantially across the wide variety of types of hybrid and electric vehicles now being commercialized including stop-start microhybrids, mild hybrids, strong or full hybrids, plug-in hybrids, and BEVs.[14–17] While the definitions of electric-drive vehicle types are not entirely consistent, a general consensus has evolved. Particularly from the point of view of understanding battery requirements, it is useful to classify HEVs and EVs according to their performance features and degree of electrification, as shown in Fig. 26A.9. In the most minimal hybrid design, the stop-start HEV, the ICE is turned off during idle and/or deceleration. In the mild HEV, the electric motor additionally provides assist during acceleration. In the strong hybrid or full HEV, the vehicle is propelled by the electric drive with the engine off over at least some speed ranges. In the PHEV, a significant all-electric range is added with the capability to recharge the battery from the utility grid. The electric vehicle (BEV) is fully powered by battery and does not include an ICE drivetrain. In all cases, regenerative braking energy can be captured to power the various electric performance features. All designs provide for improved fuel economy and the benefit generally increases with degree of electrification. The size of the batteries also increases with degree of electrification as well as the overall system complexity and HEV cost.

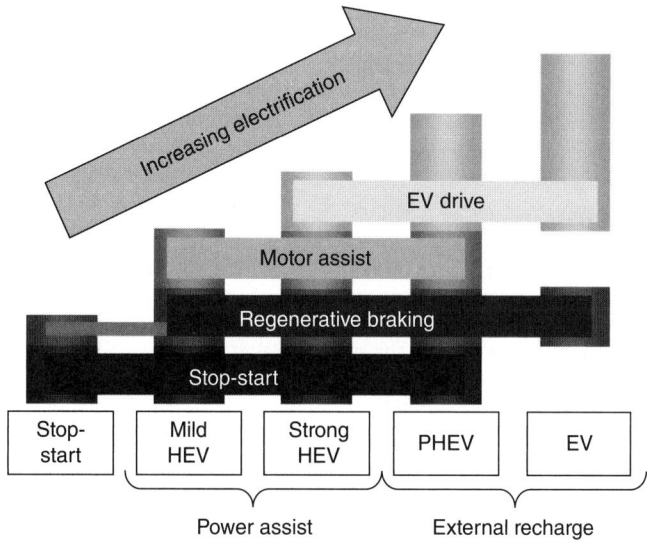

**FIGURE 26A.9** Types of hybrid electric vehicles based on performance features.

## 26A.1.5   Battery Pack Systems

EVs and hybrid vehicles are powered by battery packs that contain multiple battery modules which in turn are composed of several battery cells as illustrated in Fig. 26A.10. Battery cells operate typically in the range of 1 to 5 V. The nominal discharge voltage for nickel-metal hydride battery cells is 1.2 V. Lead-acid batteries operate at 2 V per cell. Lithium-ion batteries typically operate in the range of 3 to 4 V depending on the chemistry. This low-voltage range is insufficient to provide the tens of kilowatts required for EV applications since practical considerations with electric motors and power electronics limit current flow to less than 500 A. EVs operate generally in the range of 300 to 400 V. HEVs operate at lower voltages ranging from 12 V for SSVs to over 200 V for full power-assist HEVs. A series or series-parallel arrangement of up to hundreds of battery cells is used to power PEVs and HEVs.

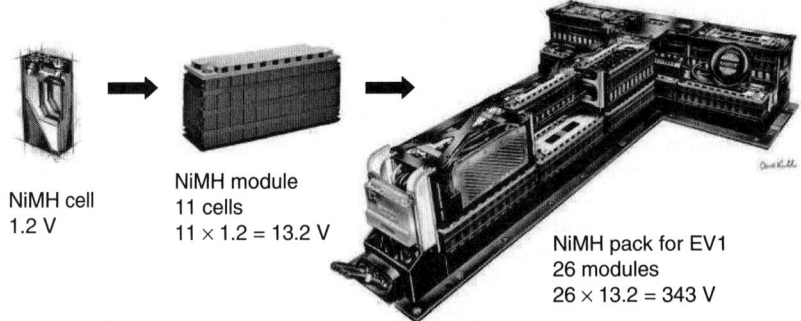

NiMH cell
1.2 V

NiMH module
11 cells
11 × 1.2 = 13.2 V

NiMH pack for EV1
26 modules
26 × 13.2 = 343 V

**FIGURE 26A.10**   Battery modules comprised of cells are the building blocks of battery packs. (*Courtesy of General Motors.*)

Multiple battery cells are combined electrically and mechanically to provide higher voltage units called battery modules. Modules are the building blocks used to assemble PEV and HEV battery packs. Battery modules typically contain 5 to 25 cells arranged so that the voltage is limited to less than 60 V (safe against electrocution hazards) with a weight of less than 50 lb for easy handling. Means for thermal management, cell voltage monitoring, and control electronics may also be incorporated into the module depending on the specific technology and product.

Battery modules are combined electrically and mechanically into battery packs to provide the full power and energy required for each application. Battery pack systems incorporate a series-connected string of modules together with power interconnections, mechanical packaging, electronic sensors and controls, and thermal management components necessary for electric propulsion operation.

## 26A.1.6   Electronic Controls

Electronic controls are used to maintain the battery pack within normal operating conditions. The battery management system (BMS) measures the pack, module, and cell voltages and battery temperatures as well as the battery pack current. The BMS inputs are used to monitor and control the state-of-charge (SOC) and also monitor the state of health throughout its life (by tracking battery capacity and resistance). Battery overdischarge is avoided by reduction and/or suspension of discharge power with cell voltage excursions below specified low-voltage limits. Battery overcharge is avoided by reduction and/or suspension of charge power with cell voltage excursions above specified voltage limits. Similarly, the BMS acts to keep the battery pack from exceeding high-temperature limits. Voltage limits are compensated for dependences on temperature and current.

PEVs are recharged with onboard and/or off-board chargers that interact with the BMS as well. Together the charger and BMS act to provide for full charge within normal cell voltage and temperature conditions. The BMS also serves to monitor the SOC of the EV battery pack during charge and driving typically by ampere-hour counting of the net discharge capacity.

For the case of lithium-ion batteries, which cannot be overcharged without safety issues and/or compromise of battery life, electronic controls are essential to keep cells within proper operational voltage limits. These electronic controls can be incorporated at both the battery module and battery pack levels. They must be capable of detecting voltage deviations between cells arising from variations in self-discharge and cell capacity degradation rates. These differences are usually due to thermal inhomogeneity within battery packs. Electronic circuits to balance the SOC between individual cells are usually included. These controls add significantly to the complexity and cost of lithium-ion battery packs, though costs have declined in recent years as these devices mature.

Electronic controls for HEV batteries depend on the type of HEV. Minimal controls are required for lead-acid batteries used in SSV batteries that employ simple integration similar to that for conventional SLI (starting/lighting/ignition) batteries. Power-assist HEV battery controls are generally more complex than controls for EV applications. Control of the SOC is more difficult because the SOC is not reset to 100% upon recharge after each drive. Neither is the battery fully discharged as the SOC limits need to be maintained for the available energy range where discharge and regen power can be maintained. SOC algorithms based on ampere-hour counting are difficult because small errors due to self-discharge and round off can cause the SOC estimation to drift into error. Many lithium-ion batteries have an adequate dependence of battery voltage on the rate to reasonably estimate SOC from battery voltage measurements. For NiMH batteries and lithium-iron phosphate batteries (both with very flat discharge curves), SOC algorithms generally require a complex strategy utilizing occasional voltage excursions near fully charged and deeply discharged states, and/or resistance estimates based on driving transients.

## 26A.1.7   Thermal Management

PEV and HEV battery packs employ a thermal management system designed to maintain a uniform battery cell temperature within the normal operating range. Ambient temperature batteries such as lead-acid, nickel-metal hydride, and lithium-ion batteries operate best in the range of 20 to 40°C; coincidentally temperatures at which human beings are comfortable. As batteries approach freezing temperatures, the power performance suffers. Elevated temperatures over 40°C can result in reduced charging efficiency as well as acceleration of failure modes leading to decreased life. Extreme elevated temperatures accelerate failure modes and aggravate safety issues.

Electric drive operation generates significant waste heat within a battery pack that must be rejected. Joule heat generation on the order of 10% of the average power draw can be generated. Average heat generation during operation can reach 2 to 5 kW in PEVs. A lower heat generation is observed in HEVs, ranging from less than 100 W in SSVs to 500 W in some power-assist HEVs. In addition to keeping the average cell temperatures within an optimal operating range, it is important to maintain a tight uniformity in cell temperatures. Variability in cell temperatures within the battery pack will lead to inhomogeneous self-discharge leading to SOC imbalance among cells of the battery pack. Methods of thermal management in PEVs and HEVs have included passive cooling utilizing natural convection, forced air cooling, and forced liquid cooling. Additionally, refrigeration of the cooling medium has been used to increase heat rejection rates. Recently, the use of phase-change composites has been developed as another thermal management technique.

Passive cooling is the simplest and least expensive cooling method. It is typically utilized for small batteries with low average power draws such as SLI batteries and 12-V batteries for SSVs. Passive cooling is limited in its heat rejection capability and has deficiencies in its ability to maintain all cells at a uniform temperature, particularly in unsymmetrical battery packs. When attempted in large PEV packs, passive cooling has often resulted in battery life issues.

Forced air cooling has been a common approach utilized for electric drive battery packs. It has the advantage of simplicity and it also typically provides a lighter and cheaper solution than can be achieved with liquid cooling. However, air has a limited heat capacity. Air flow channels are bulky and require symmetrical geometries that may not be practical in some vehicle designs. Fans consume energy and can be noisy. Finally, dirt and water must be removed from intake air to avoid ground fault issues. Forced air battery cooling was used in the GM EV1 electric vehicle as well as the Toyota Prius HEVs.

Utilizing the much higher heat capacity and thermal conductivity of liquid cooling media, significantly higher heat rejection rates can be obtained with forced liquid cooling. Liquid cooling also provides for a more uniform distribution of battery cell temperatures. A more compact battery pack design can be achieved since flow channels can be smaller. Automotive coolants comprising water-glycol formulations provide for high heat capacity and good flow characteristics under vehicle operation. Examples of electrified vehicles with forced liquid cooling include the Tesla EVs as well as the GM Chevy Volt PHEV and GM Chevy Bolt EV.

Liquid-cooled thermal management carries disadvantages in terms of weight, complexity, and cost. The liquid coolant can add complexity by introducing another conduction path that may result in ground faults. Additional discipline is required to avoid reliability issues with respect to leaks. Liquid cooling also requires a second heat exchanger such as a radiator which is then ultimately air cooled. It would be especially useful if battery packs could share coolant loops with the ICE cooling system. This is not yet practical with current battery chemistries that require substantially lower temperatures, but it may be possible with future solid-state battery technology.

Both air cooling and liquid cooling can also utilize active refrigeration systems to further accelerate heat rejection from the battery pack. For example, air inlets from an air-conditioned cabin have provided for refrigerated air to cool propulsion battery packs in Toyota Prius HEVs. The 2004 Ford Escape HEV utilized a separate HVAC duct system to condition air for its NiMH battery pack.

### 26A.1.8   Vehicle Integration

EV battery packs must also be contained physically within the vehicle in such a way as to be safe in the event of crashes and remain isolated electrically from the vehicle. This structure and containment also provides significant weight and volume burden to the battery pack. Due to the restrictive nature of volume limitations on the battery pack, the engineering of the battery packaging envelope is always a critical feature in the design of an EV. As a rule of thumb, the PEV battery pack is typically less than one-third of the vehicle weight and even a smaller fraction of the vehicle volume.

For HEVs, smaller and lighter battery packs are utilized, but due to the size and weight of the ICE drivetrain in hybrid vehicles, the packaging constraints for the battery pack are still challenging. Particularly, in the development of hybrid versions of conventional ICE vehicles, it is often challenging to find a suitable location for the battery pack that can simultaneously meet requirements of facile integration with the drivetrain, avoidance of crash safety issues, and environmental issues including exposure to water or extreme temperatures.

Electrified vehicles utilize battery pack systems for propulsion, so battery requirements are based on the power, energy, life, weight, volume, and cost of full battery pack systems. System components other than the battery cells and modules add no power or energy but can contribute significantly to the weight, volume, and cost of the battery pack system. Thus, the specific performance of the battery pack system is always less than that of the battery cells and modules. Similarly, the cost per unit energy ($/kWh) is always higher. EV and HEV battery performance targets are thus rightfully set at the battery system level.

### 26A.1.9   Battery Pack Safety Requirements

Tolerance to mechanical, thermal, and electrical abuse on a system level is an inherent safety requirement for all types of PEV and HEV battery packs. A variety of mandatory government regulations and voluntary industrial standards have been developed to guide the design and testing of battery packs to ensure vehicle safety. The USABC has published safety abuse test procedures[18] and developed a battery hazard mode and risk mitigation analysis.[19] This is based on following a similar methodology to the failure modes and effects analysis (FMEA) commonly used in the international automotive community. This disciplined systems approach to battery pack safety is important in the design of battery packs, their control systems, and their integration into vehicles.

Mechanical abuse tests typically involve a mechanical integrity or crush test; however, the test conditions can vary substantially. Most testing documents also call for a drop test to alternatively simulate maintenance accidents or in some cases as a proxy for a crash event. Another common, though not ubiquitous, test is a mechanical shock or aggressive vibration test designed to simulate a crash impulse. Thermal abuse tests include a rapid thermal rise through either simulated fire exposure or a thermal shock regime. Thermal shock testing is usually performed using controlled temperature ovens and can involve hot and cold cycles. The simulated fire exposure test typically involves direct and indirect exposure to a bed of burning liquid fuel for controlled periods of time. Standard electrical abuse tests include overcharge, overdischarge, and short circuit. Additional electrical tests may be specified depending on the application.

The European Council for R&D (EUCAR) has created a rating system to describe the safety response of a test article to an abuse test. Response score values range from 0 (no damage) to 7 (explosion). Response scores of 1 to 2 indicate very minor damage, 3 to 4 indicate venting, and 5 to 6 represent fire and rupture.

## 26A.1.10  Battery Standardization

The automotive industry depends heavily on parts standardization to lower cost and complexity in its manufacturing process through economies of scale. As commercialization of PEVs and HEVs has proceeded, interest in standardized cell sizes has grown as a strategy to lower cost. This interest has met with limited success given the diversity of designs (of both the vehicle and the battery) and fast rate of design evolution. In the last few years, three different countries, Germany, United States, and China, have indicated national preferences for automotive lithium-ion cell sizes through their national standard bodies and industrial groups. Following this effort, ISO attempted to create global standards; however, due to the challenges stated above, a standard issued in 2014 listed 63 different cell sizes to accommodate competing national interests.

## 26A.2  PLUG-IN ELECTRIC VEHICLES

### 26A.2.1  PEV Applications: Battery Electric Vehicles

The mission of the USABC has been to develop electrochemical energy-storage technologies that support commercialization of electric, hybrid, and fuel cell vehicles. An important contribution of the USABC has been the development of performance targets and detailed test procedures for electric and hybrid vehicle batteries.[20–25] The USABC established quantitative performance targets for batteries for EV applications starting in the early 1990s.

Midterm goals including 80 Wh/kg specific energy, 200 W/kg specific power, and 1000 lifetime charge-discharge cycles have been largely achieved by nickel-metal hydride EV batteries developed in the 1990s, at least on a battery module basis. This enabled the demonstration EV programs in response to the CARB ZEV mandate. However, range limitations of this midterm technology limited its commercial appeal. More significantly, battery costs were several times the $150/kWh cost target and commercialization did not proceed.

Since 2000, the focus has been on the more aggressive long-term EV battery goals aimed at widespread commercialization, utilizing higher energy density lithium-ion batteries. Table 26A.1 shows the USABC goals for

**TABLE 26A.1**  USABC Goals for Advanced Batteries for Electric Vehicles

| Characteristics at end of life | Units | System level | Cell level |
|---|---|---|---|
| Peak discharge power density, 30 s pulse | W/L | 1000 | 1500 |
| Peak specific discharge power, 30 s pulse | W/kg | 470 | 700 |
| Peak specific regen power, 10 s pulse | W/kg | 200 | 300 |
| Usable energy density, $C/3$ discharge rate | Wh/L | 500 | 750 |
| Usable specific energy, $C/3$ discharge rate | Wh/kg | 235 | 350 |
| Usable energy, $C/3$ discharge rate | kWh | 45 | N/A |
| Calendar life | Years | 15 | 15 |
| DST cycle life | Cycles | 1000 | 1000 |
| Selling price, 100 K units | $/kWh | 125 | 100 |
| Operating environment | °C | −30 to +52 | −30 to +52 |
| Normal recharge time | Hours | <7 hours, J1772 | <7 hours, J1772 |
| High rate charge | Minutes | 80% ΔSOC in 15 min | 80% ΔSOC in 15 min |
| Maximum operating voltage | V | 420 | N/A |
| Minimum operating voltage | V | 220 | N/A |
| Peak current, 30 s | A | 400 | 400 |
| Unassisted operating at low temperature | % | >70% usable energy at $C/3$ discharge rate at −20°C | >70% useable energy at $C/3$ discharge rate at −20°C |
| Survival temperature range, 24 h | °C | −40 to +66 | −40 to +66 |
| Maximum self-discharge | %/month | <1 | <1 |

advanced batteries for EVs as updated in 2015.[21] Aiming at commercialization in the 2020 timeframe, these long-term goals for performance and cost are provided at the battery system level but also translated into cell level targets assuming overheads for weight, volume, and cost.

The specific power and power density requirements are based on the peak discharge power capability at the end of a 30 s constant current power pulse. The specific power and power density are the peak power of the battery divided by the battery weight and volume, respectively. In the revised specifications, performance targets are given at the system level and at the cell level. System level requirements of 470 W/kg and 1000 W/L become even more challenging at the cell level translating to 700 W/kg and 1500 W/L since they must anticipate weight and volume burdens at the system level. Similarly, the peak specific regen power requirement of 300 W/kg at the system level is based on the peak regen charge power capability at the end of a 10 s constant current power pulse. Both the power and regen requirements must be achieved at the system level without exceeding a maximum voltage or 420 V and without dropping below a minimum value of 220 V. (This voltage range represents the practical operation range of the power electronics in driving the traction electric motor.) Furthermore, both the pulse power and regen requirements must be simultaneously met throughout the entire depth of discharge ranging from full charge to a full discharge of the available energy.

The available energy targets are based on a constant current discharge at the $C/3$ rate. At the system level, the specific power requirement is 235 Wh/kg and the energy density requirement is 500 Wh/L. At the cell level, the comparable targets of 350 Wh/kg and 750 Wh/L, respectively, are more challenging to accommodate weight and volume burdens at the system level. Furthermore, the targets must be met at the end of life so that an energy margin must be built to accommodate degradation on the order of 10% to 20% during the battery lifetime. Consequent cell gravimetric and volumetric energy density targets at the beginning of life approaching 500 Wh/kg and 1000 Wh/L will require a doubling of the energy density currently available in commercially available lithium-ion batteries.

While the USABC goals are written in intrinsic variables of specific power, power density, specific energy, and energy density, there was an implicit assumption of a 45 kWh battery pack with a pulse power discharge power capability of 90 kW based on a power-to-energy ratio of 2:1. This battery pack would weigh no more than 191 kg (or 421 lb) and occupy a volume of 90 L (or 24 gal) or less so that the density would be 2.1 kg/L just meeting both goals simultaneously. The regen power target corresponds to 38 kW.

The life requirements for the battery pack are based on the expectation that the battery will last the life of the vehicle, generally 15 years and/or 150,000 mi (240,000 km). The cycle life targets are based on deep-depth cycles under dynamic stress test (DST)-simulated driving profile conditions. The DST profile was developed to simulate a more realistic and aggressive version of the Federal Urban Driving Schedule (FUDS) in a limited number of constant power steps, as shown in Fig. 26A.11. USABC cycle life tests subject batteries to repeated deep charge-discharge cycles until the power and energy goals can no longer be met. The USABC test specification calls for full discharge of the available energy target in these deep cycles. The cycle life target of 1000 deep cycles is designed to provide for the battery to last a vehicle life of 150,000 mi (240,000 km) implicitly assuming that a range of at least 150 mi (240 km) can be achieved with DST discharge of the battery pack.[21]

Testing for battery power and energy is relatively quick and can be accomplished in some cases in a single day. However, cycle life testing can be very time consuming. Assuming 6 h charge times and 3 h discharge times, it would take about 1 year to perform 1000 charge-discharge cycles. Even accelerated testing at higher rates takes at least several months. A recent innovation is high precision coulometry developed as a potential tool to predict cycle life from measurements of inefficiency early in life.[26]

Calendar life is a separate issue from cycle life.[27,28] Some failure modes are independent of charge-discharge cycling and are driven by time-dependent chemical processes such as corrosion. A general approach is to accelerate known stress conditions. For example, failure due to temperature-dependent failure mode processes can be accelerated by life testing at elevated temperatures. Results at several temperatures can then be extrapolated back to estimate the calendar life at room temperature. However, accelerated calendar life testing to predict a battery life meeting USABC goals of 15 years or more is a somewhat speculative art, since failure modes may change with stress conditions. There is an inherent risk in basing business decisions on these unproven projections.

The selling price for a 45 kWh battery pack produced at volumes of 100,000 units per year is targeted at $5625. Even when not including the electric motor, the target battery cost is higher than the cost of the gas tank and ICE in a conventional vehicle that costs on the order of $2000 or less. However, savings in fuel cost in comparison to less expensive energy in the form of electricity may be used to justify a higher battery cost. Cost is the single most challenging target for EV batteries. This has been the key issue holding back the commercialization of EVs.

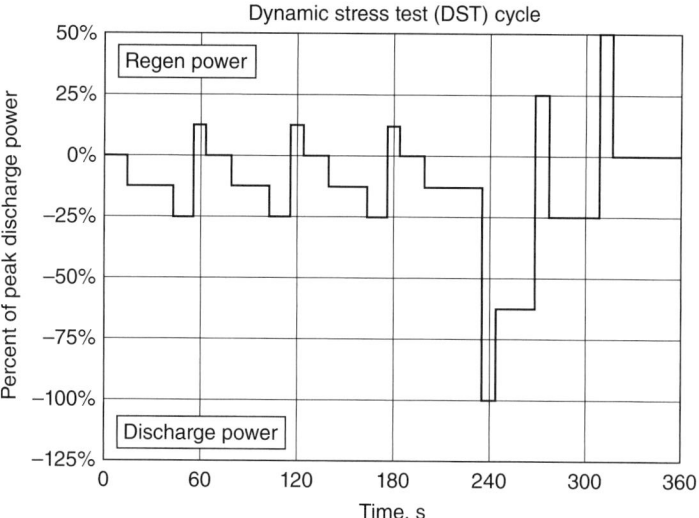

**FIGURE 26A.11**    USABC Dynamic Stress Test power profile. (*Courtesy of USABC.*)

Other USABC goals for advanced batteries for EVs in Table 26A.1 include an operational temperature range between −30 and +52°C that is to be achieved on a system level. Capability for rapid recharge on the order of 80% in 15 min is desired to supplement overnight charging at the 6 h rate. Very high charge retention is expected with the self-discharge rate limited to no more than 1% per month. Additionally, battery systems should be fully maintenance-free. Finally and importantly, battery packs must be tolerant to electrical and mechanical abuse on a system level. A series of electrical and mechanical abuse tests have been developed that battery packs must pass to ensure safe operation in EVs.[18]

### 26A.2.2    PEV Applications: Plug-In Hybrid Electric Vehicles

Plug-in hybrid vehicles have the potential to provide most of the benefits of EVs at a lower cost and without the key BEV performance disadvantages of short range and slow refueling. The very substantial fraction of ZEV operation provides high fuel economy through the efficient use of grid electric power and reduced use of petroleum-based fuel. This vehicle type, which was not a target of the original PNGV program, was developed and promoted in the 1990s by university vehicle competition teams[29,30] and more recently championed by Plug-In America and other groups for its potential to strategically reduce U.S. dependence on foreign oil.

The key feature of PHEVs is the ability to recharge the battery by plugging into the electrical grid. These vehicles utilize both grid electricity and onboard petroleum-based fuel as energy sources, with the electrical energy fraction depending on the size of the battery and the operating strategy. The usual operating strategy is to first utilize charge-depletion operation after recharge of the battery pack from the grid connection. This consists of all-electric operation with the electric drivetrain providing full traction power unless the ICE drivetrain is needed to meet power demands. Charge-depletion operation is continued until the battery pack is substantially depleted at which point charge-sustaining HEV operation is commenced. In charge-sustaining operation, the SOC is maintained at a target level by balancing the discharge of the battery with charge from regenerative braking as in the power-assist hybrid. Charge-sustaining operation is essentially the same as the operational mode for the power-assist hybrid vehicle, but typically at a lower SOC. The PHEV operational modes are illustrated by the plot of SOC in Fig. 26A.12.

PHEVs mix features of strong hybrids and electric vehicles. They can be thought of as strong hybrids with extended EV drive capability. Alternatively, they can be thought of as EVs with extended range from an ICE-driven generator. An important variable in PHEV design is the degree of electrification. Generally, the

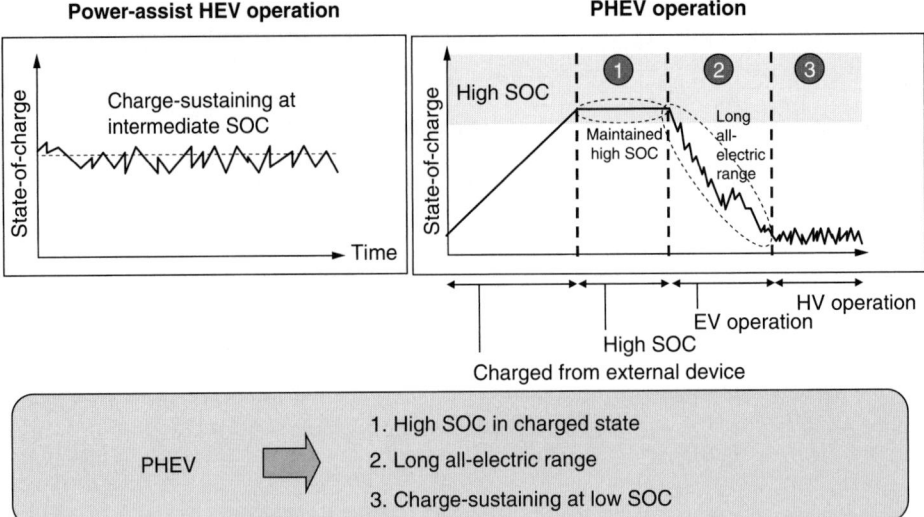

**FIGURE 26A.12** Plug-in hybrid electric vehicle battery state-of-charge during operational modes including charge, standby, charge-depletion operation, and charge-sustaining operation.

electric drivetrain of the PHEV is designed to provide full acceleration power capability without the use of the ICE drivetrain during some significant all-electric range. The all-electric range of the plug-in hybrid is an important design criterion. Most development has focused on designs with a 10 to 40 mi (16–64 km) range. The larger the all-electric range, the larger the benefit in terms of improved fuel economy and reduced emissions. However, this requires larger batteries that increase the weight and cost of the vehicle substantially.

Another key aspect of the PHEV design is the control strategy. With large battery packs providing a substantial all-electric range, the electric drivetrain may be designed to provide all traction power during the charge-depletion phase of operation. In this way, the PHEV may operate totally as a BEV unless the all-electric range is exceeded. These PHEVs may be properly described as E-REVs. There are advantages to utilizing series HEV architectures with E-REVs such as the Chevy Volt. Alternatively, to avoid the cost of large battery packs and full power electric drivetrains, some PHEVs are designed with blended-mode charge-depletion operation. In this case, the ICE drivetrain will turn on at certain power-demand or vehicle speed thresholds prior to reaching charge-sustaining operation. These PHEVs typically utilize a series-parallel HEV architecture. For comparison purposes, measurements on PHEVs with blended electric and ICE operation in the charge-depletion mode are still analyzed to provide an equivalent electric range, the range provided by the battery energy by subtracting out the range provided by the ICE drivetrain in the charge depletion mode.

Another aspect to PHEV control strategy is the state-of-charge set points for full charge and for charge-sustaining operation. To maximize all-electric range for the greatest vehicle energy efficiency and emission benefits, it would be desirable to fully charge the battery pack and drive the battery discharge almost to completion in the charge-depletion mode. However, fully charging a lithium-ion PHEV battery pack accelerates failure modes and may expose the battery pack to overcharge abuse during regenerative braking in subsequent driving. The life of the battery pack is enhanced and control is also simplified by limiting charge to about 80% of the rated capacity. Similarly, the efficiency in charge-sustaining mode is reduced if the battery pack SOC is too low. Additionally, there is risk of overdischarge under high-power demand conditions. So, the HEV efficiency and control is simplified by limiting discharge to operate in charge-sustaining mode at about 30% SOC. This has the effect of reducing the usable energy to about half of the total rated discharge energy. As a result, a vehicle with an energy consumption of about 200 Wh/mi (125 Wh/km) would require a 16 kWh battery pack to achieve a 40 mi (64 km) range.

Table 26A.2 summarizes the USABC energy-storage system performance goals for plug-in hybrid vehicles as updated in 2014.[22] Three sets of goals are listed based on different vehicle targets for all-electric range. These vehicles, PHEV-20, PHEV-40, and E-REV-50, are aimed at all-electric ranges of 20, 40, and 50 mi (32, 64, and 80 km),

**TABLE 26A.2** USABC Goals for Advanced Batteries for Plug-In HEVs

| Characteristics at end of life | Units | 20 mile PHEV | 40 mile PHEV | 50 mile E-REV |
|---|---|---|---|---|
| Commercialization timeframe | | 2018 | 2018 | 2020 |
| All-electric range (AER) | Miles | 20 | 40 | 50 |
| Peak pulse discharge power, 10 s | kW | 37 | 38 | 100 |
| Peak pulse discharge power, 2 s | kW | 45 | 46 | 110 |
| Peak regen pulse power, 10 s | kW | 25 | 25 | 60 |
| Available energy for CD (charge depleting) mode | kWh | 5.8 | 11.6 | 14.5 |
| Available energy for CS (charge sustaining) mode | kWh | 0.3 | 0.3 | 0.3 |
| Minimum round-trip energy efficiency | % | 90 | 90 | 90 |
| Cold cranking power at −30°C, 2 s − 3 pulses | kW | 7 | 7 | 7 |
| CD life/discharge throughput | Cycles/MWh | 5000/29 | 5000/58 | 5000/72.5 |
| CS HEV cycle life, 50 Wh profile | Cycles | 300,000 | 300,000 | 300,000 |
| Calendar life, 30°C | Year | 15 | 15 | 15 |
| Maximum system weight | kg | 70 | 120 | 150 |
| Maximum system volume | Liter | 47 | 80 | 100 |
| Maximum operating voltage | Vdc | 420 | 420 | 420 |
| Minimum operating voltage | Vdc | 220 | 220 | 220 |
| Maximum self-discharge | %/month | <1 | <1 | <1 |
| System recharge rate at 30°C | kW | 3.3 (240V/16A) | 3.3 (240V/16A) | 6.6 (240V/32A) |
| Unassisted operating and charging temp range | °C | −30 to +52 | −30 to +52 | −30 to +52 |
| 30°–52° | Power | 100 | 100 | 100 |
| 0° | Power | 50 | 50 | 50 |
| −10° | Power | 30 | 30 | 30 |
| −20° | Power | 15 | 15 | 15 |
| −30° | Power | 10 | 10 | 10 |
| Survival temperature range | °C | −46 to +66 | −46 to +66 | −46 to +66 |
| Max system production price, 100 k units/yr | $ | $2200 | $3400 | $4250 |

estimated to require 5.8, 11.6, and 14.5 kWh of energy available for charge-depletion operation, respectively. Note that the PHEV usable energy requirements are more than an order of magnitude larger than that required for power-assist HEVs, but a fraction of that required for BEVs with no ICE power. The power requirements for E-REV-50 are higher and comparable to those for EV battery systems. The lower-power requirements for PHEV-20 and PHEV-40 may anticipate more blended charge-depletion operation.

The goals for charge-sustaining operation are similar to those for the power-assist HEV. In addition to goals for charge-sustaining cycle life, challenging goals are added for the charge-depletion cycle life. However, the cost targets amounting to $379/kWh, $293/kWh, and $293/kWh for 20, 40, and 50 mi (32, 64, and 80 km) all-electric range, respectively, are challenging but more achievable than the aggressive EV cost goal of $125/kWh.

## 26A.2.3 Batteries for PEVs

With the introduction of EVs in the 19th century, almost all EVs were powered by lead-acid batteries. Around the turn of the century, Edison developed the nickel-iron battery to provide a battery solution with a somewhat higher energy density that was much more durable as well as very robust to electrical abuse. While it provided a significantly improved performance over lead-acid batteries, its high cost precluded widespread commercial success.

In the development of EVs in the 20th century, a variety of other batteries have also been utilized including nickel-cadmium, nickel-zinc, sodium-sulfur, sodium metal-chloride, zinc bromine, and zinc chlorine.[31] These advanced battery types all provided higher specific energy and energy density performance as well as longer life. However, several battery types raised significant toxicity and/or safety issues. All of the other batteries were several-fold more expensive than lead-acid batteries. Despite many years of advanced battery development in the 20th century, none of these eclipsed the lead-acid battery in terms of overall commercial viability. It is noteworthy that when GM introduced its revolutionary EV1 at the turn of the century, it was initially powered by deep-cycle lead-acid batteries. However, the practical range of less than 100 mi (160 km) was not satisfactory.

**TABLE 26A.3**  Performance Specifications for GM Ovonic NiMH Battery Module and EV1 Battery Pack

|  | GM Ovonic battery module (11 cells) | EV1 battery pack (26 modules) |
|---|---|---|
| Nominal voltage | 13.2 V | 343 V |
| Discharge capacity | 90 Ah | 90 Ah |
| Energy | 1.2 kWh | 30 kWh |
| Power | 3.6 kW | 94 kW |
| Weight | 18 kg | 535 kg |
| Specific power | 200 W/kg | 175 W/kg |
| Specific energy | 66 Wh/kg | 56 Wh/kg |

Successful midterm EV battery development in the 1990s focused largely on nickel-metal hydride batteries by developers including Ovonic Battery Company (previously a subsidiary of Energy Conversion Devices since acquired by BASF), its manufacturing operation GM Ovonics (previously a joint venture with GM and now known as Bosch Battery Systems), Saft, and PEVE. GM Ovonic, Saft, and PEVE all produced NiMH battery packs for EVs introduced by major automotive OEMs in California around 2000. Table 26A.3 shows representative performance achieved by the GM Ovonics nickel battery module and the GM EV1 battery pack.[32]

In this NiMH battery pack for the EV1, the battery modules were arranged in a T-shape, as illustrated in Fig. 26A.10. Thermal management was achieved by air cooling which was complicated by the asymmetry of the battery pack shape. The battery pack also included a BMS that operated effectively with module level voltage and selected temperature sensors. Safety disconnects, fuses, and power contactors were also included. The battery pack hardware added nearly 15% to the battery pack weight (and considerably more to the pack system volume).

PEVE developed 12 V NiMH battery modules with a capacity of 100 Ah, which were used in air-cooled battery packs for the Toyota RAV-4 EV and the Honda EV-Plus. Saft developed similar NiMH battery modules that were utilized in liquid-cooled battery packs for the Chrysler EPIC minivan. NiMH batteries provided typically 65 Wh/kg and 200 W/kg at the battery module level which was adequate to provide an EV range in excess of 100 mi (160 km). Additionally, a cycle life of 1000 cycles was achieved which provided for excellent durability and several times the life of lead-acid battery packs. The NiMH EV performance was generally quite good, and these vehicles stimulated by the ZEV mandate were popular.

While the energy density of the NiMH battery technology still limited the range to less than 200 mi (320 km), a bigger issue was the cost of nickel-metal hydride batteries. At the low volumes, the nickel-metal hydride battery modules cost well in excess of $1000/kWh. Pack assembly and vehicle integration roughly doubled this cost at the battery pack system level. For large EV battery packs required to provide a vehicle range of more than 100 mi (320 km), this amounted to costs of over $50,000 for the battery pack system alone. This was not a viable proposition since this component was more than twice the desired selling cost of the entire vehicle. NiMH battery developers projected production costs to drop below $300/kWh at mature volumes of 20,000 battery packs per year. Actual volumes achieved were more than an order of magnitude lower and the mass production cost potential was not realized. Consequently, the California mandate ZEV production targets were relaxed and commercialization was delayed. NiMH battery technology was not sufficient for the commercialization of PEVs.

Large lithium-ion batteries were developed by Sony and Hitachi for EV applications in the early 1990s due to their strong promise for high energy density. In those early days, the promise of high energy density was tempered with low-power performance and serious safety issues. However, the development of lithium-ion batteries for EV applications was accelerated by the massive effort of the battery industry aimed at commercialization of small lithium-ion batteries for portable electronics applications, particularly cell phones and laptop computers. Not only has the promise of high energy density been realized, but the thin electrode designs have evolved to very high-power lithium-ion batteries. The serious safety issues associated with earlier products have been greatly mitigated by improved manufacturing techniques and cell designs as well as new chemistries with reduced volatility.[33–36]

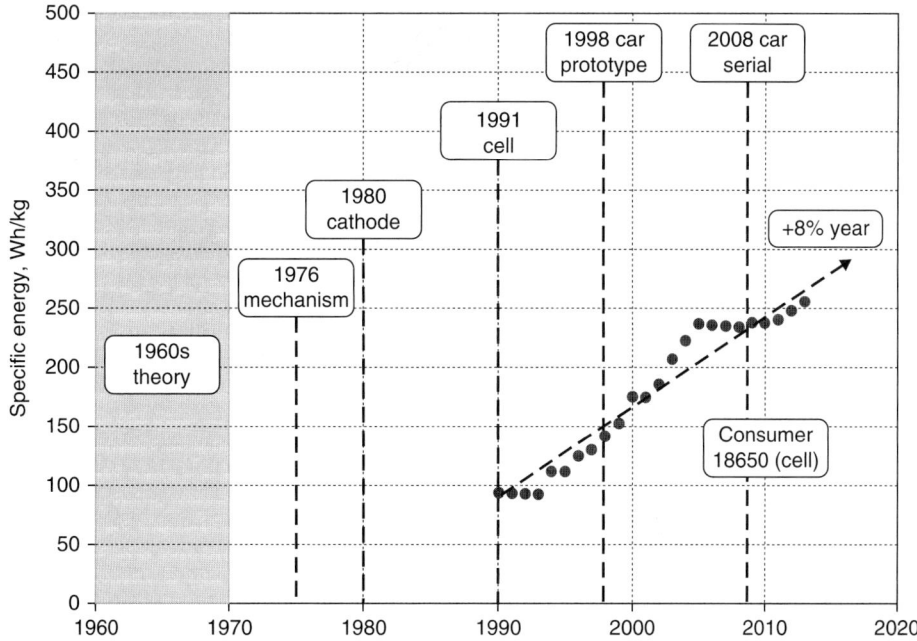

**FIGURE 26A.13**    History of Li-ion battery advancements and specific energy progression.

A multibillion dollar market for small 18650 lithium-ion cells (cylindrical cells for laptop computers with a standard size of 18 mm diameter by 65 mm long) has developed with a rapidly evolving technology with strong competition for high performance and low cost, as illustrated in Fig. 26A.13. The specific energy and energy density have improved dramatically from less than 100 Wh/kg and 200 Wh/L in 1990 to around 250 Wh/kg and 600 Wh/L, respectively, in 2015. The reduction in cost was even more dramatic. By 2010, the cost of 18650 cells was reduced from $3000/kWh to less than $200/kWh.[10]

In one innovative approach, small 18650 Li-ion cells developed for portable electronics were used directly to power a high-performance EV.[37] Tesla Motors had developed a novel battery pack for its Tesla Roadster sports car that incorporated 6831 of these cells in a series-parallel arrangement to provide 53 kWh of energy with a peak power of about 200 kW. High specific energy cobalt oxide cathode chemistry was used to provide over 2 Ah of capacity at about 3.6 V in a cell weighing less than 50 g yielding nearly 200 Wh/kg. The parts count and manufacturing complexity of a battery pack with thousands of cells was challenging from a cost point of view. However, the use of the high volume 18650 cell provided very competitive cell costs with the full effect of high volume materials production. Additionally, a variety of alternative suppliers were available to provide high quality and reliable cells from this mature market.

The potential volatility of this high-energy chemistry was mitigated by the small cell size and internal cell safety features such as a positive temperature coefficient (PTC) current limiting device and a current interrupt device (CID) incorporated into each cell. The series-parallel network provides for inherent redundancy, and the pack was designed with cell level fuses to survive the failure of multiple individual cells. Sophisticated electronics at the battery module and pack level were designed to provide for safe control of this high-energy battery pack. Thermal management of the battery pack was achieved with liquid cooling. Based on a weight of 450 kg, the battery pack system achieved an impressive specific energy of 120 Wh/kg and a specific power of 400 W/kg.

More traditionally, larger prismatic lithium-ion battery cells have been utilized in battery packs for EV applications. An example was the battery pack developed for the Mitsubishi iMiEV that utilized a single series string of 88 prismatic lithium-ion batteries developed by GS Yuasa.[38] This pack, shown in Fig. 26A.14, consisted of 22 series-connected modules each consisting of 4 series-connected cells. The cells delivered 50 Ah at a nominal

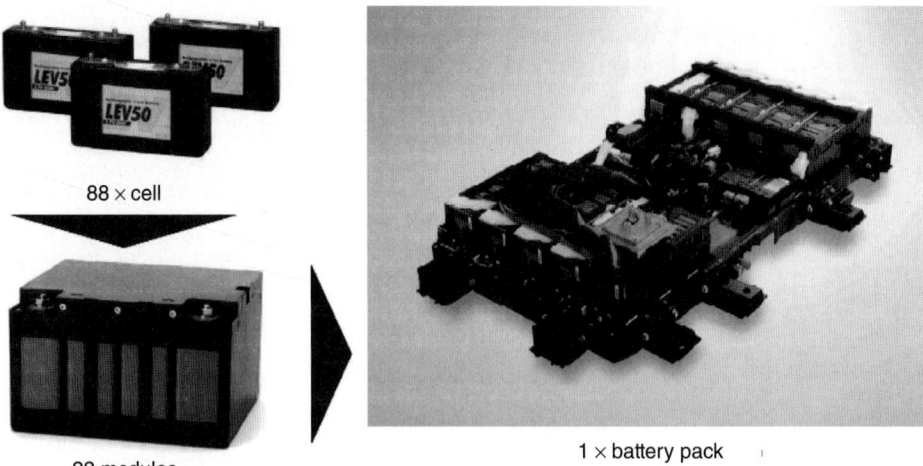

88 × cell

22 modules

1 × battery pack

**FIGURE 26A.14** Lithium-ion battery cells, module, and battery pack for the Mitsubishi iMiEV. (*Courtesy of Mitsubishi.*)

cell voltage of 3.7 V with a specific energy of 109 Wh/kg. Thermal management of this pack was achieved by forced air cooling. The battery pack, including batteries, thermal management, and control electronics, weighed 200 kg yielding a specific energy of 82 Wh/kg. Performance specifications for the cells, modules, and battery packs for the iMiEV are summarized in Table 26A.4.

Lithium-ion battery cells encompass a wide variety of chemistries and cell designs that now offer a variety of technology options for PEV developers[33-36] (see Chap. 17). The cobalt oxide cathode chemistry predominant in portable electronics applications provides high energy density, but many PEV developers have avoided this chemistry due to concerns about its volatility with respect to safety. Mixed layered oxide cathode chemistries such as the nickel-cobalt-aluminum oxide and the nickel-cobalt-manganese oxide have relatively high energy density and improved stability with respect to safety issues. The most benign chemistries from a safety point of view

**TABLE 26A.4** Performance Specifications for GS Yuasa LEV50 Cells and Modules and the Mitsubishi iMiEV Battery Pack

|  | Yuasa LEV50 battery cell | Yuasa LEV50 battery module (4 cells) | Mitsubishi MiEV battery pack (22 modules) |
|---|---|---|---|
| Nominal voltage | 3.7 V | 14.8 V | 326 V |
| Discharge capacity | 50 Ah | 50 Ah | 50 Ah |
| Energy | 185 Wh | 740 Wh | 16 kWh |
| Power | 935 W | 3740 W | 60 kW |
| Weight | 1.7 kg | 7.5 kg | 200 kg |
| Length | 171 mm | 175 mm | 1400 mm |
| Width | 44 mm | 194 mm | 700 mm |
| Height | 114 mm | 116 mm | 200 mm |
| Volume | 0.85 L | 3.9 L | 196 L |
| Specific power | 550 W/kg | 500 W/kg | 300 W/kg |
| Power density | 1100 W/L | 960 W/L | 306 W/L |
| Specific energy | 109 Wh/kg | 100 Wh/kg | 82 Wh/kg |
| Energy density | 218 Wh/L | 190 Wh/L | 83 Wh/L |

include the manganese oxide spinel and lithium iron phosphate olivine phase cathodes, albeit with some sacrifice in energy density. Yet, all of these alternative chemistries are attractive in mitigating materials cost issues with cobalt. All of the cell chemistries are offered in a variety of cell designs including cylindrical wound cells, elliptical wound cells, flat wound cells, traditional large format prismatic cells, and laminated pouch cells. The great commercial promise of lithium-ion batteries has spurred intense international competition resulting in a diverse multitude of suppliers and developers.

**FIGURE 26A.15**   A lithium-ion battery pack for Nissan Leaf EV. (*Courtesy of Nissan.*)

Even prior to 2010, lithium-ion EV battery packs met minimum goals for commercialization on specific power, power density, and cycle life; however, the calendar life was problematic, and they fell short on specific energy and energy density needed for a 300 mi (480 km) range. They were also several times too expensive. However, a variety of limited-range EVs were developed including the 2011 Nissan Leaf EV that utilized lithium-ion pouch cells. A range of about 80 mi (130 km) was achieved with a battery pack that delivered 24 kWh at 360 V. The battery pack was composed of 48 modules each with a series-parallel arrangement of four cells. Based on a battery pack weight of 295 kg, the pack specific energy was 82 Wh/kg. An innovative feature of this battery pack was a passive cooling system in a fully sealed battery pack, as shown in Fig. 26A.15. The battery pack cost and energy density limited the battery pack size and vehicle range.

Concurrently with the introduction of 100 mi (160 km) range EVs based on Li-ion batteries, GM pioneered the "extended range EV" concept as a shortcut strategy to EV commercialization.[39,40] This vehicle type also known as a plug-in HEV enables full vehicle range with a smaller and less expensive battery pack yielding a practical vehicle, even with EV battery technology falling short of long-term goals. Performance targets for PHEVs (see Table 26A.2) are similar to those for BEVs, but the requirements for specific energy, energy density, and specific cost ($/kWh) are relaxed by a factor of two, which could nearly be met by current technology.

As introduced by GM in 2011, the Chevy Volt PHEV provided a 40 mi (160 km) all-electric range and an extended range of around 400 mi (640 km) utilizing the onboard ICE generator. It utilized lithium-ion pouch cell battery technology from LG Chem in a large T-shaped battery pack, as shown in Fig. 26A.16. The battery pack was T-shaped reminiscent of the GM EV1 and favorable for vehicle handling. However, forced air thermal

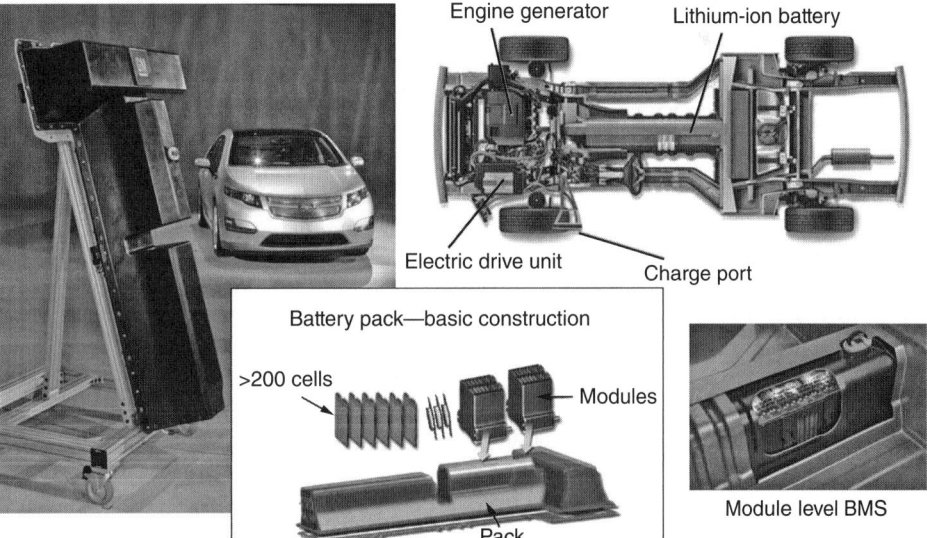

**FIGURE 26A.16**   Lithium-ion battery pack for Chevy Volt PHEV. (*Courtesy of GM.*)

management of PHEV batteries is perhaps even more challenging than for EV batteries for battery pack designs lacking in symmetry. Liquid cooling was utilized in the Volt to keep its large battery pack within the recommended operating range and to keep the battery cells at uniform temperatures.[41,42]

The 2016 Generation 2 Chevrolet Volt battery pack[43] utilized 192 Li-ion pouch cells in a 96s × 2p configuration (series connection of 96 sets of 2 cells in parallel) yielding a total energy of 18.4 kWh at 360 V. Individual cells delivered 26 Ah at a nominal voltage of 3.75 V. With a battery pack weight of 183 kg, the specific energy was 101 Wh/kg. The volumetric energy density was 119 Wh/L. The battery pack power exceeded 120 kW, equivalent to a specific power of 600 W/kg. The battery pack was liquid-cooled using cooling fins with flow channels interspersed between cells. This battery pack generally met USABC performance targets given in Table 26A.2 for the E-REV 50 extended range EV (PHEV with 50 mi or 80 km range) except for being 30% heavier and 50% larger than the target.

In 2017, GM introduced the Chevy Bolt, an affordable all-electric BEV with a range of over 200 mi (320 km). The Bolt battery pack[44] utilized 288 lithium-ion pouch cells in a 96s × 3p configuration delivering 60 kWh at 350 V. The battery pack featured a flat geometry spanning the entire floor of the vehicle from front to back, as shown in Fig. 26A.17, providing a low center of gravity for excellent vehicle handling. As with the Volt, the Bolt battery pack was liquid cooled. The battery pack weight and volume were 436 kg and 298 L, yielding a specific energy and energy density of 138 Wh/kg and 211 Wh/L, respectively. The battery pack delivered over 160 kW for propulsion, yielding a specific power and power density of at least 367 W/kg and 537 W/L, respectively.[36,44]

**FIGURE 26A.17**    Chevy Bolt EV battery pack and integration into the vehicle. (*Courtesy of GM.*)

GM achieved excellent vehicle performance in spite of significant shortfalls in battery energy density in comparison to USABC performance goals for EV batteries. However, there has been very significant and substantial progress toward cost targets, historically the biggest obstacle to EV commercialization.[10]

## *26A.3*   *HYBRID ELECTRIC VEHICLES*

### 26A.3.1    HEV Applications: Stop-Start Vehicles

Stop-start HEVs (also called microhybrids) provide for vehicle efficiency improvements of 5% to 10% by turning off the ICE instead of idling it during coast, braking, and stops.[14–17] With the engine off, the stop-start battery powers the vehicle electrical system, including accessories such as lights, sound systems, HVAC fans, and window defrosters. The engine is restarted to power acceleration and to maintain speed during driving. SSV systems are designed to provide seamless engine on-off transitions providing their greatest benefits during stop-and-go urban driving.

The simplest approach to SSVs utilizes a 14-V electrical system (alternator voltage) with an enhanced 12-V lead-acid battery. Similar to the conventional SLI operation, the battery is charged whenever the engine is operating, so it is maintained at a high SOC. Although this limits the opportunity to accept regenerative braking energy, it avoids premature sulfation failure of lead-acid batteries in partial SOC operation (see Chap. 14).

The USABC goals for advanced batteries for 12-V stop-start vehicle applications are provided in Table 26A.5 (released in 2015).[23] Superficially, the power and energy requirements for the stop-start HEV are no more challenging than for the conventional SLI application. The specific power and specific energy requirements of 600 W/kg and 25 Wh/kg, respectively, can be met with lead-acid battery technology. Cold crank capability to deliver 6 kW at −30°C is an important requirement difficult for some Li-ion batteries.

**TABLE 26A.5**   USABC Goals for Advanced Batteries for 12-V Stop-Start Vehicles

| End of life characteristics | Units | Target | |
|---|---|---|---|
| | | Under hood | Not under hood |
| Discharge pulse, 1 s | kW | 6 | |
| Max discharge current, 0.5 s | A | 900 | |
| Cold cranking power at −30°C (three 4.5-s pulses, 10 s rests between pulses at min SOC) | kW | 6 kW for 0.5 s followed by 4 kW for 4 s | |
| Min voltage under cold crank | Vdc | 8 | |
| Available energy (750 W accessory load power) | Wh | 360 | |
| Peak recharge rate, 10 s | kW | 2.2 | |
| Sustained recharge rate | W | 750 | |
| Cycle life, every 10% life RPT with cold crank at min SOC | Engine starts/miles | 450k/150k | |
| Calendar life at 30°C, 45°C if under hood | Years | 15 at 45°C | 15 at 30°C |
| Minimum round-trip energy efficiency | % | 95 | |
| Maximum allowable self-discharge rate | Wh/day | 2 | |
| Peak operating voltage, 10 s | Vdc | 15 | |
| Sustained operating voltage—max | Vdc | 14.6 | |
| Minimum operating voltage under autostart | Vdc | 10.5 | |
| Operating temperature range (available energy to allow 6 kW (1 s) pulse) | °C | −30 to +75 | −30 to +52 |
| 30°C–52°C | Wh | 360 (to 75°C) | 360 |
| 0°C | Wh | 180 | |
| −10°C | Wh | 108 | |
| −20°C | Wh | 54 | |
| −30°C | Wh | 36 | |
| Survival temperature range (24 h) | °C | −46 to +100 | −46 to +66 |
| Maximum system weight | kg | 10 | |
| Maximum system volume | L | 7 | |
| Maximum system selling price (250k units/year) | $ | $220 | $180 |

On a system level, the control electronics are simple in comparison to other hybrid vehicle types. Additionally, because of the minimal power duty cycle, passive thermal management can be utilized. However, the life requirement of several hundred thousand cycles and megawatt-hours of total energy throughput makes this application very difficult for lead-acid batteries. The multiple engine starts per mi of operation will require more than 100,000 shallow cycles in 4 to 5 years. Operation during idle periods has been estimated to average from 1 to 5 Wh to as much as 50 Wh with HVAC operation. Thus, the total energy throughput requirement is on the order of several MWh.

Given the limitation of conventional flooded lead-acid SLI batteries to provide an accumulated Ah turnover of about 150 times the nominal capacity in shallow discharge cycles, they are not likely to provide even 10,000 mi (16,000 km) of operation in this more challenging application. Enhanced flooded lead-acid batteries and VRLA batteries have been utilized providing a several-fold improvement in these SSV applications. Still the USABC life targets of 450,000 starts and operation for 15 years appear to be beyond the capability of lead-acid battery technology. This may require the use of advanced batteries making it difficult to meet full system cost targets at around \$30/kW on a cost per unit power basis (about \$600/kWh on a cost per unit energy basis).

Microhybrids have also been developed using 42-V electrical systems (42 V alternator voltages with batteries having a nominal voltage of around 36 V). In 2003, USABC provided target performance specification for 42-V stop-start batteries (these targets have since been archived). Historically, there have been intermittent attempts to transition conventional automobile electrical systems to utilize 36-V batteries. Due to system complications and costs involved, these attempts failed and the 42-V stop-start vehicle similarly appears to have no current proponents. Efforts have transitioned to the 48-V mild HEVs, which have the stop-start function as well as regen and motor-assist capability (see Sec. 26A.3.4).

## 26A.3.2   HEV Applications: Power-Assist HEVs

In power-assist HEVs, supplementary traction power is provided by the onboard electric power train. This motor assist is accomplished by using battery energy to power the HEV's electric motor and enabling it to contribute power to the driveshaft. Many different hybrid vehicle designs are under development and vary in their integration into the existing mechanical drivetrain, such as through simple clutches or more complicated power-split devices. The extent of electric power available for motor assist varies. At the low power end are mild hybrids that provide motor assist but no electric launch or all-electric propulsion. More substantial power-assist HEVs feature electric launch and engine-off electric drive. These higher-power HEVs with electric drive are called full hybrids or strong hybrids and have also been termed dual-mode hybrids since they feature the electric drive mode. All power-assist HEVs, both mild and strong hybrids, utilize regenerative braking.

Table 26A.6 summarizes FreedomCAR/USABC energy storage system performance goals for power-assist hybrid vehicles.[24] The most important criterion for HEV applications is power, both pulse discharge power for acceleration and pulse charge power for regenerative braking. The discharge and regen power must also be delivered at voltages within the operating range of the power electronics for the electric motor. So, the minimum voltage on discharge must be no lower than 55% of the maximum voltage on charge. The discharge and regen pulse power performance are specified for 10-s pulses. FreedomCAR provides goals for two representative cases. The minimum power-assist HEV targets call for 25 kW for discharge and 20 kW for regen charge. The maximum power-assist HEV targets call for 40 kW discharge and 35 kW regen charge. With the battery system weight targets of 40 and 60 kg and volume targets of 32 and 45 L for the minimum and maximum power-assist HEV cases, respectively, this represents over 600 W/kg and around 800 W/L for the specific power and power density requirements at a battery pack system level.

The power-assist HEV application target for energy is specified as the total available energy over the SOC range at which the battery system will provide the specified pulse discharge power and pulse regen charge power. This available energy, also termed usable energy, is specified as energy that can be delivered on the $C$-rate discharge from the highest SOC where the regen power goals are met to the lowest SOC where the discharge power goals are met. This is illustrated in Fig. 26A.18. The power and regen performance are determined from the hybrid pulse power characterization (HPPC) test that utilizes constant current pulses 10 s in duration at 10% depth of discharge increments in a $C$-rate discharge. The available energy is the discharge energy corresponding to the state-of-charge region where the discharge power and regen power targets are simultaneously met.

The 90% round-trip energy efficiency target is based on a simple series of discharge and charge pulses with peak discharge pulses at 24 kW and peak charge pulses at 21 kW for the maximum power-assist HEV case. The energy efficiency profile for the maximum power-assist HEV is provided in Fig. 26A.19. This profile delivers 50 Wh of discharge energy per cycle. The similar but lower power profile for the minimum power-assist HEV delivers 25 Wh of discharge energy per cycle.

The 25 Wh and 50 Wh energy efficiency cycles are also utilized for HEV cycle life testing of the minimum and maximum power-assist HEVs, respectively. Battery pack systems for the maximum and minimum power-assist HEV must be capable of delivering 300,000 HEV cycles. This is equivalent to an energy throughput of 7.5 and 15 MWh for the minimum and maximum power-assist HEVs, respectively. These goals are designed to

**TABLE 26A.6**  FreedomCAR/USABC Goals for Advanced Batteries for Power-Assist HEVs

| Characteristics at end of life | Condition | Units | Power-assist miniumum goal | Power-assist maximum goal |
|---|---|---|---|---|
| Pulse discharge power | 10 s pulse | kW | 25 | 40 |
| Peak regenerative pulse power | 10 s pulse | kW | 20 | 35 |
| Cold cranking power | 3 pulses of 2 s at −30°C | kW | 5 | 7 |
| Total available energy | 1 C discharge over DOD range where power goals are met | kWh | 0.3 | 0.5 |
| Round-trip energy efficiency | 25-Wh cycle for minimum goal 50-Wh cycle for maximum goal | % | 90 | 90 |
| HEV cycle life | 25-Wh cycle for minimum goal 50-Wh cycle for maximum goal | Cycles | 300,000 | 300,000 |
| HEV cycle life energy throughput | | MWh | 7.5 | 15 |
| Calendar life | | Years | 15 | 15 |
| Maximum weight | | kg | 40 | 60 |
| Maximum volume | | L | 32 | 45 |
| Operating voltage limits | | V | max < 400 min > 0.55 × | max < 400 min > 0.55 × |
| Self-discharge | Maximum allowable rate | Wh/day | 50 | 50 |
| Temperature range | Equipment operation | °C | −30 to +52 | −30 to +52 |
| Temperature range | Equipment survival | °C | −46 to +66 | −46 to +66 |
| System selling price | At 100,000 production volume | $ | 500 | 800 |

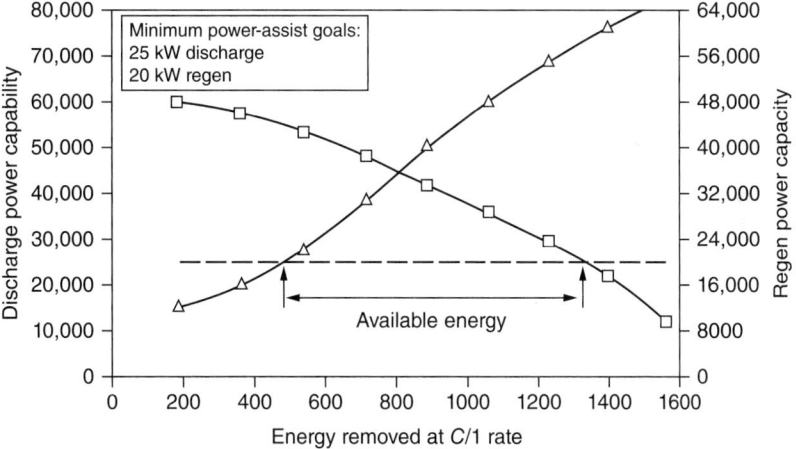

**FIGURE 26A.18**  Illustration of USABC available energy determined from pulse power and regen performance as a function of state-of-charge. (*Courtesy of USABC.*)

provide the capability for the battery to last the life of a car or 150,000 mi (240 km) and 15 years. A calendar life of 15 years is separately targeted since energy throughput and time can stress the battery differently for some failure modes. A key point for FreedomCAR/USABC life targets for hybrid vehicles is that end of life is defined as failure to meet either the power or energy performance goals. Thus, to meet the USABC life targets, HEV batteries need to be designed with excess power and energy providing a buffer so that performance targets are still met at the end of life. This is illustrated in Fig. 26A.20.

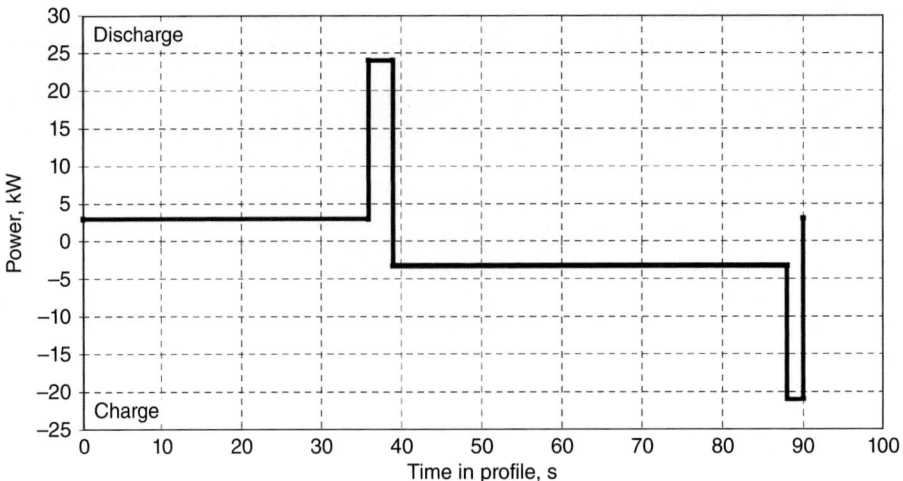

**FIGURE 26A.19**   Energy efficiency and baseline cycle life test profile for power-assist HEV. (*Courtesy of USABC.*)

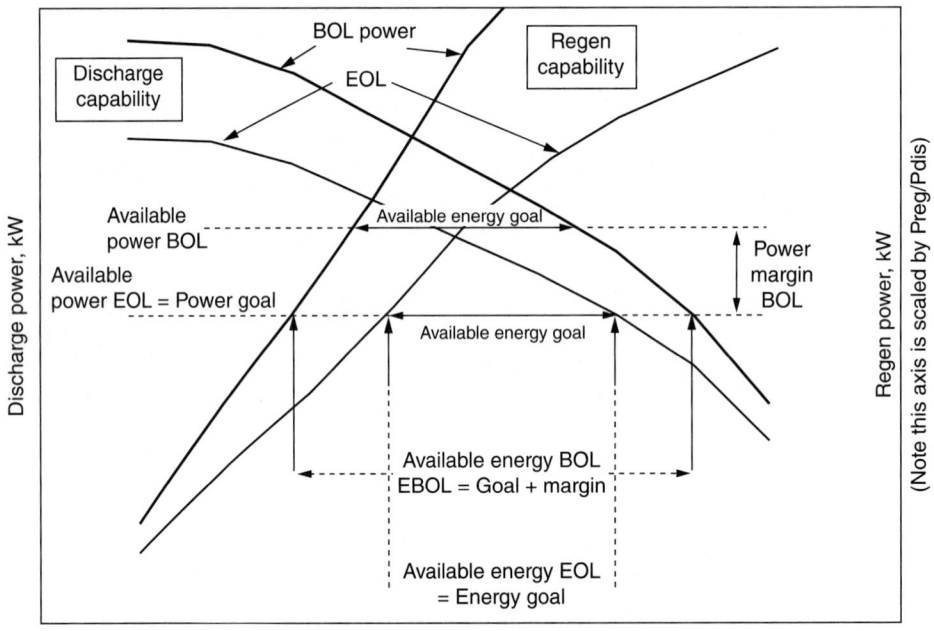

**FIGURE 26A.20**   Illustration of available energy and power margins over battery system life. (*Courtesy of USABC.*)

### 26A.3.3    HEV Applications: Strong HEVs

Strong hybrid vehicles or full HEVs, as they are sometimes called, are a subset of power-assist hybrid vehicles that are capable not only of motor-assist but also all-electric propulsion that may include electric launch. The torque and power capability of the electric drivetrain are comparable to that of the ICE. The 25 to 50 kW power capability is provided at voltage of over 200 V to maintain electrical currents at practical levels. These vehicles require very high specific power batteries, approaching 1000 W/kg or more, with the capability to accept high-power regen charge pulses as well as deliver high-power discharge pulses. The FreedomCAR goals for power-assist HEVs are aimed principally at strong HEVs, especially the maximum power-assist HEV case. Strong HEVs utilize petroleum-based fuels in ICEs as the source of energy, but use batteries to improve the efficiency of operation by limiting the operation of the ICE to its highest efficiency operational conditions, which also provides for reduced emissions of noxious pollutants. Thus, the strong or full hybrid provides for the high fuel economy improvements of 50% or more in city driving and large emission reduction opportunities. Good examples of the technical and engineering success of this approach are the Toyota Prius and the Ford Fusion Hybrid.

### 26A.3.4    HEV Applications: Mild HEVs

Mild hybrid vehicles are the lower-power subset of power-assist hybrid vehicles that are capable of motor assist, but not all-electric propulsion. This design approach addresses the cost challenges of the full hybrid and provides an alternative lower-cost design that still provides substantial fuel economy and emission reduction advantages. The torque and power capability of the electric drivetrain is significantly less than that of the ICE. The 10 to 20 kW power capability is provided at a lower voltage that still may exceed 100 V, but some mild hybrid systems have been designed even with 42-V systems. These vehicles still require very high specific power batteries, approaching 1000 W/kg or more, but smaller batteries are used, which lowers the system cost. Examples of mild hybrid vehicles produced commercially include the Honda Civic Hybrid, which utilized a 144-V NiMH battery with a 10-kW power capability, and the Saturn VUE Greenline mild hybrid, which utilized a 42-V NiMH battery pack with a power capability of less than 10 kW.

Battery packs for mild hybrid vehicles thus span a rather wide range from moderate-power performance similar to that in the minimum power-assist HEV given in Table 26A.6 down to power requirements for the stop-start hybrid, but with regen capability. There has recently been a trend to provide mild hybrid battery packs with a nominal voltage of 48 V. This minimizes the electric shock hazards that require special safety mitigation with voltages higher than 50 V.

Table 26A.7 summarizes USABC performance targets for 48-V systems for mild hybrid applications.[25] The peak pulse discharge power requirement of 9 kW for 10 s is higher than that for the 12-V stop-start application and for a longer time so as to provide motor-assist capability. The 48-V mild hybrid application also targets 11 kW regen power for 5 s so as to utilize significant regen braking energy. While the available energy for cycling is only 105 Wh, the total usable energy requirement is 313 Wh so as to provide for a 5 kW accessory load for up to 5 min. The HEV cycle requirement of 75,000 charge-sustaining 48-V cycles includes six charge-discharge cycles including stop-starts. So, the actual cycle life requirement is 450,000 charge-discharge cycles.

### 26A.3.5    Comparison of HEV Battery Performance Requirements

In Table 26A.8, the HEV battery specific performance targets derived from the USABC goals in Tables 26A.5 to 26A.7 are compared to the specific performance targets for PEV batteries in Tables 26A.1 and 26A.2. Specific performance criteria, including specific energy, specific power, power-to-energy ratio (units of W/Wh = 1/h), and cost in $/kWh and $/kW, are listed for the hybrid and electric vehicle applications in order of increasing degree of electrification. The specific energy targets for stop-start, mild hybrids, and power-assist hybrids are all quite modest and within the capability of lead-acid batteries. The specific energy requirements for plug-in hybrids are more challenging but could be met by lithium-ion batteries. The power capability required by most HEV batteries is around 600 W/kg, well within the capability of high-power nickel-metal hydride and lithium-ion batteries. The high-energy PHEV and EV applications require moderately less power. High-power nickel-metal hydride and lithium-ion batteries have been developed that can easily meet these power targets.

**TABLE 26A.7** USABC Goals for Batteries for 48-V Mild HEVs

| Characteristics at end of life | Units | Target |
|---|---|---|
| Peak pulse discharge power (10 s) | kW | 9 |
| Peak pulse discharge power (1 s) | kW | 11 |
| Peak regen pulse power (5 s) | kW | 11 |
| Available energy for cycling* | Wh | 105 |
| Minimum round-trip energy efficiency | % | 95 |
| Cold cranking power at −30°C (three 4.5-s pulses, 10 s rests between pulses at min SOC) | kW | 6 kW for 0.5 s followed by 4 kW for 4 s |
| Accessory load (2.5 min duration)* | kW | 5 |
| Charge sustaining 48 V HEV cycle life† | Cycles/MWh | 75,000/21 |
| Calendar life, 30°C | Year | 15 |
| Maximum system weight | kg | <8 |
| Maximum system volume | Liter | <8 |
| Maximum operating voltage | Vdc | 52 |
| Minimum operating voltage | Vdc | 38 |
| Minimum voltage during cold crank | Vdc | 26 |
| Maximum self-discharge | Wh/day | 1 |
| Unassisted operating temp range (power available to allow 5 s charge and 1 s discharge pulse) at min. and max. operating SOC and voltage | °C | −30 to +52 |
| 30–52°C | kW | 11 |
| 0°C | kW | 5.5 |
| −10°C | kW | 3.3 |
| −20°C | kW | 1.7 |
| −30°C | kW | 1.1 |
| Survival temperature range | °C | −46 to +66 |
| Max system production price @ 250 k units/year | $ | $275 |

*Total usable energy will include cycling energy and accessory load energy. The usable energy will be 313 Wh.
†Each individual cycle profile includes six (6) start-stop events, for a total of 450k events over the duration of the test.

**TABLE 26A.8** Comparison of Specific Performance Targets for HEVs and EVs

| | Specific energy, Wh/kg | Specific power, W/kg | P/E ratio | Energy cost, $/kWh | Power cost, $/kW |
|---|---|---|---|---|---|
| Stop-start 12 V hybrid (under hood) | 36 | 600 | 17 | $611 | $37 |
| Stop-start 12 V hybrid (not under hood) | 36 | 600 | 17 | $500 | $30 |
| 48V HEV | 39 | 1125 | 29 | $879 | $31 |
| Power-assist (minimum) | 8 | 625 | 83 | $1667 | $20 |
| Power-assist (maximum) | 8 | 667 | 80 | $1600 | $20 |
| Plug-in hybrid (PHEV-20) | 87 | 529 | 6 | $361 | $59 |
| Plug-in hybrid (PHEV-40) | 99 | 317 | 3 | $286 | $89 |
| Plug-in hybrid (E-REV 50) | 99 | 667 | 7 | $287 | $43 |
| Electric vehicle | 235 | 470 | 2 | $125 | $63 |

The key difference between the battery specific performance requirements for HEV and EV applications is the power-to-energy ratio. Hybrid vehicles need high power, but have modest energy requirements, except for PHEVs. Consequently, their specific energy requirements are lower. EVs and PHEVs need high energy, resulting in larger batteries. While the power requirement is also higher, the specific power requirement is lower for these larger batteries. As a result, the power-to-energy ratio is more than an order of magnitude higher for HEV applications except for PHEV applications that more closely resemble EVs in their requirements.

Batteries can be designed for high power or high energy. High-energy designs tend to feature a smaller number of thicker electrodes and have a higher fraction of active materials. High-power batteries tend to feature a larger number of thinner electrodes and utilize a higher fraction of conductive components at the expense of the active material fraction. An example can be provided in a comparison of nickel-metal hydride batteries for HEV and EV applications. High-energy NiMH EV batteries have been developed with a specific energy of 80 Wh/kg. They have a specific power of about 200 W/kg. High-power NiMH HEV batteries have been developed with a specific power of 1300 W/kg. However, this was achieved through design trade-offs that resulted in a lower specific energy of 45 Wh/kg in the HEV battery.

The cost requirements and criteria are different for HEV and EV applications. The cost target for EV batteries is $125/kWh. In HEV batteries, cost per unit power is the key variable. For power-assist HEVs, the cost target is $20/kW. On a cost per unit energy basis, the power-assist HEV cost target is about $1600/kWh, an order of magnitude higher than the EV battery target. Even taking into account the higher cost of the high-power HEV designs, this cost target is much more achievable. One somewhat oversimplified way to look at this is that it is easier for HEV batteries to compete with engines on the cost of providing power than it is for EV batteries to compete with gasoline tanks on the cost of providing energy.

### 26A.3.6 Batteries for Hybrid Electric Vehicles

Lead-acid batteries were initially utilized for many HEV development projects in the period from 1970 to 1990. For power-assist HEV applications, with both mild and strong hybrids, lead-acid batteries did not offer adequate durability. This is principally due to premature sulfation failure during operation at partial states of charge. Long life can be obtained with lead-acid batteries maintained at the fully charged state, but this is not feasible in HEV applications utilizing regenerative braking. The power-assist HEV battery must be operated at intermediate states of charge to provide capability to accept regenerative braking charge current.

For microhybrid SSV applications, lead-acid batteries remain the battery of choice because they perform the necessary functions at the lowest cost. However, this application requires substantially increased number of engine starts and deeper discharge operation to operate accessories than the SLI operation in conventional ICE vehicles. This requires a more robust battery and premium lead-acid technologies such as enhanced flooded batteries or VRLA batteries are preferred. Even then, the operational lifetime is limited to several years. The USABC lifetime target of 150,000 mi (240 km) over 15 years will require advanced technologies and/or hybrid systems involving lead-acid batteries combined with lithium-ion batteries or supercapacitors.

Additionally, microhybrid developers are exploring ways to accept regen braking energy while still utilizing relatively inexpensive lead-acid technology. Battery companies are also developing lead-acid batteries that have improved durability in partial SOC operation through the use of carbon additives.[45] Of particular interest is Ultrabattery developed by the Australian Commonwealth Scientific and Industrial Research Organization (CSIRO). The Ultrabattery utilizes supercapacitor carbon materials in the anode that enables partial state-of-charge operation.[46] East Penn has recently developed production facilities to offer this battery that shows promise for stop-start hybrid vehicle applications (see Chap. 14 for details).

Starting in the 1990s, high-power NiMH batteries for HEV applications were developed by a variety of battery companies including ECD Ovonics, PEVE, Sanyo, Varta, and Saft. High-power NiMH batteries were first introduced commercially in the original Toyota Prius in 1997. The high specific power and high regen capability enabled the Prius to achieve excellent fuel economy under power-assist HEV operation. Nickel-metal hydride battery technology has the capability to operate in intermediate states of charge and deliver hundreds of thousands of shallow charge-discharge cycles with a 1% to 2% state-of-charge swing. This has enabled a battery pack capable of operating for 10 years and/or 100,000 mi (160,000 km) with appropriate control strategies. This enabled the commercialization of hybrid vehicles at the turn of the century.

These first NiMH HEV batteries evolved from high-power cylindrical cells produced at high volume for consumer applications.[47] They featured the typical jelly-roll electrode winding of long thin electrodes with circular nickel end plates welded to the electrode edges to provide higher specific power. Cylindrical NiMH HEV batteries supplied by Panasonic were used to power the original Prius. These batteries were also utilized in the Honda Insight in 1999 in an air-cooled battery pack. This battery pack consisted of 120 cells packaged in 20 modules each with six cells in a string. The pack provided slightly less than 1 kWh of energy, but 10 kW of power, enabling the Insight to achieve exceptional fuel economy. At a module level, a specific power of 600 W/kg was achieved, approaching USABC power targets for mild HEV applications.

PEVE later developed higher-power prismatic NiMH batteries that were utilized first in the 2001 Toyota Prius when it was introduced in North America.[48–50] This module also consisted of six series-connected cells each 6.5 Ah in capacity, but featured a unique prismatic design. Cells in the module were placed edge to edge in a long thin package that provided for the thin cells to have a high area cooling surface, as shown in Fig. 26A.21. The prismatic PEVE NiMH HEV battery modules utilized thin electrodes and separators with current collection along the long edge of the electrodes providing for an improved specific power of 1000 W/kg. The specific power was further improved in 2003 to 1300 W/kg in large part by improving electrical power interconnections between cells.[50] These design changes for improved power were at the expense of a more modest specific energy, 45 Wh/kg at the module level in comparison to around 80 Wh/kg for EV batteries. However, these trade-offs still provided a considerable excess in specific energy over the 8 Wh/kg required for the power-assist HEV application. HEV cycle life tests confirmed capabilities well in excess of the 300,000 cycle requirement.

The packaging of PEVE prismatic battery modules into battery pack systems for the 2003 Toyota Prius HEV is also illustrated in Fig. 26A.21. The prismatic battery modules are stacked together with interlocking plastic circular nubs and ribs that hold the batteries in place and space them for optimal flow of cooling air. A cutaway view shows the module stack in a steel case that provides for an upper and lower plenum directing air flow between the modules. This arrangement provides for a more uniform cooling than achieved in the previous

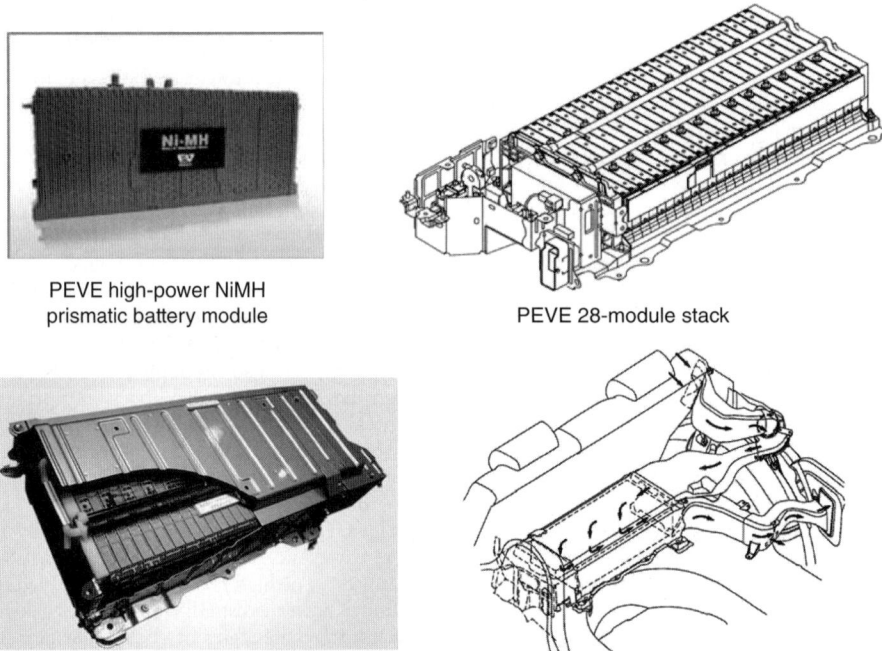

PEVE high-power NiMH
prismatic battery module

PEVE 28-module stack

Cutaway view of Prius pack

Air flow directed through battery pack

**FIGURE 26A.21**   PEVE 2003 prismatic NiMH HEV battery module and 2003 Prius battery pack. (*Courtesy of Panasonic EV Energy.*)

cylindrical cell battery pack for the Prius where air flow sequentially cooled a series of battery modules. In this design, all battery modules and cells are cooled with nearly identical parallel air flow.

The 2003 Prius utilized a battery pack comprising 28 NiMH modules, each delivering 6.5 Ah at 7.2 V and each weighing 1.04 kg. The fully integrated battery pack weighed 38 kg and occupied about 40 L of volume. At the pack level, the peak power of 39 kW and total energy of 1.3 kWh yielded a specific power of 1026 W/kg and a specific energy of 34 Wh/kg. With a proven cycle life of over 500,000 cycles, this NiMH battery pack handily met all of the USABC performance requirements for power-assist HEVs (see Table 26A.6). However, the system cost target was probably exceeded at that time. Further improvements in these PEVE NiMH HEV batteries have resulted in smaller and lighter battery packs with exceptional reliability and durability for two decades of Prius hybrid vehicles. The basic design is retained in the NiMH batteries that still power the Toyota family of Prius hybrids today.

During 2012–2013, several major auto companies including Honda, Ford, and GM transitioned their HEV battery pack technology from NiMH to Li-ion.[51] Reasons cited for this change included advantages in lithium-ion technology in power density, energy efficiency, and durability. It would have seemed that the added volume, weight, complexity, and cost of the more complicated control systems for lithium-ion batteries as well as safety concerns might have discouraged this change. However, high-energy lithium-ion batteries are clearly the technology of choice for EV and PHEV applications. Auto companies are now more comfortable with this technology and also see more opportunity for cost reduction over time. Also, NiMH suppliers have been largely consolidated with the largest supplier, PEVE (rebranded as Primearth EV Energy in 2010), acquiring the NiMH HEV battery business of Sanyo, the second largest supplier. Toyota has a controlling interest in Primearth and has continued with NiMH technology where it has made major investments in substantial manufacturing capabilities (see Chap. 15 for details).

In 2008, Mercedes Benz was the first to go into mass production with a hybrid vehicle utilizing a lithium-ion battery when it produced a mild hybrid version of the Mercedes S-class luxury sedan, the S400.[52] The battery pack, supplied by the Johnson Controls-Saft (JCS) joint venture, comprised 35 high-power cylindrical lithium-ion cells each with a capacity of 6.5 Ah delivered at a nominal voltage of 3.6 V. The 126-V battery pack was capable of 19 kW, corresponding to about 750 W/kg on a battery pack level. Mercedes utilized the high specific power and power density of the JCS lithium-ion battery to locate this battery under the hood within the same general packaging envelop of the lead-acid SLI battery. To maintain the temperature of the battery cells below 50°C in the engine compartment, the battery pack was liquid-cooled and utilized the vehicle HVAC system as needed.

A variety of strong power-assist HEVs entering the market in recent years have been powered by lithium-ion batteries. An example is the 2016 Chevrolet Malibu Hybrid.[53] The Malibu HEV battery pack[54] as shown in Fig. 26A.22 was based on prismatic Hitachi lithium-ion cells each delivering 5.2 Ah at 3.7 V. With a cell weight of 0.24 kg, the specific energy was a modest 80 Wh/kg, in a high-power design yielding 5000 W/kg. The battery

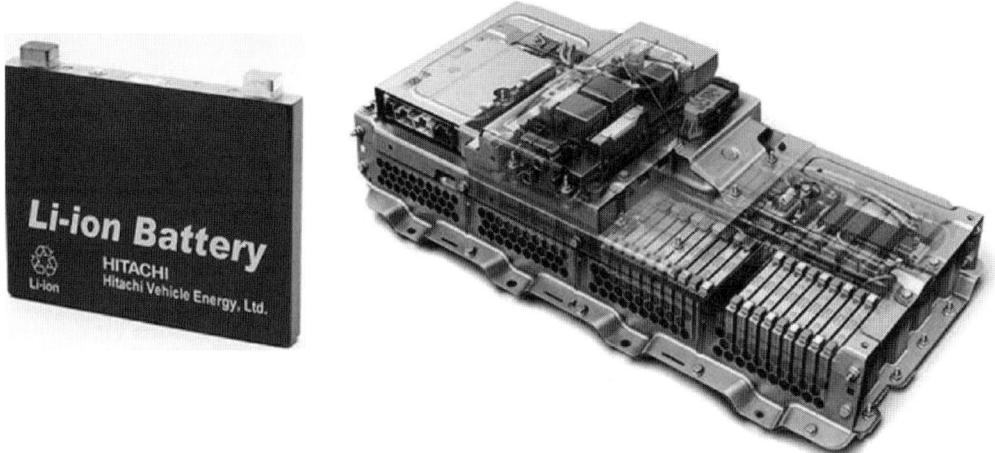

**FIGURE 26A.22**    A Li-ion battery cell and battery pack for 2016 Chevrolet Malibu Hybrid. (*Courtesy of GM.*)

pack comprised 80 series-connected lithium-ion cells in 8 modules of 10 cells each. The battery pack has a total energy of 1.5 kWh at 300 V. The usable energy is 450 Wh. The pack easily delivers the 52 kW discharge power draw required of the drivetrain and can similarly accept 65 kW of regen power with high efficiency. The air-cooled battery pack weighs 43 kg. The specific power is well in excess of 1200 W/kg and the specific energy is 10 Wh/kg based on usable energy. This pack was integrated behind the rear seats and the thermal management system utilized conditioned cabin air. Thus, the USABC power and energy targets were met. The Malibu HEV powered by Li-ion batteries is a high-performance HEV with an excellent fuel economy rating of 49 mi/gal in the city and 43 mi/gal on the highway.

## 26A.4   FUTURE TRENDS IN BATTERY TECHNOLOGY

Lithium-ion battery technology has been the principal focus of funded EV battery R&D in the United States in recent decades to exploit the promise of this high-energy technology (see Chap. 17A). Further improvements in energy density can drive the cost down by lowering materials costs. One promising approach is the development of higher-voltage cathode materials. A second approach is the development of higher specific capacity cathodes and anodes. There is particular promise with the development of durable higher-energy Si-based anodes (see Chap. 17). The development of other battery technologies exceeding the performance of lithium-ion is also being encouraged. The focus is on high energy density; however, the full set of performance criteria including power and life need to be met simultaneously. There has been a strong emphasis on lithium-air batteries given their exceptional theoretical specific energy (see Chap. 18C). However, substantial new research has also revealed serious obstacles to be overcome with this technology that utilizes challenging fuel cell technology including bifunctional air cathodes. Lithium-sulfur batteries, with the simplifications of a closed system, continue to offer promise for high energy densities and low cost (see Chap. 17B). There is renewed interest in solid-state lithium batteries that offer substantial potential advantages in energy density and safety as well as the ability to operate at higher temperatures, which could lead to smaller and less expensive thermal management systems (see Chap. 22C).

Hydrogen fuel cells have long been considered the ultimate high-energy electrochemical energy conversion device. Hydrogen has the highest specific energy electroactive material (the specific energy exceeds gasoline). However, it has a very poor volumetric energy density and hydrogen storage is a significant problem. An attraction is that fast refueling comparable with that in conventional gas-powered vehicles may be possible. In recent decades, there was substantial progress in the development of PEM fuel cell power plants that meet the power requirements for automotive propulsion. However, some significant technical and commercial barriers remain including durability under practical driving conditions, hydrogen storage system weight and size, and the cost of hydrogen fuel. A serious commercial barrier has been the cost of platinum-based PEM fuel cells that utilize noble metal catalysts and expensive proprietary membrane materials. Fuel cell vehicles also utilize high-power batteries to enable them to capture and utilize regenerative braking energy as well as to provide for peak acceleration power, which allows utilization of smaller and less expensive fuel cell stacks. Thus, the proper term for the modern fuel cell vehicle is fuel cell hybrid electric vehicle (FCHEV). Although the timing for commercial introduction now appears to be beyond 2020, optimism remains for this promising technology within the automotive community. For very long-range high-power applications, such as long haul semitrailers, no other electric propulsion system seems viable (see Chap. 19).

High-power battery technology for HEV applications has not been an urgent priority for funded R&D. This is because the 600 W/kg specific power targets for HEV applications have been handily met by both NiMH and lithium-ion technology. NiMH batteries were developed to nearly 1500 W/kg at the battery module level. A variety of battery companies, including LG Chem, SK Energy, Hitachi, AESC, Panasonic, A123, Samsung, CATL and others, have now developed very high-power lithium-ion HEV batteries in the range of 2000 to 5000 W/kg on a cell level. However, batteries do not yet meet industry cost targets for HEV applications. HEV batteries are oversized because they cannot be fully discharged and charged during HEV acceleration and regen braking events, which occur on the order of seconds. Batteries require hours or minutes for full charge-discharge. A faster electrochemical device would be more suitable to power HEV acceleration times of seconds.

Supercapacitors, also known as electrochemical double-layer capacitors or ultracapacitors, are electrochemical power devices that can be fully charged or discharged in seconds, and may have promise for HEV applications (see Chap. 20). These devices have been developed by a variety of companies throughout the world, including Maxwell and Ioxus, and have been commercially introduced in heavy-duty HEV applications. They have powered many hundreds of HEV busses in China, for example, as a viable alternative to high-power battery energy storage.

In addition to very high specific power, supercapacitors offer advantages in HEV cycle life with the capability to deliver hundreds of thousands of deep charge-discharge cycles. A strong advantage is also the excellent low-temperature performance particularly of supercapacitors with little attenuation in battery power down to −30°C in some devices. The most significant performance drawback is specific energy. The 5 Wh/kg commercially available is formally insufficient to meet the 8 Wh/kg specific energy requirements for available energy in power-assist HEV applications.

However, there has been reconsideration of this target in recent studies. Incorporation of supercapacitors into the Saturn VUE mild hybrid vehicle has provided excellent and promising results.[55] Modeling studies have also shown effective HEV operation with available energy storage at 4 Wh/kg levels and below. Not considering energy to run accessories, the USABC usable energy of 500 Wh target for strong power-assist HEVs is higher than needed for HEV propulsion. Consequently, USABC has considered lower energy density systems in their goals for high-power, lower-energy energy-storage systems (LEESS) for power-assist HEV applications. The energy window was lowered to 165 Wh in the LEESS goals, which is easily achievable by ultracapacitors.

Another development of interest is the hybrid supercapacitor that couples a battery electrode with a capacitor electrode. An example is the lithium-ion capacitor (LIC), which couples a lithium anode with an activated carbon cathode. These devices, now commercial, provide excellent power with a specific energy greater than 10 Wh/kg, meeting USABC power-assist HEV performance targets. In a recent National Renewable Energy Laboratory (NREL) study, the NiMH HEV battery pack was replaced in a Ford Fusion Hybrid by a LIC energy-storage device supplied by JSR micro which enabled equivalent energy efficiency in HEV operation.[56]

## ACKNOWLEDGMENT

We thank Matthew Shirk of Idaho National Lab for his assistance in understanding the updated USABC battery requirements and for his diligent review of the manuscript.

## REFERENCES

1. M. H. Westbrook, "The Electric and Hybrid Electric Car," Society of Automotive Engineers, Warrendale, Pennsylvania (2001).

2. C. C. Chan, "The State of the Art of Electric, Hybrid, and Fuel Cell Vehicles," *Proceedings of the IEEE* **95**:704–718 (2007).

3. I. Husain, "Electric and Hybrid Vehicles," CRC Press, Boca Raton, Florida (2003).

4. M. Ehasani, Y. Gao, S. Gay, and A. Emadi, "Modern Electric, Hybrid, and Fuel Cell Vehicles: Fundamentals, Theory, and Design," 2nd Edition, CRC Press, Boca Raton, Florida (2010).

5. J. Larminie and J. Lowry, "Electric Vehicle Technology Explained," 2nd edition, John Wiley, Hobroken, New Jersey (2012).

6. A. Emadi, "Advanced Electric Drive Vehicles," CRC Press, Boca Raton, Florida (2015).

7. M. Shnayerson, "The Car that Could: The Inside Story of GM's Revolutionary Electric Vehicle," Random House, New York, 1996.

8. "Who Killed the Electric Car," documentary film written and directed by Chris Paine, produced by Jessie Deeter, Sony Picture Classics (2006).

9. Argonne National Laboratory, Energy Systems, Light Duty Electric Drive Vehicles Monthly Sales Updates, https://www.anl.gov/es/light-duty-electric-drive-vehicles-monthly-sales-updates, accessed February 8, 2019.

10. M. Sanders, "The Rechargeable Battery Market and Main Trends 2016–2025," presented at The Battery Show, Novi, Michigan (2017).

11. John Voeker, "Hybrid Market Share Peaked in 2013—Down Since Then," Green Car Reports, http://www.greencarreports.com/news/1108483_hybrid-market-share-peaked-in-2013-down-since-then, accessed January 23, 2017.

12. E. Taubes, "Stop-Start Technology is Spreading (like it or not)," *New York Times*, April 7, 2017.

13. P. Savagian, "Driving the Volt," SAE Hybrid Vehicle Technology Conference, San Diego, California, February 2008.

14. M. Anderman, "The Challenge to Fulfill Electrical Power Requirements of Advanced Vehicles," *J. Power Sources* **127**:2–7 (2004).

15. O. Bitsche and G. Gutman, "Systems for Hybrid Cars," *J. Power Sources* **127**:8–15 (2004).

16. E. Karden, P. Shinn, P. Bostock, J. Cunningham, E. Schoultz, and D. Kok, "Requirements for Future Automotive Batteries—A Snapshot," *J. Power Sources* **144**:505–512 (2005).

17. E. Karden, S. Ploumen, B. Fricke, T. Miller, and K. Snyder, "Energy Storage Devices for Future Hybrid Electric Vehicles," *J. Power Sources* **168**:2–11 (2007).

18. D. H. Doughty and C. C. Craft, "FreedomCAR Electrical Energy Storage System Abuse Test Manual for Electric and Hybrid Vehicle Applications," Sandia National Laboratories, SAND 2005-3123, June 2005.

19. C. N. Ashtiani, "Battery Hazard Modes and Risk Mitigation Analysis," United States Advanced Battery Consortium Manual, August 2007.

20. USABC Energy Storage System Goals and Test Manuals, USABC web page, http://www.uscar.org/guest/teams/12/U-S-Advanced-Battery-Consortium-LLC, accessed February 8, 2019.

21. "Battery Test Manual for Electric Vehicles," Vehicle Technologies Program, U.S. Department of Energy, Idaho National Laboratory Report INL/EXT-15-34184, Rev 3, June 2015.

22. "Battery Test Manual for Plug-In Hybrid Electric Vehicles," Vehicle Technologies Program, U.S. Department of Energy, Idaho National Laboratory Report INL/EXT-14-32849, Rev 0, March 2017.

23. "Battery Test Manual for 12 Volt Start/Stop Vehicles," Vehicle Technologies Program, U.S. Department of Energy, Idaho National Laboratory Report INL/EXT-12-26503, Rev 1, May 2015.

24. "FreedomCAR Battery Test Manual for Power Assist Hybrid Electric Vehicles," Vehicle Technologies Program, U.S. Department of Energy, Idaho National Laboratory Report DOE/ID-11069, October 2003.

25. "Battery Test Manual for 48 Volt Mild Hybrid Electric Vehicles," Vehicle Technologies Program, U.S. Department of Energy, Idaho National Laboratory Report INL/EXT-15-36567, Rev 0, March 2017.

26. A. Smith, J. Burns, S. Trussler, and J. Dahn, "Precision Measurements of the Coulombic Efficiency of Lithium-Ion Batteries and of Electrode Materials for Lithium-Ion Batteries," *J. Electrochem. Soc.* **157**:A196–A202 (2010).

27. I. Bloom, B. Cole, J. Sohn, S. Jones, E. Polzin, V. Battaglia, and G. Henriksen, et al., "An Accelerated Calendar and Cycle Life Study of Li-Ion Cells," *J. Power Sources*, **101**:238–247 (2001).

28. J. Belt, D. Bernardi, and V. Utgikarb, "Development and Use of a Lithium-Metal Reference Electrode in Aging Studies of Lithium-Ion Batteries," *J. Electrochem. Soc.* **161**:A1116–A1126 (2014).

29. A. A. Frank, "Charge Depletion Control Method and Apparatus for Hybrid Powered Vehicles," U.S. Patent 5,842,534, December 1, 1998.

30. B. Johnston, T. McGoldrick, D. Funtson, H. Kwan, M. Alexander, F. Aliato, and N. Culaud, et al., University of California, Davis, PNGV FutureCar Technical Report, SP-1359 SAE, June 1997.

31. D. A. J. Rand, R. Woods, and R. M. Dell, "Batteries for Electric Vehicles," Society of Automotive Engineers, Warrendale, Pennsylvania (1998).

32. R. S. Stempel, S. R. Ovshinsky, P. R. Gifford, and D. A. Corrigan, "Nickel-Metal Hydride: Ready to Serve," *IEEE Spectrum* **35**:29 (1998).

33. R. Spotnitz, "Large Li Ion Battery Design Principles," Tutorial A, The 8th International Advanced Automotive Battery Conference, Tampa, Florida, May 2008.

34. M. S. Whittingham, "Lithium Batteries and Cathode Materials," *Chem. Rev.* **104**:4271 (2004).

35. G. Nazri and G. Pistoria, "Lithium Batteries: Science and Technology," Kluwer Academic Publishers, New York (2004).

36. M. Alamgir, "Lithium Has Transformed Vehicle Technology," *IEEE Electrification Magazine* **5**(1):43–52 (2017).

37. G. Berdichevsky, K. Kelty, J. B. Straubel, and E. Toomre, "The Tesla Roadster Battery System," Tesla Motors, December 2007.

38. T. Miyashita and Y. Tominga, "Development of High Energy Lithium-Ion Battery Pack for Pure EV Applications," Large Lithium-ion Battery Technology and Application Symposium, The 8th International Advanced Automotive Battery Conference, Tampa, Florida, May 2008.

39. K. Brooke, "Chevrolet Volt, Development Story of the Pioneering Electrified Vehicle," SAE International, Warrendale, Pennsylvania (2011).

40. E. Tate, M. Harpster, and P. Savagian, "The Electrification of the Automobile: From Conventional Hybrid, to Plug-In Hybrids, to Extended-Range Electric Vehicles," SAE Technical Paper No. 2008-01-0458, Society of Automotive Engineers, Warrendale, Pennsylvania (2008).

41. R. Parrish, K. Elankumaran, M. Gandhi, B. Nance, P. Meehan, D. Milburn, and S. Siddiqui, et al., "Voltec Battery Design and Manufacturing," SAE Technical Paper No. 2011-01-1360, Society of Automotive Engineers, Warrendale, Pennsylvania (2011).

42. R. Matthe, L. Turner, and H. Mettlach, "Voltec Battery System for Electric Vehicle with Extended Range," SAE Technical Paper No. 2011-01-1373, Society of Automotive Engineers, Warrendale, Pennsylvania (2011).

43. 2016 Chevrolet Volt Battery System, https://media.gm.com/content/dam/Media/microsites/product/Volt_2016/doc/VOLT _BATTERY.pdf, accessed February 8, 2019.

44. Drive Unit and Battery at the Heart of the Chevrolet Bolt, http://media.chevrolet.com/media/us/en/chevrolet/news. detail.html/content/Pages/news/us/en/2016/Jan/naias/chevy/0111-bolt-du.html, accessed December 2017.

45. W. Buiel, "Axion Power's Asymmetric Ultracapacitor/Lead-Acid Technology Applied to High-Rate Partial State of Charge HEV Cycling," Large EC Capacitor Technology and Application Symposium, The 9th International Advanced Automotive Battery Conference, Long Beach, California, June 8, 2009.

46. A. Cooper, J. Furakawa, L. Lam, and M. Kellaway, "The UltraBattery—A New Battery Design for a New Beginning in Hybrid Electric Vehicle Energy Storage," *J. Power Sources* **188**:642–649 (2009).

47. A. Taniguchi, N. Fujioka, M. Ikoma, and A. Ohta, "Development of Nickel/Metal-Hydride Batteries for EVs and HEVs," *J. Power Sources* **100**:117–124 (2001).

48. B. G. Potter, T. Q. Duong, and I. Bloom, "Performance and Cycle Life Test Results of a PEVE First-Generation Prismatic Nickel/Metal Hydride Battery Pack," *J. Power Sources* **158**:760–764 (2006).

49. M. Zolot, A. A. Pesaran, and M. Mihalic, "Thermal Evaluation of Toyota Prius Battery Pack," SAE Technical Paper No. 2002-01-1962, Society of Automotive Engineers, Warrendale, Pennsylvania, 2002.

50. M. Ohnishi, K. Ito, S. Yuasa, N. Fujioka, T. Asahina, S. Hamada, and T. Eto, "Development of Prismatic Type Nickel/ Metal-Hydride Battery for HEV," The 3rd International Advanced Automotive Battery Conference, Nice, France, June 21, 2003.

51. K. Snyder, X. G. Yang, and T. J. Miller, "Hybrid Vehicle Battery Technology – The transition from NiMH to Li-Ion," SAE Technical Paper No. 2009-01-1385, Society of Automotive Engineers, Warrendale, Pennsylvania, 2009.

52. W. Wiedemann, O. Vollrath, N. Armstrong, J. Schenk, O. Bitsche, and A. Lamm, "Advanced Energy Storage Systems for Hybrids," The 9th International Advanced Automotive Battery Conference, Long Beach, California, June 2009.

53. 2016 Chevrolet Malibu Hybrid, http://media.chevrolet.com/media/us/en/chevrolet/vehicles/malibu-hybrid/2017.html, accessed February 10, 2019.

54. 2016 Chevrolet Malibu Hybrid Batteries, http://gmauthority.com/blog/2015/06/hitachi-talks-up-its-2016-chevrolet-malibu-hybrid-batteries/, accessed February 10, 2019.

55. J. Gonder, A. Pesaran, J. Lustbader, NREL; and H. Tataria, "Fuel Economy and Performance of Mild Hybrids with Ultra-capacitors: Simulations and Vehicle Test Results," Large EC Capacitor Technology and Application Symposium, The 9th International Advanced Automotive Battery Conference, Long Beach, California, June 2009.

56. J. Gonder, J. Cosgrove, and A. Pesaran, "Performance Evaluation of Lower-Energy Energy Storage Alternatives for Full-Hybrid Vehicles," NREL (National Renewable Energy Laboratory). NREL (National Renewable Energy Laboratory [NREL], Golden, CO [United States]); SAE2014 Hybrid and Electric Vehicle Technologies Symposium, La Jolla, California, February 2014.

# SECTION B

# TRACTION AND MOTIVE POWER VEHICLES

**Ronald T. Moelker**

## 26B.0   OVERVIEW

Industrial motive power vehicle designs and duties have evolved greatly in recent years. Typical industrial class electric vehicles are listed and briefly described below. Subsequently, the available power packages (batteries) are detailed and the rationale for choosing a particular system based on required duty cycles and their equipment utilization efficiencies are analyzed. Table 26B.1 lists the various categories and generic uses for various industrial electric vehicles.

## 26B.1   BATTERY SELECTION FOR MOTIVE POWER VEHICLE APPLICATIONS

Greater demands on motive power electric vehicles has necessitated a constant search for better battery/power packages. Demands for increased equipment up-time and throughput increases in AC-driven lift trucks and 24 h/7 day a week automatic guided vehicles (AGVs) have placed new challenges on the motive power battery. Additionally, optional battery technologies, such as thin-plate pure lead (TPPL), 2-V pure lead, fuel cells, lithium batteries, and fast-charge lead-acid have created an unprecedented level of new choices for the material-handling community. Whether in warehousing, manufacturing, mining, or airline ground operations, users now have several choices beyond the typical "run and swap" battery systems that required large battery rooms, high maintenance and electric expenses, and excessive labor costs. New choices can achieve the following results while having the ability to operate 6 days a week without battery changes:

- Reduce required number of power package systems by up to 50%
- Reduce overall power package costs
- Reduce labor that was associated with the old battery swap process by up to 80%
- Reduce space requirements by up to 80%
- Reduce energy usage by 30%

Using these "new" power packages can provide improved operational efficiencies and cost benefits for the various battery application choices as described below.

### 26B.1.1   Swap-Out: Application I

*Multiple Batteries/Vehicles/Chargers.*   The traditional method of running one battery at a time until discharged and then swapping it out with another demands a storage fleet and support personnel. This process is used in less than 50% of material-handling applications and provides an 80% battery capacity allowable throughput. Table 26B.2 details the preferred options for swap-out operations.

However, this "old way" of powering motive power equipment has become better because of new battery technology. For example, very high throughput warehouses with reach trucks can now reduce battery room

**TABLE 26B.1**  Industrial Electric Vehicle Classification Chart

| Class/type | Description | Usage |
|---|---|---|
| Class 1/Electric motor rider lift vehicles | Counterbalanced stand-up rider | Standard ride-on, stand-up vehicle for dock work and palletize product movement |
| | Three-wheel sit-down rider | Standard sit-down palletized product mover with great turning radius for tight areas |
| | Four-wheel sit-down rider | Standard sit-down palletized product mover with great stability and the ability to handle heavy loads |
| Class 2/Electric motor narrow aisle lift trucks | High-lift straddle | Stand-up truck designed for narrow short-turning radius, but using load straddle arms to support the load in front of the truck instead of rear counterbalance |
| | Order picker | Product order picker with stand-up rider for multilevel vertical rack access |
| | Reach truck | Short-aisle truck with pantograph forks, sharper turning and stand-up rider |
| | Side loader-turret trucks-swing mast trucks | Specialty trucks for storage and handling of carrier loads with unique lengths and shapes |
| Class 3/Electric motor hand control or walk ride pallet movers | Low-lift platform/pallet movers | Walk behind pallet movers |
| | Tow tractors | Electric vehicle towing carts |
| | Low-lift rider and/or center control | Pallet movers where the operator can also ride on the vehicle for traveling longer distances—can handle one or two pallets end to end |
| | Pallet straddle low-lift or high-lift (various designs) | Single-pallet mover, typically walk behind, able to stack palletize product in racks |
| Automatic guided vehicles: driverless | AGC (automatic guided carts) | Very small fully automated tow or burden-carrying vehicles |
| | AGV (automatic guided vehicles) | Heart of the line large load-carrying unmanned vehicles |
| | Automatic-guided fork AGV | Standard counterbalance fork trucks equipped with controls to make them fully driverless |
| | Light load AGV | Small-load delivery in mail rooms and hospitals |
| | Assembly AGV | Product conveyance through blind assembly line used to replace assembly drag lines |
| Airlines | Tugs and tractors | Electric drive units that tow luggage and other items around the facility |
| | Container loaders | Container boxes that are loaded into the belly of the planes |
| | Pushback tugs and tractors | Large units that maneuver unpowered aircraft on the ground |
| | Belt loaders | Drivable powered conveyer units used to load and unload to and from planes and buildings |
| Floor care | Floor scrubbers and sweepers | Power facility cleaning equipment of various sizes and types (from walk behind to riding units) |
| Personnel carriers | Personnel carriers | A large variety of vehicles used for transit of personnel, supplies, and equipment in large facilities for viewing, towing, order picking, etc. High demand by design that they are always needed to be "on call" |
| Mining vehicles (zero-emission vehicles for confined/ underground spaces) | Load/haul dump | Low-profile front loader used to move rock (320 V) |
| | Locomotive | Personnel and materials hauler used in and out of mines (240 V) |
| | Scoop | Rubber tired "forklift" typical to mining industry (240 V and 128 V) |
| | Rock drill | Special-purpose hard rock drilling vehicles for blasting, reinforcing rods, etc. (240–360 V) |
| | Coal hauler | Main rubber tire haulers for material transfer from work site to outbound conveyer (240–360 V) |

**TABLE 26B.2**    Swap-Out Battery Applications

| Class 1 | Class 2 | Class 3 | AGC | AGV | AGV-fork | AGV-light | AGV-assembly | Airline | Floor care | Personnel carrier | Mining |
|---|---|---|---|---|---|---|---|---|---|---|---|
| Excellent/ very good | Very good | Good | Not advised | Good | Good | Good | Not advised | Good | Not advised | Not advised | Excellent |

space and labor dramatically by using high-output batteries (i.e., lead-acid batteries with 18–20% more specific power). Coupled with high-efficiency modular chargers, power consumption cost can be lowered by up to 25%. Also, when a fully developed *battery management system* is used in the battery change-out room, a display screen will automatically specify the most charged/most cooled battery for the operator, eliminating battery room support labor. Systems such as EZ-Select remove the guess work, adding productivity to the lift truck and extending the battery cycle life with considerably lower maintenance costs. High-output batteries with advanced square tubular battery technology allow one battery to increase work up to 35% with each discharge cycle, reducing battery count, change-out time, and storage space by up to 35%. Smaller operations may not need their second batteries at all. Where swap-out batteries are needed, battery providers are now able to offer proven high ampere-hour batteries that fit difficult applications, such as high-rise freezer warehouses, grocery warehouses, "just-in-time" manufacturing, and other very high throughput operations.

### 26B.1.2    No Changing/Opportunity Charging: Application II

***One Battery/One Vehicle/One Charger.***    High fossil fuel cost, high labor cost, and high throughput operations have led to a move away from the old standard (swap-out) battery package operational process. Standard batteries, due to limitations in design and charging equipment, only allowed one 80% cycle a day. The automotive industry and others using AGVs have forced a total rethinking of battery requirements and the "cycle/change-out" operating mode. Battery systems without a need for off-line charging or change-out would be limited only by *throughput allowed by the manufacture*. By recharging batteries with advanced chargers during work breaks or with through-the-floor charging plates, lift trucks, driverless vehicles, and AGVs might save up to $1 million a year in very large automotive plant operations.

By using high-quality batteries partnered with high frequency, precision opportunity chargers, battery change-out can be eliminated. Opportunity chargers, using high charge currents and special charging profiles (up to a 25% charge rate/100 Ah of the battery, 0.25$C$), now allow batteries to be plugged at lunch, shift changes, or at other work breaks, and run for two full shifts without any change-out. Advanced opportunity chargers, placed throughout a plant in work and break areas, allow one to use 120% of the name plate battery rating (BCI International Standard) in 24 h with a 4 to 5 years life. Compared to a swap-out battery system, this allows a throughput advantage of up to 34% from one battery in the same 24 h. While every application requires a power usage study, reputable suppliers will typically guarantee the product to meet expected life and/or equipment lease terms.

These studies measure power demand peaks, idle time, temperature profiles, break times, and total run time to properly size the battery and charger using a proven spreadsheet program. Users see the exact details and can work with power system suppliers to meet both present and future needs. While warranty/expected life may be reduced, since *one battery now does the work of two*, the overall benefits have led to use by almost 50% of all motive power users in North America, including applications in motive power trucks, AGVs, airline support equipment, and mining vehicles. Table 26B.3 details the range of applications for "no change/opportunity charge" systems.

**TABLE 26B.3**    No Changing/Opportunity Charging Battery Applications

| Class 1 | Class 2 | Class 3 | AGC | AGV | AGV-fork | AGV-light | AGV-assembly | Airline | Floor care | Personnel carrier | Mining |
|---|---|---|---|---|---|---|---|---|---|---|---|
| Very good | Very good | Very good | Good | Excellent | Excellent | Good | Very good | Very good | Very good | Good | Very good |

### 26B.1.3  Fast Charging: Application III

*One Battery/One Vehicle/One Charger.*    Fast charging (FC) is actually just opportunity charging but with higher charge rates. The concept simply specifies charge current up to a 50% charge rate/100 Ah of the battery or 0.5*C*. Typically, a 1000 Ah battery was charged at 160 A (0.16*C*) in swap-out applications. In the opportunity charge mode, as listed in Sec. 26B.1.2, a 250 A charge rate is used (0.25*C*). With fast charge, a 500 A charge rate is used (0.5*C*), allowing operating with *one truck with one battery for three full shifts without a swap out*. Once again the battery will be charged whenever/wherever down time might occur.

However, for any lead-acid battery operation, an equalizing charge of 12 h on one day a week *must* be used. The charger *must* use a reliable, fully qualified charge protocol and *must* have *active battery temperature management*. Heat readily degrades batteries and supports the statement, "The worst thing you can do to a battery is…*charge it.*" High charging rates demand the use of a high-quality charger with a highly developed charge program. Such chargers may cost up to three times a normal swap-out charger; and fast-charge batteries will also be around 30% more costly. However, with just one battery per vehicle and the deletion of all the swap-out equipment, time, and space, the overall savings with this concept are clearly proven by the return on investment (ROI) portion of the power study spreadsheet. Premium batteries, specifically designed for fast charge, should be matched up to handle the high charge rates.

A power study is essential so that the proper battery and charger can be specified to meet the throughputs demanded over the life of the product, based on the type of vehicle and operating environment. Also, since batteries lose capacity over time, the analysis must compensate to maintain the required duty cycle for the expected life.

Overall, significant savings are realized when just one battery and one charger, placed anywhere in the operation, can run uninterrupted for one week. This system eliminates swap outs, battery rooms, and a large support labor pool, all with a greatly reduced array of battery equipment. Motive power applications can now run 24/7 for up to 6 days a week. With the savings achieved from this revised mode of operation, some users have customized their charging slot allocations and added additional electric vehicles to allow 7 day a week capability and still save substantially as compared to the old swap processes. Fossil fuel lift trucks are also at a great disadvantage, based on the cost savings of fast-charged electric vehicles. Table 26B.4 details the benefits of FC in various applications.

**TABLE 26B.4**  Fast Charging Battery Applications

| Class 1 | Class 2 | Class 3 | AGC | AGV | AGV-fork | AGV-light | AGV-assembly | Airline | Floor care | Personnel carrier | Mining |
|---|---|---|---|---|---|---|---|---|---|---|---|
| Very good | Excellent | OK | Very good | Very good | Excellent | OK | Excellent | Excellent | OK | OK | Good |

### 26B.1.4  Comparison of Various Operating Modes

Figure 26B.1 provides a graphical depiction of the three operating schemes (i.e., swap out, opportunity charge, and fast charge) discussed in the sections above.

## 26B.2  NEW BATTERY TECHNOLOGIES FOR MOTIVE POWER APPLICATIONS

> *In any moment of decision, the best thing you can do is the right thing. The worst thing you can do is nothing.*
>
> —*Theodore Roosevelt*

This statement finds particular relevance in today's material-handling world. If new decisions on the powering of material-handling equipment are not made, a company's operations will suffer in the long term. However, even with great advances in lead-acid batteries that are now being applied in most of the applications, the

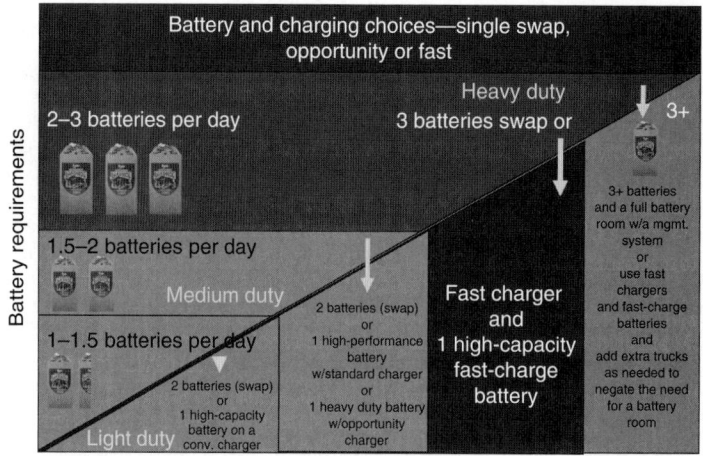

**FIGURE 26B.1**  Comparison of various motive power applications and associated battery types.

increasing impacts on time, labor, and equipment count require material handlers to consider other options. The most relevant new battery technology choices that have evolved for potential use in motive power applications in recent years are as follows:

1. TPPL blocks and 2-V pure lead (see Chap. 14)
2. Fuel cells (see Chap. 19)
3. Lithium batteries (see Chap. 17)

All of these systems will likely find a fit somewhere in most applications. All three systems greatly reduce charging time, eliminate battery "watering," reduce maintenance up to 80%, and allow for only one vehicle, one power source, and one charger, no matter what type of electric vehicle is being used. These technologies are reviewed below for use in motive power applications.

## 26B.2.1  Thin-Plate/2-V Pure Lead (TPPL) Systems

TPPL 12 V blocks or 2-V pure lead-acid technology is a lead-acid-based battery technology that provides enhanced performance. In opportunity/fast-charge concepts, TPPL has potential advantages in five areas:

1. Energy capacity
2. Counterbalance weight (no issue)
3. Cost
4. Maintenance
5. Disposal

TPPL blocks and the most recent 2-V pure lead cells provide the counterbalance weight needed and are fully compatible with all present vehicle systems. The nearly pure lead plates are *95% recyclable* (far better than any competing system and a huge environmental benefit). With increased energy and lower maintenance, cost per unit energy is improved as shown in Table 26B.5 compared to advanced technologies.

TPPL has existed for some time but was used mostly by the military because of its large power advantages, quick charging time, little to no standing loss, maintenance-free needs, and its availability in commonly used

**TABLE 26B.5**  Comparison of Various Advanced Battery Technologies

| Battery type | Battery cost |
|---|---|
| Lead acid (conventional) | $255/kWh |
| TPPL (thin-plate pure lead) | $380/kWh |
| Fuel cells | $190/kWh |
| | $2300/kWh (estimated system cost) |
| Lithium | $700–$1000 kWh |

Cost numbers based on 2017 retain product quotes.

package sizes. Released to the marketplace by EnerSys/Hawker in 2007 for the AGV, personnel carrier, and floor-cleaning equipment markets, it quickly became the power of choice *for quick charging, no changing, no maintenance, and continuous running* vehicle applications. Class 3 power pallet trucks are another key market. The slight higher cost is offset by higher power, zero maintenance, no swapping out, and the ability to be charged at rates from 0.4 to 1$C$-rate (i.e., charged in just over 1 h). TPPL block configurations are an impressive lead-acid technology and extremely viable for all small to midsize vehicles. For large vehicles, the larger 2-V pure lead cells (with charging at 0.4$C$ rates) and TPPL block packages configured in stacks (charging at 0.7$C$ to 1$C$) can be used in most material-handling vehicles today. The larger industrial 2-V pure lead cells are already in full use in Europe (large lift trucks and vehicles), and the transition to U.S. dimensions is pending. Eliminating change-outs and need for watering, shortening charging times, and increasing the power density have resulted in a true "*no–maintenance*" lead-acid product applicable to all electric industrial vehicles today. Table 26B.6 compares the use of TPPL batteries in various motive power applications.

**TABLE 26B.6**  Thin Plate and 2-V Pure Lead Battery (TPPL) Applications

| Class 1 | Class 2 | Class 3 | AGC | AGV | AGV-fork | AGV-light | AGV-assembly | Airline | Floor care | Personnel carrier | Mining |
|---|---|---|---|---|---|---|---|---|---|---|---|
| Very good | Very good | Excellent | Excellent | Excellent | Very good | Excellent | Very good | Very good | Excellent | Excellent | OK |

## 26B.2.2  Fuel Cells

Fuel cells have existed for a long time with much of the early development being government subsidized and typically aimed at the automotive industry. Both the U.S. government and the auto industry have now shifted focus, and new fuel cell engineering and development is being pushed toward the industrial lift truck industry. Current options based on past technologies and provided by a small group of manufacturers today work well. Hydrogen as a fuel is cheap and the power density is good. Hydrogen availability and generation is the bigger issue when looking at vehicle operating costs. Fuel cells, unlike batteries in swap out, are not changed out, but refilled just as fossil fuel systems. Because the hydrogen generator is actually an "engine" of sorts, there is maintenance and wear, requiring replacement over time. Fuel cells are viewed as "green," recharge in 2 to 3 minutes, and do not need to be swapped. The challenge is the cost of the infrastructure and hydrogen transportation/generation. In addition, high throughput vehicles (reach trucks) need a battery or supercapacitor to meet high ampere draw requirements. Even though the supplier list is very small, these companies are working to provide a "full package" concept to cover these challenges. Hydrogen as a converted fuel is less efficient than lead or lithium batteries. Fuel cell stack maintenance and replacement, and the supply sources with infrastructure for hydrogen are issues that must be resolved in using fuel cells. Overall, fuel cells presently are available, but must be implemented with assistance of a quality supplier. Table 26B.7 shows the best applications for fuel cells.

**TABLE 26B.7**    Fuel Cell Applications

| Class 1 | Class 2 | Class 3 | AGC | AGV | AGV-fork | AGV-light | AGV-assembly | Airline | Floor care | Personnel carrier | Mining |
|---|---|---|---|---|---|---|---|---|---|---|---|
| Very good | OK | Very good | Not advised | Not advised | OK | Not advised | OK | Very good | Not advised | Not advised | ? |

### 26B.2.3   Lithium Ion

Lithium battery packages for material-handling applications of all types are growing rapidly. The AGV world has already moved into lithium batteries, which work well in their intensely managed environments using the present small lithium battery packs that are available. The energy density of a lithium battery is exceptional with an ability to cycle for 6000 to 10,000 cycles. The technology is evolving to be more "green," including recycling. With no maintenance, little to no self-discharge, an acceptable temperature range, and very high charge rates ($1C$ is common—1 h or less to recharge), these systems are gaining favor. However, numerous variations exist (at least 6 basic chemical systems), and the early designs are typically only available in small 18650 cells assemblies from Asia (LMO [lithium metal oxide]) or in small prismatic layouts (LFP [lithium iron phosphate]) requiring hundreds of battery connections in a power package.

Overseeing this complex system of cells is a battery management system (BMS) that requires expertise not typical in the material-handling field where vehicles are used by multiple drivers with multiple skill levels. The BMS must be reliable and rugged with little need for user interface where at all possible. Newer large-format lithium battery technologies are coming quickly into motive power and automotive designs. Using nickel manganese cobalt (NMC), these batteries will be simpler to construct, safer, have much more power, and be much easier to configure for large vehicle designs. However, motive power users will quickly adopt lithium batteries because of the greater expected life, the ability to meet all power demands with no change out, and very fast recharge capability.

For large high throughput vehicles and "just-in-time" manufacturing vehicles, lithium cells are a potential option. Smaller vehicles in continuous use, where change out and availability are critical issues, will also be a good consideration for lithium. Maintenance issues will be a new challenge due to the complexity of lithium systems. The present market size will be limited by high cost, and measured by high vehicle throughput and production ability. In the 24/7 operations, ground support, and the AGV business, lithium batteries will be ideal. This electric power option dictates the use of a competent supplier with the best BMS available as key deciding factors beyond cost. Table 26B.8 lists the benefits of lithium cells in various applications.

**TABLE 26B. 8**    Lithium Battery Applications

| Class 1 | Class 2 | Class 3 | AGC | AGV | AGV-fork | AGV-light | AGV-assembly | Airline | Floor care | Personnel carrier | Mining |
|---|---|---|---|---|---|---|---|---|---|---|---|
| Excellent | Excellent | Good | Very good | Excellent | Excellent | Very good | Very good | Very good | Good | Good | ? |

## 26B.3   SUMMARY

Even with these great new technologies, there will still be many users selecting advanced lead-acid batteries for many applications due to cost considerations in the "near" future. However, the market has now finally evolved to where a range of practical choices are available to match power needs at all levels of any particular business looking to maximize operating profits.

# SECTION C

# ELECTRIC AIRCRAFT

**George E. Bye**

## *26C.0   OVERVIEW*

Boeing, the world's largest aerospace company and leading manufacturer of commercial jetliners and defense, space, and security systems, made a staggering projection in 2017 that over 637,000 new commercial pilots will be needed over the next 20 years.[1] Previously, in July 2016, Airbus had predicted the number of new airline pilots needed over the next 20 years will top 560,000.[2] According to the Federal Aviation Administration (FAA), at the end of 2016 the United States only had about 158,000 Airline Transport pilots.[3]

How did this occur? There are several reasons why the number of available U.S. commercial pilots has decreased to alarming levels, but they include the impact of a shrinking pool of military pilots, including a U.S. Air Force that is producing fewer pilots, and the significant cost of receiving a commercial pilot's license from civilian pilot schools.

In terms of training aircraft, the current fleet of flight trainers in the United States now averages five decades old. In addition, the required flight hours to obtain the proper training and experience to qualify for airline pilot capability increased substantially to 1500 flight hours—creating a large expense for new pilots hoping to become airline pilots. The challenge to train new pilots will help drive flight schools to use high tech, lower operating cost trainers. Given the convergence of these flight training anomalies, the time has come to find a sensible, workable solution.

In 2014, Bye Aerospace launched a program to develop a practical, certified electric-powered airplane to serve the aviation flight training market. The FAA-certified family of aircraft trainers being developed is called "Sun Flyer." In April 2018, the company completed its FAA certification application for the Sun Flyer—the first electric aircraft to do so. Once certified, the two-seat "Sun Flyer 2" aircraft (see Fig. 26C.1) is predicted to have 3½ h of flight endurance and a regenerative motor-generator propeller. The Sun Flyer will have a very low $3 per flight hour of electric energy operating costs (compared to over $50 per flight hour aviation fuel cost for a typical trainer), and virtually no noise or any aviation fuel-related lead and $CO_2$ emission pollutants.

**FIGURE 26C.1**    Sun Flyer 2 prototype in flight testing.

## *26C.1   TECHNICAL APPROACH*

Why is an electric-powered aircraft preferable to address flight training challenges? Over the last 10 years, battery-powered technology has advanced significantly. Driven by cell phones and other personal electronic devices, as well as the growing electric-powered automobile industry, major strides are being made with electric motors, batteries, and the software that manages electric propulsion systems. Additionally, increases in battery energy densities (energy stored per unit weight) now enable flight endurance to be measured in hours, rather than minutes.

The first key elements, efficient light-weight electric motors and controllers, are then combined with lithium-ion batteries. The battery cells are configured in packs, which can be recharged efficiently. The second element is a sleek, light, carbon fiber structure, low-drag fuselage with efficient long-wing (high "aspect ratio") advanced aerodynamics (see Fig. 26C.2).

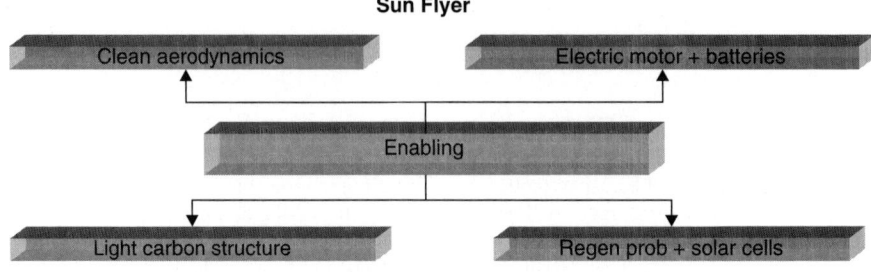

**FIGURE 26C.2**   Key elements to Sun Flyer success.

The Sun Flyer aircraft will be certified under Federal Aviation Regulations (FAR) Part 23 and will be standard category, day-night visual flight rules (VFR) with a target gross weight of less than 2000 lb. It is intended to be a sharp-looking, high-performance aircraft that will not compromise performance, given a projected speed of 135 knots and climb rate of over 1000 ft/min. Of more importance is the operating cost benefit of electric propulsion ($50 of aviation gasoline per flight hour vs. $3/h of electricity for the quiet, zero-pollution Sun Flyer).

## 26C.2   APPLICATION REQUIREMENTS

The electrical energy requirements for the Sun Flyer 2 depend on the flight profile. Missions may vary dependent on flight time, maximum attained altitudes, and the distance traveled. Energy requirements for two different flight profiles of approximately 3 h duration and a 1.2 h training flight as of 2018 are summarized in Tables 26C.1 to 26C.3.

**TABLE 26C.1**   Sun Flyer 2: 2500 ft MSL, 3 h, 240 nm Cross-Country

| Flight mission component | Time, min | Energy, kWh |
|---|---|---|
| "Start" and taxi | 5 | Negligible |
| Takeoff | 0.4 | 1 |
| Climb to 2500 ft AGL (assume 2500 ft MSL) | 3.7 | 5 |
| Cruise 80 KIAS (83 KTAS) | 165.3 | 68 |
| Descent (prop energy being regen to battery not included for conservatism) | 5 | 0 |
| Landing | 0.3 | Negligible |
| **Total** | **179.7 min** | **73 kWh**[*] |

[*]*Basis:* 86 kWh "full battery"; calculation does not include regen propeller (13–15% benefit); 73 kWh usable provides 30 min "VFR reserves," (officially TBD with FAA prior to certification).

**TABLE 26C.2**   Sun Flyer 2: 10,000 ft MSL, 2.8 h, 256 nm Cross-Country

| Flight mission component | Time, min | Energy, kWh |
|---|---|---|
| "Start" and taxi | 5 | Negligible |
| Takeoff | 0.4 | 1 |
| Climb to 1000 ft AGL (assume 10,000 ft MSL) | 14.0 | 18 |
| Cruise 80 KIAS (83 KTAS) | 131.1 | 54 |
| Descent (prop energy being regen to battery not included for conservatism) | 20 | 0 |
| Landing | 0.3 | Negligible |
| **Total** | **167.1 min** | **73 kWh** |

*Note:* The information presented in the performance tables may be superseded as the aircraft characteristics continue to mature.

**TABLE 26C.3**   Sun Flyer 2: 1.2 h Training Mission

| Flight mission component | Time, min | Energy, kWh |
|---|---|---|
| "Start" and taxi | 5 | Negligible |
| Takeoff | 0.4 | 1 |
| Climb to 2500 ft AGL (assume 2500 ft MSL) | 3.7 | 5 |
| Cruise to practice area | 12 | 7 |
| Practice area | 45 | 16 |
| Descent (prop energy being regen to battery not included for conservatism) | 5 | 0 |
| Landing | 0.3 | Negligible |
| **Total** | **71.3 min** | **29 kWh** |

*Note:* Distance from airport to practice area is assumed to be 10 nm each way, 20 nm total.

## 26C.3   APPLICATION ENGINEERING PERSPECTIVES

When Bye Aerospace was founded in 2007, it was built on understanding the marketplace breakthroughs that could be attained when projecting the trajectories of key battery technologies that would soon advance to the point where electric motors could realistically replace combustion engines in certain aircraft configurations.

Like any program, developing Sun Flyer presents unique challenges. As it turns out, the largest obstacles are the system components previously mentioned—particularly, electric motors and batteries. The challenge is to make each component as "productive," or energy efficient as possible. Batteries need to have maximum energy density and purpose-built electric motors, for electric aircraft must have proper power ranges for a particular airplane design.

The miniaturization of commercial electronics has contributed to progress with the power cells and batteries. For example, the typical D battery cells used in flashlights have been replaced with more efficient lighting sources and hence smaller battery "size, weight, and power" (SWAP). Remarkable achievements over the last few decades have increased battery energy densities, and the battery weight to meet specific application requirements has decreased substantially. As a result, the life of the product with the battery has longer endurance. As energy densities improved, a general aviation research and development project became feasible a few years ago. Since 2012, energy density has doubled, providing the basis for this first practical two-seat general aviation flight training aircraft.

Aviation use of higher energy density batteries does come with appropriate precautions. The ability to monitor the safety of each cell during operations, including under- or overcharges and circuit protection temperature extremes, is a must. As with electric vehicles, a battery management system monitors all these elements and feeds the corresponding data to the overall information management system to the pilot in the cockpit (see Fig. 26C.3). In its research, Bye Aerospace has worked with Panasonic and Dow-Kokam, but it is now using the LG Chem MJ1 lithium-ion 18650 cell in its Sun Flyer 2 prototype battery.

**FIGURE 26C.3**    Ground tests for Sun Flyer 2 prototype.

In working with the individual battery cell characteristics, it is clear that the modern battery design has made great strides in improving every aspect of battery performance and safety. However, a practical "pack" must be designed that provides energy, properly configured for the 3-½ h flight endurance of the two-seat "Sun Flyer 2." Multiple cells are needed, arranged in series and in parallel, to align with the electric motor voltage and current requirements. Multicell safety must be fully addressed with sufficient layers of redundancy to achieve the desired outcome for an FAA-certified aircraft—a significant effort!

On the whole, electric motors and batteries need to be engineered, designed, and configured for their application, and translated with all appropriate safety aspects included, to FAA regulatory standards so they can be put into practical use in general aviation.

The task of bringing the electric Sun Flyer family of aircraft to market is monumental, but certainly worth the effort. The two most critical factors for success in this effort may well be optimism and perseverance, but nothing of any value was ever accomplished without pursuing something that has never been done before.

## REFERENCES

1. 2017 Boeing Pilot & Technician Outlook.

2. Airbus "Global services forecast," July 2016.

3. U.S. Civil Airmen Statistics; https://www.faa.gov/data_research/aviation_data_statistics/civil_airmen_statistics/.

# CHAPTER 27
# BATTERIES FOR STATIONARY ENERGY STORAGE APPLICATIONS

**Babu R. Chalamala, Summer R. Ferreira, Raymond H. Byrne,
Daniel Borneo, Imre Gyuk**

## 27.0 INTRODUCTION TO THE ELECTRICITY GRID

Electrification ranks as, perhaps, one of the most important engineering achievements in history.[1] Electricity has been central to most of the recent progress in human quality of life, and it underpins the core infrastructure for our modern world. Over the last 100 years, the grid has evolved into a centralized electricity delivery system based on a model that has remained unchanged for decades. This traditional electricity system (shown in Fig. 27.1[2]) consists of large central generators (typical large power plants using coal, natural gas, or nuclear fuels), an electricity transmission system, and loads. In this model, generation and load are always balanced with very limited flexibility.

Successful operation of the electric grid rests on two key principles: reasonably predictable loads and a measure of control over generation. This approach has worked well for many decades. Electric utilities deliver reliable power at prices set by government agencies or regulators in most markets, and variable market-driven pricing in a few unregulated markets.[3] The electric grid, especially in the developed world, is reliable and efficient with electricity delivery mechanisms that are robust, and utility market structures that are transparent and cost effective.

The current grid infrastructure is stable and resilient: the number of agents that could affect the state of the system is limited; the variability of energy input to the system is small; and many of the largest loads intrinsically reduce consumption in response to a drop in system frequency. For any given area, a handful of generators and control devices can reliably maintain frequency and voltage. Grid architecture and control systems are designed with sufficient margin that disruptive events such as a loss of a generator or loss of a major load center can be swiftly isolated and remedied, thereby avoiding cascading failures.

However, the electric grid is rapidly evolving from the traditional model of largely centralized, controllable generators, and distributed independently controlled loads to a more flexible grid with a much larger number of distributed, variable generators and participatory, coordinated loads. Key trends driving the electric grid include a changing mix of types and characteristics of electric generation, growing supply- and demand-side opportunities for customers to participate in electricity systems, and the rapidly increasing role of DC loads, such as computers and electric vehicles. While there has been greater investment on the generation side, especially in large amounts of natural gas and solar and wind power, updates to the transmission and distribution (T&D) infrastructure have not kept pace. Also, with the increasing role of network communications and electronic loads, there is a greater need to make energy systems robust against malicious attacks.

The electricity grid is a complex system with a total worldwide electricity generation capability exceeding 6 TW. With significant new capacity additions in the developing world, power generation output is expected to grow to over 7.5 TW by 2020.[4] Figure 27.2[5] shows the worldwide electric energy generation by source.

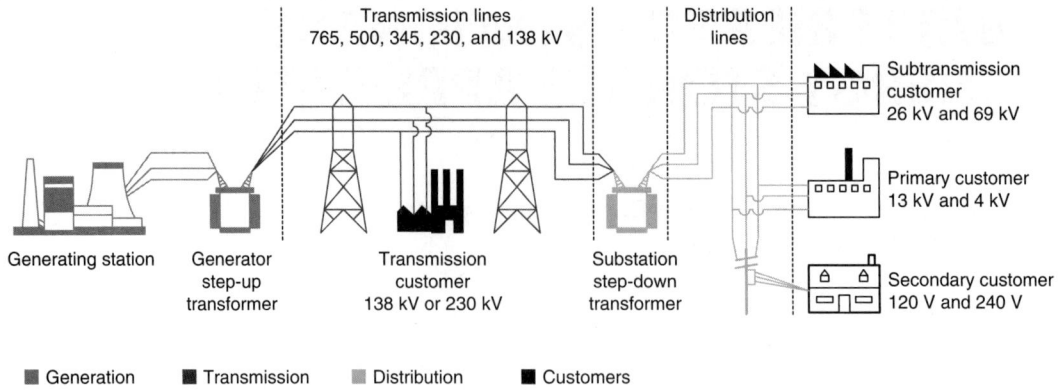

**FIGURE 27.1**  Architecture of the traditional electricity system.[2]

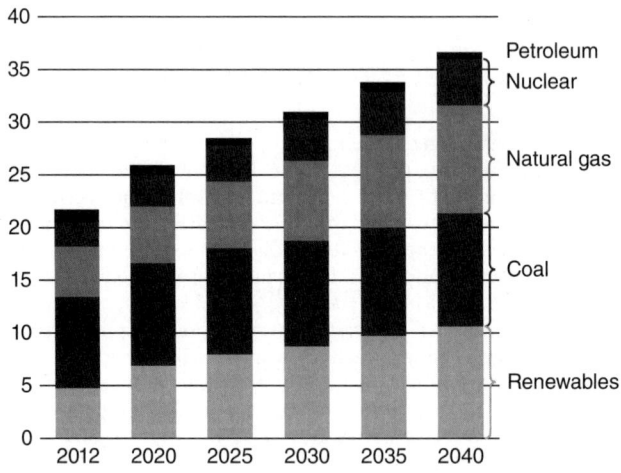

**FIGURE 27.2**  Current and projection for worldwide net electricity generation by energy source in trillion kWh (EIA, 2016).[5]

The electric grid is rapidly evolving toward a hybrid system (shown in Fig. 27.3), where the distinction between generation, transmission and distribution, and consumption is becoming less clear. Renewable energy sources are growing at a rapid rate and, according to the 2016 EIA Energy Outlook, will grow at an average of about 3.8% per year between now and 2035.[6] By 2020, total installed renewable sources will reach ~1 TW and market penetration is expected to be as high as 30% to 40% in some areas. This change in paradigm results from a combination of trends: the retirement of older coal plants; low cost natural gas; new renewables to meet mandates and policies; and growing investments in distributed generation, energy efficiency, and demand response. As the generation mix becomes much more diverse, the delivery system is challenged to meet the needs for two-way power flow and massively increased flexibility. With increasing amounts of distributed generation and stochastic loads, the need for flexible balancing resources, such as energy storage, becomes increasingly important.[7] Larger scale integration of energy storage within today's grid infrastructure can improve grid performance while reducing electricity cost. Grid operators are more often considering energy storage as a flexible resource that can be effective in managing the grid. In this chapter, only energy storage connected to the electric utility grid is described, also sometimes called "in front of the meter."

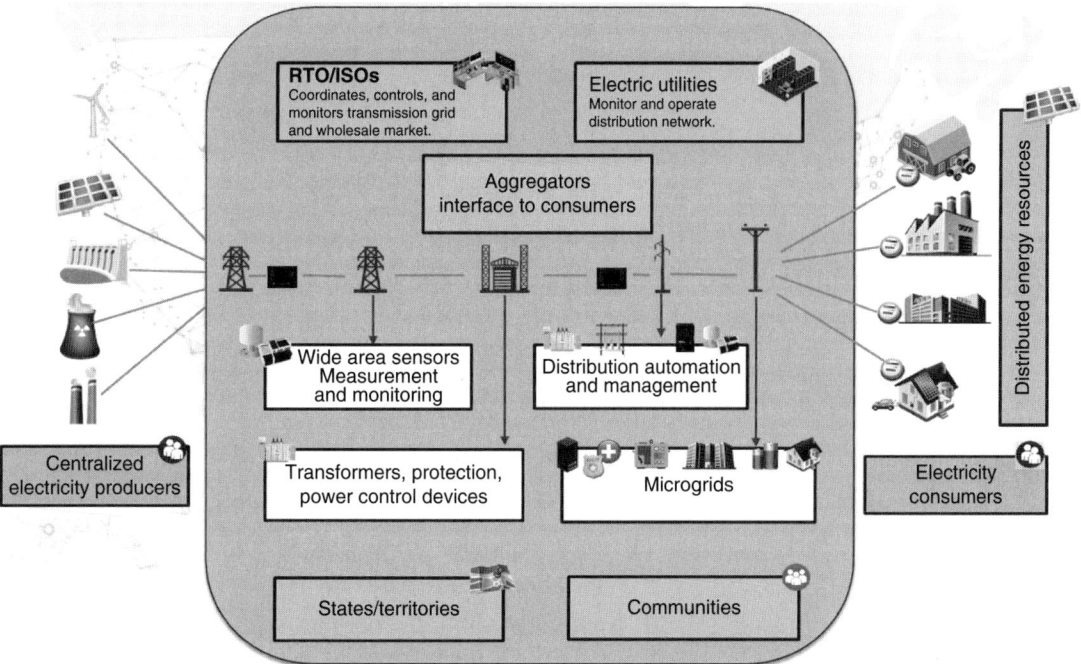

**FIGURE 27.3**   Hybrid grid architecture. (*Source: U.S. Department of Energy. Office of Electricity Delivery and Energy Efficiency, 2017.*)

## 27.1   *ENERGY STORAGE IN THE ELECTRICITY GRID*

Among all types of energy storage, battery energy storage in grid infrastructure has been growing most rapidly during the last few years. According to the DOE Global Energy Storage Database,[8] over 3 GWh of grid-tied battery storage systems were deployed from 2012 through the end of 2017. With continued reduction in the cost of battery storage systems and a greater need for energy storage in the electric grid, forecasts indicate continued acceleration in the pace of global battery energy storage deployments in the coming decade.[9]

Large-scale integration of energy storage in the electricity grid has a number of benefits, including distribution system resiliency, reducing bottlenecks in the transmission system, and improving the efficiency of the generation fleet. It is estimated that energy storage can help to defer some of the annual $100 billion in upgrades to T&D infrastructure in the United States alone.[10] Energy storage can also reduce commercial and industrial economic losses from power outages while providing backup power and islanding capabilities during natural disasters.[11]

Grid-scale energy storage can enable significant cost savings to industry while improving infrastructure reliability and efficiency.[12] Before energy storage becomes ubiquitous throughout the electric system, there needs to be improvements in the performance, cost, safety, and reliability of grid energy storage systems (ESSs). Grid ESSs are complex, because they integrate batteries with power electronics and power conversion systems, as well as advanced control and energy management systems (EMSs) that enable the grid operators to manage these systems like any other grid asset. In addition to having batteries that are safe and reliable, the systems must be robust and cost effective for the industry to become commercially viable across all markets. Simultaneously, the value and benefits of grid-tied storage must be defined and articulated.

ESSs, if suitably deployed, give grid operators more flexibility to effectively manage variability in generation and demand. System power ratings and energy capacity needs of various applications are compared to a range of energy storage technologies in Fig. 27.4. While batteries can satisfy the power and energy needs for a large number of applications, only large-scale pumped hydro storage and compressed air energy storage (CAES) can provide long-duration storage.

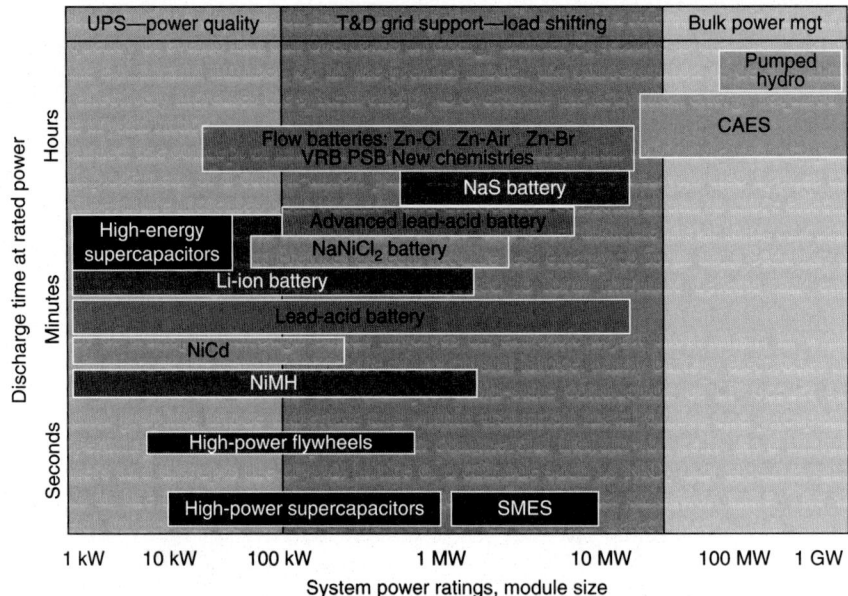

**FIGURE 27.4** Range of applications and storage system needs. (*Source: DOE/EPRI Electricity Storage Handbook in Collaboration with NRECA, 2013.*)[2]

Grid applications of energy storage are typically categorized as "energy" applications and "power applications." A summary of important grid applications of energy storage is listed in Table 27.1.

**TABLE 27.1** Summary of Storage Applications

| Energy applications | Power applications |
|---|---|
| Arbitrage | Frequency regulation |
| Renewable energy time shift | Voltage support |
| Demand charge reduction | Small signal stability |
| Time-of-use charge reduction | Frequency droop |
| T&D upgrade deferral | Synthetic inertia |
| | Renewable capacity firming |

The need for energy storage in the current and future electric grid is significant. For example, peak shaving in the territory managed by the New York Independent System Operator (NYISO) alone would require around 500 GWh of energy storage to shave 15% of the total demand (Fig. 27.5).[13] Ramp support and peak shaving in the territory managed by the California Independent System Operator (CAISO) would require at least 50 GWh of energy storage as the renewable penetration has been rapidly growing, having reached 42% in May 2017 (Fig. 27.6[14]) such that handling the ramp rates in the gigawatt scale is becoming a challenge for CASIO. T&D upgrades are expensive and energy storage can be effectively used in deferring T&D upgrades. Energy storage presents a significant opportunity for making the distribution system resilient by installing small amounts of energy storage at distribution substations. For example, in the United States alone, there are over 66,000 substations that currently have no installed backup capacity.

Currently, most of the installed capacity of energy storage on the grid is mainly provided by large pumped hydro energy storage plants.[8] The total installed energy storage capacity, including pumped hydro, is relatively

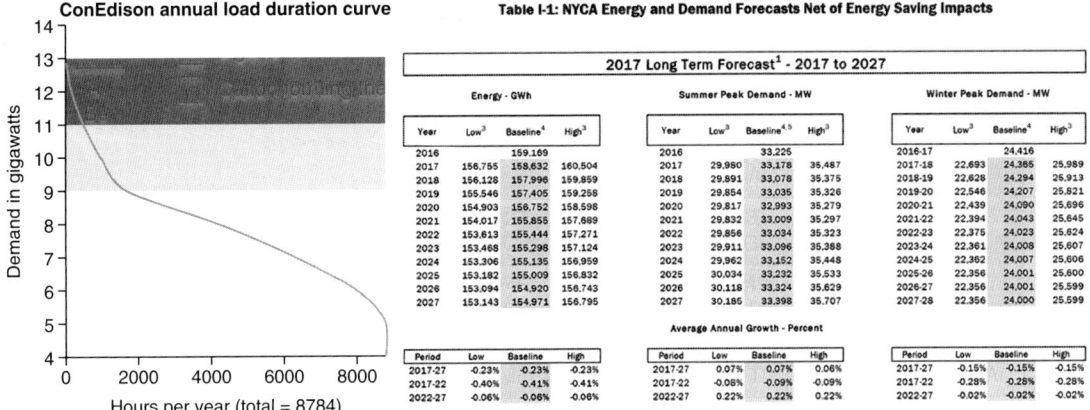

**FIGURE 27.5** Potential for peak shaving in the territory under New York Independent System Operator (NYISO). (Refer to Ref. 13 for details specified in the notes citations.)

small compared to the total generation capacity. The amount of energy storage installed (see Table 27.2) represents about 2.5% of the total generation capacity. For example, the worldwide generation capacity is estimated to be about 6 TW. In the case of the United States, the summer peak demand reached 1250 GW in 2017, and all the installed energy storage represents roughly 2.9% of the peak demand.[15] Clearly there is a big opportunity for energy storage to grow in the future to fill this need.

Key challenges for the large-scale deployment of energy storage were identified in the 2013 DOE Grid Energy Storage Strategy.[16] The report identifies key issues for the future of energy storage in the United States. These include the need for energy storage technologies to be economical and competitive with other technologies providing similar services; the development of market mechanisms that accurately recognize energy storage for its value in providing multiple benefits simultaneously; and ultimately, the need for storage technology to seamlessly integrate with existing systems and subsystems. During the last few years, major cost reductions in lithium-ion (Li-ion) battery technology have enabled initial market penetration and revenue for a small fraction of grid services such as frequency regulation, but further technology advances are needed to economically handle all utility class applications. Additionally, the values of the services that energy storage provides are not fully understood and/or monetized within the existing policy and regulatory framework.[17]

Recent growth in the deployment of energy storage is primarily in Li-ion battery ESSs. Other technologies, such as sodium-sulfur (NaS) and redox flow batteries (RFBs), are also beginning to be widely deployed. According to the U.S. DOE Global Energy Storage Database, in 2017, 1300 MW of Li-ion batteries were deployed in the United States, while other deployments included 89 MW of lead-acid (Pb-acid) technologies, 207 MW of NaS batteries, and 75 MW of RFBs, as shown in Fig. 27.7.

## 27.2   BATTERY ENERGY STORAGE TECHNOLOGY TYPES FOR GRID APPLICATIONS

Technology maturities for utility scale energy storage technologies are shown in Fig. 27.8. Many of the deployed ESSs in the grid are based on mature battery technologies such as lead acid, NaS, Li-ion, and RFBs.[18,19] Some of the early large-scale projects also used nickel-cadmium (NiCd) batteries such as the 46 MW system in Fairbanks, Alaska, which went online in 2003 and is still in operation today. The Golden Valley Electric Association Battery Energy Storage System (GVEA) was at the time the largest battery storage system in the world and had the world's highest battery voltage (operating up to 5200 V).[20] Major cost reductions in Li-ion technology have enabled initial market penetration and revenue for a small fraction of grid services such as frequency regulation, but further cost reduction is needed to economically handle all utility class applications. Among the most

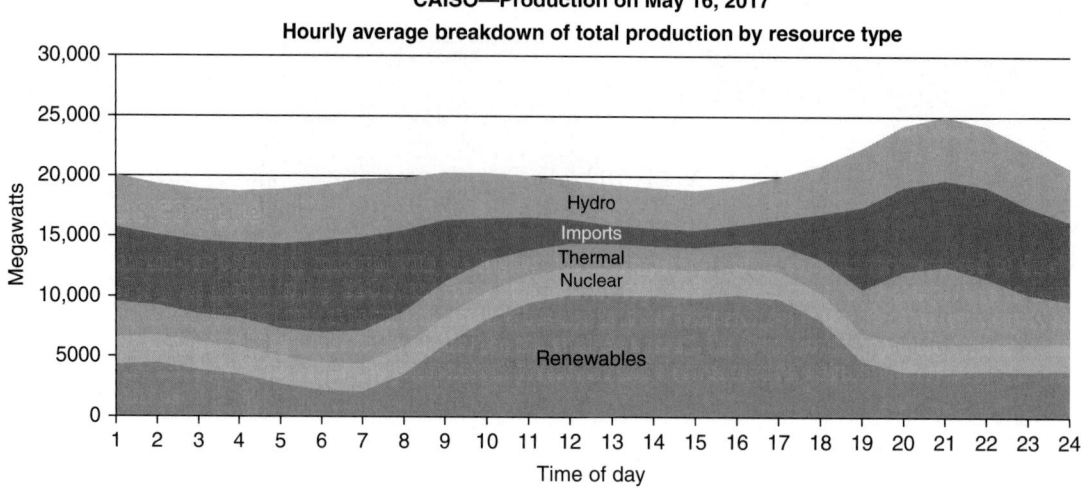

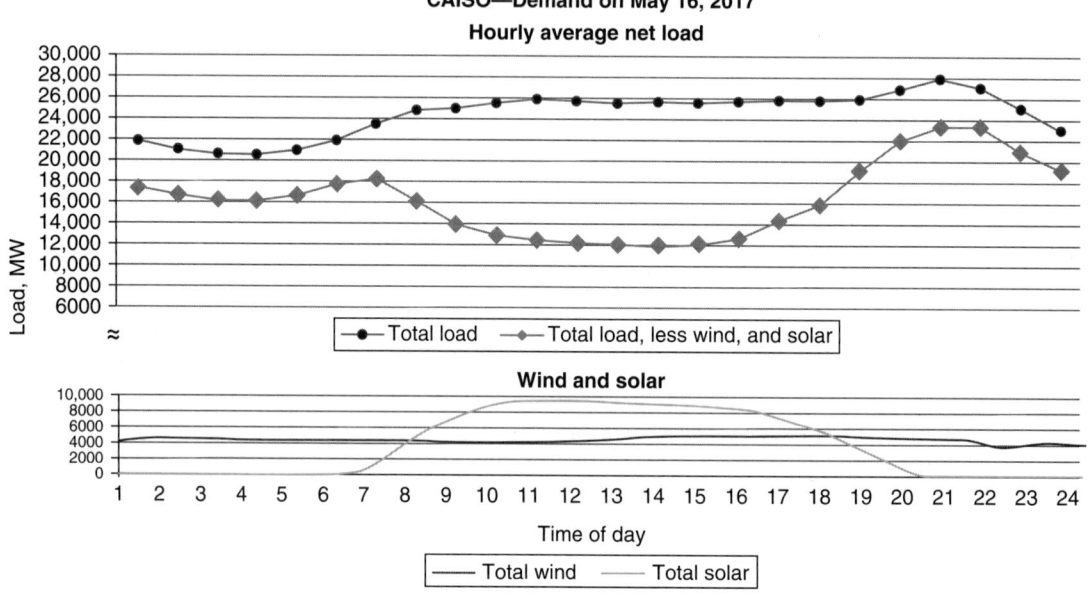

**FIGURE 27.6**    CAISO production and demand in May 2017. (*Source: California Independent System Operator, Folsom, CA.*)[14]

**TABLE 27.2**    Worldwide Utility Scale Energy Storage Installed Capacity

| Technology | Projects | Rated power (MW) |
|---|---|---|
| Electrochemical storage | 991 | 3259 |
| Pumped hydro storage | 352 | 183,800 |
| Thermal storage | 206 | 3622 |
| Electromechanical storage | 70 | 2616 |

*Source:* US DOE Global Energy Storage Database, 2017.[8]

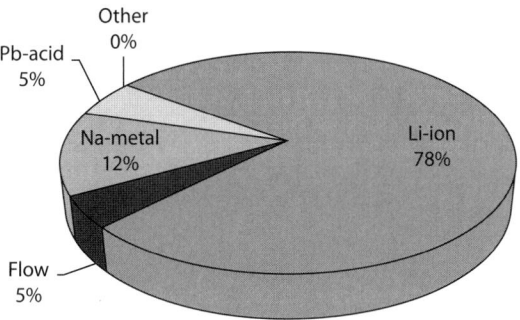

**FIGURE 27.7**    Battery energy storage deployment by technology.
(*Source: US DOE Global Energy Storage Database, 2017.*)

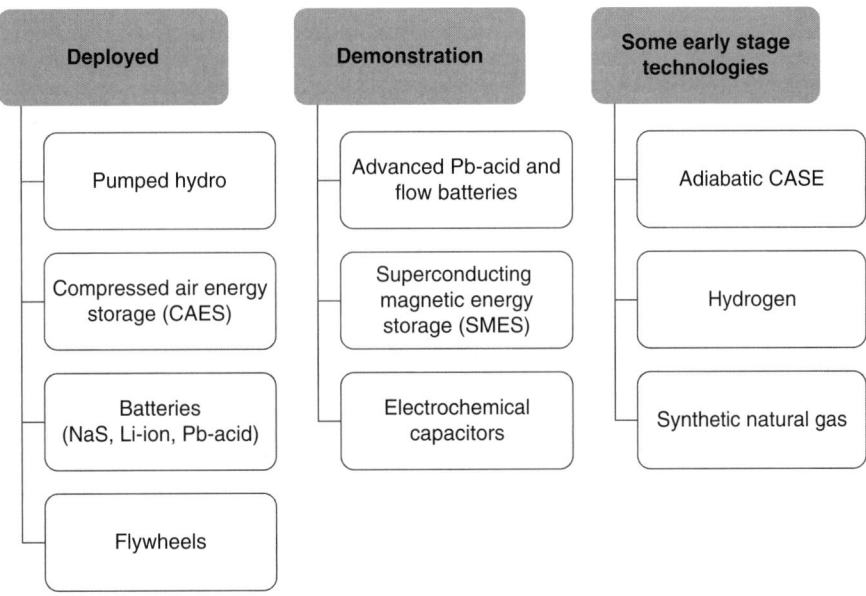

**FIGURE 27.8**    Technology maturities. (*Source: U.S. Department of Energy, "Grid Energy Storage," December 2013.*)

promising other technologies for the grid-scale ESS are RFBs, which can store a large quantity of electricity (multi-MWs/MWhs) in a relatively simple and straightforward design. Of the several RFB technologies (Chap. 22B), all-vanadium redox flow batteries (VRFBs) have received significant attention because of their excellent electrochemical reversibility, high round-trip efficiency, and negligible cross-contamination between positive and negative electrolytes.[21]

Which ESS technology is most suitable for a given application depends on the requirements in terms of power output, energy storage capacity, and expected cycle and system life. Comparisons between options can be made on the basis of these characteristics and limitations.[22] Cycle life plays heavily and is yet not well characterized, which affect large-scale adoption of battery ESSs in the electricity grid infrastructure.[23] Current trends in stationary storage show that as of 2015 close to 50% of battery energy storage was for arbitrage and most of the balance was for power quality applications.[24] By capacity, 48% of battery-based energy storage is Li-based, 15% lead acid, 4% flow batteries, and 2% NaS. By number of projects, 65% is Li-based, 14% lead acid, 8% flow batteries, and 7% NaS-based. Therefore, we focus here on these four technology categories as these are currently implemented in stationary storage, namely Li-ion, lead-acid, RFB, and NaS batteries. (Also, see Chaps. 17A, 14, 22B, and 1 for further details on these respective technologies).

### 27.2.1   Lithium Ion

Li-ion batteries are the most prevalent category of battery technology currently selected for stationary storage projects. Li-ion batteries provide long cycle-life, high efficiency, and high energy and power density. While much conventional wisdom has asserted that the high energy density of Li-ion required for mobile applications (such as electric vehicles) is not needed for stationary applications (since the cost of moving the storage technology is immaterial), in practice Li-ion is still the most commonly selected technology for stationary storage.[25] Reviews of ESSs often benchmark against Li-ion to show the favorability of other specific technologies for ESS use cases.[26,27] Nonetheless, systems that are implemented continue to utilize Li-ion batteries. This is most likely due to their maturity as a result of the commercial use in consumer electronics and electric vehicles. Reproducibility, known cycle life, and robust design are significant considerations that come with high-volume manufacturing.[28] Much of the current research on performance of electrochemical energy storage is on Li-ion batteries.[29,30] Furthermore, safety improvements have been made[31,32] and ongoing work aims to improve safety of various chemistries even further.[33]

With cost reductions of utility scale Li-battery storage, many large Li-ion-based battery systems are being deployed throughout the grid. A 30 MW/120 MWh system was deployed in 2017 by San Diego Gas and Electric Co in Escondido, California. This project was deployed as part of a multipart response to reduce dependence on natural gas fired peaker plants by the California Public Utilities Commission in May 2016.[34] Li-ion battery systems are also being deployed to improve the resiliency of the electric grid in small municipal and rural utility systems. Figure 27.9 shows a 2 MW/3.9 MWh ESS at Sterling Municipal Light Department in Sterling, MA, which provides 2 MW for 2 h to give resiliency for first responders and to maintain emergency services.[35] Li-ion battery ESSs are poised to be the dominant technology especially for power applications in the electricity markets.

**FIGURE 27.9**    2 MW/3.9 MWh ES System at Sterling Municipal Light Department, Sterling, MA. (*Source: Sterling Municipal Light Department, Sterling, MA, 2017.*)

### 27.2.2   Advanced Lead Acid Batteries

Traditional lead-acid batteries were developed in the 1800s. Their use in automotive applications as starter batteries and in uninterrupted power supply (UPS) applications is ubiquitous. As a UPS, the lead acid battery has long been used in stationary storage as a backup resource. A defining trait of lead acid in such applications is that the cell is maintained at or near top of charge, i.e., 100% state-of-charge (SoC), most of the time. This is important to keeping the performance of the batteries in good health. When a lead-acid battery is left in an intermediate charge state and not returned to a fully charged state, rapid degradation of the cell results from the formation and growth of lead sulfate ($PbSO_4$) crystals. When these become too large they are described as hard sulfation, an irreversible deposit that damages the electrodes and reduces the capacity and the cycle life of the cell. Utility applications may require short duration, high-rate, and partial SoC cycling (HRPSoC). Under HRP-SoC duty, conventional valve regulated lead acid (VRLA) cells fail prematurely from irreversible $PbSO_4$ formation within the negative plates.[36] Regular cycling to 100% SoC mitigates $PbSO_4$ crystal formation and growth. However, regularly cycling to 100% SoC is not viable for many utility energy storage applications. Large $PbSO_4$ crystals are not easily reduced back to metallic lead during HRPSoC charging, reducing cycle life. Reduced cycle life of VRLA batteries increases the operating cost, thereby limiting their practicality for utility applications. For most such uses, the battery remains in an intermediate SoC regularly. While VRLA batteries have long been brushed off as old technology that does not meet utility storage needs, they remain cost competitive and continue to be placed in stationary storage applications. This technology is still selected for demonstration projects and implemented in system-level projects as lead-acid batteries meet the low-cost needs of stationary storage.[37,38] They also have a lower hazard of failure leading to thermal runaway and fire than other stationary storage battery options. While VRLA cells use large quantities of lead, lead is highly recyclable, and when done optimally, contributes to a lower environmental impact.[39,40]

Advanced lead-acid batteries typically have a carbon additive in the negative electrode to make them more stable against hard sulfation, which increases their applicability in stationary applications.[41] In the grid today, over 50% of lead-acid battery projects use advanced lead-acid technology, and 81% of projects begun between 2010 and 2015 use advanced lead acid, showing there is continued market for lead-acid batteries.[41]

A large number of utility class lead battery installations are providing grid services including ramp rate support, peak shaving for solar power plants, and reduction of diesel usage in grid-tied and off-grid applications. For example, Fig. 27.10 shows a 1.3 MW/1.9 MWh advanced lead-acid battery system used for renewable support in Alt Daber, Germany. There are many such installations that can be found throughout the world, especially in developing countries and island markets.

**FIGURE 27.10**   1.3 MW/1.9 MWh advanced lead-acid battery system providing support for a 68 MW solar farm in Alt Daber, Germany. (*Source: BAE Batteries, 2018.*)

Partial-SoC applications can be effectively served for extended periods by advanced gel type batteries utilizing carbon as a major additive to the plate structure and by a more traditional lead-acid utilizing some carbon, tubular plates and gelled electrolyte.[42] Especially promising are advanced lead acid batteries using a carbon-containing anode, coupled with a capacitor in parallel with the advanced lead-acid battery cell design. The capacitor can absorb high power charges and discharges to help protect the electrodes of the battery cell and serve to operate longer under partial state-of-charge operations.[43] Figure 27.11 shows a 3 MW/3 MWh system performing frequency regulation services in the PJM market, which regulates the mid-Atlantic area of the United States.

**FIGURE 27.11**   A 3 MW/3 MWh advanced lead-acid battery system for utility applications. (*Source: EastPenn Manufacturing, East Lyons, PA.*)

### 27.2.3   Redox Flow Batteries

In RFBs, electrolyte solutions are stored externally to the electrochemical cell and pumped through porous electrodes on either side of the cell stack as shown in Fig. 27.12. Each half-cell is separated by an ion exchange membrane or porous separator to prevent mixing and shorting of the cell. The unique architecture of flow batteries allows them to independently scale their power and energy capacity. Power is determined by the size and geometry of the cell stack, and capacity is determined by the amount of material stored in the tanks. This feature makes RFB more flexible in design for a given application. Since only a small percent of the electrolyte is running through the stack at any moment, it reduces the vulnerability to faults and uncontrolled energy release. In general, RFBs have lower roundtrip efficiency in comparison with Li-ion or lead-acid batteries. However, RFBs often have longer cycle life (>10,000 cycles), and their cycle life is not dependent on the depth of discharge (DOD).[44]

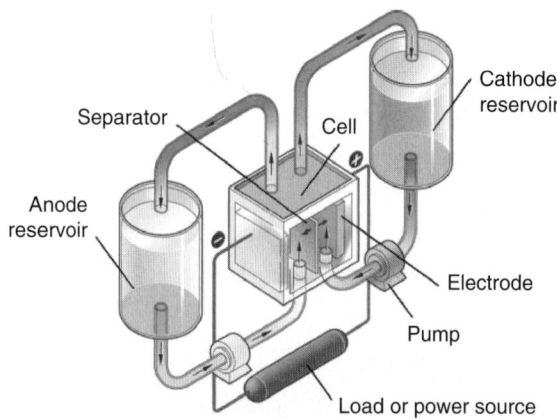

**FIGURE 27.12**   Simplified schematic of a redox flow battery system. (*Courtesy of T.M. Anderson, Sandia National Laboratories.*)

While a range of RFB chemistries are available, only vanadium RFB and zinc-bromine (Zn-Br) flow batteries are in commercial production. Vanadium-based RFBs are technically more mature and these systems are beginning to find utility scale deployments in the megawatt scale applications. One important advantage of vanadium RFB over other types of RFBs is that the two electrolytes are identical after full discharge. This feature greatly simplifies battery electrolyte management during operation, maintenance, and shipment. The first generation of vanadium flow batteries had lower energy densities and needed adequate thermal management of the electrolyte to reduce the potential for precipitation of the electrolyte at higher temperature.[45] Second-generation flow batteries with improved electrolytes have enabled large megawatt scale deployments to take place. In fact, large plants in the MWh to over 100 MWh size are being implemented for applications ranging from grid resiliency at a substation (shown in Fig. 27.13) to a large 15 MW/60 MWh flow battery plant (Sumitomo Electric, Yokohama, Japan) for supporting a large wind farm.

**FIGURE 27.13**    A 100 kW/400 kWh vanadium redox flow battery system at a substation. (*Source: Electric Power Board, Chattanooga, TN.*)

Major trends in flow battery technology are the increasing size of the stacks and improving energy densities. For example, Fig. 27.14 shows a meter size stack with a power capacity of 32 kW. Such large stack sizes improve power efficiency and reduce the overall dimensions of power conversion systems. There are considerable

**FIGURE 27.14**    A large commercial flow battery stack with a power capacity of 32 kW. (*Source: Rongke Power/UniEnergy Technologies, 2017.*)

advances in system design for larger scale flow battery plants. While smaller 1 to 10 MWh flow batteries typically used containerized architectures (as shown in Fig. 27.13), for larger systems, the plants are customized and built at the deployment site.

### 27.2.4  Sodium Sulfur (NaS) Batteries

The most common sodium battery systems are the NaS and sodium-nickel chloride ($NaNiCl_2$) batteries, which traditionally operate at 270 to 350°C.[46] In each of these cases, a molten sodium anode is separated from a molten sulfur or molten nickel chloride/aluminum chloride molten salt cathode using a ceramic, sodium-selective ion-conducting separator. In each of these systems, high-temperature operation is necessary to ensure the molten state of the anode and cathode as well as the high ionic conductivity of the ceramic separators. However, these high temperatures can increase the likelihood and consequences of undesirable or unsafe side reactions, decrease cycling capacity of the electrolytes, and limit the lifetime of the materials used in these self-contained ESSs.

NaS is a mature technology with a large number of installations, primarily in Japan. This is the most widely deployed utility scale energy storage technology in Japan with over 530 MW/3700 MWh installed capacity industrial and utility applications. A 34 MW/245 MWh grid connected NaS battery system was deployed to stabilize a 51 MW wind farm in Rokkasho, Aomori, Japan by NGK in 2017.

High-temperature sodium batteries offer the potential for cost-effective, reliable, and long-duration batteries for grid applications as they do not have the limitations of conversion chemistries like lead acid or intercalation mechanisms in Li-ion batteries. Work continues to enable lower temperature batteries utilizing sodium materials that would reduce drawbacks from high-temperature operation such as reduced long-term battery performance, reliability, and safety. Lower temperature separators for sodium systems would allow for the use of promising higher energy, long lifetime iodine, or bromine-based cathode chemistries not suitable for operation at traditional 270 to 350°C operating temperatures.[47] Significant reductions in operating temperature could also advance low-cost, non-toxic water-based battery chemistries, such as aqueous hybrid ion or emerging secondary alkaline technologies.

### 27.2.5  Other Battery Technologies

Alkaline batteries based on a $Zn/MnO_2$ system continue to be an area with tremendous promise due to the low cost of starting materials. The battery consists of a $MnO_2$ cathode and a Zn anode with a separator and alkaline electrolyte (see Fig. 27.15). These materials are known for their low cost, long shelf life, and high current density.[48]

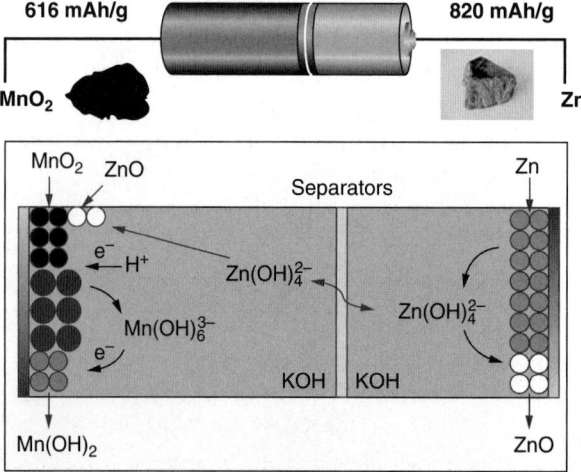

**FIGURE 27.15**  Alkaline $Zn$-$MnO_2$ batteries. (*Figure courtesy of CUNY Energy Institute, CUNY, New York, 2018.*)

**FIGURE 27.16**  2 kW/2 kWh Zn-MnO$_2$ prototype systems. (*Source: Urban Electric Power, New York, NY, 2017.*)

Current low-cost aqueous Zn-MnO$_2$ batteries are based on low DOD operation to get desired recyclability with a cycle life of 3000 to 5000 cycles. The early commercial prototypes shown in Fig. 27.16 have relatively low energy density of about 60 Wh/L. Low DOD of the active materials and low voltages have been the main reasons for the low energy densities.

Recent research has shown that the MnO$_2$ cathode can be stabilized at full DOD for an indefinite amount of cycling in the absence of zinc.[49] Extending the cycle life could mean that Zn/MnO$_2$ battery costs might someday reach the promised \$100/kWh. A number of technical challenges need to be addressed, especially with regard to the zinc anode. Zinc anodes show a large decrease in cycle ability as DOD is increased. This is due mainly to dendrite formation, electrode shape change, and ZnO passivation.[50] Understanding the zinc dissolution/deposition mechanism would provide insight into appropriate additives to enhance the anode's cycle life. Stabilizing the zinc anode will not only improve Zn/MnO$_2$ batteries, but any battery system relying on a zinc negative electrode, including zinc-air and zinc-nickel cells among others.

## 27.3   APPLICATION OF GRID STORAGE BATTERIES IN THE ELECTRICITY INFRASTRUCTURE

Most large-scale deployments of battery ESSs have been limited to select markets, where the essential services that energy storage can provide have been adequately monetized. For example, ESSs have found application for frequency regulation services where market mechanisms exist to pay for the faster response that storage assets can provide compared to other more conventional technologies. Since 2011, when energy storage was first allowed in the PJM Market, over 300 MW of new systems have been deployed. Storage can provide faster or better response than conventional resources in numerous other grid service areas, but market mechanisms are not yet in place to provide revenue for these services. Unlike frequency regulation, other applications like arbitrage, generation optimization, and renewable energy time shift are not economically viable with the relatively high cost of current technologies. Further research and development advances are needed to move lower cost technologies toward large-scale manufacturing, improve the overall system safety and reliability, and develop the engineering and market solutions to make large-scale deployments attractive for utilities in the United States and around the world. Promising use cases for energy storage descried in this chapter include

- Transmission systems (peaker replacement, T&D upgrade deferral and grid stability)
- Renewable integration and firming
- Microgrids

## 27.3.1   Role of Energy Storage in the Transmission Systems

*Peaker Replacement.*   The generation assets in an electric power grid are typically dispatched in an optimal fashion. In places where energy markets are not controlled by a single operating entity, the optimal dispatch is made through energy markets. Vertically integrated utilities, on the other hand, often utilize production cost modeling software to identify the lowest cost dispatch. Normally, the lowest cost or must-run generation units (i.e., power plants) provide the energy to meet the base load. As demand increases, additional units are brought online to meet load. In many jurisdictions due to the nature of the interconnection agreement between the local utility and the renewable generators, renewable generation must always be procured or it is curtailed when generation exceeds load. An example of a generator dispatch profile for Maui, HI is shown in Fig. 27.17.[51]

Power plants that are brought online to meet peak load (the few hours of the day with highest power demand) are often referred to as "peaker plants." These are usually flexible resources, but often have high energy production costs. Natural gas combustion turbines are commonly used. Other technologies include hydroelectric and diesel generation. In many cases, these resources are only called upon infrequently, seasonally, or even for a few hours a year. During these times of peak demand, electricity prices are often very high. To illustrate, the cumulative density function for the Electric Reliability Council of Texas (ERCOT) load data from 2016 is illustrated in Fig. 27.18.[52] Here, cumulative density represents the fraction of time in that year that the total load on the grid was below a given value. The minimum hourly load for 2016 was 25.1 GW, while the maximum hourly load was 71.1 GW. However, 90% of the time the load was below 54.9 GW. This means that approximately 16 GW of generation was deployed less than 10% of the time in 2016. This is typical of large power systems.

Even with the limited deployment, the high-energy prices during peak use often justify the investment in the peaker plant. Energy storage is being considered for peaker plant applications where there will be an economic benefit.

Another related solution is the combination of energy storage with a natural gas turbine. Southern California Edison (SCE) recently deployed such a battery-turbine combined system, developed by General Electric.[53] The system combines a 10 MW Li-ion battery ESS with a 50 MW LM6000 aero-derivative gas turbine. In addition to serving as a peaker plant, the hybrid system can also be used for frequency regulation, primary frequency response, and voltage support.[54]

*Transmission Infrastructure Upgrade Deferral.*   As load growth increases, transmission lines at some point reach their maximum transfer limits. This results in congestion (i.e., overloads), and energy from the lowest cost generators may not have a path to the demand. In this case, higher cost generation that has a different transmission path to the load must be employed. This results in higher electricity prices for the consumer. If there are no alternate paths, load cannot be met when it exceeds the transmission limits. Remedies include reducing load (e.g., through demand response), upgrading the transmission system (e.g., reconductoring or building new lines), or deploying energy storage or generation closer to the load.

The cost of upgrading a transmission line or upgrading a transformer can be very high. For example, typical transmission line construction costs are summarized in Table 27.3. These baseline costs increase significantly in urban areas (1.59 multiplier in the Western Electricity Coordinating Council [WECC]), as well as for short distances (1.5 multiplier if less than 3 miles).[55] Typical substation transformer capital costs are summarized in Table 27.4. If a transmission grid asset is reaching its limits, and expected load growth requires that the device be upgraded, employing energy storage to defer the upgrade can result in a large positive net benefit.

*Frequency Regulation.*   Frequency regulation is the second-by-second adjustment of AC generation to maintain system frequency; in the United States, this is 60 Hz. If the generation exceeds load, the system frequency will increase. Similarly, if the generation is less than load, the system frequency will decrease. An automatic generation control (AGC) signal is transmitted to all assets that provide frequency regulation. Depending on the balancing area, the signal is typically sent every 2 to 4 s. A representative signal from PJM is shown in Fig. 27.19. With the adoption of FERC Order 755 in October 2011, regional transmission operators (RTOs) and independent system operators (ISOs) are

> *required to compensate frequency regulation resources based on the actual service provided, including a capacity payment that includes the marginal units opportunity costs and a payment for performance that reflects the quantity of frequency regulation service provided by a resource when the resource is accurately following the dispatch signal.*[56]

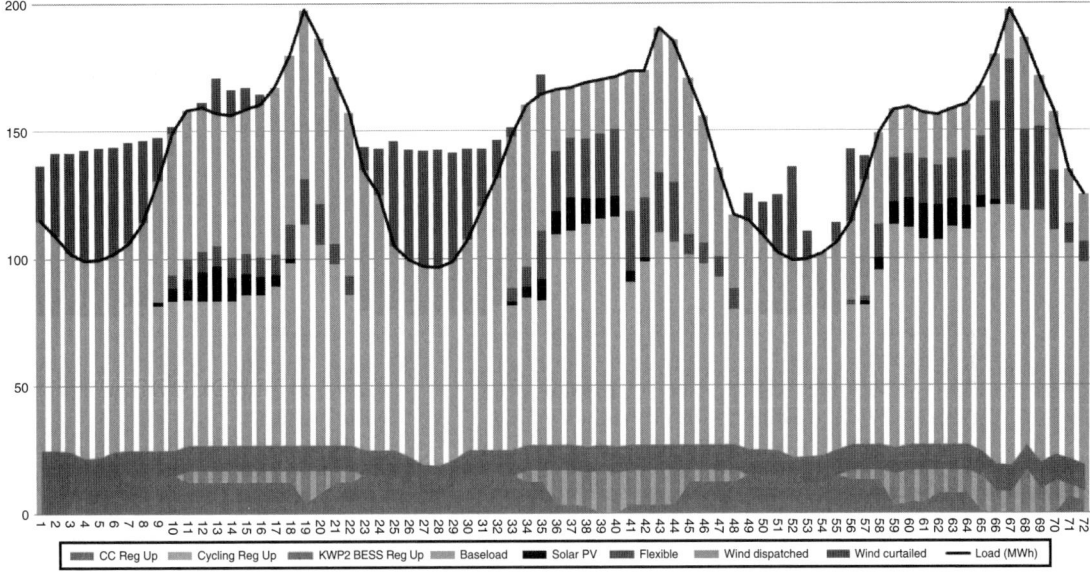

| CC Reg Up | Cycling Reg Up | KWP2 BESS Reg Up | Baseload | Solar PV | Flexible | Wind dispatched | Wind curtailed | Load (MWh) |

**FIGURE 27.17**    Maui, HI dispatch January 1 to January 3, 2015.[51] Vertical axis is megawatt of load and horizontal axis is a bar graph in hours for a total of 3 days.

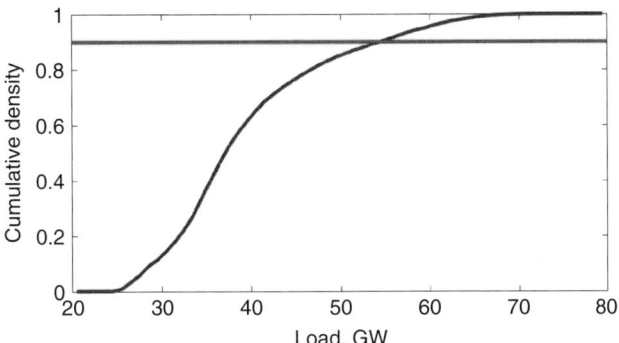

**FIGURE 27.18**    ERCOT 2016 load data, cumulative density function.[54] The plot illustrates the fraction of time the grid was at or below a given load demand during the year. The red line represents 90% cumulative density.

This has leveled the playing field for fast-responding resources like energy storage by compensating them for better performance tracking the commanded regulation signal. In addition, some ISOs have implemented separate regulation signals for fast responding resources, like the PJM dynamic regulation signal (REG-D) shown in Fig. 27.19. This figure shows a typical regulation signal that the regulation resource needs to follow within the performance requirements imposed by the system operator. Providing frequency regulation represents one of

**TABLE 27.3**    New Transmission Line Costs (2014 Data)

| Line description | New line cost ($/mile) |
| --- | --- |
| 230 kV single circuit | 959,700 |
| 230 kV double circuit | 1,536,400 |
| 345 kV single circuit | 1,343,800 |
| 345 kV double circuit | 2,150,300 |
| 500 kV single circuit | 1,919,450 |
| 500 kV double circuit | 3,071,750 |
| 500 kV HVDC bipole | 1,536,400 |
| 600 kV HVDC bipole | 1,613,200 |

Assumptions: Aluminum conductor steel reinforced (ACSR), tubular (230 kV)/lattice (345–600 kV), length > 10 miles.[23] Aluminum conducts to 600 kV.

**TABLE 27.4**    Substation Transformer Capital Costs (2014 Data)[55]

| Transformer cost ($/MVA) | 230 kV substation | 345 kV substation | 500 kV substation |
| --- | --- | --- | --- |
| 115/230 kV XFMR | $7250 | — | — |
| 115/345 kV XFMR | — | $10,350 | — |
| 115/500 kV XFMR | — | — | $10,350 |
| 138/230 kV XFMR | $7250 | — | — |
| 138/345 kV XFMR | — | $10,350 | — |
| 138/500 kV XFMR | — | — | $10,350 |
| 230/345 kV XFMR | $10,350 | — | — |
| 230/500 kV XFMR | $11,400 | — | $11,400 |
| 345/500 kV XFMR | — | $13,450 | $13,450 |

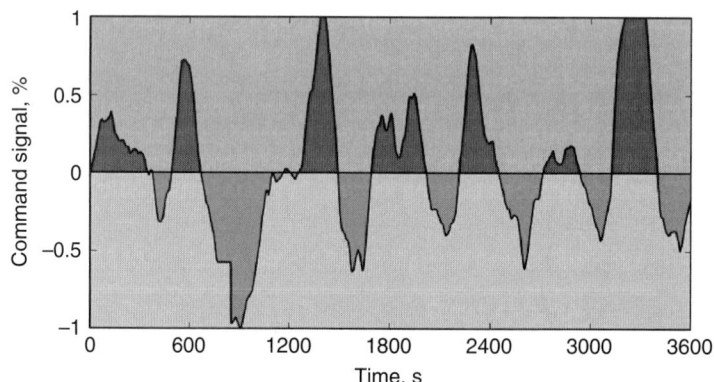

**FIGURE 27.19**    Sample automatic generation control (AGC) signal from PJM (REG-D).

the promising value streams for energy storage technologies provided that the technology is capable of the high cycling requirements for this application.

***Grid Stability.***    Power system stability can be divided into two categories: angle stability and voltage stability.[57] Angle stability refers to the ability of the interconnected synchronous machines (e.g., traditional generation) to remain synchronized. Angle stability can be broken down further into small signal stability and transient stability. Small signal stability refers to the response of a system to small disturbances, where the response can be assumed to be linear. Transient stability refers to the response of the system to large disturbances. In power systems, transient stability analysis focuses on the first swing in rotor angle after a large disturbance.[57] Voltage stability is concerned with maintaining a steady acceptable voltage at all buses in the power system. Insufficient reactive power is often the culprit when voltage stability problems occur. Energy storage can improve angle and voltage stability in power systems through the injection of real and reactive power.

*Small Signal Stability.* All large power systems with rotating generator equipment are subject to low-frequency electro-mechanical oscillations in the 0.2 to 1.0 Hz range.[58] Normally, these oscillations are well damped. However, in certain stressed situations these oscillations can become lightly damped and lead to a system shutdown. The 1996 west coast blackout in the United States was partially attributed to undamped interarea oscillations.[59] Real power injection at various locations in the electric power grid can be employed to improve damping.[60] Since the response is energy neutral, energy storage can provide the real power injections to augment small signal stability. The control mechanism is based on frequency measurements at various locations in the grid. Therefore, fast communications are required to implement a wide area damping control scheme. Because the oscillations have a relatively extended period (e.g., up to 5 s), a communications delay of up to 100 ms is tolerable, which is easily achievable with modern fiber optic links over long distances. A prototype system has been demonstrated on the Pacific DC Intertie, a high voltage DC link connecting Oregon and California.[61]

*Angle Stability.* The response of a typical power system to a loss of generation is illustrated in Fig. 27.20. The rate of change of frequency (RoCoF) is a function of the amount of inertia in the power system. The greater the inertia, the slower the rate of change. The frequency nadir, or lowest frequency after the event, is a function of the initial rate of change and the amount of frequency responsive generation in the system. Large traditional rotating generators provide most of the inertia in a power system. Smaller systems typically have less inertia, and displacing traditional generation with inverter-based renewable generation (which does not provide inertia unless a synthetic inertia has been implemented) further reduces the amount of inertia. Reduced system inertia increases the likelihood of a loss of synchronism and underfrequency load-shedding as the result of a large disturbance.

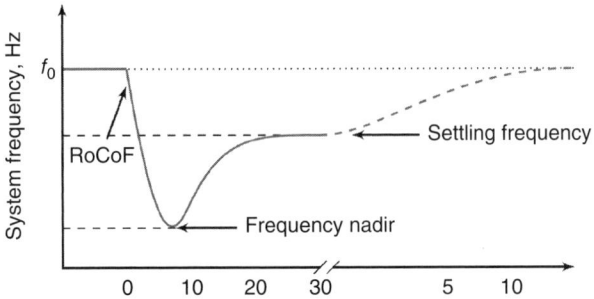

**FIGURE 27.20** System frequency after a loss of generation.

Energy storage can improve transient stability by injecting real power. A synthetic inertia control law can be employed to increase the amount of system inertia, where the output power is proportional to the RoCoF (second derivative of rotor angle). This is especially important for low inertia grids (e.g., small grids with high renewable penetrations). A study evaluating the benefits of synthetic inertia provided by energy storage for island grids has been recently published.[62,63]

Energy storage can also be deployed to improve the first swing stability of generators in the face of line trips by sourcing/sinking power to reduce the transient impact of the lost line. Energy storage can raise the frequency nadir by employing frequency droop control where the output power is proportional to the frequency error.

*Voltage Stability.* The principal causes of voltage instability in power systems are[64]:

- Overloaded transmission lines
- Voltage sources are too far from the load
- Voltage sources are too low
- Insufficient reactive power

Energy storage can help with many of these. In the case of overloaded transmission lines, energy storage can be deployed at the end of the line, close to the load. By discharging when the load is high, and charging when the load is low, energy storage can reduce the overloading of the transmission line. This is the same as transmission infrastructure upgrade deferral if the deployment of energy storage can delay the upgrade of the transmission line. Energy storage can supply reactive power provided that the inverter is properly sized and designed for this purpose. This capability is available over a wide range of SoC, and has minimal impact on the SoC. Typical control laws for reactive power control involve a proportional or proportional-integral control based on the voltage error at the node of interest. More discussion of the potential benefits of energy storage in preventing voltage collapse can be found in the literature.[65]

## 27.3.2    Energy Storage for Renewable Integration and Firming

One of the most important challenges in renewable integration is making the generated power available when it is needed. Unlike traditional generation that is "dispatchable," renewable generation is inherently less certain and variable or, in the case of solar, follows a diurnal pattern. Consider two of the most prevalent sources of renewable generation: solar and wind. Solar power can only be generated when the sun is up, and it varies greatly depending on cloud cover and other aspects of the weather and the atmosphere. Wind power generation, on the other hand, may be available at any time of day but fluctuates dramatically with wind speed and direction, can be hard to predict, and can change on a second-by-second basis. Because of this, renewable generation may not be available when it is needed. Energy storage can mitigate this challenge by storing energy when there is an excess of renewable generation and supplying it when it is required, thereby making renewable generation dispatchable. This is becoming more and more necessary as renewable penetration increases.

The inherent variability of renewable generation can cause grid integration challenges. Two noted areas of concern are the mismatch between supply and demand (load) during the course of the day and the impact of rapid fluctuations in renewable generation. The first phenomenon, often referred to as the "duck curve," is illustrated in Fig. 27.21.[66] The "belly" of the duck is caused by over-generation of solar energy compared to demand in the middle of the day. The sharp neck is caused by the drop in solar generation just at the time that load is starting to increase. This creates a high ramp rate requirement for other generation units. In California, this is being addressed through the introduction of a ramping product in the market. Energy storage can shift the energy from mid-day to early evening, thus eliminating the need for rapid ramping of generation. This capability is referred to as renewable energy time shift.

Other applications for energy storage in the integration of renewables involve "smoothing" and "firming" the power generated by renewables. Smoothing involves reducing the variability in the renewable generation so that the output is relatively smooth, just as a filter reduces the noise in a signal. Firming involves regulating the renewable generation output so that it meets certain values or follows certain profiles. These procedures make the generation more predictable and facilitate its dispatch to meet demand. Figure 27.22 shows solar power generation in Sterling, MA on cloudy day. The actual generation is shown in the dark gray curve, and the nominal generation is shown in light gray.

The second-by-second variability of renewable generation can cause voltage fluctuations that are damaging to equipment, especially tap changing transformers. In some areas like Puerto Rico and Hawaii, penalties are being placed on renewable resources that exceed ramp rate limits (e.g., a percentage of nameplate over a predefined period). These penalties can be mitigated by deploying energy storage to provide smoothing, thus limiting the observed ramp rates.

A simple control law can be used to determine the injection or absorption of power by the energy storage device to provide smoothing or ramp rate limiting. This control law is often designed as a constant value (or "gain") multiplied by the difference between the generated power and a reference signal that ensures the total power from the storage and renewable generation meet the ramp rate requirement. A block diagram for this control law is shown in Fig. 27.23. The power from the renewable source ($P_{gen}$) is filtered to obtain a signal that meets the ramp rate or other requirements ($P_{ref}$). Then, the command to the ESS ($P_f^c$) is the difference between the renewable generation and the smoothed signal.

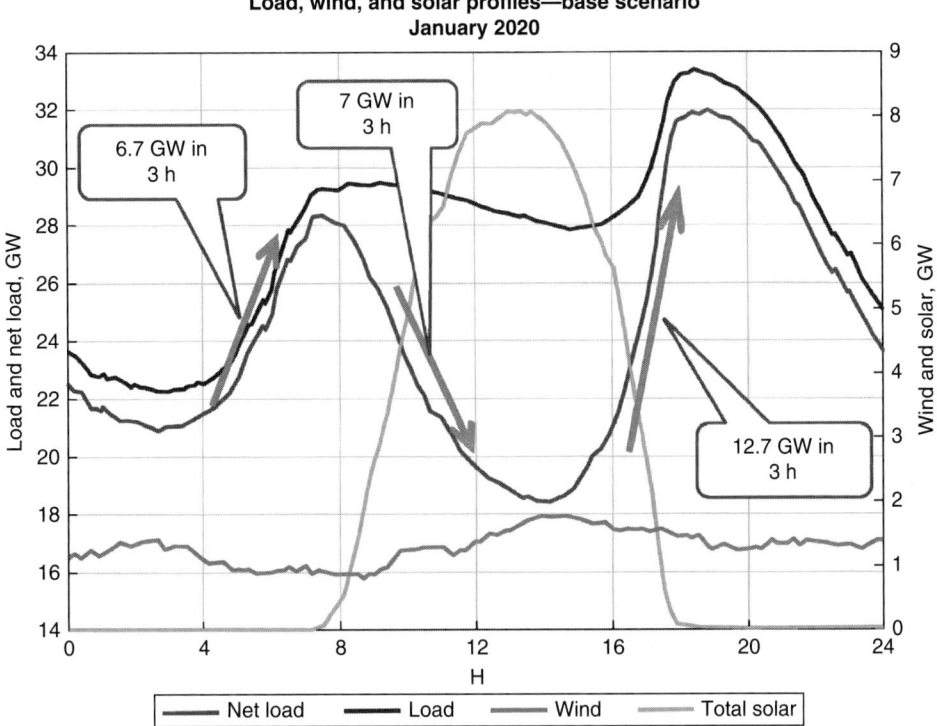

**FIGURE 27.21** Renewable energy time shift example—California 2020 load, wind, and solar profile.[66] (*Reprinted with permission from CAISO.*)

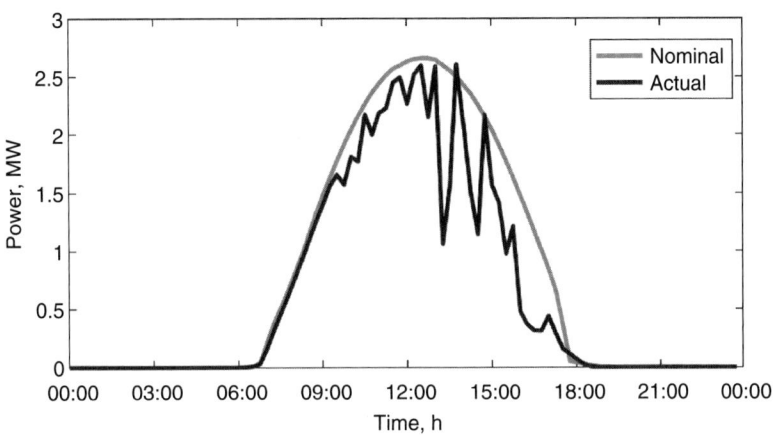

**FIGURE 27.22** Example of solar variability from Sterling, MA, on September 28, 2015.

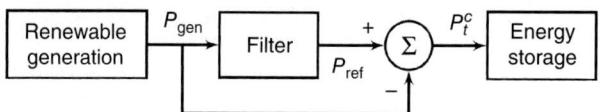

**FIGURE 27.23** Block diagram for renewable capacity firming (e.g., ramp rate limiting).

### 27.3.3 Microgrids

A microgrid can be described as a cluster of distributed generators, ESSs, and local loads, managed by an EMS. Figure 27.24 shows the schematic of a typical microgrid. A microgrid operates in two modes: grid-connected mode or islanded mode. In grid-connected mode, the grid acts as a balancing entity that absorbs the surplus power from the renewable generators (e.g., solar panels, wind turbines) or supplies deficit power to maintain the loads. When faults occur in the grid, a microgrid should be able to seamlessly island itself with minimum interruption to its load. In this islanded mode, a microgrid must self-supply all or part of its load for a certain amount of time until the grid problems are resolved.

Regardless of the operating modes of the microgrid, energy storage, with the ability to absorb and generate power, plays a crucial role in enabling a wide range of performance-enhancing and cost-cutting functionalities including the following:

- *Renewable smoothing/firming:* While renewable generators provide clean and cheap energy, their highly intermittent outputs create issues for microgrid frequency and voltage control. ESSs can reduce the intermittency by smoothing or firming the outputs of the renewable generators. Short-duration and fast-response ESSs such as electrical double-layer capacitors and flywheel ESSs are the most suitable for this functionality.

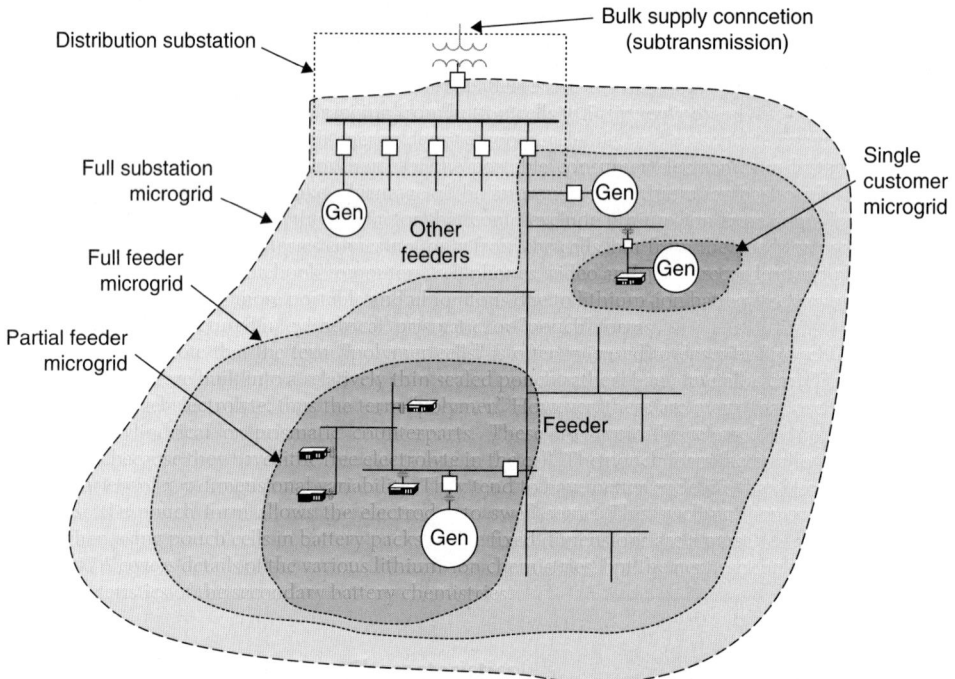

**FIGURE 27.24** Schematic of a typical microgrid system. (*Source: Sandia National Laboratories.*)

- *Reliability improvement:* In any distribution system including microgrids, the power supplies for the critical loads (e.g., hospitals, police stations) need to be maintained all the time. This requires a large amount of on-site back-up generation that significantly increases the investment and operating cost for the customers. ESSs can be a reliable source of reserve, thereby improving overall system reliability while also providing other functionalities.

- *Time shifting and peak shaving:* In a renewable microgrid, the peak loads and the peak renewable generation do not necessarily occur at the same time. For example, photovoltaic systems often output maximum power at noon while peak loads might occur at 6 PM. Therefore, the excess power at noon must be curtailed or absorbed by the grid at low energy price. In this case, energy storage can be used to store the excess renewable energy when the load is low and use that energy later to support the peak loads. This functionality often requires longer duration energy storage such as battery ESS.

- *Volt/var support:* With the recent improvement in power electronic technologies, ESSs can provide reactive power support that is essential to regulate the voltages at local nodes and reduce the overall power losses in microgrids. Since providing reactive power support does not significantly affect the SoC of the ESS, this functionality can be implemented simultaneously with other functionalities to increase the overall benefit of the ESS in a microgrid.

- *Synthetic inertia:* In island microgrids, diesel generators are often incorporated to generate power and provide the rotating mass needed for system stability. With the increase in renewable energy penetration, diesel generators are no longer the main source for power supply. However, relatively low rotating mass in the microgrids could cause instability issues when disturbances occur. In this case, ESSs could provide needed inertia via control algorithms.

Although ESSs can provide a wide range of applications, there are challenges when incorporating them in microgrids such as:

- The overall economic gains of energy storage deployments are limited by the round-trip efficiency and the degradation of the capacity and performance characteristics of the energy storage devices.

- It is difficult to coordinate multiple energy systems in a microgrid because they are often different in rating, capacity, and technology, and sophisticated distributed communication and optimization methods may be required to coordinate them effectively.

## 27.4 ENERGY STORAGE INTEGRATION AND NEEDED DEVELOPMENT

Properly integrated energy storage can transform the power grid to a more efficient, flexible, and reliable grid of the future. Unfortunately, there are several impediments to widespread deployment, including:

- Minimal modeling and valuation tools for optimal energy storage sizing, placement, and valuation.

- A regulatory structure that fairly compensates energy storage for services provided to the grid.

- Control and communications architectures for a power grid composed largely of distributed energy resources.

Applied and fundamental research in several key areas can significantly lower these barriers to the widespread deployment of energy storage. Some approaches are discussed below.

*Valuation* of energy storage is difficult. Radically different approaches are required depending on whether the device is in a market area or under a vertically integrated utility. Some ISOs are starting to introduce ramping and voltage support products that provide additional potential revenue streams for energy storage. There are also many grid functions that remain uncompensated, including carbon reduction, synthetic inertia, governor response, and small signal stability. For large-scale integration of energy storage, there needs to be well-established market mechanisms to value energy storage for all the services ESSs can provide.

*Cost modeling* is typically performed to evaluate operational savings from different energy storage scenarios in a vertically integrated utility. In addition, dynamic simulations must be performed to evaluate storage requirements to meet technical goals like a minimum frequency nadir after a generator drop. Then, costs of energy storage must be compared against other approaches for meeting the technical objectives (e.g., demand response, peaker plants, new transmission, etc.).

*Dynamic simulation tools* have been used to model and analyze the dynamic response of the power grid to ensure small signal and transient stability. Some examples are PSLF (General Electric), PSS®E (Siemens), and PowerWorld. These same tools are used in transmission planning studies. They typically aggregate load and primarily model the bulk transmission system. Distribution level modeling is performed with a different modeling tool like OpenDSS.[67] While OpenDSS can perform dynamic simulations, this capability is rarely used because in the United States, it would be extremely rare for any type of distribution event to significantly impact grid frequency. Therefore, the grid frequency is assumed to be constant and a quasi-static time-series simulation is employed. New techniques and tools need to be developed to model and analyze the distributed, flexible grid of the future that is composed largely of distributed resources controlled through communications. Development of these tools is a prerequisite to identifying viable control and communications architectures that must be codified in standards to enable widespread deployment.

*Communications and control* will be essential to a grid of the future, composed largely of distributed renewable generation, energy storage, and responsive load. With storage in the mix, this set of resources can disconnect from the grid and operate autonomously if needed, and are referred to as a microgrid. This is radically different from the grid of today because it must elegantly handle two-way power flows, and will require a yet-to-be-developed communications and control architecture to maintain the expected levels of reliability. A significant amount of research has been performed on microgrid design optimization and control algorithms, but very little research has been performed on how to most efficiently run a grid composed largely of microgrids. Therefore, a critical research area is the development of control algorithms and the associated communications infrastructure that will be required to operate a power grid composed largely of distributed renewable generation, energy storage, and responsive load. Fundamental research is required to assess the trade-offs in controls and communications complexity to arrive at a potential solution that meets technical performance requirements for a reasonable cost while leveraging as much existing infrastructure as possible.

## 27.5   SYSTEMS ENGINEERING ASPECTS OF GRID ENERGY STORAGE USING BATTERIES

A battery ESS is composed of four major components: the battery and battery management system (BMS), the power conditioning system (PCS), the balance of plant (BOP), and the EMS. The components of an ESS are summarized in Fig. 27.25 and described in more detail below.

### 27.5.1   Battery and Battery Management System (BMS)

The battery system comprises the battery cell stacks, electrolyte tanks (when used), modules, the BMS, overload and short circuit protection, and the necessary racking means. The batteries provide the electrochemical energy storage, while the modules contain the batteries conveniently arranged in appropriate combinations for the given kW and kWh rating. This allows for the batteries to be placed in racks, thereby making up the system. Depending on the kW and kWh ratings, these racks can be housed in shipping-type containers for easy transport and installation, or be placed in a building.

The BMS controls (in conjunction with the EMS) the charge and discharge of the battery. If necessary, the BMS can also provide battery trickle charge. The BMS can also monitor cell health and detect and annunciate cell and module failures.

Several battery housing configurations are shown in Fig. 27.26 through Fig. 27.29. Figure 27.26 shows the walkway through the large NEC Li-ion battery installed at Sterling (MA) Municipal Light Department. Figure 27.27 is a somewhat smaller NEC Li-ion battery system, housed in a conventional 40 ft (12.2 m) container. Figure 27.28

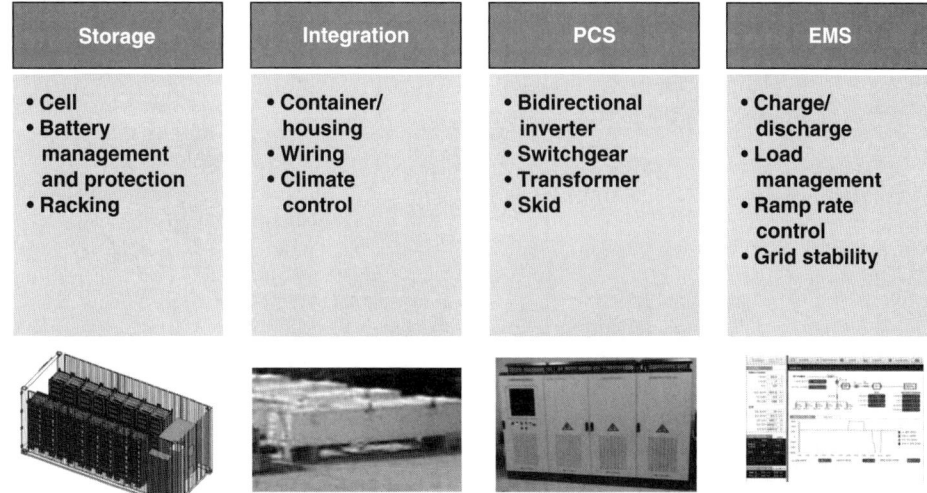

**FIGURE 27.25**  Elements of a grid energy storage system.

illustrates all the components of a Kokam 2.4 MWh Li-ion battery ESS. Figure 27.29 shows the inside of a UniEnergy flow battery installed at Sandia National Laboratories, Albuquerque, NM.

## 27.5.2  Power Conditioning System (PCS)

The purpose of the PCS is to control the DC battery output (or input) to match the AC load (or source). During discharging the PCS will convert DC from the battery into AC output power for connection to the grid. Conversely, when charging, the PCS will rectify an AC voltage to a DC voltage. The PCS thus consists of

**FIGURE 27.26**  Sterling project: internal view of container, rated at 2000 kW, 3900 kWh. (*Courtesy of NEC Energy Solutions.*)

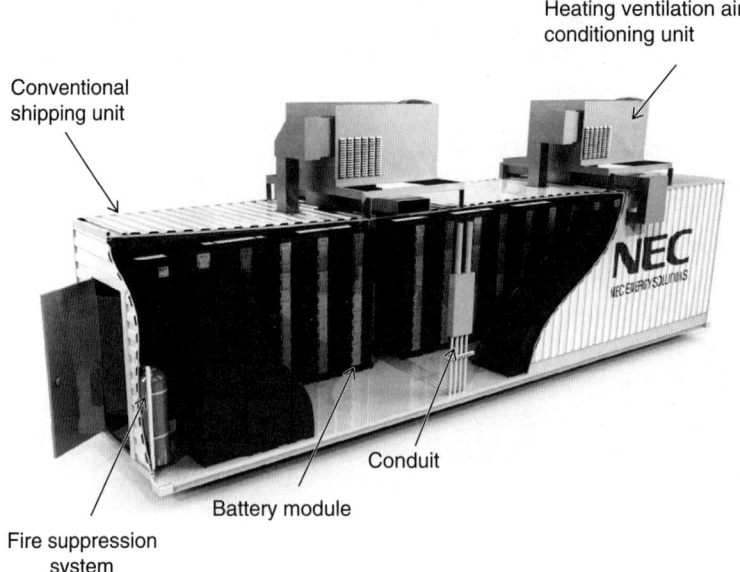

**FIGURE 27.27**   Li-ion ESS in a 40-ft (12.2-m) long container, rated 2800 kW and 2800 kWh.

bidirectional inverters[a] that both convert and rectify AC and DC voltages, and may have an isolation and/or step-up/step-down transformer. The PCS will also include harmonic filtration, distribution switchgear with protection-coordinated circuit breakers (overload and short circuit), and any automatic switchgear control.

A battery, as seen by the PCS, behaves as a voltage source, and therefore a PCS inverter will be rated in both power and voltage, depending on the application. Inverters are rated in kW and presently range in sizes up to 500 kW. AC voltages on the line side of the inverter would normally not be greater than 480 V/3-phase AC and 600 to 900 V DC on the DC bus of the inverter. However, inverters are now capable of handling higher AC voltage with the new IGBT (insulated-gate bipolar transistor) and SiC technologies. Another attribute of the inverter system is whether it is single phase or three-phase. Smaller systems for residential use are single phase. Commercial and industrial as well as utility systems would most likely be three-phase. Modern PCS components are becoming more and more efficient.[69] Several examples of PCS subsystems, a bidirectional inverter and a three-phase transformer, are shown in Figs. 27.30 and 27.31, respectively.

### 27.5.3   Balance of Plant (BOP)

The BOP system consists of any components that are not directly part of the battery system, but are critical in the safe, reliable operation of the system. These items vary for the different battery technologies, but could include but not be limited to containers (building, skid, or shipping container), air conditioning and heating units, pumps, lighting, and fire protection. Figure 27.32 shows the cooling system for the bidirectional inverter in Fig. 27.30.

An example of a pump system is shown in Fig. 27.33. On the left-hand side of the container are two long skinny cylinders, which are the two flow pumps, anolyte and catholyte, on a vanadium RFB system. Depending on the ambient temperature of the installation, there could be air conditioning units, fans, or humidification systems, or all three. Another component that could be included in BOP is a leak detection system. This is

---

[a]A power inverter, or inverter, is an electronic device or circuitry that converts direct current (DC) to alternating current (AC) and rectifies AC to DC.

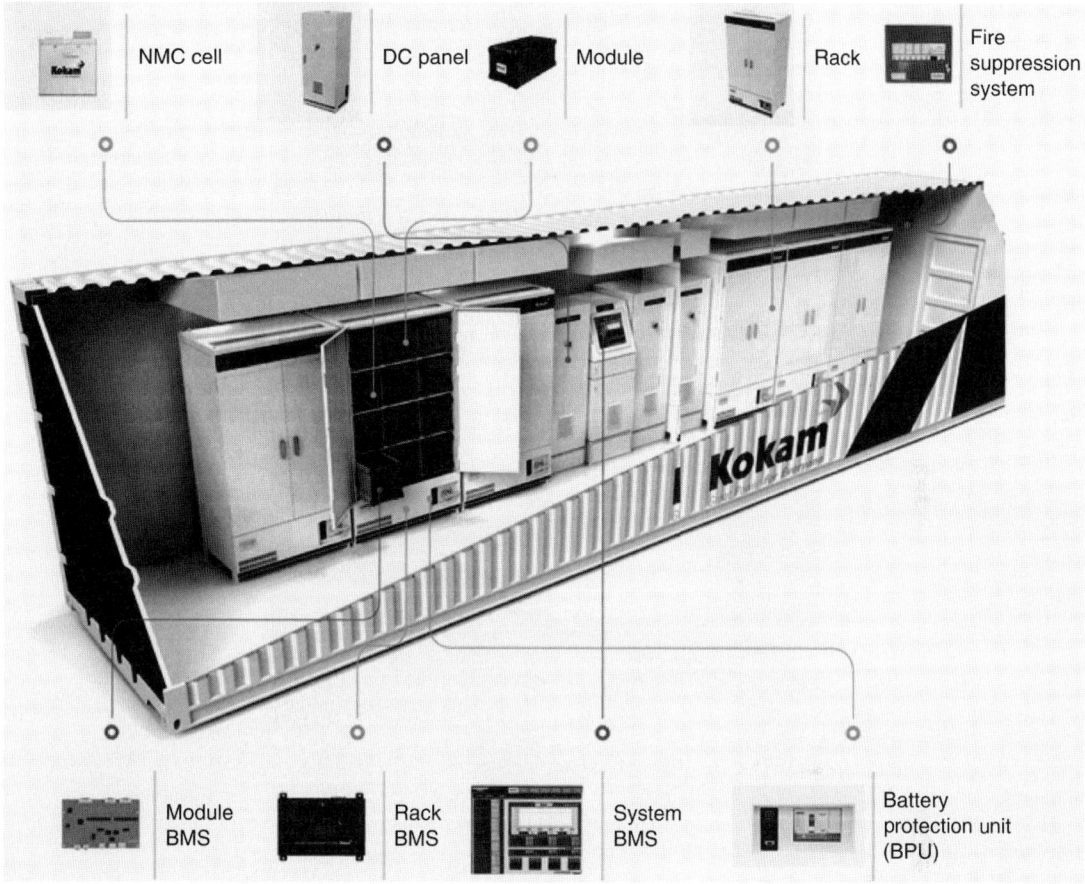

**FIGURE 27.28**    Kokam 2.4 MWh LiMNC battery energy storage system.[68]

important in a flow battery system, as it will detect any leaks that might occur. Usually this system will only be placed in the stack section of the ESS enclosure, as this area is the most vulnerable to leaks. However, the containers that house the anolyte and catholyte are usually built in such a way and with adequate containment that leak detection is not required. For other battery technologies, a critical component would be a fire detection system that would shut down the system and alarm operations personnel in the event of smoke or fire.

### 27.5.4    Energy Management System (EMS)

The ESS needs to be controlled in such a way that it can provide beneficial service to the owner. This operation is performed by the EMS, which is computer-based. This system has an important role in managing the battery's interface with the grid. It looks at information on the grid, such as time-of-use rates, weather forecast, load profile and criticality, and grid stability. It also monitors the battery SoC, load profile constraints, ramp rates, and other variables that could impact battery performance, such as cumulative cycles. With some of the more advanced technology, the EMS can make smart decisions concerning how the battery should operate and optimize the operation to increase the benefits of ESS.

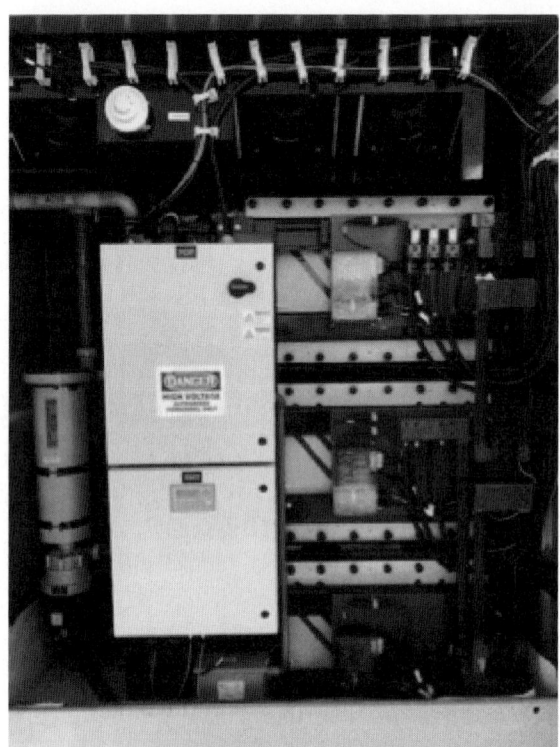

**FIGURE 27.29**   Battery management system. Sandia National Laboratories 1 MWh vanadium flow battery installation. (*Courtesy of UniEnergy.*)

Cooling system

Connection point and fusing

Inverter

Inverter cabinet

**FIGURE 27.30**   Two views of the 125 kW AC/DC bidirectional inverter. (*Courtesy of EPC Power.*)

**FIGURE 27.31**    150 kVA 480 V delta/wye three-phase transformer.

## 27.6   *DESIGNING THE BATTERY ENERGY STORAGE SYSTEM*

When designing for the installation of a battery ESS to be interfaced with a utility power distribution system, certain items need to be considered. The first order of business is to conduct an analysis to determine which application(s) will provide the most benefit for the given market and/or need. There are numerous applications, as discussed previously,[70] with all of them providing either primarily a power benefit or an energy benefit. These categories are roughly defined by the optimal discharge time of the battery. Power applications typically operate for less than 15 min, while energy applications generally require greater than 15 min of operation. Some technologies fall into both categories. Common applications are summarized in Table 27.5.

To determine the most suitable applications for an ESS, the utility market rate structure where the system will be operated should be known and understood. Also, the kW and kWh ratings for the considered application will need to be calculated. One common way to do this calculation is to develop a power load profile for the given application. Load profiles tell a system designer the maximum demand (power load), the duration, and the cycles. The maximum demand indicates the kW rating of the system, while the duration indicates the kWh rating. The cycle data are important in determining how many cycles the system will have to perform at max load and for duration for the life of the project. Another important element in selecting a battery is the ambient temperature where the system will be located. This information is important to determine if any additional

**FIGURE 27.32** On board cooling system for EPC 125 kW bidirectional inverter—Sandia energy storage installation. (*Courtesy of EPC Power.*)

cooling or heating will be required for the battery to operate as intended. The above information will serve to provide battery vendors with a bid specification. Examples of load profiles for testing various applications can be found in Figs. 27.34 and 27.35.[71]

Some other considerations to be addressed in the design phase include the following:

1. Who will own and operate the system—power purchase agreement (PPA) or owner owned and operated? The pros and cons are listed in Table 27.6.

2. How will the system be contracted for design and construction? Several approaches are presented in Table 27.7.

## 27.7 INSTALLATION

When installing an ESS the following should be considered:

*Building and fire codes:* As of this writing, installation codes are just catching up with the ESS installations, and in some locales, there are little or no codes of which to speak. Also, with certain technologies there is increased concern for fire safety. Therefore, for any system of size (>5 kW) it is imperative to work with the

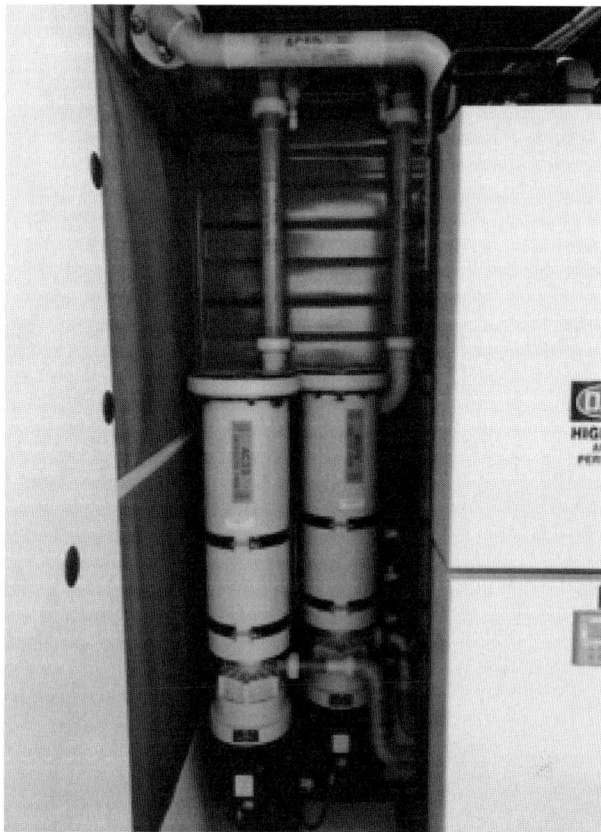

**FIGURE 27.33**    Example of balance of plant components—flow pumps on a flow battery. (*Courtesy of UniEnergy Technologies.*)

**TABLE 27.5**    Energy Storage Applications

| Application | Power <15 min | Energy >30 min |
|---|:---:|:---:|
| Renewable firming | × | × |
| T&D upgrade deferral | | × |
| Demand reduction | | × |
| Energy shifting | | × |
| Power quality/reliability | × | × |
| Spinning reserve | | × |
| Frequency regulation | × | |
| Capacity | | × |

local code official and fire department to understand the requirements for installation. For instance, depending on the location of the system, building type or city, the installer may need to follow clearance requirements, as well as containment, and installation material requirements. In addition, depending on the battery chemistry, there could be fire suppression requirements that the authority having jurisdiction (AHJ) may want incorporated. Many codes are just now being developed based on lessons learned from

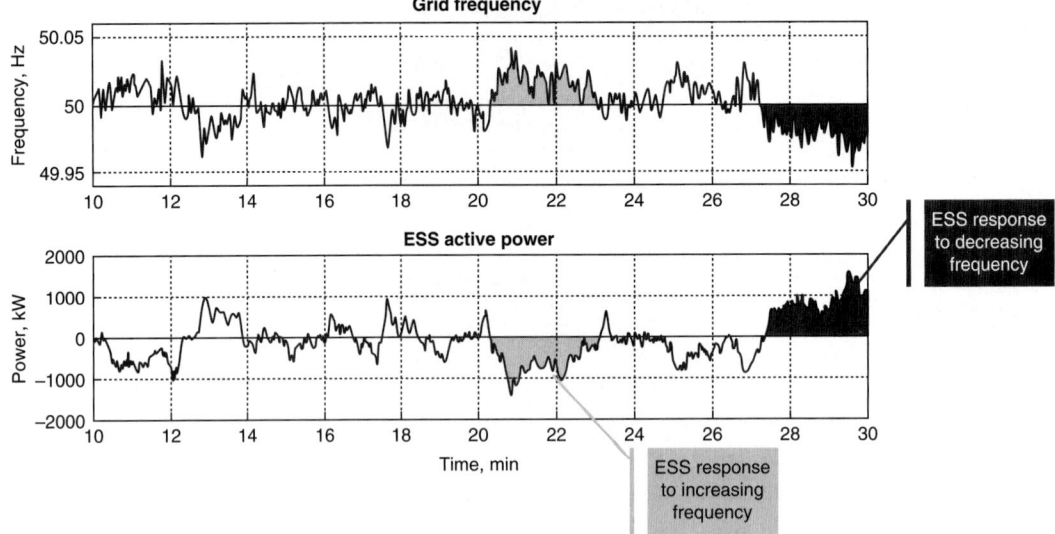

**FIGURE 27.34**   Frequency regulation load profile.

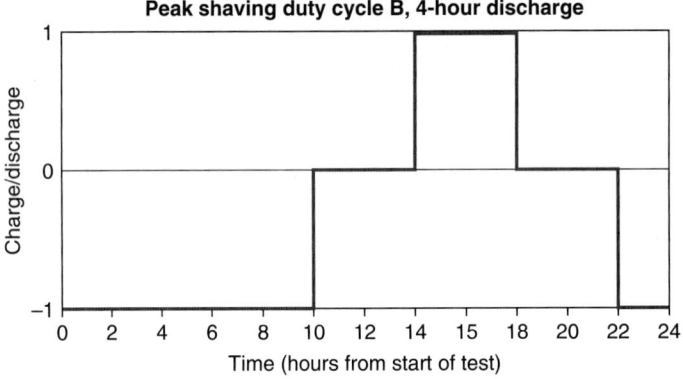

**FIGURE 27.35**   Peak Shaving Duty Cycle B, 4-h discharge load profile.

previous installations. Therefore, codes could change quickly, which is why it is important to work closely with the AHJ. *Specific comments on battery safety are addressed in Sec. 27.10.*

*Interconnection agreements with the utility:* Since the utilities are concerned about what generation is being attached to their grid, and a battery system is considered a generating source when discharging to the grid, permission will be required to connect the ESS to the grid. This is primarily driven by a safety concern that the utility lineman will not work on a circuit believed to be "dead" or not energized, only to find out it is energized by an unknown battery connected to the grid. Interconnection agreements should inform the utility of the size, location of the ESS, and the applications that the system will be used for. If islanding will be done, then the utility will need to ensure that the system installer has the disconnection means that will isolate it from the grid.[72]

**TABLE 27.6** Characteristics of Different Ownership Models

| Ownership | Pros | Cons |
|---|---|---|
| Power purchase agreement (PPA) project: developer/operator builds and operates ESS. Owner pays for kWh delivered. | Performance risk is placed onto developer as owner only pays for kWh delivered. Maintenance by developer/operator. This is valuable in projects where the owner does not have support staff. | Lack of ownership. May be locked into operating load profiles and/or applications that become inconsistent with market needs. |
| Owner owned and operated: owner pays for developer to build system. Owner will own and operate system once commissioned. | Complete control of system installation and operation. Ability to adjust operating load profiles and applications as markets warrant. | Owner assumes risk. If system does not perform as specified, would only have contract requirements, warranty, or O&M agreements to solve operational issues. Will need access to maintenance support for minor inspections/adjustments not covered by warranty or maintenance agreements. |

**TABLE 27.7** Characteristics of Different Contracting Models

| Contracting strategy | Description | Comments |
|---|---|---|
| Design/build (D/B) | In this strategy a firm is hired by the owner to design and build the EES project. This is sometimes called a turnkey system. | This is a convenient strategy when the owner has limited engineering and/or construction management resources. The D/B firm can be contracted to be all-inclusive and can also act as the owner representative. |
| Design/bid/build (DBB) or engineer/procure/contract (EPC) | Using this strategy the owner will place a contract with a design firm, and then once the design is complete, the design is put out to bid to an installer. | When the owner has adequate staff, this strategy allows the owner more control, as they can act as the gate between design and construction. |

*Application considerations:* What applications will be utilized and are the right distribution components installed? For example, if the system will be used in an islanding application, the ESS inverter must have black start capability.[b]

*Resiliency:* When installing an ESS as part of a resiliency effort, care should be taken to install the ESS in an area that is less prone to flooding, rain and wind damage, and fire damage. Also, the distribution system to which the ESS connects should be selected or designed in such a way that minimizes failure due to events of nature or otherwise. For instance, distribution wiring should be installed underground, with the switchgear preferably located above the flood line.

Finally, to avoid unnecessary cost, the ESS should be located near a distribution system that will allow for an effective connection point.

---

[b]An inverter has two modes of operation: (1) Grid following—the inverter provides current while following grid voltage. (2) Grid forming—the inverter provides voltage as a standby generator would. Black start capability is available when the inverter is in grid forming mode, and has a power source to energize the PCS controls during a power outage. With this capability, the inverter can provide voltage to the grid if the ESS has sufficient charge.

## *27.8   UTILITY INTERCONNECTION*

Utility interconnection can be as simple as a connection to a breaker in a branch panel, or as elaborate as installing separate switchgear with islanding control. Islanding was explained earlier in Sec. 27.3.3. In addition to the physical connection, most utilities also require a monitoring or communication connection to the utility management system. This could be operated through the EMS discussed previously.

Figure 27.36 shows different interconnection/distribution panels. Figure 27.37 shows typical service entrance switchgear, while Fig. 27.38 shows a step-up transformer. These are all located at the Sandia National Laboratories ESS test facility.

**FIGURE 27.36**   A 300-A panel and a 200-A panel for connecting to the distribution grid.

Not every ESS utility system is designed for islanding, but when it is, islanding control includes not only the power components, e.g., motor-operated breakers or static switches, but also the actual computer-based controller that will operate the power devices. Islanding may also require synchronization control gear that allows the battery to be in phase with the grid when grid power returns. This synchronization allows for a seamless return to grid. One important aspect of interconnection is that the system protection scheme needs to have the ability to protect the system with the ESS as a load and as a source. In most modern protection relaying systems this is accomplished by dual protection relay settings that adjust, depending on if the battery is charging or discharging.

## *27.9   COMMISSIONING*

Commissioning is an important step in the overall ESS installation process. Before declaring a system operational, this activity is needed to determine and ensure that the system was installed as designed, can be safely operated, and will perform the applications as required. The commissioning program contains design verification checklists, safety system checks, installation to codes and standards verification, application testing, and baseline data monitoring.

Figure 27.39 shows the phases of commissioning required. As illustrated, this does not all happen at the end of construction and installation. Rather, start-up procedures, test procedures, routine operation, and maintenance procedures and emergency operations planning all need to be developed while system building is underway. Testing protocols[73] can be found in the Sandia National Laboratories/Pacific Northwest National Laboratory protocol handbook. The project sponsor determines detailed requirements for the factory witness test, operational acceptance test, and functional acceptance test.

**FIGURE 27.37**    Typical service entrance switchgear (SES) for commercial/industrial/utility installations (Sandia National Laboratories).

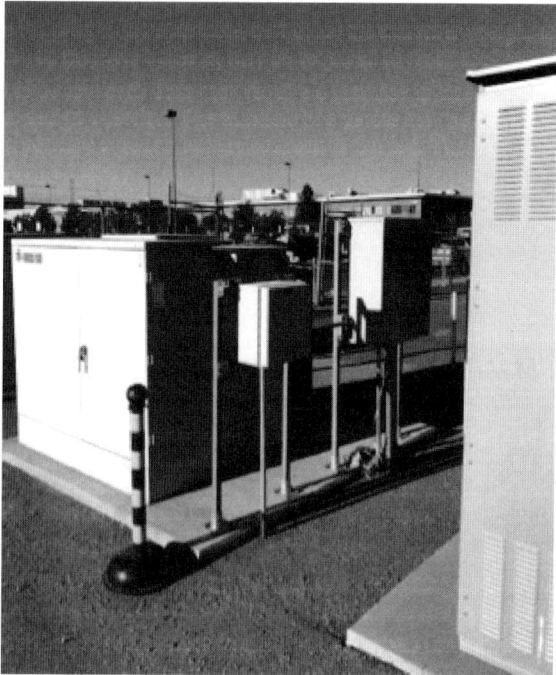

**FIGURE 27.38**    1.5 MVA, 12470/480-277 three-phase step-up transformer for connection to grid (Sandia National Laboratories).

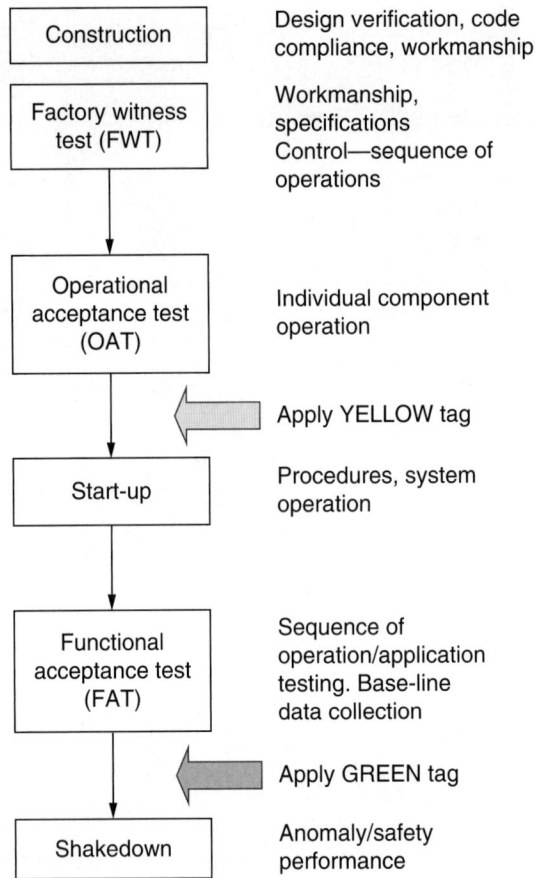

**FIGURE 27.39**   Phases of the commissioning process. Tags mean the following: Yellow tag—contractor owned, owner-operated, not transitioned. Green tag—transition (hand-off) to operation.

## 27.10   *SAFETY AND RELIABILITY OF LARGE BATTERY ENERGY STORAGE SYSTEMS*

Battery safety is of increasing concern to both the public and the stakeholders within the utility grid energy storage industry. Increasing battery prevalence and the scale of stationary ESSs drives safety concerns. These are exacerbated by high-profile, public incidence of failures in consumer electronics, aviation, and electric vehicle battery technologies that create reticence to allow adoption of storage in some instances. This is particularly true in urban environments, where adoption of energy storage requires robust safety codes and protocols. Safety research encompasses identifying and mitigating risks. Internally initiated failures can result from internal shorts from manufacturing defects, aging effects leading to early degradation, abuse conditions, power electronics faults, etc. External events can be fire in an adjacent room or mechanical impact that compromises a battery. Safety research is further separated into materials and cell-level topics, and those that impact system-level scales. In 2014, the U.S. Department of Energy issued a strategy document outlining the research needs in this area.[74] Critical areas are highlighted below.

*Risk determination* is of immediate concern for identifying failure scenarios and evaluating the consequence for ESS. To better mitigate severe consequences, a better understanding of failures that relate to thermal runaway, burn behavior, vent gas, fire modeling/analysis, and suppression testing are the highest priority areas of R&D for system-level energy storage. Stranded energy should also be addressed. Tools to assess when stranded energy presents a danger and best practices for addressing stranded energy after a failure must be developed. Additionally, advanced monitoring and controls for predicting and mitigating failure are critical to risk mitigation. Evaluation of the performance and aging of cells as well as methods to identify potential failure through electrochemical signatures, or other sensor techniques is a significant area of research to increase the resilience of ESS.

*Codes, standards, and regulations* (CSR), as they pertain to safety concerns with ESS, must also be developed. This includes identifying gaps and contradiction in the CSR environment and where R&D is needed to elucidate best practices. These efforts must also be coupled with rigorous risk assessment processes such as failure mode and effects analysis (FMEA). Regardless of the general areas of safety consideration as covered by existing codes and standards, the following areas, as outlined in Energy Storage System Guide for Compliance,[75] need to be addressed:

- Smoke and fire detection, fire containment and suppression, fire extinguishing
- Ventilation, thermal management
- Egress and access, security, illumination, signage
- Electrical safety on the customer or utility side of the meter
- Anchoring and protection from natural disaster and the elements
- Spill containment, neutralization, and disposal; personnel protection; hazmat equipment
- Communications and battery management

The relevant codes and standards for each of these categories are found in Refs. 76, 77, and 78.

*Battery design* can mitigate failure through several avenues such as developing inherently safe chemistries. This can also include cell-level safety devices and improved active and inactive materials, ranging from separators and electrolyte materials to plastic housings. Modification of standard lithium ion design should include engineered flame-resistant electrolytes. Cell-level thermal propagation experiments and thermal modeling must be conducted to inform heat management strategies for mitigating propagation of thermal runaway. Vent gas analysis and modeling is needed to understand fuel air mixtures and the gas products that result from battery failure for conflagration studies and human risk. Significant areas for improvement lie in power electronics research and software design to prevent failures by identifying conditions before catastrophic failure can occur.

*Online repository* of data, test plans, and discussion is critically needed to consolidate information related to safety R&D for ESS, and to coordinate collaborations, share findings in a timely manner, and elucidate the critical paths to improving risk management of ESS. This should be complemented by coordinated meetings such as the ESS Safety and Reliability Forum.

Safety and reliability of ESS have been considered the largest unknown variables in cost forecasting, and ESS safety is a critical area needing significant research.[79] Determining the risk of a system is needed for insurance purposes and to develop the most effective engineered safety when installing a system. Significant focus on the use and risks of energy storage concentrates on Li-ion batteries.[80] This is understandable because the prevalence and high energy density of Li-ion batteries combine to add likelihood and consequence together into a growing risk.

*Reliability and life cycle predictions* are difficult even in a laboratory setting at the cell level. At the system scale, reliability is further complicated by many factors including:

- Cell chemistry and associated battery management protocols
- Projected usage case profiles and variations in the profile
- Handling/mishandling on install
- Reliability of lead connections
- Sensor placement and algorithms to incorporate cell diagnostics using available electrochemical signals
- BMS, data management, data storage
- Cell temperature, thermal management (both passive and active), HVAC design, hot spot detection/prevention
- Power issues, inverter response, and more

It has been reported that installation of energy storage is delayed in some areas due to unfamiliarity with ESS and uncertainty of risks and mitigation strategies. A recently commissioned study by NYSERDA was carried out on various battery energy storage technologies under abuse conditions, specifically intended to alleviate concerns that were inhibiting installation of energy storage.[81] Continued uncertainty exists by installers about both the safety and the reliability of storage, which impact the understanding of risk and the levelized cost of ownership for using storage. More robust understanding of safety and reliability are therefore important to implementation of storage in a number of use cases and markets.

Power electronics are an important component of the ESS, providing conversion between AC and DC power, among other functions. Power electronics control aspects contribute to the safety and reliability of the system.[82,83] In turn, power electronics themselves represent a safety risk, predominantly due to the potential failure of electrolytic capacitors.[84] Thermal management is an equally critical component to managing risk and improving safety of systems.[85] Assessing the potential for thermal runaway[86] relies heavily on modeling[87] as cell failure moves to string[88] and then system failure.

Effects of failures include voltage, arc-flash/blast, fire, combustion, toxicity, stranded energy, and reignition. Fire is the most commonly considered and discussed concern with storage system failures.[89] Standardized ways of conducting failure testing is needed, and confidence developed to assist in quantifying the risk of batteries is needed. This has been addressed by some reports with recommended practices for performance and safety testing.[90,91] To date, several failures resulting in fires with systems have been reported. Due to the proprietary nature of demonstration projects and commercial systems, any close calls and lessons learned not resulting in fire have likely not been reported. Of the known fires, at least one was attributed to power electronics failure, illustrating that safety considerations cannot be limited to the battery alone when dealing with systems (Table 27.8[92–95]).

**TABLE 27.8**    Summary of Energy Storage Failures Resulting in Fire

| Name | Technology | Size (kW) | Year | Notes |
|---|---|---|---|---|
| Port Angeles Landing Mall | Lithium iron phosphate | 75 | 2013[92] | |
| Kahuku Wind-Energy Storage Farm | Advanced lead acid | 15,000 | 2012[93] | Fires were attributed to ECI capacitors in inverters |
| Tsukuba Plant | Sodium sulfur (NGK) | 2000 | 2011[94] | |
| PREPA BESS 2 | Lead acid | 20,000 | 2006[95] | |

## 27.11    ENERGY STORAGE IN THE FUTURE GRID

The current trends and research and development projects in grid modernization point toward a future grid with advancements and disruptive changes from generation and transmission all the way to how we consume and pay for energy. Some of these disruptive changes include bidirectional power flows in both generation and distribution, power electronics enabled infrastructure across the grid, system-wide coordinated sensing and control, asynchronous networks, and a diversity of energy products and services. The widespread adoption and integration of energy storage will affect every level of the grid and will play a key role in addressing the challenges of the future electric grid. However, to fully realize the benefits of energy storage in the future, grid scale energy storage will be required to have lower capital cost (on the order of $100/kWh) as well as longer cycle life (perhaps up to 10,000 deep cycles) and deeper discharge to serve multiple grid applications.

One of the challenges for which energy storage will play a critical role is in the integration of large amounts of renewable generation in the T&D grids. Of the 6 TW of current worldwide generation capacity, renewables are reaching 20% penetration in many markets, and that percentage is rapidly approaching 30% to 40% in some markets. In these areas, handling variability and intermittency is crucial. Energy storage will be used to firm renewable generation capacity and make it competitive with fossil fuels and nuclear generation. For this to become reality, more advanced power electronics and EMSs will be required.

While there has been significant progress in research and development of energy storage technologies, several areas require further research: (1) electrical models that fully reflect the performance and cycle life characteristics of battery systems, (2) improved operational optimization approaches for stacking the values of multiple grid applications, (3) improved decentralized and distributed optimization approaches for the integration of energy storage as a distributed energy resource, (4) improved grid simulation tools that combine the T&D networks to assess the impact of installation of resources like energy storage across the entire system, and (5) the development of energy/power management systems that enable the integration of massive distributed energy resources.

Any ESS is a complex system consisting of the energy storage technology, power conversion system, and grid interconnect. Materials R&D continues to evolve and significant strides in understanding the materials origins of battery failures have been made over several decades. These new materials can significantly expand the design space and new topologies can be developed beyond current system designs. By investing strategically into novel, low-cost, reliable, and safe technologies that are specific in their ability to address the utility grid needs and challenges, market penetration can be maximized. Newer technologies include redox flow, high temperature, and alkaline battery types. Power electronics are a universal piece of ESSs that have not received a commensurate amount of R&D focus, even as power electronics represents a significant cost within a storage system and contribute significantly to system efficiency. Safety is another universal concern that has variations in its risks depending on factors such as technology, deployment location, and use case, but that contribute to cost through reduced understanding of risk that must be compensated through other means. Further uncertainty in risks leads to lag times in approvals for systems that would otherwise move forward. Mitigating risks before some potential future incident(s) is important to enabling the current growth of ESS. Finally, integration issues must be approached in a more systematic manner to better understand and maximize the value of ESS.

## ACKNOWLEDGMENTS

This work was funded by the U.S. Department of Energy Office of Electricity Energy Storage Program. The authors thank Susan Schoenung, Tu Nguyen, and David Copp for critical review and edits of the manuscript. Sandia National Laboratories is a multimission laboratory managed and operated by National Technology and Engineering Solutions of Sandia, LLC, a wholly owned subsidiary of Honeywell International, Inc., for the U.S. Department of Energy's National Nuclear Security Administration under contract DE-NA-0003525.

## REFERENCES

1. G. Constable and B. Somerville (eds.), *A Century of Innovation: Twenty Engineering Achievements That Transformed Our Lives, National Academy of Engineering*, Joseph Henry Press, Washington, DC, 2003.

2. DOE/EPRI Electricity Storage Handbook, 2013.

3. J. Makansi, *An Investor's Guide to the Electricity Economy*, John Wiley & Sons, New York, NY, 2002.

4. Key World Energy Statistics, International Energy Agency (IEA), 2016.

5. U.S. Energy Information Administration (EIA) Energy Outlook, 2016.

6. Annual Energy Outlook 2016 with Projections to 2040, U.S. Energy Information Administration, August 2016. Report: DOE/EIA-0383(2016).

7. D. Stenclik, P. Denholm, and B. Chalamala, "Maintaining Balance: The Increasing Role of Energy Storage for Renewable Integration," *IEEE Power Energy Mag.* **15**(6):31–39 (November to December 2017). doi: 10.1109/MPE.2017.2729098.

8. U.S. Department of Energy, Global Energy Storage Database at: http://www.energystorageexchange.org/ (accessed 2018).

9. International Energy Agency, Country Forecasts for Grid-Tied Energy Storage, 2015.

10. Grid 2030: A National Vision for Electricity's Second 100 Years, United States Department of Energy, Office of Electric Transmission and Distribution, July 2003.

11. K. H. LaCommare and J. Eto, "Understanding the Cost of Power Interruptions to U.S. Electricity Consumers," Lawrence Berkeley National Laboratory, 2004. https://escholarship.org/uc/item/1fv4c2fv.

12. U.S. Department of Energy, Quadrennial Technology Review: An Assessment of Energy Technologies and Research Opportunities, U.S. Department of Energy, Washington, DC, September 2015.

13. New York Independent System Operator (NYISO), 2017 Load & Capacity Data Report Gold Book, April 2017, NYISO, Albany, NY.

14. California Independent System Operator, Folsom, CA. http://www.caiso.com/market/Pages/ReportsBulletins/Default.aspx.

15. North American Electric Reliability Corporation (NERC), Atlanta, GA, 2017 Summer Reliability Assessment Report. Available at: http://www.nerc.com/pa/RAPA/ra/.

16. U.S. Department of Energy, Grid Energy Storage, December 2013.

17. R. D. Masiello, B. Roberts, and T. Sloan, "Business Models for Deploying and Operating Energy Storage and Risk Mitigation Aspects," *Proc. IEEE* **102**(7):1052–1064 (July 2014). doi: 10.1109/JPROC.2014.2326810.

18. Y. Zhang, V. Gevorgian, C. Wang, X. Lei, E. Chou, R. Yang, Q. Li et al., "Grid-Level Application of Electrical Energy Storage: Example Use Cases in the United States and China," *IEEE Power Energy Mag.* **15**(5):51–58 (2017).

19. B. Dunn, H. Kamath, and J.-M. Tarascon, "Electrical Energy Storage for the Grid: A Battery of Choices," *Science* **334**:928–935 (2011).

20. B. Roberts and J. McDowall, "Commercial Successes in Power Storage," *IEEE Power Energy Mag.* **3**(2):24–30 (2005). doi: 10.1109/MPAE.2005.1405867.

21. B. R. Chalamala, T. Soundappan, G. R. Fisher, M. R. Anstey, V. V. Viswanathan, and M. L. Perry, "Redox Flow Batteries: An Engineering Perspective," *Proc. IEEE* **102**(6):976–999 (June 2014). doi: 10.1109/JPROC.2014.2320317.

22. H, Ibrahim, A. Ilinca, and J. Perron, "Energy Storage Systems—Characteristics and Comparisons," *Renew. Sust. Energy Rev.* **12**(5):1221–125 (2005).

23. B. Zakeri and S. Syri, "Electrical Energy Storage Systems: A Comparative Life Cycle Cost Analysis," *Ren. Sust. Energy Rev.* **42**:569–596 (2015).

24. E. Telaretti and L. Dusonchet, "Stationary Battery Technologies in the U.S.: Development Trends and Prospects," *Ren. Sust. Energy Rev.* **75**:380–392 (2017).

25. J. Araiza, J. Hambrick, J. Moon, M. Starke, and C. Vartanian, "Grid Energy-Storage Projects: Engineers Building and Using Knowledge in Emerging Projects," *IEEE Electrification Magazine* **6**(3):14–19 (2018).

26. H. Chen, T. N. Cong, W. Yang, C. Tan, Y. Li, and Y. Ding, "Progress in Electrical Energy Storage System: A Critical Review," *Prog. Nat. Sci.* **19**(3):291–312 (2009).

27. X. Luo, J. Wang, M. Dooner, and J. Clarke, "Overview of Current Development in Electrical Energy Storage Technologies and the Application Potential in Power System Operation," *Appl. Energy* **137**:511–536 (2015).

28. K. Wu, Y. Zhang, Y. Zeng, and J. Yang, "Safety Performance of Lithium-Ion Battery," *Prog. Chem.* **23**:401–409 (2011).

29. A. Eddahech, O. Briat, and J.-M. Vinassa, "Performance Comparison of Four Lithium–Ion Battery Technologies under Calendar Aging," *Energy* **84**:542–550 (2015).

30. A. Barré, B. Deguilhem, S. Grolleau, M. Gérard, F. Suard, and D. Riu, "A Review on Lithium-Ion Battery Ageing Mechanisms and Estimations for Automotive Applications," *J. Power Sourc.* **241**:680–689 (2013).

31. S. Y. Chen, Z. Wang, Z. Hailei, and L. Chen, "Safety-Enhancing Additives for Lithium Ion Batteries," *Prog. Chem.* **21**(4):629–636 (2009).

32. H. M. Barkholtz, A. Fresquez, B. R. Chalamala, and S. R. Ferreira, "A Database for Comparative Electrochemical Performance of Commercial 18650-Format Lithium-Ion Cells," *J. Electrochem. Soc.* **164**(12):A2697–A2706, 2017.

33. B. Scrosati, J. Hassoun, and Y.-K. Sun, "Lithium-Ion Batteries: A Look into the Future," *Energy Environ. Sci.* **4**(9):3287–3295 (2011).

34. California Energy Commission, Tracking Progress in Energy Storage, November 2017. http://www.energy.ca.gov/renewables/tracking_progress/documents/energy_storage.pdf.

35. R. H. Byrne, S. Hamilton, D. R. Borneo, T. Olinsky-Paul, and I. Gyuk, "The Value Proposition for Energy Storage at the Sterling Municipal Light Department," 2017 IEEE Power & Energy Society General Meeting, Chicago, IL, 2017, 1–5. doi: 10.1109/PESGM.2017.8274631.

36. L. Lam, N. Haigh, C. Phyland, and A. Urban, "Failure Mode of Valve-Regulated Lead-Acid Batteries under High-Rate Partial-State-of-Charge Operation," *J. Power Sources* **133**:126–134 (2004). https://doi.org/10.1016/j.jpowsour.2003.11.048.

37. G. L. Soloveichik, "Battery Technologies for Large-Scale Stationary Energy Storage," *Ann. Rev. Chem. Biomol. Eng.* **2**(1):503–527 (2011).

38. B. Zakeri and S. Syri, "Electrical Energy Storage Systems: A Comparative Life Cycle Cost Analysis," *Renew. Sust. Energy Rev.* **42:**569–596 (2015).

39. C. J. Rydh, "Environmental Assessment of Vanadium Redox and Lead-Acid Batteries for Stationary Energy Storage," *J. Power Sources* **80**(1):21–29 (1999).

40. L. Unterreiner, V. Jülch, and S. Reith, "Recycling of Battery Technologies–Ecological Impact Analysis Using Life Cycle Assessment (LCA)," *Energy Procedia* **99:**229–234 (2016).

41. M. Perrin, Y. M. Saint-Drenan, F. Mattera, and P. Malbranche, "Lead–Acid Batteries in Stationary Applications: Competitors and New Markets for Large Penetration of Renewable Energies," *J. Power Sources* **44**(2):402–410 (2005).

42. M. Shiomi, T. Funato, K. Nakamura, and T. Takahashi, "Effects of Carbon in Negative Plates on Cycle-Life Performance of Valve-Regulated Lead/Acid Batteries," *J. Power Sources* **64**(1–2):147–152 (1997).

43. B. B. McKeon, J. Furukawa, and S. Fenstermacher, "Advanced Lead–Acid Batteries and the Development of Grid-Scale Energy Storage Systems," *Proc. IEEE* **102**(6):951–963 (June 2014). doi: 10.1109/JPROC.2014.2316823.

44. B. R. Chalamala, T. Soundappan, G. R. Fisher, M. R. Anstey, V. V. Viswanathan, and M. L. Perry, "Redox Flow Batteries: An Engineering Perspective," *Proc. IEEE* **102**(6):976–999 (June 2014). doi: 10.1109/JPROC.2014.2320317.

45. Z. Yang, J. Zhang, M. C. W. Kintner-Meyer, X. Lu, D. Choi, J. P. Lemmon, and J. Liu, "Electrochemical Energy Storage for Green Grid," *Chem. Rev.* **111**(5):3577–3613 (2011). doi: 10.1021/cr100290v.

46. J. L. Sudworth, "The Sodium/Nickel Chloride (ZEBRA) Battery," *J. Power Sources* **100:**149–163 (2001).

47. L. J. Small, A. Eccleston, J. Lamb, A. C. Read, M. Robins, T. Meaders, D. Ingersoll, et al., "Next Generation Molten NaI Batteries for Grid Scale Energy Storage," *J. Power Sources* **360**(6):569–574, 2017.

48. N. D. Ingale, J. W. Gallaway, M. Nyce, A. Couzis, and S. Banerjee, "Rechargeability and Economic Aspects of Alkaline Zinc-Manganese Dioxide Cells for Electrical Storage and Load Leveling," *J. Power Sources* **276:**7–18 (2015).

49. G. G. Yadav, J. W. Gallaway, D. E. Turney, M. Nyce, J. Huang, X. Wei, and S. Banerjee, "Regenerable Cu-Intercalated $MnO_2$ Layered Cathode for Highly Cyclable Energy Dense Batteries," *Nat. Commun.* **8:**1–9 (2017). doi:10.1038/ncomms14424.

50. F. R. McLarnon and E. J. Cairns, "The Secondary Alkaline Zinc Electrode," *J. Electrochem. Soc.* **138**(2):645–656 (1991).

51. J. Ellison, D. Bhatnagar, and B. Karlson, "Maui Energy Storage Study," Sandia National Laboratories, SAND2012-10314, Albuquerque, NM 87185, 2012.

52. ERCOT, "2016 ERCOT Hourly Load Data," December 2017. (Online.) Available at: www.ercot.com (accessed December 1, 2017).

53. Editors of Power Engineering, "GE, Southern California Edison Reveal World's First Battery-Gas Turbine Hybrid," Power Engineering, April 17, 2017.

54. General Electric, "LM6000 Hybrid EGT," General Electric, 2017. (Online.) Available: https://www.gepower.com/services /gas-turbines/upgrades/hybrid-egtv (accessed December 1, 2017).

55. R. Pletka, J. Khangura, A. Rawlins, E. Waldren, and D. Wilson, *Capital Costs for Transmission and Substations, Updated Recommendations for WECC Transmission Expansion Planning*, Western Electricity Coordinating Council (WECC), Salt Lake City, UT, 2014.

56. U.S. Federal Energy Regulatory Commission (FERC), "Final Rule Order No. 755: Frequency Regulation Compensation in the Organized Wholesale Power Markets," U.S. Federal Energy Regulatory Commission (FERC), Washington, DC, 2011.

57. P. Kundur, *Power System Stability and Control*, McGraw-Hill, Inc., New York, NY, 1994.

58. D. Trudnowski and J. Pierre, "Signal Processing Methods for Estimating Small-Signal Dynamic Properties from Measured Responses," in *Inter-Area Oscillations in Power Systems*, Springer, New York, New York, 2009.

59. North American Electric Reliability Corporation (NERC), *1996 System Disturbances: Review of Selected 1996 Disturbances in North America*, North American Electric Reliability Corporation (NERC), Princeton, NJ, 2002.

60. D. J. Trudnowski, D. Kosterev, and J. Undrill, "PDCI Damping Control Analysis for the Western North American Power System," in IEEE Power and Energy Society General Meeting, Vancouver, BC, 2013.

61. D. A. Schoenwald, B. J. Pierre, F. Wilches-Bernal, and D. J. Trudnowski, "Design and Implementation of a Wide-Area Damping Controller Using High Voltage DC Modulation and Synchrophasor," *IFAC-PapersOnLine* **50**(1):67–72 (2017).

62. B. J. Pierre, F. Wilches-Bernal, D. A. Schoenwald, R. T. Elliott, J. C. Neely, R. H. Byrne, and D. J. Trudnowski, "Open-Loop Testing Results for the Pacific DC Intertie Wide Area Damping Controller," IEEE Manchester PowerTech, Manchester, England, 2017.

63. G. Delille, B. François, and G. Malarange, "Dynamic Frequency Control Support by Energy Storage to Reduce the Impact of Wind and Solar Generation on Isolated Power System's Inertia," *IEEE Trans. Sust. Energy* **3**(4):931–939 (2012).

64. T. Van Cutsem and C. Vournas, *Voltage Stability of Electric Power Systems*, Springer-Verlag, 1998.

65. J. A. Diaz de Leon II, and C. W. Taylor, "Understanding and Solving Short-Term Voltage Stability Problems," IEEE Power Engineering Society Summer Meeting, Chicago, IL, 2002.

66. R. W. Cummings, "Energy Storage and Reliability," in *Energy Storage Workshop Southwest Public Utility Regulatory Commissioners*, Albuquerque, NM, 2016.

67. Electric Power Research Institute. OpenDSS. Available at: http://smartgrid.epri.com/SimulationTool.aspx.

68. http://microgridmedia.com/kepco-installs-worlds-largest-frequency-regulation-bess/.

69. S.A. "Role of WBG Power Electronics and Power Conversion Systems in Grid-Tied Energy Storage," Sandia National Laboratories, Report # SAND2016-11282C, 2016.

70. Energy Storage for the Electricity Grid: Benefits and Market Potential Assessment Guide A Study for the DOE Energy Storage Systems Program, p. 6, Jim Eyer Garth Corey. http://www.sandia.gov/ess/publications/SAND2010-0815.pdf.

71. http://www.sandia.gov/ess/publications/SAND2016-3078R.pdf.

72. According to IEEE 1547.

73. http://www.sandia.gov/ess/publications/SAND2016-3078R.pdf.

74. Energy Storage Safety Strategic Plan, U.S. Department of Energy Office of Electricity Delivery and Energy Reliability, December 2014. http://www.sandia.gov/ess/publication/doe-office-of-electricity-doe-publications-2/.

75. P. Cole and D. R. Conover, Energy Storage System Guide for Compliance with Safety Codes and Standards, Pacific Northwest National Laboratory and Sandia National Laboratory, Richland, WA, June 2016.

76. D. Conover, Inventory of Safety-Related Codes and Standards for Energy Storage Systems with Some Experiences Related to Approval and Acceptance, Pacific Northwest National Laboratory, 2014.

77. R. Schubert, "Code Compliance for Stationary Battery Systems," in IEEE Power and Energy Society: Energy Storage and Stationary Battery Committee, Energy Storage Tutorial, June 2017. Available at: IEEE PES Resource Center (resourcecenter.ieee-pes.org; sites.ieee.org/pes-essb/files/2017/07/2017-SM-Energy_Storage_Tutorial.pdf).

78. D. Conover, Overview of Development and Deployment of Codes, Standards and Regulations Affecting Energy Storage System Safety in the United States, Pacific Northwest National Laboratory, 2014.

79. S. Whittingham, "The Role of the Materials Scientist in Battery Safety," *MRS Bull.* **42**(6):413–413 (2017).

80. C. Mikolajczak, M. Kahn, K. White, and R. T. Long, *Lithium-Ion Batteries Hazard and Use Assessment*. Springer Science & Business Media, 2012.

81. D. Hill, N. Warner, W. Kovacs III, B. Reichborn-Kjennerud, *Considerations for ESS Fire Safety: Consolidated Edison and NYSERDA New York, NY*. 2017, DNV-GL: Dublin, OH. p. 97.

82. M. G. Molina, "Energy Storage and Power Electronics Technologies: A Strong Combination to Empower the Transformation to the Smart Grid," *Proc. IEEE* **105**(11):2191–2219 (November 2017). doi: 10.1109/JPROC.2017.2702627.

83. X. Hu, C. Zou, C. Zhang, and Y. Li, "Technological Developments in Batteries: A Survey of Principal Roles, Types, and Management Needs," *IEEE Power Energy Mag.* **15**(5):20–31 (September–October 2017). doi: 10.1109/MPE.2017.2708812.

84. J. R. Miller and A. F. Burke, "Electrochemical Capacitors: Challenges and Opportunities for Real-World Applications," *Electrochem. Soc. Interface* **17**(1):53 (2008).

85. Z. An, L. Jia, Y. Ding, C. Dang, and X. Li, "A Review on Lithium-Ion Power Battery Thermal Management Technologies and Thermal Safety," *J. Therm. Sci.* **26**(5):391–412 (October 2017).

86. T. M. Bandhauer, S. Garimella, and T. F. Fuller, "A Critical Review of Thermal Issues in Lithium-Ion Batteries," *J. Electrochem. Soc.* **158**(3):R1–R25 (2011).

87. P. T. Coman, E. C. Darcy, C. T. Veje, and R. E. White, "Numerical Analysis of Heat Propagation in a Battery Pack Using a Novel Technology for Triggering Thermal Runaway," *Appl. Energy* **203**:189–200 (2017).

88 J. Lamb, C. J. Orendorff, L. M. Steele, and S.W. Spangler, "Failure Propagation in Multi-Cell Lithium Ion Batteries," *J. Power Sources* **283**:517–523 (2015).

89. D. Rosewater and A. Williams, "Analyzing System Safety in Lithium-Ion Grid Energy Storage," *J. Power Sources* **300**:460–471 (2015).

90. D. H. Doughty and C. C. Crafts, FreedomCAR Electrical Energy Storage System Abuse Test Manual for Electric and Hybrid Electric Vehicle Applications. 2006, Sandia National Laboratories, Albuquerque, NM.

91. J. Lamb and C. J. Orendorff, "Evaluation of Mechanical Abuse Techniques in Lithium Ion Batteries," *J. Power Sources* **247:**189–196 (2014).

92. https://www.energystorageexchange.org/projects/71 (accessed December 2017); http://www.peninsuladailynews.com/news/wind-power-battery-reignites-at-port-angeles-landing-mall/ (accessed December 2017).

93. https://www.greentechmedia.com/articles/read/battery-room-fire-at-kahuku-wind-energy-storage-farm (accessed December 2017).

94. https://www.greentechmedia.com/articles/read/Exploding-Sodium-Sulfur-Batteries-From-NGK-Energy-Storage (accessed December 2017); https://www.energystorageexchange.org/projects/119 (accessed December 2017).

95. https://www.energystorageexchange.org/projects/1445 (accessed December 2017); http://www.energy.ca.gov/research/notices/2005-02-24_workshop/06%20Farber-deAnda022405.pdf (accessed December 2017).

# CHAPTER 28
# IMPLANTABLE MEDICAL CELLS

**Steven M. Davis, Christopher R. Feger, Timothy R. Marshall,
Michael J. Root, Thomas F. Strange
(Emeritus Contributors: Randolph A. Leising, Nancy R. Gleason,
Barry C. Muffoletto, Curtis F. Holmes)**

## 28.1  APPLICATIONS AND REQUIREMENTS FOR IMPLANTABLE MEDICAL DEVICE BATTERIES

Batteries for medical applications have unique sets of requirements that are as varied as the many devices used to treat a wide range of medical conditions. Over the past 50 years considerable research and development has been devoted to the materials, chemistry, design, and manufacturing of batteries for these specialized applications. In this section, selected applications and general requirements for the batteries that power implantable medical devices are described.

### 28.1.1  Implantable Medical Device Batteries

The number of devices used to treat various medical conditions is expanding rapidly every year. Examples of battery-powered implantable medical devices are presented in this section, and summarized in Table 28.1 along with some of the medical conditions treated or monitored by these devices. A wide variety of cell chemistries have been used in battery-powered implantable medical devices. A list of currently used battery chemistries is summarized in Table 28.2.

A battery chemistry selected for an implantable device must exhibit a predictable voltage and internal resistance profile throughout its service life. The battery is usually discharged at a lower device background load as well as at intermittent, and often variable, higher current levels for therapy, wireless communication, and monitoring as required by the specific application. Implantable device batteries typically must also survive multiple years of continuous exposure at or near body temperature, dependent on the implant site. The appropriate battery chemistry and design is determined by the total energy, power demands, size requirements, and form factor of the device, as well as by the desired time over which therapy is to be provided.

*Implantable Cardiac Pacemakers.*  In Stockholm, Sweden, on October 8, 1958, heart surgeon Dr. Ake Senning teamed with electrical engineer Rune Elmquist to implant a cardiac pacemaker in Arne Larsson, which marked the first application of an implantable battery-powered medical device for human use. The device lasted only a few hours while its replacement lasted just a few weeks.[1] At the same time in the United States, Wilson Greatbatch was developing his own implantable cardiac pacemaker. The Chardack-Greatbatch pacemakers were implanted in animals in 1958 and in 10 patients in 1960. These pacemakers had greater longevity, but still suffered from poor performance that was commonly attributed to the mercury-zinc batteries that were used. However, it has been suggested that there may have been other causes for the device failures.[2] While some

**TABLE 28.1**    Examples of Battery-Powered Implantable Medical Devices

| Condition | Device | Typical battery discharge rate |
|---|---|---|
| Bradycardia | Pacemaker | Low |
| Tachycardia | Implantable cardioverter defibrillator (ICD) Subcutaneous implantable cardioverter defibrillator (S-ICD) | Low, medium, and high |
| Heart failure | Cardiac resynchronization therapy defibrillator (CRT-D) Cardiac resynchronization therapy pacemaker (CRT-P) | Low, medium, and high |
| Syncope | Implantable cardiac monitor | Medium |
| End stage heart failure | Left ventricular assist devices (LVAD) Total artificial heart (TAH) | High |
| Chronic pain | Neurostimulator | Medium |
| Epilepsy | Neurostimulator | Medium |
| Hearing loss | Neurostimulator Electromechanical | Medium |

**TABLE 28.2**    Battery Chemistries Currently Used in Implantable Devices

| Battery chemistry | Device application |
|---|---|
| Li/iodine | Pacemaker |
| Li/CF$_x$ | Pacemaker, CRT-P, neurostimulator, cardiac monitor, drug pumps |
| Li/SOCl$_2$ | Cardiac monitor |
| Li/SVO | Neurostimulator, ICD |
| Li/(SVO + CF$_x$) | Neurostimulator, pacemaker, ICD |
| Li/MnO$_2$ | Middle ear implant, pacemaker, CRT-P, ICD, S-ICD, CRT-D |
| Li-ion | Neurostimulator |
| Li-polymer rechargeable | Spinal cord oscillatory field stimulator |

pacemakers circumvented short device longevities by using rechargeable nickel-cadmium batteries or nuclear power sources, Greatbatch went on to develop a more robust primary battery power source for the pacemaker, leading to the development of the lithium/iodine battery.[3–5] The first human use of an implantable cardiac pacemaker powered by a Li/I$_2$ battery occurred in 1972 and was one of the first successful commercial uses of primary lithium batteries. The success of this battery technology contributed to the growth of pacemakers as an important treatment option for cardiac patients with an estimated 1,000,000+ pacemakers implanted each year worldwide.[6]

Incidentally, Larsson (the first recipient of a pacemaker) lived another 43 years to the age of 86 after the first pacemaker implant. During that time, he received 26 implantable pacemakers.

A pacemaker is designed to treat bradycardia, a slow heartbeat, by stimulating heart tissue to contract using low energy electrical pulses. A pacemaker (Fig. 28.1) consists of a battery and the electrical circuits that generate the pacing pulses. The device also monitors the heart rate of the patient. A pacemaker, sensing an inadequate heart rate, will deliver electrical pulses that sustain a healthy pacing rate. Pacing leads, connected to the pulse generator, convey the low voltage (1 to 5 V) pulses to the heart. Pacing of a single heart chamber is accomplished using a single lead placed in either the right ventricle or the right atrium. Dual chamber pacing requires a pacing lead located in the right ventricle and another in the right atrium. The sizes of pacemakers today are typically in the 8 to 15 cm$^3$ range. The recent development of transcatheter pacemakers, also known as leadless pacemakers, has decreased the size to approximately 1 cm$^3$, as illustrated in Fig. 28.2, although with functions that are more limited compared to conventional pacemakers with transvenous leads. These devices are directly implanted into the right ventricle of the heart by means of a catheter inserted through the femoral vein.[7]

Battery longevity in pacemakers is determined by a number of factors, including the frequency and intensity of pacing required by the patient, the efficiency of the electrical circuits in the device, and the size and design of

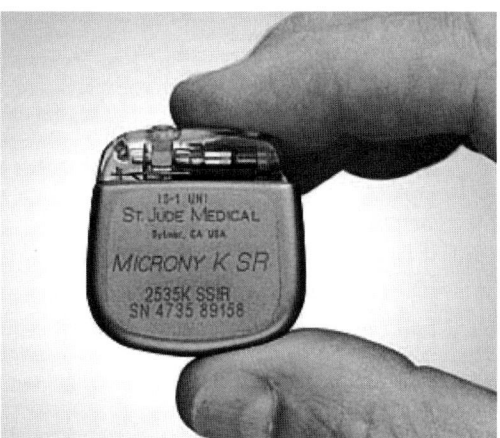

**FIGURE 28.1** An implantable cardiac pacemaker designed for pediatric patients. (*Courtesy of Abbott Laboratories.*)

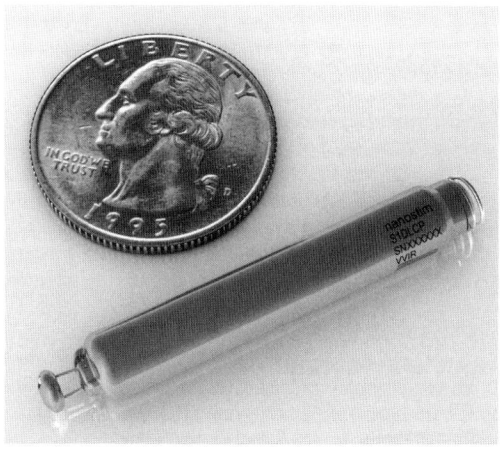

**FIGURE 28.2** Transcatheter (leadless) implantable cardiac pacemaker. (*Courtesy of Abbott Laboratories.*) (*Nanostim™ leadless pacemaker and St. Jude Medical are trademarks of St. Jude Medical, LLC, or its related companies. Reproduced with permission of St. Jude Medical, ©2017. All rights reserved.*)

the battery. Typical power consumption for a conventional cardiac pacemaker is on the order of 10 to 100 μW. Because of this relatively low power requirement, $Li/I_2$ batteries were well suited for the original pacemakers with leads. However, pacemakers now typically incorporate wireless telemetry features that greatly improve patient monitoring functions, although at the cost of increased power consumption of the device. In many cases, a battery capable of intermittently delivering several milliwatts of power is needed by these advanced devices (Table 28.2).

When the device is approaching the end of its service life, the patient and physician should be alerted sufficiently in advance such that surgery to replace the pacemaker can be scheduled. This may be accomplished by means of an auditory or vibratory alert originating from the device, an alert sent by wireless communication through a home monitoring station, or in the near future, a wireless transmission to the patient's mobile phone.

***Implantable Cardioverter-Defibrillators.*** The implantable cardioverter-defibrillator (ICD) was invented by Dr. Michel Mirowski in 1979. Mirowski witnessed repeated fainting spells and the eventual death of a colleague who suffered from ventricular tachycardia, a rapid heartbeat that can be in the 160 to 240 beats per minute range. After this experience, Mirowski collaborated with Dr. Martin Mower and Dr. Stephen Heilman to develop the first ICD; this first prototype was built using two Mallory zinc-mercuric oxide cells.[8] Mirowski took his idea for an implantable defibrillator from external defibrillators, first developed in the 1940s. These devices were successful in converting ventricular tachycardia into a normal heart rhythm.

Ventricular tachycardia is a life threatening rapid heart rhythm that can impede the heart from efficiently pumping blood and can lead to ventricular fibrillation. Ventricular fibrillation is a condition where the electrical activity of the heart becomes chaotic. The ventricles contract in a rapid, unsynchronized way, and the heart pumps little or no blood. An ICD detects these conditions and applies a high-voltage shock to the heart to temporarily stop the heart and interrupt the arrhythmia, after which the heart usually recovers and regains a normal rhythm. Additionally, ICDs function as pacemakers, providing the low current pulses required to correct slow heart rhythms (bradycardia). An image showing the main internal components of an ICD is shown in Fig. 28.3. The efficacy of ICDs has been demonstrated in a number of clinical studies. For example, in a comparison of over 1000 patients, ICDs were found to be superior to anti-arrhythmic drugs for increasing overall survival.[9]

Batteries are relatively low-voltage electrochemical power sources that alone are unable to meet the high voltage (600 to 800 V) requirements of the ICD to defibrillate the heart. Rather, the ICD batteries charge two to three high-voltage capacitors, typically aluminum or tantalum electrolytic capacitors, which subsequently

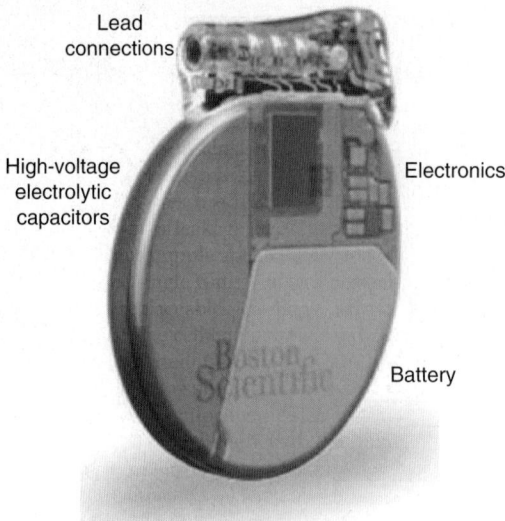

**FIGURE 28.3**   Implantable cardioverter-defibrillator showing the main internal components. (*Courtesy of Boston Scientific.*)

provide a high-voltage shock to the heart. The additional components needed for the ICD result in a larger device size compared to pacemakers. The first ICDs used in the mid-1980s were quite large, over 160 cm³ in volume.[10] The large size of the device required abdominal implantation. The leads that delivered the high-voltage shocks were attached to the outside of the heart. As the technology continued to improve, the size of the device became smaller, reaching about 60 cm³ in the mid-1990s and down to the 30 to 40 cm³ volume range for modern ICDs. Examples of contemporary ICDs are shown in Fig. 28.4. The volume of the larger ICD shown in Fig. 28.4 is 31 cm³ and the smaller ICD can be as small as 26.5 cm³, depending on the number of lead connections needed. The decrease in ICD sizes enabled pectoral implantation of the ICD, which is a more desirable position for device placement, due to a simpler implant procedure. Another advance was the development of endocardial leads that are positioned inside the right atrium and right ventricle of the heart.

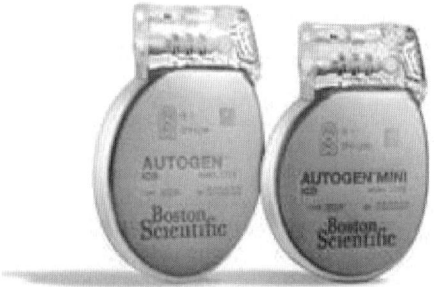

**FIGURE 28.4**   Implantable cardioverter defibrillators. (*Courtesy of Boston Scientific.*)

More recently, a subcutaneous implantable cardioverter defibrillator (S-ICD) was introduced (Fig. 28.5). Contrary to ICD and cardiac resynchronization therapy defibrillator (CRT-D) devices, the S-ICD electrode that delivers the high-voltage, high-energy shocks is implanted with no direct contact to the heart.

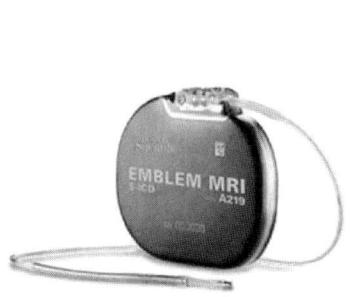
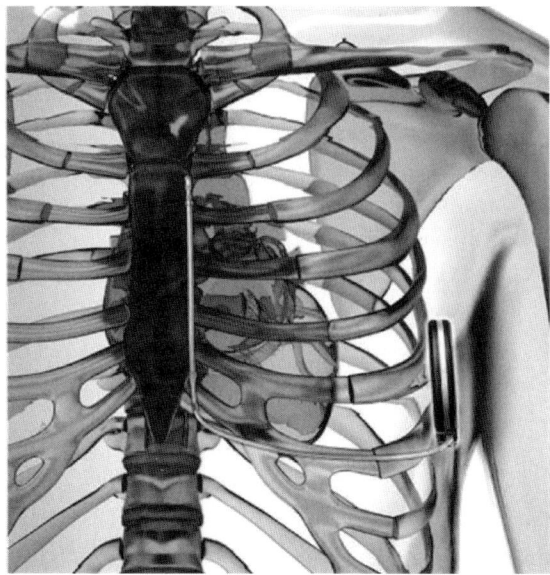

**FIGURE 28.5**    A subcutaneous ICD (S-ICD) showing the device and lead on the left, with typical anatomical placement on the right. (*Courtesy of Boston Scientific.*)

The battery chemistries of choice for ICDs are lithium/silver vanadium oxide (SVO), lithium/silver vanadium oxide-carbon monofluoride (SVO-CF$_x$), and lithium/manganese dioxide. The earliest ICDs used lithium/vanadium oxide batteries, but exhibited performance challenges and were quickly replaced by SVO batteries. Lithium batteries using a hybrid cathode combining silver vanadium oxide with carbon monofluoride now have largely supplanted SVO batteries. This hybrid battery system is divided into two types, mixed and layered. The mixed system consists of the SVO blended with CF$_x$ making a homogeneous cathode. The layered concept consists of discrete SVO and CF$_x$ layers that are consolidated to form a cathode. Other ICDs and the S-ICDs use Li/MnO$_2$ batteries. Each of these battery types meets the diverse power requirements for ICDs:

- Low power for monitoring and pacing functions
- Medium power for wireless communication (radio frequency telemetry)
- High pulse current on demand to rapidly charge the high-voltage capacitors

The cardiac pacing and monitoring functions of an ICD typically consume low power, normally between 10 and 100 μW. The pulse currents that charge the high-voltage capacitors, on the other hand, can be in the range of 2 to 4 A, with pulse durations generally lasting 5 to 15 s. While one pulse is often sufficient to restore the patient's normal heart rhythm, the device is programmed to provide several sequential pulses if required during an individual cardiac episode. In addition to wireless communication through a home monitoring station, some devices are equipped to provide an audible or vibratory alarm to alert the patient when the battery needs to be replaced in advance of complete battery depletion.[11]

***Implantable Cardiac Resynchronization Therapy Defibrillators.***    Contemporary ICDs, in addition to supplying defibrillation therapy, also function as pacemakers, including single and dual chamber pacing.[12] Pacing leads are placed in either the right atrium or right ventricle for single chamber pacing. A dual chamber defibrillator has a pacing lead in the right atrium and a pacing and defibrillation lead placed in the right ventricle. A three chamber heart failure pacing system utilizes yet another pacing lead that is placed in a cardiac vein outside the left ventricle.[12] In addition to providing the pacemaker and defibrillator functions of an ICD, implantable CRT-Ds utilize ventricular resynchronization to treat heart failure in certain patients. CRT-Ds are similar to

ICDs in shape and size. Less commonly implanted are cardiac resynchronization therapy pacemakers (CRT-Ps) that function as heart failure devices and pacemakers without defibrillation capability.

Heart failure is a condition where the heart cannot pump sufficient blood to the body's other organs. The weakened, often enlarged, heart muscle cannot pump all of the blood it receives, which causes pulmonary edema (fluid in the lungs). Heart failure can result from a number of factors[13]:

- Coronary artery disease
- Scar tissue from past heart attack
- High blood pressure
- Heart valve disease
- Disease of the heart muscle (cardiomyopathy)
- Congenital heart defects
- Infection of the heart valves and muscle

While the battery requirements for CRT devices are comparable to ICDs, providing high-power pulse currents for defibrillation, the lower power pacing requirements are significantly higher, up to five times greater, for the CRT versus a single or dual chamber pacing ICD or pacemaker. Thus, a CRT device may require pacing currents in the 50 to 100 µA range.

***Implantable Cardiac Monitors.***   Implantable cardiac monitors are devices that are implanted subcutaneously in the chest area. These devices continuously monitor the cardiac output of patients who have displayed syncope, or fainting spells, and record their electrocardiograms. Syncope results from a temporary reduction in blood flow producing a shortage of oxygen to the brain. However, there can be different causes for this condition (including many noncardiac causes) so continuous cardiac monitoring, especially during a syncopal episode, can be critical in determining a correct diagnosis of the condition.

These devices are small in size, historically on the order of 6 to 10 cm$^3$, so a small battery (originally Li/SOCl$_2$) is required for the device. New designs are typically 1 to 2 cm$^3$ and use Li/CF$_x$ batteries. A comparison of older and newer models is shown in Fig. 28.6. The monitoring function requires currents in the microampere range; telemetry requires the battery to produce milliampere level currents on an intermittent basis. The typical lifetime of the implanted device is 3 years.

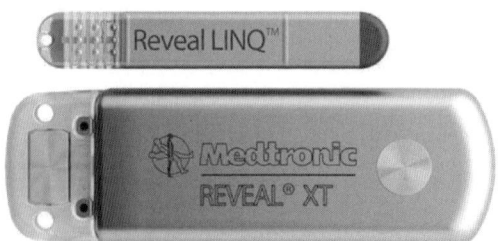

**FIGURE 28.6**   Implantable cardiac monitor, newer and older generation. (*Reproduced with permission of Medtronic, Inc.*)

***Neurostimulation or Neuromodulation Devices.***   Neurostimulators are implanted devices that provide electrical stimulation to treat a wide variety of neurological conditions, from movement disorders to bladder control to pain management. This field continues to expand and offers alternatives to surgery or drug therapy. Pain management by electrical spinal cord stimulation is accomplished by means of a lead placed in the spinal column that is connected to the device implanted in the lower back. Another neurostimulation application is deep brain stimulation wherein leads inserted into the brain can be used to treat various conditions, including epilepsy, Parkinson's disease, dystonia, essential tremors, and more. The deep brain neurostimulation device is implanted pectorally. Conditions that are potentially treated by neurostimulation are listed in Table 28.3.

**TABLE 28.3**    Examples of Potential Applications of Neurostimulation

| Stimulated area | Condition treated |
| --- | --- |
| Bone | Bone fractures |
| Cochlea | Severe to profound hearing loss |
| Deep brain | Parkinson's disease, essential tremor, dystonia, Alzheimer's disease, obsessive compulsive disorder, Tourette's syndrome |
| Gastric nerve | Obesity |
| Gracilus muscle | Fecal incontinence |
| Middle ear | Mild to severe sensorineural hearing loss |
| Occipital nerve | Chronic migraine headaches |
| Optic nerve | Blindness |
| Phrenic nerve | Diaphragm pacing to restore breathing |
| Sacral nerve | Incontinence |
| Spinal cord | Chronic pain |
| Stomach muscle | Gastroparesis |
| Vagus nerve | Epilepsy, depression, obesity |

Implantable neurostimulators are typically between the size of implantable pacemakers and defibrillators. Power requirements for neurostimulators are typically medium rate, with pulse currents greater than that of pacemakers, often approaching 1 mA or more. Devices can utilize either a primary or secondary battery, depending on size and overall energy required, as well as the patient's ability to recharge the device. The primary batteries used for this application include $Li/SOCl_2$, $Li/CF_x$, and combined $Li/SVO-CF_x$, while the secondary systems employ Li-ion technologies.

***Implantable Drug Pumps.***    Implantable drug pumps have been used to deliver therapeutic medications into the body, and often employed $Li/CF_x$ batteries due to their high capacity and low self-discharge. In some devices, pain medication can be administered directly into the problem area. The main issue with such devices has been the replenishment of the medication through a reservoir fill port. Additionally, the long-term stability of the medication can be affected by long-term storage at body temperature. External drug pumps are frequently used in place of these implantable devices.

***Implantable Hearing Assist Devices.***    A completely implantable hearing device has been approved by the Food and Drug Administration (FDA). This device is indicated for use with nerve-related (sensorineural) hearing loss, and operates by converting sound into electrical signals that are subsequently converted into mechanical vibrations that stimulate the inner ear (Fig. 28.7). Unlike traditional hearing aids that amplify sounds, this device provides a more natural sound to the patient. Due to the low current requirements for therapy, a $Li/I_2$ battery is utilized which provides typical device life of 4.5 to 9 years.[14,15]

## 28.1.2    Externally Powered Medical Device Batteries

In addition to batteries used to power implantable medical devices, there are many different primary and secondary batteries used to power externally worn or nonimplanted medical devices. Some devices are implanted, but the battery is external to the body. One major important difference with these systems is that the power source can be replaced without having to explant the medical device. As a result, the bulk of these devices utilize common chemistries, form factors, and sizes found in commercial batteries. As these devices do use available battery technologies, the advancement in these devices is on therapy rather than energy storage. External medical devices in use today include:

- Automated external defibrillators
- Hearing assist devices (hearing aids)
- External drug pumps

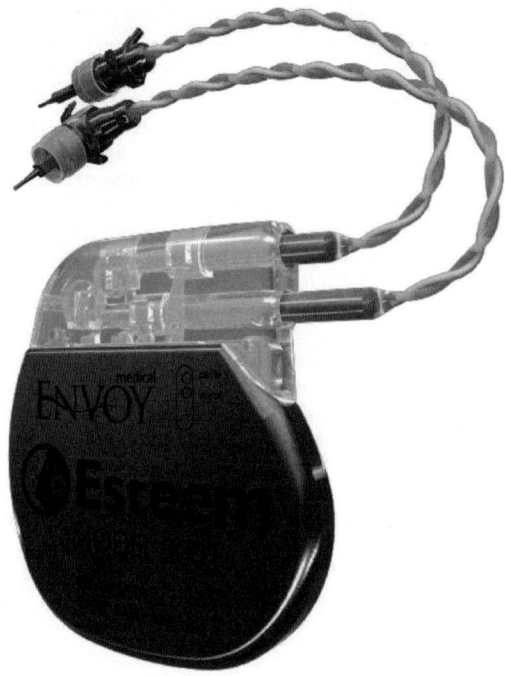

**FIGURE 28.7**   Sound processor with leads for implantable hearing assist device. (*Reproduced with permission of Envoy Medical Corporation—Esteem® model photograph is copyright protected property of Envoy Medical Corporation.*)

Implanted medical devices using external batteries include:

- Left ventricular assist devices (LVADs)
- Total artificial hearts (TAH)
- Hearing assist devices (cochlear implants)

## 28.2   SAFETY AND RELIABILITY CONCERNS FOR IMPLANTABLE MEDICAL DEVICE BATTERIES

Batteries for implantable devices are used in applications where the failure of the cell to perform to specification can often result in direct physical harm to the patient. Biomedical cells are almost always in close or direct contact with the patient, and must perform safely under all conditions. Even when cells perform as predicted, replacement of the device due to battery depletion presents surgical risk and financial cost.[16,17] For implantable cells, reliability and patient safety cannot be separated. Cells produced for implantable biomedical applications are regulated by a number of national and international bodies as hazardous materials, electrical components, and medical device components. In addition to safety testing required by regulatory agencies, cells typically undergo qualification and design assurance testing. Once in production, sampling followed by destructive and nondestructive testing is used to track and verify safe, consistent, and predictable performance.

## 28.2.1    Safety Standards

Historically, the most demanding biomedical applications have utilized primary lithium anode chemistries, with some newer applications utilizing secondary lithium-ion-based cells. Lithium-based systems are inherently power and energy dense, but are unsafe at temperatures near and above 180°C, the melting point of the highly reactive lithium metal component. Lithium-ion cells, which utilize a carbon anode containing lithium ions, have been the subject of multiple highly publicized safety issues, including the complete ban of a popular portable electronic device from civilian aviation carriers due to fire risk.[18,19] This risk was ultimately traced to two root causes at two different manufacturers. In the first case, the design exhibited a flaw that could produce a short circuit, whereas the second supplier had a fabrication issue that caused a welding defect. The two separate issues both resulted in short circuit conditions with a possibility of battery ignition.[20] Lithium-based cells have historically been considered hazardous goods and materials, but they present such a unique risk profile and are so commonly transported that separate regulations have been developed for transport via land, sea, and air. These safety regulations and standards are maintained throughout the world by a number of regulatory organizations as shown in Tables 28.4 and 28.5.

Perhaps the most commonly discussed safety standards for lithium cells are transportation standards. These include United Nations 38.3 standard for lithium cells, commonly referred to as UN/DOT 38.3, as well as IEC 62281.[21]

**TABLE 28.4**    Standards Organizations

| Abbreviation | Organization | Website |
|---|---|---|
| IATA | International Air Transport Association Dangerous Goods Regulations | www.iata.org |
| ICAO | International Civil Aviation Organization Technical Instructions | www.icao.int |
| IMDG | International Maritime Organization | www.imo.org |
| UN | United Nations | www.un.org |
| FDA | U.S. Food and Drug Administration | www.fda.gov |
| PMHSA | U.S. Pipeline and Hazardous Materials Safety Administration | www.phmsa.dot.gov |
| DOT | U.S. Department of Transportation | www.transportation.gov |

**TABLE 28.5**    Lithium Battery Safety Standards and Regulations

| Organization | Standard | Description | FDA consensus standard |
|---|---|---|---|
| IEC | 60086-4 | Primary batteries—Part 4: Safety of lithium batteries | Yes |
| UL | 1642 | Lithium batteries | Yes |
| IEC | 61960 | Secondary lithium cells and batteries for portable applications | No |
| IEC | 62281 | Safety of primary and secondary lithium cells and batteries during transport | No |
| IEC | 62133 ed. 2.0 | Secondary cells and batteries containing alkaline or other nonacid electrolytes—safety requirements for portable sealed secondary cells, and for batteries made from them, for use in portable applications | Yes |
| ANSI | C18.3M Part 2 | National standard for portable lithium primary cells and batteries—safety standard | No |
| UN | 38.3 | Transportation testing for lithium batteries | No |
| IEEE | 1679.1 | Characterization and evaluation of lithium | No |
| IATA | DGR, 54th ed. | Dangerous goods regulations | No |
| IACO | 9284 | Technical instruction for the safe transport of dangerous goods by air | No |
| PMHSA | 49 CFR 173.185 | Lithium cells and batteries | No |

These tests must be passed, even outside of more highly restrictive hazardous goods restrictions, in order to ship lithium-based cells to most locations. Different modes of transportation have different rules for the size and number of cells which can be transported, as well as the packaging requirements. It is important to note that these standards have not been developed specifically with biomedical applications in mind, but rather as general standards for lithium battery safety. All applicable local and international regulations must be consulted prior to shipping or utilizing lithium cells in any application.

In addition to transportation standards, several general standards have been developed for lithium-based cells. In the United States, the FDA currently recognizes some UL and IEC standards, known as "consensus standards," in order to "support a reasonable assurance of safety and/or effectiveness for many applicable aspects of medical devices."[22] As of August 2017, the United States Food and Drug Administration database lists 10 "consensus standards" regarding the keyword "batteries."[23] These standards include both tests for intended use as well as for foreseeable misuse or abuse. The FDA does acknowledge that at least some of these standards are not intended for and do not apply to implantable medical device batteries.[24,25] The FDA database is updated regularly and should be consulted for the exact consensus standard cited before use. Actual written standards can be downloaded and/or purchased from the standards agencies.

Lithium-based chemistries present several unique safety challenges. First among these is the fact that the molten lithium metal is highly reactive. Excursions beyond this temperature can result in immediate violent venting of the battery with fire. In December 2016, the U.S. FDA issued a letter to healthcare facility administrators warning of reports of explosion and fire from battery powered medical carts.[26] Incidents with rechargeable lithium battery-powered electronic cigarettes and vaporizers have also been reported.[27]

## 28.2.2    Reliability Systems

As previously stated, ensuring patient safety for a biomedical battery application requires more than normal safety standards. In the case of consumer electronics, safe early failure of the device typically causes only an inconvenience to the user. Failure of a life-sustaining biomedical device to perform as predicted can result in death or adverse health effects to the patient, including those incurred from premature device replacement. Prior work by Levy on the reliability of lithium anode based systems emphasized lot analysis, individual cell reliability, and postmortem analysis.[28–31]

In 2013, the FDA added a focus on batteries and battery-powered devices to its "case for quality" program, which emphasizes those aspects of the device design and production process that are important to ensure device quality and ultimately patient safety.[32] A quality inspection pilot program for batteries was launched in 2013. The implementation report from this program was released in 2015, making specific recommendations for assuring battery reliability in three areas: corrective and preventative actions, design controls, and process controls.[33] The first area includes a recommendation to trace-suspected battery-related problems from known complaints, quality systems data, and design verification and validation to underlying causes, followed by implementation of appropriate corrective actions. Multiple recommendations for the design controls subsystem are also identified, including a recommendation to understand actual device requirements and verify reliability through prediction models, electrical stress analysis, and exploratory testing. Chemistries that can potentially experience high resistance at middle of life such as silver vanadium oxide must be designed properly to avoid this failure mechanism. Several failure modes specific to rechargeable lithium-ion chemistry are noted. Accelerated life testing is allowed to support reliability claims, but at least 3 years of battery data are required to support a longevity claim of 9 to 10 years.[33] Process control recommendations include maintaining appropriate environmental controls, avoiding manufacturing defects, validating all processes including all welds, and ensuring specification conformance through appropriate process controls, material controls, tool controls, and maintenance controls, both in house and at all suppliers.

In addition to the FDA, industry groups have also developed recommendations for successful use of battery-powered devices.[34]

## 28.2.3    Design Verification and Validation

Biomedical devices must pass qualification protocols based on the specific device application. Although exact protocols are specific to each manufacturer, Visbisky et al.[35] provided an early example of an ICD qualification protocol.

Over time, these protocols have expanded to include understanding of newly discovered failure modes. Well-established quality tools (such as failure modes and effects analysis [FMEA]; failure modes, effects, and criticality analysis [FMECA]; and fault tree analysis [FTA]) are used to identify areas where additional risk control measures or process improvements may be required. New manufacturing processes and any changes by suppliers must be fully reviewed and validated to ensure battery (and device) reliability.

*Destructive Testing.*    In the most extreme case, destructive physical analysis (DPA) indicates the careful disassembly of the cell for the purpose of analyzing individual components. These components are inspected visually, as well as via basic analytical techniques such as SEM/EDS (scanning electron microscopy/energy dispersive x-ray spectroscopy) and ICP-AES (inductively coupled plasma-atomic emission spectrometry). This approach is useful both for lot sampling and release purposes as well as for analysis of returned failed device battery components.

Life testing refers to electrochemical discharge under conditions similar to actual device usage, consuming all cell capacity available for therapy. These tests are also usually performed on a lot sample basis, and run concurrently with cells in the field. When originally implemented for $Li/I_2$ cells, the life test program was reported to serve three purposes. First, it provides a real time verification of accelerated battery performance projections. Second, it provides a measure of cell-to-cell variability and failure rates. Lastly, it should provide a warning of any unexpected cell behavior.[35]

*Nondestructive Testing.*    There are several nondestructive analysis techniques that can be used to estimate the reliability of a given battery system. All cells are typically subjected to some discharge and pulse protocol prior to use. This testing consumes a minimal amount of cell capacity, but can provide important reliability information. Open circuit voltage readings may reveal the presence of unwanted electroactive species, or the presence of an internal or external short circuit condition.[36] Pulse testing reveals information on the beginning of life internal resistance of the cell, which must be controlled in order to ensure proper device operation throughout discharge. Dimensional and weight measurements are also typically taken, as well as x-ray measurements. Leak testing equipment can be used to check that cells have remained hermetic following the production process. This is a specific requirement for biomedical cells, which are sealed into devices. Any leaking of cell materials into the device can damage circuitry and other components, compromising device performance.

The use of microcalorimetry in the analysis of battery self-discharge was first developed in the late 1970s.[37-39] These instruments are sensitive to the microwatt level and can help understand the self-discharge properties of battery chemistries and designs intended to be implanted over several years. Cell heat dissipation can be related to internal cell reactions and if the cell reactions are well understood, can be expressed in terms of the total capacity of the cell. This approach can also analyze failed batteries. The presence of a small internal short can be characterized by elevated heat dissipation.

Electrochemical impedance spectroscopy (EIS) can be used to effectively assess the condition of a cell in a nondestructive manner.[40] This technique is used frequently to identify the internal resistance contribution of the cathode, anode, separator, and electrolyte components.

## 28.3   CHARACTERISTICS OF BATTERIES FOR IMPLANTABLE BIOMEDICAL DEVICES

Implantable medical device batteries often use chemistries that may be available commercially in other form factors, such as coin and cylindrical cells. However, implantable medical devices often require battery designs, features, and manufacturing processes that are specifically devised for just that single application. Among the key characteristics that implantable medical device battery designers and developers must consider are the following:

- All power requirements of the target application
- High level of safety and reliability
- Predictable discharge voltage and internal resistance throughout discharge
- Low self-discharge rate
- Hermetic seals

Selected battery design features that support the key characteristics listed above are described in the following sections.

### 28.3.1  Power Output

The power requirements for implantable battery-powered medical devices are often divided into three broad groups that are defined by the maximum power output that may be required by the device, either continuously or intermittently:

- Continuous or frequent low rate—tens to hundreds of microwatts
- Intermittent medium rate—milliwatts
- Infrequent high rate—one to ten watts

Low-rate discharge of implantable batteries commonly supports device operations that may be frequent, for example, cardiac sensing and pacing or neuromodulation therapy, or continuous, such as maintaining power to the electronic circuits. Radio frequency telemetry is a function that requires medium rate power from the battery. This is a feature that is now commonplace in implantable cardiac pacemakers and defibrillators. High-rate discharge may be relatively short and infrequent as for cardiac defibrillation (5–15 s pulses every few months to years), or continuous as with LVAD and TAH systems.

For devices like implantable defibrillators that require high-power pulses, battery designs may include thin anodes and cathodes with high superficial electrode areas to reduce the internal resistance.[41,42]

### 28.3.2  Chemistry

In addition to the mechanical design of the battery, selection of an appropriate cell chemistry plays a key role in the power output capability of a battery. Many of the basic primary and secondary battery chemistries used in implantable medical devices are rather similar to those used in consumer, military, and OEM batteries, including $Li/MnO_2$, $Li/CF_x$, and lithium ion (see Part IV, Chaps. 11 to 22). These chemistries may be integrated into custom-designed batteries that incorporate other attributes that are essential for use in implantable medical devices, such as high-rate electrode configurations (see Sec. 28.3.1) and hermetic enclosures.

However, there are a few cell chemistries that have been used nearly exclusively in various medical devices. Specialized cell chemistries first used in implantable cardiac rhythm devices (pacemakers and defibrillators) are described in the following sections.

***Lithium/Iodine.***    Lithium/iodine ($Li/I_2$) is a low-rate battery that was developed for implantable cardiac pacemakers. The first pacemakers used $Zn/HgO$ batteries, but these devices had relatively short service lives, about 3 to 5 years. Rechargeable NiCd batteries and even highly radioactive nuclear power sources were introduced to improve device longevity. Neither was widely used, though.

The first human implant of a pacemaker powered by a $Li/I_2$ occurred in 1972.[43] Since then, millions of these batteries have been implanted. Other pacemaker battery chemistries were also developed and used, such as lithium/silver chromate, lithium/copper sulfide, and lithium/thionyl chloride. These chemistries are no longer available for pacemakers and only $Li/I_2$ batteries are still produced for implantable pacemakers.[44]

The standard design for a $Li/I_2$ cell includes a central lithium anode that is corrugated to increase its superficial surface area (Fig. 28.8). The cell is filled with a mixture of $I_2$ and poly-2-vinylpyridine (PVP) that envelops the lithium anode and contacts the stainless steel case to form a case-positive cell. PVP is also coated onto the lithium anode as a thin film to reduce the internal resistance of the cell throughout discharge,[45] as illustrated in Fig. 28.9.

The discharge reaction of a $Li/I_2$ cell is

$$Li + \tfrac{1}{2}I_2 \rightarrow LiI$$

The kinetics of the reaction are improved through the use of charge-transfer complexes between $I_2$ and an organic donor like PVP. The first report of the use of an $I_2$ charge transfer complex was made by

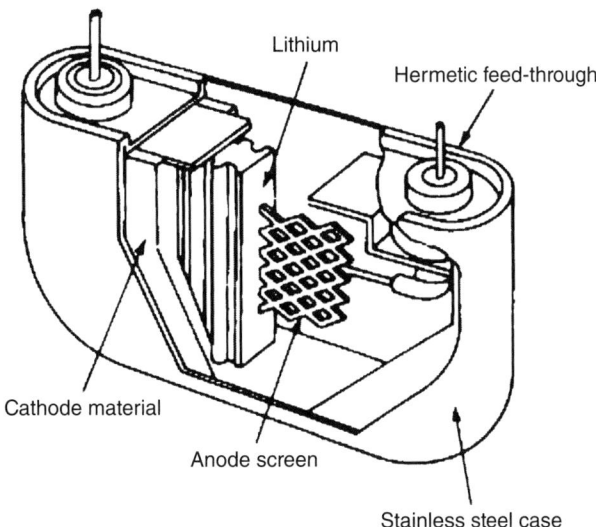

**FIGURE 28.8**    Cutaway view of typical case-positive Li/iodine cell. (*Courtesy of Greatbatch, Inc.*)

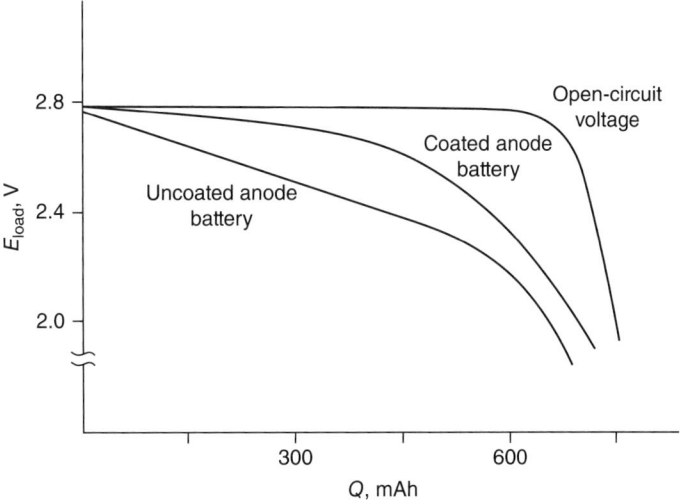

**FIGURE 28.9**    Loaded voltage versus discharge state for uncoated and PVP-coated anode Li/iodine batteries discharged at 6.7 μA/cm$^2$ at 37°C. (*Courtesy of Medtronic.*)

Gutman et al. in 1967.[3] This was followed by a 1972 patent that described a solid-state primary cell having a lithium anode, an iodine cathode, and a solid lithium halide electrolyte.[4] The thermal reaction between $I_2$ and PVP to form a $I_2$/PVP cathode material was disclosed in a 1973 patent,[46] and $I_2$/PVP charge transfer complex materials continue to be used in the manufacture of Li/$I_2$ pacemaker batteries today.

The reaction product, lithium iodide, forms in situ as the cell is discharged. LiI acts as both the separator and the solid electrolyte for the battery. The LiI layer that forms between the cathode and anode increases in thickness as the cell discharges causing a gradual increase in the internal impedance of the battery. The discharge voltage curve for a Li/I$_2$ cell along with the resistance curve is plotted as a function of discharged capacity in Fig. 28.10. The logarithm of the increase in resistance is linear with discharged capacity. This gradual increase in internal resistance can be used to predict the remaining service life of the battery.

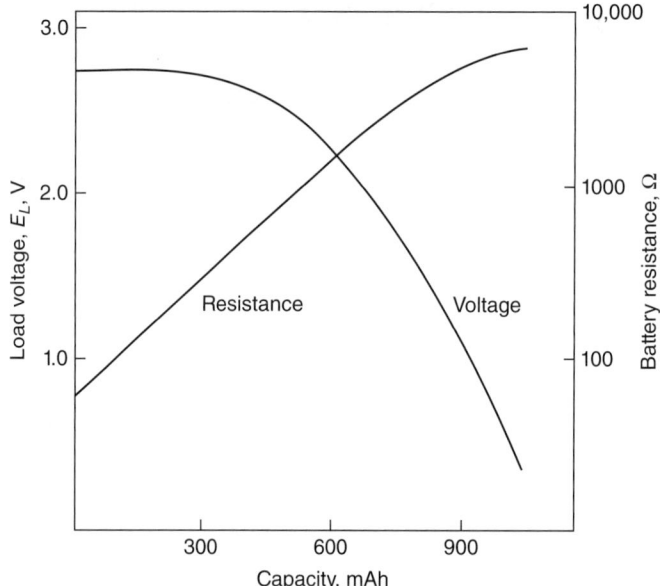

**FIGURE 28.10**   Loaded voltage and battery resistance for coated-anode Li/iodine battery discharged at 100 μA at 37°C. (*Courtesy of Medtronic.*)

The formation of the solid LiI electrolyte interface between I$_2$/PVP and the Li anode during discharge becomes a self-healing separator; an internal short between cathode material and anode generates LiI at the site of the short thus forming a layer of solid electrolyte separator thereby terminating the short circuit.

Self-discharge of the Li/I$_2$ battery system occurs by the direct reaction of Li with I$_2$ that has diffused through the LiI layer. The amount of I$_2$ that is able to diffuse through the solid electrolyte separator layer is dependent on the thickness of that separator film. Hence, self-discharge is greater early in the discharge reaction of the battery when the LiI layer is thinnest. A plot of heat dissipation data measured by microcalorimetery made at open circuit as a function of cell capacity is illustrated in Fig. 28.11. From this figure, it is evident that a large majority of the battery self-discharge process occurs during the first 25% of the battery discharge.

More recently, pacemakers have been introduced that can be programmed and communicated remotely using radio frequency telemetry. Li/I$_2$ batteries are unable to provide the medium power levels required for this function. As a result, battery designers incorporated other battery chemistries and designs with higher rate capabilities, including Li/CF$_x$, Li/MnO$_2$, and Li/SVO-CF$_x$ pacemaker batteries (see below).

***Lithium/Silver Vanadium Oxide.***   When ICDs were first implanted in the mid-1980s, they were powered by lithium/vanadium pentoxide (Li/V$_2$O$_5$) batteries.[47] However, device manufacturers sought to improve their performance with a different chemistry. Lithium/silver vanadium oxide (Li/Ag$_2$V$_4$O$_{11}$ or Li/SVO) batteries were originally developed for high-temperature commercial applications.[48] However, realization that Li/SVO could improve the performance of ICDs led to its adoption as the chemistry of choice for this type of device since its first use in 1987. Later, Li/SVO would also be implemented in implantable drug pumps.

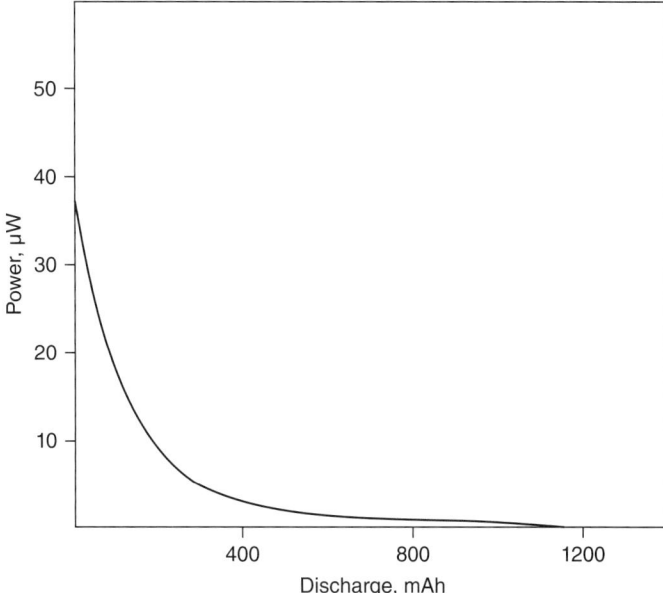

**FIGURE 28.11**    Power (heat) loss due to self-discharge versus discharge state for typical Li/iodine battery (calorimetric measurements made at open-circuit voltage). (*Courtesy of Medtronic.*)

The SVO ε-phase used in batteries comprises edge-shared distorted octahedra in $V_4O_{11}$ clusters, which then form corner-shared vanadium oxide strings in a layered structure.[49] Silver is located between the $V_4O_{11}$ layers. The cell reactions for Li/SVO discharge are as follows:

*Anode reaction:*          $Li \rightarrow Li^+ + e^-$

*Cathode reaction:*      $Ag_2V_4O_{11} + 7Li^+ + 7e^- \rightarrow Li_7Ag_2V_4O_{11}$

*Overall cell reaction:*    $7Li + Ag_2V_4O_{11} \rightarrow Li_7Ag_2V_4O_{11}$

The electrolyte is formulated using organic solvents (propylene carbonate and dimethoxyethane) to which a lithium salt such as $LiAsF_6$ has been added.

The Li/SVO discharge reactions have been studied in detail by a combination of physical, chemical, and electrochemical methods.[50–52] The first reduction step of SVO primarily forms silver metal ($Ag^+ + e^- \rightarrow Ag^0$) with $Li^+$ displacing the silver between the vanadium oxide layers.[53] The silver is extruded from the vanadium oxide structure.[54]

The silver metal product that forms greatly increases the conductivity of the cathode material that contributes to the high current-carrying capability of the Li/SVO system. Some V(V) is also reduced during the first discharge step.[55] As the discharge reaction continues, V(V) is reduced to V(IV) and V(III). These reactions occur in a stepwise manner that exhibit multiple voltage plateaus during discharge (Fig. 28.12). The stepped discharge voltage curve provides a rather predictable state-of-charge indication for the battery that can be used to estimate the remaining lifetime of the implantable device.

One of the characteristics of SVO discharge is an increase in the internal resistance in the middle of battery service life.[56–60] This behavior is innate to SVO and can present battery performance challenges. When the internal resistance of a defibrillator battery increases, delivery of life-saving defibrillation therapy is delayed.

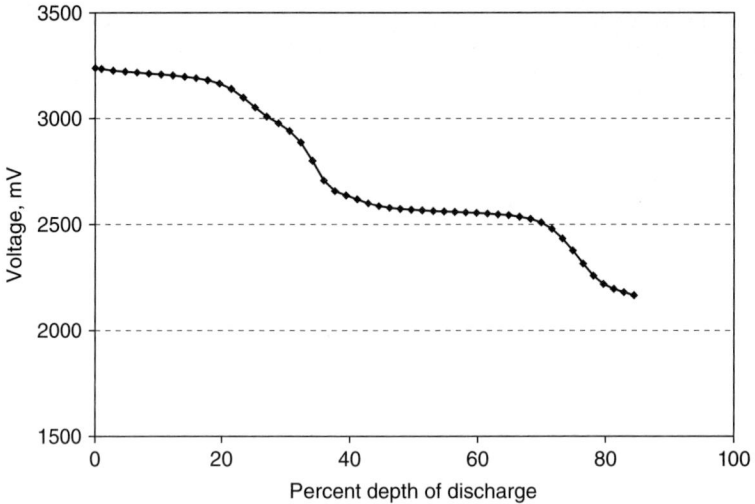

**FIGURE 28.12**   Discharge of a Li/SVO battery under a 100 kΩ constant resistance load at 37°C, with pulse loads applied every 30 days. The total test time for the discharge was 4 years. (*Courtesy of Greatbatch, Inc.*)

The increased resistance is manifest as a voltage delay observed during high current pulse discharge. The voltage delay was shown to result from film formation on the lithium anode.[56] Figure 28.13 displays the discharge of a Li/SVO battery under a 7-year test regime. Voltage delay (lower pulse minimum voltage at the beginning of the pulse) can be seen in the 40% to 50% depth of discharge (DOD) region of the discharge.

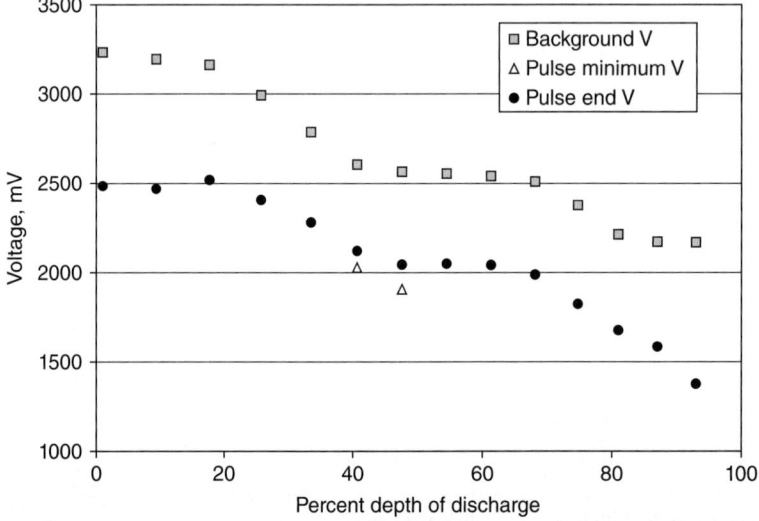

**FIGURE 28.13**   Discharge of a Li/SVO battery under a 7-year test regime at 37°C. The cell was discharged under a 80 kΩ constant resistance load, with pulse loads applied every 180 days. (*Courtesy of Greatbatch, Inc.*)

Under ordinary use conditions, the anode film may be removed and the internal resistance of the battery temporarily lowered by discharging the cell with high current pulses. If cells are not pulsed at an adequate level and duration, the film on the anode is not sufficiently removed and a permanent increase in cell resistance results. A number of studies have been done to explore the effectiveness of electrolyte additives on reducing voltage delay by modifying the film formed on the lithium anode. The addition of $CO_2$ or substances such as dibenzyl carbonate or benzyl succinimidyl carbonate have been reported to reduce anode passivation. These materials are believed to operate by lowering the impedance of the solid electrolyte interface (SEI) layer that naturally forms on the surface of the lithium anode.[56,57,61]

An example of a prismatic implantable Li/SVO cell designed for use in an ICD is shown in Fig. 28.14.

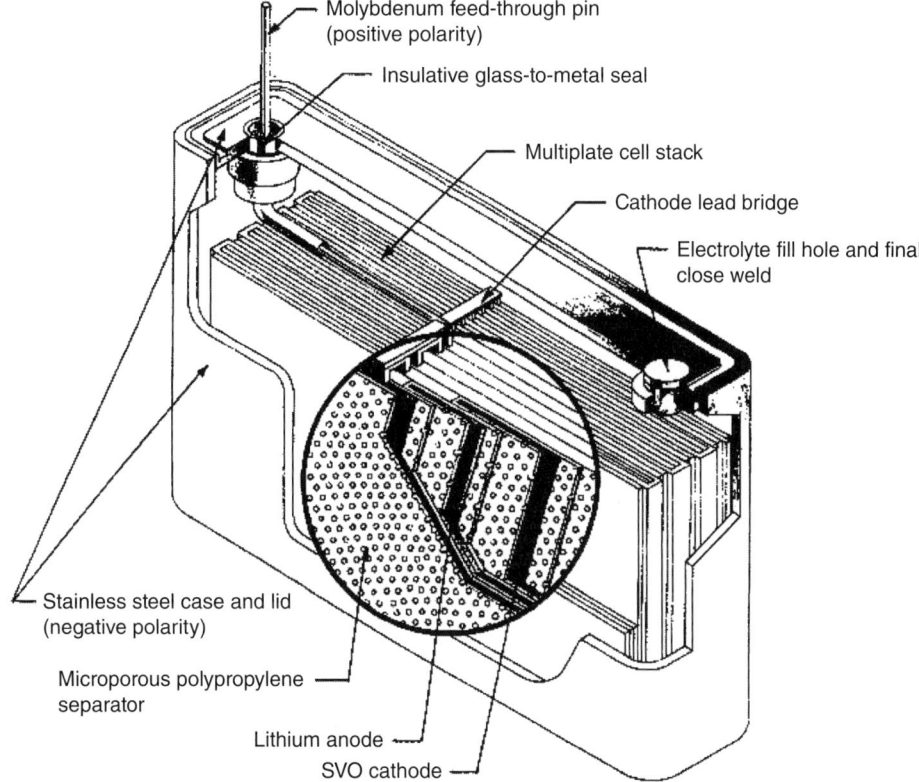

**FIGURE 28.14**  Construction of a Li/SVO cell designed for an ICD. (*Courtesy of Greatbatch, Inc.*)

***Lithium/Silver Vanadium Oxide-Carbon Monofluoride.***    More recently, other ICD and CRT-D battery chemistries have been implemented that overcome the performance drawbacks of Li/SVO related to the internal resistance increase during discharge. These chemistries are Li/MnO$_2$ and Li/SVO-CF$_x$ and they have largely replaced Li/SVO defibrillator batteries.[62]

One strategy for overcoming the internal resistance increase experienced by Li/SVO batteries is to replace SVO entirely with another cathode material like MnO$_2$. Li/MnO$_2$ batteries can provide the intermittent high power as well as the continuous and frequent low power load demands of implantable defibrillators. Further, the internal resistance Li/MnO$_2$ batteries does not increase in the middle of discharge as with Li/SVO batteries.[63]

Another strategy is to reduce the amount of SVO, a high-rate cathode material, by combining it with another material to eliminate the internal resistance increase during the middle of discharge. Carbon monofluoride was chosen for this role to yield a Li/SVO-CF$_x$ hybrid cathode battery. This approach takes advantage of both cathode materials by combining the high-power capability of SVO with the high energy density of $CF_x$. This is particularly important when considering the disparate power requirements for ICDs and CRT-Ds.

Li/SVO and Li/$CF_x$ batteries have been utilized in implantable medical applications successfully for up to three decades. A comparison of Li/SVO and Li/$CF_x$ battery technologies is shown in Table 28.6.[64] The Li/SVO system has much higher rate capability and lower internal resistance, while the Li/$CF_x$ system has much higher energy density (nearly 300 Wh/L higher than Li/SVO) and has a low self-discharge rate. Throughout much of discharge, the voltage of the Li/$CF_x$ couple is greater than the Li/SVO couple. This means that the Li/$CF_x$ is powering the lower rate functions of the device. When a patient needs defibrillation therapy, the Li/SVO couple delivers the higher current pulse that is necessary.

**TABLE 28.6**    Comparison of Li/SVO and Li/$CF_x$ Implantable Battery Technologies

| Category | Li/SVO | Li/$CF_x$ |
|---|---|---|
| Typical running voltage, V | 2.7 | 2.9 |
| Energy density, Wh/L | 730 | 1000 |
| Typical current density, mA/cm$^2$ | 35 | 1 |
| Typical internal resistance (BOL),* $\Omega$ | 0.250 | 40 |
| Time-dependent internal resistance increase, % | 40–60 DOD | None |
| Self-discharge, %/yr | 1 | <1 |
| Stepped discharge curve | Yes | No |

*BOL: beginning of life.

The combination of these two materials into one cathode system was introduced to improve the performance of batteries for implantable defibrillators, pacemakers, and other devices. One form of this system uses an admixture of SVO with $CF_x$ to form a hybrid cathode,[65] while another form uses discrete layers of SVO contacting a $CF_x$ layer, with a current collector between.[66] The use of two different battery chemistries within a single implantable device has precedence. Two energy sources, a high-rate Li/SVO and a low-rate Li/$I_2$ battery, have been used to power ICDs.[67] Here, the Li/SVO cell was used to provide high power for the defibrillation shocks and the Li/$I_2$ cell was used for the low power pacing and monitoring functions.

Li/SVO-$CF_x$ hybrid cathode batteries are used to power a number of implantable devices, including pacemakers, hemodynamic monitors, drug-delivery devices, and pulse generators that treat atrial fibrillation, and to provide cardiac resynchronization therapy.[67] The ratio of SVO to $CF_x$ can be varied in these batteries to meet application requirements. A greater proportion of $CF_x$ enhances the low-rate discharge capacity, while a greater proportion of SVO improves the power capability and end-of-service detection characteristics. These batteries matched Li/$I_2$ cells in energy density at about 1 Wh/cm$^3$.[68]

Extensive modeling studies have been conducted on the Li/SVO-$CF_x$ hybrid cathode batteries.[68] The physically based models show good agreement with data collected for the hybrid batteries with varying cathode thickness, geometric area, and SVO-$CF_x$ mix ratio. An advantage of the SVO-$CF_x$ system is the two voltage level changes that serve as early indicators that the end of the battery service life is approaching (Fig. 28.15).[64]

***Lithium Ion.***    Some implantable medical devices use secondary batteries because of the high power demands required by certain applications, such as neuromodulation pain management and ventricular assist devices. The ability to recharge the battery can extend service life and enable a smaller sized device compared to those that use primary batteries. Both longevity and size are important considerations for implantable medical device design. While other secondary battery chemistries are still used in some medical devices, lithium ion batteries are widespread, particularly in implantable medical devices.

One of the challenges with using secondary batteries in implantable medical devices is making sure they are recharged so that therapy is available as needed. If a patient should neglect to recharge the battery, then either the device or self-discharge continues to deplete the battery such that the battery becomes over-discharged as the voltage drops below the lower operating voltage limit. When this occurs, battery performance can be diminished and may even lead to safety concerns.

The recharging of an implanted secondary battery is accomplished by means of "transcutaneous energy transmission," a method of inductive charging. The external charger uses an induction coil to produce an

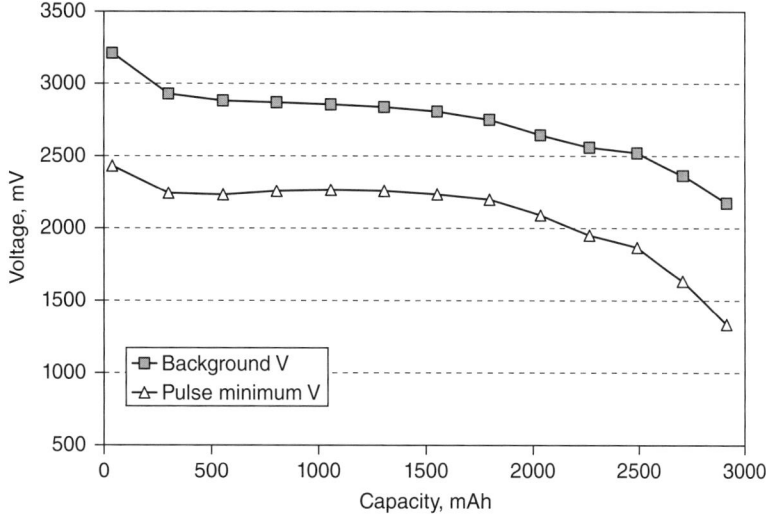

**FIGURE 28.15**   Discharge profile for an implantable high-rate Li/SVO-CF$_x$ sandwich cell. Cell pulsed with 3000 mA current for 10 s every 60 days, with a background load of 20 kΩ at 37°C. (*Courtesy of Greatbatch, Inc.*)

alternating electromagnetic field, inducing a current in a secondary, typically smaller coil in the implanted device. The efficiency of the energy transfer can be improved by utilizing a resonant circuit in the implanted device. The distance between the external charger and the secondary battery is minimized during the recharging process by holding the external charger in place above the implanted device, using a belt or other restraint. A balance must be struck between the desire to reduce charge time in order to improve patient compliance and comfort, and the need to limit power transmission across the cutaneous barrier in order to prevent heating effects in the skin, charger, or device.[69] Despite these limitations, devices powered by rechargeable cells, such as neurostimulators, are currently on the market from major device manufacturers.

One reason for compromised cell performance derives from the copper negative electrode current collector used in typical lithium ion cells.[70] When the cell voltage is too low, the negative electrode potential could reach the potential at which the copper current collector corrodes (Fig. 28.16). Using a different metal with a higher

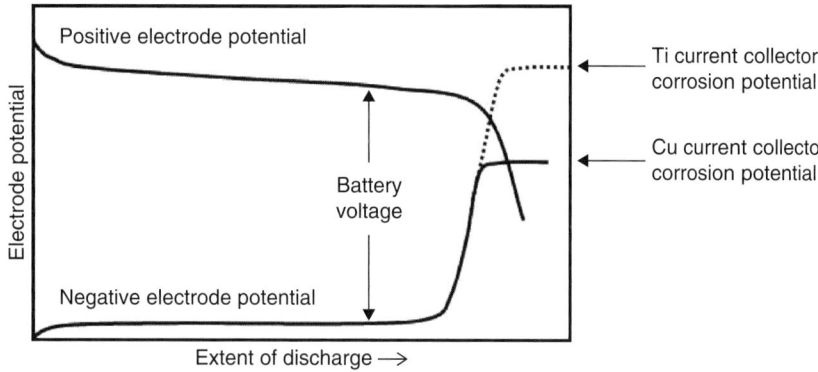

**FIGURE 28.16**   Electrode discharge potentials and battery voltage with extent of discharge showing the corrosion potentials for different negative electrode current collectors.

corrosion potential as the negative current collector, such as titanium, can eliminate corrosion-related performance issues when the battery is allowed to discharge to low voltages. Another solution is to replace the positive electrode material with one that discharges at a lower potential[71] or a negative electrode material that discharges at a higher potential.[72]

## ACKNOWLEDGMENTS

Major content for this chapter was provided in whole or in part by R. A. Leising, N. R. Gleason, B. C. Muffoletto, and C. F. Holmes, Chap. 31, Batteries for Biomedical Applications, *Linden's Handbook of Batteries*, 4th ed., T. B. Reddy and D. Linden, eds., McGraw-Hill, 2011.

## REFERENCES

1. V. S. Mallela, V. Ilankumaran, and N. S. Rao, "Trends in Cardiac Pacemaker Batteries," *Indian Pacing Electrophysiol. J.* **4:**201 (2004).

2. B. Parker, "Obituary: A Vindication of the Zinc-Mercury Pacemaker Battery" *Pacing Clin. Electrophysiol.* **1:**148 (1978).

3. F. Gutmann, A. M. Hermann, and A. Rembaum, "Solid-State Electrochemical Cells Based on Charge Transfer Complexes," *J. Electrochem. Soc.* **114:**323 (1967).

4. J. R. Moser, "Solid State Lithium-Iodine Primary Battery," U.S. Patent 3,660,163, May 2, 1972.

5. R. T. Mead, C. F. Holmes, and W. Greatbatch, "Design Evolution of the Lithium Iodine Pacemaker Battery," *Proc. Electrochem. Soc.* **79**(1):327 (1979).

6. H. G. Mond and A. Proclemer, "The 11th World Survey of Cardiac Pacing and Implantable Cardioverter-Defibrillators: Calendar Year 2009—A World Society of Arrhythmia's Project," *Pacing Clin. Electrophysiol.* **34:**1013 (2011).

7. J. Sperzel, H. Burri, D. Gras, F. V. Tjong, R. E. Knops, G. Hindricks, C. Steinwender et al., "State of the Art of Leadless Pacing," *Europace* **17:**1508 (2015).

8. H. F. Clemo and K. A. Ellenbogen, Chapter 4 in A. W. C. Chow and A. E. Buxton (eds.), *Implantable Cardiac Pacemakers and Defibrillators*, Blackwell Publishing, Malden, MA, 2006.

9. D. P. Zipes, D. G. Wyse, P. L. Friedman, A. E. Epstein, A. P. Hallstrom, H. L. Greene, E. B. Schron et al., "A Comparison of Antiarrhythmic-Drug Therapy with Implantable Defibrillators in Patients Resuscitated from Near-Fatal Ventricular Arrhythmias," *N. Eng. J. Med.* **337:**1576 (1997).

10. R. S. Nelson, Chapter 12 in M. W. Kroll and M. H. Lehmann (eds.), *Implantable Cardioverter Defibrillator Therapy: The Engineering-Clinical Interface*, Kluwer Academic Publishers, Norwell, MA, 1996.

11. K. E. Ellison, Chapter 6 in A. W. C. Chow and A. E. Buxton (eds.), *Implantable Cardiac Pacemakers and Defibrillators*, Blackwell Publishing, Malden, MA, 2006.

12. M. Kirk, Chapter 1 in A. W. C. Chow and A. E. Buxton (eds.), *Implantable Cardiac Pacemakers and Defibrillators*, Blackwell Publishing, Malden, MA, 2006.

13. American Heart Association Website, www.americanheart.org (accessed August 9, 2017).

14. S. J. Marzo, J. M. Sappington, and J. A. Shohet, "The Envoy Esteem Implantable Hearing System," *Otol. Clin. N. Am.,* **47:**941 (2014).

15. FDA Website, https://www.accessdata.fda.gov/cdrh_docs/pdf9/p090018c.pdf (accessed December 1, 2017).

16. M. R. Sohail, C. A. Henrikson, M. J. Braid-Forbes, K. F. Forbes, and D. J. Lerner, "Mortality and Cost Associated with Cardiovascular Implantable Electronic Device Infections," *Arch. Intern. Med.* **171:**1821 (2011).

17. M. R. Sohail, C. A. Henrikson, M. J. Braid-Forbes, K. F. Forbes, and D. J. Lerner, "Increased Long-Term Mortality in Patients with Cardiovascular Implantable Electronic Device Infections," *Pacing Clin. Electrophysiol.* **38:**231 (2015).

18. DOT Bans All Samsung Galaxy Note7 Phones from Airplanes, https://www.transportation.gov/briefing-room/dot-bans-all-samsung-galaxy-note7-phones-airplanes/ (accessed September 12, 2017).

19. Samsung Recall Support Note7 Investigation, https://img.us.news.samsung.com/us/wp-content/uploads/2017/01/22201435/EXPONENT-Galaxy-Note7-Press-Conference.pdf (accessed September 12, 2017).

20. Samsung Says Two Separate Battery Issues Were to Blame for All of Its Galaxy Note 7 Problems, https://www.recode.net/2017/1/22/14330404/samsung-note-7-problems-battery-investigation-explanation (accessed October 16, 2017).

21. Recommendations on the Transport of Dangerous Goods: Manual of Tests and Criteria, 5th Revised ed., Amendment 1, Section 38.3, UN/DOT.

22. FDA website, https://www.fda.gov/downloads/MedicalDevices/DeviceRegulationandGuidance/GuidanceDocuments/ucm077295.pdf (accessed September 12, 2017).

23. FDA website, https://www.accessdata.fda.gov/scripts/cdrh/cfdocs/cfStandards/search.cfm (accessed September 12, 2017).

24. FDA website, https://www.accessdata.fda.gov/scripts/cdrh/cfdocs/cfstandards/detail.cfm?id=32330 (accessed October 18, 2017).

25. FDA website, https://www.accessdata.fda.gov/scripts/cdrh/cfdocs/cfStandards/detail.cfm?standard__identification_no=32632 (accessed October 18, 2017).

26. FDA website, https://www.fda.gov/MedicalDevices/ResourcesforYou/HealthCareProviders/ucm534566.htm (accessed September 12, 2017).

27. FDA website, https://www.fda.gov/TobaccoProducts/Labeling/ProductsIngredientsComponents/ucm539362.htm (accessed September 12, 2017).

28. K. Fester and S. C. Levy, Chapter 4 in B. B. Owens (ed.), *Batteries for Implantable Biomedical Devices*, Plenum Press, New York, 1986.

29. S. C. Levy and P. Bro, "Reliability Analysis of Lithium Cells," *J. Power Sources* **26**:223 (1989).

30. P. Bro and S. C. Levy, *Quality and Reliability Methods for Primary Batteries*, Wiley-Interscience, New York, 1990.

31. S. C. Levy, "Safety and Reliability Considerations for Lithium Batteries," *J. Power Sources* **68**:75 (1997).

32. FDA website, https://www.fda.gov/MedicalDevices/Device Regulation and Guidance/Medical Device Quality and Compliance/ucm378185.htm (accessed September 12, 2017).

33. FDA website, https://www.fda.gov/downloads/MedicalDevices/Device Regulation and Guidance/Medical Device Quality and Compliance/UCM469128.pdf (accessed September 12, 2017).

34. Successful Practices for Battery Powered Medical Devices, AdvaMed website, https://www.advamed.org/resource-center/successful-practices-battery-powered-medical-devices (accessed September 12, 2017).

35. M. Visbisky, R. C. Stinebring, and C. F. Holmes, The Reliability Evaluation of Medical Implantable Batteries, *Proceedings of the 3rd Annual Battery Conference on Applications and Advances*, California State University, Long Beach, 1988.

36. NASA Technical Reports Servers, https://ntrs.nasa.gov/search.jsp?R=20060013441 (accessed September 12, 2017).

37. L. D. Hansen and R. M. Hart, "The Characterization of Internal Power Losses in Pacemaker Batteries by Calorimetry," *J. Electrochem. Soc.* **125**:842 (1978).

38. D. F. Untereker, "The Use of a Microcalorimeter for Analysis of Load-Dependent Processes Occurring in a Primary Battery," *J. Electrochem. Soc.* **125**:1907 (1978).

39. W. Greatbatch, R. McLean, W. Holmes, and C. Holmes, "A Microcalorimeter for Nondestructive Analysis of Pacemakers and Pacemaker Batteries," *IEEE Trans. Biomed. Eng.* **26**:309 (1979).

40. M. E. Orazem and B. Tribollet, *Electrochemical Impedance Spectroscopy*, Wiley, Hoboken, NJ, 2008.

41. E. S. Takeuchi, "Reliability Systems for Implantable Cardiac Defibrillator Batteries," *J. Power Sources* **54**:115–119 (1995).

42. M. J. O'Phelan, T. G. Victor, B. J. Haasl, L. D. Swanson, R. J. Kavanagh, A. G. Barr, and R. M. Dillon, U.S. Patent 7,479,349, 2009.

43. G. Antonioli, F. Baggioni, F. Consiglio, G. Grassi, R. LeBrun, and F. Sanardi, "Stimulatore Cardiaco Impiantabile con Nuova Battaria a Stato Solido al Litio," *Minerva Med.* **64**:2298 (1973).

44. C. F. Holmes, "The Role of Electrochemical Power Sources in Modern Health Care," *Interface* **8**:32–34 (1999).

45. R. T. Mead, W. Greatbatch, and F. W. Rudolph, "Lithium-Iodine Battery Having Coated Anode," U.S. Patent 3,957,533, 1976.

46. R. T. Mead, "Solid State Battery," U.S. Patent 3,773,557, 1973.

47. C. F. Holmes, Chapter 10 in M. W. Kroll and M. H. Lehmann (eds.), *Implantable Cardioverter Defibrillator Therapy: The Engineering-Clinical Interface*, Kluwer Academic Publishers, Norwell, MA, 1996.

48. C. C. Liang, M. E. Bolster, and R. M. Murphy, "Metal Oxide Composite Cathode Material for High Energy Density Batteries," U.S. Patent 4,391,729, 1983.

49. M. Onoda and K. Kanbe, "Crystal Structure and Electronic Properties of the $Ag_2V_4O_{11}$ Insertion Electrode," *J. Phys. Condens. Matter* **13**:6675 (2001).

50. R. A. Leising, W. C. Thiebolt, and E. S. Takeuchi, "Solid-State Characterization of Reduced Silver Vanadium Oxide from the Li/SVO Discharge Reaction," *Inorg. Chem.* **33**:5733 (1994).

51. P. M. Skarstad, Lithium/Silver Vanadium Oxide Batteries for Implantable Cardioverter-Defibrillators, *Proceedings of the Twelfth Annual Battery Conference on Applications and Advances* (IEEE 97th 8226), IEEE, 1997, p. 151.

52. R. P. Ramasamy, C. Feger, T. Strange, and B. N. Popov, "Discharge Characteristics of Silver Vanadium Oxide Cathodes," *J. Appl. Electrochem.* **36**:487 (2006).

53. M. Morcrette, P. Martin, P. Rozier, H. Vezin, F. Chevallier, L. Laffont, P. Poizot et al., "$Cu_{1.1}V_4O_{11}$: A New Positive Electrode Material for Rechargeable Li Batteries," *Chem. Mater.* **17**:418 (2005).

54. N. R. Gleason, R. A. Leising, M. Palazzo, E. S. Takeuchi, and K. J. Takeuchi, Microscopic Study of the First Voltage Plateau in the Discharge of SVO and the Consequences on Electrical Conductivity, 208th Meeting of the Electrochemical Society, Los Angeles, October 21, 2005.

55. N. D. Leifer, A. Colon, K. Martocci, S. G. Greenbaum, F. M. Alamgir, T. B. Reddy, N. R. Gleason, et al., "Nuclear Magnetic Resonance and X-Ray Absorption Spectroscopic Studies of Lithium Insertion in Silver Vanadium Oxide Cathodes," *J. Electrochem. Soc.* **154**:A500 (2007).

56. H. Gan and E. S. Takeuchi, "Lithium Electrodes with and Without $CO_2$ Treatment: Electrochemical Behavior and Effect on High Rate Lithium Battery Performance," *J. Power Sources* **62**:45 (1996).

57. H. Gan and E.S. Takeuchi, "Correlation of Anode Surface Film Chemical Composition and Voltage Delay in Silver Vanadium Oxide Cell System," *198th Meeting of the Electrochemical Society*, Phoenix, October 22, 2000.

58. A. Crespi, C. Schmidt, J. Norton, K. Chen, and P. Skarstad, "Modeling and Characterization of the Resistance of Lithium/SVO Batteries for Implantable Cardioverter Defibrillators," *J. Electrochem. Soc.* **148**:A30–A37 (2001).

59. K. Syracuse, N. Waite, H. Gan, E. S. Takeuchi, U.S. Patent 6,930,468, 2005.

60. M. J. Root, "Resistance Model for Lithium-Silver Vanadium Oxide Cells," *J. Electrochem. Soc.* **158**:A1347–A1353 (2011).

61. H. Gan and E.S. Takeuchi, U.S. Patent 5,753,389, 1998.

62. M. J. Root, "Implantable Cardiac Rhythm Device Batteries," *J. Cardiovasc. Trans. Res.* **1**:254 (2008).

63. M. J. Root, "Lithium–Manganese Dioxide Cells for Implantable Defibrillator Devices—Discharge Voltage Models," *J. Power Sources* **195**:5089 (2010).

64. H. Gan, R. Rubino, and E. Takeuchi, "Dual-Chemistry Cathode System for High-Rate Pulse Applications," *J. Power Sources* **146**:101 (2005).

65. C. L. Schmidt and P. M. Skarstad, "The Future of Lithium and Lithium-Ion Batteries in Implantable Medical Devices," *J. Power Sources* **97–98**:742 (2001).

66. H. Gan and E. Takeuchi, "Novel Electrode Design for High Rate Implantable Medical Cell Application," Abstract 219, *204th Meeting of the Electrochemical Society*, October 12–16, 2003.

67. K. Chen, D. R. Merritt, W. G. Howard, C. L. Schmidt, and P. M. Skarstad, "Hybrid Cathode Lithium Batteries for Implantable Medical Applications," *J. Power Sources* **162**:837 (2006).

68. P. M. Gomadam, D. R. Merritt, E. R. Scott, C. L. Schmidt, P. M. Skarstad, and J. W. Weidner, "Modeling Lithium/Hybrid-Cathode Batteries," *J. Power Sources* **174**:872 (2007).

69. C. Niu, H. Hao, L. Li, B. Ma, and M. Wu, The Transcutaneous Charger for Implanted Nerve Stimulation Device, *Proceedings of the 28th IEEE EMBS Annual International Conference.*

70. C. Kishiyama, M. Nagata, T. Piao, J. Dodd, P. Lam, and H. Tsukamoto, "Improvement of Deep Discharge Capability for Lithium Ion Batteries," *Proc. Electrochem. Soc.* **28**:352 (2004).

71. H. Tsukamoto, C. Kishiyama, M. Nagata, H. Nakahara, and T. Piao, U.S. Patent 6,596,439, 2003.

72. E. R. Scott, W. G. Howard, and C. L. Schmidt, U.S. Patent 7,811,705, 2010.

# CHAPTER 29

# RESERVE CELLS

R. David Lucero, Alexander P. Karpinski, Benjamin M. Meyer
(Emeritus Contributors: David L. Chua, William J. Eppley,
Jeffrey A. Swank, Michael Ding, Charles M. Lamb)

## 29.0  INTRODUCTION

Reserve cells and batteries are specifically designed to withstand long shelf life periods by constructing cells in an inactive state. This is typically accomplished by storing the electrolyte in a separate container, or in the case of thermal reserve batteries employing an electrolyte that is solid at room temperature. These cells are considered to be inactive and will not deliver current until activated. Activation can be accomplished through various means to bring the electrolyte into direct contact with the electrodes in the cell. In the case of thermal batteries, heat is generated to melt the solid electrolyte salt creating an ionically conductive system.

In this chapter the following reserve battery systems are reviewed:

- Water-activated magnesium anode
- Water-activated zinc/silver oxide
- Water-activated aluminum/silver oxide
- Lithium anode (nonaqueous, ambient temperature cells)
- Spin-dependent reserve batteries

While a brief introduction to thermal batteries is addressed in this chapter, the reader is referred to Chap. 21 for a more thorough discussion of thermal battery electrochemistry. In addition, further details on the various electrochemistries used for reserve cells as discussed in this chapter can be found in the chapters covering the various primary cells (see Chaps. 11B, 12C, and 13 for relevant details).

Also, the batteries covered in this chapter are primary (single use) systems. While secondary electrochemistries could (with proper design and test) be viable alternatives for use in reserve cells, the typical reserve cell application does not require recharge ability. One notable exception is a type of lead acid cell, known as "dry charge," produced as a precharged battery that is shipped without electrolyte but activated when installed into an automobile. This technique is used to eliminate problems with dead batteries from excessive periods of high-temperature storage prior to use (see Chap. 14 for details on lead-acid cells).

## 29.1    OVERVIEW OF TYPICAL RESERVE BATTERIES

### 29.1.1    Water-Activated Reserve Batteries

***General Characteristics of Reserve Magnesium Batteries.***    The water-activated battery was first developed in the 1940s to meet a need for a high-energy-density, long-shelf-life battery, with good low-temperature performance, for military applications.

The battery is constructed dry, stored in the dry condition, and activated at the time of use by the addition of water or an aqueous electrolyte. Most water-activated batteries use magnesium as the anode material. Several cathode materials have been used successfully in different types of designs and applications.

The magnesium/silver chloride seawater-activated battery was developed by Bell Telephone Laboratories as the power source for electric torpedoes.[1] This work resulted in the development of small high-energy-density batteries readily adaptable for use as power sources for sonobuoys, electric torpedoes, weather balloons, air-sea rescue equipment, pyrotechnic devices, marine markers, and emergency lights.

The magnesium/cuprous chloride system became commercially available in 1949.[2,3] Compared with the magnesium/silver chloride battery, this system has lower energy density, lower rate capability, and less resistance to storage at high humidity, but its cost is significantly lower. Although the magnesium/cuprous chloride system can be used for the same purposes as the magnesium/silver chloride battery, its major application was in airborne meteorological equipment, where the use of the more expensive silver chloride system was not warranted. The cuprous chloride system does not have the physical or electrical characteristics required for use as the power source for electric torpedoes. More recently, the magnesium/cuprous chloride chemistry has been developed for aviation and marine life jacket lights.

Because of the high cost of silver and the impracticality of recovery after use, other nonsilver, water-activated batteries were developed, primarily as the power source for antisubmarine warfare (ASW) equipment.

The systems that have been developed and used successfully are magnesium/lead chloride,[4] magnesium/cuprous iodide-sulfur,[5–7] magnesium/cuprous thiocyanate-sulfur,[8] and magnesium/manganese dioxide utilizing an aqueous magnesium perchlorate electrolyte.[9–11] None of these systems can compete with the magnesium/silver chloride system in most attributes except cost.

Magnesium seawater-activated batteries, using dissolved oxygen in the seawater as the cathode reactant, also have been developed for application in buoys, communications, and underwater propulsion.

Another seawater battery system considered for low-rate, long-duration undersea vehicle applications consists of a magnesium anode, a palladium- and iridium-catalyzed carbon paper cathode, and a solution-phased catholyte of seawater, acid, and hydrogen peroxide. The magnesium/hydrogen peroxide system has a voltage of 2.12 V and has been estimated to be capable of more than 500 Wh/kg when configured for large-scale unmanned undersea vehicles.[12]

The advantages and disadvantages of the various water-activated magnesium anode batteries are given in Table 29.1.

***General Characteristics of Zinc/Silver Oxide Reserve Batteries.***    An important reserve battery, particularly for missile and torpedo applications, is the zinc/silver oxide electrochemical system, which is noted for its high-rate capability and high energy density. The cells of the battery are designed with thin plates and large-surface-area electrodes, which augment its high-rate and low-temperature capability and provide a flat discharge characteristic. This design, however, reduces the activated or wet shelf life of the battery, necessitating the use of a reserve-battery design to meet storage requirements.

The zinc/silver oxide electrochemical system was conceived by the work of Alessandro Volta who demonstrated the possibility of using dissimilar metals in a "pile-type" multicell construction to obtain a substantial electric voltage. The system existed somewhat as a laboratory device until Professor Henri André designed a practical secondary cell early in World War II.

Subsequent to World War II, the U.S. military became interested in a dry-charged primary version for use in airborne electronics, missiles, and torpedoes because of its very high energy output per unit weight and volume and high-rate capability. The ultimate result of this interest was the development of lightweight batteries for the aerospace industry, both military and civilian. The entire manned space program was keyed to zinc/silver oxide reserve batteries as the power sources for the various flight vehicles.

**TABLE 29.1**   Comparison of Silver and Nonsilver Cathode Batteries

| Advantages | Disadvantages |
|---|---|
| Silver chloride cathodes | |
| Reliable | High raw material costs |
| Safe | High rate of self-discharge after activation |
| High power density | |
| High energy density | |
| Good response to pulse loading | |
| Instantaneous activation | |
| Long unactivated shelf life | |
| No maintenance | |
| Nonsilver cathodes | |
| Abundant domestic supply | Requires supporting conductive grid |
| Low raw-material cost | Operates at low current densities |
| Instantaneous activation | Low energy density compared to silver |
| Reliable, safe | High rate of self-discharge after activation |
| Long unactivated shelf life | |
| No maintenance | |

Zinc/silver oxide reserve batteries are divided into two classes, manually activated and remotely activated. In general, the manually activated types are used for space launch vehicles and accessible terrestrial applications and are usually packaged in more conventional configurations. The remote or automatically activated types are used principally for torpedo and missile systems. This use requires a long period of readiness (in storage), a means for rapid remote activation, and an efficient discharge at high discharge rates, typically from about 10 s to over 4 h, inclusive of open-circuit wet-stand periods. The specific energy and energy density of manually activated types range from about 60 to 220 Wh/kg and 120 to 550 Wh/L. For remotely activated types, the specific energy and energy density are reduced because of the self-contained activating device and range from about 11 to 88 Wh/kg and 24 to 320 Wh/L.

***General Characteristics of Aluminum/Silver Oxide Reserve Batteries.***   The aluminum/silver oxide aqueous battery was developed by the U.S. Navy for torpedo propulsion in the early 1970s.[13] As with the reserve magnesium batteries, the battery is constructed dry, stored in the dry condition, and activated at the time of use by the addition of seawater or an aqueous electrolyte. This system is capable of delivering very high rates in excess of 1600 mA/cm$^2$ while maintaining system level energy densities over 200 Wh/kg and 264 Wh/L.

## 29.1.2   Nonaqueous, Ambient Temperature Lithium Reserve Cells

The use of lithium metal as an anode in reserve batteries provides a significant energy advantage over traditional reserve batteries because of the high potential and high capacity per mass (3.86 Ah/g) of lithium. A lithium reserve battery can operate at a voltage close to twice that of the conventional aqueous types. Due to the reactivity of lithium in aqueous electrolytes, with the exception of the special lithium/water and lithium/air batteries, lithium batteries must use a nonaqueous electrolyte with which lithium is nonreactive. Table 29.2 lists typical features of lithium reserve batteries.

The various ambient-temperature active (nonreserve) lithium primary batteries demonstrating higher energy density and rate capability are $Li/SO_2$, $Li/V_2O_5$, $Li/SOCl_2$, $Li/SO_2Cl_2$, and $Li/Li_xCoO_2$. The discharge characteristics of these batteries are shown in Fig. 29.1 and are the predominate electrochemical systems employed in reserve-type configurations.

In the lithium reserve battery construction, the electrolyte is stored in a separate reservoir isolated from the electrode active materials until the battery is activated. This design feature provides a capability of

**TABLE 29.2**   Typical Characteristics of Lithium Anode Reserve Batteries

Operating temperature range −55 to 70°C
10- to 20-year unactivated storage life
Hermetically sealed
High energy density
High reliability
Low electrical noise
Flat discharge voltage profile
Rapid voltage rise after initiation
Mechanical environmental capability:
   Acceleration shocks up to 20,000 *g*
   High spin up to 20,000 rpm
   Transportation and deployment vibration levels
Operating life from several seconds up to 1 year

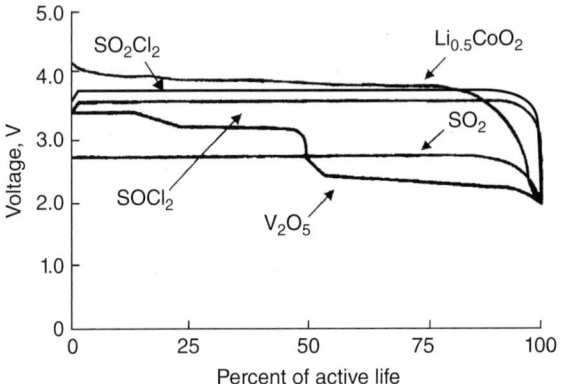

**FIGURE 29.1**   Performance comparison of lithium anode primary systems at 20°C. Thionyl chloride ($SOCl_2$)—3.6 V; vanadium pentoxide ($V_2O_5$)—3.4 V; sulfur dioxide ($SO_2$)—2.9 V; precharged lithiated cobalt oxide ($Li_xCoO_2$, $x = 0.4$–$0.5$) 4.0 V; sulfuryl chloride ($SO_2Cl_2$)—3.8 V.

essentially undiminished output even after storage periods in the inactive state of over 20 years. However, the use of a reservoir system results in an energy density penalty of as much as 50% compared with the active lithium primary batteries. Key contributors to this penalty are the activation device and the electrolyte reservoir.

In designing a lithium anode reserve battery, important considerations include the physical properties of the electrolyte solution and performance as a function of the discharge conditions as well as factors such as the stability and the compatibility of the electrolyte with the materials of construction of the electrolyte reservoir. Use of environmentally friendly systems has additional importance, leading to an emerging interest in pursuing organic-based electrolyte solutions in the development of new cell systems.

## 29.1.3   Spin-Dependent Reserve Cells

Various military and a few civilian electrical power applications with long shelf-life requirements require the use of reserve batteries. This is particularly true when the system requires that the power supply be integrally packaged with the electronics and not replaced throughout the storage life of the system. Typical of such applications are fuzing, control, and arming systems for artillery and other spin-stabilized projectiles.

High spin forces, such as those encountered in artillery projectiles, may produce a difficult environment for many battery designs. However, special designs for liquid-electrolyte reserve batteries have evolved that take advantage of spin to bring about their activation and then keep the electrolyte within the cell structure.

A typical spin-dependent reserve battery is illustrated in Fig. 29.2. The electrode stack consists of electrodes and cell spacers of an annular configuration packaged dry and therefore capable of long-term storage. A metal ampoule, inserted in the center hole of the stack, houses the electrolyte. Upon firing of the gun, the ampoule opens; the electrolyte is released and is then distributed into the annular-shaped cells centrifugally, thereby causing the battery to become active.

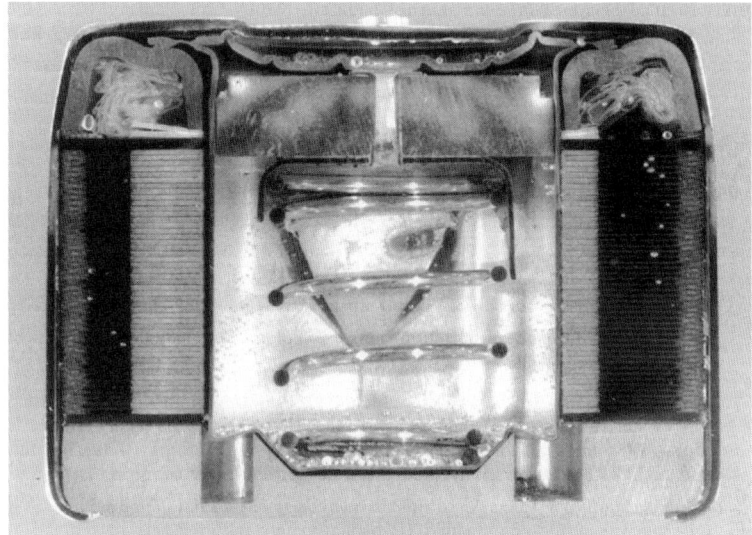

**FIGURE 29.2**  Cross section of a lead/fluoboric acid/lead dioxide multicell reserve battery showing a "dashpot" cutter for a copper ampoule. (*Courtesy of U.S. Department of the Army.*)

For several decades, the chemistry most commonly employed in spin-dependent liquid-electrolyte reserve batteries has been the lead/fluoboric acid/lead dioxide cell represented by the following simplified reaction:

$$Pb + PbO_2 + 4HBF_4 \rightarrow 2Pb(BF_4) + 2H_2O$$

Fluoboric acid, rather than the more common sulfuric acid electrolyte, is used because it performs better at the very low temperatures required for these military applications. This low-temperature performance is due in part to the absence of insoluble reaction products as the reserve battery discharges. In the early 1990s, the last two facilities that were capable of producing the specialized electrode material used in these batteries were decommissioned, primarily for business reasons, and the availability of this technology essentially ended. However, fuzes powered by this electrochemistry remain in the U.S. Army's inventory.

To replace the lead/fluoboric acid/lead dioxide system, spin-dependent liquid-electrolyte reserve batteries employing lithium anodes have been developed. The most common system today is based on thionyl chloride, which serves in the dual role of electrolyte carrier and active cathodic depolarizer. The accepted cell reaction for this system is

$$4Li + 2SOCl_2 \rightarrow 4LiCl + S + SO_2$$

At one time, the zinc/potassium hydroxide/silver oxide system was also employed in spin-dependent reserve batteries. More frequently, this reserve system has been used in nonspin applications, such as missiles, where the electrolyte is driven into place by a gas generator or some other activation method. This system is again finding favor in some applications where the potential hazards of lithium-based systems can create safety problems. The chemistry of the zinc/silver oxide couple can be represented by either of two reactions, depending on the oxidation state of the silver oxide:

$$2AgO + Zn \rightarrow Ag_2O + ZnO$$

$$Ag_2O + Zn \rightarrow 2Ag + ZnO$$

In the early 1970s, thermal batteries using the $Ca/LiCl-KCl-CaCro_4/Fe$ system that could operate at high spin rates (300 rps) were developed and successfully demonstrated by Sandia National Laboratories. In the early 1990s, this process was repeated for the now standard lithium (alloy)/iron disulfide couple employed in most thermal batteries.

### 29.1.4   Thermal Reserve Batteries

Thermal batteries are primary reserve batteries that employ inorganic salt electrolytes. These electrolytes are relatively nonconductive solids at ambient temperatures. Integral to the thermal battery are pyrotechnic materials scaled to supply sufficient thermal energy to melt the electrolyte. The molten electrolyte is highly conductive, and high currents may then be drawn from the cells.

The activated life of a thermal battery depends on several factors involving cell chemistry and construction. Once activated, and as long as the electrolyte remains molten, thermal batteries may supply current, discharging the active materials to the point of functional exhaustion. On the other hand, even with excess active materials present, the batteries will eventually cease functioning due to the loss of internal heat and subsequent resolidification of the electrolyte. Hence, two of the primary factors behind thermal battery active life are:

1. Compositions and masses of the active cell-stack materials (i.e., anodes and cathodes)
2. Other construction details, including the overall battery shape and the types and amounts of thermal insulation

Depending on the battery design, which is ultimately determined by the specific requirements of the application, the activated thermal battery may supply electric power for only a few seconds or may function for over an hour.

Initiation of a thermal battery is normally provided by an energy impulse from an external source to a built-in initiator. The initiator, typically an electric match, an electroexplosive device (squib), or a percussion primer, ignites the cell-stack pyrotechnics. Rise time, the time interval between the initiation impulse and that time at which the battery can sustain a current at voltage, varies as a function of battery size, design, and chemistry. Rise times of several hundred milliseconds are not uncommon for large units. Small batteries have been designed to reliably achieve operating conditions within 10 to 20 ms.

The shelf life of an unactivated thermal battery is typically 10 to 25 years, depending upon the design. Once activated and discharged, though, they are not reusable or rechargeable.

Current developments in extending the activated life capabilities of thermal batteries have widened their suitability and application potential in new military as well as industrial/civilian systems.

Thermal batteries were first developed in Germany in the 1940s, and were used primarily for weapons applications.[14-16] Batteries containing multiple cells and integral pyrotechnic materials have been produced since 1947.[17] Because of their high reliability and long shelf life, thermal batteries are ideally suited for military ordnance purposes. Consequently, they have been widely used in missiles, bombs, mines, decoys, jammers, torpedoes, space exploration systems, emergency escape systems, and similar applications.

A more detailed description of thermal battery technology, construction, and performance is discussed in Chap. 21.

## 29.2   *CELL AND RESERVE BATTERY SYSTEMS/TYPES*

### 29.2.1   **Water-Activated Batteries**

Water-activated batteries are manufactured in the following basic types:

1. *Immersion* batteries are designed to be activated by immersion in the electrolyte. They have been constructed in sizes to produce 1.0 V to several hundred volts at currents up to 50 A. Discharge times can vary from a few seconds to several days. A typical immersion-type water-activated battery is shown in Fig. 29.3.

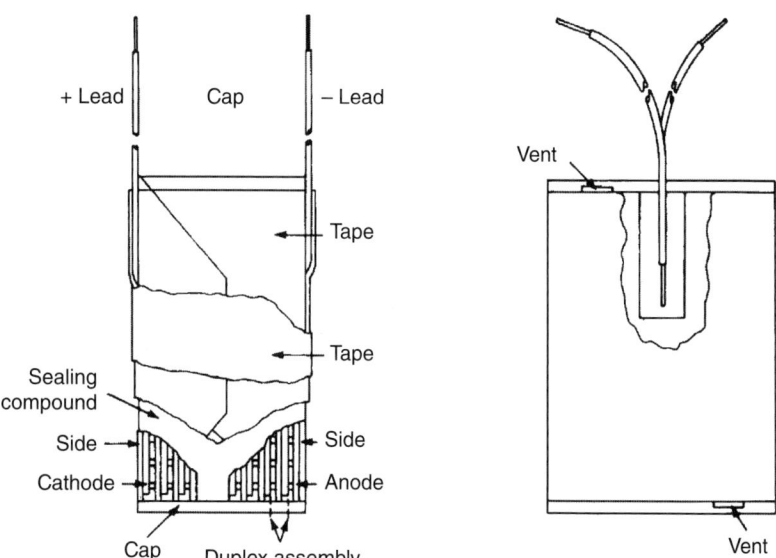

**FIGURE 29.3**   A seawater battery, immersion type.

2. *Forced-flow* batteries are designed for use as the power source for electric torpedoes. The name is derived from the fact that seawater is forced through the battery as the torpedo is driven through the water. Because of the heat generated during discharge and electrolyte recirculation, these systems can perform at current densities over 500 mA/cm$^2$ of the cathode surface area. Batteries containing 118 to 460 cells that will produce 25 to 460 kW of power have been developed. Discharge times are typically about 10 to 15 min but can exceed an hour depending on the power requirements. A diagrammatic representation of a torpedo battery and a torpedo battery with recirculation voltage control is shown in Figs. 29.4a to 29.4c.

3. *Dunk-type* batteries are designed with an absorbent separator between the electrodes and are activated by pouring the electrolyte into the battery, where it is absorbed by the separator. Batteries of this type have been designed to produce 1.5 to 130 V at currents up to about 10 A. Lengths of discharge vary from about 0.5 to 15 h. Figure 29.5 is a diagrammatic representation of a magnesium/cuprous chloride battery used in radiosonde applications. A pile-type construction is used. A sheet of magnesium is separated from the cuprous chloride cathode by a porous separator, which also serves to retain the electrolyte. The cathode is pasted type made by applying a paste of powdered cuprous chloride and a liquid binder onto a copper grid or screen. The assembly is taped together to form the battery. The batteries are also made in spiral or jelly-roll design. Other cathode materials such as silver chloride can also be used in this configuration.

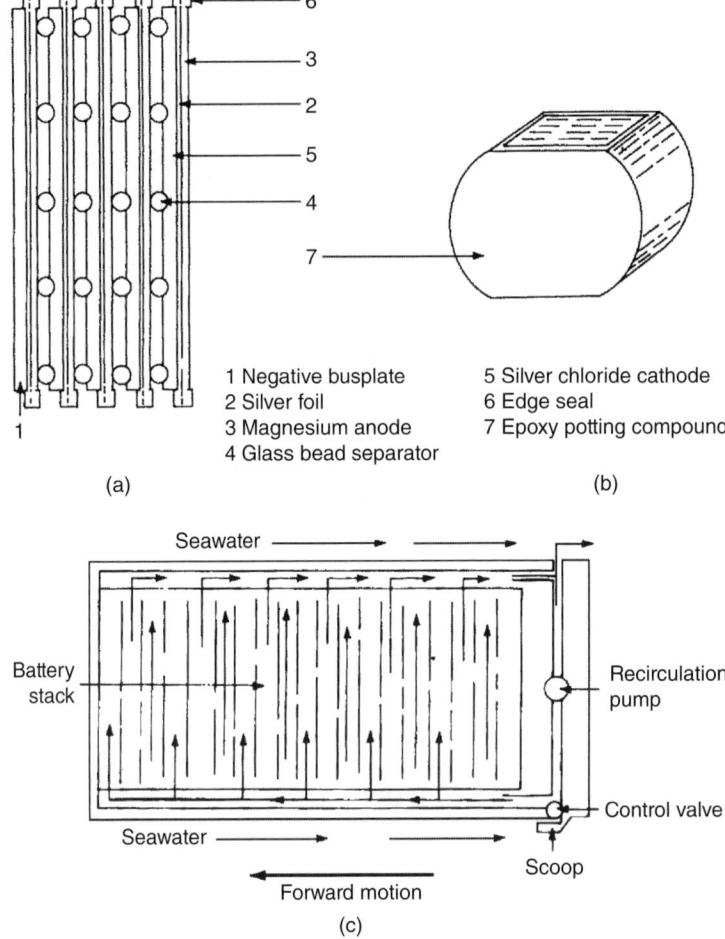

1 Negative busplate
2 Silver foil
3 Magnesium anode
4 Glass bead separator

5 Silver chloride cathode
6 Edge seal
7 Epoxy potting compound

(a)                                                          (b)

(c)

**FIGURE 29.4**   Diagrammatic representation of a torpedo battery construction. (*a*) Cell construction. (*b*) Battery configuration. (*c*) Recirculation voltage control.

## 29.2.2   Lithium Reserve Batteries

A variety of lithium anode reserve cell designs are available:

- Glass ampoule electrolyte storage
- Single cells with bellows electrolyte storage
- Multicell batteries with either glass ampoule or electrolyte reservoirs

Many typical lithium battery chemistries may be feasible for use, including both liquid and solid cathodes with a broad range of organic and inorganic electrolytes. Key features for lithium reserve cells include extremely low levels of moisture contamination and the use of hermetic enclosures. Both requirements make these types of cells a challenge to design and produce. Some examples of production lithium reserve cells include the following military applications:

- MOFA for prox-fuzed artillery
- M762/M767 electronic time fuzes

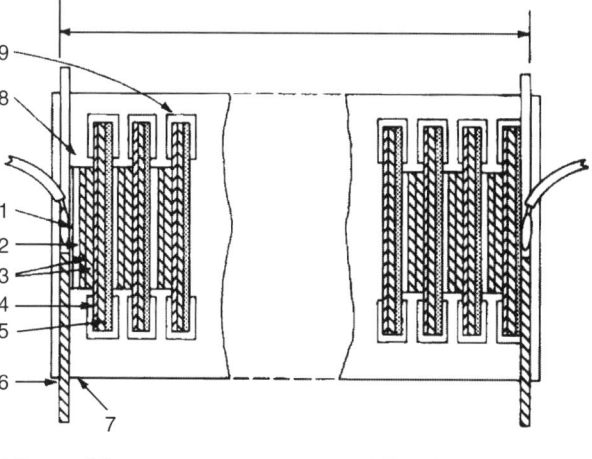

1 Copper foil
2 Cuprous chloride and cotton gauze
3 Cotton Webril
4 Paper separator
5 Magnesium

6 Formica case
7 Varnish-coated paper
8 Void (for electrolyte)
9 Tape

**FIGURE 29.5**   Diagrammatic representation of a magnesium/cuprous chloride dunk-type battery. (*Courtesy of Eagle-Picher Technologies.*)

- Excalibur 155 mm round for prelaunch battery
- Self-destruct fuze (SDF)

### 29.2.3   Spin-Dependent Reserve Batteries

*Electrode-Stack Arrangement.*    The electrode stack may be arranged in two ways. One favors a high-voltage output and the other a high-current output. The former generally uses bipolar electrodes—that is, electrodes wherein anodic and cathodic materials, respectively, are applied to the opposite sides of a metal substrate. Such bipolar electrode plates are stacked in a pile or series configuration, making automatic contact from one cell to the next. The voltage output of such a stack is the sum of all the cells. In the high-current configuration, electrode plates coated with anodic material on both sides of the substrate are stacked alternately with plates coated with cathodic material on both sides. All anodic plates are connected electrically in parallel through tabs. All cathodic plates are similarly connected. The two electrical connections constitute the effective terminals of the battery. This type of parallel stack is, in effect, a single electrochemical cell with a larger electrode area. Where required by the application, multiple series stacks can be connected in parallel, thereby yielding both high-voltage and high-current outputs (see Fig. 29.6).

*Electrolyte Volume Optimization.*    The electrolyte capacity of the ampoule must be matched to the composite volume of all the cells in the battery. A parallel-construction battery is reasonably tolerant of electrolyte flooding or starvation since it is a single cell. A series configuration, however, can be greatly degraded by flooding since that condition produces intercell short circuits in the electrolyte fill channel or manifold. The opposite condition—that is, insufficient electrolyte—may leave one or more cells empty and therefore fail to provide continuity throughout the cell stack.

Since temperature extremes have a greater effect on the expansion and contraction of the liquid electrolyte than on the volume of the cells, an electrolyte volume that ensures that the cells will be reasonably full at low temperatures usually leads to an excess of electrolyte at higher temperatures. In the lead/fluoboric acid/lead dioxide system, this excess must be accommodated in the design of the battery by the use of a "sump," which is placed at the end of the path the electrolyte follows to fill the cells. The sump is used to catch not only any excess of electrolyte created by thermal expansion, but also that resulting from eccentric

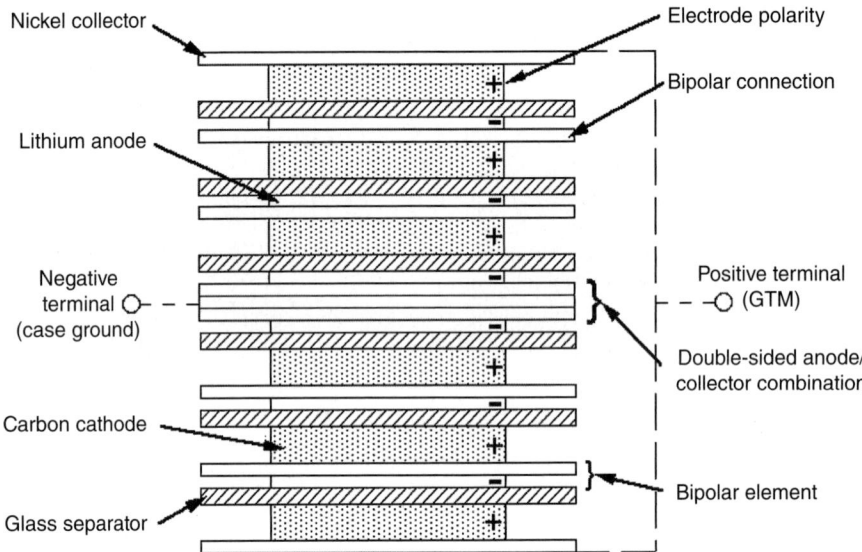

**FIGURE 29.6**    Quarter cross-sectional view of the cell stack of a lithium/thionyl chloride reserve battery showing an example of series and parallel construction. (*Courtesy of EnerSys Advanced Systems.*)

spin conditions. Lithium/thionyl chloride batteries may be more tolerant of cell flooding, depending on the particulars of the individual design. In some short-life batteries, a match is established at high temperature with the recognition that cells will be less than full at lower temperatures. To ensure that some electrolyte enters each cell (so that continuity can be maintained), leveling holes may be provided from cell to cell. Though kept very small to reduce the effect of inevitable intercell short circuits, these holes do dissipate some of the capacity of the battery.

***Cell Sealing.***    Since the individual cells of a spin-dependent liquid-electrolyte reserve battery are generally annular in shape and are filled by centrifugal force, the periphery of the cell must be sealed to keep electrolyte from leaking out. This sealing is typically accomplished by a plastic barrier formed around the outside of the electrode-spacer stack. For lead/fluoboric acid/lead dioxide batteries, this barrier is formed by fish paper (a dense, impervious paper) coated with polyethylene that melts at a relatively low temperature (similar to that used on milk cartons). Cell spacers are punched from the coated fish paper and placed between the electrodes. The stack is then clamped together and heated in an oven at a temperature sufficient to fuse the polyethylene, which then acts as an adhesive and sealer between the electrodes. Cell sealing is critically important in this system because the electrolyte is highly conductive and any leakage that leads to intercell shorting will rapidly deplete the capacity of the battery.

Cell sealing can be somewhat less critical in some lithium/thionyl chloride batteries, as in the moderately powerful batteries typically found in modern radar proximity fuzes used in artillery applications. In a recently produced multicell battery (Fig. 29.7), the cell parts were designed so that the annular electrodes and separators could more easily be inserted into a Tefzel® cell cup to create the cell stack. This was done to simplify automated assembly. The intercell shorting that resulted from the slightly relaxed fit at the outside diameter of the cell stack was reduced by the less conductive (relative to the fluoboric acid used in the previous system) catholyte, and somewhat offset by the ability to easily package excess capacity into that particular design. For the higher-power and energy batteries required by missile applications, cell sealing and isolation remain critically important.

***Ampoules.***    Early designs of liquid-electrolyte reserve batteries used glass ampoules to house the electrolyte, and in fact, some modern batteries still use such ampoules. These ampoules are generally smashed by the acceleration force of gunfire or by the explosive output of a primer or squib. Although these forces are ample, there

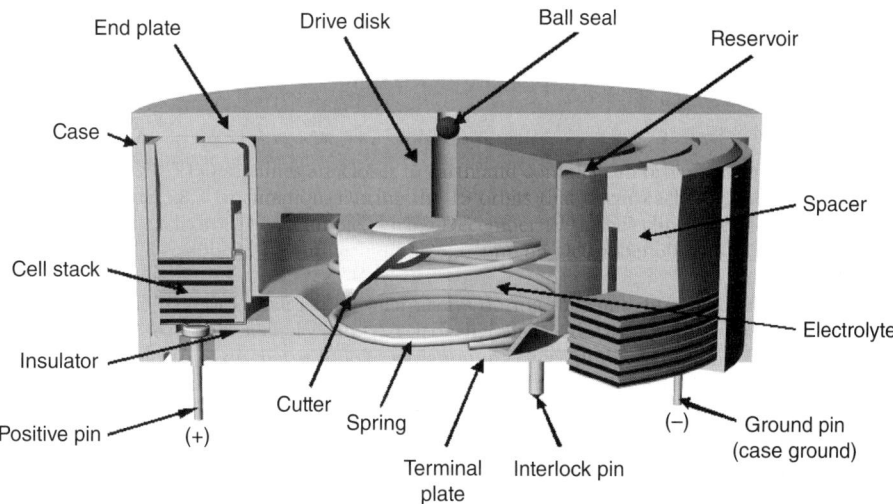

**FIGURE 29.7** Cross section of a lithium/thionyl chloride multicell reserve battery. (*Courtesy of EnerSys Advanced Systems.*)

is also a tendency for rough handling or a drop on a hard surface to cause inadvertent glass ampoule breakage. This destroys the battery due to the premature leakage of electrolyte into the cells.

A major advance in battery ruggedness resulted from the design of metal (usually copper in lead/fluoboric acid/lead dioxide batteries and stainless steel in the lithium/thionyl chloride designs) ampoules (or reservoirs) with internal cutting mechanisms. One version employs a cutter that is activated by a combination of spin and acceleration (Fig. 29.8), both provided by the act of gunfire. Other versions rely on a dashpot cutter mechanism (Figs. 29.2 and 29.7 referenced previously, and Fig. 29.9 shown below). This mechanism requires a sustained acceleration (several milliseconds experienced in gunfire), but it will not function when subjected to the much shorter (a portion of a millisecond) shock pulse resulting from being dropped on a hard surface. The use of these "intelligent" ampoules, which are capable of discriminating between the forces of gunfire and those of rough handling, has resulted in a substantial improvement in battery reliability and safety.

An assembly diagram for a typical lead/fluoboric acid/lead dioxide multicell battery is shown in Fig. 29.9.

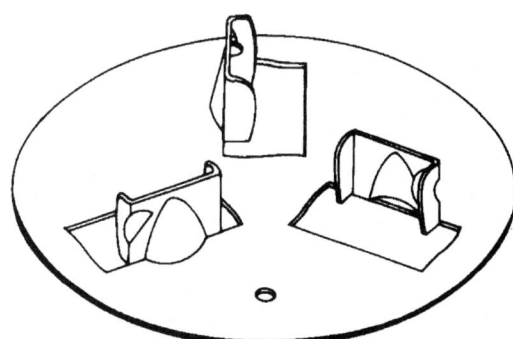

**FIGURE 29.8** A three-bladed cutter for a copper ampoule requiring spin and acceleration for activation. (*Courtesy of U.S. Department of the Army.*)

***Safety in Lithium-Based Batteries.*** For at least the past 20 years, reserve single-cell batteries that use various lithium-based electrochemistries have been employed in a number of fuzing applications. These cells are normally constructed with an anode-separator-cathode assembly spirally wound around a centrally located glass electrolyte ampoule. The electrolyte ampoule is normally broken upon gunfire or as the result of the bottom of the cell case being struck with a squib- or spring-driven device. Since these are single-cell devices, there is no chance for intercell short-circuiting and subsequent safety problems.

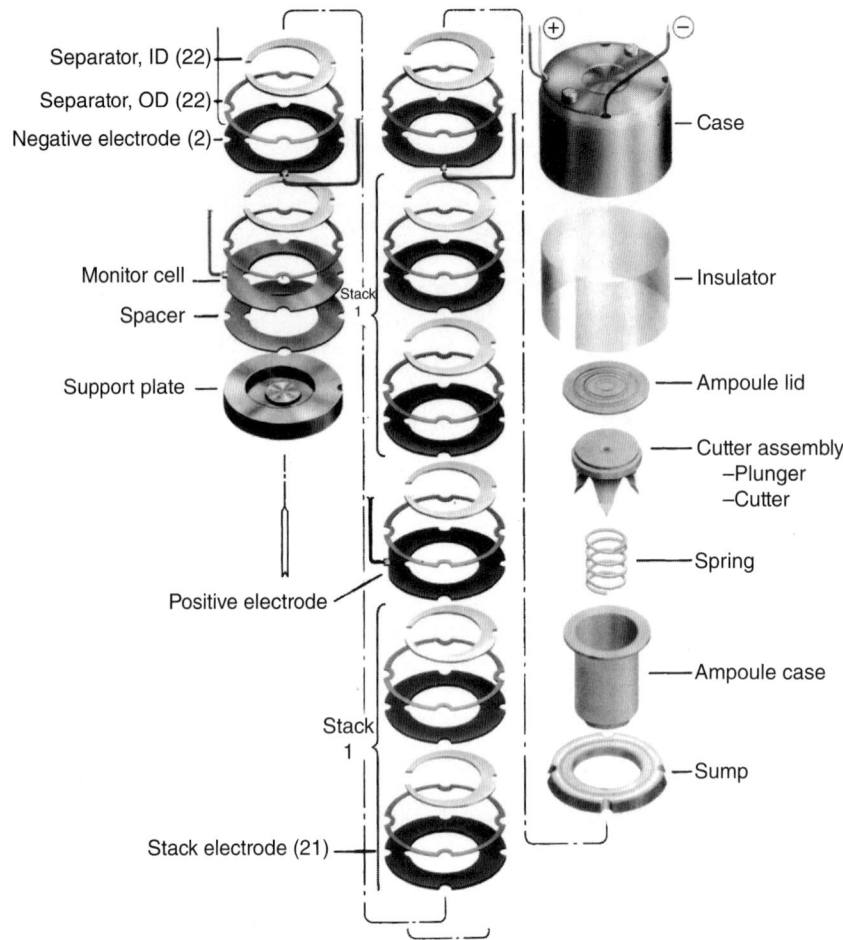

**FIGURE 29.9**    Component parts of a lead/fluoboric acid/lead dioxide multicell reserve battery, PS 416 power supply. (*Courtesy of U.S. Department of the Army.*)

However, in multicell pile-configuration reserve batteries, there is a considerable chance for intercell short-circuiting in the common electrolyte manifold. This intercell short-circuiting not only dissipates the capacity of the cells, but it can also allow for dendritic growth, which can lead to electronic short-circuiting of the cells with catastrophic results. Experience has shown that such dendritic growth can be minimized or eliminated if all interior metallic surfaces of the battery have a nonconductive (usually Teflon-based) coating.

## 29.3   CONSTRUCTION AND PERFORMANCE OF WATER-ACTIVATED BATTERIES

### 29.3.1   Magnesium Anode Cells

***Construction.***    Water-activated cells consist of an anode, a cathode, a separator, terminations, and some form of encasement. A battery consists of a multiplicity of cells connected in series or series-parallel. Such an assembly requires a method to connect the cells in the desired configuration plus a method to control leakage currents (i.e., internal electrical short circuits through electrolyte pathways). The voltage of a cell depends primarily on the electrochemical system involved. To increase voltage, a number of cells must be connected in series. The capacity of a cell in ampere-hours is primarily dependent on the quantity of active material in the electrodes. The ability of a cell to produce a given current at a usable voltage depends on the area of the electrode. To decrease current density so as to increase load voltage, the electrode area must be increased. Power output depends on the temperature and salinity of the electrolyte. Power output can be increased by increasing the temperature or the salinity of the electrolyte.

The basic components of a single cell, a duplex assembly for connecting cells in series, and a finished battery are illustrated in Fig. 29.10 and Figs. 29.11a and 29.11b, respectively.[12,18–21] The illustrations represent batteries designed for use by immersion in the electrolyte as contrasted to a dunk-type (radiosonde) battery, which is activated by pouring the electrolyte into the battery, or a forced-flow electric torpedo battery. The construction principles with slight variations are similar in all cases.

*Components.* A more detailed description of the various cell and battery components and construction elements follows.

ANODE (NEGATIVE PLATE). The anode is made from sheet magnesium. Magnesium AZ61A is preferred because it tends to sludge and polarize less. In some cases AZ31B alloy is used; however, this alloy gives slightly lower voltage, polarizes at high current densities,

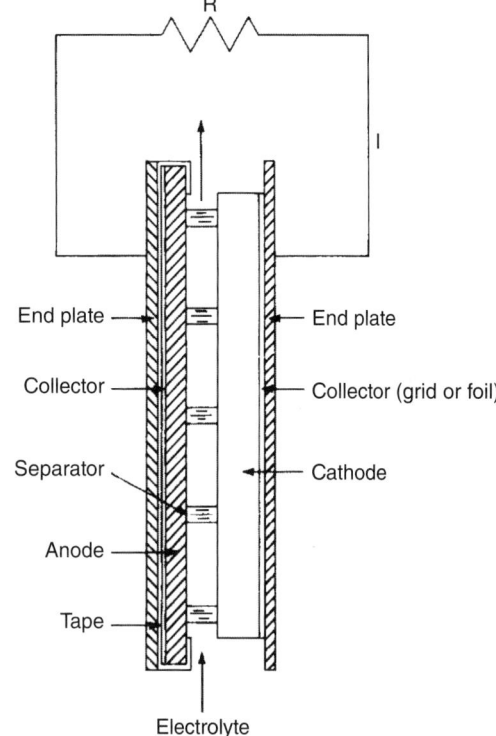

**FIGURE 29.10**    A basic water-activated cell.

and deposits more sludge. In recent years, magnesium alloys AP65 and MTA75 have been developed and evaluated. These are high-voltage alloys giving load voltages of 0.1 to 0.3 V higher than AZ61A. MTA75 is a higher-voltage alloy than AP65. These alloys foul more; under some forced-flow discharge conditions, the sludging problem may be controlled. These alloys are not used extensively in the United States; they are used in the United Kingdom and Europe in electric torpedo batteries. Composition ranges of these alloys are shown in Table 29.3.

CATHODE (POSITIVE PLATE). The cathode consists of a depolarizer and a current collector. These depolarizers are powders and are nonconductive. In order for the depolarizer to function, a form of carbon is added to impart conductivity; a binder is added for cohesion, and a metal grid is used as a current collector, a base for the cathode, to facilitate intercell connections and battery terminations. Possible cathode formulations are shown in Table 29.4.

Silver chloride is a special case. Silver chloride can be melted, cast into ingots, and rolled into sheet stock in thicknesses from about 0.08 mm up. Since this material is malleable and ductile, it can be used in almost any configuration. Silver chloride is nonconductive and is made conductive by superficially reducing the surface to silver by immersion in a photographic developing solution. No base grid need be used with silver chloride.

Nonsilver cathodes are usually prismatic in shape and are flat. Silver chloride cathodes are used flat and corrugated in many configurations.

SEPARATORS. Separators are nonconductive spacers placed between the electrodes of immersion- and forced-flow-type batteries to form a space for free ingress of electrolyte and egress of corrosion products. Separators in the form of disks, rods, glass beads, or woven fabrics may be used.[18,19]

Dunk-type batteries utilize a nonwoven, absorbent, nonconductive material for the dual purpose of separating the electrodes and absorbing the electrolyte.

INTERCELL CONNECTIONS. In a series-arranged battery of pile construction, the anode of one cell is connected to the cathode of the adjacent cell. To accomplish this without producing a short-circuited cell, an

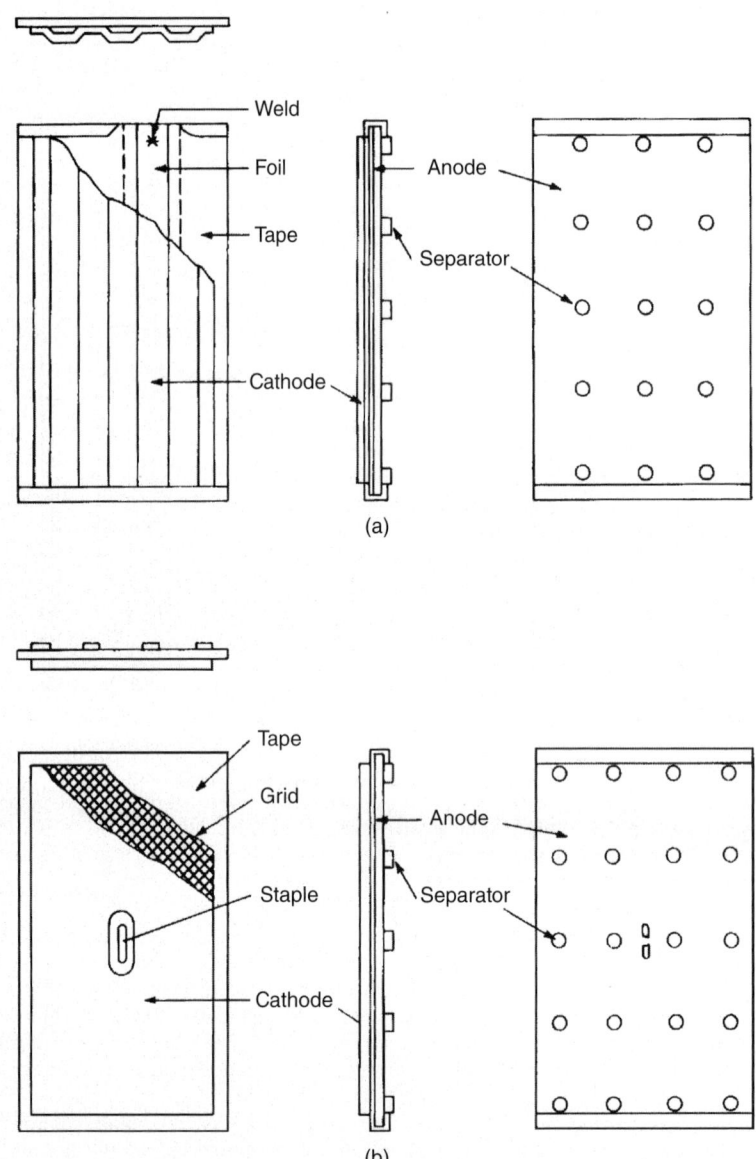

**FIGURE 29.11**  Duplex electrode assemblies. (*a*) Silver. (*b*) Nonsilver.

insulating tape or film is placed between the electrodes on nonsilver batteries. For silver batteries, silver foil is used alone or in conjunction with an insulating tape.

For nonsilver cells, the connection is made by stapling the electrodes together through the insulator.[20] For silver cells, the silver chloride, surface reduced to silver, is heat-sealed to silver foil, which has been previously bonded to the anode. Where large surface areas are involved, contact between silver and silver foil can be made by pressure alone.

**TABLE 29.3** Composition Range for Battery Plate Alloys

| | AZ31 | | AZ61 | | AP65 | | MELMAG 75 | |
|---|---|---|---|---|---|---|---|---|
| Element | %Min. | %Max. | %Min. | %Max. | %Min. | %Max. | %Min. | %Max. |
| Al | 2.5 | 3.5 | 5.8 | 7.2 | 6.0 | 6.7 | 4.6 | 5.6 |
| Zn | 0.6 | 1.4 | 0.4 | 1.5 | 0.5 | 1.5 | — | 0.3 |
| Pb | — | — | — | — | 4.4 | 5.0 | — | — |
| Tl | — | — | — | — | — | — | 6.6 | 7.6 |
| Mn | 0.15 | 0.7 | 0.15 | 0.25 | 0.15 | 0.30 | — | 0.25 |
| Si | — | 0.1 | — | 0.05 | — | 0.3 | — | 0.3 |
| Ca | — | 0.04 | — | 0.3 | — | 0.3 | 0.3 | — |
| Cu | — | 0.05 | 0.05 | 0.05 | 0.005 | — | — | — |
| Ni | — | 0.005 | — | 0.005 | — | 0.005 | — | 0.005 |
| Fe | — | 0.006 | — | 0.006 | — | 0.010 | — | 0.006 |

**TABLE 29.4** Cathode Compositions

| | Silver chloride[1] | Cuprous iodide[5,6] | Cuprous thiocyanate[8] | Lead chloride[4] | Cuprous chloride |
|---|---|---|---|---|---|
| Depolarizer, %/w | ++100 | 73 | 75–80 | 80.7–82.5 | 95–100 |
| Sulfur, %/w | — | 20 | 10–12 | — | — |
| Additive, %/w | — | — | 0–4 | 2.3–4.4 | — |
| Carbon, %/w | — | 7 | 7–10 | 9.6–9.8 | — |
| Binder, %/w | — | — | 0–2 | 1.5–1.6 | 0–5 |
| Wax, %/w | — | — | — | 3.8 | — |

TERMINATIONS.    For silver chloride cathodes, the lead is soldered directly to silver foil, which has been heat-sealed to one surface of the silver chloride. Leads are soldered directly to the collector grid of nonsilver cathodes or soldered to a piece of copper foil, which has been stapled to the collector grid.

The anode connection is made by soldering the lead to silver foil that has been bonded to the anode or by welding directly to the anode.

ENCASEMENT.    The battery encasement must effectively rigidize the battery and provide openings at opposite ends to allow free ingress and egress of electrolyte and corrosion by-products.

The periphery of the battery must be sealed in such a manner that the cells contact the external electrolyte only at the openings provided at the top and bottom of the battery. The encasement can be accomplished by using premolded pieces, caulking compounds, epoxy resins, an insulating sheet, or hot-melt resins.[18–21] For single batteries, these precautions are not necessary.

*Leakage Current.*    All the cells in the immersion- and forced-flow–type batteries operate in a common electrolyte. Since the electrolyte is conductive and continuous from cell to cell, conductive paths exist from each point in a battery to every other point. Current will flow through these conductive paths to points of different potential. This current is referred to as "leakage current" and is in addition to the current flowing through the load. Electrodes must be designed to compensate for these leakage currents.

Leakage currents for a small number of cells can be reduced by increasing the resistance path from a cell to the common electrolyte or that of the common electrolyte between adjacent cells. Leakage currents for a large number of cells can be reduced by increasing the resistance of the common electrolyte external to the individual cells.

During construction, the conducting paths from cell to cell are made as long as possible. In many instances, the negative or positive of the battery is connected to an external metal surface. Leakage currents flow from the battery to this surface. These leakage currents are controlled by placing a cap containing a slot over the battery openings. If one terminal is connected to an external conductive surface, the slot in the cap is opened to the electrolyte only on that side of the battery. Where neither terminal is connected to an external conductive surface, either end of the cap may be opened, but only on one side of the battery.

The resistance (ohms) of the slot in the cap may be calculated using the formula

$$R = p\frac{l}{a}$$

where $R$ = resistance, $\Omega$
    $l$ = length of slot, cm
    $a$ = cross-sectional area of slot, cm$^2$
    $p$ = resistance of electrolyte for temperature and salinity in which battery is operating, $\Omega \cdot$ cm

For dunk-type batteries, the electrolyte continuity from cell to cell is broken when the electrolyte is absorbed in the separator. The excess is poured off the battery or spun away from the cells by some external force applied to the battery.

*Electrolyte.*    Seawater-activated batteries are designed to operate in an infinite electrolyte, namely, the oceans of the world. However, for design, development, and quality-control purposes, it is not practical to use ocean water. Thus, it is common practice throughout the industry to use simulated ocean water. A commercial product, composed of a blend of all the ingredients required, simplifies the manufacture of simulated ocean water test solutions.

Dunk-type batteries, activated by pouring the electrolyte into the battery where it is absorbed by the separa-tor, can utilize water or seawater when the temperature is above the freezing point. At lower temperatures spe-cial electrolytes can be used. The use of a conducting aqueous electrolyte will result in faster voltage buildup. However, the introduction of salts in the electrolyte will increase the rate of self-discharge.

**Performance Characteristics.**    A summary of the performance characteristics of the major water-activated batteries currently available is given in Table 29.5.

*Voltage versus Current Density.*    Figures 29.12 and 29.13 are representative of voltage versus current density curves for several water-activated battery systems at 35 and 0°C, respectively, using a simulated ocean water electrolyte.

**TABLE 29.5**    Performance Characteristics of Water-Activated Batteries

| Cathode | Silver chloride | Lead chloride | Cuprous iodide | Cuprous thiocyanate | Cuprous chloride[a] |
|---|---|---|---|---|---|
| Anode | Magnesium | | | | |
| Electrolyte | Tap water, seawater, or other conductive aqueous solutions | | | | |
| Open-circuit voltage, V[b] | 1.6–17 | 1.1–1.2 | 1.5–1.6 | 1.5–1.6 | 1.5–1.6 |
| V per cell at 5 mA/cm$^2$ | 1.42–1.52 | 0.90–1.06 | 1.33–1.49 | 1.24–1.43 | 1.2–1.4 |
| Activation, s: | | | | | |
|   35°C[c] | <1 | <1 | <1 | <1 | |
|   RT[d] | — | — | — | — | 1–10 |
|   0°C[e] | 45–90 | 45–90 | 45–90 | 45–90 | |
| Internal resistance, $\Omega$[f] | 0.1–2 | 1–4 | 1–4 | 1–4 | 2 |
| Ah/g cath. theor.[g] | 0.187 | 0.193 | 0.141 | 0.220 | 0.271 |
| Usable capacity, % of theoretical | 60–75 | 60–75 | 60–75 | 60–75 | 60–75 |
| Wh/kg | 100–150 | 50–80 | 50–80 | 50–80 | 50–80 |
| Wh/L | 180–300 | 50–120 | 50–120 | 50–120 | 20–200 |
| Operating temperatures, °C[h] | −60 to +65 | | | | |

[a]All but cuprous chloride are immersion type. Cuprous chloride is dunk type.
[b]See voltage versus current density curves.
[c]Battery preconditioned at +55°C, then immersed in simulated ocean water of 3.6 wt %.
[d]Electrolyte at room temperature poured into battery and absorbed by separator.
[e]Battery preconditioned at −20°C, then immersed in simulated ocean water of 1.5 wt %.
[f]Depends on battery design.
[g]100% active material.
[h]Following activation at room temperature.

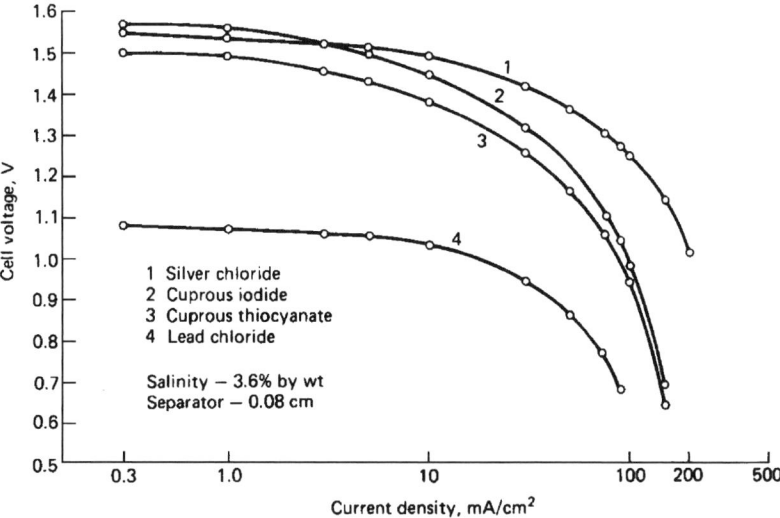

**FIGURE 29.12**    Representative cell voltages versus current density at 35°C.

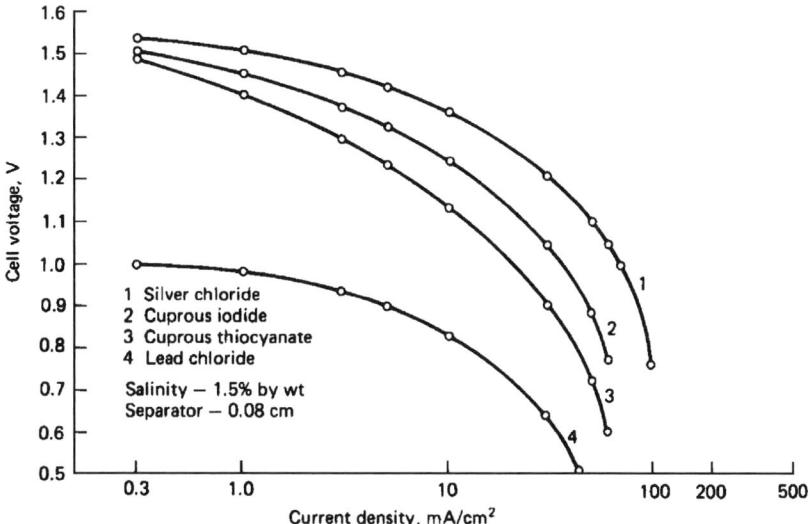

**FIGURE 29.13**    Representative cell voltages versus current density at 0°C.

*Discharge Curves.*    Discharge curves of the magnesium/silver chloride, magnesium/cuprous thiocyanate-sulfur, magnesium/cuprous iodide-sulfur, and magnesium/lead chloride electrochemical systems, discharged continuously through various resistances in simulated ocean water at high and low temperatures and salinities, are shown in Figs. 29.14 to 29.21. These data show the advantageous performance of the silver chloride system.

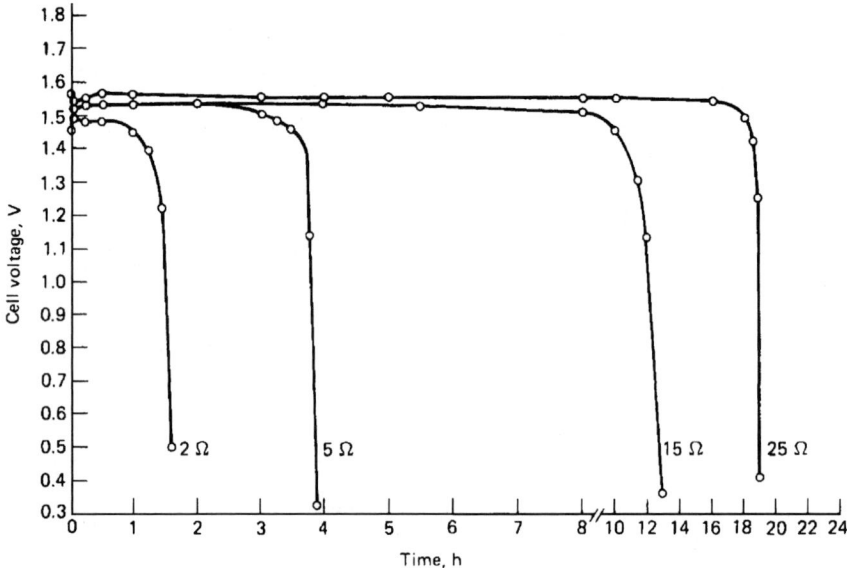

**FIGURE 29.14** A magnesium/silver chloride seawater-activated cell discharged continuously at 35°C in simulated ocean water, 3.6% salinity.

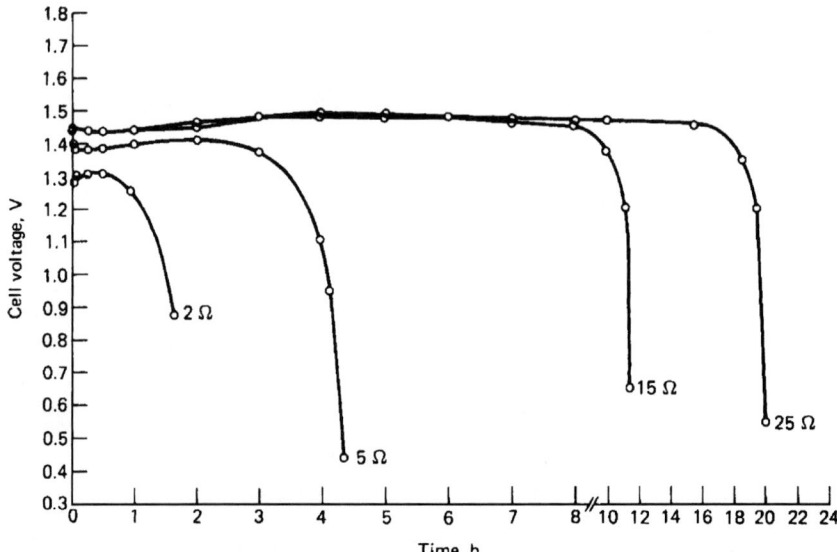

**FIGURE 29.15** A magnesium/silver chloride seawater-activated cell discharged continuously at 0°C in simulated ocean water, 1.5% salinity.

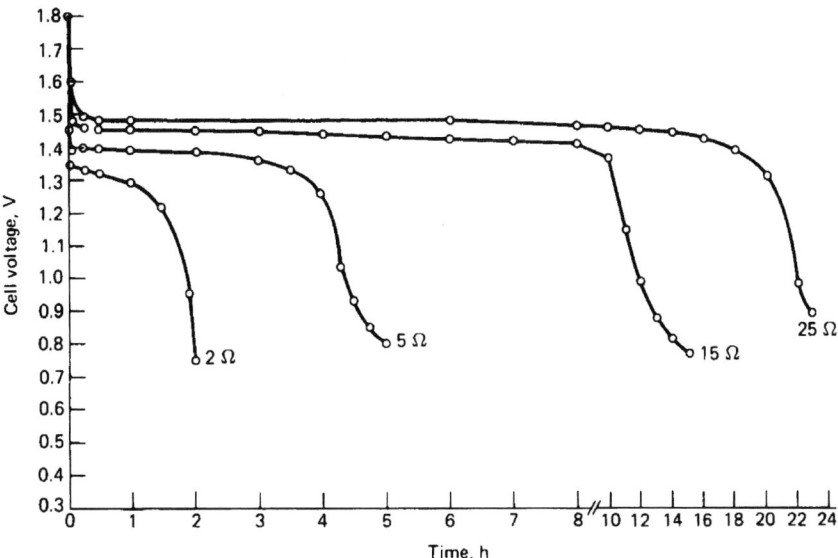

**FIGURE 29.16**  A magnesium/cuprous thiocyanate seawater-activated cell discharged continuously at 35°C in simulated ocean water, 3.6% salinity.

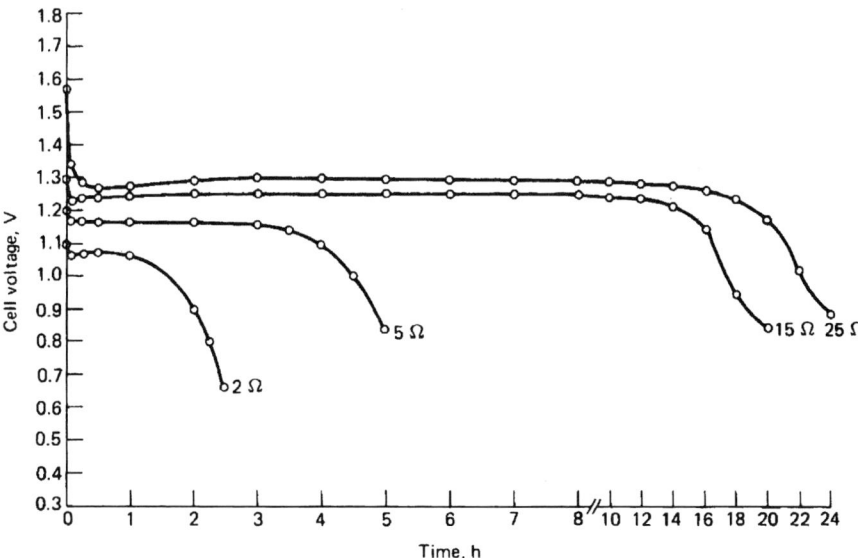

**FIGURE 29.17**  A magnesium/cuprous thiocyanate seawater-activated cell discharged continuously at 0°C in simulated ocean water, 1.5% salinity.

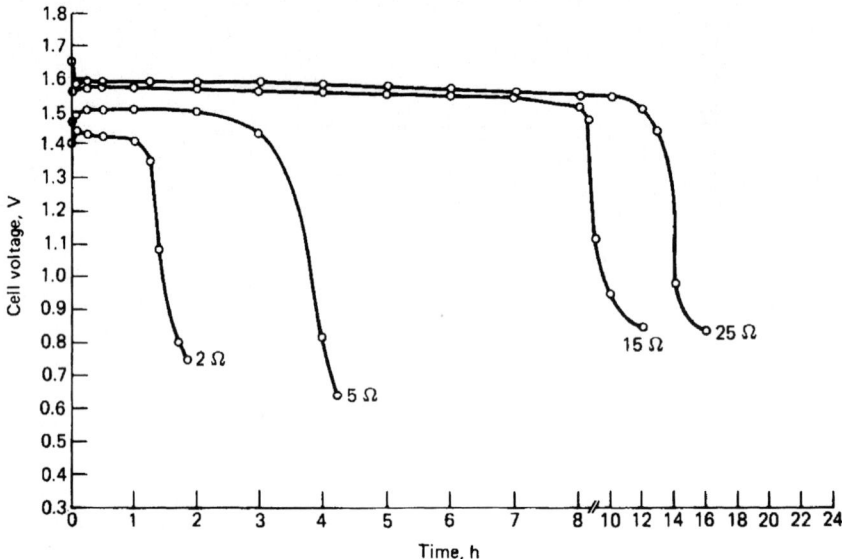

**FIGURE 29.18** A magnesium/cuprous iodide seawater-activated cell discharged continuously at 35°C in simulated ocean water, 3.6% salinity.

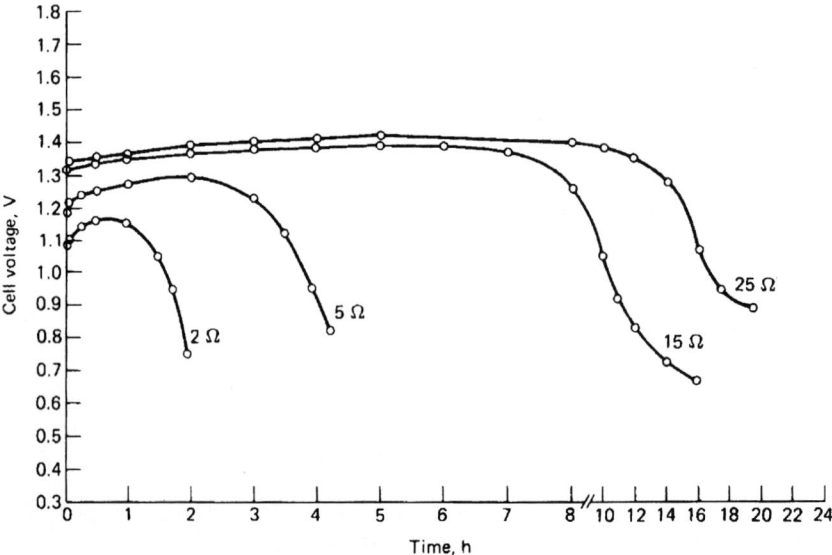

**FIGURE 29.19** A magnesium/cuprous iodide seawater-activated cell discharged continuously at 0°C in simulated ocean water, 1.5% salinity.

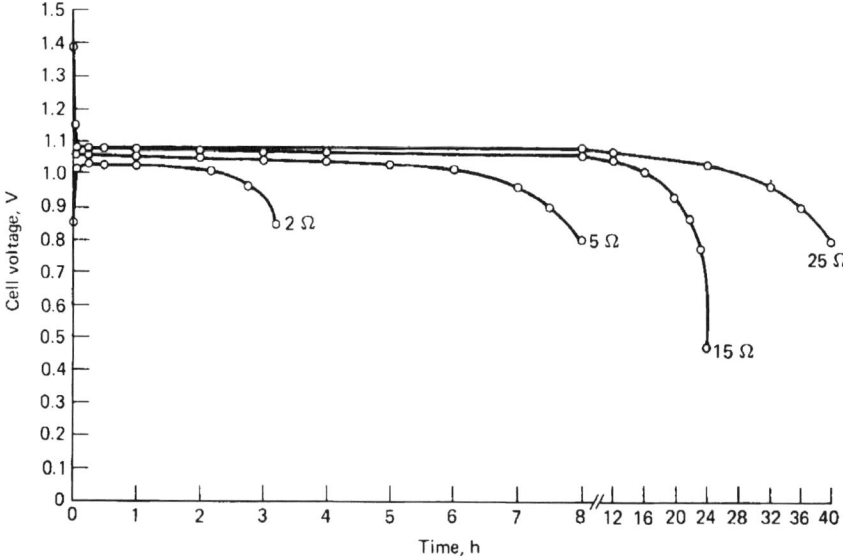

**FIGURE 29.20**  A magnesium/lead chloride seawater-activated cell discharged continuously at 35°C in simulated ocean water, 3.6% salinity.

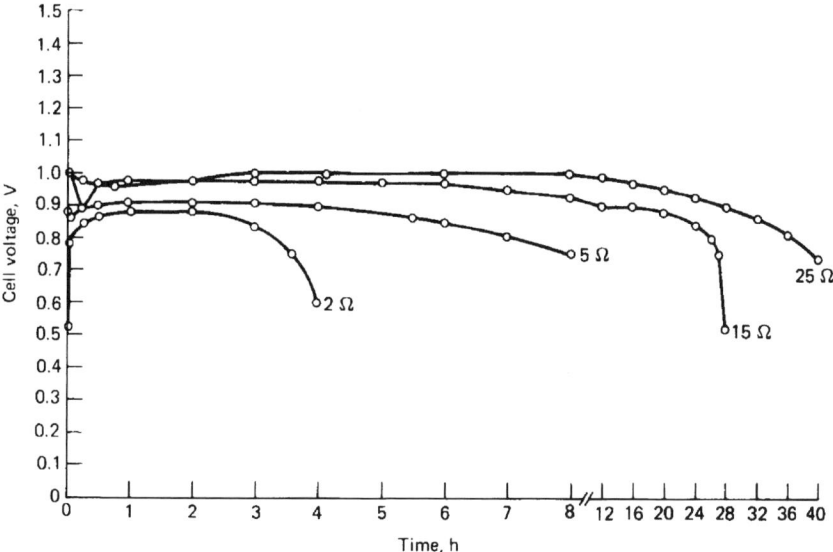

**FIGURE 29.21**  A magnesium/lead chloride seawater-activated cell discharged continuously at 0°C in simulated ocean water, 1.5% salinity.

*Service Life.*     The capacities per unit of weight versus the average power output of these same electrochemical systems, at high and low temperatures and salinities, are shown in Figs. 29.22 and 29.23.

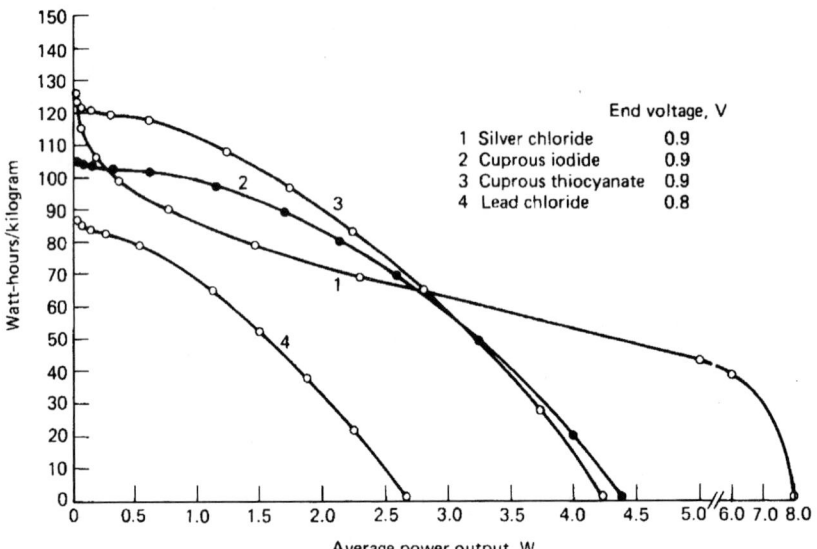

**FIGURE 29.22**   Capacity versus power output of seawater-activated cells discharged continuously at 35°C in simulated ocean water, 3.6% salinity.

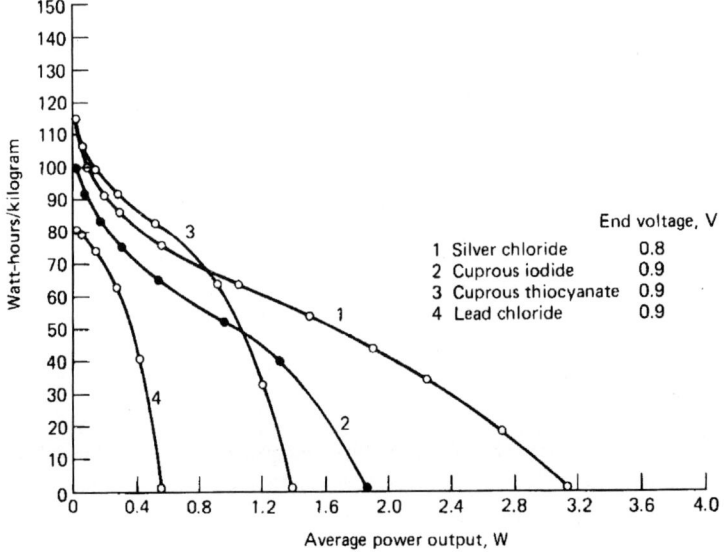

**FIGURE 29.23**   Capacity versus power output of seawater-activated cells discharged continuously at 0°C in simulated ocean water, 1.5% salinity.

*Immersion-Type Batteries.*    The performance of these same systems, designed as immersion-type batteries to meet the physical, electrical, and environmental specifications listed in Table 29.6, are shown in Figs. 29.24 to 29.26. The performance characteristics are summarized in Table 29.7.

**TABLE 29.6**    Performance Specifications for a Seawater-Activated Battery

| Load | $80 \pm 2\ \Omega$ | |
| --- | --- | --- |
| Life | 9 h | |
| Voltage | 15.0 V: Initial from 90 s to 9 h | |
| | 19.0 V max. | |
| Activation* | 60 s to 13.5 V | |
| | 90 s to 15.0 V | |
| Battery size: | Silver | Nonsilver |
| Height, cm | 7.7 max. | 10.6 max. |
| Width, cm | 5.7 max. | 7.6 max. |
| Thickness, cm | 4.2 max | 5.7 max. |
| Weight, g | $255 \pm 14$ | $482 \pm 85$ |
| Environmental: | | |
| Storage | From –60 to +70°C for 5 years† | |
| | 90 days at –50 to +40°C at 90% RH | |
| | 10 days per MIL-T-5422E | |
| Vibration, Hz | 5–500 | |
| Electrolyte: | | |
| Low temperature | Ocean water of 1.5% salinity by weight at $0 \pm 1$°C | |
| High temperature | Ocean water of 3.6% salinity by weight at $+34 \pm 1$°C | |

*Battery preconditioned at –20°C prior to immersion in ocean water of 1.5% salinity by weight at $0 \pm 1$°C.

†In equipment packed in sealed plastic container with appropriate desiccant.

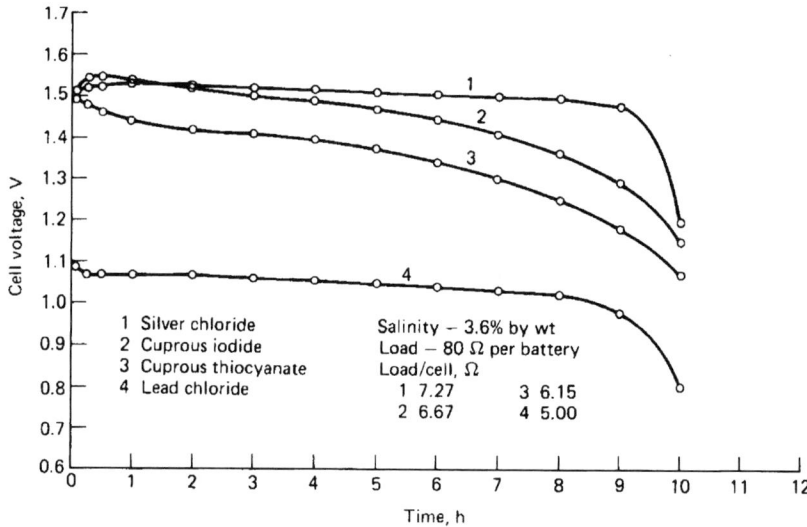

**FIGURE 29.24**    Discharge curves of seawater-activated batteries at 35°C.

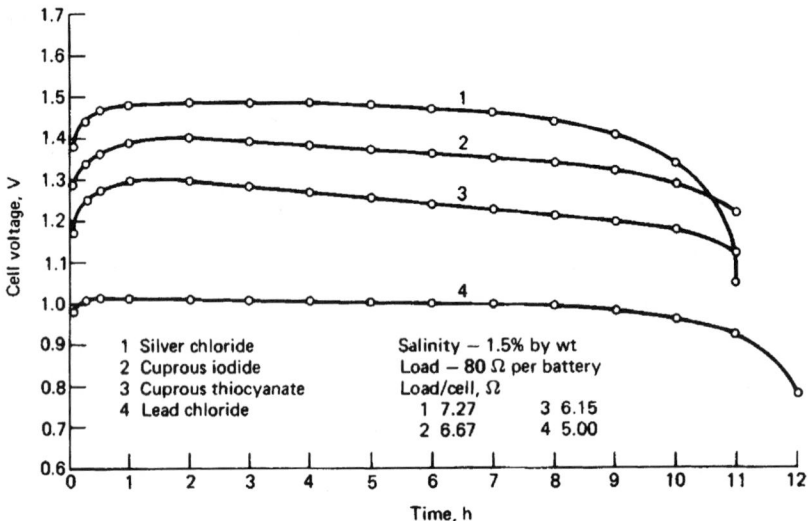

**FIGURE 29.25**    Discharge curves of seawater-activated batteries at 0°C.

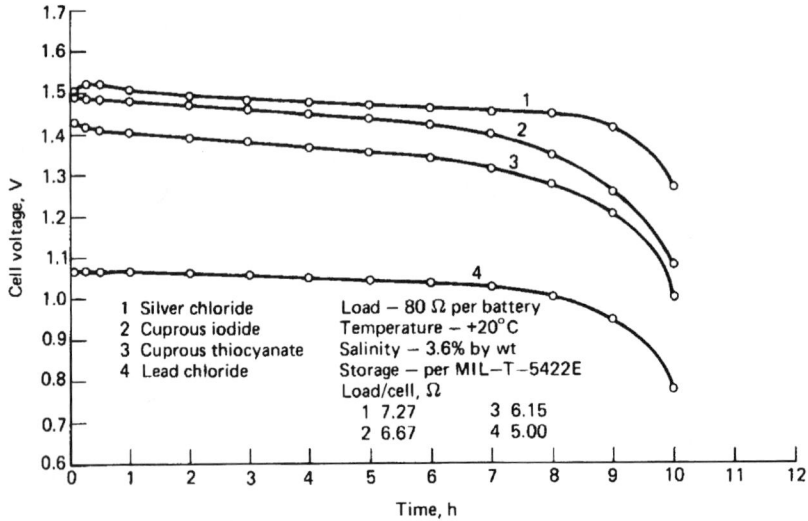

**FIGURE 29.26**    Discharge curves of seawater-activated batteries, 10-day humidity at 90% RH.

*Forced-Flow Batteries.*    With the development of the recirculation system in which the inflow of fresh electrolyte can be controlled, thereby maintaining the temperature and conductivity of the electrolyte, the performance of electric torpedo batteries has been improved markedly. With recirculation and flow control, a recirculation pump and a voltage-sensing mechanism are added to the battery system. By this method, the temperature of the battery and the conductivity of the seawater electrolyte increase. Since battery voltage increases directly with temperature and conductivity, it is possible to control the output of the battery by controlling the intake of electrolyte by means of the voltage-sensing mechanism.

**TABLE 29.7**  Performance Summary of Seawater-Activated Batteries

| | Silver chloride | Cuprous iodide | Cuprous thiocyanate | Lead chloride |
|---|---|---|---|---|
| Number of cells | 11 | 12 | 13 | 16 |
| Battery dimensions: | | | | |
|    Height, cm | 7.5 | 9.8 | 10.2 | 10.5 |
|    Width, cm | 5.5 | 7.6 | 7.4 | 7.5 |
|    Thickness, cm | 3.9 | 4.4 | 5.7 | 4.5 |
|    Weight, g | 252 | 516 | 478 | 458 |
| Activation: | | | | |
| Low temp.: | | | | |
|    To 13.5 V, s | <15 | <15 | <15 | <15 |
|    To 15.0 V, s | 60 | 60 | 60 | 15 |
| High temp.: | | | | |
|    To 15.0 V, s | <1 | <1 | <1 | <1 |
| Life: | | | | |
|    High temp., h | 9.67 | 9.4 | 9.3 | 9.5 |
|    Low temp., h | 9.80 | 10.3 | 10.3 | 10.7 |
| Load resistance (per cell), $\Omega^*$ | 7.27 | 6.67 | 6.15 | 5.0 |
| Cutoff voltage (per cell), $V^*$ | 1.364 | 1.25 | 1.154 | 0.9375 |
| Average current, A | 0.206 | 0.220 | 0.236 | 0.219 |
| Average volts per cell, $V^*$ | 1.497 | 1.463 | 1.378 | 1.048 |
| Wh/L | 204 | 110 | 90 | 100 |
| Wh/kg | 130 | 70 | 79 | 75 |

$^*$As each battery system contains a different number of cells, cell load resistances and cell voltages are different for each battery.

The performance of one type of torpedo battery with and without recirculation voltage control is shown in Fig. 29.27.[22] The blocked-in area represents the limits within which an electric torpedo battery with recirculation and flow control will perform when discharged under any of the conditions shown by the three individual curves. All voltages pertinent to the start and finish of the battery are shown by the three individual curves.

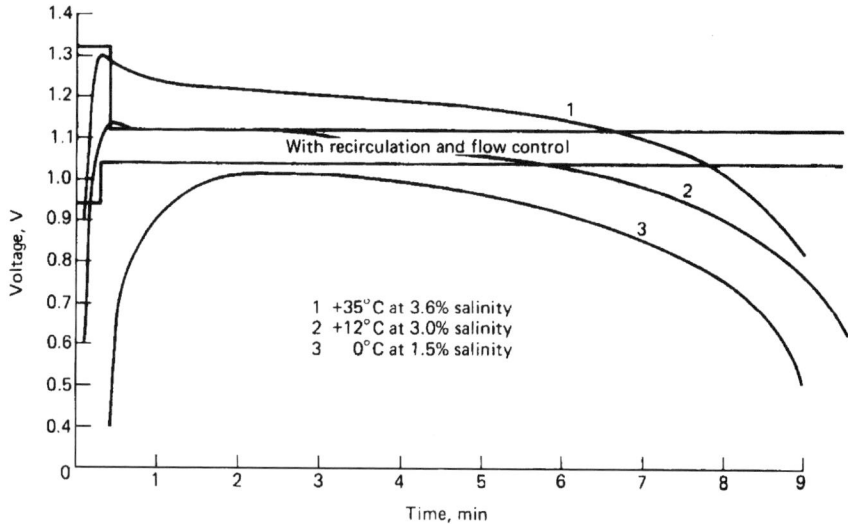

**FIGURE 29.27**  Discharge curves of a torpedo battery—effect of recirculation and flow control.

*Dunk-Type Batteries*

MAGNESIUM/CUPROUS CHLORIDE BATTERIES.    The magnesium/cuprous chloride battery was widely used in applications requiring low-temperature performance, such as radiosondes, having replaced the more expensive magnesium/silver chloride system in applications where weight and volume are not critical. Figure 29.28 illustrates a typical magnesium/cuprous chloride battery. The pile-type construction shown in Fig. 29.5 is used.

The battery is activated by filling it with water, and full voltage is reached within 1 to 10 min. The battery is best suited for discharge at about the 1 to 3 h rate at temperatures from +60 to –50°C after activation at room temperature. Overheating and dry-out will occur on high current drains, and self-discharge limits the life after activation. For best service, these batteries should be put into use soon after activation. The heat that is developed during discharge can be used to advantage in batteries that are operated at low temperatures; therefore, the energy output varies little with decreasing temperature. Figure 29.29 shows the discharge curve for this battery at various temperatures. Figure 29.30 gives some typical discharge curves for this type of battery with a similar design at various discharge loads.

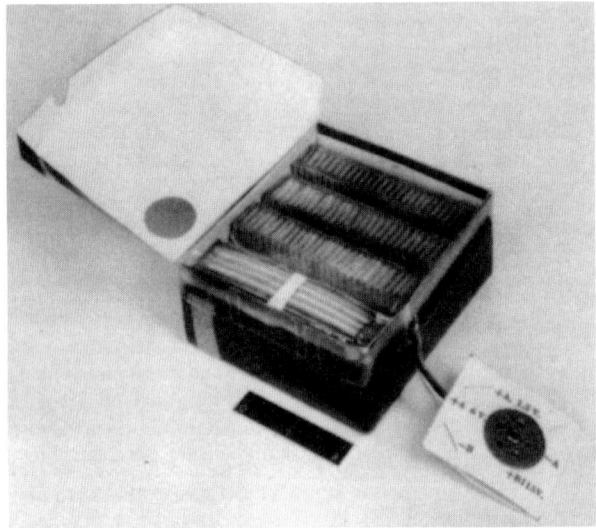

**FIGURE 29.28**  A magnesium/cuprous chloride radiosonde battery. Size: 10.2, 11.7, 1.9 cm; weight: 450 g; rated capacity: $A_1$ section—1.5 V, 0.3 Ah; $A_2$ section—6.0 V, 0.4 Ah; B section—115 V, 0.08 Ah.

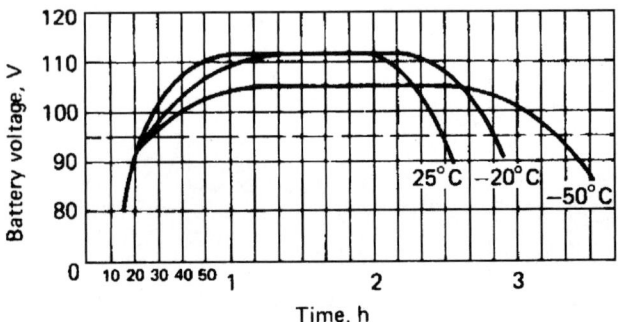

**FIGURE 29.29**  Discharge curves of a magnesium/cuprous chloride radiosonde battery, 115 V section; discharge load: 3050 W.

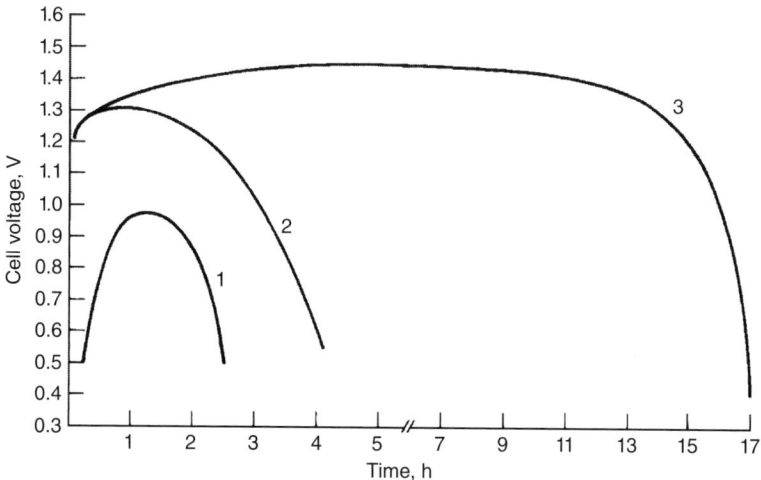

**FIGURE 29.30**  Discharge curves of magnesium/cuprous chloride water-activated batteries at 20°C; electrolyte: tap water.

| Cell no. | Load, Ω | Dimensions, cm | | | |
|---|---|---|---|---|---|
| | | Volume | Length | Height | Thickness |
| 1 | 2.5 | 10.2 | 8.2 | 2.5 | 0.5 |
| 2 | 8.0 | 2.5 | 2.2 | 3.8 | 0.3 |
| 3 | 125 | 1.3 | 2.0 | 2.0 | 0.3 |

*Magnesium/Manganese Dioxide Battery.*    This reserve battery consists of a magnesium anode and a manganese dioxide cathode.[10,23] It is activated by pouring an aqueous magnesium perchlorate electrolyte into the cells of the battery, where it is absorbed by the separators. Electrolyte absorption occurs within a few seconds at 0°C or above, but 3 min or more are required at −40°C due to the viscosity of the electrolyte.

The battery can deliver between 80 and 100 Wh/kg over the temperature range of −40 to +45°C at the 10 to 20 h discharge rate. Over 75% of the battery's fresh capacity is available after 7 days' activated stand at 20°C and 4 days' storage at 45°C. Typical discharge curves are shown in Figs. 29.31a and 29.31b for a five-cell 10 Ah battery, weighing about 1 kg with a volume of 655 cm².

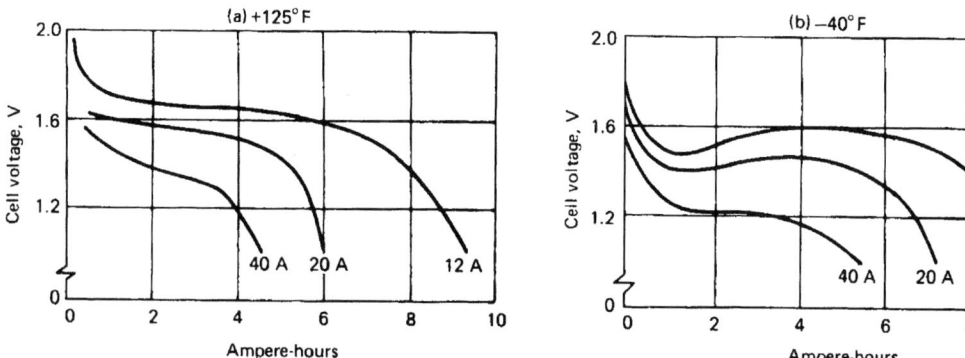

**FIGURE 29.31**  Typical discharge curves of a magnesium/manganese dioxide cell, 10 Ah size. (*Courtesy of Eagle-Picher Industries.*)

### 29.3.2 Zinc/Silver Oxide Batteries

***Construction.*** A typical assembly of the manually activated reserve zinc/silver oxide cell is shown in Fig. 29.32. These batteries are designed to be filled with electrolyte just before use. The conventional cell design is a prismatic thermoplastic container with positive and negative terminals and a combination fill/vent cap. Batteries are formed by connecting single cells in series and packaging them in a unit container. Batteries used in space programs utilize thin-gauge stainless steel, aluminum, titanium, magnesium, or composite containers to minimize weight.

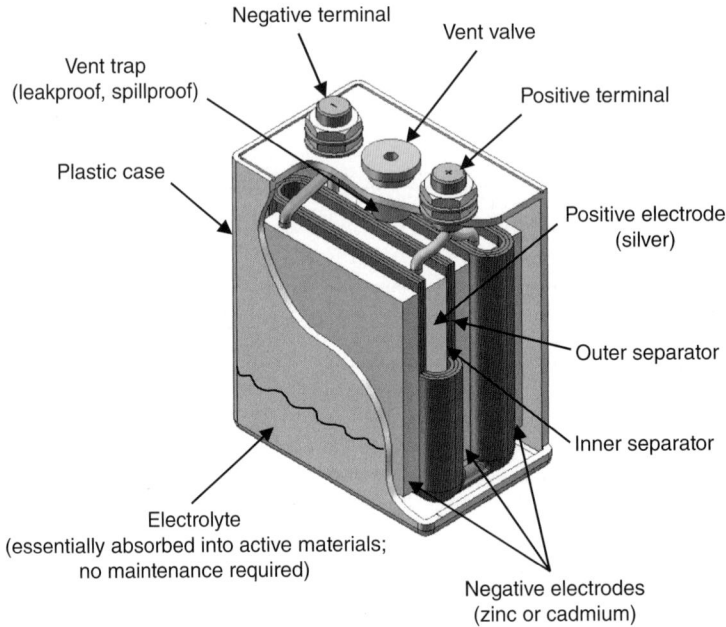

**FIGURE 29.32** Typical construction of a primary reserve zinc/silver oxide cell. (*Courtesy of Yardney Technical Products, Inc.*)

*Cell Components.* The components of a reserve zinc/silver oxide cell consist of the positive plates (silver), the negative plates (zinc), and the separators. The components are assembled such that each negative plate is protected from direct contact with the adjacent positive plate by a separator. The cell components are assembled and packaged in a container; the plates can be prepared in either a dry and charged condition or dry and uncharged condition. The latter requires formation by the user.

An alternate construction technique that has been successfully used on various applications is pile batteries. The major incentive for use of bipolar cells in a pile configuration is the elimination of intercell connectors, other heavy current-carrying structural elements, and some of the cell containment features. This allows for a substantial weight reduction, increasing the power and energy density of the battery. Some of the disadvantages of the bipolar electrode are that the electrode can only react on one surface and the potential for intercell electrolyte leakage, which could create parasitic currents between some cells.

The construction usually consists of a bipolar or duplex matrix where both positive and negative electrodes are built on a common current collector. These are stacked in a pile configuration with a separator or polymeric standoffs placed between each bipolar electrode. This results in a high-voltage, multiple-cell battery. Each cell consists of the positive side of a bipolar electrode, the negative side of the next bipolar electrode, the separator in between, and a plastic frame. The common current collector has two main functions. It separates the positive and negative portions of each bipolar electrode and serves as an intercell connector and electrolyte barrier.

POSITIVE PLATES.    The positive plates are prepared by applying silver or silver oxide powder to a metallic grid. Copper, nickel, and silver have all been used for grid material, with silver being the most prevalent for reasons of electrochemical stability and conductivity. After the silver powder is pressed or sintered to the grid, the plates are electroformed in an alkaline solution, then rinsed thoroughly and air-dried at a moderate temperature (usually 20–50°C). The nominally divalent oxide thus formed is relatively stable at ambient temperatures but tends to lose oxygen and degrade to the monovalent state with increasing temperatures and time. Continuous exposure to high temperatures (70°C) causes reduction to the monovalent oxide in a few months.

NEGATIVE PLATES.    The negative plates may be prepared by pasting or pressing zinc powder or zinc oxide onto a grid or by electroplating zinc from an alkaline bath to form a very active spongy zinc deposit.

Both positive and negative electrodes may vary in thickness from 0.12 mm as a practical minimum to 2.5 mm maximum for positives and 3.5 mm maximum for negatives. The extremely thin plates are utilized for very short-life, high-discharge-rate automatically activated batteries; the thick plates are employed in manually activated batteries designed for continuous discharge over several months at very low currents.

SEPARATOR MATERIALS.    Typical separator materials used in zinc/silver oxide cells include regenerated cellulose films (cellophane, fiber-reinforced, or silver-treated cellophane), and woven nylon or nonwoven synthetic fiber mats of nylon, hemp tissue, polyvinyl alcohol, rayon, or high alpha cellulose content papers. The woven nylon, paper, or synthetic fiber mats are frequently placed adjacent to the positives to protect the cellophane from the highly oxidizing influence of that material. The cellophane, a semipermeable film, prevents buildup of particles between plates (while allowing ionic transfer), thus preventing interplate short circuits. The mats also absorb the electrolyte solution and distribute it over the electrode surfaces. Cells intended for automatic activation normally are not designed with the film separators for they take too long for complete wetting. The open-mat separators provide sufficient protection from interplate short-circuiting for several minutes to hours.

Separator materials are necessary for cell operation because they prevent short circuits, but they also impede current flow, causing an *IR* drop within the cell. Very high-discharge-rate cells must have very low internal impedance, hence a minimum of separator material. As a result, this type of cell is restricted to very short wet-life applications. The semipermeable film is the separator that contributes most to *IR* drop and also to protection against short circuits. Long-life cells may contain five or six layers of cellophane. They are therefore better suited to medium or low discharge rates.

ELECTROLYTE.    The electrolyte used for reserve zinc/silver oxide cells is an aqueous solution of potassium hydroxide. High and medium discharge rate cells use a 31% by weight electrolyte solution because this composition has the lowest freezing point and is close to the minimum resistance, which occurs at 28 wt.%. Low-rate cells may use a 40% to 45% solution since lower rates of hydrolysis of cellulosic separators occur with the higher KOH concentrations.

*High- and Low-Rate Designs.*    A battery intended to be discharged at a 5 to 60 min rate is considered a high-rate design. These cells are designed primarily to deliver high current and require a large plate surface area. They therefore contain many very thin plates. These separators also must have the lowest possible impedance, that is, one or two layers of cellophane versus five or six layers for low-rate cells. The 31% potassium hydroxide electrolyte has a high conductivity and is, therefore, employed in high-rate cells.

Low-rate batteries are in a class intended for discharge at rates ranging from 10 to 1000 h with emphasis on high specific energy and energy density. The plates are thick (2 mm), and relatively high-impedance separator wraps are used. A higher concentration of electrolyte (40%), which permits a greater ampere-hour capacity, can also be used. This design configuration also gives a substantial improvement in the activated or wet-stand capability of the cell.

*Automatically Activated Types.*    The automatically activated-type battery is a class of reserve battery intended for quick preparation for use after an undetermined period subsequent to installation. The very high power output of the primary zinc/silver oxide system and the use of an integrally designed system for injecting the electrolyte into the cells combine to provide an efficient power source for weapons and other systems requiring a long-term ready state. Figure 29.33*a* shows typical automatically activated batteries for military applications. Figure 29.33*b* shows a typical automatically activated battery for a torpedo application.

Four kinds of activation systems have been utilized in this type of battery for transferring electrolyte from a reservoir to the cells. All the systems depend on gas pressure above ambient (e.g., 4 to 1 ratio) to move the electrolyte, and the most conventional source of gas is a pyrotechnic device.

(a)

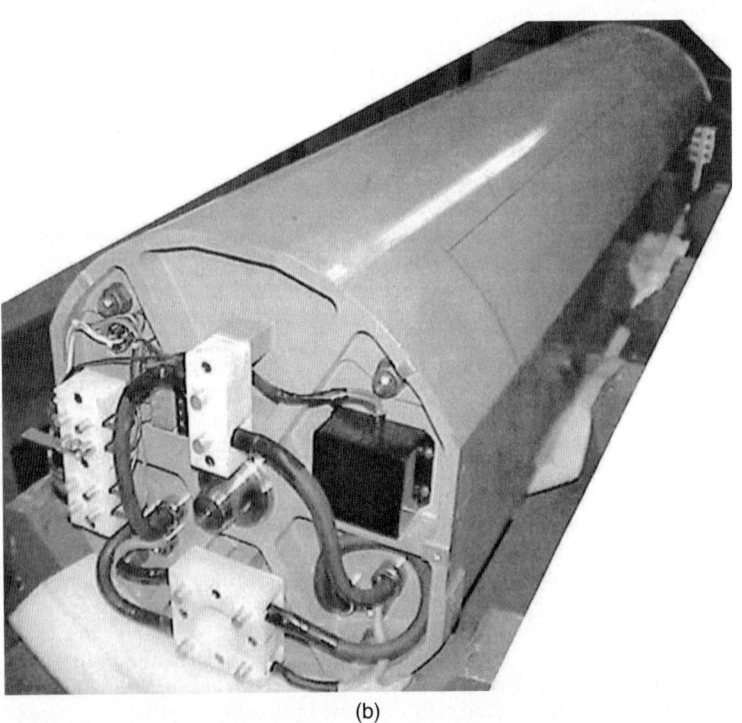

(b)

**FIGURE 29.33**    (*a*) Zinc/silver oxide primary reserve batteries designed for automatic activation. (*b*) An automatically activated zinc/silver oxide torpedo battery. (*Courtesy of Yardney Technical Products, Inc.*)

The "gas generator" is a small cartridge that contains an ignitable propellant material and an electrically fired ignitor or "match." Figures 29.34a to 29.34d show the four types of battery designs. In some instances high pressure tanks have also been used for larger torpedo batteries.

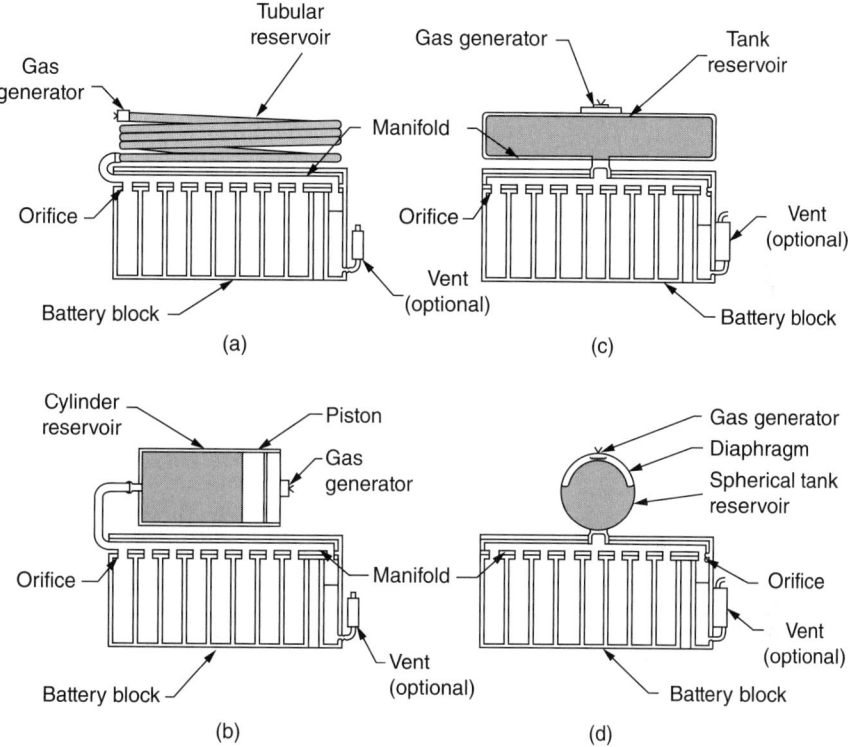

**FIGURE 29.34**  Schematic drawings of four types of activation systems used in automatically activated batteries. (*a*) Tubular reservoir. (*b*) Piston activator. (*c*) Tank activator. (*d*) Tank-diaphragm activator. (*Courtesy of Yardney Technical Products, Inc.*)

The tubular reservoir (Fig. 29.34a) can assume many forms. It is usually coiled around the battery, as shown in Fig. 29.35 (an assembly of a battery with a tubular reservoir), but it can also be formed with 180° bends into a flat shape, or it can be configured to fit into available nonstandard volumes into which a missile battery is often mounted. The tubular reservoir is fitted with foil diaphragms at each end. For activation, the gas generator located at one end can be electrically ignited; the gas causes the diaphragms to break, and the electrolyte is forced into a manifold which distributes it to the cells of the battery. The piston activator (Fig. 29.34b) operates by pushing the electrolyte out of a cylindrical reservoir when a gas generator is fired behind it. The tank activator (Fig. 29.34c) contains the electrolyte in a variable-geometry tank with a gas generator located at the top. When the gas enters at the top, the electrolyte is forced out through an aperture at the bottom. The system is position-sensitive and will operate properly only when in an upright position relative to the components. The tank-diaphragm activator (Fig. 29.34d) uses a sphere or spheroid tank with a diaphragm attached internally at the major circumference. When the gas generator is fired, the diaphragm moves to the opposite side, forcing the electrolyte out through an aperture in the reservoir side of the tank.

Of the four systems, the tubular system is the most versatile, but in simple battery shapes it may be heavier. The piston and diaphragm systems have moving parts and thus can be less reliable; they are also less adaptable to special shapes. The tank is efficient but position-sensitive.

**FIGURE 29.35** Assembly of an automatically activated zinc/silver oxide primary battery with a tubular coil reservoir. (*Courtesy of Eagle-Picher Technologies.*)

Depending on the battery design and intended applications, the battery can be configured to externally vent or retain any internal gases due to the activation system and normal gassing of the cells.

The operating sequence of an automatically activated battery involves (1) application of ignition current, (2) gas generator propellant burning and associated gas production, (3) rupture of a diaphragm, (4) movement of the electrolyte out of the reservoir into the distribution manifold, and (5) filling of the cells with the electrolyte. In a typical operation, the total sequence involves less than 1 s. In many applications the electrical load is wired directly to the battery, and so the battery activates under load. Figure 29.36 shows the rise times of the voltage under load (A) and under no-load (B) condition for a battery not used until 6 h later. The delayed-use battery has film separators, and the slower wetting is reflected by the longer rise time.

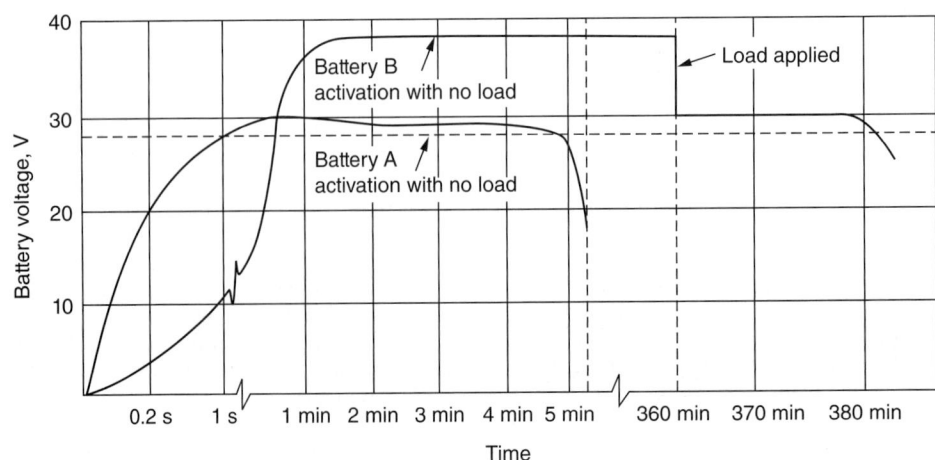

**FIGURE 29.36** Voltage rise time for automatically activated zinc/silver oxide batteries at 25°C. (*Courtesy of Eagle-Picher Technologies.*)

In some applications, reserve batteries include multiple power taps and can utilize cells with different capacity ratings within the same battery container.

Automatically activated batteries suffer a weight and volume penalty compared with manually activated types, but the design permits the use of a high-performance battery when there is no time available to activate manually or the unit is inaccessible. In many applications, both conditions exist. The volume penalty is usually about 2 times and the weight penalty about 1.6 times the basic battery. Most automatically activated battery designs utilize an integral electric heater. The heater maintains the electrolyte at about 40°C or at a temperature that will raise a cold battery to 40°C when activation occurs. The use of heaters permits the design of batteries that can meet close voltage tolerances, thus improving the capability of the weapons' electric and electronic systems when operating over a wide temperature range.

***Performance Characteristics.***    Zinc/silver oxide reserve batteries as a class are somewhat unique in that they are almost entirely committed to specific applications. These applications require the flat voltage profile and the high specific energy and energy density available from this system, and they often demand a special design for each requirement. If a low-temperature environment is involved, battery heaters are used. If the discharge requires a wide range of current with only a small voltage variation, many very thin plates are used. A very high capacity requirement at low rates requires the use of thicker plates and more concentrated electrolyte. There is no standard design or size because there is no typical application. The applications always demand the maximum from the battery design in capacity and voltage regulation at a minimum weight and volume. Multiple batteries are commonly packaged as a single unit, providing a range of load current and ampere-hour capacity in one convenient package.

*Voltage.*    The open-circuit voltage of the zinc/silver oxide cell will range from 1.6 to 1.85 V per cell. The nominal load voltage is 1.5 V, and typical end voltages are 1.35 V for low-rate cells and 1.2 V for high-rate cells. At high rates, such as a 5 to 10 min discharge rate, the output voltage would be about 1.3 to 1.4 V per cell, whereas the 2 h rate discharge voltage would be slightly above 1.5 V. Figure 29.37 shows a family of discharge curves at four different current densities. The voltage level is inversely related to the current density (calculated from the area of the active plate surface). Thus, based on 100 cm$^2$ of positive-plate surface, a 10 A discharge rate would be a 0.1 A/cm$^2$ current density. If the discharge rate is doubled to 0.2 A/cm$^2$, the voltage level would decrease, and if the rate is lowered to 0.05 A/cm$^2$, the voltage level would increase.

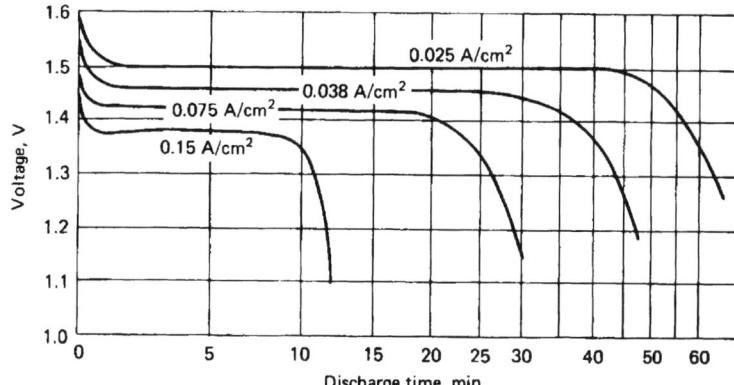

**FIGURE 29.37**    Effect of changing current density on battery voltage at 25°C. (*Courtesy of Eagle-Picher Technologies.*)

In cell design, the ampere-hour capacity of the cell is determined by the amount of silver oxide active material present. Zinc active material is provided in excess because of the cost relationship to silver. Voltage is determined by the current density. In a fixed volume, higher discharge rates can be obtained without lowering the battery voltage by using thinner plates (thus providing more plates per cell element and lowering the current density), but with a reduction of capacity. The lower the current density at which a battery can operate, the better the voltage regulation with changing rates of discharge.

*Discharge Curves.*    A set of discharge curves for high-rate batteries is shown in Fig. 29.38 and for low-rate batteries in Fig. 29.39. The designs for these two types of batteries are quite different, with the principal difference being the thickness of the plates. The thin plates used in high-rate cells provide more surface area for lower current density, thus better voltage control and also more efficient utilization of the active material. At lower rates of discharge, as in Fig. 29.39, the voltage level is higher and active material utilization is also excellent, both because of lower current density. It will be noted that the low-rate discharge curves are above 1.6 V for a period of time. This is the effect of the divalent oxide, which affects voltage only at low rates. Most of the divalent capacity is obtained at high rates, but its voltage is decreased by the higher current density imposed.

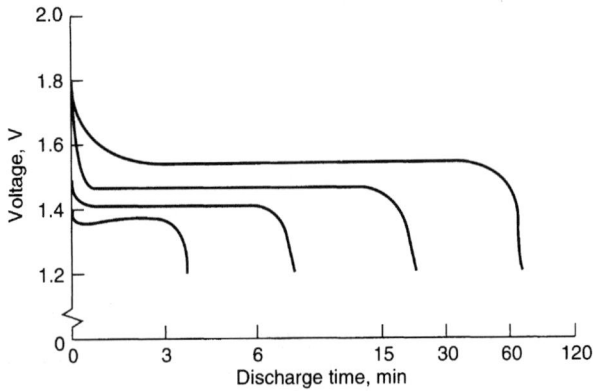

**FIGURE 29.38**    Discharge curves for high-rate zinc/silver oxide batteries at 25°C. (*Courtesy of Eagle-Picher Technologies.*)

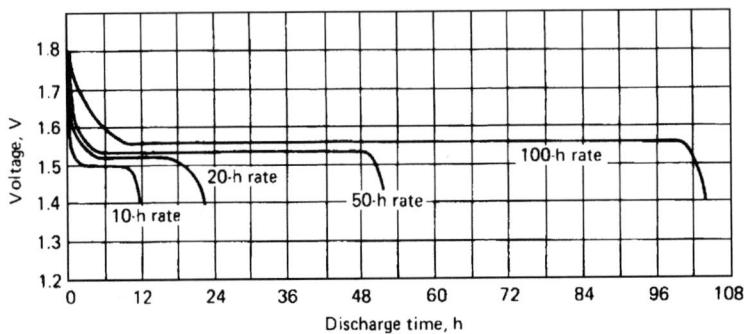

**FIGURE 29.39**    Discharge curves for low-rate zinc/silver oxide batteries at 25°C. (*Courtesy of Eagle-Picher Technologies.*)

*Effect of Temperature.*    The family of curves shown in Fig. 29.40 illustrates the performance obtained from a high-rate battery when discharged over a range of temperatures. It should be understood that the change in voltage levels caused by temperature is closely related to the changes caused by current density. Thus the adverse effect of cold temperature can be improved by reducing the current density of the cell, and the voltage and capacity of batteries discharged at high current densities can be improved by increasing their operating temperature. Figure 29.41 shows a family of curves for low-rate batteries discharged at various temperatures. The two sets of curves show that the zinc/silver oxide system is significantly affected at temperatures below 0°C and thus is not recommended for applications in this environment without heaters.

*Impedance.*    Figure 29.42 shows the dynamic internal resistance (DIR) of a high-rate cell at various stages of discharge and temperature. These curves show a declining ($\Delta V/\Delta I$) ratio until the end of discharge, at which

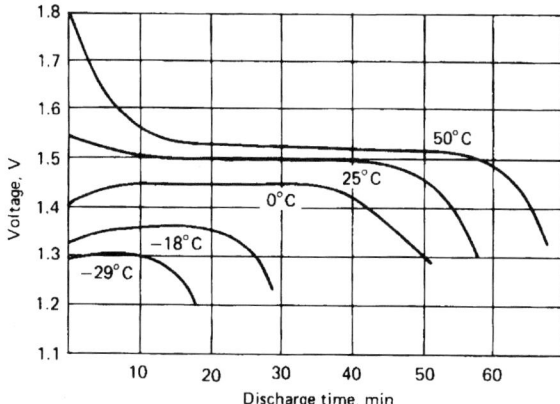

**FIGURE 29.40**   Effects of temperature on high-rate zinc/silver oxide primary batteries discharged at the 1 h rate. (*Courtesy of Eagle-Picher Technologies.*)

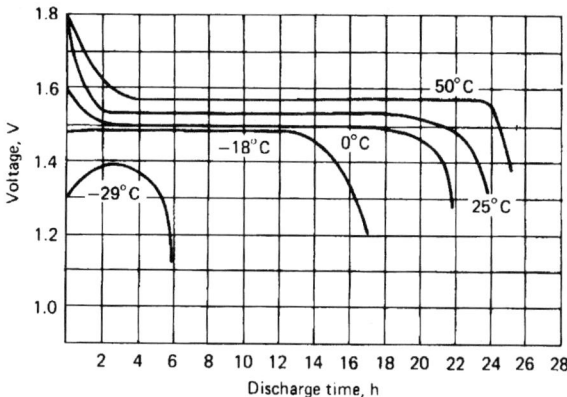

**FIGURE 29.41**   Effects of temperature on low-rate zinc/silver oxide primary batteries discharged at the 24 h rate. (*Courtesy of Eagle-Picher Technologies.*)

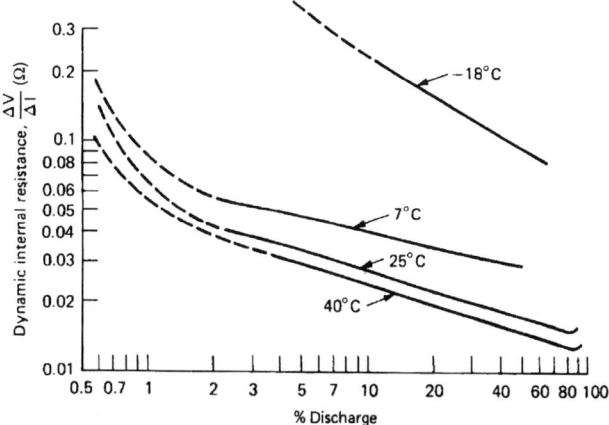

**FIGURE 29.42**   Dynamic internal resistance of zinc/silver oxide primary batteries. (*Courtesy of Eagle-Picher Technologies.*)

time the dynamic resistance rises rapidly. The declining impedance is caused by an improvement in the positive-plate conductivity and a temperature rise during the discharge. This feature can vary considerably, depending on cell design, ambient temperature of the discharge, and the point in time after the change of discharge rate when the voltage change is observed.

*Service.* The performance of zinc/silver oxide batteries in amperes per unit weight and volume versus service time is given in Fig. 29.43. It can be noted, again, that this battery system is particularly sensitive to temperatures below 0°C. These data are applicable, within reasonable accuracy, for both high- and low-rate designs.

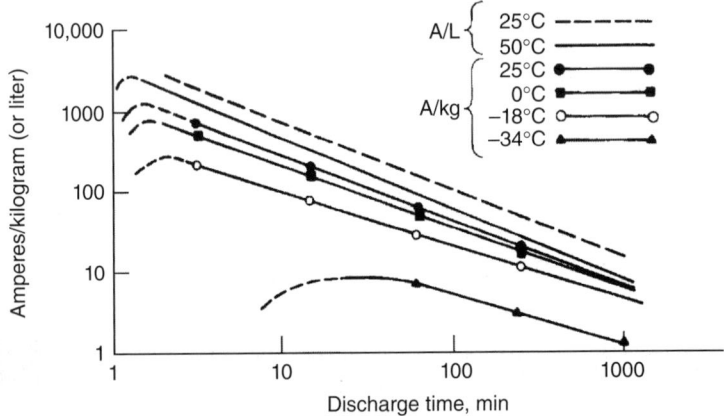

**FIGURE 29.43** Service life of zinc/silver oxide primary batteries. (*Courtesy of Eagle-Picher Technologies.*)

*Shelf Life.* The dry shelf life of the zinc/silver oxide battery is shown in Fig. 29.44 that gives storage data at 25, 50, and 74°C for periods of up to 2 years. The losses shown are based on the assumption that the positive active material is divalent silver oxide, which slowly degrades to monovalent oxide at temperatures above about 20°C. Degradation of the negative plate is minimal. It is expected that the monovalent oxide level would be reached in about 30 months when the storage temperature is 50°C. Experience has shown that batteries stored at average ambient temperatures of 25°C or lower retain capacity at or above the monovalent oxide level for a period of 25 years or longer.

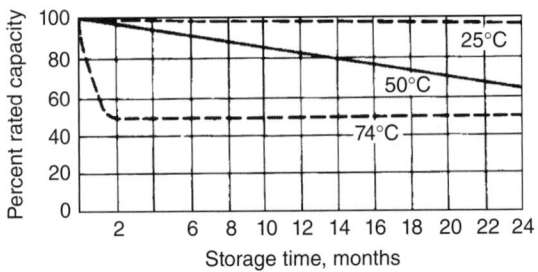

**FIGURE 29.44** Dry storage of zinc/silver oxide primary batteries. (*Courtesy of Eagle-Picher Technologies.*)

The wet shelf life of the zinc/silver oxide battery varies considerably with design and method of manufacture. Figure 29.45 provides a guide to the expected performance of most designs. The wet shelf-life degradation is caused principally by loss of negative-plate capacity (dissolution of the sponge zinc in the electrolyte) or development of short circuits through the cellulosic separators.

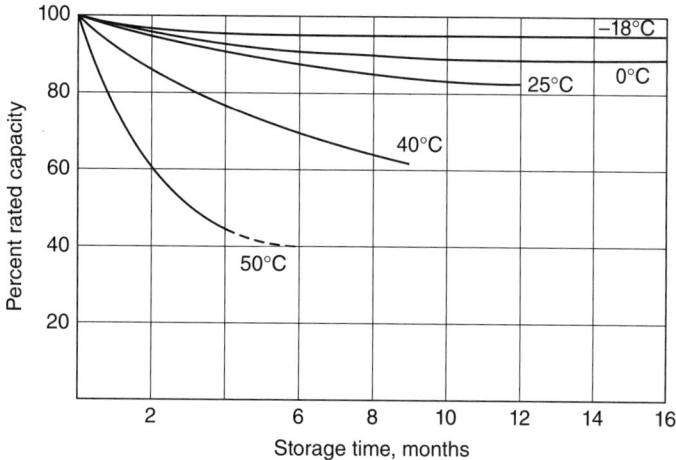

**FIGURE 29.45**    Wet (activated) storage of zinc/silver oxide primary batteries. (*Courtesy of Eagle-Picher Technologies.*)

*Special Features and Handling.*    Both manually and automatically activated zinc/silver oxide batteries were developed to meet highly stringent requirements with regard to performance and reliability. The time and temperature of storage prior to use are of importance, and records should be maintained to ensure use within allowable limits. Special care must be exercised to ensure that the proper amount of the specified type of electrolyte is added to each cell of a manual-type battery and that, after activation, the unit is discharged within the shelf-life limitation at the proper temperature. Some battery containers have pressure-relief valves or heaters, or both, and these must be carefully maintained and monitored.

Automatically activated batteries require special preinstallation check out of gas generator ignitor circuits, heater circuits, and vent fittings. For long-term installations, there should be monitoring of the ambient temperature to prevent degradation caused by exposure to high temperatures. Periodic checks should be made to ensure that the ignitor circuits are intact, as some circuits are sensitive to electromagnetic fields. After activation, if the battery is not discharged within the specified time, it must be replaced.

The proper electrical performance of these batteries is best ensured by operating temperatures at or slightly above room temperature. Temperatures below 15°C can adversely affect the voltage regulation of high-rate batteries, and below 0°C there is also considerable loss of capacity for both types.

*Cost.*    The cost of high-performance primary zinc/silver oxide batteries is dependent on the specifications to which they are built and the quantity involved. Manual-type batteries may cost anywhere from $5 to $15 per watt-hour; remote-activated types will cost about $15 to $20 per watt-hour. When the price of silver is high, material cost becomes one of the chief disadvantages of these batteries. There are many applications, however, in which no other technology can meet the high energy density of the zinc/silver oxide primary system.

### 29.3.3    Aluminum/Silver Oxide Batteries

Figure 29.46 illustrates the voltage potential versus current density for the three most common aqueous systems using silver as the cathode. In the case of the aluminum/silver oxide and magnesium/silver chloride batteries the construction represents a bipolar design while the silver zinc is monopolar in construction.

*Aluminum/Silver Oxide System Construction.*    Unlike conventional batteries, the aluminum/silver oxide system requires a continuous circulation of the electrolyte through the cell stack or cartridge. The cell stack consists of bipolar silver oxide cathodes and aluminum alloy anodes coupled with the electrolyte management system which represents the balance of plant. The electrolyte management system carefully manages the electrolyte properties for the aluminum/silver oxide cell stack during operation.[24] This system remains inert until the introduction

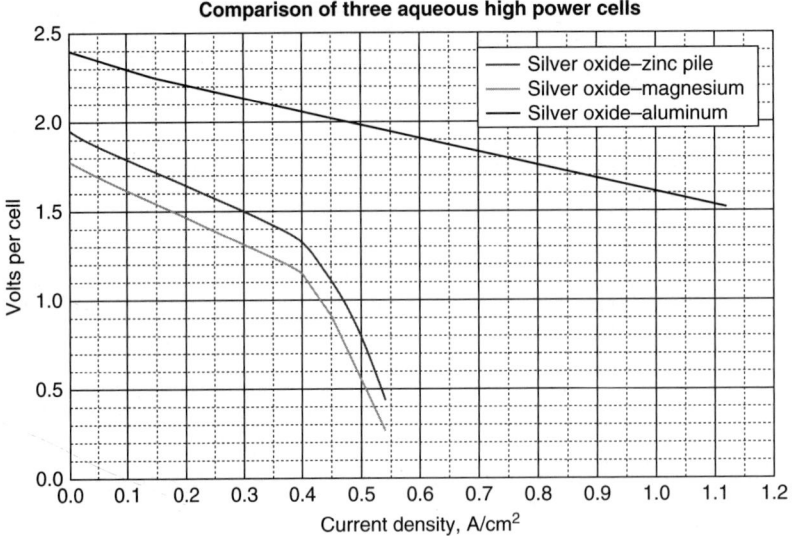

**FIGURE 29.46**    Comparison of three aqueous high power cells. (*Courtesy of Eagle Picher.*)

of seawater. Figure 29.47 is a schematic of the energy section, which includes the aluminum/silver oxide cell stack. The operation consists of system activation by the thermal batteries until the activation valve and electrolyte pump are operational. Once seawater enters a mixing chamber, the electrolyte is formed and thermally metered through the cell stack. A gas separator is included to remove any insoluble hydrogen from the electrolyte.

*Chemistry.*    The following reactions occur in alkaline electrolyte solutions:

| | | |
|---|---|---|
| *Overall:* | $2Al + 3AgO + 2OH^- \rightarrow 2AlO_2^- + 3Ag + H_2O$ | $E^0 = 2.69$ V |
| *Anode:* | $2Al + 8OH^- \rightarrow 2AlO_2^- + 4H_2O + 6$ electrons | $E^0 = -2.34$ V |
| *Cathode:* | $3AgO + 3 H_2O + 6$ electrons $\rightarrow 3$ Ag $+ 6$ OH | $E^0 = 0.35$ V |
| *Corrosion:* | $2Al + 2H_2O + 2OH^- \rightarrow 2AlO_2^- + 3H_2$ | |

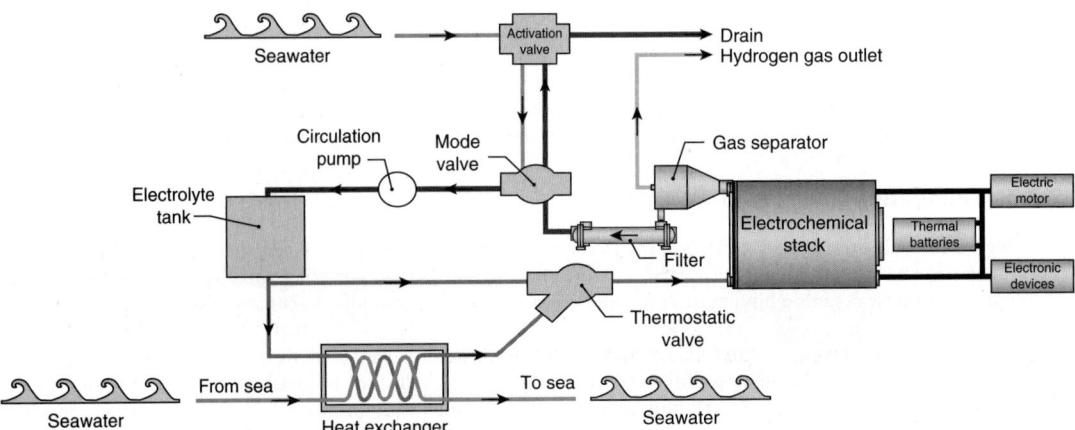

**FIGURE 29.47**    Energy section of an aluminum/silver oxide reserve cell stack. (*Courtesy of Eagle Picher.*)

As noted in these reactions, in addition to consuming the anode and the electrolyte, the corrosion generates a substantial amount of heat (100 kcal/gmol Al) and an insoluble gas species ($H_2$), which needs to be managed within the electrolyte management system. These side effects are usually controlled with a heat exchanger integrated within the torpedo shell and a cyclonic gas separator that is in-line with the electrolyte flow domain.

***Cell Components and Assembly.***     Aluminum/silver oxide battery systems are pile configured while electrically connected in series using bipolar electrodes. The shape of the electrodes is generally circular which facilitates assembly within a cylindrical shell. In a bipolar design, as the number of electrode elements increases (e.g., one anode and one cathode for each cell) the corresponding stack voltage cumulatively increases to reflect the total number of electrode pairs. Conversely in the monopolar configuration all the electrode pairs are in parallel. The cell voltage in this example would be the potential across the anode and the cathode. The biggest advantages of the bipolar configuration versus the traditional monopolar layout are (1) reduction of weight and volume due to the limited number of cell containment features, and (2) lower internal resistance (IR drops) due to elimination of cell-to-cell current conductors, which also favorably impact weight and volume.

Figure 29.48 is a comparison of a monopolar cell versus a bipolar stack configuration.

The hydraulic configuration of the aluminum/silver oxide cell stack is critical to the uniform distribution of electrolyte to each cell. Under normal circumstances it is desirable to limit the static pressure variation among individual cells to less than 1% of the cartridge drop. Finite element analysis (FEA) modeling has been successfully utilized to define optimum geometries and layouts which limit static pressure variations associated with velocity gradients in the electrolyte distribution plenum. Unequal flow distribution will result in temperature gradients from cell to cell, which in turn results in inefficient cell material utilization and reduced energy yield.

Another consequence of a pile design is leakage or shunt currents within individual or groups of cells within the cartridge. The design of the electrolyte delivery system to the cells is critical in maintaining the lowest possible shunt currents which impact voltage and energy output. This is a function of the manifold geometry between the cells and the flow bar geometry on the inlet and outlet side of each cell while limiting excessive electrolyte pressure drop through the cartridge. Using a 250 cell stack as an example, Fig. 29.49 is a hydraulic model that predicts leakage currents at two opposing scenarios. One is a common sump for all 250 cells, while the second is minimizing the common reservoir to only a limited number of distinct cell modules within the stack.

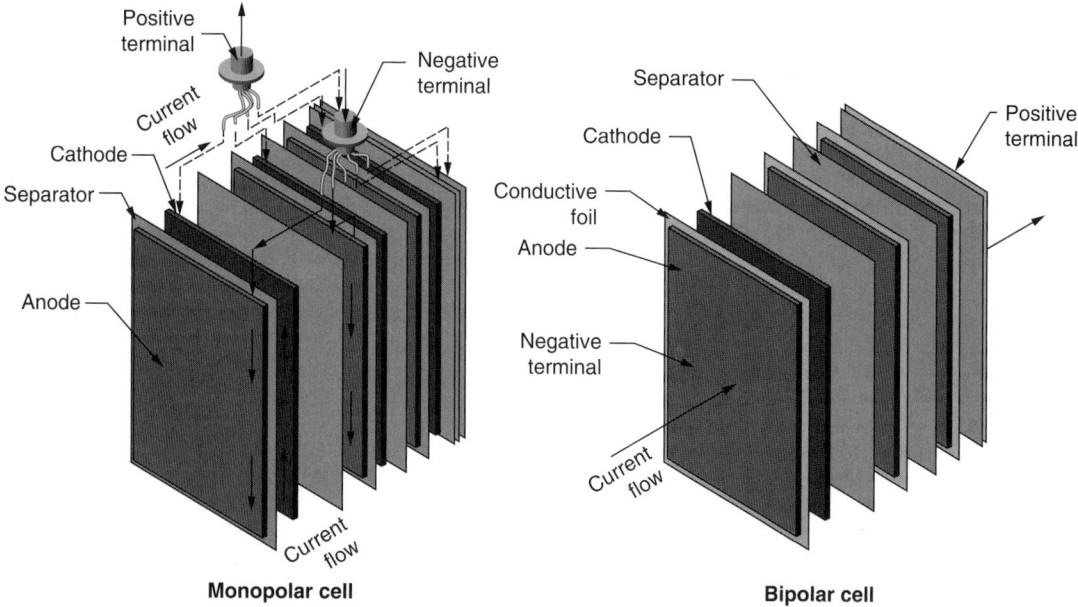

**FIGURE 29.48**   Bipolar and monopolar stack configurations. (*Courtesy of Eagle Picher.*)

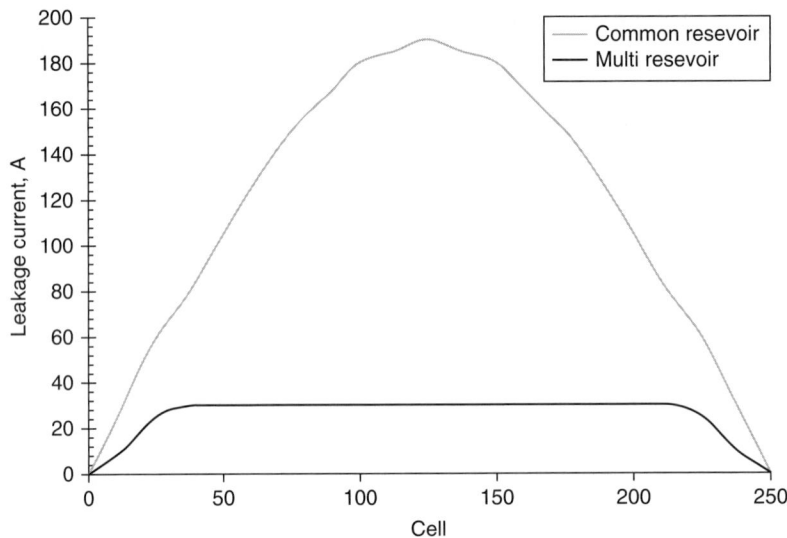

**FIGURE 29.49**    Leakage current simulation. (*Courtesy of Eagle Picher.*)

*Positive.*    The silver oxide cathode design is very similar to those used in other silver-based chemistries such as the silver zinc system (Chap. 16). These are designed around an energy output of 0.45 ampere-hours per gram of silver. Construction can be either polymer bonded (as AgO) or sintered electrodes, with or without a grid. In some cases, a thin sheet of electrically conductive foil (e.g., copper or silver) may be laminated on one side to serve as an intercell connector to the aluminum anode of the adjacent cell.

*Negative.*    The anode used in the aluminum/silver oxide system is centered on a high purity aluminum metal that is doped with various alloy elements to facilitate higher cell voltage and the reduction of hydrogen gas. The energy output of aluminum is designed around 2.98 ampere-hours per gram of aluminum alloy and has typically exhibited high coulombic efficiencies of greater than 95%. The initial development of aluminum as an anode in batteries was by Zaromb.[25] Specialty alloys used for aluminum/silver oxide battery systems were further developed in the 1970s by Reynolds and Alcoa followed by ALCAN in the 1980s.[26] As an alternative to the intercell connection feature on the cathode, one side of the aluminum anode may similarly include a thin electrically conductive foil laminated on one side.

*Separator.*    The separators in an aluminum/silver oxide pile system must establish a uniform gap between the anode and cathode and prevent any shorting between the electrodes. The formed gap defines the flow domain within each cell to facilitate the efficient chemical reaction within the cell. Separation techniques come in various forms such as polymer disks (typically called picots) or glass beads. These are sized and numbered to facilitate the lowest impact to the cells surface area (<10%), electrical resistance (defined by gap height), while resulting in a lowest possible pressure drop across the cell (defined by separator diameter). Glass beads are normally compressed within the silver cathode matrix. This is similarly used in magnesium/silver chloride seawater batteries while polymer disks are normally laminated to the aluminum anode. In either case, the separators require sufficient compressive strength to withstand the cell stack preload while not damaging the electrodes.

*Electrolyte.*    The aluminum/silver oxide system is activated with a mixture of an anhydrous alkaline compound, which dissolves in seawater to form a liquid electrolyte. Sodium hydroxide (NaOH), in pellet or powder form with additional additives, is commonly used in this system due to its lower weight and higher exothermic reaction which is beneficial during start-up operation. Refer to Table 29.8, which highlights a comparison between potassium hydroxide and sodium hydroxide as an electrolyte.

There have been many studies that focused on electrolyte operating conditions that include careful control of thermal, hydraulic, and chemical operating conditions for a wide variety of operating scenarios.[27] Electrolyte concentrations as high as 9 N have been used to provide the optimum thermal conditions to activate the aluminum anode without the risk of passivation, which typically occurs below 46°C. In conjunction with tailoring the

**TABLE 29.8**  Comparison of Sodium Hydroxide and Potassium Hydroxide Electrolyte

| Characteristics | Sodium hydroxide | Potassium hydroxide |
|---|---|---|
| Weight | Lower | Higher |
| Electrical conductivity | Lower | Higher |
| Exothermic reaction | Higher | Lower |
| Solubility | Lower | Higher |
| Boiling point | Higher | Lower |
| Cost | Lower | Higher |

flow rate, solute concentration, and temperature, the system start-up time and power output can be optimized during operation. As a consequence, some electrolyte extremes can accelerate corrosion of the aluminum anode which may impact run time.

**_Discharge Characteristics._**    Since the 1990s the aluminum/silver oxide system has been fielded in the lightweight torpedo (MU90) and later scaled up to the heavyweight configuration (Black Shark) in 2000. Performance is primarily influenced by the electrolyte characteristics such as normality, flow rate, and temperature. These can be adjusted to facilitate low power, long endurance runs or provide very high power, for short bursts of high speed. Figures 29.50 and 29.51 summarize a typical discharge for a full-scale discharge at low power (>40 kW) and a high power (>240 kW).

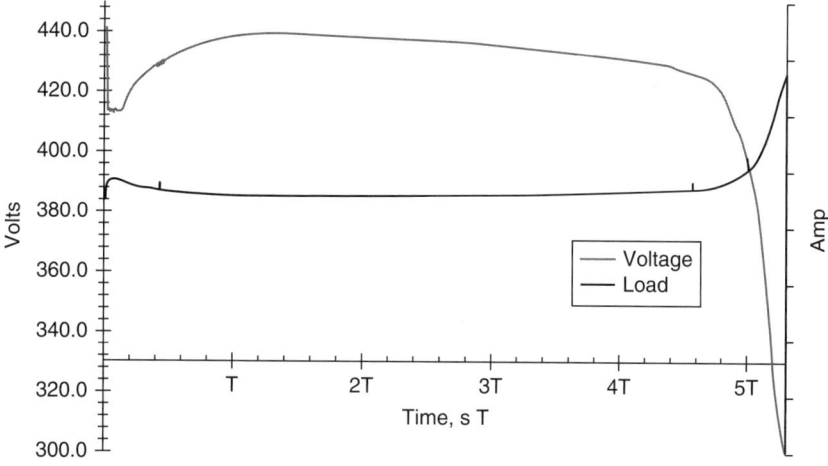

**FIGURE 29.50**   Low-power discharge. (*Courtesy of Eagle Picher.*)

## 29.4  AMBIENT TEMPERATURE LITHIUM

### 29.4.1  Construction

Lithium anode reserve batteries are basically composed of three major components:

1. Activation and electrolyte delivery system
2. Electrolyte reservoir
3. Cell and/or battery unit

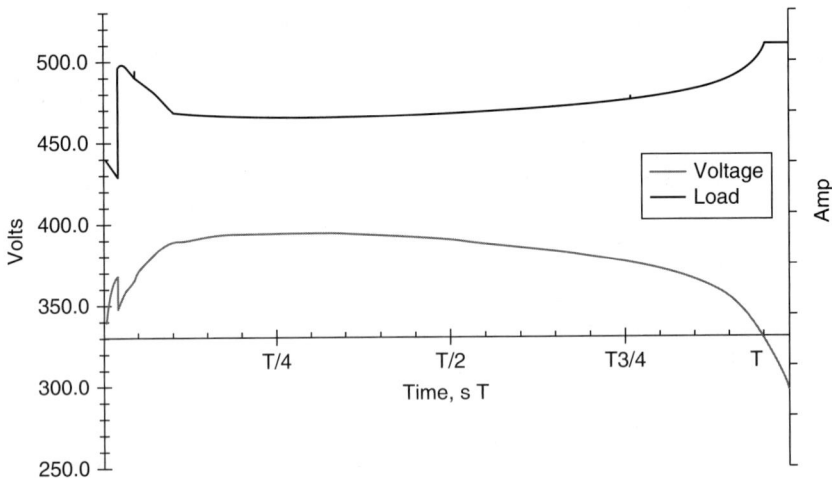

**FIGURE 29.51** High-power discharge. (*Courtesy of Eagle Picher.*)

However, the actual design can vary widely, depending on the application. The design can vary from a simple, small, single cell with an ampoule manually activated, to a very large, complex, multicell battery with an automatic electric initiation mechanism to transfer the electrolyte from the reservoir chamber to a high-voltage multicell battery stack. Both the electrodes and the hardware components are essentially the same as the primary active units, but with allowances made for electrolyte storage and electrolyte delivery into the cells at the time of activation. In addition, the electrochemical and hardware components must be constructed of a rugged maintenance-free design to survive severe environmental and performance requirements as most are used in military or special applications. Table 29.2 (see Sec. 29.1.2) lists typical requirements of lithium reserve batteries, which illustrates the reason for many of their unique construction and design features.

Some common construction features are used in the design of lithium reserve batteries. The outer case is generally made of a 300-series stainless steel since it offers the corrosion resistance against both the internal system and the external environment during its long-term use. Various welding techniques such as laser, tungsten inert gas (TIG), resistance, and electron beam can be applied to the 300-series stainless steels. Thus the outer case provides a true 20-year reliability, capable of maintaining the hermeticity required for reserve lithium batteries. The electrical terminals used are generally glass-to-metal types, which also provide the hermeticity required for long-term storage.

### 29.4.2 Types of Lithium Anode Reserve Batteries

Three basic lithium reserve battery types are being manufactured at the present time:

1. Single-cell battery with electrolyte stored in a glass ampoule
2. Multiple single cells using bellows for the electrolyte storage reservoir
3. Multicells of bipolar construction with either a glass ampoule or a reservoir for electrolyte storage

*Ampoule Type.*  Single-cell reserve types using an ampoule as the electrolyte storage reservoir are the most reliable of the reserve designs due to their simple construction and lack of intercell leakage problems associated with multicell batteries. One group of these cells is sized to the American National Standards Institute (ANSI) specifications, and the other group consists of those cells built for special-purpose applications that are not sized to the ANSI specifications. Both groups, however, are very similar in construction.

Figure 29.52 shows the cross section of a reserve lithium anode cell in an A-size configuration of about 1 Ah, using the Li/SOCl$_2$ system.[28] The cell consists of concentrically arranged components. A lithium anode is swaged against the inner wall of a stainless steel cylindrical can. A nonwoven glass separator is located adjacent to the anode. The Teflon®-bonded carbon cathode is inserted against the separator. A cylindrical nickel current collector provides the electrical contact to the positive terminal and houses the hermetically sealed glass ampoule. The ampoule is held firmly in place by upper and lower insulating supports, which protect it from premature breakage while permitting transmission of a direct force at the bottom of the case to shatter the ampoule at the time of activation. The unit is sealed hermetically to ensure long shelf life in the unactivated condition. Activation is achieved by applying a sharply directed force at the bottom of the cell case to shatter the glass ampoule. The electrolyte is absorbed by the porous cathode and the glass separator, thereby activating the battery.

Another design has been developed for mine and fuze applications, using both the Li/V$_2$O$_5$ and the Li/SOCl$_2$ systems, in the capacity range of 100 to 500 mAh.[29] The cross sections of these two cells are shown in Figs. 29.53 and 29.54, respectively. Both cells are similar with respect to the external hardware and the internal arrangement of the components. The case and header assembly are projection-welded together at the case flange. The header serves as the cover for the cells and incorporates a glass-to-metal seal for the center terminal pin made of nickel-iron Alloy 52. The terminal pin has negative polarity (both

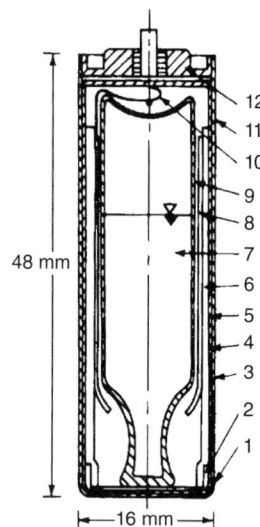

| | |
|---|---|
| 1 Insulator | 7 Electrolyte |
| 2 Bottom separator | 8 Current collector |
| 3 Cell can | 9 Glass ampoule |
| 4 Lithium anode | 10 Positive terminal tab |
| 5 Separator | 11 Top spacer |
| 6 Carbon cathode | 12 Cell cover |

**FIGURE 29.52**  Cross section of Li/SOCl$_2$ A-size reserve cell. (*Courtesy of Tadiran Industries, Ltd.*)

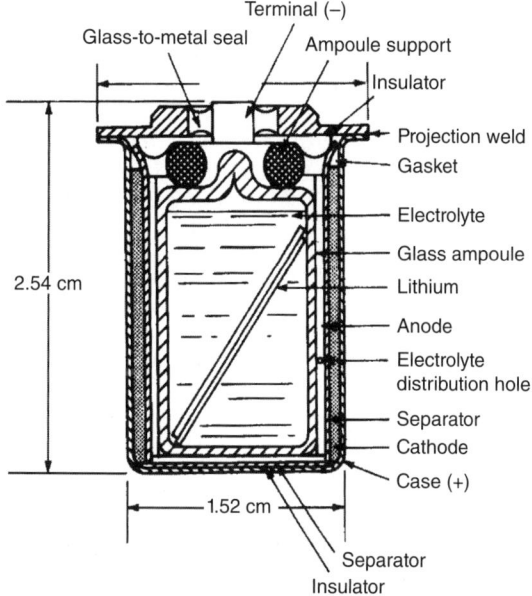

**FIGURE 29.53**  Cross section of Li/V$_2$O$_5$ reserve cell. Alliant model G2659. (*Courtesy of Alliant Techsystems, Inc.*)

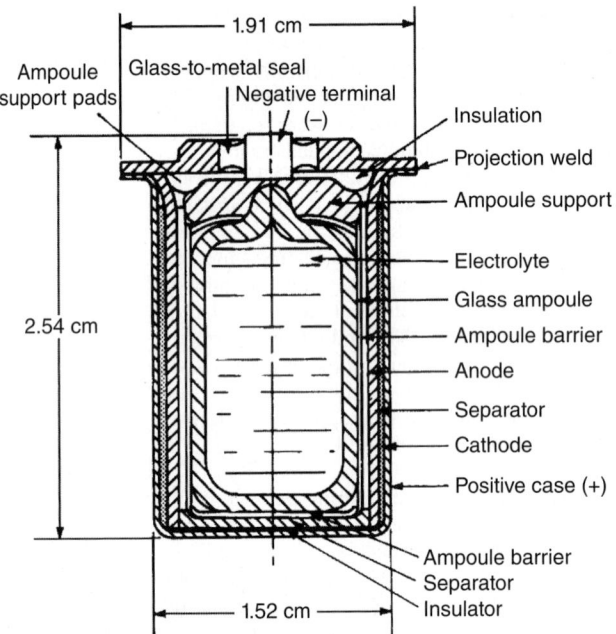

**FIGURE 29.54**   Cross section of a LiSOCl₂ reserve cell. Alliant model G2659B1. (*Courtesy of Alliant Techsystems, Inc.*)

cell designs), and the balance of the header and case surface have positive polarity. The hermetically sealed hardware in conjunction with the reserve feature of the design makes it possible to achieve storage times in excess of 20 years.

The internal arrangement of the components consists of annularly located electrodes about a central glass ampoule used as the electrolyte solution reservoir. In addition, there are various insulating components in the upper and lower portions of the cell, used to prevent internal short-circuiting.

Several features account for most of the design differences between these two cells. In the Li/SOCl₂ reserve cell, the glass ampoule also contains the cathode oxidant, SOCl₂, while the cathode oxidant of the Li/V₂O₅ reserve cells is contained in the cathode structure. Directly adjacent to the Li/SOCl₂ cell case is the Teflonated carbon, while in the case of the Li/V₂O₅ cell, the cathode is molded from a dry mixture of V₂O₅ and graphite. The Teflonated carbon cathode for the reduction of SOCl₂ is made in sheet form and is attached to a metal grid, rolled to shape, and inserted against the inside wall of the case. Another difference is the way the electrical connection is made for the two cathodes. The V₂O₅ connection is made by the direct-pressure contact of the molded cathode, whereas with the SOCl₂ system the cathode lead is welded to the case at the time the cover is welded. The lithium anode structure consists of pure lithium metal, which is pressed onto an expanded metal grid of 316L stainless steel. One end of a flat 316L stainless steel lead is spot-welded to the pin of the glass-to-metal seal. Rolled into a cylinder, the anode is inserted into the cell next to the separator. Both cells are provided with an ampoule support in order to survive the shock environment specified. In the Li/SOCl₂ system, Tefzel and glass have been found to be chemically stable for use as insulators, separators, and supports. The Li/V₂O₅ system allows more flexibility because many rubbers and plastics can be used.

***Multicell Single-Activator Design.***   For those applications where higher than single-cell voltages are required, a battery is constructed of two or more cells, depending, of course, on the voltage needed. Typical voltages are 12 and 28 V, and for lithium anode cells with a 2.7 to 3.3 V operating voltage, this would require anywhere from 4 to 10 cells for each battery. This family of batteries is unique with respect to the method of cell activation and the containment of electrolyte in multiple cells initiated from a single self-contained reservoir of electrolyte.

Batteries of this design are used in preference to the bipolar type to achieve higher cell capacities and to allow discharge times up to 1 year or more, through the tight control of intercell leakage. The leakage currents are controlled and limited to usually less than several percent of the discharge current. This feature, however, limits these batteries from being miniaturized, which is possible with many bipolar designs.

An example of this design approach is the $Li/SO_2$ reserve battery illustrated in Fig. 29.55. The battery is cylindrical and contains three main components: (1) the electrolyte storage reservoir section, (2) the electrolyte manifold and activation system, and (3) the reserve cell compartment. About one-half of the internal battery volume contains the electrolyte reservoir. The reservoir section consists primarily of a collapsible bellows in which the electrolyte solution is stored. Surrounding the bellows, between it and the outer battery case, is a space that holds a specific amount of gas/liquid. The gas is selected such that its vapor pressure always exceeds that of the electrolyte, thereby providing the driving force for eventual liquid transfer into the cell chamber section once the battery has been activated.

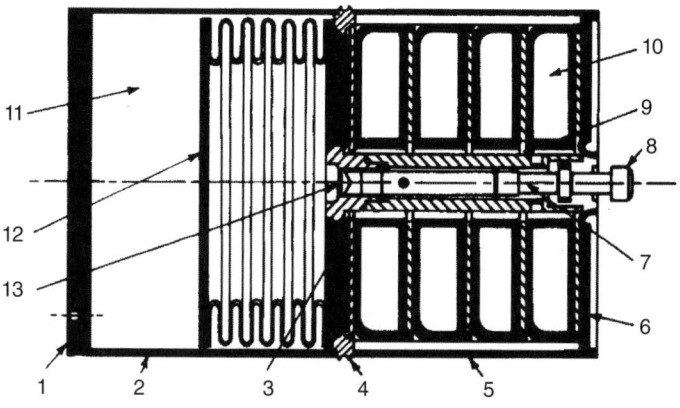

Legend:

1 Battery top bulkhead
2 Upper battery case
3 Bulkhead
4 Intermediate bulkhead ring
5 Lower battery case
6 Battery bottom bulkhead
7 Activation manifold

8 Activation stud
9 Intercell insulation
10 Single 20 Ah cell
11 Freon backfill volume
12 Electrolyte storage bellows
13 Manifold diaphragm

**FIGURE 29.55**    Cross section of a 20 Ah $Li/SO_2$ multicell battery.

In the remaining half of the battery volume there is the centrally located electrolyte manifold and activation system housed in a 1.588-cm-diameter tubular structure plus the series stack of four toroidal-shaped cells that surround the manifold/activation system.

The manifold and cells are separated from the reservoir by an intermediate bulkhead. In the bulkhead there is a centrally positioned diaphragm of thin section to be pierced by the cutter contained within the manifold. In fabrication, the diaphragm is assembled as part of the tubular manifold which, in turn, is welded as a subassembly to the intermediate bulkhead. Figure 29.56 is a more detailed cross-sectional view of the electrolyte manifold and the activation system with the major components identified.

The activating mechanism consists of a cutter that is manually moved into the diaphragm, cutting it and thereby allowing electrolyte to flow. The movement of the cutter is accomplished by the turning of an external screw that is accessible in the bottom base of the battery. The cutter section and the screw mechanism are isolated from one another by a small collapsible metal cup that is sealed hermetically between the two sections. This prevents external electrolyte leakage. The manifold section is a series of small nonconductive plastic tubes connected to one end of the central cylinder and to each of the individual cells at the other end. The long length

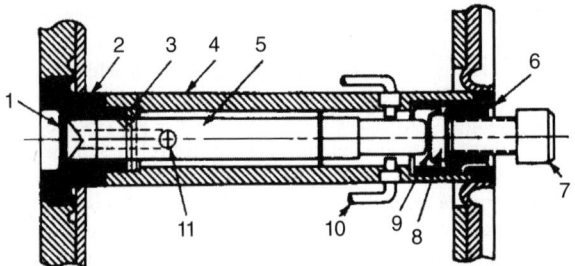

Legend:

1 Diaphragm
2 Top bushing
3 Shear pin
4 Center tube
5 Cutter
6 Bottom bushing

7 Activator stud
8 Drive disk
9 Collapsing cup
10 Electrolyte supply tube
11 Electrolyte entry flow

**FIGURE 29.56**   Cross section of an electrolyte manifold and an activation system.

and small cross-sectional area of the tubes minimize intercell leakage losses during the period of time that electrolyte is present in the manifold structure.

In this application, four individual cells are required to meet the voltage requirement. (The number of cells is, of course, adjustable with minor modification to meet a wide range of voltage needs.) Each cell contains flat circular anodes and cathodes that are separately wired in parallel to achieve the individual cell capacity and plate area needed for a given set of requirements. To fabricate, the components, with intervening separators, are alternately stacked around the cell center tube, after which the parallel connections are made. The cells are individually welded about the inner tube and outer perimeter to form hermetic units ready for series stacking within the battery. Connections from the cells are made to external terminals, which are located in the bottom bulkhead of the battery.

Figure 29.57 shows the major battery components prior to assembly. The components shown are fabricated primarily from 321 stainless steel, and the construction is accomplished with a series of TIG welds. The hardware shown is designed specifically for use with the lithium/sulfur dioxide electrochemical system; however, it is adaptable, with minor modifications, to other liquid and solid oxidant systems. The battery can also be adapted to electrical rather than manual activation.

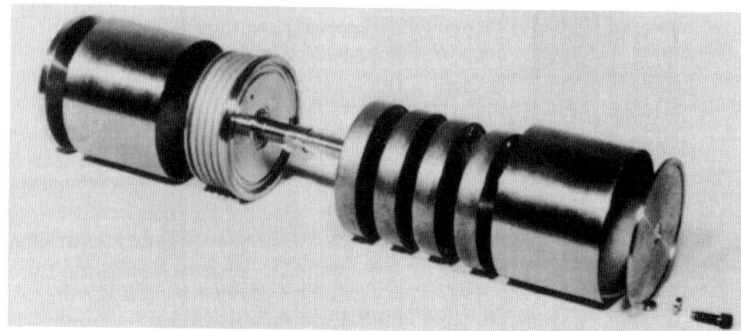

**FIGURE 29.57**   Pictorial view of a 20 Ah Li/SO$_2$ multicell battery.

An example of this reserve design approach being used with the lithium/thionyl chloride chemistry is shown in Fig. 29.58*a*. This high-power reserve battery, designated by the U.S. Navy as battery BA-6511 SLQ, was developed to provide electric power for a family of ocean buoys.[30] The reserve battery was selected for this application not only to eliminate the problems of self-discharge and passivation associated with extended stand of an active battery, but also for safety as the electrolyte is stored separately from the battery until activation.

The battery weighs about 145 lb and is contained in a package that is 29.2 cm in diameter and 43.2 cm long. The battery contains 21 cells; 18 cells compose a 56 V section, delivering 4 kW, and rated at 65 Ah; 3 cells are in a 10 V section, delivering 7 A, and rated at 57 Ah. The electrolyte is stored in a reservoir and is distributed to the 21 cells via a unique manifold system. Activation is initiated by an explosive squib, and a stored energy system within the reservoir provides the motive power. The cell design used for this battery (Fig. 29.58*b*) is a circular wafer with a hole through the center to provide a channel for electrical and tubing connections. The two types of cells used in the battery are physically similar, differing only in height and capacity as a result of one less set of electrodes. The cells used in the high-voltage, high-rate section contain five anodes and six cathodes. Anodes are single-sided with lithium pressed onto expanded nickel grids. The cathode is cut from coated stock of Teflonated carbon on a nickel screen, as shown in Fig. 29.58*c*. Nonwoven glass separators are used. The specifications for the two cells are listed in Table 29.9.

Another example of this reserve design using a precharged $Li_xCoO_2$ $(0.5 \leq x < 1)$ chemistry is shown in Fig. 29.59.[31] This reserve battery consists of three hermetically welded cell cases with a central reservoir enclosed in a stainless steel battery housing. Such a battery was developed to power the hand-emplaced wide area munitions (HWAM).

Figure 29.60 shows another multicell battery designed for lightweight missile applications. Based on the Li/oxyhalide technology, it used advanced thin electrode technology that has high energy utilization and low electrical impedance. A composite separator was also used that combines high electrolyte absorption and mechanical integrity. This type of high-power design sees applications such as the Theater High Altitude Area Defense (THAAD) in a Kill Vehicle for the Ground Based Interceptor (GBI) program.[32] Other advantages of this lightweight, high-power battery are (1) reduction in battery weight over the thermal or silver-zinc system, (2) specific energy greater than 250 Wh/kg, (3) gain in weight advantage as the mission time

(a)

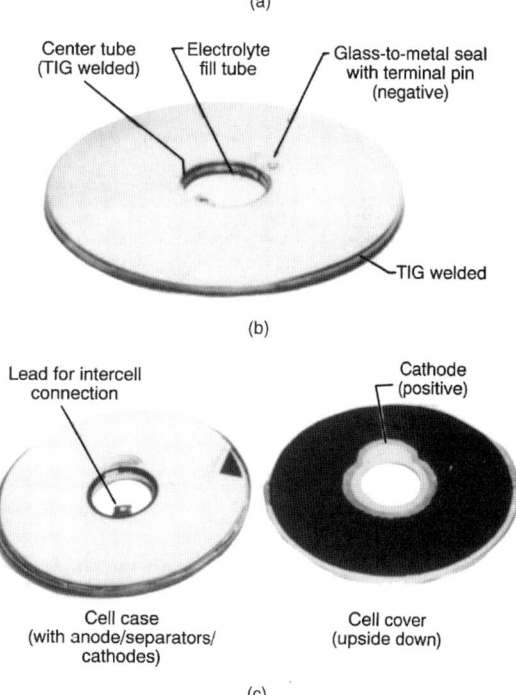

(b)

(c)

**FIGURE 29.58** A high-power reserve battery BA-6511/SLQ. (*a*) A Li/SOCl₂ reserve battery. (*b*) High-power cells. (*c*) A high-power cell case and an electrode assembly. (*Courtesy of Alliant Techsystems, Inc.*)

**TABLE 29.9**    Characteristics of Li/SOCl₂ Reserve Cells Model G3070A2

| | Low-rate reserve cell | High-rate reserve cell |
|---|---|---|
| Performance: | | |
| Open-circuit voltage (activated) | 3.67 V | 3.67 V |
| Voltage under load | 3.40 V, 7 A at 20°C | 3.10 V, 72 A at 20°C |
| Rated capacity | 57 Ah at 7 A to 2.67 V at 20°C | 65 Ah at 72 A to 2.63 V at 20°C |
| Physical characteristics: | | |
| Max. diameter, OD | 28.5 cm | 28.5 cm |
| Max. diameter, ID | 6.7 cm | 6.7 cm |
| Max. height | 0.89 cm | 1.04 cm |
| Cell weight with electrolyte | 1310 g | 1485 g |
| Case material | Stainless steel | Stainless steel |

*Source:* Alliant Techsystems, Inc.

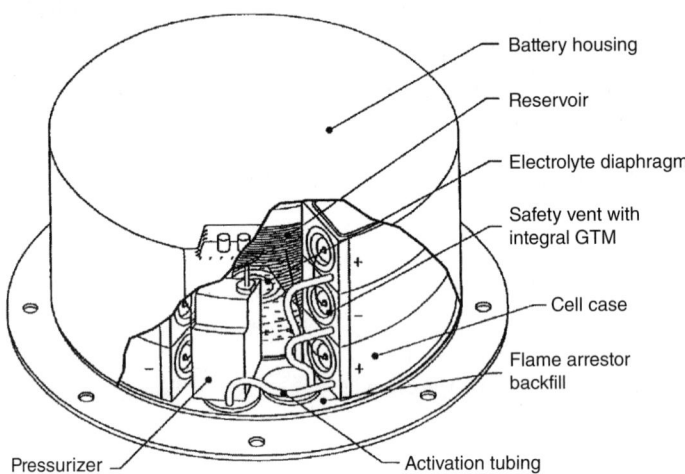

**FIGURE 29.59**   Design of a reserve Li/Li$_x$CoO$_2$ battery for hand-emplaced wide area munitions (HWAM). (*Courtesy of Alliant Techsystems, Inc.*)

increases or as the energy to power ratio increases, (4) high-power delivery even after 10 years of storage at temperature below 32°C, and (5) low operating temperature allowing locations near heat-sensitive electronics.

***Multicell Bipolar Construction with Single-Activator Reservoir.***    Lithium anode reserve batteries, using bipolar construction, are relatively few in number and always developed for specific applications. The bipolar construction—one component used as both the anode collector of one cell and the cathode collector of the next cell in the stack—is not unique to the lithium reserve battery, but an adaptation of techniques used in other types of batteries. There are several advantages of the bipolar construction:

- Very high energy and power density for high-voltage batteries
- Rugged construction to withstand spin and setback forces from artillery firing
- Flexibility to adjust voltages in the cell stack
- Adaptability to varying energy and power requirements

Figure 29.61 is an illustration of a reserve lithium/thionyl chloride battery using a bipolar plate construction. This battery weighs approximately 5.4 kg and has a volume of 2000 cm³.

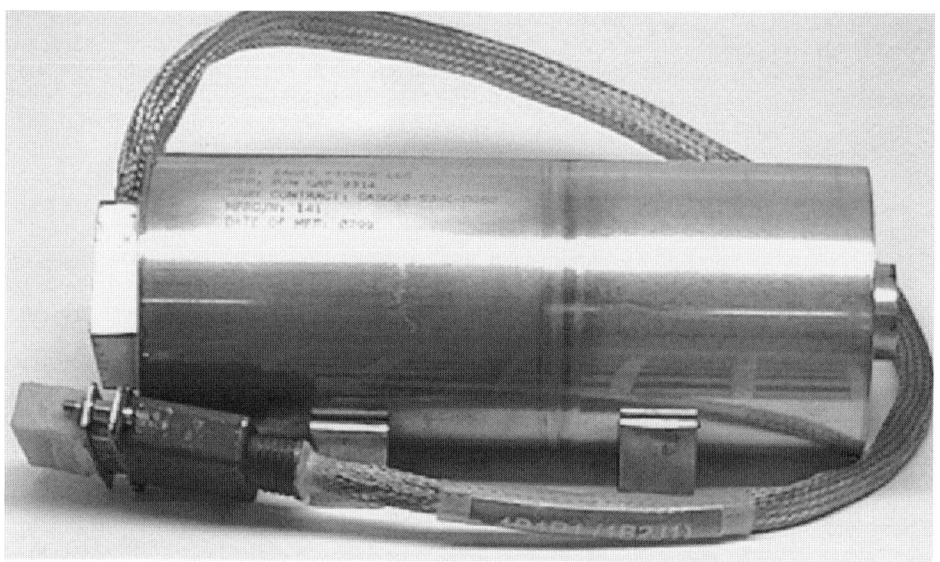

**FIGURE 29.60** A high-power 1 kW Li/oxychloride reserve battery. (*Courtesy of Eagle-Picher Technologies.*)

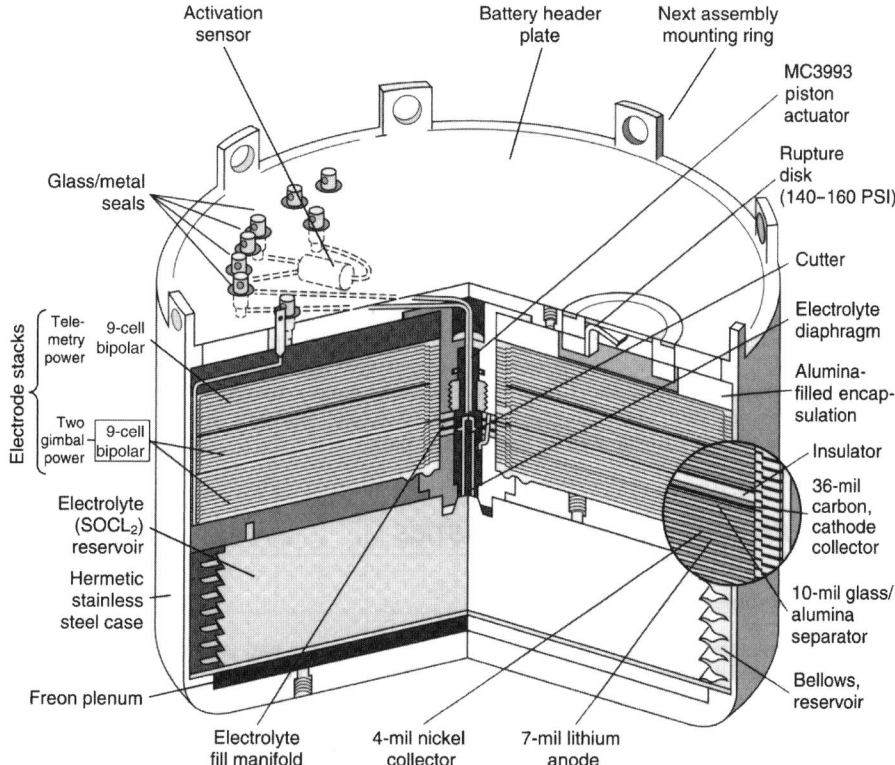

**FIGURE 29.61** Sandia National Laboratories Li-SOCl₂ reserve battery model MC3945.

Activation of the reserve battery is accomplished by supplying an electric pulse to the battery by firing an electric squib or actuator or by some mechanical means. This type of reserve battery has been used chiefly in artillery shells for electronic fuze power supplies and in missiles for the electronic power supply. Therefore, the electric pulse can be supplied prior to firing or at the time of launch. However, for artillery fuze power supplies, the battery is usually activated by the launch acceleration (setback) and/or the spin forces. The acceleration force of the artillery shell releases a firing pin that strikes and fires a primer. The primer can ignite a gas generator or directly release a stored gas by breaking open a metal diaphragm.

Once the battery has been initiated as described, the gas pressure (such as from a gas generator, stored gas/liquid, or $CO_2$) forces the electrolyte into each of the cells through a manifold (electrolyte distribution network).

The electrolyte reservoir is generally made using a collapsible cup, a bellows, or a wound tubing design. These serve to hold the electrolyte during the long inactive storage period and act as the delivery mechanism during activation. Each reservoir has some type of diaphragm, which is broken with high pressure or mechanical means to allow the electrolyte to enter the cell-stack part of the battery hardware.

The bipolar cell stack with the electrolyte distribution manifold in the center comprises the battery section. When electrolyte enters the center manifold, it is distributed to each cell through holes or passageways in the housing encompassing the battery. The design of the manifold is the key to controlling intercell leakage. For bipolar batteries, life requirements are relatively short (seconds to several hours); therefore, the manifold is relatively simple. But when longer life is needed, the parasitic leakage currents are controlled by the length and area of the leakage path.

Another battery of this design was developed as a power source for the extended-range guided missile (ERGM).[33] Ultimately, this design would be similar to the design shown in Fig. 29.60. However, the development test fixture used is shown in Fig. 29.62. It uses the $Li/SOCl_2$ chemistry but with a special lithium tetrachlorogallate electrolyte optimized for performance. The test fixture employs gas pressure to compress the bellows, rupturing the diaphragm and activating the battery. In actual use, setback forces on launch perform this function. The specifications for the ERGM battery are given in Table 29.10.

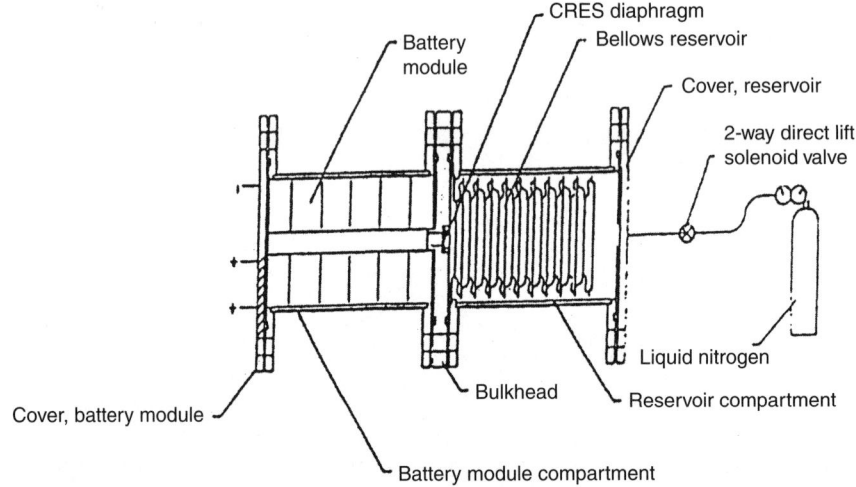

**FIGURE 29.62** Laboratory test fixture for activation of an extended-range guided missile (ERGM) battery using a $Li/SOCL_2$ system. (*Courtesy Yardney Technical Products, Inc.*)

### 29.4.3 Performance

***Ampoule-Type Batteries.*** Voltage characteristics at the "time of activation" are unique and an important feature of reserve batteries. This is especially true for military applications, where reserve batteries must normally be designed to meet operational voltage in less than 1 s and in many cases even less than 500 ms. For nonmilitary use, activation times to operating voltage level are less critical. However, for a given reserve battery design and the electrochemical couple used, the activation time is dependent on the discharge rate and temperature.

**TABLE 29.10** Specifications for the ERGM Battery

| Description | Specifications |
| --- | --- |
| 12 V section regulation | • Load voltage range: 9.5 to 16.0 V<br>• 60 W applied continuously |
| 28 V section regulation | • Load voltage range: 24 to 40 V<br>• 15 W applied continuously<br>• 125 pulses, 8 A, 0.1 s<br>• Duration, evenly distributed |
| Operating life | • 480 s minimum |

In general, the voltage rise times for both $Li/SOCl_2$ and $Li/V_2O_5$ have similar characteristics. Figure 29.63 shows the rise-time characteristics for the $Li/SOCl_2$ battery (illustrated in Fig. 29.54) at five temperatures at a current density of 0.1 mA/cm (approximately a $C/500$ rate). Rise times are typically below 20 ms at ambient (24°C) and higher temperatures, but increase up to 500 ms at the lower temperatures. The ability to activate rapidly is primarily due to the cell design, which allows the electrolyte to penetrate and wick into the porous electrode and separator at the instant of ampoule breakage.

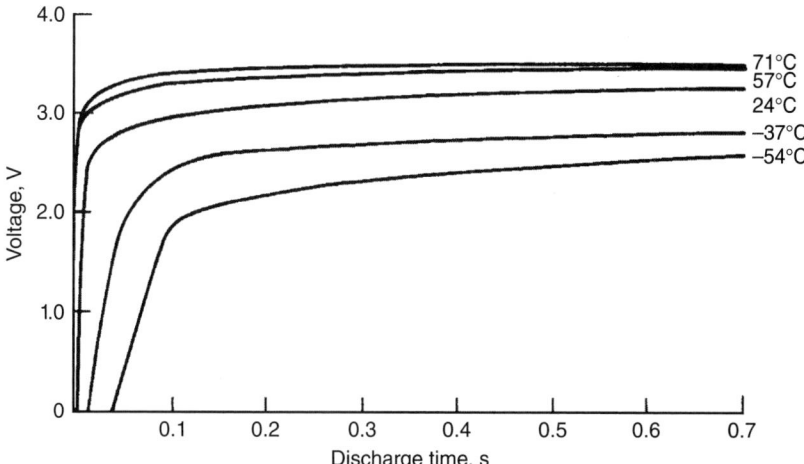

**FIGURE 29.63** Rise-time characteristics after activation of the $Li/SOCl_2$ reserve battery, Alliant model G2659B1; load = 4.35 kW. (*Courtesy of Alliant Techsystems, Inc.*)

The voltage levels of both the $Li/V_2O_5$ and the $Li/SOCl_2$ systems (batteries illustrated in Figs. 29.53 and 29.54) under steady-state discharge conditions are shown in Fig. 29.64. These two systems are very close in voltage at the lower temperatures, ranging from 3.3 to 3.0 V at current densities of less than 1 mA/cm². At higher temperatures, ambient and above up to 74°C, the $Li/SOCl_2$ battery operates above 3.5 V, whereas the $Li/V_2O_5$ battery normally operates between 3.2 and 3.4 V. The higher voltage and increased capacity account for the significant increase in energy density of the $SOCl_2$ battery over the $V_2O_5$ battery. As shown in Fig. 29.64, the $V_2O_5$ system has very little change in capacity over the wide temperature range but is still much lower in capacity than the $SOCl_2$ battery when discharged at the same rate, namely 0.1 mA/cm². Although the capacity and voltage of the $Li/SOCl_2$ battery are lower at cold temperatures, its output is still higher than that of most other systems and its voltage profile is characterized by a flat single-step plateau. The high-temperature curve is also extremely flat and typically discharges above 3.6 V at a current density of 0.1 mA/cm². The voltage characteristics on ambient temperature discharges are similar to those at high temperature except for a slightly lower load voltage when discharged at the same rate, averaging 3.5 V. Table 29.11 compares the output

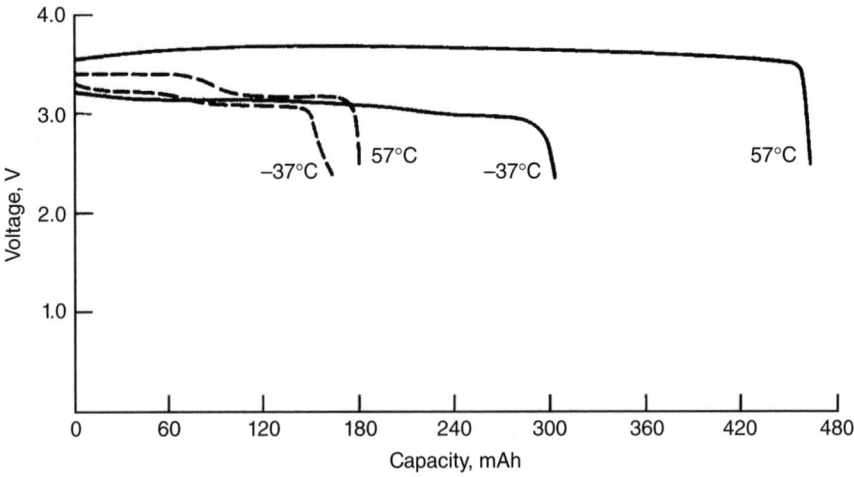

**FIGURE 29.64**   Comparison of discharge profiles for reserve-type Li/V$_2$O$_5$ (— —) batteries and (—) Li/SOCl$_2$. Current density = 0.1 mA/cm$^2$.

**TABLE 29.11**   Performance Comparison between Li/SOCl$_2$ and Li/V$_2$O$_5$ Systems

| System | Temperature, °C | Cell voltage, V | Capacity, mAh | Cell volume, cm$^3$ | Cell weight, g | Specific energy, Wh/kg | Energy density, Wh/L |
|---|---|---|---|---|---|---|---|
| Li/V$_2$O$_5$[*] | −37 | 3.15 | 160 | 5.1 | 10 | 50.4 | 98.8 |
|  | 57 | 3.30 | 180 | 5.1 | 10 | 59.4 | 116.5 |
| Li/SOCl$_2$[†] | −37 | 3.05 | 300 | 5.1 | 10.5 | 87.1 | 179.4 |
|  | 57 | 3.60 | 450 | 5.1 | 10.5 | 154.3 | 317.6 |

[*]Alliant model G2659.
[†]Alliant model G2659B1.

parameters of the two systems with identical hardware and shows the superior performance of the Li/SOCl$_2$ battery. The similarity in voltage and the fact that the same hardware is used for both systems permits a one-for-one replacement.

Figure 29.65 shows the effect of inactive storage of up to 12 months at 71°C on the Li/SOCl$_2$ battery performance over the temperature range of −54 to 71°C. No significant effect on performance was found as a result of the storage. The slightly lower voltages during discharge or the voltage delays when the load is first applied on active (nonreserve) batteries were not present with the reserve batteries. Figure 29.65 also gives a summary of the performance of fresh batteries at various discharge loads and temperatures.

Figure 29.66 shows the discharge curves of the Li/SOCl$_2$ reserve A-size battery (illustrated in Fig. 29.55). The current drain capability of a reserve system significantly exceeds that of the corresponding active primary battery. Figure 29.66 shows the discharge characteristics at 1.25 kΩ (about 3 mA) or 0.15 mA/cm$^2$. Currents higher than 1.5 A (current density of 100 mA/cm$^2$) can be obtained at voltages higher than 2.0 V for several minutes at −10°C. The specific energy of the cells, to a cutoff voltage of 2.0 V, as a function of the discharge current at various temperatures is shown in Fig. 29.67. The performance at higher temperatures is close to that for 25°C.

***Multicell Battery Design.***   The performance characteristics of the Li/SO$_2$ multicell single-activation design battery are shown in Fig. 29.68. The activation and discharge profiles for a 12 V, 100 Ah battery using a LiAsF$_6$ in AN-SO$_2$ electrolyte are illustrated. Because the battery is activated manually, the slower voltage rise time is expected because the cutting of the diaphragm in the center bulkhead requires several turns on the activation bolt. The battery could easily be activated with an electric or mechanical input to a piston actuator or squib to cut the

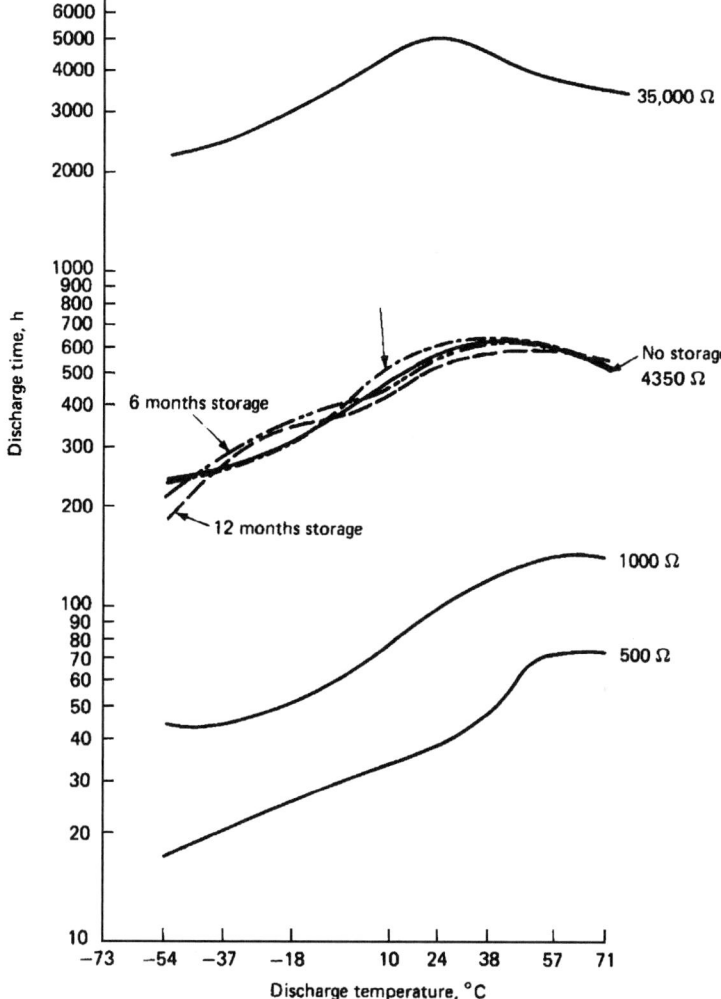

**FIGURE 29.65**   Effect of the discharge rate at various loads and inactive storage on a Li/SOCl₂ reserve battery, Alliant model G2659B1. (*Courtesy of Alliant Techsystems, Inc.*)

diaphragm to improve the voltage rise time. Although the operational life of the battery can be very short, the data illustrate its capability for a long-term discharge at low discharge rates if a stable electrolyte is used.

Typical discharge curves at several temperatures for the two types of single-cells used in the Li/SOCl₂ reserve battery illustrated in Fig. 29.58 are shown in Fig. 29.69.

## 29.5   *SPIN-DEPENDENT DESIGNS*

### 29.5.1   **Energy and Power Density**

Liquid-electrolyte reserve batteries are not normally rated in terms of energy or power per unit weight or per unit volume. Because of the need to provide double the volume for the electrolyte (one volume in an ampoule, the other in the cells themselves), such batteries are not highly space efficient. Space is also consumed by the

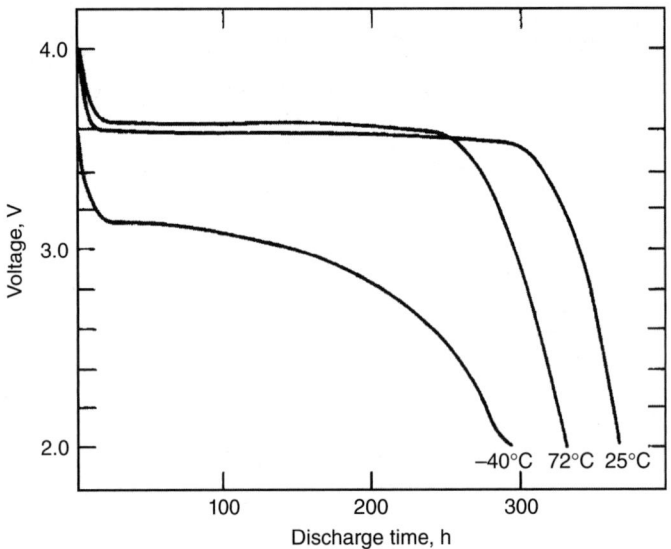

**FIGURE 29.66** Typical discharge curves on 1.25 kΩ of a Li/SOCl₂ reserve battery, Tadiran model TL-5160. (*Courtesy of Tadiran Industries, Ltd.*)

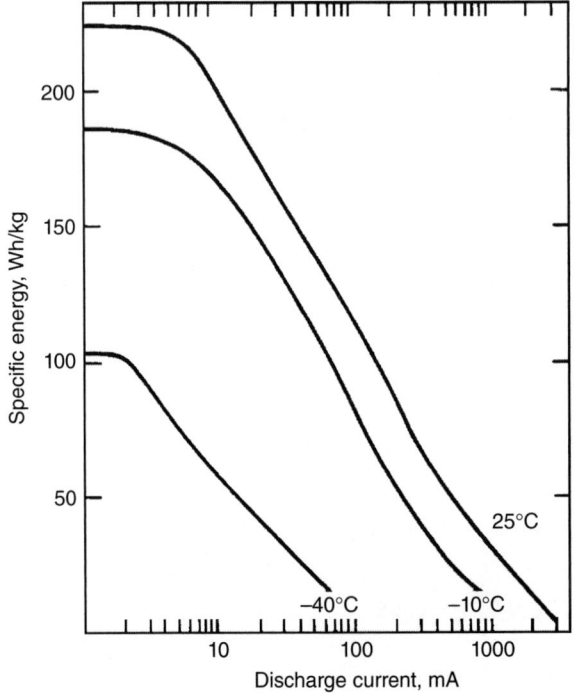

**FIGURE 29.67** Specific energy as a function of discharge current and temperature for a Li/SOCl₂ reserve battery, Tadiran model TL-5160. (*Courtesy of Tadiran Industries, Ltd.*)

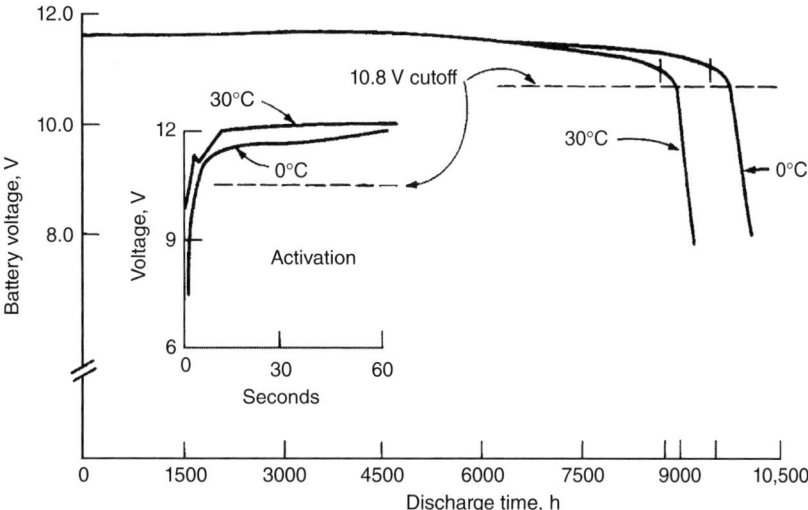

**FIGURE 29.68**  Activation and discharge voltage profile for a 12 V, 100 Ah Li/SO$_2$ battery; electrolyte: LiAsF$_6$ in AN-SO$_2$.

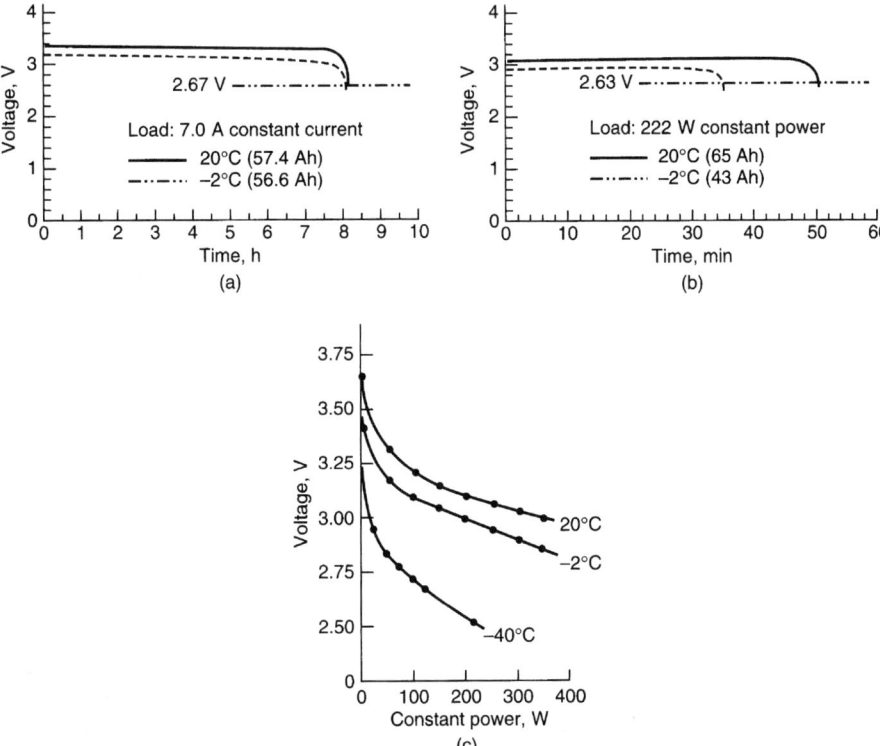

**FIGURE 29.69**  Discharge characteristics of Li/SOCl$_2$ reserve batteries, Alliant model G3070A2. (*a*) Low-rate cell discharge profile. (*b*) High-rate cell discharge profile. (*c*) Voltage characteristics of a high-rate cell at various temperatures and power levels. (*Courtesy of Alliant Techsystems, Inc.*)

ampoule-opening mechanism and the cell-sealing material. Furthermore, the cell area is sometimes not exposed to the electrolyte because of the spin eccentricity of the projectile, which houses the battery. Finally, such batteries are generally designed for short-lifetime applications, such as the flight time of an artillery projectile (approximately 3 min).

### 29.5.2  Operating Temperature Limits

Like most other batteries, the performance of liquid-electrolyte reserve batteries is affected by temperature. Military applications frequently demand battery operations at all temperatures between −40 and 60°C, with storage limits of −55 to 70°C. These requirements are routinely met by the lead/fluoboric acid/lead dioxide systems and, with some difficulty at the low-temperature end, by the lithium/thionyl chloride and zinc/potassium hydroxide/silver oxide systems. Provision is occasionally made to warm the electrolyte prior to the activation of the two latter systems.

### 29.5.3  Voltage Regulation

Since the voltage sustained by a liquid-electrolyte reserve battery at low temperatures and under heavy electric loading is much lower than that which it delivers at high temperatures, a serious problem of voltage regulation frequently results. In some situations, the ratio of high- to low-temperature voltage may be as much as 2:1. This problem may be avoided by the use of thermal batteries (Chap. 21), which provide their own pyrotechnically induced operating temperature, irrespective of the ambient temperature. Until recently, thermal batteries were extremely ineffective at high spin rates, but progress has been made in this field and thermal batteries capable of withstanding spin rates of 300 rev/s are now available.

### 29.5.4  Shelf Life

The shelf life of liquid-electrolyte reserve batteries is highly dependent on the storage temperature, with high temperatures being the more deleterious. Zinc/silver oxide cells are probably the most vulnerable of the generally used systems because of the reduction of silver oxide and the passivation of zinc. Ten-year storage life is probably the best that can be expected unless the battery is substantially overdesigned. Lead/fluoboric acid/lead dioxide batteries also degrade with time, in both the loss of capacity and the lengthening of activation time. However, if objectionable organic materials are avoided in battery construction and the battery is designed with some safety factors, 20 to 25 years of shelf life may be realized. Lithium/thionyl chloride reserve systems employing a neutral catholyte formulation stored in a glass ampoule, as shown in Fig. 29.54, have demonstrated excellent shelf life, exceeding 16 years in a real-time study. Other lithium/thionyl chloride batteries (Fig. 29.7) have incorporated formed and/or welded stainless steel (304L or 316L) reservoirs filled with nonneutral catholytes that may also contain additives to improve certain aspects of performance. Predicting shelf life in this circumstance is less straightforward because of the highly corrosive nature of the catholytes and the potential interaction with/of the various additives, but several studies have projected long storage capability for properly (dry) built and sealed batteries.

### 29.5.5  Linear and Angular Acceleration Limits

Since spin-activated batteries are normally expected to be used in environments where guns are used, they must be built to withstand the forces of gunfire. With the development of the ampoules and the construction methods described, such batteries can withstand linear acceleration to the 20,000 to 30,000 $g$ level and spin rates as great as 30,000 rev/min. The sizes intended for small-caliber (20 to 40 mm) projectiles will withstand linear $g$ levels two to five times that high.

As an assist in withstanding these forces, the battery assembly is sometimes encapsulated in a supporting plastic. A popular design involved a molded plastic cup to house the stack and ampoule assembly, which was locked in place with an epoxy resin. More recently, the stack and ampoule assembly was encapsulated in situ in a RIM (reaction impingement molding) process using a high-impact polyurethane foam, a process that allows demolding in just minutes. These two types of support are shown in Fig. 29.70.

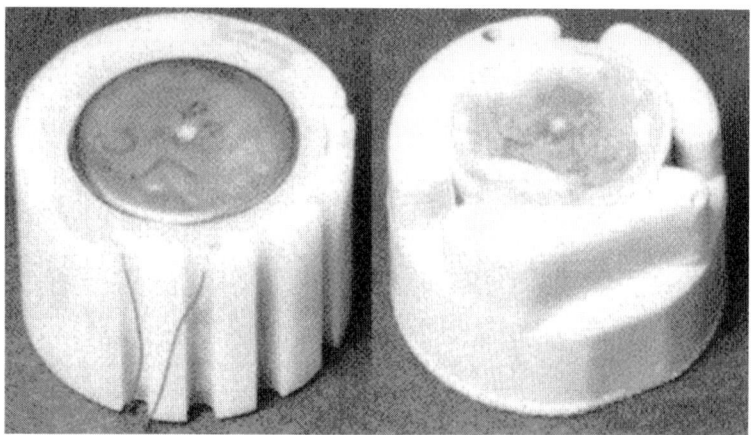

**FIGURE 29.70**  The stack and ampoule assembly of a lead/fluoboric acid/lead dioxide reserve battery supported by potting in epoxy in a molded case (*left*) and by in situ molding using a reaction impingement molded polyurethane foam (*right*). (*Courtesy of U.S. Department of the Army.*)

### 29.5.6  Activation Time

The time from initiation of the battery to the point at which it delivers and sustains a requisite level of voltage across a specified electric load is defined as the activation time. For a spin-dependent liquid-electrolyte reserve battery, this time would include the times for ampoule opening, electrolyte distribution, clearing of electrolyte short circuits in the filling manifold, depassivation of electrodes, and elimination of any form of polarization. Activation times are usually longest at low temperatures, where increased viscosity of the electrolyte and decreased ion mobility are most significant.

The application normally establishes the maximum allowable activation time, and reserve batteries are frequently designed to reach 75% or 80% of their peak voltage within this required time. A typical application requiring a very short activation time, perhaps less than 100 ms, would be a time fuze for an artillery projectile. Battery power is required to start the timer. Hence a stretch-out or uncertainty of time to reach timer voltage could result in a serious timing error, with a corresponding ineffectiveness of gunfire. In some cases, safety can be adversely affected by a timing error. In less critical situations, 0.5 to 1.0 s is allowed for activation.

The physical and electrical characteristics of several spin-dependent reserve batteries are presented in Table 29.12.

**TABLE 29.12**  Typical Spin-Dependent Reserve Batteries

| Reference | Electrochemical system | Height, cm | Diameter, cm | Weight, g | Nominal voltage, V | Nominal energy, Wh |
|---|---|---|---|---|---|---|
| Fig. 29.2 | $Pb/HBF_4/PbO_2$ | 4.1 | 5.7 | 280 | 35 | 0.5 |
| Fig. 29.7 | $Li/SOCl_2$ | 1.67 | 3.8 | 70 | 9 | 0.37 |
|  | $Zn/KOH/AgO$ | 1.3 | 5.1 | 80 | 1.4 | 0.65 |

## 29.6  SUMMARY AND EXAMPLES OF VARIOUS TYPES OF RESERVE BATTERIES

### 29.6.1  Water-Activated Battery Applications

Water-activated batteries can be viable candidates as the power source for many types of equipment. The choice of battery to use becomes one of economics. With proper design, all will perform similarly. Where high current densities are required and cost is secondary, the magnesium/silver chloride system is best. All can be used as

immersion or dunk-type batteries; all but the magnesium/cuprous chloride system will withstand long storage times at high temperatures and high humidity. At the present state of the art, only the magnesium/silver chloride system is suitable for use in forced-flow batteries.

***Water-Activated Batteries for Aviation and Marine Life Jacket Lights.*** The magnesium/cuprous chloride water-activated battery system is being used in Federal Aviation Administration (FAA) and U.S. Coast Guard approved aviation and marine life jacket lights. A typical light is shown in Fig. 29.71.

The single-cell battery has a cathode approximately 5 mm thick with a footprint of 7.25 by 2 mm. Table salt is added to the cathode mix[19] to obtain an adequate voltage in freshwater. The holes in the battery case are optimized to maintain electrolyte salinity while allowing flushing of discharge products. After being mixed with heating, and then cooled and rechopped, the powder is pressed and reheated in an automatic hydraulic press. The cathode is pressed onto a titanium wire current collector, which is wire brushed before manufacture to remove oxide buildup.

The cell is constructed with two anodes, each with the same footprint as the cathode, connected in parallel and placed on either side of the cathode. The anodes are AZ61 electrochemical magnesium sheets.

Typical cell voltage at a 220 to 240 mA (C/12) discharge (against a miniature incandescent lamp) starts at 1.77 V in salt-

**FIGURE 29.71** Life jacket light, using a magnesium/cuprous chloride water-activated battery. (*Courtesy of Electric Fuel Ltd.*)

water and goes down gradually to about 1.65 V before a sharp voltage drop signaling the end of discharge. Voltages in freshwater are about 0.1 V lower. Total capacity is about 3000 mAh.

A battery, with two cells wired in series for international marine applications, uses an AT61 sheet because of the requirement for higher voltage. In saltwater, the cell voltage at a 340 mA (C/8) discharge (against a highly efficient gas-filled miniature lamp) is as high as 1.87 V early in the discharge and drops to about 1.8 V after 8 h. Again, the freshwater voltage is about 100 mV less per cell. This discharge is shown in Fig. 29.72.

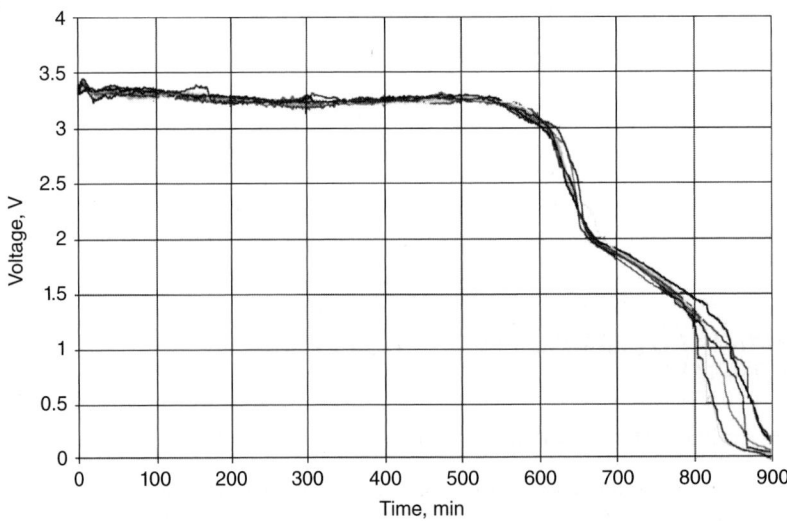

**FIGURE 29.72** Typical discharge of 6 WAB-MX8 batteries in fresh tap water at 330 mA. (*Courtesy of Electric Fuel Ltd.*)

Because the salt added to the cathode makes the cathode even more hygroscopic than it would otherwise be, the batteries are preferably stored with a removable pull-plug used to seal the holes in the battery case.

The characteristics of the life jacket lights are given in Table 29.13.

**TABLE 29.13**  Characteristics of Life Jacket Lights

| Electric fuel model no. | Nominal voltage, V | Nominal size, cm | | | Nominal discharge capacity | | |
|---|---|---|---|---|---|---|---|
| | | Length | Width | Height | Time | Wh | Normal usage mode |
| WAB-H12 | 1.7 | 2.9 | 1.6 | 9.3 | 12h | 4.4 | Aviation/marine life jacket light |
| WAB-H18 | 1.7 | 2.9 | 1.6 | 9.3 | 8h+ | 3.3 | Aviation life jacket light |
| WAB-MX8 | 3.6 | 3.1 | 3.3 | 9.5 | 8h+ | 10.7 | Marine life jacket light |

*Source:* Electric Fuel Ltd.[20]

***Applications for Magnesium/Silver and Cuprous Chloride Batteries.***   Figure 29.73 illustrates two of the magnesium/silver chloride batteries currently manufactured. These batteries are used in the following types of applications:

- Lifeboat emergency equipment on commercial airlines
- Sonobuoys
- Radio and light beacons
- Underwater ordnance
- Radiosonde units—balloon transport equipment; high altitude, low ambient temperature operation

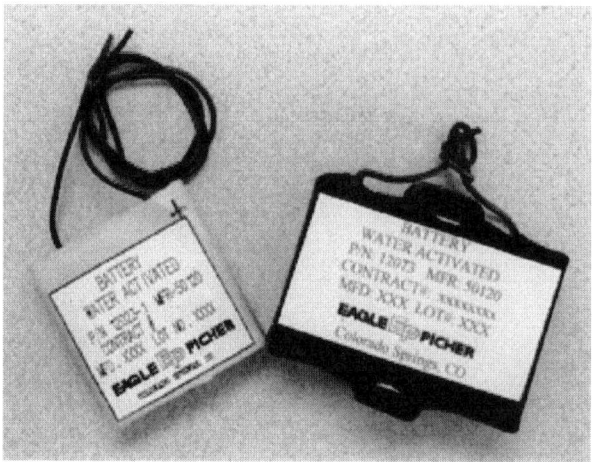

**FIGURE 29.73**   Magnesium/silver chloride batteries; 12023-1 and 12073.

Although "standard lines" of water-activated batteries were once manufactured, most batteries now are designed and manufactured for specific applications. Tables 29.14 and 29.15 list some of the standard and special-purpose magnesium/cuprous chloride and magnesium/silver chloride batteries that were manufactured.

***Zinc/Silver Oxide Battery Types and Sizes for Various Applications.***   Single-cell units of the reserve zinc/silver oxide type are available in sizes from about 0.375 Ah as a minimum up to about 775 Ah. Tables 29.16 and 29.17 provide the specifications for a series of high-rate cells ranging in capacity from 1 to 250 Ah and a series of low-rate cells ranging in capacity from about 2 to 2680 Ah. These are all manually activated.

Table 29.18 lists a number of automatically activated batteries that have been designed to meet various specific applications. Most of these batteries are high-rate with a short wet life. The weight and volume of this type are more a function of the load requirements and the space envelope provided than voltage and capacity.

**TABLE 29.14**  Magnesium/Cuprous Chloride Water-Activated Batteries

| E-P number | Other designation | Nominal voltage/ selection, V | Nominal size, cm | | | Nominal discharge capacity | | Normal usage mode |
|---|---|---|---|---|---|---|---|---|
| | | | Length | Width | Height | Time | Wh | |
| MAP-12037 | PIBAL | 3.0 | 1.3 | 3.2 | 5.1 | 30 min | 0.8 | Airborne, lighting type |
| MAP-12051 | — | 18.0 | 6.8 | 3.8 | 5.7 | 120 min | 2.16 | Airborne, radiosonde |
| MAP-12053 | BA-259 | A: 1.5 B: 6.0 C: 115.0 | 11.7 | 10.2 | 6.0 | A: 90 min B: 90 min C: 90 min | 0.34 1.89 650.4 | Airborne, radiosonde |
| MAP-12060 | — | 18.0 | 5.1 | 5.4 | 5.1 | 120 min | 5.4 | Airborne, radiosonde |
| MAP-12061 | — | 22.5 | 5.1 | 7.0 | 5.1 | 90 min | 7.59 | Airborne, radiosonde |
| MAP-12064 | BA-253 | 6.0 | 10.2 | 3.8 | 3.8 | 45 min | 2.25 | Airborne, lighting type |
| MAP-12071 | — | 20.0 | 6.3 | 7.6 | 16.0 | 8.1 h | 53.46 | Submerged, buoy system |

*Source:* Eagle-Picher Technologies, LLC.[21]

**TABLE 29.15**  Magnesium/Silver Chloride Water-Activated Batteries

| E-P number | Other designation | Nominal voltage, V | Approximate size, cm | | | Nominal discharge capacity | | |
|---|---|---|---|---|---|---|---|---|
| | | | Length | Width | Height | Time | Wh | Ah |
| MAP-2023-1 | Squib firing battery | 5.5 | 5.1 | 2.5 | 5.4 | 1 min | 0.315 | 0.0572 |
| MAP-12062 | — | 48 | 12.1 | dia. | 33 | 20 min | 400 | 8.33 |
| MAP-12065 | — | 4.5 | 6.3 | 6.7 | 13.9 | 50 h | 157.5 | 35 |
| MAP-12066 | — | 7.5 | 5.1 | 5.1 | 16.5 | 14 h | 138 | 18.4 |
| MAP-12067 | MK-72 Squib firing battery | 0.75 | 2.8 | dia. | 2.5 | 13 s | 0.0010 | 0.0014 |
| MAP-12069 | — | 10 | 7.6 | 2.5 | 8.9 | 6 h | 1.5 | 0.15 |
| MAP-12070 | — | 12 | 5.1 | 2.6 | 10 | 9 h | 53.2 | 4.44 |
| MAP-12073 | — | 14.5 | 7.6 | 2.8 | 5.1 | 15 h | 14.55 | 1 |
| MAP-12074 | — | 10.5 | 4.1 | 5.1 | 25 | 48 h | 95.35 | 9 |
| 1473132 | MK61 | 130 | 9.88 | dia. | 12.875 | 6 min. | 2,925 | 20 |

*Source:* Eagle-Picher Technologies, LLC.

**TABLE 29.16**  Zinc/Silver Oxide Manually Activated Batteries

| High-rate cells, 15 min rate | | | | | Low-rate cells, 20 h rate | | | | | Physical dimensions, cm | | |
|---|---|---|---|---|---|---|---|---|---|---|---|---|
| Cell type* | Cap., Ah | Specific energy, Wh/kg | Energy density, Wh/L | Wt, g | Cell type* | Cap., Ah | Specific energy, Wh/kg | Energy density, Wh/L | Wt, g | Length | Width | Height |
| SZH 1.0 | 1.0 | 57 | 104 | 25 | SZL 1.7 | 1.7 | 84 | 171 | 30 | 1.09 | 2.69 | 5.16 |
| SZH 1.6 | 1.6 | 66 | 110 | 35 | SZL 2.8 | 2.8 | 88 | 201 | 50 | 1.25 | 3.07 | 5.72 |
| SZH 2.4 | 2.4 | 66 | 116 | 55 | SZL 4.5 | 4.5 | 92 | 220 | 75 | 1.42 | 3.50 | 6.32 |
| SZH 4.0 | 4.0 | 66 | 128 | 90 | SZL 7.5 | 7.5 | 97 | 250 | 120 | 1.63 | 4.00 | 7.09 |
| SZH 7.0 | 7.0 | 66 | 134 | 160 | SZL 16.8 | 16.8 | 106 | 305 | 240 | 2.00 | 4.95 | 8.48 |
| SZH 16.0 | 16.0 | 66 | 140 | 370 | SZL 43.2 | 43.2 | 125 | 397 | 520 | 2.54 | 6.27 | 10.39 |
| SZH 68.0 | 68.0 | 80 | 196 | 1290 | SZL 160.0 | 160.0 | 187 | 470 | 1330 | 3.73 | 9.27 | 15.09 |
| SZH 250.0 | 250.0 | 154 | 410 | 2450 | — | — | — | — | — | 4.32 | 9.45 | 22.43 |
| — | — | — | — | — | SZL 410.0 | 410.0 | 210 | 560 | 3000 | 4.22 | 13.84 | 19.35 |
| — | — | — | — | — | SZL 775.0 | 775.0 | 276 | 957 | 4380 | 6.96 | 8.36 | 21.70 |

*Eagle-Picher Technologies.

**TABLE 29.17** Zinc/Silver Oxide Manually Activated Primary Batteries*

| Cell model[†] | Type[‡] | Voltage[§] | Capacity, Ah | Specific energy, Wh/kg | Energy density, Wh/L | Weight, g | Dimensions, cm | | | Volume, L |
|---|---|---|---|---|---|---|---|---|---|---|
| | | | | | | | Height | Width | Depth | |
| PM1 | HR | 1.42 | 2.0 | 92 | 147 | 31 | 5.13 | 2.74 | 1.37 | 0.019 |
| PMV2 | HR | 1.48 | 5.3 | 103 | 184 | 76 | 6.42 | 4.37 | 1.52 | 0.043 |
| PM3 | HR | 1.41 | 6.4 | 106 | 187 | 85 | 7.26 | 4.37 | 1.52 | 0.048 |
| PML4 | HR | 1.42 | 8.3 | 113 | 208 | 104 | 8.53 | 4.37 | 1.52 | 0.057 |
| PM5 | MR | 1.49 | 9.9 | 119 | 187 | 124 | 7.36 | 5.28 | 2.03 | 0.079 |
| PMC5 | MR | 1.48 | 12.3 | 141 | 231 | 129 | 7.36 | 5.28 | 2.03 | 0.079 |
| PMC10 | MR | 1.48 | 28 | 152 | 312 | 272 | 12.00 | 5.89 | 1.88 | 0.133 |
| PM15 | HR | 1.42 | 19 | 92 | 180 | 292 | 12.55 | 5.89 | 2.03 | 0.150 |
| PMV16 | HR | 1.47 | 18 | 72 | 141 | 365 | 15.57 | 5.84 | 2.06 | 0.187 |
| PM30 | HR | 1.44 | 41 | 98 | 169 | 600 | 16.64 | 8.28 | 2.54 | 0.350 |
| PM58 | HR | 1.42 | 56 | 85 | 162 | 938 | 18.42 | 8.26 | 3.23 | 0.491 |
| PML100 | LR | 1.50 | 118 | 180 | 376 | 982 | 13.74 | 9.70 | 3.53 | 0.470 |
| PML140 | LR | 1.49 | 165 | 197 | 439 | 1,250 | 16.36 | 9.70 | 3.53 | 0.560 |
| PML170 | LR | 1.48 | 200 | 197 | 469 | 1,500 | 18.44 | 9.70 | 3.53 | 0.631 |
| PML400 | LR | 1.47 | 375 | 218 | 566 | 2,525 | 16.10 | 15.27 | 3.96 | 0.974 |
| PML2500 | LR | 1.48 | 2680 | 221 | 721 | 17,960 | 47.90 | 10.72 | 10.72 | 5.505 |

*These batteries are normally used as primaries. However, they all can be recharged (typically 3 to 10 cycles).
[†]Yardney Technical Products.
[‡]HR = High rate, MR = Medium rate, LR = Low rate.
[§]HR = 15 min rate, MR = 1 h rate, LR = 5 h rate.

**TABLE 29.18** Zinc/Silver Oxide Automatically Activated Batteries

| Part number* | Application | Weight, kg | Volume, L | Voltage, V | Current, A | Capacity, Ah | Specific energy and energy density | |
|---|---|---|---|---|---|---|---|---|
| | | | | | | | Wh/kg | Wh/L |
| EPI 4331 | AIM-7 | 1.0 | 0.45 | 26 | 10.0 | 0.8 | 21 | 46 |
| EPI 4568 | Peacekeeper | 3.3 | 1.89 | 30 | 2.0 | 3.8 | 35 | 60 |
| EPI 4500 | Patriot | 3.6 | 1.61 | 51 | 18.0 | 1.5 | 21 | 48 |
| YTP 15148 | Trident I (C-4) | 5.0 | 1.20 | 28 | 6.0 | 12.0 | 65 | 284 |
| EPI 4567 | Peacekeeper | 6.2 | 3.46 | 30 | 11.0 | 16.0 | 77 | 139 |
| EPI 4470 | Harpoon | 8.6 | 3.5 | 28 | 27, 40 | 8, 12 | 65 | 160 |
| EPI 4445 | Torpedo | 9.3 | 4.8 | 28 | 30 | 20 | 60 | 117 |
| YTP 15066 | Trident I | 14.5 | 3.8 | 30, 31 | 15, 23 | 4, 10 | 30 | 112 |
| YTP 5659985 | Trident II (D-5) | 30.0 | 31.2 | 34, 32 | 113, 10 | 15, 7 | 21.7 | 20.83 |
| YTP P-530 | Minuteman | 0.77 | 0.36 | 30 | 10.0 | 0.46 | 17.9 | 38.3 |
| YTP P-515 | Sparrow | 0.99 | 0.45 | 24 | 11.0 | 0.45 | 10.9 | 24.0 |
| YTP P-512 | NMD | 0.86 | 0.30 | 30 | 13.0 | 0.30 | 10.5 | 30.0 |
| YTP P-468 | AGM130 | 7.03 | 2.70 | 28 | 30 | 12.08 | 45.0 | 117.0 |
| YTP P-471 | Peacekeeper | 19.5 | 12.1 | 76, 31 | 16.7, 40 | 5.46, 40.90 | 86.3 | 139.3 |
| YTP P-329[†] | Hawk | 3.18 | 1.64 | 59, 25, 19, 12, −13, −59 | 2.2, 0.8 | 0.5 | 5.7 | 11.0 |
| YTP 17511 | MK37 Torpedo | 120 | 96.6 | 85, 76 | 900, 450 | 79, 22 | 67.5 | 83.9 |
| YTP 19580 | SST-4 Torpedo | 408 | 467 | 210, 115 | 480, 525 | 29, 110 | 54.3 | 47.4 |
| YTP[‡] | Tigerfish | 583 | 367 | 45/60 | 1200/750 | 240 | 54.3, 47.6 | 75.61 |

*EPI—Eagle-Picher Technologies; YTP—Yardney Technical Products, Inc.
[†]6 V taps.
[‡]Consists of one forward and one aft battery.

### 29.6.2  Spin-Dependent Reserve Systems

The physical and electrical characteristics of several typical spin-dependent reserve batteries were presented above in Table 29.13. Further application details for the various listed cell types are discussed below.

*Lead/Fluoboric Acid/Lead Dioxide Battery.*   Discharge curves for a typical lead/fluoboric acid/lead dioxide liquid-electrolyte reserve battery employed to power the proximity fuze of an artillery shell are given in Fig. 29.74. The slight rise in the low-temperature curve is due to its gradual rise in temperature in a room-temperature spinning tester. Similarly, the high-temperature curve is falling faster than it would in a true iso-thermal situation.

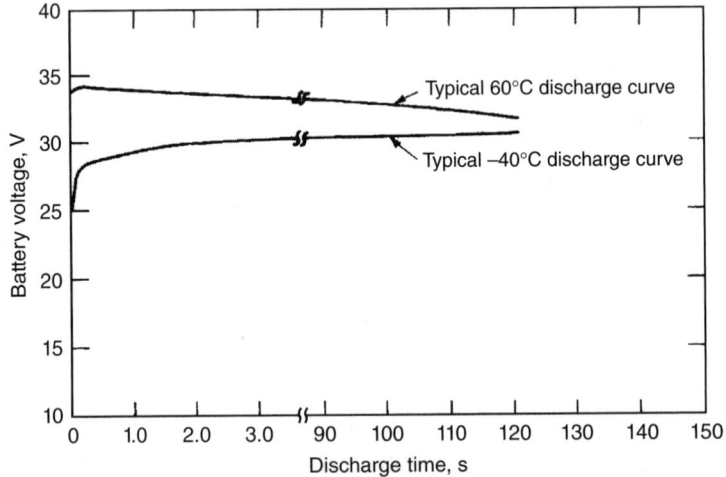

**FIGURE 29.74**   Discharge curves of a spinning lead/fluoboric acid/lead dioxide series-configuration reserve battery. Current density: 100 mA/cm². (*Courtesy of U.S. Department of the Army.*)

*Lithium/Thionyl Chloride Battery.*   Discharge curves for the multicell lithium/thionyl chloride liquid-electrolyte reserve battery shown in Fig. 29.7, which was designed to power the proximity fuze of an artillery shell, are shown in Figs. 29.75 and 29.76 and indicate the effect of operating temperature on the output voltage level, deliverable capacity, and rise time of the battery.

In addition to simply replacing the lead/fluoboric acid/lead dioxide system, the advent of lithium/thionyl chloride batteries introduced some operational advantages. With the former system, spin-dependent batteries were expected to function for short periods of time and only under sustained spin (necessary to keep the electrolyte within the cells). New applications have arisen that require a battery capable of withstanding artillery fire and spinning for a short time followed by some substantial operating time in a nonspin mode. Such applications include artillery delivery of mines or communication jammers intended to function after impact with the ground, or projectiles and submunitions that are operative while being slowed down by a parachute.

The lithium-based liquid-electrolyte reserve battery is capable of fulfilling this difficult combination of requirements. A typical cell, as illustrated in Fig. 29.77, incorporates an absorbing separator such as a nonwoven glass mat between the electrodes, and a long, high-resistance electrolyte filling path. After cell filling under spin, the absorbing material causes the electrolyte to be retracted away from the manifold and retained within the cell after cessation of spin. These design features, coupled with the long wet-stand capability of the lithium/thionyl chloride system, have paved the way for reserve batteries in applications that previously had

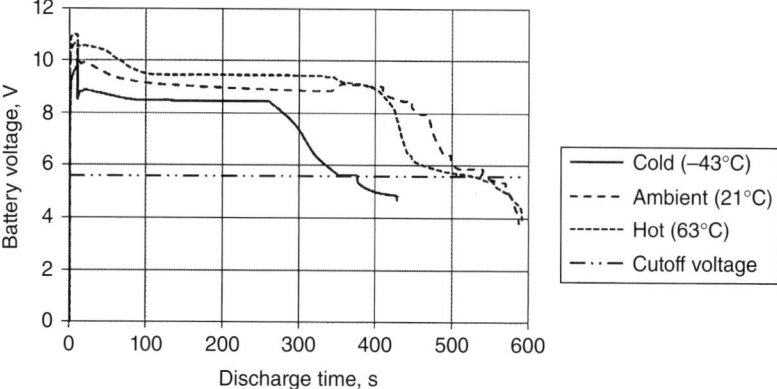

**FIGURE 29.75** Discharge curves of a spinning (80 rev/s) lithium/thionyl chloride reserve battery at various temperatures. Discharge current density is 2 mA/cm² until 10 s have elapsed, after which it is 35 mA/cm². (*Courtesy of U.S. Department of the Army.*)

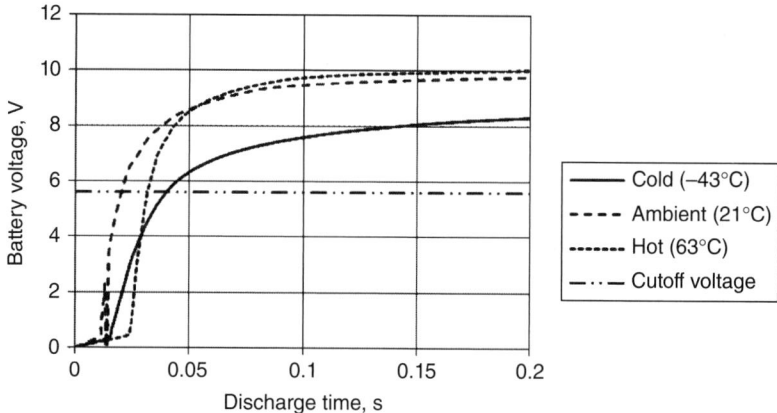

**FIGURE 29.76** Rise-time curves of the spinning (80 rev/s) lithium/thionyl chloride reserve battery shown in Fig. 29.75 at various temperatures. Discharge current density is 2 mA/cm². (*Courtesy of U.S. Department of the Army.*)

to depend on the use of active batteries with relatively shorter storage capability. The discharge curves for such a multicell, liquid-electrolyte reserve battery that demonstrate this long wet-stand capability are given in Fig. 29.78.

***Spin-Capable Thermal Batteries.*** Because of their lower susceptibility to temperature extremes and known superior shelf life (without degradation), thermal batteries have been desired as an alternative to liquid-electrolyte reserve batteries for some time. The primary failure mode for thermal batteries in a high-spin environment had always been intercell short-circuiting at the cell stack edges due to the leakage of molten conductive materials at battery operating temperatures. Novel construction techniques, electrochemistries that allow for higher electrolyte binder contents, and lithium alloy anodes that prevent migration of the anode material have made spin-capable thermal batteries practical.

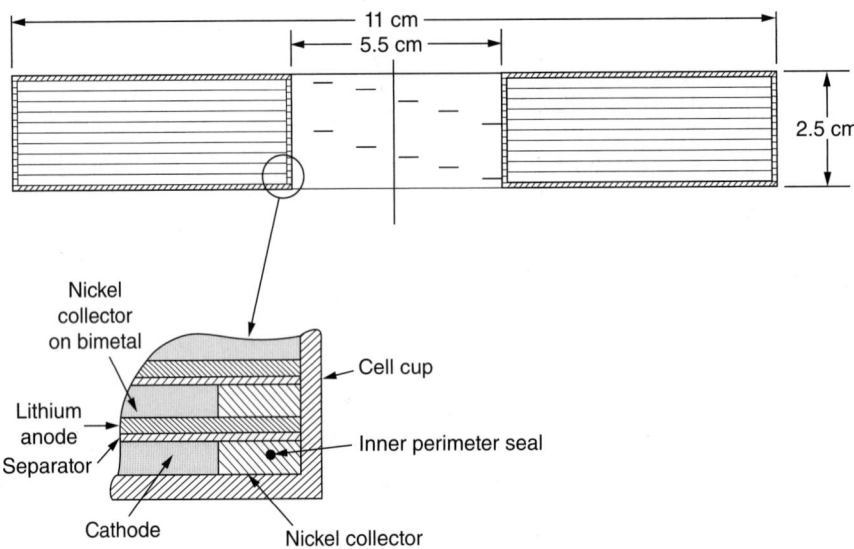

**FIGURE 29.77**   A lithium anode/carbon cathode cell stack for a thionyl chloride liquid-electrolyte reserve battery. (*Courtesy of EnerSys Advanced Systems.*)

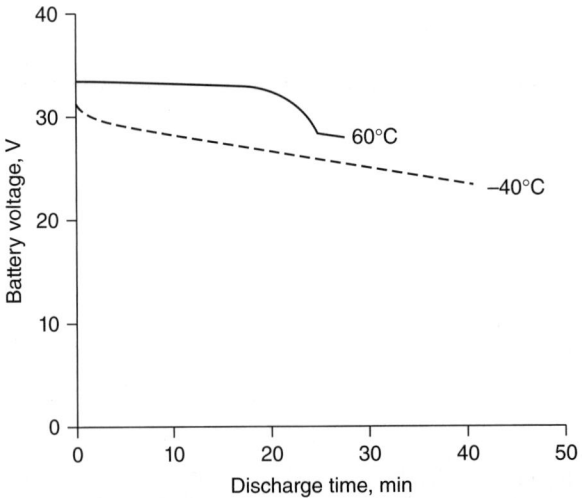

**FIGURE 29.78**   Discharge curves of a lithium/thionyl chloride series-configuration reserve battery. Current density is 50 mA/cm². (*Courtesy of U.S. Department of the Army.*)

## ACKNOWLEDGMENTS

Major content for this chapter was provided in whole or in part by R. David Lucero and Alexander P. Karpinski (Chap. 34, Reserve Magnesium Anode and Zinc/Silver Oxide Batteries); David L. Chua, Benjamin M. Meyer, William J. Eppley, Jeffrey A. Swank, and Michael Ding (Chap. 35, Reserve Military Batteries); and Charles M. Lamb (Chap. 36, Thermal Batteries), *Linden's Handbook of Batteries*, 4th ed., T. B. Reddy and D. Linden, eds., McGraw-Hill, 2011.

## REFERENCES

1. National Defense Research Committee, Final Report on Seawater Batteries, Bell Telephone Laboratories, New York, 1945.

2. L. Pucher, "Cuprous Chloride-Magnesium Reserve Battery," *J. Electrochem. Soc.* **99**:203C (1952).

3. B. N. Adams, "Batteries," U.S. Patent 2,322,210, 1943.

4. H. N. Honer, F. P. Malaspina, and W. J. Martini, "Lead Chloride Electrode for Seawater Batteries," U.S. Patent 3,943,004, 1976.

5. H. N. Honor, "Deferred Action Battery," U.S. Patent 3,205,896, 1965.

6. N. Margalit, "Cathodes for Seawater Activated Cells," *J. Electrochem. Soc.* **122**:1005 (1975).

7. J. Root, "Method of Producing Semi-Conductive Electronegative Element of a Battery," U.S. Patent 3,450,570, 1969.

8. R. F. Koontz and L. E. Klein, "Deferred Action Battery Having an Improved Depolarizer," U.S. Patent 4,192,913, 1980.

9. E. P. Cupp, "Magnesium Perchlorate Batteries for Low Temperature Operation," *Proc. 23d Annual Power Sources Conf.*, Electrochemical Society, Pennington, NJ, 1969, p. 90.

10. N. T. Wilburn, "Magnesium Perchlorate Reserve Battery," *Proc. 21st Annual Power Sources Conf.*, Electrochemical Society, Pennington, NJ, 1967, p. 113.

11. W. A. West-Freeman and J. A. Barnes, "Snake Battery; Power Source Selection Alternatives," NAVSWX TR 90-366, Naval Surface Warfare Center, Carderock Div. 1990.

12. M. G. Medeiros and R. R. Bessette, "Magnesium-Solution Phase Catholyte Seawater Electrochemical System," *Proc. 39th Power Sources Conf.*, Cherry Hill, NJ, June 2000, p. 453.

13. Patent no. 3,953,239, Al-AgO Primary Battery dated April 27, 1976, issued to George Perkons, assigned to the U.S. Navy.

14. G. O. Erb, "Theory and Practice of Thermal Cells," *Publication BIOS/Gp 2/HEC 182 Part II*, Halstead Exploiting Centre, Jun. 6, 1945.

15. O. G. Bennett et al., U.S. Patent 3,575,714, Apr. 20, 1971.

16. B. H. van Domelen and R. D. Wehrle, "A Review of Thermal Battery Technology," *Intersoc. Energy Convers. Conf.*, 1974.

17. F. Tepper, "A Survey of Thermal Battery Designs and Their Performance Characteristics," *Intersoc. Energy Convers. Conf.*, 1974.

18. M. E. Wilkie and T. H. Loverude, "Reserve Electric Battery with Combined Electrode and Separator Member," U.S. Patent 3,061,659, 1962.

19. K. R. Jones, J. L. Burant, and D. R. Wolter, "Deferred Action Battery," U.S. Patent 3,451,855, 1969.

20. H. N. Honor, "Seawater Battery," U.S. Patent 3,966,497, 1976.

21. H. N. Honer, "Multicell Seawater Battery," U.S. Patent 2,953,238, 1976.

22. J. F. Donahue and S. D. Pierce, "A Discussion of Silver Chloride Seawater Batteries," Winter Meeting, American Institute of Electrical Engineers, New York, 1963.

23. H. R. Knapp and A. L. Almerini, "Perchlorate Reserve Batteries," *Proc. 17th Annual Power Sources Conf.*, Electrochemical Society, Pennington, NJ, 1963, p. 125.

24. Patent No. 5,506,065, Electrolyte Activated Battery dates April 9, 1996, issued to Silvano Tribioli et al, Whitehead Alenia Sistemi Subacquei.

25. Zaromb, S. The use and behavior of aluminum anodes in alkaline primary batteries. *J. Electrochem. Soc.*, 1962, 109, 1125–1130.

26. Patent no. 4,942,100, Aluminum Batteries dated 17 July 1990, issued to John Hunter et al, Alcan International Limited.

27. Eric Dow, The Development of Aluminum Aqueous Batteries for Torpedo Propulsion, NUWC-Newport Division.

28. M. Babai, U. Meishar, and B. Ravid, "Modified Li/SOCl$_2$ Reserve Cells with Improved Performance," *Proc. 29th Power Sources Conf.*, Jun. 1980.

29. W. J. Eppley and R. J. Horning, "Lithium/Thionyl Chloride Reserve Cell Development," *Proc. 28th Power Sources Symp.*, 1978.

30. J. Nolting and N. A. Remer, "Development and Manufacture of a Large Multicell Lithium-Thionyl Chloride Reserve Battery," *Proc. 35th International Power Sources Symp.*, 1992.

31. C. Kelly, "Development of HWAM Li$_x$CoO$_2$ Reserve Battery," Report No. NSWCCD-TR-98/005, Apr. 1997.

32. S. McKay, M. Peabody, and J. Brazzell, *Proc. 39th Power Sources Conf.*, pp. 73–76, 2000.

33. P. G. Russell, D. C. Williams, C. Marsh, and T. B. Reddy, *Proc. 6th Workshop for Battery Exploratory Development*, pp. 277–281, 1999.

## *BIBLIOGRAPHY*

Askew, B. A. and R. Holland: "A High Rate Primary Lithium-Sulfur Battery," *Power Sources* **4** (1972).

Baird, M. D., A. J. Clark, C. R. Feltham, and L. H. Pearce: "Recent Advances in High Temperature Primary Lithium Batteries," *Power Sources*, **7** (1978).

Bauer, P.: *Batteries for Space Power Systems*, U.S. Government Printing Office, Washington, D.C., 1968.

Benderly, A. A.: "Power for Ordnance Fuzing," *National Defense*, Mar.–Apr. 1974.

Biggar, A. M.: "Reserve Battery Requiring Two Simultaneous Forces for Activation," *Proc. 24th Annual Power Sources Symp.*, Electrochemical Society, Pennington, NJ, pp. 39–41, 1970.

Biggar, A. M., R. C. Proestel, and W. H. Steuernagel: "A 48-Hour Reserve Power Supply for a Scatterable Mine," *Proc. 26th Annual Power Sources Symp.*, Electrochemical Society, Pennington, NJ, pp. 126–129, 1974.

Birt, D., C. Feltham, G. Hazzard, and L. Pearce: "The Electrochemical Characteristics of Iron Sulfide and Immobilized Salt Electrolytes," *Power Sources*, **7** (1978).

Bowser, G. C., et al.: U.S. Patent 3,891,460, Jun. 24, 1975.

Bowser, G. C., et al.: U.S. Patent 3,930,888, Jan. 1976.

Bush, D. M. and D. A. Nissen: "Thermal Cells and Batteries Using the Mg/FeS$_2$ and LiAl/FeS$_2$ Systems," *Proc. 28th Power Sources Symp.*, 1978.

Cahoon, N. C. and G. W. Heise: *The Primary Battery*, Wiley, New York, 1969.

Chubb, M. F. and J. M. Dines: "Electric Battery," U.S. Patent 3,022,364.

Cieslak, W. R., F. M. Delnick, and C. C. Crafts: "Compatibility Study of 316L Stainless Steel Bellows for XMC3690 Reserve Lithium/Thionyl Chloride Battery," Sandia Report *SAND85-1852*, February 1986.

Collins, W. H.: U.S. Patent 1,482,738, Aug. 10, 1977.

De Gruson, J. A.: "Improved Thermal Battery Performance," *AFAPL-TR-79-2042*, Eagle-Picher Industries, 1979.

Delnick, F. M., R. A. Guidotti, and D. K. McCarthy: "Chromium (V) Compounds as Cathode Materials in Electrochemical Power Sources," U.S. Patent 4,508,796, Apr. 2, 1985.

Doddapaneni, N., D. L. Chua, and J. Nelson: "Development of a Spin Activated, High Rate, Li/SOCl$_2$ Bipolar Reserve Battery," *Proc. 30th Annual Power Sources Symp.*, Electrochemical Society, Pennington, NJ, pp. 201–204, 1982.

Eagle-Picher Technologies, Joplin, MO, website: www.eagle-picher.com.

Fleiseher, A. and J. J. Lander: *Zinc Silver Oxide Batteries*, Wiley, New York, 1971.

Grothaus, K. R.: "Thermal Battery for Artillery," *Proc. 26th Power Sources Conference*, U.S. Army CECOM/ARL, pp. 141–144, Apr. 29–May 2, May 1974.

Guidotti, R. A. and F. W. Reinhardt: "Lithiation of FeS$_2$ for Use in Thermal Batteries," *Proc. 2nd Annual Battery Conf. on Applications and Advances*, paper 87DS-3, 1987.

Guidotti, R. A., F. M. Reinhardt, and W. F. Hammeter: "Screening Study of Lithiated Catholyte Mix for a Long-Life Li(Si)/FeS$_2$ Thermal Battery," *SAND 85-1737*, 1988.

Hansen, M.: *Constitution of Binary Alloys*, McGraw-Hill, New York, 1958.

Harney, D. E.: U.S. Patent 4,221,849, Sept. 9, 1980.

Hollman, F. G., et al.: "Silver Peroxide Battery and Method of Making," U.S. Patent 2,727,083.

Jasinski, R.: *High Energy Batteries*, Plenum, New York, 1967.

Kuper, W. E.: "A Brief History of Thermal Batteries," *Proc. 36th Power Sources Conf.*, Cherry Hill, NJ, June 1994.

Morganstein, M. and A. B. Goldberg: "Reaction Impingement Molding (RIM) Encapsulation of a Fuze Power Supply," *Proc. of the 4th International SAMPE Electronics Conference*, Society for the Advancement of Material and Process Engineering, Covina, CA, pp. 753–764, 1990.

Quinn, R. K., A. R. Baldwin, J. R. Armijo, P. G. Neiswander, and D. E. Zurawski: "Development of a Lithium Alloy/Iron Disulfide 60-Minute Primary Thermal Battery," *SAND79-0814*, 1979.

Schisselbauer, P. F. and D. P. Roller: "Reserve g-Activated, Li/SOCl$_2$ Primary Battery for Artillery Applications," *Proc. 37th Annual Power Sources Conference*, U.S. Army CECOM/ARL, Cherry Hill, NJ, pp. 357–360, 1996.

Schneider, A. A., et al.: U.S. Patent 4,119,796, Oct. 10, 1978.

Searcy, J. Q. and J. R. Armijo: "Improvements in Li(Si)/FeS$_2$ Thermal Battery Technology," *SAND82-0565*, 1982.

Szwarc, R.: "Study of Li-β Alloy in LiCl-KCl Eutectic Thermal Cells Utilizing Chromate and Iron Disulfide Depolarizer," Gepp-TM-426, General Electric Co., Neut. Dev. Dept., 1979.

Turrill, F. G. and W. C. Kirchberger: "A One-Dollar Power Supply for Proximity Fuzes," *Proc. 24th Annual Power Sources Symp.*, Electrochemical Society, Pennington, NJ, pp. 36–39, 1970.

Yardney Technical Products, Pawcatuck, CT, website: www.yardney.com.

# P · A · R · T · 6

# BATTERY INDUSTRY INFRASTRUCTURE

# CHAPTER 30
# OVERVIEW OF CELL AND BATTERY MANUFACTURING

**Anthony Sudano**

## 30.1   INTRODUCTION

Many of the prior battery technology chapters have covered general details and various specifics regarding cell construction and battery pack assembly. This chapter builds upon the prior discussions to better explain the overall generic step-by-step process used to set up and commission a battery manufacturing plant.

## 30.2   OVERALL PROCESS

Figure 30.1 outlines the overall manufacturing process typical to lithium ion pouch-cell production. Details will vary for various electrochemistries and cell formats. Other cells, for example, may have electrodes that are either stacked or wound and fitted into metallic cylindrical cans. Primary batteries and even a few secondary cells do not require any type of electrical precharge or often any discharge testing, either.

While each of the various elements may be modified, deleted, or rearranged for different battery systems, this block diagram provides an example of the type of planning that is needed in establishing a battery manufacturing operation. Specifications for each process step and all related equipment must be determined to ensure success (high quality, high output rate, low cost, high efficiency, etc.).

## 30.3   CELL PROCESSING STEPS

### 30.3.1   Mixing of Electrode Slurries/Pastes

As with any recipe, quality batteries come from the proper selection, dosing, and mixing of ingredients. Regardless of the raw materials used, proper mixing is needed to obtain a homogeneous dispersion and consistent processing. The following sections describe the overall manufacturing process outlined in Fig. 30.1 for a typical lithium battery manufacturing operation.

*Materials.*   Electrode slurries, both anode and cathode, are individually mixed in planetary mixers with high shear dispersing capability in order to break down any clumps or agglomerates from the solid components. A solvent, such as NMP (*N*-methyl-2-pyrrolidone), is used into which polymer binder, solid energetic materials (i.e., electrochemically active electrode ingredients), and conductive carbons are added. The mixing order may vary.

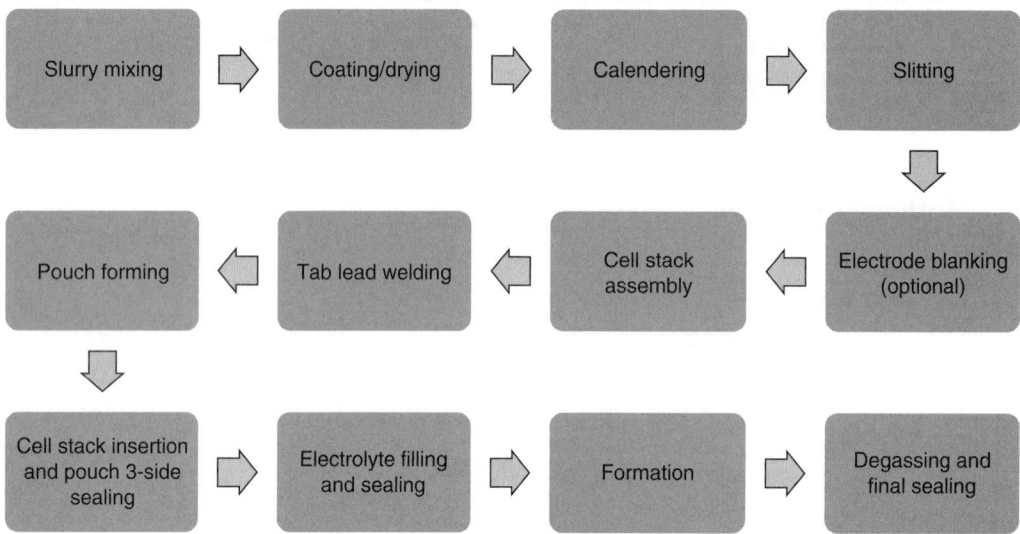

**FIGURE 30.1**   Manufacturing steps.

***Equipment.***   To avoid any risk of cross contamination, it is paramount to have dedicated mixing equipment for anode and cathode. Equipment cost for mixers is not negligible; however, the material handling, controls system, solvent and powder compatible devices, and in-place cleaning of vessels, pumps, and pipes add an even more significant amount of cost to the mixing operation.

These materials are mixed together in a batch process for several hours in vessels as large as 1000 L, until a homogeneous dispersion is obtained. Preparing sufficient mixed slurries in large facilities can require several mixers as seen in Fig. 30.2. The key objective is to assure that the conductive and solid particles form an evenly

**FIGURE 30.2**   High shear, double planetary mixers. (*Courtesy of Charles Ross and Son Company.*)

distributed network through which electrons and ions can easily and quickly migrate. A cooling jacket on the outside of the mixing vessel helps keep slurry temperatures under control to avoid degradation. Vacuum applied to the vessel contents helps eliminate air bubbles from the slurry and extract gases that are formed during the mixing process.

Although traditionally a batch process, recent developments of continuous mixing processes have emerged with obvious benefits of fewer vessels, shorter mixing time, reduced equipment capital, and floor space (see Fig. 30.3). This method has recently been adopted by only a few battery manufacturers, although given its economic advantages, this method is certainly worth investigating as an option.

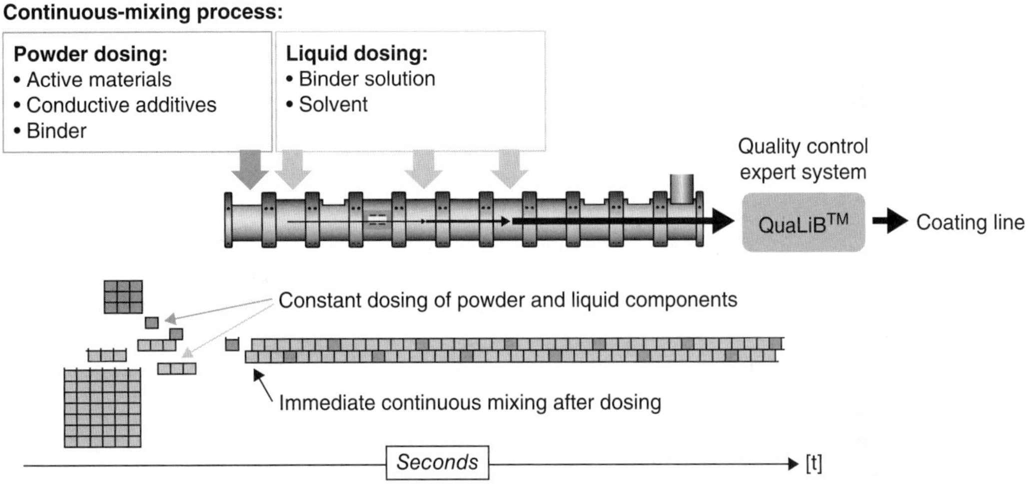

**FIGURE 30.3**  Continuous mixing. (*Courtesy of Bühler AG.*)

## 30.3.2  Electrolytes and Miscellaneous Items

*Electrolytes.*  Electrolyte solutions used in various electrochemical systems require very stringent controls on composition and contaminate levels. However, lithium cells require an even more extreme degree of purity. Dissolved salts and gases, metal particles, moisture, or other contaminants are very harmful to long-term stability. The raw materials and equipment used for electrolyte preparation are also very expensive, and electrolytes are often prepared by specialty chemical producers. Further details on electrolyte systems are provided in Chap. 3B.

*Piece Part Manufacture.*  Various mechanical components (current collectors/insulators/tabs/packaging parts/etc.) go into finished cells. These consist of items from various specialized industries including

- Thin metallic aluminum and copper foils used as current collectors for the electrodes
- Porous polymeric films for the cell separator providing electrical isolation between the anode and cathode while allowing a pathway for the ions to flow through the liquid electrolyte that fills the pores of the separator film
- Metal strips (tabs) with bonded sealants acting as the current conducting channel between the cell stack inside the packaging and the external electrical circuit
- Multilayer metalized plastic laminate films for cell packaging

### 30.3.3  Electrodes

***Fabrication of Electrodes: Coating/Drying.***    In order to satisfy battery performance requirements, both anode and cathode electrodes need to be separately coated onto both sides of thin current carrying foils. Typically, the finished electrodes range in overall thickness from 100 to 200 μm and contain active and conductive materials with precise control of thickness and porosity. The anode is coated on copper foil, whereas the cathode is coated on aluminum foil. The coatings are applied to both sides of the respective foils to improve overall specific energy once assembled as cells.

The coating process is integrated with in-line drying to evaporate the solvent. Slot die coating technology is typically used in lithium-ion battery electrode operations to precisely apply the slurry onto the foil. The mixed slurry is pumped through the slot die onto the moving foil web. Slot die coating technique allows for better thickness control and the possibility of continuous or intermittent coating, also known as skip coating, and either stripe or full web width coating. The coating pattern is a function of the selected cell configuration and as required by the downstream cell assembly process. Web and coating widths can vary from 600 mm to over 1 m wide as per Fig. 30.4. Historically, one side of the web is coated and dried at a time; however, simultaneously coating both sides of the web prior to entering the drying oven is also possible and gaining popularity (see Fig. 30.5).

**FIGURE 30.4**    Coating line with integrated drying oven. (*Courtesy of Durr MEGTEC.*)

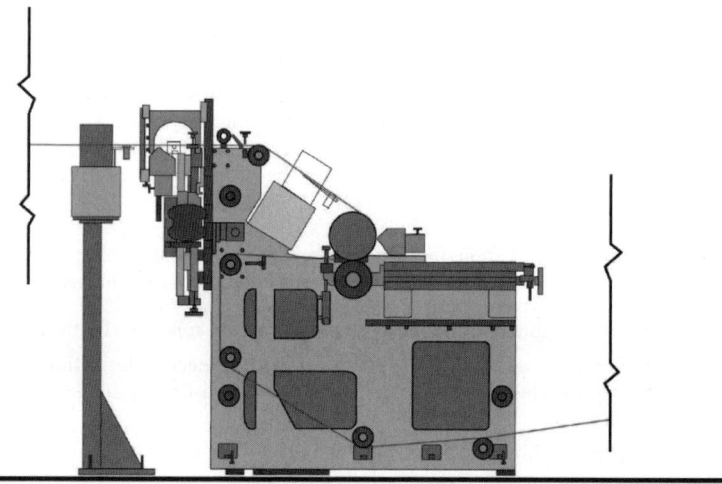

**FIGURE 30.5**    Simultaneous double-sided coating. (*Courtesy of Durr MEGTEC.*)

The coated foil enters a drying oven where hot air is directed over the moving web to heat it up bringing the solvent to its evaporation point and evacuated from the dryer via exhaust fans. The length of the dryer is essentially proportional to the web speed for a given set of coating conditions. Drying ovens on the order of 20 to 40 meters long with web speeds of 15 to 30 m/min are normally found in battery electrode manufacturing facilities, although this depends heavily on coating thickness and solid concentration in the slurry.

To protect the environment and for cost considerations, the evacuated solvent-laden air is treated either by incinerating and scrubbing or more commonly by recovering the solvent out of the airstream as shown in Fig. 30.6. In order to recover the solvent, the solvent-laden exhaust stream is cooled in multiple steps to condense the liquid out of the gas stream. The reclaimed solvent can be filtered or distilled to purify for the purpose of reutilizing in the mixing process.

***Processing of Electrodes: Calendering/Slitting.***    Once both sides of the web are coated and dried, the rolls of flexible electrode are then moved to a roll-to-roll pressing or calendering operation (see Fig. 30.7). The coating/drying process creates electrodes that are porous in nature once dried. However, for proper cell performance and for maximizing the amount of active material in the container, the thickness of the electrodes needs to be reduced while maintaining sufficient porosity for the liquid electrolyte. The web is run through a set of high-pressure rollers, which are sometimes heated to help the material compress more easily.

The coated and calendered electrodes are then slit into finished widths as required by the cell size and configuration. Rotating shear cutting knives and anvils are used to slit the moving web prior to being wound into finished width reels as shown in Fig. 30.8. In some instances, the slitting and calendering operations are

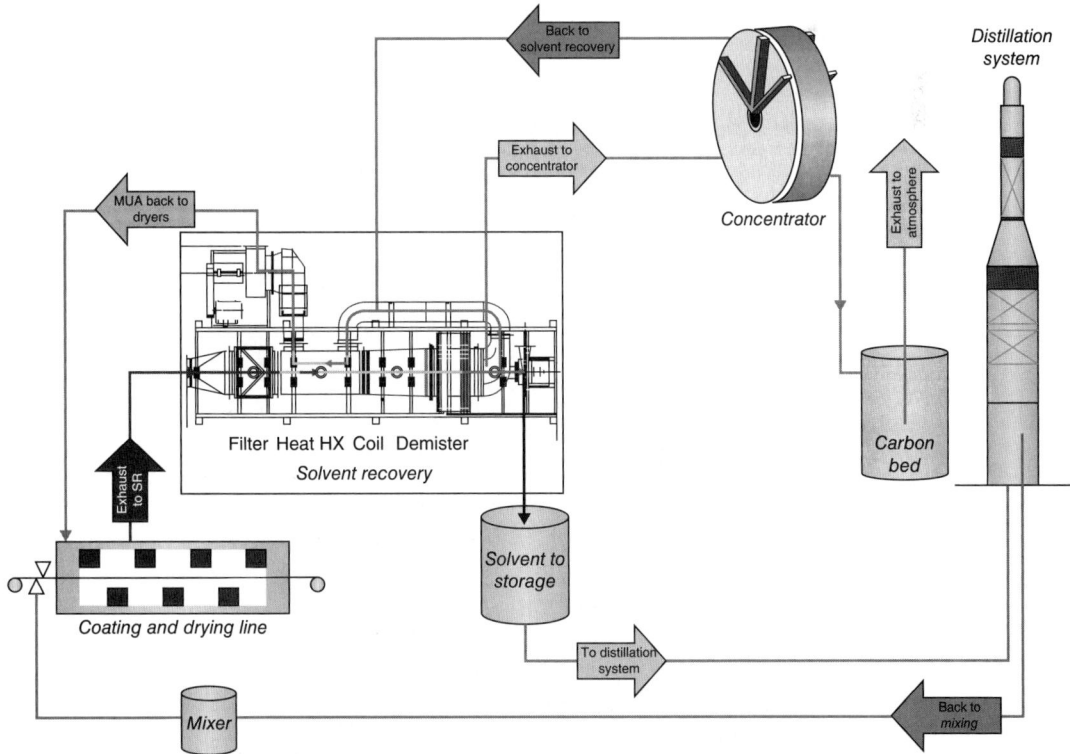

**FIGURE 30.6**    Solvent recovery and purification system. (*Courtesy of Durr MEGTEC.*)

**FIGURE 30.7** Electrode calendering press. (*Courtesy Innovative Machine Corporation.*)

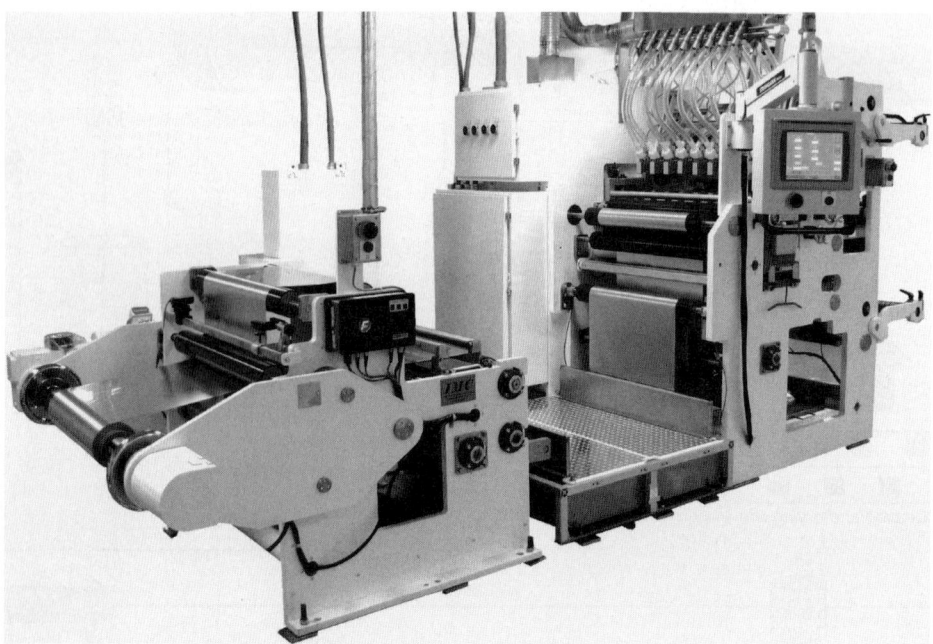

**FIGURE 30.8** Electrode slitting machine. (*Courtesy Innovative Machine Corporation.*)

inversed, although this leads to greater calendering equipment needs since greater linear length of electrodes must be processed and since line speed is a limiting factor. Details on other types of electrode manufacturing technologies are presented in Chap. 3A.

### 30.3.4   Cell Assembly

*Electrode Stacking.*   Although cylindrical format cells are quite common, this chapter will focus on prismatic pouch cells such as seen in Fig. 30.9. In order to preserve low moisture of materials going into a finished cell, all assembly processes are typically performed in humidity-controlled dry rooms. Humidity levels on the order of 0.5% RH or −40°C dew point are usually maintained. To assure such low moisture conditions, production rooms require expensive air handling equipment, not to mention high operating (energy) costs.

**FIGURE 30.9**   Examples of prismatic pouch cells. (*Courtesy of Manz AG.*)

There are various means by which the flat rectangular cell format can be achieved. Common methods include flat winding, Z-fold stacking, pure stacking, and stack winding. Most versatile and easiest to set up is Z-fold stacking, as this method can easily transition and scale up from pilot to high-volume manufacturing. The electrodes need to be notched, to create the bare foil extension that serves for tabbing, and blanked, to yield the specific rectangular size required by the cell size. The electrode cutting is achieved by using a punch and die set as shown in Fig. 30.10 or by laser cutting as per Fig. 30.11. The cutting operation can be done off-line such that the cut electrode sheets are presented to the Z-fold stacking machine in magazines, or it can be done by cutting in-line along with the cell assembly operation.

At the Z-fold stacking machine, pick-up heads alternately pick anode and cathode electrodes from their respective magazines, if not cut in-line, and place each onto a common stacking platform between successive folds of a continuous band of separator material as detailed in Fig. 30.12. The separator width is selected to cover the active area of the electrodes, assuring electrical isolation between the electrodes, while allowing the uncoated bare foil leads to extend beyond the edge of the separator with the electrode foils properly oriented and aligned to ensure proper polarity and to facilitate downstream tabbing of electrode in the cell stack. The appropriate number of electrode sheets is stacked to suit the cell design. The separator is used to form a final wrap around the stacked cell and taped to help immobilize the cell stack.

The stack-winding method, whether using either in-line cutting or pick and place of precut electrodes, may also incorporate heat-bonding of electrodes onto a continuous strip of modified (i.e., heat-activated) separator using a hot lamination process. In this process the spacing between electrodes is increased to allow for the

**FIGURE 30.10**    High-speed electrode notching. (*Courtesy of Manz AG.*)

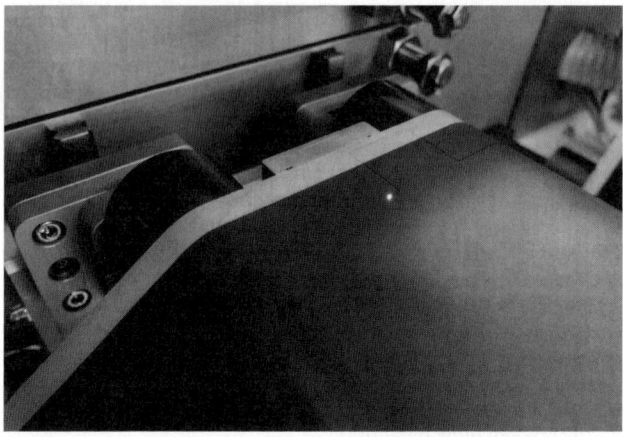

**FIGURE 30.11**    Electrode laser notching. (*Courtesy of Manz AG.*)

increase in cell thickness during winding of each individual cell. The continuous strip separator with the adhered electrodes is typically wound around a flat spindle at high speed to complete the cell stack. This method is generally a higher speed process than Z-fold stacking. However, to avoid any electrode shifting during winding more controlled, in-line functions are required for cutting and placing the electrodes as well as for the thermal bonding of the electrodes to the separator (see Fig. 30.13). The modified separator is more expensive than traditional separator, although the additional cost is offset by increased manufacturing throughput (i.e., higher line speed and improved cell quality and yield imparted by the use of electrodes that are less likely to shift within the cell during and after assembly).

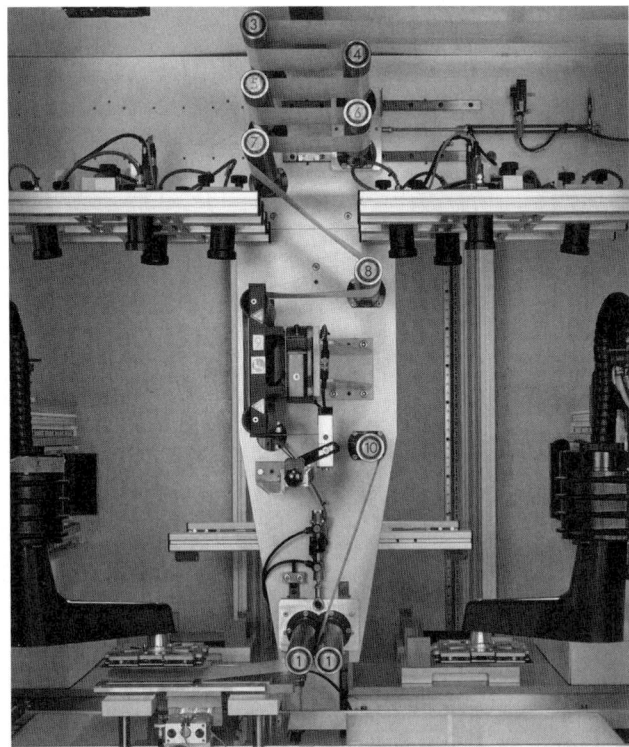

**FIGURE 30.12**   Z-fold stacking. (*Courtesy of Manz AG.*)

**FIGURE 30.13**   Stack winding machine. (*Courtesy of Manz AG.*)

The pure stacking process can also incorporate in-line cutting of the electrodes as well as the separator. Cut sheets of electrodes and separators are alternatively picked and stacked by a high-speed mechanism and held in place with hold-down fingers until the entire stack is completed. Pieces of tape are applied to the finished stack to keep the sheet components from moving with respect to each other (see Fig. 30.14).

**FIGURE 30.14**    Pure stacking mechanism. (*Courtesy of Manz AG.*)

***Lead Attachment.***    Since the individual cell stacks need to be hermetically packaged to contain the liquid electrolyte and to prohibit moisture or oxygen infiltration, a means of providing a hermetic electrical pass-thru between the stacked cell inside the packaging and the outside world needs to be provided. With pouch cells, conductive metallic electrical leads or tabs are ultrasonically welded to each set of bare foil extension stacks forming the positive and negative leads. The bare foil electrode extensions are first compressed together, ultrasonically welded into a nugget, and trimmed to a common length. The metallic tab is then positioned over the nugget and ultrasonically welded to it as per Fig. 30.15. This process is then repeated for the opposite polarity or both polarities can also be welded simultaneously.

***Pouch Forming and Cell Stack Insertion.***    The packaging material containing the cell stack is typically composed of a multilayer composite to assure a moisture and oxygen resistant barrier, to provide a leak-tight enclosure to contain the liquid electrolyte, and to maintain electrical isolation between the cell stack and the outside world. The packaging material is normally made up of aluminum foil with a polypropylene inner layer and a protective nylon outer layer. The composite film is formed to the cell stack dimensions and depth using a cold-drawn punch and die. Depending on the thickness of the cell stack, one or both sides of the pouch can be formed. The depth of cavity is limited by the ability of the aluminum layer to withstand rupture or tearing during the drawing process—typically in the range of 5 mm deep.

With the pouch packaging formed, the cell stack is placed into one of the formed packaging halves. If the cell stack is not too thick, then the second half of the packaging can be a simple flat layer of packaging material and either folded over or applied as a separate layer. Alternatively, for thicker cell stacks, two formed packaging halves are required. Once the cell stack is placed inside the packaging, the outer perimeter of the packaging, with the exception of one side for subsequent electrolyte filling, is sealed using heat seal bars.

**FIGURE 30.15**   Tab welding. (*Courtesy of Manz AG.*)

One of the sealed edges of the cell stack contains the metallic electrical tabs. In order to provide a leak-tight seal, the metallic tabs contain polymer adhesive sealing strips on each side of the tab and are positioned to align with the sealing area of the packaging. The sealing strips are bonded to the metallic tab using primers and other means to assure a leak-tight adhesion such that no liquid electrolyte, moisture, or air can make its way between the metallic tab and the sealant strips. The sealant strips are a polymeric material similar to the inner layer of the cell packaging material, which helps assure a solid plastic weld between the inner layers of the packaging material and the sealant strips of the tabs.

Note that alternatively, rigid metal cans instead of multilayer flexible laminate packaging are often used for more abusive environments and longer life (see Chap. 3E for further details). The assembly steps are similar but the closure (crimping, welding, etc.) and electrolyte fill (see "Electrolyte Filling and Cell Closure" below) techniques can be considerably more complex.

***Electrolyte Filling and Cell Closure.***    With three of the four edges of the pouch already sealed, the remaining open edge is used for filling the cell with liquid electrolyte. The cell is vertically oriented with its open edge up. A filling nozzle is inserted into the open edge of the pouch and a precise amount of liquid electrolyte is injected into the cell packaging. A precision metering pump accurately doses the amount of electrolyte to be injected. This process can be done either at atmospheric pressure, or in some instances, the filling process is performed in a vacuum chamber whereby high vacuum is drawn on the cell contents thereby evacuating the electrodes and separator of air trapped within their pores. Once the air is evacuated, the electrolyte is then injected into the cell where the negative pressure more quickly draws the liquid electrolyte into the void pores and accelerates wetting of the materials.

Another method injects liquid electrolyte while the cell is at atmospheric pressure. Then the cell is subjected to vacuum and sealed in a separate operation. Filling or even simply sealing a filled cell under vacuum requires a vacuum chamber, pump, piping, and controls. Despite the additional vacuum equipment cost and processing time, the result is considered by some battery manufacturers to be superior since the pores in both the electrodes and separator are more easily and evenly filled with the liquid electrolyte. Filling under atmospheric pressure, on the other hand, assumes that the electrolyte will gradually but fully soak into the pores and displace the gases trapped deep in the matrix.

Conversely, others consider drawing a vacuum at this stage to be a futile and costly exercise since the cell undergoes a soaking and degassing process downstream where all residual gases and gaseous by-products from the cell activation reactions are extracted from the cell prior to a final sealing process. From an overall processing time standpoint, both variants turn out to be similar, where the vacuum time is replaced by the soaking time.

### 30.3.5   Cell Finishing

***Formation.***    The formation process is required to form a protective layer on both electrodes via the initial charge and discharge of the cell. The first few charge/discharge cycles also serve to finalize wetting of the electrodes and separator by the electrolyte.

The formation process is achieved by subjecting each individual cell to a controlled number of electrical charge, discharge, and rest periods, some of these being at elevated temperature. The overall process can take upwards of 1 week where the majority of this time period is spent by the cell resting at either room or elevated temperature. Nonetheless, the charge/discharge equipment is very costly both from an equipment standpoint as well as the storage area required to house numerous cells throughout the duration of the lengthy process. Figure 30.16 shows a typical cell formation facility.

**FIGURE 30.16**    Cell formation and aging warehouse. (*Image courtesy of PEC.*)

Upwards of 20 handling steps between the different work stations can be involved in the formation process, and given the number of cells, this can only be accomplished effectively by an automated system. To facilitate the frequent relocation of the cells between the various conditioning stations, an Automated Storage and Retrieval System (ASRS) is employed to shuttle pallets containing several cells between stations as seen in Fig. 30.17.

Near the very end of the overall formation process, the pouch cells are placed in a vacuum chamber, where the pouch is pierced in order to extract any gases which formed during the process. The pouch is then sealed for a last time and any excess packaging material is trimmed to complete the cell.

***Test and QC.***    Throughout the assembly and formation processes, several measurements are taken to determine the quality of the cells. These measurements include short-circuit tests, open-circuit voltage, voltage loss over time, and impedance. These are used to either discard bad cells or categorize (grade) each cell according to predetermined set of quality parameters. This grading system serves to match similar quality cells, as is necessary in future module and pack assembly processes.

***Battery Pack Assembly/Test.***    Once cells are completed and tested, they often get assembled into modules and packs. The end application may vary from small packs for consumer or e-bike applications to larger applications such as electric vehicles (EVs) or very large stationary energy storage systems (ESSs). (See Chaps. 23 through 27 for further details on these respective applications.) Depending on the application and electrical requirements such as voltage, energetic capacity, and power, multiple cells are connected in a series-parallel arrangement first

**FIGURE 30.17**   ASRS. (*Image courtesy of PEC.*)

within modules and then into packs. Electrical interconnection means can be via mechanical connectors or by welded connections (Chap. 3D). Where necessary, either air or liquid thermal management systems are added to control the temperature of the cells during use. Additionally, controls, inverter, and battery management system are added to complete the functionality of the system (see Chap. 31). This is particularly the case for large ESSs as well as for EVs.

## 30.4   SUMMARY

Several factors can influence overall cost of setting up a cell manufacturing facility: equipment quality, cell size, labor costs, etc. Generally speaking, for very large cell manufacturing facilities of >1 GWh yearly capacity, investment and direct labor can be broken down as seen in Table 30.1. Capital investment of $100 million should be expected for this size of a facility. These costs do not include module or pack assembly as this varies depending on cell size and configurations and final battery pack size and features.

**TABLE 30.1** Cost Breakdown for a Typical Gigawatt Battery Factory

| Processing area | Capital cost | Labor (by operator hours) |
|---|---|---|
| Electrode preparation | 33% | 40% |
| Cell assembly—including dryroom | 33% | 40% |
| Cell formation | 33% | 20% |
| Total | 100% | 100% |

This brief, high-level description of the steps involved in fabricating battery electrodes and assembling cells has been presented to depict the complexities in constructing a typical battery manufacturing facility. Many assembly possibilities and permutations exist at all levels, and are often governed by individual manufacturer's preference and end product. As demand grows for advanced battery technologies, it is imperative that the manufacturing technology keeps pace with the trends.

# CHAPTER 31
# BATTERY CHARGERS, CONTROL, AND SAFETY ELECTRONICS

**David Simm**

## 31.0 INTRODUCTION

The two battery types, primary and secondary, have common as well as uniquely different monitoring and control requirements. Secondary batteries require a charging system, and primary batteries typically require protection from charging. Both require safety features to minimize damage from abuse and abnormal environmental exposure. Solutions to these requirements take a variety of forms, ranging from separate battery pack and charger implementations to combined configurations as appropriate for larger batteries. Nevertheless, the requirements driving these designs have mostly common elements, varying with regards to chemistry, size, and application. Figure 31.1 depicts the circuitry required for these two types.

## 31.1 REQUIREMENTS FOR BATTERY PACKS AND CHARGERS

The following sections discuss the various considerations for proper operation and safety of the overall charger and battery system.

### 31.1.1 Charge and Discharge Control

Various cell chemistries have different charge control requirements. Electrochemical couples such as nickel cadmium, nickel metal hydride, lead acid, and to a certain extent nickel hydrogen are relatively tolerant of voltage excursions past their recommended charge voltage limits. Their ability to accept a limited overcharge, due to the electrolysis of water at higher voltages, mitigates cell imbalance while operating within a prescribed temperature range. However, the evolution of gases, hydrogen in particular, requires provisions for cell venting to reduce excessive pressure buildup.

Lithium chemistries, which lack any mechanism to absorb the overcharge energy, generally have much more stringent cell voltage requirements for charge operation. Individual cell voltages must be monitored to prevent exceeding cell voltage limits. When any cell's voltage reaches the maximum allowed, charging current must be reduced or terminated. However, high charge rates, within allowable maximums, can increase individual cell voltages due to resistance increases possibly terminating charge prematurely, limiting total recharge capacity. The same concerns apply to discharge control except that discharge must be terminated to prevent any individual cell voltage exceeding its lower limit.

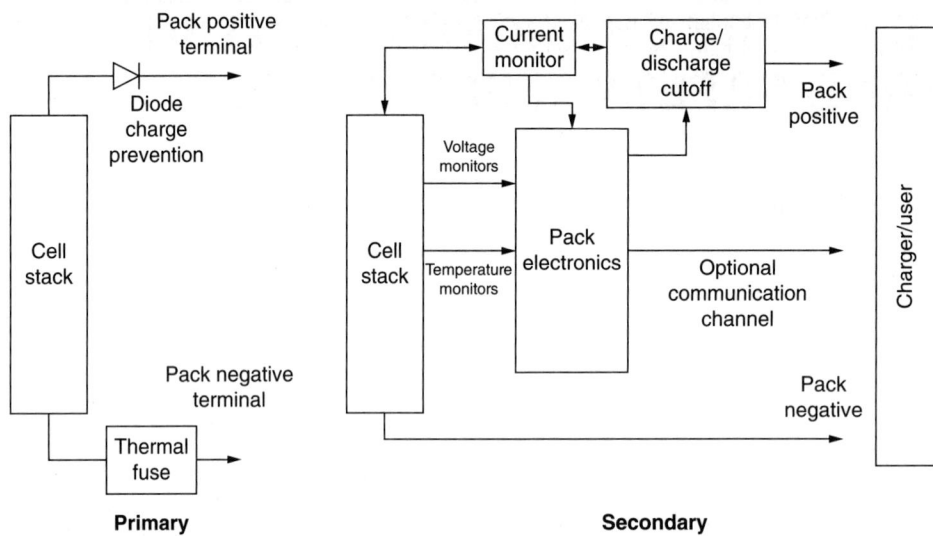

**FIGURE 31.1**    General battery electronics types.

Additionally, excessively high discharge currents, such as those occurring from a short circuit on the battery output, must enable a termination of discharge. The battery must also be able to resume normal operation once the high-rate load is removed.

## 31.1.2    Cell Balancing

In order for series-connected cells to reach full charge at the same time, each cell must begin the process at an equivalent state-of-charge (SOC) and have equal temperature and charge/discharge performance characteristics. While these conditions may be close to reality in controlled situations and at the beginning of life, as cycle life progresses this becomes increasingly unlikely.

For chemistries such as nickel cadmium and lead acid, cells reaching full charge before others in the series string will continue to accept additional charge (into overcharge) while lesser charged cells "catch up" and attain full charge. Overcharge results in heat generation from various side reactions, whose effects can be mitigated by limiting charging current.

Lithium ion cells must not be overcharged (or over discharged for that matter), thus preventing a SOC imbalance that would accumulate over cycle life, significantly limiting battery capacity. In the case of lithium ion, maximum charge and discharge voltages for the individual cells in a series string must not be exceeded. Differences in individual cell resistance accentuates cell voltage differences under charge and can cause cells to reach charge voltage limits prematurely, causing charge termination for the entire series string. One approach to deal with this problem is to monitor cell voltage differentials and limit charge current to keep them within established limits in order for string members to reach the highest charge state possible without going over charge voltage limits. While this method has limited effect on SOC imbalance, at least the imbalance due to the cell's DC resistance may be mitigated. DC resistance is the term used in this chapter instead of cell impedance. Impedance is measured by applying an AC signal to a battery or cell and measuring its variable response. The resistance effect discussed here is a linear function describing an IR (current times resistance) voltage drop as battery currents change.

Cell DC resistance has a similar effect on discharge, which can trigger early discharge termination as a result of a high resistance cell's voltage dropping to the discharge voltage cutoff limits due to IR voltage effects, thus limiting battery discharge capacity. These voltage differentials are accentuated by increased charge and discharge currents, thus becoming a significant issue in high-rate batteries. As a practical matter, allowable cell voltage differentials should be evaluated and controlled to minimize impact on system capacity. Both active and passive methods are available to ameliorate cell imbalance. Approaches to both of these methods will be addressed in later sections.

### 31.1.3  Communications

Batteries that indicate status to the user (i.e., participate in the charge/discharge control process or communicate status and operational data, either real time or historical) require a communication capability. Signals can be as simple as a visual indicator on the battery pack itself; a dedicated analog signal lines, typically to an external charger; or a complex command and data system (Fig. 31.2).

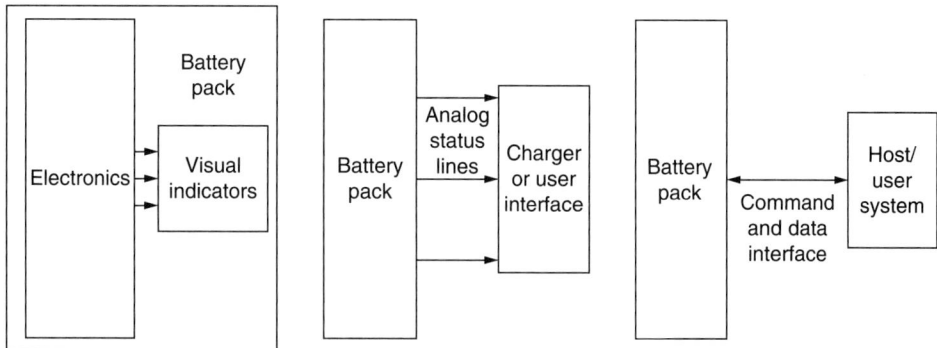

**FIGURE 31.2**    Battery communication options.

Visual indicators on the pack can signal battery status or SOC. This kind of display can be supported by dedicated analog or mixed signal semiconductors or by more complex microcontroller systems and may be implemented as an LED or LCD display.

Analog parameters such as thermistor outputs can be conducted outside the pack on a dedicated connector contact. This approach is often used with NiCd power tool chargers for charge control to prevent pack overheating. Due to typical limitations on the number of contacts on battery packs, this method is usually limited to one or maybe two parameters.

System parameters can also be transmitted from the pack to charger and/or user in the form of a simplex serial data stream, either clocked or asynchronous. (A simplex stream indicates that data flows only in one direction.) Parallel data transmission could also be used but would require several data lines and associated contacts, an approach that is generally not favored. Instead, the data can be collected and formatted, either by a dedicated purpose mixed signal semiconductor or by a microcontroller. Devices are readily available that support serial communication, data collection, and processing. This topic will be discussed in later sections.

Serial data transmission can also allow systems outside the battery pack to communicate and control operation of pack electronics. This technique is called a half-duplex serial transmission, meaning that data flows in two directions, but one direction at a time. Thus, user (charger) and battery pack share a signal line. This method can be used for mode control, stored parameter modification, data collection, as well as system initialization at manufacture. A well-defined data structure is required, particularly in the microcontroller case as it needs to be incorporated into system software. Dedicated semiconductors typically have their communication structure fully defined. Typical examples are listed in Table 31.1.

When selecting a system data rate, or rates, provisions to allow enough time for collection, conversion processing, and control are necessary. Relatively modest data rates allow sufficient time for system processing and generally do not impact overall system precision. More data are not necessarily better.

Larger systems such as motive power traction batteries may benefit from the use of "Bluetooth" hardware and software for data interaction, which would be particularly useful for the management of groups of tow vehicles and/or forklifts. This would allow development of smartphone "apps" for battery system interaction.

**TABLE 31.1** Battery/Charger Chips and Their Communication Structure

| Semiconductor OEM | Device | Communication |
|---|---|---|
| Texas instruments | BQ24721 | Similar to SMBus |
| Microchip | PIC microcontrollers | UART, SPI, SDA |
| Maxim | DS2438 | 1-Wire w/CRC |
| Linear technology | LTC4100 | SMBus |
| ST micro | STC3117 | I²C |

*Note:* 1-Wire system originally developed by Dallas Semiconductor.

### 31.1.4 Diagnostic Information

The most commonly collected types of information are individual cell voltages. These values are sampled as analog data and either utilized in that form or digitized and processed. Digitized data are required for storing historical data as analog data usually exist only in latest value form. Stored data also allow for computation of differentials, which can be useful for performance trend analysis, limited only by how much storage is available. Temperature data can be collected and handled in the same way as cell voltages.

Techniques such as Kalman filtering and other statistical processing can be employed, given enough computational power to capture the trend characteristics of data to minimize the actual stored data volume.

In larger or more critical systems, measuring of actual battery impedance (AC) may be useful. In almost all cases, however, the required hardware makes this impractical for the limited value of the result. Effective DC resistance can be useful for diagnostic trending and as an aid in SOC assessment (fuel gauging). Resistance can be derived from the change in cell voltage between periods of charge/discharge and rest.

Another important feature is the ability to sense and measure charge and discharge current. Current should preferably be sensed on the battery's "high side," that is, the conductor connecting the positive terminal of the battery to the charge/discharge connection. This allows easy measurement of current in both directions: charge and discharge. Low-side measurement might seem easier but due to reference voltage considerations, only discharge current can be easily measured.

### 31.1.5 State-of-Charge Monitoring

Also known as "fuel gauging," SOC monitoring would seem to be very useful and straightforward. However, using a water and bucket analogy the bucket capacity is fixed while a battery's "volume" is not. Effective capacity varies with temperature and the rate of charge and discharge. Thus, integration of charge and discharge current is not particularly effective, even though the method is widely touted.

Perhaps the best indicator of battery SOC, across various electrochemistries, is open-circuit voltage. However, in a dynamic system voltage levels are constantly in flux.

System designers need to first determine the importance of knowing the actual SOC. What will the user do with this information? Does it really make any difference? Objective answers will dictate the complexity, cost, and utility of the resulting design. For instance, what difference does it make if the SOC is 17% or 22%? Does the user care if the "anticipated run time" at present conditions is 13 min, knowing that present conditions may not be future conditions? What does the user, or intelligent system, really need to know about SOC?

### 31.1.6 Unique Safety Requirements for Primary Batteries

Primary, or nonrechargeable, cells and batteries must not be allowed to recharge. Consequences of charging can vary from insignificant to catastrophic, depending on chemistry. Lithium primary chemistries are perhaps the most dangerous. Small alkaline cells might generally present few dangerous risks, except for performance degradation and perhaps leakage. Since cells can be connected in various series–parallel arrays, care needs to be taken when paralleling series strings to prevent one string from charging a neighboring string. For greater negative consequences, more caution should be exercised. For chemistries capable of particularly high discharge rates, the battery should be protected by a discharge control device activated by high temperature. The sensor for this circuit needs to be installed at a location within the battery pack that experiences the greatest amount of heating while in discharge. (These circuits are also useful for secondary batteries.) Thermal fuses are commonly used for this purpose, both as the only protective device and as a backup to "smart" battery systems. Due to self-heating effects in the thermal fuse, these devices are useful for discharge rate protection as well.

Some primary batteries, like the ubiquitous military BA-5590, provide the capability to completely drain the battery capacity to eliminate the danger of lithium metal release after the battery is discarded. The circuit needs to be designed for an adequate discharge rate that does not overheat the pack. It is necessary before using this circuit to have the subject battery at a sufficiently low SOC to be safely drained to prevent overheating and possible fire. In this and most other cases, consideration of unlikely events is an important exercise.

## 31.2   *BATTERY ELECTRONICS DESIGN APPROACHES*

There are several elements to consider in the design of batteries containing electronics, which are discussed in the following sections.

### 31.2.1   Pack Electronics Architecture

Very simple batteries, like those for nickel cadmium power tools, generally require little if any pack electronics. The exception might be a thermistor whose output is used by the charger to terminate charge in case of excessive pack temperature. Lithium batteries (and perhaps other chemistries) generally require a safety system typically composed of at least a thermal fuse and perhaps a diode in the case of a primary battery. "Smart" battery packs will have additional circuitry for data monitoring, control, and communication. These circuits are often used in addition to basic safety circuits.

The need to design for manufacture, test, and initialization is an important early decision in the design process. Manufacture, test, and initialization of a few prototypes might be easily accomplished, but hundreds, thousands, even ten thousands batteries present an entire set of new problems. Design of the electronics should incorporate the minimum number of components on a circuit substrate that is most economical and easy to handle in automated processing. This usually indicates fiberglass (FR4 or similar) circuit boards. Flex circuits in lieu of wired daughter boards sometimes make sense versus hand-installed cabling. These films may initially be more expensive, but save labor during assembly, limiting overall cost. A typical test set-up is shown in Fig. 31.3.

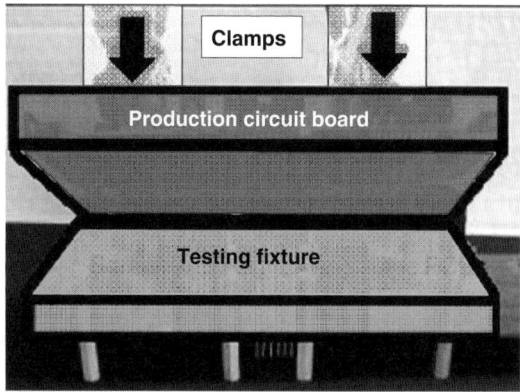

**FIGURE 31.3**   Sample test jig for electronic PCBs.

Bulk programming before assembly is one approach to initializing programmable components, eliminating the need for initialization of the entire assembly. When using EEPROM or Flash devices, this approach limits the ability of reprogramming. Circuits need to be manually reworked to replace the old chip. It is much better to provide an "in-circuit" programming capability. Thus, parameters can be changed without rework, and by using Flash devices, entire operating programs can be modified and updated without hardware changes.

In order to efficiently test large numbers of circuit assemblies, they need to be designed so that reference configurations can be commanded while inserted in fixtures that simulate battery parameter inputs so that correct operation can be observed and verified. The test process can be easily automated, only requiring the circuit to be inserted into the test fixture. A log of test results can also be generated from this process.

### 31.2.2   Monitoring and Measurement

Many applications require charge or discharge to be terminated when an individual cell voltage reaches a certain value, or falls below another value. Semiconductor operational amplifiers, or "op amps," can be configured to perform this function (called a window comparator). As long as the cell voltage is maintained within the defined window, the op amp output remains high (or low depending on design). Values outside this window cause the op amp output to switch states. This output is coupled to the charge/discharge current cutoff device. This approach can be used for a single cell, or "stacked" to monitor a series string of cells as long as the op amp supply voltage requirements are not exceeded by the battery voltage. Op amp outputs are digitally monitored to control cutoff.

A design approach utilizing one differential op amp and one A/D channel involves using a multiplexer. This chip provides an analog value at its output corresponding to one of several analog inputs depending on a selection address. This allows all cell voltages (as many as eight) to be sampled by the same circuitry, minimizing errors between samples. If a known voltage reference is provided for sampling, the analog process can be calibrated.

Analog values may not be a useful measure. Thermistor output is approximately linear over only a limited range of values and needs to be calibrated in order to be useful. This is easily done by applying a calibration polynomial, which can have as many terms as necessary. The simplest, the linear, has one term and a constant. Even the most precise calibrations are adequately done with five terms.

Battery current measurement for both charge and discharge can be implemented utilizing a sense resistor. Typically of very low value on the order of $0.10\ \Omega$, the voltage drop across this resistor is measured and used to compute a current value. Sense resistors in the battery positive conductor, or "high side," can sense both charge and discharge currents due to the polarity of the differential with respect to system reference, which is the negative pole of the battery. Sense resistors in the negative battery line can only sense discharge because the differential voltage generated by a charge current has a polarity less than the battery negative terminal, and therefore not within the possible range of measurement.

Smart battery systems require A/D converters to convert sampled analog values to digital values. Among an A/D's important characteristics is the precision of digital output, typically 8, 10, 12, or more bits. This means that the analog sampled value is scaled into 256, 1024, or 4096 parts (with respect to the A/D reference voltage) depending on the number of bits. Lesser precision makes design and operation easier, so precision should be limited to a meaningful value. A precision of $0.1\ mV$ is generally not practical for reasons such as noise, stability, and computational ease.

When designing a high-voltage battery, it may be necessary to divide the data and control sections of the pack electronics to prevent excessive voltages being applied to semiconductors. This will require galvanic isolation of some sort. Providing an isolated bus within the pack itself may be required to support communication across large terminal voltages. Likewise, charge and discharge control may require isolation to prevent component overvoltage. These various designs are shown in Fig. 31.4.

### 31.2.3   Safety Devices

Thermistors are resistors whose resistance varies with temperature. They are specified for operation over a specific range of temperatures but have a very nonlinear response over their entire range. Small regions of interest may be sufficiently linear for direct use, but large operating ranges will require output conversion and calibration. There are two thermistor types: negative temperature coefficient (NTC) and positive temperature

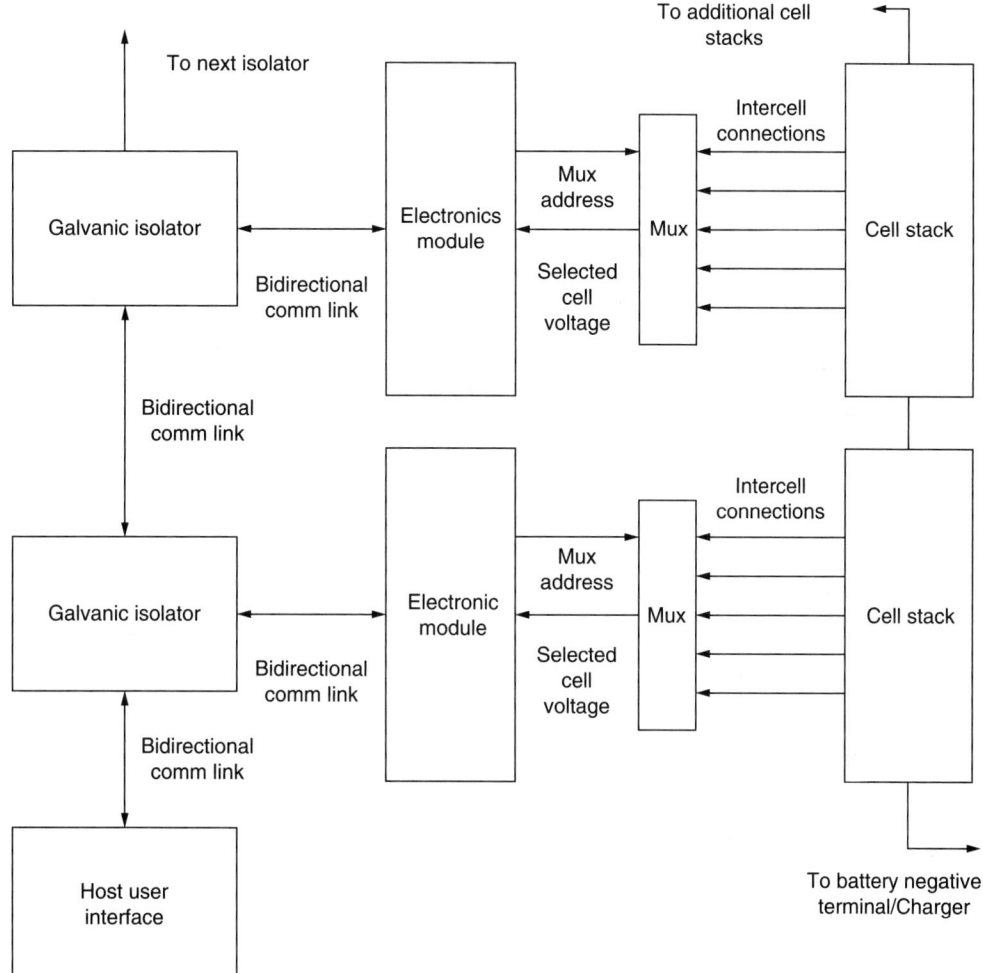

**FIGURE 31.4**   Stacked multiplexers for high-voltage batteries with communications.

coefficient (PTC). NTC thermistors are the most commonly used type. NTC output falls with rising temperature and PTC output rises with rising temperature. It should be noted that thermistors are used in a voltage divider circuit that can draw more than a negligible amount of system current (always a consideration in pack electronics design). In cases where the thermistor is connected remotely to the charger, this current consumption is generally not a problem.

Another approach uses semiconductor temperature sensors whose voltage output varies directly with temperature. Some of these devices have calibrated outputs on the order of 0.1 V per °C. Others have linear outputs that are easily converted for use. These devices are easy to use and have very low quiescent current on the order of tens of microamperes.

There are two fundamental types of fuses: single use and resettable. Single-use fuses limit the magnitude of current and are generally sized to prevent catastrophic failure of the battery when recovery from overload would not be likely. Resettable fuses open when either current or temperature limits are exceeded but can resume operation when fault conditions are removed. Applications of any of these devices involve philosophical as well as risk assessment considerations.

### 31.2.4   Cell Balancing

The concept of cell balancing refers to the equalization of individual cell SOC in a battery. A variety of factors cause the SOC of cells to diverge from one another under charge and discharge cycling, none of which are topics of discussion in this section. What will be discussed are some techniques that may be useful in minimizing divergence.

There are proponents of a scheme referred to as passive balancing, which takes no direct action to balance cells but relies on cell uniformity and electrical arrangement to minimize the "disturbance forces" driving imbalance. For example, a group of cells connected in parallel will "clamp" each other's voltage by virtue of their electrical connection. Given that we measure SOC by a cell's voltage, all these cells' SOC percentages will be the same regardless of their actual capacity. Series-connected cells all "see" the same charge and discharge currents, but may have different voltages and SOC. Judicious design of series/parallel connections of batteries with relatively large number of cells can result in a battery whose cell voltage diversion is minimized.

The flip side of passive balancing is active balancing of which there are two types: charge conservation and charge dissipation. Charge conservation requires the draining of charge from high-capacity cells, moving that charge to lower charge cells. This can be accomplished by what's called the "flying capacitor" method or by any number of esoteric magnetic converter methods. The question quickly becomes, "How much battery do you want with your electronics?"

Dissipative methods involve connecting a small resistor to the highly charged cells to bring their charge in line with the other cells. This is done under electronic control, either data driven or analog. It has been found that the specific point in the charge cycle where balancing is done is important to maintaining system stability. Arguments have been made that this reduces the battery capacity, but this energy would not be accessible anyway and the resulting balance prepares for more effective charging on the next cycle. Others argue that the money saved on electronics could be used to make larger batteries, achieving similar results with less "heartburn."

### 31.2.5   Fuel Gauging

Conventional wisdom regarding fuel gauging, or SOC determination, is that integration of charge and discharge is the best or only approach. Over limited periods this may be true, but this method, like all integrative processes, quickly accumulates error. Hence, an additional SOC reference value, collected periodically, will help to correct the integrated value. The best available optional SOC measure is open-circuit voltage. If charge and discharge currents are measured, it is straightforward to determine periods when cell voltages are equal to open-circuit voltages. By applying a conversion algorithm or consulting a lookup table, a SOC correction could be obtained. These estimates are all conditioned on a variety of cell parameters that make accurate assessment difficult.

## 31.3   MICROCONTROLLER HARDWARE AND SOFTWARE CONSIDERATIONS

Microcontrollers are single-chip computer systems containing ROM memory, RAM memory, I/O devices, and a core processor. Usually targeted for a single or small group of applications, peripherals can be included such as A/D converters, timers, and serial data peripherals.

The program that controls operation resides in ROM memory in a microcontroller, which can be implemented in EEPROM, EPROM, or Flash hardware. EEPROM and Flash memory can be loaded or modified while the physical device is installed in the circuit board. Application notes and support hardware and software are typically provided by each manufacturer, such as Motorola, National Semiconductor, Maxim, Microchip, and others.

Microcontroller RAM is used to store variable parameters and is typically limited to several thousand bytes, not suited to large data arrays. Compilers for microcontrollers are available utilizing several languages that will assist in establishing RAM and ROM data requirements as well as providing implementation of floating point arithmetic algorithms.

Data and command management is an important element of smart battery software. It is important to not only provide for all needed and expected commands and data telemetry, but also to make provision for modular expansion, if necessary. During system development, changes occur on a regular basis, and having means for accommodating revisions will make the effort easier.

Battery electronics need to operate on as little electrical current as possible to minimize impact on overall energy storage. Because microcontroller current draw is proportional to basic operating frequency, its selection is important. Many controllers have a built-in sleep mode in which the program is held static while the chip enters a low-power sleep mode. The flip side is, of course, that the program needs to "wake up" at some time to operate correctly. This is typically done utilizing on-chip timers and processors having a corresponding interrupt service.

Most microcontrollers also have a watchdog capability based on a time that needs to be periodically reset by a properly operating program. If it is not reset, the timer "times out" and resets the controller program execution. This feature prevents system failure from program "lockup."

All smart battery applications will require some type of a communication interface, which is typically serial. Serial data can be "clocked" or "unclocked." Clocked data are utilized by I2C (Inter Integrated Circuit—of which SMBus is one type) and SPI (Serial Peripheral Interface). Both of these services require a "master" to generate the clock signal, although SMBus can be operated as a multi-master service. The well-known KISS (keep it simple . . .) principle would indicate avoiding multi-master mode if possible.

Unclocked serial data, or asynchronous data, does not require a separate clock line but maintains bit definition by timing. Asynchronous serial systems can be done with one or two conductors. Two conductor asynchronous serial can be provided by a universal asynchronous receiver-transmitter (UART) peripheral on the microcontroller. (It can also be manually generated, or "bit banged," but requires great care to maintain correct timing.) One wire serial or DQ/HDQ can also be generated either way, but is most often used in dedicated specialized chips. As mentioned before, the design and use of any of these semiconductors are well supported with manufacturer hardware, software, and application notes. Refer to Fig. 31.5.

### 31.3.1  SMBus Implementations

The System Management Bus Specification (SMBus)[1] was developed by the Intel and Duracell Corporations. It is composed of a set of commands and data that can be exchanged among battery pack, charger, and system user and is intended to improve system performance and component interchangeability. While most of the parameters utilized are predefined, there are user configurable commands and data available. The hardware and bit timing of the specification is based on Philips Semiconductor I2C system. It can be operated in a single master mode, where only one node sends commands and the other node(s) respond, or in a multi-master mode in which all nodes can originate commands.

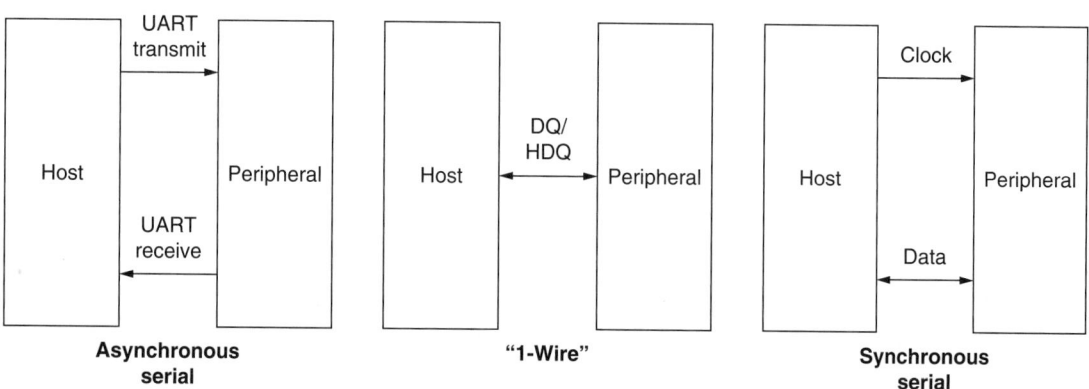

**FIGURE 31.5**  Serial data configurations.

SMBus systems can be implemented using programmable processors or with single chip integrated mixed signal dedicated semiconductors, in which case no programming is required for battery pack development. SMBus does not directly address safety concerns of certain chemistries with specialized current and voltage limits; additional safety circuits are required in these cases. SMBus is more generally directed toward cell chemistry and pack identification plus fuel gauging. A fully implemented SMBus system allows chargers to safely charge batteries of different configurations and chemistries.

### 31.3.2   Charger Specific Requirements

When designing a charger, the nature and source of charging current need to be identified. Will the system have a DC or AC input, or both? What voltage? Current requirements? If AC, what frequency? If it is a military system, is it automotive, marine, or aircraft? Commercial systems also need to utilize proper, allowable power systems, be it 120 V AC household or data center power.

Power supply architecture needs to be determined. Most systems utilize switch mode converters, but in special applications, linear regulation provides benefits in electrical noise reduction. If the charger is to communicate with the battery pack or provide system data to the user, hardware and software to implement those communication channels need to be defined and provided.

System control is also required. Relatively simple circuits may suffice for simple battery systems like lead acid, nickel cadmium, and nickel metal hydride, while a full functional "smart" battery will require a data processor to communicate with the battery and user while maintaining charge. Various battery builders have developed proprietary charging techniques, some using methods such as pulse charging and short periodic discharge. Control of these activities requires a data processor and control system.

### 31.3.3   Charger Deployment

Relatively small battery systems, such as those used in power tool, household appliances, and electronic devices, generally utilize chargers separated from the battery. This minimizes appliance weight and allows chargers to support a number of batteries. After all, what is the point of having a portable, battery-powered device if a charger must be carried along with the equipment?

Larger battery systems may include charge electronics, either in part or completely, in the battery case itself. Stationary batteries such as those in data centers would be a primary example. Forklift batteries typically have separate chargers to minimize vehicle weight and allow continuous, sequential recharge of multiple vehicles. Devices like cell phones, which charge their battery from a USB connection, need to have some charge control functions in the phone as single-cell lithium batteries are typically charged to 4.1 V or less.

In the end, the system designer needs to consider how best to separate functions along with both battery and charger system volume and weight.

### 31.3.4   Charger Electronics

Fundamentally, chargers are DC power supplies with output controls to limit current and control voltage. In the case of an SMBus system, it also contains a communication interface which can implement some controls on system output. In a more general case, a smart battery charger would also communicate to the user.

Battery/charger original equipment manufacturers (OEMs) have numerous component and subsystem choices when considering new systems. Once operational requirements have been defined, design options can be explored. Rather than build a DC power supply in-house, the OEM may go to a power supply design and manufacturing house for that portion of the charger. At suitable volumes and costs, manufacturers may modify standard products for the OEM. If developing in-house, the OEM has a range of components available for their design. Contract manufacturers are also available to manufacture subsystems or entire assemblies.

Battery pack electronics can be developed in a similar manner, although complete pack electronic assemblies are not as readily available as voltage converters. Specialty firms exist and design support is readily available from semiconductor manufacturers; so in-house development need not be prohibitive.

Charger weight, volume, and cost, as might be expected, are proportional to power capability. Chargers based on linear regulation are often cheaper, but larger, than switch-mode versions. They are also generally electrically quieter (i.e., electrical noise) than switch mode.

### 31.3.5  Charger Failure and Failsafe Issues

When designing a charger that will be connected to the mains for power, it is important to assure safe operation under both normal and extreme conditions. Obviously, the mains "hot" input must be protected from user contact. In chargers without a transformer to provide galvanic isolation, it is possible that there is continuity from the "hot" side of the supply to the output. This is a good reason not to use electronics/chargers in the bathroom or around water. Some OEMs provide a polarized plug to minimize this danger, but if the wall outlets are wired "backwards," or connected through a nonpolarized extension cord or adapter, the safety plug's function is thwarted. Sometimes the "ground" or common connection of a device can be accessed with similar results. Most of the time these issues are not significant, but problems can occur when connecting standalone systems together electrically.

Perhaps it should go without saying that a poorly designed charger will not necessarily function as desired in use. It might exceed a battery's charge voltage, maximum charging current, ignore over temperature, etc., thus damaging battery life and safety. The larger the battery and the more dangerous the battery chemistry, the more dramatic will be the potential safety consequences of a poorly designed charger.

There are charger/battery interactions that are hazardous as well. Pack electronics in lithium batteries should, in many developers' opinions, limit the individual cell discharge voltage cutoff. Excessive discharge can damage the battery cells, causing them to reach full charge at lower and lower voltages. The charger does not "know" this and continues charging as designed, which is now an overcharge, causing excessive heating and cell failure. In the case of internal cell shorts or other cell failures, the charger generally cannot "know" what happened and continues charging. A thermal feedback from the battery would indicate a failure, but too late to affect the outcome, particularly in the case of fire or cell venting.

Failures receiving recent attention include "Hoverboard" fires, Boeing 777 battery fires, and Samsung cell phone battery failures. These are all primarily the result of internal cell shorting and other faults. Very early in the "failure incidents" of the systems mentioned above, battery electronics might have been able to respond in some way, but not to any appreciable degree.

## 31.4  OVER-ARCHING ELECTRONICS DESIGN CONSIDERATIONS

### 31.4.1  Strategies

The paragraphs below detail some facts and features of the battery electronics system (i.e., the battery pack circuitry and charger). Some initial caveats are as follows:

1. Battery electronics can only measure parameters and indicate system mode or status, terminate charge/discharge, communicate with a charger or user, and provide a few basic electrical/thermal safeguards.

2. It cannot limit the effect of poor design and construction.

3. Misunderstanding of thermal characteristics cannot be corrected.

4. Addition of electronics does not increase battery capacity no matter how many components are added.

Listed below are some battery/charger system design steps based on the following scenario:

*A battery is needed because a device can't be plugged into the mains.*

a. Primary or secondary? How long does the battery need to last? If a primary battery will last a long time, or the device is single use (like some rocket boosters), no need to bother with a secondary and its charger. For example, with a primary lithium camera battery the need for an auxiliary charger is eliminated.

**b.** If a primary battery is used, what chemistry? Are safety devices needed? What is the system voltage and current design? Have thermal issues been assessed? Will the battery experience excessive heating? Are thermal fuses required? Must the battery be protected from charging (depending on chemistry)? Have additional considerations like terminal protection and end of life issues been addressed?

**c.** If secondary, what chemistry? Are safety devices required? Is the charge/discharge control decided? (There are developers that believe some discharge limits do not need protection—a decision that is ignored at one's own peril.)

**d.** Is a dedicated or universal charger required? A dedicated charger is designed for a single battery application, while a universal one must accommodate several different batteries. Are economies really accomplished with a universal charger or are the complications prohibitive?

**e.** Is there some advantage to having a smart battery? (This can be a deceptively tricky question.) Are the things a smart battery can do really important for the specified goals?

## 31.4.2  Summary

There are many options for the design of battery pack electronics and chargers, ranging from the most basic pack protection and unregulated charging to very complex data-driven systems hosting data communication for the control of charging and user interaction. Please again note the list of things battery electronics *cannot* do from Sec. 31.4.1.

The highest level of consideration is, of course, safety, but system performance for maximum useful life is important as well. The system designer needs to evaluate performance against cost and reliability. Low reliability presents issues of safety and utility. It could be argued that a less complex, more reliable system is a better value than an involved design. It is important to remember that "more is not always better" with respect to electronics and at some point the extra cost of electronics would be better used increasing cell size or quality.

## REFERENCE

1. www.smbus.org/specs.

# CHAPTER 32
# ANCILLARY SERVICES IN THE BATTERY INDUSTRY

## 32.0 OVERVIEW

The "mechanics" of the electrochemical energy storage industry were delineated in Chaps. 1 through 31. However, without a supporting structure of auxiliary or ancillary services and activities, such as industry-specific legal and financial advice, diagnostic and analytical software, test equipment, logistics, IT, market and technical databases, and other such indirect support networks, the growth and efficiency of the industry would be compromised. Chapter 32 provides a few insights into this world of activity. Components of the service industry may come and go, but the need for consultants, analysts, and a service sector will not change greatly. The sections in this chapter include the following:

Chapter 32A: Intellectual Property

Chapter 32B: Failure Analysis

Chapter 32C: Battery Test Equipment

Chapter 32D: Safety and Performance Testing

Chapter 32E: Business and Financing Strategies

Chapter 32F: Strategic Market Analyses

This list of services is certainly not complete. Other topics that might have been discussed, if space allowed, include packaging and shipping,[1] hazardous waste disposal,[2] automated battery build/test systems,[3] advanced analytical equipment,[4] quality systems,[5] etc. The key for success in any of these service-related businesses is, of course, the ability to add value to the product offering. Adapting existing know-how to the benefit of the battery industry is one approach to success. Developing new service models and analysis techniques is another option. And as emphasized throughout this Handbook, one example of a potentially high-value service may be an industry services sector that helps make wise decisions on the energy efficiency of battery raw material and component sourcing; product design and battery assembly methods; and application engineering and system deployment/utilization concepts.[6]

## References

1. B. Richard, The challenges of shipping damaged lithium batteries, 35th Annual International Battery Seminar and Exhibit, Ft. Lauderdale, FL, March 26–29, 2018.

2. G. Kerchner, Understanding the complexities of shipping new, refurbished, and waste lithium batteries, 35th Annual International Battery Seminar and Exhibit, Ft. Lauderdale, FL, March 26–29, 2018.

3. D. Strand, Accelerating development of high nickel NMC cathodes, 35th Annual International Battery Seminar and Exhibit, Ft. Lauderdale, FL, March 26–29, 2018.

4. M. Costello, R. Sterbenz, New battery test capability maximizing test coverage, 35th Annual International Battery Seminar and Exhibit, Ft. Lauderdale, FL, March 26–29, 2018.

5. B. Miller, Quality philosophy in the manufacture of lithium ion batteries, 35th Annual International Battery Seminar and Exhibit, Ft. Lauderdale, FL, March 26–29, 2018.

6. J. Spangenberger, L. Gaines, Q. Dai, Comparison of lithium-ion battery recycling processes using the ReCell Model, 35th Annual International Battery Seminar and Exhibit, Ft. Lauderdale, FL, March 26–29, 2018.

# SECTION A

# INTELLECTUAL PROPERTY STRATEGIES IN THE ADVANCED BATTERY INDUSTRY

**Matthew Rappaport, Daniel Abraham**

## 32A.0 PREFACE

Intellectual property (IP) considerations are important for emerging battery technologies such as advanced lithium ion, solid state, flow batteries, and others. The authors are not attorneys. A qualified attorney should be consulted for any legal advice. This section is directed to IP aspects of business strategy for technical and business professionals.

## 32A.1 INTRODUCTION

High-performance batteries are enabling key transformations in society such as electrifying transportation, autonomous vehicles, renewable energy utilization, and smart grid. Innovation and development are core drivers for increasing energy and power density, and behind any innovation lies the underlying technology ownership rights or IP. As such, IP represents a key value driver in many emerging technologies. Developing unique innovations is not always sufficient for successful implementation. IP can create a lawful barrier to entry thereby enhancing commercial success.

Published patent information provides a window into the competitive technology development around the world. Not only does IP show who is patenting what, but also when and in what jurisdictions along with many other criteria that are valuable to understanding a market. For example, IP filings in advanced batteries have increased rapidly since 2010 as shown in Fig. 32A.1. Furthermore, Toyota, Samsung, Panasonic, and LG consistently file more battery patents than any other entity; however, other large companies are not far behind (Fig. 32A.2).

## 32A.2 TRENDS IN PATENT STRATEGIES

Robust innovation leads to more IP that generally enables development and commercialization. Dense collections of patents protecting overlapping aspects of a technology are likely to emerge in the overall patent landscape. These so-called patent *thickets* have already emerged in several areas related to advanced lithium ion batteries. For example, substantial patent thickets have emerged for LFP cathodes, electrolytes, and silicon-based anodes. As shown in Fig. 32A.3, patents in silicon-based anodes can be parsed to show different aspects of the technology and where thickets are likely to emerge.

A thicket owned by a single entity can present a substantial barrier to entry thereby securing a competitive edge. However, patent thickets can be problematic if distributed over multiple entities leading to negative outcomes such as patent litigation, lengthy development timelines, and other unintended drags on market adoption. As shown in Fig. 32A.4, silicon-based anodes are under development by many companies who may be developing overlapping aspects of the technology.

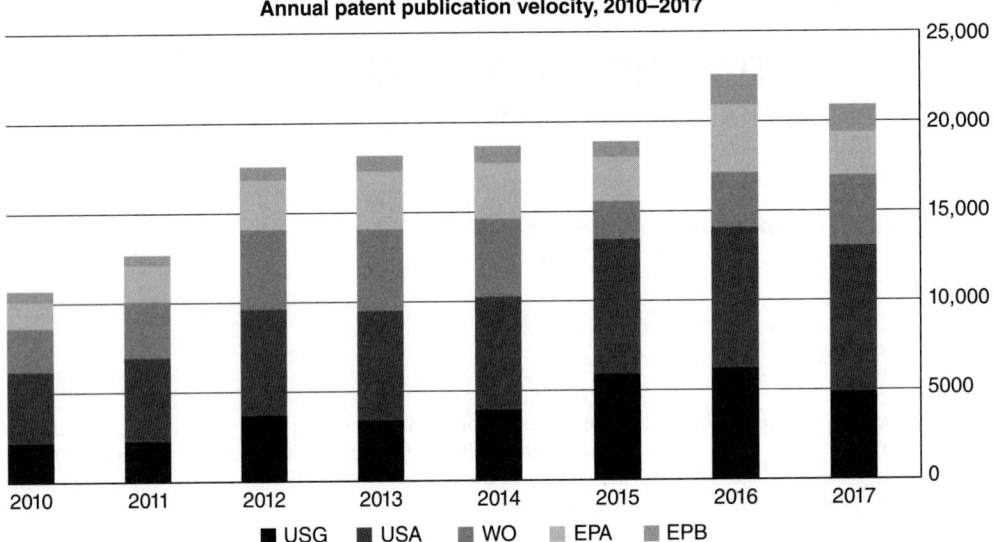

**FIGURE 32A.1** Worldwide patent filing activity from 2010 to 2017. (*From the Advanced Battery and Capacitor PatentEdge*™; www.abcpatentedge.com.)

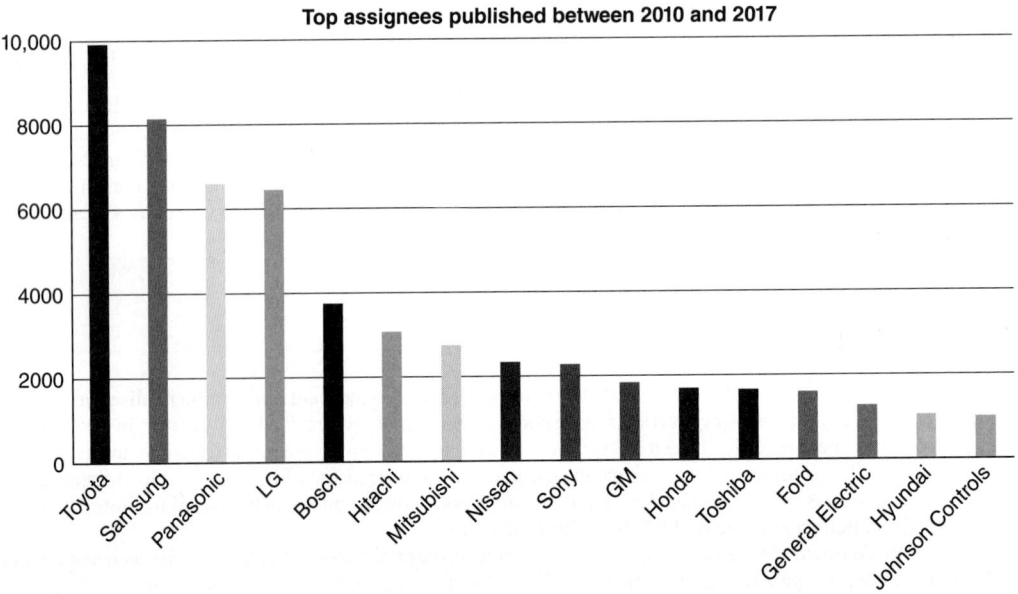

**FIGURE 32A.2** Top assignees in advanced batteries from 2010 to 2017. (*From the Advanced Battery and Capacitor PatentEdge*; www.abcpatentedge.com.)

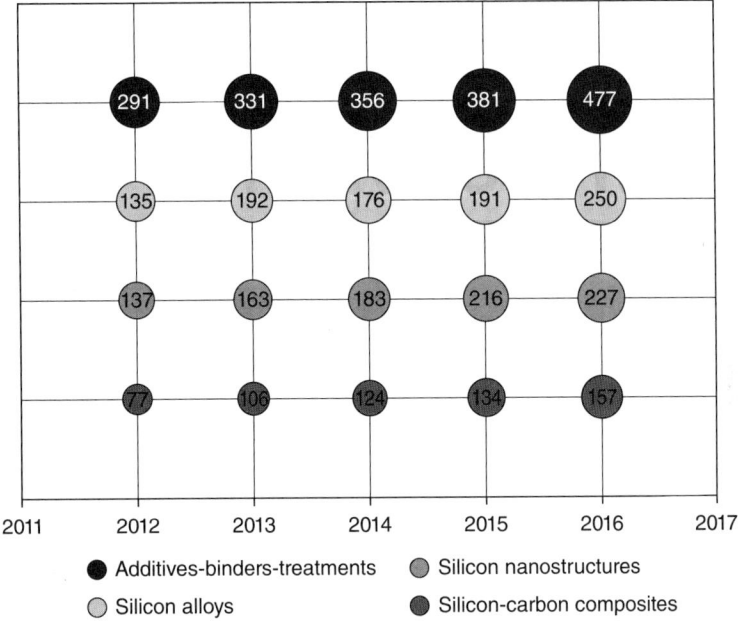

**FIGURE 32A.3** Patent filings for different aspects of silicon-based anodes. (*From the Advanced Battery and Capacitor PatentEdge*; www.abcpatentedge.com.)

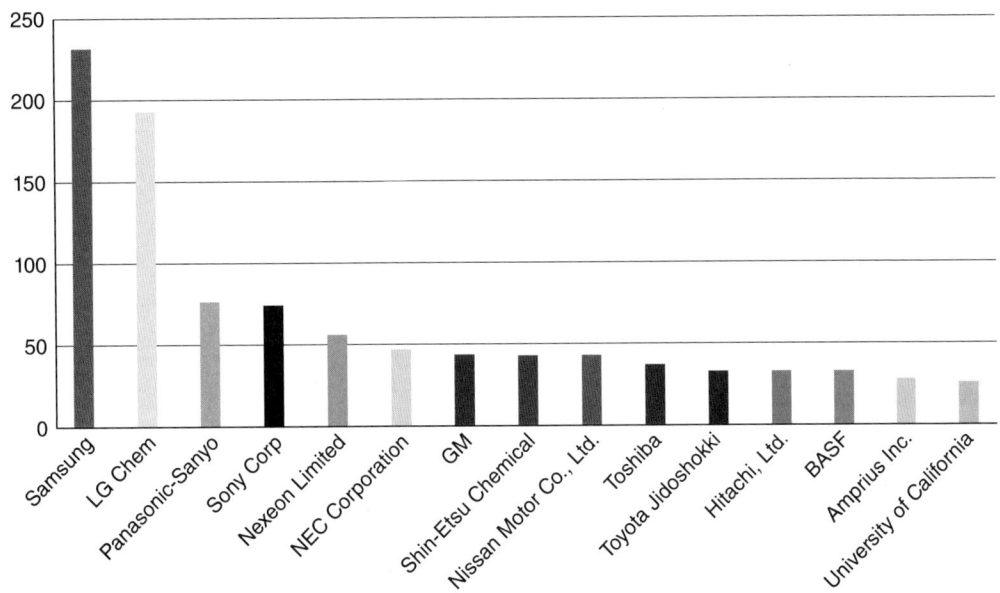

**FIGURE 32A.4** Top patent owners in Silicon anodes based on patents filed between 2012 and 2016. (*From the Advanced Battery and Capacitor PatentEdge*; www.abcpatentedge.com.)

Technology adoption in other areas, like smartphones and LEDs, has shown that patent thickets can present challenges such as increasing costs, lowering profits, reducing incentives to innovate, and creating confusion in the market. While patent thickets represent significant creative activity, negotiating thickets can be a complex problem. As evidenced by other industries, several pathways are possible.

## 32A.3    PATENT LITIGATION

From a global market perspective, the solution of last resort is patent litigation where parties enforce their patents over alleged infringing products. Court battles result in heavy costs for both plaintiffs and defendants as legal expenses pile up over protracted proceedings. Resources end up on discovery, depositions, and legal wrangling instead of new research and development. All of these have occurred in abundance during the last decade in the smartphone patent space creating market chaos as well as misplaced ire toward patents. Already in the lithium ion battery space, lawsuits are on the rise as shown in Fig. 32A.5. A key nascent industry can ill-afford these problems.

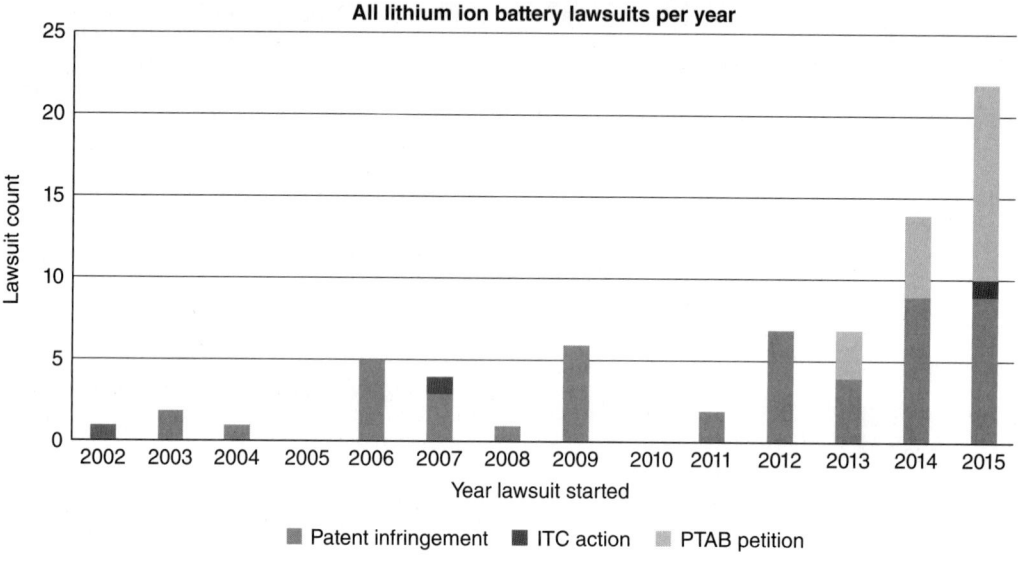

**FIGURE 32A.5**    Patent lawsuits related to lithium ion batteries. (*Data compiled by MPEG LA LLC.*)

Alternatively, the light emitting diode (LED) industry resolved patent thicket issues related to white LEDs by establishing multiple cross-licensing deals involving the largest patent holders. It took years and millions of dollars involving negotiations between companies with diverse interests. Cross-licensing helped free the gridlock, resulting in increased technology development and product cost reductions. Still many smaller companies ended up paying relatively high license fees to avoid patent infringement.

## 32A.4    IP STRATEGY

An organic approach to navigating patent thickets involves R&D and market analysis informing a thoughtful IP strategy. This approach requires companies to diligently review their own technical innovations with respect to the competitive patent and market landscapes to choose the most fruitful opportunities. This is accomplished in part through analysis of the "white space" or relatively sparse areas within the patent landscape. Dedicated

## MPEG LA® licensing model

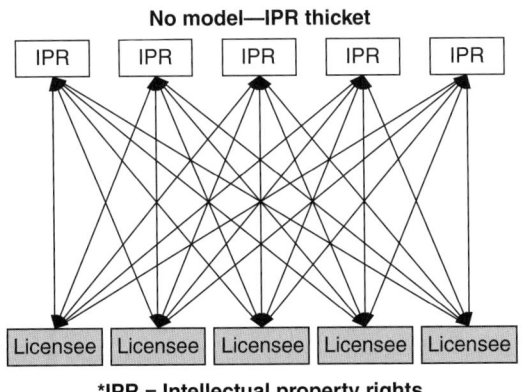

FIGURE 32A.6    "Many-to-many" patent pool licensing model . (*From MPEG LA LLC.*)

R&D to "design around" others' technologies coupled with acquisitions may help to better negotiate the thicket. Leveraging this research-based approach can result in the development of a harmonized IP strategy that is aligned with a company's R&D and business development objectives. Although this solution results in much less friction, it also requires a sophisticated and coordinated approach among the R&D, business development, and legal departments to ensure all internal organizations are operating in lock step.

An emerging model to avoid patent thickets is through IP aggregation or patent pools. As shown in Fig. 32A.6, an objective third party can either acquire the patents outright (aggregator) or administer a collection of overlapping rights (patent pool). Traditionally, patent pools have been most effective when de facto technology standards comprise multiple patent owners and multiple patent users. The pool provides an efficient market solution to ensure licensees (those making, using, and/or selling products utilizing IP rights) can access intellectual property rights (IPR) in an equitable way, while the licensors (those committing IP rights to the pool) are compensated at a fair market value by the pool.

The pool model offers an effective market-based solution to the patent thicket problem for the advanced battery industry. One sign that a technology area is ripe for a pool is a distribution of patents across a variety of technology and product developers. For example, as shown previously in Fig. 32A.4, it appears, from patent ownership statistics for silicon anodes in lithium ion batteries, that a patent pool may benefit adoption and commercialization. Indeed, there may be many areas in the battery space that could benefit from a patent pool as conceptualized in Fig. 32A.7.

## 32A.5    SUMMARY

Battery energy storage plays a crucial role in key enabling industries such as mobile electronics, electric and autonomous transportation, and renewable energy utilization. To affect these changes, innovation and development require efficient conduits to commercialization. Patents can help bring innovation to market, but in some cases patent thickets can present challenges of their own. Effective IP strategies can navigate thickets by various approaches gleaned from other industries. Among those, patent analytics and pools represent long-term market-efficient strategies that will help enable battery energy storage in the future.

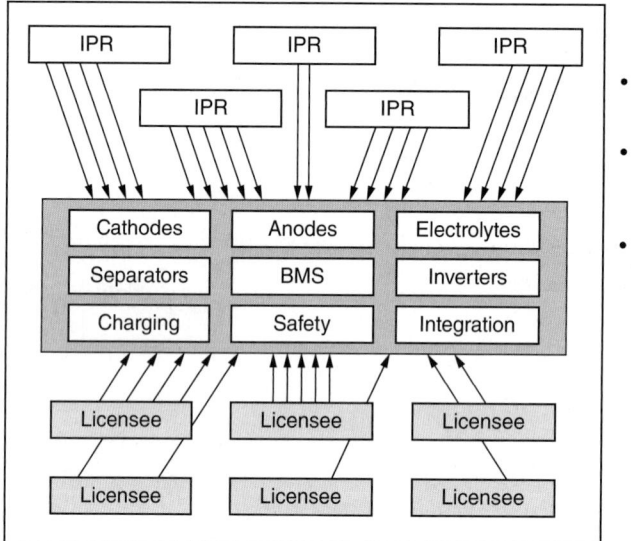

**FIGURE 32A.7**    Potential patent pools for the advanced battery industry. (*From MPEG LA LLC.*)

# SECTION B

# FAILURE ANALYSIS
# (QUALITY, DESIGN ASSESSMENT, AND
# ROOT CAUSE ANALYSIS OF BATTERIES)

**Vidyu Challa, Michael A. Howard, Seth Ayliffe Binfield,
Lawrence Edward Weinstein**

## 32B.1   INTRODUCTION—BATTERY MARKET TRENDS AND RISK OF FIELD FAILURES

In recent years the battery market has seen significant changes, which may elevate both the risk and impact of battery failures for the consumer. The trend toward higher energy batteries, both rechargeable and primary, a diffuse supply chain, and their use on or close to the body are significant contributors.

The market share of primary batteries has declined while the use of rechargeable lithium-ion batteries has risen in traditional, small form factor battery markets such as communications and consumer electronics. New market areas such as industrial, automotive, and stationary segments are also emerging. Major battery market trends may be summarized as follows:

1. Technology advancements have made many power-hungry devices, like incandescent flashlights that used many disposable alkaline batteries, less prevalent. Replacements like LED flashlights are increasingly powered with rechargeable batteries or lithium primary batteries.[1]

2. The availability of cheaper and better performing rechargeable batteries has resulted in a growing end-user preference for rechargeable options. Growing environmental awareness may also contribute to the shift in consumer preference toward rechargeable batteries.

3. As the primary battery market continues to dwindle, and the rechargeable battery market continues to expand, lithium-ion batteries are becoming the preferred choice for many applications because of their high energy density. Most individuals cannot get through a single day without using a device that incorporates a lithium-ion battery. These batteries are widely used in wearable and other devices that are worn on or in close proximity to the human body, leading to elevated impact in the event of catastrophic failures.

4. With the recent emergence of IOT (Internet of things) and M2M (machine to machine) applications, many smaller device manufacturers that do not have core battery expertise have started to buy and use COTS (commercial off the shelf) lithium-ion batteries. While large Japanese companies have historically dominated lithium-ion battery manufacturing, competitors have emerged in China, South Korea, and other Asian countries, while specialty battery companies have come into play in the United States and Europe. Many smaller players, some with questionable manufacturing quality, have entered the fray. The battery supply chain has also become more convoluted, with many companies selling batteries under their brand name that were made by other companies—sometimes fraudulently claiming higher performance than is warranted. Historically, device manufacturers could rely on a few key battery suppliers to deliver reliable products typically with nonlithium-ion chemistries requiring less stringent safety precautions than lithium-ion batteries. By contrast, safely using lithium-ion batteries today requires insight into the cradle-to-grave battery life cycle, including manufacturing, transportation, warehousing, device integration, use, and recycling.

5. There has also been increased competition within the lithium-ion battery industry. At the high end of the market, there is a trend toward increased capacities and power ratings. Increases in capacity have largely

come from increased active material content within batteries, sometimes accompanied by thinner separators. This places greater mechanical stress on the cell assembly and relies on tighter manufacturing tolerances, increasing the risk of failure. At the lower end of the market, the pressure has been on reducing cost. This has resulted in lower quality cells being pushed into the marketplace.

Within the primary battery industry, there has also been a move to increased price competition at the low end of the market, with a consequent drive to minimize cost at the possible expense of reliability. At the higher end of the market, some applications require higher current drain and/or greater reliability than has previously been needed. And regulations have required the elimination of heavy metals such as mercury and lead from alkaline button cells; these heavy metals act to reduce corrosion in zinc, and their removal requires strategies to maintain shelf life.

Lithium-ion batteries require much more stringent manufacturing process controls than other kinds of batteries. This fact, combined with the market trends noted above, has resulted in most battery field failures coming from lithium-ion batteries. Indeed, the proportion of battery failures stemming from lithium-ion chemistry has become so high that most battery failures are automatically associated with this specific chemistry. Because of their high energy density and the use of a flammable electrolyte, lithium-ion batteries can pose safety issues when they fall outside their operating window in their lifetime. The increasing use of lithium-ion batteries (particularly in wearable and IOT applications that are used in close proximity to the body), combined with a push toward higher energy densities, and a diffuse supply chain, all contribute to the increasing risk of field failures.

## 32B.2    BATTERY FAILURES

Batteries can fail energetically, or in a more benign manner. The vast majority of battery failures are nonenergetic and result in loss of performance that does not pose safety issues. Energetic failures are more prevalent with lithium-based chemistries than for those that use aqueous electrolytes. The severity of failure depends on the specific battery chemistry and on the size and physical construction of the cell. For instance, an electrical short in an alkaline battery will have a lower impact than one in a lithium-ion battery that uses a flammable electrolyte and has much higher power density. Likewise, for a given chemistry, a short in a cell with a bobbin construction will have less impact than in a cell that uses a higher surface area jellyroll construction. Smaller cells with capacities of an amp hour or less release less energy and fail less spectacularly than cells with a capacity of several ampere hour or more.

While the rate of battery field failures is statistically low at roughly between 1 and 10 ppm, the impact of energetic failures can be so severe as to be life changing. Explosions and fires caused by lithium-ion battery failures have resulted in billions of dollars' worth of damage, and even deaths. A 2017 report from the U.S. Consumer Products Safety Commission (CPSC) reports three deaths from battery fires in hoverboard applications.[2] Violent explosions from e-cig (electronic cigarette) failures have been reported to cause loss of body parts, disfigurement, and third-degree burns.[3] The first e-cig–related death was reported in 2018.[4]

The economic impact of battery failures for the companies involved has also been massive. Samsung recalled 2.5 million Galaxy Note 7 phones in 2016 eventually scrapping the entire product line, with a total loss of $5 billion.[5] Nokia recalled 46 million cell phone batteries in 2007 due a manufacturing defect that created a risk of explosion.[6] Sony's 2006 battery failures stemming from a manufacturing defect caused it to recall 10 million laptop batteries.[7]

High-profile battery failures such as those in Samsung's 2016 Galaxy Note 7,[8] Boeing's 2012 failures,[9] Nokia's 2007 failures, and Sony's 2006 failures[7] highlight manufacturing defects as one of the top causes of field failures in lithium-ion batteries.[10,11] A second cause of failures, based on field failure data as well as experience by DfR Solutions in dealing with customer failures, result from improper battery integration into the host device.[3,8] By not providing sufficient clearance to accommodate the battery as it ages, undue mechanical stress is created on the battery, which can lead to cell failure. Surrounding electronics or other mechanical features in the product can impinge on expanding lithium-ion pouch cells, increasing gassing and the risk of electrical shorts.

A third cause of field failures is due to lack of proper handling procedures during the life cycle of the lithium-ion battery. DfR Solutions has observed several battery failures from swelling due to improper storage and recharging procedures, which resulted in deep discharge of the batteries. A fourth category of lithium-ion battery field failures is inadequate battery protection design and implementation. While users might inadvertently use a wrong charger, resulting in a thermal runaway, "design for reliability" principles call for the use of a custom connector that prevents such incidents. Battery safety and reliability assurance is a system-level activity involving many job functions, including manufacturing, product engineering, quality, reliability, compliance, and logistics.

Therefore, it is necessary to conduct a design and quality assessment whenever integrating lithium-ion batteries into a device to prevent field failures. Evaluations by DfR Solutions reveal that failures often occurred when a device manufacturer did not know the manufacturing quality of the cells being used, and therefore could not appropriately assess the impact of the product application on the battery. This chapter therefore focuses on techniques that can be used for design and quality assessment as well as root cause analysis of battery failures. Lithium-ion batteries will be the major focus because of the high impact of their failure and their sensitivity to manufacturing defects. Some analyses on primary batteries are also included.

## 32B.3  CHARACTERIZATION AND ROOT CAUSE ANALYSIS TECHNIQUES

The following techniques may be used to diagnose battery failure root cause as well as for initial quality and design assessment.

### 32B.3.1  Computed Tomography (CT) Scans

Computed tomography (CT) scanning allows the detection of many battery quality and design issues by producing "virtual slices" of an object. To produce the CT scan, hundreds or thousands of x-ray images of a sample are collected while the object is rotated, and then reconstructed into a three-dimensional representation of the object. In a two-dimensional x-ray, the entire volume of the battery is integrated in one image, and therefore defects or features that do not create a significant density change are masked. A significant benefit of CT scanning is that the technique is nondestructive and noninvasive. Thus, CT scanning does not require the special precautions necessary to disassemble cells and allows safe analysis of fully charged cells.

Figure 32B.1 shows a deeply discharged commercial cellphone battery with the protection circuit activated and an open circuit voltage of 0 V. Gas pockets are visible within the flat-wound cell where they have forced the electrodes apart. When a lithium-ion battery goes into a deeply discharged state (usually below 2 V), the solid electrolyte interphase (SEI) layer is disrupted and reformed along with potential dissolution of the copper negative electrode current collector. The SEI disruption causes gas formation and swelling within the battery.[12] The battery in Fig. 32B.1 was not visibly deformed because the metallic can enclosure is able to withstand much higher internal pressure than pouch cells encased in a polymer laminate material. The higher internal pressure generated by the gas pockets is evident in the CT scan image. Nonetheless, the gas buildup was clearly able to displace the battery windings. Periodically recharging lithium-ion batteries will prevent cells from going into a deeply discharged state, preventing failures similar to those in the figure.

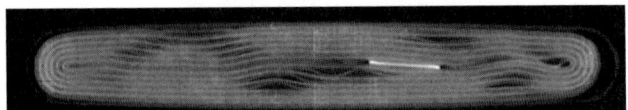

**FIGURE 32B.1**   Prismatic 2.96 Wh battery that underwent deep discharge. (*Image courtesy of DfR Solutions.*)

Figure 32B.2, reproduced from Finegan et al.,[13] shows CT scans of commercial 18650 cells that were fully charged to 4.2 V and then subjected to elevated temperatures (>250°C). The CT images were captured after cell venting to illustrate the need for internal cylindrical support, as used in Fig. 32B.2*a*, which prevented jellyroll collapse. Jellyroll deformation, as seen in Fig. 32B.2*b*, results from stress concentrations in certain regions of the jellyroll, increasing the risk of internal shorting and a thermal event.

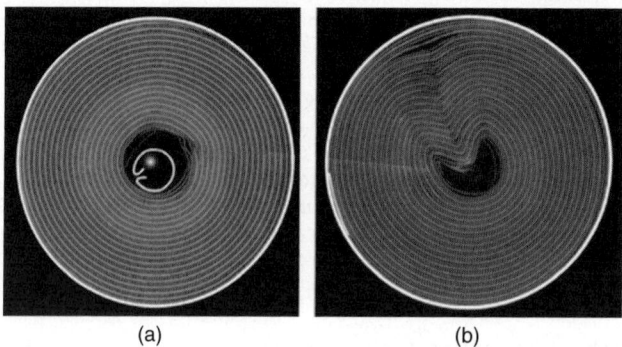

(a)                                    (b)

**FIGURE 32B.2**    CT scans of spirally wound cells from Finegan et al.[13] (*a*) with internal support and (*b*) without internal support. (*Reproduced under Wiki Commons License.*)

CT scan images of a commercial zinc-silver oxide button cell for a watch application are shown in Fig. 32B.3. The smaller cup shaped compartment is the negative electrode and is filled with zinc particles. The larger compartment houses a pellet of silver oxide, a conductive additive, and PTFE binder that aids in pelletization. In this battery chemistry, cells are designed to be anode-limited with at least 5% to 10% more cathode capacity, as evident in the CT scans. This is done to prevent the hydrogen gassing associated with cathode-limited designs. Cracks in the pellet are seen in the side view slice to the far right of Fig. 32B.3. While this is not necessarily a problem for low drain applications, these images demonstrate the sensitivity of CT scans to manufacturing and design flaws.

Figure 32B.4 shows CT scans of a commercial zinc-air hearing aid button cell. Notice here that in contrast to Fig. 32B.3, most of the cell volume is filled up by the negative electrode, which is composed of zinc particles surrounded by electrolyte gel. The cathode uses oxygen from the air, and the cell is activated just before use by removing a sealing tab that covers vent holes on the cathode side. The thin multilayered cathode structure consists of gas diffusion membranes and catalytic materials, along with a current collector mesh. By using most of the allowable cell volume for the anode active material and relying on ambient air for the cathode reaction, this cell chemistry is able to achieve very high energy density. Note the fairly loose packing of zinc particles inside the anode compartment, with significant void volumes. This is intentionally done to accommodate an increase in both mass and volume during the battery discharge reaction where zinc is converted to zinc oxide.

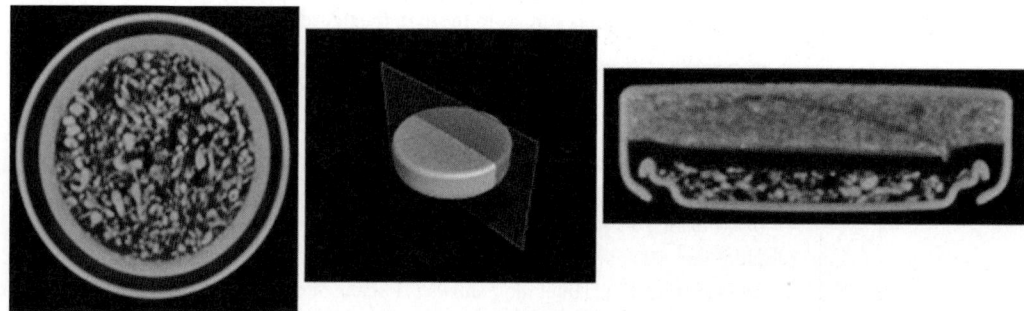

**FIGURE 32B.3**    CT scans of commercial zinc-silver oxide 364 watch battery. (*Images courtesy of DfR Solutions.*)

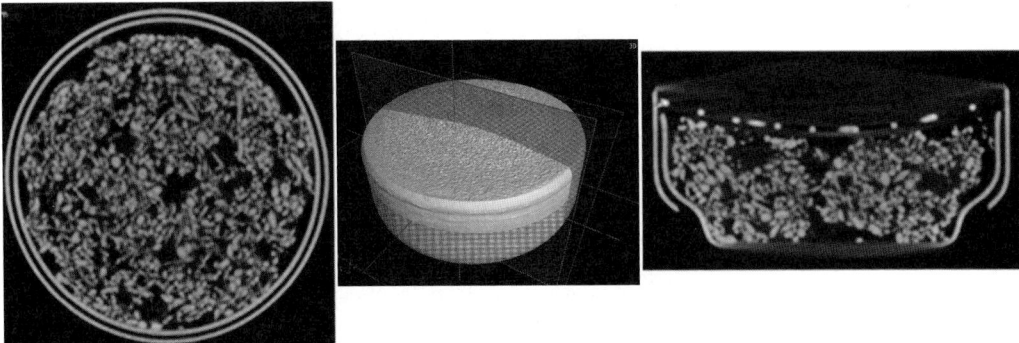

**FIGURE 32B.4**    CT scan of commercial zinc-air 312 battery. (*Image courtesy of DfR Solutions.*)

## 32B.3.2    Cell Teardowns

Cell teardowns are useful not only for root cause analysis of failures, but also as a way of assessing initial cell quality. This is particularly true with defects like electrode delamination or separator defects, which are harder to detect using practical resolutions with CT scans. While a full description of how to conduct lithium-ion battery teardowns, the associated hazards, along with the safety procedures to do so, can be found elsewhere,[14] the basic steps below must be followed. Nondestructive analyses must always precede any destructive analyses.

Lithium-ion battery teardown must only be conducted by qualified personnel and with adequate engineering controls. A safety data sheet must be reviewed, safety risks should be identified, and mitigation procedures, including use of adequate engineering controls and personal protective equipment (PPE), should be in place before any battery teardowns.

1. The external surface condition of the battery should be documented through optical microscopy or photographs, and any defects noted. Examples of defects include leaks, corrosion, and any deformation.

2. Background information on failure symptoms, product usage history, and any history on manufacturing date codes should be recorded.

3. Basic electrical characterization such as open circuit voltage and cell internal resistance should be measured. Operational characteristics of the battery may be obtained from product specification sheets.

4. The location of the initial cutting operation may be identified through CT scans or 2D x-ray. Figure 32B.5 illustrates how a safe cutting position that does not create shorting of the electrodes may be determined.

5. Cells should be discharged to below 3 V for lithium cobalt oxide chemistry for safety reasons. (For other lithium-ion chemistries, cells should be discharged to cutoff voltage specified in the data sheet.) This significantly reduces the cell energy since there is very little residual capacity below this voltage. Reducing the amount of energy stored decreases the risk of an energetic event.

6. For safety reasons, it is recommended that lithium-based cells be disassembled inside an argon-filled glovebox with $O_2$ and $H_2O$ content maintained below 5 ppm. The chemical state of the electrodes can also change in contact with air, and disassembly in an inert environment preserves their states and any failure signatures.

7. The initial cut should be made in a way that avoids shorting the two electrodes at a location identified by CT scan or two-dimensional x-ray in step 4. The metallic can or pouch is then gently peeled while ensuring that the jellyroll or stack is not mechanically deformed during the disassembly process.

8. The electrodes and separator may be separated and further analysis techniques may be used.

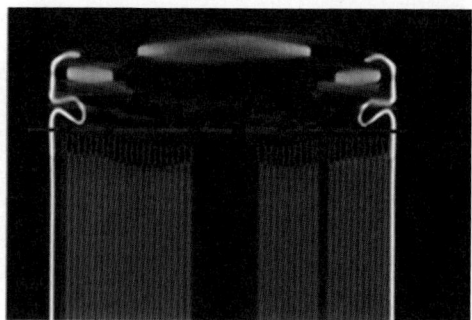

**FIGURE 32B.5**   CT scan depicting safe cutting position that prevents electrode shorting. (*Image courtesy of DfR Solutions.*)

## 32B.4   SAMPLE PREPARATION AND CHARACTERIZATION TECHNIQUES

A detailed description of characterization techniques that may be used can be found elsewhere.[15] The most common techniques that provide useful information are listed below.

### 32B.4.1   Visual Examination

After disassembly, the electrodes and separator layers can be visually inspected after gently unraveling the jellyroll (or de-stacking in case of a prismatic stacked cell). Negative electrode areas having thinned coatings (which increase the risk of lithium plating), micro-shorting on separator/electrodes, weld defects, particulate contaminants, and evidence of lithium plating, are some examples of defects that can be visually identified, without any need for further analytical techniques. Figure 32B.6 shows lithium plating on the negative electrode

**FIGURE 32B.6**   Lithium plating on the graphite negative electrode. (*Image reproduced with permission, Matthew Gantner, Lithium-ion manufacturing seminar, April 25, 2018, Rochester, NY.*)

of a prototype lithium-ion pouch cell. Good electrode design principles were not followed in this cell, creating a situation where the cathode supplied more lithium ions than the anode could accommodate. Lithium plating is an insidious failure mode, which has been implicated in many field failures.[10]

## 32B.4.2 Scanning Electron Microscopy (SEM) with (Energy Dispersive X-Ray Spectroscopy) EDS

Scanning electron microscopy (SEM) may be used to analyze the morphology of the electrodes, the shape and size of the active particles, conductive additives, and the condition of the separator. The chemical composition of battery materials and contamination may be assessed and mapped simultaneously using energy dispersive x-ray spectroscopy (EDS). Figure 32B.7 shows cracking of cathode active particles in a lithium-ion battery due to cycling. Aging effects such as active material cracking due to charge-discharge effects are described in detail by Vetter et al.[16]

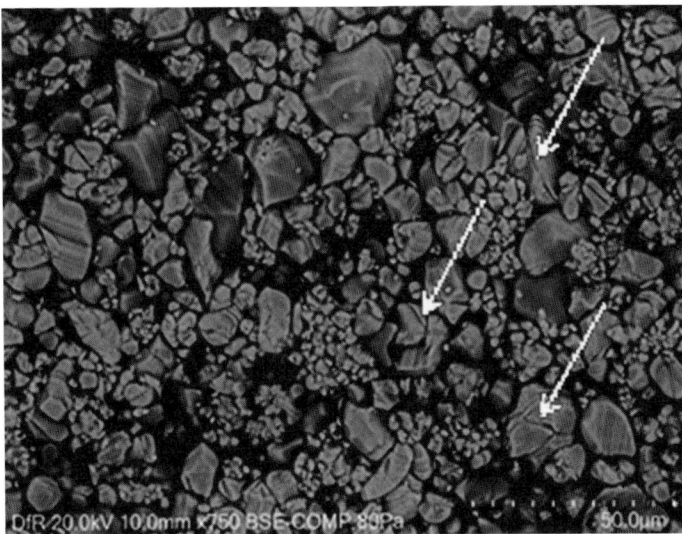

**FIGURE 32B.7** Cracking of lithium cobalt oxide particles due to charge-discharge cycling. (*Image courtesy of DfR Solutions.*)

## 32B.4.3 Cross Sectioning

Cross sectioning begins with extracting the electrolyte from the cell and filling the empty cell volume with an epoxy potting resin. The resin is allowed to cure, holding the internal components together. Mechanical grinding and polishing exposes internal structures of the battery for analysis. Cross sections can reveal shorts across the separator, active particle migration, and leaks or improper seals in the crimp area.

## 32B.4.4 X-Ray Fluorescence (XRF)

X-ray fluorescence (XRF) uses an x-ray source to bombard the sample surface. Secondary x-ray emissions are detected, and each chemical element has unique x-ray emission energies. Thus, XRF can be used to nondestructively determine the composition of a sample and to establish whether a metal sample is of uniform composition or is a coated material such as plated metal. XRF can also be used to measure the thickness of thin coatings on a substrate. To make thickness measurements, engineering standard references are made with the exact same coating and substrate stack up, with known and varying thicknesses of the relevant coating. The ratios of the detected elements are used to create a calibration curve. Multiple locations of a sample can be measured to assess the uniformity of the coating.

Plated materials are sometimes used in battery connectors. For example, nickel-plated steel and copper battery terminals are commercially available. These combine the joinability and corrosion resistance of nickel with the electrical conductivity of copper or the mechanical properties and the lower cost of steel. XRF can be used to quickly distinguish between pure nickel battery tabs or terminals and nickel-plated materials. In some applications, such as pack building, pure nickel tabs are preferred over nickel-plated steel but are more expensive. With appropriate standards as described above, XRF can also be used to measure plating thickness and uniformity.

Rigid battery cases are also often made out of plated materials—nickel-plated steel in particular. Nickel-plated steel can be used for the can in both alkaline battery chemistries and in lithium-ion chemistries. XRF can be used to nondestructively assess the thickness of metal plating using standards as described above; it can also detect locally thin areas. The nickel plating is used to protect the underlying steel from corrosion, and if it is too thin, it will not be effective.

### 32B.4.5    Fourier Transform Infrared (FTIR)

Fourier transform infrared (FTIR) spectroscopy is a common analytical instrument in chemistry laboratories. By detecting the vibrations of chemical bonds, for example, carbon-oxygen bonds, FTIR spectroscopy allows analysis of many organic and some inorganic materials. Each material has a unique FTIR spectrum. Identification of the polymers used in a battery can help establish parameters such as the temperature or pressure where a given battery vents, and FTIR can readily show when a manufacturer has changed materials and/or help prove that a component is counterfeit. Organics used in batteries include polyethylene, cellulose, and polypropylene for separators, nylon and high-density polypropylene for the gasket, and various materials applied to gaskets to act as crack fillers. Polypropylene and heat-resistant materials such as nylon are sandwiched around aluminum foil in pouch cell laminates, while heat sealable polymers such as modified polypropylene are sandwiched around a higher melting polymer core for pouch cell battery tabs.

### 32B.4.6    Particle Size Measurement Techniques

Particle shape, surface area, tap density, and size distribution of the battery electrode raw material powders have major influences on slurry properties, electrode coating, and ultimately cell performance. Finer particles are much harder to disperse and need high energy mixing and higher solvent content during electrode slurry preparation. This not only lowers the solids content, and subsequently the cell energy density, but also poses safety concerns. From a safety perspective, the surface area should be minimized and the fraction of submicron particles controlled to mitigate the hazard from exothermic reactions between electrolyte and charged electrode material at high temperatures. High power cells have greater electrode void volumes rather than a high fines content for safety and economic reasons. Greater available void volume in the electrodes for electrolyte enhances ionic conductivity and increases cell power output, but at the expense of capacity.

Owing to their significant impact on cell safety and performance, particle size distribution (PSD) measurements of incoming materials are critical and are necessary for dependable and consistent products. Many techniques for measuring PSD, including sieving, laser diffraction, dynamic light scattering, and image analysis, are applicable for measuring the PSD relevant to anode, cathode, and conductive powders.

The SEM image below (Fig. 32B.8) shows shape and qualitative PSD of $Ni(OH)_2$ particles from two different commercial vendors. Shape, size, and PSD impact tap density and packing density of the electrode. Higher packing density enables higher energy density cells. The more spherical particles in Fig. 32B.8*b* resulted in higher tap density in this instance, while the irregular shaped particles in the image (Fig. 32B.8*a*) led to lower packing and energy density. In some cases, however, smaller irregular shaped particles can fill up gaps between larger particles and lead to higher packing density. PSD using laser light scattering in Fig. 32B.9 correlates with the SEM quantitative measurements.

(a)          (b)

**FIGURE 32B.8** Comparison of (*a*) irregular shaped nickel hydroxide active particles versus (*b*) spherical particles with a narrow particle size distribution that allow for greater packing density and higher capacity. (*Images courtesy of Flexel LLC.*)

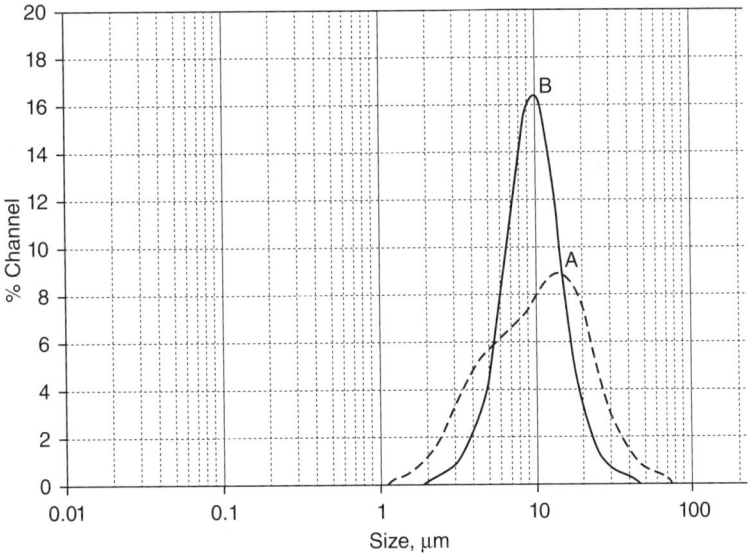

**FIGURE 32B.9** Particle size distribution using laser light scattering method showing narrow particle size distribution for vendor B versus bimodal and broader particle size distribution for vendor A. (*Image courtesy of Flexel LLC.*)

## 32B.5  *APPLICATION OF ROOT CAUSE ANALYSIS TECHNIQUES*

It is beyond the scope of this chapter to discuss battery failure symptoms and the root cause failure mechanisms in detail. These can be found elsewhere in the literature.[12,16–18] However, a few examples of failure analyses and quality/design assessments are presented here, using some of the techniques and methods discussed above.

### 32B.5.1    Case Study: Fraudulent Cells

In recent years, the practice of buying spare e-cig batteries online has increased, with users seeking higher ampere-hour power batteries for a better vaping experience. However, not all lithium-ion batteries sold online are genuine, with some having grossly overstated capacity and discharge rate. Because lithium-ion batteries have been reported to cause thermal events and explosions in e-cig applications, mislabeled batteries are a major concern. Mislabeled batteries include fraudulent batteries, counterfeit batteries, and rewrapped batteries. Counterfeit batteries are those that are intentionally trying to masquerade as ones from a reputable brand, while fraudulent batteries are those that make false representations of capacity and electrical performance.

Rewrapped batteries are batteries that were made by one company and are sold by another company under a different brand name. As the name implies, this involves the application of a plastic label (wrapping) over the cell with a different name than the original manufacturer. Some rewrapped batteries are simply safe, lower grade cells from major manufacturers sold as budget cells. Others, however, are marked with higher capacity or discharge rate capability than is warranted, or are even defective, unsafe cells that were headed for recycling.

DfR Solutions recently came across fraudulent cells that were sold on a major online retailer's website. An inspection of markings on the cell case (Fig. 32B.10) revealed misspelled words, often a telltale sign of fraudulent, rewrapped, or counterfeit batteries. These cells were also measurably lighter than typical 18650 cells. 18650 cells typically weigh at least 42 g, while the suspect fraudulent cells weighed 34 g, indicating less active material and lower capacity than a typical 18650.

**FIGURE 32B.10**    Fraudulent battery with misspelled words. (*Image courtesy of DfR Solutions.*)

CT scans of the suspect cells are compared with known high-quality 18650 cells in Fig. 32B.11. Notice the shorter length of the jellyroll in the suspect cell compared to the "good" cell. This would translate into a lower cell capacity and discharge rate capability. Deconstruction of the suspect cell (Fig. 32B.12) also shows heavily delaminated electrodes that demonstrate poor coating adhesion and low cell quality. Furthermore, no protection circuit was found in or on the battery as was claimed. One online forum user reports opening up an 18650 lithium-ion cell of this brand name and finding a smaller cell on the inside.[19]

Since lithium-ion battery safety is a function of cell manufacturing quality, protection adequacy, application integration, and user behavior, safety can be compromised by sourcing lithium-ion batteries online. This is of particular importance for high-rate e-cig applications, since the manufacturing quality, specific lithium-ion battery chemistry, and protection circuit adequacy are often unknown.

### 32B.5.2    Case Study: Lithium-Ion Cell Quality Assessment

Internal shorts in lithium-ion batteries are a source of significant concern because they are not effectively mitigated by battery safety systems and are not addressed by standards-based tests, which are focused on abuse tolerance. Optimum control of the manufacturing process is essential to preventing and mitigating manufacturing defects. Contamination, physical damage to the electrodes, and burrs on the foils/electrodes are dangerous because they can penetrate the separator and create internal short circuits in the cell.

CT scans can provide a quick and nondestructive way of detecting many of these battery quality and design issues. In the example given in Fig. 32B.13, the negative electrode windings are seen in dark gray, while the

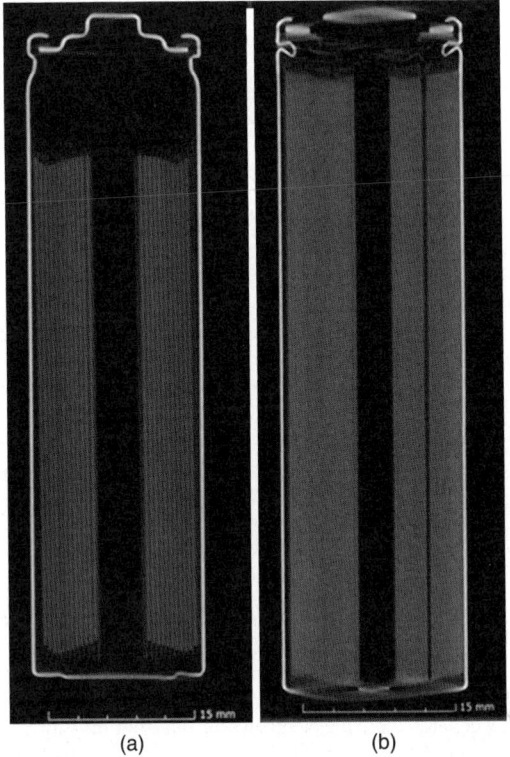

(a)                    (b)

**FIGURE 32B.11**    Fake battery (*a*) that has approximately 20% shorter jellyroll compared to (*b*) genuine battery. (*Image courtesy of DfR Solutions.*)

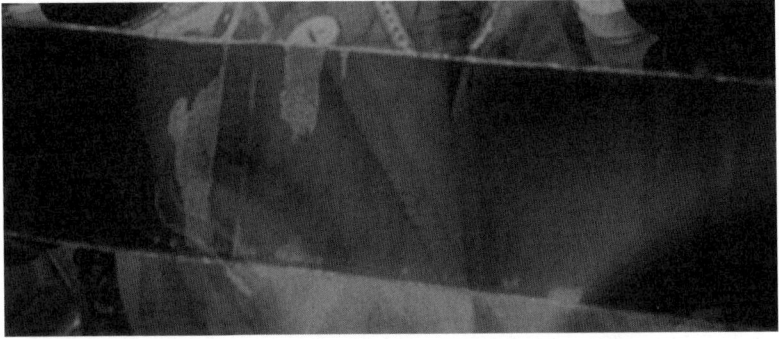

**FIGURE 32B.12**    Poor electrode quality showing anode delamination and copper current collector exposure on fake cell in pristine state. (*Image courtesy of DfR Solutions.*)

positive electrode windings are light gray. The Institute of Electrical and Electronics Engineers (IEEE) standards 1625[20] and 1725[21] for lithium-ion batteries call for the negative electrode windings to overlap the positive by a minimum of 0.1 mm, plus manufacturing tolerance. These guidelines are designed to greatly minimize the risk of short circuits created by dendritic growth from lithium ions. Areas where the cathode overhangs the anode can develop lithium dendrites, since the lithium within the battery starts out in the cathode and can locally supply more lithium than the anode can accommodate, resulting in lithium plating. The cell in Fig. 32B.13b fails the 0.1 mm overlap criterion. The cell in Fig. 32B.13a shows good electrode overlap.

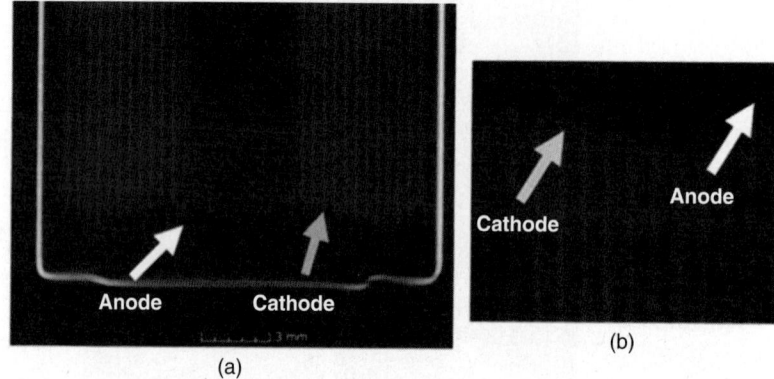

(a)　　　　　　　　　　(b)

**FIGURE 32B.13**  CT scans demonstrating anode versus cathode overlap. (*Images courtesy of DfR Solutions.*)

The SEM image in Fig. 32B.14 below shows a copper particle on the graphite anode of a lithium-ion battery. Metallic particles can be generated during electrode slitting from dull cutting wheels. These can lead to shorting across the separator, or block the separator, leading to anode over-lithiation and lithium plating.

**FIGURE 32B.14**  Copper particle on graphite anode in lithium ion battery generated from electrode slitting operation. (*Image courtesy of DfR Solutions.*)

### 32B.5.3 Case Study: Comparative Supplier Design Assessment

In this case study two different commercial CR 2032-coin cells for a memory backup application were assessed using CT scans and cell teardowns. For this particular application, cell failure has major implications for the larger system, necessitating a design assessment.

Cell design A (Fig. 32B.15a) shows the use of an expanded metal current collector that is welded to the positive cup and contacting the cathode pellet. Cell design B (Fig. 32B.15b) does not use such a current collector; the cathode pellet directly contacts the positive can. Design B is more prone to loss of electrical contact, internal resistance increase, and rate capability loss as cell discharge proceeds and electrode volumes change. This is clearly demonstrated in Fig. 32B.16, where penetrating cracks, such as those seen in CT slice 2, will electrically cut off significant portions of the active material. Figure 32B.17 shows a cell with a perforated metal current collector, similar to the current collector in design A, that is then welded to the can allowing for better cathode active material utilization and uniform current density.

(a)                                           (b)

**FIGURE 32B.15** (*a*) CR 2032 cell design showing use of an expanded metal cathode current collector that was welded to the positive cup (reproduced under Wiki commons license). (*b*) CR 2032 design without the use of a separate current collector. (*Image courtesy of DfR Solutions.*)

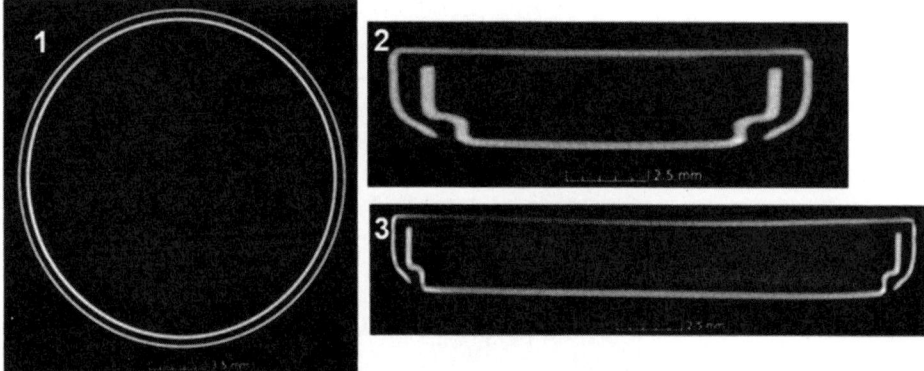

**FIGURE 32B.16** Cell design A (Fig. 32B.15a) showing through-cracks in pellet (image 2) leading to an increased likelihood of contact loss. (*Image courtesy of DfR Solutions.*)

**FIGURE 32B.17**   CT scan showing the use of a perforated current collector on positive electrode. (*Image courtesy of DfR Solutions.*)

### 32B.5.4   Case Study: Lithium-Ion Battery Swelling from Deep Discharge

Figure 32B.18 shows a swollen commercial battery from a smartphone. The CT scan (Fig. 32B.19) shows severe jellyroll distortion from internal gassing.

**FIGURE 32B.18**   Swollen lithium-ion battery from a smartphone. (*Image courtesy of DfR Solutions.*)

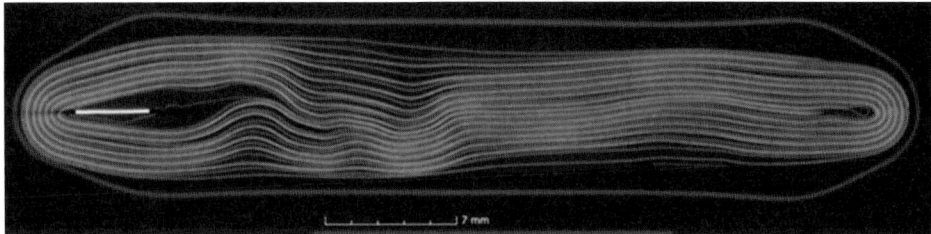

**FIGURE 32B.19**    CT scan of battery from deep discharge swelling. (*Image courtesy of DfR Solutions.*)

Lithium-ion batteries can swell due to many causes. These include

1. Overcharge conditions from elevated voltage that accelerate parasitic reactions between the electrodes and electrolyte. A hallmark of an overcharge condition is heat and gas release (often with lithium plating on the anode).

2. Poor cell quality and design with low anode to cathode stoichiometric ratios, cathode electrode overhang (which increases risk of lithium plating and gassing), and particulate contamination (self-discharge and gassing).

3. Mechanical damage to electrodes induced either during cell assembly or from the product application can both cause gassing.

4. Excessive temperatures can also cause gassing from electrolyte decomposition.

5. Deep discharge of cells.

Deep discharge or failure to periodically charge lithium-ion batteries is an often overlooked cause of battery swelling. This condition can also create safety hazards. When a lithium-ion cell goes into deep discharge, it is in a highly de-intercalated state that causes loss of the SEI layer and reformation of a new SEI, leading to gas formation. The copper current collector on the negative electrode also starts to dissolve as shown in Fig. 32B.20.

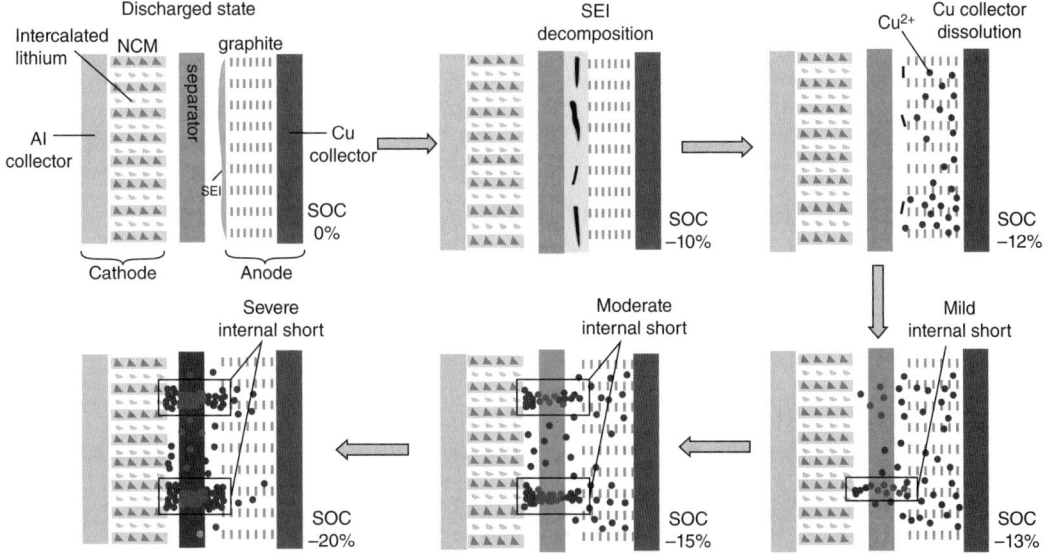

**FIGURE 32B.20**    Mechanism of deep discharge in a lithium-ion battery. (*Reproduced under Wiki Commons license.*)

**FIGURE 32B.21**    Copper deposits on graphite anode for a deeply discharged lithium-ion battery. (*Image courtesy of DfR Solutions.*)

Upon recharge, the dissolved copper redeposits on electrode or separator surfaces, potentially leading to a short and the risk of thermal runaway. Figure 32B.21 is an SEM image showing copper deposits in light gray on the darker gray graphite negative electrode.

Alternatively, the dissolved copper current collector will not be replated onto the anode's graphite surfaces in cells that are not deep discharged. A deeply discharged battery should not be recharged and a deep discharge voltage cutoff must be implemented in the battery protection system. This can include an initial low-rate charge when recharging the battery; if a certain threshold voltage is not reached within a given time, the battery would be determined to be damaged and charging would be prevented. If the battery is safe, the protection circuit would increase charging current for the rest of the charge cycle.

### 32B.5.5    Case Study: Leaking Batteries and Cell Closure Design

Cell closure design is critical to maintaining the integrity of a battery package. The battery package must allow electrical contact between the electrodes and external terminals on the package; this requires that the terminals extend through the battery, so electrical insulation around one or both terminals is needed. The packaging must also seal the contents of the battery from the outside environment. Once a battery closure is breached and electrolyte leaks, various unwanted reactions start to take place, paving the way for battery failures.

Battery packaging is also generally designed to vent intentionally at a moderate pressure; this prevents a more catastrophic spontaneous disassembly at higher pressure, cools the battery, and allows volatile electrolyte species to escape, slowing or stopping any undesired exothermic reactions. Failure to vent properly can result in a more catastrophic failure than would otherwise occur.

Defects, however, are possible when the terminal, gasket, and/or crimping die are misaligned. The x-ray image in Fig. 32B.22 shows the can crimped around a gasket (not visible in the x-ray due to low density), which holds a terminal (the disc) in place. Because of the misalignment of the terminal in the crimped seal, an uneven pressure is developed in the crimp area. Therefore, some areas are not properly sealed, allowing the cell to leak. This is manifested electrically as a drop in cell open circuit voltage.

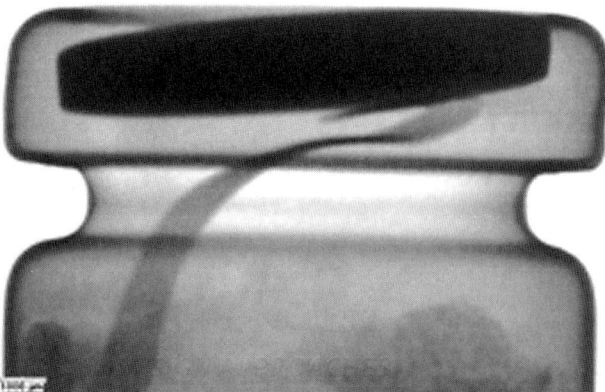

**FIGURE 32B.22**    2D x-ray showing improper sealing from poor gasket seating and consequent cap placement. (*Image courtesy of Flexel LLC.*)

The corrosion protection of metallic battery enclosures is also important for good battery sealing. Corrosion of a battery package—either from the outside or the inside—can cause the package to leak, resulting in battery failure. Corrosion is prevented either by making the entire casing out of a corrosion resistant material, or by applying a corrosion-resistant coating (such as plating or cladding) to a corrosion-prone material. Stainless steel casings, for example, are often used in lithium and lithium-ion batteries; stainless steel is resistant to corrosion inside and outside the battery.

Nickel plating applied on top of carbon steel can be used as a battery case or terminal material for alkaline, nickel-metal hydride, nickel-cadmium, and lithium-ion batteries. Carbon steel, while much less corrosion resistant than nickel or stainless steel, is strong and inexpensive. The specifics of the nickel plating required depend on the chemistry chosen. In general, excess exposed iron—even if only in patches—would result in corrosion, and potential cell leakage. Exposure of iron generally occurs because cans are stamped from plated steel, rather than being plated after stamping.

EDS mapping under an SEM can readily establish the thickness and continuity of a plating layer on a battery component with greater resolution than is possible using XRF, as seen in Fig. 32B.23.

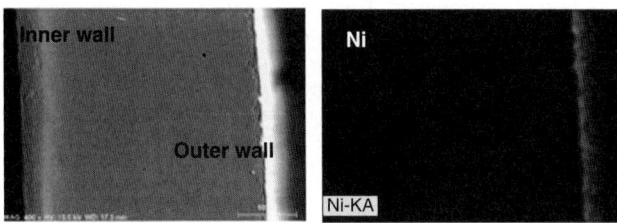

**FIGURE 32B.23**    SEM and EDS images showing nickel plating on steel battery can. (*Image courtesy of Flexel LLC.*)

The EDS map of the can shows insufficient nickel plating on the inner wall of an alkaline cell container.

## 32B.5.6   Case Study: Battery Protection Circuit Failures

Even if the battery is designed in a robust manner and constructed with sufficient controls, failure of the battery protection circuit can also initiate unexpected thermal events. Battery protection circuits typically consist of two systems, a protection circuit module or PCM and a smart battery charger (SBC). Both systems share many of the attributes of other electronic designs and are therefore susceptible to similar failure mechanisms.

Because of the need to meet several performance requirements (size, cost, safety, electrical, etc.), failure mechanisms in battery protection circuits can sometimes have multiple drivers. For instance, one failure observed was due to the presence of a potential counterfeit part. This is a significant risk with all electronic designs, but it is especially concerning with battery protection circuits owing to the higher impact of failure. In this particular case study, an OEM (original equipment manufacturer) reported issues with a p-channel MOSFET (metal oxide semiconductor field effect transistor). A simple comparison of the markings on the known good part and the suspect parts showed clear differences in the font thickness, even though both parts were supposedly manufactured in 2010 (see Fig. 32B.24).

**FIGURE 32B.24**   Good device on the *left*, suspect device on the *right*.

Further physical characterization identified differences in wire bond material (copper vs. gold) and, after decapsulation, differences in die markings. The suspect part appeared to be a Fairchild die fabricated in 2000.

Another case of multiple failure drivers within a battery protection circuit was seen with a battery charger being used in an industrial environment. The failure mode was overheating of the batteries. During investigation of some failed units, evidence of corrosion was noted across the PCB in areas where the solder mask did not provide complete coverage. Interviews with the customer and elemental analysis determined that the corrosion might have been due to sulfur coming from a natural gas processing facility (see Fig. 32B.25).

Normally, corrosion in electronics would be a clear-cut root cause. However, copper sulfide, which was the corrosion product, has limited growth and migration. If separation distances are sufficient, sulfur corrosion would not necessarily induce an electrical short and a corresponding thermal event. But, with this design, the

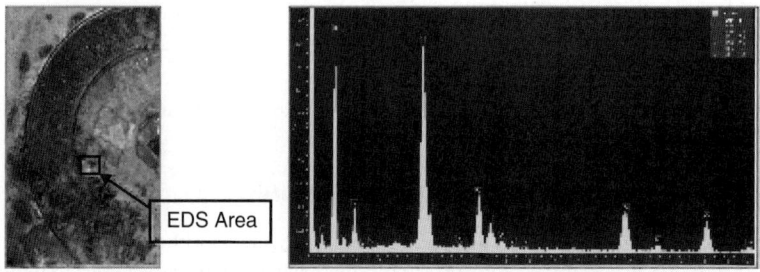

**FIGURE 32B.25**   EDS spectrum with a high amount of sulfur detected.

ground plane was placed in close proximity to the battery leads (see Fig. 32B.26). This insufficient clearance was an additional root cause of failure and subsequent mitigations included use of conformal coating and a redesign of the board layout.

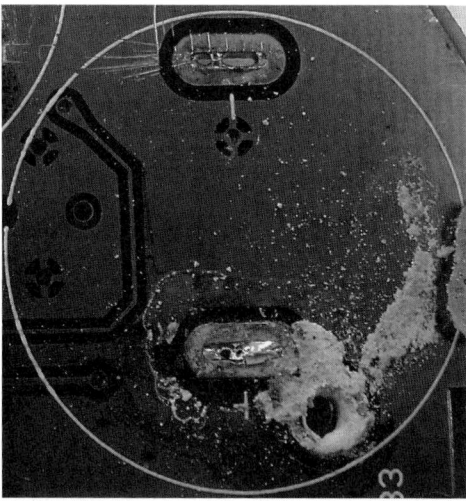

**FIGURE 32B.26**   Photograph of a battery protection circuit board with corrosion.

## 32B.6   SUMMARY

Recently, a number of market factors have combined to make battery failures much more of a high-profile issue than they were historically. Consumer preferences have driven a shift away from well-understood, safer primary batteries to secondary lithium-ion batteries, while emerging applications such as electronic cigarettes place batteries under greater loads than they have seen previously. At the same time, commoditization and competition have driven battery companies to cut corners in manufacturing to reduce price at the low end of the market and to pack more active materials to increase capacity at the high end of the market. Both of these trends can result in failure-prone batteries, and lithium-ion battery failures can be much more dramatic than, for instance, alkaline battery failures.

Furthermore, even high-quality lithium-ion batteries can fail if improperly integrated into a device, and many OEMs lack sufficient in-house battery expertise to properly select and integrate lithium-ion batteries into their products. Lithium-ion battery failures can stem from improper understanding and management of the cradle-to-grave life cycle. This life cycle includes manufacturing controls, application integration, battery protection circuit adequacy, and usage and handling of the battery. Well-established analytical techniques can detect defects in battery design and manufacturing that would potentially lead to field failures. And a quality and design assessment of the battery and protection system can prevent or mitigate many potential battery failures.

## REFERENCES

1. Anonymous, "Out of Juice," January 18, 2014. (Online.) Available at https://www.economist.com/business/2014/01/18/out-of-juice (accessed May 21, 2018).

2. U.S. Consumer Product Safety Commission, "Lithium-Ion Battery Safety Standards for Consumer Product Import into the United States," May 16, 2017. (Online.) Available at https://www.cpsc.gov/s3fs-public/3LeeCPSC.En_.pdf?QMvz78vcq0web.KaXE_TJD.dpk7DbADF (accessed May 21, 2018).

3. L. A. McKenna, "Electronic Cigarette Fires and Explosions in the United States 2009–2016," July 2017. (Online.) Available at https://www.usfa.fema.gov/downloads/pdf/publications/electronic_cigarettes.pdf (accessed May 21, 2018).

4. E. Rosenberg, "Exploding Vape Pen Caused Florida Man's Death, Autopsy Says," May 17, 2018. (Online.) Available at https://www.washingtonpost.com/news/to-your-health/wp/2018/05/16/man-died-after-a-vape-pen-exploded-and-embedded-pieces-into-this-head-autopsy-says/?noredirect=on&utm_term=.15bb66e2568c (accessed May 25, 2018).

5. J. Mullen and M. Thompson, "Samsung takes $10 billion hit to end Galaxy Note 7 fiasco," October 11, 2016. (Online.) Available at http://money.cnn.com/2016/10/11/technology/samsung-galaxy-note-7-what-next/index.html (accessed May 25, 2018).

6. "Nokia Recalls 46 Million Cell-Phone Batteries," August 15, 2007. (Online.) Available at: http://www.foxnews.com/story/2007/08/15/nokia-recalls-46-million-cell-phone-batteries.html (accessed May 25, 2018).

7. J. Christman, "The Case of Burning Laptops," *J. Case Stud.* **30**(1):88–97 (June 2012).

8. M. Humrick, "Samsung Reveals Root Cause of Galaxy Note7 Battery Fires," January 23, 2017. (Online.) Available at: https://www.anandtech.com/show/11060/samsung-reveals-root-cause-of-galaxy-note7-battery-fires (accessed April 13, 2018).

9. U.S. National Transportation Safety Board, "Auxiliary Power Unit Battery Fire Japan Airlines Boeing 787-8, JA829J," November 21, 2014. (Online.) Available at: https://www.ntsb.gov/investigations/AccidentReports/Pages/AIR1401.aspx (accessed April 13, 2018).

10. Z. J. Zhang, P. Ramadass, and W. Fang, "Safety of Lithium-Ion Batteries," in G. Pistoia (ed.), *Lithium-Ion Batteries, Advances and Applications*, 1st ed., Elsevier, 2014, pp. 409–435.

11. V. Challa, "Top Causes of Lithium-Ion Battery Field Failures," in *Design for Reliability Conference*, Baltimore, MD, 2018.

12. R. Guo, L. Lu, M. Ouyang, and X. Feng, "Mechanism of the Entire Overdischarge Process and Overdischarge-Induced Internal Short Circuit in Lithium-Ion Batteries," *Sci. Rep.* **6** (2016).

13. D. P. Finegan, M. Scheel, J. B. Robinson, B. Tjaden, I. Hunt, T. J. Mason, J. Millichamp et al., "In-Operando High-Speed Tomography of Lithium-Ion Batteries During Thermal Runaway," *Nat. Commun.* 6 (2015).

14. B. Sood, L. Severn, M. Osterman, M. Pecht, A. Bougaev, and D. McElfresh, "Lithium-Ion Battery Degradation Mechanisms and Failure Analysis Methodology," *ISTFA 2012: Proceedings from the 38th International Symposium for Testing and Failure Analysis*, Phoenix, AZ, 2012.

15. T. Waldmann, A. Iturrondobeitia, M. Kasper, N. Ghanbari, F. Aguesse, E. Bekaert, L. Daniel et al., "Review—Post-Mortem Analysis of Aged Lithium-Ion Batteries: Disassembly Methodology and Physico-Chemical Analysis Techniques," *J. Electrochem. Soc.* **163**(10): A2149–A2164 (2016).

16. J. Vetter, P. Novák, M. Wagner, C. Veit, K.-C. Möller, J. Besenhard, M. Winter et al., "Ageing Mechanisms in Lithium-Ion Batteries," *J. Power Sources* **147**(1–2): 269–281 (September 9, 2005).

17. C. Mikolajczak, M. Kahn, K. White, and R. T. Long, "Lithium-Ion Batteries Hazard and Use Assessment," July 2011. (Online.) Available at http://www.prba.org/wp-content/uploads/Exponent_Report_for_NFPA_-_20111.pdf (accessed May 22, 2018).

18. Q. Liu, C. Du, B. Shen, P. Zuo, X. Cheng, Y. Ma, G. Yin et al., "Understanding Undesirable Anode Lithium Plating Issues in Lithium-Ion Batteries," *RSC Adv.* **91** (2016).

19. Kronological, "UltraFire Batts Meet OPUS Meet Pellet Gun—Teardown Photos Added to OP," March 9, 2015. (Online.) Available at http://budgetlightforum.com/node/38133 (accessed May 22, 2018).

20. ANSI, *ANSI/IEEE 1625-2008—IEEE Standard for Rechargeable Batteries for Multi-Cell Mobile Computing Devices*, American National Standards Institute, 2009.

21. IEEE, *IEEE Std 1725-2011(Revision to IEEE Std 1725-2006)—IEEE Standard for Rechargeable Batteries for Cellular Telephones*, IEEE, 2011.

# SECTION C

# BATTERY TEST EQUIPMENT

**Miguel Sandoval**

## 32C.0   OVERVIEW

Battery test equipment is used for a wide range of applications such as materials research, portable electronic device testing, quality control, R&D, battery pack performance verification, super-capacitor/fuel cell/thermal battery testing, etc. As detailed in Chap. 32D, advanced battery testers are key to the development, qualification, and quality of any battery system.

## 32C.1   TEST EQUIPMENT HARDWARE

A typical automated, computerized test system requires high accuracy and excellent time resolution. A battery tester ideally will have multiple independent channels that allow for a wide range of multistep tests, including pulse tests, with fixed current, fixed power, fixed voltage, fixed resistance, voltage ramp (cyclic voltammetry), or other protocols. The system will record data at operator-specified intervals of time, voltage, current, etc., and the system will be calibrated to NIST traceable standards and require minimal calibration. Various optional hardware may often be added (i.e., reference voltage electrodes, pH electrodes, auxiliary inputs, SMB [system management bus] communications, TTL [transistor–transistor logic] inputs/outputs, external controller interfaces, environmental chamber controllers, etc.). And it is essential that the testers are available as turn-key installations with upgrade ability. A typical battery test unit is shown in Fig. 32C.1. These testers are usually arranged in large arrays adjacent to the battery test room, so that the electronics are protected from any failures by a secure barrier as shown in Fig. 32C.2.

These single-cell testers will range from 150 $\mu$A (full scale $\pm$0.03 $\mu$A) to 5000 mA (full scale $\pm$1.0 mA) or even to 2000 A in custom built units. The accuracy of control in these systems is 300 $\eta$A to 5 A. Battery voltage ranges from $-2$ to $+10$ V with $\pm$0.02% full scale range. Programmable time intervals are 10 ms (1 and 5 ms for custom units). For large-scale, full size battery testing, other model units are available. Specifications for one such battery pack tester are detailed in Table 32C.1.

## 32C.2   TEST EQUIPMENT SOFTWARE

Battery and energy storage device automated test systems require software to program and control tests, record data, and process the test data locally or remotely. Most software systems now use standard operating systems (i.e., MS Windows). Once an operator connects test cells and programs the test protocols, the systems will run automatically for thousands of cycles or months of unattended operation.

Programs typically use a grid layout and menu-driven cells with dozens of possible steps that direct the sequential operation of the test system. Features include:

- Subroutine that allow unlimited test loops
- Constant current or voltage with taper limits or other cut-off criteria
- Pulse regimes to simulate digital signaling or other complex procedures
- Ability to use temperature and pressure sensor inputs

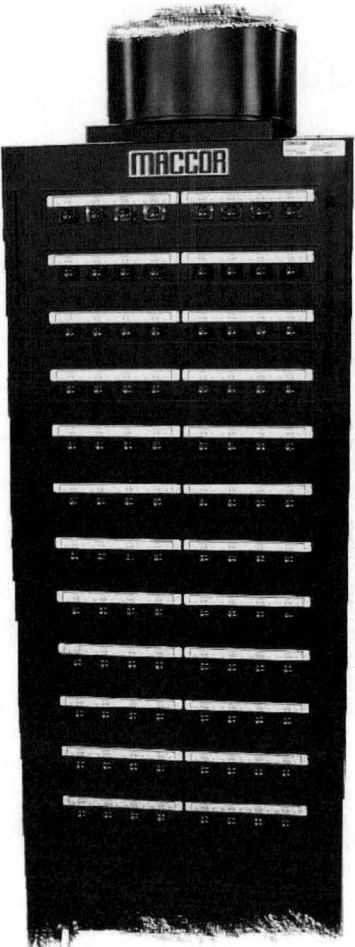

**FIGURE 32C.1**  Typical 96 channel single cell battery tester. (*Image courtesy of Maccor, Inc.*)

Once a test program has finished, built-in software or standard spreadsheet programs (i.e., MS Excel) can be used to consolidate and manipulate the data output for specific needs. These may include voltage and current logs, charge and discharge capacity and energy, and various other calculated values (i.e., internal resistance). And of course, data may be presented either in table form or as a graph.

One final key feature for any test system is to have a fully automatic, safe, and secure data storage for multiple battery test systems over a local area network for backup and further data processing. By backing up test data as well as critical configuration and calibration files from each test, systems data may be analyzed offline and recovered or restored after any events. Figure 32C.3 shows a typical battery test programing screen and Fig. 32C.4 depicts some of the output graphics available.

**FIGURE 32C.2**    Battery test facility. (*Image courtesy of Maccor, Inc.*)

**TABLE 32C.1**    Technical Specifications Model Series 8500

| | |
|---|---|
| Maximum voltage range | Customer specified from 48–1500 V |
| Minimum voltage | 5% of full-scale voltage |
| Maximum current range | Customer specified from 5–5000 A |
| Minimum current | 2% of full-scale current |
| Maximum continuous power | 10 kW–1 MW |
| Measurement accuracy | ±0.05% of full scale |
| Control accuracy | ±0.10% of full scale |
| Resolution | 16 bit (1 part in 65,536) |
| Maximum number of test cycles | $2^{32}$ |
| Maximum number of waveform steps | $2^{16}$ |
| Maximum number of test steps | 128 |
| Minimum step time | 10 ms (faster speeds available as option) |
| Minimum data sampling rate | 10 ms (faster speeds available as option) |
| Input power supply | 380/440/480 VAC, +10% to 15%, three-phase, 50/60 Hz |
| Total harmonic distortion | <3.0% |
| Dimensions and weight will vary with system voltage and power. Contact Maccor for details | |

*Source:* Courtesy of Maccor, Inc.

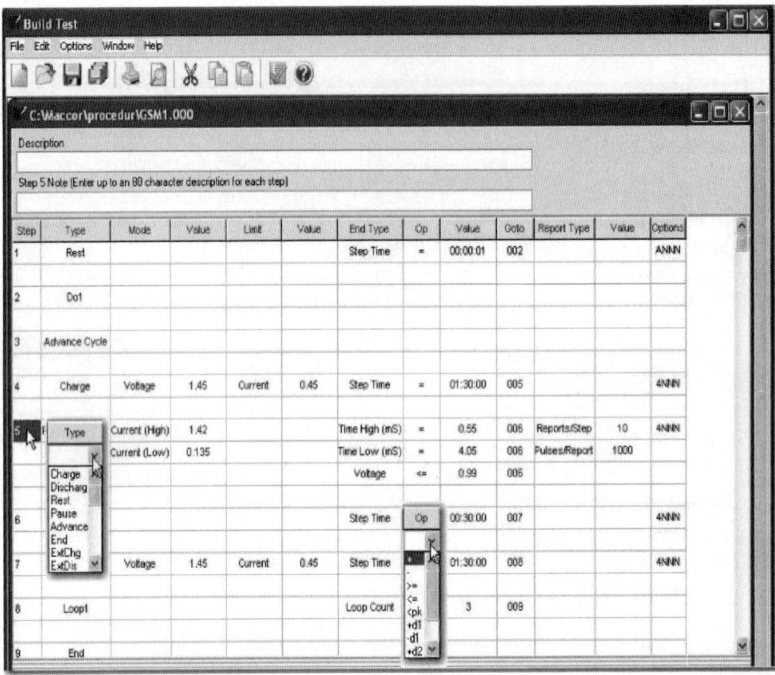

**FIGURE 32C.3** Typical test program screen display. (*Courtesy of Maccor, Inc.*)

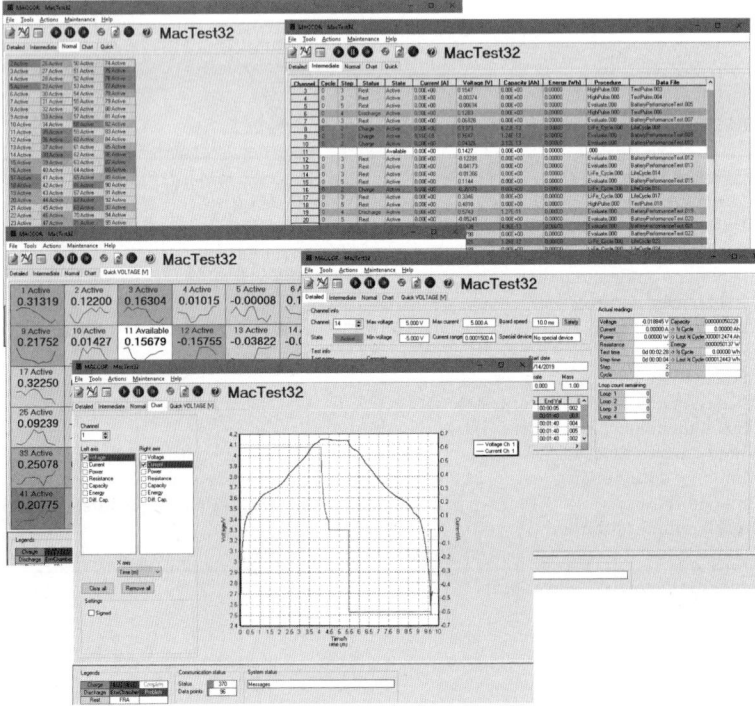

**FIGURE 32C.4** Various data output displays. (*Courtesy of Maccor, Inc.*)

## 32C.3   TEST EQUIPMENT SPECIFICATION GUIDELINES

When selecting equipment, the accuracy, resolution, and precision must be defined. Accuracy relates to the "real" value of something while resolution depicts the "fineness" and precision references "repeatability." All three are important but accuracy is the most valuable.

For example, a weight of precisely 100 kg may vary on different scales but the most accurate will be most trusted. A high resolution scale may provide a reading with many decimal places but that number may be very wrong. Finally, a high precision scale provides the same weight each time but may also be wrong. Having high accuracy is most crucial to insure the number is most correct, even if it is a gross value and not the most consistent. By using repeated measurements, the measured value will be very accurate (i.e., most correct). The typical steps for battery tester equipment selection are as follows:

1. Specify the number of test channels needed and make sure the unit is upgradeable in the field (different current or voltage ranges, etc.)
2. Set maximum charge voltage. Typically lithium cells use 4.2 VDC but new chemistries may need 4.6 VDC. Also, allow for voltage drop under high current and with long cables (i.e., high resistance leads).
3. Select the minimum discharge voltage. Lithium cells are limited to around 2.5 VDC but other chemistries may need the cells to be discharged down to 0.5 VDC or even lower.
4. Determine the maximum current needed in both charge and discharge. Some testers have set ranges (i.e., 5 A, 150 mA, 5 mA, and 150 µA), but others may be flexible/adjustable.
5. Choose any accessories that may be required during the testing.

The ultimate choice will depend on pricing and logistics, but additionally, the costs of dedicated, properly trained personnel to run the test laboratory should also be factored into the decisions and justifications.

# SECTION D

# BATTERY SAFETY AND PERFORMANCE TESTING

**Miguel Sandoval**

## 32D.0   OVERVIEW

Battery testing is an import and required task in developing and commercializing not only batteries but the products that use batteries. Two major categories of battery testing are safety and performance. These test categories may have some overlap, but serve different purposes.

Battery companies may test many things but most always use battery testers to provide safety and performance data to users of batteries, capacitors, and fuel cells. After purchase, the users may then want to confirm the capabilities relevant to their specific use. Electronic equipment companies that sell battery-powered devices will also want to simulate the discharge and charge profiles in their specific device. This testing will help select the best battery for the device. Safety and performance testing is necessary to confirm the limits of the batteries and assess damage if abused. Overall, battery testing is done by material suppliers, battery manufacturers, device/accessory manufacturers, and end users. Safety regulators, shippers, and approving/certification agencies will also play a major role in testing batteries and associated devices.

## 32D.1   SAFETY TESTING

Safety testing is performed to avert danger, injury, or loss by any user of the battery or fuel cell, including, but not limited to, end-use consumers of products that utilize batteries and fuel cells. Safety testing has evolved over the years in response to two controlling ideas: standards and regulations. Safety testing verifies that the battery will perform safely in its intended application. While standards and regulations may be closely related, it is important to understand the difference between the two.

### 32D.1.1   Standards

Engineers and experts within an industry reach a consensus to develop usual and customary principles that become the model for comparison, also known as *standards*, which are voluntary. There is no shortage of standards and issuing organizations in the field of batteries. Some common standards organizations include Underwriters Laboratory (UL), International Electrotechnical Committee (IEC), National Electrical Manufacturers Association (NEMA), Institute of Electrical and Electronics Engineers (IEEE), SAE International (formerly known as the Society of Automotive Engineers, or SAE), Japanese Industrial Standards (JIS), and Battery Safety Organization (BATSO). Many of these organizations' standards overlap.

### 32D.1.2   Regulations

*Regulations* are legally binding requirements issued by governments. Regulations often start out as standards that are eventually codified by local, national, or worldwide governmental bodies.

The United Nations' Manual of Tests and Criteria (UN Manual) specifies criteria, test methods, and procedures for the classification of dangerous goods that may be transported and is considered the most widely accepted regulatory guidelines for battery testing with regard to safety. The UN Manual, originally developed by the United Nations' Economic and Social Council's Committee of Experts on the Transport of Dangerous Goods, was first adopted in 1984. Since 2001, the Committee of Experts on the Transport of Dangerous Goods and on the Globally Harmonized System of Classification and Labelling of Chemicals has been responsible for updating the UN Manual. The Sixth Revised Edition, the most current edition, includes all amendments through 2014.

Section 38.3 of Part III of the UN Manual covers lithium metal and lithium ion batteries. This section of the UN Manual provides detailed specification on tests, procedures, and criteria applicable to lithium cells and batteries. Nearly all lithium batteries are required to pass section 38.3 of the UN Transport of Dangerous Goods Manual of Tests and Criteria. There are eight (8) tests outlined in Sec. 38.3 of the manual. These tests are as follows:

T1—Altitude simulation (primary and secondary cells and batteries)

T2—Thermal test (primary and secondary cells and batteries)

T3—Vibration (primary and secondary cells and batteries)

T4—Shock (primary and secondary cells and batteries)

T5—External short circuit (primary and secondary cells and batteries)

T6—Impact (primary and secondary cells)

T7—Overcharge (secondary batteries)

T8—Forced discharge (primary and secondary cells)

Depending on the type of cell (primary, secondary, single-cell battery, loose cells, etc.), different combinations of testing are required. The number of test cells and test sequences also may vary according to the cell types and specified tests. The criteria for these tests are listed in Table 32D.1.

**TABLE 32D.1**    Criteria for UN 38.3 Battery Testing

| Applicable test series | Criteria (result) |
| --- | --- |
| 1, 2, 3, 4 | Leakage/venting |
| 1, 2, 3, 4, 5, 6, 7, 8 | Disassembly/rupture |
| 1, 2, 3, 4, 5, 6, 7, 8 | Fire |
| 1, 2, 3, 4 | Voltage criteria |
| 5, 6 | Temperature limits |

These tests can be performed by many different laboratories around the world on many different types of equipment. When selecting lithium cell or battery test equipment, it is important to consider which safety regulations may be enforced and the reliability of the test equipment. Failure to abide by safety guidelines can result in financial loss, injury, or even loss of life.

## 32D.2   PERFORMANCE TESTING

Unlike the safety testing, performance testing is not regulated. Performance testing is used to measure the characteristics of the cell or battery. Similar tests are implemented on individual components that make up a battery such as anode, cathode, electrolyte, etc., as well as cells, modules, and packs.

Battery testers are available from various suppliers around the world. Maccor's battery testers are standard within the battery industry. Although many different types of tests can be performed, the single most important feature of each of these tests is the output, or data. The data obtained from the testers can be invaluable to expanding or determining further development efforts. With proper testing, materials or cells can be evaluated and compared. Performance data, such as capacity, energy, cycle life, fade rate, etc., are mainly obtained through battery testers. These characteristics will change with temperature (high and low), rate of charge or discharge, and after abuse (e.g., drop or vibration), requiring tests to be performed at various conditions.

A battery tester needs to be able to source (charge) and sink (discharge) current at a wide range of values. Although many tests are performed at constant current, a fixed value of current, other modes need to be available such as constant voltage charge or discharge, constant power, and constant resistance.

A battery tester must allow the operator to program a series of test steps with several safety conditions to prevent the battery under test from damage. As battery-powered devices become more sophisticated, so must battery testers. Battery testers must accurately simulate the conditions of battery usage. A battery that performs well in one device may not perform well in a different device.

Also, as the number of cells in a battery pack increases, it becomes even more crucial to perform tests on these packs. Tests will assist in finding hot spots, poor interconnections between cells, and will greatly assist in creating a superior product.

Lastly, the battery tester should allow for destructive tests such as overcharge or reverse voltage.

## 32D.3  SUMMARY

While the descriptions in this section pertain mostly to lithium-based cells, similar criteria apply to other electrochemistries. Acid electrolytes are especially troubling for use or shipment on aircraft due to corrosion problems and all failure conditions must be assessed. Also, various testing laboratories and OEM manufacturers will have their own procedures and separate test goals that should be referenced (see Chap. 23 and specific battery technology or application chapters for more details).

# SECTION E

# BUSINESS AND FINANCING STRATEGIES

**Kirby W. Beard**

## 32E.1 INTRODUCTION

No discussion on advanced battery technology is relevant without consideration of the financing structure used to achieve success. Dozens of methods have been used with varying degrees of success to achieve both technical and commercial successes. No one model is perfect and as with most new business venture, the failures greatly outweigh the successes. Hence, the statistics for declaring winning formulae are highly suspect. Typical statistics cite about a 1% success rate for new business start-ups, but even those numbers are questionable as both unsuccessful and successful entities get bought, absorbed, and dissolved without notice.

On the other hand, large, well-functioning, and profitable corporations also manage to bring new technologies to market, alone or with government assistance or other partnerships. These successes are not as visible due to the more secretive nature of their undertakings.

Philosophically, if the only goal were to advance the state-of-the-art, then the situation might be considered quite good. But when financial resources are not commensurate with the needs and the returns on the investment are risky, the overall health of the industry will ultimately suffer, driving entrepreneurs, engineers, scientists, investors, and other stakeholders to pursue opportunities with quicker, more direct, and more lucrative payouts.

Electrochemistry and battery technology are not simple. Development costs are high and barriers to entry nearly prohibitive. A cell phone app, such as where to find your parked car or the nearest dry cleaner, will clearly be a better investment for most investors, despite the irony that the app requires a high performance, safe battery to function.

There are many examples of disconnects between funding and success. A123, one of the best funded and most applauded start-ups of all time, could not survive intact. Was financing, technical resources, or market clout to blame? Or does this example show that it is the nexus of technology/IP, market fit/timing, business organization, etc., as well as financing that needs to be strategically analyzed in a precise, methodical manner?

Thomas Edison, one of the greatest entrepreneurs ever, with the backing of one of the most powerful financiers of the time (J. P. Morgan), failed to succeed at electrifying the world based on a direct current, battery-powered system (Edison's nickel-iron battery). Nicola Tesla and George Westinghouse succeeded using a more fundamentally sound concept (i.e., the alternating current or AC induction motor). Determining a reasonable business strategy is not a guarantee of technical or financial success. Hence, being resolute, but flexible in business planning may be the best advice. However, as discussed below, it is also critical to then at least be aware of the potential business strategies that may be appropriate for any given situation.

## 32E.2 BUSINESS MODELS FOR IMPLEMENTING NEW BATTERY TECHNOLOGY

Figure 32E.1 details the initial potential entry points for developing a hypothetical new superior battery technology, and the steps for establishing a commercially viable business based on a unique battery system or related components. Figure 32E.2 shows some potential models for developing, financing, and growing a battery technology/business venture, based on an assumed new, unproven technology. These charts are not considered all-inclusive or definitive and are based on anecdotal information. Further analysis and professional expertise are required for establishing a proper strategic plan.

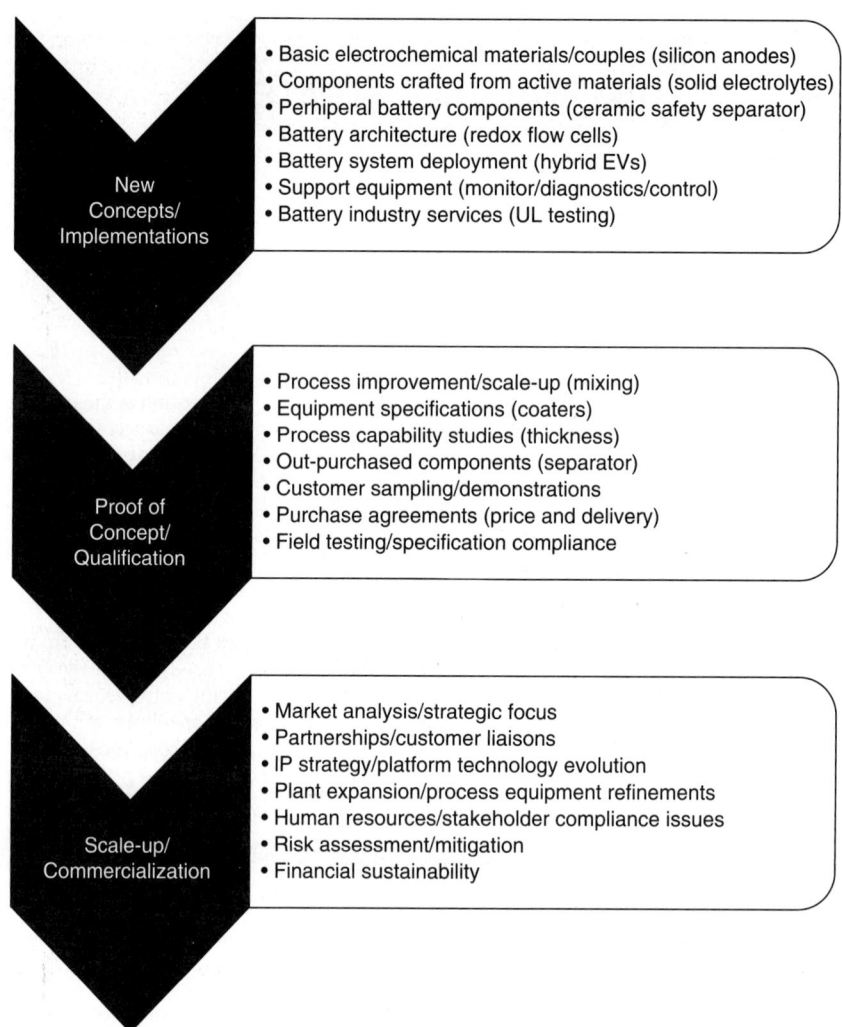

**FIGURE 32E.1** Entrepreneurial opportunities in the battery industry.

## 32E.3 RECENT STUDIES

A recent paper highlighted the special challenges facing the battery industry.[1] The authors studied start-up company success rates in the pharmaceutical industry and hypothesized that the use of such an alternative business model might be beneficial to the battery industry. The main impediments to success were noted to be the high capital needs, delayed payback, and high technology risk. One report[2] stated that only 39 companies (presumably in the United States only) had received more than $500,000 in funding since 2000 and only two from that group returned profit. The use of corporate partnerships and joint development efforts was suggested—a tenet of Fourth Generation R&D.[3]

Another recent presentation offered another approach for start-up company success.[4] In addition to describing various opportunities for advancements and commercialization, the paper also described the massive array

| Development of New Technologies and Business Concepts | Proof of Initial Concepts/Strategies | Establishment of a Successful, On-going Commercial Venture |
|---|---|---|
| • Costs: $10k–$10M<br>• Funding Sources:<br>  • self-funded<br>  • friends & family<br>  • misc. loans/grants<br>  • corporate funds<br>  • skunk-works<br>  • venture capitalists<br>• Concept Developers:<br>  • individuals<br>  • academia<br>  • government labs<br>  • trade associations<br>  • corporate labs<br>• Facilities/Equipment:<br>  • borrowed/loaned<br>  • shared/cooperative<br>  • government<br>  • corporation<br>  • used/repurposed | • Costs: $10M–$100M<br>• Funding Sources:<br>  • venture capitalists<br>  • corporations<br>  • government grants<br>  • government loans<br>  • wealthy benefactors<br>   and foundations<br>  • boutique financing<br>• Leadership:<br>  • founders/inventors<br>  • entrepreneurial<br>   professionals<br>  • corporate entities<br>• Facilities/Equipment:<br>  • partnerships<br>  • contract suppliers<br>  • leased<br>  • used/repurposed<br>  • brand new | • Costs: $100M–$1B⁺<br>• Funding Sources:<br>  • banks/financial<br>   corporation<br>  • loans/convertible<br>   debt/equity stakes<br>  • corporate funds<br>  • venture capitalist<br>   (less typically)<br>  • IPO/stock issuance<br>• Governance:<br>  • board of directors<br>  • corporate officers<br>  • consortiums<br>  • Individual owner/<br>   founder (atypically)<br>• Facilities/Equipment:<br>  • dedicated/special<br>   purpose/proprietary<br>  • supply chain<br>   partnerships |

**FIGURE 32E.2**   Strategic models applicable to the battery industry.

of obstacles that come from numerous sources. As a result, the development of simple and basic cell components or additives (i.e., new surface treatments for separators or lightweight electroplated copper web current collectors) was suggested as a more viable option than commercializing a full cell. And subsequently, the path forward might include licensing or material supply, whichever option is best supported by the financial returns.

## 32E.4   SUMMARY

The pathway to success in the battery industry is not fundamentally different than many other industries. However, the scope and complexity of the technology pertinent to the battery industry are such that advancements are slow, prodding, and difficult. If a battery developer is not methodical, any number of trivial details or minor missteps can result in catastrophe. These situations can never be fully anticipated by financial investors or by the best collective advice of any advisors.

Some companies have had the technical expertise and financial backing to deal with battery failures and survive. Most start-up companies do not have that luxury. A single technical, marketing, or financial setback is more than ample cause to shut down companies with even fundamentally sound technology and financing. The Sony lithium ion battery (and perhaps, the entire industry as a whole) survived several rather severe early failures when the first commercial lithium ion cells were introduced in the early 1990s, but probably only survived as a result of Sony's size and determination to succeed.

On another front, the ill effects of patent infringement lawsuits (see Chap. 32A) can also financially overwhelm any company with even excellent financial backing. Whether warranted or not, whether defensive or offensive, many lawsuits, especially in the United States, often impact the survivability of the damaged or infringed party. Additionally, IP protections are nearly nonexistent in certain countries. Unless a company has been very clever in disguising, protecting, or solidifying their IP, investors and shareholders will not see their rightful return on their investment. For the industry to prosper as a whole, regardless of which party is in the right, alternatives to lengthy legal battles are needed. Financing of battery ventures is not a purely economic activity. IP protections and associated judicial procedures can impact any venture.

## REFERENCES

1. https://www.cambridge.org/core/journals/mrs-energy-and-sustainability/article/applying-insights-from-the-pharma-innovation-model-to-battery-commercializationpros-cons-and-pitfalls/DE3F5D3D608E00854A2178115C289F07 (web: June 6, 2018).

2. C. Morris (https://chargedevs.com/author/charles-morris/), filed under Newswire (https://chargedevs.com/category/newswire/), The Tech (https://chargedevs.com/category/newswire/the-tech/), posted October 1, 2017.

3. W. L. Miller and L. Morris, *Fourth Generation R&D—Managing Knowledge, Technology, and Innovation*, John Wiley & Sons, Inc., New York, 1999.

4. C. Renn, Commercialization and manufacturing of advanced battery materials, 35th Annual International Battery Seminar and Exhibit, Ft. Lauderdale, FL, March 26–29, 2018.

# SECTION F

# BUSINESS PLANNING AND STRATEGIC ANALYSIS (A CASE STUDY: HOW BATTERIES CAN REALIZE A THIRD GREAT ENERGY REVOLUTION)

**Andreas de Vries, Salman Ghouri**

## 32F.0   OVERVIEW

Under the influence of technological innovation, the global energy industry underwent a remarkable transformation during the 19th and 20th centuries. Innovation in the area of battery technology could cause a similar transformation over the coming 20 years and lead to a Third Great Energy Revolution.

## 32F.1   THE HISTORY OF ENERGY REVOLUTIONS

The world underwent its First Great Energy Revolution following the invention of the steam engine during the 18th century. Before that moment, wood, wind, and water (in the form of the classic wind and water mills) provided the little energy the global economy used. However, as the steam engine revolutionized industry and mass transportation during the 19th century, it not only greatly increased the energy consumed by the global economy, but also boosted demand for the required fuel sources, wood and coal. Consequently, by the end of the 19th century, wood and coal had become the dominant sources of energy, both supplying approximately 50% of global energy consumption.[1]

At the beginning of the 20th century, Henry Ford's decision to use the recently invented conveyor belt[2] to make a car powered by the internal combustion engine, another invention of that time,[3] would cause a Second Great Energy Revolution. Mass production lead the internal combustion engine to surpass the steam engine not only in terms of power, range, ease of operation and maintenance, but also in affordability. Consequently, transportation was transformed again. In personal transportation, the automobile replaced the horse-drawn carriage, and in mass transportation, the locomotive was replaced by the diesel train; the steamship, by the motorized vessel; and the zeppelin, by the airplane. This rise of the internal combustion engine fundamentally changed lifestyles around the globe, as it enabled humanity to travel farther, faster, and cheaper than ever before. Its liquid fuel demand also led to crude oil becoming the "transportation fuel of choice," as a result of which crude oil became the most important energy source.

The rise of oil had a dramatic effect on the demand for wood, and had it not been for another invention, Thomas Edison's lightbulb, coal would have suffered the same fate. But the lightbulb's superiority over the oil lamp led to invention of the modern electricity grid,[4] which in turn led to the wholesale electrification of day-to-day life. As coal was the ideal fuel to power this electricity grid, electrification effectively breathed new life into coal demand.

Initially, this was bad news for natural gas. From the 19th century to the early 20th century, street lighting primarily used natural gas.[5] The coal-fired electricity grid enabled the lightbulb to comprehensively outperform natural gas in this area, however, providing a brighter, more reliable and also cheaper source of light. Coal's reemergence as the "electricity fuel of choice" thus led to a drop in natural gas' share in the global energy mix.

But again technological innovation changed the situation (and natural gas' fortunes). After World War II, innovations in welding, pipe rolling, and metallurgy made it economically feasible to transport natural gas over long distances,[6] providing it an opportunity to take on coal in the area of electricity generation. An abundance of natural gas production globally led natural gas to grow its share in electricity generation steadily during the latter half of the 20th century, primarily at the expense of coal.

As a result of all these events, at the end of the 20th century, wood's share in the global energy mix was devastated—from 50% a hundred years earlier to supplying just 11%. Virtually all of the increase in global energy consumption during the 20th century was derived from coal, oil, and natural gas. Coal's relative share in the energy mix had also declined to 29%, despite growing in absolute terms. Spectacular growth in oil and natural gas supply and consumption lead these two energy sources to provide 44% and 26%, respectively, of global energy demand.[7] Figure 32F.1 details the worldwide energy use over the past two centuries.

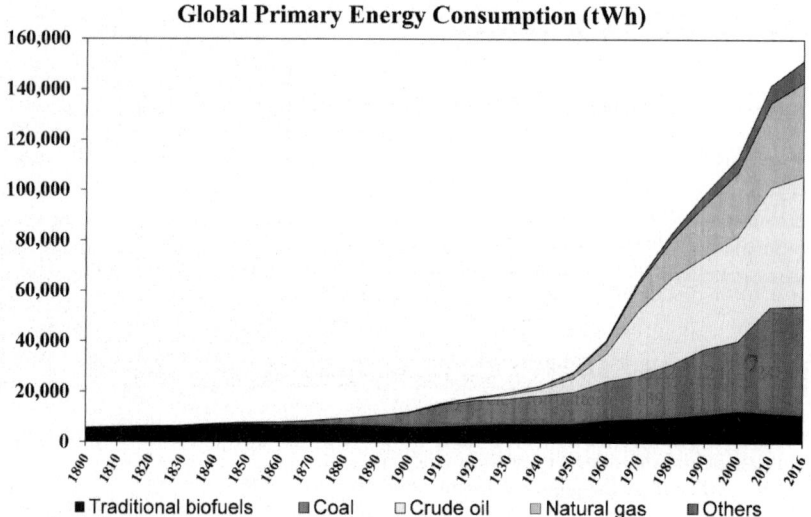

**Global Primary Energy Consumption (tWh)**

■ Traditional biofuels     ■ Coal     □ Crude oil     ■ Natural gas     ■ Others

**FIGURE 32F.1**    Global primary energy consumption since 1800.[8]

## 32F.2   CURRENT TRENDS IN ENERGY: THE DAWN OF A THIRD GREAT ENERGY REVOLUTION?

Almost 20 years into the 21st century, a new wave of technological innovation is again changing the energy landscape. In the transportation industry, for example, the electric drivetrain is now challenging the "good old" internal combustion engine. In part, this is due to one of the defining trends of the 21st century so far, which is increased environmental awareness and concern among consumers. These sentiments have not only led to adoption and continued tightening of emission control regulations by governments around the world, they have also created a realization by many businesses that "environmental awareness" can be an advantage in the global marketplace. The first concern (reducing emissions) is pushing the leading companies in the automobile industry to investigate replacement of the internal combustion engine in their vehicles with electric drivetrains.[8] The latter factor (green energy) is pulling businesses to do so, as newcomers in the industry specializing in vehicles designed around the electric drivetrain have proven massive consumer interest in such vehicles.[9]

A simple switch from the internal combustion engine to the electric drivetrain would, of course, have far-reaching consequences for oil demand, since passenger and commercial vehicles use approximately 42 million barrels of crude oil per day, over 40% of daily global oil demand.[10]

At the same time, such a switch would appear a boon for natural gas and coal, since they presently power the bulk of the electricity grids around the world. However, electricity generation is also experiencing disruption driven by technology innovation. Due to various technological advances, and a resulting increase in the scale of manufacturing, the cost to produce electricity using solar panels and wind turbines has decreased dramatically since 2009, by 85% and 66%, respectively.[11] Consequently, in certain environments, solar- and wind-powered electricity is now cheaper than electricity from conventional coal or natural gas plants. Together with the popularity of environmentally friendly solutions among consumers, these improved economics have led to a massive increase in investment for solar- and wind-based electricity generation capacity at the expense of investment in electricity generation capacity that is based on coal or natural gas.[12]

These trends raise the prospect of a future "double whammy" for the energy industry, with cars switching to electricity instead of crude oil, while electricity switches to solar rays and wind instead of natural gas and coal.[13]

## 32F.3   FROM TREND TO REVOLUTION: BATTERY TECHNOLOGY INNOVATION

It remains uncertain whether or not this particular future will indeed materialize, however. While the newcomers in the transportation industry have proven that the electric drivetrain can compete with the internal combustion engine in terms of range,[14] and can outperform it in cost or operation (fuel, maintenance, and repairs)[15] and driving comfort, they have also found that manufacturing electric vehicles (EVs)—be it passenger cars, trucks, or busses—is substantially more expensive.[16]

The main and quite possibly only reason for this issue is the cost of an EVs battery pack. Despite battery costs falling from approximately USD 1000/kWh in 2010 to USD 227/kWh in 2016,[17] much faster than expected,[18] the current cost of an EVs battery pack remains too high.[19] Past and projected cost trends for EV batteries are shown in Fig. 32F.2, while Fig. 32F.3 shows the associated cost breakdown for a typical EV.

### Lithium Ion Battery Cost (US$/kWh)

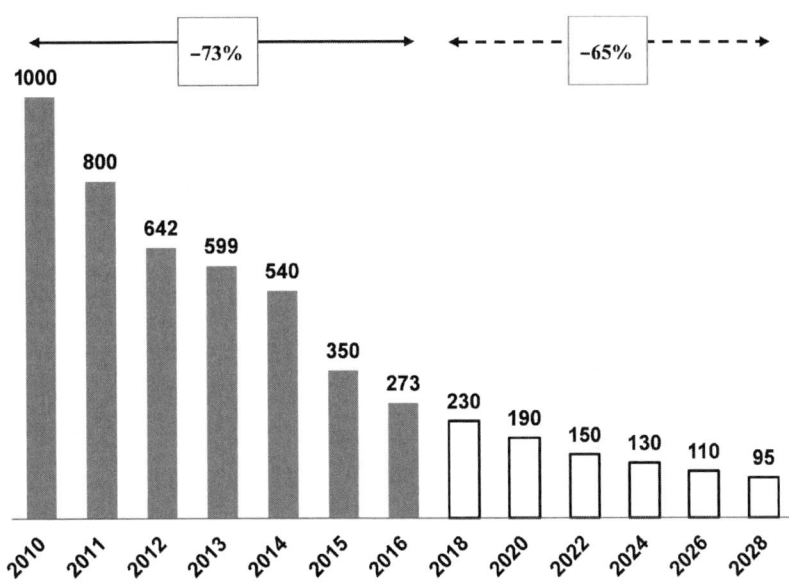

**FIGURE 32F.2**   The cost of lithium-ion batteries in USD/kWh has decreased 73% since 2010. Extrapolation of this trend sees the price decrease another 65% by 2028.[20]

Solar- and wind-powered electricity also struggle with an important disadvantage compared to coal and natural gas, which is their dependency on the weather. Unlike coal- and natural gas-powered turbines, solar panels and wind turbines cannot be "turned on" whenever there is a need for electricity. Rather, their supply to the grid is dependent on whether the sun shines or the wind blows.

**Electric Vehicle Manufacturing Cost (US$000's, 2016)**
**Compared to Internal Combustion Engine Vehicle Baseline**

**FIGURE 32F.3**    The battery pack still represents approximately 48% of the total cost to manufacture an electric vehicle, making them too expensive to be competitive with internal combustion engine-powered vehicles. But this is forecasted to change during the 2020s.[20]

Here, too, the cost of battery technology is the main driver behind this important disadvantage of solar- and wind-based electricity generation. In theory, batteries could resolve the "intermittency challenge" of solar- and wind-based electricity generation, as they could store the electricity generated by nature when it is not needed and then release it to the grid when it is needed. Although the first projects that couple batteries with solar- and wind-based electricity generation at macro grid-scale have been completed already,[20] in most environments this remains an expensive solution when compared with coal- or natural gas-based power generation.[21]

The overall implication is that batteries are key for realizing a Third Great Energy Revolution. Battery technology innovation is still necessary for bringing about EVs that outperform internal combustion-powered vehicles comprehensively (i.e., for all metrics of importance to customers), and for allowing an electricity grid that is powered by solar rays and wind that then fuels these vehicles.

## 32F.4    THE THIRD GREAT ENERGY REVOLUTION: TIMING AND IMPACT

Electrification of transportation would cause a real Third Great Energy Revolution, since this event would lead to a fundamental change in energy production, storage, and usage, yet again. Specifically, electrification would enable a switch to driverless transport,[22] which would, among other things, remove the need for people to live (relatively) close to work. In other words, it would halt the trend of urbanization.

The battery technology innovation necessary for realizing a Third Great Energy Revolution falls well within the realms of the possible. Simple extrapolation of the rate of improvement over recent years would already bring the cost of batteries down to the USD 100/kWh level that is necessary for the manufacturing of EVs to reach cost parity with internal combustion engine (ICE) vehicles by the middle of the 2020s (see Fig. 32F.3). However, the increasing interest in batteries is already driving up the price of critical components such as lithium and cobalt,[23] which could delay battery technology reaching the tipping point needed for the Third Great Energy Revolution, or even prevent it from occurring.[24]

But what would happen when battery technology reaches the tipping point? Conservative estimation, assuming the recent trend figures of 60% annual growth of EV sales,[25] predicts that crude oil demand would peak during the 2020s and begin to shrink during the 2040s.[26] At the same time, continuation of the current rate of improvement in battery technology would bring investment in coal- and natural gas-powered electricity to a halt during the 2020s, and with it growth in coal and natural gas demand. Meanwhile, solar and wind power, coupled with battery storage, would outperform these conventional sources of electricity even on full life cycle basis (building and operating).[27]

Some foresee even more far-reaching implications during the next 10 to 15 years. For once battery technology innovation may allow EVs to be produced at a cost that is lower than that of conventional vehicles. But also, the cost of *building and operating* integrated solar-wind battery storage facilities could potentially drop below that of *operating* conventional coal- and natural gas-powered electricity generation.

In this scenario, internal combustion-powered vehicles and coal and natural gas power plants would effectively become "stranded assets"—cheaper to discard and replace with what the Chinese call "New Energy" solutions than to keep in operation. In the future, predicted by Tony Seba, fossil fuel demand would not just peak during the 2020s but essentially disappear altogether.[28]

The social, environmental, and financial implications of either scenario would of course be tremendous. And the world would have battery technology to thank for it.

## REFERENCES

1. https://ourworldindata.org/energy-production-and-changing-energy-sources.

2. https://en.wikipedia.org/wiki/Conveyor_belt.

3. https://en.wikipedia.org/wiki/History_of_the_internal_combustion_engine.

4. http://instituteforenergyresearch.org/history-electricity/.

5. https://en.wikipedia.org/wiki/Street_light.

6. http://naturalgas.org/overview/history/.

7. Ibid. note 1, as of 1999.

8. https://phys.org/news/2018-01-daimler-struggling-european-emissions-standards.html.

9. http://www.thedrive.com/sheetmetal/13007/over-a-half-million-people-have-reserved-a-tesla-model-3.

10. https://oilprice.com/Energy/Energy-General/Can-We-Expect-Oil-Demand-To-Slow-Anytime-Soon.html.

11. http://beta.energyintel.com/world-energy-opinion/should-energy-security-go-green/.

12. https://www.theguardian.com/environment/2017/oct/04/solar-power-renewables-international-energy-agency.

13. https://www.elektormagazine.com/news/thinking-the-unthinkable-strategy-options-for-an-age-of-disruption-in-the-energy-industry.

14. https://en.wikipedia.org/wiki/Tesla_Model_S.

15. https://cleantechnica.com/2017/09/05/10492-tesla-model-s-maintenance-charging-costs-300000-miles/.

16. https://insideevs.com/ghosn-ev-sales-are-driven-by-mainly-state-and-company-incentives/.

17. https://electrek.co/2017/01/30/electric-vehicle-battery-cost-dropped-80-6-years-227kwh-tesla-190kwh/.

18. https://www.mckinsey.com/business-functions/sustainability-and-resource-productivity/our-insights/battery-technology-charges-ahead.

19. "Lithium-Ion Battery Costs and Market," Claire Curry, Bloomberg New Energy Finance, https://data.bloomberglp.com/bnef/sites/14/2017/07/BNEF-Lithium-ion-battery-costs-and-market.pdf.

20. https://www.reuters.com/article/us-australia-power-tesla/tesla-switches-on-giant-battery-to-shore-up-australias-grid-idUSKBN1DV3VRandhttps://www.technologyreview.com/s/603531/tesla-just-added-a-huge-stack-of-batteries-to-the-california-power-grid/.

21. https://www.technologyreview.com/s/608273/grid-batteries-are-poised-to-become-cheaper-than-natural-gas-plants-in-minnesota/.

22. http://energyfuse.org/the-second-automotive-revolution-implications-for-the-oil-industry/.

23. https://www.platts.com/latest-news/metals/dublin/lithium-supply-to-outweigh-demand-by-2018-cobalt-26720886.

24. http://energypost.eu/can-renewables-avoid-fate-nuclear-power/.

25. https://www.iea.org/publications/freepublications/publication/GlobalEVOutlook2017.pdf.

26. http://energypost.eu/wake-call-oil-companies-electric-vehicles-will-bigger-impact-oil-demand-think.

27. https://www.lazard.com/perspective/levelized-cost-of-energy-2017/.

28. https://www.rethinkx.com/transportation.

# APPENDICES

# APPENDIX A
# DEFINITIONS

**Accumulator**   *See* Secondary Battery.

**Activated Stand Life**   The period of time, at a specified temperature, that a battery can be stored in the charged condition before its capacity falls below a specified level.

**Activation**   The process of making a reserve battery functional, either by introducing an electrolyte, by immersing the battery into an electrolyte, or by other means.

**Activation Polarization**   That part of electrode or battery polarization arising from the charge-transfer step of the electrode reaction. (*See also* Polarization.)

**Active Cell or Battery**   A cell or battery containing all components and in a charged state ready for discharge (as distinct from a Reserve Cell or Battery).

**Active Material**   The material in the electrodes of a cell or battery that takes part in the electrochemical reactions of charge or discharge.

**Adsorption**   The taking up or retention of one material or medium by another by chemical or molecular action.

**Aging**   Permanent loss of capacity due either to repeated use or to passage of time.

**Ambient Temperature**   The average temperature of the surroundings.

**Ampere-Hour Capacity (also Amp-Hour Capacity)**   The quantity of electricity measured in ampere-hours (Ah) that may be delivered by a cell or battery under specified conditions.

**Ampere-Hour Efficiency (also Amp-Hour Efficiency)**   The ratio of the output of a secondary cell or battery, measured in ampere-hours, to the input required to restore the initial state-of-charge, under specified conditions (also coulombic efficiency).

**Anion**   Ion in the electrolyte carrying a negative charge.

**Anode**   The electrode in an electrochemical cell where oxidation takes place. During discharge, the negative electrode of the cell is the anode. In a rechargeable battery, during charge, the situation reverses and the positive electrode of the cell is the anode.

**Anolyte**   The portion of the electrolyte in a galvanic cell adjacent to the anode; if a diaphragm is present, the electrolyte on the anode side of the diaphragm.

**Aprotic Solvent**   A nonaqueous solvent that does not contain any reactive protons although it may contain hydrogen atoms in the molecule.

**Available Capacity**   The total capacity (amp-hours) that will be obtained from a cell or battery at defined discharge rates and other specified discharge or operating conditions.

**Battery**   One or more electrochemical cells electrically connected in an appropriate series/parallel arrangement to provide the required operating voltage and current levels including, if any, monitors, controls and other ancillary components (fuses, diodes), case, terminals, and markings.

**Bipolar Plate**   An electrode construction where positive and negative active materials are on opposite sides of an electronically conductive plate.

**Bobbin**  A cylindrical electrode (usually the positive) pressed from a mixture of the active material, a conductive material, such as carbon black, the electrolyte, and/or binder with a centrally located conductive rod or other means for a current collector.

**Boost Charge**  Charging of batteries in storage to maintain their capacity and counter the effects of self-discharge.

**Boundary Layer**  The volume of electrolyte solution immediately adjacent to the electrode surface in which concentration changes occur due to the effects of the electrode process.

**C Rate**  The discharge or charge current, in amperes, expressed as a multiple of the rated capacity in ampere-hours:

$$I = M \times C_n$$

where   $I$ = current, A
      $C$ = numerical value of rated capacity of a battery in ampere-hours (Ah)
      $n$ = time in hours for which rated capacity is specified
      $M$ = multiple or fraction (of $C$)

For example, the $0.05C$ or $C/20$ discharge current for a battery rated at 5 Ah at the $0.2C$ or $C/5$ rate is 250 mA:

$$I = M \times C_{0.2} = (0.05)(5) = 0.250 \text{ A}$$

Conversely, a battery rated at 300 mAh at the $0.5C$ or $C/2$ rate, discharged at 30 mA, is discharged at the $0.1C$ or $C/10$ rate, which is calculated as follows:

$$M = \frac{I}{C_{0.5}} = \frac{0.030}{0.300} = 0.1 \text{ or } C/10$$

**Capacitance Current**  The fraction of the cell current consumed in charging the electrical double layer.

**Capacity**  The total number of ampere-hours (Ah) that can be withdrawn from a fully charged cell or battery under specified conditions of discharge. (*See also* Available Capacity, Rated Capacity.)

**Capacity Fade**  Gradual loss of capacity of a secondary battery with cycling.

**Capacity Retention**  The fraction of the full capacity available from a battery under specified conditions of discharge after it has been stored for a period of time.

**Cathode**  The electrode in an electrochemical cell where reduction takes place. During discharge, the positive electrode of the cell is the cathode. In a rechargeable battery, during charge, the situation reverses, and the negative electrode of the cell is the cathode.

**Catholyte**  The portion of an electrolyte in a galvanic cell adjacent to a cathode; if a diaphragm is present, the electrolyte on the cathode side of the diaphragm.

**Cation**  Ion in the electrolyte carrying a positive charge.

**Cell**  The basic electrochemical unit providing a source of electrical energy by direct conversion of chemical energy. The cell consists of an assembly of electrodes, separators, electrolyte, container, and terminals.

**Charge**  The conversion of electrical energy, provided in the form of a current from an external source, into chemical energy within a cell or battery.

**Charge Acceptance**  Ability of a battery to accept charge. May be affected by temperature, charge rate, and state-of-charge.

**Charge Control**  Techniques for effectively terminating the charging of a rechargeable battery.

**Charge Efficiency**  *See* Efficiency.

**Charge Rate**   The current applied to a secondary cell or battery to restore its capacity. This rate is commonly expressed as a multiple of the rated capacity of the cell or battery. For example, the $C/10$ charge rate of a 500 Ah cell or battery (rated at the 0.2 rate) is expressed as

$$\frac{C_{0.2}}{10} = \frac{500 \text{ Ah}}{10} = 50 \text{ A}$$

**Charge Retention**   *See* Capacity Retention.

**Closed-Circuit Voltage (CCV)**   The potential or voltage of a cell or battery when it is discharging, normally under a specified load.

**Concentration Polarization**   Polarization caused by the depletion of ions in the electrolyte at the surface of the electrodes, leading to concentration gradients of battery reactants and products, caused by the passage of current. (*See also* Polarization.)

**Conditioning**   Cycle charging and discharging of a battery to ensure that it is fully formed and fully charged. Sometimes indicated when a battery is first placed in service or returned to service after prolonged storage. (*See also* Formation.)

**Constant-Current Charge**   A method of charging the battery using a current having little variation.

**Constant-Voltage Charge**   A method of charging the battery by applying a fixed voltage, and allowing variations in the current. Also called constant potential charge.

**Continuous Test**   A test in which a battery is discharged to a prescribed end-point voltage without interruption.

**Coulometer**   Electrochemical or electronic device, capable of integrating current-time curves, used for charge control and for measurement of charge inputs and discharge outputs. Results usually reported in ampere-hours.

**Counter Electromotive Force**   A voltage of an electrochemical cell opposite to the applied external voltage. Also referred to as back EMF.

**Couple**   Combination of anode and cathode materials that engage in electrochemical reactions that will produce current at a voltage defined by the reactions.

**Creepage**   The movement of electrolyte onto surfaces of electrodes or other components of a cell with which it is not normally in contact.

**Current Collector**   An inert member of high electrical conductivity used to conduct current from or to an electrode during discharge or charge.

**Current Density**   The current per unit active area of the surface of an electrode.

**Cutoff Voltage**   The battery voltage at which the discharge is terminated. Also called end voltage.

**Cycle**   The discharge and subsequent or preceding charge of a secondary battery such that it is restored to its original conditions.

**Cycle Life**   The number of cycles under specified conditions that are available from a secondary battery before it fails to meet specified criteria as to performance.

**Cycle Service**   A duty cycle characterized by frequent and usually deep discharge-charge sequences, such as motive power applications.

**Deep Discharge**   Withdrawal of at least 80% of the rated capacity of a battery.

**Density**   The ratio of a mass of material to its own volume at a specified temperature.

**Depolarization**   A reduction in the polarization of an electrode.

**Depolarizer**   A substance or means used to prevent an increase in polarization. The term "depolarizer" is often used to describe the positive electrode or cathode of a primary cell.

**Depth of Discharge (DOD)**   The ratio of the quantity of electricity (usually in ampere-hours) removed from a cell or battery on discharge to its rated capacity.

**Desorption**    The opposite of absorption, whereby the material retained by a medium is released.

**Diaphragm**    A porous or permeable material for separating the positive and negative electrode compartments of an electrochemical cell and preventing mixing of catholyte and anolyte.

**Diffusion**    The movement of species under the influence of a concentration gradient.

**Discharge**    The conversion of the chemical energy of a cell or battery into electrical energy and withdrawal of the electrical energy into a load.

**Discharge Rate**    The rate, usually expressed in amperes, at which electrical current is taken from the cell or battery.

**Double Layer**    The region in the vicinity of an electrode-electrolyte interface where the concentration of mobile ionic species has been changed to values differing from the bulk equilibrium value by the potential difference across the interface.

**Double-Layer Capacitance**    The capacitance of the electrical double layer at an electrode-electrolyte interface.

**Dry Cell**    A cell with immobilized electrolyte. The term "dry cell" is often used to describe the Leclanché cell.

**Dry-Charged Battery**    A battery in which the electrodes are in a charged state, ready to be activated by the addition of the electrolyte.

**Duplex Electrode or Plate**    *See* Bipolar Plate.

**Duty Cycle**    The operating regime of a cell or battery including factors such as charge and discharge rates, depth of discharge, cycle length, and length of time in the standby mode.

**E-Rate**    The discharge or charge power, in watts, expressed as a multiple of the rated energy, in watt-hours:

$$P = M \times E_n$$

where   $P$ = power, W
   $E$ = numerical value of the rated energy of a battery in watt-hours (Wh)
   $n$ = time in hours at which the battery was rated
   $M$ = multiple or fraction (of $E$)

For example, the $0.05E$ or $E/20$ discharge power for a battery rated at 5 h at the $0.2E$ or $E/5$ rate is 250 mW:

$$P = M \times E_{0.2} = (0.05)(5) = 0.250 \text{ W}$$

Conversely, a battery rated at 300 mWh at the $0.5E$ or $E/2$ rate, discharged at 30 mW, is discharged at the $0.1E$ or $E/10$ rate, which is calculated as follows:

$$M = \frac{I}{E_{0.5}} = \frac{0.030}{0.300} = 0.1$$

**Efficiency**    The ratio of the output of a secondary cell or battery on discharge to the input required to restore it to the initial state-of-charge under specified conditions. (*See also* Ampere-Hour Efficiency, Energy Efficiency, Voltage Efficiency, and Watt-hour Efficiency.)

**Electrical Double Layer**    *See* Double Layer.

**Electrocapillarity**    The surface tension between liquid mercury and an electrolyte solution is modified by the potential difference across the interface. The effect is termed electrocapillarity.

**Electrochemical Cell**    A cell in which the electrochemical reactions are caused by supplying electrical energy or which supplies electrical energy as a result of electrochemical reactions. If the first case only is applicable, the cell is an electrolysis cell; if the second case only, the cell is a galvanic cell.

**Electrochemical Couple**    *See* Couple.

**Electrochemical Equivalent**    Weight of one equivalent of a substance being electrolyzed, which is its gram atomic weight or its gram molecular weight divided by the number of electrons in the electrode reaction. (*See also* Faraday.)

**Electrochemical Series**    A classification of the elements according to the values of the standard potentials of specified electrochemical reactions.

**Electrode**    The site, area, or location at which electrochemical processes take place.

**Electrode Potential**    The voltage developed by a single plate, either positive or negative, against a standard reference electrode, typically the standard hydrogen electron. The algebraic difference in voltage of any two electrodes equals the cell voltage.

**Electroformation**    A term applied to the conversion of the material in both the positive and negative plates to their respective active materials. Also referred to as formation.

**Electrolyte**    The medium that provides the ion transport mechanism between the positive and negative electrodes of a cell.

**Electromotive Force (EMF)**    The standard potential of a specified electrochemical action.

**Electromotive Series**    *See* Electrochemical Series.

**Electron**    The elemental particle of an atom having a negative charge.

**Element**    The negative and positive electrodes together with the separators of a single cell. It is used almost exclusively in describing lead-acid cells and batteries.

**End Voltage**    The prescribed voltage at which the discharge (or charge, if end-of-charge voltage) of a battery may be considered complete (also called cutoff voltage).

**Energy Density**    The ratio of the energy available from a battery to its volume (Wh/L). (*See also* Specific Energy.)

**Energy Efficiency**    *See* Watt-hour Efficiency.

**Equalization**    The process of restoring all cells in a battery to an equal state-of-charge.

**Equilibrium Electrode Potential**    The difference in potential between an electrode and an electrolyte when they are in equilibrium for the electrode reaction that determines the electrode potential.

**Equivalent Circuit**    An electrical circuit that models the fundamental properties of a device (e.g., a cell) or a circuit.

**Exchange Current**    Under open-circuit conditions, the forward and backward current of an electrochemical process are equal and opposite. This equilibrium current in one direction is defined as the exchange current.

**Faraday**    One gram equivalent weight of matter is chemically altered at each electrode of a cell for each 96,494 international coulombs, or one Faraday, of electricity passed through the electrolyte.

**Fast Charge**    A rate of charging, which returns capacity to a rechargeable battery, usually within a few hours.

**Fauré Plate**    *See* Pasted Plate.

**Flash Current**    *See* Short-Circuit Current.

**Flat-Plate Cell**    A cell fabricated with rectangular flat-plate electrodes (also called a prismatic cell).

**Float Charge**    A method of maintaining a battery in a charged condition by continuous, long-term constant-voltage charging, at a level sufficient to balance self-discharge.

**Flooded Cell**    A cell design that incorporates an excess amount of electrolyte.

**Forced Discharge**    Discharging a cell or battery below 0 V into voltage reversal.

**Formation**    Electrochemical processing of a battery electrode that transforms the active materials into their usable form.

**Fuel Cell**    A galvanic cell in which the active materials are continuously supplied from a source external to the cell and the reaction products continuously removed, converting chemical energy to electrical energy.

**Galvanic Cell**    An electrochemical cell that converts chemical energy into electrical energy by electrochemical action.

**Gas Recombination**    Method of suppressing hydrogen generation during charging by recombining oxygen gas on the negative electrode as the cell approaches full charge. Batteries using this method normally contain excess capacity in the negative electrode.

**Gassing**   The evolution of gas from one or more of the electrodes in a cell. Gassing commonly results from local action (self-discharge) or from the electrolysis of the electrolyte during charging.

**Grid**   A framework for a plate or electrode that supports or retains the active materials and acts as a current collector.

**Group**   An assembly of positive or negative plates that fit into a cell.

**Half-Cell**   An electrode (either the anode or cathode) immersed in a suitable electrolyte.

**Hourly Rate**   A discharge rate, in amperes, of a battery that will deliver the specified hours of service to a given end voltage.

**Hydrogen Electrode**   An electrode of platinized platinum saturated by a stream of pure hydrogen, immersed in an electrolyte of known acidity (pH).

**Hydrogen Overvoltage**   The activation overvoltage for hydrogen evolution on an electrode.

**Initial (Closed-Circuit) Voltage**   The on-load voltage at the beginning of a discharge under a specified load.

**Inner Helmholtz Plane**   The plane of closest approach to an electrode of ions in solution. It corresponds to the plane that contains the adsorbed ions and the innermost layer of water molecules.

**Intermittent Test**   A test during which a battery is subjected to alternate periods of discharge and rest according to a specified discharge regime.

**Internal Impedance**   The opposition or resistance of a cell or battery to an alternating current of a particular frequency.

**Internal Resistance**   The opposition or resistance to the flow of an electric current within a cell or battery; the sum of the ionic and electronic resistances of the cell components.

**Ion**   A particle in solution that carries a negative or positive charge.

**IR Drop**   A voltage that is the product of the electrical resistance ($R$) of a cell or battery and the current ($I$). The value is the product of the resistance in ohms and the current in amperes.

**Life**   For rechargeable batteries, the duration of satisfactory performance, measured in years (float life) or in the number of charge/discharge cycles (cycle life).

**Load**   A term used to indicate the current drain on a battery, either directly imposed or through a resistance.

**Local Action**   Chemical reactions within a cell that convert the active materials to a discharged state without supplying energy through the battery terminals (self-discharge).

**Luggin Capillary**   The bridge from an external reference electrode to a cell solution often has a capillary tip. The capillary that is often situated close to the working electrode to minimize the IR drop is called a Luggin capillary.

**Maintenance-Free Battery**   A secondary battery that does not require periodic "topping up" to maintain electrolyte volume.

**Maximum-Power Discharge Current, $I_{mp}$**   Discharge rate at which maximum power is transferred to the external load. This is the discharge rate when the discharge voltage is approximately equal to one-half of the open circuit voltage if the discharge is purely ohmic.

**Mechanical Recharging**   Restoring the capacity of a cell by replacing a spent or discharged electrode with a fresh one.

**Memory Effect**   A phenomenon in which a cell, operated in successive cycles to the same, but less than a full, depth of discharge experiences a depression of its discharge voltage and temporarily loses the rest of its capacity at normal voltage levels (see Chap. 15).

**Midpoint Voltage**   The voltage of a battery midway in the discharge between the fully charged state and the end voltage.

**Migration**   The movement of a charged species under the influence of a potential gradient.

**Motive Power Battery**   *See* Traction Battery.

**Negative Electrode**   The electrode acting as an anode when a cell or battery is discharging.

**Negative-Limited**   The operating characteristics (performance) of the cell is limited by the negative electrode.

**Nominal Voltage**   The characteristic operating voltage or rated voltage of a battery (as distinct from Midpoint Voltage, Working Voltage, etc.).

**Off-Load Voltage**   *See* Open-Circuit Voltage.

**Ohmic Overvoltage**   Overvoltage caused by the ohmic drop in an electrolyte.

**Ohmic Polarization**   That part of electrode or battery polarization arising from current flow through ohmic resistances within an electrode or battery. (*See also* Polarization.)

**On-Load Voltage**   The difference in voltage between the terminals of a cell or battery when it is discharged under a specified load.

**Open-Circuit Voltage (OCV)**   The difference in voltage between the terminals of a cell or voltage when the circuit is open (no-load condition).

**Outer Helmholtz Plane**   The plane of closest approach of those ions that do not contact-absorb but approach the electrode with a sheath of solvated water molecules surrounding them.

**Overcharge**   The forcing of current through a battery after all the active material has been converted to the charged state. In other words, charging continued after 100% state-of-charge is achieved.

**Overdischarge**   Discharge past the point where the full capacity of the battery has been obtained.

**Overvoltage**   The potential difference between the equilibrium potential of an electrode and that of the electrode under an imposed polarization current.

**Oxygen Recombination**   The process by which oxygen generated at the positive plate during charge is reacted at the negative plate.

**Paper-Lined Cell**   Construction of a cell where a layer of paper, wetted with electrolyte, acts as the separator.

**Parallel**   Term used to describe the interconnection of cells or batteries in which all of the like terminals are connected together. Parallel connections increase the capacity of the resultant battery as follows:

$$C_p = n \times C_u$$

where $C_p$ = the resultant capacity
$n$ = the number of cells or batteries connected in parallel
$C_u$ = capacity of the unconnected cell or battery

**Passivation**   The phenomenon by which an electrode, although in conditions of thermodynamic instability, remains unattacked because of its surface condition.

**Paste**   Mixtures of various compounds that are applied to positive and negative grids of lead batteries. These pastes are then converted to positive and negative active materials. (*See also* Formation.)

**Paste-Lined Cell**   Leclanché cell constructed so that a layer of gelled paste acts as the separator.

**Pasted Plate**   A plate manufactured by coating a grid or support strip with active materials in paste form.

**Planté Plate**   A plate for a lead-acid battery in which the active materials are formed directly from a lead substrate by electrochemical processing.

**Plate**   A structure containing active materials held firmly to a grid or conductor.

**Pocket Plate**   A plate for a secondary battery in which active materials are held in perforated metal pockets on a support strip.

**Polarity**   Denoting either positive or negative potential.

**Polarization**   The change of the potential of a cell or electrode from its equilibrium value caused by the passage of an electric current, includes phenomena such as activation, concentration, and ohmic polarization.

**Positive Electrode**   The electrode acting as a cathode when a cell or battery is discharging.

**Positive Limited**   The operating characteristics (performance) of the cell is limited by the positive electrode.

**Power Density**    The ratio of the power available from a battery to its volume (W/L). (*See also* Specific Power.)

**Primary Cell or Battery**    A cell or battery that is not intended to be recharged and is discarded when it has delivered all its electrical energy.

**Prismatic Cell**    *See* Flat-Plate Cell.

**Rate Constant**    At equilibrium, the forward and backward Faradic currents of an electrode process are equal and referred to as the exchange current. This exchange current can be defined in terms of a rate constant called the standard heterogeneous rate constant for the electrode process.

**Rated Capacity**    The number of ampere-hours a battery can deliver under specific conditions (rate of discharge, end voltage, temperature); usually the manufacturer's rating.

**Recharge**    *See* Charge.

**Rechargeable Battery**    *See* Secondary Battery.

**Recombination**    A term used in a sealed cell construction for the process whereby internal pressure is relieved by reaction of oxygen with the negative active material.

**Recovery**    *See* Recuperation.

**Recuperation**    The lowering of the polarization of a cell during rest periods.

**Redox Cell**    A secondary cell in which two soluble ionic reactants, separated by a membrane, form the active materials.

**Reference Electrode**    A specially chosen electrode that has a reproducible potential against which other electrode potentials may be measured. (*See also* Hydrogen Electrode.)

**Reserve Cell or Battery**    A cell or battery that may be stored in an inactive state and made ready for use by adding electrolyte, another cell component, or, in the case of a thermal battery, melting a solidified electrolyte.

**Reversal**    The changing of the normal polarity of a cell or battery.

**Secondary Battery**    A galvanic battery which, after discharge, may be restored to the charged state by the passage of an electric current through the cell in the opposite direction to that of discharge.

**Self-Discharge**    The loss of useful capacity of a cell or battery due to internal chemical action (local action).

**Semipermeable Membrane**    A film that will pass selected ions.

**Separator**    An ion-permeable, electronically nonconductive spacer or material that prevents electronic contact between electrodes of opposite polarity in the same cell.

**Series**    The interconnection of cells or batteries in such a manner that the positive terminal of the first is connected to the negative terminal of the second, and so on. Series connections increase the voltage of the resultant battery as follows:

$$V_s = n \times V_u$$

where   $V_s$ = the resultant voltage
         $n$ = the number of cells or batteries connected in series
         $V_u$ = voltage of the unconnected cell or battery

**Service Life**    The period of useful life of a primary battery before a predetermined end point voltage is reached.

**Shallow Discharge**    A discharge on a secondary battery equaling only a small part of its total capacity.

**Shape Change**    Change in shape of an electrode due to migration of active material during charge/discharge cycling.

**Shedding**    The loss of active material from a plate during cycling.

**Shelf Life**    The duration of storage under specified conditions at the end of which a cell or battery still retains the ability to give a specified performance.

**Short-Circuit Current**    The initial value of the current obtained from a battery in a circuit of negligible resistance.

**Sintered Electrode**    An electrode construction in which active materials are deposited in the interstices of a porous metal matrix made by sintering metal powder.

**SLI Battery**    A battery designed to start internal combustion engines and to power the electrical systems in automobiles when the engine is not running (starting, lighting, ignition). Typically this is a lead-acid battery.

**Specific Energy**    The ratio of the energy output of a cell or battery to its weight (Wh/kg). (*See also* Energy Density.)

**Specific Gravity**    The specific gravity of a solution is the ratio of the weight of the solution to the weight of an equal volume of water at a specified temperature.

**Specific Power**    The ratio of the power delivered by a cell or battery to its weight (W/kg). (*See also* Power Density.)

**Spirally Wound Cell**    A cylindrical cell that uses an electrode structure made by winding the electrodes and separators into a cylindrical "jelly-roll" construction.

**Standard Electrode Potential**    The equilibrium value of an electrode potential when all the constituents taking part in the electrode reaction are in the standard state.

**Standby Battery**    A battery designed for emergency use in the event of a main power failure.

**Starved-Electrolyte Cell**    A cell containing little or no free fluid electrolyte. This enables gases to reach electrode surfaces during charging and facilitates gas recombination.

**State-of-Charge (SOC)**    The available capacity in a battery expressed as a percentage of rated capacity.

**Stationary Battery**    A secondary battery designed for use in a fixed location.

**Storage Battery**    *See* Secondary Battery.

**Storage Life**    *See* Shelf Life.

**Sulfation**    Process occurring in lead batteries that have been stored and allowed to self-discharge for extended periods of time. Large crystals of lead sulfate grow that interfere with the function of the active materials.

**Taper Charge**    A charge regime delivering moderately high-rate charging current when the battery is at a low state-of-charge and tapering the charging current to lower rates as the battery is charged.

**Thermal Runaway**    A condition whereby a battery on charge or discharge will overheat and destroy itself through internal heat generation caused by high overcharge or overdischarging current or other abusive condition.

**Traction Battery**    A secondary battery designed for the propulsion of electric vehicles or electrically operated mobile equipment operating in a deep-cycle regime.

**Transfer Coefficient**    The transfer coefficient determines what fraction of the electrical energy of a system results from the displacement of the potential from the equilibrium value that affects the rate of electrochemical transformation (see Chap. 4).

**Transference Number**    The fraction of the total cell current carried by the cation of an electrolyte solution is called the "cation transference number." Similarly, the fraction of the total current carried by the anion is referred to as the "anion transference number."

**Transition Time**    The time of an electrode process from the initiation of the process at constant current to the moment an abrupt change in potential occurs, signifying that a new electrode process is controlling the electrode potential.

**Trickle Charge**    A charge at a low rate, balancing losses through a local action and/or periodic discharge, to maintain a battery in a fully charged condition.

**Tubular Plate**    A battery plate in which an assembly of perforated metal or polymer tubes holds the active materials.

**Unactivated Shelf Life**    The period of time, under specified conditions of temperature and environment, that an unactivated or reserve battery can stand before deteriorating below a specified capacity.

**Vent**    A normally sealed mechanism that allows for the controlled escape of gases from within a cell.

**Vented Cell**    A cell design incorporating a vent mechanism to relieve excessive pressure and expel gases that are generated during the operation or abuse of the cell.

**Voltage Delay**    Time delay for a battery to deliver the required operating voltage after it is placed under load.

**Voltage Depression**    An abnormal low voltage, below the expected value, during the discharge of a battery.

**Voltage Efficiency**    The ratio of average voltage during discharge to average voltage during recharge under specified conditions of charge and discharge.

**Watt-hour Capacity**    The quantity of electrical energy, measured in watt-hours, that may be delivered by a cell or battery under specified conditions.

**Watt-hour Efficiency**    The ratio of the watt-hours delivered on discharge of a battery to the watt-hours needed to restore it to its original state under specified conditions of charge and discharge. Also called energy efficiency.

**Wet Shelf Life**    The period of time that a battery can stand in the charged or activated condition before deteriorating below a specified capacity.

**Working Voltage**    The typical voltage or range of voltages of a battery during discharge.

# APPENDIX B
# STANDARD REDUCTION POTENTIALS

**TABLE B.1** Standard Reduction Potentials of Electrode Reactions at 25°C

| Electrode reaction | $E^0$, V |
|---|---|
| $Li^+ + e \rightleftharpoons Li$ | −3.01 |
| $Rb^+ + e \rightleftharpoons Rb$ | −2.98 |
| $Cs^+ + e \rightleftharpoons Cs$ | −2.92 |
| $K^+ + e \rightleftharpoons K$ | −2.92 |
| $Ba^{2+} + 2e \rightleftharpoons Ba$ | −2.92 |
| $Li^+ + 6C + e \rightleftharpoons LiC_6$ | −2.90 |
| $Sr^2 + 2e \rightleftharpoons Sr$ | −2.89 |
| $Ca^{2+} + 2e \rightleftharpoons Ca$ | −2.84 |
| $Na^+ + e \rightleftharpoons Na$ | −2.71 |
| $Mg(OH)_2 + 2e \rightleftharpoons Mg + 2OH^-$ | −2.67 |
| $Mg^{2+} + 2e \rightleftharpoons Mg$ | −2.38 |
| $Al(OH)_3 + 3e \rightleftharpoons Al + 3OH^-$ | −2.34 |
| $Ti^{2+} + 2e \rightleftharpoons Ti$ | −1.75 |
| $Be^{2+} + 2e \rightleftharpoons Be$ | −1.70 |
| $Al^3 + 3e \rightleftharpoons Al$ | −1.66 |
| $Zn(OH)_2 + 2e \rightleftharpoons Zn + 2OH^-$ | −1.25 |
| $Mn^{2+} + 2e \rightleftharpoons Mn$ | −1.05 |
| $Fe(OH)_2 + 2e \rightleftharpoons Fe + 2OH^-$ | −0.88 |
| $2H_2O + 2e \rightleftharpoons H_2 + 2OH^-$ | −0.83 |
| $H^+ + M + e \rightleftharpoons MH$ | −0.83 |
| $Cd(OH)_2 + 2e \rightleftharpoons Cd + 2OH^-$ | −0.81 |
| $Zn^{2+} + 2e \rightleftharpoons Zn$ | −0.76 |
| $Ni(OH)_2 + 2e \rightleftharpoons Ni + 2OH^-$ | −0.72 |
| $Ga^{3+} + 3e \rightleftharpoons Ga$ | −0.52 |
| $S + 2e \rightleftharpoons S^{2-}$ | −0.48 |
| $Fe^{2+} + 2e \rightleftharpoons Fe$ | −0.44 |
| $Cd^{2+} + 2e \rightleftharpoons Cd$ | −0.40 |
| $PbSO_4 + 2e \rightleftharpoons Pb + SO_4^{2-}$ | −0.36 |
| $In^{3+} + 3e \rightleftharpoons In$ | −0.34 |
| $Tl^+ + e \rightleftharpoons Tl$ | −0.34 |
| $Co^{2+} + 2e \rightleftharpoons Co$ | −0.27 |
| $Ni^{2+} + 2e \rightleftharpoons Ni$ | −0.23 |
| $Sn^{2+} + 2e \rightleftharpoons Sn$ | −0.14 |

*(Continued)*

**TABLE B.1** Standard Reduction Potentials of Electrode Reactions at 25°C (*Continued*)

| Electrode reaction | $E^0$, V |
|---|---|
| $Pb^{2+} + 2e \rightleftharpoons Pb$ | $-0.13$ |
| $O_2 + H_2O + 2e \rightleftharpoons HO_2^- + OH^-$ | $-0.08$ |
| $D^+ + e \rightleftharpoons \frac{1}{2}D_2$ | $-0.003$ |
| $H^+ + e \rightleftharpoons \frac{1}{2}H_2$ | $0.000$ |
| $HgO + H_2O + 2e \rightleftharpoons Hg + 2OH^-$ | $0.10$ |
| $CuCl + e \rightleftharpoons Cu + Cl^-$ | $0.14$ |
| $AgCl + e \rightleftharpoons Ag + Cl^-$ | $0.22$ |
| $\gamma\text{-}MnO_2 + H_2O + e \rightleftharpoons \alpha\text{-}MnOOH + OH^-$ | $0.30$ |
| $Cu^{2+} + 2e \rightleftharpoons Cu$ | $0.34$ |
| $Ag_2O + H_2O + 2e \rightleftharpoons 2Ag + 2OH^-$ | $0.35$ |
| $\gamma\text{-}MnO_2 + H_2O + e \rightleftharpoons \lambda\text{-}MnOOH + OH^-$ | $0.36$ |
| $\frac{1}{2}O_2 + H_2O + 2e \rightleftharpoons 2OH^-$ | $0.40$ |
| $NiOOH + H_2O + e \rightleftharpoons Ni(OH)_2 + OH^-$ | $0.45$ |
| $Cu^+ + e \rightleftharpoons Cu$ | $0.52$ |
| $I_2 + 2e \rightleftharpoons 2I^-$ | $0.54$ |
| $2AgO + H_2O + 2e \rightleftharpoons Ag_2O + 2OH^-$ | $0.57$ |
| $LiCoO_2 + 0.5e \rightleftharpoons Li_{0.5}CoO_2 + 0.5Li^+$ | $\sim0.70$ |
| $Hg^{2+} + 2e \rightleftharpoons 2Hg$ | $0.80$ |
| $Ag^+ + e \rightleftharpoons Ag$ | $0.80$ |
| $O_2 + 4H^+(10^{-7}\ M) + 4e \rightleftharpoons 2H_2O$ | $0.82$ |
| $Pd^{2+} + 2e \rightleftharpoons Pd$ | $0.83$ |
| $Ir^{3+} + 3e \rightleftharpoons Ir$ | $1.00$ |
| $Br_2 + 2e \rightleftharpoons 2Br^-$ | $1.08$ |
| $O_2 + 4H^+ + 4e \rightleftharpoons 2H_2O$ | $1.23$ |
| $MnO_2 + 4H^+ + 2e \rightleftharpoons Mn^{2+} + 2H_2O$ | $1.23$ |
| $Cl_2 + 2e \rightleftharpoons 2Cl^-$ | $1.36$ |
| $PbO_2 + 4H^+ + 2e \rightleftharpoons Pb^{2+} + 2H_2O$ | $1.46$ |
| $PbO_2 + SO_4^{2-} + 4H^- + 2e \rightleftharpoons PbSO_4 + 2H_2O$ | $1.69$ |
| $F_2 + 2e \rightleftharpoons 2F^-$ | $2.87$ |

# ELECTROCHEMICAL EQUIVALENTS OF BATTERY MATERIALS

**TABLE C.1** Electrochemical Equivalents of Battery Materials

| Material | Symbol | Atomic no. | Atomic wt., g | Density, $g/cm^3$ | Valence change | Electrochemical equivalents Ah/g | g/Ah | $Ah/cm^3$ |
|---|---|---|---|---|---|---|---|---|
| | | | Elements | | | | | |
| Aluminum | Al | 13 | 26.98 | 2.699 | 3 | 2.98 | 0.335 | 8.05 |
| Antimony | Sb | 51 | 121.75 | 6.62 | 3 | 0.66 | 1.514 | 4.37 |
| Arsenic | As | 33 | 74.92 | 5.73 | 3 | 1.79 | 0.559 | 10.26 |
| Barium | Ba | 56 | 137.34 | 3.78 | 2 | 0.39 | 2.56 | 1.47 |
| Beryllium | Be | 4 | 9.01 | — | 2 | 5.94 | 0.168 | — |
| Bismuth | Bi | 83 | 208.98 | 9.80 | 3 | 0.385 | 2.59 | 3.77 |
| Boron | B | 5 | 10.81 | 2.54 | 3 | 7.43 | 0.135 | 18.87 |
| Bromine | Br | 35 | 79.90 | — | 1 | 0.335 | 2.98 | — |
| Cadmium | Cd | 48 | 112.40 | 8.65 | 2 | 0.477 | 2.10 | 4.15 |
| Cesium | Cs | 55 | 132.91 | 1.87 | 3 | 0.574 | 1.74 | 1.07 |
| Calcium | Ca | 20 | 40.08 | 1.54 | 2 | 1.34 | 0.748 | 2.06 |
| Carbon (graphite) | C | 6 | 12.01 | 2.25 | 4 | 8.93 | 0.112 | 20.09 |
| Chlorine | Cl | 17 | 35.45 | — | 1 | 0.756 | 1.32 | — |
| Chromium | Cr | 24 | 52.00 | 6.92 | 3 | 1.55 | 0.647 | 10.72 |
| Cobalt | Co | 27 | 58.93 | 8.71 | 2 | 0.910 | 1.10 | 7.93 |
| Copper | Cu | 29 | 63.55 | 8.89 | 2 | 0.843 | 1.19 | 7.49 |
| | | | | | 1 | 0.422 | 2.37 | 3.75 |
| Fluorine | F | 9 | 19.00 | — | 1 | 1.41 | 0.709 | — |
| Gold | Au | 79 | 197.00 | 19.30 | 1 | 0.136 | 7.36 | 2.62 |
| Hydrogen | H | 1 | 1.008 | — | 1 | 26.59 | 0.0376 | — |
| Indium | In | 49 | 114.82 | 7.28 | 3 | 0.701 | 1.43 | 5.10 |
| Iodine | I | 53 | 126.90 | 4.94 | 1 | 0.211 | 4.73 | 1.04 |
| Iron | Fe | 26 | 55.85 | 7.85 | 2 | 0.96 | 1.04 | 7.54 |
| | | | | | 3 | 1.44 | 0.694 | 11.30 |
| Lead | Pb | 82 | 207.20 | 11.34 | 2 | 0.259 | 3.87 | 2.94 |
| Lithium | Li | 3 | 6.94 | 0.534 | 1 | 3.86 | 0.259 | 2.06 |
| Magnesium | Mg | 12 | 24.31 | 1.74 | 2 | 2.20 | 0.454 | 3.83 |
| Manganese | Mn | 25 | 54.94 | 7.42 | 2 | 0.976 | 1.02 | 7.24 |
| Mercury | Hg | 80 | 200.59 | 13.60 | 2 | 0.267 | 3.74 | 3.63 |
| Molybdenum | Mo | 42 | 95.94 | 10.20 | 6 | 1.67 | 0.597 | 17.03 |

*(Continued)*

**TABLE C.1** Electrochemical Equivalents of Battery Materials (*Continued*)

| Material | Symbol | Atomic no. | Atomic wt., g | Density, g/cm³ | Valence change | Electrochemical equivalents | | |
|---|---|---|---|---|---|---|---|---|
| | | | | | | Ah/g | g/Ah | Ah/cm³ |
| | | | | Elements | | | | |
| Nickel | Ni | 28 | 58.71 | 8.60 | 2 | 0.913 | 1.09 | 7.85 |
| Nitrogen | N | 7 | 14.01 | — | 3 | 5.74 | 0.174 | — |
| Oxygen | O | 8 | 16.00 | — | 2 | 3.35 | 0.298 | — |
| Platinum | Pt | 78 | 195.09 | 21.37 | 4 | 0.549 | 1.82 | 11.73 |
| Potassium | K | 19 | 39.10 | 0.87 | 1 | 0.685 | 1.46 | 0.59 |
| Silver | Ag | 17 | 107.87 | 10.50 | 1 | 0.248 | 4.02 | 2.60 |
| Sodium | Na | 11 | 22.99 | 0.971 | 1 | 1.17 | 0.858 | 1.14 |
| Sulfur | S | 16 | 32.06 | 2.00 | 2 | 1.67 | 0.598 | 3.34 |
| Tin | Sn | 50 | 118.69 | 7.30 | 4 | 0.903 | 1.11 | 6.59 |
| Vanadium | V | 23 | 50.95 | 5.96 | 5 | 2.63 | 0.380 | 15.67 |
| Zinc | Zn | 30 | 65.38 | 7.10 | 2 | 0.820 | 1.22 | 5.82 |
| Zirconium | Zr | 40 | 91.22 | 6.44 | 4 | 1.18 | 0.851 | 7.60 |
| | | | | Compounds | | | | |
| Bismuth trioxide | $Bi_2O_3$ | | 466 | 8.5 | 6 | 0.345 | 2.90 | 2.97 |
| Bismuth trifloride | $BiF_3$ | | 265.9 | — | 3 | 0.302 | 3.31 | — |
| Calcium chromate | $CaCrO_4$ | | 156.1 | — | 2 | 0.34 | 2.90 | — |
| Carbon monofluoride | $CF_x$ | | 31 | 2.7 | 1 | 0.862 | 1.16 | 2.32 |
| Cobalt difluoride | $CoF_2$ | | 96.9 | — | 2 | 0.553 | 1.81 | — |
| Cuprous chloride | CuCl | | 99 | 3.5 | 1 | 0.27 | 3.69 | 0.95 |
| Cupric chloride | $CuCl_2$ | | 134.5 | 3.1 | 2 | 0.40 | 2.50 | 1.22 |
| Cupric fluoride | $CuF_2$ | | 101.6 | 2.9 | 2 | 0.528 | 1.89 | 1.52 |
| Cupric oxide | CuO | | 79.6 | 6.4 | 2 | 0.67 | 1.49 | 4.26 |
| Cupric sulfate | $CuSO_4$ | | 159.6 | 3.6 | 2 | — | — | — |
| Cupric sulfide | CuS | | 95.6 | 4.6 | 2 | 0.56 | 1.79 | 2.57 |
| Iron monosulfide | FeS | | 87.9 | 4.84 | 2 | 0.61 | 1.64 | 2.95 |
| Iron disulfide | $FeS_2$ | | 119.9 | 4.87 | 4 | 0.89 | 1.12 | 4.35 |
| Iron trifluoride | $FeF_3$ | | 112.8 | — | 3 | 0.712 | 1.40 | — |
| Lead bismuthate | $Bi_2Pb_2O_5$ | | 912 | 9.0 | 10 | 0.29 | 3.41 | 2.64 |
| Lead chloride | $PbCl_2$ | | 278.1 | 5.8 | 2 | 0.19 | 5.18 | 1.12 |
| Lead dioxide | $PbO_2$ | | 239.2 | 9.3 | 2 | 0.22 | 4.45 | 2.11 |
| Lead iodide | $PbI_2$ | | 461 | 6.2 | 2 | 0.12 | 8.60 | 0.72 |
| Lead oxide | $Pb_3O_4$ | | 685 | 9.1 | 8 | 0.31 | 3.22 | 2.85 |
| Lead sulfide | PbS | | 239.3 | 7.5 | 2 | 0.22 | 4.46 | 1.68 |
| Lithiated carbon | $LiC_6$ | | 79 | — | 1 | 0.372* | 2.69* | — |
| Lithium cobalt oxide | $LiCoO_2$ | | 98 | 5.05 | 0.55 | 0.150 | 6.67 | 0.757 |
| Lithium iron phosphate | $LiFePO_4$ | | 117.7 | 3.60 | 1 | 0.160 | 6.25 | 0.576 |
| Lithium manganese oxide (spinel) | $Li_{1.1}Mn_{1.9}O_2$ | | 144.0 | 4.18 | 1 | 0.120 | 8.33 | 0.502 |
| Lithium nickel manganese cobalt oxide (NMC) | $Li(Ni_{1/3}Mn_{1/3}Co_{1/3})O_2$ | | 96.4 | 4.77 | 0.59 | 0.163 | 6.13 | 0.777 |
| Manganese dioxide | $MnO_2$ | | 86.9 | 5.0 | 1 | 0.31 | 3.22 | 1.54 |
| Manganese trifluoride | $MnF_3$ | | 111.9 | — | 3 | 0.719 | 1.39 | — |
| Mercuric oxide | HgO | | 216.6 | 11.1 | 2 | 0.247 | 4.05 | 2.74 |
| Molybdenum trioxide | $MoO_3$ | | 143 | 4.5 | 1 | 0.19 | 5.26 | 0.84 |
| Nickel difluoride | $NiF_2$ | | 96.7 | — | 2 | 0.554 | 1.80 | — |

*Based on weight of carbon only.

**TABLE C.1**    Electrochemical Equivalents of Battery Materials (*Continued*)

| Material | Symbol | Atomic no. | Atomic wt., g | Density, g/cm$^3$ | Valence change | Electrochemical equivalents | | |
|---|---|---|---|---|---|---|---|---|
| | | | | | | Ah/g | g/Ah | Ah/cm$^3$ |
| Compounds | | | | | | | | |
| Nickel oxide | NiOOH | | 91.7 | 7.4 | 1 | 0.29 | 3.42 | 2.16 |
| Nickel sulfide | Ni$_3$S$_2$ | | 240 | — | 4 | 0.47 | 2.12 | — |
| Silver chloride | AgCl | | 143.3 | 5.56 | 1 | 0.19 | 5.26 | 1.04 |
| Silver chromate | Ag$_2$CrO$_4$ | | 331.8 | 5.6 | 2 | 0.16 | 6.25 | 0.90 |
| Silver oxide (monovalent) | Ag$_2$O | | 231.8 | 7.1 | 2 | 0.23 | 4.33 | 1.64 |
| Silver oxide (divalent) | AgO | | 123.9 | 7.4 | 2 | 0.43 | 2.31 | 3.20 |
| Sulfur dioxide | SO$_2$ | | 64 | 1.37 | 1 | 0.419 | 2.39 | — |
| Sulfuryl chloride | SO$_2$Cl$_2$ | | 135 | 1.66 | 2 | 0.397 | 2.52 | — |
| Thionyl chloride | SOCl$_2$ | | 119 | 1.63 | 2 | 0.450 | 2.22 | — |
| Vanadium pentoxide | V$_2$O$_5$ | | 181.9 | 3.6 | 1 | 0.15 | 6.66 | 0.53 |

# APPENDIX D
# STANDARD SYMBOLS AND CONSTANTS

**TABLE D.1**  SI Base Units

| Quantity | Unit | Symbol |
|---|---|---|
| Length | Meter | m |
| Mass | Kilogram | kg |
| Time | Second | s |
| Electric current | Ampere | A |
| Thermodynamic temperature* | Kelvin | K |
| Amount of substance | Mole | mol |
| Luminous intensity | Candela | cd |

*Celsius temperature is, in general, expressed in degrees Celsius (symbol °C).

*Source:* From D. G. Fink and W. Beaty (eds.), *Standard Handbook for Engineers,* 12th ed., McGraw-Hill, NY, 1987; reproduced from IEEE Standard 268–1982, by permission.

**TABLE D.2**  SI Prefixes Expressing Decimal Factors

| Factor | Prefix | Symbol | Factor | Prefix | Symbol |
|---|---|---|---|---|---|
| $10^{18}$ | Exa | E | $10^{-1}$ | Deci | d |
| $10^{15}$ | Peta | P | $10^{-2}$ | Centi | c |
| $10^{12}$ | Tera | T | $10^{-3}$ | Milli | m |
| $10^{9}$ | Giga | G | $10^{-6}$ | Micro | $\mu$ |
| $10^{6}$ | Mega | M | $10^{-9}$ | Nano | n |
| $10^{3}$ | Kilo | k | $10^{-12}$ | Pico | p |
| $10^{2}$ | Hecto | h | $10^{-15}$ | Femto | f |
| $10^{1}$ | Deka | da | $10^{-18}$ | Atto | a |

*Source:* From D. G. Fink and W. Beaty (eds.), *Standard Handbook for Engineers,* 12th ed., McGraw-Hill, NY, 1987; adapted from IEEE Standard 268–1982, by permission.

**TABLE D.3**  Greek Alphabet

| Greek letter | | Greek name | English equivalent | Greek letter | | Greek name | English equivalent |
|---|---|---|---|---|---|---|---|
| A | $\alpha$ | Alpha | A | N | $\nu$ | Nu | N |
| B | $\beta$ | Beta | B | $\Xi$ | $\xi$ | Xi | X |
| $\Gamma$ | $\gamma$ | Gamma | G | O | o | Omicron | $\breve{\text{O}}$ |
| $\Delta$ | $\delta$ | Delta | D | $\Pi$ | $\pi$ | Pi | P |
| E | $\varepsilon$ | Epsilon | $\breve{\text{E}}$ | P | $\rho$ | Rho | R |
| Z | $\xi$ | Zeta | Z | $\Sigma$ | $\sigma$ | Sigma | S |
| H | $\eta$ | Eta | $\bar{\text{E}}$ | T | $\tau$ | Tau | T |
| $\Theta$ | $\theta$ | Theta | Th | Y | $\upsilon$ | Upsilon | U |
| I | $\iota$ | Iota | I | $\Phi$ | $\phi$ | Phi | Ph |
| K | $\kappa$ | Kappa | K | X | $\chi$ | Chi | Ch |
| $\Lambda$ | $\lambda$ | Lambda | L | $\Psi$ | $\psi$ | Psi | Ps |
| M | $\mu$ | Mu | M | $\Omega$ | $\omega$ | Omega | $\bar{\text{O}}$ |

**TABLE D.4**  Standard Symbols for Units

| Unit | Symbol | Notes |
|---|---|---|
| Ampere | A | SI unit of electric current |
| Ampere-hour | Ah | |
| Angstrom | Å | $1\ \text{Å} = 10^{-10}\ \text{m}$ |
| Atmosphere, standard | atm | $1\ \text{atm} = 101{,}325\ \text{N/m}^2$ or Pa |
| Atmosphere, technical | at | $\text{at} = \text{kg}_f/\text{cm}^2$ |
| Atomic mass unit (unified) | u | The (unified) atomic mass unit is defined as one-twelfth of the mass of an atom of the $^{12}C$ nuclide. Use of the old atomic mass unit (amu), defined by reference to oxygen, is deprecated |
| Atto | a | SI prefix for $10^{-18}$ |
| Bar | bar | $1\ \text{bar} = 100{,}000\ \text{N/m}^2$ |
| Barn | b | $1\ \text{b} = 10^{-28}\ \text{m}^2$ |
| Barrel | bbl | $1\ \text{bbl} = 9702\ \text{in}^3 = 0.15899\ \text{m}^3$ This is the standard barrel used for petroleum, etc. A different standard barrel is used for fruits, vegetables, and dry commodities |
| British thermal unit | Btu | |
| Calorie (International Table calorie) | $\text{cal}_{rr}$ | $1\ \text{cal}_{rr} = 4.1868\ \text{J}$ The 9th Conférence Générale des Poids et Mesures adopted the joule as the unit of heat. Use of the joule is preferred |
| Calorie (thermochemical calorie) | cal | $1\ \text{cal} = 4.1840\ \text{J}$ (see note for International Table calorie) |
| Centi | c | SI prefix for $10^{-2}$ |
| Centimeter | cm | |
| Coulomb | C | SI unit of electric charge |
| Cubic centimeter | $\text{cm}^3$ | |
| Cycle | c | |
| Cycle per second | Hz, c/s | See hertz. The name "hertz" is internationally accepted for this unit; the symbol Hz is preferred to c/s |
| Day | d | |
| Deci | d | SI prefix for $10^{-1}$ |
| Decibel | dB | |
| Degree (temperature): | | |
| Degree Celsius | °C | Note that there is no space between the symbol ° and the letter. The use of the word *centigrade* for the Celsius temperature scale was abandoned by the Conférence Générale des Poids et Mesures in 1948 |
| Degree Fahrenheit | °F | |
| Kelvin | K | See Kelvin |
| Degree Rankine | °R | |

**TABLE D.4**  Standard Symbols for Units (*Continued*)

| Unit | Symbol | Notes |
|---|---|---|
| Deka | da | SI prefix for 10 |
| Dyne | dyn | |
| Electron | e | This symbol is used in this Handbook to denote an electron. More conventionally shown as e⁻ |
| Electronvolt | eV | |
| Erg | erg | |
| Farad | F | SI unit of capacitance |
| Femto | f | SI prefix for $10^{-15}$ |
| Gauss | G | The gauss is the electromagnetic CGS unit of magnetic flux density. Use of SI unit, the tesla, is preferred |
| Giga | G | SI prefix for $10^9$ |
| Gilbert | Gb | The gilbert is the electromagnetic CGS unit of magnetomotive force. Use of the SI unit, the ampere (or ampere turn), is preferred |
| Gram | g | |
| Gram per cubic centimeter | g/cm³ | |
| Hecto | h | SI prefix for $10^2$ |
| Henry | H | SI unit of inductance |
| Hertz | Hz | SI unit of frequency |
| Hour | h | |
| Joule | J | SI unit of energy |
| Joule per kelvin | J/K | SI unit of heat capacity and entropy |
| Kelvin | K | In 1967 the CGPM gave the name "kelvin" to the SI unit of temperature, which had formerly been called "degree Kelvin" and assigned it the symbol K (without the symbol °) |
| Kilo | k | SI prefix for $10^3$ |
| Kilogram | kg | SI unit of mass |
| Kilogram-force | kg_f | In some countries the name *kilopond* (kp) has been adopted for this unit |
| Kilohm | kΩ | |
| Kilometer | km | |
| Kilometer per hour | km/h | |
| Kilovolt | kV | |
| Kilowatt | kW | |
| Kilowatt-hour | kWh | |
| Liter | L | $1 \text{ L} = 10^{-3} \text{ m}^3$ |
| Liter per second | L/s | |
| Lumen | lm | SI unit of luminous flux |
| | | SI unit of illuminance |
| Maxwell | Mx | The maxwell is the electromagnetic cgs unit of magnetic flux. Use of the SI unit, the weber, is preferred |
| Mega | M | SI prefix for $10^6$ |
| Megohm | MΩ | |
| Meter | m | SI unit of length |
| Mho | mho | CGPM has adopted the name "siemens" (S) for this unit |
| Micro | μ | SI prefix for $10^{-6}$ |
| Microampere | μA | |
| Microgram | μg | |
| Micrometer | μm | |
| Micron | μm | See micrometer. The name "micron" was abrogated by the Conférence Générale des Poids et Mesures, 1967 |
| Microsecond | μs | |
| Microwatt | μW | |

(*Continued*)

**TABLE D.4** Standard Symbols for Units (*Continued*)

| Unit | Symbol | Notes |
|---|---|---|
| Milli | m | SI prefix for $10^{-3}$ |
| Milliampere | mA | |
| Milligram | mg | |
| Milliliter | ml | |
| Millimeter | mm | |
| Conventional millimeter of mercury | mmHg | 1 mmHg = 133:322 $N/m^2$ |
| Millimicron | nm | Use of the name "millimicron" for the nanometer is deprecated |
| Millisecond | ms | |
| Millivolt | mV | |
| Milliwatt | mW | |
| Minute (time) | min | Time may also be designated by means of superscripts as in the following example: $9^h46^m20^s$ |
| Mole | mol | SI unit of amount of substance |
| Nano | n | SI prefix for $10^{-9}$ |
| Nanoampere | nA | |
| Nanometer | nm | |
| Nanosecond | ns | |
| Newton | N | SI unit of force |
| Newton meter | $N \cdot m$ | |
| Newton per square meter | $N/m^2$ | SI unit of pressure or stress; see pascal |
| Newton second per square meter | $N \cdot s/m^2$ | SI unit of dynamic viscosity |
| Oersted | Oe | The oersted is the electromagnetic cgs unit of magnetic field strength. Use of the SI unit, the ampere per meter, is preferred |
| Ohm | $\Omega$ | SI unit of resistance |
| Pascal | Pa | $Pa = N/m^2$ |
| | | SI unit of pressure or stress. This name accepted by the 14th Conférence Générale des Poids et Mesures |
| Pico | P | SI prefix for $10^{-12}$ |
| Picowatt | pW | |
| Revolution per second | r/s | |
| Second (time) | s | SI unit of time |
| Siemens | S | $S = \Omega^{-1}$ |
| | | SI unit of conductance. This name and symbol were adopted by the 14th Conférence Générale des Poids et Mesures. The name "mho" is also used for this unit in the United States |
| Square meter | $m^2$ | |
| Tera | T | SI prefix for $10^{12}$ |
| Tesla | T | SI unit of magnetic flux density |
| Tonne | T | 1 t = 1000 kg (in USA: ton, metric) |
| (Unified) atomic mass unit | u | The (unified) atomic mass unit is defined as one-twelfth of the mass of an atom of the $^{12}C$ nuclide. Use of the old atomic mass unit (amu), defined by reference to oxygen, is deprecated |
| Volt | V | SI unit of voltage |
| Volt per meter | V/m | SI unit of electric field strength |
| Voltampere | VA | IEC name and symbol for the SI unit of apparent power |
| Watt | W | SI unit of power |
| Watt per meter kelvin | $W/(m \cdot K)$ | SI unit of thermal conductivity |
| Watt-hour | Wh | |

***Source:*** From D. G. Fink and W. Beaty (eds.), *Standard Handbook for Engineers*, 12th ed., McGraw-Hill, NY, 1987; adapted from ANSI/IEEE Standard 260–1982.

# APPENDIX E
# CONVERSION FACTORS

**TABLE E.1** Length Conversion Factors*

### A. Length units decimally related to 1 meter

| | Meters (m) | Kilometers (km) | Decimeters (dm) | Centimeters (cm) | Millimeters (mm) | Micrometers (μm) | Nanometers (nm) | Angströms (Å) |
|---|---|---|---|---|---|---|---|---|
| 1 meter = | 1 | 0.001 | 10 | 100 | 1000 | 1,000,000 | $10^9$ | $10^{10}$ |
| 1 kilometer = | 1000 | 1 | 10,000 | 100,000 | 1,000,000 | $10^9$ | $10^{12}$ | $10^{13}$ |
| 1 decimeter = | 0.1 | 0.0001 | 1 | 10 | 100 | 100,000 | $10^8$ | $10^9$ |
| 1 centimeter = | 0.01 | 0.00001 | 0.1 | 1 | 10 | 10,000 | $10^7$ | $10^8$ |
| 1 millimeter = | 0.001 | $10^{-6}$ | 0.01 | 0.1 | 1 | 1000 | 1,000,000 | $10^7$ |
| 1 micrometer (micron) = | $10^{-6}$ | $10^{-9}$ | 0.00001 | 0.0001 | 0.001 | 1 | 1000 | 10,000 |
| 1 nanometer = | $10^{-9}$ | $10^{-12}$ | $10^{-8}$ | $10^{-7}$ | $10^{-6}$ | 0.001 | 1 | 10 |
| 1 angstrom = | $10^{-10}$ | $10^{-13}$ | $10^{-9}$ | $10^{-8}$ | $10^{-7}$ | 0.0001 | 0.1 | 1 |

### B. Nonmetric length units less than 1 meter

| | Meters (m) | Yards (yd) | Feet (ft) | Inches (in) | Mils (mil) | Microinches (μin) |
|---|---|---|---|---|---|---|
| 1 meter = | 1 | 1.09361330 | 3.28083939 | 39.3700787 | $3.93700787 \times 10^4$ | $3.93700787 \times 10^7$ |
| 1 yard = | 0.9144 | 1 | 3 | 36 | 36,000 | $3.6 \times 10^7$ |
| 1 foot = | 03048 | 1/3 = 0.3333 | 1 | 12 | 12,000 | $1.2 \times 10^7$ |
| 1 inch = | 0.0254 | 1/36 = 0.0277 | 1/12 = 0.0833 | 1 | 1000 | 1,000,000 |
| 1 mil = | $2.54 \times 10^{-5}$ | $2.777 \times 10^{-5}$ | $8.333 \times 10^{-5}$ | 0.001 | 1 | 1000 |
| 1 microinch = | $2.54 \times 10^{-8}$ | $2.777 \times 10^{-8}$ | $8.333 \times 10^{-8}$ | $10^{-6}$ | 0.001 | 1 |

### C. Nonmetric length units greater than 1 meter (with equivalents in feet)

| | Meters (m) | Rods (rd) | Statute miles (mi) | Nautical miles (nmi) | Astronomical units (AU) | Parsecs (pe) | Feet (ft) |
|---|---|---|---|---|---|---|---|
| 1 meter = | 1 | 0.19883378 | $6.21371192 \times 10^{-4}$ | $5.39956904 \times 10^{-4}$ | $6.68449198 \times 10^{-12}$ | $3.24073317 \times 10^{-17}$ | 3.28083989 |
| 1 rod = | 5.0292 | 1 | 0.003125 | $2.71555076 \times 10^{-3}$ | $3.36176471 \times 10^{-11}$ | $1.62982953 \times 10^{-16}$ | 16.5 |
| 1 statute mile = | 1609.344 | 320 | 1 | 0.86897624 | $1.07576471 \times 10^{-8}$ | $5.21545450 \times 10^{-14}$ | 5280 |
| 1 nautical mile = | 1852 | 368.249423 | 1.15077945 | 1 | $1.23796791 \times 10^{-8}$ | $6.00183780 \times 10^{-14}$ | 6076.11548 |
| 1 astronomical unit† = | $1.496 \times 10^{11}$ | $2.97462817 \times 10^{10}$ | 92,957,130.3 | 80,777,537.8 | 1 | $4.84813682 \times 10^{-6}$ | $4.90813648 \times 10^{11}$ |
| 1 parsec = | $3.08572150 \times 10^{16}$ | $6.13561102 \times 10^{15}$ | $1.91737844 \times 10^{13}$ | $1.66615632 \times 10^{13}$ | 206,264,806 | 1 | $1.01237582 \times 10^{17}$ |
| 1 foot = | 0.3048 | 0.060606 | $1.893939 \times 10^{-4}$ | $1.64578833 \times 10^{-4}$ | $2.03743316 \times 10^{-12}$ | $9.87775472 \times 10^{-18}$ | 1 |

## D. Other length units

| | | |
|---|---|---|
| 1 cable = **720** feet = **219.456** meters | 1 hand = **4** inches = **0.1016** meter | 1 millimicron = **1** nanometer = **10⁻⁹** meter |
| 1 cable (U.K.) = **608** feet = **185.3184** meters | 1 league (international nautical) = 3 nautical miles = **5556** meters | 1 myriameter = **10,000** meters |
| 1 chain (engineers') = **100** feet = **30.48** meters | 1 league (statute) = 3 statute miles = **4828.032** meters | 1 nautical mile (U.K.) = 1853.184 meters |
| 1 chain (surveyors') = **66** feet = **20.1168** meters | 1 league (U.K. nautical) = 5559.552 meters | 1 pale = **1** rod = **5.0292** meters |
| 1 fathom = **6** feet = **1.8288** meters | 1 light-year = 9.4608952 × 10¹⁵ meters (= distance traveled by light in vacuum in one sidereal year) | 1 perch (linear) = **1** rod = **5.0292** meters |
| 1 fermi = **1** femtometer = **10⁻¹⁵** meter | 1 link (engineers') = **1** foot = **0.3048** meter | 1 pica = **1/6** inch (approx.) = 4.217518 × 10⁻³ meter |
| 1 foot (U.S. Survey) = 0.3048006 meter | 1 link (surveyors') = **7.92** inches = **0.201168** meter | 1 point = **1/72** inch (approx.) = 3.514598 × 10⁻⁴ meter |
| 1 furlong = **660** feet = **201.168** meters | 1 micron = **1** micrometer = **10⁻⁶** meter | 1 span = **9** inches = **0.2286** meter |

*Exact conversions are shown in boldface type. Repeating decimals are <u>underlined</u>. The SI unit of length is the meter.
†As defined by the International Astronomical Union, 1964.
**Source:** D. G. Fink and W. Beaty (eds.), *Standard Handbook for Electrical Engineers* (12th ed.), McGraw-Hill, New York, 1987.

**TABLE E.2**   Area Conversion Factors[*]

### A. Area units decimally related to 1 square meter

| | Square meters (m²) | Square kilometers (km²) | Hectares (square hectometers) (hm²) | Square centimeters (cm²) | Square millimeters (mm²) | Square micrometers (µm²) | Bams (b) |
|---|---|---|---|---|---|---|---|
| 1 square meter = | **1** | $10^{-6}$ | **0.001** | **10,000** | **1,000,000** | $10^{12}$ | $10^{28}$ |
| 1 square kilometer = | **1,000,000** | **1** | **100** | $10^{10}$ | $10^{12}$ | $10^{18}$ | $10^{34}$ |
| 1 hectare = | **10,000** | **0.01** | **1** | $10^{8}$ | $10^{10}$ | $10^{16}$ | $10^{32}$ |
| 1 square centimeter = | **0.001** | $10^{-10}$ | $10^{-8}$ | **1** | **100** | $10^{8}$ | $10^{24}$ |
| 1 square millimeter = | $10^{-4}$ | $10^{-12}$ | $10^{-10}$ | **0.01** | **1** | $10^{6}$ | $10^{22}$ |
| 1 square micrometer = | $10^{-12}$ | $10^{-18}$ | $10^{-16}$ | $10^{-8}$ | $10^{-6}$ | **1** | $10^{16}$ |
| 1 barn = | $10^{-28}$ | $10^{-34}$ | $10^{-32}$ | $10^{-24}$ | $10^{-22}$ | $10^{-16}$ | **1** |

### B. Nonmetric area units (with square meter equivalents)

| | Square meters (m²) | Square statute miles (mi²) | Acres (acre) | Square rods (rd²) | Square yard (yd²) | Square feet (ft²) | Square inches (in²) | Circular mils (cmil) |
|---|---|---|---|---|---|---|---|---|
| 1 square meter = | **1** | $3.86102159 \times 10^{-7}$ | $2.47105382 \times 10^{-4}$ | $3.95368610 \times 10^{-2}$ | 1.19599005 | 10.7639104 | 1550.00310 | $1.97342524 \times 10^{9}$ |
| 1 square statute mile = | 2,589,988.1 | **1** | **640** | **102,400** | **3,097,600** | **27,878,400** | $4.01448960 \times 10^{9}$ | $5.11140691 \times 10^{15}$ |
| 1 acre = | 4046.85641 | **1/640** − 0.0015625 | **1** | **160** | **4840** | **43,560** | **6,272,640** | $7.98657330 \times 10^{12}$ |
| 1 square rod = | 25.2928526 | $9.765625 \times 10^{-6}$ | **1/160** − 0.00625 | **1** | **30.25** | **272.25** | **39,204** | $4.99160831 \times 10^{10}$ |
| 1 square yard = | **0.83612736** | $3.22830579 \times 10^{-7}$ | $2.06611570 \times 10^{-4}$ | $3.30578512 \times 10^{-2}$ | **1** | **9** | **1296** | $1.65011845 \times 10^{9}$ |
| 1 square foot = | **0.09290304** | $3.58700643 \times 10^{-8}$ | $2.29568411 \times 10^{-5}$ | $3.67309458 \times 10^{-3}$ | **1/9** = 0.11111$\underline{1}$ | **1** | **144** | $1.83346495 \times 10^{8}$ |
| 1 square inch = | $6.4516 \times 10^{-4}$ | $2.49097669 \times 10^{-10}$ | $1.59422508 \times 10^{-7}$ | $2.55076013 \times 10^{-5}$ | $7.71604938 \times 10^{-4}$ | **1/144** = 0.0069444$\underline{4}$ | **1** | $1.27323955 \times 10^{6}$ |
| 1 circular mil = | $5.06707479 \times 10^{-10}$ | $1.95640851 \times 10^{-16}$ | $1.25210145 \times 10^{-13}$ | $2.00336232 \times 10^{-11}$ | $6.06017101 \times 10^{-10}$ | $5.45415391 \times 10^{-9}$ | $7.85398163 \times 10^{-7}$ | **1** |

Exact conversions are

1 acre = **4046.8564224** square meters

1 square mile = **2,589,988.110336** square meters

### C. Other area units

1 are = 100 square meters

1 centiare (centare) = 1 square meter

1 perch (area) = **1** square rod = **30.25** square yards = 25.2928526 square meters

1 rood = **40** square rods = 1011.71411 square meters

1 section = **1** square statute mile = 2,589,988.1 square meters

1 township = **36** square statute miles = 93,239,572 square meters

[*]Exact conversions are shown in boldface type. Repeating decimals are underlined. The SI unit of area is the square meter.
**Source:** D. G. Fink and W. Beaty (eds.), *Standard Handbook for Electrical Engineers* (12th ed.), McGraw-Hill, New York, 1987.

**TABLE E.3** Force Conversion Factors[a]

| | Newtons (N) | Kips (kip) | Slugs-force (slug) | Kilograms-force (kg) | Avoirdupois pounds-free (lb$_f$ avdp) | Avoirdupois ounces-force (oz$_f$ advp) | Poundals (pdl) | Dynes (dyn) |
|---|---|---|---|---|---|---|---|---|
| 1 newton = | **1** | $2.24808943 \times 10^{-4}$ | $6.98727524 \times 10^{-3}$ | 0.10197162 | 0.22480894 | 3.59694309 | 7.2330142 | **100,000** |
| 1 kip = | 4448.22162 | **1** | 31.080949 | 453.592370 | **1000** | **16,000** | 32,174.05 | 444,822,162 |
| 1 slug-force = | 143.117305 | 0.03217405 | **1** | 14.593903 | 32.17405 | 514.78480 | 1035.1695 | 14,311,730 |
| 1 kilogram-force = | **9.806650** | $2.20462262 \times 10^{-3}$ | $6.8521763 \times 10^{-2}$ | **1** | 2.20462262 | 35.2739619 | 70.9316384 | **980,665** |
| 1 avdp pound-force = | 4.44822162 | 0.001 | $3.10809488 \times 10^{-2}$ | 0.45359237 | **1** | **16** | 32.17405 | 444,822.162 |
| 1 avdp ounce-force = | 0.27801385 | **1/16,000 = 0.0000625** | $1.94255930 \times 10^{-3}$ | $2.834952 \times 10^{-2}$ | **1/16 = 0.0625** | **1** | 2.01087803 | 27,801.385 |
| 1 poundal = | 0.13825495 | $3.1080949 \times 10^{-5}$ | $9.6602539 \times 10^{-4}$ | 0.14098081 | 0.03108095 | 0.49729518 | **1** | 13,825.495 |
| 1 dyne = | **0.00001** | $2.24808943 \times 10^{-8}$ | $6.98727524 \times 10^{-8}$ | $1.01971621 \times 10^{-6}$ | $2.24808943 \times 10^{-6}$ | $3.59694310 \times 10^{-5}$ | $7.2330142 \times 10^{-5}$ | **1** |

The exact conversion is: 1 avdp pound-force = **4.448221652605** newtons

[a]Exact conversions are shown in boldface type. The SI unit of force is the newton (N).

**Source:** D. G. Fink and W. Beaty (eds.), *Standard Handbook for Electrical Engineers* (12th ed.), McGraw-Hill, New York, 1987.

**TABLE E.4**  Volume and Capacity Conversion Factors

### A. Volume units decimally related to one cubic meter

| | Cubic meters (steres) (m³) | Cubic decimeters (dm³) | Cubic centimeters (cm³) | Liters (L) | Centiliters (cL) | Milliliters (mL) | Microliters (µL) |
|---|---|---|---|---|---|---|---|
| Cubic meter = | 1 | 1000 | 1,000,000 | 1000 | 100,000 | 1,000,000 | $10^9$ |
| Cubic decimeter = | 0.001 | 1 | 1000 | 1 | 100 | 1000 | 1,000,000 |
| Cubic centimeter = | 0.000001 | 0.001 | 1 | 0.001 | 0.1 | 1 | 1000 |
| Liter = | 0.001 | 1 | 1000 | 1 | 100 | 1000 | 1,000,000 |
| Centiliter = | 0.00001 | 0.01 | 10 | 0.01 | 1 | 10 | 10,000 |
| Milliliter = | 0.000001 | 0.001 | 1 | 0.001 | 0.1 | 1 | 1000 |
| Microliter = | $10^{-9}$ | 0.000001 | 0.001 | 0.000001 | 0.0001 | 0.001 | 1 |

### B. Nonmetric volume units (with cubic meter and liter equivalents)

| | Cubic meters (steres) (m³) | Liters (L) | Cubic inches (in³) | Cubic feet (ft³) | Cubic yards (yd³) | Barrels (U.S.) (bbl) | Acre-feet (acre-ft) | Cubic miles (mi³) |
|---|---|---|---|---|---|---|---|---|
| 1 cubic meter = | 1 | 1000 | $6.10237441 \times 10^4$ | 35.314666 | 1.30795062 | 6.2898109 | $8.10713194 \times 10^{-4}$ | $2.39912759 \times 10^{-10}$ |
| 1 liter = | 0.001 | 1 | 61.10237441 | 0.03531466 | $1.30795062 \times 10^{-3}$ | $6.28981097 \times 10^{-3}$ | $8.10713193 \times 10^{-7}$ | $2.39912759 \times 10^{-13}$ |
| 1 cubic inch = | $1.6387064 \times 10^{-5}$ | $1.6387064 \times 10^{-2}$ | 1 | $1/1728 = 5.78703703 \times 10^{-4}$ | $1/46656 = 2.14334705 \times 10^{-5}$ | $1.03071532 \times 10^{-4}$ | $1.32852090 \times 10^{-8}$ | $3.93146573 \times 10^{-15}$ |
| 1 cubic foot = | $2.83168466 \times 10^{-2}$ | 28.3168466 | 1728 | 1 | $1/27 = 0.037037$ | 0.17810761 | $1/43560 = 2.29568411 \times 10^{-5}$ | $6.79357278 \times 10^{-12}$ |
| 1 cubic yard = | 0.76455486 | 764.554858 | 46,656 | 27 | 1 | 4.80890538 | $6.19834711 \times 10^{-4}$ | $1.83426465 \times 10^{-10}$ |
| 1 barrel (U.S.) = | 0.15898729 | 158.987294 | 9702 | 5.61458333 | 0.20794753 | 1 | $1.28893098 \times 10^{-4}$ | $3.81430805 \times 10^{-11}$ |
| 1 acre-foot = | 1233.48184 | $1.23348184 \times 10^6$ | $7.52716800 \times 10^7$ | 43,560 | 1613.33333 | 7758.36734 | 1 | $2.95928030 \times 10^{-7}$ |
| 1 cubic mile = | $4.1681183 \times 10^9$ | $4.1681183 \times 10^{12}$ | $2.54358061 \times 10^{14}$ | $1.47197952 \times 10^{11}$ | $5.451776 \times 10^9$ | $26.2170749 \times 10^9$ | 3,379,200 | 1 |

Exact conversion: 1 cubic foot = **28.316846592 liters**

### C. U.S. liquid capacity measures (with liter equivalents)

| | Liters (L) | Gallons (U.S. gal) | Quarts (U.S. qt) | Pints (U.S. pt) | Gills (U.S. gi) | Fluid ounces (U.S. floz) | Fluidrams (U.S. fldr) | Minims (U.S. minim) |
|---|---|---|---|---|---|---|---|---|
| 1 liter = | 1 | 0.26417205 | 1.056688 | 2.113376 | 8.453506 | 33.814023 | 270.51218 | 16,230.73 |
| 1 gallon, U.S. = | 3.7854118 | 1 | 4 | 8 | 32 | 128 | 1024 | 61,440 |
| 1 quart, U.S. = | 0.9463529 | $1/4 = 0.25$ | 1 | 2 | 8 | 32 | 256 | 15,360 |
| 1 pint, U.S. = | 0.4731765 | $1/8 = 0.125$ | $1/2 = 0.5$ | 1 | 4 | 16 | 128 | 7680 |
| 1 gill, U.S. = | 0.1182941 | $1/32 = 0.03125$ | $1/8 = 0.125$ | $1/4 = 0.25$ | 1 | 4 | 32 | 1920 |
| 1 fluid ounce, U.S. = | $2.957353 \times 10^{-2}$ | $1/128 = 0.0078125$ | $1/32 = 0.03125$ | $1/16 = 0.0625$ | $1/4 = 0.25$ | 1 | 8 | 480 |
| 1 fluidram, U.S. = | $3.6966912 \times 10^{-3}$ | $1/1024 = 9.765625 \times 10^{-4}$ | $1/256 = 3.90625 \times 10^{-3}$ | $1/128 = 0.0078125$ | $1/32 = 0.03125$ | $1/8 = 0.125$ | 1 | 60 |
| 1 minim, U.S. = | $6.161152 \times 10^{-5}$ | $1/61440 = 1.62760416 \times 10^{-5}$ | $1/15360 = 6.51041666 \times 10^{-5}$ | $1/7680 = 1.30208333 \times 10^{-4}$ | $1/19200 = 5.2083333 \times 10^{-4}$ | $1/480 = 2.0833333 \times 10^{-3}$ | $1/60 = 0.0166666$ | 1 |

Exact conversion: 1 liquid quart, U.S. = **0.946352946 liter**

D. British Imperial liquid capacity measures (with liter equivalents)

| | Liters (L) | Gallons (U.K.) | Quarts (U.K. qt) | Pints (U.K. pt) | Gills (U.K. gi) | Fluid ounces (U.K. floz) | Fluidrams (U.K. fldr) | Minims (U.K. minim) |
|---|---|---|---|---|---|---|---|---|
| 1 liter = | 1 | 0.2199692 | 0.8798766 | 1.759753 | 7.039018 | 35.19506 | 281.5605 | 16,893.63 |
| 1 gallon, U.K. = | 4.546092 | **1** | **4** | **8** | **32** | **160** | **1280** | **76,800** |
| 1 quart, U.K. = | 1.136523 | **1/4 = 0.25** | **1** | **2** | **8** | **40** | **320** | **19,200** |
| 1 pint, U.K. = | 0.5682615 | **1/8 = 0.125** | **1/2 = 0.5** | **1** | **4** | **20** | **160** | **9600** |
| 1 gill, U.K. = | 0.1420654 | **1/32 = 0.03125** | **1/8 = 0.125** | **1/4 = 0.25** | **1** | **5** | **40** | **2400** |
| 1 fluid ounce, U.K. = | $2.841307 \times 10^{-2}$ | **1/160 = 0.00625** | **1/40 = 0.025** | **1/20 = 0.05** | **1/5 = 0.2** | **1** | **8** | **480** |
| 1 fluidram, U.K. = | $3.551634 \times 10^{-3}$ | **1/1280 = $7.8125 \times 10^{-4}$** | **1/320 = 0.003125** | **1/160 = 0.00625** | **1/40 = 0.025** | **1/8 = 0.125** | **1** | **60** |
| 1 minima, U.K. = | $5.919391 \times 10^{-5}$ | 1/76800 = $1.30209333 \times 10^{-5}$ | 1/19200 = $5.2083333 \times 10^{-5}$ | 1/9600 = $1.04166666 \times 10^{-4}$ | 1/2400 = $4.16666666 \times 10^{-4}$ | 1/480 = $2.08333333 \times 10^{-3}$ | 1/60 = 0.01666666 | **1** |

E. U.S. and British dry capacity measures (with liter equivalents)

| | | U.S. dry measures | | | | British dry measures | | | |
|---|---|---|---|---|---|---|---|---|---|
| | Liters (L) | Bushels (U.S. bu) | Pecks (U.S. peck) | Quarts (U.S. qt) | Pints (U.S. pt) | Bushels (U.K. bu) | Pecks (U.K. peck) | Quarts (U.K. qt) | Pints (U.K. pt) |
| 1 liter = | 1 | 0.02837759 | 0.11351037 | 0.90808299 | 1.81816598 | 0.0274961 | 0.1099846 | 0.8798766 | 1.7597534 |
| 1 bushel, U.S. = | 35.239070 | **1** | **4** | **32** | **64** | 0.9689387 | 3.8757549 | 31.00604 | 62.01208 |
| 1 peck, U.S. = | 8.8097675 | **1/4 = 0.25** | **1** | **8** | **16** | 0.2422347 | 0.9689387 | 7.751509 | 15.50302 |
| 1 quart, U.S. = | 1.1012209 | **1/32 = 0.03125** | **1/8 = 0.125** | **1** | **2** | 0.03027934 | 0.1211173 | 0.9689387 | 1.937878 |
| 1 pint, U.S. = | 0.5506105 | **1/64 = 0.015625** | **1/16 = 0.0625** | **1/2 = 0.5** | **1** | 0.01513967 | 0.06055867 | 0.4844693 | 0.9689387 |
| 1 bushel, U.K. = | 36.36873 | 1.032057 | 4.128228 | 33.02582 | 66.95165 | **1** | **4** | **32** | **64** |
| 1 peck, U.K. = | 9.092182 | 0.2580143 | 1.032057 | 8.256456 | 16.51291 | **1/4 = 0.25** | **1** | **8** | **16** |
| 1 quart, U.K. = | 1.136523 | 0.03225178 | 0.1290071 | 1.032057 | 2.0641142 | **1/32 = 0.03125** | **1/8 = 0.125** | **1** | **2** |
| 1 pint, U.K. = | 0.5682614 | 0.01612589 | 0.0645036 | 0.5160184 | 1.032057 | **1/64 = 0.015625** | **1/16 = 0.0625** | **1/2 = 0.5** | **1** |

Exact conversion: 1 dry pint, U.S. = **33.6003125** cubic inches

F. Other volume and capacity units

1 barrel, U.S. (used for petroleum, etc.) = **42** gallons = 0.1589987296 cubic meter
1 barrel (old barrel) = **31.5** gallons = 0.119240 cubic meter
1 board foot = **144** cubic inches = $2.359737 \times 10^{-3}$ cubic meter
1 cord = **128** cubic feet = 3.624556 cubic meters
1 cord foot = **16** cubic feet = 0.4530695 cubic meter
1 cup = **8** fluid ounces, U.S. = $2.365882 \times 10^{-4}$ cubic meter

1 gallon (Canadian liquid) = $4.546090 \times 10^{-3}$ cubic meter
1 perch (volume) = 24.75 cubic feet = 0.700842 cubic meter
1 stere = **1** cubic meter
1 tablespoon = **0.5** fluid ounce, U.S. = $1.478677 \times 10^{-5}$ cubic meter
1 teaspoon = **1/6** fluid ounce, U.S. = $4.928922 \times 10^{-6}$ cubic meter
1 ton (register ton) = **100** cubic feet = 2.83168466 cubic meters

*Exact conversions are shown in boldface type. Repeating decimals are underlined. The SI unit of volume is the cubic meter.
**Source:** D. G. Fink and W. Beaty (eds.), *Standard Handbook for Electrical Engineers* (12th ed.), McGraw-Hill, New York, 1987.

**TABLE E.5**    Mass Conversion Factors[*]

A. Mass units decimally

| | Kilograms (kg) | Tonnes (metric tons) (t) | Grams (g) |
|---|---|---|---|
| 1 kilogram = | **1** | 0.01 | 1000 |
| 1 tonne = | 1000 | **1** | 1,000,000 |
| 1 gram = | 0.001 | 0.000001 | **1** |
| 1 decigram = | 0.0001 | $10^{-7}$ | 0.1 |
| 1 centigram = | 0.00001 | $10^{-8}$ | 0.01 |
| 1 milligram = | 0.000001 | $10^{-9}$ | 0.001 |
| 1 microgram = | $10^{-9}$ | $10^{-12}$ | 0.000001 |

B. Nonmetric mass units less than one

| | Grams (g) | Avoirdupois ounces-mass ($oz_m$, avdp) | Troy ounces-mass ($oz_m$, troy) |
|---|---|---|---|
| 1 gram = | **1** | 0.035273962 | 0.032150747 |
| 1 avdp ounce-mass = | 38.3495231 | **1** | 0.91145833 |
| 1 troy ounce-mass = | 31.1031768 | 1.09714286 | **1** |
| 1 avdp dram = | 1.77184520 | **1/16 = 0.0625** | 0.056966.15 |
| 1 apothecary dram = | 3.88793458 | 0.137142857 | **1/8 = 0.125** |
| 1 pennyweight = | 1.55517383 | 0.054863162 | **1/20 = 0.05** |
| 1 grain = | **0.06479891** | $1/437.5 = 2.28571429 \times 10^{-3}$ | **1/480** = 0.002083<u>33</u> |
| 1 scruple = | 1.29597820 | $4.57142858 \times 10^{-2}$ | **1/24** = 0.04166<u>66</u> |

C. Nonmetric mass units of one pound-mass

| | Kilograms (kg) | Long tons (long ton) | Short tons (short ton) |
|---|---|---|---|
| 1 kilogram = | **1** | $9.842065 \times 10^{-4}$ | $1.10231131 \times 10^{-3}$ |
| 1 long ton = | 1016.0469 | **1** | **1.12** |
| 1 short ton = | **907.18474** | $200/224 = 0.89285714$ | **1** |
| 1 long hundredweight = | 50.802.3454 | **0.05** | **0.056** |
| 1 short hundredweight = | **45.359237** | $10/224 = 0.04464286$ | **0.05** |
| 1 slug = | 14.593903 | 0.01436341 | 0.01608702 |
| 1 avdp pound-mass = | **0.45359237** | $1/2240 = 4.46428571 \times 10^{-4}$ | **0.0005** |
| 1 troy pound-mass = | 0.37324172 | $3.67346937 \times 10^{-4}$ | $4.11428570 \times 10^{-4}$ |

Exact conversions: 1 long ton = **1016.0469088** kilograms

1 troy pound-mass = **0.3732417216** kilogram

D. Other

1 assay ton = 29.166667 grams
1 carat (metric) = **200** milligrams
1 carat (troy weight) = **31/6** grains = 205.19655 milligrams
1 mynagram = **10** kilograms
1 quintal = **100** kilograms
1 stone = **14** pounds, advp = **6.35029328** kilograms

[*]Exact conversions are shown in boldface type. Repeating decimals are <u>underlined</u>. The SI unit of mass is the kilogram.
**Source:** D. G. Fink and W. Beaty (eds.), *Standard Handbook for Electrical Engineers* (12th ed.), McGraw-Hill, New York, 1987.

related to kilogram

| Decigrams (dg) | Centigrams (cg) | Milligrams (mg) | Micrograms (μg) |
|---|---|---|---|
| 10,000 | 100,000 | 1,000,000 | $10^9$ |
| $10^7$ | $10^8$ | $10^9$ | $10^{12}$ |
| 10 | 100 | 1000 | 1,000,000 |
| 1 | 10 | 100 | 100,000 |
| 0.1 | 1 | 10 | 10,000 |
| 0.01 | 0.1 | 1 | 1000 |
| 0.00001 | 0.0001 | 0.001 | 1 |

pound-mass (with gram equivalents)

| Avoirdupois drams (dr avdp) | Apothecary drams (dr apoth) | Pennyweights (dwt) | Grains (grain) | Scruples (scruple) |
|---|---|---|---|---|
| 0.56438339 | 0.25720597 | 0.64301493 | 15.4323584 | 0.77161792 |
| **16** | 7.29166666 | 18.2271667 | 437.5 | 21.875 |
| 17.5542857 | **8** | **20** | 480 | 24 |
| **1** | 0.45572917 | 1.13932292 | 27.34375 | 1.3671875 |
| 2.19428570 | **1** | **2.5** | 60 | 3 |
| 0.87771428 | **1/2.5 = 0.4** | **1** | 24 | 1.2 |
| $\mathbf{3.65714285 \times 10^{-2}}$ | **1/60 = 0.010666666** | **1/24 = 0.04166666** | **1** | 0.05 |
| 0.73142857 | **1/3 = 0.33333333** | **5/6 = 0.83333333** | 20 | 1 |

and greater (with kilogram equivalents)

| Long hundredweights (long cwt) | Short hundredweights (short cwt) | Slugs (slug) | Avoirdupois pounds-mass (lb$_m$, avdp) | Troy pounds-mass (lb$_m$, troy) |
|---|---|---|---|---|
| $1.96841131 \times 10^{-2}$ | $2.20462262 \times 10^{-2}$ | 0.06852177 | 2.20462262 | 2.67922889 |
| **20** | **22.4** | 69.621329 | **2240** | 2722.22222 |
| **400/224 =** 17.8571429 | **20** | 62.161901 | **2000** | 2430.55555 |
| **1** | **1.12** | 3.4810664 | **112** | 136.111111 |
| **100//112 =** 0.89285714 | **1** | 3.1080950 | **100** | 121.527777 |
| 0.2872683 | 0.3217405 | **1** | 32.17405 | 39.100406 |
| **1/112 = $8.92857143 \times 10^{-3}$** | **0.01** | $3.1080950 \times 10^{-2}$ | **1** | 1.215277777 |
| $7.34693879 \times 10^{-3}$ | $8.22857145 \times 10^{-3}$ | 0.02557518 | 0.82285714 | **1** |

mass units

**TABLE E.6**   Pressure/Stress Conversion Factors

### A. Pressure units decimally related to one pascal

|  | Pascals (Pa) | Bars (bar) | Decibars (dbar) | Millibars (mbar) | Dynes per square centimeter (dyn/cm²) |
|---|---|---|---|---|---|
| 1 pascal = | 1 | 0.00001 | 0.0001 | 0.01 | 10 |
| 1 bar = | 100,000 | 1 | 10 | 1000 | 1,000,000 |
| 1 decibar = | 10,000 | 0.1 | 1 | 100 | 100,000 |
| 1 millibar = | 100 | 0.001 | 0.01 | 1 | 1000 |
| 1 dyne per second centimeter = | 0.1 | 0.000001 | 0.00001 | 0.001 | 1 |

### B. Pressure units decimally related to one kilogram-force per square meter (with pascal equivalents)

|  | Kilograms-force per square meter (kg/m²) | Kilograms-force per square centimeter (kg/cm²) | Kilograms-force per square millimeter (kg/mm²) | Grams-force per square centimeter (g/cm²) | Pascals |
|---|---|---|---|---|---|
| 1 kilogram-force per square meter = | 1 | 0.001 | 0.000001 | 0.1 | 9.80665 |
| 1 kilogram-force per square centimeter = | 10,000 | 1 | 0.01 | 1000 | 98,066.5 |
| 1 kilogram-force per square millimeter = | 1,000,000 | 100 | 1 | 100,000 | 9,806,650 |
| 1 gram-force per square centimeter = | 10 | 0.001 | 0.00001 | 1 | 98.0665 |
| 1 pascal = | 0.10197162 | $1.0197162 \times 10^{-5}$ | $1.0197162 \times 10^{-7}$ | $1.0197162 \times 10^{-2}$ | 1 |

NOTE: 1 atmosphere (technical) = 1 kilogram-force per square centimeter = 98,066.5 pascals.

### C. Pressure units expressed as heights of liquid (with pascal equivalents)

|  | Millimeters of mercury at 0°C (mmHg, 0°C) | Centimeters of mercury at 60°C (cmHg, 60°C) | Inches of mercury at 32°F (inHg, 32°F) | Inches of mercury at 60°F (inHg, 60°F) | Centimeters of water at 4°C (cmH₂O, 4°C) | Inches of water at 60°F (inH₂O, 60°F) | Feet of water at 39.2°F (ftH₂O, 39.2°F) | Pascals (Pa) |
|---|---|---|---|---|---|---|---|---|
| 1 millimeter of mercury, 0°C = | 1 | 0.100282 | 0.0393701 | 0.0394813 | 1.359548 | 0.5357756 | 0.0446046 | 133.3224 |
| 1 centimeter of mercury, 60°C = | 9.971830 | 1 | 0.3925919 | 0.3937008 | 13.55718 | 5.342664 | 0.4447895 | 1329.468 |
| 1 inch of mercury, 32°F = | 25.4 | 2.547175 | 1 | 1.0028248 | 34.53252 | 13.60870 | 1.132957 | 3386.389 |
| 1 inch of mercury, 60°F = | 25.32845 | 2.54 | 0.9971831 | 1 | 35.43525 | 13.57037 | 1.129765 | 3376.85 |
| 1 centimeter of water, 4°C = | 0.735539 | 0.073762 | 0.028958 | 0.0290400 | 1 | 0.3940838 | 0.0328084 | 98.0638 |
| 1 inch of water, 60°F = | 1.866453 | 0.187173 | 0.073482 | 0.0736900 | 2.537531 | 1 | 0.0832524 | 248.840 |
| 1 foot of water, 39.2°F = | 22.4192 | 2.248254 | 0.882646 | 0.885139 | 30.47998 | 12.01167 | 1 | 2988.98 |
| 1 pascal = | $7.500615 \times 10^{-3}$ | $7.521806 \times 10^{-4}$ | $2.952998 \times 10^{-4}$ | $2.96134 \times 10^{-4}$ | $1.01974 \times 10^{-2}$ | $4.01865 \times 10^{-3}$ | $3.34562 \times 10^{-4}$ | 1 |

## D. Nonmetric pressure units (with pascal equivalents)

| | Atmospheres (atm) | Avoirdupois pounds-force per square inch (psi) | Avoirdupois pounds-force per square foot (lb/ft², avdp) | Poundals per square foot (pd/ft²) | Pascals (Pa) |
|---|---|---|---|---|---|
| 1 atmosphere = | **1** | 14.69595 | 2116.217 | 68,087.24 | **101,325** |
| 1 avdp pound-force per square inch = | $6.80460 \times 10^{-2}$ | **1** | **144** | 4633.063 | 6894.757 |
| 1 avdp pound-force per square foot = | $4.725414 \times 10^{-4}$ | **1/144** = 0.006944 | **1** | 32.17405 | 47.88026 |
| 1 poundal per square foot = | $1.468704 \times 10^{-3}$ | $2.158399 \times 10^{-4}$ | 0.0310809 | **1** | 1.488165 |
| 1 pascal = | $9.869233 \times 10^{-6}$ | $1.450377 \times 10^{-4}$ | 0.0208854 | 0.6719689 | **1** |

NOTE: 1 normal atmosphere = 760 torr = **101,325** pascals.

*Exact conversions are shown in boldface type. Repeating decimals are underlined. The SI unit of pressure or stress is the Pascal (Pa).
**Source:** D. G. Fink and W. Beaty (eds.), *Standard Handbook for Electrical Engineers* (12th ed.), McGraw-Hill, New York, 1987.

**TABLE E.7**  Energy/Work Conversion Factors*

### A. Energy/work units decimally related to 1 joule

| | Joules (J) | Megajoules (MJ) | Kilojoules (kJ) | Millijoules (mJ) | Microjoules (µJ) | Ergs (erg) |
|---|---|---|---|---|---|---|
| 1 joule = | **1** | **0.000001** | **0.001** | **1000** | **1,000,000** | **$10^7$** |
| 1 megajoule = | **1,000,000** | **1** | **1000** | **$10^9$** | **$10^{12}$** | **$10^{13}$** |
| 1 kilojoule = | **1000** | **0.001** | **1** | **$10^6$** | **$10^9$** | **$10^{10}$** |
| 1 millijoule = | **0.001** | **$10^{-9}$** | **$10^{-6}$** | **1** | **1000** | **10,000** |
| 1 microjoule = | **0.000001** | **$10^{-12}$** | **$10^{-9}$** | **0.001** | **1** | **10** |
| 1 erg = | **$10^{-7}$** | **$10^{-13}$** | **$10^{-10}$** | **0.0001** | **0.1** | **1** |

NOTE: 1 watt-second = 1 joule.

### B. Energy/work units less than 10 joules (with joule equivalents)

| | Joules (J) | Foot-poundals (ft · pdl) | Foot-pounds-force (ft · lbf) | Calories (International Table) (cal, IT) | Calories (thermochemical) (cal, thermo) | Electronvolts (eV) |
|---|---|---|---|---|---|---|
| 1 joule = | 1 | 23.73036 | 0.7375621 | 0.2388459 | 0.2390057 | $6.24146 \times 10^{18}$ |
| 1 foot-poundal = | $4.2104011 \times 10^{-2}$ | 1 | $3.108095 \times 10^{-2}$ | $1.006499 \times 10^{-2}$ | $1.007173 \times 10^{-2}$ | $2.63016 \times 10^{17}$ |
| 1 foot-pound-force = | 1.355818 | 32.17405 | 1 | 0.3233816 | 0.3240483 | $8.46228 \times 10^{18}$ |
| 1 calorie (Int. Tab.) = | **4.1868** | 99.35427 | 3.088025 | 1 | 1.000669 | $2.61317 \times 10^{19}$ |
| 1 calorie (thermo) = | **4.184** | 99.28783 | 3.085960 | 0.9993312 | 1 | $2.61143 \times 10^{19}$ |
| 1 electronvolt = | $1.60219 \times 10^{-19}$ | $3.80205 \times 10^{-18}$ | $1.18171 \times 10^{-19}$ | $3.82677 \times 10^{-20}$ | $3.82933 \times 10^{-20}$ | 1 |

### C. Energy/work units greater than 10 joules (with joule equivalents)

| | Joules (J) | British thermal units, International Table (Btu, IT) | British thermal units, thermochemical (Btu, thermo) | Kilowatthours (kWh) | Horsepower-hours, electrical (hp h, elec) | Kilocalories, International Table (kcal, IT) | Kilocalories, thermochemical (kcal, thermo) |
|---|---|---|---|---|---|---|---|
| 1 joule = | 1 | $9.478170 \times 10^{-4}$ | $9.4845165 \times 10^{-4}$ | **1/3.6 × 10⁶** = $2.777 \times 10^{-7}$ | $3.723562 \times 10^{-7}$ | $2.388459 \times 10^{-4}$ | $2.3900574 \times 10^{-4}$ |
| 1 British thermal unit, Int. Tab. = | 1055.056 | 1 | 1.000669 | $2.9307111 \times 10^{-4}$ | $3.928567 \times 10^{-4}$ | 0.2519958 | 0.2521644 |
| 1 British thermal unit (thermo) = | 1054.35 | 0.999331 | 1 | $2.928745 \times 10^{-4}$ | $03.925938 \times 10^{-4}$ | 0.2518272 | 0.2519957 |
| 1 kilowatt-hour = | **3,600,000** | 3412.141 | 3414.426 | 1 | **1/0.746** = 1.3404826 | 859.8452 | 860.4207 |
| 1 horsepower hour, electrical = | **2,685,600** | 2545.457 | 2547.162 | **0.746** | 1 | 641.4445 | 641.8738 |
| 1 kilocalorie, Int. Tab. = | **4,186.8** | 3.968320 | 3.970977 | 0.001163 | $1.558981 \times 10^{-3}$ | 1 | 1.000669 |
| 1 kilocalorie, thermochemical = | **4184** | 3.965666 | 3.968322 | 0.0011622 | $1.5579386 \times 10^{-3}$ | 0.999331 | 1 |

The exact conversion is 1 British thermal unit, International Table = **1055.05585262** joules.

*Exact conversions are shown in boldface type. Repeating decimals are underlined. The SI unit of energy and work is the joule (J).
*Source:* D. G. Fink and W. Beaty (eds.), *Standard Handbook for Electrical Engineers* (12th ed.), McGraw-Hill, New York, 1987.

**TABLE E.8** Power Conversion Factors*

**A. Power units decimally related to 1 watt**

| | Watts (W) | Megawatts (MW) | Kilowatts (kW) | Milliwatts (mW) | Microwatts (μW) | Picowatts (pW) | Ergs per second (ergs/s) |
|---|---|---|---|---|---|---|---|
| 1 watt = | 1 | 0.000001 | 0.001 | 1000 | 1,000,000 | $10^9$ | $10^7$ |
| 1 megawatt = | 1,000,000 | 1 | 1000 | $10^9$ | $10^{12}$ | $10^{15}$ | $10^{13}$ |
| 1 kilowatt = | 1000 | 0.001 | 1 | 1,000,000 | $10^9$ | $10^{12}$ | $10^{10}$ |
| 1 milliwatt = | 0.001 | $10^{-9}$ | 0.000001 | 1 | 1000 | 1,000,000 | 10,000 |
| 1 microwatt = | 0.000001 | $10^{-12}$ | $10^{-9}$ | 0.001 | 1 | 1000 | 10 |
| 1 picowatt = | $10^{-9}$ | $10^{-15}$ | $10^{-12}$ | 0.000001 | 0.001 | 1 | 0.01 |
| 1 erg per second = | $10^{-7}$ | $10^{-13}$ | $10^{-10}$ | 0.0001 | 0.1 | 100 | 1 |

NOTE: 1 watt = 1 joule per second (J/s).

**B. Nonmetric power units (with watt equivalents)**

| | British thermal units (International Table) per hour (Btu/h, IT) | British thermal units (thermochemical) per minute (Btu/min, thermo) | Avoirdupois foot pounds force per second (ft lb$_f$/s avdp) | Kilocalories per minute (thermochemical) (kcal/min, thermo) | Kilocalories per second (International Table) (kcals, IT) | Horsepower [electrical] (hp, elec) | Horsepower (mechanical) (hp, mech) | Watts (W) |
|---|---|---|---|---|---|---|---|---|
| 1 British thermal unit (Int. Tab.)-per hour = | 1 | 0.0166778 | 0.2161581 | $4.2027405 \times 10^{-3}$ | $6.9998831 \times 10^{-5}$ | $3.9285670 \times 10^{-4}$ | $3.930148 \times 10^{-4}$ | 0.2930711 |
| 1 British thermal unit (thermo) per minute = | 59.959853 | 1 | 12.960810 | 0.2519957 | $4.1971195 \times 10^{-3}$ | 0.0235556 | 0.0235651 | 17.57250 |
| 1 foot-pound-force per second = | 4.6262426 | 0.0771557 | 1 | 0.0194429 | $3.2383157 \times 10^{-4}$ | $1.8174504 \times 10^{-3}$ | **1/550** $=1.818181 \times 10^{-3}$ | 1.355818 |
| 1 kilocalorie per minute (thermo) = | 237.93998 | 3.9683217 | 51.432665 | 1 | 0.0166555 | 0.0934763 | 0.0935139 | 69.733333 |
| 1 kilocalorie per second (Int. Tab.) = | 14,285.953 | 238.25864 | 3088.0251 | 60.040153 | 1 | 5.6123324 | 5.6145911 | **4186.800** |
| 1 horsepower (electrical) = | 2545.4574 | 42.452696 | 550.22134 | 10.697898 | 0.1781790 | 1 | 1.0004024 | **746** |
| 1 horsepower (mechanical) = | 2544.4334 | 42.435618 | **550** | 10.693593 | 0.1781074 | 0.9995977 | 1 | 745.6999 |
| 1 watt = | 3.4121413 | 0.0569071 | 0.7375621 | 0.0143403 | $2.3884590 \times 10^{-4}$ | **1/746** $= 1.3404826 \times 10^{-3}$ | $1.3410220 \times 10^{-3}$ | 1 |

NOTE: The horsepower (mechanical) is defined as a power equal to **550** foot-pounds-force per second.

Other units of horsepower are

1 horsepower (boiler) = 9809.40 watts
1 horsepower (metric) = 735.499 watts
1 horsepower (water) = 746.043 watts
1 horsepower (U.K.) = 745.70 watts
1 ton (refrigeration) = 3516.8 watts

*Exact conversions are shown in boldface type. Repeating decimals are underlined. The SI unit of power is the watt (W).
**Source:** D. G. Fink and W. Beaty (eds.), *Standard handbook for Electrical Engineers* (12th ed.), McGraw-Hill, New York, 1987.

**TABLE E.9**    Temperature Conversions*

| Celsius (°C) $°C = 5(°F − 32)/9$ | Fahrenheit (°F) $°F = [9(°C)/5] + 32$ | Absolute (K) $K = °C + 273.15$ |
|---|---|---|
| **−273.15** | **−459.67** | **0** |
| **−200** | **−328** | 73.15 |
| **−180** | **−292** | 93.15 |
| **−160** | **−256** | 113.15 |
| **−140** | **−220** | 133.15 |
| **−120** | **−184** | 153.15 |
| **−100** | **−148** | 173.15 |
| **−80** | **−112** | 193.15 |
| **−60** | **−76** | 213.15 |
| **−40** | **−40** | 233.15 |
| **−30** | **−22** | 243.15 |
| **−20** | **−4** | 253.15 |
| −17.7_7_ | **0** | 255.37_2_ |
| **−10** | 14 | 263.15 |
| −6.6_6_ | **20** | 266.48_3_ |
| **0** | **32** | 273.15 |
| **5** | **41** | 278.15 |
| **10** | **50** | 283.15 |
| **15** | **59** | 288.15 |
| **20** | **68** | 293.15 |
| **25** | **77** | 298.15 |
| **30** | **86** | 303.15 |
| **35** | **95** | 308.15 |
| **40** | **104** | 313.15 |
| **45** | **113** | 318.15 |
| **50** | **122** | 323.15 |
| **55** | **131** | 328.15 |
| **60** | **140** | 333.15 |
| **65** | **149** | 338.15 |
| **70** | **158** | 343.15 |
| **75** | **167** | 348.15 |
| **80** | **176** | 353.15 |
| **85** | **185** | 358.15 |
| **90** | **194** | 363.15 |
| **95** | **203** | 368.15 |
| **100** | **212** | 373.15 |
| **105** | **221** | 378.15 |
| **110** | **230** | 383.15 |
| **115** | **239** | 388.15 |
| **120** | **248** | 393.15 |
| **140** | **284** | 413.15 |
| **160** | **320** | 433.15 |
| **180** | **356** | 453.15 |
| **200** | **392** | 473.15 |
| **250** | **482** | 523.15 |
| **300** | **572** | 573.15 |
| **350** | **662** | 623.15 |
| **400** | **752** | 673.15 |
| **450** | **842** | 723.15 |
| **500** | **932** | 773.15 |
| **1000** | **1832** | 1273.15 |
| **5000** | **9032** | 5273.15 |
| **10,000** | **18,032** | 10,273.15 |

*Conversions in boldface type are exact. Continuing decimals are <u>underlined</u>. Temperature in kelvins equals temperatures in degrees Rankine divided by 1.8 [K = °R/1.8].

**Source:** D. G. Fink and W. Beaty (eds.), *Standard Handbook for Electrical Engineers* (12th ed.), McGraw-Hill, New York, 1987.

# INDUSTRY/GOVERNMENT/UNIVERSITY BATTERY ORGANIZATIONS[a]

**Vaidevutis Alminauskas**

## AUSTRIA

**Institut fur Chemische Technologie[b]**
Technischen Universitat
A-8010 Graz
*www.ictos.tugraz.at*

## AUSTRALIA

**CSIRO Energy Technology[b]**
Locked Bag 10
Clayton South VIC 3169
Tel: +61 3 9545 2176
Fax: +61 3 9545 2175
*www.det.csiro.au*

## BELGIUM

**Association of European Storage Battery Manufacturers[c]**
Avenue Jules Bordet 142
B-1140 Brussels
Tel: +32 2 761 1653
Fax: +32 2 761 1699
*www.eurobat.org*

**European Portable Battery Association[c]**
Avenue Jules Bordet 142
B-1140 Brussels
Tel: +32 2 761 1602
Fax: +32 2 761 1699
*www.epbaeurope.net*

## BULGARIA

**Institute of Electrochemistry and Energy Systems (IEES) Bulgarian Academy of Sciences[b]**
Acad. G, Bonchev Street, Block 10, Sofia
Tel: +359 2 872 25 45
Fax: +359 2 872 25 44
*www.bas.bg/cleps/*

## CANADA

**Dalhousie University, Depts. of Physics and Chemistry[d]**
6300 Coburg Road
Halifax, Nova Scotia, B3H 3J5
Tel: (902) 494 2991
*www.dal.ca*

**Institut de Recherché d'Hydro-Quebec[b]**
1800 Montée Ste-Julie
Varennes, Quebec J0L 2P0
Tel: (514) 652 8011
Fax: (450) 652 8990
*www.ireq.ca*

---

[a]See also Chap. 8 for battery standards organizations.
[b]Government laboratory.
[c]Battery trade group or association.
[d]University/university affiliated.

## CHINA (PRC)

**China Industrial Association of Power Sources**[c]
No. 18, Lizhuangzi Road
Nankai District
Tianjin 300381
Tel: +86 22 23959268
Fax: +86 22 23380938
*www.cibf.org.cn*
*www.chinabatteryonline.com*

**China National Battery Industry Information Center**[c]
No. 1 Yangtianhu Xincun, Changsha
Hunan 410007
Tel: +86 731 5141901
*www.batterypub.com*

**Shenyang Storage Battery Research Institute**[b]
No. 33 Beier Zhong Road, Tie-xi, Shenyang
Tel: +86 24 85610109
Fax: +86 24 85610109

**Tianjin Institute of Power Sources**[b]
No. 18, Lingzhuangzi Road, Nankai
Tianjin 300381
Tel: +86 22 2399396
Fax: +86 22 23383783

## ISRAEL

**Bar-Ilan University**[d]
Ramat Gan, 52900
Tel: +972 3 531 8111
*www1.biu.ac.il*

**Tel Aviv University**[d]
P.O. Box 39040
Tel Aviv 69978
Tel: +972 3 640 8111
*www.tau.ac.il*

## ITALY

**Università degli Studi di Roma "La Sapienza"**[d]
Dipartimento di Chimica
Piazzale Aldo Moro 5, 00185 Roma
Tel: +39 06 446 2866
Fax: +39 06 491769
*www.uniroma1.it*

## JAPAN

**Japan Batteries Industry Association**[c]
Kikai Shinkokaikan Building, 3-5-8
Shiba-koen, Minato-ku, Tokyo 105-0011
Tel: +81 3 343 40261
Fax: +81 3 343 42691
*www.baj.or.jp*

## POLAND

**Centralne Laboratorium Akumulatorow I Ogniw**[b]
ul. Forteczna 12
Poznan 61-362
Tel: +48 61 8790517
Fax: +48 61 8793012
*www.claio.poznan.pl*

## RUSSIA

**International Assn. of Chemical Power Sources Manufacturers INTERBAT**[c]
1-St.Smolensky Byst 7
123099 Moscow
Tel: +7 095 244 0735
Fax: +7 095 244 0369
*www.interbat.ru*

## SOUTH KOREA

**Korea Electrotechnology Research Institute**[b]
28-1 Seongju-dong, Changwon-si
Gyeongsangnam-do, 641-120
Tel: +82 55 280 1114
Fax: +82 55 280 1216
*www.keri.re.kr*

## UNITED KINGDOM

**British Battery Manufacturers Association**[c]
26 Grosvenor Gardens
London SW1W 0TG
Tel: +44 20 7838 4800
Fax: +44 20 7838 4871
*www.bbma.co.uk*

**DERA Tech. Management Services**[b]
Aquila, Golf Road
Bromley
Kent BR1 2JB
Tel: +44 20 8285 7127
Fax: +44 20 8285 7346
*www.dera.gov.uk*

**Electrochemical Power Sources Centre**[b]
DERA Haslar
Gosport Hampshire PO12 2AG
Tel: +44 1705 335358
Fax: +44 1705 335102
*www.dera.gov.uk*

## USA

**Air Force Wright Aeronautical Laboratory**[b]
Aerospace Power Division
Wright-Paterson AFB, OH 45435
Tel: (937) 255 7770
Fax: (937) 656 7529

**Argonne National Laboratory**[b]
Chemical Sciences and Engineering Division
9700 South Cass Avenue
Argonne, IL 60439
Tel: (630) 252 4383
Fax: (630) 252 5528
*www.cse.anl.gov*

**Battery Council International**[c]
401 North Michigan Avenue
Chicago, IL 60611
Tel: (312) 644 6610
Fax: (312) 527 6640
*www.batterycouncil.org*

**Electric Power Research Institute**[c]
3420 Hillview Avenue
Palo Alto, CA 94304
Tel: (650) 855 2121
*www.epri.com*

**Hunter College Physics Dept., City University of New York**[d]
695 Park Ave
New York, NY 10065
Tel: (212) 772-4973
Fax: (212) 772 5390
*www.hunter.cuny.edu/physics/faculty/greenbaum/research*

**Jet Propulsion Laboratory**[b]
Electrochemical Technologies Group
California Institute of Technology
4800 Oak Grove Drive
Pasadena, CA 91109
Tel: (818) 354 4321
Fax: (818) 393 6951
*www.jpl.nasa.gov*

**Lawrence Berkley Laboratory**[b]
Advanced Energy Technologies Department
Environmental Energy Technologies Division
1 Cyclotron Road, MS 70R0108B
Berkeley, CA 94720
Tel: (510) 486 4202
Fax: (510) 486 7303
*eetd.lbl.gov/aet/batteries.html*

**NanoPower Research Laboratories**[d]
Rochester Institute of Technology
85 Lomb Memorial Drive
Rochester, NY 14623
Tel: (585) 475 2480
Fax: (585) 475 7890
*www.sustainability.rit.edu/nanopower*

**National Alliance for Advanced Technology Batteries (NAATBATT)**[c]
122 South Michigan Avenue, Suite 1700
Chicago, IL 60603
Tel: (312) 588 0477
*www.naatbatt.org*

**NASA Glenn Research Center**[b]
21000 Brookpark Road
Cleveland, OH 44135
Tel: (216) 433 4000
*www.nasa.gov/centers/glenn*

**Naval Sea Systems Command, Carderock Division**[b]
Power Systems Branch, Code 643
9500 MacArthur Boulevard
West Bethesda, MD 20817-5700
Tel: (301) 227-5681
Fax: (301) 227-4733
*www.dt.navy.mil/sur-str-mat/fun-mat/pow-sys-bra/*

**Naval Sea Systems Command, Crane Division**[b]
Power and Circuit Board Technologies Division, Code GXS
Crane, IN 47522
Tel: (812) 854 1593
Fax: (812) 854 3589
*www.crane.navy.mil/whatwedo/PowerSystems.asp*

**Portable Rechargeable Battery Association**[c]
1776 K Street
Washington, DC 20006
Tel: (202) 719 4978
*www.prba.org*

**Rutgers Energy Storage Research Group (ESRG)**[b,d]
Department of Materials Science and Engineering
607 Taylor Road
Piscataway, NJ 08854
Tel: (732) 932 6850
*mse.rutgers.edu*

**Sandia National Laboratories**[b]
Energy Storage Systems
P. O. Box 5800
Albuquerque, NM 87185
Power Sources Technology Group
Tel: (505) 844 0452
Fax: (505) 844 5824
*www.sandia.gov*

**Underwriters Laboratories, Inc.**[c]
333 Pfingsten Road
Northbrook, IL 60062
Tel: (847) 272 8800
Fax: (847) 272 8129
*www.ul.com*

**University of South Carolina**[d]
Department of Chemical Engineering
College of Engineering and Computing
301 Main Street
Columbia, SC 29208
Tel: (803) 777 3270
Fax: (803) 777.8265
*www.che.sc.edu*

**U.S. Army RDECOM CERDEC C2D**[b]
Army Power Division, RDER-CCA
328 Hopkins Road, Building 1105
Aberdeen Proving Ground, MD 21005
Tel: (410) 278 9229
Fax: (410) 278 8990
or
Fort Monmouth, NJ 07703
Tel: (732) 532 9000
*www.cerdec.army.mil/directorates/c2d_army_power.asp*

**U.S. Army Research Laboratory**[b]
Sensors and Electronic Devices Directorate
2800 Powder Mill Road
Adelphi, MD 20783
Tel: (301) 394 5429
*www.arl.army.mil*

# INDEX